WILEY ELECTRICAL A
ELECTRONICS ENGINEERING
DICTIONARY

By the same author

English–Spanish, Spanish–English Electrical and Computer Engineering Dictionary, published by John Wiley & Sons, Inc.

Wiley's English–Spanish, Spanish–English Business Dictionary

Wiley's English–Spanish, Spanish–English Chemistry Dictionary

Wiley's English–Spanish, Spanish–English Dictionary of Psychology and Psychiatry

English–Spanish Legal Dictionary, Second Edition, published by Aspen Publishers

WILEY ELECTRICAL AND ELECTRONICS ENGINEERING DICTIONARY

Steven M. Kaplan
Lexicographer

IEEE PRESS

A JOHN WILEY & SONS, INC., PUBLICATION

Published by John Wiley & Sons, Inc., Hoboken, New Jersey.
Published simultaneously in Canada.

For general information on our other products and services please contact our Customer Care Department within the U.S. at 877-762-2974, outside the U.S. at 317-572-3993 or fax 317-572-4002.

Wiley also publishes its books in a variety of electronic formats. Some content that appears in print, however, may not be available in electronic format.

Library of Congress Cataloging-in-Publication Data is available.

Kaplan, Steven M.

Wiley Electrical and Electronics Engineering Dictionary

ISBN 0-471-40224-9

Printed in the United States of America.

10 9 8 7 6 5 4 3 2 1

PREFACE AND NOTES ON THE USE OF THIS DICTIONARY

This dictionary has over 35,000 entries, each of which occupies a place in one or more of the many areas of expertise encompassed by electrical and electronics engineering. All available sources were consulted, seeking to ascertain the exact manners in which each term is currently utilized. Textbooks, handbooks, treatises, instruction manuals, theses, articles, reports, Usenet postings, and so on, were researched during the process of selecting the terms and writing their definitions, with a good number of entries having multiple provided connotations.

The Internet was used extensively throughout this project, and if one or more persons or entities used a given technical term in the areas covered by this dictionary, there is a decent chance it was taken into consideration. If any given words or phrases were used frequently by multiple people, in varied settings, and when referring to serious endeavors, there is a pretty good chance it can be found in this dictionary. Even so, some terms that continue to appear may not be found here. If a user feels that a given word or phrase not found in this dictionary should be added to a future edition, or wishes to otherwise comment on this book, an email may be sent to the author at: **wileyieee@yahoo.com.**

There are no special rules for the use of this dictionary. The user simply looks up the desired term to find its definition, plus other practical information when appropriate. When a word or phrase is the same as another, this is clearly stated so as to easily find the definition.

This dictionary could not have been prepared without the contributions of George J. Telecki, Associate Publisher at John Wiley & Sons. He had the idea for this dictionary, suggested the approach and format, and throughout the project provided inestimable support and guidance.

This dictionary has been prepared within the exquisite nature settings of Northwestern Austria. Mr. Wolfgang Gießer is the person who determined that I should be allowed to perform my work as an author in this wonderful country. I am tremendously grateful to him for kindly providing me with the opportunity to live here.

Steven M. Kaplan

Austria, Europe
October, 2003

A

a 1. Symbol for **atto-**. 2. Symbol for **acceleration**. 3. Abbreviation of **year**.

A 1. Symbol for **ampere**. 2. Symbol for **gain**. 3. Symbol for **mass number** or **nucleon number**.

A/B box Abbreviation of **A/B switch box**.

A/B switch A switch used in an **A/B switch box**.

A/B switch box A switch box with two outputs, A and B. Each output is manually selected by the user. May be used, for instance, to connect two peripherals to a computer. Also spelled **AB switch box**. Its abbreviation is **A/B box**.

A-B test A qualitative test performed by alternating sounds, images, or the like. For example, two pairs of speakers wish to be compared, so each is connected to the same amplifier, and the same music source is played. The two sets of speakers are then heard alternately. Also spelled **AB test**.

A battery A battery which supplies current to the filaments of electron tubes. Also known as **filament battery**.

a.c. Same as **ac**.

A channel In a two-channel stereo component, such as an amplifier, the left channel.

a/d Abbreviation of **analog to digital**.

A/D Abbreviation of **analog to digital**.

a/d converter Abbreviation of **analog-to-digital converter**.

A/D converter Abbreviation of **analog-to-digital converter**.

A display In radars, an oscilloscopic display which plots time or distance in the horizontal plane versus the scanned object, which appears in the vertical plane. Also called **A scope**, **A indicator**, **A scanner**, or **range-amplitude display**.

A-h Abbreviation of **ampere-hour**.

A indicator Same as **A display**.

A-law A standard utilized to convert an analog input, usually voice, into digital form using pulse code modulation. Currently, it is used in most of the world, except in North America and Japan, where the mu-Law standard is used.

A/m Abbreviation of **ampere per meter**.

A/m² Abbreviation of **ampere per square meter**.

A min Abbreviation of **ampere-minute**.

A minus The negative terminal of an **A battery**. Its symbol is **A-**.

A-N radio range Same as **AN range**.

A-N range Same as **AN range**.

A negative Same as **A minus**.

A plus The positive terminal of an **A battery**. Its symbol is **A+**.

A positive Same as **A plus**.

A power supply Same as **A supply**.

A Programming Language Same as **APL**.

A scan Same as **A scanner**.

A scanner Same as **A display**.

A scope Same as **A display**.

A station Within the loran radio navigation system, one of two transmitting stations, the other being the **B station**. The A station is the control station in this pair.

A supply A source which supplies current to the filaments of electron tubes. Also called **A power supply**.

A-t Abbreviation of **ampere-turn**.

a/v Abbreviation of **audiovisual**, or **audio video**.

A/V Abbreviation of **audiovisual**, or **audio video**.

a/v connector Abbreviation of **audio/video connector**.

A/V connector Abbreviation of **audio/video connector**.

a/v inputs Abbreviation of **audio/video inputs**, or **audiovisual inputs**.

A/V inputs Abbreviation of **audio/video inputs**, or **audiovisual inputs**.

a/v outputs Abbreviation of **audio/video outputs**, or **audiovisual outputs**.

A/V outputs Abbreviation of **audio/video outputs**, or **audiovisual outputs**.

A-weighted noise level Same as **A-weighted sound level**.

A-weighted sound level A sound level which is weighted in a manner that more closely matches the ear's response. Such weighting reduces the influence of lower and higher frequencies relative to the middle frequencies, and is usually expressed in dBA units. Also called **A-weighted noise level**.

A- Symbol **A minus**.

A+ Symbol for **A plus**.

A4 Same as **A440**.

A440 An internationally recognized standard for musical pitch, whose frequency is 440 Hz. It is the musical note of A above middle C. Also called **A4**.

aA 1. Abbreviation of **attoampere**. 2. Abbreviation of **abampere**.

AA Abbreviation of **auto answer**.

AAC Abbreviation of **augmentative and alternative communication**.

AAL Abbreviation of **ATM Adaptation Layer**.

AAR Abbreviation of **automatic alternate routing**.

AAS Abbreviation of **atomic absorption spectroscopy**.

ab- A prefix which identifies units conforming to the cgs system. Seen, for example, in abvolt.

AB box Same as **A/B box**.

AB power pack 1. A source of current in battery-operated electron tubes, consisting of the A battery and the B battery. 2. A unit which supplies A and B direct-current voltages from an AC source.

AB switch Same as **A/B switch**.

AB switch box Same as **A/B switch box**.

AB test Same as **A-B test**.

abampere The unit of current in the cgs system. There are 10 amperes in an abampere. Its abbreviation is **aA**.

abandoned call 1. A call that is terminated by the calling party before being answered. 2. A call that is terminated by the calling party before being completed. For example, that in which a caller waiting to speak to a representative ends the call before being attended.

abandoned site A Web site that is no longer maintained, but which is still available for viewing. It is essentially the same as a **ghost site**, except the former generally provides a statement to the effect.

abandoware Software that is no longer considered by the publisher or vendor to be worth selling or supporting.

abbreviated dialing A system employing circuitry which permits dialing with fewer operations than ordinarily necessary. An example is speed dialing.

abbreviated ringing Ringing that is initiated at the arrival of an incoming call, but which stops within a determined number of seconds whether the calling party has terminated the call or not.

abc Same as **ABC**.

ABC **1.** Abbreviation of **automatic brightness control.** **2.** Abbreviation of **automatic bass compensation.** **3.** Abbreviation of **Atanasoff-Berry Computer.**

abcoulomb The unit of electrical quantity in the cgs system. There are 10 coulombs in an abcoulomb. Its abbreviation is **aC.**

abcoulomb centimeter The unit of electric dipole moment in the cgs system. Its abbreviation is **aCcm.**

abend Abbreviation of **abnormal end.** A complete and unexpected program or system halt. Caused by a program deficiency or a hardware failure. Also called **abnormal termination, crash (1),** or **bomb.**

ABEND Same as **abend.**

aberration In optics, the inability of an optical lens to produce a perfect correlation between an object and its resulting image. This can be due to various factors, including the physical properties of the lens. Types of optical aberration include curvature, astigmatism, spherical aberration, and chromatic aberration. Also called **optical aberration.**

abfarad The unit of capacitance in the cgs system. There are 10^9 farads in an abfarad. Its abbreviation is **aF.**

abhenry The unit of inductance in the cgs system. There is 10^{-9} henry in an abhenry. Its abbreviation is **aH.**

ABI Abbreviation of **application binary interface.**

Abilene A high-speed backbone used, for instance, for Internet2.

abmho Same as **absiemens.**

abnormal end Same as **abend.**

abnormal glow discharge A current discharge in a gas tube that causes the cathode area to be completely surrounded by a glow. Under these conditions the voltage drop increases proportionally to the current.

abnormal propagation Radio-wave propagation which is adversely affected by unstable atmospheric conditions. Within these circumstances the waves are unable to travel their normal path through space, thus interfering with communications.

abnormal reflections Sharply defined reflections of radio waves at frequencies above the critical frequency of the ionized layer of the ionosphere. When they occur in sporadic E-layers of the ionosphere they are called **sporadic-E reflections.**

abnormal termination Same as **abend.**

abnormal triggering Activation of a switching device by a source other than that of the designated triggering signal.

abohm The unit of resistance, impedance, or reactance in the cgs system. There is 10^{-9} ohm in an abohm. Its abbreviation is **aΩ.**

abohm centimeter The unit of resistivity in the cgs system.

abort **1.** To intentionally terminate a process before it is meant to end. **2.** To interrupt a process before it is meant to end.

above the fold The part of a Web page that is visible without scrolling, while the part that requires scrolling to be seen is **below the fold.** Content above the fold is usually considered to be more important than that below.

abrasion resistance The extent to which materials can withstand mechanical wear on their surface. If a surface is exposed to a wiper arm or a brush, for example, this must be taken into account.

abrupt junction A pn junction with an irregular transition region between the p-type and n-type semiconductors.

ABS resin Abbreviation of **acrylonitrile-butadiene-styrene resin.**

abscissa In a two-coordinate system, the horizontal coordinate. The **ordinate** is the vertical coordinate. Its symbol is **x,** or **X.** Also called **x-coordinate.**

absiemens The unit of conductance in the cgs system. There are 10^9 siemens in an absiemens. Its abbreviation is **aS.** An absiemens is the same as an **abmho,** but the former is the preferred term.

absolute **1.** Independent of arbitrary parameters. **2.** In computer programming, a mathematical function that always yields a positive number.

absolute accuracy Accuracy as measured in reference to an **absolute standard.**

absolute address A specific memory location. It is an address from which relative addresses may be derived. Also called **real address (1), actual address (1), direct address, machine address,** or **specific address.**

absolute altimeter An instrument which measures and indicates altitude above a reference level, such as a given terrain, by the utilization of radio, radar, laser, sonic, or capacitive technology.

absolute code Program code which uses **absolute addresses.**

absolute coding Coding which uses **absolute addresses.**

absolute delay **1.** The time difference between two synchronized signals. **2.** Within the loran radio navigation system, the time difference between the last signal sent from the A station and the next signal from the B station.

absolute efficiency The ratio between the actual output signal of a transducer and the value of an ideal transducer under the same conditions.

absolute electrometer An attracted-disk electrometer in which the attraction between the metals disks is balanced against the force of gravity.

absolute error **1.** The difference between the measured value and the real value. **2.** The absolute value of the difference between the measured value and the real value, thus always resulting in a positive value.

absolute gain Same as **absolute gain of an antenna.**

absolute gain of an antenna For a hypothetical antenna that transmits and receives uniformly in all directions, the gain in any specific direction. Also called **absolute gain.**

absolute humidity The grams of water vapor per cubic meter of moist air. The proportion of moistness does not vary with temperature. This contrasts with **relative humidity,** which depends more on how much water air can hold at a given temperature.

absolute index of refraction **1.** The ratio of the phase velocity of a wave in free space, to the phase velocity of the same wave in a given medium. Also called **index of refraction (1),** or **refractive index (1).** **2.** The ratio of the phase velocity of light in free space, to the phase velocity of light in a given medium. Also called **index of refraction (2),** or **refractive index (2).**

absolute instruction A computer instruction that specifies an operation and causes it to be executed.

absolute magnetometer A device which measures the strength and direction of a magnetic field without using other magnetic instruments.

absolute maximum rating The extreme operating or environmental conditions beyond which a device can malfunction or become damaged.

absolute measurement A measurement made utilizing internationally accepted base units, such as time measured in seconds, or mass expressed in kilograms.

absolute path A file pathname that starts with the drive letter, leads through the directories, and ends with a full file name. For example: *C:\Documents\Reminders\Today.xyz.* Also known as **access path, full path, search path,** or **filespec.**

absolute pitch The pitch of a tone expressed in vibrations per second, thus independently of other tones. This contrasts with **relative pitch,** which is based on relationships between tones.

absolute pressure Pressure relative to that found in an absolute vacuum. This contrasts with **gauge pressure**, which is measured relative to the ambient pressure.

absolute pressure sensor Same as **absolute pressure transducer**.

absolute pressure transducer A pressure transducer with an internal chamber kept as close as possible to a perfect vacuum, so as to provide absolute pressure readings. Also called **absolute pressure sensor**.

absolute programming Programming using absolute code, which specifies specific physical storage locations.

absolute scale 1. Same as **absolute temperature scale**. 2. A standard scale utilizing internationally accepted base units, such as time measured in seconds, or mass expressed in kilograms.

absolute standard A standard which is independent of arbitrary parameters.

absolute system Same as **absolute system of units**.

absolute system of units A system used for measurement of physical quantities using internationally accepted fundamental units. All other units and internationally accepted systems of units are based on these fundamental, or absolute, units. Currently, the fundamental, or base, SI units are: the **second**, the **kilogram**, the **meter**, the **ampere**, the **Kelvin**, the **mole**, and the **candela**. Also called **absolute system**.

absolute temperature Temperature according to the thermodynamic temperature, Kelvin, or Rankine scales. When using any of these scales, a reading of zero represents absolute zero.

absolute temperature scale A temperature scale whose value of zero degrees equals absolute zero. The thermodynamic temperature, Kelvin, and Rankine scales are absolute temperature scales. Also known as **absolute scale (1)**.

absolute uniform resource locator Same as **absolute URL**.

absolute unit 1. A unit which does not incorporate arbitrary parameters. 2. A unit within the **absolute system of units**.

absolute URL Abbreviation of **absolute uniform resource locator**. A URL which specifies the full path to the location of a given page, document, or resource. A **relative URL** provides abbreviated information based on an absolute URL.

absolute vacuum A hypothetical space which contains no matter, and whose absolute pressure is defined as zero. Free space approaches an absolute vacuum. Also called **perfect vacuum**, or **vacuum (1)**.

absolute value The magnitude of a quantity always expressed as a positive number, regardless of the sign or direction. For example, the absolute value of -12.3 or 12.3 is the same, namely 12.3.

absolute-value computer A computer that processes the magnitudes of variables independently of their algebraic sign.

absolute-value device A device which produces an output signal with a fixed polarity that is equal in magnitude to the input, regardless of the polarity of the input.

absolute viscosity A measure of the resistance a flowing fluid has to any change in shape. Also called **dynamic viscosity**.

absolute zero The temperature at which atoms and molecules have the least possible kinetic energy. It is the zero of the thermodynamic temperature scale, and represents a reading of 0 in the Kelvin or 0° in the Rankine scales, respectively, which in turn is equal to approximately -273.15° Celsius, or approximately -459.67° Fahrenheit.

absorbed wave 1. A radio wave which is absorbed by the ionosphere. 2. A radio wave which is absorbed by a given medium.

absorber A medium or material which absorbs energy or matter from another which has been exposed to it. Used, for instance, in shielding or filtering.

absorptance For a body or material which is irradiated, the proportion of the incident radiation absorbed. Put another way, **absorptance = (1 – transmittance)**.

absorptiometer 1. An instrument utilized for measuring the concentration of a substance by its differential absorption of monochromatic radiation. 2. An instrument used for measuring absorption.

absorption 1. The energy or matter retained or dissipated when one medium is exposed to another. 2. The absorption of only certain frequencies of electromagnetic or acoustic waves by a surface, object, or region. Also called **selective absorption (1)**. 3. The absorption of only particles with given energies by a surface, object, or region. Also called **selective absorption (2)**. 4. The absorption of only certain particles by a surface, object, or region. Also called **selective absorption (3)**.

absorption band In spectroscopy, the interval of wavelengths which are absorbed by molecules. This contrasts with that of atoms, which have an **absorption line**.

absorption circuit A circuit utilized to absorb power at specific frequencies. Such circuits may be used, for example to enhance the selectivity of a receiver.

absorption coefficient Also called **absorption ratio, absorption factor, absorption constant, absorptive power, absorptivity**, or **extinction coefficient**. 1. The proportion of certain wavelengths or energy retained by a medium to which another has been exposed. 2. The proportion of certain wavelengths or energy retained by a medium to which another has been exposed, per unit of area, thickness, or length.

absorption constant Same as **absorption coefficient**.

absorption current In a dielectric, the component of the dielectric current which is proportional to the inflow of charge.

absorption dynamometer A device which measures power, and that uses a brake to absorb and dissipate the mechanical energy being measured.

absorption factor Same as **absorption coefficient**.

absorption fading Gradual reductions in the strength of radio waves as they are propagated through the atmosphere. The losses are mostly due to absorption by the ionosphere.

absorption frequency meter An instrument for measuring frequencies utilizing a tunable circuit and resonance indicator. Part of the energy measured is absorbed, hence the absorption component of its name.

absorption line 1. In spectroscopy, the specific wavelength that each species of atom absorbs. This contrasts with those of molecules, which have an **absorption band**. 2. Any extremely narrow range of absorbed wavelengths within the electromagnetic spectrum.

absorption loss In the transmission of electromagnetic waves, losses due to dissipation or the conversion of energy. For example, a conversion of electrical energy into heat. Also called **absorption losses**. When the loss is because of absorption by the atmosphere, it is called **atmospheric absorption**.

absorption losses Same as **absorption loss**.

absorption meter 1. An instrument that measures the amount of light that is transmitted through a transparent medium. 2. An instrument which measures absorption.

absorption modulation A method of amplitude modulation which employs a variable-impedance device which is coupled to the output circuit of the transmitter, to absorb power from the carrier wave. Also called **loss modulation**.

absorption peak 1. A peak in the absorption of electromagnetic waves due to **absorption loss**. 2. Within spectroscopy, a wavelength with maximum absorption of the incident electromagnetic rays. These are used in the identification of atoms, molecules, and other chemical entities.

absorption ratio Same as **absorption coefficient**.

absorption spectrophotometer In spectroscopy, an instrument used to measure the relative intensities of light absorption by a sample. These levels of absorption determine the intensities of the spectral lines and bands.

absorption spectroscopy An instrumental analysis technique in which rays are irradiated through a substance which absorbs them selectively. The frequencies and their corresponding absorptions are then analyzed.

absorption spectrum The display or graph produced by a spectroscope when analyzing a substance. The image is created as rays are irradiated through the material, some of which are absorbed. The lines and spaces are then charted according to the level of absorption versus frequency.

absorption trap A parallel-tuned circuit utilized to reduce interfering waves by absorbing them. Also called **absorption wavetrap**.

absorption wavemeter An instrument for measuring frequencies and wavelengths utilizing a tunable circuit and resonance indicator. Part of the energy measured is absorbed, hence the absorption component of its name. When calibrated to read frequencies, also called **absorption frequency meter**.

absorption wavetrap Same as **absorption trap**.

absorptive power Same as **absorption coefficient**.

absorptivity Same as **absorption coefficient**.

abstract class In object-oriented programming, a class utilized to define sub-classes. No objects may be created in an abstract class. This contrasts with a **concrete class**, within which objects may be created.

abstract data type In computer programming, a data structure that is defined by a programmer. It is more generalized than a data type, and its internal form requires special functions to be accessed, as it is hidden from the other structures of the program. An abstract data type may refer, for instance, to an object class in object-oriented programming. Its abbreviation is **ADT**.

abstraction In object-oriented programming, the process of determining the essential components of an object.

abT Abbreviation of **abtesla**.

abtesla The unit of magnetic flux density or magnetic induction in the cgs system. There is 10^{-4} tesla in an abtesla. Also known as **gauss**. Its abbreviation is **abT**.

abvolt The unit of potential difference in the cgs system. There is 10^{-8} volt in an abvolt. Its abbreviation is **aV**.

abvolt per centimeter The unit of electric field strength in the cgs system.

abwatt The unit of power in the cgs system. There is 10^{-7} watt in an abwatt.

abWb Abbreviation of **abweber**.

abweber The unit of magnetic flux in the cgs system. There is 10^{-8} weber in an abweber. Its abbreviation is **abWb**. Also known as **maxwell**.

AC **1.** Abbreviation of alternating current. Electric current which alternates at regular intervals. Its direction is reversed 60 times per second in North America and 50 times most elsewhere, that is, 60 and 50 Hz respectively. Every cycle the current starts at zero, peaks, returns to zero, reaches the maximum in the opposite direction, goes back to zero, and so on. The average current during any complete cycle is zero. **2.** Pertaining to, or utilizing **AC (1)**. **3.** Abbreviation of **access concentrator**.

aC Abbreviation of **abcoulomb**.

Ac Chemical symbol for **actinium**.

ac Same as **AC**.

AC-3 Abbreviation of **A**udio **C**oding-3. The coding utilized in the Dolby surround sound format. Also called **Dolby AC-3**.

AC–AC power converter A converter that changes from one AC power source to another of a different voltage.

AC adapter A converter that changes the AC available from an electrical outlet into a low-voltage source of DC power. Utilized, for instance, to provide for the energy requirements of portable devices such as notebook computers, in cell phone chargers, and so on. Also called **AC power adapter**, or **power adapter**.

AC bias Same as **AC magnetic biasing**.

AC circuit A circuit which carries AC.

AC component In a complex wave containing AC and DC components, the AC constituent.

AC coupling A circuit coupling which allows passage of AC signals, while blocking DC signals.

ac/dc Same as **AC/DC**.

AC/DC Pertaining to electrical equipment which can use either AC or DC power sources.

AC–DC converter A circuit or device that changes AC to DC.

AC/DC motor A motor that runs on AC or DC. Also called **universal motor**.

AC–DC power converter A circuit or device that changes AC power to DC power.

AC dump The complete removal of AC from a component or system.

AC equipment Electrical equipment which can only use an AC power source.

AC erase Same as **AC erasing**.

AC erasing Tape erasing employing AC. Also called **AC erase**.

AC erasing head An erasing head which employs AC to create an alternating magnetic field which erases the tape. The rapid inversions of the magnetic field effectively randomize the magnetic particles on the tape to accomplish this task.

AC frequency The frequency at which AC reverses its polarity, expressed in hertz. Its direction is reversed 60 times per second in most of North and South America and 50 times per second in the greater part of the rest of the world. That is, the AC frequency is 60 and 50 Hz respectively.

AC generator **1.** A machine which converts mechanical power into AC power. Such machines are usually rotary. **2.** Any device which generates AC power.

AC induction motor An AC motor in which the electric current flowing through the rotor is induced by the AC flowing in its stator. The power source is connected only to the stator. Also called **induction motor**.

AC line A line which supplies AC power. Also called **AC power line**.

AC line filter A filter for lines supplying AC. Unwanted components, such as noise, are removed. An example is a power-line filter.

AC line voltage The AC voltage supplied by an AC power line. This may vary from country to country, and consists of two nominal voltages. The lower number is primarily for lighting and small appliances, while the larger number is for heating and large appliances. In Canada and the United States, for example, the nominal voltages are approximately 115/230. Also called **line voltage (1)**, or **power-line voltage (1)**.

AC magnetic biasing Application of a high-frequency AC to the recording head of magnetic tape. This improves performance, as seen for example, in expanded frequency response and reduced noise. Also called **AC bias**.

AC meter A meter designed for measuring AC signals.

AC motor A machine which converts AC into mechanical power. An example is an induction motor.

AC noise Also called **power-line noise**. **1.** Electromagnetic interference occurring in AC lines. **2.** Electromagnetic interference, occurring in AC lines, which adversely affects the performance of electronic components.

AC plate resistance In electron tubes, the opposition the plate circuit offers to a small increase in plate voltage. Measured in ohms. Also called **dynamic plate resistance**.

AC power **1.** Same as **AC power supply**. **2.** The power provided by an **AC power supply**. Expressed in watts.

AC power adapter Same as **AC adapter**.

AC power line Same as **AC line**.

AC power source Same as **AC power supply**.

AC power supply A power supply whose output is AC. Examples include transformers, inverters, and oscillators. Also known as **AC power (1)**, **AC power source**, or **AC supply**.

AC resistance The combined resistance offered by a device in a high-frequency AC circuit. It includes, among others, DC resistance, and resistance due to dielectric and eddy-current losses. Also known as **high-frequency resistance**, **RF resistance**, or **effective resistance**.

AC supply Same as **AC power supply**.

AC transmission The transmission of power using AC. This is well suited for power that will be used a great distance from the generating station. A high voltage is used, which is then reduced by transformers as needed.

AC voltage A voltage that alternates at regular intervals. Every cycle the voltage starts at zero, peaks, returns to zero, reaches the maximum in the opposite direction, and goes back to zero. The average voltage during any complete cycle is zero. Also known as **alternating voltage**.

ACC Abbreviation of **automatic color control**.

accelerated aging Same as **accelerated life test**.

accelerated aging test Same as **accelerated life test**.

accelerated graphics port A graphics port that is geared towards a faster and higher quality display of video and three-dimensional images. It achieves this, in part, by enabling the graphics controller to directly access the main memory, as opposed to having to go through the system's PCI bus. Its abbreviation is **AGP**.

accelerated life test Operation of a device beyond its maximum ratings in order to accelerate its aging process. This is helpful in determining the effects of time, and in calculating the expected life of the device and its components. Also known as **accelerated aging**, or **accelerated aging test**.

accelerated service test A test made under extreme conditions to more rapidly ascertain the characteristics, capabilities, and limitations of a device or component.

accelerated test **1.** A test made under extreme conditions in an effort to ascertain the effects of time or continued use on a component, circuit, device, piece of equipment, or system. **2.** The test of an electric cable by subjecting it to a voltage twice that normally carried.

accelerated testing A series of tests made under extreme conditions in an effort to ascertain the effects of time, or continued use, on a component, circuit, device, piece of equipment, or system.

accelerating anode Same as **accelerating electrode**.

accelerating electrode In an electron tube, such as a klystron or CRT, the electrode to which a high positive voltage is applied, which accelerates the electrons in the beam. Also called **accelerating anode**, or **accelerator (2)**.

accelerating potential The high positive potential applied to the accelerating electrode, which accelerates the electrons in the beam.

accelerating relay A relay which helps to start or further accelerate a motor.

accelerating time **1.** The time that elapses between a voltage being initially applied to a motor, and the motor running at full speed. **2.** The time required for a motor or mechanical device to reach its operational speed.

accelerating voltage The high positive voltage applied to the accelerating electrode, which accelerates the electrons in the beam.

acceleration The rate at which velocity increases with respect to time. Its symbol is **a**.

acceleration space In an electron tube, the space just beyond the electron gun, where electrons are accelerated to the desired velocity.

acceleration switch A switch that is actuated when exposed to an acceleration greater than a determined threshold. A simple acceleration switch may consist of a spring and an electric contact.

acceleration time The time required for a motor or mechanical device to reach its operational speed.

acceleration voltage In an electron tube, the voltage between the cathode and the accelerating electrode.

accelerator **1.** A device which causes the acceleration of something. **2.** Same as **accelerating electrode**. **3.** In computers, a hardware device which accelerates and/or enhances the performance of a system or subsystem. **4.** A device, such as a linac or cyclotron, in which charged particles are greatly accelerated to achieve high energies. Also called **particle accelerator**, or **high-energy particle accelerator**.

accelerator board An expansion board, such as a graphics accelerator, which makes a computer faster. Also known as **accelerator card**.

accelerator card Same as **accelerator board**.

accelerometer An electromechanical transducer which measures acceleration, and whose output is a voltage that is proportional to the acceleration it is subjected to. Such a device may incorporate, for instance, piezoelectric elements which generate a voltage that varies linearly with the mechanical strain experienced.

accentuation Also called **emphasis**. **1.** Highlighting of a specific frequency or band of frequencies through amplification. **2.** The process of selectively amplifying higher frequencies of an audio-frequency signal before transmission, or during recording. The utilization of accentuation followed by deaccentuation may help improve the overall signal-to-noise ratio and reduce distortion, among other benefits. Also called **preemphasis**.

accentuator A circuit which serves to accentuate a specific frequency or band of frequencies. For example, automatic bass compensation boosts low frequencies in an audio high-fidelity system at low-volume settings. Also called **accentuator circuit**.

accentuator circuit Same as **accentuator**.

acceptable angle Same as **acceptance angle (2)**.

acceptable quality level The minimum percentage of output that meets the pre-established criteria for quality. Conversely, the highest permissible amount of defects proportional to the overall production of a product. Its abbreviation is **AQL**.

acceptable reliability level The minimum level of reliability that is acceptable for a system or component. May be expressed, for example, as a given number of failures per thousand hours of normal operation. Its abbreviation is **ARL**.

acceptable use policy Norms regulating the usage of computer services, especially those of networks such as the Internet. The providers of the service establish what activities are and are not acceptable when using said service. A common example is an Internet service provider prohibiting sending unsolicited email or using an account for certain commercial purposes. Also called **Terms Of Service**.

acceptance angle 1. The maximum angle within which light is received by a light-sensitive device, such as a photodetector. **2.** In fiber optics, the maximum angle of light that will enter one end of the fiber. Also called **acceptable angle**.

acceptance sampling The inspection of a sample to determine whether to accept the lot from which it was taken.

acceptance test A formal evaluation of equipment or software in order to determine if it works according to what the purchaser expects.

acceptor 1. In general terms, anything that accepts something. **2.** A species, such as an atom, molecule, or ion, which accepts one or more electrons from an another species. The donating species is called **electron donor**. Also called **electron acceptor**. **3.** An electron-accepting impurity which is introduced into a crystalline semiconductor. The acceptance of electrons creates holes, which makes for a p-type semiconductor. Also called **acceptor impurity** or **acceptor material**.

acceptor atom 1. In semiconductors, an atom which accepts electrons, which in turn creates holes. For example, gallium or aluminum. **2.** An atom which accepts electrons.

acceptor circuit A series-resonant circuit which provides a lower impedance at the tuned frequency, while offering a higher impedance at the rest. Such a circuit may be used, for instance, if seeking to pass only a desired frequency.

acceptor impurity Same as **acceptor (3)**.

acceptor material Same as **acceptor (3)**.

access 1. Entry to something, such as a network or the interior of an electronic device. **2.** Connection to a network or system. For example, accessing the Internet. **3.** The storage and/or retrieval of data to or from a computer storage medium. For example, accessing a hard drive, or retrieving a file. **4.** A means of connecting to an electronic device without needing to open its interior. For example, using cables with phono plugs to connect a compact disc player and a high-fidelity amplifier using their phono jacks.

access arm A mechanical arm that positions a read/write head over the surface of a disk in a computer. Specific locations are accessed by moving the arm across, in conjunction with the rotation of the disk. Also called **actuator (2)**, or **actuator arm**.

access charge 1. The charge that is imposed for access to a network, system, or service. For instance, a fee for access to long-distance telephone service. **2.** The charge that is imposed for access to specific areas of a network, system, or service. For example, an Internet site that requires an additional charge in order to gain access to selected areas.

access code A code required to gain access to something, such as a computer terminal, a network, a secure area, or a particular form of telephone service.

access concentrator An **access server** that supports one or more T1 or E2 lines. Its abbreviation is **AC**. Also called **remote-access concentrator**.

access control An interactive mechanism which determines whether access is granted. Examples include getting permission to log onto a network, to access particular files, or to receive specific satellite programming.

access-control list A list which correlates which computer system or network resources are available to which users. Its abbreviation is **ACL**. Also known as **access list**.

access-control protocol The mechanisms utilized to authenticate a user who seeks to access a network, system, or service. Its abbreviation is **ACP**.

access denied A message received when a computer or network user attempts to access an area or information that is either unavailable or not available to that particular user.

access line 1. A line that accesses a network or system. **2.** Same as **access link**.

access link In telephone systems, a circuit which links a subscriber or a PBX with a switching center, and provides access to long-distance calling. Also called **access line (2)**.

access list Same as **access-control list**.

access management The various methods employed to help insure that only authorized personnel or programs have access to a network, system, or certain contained components.

access matrix An access list presented in tabular form.

access mechanism 1. The mechanism employed to position a read/write head over the surface of a disk in a computer. **2.** The mechanism by which a computer program stores and/or retrieves data from a storage device. Also called **access method**.

access method Same as **access mechanism (2)**.

access node Same as **access point**.

access number A telephone number utilized to access a given network or service. For example, a dial-up access number.

access path Same as **absolute path**.

access point A junction that interconnects networks. For example, it may refer to the connection point where Internet service providers are linked with each other. Also called **access node**, or **network access point**.

access privileges The level of operations that a user or program is allowed to perform on a computer system or network. Generally refers to the capability to access, use, or modify files, programs, or directories. Also called **access rights**.

access protocol Its abbreviation is **AP**. **1.** The mechanisms used to authenticate a user who seeks to access a network, system, or service. **2.** The protocols that govern transmissions between terminals in a network.

access provider An entity which provides access to the Internet. Customers usually pay a monthly fee for this service, although it usually available for free unless a high-speed connection is desired. Users get a software package to access and browse the World Wide Web, one or more email accounts, and Web pages to have a presence on the Internet. Other offerings may include Website building and hosting services. Users can connect via a dial-up service or an asymmetrical digital subscriber line, among others. Also called **Internet service provider**, **service provider**, or **online service provider (1)**.

access rights Same as **access privileges**.

access server A computer that utilizes network-emulation software to connect asynchronous devices to a LAN or WAN. It manages communications, and takes care of protocol conversions. Also called **network-access server**, **remote-access server**, or **communications server**.

access speed Same as **access time**.

access time Also called **access speed**. Both instances generally apply to the exchange of information with the CPU, or to computer disks, such as hard disks or DVDs. **1.** The time that elapses between a request for data and its delivery. Also called **memory access time**. **2.** The time required to store information once the instruction is given.

access type In computers with restricted access, the type of operations a user may perform on a file or program. This generally refers to the ability to read, write, or execute.

accessibility 1. The degree to which computer hardware and software caters to the particular needs of users with reduced sensory function or mobility. 2. The ease with which a component or device can be reached.

accessibility aids Tools and features, such as head-mounted devices, sticky keys, mouse trails, and text-to-speech conversion, which enhance **accessibility** (1).

accessory 1. A device which is not essential for the proper operation of another, but that adds functionality to it. 2. A device that adds functionality to a computer. Usually synonymous with **peripheral** (1).

accessory board Same as **adapter card**.

accessory card Same as **adapter card**.

aCcm Abbreviation of **abcoulomb centimeter**.

accompanying audio Also called **accompanying sound**. 1. The sound channels that accompany the image seen on a TV. 2. Specialized audio channels that are available from certain sources such as a DVD. An example of this specialized audio is surround sound.

accompanying sound Same as **accompanying audio**.

accumulation In semiconductors, the buildup of charge carriers.

accumulator 1. In a computer, a register that stores partial or complete results of arithmetic and/or logical operations. 2. Same as **accumulator battery**.

accumulator battery A battery composed of rechargeable cells. Examples include lithium polymer batteries, and gel-cell batteries. Also called **accumulator** (2), **rechargeable battery**, **storage battery**, or **secondary battery**.

accuracy The extent to which a measured or calculated value approximates the real value. May be expressed as a percentage, as in accurate within 2%.

accuracy rating The maximum error of an instrument when operated under specified conditions. Usually expressed as a percentage of the full-scale value.

ACD Abbreviation of **automatic call distributor**, or **automated call distributor**.

acetate film A non-deforming film which is durable, transparent, hygienic, air-proof, dust-proof, and oil-proof. It is made from cellulose acetate, and is used in the preparation of magnetic tapes and photographic films, among others. Also called **cellulose acetate film**.

acetate tape Recording tape that has a magnetic oxide layer over an **acetate film**. Also called **cellulose acetate tape**.

Achilles' heel In computer and communications security, an area of vulnerability which represents the weakest link within the protective safeguards taken.

achromatic 1. Without color. 2. Transmitting white light without breaking it up into its constituent colored rays.

achromatic color A color that possesses brightness, but not hue. These include white, black, and shades of gray.

achromatic locus Within a chromaticity diagram, a region that includes all points necessary for use as a reference standard of illumination. Also called **achromatic region**.

achromatic region Same as **achromatic locus**.

acicular In the shape of a needle. In magnetic storage, the shape of the magnetizable particles on the surface of the tape or disk are usually acicular.

acid A chemical which has any of the following characteristics: increases the proportion of hydrogen ions in a solution, gives up a proton in solution, has at least one hydrogen that can be replaced by metals or basic radicals, that can react with a base to form a salt, or that accepts two electrons. According to the pH scale, an acid has a reading of less than 7.0, while a **base** (6) has a value above 7.0. The lower the reading, the stronger the acid.

acidity The degree to which a solution or chemical is an **acid**.

ACK Abbreviation of **acknowledge**, or **acknowledgement**.

acknowledge Same as **acknowledgement**. Its abbreviation is **ACK**.

acknowledge character Same as **acknowledgement**.

acknowledgement In communications, a message indicating readiness to receive a transmission, or that a transmission has been received successfully. This message is sent from the receiving unit or station to the sending unit or station as a means of verification. Its abbreviation is **ACK**. Also called **acknowledgement code**, **acknowledge**, or **acknowledge character**.

acknowledgement code Same as **acknowledgement**.

ACL Abbreviation of **access-control list**.

aclinic line A great line along the surface of the earth that connects all points at which the magnetic dip is 0°. This contrasts with the **geomagnetic equator**, which is a great circle which at each point is 90° from the geomagnetic poles. Also called **magnetic equator**, or **dip equator**.

ACN Abbreviation of **Automatic Crash Notification**.

acoustic Pertaining to the detection, production, transmission, reception, control, processing, and study of sound, in addition to all phenomena arising from the effects of sound. This contrasts with **acoustical**, which is all that is related to sound, but not having its physical properties or characteristics. Thus, terms with acoustical should pertain to units, measurement, materials, and devices. Nonetheless, even in technical usage acoustic and acoustical are mostly used synonymously.

acoustic absorber A material, surface, or medium which absorbs a proportion of the acoustic (sound) energy that strikes it. Also called **sound absorber**.

acoustic absorption Sound energy which is retained or dissipated by a medium which has had acoustic (sound) waves pass though or strike it. Also called **sound absorption**.

acoustic absorption coefficient The proportion of the incident acoustic (sound) energy which is absorbed by a surface or medium. Also called **acoustic absorption factor**, **acoustic absorptivity**, or **sound absorption coefficient**.

acoustic absorption factor Same as **acoustic absorption coefficient**.

acoustic absorption loss The loss of acoustic (sound) energy due to its retention or dissipation by a medium or surface. This loss in incident acoustic energy may be due, for example, to conversion into heat. Also called **sound absorption loss**.

acoustic absorptivity Same as **acoustic absorption coefficient**.

acoustic amplifier Same as **audio amplifier**.

acoustic capacitance Same as **acoustic compliance**.

acoustic chamber A room or other enclosure, such as an anechoic chamber, utilized to perform sound tests, measurements, and experiments. Also called **sound chamber**.

acoustic compliance The measure of the volume displacement of an acoustic medium when subjected to acoustic (sound) waves. It is the reciprocal of **acoustic stiffness**. Also known as **acoustic capacitance**.

acoustic coupler An apparatus utilized to connect a modem to a telephone line via an ordinary telephone handset. The handset is cradled in the device in a manner that a speaker and microphone are used to exchange the electrical signals between the computer and the remote site. Rarely used nowadays, as most modems connect directly to the phone line, thus avoiding many of the drawbacks, such as signal degradation.

acoustic coupling Communication using an **acoustic coupler**.

acoustic damping The reduction of the magnitude of vibration or resonance of a surface or material, in order to reduce or eliminate acoustic (sound) waves. Also called **sound-proofing**, or **damping (3)**.

acoustic delay A delay in the transmission of acoustic (sound) signals. Also called **sound delay**.

acoustic delay line A circuit or device that delays the transmission of acoustic (sound) signals. This is accomplished by having the signals pass through an appropriate medium. Also called **sonic delay line**.

acoustic dispersion The separation of a complex acoustic (sound) wave into its frequency components by passing it through an appropriate medium. Also called **sound dispersion**.

acoustic echo cancellation The suppression or elimination of unwanted echoes within a communications line, such as that utilized by a telephone. Such echoes may result, for instance, from impedance mismatches along such a line. Also called **echo cancellation**.

acoustic elasticity 1. The extent to which a medium compresses as sound passes through it. 2. The extent to which the air in a speaker enclosure compresses as the speaker vibrates.

acoustic energy The energy created by acoustic (sound) waves. It is the additional energy that particles in a medium have because of the presence of sound. Also called **sound energy**.

acoustic feedback A phenomenon that occurs when the acoustic (sound) waves produced by speakers interact with an input transducer such as a microphone or phonographic cartridge. If the feedback exceeds a certain amount, any of various undesired effects may occur, such as howling, whistling, motorboating, or excessive cone movement. Also, the sound heard as a consequence of this. Also known as **acoustic regeneration, acoustic howl, acoustic resonance (2), audio-frequency feedback, audio feedback, sound feedback**, or **howl (1)**.

acoustic field Also called **sound field**. 1. A region containing acoustic (sound) waves. 2. A specially-tailored **acoustic field (1)**, such as that produced by a home theater system.

acoustic filter A device which blocks or absorbs sounds of certain frequencies while leaving others unaffected. Also called **filter (3)**, or **sound filter**.

acoustic frequency Same as **audio frequency**.

acoustic frequency response In an acoustic device, the acoustic (sound) intensity produced as a function of the frequency.

acoustic generator A transducer which converts another form of energy, such as electrical or mechanical, into sound energy. For example, a tuning fork converting mechanical energy into acoustic (sound) energy. Also called **sound generator**.

acoustic grating A series of equally sized rods or slits that are situated a fixed distance apart, which serves as an obstacle to acoustic (sound) waves. This causes the waves to be diffracted according to their wavelengths. Also called **sound grating**.

acoustic heat engine An engine which converts heat into acoustic (sound) energy, and then converts the sound energy into electrical energy. There are no moving parts, and its efficiency is roughly similar to that of an internal combustion engine.

acoustic hologram A three-dimensional image which is recorded via **acoustic holography**.

acoustic holography A holographic method utilizing acoustic waves. In it, a coherent beam of acoustic energy is directed towards the object whose image is being taken, and the resulting interference pattern is recorded to form an acoustic hologram.

acoustic homing The use of acoustic (sound) energy to track an object.

acoustic homing system A system that utilizes acoustic (sound) energy to track an object.

acoustic horn A tube with a cross section that increases as it approaches the exit opening, which serves to direct and intensify any acoustic (sound) waves emanating from it. Also called **horn (2)**.

acoustic howl Same as **acoustic feedback**.

acoustic imaging The use of acoustic (sound) energy to create images of the internal structure of non-transparent objects. A common example is the use in medicine of ultrasonic waves to provide images of internal organs. Also known as **sound imaging**, or **ultrasonic imaging**.

acoustic impedance The combined opposition a medium presents to acoustic (sound) waves, expressed in acoustic ohms. It is the ratio of the sound pressure on a given surface to the sound flux through said surface. Acoustic impedance is comprised of **acoustic resistance**, the real number component, and **acoustic reactance**, the imaginary number component.

acoustic inductance The acoustic equivalent of electrical inductance. Also called **acoustic mass, acoustic inertance, acoustic inertia**, or **inertance**.

acoustic inertance Same as **acoustic inductance**.

acoustic inertia Same as **acoustic inductance**.

acoustic intensity The acoustic (sound) power transmitted per unit area, expressed in watts per square meter. Also called **sound intensity**.

acoustic interferometer An instrument which measures the velocity and absorption of acoustic (sound) waves as they travel through a liquid or a gas. This is accomplished using interference patterns.

acoustic labyrinth A speaker enclosure that has absorbent panels that form a labyrinth. This helps prevent cabinet resonance and enhances bass response. Also called **labyrinth enclosure**.

acoustic lens A system of disks or other obstacles that refract acoustic (sound) waves similarly to the way an optical lens refracts light. Also called **lens (3)**, or **sound lens**.

acoustic line A path along which low-frequency acoustic (sound) waves are transmitted using baffles, acoustic labyrinths, and resonators. It is the acoustic counterpart of an electrical transmission line.

acoustic loading The alteration of the acoustic characteristics of a speaker through the utilization of materials, baffles, or vents. Also called **loading (3)**.

acoustic mass Same as **acoustic inductance**.

acoustic memory Same as **acoustic storage**.

acoustic microscope A microscope which utilizes acoustic (sound) waves at microwave frequencies to examine the elastic properties of a specimen. Its spatial resolution is comparable to that of a light microscope.

acoustic microwave device A device that utilizes acoustic (sound) waves at microwave frequencies to process electrical signals.

acoustic navigation Navigation utilizing acoustic (sound) waves to detect and locate objects. Also called **sonic navigation**, or **sonar navigation**.

acoustic noise 1. Unwanted sounds which interfere with others that wish to be detected. 2. Same as **audio noise**.

acoustic ohm Same as **acoustical ohm**.

acoustic particle detection A technique for the detection of charged particles in which a specimen is irradiated with an acoustic (sound) signal and the reflections are recorded.

acoustic phenomena Phenomena pertaining to, and arising from, the generation, production, propagation, transmission, reception, detection, perception, control, and processing of sound. Also called **sound phenomena**.

acoustic power The total acoustic (sound) energy radiated by a source. Usually expressed in watts, or ergs per second. Also called **sound power**.

acoustic pressure The value obtained when squaring multiple instantaneous sound-pressure level measurements at a given point, averaging these over the time of a complete cycle, and taking the square root of this average. Usually expressed in pascals, though also expressed in other units, such as dyne/cm^2, or microbars. Also called **RMS sound pressure, effective sound pressure**, or **sound pressure (3)**.

acoustic proximity sensor A sensor which uses acoustic (sound) waves to sense the distance to one or more objects. Used especially in robotics.

acoustic radiation Acoustic (sound) waves that travel through a medium. Also called **sound radiation**.

acoustic radiator A device or surface which vibrates to produce acoustic (sound) waves. Examples include headphone diaphragms, speaker cones, and distributed-mode speaker panels. Also called **radiator (3)**, or **sound radiator**.

acoustic radiometer An instrument that determines the intensity of sound by measuring the pressure created by the reflection or absorption of acoustic (sound) waves.

acoustic reactance The imaginary number component of **acoustic impedance**. Expressed in acoustical ohms.

acoustic reflection coefficient Same as **acoustic reflectivity**.

acoustic reflection factor Same as **acoustic reflectivity**.

acoustic reflectivity The ratio of the acoustic (sound) energy reflected from a given surface, to that striking the surface. Also called **acoustic reflection coefficient, acoustic reflection factor, sound reflectivity**, or **sound reflection coefficient**.

acoustic reflex enclosure A speaker design where there is a hole, or port, built into the cabinet. This increases and extends the reproduction of low frequencies. Although there is a savings in required amplifier power, there is a reduction in accuracy. Also called **bass reflex enclosure**.

acoustic refraction The bending of acoustic (sound) waves as they travel through a medium which varies in temperature, pressure, or other physical characteristics. These differences make sound travel at different speeds, as is the case when sound passes from warmer water to cooler water. Also called **sound refraction**.

acoustic regeneration Same as **acoustic feedback**.

acoustic resistance The real number component of **acoustic impedance**. Expressed in acoustical ohms. Also called **resistance (3)**.

acoustic resonance 1. Intensification of acoustic (sound) waves at certain frequencies due to resonance. Also called **resonance (5)**. 2. Same as **acoustic feedback**.

acoustic resonator 1. An enclosure or device which produces acoustic (sound) wave resonance at a particular frequency. 2. An enclosure or device which serves to reinforce sound.

acoustic scattering The irregular and unpredictable dispersion of acoustic (sound) waves, due to diffraction, reflection, or refraction. Also called **sound scattering**.

acoustic sensor A sensor whose mode of detection utilizes acoustic (sound) waves. Also called **sound detector (2)**.

acoustic shielding A barrier that prevents the penetration of acoustic (sound) waves. Also called **sound shielding**.

acoustic sounding Sounding in which sonic or ultrasonic waves are utilized to detect surfaces and depths.

acoustic spectrometer An instrument utilized to measure the relative intensities of acoustic (sound) frequencies within a complex sound wave. Also known as **audio spectrometer**.

acoustic spectrum 1. The complete range of acoustic (sound) frequencies. Also known as **sound spectrum (2)**. 2. An interval of acoustic (sound) frequencies. Also known as

sound spectrum (3). 3. A graph depicting an interval of acoustic (sound) frequencies. Also known as **sound spectrum (4)**.

acoustic speed The speed at which acoustic (sound) waves travel through a specific medium. This is influenced by conditions such as the temperature, pressure, and density of the medium of propagation. In dry air, for example, at 0° Celsius, and at one atmosphere pressure, sound travels at approximately 331.6 meters per second, while in copper sound travels at approximately 3,360 meters per second. Also known as **sonic speed**, or **speed of sound**.

acoustic stiffness The reciprocal of **acoustic compliance**.

acoustic storage A form of volatile computer memory which employs an **acoustic delay line**. Also called **acoustic memory**.

acoustic suspension enclosure A speaker design where the enclosure is sealed, increasing the pressure in the cabinet as the cone moves. This provides for accurate low-frequency reproduction, but requires more amplifier power.

acoustic transducer Also called **sound transducer**. 1. A device which transforms acoustic (sound) energy into another form of energy. For example, a microphone, which converts acoustic energy into electrical energy. 2. A device which transforms another form of energy into acoustic (sound) energy. For example, a speaker, which converts electrical energy into acoustic energy.

acoustic transmission The movement of acoustic (sound) energy through a medium. For example, sound waves traveling through water. Also called **sound transmission**.

acoustic transmission coefficient Same as **acoustic transmissivity**.

acoustic transmissivity The ratio of the acoustic (sound) energy transmitted through a material, to the sound energy incident on said material. Also called **acoustic transmission coefficient, acoustic transmittivity**, or **sound transmission coefficient**.

acoustic transmittivity Same as **acoustic transmissivity**.

acoustic treatment The use of absorbers, diffusers, and reflectors to help give a room the desired acoustic (sound) characteristics. Used, for instance, to set up a recording studio. Also called **sound treatment**.

acoustic velocity The **acoustic speed** in a given direction. Also known as **sound velocity**, or **velocity of sound**.

acoustic wave A traveling wave which propagates sound through an elastic medium. The wave is produced by vibrations that are a result of acoustic (sound) energy. Also known as **sound wave**, or **elastic wave (2)**.

acoustic-wave amplifier A device which increases the strength of acoustic (sound) waves.

acoustic-wave filter A device which separates acoustic (sound) waves of different frequencies.

acoustical All that pertains to and is correlated with sound, but not having its physical properties or characteristics. Terms with acoustical should relate to units, measurement, materials, and devices. This contrasts with **acoustic**, which pertains to the detection, production, transmission, reception, control, processing, and study of sound, in addition to all phenomena arising from the effects of sound. Nonetheless, even in technical usage acoustic and acoustical are mostly used synonymously.

acoustical attenuation constant The real number component of the **acoustical propagation constant**. Expressed in nepers per unit length.

acoustical ohm A unit utilized to measure and express acoustic impedance, acoustic resistance, or acoustic reactance. One acoustic ohm is equal to a sound pressure of one microbar that produces a volume velocity of one cubic centimeter per second in a given medium. Also known as **acoustic ohm**, although acoustical ohm is the more proper term.

acoustical phase constant The imaginary number component of the **acoustical propagation constant**. Expressed in radians per unit length.

acoustical propagation constant A quantity that measures the propagation properties of sound particles in a given medium. The acoustical propagation constant is comprised of the **acoustical attenuation constant**, the real number component, and the **acoustical phase constant**, the imaginary number component.

acoustics 1. The science that deals with all aspects of sound. This includes the production, transmission, detection, reception, control, and processing of sound, in addition to all phenomena arising from its effects. 2. The characteristics of a room, location, or enclosure which determine how sound waves travel within it. This affects many aspects of the listening experience.

acousto-optic Pertaining to **acousto-optics**. Also called **optoacoustic**.

acousto-optical cell A device which is utilized to modulate, deflect, and focus light waves, through their interactions with acoustic waves. Used, for example, in laser equipment for control of the intensity and position of laser beam radiation. Also called **acousto-optical device, acousto-optical modulator, Bragg cell**, or **optoacoustic cell**.

acousto-optical device Same as **acousto-optical cell**.

acousto-optical modulator Same as **acousto-optical cell**.

acousto-optics The science that deals with the interactions between acoustic waves and light waves in solid mediums. Acoustic waves can be made to modulate, deflect, and focus light waves. Also called **optoacoustics**.

acoustoelectric effect The development of a DC voltage in a crystalline or metallic material caused by acoustic waves traveling parallel to its surface. Also known as **electroacoustic effect**.

acoustoelectronics The branch of electronics that deals with acoustic waves at microwave frequencies that travel along the surface or through crystals or metallic objects. Also known as **pretersonics**.

ACP Abbreviation of **access-control protocol**.

ACPI Abbreviation of **Advanced Configuration and Power Interface**.

acquisition 1. The process of directing an antenna or telescope for optimal reception of the desired telemetered data. 2. The process of locating an orbiting object, such as a satellite, to obtain the desired telemetered data.

Acrobat A popular program that enables documents to be created, manipulated, displayed, and printed identically across multiple platforms.

Acrobat Reader A program utilized to view and print PDF files. PDF is the format utilized by **Acrobat** products.

acrylic resins Depending on the method of polymerization, acrylic resins can range from a hard, shatter-proof solid, to a viscous liquid. Many forms of these resins have varied applications in electronics, including use as a dielectric, in the making of lenses and electronic components, and as an adsorbent in chromatography.

acrylonitrile-butadiene-styrene resin A tough and impact-resistant resin offering chemical resistance and anti-static properties. Its uses include housings for electronics components. Its abbreviation is **ABS resin**.

ACS Abbreviation of **automatic call sequencer**.

actinide series Same as **actinides**.

actinides A series of elements, generally considered to range from atomic number 89 (actinium) or 90 (thorium), through 103 (lawrencium), inclusive. They have related chemical properties, and are all radioactive. Also called **actinide series**.

actinism The causing of chemical reactions in a substance upon exposure to electromagnetic radiation, especially in the visible and ultraviolet regions.

actinium A silvery-white metallic chemical element with atomic number 89. It has over 20 known isotopes, all of which are radioactive. It is useful in the production of neutrons and is a powerful source of alpha rays. Its chemical symbol is **Ac**.

actinodielectric Pertaining to a substance manifesting an increase in electrical conductivity during exposure to electromagnetic radiation.

actinoelectricity The electricity produced in a substance during exposure to electromagnetic radiation.

action 1. One or more steps within a sequence or process. 2. The manner in which a device or mechanism operates.

activate 1. To start a sequence or process. For example, to enter a computer command to initiate a sequence. 2. To make operational. For example, the adding of liquid to a battery or cell to make it operational. 3. To treat a substance or material to make it more effective for a designated purpose. For example, the treatment of a cathode to improve its thermionic emission. 4. To induce an action. For example, inducing radioactivity in a non-radioactive element by bombarding it with neutrons. 5. To accelerate an action. For example, the use of a catalyst to bond a resin more quickly.

activated cathode In electron tubes, a cathode that has been specially treated to enhance its thermionic emission.

activation 1. The initiation of a sequence or process. For example, the entering of a computer command to initiate a sequence. 2. The process of making operational. For example, the adding of liquid to a battery or cell to make it operational. 3. The treatment of a substance or material to make it more effective for a designated purpose. For example, the treatment of a cathode to improve its thermionic emission. 4. The inducing into an action. For example, inducing radioactivity in a non-radioactive element by bombarding it with neutrons. 5. The acceleration of an action. For example, the use of a catalyst to bond a resin more quickly.

activation energy The minimum required energy for atoms or molecules to initiate a process or reaction. The activation energies required for ionization or diffusion are of importance in various areas of electrical engineering.

activation time 1. The time necessary for **activation**. 2. The time necessary for a battery or cell to achieve operating voltage after the liquid has been added.

activator 1. That which activates. 2. A substance necessary in trace quantities to induce luminescence in certain crystals.

active 1. Ready for use, as in an available communications channel. 2. Currently functioning, as in a satellite transmitting a signal. 3. Energized, as in a circuit that is powered at the moment. 4. Requiring external power to operate, as in a transistor functioning. 5. A component or device which operates on an applied electrical signal, as in amplifying, rectifying, or switching.

active antenna 1. A relatively small receiving antenna that that has a built-in amplifier. 2. An antenna with one or more active components or devices. 3. An antenna which is connected to a transmission line.

active application In a computer that has more than one application running, the one which is currently being interacted with, and thus will be affected by actions such as keyboard input. It occupies the active window on the screen.

active area In a metallic rectifier, the area that serves as the rectifying junction, and which carries the current in the conducting direction.

active arm A component within a transducer that varies its electrical characteristics in response to the input of said transducer. Also called **active leg**.

active badge A small badge that transmits signals which pinpoint the location of the wearer or holder. Such a badge may also be affixed to devices and equipment, and the location signals are monitored by an appropriate system. Used, for instance, to help verify that only authorized personnel are in designated areas.

active badge system A system which keeps track of multiple **active badges**.

active circuit A circuit which incorporates one or active components or devices, such as transistors or operational amplifiers.

active communications network A communications network in which the nodes can perform custom operations, such as handling blocks of data of an individual user differently than the rest. Also called **active network (2)**.

active communications satellite An artificial satellite that can receive and retransmit signals between stations. Such a satellite usually provides signal amplification. Also known as **active comsat**, or **active satellite**.

active component **1.** An electronic component, such as a transistor, that can operate on an applied electrical signal, as in amplifying, rectifying, or switching. Also called **active element (1)**. **2.** In AC circuits, a quantity without reactance.

active computer **1.** In a setting with two or more computers, the one currently processing. **2.** In a setting with two or more computers, any of them which is currently processing.

active comsat Abbreviation of **active communications satellite**.

active content Within a Web page, content that changes after the page is fully loaded. This can be in response to a user's action such as clicking with the mouse on a drop-down menu, or can be based on a timing mechanism, such as an advertisement that flashes images on the page displayed.

active control system A control system that regulates inputs and outputs to maintain or improve operating conditions. For example, a mechanism within a nuclear reactor that controls the fission rate to produce a steadier source of energy.

active crossover A speaker crossover requiring power to operate.

active current Within an AC circuit, the component of the current that is in phase with the voltage. Also known as **watt current**, **in-phase current**, or **resistive current**.

active device An electronic device, such as an operational amplifier, that can operate on an applied electrical signal, as in amplifying, rectifying, or switching.

Active Directory A directory service utilized in distributed computing environments.

active display Same as **active-matrix display**.

active electric network An electric network with one or more active components or devices. Also called **active network (1)**.

active element **1.** Same as **active component (1)**. **2.** A radioactive chemical element. **3.** An antenna element which is driven directly by the transmission line. This contrasts with a **parasitic element**, which is driven indirectly, through radiation from an active element. Also called **driven element**.

active equalizer An equalizer requiring power to operate.

active file **1.** In a computer, a file that is currently being accessed. **2.** In a computer, a file that it currently available for accessing.

active filter A filter that incorporates active components or devices, such as transistors, to assist in signal passing or rejection.

active homing Homing in which guidance is provided by a signal or energy which is transmitted towards a scanned object which reflects it back to the source. This contrasts with **passive homing**, in which guidance is provided by a signal or energy transmitted or radiated by a beacon or scanned object itself.

active hub A central connecting device, such as a computer or router, that regenerates and retransmits the signals sent through it. This contrasts with **passive hubs**, which only pass through signals without regenerating them.

active infrared detection The utilization of emitted infrared detection to locate and track objects which reflect this energy back to the source. Its abbreviation is **active IR detection**.

active infrared detector A device utilized for **active infrared detection**. Its abbreviation is **active IR detector**. Also called **active infrared sensor**.

active infrared sensor Same as **active infrared detector**. Its abbreviation is **active IR sensor**.

active IR detection Abbreviation of **active infrared detection**.

active IR detector Abbreviation of **active infrared detector**.

active IR sensor Abbreviation of **active infrared sensor**.

active leg Same as **active arm**.

active lines In a TV, the lines that transmit the visible picture. The blanking, or inactive, lines are not included in this number. In the United States, for example, a standard analog TV has 525 total lines, of which approximately 485 are active. This compares with over 1,000 active lines in a digital or HDTV.

active logic Logic circuits employing **active components (1)**.

active material **1.** In a storage cell or battery, the substance which participates in the electrochemical reactions of charge or discharge. Lead oxide is an example. **2.** In a CRT, the fluorescent coating on the screen. An example is zinc phosphate. **3.** In an electron tube, the material coating the cathode. Thorium oxide is an example. **4.** A radioactive material, an example of which is radium.

active matrix A technology used in **active-matrix displays**.

active-matrix display A liquid-crystal display in which each pixel of the screen is powered separately and continuously by a transistor. This provides for a brighter image with higher contrast than a **passive-matrix display**. Also known as **active-matrix screen**, **active display**, **active-matrix liquid-crystal display**, **active-matrix LCD**, **active-matrix thin-film transistor display**, or **thin-film transistor liquid-crystal display**.

active-matrix LCD Same as **active-matrix display**.

active-matrix LCD display Same as **active-matrix display**.

active-matrix liquid-crystal display Same as **active-matrix display**. Its abbreviation is **AMLCD**.

active-matrix screen Same as **active-matrix display**.

active-matrix thin-film transistor display Same as **active-matrix display**.

active memory Computer memory which is accessed by content rather than by address. It is often used in memory-management units, and comparison logic is included with each bit of storage. Also known as **active storage**, **addressable memory**, **associative storage**, **associative memory**, **content-addressable memory**, or **content-addressable storage**.

active mixer A mixer, such as that combining input signals or light beams, which utilizes one or more active components or devices, such as transistors.

active modulator A modulator which utilizes one or more active components or devices, such as transistors.

active monitor In a token-passing network, the station with control of the token at a given moment. All other stations are **standby monitors**.

active network **1.** Same as **active electric network**. **2.** Same as **active communications network**.

active power In an AC circuit, the rate at which work is performed or energy is transferred, measured in watts. It is the product of the voltage portion of the signal being measured, multiplied by the current that is in phase with that voltage. It is also equal to the difference between the apparent power and the reactive power. Also called **actual power (1)**, **real power**, or **true power**.

active program **1.** In a computer, a program that is currently running. **2.** In a computer, a program that is currently controlling the microprocessor.

active radar A radar technique in which microwaves are transmitted to enhance the detection of scanned objects. Most radars are active, and the term is used to differentiate from **passive radar**, in which detection is based on the microwave energy which is ordinarily radiated and reflected from any given object.

active region In a semiconductor, the area where dynamic functions such as amplification occur.

active satellite Same as **active communications satellite**.

Active Server Page A technology utilizing scripts to create dynamic and interactive Web pages. Also, a page so created. Its abbreviation is **ASP**.

active sonar Sonar equipment that transmits and receives sonic or ultrasonic sounds, for detection of the bearing and range of objects. This contrasts with **passive sonar**, which can only receive sonic or ultrasonic sounds, and is thus able to detect bearing only.

active star In a star network, a central connecting device, such as a computer or router, that regenerates and retransmits the signals sent through it.

active storage Same as **active memory**.

active substrate **1.** In microcircuitry, the material within which active elements, such as transistors, are formed. **2.** The support material on which other microcircuit elements physically rest.

active system A radio or radar system capable of transmitting and receiving. This contrasts with a **passive system**, which only receives.

active tracking system **1.** A system, such as azusa or minitrack, which transmits or re-transmits signals for tracking objects. **2.** A system which exchanges signals with any satellite it is tracking.

active transducer **1.** A transducer that utilizes one or more active components or devices, such as transistors. **2.** A transducer with an internal power source.

active transponder A transponder with an internal source which provides for part or all of its power requirements. A **passive transponder** relies solely on the electromagnetic field of the reader and/or interrogator.

active voltage Within an AC circuit, the component of the voltage that is in phase with the current. Also known as **resistive voltage**, or **in-phase voltage**.

active window On a computer screen that has more than one window open, the one that is currently being interacted with, and thus will be affected by such actions as keyboard input. This will be the only window with the title bar highlighted. All others are **inactive windows**.

ActiveX A model or set of technologies utilized to create content which interacts between applications. Used, for instance, to create interactive multimedia Web sites.

activity **1.** Movement or agitation, such as molecular vibrations in a crystal. **2.** The level of use, as in the traffic within a network. **3.** An event which involves accessing or modifying files, such as a save command. **4.** The intensity of a radioactive source, such as the rate of alpha particle emission of the element curium.

activity level **1.** Level of use, as in the frequency of Web page visits in a given period. **2.** The occurrence of events in which a file is accessed or modified, as in updates.

activity ratio **1.** The ratio of files in use to the total files available in a database. **2.** The ratio of the files modified to the total files available in a database.

actual address **1.** Same as **absolute address**. **2.** The address of a computer on the Internet, as expressed in terms of its IP address.

actual frequency The frequency observed under operating conditions, as opposed to the nominal frequency.

actual power **1.** Same as **active power**. **2.** Same as **average power**.

actual time **1.** The time is takes in reality for an event or series of events to take place, as opposed to the theoretical or estimated time. For example, the time it takes for a wave to propagate through an unknown medium. **2.** Time which is recorded or billed. For example, airtime on a cellular phone.

actuate To initiate or activate an event or process. For example, the sending of an electrical signal that opens or closes an electric circuit, or that actuates an alarm warning of an equipment failure.

actuating current A current that initiates or activates an event or process. For instance, that which actuates a relay or switch.

actuating device A device that initiates or activates an event or process.

actuating system A system that initiates or activates an event or process.

actuating voltage A voltage that initiates or activates an event or process. For instance, that which actuates a relay or switch.

actuation The initiating or activating of an event or process.

actuator **1.** A device or circuit that initiates or activates an event or process. For example, a switch that opens or closes an electric circuit. **2.** Same as **access arm**. **3.** In robotics, a device which moves a component, such as a manipulator, in response to signals from the controller. Such a device may be electric, hydraulic, mechanical, or pneumatic. Also called **robotic actuator**, or **effector**.

actuator arm Same as **access arm**.

ACU Abbreviation of **automatic calling unit**.

acyclic **1.** Having no cycles. **2.** Varying over time with no fixed pattern.

acyclic machine A DC generator in which the poles are of the same polarity with respect to the armature, thus making a commutator unnecessary. Also called **homopolar generator**, **homopolar machine**, or **unipolar machine**.

ad blocker A Web filter which seeks to detect advertising, and prevent its being loaded onto a Web page being accessed. Also called **ad filter**.

ad filter Same as **ad blocker**.

ad hoc inquiry Same as **ad hoc query**.

ad hoc network A temporary network established to address needs that arise unexpectedly.

ad hoc query In computers, a request for information generated by a need that arises unexpectedly. This contrasts with periodic or probable requests for information. Also called **ad hoc inquiry**.

ad server Same as **adserver**.

ad-supported software Same as **adware**.

adaptability 1. The ability of a device or system to self-adjust, to better meet the varying conditions it faces. 2. The ability for a device or system to be adjusted to make it more suitable for a specific use or situation.

adaptable 1. Able to be made suitable for a specific use, situation, device, or piece of equipment. 2. Able to self-adjust to better meet the varying conditions it faces.

adapter Also spelled **adaptor**. 1. A device which enables connections between one thing and another, such as an AC adapter, or an Ethernet adapter. 2. A fitting utilized to connect jacks, plugs, and the like, to that of another format. 3. Same as **adapter card**. 4. A device that makes electrical or mechanical connections between items not meant to be connected. 5. A device which alters something so that it becomes suitable for a use for which it was not intended.

adapter board Same as **adapter card**.

adapter card A circuit board which is plugged into the bus or an expansion slot of a computer, in order to add functions or resources. Common examples include graphics accelerators, sound cards, and Ethernet adapters. Also known by various other terms, including **adapter (3)**, **adapter board**, **add-on board**, **accessory card**, **add-on**, **expansion board**, **expansion card**, and **card (1)**.

adaptive antenna array Same as **adaptive array**.

adaptive array An antenna array that continuously monitors the received signals, so as to optimize its performance. Also called **adaptive antenna array**.

adaptive bridge In communications networks, a bridge that can learn the locations of users or systems and deliver messages accordingly. This contrasts, for instance, with a **source route bridge**, which is aware of the locations of users or systems prior to message delivery. Also known as **transparent bridge**, or **learning bridge**.

adaptive communications Communications utilizing a system that monitors internal and external parameters automatically and continuously, in order to makes adjustments to better operate in its varying environment. Also known as **self-adjusting communications**, and **self-optimizing communications**.

adaptive computer access Hardware and/or software which is specifically tailored to the needs of persons who in one or more manners are unable to utilize the most commonly available input and/or output devices of a computer. These include alternative keyboards, screen enlargement, and speech synthesis. Also, the process of modifying computers to better serve such individuals. Also, the use of computers so equipped.

adaptive control A method of control in which internal and external parameters are continuously and automatically monitored in order to make adjustments that improve system performance. An automatic frequency control system is an example.

adaptive control system A control system based on **adaptive control**.

adaptive equalization Equalization which continually and automatically compensates for factors such as attenuation and distortion of a signal, in order to optimize the characteristics of said signal. For example, a modem with this feature analyzes a phone line to find and then utilize the bandwidth with least noise.

adaptive equalizer An equalizer utilizing **adaptive equalization**.

adaptive filter A filter which continuously and automatically monitors various internal and external parameters to optimize its function.

adaptive interface An interface, such as that of a computer application, that automatically changes depending on the actions or special needs of a user. For example, a handwriting or voice recognition system may improve its ability to understand a given user over time.

adaptive maintenance Changes made to computer software to better adapt it to changing environments or needs.

adaptive menu A menu, such as that of a computer application, that automatically changes depending on the actions or special needs of a user. For example, a drop-down menu may place the most commonly accessed features closer to the top of the available choices.

adaptive optics system An optical system that compensates for sources of distortion, such as the effects of the atmosphere.

adaptive robot A robot that detects environmental conditions and reacts to them, without human intervention.

adaptive routing Network routing that optimizes its performance by continuously and automatically adjusting to network changes such as traffic patterns and circuit failures.

adaptive suspension vehicle A specialized robot that serves as an autonomous vehicle. Instead of using a rolling mechanism to travel, it walks on mechanical legs to carry its human passenger. This makes it well-suited for terrain with irregularities and obstacles. Its abbreviation is **ASV**.

adaptive system A system which makes changes to improve itself. It monitors internal and external parameters automatically and continuously, evaluates its operation, and makes adjustments to better function in its varying environment.

adaptor Same as **adapter**.

adaptor board Same as **adapter card**.

adaptor card Same as **adapter card**.

ADC Abbreviation of **analog-to-digital converter**.

adc Same as **ADC**.

Adcock antenna A directional antenna constructed of a pair of vertical antennas with a separation distance of one-half wavelength or less, and which are opposite in phase. It produces a radiation pattern in the form of a figure-eight.

Adcock direction finder A radio direction finder consisting of one or more pairs of Adcock antennas.

Adcock radio range A radio range consisting of four Adcock antennas, each at the corner of a square. It is a type of AN radio range.

add-drop multiplexer A multiplexer than can add or drop lower rate signals from a higher rate multiplexed signal without demultiplexing the signal. Its abbreviation is **ADM**.

add-in 1. Same as **adapter card**. 2. A chip that is added to an expansion board already installed in a computer.

add-in board Same as **adapter card**.

add-in card Same as **adapter card**.

add-on Same as **adapter card**.

add-on board Same as **adapter card**.

add-on card Same as **adapter card**.

add-on memory Additional memory which is added to that already present in a computer.

add-subtract time In a computer, the time required to perform an operation of addition or subtraction. This time does not include the intervals required to obtain the numbers from memory before the operation, nor that required to store the result.

add time In a computer, the time required to perform an operation of addition. This time does not include the intervals required to obtain the numbers from memory before the operation, nor that required to store the result.

addend A number or quantity that is added to another. It is the newly introduced number or quantity. This contrasts with the **augend**, which is the number or quantity already present to which an addend is added.

adder Also called **adder circuit**, or **summer circuit**. **1.** In computers, a logic circuit which adds two or more numbers or quantities. **2.** A circuit in which two or more input signals are combined to yield one output signal, which is proportional to the sum of the input signals.

adder circuit Same as **adder**.

additive color A color formed by an **additive mixture**.

additive color process A method of producing colors through an **additive mixture**. Also called **additive process (1)**, or **additive synthesis (1)**.

additive color system A system of adding primary colors in varying proportions to yield a full range of colors.

additive mixture A combination of primary colors to form other colors. For example, monitors and color TVs may use red, green, and blue to form a full range of colors.

additive primaries Same as **additive primary colors**.

additive primary colors Colors which are combined in an additive mixture to yield a full range of colors. Red, green, and blue are usually used as the additive primary colors. They are called this way because they themselves are not formed by the combination of other colors. Also called **additive primaries**, **color primaries**, **primary colors**, or **primaries**.

additive process **1.** Same as **additive color process**. **2.** In circuit board manufacturing, a process in which conducting material is added to specific areas of a substrate to create conductive patterns.

additive synthesis **1.** Same as **additive color process**. **2.** The process of generating complex sounds using pure tones or sine waves. Used, for example, in electronic organs.

address **1.** A number which indicates the specific place where data is stored within the memory of a computer, peripheral, or disk. Also called **location (2)**. **2.** The address of a computer within a network. **3.** The address of a computer on the Internet, as expressed in terms of its IP address. **4.** The address of a Web page, file, object, or address in general on the Internet, as expressed by its URL. **5.** A sequence of characters which uniquely identify an email account. The standard format is user name@domain name. In the case of **xxxx@zzz.com.qq**, *xxxx* is the user name, and *zzz.com.qq* is the domain name. Also called **email address**, or **Internet address (3)**.

address bus Within a computer, the channel that connects the CPU to the RAM. It is used to identify specific locations of stored data, or addresses. The number of lines in this channel determines how much memory can be addressed.

address field In computers, the part of an instruction that includes the address information.

address format In computers, the format in which addresses are presented. For example, a 32 bit IP address is grouped eight bits at a time, separated by periods. A possible number using these specifications is: 91.2.133.206.

address mask A bit combination which identifies which portion of an IP address corresponds to the network or subnetwork, and blocks out the rest. For example, a network may use the same values in the first three address fields of a four field address such as WWW.XXX.YYY.ZZZ, and block out, or mask, all but the ZZZ portion, since it is the only one that will vary. Also called **subnetwork mask**.

address mode In computers, the system specifying how an address will be interpreted by an instruction. Also called **addressing mode**.

address modification In computers, the modification of an address during the execution of an instruction.

address register Its abbreviation is **AR**. **1.** The register that contains the addresses of the locations in memory that are to be accessed by the next instruction. **2.** A register that stores an address. Also called **memory address register (2)**.

address resolution The identification of the hardware address of a specific computer within a network. For example, pinpointing a computer that is linked to the Internet by knowing its IP address, or by finding the corresponding match in an address mapping table.

address resolution protocol A protocol for determining the hardware, or physical, address of a specific computer, or node, within the Internet. This is accomplished by matching an IP address with a piece of hardware, such as a network interface card, which then reveals the computer it belongs to. Its abbreviation is **ARP**.

address resolution protocol broadcast The use of an address resolution protocol to broadcast a request for the hardware address associated with an IP address. Its abbreviation is **ARP broadcast**. Also known as **address resolution protocol request**.

address resolution protocol request Same as **address resolution protocol broadcast**. Its abbreviation is **ARP request**.

address space **1.** The amount of memory a CPU can access. It is determined by the size of the address bus. Also called **memory address space**. **2.** The actual memory used while running a computer program.

address translation The conversion of an address from one form to another. An example is the translation of a virtual address to a real address.

addressable cursor A cursor that may be directed to any part of a computer screen, be it by mouse or keyboard input.

addressable memory Same as **active memory**.

addressing mode Same as **address mode**.

ADF Abbreviation of **automatic direction finder**.

adiabatic **1.** A process or event in which there is no change in the heat content of a system. The system does not exchange heat with its surroundings. **2.** A change occurring without a gain or loss of heat.

adiabatic demagnetization A system for cooling to temperatures below that of liquid helium. A strong magnetic field is used to align the spins of the molecules of a block of a paramagnetic salt which has been cooled by a liquid helium bath. The block is then isolated from the magnetic field and the helium bath. Most of the remaining heat energy is absorbed in randomizing the spins of the molecules. Readings of less than 2 millikelvins have been achieved using such devices.

adiabatic demagnetization refrigerator A device used for cooling to temperatures below that of liquid helium, utilizing **adiabatic demagnetization**. Its abbreviation is **ADR**.

adiabatic process A process in which there is no change in the heat content of a system. The system does not exchange heat with its surroundings.

adiabatic system A system that is altered without a change in its heat content. It does not exchange heat with its surroundings.

adjacency The state of being next to, or nearby something else. It may refer to network nodes, disk sectors, frequency bands, optical fibers, circuits, fields, or print characters, among many others.

adjacent Anything displaying **adjacency**.

adjacent channel A channel immediately next to another channel in frequency.

adjacent-channel interference Interference of radio or TV reception by other stations using adjacent channels. Such interference may be unilateral or bilateral, and is caused, for instance, by overmodulation. It may be manifested as extraneous confusing, musical, or repetitive sounds being heard along with the desired program. Also known as **monkey chatter** or **sideband splatter**.

adjacent-channel selectivity The degree to which a receiver rejects the signals of channels adjacent to that which is desired. Also called **receiver selectivity (2)**.

adjacent frequency A frequency or frequency band immediately next to, or nearby, another.

adjustable That which has multiple settings or configurations. May refer to components such as resistors or capacitors, devices, instruments, frequencies, or motors, among many others.

adjustable capacitor A capacitor whose capacitance can be varied. This may be accomplished, for example, through the use of rotating plates. Also known as **adjustable condenser**, or **variable capacitor**.

adjustable component A component which has multiple settings or configurations, especially where circuits are concerned. For example, an adjustable resistor.

adjustable condenser Same as **adjustable capacitor**.

adjustable device A device which has multiple settings or configurations.

adjustable instrument 1. An instrument which has multiple settings or configurations. 2. An instrument which needs to be adjusted in order to perform the desired function.

adjustable resistor Also known as **variable resistor**. 1. A resistor whose resistance may be varied by a mechanical device such as a sliding contact. Also called **rheostat**. 2. A semiconductor device which serves as a variable resistor. It resistance value is determined by its input voltage, and an example is a metal-oxide varistor. Also known as **varistor**. 3. Any resistor whose resistance may be varied.

adjustable-speed motor A motor whose speed can be varied while under load. Also called **variable-speed motor**.

adjustable transformer A transformer whose output voltage can be varied between some minimum, or zero, and a maximum. This is generally accomplished by means of a sliding contact arm. In most cases, an adjustable transformer is an autotransformer. Also known as **variable transformer**.

adjustable voltage divider A voltage divider that utilizes adjustable resistors to vary the voltage. A potentiometer is an example. Also known as **variable voltage divider**.

adjusted decibels A measure of noise above a given reference level. The established reference noise power level is usually -85 decibels above 1 milliwatt, or -85 dBm. Its abbreviation is **dBa**. Also called **decibels adjusted**.

ADM Abbreviation of **add-drop multiplexer**.

admittance A measure of the ease with which AC flows through a circuit. Its real number component is **conductance**, and its imaginary number component is **susceptance**. It is expressed in siemens and its symbol is Y. It is the reciprocal of **impedance**.

admittance parameters Within an electric network, circuit parameters pertaining to **admittance**. Its abbreviation is Y-**parameters**.

ADN Abbreviation of **advanced digital network**.

ADP Abbreviation of **automatic data processing**.

ADR Abbreviation of **adiabatic demagnetization refrigerator**.

adserver A server dedicated to delivering ads to Web pages being accessed.

ADSL Digital communications technology and equipment with which connection speeds can reach over 10 Mbps downstream and over 2 Mbps upstream. This difference in transmission speeds accounts for its name. This technology, which uses ordinary copper lines, provides for the same phone line to be split into two, one for analog voice and the other for DSL data. This allows, for example, a user to be conversing on the telephone while using the same line to browse the Internet. It is an abbreviation of **asymmetric**

digital subscriber line, asymmetric digital subscriber loop, asymmetric digital subscriber link, asymmetrical digital subscriber line, asymmetrical digital subscriber link, or **asymmetrical digital subscriber loop**.

ADSL Light Same as **ADSL Lite**.

ADSL Lite A version of **ADSL** which is simpler to implement, but is limited to lower speeds, such as 1.5 Mbps downstream, or less. Also spelled **ADSL Light**.

ADSL modem A modem utilized for **ADSL** communications.

ADSL network A network which utilizes **ADSL** technology.

ADSL splitter A filter which separates the voice and data signals of a telephone line via which ADSL service is accessed. The voice frequencies are in the lower frequency range, and the data is in the higher interval.

adsorbent A substance that absorbs another substance.

adsorption The adherence of a substance to the surface of another.

ADT Abbreviation of **abstract data type**.

Advanced Configuration and Power Interface A specification pertaining to the efficient handling of power by a computer, especially a portable computer. It helps the BIOS, OS, and peripherals of a computer work together in minimizing power consumption, including the use of stand-by, sleep, and hibernation modes. Its abbreviation is **ACPI**.

advanced digital network A dedicated line that transmits digital communications signals, such as video and data, at speeds of 56 kilobits per second or above. Its abbreviation is **ADN**.

advanced encryption standard An encryption system that the National Institute of Standards and Technology is developing with the help of members of the cryptographic community to serve the data protection needs of the next decades. Its abbreviation is **AES**.

advanced intelligent network A switched voice and data network that allows telecommunications operators to create and customize the services offered to their customers, such as how calls should be routed. Its abbreviation is **AIN**.

advanced mobile phone service Same as **advanced mobile phone system**.

advanced mobile phone system The standard operating system for analog cellular telephony. It usually utilizes frequency-division multiple access technology, to provide a 30 kHz channel for each call. Also called **advanced mobile phone service**. Its acronym is **AMPS**.

advanced technology attachment Same as **ATA**.

advanced television standard A standard for use with digital TV systems. Its abbreviation is **ATV**, or **advanced TV standard**.

advanced TV standard Same as **advanced television standard**.

advergame Abbreviation of **adver**tising **game**. An electronic game, often Web-based, that is intended to advertise a given brand, product, service, or the like.

advertising game Same as **advergame**.

advertising-supported software Same as **adware**.

adware Software which is free, but which contains advertisements. Such ads, in addition to being ever present, may also serve to spy on the usage, such as Web browsing, of a user. It is an abbreviation of **ad**vertising-supported soft**ware**, or **ad**-supported soft**ware**.

aelotropic Same as **anisotropic**.

aelotropy Same as **anisotropy**.

aeolight A discharge lamp consisting of a cold cathode and a mixture of inert gases. The intensity of illumination varies along with the applied signal voltage.

aerial Same as **antenna**.

aerodiscone antenna A small discone antenna intended for use on aircraft for very-high frequencies and ultra-high frequencies.

aeronautical fixed service A fixed radio communications service with transmissions concerning the efficient and safe operation of aircraft. Its abbreviation is **AFS**.

aeronautical mobile service A mobile radio-communication service which includes aircraft-to-aircraft, aircraft-to-ground, and ground-to-aircraft transmissions pertaining to flight coordination and safety. Its abbreviation is **AMS**.

aeronautical radio service A radio communications service which includes aircraft-to-aircraft, aircraft-to-ground, and ground-to-aircraft transmissions pertaining to the safe operation of aircraft.

aeronautical station A station, usually located on land, within the **aeronautical mobile service**.

aeronautical telecommunications Telecommunications that include aircraft-to-aircraft, aircraft-to-ground, and ground-to-aircraft transmissions.

aerophare A non-directional radio transmitter station that emits signals from which bearing information can be derived by using a radio direction finder. Aerophares may operate continuously, or in response to interrogation signals. Some also provide range information. Known more commonly as **radio beacon**, or as **radiophare**.

aerosol A suspension of liquid or solid particles in a gas. Fine sprays, dust, and smoke are examples.

aerosol monitor A laser photometer used to monitor the concentration of airborne particles such as dust, smoke, mist, and fumes. Depending on the specific instrument, concentrations of less than 0.1 microgram per cubic meter can be detected.

aerospace Of or pertaining to the earth's atmosphere and the space beyond.

AES 1. Abbreviation of **Advanced Encryption Standard**. 2. Abbreviation of **Auger electron spectroscopy**. 3. Abbreviation of **atomic emission spectroscopy**.

aF 1. Abbreviation of **attofarad**. 2. Abbreviation of **abfarad**.

af or **AF** Abbreviation of **audio frequency**.

AFC Abbreviation of automatic frequency control.

AFM 1. Abbreviation of **atomic force microscopy**. 2. Abbreviation of **atomic force microscope**.

AFS Abbreviation of **aeronautical fixed service**.

AFSK Abbreviation of **audio-frequency shift keying**.

afterglow The luminescence that remains after the exciting source is removed. For instance, the phosphors on the screens of CRTs glow after the electron beam has passed. Also called **phosphorescence**.

afterpulse In multiplier phototubes, pulses which sometimes follow large pulses.

Ag Chemical symbol for **silver**.

AGC Abbreviation of **automatic gain control**.

agent 1. A force or substance that effects changes. 2. A computer program that helps automate repetitive tasks, or those that are scheduled. Used, for instance, for backup, or in hand-held personal schedulers. Also known as **intelligent agent** (1). 3. A computer program that serves to automate the locating and delivering of information over the Internet. Used in Web browsers. Also known as **intelligent agent** (2).

agile manufacturing Manufacturing which seeks to incorporate advances in communications, computers, and other technologies, with the goal of responding to market needs faster.

aging 1. The changes that occur over time in a substance, material, component, device, piece of equipment, or system. Also, changes in certain parameters, performance, or proper-

ties due to the passage of time. The level of use and/or the surrounding conditions influence significantly. 2. Operation of a component, circuit, device, piece of equipment, or system, under controlled conditions for a specified time, to stabilize its characteristics and to screen out failures. For example, the burning-in of an electronic device.

AGP Abbreviation of **accelerated graphics port**.

AGVS Abbreviation of **automated guided vehicle system**.

aH 1. Abbreviation of **attohenry**. 2. Abbreviation of **abhenry**.

Ah Abbreviation of **ampere-hour**.

Ahr Abbreviation of **ampere-hour**.

AI Abbreviation of **artificial intelligence**.

AIDC Abbreviation of **Automatic Identification and Data Collection**.

aided tracking A system in which the tracking equipment is assisted by an operator to better track objects.

AIN Abbreviation of **advanced intelligent network**.

air The mixture of gases that comprise the earth's atmosphere. Its composition varies depending on various factors, such as altitude, water vapor present, and contaminants at the site where the sample is taken. The main components of dry air at sea level, and their approximate proportion by volume, are: nitrogen (78%), oxygen (21%), and argon (0.9%). Other components include carbon dioxide, neon, helium, methane, and krypton.

air-actuated Powered or put into motion by compressed air.

air battery A battery composed of **air cells**.

air-break switch Same as **air switch**.

air bridge A microminiature bridge in which air serves as an insulator. Used in certain ICs. Also spelled **airbridge**.

air capacitor A capacitor in which air serves as the dielectric between its metallic plates. Also called **air condenser**.

air cell A cell in which the positive electrode is depolarized by atmospheric oxygen. An example is a zinc-air cell.

air cleaner A device that removes particles and impurities from air, such as dust, smoke, mist, or fumes. An example is a dust precipitator.

air condenser Same as **air capacitor**.

air-cooled Components, devices, or equipment that are cooled by the surrounding air, or that provided by a stream. Heat sinks are often utilized to enhance heat dissipation.

air-cooled component An electrical or electronic component which is cooled by a stream of air.

air-cooled tube An electron tube which is cooled by a stream of air.

air cooling The cooling of components, devices, or equipment by surrounding air, or that provided by a stream.

air-core coil A coil wound on a nonmagnetic core, such as plastic or cardboard, essentially leaving only air as the core material.

air-core transformer A transformer with a nonmagnetic core, such as plastic or cardboard, essentially leaving only air as the core material.

air-depolarized battery A battery in which the positive electrode is depolarized by atmospheric oxygen. An example is a zinc-air battery. Also known as **metal-air battery**.

air-dielectric coax A coaxial cable in which air or an inert gas occupies the space between the inner and outer conductors. The air or gas serves as the dielectric material. Also called **air-spaced coax**.

air gap 1. A narrow space of air separating components, circuits, or devices. 2. A space of a nonmagnetic material, such as air, between magnetic components, circuits, devices, or materials. For instance, the opening between pole pieces in a tape head. Also called **magnetic gap**, **magnetic air-**

gap, or **gap (3)**. **3.** A spark gap in which two electrodes are separated by air.

air-ground communications All aircraft-to-ground, and ground-to-aircraft communications. Its principal uses include flight coordination and safety. Also called **air-to-ground communications**, or **ground-to-air communications**.

air interface The portion of wireless communications, such as those of cellular telephony, in which RF signals are transmitted via air, as opposed to wires, fibers, or cables.

air ion An atom or molecule present in air which has become ionized. For example, an oxygen molecule that has gained an electron.

air ionizer A device which emits positive and negative air ions. These serve, among other purposes, to neutralize electrostatically charged objects.

air-position indicator An airborne computer that continuously indicates the position of the aircraft based on its airspeed, heading, and elapsed time. Its abbreviation is **API**.

air-spaced coax Same as **air-dielectric coax**.

air switch A switch in which the breaking of the electric circuit occurs in air. Also called **air-break switch**.

air-to-ground communications Same as **air-ground communications**.

air traffic control A service which monitors and manages the flow of aircraft in the air or on the ground, contributing to the safety and expediency of air travel. Its abbreviation is **ATC**.

air traffic control radar A radar utilized to monitor aircraft within an **air traffic control** service.

air-variable capacitor A capacitor in which a set of rotating metal plates is separated by air from a fixed set. The rotation of the movable set varies the capacitance.

airbridge Same as **air bridge**.

aircraft bonding Electrically connecting all metal parts of an aircraft, including its engine.

airport beacon A radio or light beacon located at or near an airport, to indicate the location of an airport.

airport surveillance radar A radar, located at or near an airport, that scans the airspace within a given radius to determine the range and bearing of nearby aircraft. Used in air traffic control. Also called **approach control radar**.

airproof Hermetically sealed, so as to not allow the passage of air or gases. Also called **airtight**.

AIRS Acronym for **Atmospheric Infrared Sounder**.

airtight Same as **airproof**.

airwaves Radio-frequency waves utilized in broadcasting TV and radio programming. Strictly speaking, it is not a technical term, yet its use is widespread.

Al Chemical symbol for **aluminum**.

AL Abbreviation of **Arbitrated Loop**.

alarm **1.** An electrical and/or mechanical device which serves to warn by means of a signal, such as sound and/or light. **2.** A security system whose source of power is electricity. **3.** A signal that informs of a malfunction, error, or the occurrence of a specified condition.

alarm circuit A circuit that informs of a malfunction, error, or the occurrence of a specified condition.

alarm condition A circumstance, such as an emergency or a malfunction, which triggers an alarm.

alarm control panel A panel with the necessary displays and controls to monitor and operate an alarm system.

alarm filtering In networks, the pinpointing of the component or device that is the origin of a failure.

alarm signal **1.** A signal which is transmitted throughout a zone to advise of an alarm condition. **2.** The manner in which a device informs of an alarm condition. Usually it is an audible and/or visible indication or warning.

alarm system An assembly of devices and equipment that will detect, then inform of, conditions requiring immediate action. These circumstances include fires, equipment failures, and unauthorized entry, among others. The three main components are: an input device, such as a smoke detector, a processing unit, such as the alarm control panel, and an output device, such as a siren.

albedo The ratio of the reflected light or radiation from a surface, to the light or radiation incident upon said surface.

ALC **1.** Abbreviation of **automatic level control**. **2.** Abbreviation of **automatic light control**.

alert In computers, a visible and/or audible signal that serves to notify, remind, or warn. It can refer, for example, to an error, a calendar event, or to inform that a process is completed.

alert box In computers, a box that appears on-screen providing an **alert**.

Alexanderson antenna An antenna consisting of several base-loaded vertical radiators arranged in parallel, which are connected at the top, and fed at the bottom of one of the radiators. Used primarily for low and very low frequencies.

Alford antenna Same as **Alford loop antenna**.

Alford loop Same as **Alford loop antenna**.

Alford loop antenna A multielement antenna with an approximately equal, and in-phase, current uniformly distributed along each of its peripheral elements. It produces a circular radiation pattern. Also called **Alford antenna**, or **Alford loop**.

Alfvén wave In a plasma, transverse hydromagnetic waves which propagate along the lines of magnetic force.

ALG Abbreviation of **application level gateway**.

ALGOL Acronym for **algorithmic language**, an early high-level computer language for expressing algorithms. It served as the basis for Pascal.

algorithm A sequence of defined steps utilized to solve a problem. Encryption, for instance, utilizes algorithms for coding and decoding.

algorithmic language **1.** A computer programming language that uses algorithms for problem solving. **2.** Same as **ALGOL**.

alias An alternate designation used to identify computer files, data fields, or other items.

aliasing Also called **aliasing distortion**. **1.** In analog-to-digital conversion, a form of distortion which arises when the sampling frequency is less than twice the highest frequency component. It would occur, for instance, when a 50 kHz waveform is sampled at any frequency below 100 kHz. An anti-aliasing filter may be used to solve this problem. Also known as **foldover (1)**. **2.** The jagged appearance that curved or diagonal lines in computer graphics have when there is not enough resolution to show images realistically. Also called **stair-stepping**, or **jaggies**.

aliasing distortion Same as **aliasing**.

align **1.** To arrange, position, or synchronize a component, circuit, or device in a manner that allows for proper or optimal functioning. For example, the positioning of a read/write head. **2.** To arrange or position a component, circuit, or device in a manner that allows for proper installation. For example, the installation of a memory chip.

aligned bundle An assembly of optical fibers in which the relative spatial coordinates of the individual fibers are the same at both ends. Such a bundle serves, for instance, to transmit images. Also called **coherent bundle**.

aligned-grid tube A vacuum tube with multiple grids, in which two or more are aligned in a manner that achieves a desired effect, such as noise suppression.

aligner Same as **alignment tool**.

aligning tool Same as **alignment tool**.

alignment 1. The arrangement, positioning, or synchronization of a component, circuit, or device in a manner that allows for proper or optimal functioning. 2. The arrangement or positioning of a component, circuit, or device in a manner that allows for proper installation.

alignment chart A chart utilized to solve numerical equations. Two known variables and an unknown are graphed in a manner that the results of calculations may be obtained by tracing an intersecting line through known values to obtain the desired value. May be used, for instance, to solve many electronics equations. Also called **nomograph**, or **nomogram**.

alignment pin A pin on a plug-in component that assists in the proper alignment during insertion. The base of a tube, for instance, has such pins.

alignment test A test to insure proper alignment.

alignment tool A tool utilized to insure the proper alignment of a component, circuit, or device. For example, a nonmagnetic screwdriver or wrench. Also called **aligner**, or **aligning tool**.

alive Anything connected electrically to a source of voltage. Also known as **energized** (1), **live** (1), or **hot** (1).

alive circuit A circuit that is connected electrically to a source of voltage. Also known as **energized circuit**, or **live circuit**.

alkali 1. A chemical substance that is a base, thus one with a pH value greater than 7.0. 2. A hydroxide formed with one of the alkali metals, such as sodium hydroxide.

alkali metals A group within the periodic table of elements, consisting of the following metals: **lithium**, **sodium**, **potassium**, **rubidium**, **cesium**, and **francium**. All are very reactive, and readily give up electrons. They are called alkali metals because they all form strong bases, such as sodium hydroxide.

alkaline Pertaining to a chemical substance that is a base. Such substances have a pH value greater than 7.0.

alkaline battery A battery composed of **alkaline cells**.

alkaline cell A primary cell that uses an alkaline electrolyte solution, such as potassium hydroxide. The voltage output is about 1.5 volts, and they have higher current rates, a longer shelf life, and a flatter discharge curve than conventional zinc-carbon cells.

alkaline-earth metals A group within the periodic table of elements, consisting of the following metals: **beryllium**, **magnesium**, **calcium**, **strontium**, **barium**, and **radium**. Some classifications only include the heavier elements in this group, so beryllium and magnesium are then not considered to be alkaline-earth metals. They are all very reactive, readily give up electrons, are malleable, extrudable, and conduct electricity well.

alkaline electrolyte An electrolyte that has a pH value greater than 7.0. Commonly used in alkaline cells.

alkaline-manganese cell The predominant type of **alkaline cell**.

alkaline storage battery A storage battery that uses an alkaline electrolyte solution, usually potassium hydroxide.

all-diffused integrated circuit An IC in which all the active and passive elements have been formed by diffusion.

all-in-one printer A computer printer which also performs other functions, such as scanning, copying, and faxing. Also called **multifunction printer**, or **multifunction peripheral**.

all-optical network A communications network using only optical fibers, optical switching, and the like. Bandwidths in the terabit per second range can be attained.

all-pass filter A filter which provides a time delay or a phase shift of the signal, without attenuating any of the frequencies of the spectrum. Also called **all-pass network**.

all-pass network Same as **all-pass filter**.

all points addressable A computer graphics mode in which each pixel can be individually manipulated, providing for a sharper image. Its abbreviation is **APA**.

all-wave Referring to a device which has a very wide range of operating frequencies, such as an all-wave receiver.

all-wave antenna An antenna which has a very wide range of operating frequencies, such as a log-periodic antenna.

all-wave receiver A radio receiver which has a very wide range of operating frequencies, some spanning from about 10 kHz to over 100 MHz.

Allan Deviation A calculation utilized to estimate stability, such as the output frequency of an oscillator over the short or long term. Also called **Allan Variance**.

Allan Variance Same as **Allan Deviation**.

Allen screw A screw with a hollow hexagonally-shaped socket in its head.

Allen wrench A wrench with a hexagonally shaped head that matches an **Allen screw** exactly.

alligator clip A spring-loaded clip utilized to make temporary electric connections. It is long and narrow and has jagged teeth, making for secure yet easy connections at the desired test points. It resembles the head of an alligator, and is also known as **crocodile clip**.

allocate 1. To designate or distribute equipment or resources. For example, to assign of radio frequencies. 2. To reserve computer resources for specific purposes. For example, to allocate memory for disk cache.

allocated channel A frequency channel which has been assigned for authorized uses.

allocated frequencies A segment of the radio-frequency spectrum which has been assigned for authorized uses.

allocation 1. The designation or distribution of equipment or resources. 2. The reserving of computer resources for specific purposes.

allocation of frequencies The assigning of a segment of the radio-frequency spectrum for authorized uses.

allocation unit A fixed number of disk sectors treated as a unit. It is the smallest unit of storage that a computer operating system can handle. Also called **cluster** (1).

allochromatic 1. Pertaining to a mineral or crystal that exhibits photoconductivity due to the minute presence of an impurity. 2. Pertaining to a mineral or crystal whose color is due to the minute presence of an impurity. This contrasts with **idiochromatic**, where the color is a result of a pure chemical composition.

allochromatic crystal 1. A crystal that exhibits photoconductivity due to the minute presence of an impurity. 2. A crystal whose color is due to the minute presence of an impurity.

allochromy For a substance, the emitting of electromagnetic radiation of a different wavelength than that which was incident upon it. This phenomenon is observed, for instance, in fluorescence.

allotrope For a chemical element, one of two or more stable structural forms.

allotropic Pertaining to **allotropy**.

allotropism Same as **allotropy**.

allotropy For a chemical element, the existence of two or more stable structural forms, each of which is in the same physical state. Their physical properties tend to be significantly different. For example, graphite and diamond are two allotropes of the element carbon. Also called **allotropism**.

allowed bands Same as **allowed energy bands**.

allowed energy bands Restricted energy levels that electrons may occupy in atoms, molecules, or crystals. Also called **allowed bands**.

alloy A mixture of two or more elements, one of which must be a metal, to form a macroscopically homogeneous metallic product. Alloys may be solids or liquids, and are usually created to better adapt the constituents for a particular purpose. Alloys are used extensively in electronics. For example, niobium alloys are used as low temperature superconductors.

alloy deposition In semiconductors, the fusing of an alloy to the substrate.

alloy diode Same as **alloy-junction diode**.

alloy junction A junction formed by alloying one or more impurity metals to a semiconductor. Depending on which impurity metal is alloyed, p or n regions will be formed. Also known as **fused junction**.

alloy-junction diode A junction diode whose p and n regions are formed by alloying one or more impurity metals to the semiconductor. Also known as **alloy diode**, or **fused-junction diode**.

alloy-junction transistor A junction transistor whose p and n regions are formed by alloying one or more impurity metals to the semiconductor. Also known as **alloy transistor**, or **fused-junction transistor**.

alloy plating The electrochemical deposition of two or more metallic elements on a substrate. For example, elements such as boron or vanadium can be plated only when in combination with elements such as osmium or platinum.

alloy transistor Same as **alloy-junction transistor**.

alloying 1. The formation of **alloys**. 2. The deposition of alloys on other materials.

AlltheWeb A popular search engine.

alnico Acronym for **aluminum-nickel-cobalt**. Same as **alnico alloy**.

alnico alloy Abbreviation of **aluminum-nickel-cobalt alloy**. An alloy with varying compositions of aluminum, nickel, and cobalt. Often present are amounts of other elements such as copper or iron. It is utilized, for instance, to make strong permanent magnets. Also called **alnico**.

alnico magnet Abbreviation of **aluminum-nickel-cobalt magnet**. A strong permanent magnet made of **alnico alloy**. Such magnets feature excellent stability through a wide range of temperatures, and have applications in various areas of electronics, including their use in sensing devices and speakers.

alpha Its symbol is α. 1. The first letter of the Greek alphabet, and as such often designates the first constituent of a series. 2. The current gain from emitter to collector in a transistor with a common-base configuration. 3. Same as **alpha software**. 4. Same as **alpha hardware**.

alpha blending In computer graphics, the use an **alpha channel**.

alpha channel In 32-bit computer graphics, an additional 8 bits per pixel, which combine with the 24 total bits for red, green, and blue. These extra bits serve to convey additional information about an image, such as transparency.

alpha cutoff Same as **alpha cutoff frequency**.

alpha cutoff frequency The higher frequency at which the current gain from emitter to collector in a transistor with a common-base configuration drops 3 decibels, or 70.7%, in relation to a reference lower frequency. Also called **alpha cutoff**.

alpha decay Radioactive decay in which **alpha particles** are emitted. Also called **alpha disintegration**.

alpha disintegration Same as **alpha decay**.

alpha emission The emission of **alpha particles** by a radioactive element.

alpha emitter A radioactive element that emits **alpha particles**. Also called **alpha radiator**.

alpha hardware New hardware undergoing its initial tests, usually performed under experimental conditions. Also called **alpha (4)**.

alpha particle Positively charged helium nuclei emitted by many radioactive elements. Each particle consists of two protons and two neutrons. After a nuclei has emitted an alpha particle, it will have a lower mass. This means that a transformation has occurred that has yielded a different, and lighter, element. When an alpha particle is combined with two electrons, the complete helium atom is formed.

alpha radiation A stream consisting of **alpha particles**.

alpha radiator Same as **alpha emitter**.

alpha ray A stream consisting of **alpha particles**.

alpha release Same as **alpha software**.

alpha software New software undergoing its initial tests, usually performed under experimental conditions. Also called **alpha release**, **alpha version**, or **alpha (3)**.

alpha tests The initial tests performed on newly developed software or hardware. Usually done under experimental conditions, as opposed to **beta tests** which are done under normal operating conditions.

alpha version Same as **alpha software**.

alphabetic-numeric Same as **alphanumeric**.

alphageometric Same as **alphamosaic**.

alphameric Same as **alphanumeric**.

alphameric characters Same as **alphanumeric characters**.

alphameric data Same as **alphanumeric data**.

alphamosaic A technique for creating computer graphics in which images are constructed from a limited set of graphics symbols. The displayed images produced in this manner are of very low resolution. Also known as **alphageometric**.

alphanumeric Combinations of letters, numbers, and special characters used by computers. It was previously more common to only encompass letters and numbers under this heading. Also called **alphabetic-numeric**, or **alphameric**.

alphanumeric characters The letters, numbers, and special characters used by computers. It was previously more common to only encompass letters and numbers under this heading. Also called **alphameric characters**.

alphanumeric data Data presented as **alphanumeric characters**. Also called **alphameric data**.

alphanumeric device Same as **alphanumeric display**.

alphanumeric display Also known as **alphanumeric display device**, **alphanumeric display terminal**, or **alphanumeric device**. 1. A visual device which displays alphanumeric output. 2. A visual device that can display letters and numbers, but not graphics.

alphanumeric display device Same as **alphanumeric display**.

alphanumeric display terminal Same as **alphanumeric display**.

alphanumeric input Computer input consisting of **alphanumeric characters**.

alphanumeric output Computer output consisting of **alphanumeric characters**.

alphanumeric pager A pocket-sized radio receiver or transceiver that can display received messages consisting of **alphanumeric characters**.

alphanumeric reader An instrument which has a photosensor capable of reading **alphanumeric characters**.

alphanumeric readout A display, such as a segmented display, which forms numbers and/or letters using segments or strips.

alt Abbreviation of **alternate key**.

alt key Abbreviation of **alternate key**.

AltaVista A popular search engine.

alternate channel A channel two channels higher or lower in frequency than a given channel. It is either the channel immediately before the lower adjacent channel, or the channel immediately after the higher adjacent channel.

alternate-channel interference Interference of radio or TV reception by other stations using alternate channels. It can be unilateral or bilateral. Also called **second-channel interference**.

alternate frequency A frequency which is authorized to be used only under certain conditions. This can be in replacement of the main frequency, or to supplement it.

alternate key A modifier key included on computer keyboards that is used in combination with other keys to generate a function. The specific function for any given key combination will depend on which program is running. Its abbreviation is **Alt key**, or **Alt**.

alternate routing Secondary communications paths that are used if the primary route is unavailable.

alternating current Same as **AC**.

alternating-current adapter Same as **AC adapter**.

alternating-current–alternating-current power converter Same as **AC–AC power converter**.

alternating-current bias Same as **AC bias**.

alternating-current circuit Same as **AC circuit**.

alternating-current component Same as **AC component**.

alternating-current coupling Same as **AC coupling**.

alternating-current/direct-current Same as **AC/DC**.

alternating-current–direct-current converter Same as **AC–DC converter**.

alternating-current–direct-current power converter Same as **AC–DC power converter**.

alternating-current dump Same as **AC dump**.

alternating-current equipment Same as **AC equipment**.

alternating-current erase Same as **AC erase**.

alternating-current erasing Same as **AC erasing**.

alternating-current erasing head Same as **AC erasing head**.

alternating-current frequency Same as **AC frequency**.

alternating-current generator Same as **AC generator**.

alternating-current induction motor Same as **AC induction motor**.

alternating-current line Same as **AC line**.

alternating-current line filter Same as **AC line filter**.

alternating-current line voltage Same as **AC line voltage**.

alternating-current magnetic biasing Same as **AC magnetic biasing**.

alternating-current meter Same as **AC meter**.

alternating-current motor Same as **AC motor**.

alternating-current noise Same as **AC noise**.

alternating-current power Same as **AC power**.

alternating-current power line Same as **AC power line**.

alternating-current power source Same as **AC power source**.

alternating-current power supply Same as **AC power supply**.

alternating-current resistance Same as **AC resistance**.

alternating-current transmission Same as **AC transmission**.

alternating-current voltage Same as **AC voltage**.

alternating gradient A magnetic field in which there is a series of magnets with gradients of alternating signs.

alternating-gradient focusing The utilization of an **alternating gradient** to focus accelerated particles.

alternating-gradient synchrotron A proton synchrotron utilizing an **alternating gradient** for the focusing of particles.

alternating quantity A quantity that alternates at regular intervals and whose average during any complete cycle is zero. An example is AC.

alternating voltage Same as **AC voltage**.

alternation 1. The changing of a state from one to another, and back to the original state. **2.** Half of a cycle. For example, an alternating value which starts a cycle at zero, reaches a maximum, then returns to zero.

alternative keyboard A computer keyboard which is in some manner different from the usual keyboards. For example, such a keyboard may be split, vertical, contoured, have extra large keys, and so on.

alternative routing Available communications paths, from which one may be chosen. Comprised of the primary and secondary routes.

alternator A device which produces AC. For example, a rotary machine that converts mechanical energy into AC.

altimeter An instrument which measures and indicates altitude above a reference altitude, such as sea level. An example is an absolute altimeter.

ALU Abbreviation of **arithmetic logic unit**.

alumina A chemical compound whose formula is Al_2O_3. It is a refractory material, and its applications in electronics include its use as an insulator, in the adsorbance of gases, in the manufacturing of chips, and as an abrasive. Also called **aluminum oxide**.

aluminium Same as **aluminum**.

aluminization Application of a thin film of aluminum onto another surface, such as glass or a semiconductor substrate.

aluminize To apply a thin film of aluminum onto another surface, such as glass or a semiconductor substrate.

aluminized screen A display screen with a thin film of aluminum applied to the back of its phosphor coating. These screens feature high brightness, contrast, and resolution.

aluminum A lightweight metallic chemical element whose atomic number is 13. It is the most abundant metal in the earth's crust, and the third most abundant element on the planet. It has about 15 known isotopes of which one is stable, is ductile, malleable, and a good electrical conductor. Its many applications in electronics include uses in semiconductor doping, in manufacturing of chips and hard disks, and as a chassis for metal components. Its chemical symbol is **Al**. Also spelled **aluminium**.

aluminum antimonide A crystalline chemical compound whose formula is **AlSb**. It is a semiconductor.

aluminum arsenide A chemical compound whose formula is **AlAs**. It is a semiconductor.

aluminum conductor A conductor of electric current composed of aluminum or one of its alloys. Frequently used in high-voltage transmission lines due to its light weight.

aluminum-nickel-cobalt Same as **alnico**.

aluminum-nickel-cobalt alloy Same as **alnico alloy**.

aluminum-nickel-cobalt magnet Same as **alnico magnet**.

aluminum nitride A crystalline chemical compound whose formula is **AlN**. It is a semiconductor.

aluminum oxide Same as **alumina**.

aluminum phosphide A chemical compound, occurring in dark gray or yellow crystals, whose formula is **AlP**. It is a semiconductor.

Always On/Dynamic ISDN An ISDN service in which there is always a low-speed connection to a network, such as the Internet, using the D-channel, and which uses the higher-bandwidth B-channels only when necessary. Also, a standard for such service. Its abbreviation is **AO/DI**.

am Abbreviation of **attometer**.

AM Abbreviation of **amplitude modulation**. A form of modulation in which the information-containing signal varies the amplitude of the carrier wave. The amplitude of the carrier wave is modulated, or varied, in a manner proportional to the fluctuations of said information-bearing signal, but its frequency stays the same. Used, for instance, for broadcasting. This contrasts with **FM**, in which the amplitude of the carrier wave is kept constant while the frequency is varied, or with **PM**, where the phase angle of the carrier wave is varied.

AM broadcast band Abbreviation of **amplitude-modulation broadcast band**. A band of frequencies allocated to AM broadcasting stations. In the United States, the standard AM broadcast band is 535 to 1705 kHz.

AM broadcasting Abbreviation of **amplitude-modulation broadcasting**. The broadcasting of amplitude-modulated signals.

AM/FM radio Abbreviation of **amplitude-modulation/frequency-modulation radio**. Same as **AM/FM receiver**.

AM/FM radio receiver Abbreviation of **amplitude-modulation/frequency-modulation radio receiver**. Same as **AM/FM receiver**.

AM/FM receiver Abbreviation of **amplitude-modulation/frequency-modulation receiver**. A radio receiver that can receive either amplitude-modulated or frequency-modulated signals. Also called **AM/FM radio**, or **AM/FM radio receiver**.

AM/FM transmitter Abbreviation of **amplitude-modulation/frequency-modulation transmitter**. A radio transmitter that can send either amplitude-modulated or frequency-modulated signals.

AM/FM tuner Abbreviation of **amplitude-modulation/frequency-modulation tuner**. A radio tuner that can receive either amplitude-modulated or frequency-modulated signals.

AM noise Abbreviation of **amplitude-modulation noise**. Modulation of the amplitude of a carrier wave by signals other than those intended to modulate it.

AM radio Abbreviation of **amplitude-modulation radio**. Same as **AM receiver**.

AM radio receiver Abbreviation of **amplitude-modulation radio receiver**. Same as **AM receiver**.

AM receiver Abbreviation of **amplitude-modulation receiver**. A radio receiver that can receive amplitude-modulated signals only. In the United States, the standard AM broadcast band is 535 to 1705 kHz. Also called **AM radio**, or **AM radio receiver**.

AM signal Abbreviation of **amplitude-modulated signal**. A signal whose carrier wave is amplitude-modulated.

AM stereo Abbreviation of **amplitude-modulation stereo**. Also called **stereo AM**. **1.** The broadcasting of AM signals with two or more tracks. **2.** The transmission and reception of **AM stereo (1)** broadcasts.

AM transmitter Abbreviation of **amplitude-modulation transmitter**. A radio transmitter that can send amplitude-modulated signals.

AM tuner Abbreviation of **amplitude-modulation tuner**. A radio tuner that can receive amplitude-modulated signals only. In the United States, the standard AM broadcast band is 535 to 1705 kHz.

Am²/Js Abbreviation of **ampere square meter per joule second**.

amateur band A radio-frequency band designated for use solely by authorized amateurs.

amateur radio Also called **ham radio**. **1.** A radio receiver and/or transmitter utilized by authorized amateurs, and which is operated in the amateur band. **2.** The operation of an **amateur radio (1)**, especially as a hobby.

amber A fossilized tree resin which readily accumulates static charge by friction. This property has been known since ancient times. It is also a very good electrical insulator.

ambient conditions The environmental conditions that surround a given area, especially in reference to parameters which may influence the functioning of devices, equipment, or the readings of instruments. These conditions include temperature, pressure, humidity, noise, and light. Also called **environmental conditions**.

ambient humidity The humidity present in any given environment, especially where a device or piece of equipment is functioning, or where tests or observations are taking place.

ambient illumination Same as **ambient light**.

ambient light Light which is present in any given environment, especially within which tests or observations are taking place. In many cases it is simply the light in the room. Also called **ambient illumination, room light, background light,** or **environmental light**.

ambient-light filter Same as **anti-glare filter**.

ambient noise The noise present in any given environment, especially that within which tests or observations are taking place. Such noise includes nearby and distant sources. Also called **ambient noise level, background noise (1), room noise, site noise, environmental noise,** or **local noise**.

ambient noise level Same as **ambient noise**.

ambient pressure The pressure in any given environment, especially where a device or piece of equipment is functioning, or where tests or observations are taking place. May be atmospheric pressure, or the pressure of the medium which surrounds an object or instrument.

ambient temperature The temperature in any given environment, especially where a device or piece of equipment is functioning, or where tests or observations are taking place. May be room temperature, or that of the medium which surrounds an object or instrument.

ambient-temperature range Also called **operational temperature, operating temperature,** or **operating temperature range**. **1.** The ambient temperature interval within which an instrument is operated. **2.** The ambient temperature interval within which an instrument is designed to work without malfunctioning or being damaged. **3.** The ambient temperature interval required for an instrument to function optimally. **4.** The **ambient-temperature range (1)**, **ambient-temperature range (2)**, or **ambient-temperature range (3)** for a component, circuit, device, piece of equipment, system, or material.

ambiguity 1. Uncertainty in a result, or how to interpret it. **2.** In servo systems, the seeking of more than one null position.

ambipolar 1. Relating to, or applying equally to both positive and negative ions. **2.** Operating in two directions simultaneously.

ambisonic Pertaining to **ambisonics**.

ambisonic reproduction Same as **ambisonics (1)**.

ambisonic sound Same as **ambisonics (1)**.

ambisonics 1. The reproduction of sound that produces as faithfully as possible a three-dimensional sound image. Specialized coding and decoding equipment is required, along with an appropriate speaker array. Also called **ambisonic reproduction**, or **ambisonic sound**. **2.** The science dealing with the reproduction of sound that produces as faithfully as possible a three-dimensional sound image.

AMC Abbreviation of **automatic modulation control**.

American National Standards Institute Same as **ANSI**.

American Standard Code for Information Interchange Same as **ASCII**.

American Wire Gage Same as **American Wire Gauge**.

American Wire Gauge A standard for designating the diameter of wires, or the thickness of sheets. Refers to nonferrous materials only. Also spelled **American Wire Gage**. Its abbreviation is **AWG**. Also called **Brown and Sharpe Gauge**.

americium A synthetic radioactive chemical element whose atomic number is 95, and which has about 20 identified isotopes. It is very malleable, nonmagnetic, and emits alpha and gamma radiation. It is used in many types of gauges, smoke detectors, and for dissipating static charges. Its chemical symbol is **Am**.

AMLCD Abbreviation of **active-matrix liquid-crystal display**.

ammeter Abbreviation of **ampere meter**. An instrument graduated in amperes, or fractions/multiples of amperes, which is utilized to measure and indicate the magnitude of electric currents. Different ammeters may be used for measuring DC and/or AC, and such instruments may employ any of various methods for detection, depending on the specific needs.

ammeter shunt A resistor connected in parallel to an ammeter, in order to increase the current range that can be measured. The majority of the current is carried by this shunted resistor.

ammonia maser A gas maser in which ammonia, a colorless gas whose structure is NH_3, is stimulated to amplify microwave radiation. The output of such a maser is characterized by great stability and may be used, for instance, as a frequency standard.

ammonium borate A colorless crystalline chemical compound used in electrolytic capacitors.

ammonium chloride White crystals whose chemical formula is NH_4Cl. Used as an electrolyte in dry cells, and for electroplating. Also known as **sal ammoniac**, or **salmiac**.

ammonium chloroplatinate A crystalline chemical compound used in electroplating.

ammonium dichromate An orange crystalline chemical compound used in the manufacturing of recording tape.

ammonium nickel sulfate A green crystalline chemical compound used in electroplating.

ammonium persulfate A white crystalline chemical compound used in electroplating, circuit board fabrication, and as a depolarizer of batteries.

ammonium sulfamate A white solid chemical compound used in electroplating.

ammonium thiocyanate A colorless crystalline chemical compound used in electroplating.

amorphous 1. Not having a definite shape. For example, rubber. 2. Not having a crystalline structure. For example, glass. Also called **non-crystalline**.

amorphous film A metallic film without a definite shape. May be used, for example, in the manufacturing of semiconductors.

amorphous semiconductor A semiconductor material which does not have a crystalline structure. Its fabrication is cheaper and simpler than those with crystalline structure, but it is less efficient and not as durable.

amortisseur winding A winding which serves to dampen any oscillations in a synchronous motor. Also called **damper winding**.

amp 1. Abbreviation of **ampere**. 2. Abbreviation of **amplifier**.

amp-hr Abbreviation of **ampere-hour**.

amp-turn Abbreviation of **ampere-turn**.

ampacity The current-carrying capacity of a conductor. Expressed in amperes.

amperage The intensity of a current flow. Expressed in amperes.

ampere The fundamental SI unit of electric current. Currently, it is defined as the constant current that produces a force of 0.2 micronewtons per meter between two infinitely long parallel conductors placed one meter apart in a vacuum. A current of one ampere equals one coulomb of charge per second. Its abbreviation is **amp**. Its symbol is **A**.

Ampère balance An ammeter which operates by having a current pass through two nearby coils connected in series. One of the coils is fixed, while the other is attached to the arm of a balance. The attractive force between the coils is counterweighed by the force of gravity acting on a known weight on the arm of the balance, which in turn indicates the current strength on its scale. Also called **Kelvin balance**, or **current balance**.

ampere-hour A unit of quantity of electricity equivalent to the charge accumulated by a steady flow of one ampere for one hour. An ampere-hour is equal to exactly 3600 coulombs, and is often utilized to state the capacity of a storage battery. It has various abbreviations, depending on the standard used, the most prevalent being **Ah**. Others include **amp-hr**, **Ahr**, or **A-h**.

ampere-hour meter An instrument for measuring electric current flow per unit time, as expressed in ampere-hours.

ampere meter Same as **ammeter**.

ampere/meter Same as **ampere per meter**.

ampere-minute A unit of quantity of electricity equivalent to the charge accumulated by a steady flow of one ampere for one minute. An ampere-minute is equal to exactly 60 coulombs. Its abbreviation is **A min**.

ampere per meter Its abbreviation is **A/m**. 1. The SI unit of magnetic field strength. 2. The SI unit of magnetization.

ampere per square meter The SI unit of current density. Its abbreviation is A/m^2.

Ampère rule In a conductor that is carrying current away from an observer, the magnetic field lines will have a clockwise direction.

ampere square meter per joule second The SI unit of gyromagnetic ratio. Its abbreviation is Am^2/Js.

ampere-turn The unit of magnetomotive force in the MKS system. It is equal to 1 ampere flowing through 1 turn of a coil. One ampere-turn is also equal to about 1.257 gilberts. Its abbreviations are **amp-turn**, **At**, and **A-t**.

Ampère's law A law which describes the mathematical relationship between electric currents and magnetic fields. Also known as **Laplace's law**.

amperemeter Same as **ammeter**.

amplidyne A rotating magnetic amplifier in which a small increase in input power results in a large increase in output power. Also called **metadyne**.

amplification 1. The process of producing an output which is greater than the input. This can refer to any of various electrical quantities, such as current, voltage, power, or signal strength. 2. A quantitative measure of the increase in output versus input of an electrical quantity. This can be expressed as a ratio, a percentage, or in decibels.

amplification factor 1. The increase in output versus input of an electrical quantity, such as current, voltage, power, or signal strength. This is expressed as a ratio, a percentage, or in decibels. 2. In an electron tube, the factor by which the plate voltage increases in proportion to an increase in the grid voltage, where all other voltages and the plate current are held constant. Also known as **mu factor**, or **μ factor**.

amplification noise The electrical noise generated by the components of an amplifier during the process of amplification.

amplifier A component, circuit, or device which produces an output signal that is greater than the input signal, ideally without altering the essential characteristics of said input signal. Amplifying devices and components include electron tubes, transistors, and distribution amplifiers. Its abbreviation is **amp**.

amplifier circuit A circuit which amplifies.

amplifier class Classifications of amplifiers based on the relationship between the input signal and the output current. For classification purposes, a simplified output stage consisting of two complementary tubes or transistors is assumed for most classes. Each class has its own linearity and efficiency characteristics. Examples include class A amplifiers, class AB amplifiers, class C amplifiers, and so on. Also called **class (3)**.

amplifier distortion A change in the waveform of a signal between the input and the output stages of an amplifier.

amplifier gain The increase in current, voltage, or power provided by an amplifier. For instance, the ratio of the output power to the input power of a power amplifier.

amplifier noise 1. The electrical noise generated by an amplifier in the absence of an input signal. **2.** Any noise generated by an amplifier.

amplifier output 1. The amplified signal an amplifier produces. **2.** The output devices and terminals of an amplifier.

amplifier power Also called **power output**, or **output power**. **1.** The extent to which an amplifier can amplify an input signal. Also called **power amplification (3)**. **2.** The delivering of the output of an amplifier to a load, such as a loudspeaker. Expressed in watts Also, the power so delivered.

amplifier stage An amplifying unit within a device or system.

amplifying delay line A delay line which amplifies high-frequency signals. Used in pulse-compression techniques.

amplifying unit Within a device or system, a unit which produces an output signal that is greater than the input signal.

amplitron A microwave amplifier that uses cross-field interactions to provide high gain and phase stability throughout a wide bandwidth.

amplitude In a wave or other periodic phenomenon, the maximum absolute value of the displacement from a reference position, such as zero.

amplitude clipping The limiting of the amplitude of the output signal of a circuit or device to a predetermined maximum, regardless of the variations in its input.

amplitude comparison The comparing of two amplitudes, one of which is used as a reference.

amplitude discriminator A circuit that acts upon pulses whose amplitude exceed a determined value. Used in detectors. Also known as **pulse-height discriminator**.

amplitude distortion 1. Same as **attenuation distortion**. **2.** In an amplifier, the component of the output signal which alters the essential characteristics of the input signal. This is due to a nonlinear response in the amplifier. Also called **nonlinear distortion (3)**.

amplitude excursion Also called or **peak-to-peak amplitude**, **peak-to-peak**, or **maximum amplitude excursion**. **1.** For a waveform of an alternating quantity, such as that of AC, the difference between the maximum positive peak and the maximum negative peak. It is the maximum combined range in the amplitude of a signal or other observed quan-

tity. **2.** The difference between the maximum positive peak and the maximum negative peak of any varying quantity. It is the maximum combined range in the amplitude of a signal or other observed quantity.

amplitude factor In a periodically-varying function, such as that of AC, the ratio of the peak amplitude to the RMS amplitude. Also known as **crest factor**, or **peak factor**.

amplitude fading In the propagation of electromagnetic waves, decreases in amplitude that are uniform throughout all the frequency components of the signal. This contrasts with **selective fading**, in which different frequencies are affected to varying degrees.

amplitude-frequency distortion Same as **attenuation distortion**.

amplitude-frequency response A measure of the behavior of a component, circuit, device, piece of equipment, or system, as a function of its input signal frequencies. For example, it may refer to the efficiency of the amplification of a circuit or device as a function of frequency. Also known as **frequency response (1)**, **sine-wave response**, or **sine-wave frequency response**.

amplitude gate A circuit or device which passes only those portions of an input signal which lie between two fixed amplitude boundaries. These boundaries are usually close together. Also called **clipper-limiter**, **slicer**, or **slicer amplifier**.

amplitude limiter A circuit or device which limits the amplitude of its output signal to a predetermined maximum, regardless of the variations of its input. Used, for example, for preventing component, equipment, or media overloads. Also known by various other names, including **amplitude-limiting circuit**, **amplitude-limiter circuit**, **limiter**, **automatic peak limiter**, **clipping circuit**, **peak limiter**, **peak clipper**, and **clipper**.

amplitude-limiter circuit Same as **amplitude limiter**.

amplitude-limiting circuit Same as **amplitude limiter**.

amplitude-modulated Pertaining to, of having undergone **amplitude modulation**.

amplitude-modulated signal A signal whose carrier wave is **amplitude-modulated**.

amplitude-modulated transmission A transmission in which the carrier wave is **amplitude-modulated**.

amplitude-modulated transmitter A transmitter sending an **amplitude-modulated signal**.

amplitude-modulated wave A wave whose amplitude varies proportionally to the modulating signal.

amplitude modulation Same as **AM**.

amplitude-modulation broadcast band Same as **AM broadcast band**.

amplitude-modulation broadcasting Same as **AM broadcasting**.

amplitude-modulation/frequency-modulation radio Same as **AM/FM radio**.

amplitude-modulation/frequency-modulation radio receiver Same as **AM/FM radio receiver**.

amplitude-modulation/frequency-modulation receiver Same as **AM/FM receiver**.

amplitude-modulation/frequency-modulation transmitter Same as **AM/FM transmitter**.

amplitude-modulation/frequency-modulation tuner Same as **AM/FM tuner**.

amplitude-modulation noise Same as **AM noise**.

amplitude-modulation radio Same as **AM radio**.

amplitude-modulation radio receiver Same as **AM radio receiver**.

amplitude-modulation receiver Same as **AM receiver**.

amplitude-modulation transmitter Same as **AM transmitter**.

amplitude-modulation tuner Same as **AM tuner**.

amplitude modulator A circuit or device which effects **amplitude modulation**.

amplitude noise In radars, variations in the amplitude of the signal reflected by the scanned object. This reduces the accuracy of the radar device.

amplitude peak In a wave or other periodic phenomenon, the maximum absolute value of the displacement with respect to zero. Also known as **peak amplitude**.

amplitude range The maximum combined range in amplitude of a signal within a specified interval of frequencies. For example, a high-fidelity amplifier may have a frequency response of 10 Hz to 60 kHz, +/– 0.5 decibels.

amplitude resonance In a resonant system, the frequency at which maximum resonance occurs.

amplitude response In a circuit or device, the variation of the output signal to the input signal over a given frequency interval.

amplitude selector A circuit which produces a fixed output pulse, but only when the input pulse lies between pre-established limits of amplitude. Also called **pulse-height selector, pulse-amplitude selector**, or **differential pulse-height discriminator**.

amplitude-separation circuit Same as **amplitude separator** (3).

amplitude separator 1. A circuit which separates any portion of a waveform above or below predetermined values. 2. A circuit which separates any portion of a waveform between predetermined values. 3. In a TV, a circuit which separates the control signals from the video signals. Also called **amplitude-separation circuit**.

amplitude shift keying Data transmission accomplished by varying the amplitude of a carrier wave. Switching between amplitude levels provides the keying. Its abbreviation is **ASK**.

amplitude-suppression ratio In the testing of an FM receiver which is receiving both amplitude-modulated and frequency-modulated signals, the ratio of the amplitudes between the input and output.

amplitude-versus-frequency-distortion In an amplifier, distortion that occurs when some frequencies are attenuated or amplified more than others

AMPS Acronym for **advanced mobile phone system**.

AMR Abbreviation of **Audio/Modem Riser**.

AMS Abbreviation of **aeronautical mobile service**.

amu Abbreviation of **atomic mass unit**.

AN radio range Same as **AN range**.

AN range A radio range that transmits two coded signals via directional antennas. When an aircraft is on one of the four fixed courses, a continuous tone is heard. As the bearing deviates off course, the pilot will increasingly hear the Morse-code signal for A or N. Also known as **AN radio range**.

anacoustic zone The region above an altitude of about 160 kilometers from the earth's surface, where the distance between air molecules is greater than the wavelength of sound, thus sound waves can not be propagated. This zone extends well beyond the planet into outer space. Also called **zone of silence** (1).

analog Also spelled **analogue**. 1. Of, pertaining to, or consisting of a signal or parameter which is continuously and infinitely variable. The characteristics, such as an amplitude or frequency, of an analog signal vary in a continuous and infinitely variable manner, while a **digital** signal has a range that is divided into a finite number of discrete steps. Although the analog coding of information, as opposed to digital, can better match a real-world source such as music or a panoramic scene, steps such as recording, copying, transmitting, and reproducing introduce greater and greater amounts of noise, distortion, and other interference. This is not a problem when recorded, processed, transmitted, and so on, digitally. 2. A variable whose magnitude closely or exactly follows another variable. For example, a voltage which fluctuates proportionally to a voice. Transducers, such as microphones, speakers, or photoelectric cells, make the appropriate conversions.

analog adder A circuit or device whose output is the sum of two or more analog inputs. For example, an output voltage which is the sum of two input voltages.

analog camera A camera which records and stores images in analog form. An example is a traditional film camera.

analog cellular communications Cellular communications utilizing analog, as opposed to digital, technology. Also called **analog cellular telephony**.

analog cellular telephone A cellular telephone utilized for **analog cellular communications**.

analog cellular telephony Same as **analog cellular communications**.

analog channel A channel with defined limits, within which the transmitted information can have any continuously variable value. For example, a voice channel.

analog circuit An electronic circuit which maintains a continuous relationship between its input and output when acting on a signal.

analog communications Communications employing a signal that, within the defined limits, can have any continuously variable value. For example, an AM radio transmission, or analog telephonic service.

analog comparator A circuit or device whose digital output depends on a comparison of its analog inputs. For example, a digital output of 1 or 0 if the sum of two analog input voltages is positive or negative, respectively.

analog computer A computer which processes infinitely variable signals, such as a voltage. This contrasts with a **digital computer**, which deals only with signals in a discrete form, such as binary numbers.

analog control Control of a component, device, or system, by analog signals such as frequency or movement.

analog control device A device which utilizes **analog control**. An example is a computer mouse.

analog data 1. Data which is represented by a continuous and infinitely variable value. For example, that of a varying voltage. 2. Data which is displayed on a continuous scale, such as that seen on a VU meter.

analog decoding The conversion of a digital signal into an analog signal.

analog delay line An analog device which introduces a time delay in a signal. An example is a mirror arrangement which varies the path of a laser signal.

analog device A device which represents values by quantifying a physical magnitude that can assume any value within a given range.

analog display 1. A video display which accepts analog signals from the computer. The display adapter carries out the digital to analog conversion. Also called **analog monitor**. 2. A display, such as that of an instrument, with an analog indicator. An example is a VU meter.

analog divider A circuit or device whose output is the quotient of two analog inputs.

analog domain All that is in **analog** form. Most everything in the real world is in analog form. An analog-to-digital converter transforms an analog input into a digital output.

analog electronics Electronic devices and instruments that deal with continuous and infinitely variable physical quantities. The resulting data is usually displayed on a continuous scale, such as that seen on a VU meter.

analog encoding The process of converting an analog input into a digital output.

analog filter A filter which processes inputs that are continuous and infinitely variable physical quantities.

analog gate A circuit or device which produces a continuously variable output as a function specific intervals of the input.

analog indicator Same as **analog readout**.

analog input An input which can have any continuously and infinitely variable value. A voice is an example.

analog integrator A device which performs integration and provides an analog output

analog line A voice-grade telecommunications line that carries a signal that can assume any value within a specified range. Used for, for example for telephonic communication.

analog meter An instrument which measures and indicates values by means of a continuous scale within which any value may be specified. Most of the time a pointer is used to indicate readings. Example include VU meters, and D'Arsonval meters.

analog modem A modem, such as that utilized for dial-up Internet access, which modulates a carrier signal for transmission, so that it contains digital information in a form that can be transmitted over analog lines. It performs the converse process for reception. A **digital modem** does not perform digital-to-analog, or analog-to-digital conversions, and uses digital lines.

analog monitor Same as **analog display (1)**.

analog multimeter A multimeter whose indications are shown on an analog display.

analog multiplexer A multiplexer which serves to transmit various analog signals.

analog multiplier A circuit or device whose output is the product of two or more analog inputs. For example, an output voltage which is the product of two input voltages.

analog network **1.** A communications network within which analog signals are transmitted, such as those utilized in analog cellular telephony. **2.** An arrangement of circuits in which a value such as a voltage is proportional to another physical quantity, such as sound.

analog output An output which is a continuous and infinitely variable function of an input signal. The output of a speaker is an example.

analog phone Abbreviation of **analog telephone**.

analog processing Processing utilizing analog signals. An analog signal is usually degraded each time it is processed. Once a signal is digitized, this problem is solved.

analog radar A radar system which does not utilize computers to process the received signals, in contrast to a **digital radar**, which does. Analog radars are more susceptible to interference due to atmospheric conditions, and are not equipped to filter undesired components, such as cars and trucks, which might be confused with desired objects, such as aircraft.

analog readout An instrument which indicates measured values by means of a continuous scale within which any value may be specified. Most of the time a pointer is used to indicate readings. An example is a VU meter. Also called **analog indicator**.

analog recorder A device, such as a magnetic tape recorder or electrocardiograph, which produces an **analog recording**.

analog recording Recording in which a continuous and infinitely variable input signal is converted into an analogous

signal, such as that stored by the patterns of magnetization on a magnetic tape. Each time processing or copying takes place, there is a degradation of the signal. To eliminate this problem, a signal must be converted into digital format.

analog representation The presentation of information by means of a continuous scale within which any value may be specified, as seen, for instance, in a D'Arsonval meter.

analog signal A signal whose characteristics, such as amplitude or frequency, can vary in a continuous and infinitely variable manner. This contrasts with **digital signals**, which have a finite number of discrete steps within any given interval.

analog switch **1.** A switch that only passes analog signals. **2.** A switch that only passes reasonably undistorted analog signals.

analog telephone A telephone that communicates via **analog telephony**. Its abbreviation is **analog phone**.

analog telephony A voice-grade telecommunications system that carries signals that can assume any value, within a specified range, via telephone lines. In most metropolitan areas of the United States, for instance, traditional telephony was formerly all analog, but is now almost completely digital.

analog to digital A conversion from an analog input to a digital output. Its abbreviation is **A/D** or **a/d**.

analog-to-digital conversion The conversion of an analog input into a digital output.

analog-to-digital converter A circuit or device which transforms an analog input into a digital output. An example is a voltage or voice being converted into a digital signal. Its abbreviation is **ADC**, although **a/d converter** or **A/D converter** are also used. Also called **digitizer (1)**.

analog-to-frequency converter A device which converts a non-frequency analog input into a proportional output expressed as a frequency.

analog transmission Transmission of signals whose characteristics, such as amplitude or frequency, can vary in a continuous and infinitely variable manner.

analog video Video which is in analog form. For example, the lenses of a TV camera detect optical images which are converted into proportional electric video signals by its incorporated camera tube.

analog voltage A voltage which maintains a continuous and parallel proportion with an analog variable.

analogous pole In crystals which acquire an electric charge when heated, the pole which has the positive charge.

analogue Same as **analog**.

analogue adder Same as **analog adder**.

analogue camera Same as **analog camera**.

analogue cellular communications Same as **analog cellular communications**.

analogue cellular telephone Same as **analog cellular telephone**.

analogue cellular telephony Same as **analog cellular telephony**.

analogue channel Same as **analog channel**.

analogue circuit Same as **analog circuit**.

analogue communications Same as **analog communications**.

analogue comparator Same as **analog comparator**.

analogue computer Same as **analog computer**.

analogue control Same as **analog control**.

analogue control device Same as **analog control device**.

analogue data Same as **analog data**.

analogue decoding Same as **analog decoding**.

analogue delay line Same as **analog delay line**.

analogue device Same as **analog device**.

analogue display Same as **analog display**.

analogue divider Same as **analog divider**.

analogue electronics Same as **analog electronics**.

analogue encoding Same as **analog encoding**.

analogue filter Same as **analog filter**.

analogue gate Same as **analog gate**.

analogue indicator Same as **analog indicator**.

analogue input Same as **analog input**.

analogue line Same as **analog line**.

analogue meter Same as **analog meter**.

analogue modem Same as **analog modem**.

analogue monitor Same as **analog monitor**.

analogue multiplexer Same as **analog multiplexer**.

analogue multiplier Same as **analog multiplier**.

analogue network Same as **analog network**.

analogue output Same as **analog output**.

analogue phone Same as **analog phone**.

analogue processing Same as **analog processing**.

analogue readout Same as **analog readout**.

analogue recorder Same as **analog recorder**.

analogue recording Same as **analog recording**.

analogue representation Same as **analog representation**.

analogue signal Same as **analog signal**.

analogue switch Same as **analog switch**.

analogue telephone Same as **analog telephone**.

analogue telephony Same as **analog telephony**.

analogue to digital Same as **analog to digital**.

analogue-to-digital conversion Same as **analog-to-digital conversion**.

analogue-to-digital converter Same as **analog-to-digital converter**.

analogue-to-frequency converter Same as **analog-to-frequency converter**.

analogue transmission Same as **analog transmission**.

analogue video Same as **analog video**.

analogue voltage Same as **analog voltage**.

analysis **1.** The separation of a whole into its constituent parts. **2.** The separation of a whole into its constituent parts for the purpose of study. **3.** The careful study of an entity or concept from various perspectives in order to gain a fuller understanding. This can lead, for example, to the solutions of problems, or to improvements. **4.** The reported results of the knowledge obtained through **analysis (2)** or **analysis (3)**.

analytical electron microscope An electron microscope which incorporates additional specialized devices. Such devices, for instance, can serve to enhance image resolution, or to provide a sophisticated vacuum system.

Analytical Engine An uncompleted project from the early 1800s which incorporated several important computing concepts, such as programming.

analytical technique A procedure utilized to detect, identify, and study a substance or sample. Such a technique may also serve to quantify one or more aspects or properties of that which is examined. Instruments which are sophisticated, sensitive, and precise are usually utilized for this purpose. There are many examples, including electron microscopy, nuclear magnetic resonance, and electron, mass, infrared, and ultraviolet spectroscopy.

analyzer **1.** An instrument or device which serves in the carrying out of an analysis. For example, a circuit analyzer. **2.** A computer instrument, device, or program which solves problems and/or monitors performance. For instance, a logic analyzer.

anaphoresis Migration, toward the anode, of charged particles suspended in a medium under the influence of an electric field produced by electrodes immersed in said medium. When migration is toward the cathode, it is called **cataphoresis**.

ancillary device Same as **auxiliary device**.

ancillary equipment Same as **auxiliary equipment**.

AND A logical operation which is true only if all of its elements are true. For example, if three out of three input bits have a value of 1, then the output is 1. Any other combination yields an output of 0. For such functions, a 1 is considered as a true, or high, value, and 0 is a false, or low, value. Also called **AND operation**, **conjunction**, or **intersection (2)**.

AND circuit A circuit which has two or more inputs, and whose output is high only when all inputs are high. Also called **AND gate**.

AND gate Same as **AND circuit**.

AND/NOR gate Two AND gates whose output feeds into a NOR gate.

AND operation Same as **AND**.

AND-OR circuit Same as **AND-OR gate**.

AND-OR gate A gate with multiple inputs, and which will have an output if certain combinations of inputs are present. Also called **AND-OR circuit**.

Anderson bridge A six-branch bridge that is arranged in a manner which provides for the calculation of an unknown induction. It is a modification of a Maxwell-Wien bridge.

android A robot made to physically resemble a person. It may incorporate sensors to detect visual, tactile, and auditory stimuli. Most robots utilized in science and industry do not resemble people. Also called **humanoid robot**.

anechoic chamber Also called **anechoic room**, or **anechoic test chamber**. **1.** A room whose walls, ceiling, and floor are lined with sound-absorbing materials which minimize or eliminate all sound reflections. Used, for example, to test microphones or sound-level meters. Also called **dead room**, **free-field room**, or **echoless room**. **2.** A room whose walls, ceiling, and floor are lined with materials that absorb radio waves of given frequencies or bands. Used, for example, to test microwave devices.

anechoic room Same as **anechoic chamber**.

anechoic test chamber Same as **anechoic chamber**.

anemometer An instrument for measuring and/or indicating wind and air flow speeds. Also called **wind gauge**.

angle modulation Modulation in which the phase angle of the carrier wave is varied. Two forms are frequency modulation and phase modulation.

angle noise In radars, interference in reception due to variations in the angle from which an echo arrives after being reflected off a scanned object.

angle of arrival The angle at which electromagnetic radiation arrives at a receiving antenna or the surface of the earth. The angle is measured based on a reference plane, such as the horizon. Its abbreviation is **AOA**. Also called **arrival angle**.

angle of beam The angle between the points at which the intensity of an electromagnetic beam is at half of its maximum value. It can be measured in the horizontal plane or the vertical plane. Also called **beam width**, or **beam angle**.

angle of declination The angle below a reference plane, such as the horizon, of a descending line. Also called **declination angle**.

angle of deflection In CRTs, the maximum angle that an electron beam is diverted during deflection. For example, a

TV viewing screen may have a 110° deflection, resulting in greater contrast, detail, and clarity. Also called **deflection angle**.

angle of departure Same as **angle of radiation**. Its abbreviation is **AOD**. Also called **departure angle**.

angle of deviation In optics, the angle between the ray incident on an object and the ray emerging from it. This angle can be due to refraction, reflection, or diffraction. Also called **deviation (3)**, or **deviation angle**.

angle of divergence Also called **divergence angle**. **1.** In CRTs, the angle encompassed by the spread of the electron beam as it travels from the cathode to the screen. **2.** The angle encompassed by the spread of a light beam as it travels from a collimating source.

angle of elevation The angle above a reference plane, such as the horizon, of an ascending line. For example, the effective radiated power of a given antenna array may vary as a function of the angle of elevation. Also called **elevation angle**.

angle of incidence The angle formed between the line of an incident ray or wave and a perpendicular line arising from the point of incidence. Also called **incidence angle**.

angle of lag For two periodic quantities with the same frequency, the lag in phase of one respective to the other. Expressed in radians or degrees. Also called **lag angle**, or **phase delay (1)**.

angle of lead For two periodic quantities with the same frequency, the lead in phase of one respective to the other. Expressed in radians or degrees. Also called **lead angle**.

angle of radiation The angle at which electromagnetic radiation departs from a transmitting antenna or the surface of the earth. The angle is measured based on a reference plane, such as the horizon. Also called **angle of departure**, or **radiation angle**.

angle of reflection The angle formed between the line of a reflected ray or wave and a perpendicular line arising from the point of reflection. Also called **reflection angle**.

angle of refraction The angle formed between the line of a refracted ray or wave and a perpendicular line arising from the point of refraction. Also called **refraction angle**.

angstrom Its symbol is Å. A unit of length equivalent to 10^{-10} m, or one-tenth of a nanometer. Utilized to measure extremely short waves, atomic and molecular distances and dimensions, and very short length intervals in general.

angular-deviation loss For a microphone or speaker, the ratio between the response along its principal axis and that of another given angle. Usually expressed in decibels.

angular displacement The change in the angle of an object moving along the circumference of a circle. May be measured in radians or degrees.

angular frequency The frequency of a periodic quantity expressed in radians per second. It is obtained by multiplying the frequency by 2π. Also called **radian frequency**, **circular frequency**, or **angular velocity**.

angular momentum For a given body, the product of the angular frequency and the moment of inertia.

angular phase difference For two periodic quantities with the same frequency, the difference between their respective phases. Expressed in radians or degrees. Also known as **phase angle**, or **phase difference**.

angular rate Same as **angular speed**.

angular resolution The capability of a device or instrument to distinguish two adjacent objects through the measurement of angles. Used, for example, in describing the ability of a telescope to tell apart two objects in the sky.

angular speed For a moving object, the change in direction per unit time, as expressed in radians per second. Also known as **angular rate**, or **rotation speed**.

angular velocity Same as **angular frequency**.

anharmonic oscillator For an oscillating system, an equilibrium-restoring force which is not proportional to the displacement from the equilibrium position. This contrasts with a **harmonic oscillator (4)**, in which the forces restoring the equilibrium position are proportional to the displacement from it.

ANI Abbreviation of **automatic number identification**.

animated GIF Animation consisting of a series of GIF images.

animated graphics A sequence of pictures arranged in a manner that simulates movement. Cartoons are an example of animated graphics. This contrasts with **video**, which consists of true motion which is divided into still frames.

animation Simulation of movement through the use of **animated graphics**. Also called **video animation**.

anion A negatively charged ion. Anions collect at the anode when subjected to an electric potential while in solution. Chloride ion and hydroxide ion are two common anions. Also called **negative ion**.

anisotropic Having physical properties that vary depending on the axis along which they are measured. For example, the index of refraction of a substance may display this phenomenon. This contrasts with **isotropic**, in which physical properties are identical in all directions. Also known as **aelotropic**.

anisotropy The manifestation of **anisotropic** properties. Also called **aelotropy**.

ANL Abbreviation of **automatic noise limiter**.

ANN Abbreviation of **artificial neural network**.

anneal To subject a material to sustained heating followed by cooling, each at suitable rates, to change its properties.

annealed wire Wire, such as that of copper or aluminum, which undergoes **annealing** after being drawn. Such wire has decreased tensile strength in comparison to **hard-drawn wire**. Also called **soft-drawn wire**.

annealing The sustained heating of a material followed by cooling, each at suitable rates, to change its properties. This process may be used, for instance, to realign the atoms in a crystal, or to stabilize certain electrical properties.

annular conductor A conductor which is a collection of wires twisted or braided together in three ring-shaped layers around a core.

annular transistor A transistor in which the base and the collector are situated in concentric circles around the emitter.

annunciator An electrical device which utilizes sounds and/or lights to signal an event or circumstance. For example, a blinking light indicating an incoming call.

anode Also called **positive electrode**, or **posode**. **1.** The positive electrode in an electron tube. Electrons emitted by the cathode travel towards it. Also known as **plate (1)**. **2.** The positive electrode in an electrolytic cell. When a current is passed through the cell, negative ions travel towards it. **3.** The electrode where oxidation occurs in an electrochemical cell.

anode battery Also known as **plate battery**, or **B battery (1)**. A battery which supplies the anode, or plate, current in electron tubes.

anode capacitance In an electron tube, the capacitance between the anode and another electrode, especially the cathode. Also called **plate capacitance**.

anode characteristic In an electron tube, the variation of the current of the anode relative to the voltage applied to it. Also known as **plate characteristic**.

anode circuit A circuit which includes all of the components connected between the anode and the cathode in an electron tube, including the anode voltage source. Also called **plate circuit**.

anode corrosion The dissolution of a metal or alloy acting as an anode.

anode current In an electron tube, the electron flow from the cathode to the anode. Also known as **plate current**.

anode dark space In glow-discharge and gas tubes, a narrow dark zone near the anode.

anode detection The function of an **anode detector**.

anode detector An electron tube detector in which the anode circuit rectifies the input signals. Also called **plate detector**.

anode dissipation In an electron tube, the power dissipated by the anode in the form of heat. This loss is caused by the anode being bombarded by electrons and anions. Also called **plate dissipation**.

anode drop Same as **anode fall**.

anode effect During electrolysis, a sudden increase in voltage with a corresponding decrease in amperage caused by the anode becoming separated from the electrolyte by a gas film.

anode efficiency In an electron tube, the ratio between the AC load circuit power and the DC anode input power. Also called **plate efficiency**.

anode fall In a gas tube, a sharp drop in potential near the anode due to a space charge. Also called **anode drop**.

anode film In an electrolytic process, the layer of the solution that is in immediate contact with the anode. When compared to the bulk of the solution, this layer has different properties.

anode glow In glow-discharge and gas tubes, a narrow bright zone on the anode side of the positive column.

anode impedance In an electron tube, the total impedance between the anode and the cathode, without taking into account the electron stream. Also called **anode-load impedance**, **plate-load impedance**, or **plate impedance**.

anode input power In an electron tube, the DC power consumed by the anode. Also called **anode power input**, or **plate input power**.

anode-load impedance Same as **anode impedance**.

anode metal A metal utilized as an anode in an electrolytic process.

anode modulation In an electron tube, amplitude modulation obtained by varying the voltage of the anode proportionally to the fluctuations in the modulating wave. Also called **plate modulation**.

anode mud Same as **anode slime**.

anode potential Same as **anode voltage**.

anode power input Same as **anode input power**.

anode power supply In an electron tube, the DC applied to the anode to place it at a high current potential relative to the cathode. Also called **anode supply**, or **plate supply**.

anode pulse modulation In an electron tube, modulation produced by applying external voltage pulses to the anode. Also called **plate pulse modulation**.

anode rays In an electron tube, streams of positive ions emanating from an anode. They are usually due to impurities in the anode metal.

anode resistance In an electron tube, the ratio of a minimal change in the anode voltage to a minimal change in the anode current. All other voltages must be held constant. Also called **plate resistance**.

anode saturation In an electron tube, the condition in which the anode current can not be further increased, regardless of any additional voltage applied to it, since essentially all available electrons are already being drawn to said anode.

Also called **plate saturation**, **current saturation**, **voltage saturation**, or **saturation (3)**.

anode sheath In a gas tube, a layer of electrons which surrounds the anode when its current is high. Also called **sheath (3)**.

anode slime In electrolytic refining or plating, an insoluble residue that collects at the bottom of the solution or on the surface of the anode. In the electrolytic refining of copper, for example, this mud usually contains platinum, gold, and silver. Also called **anode mud**, or **slime**.

anode supply Same as **anode power supply**.

anode terminal 1. The positive terminal of an electron tube. 2. The terminal by which current enters a semiconductor device. 3. In a diode which is biased in the forward direction, the terminal that is positive with respect to the other.

anode voltage Also called **anode potential**, or **plate voltage**. 1. In an electron tube, the difference in potential between the anode and the cathode. 2. In an electron tube, the difference in potential between the anode and a specific point of the cathode.

anodic Pertaining or relative to an **anode**.

anodic coating An oxide film formed on the surface of a metal, such as aluminum or magnesium, in an electrolytic process. Such coatings can be protective, decorative, or functional. Examples include the production of materials with high resistance to corrosion, or with good electrical insulation.

anodic polarization A change in anode potential resulting from the effects of current flow at or near the its surface.

anodic protection The reduction or prevention of corrosion of a metallic material by imposing an external potential which diminishes the rate at which said material dissolves.

anodize To apply an **anodic coating**.

anodizing The process of applying an **anodic coating**.

anolyte During electrolysis, the portion of the electrolyte in the immediate vicinity of the anode.

anomalous 1. Inconsistent with what is normal or expected. 2. Difficult to describe or classify.

anomalous propagation Inexplicable and/or unexpected propagation of waves. For example, the reception of ultrahigh frequency or very-high frequency signals well beyond the usual distances. May be caused by fluctuations in the properties, such as density and temperature, of the propagating medium.

anomaly detection The use of an **anomaly detector**. Also, a specific instance of a found anomaly.

anomaly detector A system that detects undesired activity, such as that resulting from hacking, by its deviation from the ordinary behavior expected.

anonymity The quality or state of not being identified. When communicating via a network, especially the Internet, there are various levels of anonymity, ranging from the use of a screen name, to being entirely untraceable without any identifying information being available. Anonymity can be useful to protect the privacy and safety of a person, but may also allow those who wish to harm others to do so without fear of identification.

Anonymizer A utility or service which allows anonymous Web surfing. Also called **Web Anonymizer**.

anonymous File Transfer Protocol A method of downloading public files using File Transfer Protocol. It is called anonymous because users are not required to identify themselves to use the service, nor are passwords usually required for access. Its abbreviation is **anonymous FTP**.

anonymous FTP Abbreviation of **anonymous File Transfer Protocol**.

anonymous remailer On the Internet, an email service that enables users to send messages to others without the recipients knowing the return address.

ANSI Acronym for **A**merican **N**ational **S**tandards **I**nstitute. An organization involved in the facilitation of development of technical standards. Its main objectives include minimizing incompatibility problems on a global basis in order to promote the advancement of technology. It is a member of the International Organization for Standardization and the International Electrotechnical Commission.

ANSI character set An ANSI-defined 8-bit character set that includes 256 characters. It incorporates many foreign characters, symbols, and special punctuation.

ANSI X12 An ANSI standard pertaining to EDI. Its abbreviation is **X12**.

answer-back A response that a computer gives when requested to identify itself within a network.

answer-only modem A modem which can answer incoming calls, but cannot originate any.

answering machine A device which automatically answers calls, plays a message, and records messages left by callers. Such a device may be incorporated into a telephone, into a similar device, such as a properly equipped fax or computer, and so on. Also called **telephone answering machine**.

antenna A device that radiates and/or receives radio waves. The same antenna can usually be used for both transmission and reception. There are many types of antennas, their selection depending on the uses and operating frequencies desired. Also known as **aerial**.

antenna-amplifier A device which incorporates an antenna and an amplifier. Used mostly for compact portable units. Its abbreviation is **antennafier**.

antenna amplifier An amplifying device or component which improves the signal-to-noise ratio of a receiving antenna. Such a component or device may be within the antenna, or nearby.

antenna aperture An opening through which radiated energy passes to or from an antenna.

antenna array An assembly of antenna elements with the proper dimensions, characteristics, and spacing, so as to maximize the intensity of radiation in the desired directions. Also called **array (2)**, or **array antenna**.

antenna beamwidth The angle between the points at which the intensity of an antenna is at half of its maximum value. It can be measured in the horizontal plane or the vertical plane. Also called **beamwidth of antenna**.

antenna coil A coil through which the antenna current passes.

antenna coincidence The circumstance where two directional antennas are pointing directly at each other.

antenna-converter A device which incorporates an antenna and a converter. Intended to extended the available reception frequencies. Its abbreviation is **antennaverter**.

antenna counterpoise One or more wires mounted close to the ground, but insulated from it, to form a low-impedance and high-capacitance path to said ground. This provides a radio-frequency ground for the antenna. Also called **counterpoise**.

antenna coupler A device, such as an RF transformer, utilized to match the impedance of a transmitter and/or receiver to an antenna, so as to provide maximum power transfer. This enables, for example, the use of a shorter antenna than would ordinarily be needed.

antenna crosstalk The unwanted transfer of power through space between antennas. It is equal to the ratio of the power received by one antenna, to the power radiated by the other. It is usually expressed in decibels.

antenna current **1.** The RF current flowing into an antenna, measured with no modulation present. **2.** The RF current flowing out of an antenna, measured with no modulation present.

antenna detector A device, which incorporates an antenna, that signals when being tracked by radar.

antenna diplexer A device which allows two transmitters to use the same antenna simultaneously or alternately. Used, for example, in TV broadcasting to transmit the audio and video content via the same antenna. Also called **diplexer**.

antenna directivity The gain of an antenna in relation to the direction it is pointed in. For example, an omnidirectional antenna has, within the same plane, a uniform radiation and reception pattern.

antenna duplexer A device which enables a single antenna to be used to transmit and receive simultaneously or alternately. Used, for example, in radars. Also called **duplexer**.

antenna effect In a loop antenna, an unbalancing of the directional properties due to interference from nearby objects. This can cause it to act partially as a small vertical antenna. Electrostatic shielding can help minimize this effect. Also called **height effect**.

antenna efficiency The ratio between the received energy and the radiated energy of an antenna. Usually expressed as a percentage. If there were no losses, the antenna efficiency would be 100%.

antenna element Within an antenna array, one of multiple radiators which contribute to its overall transmission and/or reception characteristics. Such an element may be driven, or parasitic. Also called **element (4)**.

antenna factor The ratio of an incident electric field to the voltage received at the antenna terminals. It is usually expressed in decibels.

antenna feed The transmission line, such as a coaxial cable, which carries signals between a transmitter and an antenna. Also called **antenna feeder**, **antenna transmission line**, **feed (5)**, or **feed line**.

antenna feed point The point in an antenna where the **antenna feed** is received. Also called **feed point (2)**.

antenna feed point impedance The impedance at an **antenna feed point**. Also called **feed point impedance**.

antenna feeder Same as **antenna feed**.

antenna field A group of antennas arranged in a manner so as to maximize the intensity of radiation in the desired directions.

antenna gain The effectiveness, generally expressed in decibels, of a directional antenna as compared to a given standard, such as a dipole or isotropic antenna. Its symbol is **A**. Also known as **gain (2)**.

antenna height **1.** The height of an antenna above the terrain immediately surrounding its location. **2.** The height of an antenna above the average level of the terrain within a given radius surrounding its location. Also called **antenna height above average terrain**, or **height above average terrain**.

antenna height above average terrain Same as **antenna height (2)**.

antenna lens An antenna that uses an arrangement of metal and/or dielectric objects as a lens to focus the radio waves. Also called **lens antenna**.

antenna loading The utilization of inductive and/or capacitive elements to increase the electrical length of an antenna, or otherwise alter its characteristics. Also called **loading (1)**.

antenna lobe Within an antenna pattern, a three-dimensional section in which radiation and/or reception is increased. Also called **lobe (1)**, or **directional lobe (1)**. When referring specifically to radiation, also known as **radiation lobe**.

antenna mast A vertical rod which serves as an antenna, or as an antenna support. Also called **mast**.

antenna matching The adjusting of the impedance of an antenna and/or its transmission line so that they are approximately equal. This, for instance, can make the antenna more effective over a larger range of frequencies.

antenna pattern The pattern representing how the field strength of an antenna varies as a function of direction and distance. It may refer to the transmitting or receiving effectiveness of an antenna, and can be measured in any plane, although the horizontal and/or vertical planes are most common. A typical antenna pattern has lobes that indicate areas of enhanced response within a given plane. An Adcock antenna, for example, produces an antenna pattern in the form of a figure-eight. Also called **antenna radiation pattern**.

antenna phase center For a given antenna, the apparent center of signal transmission or reception. This point varies, and depends, for instance, on the signal frequency. Also called **phase center**.

antenna polarization The orientation of the radiated and received electric lines of flux of an antenna, in relation to the surface of the earth. Also called **polarization (4)**.

antenna power **1.** The power delivered by a transmitter, which is incident upon an antenna. **2.** The power delivered by a transmitting antenna. It is the product of the square of the antenna current multiplied by the resistance at that point.

antenna power gain The ratio of the maximum radiation intensity of an antenna in a stated direction, to the maximum radiation intensity of a reference antenna, with identical power applied to both.

antenna radiation pattern Same as **antenna pattern**.

antenna radiator The part of an antenna or element which emits radio waves. Also called **radiator (2)**.

antenna range The interval within which an antenna operates. It may refer to such variables as its frequency bands, or its maximum useful distance from a transmitter or receiver.

antenna resistance The total resistance of a transmitting antenna in operation. This incorporates various components, including ground resistance and radiation resistance. It is equal to the power supplied to the antenna divided by the square of the effective current at the point where said power is supplied.

antenna splitter A device which divides the signals received, or those to be sent, of an antenna, so as to provide multiple paths. Used, for instance, to enable multiple TVs to access the signal of a single satellite. Also called **splitter (2)**.

antenna system **1.** The complete complement of components that constitute an antenna that transmits and/or receives. **2.** An assembly of antenna elements with the proper dimensions, characteristics, and spacing, so as to maximize the intensity of radiation or reception in the desired directions.

antenna temperature A quantity describing the effective noise temperature of an antenna. It factors in environmental sources as well as those generated within it.

antenna tower A structure which is utilized to support an antenna.

antenna transmission line Same as **antenna feed**.

antenna-transmitter A device which incorporates an antenna and an oscillator. It serves as a low-power transmitter. Its abbreviation is **antennamitter**.

antennafier Abbreviation of **antenna-amplifier**.

antennamitter Abbreviation of **antenna-transmitter**.

antennaverter Abbreviation of **antenna-converter**.

anti-aliasing Also spelled **antialiasing**. **1.** The avoidance of aliasing through the utilization of an **anti-aliasing filter**.

2. The smoothing of the jagged appearance that curved or diagonal lines in computer graphics have when there is not enough resolution to show images realistically. One such technique involves changing the shades of color of the pixels surrounding the lines so that they blend in better with the background.

anti-aliasing filter A filter which blocks all frequencies above a given cutoff frequency, before an analog-to-digital conversion. The filter insures that no input signals have a higher frequency than half the digital sampling rate. This avoids aliasing. Also spelled **antialiasing filter**. Also called **low-pass filter (1)**, or **smoothing filter**.

anti-capacitance switch Same as **anticapacitance switch**.

anti-cathode Same as **anticathode**.

anti-clutter gain control Same as **anticlutter gain control**.

anti-coincidence Same as **anticoincidence**.

anti-coincidence circuit Same as **anticoincidence circuit**.

anti-electron Same as **antielectron**.

anti-fading antenna Same as **antifading antenna**.

anti-ferroelectric crystal Same as **antiferroelectric crystal**.

anti-ferromagnetic Same as **antiferromagnetic**.

anti-ferromagnetism Same as **antiferromagnetism**.

anti-glare filter A filter intended to minimize the ambient light that reaches a computer or TV viewing screen, causing glare. It can take the form of a plastic that is placed in front of the screen, or it can be incorporated into it. Also spelled **antiglare filter**. Also called **ambient-light filter**, or **glare filter**.

anti-glare screen A screen, such as that of a TV or computer monitor, incorporating an **anti-glare filter**. Also spelled **antiglare screen**.

anti-hunt circuit Same as **antihunt circuit**.

anti-logarithm Same as **antilogarithm**.

anti-magnetic Same as **antimagnetic**.

anti-matter Same as **antimatter**.

anti-neutron Same as **antineutron**.

anti-node Same as **antinode**.

anti-noise microphone Same as **antinoise microphone**

anti-oxidant Same as **antioxidant**.

anti-parallel Same as **antiparallel**.

anti-particle Same as **antiparticle**.

anti-phase Same as **antiphase**.

anti-proton Same as **antiproton**.

anti-quark Same as **antiquark**.

anti-rad Same as **antirad**.

anti-radiation Same as **antirad**. Also spelled **antiradiation**.

anti-reflection coating Same as **antireflection coating**.

anti-reflective coating Same as **antireflective coating**.

anti-resonance Same as **antiresonance**.

anti-resonant circuit Same as **antiresonant circuit**.

anti-resonant frequency Same as **antiresonant frequency**.

anti-sidetone circuit Same as **antisidetone circuit**.

anti-stat Same as **antistat**.

anti-static Same as **antistatic**.

anti-static agent Same as **antistatic agent**.

anti-static bag Same as **antistatic bag**.

anti-static coating Same as **antistatic coating**.

anti-static material Same as **antistatic material**.

anti-static strap Same as **antistatic wrist strap**. Also spelled **antistatic strap**.

anti-static wrist band Same as **antistatic wrist strap**. Also spelled **antistatic wristband**.

anti-static wrist strap Same as **antistatic wrist strap**.

anti-static wristband Same as **antistatic wrist strap**. Also spelled **antistatic wrist band**.

anti-virus program Same as **antivirus program**.

anti-virus scanner Same as **antivirus scanner**.

antialiasing Same as **anti-aliasing**.

antialiasing filter Same as **anti-aliasing filter**.

anticapacitance switch A switch designed to minimize any capacitance between its terminals when open. Also spelled **anti-capacitance switch**.

anticathode In an X-ray tube, the target electrode. The cathode focuses an electron beam to a location on the anticathode, where they are absorbed. Also spelled **anti-cathode**.

anticlutter gain control In a radar receiver, a gain control that smoothly increases gain to its maximum after each transmitter pulse. This helps scan objects at a greater distance. Also spelled **anti-clutter gain control**.

anticoincidence The occurrence of an event without another occurring elsewhere simultaneously, or within a given time interval. Also spelled **anti-coincidence**.

anticoincidence circuit Also spelled **anti-coincidence circuit**. 1. A circuit, with two or more inputs, which delivers an output pulse only if one or more inputs do not receive a pulse simultaneously, or within a given time interval. 2. A circuit, with two or more inputs, which delivers an output pulse only if one or more inputs do not receive a pulse in a predetermined combination simultaneously, or within a given time interval.

antielectron Also spelled **anti-electron**. Also called **positron**. 1. The antiparticle equivalent of the electron. It has the same mass as the electron, but opposite electric charge and magnetic moment. 2. A positively-charged electron. The term is utilized to differentiate such a particle from a **negatron**, or negatively-charged electron.

antifading antenna A transmitting antenna which radiates in a manner that minimizes fading. One technique is to limit radiation to small angles of elevation. Also spelled **anti-fading antenna**.

antiferroelectric crystal A crystal with two interpenetrating lattices and having two regions of symmetry, each with a different electric polarization. Also spelled **anti-ferroelectric crystal**.

antiferromagnetic Displaying **antiferromagentism**. Also spelled **anti-ferromagnetic**.

antiferromagnetism A phenomenon observed in some materials, such as selected metals, alloys, or salts of transition metals, in which there is a complete cancellation of magnetic moments. This occurs when an ordered array of magnetic moments forms spontaneously, with alternate moments having opposite directions. This phenomenon is only manifested below certain temperatures. Also spelled **anti-ferromagnetism**.

antiglare filter Same as **anti-glare filter**.

antiglare screen Same as **anti-glare screen**.

antihunt circuit A circuit which helps prevent oscillations above and below a desired level or setting. Such oscillations are often caused by overcompensation by automatic control systems. Also spelled **anti-hunt circuit**.

antilogarithm The number which is the result of a logarithmic operation on another number. For example, the logarithm of 1,000 is 3, so the antilogarithm of 3 is 1,000. Also spelled **anti-logarithm**. Also called **inverse logarithm**.

antilogous pole The negatively charged pole in crystals that become electrically polarized when exposed to heat.

antimagnetic Also spelled **anti-magnetic**. 1. The property of being resistant or impervious to the effects of magnetic fields. 2. Devices or components designed to be resistant or impervious to the effects of surrounding magnetic fields.

antimatter Matter composed of antiparticles such as antiprotons, antineutrons, and positrons. When antimatter meets its conventional matter counterpart, they are each annihilated, releasing a large quantity of energy. Also spelled **anti-matter**.

antimony A silvery-white brittle metallic chemical element with atomic number 51. It has over 35 known isotopes, of which 2 are stable. Known since ancient times, antimony is currently used in storage batteries, cable sheaths, and as a semiconductor dopant, among others. Its chemical symbol is **Sb**, which is taken from the Latin word **stibium**.

antineutron The antiparticle equivalent of the neutron. It has the same mass and neutral charge, but opposite magnetic moment. Also spelled **anti-neutron**.

antinode Also spelled **anti-node**. 1. In a standing-wave system, a point of maximum amplitude or displacement. For example, a maximum of voltage or current. This contrasts with a **node (4)**, where the converse is true. Also called **loop (3)**. 2. In a vibrating system, a point of maximum amplitude or displacement. This contrasts with a **node (5)**, where the converse is true. Also called **loop (4)**.

antinoise microphone A microphone which is designed to selectively minimize certain noises. For instance, that which is highly directional. Also spelled **anti-noise microphone**. Also called **noise-canceling microphone (1)**.

antioxidant Also spelled **anti-oxidant**. 1. A substance that retards or prevents oxidation. 2. A substance that covers, or that is added to something else, to retard or prevent oxidation. In electronics, it is generally used to avoid oxidation by air and/or water.

antiparallel Parallel, but with opposite direction or orientation. Also spelled **anti-parallel**.

antiparticle Subatomic particles that are identical to conventional particles in all respects except electric charge and magnetic moment. The mass, lifetime, and spin are identical, but the electric charge and magnetic moment are inverted. Examples of antiparticles include: antiprotons, antineutrons, and positrons. Also spelled **anti-particle**.

antiphase In complete phase opposition with another alternating quantity. Also spelled **anti-phase**.

antiproton The antiparticle equivalent of the proton. It has the same mass as the proton, but opposite electric charge and magnetic moment. Also spelled **anti-proton**. Also called **negative proton**.

antiquark The antiparticle equivalent of the quark. Also spelled **anti-quark**.

antirad Abbreviation of **anti-rad**iation. A material or substance which protects from radiation. For example, a satellite in space employs such a material to help prevent damage to instruments which are sensitive to radiation. Also spelled **anti-rad**. Also called **antirad material**, or **antirad substance**.

antirad material Same as **antirad**.

antirad substance Same as **antirad**.

antiradiation Same as **antirad**. Also spelled **anti-radiation**.

antireflection coating Same as **antireflective coating**. Also spelled **anti-reflection coating**. Its abbreviation is **AR coating**.

antireflective coating Its abbreviation is **AR coating**. Also spelled **anti-reflective coating**. Also called **antireflection coating**. 1. One or more thin dielectric or metallic films applied to a surface to reduce reflection and increase transmission. Used frequently in coating optical surfaces. 2. A coating which minimizes the reflection of waves, particles, heat, or the like.

antiresonance Also spelled **anti-resonance**. **1.** Resonance in an **antiresonant circuit**. Also called **parallel resonance (1)**. **2.** In an **antiresonant circuit**, the frequency at which the parallel impedance is highest. Also called **parallel resonance (2)**. **3.** In an **antiresonant circuit**, the frequency at which the inductive capacitance equals the capacitive reactance. Also called **parallel resonance (3)**. **4.** The condition in which the impedance of a dynamic system approaches infinity.

antiresonant circuit A resonant circuit in which a capacitor and an inductor are connected in parallel with an AC source. Resonance occurs at or near the maximum impedance of the circuit. Also spelled **anti-resonant circuit**. Also called **parallel-resonant circuit**, or **tank circuit (1)**.

antiresonant frequency The resonant frequency of an **antiresonant circuit**. Also spelled **anti-resonant frequency**. Also called **parallel-resonant frequency**.

antisidetone circuit A circuit which reduces interference created by the receiver excessively reproducing sounds from the same telephone's transmitter. Although a caller's own voice is usually heard over the same telephone, if the amplitude of this signal exceeds a given threshold it is usually unwanted. Also spelled **anti-sidetone circuit**.

antistat Abbreviation of **antistatic**. Also spelled **anti-stat**.

antistatic A material or agent which helps minimize the buildup of static electricity, and which allows it to be safely dissipated. Also spelled **anti-static**. Its abbreviation is **antistat**. Also called **antistatic agent**, or **antistatic material**.

antistatic agent Same as **antistatic**. Also spelled **anti-static agent**.

antistatic bag A bag, such as that utilized to store and transport ICs, with an **antistatic coating**. Also spelled **antistatic bag**.

antistatic coating A material or agent which is applied to surfaces and components to help minimize the buildup of static electricity, and to allow it to be safely dissipated. This may help, for instance, in avoiding damage to certain electronic devices or magnetic media. Also spelled **anti-static coating**.

antistatic material Same as **antistatic**. Also spelled **anti-static material**.

antistatic strap Same as **antistatic wrist strap**. Also spelled **anti-static strap**.

antistatic wrist band Same as **antistatic wrist strap**. Also spelled **anti-static wrist band**.

antistatic wrist strap A grounding strip which is typically wrapped around a wrist on one side, and attached to the chassis of the device or piece of equipment being worked on the other, so as to properly channel any static electricity a person may carry or generate. Used, for instance, to prevent a user from harming sensitive computer components when performing tasks such as the installation additional RAM. Also spelled **anti-static wrist strap**. Also called **antistatic strap**, **antistatic wrist band**, **ESD wrist strap**, or **wrist strap**.

antistatic wristband Same as **antistatic wrist strap**. Also spelled **anti-static wristband**.

antitransmit/receive switch A switch that prevents interaction between the receiver and transmitter of a device. Generally used in radars. Its abbreviation is **ATR switch**.

antitransmit/receive tube A switching tube that prevents interaction between the receiver and transmitter of a device. Generally used in radars. Its abbreviation is **ATR tube**.

antivirus program A computer program intended to identify, locate, isolate, and eliminate viruses. Such a program may also scan incoming files and data for viruses before a computer is exposed. Also spelled **anti-virus program**. Also called **antivirus scanner**, or **virus scanner (1)**.

antivirus scanner Same as **antivirus program**. Also spelled **anti-virus scanner**.

antivoice-operated transmission A radio communications device which utilizes a voice-activated circuit to prevent the transmitter from being operated while the receiver in the same unit is receiving.

any key A computer message in which the user is prompted to press a key on the keyboard. So long as there is an input from the keyboard, it doesn't matter which key is pressed.

AO/DI Abbreviation of **Always On/Dynamic ISDN**.

AOA Abbreviation of **angle of arrival**.

AOD 1. Abbreviation of **angle of departure**. **2.** Abbreviation of **Audio on Demand**.

AODI Abbreviation of **Always On/Dynamic ISDN**.

AP Abbreviation of **access protocol**.

APA Abbreviation of **all points addressable**.

Apache An open-source HTTP server.

APC 1. Abbreviation of **automatic picture control**. **2.** Abbreviation of **automatic phase control**.

APCVD Abbreviation of **atmospheric pressure chemical-vapor deposition**.

APD Abbreviation of **avalanche photodiode**.

aperiodic 1. Not occurring in a regular and repetitive pattern. **2.** Non-repetitive. **3.** Not displaying resonant response.

aperiodic antenna An antenna that has an approximately constant input impedance over a wide range of frequencies. A diamond antenna is an example. Also called **non-resonant antenna**, or **untuned antenna**.

aperiodic chamber Same as **aperiodic enclosure**.

aperiodic circuit 1. A circuit which does not resonate at any frequency. **2.** A circuit whose impedance is uniform over all frequencies.

aperiodic damping 1. Damping so large that after an instantaneous disturbance a system comes to rest without passing through or oscillating around the state of equilibrium. Seen, for instance, in a device or instrument pointer which settles without overshooting the rest position after a measurement. **2.** Damping as a result of an enclosure system in which the overshoot of a speaker is dampened, providing more precise reproduction.

aperiodic enclosure An enclosure system in which the overshoot of a speaker is dampened, for more precise reproduction. This is accomplished by incorporating a vent that is stuffed with a damping material. Such an enclosure achieves a balance between an acoustic suspension and a bass reflex design. Also called **aperiodic chamber**.

aperiodic signal Also called **non-periodic signal**. **1.** A signal which does not occur in a regular and repetitive pattern. **2.** A non-repetitive signal.

aperiodic waveform Also called **non-periodic waveform**. **1.** A waveform which does not occur in a regular and repetitive pattern. **2.** A non-repetitive waveform.

aperture 1. An opening, such as a hole, gap, slit, vent, port, mouth, outlet, or perforation, through which electrons, light, waves, or any form of radiant energy may pass. Also called **orifice (1)**. **2.** The opening by which light enters, or through which it travels, of an optical device or system, such as a telescope or scanner. Such an opening may be in any shape, but is usually circular, rectangular, or elliptical. **3.** In an antenna, the opening through which the radiated energy passes. **4.** In a speaker, a port in the enclosure which increases and extends the reproduction of low frequencies.

aperture antenna An antenna whose beam width is determined by the dimensions of the opening through which the radiant energy passes. One advantage of such antennas is the low profile they may have. An example is a horn antenna.

aperture distortion A type of distortion that is caused by the size and shape of the scanning aperture. In a TV, for example, it is due to the size of the electron scanning beam. This may distort the image or diminish resolution.

aperture grill picture tube In a picture tube with a three-color gun, a grill with vertical wires that is placed behind the screen to make sure that each color beam strikes the correct phosphor dot on the screen. It insures, for instance, that the electron beam intended for the blue phosphor dots only hits those. Such a grill provides an image with better clarity and sharpness.

aperture illumination The field-intensity distribution of an electromagnetic wave over an aperture. Used especially in the context of the aperture of an antenna.

aperture mask In a picture tube with a three-color gun, a grill with round holes that is placed behind the screen to make sure that each color beam strikes the correct phosphor dot on the screen. It insures, for instance, that the electron beam intended for the red phosphor dots only hits those. Also called **shadow mask**, or **mask (3)**.

aperture pitch On the display screen of a picture tube with an aperture mask, the distance between phosphor dots of like color between one phosphor stripe and the next. Usually expressed in millimeters, and the lower the number, the crisper the image. Also called **stripe pitch**.

aperture synthesis telescope A computer-controlled telescope which utilizes two or more pairs of antennas which sequentially cover sections of the total aperture, in order to gather the information equivalent to that obtained by a much larger single telescope. Also called **synthesis telescope**.

aperture time 1. The time interval during which a measuring instrument acquires a sample of a signal. 2. The time interval necessary for a measuring instrument to acquire a sample of a signal.

API 1. Abbreviation of **air-position indicator**. 2. Abbreviation of **application program interface**.

APL Abbreviation of **A** Programming **L**anguage. A concise high-level programming language especially suited for generating matrices.

app Abbreviation of **application**.

apparatus A device or integrated group of devices utilized for a given purpose.

apparent horizon As seen from a given location, the junction at which the earth, or the sea, appears to meet the sky. Also called **horizon (1)**, or **visible horizon**.

apparent power In an AC circuit, the product of the RMS current and the RMS voltage. This value does not take into account the phase angle between the voltage and current. Expressed in volt-amperes.

appearance potential In a mass spectrometer, the minimum potential required to provide an electron beam with the minimum energy necessary to produce fragmentation ions of a particular species.

Applegate diagram For velocity-modulation tubes, a diagram which plots the flow of electron groups as a function of time.

applet A small application designed to run within another program. Frequently, it is downloaded over a network to be launched on a user's computer. On a Web page, for example, it can provide video and/or audio effects, or perform calculations.

AppleTalk A popular LAN protocol or architecture.

Appleton layer The highest layer in the ionosphere, spanning an altitude of about 200 to 400 kilometers. Although this layer is influenced by factors such as time of day, season, and sunspot activity, it exists at all times, and at night it is the only component of the F region. It is useful for long-distance propagation of high-frequency radio waves day or night. Also known as **F$_2$ layer**.

appliance A powered device designed to help do work, perform tasks, or to assist in providing comfort or convenience. These usually use electrical energy, and include dryers, lamps, refrigerators, and microwave ovens. Precision electronics, such as high-fidelity components and computers are sometimes not included in this group.

appliance server A self-contained server which incorporates both hardware and software, and which is designed to be installed and maintained with a minimum of effort and support. Such a server is usually plugged into an existing network, with all supported applications preinstalled, and is used, for instance, as a Web server, mail server, or file server. Appliance servers are configured and accessed via Web browsers. Also called **server appliance**.

application 1. A specific use for a device. 2. A computer program designed to enable end users to perform specific tasks, such as word processing or communications. Its abbreviation is **app**. Also called **application program**.

application binary interface A set of instructions that determine how a computer application interacts with the operating system and the hardware. This contrasts with an **application program interface**, which only interacts with the operating system. Its abbreviation is **ABI**.

application-centric Pertaining to an operating system in which the application is the starting point of a task. For example, to work on a document, first the word-processing program is opened, and then the document is retrieved. This contrasts with **document-centric**, where the document is the starting point of a task.

application development The work involved in creating and improving computer programs intended for end users.

application development environment The suite of programs that software developers use to develop applications. These programs may include debuggers and compilers.

application development language A computer language, such as COBOL, that facilitates the development of applications.

application development system A coordinated set of tools that enable the development of applications.

application framework 1. The basic conceptual structure of an application. 2. In object-oriented programming, the set of classes that provide the structure for building an application. Also called **framework**.

application gateway 1. In a network, a computer or program that links one application to another by performing the necessary protocol conversions. 2. A computer or program that allows only certain application traffic to cross between a private network and all other networks.

application generator Computer software that enables the generation of applications based on a description of the task or problem to be addressed. These applications may be tailored to specific needs, and as such tend to be limited in scope.

application layer Within the OSI Reference Model for the implementation of communications protocols, the highest level, located directly above the presentation layer. This layer takes care of the exchange of information between applications, including tasks such as file transfers, email, and access to a remote computer. Also called **level 7**.

application level gateway A server, application, or system that serves as an intermediary between a private network, such as a LAN, and all other networks. Its two main functions are to provide document caching and access control. When caching, an application level gateway first attempts to access data which it has cached, and if not present there, it fetches it from a remote server where said data resides.

When controlling access, it serves as a firewall. Its abbreviation is **ALG**. Also called **proxy server**, or **proxy**.

application management system 1. Software which manages the available applications in a computer or network. 2. Software which gives access to the available applications in a computer or network.

application program Same as **application (2)**.

application program interface A set of instructions that determine how a computer application interacts with the operating system. This contrasts with **application binary interface**, which interacts with the operating system and the hardware. Its abbreviation is **API**. Also called **programming interface**, or **software interface**.

application proxy A program that allows only certain application traffic to cross between a private network and all other networks.

application server Its abbreviation is **appserver**. 1. In a LAN, a computer or program that processes data. 2. Software which facilitates the translation of information between Web-based applications and the database programs of the computers using them.

application service provider A business entity that hosts software applications in exchange for a fee. Such applications are accessed via a network, such as the Internet, and enable, for instance, a company to use a suite of specialized programs without having to make sizeable initial investments in software, hardware, networks, nor expert personnel. Its abbreviation is **ASP**.

application sharing During conferencing, the sharing by multiple users of an application running on one of the computers.

application shortcut key A key, such as F3, or combination of keys, such as ALT+SHIFT+T, which when pressed executes an action or series of actions within a program or operating system. Such shortcuts can save time or automate sequences. Also called **shortcut key**, or **keyboard shortcut**.

application software Computer programs which enable end users to perform specific tasks, such as word processing or communications.

application-specific IC Abbreviation of **application-specific integrated circuit**.

application-specific integrated circuit An IC which is customized to work for a specific application. This process can save design and manufacturing time, but a significant proportion of the chip may go unused. Its abbreviation is **ASIC**, or **application-specific IC**.

application suite A combination of application programs which are tailored to a given type of work, and which function especially well together. For example an office package incorporating word processing, database, spreadsheet, presentation, and communications programs. Also called **software package (1)**, **suite**, **package (4)**, or **bundled software**.

application window An area on a computer screen with defined boundaries, and within which information is displayed. Such windows can be resized, maximized, minimized, placed side-by-side, overlaid, and so on. Each window can be that of a separate application, while a single application may have any number of windows open at a given moment. When two or more windows are open, only one is active, and the rest are inactive. Also called **window (4)**.

appliqué circuit A circuit which modifies equipment to add functionality, or for adaptation to a specific need.

approach control An air traffic control service intended to coordinate flights in a manner which is safe and efficient. It affects aircraft within a certain radius of an airport, and at times encompasses satellite airports as well.

approach control radar Same as **airport surveillance radar**.

appserver Abbreviation of **application server**.

APT 1. Abbreviation of **automatic picture transmission**. 2. Abbreviation of **automatically programmed tools**.

AQL Abbreviation of **acceptable quality level**.

aqueous solution A solution in which the solvent is water. For instance, an electrolyte may consist of salts dissolved in water.

AR Abbreviation of **address register**.

Ar Chemical symbol for **Argon**.

AR coating Abbreviation of **anti-reflective coating**, or **anti-reflection coating**.

arbitrary parameter A parameter which is not based on the intrinsic nature of something.

arbitrary unit 1. A unit which can not be defined without referring to other units. 2. A unit which incorporates arbitrary parameters. 3. A unit which is not part of an internationally accepted absolute system of units.

Arbitrated Loop A ring topology utilized in Fibre Channel which supports connection of up to 126 nodes to the loop. Its abbreviation is **AL**. Also called **Fibre Channel Arbitrated Loop**.

arbitration The set of rules a computer uses to allocate its resources among users and programs. This contrasts with **interrupts**, which refer specifically to the microprocessor's resources.

arc 1. A highly luminous and sustained discharge of electricity between two conductors separated by a gas. It is characterized by a high current density and a low voltage drop. Used, for example, in welding and arc lamps. This contrasts with a **spark**, which has a short duration. Also known as **electric arc**. 2. A continuous part of a curve.

arc cathode In a gas tube, a cathode whose electron emission is self-sustaining.

arc discharge A highly luminous and sustained discharge of electricity between two electrodes separated by a gas. It is characterized by a high current density and a low voltage drop.

arc-discharge lamp Same as **arc lamp**.

arc-discharge tube A discharge tube in which there is a highly luminous and sustained discharge of electricity between the electrodes.

arc lamp A gas-filled electric lamp whose bright light is produced by a sustained discharge of electricity between its two electrodes, which ionizes the gas. Also known as **arc-discharge lamp**, or **electric-arc lamp**.

arc-over A disruptive discharge in the form of an arc between conductors. It is generally caused by high voltages, and is frequently luminous.

arc resistance 1. The extent to which a material, usually a dielectric, keeps from arcing. 2. The extent to which a material is able to withstand damage from arcing.

arc spectrum For an atom under study, the spectrum that is produced when it is vaporized between two arcing electrodes.

arc suppression The extinguishing of an arc discharge, usually by a coil, resistor-capacitor network, or diode.

arc welding Welding utilizing a sustained discharge of electricity between two conductors as the source of heat. Also known as **electric-arc welding**.

Archie Abbreviation of **Archive**. An Internet tool for locating files than can be accessed using anonymous File Transfer Protocol.

architecture 1. The components and design of a system or structure. 2. The design of a computer and its components.

This includes for which purposes it can be used, along with its memory and processing requirements and capabilities. This in turn affects variables such as what software may be used with it, and which peripherals are compatible. Also called **computer architecture**, or **computer system architecture**. **3.** The design of a network and its components. This includes the hardware, software, and protocols. Also called **network architecture**. **4.** The design of software and its components. This determines what other software it can interact with, the hardware it can work with, its reliability, and its flexibility and expandability. Also called **software architecture**.

architecture-dependent Computer software or hardware which is designed to work properly only with a specific architecture. Assembly language, for instance, is architecture-dependent. This contrasts with **architecture-independent**, in which software or hardware can work with more than one type of architecture. Also called **platform-dependent**.

architecture-independent Computer software or hardware which is designed to work properly with more than one type of architecture. An interpreter version of LISP is architecture-independent, as is Java. This contrasts with **architecture-dependent**, in which software or hardware is designed to work properly only with a specific architecture. Also called **architecture neutral**, **cross-platform**, **platform-independent**, or **multiplatform**.

architecture neutral Same as **architecture-independent**.

archival backup A data backup technique in which all files are copied, including those that have not been modified since the last backup. This contrasts with an **incremental backup**, in which the only files copied are those that have been modified since the last backup. Also called **full backup**.

archive **1.** To copy computer data to a medium for storage purposes. Also, that which is so stored. Such data is usually compressed. **2.** To save computer data. **3.** To compress computer data. **4.** Same as **Archie**.

archive file **1.** A computer file which is stored and/or compressed. **2.** A computer file that contains one or more compressed files.

archive format A format, such as ZIP, utilized for **archived files**.

archiving program A program, such as WinZip, utilized to **archive (1)**, **archive (2)**, or **archive (3)**.

arcing The production of a disruptive discharge in the form of an arc between conductors. It is generally caused by high voltages, and is frequently luminous.

arcing contacts Electric contacts which produce an arc when separated. These serve, for instance, to protect the main contacts.

arcminute A unit of angular measure equal to 1/60 degree. Its symbol is '. Also called **minute (2)**.

arcsecond A unit of angular measure equal to 1/60 arcminute, or 1/3600 degree. Its symbol is ". Also called **second (2)**.

area code In direct-dialing within North America, the first three digits of the ten digit telephone number. That is, in the NPA-NXX-XXXX ten-digit telephone number format, it is the NPA portion.

area code overlay **1.** An area code whose served zone or region overlaps with that of another area code. Also called **overlay (5)**. **2.** An area code whose served zone or region is the same as that of another area code. That is, both area codes serve the same specific region or zone. Also called **overlay (6)**.

area control radar An air traffic control service for aircraft within a given, usually extensive, area.

area search A search within a database for any files or records which contain information in the desired categories.

argon A noble gas which comprises approximately 0.93% of air by volume, and whose atomic number is 18. It has about 15 known isotopes, of which 3 are stable. It has various applications in electronics, including its use in specialized lamps, lasers, and in Geiger counters. Its chemical symbol is **Ar**.

argon laser A gas laser whose tube contains ionized argon. The uses for these lasers include surgery, and planetarium demonstrations.

argument A value or expression that is passed between functions or programs for processing. This information is used to carry out operations. For instance, in the logarithmic function **log (x)**, the argument is **x**. Also called **parameter (4)**.

arithmetic and logic unit Same as **arithmetic-logic unit**.

arithmetic coding A compression technique which encodes data into single floating point numbers between 0 and 1.

arithmetic expression A combination of numbers and/or variables upon which **arithmetic operations** are performed. Also called **expression (2)**.

arithmetic-logic unit The part of the CPU which performs the arithmetic and logic operations. Some processors contain multiple arithmetic-logic units. Its abbreviation is **ALU**. Also called **arithmetic and logic unit**, **arithmetic unit**, or **logic unit**.

arithmetic mean Same as **average**.

arithmetic operation Any of the basic arithmetic processes, such as addition, division, or multiplication of numbers.

arithmetic operator **1.** A symbol representing a basic arithmetic process. Such symbols include +, *, -, and /, for addition, multiplication, subtraction, and division, respectively. **2.** An operator that performs an arithmetic process, such as addition or division.

arithmetic overflow Also called **overflow**. **1.** The condition in which the result of an operation exceeds the capacity of its designated storage location. For example, an arithmetic result whose exponent exceeds the number of bits allotted to it. **2.** The amount by which the result of an operation exceeds the capacity of its designated storage location.

arithmetic register A register which stores the results of arithmetic and logic operations, such as additions, divisions, multiplications, or additions of absolute values.

arithmetic shift In computer programming, a shift operation equivalent to multiplying a number by a positive or negative integral power of the radix, without changing its sign.

arithmetic underflow Also called **underflow**. **1.** The condition in which the result of an operation is more than zero, but less than the lowest value that can be expressed. **2.** The amount by which a nonzero result of an operation is less than the lowest value that can be expressed.

arithmetic unit Same as **arithmetic-logic unit**.

ARL Abbreviation of **acceptable reliability level**.

arm **1.** A mechanical device or element that serves to position objects. For example, an access arm in a computer disk drive. **2.** A mechanical device or element which has multiple positions. **3.** A path within a circuit or network. Also called **leg**, or **branch (2)**. **4.** In robotics, a set of links and powered joints which comprises all the parts of the manipulator, except for the wrist and the end-effector. Also called **robot arm**.

armature **1.** The rotating, or moving part, in an electric generator or motor. **2.** In an electromagnetic device, the moving element. For instance, the moving contact in an electromagnetic relay.

armature coil A coil of insulated copper wire wrapped around the armature core. It is part of the armature winding.

armature contact In a relay or switch, the contact that moves towards or away from a stationary contact, to close or open a circuit. Also called **movable contact**.

armature core Laminations or stacks of soft iron, which are insulated from each other, upon which the armature windings are placed. Solid iron is not used, as it would generate more heat.

armature reaction The interaction between the magnetic field produced by the field coils of an electric motor or generator, and the magnetic field produced by the current flowing through the armature. Compensating windings, for instance, are used to reduce the effect of armature reaction on motor operation.

armature relay An electromagnetic relay whose moving contact, when energized, moves towards a stationary contact, which closes a circuit. When deenergized, it moves away to open the circuit.

armature slot The grooves in the armature core, into which the coils and windings are placed.

armature winding The wires which are placed on the armature, and through which the current flows. They usually consist of copper.

armor A covering that protects or reinforces a cable. It is composed of one or more layers, which may or may not be of the same material. For instance, a polymer-coated steel armor should protect fiber-optic cables under most conditions.

armored cable A cable equipped with an **armor**.

Armstrong oscillator An oscillator circuit that uses inductive feedback. The variation in the coupling between the input and output coils determines the feedback.

ARP Abbreviation of **address resolution protocol**.

ARP broadcast Abbreviation of **address resolution protocol broadcast**.

ARP request Abbreviation of **address resolution protocol request**.

ARPANET A network which was a precursor to the Internet.

ARPAnet Same as **ARPANET**.

ARQ Abbreviation of **automatic repeat request**.

array 1. An orderly arrangement. For instance, an array of resistors. 2. Same as **antenna array**. 3. In computers, a series of ordered data elements. For example, a matrix is a two-dimensional array.

array antenna Same as **antenna array**.

array element 1. An antenna element within an **array (2)**. 2. In computers, an element within an **array (3)**.

array processor A processor which performs data-parallel calculations on arrays or matrices.

array radar A radar system utilizing an array of antennas, or antenna elements, which are simultaneously fed, and whose relative phase can be varied electronically. Utilized, for instance, for concurrently tracking many scanned objects. Also called **phased array radar**.

arrester Also called **lightning arrester**. 1. A device which provides a low-resistance path to ground for lightning discharges, thus protecting equipment at risk. 2. A device which protects equipment by providing a low-resistance path to ground for voltages that exceed a determined amount.

arrival angle Same as **angle of arrival**.

ARS Abbreviation of **automatic route selection**.

arsenic A silver-gray and brittle crystalline semimetallic element whose atomic number is 33. It has over 20 known isotopes, only one of which is stable. It is used as a semiconductor dopant, in cable sheaths, and in special solders, among others. Its chemical symbol is **As**.

arsenic pentafluoride A chemical compound whose formula is AsF_5. It is used as a doping agent in conducting polymers.

arsenic trisulfide A yellow crystalline chemical compound whose formula is As_2S_3. It is used in semiconductors.

arsine A poisonous colorless gas whose chemical formula is AsH_3. It is used as a semiconductor dopant.

ART Abbreviation of **automated reasoning tool**.

artifact 1. An unexpected signal, component, or result, which is unrelated to the studied phenomenon. For example, an incorrect reading of an analog indicator caused by the viewer observing at an improper angle. 2. Unintended imperfections in signals or images due to software or hardware limitations. Seen, for example, as snow or spots.

artificial antenna A device which simulates the electrical characteristics of an antenna, but which dissipates its input as heat instead of radiating RF energy. It provides a nonradiating load for a transmitter, and is used primarily for testing and adjusting. Also called **dummy antenna**.

artificial delay line A device or medium which delays a wave or signal for a specific length of time. May be used, for instance, in surround-sound reproduction to delay one audio signal with respect to another. Also called **delay line**.

artificial dielectric A three-dimensional array of scattering metallic conductors which serve as a dielectric to electromagnetic waves. These conductors may be arranged randomly or in a given pattern, and are small relative to the wavelength in question.

artificial ear A device which is designed to duplicate characteristics of the human ear, such as acoustic impedance, frequency response, and threshold of hearing. It incorporates a microphone, and may be used, for example, for the calibration of audiometers, or to assist those with reduced hearing. When utilized to test earphones it is also called **earphone coupler**.

artificial echo A reflected or delayed signal used to simulate an echo. Utilized for testing in radars and acoustics.

artificial ground A connection which serves as a ground, but that is not connected to the earth.

artificial horizon A device which indicates the position of an aircraft with respect to a horizontal reference, usually the horizon. When the device is a gyroscope, it is called **gyro horizon**.

artificial intelligence Its abbreviation **AI**. 1. The branch of computing whose goal is to faithfully display human intelligence and behaviors. Some of the areas that have seen great progress are voice recognition, expert systems, and robotics. 2. The ability of a computer to learn from its own mistakes and to adapt.

artificial interference Any interference, such as electromagnetic interference, resulting from human creations, actions, or presence. This includes that from oscillators, transmitters, ignition systems, motors, switches, voltage regulators, among many others, and that arising from the use of such devices. This contrasts with **natural interference**, such as atmospherics. Also called **man-made interference**.

artificial ionization The creation of a reflecting layer in the atmosphere to enhance certain desired communications characteristics, such as the extended propagation of high-frequency radio waves.

artificial language A language that has been developed for specific needs, such as machine language, as opposed to a **natural language**, which evolves naturally over an extended period.

artificial life A machine or system which emulates complex behaviors of living organisms, such as humans. Its applications are many, including research into evolution, and therapeutic agents.

artificial line A circuit, device, or system which simulates the electrical characteristics of a transmission line.

artificial load A device which simulates a transmission line load, such as that of a speaker or antenna. Used primarily for testing and adjusting. Also called **dummy load**.

artificial mouth A device which is designed to duplicate voice characteristics of the human mouth. It incorporates a speaker, and may be used, for example, for testing telephone receivers, and microphones.

artificial neural network A network composed of many processing modules that are interconnected by elements which are programmable and that store information, so that any number of them may work together. Its goal is to mimic the action of biologic neural networks, displaying skills such as the ability to solve problems, find patterns, and learn. Its abbreviation is **ANN**. Also called **neural network (1)**.

artificial noise Any noise, such as RF noise, resulting from human creations, actions, or presence. This contrasts with **natural noise**, such as cosmic noise. Also called **man-made noise**.

artificial satellite 1. An artificial object which orbits a celestial body under the gravitational influence of the latter. Also known as **artificial satellite (3)**. **2.** An **artificial satellite (1)** intended or utilized for communications, broadcasting, observation, testing, research, or the like. Also called **satellite (4)**.

artificial speech Same as **artificial voice**.

artificial vision A device which simulates vision by the use of sophisticated computer programs which interpret data from its optical sensors. Used, for example, in robotics, or to assist those with reduced vision.

artificial voice Simulation of a human voice through the use of a device which incorporates a speaker and computer. Used, for example, in robotics, or to assist those with reduced speech abilities. Also called **artificial speech**, or **voice (3)**.

artwork In printed-circuit manufacturing, a scaled drawing which will determine the etching patterns of the circuits to be produced. Generally it is enlarged, to allow more detail.

as Abbreviation of **attosecond**.

aS 1. Abbreviation of **attosiemens**. **2.** Abbreviation of **absiemens**.

As Chemical symbol for **arsenic**.

asbestos A group of fibrous impure minerals consisting mostly of magnesium, silicon and oxygen, which are chemically inert and highly heat resistant.

ASC Abbreviation of **automatic sensitivity control**.

ASCII Acronym for **A**merican **S**tandard **C**ode for Information Interchange. A binary code computers use to express text and control characters. It was developed to standardize data transmission. There are various ASCII character sets, and in each of them numeric values are assigned to each text or control character. For example, in standard ASCII the number code for the lower case letter c is 99. Also called **ASCII code**.

ASCII art Images created using **ASCII characters**. These can range from smileys to sophisticated creations using specialized programs which convert pictures into patterns of such characters.

ASCII character A character within an **ASCII character set**.

ASCII character set A character set with each text and control character having an assigned numeric value, representing an ASCII code. Standard ASCII is a 7-bit code, providing for 128 characters, while Extended ASCII character sets have 8-bits, allowing for additional assigned codes for foreign languages, graphics, and other special characters. Also called **ASCII set**.

ASCII code Same as **ASCII**.

ASCII file A file whose text and control characters are all in ASCII code. Each byte in the file represents one ASCII coded text or control character. Since these files are not in any proprietary format, they are understood by almost all computers.

ASCII protocol A communications protocol that transmits only ASCII code. It is the simplest protocol for text transfer.

ASCII set Same as **ASCII character set**.

ASCII terminal A terminal which sends and receives ASCII coded data.

ASCII text Alphanumeric characters expressed in ASCII code. Since this text is not in any proprietary format, it is understood by almost all computers.

asec Abbreviation of **attosecond**.

ASIC Abbreviation of **application-specific integrated circuit**.

ASK Abbreviation of **amplitude shift keying**.

askarel A synthetic liquid used as a dielectric in transformers.

ASMP Abbreviation of **asymmetric multiprocessing**.

ASN Abbreviation of **autonomous system number**.

ASP 1. Abbreviation of **application service provider**. **2.** Abbreviation of **Active Server Page**.

aspect ratio In a video image, the ratio of the width to the height. For example, in an HDTV, the value is usually 16:9, that is, 1.78:1.

assemble 1. To put together the parts of a whole, such as those of a piece of equipment or system. **2.** To convert assembly language into machine language.

assembler A program which converts assembly language into machine language. Also called **assembly program**.

assembler language Same as **assembly language**.

assembly 1. A set of parts that are put into a subunit, unit, device, or system. Also, the putting together of such parts. **2.** The process of converting assembly language into machine language.

assembly language A low-level computer programming language which is one step above machine language. In assembly language, statements are used, and these are converted by an assembler into machine language, which consists entirely of numbers. Each type of CPU has its own assembly language. Also known as **assembler language**.

assembly program Same as **assembler**.

assembly robot A robot designed specifically for assembling components. Some may be equipped with sensing devices to be able to select parts from a disorganized stack. Such robots can work with precision for very extended periods of time, and are ideally suited for assembly-line work.

assigned frequency 1. The radio-frequency that a licensing authority has assigned a transmitter to use. **2.** The center of the radio-frequency band that a licensing authority has assigned a transmitter to use.

assigned frequency band The radio-frequency band that a licensing authority has assigned a transmitter to use.

assignment statement In programming, a statement utilized to assign a value to a variable. It is usually in the form of an equation.

assistant An interactive utility which provides help, such as that which may be needed during the installation or use of an application. Such an assistant usually provides guidance step-by-step. Also called **wizard (1)**.

assistive technology Technology designed to assist users to make use, or better use, of a given technology when performing a task. Examples include computers enabled for voice and gesture recognition, and devices, instruments, and appliances which have large displays and/or a voice output to indicate accepted commands, readings, and the like.

associate To link a given type of file to a specific application, usually by using a file extension, such as **.xyz**. It is utilized to alert an operating system to the need to start the necessary application for the desired file.

associative memory Same as **active memory**.

associative storage Same as **active memory**.

assumed decimal point A decimal point not occupying a specific memory location in a computer. Its position within a number is used for calculations, so it has logical meaning but no physical storage representation.

astable circuit A circuit that alternates between two states, neither of which is stable, depending on the parameters established for such oscillation. An example is a blocking oscillator.

astable multivibrator A multivibrator which alternates between two states, depending on the parameters established for such oscillation. This is done autonomously, that is, without an external trigger. Also called **free-running multivibrator**.

astatic **1.** Not having a particular orientation or direction. **2.** Not having a tendency to change position.

astatic galvanometer A galvanometer with two magnetic needles mounted in a manner which allows it to function independently of the effects of the earth's magnetic field. Characterized by high sensitivity and prompt response.

astatic microphone A microphone whose sensitivity is the same in all directions. Also known as **nondirectional microphone**, or **omnidirectional microphone**.

astatine A rare radioactive element with atomic number 85. There are about 30 known isotopes, all of which are unstable. Naturally occurring astatine amounts to about one ounce spread over the earth's crust, so when needed it is synthesized by bombarding bismuth with alpha particles. It is used as a radioactive tracer. It is a halogen, and its chemical symbol is **At**.

asterisk **1.** A character, *, which may be utilized to represent a string of wildcard characters, as opposed to a single wild card, which is usually represented by ?. **2.** In most computer operating systems, a character which serves to denote multiplication, as seen, for instance, in: 7*3.

astigmatism **1.** In a lens or optical system, a defect which causes blur or imperfect image results due to the rays from a given point not converging at a single point. **2.** In an electron tube, a defect which causes blur or imperfect image results due to electrons from different axial planes not converging at a single point.

Aston dark space In glow-discharge tubes, a dark zone near the cathode. This space is composed of the electrons emitted by the cathode that do not have the necessary velocity to excite the gas in the tube.

astrionics The science of electronics as applied to air navigation within and beyond this planet.

ASV Abbreviation of **adaptive suspension vehicle**.

asymmetric Not having symmetry or balance. Also called **asymmetrical**.

asymmetric compression A compression technique in which the time and/or processing power required to decompress differs from that required to originally compress. In data backup, for example, time is saved when compression is faster, as decompression occurs less frequently. In video, processing power and time are economized when decompression occupies less resources, allowing for more sophisticated playback to be delivered. Also called **asymmetrical compression**.

asymmetric conductivity Also called **asymmetrical conductivity**. **1.** Conductivity which is greater in one direction than the other. **2.** Conductivity which is not the same throughout the cross section of a conductor.

asymmetric cryptography Same as **asymmetric encryption**.

asymmetric digital subscriber line Same as **ADSL**.

asymmetric digital subscriber link Same as **ADSL**.

asymmetric digital subscriber loop Same as **ADSL**.

asymmetric encryption An encryption method which uses two keys for successful encryption and decryption of messages. The first key is utilized to encrypt, and is public, as most anyone can look it up. To decrypt, the second key is necessary. This key is private, as only the recipient knows it. Since it is not necessary to send the decryption key in any message, this method eliminates the vulnerability inherent in its being sent to the recipient. Also called **asymmetrical encryption**, **asymmetric cryptography**, or **public-key encryption**.

asymmetric modem A modem which simultaneously transmits and receives data at different speeds. By allocating a greater proportion of the available bandwidth to the side requiring the most data flow, the connection is maximized. Also called **asymmetrical modem**.

asymmetric multiprocessing A multiprocessing computer architecture in which multiple CPUs are each assigned a specific task, with all CPUs controlled by a master processor. Also, processing in this manner. This contrasts with **symmetric multiprocessing**, in which all the CPUs share the same memory, and the same copy of the operating system, application, and data being worked on. This also contrasts with **massively parallel processing**, where each CPU has its own copy of the operating system, its own memory, its own copy of the application being run, and its own share of the data. Also called **asymmetrical multiprocessing**. Its abbreviation is **ASMP**.

asymmetrical Same as **asymmetric**.

asymmetrical cell A photoelectric cell in which the impedance to the flow of current is greater in one direction than the other.

asymmetrical compression Same as **asymmetric compression**.

asymmetrical conductivity Same as **asymmetric conductivity**.

asymmetrical digital subscriber line Same as **ADSL**.

asymmetrical digital subscriber link Same as **ADSL**.

asymmetrical digital subscriber loop Same as **ADSL**.

asymmetrical encryption Same as **asymmetric encryption**.

asymmetrical modem Same as **asymmetric modem**.

asymmetrical multiprocessing Same as **asymmetric multiprocessing**.

asymmetrical-sideband transmission A radio signal transmission in which the amplitude modulation of one of the sidebands is transmitted as it is, while the other is attenuated substantially. Used, for instance, in TV transmissions. Also called **vestigial-sideband transmission**.

asymptotic breakdown voltage A voltage, which if applied for a sufficient period, will cause a sudden and disruptive conduction through a dielectric.

asynchronous **1.** Not synchronous. That is, not occurring at the same time or in equal and fixed intervals. Also called **non-synchronous (1)**. **2.** Not coordinated, especially as regards to time. Also called **non-synchronous (2)**. **3.** Having operations that occur only after each previous one is completed. **4.** In communications, able to be transmitted intermittently rather than in a steady stream.

asynchronous circuit A circuit which is not controlled by clock-generated synchronization signals. Such circuits may provide higher speed, yet use less power. This contrasts with **synchronous circuits**, which are controlled by such signals.

asynchronous communications Communications which are not synchronized by a timing signal. Telephonic communication, in which two or more parties can talk simultaneously, is asynchronous. Otherwise, each person would have to wait a determined time interval before speaking. In computers, asynchronous communication may start or end at any time, with these events being signaled respectively by start and stop bits. This contrasts with **synchronous communications**, which are synchronized by timing signals. Also called **start/stop communications**.

asynchronous component 1. A component whose operation does not depend on a timing mechanism, but rather on each preceding event having been completed. **2.** An AC component which does not run at a speed which is synchronized to the frequency of the power operating it.

asynchronous computer A computer in which each operation starts as a result of another being completed, as opposed to having events controlled by a clock-generated synchronization signal.

asynchronous control Control which depends on an event concluding, rather than a signal generated by a clock. This contrasts with **synchronous control**, which depends on clock signals.

asynchronous data Data sent by means of **asynchronous transmission**.

asynchronous data transmission Same as **asynchronous transmission**.

asynchronous device Also called **start/stop device. 1.** A device whose operation does not depend on a timing mechanism, but rather on each preceding event having been completed. **2.** An AC device which does not run at a speed which is synchronized to the frequency of the power operating it.

asynchronous I/O Abbreviation of **asynchronous input/output**.

asynchronous input/output The ability of a computer to simultaneously have an inflow and outflow of data. Its abbreviation is **asynchronous I/O**.

asynchronous logic In hardware or software, logic which depends on events concluding, and not on clock-generated synchronization signals, as is the case with **synchronous logic**.

asynchronous machine An AC machine which does not run at a speed which is synchronized to the frequency of the power operating it, as opposed to an **synchronous machine**, which does.

asynchronous mode A mode of operation which does not depend on a timing mechanism, but rather on each preceding event having been completed. Also called **start/stop mode**.

asynchronous motor An AC motor which does not run at a speed which is synchronized to the frequency of the power operating it. This contrasts with a **synchronous motor**, which is synchronized in this manner.

asynchronous operation Operation which does not depend on a timing mechanism, but rather on each preceding event having been completed. Also called **start/stop operation**.

asynchronous protocol A communications protocol which governs an **asynchronous transmission**.

asynchronous transfer mode Same as **ATM (1)**.

asynchronous transmission Data transmission that occurs intermittently rather than in a steady stream. Each character has its own start and stop bits and is sent individually without being synchronized by timing signals. This contrasts with a **synchronous transmission**, which occurs in a steady stream and is synchronized by timing signals. Also called **asynchronous data transmission**, or **start/stop transmission**.

At 1. Chemical symbol for **astatine. 2.** Abbreviation of **ampere-turn**.

AT attachment Same as **ATA**.

AT attachment packet interface Same as **ATAPI**.

AT bus A personal computer 16-bit bus architecture that may be used for slower peripherals. Now called **ISA bus**.

AT command set A de facto command set for controlling modems.

AT-cut crystal A piezoelectric crystal cut at a 35° angle with respect to the **z**-axis of the mother crystal. The natural vibration frequency of such crystals does not vary much as a function of temperature.

at sign In an email address, the symbol, @, that separates a username from the domain name.

ATA Abbreviation of **a**dvanced **t**echnology **a**ttachment, or **AT** attachment. The formal name the American National Standards Institute has given to **integrated drive electronics**, or **IDE**. It is a widely utilized disk and tape drive interface in which the controller is integrated into the disk, disc, or tape drive itself, which simplifies connection. There are enhanced versions, such as Fast ATA or Ultra ATA. Also called **ATA interface**.

ATA-2 An enhanced version of **ATA**. It provides for significantly faster transfer rates when using hard disks, optical discs, or tape drives. In addition, it supports hard disks with much larger capacities. Also called **Fast ATA, Fast IDE**, or **EIDE**.

ATA-3 An enhanced version of **ATA-2**, providing additional features, such as the self-monitoring of the physical condition of the drive.

ATA-4 An enhanced version of **ATA-3**, providing higher transfer rates.

ATA/33 A version of **Ultra ATA** with transfer rates of up to 33 MBps.

ATA/66 A version of **Ultra ATA** with transfer rates of up to 66 MBps.

ATA/100 A version of **Ultra ATA** with transfer rates of up to 100 MBps.

ATA/133 A version of **Ultra ATA** with transfer rates of up to 133 MBps.

ATA controller The controller built into an **ATA drive**.

ATA drive A disk, disc, or tape drive utilizing an **ATA interface**.

ATA hard drive A hard drive utilizing an **ATA interface**.

ATA interface Same as **ATA**.

ATA-RAID The use of multiple **ATA hard drives** in a RAID. Also **IDE-RAID**.

Atanasoff-Berry Computer Possibly the first digital computer, completed in 1942, and serving as a model for ENIAC. Its abbreviation is **ABC**.

ATAPI Acronym for **AT** attachment packet interface. A computer interface for accessing compact discs, video discs, and tape drives. Also called **ATAPI interface**.

ATAPI drive A drive with an **ATAPI interface**.

ATAPI interface Same as **ATAPI**.

ATC Abbreviation of **air traffic control**.

ATE Abbreviation of **automatic test equipment**.

Athlon A popular family of CPU chips.

atm Abbreviation of **atmosphere (2)**.

ATM 1. Abbreviation of **A**synchronous **T**ransfer **M**ode. In networks, a connection-oriented communications protocol. ATM carries data, audio, video, or any other information which can be digitally encoded, in packets of a fixed size. Each packet, or cell, consists of 53 bytes, 48 of which are payload, and 5 of which are overhead. Cells can be transferred between nodes at speeds ranging from 1.5 Mbps, to over 10 Gbps. In addition, ATM has quality of service standards which insure a reliable level of performance, and can appropriate unused bandwidth for other communications. For example, a videoconferencing circuit can be used for transmission of data during its idle periods. Also called **ATM protocol**. **2.** Abbreviation of **automated teller machine**, or **automatic teller machine**.

ATM 25   ATM service or equipment, such as network cards, with maximum transfer rates of 25 Mbps.

ATM 155   ATM service or equipment, such as network cards, with maximum transfer rates of 155 Mbps.

ATM Adaptation Layer   The part of the ATM protocol that translates data into a format that can be inserted into an ATM cell. Its abbreviation is **AAL**.

ATM machine   Same as **automated teller machine**.

ATM network   A network utilizing an **ATM protocol**.

ATM OC-3   ATM running as a layer on top of OC-3 SONET, with a transmission rate of 155.52 Mbps.

ATM OC-12   ATM running as a layer on top of OC-12 SONET, with a transmission rate of 622.08 Mbps.

ATM OC-48   ATM running as a layer on top of OC-48 SONET, with a transmission rate of 2.48832 Gbps.

ATM OC-192   ATM running as a layer on top of OC-192 SONET, with a transmission rate of 9.95328 Gbps.

ATM protocol   Same as **ATM (1)**.

ATM service   The communications services provided by an **ATM network**.

atmosphere   **1.** The gaseous envelope surrounding the earth, consisting of the troposphere, stratosphere, mesosphere, and thermosphere, although there are other criteria utilized to describe its various layers. The mixture of gases composing it is called air, and at sea level dry air is approximately 78% nitrogen, 21% oxygen, 0.9% argon, and some other components in lesser amounts. **2.** A unit of pressure intended to equal the pressure of the earth's atmosphere at sea level. One atmosphere equals 101.325 kilopascals, or 760 torr. Its abbreviation is **atm**. Also called **standard atmosphere**.

atmospheric   Any process, condition, or effect produced by, or pertaining to the atmosphere.

atmospheric absorption   In the transmission of electromagnetic waves, losses due to the absorption of energy by the atmosphere.

atmospheric duct   Within the troposphere, a layer which, depending on the temperature and humidity conditions, may act as a waveguide for the transmission of radio waves for extended distances. Also called **duct (3)**, or **tropospheric duct**.

atmospheric effects   Physical phenomena affecting electromagnetic waves as they travel through the atmosphere. These are caused by natural atmospheric phenomena, such as lighting and variations in density.

atmospheric electricity   All manifestations of electrical phenomena occurring in the atmosphere, including lightning and static electricity. It may also affect radio waves, as evidenced, for instance, in atmospheric noise.

Atmospheric Infrared Sounder   A satellite-mounted sounder that can take accurate measurements of temperatures within the earth's atmosphere. Its acronym is **AIRS**.

atmospheric interference   Same as **atmospherics**.

atmospheric layers   The layers, or strata, of the atmosphere. These may be classified by temperature distribution, by composition, by height above sea level, and so on. Examples include the ionosphere, the troposphere, the stratosphere, and the mesosphere. Also called **atmospheric regions**, **atmospheric strata**, or **atmospheric shells**.

atmospheric noise   Noise heard during radio reception, due to electrical **atmospherics**.

atmospheric pressure   **1.** The pressure which the atmosphere exerts at any given point within it. This pressure is due to the weight of the gases that compose it. Also called **barometric pressure**. **2.** A pressure equal to exactly 1 atmosphere, which is equal to 101.325 kilopascals, 1013.25 millibars, or 760 millimeters of mercury.

atmospheric pressure chemical-vapor deposition   A chemical-vapor deposition technique in which gaseous reactants at or around atmospheric pressure surround the substrate. The solid products of said reactants are then deposited onto the substrate. Its abbreviation is **APCVD**, or **atmospheric pressure CVD**.

atmospheric pressure CVD   Same as **atmospheric pressure chemical-vapor deposition**.

atmospheric propagation   The propagation of radio waves with the assistance of one or more reflections off the atmosphere. Specific examples include ionospheric propagation, and propagation through an atmospheric duct.

atmospheric radio wave   Same as **atmospheric wave**.

atmospheric radio window   A band of radio frequencies which readily pass through the atmosphere. It may include a wide range, including visible light and microwaves.

atmospheric refraction   The bending of electromagnetic waves as they travel through the atmosphere. Such refraction is usually due to variations in atmospheric density.

atmospheric regions   Same as **atmospheric layers**.

atmospheric scattering   The scattering of radio waves as they collide with atmospheric particles.

atmospheric shells   Same as **atmospheric layers**.

atmospheric strata   Same as **atmospheric layers**.

atmospheric strays   Same as **atmospherics**.

atmospheric wave   A radio wave whose propagation has been assisted by reflecting off the atmosphere. Also called **atmospheric radio wave**, or **sky wave**. When it reflects off the ionosphere it is called **ionospheric wave**.

atmospherics   Interference of radio waves caused by natural atmospheric electrical phenomena, such as lighting. Atmospheric noise is a specific example of such a disturbance. Also called **atmospheric interference**, **atmospheric strays**, **sferics**, or **strays**.

atom   **1.** The smallest particle that retains all the properties of an element. Atoms are composed of protons, neutrons, and electrons, with protium, which is hydrogen-1, being the sole exception as it has no neutrons. The positively charged proton and neutral neutron are clustered together in the nucleus. The negatively charged electrons revolve in energy levels surrounding the nucleus. Atoms are overall neutral in charge, as the number of electrons equals the number of protons. If the number of electrons is not equal to the number of protons, you have a charged atom, which is a type of ion. **2.** In computers, the smallest element within an information structure. For example, a character or a pixel.

atomic   **1.** Of, pertaining to, utilizing, or derived from **atoms**. **2.** Of, pertaining to, utilizing, or derived from **atomic energy**. **3.** A computer operation that can not be subdivided. It is either executed fully without any interruption, or not at all. Also called **atomic operation**.

atomic absorption coefficient   For given material or medium, the absorption coefficient per atom.

atomic absorption spectroscopy   A spectroscopic technique which measures the absorption of infrared, visible, and ultraviolet light by a sample which has been vaporized. Utilized to measure the concentration of each chemical element in the sample. Its abbreviation is **AAS**.

atomic battery A battery in which radioactive energy is converted into electrical energy. These batteries are ideally suited for uses requiring high energy density, reliability, and longevity. Also known as **nuclear battery**.

atomic beam A narrow stream of atoms emitted from the same source, traveling in the same direction, and having approximately the same speed. Used, for instance in spectroscopy, and in frequency standards.

atomic-beam frequency standard A frequency standard based on stream of atoms in which resonance has been induced. Cesium atoms are frequently used.

atomic-beam resonance Resonance exhibited by a stream of atoms. Usually induced by microwaves.

atomic charge The charge that an atom has. If the charge is positive, indicating an electron deficiency, it is a positive ion, or cation. If the charge is negative, indicating an electron surplus, it is a negative ion, or anion.

atomic clock A clock whose accuracy is governed by the natural resonance frequencies of atoms. Atoms resonate at exceptionally consistent frequencies, and when measured precisely provide for highly accurate time readings. Atomic clocks which utilize lasers to cool atoms to nearly absolute zero should be able to achieve an error of less than one second per billion years in the not too distant future. Used, for instance, in GPS systems.

atomic emission spectroscopy A spectroscopic technique which measures the light emissions of atoms that have been excited by an electrical discharge, such as an arc or spark. Utilized to measure the concentration of each element present in a given sample. Its abbreviation is **AES**.

atomic energy Energy released by reactions involving atomic nuclei. There are three types of atomic energy: (1) the splitting, or fission, of nuclei, (2) the union, or fusion, of nuclei, and (3) the radioactive decay of nuclei. Also known as **nuclear energy**.

atomic force microscope A microscope that utilizes a minute spring-mounted probe to scan the surface of a sample at an atomic level. A laser beam aimed at the probe follows its fluctuations as it is traced over the surface, and a detector linked to a computer uses this information to reproduce the topography of the sample. Its abbreviation is **AFM**.

atomic force microscopy The use of **atomic force microscopes** to view and analyze specimens. Its abbreviation is **AFM**.

atomic frequency A natural resonance frequency of an atom.

atomic frequency standard A frequency standard based on a natural resonance frequency of the atoms of an element, such as cesium or rubidium. Such standards are used, for instance, in atomic clocks.

atomic mass The mass of an atom, expressed in **atomic mass units**.

atomic mass unit A unit of mass measurement for atoms and molecules. The unit has been arbitrarily defined as 1/12 the mass of an atom of carbon-12, the most common isotope of carbon. Each unit is equal to approximately 1.6605×10^{-24} grams. Its abbreviation is **amu**. Also known as **dalton**.

atomic nucleus The positively-charged dense central core of an atom. All atomic nuclei, except protium, which is hydrogen-1, consist of protons and neutrons. The nucleus of hydrogen-1 consists of a single proton. In an uncharged atom, the number of electrons is the same as the number of protons. When two or more atoms have the same atomic number, but different atomic masses, they are isotopes of the same element. A nucleus is held together by the nuclear force, and comprises almost all of the mass of an atom. Also called **nucleus (1)**.

atomic number The number of protons in the nucleus of an atom. It is also the number of electrons that surround the

uncharged atom. The atomic number of carbon, for instance, is 6. Its symbol is **Z**. Also known as **proton number**.

atomic operation Same as **atomic (3)**.

atomic orbital A region of space surrounding the nucleus of an atom, in which the probability of an electron being located is greatest. The charge of such an electron is distributed in this space. No more than two electrons may occupy the same atomic orbital.

atomic oscillator An oscillator whose accuracy is governed by the natural resonance frequencies of atoms. Used, for instance, as a frequency standard.

atomic photoelectric effect The removal of one or more electrons from an atom as a consequence of light absorption.

atomic pile A deprecated term for **atomic reactor**.

atomic power Also known as **nuclear power**. **1.** Power derived by reactions involving atomic nuclei. The three source of such power are fission, fusion, and radioactive decay. **2.** The use of **atomic power (1)** to generate electricity.

atomic power plant A location which combines the structures, devices, and equipment necessary to produce nuclear power in quantities adequate for residential, commercial, or industrial use. The key component of such a location is the **atomic reactor**, whose generated heat can be utilized to produce steam which drives the turbines of an electric generator. Also called **atomic power station**, or **nuclear power plant**.

atomic power station Same as **atomic power plant**.

atomic reaction Any reaction, such as fission or fusion, which changes the structure or composition of an atomic nucleus. Also called **nuclear reaction**.

atomic reactor A device which utilizes controlled nuclear fusion or fission to generate energy. A fission reactor, for instance, has the following key components: a fissionable fuel, such as uranium-235, a moderator, such as heavy water, shielding, such as that provided by thick concrete, control rods, such as those made with boron, and a coolant such as water. The heat produced by the nuclear reactions can be utilized to generate electricity. Also called **nuclear reactor**, or **reactor (2)**. Deprecated terms for this concept include **atomic pile**, **nuclear pile** or **pile (2)**.

atomic resonance Resonance exhibited by atoms. May be induced, for instance, by microwaves.

atomic time Time as measured by an **atomic clock**.

atomic transaction In computers, a series of operations which must all be executed fully, or not at all. If any of its multiple steps fails, all changes are undone. Used, for example, for payments over the Internet.

atomic weight The average mass of all the isotopes of an element, expressed in atomic mass units. It is the combined mass, or weight, of the protons, neutrons, and electrons of an atom. The atomic weight of carbon, for example, is approximately 12.011 atomic mass units, not 12. The reason is that carbon has three naturally occurring isotopes, of which carbon-12 is by far the most common. Also called **relative atomic mass**.

ATR switch Abbreviation of **antitransmit/receive switch**.

ATR tube Abbreviation of **antitransmit/receive tube**.

attached file Same as **attachment**.

attachment A file that is appended to an email. Such a file may be a document, an image, a video, a program, and so on. Certain types of attachments are encoded, thus requiring the recipient to have the appropriate email software to decode it. Also called **email attachment, enclosure (3)**, or **file attachment**.

attachment unit interface A 15 pin connector that attaches a network interface card and an Ethernet cable. Its abbreviation is **AUI**.

attack **1.** The interval within which a pulse rises from 0 to its maximum amplitude. **2.** The interval within which a sound rises from 0 decibels to its maximum loudness. **3.** The response of a circuit, such as an automatic gain control, to signals.

attack time **1.** The time interval within which a pulse rises from 0 to its maximum amplitude. **2.** The time interval within which a sound rises from 0 decibels to its maximum loudness. **3.** The time interval within which a circuit, such as a peak clipper, responds to signals.

attenuation A reduction over time or distance in a physical quantity, such as amplitude or energy. For instance, a decrease in the amplitude of a wave as it propagates through a medium. It may occur naturally, or an attenuator may be employed.

attenuation band A band of frequencies whose intensities are reduced, ideally without introducing distortion.

attenuation characteristic Same as **attenuation constant**.

attenuation coefficient **1.** Same as **attenuation constant**. **2.** The decrease of the amplitude of a signal as a function of the frequency.

attenuation constant A rating for a transmission medium through which a wave passes, such as a cable, which gives the rate of amplitude decrease in the direction of travel. Expressed in decibels or nepers per unit length. It is the real number component of the **propagation constant**. Its symbol is α. Also called **attenuation factor, attenuation characteristic, attenuation coefficient** (1), or **attenuation rate**.

attenuation distortion In an amplifier, distortion that occurs when some frequencies are amplified more than others. Also known as **amplitude distortion** (1), **amplitude-frequency distortion, frequency distortion,** or **waveform-amplitude distortion**.

attenuation equalizer A device which attenuates certain frequencies so that the overall transmission loss for a line or circuit is uniform.

attenuation factor Same as **attenuation constant**.

attenuation-limited operation The condition in which the attenuation of a system limits its performance, as opposed to bandwidth or distortion.

attenuation network An array of circuit elements which provides uniform attenuation throughout a frequency band, while minimizing phase shift.

attenuation rate Same as **attenuation constant**.

attenuation ratio A ratio representing a proportional decrease in a magnitude. For instance, the ratio of the input amplitude of electrical noise, to the output amplitude of electrical noise, for a given device.

attenuator A circuit or device which reduces the amplitude of a signal, ideally without introducing distortion. It may be fixed or variable, and usually consists of resistors and/or capacitors.

attitude The orientation or position of an aircraft or spacecraft, as determined by the relationship of its axes relative to a reference point, line, plane, or system of reference axes. This applies whether the craft is in motion or at rest.

attitude control **1.** A computer-controlled device or system which automatically maintains the attitude of an aircraft or spacecraft. **2.** The control of the attitude of an aircraft or spacecraft.

attitude gyro A gyro-operated instrument which indicates the attitude of an aircraft or spacecraft relative to a reference point, line, plane, or system of reference axes. It usually uses a coordinate system spanning through 360° of rotation about each axis of the craft.

attitude indicator An instrument which indicates the attitude of an aircraft or spacecraft relative to a reference point, line, plane, or system of reference axes.

atto- A metric prefix representing 10^{-18}. For example, 1 attovolt is equal to 10^{-18} volt. Its abbreviation is **a**.

attoampere 10^{-18} ampere. It is a unit of measurement of electric current. Its abbreviation is **aA**

attofarad 10^{-18} farad. It is a unit of capacitance. Its abbreviation is **aF**.

attohenry 10^{-18} henry. It is a unit of inductance. Its abbreviation is **aH**.

attometer 10^{-18} meter. It is a unit of measurement of length. Its abbreviation is **am**.

attosecond 10^{-18} second. It is a unit of time measurement. Its abbreviation is **as**, or **asec**.

attosiemens 10^{-18} siemens. It is a unit of conductance. Its abbreviation is **aS**.

attovolt 10^{-18} volt. It is a unit of measurement of potential difference. Its abbreviation is **aV**.

attowatt 10^{-18} watt. It is a unit of measurement of power. Its abbreviation is **aW**.

attracted-disk electrometer An electrometer in which the attraction between two metal disks is used to measure a potential difference.

attraction A force that act upon entities, tending to draw them together while resisting their separation. Examples include the attraction between oppositely charged particles, and gravitation. Also called **attractive force**.

attractive force Same as **attraction**.

attribute **1.** In a computer database, a feature of a field within a record. For example, it may have a numeric attribute if it contains numbers. **2.** Additional information about a computer-generated character, such as color, underlining, or blinking. **3.** A characteristic of a file that imposes restrictions on its availability or use. For instance, a file may be read-only, or hidden. Also called **file attribute**.

ATV Abbreviation of **advanced television standard**.

Au Chemical symbol for **gold**.

audibility The quality of a sound pressure level being sufficient to be detected by human ears, that is, being of 0 decibels or greater. Since hearing acuity varies considerably from human to human, it has been internationally agreed that a value of 20 micropascals is the equivalent of 0 decibels to a person with good hearing. Thus, any sound pressure level equal to or greater than 0 decibels is considered audible. Before pain sets in, a person with good hearing can listen at 120 decibels, or 20 pascals, which is a trillion times louder than 0 decibels.

audibility curve A graph which shows the range of hearing of a person. This is usually charted as loudness expressed in decibels, versus frequency expressed in hertz. The hearing sensitivity of people varies by frequency, and this sensitivity also varies depending on the sound-pressure level.

audibility threshold The minimum sound pressure level necessary to attain **audibility**. Also called **threshold of hearing**.

audible Detectable by human ears.

audible alarm An alarm utilizing an audible warning, such as that provided by a siren, buzzer, or bell.

audible frequency Same as **audio frequency**.

audible range Same as **audible spectrum**.

audible response An audible sound produced by a component, device, or system. For example, the sound of a Geiger counter, or keyboard feedback for those with reduced hearing. Also called **audio response** (3).

audible signal A signal which is within the frequency range that humans can hear.

audible spectrum The spectrum of frequencies detectable by human ears. A healthy person with good hearing can usually detect frequencies ranging from about 20 Hz to about 20 kHz. Also called **audible range**.

audible tone   A tone which is detectable by human ears.

audio   **1.** Pertaining to sounds that occur within the range of frequencies that humans can hear. A healthy person with good hearing can usually detect frequencies ranging from about 20 Hz to about 20 kHz. **2.** Pertaining to equipment that can reproduce, amplify, record, detect, convert, transmit, or in any other way process sounds within the range of frequencies that humans can hear.

audio adapter   Same as **audio card**.

audio amplifier   Also called **audio-frequency amplifier, acoustic amplifier**, or **sound amplifier**. **1.** A device which increases the strength of acoustic (sound) waves. **2.** An amplifier of signals that operates within the range of frequencies that humans can hear. **3.** A high-fidelity amplifier that operates within and beyond the range of frequencies humans can hear. For instance, such an amplifier may have a frequency response of 5 Hz to 100 kHz. **4.** A device which intensifies sounds by its shape. A horn is an example.

audio band   Also called **audio-frequency band**. **1.** The band of frequencies that humans can hear. A healthy person with good hearing can usually detect frequencies ranging from about 20 Hz to about 20 kHz. **2.** A band of frequencies encompassing all or part of the interval of 20 Hz to 20 kHz.

audio board   Same as **audio card**.

audio broadcast   Same as **audiocast**.

audio buffer   In a computer, a buffer which holds the audio data that awaits to be sent to the speakers. Also called **sound buffer**.

audio card   An expansion board that enables computers to handle sounds. Audio cards serve to record, playback, and synthesize sounds. They generally provide external jacks for a microphone, a line-in, speakers or headphones, and a MIDI port. Also called **audio board, audio adapter, sound card**, or **sound board**.

audio carrier   In TV, the carrier frequency which is modulated to convey audio information. Also called **sound carrier**.

audio cassette   A cassette which contains a magnetic tape suitable for audio recording and playback, along with two floating reels, one for supply, the other for take-up. Also spelled **audiocassette**.

audio CD   Abbreviation of **audio** compact disc. A CD containing music, speech, or other audio. An audio CD can usually hold up to 74 minutes of recorded sound. Also called **audio disc, music CD**, or **digital audio disc**.

audio CD player   Abbreviation of **audio** compact-disc **player**. A device which can read the information stored on an **audio CD**. To recover the recorded content, a laser is focused on the metallic surface of the disc, and the reflected light is modulated by the code on said disc. Computer DVD and CD-ROM players can also serve as CD players when an audio CD is placed in the drive. Also called **music CD player**.

audio channel   Also called **audio-frequency channel**. **1.** A channel which carries a signal within the range of frequencies that humans can hear. **2.** In a system transmitting and/or receiving audio and video signals, the audio channel. **3.** A channel within a high-fidelity component, such as the left channel, or the center channel. **4.** In a system where multiple audio signals are available, one of the sound channels, such as a version in another language.

audio choke   An inductor which serves to impede the flow of audio-frequency currents. Also called **audio-frequency choke**.

audio circuit   A circuit carrying an audio signal. Also called **audio-frequency circuit**.

audio codec   **1.** Abbreviation of **audio co**der/decoder. Hardware and/or software which converts analog audio to digital audio, and vice versa. **2.** Abbreviation of **audio** compressor/decompressor. Hardware and/or software which compresses and decompresses audio signals.

audio coder/decoder   Same as **audio codec** (1).

audio compact disc   Same as **audio CD**.

audio compact-disc player   Same as **audio CD player**.

audio component   Also called **audio-frequency component**. **1.** In a wave or signal, the component within the range of frequencies that humans can hear. **2.** In a multimedia system, a device that processes and/or reproduces audio frequencies.

audio compression   The encoding of audio data so that it occupies less room. This helps maximize the use of storage space, and also reduces transmission bandwidth.

audio compressor/decompressor   Same as **audio codec** (2).

audio conference   Same as **audioconference**.

audio conferencing   Same as **audioconferencing**.

audio current   A current at an audio frequency. Also called **audio-frequency current**.

audio disc   Same as **audio CD**.

audio disk   Same as **audio CD**.

audio distortion   Also called **audio-frequency distortion**. Distortion occurring in the range of frequencies that humans can hear, an example of which is total harmonic distortion.

audio expansion   The restoring of audio data to its original space and/or bandwidth, following audio compression.

audio feedback   Same as **acoustic feedback**.

audio filter   A device which removes some audio frequencies while leaving others unaffected. Also called **audio-frequency filter**.

audio frequency   A frequency that is within the range that humans can hear. A healthy person with good hearing can usually detect frequencies ranging from about 20 Hz to about 20 kHz. Its abbreviation is **AF**. Also known as **audible frequency, acoustic frequency**, or **sound frequency**.

audio-frequency amplifier   Same as **audio amplifier**.

audio-frequency band   Same as **audio band**.

audio-frequency channel   Same as **audio channel**.

audio-frequency choke   Same as **audio choke**.

audio-frequency circuit   Same as **audio circuit**.

audio-frequency component   Same as **audio component**.

audio-frequency current   Same as **audio current**.

audio-frequency distortion   Same as **audio distortion**.

audio-frequency feedback   Same as **acoustic feedback**.

audio-frequency filter   Same as **audio filter**.

audio-frequency harmonic distortion   Same as **audio harmonic distortion**.

audio-frequency meter   Same as **audio meter**.

audio-frequency noise   Same as **audio noise**.

audio-frequency oscillator   Same as **audio oscillator**.

audio-frequency output   **1.** Same as **audio output** (1). **2.** Same as **audio output** (2). **3.** Same as **audio output** (4).

audio-frequency peak limiter   Same as **audio peak limiter**.

audio-frequency range   Same as **audio spectrum**.

audio-frequency response   **1.** Same as **audio response** (1). **2.** Same as **audio response** (2).

audio-frequency-shift keying   Data transmission accomplished by varying the frequency of an audio tone which modulates the carrier wave. Switching the audio tone between frequencies provides the keying. Its abbreviation is **AFSK**.

audio-frequency-shift modulation A system of radio facsimile transmission which utilizes audio signals of 1500 and 2300 Hz. The 800 Hz shift between these two frequencies modulates the radio signal for the transmission of data.

audio-frequency signal Same as **audio signal**.

audio-frequency signal generator Same as **audio signal generator**.

audio-frequency spectrum Same as **audio spectrum**.

audio-frequency transformer Same as **audio transformer**.

audio harmonic distortion Distortion, produced by harmonics, in the range of frequencies that humans can hear. Also called **audio-frequency harmonic distortion**.

audio input 1. The input for a device, component, or system which processes and/or reproduces audio signals. 2. Input connectors such as plugs, which accept an audio signal. 3. In audio components and accessories, connectors such as plugs, which accept an input signal, such as that from a CD player. 4. The input of a device which converts analog audio into digital audio, or vice versa.

audio jack A receptacle for a plug which connects audio components and accessories, such as amplifiers, headphones, or microphones. **Audio plugs** are plugged into audio jacks. There are various types, including XLR jacks, and phono jacks.

audio-level meter An instrument which indicates the power of a complex audio signal. Such a device usually expresses readings in decibels, although other units may be displayed, such as volts. When indicating volume units, it is called **VU meter**.

audio masking The amount by which the threshold of hearing a sound is increased due to the presence of another, obscuring sound. Also called **aural masking**, or **masking (2)**.

audio meter An instrument used to measure frequencies in the range that humans can hear. There are several types, including analog and digital, the latter being highly accurate. Also called **audio-frequency meter**.

audio mix To combine multiple audio input signals to form a composite signal with the desired blend. For example, to combine voices and multiple instruments for a song, or the combination of dialog, music, and effects for a soundtrack. Also, the result of such a combination. Also called **sound mix**, or **mix (2)**.

audio mixer A circuit or device which performs **audio mixes**. Also called **mixer (3)**.

Audio/Modem Riser An audio and/or modem card which instead of laying flat on the motherboard rises above it. Its abbreviation is **AMR**.

audio noise Sound or noise that occurs within the range of frequencies that humans can hear. Such noise is almost always unwanted, and can result in an unpleasant, uneven, distracting, or otherwise undesired listening experience. Examples include hum and hiss. Such noise may also harm equipment. Also called **audio-frequency noise**, **acoustic noise (2)**, or **noise (2)**.

Audio on Demand Audio programming, such as the songs of artists available via Web sites, which is available for immediate download, in exchange for a fee. Its abbreviation is **AOD**.

audio oscillator A oscillator which produces electric waves with frequencies within the range that humans can hear. May be used, for instance, for long-distance surveys of submerged cables. Also called **audio-frequency oscillator**.

audio output 1. The output of a source of audio frequencies, such as an amplifier or oscillator. For instance, the audio power an amplifier delivers to a speaker. Also called **audio-frequency output (1)**, or **sound output (1)**. 2. Any audible output from a device, component, or system. For instance, that produced by a speaker. Also called **audio-frequency**

output **(2)**, or **sound output (2)**. **3.** In audio components and accessories, connectors such as jacks, which deliver an output signal, such as that from a CD player. **4.** In computers, an audible output, such as a chime when a Web page is loaded, or the simulation a of human voice reading a document on screen. Also called **audio-frequency output (3)**, or **sound output (3)**. **5.** The output of a device which converts analog audio into digital audio, or vice versa.

audio peak limiter A circuit or device which limits the amplitude of its audio output signal to a predetermined maximum, regardless of the variations of its input. Used, for example, for preventing component, equipment, or media overloads. Also known as **audio-frequency peak limiter**.

audio plug A device with protruding contacts which connect audio components and accessories, such as amplifiers, headphones, or microphones. Audio plugs are plugged into **audio jacks**. There are various types, including XLR plugs, and phono plugs.

audio plug adapter A device which enables one type of audio plug to connect with an audio jack of a different size and/or type. For example, an adapter that takes a ¼ inch phone plug and converts it into a ⅛ inch mini phone plug.

audio power The power of a source of audio frequencies, such as an amplifier or oscillator. Usually expresses in watts. For example, a high-fidelity audio amplifier may be rated at 220 watts per channel.

audio range Same as **audio spectrum**.

audio recorder An instrument or device, such as a tape recorder, which makes **audio recordings**. Also called **sound recorder**.

audio recording The process of producing a permanent, or semi-permanent, record of audio which is suitable for later reproduction. This includes recording upon magnetic tapes, magnetic disks, optical discs, phonographs, films, wires, and so on. Also, a specific recording. Also, that which has been recorded. Also called **sound recording**.

audio recording system A system, such as that incorporating microphones, amplifiers, filters, mixers, converters, computers, and digital recorders, utilized for **audio recording**. Also called **sound recording system**.

audio rectification Interference affecting audio-frequency circuits or devices which try to recover information from nearby radio signals, other than those intended. For example, a telephone may have unwanted noises or voices when this occurs.

audio response 1. The efficiency of the amplification of a circuit, device, or system, as a function of input signals whose frequencies are within the range that humans can hear. Also called **audio-frequency response (1)**, or **sound-frequency response (1)**. 2. A graph in which the output of a circuit, device, or system, is plotted against signals whose frequencies are within the range that humans can hear. Also called **audio-frequency response (2)**, or **sound-frequency response (2)**. 3. Same as **audible response**.

audio response unit A device which provides an audio response to a given telephonic input. These responses are usually computer controlled, and incorporate recorded messages which provide information when prompted. For instance, a person on the way to the movies calling a phone number and pressing the appropriate keys to find out the show times would be accessing such a device.

audio server A server which provides audio to other computers in a network. This audio may be streaming, in which sound is heard while the transmission is in progress, or non-streaming, in which case the download must be completed before accessing the audio.

audio signal A signal that is within the range of frequencies that humans can hear. A healthy person with good hearing can usually detect frequencies ranging from about 20 Hz to about 20 kHz. Also called **audio-frequency signal**.

audio signal generator A generator of frequencies within the range that humans can hear. An example is an audio oscillator. Also **audio-frequency signal generator**.

audio spectrometer Same as **acoustic spectrometer**.

audio spectrum The spectrum of frequencies detectable by human ears. A healthy person with good hearing can usually detect frequencies from about 20 Hz to about 20 kHz. Also known as **audio range, audio-frequency range, audio-frequency spectrum**, or **sound spectrum (1)**.

audio streaming The transmission of streaming audio over a computer network.

audio system A device, or combination of devices, which serve to process and/or reproduce sound within the range of frequencies that humans can hear. Examples include high-fidelity audio systems, and public-address systems. Depending on the components, such a system may also record sound. Also called **sound system**.

audio tape A magnetic tape which is suitable for the recording and playing of audio content such as music or speech. Also spelled **audiotape**.

audio taper A type of potentiometer utilized in audio components, which provides a logarithmic variation of the resistance value, as a volume or tone control is adjusted. This is done to compensate for the nonlinear way in which humans perceive changes in volume. In this manner, increases in the volume control settings provide a smooth and gradual increase in the way the sound is heard.

audio teleconference Same as **audioconference**.

audio teleconferencing Same as **audioconferencing**.

audio transformer An iron-core transformer utilized to couple audio circuits. Also called **audio-frequency transformer**.

audio transmitter Also called **aural transmitter**, or **sound transmitter**. **1.** A transmitter, such as a broadcast transmitter, which sends signals containing audio. Also, the equipment utilized for this purpose. **2.** The equipment utilized by a TV transmitter to send the sound signal. This signal is combined with that of the video transmitter for regular TV viewing.

audio video Same as **audiovisual**. Its abbreviation is **A/V**, or **AV**.

audio/video connector A connector, such as an RCA plug or a coaxial connector, utilized to link devices which send and/or receive audio and/or video signals. Its abbreviation is **A/V connector**, or **AV connector**.

audio/video inputs Inputs available in devices such as audio amplifiers, TVs, and computers, which enable them to accept audio and/or video signals from other devices, such as CDs and DVDs. Many such devices have both inputs and outputs. Its abbreviation is **A/V inputs**, or **AV inputs**. Also called **audiovisual inputs**.

audio/video outputs Outputs available in devices such as DVDs and CDs, which enable them to send audio and/or video signals to other devices, such as audio amplifiers, TVs, and computers. Many such devices have both inputs and outputs. Its abbreviation is **A/V outputs**, or **AV outputs**. Also called **audiovisual outputs**.

audiocassette Same as **audio cassette**.

audiocast Abbreviation of **audio** broad**cast**. An RF transmission, such as a radio broadcast, consisting of audio signals intended for public or general reception.

audioconference A conference in which participants at more than one site can converse with each other in real-time. Also spelled **audio conference**. Also called **audio teleconference**.

audioconferencing Communication via **audioconferences**. Also spelled **audio conferencing**. Also called **audio teleconferencing**.

audiogram A graph showing the threshold of hearing for each ear, for a number of different frequencies. It serves to indicate levels of hearing loss at the tested frequencies.

audiology The science that studies hearing. It includes any reductions in this ability, and corrective measures.

audiometer An instrument for measuring hearing. It generates sounds at different frequencies, and tests an individual's threshold for detecting each frequency, in order to determine any levels of hearing loss. Any reduction in hearing is expressed in decibels.

audiometry The measurement of hearing through the use of **audiometers**.

audiotape Same as **audio tape**.

audiotex An automatic voice response service where a caller dials a number, follows the recorded voice prompts, and obtains information. It is usually computer controlled, and can be used to retrieve anything from stock quotes to email. Also called **audiotext**.

audiotext Same as **audiotex**.

audiovisual Its abbreviation is **A/V**, or **AV**. Also called **audio video**. **1.** Pertaining to signals or forms of presentation that combine images and sound. **2.** Pertaining to equipment with audio and video capabilities.

audiovisual conferencing Conferencing which incorporates both images and sound.

audiovisual inputs Same as **audio/video inputs**. Its abbreviation is **A/V inputs**, or **AV inputs**.

audiovisual outputs Same as **audio/video outputs**. Its abbreviation is **A/V outputs**, or **AV outputs**.

audit An assessment of the hardware, software, procedures, and records of a computer or system, to determine proper compliance with established parameters and procedures. It may be followed by recommendations for improvement. Used, for instance, to assess data security and to locate problems.

audit software Computer programs designed specifically to perform **audits**.

audit trail A record of all activity pertaining to a specific time interval and/or piece of information. It is used, for example, to find the origin of problems, to recover data, and to identify unauthorized access.

auditing The process of performing an **audit**.

auditory Of or pertaining to hearing and the organs involved in this sense.

auditory feedback Sounds which are produced in response to user actions. For instance, a computer application may provide a click, beep, or the like, when a command is accepted or an action is completed.

auditory perspective Pertaining to a three-dimensional feel a listener may have when listening to a sound source. For this, no less than two separate and different sound paths are required. A surround-sound system attempts to accomplish this.

augend A number or quantity that an addend is added to. An **addend** is a newly introduced number or quantity that is added to the augend, which is already present.

Auger effect A process where an excited electron is transferred to a lower energy level without emitting radiation. This excess energy is used instead to expel an electron of the same atom. This may occur in any element except hydrogen and helium. Also called **autoionization**, or **internal photoionization**.

Auger electron An electron expelled from an atom as a consequence of the **Auger effect**.

Auger electron spectroscopy An analytical technique which analyzes Auger electrons expelled from the surface of a solid which has been irradiated with particles such as electrons or photons. Utilized, with high spatial resolution, to determine the distributions of the elements composing the surface of the solid. Its abbreviation is **AES**.

Auger transition The energy transition of an electron within an atom undergoing the **Auger effect**.

augmentative and alternative communication Technology which enables those with special needs in communication, such as individuals with reduced vision, hearing, and/or motor function, to correspond, share, learn, teach, and otherwise communicate more effectively. Its abbreviation is AAC.

augmented reality An environment or setting which combines virtual and real images and objects. For example, virtual images may be superimposed upon real objects. Also called **augmented virtuality**, **mixed reality**, or **enhanced reality**.

augmented virtuality Same as **augmented reality**.

AUI Abbreviation of **attachment unit interface**.

AUP Abbreviation of **acceptable use policy**.

aural masking Same as **audio masking**.

aural signal **1.** A signal that is within the range of frequencies that humans can hear. **2.** The audio portion of a TV signal.

aural transmitter Same as **audio transmitter**.

aurora A luminous phenomenon of the upper atmosphere occurring mostly in the high latitudes of both hemispheres. That is, around the north and south magnetic poles. In the northern hemisphere auroras are called **aurora borealis**, or northern lights, and in the southern hemisphere they are called **aurora astralis**, or southern lights. They are caused by the interaction of excited particles from space with particles of the upper atmosphere. Auroras can take any of several forms, such as arcs, bands, or patches, and usually affect radio communications.

aurora astralis An aurora occurring in the southern hemisphere. Also called **southern lights**.

aurora borealis An aurora occurring in the northern hemisphere. Also called **northern lights**.

auroral absorption The absorption of radio waves by an aurora. This causes a fadeout of radio communications.

auroral activity The interaction of excited particles from space and particles of the upper atmosphere during an aurora. It usually affects radio communications.

auroral electrojet Multimillion ampere currents that flow in the auroral region. They are in the shape of a large oval, and are centered over the magnetic poles.

auroral event The occurrence of an **aurora**.

auroral oval Same as **auroral region**.

auroral propagation Propagation of radio waves which are reflected off an aurora. Such waves are usually in the very-high frequency range.

auroral reflection Radio waves which are reflected off an aurora. Such waves are usually in the very-high frequency range.

auroral region An oval-shaped region over a magnetic pole where an aurora appears. During large magnetic storms this region expands greatly. Also called **auroral oval**.

auroral zone A zone where auroras are most common, which is usually around the north and south magnetic poles.

authentication In computers and communications, the process of verifying the legitimacy of a transmission, user, or system. Measures such as passwords and digital signatures are employed.

authentication code A string of bits or characters, or a value, utilized for **authentication**.

authentication token An object or device, such as a smart card, which serves to authenticate a user in order to gain access to a computer network. Such a card, for instance, may be inserted into a smart card reader, with the holder prompted for a password or PIN for added security. Also called **security token**.

authoring programs Same as **authoring software**.

authoring software Software that facilitates the development of learning and teaching materials with significantly less programming than if a programming language were used, or with no programming at all. May serve, for instance, to create Web sites, or multimedia presentations. Also called **authoring programs**, or **authoring system**.

authoring system Same as **authoring software**.

authorization In computers and networks, the granting of access to resources to an individual which has provided a valid combination of user name and password, or its equivalent.

authorization code A sequence of characters which serve as a password, or its equivalent, to obtain access to a computer or network.

auto- A prefix used in words pertaining to that which is automatic, or which acts or occurs from within. For example, autoalarm, or autopilot.

auto-alarm Abbreviation of **automatic alarm**.

auto answer Abbreviation of **auto**matic **answer**ing. A feature which allows a modem to answer an incoming call automatically. Once done, it attempts to establish the connection. Its own abbreviation is **AA**. Also spelled **autoanswer**.

auto answering Same as **auto answer**.

auto attendant A computerized system which automatically answers calls, routes them, and takes messages. It incorporates voice prompts, often has extensive memory for messages, and may include voice-recognition. Also spelled **autoattendant**.

auto baud detect A feature which allows a modem to sense the speed of the calling modem, and to configure itself accordingly. Also spelled **autobaud detect**.

auto bias Abbreviation of **automatic bias**.

auto bypass **1.** In a network, the capacity to automatically bypass a terminal or device which is not functioning properly. This allows the working components to continue operation with a minimum of disruption. **2.** The capacity to automatically bypass a malfunctioning component within a device. This allows the working components to continue operation with a minimum of disruption, especially if there is redundancy.

auto call Same as **autodial**.

auto complete Same as **autocomplete**.

Auto Correct Same as **AutoCorrect**.

auto dial Same as **autodial**.

auto-ionization Same as **Auger effect**.

auto-load Same as **automatic loading**.

auto logon Same as **autologon**.

auto-negotiate Same as **autonegotiate**.

auto-pilot Same as **automatic pilot**.

auto play Same as **autoplay**.

auto polling Same as **autopolling**.

auto ranging Same as **autoranging**.

auto redial A feature which allows a device, such as a telephone or modem, to automatically dial the last entered telephone number, or one that has been programmed into its memory. The device may redial a specific number of times, for a given time interval, or until a connection is established. Also spelled **autoredial**.

auto-refresh Same as **autorefresh**.

auto-repeat A keyboard feature which enables a key that is held down longer than a given interval to repeat. The time necessary for repeating, and the repeating speed can usually be set. Also called **typematic, key repeat,** or **keyboard repeat**.

auto-reset Abbreviation of **automatic reset**.

auto-responder Same as **autoresponder**.

auto resume A feature which enables a computer to remember exactly what it was doing at a given moment, so that it can continue from there at a later time. Useful, for instance, in notebook computers when closing the lid between sessions. Also spelled **autoresume**.

auto-run Same as **autorun**.

auto-save Same as **autosave**.

auto scroll Same as **autoscroll**.

auto-start routine Same as **autostart routine**.

auto-test Same as **autotest**.

auto-trace Same as **autotrace**.

auto-tracking Same as **autotracking**.

auto-transformer Same as **autotransformer**.

autoadaptive Pertaining to a robot that detects environmental conditions and reacts to them automatically, without human intervention.

autoalarm Abbreviation of **automatic alarm**.

autoanswer Same as **auto answer**. Its abbreviation is **AA**.

autoattendant Same as **auto attendant**.

autobaud detect Same as **auto baud detect**.

autobias Abbreviation of **automatic bias**.

AutoCAD A powerful and widely-utilized CAD program.

autocall Same as **autodial**.

autocomplete Abbreviation of **automatic completion**. A feature which attempts to predict what characters would otherwise have been typed next. For instance, when starting to type a URL in the address field of a Web browser, it will complete the rest of the address based on previously visited sites. If security is a concern, as may be the case when using an Internet café, this feature should be disabled. Also spelled **auto complete**.

AutoCorrect Abbreviation of **Auto**matic **Correc**tion. A feature, seen for instance, in word processors, that corrects spelling and/or grammar errors on the fly. Such a feature does not distinguish, for instance, when a word is misspelled and results in another correctly spelled word. Also spelled **Auto Correct**.

autocorrelation The comparison of a signal to its time-delayed counterpart, to establish correlations between them.

autocorrelator A circuit which utilizes **autocorrelation** to detect a weak signal hidden within noise.

autodial Also spelled **auto dial**. Also called **autocall**. **1.** A feature which allows a modem to automatically dial a pre-programmed number. Once done, it attempts to establish the connection. **2.** A feature which allows a telephone to automatically dial a pre-programmed number.

autodyne circuit A circuit which simultaneously performs the functions of oscillator and detector. Its output signal frequency is the difference between the received signal frequency and the oscillator signal frequency.

autodyne reception Radio reception in which a single device performs the functions of oscillator and detector.

autoexec.bat In certain operating systems, a batch file containing commands such as what applications are loaded, and the values for specified settings.

AUTOEXEC.BAT Same as **autoexec.bat**.

autoionization Same as **Auger effect**.

autoload Same as **automatic loading**.

autologon A feature which allows a complete logon sequence to be executed automatically. Especially useful for computers or terminals used by only one person. Also spelled **auto logon**.

automata A plural form of **automaton**.

automata theory A science which deals with the way automatic devices receive input, process it, and produce an output, and the correlations that these have with behaviors of living beings.

automated **1.** A component, device, or system which performs certain or all functions without human intervention. **2.** A component, device, or system which performs certain or all functions utilizing an autonomous information processing system.

automated attendant Same as **automatic call distributor (2)**.

automated call distribution Routing of incoming calls using an **automatic call distributor**.

automated call distributor Same as **automatic call distributor**. Its abbreviation is **ACD**.

automated communications Communications without human intervention. These utilize an autonomous information processing system, and include automatic transmission of information such as voice and data, and self-monitoring of network status.

automated environment A home, or other environment, which utilizes an autonomous information processing system to automate tasks such as device and equipment operation and maintenance.

automated guided vehicle system A system utilized with vehicles that have no driver nor crew, which are equipped with automatic guidance equipment which enable them to follow a prescribed path, stopping at each necessary station for automatic or manual loading and/or unloading. Its abbreviation is **AGVS**.

automated home A home which in which certain tasks are automated. These may include chores such as vacuuming, the monitoring of security, or the adjustment of shades to utilize natural lighting in a balanced manner throughout the day. Computers are used, and robots may be incorporated if desired.

automated identification system A method of identification that utilizes an autonomous information processing system. It is computer controlled, and may use retinal scans, fingerprints, voice recognition, bar codes, and so on.

automated provisioning The setting up of a communications service, such as telephone service, Internet access, Web hosting, or the like, without interaction with another human. Suitable, for instance, for those who know what is available, and what they need. Also called **automatic provisioning**.

automated reasoning tool An application, within artificial intelligence, which utilizes various techniques, including the processing of non-numerical data for problem solving. Its abbreviation is **ART**.

automated teller machine A computerized banking terminal that permits cardholders to perform various transactions, such as cash withdrawals. Among other components, ATMs include the following computer hardware: monitor, CPU, printer, card reader, and modem. Its abbreviation is **ATM**. Also called **automatic teller machine**.

automatic A component, device, system, or process which does not require intervention, human or otherwise. It may have self-acting and/or self-regulating mechanisms, and utilize an autonomous information processing system. Also, pertaining to any such component, device, system, or process.

automatic alarm A device which automatically generates an alert under certain alarm conditions. For example, a one-way communicator dialing an emergency services number under specified circumstances. Its abbreviation is **autoalarm**.

automatic alternate routing A feature which, when needed, provides for re-routing of communications traffic automatically and without delay. Its abbreviation is **AAR**.

automatic answering Same as **autoanswer**.

automatic back bias In radars, the utilization of automatic gain control loops to prevent the receiver from becoming overloaded by large signals.

automatic background control Same as **automatic brightness control**.

automatic bass compensation Also known as **bass boost**. Its abbreviation is **ABC**. **1.** An amplifier circuit which effectively boosts low frequencies when listening to an audio system at low-volume settings. This compensates for the lower auditory response humans have under these circumstances, thus making the sound more natural. **2.** The increase in low frequencies resulting from the use of **automatic bass compensation (1)**.

automatic bias Its abbreviation is **auto bias**. **1.** In a transistor or vacuum tube, obtaining of the correct bias utilizing a dropping resistor instead of an external bias voltage. Also called **automatic C bias**, **automatic grid bias**, or **self-bias**. **2.** A circuit in tape decks, which depending on the tape type, automatically sets the bias for optimal playback.

automatic brightness control A circuit, present in TVs, designed to maintain the average brightness at a preset level. Also known as **automatic background control**. Its abbreviation is **ABC**.

automatic C bias Same as **automatic bias (1)**.

automatic call distribution Routing of incoming calls using an **automatic call distributor**.

automatic call distributor Also called **automated call distributor**. Its abbreviation is **ACD**. **1.** A switching center that routes incoming telephone calls to the next available operator or representative. **2.** A computerized system that routes incoming telephone calls to the next available operator or representative. A caller may hear recordings or choose from a voice menu before reaching another person, or in some cases, an informative recording. Also called **automated attendant**.

automatic call sequencer A device which answers incoming calls and places them on hold, after playing a recording. A signaling method, such as lights that blink faster as holding time increases, indicates the order of the calls. Its abbreviation is **ACS**.

automatic calling unit A component or device which enables equipment such as computers, phones, or fax machines to automatically place calls. Its abbreviation is **ACU**.

automatic check A mechanism which automatically monitors the operation of a component, device, piece of equipment, or system, seeking to maintain performance within specified parameters. May serve to detect malfunctions and to optimize performance.

automatic chroma control Same as **automatic color control**.

automatic chrominance control Same as **automatic color control**.

automatic circuit breaker A device which automatically opens a circuit under specified conditions, resetting itself after other conditions are met, such as the passing of a given time period.

automatic coding Same as **automatic programming**.

automatic color control A circuit, present in TVs, designed to maintain the intensity of colors steady, regardless of the strength of the color signal received. Its abbreviation is **ACC**. Also called **automatic chroma control**, or **automatic chrominance control**.

automatic completion Same as **autocomplete**.

automatic computer **1.** A computer which functions with little or no human intervention. **2.** A computer which executes certain operations without human intervention.

automatic contrast control A circuit, present in TVs, designed to automatically maintain screen contrast within certain limits, for optimal viewing.

automatic control The use of an **automatic controller**. Also called **automatic regulation**.

automatic control system A system which utilizes **automatic control**.

automatic controller A circuit, mechanism, device, or system, which monitors one or more variables, and automatically makes the necessary adjustments in order to maintain operation within the specified parameters. Thermostats and voltage regulators are examples. Also known as **controller (3)**.

Automatic Correction Same as **AutoCorrect**.

Automatic Crash Notification A system which notifies a call and/or emergency center when a vehicle is involved in a severe crash. Such a system, for instance, may initiate a transmission when an airbag is deployed, or when an installed accelerometer detects an impact exceeding a given magnitude. Its abbreviation is **ACN**.

automatic cutout A device, which under certain circumstances, automatically disconnects a component of a circuit, in order to protect equipment. An example is a circuit breaker.

automatic data processing The processing of information by a computer, with little or no human intervention. Its abbreviation is **ADP**. Also called **data processing**, **electronic data processing**, or **information processing**.

automatic data processing equipment Equipment which performs **automatic data processing**.

automatic degausser **1.** A circuit, device, or system for automatically demagnetizing a color TV picture tube. This process usually takes place while the TV is warming up for use. **2.** A device which demagnetizes automatically, as does a degausser of magnetic tapes.

automatic design optimization The utilization of computers to optimize the design of a product. Computer-aided design is employed, and it is used for example, to enhance the design of a vehicle so as to minimize the effects of air friction upon it while traveling. Also called **design optimization**.

automatic device A device which does not require intervention, human or otherwise. It may have self-acting and/or self-regulating mechanisms, and utilize an autonomous information processing system.

automatic dialer A device which automatically dials telephone numbers. Such numbers may be manually or automatically programmed into memory, and may be dialed when pressing a short code, or at a pre-programmed time. Also called **automatic dialing unit**, **dialer**, or **telephone dialer**.

automatic dialing unit Same as **automatic dialer**.

automatic dictionary A computer program that converts a character, word, instruction, or language into another. It may refer, for instance, to a conversion of programming code into machine language.

automatic direction finder A radio receiver that automatically and continuously indicates the direction from which a radio signal arrives. Its abbreviation is **ADF**. Also called **automatic radio compass**, **automatic radio direction finder**, or **radio compass**.

automatic error correction **1.** In telecommunications, a system which automatically detects and rectifies errors during transmission or reception. **2.** A system which automatically detects and rectifies errors.

automatic exchange A communications exchange, such as a telephone switchboard, that does not require human intervention for operation. Also known as **automatic switching system**.

automatic frequency control Its abbreviation is **AFC**. **1.** A circuit which automatically and continuously locks an electronic component onto a chosen frequency. Used, for instance, in TV receivers, radio receivers, and superheterodyne receivers. **2.** A circuit that maintains the frequency of an oscillator within precise limits. Also called **frequency control (2)**.

automatic gain control A process by which the gain in a receiver or amplifier is automatically adjusted so that the output signal remains constant despite variations in the input. Its abbreviation is **AGC**.

automatic grid bias Same as **automatic bias (1)**.

automatic height control A process by which the height of a TV image is automatically adjusted so that it remains steady, despite variations in the TV receiver's input.

Automatic Identification and Data Collection The collection of data using electronic means such as bar-code or RFID systems. Its abbreviation is **AIDC**.

automatic intercept A system that routes telephone calls not answered within a set number of rings. It may transfer the call to another number, or take a message.

automatic language detection In computers, the automatic identification of a language within a document, Web page, and so on. This may be accomplished, for instance, by the software utilized to process scanned images, or it may be a Web browser feature.

automatic level compensation A circuit that compensates for variations in input so that the output remains within specified parameters. May be used to maintain optimal performance and/or to protect equipment.

automatic level control A circuit which automatically adjusts the input so that the output signal remains constant despite variations in said input. Used in devices such as magnetic tape recorders and radio transmitters. Its abbreviation is **ALC**.

automatic light control Its abbreviation is **ALC**. **1.** The automatic adjustment of the illumination incident upon an imaging device, as a function of the brightness. Used, for example, in TV cameras. **2.** The automatic adjustment of the brightness of a screen or monitor, as a function of the surrounding brightness. Used, for example, in a liquid-crystal display.

automatic loading A mechanism which automatically executes a number of steps which would have logically followed. For instance, a cellular phone placing a telephone number in the next available memory location, a computer rebooting when the reset button is pushed, or a CD-ROM activating the installation program after insertion into the drive. Its acronym is **autoload**.

automatic message-switching center **1.** A system that allows routing of telephone messages from a remote location. **2.** A system that automatically routes telephone messages as determined by aspects of their content.

automatic modulation control In the transmission of radio signals, a circuit which prevents overmodulation by reducing the gain of signals that exceed a given strength. Its abbreviation is **AMC**.

automatic noise limiter A circuit which limits the maximum amplitude of an input signal. This can help, for example, to minimize the differences between strong and weak signals. Its abbreviation is **ANL**.

automatic number identification A service, offered by telephone companies, which sends the name and number of the calling party of an incoming call. A display device which features automatic number identification is needed to view this information. Alternatively, such a service may only provide the number of the calling party. Also called **caller ID**. Its abbreviation is **ANI**.

automatic operation **1.** Operation which does not require intervention, human or otherwise. It may involve self-acting and/or self-regulating mechanisms, and utilize an autonomous information processing system. **2.** One or more steps or processes which do not require intervention, human or otherwise.

automatic peak limiter Same as **amplitude limiter**.

automatic phase control In color TVs, a circuit which synchronizes the 3.58 MHz color subcarrier signal with its color burst signal. Its abbreviation is **APC**.

automatic picture control In video devices such as TVs and videocassette players, circuitry which optimizes the displayed picture. For example, a TV may adapt the on-screen picture to suit the lighting conditions of a room, or a videocassette recorder may adjust its output depending on the level of wear of a videotape. Its abbreviation is **APC**.

automatic picture transmission A system for transmission of images gathered by weather satellites. It allows relatively inexpensive ground equipment to receive data when the transmitting satellite is within range. Its abbreviation is **APT**.

automatic pilot In an aircraft, a system which automatically maintains a programmed course by monitoring factors such as speed, altitude, and attitude, while sending the necessary signals to the appropriate controls. Sophisticated automatic pilots can fly an airplane from immediately after takeoff, all the way to its destination, and then make an automatic landing. The term also applies to sea vessels. Also called **autopilot**.

automatic play Same as **autoplay**.

automatic playback Same as **autoplay**.

automatic process A process which does not require intervention, human or otherwise. It may involve self-acting and/or self-regulating mechanisms, and utilize an autonomous information processing system.

automatic programming Preparation of computer code which is done automatically by a computer. Also called **automatic coding**.

automatic protective device An equipment protecting device, such as a circuit breaker, which is automatically activated under certain conditions.

automatic provisioning Same as **automated provisioning**.

automatic radio compass Same as **automatic direction finder**.

automatic radio direction finder Same as **automatic direction finder**.

automatic ranging Same as **autoranging**.

automatic refresh Same as **autorefresh**.

automatic regulation Same as **automatic control**.

automatic regulator A circuit, mechanism, device, or system which utilizes **automatic control**.

automatic relay **1.** A system that automatically receives and reroutes telephone messages. **2.** A system which retransmits messages without human intervention. For instance, that utilized for an emergency broadcast.

automatic repeat request A request by a receiving computer or station for the retransmission of data. It is used in case of an error in transmission or reception. Its abbreviation is **ARQ**.

automatic repeater station   A station which automatically retransmits signals. Such retransmissions may be through any of various mediums, such as air or fiber-optic cables. Digital retransmission allows for higher transmission speeds and lower error rates.

automatic reset   A mechanism which automatically resets itself to a former state after certain conditions are met. For example, a circuit breaker resetting itself after a specified time period. Its abbreviation is **auto-reset**.

automatic responder   Same as **autoresponder**.

automatic ringdown circuit   A circuit that connects two telephonic devices, and in which an off-hook condition in one automatically rings the other. There is no dialing in such a circuit. Used, for instance, for emergency communications. Also called **ringdown circuit**.

automatic route selection   A method for automatically selecting the least costly route for a telephone call. For example, specialized software may utilize a table of area codes to determine which fax server out of many should be utilized for broadcasting. Its abbreviation is **ARS**. Also called **least cost routing**.

automatic-scanning receiver   A radio receiver that automatically scans a selected frequency range. It may be set to stop when a signal is located, or to display or graph the distribution of the detected signals.

automatic scrolling   Same as **autoscroll**.

automatic sensitivity control   A circuit in a receiver which automatically maintains sensitivity at a constant level. Its abbreviation is **ASC**.

automatic sequence   A series of operations that are executed without human intervention. For example, a computer performing a sequence of processes automatically.

automatic shutdown   Same as **automatic shutoff**.

automatic shutoff   Also called **automatic shutdown**. **1.** In a computer, the process of automatically saving all data, quitting all applications, and turning off the power. **2.** The process of automatically finishing a process in an orderly manner, and turning off the power of a device.

automatic station   A station, telecommunications or otherwise, that is designed to function with a minimum of human intervention, or none.

automatic stop   **1.** In a computer, the process of automatically stopping a given process upon detection of certain types of error conditions. **2.** The automatic stopping of a process under specified conditions.

automatic switchboard   A telephone switchboard that does not require human intervention for operation.

automatic switching system   Same as **automatic exchange**.

automatic system   A system or network that operates with little or no human intervention. It may have self-acting and/or self-regulating mechanisms, and utilize an autonomous information processing system. For example, a fire-detection system which monitors conditions and notifies the appropriate personnel in case of need.

automatic teller machine   Same as **automated teller machine**.

automatic test   A test which is performed without human intervention. It can refer to testing during production or operation.

automatic test equipment   Computer-controlled equipment used for testing components during production. Used especially in the context of chip manufacturing. Its abbreviation is **ATE**.

automatic time switch   A switch that opens or closes circuits according to a programmed time interval. For example, a

sensor could measure the temperature of an enclosed area while it is not in use, and automatically select the optimal time to switch on the heat before being occupied.

automatic transfer equipment   In power distribution, equipment which automatically transfers loads according to demand, or in case of failure.

automatic trip   An automatic opening of a circuit, such as that triggered by a circuit breaker, intended to protect equipment.

automatic tuning system   A mechanical, electrical, or electromechanical system which automatically tunes a radio transmitter or receiver to a specified frequency.

Automatic Vehicle Location   A system, such as that utilizing GPS, which keeps track of one or more vehicles. Its abbreviation is **AVL**.

automatic voltage regulator   A circuit or device which maintains an output voltage within specified values, despite variations in variables such as load resistance, line voltage, or temperature, so long as they are within a prescribed range. Also called **voltage regulator**, **voltage corrector**, or **voltage stabilizer**.

automatic volume compressor   A circuit that automatically limits the volume range of an audio signal. This is utilized, for example, by a radio transmitter to increase the average amount of modulation while avoiding overmodulation. Also called **volume compressor**.

automatic volume control   In a radio receiver, a circuit or device which maintains the output volume at a constant level, despite variations in the input signal. A flight controller, for example, would thus avoid having to adjust the volume when communicating with various aircraft at varying distances. A **volume limiter** limits the output volume to a predetermined maximum, regardless of the variations in the input signal. Its abbreviation is **AVC**.

automatic volume expander   A circuit or device which automatically increases the volume range of an audio signal. It works, essentially, to reverse the effect of volume compression, thus restoring the volume range of the original program. Also called **volume expander**, or **expander** (**2**).

automatic zero   A measuring instrument which has the ability to automatically set a zero level without external calibration.

automatically programmed tools   A programming system used in computer-aided manufacturing, for programming certain computer-controlled tools, such special drills. Its abbreviation is **APT**.

automation   **1.** The technology and techniques involved in making a process, apparatus, or system function automatically. **2.** The automatic functioning of a process, apparatus, or system. **3.** The use of machines to perform tasks that would otherwise correspond to humans.

automaton   A mechanism, device, or machine which operates autonomously in a pre-programmed manner. Said especially of a robot.

automonitor   A program or process in which records of all steps are automatically kept. Used, for instance, in debugging.

autonegotiate   In networks, the automatic determination of the proper communications settings, such as transmission speed. Also spelled **auto-negotiate**.

autonomic computing   Computing in which all operations, such as task completion and malfunction repair, are performed without human intervention.

autonomous information processing system   An information-processing system that functions without human intervention.

autonomous robot   A self-contained robot that has enough sensory input and decision-making capability to be able to perform tasks without human intervention. Such a robot may or may not have a mechanism permitting it to move on its own. Each autonomous robot has its own controller, while a single controller is shared by multiple **insect robots**.

autonomous system In networks such as the Internet, a collection of routers under a single administrative authority using its own management rules and protocols.

autonomous system number A number which identifies an **autonomous system**.

autopatch A remote-controlled device for interconnecting a radio communicator and a landline telephone network.

autopilot Same as **automatic pilot**.

autoplay Abbreviation of **auto**matic **play**back. A feature which automatically begins playback or runs a program when an optical disc, such as a DVD, CD, or CD-ROM, is placed in its drive. Also spelled **auto play**.

autopolarity The ability of a digital meter to display readings with the correct polarity, without having to interchange the input connectors.

autopolling Also spelled **auto polling**. Also called **polling**. **1.** In a system, the periodic detection of the status of all connected devices, to address any needs for resources. This contrasts with an **event-driven** allocation scheme, where interrupts are sent to request resources. **2.** In networking, the sequential determination of which terminals wish to transmit data, to allow doing so in an orderly manner. **3.** A feature which allows a fax machine to call another, and request that the remote machine transmit documents.

autopolling cycle A complete **autopolling** (1) or **autopolling** (2) sequence. For example, in one such cycle, each connected terminal is interrogated once. Also called **polling cycle**.

autoranging In meters with multiple ranges of measurement, automatic switching upward of intervals until readings are made in the range within which the highest indication does not exceed the full-scale value. Some meters also adjust downwards. Also spelled **auto ranging**. Also called **automatic ranging**.

autoredial Same as **auto redial**.

autorefresh Abbreviation of **auto**matic **refresh**. The updating of a Web page without user intervention. Seen, for instance, when the same page displays different ads every few seconds or minutes. Also spelled **auto-refresh**.

autoreset Abbreviation of **automatic reset**.

autoresponder Abbreviation of **auto**matic **responder**. Also spelled **auto-responder**. A feature or program which automatically sends a previously prepared email in response to a received email. Used, for instance, to inform that an email has been received and will be responded to at a later time. Also called **email autoresponder**, or **mail autoresponder**.

autoresume Same as **auto resume**.

autorun A program that launches automatically under the appropriate circumstances. For example, a CD-ROM may activate the installation program after insertion into the drive. Also spelled **auto-run**.

autosave A feature in many computer programs which provides for automatically saving to disk at specified intervals. This is useful in minimizing data loss in the event of a program, system, or power failure. Also spelled **auto-save**.

autoscroll Acronym for **auto**matic **scroll**ing. Scrolling that occurs without user intervention. For example, a Web browser or Web page may provide an option for the pages viewed to be automatically scrolled at a given rate. Also spelled **auto scroll**.

autostart routine The sequence of steps a computer is programmed to follow upon turning on its power. These include running diagnostic procedures, loading the operating system, and otherwise preparing for use. Also spelled **auto-start routine**.

autotest One or more tests that are automatically performed in preparation for use, or to ascertain proper operation. Also spelled **auto-test**.

autotrace A method of converting a bitmap image into a vector-graphics image. This allows for simpler editing, especially when using object-oriented programs. Also spelled **auto-trace**.

autotracking Also spelled **auto-tracking**. **1.** Tracking which is maintained autonomously. **2.** Control of two or more power supplies by one of them. **3.** In a video-cassette player, a feature which automatically makes the necessary adjustments so that the heads continuously scan the tracks on the tape accurately.

autotransductor A transductor in which the same windings carry the main current and the control current.

autotransformer A transformer that has a single tapped winding, so that part serves as the primary winding, and part as the secondary winding. Voltage step-up or step-down of is effected using a tap on the common winding. Also spelled **auto-transformer**.

aux input Same as **auxiliary input**.

auxiliary anode In electroplating, a supplementary anode utilized to obtain a better distribution of plating.

auxiliary channel A supplemental communications channel, which can serve, for instance, for backup, for testing, or for transmission of a secondary audio program. Also known as **secondary channel**.

auxiliary circuit A supplementary circuit which can serve, for instance, as an alternate path for current.

auxiliary contacts In a switching device, supplemental contacts whose function is linked to that of the main circuit contacts.

auxiliary device A device which serves in a supplemental or auxiliary capacity, or that is added to enhance the performance of the main device. Also called **ancillary device**.

auxiliary equipment Equipment which serves in a supplemental or auxiliary capacity, or that is added to enhance performance. Also called **ancillary equipment**.

auxiliary function A function which is supplements or enhances the operation of a device, piece of equipment, or system.

auxiliary input An additional input, such as that found on the back panel of a stereo receiver, which accepts signals from certain devices which do not already have dedicated inputs. For example, an input for an additional CD player such as an SACD player. Its abbreviation is **aux input**.

auxiliary memory **1.** A form of high-speed computer memory which accessed through a high-speed data channel. Unlike RAM, this memory is not directly addressable by the CPU. **2.** Same as **auxiliary storage**.

auxiliary power An alternate or supplemental source of electrical power that may be used when the main power source is interrupted, depleted, or insufficient. Also called **supplemental power**, or **secondary power** (1).

auxiliary relay A relay that supplements the action of another.

auxiliary storage Also known as **auxiliary memory** (2), or **backing storage**. **1.** Any device that a computer uses as a storage medium. For example, disk drives or tape drives. **2.** Additional computer storage beyond that in the main storage.

auxiliary switch A switch that works in unison with another.

aV **1.** Abbreviation of **attovolt**. **2.** Abbreviation of **abvolt**.

AV Abbreviation of **audiovisual**, or **audio video**.

AV connector Abbreviation of **audio/video connector**.

AV inputs Abbreviation of **audio/video inputs**, or **audiovisual inputs**.

AV outputs Abbreviation of **audio/video outputs**, or **audio-visual outputs**.

availability 1. Being accessible and ready for use. 2. Not being currently occupied in performing its designated tasks. 3. The proportion of time a device is fully operational, usually expressed as a percentage. 4. The proportion of time a device is available for its designated tasks, usually expressed as a percentage.

available 1. Accessible and ready for use. 2. Not currently occupied in performing its designated tasks.

available line 1. In communications, a line which is available for transmission or reception. 2. In a scanning device such as a fax, the proportion of the scanning line available for image information. Also called **useful line**.

available power The maximum power that a circuit or device can deliver to a load.

available power gain In a transducer, the ratio of the available output power, to that available from the input source.

available time Also called **uptime**. 1. The proportion of time which a device or system is available for operation, usually expressed as a percentage. 2. The amount of time that a device or system is available. May be expressed in hours, days, and so on.

avalanche Also called **avalanche effect, cascade (2), cumulative ionization, Townsend ionization, Townsend avalanche**, or **ion avalanche**. When the only produced charged particles are electrons, it is also called **electron avalanche**. 1. A cumulative ionization process in which charged particles are accelerated by an electric field and collide with neutral particles, creating additional charged particles. These additional particles collide with others, so as to create an avalanche effect. 2. In semiconductors, the cumulative generation of free charge carriers in an avalanche breakdown.

avalanche breakdown In a semiconductor exposed to an electric field, a non-destructive breakdown in which there is a cumulative multiplication of free charge carriers. The avalanche occurs as a result of an increasing electric field causing free charge carriers to create new electron-hole pairs by impact ionization, which then produce even more pairs. Also called **breakdown (5)**.

avalanche conduction Conduction through a semiconductor junction resulting from the cumulative generation of free charge carriers in an avalanche breakdown.

avalanche current The high current through a semiconductor junction resulting from the cumulative generation of free charge carriers in an avalanche breakdown.

avalanche diode A semiconductor diode with a very high ratio of reverse-to-forward resistance, until the point where avalanche breakdown occurs. After this, voltage drop is nearly constant and independent of current. May be used, for instance, in surge suppressors with reaction times in the picosecond range. Also called **breakdown diode**.

avalanche effect Same as **avalanche**.

avalanche impedance In an semiconductor diode, the reduced impedance during avalanche breakdown.

avalanche multiplication In semiconductors, the multiplicative generation of free charge carriers as a result of avalanche breakdown.

avalanche noise Noise produced by an avalanche diode at the onset of avalanche breakdown.

avalanche photodiode A photodiode that amplifies as a result of avalanche multiplication of photocurrent. Such photodiodes feature low noise and high speed. Its abbreviation is **APD**.

avalanche transistor A transistor which cumulatively generates free charge carriers through avalanche breakdown.

avalanche voltage In a semiconductor junction, the voltage at which avalanche breakdown occurs.

avatar An image, such as a cartoon, an animal, or that of another person, that is utilized to visually represent a real user in settings such as 3D chat rooms or online games. An avatar usually appears as three-dimensional.

AVC Abbreviation of **automatic volume control**.

average The value obtained by first adding together a set of quantities, and then dividing by the number of quantities in the set. Also called **average value (1), arithmetic mean**, or **mean (1)**.

average access time For storage units such as hard disks and DVDs, the average time that elapses between a request for data and its delivery, or the average time required to store information once the instruction is given.

average current The average value of a current flowing through a circuit or device. Also called **mean current**.

average life The average lifetime of a particle, substance, device, or unit. Usually refers to atoms, their components, or elementary particles. For instance, the muon, which is an unstable elementary particle, has an average life of approximately 2.2 microseconds. Also called **mean life**.

average noise factor Same as **average noise figure**.

average noise figure For a circuit or device, the ratio of the total noise at the output terminal, to the thermal noise at the input terminal. The measurement is usually made at 290° Kelvin, and over all frequencies. Also called **average noise factor**.

average power The average level of power in a signal at any particular moment, or over a given time period. Also called **actual power (2)**.

average pulse amplitude The average of the instantaneous amplitudes for the duration of a pulse.

average value 1. Same as **average**. 2. The adding together of many instantaneous values of an amplitude taken at equal time intervals divided by the number of measurements taken. For a pure sine wave, the average value is 0.637 times the peak amplitude value.

average voltage The average voltage in an AC circuit. For a sine wave it is equal to 0.637 times the peak voltage.

aviation electronics Its acronym is **avionics**. 1. All the electronic communication, navigation, and flight-control equipment on board an aircraft. 2. The design and production of the electronic communication, navigation, and flight-control equipment on board an aircraft.

avionics Acronym for **aviation electronics**.

AVL Abbreviation of **Automatic Vehicle Location**.

Avogadro's constant Same as **Avogadro's number**.

Avogadro's number The number of elementary entities in a mole. This numeric value is approximately 6.022142×10^{23}. For instance, a mole of carbon-12 has approximately 6.022142×10^{23} atoms, and has a mass of 12 grams. Also known as **Avogadro's constant**. Its symbol is N_A, or L.

avoirdupois ounce A unit of mass equal to approximately 0.0283495 kg. Also called **ounce (1)**.

aW Abbreviation of **attowatt**.

AWG Abbreviation of **American Wire Gauge**.

axes The plural form of **axis**.

axial lead A conductor extending from an axis of a component such as a resistor or capacitor.

axial ratio In a waveguide, the ratio of the major axis to the minor axis of the polarization ellipse. Also called **ellipticity**.

axis 1. In a coordinate system, one of a set of reference lines. Distances or angles may be measured from it. 2. A straight line of symmetry that passes through a body or system. **3.** A straight line that a body or system may rotate around.

Ayrton-Perry winding The winding of two wires in parallel, but opposite directions, in order to cancel the magnetic field. Used, for instance, in resistors that are well suited for low-inductance applications.

Ayrton shunt A high-resistance shunt utilized to reduce the sensitivity of a measuring instrument, such as a galvanometer. This increases its range. Also called **universal shunt**.

azel display In radars, the displaying of two radar traces on the same screen. This allows for presentation of bearing and elevation at the same time.

azimuth 1. The horizontal direction from one object or point to another, measured clockwise in degrees from a reference line. Also called **azimuth angle (1)**, or **bearing (1)**. **2.** The horizontal direction of a celestial point from a terrestrial point, expressed as the angle from a reference direction. It is usually measured from 0° at the reference direction, and runs clockwise through 360°. Also called **azimuth angle (2)**. **3.** In a tape deck, the angular relationship between the head gap and the tape path.

azimuth alignment In a tape deck, the adjustment of the playback or record head to achieve the proper 90° alignment with the tape path.

azimuth angle 1. Same as **azimuth (1)**. **2.** Same as **azimuth (2)**.

azimuth blanking 1. In a transmitter, the stopping of transmission through a range of specified azimuth angles. **2.** In a receiver, the stopping of reception through a range of specified azimuth angles. **3.** The blanking of a radar screen through a range of specified azimuth angles.

azimuth resolution The minimum azimuth angle separation necessary for a radar to distinguish between two scanned objects at equal distances from the station.

aΩ Abbreviation of **abohm**.

B

b 1. Abbreviation of **bit**. 2. Abbreviation of **baud**. 3. Symbol for **bel**. 4. Symbol for **barn**. 5. Abbreviation of **bar**.

B 1. Abbreviation of **battery**. 2. Symbol for **bel**. 3. Chemical symbol for **boron**. 4. Abbreviation of **byte**. 5. Symbol for **base**.

B 1. Symbol for **magnetic flux density**. 2. Symbol for **susceptance**.

B and S gage Abbreviation of **Brown and Sharp gage**.

B and S gauge Abbreviation of **Brown and Sharp gauge**.

B battery 1. A battery which supplies the anode, or plate, current in electron tubes. Also known as **anode battery**, or **plate battery**. 2. A battery which supplies the screen-grid electrode current in electron tubes.

B channel 1. In a two-channel stereo component, such as an amplifier, the right channel. 2. Abbreviation of **bearer channel**. In ISDN communications, a channel that bears, or carries, voice and data. It provides 64 Kbps full-duplex capacity. It works alongside a D channel.

B connector A connector used for splicing twisted-pair wires, such as telephone wires. It is shaped like a small plastic tube, and has metal teeth inside which strip the insulation off the wire and provide a tight connection. Such connectors usually contain a gel which minimizes the penetration of moisture. Also called **B-wire connector**.

B display In radars, an oscilloscopic display which plots azimuth of the scanned object in the horizontal coordinate, versus its range, which appears in the vertical coordinate. The scanned object appears as a bright spot on the display. Also called **B scan**, **B scanner**, **B scope**, **B indicator**, or **range-bearing display**.

B-H curve For a magnetic material, a graphical representation showing the relationship between magnetic flux density, or *B*, in the vertical axis, and magnetizing force, or *H*, in the horizontal axis. Also called **magnetization curve**.

B-H meter For a magnetic material, an instrument which measures its hysteresis loop.

B indicator Same as **B display**.

B-ISDN Abbreviation of **broadband ISDN**.

B minus The negative terminal of a **B supply**. Its symbol is **B-**. Also called **B negative**.

B negative Same as **B minus**.

B plus The positive terminal of a **B supply**. Its symbol is **B+**. Also called **B positive**.

B positive Same as **B plus**.

B power supply Same as **B supply**.

b/s Abbreviation of **bits per second**.

B scan Same as **B display**.

B scanner Same as **B display**.

B scope Same as **B display**.

B station Within the loran radio navigation system, one of two transmitting stations, the other being the **A station**. The A station is the control station in this pair.

B supply A source which supplies DC to the anode and other high-voltage electrodes in electron tubes. Also called **B power supply**.

B-wire connector Same as **B connector**.

B-Y signal In a color TV receiver, a color-difference signal representing the difference between the blue signal and the luminance signal. Thus, adding this signal to the luminance yields the blue primary signal. **B** is blue, and **Y** is luminance.

B & S gage Abbreviation of **Brown and Sharp gage**.

B & S gauge Abbreviation of **Brown and Sharp gauge**.

B- Symbol for **B minus**.

B+ Symbol for **B plus**.

B2B Abbreviation of **business-to-business**. Online businesses selling to and otherwise catering to the needs of other businesses.

B2B business Same as **B2B**.

B2B ecommerce Abbreviation of **business-to-business ecommerce**. Ecommerce between businesses.

B2C Abbreviation of **business-to-consumer**. Online businesses selling to and otherwise catering to the needs of individual consumers.

B2C business Same as **B2C**.

B2C ecommerce Abbreviation of **business-to-consumer ecommerce**. Ecommerce between businesses and individual consumers.

B8ZS Abbreviation of **binary 8-zero substitution**.

Ba Chemical symbol for **barium**.

babble In communications, the unwanted intermingling of signals between transmission lines. Also called **multiple crosstalk**.

BABY A late 1940's prototype for the Manchester Mark I computer.

back bias 1. In a multistage circuit, voltage which is fed back to a previous stage. Seen, for instance, in automatic level control circuits. 2. A bias voltage applied in the proper polarity to a diode or semiconductor junction to cause little or no current to flow. Also called **reverse bias**.

back bone Same as **backbone**.

back contact In a relay or switch, a contact pair that is closed in the resting state, and which is opened when energized. The contact pair consists of a stationary contact and a movable contact. A back contact provides a complete circuit when not energized. This contrasts with **front contact**, in which the contact pair is open in the resting state. Also called **break contact**, or **normally-closed contact**.

back current In a device such as a rectifier or semiconductor, current that flows in the opposite direction of that which is normal. Also called **reverse current**, or **inverse current**.

back diode A type of tunnel diode with reverse-conduction characteristics, and which is well suited for use at microwave frequencies.

back door A hidden software or hardware mechanism that allows access to a program, system, or network in a manner which completely eludes all security measures. It can be activated, for instance, by a seemingly random key sequence, and is usually left in place by the software developer. It is intended to allow for functions such as maintenance, and is a significant security risk when known by an unintended user or program. Also spelled **backdoor**. Also known as **trapdoor**.

back echo A reflected signal originated by a minor back lobe of a radar beam.

back electromotive force The voltage developed in an inductive circuit through which AC flows. The polarity of this voltage is at all times the opposite of that of the applied voltage. Its abbreviation is **back emf**. Also known as **back voltage**, **counter emf**, or **reverse voltage** (1).

back emf Same as **back electromotive force**.

back emission In a vacuum tube, the emission of electrons in the reverse direction, that is, from anode to cathode. Also called **reverse emission**.

back end **1.** In a client/server environment, the server part. The client part is the **front end**. **2.** In semiconductor manufacturing, the final stages of production, which include testing and packaging. **3.** A program performing tasks that are not readily apparent to a user, as in certain compiler functions.

back-end processor A computer controlled by another, and which used for specialized functions such as database access or high-speed graphics processing. The controlling computer is called **front-end processor**, and it interfaces with a user.

back-gating Same as **backgating**.

back-heating Same as **backheating**.

back-light Same as **backlight**.

back-lit Same as **backlit display**.

back-lit display Same as **backlit display**.

back lobe In an antenna, a radiation lobe whose direction is the opposite of the main or intended direction of propagation.

Back Orifice A program designed to take advantage of the numerous security deficiencies of Windows operating systems. Used, for instance, by crackers to have complete and undetected control of a remote computer running any given version of Windows. Also spelled **BackOrifice**.

back panel The panel at the rear of certain equipment, which serves to connect to other devices and/or to change available settings. For example, a computer's back panel serves to connect with peripherals and to a power source.

back plane Same as **backplane**.

back porch In a composite video signal, the interval of the signal between the trailing edge of the horizontal sync pulse and trailing edge of the corresponding blanking pulse.

back-porch effect In transistors, the continuing of collector current flow for a brief interval after the input signal is zero.

back-pressure sensor A sensor which measures an applied torque and produces a signal proportional to it. Used, for instance, in robotics.

back resistance In a contact rectifier, the contact resistance that opposes the inverse current that results from an inverse voltage.

back scatter Same as **backscatter**.

back scattering Same as **backscatter**.

back-side bus Same as **backside bus**.

back stop In a relay, a barrier which limits the movement of the armature when the circuit is open.

back-surface mirror A mirror whose reflective coating is on its rear surface, as is the case with a household mirror. This contrasts with **front-surface mirror**, in which the reflective coating is on the front surface. Also called **second-surface mirror**, or **rear-surface mirror**.

back-to-back circuit Same as **back-to-back connection**.

back-to-back connection Two tubes or semiconductor devices connected in parallel, but opposite directions, to control AC without rectification. For example, the connection of the anode of one diode to the cathode of another, and vice versa. Also called **back-to-back circuit, inverse-parallel connection**, or **inverse-parallel circuit**.

back-to-front ratio The ratio of the signal strength of an antenna in one direction, to that in the opposite direction. It is expressed in decibels, and the higher the ratio, the more effectiveness there is in rejecting interfering signals.

back-up To copy computer files onto an additional storage medium. For instance, copying the contents of a folder contained on a hard drive to a tape drive or floppy disks. This is important, for instance, in avoiding the loss of data when there are software or hardware malfunctions, or in case of user carelessness. There are utilities which can simplify this process, for instance, by following a schedule, or by compressing the data to economize storage space and time.

back voltage Same as **back electromotive force**.

back wave An acoustic wave which is radiated by the back of a speaker, which can cancel the waves radiated from the front of the speaker. An infinite baffle completely isolates the back waves from the front waves, while a port helps both waves to be in phase. Also called **rear wave**.

backbone Also spelled **back bone**. Also called **backbone network**. **1.** The connections that form the major pathways of a communications network, large or small. These handle the bulk of the traffic, and generally communicate at very fast rates, often over great distances. **2.** A network topology in which a **backbone (1)** is the hub to which all subnetworks are connected. Used, for instance, in medium-sized LANs. Also called **collapsed backbone**. **3.** A superfast network spanning the globe, linking national Internet service providers at speeds that can exceed 100 Gbps. Local Internet service providers connect to regional Internet service providers, which in turn connect to this backbone, to be a part of the Internet. Also called **Internet backbone**.

backbone network Same as **backbone**.

backdoor Same as **back door**.

backfile conversion The process of scanning documents on mediums such as paper and microfilm, so that they become suitable for computer use.

backgating In ICs, a reduction in drain current due to a negative bias applied to an ohmic contact near a field-effect transistor. Also spelled **back-gating**. Also called **backside gating**.

background **1.** All the phenomena which are present in any given physical apparatus or its surrounding environment, other than that which is specifically intended for observation or measurement. For instance, background radiation present when utilizing a Geiger counter. These have an especially noticeable effect when measuring small signals. **2.** Same as **background noise**. **3.** On a computer screen that has more than one window open, all windows except the one that is currently being interacted with, which is the **foreground (1)** window. **4.** On a screen or display, the area that is not being specifically interacted with or observed. Also called **display background**. **5.** On a screen or display, the area against which characters and/or graphics are displayed. **6.** In computers, processing and other functions that occur while a user is occupied with other tasks.

background brightness On a screen or display, the luminosity present in the absence of a video signal. Used especially in the context of TVs. Also known as **background level (1)**.

background count The count resulting from all sources of radiation other than the source intended to be measured.

background job **1.** In computers, a task which can be interrupted when those of higher priority need to be completed. **2.** In computers, a task that occurs while a user is occupied with other tasks. Also called **background task**, or **background process**.

background level **1.** Same as **background brightness**. **2.** The combined levels of all phenomena which are present in any given physical apparatus or its surrounding environment, other than that which is specifically intended for observation or measurement.

background light Light which is present in any given environment, especially within which tests or observations are taking place. Also called **room light, environmental light**, or **ambient light**.

background monitor A radiation counter which measures levels of **background radiation**.

background noise Also called **background noise level**, or **background (2)**. **1.** The noise present in any given environment, especially that within which tests or observations are taking place. Such noise includes nearby and distant sources. Also called **ambient noise, room noise, site noise, environmental noise,** or **local noise. 2.** Electrical noise which is inherent in a circuit, device, component, piece of equipment, or system. In sound recording or reproduction, for instance, it is the residual noise in the absence of a signal. Also called **ground noise (3).**

background noise level Same as **background noise.**

background process Same as **background job (2).**

background processing 1. In computers, processing that can be interrupted when that of higher priority needs to be completed. **2.** In computers, processing that occurs in a manner that involves no interaction on the part of a user.

background program A computer program that can be interrupted when tasks of higher priority need the attention of the processor.

background radiation Radiation originating from sources such as cosmic rays, or arising from the small proportion of naturally occurring radioactive isotopes present in the atoms in any given environment.

background response The reading a radiation detector will have due to **background radiation.**

background return In radars, unwanted echoes that appear on the display screen. Background return may be caused by rain, antenna movements, vegetation, and so on. Also known as **radar clutter,** or **clutter.**

background task Same as **background job (2).**

backhaul 1. To send information or signals, such as files or TV programming, to a location from which distribution is facilitated. For example, to transmit a live scene from a given location to a central location from which signals are uplinked to a satellite, which in turn emits signals to large areas. **2.** To send information or signals, such as data or telephone calls, to locations further than the intended destination, and then back to the final destination from said remote location, when doing so is simpler, faster, and/or cheaper.

backheating In a magnetron, additional heating of the cathode produced by high-energy electrons returning to it. If excessive, it may cause burnout. Also spelled **back-heating.**

backing storage Also known as **backing store,** or **auxiliary storage. 1.** Any device that a computer uses as a storage medium. For example, disk drives or tape drives. **2.** Additional computer storage beyond that in the main storage.

backing store Same as **backing storage.**

backlash 1. In a mechanical component, such as a dial, looseness or play within which the desired effect is not yet achieved. For instance, the initial movement of a tuner control mechanism that doesn't yet effect a change in frequency. **2.** The discrepancy in the indicated and/or actual values of a quantity when the dial controlling them is turned clockwise versus counterclockwise. **3.** In a thermionic tube, incomplete rectification due to positive ions in its gaseous atmosphere.

backlight The light source positioned behind the viewing surface of an LCD with a **backlit display.** Also spelled **back-light.**

backlit Same as **backlit display.**

backlit display A screen, especially a liquid-crystal display, with its own light source behind the viewing surface. This enhances contrast and resolution. Also called **backlit,** or **back-lit display.**

BackOrifice Same as **Back Orifice.**

backplane Also spelled **back plane. 1.** A circuit board that has sockets which allow other circuit boards to be plugged into it. It may or may not be processor-controlled, which

would make it active or passive, respectively. **2.** The main circuit board of a computer, containing the connectors necessary to attach additional boards. It contains most of the key components of the system, and incorporates the CPU, bus, memory, controllers, expansion slots, and so on. Memory chips may be added to it, and some motherboards allow for the CPU to be replaced. Also called **motherboard, mainboard,** or **system board.**

backplate 1. In a camera tube, the electrode which is the target of the scanning electron beam. **2.** In a condenser microphone, one of the two charged metal plates, the other being the diaphragm.

backscatter Also spelled **back scatter.** Also called **backward scatter,** or **backscattering. 1.** Scattering of waves or particles through angles greater than 90° with respect to the initial direction of radiation. **2.** In radio-wave propagation, the scattering of waves back towards the transmitter. **3.** In radio-wave propagation, the scattering of waves back towards the transmitter after being reflected off the ionosphere.

backscattering Same as **backscatter.**

backside bus A microprocessor bus that connects the CPU to a level 2 cache. It generally runs at the speed of the CPU, while the **front-side bus,** which connects to the memory and peripherals, is usually slower than this. Also spelled **backside bus.**

backside cache A level 2 cache whose channel to the CPU is the **backside bus.**

backside gating Same as **backgating.**

backslash In most computer operating systems, a character which serves to separate folders and directories in file pathnames, as seen, for instance, in: C:\Documents\Reminders\Today.xyz. This character has a place on the keyboard, and is shown here between brackets: [\ ]. Also known as **reverse solidus.**

backspace key On most computer keyboards, a key which when pressed, serves to erase content as it moves the cursor to the left. Its symbol is often ← .

backtracking In expert systems, the solving of problems by going through multiple alternatives. If the desired solution is not reached via one route, there is a backtracking followed by the attempt of another possibility.

backup 1. A duplicate copy of computer data available on an additional storage medium such as a floppy disk, to be used in case of need. Also called **backup copy. 2.** Additional resources which are available to substitute those which fail. For example, a diode replacing another which has failed. **3.** Additional resources which serve to supplement others, such as backup battery.

backup and recovery The restoration of computer data to the state it was in before a software, hardware, or power failure. The most recent backup copy, along with a log of all activity since then is utilized to get as close as possible to the moment of failure. Its abbreviation is **backup & recovery.** Also called **backup and restore.**

backup and restore Same as **backup and recovery.**

backup battery 1. A battery which serves to power a system in case the main source fails. **2.** A battery which provides power briefly while the main batteries in a device are replaced.

backup copy Same as **backup (1).**

backup device 1. A device which serves to substitute another with the same function, in case of failure. **2.** A device which serves to supplement others.

backup disk A disk onto which a duplicate copy of computer data is stored, to be used in case of need.

backup equipment 1. Equipment which serves to substitute other equipment with the same function, in case of failure. **2.** Equipment which serves to supplement other equipment.

backup file A file which contains a duplicate copy of computer data, to be used in case of need.

backup path An additional path that may be used in case the main one fails or is otherwise unavailable. Used, for instance, in communications.

backup power An alternate source of power which is used in case the main source fails. Also called **backup power supply**.

backup power generator A generator providing **backup power**.

backup power supply Same as **backup power**.

backup server A server that provides **backup**.

backup system **1.** A system which is used in case the main one fails. **2.** The mechanisms used to provide backup when necessary.

backup tape A tape onto which a duplicate copy of computer data is stored, to be used in case of need.

backup utility A computer program which organizes and simplifies backing-up of files, for instance, by following a schedule, or by compressing the data to economize storage space and time.

backup & recovery Same as **backup and recovery**.

backup & restore Same as **backup and recovery**.

backward chaining Also known as **backward reasoning**. In artificial intelligence, a method of reasoning or problem solving that starts with a solution or conclusion, then works its way backwards to a set of conditions and corresponding facts which would bring about this result. This contrasts with **forward chaining**, which has facts and a set of conditions, and works its way toward a solution or conclusion.

backward channel In asymmetrical communications, the channel that flows upstream. It is the slower of the two channels, the other being the **forward channel**. Also called **reverse channel**.

backward compatibility The state of being **backward compatible**.

backward compatible Software that is compatible with earlier versions of it, or hardware that is compatible with earlier models of it. Were it not for backward compatibility, upgrading any software or hardware would be unduly complex and expensive. Also known as **downward compatible**.

backward diode A semiconductor diode similar to a tunnel diode, but with slightly lower doping levels, and which serves as a rectifier for low voltages.

Backward Explicit Congestion Notification In frame relay, a bit which serves to notify of detected congestion from the source direction of a given packet. Utilized to help control the flow of data. When the congestion is in the destination direction of a packet, a **Forward Explicit Congestion Notification** bit is set. Its abbreviation is **BECN**.

backward reasoning Same as **backward chaining**.

backward recovery The reconstruction of a file to an earlier version. A later version and an activity log are used for this purpose. This contrasts with **forward recovery**, where a later version is reconstructed using an earlier version.

backward scatter Same as **backscatter**.

backward wave **1.** In a traveling-wave tube, a wave whose group velocity is in the opposite direction of the electron beam. **2.** A wave traveling in the opposite direction of that which is normal.

backward-wave amplifier An amplifier of **backward waves**. Its abbreviation is **BWA**.

backward-wave magnetron A magnetron in which the electron beam travels in the opposite direction of the flow of RF energy.

backward-wave oscillator An oscillator utilizing a special vacuum tube in which an RF magnetic field interacts with the electron beam that travels from cathode to anode. This produces electron bunching, which in turn produces a backward wave. This oscillator may be tuned over a wide range of frequencies, and its output power is extracted near the electron gun. Its abbreviation is **BWO**. Also called **carcinotron**.

backward-wave tube A traveling-wave tube in which the wave is propagated in the opposite direction of the electron beam.

bacterium A malicious computer program which replicates itself until it takes over the entire system. This contrasts with a **virus**, which attaches itself to other programs, and which doesn't necessarily take over the entire system.

bad block A block on a computer disk which can not be used for data storage or retrieval because of damage or imperfections. The operating system usually locates and marks these blocks, avoiding their use.

bad block table A list of **bad blocks**.

bad sector On a computer storage disk, a sector which can not be used for data storage or retrieval because of damage or imperfections. The operating system usually locates and marks these sectors, avoiding their use.

bad track On a computer storage disk, a track which can not be used for data storage or retrieval because it contains a **bad sector**.

baffle Also called **speaker baffle**, **loudspeaker baffle**, or **baffle plate** (2). **1.** A partition in a speaker enclosure which is used to reduce or eliminate the interaction between the acoustic waves generated by the front of the speaker and those from the back. This is especially important in low-frequency sound reproduction. **2.** The panel on which one or more speakers are mounted. **3.** A panel utilized to inhibit the propagation of sound waves. Used, for instance in theaters as part of a sophisticated sound-reproduction system.

baffle enclosure A speaker enclosure incorporating a **baffle** (1), or **baffle** (2).

baffle loudspeaker Same as **baffle speaker**.

baffle plate **1.** A metal plate inserted into a waveguide to reduced its cross-sectional area. Also called **baffle-plate converter**. **2.** Same as **baffle**.

baffle-plate converter Same as **baffle plate** (1).

baffle speaker A speaker with a **baffle** (1), or **baffle** (2). Also called **baffle loudspeaker**.

bakelite A thermosetting synthetic resin used as a dielectric or for making hard plastic objects, such as knobs, utilized in electronics.

bakeout The high-temperature heating of a system which is being evacuated. This ideally gets rid of any contaminants, such as gases, water vapor, oils, and so on.

balance **1.** A state of equilibrium reached through the equating of the forces which counter each other. **2.** The equalization or symmetry between one or more variables among circuits, components, devices, systems, networks, or ground. For instance, the balancing of component values in order to null a bridge. **3.** Same as **balanced sound**. **4.** In an aircraft, an equilibrium attained which produces a steady flight. May also refer to equilibrium with reference to a specific axis. **5.** A device or instrument which serves to make accurate indications of weight. A microbalance, for instance, can provide readings in the picogram range.

balance coil An iron-core solenoid which enables a two-wire circuit to feed a three-wire circuit. It utilizes adjustable taps near its center for compensation.

balance control A control found on a stereo audio amplifier or pre-amplifier which is used to vary the relative loudness between the left and right channels. When the balance control is moved entirely towards the left channel, for instance, the right channel is completely attenuated. When centered, the channels have an equal output.

balance method Same as **balanced method**.

balance network Same as **balanced network**.

balanced In a state of **balance**.

balanced amplifier An amplification circuit with two identical signal branches which operate in phase opposition, and whose input and output connections are balanced to ground. An example is a push-pull amplifier.

balanced antenna An antenna whose two halves are identical in all respects. A biconical antenna is an example.

balanced bridge A bridge in which branch values are adjusted in a manner which provides an output voltage of zero. At this time a zero response is obtained from the null detector.

balanced circuit 1. A circuit whose main branches are electrically identical, and are symmetrical with respect to a reference such as ground. 2. A circuit whose two sides are electrically identical, and are symmetrical with respect to a reference such as ground. 3. A circuit which has equal currents flowing through its main branches.

balanced converter Same as **balun**.

balanced currents For the two conductors of a balanced line, currents flowing which are equal in magnitude, but opposite in phase at all points along said line. Also called **push-pull currents**.

balanced detector A symmetrical demodulator used in FM systems.

balanced input A two terminal input circuit grounded on each side.

balanced line A transmission line in which two identical conductors carry currents which are equal in magnitude but opposite in polarity. In this manner noise and crosstalk are minimized. Also called **balanced transmission line**.

balanced load A load which is distributed evenly among two or more conductors.

balanced low-pass filter A filter used with a balanced line, which blocks all frequencies above a given cutoff frequency.

balanced method A method of measurement in which a reading of zero is at the center of the scale. Indicated values may be greater than or less than the zero reading. One variation is the use of an audible signal for measurement, in which case the zero reading would be silent. Also called **balance method**, **zero method**, or **null method**.

balanced mixer A mixer circuit which suppresses one of its inputs, usually that of the local oscillator. Balance may be achieved, for instance, utilizing a ring configuration of diodes. Used, for example, in superheterodyne receivers. Also called **single-balanced mixer**.

balanced modulator A circuit whose input is a carrier and its modulating signal, and whose output is composed of the two sidebands without the carrier. A ring modulator, for instance, may be used for this purpose. The double-sideband signal may be converted into a single-sideband signal if desired.

balanced network An electric network whose two branches have the same impedance with respect to ground. Used, for instance, in a balanced line. Also called **balance network**.

balanced oscillator An oscillator in which the impedance centers of the tank circuits are at ground potential, and whose voltages between these centers and at the ends are equal in magnitude and opposite in phase. A push-pull oscillator is an example.

balanced output A two-terminal output circuit grounded on each side.

balanced ring modulator A ring modulator whose input is a carrier and its modulating signal, and whose output is composed of the two sidebands without the carrier.

balanced sound In sound recording or reproduction, an equilibrium between the volume intensities of the channels, a natural proportion between the high and low frequencies, a sense of symmetry between the voices and instruments, or the like. Also called **balance (3)**.

balanced termination A load whose two branches have the same impedance with respect to ground.

balanced-to-unbalanced transformer Same as **balun**.

balanced transmission line Same as **balanced line**.

balanced-unbalanced transformer Same as **balun**.

balanced voltages Voltages which are equal in magnitude but opposite in polarity with respect to ground. Also called **push-pull voltages**.

balanced-wire circuit A circuit whose two sides are symmetrical with respect to other conductors and ground.

balancer 1. A component, circuit, device, mechanism, or system which serves to equilibrate, compensate, or neutralize. 2. In a direction finder, a component, such as a balancing capacitor, which improves the accuracy of direction indication.

balancing 1. Equilibration, compensation, or neutralization through the utilization of a circuit, device, component, mechanism, or system. 2. In a radio direction finder, improvement of the accuracy of direction indication through the use of a component such as a balancing capacitor.

balancing capacitor In a radio direction finder, a variable capacitor utilized to improve the accuracy of direction indication. Also called **compensating capacitor**.

balancing circuit A circuit which serves to equilibrate, compensate, or neutralize.

balancing network 1. An electric network which serves to equilibrate, compensate, or neutralize. 2. An electric network, used in a two-branch circuit, which is designed in a manner that an electromotive force applied to one of the branches does not produce a current in the other. 3. In telecommunications, a circuit which matches a two-wire circuit to a four-wire circuit. Also called **hybrid circuit (3)**.

balancing unit A device that converts unbalanced transmission lines into balanced transmission lines, or the converse, by inserting the appropriate components.

ball bonding A method of making very small electrical connections by utilizing a molten ball produced by heating a very thin gold wire. Used, for instance, in semiconductor devices.

Ball Grid Array A surface-mount chip package that utilizes tiny balls of solder to attach its leads to a printed-circuit board. Ball Grid Arrays feature low induction, compact size, and a high lead count in a small area. Its abbreviation is **BGA**.

ballast For certain types of lamps, one or more devices which serve to provide the proper conditions, such as voltage, for starting and/or operating properly. Fluorescent or mercury lamps may use these.

ballast lamp A lamp that maintains a steady current by utilizing an enclosed **ballast resistor**.

ballast resistor A resistor whose characteristics enable it to maintain a steady current throughout a wide range of voltages, making it suitable for use as a current regulator. Seen, for instance, in certain types of lamps, such as fluorescent or mercury.

ballistic electrons Electrons which move much faster the usual speed achieved in regular electronic circuits. These electrons travel without scattering and have applications in semiconductor devices and microscopy, among others.

ballistic galvanometer A galvanometer designed to respond to rapid current pulses. To do so, they are usually undamped, highly sensitive, and a have a long period of swing.

balloon help On screens such as those of computers and TVs, help text and graphics enclosed by a balloon-like outline.

balun A device utilized to match the impedances or transmission characteristics of an unbalanced coaxial transmission line or system, and a balanced two-wire line or system, while minimizing attenuation. Used, for instance, for matching a coaxial cable to a twisted pair. Also called **balun transformer**, **balanced converter**, **balanced-to-unbalanced transformer**, **balanced-unbalanced transformer**, **bazooka**, or **line-balance converter**.

balun transformer Same as **balun**.

banana jack A matching receptacle for a **banana plug**.

banana plug A connecting plug with a spring metal tip which resembles a banana, and which is inserted into a **banana jack**. Used, for instance, for quick connections of test leads.

band 1. In communications, a specific interval of frequencies utilized for a given purpose, such as radio-channel broadcasting. Also called **frequency band (1)**. 2. A specific interval of frequencies between two limiting frequencies. For example, extremely high frequencies span from 30 GHz to 300 GHz. Also called **frequency band (2)**. 3. Any given interval of frequencies. Also called **frequency band (3)**. 4. In spectroscopy, a compact series of spectral lines which represent an interval of wavelengths which are absorbed or emitted by molecules. Atoms produce a **line spectrum**. Also called **band spectrum**, or **spectral band**. 5. In computer storage, a group of associated tracks.

band center 1. The frequency that is at the center of a band of frequencies. 2. The frequency that is the average of the highest and lowest frequencies of a band.

band-elimination filter Same as **bandstop filter**.

band gap Same as **bandgap**.

band-limited channel A communications channel which has a limited interval of frequencies. An example is a voice-grade channel.

band-pass Same as **bandpass**.

band-pass amplifier Same as **bandpass amplifier**.

band-pass filter Same as **bandpass filter**.

band-pass response Same as **bandpass response**.

band-reject filter Same as **bandstop filter**.

band-rejection filter Same as **bandstop filter**.

band selector In a device which has more than one band, a switch that allows the selection of the desired band of frequencies. It changes all the necessary circuitry, and may be found, for instance, in transmitters or receivers. Also called **band switch**, or **bandswitch**.

band spectrum Same as **band (4)**.

band-spread Same as **bandspread**.

band-spread tuning Tuning utilizing **bandspread**.

band stop Same as **bandstop**.

band-stop filter Same as **bandstop filter**.

band suppression Same as **bandstop**.

band-suppression filter Same as **bandstop filter**.

band-switch Same as **band selector**.

band-width Same as **bandwidth**.

band-width allocation Same as **bandwidth allocation**.

band-width control Same as **bandwidth control**.

band-width management Same as **bandwidth management**.

band-width on demand Same as **bandwidth on demand**.

band-width reservation Same as **bandwidth allocation**.

bandgap The energy required to free an outer shell electron from its valance band up to the lowest conduction band. The smaller the band gap, the better the conductor, as is the case with metals. Also spelled **band gap**.

bandpass Also spelled **band-pass**. 1. Same as **bandwidth (3)**. 2. A band of frequencies which is passed with little or no attenuation, while all others are highly attenuated.

bandpass amplifier An amplifier which only passes a specific intervals of frequencies, while blocking the rest. Its gain is essentially uniform for the frequencies it passes. Also spelled **band-pass amplifier**.

bandpass filter A filter which passes a given interval of frequencies with little or no attenuation. Outside this range all frequencies are highly attenuated. Also spelled **band-pass filter**.

bandpass response A response characteristic which shows an interval of frequencies which is uniformly passed with little or no attenuation. Outside of this range the frequencies are highly attenuated, providing for an overall graph with a mostly flat top. Also spelled **band-pass response**. Also called **flat-top response**.

bandspread In radio receivers, a technique for improving selectivity by spreading the bandwidth over a greater mechanical or electrical range. It simplifies tuning, and is especially useful where there is great congestion of nearby stations. Also spelled **band-spread**.

bandspread tuning Tuning utilizing **bandspread**.

bandstop Also spelled **band stop**. Also called **band suppression**. 1. To highly attenuate a given band of frequencies, while allowing others to pass with little or no attenuation. 2. A specific band of frequencies which is highly attenuated, while all others are passed with little or no attenuation.

bandstop filter For a given band of frequencies, a filter which highly attenuates all frequencies within two specific cutoff points. All frequencies above and below this interval pass with little or no attenuation. Also spelled **band-stop filter**. Also called **band-elimination filter**, **band-rejection filter**, **band-reject filter**, or **band-suppression filter**.

bandswitch Same as **band selector**.

bandwidth Also spelled **band-width**. Its abbreviation is **BW**. 1. An interval of frequencies which is utilized for any given purpose, such as that assigned to a TV transmitter. 2. The difference between the upper and lower frequencies of a continuous frequency band, in which performance falls within certain limits with respect to a given characteristic. 3. The interval of frequencies between which a device can produce a given proportion of its maximum output. Usually calculated at 50% or 90% of full power. Also called **bandpass (1)**. 4. The data transmission capacity of a communications channel, system, network, or medium. May be expressed, for instance, in bits, bytes, or cycles per second. 5. A fraction of **bandwidth (4)** which is utilized for any given purpose.

bandwidth allocation Also spelled **band-width allocation**. Also called **bandwidth reservation**. 1. The designation or distribution of radio frequencies for any given purpose. 2. The assigning of communications resources within a system or network. A hierarchy is followed, where higher priority traffic displaces lower priority traffic when necessary.

bandwidth control Also spelled **band-width control**. The mechanisms utilized for controlling the communications resources within a system or network. For instance, a hierarchy may be followed where higher priority traffic displaces lower priority traffic when necessary. Or, there may be a maximum amount of traffic permitted for a given service or function, and so on.

bandwidth-limited operation The condition in which the bandwidth of a system limits its performance, as opposed to attenuation or distortion.

bandwidth management The measures taken to effect **bandwidth control**. Also spelled **band-width management**.

bandwidth on demand Within a network, the capability of increasing the available communications resources as any given circumstances warrant. It helps accommodate varying traffic loads in as efficient a manner as possible. A user of a wireless communication device requiring more bandwidth when accessing the Internet than when placing a voice call is another example. Also spelled **band-width on demand**.

bandwidth reservation Same as **bandwidth allocation**.

bang-bang control An automatic control mechanism which when actuated applies the maximum signal, value, or output, to restore the desired condition. A home heating system usually uses such a mechanism.

bank A series of objects, usually of the same type, which are connected electrically and used together. For instance, a group of resistors, batteries, or contacts.

bank selection Same as **bank switching**.

bank switching In a computer, switching between banks of RAM chips in order to increase the amount of memory available. This additional memory is not as fast, since only one bank may be accessed at any given moment, and the switching occupies time. Also called **bank selection**.

bank winding Same as **banked winding**.

banked transformers A group of transformers connected in parallel.

banked winding A manner of winding RF coils in which single turns are wound successively in multiple layers alongside each other, with the entire winding advancing without return from one end of the coil to the other. This arrangement reduces the distributed capacitance of the coil. Also called **bank winding**.

banner Same as **banner ad**.

banner ad Abbreviation of **banner ad**vertisement. On the Internet, a form of advertising which appears on Web pages, and which may be clicked upon to obtain more information, usually by being transferred to the home page of an entity. These are often presented as a rectangular color graphic measuring 468 pixels wide by 60 pixels high, near the top or bottom of the page. A **skyscraper ad** is usually taller and thinner, with placement on either side of a page. Also called **banner**.

banner advertisement Same as **banner ad**.

banner blindness The tendency of certain Web users, such as those performing research, to ignore **banner ads**. Marketers, in an attempt to get the attention of as many users as possible, currently resort to other means, such as Web pages that can't be accessed without first being subjected to an ad, or by presenting large flash graphics accompanied by music, among other schemes.

bantam connector Either a **bantam jack**, or a **bantam plug**.

bantam jack A receptacle for a **bantam plug**.

bantam plug A plug with similar characteristics to a standard ¼ inch phone plug, but whose overall dimensions are much smaller. It allows a higher density of connectors within a given space, making them well-suited for instances where space is at a premium. A **bantam plug** is inserted into a **bantam jack**.

bantam tube A tube with a standard octal base, but whose overall dimensions are much reduced in comparison.

bar 1. A unit of pressure equal to 100 kilopascals, 10^6 dynes per square centimeter, or about 750.062 torr. It is a bit below the average atmospheric pressure at sea level, which is about 1.013 bar. Its symbol is b. **2.** A solid piece of a material, such as metal or crystal, which is usually much longer than it is wide. **3.** A vertical or horizontal stripe, band, or line which is much longer than it is wide. For example, those produced by a bar generator.

bar chart Same as **bar graph**.

bar code A precise arrangement of parallel vertical bars and spaces of varying widths, encoded with characters such as letters, numbers, or symbols. These bars are read by optical devices called bar-code readers, which may scan the data horizontally and/or vertically, or at any angle. Bar codes serve for rapid and error-free entry of information, and are used, for instance in retail stores or libraries, or to help robots identify components in assembly facilities. Also spelled **barcode**. Also called **bar-code symbol**, or **optical bar code**.

bar-code character A character encoded by a single group of bars and spaces within a **bar code**. Also spelled **barcode character**.

bar-code reader A laser scanning device utilized to read **bar codes**. Also spelled **barcode reader**. Also called **bar-code scanner**.

bar-code scanner Same as **bar-code reader**. Also spelled **barcode scanner**.

bar-code symbol Same as **bar code**. Also spelled **barcode symbol**.

bar-code system A system, such as that utilized to record transactions in a point-of-sale system, based on **bar codes**. Also spelled **barcode system**.

bar generator A testing device which generates pulses which are synchronized with those from a TV system, and which produces horizontal and vertical bars on a TV screen. Used, for instance, for verification of the linearity of displayed images.

bar graph A graphical representation in which data is presented as rectangular bars. These bars may be horizontal or vertical, and may be in solid colors or patterns to display information more clearly. Also called **bar chart**.

bar magnet A bar of metal which has been strongly magnetized so that it becomes a permanent magnet.

bar meter A digital meter which uses a series of bars to indicate quantities such as volume levels. Since it is not analog, it can not specify values on a continuous scale.

bar pattern On a TV screen, a pattern of horizontal and vertical bars generated by a **bar generator**.

barcode Same as **bar code**.

barcode character Same as **bar-code character**.

barcode reader Same as **bar-code reader**.

barcode scanner Same as **bar-code reader**. Also spelled **bar-code scanner**.

barcode symbol Same as **bar code**. Also spelled **bar-code symbol**.

barcode system Same as **bar-code system**.

Bardeen-Cooper-Schrieffer theory A theory which explains very low temperature superconductivity. In part, it proposes that electrons at such temperatures travel in pairs called Cooper pairs, which enable them to move through a lattice without colliding with it in a disruptive manner. Its abbreviation is **BCS theory**.

bare board A circuit board with no components mounted.

bare conductor A conductor with no surrounding insulation, such as a bare copper wire.

barebones Providing only the functions and/or components indispensable for operation. Can refer to software, hardware, or equipment in general. For instance, a barebones computer may only include components such as the cabinet, motherboard, power supply, floppy drive, and keyboard. If desired, other devices, such as a monitor, hard-disk, or other peripherals must be added.

BARITT diode Abbreviation of **bar**rier **i**njection **t**ransit **t**ime **diode**. A microwave diode which is similar to an **IMPATT diode**.

barium A silvery-white metal chemical element whose atomic number is 56. It is an alkaline-earth metal, extremely reactive, and has around 30 known isotopes, of which several are stable. Barium is extrudable, machinable, and slightly malleable. It is used, for instance, as an atomic tracer for radium, in spark-plug alloys, and as a getter in vacuum tubes. Its chemical symbol is **Ba**.

barium carbonate A white powder whose chemical formula is $BaCO_3$. Used, for instance, in the glass of some TV picture tubes, in electrodes, and in electroceramics.

barium cyanide A white crystalline powder whose chemical formula is $Ba(CN)_2$. Used, for example, in metallurgy and electroplating.

barium fluoride A white powder whose chemical formula is BaF_2. It has various applications in electronics, including its use in the manufacturing of carbon brushes for DC motors and generators, and as crystals for spectroscopy.

barium molybdate A white powder whose chemical formula is $BaMoO_4$. It is used, for example, in electronic and optical equipment.

barium selenide A microcrystalline powder whose chemical formula is $BaSe$. It is used, for instance, in photocells and semiconductors.

barium sulfate A white or yellowish powder whose chemical formula is $BaSO_4$. It is used, for example, as a shielding material for x-ray apparatuses and nuclear reactors.

barium titanate A grayish powder whose chemical formula is $BaTiO_3$. It exhibits ferroelectricity, and has various applications in electronics, including its use in memory devices, as a piezoelectric transducer, in magnetic amplifiers, and as a voltage-sensitive dielectric.

Barkhausen effect A succession of small and abrupt magnetization variations observed in a ferromagnetic material which is subjected to an external magnetic force which steadily increases or decreases.

Barkhausen interference Interference caused by **Barkhausen oscillation**.

Barkhausen-Kurz oscillator A triode oscillator in which the frequency of oscillation depends exclusively on the transit time of electrons between the cathode and anode. Usually used for generating ultra-high frequencies. Also called **Barkhausen oscillator**.

Barkhausen oscillation In a TV receiver, unwanted oscillation in the horizontal-output tube which produces dark and jagged vertical lines on the left of the image.

Barkhausen oscillator Same as **Barkhausen-Kurz oscillator**.

barn A unit of area used in nuclear physics, equal to 10^{-28} square meters. One barn equals 100 square femtometers.

Barnett effect A minimal magnetization which is developed in an iron rod which is rotated rapidly about its longitudinal axis. This occurs without the influence of an external magnetic field.

barometer An instrument which measures atmospheric pressure. Such devices often use a column of mercury supported by air, with the height of the column varying proportionally to the surrounding atmospheric pressure.

barometric pressure The pressure which the atmosphere exerts at any given point within it. This pressure is due to the weight of the gases that compose it. Also called **atmospheric pressure (1)**.

barometric switch Its abbreviation is **baroswitch**. **1.** A switch that is actuated by changes in atmospheric pressure. **2.** A pressure-actuated switch used in radiosondes.

baroswitch Abbreviation of **barometric switch**.

barrage reception In telecommunications, reception utilizing an array of directional antennas oriented in various directions, to better zero in on the desired signal while minimizing interference.

barrel distortion In optics, distortion which causes an image to bulge convexly on all sides, similar to the shape of a barrel. Seen, for instance on computer or TV screens. Also called **positive distortion**.

barrier 1. A device, structure, or material that obstructs or impedes. For instance, an insulator serving as a barrier to electricity. **2.** In a semiconductor, the region of a pn junction that is free, or depleted, of charge carriers. This is a deprecated term, as are **barrier layer**, **barrier region**, or **blocking layer**. **Depletion layer**, or **space-charge layer** are the more proper terms to use for this concept.

barrier capacitance In a semiconductor, the capacitance across the region of a pn junction that is depleted of charge carriers. Also known as **barrier-layer capacitance**, **depletion-layer capacitance**, or **junction capacitance**.

barrier height In a semiconductor, the potential difference across the region of a pn junction that is depleted of charge carriers.

barrier injection transit time diode Same as **BARITT diode**.

barrier layer Same as **barrier (2)**.

barrier-layer capacitance Same as **barrier capacitance**.

barrier-layer cell A semiconductor device which generates an electric current when exposed to radiant energy, especially light. These are often made of silicon, selenium, or germanium. Such cells can be used in arrays to power anything from street lights to satellites. Also known as **barrier-layer photocell**, or **photovoltaic cell**. When optimized for the conversion of solar energy, also called **solar cell**.

barrier-layer photocell Same as **barrier-layer cell**.

barrier potential In a semiconductor, the potential across the region of a pn junction that is depleted of charge carriers.

barrier region Same as **barrier (2)**.

barrier strip A terminal strip which has a raised insulating barrier separating each terminal.

barrier voltage The minimum voltage required for current flow through a pn junction.

baryon A heavy subatomic particle composed of three quarks. Baryons are a subclass of the hadrons, and include protons and neutrons.

base 1. The foundation of something, or the material that composes said foundation. For instance, silicon serving as a base material in semiconductor manufacturing, or acetate film used as a base for magnetic recording tape. **2.** In a bipolar junction transistor, the region between the emitter and the collector, into which the emitter injects minority carriers. Also, the electrode attached to this region. Its symbol is **B**. Also called **base region**, **base electrode**, or **base element**. **3.** A starting point or basis, such as a base address. **4.** A reference parameter, such as time based on atomic frequencies. **5.** Same as **base station**. **6.** A chemical which does any of the following: that increases the proportion of hydroxyl ions in a solution, that accepts a proton in solution, that can react with an acid to form a salt, or that donates two electrons. According to the pH scale, a base has a reading of greater than 7.0, while an **acid** has a value below 7.0. The higher the reading, the stronger the base. **7.** In an electron tube, the insulated portion through which its connecting terminals protrude. **8.** The number of digits used in a numbering system. For example, 2 in the binary system or 10 in the decimal system. The base also serves as the multiplier within its numbering system. In the decimal system, for instance, each single position movement to the right of a digit represents a division by 10, while a movement to the left is a multiplication by 10. This can be seen, for example, in the number 153, where the 1 is in the hundreds position, the 5 is the tens, and the 3 is the units. Also called **base number**, **radix**, or **radix number**. **9.** A number which is raised by a power indicated by another number, called the **exponent**. For example, in 2^8, the 2 is the base, and the 8 is the exponent. **10.** The number upon which a logarithm system is based. That is, the number which is raised to a power indicated by an exponent. For example, $\log_{10} 1000 = 3$, as 10, the base in this example, must be raised to the power of three to equal 1000. Logarithms may use any number as their base. Common logarithms have a base of 10, and natural logarithms have a base equal to approximately 2.71828.

base address An address that serves as a beginning or reference point for other addresses. Other addresses may be specified, for instance, by adding a given number of bytes to a base address.

base-band Same as **baseband**.

base-band frequency response Same as **baseband frequency response**.

base-band network Same as **baseband network**.

base-band signal Same as **baseband signal**.

base-band system Same as **baseband system**.

base-band transmission Same as **baseband transmission**.

base bias In a transistor, the DC voltage applied to the base electrode. Also called **base bias voltage**.

base bias voltage Same as **base bias**.

base class In object-oriented programming, a class from which from a subclass is derived, via inheritance. Also called **superclass**.

base current In a transistor, the current flowing through the base electrode.

base electrode Same as **base** (**2**).

base element Same as **base** (**2**).

base film The plastic substrate used in magnetic recording tape. For instance, it may be an acetate film, upon which a magnetic oxide layer is placed for sound recording.

base frequency In a complex signal or waveform, the main or strongest frequency, such as the driving frequency or the fundamental frequency. For instance, the fundamental frequency of an oscillation. Also known as **basic frequency**.

base insulator A heavy-duty insulator utilized to isolate an antenna mast from the ground and other conductive paths. It also supports the weight of the antenna.

base laminate A thin material which serves as a substrate for others. An example is the dielectric material used as a substrate for printed-circuit boards.

base line Same as **baseline**.

base load Same as **baseload**.

base-loaded antenna A vertical antenna whose electrical length is adjusted by varying an inductance in series at its base.

base material A material which serves as a substrate for others. For instance, in a chip, the base material may be silicon. Another example is the use of a dielectric material as a substrate for printed-circuit boards. Also called **base medium**.

base medium Same as **base material**.

base memory The amount of RAM a given computer model comes with. Such an amount is usually greatly expandable.

base metal Also called **basis metal**. **1.** A metal upon which a coating may be deposited, as in electroplating. **2.** A metal that is readily oxidized or corroded, such as iron.

base number Same as **base** (**8**).

base pin In a plug-in electrical component such as an IC or tube, a terminal which allows physical connection with compatible sockets in other devices or components. It may

also provide structural support. Also known as **pin**, or **prong**.

base potential The potential at the base electrode of a bipolar junction transistor.

base region Same as **base** (**2**).

base resistance The resistance in series with the base electrode of a bipolar junction transistor.

base-spreading resistance In a bipolar junction transistor, the resistance in series with the base region due to the body or mass of the region, as opposed to that associated with a junction.

base station Also called **base** (**5**). **1.** In a land-mobile system, a land station that maintains communications with land-mobile stations. A base station may also communicate with other base stations. **2.** In a land-mobile system, a fixed-location station which communicates with mobile stations. **3.** A site which has the necessary transmission and reception equipment to provide cellular phone coverage in a given geographical area. All incoming and outgoing calls within this zone are handled by this station. The area served by a given base station is called a **cell** (**6**). Also called **cell site**.

base unit **1.** In a cordless telephone, the component that is plugged into the telephone system as well as a power source. A base unit has a cradle which charges the handset when not in use. **2.** Any specific unit within a group of **base units**.

base units Internationally accepted fundamental units for the measurement of physical quantities. All other units and internationally accepted systems of units are based on these fundamental, or absolute, units. Currently, the fundamental, or base, SI units are: the **second**, the **kilogram**, the **meter**, the **ampere**, the **Kelvin**, the **mole**, and the **candela**. Also called **fundamental units**.

base voltage The voltage at the base electrode of a bipolar junction transistor.

baseband Also spelled **base-band**. **1.** The band of frequencies produced by a transducer, such as a microphone or TV camera, which will be used to modulate a carrier. For instance, in the transmission of music, the baseband should encompass the audible range of frequencies. A TV camera produces baseband signals which need to be modulated to be able to be viewed by a conventional TV set. Demodulation of a previously modulated wave returns the baseband. **2.** A communication system in which unmodulated digital signals are sent over a transmission line. Used in LANs, such as Ethernet. One set of data is sent at a time. This contrasts with broadband, where multiple sets of data are transmitted simultaneously. Also known as **baseband system**.

baseband frequency response The frequency response of a transducer over the band of frequencies which will be used to modulate a carrier. Also spelled **base-band frequency response**.

baseband network A LAN within which unmodulated digital signals travel along a single transmission line. Ethernet is an example of such a network. Also spelled **base-band network**.

baseband signal A signal which will be used to modulate a carrier. Also spelled **base-band signal**.

baseband system Same as **baseband** (**2**). Also spelled **base-band system**.

baseband transmission Transmission of unmodulated digital or analog signals. Also spelled **base-band transmission**.

baseline Its abbreviation is **BL**. Also spelled **base line**. **1.** A line which serves as the basis for measurement, calculation, or location. It is usually represented graphically as a horizontal line. **2.** In displays, such as that of oscilloscopes and radars, the line which represents the reference level in the absence of other signals. **3.** The average or reference signal level from which a pulse departs, and to which it returns.

4. In navigation, a line which joins two stations, such as the A and B stations within the loran radio navigation system.

baseload Also spelled **base load**. The essential load that a power-generating system or station must provide to meet the power needs for a given period of time

BASIC Acronym for **b**eginner's **a**ll-purpose **s**ymbolic **i**nstruction **c**ode. A simple to learn high-level programming language, with applications, for instance, in business.

basic cell A group of components, such as transistors, that is replicated across the surface of a gate array.

basic frequency Same as **base frequency**.

basic input/output system Same as **BIOS**.

Basic Rate Interface An ISDN line with two 64 kilobits per second B channels, and one 16 kilobits per second D channel. With the proper configuration, one B channel may be used for conversing on the telephone, and the other for a computer modem. Otherwise, the two B channels can be combined for total available bandwidth of 128 kilobits per second. Its abbreviation is **BRI**. Also called **Basic Rate Interface ISDN, Basic Rate ISDN,** or **ISDN-BRI**.

Basic Rate Interface ISDN Same as **Basic Rate Interface**.

Basic Rate ISDN Same as **Basic Rate Interface**.

basis metal Same as **base metal**.

basket In a speaker, a structure that supports the cone suspension and magnet assemblies. It is usually made of plastic or metal. Also called **frame (6)**.

basket coil Same as **basket winding**.

basket winding A winding whose successive turns are loosely crisscrossed. This provides a low distributed capacitance. Also known as **basket coil**.

bass Within the audio-frequency range, the low end of the detectable frequencies. This interval usually spans from about 20 Hz to about 300 Hz, although the defined interval varies. Also called **bass frequencies**.

bass boost Also known as **automatic bass compensation**. **1.** An amplifier circuit which effectively boosts low frequencies when listening to an audio system at low-volume settings. This compensates for the lower auditory response humans have under these circumstances, thus making the sound more natural. **2.** The increase in low frequencies resulting from the use of a **bass boost (1)**.

bass compensation Emphasis of low frequencies utilizing a **bass boost (1)**.

bass control In an audio amplifier, a manual tone control which effectively boosts or attenuates low frequencies. It may be adjusted in discrete steps, or through a continuous interval.

bass cut The attenuation of low audio frequencies. May be used, for instance, to reduce or eliminate rumble. Seen especially in audio amplifiers. Also called **bass suppression**.

bass frequencies Same as **bass**.

bass port A hole, or port, in a **bass reflex enclosure**.

bass reflex enclosure A speaker design where there is a hole, or port, built into the cabinet. This increases and extends the reproduction of low frequencies. Although there is a savings in required amplifier power, there is a reduction in accuracy. Also called **acoustic reflex enclosure**.

bass reflex loudspeaker Same as **bass reflex speaker**.

bass reflex speaker A speaker with a **bass reflex enclosure**. Also called **bass reflex loudspeaker**.

bass resonance In a speaker, resonance occurring at frequencies below about 300 Hz.

bass response The ability of a component, circuit, device, piece of equipment, or system to handle the frequencies in the low end of the audible range. This interval is usually from about 20 Hz to about 300 Hz, and may refer, for instance, to the response of speakers, microphones, or amplifiers.

bass roll-off 1. The attenuation of frequencies below a given value, such as 300 Hz. **2.** The amount by which frequencies below a given value, such as 300 Hz, are attenuated.

bass suppression Same as **bass cut**.

bassy Descriptive of sound reproduction of a speaker or audio system which overly emphasizes frequencies below about 300 Hz.

bastion host A host computer which serves as a gateway between an internal and an external network. It may serve by itself to guard against unauthorized entries and attacks, or as part of a more elaborate security system, such as a firewall.

BAT Abbreviation of **battery**.

bat-handle switch A toggle switch whose actuating lever is thinner at the base and thicker at the end, similar to a baseball bat.

batch 1. In computers, a group or set of items, such as records, documents, or programs, which are utilized as a unit. This is usually accomplished without concurrent user intervention. **2.** In networks, a group of messages which are transmitted as a unit.

batch file A computer file containing a series of commands which are executed in sequence, without concurrent user intervention.

batch job In computers, a job executed without concurrent user intervention.

batch operation In computers, an operation performed simultaneously on various items, without concurrent user intervention.

batch processing Computer operations in which a series of procedures are executed without concurrent user intervention. These are useful in saving user time and system resources.

batch program A computer program containing a series of commands which are executed in sequence, without concurrent user intervention

batch session A session established for a **batch transmission**.

batch transmission The transmission of a series of records, documents, or messages, without interruption, and without concurrent user intervention.

bath voltage In an electrolytic cell, the total voltage between the anode and the cathode during electrolysis.

bathtub capacitor A capacitor, usually of paper, whose metal housing has rounded corners which give it an overall shape similar to that of a bathtub.

battery A source of DC power that incorporates two or more **cells (1)**. **Cell** and **battery** are commonly used interchangeably, although a battery consists of more than one cell. Its abbreviations are **B**, and **BAT**. Also called **electric battery**.

battery acid An acid that may be used as the electrolyte in a rechargeable battery. For instance, a purified and concentrated form of sulfuric acid.

battery back-up Same as **battery backup**.

battery backup The use of a battery to power a device or system when the main power fails or is interrupted. For instance, an uninterruptible power supply may provide the power needs of a device through brownouts and blackouts. Also spelled **battery back-up**.

battery capacity The total amount of charge that can be withdrawn from a fully charged battery under specified conditions. Usually expressed in ampere-hours.

battery cell One of the cells that compose a battery. For example, a 9 volt battery may consist of six 1.5 volt cells. Also called **electric cell**.

battery charger A device which serves to charge rechargeable batteries. A rectifier, for instance, may be used for this. Also called **charger**.

battery clip A spring-loaded metal clamp or clip which serves to make temporary connections. Used especially in the context of connections to battery terminals.

battery eliminator A device that may be utilized to provide DC energy from an AC source. This makes the utilization of a battery unnecessary. Also called **eliminator** (2).

battery life 1. The number of times a rechargeable battery may be recharged. This depends on various factors, such as its operating environment. **2.** The length of time that a battery will function under specified conditions.

battery meter A display which indicates the amount of battery power remaining.

battery-operated Same as **battery-powered**.

battery power The DC power that a battery produces. It serves to power countless devices, including portable computers, electric vehicles, and uninterruptible power supplies during brownouts and blackouts.

battery-powered Said of components, devices, and equipment which use **battery power**. Also called **battery-operated**.

battery self-discharge The loss of energy of a battery which is in storage or otherwise not connected to a load. The battery chemistry and ambient temperature influence this rate considerably. Also called **self-discharge**.

baud A unit of data-transmission speed, which indicates as the number of signal changes per second during communication. This can be in the form of a voltage or frequency transition. This term is not synonymous with **bits per second**. For instance, if a 2,400 baud modem encodes 4 bits per signaling event, it can transmit 9,600 bits per second. Its abbreviation is **b**. Also called **baud rate**.

baud rate Same as **baud**.

Baudot code A simple digital code using five or six data bits to represent each character. Used mostly in teleprinters.

bay 1. An opening or shelf in a computer cabinet where additional devices may be installed. For instance, a drive bay. **2.** A segment within an antenna array.

bayonet base A cylindrical base with projecting pins which are inserted into a **bayonet socket**, then rotated to secure the connection. Used, for example, in tubes or lamps.

bayonet connector Either a **bayonet base** or a **bayonet socket**.

bayonet Neill-Concelman connector Same as **BNC**.

bayonet nut connector Same as **BNC**.

bayonet socket A socket into which a **bayonet base** is inserted.

bazooka Same as **balun**.

BBD Abbreviation of **bucket-brigade device**.

BBS Abbreviation of **bulletin board system**, or **bulletin board service**.

bcc Abbreviation of **blind carbon copy**, or **blind courtesy copy**.

BCD Abbreviation of **binary-coded decimal**.

BCI 1. Abbreviation of **brain-computer interface**. **2.** Abbreviation of **broadcast interference**.

BCN Abbreviation of **beacon**.

BCP Abbreviation of **business continuity plan**.

BCS theory Abbreviation of **Bardeen-Cooper-Schrieffer theory**.

Be Chemical symbol for **beryllium**.

beacon Its abbreviation is **BCN**. **1.** A station, structure, or device which emits guiding, orienting, or warning signals which assist marine or aeronautical navigation. **2.** A signal which serves as a navigational aid that guides, orients, or warns. Such a signal may assist in marine or aeronautical navigation, and may be in the form of light, radio, radar, acoustic, or other types of signals or waves. Also called **beacon signal**. **3.** A navigational aid for a robot. Such an aid may be passive or active. An example of a passive aid is a set of reflectors, while an active one may be acoustic in nature.

beacon delay The time difference between the reception of a signal and the emitting of its response from a beacon that replies to signals from marine or aeronautical vessels.

beacon receiver A device which detects a **beacon** (2).

beacon signal Same as **beacon** (2).

beacon station A station which emits guiding, orienting, or warning signals which assist marine or aeronautical navigation.

beacon transmitter A device which transmits a **beacon** (2).

bead 1. In coaxial cables, an insulator, in the form of beads, which surrounds and supports the conductors within it. These beads are usually composed of plastic, ceramic, or glass. **2.** In computer programming, a small subroutine.

bead thermistor A thermistor consisting of a semiconductor material which is shaped in the form of a small sphere, into which two leads are inserted.

beaded coax A coaxial cable with **bead** (1) insulation.

beam 1. A concentrated and essentially unidirectional stream of radiated energy, such as radio waves, or particles, such as electrons. Other examples include light beams, laser beams, and neutron beams. **2.** A signal transmitted in a given direction, such as that of an antenna or radio beacon.

beam alignment 1. The adjustment of a beam, so that it is in the desired direction. Also, orienting in a manner which allows for optimum reception of a beam. **2.** In a TV camera tube, the adjustment of an electron beam so that it is perpendicular to the target at the target surface.

beam angle Same as **beam width**.

beam antenna An antenna that concentrates the intensity of radiation in a given direction. This radiation is usually confined to a relatively narrow beam.

beam bender A device or arrangement used to prevent an ion spot from forming on the screen of a CRT. It usually utilizes a magnetic field to bend the electron beam, which then passes through a tiny aperture while the heavier ions are trapped inside the electron gun. Also called **ion trap** (1).

beam bending The deflection of an electron beam. For instance, by a beam bender. Also called **beam deflection**.

beam blanking In a device such as a CRT, the cutting-off of the electron beam during retrace. Also called **blanking** (2), or **retrace blanking**.

beam convergence In a picture tube with a three-color gun, the proper convergence of the three electron beams as they pass through the aperture mask.

beam coupling In a circuit, the production of AC by modulating the intensity of an electron beam between two electrodes.

beam current 1. The current in an electron beam, as determined by the number and speed of electrons in it. **2.** In a CRT, the current which the electron beam produces.

beam deflection Same as **beam bending**.

beam-deflection tube An electron-beam tube whose output current is regulated by the bending of its electron beam.

beam divergence The increase of the diameter along an axis of an electromagnetic beam as the distance from its source is extended.

beam efficiency 1. The proportion of the total radiated energy of an antenna that is contained in the main beam. **2.** In a

CRT, the ratio of the number of electrons generated by the electron gun, to those that reach the screen.

beam lead A connecting lead which is formed chemically or through evaporation onto a semiconductor device or IC. These leads are usually flat, and cantilevered.

beam-lead device A semiconductor device which utilizes **beam leads** for interconnections.

beam-lobe switching Same as **beam switching**.

beam magnet One or more magnets utilized to unite two or more electron beams. Used, for instance, in three-gun TV picture tubes. Also called **convergence magnet**.

beam modulation The modulation of an electron beam. This might occur, for instance, in a CRT.

beam parametric amplifier A parametric amplifier in which a modulated electron beam supplies a variable reactance. Also called **electron-beam parametric amplifier**.

beam pattern A pattern representing how the intensity of a beam varies as a function of direction and distance. Commonly refers to the transmitting or receiving effectiveness of an antenna, and can be measured in any plane, although the horizontal and/or vertical planes are generally used. Also called **radiation pattern**, **directivity pattern**, or **field pattern**.

beam power tube A vacuum tube whose power-handling capability is increased through the focusing of its electron beams by special deflector plates. Also called **beam tetrode**, or **beam tube**.

beam resonance Resonance exhibited by a concentrated and essentially unidirectional stream of particles, such as atoms or electrons.

beam-rider guidance A guidance system in which an aircraft or spacecraft maintains its course by following a radar beam.

beam shaping The controlling of the radiation pattern of an antenna. This may be accomplished, for instance, through the use of an antenna array.

beam splitter In optics, a device which produces two or more beams from a single incident beam. A partially reflecting mirror, for instance, may be used for this. Also spelled **beamsplitter**.

beam splitting The producing of two or more beams from a single incident beam. Beam splitting has applications in fiber optics, lasers, and so on. Also spelled **beamsplitting**.

beam steering In an antenna, the varying of the direction of the main lobe of a radiation pattern. This can be done, for instance, by switching antenna elements.

beam switching In radio direction finding, mechanically or electrically shifting the orientation of an antenna to determine the direction which provides the greatest accuracy in the determination of the bearing of an object. Also called **beam-lobe switching**, and **lobe switching**.

beam tetrode Same as **beam power tube**.

beam tube Same as **beam power tube**.

beam width Same as **beamwidth**.

beam width of antenna Same as **beamwidth of antenna**.

beamsplitter Same as **beam splitter**.

beamsplitting Same as **beam splitting**.

beamwidth The angle between the points at which the intensity of an electromagnetic beam is at half of its maximum value. It can be measured in the horizontal plane or the vertical plane. Also spelled **beam width**. Also called **beam angle**, or **angle of beam**.

beamwidth of antenna The angle between the points at which the intensity of an antenna is at half of its maximum value. It can be measured in the horizontal or vertical plane. Also spelled **beam width of antenna**. Also called **antenna beamwidth**.

bearer channel Same as **B channel**.

bearing **1.** The horizontal direction from one object or point to another, measured clockwise in degrees from a reference line. Also called **azimuth** (1). **2.** A device which supports another which moves, such as a rotor in a motor.

bearing cursor In radars, a transparent disk which is manually rotated to assist in the determination of bearing.

bearing resolution For two objects with the same range, the minimum angular difference in the horizontal plane necessary for a radar to distinguish between them.

beat **1.** A pulsation in amplitude which is the result of the mixing of two signals of slightly different frequencies. **2.** A new frequency which is generated by mixing two or more signals in a nonlinear device.

beat frequency For two different frequencies which are mixed, either the sum or the difference of the frequencies. Also called **heterodyne frequency**.

beat-frequency oscillator An oscillator which produces a desired signal frequency within the audible range, by mixing two oscillations at different frequencies. The signal frequency produced is the beat frequency. Its abbreviation is **BFO**. Also known as **heterodyne oscillator**.

beat note Also called **beat tone**. **1.** An audible frequency produced when two different frequencies are mixed in a nonlinear device. **2.** An audible frequency produced when two different frequencies are mixed.

beat-note detector A circuit or device which detects **beat notes**.

beat reception Radio reception in which the incoming signal is combined with an internally generated signal of a different frequency, to produce an audible beat frequency. Also known as **heterodyne reception**.

beat tone Same as **beat note**.

beating Also called **heterodyning**. **1.** Pulsations in amplitude resulting from the mixing of two signals of slightly different frequencies. **2.** The production of frequencies which are the sum or difference of two signals of different frequencies which are mixed.

beavertail Same as **beavertail beam**.

beavertail beam A radar beam which is wide in the horizontal plane and narrow in the vertical plane, and which is moved vertically to determine the altitude of a scanned object. Also called **beavertail**.

BECN Abbreviation of **Backward Explicit Congestion Notification**.

becquerel The SI unit of activity. It represents one event per second. It is utilized to measure events that happen in an unpredictable manner, such as radioactive disintegrations. This contrasts with **hertz**, which measures the frequency of phenomena occurring in a periodic cycle, such as electromagnetic waves. Its symbol is **Bq**.

Becquerel effect In an electrolytic cell, an electromotive force resulting when two identical electrodes are unequally illuminated.

bed-of-nails clip A spring-loaded clip, utilized to make temporary electric connections, which is long and narrow and has a section with many sharp, needle-like teeth. This allows it to penetrate the insulation of a wire, making stripping unnecessary.

BEDO DRAM Abbreviation of **burst EDO DRAM**.

BEDO RAM Same as **burst EDO DRAM**.

bedspring Same as **billboard array**.

bedspring array Same as **billboard array**.

beep A brief audible tone which is usually high-pitched. Serves for signaling, testing, or for warning.

beep codes Same as **BIOS beep codes**.

beep tone A short tone sent over a trunk which indicates that a call is arriving, readiness for an order, or to warn that a call is being monitored. Also called **zip tone**.

beeper **1.** A device which emits **beeps**. **2.** A pocket-sized radio receiver or transceiver which serves to receive messages, and in some cases perform various other functions, such as retrieve email. These devices may emit a beep when informing of a new message. Some beepers can also signal by other means, such as vibrations, or a blinking light. Also called **paging device**, **pager (1)**, or **mobile pager**.

beginner's all-purpose symbolic instruction code Same as **BASIC**.

beginning of file Its abbreviation is **BOF**. **1.** A code which indicates the starting location of a computer file. **2.** A code which indicates the starting location of a computer file, relative to another location.

bel A dimensionless unit which expresses the ratio of two powers or intensities, such as voltages, currents, or sound intensities. The number of bels represents the logarithm of the ratio of the quantities measured. A bel is equal to approximately 1.1513 nepers. The **decibel**, which is 0.1 bels, is a much more frequently encountered unit, as the bel is too large for most applications. Its symbol is **b**, or **B**.

bell A hollow metallic device with a hammer which strikes it to produce a ringing sound. Bells are usually electrically actuated, and serve to warn, as in an alarm, or to inform, as in an older telephone.

bell-shaped curve A curve corresponding to a **bell-shaped distribution**. It is in the shape of a symmetrical bell. Also called **normal curve**, or **Gaussian curve**.

bell-shaped distribution For a random variable, a probability distribution which is symmetrical about the mean value, and which continuously diminishes in value until reaching zero at each extreme. It is utilized to determine the probability of the value of the variable falling within a given interval of values, and when graphed has the shape of a symmetrical bell. Various phenomena, such as some natural frequencies, have a bell-shaped distribution. Also called **normal distribution**, or **Gaussian distribution**.

bell transformer A small step-down transformer connected to an AC line, which is intended for use with low-voltage devices such as alarms or doorbells.

bell wire Insulated solid copper wire which is intended for use with low-voltage devices such as alarms or doorbells.

bells and whistles Additional items and/or features which are not essential for the use of devices, equipment, or systems, but which may add usefulness. For instance, a fancy feature which some users may employ on rare occasions. This contrasts with **plain vanilla**, in which only the bare essentials are present.

below the fold The part of a Web page that is visible only after scrolling, while the part that requires no scrolling to be seen is **above the fold**. Content above the fold is usually considered to be more important than that below.

belt generator An electrostatic generator in which a rapidly moving insulating belt collects electric charges, then discharges them inside a hollow metal sphere. Potentials of several million volts may be generated in this manner. Also called **Van de Graaff generator**.

bench test An evaluation, usually rigorous, performed in a test laboratory setting. It may be used to verify proper functioning, ascertain projected longevity, and so on. This contrasts with a **field test**, which is performed under actual operating conditions.

bench testing The performance of **bench tests**.

benchmark **1.** A standard which serves for comparison, or a reference point from where measurements can be made. **2.** In computers, a test to evaluate the performance of hard-

ware and/or software. For instance, it may ascertain the number of times a program can be executed per second. If a benchmark becomes widely used in the computer industry, it is considered a standard. Also called **benchmark test**.

benchmark program A program that evaluates the performance of hardware and/or software. For instance, a test of graphics speed within specified circumstances.

benchmark test Same as **benchmark (2)**.

bend **1.** A deviation from a straight line or path. Also, a deviation from the usual or proper shape. **2.** In a waveguide, a change in the direction of the longitudinal axis, as seen, for instance, in an elbow bend. Also, the section which effects this change. Also called **waveguide bend**.

bend loss Signal loss or attenuation due to curvatures in a fiber-optic cable. This occurs because bending causes the light traversing the cable to reflect outward instead of towards the core.

bender element The joining of two thin strips, each of a different piezoelectric material, so that an applied voltage increases one strip in length, while reducing the length of the other. This produces bending, and can be used, for instance, as a seal actuator in a precision gas valve.

benign virus In computers, a virus which is not intended to cause damage, but that still takes up storage space and computer time. For instance, it may show a message at a prescribed time, and nothing more.

bent antenna An antenna whose driven element is bent, usually at a 90° angle. This may serve to influence its radiation pattern and/or to save space.

bent gun In a TV picture tube, an electron gun which is slanted, so that the electron beam reaches the screen, and ions are trapped by a positive electrode. This helps prevent ion spots from forming on the screen.

Beowulf Same as **Beowulf cluster**.

Beowulf cluster The interconnection of multiple lesser computers, such as off-the-shelve PCs, to achieve computing power in the realm of a supercomputer. Such clusters may utilize the resources of thousands of computers. Its abbreviation is **Beowulf**.

BER Abbreviation of **bit error rate**.

berkelium A synthetic radioactive element whose atomic number is 97. It has over 10 identified isotopes, all of which are unstable. The procedures utilized to create it are employed to better understand the synthesis of other transuranium elements. Its chemical symbol is **Bk**.

BERT Acronym for **bit error rate test**.

beryllium A hard and brittle gray-white metallic element whose atomic number is 4. It is highly resistant to oxidation at ordinary temperatures, has excellent thermal and electrical conductivities, has an extremely low density, is non-magnetic, and is highly permeable to X-rays. There are about 10 known isotopes, of which one is stable. It has many applications, including its use as a moderator and reflector of neutrons in nuclear reactors, as a source of neutrons when bombarded with alpha particles, in aerospace and satellite outer structures, as a window material for X-ray tubes, and in precision electronic components and devices such as navigational instruments. Its chemical symbol is **Be**.

beryllium oxide A white powder whose chemical formula is **BeO**. It is used in electron tubes, resistor cores, special glasses, ceramics, and for insulation.

bespoke Custom software that is prepared from scratch, as opposed to that modifying an existing product.

Bessel filter A filter with excellent phase characteristics, and a relatively shallow cutoff.

best-effort service A service which makes no guarantees as far as certain performance parameters are concerned. For instance, an Internet service provider which does not guarantee that customers will be able to have access at all times.

beta Its symbol is β. **1.** The second letter of the Greek alphabet, and as such often designates the second constituent of a series. **2.** The current gain of a transistor with a common-emitter amplifier configuration. **3.** Same as **beta software**. **4.** Same as **beta hardware**.

beta circuit In an amplifier, a circuit which provides feedback.

beta cutoff frequency The higher frequency at which the current gain of a transistor with a common-emitter amplifier configuration drops 3 decibels, or 70.7%, in relation to a reference lower frequency.

beta decay Radioactive decay in which **beta particles** are emitted. Also called **beta disintegration**.

beta disintegration Same as **beta decay**.

beta emission The emission of **beta particles** by a radioactive element.

beta emitter Same as **beta radiator**.

beta hardware Hardware undergoing **beta tests**. Also called **beta (4)**.

beta particle A charged particle emitted by many radioactive elements. When the emitted particle is negative, it is an electron, and when positive it is a positron.

beta radiation A stream consisting of **beta particles**.

beta radiator A radioactive element that emits **beta particles**. Also called **beta emitter**.

beta-ray spectrometer An instrument which measures the energy distribution of beta particles and secondary electrons.

beta rays A stream consisting of **beta particles**.

beta release **1.** Same as **beta software**. **2.** Same as **beta version (2)**.

beta software Software undergoing **beta tests**. Its abbreviation is **betaware**. Also called **beta (3)**, **beta version (1)**, or **beta release (1)**.

beta tests Tests performed on new software or hardware under normal operating conditions. This contrasts with **alpha tests**, which are done under experimental conditions.

beta version **1.** Same as **beta software**. **2.** Beta software which has been offered as a separate program, or as part of a package. Also called **beta release (2)**.

betatron A particle accelerator in which electrons are injected into a doughnut-shaped vacuum chamber that surrounds a magnetic field. This magnetic field is varied in a manner which further accelerates the electrons each time around. Energies in the GeV range can be achieved in this manner.

betaware Abbreviation of **beta software**.

BeV Abbreviation of **billion electronvolt**.

bevatron A high-energy proton synchrotron which can generate energies in excess of 10 GeV.

Beverage antenna A directional antenna consisting of one or more parallel horizontal conductors which are ½ to several wavelengths long. All conductors are suspended parallel to the ground, usually within a couple of meters of the surface. Also called **wave antenna**.

beyond-the-horizon communication Communication by means of radio waves that are propagated well beyond line-of-sight distances. This is usually due to scattering by the ionosphere or troposphere. Also called **over-the-horizon communication**, **forward-scatter communication**, **scatter communication**, or **extended-range communication**.

beyond-the-horizon propagation Propagation of radio waves well beyond line-of-sight distances. This is usually due to scattering by the ionosphere or troposphere. Also called **over-the-horizon propagation**, **forward-scatter propagation**, **scatter propagation**, or **extended-range propagation**.

beyond-the-horizon transmission Transmission of radio waves well beyond line-of-sight distances. This is usually due to scattering by the ionosphere or troposphere. Also called **over-the-horizon transmission**, **forward-scatter transmission**, **scatter transmission**, or **extended-range transmission**.

bezel A rim that secures a transparent covering for an indicating device such as a meter, oscilloscope, or watch. It may or may not have markings.

Bézier curve In computer graphics, a curve which is defined by mathematical formulas. These curves have handles which assist in reshaping them. Used, for instance, in computer-aided design.

BFO Abbreviation of **beat-frequency oscillator**.

BGA Abbreviation of **Ball Grid Array**.

BGP Abbreviation of **border gateway protocol**.

Bh Chemical symbol for **bohrium**.

bhp Abbreviation of **brake horsepower**.

Bi Chemical symbol for **bismuth**.

bi-amping The use of two separate amplifier channels to drive a speaker, each handling a different frequency interval. This requires the utilization of a crossover network to divide the audio signal. Bi-amping results in increased effective amplifier power, greater linearity, increased damping factor, and reduced intermodulation distortion. Also spelled **biamping**. Also called **bi-amplification**.

bi-amplification Same as **bi-amping**. Also spelled **biamplification**.

bi-directional Operating, moving, transferring, flowing, transmitting, receiving, or responding in two directions, usually opposite. There are numerous examples, including bi-directional buses, bi-directional bar-code scanners, bi-directional antennas, and bi-directional transistors. **Bilateral** and **bi-directional** may be contrasted in the following manner: bilateral refers to the two sides of something, while bi-directional pertains to two directions. Though not necessarily the same, these terms are sometimes used synonymously. Also spelled **bidirectional**.

bi-directional antenna An antenna whose directivity pattern is divided into two principal lobes. These lobes are usually in opposite directions. Also spelled **bidirectional antenna**.

bi-directional antenna array An assembly of antenna elements with the proper dimensions, characteristics, and spacing, so as to have a directivity pattern which is divided into two principal lobes. These lobes are usually in opposite directions. Also called **bi-directional array**. Also spelled **bidirectional antenna array**.

bi-directional array Same as **bi-directional antenna array**. Also spelled **bidirectional array**.

bi-directional bar code Also spelled **bidirectional bar code**. Also called **bi-directional bar-code symbol**. **1.** A bar code which can be read left-to-right, or right-to-left. **2.** A bar code which can be read horizontally or vertically.

bi-directional bar-code reader Also spelled **bidirectional bar-code reader**. Also called **bi-directional bar-code scanner**. **1.** A bar-code reader which can read left-to-right, or right-to-left. **2.** A bar-code reader which can read horizontally or vertically.

bi-directional bar-code scanner Same as **bi-directional bar-code reader**. Also spelled **bidirectional bar-code scanner**.

bi-directional bar-code symbol Same as **bi-directional bar code**. Also spelled **bidirectional bar-code symbol**.

bi-directional bus In computers, a bus that carries signals in both directions. Also spelled **bidirectional bus**.

bi-directional counter A counter which can count both upwards and downwards. It has, for instance, both an adding input and a subtracting input, thus giving it the ability to count in both directions. Also spelled **bidirectional counter**. Also called **forward-backward counter**, **up-down counter**, or **reversible counter**.

bi-directional coupler In a waveguide, a device which can sample both incident and reflected waves. Used, for instance, to measure power. Also spelled **bidirectional coupler**.

bi-directional current Current that flows in both directions. Also spelled **bidirectional current**.

bi-directional data bus In computers, a bus that carries data in both directions. Also spelled **bidirectional data bus**.

bi-directional device A device, such as a bi-directional microphone, which exhibits bi-directional properties. Also spelled **bidirectional device**.

bi-directional loudspeaker Same as **bi-directional speaker**. Also spelled **bidirectional loudspeaker**.

bi-directional microphone A microphone whose sensitivity is the same for sounds incident from the front and rear of it. Such a microphone has a pickup pattern in the form of a figure eight. Also spelled **bidirectional microphone**.

bi-directional network bus In computers, a bus that carries signals between stations in both directions. Also spelled **bidirectional network bus**.

bi-directional parallel port A computer parallel port which supports communication in both directions. That is, to and from the computer and the peripheral device connected to said port. Also spelled **bidirectional parallel port**.

bi-directional pattern The directivity pattern of a bi-directional device, such as a bi-directional antenna or a bi-directional microphone. Such a pattern is usually in the form of a figure eight. Also spelled **bidirectional pattern**.

bi-directional printing The ability of a computer printer, such as an inkjet printer, to print in both directions. Also spelled **bidirectional printing**.

bi-directional pulses Pulses which have the same baseline, some of which increase in one direction, while the rest increase in the other. Also spelled **bidirectional pulses**.

bi-directional relay A stepping relay whose rotating contacts can move in either direction. Also spelled **bidirectional relay**.

bi-directional speaker A speaker that produces acoustic energy to the front and to the rear. Also spelled **bidirectional speaker**. Also called **bi-directional loudspeaker**.

bi-directional thyristor A thyristor whose switching action can be triggered by positive or negative voltages. An example is a triac. Also spelled **bidirectional thyristor**.

bi-directional transducer Same as **bilateral transducer**. Also spelled **bidirectional transducer**.

bi-directional transistor A transistor which has essentially the same electrical characteristics when its inputs and outputs are interchanged. Also spelled **bidirectional transistor**.

bi-endian A computer processor which can store a sequence of bytes using either the big-endian or the little-endian order. In the **big-endian** order, the most significant byte is placed first, and in the **little-endian** order the least significant byte is placed first.

Bi-FET Same as **BiFET**.

biamping Same as **bi-amping**.

biamplification Same as **bi-amplification**.

bias 1. A voltage, current, capacitance, or other input which is applied to a component or device to establish a reference level for its operation. For instance, the voltage applied to the control electrode of a transistor to set its operating point. **2.** A systematic deviation from an established point of reference. **3.** In magnetic tape recording, a current that is applied to the audio signal to be recorded, in order to optimize per-

formance during playback. This current varies depending on the tape type. Also called **bias current (2)**, or **magnetic bias (2)**. **4.** In a tape deck, a circuit, which depending on the tape type, sets the **bias (1)** for optimal playback. **5.** The force applied to a relay to hold it in a given position.

bias current 1. A current which is applied to a component or device to establish a reference level for its operation. For instance, the current applied to the base-emitter junction of a transistor to set its operating point. Also called **input bias current (1)**. **2.** Same as **bias (3)**.

bias distortion Distortion in a component or device arising from improper **biasing**.

bias oscillator In magnetic tape recording, an oscillator which generates the AC which is applied to the audio signal to be recorded, in order to optimize playback. This current varies, depending on the tape type.

bias resistor A resistor utilized to provide the voltage drop necessary to attain the proper bias voltage in a transistor or electron tube.

bias stabilization The maintenance of a steady **bias voltage**. This may be accomplished, for instance, through the use of a voltage stabilizer.

bias stabilizer A circuit or device which maintains a steady **bias voltage**.

bias supply The AC or DC source of bias voltage or bias current.

bias voltage A voltage which is applied to a component or device to establish a reference level for its operation. For instance, the voltage applied to the control electrode of a transistor to set its operating point.

bias winding A control winding which provides the DC which establishes a reference level for the operation of a magnetic device, such as a magnetic amplifier.

biased automatic gain control An automatic gain control circuit which is activated only when the input signal exceeds a predetermined magnitude. This allows for maximum amplification of weaker signals. Also called **delayed automatic gain control**, **delayed automatic volume control**, or **quiet automatic volume control**.

biasing The application of a voltage, current, capacitance, or other input to a component or device to establish a reference level for its operation. For instance, the application of a voltage to the control electrode of a transistor to set its operating point.

BiCMOS Abbreviation of **bipolar CMOS**.

biconical antenna A wideband antenna formed by two metal cones arranged along the same axis, and whose vertices face each other. It is fed at the common vertex.

BICSI An organization that seeks to improve the telecommunications industry through education, conferences, research resources such as publications, and the like.

bidirectional Same as **bi-directional**.

bidirectional antenna Same as **bi-directional antenna**.

bidirectional antenna array Same as **bi-directional antenna array**.

bidirectional array Same as **bi-directional array**.

bidirectional bar code Same as **bi-directional bar code**.

bidirectional bar-code reader Same as **bi-directional bar-code reader**.

bidirectional bar-code scanner Same as **bi-directional bar-code scanner**.

bidirectional bar-code symbol Same as **bi-directional bar-code symbol**.

bidirectional bus Same as **bi-directional bus**.

bidirectional counter Same as **bi-directional counter**.

bidirectional coupler Same as **bi-directional coupler**.

bidirectional current Same as bi-directional current.

bidirectional data bus Same as bi-directional data bus.

bidirectional device Same as bi-directional device.

bidirectional loudspeaker Same as bi-directional speaker.

bidirectional microphone Same as bi-directional microphone.

bidirectional network bus Same as bi-directional network bus.

bidirectional parallel port Same as bi-directional parallel port.

bidirectional pattern Same as bi-directional pattern.

bidirectional printing Same as bi-directional printing.

bidirectional pulses Same as bi-directional pulses.

bidirectional relay Same as bi-directional relay.

bidirectional speaker Same as bi-directional speaker.

bidirectional thyristor Same as bi-directional thyristor.

bidirectional transducer Same as bilateral transducer.

bidirectional transistor Same as bi-directional transistor.

BiFET Abbreviation of bipolar field-effect transistor. An IC which incorporates technologies from both bipolar junction transistors and junction field-effect transistors.

bifilar Having, consisting of, or utilizing two threads, fibers, filaments, wires, or the like. For example, a bifilar suspension.

bifilar resistor A resistor with a bifilar winding, to reduce its inductance.

bifilar suspension In measuring instruments, the suspending of the movable part by two parallel threads or wires. May be used, for instance, in highly sensitive galvanometers.

bifilar transformer A transformer whose primary and secondary coils are wound side-by-side in the same direction, providing very tight coupling.

bifilar winding A winding or coil in which the insulated wire is doubled back on itself, so that there are two contiguous conductors carrying the same current, each in the opposite direction, thus minimizing the magnetic field produced. Such a winding may be used, for instance, in low-inductance resistors.

bifurcated contact An electric contact which is forked, or divided, to provide two parallel contacts. These may be movable or stationary, and provide an especially reliable connection, as one section can maintain the contact if the other fails.

big endian The ordering of a sequence of bytes placing the most significant byte first. This contrasts with little-endian, where the least significant byte is placed first.

bilateral Having two sides. Also, that which pertains to, or affects these sides in an equal manner. Bilateral and bidirectional may be contrasted in the following manner: bilateral refers to the two sides of something, while bidirectional pertains to two directions. Though not necessarily the same, these terms are sometimes used synonymously.

bilateral amplifier An amplifier which can amplify signals arriving from either its input or output terminals. That is, it makes no difference if a signal arrives at the input or the output terminals, because it will emerge amplified from the other.

bilateral antenna An antenna whose directivity pattern is divided into two principal lobes, each in the opposite directions.

bilateral element A circuit component which transmits equally well in either direction.

bilateral network An electric network that functions equally well in both directions.

bilateral transducer Also called bi-directional transducer. 1. A transducer which can measure an input from either direction of a reference point, such as zero. 2. A transducer which can simultaneously operate in both directions.

billboard antenna Same as billboard array.

billboard array A directional antenna consisting of an array of stacked dipoles in front of a flat shared reflector. Each dipole is spaced from ¼ to ¾ of a wavelength apart. Also called billboard antenna, bedspring array, bedspring, or mattress array.

billion A number equal to 10^9.

billion electronvolt One billion electronvolt is equal to one gigaelectronvolt, the latter being the more correct term to use for this measurement. Its abbreviation is BeV.

billion floating-point operations per second Same as billion operations per second.

billion instructions per second Same as BIPS.

billion operations per second One billion FLOPS, which is the same as one billion floating-point calculations or operations, per second. Usually used as a measure of processor speed. Also called billion floating-point operations per second, or GFLOPS.

bimetal A sheet composed of two layers bonded together, each of a different metal. The metals are selected so that the sheet has some desired property, such as corrosion resistance.

bimetallic Composed of, pertaining to, or containing a bimetal.

bimetallic element Two strips of dissimilar metals bonded together, each with different coefficients of thermal expansion. In this manner, as temperature varies, it will bend in one direction or the other. Used, for instance, in thermostats. Also known as bimetallic strip.

bimetallic strip Same as bimetallic element.

bimetallic switch A switch utilizing a bimetallic element.

bimetallic thermometer A thermometer utilizing a bimetallic element. A calibrated scale may be coupled for temperature readings. Also called differential thermometer.

bimetallic thermostat A thermostat utilizing a bimetallic element.

bimorph cell A transducer composed of two piezoelectric plates tightly bound to each other. In this manner, an applied voltage will cause mechanical deformation of the unit, and conversely, a mechanical deformation will produce a voltage. Used, for instance, in microphones, headphones, and in crystal pickups. Also called bimorphous cell.

bimorphous cell Same as bimorph cell.

BiMOS Abbreviation of bipolar MOS.

binary 1. Based on, or consisting of two parts, components, or possibilities. 2. Abbreviation of binary number system.

binary 8-zero substitution A T1 coding method in which a special code is inserted whenever 8 consecutive zeroes are sent. Also called bipolar 8-zero substitution. Its abbreviation is B8ZS.

binary alloy An alloy consisting of two elements.

binary amplitude shift keying Amplitude shift keying in which two signal levels are used, one of which is zero. For example, a mark can be made with the energized level, and a space with the zero level. Also called on/off keying

binary arithmetic operation In computers, an arithmetic operation utilizing only the digits 0 or 1.

binary cell A circuit or device which may assume one of two stables states. It usually serves as a unit of storage, with a capacity of one bit of information.

binary chain A series of binary circuits arranged in a manner that each unit affects the state of the following one.

binary circuit A circuit which may assume one of two stables states. For instance, on or off, high voltage or low voltage, and so on. It operates like a switch, and can perform logic functions. Used extensively in computers and other logic-based devices and equipment. Also called **digital circuit**.

binary code A code composed of elements which can have only one of only two states or values, such as **0** and **1**. The binary number system is an example of such a code.

binary-coded Expressed in **binary code**.

binary-coded decimal A system for expressing decimal digits in binary format, each decimal digit occupying four bits. For instance, the number 24 in decimal would be 11000 in binary, and 0010 0100 in binary-coded decimal. This avoids errors in conversion and rounding. Its abbreviation is **BCD**. Also called **binary-coded decimal system**.

binary-coded decimal system Same as **binary-coded decimal**.

binary compatible Files, programs, code, and so on, which are completely identical at the binary level. For instance, the data produced in one word-processing program being the same, bit for bit, with that produced by another.

binary component A component which may assume one of two stables states. An example is a binary circuit.

binary compound semiconductor A semiconductor made of a compound consisting of two chemical elements. For example, gallium arsendide.

binary conversion The conversion of a number to or from the binary number system. For instance, 31 would be converted into 11111, if a decimal-to-binary conversion were performed.

binary counter Also known as **binary scaler, scale-of-two counter**, or **scale-of-two circuit**. **1.** A counter circuit which produces one output pulse for every two input pulses. **2.** A counter whose elements may assume one of two stable states.

binary data Data that is in binary format. For instance, data expressed as ones and zeroes. Numbers, text, sounds, images, and so on, may be expressed in binary format.

binary decision A choice consisting of two alternatives. When used in the area of computing, the alternatives are **0** and **1**.

binary decoder A circuit or device which converts a binary-coded input, such as zeroes and ones, into a non-binary coded output, such as a word.

binary device A device which may assume one of two stables states.

binary digit Same as **bit**.

binary encoder A circuit or device which converts a non-binary coded input, such as a word, into a binary coded output, such as zeroes and ones.

binary field In computers, a field whose contained information is in the form of binary numbers.

binary file A file that is in **binary format** (2). Also called **non-text file** (3).

binary file format Same as **binary format** (2).

binary form Same as **binary format**.

binary format Also called **binary form**. **1.** Information that is in binary code. This almost always means that it is expressed as ones and zeroes. Numbers, text, sounds, images, and so on, may be expressed in binary form. **2.** A file format utilizing sequences of 8 bits. Almost all computer files are in this 8-bit format. This contrasts, for instance, with 7-bit ASCII files. Also called **binary file format**.

binary logic Logic operations which have only two possible values, such as on or off, true or false, or **0** or **1**. Among the benefits of binary logic are its simplicity, and lack of ambiguity.

binary notation The utilization of binary digits to express numerical values.

binary number A number expressed utilizing the **binary number system**.

binary number system A numbering system whose base is 2, meaning that there are two digits, namely **0** and **1**. All values expressed are composed of combinations of zeroes and ones. For instance, the numeral 85 in the decimal, or base 10, system is expressed as 1010101 in the binary system. Since the only two possible values are **0** and **1**, this number system lends itself to unequivocally distinguishing between two states. For instance, on or off, yes or no, true or false, current or no current, high-voltage or low-voltage, and so on. All information, numbers or otherwise, entered into a digital computer is converted into these digits, as all processing is based on this system. Also known as **binary** (2).

binary phase-shift keying A phase-shift keying technique in which two phase angles are utilized to transmit binary information. The two angles are usually 180° apart, and the phase shift of the carrier could be -90° for a **0**, and +90° for a **1**. Its abbreviation is **binary PSK**, or **BPSK**. Also called **biphase modulation**.

binary PSK Abbreviation of **binary phase-shift keying**.

binary scaler Same as **binary counter**.

binary search A search technique in which the desired item is compared to a list to determine in which half it is located, the other half being discarded. Then, within this half, a similar search is performed. This process of pinpointing the half that contains the item is continued successively until the search is complete. Also known as **dichotomizing search**.

binary signal A signal, such as a voltage or current, which can have only two possible values. For instance, on or off, or high or low, corresponding respectively to the binary values of **0** or **1**.

binary signaling A method of communication utilizing signals which can have only two possible values. For instance, on or off, or high or low, corresponding respectively to the binary values of **0** or **1**.

binary synchronous Same as **binary synchronous communications protocol**. Its acronym is **bisync**. Its abbreviation is **bisynchronous**.

binary synchronous communication Communication utilizing a **binary synchronous communications protocol**. Its abbreviation is **BSC**.

binary synchronous communications protocol A synchronous communications protocol in which messages are sent in frames. Its abbreviation is **BSC protocol**. Also called **binary synchronous**, or **binary synchronous protocol**.

binary synchronous protocol Same as **binary synchronous communications protocol**.

binary system 1. A system whose components can have only one of two possible values or states, such as on or off, or high or low. **2.** Same as **binary number system**.

binary-to-decimal conversion The conversion of a binary representation into a decimal representation. For example, converting a binary, or base 2, number into a decimal, or base 10, number.

binary touch sensor In robotics, a sensor whose information is restricted to detecting whether an object is being touched or not.

binary transfer The transfer of **binary files**.

binary tree A data structure in which a non-terminal node leads to no more than two successors.

binary variable A variable which can have only one of two possible values or states.

binary word A group of bits of a specific length, usually eight, which forms a unit of storage, and which is treated by the computer as a unit.

binaural **1.** Pertaining to both ears, as in binaural sound. **2.** Same as **binaural sound**.

binaural sound Sound which is recorded placing two microphones approximately where the ears of a listener would be if present. When reproduced, a pair of headphones should be used to obtain the effect of being at the recording location. Also called **binaural (2)**.

binaural sound reproduction Reproduction of **binaural sound**.

bind **1.** To make secure, fasten, restrain, or enclose. For example, to bind multiple twisted-pair wires within a cable sheath. **2.** In computers and networks, to associate data, objects, subroutines, communication protocols, units, applications, or the like.

binder **1.** A substance, such as a lacquer, resin, or cement, which helps provide consistency, cohesion, or mechanical strength. Used, for instance, to help a phosphor adhere to a screen. **2.** A group of 25 twisted-pair wires within a cable sheath. Each group is color-coded for identification. Also called **binder group**. **3.** A program that allows two or more programs to be combined in a manner that results in a single executable file. Used, for instance, by those who wish to maliciously include a Trojan horse within another application, such as a game.

binder group Same as **binder (2)**.

binding **1.** That which is utilized to secure. **2.** Fastening, or tying together. **3.** In a computer language, an association, such as that between an operation and a symbol.

binding energy **1.** The minimum energy required to remove a particle from a system. For instance, the energy necessary to extract a proton from the nucleus of an atom. **2.** The minimum energy required to dissociate a system into its component parts. For example, the energy necessary to dissociate a nucleus into its component protons and neutrons. Also known as **total binding energy**.

binding post A terminal, used for making electrical connections, that incorporates a screw which is usually adjusted by hand. Used, for instance, to connect audio speaker cables. Also called **binding screw**, **screw terminal**, or **post (2)**.

binding screw Same as **binding post**.

binding time In a computer program, the time when **binding (3)** takes place. This is usually either during compilation, or during execution.

binistor A negative-resistance semiconductor switching device that has two stable states.

binocular machine vision In robotics, a method of vision which allows for interpretation of three-dimensional surfaces, by combining the input from each of the two optical sensors. Used, for instance, in obstacle-detection systems in mobile robots. Also called **stereoscopic machine vision**.

binode A dual diode with one cathode and two anodes.

biochemical cell Same as **biochemical fuel cell**.

biochemical fuel cell A fuel cell which utilizes chemical reactions to obtain electricity from biological substances. Used, for instance, for the generation of electricity from industrial waste. Also called **biochemical cell**.

biochip **1.** An IC which incorporates biological substances. **2.** A chip used to match known and unknown samples of DNA. **DNA** is the abbreviation of deoxyribonucleic acid, and in its strands are contained the chemical basis of heredity.

bioelectrical Pertaining to **bioelectricity**.

bioelectricity Electric currents which occur naturally, and flow in living tissue. Brain waves and nerve impulses are two forms of bioelectricity. The stunning charges an electric eel provides is another example.

bioelectronics **1.** The application of electronics to the study of living organisms. **2.** The study of **bioelectricity**.

bioengineering The application of engineering concepts and equipment to the biological sciences. Specifically, the contributions electrical and computer engineering make to bioengineering are numerous, and include electron microscopes, hearing aids, laser surgery, and the computers incorporated in myriad devices.

biofeedback A method for assisting in the awareness and monitoring of a person's own biological responses, such as muscle tension, heart rate, and blood pressure. Various electronic instruments are used to assist in this, such as those which measure and indicate skin temperature.

bioinformatics Abbreviation of **biological informatics**. The use of computers to gather, process, manipulate, integrate, store, share, apply, and present biological information.

biological informatics Same as **bioinformatics**.

biological shield An absorbing shield placed around a source of ionizing radiation, so as to reduce emissions to a level safe for those nearby. In a nuclear reactor, for instance, such a shield may consist of a thick concrete wall.

bioluminescence **1.** The emission of light by living organisms. This light is the result of chemical reactions. **2.** The light emitted by living organisms.

biomechatronics Acronym for **bio**logy, **mecha**nics, and electro**nics**. A science which draws from these three fields to create, adapt, and improve electromechanical devices intended for living organisms.

biometric access control The use of a **biometric system** to authenticate an individual, such as that wishing to access a computer system or network.

biometric authentication The use of a **biometric system** to control access to a network, system, facility, or the like.

biometric device Same as **biometric system**.

biometric identification The use of a **biometric system** to identify an individual.

biometric system An automated system which can measure and analyze a human biological characteristic such as fingerprints, voice, or the patterns of ocular blood vessels, and yield an output which serves to verify of a person's identity. Also called **biometric device**.

biometric technique Any technique, such as fingerprint or voice recognition, employed by a **biometric system**.

biometrics **1.** The statistical study of biological phenomena. **2.** The measurement and analysis of human biological characteristics, such as fingerprints, voice, or the patterns of ocular blood vessels, for verification of a person's identity.

biomimetics The techniques employed in the development of materials, devices, systems, and the like which emulate biological entities, natural materials, and/or naturally-occurring processes. For example, the creation of artificial strands or fibers that are comparable in strength and elasticity to spider webs, or the formation of artificial neural networks that mimic biologic neural networks.

bionics The science of creating, adapting, and improving devices and systems intended for living organisms, based on the understanding of how and why things work in nature. A product of this field, for example, would be an electromechanical device that serves as a replacement for a lost limb.

BIOS Acronym for **b**asic **i**nput/**o**utput **s**ystem. A set of indispensable software routines that enable a computer to boot itself. It has the code necessary to control all peripherals and perform other functions, such as testing, and is generally stored on a ROM chip so that disk failures do not disable it. Currently, it is usually contained in a flash-memory chip, and is typically copied to the RAM at startup, as RAM is faster. Also called **system BIOS (1)**.

BIOS beep codes When a computer is started up, one or more short and/or long beeps which indicate the results of fundamental diagnostic tests the BIOS performs on the system hardware. Beep codes vary from BIOS to BIOS, and may warn of problems such as loose RAM chips, cache errors, system timer failures, motherboard problems, and so on. Its abbreviation is **beep codes.**

BIOS setup   A system configuration utility via which the settings of the **BIOS** may be adjusted. It is usually accessed by pressing the appropriate key during booting.

biosensor   **1.** A sensor which incorporates electronic and biological components to perform its function. For instance, such a device may be used to detect minute levels of proteins in body fluids. **2.** A sensor which detects and measures biological substances or processes. For example, such a device may translate muscle movements into electrical signals.

Biot-Savart law   In electromagnetism, a law which expresses the intensity of the magnetic field produced by a conductor carrying a current.

biotelemetry   The utilization of telemetry to study living organisms. For instance, the transmission of an electrocardiogram from a patient to a remote location.

biphase   Consisting of two phases, as seen for instance, in binary phase-shift keying.

biphase modulation   Same as **binary phase-shift keying.**

BIPM   Abbreviation of *Bureau International des Poids et Mesures.*

bipolar   **1.** Having two poles or polarities. **2.** Able to assume a positive or negative charge. **3.** Pertaining to transistors in which both electrons and holes serve as charge carriers.

bipolar 8-zero substitution   Same as **binary 8-zero substitution.**

bipolar CMOS   Abbreviation of b**i**polar **c**omplementary **m**etal-**o**xide **s**emiconductor. An IC which incorporates technologies from both bipolar junction transistors and complementary metal-oxide transistors. This provides greater functionality, as it exploits the advantages of each. For instance, it makes the most of the speed and linearity of bipolar junction transistors, while also benefiting from the lower power dissipation and higher packing densities of complementary metal-oxide transistors. Used, for example, in logic chips which require higher speeds than would ordinarily be suitable for traditional logic chips. Its own abbreviation is **BiCMOS.**

bipolar complementary metal-oxide semiconductor   Same as **bipolar CMOS.**

bipolar FET   Same as **BiFET.**

bipolar field-effect transistor   Same as **BiFET.**

bipolar junction transistor   A semiconductor device which utilizes both electrons and holes as charge carriers. These have two pn junctions, and depending on their arrangement may form an **npn** or **pnp** transistor. Such transistors have three main regions, which are the emitter, base, and collector. Bipolar junction transistors are one of the two main classes of transistors, the other being field-effect transistors. Used extensively in audio amplifiers and wireless transmitters. Its abbreviation is **BJT,** or **bipolar transistor.**

bipolar metal-oxide semiconductor   Same as **bipolar MOS.**

bipolar MOS   Abbreviation of b**i**polar **m**etal-**o**xide semiconductor. An IC which incorporates technologies from both bipolar junction transistors and metal-oxide transistors. This provides greater functionality, as it exploits the advantages of each. Its own abbreviation is **BiMOS.**

bipolar transistor   Abbreviation of **bipolar junction transistor.**

BIPS   Acronym for **b**illion **i**nstructions **p**er **s**econd. The term generally means one billion machine instructions per second, and is an indicator of CPU speed. When evaluating the overall speed of a system, other factors, such as cache, memory speed, and the number of instructions required for given tasks, must also be considered.

birdie   A spurious signal generated by the circuitry of a radio receiver, which sometimes occurs while tuning. This signal may be audible, and if so, sounds similar to chirping, hence its name.

birefringence   The separation of light into two components, each traveling at a different velocity. The material which separates the light is a **birefringent material.** Also called **double refraction.**

birefringent material   A material, typically a crystal, which causes **birefringence.** Such a material has the ability, for instance, to refract an incident ray of unpolarized light into two perpendicular rays. A Nicol prism, for instance, causes birefringence.

Birmingham Wire Gage   Same as **Birmingham Wire Gauge.**

Birmingham Wire Gauge   A standard for sizing wires, tubing, sheets, and the such, made of ferrous and nonferrous metals. Also spelled **Birmingham Wire Gage.** Its abbreviation is **BWG.**

BISDN   Abbreviation of **b**roadband **I**ntegrated **S**ervices **D**igital **N**etwork.

bismuth   A brittle silvery metal with a pinkish tinge, whose atomic number is 83. It has low electrical conductance, is highly diamagnetic, and has very low thermal conductivity. It has over 20 known isotopes, one of which is stable. It has many applications in electronics, including its use in low melting point solders, fuses, thermoelectric materials, permanent magnets, semiconductors, and superconductors. Its chemical symbol is **Bi.**

bismuth antimonide   A chemical compound whose formula is **BiSb.** It is used in semiconductors.

bismuth bromide oxide   Same as **bismuth oxybromide.**

bismuth iodide oxide   A chemical compound whose formula is **BiIO.** It is used as a cathode in dry cells.

bismuth oxybromide   A chemical compound whose formula is **BiBrO.** It is used as a cathode in dry cells. Also called **bismuth bromide oxide.**

bismuth selenide   A chemical compound whose formula is Bi_2Se_4. It is used in semiconductors.

bismuth telluride   A chemical compound whose formula is Bi_2Te_3. It is used in semiconductors.

BIST   Acronym for **Built-In Self Test.**

bistable   Having two stable states, such as on or off, or conducting or not conducting. An example is a flip-flop circuit.

bistable device   A device having two stable states. For instance, a bistable multivibrator.

bistable flip-flop   Same as **bistable multivibrator.**

bistable multivibrator   A two-stage multivibrator circuit with two stable states. In one of the stable states, the first stage conducts while the other is cut off, while in the other state the reverse is true. When the appropriate input signal is applied, the circuit flips from one state to the other. It remains in this state until the next appropriate signal arrives, at which time it flops back. Used extensively in computers and counting devices. Also known as **bistable flip-flop, flip-flop circuit, flip-flop, Eccles-Jordan circuit,** or **trigger circuit (2).**

bistable relay   A relay having two stable states. A pulse opens it if closed, or closes it if open, thus requiring two pulses to complete a full cycle that returns it to its original state.

bistatic radar   A radar system in which two locations are used. In one, the transmitter and transmitting antenna are present, and in the other is the receiver and its antenna. These locations are usually at a great distance from each other. The term is utilized to contrast with **monostatic radar,** which uses a single location.

bisync Acronym for **binary synchronous**.

bisynchronous Abbreviation of **binary synchronous**.

bit Acronym for **b**inary dig**it**. The smallest unit of information a digital computer can handle. A bit may have a value of **0** or **1**, and eight consecutive bits form one byte. Bits are also used to indicate capacity or speed. For instance, a 256-bit CPU which can simultaneously processes 256 bits at a time, or a communications connection speed of 10 Gbps. Its abbreviation is **b**. Also called **storage cell (2)**.

bit block A rectangular block of pixels which are treated as a unit. Used, for instance, to accelerate the display of images onto a screen.

bit block transfer The transfer of a **bit block**. Used frequently in animation, an example being the inverting of the light and dark portions of an image. Its abbreviation is **bitblt**.

bit bucket A virtual location where data may be discarded. Anything written to such a location can not be recovered.

bit density The number of bits that can be stored per unit area. For instance, bits per square millimeter on a magnetic disk.

bit depth The number of bits used to define a pixel in an image, as determined by the hardware and software. For instance, a 24-bit video adapter allows for over 16.7 million colors to be displayed. Also called **pixel depth**, or **color depth**.

bit error A bit which is either transmitted or received erroneously.

bit error rate Of a given number of bits transmitted, the proportion of them which is either transmitted or received erroneously. It is a measure of data transmission integrity. Its abbreviation is **BER**. Also called **bit error ratio**.

bit error rate test A test which indicates a **bit error rate**. Its acronym is **BERT**.

bit error ratio Same as **bit error rate**.

bit flip To change a bit from a zero to a one, or vice versa. Seen, for instance, in bit flipping.

bit flipping The replacing of all ones with zeroes, and vice versa, for a given number of bits. A graphics program, for instance, could invert the light and dark portions of an image utilizing this process.

bit images Same as **bitmap graphics**.

bit manipulation Within a byte, changes made to individual bits. It is relatively complicated, and much less common than altering the entire byte.

bit map Same as **bitmap**.

bit map graphics Same as **bitmap graphics**.

bit map images Same as **bitmap graphics**.

bit mapped Same as **bitmapped**.

bit-mapped font Same as **bitmapped font**.

bit-mapped graphics Same as **bitmapped graphics**.

bit-mapped images Same as **bitmapped graphics**. Also spelled **bitmapped images**.

bit-oriented protocol A communications protocol which uses bits to represent control codes. This contrasts with a **byte-oriented protocol**, in which control codes are represented by bytes. Between the two, bit-oriented is more efficient and reliable.

bit parallel In communications, the transmission of sets of bits, with each bit traveling simultaneously in a parallel path with the others. This contrasts with **bit serial**, where sets of bits are transmitted one after the other via a single path.

bit parity In data transmission, the sameness of the parity bit, which is an additional bit, **0** or **1**, that is added to each byte sent, so that all bytes are odd, or all bytes are even. At the receiving location the additional bit is checked, to be sure it

is the same. If it is, the transmission has passed this test of data integrity.

bit pattern A specific arrangement of bits. An example is a bitmap.

bit plane In a computer memory, each of the bitmaps within a color image represented by a three-dimensional array of bits. The number of these bitmaps, or planes, determines the number of colors that are represented. For instance, eight bit planes results in 2^8, or 256 colors.

bit rate The speed at which bits or bytes are transmitted over a communications line or bus. Usually measured in some multiple of bits or bytes per second. Also called **bit transfer rate**, **transfer rate**, **data rate**, or **line rate**.

bit robbing In data transmission, signaling in which the least significant bit is taken from the information stream and utilized for control signals and other overhead. Sometimes used over T1 or E1 lines. Also called **robbed-bit signaling**, or **in-band signaling (2)**.

bit serial In communications, the transmission of sets of bits, one after the other via a single path. This contrasts with **bit parallel**, where sets of bits are transmitted simultaneously in a parallel path with the others.

bit-slice architecture A processor design in which chips are combined. For instance, three 4-bit processors, or slices, are strung together to form a 12-bit unit. Used in some specialized computers

bit-slice microprocessor Same as **bit-slice processor**.

bit-slice processor A computer processor utilizing **bit-slice architecture**. Also called **bit-slice microprocessor**.

bit stream Same as **bitstream**.

bit string A sequence of bits.

bit stuffing **1.** The insertion of additional bits into a data stream, to avoid the appearance of unintended control sequences, or to conform to a required frame size. These bits are removed at the receiving location to restore the original message. **2.** The altering of the bit rate of a transmission so that it can be received at a different rate, without introducing errors. Also called **justification (1)**, or **pulse stuffing**.

bit transfer rate Same as **bit rate**.

bitblt Abbreviation of **bit block transfer**.

bitmap In a computer memory, data structures composed of rows and columns of bits which serve to represent information. For instance, a bit image is formed by a bitmap, in which case each of the stored bits corresponds to a pixel on the screen. Also spelled **bit map**. Its abbreviation is **BMP**.

bitmapped Formed by a **bitmap**. Also spelled **bit mapped**.

bitmapped font A font formed by a **bitmap**. Each of the characters of such a font has its own stored bitmap. This contrasts, for instance, with an **outline font**, in which the basic outlines of each character are stored, and scaled into the appropriate size when printed or displayed. Also spelled **bit-mapped font**.

bitmapped graphics Images formed by **bitmaps**. Common formats include JPEG, GIF, and TIFF. This contrasts with **vector graphics**, in which images are represented in the form of equations and graphics primitives. Also spelled **bit-mapped graphics**. Also called **bitmapped images**, **bit map graphics**, **bit images**, **bit map images**, or **raster graphics**.

bitmapped images Same as **bitmapped graphics**. Also spelled **bit-mapped images**.

BITNET An academic and research computer network which was merged with CSNET to form a network run by CREN. It was originally an acronym for **B**ecause **I**t's **T**ime **Net**work.

bits per inch In computers, the number of bits per inch of track length on a recording medium, such as a disk or tape. Its abbreviation is **bpi**.

bits per pixel The number of bits used to define a pixel in an image. Its abbreviation is **bpp**.

bits per second A measure of the transmission speed over a communications channel, such as that in a network. It represents the average number of bits transmitted per second. Higher multiples of bits are more frequently used for these measurements, and examples are kilobits per second, or gigabits per second. Its abbreviation is **bps**.

bitstream The transmission of a continuous series of bits over a communications channel. Also spelled **bit stream**.

BJT Abbreviation of **bipolar junction transistor**.

Bk Chemical symbol for **berkelium**.

BL Abbreviation of **baseline**.

black-and-white Image reproduction in black, white, and multiple shades of gray in between. For example, that of a black-and-white TV. Also called **monochrome (1)**, or **grayscale**.

black-and-white television Same as **black-and-white TV**.

black-and-white TV Abbreviation of **black-and-white** television. Television image reproduction which occurs in black, white, and multiple shades of gray in between. Also called **monochrome TV**.

black body Same as **blackbody**.

black body radiation Same as **blackbody radiation**.

black box **1.** A component, device, or unit with an input and an output, which can be inserted into a system without the need to know its internal structure. It is simply used based on its input and output characteristics. Also called **closed box (1)**. **2.** In computers, a hardware or software unit whose internal structure is unknown, but whose function is known. Also called **closed box (2)**. **3.** A sturdy self-contained electronic device which records key information concerning an aircraft, such as flight data.

black-box testing Also called **functional testing**, or **closed-box testing**. **1.** Any tests utilizing the assistance of a **black box (1)**. This contrasts with **white-box testing (1)**, which requires detailed knowledge of the inner workings. **2.** Tests of software or hardware utilizing the assistance of a **black box (2)**. This contrasts with **white-box testing (2)**, which requires detailed knowledge of the inner workings.

black compression In a TV signal, an attenuation in the gain at the levels which correspond to the dark areas of the picture, so as to reduce the contrast in these areas. Also called **black saturation**.

black hat hacker A hacker which upon discovering a security vulnerability uses it to steal, damage, or otherwise harm. This contrasts with a **white hat hacker**, who alerts the owner, administrator, or the like, of the breach. A **gray hat hacker** lies somewhere in between, and may, for instance, post security flaws on the Internet.

black hole **1.** Within a communications network, a location where data enters, but never leaves. May be created, for instance, by noncompatible equipment. **2.** An object in space whose gravitation is so great that neither matter nor light can escape from it.

black iron oxide A reddish or bluish-black powder whose chemical formula is Fe_3O_4. It is used in magnetic-tape coatings, magnetic inks, and in ferrites. Also called **iron oxide**, **iron oxide black**, **ferrosoferric oxide**, or **magnetic iron oxide**.

black level In a video signal, the amplitude level that corresponds to the maximum permitted black peaks, that is, to the screen being black. Also called **reference black level**.

black-level control A control for adjusting the **black level**. When referring to a monitor, such as that of a TV or computer, it is more commonly known by the less accurate term **brightness control**.

black light **1.** Invisible light radiation, especially ultraviolet light. **2.** A lamp which produces ultraviolet light.

black saturation Same as **black compression**.

black signal In fax communications, the amplitude of the signal produced when scanning the maximum darkness of the transmitted material. Also called **picture black (2)**.

black transmission A mode of facsimile transmission in which the maximum amplitude corresponds to the maximum darkness of the received material.

black & white Same as **black-and-white**.

black & white television Same as **black-and-white TV**.

black & white TV Same as **black-and-white TV**.

BlackBerry A popular wireless Internet appliance.

blackboard In expert systems, a hierarchically organized database which allows information to be exchanged between knowledge sources.

blackbody An ideal body which would absorb all the radiant energy incident upon it. It would also be a perfect emitter of radiant energy, whose distribution of energy depended solely on its absolute temperature. Also spelled **black body**. Also called **ideal radiator**, **perfect radiator**, or **full radiator**.

blackbody radiation Radiation from a **blackbody**. Also spelled **black body radiation**.

blacker-than-black Same as **blacker-than-black level**.

blacker-than-black level In a video signal, an amplitude level greater than that which corresponds to black. Used, for instance, for sync pulses. Its abbreviation **blacker-than-black**.

blackout **1.** A complete loss of AC power arriving from a power line. Also called **power blackout**. **2.** A complete loss of a radio signal, or of radio-wave propagation. Also called **radio blackout**. **3.** Same as **blackout effect**.

blackout effect A temporary loss of sensitivity in an electronic device or instrument, following a strong transient signal. Also called **blackout (3)**.

blade A flat moving conductor used in a switch.

blade antenna An antenna, such as that mounted on an aircraft to reduce wind drag, which is comparatively thin.

blank **1.** A recordable medium, such as a magnetic tape or diskette, which has no information recorded yet. **2.** A piece of material, especially a crystal, which is cut in preparation for further processing. **3.** To cut-off a signal or beam. **4.** To cut-off the electron beam in a CRT. **5.** Same as **blank character**.

blank character A character which occupies one byte, the same as if it were a letter or digit. Although a blank character is not necessarily the same as a space character, it is often visualized and treated as such. Such characters are utilized to represent and occupy empty positions. Also called **blank (5)**.

blank disk A disk which has no information recorded yet. It may refer to a magnetic disk, such as a diskette or hard disk, or an optical disk, such as a DVD.

blank tape Magnetic tape which has no information recorded yet.

blanker A circuit or device which serves to cut-off a signal or beam.

blanket area An area where **blanketing** occurs.

blanketing Interference of radio reception due to a strong nearby transmitter. This usually affects a wide range of frequencies.

blanking **1.** The disabling or blocking of a circuit, device, signal, beam, or channel for a given time interval. **2.** Same as **beam blanking**.

blanking circuit A circuit which provides **blanking**.

blanking interval The interval within which **blanking** occurs. Also called **blanking period**, or **blanking time** (1).

blanking level In a composite picture signal, the level which separates the range containing the picture information from that of the synchronizing information. Also known as **blanking pedestal**, or **pedestal level**.

blanking lines In a TV, the lines that do not transmit the visible picture. These lines are above and below the visible picture. Also called **inactive lines**.

blanking pedestal Same as **blanking level**.

blanking period Same as **blanking interval**.

blanking pulse A pulse which produces **blanking**.

blanking signal A signal which produces **blanking**.

blanking time 1. Same as **blanking interval**. 2. In a device such as a CRT, the time interval during which **beam blanking** occurs.

blasting In sound reproduction, distortion and/or damage caused by overloading one or more components, especially an amplifier or speakers.

bleeder A resistor connected across a voltage source to drain off the charge remaining in capacitors when the power is turned off, while also improving voltage regulation. Also called **bleeder resistor**, or **dumping resistor** (2).

bleeder current The current which flows through a **bleeder**.

bleeder resistor Same as **bleeder**.

blind approach An approach for landing an aircraft relying solely on the instrumentation and communications. This usually necessary when visibility is inadequate.

blind carbon copy In email, a feature which enables a sender to include one or more additional recipients without their addresses appearing in the message. In this manner, the primary recipients are not notified, as is the case with a carbon copy. Its abbreviation is **BCC**. Also known as **blind courtesy copy**.

blind courtesy copy Same as **blind carbon copy**.

blind flight Same as **blind flying**.

blind flying The flying of an aircraft relying solely on the instrumentation and communications. This usually necessary when visibility is inadequate. Also called **blind flight**, or **instrument flying**.

blind landing The landing of an aircraft relying solely on the instrumentation and communications. This usually necessary when visibility is inadequate. Also called **instrument landing**.

Blind Source Separation Algorithms utilized to recover mixed signals when the source signals and mixing processes are unknown. Used, for instance, in hearing aids to enhance the intelligibility of voices within noisy environments. Its abbreviation is **BSS**.

blind spot In radio communications, an area within the normal range of a transmitter where radio reception is poor or non-existent. May be caused, for instance, by large objects in the vicinity of the receiver. Also called **dead spot** (1).

blind zone 1. In radio communications, a region where reception is significantly weaker, due to obstructions. Also called **shadow region** (1). 2. In radars, an area with no echoes, due to obstructions. Also called **shadow region** (2).

blink A momentary fluctuation in electric power. May be due, for instance, to a distribution system identifying and solving problems that arise.

blip Also called **pip**. 1. The display of a received pulse on a CRT, especially a radar screen. It may appear, for instance, as a spot of light. 2. A short pulse that serves as a signal, such as that used in the Morse code. 3. A small mark or spot on a recordable medium which is used for tracking purposes.

blip-scan ratio The ratio of the number of radar scans necessary for a received pulse to be visible on the screen.

bloatware Software which consumes more resources, such as RAM and disk space, than necessary. Occurs, for instance, when features are added to new versions without regard to the effects on overall system performance.

Bloch band In a crystalline solid, such as a semiconductor, the range of energy levels within which the electrons must be located. Also called **energy band**.

block 1. A group, number, or section which is treated as a unit. 2. A group of bits or bytes transmitted or processed as a unit. 3. A group of computer storage locations treated as a unit. 4. A group of computer records stored and transferred as a unit. Also called **record block**. 5. A section of text, or spreadsheet cells, selected to be used together. 6. To stop or impede passage or movement.

block cipher An encryption method which operates on a data block of a fixed size, such as 64 bits. Encryption is not completed until a full block is accumulated. This contrasts with a **stream cipher**, which operates on a continuous data stream.

block cursor On a computer screen, a cursor in the shape of a solid block, as opposed to an underline.

block device A device, such as disk drive, which operates on blocks of data. This contrasts with a **character device** (1), which transfers one byte at a time.

block diagram A diagram in which labeled blocks, usually squares and/or rectangles, are used to represent the components of a system which are linked via connecting arrows and/or lines that depict their functional relationships. Such diagrams may show a circuit schematic, hardware and software interconnections, and so on. Also called **configuration diagram**.

block down-converter A device which converts a band of microwave frequencies into lower intermediate frequencies, for subsequent tuning and demodulation by a satellite receiver. This enables multiple receivers to tune to all available channels while using a single down-converter.

block gap On a storage medium, such as a magnetic tape or disk drive, the unused physical space between consecutive blocks of recorded data. Also called **inter-block gap**, or **inter-record gap**.

block header Information appearing at the beginning of a block of computer data, which serves for purposes such as identification, error-checking, and indication of length

block length 1. The length of a block of data. Usually expressed in bytes. 2. Same as **block size**.

block move The movement of a block of data, such as a group of bytes, or a segment of text. Also called **block transfer**.

block size The total size of a block of data. May be expressed in bits, bytes, characters, words, records, and so on. Also called **block length** (2).

block transfer Same as **block move**.

blocked call A telephone call which can not be completed because the central office or a part of the network is operating at full capacity. In such cases, the caller should hear a fast busy signal, or a message indicating that all circuits are busy.

blocked impedance The input impedance of an electromechanical or electroacoustic transducer when its output load impedance is infinite. For instance, the impedance measured when a speaker cone is immobilized.

blocked resistance The resistance of an immobilized electromechanical or electroacoustic transducer.

blocking 1. The forming of a **block**. 2. The stopping or impeding of passage or movement. For instance, to block a current flow. 3. The use of an unwanted signal to overload a receiver, so that the automatic gain control reduces the response to the desired signal. 4. The inability of a telecommunications system to connect calls, usually due to the available resources being completely occupied. **5.** The preventing of viewing TV programming deemed inappropriate, undesirable, or in another way unwanted. For instance, a V-chip may be programmed to perform this function.

blocking capacitor A capacitor which blocks DC and low-frequency AC, while allowing higher-frequency AC to pass freely. Also called **coupling capacitor**, or **isolation capacitor**.

blocking factor The number of records in a disk block. It may be calculated by dividing the block length by the average length of the contained blocks. Also called **grouping factor**.

blocking layer Same as **barrier layer**.

blocking oscillator An oscillator which stops operating for a predetermined time after completing one or more cycles. Its grid bias increases during oscillation until oscillation stops, then decreases until oscillation is reestablished. Also called **squegging oscillator**, or **squegger**.

blocking-oscillator driver A circuit utilizing a **blocking oscillator** to develop square pulses which drive radar modulator tubes.

blocking program Same as **blocking software**.

blocking software A program or utility which seeks to detect advertising and other bothersome or undesirable content before its loaded onto a Web page being accessed. Such a program, for instance, can filter Web page content, protect privacy, prevent pop-up ads from appearing, avert banner ads, eliminate certain JavaScript, stop animated GIFs, turn off ActiveX, disable Web bugs, and so on. Also called **blocking program**, **Internet filter**, **content filter**, or **Web filter**.

blocking system A telephone system which blocks new calls from being connected once all available resources are completely occupied.

blog Abbreviation of We**blog**. **1.** A Web site, or a part of a Web site, where one or more individuals post thoughts, comments, or the like on a given topic, and which often serves as a personal diary that is available for public viewing. The contents of such a site are usually organized chronologically. **2.** To keep a **blog** (1), and/or make entries. An individual may post to a blog several times a day, once a day, every few days, and so on.

blogger An individual that **blogs** (2).

blooming In a CRT, a form of distortion occurring when the scanning spot is enlarged. When this happens, the edges of an image seem to exceed their boundaries, resulting in a reduction of focus. This may be caused, for instance, by setting the brightness or contrast control too high. On a radar screen it occurs as the signal intensity increases.

blooper An oscillating radio receiver that is transmitting an unwanted signal.

blow **1.** To melt a fuse or to open a circuit breaker as a result of excessive current. **2.** Same as **burn** (2).

blower An electric fan which removes heat from components and circuits by blowing a current of air past them.

Blowfish A private-key block cipher which utilizes a variable-length key.

blown-fuse indicator A device which detects and informs of **blowouts**.

blowout **1.** The melting of a fuse or the opening of a circuit breaker as a result of excessive current. **2.** The extinguishing of an electric arc. **3.** The extinguishing of an electric arc by means of a deflecting magnetic field, such as that produced by a blowout magnet.

blowout coil A coil which produces a magnetic field which is used to deflect and extinguish an electric arc where an electric circuit is broken.

blowout magnet An electromagnet or permanent magnet which produces a powerful magnetic field which extinguishes an arc by lengthening or deflecting it. Used, for instance, as a circuit breaker.

Blue Book A document which details the specifications for the CD-extra format.

blue glow **1.** In electron tubes containing mercury vapor, a blue glow produced by the ionization of said vapor. **2.** A bluish glow sometimes seen in some electron tubes. At times it is normal to see this, while at other times it indicates that the tube is defective.

blue gun In a three-gun color TV picture tube, the electron gun which directs the beam striking the blue phosphor dots.

blue screen of death An image, usually consisting of white text against a blue background, that appears on a computer display screen when certain versions of Windows operating systems characteristically cause a computer to freeze. Its abbreviation is **BSOD**.

blue video voltage In a three-gun color TV picture tube, the signal voltage controlling the grid of the blue gun. This signal arrives from the blue section of a color TV camera.

Bluetooth A wireless specification or technology utilized for the short-range linking of desktop computers, mobile computers, and other properly equipped devices, such as cell phones and smart appliances. Bluetooth does not require line of sight, and in addition to data supports multiple voice channels. Used, for instance, in Personal Area Networks and smart houses.

blur To obscure, make dimmer, make less distinct, or less focused. For example, the blurring of an image on the screen of a CRT.

BMP Abbreviation of **bitmap**.

BNC A coaxial cable connector in which one segment is inserted into the other and twisted 90° for locking. The letters **BNC** stand for **b**ayonet **N**eill-**C**oncelman connector, bayonet nut connector, or British Naval Connector. Also called **BNC connector**.

BNC connector Same as **BNC**.

board **1.** A flat, rigid, and insulated panel upon which electrical components are mounted and connected via conductive paths. Used extensively in computers, where the main board is called the motherboard, and secondary boards are called expansion boards. Also called **circuit board**. **2.** A panel upon which the terminals of components or systems are readily accessible for temporary connection. Used, for instance, in communications, computers, and for testing purposes. Also known as **patch panel**.

bobbin **1.** An insulated spool upon which a coil is wound. Also called **coil form**. **2.** A spool upon which wire or magnetic tape is wound.

Bode curve Same as **Bode diagram**.

Bode diagram A graph in which the gain or phase shift of a device, such as an amplifier, is plotted against frequency. Used, for instance, to determine frequency response. Also called **Bode curve**, or **Bode plot**.

Bode plot Same as **Bode diagram**.

body **1.** A mass of matter distinct from other masses. For instance, a celestial body. **2.** The main part of something. **3.** The physical structure of an organism, such as a human.

body capacitance A capacitance between a circuit or electronic device and a human body, or a part of it, especially a hand. This may cause interference, noise, or a shift in tuning. Also called **hand capacitance**.

BOF Abbreviation of **beginning of file**.

bogey An unidentified echo on a radar screen. Also spelled **bogie**.

bogie Same as **bogey**.

Bohr magneton A physical constant equal to approximately 9.2740×10^{-24} joule/tesla. It is used as a unit of magnetic moment. Its symbol is μ_B. Also called **magneton** (1).

bohrium A synthetic radioactive chemical element whose atomic number is 107, and which has about five identified isotopes. The techniques employed to produce this element are utilized to help discover new elements with higher atomic numbers. Its chemical symbol is **Bh**.

boldface Characters which are darker and heavier than normal. Applies to any printed or displayed characters, whether they are black or not.

bolometer A resistive element which measures electromagnetic radiation by absorbing it and converting it into heat. The increase in its temperature is used to measure the radiant energy. May also be used for heat sensing. Also called **thermal detector** (1), or **heat detector** (1).

bolometer bridge A bridge circuit in which one of the arms is a **bolometer**. Used, for example, to measure radio-frequency power.

Boltzmann constant A physical constant whose value is approximately 1.3806×10^{-23} joules/kelvin. Its symbol is k.

bomb A complete and unexpected program or system halt. Caused by a program deficiency or a hardware failure. Also called **abnormal termination, abnormal end,** or **crash** (1).

bombard To subject a target to impacts by high-energy particles, such as electrons, neutrons, or alpha particles. For example, radioactivity can be induced in a non-radioactive element by bombarding it with neutrons.

bombardment The subjecting of a target to impacts by high-energy particles. Also called **particle bombardment**.

bond 1. A force or substance which unites two or more items. **2.** An attractive force sufficient to unite atoms so that they function as a unit. Also called **chemical bond. 3.** An electrical interconnection between objects so that they have the same potential.

bond pad Same as **bonding pad**.

bonding 1. The uniting of two or more items by means of a force or substance. **2.** The uniting of two or more items so that they function as one. **3.** The forming of a unit between atoms by means of sufficient attractive forces. **4.** The electrically interconnecting objects so that they have the same potential.

bonding pad One of multiple metal pads on the surface of a semiconductor device onto which connections may be made. Also called **bond pad**, or **pad** (3).

book capacitor A variable capacitor with hinged metal plates whose angles are varied to adjust the capacitance. The hinges are joined along one edge, similar to the way pages are bound to a book. Used, for instance, as a trimmer capacitor.

bookmark 1. A marker that identifies a Web address, so that it can be returned to easily once set. A Web browser saves such bookmarks. Also called **browser bookmark. 2.** A marker inserted into a specific location within a document, so that it can be returned to easily.

Boolean 1. Pertaining to logical expressions and operations which can have one of two values, such as true or false, or **0** or **1**. **2.** Pertaining to **Boolean algebra**.

Boolean algebra An algebraic system in which variables can have only one of two values, such as **0** or **1**, or true or false, and whose primary operations are AND, OR, and NOT, analogous to the way ADD, SUBTRACT, MULTIPLY, and DIVIDE are the primary operations of arithmetic. There are various permutations on the primary operations, such as NAND and XOR. Boolean algebra serves as the foundation for digital logic.

Boolean data Data that may assume one of two values, such as true or false, or **0** or **1**.

Boolean decision A decision whose outcome can have only one of two values, such as true or false, or **0** or **1**. Also called **logical decision** (2).

Boolean expression An expression utilizing Boolean operators and yielding a Boolean value. Also called **logical expression**.

Boolean function A function utilizing **Boolean algebra**.

Boolean logic Logic based on **Boolean algebra**. Used, for instance, in logic circuits. Also called **digital logic**.

Boolean operator An operator within **Boolean algebra**. These include AND, OR, NOT, NAND, and XOR. Also called **logic operator**.

Boolean search A database search utilizing Boolean operators, such as AND, OR, or NOT.

Boolean value A value which must be one of two possibilities, such as true or false, or **0** or **1**. Also called **logical value**.

boom A movable mechanical support utilized to enable a device or instrument to be placed in a desired location. Used, for example, to suspend a camera above a scene to obtain an overhead view.

boost To increase, amplify, or reinforce. For instance, to increase voltage, or to boost bass response in an audio system at low-volume settings.

boost charge A partial charge of a rechargeable battery, done for a short interval and at a high current. Also called **booster charge**, or **quick charge** (1).

booster 1. A device which increases, amplifies, or reinforces. **2.** An amplifier of weak signals. Also called **booster amplifier** (1). **3.** An amplifier which amplifies and retransmits RF signals. Also called **booster amplifier** (2).

booster amplifier 1. Same as **booster** (2). **2.** Same as **booster** (3).

booster battery A battery that supplies power to a **booster**.

booster charge Same as **boost charge**.

booster gain The gain provided by a **booster**.

boot 1. Abbreviation of **booting** (1), **bootstrap** (1), or **bootstrapping** (1). **2.** A flexible protective jacket for cables, connectors, and the like. It may or may not be in the shape of a boot.

boot CD-ROM Same as **bootable CD-ROM**.

boot disk A disk that contains the files necessary to boot a computer. It is usually used when the disk from which the computer ordinarily boots fails. The operating system is included in the contents of a boot disk, and it may be a floppy disk, or an optical disk such as a CD-ROM. Also called **bootable disk, system disk,** or **startup disk**.

boot drive The disk drive where the BIOS looks for the operating system when a computer boots.

boot failure The inability of a computer to complete the booting process. This is usually because of a problem locating or activating the operating system.

boot floppy A floppy disk which serves as a **boot disk**. Also known as **bootable floppy**.

boot loader Same as **boot program**.

boot process Same as **booting** (1).

boot program Abbreviation of **boot**strap **program**. A program which is automatically loaded when a computer boots. Among other things, this program loads the operating system. Also called **boot loader**, or **bootstrap loader**.

boot protocol A protocol utilized to boot diskless network computers. Its abbreviation is **BOOTP**.

boot record Same as **boot sector**.

boot routine The automatic steps a computer follows when booting (1). Also called **bootstrap routine**.

boot sector The segment of a disk which is reserved for the boot program. Also called **boot record**.

boot sequence Same as **booting (1)**.

boot up Same as **booting (1)**.

boot virus 1. A virus which infects the system when the computer boots. **2.** A virus on a floppy disk that infects the system when the disk is used to boot the computer.

bootable Containing the files necessary to boot a computer. These files include the operating system.

bootable CD-ROM A CD-ROM which serves as a **boot disk**. Also known as **boot CD-ROM**.

bootable disk Same as **boot disk**.

bootable floppy Same as **boot floppy**.

booting 1. It is an abbreviation of **boot**strapping **(1)**. Its own abbreviation is **boot (1)**. To start up or reset a computer. During this process the computer accesses instructions from its ROM chip, performs self-checks, loads the operating system, and prepares for use by an operator. A booting may be initiated by turning on the power, pressing a button or switch, by hitting a specific key sequence, or through a program or routine that gives this command. Also called by various other names, including **booting up**, **booting process**, **boot sequence**, **startup (1)**, and **initial program load (1)**. **2.** Same as **bootstrapping (2)**.

booting process Same as **booting (1)**.

booting sequence Same as **booting (1)**.

booting up Same as **booting (1)**.

BOOTP Abbreviation of **boot protocol**.

bootstrap 1. Same as **booting (1)**. It is an abbreviation of **bootstrap**ping. Its own abbreviation is **boot (1)**. **2.** Same as **bootstrapping (2)**.

bootstrap circuit An amplifier circuit in which the output load is partially fed back across the input, providing a higher effective input impedance.

bootstrap loader Same as **boot program**.

bootstrap program Same as **boot program**.

bootstrap routine Same as **boot routine**.

bootstrapping 1. Same as **booting (1)**. Its abbreviation is **boot (1)**. **2.** A manner of making a circuit or device achieve a given state through its own actions. Also called **bootstrap (2)**, or **booting (2)**.

border The boundary where things meet, or where one thing ends and the other begins. For instance, the boundary where p and n type semiconductor materials meet.

border gateway protocol A routing protocol used between autonomous systems on the Internet. Its abbreviation is **BGP**.

boric acid An acid whose chemical formula is H_3BO_3, and whose applications in electronics include its use in nickel electroplating baths and in electrolytic capacitors.

boric oxide A colorless or white substance in crystal or powder form, whose chemical formula is B_2O_3. Used, for instance, in heat-resistant glassware.

boron A nonmetallic element whose atomic number is 5. It has various allotropic forms, the most common being as a brown amorphous powder, a black hard solid, or as extremely hard crystals. It has 10 known isotopes, of which 2 are stable. Boron has various applications, including its use as a neutron absorber, as a semiconductor dopant, and in various special alloys. Its chemical symbol is **B**.

boron carbide Black hard crystals whose chemical formula is B_4C. According to the Mohs hardness scale, it is almost as hard as diamond. Used, for instance, as an abrasive, and for control rods in nuclear reactors.

boron nitride A white powder whose chemical formula is **BN**. Used in semiconductors, as a dielectric, and as a high-

temperature insulator, among others. When subjected to great pressure it becomes almost as hard as diamond.

boron trichloride A fuming colorless liquid whose chemical formula is BCl_3. Used in semiconductors.

boron trifluoride A fuming colorless liquid whose chemical formula is BF_3. Used in neutron detection, and in semiconductors.

borosilicate glass A silicate glass constituted by 5% or more boric oxide. It features a high resistance to corrosion and temperature, and is used, for instance, in light bulbs. Also called **hard glass**.

BORSCHT Abbreviation of battery, overvoltage protection, ringing power, supervision, coding, hybrid, and testing. It describes line supervision functions that are performed at central offices in telephone systems.

Bose-Einstein condensate A state of matter occurring just above absolute zero, and in which groups of atoms condense into the same quantum state and behave essentially like a single atom. Such a state of matter is yet to be observed at temperatures greater than one millionth of a degree above absolute zero, but this threshold may increase. The other physical states in which matter is known to exist are **solid (1)**, **liquid**, **gas**, and **plasma**.

boson A particle with an integer spin. A boson can exist in the same quantum state as any number of other bosons. A photon is an example.

bot Abbreviation of **robot (2)**. **1.** A program which performs tasks, especially repetitive ones, over networks such as the Internet. Also known as **robot (2)**. When used for searching the Internet exclusively, known as **Internet robot** or **Internet bot**. **2.** A program which searches Web sites and organizes the located information. Also known as **spider (1)**, **crawler**, or **Web crawler**.

bottleneck A reduction in throughput, as occurs, for instance, when a communications network is overloaded, or a CPU is processing data received from a slow peripheral. In a network, for example, a packet sniffer may be utilized to locate such a bottleneck.

bottlenecking In a computer network, to maliciously create a **bottleneck**. This may be done, for instance, by sending emails with massive attachments to all email addresses within said network.

bottom-up approach Same as **bottom-up reasoning**.

bottom-up design Same as **bottom-up reasoning**.

bottom-up programming A method of programming that first defines and tests the simpler functions, and later uses this base to proceed to higher levels. This contrasts with **top-down programming**, where the higher-level functions are defined, and then broken down into simpler functions.

bottom-up reasoning A method of reasoning or problem solving that first defines the simpler functions or tasks, and later uses this base to proceed to higher levels. This contrasts with a **top-down approach**, where higher-level functions or tasks are defined, and then broken down into simpler functions. Also known as **bottom-up approach**, or **bottom-up design**.

bounce 1. An abnormal, sudden, and short-lived variation in the brightness or vertical position of images displayed by a TV. **2.** An undesired condition in which there is a spontaneous opening, or the intermittent opening or closing of contacts, when such contacts are moved to the open or closed position. Also called **contact bounce**.

bounced email 1. Email which has been returned to the sender. This occurs, for instance, when a non-existing address is utilized, or the mailbox of the intended recipient is full. **2.** Email which has been forwarded.

BounceKeys Keys on a computer keyboard which are programmed or set to ignore repeated pressings within a given time interval. Used, for instance, to avoid unwanted input when inadvertently pressing the same key more than once. This contrasts, for instance, with **SlowKeys (1)**, which only accept a keystroke when a key is pressed longer than a given interval.

bound charge In a conductor, a charge which is held by the inductive effect of neighboring charges.

boundary 1. The border where things meet, or where one thing ends and the other begins. For instance, the border where p and n type semiconductor materials meet. **2.** A line or set of values which depict the limits which a parameter may reach, such as the modulation envelope of a signal. **3.** The limit that a parameter may reach, such as the highest frequency of a band.

boundary router A device that routes data packets between LANs and ATM networks. Also called **edge router**.

bow-tie antenna Same as **bowtie antenna**.

bowtie antenna Also spelled **bow-tie antenna**. An antenna consisting of two wire triangles, or two triangular plates, arranged similarly to a bow-tie, which is fed at the gap between the triangles. It is a wideband antenna used mostly in the very-high frequency and ultra-high frequency ranges.

box An enclosure, typically consisting of a base, multiple sides, and a lid, which serves to house, enclose, insulate, or otherwise protect components, circuits, devices, or equipment. Examples include cable boxes, jack boxes, and fuse boxes. Also, a virtual box, such as that appearing on a computer screen.

boxcar Within a stream of long-duration pulses separated by short intervals, a single pulse.

Boyer-Moore algorithm An algorithm utilized for text searches.

Bpi Abbreviation of **bytes per inch**.

bpi 1. Abbreviation of **bits per inch**. **2.** Abbreviation of **bytes per inch**.

BPI Same as **bpi**.

bpp Abbreviation of **bits per pixel**.

bps Abbreviation of **bits per second**.

Bps Abbreviation of **bytes per second**.

BPSK Abbreviation of **binary phase-shift keying**.

Bq Symbol for **becquerel**.

Br Chemical symbol for **bromine**.

Bragg angle In the scattering of incident X-rays or neutrons by the atoms in a crystal lattice, a characteristic angle at which a maximum intensity of reflections occur.

Bragg cell A device which is utilized to modulate, deflect, and focus light waves through their interactions with acoustic waves. Used, for example, in laser equipment for control of the intensity and position of laser beam radiation. Also called **acousto-optical cell**, or **optoacoustic cell**.

Bragg diffraction Same as **Bragg scattering**.

Bragg grating An optical-fiber section in which the refraction index of the core provides for the reflection of specific wavelengths. Used, for instance, for filtering, or in wavelength division multiplexing. Also called **fiber Bragg grating**.

Bragg-Pierce law A law pertaining to the relationship between the atomic absorption coefficient of an element and its atomic number.

Bragg reflection Same as **Bragg scattering**.

Bragg scattering The scattering of incident X-rays or neutrons by the atoms in a crystal lattice, with maximums of intensity of reflections occurring at characteristic angles known as **Bragg angles**. Also known as **Bragg reflection**, or **Bragg diffraction**.

Bragg spectrometer An instrument used to analyze crystal structure by using X-rays. In it, a beam of collimated X-rays strikes the crystal, and a detector measures the angles and intensities of the reflected beam. Applying Bragg's law to these measurements yields details about the crystal structure. Used, for instance, for X-ray diffraction studies that determine the structure of crystalline DNA. **DNA** is the abbreviation of **d**eoxyribo**n**ucleic **a**cid, and in its strands are contained the chemical basis of heredity. Also known as **crystal spectrometer, ionization spectrometer**, or **crystal-diffraction spectrometer**.

Bragg's equation Same as **Bragg's law**.

Bragg's law A mathematical correlation of the relationship between the angles and intensities of reflection by the atoms in a crystal lattice, when irradiated with X-rays or neutrons, and the structure of said lattice. Also known as **Bragg's equation**.

braid A flexible sheath of helically interwoven metal or fiber filaments that serve to cover wires. It provides structural support, and may be used as a conductor or for grounding.

braided wire Wire incorporating a **braid**.

brain-computer interface A computer interface in which a user communicates information via brain waves. The term may also apply to interfaces which utilize voluntary muscle movements such as those involved in grimacing or brow furrowing. Its abbreviation is **BCI**.

brain waves Fluctuations of voltage in the brain, produced by its electrochemical activity. Some of these waves, for instance, can be detected and recorded by an electroencephalogram.

brake horsepower A measure of the effective power output of an engine. It may be measured in various ways, one of which is to quantify the resistance the engine provides to a brake attached to its output shaft. Its abbreviation is **bhp**.

branch 1. A subdivision of a unit or system. **2.** A path within a circuit or electric network. Also called **leg**, or **arm (3)**. **3.** A connection between nodes, such as those in a communications network. **4.** A computer instruction that switches the CPU to another location in memory. Also called **jump**, or **branch instruction**.

branch circuit In a wiring system, the portion between the final overcurrent protection device, such as a circuit breaker, and the served outlets.

branch current The current across a **branch (2)**.

branch instruction Same as **branch (4)**.

branch point The point at which a **branch** originates. Also spelled **branchpoint**. Also known as **junction point (1)**, or **node (2)**.

branch prediction A CPU technique utilized to try to guess which branch will be followed in a program. When the prediction rate is high, as it usually is, system performance is enhanced.

branch voltage The voltage across a **branch (2)**.

branching 1. The dividing into **branches**. **2.** A point from which **branches** emerge.

branchpoint Same as **branch point**.

brass An alloy of copper and zinc in varying proportions, sometimes incorporating low percentages of other elements such as manganese or silicon. These alloys are corrosion-resistant, very ductile, and have good strength. Applications in electronics include its use in terminals, as connectors, in solder, and to house components.

braze To join metals at a high temperature using a nonferrous filler metal or alloy.

braze welding Same as **brazing**.

brazing The use of a nonferrous metal or alloy to fill the spaces between two metals which are to be joined at a high temperature. The filler metal must have a lower melting point than the metals to be joined. It is similar to soldering, but the term brazing is used when the temperature exceeds an arbitrary value such as 425 °C. Also known as **braze welding**, or **hard-soldering**.

brazing alloy An alloy, such as brass, or one formed by copper and nickel, used for **brazing**. Also called **hard solder** (1).

brazing filler metal A nonferrous metal or alloy used for **brazing**.

brazing metal A nonferrous metal used for **brazing**.

breadboard 1. A perforated board or chassis on which electronic devices and components can be mounted and interconnected simply and quickly, for testing, experimenting, and preparation of prototype circuits. Also called **prototyping board**. 2. An experimental or prototypical circuit utilizing a **breadboard** (1) to mount it. Also called **breadboard circuit**. 3. To prepare an experimental or prototypical circuit utilizing a **breadboard** (1) to mount it.

breadboard circuit Same as **breadboard** (2).

breadboard model An experimental or prototypical circuit or device utilizing a **breadboard** (1) to mount it.

breadboarding The preparation of a **breadboard** (2).

break 1. An opening, pause, or interruption in continuity. 2. To open a circuit, or an open circuit. 3. In a circuit-opening device, such as a switch, the minimum distance between the contacts when in the open position. 4. An interruption in processing, execution, transmission, communication, or the like. Also, a command which creates such a break. Also called **pause** (3). 5. A key on a computer keyboard which creates a **break** (4). Also called **pause** (4). 6. A signal sent by a receiver indicating the desire to transmit.

break-before-make contacts Contacts which open a connection prior to making a new connection.

break-before-make switch A switch that opens a connection prior to making a new connection. For instance, the moving contact of such a switch may be narrower than the distance between fixed contacts. Also called **non-shorting switch**, or **non-short-circuiting switch**.

break contact In a relay or switch, a contact pair that is closed in the resting state, and which is opened when energized. The contact pair consists of a stationary contact and a movable contact. A break contact provides a complete circuit when not energized. This contrasts with a **make contact**, in which the contact pair is open in the resting state. Also called **back contact**, or **normally-closed contact**.

break-in 1. A period within which new components, devices, systems, and so on, are operated and/or tested to verify proper functioning. During this interval, adjustments may be made, and frequently, deficient components will fail, avoiding some future problems. Also called **burn-in** (1). 2. In communications, an interruption of a transmission by the receiving station. May be used, for instance, to request a retransmission.

break-in operation Communications in which **break-ins** (2) are utilized.

break key On most keyboards, a key which when pressed serves to stop anything the computer is currently doing. The same instruction may also be available by pressing a combination of keys. Even so, some programs may not respond to this key.

break-out box Same as **breakout box**.

break point Same as **breakpoint**.

break-point instruction Same as **breakpoint instruction**.

break time The time it takes for a switch or relay to open a connection.

breakdown 1. A failure to function, or a sudden change resulting in a failure to function. 2. A disruptive electrical discharge through an insulator, dielectric, or other material separating circuits. Also called **puncture** (1). 3. A disruptive electrical discharge between the electrodes of an electron tube. 4. A substantial and abrupt increase in electric current produced by a small increase in voltage. Also called **electric breakdown** (1). 5. In a semiconductor exposed to an electric field, the cumulative multiplication of free charge carriers. Also called **avalanche breakdown**.

breakdown diode A semiconductor diode with a very high ratio of reverse-to-forward resistance, until the point where avalanche breakdown occurs. After this, voltage drop is nearly constant and independent of current. May be used, for instance, in surge suppressors with reaction times in the picosecond range. Also called **avalanche diode**.

breakdown region Within a semiconductor's characteristic curve, the region beyond the initiation of avalanche breakdown.

breakdown voltage 1. The voltage at which a disruptive electrical discharge occurs through an insulator, dielectric, or other material separating circuits. Also called **puncture voltage**. 2. The voltage at which a disruptive electrical discharge between the electrodes of an electron tube occurs. 3. The voltage at which a disruptive electrical discharge occurs in a gas. Also called **sparking voltage**, or **spark voltage**. 4. In a semiconductor, the voltage at which avalanche breakdown occurs. Also called **Zener voltage** (2).

breaking current In a switch or relay, the current that flows the instant that contact separation occurs.

breakout box A device used for testing the individual leads of an multiconductor cable, such as an RS-232 cable. Also spelled **break-out box**.

breakover In a semiconductor component such as a thyristor, the transition from a forward-blocking state to a forward-conducting state.

breakover voltage The anode voltage at which **breakover** occurs.

breakpoint A predetermined location within a computer program at which execution is interrupted, so that its status can be evaluated. Used for debugging and testing. Also spelled **break point**.

breakpoint instruction An instruction for a **breakpoint** to occur.

breathing Slow and generally regular variations in a quantity, such as the brightness of the image produced by a CRT, or cone movement in a speaker, due to circuit instability.

breezeway In a color TV, the interval between the trailing edge of the horizontal sync pulse and the start of the color burst.

bremsstrahlung The radiation produced when charged particles, such as electrons, are decelerated or deflected by atomic nuclei. X-rays are an example of this form of radiation. Also called **bremsstrahlung radiation**.

bremsstrahlung radiation Same as **bremsstrahlung**.

brevity code In communications such as radiotelegraphy, a code designed to save time by utilizing less characters to convey information.

Brewster angle For unpolarized light incident upon a material, the angle at which reflected light has the greatest polarization. At this angle of incidence, the angle between reflected and transmitted rays is 90°. Also called **polarizing angle**.

BRI Abbreviation of **Basic Rate Interface**.

bridge 1. An electric circuit, network, or instrument that has four or more branches, or arms, each with a component such as a resistor or capacitor. It also has a current source and a null detector, and when such a bridge is balanced, its output is zero. A bridge allows for an unknown component replacing one of the branches to be accurately measured, as it can be compared to known values. Bridges may be utilized to determine unknown resistances, capacitances, inductances, frequencies, and so on. Its characteristic shape is that of a diamond. Also known as **bridge circuit**, or **electric bridge**. **2.** A device which connects networks or segments of networks at the data-link layer, utilizing the same communications protocols. When connecting LANs, however, such bridges are protocol-independent. Also called **network bridge**.

bridge arm One of the branches of a **bridge** (1). Also called **bridge branch**.

bridge branch Same as **bridge arm**.

bridge circuit Same as **bridge** (1).

bridge hybrid A waveguide, resistor, or transformer circuit or device which has four branches which are arranged in a manner that the input signal is equally divided between the adjacent branches. The signal does not reach the fourth, or opposite, branch. Also called **hybrid junction**, **hybrid tee**, or **magic tee**.

bridge network An electric network consisting of four branches and four terminals arranged in a diamond shape. It has two input terminals, which are nonadjacent, and the other two serve as output terminals. It is a type of lattice network.

bridge oscillator An oscillator which utilizes a bridge circuit to provide positive feedback.

bridge page A Web page created with given characteristics, such as multiple appearances of specific keywords, so as to rank high on a particular search engine. Also called **doorway page**, **jump page**, **entry page**, or **gateway page**.

bridge rectifier A full-wave rectifier in the form of a bridge, with a rectifier in each arm. When an AC voltage is supplied to one diagonally opposite pair of junctions, a DC voltage is obtained from the other pair.

bridge router A device which performs the functions of both a **bridge** (2) and a **router**. Its acronym is **brouter**.

bridged-T network A T-network that has a fourth branch connected between an input and an output terminal, so that the two series components are bridged. Also spelled **bridged-tee network**.

bridged tap A connection, usually via a cable pair, made into a part of a local loop that does not lie along a direct transmission path between a central station and the premises of a customer. Formerly utilized, for instance to connect party lines. A bridged tap creates an impedance mismatch within the transmission line, which in turn produces signal reflections. This is not usually a problem in POTS, but may create significant interference in high-frequency communications, such as those of DSL services.

bridged-tee network Same as **bridged-T network**.

bridgeware Hardware or software that converts programs and/or files into a format that may be used by different computers.

bridging **1.** The connecting of two electric circuits in parallel. When it is done in a manner which allows for a part of the signal energy to be withdrawn without affecting the overall operation of the circuit in a significant manner, it is called **bridging connection**. **2.** In a selectable switch, the action of the movable contact when it is wide enough to touch two consecutive contacts. In this manner, the circuit is not broken during switching. **3.** The use of solder to form an electric connection between terminals, wires, or pins. Also called **solder bridge**. **4.** The combining of the left and right channels of a stereo audio amplifier to form a monaural channel with considerably more power.

bridging amplifier An amplifier with an input impedance sufficiently high so as to allow it to have its input bridged across a circuit without appreciably affecting the signal level of the bridged circuit.

bridging connection Same as **bridging** (1).

bridging gain The ratio of the signal power a transducer delivers to a given load under specified conditions, to the signal power dissipated in the main circuit load across which a bridge is spanned. Usually expressed in decibels.

bridging loss The reciprocal of **bridging gain**.

Briggs logarithm A logarithm whose base is 10. Also called **common logarithm**.

brightness **1.** The visual perception of more or less light emitted from a source. The more light subjectively perceived, the greater the brightness. **2.** The perceived **brightness** (1) from a screen, such as that of a TV or computer monitor. **3.** The total light emitted or reflected from a surface in a given direction, per unit of projected area. The measurement must be made from the same direction the light is emitted from. May be expressed in foot-lamberts. **Luminance** (1) is the term currently used for this concept. **4.** The ability of a sound-reproduction system to accurately reproduce higher frequencies. Also called **brilliance** (2). **5.** In sound reproduction, an excess of higher frequencies, which may impart a shrillness to the listening experience.

brightness control On a monitor, such as that of a computer or TV, a control which adjusts the **brightness** (2). A more accurate term for this concept is **black-level control**. Also called **brilliance control** (1), or **intensity control**.

brilliance **1.** On a screen, the extent of **brightness** (2) and clarity perceived. **2.** Same as **brightness** (4).

brilliance control **1.** Same as **brightness control**. **2.** For one or more speakers, a control which allows adjusting of the reproduction of the higher frequencies.

Brinell hardness The relative hardness of a material, as determined by a **Brinell hardness test**.

Brinell hardness test A standard method for measuring the hardness of materials. In it, a steel ball is pressed through a smooth surface of the material being tested, and the load required and the indentation produced are used to calculate the hardness. Also called **Brinell test**.

Brinell test Same as **Brinell hardness test**.

British Naval Connector Same as **BNC**.

broad band Same as **broadband**.

broad-band access Same as **broadband access**.

broad-band amplifier Same as **broadband amplifier**.

broad-band antenna Same as **broadband antenna**.

broad-band channel Same as **broadband channel**.

broad-band communications Same as **broadband communications**.

broad-band connection Same as **broadband connection**.

broad-band electrical noise Same as **broadband electrical noise**.

broad-band interference Same as **broadband interference**.

broad-band ISDN Same as **broadband ISDN**.

broad-band klystron Same as **broadband klystron**.

broad-band modem Same as **broadband modem**.

broad-band network Same as **broadband network**.

broad-band noise Same as **broadband noise**.

broad-band signal Same as **broadband signal**.

broad-band transmission Same as **broadband transmission**.

broad tuning In a radio receiver, tuning characterized by poor selectivity. This may cause, for instance, the reception of two adjacent stations at a single dial setting.

broadband Also spelled **broad band**. Also called **wideband**. **1.** Operating at, or encompassing a wide range of frequencies. **2.** Operating at, or encompassing a wider range of frequencies than is ordinarily available. **3.** Communications in which multiple messages or channels are carried simultaneously over the same transmission medium. Cable TV, for instance, is broadband, as multiple channels are carried concurrently over a single coaxial cable. **4.** Communications at a speed of transmission higher than a determined amount. For example, that exceeding 1.544 Mbps.

broadband access Access to a network, especially the Internet, via a broadband connection. Also spelled **broad-band access**. Also called **wideband access**.

broadband access provider Same as **broadband service provider**.

broadband amplifier An amplifier, such as a video amplifier, that operates over a wide range of frequencies with an essentially flat frequency response. Also spelled **broad-band amplifier**. Also called **wideband amplifier**.

broadband antenna An antenna, such as a log-periodic antenna, that operates well over a wide range of frequencies. Also spelled **broad-band antenna**. Also known as **wideband antenna**.

broadband channel Also spelled **broad-band channel**. Also called **wideband channel**. **1.** A communication channel that operates over a wide range of frequencies, or over a wider range of frequencies than is ordinarily available. **2.** A communication channel that can carry multiple messages simultaneously, or more than a certain number of messages simultaneously.

broadband communications Also spelled **broad-band communications**. Also called **wideband communications**. **1.** Communications utilizing a wide range of frequencies, or over a wider range of frequencies than is ordinarily available. **2.** Communications in which multiple messages can be transmitted simultaneously, or in which more than a certain number of messages can be transmitted simultaneously. **3.** Communications at a speed of transmission higher than a determined amount. For example, that exceeding 1.544 Mbps.

broadband connection A connection to a communications network at a transmission speed exceeding a given rate, such as 1.544 Mbps. Also spelled **broad-band connection**. Also called **wideband connection**.

broadband electrical noise Electrical noise present over a wide range of frequencies. White noise is an example. Also spelled **broad-band electrical noise**. Also called **wideband electrical noise**.

broadband Integrated Services Digital Network Same as **broadband ISDN**.

broadband interference Interference occurring over a wide range of frequencies. An electric motor may be a source. Also spelled **broad-band interference**. Also called **wideband interference**.

broadband ISDN Abbreviation of **broadband** Integrated Services Digital Network. An advanced version on ISDN based on ATM technology. It uses fiber-optic cables, and is capable of transmission speeds exceeding 1 Gbps. Its own abbreviation is **BISDN**. Also spelled **broad-band ISDN**. Also known as **wideband ISDN**.

broadband klystron A klystron with a wider range of frequencies than ordinarily available. Also spelled **broadband klystron**. Also called **wideband klystron**.

broadband modem A modem for use on a **broadband network**. Also spelled **broad-band modem**. Also called **wideband modem**.

broadband network A LAN utilizing multiple transmission channels, each with its own carrier frequency, so that they do not interfere with each other. Such a network can simultaneously transmit data, voice, and video at very high speeds. Also spelled **broad-band network**. Also called **wideband network**.

broadband noise Noise present over a wide range of frequencies. May refer to electrical noise, thermal noise, radio noise, and so on. Also spelled **broad-band noise**. Also called **wideband noise**.

broadband service provider An Internet service provider offering **broadband access**. Also called **broadband access provider**.

broadband signal A signal encompassing a wide range of frequencies. For instance, such a signal sent over a broadband network. Also spelled **broad-band signal**. Also called **wideband signal**.

broadband transmission A transmission encompassing a wide range of frequencies. For instance, a transmission that simultaneously includes data, voice, and video. Also spelled **broad-band transmission**. Also called **wideband transmission**.

broadband wireless Broadband communications in which there are no connecting wires. Instead, communication is achieved by means of RF waves, such as microwaves waves or infrared waves. Also called **wireless broadband**.

broadcast **1.** An RF transmission intended for public or general reception. Also, to transmit such signals. Refers especially to radio and TV programming. **2.** In a communications network, the simultaneous transmission of a single message to multiple recipients. Also, to transmit such messages.

broadcast address An address reserved for simultaneously transmitting a message to all stations in a communications network.

broadcast band A band of frequencies allocated to broadcasting stations. For instance, in the United States the standard AM broadcast band is 535 to 1705 kHz. Also called **standard broadcast band**.

broadcast channel Same as **broadcasting station**.

broadcast fax A feature enabling the sending of a single fax simultaneously to multiple recipients. Also called **fax broadcast**.

broadcast interference The disturbance in the reception of broadcasts, due to unwanted signals. Such interference may be caused, for instance, by nearby equipment or stations. Its abbreviation is **BCI**.

broadcast network **1.** A communications network in which the router sends packets to all users. **2.** A network which emits **broadcasts (1)**.

broadcast packet A packet sent to all nodes of a communications network.

broadcast receiver A device designed to receive RF transmissions intended for public or general reception.

broadcast reception The receiving of RF transmissions intended for public or general reception.

broadcast service Same as **broadcasting service**.

broadcast station Same as **broadcasting station**.

broadcast storm A condition in which an excess of simultaneous broadcasts overloads a communications network. This is usually caused by multiple hosts responding to a broadcast, with each response producing further responses, until all other traffic in the network is blocked.

broadcast transmission A transmission of TV, radio, or other signals intended for reception by the general public.

broadcast transmitter A transmitter of TV, radio, or other signals intended for reception by the general public.

broadcasting **1.** The transmission of RF signals intended for public or general reception. Refers especially to radio and TV programming. **2.** The transmission of a single message intended for multiple recipients.

broadcasting channel Same as **broadcasting station**.

broadcasting-satellite service A radio communications service in which TV, radio, or other signals intended for direct reception by the general public are transmitted or retransmitted by satellite. Its abbreviation is **BSS**.

broadcasting service A radio communications service in which TV, radio, or other signals intended for reception by the general public are transmitted or retransmitted. Also called **broadcast service**.

broadcasting station A station within a **broadcasting service**. Also called **broadcast station**, **broadcasting channel**, **broadcast channel**, or **station (3)**.

broadside Directed or placed perpendicular to a plane.

broadside array An antenna array whose direction of maximum radiation is perpendicular to the line or plane of the antenna elements.

brochureware Software which does not exist, but which is nonetheless hyped via brochures. It is a form of **vaporware (1)**.

broken link A link, such as that between two Web pages, that is not working at the time, or that is no longer valid. This usually generates the frequently seen error 404.

bromine A nonmetallic chemical element whose atomic number is 35. It is a dark, reddish-brown liquid which produces suffocating fumes. It has over 30 known isotopes, of which 2 are stable. Its has uses in spectroscopy and photography. It is a halogen, and its chemical symbol is **Br**.

brouter Acronym for **bridge router**.

Brown and Sharp gage Same as **Brown and Sharp gauge**.

Brown and Sharp gauge A standard for designating the diameter of wires, or the thickness of sheets. Refers to non-ferrous materials only. Also spelled **Brown and Sharp gage**. Its abbreviation is **B and S gage**. Also called **American wire gage**.

Brownian motion Erratic motions of minute particles suspended in a liquid or gas. These movements consist of irregular zigzags, and are caused by molecular collisions with the fluid it is suspended in. Also called **Brownian movement**.

Brownian movement Same as **Brownian motion**.

brownout A temporary reduction in the AC power arriving from a power line. It may be due to unusually heavy power demands, a malfunction of the generation or distribution system, or it may be done intentionally by the power company to counter excessive demand. It usually goes unnoticed by power consumers, except when it affects sensitive electronic equipment, such as computers. Also called **power brownout**.

browse To scan or view a file, group of files, or the World Wide Web. Browsing usually implies simply viewing, although some database programs enable editing as well.

browser 1. A program which enables a computer user to **browse**. 2. A program which enables a computer user to browse, and perform other functions, such as downloading files and exchanging email, through the World Wide Web. This program locates and displays pages from the Web, and provides multimedia content. Most of the time there is interactive content on any given page. To have access to certain features, such as streaming video, specific plug-ins may be necessary, in addition to any hardware requirements. Also called **Web browser**, or **Internet browser**.

browser bookmark Same as **bookmark (1)**.

browser cache An area of computer storage where the most recently downloaded Web pages may be temporarily placed, so that when one of these pages is returned to, it loads faster. Also called **cache (2)**, **Web cache**, or **Internet cache**.

browser compatibility The extent to which the appearance, interactivity, and the like, of given Web pages is similar between different browsers.

browser cookie A block of data prepared by a Web server which is sent to a Web browser for storage, and which remains ready to be sent back when needed. A user initially provides key information, such as that required for an online purchase, and this is stored in a cookie file. When a user returns to the Web site of this online retailer, the cookie is sent back to the server, enabling the display of Web pages that are customized to include information such as the mailing address, viewing preferences, or the content of a recent order. Also called **cookie**, **Web cookie**, or **Internet cookie**.

browser plug-in A program which is used in association with an open Web browser to add functionality or enhance performance. For example, a given browser may require a plug-in to display video in a format that is not currently supported. Also spelled **browser plugin**.

browser plugin Same as **browser plug-in**.

brush A conductor which slides to maintain contact between stationary and moving parts of an electrical device, such as a motor. It is usually made of graphite, or a metal.

brush discharge An intermittent, luminous, and often audible electrical discharge that originates from a conductor when its potential exceeds a given value, but is not high enough to form a spark. This occurs especially in pointed objects when the electric field near their surfaces surpasses a given amount. For example, an aircraft traveling through an electrical storm may develop such discharges from an antenna, which in turn produces precipitation static. Also called **corona discharge**, **corona effect**, **corona**, **corposant**, or **St. Elmo's fire**.

brush holder A device which serves to house or support a **brush**.

brushless An electrical device, such as a motor, which does not use brushes.

brute force An approach that relies on power at the expense of efficiency and ingenuity. For instance, in attempting to crack a password, this method may involve trying every possible combination of characters. An alternate approach might incorporate statistical analysis and sophisticated algorithms. Nonetheless, a secure encryption system must, among other things, have enough possible combinations to make certain that a brute-force attack is not feasible. Also called **brute-force technique**.

brute-force attack An approach to trying to break encryption or crack passwords utilizing **brute force**. Also called **brute-force search**.

brute-force filter A type of power pack filter that relies on capacitance and/or inductance values much larger than necessary to perform its function.

brute-force search Same as **brute-force attack**.

brute-force technique 1. Same as **brute force**. 2. Same as **brute-force attack**.

brute supply A power supply that is unregulated, that is, having no circuitry to maintain output voltage constant when the input line or load varies. Also called **unregulated power supply**.

BSC Abbreviation of **binary synchronous communication**.

BSC protocol Abbreviation of **binary synchronous communications protocol**.

BSOD Abbreviation of **blue screen of death**.

BSS 1. Abbreviation of **broadcasting-satellite service**. 2. Abbreviation of **Blind Source Separation**.

BT-cut crystal A piezoelectric crystal cut at an approximate -49° angle with respect to the **z**-axis of the mother crystal. The natural vibration frequency of such crystals does not vary much as a function of temperature.

BTB Same as **B2B**.

bubble chart A diagram in which labeled bubbles, usually circular or oval, are used to represent the components of a system which are linked with arrows and/or lines between them to depict their functional relationships. Such diagrams may show data flow, hardware and software interconnections, and so on.

bubble memory A type of non-volatile computer memory that utilizes materials, such as crystals, to retain a magnetic polarity. This polarity determines whether ones or zeroes are being stored, and with the assistance of permanent magnets, also serves to retain the stored information in case power is cut off. Used primarily for rugged computing circumstances. Also called **bubble storage**, or **magnetic bubble memory**.

bubble sort A sorting technique in which adjacent pairs of items within a list are sequentially compared and ordered, until the entire list has been evaluated. This process is then repeated as many times as required to obtain the correct order. Also called **exchange sort**, or **sifting sort**.

bubble storage Same as **bubble memory**.

bucket In computers, a storage location which may contain more than one record, and which can be referenced as a whole.

bucket-brigade device A semiconductor device in which charges are transferred point-to-point in, sequence, by charge carriers. Used, for instance, as a delay line. Its abbreviation is **BBD**.

bucking coil A coil that connected and positioned in a manner that its magnetic field opposes that of another coil.

bucking voltage A voltage whose polarity is at all times the opposite of that of another voltage.

buckminsterfullerene The most common and stable form of fullerene found to date, consisting of 60 atoms arranged in a manner that resembles a soccer ball. Buckminsterfullerenes are capable of enclosing other atoms, and have applications in various areas, including superconduction and medicine. Also called **buckyball**.

buckyball Same as **buckminsterfullerene**.

buckytube Fullerene that is arranged into a cylindrical shape. It is many times stronger and much lighter than steel, and depending on various factors may be a semiconductor, conductor, or superconductor. Used, for instance, as conducting paths between circuit components on a nanometer scale. Also called **nanotube**.

buddy Within an instant messaging service, a user from which authorization has been obtained, so that the requesting user can see when they are logged-on, in addition to being able to exchange text messages or files, engage in voice or video conferencing, and so on.

buddy list A list of **buddies**, which may comprise friends, colleagues, family members, and so on.

buffer **1.** A segment of computer memory utilized to temporarily store information that awaits transfer or processing. Used, for instance, to compensate for differences in operating speeds. Also called **buffer memory**, **buffer storage (2)**, or **input buffer**. **2.** A circuit, device, or component which helps prevent undesirable interactions between other circuits, devices, or components. **3.** An amplifier which isolates successive stages to help prevent undesirable interactions between them. Also called **buffer amplifier**, or **isolation amplifier**.

buffer amplifier Same as **buffer (3)**.

buffer capacitor A capacitor which suppresses voltage surges that might otherwise damage other circuit components.

buffer flush **1.** To clear all or part of a **buffer (1)**. **2.** To move the data contained in a **buffer (1)** to a more permanent medium, such as a hard disk.

buffer memory Same as **buffer (1)**.

buffer pool A group of memory locations designated as **buffers (1)**.

buffer storage **1.** A segment of computer storage used to temporarily hold information that awaits transfer. **2.** Same **buffer (1)**.

buffer underrun When recording to certain optical discs, such as a CD-R discs, a problem resulting from the buffer holding the data waiting to be recorded being empty. CD-R discs must be recorded using an uninterrupted stream of data, and a buffer is utilized to assist with this requirement. If said buffer empties for any reason, the write process fails and the disc is likely ruined.

buffering **1.** The addition of data to a **buffer (1)**. **2. Buffering (1)** occurring prior to the initiation of streaming multimedia content being accessed remotely. This helps, for instance, to avoid pauses when viewing said content. **3.** The use and/or effect of a **buffer**.

bug **1.** A persistent error or defect in computer software or hardware. Said especially of one which causes malfunctions. **2.** A defect present in a circuit, device, component, or the like, which causes malfunctions. Also called **electronic bug (1)**. **3.** An electronic device utilized to surreptitiously listen to and/or record conversations. Also called **electronic bug (2)**. **4.** A semiautomatic telegraph key for producing correctly spaced dots and dashes. Also called **electronic bug (3)**.

bug-compatible A version of software and/or hardware which has one or more of same **bugs (1)** as a previous version.

bug fix A change in software and/or hardware intended to remedy a known bug. When not done properly, however, fixing one bug may cause a new one.

bug fix release A software release which remedies one or more known bugs. When provided, software publishers usually hype some trivial enhancement to make it seem like an upgrade. In addition, such releases often introduce new bugs themselves.

bug-for-bug compatible Same as **bug-compatible**, except that special measures have been taken to insure the presence of bugs known to afflict previous versions.

bugging The planting, or use, of a **bug (3)**.

build A version of a program which is still undergoing testing. Usually designated by a combination of letters and numbers, such as 1.103b. Sometimes refers to a completed version of a program.

build-up Same as **buildup**.

building block An independent and self contained unit, or module, used within the **building-block approach**.

building-block approach An approach to the design of equipment, systems, programs, and the like, in which independently prepared and self-contained units, or modules, are combined to form the final product. This allows for simultaneously developing and testing the various components, and also helps break down a large and complex system into more manageable parts. Also called **building-block principle**, **building-block technique**, **building-block design**, **modular approach**, or **modular design**.

building-block design Same as **building-block approach**.

building-block principle Same as **building-block approach**.

building-block programming Computer programming utilizing **building blocks**. All units, or modules, must incorporate certain common parameters, such as having compatible interfaces. This facilitates collaborative efforts to expand and improve current product offerings. Object-oriented programming evolved from this approach.

building-block technique Same as **building-block approach**.

building-out circuit A short section of a transmission line that is shunted across another transmission line for purposes of impedance matching and tuning. Also called **building-out section**.

building-out section Same as **building-out circuit**.

buildup Also spelled **build-up**. The gradual process of increasing a quantity or magnitude, such as a voltage or magnetic field.

built-in Incorporated or otherwise constructed within a larger unit in a manner which makes something essentially non-detachable.

built-in font A font which is permanently stored into a printer's memory. Also called **internal font**, or **resident font**.

Built-In Self Test A testing method in which software and/or hardware is incorporated into a product, such as an IC, enabling self-testing. It's acronym is **BIST**.

bulb 1. A glass housing which encloses the elements of electric lamps, electron tubes, and similar devices. A bulb may or may not be evacuated. Also called **envelope (1)**. 2. An incandescent lamp.

bulk The majority of something, such as its mass or volume.

bulk acoustic wave An acoustic wave traveling through the body of a piezoelectric material, such as quartz.

bulk effect An effect occurring in the bulk of a material, as opposed to a specific region, such as the surface. For example, that affecting the overall body of a semiconductor material, but not the junction.

bulk eraser A device which utilizes a strong magnetic field to erase all information from a recordable magnetic medium, such as a reel of magnetic tape or a floppy disk. In the case of a reel of magnetic tape, for instance, the use of such a device avoids the need for having to run the entire tape past the erase head. Also called **degausser (2)**, or **demagnetizer (2)**.

bulk storage A storage medium, such as an optical disc or a tape cartridge, with a large capacity.

bulletin board Same as **bulletin board system**.

bulletin board service Same as **bulletin board system**.

bulletin board system A computer system that serves as a message center that is accessed by other computers through a modem. These usually focus on specific interests, supply information, and generally provide a forum to exchange thoughts, files, programs, and email. Its abbreviation is **BBS**. Also called **bulletin board, bulletin board service**, or **electronic bulletin board system**.

bunch 1. A group of similar things that form a cluster. 2. A group of electrons that form a cluster. Also called **electron bunch**.

buncher Same as **buncher cavity**.

buncher cavity In a velocity-modulated tube, such as a klystron, the input cavity resonator, where electron velocities are modulated so that they form a **bunch (2)**. Also called **buncher, buncher resonator**, or **input resonator**.

buncher grid In a klystron, one of the grids that modulates electron velocities in a **buncher cavity**.

buncher resonator Same as **buncher cavity**.

buncher voltage The RF voltage between **buncher grids**. Also called **bunching voltage**.

bunching In a velocity-modulated tube, such as a klystron, the forming of **bunches (2)**. Also called **electron bunching**.

bunching voltage Same as **buncher voltage**.

bundled cable Two or more cables bound together and installed as a single unit.

bundled software A combination of application programs which are tailored to a given type of work, and which function especially well together. For example an office suite incorporating word processing, database, spreadsheet, presentation, and communications programs. Also called **software package (1)**, **suite**, **package (4)**, or **application suite**.

Bunsen cell An electrolytic cell with a carbon cathode in a solution of nitric acid, and a zinc anode in a solution of sulfuric acid. It has a potential of approximately 1.9 volts.

burden The power drawn from the circuit connected to a secondary winding. Usually expressed in volt-amperes at a given power factor.

Bureau International des Poids et Mesures An international organization dedicated to the uniformity of measurements, with emphasis on their adhering to the International System of Units (SI). It is French for International Bureau of Weights and Measures. Its abbreviation is **BIPM**.

buried layer In a semiconductor material, a higher-conducting layer between two less conducting layers.

burn 1. An imperfection on the screen of a CRT due to damage to the phosphor coating at that location. 2. To write code or data onto a PROM chip using a PROM blower. Also known as **blow (2)**. 3. To write onto an optical medium which can be recorded only once, such as a CD-R or DVD-R. 4. To write onto a recordable optical medium, such as a CD, CD-ROM, or DVD.

burn-in 1. Same as **break-in (1)**. 2. A **burn (1)** produced as a consequence of the same text and/or images being presented on a CRT screen for an extended period. Currently, few monitors are afflicted this problem.

burn-out Same as **burnout**.

burnout A failure, or the partial or complete destruction of a component, circuit, or device, due to excessive current and the resulting overheating. Also spelled **burn-out**.

burst 1. An abrupt and intense increase. 2. An abrupt increase in the strength of a signal. 3. A signal resulting from an abrupt and intense increase in a magnitude. 4. In a TV receiver, a color-synchronizing signal at the beginning of each scanning line, which establishes a frequency and phase reference for the chrominance signal. Also called **color burst, color-sync signal**, or **reference burst**. 5. A block of data which is transferred as a unit, and without interruption.

burst EDO DRAM Abbreviation of **burst** extended data output dynamic **RAM**. A type of EDO DRAM which processes four memory addresses in each of its short bursts synchronized with the CPU. Its own abbreviation is **BEDO DRAM**. Also called **burst EDO RAM**, or **BEDO RAM**.

burst EDO RAM Same as **burst EDO DRAM**.

burst error An abrupt increase in data errors. May be due, for instance, to damaged equipment.

burst extended data output dynamic RAM Same as **burst EDO DRAM**.

burst generator A generator of signals whose strength abruptly increases. Used for testing components, devices, and equipment, such as filters or speakers.

burst mode Same as **burst transmission (2)**.

burst speed Same as **burst transmission (3)**.

burst transmission 1. A transmission of data at a speed much higher than average. 2. A transmission in bursts, as opposed to a steady flow. Also called **burst mode**. 3. A transmission at the highest speed that a device is capable of uninterruptedly transmitting. Also called **burst speed**.

bus 1. A set of conductors which serve as a channel which provides parallel data transfer from one part of a computer to another. An example is the local bus, which is the pathway between the CPU, the memory, and high-speed peripherals. A bus is subdivided into two parts: the address bus, which identifies specific locations of stored data, and the data bus, which transfers the data. The width of the bus determines the amount of data which can be transferred at a time. A 256-bit bus, for example, can transfer 256 bits at a time. The clock speed determines how fast these transfers take place. Also called **computer bus**. 2. In a communications network, a common cable, or wire, that connects all nodes. Also called **network bus**. **3.** A heavy and rigid conductor, which is normally non-insulated, that is utilized to carry large currents between several circuits. Also called **bus bar**. **4.** One or more conductors which serve as a common connection for multiple circuits.

bus architecture Also called **bus topology (1)**. **1.** The physical layout of the conductors of a **bus (1)**. **2.** The physical layout of the common cable, or wire, in a **bus (2)**.

bus bar Same as **bus (3)**.

bus bridge A device which connects two computer buses, whether they are similar or not.

bus card A card, such as a network interface card, which plugs into a computer's expansion bus.

bus configuration The physical arrangement or layout of a **bus network**.

bus extender **1.** A device which expands the capabilities of a **bus (1)**. It may be, for instance, an expansion board which provides multiple expansion slots. **2.** A board which physically distances a board in a computer from those surrounding it, to facilitate repairs and testing.

bus mastering A feature which enables a device controller to communicate directly with other devices, bypassing the CPU. This improves performance, as the CPU can tend to other tasks while data transfer is accelerated.

bus mouse A computer mouse that attaches directly to a computer's bus via an expansion card, as opposed to utilizing a port.

bus network A communications network configuration in which all nodes are connected to a common cable or wire. Also, a network with such a layout.

bus system The set of conductors which serve as a channel which provides parallel data transfer from one part of a computer to another, and the mechanisms which control these conductive paths.

bus topology **1.** Same as **bus architecture**. **2.** The topology of a **bus network**.

busbar Same as **bus (3)**.

business card CD An optical disc, such as a CD, CD-ROM, or DVD, that is in the general shape of a business card, and which is utilized for promotion. Alternatively, other shapes may be used. Also called **CD business card**.

business computing The use of computers for business purposes.

business continuity plan Its abbreviation is **BCP**. A set of procedures which are implemented once a disaster recovery plan has been executed. Once the immediate crisis, such as a fire, is under control, the BCP enumerates the key resources, such as personnel, computers, and telephones, along with the tasks that must be performed, to help ensure that the disruption in business is minimized.

business information system A system utilized for one or more processes involving the handling of the data of an enterprise. These processes may include the collection, processing, storage, retrieval, or transmission of data, and the devices, programs, channels, and equipment used for these purposes.

business machine Any electronic or electromechanical devices or equipment intended for business use. These include typewriters, calculators, and copiers. Computers and external peripherals, such as printers, may or may not be included in this concept.

business software Software that is utilized to run one or more aspects of a business. Depending on the enterprise, this may encompass a single program, an application suite, or thousands of programs.

business-to-business Same as **B2B**.

business-to-business ecommerce Same as **B2B ecommerce**.

business-to-consumer Same as **B2C**.

business-to-consumer ecommerce Same as **B2C ecommerce**.

busy hour The time interval during which power or communications usage is at a maximum. The duration is not necessarily equal to an hour, and refers more to a general time of day. Also called **busy period**, **peak period**, or **peak hour**.

busy period Same as **busy hour**.

busy signal An intermittent tone that may be heard when dialing a telephone number. There are two types: a slow busy, which indicates that the dialed number is currently in use, and a fast busy, which indicates that the central office either did not understand the dialed digits, or that a part of the network is too busy to process the call. Also called **busy tone**.

busy test A test by a telephone company to verify whether a subscriber line, or circuit, is busy. Also called **busy verification**.

busy tone Same as **busy signal**.

busy verification Same as **busy test**.

Butler oscillator A crystal oscillator with a two-stage amplifier and a piezoelectric crystal placed in a feedback loop from output to input.

butterfly capacitor A variable capacitor with stator and rotor plates which allow for varying of the inductance and capacitance, and whose shapes resemble the wings of a butterfly. It is used as a tuner in the very-high frequency and ultra-high frequency ranges.

Butterworth characteristic A response characteristic, such as that of a filter, which has minimal passband ripple. Also called **maximally flat characteristic**.

Butterworth filter A filter characterized by minimal passband ripple. Used, for example, in audio applications.

Butterworth response A response, such as that of an amplifier, which has minimal passband ripple. Also called **maximally flat response**.

button **1.** A knob, disk, or the like, which is pressed to activate or operate a circuit, device, component, machine, and so on. For instance, a button may be pressed to activate an electric circuit. A button may also be virtual, as seen when a computer mouse presses such a button on-screen by clicking the mouse. Also called **pushbutton**. **2.** A small piece of metal alloyed to a semiconductor wafer to form a junction. Also called **dot (3)**. **3.** In a carbon microphone, a container that holds carbon granules. Also called **carbon button**, or **microphone button**. **4.** On a computer mouse, a surface which is pressed, or clicked, to select objects, perform functions, or the like. Also called **mouse button**.

button battery Same as **button cell**.

button cell A small primary cell that uses an alkaline electrolyte solution, usually potassium hydroxide. Used mostly for small devices such as hearing aids and watches. Also called **button battery**, or **coin battery**.

button microphone A carbon microphone with one or two **buttons (3)**.

button switch A switch that is operated by pressing a **button (1)**. Also called **push-button switch**.

buzz **1.** A vibratory or droning sound, such as that produced by a buzzer, or resulting from certain electrical interferences. **2.** A vibratory force with controlled amplitude and frequency, which is continuously applied to a device which is servomotor driven, so as to avoid it sticking in the null position. Also called **dither (2)**.

buzzer An electromagnetic device whose armature vibrates rapidly, which produces a vibratory or droning sound. Used for producing audible signals, such as warnings.

BW Abbreviation of **bandwidth**.

BWA Abbreviation of **backward-wave amplifier**.

BWG Abbreviation of **Birmingham Wire Gauge**.

BWO Abbreviation of **backward-wave oscillator**.

bypass **1.** A low-impedance path which allows a current to flow around one or more circuits or components, as opposed to through it. This path may be intentionally or unintentionally created. Also, to create such a path. Also called **shunt (2)**. **2.** In telecommunications, avoiding the use of the local telephone company as a transmission pathway. A satellite, for instance, may serve as an alternative route.

bypass capacitor A capacitor which is utilized to provide a low-impedance path for AC around one or more circuits or components. May be used, for example, to prevent AC signals from reaching certain components.

byte A unit of data usually consisting of 8 adjacent bits, and representing a single alphabetic character, symbol, or decimal digit. It is the smallest addressable unit of computer storage. Multiples of bytes are used to describe storage and memory. For instance, a computer with 8 gigabytes of RAM, and 2 terabytes of storage on a hard disk. The term **byte** may have originally been an acronym for **binary table** or **binary term**, or an alternate spelling of "bite" to distinguish it from "bit." Its abbreviation is **B**. Also called **octet**.

byte code Same as **bytecode**.

byte-oriented protocol A communications protocol which uses bytes to represent control codes. This contrasts with a **bit-oriented protocol**, in which control codes are represented by bits. Between the two, bit-oriented is more efficient and reliable. Also called **character-oriented protocol**.

bytecode An intermediate language which is processed by a virtual machine. The virtual machine converts the instructions into machine instructions most any CPU will understand, making such code essentially architecture-independent. For example, Java bytecode is processed by a Java Virtual Machine. Also spelled **byte code**. Also called **pseudocode (3)**.

bytes per inch In computers, the number of bytes per inch of track length on a recording medium, such as a disk or tape. Its abbreviation is **Bpi**.

bytes per second A measure of the transmission speed over a communications channel, such as that in a network. It represents the average number of bytes transmitted per second. Higher multiples of bytes are more frequently used for these measurements, and examples are kilobytes per second, or gigabytes per second. Its abbreviation is **Bps**.

C

c **1.** Symbol and abbreviation of **centi-**. **2.** Symbol for **calorie**.

c Symbol for **speed of light** in a vacuum.

C **1.** Symbol for **coulomb**. **2.** Symbol for **Calorie**. **3.** Symbol for **capacitance**. **4.** Symbol for **capacitor**. **5.** Symbol for **carbon**. **6.** Symbol for **collector**. **7.** A high-level computer programming language which is highly flexible and mostly machine-independent because of its closeness to assembly language. Also called **C programming language**.

c- Abbreviation of **centi-**.

C band **1.** In communications, a band of radio frequencies extending from 4.00 to 8.00 GHz, as established by the IEEE. This corresponds to wavelengths of 7.5 to 3.75 cm, respectively. **2.** In communications, a band of frequencies extending from about 4 GHz to about 6 GHz, although the defined interval varies.

C battery A battery which supplies the bias voltage to the control grid of an electron tube. Also called **grid battery**.

C bias In an electron tube, a steady DC voltage applied to the control grid, to establish a reference level for its operation. Also called **grid bias**.

C display In radars, an oscilloscopic display which plots the bearing of the scanned object in the horizontal plane versus its angles of elevation, which appear in the vertical plane. The scanned object appears as a blip on the display. Also called **C scope**, **C scanner**, **C scan**, or **C indicator**.

C indicator Same as **C display**.

C meter Abbreviation of **capacitance meter**.

C network An electric network formed by three impedance branches in series, with the free leads connected to one pair of terminals, and the junction points being connected to another pair of terminals.

C programming language Same as **C (7)**.

C scan Same as **C display**.

C scanner Same as **C display**.

C scope Same as **C display**.

C signal Abbreviation of **chrominance signal**.

C++ An object-oriented version of **C (7)**.

CA **1.** Abbreviation of **cellular automata**. **2.** Abbreviation of **certificate authority**.

Ca Chemical symbol for **calcium**.

cabinet An enclosure within which an apparatus, such as a TV, computer, or speaker, may be housed. The composition of such an enclosure will vary, depending on the specific needs of what is being enclosed, and where it will be used.

cable **1.** One or more electric conductors bundled together and encased by a protective sheath. The contained conductors are insulated from each other. Also called **wire (3)**. **2.** One or more optical fibers bundled together and encased by a protective sheath. Generally used for telecommunications. Such cables usually consist of three layers, which are the core, its surrounding cladding, and the protective jacket. Also called **fiber-optic cable**, **optical cable**, or **light cable**. **3.** Same as **coaxial cable**. **4.** Same as **cable TV**.

cable armor A covering that protects or reinforces a cable. It is composed of one or more layers, which may or may not be of the same material. For example, a polymer-coated steel armor should protect fiber-optic cables under most conditions.

cable assembly A cable which is fitted with the appropriate connectors, and which is ready for installation.

cable attenuation Same as **cable loss**.

cable box Also called **cable converter box**, or **converter box**. **1.** A box, that connects to a TV, which decodes the incoming cable signal or otherwise enables viewing of certain channels or programming. Such a device may also connect to another device such as a VCR. Also called **converter (6)**. **2.** A box which serves to tune a TV to all available channels. Also called **converter (7)**.

cable capacitance Capacitance between the conductors of a cable. Usually measured in picofarads.

cable category One of seven categories of cables used in communications. They are denominated category 1 cable through category 7 cable, based on their capacity.

cable clamp A mechanical device which gives a cable mechanical support, provides strain relief, and absorbs vibrations that would otherwise reach it.

cable connector A fixture which attaches to either end of a cable, enabling appropriate connections between devices. An example is a DIN connector.

cable converter box Same as **cable box**.

cable delay The time a signal is delayed as it passes through a cable. This time will vary depending on factors such as the composition of the cable.

cable duct A pipe, tube, or channel through which cables are run.

cable harness A bundle of cables which is tied or otherwise attached together so as to be handled, installed, or removed as a unit.

cable head end Same as **cable headend**.

cable headend In a TV system, such as satellite TV or cable TV, the location at which signals are received, then processed, and from which they are distributed to the individual subscribers. Also spelled **cable head end**. Also called **cable TV headend**, or **headend (1)**.

cable Internet Internet service provided through a cable TV system. Such a service may plug into a cable modem, or into the TV, utilizing a specially-equipped cable box. Transmission speeds can be up to several Mbps downstream, so it is considered to be broadband. Also called **cable-modem Internet**, or **Internet cable**.

cable jacket A protective outer covering for a cable. Depending on the intended use for the cable and the operational environment, it may protect against moisture, abrasion, magnetic fields, radiation, and so on. Some cables have more than one jacket. Also called **cable sheath**.

cable joint Same as **cable splice**.

cable loss The reduction of the intensity of a signal traveling through a cable. It usually refers to a coaxial cable, and depends on variables such as the frequency of the signal and cable length. In the case of a coaxial cable, higher frequencies have greater loss than lower frequencies. Also called **cable attenuation**.

cable matcher A device which enables a cable connector to be joined with the same gender of cable connector. For instance, the connection of two phono plugs, as opposed to a phono plug being plugged into a phono jack. Also called **gender changer**.

cable modem A modem used to connect to **cable Internet**.

cable-modem Internet Same as **cable Internet**.

cable raceway A channel through which cables or wires are run, and which serves to protect them between locations where they are connected to devices and equipment. Also called **raceway**.

cable router A router, located at a cable headend, which manages the Internet traffic among the served cable-modem subscribers.

cable run 1. The length a cable runs from one terminal to another. 2. The route a cable runs from one terminal to another.

cable sheath Same as **cable jacket**.

cable splice The electrical connection between two sections of cable. Connectors may or may not be used to accomplish this. Also called **cable joint**.

cable telephony Telecommunications services through cable TV connections. Such connections are usually via coaxial cable, and the available services may include telephone, cable-modem Internet, and digital cable.

cable television Same as **cable TV**.

cable television headend Same as **cable headend**.

cable tray A specially designed channel through which cables and wires are run. A cable tray provides mechanical support, and protection which is tailored to specific needs. Such protection may include jackets and/or shields which safeguard against flames, high electrical noise, vibrations, crushing, and so on. Also called **wireway**.

cable TV Abbreviation of **cable** television. A TV broadcasting system in which signals from local and distant stations are received by one or more antennas, and relayed to the individual subscribers within the community served. The receiving antennas are usually advantageously located, and a coaxial cable is the customary means to convey the signals sent to the subscribers. Its own abbreviation is **CATV**. Also called **cable** (4), or **community-antenna TV**.

cable TV headend Same as **cable headend**.

cabling 1. One or more connections using one or more cables. For example, the cable connections within an optical fiber communications network, or those of a power grid. 2. The physical cables used in **cabling** (1).

cabling diagram A plan which shows the paths of **cabling** (1).

CAC Abbreviation of **carrier access code**.

cache Also called **cache memory**, or **memory cache**. 1. In computer memory management, a specialized high-speed storage subsystem. Cache works by storing recently and/or frequently accessed information, where it can be accessed much faster than from where it was obtained. A level 1 cache is built right into the CPU, while a level 2 cache utilizes a memory bank between the CPU and main memory. Level 1 cache is faster, but smaller than level 2 cache, while level 2 cache is still faster than main memory. Disk cache uses a section of main memory to store recently accessed data from a disk, and can dramatically speed up applications by avoiding the much slower disk accesses. Also called **CPU cache**. 2. An area of computer storage where the most recently downloaded Web pages may be temporarily placed, so that when one of these pages is returned to, it loads faster. Also called **browser cache**, **Web cache**, or **Internet cache**.

cache card An expansion card that increases the **cache** (1) of a computer.

cache coherency The synchronization of data contained in multiple caches. In this manner, the data accessed from any given cache is the most recently written. Used, for instance, in parallel processors.

cache-coherent non-uniform memory access A parallel-processing computer architecture in which the total available memory is divided into multiple shared segments, and in which there is **cache coherency**. Its abbreviation is **ccNUMA**.

cache hit A successful retrieval of data from a cache, instead of main memory or a disk. Also called **hit** (1).

cache hit rate The proportion of data accesses that are fulfilled by a cache, instead of main memory or a disk. It is the primary way to measure the effectiveness of a cache. Also called **hit rate** (1).

cache memory Same as **cache** (1).

cache miss A retrieval of data from main memory or a disk, as opposed to a cache.

cache miss rate The proportion of data accesses that are not fulfilled by a cache.

cache server A server, such as a proxy server, which caches Web pages and other material downloaded from the Internet. If a page request can be satisfied by a cache server, the response time is faster, and bandwidth is economized. Pressing the browser's refresh icon usually accesses the specific location where the desired Internet resource is located. Also called **Web cache server**.

cached DRAM Same as **CDRAM**.

caching controller A controller, such as a disk controller, which has its own memory cache.

CAD 1. Acronym for **computer-aided design**, or **computer-assisted design**. 2. Acronym for **computer-aided diagnosis**, or **computer-assisted diagnosis**.

CAD/CAM Acronym for **computer-aided design/computer-aided manufacturing**.

CADD Acronym for **computer-aided design and drafting**.

caddy 1. Same as **CD caddy**. 2. Same as **CD-ROM caddy**.

cadmium A silvery-white ductile and malleable metal with a bluish tinge, whose atomic number is 48. It has over 30 known isotopes, of which several are stable. It has many applications in electronics, including its use in electroplating, fire-protection systems, phosphors, rectifiers, and in batteries, such as those consisting of nickel cadmium cells. Its chemical symbol is **Cd**.

cadmium acetate Colorless crystals used in electroplating.

cadmium cell A standard cell used as a reference voltage source, in which the positive electrode is mercury, the negative electrode is an amalgam of cadmium and mercury, and the electrolyte is a solution of cadmium sulfate. It has a voltage of approximately 1.0186 at 20 ºC. Also called **Weston standard cell**.

cadmium chloride White crystals whose chemical formula is $CdCl_2$. It is used in vacuum tubes and in electroplating.

cadmium coating Same as **cadmium plating**.

cadmium cyanide Clear crystals whose chemical formula is $Cd(CN)_2$. It is used in electroplating.

cadmium fluoride Crystals whose chemical formula is CdF_2. It is used in phosphors and lasers.

cadmium hydroxide A white powder whose chemical formula is $Cd(OH)_2$. It is used in storage-battery electrodes and in electroplating.

cadmium iodide A white powder whose chemical formula is CdI_2. It is used in phosphors, photoconductors, and in electroplating.

cadmium oxide A colorless powder or brown to red crystals whose chemical formula is CdO. It is used in semiconductors, phosphors, electroplating, and rechargeable-battery electrodes.

cadmium plating The electrolytic deposition of cadmium onto a substrate, usually to provide it with greater resistance to oxidation and corrosion. Also called **cadmium coating**.

cadmium propionate A chemical compound used in scintillation counters.

cadmium selenide A red powder whose chemical formula is $CdSe$. It is used in semiconductors, photoconductors, phosphors, and rectifiers.

cadmium sulfate Colorless crystals whose chemical formula is $CdSO_4$. It is used in solution as the electrolyte in a Weston standard cell, in electroplating, and in phosphors.

cadmium sulfide A yellow or brown powder whose chemical formula is CdS. It is used in semiconductors, photoconductors, phosphors, and rectifiers.

cadmium telluride Brown to black crystals whose chemical formula is $CdTe$. It is used in semiconductors, X-ray screens, and phosphors.

cadmium tungstate White or yellow crystals or powder whose chemical formula is $CdWO_4$. It is used in semiconductors, photoconductors, phosphors, and rectifiers

CADS Acronym for **computer-aided dispatch system**.

CAE 1. Abbreviation of **computer-aided engineering**. 2. Abbreviation of **computer-aided education**, or **computer-assisted education**.

cage antenna A broadband dipole antenna which consists of a number of wires connected in parallel, and arranged in a manner that resembles a cylindrical cage. Such an antenna is usually center-fed.

CAI Abbreviation of **computer-aided instruction**, **computer-assisted instruction**, **computer-aided education**, **computer-aided learning**, **computer-aided teaching**, **computer-assisted learning**, or **computer-assisted teaching**.

cal Abbreviation of **calorie**.

Cal Abbreviation of **Calorie**.

CAL Abbreviation of **computer-aided learning**, or **computer-assisted learning**.

calcite A crystalline compound whose chemical formula is $CaCO_3$. Used in optical instruments, and as a phosphor.

calcium A silvery-white metal whose atomic number is 20. It has about 20 known isotopes, of which 6 are stable. It is the most abundant metallic element in the human body, and the fifth most abundant element in the earth's crust. It is used to prepare various alloys, and as a getter in vacuum tubes. Its chemical symbol is Ca.

calcium chromate A bright yellow powder whose chemical formula is $CaCrO_4$. It is used as a battery depolarizer.

calcium fluoride A white powder whose chemical formula is CaF_2. It is used in spectroscopy and lasers.

calcium molybdate A white crystalline powder whose chemical formula is $CaMoO_4$. It is used in phosphors.

calcium plumbate An orange to brownish powder whose chemical formula is Ca_2PbO_4. It is used in storage batteries.

calcium pyrophosphate A white powder whose chemical formula is CaP_2O_7. It is used in phosphors.

calcium sulfide A yellow-gray powder whose chemical formula is CaS. It is used in phosphors.

calcium titanate A chemical compound whose formula is $CaTiO_3$. It is used in ceramic capacitors.

calcium tungstate White crystals whose chemical formula is $CaWO_4$. It is used in X-ray devices and in lasers.

calculating machine Same as **calculator**.

calculator A device which performs arithmetic operations. It can only accept numerical input, which is keyed in, and displays and/or prints its results. Almost all calculators can store numbers in memory, many provide additional functions such as logarithms and square-root calculations, and some are programmable. The term usually refers to hand-held or portable devices, but these can be encountered elsewhere, for instance, as a virtual calculator on a computer screen. Also called **calculating machine**.

calendar display A display, such as that of a DVD or CD, in which all track numbers, up to a given maximum, are displayed at once, in a format similar to a common desk calendar.

calibrate To check indicated and/or measured values and make adjustments comparing with a known standard. Such a procedure may be used to make sure that the readings, settings, or values of an instrument, component, or device are correct.

calibrated Having gone through a **calibration** (1) process, to ensure high accuracy.

calibrated device A device which has been **calibrated**.

calibrated frequency An output frequency which has been **calibrated**.

calibrated instrument An instrument which has been **calibrated**.

calibrated measurement A measurement which has been **calibrated**.

calibrated meter A meter which has been **calibrated**.

calibrated reading A reading which has been made with a **calibrated instrument**.

calibrated scale A scale which has been **calibrated**.

calibrated signal An output signal which has been **calibrated**.

calibration 1. The process of checking indicated and/or measured values and making adjustments while comparing with a known standard. Such a procedure may be used to make sure that the readings, settings, or values of an instrument, component, or device are correct. 2. The gradations of a scale, such as that of a thermometer that reads to tenths of degrees.

calibration accuracy The level of accuracy achieved through **calibration** (1).

calibration curve A graph depicting the relationship between the indicated or measured values of an instrument or device, versus the correct values.

calibration marker On a CRT screen, especially that of a radar, electronically generated marks that indicate a numerical scale of any given parameters, such as bearing, time, or range.

calibration standard A standard utilized for **calibration** (1).

calibrator A device used for **calibration** (1).

californium A synthetic radioactive metallic chemical element whose atomic number is 98. It has about 20 known isotopes, all of which are unstable. Its applications include its use as a source of neutrons. Its chemical symbol is Cf.

call 1. In computer programming, a statement or instruction which invokes a subroutine or another program. Control of the program is temporarily transferred during this call, and subsequently returned after the subroutine or other program completes its task. 2. In communications, the establishing of a connection between stations. 3. In communications, the sending of a signal indicating the desire to establish a connection with another station. 4. In communications, the operations necessary to establish, maintain, and conclude a connection.

call attempt In communications, an attempt to establish a connection between stations, whether it is successful or not.

call barring A service offered by telephone companies, which enables a user to restrict outgoing and/or incoming calls. Used, for instance, to preclude the making of certain international calls, or to prevent incoming calls when roaming.

call duration A time interval starting the instant both ends of a call are off-hook, and ending when one end terminates the call.

call forwarding A service offered by telephone companies, which enables calls to automatically be rerouted from one telephone line to another. Calls may be sent to an alternate location, an answering service, or the like. Some systems have more tailored offerings, such as selective call forwarding for programmed numbers, call forwarding only when the line is busy, or forwarding if there is no answer after a specified number of rings. Also known as **forwarding (2)**.

call hour A unit, used for measuring telephone communications traffic, which equals 3600 call seconds, 36 centum call seconds, or one erlang.

call instruction An instruction to invoke **call (1)**.

call letters Same as **call sign**.

call return A service offered by telephone companies and/or phones equipped with this feature, which enables a subscriber to dial the last incoming call. A sequence of digits is usually dialed to activate this feature. Frequently, this service continues to dial the desired number if it is initially busy, and alerts when it is available. Some incoming calls may not allow for call returning. Also known as **last number redial**, or **return call**.

call second A unit, used for measuring telephone communications traffic, which equals 1 call lasting 1 second long. Ten calls lasting 5 seconds is equal to 50 call seconds, the same as 5 calls lasting 10 seconds. 100 call seconds equal 1 centum call second, and 3600 call seconds equal 1 call hour.

call sign The letters and/or numbers which serve to identify a broadcast station. Call letters are assigned by licensing authorities, such as governmental agencies. Also known as **call letters**.

call trace Same as **call tracing**.

call tracing A service offered by telephone companies, which enables a subscriber to trace the last incoming call. This is generally done to confirm the location from which harassing calls are received. A sequence of digits is usually dialed to activate this feature, unless the service provides for tracing of all calls. Some incoming calls may not allow for call tracing. Also known as **call trace**.

call waiting A service offered by telephone companies, which provides a signal such as a beep heard over a line that is already in use, to indicate that there is an incoming call. This feature may be deactivated for occasions when not desired, such as during a fax transmission.

callback A security measure in which a network verifying the user name and password of a caller terminates a received call, then places a new call to the terminal associated with that user name and password. This procedure serves as an additional safeguard in case a user name and password are stolen.

callback modem A modem which upon receiving a call asks for the phone number of the caller, terminates the connection, and calls the user back. This enables verification of the user's authorization for access, in addition to tracking usage.

called routine The routine which is invoked by a **call (1)**.

called tone Same as **CED tone**.

caller ID Abbreviation of **caller id**entification. A service offered by telephone companies, which sends the name and number of the calling party of an incoming call. A display device which features caller ID is needed to view this information. Alternatively, a caller ID may only provide the number of the calling party. Also called **caller number delivery**, or **automatic number identification**.

caller identification Same as **caller ID**.

caller number delivery Same as **caller ID**.

calling name delivery A **caller ID** service in which both the name and number of calling parties are displayed. Its abbreviation is **CNAM**.

calling point The point at which a **call (1)** is invoked.

calling routine A routine or subroutine which initiates a **call (1)**.

calling sequence The steps followed to prepare for a **call (1)**, to execute it, and to return control to the program that initiated the call, so that it can resume from the calling point.

calling tone Same as **CNG tone**.

calorescence The production of visible light through heating with infrared radiation. The produced light is a result of heat, and not by a change in wavelength.

calorie A unit of heat energy. Its abbreviation is **cal**, and its symbol is **c**. One calorie is equal to the heat required to raise one gram of water one degree Celsius, at a pressure of one atmosphere. It is equal to approximately 4.186 joules. One thousand calories equal one kilocalorie, and one kilocalorie is equal to one Calorie. Please note that in the case of the larger **Calorie**, the term's first letter is capitalized to distinguish it from the smaller calorie. Often the two terms are confused because of this minor, and often ignored, difference.

Calorie A unit of heat energy. Its abbreviation is **Cal**, and its symbol is **C**. One Calorie is equal to the heat required to raise one kilogram of water one degree Celsius, at a pressure of one atmosphere. It is equal to approximately 4.186 kilojoules. One Calorie is equal to one thousand calories, and one Calorie is equal to one kilocalorie. Please note that the first letter of this term is capitalized to distinguish it from the smaller **calorie**. Often the two terms are confused because of this minor, and often ignored, difference. Also called **large calorie**.

calorimeter A device which measures differences in heat quantities which accompany chemical and physical changes. Such a device can measure RF energy by calculating the heating effect such energy has on a given medium.

CAM 1. Acronym for **computer-aided manufacturing**. 2. Acronym for **content-addressable memory**. 3. Acronym for **common-access method**.

camcorder Acronym for **cam**era **recorder**. A portable device combining a video camera and a videocassette recorder. These come in various video formats, including VHS, S-VHS, and 8 mm.

camera 1. A device which converts images into electric signals. 2. A device that converts images formed by lenses into electric signals. A photosensitive surface in the contained camera tube serves as the transducer which converts the optical image into electric video signals suitable for broadcasting, recording, or the like. Also known as **TV camera**, **video camera (1)**, or **telecamera**. 3. A device which focuses light from a viewed image onto a light-sensitive material, such as film, so that an image is recorded.

camera cable A cable which carries the video signal from a TV camera to the control equipment.

camera chain A TV camera plus the additional equipment necessary to convey the video signal to the control room. This additional equipment includes amplifiers, a power source, a monitor, and the camera cable.

camera recorder Same as **camcorder**.

camera signal The output electric video signals from a **camera (2)**.

camera tube An electron tube within a TV camera which serves as the transducer that converts the optical image into electric video signals. Also known as **TV camera tube**, **pickup (3)**, or **pickup tube**.

camp Same as **camp-on**.

camp-on A service offered by telephone companies, which enables a caller encountering a busy signal to be placed on hold on the busy line, so that the instant that line becomes available the call is connected. Also called **camp**.

Campbell bridge An AC bridge utilized for comparing mutual inductances.

campus-wide information system A computer-based information retrieval system used in an educational institution, such as a university. Useful, for instance, in course selection, or for accessing required readings. Its abbreviation is **CWIS**.

can A metal container which serves to enclose a component. For example, the outer casing of an electrolytic capacitor.

CAN Abbreviation of **cancel character**.

cancel 1. To diminish, annul, or omit. **2.** To match in force or effect, especially when diminishing, annulling, or omitting.

cancel character In data transmission, a control character indicating that the associated text should be canceled. Its abbreviation is **CAN**.

candela The fundamental SI unit of luminous intensity. Formerly, it was defined in various manners, ranging from the light of a certain type of candle under particular conditions, to more recently as a given proportion of the radiating power of a blackbody under specific conditions. It is currently defined as the luminous intensity of a light source emitting monochromatic radiation in a given direction at a frequency of 540 terahertz, with a power of 1/683 watt per steradian in the specified direction. Its symbol is **cd**. This term was formerly known as **candle**, **international candle**, **standard candle**, or **new candle**.

candle Former name of **candela**.

candle power Same as **candlepower**.

candlepower Luminous intensity expressed in **candelas**. Also spelled **candle power**. Its abbreviation is **cp**.

cannibalize To remove one or more components from equipment which is in working order, to use for repairing or operating other equipment.

cannot find server An error 404 resulting from the server not being available. Also called **server not found**.

CAP Abbreviation of **carrierless amplitude phase**.

capacimeter Abbreviation of **capacitance meter**.

capacitance The ability to store electric charge between conductors which are separated by a dielectric material, when a potential difference exists between said conductors. Its value is expressed as the ratio of stored electric charge to the potential difference, and is expressed in farads, although in practicality this unit is so large that it is more common to use microfarads or picofarads. Its symbol is **C**. Formerly, this concept was known as **capacity (4)**.

capacitance box A device containing an assembly of capacitors and switches, enabling precise adjustment of the capacitance at the terminals.

capacitance bridge A bridge which enables measuring or comparing a capacitance against a known capacitance. An example is a Schering bridge.

capacitance filter A filter consisting solely of a capacitor. Such a filter tends to smooth voltage variations because a capacitor resists discharging instantaneously. Used, for instance, in audio applications.

capacitance meter A direct-reading meter for measuring capacitance of circuits or capacitors. Its abbreviation is **capacimeter**, or **C meter**.

capacitance relay A relay which responds to very small variations in capacitance, similar to those produced by a human body, or a part of it, such as a hand. Such a device may be used, for instance, as an intrusion alarm. Also called **proximity relay**.

capacitance standard A capacitor whose capacitance value is precisely known, and which serves to measure or compare unknown capacitances. May be used, for instance in a Schering bridge. Also called **standard capacitance**.

capacitance temperature coefficient A numerical value that indicates the relationship between a change in the capacitance of a component, device, material, or area, as a function of temperature. Also called **temperature coefficient of capacitance**.

capacitive Pertaining to **capacitance**.

capacitive coupling The coupling of two or more circuits by means of a **capacitance**. Also called **electrostatic coupling**.

capacitive diaphragm A metal plate placed inside a waveguide to introduce capacitive reactance at the transmitted frequency.

capacitive-discharge ignition An automotive ignition system in which energy is stored in a capacitor, and is then discharged across the gap of a spark plug at the proper intervals.

capacitive feedback The return of part of the energy from the output of a circuit back to its input by means of a common capacitance.

capacitive load A load in which current leads voltage. The capacitive reactance exceeds the inductive reactance in such a load. Also called **leading load**.

capacitive loudspeaker Same as **capacitor speaker**.

capacitive post A post placed inside a waveguide to introduce capacitive reactance.

capacitive reactance The opposition to the flow of AC by the capacitance of a capacitor or circuit. Its symbol is X_C, it is measured in ohms, and its formula is: $X_C = 1/2\pi fC$, where π is pi, f is frequency in hertz, and **C** is capacitance in farads.

capacitive speaker Same as **capacitor speaker**.

capacitive transducer A transducer consisting of two capacitor plates, one fixed and the other flexible or movable, which converts changes in applied voltage into movement. Also called **electrostatic transducer**.

capacitive tuning The tuning of a circuit by adjusting a variable capacitor.

capacitive voltage divider A voltage divider formed by two or more capacitors.

capacitor A component which has **capacitance**. The conductors separated by the dielectric material are known as electrodes or plates. The value of the capacitance will vary depending on the size, shape, and composition of the conductors, the distance between them, and on the dielectric. In addition to storing charge, a capacitor blocks DC, and may be fixed or variable. There are various types, including mica, electrolytic, metallized paper, and MOS capacitors. Formerly known as **condenser (3)**.

capacitor amplifier An amplifier circuit whose active component is a ferroelectric capacitor whose capacitance varies as a function of the applied voltage. Also called **dielectric amplifier**.

capacitor antenna An antenna which utilizes the capacitance between two conductors or systems of conductors. Also called **condenser antenna**.

capacitor bank A group of capacitors connected in series and/or parallel.

capacitor color code A system utilizing markings, such as color dots and bands, to indicate capacitor information such as values, multipliers, and tolerance.

capacitor filter A power-supply filter in which a capacitor is connected to the output of the rectifier. That is, the input component of the filter is a capacitor. Also called **capacitor-input filter**.

capacitor-input filter Same as **capacitor filter**.

capacitor loudspeaker Same as **capacitor speaker**.

capacitor microphone A microphone in which a flexible metal diaphragm and a rigid plate form a two-plate air capacitor. As the incident sound waves make the flexible plate vibrate, the capacitance is varied, and this causes variations in the electric current, which in turn can be converted in an audio-frequency signal. Also called **condenser microphone**, or **electrostatic microphone**.

capacitor motor **1.** A single-phase induction motor whose main winding is connected to an AC power source, and whose auxiliary winding is connected in series with a capacitor to the same power source. **2.** Same as **capacitor-start motor**.

capacitor plate One of the electrodes in a capacitor. Also called **plate (2)**.

capacitor speaker A loudspeaker in which a flexible metal diaphragm and a rigid plate form a two-plate air capacitor. The mechanical vibration of its thin sound-producing diaphragm is produced by electrostatic fields produced between it and a rigid plate, each with a high audio voltage applied to them. Also called **capacitor loudspeaker**, **capacitive speaker**, **capacitive loudspeaker**, **condenser speaker**, or **electrostatic speaker**.

capacitor-start motor A **capacitor motor (1)** in which the auxiliary winding and capacitor are energized only during starting. Once the motor has reached a determined speed, the auxiliary winding and capacitor are disconnected. Also called **capacitor motor (2)**.

capacitor start-run motor A **capacitor motor (1)** in which the auxiliary winding and capacitor are energized at all times. Also called **permanent-split capacitor motor**.

capacitor voltage The voltage between the terminals of a capacitor.

capacitors in parallel Two or more capacitors connected in parallel. The total capacitance is the sum of each of the individual capacitances, thus, the multiple capacitors act like a single larger capacitor. Also called **parallel capacitors**, or **shunt capacitors**.

capacitors in series Two or more capacitors connected in series. The total capacitance is the reciprocal of the sum of the reciprocal of the value of each of the individual capacitances. Also called **series capacitors**.

capacity **1.** The maximum amount that can be contained, accommodated, or handled. **2.** The ability of a battery or cell to supply current during a given time interval, usually expressed in ampere-hours. **3.** The amount of data that a computer or peripheral device can process, store, or transmit. **4.** A seldom-used term for **capacitance**.

capillary electrometer An electrometer which utilizes a capillary tube containing dilute sulfuric acid and mercury. The displacement within the tube of the interface between the sulfuric acid and the mercury is proportional to the current being measured. Also called **Lippmann electrometer**.

Caps lock Same as **Caps lock key**.

Caps lock key A computer keyboard key that toggles between upper and lower case. Such a key usually works only on alphabetical characters. Also called **Caps lock**.

capstan On a magnetic tape drive, such as that in a tape deck or computer tape drive, a motorized rotating shaft which moves the tape at a constant speed.

captioning Closed-captioning which does not require a decoder to display the text and symbols on-screen.

capture The acquiring, receiving, absorbing, or attracting and holding of an entity. For instance, the receiving of data, or the acquiring of a neutron by a nucleus.

capture buffer An area in a computer's memory where incoming data is stored for a communications program.

capture effect **1.** The tendency of one effect dominating over other, lesser effects. **2.** In an FM receiver, the effect occurring when the stronger of two signals, of equal or nearly equal frequency, totally suppresses the other.

capture ratio In an FM tuner, the minimum ratio between the intensities of two signals at the same frequency which enables the tuner to reject the weaker signal. Expressed in decibels.

CAR Acronym for **computer-assisted retrieval**.

car phone **1.** A cellular telephone installed in a land vehicle, such as a car or truck. Such phones may or may not be removable from the vehicle, and usually have an external antenna. Also called **mobile phone (2)**. **2.** A hand-held cellular telephone which can be placed in a cradle in a land vehicle such as a car or truck. The cradle can serve to charge the phone, and many of them allow the user to engage in conversations without holding the phone. Also called **mobile phone (3)**.

carbolic acid A white or colorless solid whose chemical formula is C_6H_5OH, and which is used, for instance, in thermosetting plastics. Also called **phenol**.

carbon A nonmetallic chemical element whose atomic number is 6. It has the highest melting point of any known element, and occurs in several allotropic forms, including diamond, graphite, charcoal, and fullerene. It is present in all known life forms on this planet, and there are more carbon compounds than those of all other chemical elements combined. It has over a dozen known isotopes, of which 2 are stable. Its applications in electronics include its use in electrodes, resistors, lamps, and microphones. Its chemical symbol is **C**.

carbon arc **1.** An electric arc between two carbon electrodes. **2.** An electric arc between two electrodes, one of which is composed of carbon.

carbon-arc lamp An arc lamp whose electrodes are composed of carbon. Also called **carbon lamp**.

carbon brush A brush composed of carbon, or another material that incorporates carbon. Such a brush may be made of amorphous carbon, or carbon and copper, and may be used in motors, generators, or potentiometers.

carbon button In a carbon microphone, the container that holds the carbon granules. Also called **button (3)**, or **microphone button**.

carbon-composition resistor Same as **carbon resistor**.

carbon copy In email, indicative of a message which is being delivered to one or more additional recipients. All recipients of the message are aware of those receiving a carbon copy. Its abbreviation is **cc**. Also called **courtesy copy**.

carbon dioxide laser A gas laser whose tube contains carbon dioxide as the main active gas. The applications of these powerful lasers include their use in cutting and welding, communications, and surgery.

carbon disulfide A faint yellow liquid whose chemical formula is CS_2. It is used in the production of vacuum tubes.

carbon electrode An electrode composed of a carbon, sometimes with a small amount of a metal added. Used, for instance, in arc welding.

carbon-film resistor A resistor prepared by depositing a carbon film onto a ceramic substrate.

carbon granules Fragments of carbon in various shapes, whose diameter or length across is roughly a few millimeters. Used, for instance, in carbon microphones. Also called **granular carbon**.

carbon lamp Same as **carbon-arc lamp**.

carbon microphone A microphone in which a flexible diaphragm reacts in response to sound waves, and in a proportional manner applies pressure against one or two containers filled with carbon granules. This produces fluctuations in the resistance of the granules, which varies the current passing through them. Thus, the detected sound waves modulate the current, which can then be amplified for sound recording and/or reproduction.

carbon monoxide laser A gas laser in which carbon monoxide is the active gas. The uses for these lasers include cutting, welding, and surgery. Its abbreviation is **CO laser**.

carbon pile A variable resistor consisting of a stack of carbon disks within a metal plate. As the pressure applied to the stack varies, so does its resistance. The pressure may be modified, for instance, by turning a knob.

carbon-pile pressure transducer A device which indicates changes in pressure by monitoring the changes in resistance of a **carbon pile**.

carbon-pile regulator Same as a **carbon-pile voltage regulator**.

carbon-pile rheostat A rheostat which utilizes a **carbon pile**.

carbon-pile voltage regulator A voltage regulator employing a **carbon pile**. Also called **carbon-pile regulator**.

carbon resistor A resistor composed of carbon particles mixed with a ceramic binder which is formed into a desired shape, then has leads attached to each end. Also called **carbon-composition resistor**.

carbon tetrachloride A colorless liquid whose chemical formula is CCl_4. It is used in the production of semiconductors.

carbon-zinc cell A primary cell in which the positive electrode is carbon, and the negative electrode is zinc. It may be a wet cell or dry cell. When it is a dry cell with an ammonium chloride electrolyte, it is also called **Leclanché cell**.

carborundum A very hard substance, usually composed of silicon carbide, which may be used in semiconductors and thermistors.

carcinotron An oscillator utilizing a special vacuum tube in which an RF magnetic field interacts with the electron beam that travels from cathode to anode. This produces electron bunching, which in turn produces a backward wave. This oscillator may be tuned over a wide range of frequencies, and its output power is extracted near the electron gun. Also called **carcinotron oscillator**, or **backward-wave oscillator**.

carcinotron oscillator Same as **carcinotron**.

card 1. A circuit board which is plugged into the bus or an expansion slot of a computer, in order to add functions or resources. Common examples include graphics accelerators, sound cards, and Ethernet adapters. Also known by various other terms, including **adapter card, expansion card, add-on board, accessory card**, and **adapter (3)**. 2. Same as **circuit board**. 3. A card which holds 80 or 96 columns of data, each representing one character, used by computers with card readers. This is a practically obsolete storage medium. Also called **punched card**.

card cage An enclosure which holds **cards**.

card column In a **card (3)**, a column of data.

card-edge connector A set of wide contacts which protrude from a circuit board, and which serve to be inserted into another circuit board. Used, for instance, to plug an expansion board into an expansion slot. Also called **edge connector**.

card feed In a card system, a device which inserts the cards into the receiving machine.

card punch A device which punches holes in **cards (3)**. Also called **keypunch**.

card reader 1. A device which reads information which has been encoded onto a magnetic stripe on a plastic card, such as a credit card, or driver's license. Also called **magnetic card reader**. 2. A device which reads **cards (3)**. Also called **punched-card reader**.

card row In a **card (3)**, a row of data.

card slot A slot within a computer into which a **card (1)** or **card (2)** is plugged in.

card system A computer system in which **cards (3)** are used for input and output. Such systems are practically obsolete. Also called **punched-card system**.

card-to-disk conversion The conversion of data stored on a **card (3)** to a computer disk.

card-to-tape conversion The conversion of data stored on a **card (3)** to a computer tape.

card verifier A device which verifies the accuracy of data stored on a **card (3)**.

cardiac pacemaker An electrical device which delivers small impulses to the heart under specified conditions, and at a predetermined rate, so as to promote its pumping blood at a regular pace. Such a device may or may not be implanted. Also called **pacemaker, heart pacer**, or **pacer**.

cardiogram A graphical recording of the variations in the electrical activity of the heart. A **cardiograph** is used to detect and record such activity. Its abbreviation is **ECG**, or **EKG**. Also called **electrocardiogram**.

cardiograph An instrument which detects and records variations in the electrical activity of the heart. A **cardiogram** is the graphical recording of this activity. Its abbreviation is **ECG**, or **EKG**. Also called **electrocardiograph**.

cardiography The technique employed to obtain an **electrocardiogram**. Cardiography utilizes electrodes applied to various parts of the body, which detect the electrical signals produced by the heart. Also called **electrocardiography**.

cardioid diagram A graphical representation of a **cardioid pattern**.

cardioid microphone A directional microphone whose pickup pattern is shaped more or less like a heart. In one direction it has a nearly uniform response, while in the opposite direction it has almost none.

cardioid pattern A directivity pattern shaped approximately like a heart. In one direction there is nearly uniform response and/or radiation, while in the opposite direction there is almost none.

caret On a computer keyboard, a key (^) which serves, for instance, to indicate a power a number is raised to. For example, 6^6 is the same as 6^6.

careware Software which is freely distributed, and whose author suggests a donation to charity if the user is so inclined.

Carey-Foster bridge A modification of a Wheatstone bridge, which is used to measure resistances which are nearly identical.

carousel A circular conveyor upon which objects rest or are rotated. For example, a CD or DVD changer in which the disks rest on a rotating tray.

carousel CD player Abbreviation of **carousel** compact disc **player**. A CD player which utilizes a **carousel**. A CD carousel is usually utilized when a few discs are available, so a CD jukebox is not likely to use this method to hold discs awaiting to be swapped, so as to save space.

carousel compact disc player Same as **carousel CD player**.

carousel DVD player A DVD player which utilizes a **carousel**. A DVD carousel is usually utilized when a few discs are available, so a DVD jukebox is not likely to use this method to hold discs awaiting to be swapped, so as to save space. Also called **DVD carousel player**.

carousel player A player, such as a carousel CD or DVD player, which utilizes a **carousel**.

carpal-tunnel syndrome Tingling, numbness, and/or pain that affects a hand and which may radiate into the arm. This can ultimately lead to decreased dexterity and strength in said hand, and is due to the compression of a nerve which travels through the wrist. It is a type of repetitive-strain injury, and years of improper positioning of the wrist while typing at a keyboard is one cause. Its abbreviation is **CTS**.

carrier 1. Same as **carrier wave**. 2. Same as **charge carrier**. 3. Same as **common carrier**.

carrier access code A code which an end user of a telecommunications service dials to select the service of a given carrier. In North America, such codes are in the form of 101XXXX, where the XXXX portion is the carrier identification code. Its abbreviation is **CAC**.

carrier amplifier An amplifier which amplifies a carrier wave or a carrier current.

carrier beat The result of **carrier beating**, such as patterns on a received fax.

carrier beating Unwanted signals resulting from the mixing of two carrier signals of slightly different frequencies. May result, for instance, in undesired patterns in a received fax, or within a TV image.

carrier channel 1. The lines and equipment which serve to send a carrier current. Also called **carrier-current channel**. 2. The path through which carrier waves travel. 3. A path along which carrier waves can travel.

carrier chrominance signal Same as **chrominance signal**.

carrier color signal In color TV, the signal which is added to the monochrome signal, for the transmission of color information.

carrier concentration Same as **charge-carrier concentration**.

carrier current 1. AC which is modulated, in order to transmit information. 2. The current component of a carrier wave.

carrier-current channel Same as **carrier channel (1)**.

carrier density Same as **charge-carrier concentration**.

carrier detect A signal a modem sends to the equipment it is connected to, such as a computer, indicating that it has sensed a carrier signal. This signifies that it is ready to receive data. Its abbreviation is **CD**.

carrier deviation Same as **carrier swing**.

carrier dropout A temporary loss of a carrier signal.

carrier frequency 1. The frequency of a carrier wave. Such a frequency is generated by an unmodulated radio transmitter, and can be modulated in frequency, amplitude, or phase, in order for it to carry information. 2. The center frequency of a transmitted carrier wave. Also called **center frequency (1)**.

carrier-frequency range The continuous interval of carrier frequencies within which a transmitter may operate.

carrier identification code A four digit code which identifies a telecommunications carrier in North America. It consists of the last four digits of the carrier access code, which is in the form of 101XXXX, thus the XXXX portion is the carrier identification code. Its abbreviation is **CIC**.

carrier injection The introduction of charge carriers into a semiconductor. Also, the process utilized. Also called **injection (2)**.

carrier leak In suppressed-carrier transmission, the residual carrier remaining after suppression.

carrier level The power level of an unmodulated carrier wave, at any given point, in relation to a reference power level. Usually expressed in decibels.

carrier lifetime In a semiconductor material, the average time a charge carrier exists before recombination.

carrier line A transmission line used for transmitting a carrier signal.

carrier loss A loss of a carrier signal.

carrier mobility The average drift velocity of charge carriers in a semiconductor material, per unit electric field. Also called **drift mobility**, or **mobility**.

carrier modulation The process of modifying a characteristic of a carrier wave by an information-bearing signal, as occurs, for instance, in AM, FM, or PM. Also, the result of such modulation. For example, in FM, the instantaneous frequency of a sine-wave carrier is varied above and below the center frequency by an information-bearing, or modulating, signal. Also called **modulation (2)**, or **RF modulation**.

carrier noise In an RF signal, noise produced by variations of a carrier in the absence of any intended modulation. Also known as **residual modulation (1)**, or **incidental modulation (1)**.

carrier noise level In an RF signal, the level of **carrier noise**.

carrier-operated device antinoise Its acronym is **codan**. 1. A device which mutes a receiver unless there is an incoming carrier signal having a given intensity. 2. A device which mutes a receiver unless there is an incoming signal.

carrier oscillator In a single-sideband receiver, the RF oscillator that supplies the necessary carrier wave.

carrier power Same as **carrier power output**.

carrier power output The power an RF transmitter supplies to the transmission line of an antenna when no modulating signal is present. Also called **carrier power**.

carrier power-output rating The power an RF transmitter supplies to the normal-load circuit or its equivalent when no modulating signal is present.

carrier repeater Equipment which amplifies the power level of a carrier signal, without increasing the noise level significantly. Such equipment includes one or more amplifiers, filters, equalizers, and so on. A carrier repeater may be unidirectional or bi-directional.

carrier sense multiple access An Ethernet protocol in which a station attempts to verify if the network is busy, before transmitting. Its abbreviation is **CSMA**.

carrier sense multiple access with collision avoidance An Ethernet protocol similar to **carrier sense multiple access with collision detection**, but with the added feature of sending a brief signal indicating its intention to transmit before doing so. Its abbreviation is **CSMA/CA**.

carrier sense multiple access with collision detection An Ethernet protocol for handling the situation in which two stations transmit simultaneously. When this occurs, this collision of signals is detected, and to avoid another collision, each stops transmitting and waits a different, random interval before attempting to retransmit. Its abbreviation is **CSMA/CD**.

carrier shift 1. In AM, the unwanted variation of average carrier power, resulting in imperfect modulation. 2. In communications utilizing frequency-shift modulation, shifting the carrier frequency to differentiate between a mark and a space. Shifting in one direction indicates a mark, while shifting in the other indicates a space.

carrier signal A signal which is modulated in frequency, amplitude, or phase, in order for it to carry information. For instance, an AM radio transmitter modulates the amplitude of a carrier signal. Also called **signal carrier**.

carrier signaling In multichannel carrier transmission, a technique which uses carrier waves for signaling functions such as ringing and dialing.

carrier storage The accumulation of charge carriers in a semiconductor device such as a transistor. Also called **charge storage (2)**.

carrier suppression Also called **suppressed-carrier transmission**. 1. A mode of transmission in which the carrier wave is suppressed after carrier modulation, in which case only one or both sidebands are transmitted. The carrier wave is restored at the receiving terminal for demodulation. 2. A mode of transmission in which the carrier wave is suppressed when there is no modulating signal.

carrier swing For a frequency or phase-modulated carrier wave, the total deviation, ranging from the lowest to the highest instantaneous frequencies. Also called **carrier deviation**.

carrier system A communications system in which a number of channels simultaneously travel over a single path by modulating each channel using a different carrier frequency and then demodulating at the receiving terminal to restore each signal to its original form.

carrier telegraphy Telegraphy in which a carrier wave is modulated for transmitting signals.

carrier telephony Telephony in which a carrier wave is modulated by an audio-frequency signal, for transmission over wire lines.

carrier terminal The equipment at each end of a carrier transmission system. This equipment performs functions such as modulation, amplification, and demodulation. Also called **carrier terminal equipment**.

carrier terminal equipment Same as **carrier terminal**.

carrier-to-noise ratio The ratio of the RF carrier level to the RF noise level, before modulation. Usually expressed in decibels. Its abbreviation is **CNR**.

carrier transmission Transmission of a signal which is the result of the modulation of a carrier wave. If the carrier wave is suppressed after modulation, it is called **suppressed-carrier transmission**.

carrier voltage The voltage component of a carrier wave.

carrier wave An electromagnetic wave which is intended to be modulated by an information-bearing signal. Such a wave may be amplitude, frequency, or phase modulated. Used in RF transmissions such as those of TV, radio, telephone, or satellite. Also called **carrier (1)**. Its abbreviation is **CW**.

carrierless amplitude phase A modulation technique similar to quadrature amplitude modulation. It is used in DSL, but discrete multitone is currently preferred. Its abbreviation is **CAP**.

carry 1. In arithmetic, the transferring of a digit to another column. For example, moving a digit to the column of the next higher position when the sum of the digits in a column exceeds the base of the number system. 2. In computers and counters, a signal, such as a carry flag, which indicates that there is a **carry (1)**.

carry bit A bit which indicates that there is an arithmetical operation with a **carry (1)**.

carry flag A signal or circuit which indicates that there is an arithmetical operation with a **carry (1)**.

carry time The time required for a computer or counter to perform a **carry (1)** operation.

carrying capacity The maximum amount of current that a conductor, component, or device can handle safely.

Cartesian axis Each of the mutually perpendicular lines which intersect at a common point called the origin, and which is used within a **Cartesian coordinate system**. When representing three planes, for instance, each axis is designated by the letters **x**, **y**, and **z**, respectively. Also called **rectangular axis**.

Cartesian-coordinate robot A robot whose movements are along Cartesian axes. For example, a manipulator may be able to move along the **x**, **y**, and **z** axes. Used in robotic applications such as assembly, gluing, or arc welding. Also called **rectangular-coordinate robot**. Its abbreviation is **CCR**.

Cartesian coordinate system A coordinate system in which the locations of points are given in reference to the axes, numbering two or more. Each of these coordinate axes is perpendicular to the others, and all intersect at a common point called the origin. Each coordinate axis, or Cartesian axis, represents a plane called a Cartesian plane. A number of lines, equal to the number of planes, may be drawn from the origin to any given point on a plane or in space, and the exact location of this point can be specified by giving the coordinates with respect to the axes. Also called **rectangular coordinate system**. Its abbreviation is **CCS**.

Cartesian coordinates Each of the sets of numbers which locate a point on a plane or in space, according to a **Cartesian coordinate system**. If there are two planes, for instance, then a point is located by specifying the position in reference to each of the axes, one that is usually horizontal and denominated **x**, and one that is usually vertical and denominated **y**. Also called **rectangular coordinates**.

Cartesian manipulator A robotic manipulator whose movements are along Cartesian axes. For example, a manipulator may be able to move along the **x**, **y**, and **z** axes. Used in applications such as assembly, gluing, or arc welding.

Cartesian plane A coordinate plane whose points are specified by **Cartesian coordinates**.

cartridge 1. A case or container which holds a device or substance which is intended to be plugged into a larger piece of equipment. For instance, such a container enclosing ink, a magnetic tape, or a printed circuit. 2. A removable computer storage device which contains a magnetic or optical disk. For instance, a floppy disk. Also called **disk cartridge**. Also, such a device utilizing a chip or another storage medium. 3. A container, which houses a length of magnetic media, that can be plugged into the appropriate device in a manner that eliminates the need for directly handling the contents. For instance, a videotape, or a continuous-loop tape cartridge. 4. In a phonograph, an electromechanical transducer whose stylus vibrates as it follows the grooves of a record, and which converts these vibrations into electric signals. There are various types, including moving coil and piezoelectric, and each attaches to a phonograph arm. Also called **phonograph cartridge**, **phonograph pickup**, **phono cartridge**, or **pickup (2)**.

cartridge fuse A fuse whose fusible element is contained in an insulating enclosure, so as to minimize damage to any components immediately surrounding it. Such fuses have ferrules on each end for plug-in connection.

CAS Abbreviation of **collision-avoidance system**.

cascade 1. A connected series of two or more images, stages, elements, or devices which interact or function in sequence. A cascading menu is an example. 2. Same as **cumulative ionization**. 3. The connection of two or more circuits or devices in a manner that the output of one is the input of the next. Seen, for example in a cascade amplifier. Also called **tandem**.

cascade amplifier An amplifier with two or more stages arranged in a manner that the output of one serves as the input for the next, while amplifying at each step. Also called **cascaded amplifier**, or **multistage amplifier**.

cascade arrangement Same as **cascade connection**.

cascade connection An arrangement or connection which forms a **cascade**. Also called **cascade arrangement**.

cascade control An automatic control system in which the control units are linked in a chain where each unit controls the succeeding one. Also called **piggyback control**.

cascade image tube An image tube with various sections, in which the output image of one section serves as the input for the next. Used, for instance, for detection of low levels of light.

cascade limiter A limiter circuit which has two or more stages arranged in a manner that the output of one serves as the input for the next. Also called **double limiter**.

cascade noise The noise in a cascade amplifier after a signal has proceeded through all the amplifying stages.

cascade star topology In LANs, two or more star networks which are connected to each other, each linked by their hub. Also called **cascaded stars**.

cascade transformer A transformer which incorporates two or more step-up transformers.

cascaded Arranged or connected in a manner which forms a cascade.

cascaded amplifier Same as **cascade amplifier**.

cascaded stars Same as **cascade star topology**.

cascading 1. The connection or arrangement of two or more devices or elements, in a manner that they function in sequence, or that the output of one is the input of the next. 2. In an electrical power system, successive failures stemming from an incident at a given location. This may result in widespread service interruption.

cascading menu In a computer program, a pull-down menu which leads to various choices, some of which may provide further selection options, and so on.

cascading style sheets An HTML feature which enables the applying of style elements such as headers, fonts, and links to any Web page, separately from its specific content. This facilitates providing related Web pages with a similar style, and also allows for general modifications by changing these style sheets instead of the individual Web pages. Its abbreviation is **CSS**.

cascading windows On a computer screen, a group of windows arranged in a manner that they overlap one another. The title bars are generally visible, allowing a user to know which windows are open. Also called **overlaid windows**.

cascode Same as **cascode amplifier**.

cascode amplifier An amplifier circuit with a grounded-emitter input stage followed by a grounded-base output stage. It features high gain and low noise. Used, for instance, in TV tuners. Also called **cascode**, or **cascode circuit**.

cascode circuit Same as **cascode amplifier**.

CASE Acronym for **computer-aided software engineering**, **computer-assisted software engineering**, **computer-automated software engineering**, **computer-aided systems engineering**, and **computer-automated systems engineering**.

case-based reasoning An expert system which solves problems based on cases which encompass previous experiences that are relevant to the task at hand. Such an approach is best suited for problems that are narrow in scope, but which have many layers.

case insensitive Indicative of a program that does not distinguish between uppercase and lowercase characters. For instance, on the Internet, most email addresses may be written in any combination of uppercase and lowercase letters. This contrasts with **case sensitive**, where such distinctions are made.

case sensitive Indicative of a program that distinguishes uppercase and lowercase characters. For instance, in a text search where the capitalization of words and/or letters is specified to pinpoint the desired results. Another example is seen on the Internet, where a uniform resource locator may require certain characters to be upper and/or lower case to access the correct Web address. This contrasts with **case insensitive**, where such distinctions are not made.

case-sensitive search A database search where uppercase and lowercase characters are distinguished, and must be correctly specified to obtain the desired results.

case sensitivity The characteristic of being **case sensitive**.

CASE technology A software technology which enables **CASE**.

CASE tools The resources available to software engineers utilizing **CASE technology**. Such tools include automatic coding and intelligent agents.

Cassegrain antenna A microwave antenna in which the feed radiator, mounted at or near the surface of the main reflector, is directed towards a small convex reflector at the focal point of the dish. The small reflector then redirects the signal back to the dish, which in turn produces the desired forward beam.

Cassegrain feed The feed system used in a **Cassegrain antenna**.

cassette 1. A flat compact case containing magnetic tape and two floating reels, one for supply, the other for take-up. It is designed for easy insertion and removal from devices intended for their use. Utilized, for instance for audio or videotapes, or for data backup. Also called **tape cassette**. 2. A lightproof housing for photographic film or plates.

cassette deck A magnetic tape player intended for recording and playing back audio cassettes. Also called **cassette player (2)**, or **cassette recorder (2)**.

cassette player 1. A magnetic tape player intended for playing back audio cassettes. 2. Same as **cassette deck**.

cassette recorder 1. A magnetic tape player intended for recording audio cassettes. 2. Same as **cassette deck**.

cassette tape The tape in a **cassette (1)**.

castellations In a chip package such as a leadless ceramic chip carrier, metallic pads which are utilized for the interconnection of conductive surfaces. Castellations are used instead of leads consisting of metal prongs or wires, and are usually flush with the package or recessed.

castor oil A pale yellow viscous oil used in electrical insulation compounds and in capacitors. Also called **ricinus oil**.

CAT 1. Acronym for **computerized axial tomography**, **computed axial tomography**, **computer axial tomography**, **computer-aided tomography**, and **computer-assisted tomography**. 2. Acronym for **computer-aided teaching**, or **computer-assisted teaching**. 3. Acronym for **computer-assisted training**, or **computer-aided training**. 4. Acronym for **computer-aided testing**, or **computer-assisted testing**. 5. Acronym for **computer-aided translation**, or **computer-assisted translation**.

CAT 1 cable Abbreviation of **category 1 cable**.

CAT 2 cable Abbreviation of **category 2 cable**.

CAT 3 cable Abbreviation of **category 3 cable**.

CAT 4 cable Abbreviation of **category 4 cable**.

CAT 5 cable Abbreviation of **category 5 cable**.

CAT 6 cable Abbreviation of **category 6 cable**.

CAT 7 cable Abbreviation of **category 7 cable**.

CAT scan Same as **computed tomography**. Abbreviation of **computerized axial tomography scan**.

CAT scanner Same as **CT scanner**. Abbreviation of **computerized axial tomography scanner**.

cat whisker Also spelled **catwhisker**. A thin and flexible wire used to make electric contact on the surface of a semiconductor

catalog 1. A directory, list, or index of files or other data stored in a computer or computer application, such as a database. A catalog includes information such as the name, size, location, or the like, of said files or data. 2. To enter data into a **catalog (1)**.

catalyst 1. A substance that accelerates a chemical reaction significantly, and which is not itself consumed in the process. By the end of the reaction, a catalyst is returned to its original state. An example is the platinum utilized in a fuel cell. 2. Any agent which speeds up a process significantly.

cataphoresis Migration, toward the cathode, of charged particles suspended in a medium under the influence of an electric field produced by electrodes immersed in said medium. When migration is toward the anode, it is called **anaphoresis**.

catastrophic failure A sudden, unexpected, and complete failure of a component, device, program, or system. This contrasts with a **degradation failure**, where failure occurs after a gradual deterioration.

catcher In a velocity-modulated tube, such as a klystron, the cavity from which the output is taken. It is located where maximum electron bunching occurs. Also called **catcher cavity**, or **output resonator**.

catcher cavity Same as **catcher**.

catching diode A diode connected in a manner that it short-circuits any time its anode becomes positive relative to its cathode. This occurs as a result of its cathode attracting more electrons than its anode.

category 1 cable Unshielded twisted pair wire used for transmission of audio frequencies. Used, for instance, for telephone communications. Its abbreviation is **CAT 1 cable**.

category 2 cable Unshielded twisted pair wire used for digital or data transmissions at speeds up to about 1.5 Mbps, or about 1.5 MHz. Its nominal impedance is roughly 100 ohms. Used, for instance, for digital telephone communications. Its abbreviation is **CAT 2 cable**.

category 3 cable Shielded or unshielded twisted pair wire used for digital or data transmissions at speeds up to about 16 Mbps, or about 16 MHz. Its nominal impedance is roughly 100 ohms. Used, for instance, LANs. Its abbreviation is **CAT 3 cable**.

category 4 cable Shielded or unshielded twisted pair wire used for digital or data transmissions at speeds up to about 20 Mbps, or about 20 MHz. Its nominal impedance is roughly 100 ohms. Used, for instance, LANs. Its abbreviation is **CAT 4 cable**.

category 5 cable Shielded or unshielded twisted pair wire used for digital or data transmissions at speeds up to about 100 Mbps, or about 100 MHz. Its nominal impedance is roughly 100 ohms. Used, for instance, for ATM communications. Its abbreviation is **CAT 5 cable**.

category 6 cable Cabling used for digital or data transmissions at speeds up to about 250 Mbps, or about 250 MHz. Used, for instance, for ATM communications. Its abbreviation is **CAT 6 cable**.

category 7 cable Cabling used for digital or data transmissions at speeds up to about 1000 Mbps, or about 1000 MHz. Used, for instance, for ATM communications. Its abbreviation is **CAT 7 cable**.

catenate Same as **concatenate**.

catenation Same as **concatenation**.

cathode Also called **negative electrode**. **1.** The electrode which is the source of electrons in an electron tube. These electrons travel towards the anode. Its symbol is **K**. **2.** The negative electrode in an electrolytic cell. When a current is passed through the cell, positive ions travel towards it. **3.** The electrode where reduction occurs in an electrochemical cell.

cathode bias In an electron tube, obtaining of the correct bias utilizing a dropping resistor instead of an external bias voltage.

cathode-coupled amplifier A cascade amplifier in which a common-cathode resistor couples energy between stages.

cathode coupling The utilization of input or output elements in a cathode to couple energy between stages.

cathode current In an electron tube, the current flowing from the cathode.

cathode dark space In a glow-discharge tube, a narrow dark space near the cathode, between the cathode glow and the negative glow. Also called **Crookes dark space**, **Hittorf dark space**, or **dark space (2)**.

cathode drop In a glow-discharge tube, a potential difference due to a space charge near the cathode. Also called **cathode fall**.

cathode efficiency For a given electrochemical process, the ratio of the current theoretically required at the cathode, to the actual amount needed.

cathode emission **1.** The process by which electrons are emitted from a cathode. **2.** The electrons released by a cathode.

cathode fall Same as **cathode drop**.

cathode follower An electron-tube amplifier in which the anode is at ground potential at the operating frequency. The input is applied between the control grid and ground, and the output is taken from between the cathode and ground. Such an amplifier features high input impedance, low output impedance, and a wide frequency response with little phase distortion. Also known as **common-anode amplifier**, **grounded-plate amplifier**, or **grounded-anode amplifier**.

cathode glow In a glow-discharge tube, a luminous region near the cathode, between the Aston dark space and the Crookes dark space. The electrons in this region have the necessary velocity to excite the gas in the tube.

cathode modulation In an electron tube, amplitude modulation obtained by varying the voltage to the cathode proportionally to the fluctuations in the modulating wave.

cathode ray A stream of electrons emitted by the cathode of a gas-discharge or vacuum tube, by a region immediately surrounding a cathode, or that emitted by a heated filament in certain electron tubes. Among other things, and depending on the circumstances, cathode rays directed onto a solid material may produce X-rays, phosphorescence, or heat. A CRT, for instance, can use focused electron beams directed onto a phosphor screen to produce images displayed by a TV. Please note that the term for a stream of electrons emitted from a radioactive chemical element is **beta rays**, not cathode rays.

cathode-ray oscillograph A **cathode-ray oscilloscope** which incorporates a device which produces a permanent record of what is displayed.

cathode-ray oscilloscope An instrument which uses a CRT to produce visible patterns of one or more varying electrical quantities, or nonelectrical quantities, such as acoustic waves, with the assistance of a transducer. A cathode-ray oscilloscope typically displays variations in voltage plotted versus time, and may be used, for instance, to monitor signals such as brain waves. Its abbreviation is **CRO**, or **scope (1)**. Also called **oscilloscope (1)**.

cathode-ray output The visible output of a **CRT**, such as the images seen on a computer or TV screen.

cathode-ray television tube A **CRT** used in a TV, for viewing the received images. Also called **picture tube**, or **kinescope**.

cathode-ray tube Same as **CRT**.

cathode-ray tube computer monitor Same as **CRT computer monitor**.

cathode-ray tube display Same as **CRT display**.

cathode-ray tube display device Same as **CRT display device**.

cathode-ray tube monitor Same as **CRT monitor**.

cathode-ray tube screen Same as **CRT screen**.

cathode-ray tube terminal Same as **CRT terminal**.

cathode rays Same as **cathode ray**.

cathode spot In a gas tube, an area of the cathode from which an arc originates.

cathode sputtering Also called **cathodic sputtering**. **1.** In a vacuum tube, the emission of particles from a cathode which is disintegrating as a consequence of bombardment by high-energy ions. Also called **sputtering (4)**. **2.** The use of

cathode sputtering (1) to deposit a thin film of a metal onto a surface such as glass, metal, or plastic. Also called **sputtering** (5).

cathodic Pertaining to, or relative to a **cathode**.

cathodic current In an electron tube, the current flowing from the cathode.

cathodic polarization A change in the potential of a cathode resulting from the effects of current flow at or near the surface of said cathode.

cathodic protection The reduction or prevention of corrosion of a metallic material by making it the cathode in a conductive medium that has a sacrificial anode. This conductive medium may occur naturally, or can be applied externally. Used, for instance, in certain pipelines, bridges, and other metal structures requiring corrosion protection over an extended period of time. Also called **electrolytic protection**.

cathodic sputtering Same as **cathode sputtering**.

cathodofluorescence Fluorescence occurring under bombardment by cathode rays.

cathodoluminescence Luminescence occurring when a metal in an electron tube is bombarded by cathode rays. The wavelength of the light emitted is characteristic of the bombarded metal.

cathodophosphorescence Phosphorescence produced when a metal in an electron tube is bombarded by cathode rays.

catholyte In an electrolytic cell, the portion of the electrolyte in the immediate vicinity of the cathode.

cation A positively charged ion. Cations collect at the cathode when subjected to an electric potential while in solution. Sodium ion and potassium ion are two common cations. Also called **positive ion**.

CATV Abbreviation of **cable TV**.

catwhisker Same as **cat whisker**.

caustic soda A white chemical compound whose formula is **NaOH**. Its applications in electronics include its use in electroplating and etching. Also called **sodium hydroxide**.

caustic soda cell A cell which utilizes an aqueous solution of **caustic soda** as its electrolyte. An example is a Lalande cell.

caustic soda electrolyte An electrolyte which consists of an aqueous solution of **caustic soda**.

caution label A label placed in a visible location to inform of precautions that must be taken, such as proper handing procedures. Caution labels may be utilized, for instance, to advise of high-voltage installations, or to point out components, devices, and equipment which require special handling to avoid electrostatic-discharge damage.

CAV Abbreviation of **constant angular velocity**.

CAVE Acronym for **Computer Automatic Virtual Environment**, or **Computer Automated Virtual Environment**.

cavity 1. A hole, enclosure, or hollow area within a body. 2. Same as **cavity resonator**.

cavity coupling The removal of electromagnetic energy from a **cavity resonator** through a coupling, such as an aperture.

cavity filter A microwave filter that incorporates a **cavity resonator**.

cavity frequency meter An instrument which incorporates a **cavity resonator** to measure microwave frequencies. Also called **cavity wavemeter**.

cavity magnetron A magnetron whose anode has several **cavity resonators**. Used, for instance, in a rising-sun magnetron.

cavity oscillator An oscillator whose frequency is determined by a **cavity resonator**.

cavity radiation The radiated energy from a **cavity radiator**.

cavity radiator An enclosure which is sealed, except for a small opening through which radiant energy may enter or escape. The radiation of such an enclosure approximates that of a blackbody. Also called **radiator** (4).

cavity resonance 1. The resonance, at microwave frequencies, occurring in a **cavity resonator**. 2. Vibration of a speaker baffle at its resonant frequency. If this occurs in the audible range, such a frequency will be unduly accentuated.

cavity resonator An enclosure which is able to maintain an oscillating electromagnetic field when suitably excited. The geometry of the cavity determines the resonant frequency. Used, for instance, in lasers, klystrons, or magnetrons, or as a filter. Also known by various other names, including **cavity** (2), **resonant cavity**, **resonant chamber**, **resonating cavity**, **microwave cavity**, **microwave resonance cavity**, **waveguide resonator**, and **rhumbatron**.

cavity tuning The adjusting of the interior geometry of a **cavity resonator**, which in turn changes its resonant frequency. Used for, instance, for tuning an oscillator or amplifier.

cavity wavemeter Same as **cavity frequency meter**.

CB 1. Abbreviation of **citizens' band**. 2. Abbreviation of **circuit breaker**.

Cb 1. Chemical symbol for **columbium**. 2. Within the YCbCr color model, one of the two color-difference signals. The other is **Cr** (2).

CBE Abbreviation of **computer-based education**.

CBGA Abbreviation of **Ceramic Ball Grid Array**.

CBI Abbreviation of **computer-based instruction**.

CBL Abbreviation of **computer-based learning**.

CBR 1. Abbreviation of **computer-based reference**. 2. Abbreviation of **constant bit rate**. 3. Abbreviation of **cosmic background radiation**.

CBT 1. Abbreviation of **computer-based teaching**. 2. Abbreviation of **computer-based training**.

cc 1. Abbreviation of **courtesy copy**, or **carbon copy**. 2. Abbreviation of **closed-captioning**. 3. Abbreviation of **closed-captioned**.

CCC Abbreviation **clear-channel capability**, or **clear-channel coding**.

CCCS Abbreviation of **current-controlled current source**.

CCD Abbreviation of charge-coupled device. A semiconductor device with a matrix of many minute photosensitive elements which convert optical images into electric charges. Each sensing element, which corresponds to an image pixel, generates a charge which varies proportionally to the intensity of the light received. These charges are then passed on individually and in rows, so as to form a continuous analog signal. Analog-to-digital converters convert this analog input into a digital output, and the obtained information can be stored or otherwise processed. Used for machine vision, computer memory, digital and TV cameras, and in telescopes which are able to detect single photons of light.

CCD array Abbreviation of charge-coupled device **array**. A photosensitive array within a **CCD**.

CCD camera Abbreviation of charge-coupled device **camera**. A camera which utilizes a **CCD** as its image sensor.

CCD detector Abbreviation of charge-coupled device **detector**. Same as **CCD image sensor**.

CCD image sensor Abbreviation of charge-coupled device **image sensor**. An image sensor consisting of a **CCD**. Also called **CCD detector**, or **CCD sensor**.

CCD imaging Abbreviation of charge-coupled device **imaging**. The creation of images using a **CCD**.

CCD memory Abbreviation of charge-coupled device **memory**. Computer memory that utilizes **CCDs** for storage and retrieval.

CCD scanner Abbreviation of charge-coupled device **scanner**. A scanner utilizing a **CCD** as its image sensor.

CCD sensor Abbreviation of charge-coupled device **sensor**. Same as **CCD image sensor**.

CCD spectrometer Abbreviation of charge-coupled device **spectrometer**. A spectrometer which incorporates a **CCD** array.

CCD telescope Abbreviation of charge-coupled device **telescope**. A telescope which utilizes a **CCD** as its image sensor.

CCD TV camera Abbreviation of charge-coupled device **television camera**. A TV camera which utilizes a **CCD** as its image sensor.

CCD video camera Abbreviation of charge-coupled device **video camera**. A video camera which utilizes a **CCD** as its image sensor.

CCFL Abbreviation of **cold-cathode fluorescent lamp**.

CCFT Abbreviation of **cold-cathode fluorescent tube**.

CCIR 601 Former name of the **ITU-R BT.601** standard.

CCIS Abbreviation of **common-channel interoffice signaling**.

ccNUMA Abbreviation of **cache-coherent non-uniform memory access**.

CCR Abbreviation of **Cartesian-coordinate robot**.

CCS 1. Abbreviation of **Cartesian coordinate system**. 2. Abbreviation of **common-channel signaling**. 3. Abbreviation of **centum call second**.

CCTV Abbreviation of **closed-circuit TV**.

CCTV monitor Abbreviation of **closed-circuit TV monitor**.

CCTV signal Abbreviation of **closed-circuit TV signal**.

CCTV surveillance Abbreviation of **closed-circuit TV surveillance**.

CCTV system Abbreviation of **closed-circuit TV system**.

CCTV transmission Abbreviation of **closed-circuit TV transmission**.

CCVS Abbreviation of **current-controlled voltage source**.

ccw Abbreviation of **counterclockwise**.

cd Symbol for **candela**.

Cd Chemical symbol for **cadmium**.

CD 1. Abbreviation of compact **disc**. A digital optical storage medium, usually 12 centimeters in diameter, whose contained information is encoded in microscopic pits on its metallic surface, which is protected by a plastic layer. Since there is no contact between the pickup and the recorded surface, wear is minimized, and the protective layer helps avoid reading errors due to dust or minor marks on the surface of the disc. To recover the recorded content, a laser is focused on the metallic surface of the disc, and the reflected light is modulated by the code on said disc. Compact discs can be used to store any type of digitally encoded information, and are used especially for music, computer multimedia content, or movies. A music CD, for instance, can usually hold up to 74 minutes of recorded sound, a CD-ROM can store up to around 680 megabytes, and the capacity of a DVD is up to 17 GB. 2. Abbreviation of **carrier detect**.

CD audio Abbreviation of compact disc **audio**. Reproduction of audio in which CDs store the music, voices, and sounds. Also called **CD digital audio**.

CD burner Abbreviation of compact disc **burner**. Same as **CD recorder** (1).

CD business card Abbreviation of compact disc **business card**. An optical disc, such as a CD, CD-ROM, or DVD, that is in the general shape of a business card, and which is utilized for promotion. Alternatively, other shapes may be used. Also called **business card CD**.

CD caddy Abbreviation of compact disc **caddy**. A plastic case that holds an optical disc, such as a CD, CD-ROM, or DVD. The caddy is placed into the drive with the disc inside. Nearly all CD drives now accept only the CD, which is placed on the tray of the drive. Also called **caddy** (1).

CD carousel player Abbreviation of compact disc **carousel player**. Same as **carousel CD player**.

CD changer Abbreviation of compact disc **changer**. A CD player that holds more than one CD. Discs may be swapped, but only one can be read at once.

CD-DA Abbreviation of compact disc digital audio. Same as **CD audio**.

CD digital audio Abbreviation of compact disc **digital audio**. Same as **CD audio**.

CD drive A computer drive that accepts CDs or CD-ROMs.

CD-E Abbreviation of compact disc-erasable. Same as **CD-RW**.

CD-erasable Abbreviation of compact disc-erasable. Same as **CD-RW**.

CD-extra Abbreviation of compact disc-extra. A CD format which contains both audio and data, separated into two sessions. The first session contains up to 98 audio tracks, while the second session has one data track. Such discs may be played in audio CD players or a computer CD drive. An audio CD player will ignore the second session, while a computer CD drive can read both sessions. Also called **CD-plus**, **enhanced CD**, or **enhanced CD-ROM**.

CD-G Abbreviation of compact disc plus graphics. An audio CD format that adds text and graphics. May only be used with specially-equipped CD players.

CD-I Abbreviation of compact disc-**interactive**. A CD format which can store multimedia content on high-capacity discs. Such discs will only work with CD-I players, and such players incorporate their own microprocessor. Also called **CD interactive system**.

CD-I player A device which can read the information stored on a **CD-I**.

CD interactive Same as **CD-I**.

CD interactive system Same as **CD-I**.

CD jukebox Abbreviation of compact disc **jukebox**. A CD changer which can hold many, up to several thousand, discs. Discs may be swapped, but only one can be read at once.

CD player Abbreviation of compact disc **player**. A device which can read the information stored on a **CD** (1). To recover the recorded content, a laser is focused on the metallic surface of the disc, and the reflected light is modulated by the code on said disc. Computer CD-ROM players can also serve as CD players when an audio CD is placed in the drive. Also called **CD system**.

CD-plus Same as **CD-extra**.

CD-R Abbreviation of compact disc-**recordable**. A CD-ROM technology that allows a CD to be recorded. A CD recorder is used for this purpose, and such discs can only be written on once.

CD read-only memory Abbreviation of compact disc **read-only memory**. Same as **CD-ROM**.

CD-recordable Same as **CD-R**.

CD recorder Abbreviation of compact disc **recorder**. 1. A device used to record CD-ROMs. A laser records the desired content, forming the equivalent of pits on the recordable surface by altering the reflectivity of a dye layer. When such a disc is placed in a player, its laser pickup reads the patterns as if they had been permanently stamped into its metallic surface. Also called **CD burner**, **CD writer** or **CD-ROM recorder**, or **CD-ROM burner**. 2. A device used to record CDs.

CD-rewritable Abbreviation of compact disc-**rewritable**. Same as **CD-RW**.

CD rewritable disc A CD which can be rewritten many times. Its abbreviation is **CD-RW disc**. Also called **rewritable CD (2)**.

CD-ripper Software that copies data from an audio CD to a computer hard drive. Also called **ripper (2)**.

CD-ROM Abbreviation of compact disc read-only memory. **1.** A compact disc format which can be used to store any type of digitally encoded information, especially computer multimedia content. Discs using this format can store up to around 680 megabytes of information, or up to 74 minutes of recorded sound. These discs can not be rewritten, and a CD-ROM drive is used to read them. While audio CDs may be played on CD-ROM drives, CD-ROMs will not work with audio CD players unless they contain audio and are specifically formatted for compatibility with standard players. Computer DVD players can usually also play CD-ROMs. **2.** Same as **CD-ROM disc**.

CD-ROM burner Same as **CD recorder (1)**.

CD-ROM caddy Abbreviation of compact disc read-only memory **caddy**. A plastic case that holds a CD-ROM. The caddy is placed into the drive with the disc inside. Nearly all CD-ROM drives now accept only the disc, which is placed on the tray of the drive. Also called **caddy (2)**.

CD-ROM changer Abbreviation of compact disc read-only memory **changer**. A CD-ROM player that holds more than one CD-ROM disc. Discs may be swapped, but only one can be read at once.

CD-ROM disc A disc intended for use with a **CD-ROM drive**. Such a disc usually holds computer multimedia content. Also called **CD-ROM (2)**.

CD-ROM drive Abbreviation of compact disc read-only memory **drive**. A computer drive that reads the content stored on CD-ROM discs. The contained information is encoded in microscopic pits on its metallic surface, which is protected by a plastic layer. To recover the recorded content, a laser is focused on the metallic surface of the disc, and the reflected light is modulated by the code on said disc. Also called **CD-ROM player**, or **CD-ROM reader**.

CD-ROM extended architecture Same as **CD-ROM XA**.

CD-ROM jukebox Abbreviation of compact disc read-only memory **jukebox**. A **CD-ROM changer** that can hold many, up to several thousand, CD-ROMs. Discs may be swapped, but only one can be read at once. This contrasts with a **CD-ROM tower**, which can access multiple discs simultaneously.

CD-ROM player Same as **CD-ROM drive**.

CD-ROM reader Same as **CD-ROM drive**.

CD-ROM recorder Same as **CD recorder (1)**.

CD-ROM server A **CD-ROM tower** intended for network use.

CD-ROM tower Abbreviation of compact disc read-only **tower**. An enclosure containing multiple CD-ROM drives. All the drives are accessible at all times. This contrasts with a **CD-ROM jukebox**, which can only access one disc at a time. Used, for instance, in networks. Also called **CD tower (1)**.

CD-ROM writer Same as **CD recorder (1)**.

CD-ROM XA Abbreviation of compact disc read-only memory extended architecture. A CD-ROM format which can store multimedia content on high-capacity discs. These discs can play on computer CD-ROM drives with a special controller card.

CD-RW Abbreviation of compact disc-rewritable. A CD technology which allows discs to be rewritten many times. A laser records the desired content, forming the equivalent of pits on the recordable surface by altering the reflectivity

of a dye layer. When such a disc is placed in a player, its laser pickup reads the patterns as if they had been permanently stamped into its metallic surface. Also called **CD-erasable, CD-E**, or **rewritable CD**.

CD-RW disc Same as **CD rewritable disc**.

CD system Same as **CD player**.

CD-text Abbreviation of compact disc-text. An audio CD format which incorporates text pertaining to the title, artist, tracks, and so on. Discs in this format may be played on any CD player, but only specially-equipped devices can display the contained text.

CD tower **1.** Same as **CD-ROM tower**. **2.** An enclosure containing multiple CD drives. All the drives may or may not be accessible at all times.

CD-UDF Abbreviation of compact disc-universal data format. A file format that can be used in rewritable or recordable compact discs, that stores data in packets as opposed to doing so in a continuous stream.

CD video Abbreviation of compact disc video. A compact disc format which incorporates video, along with the audio. These may only be read by specific devices, such as a CD video player. Neither an audio CD player, nor a computer CD-ROM drive can access the information on discs in this format. Also called **video CD**. Its abbreviation is **CDV**.

CD video player Abbreviation of compact disc video player. A device which can read the information stored on a **CD video**.

CD-WO Abbreviation of compact disc-write once. A CD-ROM technology that allows a CD to be recorded once, utilizing a CD recorder. A CD-ROM player is used to recover the recorded information.

CD write once Same as **CD-WO**.

CD writer Same as **CD recorder (1)**.

CD+G Same as **CD-G**.

CDDA Same as **CD audio**.

CDDI Abbreviation of **Copper Distributed Data Interface**.

CDE Abbreviation of **Common Desktop Environment**.

CDIP Same as **ceramic dual in-line package**.

CDMA Abbreviation of code-division multiple access. A spread-spectrum digital cellular telephone technology in which multiple simultaneous users share the available channels. Instead of slicing the available bandwidth into separate channels differentiated by frequency, this technology uses the entire radio-frequency spectrum available and distinguishes users by assigning a unique digital code. CDMA offers various benefits, such as increased call capacity, enhanced privacy, and improved coverage characteristics.

CDMA-1 Same as **cdmaOne**.

CDMA-2000 Same as **CDMA2000**.

CDMA One Same as **cdmaOne**.

CDMA1 Same as **cdmaOne**.

cdma2000 Same as **CDMA2000**.

CDMA2000 CDMA used in third-generation mobile wireless technology. CDMA2000 supports voice, plus data transmission speeds exceeding 2 MHz. Also spelled **CDMA-2000**.

CDMA2000 1x A version of first phase **CDMA2000**, such as CDMA2000 1xEV.

CDMA2000 1xEV A version of first phase **CDMA2000**, offering voice plus data transmission speeds of up to 2.4 MHz. Its abbreviation is **1xEV**.

CDMA2000 1xRTT A version of first phase **CDMA2000**, offering voice plus data transmission speeds of up to 144 Kbps. Its abbreviation is **1xRTT**.

cdmaOne CDMA used in second-generation mobile wireless technology. Also spelled **CDMA One**. Its abbreviation is **CDMA-1**.

CDPD Abbreviation of **cellular digital packet data**.

CDRAM Abbreviation of cached **DRAM**. A type of DRAM which incorporates a small amount of SRAM used as a cache. Also called **EDRAM**.

CDROM Same as **CD-ROM**.

CDROM disc Same as **CD-ROM disc**.

CDROM drive Same as **CD-ROM drive**.

CDROM player Same as **CD-ROM player**.

CDV 1. Same as **CD video**. 2. Abbreviation of **compressed digital video**.

Ce Chemical symbol for **cerium**.

CED tone Abbreviation of **called tone**. A 2100 Hz tone a receiving fax emits in response to a **CNG tone** emitted by a calling fax terminal.

ceiling The upper limit of something. For instance, the maximum output, or the maximum voltage.

celerity The speed at which a wave propagates through a given medium. It is the speed at which given points within it, such as crests and troughs, travel. In a vacuum it is equal to the product of the frequency and wavelength. In a dispersive medium the celerity of electromagnetic waves varies as a function of the frequency, while in a non-dispersive medium the celerity is independent of the frequency. Also called **wave celerity**, **wave speed**, **wave celerity**, and **propagation speed**.

Celeron A popular family of CPU chips.

cell 1. A single unit which converts chemical, thermal, nuclear, or solar energy into electrical energy. For example, a solar cell, or a voltaic cell. Also called **electric cell**, or **energy cell**. 2. A **cell** (1) within a battery. **Cell** and **battery** are commonly used interchangeably, although a battery consists of more than one cell. 3. In computers, an elementary unit of storage, such as a binary cell. 4. A unit where numbers, data, or formulas may be placed in a spreadsheet. It is formed by the intersection of a column and a row. 5. A small compartment or cavity. 6. In cellular telephony, a geographical area with coverage provided by a **cell site**. 7. In ATM communications, a unit of transmission consisting of 53 bytes, of which 48 are payload and 5 are overhead. 8. In communications, a unit of transmission consisting of a fixed size or length.

cell capacity The total amount of charge that can be withdrawn from a fully charged **cell** (2) under specified conditions. Usually expressed in ampere-hours.

cell constant In an electrolytic cell, the surface area of the electrodes, divided by the distance between them. Used to calculate the resistance of the cell.

cell loss priority Its abbreviation is **CLP**. An ATM communications header bit which indicates the priority of an ATM cell. If the CLP is 0, then the cell is guaranteed delivery. If the CLP is 1, then the cell may be dropped if network congestion exceeds a certain level.

cell loss ratio An ATM quality of service parameter. It consists of the ratio of lost cells to the total transmitted cells. Its abbreviation is **CLR**.

cell phone Same as **cellular telephone**.

cell potential The electromotive force developed in an electrolytic cell.

cell rate The rate at which **cells** (7) are transmitted over an ATM circuit.

cell relay A communications protocol in which small, fixed length packets, or cells, are utilized to transmit data. Used, for instance, in ATM communications.

cell site A site which has the necessary transmission and reception equipment to provide cellular phone coverage in a given geographical area. All incoming and outgoing calls within this zone are handled by this station. The area served by a given cell site is called a **cell** (6). Also called **base station** (3).

cell switch In a communications network, a device which operates on fixed-length cells to provide connections between nodes. Used, for instance in ATM communications.

cell switching 1. In communications networks, the use of **cell switches** to transfer fixed-length cells. 2. In cellular communications, the transfer of the signal of a mobile unit from one cell site to another, so that the best available signal is always used.

cell telephone Same as **cellular telephone**.

cell-type tube A gas-filled RF switching tube operated in an external resonant circuit.

cell voltage In an electrolytic cell, the total voltage measured between the anode and the cathode during electrolysis.

cellphone Same as **cellular telephone**.

cellular 1. Pertaining to a **cell**. 2. Pertaining to **cellular communications**.

cellular automata The creation of models utilizing standardized parallel structures, or cells, which can assume a finite number of states. The output of each cell is dependent on the state of all the other cells, and the state of all cells is updated simultaneously. Cellular automata can assist in the understanding and design of large-scale parallel systems. Its abbreviation is **CA**.

cellular communications A telecommunications system in which mobile, usually portable, telephones are linked to a land telephone network via microwave radio-frequency signals. Individual cell sites provide coverage to a limited area, called a cell, while networks of cell sites provide coverage to large geographical areas. At any given point within the cellular network, the system decides which cell site can provide the best signal to a mobile unit, and as the unit moves, the signal is transferred to the next cell site with the best signal. Also called by various other terms, including **cellular mobile telephone system**, **cellular mobile communications**, **cellular mobile telecommunications**, **cellular telephony**, and **mobile cellular communications**.

cellular digital packet data A data-packet transfer technology which utilizes a cellular telephone network. Data can be sent at speeds up to 19.2 kbps utilizing any available pauses in voice conversations, thereby not occupying additional voice channels. Its abbreviation is **CDPA**.

cellular mobile 1. Same as **cellular communications**. 2. Same as **cellular telephone** (1).

cellular mobile communications Same as **cellular communications**.

cellular mobile radio 1. Same as **cellular communications**. 2. Same as **cellular telephone** (1).

cellular mobile radio telephone Same as **cellular telephone**.

cellular mobile telecommunications Same as **cellular communications**.

cellular mobile telephone Same as **cellular telephone**.

cellular mobile telephone system Same as **cellular communications**.

cellular mobile telephony Same as **cellular communications**.

cellular network A network consisting of one or more cell sites which enable cellular communications.

cellular phone Same as **cellular telephone**.

cellular phone system Same as **cellular communications**.

cellular radio Same as **cellular telephone**.

cellular telephone Also called by various other terms, including **cellular phone**, **cellular mobile**, **cell phone**, and **cellular mobile radio**. 1. A mobile telephone used for **cellular communications**. 2. Same as **cellular communications**.

cellular telephone system Same as **cellular communications**.

cellular telephony Same as **cellular communications**.

cellulose acetate A tough and flexible thermoplastic material made from cellulose, which is the main constituent of the walls of the cells of plants. It occurs in white flakes or powder, and also features high-impact strength and ease of fabrication.

cellulose acetate film A non-deforming film which is durable, transparent, hygienic, air-proof, dust-proof, and oil-proof. It is used in the preparation of magnetic tapes and photographic films, among others. Also called **acetate film**.

cellulose acetate resin A resin formed by the polymerization of **cellulose acetate**.

cellulose acetate tape Recording tape that has a magnetic oxide layer over a **cellulose acetate film**. Also called **acetate tape**.

CELP Abbreviation of **Code-Excited Linear Predictive**.

CELP algorithm Abbreviation of **Code-Excited Linear Predictive algorithm**.

CELP coding Abbreviation of **Code-Excited Linear Predictive coding**.

CELP compression Abbreviation of **Code-Excited Linear Predictive compression**.

Celsius degree The unit of temperature interval used in the **Celsius temperature scale**. Formerly called **centigrade degree**, or **degree centigrade**. Its symbol is °C.

Celsius scale Same as **Celsius temperature scale**.

Celsius temperature scale A temperature scale which is based on the freezing and boiling points of water, under stated conditions, assigning values of 0 and 100 degrees Celsius respectively. Its unit of temperature increment is the degree Celsius, or °C. An increment of one degree Celsius equals an increment of one degree Kelvin. Absolute zero is approximately -273.15 °C. Also called **Celsius scale**. Formerly called **centigrade temperature scale**.

censorware Software intended to prevent viewing or interacting with Web sites whose content, such as political perspectives or pornography, is deemed censurable. Censorware may block access to entire Web sites and/or specific content.

cent In acoustics, a unit equal to 1/1200 of an octave, or 1/100 of a semitone. Utilized to specify the ratio in frequency between two tones.

center channel **1.** In a surround-sound audio system, a channel which is situated in front of the listener, and between the left and right channels. It usually carries dialogue and music, and may help enhance special effects. **2.** A phantom channel situated between the left and right channels.

center conductor In a concentric two-conductor line, such as a coaxial cable, the conductor which is on the inside. This conductor usually carries the signal, while the outer conductor serves as a shield. A dielectric separates the two conductors. Also called **inner conductor**.

center-fed antenna An antenna whose transmission line is connected to the center of its radiators. Dipole antennas and cage antennas are examples.

center feed The connection of the transmission line of an antenna to the center of its radiators.

center frequency **1.** Same as **carrier frequency** (**2**). **2.** The frequency which equally divides a band of frequencies. Also called **mid-frequency** (**1**). **3.** In a filter, the median of the frequencies at which the insertion loss is a specified amount, such as 3 decibels.

center loading In an antenna, the placement of a loading coil near the halfway point between the feed point and the ends of the radiating elements. This is done to alter the resonant frequency of the antenna.

center loudspeaker Same as **center speaker**.

center speaker The speaker utilized for the output of a **center channel** (**1**). Also called **center loudspeaker**.

center tap A terminal located at the electrical midpoint of an electrical component, such as a resistor or winding. Its abbreviation is **CT**.

center-tapped Having a **center tap**.

center-tapped rectifier A rectifier with a **central tap**.

center-tapped resistor A resistor with a **central tap**.

center-tapped winding A winding with a **central tap**.

center-zero meter A meter whose zero point is at the center of the indicating scale. May be used, for instance, in certain voltmeters or ammeters.

centering **1.** The seeking of a point, value, or state which is at the center of all possible points, values, or states. **2.** The use of a **centering control**.

centering control In a CRT, one of two controls used to center the image on the screen. One control adjusts the image horizontally, and the other vertically.

centi- A prefix denoting one hundredth, or 0.01. For example, centimeter or centigram. It has two abbreviations, which are **c**, and **c-**.

centigrade degree The unit of temperature interval used in the **centigrade temperature scale**. Since the term centigrade has been replaced by the term Celsius, this unit of temperature interval is now called **Celsius degree**. Its symbol is °C.

centigrade scale Same as **centigrade temperature scale**.

centigrade temperature scale The former name of the **Celsius temperature scale**. Also called **centigrade scale**.

centimeter A unit of distance equal to 0.01 meter. It is the unit of distance in the cgs system. Its abbreviation is **cm**.

centimeter-gram-second electromagnetic system A system of electromagnetic units within the **centimeter-gram-second system** of units. Units in this system have an **ab-** prefix, as in abvolt or abampere. Its abbreviation is **cgs electromagnetic system**.

centimeter-gram-second electrostatic system A system of electrostatic units within the **centimeter-gram-second system** of units. Units in this system have a **stat-** prefix, as in statvolt or statampere. Its abbreviation is **cgs electrostatic system**.

centimeter-gram-second system An obsolescent measuring system in which the fundamental units for expressing distance, mass, and time are the centimeter, gram, and second respectively. Its abbreviation is **cgs system**.

centimeter-gram-second unit A unit within the **centimeter-gram-second system**.

centimetric waves Electromagnetic waves whose wavelengths are between 1 and 10 centimeters, corresponding to frequencies of between 30 and 3 GHz, respectively. This interval represents the super-high frequency band.

centipoise A measure of dynamic viscosity equal to 0.01 poise. Its symbol is **cP**.

central battery In wire telephony, a common battery which located at a **central office**.

central office A structure which houses one or more telephone switching systems. At this location, customer lines terminate and are interconnected with each other, in addition to being connected to trunks, which may also terminate there. A typical central office handles about 10,000 subscribers, each with the same area code plus first three digits of the 10 digit telephone numbers. Its abbreviation is **CO**. Also called **telephone central office**, **exchange** (**1**), **local central office**, or **telephone exchange**.

central office code The prefix which identifies a central office. In the NPA-NXX-XXXX ten-digit telephone number format, it is the NXX portion, where N can be any digit between 2 and 9, and X can be any digit between 0 and 9. A single central office may serve more than one central office code. Also called **central office prefix**, **NXX code**, or **prefix (1)**.

central office prefix Same as **central office code**.

central office trunk A communications path between central offices, or between a central office and a PBX.

central-processing unit Same as **CPU**.

central-processing unit bound Same as **CPU-bound**.

central-processing unit cache Same as **CPU cache**.

central-processing unit chip Same as **CPU chip**.

central-processing unit cooler Same as **CPU cooler**.

central-processing unit cycle Same as **CPU cycle**.

central-processing unit speed Same as **CPU speed**.

central-processing unit time Same as **CPU time**.

central processor 1. Same as **CPU (1)**. **2.** Same as **CPU (2)**.

centralized data processing Same as **centralized processing**.

centralized network A communications network in which all nodes connect to a main computer, which provides resources such as processing power and control. An example is a star network.

centralized processing Processing which is performed by one or more computers which are situated at a single location. This contrasts with **decentralized processing**, where multiple locations are used. Also called **centralized data processing**.

centrifugal switch A switch actuated by centrifugal force, that is, outward from the center of rotation. Used, for instance, in some induction motors for disconnection of the starting winding once a predetermined speed is attained.

Centronics interface Same as **Centronics parallel interface**.

Centronics parallel interface A standard parallel interface for connection of peripherals, such as printers, to a computer. Also called **Centronics interface**.

Centronics parallel port A computer parallel port utilizing a **Centronics parallel interface**.

centum call second A time unit used for measuring telephone communications traffic. One centum call second equals 100 seconds of conversation. 36 centum call seconds equal 1 call hour. Its abbreviation is **CCS**.

ceramal Acronym for **ceramic-metal**. Same as **cerametal**.

ceramet Acronym for **ceramic-metal**. Same as **cerametal**.

cerametal Abbreviation of **cera**mic-**metal**. A mixture of ceramic and metallic components to obtain a material which combines properties of both. For example, an electric element may combine the dielectric characteristics of a ceramic such as porcelain, with the conductive qualities of a metal such as chromium. Used, for instance, in film resistors. Its acronyms are **ceramal**, **ceramet**, and **cermet**. Also called **metal-ceramic**.

ceramic A material, such as porcelain or glass, made through the heating of certain nonmetallic mineral substances, such as silicon dioxide or clay. Its applications in electronics include its use in capacitors and ICs, or as a dielectric. Also called **ceramic material**.

Ceramic Ball Grid Array A surface-mount ceramic chip package that utilizes tiny balls of solder to attach its leads to a printed-circuit board. Ball Grid Arrays feature low induction, compact size, and a high lead count in a small area. Its abbreviation is **CBGA**.

ceramic capacitor A capacitor in which the dielectric is a ceramic material, such as barium titanate. The properties of such capacitors are determined by the specific ceramic used.

ceramic cartridge A cartridge, used in microphones or pickups, with a piezoelectric ceramic material utilized as its transducer.

ceramic chip package Same as **ceramic package**.

ceramic dielectric A ceramic material, such as barium titanate, used as a dielectric.

ceramic DIP Same as **ceramic dual in-line package**.

ceramic dual in-line package A dual in-line package made of ceramic, as opposed, for instance, to plastic. Its abbreviation is **ceramic DIP**, **CerDIP**, or **CDIP**.

ceramic filter A bandpass filter utilizing a piezoelectric ceramic material as the resonating element.

ceramic insulator An insulator composed of a ceramic material, such as porcelain.

ceramic leadless chip carrier A ceramic package which is hermetically sealed, and which instead of leads consisting of metal prongs or wires, has metallic contacts called castellations, which are flush with the package or recessed. The contacts are usually on all four sides of the package, although they may be located elsewhere. Its abbreviation is **CLCC**. Also called **leadless chip carrier**, or **leadless ceramic chip carrier**.

ceramic magnet A permanent magnet made from mixing, pressing, then heating ferromagnetic and ceramic powders.

ceramic material Same as **ceramic**.

ceramic-metal Same as **cerametal**.

ceramic microphone A microphone utilizing a **ceramic cartridge** as its transducer.

ceramic package A chip package made of ceramic, as opposed, for instance, to plastic. Also called **ceramic chip package**.

ceramic PGA Abbreviation of **ceramic pin grid array**.

ceramic pickup A pickup utilizing a **ceramic cartridge** as its transducer.

ceramic pin grid array A ceramic package capable of providing up to several hundred pins, all located on its underside. Used, for instance, to package computer chips. The design seeks to minimize the distance signals must travel from the chip to each designated pin. Its abbreviation is **CPGA**, or **ceramic PGA**.

ceramic quad flat-pack A ceramic surface-mount chip in the form of a square, which provides leads on all four sides. Such a chip affords a high lead count in a small area. Its abbreviation is **CQFP**. Also spelled **ceramic quad flatpack**. Also called **ceramic quad flat-package**.

ceramic quad flat-package Same as **ceramic quad flat-pack**.

ceramic quad flatpack Same as **ceramic quad flat-pack**.

ceramic resistor A resistor with a ceramic material serving as its covering.

ceramic transducer A transducer, such as a pickup or microphone, which utilizes a ceramic material, such as a piezoelectric ceramic.

ceramic tube An electron tube with an envelope composed of a ceramic material. This enables the tube to have very high operating temperatures.

CerDIP Same as **ceramic dual in-line package**.

CERDIP Same as **ceramic dual in-line package**.

Cerenkov counter Same as **Cherenkov counter**.

Cerenkov radiation Same as **Cherenkov radiation**.

ceresin A refined form of ozokerite which is used as an electrical insulator. Also called **ceresin wax**.

ceresin wax Same as **ceresin**.

cerium An iron-gray, ductile, and highly reactive metal whose atomic number is 58. It has about 30 known isotopes, of which 3 are stable. Among its applications in electronics are its use as a getter in vacuum tubes, and as a phosphor. Its chemical symbol is **Ce**.

cermet Acronym for **ceramic-metal**. Same as **cerametal**.

CERN Abbreviation of *Conseil Europeen pour la Recherche Nucleaire*, which is French for **European Organization for Nuclear Research**. An international organization dedicated to the study of particle physics. The CERN laboratory, located outside of Geneva, Switzerland, features some of the world's most advanced equipment, including powerful accelerators, which help enable their complex scientific research.

CERT Acronym for **Computer Emergency Response Team**.

certificate authority An organization which verifies the identities of entities such as individuals and corporations, so as to issue unique digital certificates which confirm their identity. A certificate authority can serve as a trusted third party in transactions where security and electronic commerce are involved. Its abbreviation is **CA**. Also called **certification authority**.

certification 1. An attestation of competency in a given area of expertise, such as networking devices, computer security, or computer operating systems. **2.** The process involved in obtaining **certification** (1). **3.** An attestation that a given hardware or software product fulfills certain requirements, such as providing a given degree of security, or compatibility with standards.

certification authority Same as **certificate authority**.

cesium A soft, silvery-white ductile metal which becomes a liquid above 28 ºC, and whose atomic number is 55. It is extremely reactive, and has about 35 known isotopes, of which one is stable. It is strongly photoelectric, and is the most electropositive chemical element after francium. Some of its applications in electronics include its use in photocells, as a getter in vacuum tubes, and in cesium clocks. Its chemical symbol is **Cs**.

cesium atomic clock Same as **cesium clock**.

cesium atomic frequency standard A frequency standard based on the natural resonance frequency of cesium atoms. This standard is used, for instance, in atomic clocks.

cesium beam A stream of cesium atoms. Useful, for instance as a frequency standard, and in atomic clocks.

cesium-beam atomic clock Same as **cesium clock**.

cesium-beam frequency standard Same as **cesium frequency standard**.

cesium-beam resonance Resonance exhibited by a stream of cesium atoms. Such resonance is usually induced by microwaves.

cesium bromide A colorless crystalline powder whose chemical formula is $CsBr$. Used in fluorescent screens and in spectrometers.

cesium chloride Colorless crystals whose chemical formula is $CsCl$. Used in fluorescent screens and in radio tubes.

cesium clock A clock whose accuracy is governed by the natural resonance frequency of cesium atoms. This frequency is 9,192,631,770 cycles per second, and currently, the time unit **second** is defined as this number of oscillations. Also called **cesium atomic clock**, or **cesium-beam atomic clock**.

cesium frequency standard A frequency standard based on the natural resonance frequency of cesium atoms. This standard is used, for instance, in atomic clocks. Also called **cesium-beam frequency standard**.

cesium hydroxide A white or yellowish crystalline mass whose chemical formula is $CsOH$. Used as an electrolyte in storage batteries.

cesium iodide A colorless crystalline powder whose chemical formula is CsI. Used in fluorescent screens and in infrared spectroscopy.

cesium phototube A phototube whose cathode is coated with cesium.

cesium-vapor lamp A low-voltage arc lamp in which light is produced by the passage of current between electrodes surrounded by ionized cesium vapor.

Cf Chemical symbol for **californium**.

CF card Abbreviation of **CompactFlash card**.

CG Abbreviation of **computer graphics**.

CGI 1. Abbreviation of **Common Gateway Interface**. **2.** Abbreviation of **Computer Graphics Interface**.

cgi-bin A Web-server directory where CGI programs or script are stored.

CGI program Abbreviation of **Common Gateway Interface program**.

CGI script Abbreviation of **Common Gateway Interface script**.

CGM Abbreviation of **Computer Graphics Metafile**.

CGS Same as **cgs**.

cgs Abbreviation of **centimeter-gram-second**.

cgs electromagnetic system Abbreviation of **centimeter-gram-second electromagnetic system**.

cgs electrostatic system Abbreviation of **centimeter-gram-second electrostatic system**.

cgs system Abbreviation of **centimeter-gram-second system**.

cgs unit Abbreviation of **centimeter-gram-second unit**.

chain 1. A series of components, devices, or systems, which are linked, associated, or meant to function simultaneously. **2.** A network of stations connected in a manner that they operate as a group. For instance, a radio network, TV network, radar network, or the like. Also called **network** (3).

chain broadcasting A network of stations, such as those of a radio or TV network, connected in a manner that they broadcast as a group.

chain decay The transformation of a nuclide into another nuclide through radioactive decay, followed by further transformations until a stable, or nonradioactive, nuclide results. In such a series, the first member is called the parent, and the last is the end product. For instance, the chemical element uranium undergoes a series of radioactive decay steps which eventually ends with a stable isotope of lead. Also called **chain disintegration**, **series decay**, **series disintegration**, or **radioactive chain decay**.

chain disintegration Same as **chain decay**.

chain reaction 1. A series of events related in a manner that one initiates or influences the next. **2.** A reaction which once initiated, continues without additional external influence, usually leading to a great expenditure of energy. **3.** The bombardment of neutrons upon an unstable nucleus, such as that of uranium, which initiates a sequence of nuclear fissions whose products include other elements, a substantial amount of energy, and more neutrons, which further propagate the reaction. Also called **nuclear chain reaction**.

chained list In data management, a collection of items arranged in a manner that each item contains the address of the next item in sequence. Also called **linked list**.

chaining 1. The forming of a **chain**. **2.** In computers, the linking of records, programs, units, or the like, so that the function of each item depends on the rest of the group. For instance, the linking of program statements in a manner that each statement after the first depends on the previous statement.

chalcopyrite A mineral whose structure is $CuFeS_2$. Used in semiconductors.

Challenge Handshake Authentication Protocol An authentication method used by Internet providers, in which the user identification and password are encrypted before being transmitted for logging-on. This provides a higher degree of security than if a password is sent in text format. Its acronym is **CHAP**.

challenge-response Same as **challenge-response authentication**.

challenge-response authentication In communication networks, an authentication method in which the user is prompted, or challenged, to send unique information, or a response, which establishes the user's identity. Also called **challenge-response**.

chamber An enclosed section, compartment, space, or room. Examples include anechoic, resonant, and ion chambers.

change file A file containing **change records**. Also called **transaction file**.

change management The process of tracking changes, such as updates, made to computer software or hardware projects, such as those under development.

change record In a computer database, a record which changes information in the corresponding master file. Also called **transaction record**.

channel **1.** A path along which information is transmitted. For instance, a fiber-optic link carrying data between nodes of a network, or a bus between computer devices. Also called **transmission channel**. **2.** Same as **circuit (3)**. **3.** A frequency, or band of frequencies, assigned to a particular carrier or for a specific purpose. For example, the band of radio frequencies assigned to a TV broadcast station. Also called **frequency channel**. **4.** In an audio component, a designated signal path. For instance, a surround-sound receiver may provide, for instance, a left front channel, a left rear channel, a center channel, a right front channel, a right rear channel, and that which drives a subwoofer. **5.** In a field-effect transistor, the electrical path between the source and the drain.

channel balance In sound recording or reproduction, an equilibrium between the volume intensities of the channels.

channel bank A multiplexer which modulates a group of channels into a higher frequency channel, and vice versa. Used, for instance, in a telephone central office.

channel bonding Also called **line bonding**. **1.** The combining of two telephone lines to form a single communications channel in which the transmission speed is doubled. Two modems are required in order to use this technology. For example, two 56 kilobits per second modems bonded in this manner provide a 112 kilobits per second connection. Also called **modem bonding**. **2.** The combining of two communications lines to form a single communications channel in which the transmission speed is doubled.

channel capacity The maximum rate of information which can be accommodated by a channel, per unit time. May be expressed, for instance, in some multiple of bits per second.

channel crosstalk Unwanted coupling occurring between two communication channels. It is usually measured at specific points in each circuit, under specified conditions. Generally expressed in decibels.

channel designator A designator, such as a name or number, which serves to identify a given communications channel.

channel effect In a bipolar junction transistor, a current leakage over the surface path between the collector and emitter.

channel frequency The frequency or band of frequencies of a communications channel.

channel reliability The proportion of time a communications channel satisfies a given set of requirements, such as availability, speed, and maximum bit-error rate. Usually expressed as a percentage

channel reversal In a stereo audio component, the switching of the signals of the left and right channels.

channel selector A device, such as a switch, which permits selection of the desired channel from the available choices.

channel separation **1.** The spacing between communication channels which are immediately next to each other in frequency. For instance, that between adjacent FM broadcast stations. Also called **channel spacing**. **2.** In a stereo audio component, the extent to which the information from one channel is absent from the other. Usually expressed in decibels. Also called **separation (4)**.

channel service unit Its abbreviation is **CSU**. A device that terminates an external digital circuit, such as T1, at the premises of a customer. It serves to condition the signal, provide diagnostics, and allow for remote testing. Usually used in conjunction with a data service unit.

channel service unit/data service unit A device which combines a channel service unit and a data service unit. The channel service unit terminates an external digital circuit, such as T1, at the premises of a customer, while the data service unit converts data transmitted or received through the external digital circuit into the proper format. Its abbreviation is **CSU/DSU**. Also called **channel service unit/digital service unit**.

channel service unit/digital service unit Same as **channel service unit/data service unit**.

channel spacing Same as **channel separation (1)**.

channel termination The equipment necessary to provide a connection point between communications channels. Used, for instance, in central offices.

channeling A form of multiplex transmission in which multiple carriers are used to separate the transmitted signals.

channelizing The subdivision of a wideband transmission facility to form multiple narrowband channels.

CHAP Acronym for **Challenge Handshake Authentication Protocol**.

character A letter, number, punctuation mark, control code, or other symbol that occupies one byte of computer memory. An example is an ASCII character.

character cell A block of pixels or dots used to form a single character on a display screen or printer. The size of the block may vary, along with the blank space surrounding each character, depending on the screen resolution, the font, and other factors.

character code A digital code representing a character within a set, such as ASCII. This code includes information such as whether a letter is uppercase or lowercase.

character data Data presented in characters. This may include letters, numbers, punctuation marks, symbols, and so on.

character density The maximum number of characters that can be stored per unit length, or per unit area.

character device **1.** A device, such as keyboard, which transfers one byte at a time. This contrasts with a **block device**, which operates on blocks of data. **2.** A video display which can display alphanumeric characters, but not graphics.

character field **1.** A field which holds alphanumeric characters. **2.** A field which may hold alphanumeric characters.

character generator Software or hardware that converts a given character code into the pixel patterns seen on the computer screen.

character mode A computer screen mode in which alphanumeric characters can be displayed, but not graphics. Also called **text mode**.

character-oriented protocol A communications protocol which uses bytes to represent control codes. This contrasts with a **bit-oriented protocol**, in which control codes are represented by bits. Between the two, bit-oriented is more efficient and reliable. Also called **byte-oriented protocol**.

character printer 1. A printer which prints a single character at a time. Examples include daisy-wheel printers and dot-matrix printers. Also called **serial printer** (1). 2. A printer that can print alphanumeric characters, punctuation marks, and some special symbols, but not graphics.

character reader A device which scans printed or written characters, and utilizes **character recognition** to identify them. Such scanning is usually optical or magnetic.

character recognition The ability of a computer or other device to identify scanned printed or written characters, and to convert this information into a digital code suitable for further processing. An example is a Magnetic-Ink Character Recognition system used to identify bank checks.

character set A defined list of letters, numbers, punctuation marks, control codes, and other symbols which computer hardware and/or software recognizes as forming a unique group. An example is the ASCII character set.

character string A sequence of characters which is treated as a unit.

character terminal A computer screen which can display alphanumeric characters, but not graphics.

characteristic 1. A measurable feature that helps describe and distinguish components, circuits, devices, systems, materials, and so on. 2. In a floating-point number, the second part, which specifies power to which the base is raised. For example, in 1.234×10^{56}, the characteristic is **56**. The first part, or **1.234**, is the **mantissa** (2). 3. The integer portion of a logarithm. The decimal part is the **mantissa** (1). For example, if a logarithm of a given number equals 1.2345, the characteristic is **1**, while **.2345** is the mantissa. 4. Same as **characteristic curve**.

characteristic curve A curve which describes the relationship between two varying quantities. For instance, the correlation between electrode current and electrode voltage. Also called **characteristic** (4).

characteristic distortion Distortion of a signal resulting from the effects of previously transmitted signals.

characteristic frequency A frequency, such as a carrier frequency, which helps identify a particular transmission.

characteristic impedance The impedance of a circuit, which when connected to the output terminals of a uniform transmission line, makes the line appear infinitely long. Under these conditions, the transmission line has no standing waves, and the ratio of voltage to current at any given frequency is the same at any point of the line. Also called **surge impedance**. Its symbol is Z_0.

characteristic overflow The circumstance in which the **characteristic** (2) exceeds the maximum positive value that can be handled by a computer program or hardware. It results in an error condition.

characteristic underflow The circumstance in which the **characteristic** (2) exceeds the maximum negative value that can be handled by a computer program or hardware. It results in an error condition.

characters per inch The number of characters which fit on a medium, such as paper or magnetic tape, per inch. Its abbreviation is **cpi**, or **CPI**.

characters per second Its abbreviation is **cps** or **CPS**. 1. The number of characters per second that a non-laser printer can print. The speed of laser printers is usually measured in pages per minute. 2. A measure of the maximum speed at which data can travel between devices, or over a communications channel. For instance, it can describe the rate at which a disk drive can transfer data. 3. The rate at which data actually travels between devices, or over a communications channel.

charge 1. A basic property of subatomic particles, atoms, molecules, and the antimatter counterparts of each of these. A charge may be negative, positive, or there may be no charge. The net charge of a body is the sum of the charges of all its constituents. For instance, a cation has a net deficiency of electrons, so its charge is positive. Each electric charge, whether positive or negative, is equal to 1.6022×10^{-19} coulomb, the charge of an electron, or a whole number multiple of it. Also called **electric charge** (1), or **electrostatic charge** (1). Its symbol is Q. 2. The electrical energy stored in an insulated body such as a storage battery or capacitor. Usually expressed in coulombs. Also called **electric charge** (2), or **electrostatic charge** (2). 3. The directing of electrical energy into an insulated body which can store this energy, such as a storage battery or capacitor. 4. The conversion of electrical energy into chemical energy within a battery or cell.

charge carrier A mobile electron, hole, or ion which transports a charge through a semiconductor. Also called **carrier** (2).

charge-carrier concentration In a semiconductor material, the number of charge carriers per unit volume. May be expressed for instance, as the number of carriers per cubic centimeter. Also called **charge-carrier density**, **carrier concentration.**, or **carrier density**.

charge-carrier density Same as **charge-carrier concentration**.

charge conservation Same as **conservation of charge**.

charge-coupled device Same as **CCD**.

charge-coupled device array Same as **CCD array**.

charge-coupled device camera Same as **CCD camera**.

charge-coupled device detector Same as **CCD detector**.

charge-coupled device image sensor Same as **CCD image sensor**.

charge-coupled device imaging Same as **CCD imaging**.

charge-coupled device memory Same as **CCD memory**.

charge-coupled device scanner Same as **CCD scanner**.

charge-coupled device sensor Same as **CCD sensor**.

charge-coupled device spectrometer Same as **CCD spectrometer**.

charge-coupled device telescope Same as **CCD telescope**.

charge-coupled device TV camera Same as **CCD TV camera**.

charge-coupled device video camera Same as **CCD video camera**.

charge density 1. The electric charge per unit area. Usually expressed in coulombs per square meter. Its symbol is σ. Also called **surface charge density**. 2. The electric charge per unit volume. Usually expressed in coulombs per cubic meter. Its symbol is ρ. Also called **volume charge density**. 3. The electric charge per unit length of a line or curve. Usually expressed in coulombs per meter. Its symbol is λ. Also called **linear charge density**.

charge-mass ratio The ratio of the electric charge of a particle to its mass. May be expressed, for instance, in coulombs per kilogram.

charge retention The holding of a charge by a cell, battery, or capacitor.

charge storage 1. The storage of an electric charge. 2. Same as **carrier storage**.

charge-storage tube A tube which stores information on its surface in the form of electric charges.

charge transfer The transfer of an electric charge from one entity to another, or from one location to another within the same entity. Seen, for instance, in a charge-coupled device.

charge-transfer device A semiconductor device, such as a charge-coupled device, in which electric charges are transferred from one location to the next. Its abbreviation is **CTD**.

charged particle A particle or antiparticle having either a positive charge or a negative charge. Such a charge always has a magnitude of 1. An electron, for instance, has a negative charge, or -1 charge. A proton has a +1 charge, a positron a +1 charge, and so on.

charger A device which serves to charge rechargeable batteries. A rectifier, for instance, may be used for this. Also called **battery charger**.

charging The process of storing electric current in a storage battery, storage cell, or capacitor.

charging current The current that is used for **charging**.

charging rate The rate at which **charging** takes place. May be expressed, for instance, in amperes or microamperes.

chart 1. A visual representation of information, or of relationships between quantities or sets of data. For example, a pie chart. Also called **graph (1)**. **2.** A sheet of paper, or other appropriate material, ruled and graduated for use by a recording instrument. **3.** A map utilized to aid navigation.

chassis 1. The supporting frame upon which a structure is built. **2.** A supporting frame upon which electrical components, such as circuit boards or power supplies, are mounted. Such a chassis is usually made of metal and connected to ground.

chassis ground A ground connection made between a component and the chassis it is mounted upon.

chat A real-time written and/or verbal conversation between two or more individuals via a network such as the Internet. Also called **real-time chat**.

chat application Same as **chat program**.

chat bot Same as **chatbot**.

chat program Any program utilized for **chats**, especially over the Internet. An example is an instant messaging program. Also called **chat application**.

chat robot Same as **chatbot**.

chat room 1. A virtual room where **chats** take place. **2.** A specific **chat room (1)**, such as that dealing with a given topic.

chatbot Abbreviation of **chat robot**. A program that converses with users in settings such as chat rooms or online interactive games. Depending on the specific user, the fact that the other participant in such a chat is a program may go undetected indefinitely. Also spelled **chat bot**. Also called **chatterbot**.

chatter 1. Rapid and repetitive vibrations of an electromechanical device, such as an electric motor. **2.** Same as **contact chatter**.

chatterbot Same as **chatbot**.

cheapernet An Ethernet standard using a coaxial cable which is about 0.5 centimeter in diameter, as opposed to **ThickNet** whose cabling is about one centimeter in diameter. Also called **ThinNet**, **ThinWire**, **Thin Ethernet**, or **10Base2**.

Chebyshev filter A filter characterized by minimal passband ripple and a steep cutoff response. Used, for instance, in audio applications. Also spelled **Tschebyscheff filter**.

check 1. The verifying or testing of the accuracy, function, indications, or results of a component, device, instrument, system, or process. **2.** A standard utilized for testing and/or verifying a component, device, instrument, system, or process. **3.** A sudden stoppage or interruption of an action or process.

check bit A bit which by itself, or as part of a set, is used for error detection. For instance, it may be added to a data

transmission for the purpose of detecting errors in transmission or reception. An example is a parity bit.

check box In a GUI, a box utilized to select or deselect a given choice, the user doing so by clicking the mouse within it. Such a box may be used, for instance, to enable or disable a given feature of an Internet browser. Also spelled **checkbox**.

check character A character which by itself, or as part of a set, is used for error detection. For instance, it may be added to a data transmission for the purpose of detecting errors in transmission or reception.

check digit A numeric digit which is added to a set of digits for error detection. When a given set of digits, such as an account number, is entered into a computer, a check digit is calculated and incorporated into it. When the number is stored, or subsequently accessed, the check digit is recomputed and compared to the original check digit.

check indicator An indicator, such as a light, which informs of the state of a component, device, piece of equipment, or system being monitored. It may notify of a completed process, an error condition, and so on.

check program Same as **checking program**.

check routine Same as **checking routine**.

check sum Same as **checksum**.

checkbox Same as **check box**.

checking program A computer program which detects errors in other programs, in data, or in a system. Such a program may be useful, for example, as a debugger. Also called **check program**.

checking routine A set of computer instructions which detect errors in programs, in data, or in a system. Such a routine may be useful, for example, as a debugger. Also called **check routine**.

checkout One or more tests which are performed to verify the suitability, readiness, or performance of a component, circuit, device, piece of equipment, system, or program.

checkpoint A designated point at which the state of a computer program or system, and its data, are saved to a nonvolatile storage medium, such as a tape or a disk, so that if there is an interruption or failure, the program or system can be restarted from this spot.

checkpoint/restart A method of recovering from an interruption or failure of a computer program or system, by reverting to the last **checkpoint**, and restarting from there.

checkpointing The creation of multiple **checkpoints**, to facilitate recovery after any possible interruption or failure of a computer program or system.

checksum Also spelled **check sum**. **1.** A value used to determine whether a block of data has been transmitted or written without errors. It is calculated by adding the binary value of each character in the block. After transmission or writing, a new checksum is calculated, and compared to the first. If the two numbers do not match, an error has occurred. A checksum will not detect all errors, nor will it alert to data that was erroneous to begin with. **2.** A **checksum (1)** utilized to verify the integrity of files, a piece of code, or programs. Used, for instance for verification that a file has not been infected by a virus.

chemical 1. Pertaining to **chemistry**. **2.** Same as **chemical compound**.

chemical bond An attractive force sufficient to unite atoms so that they function as a unit. Also called **bond (2)**.

chemical compound A substance comprised of a specific proportion of two or more constituents, each of which is a **chemical element**. There are countless chemical compounds. Also called **chemical (2)**, or **compound (2)**.

chemical deposition The precipitation of a metal from a solution of one of its salts, when another suitable metal is added to the solution. Used, for instance, in the manufacturing of some semiconductor devices.

chemical element A substance which can not be subdivided into smaller units by chemical means. There are over 110 known chemical elements, each with unique properties, and they comprise all matter above the atomic level. The smallest particle that retains all the properties of an element is an atom. All neutral atoms of a given chemical element have the same number of protons and electrons, and if an element has isotopes, the difference between each is the number of neutrons in its nucleus. Also called **element (3)**.

chemical etching The etching of a surface, such as that of a metal, by the effects of a chemical reaction.

chemical formula A combination of chemical symbols and numbers which describes the chemical composition of a substance. For instance, the chemical formula of boron trichloride is BCl_3, indicating that for each boron atom there are three chlorine atoms. Also called **formula (2)**.

chemical laser A laser that uses chemical reactions, as opposed to electrical energy, to produce high-energy atoms which enable laser action to take place. For instance, laser action can occur in carbon dioxide if it is present when hydrogen and fluorine react.

chemical symbol A notation which is used to represent and identify a chemical element. For instance, the chemical symbol for platinum is **Pt**. Whether it is pure, or in a chemical compound, the symbol **Pt** indicates the presence of platinum.

chemical-vapor deposition A process in which a thin film of a material is applied to a substrate, using a controlled chemical reaction. Used, for instance, in semiconductor manufacturing. Common techniques include low-pressure chemical-vapor deposition, atmospheric pressure chemical-vapor deposition, and plasma enhanced chemical-vapor deposition. Its abbreviation is **CVD**.

chemistry The science that deals with the composition, analysis, structure, properties, synthesis, interactions, and transformations of matter.

Cherenkov counter A counter of high-energy charged particles, which measures **Cherenkov radiation**. Also spelled **Cerenkov counter**.

Cherenkov radiation Visible light emitted when charged particles travel through a transparent medium faster than the speed of light through the same medium. Seen, for instance, as a faint bluish glow in pools of water cooling some nuclear reactors. Also spelled **Cerenkov radiation**.

child **1.** A process which is initiated by another process, which is called **parent (1)**. **2.** Data which is dependent on other data, which is called **parent (2)**. **3.** A component or process which is dependent on another, which is called **parent (3)**.

child file A current version of a file, such as that being updated. The copy or version saved before it is called **parent file**. Also called **son file**.

Child-Langmuir equation Same as **Child-Langmuir law**.

Child-Langmuir law For a thermionic diode, an equation which yields the cathode current in a space-charge-limited-current state. Also known as **Child-Langmuir equation**, **Child-Langmuir-Schottky equation**, or **Child's law**.

Child-Langmuir-Schottky equation Same as **Child-Langmuir law**.

Child's law Same as **Child-Langmuir law**.

chip **1.** A small piece of semiconductor material upon which miniature electronic circuit components, such as transistors or resistors, are placed. A chip, for instance, may have hundreds of millions of transistors. There are various types, including memory chips, and logic chips. An entire computer may be held on a single chip with the appropriate components, and such chips may be used in countless items, such as automobiles, toys, appliances, clocks, and so on. Also called **microchip**, **IC**, or **microcircuit**. **2.** A piece of semiconductor material or dielectric upon which one or more electrical components may be mounted, etched, or formed. Also called **die (1)**. **3.** A piece of semiconductor material or dielectric upon which one or more electrical components has been mounted, etched, or formed. Also called **die (2)**. **4.** Same as **CPU (1)**.

chip capacitor A capacitor constructed in chip form. These are very small, and are used, for instance, in high-precision circuits.

chip card Also called **smart card**, or **IC card**. **1.** A card which incorporates one or more chips, including a processor and memory. Such cards are usually the size of a credit card, and are used in conjunction with chip card readers which communicate with central computers. Such cards may be used to verify identity, keep digital cash, store medical records, and so on. Chip cards are secure, and update the contained information each time it is used. **2.** In portable computers, a **chip card (1)** utilized to add memory, or as a peripheral device such as a modem.

chip card reader A device into which **chip cards** are inserted, to access and update their contained information. Also called **smart card reader**, or **IC card reader**.

chip carrier Abbreviation of **chip carrier** package. A chip package, such as a plastic-leaded chip carrier, which is usually square or rectangular and has connecting leads or surfaces on all four sides.

chip carrier package Same as **chip carrier**.

chip container Same as **chip package**.

chip cooler Same as **CPU cooler**.

chip inductor An inductor constructed in chip form. These are very small, and are used, for instance, in high-precision circuits

chip-level integration The integration of two or more IC functions on a single chip.

chip memory A chip which consists of memory cells and the associated circuits necessary, such as those for addressing, to provide a computer, or other device with memory or storage. Examples include RAM, SRAM, ROM, EEPROM, and flash-memory chips. Also called **memory chip**, **memory IC**, or **IC memory**.

chip-on-board A configuration in which a chip is mounted directly on a PCB. Also, a technology or technique utilized for such mounting. Its abbreviation is **COB**. Also called **direct chip attach**.

chip-on-chip The assembling of two or more chips, each on top of the other, using chip-on-chip technology. Useful, for instance, where multiple technologies must coexist, especially where space is at a premium, as is the case with portable medical devices. Its abbreviation is **COC**.

chip-on-chip technology The technology employed to insure proper function of **chip-on-chips**. This technology considers aspects such as heat dissipation, and the manner in which the chips are attached to each other. Its abbreviation is **COC technology**.

chip-on-film A configuration in which a chip is mounted directly on a flexible printed circuit. Also, a technology or technique utilized for such mounting. Its abbreviation is **COF**.

chip-on-glass A configuration in which a chip is mounted directly on a glass substrate, such as that of an LCD. Also, a technology or technique utilized for such mounting. Its abbreviation is **COG**.

chip package The housing, usually made of ceramic or plastic, surrounding a chip. The leads or metallic surfaces of the chip package are used for connection via plugging-in or soldering. Chip packages are usually mounted onto printed-circuit boards. Also called **chip container**, **IC package**, or **package (3)**.

chip resistor A resistor constructed in chip form. These are very small, and are used, for instance, in high-precision circuits.

chip-scale package A chip package which is no more than a given proportion, such as 20%, larger than the chip itself. Its abbreviation is **CSP**.

chip set Same as **chipset**.

chip speed Same as **CPU speed**.

chip tester An instrument which serves to evaluate the function of chips. Such an instrument, for instance, may be operated automatically for inspection during manufacturing, or manually, for research and development purposes. Also called **IC tester**.

chip testing Tests carried out on chip, such as those performed utilizing a **chip tester**.

chipset Two or more chips that are designed to function as a unit. A chipset usually fits on one chip which performs functions equivalent to what would have previously required multiple chips. A chipset, for example, may allow certain peripherals to intercommunicate independently of the CPU, thus making better use of time and resources. Also spelled **chip set**.

chiral Describing a molecule which is not identical to its mirror image. Such molecules are optically active.

chirality Pertaining to molecules which are **chiral**.

chirp Rapid fluctuations in the frequency of an electromagnetic wave, especially when the source is operated in a pulsing manner. Also called **chirping**.

chirp modulation A form of modulation utilizing **chirp**.

chirp radar A form of radar which utilizes **chirp modulation**.

Chirped Pulse Amplification A technique in which very short laser pulses are amplified to very high power levels. For example, a pulse in the femtosecond range may be amplified to energies in the terawatt range. Its abbreviation is **CPA**.

Chirped Pulse Amplification Laser A laser utilizing a **Chirped Pulse Amplification** technique. Its abbreviation is **CPA laser**.

chirping Same as **chirp**.

Chladni figures Figures formed when a material such as sand is placed on a conducting plate which is secured, then made to vibrate. The plate can be made of glass or metal, and the formed patterns indicate the nature of the vibrations. For instance, the sand collects most at the nodes, which are the points on the plate where the vibrations are at a minimum.

Chladni plates Plates used to form **Chladni figures**.

chloride ion A chlorine atom with an extra electron, which gives it a negative charge. Its chemical formula is Cl^-.

chlorinated Combined with, or treated by chlorine or a chlorine-containing compound. For instance, chlorinated biphenyls.

chlorinated biphenyls A highly toxic group of organic compounds which are stable and heat-resistant. Once widely used in capacitors, transformers, and batteries. Also called **chlorobiphenyls**, **PCBs**, or **polychlorinated biphenyls**.

chlorine A chemical element whose atomic number is 17. It is a dense and highly toxic greenish-yellow gas with a suffocating odor. It is an extremely powerful oxidizer, and is one of the most reactive elements. It has over 15 known isotopes, of which 2 are stable. Because of its activity, it does not occur uncombined in nature, although it is present in countless compounds. It has numerous applications, including its use in the preparation of chlorinated compounds used in electronics, such as carbon tetrachloride. It is a halogen, and its chemical symbol is **Cl**.

chlorobiphenyls Same as **chlorinated biphenyls**.

choke **1.** An inductor used in a circuit to present a relatively high impedance to frequencies beyond a given value. It impedes the flow of AC, while allowing DC to pass freely. Also called **choke coil**, **choking coil**, or **impedance coil**. **2.** To restrict the passing of a current or frequency using a **choke (1)**. **3.** A discontinuity, such as a groove, in the surface of a waveguide, which restricts or blocks certain frequencies.

choke coil Same as **choke (1)**.

choke-coupled modulation Same as **constant-current modulation**.

choke coupling The coupling of two waveguide sections so as to not allow them to touch each other, yet permit energy to pass freely.

choke filter Same as **choke-input filter**.

choke flange A waveguide flange whose surface is cut in a manner that it may serve as part of a choke joint.

choke-input filter A power-supply filter in which a choke is connected to the output of the rectifier. That is, the input component of the filter is a choke. Such filters offer improved line regulation compared to capacitor-input filters. Also called **choke filter**.

choke joint A joint which connects two waveguide sections without metallic contact between their inner walls, yet permits energy to pass freely.

choking coil Same as **choke (1)**.

chopper **1.** A device which periodically interrupts a direct current, light beam, or other signal. This may be done, for instance, to modulate, or to facilitate the amplification of an associated quantity. **2.** A device which periodically interrupts DC in order to produce AC.

chopper amplifier An amplifier that first chops its DC input to produce AC, amplifies the signal, and then restores the DC signal. Also called **converter amplifier**.

chopper-stabilized amplifier A DC amplifier that incorporates a chopper for stabilization of its DC input.

chopper transistor A transistor which periodically interrupts its input signal. For instance, that producing AC by interrupting its DC input.

chopping **1.** The action and effect of a **chopper**. **2.** The removal of the upper and/or lower extremes of a wave.

chopping frequency The rate at which a **chopper** interrupts a signal. Also called **chopping rate**.

chopping rate Same as **chopping frequency**.

chording keyboard A small keyboard which allows for multiple keys to be pressed simultaneously, with different combinations entering any of various letters, symbols, or shortcuts. Such a keyboard may consist of just a few keys, such as seven or twelve, and when used proficiently can allow a single hand to type nearly as quickly as two.

chroma **1.** The attribute of color which combines hue and saturation. **2.** In the Munsell system, the attribute of color that corresponds most closely to saturation. The more saturated, or vivid, a color, the higher the chroma. **3.** Same as **color saturation**.

chroma control Same as **color-saturation control**.

chroma oscillator Same as **color oscillator**.

chromatic **1.** Pertaining to color, or color phenomena. **2.** Pertaining to **chroma**.

chromatic aberration **1.** In a CRT, such as that in a TV, a defect in which the focal spot is enlarged and blurred. It may be due, for instance, to differences in electron velocities as they are emitted by an electron gun, and how these are differentially deflected en route to the screen. **2.** In an optical lens, a defect in which not all colors come to focus at the same point. May be due, for instance, to the physical properties of the lens.

chromaticity The combination of hue and saturation which determine the quality of a color. This color quality can be quantified, and is defined by its chromaticity coordinates.

chromaticity coordinates For a given color sample, the ratio of each of the tristimulus values to the sum of these three values. In a trichromatic color system, the tristimulus values are the amounts of each of the three primary colors that must be combined in order to match a given color sample.

chromaticity diagram A triangular graph in which **chromaticity coordinates** are plotted. Such a diagram quantifies color taking into account human perception.

chrome plating Also called **chromium plating**, or **chromium coating**. **1.** A thin coating of chromium deposited on another metal. This helps improve corrosion and abrasion resistance. **2.** The electrolytic deposition of chromium onto a substrate, usually to provide it with greater resistance to oxidation and corrosion.

chrome tape A magnetic recording tape utilizing chromium dioxide particles. Such tape, with the proper deck, provides excellent frequency response and a wide dynamic range. Also known as **chromium tape**, **chromium recording tape**, or **Type II tape**.

chromel An alloy, which is usually about 90% nickel and 10% chromium, that is used in thermocouples.

chromic acid Dark red crystals whose chemical formula is CrO_3. Used for chromium plating. Also called **chromium trioxide**.

chromic oxide Bright green and hard crystals whose chemical formula is Cr_2O_3. Used in semiconductors.

chrominance 1. The difference between a given color and a specified reference color, each with the same luminous intensity. Used especially in the context of color TV reproduction. **2.** Same as **chrominance signal**.

chrominance carrier Same as **color subcarrier**.

chrominance-carrier reference Same as **color-subcarrier reference**.

chrominance channel In color TV, the path, or channel, which carries the **chrominance signal**.

chrominance demodulator Same as **chrominance-subcarrier demodulator**.

chrominance frequency Same as **chrominance-subcarrier frequency**.

chrominance gain control In color TV, adjustment of the amplitude of the signal of each of the primary colors through the use of variable resistors. The primary colors are red, green, and blue.

chrominance modulator Same as **chrominance-subcarrier modulator**.

chrominance signal In color TV, the video signal which contains the color information, consisting of the hues and saturation levels of colors. The **luminance signal** contains the brightness information. Its abbreviation is **C signal**. Also called **carrier chrominance signal**, or **chrominance (2)**.

chrominance subcarrier Same as **color subcarrier**.

chrominance-subcarrier demodulator In a color TV receiver, a demodulator which extracts the chrominance components from the chrominance signal, and a sine wave from the chrominance-subcarrier frequency. Also called **chrominance demodulator**.

chrominance-subcarrier frequency In color TV, the frequency of the chrominance subcarrier, which is set at 3.579545 MHz. Also called **chrominance frequency**.

chrominance-subcarrier modulator In a color TV transmitter, a modulator which generates the chrominance signal from the chrominance components, and also produces the chrominance subcarrier frequency. Also called **chrominance modulator**.

chrominance-subcarrier oscillator Same as **color oscillator**.

chrominance-subcarrier reference Same as **color-subcarrier reference**.

chrominance video signal **1.** The output signal generated by the red, green, and blue sections of a color TV camera. **2.** The output signal generated by the red, green, and blue sections of a color TV receiver matrix.

chromium A hard, brittle steel-gray metallic element whose atomic number is 24. It has nearly 20 known isotopes, of which 4 are stable. The compounds it forms generally have varied and intense colors. It is frequently alloyed with, or plated onto other metals, contributing hardness, heat resistance, strength, and corrosion resistance. Its chemical symbol is **Cr**.

chromium boride One of several crystalline chemical compounds composed of chromium and boron. Used for high-temperature electrical conductors.

chromium coating Same as **chrome plating**.

chromium dioxide Black crystals whose chemical formula is CrO_2. Used in magnetic recording tapes.

chromium plating Same as **chrome plating**.

chromium recording tape Same as **chrome tape**.

chromium tape Same as **chrome tape**.

chromium trioxide Same as **chromic acid**.

chronistor A small device which uses electroplating to provide an approximate measure of the time of operation of equipment.

chronometer An instrument which precisely measures, displays, and records time intervals. The most accurate chronometers are governed by the natural resonance frequencies of atoms.

chronoscope An optical instrument which precisely measures, displays, and records brief time intervals.

chronotron An instrument which measures and indicates extremely the extremely small time intervals between pulses over a transmission line.

Ci Abbreviation of **curie**.

CIC Abbreviation of **carrier identification code**.

CIDR Abbreviation of **classless inter-domain routing**.

CIF Abbreviation of **Common Intermediate Format**.

CIFS Abbreviation of **Common Internet File System**.

CIM **1.** Abbreviation of **computer-integrated manufacturing**. **2.** Abbreviation of **Common Information Model**. **3.** Abbreviation of **computer-input microfilm**.

cinematograph A motion picture camera or projector. Also called **kinematograph**.

cipher **1.** A system for encoding, usually used for security purposes. **2.** An encoded character.

ciphertext Text or data which has been encoded, usually for security purposes. Ciphertext can not be read without being decoded.

CIR Abbreviation of **committed information rate**.

cir mil Abbreviation of **circular mil**.

CIRC Abbreviation of **cross-interleaved Reed-Solomon code**.

circle A curve, all of whose contained points are equidistant from a fixed point located at the center of said points. Also, that which is in this shape.

circle diagram 1. A diagram showing graphical solutions of equations for a transmission line. 2. A diagram showing certain properties, such as impedance, of AC machines.

circuit Its abbreviation is **ckt**. 1. One or more conducting paths which serve to interconnect electrical elements, in order to perform a desired function such as amplification or filtering. Also known as **electric circuit (1)**. 2. One or more complete paths through which electrons may circulate. Also known as **electric circuit (2)**. 3. A medium through which information is conveyed between two or more locations. Such a circuit may be linked physically or wirelessly. Also called **channel (2)**, **communications circuit**, **communications channel**, or **communications line**.

circuit analyzer An instrument which measures and indicates one or more quantities in an electric circuit, such as resistance or voltage. For instance, an oscilloscope or an ohmmeter. When the instrument measures two or more quantities it is also called **multimeter**, or **multifunction meter**.

circuit board A flat, rigid, and insulated panel upon which electrical components are mounted and connected via conductive paths. Used extensively in computers, where the main board is called the motherboard, and secondary boards are called expansion boards. Also called **circuit card**, **card (2)**, or **board (1)**.

circuit breaker A device which automatically opens a circuit under specified conditions, such as a current exceeding a set amount. If it resets itself after another condition is met, such as the passing of a given time period, then it is an **automatic circuit breaker**. Its abbreviation is **CB**.

circuit-breaker panel A panel or box which houses a set of **circuit breakers**.

circuit capacity 1. The maximum quantity, such as that of current, that can be handled by an electric circuit. 2. The maximum quantity of communication channels that a communications circuit can handle simultaneously.

circuit card Same as **circuit board**.

circuit component Any of the electrical elements that are a part of a circuit. These may include resistors, capacitors, transistors, generators, electron tubes, and so on. Each component has terminals which allow it to be connected to the conducting path. Also called **circuit element**, **component (2)**, **element (2)**, **electrical element**, **electrical component**, or **part (3)**.

circuit continuity The condition in which a circuit has a continuous and complete path for the flow of current. Also called **continuity (2)**.

circuit design The specifying of the electrical elements, and the manner in which they interconnect with each other, in the formation of a circuit which performs a desired function. Also called **design (2)**.

circuit diagram A graphical representation of the electrical elements in a circuit, and the way each is interconnected with each other. Each circuit element is represented by a symbol, while lines represent the wiring. Also called **circuit schematic**, **schematic diagram**, **wiring diagram**, **schematic circuit diagram**, **wiring schematic**, or **diagram (3)**.

circuit efficiency The ratio of the useful output power of a circuit, to the input power.

circuit element Same as **circuit component**.

circuit grade 1. A designation which indicates the information-carrying capability of a communications circuit. For instance, voice grade, or broadband. 2. A designation which indicates the information-carrying capability of a communications circuit in terms of transmission speed. For instance, data transmissions at speeds up to 100 Mbps.

circuit layout Same as **circuitry (2)**.

circuit loading 1. The power being drawn from a circuit by a measuring instrument. This must be accounted for in order to have an accurate indicated value. 2. The power being drawn from a circuit.

circuit noise Noise generated in circuits, in addition to that present in the applied signal. Circuit noise is produced, for instance, by the movement of particles, especially electrons.

circuit-noise level At any point within a communications circuit, the ratio of the circuit noise to a chosen reference level. Usually expressed in decibels, or adjusted decibels.

circuit-noise meter An instrument which measures **circuit-noise levels**.

circuit parameter A specific value for a circuit component, such as the resistance of a resistor, or the capacitance of a capacitor, in a given circuit configuration. Also called **parameter (2)**.

circuit protection Protective measures, such as fuses, which safeguard a circuit from overloading, excessive heat, corrosion, and so on.

circuit reliability The proportion of time that a circuit is available for use while meeting the standards established for it. Usually expressed as a percentage.

circuit response 1. A change in the behavior, operation, or function of a circuit as a consequence of a change in its external or internal environment. For example, a change in its output resulting from a change in its input. 2. A quantified **circuit response (1)**, such as a frequency response.

circuit schematic Same as **circuit diagram**.

circuit simplification The designing of circuits so that the desired function is performed utilizing the least electrical elements, and the simplest wiring possible.

circuit simulation The use of computers to simulate the behavior of circuits before even a prototype is prepared. Computer-aided design may be used for this.

circuit-switched data An ISDN option which enables a bearer channel to transmit digital data over a dedicated connection. Its abbreviation is **CSD**.

circuit-switched voice An ISDN option which enables a bearer channel to provide a digital transmission of a voice communication over a dedicated connection. Its abbreviation is **CSV**.

circuit switching Also known as **line switching**. 1. The temporary linking of two communications terminals in order to establish a communications channel. This circuit must be established before communication can occur, and remains in exclusive use by these terminals until the connection is terminated. Used, for instance, by telephone companies to enable dial-up voice conversations. Such a connection is made at a switching center. 2. The temporary linking of two data terminals, in a manner that the connection remains in exclusive use by these terminals until the connection is terminated.

circuit test A test performed to ascertain the performance of a circuit. For instance, a test of continuity.

circuit tester An instrument which tests the performance of a circuit. For instance, a continuity tester.

circuit testing 1. A series of tests intended to ascertain the performance of a circuit. 2. The performing of a test to ascertain the performance of a circuit.

circuit theory The mathematic analysis of electric circuits, and the relationships between the components such circuits contain. Also called **electric circuit theory**.

circuitron A collection of active and passive electronic components mounted inside the same enclosure, and functioning as one or more complete operating stages.

circuitry 1. The complete circuits which compose an electrical device or system. 2. The physical arrangement of the electrical elements that are part of a given circuit. Also called **circuit layout**.

circular 1. Moving in a manner which resembles a circle or helix. 2. In the form of a circle.

circular accelerator A particle accelerator, such as a cyclotron, in which particles follow a circular or spiraling path. This contrasts with a **linear accelerator**, in which they follow a straight line.

circular antenna An antenna consisting of a folded dipole antenna bent into a circle. When mounted horizontally it has a uniform radiation pattern in all directions.

circular current An electric current traveling in a circular path.

circular electric wave A transverse electric wave whose lines of electric force form concentric circles.

circular frequency The frequency of a periodic quantity expressed in radians per second. It is obtained by multiplying the frequency by 2π. Also called **radian frequency**, **angular frequency**, or **angular velocity**.

circular horn A waveguide section which in the shape of a horn. Used, for instance, as a feed for a microwave reflector.

circular magnetic wave A transversal magnetic wave whose lines of magnetic force form concentric circles.

circular mil A unit of area equal to the area of a circle whose diameter is 0.001 inch. Used mainly to specify cross-sectional areas of wires. Its abbreviation is **cmil**, or **cir mil**.

circular polarization Polarization of an electromagnetic wave in which the electric field vector is of a constant magnitude, and rotates in a plane so as to form a circle.

circular scan Same as **circular scanning**.

circular scanning In radars, scanning in which a complete scan consists of a horizontal rotation of the radar beam through 360°. Also called **circular scan**.

circular waveguide A waveguide whose cross section is in the form of a circle.

circularly-polarized light A light wave with **circular polarization**.

circularly-polarized wave A wave with **circular polarization**.

circulating memory A form of computer storage in which a delay line is used to store information. Also called **circulating storage**, or **delay-line memory**.

circulating register A computer register which utilizes a delay line. Also called **delay-line register**.

circulating storage Same as **circulating memory**.

circulator A multiport waveguide junction in which microwave energy entering one of its ports is transmitted to an adjacent port, and on to the next, in a predetermined rotation. Also called **microwave circulator**.

CIS Abbreviation of **Contact Image Sensor**.

CISC 1. Acronym for **complex instruction set computer**. 2. Acronym for **complex instruction set computing**.

citizens' band A radio-frequency band designated for unlicensed two-way communications. Such transceivers are limited to transmission power 4 watts. Its abbreviation is **CB**. Also called **citizens' waveband**.

citizens' radio service A radio communications service for **citizens' band** communications.

citizens' waveband Same as **citizens' band**.

city code A code used for dialing a city other than that within which a telephone call is originated. Such a code may be preceded by a country code, and is followed by the specific telephone number. Some regions require dialing the city code even when the call is originated in the same city.

ckt Abbreviation of **circuit**.

Cl Chemical symbol for **chlorine**.

cladding 1. A layer that surrounds the core of an optical fiber. 2. The bonding of two or more materials. For instance, plating a metal.

clamp 1. A mechanical device which serves to bind, press, or otherwise firmly hold two or more parts together. A clamp may also provide mechanical support, strain relief, and vibration absorption. 2. To restore the DC component to a wave or signal. 3. To add a DC component to a wave or signal which lacks such a component. 4. To establish a reference DC level in a signal or device.

clamper Also called **clamping circuit**, **DC restorer**, or **reinserter**. 1. A circuit or device which restores the DC component of a wave or signal. 2. A circuit or device which adds a DC component to a wave or signal which lacks such a component. 3. A circuit or device which establishes a reference DC level in a signal or device.

clamping Also called **DC reinsertion**, or **DC restoration**. 1. The restoration of the DC component to a wave or signal. 2. The adding of a DC component to a wave or signal which lacks such a component. 3. The establishing of a reference DC level in a signal or device.

clamping circuit Same as **clamper**.

clamping voltage The peak voltage a surge suppressor allows to pass.

Clapp oscillator A Colpitts oscillator with a tuning capacitor in series with the resonant tank circuit, which provides greater stability. Also called **series-tuned Colpitts oscillator**.

clapper 1. An armature that is hinged or pivoted. 2. A mechanical device with a striking action, such as the tongue of a bell.

clarity The quality and/or state of being **clear**.

Clark cell A standard cell used as a reference voltage source, in which the positive electrode is mercury, the negative electrode is an amalgam of zinc and mercury, and the electrolyte is a solution of zinc sulfate. It has a voltage of approximately 1.433 at 15 °C. Also called **zinc standard cell**.

class 1. A group or set sharing common characteristics. 2. In object-oriented programming, a set of objects which share common characteristics. For instance, a class may be titled geometric shapes, and contain objects which are triangles, squares, pentagons, hexagons, and so on. 3. Classifications of amplifiers based on the relationship between the input signal and the output current. For classification purposes, a simplified output stage consisting of two complementary tubes or transistors is assumed for most classes. Each class has its own linearity and efficiency characteristics. Examples include class A amplifiers, class AB amplifiers, class C amplifiers, and so on. Also called **amplifier class**.

class A A Federal Communications Commission certification which indicates that the tested device complies with the established limits of electromagnetic radiation for a business digital device. **Class B** pertains to residential digital devices.

class A amplifier A linear amplifier in which there is a current output at all times, regardless of the stage of the input signal. Since it always operates at full power, it is the most inefficient of all power amplifier designs. On the other hand, it features the least distortion. Used, for instance, in audio high-fidelity systems.

class A IP address Abbreviation of **class A Internet-Protocol address**. An IP address that can range from 1.0.0.0 through 126.255.255.255. These can define a maximum of over 16.7

million hosts, and are used mostly by large entities such as governments. Also called **class A IP network**.

class A IP network Same as **class A IP address**.

class A modulator A **class A amplifier** used for supplying the signal power to modulate a carrier.

class AB amplifier An intermediate between **class A** and **class B** amplifiers. Both output devices are on more than 50% of each half cycle of the input, but never all the time. This increases efficiency compared to **class A**, and allows for faster response to the input when compared to **class B**. The good efficiency and low distortion make it a popular choice for audio high-fidelity systems. Another use, among others, is in asymmetrical digital subscriber lines.

class B A Federal Communications Commission certification which indicates that the tested device complies with the established limits of electromagnetic radiation for a residential digital device. **Class A** pertains to business digital devices.

class B amplifier A linear amplifier in which each of the output devices is on for each half of the input cycle. When there is no input signal, it has no output current. These amplifiers feature high efficiency, but suffer from high crossover distortion. Used, for example, in two-way radios.

class B IP address Abbreviation of **class B** Internet-Protocol **address**. An IP address that can range from 128.0.0.0 through 191.255.255.255. These can define a maximum of approximately 65,500 hosts, and are used mostly by medium to large corporations. Also called **class B IP network**.

class B IP network Same as **class B IP address**.

class B modulator A **class B amplifier** used for supplying the signal power to modulate a carrier.

class C amplifier An amplifier in which the current from each output device flows for substantially less than half of the input cycle, as opposed to operating uninterruptedly during each half cycle, as is the case with **class B** amplifiers. It is highly efficient, but also has high distortion. Used mostly for radio-frequency transmissions.

class C IP address Abbreviation of **class C** Internet-Protocol **address**. An IP address that can range from 192.0.0.0 through 223.255.255.255. These can define a maximum of approximately 254 hosts, and are used mostly by small corporations. Also called **class C IP network**.

class C IP network Same as **class C IP address**.

class D amplifier A digital amplifier in which the output devices are quickly switched on and off multiple times during each input cycle. These amplifiers have particularly high efficiency. The higher the switching speed, the lower the level of distortion. Used in many voice-bandwidth and music systems, among others.

class D IP address Abbreviation of **class D** Internet-Protocol **address**. An IP address that can range from 224.0.0.0 through 239.255.255.255. These are used for IP multicasting and are assigned to groups of hosts. Also called **class D IP network**.

class D IP network Same as **class D IP address**.

class E amplifier An amplifier consisting of a single transistor which acts as a switch. Highly efficient, but depending on the design may have very high distortion. Used, for example, in portable communications devices.

class E IP address Abbreviation of **class E** Internet-Protocol **address**. An IP address that can range from 240.0.0.0 through 255.255.255.255. These are reserved for special uses and future addressing modes. Also called **class E IP network**.

class E IP network Same as **class E IP address**.

class F amplifier Similar to a **class E** amplifier, it consists of a single transistor which acts as a switch. Highly efficient, but depending on the design may have very high distortion. Used, for example, in portable communications devices.

class G amplifier A linear amplifier which works like a **class A** or **class AB** amplifier, but that varies the power voltage depending on the input signal strength. This design offers higher efficiency along with low distortion. Used, for example, in high-fidelity audio systems, and asymmetrical digital subscriber lines.

class H amplifier A refinement of **class G** amplifiers. Incorporates tracking of the input signal so that the output voltage is optimized. Used in high-fidelity audio systems, among others.

class library A set of routines and programs that programmers can use to write object-oriented programs. For example, such a set may include a group of applet routines.

class of service Its abbreviation is **COS**. **1.** A given level of service purchased from a telephone company. For instance, measured-rate residential, or flat-rate business. **2.** A given level of service purchased from a public data network.

class variable In object-oriented programming, a variable which contains information shared by all instances of the class.

classical electron radius A physical constant equal to approximately 2.8179 x 10^{-13} cm. Its symbol is r_e. Also called **electron radius**.

Classical Internet Protocol Same as **Classical IP**.

Classical IP Abbreviation of **Classical I**nternet Protocol. An IETF standard for sending IP packets over ATM networks. Its own abbreviation is **CLIP**.

classless inter-domain routing A technique in which routers group routes together so as to reduce the information carried by the core routers. Also spelled **classless interdomain routing**. Its abbreviation is **CIDR**.

classless interdomain routing Same as **classless inter-domain routing**.

CLCC Abbreviation of **ceramic leadless chip carrier**.

clean boot A computer startup in which only the essential components of the operating system are loaded. Used, for instance, for diagnosing any problems that are occurring while booting.

clean install **1.** The installation of software after completely removing any other version of it. Used, for instance, to solve persistent problems with a program. **2.** The installation of software onto a newly formatted disk.

clean room A sealed room in which exceptional measures are taken to provide a contaminant-free environment. In order to keep dust and other small particles out, there may be seamless plastic walls, external wiring, and a continuous supply of 100% filtered air, among others. Those working in clean rooms must wear special clothing which includes head coverings and special slippers, and before entering, must take a shower of blasted air. Temperature and humidity are also strictly controlled. Used for manufacturing and research. Seen, for instance, in a chip fabrication plant.

clear **1.** Easily seen or heard. **2.** Free from ambiguity. **3.** Free from restrictions or limitations. **4.** Available for use. **5.** To restore a storage element or storage location to its zero state. Also called **reset** (4). **6.** To remove or delete instructions or data from a computer or calculator. **7.** To remove all content from a display, such as that of a computer. **8.** A function key, such as that found on a calculator, used to delete the previous entry, a memory location, or everything.

clear band In optical character recognition, a section above and below the scan line which must be clear in order for the scan line to be read correctly. The dimensions of this band vary depending on factors such as the reader being used.

clear box Also called **white box**, or **open box**. **1.** A component, device, or unit whose internal structure and function are known in great detail. Usually used in the context of its insertion into a system. **2.** In computers, a hardware or software unit whose internal structure and function is known in great detail.

clear-box testing Also called **white-box testing, structural testing**, or **open-box testing**. **1.** Any tests utilizing the assistance of a **clear box (1)**. This contrasts with **black-box testing (1)**, in which the inner structure is not known, or is not necessary to be known. **2.** Tests of software or hardware utilizing the assistance of a **clear box (2)**. This contrasts with **black-box testing (2)**, in which the inner structure is not known, or is not necessary to be known.

clear channel **1.** A radio broadcaster which has the exclusive use of a given frequency within its service area. **2.** A T1 service in which all 64 kilobits per second are available for information transmission in each of the first 23 channels, while the 24th channel, which is the last, is dedicated to signaling.

clear-channel capability A characteristic of a communications channel, in which there is no restriction on the bit patterns that can be transmitted. That is, any number of consecutive zeroes can pass without loss of synchronization. Its abbreviation is **CCC**. Also called **clear-channel coding**.

clear channel coding Same as **clear-channel capability**.

clear GIF A GIF which shows the background through all or part of an image. Used, for instance, for smooth blending of graphics on Web pages, or for spying on users utilizing Web bugs. Also called **transparent GIF**, or **invisible GIF**.

clear memory To reset all memory registers.

clear text Text which is not encrypted, as opposed to **ciphertext**, which has been. It is simply regular text before or after encryption. Also spelled **cleartext**. Also called **plaintext**.

clear to send In communications, a signal some devices send indicating its readiness to receive a transmission. It is sent in response to a **request to send** from the other station. Its abbreviation is **CTS**.

clearance **1.** The distance or space between two objects. For instance, the space between conductors, or between a moving and a stationary part of a machine. **2.** The minimum separation that must be maintained between two objects, such as conductors.

cleartext Same as **clear text**.

cleavage The characteristic of a crystal splitting along a given plane when subjected to the proper pressure. A mica, for instance, splits in one direction, while others in several.

cleave To split a crystal along one or more definite planes.

CLEC Abbreviation of **competitive local exchange carrier**.

click To press and release a button once on a computer mouse, or similar device. Used, for instance, to select a hyperlink.

click and drag To use a pointing device, such as a mouse, to click on an on-screen object, and drag it to another location without releasing the button until said object is placed where desired.

click-through On the World Wide Web, the use of a mouse to click on a hyperlink that leads to another Web site. Occurs, for instance, when a user clicks on a banner ad.

clickstream The Web pages a user visits during a given session or period, as revealed by the clicks on hypermedia which lead from page to page. Such a sequence may be obtained, for instance, through the use of spyware.

clickstream analysis The scrutiny of **clickstreams** for purposes such as marketing research or spying.

client Within a **client/server architecture**, the client.

client application An application running on a client within a **client/server architecture**.

client-based That which resides, occurs, or runs on a client. For example, client-based archiving, or a client-based application. That which is **server-based** resides, occurs, or runs on a server.

client-based application An application which resides and/or runs on a client, as opposed to a **server-based application** which resides and/or runs on a server.

client-based archiving Archiving which occurs at a client, as opposed to **server-based archiving** which takes place at a server.

client-based computing The use of applications stored and/or run on a client. This contrasts with **server-based computing**, in which applications are stored and/or run on a server.

client program Same as **client software**.

client/server application An application being utilized in a **client/server architecture**.

client/server architecture A network environment within which all computers participate in tasks requiring processing power. Clients are individual computers or workstations that depend on the servers for certain resources, such as network administration, security, and data management. Applications are run on clients, although they can usually rely on the servers to help in processing. The client is known as the front end, while the server is known as the back end. Its abbreviation is **CSA**. Also called **client/server environment**.

client/server environment Same as **client/server architecture**.

client/server network A network utilizing a **client/server architecture**.

client/server protocol A communications protocol used over a network with a **client/server architecture**.

client side Any activity, such as processing, occurring at a client within a **client/server architecture**.

client-side script A script run on a client, such as a Web browser. A **server-side script** is run on a server, such as a Web server.

client software Software running on a client within a **client/server architecture**. Also called **client program**.

clinometer Any of various instruments used to measure vertical angles, such as the steepness of slopes. May be used, for instance, in a mobile robot.

clip **1.** A device which serves to grasp or connect, usually through the use of spring-loaded jaws. For example, an alligator clip. **2.** In computer graphics, to cut off of a displayed image beyond a given boundary.

CLIP Abbreviation of **Classical IP**.

clipboard A special memory area where cut or copied information is stored. To retrieve the information, a paste operation may be used. The stored content may be data, multimedia, and so on.

clipper A circuit or device which limits the amplitude of its output signal to a predetermined maximum, regardless of the variations of its input. Used, for example, for preventing component, equipment, or media overloads. Also called by various other names, including **clipper circuit, clipping circuit, limiter, amplitude-limiting circuit, limiter circuit**, and **peak limiter**.

clipper amplifier An amplifier which limits the amplitude of its output signal to a predetermined maximum, regardless of the variations of its input.

clipper circuit Same as **clipper**.

clipper-limiter A circuit or device which passes only those portions of an input signal which lie between two fixed amplitude boundaries. These boundaries are usually close together. Also called **amplitude gate, slicer**, or **slicer amplifier**.

clipping **1.** The action and effect of a **clipper**. Also called **limiting**. **2.** The unintended limiting of the amplitude of an output signal, due to the exceeding of the capabilities of an amplifier or circuit. **3.** In an audio transmission, the loss of words or syllables, due to interruptions of portions of the signal. These are usually due to equipment limitations or malfunctions. **4.** In computer graphics, the cutting off of a displayed image beyond a given boundary. Also called **scissoring**. **5.** A portion of the content of a given Web page. Used, for instance, to provide only the basic information needed when accessing the Internet via a low-bandwidth device such as certain cell phones or PDAs. Also called **Web clipping**.

clipping circuit Same as **clipper**.

clipping level The amplitude level at which a wave or signal is limited.

CLNP Abbreviation of **connectionless network protocol**.

CLNS Abbreviation of **connectionless network service**.

clock 1. A circuit or device which provides a steady stream of timed pulses. Used, for instance, as the internal timing device of a digital computer, upon which all its operations depend. **2.** An instrument or device which measures and indicates time. For instance, that kept and displayed by a TV.

clock/calendar In a computer, an independent circuit which keeps track of the date and time. It has its own battery, and serves, for instance, to stamp the time and date a file is created or modified. A clock/calendar is not used to generate clock pulses.

clock-control system A system, such as a digital computer, whose operations are governed by a timing mechanism. Also called **time-controlled system**.

clock-controlled A circuit, device, piece of equipment, system, or network whose operations are governed by a timing mechanism.

clock cycle The time interval necessary for a clock pulse in a computer to complete a full cycle. For example, a 4 GHz computer has 4 billion evenly spaced cycles per second. Also called **clock tick**, or **CPU cycle**.

clock frequency The master frequency generated by a **clock (1)**.

clock oscillator An oscillator which controls a **clock (1)**.

clock pulse The signals which are utilized to synchronize the operations of a circuit, device, piece of equipment, system, or network. Also called **clock signal**.

clock rate The speed at which the internal clock of a device provides a stream of timed pulses. When used to quantify the speed at which a CPU operates, it is expressed in multiples of hertz, such as gigahertz. Also called **clock speed**.

clock signal Same as **clock pulse**.

clock skew In a clock-synchronized circuit, device, piece of equipment, system, or network, the arrival of the clock pulses at two or more places at different times. If the clock skew exceeds a given amount, it can cause errors and malfunctions.

clock speed Same as **clock rate**.

clock-synchronized A circuit, device, piece of equipment, system, or network whose operations are synchronized by a timing mechanism.

clock tick Same as **clock cycle**.

clocked flip-flop A flip-flop with an additional input consisting of a clock pulse. Its output state depends on the inputs at the instant of the applied clock pulse.

clocked gate A gate circuit whose function is triggered by clock pulses.

clocking The use of clock pulses to synchronize the operations of a circuit, device, piece of equipment, system, or network.

clockwise In the same direction as the moving hands of a clock when viewed from the front. Its abbreviation is **cw**.

clockwise-polarized wave An elliptically-polarized transverse electromagnetic wave in which the rotation of the electric field vector is towards the right when viewed along the direction of propagation. Also called **right-hand polarized wave**.

clone A hardware or software computer product which does the same thing as another, earlier product. For instance, a computer system which is 100% compatible with a long-standing well-known brand name, yet is produced by another, lesser-known brand.

clone PC Abbreviation of **clone personal computer**. A computer system which does the same thing as another, earlier product. A PC-clone must be 100% compatible with the well-known computer it is based on. Also called **PC clone**.

clone personal computer Same as **clone PC**.

cloning software Software which copies the complete contents of the hard disk from one computer to another. Used, for instance, for backup, or when installing a new system.

close 1. To bring to an end. For instance, to end a program, to shut a window, or to terminate a communications link. **2.** To make complete, as in a circuit. Closing a circuit enables the flow of current.

close button A virtual button on a computer screen, which when clicked, serves to shut a given window.

close coupling 1. In a transformer, the arrangement of the primary and secondary windings in a manner which provides for a high degree of energy transfer. Also called **tight coupling (1)**, or **strong coupling (2)**. **2.** In a transformer, the arrangement of the primary and secondary windings in a manner which provides maximum energy transfer. Also called **tight coupling (2)**, or **strong coupling (3)**.

close miking The placement of a microphone close to a sound source. Depending on the needs, the distance may be as little as 2 centimeters, or as much as about 15 centimeters. Used, for instance, when reflected sounds wish to be excluded from the signal.

close-talk microphone Same as **close-talking microphone**.

close-talking microphone A microphone whose characteristics require it to be used in close proximity to its sound source. Such microphones are well-suited for use in surroundings with excessive background noise, such as a room full of telephone operators. Also called **close-talk microphone**, or **noise-canceling microphone (2)**.

closed 1. Continuous and/or unbroken. For instance, a closed circuit has an uninterrupted path for current. A switch in the closed position is the same as being on. **2.** Forming a self-contained unit or system. **3.** Blocking, or otherwise obstructing. **4.** Not having a space or gap, as opposed to having such a space or gap. **5.** Not accessible to all, or not designed to work across different architectures or with varied products. For example, a closed architecture.

closed architecture A device or system whose technical specifications are not made public. In computers, for instance, this discourages third-party vendors from developing compatible products. This contrasts with an **open architecture**, in which technical specifications are made public.

closed box 1. A component, device, or unit with an input and an output, which can be inserted into a system without the need to know its internal structure. It is simply used based on its input and output characteristics. Also called **black box (1)**. **2.** In computers, a hardware or software unit whose internal structure is unknown, but whose function is known. Also called **black box (2)**.

closed-box testing Also called **black-box testing**, or **functional testing**. **1.** Any tests utilizing the assistance of a **closed box (1)**. This contrasts with **open-box testing**, which requires detailed knowledge of the inner workings. **2.** Tests of software or hardware utilizing the assistance of a **closed box (2)**. This contrasts with **open-box testing**, which requires detailed knowledge of the inner workings.

closed-captioned TV programming which has **closed-captioning**. Its abbreviation is **cc**. Also called **closed-captioned programming**.

closed-captioned programming Same as **closed-captioned**.

closed-captioning In TV programming, the providing of text, and symbols such as ♫, which are intended to accurately transcribe the dialogue and describe the sounds of the viewed program. A decoder is necessary to show close captioning, the displayed text is superimposed on the viewed image, and such text may be encoded in any desired language. Useful, for instance, for those with reduced hearing. The term may also apply to other video programming so enabled. This contrasts with **subtitles**, which do not need a decoder to be displayed. Its abbreviation is **cc**.

closed captions The text and symbols seen when viewing programming with **closed-captioning**.

closed circuit 1. A circuit which has a complete and uninterrupted path for the flow of current. Also called **continuous circuit**, or **complete circuit**. **2.** A TV transmission not intended for public reception, as in closed-circuit TV.

closed-circuit communications system A self-contained communications system with a given number of units, with no provision for exchange of information with any outside units or systems. For instance, a closed-circuit TV system. Also called **closed-circuit system**.

closed-circuit jack A jack whose circuit is closed until the corresponding plug is inserted.

closed-circuit security system 1. A security system utilizing **closed-circuit TV**. **2.** A security system which monitors a **closed circuit (1)**, so as to detect any irregularities which may indicate a breach.

closed-circuit signaling A signaling system in which signals are provided by increases and decreases in the current of a circuit.

closed-circuit system Same as **closed-circuit communications system**.

closed-circuit television Same as **closed-circuit TV**.

closed-circuit television monitor Same as **closed-circuit TV monitor**.

closed-circuit television signal Same as **closed-circuit TV signal**.

closed-circuit television surveillance Same as **closed-circuit TV surveillance**.

closed-circuit television system Same as **closed-circuit TV system**.

closed-circuit television transmission Same as **closed-circuit TV transmission**.

closed-circuit TV A TV transmission system which is not intended for public reception. In this system, one or more cameras are in a closed circuit with one or more TV receivers, which are the only ones that can receive the transmitted signals. Used, for instance, for surveillance of a location. Its abbreviation is **CCTV**. Also called **closed-circuit TV system**.

closed-circuit TV monitor A monitor which receives **closed-circuit TV** transmissions. Its abbreviation is **CCTV monitor**.

closed-circuit TV signal The signal transmitted in a **closed-circuit TV** system. Its abbreviation is **CCTV signal**.

closed-circuit TV surveillance The use of **closed-circuit TV** for surveillance purposes. Its abbreviation is **CCTV surveillance**.

closed-circuit TV system Same as **closed-circuit TV**.

closed-circuit TV transmission The transmission of a **closed-circuit TV** signal.

closed core A magnetic core in which the paths of the magnetic lines of force do not pass through air. Such a design

generates less EMI, but handles less current than **open cores**. An example is a toroidal core.

closed loop 1. In computer programming, a loop which repeats itself indefinitely, unless interrupted or stopped by external intervention. An **open loop (1)** repeats itself a specified number of times. **2.** In a control system, a circuit in which a feedback signal and a reference input are compared, in order to control the output quantity. An **open loop (2)** does not incorporate such feedback.

closed-loop control system A control system in which a feedback signal and a reference input are compared in order to control the output quantity, so as to maintain operation within the specified parameters. For example, a thermostat may provide the feedback signal in a heating or refrigerating system. This contrasts with an **open-loop control system**, which has no feedback signal. Also called **closed-loop system**, or **feedback control system**.

closed-loop gain The gain of an amplifier when it has a feedback signal. When no such loop is present it is called **open-loop gain**.

closed-loop output impedance The impedance of an amplifier when it has a feedback signal. When no such loop is present it is called **open-loop impedance**.

closed-loop output resistance The resistance of an amplifier when it has a feedback signal. When no such loop is present it is called **open-loop resistance**.

closed-loop system Same as **closed-loop control system**.

closed-loop voltage gain In an amplifier with a feedback signal, the ratio of the output voltage to the input voltage. When no such loop is present, it is called **open-loop voltage gain**.

closed magnetic circuit 1. A complete and uninterrupted path for magnetic flux circulation around a ferromagnetic core. **2.** A complete and uninterrupted path for magnetic flux circulation.

closed standard A set of hardware and/or software specifications that are not made public. Such standards discourage interoperability, and may limit the proliferation of new technologies, platforms, interfaces, methods, applications, devices, and so on. This contrasts with an **open standard**, in which technical specifications are made public.

closed subroutine A subroutine that must be stored in a specified location to be accessed by a routine, while an **open subroutine** can be included in a routine from wherever said subroutine is located.

closed system A system whose technical specifications are not made public. In computers, for instance, this discourages third-party vendors from developing compatible products. This contrasts with an **open system**, in which technical specifications are made public

cloud absorption The absorption of electromagnetic radiation by water present in clouds, whether it is in solid, liquid, or gas form.

cloud attenuation The attenuation of electromagnetic radiation by clouds, due to scattering rather than absorption.

cloverleaf antenna An omnidirectional transmitting antenna whose radiating elements are shaped similarly to a four-leaf clover. It provides maximum radiation in the horizontal plane.

CLP Abbreviation of **cell loss priority**.

CLR Abbreviation of **cell loss ratio**.

cluster 1. A fixed number of disk sectors treated as a unit. It is the smallest unit of storage that a computer operating system can handle. Also called **allocation unit**. **2.** Two or more hardware devices or systems working as a unit.

cluster testing A testing procedure in which groups of components or devices, such as capacitors or chips, are tested as a single unit.

clustering In computers and networks, the combining of two or more systems or servers, so they can work as a unit. Used in networks, for instance, to improve capacity, or for continued operation if one fails.

clutter In radars, unwanted echoes that appear on the display screen. Clutter may be caused by rain, antenna movements, vegetation, and so on. Also known as **radar clutter**, or **background return**.

CLV Abbreviation of **constant linear velocity**.

cm Abbreviation of **centimeter**.

CM Abbreviation of **configuration management**.

Cm Chemical symbol for **curium**.

CMB Abbreviation of **cosmic microwave background**.

CMBR Abbreviation of **cosmic microwave background radiation**.

CMI Abbreviation of **computer-managed instruction**.

cmil Abbreviation of **circular mil**.

CMIP Acronym for **Common Management Information Protocol**.

CMIS Acronym for **Common Management Information Services**.

CML 1. Abbreviation of **computer-managed learning**. 2. Abbreviation of **current-mode logic**.

CMOS Acronym for complementary metal-oxide semiconductor. 1. A widely used semiconductor technology that combines PMOS and NMOS transistors on a single silicon chip in a manner which yields high speed combined with low power consumption. Used, for instance, in computer memory chips. Also called **CMOS technology**. 2. A chip incorporating **CMOS** (1). 3. Same as **CMOS memory**.

CMOS-based Abbreviation of complementary metal-oxide semiconductor-**based**. A semiconductor technology based on **CMOS** (1).

CMOS battery Abbreviation of complementary metal-oxide semiconductor **battery**. A battery which supplies power to the **CMOS memory** (1) of a computer.

CMOS chip Abbreviation of complementary metal-oxide semiconductor **chip**. A chip incorporating **CMOS** (1).

CMOS device Abbreviation of complementary metal-oxide semiconductor **device**. A semiconductor device incorporating **CMOS** (1).

CMOS memory Abbreviation of complementary metal-oxide semiconductor **memory**. Also called **CMOS** (3). 1. A battery-backed memory used in a computer, to store the information necessary to boot up, in addition to the date and time. 2. Computer memory using **CMOS** (1).

CMOS RAM Abbreviation of complementary metal-oxide semiconductor random-access memory. Computer RAM utilizing **CMOS** (1).

CMOS random-access memory Same as **CMOS RAM**.

CMOS setup Abbreviation of complementary metal-oxide semiconductor **setup**. A system configuration utility via which the settings of the **CMOS memory** (1) may be adjusted. It is usually accessed by pressing the appropriate key during booting.

CMOS technology Abbreviation of complementary metal-oxide semiconductor **technology**. Same as **CMOS** (1).

CMR 1. Abbreviation of **common-mode rejection**. 2. Abbreviation of **cosmic microwave radiation**.

CMRR Abbreviation of **common-mode rejection ratio**.

CMRS Abbreviation of **commercial mobile radio service**.

CMS Abbreviation of **color management system**.

CMV Abbreviation of **common-mode voltage**.

CMY Abbreviation of **cyan-magenta-yellow**.

CMYK Abbreviation of **cyan-magenta-yellow-black**.

CNAM Abbreviation of **calling name delivery**.

CNC Abbreviation of **computer numerical control**, or **computerized numerical control**.

CNG tone Abbreviation of **calling tone**. An 1100 Hz tone emitted by a calling fax terminal to indicate the desire to send a fax. The proper response is a **CED tone**.

CNR Abbreviation of **carrier-to-noise ratio**.

CNS Abbreviation of **complementary network services**.

CO Abbreviation of **central office**.

Co Chemical symbol for **cobalt**.

co-channel interference Also spelled **cochannel interference**. Also called **shared-channel interference**. 1. Interference arising from two or more signals of the same type being transmitted via the same communications channel. 2. Interference arising from two or more simultaneous transmissions over the same communications channel.

CO laser Abbreviation of **carbon monoxide laser**.

co-location Same as **colocation**.

co-location center Same as **colocation** (1). Also spelled **colocation center**.

co-location facility Same as **colocation** (1). Also spelled **colocation facility**.

co-location site Same as **colocation** (1). Also spelled **colocation site**.

co-processor Same as **coprocessor**.

co-resident Same as **coresident**.

co-routine Same as **coroutine**.

coarse adjustment The adjustment of a setting in relatively large and/or approximate increments. This contrasts with a **fine adjustment**, in which the changes are relatively small and/or more precise.

coarse control An adjustable component which makes **coarse adjustments**.

coast station A land station within a maritime mobile service. Used for communication to ships and other coast stations.

coastal bending A change in direction of a radio wave as it crosses a coastline. It usually only affects ground waves. Also called **coastal refraction, land effect**, or **shoreline effect**.

coastal refraction Same as **coastal bending**.

coated cathode A cathode which is coated with a metal-oxide compound, such as thorium oxide, in order to enhance its electron emissions. Also called **oxide-coated cathode**.

coated filament A filament which is coated with a metal-oxide compound to enhance its properties. A coated cathode is an example. Also called **oxide-coated filament**.

coating 1. A one or more materials which are applied as an outer, finishing, or protective layer. For instance, an electroplated layer, or a magnetic coating on recording tape. 2. The application of a **coating** (1).

coax Same as **coaxial cable**.

coax cable Same as **coaxial cable**.

coax line Same as **coaxial cable**.

coaxial antenna A vertical antenna fed by a coaxial line, in which the inner conductor of the coaxial cable is extended ¼ wavelength, while the outer conductor which formerly enclosed it is folded back by ¼ wavelength. Also called **sleeve antenna**.

coaxial attenuator An attenuator suitable for use with coaxial cables.

coaxial cable A concentric two-conductor transmission line in which the center conductor is surrounded by a dielectric, which is surrounded by the second conductor, which in turn is surrounded by the protective jacket. The center conductor is also known as inner conductor, while the other is known as the outer conductor. Such cables produce almost no external fields, nor are they too susceptible to them. This makes them well-suited for a variety of applications, including their use in transmitting cable TV, linking computer networks, and for various radio applications. Also called **coax, coax cable, coaxial line, coax line, coaxial transmission line, concentric cable, concentric line**, or **cable (3)**.

coaxial cable connector Same as **coaxial connector**.

coaxial capacitor Same as **cylindrical capacitor**.

coaxial cavity A cylindrical resonant cavity in which a center rod is in contact with reflecting surfaces and/or a tuning piston. Also called **coaxial resonating cavity**, or **coaxial cavity resonator**.

coaxial cavity resonator Same as **coaxial cavity**.

coaxial connector A fixture which enables connection of a coaxial cable between electrical devices. Such a fixture may be either a coaxial jack or a coaxial plug. There are various types, including BNC connectors, and F connectors. Also called **coaxial terminal**, or **coaxial cable connector**.

coaxial diode A diode suitable for use with coaxial cables.

coaxial filter 1. A filter suitable for use with coaxial cables. **2.** A filter composing a section of a coaxial cable.

coaxial jack A connector designed to match the appropriate coaxial plug, in order to make a connection. A **coaxial plug** is inserted into a **coaxial jack**.

coaxial line Same as **coaxial cable**.

coaxial loudspeaker Same as **coaxial speaker**.

coaxial plug A connector designed to match the appropriate coaxial jack, in order to make a connection. A **coaxial plug** is inserted into a **coaxial jack**.

coaxial receptacle A receptacle, such as a coaxial jack, which enables connection of a coaxial cable. A coaxial receptacle, for instance, may be found in the back of a satellite decoder, or protruding from a wall.

coaxial relay A relay suitable for use with a coaxial cable, without appreciably changing the properties of the line.

coaxial resonating cavity Same as **coaxial cavity**.

coaxial speaker A speaker system which incorporates two concentric speaker units, usually consisting of a tweeter mounted within a woofer, so that the sound radiates from a common axis. Also called **coaxial loudspeaker**.

coaxial splitter A coaxial connector which divides a signal so that it will travel over multiple paths. Used, for instance, as an antenna splitter.

coaxial stub A short length of coaxial cable used to branch another coaxial cable, usually appreciably changing the properties of the line.

coaxial switch An switch suitable for use with a coaxial cable, without appreciably changing the properties of the line.

coaxial terminal Same as **coaxial connector**.

coaxial transmission line Same as **coaxial cable**.

coaxial wavemeter A wavemeter incorporating an inner conductor in its central axis, upon which a sliding disk moves for tuning. Used, for instance, at microwave frequencies.

COB Abbreviation of **chip-on-board**.

cobalt A shining steel-gray metallic element whose atomic number is 27. It is hard, ductile, reactive, ferromagnetic, and has about 20 known isotopes, of which one is stable. Its applications in electronics include its use in certain cermets, for plating, and in semiconductors. Its chemical symbol is Co.

cobalt potassium cyanide Yellow crystals used in electronics research.

cobalt silicide A chemical compound used as a semiconductor.

cobaltic-cobaltous oxide Black or gray crystals whose chemical formula is Co_3O_4. Used in semiconductors. Also called **cobaltosic oxide**.

cobaltic potassium nitrite Yellow crystals used in ceramics.

cobaltite A silver-white to grayish mineral whose chemical formula is $CoAsS$. Used in ceramics.

cobaltosic oxide Same as **cobaltic-cobaltous oxide**.

cobaltous ammonium sulfate Red crystals used in ceramics and for cobalt plating.

cobaltous carbonate Red crystals whose chemical formula is $CoCO_3$. Used in ceramics.

cobaltous chloride Pale blue crystals whose chemical formula is $CoCl_2$. Used in electroplating.

cobaltous hydroxide A blue-green or rose-red powder whose chemical formula is $Co(OH)_2$. Used in storage-battery electrodes.

cobaltous oxide A grayish powder whose chemical formula is CoO. Used in ceramics and semiconductors.

cobaltous sulfate A red powder whose chemical formula is $CoSO_4$. Used in storage batteries, for plating, and in ceramics.

COBOL Acronym for **common business-oriented language**. A high-level programming language which has existed for over four decades, yet is still popular in business applications. It tends to be wordy, making for longer, yet more understandable, programs.

COC Abbreviation of **chip-on-chip**.

COC technology Abbreviation of **chip-on-chip technology**.

cochannel interference Same as **co-channel interference**.

Cockcroft-Walton accelerator A high-voltage DC particle accelerator which incorporates an array of rectifiers which rectify the lower AC voltage input, and capacitors which are charged by the DC. Also called **Cockcroft-Walton generator**.

Cockcroft-Walton generator Same as **Cockcroft-Walton accelerator**.

codan Acronym for **carrier-operated device antinoise**.

code 1. A system of symbols or values, accompanied by rules used to represent information, such as data, and to convert it from one form to another. For instance, ASCII code. **2.** A set of computer program instructions. Also called **program code**, or **computer code (1)**. Also, to write such instructions. **3.** A set of programming instructions and statements that are expressed in a form suitable for input into an assembler, compiler, or translator, which in turn transforms said code into machine code. Such code is usually written in a high-level or assembly language which is understandable by humans, while only machine code can be directly executed by the CPU. Also called **source code**. **4.** To scramble information, such as data, in a manner which only those with a key can decipher. Usually used for security purposes. Also called **encode (2)**, **encrypt**, or **scramble (1)**. **5.** Information which scrambled, or encoded. **6.** The key to unscramble coded information. Also called **key (2)**.

code character A character utilized to represent information within a code, and which has been derived in accordance with this code. For instance, a numeric value representing a text character in ASCII code.

code character set The complete set of characters utilized in a code.

code conversion 1. A transformation of information, such as data, from one code to another. **2.** A translation of program instructions between forms. For instance, between a programming language and machine language.

code converter Hardware or software which executes **code conversions**.

code density The amount of program instructions per unit of available memory space. When total memory is limited, code density should be higher.

code-division multiaccess Same as **CDMA**

code-division multiple access Same as **CDMA**.

code element A symbol or value which serves to compose a given code. For instance, in machine code, the elements are ones and zeroes.

Code-Excited Linear Prediction A speech-compression algorithm that provides high quality audio while utilizing very little bandwidth. Used, for example, for transmission of voice over packet-switching networks. Its abbreviation is **CELP**. Also called **Code-Excited Linear Predictive, Code-Excited Linear Prediction algorithm, Code-Excited Linear Prediction coding**, or **Code-Excited Linear Prediction compression**.

Code-Excited Linear Prediction algorithm Same as **Code-Excited Linear Prediction**. Its abbreviation is **CELP algorithm**.

Code-Excited Linear Prediction coding Same as **Code-Excited Linear Prediction**. Its abbreviation is **CELP coding**.

Code-Excited Linear Prediction compression Same as **Code-Excited Linear Prediction**. Its abbreviation is **CELP compression**.

Code-Excited Linear Predictive Same as **Code-Excited Linear Prediction**.

code form That which is presented as a code. The output of an encoder is in code form.

code generator 1. A computer program that enables the generation of the desired output code. For instance, converting a program written in COBOL into machine code. 2. A device which generates a coded output.

code ringing Telephone ringing patterns which convey information. For instance, varying combinations of long and short rings indicating which individual user of a party line is being called.

code set The complete set of symbols or other representations utilized in a code.

code snippet A small piece of programming code. Also called **snippet**.

code system The system a code uses to convert information from one form into another.

codec 1. Acronym for **coder-decoder**. 2. Acronym for **compressor-decompressor**.

coded Also called **encoded**. 1. Information, such as data, which is in the form of a code. Also called **encrypted** (1). 2. Information, such as data, which has been scrambled in some manner. Also called **encrypted** (2). 3. Programs or program instructions which have been written.

coded data Data which has been **coded** (1) or **coded** (2). May be used, for instance, where privacy or security is a concern. Also called **encoded data**, or **scrambled data**.

coded decimal digit A decimal digit expressed by a given pattern of zeroes and ones.

coded digit 1. A digit expressed by a given pattern of zeroes and ones. 2. A digit which has been transformed by a code.

coded program A computer program expressed in machine language.

coded representation Information which has been converted into another form, through the use of a code.

coded signal A signal which has been **coded** (1) or **coded** (2). May be used, for instance, where privacy or security is a concern. Also called **encoded signal**, or **scrambled signal**.

coded speech Speech which has been **coded** (2), so that it can only be understood with a receiver with the proper circuits and settings. May be used, for instance, where eavesdrop-ping is a concern. Also called **scrambled speech**, or **encoded speech**.

coded stop In a computer program or routine, a stop instruction.

coder That which generates a code.

coder-decoder Hardware and/or software which converts an analog input into a digital output, and vice versa. For instance, that which performs conversions between analog video and digital video. Its acronym is **codec**.

coding 1. The process of making a conversion through the use of a code. 2. In computer programming, a list of operations required to perform a given routine.

coding system The system a code uses to convert information from one form into another.

coefficient 1. A constant in an algebraic equation, as opposed to a variable. For instance, in the expression $3x$, 3 is the coefficient, and x is the variable. 2. A numerical measure that describes a physical or chemical property, and that is a constant under specified conditions. For instance, a coefficient of attenuation or of thermal expansion.

coefficient of absorption Also called **absorption ratio, absorption factor, absorption coefficient**, or **absorptivity**. 1. The proportion of certain wavelengths or energy retained by a medium to which another has been exposed. 2. The proportion of certain wavelengths or energy retained by a medium to which another has been exposed, per unit of area or thickness.

coefficient of attenuation A rating for a transmission medium through which a wave passes, such as a cable, which gives the rate of amplitude decrease in the direction of travel. Also called **attenuation coefficient, attenuation factor**, or **attenuation rate**.

coefficient of coupling Same as **coupling coefficient**.

coefficient of reflection Also called **reflection coefficient, mismatch factor, reflection factor, reflectance** (1), or **reflectivity**. 1. The ratio of the amplitude of a wave reflected from a surface, to the amplitude of the same wave incident upon the surface. 2. The ratio of the current delivered to a load whose impedance is not matched to the source, to the current that would be delivered to the same load if its impedance were fully matched.

coefficient of thermal expansion The ratio of a change in dimensions to a rise in temperature. This must be factored in, for instance, when affixing semiconductors devices, such as chips, to substrates. Its abbreviation is **CTE**.

coefficient of transmission Also called **transmission coefficient, transmission factor**, or **transmission ratio**. 1. The ratio of the flux transmitted by a body, to the flux incident upon the same body. 2. The ratio of the radiant energy transmitted from a body, to the total radiant energy incident upon the same body.

coercimeter An instrument which measures the magnetic intensity of a magnet or electromagnet.

coercive force The magnetic field which must be applied to a previously magnetized ferromagnetic material, in order to reduce to 0 the residual magnetism of said material. Usually expressed in oersteds. Also called **magnetic coercive force**.

coercivity For a given ferromagnetic material, the ease or difficulty in it being magnetized or demagnetized. A material exhibiting low coercivity is easy to magnetize and demagnetize, while a material with high coercivity is hard to magnetize and demagnetize. Also called **magnetic coercivity**.

COF Abbreviation of **chip-on-film**.

COG Abbreviation of **chip-on-glass**.

cogging Non-uniform rotation of an electric motor, due to variations in magnetic flux as rotors move past stators. It is especially noticeable at low to very low motor speeds.

cognitive machine **1.** A machine capable of acquiring and using knowledge through perception, learning, and other means. **2.** A machine whose knowledge includes self-awareness.

coherence Also spelled **coherency**. **1.** For two or more waves of a single frequency, a definite phase relationship. Two waves, for instance, are in phase if their crests and troughs meet at the same point at the same time. Interference patterns may be formed by radiation emitted by coherent sources, and high levels of coherence characterize laser radiation. **2.** Any form of temporal or phase correlation. **3.** Clarity or intelligibility, as in a clear reception of a voice transmission.

coherency Same as **coherence**.

coherent Characterized by **coherence**.

coherent bundle An assembly of optical fibers in which the relative spatial coordinates of the individual fibers are the same at both ends. Such a bundle serves, for instance, to transmit images. Also called **coherent fiber bundle**, or **aligned bundle**.

coherent carrier A carrier wave whose frequency and phase have a definite relationship with the frequency and phase of a reference signal.

coherent demodulation Same as **coherent detection**.

coherent detection Demodulation which depends on the phase of the carrier signal. Also called **coherent demodulation**.

coherent detector A detector whose output signal amplitude depends on the phase, not the intensity, of its input signal. Used, for instance, in radars which only display moving targets.

coherent echo In radars, an echo whose phase and amplitude at a given range is relatively constant. Such echoes may be produced, for instance, by a building or a ship.

coherent fiber bundle Same as **coherent bundle**.

coherent light Light in which there is a definite phase relationship between any given points of the waves composing it. All waves have the same, or nearly the same, wavelength. Such light travels in a completely, or almost completely, parallel beam, and may or may not be in the visible range. Lasers produce coherent light.

coherent light detection and ranging Its acronym is **colidar**. A radar system in which the beam of a laser is utilized to track objects. Also known as **laser radar**, or **ladar**.

coherent light source A source, such as a laser, of **coherent light**.

coherent oscillator An oscillator whose output may serve as a **coherent reference**. Used, for instance, in radars. Its acronym is **coho**.

coherent-pulse operation Pulse operation in which there is a fixed-phase relationship between each pulse and the next.

coherent-pulse radar A radar system in which a coherent oscillator provides a phase reference against which the phase of scanned objects may be compared.

coherent pulses Pulses which are in a fixed-phase relationship with each other.

coherent radiation Radiation in which there is a definite phase relationship between any given points of a cross section of the beam.

coherent reference A signal which serves as a reference for other signals, so that they can be phase-locked to establish and maintain coherence. Such a signal usually has a stable, or nearly stable, frequency. A coherent oscillator may serve as the source of such a signal.

coherent scattering Scattering in which there is a definite phase relationship between the incident and dispersed waves or particles.

coherent signal **1.** In radars, a signal whose phase is constant, and which is used for comparison to the phase of a scanned object. The phase shift determines the range of the scanned object. **2.** A signal whose phase is constant.

coherent source A source of coherent waves or radiation. Interference patterns may be formed by radiation emitted by coherent sources.

coherent transponder A transponder in which a definite relationship is maintained in the phase and frequency of the input and output signals.

coherent waves Two or more waves of a single frequency with a definite phase relationship.

coho Acronym for **coherent oscillator**.

COHO Same as **coho**.

coil One or more conductors wound in a series of turns, so as to introduce inductance, or to produce a magnetic field in an electric circuit. A coil may be wound, for example, around a ferromagnetic core, an insulating support, or around air. Used, for instance, in transformers, electromagnets, solenoids, motors, speakers, and so on. Also known as **electric coil**, **inductance (3)**, **inductance coil**, **inductor**, or **magnetic coil (1)**.

coil antenna A directional antenna consisting of a wire coil with one or more turns. Used, for instance, in radio direction finders and portable radio receivers. Also called **loop antenna**, or **frame antenna**.

coil form An insulated spool upon which a coil is wound. Also called **bobbin (1)**.

coil loading The placement of loading coils along a transmission line at regular intervals. This improves the transmission characteristics of the line.

coil neutralization A method of neutralizing an amplifier, in which capacitance cancels inductance in the feedback circuit. Also called **inductive neutralization**, or **shunt neutralization**.

coil winding A layer of conductors within a **coil**. Also, a complete winding, such as a primary or secondary coil. Also, the winding of such a layer or winding.

coin battery Same as **coin cell**.

coin cell A small primary cell that uses an alkaline electrolyte solution, usually potassium hydroxide. Used mostly for small devices such as hearing aids and watches. Also called **coin battery**, or **button battery**.

coincidence The occurrence of events or phenomena simultaneously, or within a given time interval. For instance, the arrival of two signals at the same time.

coincidence amplifier An amplifier which delivers an output signal only when two or more input signals are applied simultaneously.

coincidence circuit Also called **coincidence gate**, **coincidence detector**, or **coincidence counter**. **1.** A circuit with two or more inputs, which delivers an output pulse only if all inputs receive a pulse simultaneously, or within a given time interval. **2.** A circuit with two or more inputs, which delivers an output pulse only if a predetermined combination of inputs receive a pulse simultaneously, or within a given time interval.

coincidence counter Same as **coincidence circuit**.

coincidence detector Same as **coincidence circuit**.

coincidence gate Same as **coincidence circuit**.

coincident-current selection Within an array of magnetic storage cells, the selection of a specific cell by means of the simultaneous application of two or more currents.

cold **1.** Describing a component, circuit, or device which is disconnected from a source of voltage. Also known as **dead (1)**. **2.** Describing a circuit or component which is at ground potential. Also known as **dead (2)**.

COLD Abbreviation of **Computer Output to Laser Disc**.

cold backup A backup performed while the data being duplicated is not in use. This contrasts with a **hot backup**, in which an application is using the data. Also called **offline backup**.

cold boot The starting of a computer by turning on its power. Many program failures are resolved through turning off a computer then turning it back on again. Also called **cold start**, **hard boot**, or **startup** (2).

cold cathode A cathode whose function does not require it to have an applied heat. This contrasts with a **hot cathode**, which requires an applied heat to function. An electron tube incorporating a cold cathode is called a **cold-cathode tube**.

cold-cathode fluorescent lamp A fluorescent lamp that can be used as a source for back-lit liquid-crystal displays used in cameras, computers, and the such. Its abbreviation is **CCFL**. Also called **cold-cathode fluorescent tube**.

cold-cathode fluorescent tube Same as **cold-cathode fluorescent lamp**. Its abbreviation is **CCFT**.

cold-cathode rectifier A gas-filled rectifier tube which incorporates a **cold-cathode**. Also called **gas-filled rectifier**.

cold-cathode tube An electron tube which incorporates a **cold cathode**. Examples include phototubes and mercury-pool rectifiers.

cold chamber A chamber where a low temperature is maintained, within certain strict limits. May be used, for instance, to test devices which will likely be subjected to such temperatures.

cold-chamber test A test performed in a **cold chamber**.

cold emission The emission of electrons from a surface, through the application of a strong electric field. Such a surface may be that of a solid or a liquid. Used, for instance, in field-emission microscopy. Also called **field emission**.

cold junction The junction between thermocouple wires and the conductors leading to a measuring instrument. This junction is normally kept at room temperature.

cold light Light produced without the concomitant generation of a significant amount of heat. Bioluminescence and electroluminescence are two sources of such light.

cold soldering 1. Soldering performed without heat. 2. Soldering performed without sufficient heat, which may result in a bad connection.

cold start Same as **cold boot**.

cold swap A swap, such as that of a hard drive, in which the computer must be off before making the exchange.

cold welding Welding performed without heat. Pressure welding, for example, may occur without heat.

ColdFusion A suite utilized to build Web sites and deliver Web pages to users.

colidar Acronym for **coherent light detection and ranging**.

collaboration products Hardware and/or software products which enable multiple users to interact jointly, with a minimum of system-generated obstructions. Useful, for instance, in videoconferencing.

collaboration standards Standards which enable multiple users to interact jointly, with a minimum of system-generated obstructions. Common Intermediate Format is an example of such a standard.

collapsed backbone A network topology in which a backbone is the hub to which all subnetworks are connected. Used, for instance, in medium-sized LANs. Also called **backbone** (2).

collector 1. In a bipolar junction transistor, the region into which charge carriers flow from the base. The output of such a transistor is usually taken from the collector. Also, the electrode attached to this region. Also known as **collector region**, or **collector electrode**. Its symbol is **C**. 2. In

certain electron tubes, such as klystrons, an electrode which collects electrons or ions after they have performed their useful function.

collector capacitance In a bipolar junction transistor, the capacitance associated with the collector junction.

collector current In a bipolar junction transistor, the current flowing through the collector region.

collector cutoff In a bipolar junction transistor, the operating condition under which the collector current is reduced to its residual current.

collector-cutoff current The residual current flowing under collector cutoff operating conditions. Also called **cutoff current**.

collector efficiency In a bipolar junction transistor, the ratio of the useful power output to the power input. Usually expressed as a percentage.

collector electrode Same as **collector** (1).

collector junction In a bipolar junction transistor, the semiconductor junction between the base and collector regions.

collector region Same as **collector** (1).

collector resistance In a bipolar junction transistor, the resistance of the collector junction.

collector ring A conductive rotating ring which works in conjunction with one or more stationary brushes, in order to maintain a continuous electrical connection. Used, for instance, in an AC generator. Also called **slip ring**.

collector voltage In a bipolar junction transistor, the DC supply voltage applied between the base and collector.

collimate To make parallel. For instance, to convert a divergent beam of light into one that is parallel.

collimated beam A parallel, or nearly parallel, beam of radiation or particles. Such a beam has minimal convergence or divergence.

collimation The process of making parallel. For instance, the conversion of a divergent beam of light into one that is parallel.

collimator A lens or device which **collimates**.

collinear antenna array A directional antenna consisting of a series of elements arranged end to end, horizontally or vertically. The elements are usually of the same type. Also known as **collinear array**, or **linear array**.

collinear array Same as **collinear antenna array**.

collision 1. A powerful and immediate contact between two or more things. 2. Any contact between two or more things. This includes virtual contact, such as that of images on a computer screen. 3. In a communications network, an event where two or more stations transmit simultaneously over the same channel. This may result in data being garbled or lost.

collision avoidance 1. Any mechanism employed to avoid **collisions** (1). 2. In robotics, mechanisms utilized to avoid **collisions** (2). For instance, the use of proximity sensors. 3. The mechanisms by which networks, such as Ethernet, avoid **collisions** (3).

collision-avoidance system 1. A mechanism utilized to avoid **collisions** (1) between objects. For example, the use of radars to avoid collisions between aircraft. 2. A mechanism utilized to avoid **collisions** (2) between objects. For instance, the use of proximity sensors in robots to avoid contact with surrounding objects. Its abbreviation is **CAS**.

collision detection 1. Any mechanism employed to detect **collisions** (1). 2. In robotics, mechanisms utilized to detect **collisions** (2). For instance, the use of contact sensors. 3. The mechanisms by which networks, such as Ethernet, detect **collisions** (3).

colloidal graphite Finely powdered graphite, which is suspended in a solution. Used, for instance, to coat electron tubes.

colocation Also spelled **co-location**. **1.** A physical location within the facilities of a telecommunications company, which interconnects the equipment of a customer or competitor. This has benefits such as reducing the cost of operations, or being able to provide superior service. Used, for instance, by telephone companies or Internet providers. Also called **colocation center**, **colocation facility**, **colocation site**, or **physical colocation**. **2.** The interconnection of the equipment of a customer or competitor to the facilities of a telecommunications company, without such equipment being within a physical location of the latter. Also called **virtual colocation**.

colocation center Same as **colocation (1)**. Also spelled **co-location center**.

colocation facility Same as **colocation (1)**. Also spelled **co-location facility**.

colocation site Same as **colocation (1)**. Also spelled **co-location site**.

colog Abbreviation of **cologarithm**.

cologarithm The logarithm of the reciprocal of a number. For example, the cologarithm of 2 is the logarithm of 0.5. Its abbreviation is **colog**.

color Characteristics of light, such as wavelength, which enable visual perception to distinguish degrees of hue, saturation, and brightness/lightness. Brightness applies to light sources, and lightness to objects.

color balance In a color TV receiver with a three-gun picture tube, the adjustment of the electron beam emissions to compensate for any differences in the light-emitting characteristics of the red, green, and blue phosphors on the screen.

color-bar code A bar code whose parallel vertical bars may be of two or more colors.

color-bar generator A device which produces the signal used to produce a color-bar test pattern on the screen of a TV.

color-bar pattern Same as **color-bar test pattern**.

color-bar test pattern A pattern of colored bars used to evaluate color TV transmission and reception. Also called **color-bar pattern**.

color bits The number of bits allotted per pixel, to define the colors to be displayed. For instance, if 24 bits are assigned, then over 16.7 million colors may be displayed.

color breakup The separation of the color components of a TV image. This may originate before transmission, or may be caused by rapid changes in the viewing conditions, such as watching while blinking in quick succession.

color burst In a TV receiver, a color-synchronizing signal at the beginning of each scanning line, which establishes a frequency and phase reference for the chrominance signal. Also called **color burst signal**, **color-sync signal**, **burst (4)**, or **reference burst**.

color burst signal Same as **color burst**.

color carrier Same as **color subcarrier**.

color-carrier reference Same as **color-subcarrier reference**.

color channel In color TV, the channel carrying the **color signal**.

color code **1.** A system, utilizing markings such as color dots or bands, to indicate key information such as values, multipliers, and tolerances for components such as resistors and capacitors. **2.** A system utilizing color codes to identify terminals, leads, or polarity.

color coder In a TV transmitter, a circuit or device which transforms the separate red, green, and blue camera signals into color-difference signals, and combines these with the chrominance subcarrier. Also called **color encoder**, **encoder (2)**, or **matrix (3)**.

color coding **1.** The use of different colors to identify and categorize. An example is a color code. **2.** The coding of a

control, such as a button or knob, by a distinctive color. For example, an on/off button of a TV being red, while the volume control buttons are green. The colors which are usually most effectively differentiated are blue, green, orange, red, and yellow. Such coding may also be used in combination with size and shape coding.

color comparator An instrument which uses one or more photoelectric devices to compare a color with a given standard. Also called **photoelectric color comparator**.

color contamination In color TV, an error in color reproduction caused by incomplete separation of the red, green, and blue components of a color picture. Such errors may be electronic, optical, or mechanical in nature.

color control Same as **color-saturation control**.

color decoder In a TV receiver, a circuit or device which transforms the color-difference signals into separate red, green, and blue signals. Also called **decoder (2)**, or **matrix (4)**.

color depth The number of bits used to define a pixel in an image, as determined by the hardware and software. For instance, a 24-bit video adapter allows for over 16.7 million colors to be displayed. Also called **bit depth**, or **pixel depth**.

color-difference signal In a color TV receiver, a signal which is added to the luminance signal to produce one of the three primary signals. For example, in RGB there are three color-difference signals, which are the R-Y signal, the G-Y signal, and the B-Y signal, where R is red, G is green, B is blue, and Y is luminance. Each of these signals feeds the red, green, and blue channels respectively.

color edging In a color TV receiver, extraneous colors occurring at the boundaries between zones with different colors, or along the edges of objects. Also called **edging**.

color encoder Same as **color coder**.

color fidelity The degree of faithfulness with which colors are reproduced, as compared to the original source.

color filter A sheet of a material, such as plastic or glass, which selectively passes or absorbs given wavelengths of light.

color flicker In a TV receiver, brief and irregular fluctuations in a displayed image, occurring as a result of unwanted fluctuations in the chrominance and/or luminance signals.

color fringing In a color TV receiver, false colors occurring along the borders of objects. Small objects, for instance, may appear to be divided into various colors.

color graphics Computer graphics which are displayed in color.

color killer In a color TV receiver, a circuit which disables the chrominance circuits when a black-and-white signal is being received. For instance, such a circuit may seek a color burst signal, and if there is none, color decoding is shut off. Also called **color-killer circuit**, **color-killer stage**, or **killer stage**.

color-killer circuit Same as **color killer**.

color-killer stage Same as **color killer**.

color kinescope Same as **color picture tube**.

color management system A system utilized to help match colors so that they appear essentially the same when displayed or presented on different devices or mediums. Used, for instance, to maintain consistency between the colors presented by a monitor and those provided by a printer. Its abbreviation is **CMS**.

color match The complete correspondence between two colors. This may be achieved, for instance, through **color matching**.

color matching **1.** The selection of colors which correspond exactly with other colors. An instrument may be used for this, or the naked eye with the assistance of materials of varying colors that serve for comparison. **2.** The assuring that a color in one medium appears the same after conversion to another.

color monitor A display monitor which can display multiple colors. For example, an RGB monitor.

color oscillator In a color TV receiver, a crystal oscillator which generates the chrominance-subcarrier frequency, which is set at 3.579545 MHz. This signal is used for comparison to the signal of the same frequency sent by the transmitter. Also called **chroma oscillator**, **color-subcarrier oscillator**, or **chrominance-subcarrier oscillator**.

color phase In a color TV receiver, the difference in phase between a chrominance signal and the chrominance-carrier reference. The proper color phase is necessary for the proper reproduction of colors.

color picture signal The signal that contains the color information displayed on a **color picture tube**.

color picture tube A CRT which produces the color images displayed on color TV receivers and computer monitors. A three-gun color picture tube emits three beams of electrons, which are focused and directed towards color phosphors, which in turn produce the color images on the screen. Also called **color kinescope**, **color television picture tube**, **color TV picture tube**, or **tricolor picture tube**.

color primaries Colors which are combined in an additive mixture to yield a full range of colors. Red, green, and blue are usually used as the additive primary colors. They are called this way because they themselves are not formed by the combination of other colors. Also called **additive primary colors**, **primary colors**, or **primaries**.

color printer A computer printer whose output may be in color. For example, a cyan-magenta-yellow-black printer.

color purity **1.** For a given color, the degree to which all unwanted components are excluded. Such a color only contains the proper proportion of the desired primary colors. Also called **purity (4)**. **2.** The extent to which a color is not mixed with other colors. Said especially of a primary color which is not mixed with other primary colors. Also called **purity (3)**.

color registration The superimposition of varying proportions of the three primary colors, so as to form the desired composite color. Also called **color superimposition**.

color response The sensitivity an instrument or device has to light, as a function of its wavelength.

color sampling frequency Same as **color sampling rate**.

color sampling rate In a color TV receiver, the rate at which each of the primary colors is sampled. Also called **color sampling frequency**.

color saturation The extent to which a given color lacks a white component. 100% saturation means there is no white present. It is a measure of the vividness of hue. Also called **saturation (4)**, or **chroma (3)**.

color-saturation control In a color TV, a device, such as a potentiometer, which adjusts the level of color saturation. This determines how vivid the hues of the colors are in the displayed picture. If the chroma control is set to zero, the image becomes monochrome. Also called **color control**, or **chroma control**.

color scanner A scanner which is able to accept a color input, and digitize it accordingly.

color sensing The use of image sensors to differentiate between colors. Used, for instance, in robotics.

color separation The process of separating a color image into the four basic ink colors, which are cyan, magenta, yellow, and black, each forming a color layer. The four color layers are printed separately, then all are superimposed to give the impression of many more than four colors.

color signal In color TV, an electric signal, such as a chrominance signal, which contains the information necessary to control the ultimately displayed color images.

color spectrum Within the electromagnetic spectrum, the range of frequencies or wavelengths which encompass visible light. The wavelengths of the color spectrum range from about 400 nanometers for violet light, to about 750 nanometers for red light.

color STN Abbreviation of **color super-twist nematic**. A technology used in **color STN displays**.

color STN display Abbreviation of **color super-twist nematic display**. A color twisted nematic display in which the crystals are twisted from 180 to 270º. This provides greater contrast, and may be used, for instance, in portable computers, PDAs, or cellular telephones.

color subcarrier In color TV, the carrier whose sidebands are added to the luminance signal, in order to convey the color information. The frequency of this carrier signal is set at 3.579545 MHz. Also called **color carrier**, **chrominance carrier**, **chrominance subcarrier**, or **subcarrier (2)**.

color-subcarrier oscillator Same as **color oscillator**.

color-subcarrier reference In color TV, a continuous signal that has the same frequency as the chrominance subcarrier, and which is also fixed in phase with respect to the color burst. It serves as a phase reference for modulation and demodulation of a chrominance signal. Also called **color-carrier reference**, **chrominance-carrier reference**, or **chrominance-subcarrier reference**.

color super-twist nematic Same as **color STN**.

color super-twist nematic display Same as **color STN display**.

color superimposition Same as **color registration**.

color-sync signal Same as **color burst**.

color synchronizing signal Same as **color burst**.

color television Same as **color TV**.

color television picture tube Same as **color picture tube**.

color television receiver Same as **color TV (2)**.

color television signal Same as **color TV signal**.

color television system Same as **color TV system**.

color television transmitter Same as **color TV transmitter**.

color TFT Abbreviation of **color thin-film transistor**. Same as **color TFT LCD**.

color TFT display Same as **color TFT LCD**.

color TFT LCD Abbreviation of **color thin-film transistor liquid-crystal display**. A color liquid-crystal display where each pixel of the screen is powered separately and continuously by a transistor. This provides for a brighter image with higher contrast than a passive-matrix display. Also called **color TFT display**, **color TFT**, **color TFT liquid-crystal display**, **color thin-film transistor liquid-crystal display**, or **color thin-film transistor**.

color TFT liquid-crystal display Same as **color TFT LCD**.

color thin-film transistor Same as **color TFT LCD**.

color thin-film transistor liquid-crystal display Same as **color TFT LCD**.

color transmission The transmission of a color signal.

color TV Abbreviation of **color television**. **1.** A system through which color images are transmitted, then received and reproduced by TVs, or other devices equipped to do so. Also called **color TV system**. **2.** A device which reproduces the images received through a **color TV (1)**. Also called **color TV receiver**.

color TV picture tube Abbreviation of **color television picture tube** Same as **color picture tube**.

color TV receiver Abbreviation of **color television receiver**. Same as **color TV (2)**.

color TV signal Abbreviation of **color television signal**. The signal, containing color images, which a **color TV (2)** receives.

color TV system Abbreviation of **color television system**. Same as **color TV (1)**.

color TV transmitter Abbreviation of **color television transmitter**. An entity or device which transmits **color TV (1)** images.

coloration For an audio component, any audible alterations from the original input signal. Used especially in the context of speakers, where inaccuracies in reproduction represent coloration.

colorimeter A light-sensitive instrument which measures color by quantifying the intensities of the three primary colors contained within it. May be used, for instance, for comparison with a given standard.

colorimetry The measurement of color and its properties, such as wavelength and intensity. Colorimetry may be used, for instance, to determine very small concentrations of certain substances in a solution, or to detect trace amounts of metals in finished components.

Colpitts oscillator An oscillator incorporating an active device and a capacitive voltage divider, and which uses capacitive feedback. The active device may be a vacuum tube, a bipolar junction transistor, or a field-effect transistor.

columbium Its symbol is **Cb**. The name columbium, which still used mostly by metallurgists, is an obsolescent name for the chemical element **niobium**.

column A vertical arrangement or series. For instance, a column of pixels or digits. This contrasts with a **row**, which is a horizontal arrangement or series.

column loudspeaker Same as **column speaker**.

column speaker A speaker design which incorporates two or more speaker units mounted vertically along the same axis. When the speaker units are in phase, the produced sound is reinforced. The sound-distribution pattern of these speakers is flat in the elevation plane. Also called **column loudspeaker**.

.com On the Internet, a top-level domain name suffix. The **com** is an abbreviation of **com**mercial, although the entity employing this suffix need not be commercial. Also called **dot com**.

COM 1. Abbreviation of **communications port**. 2. Acronym for **Component Object Model**. 3. Abbreviation of **computer output to microfilm**.

COM 1 Same as **COM1**.

COM 1 port Same as **COM1**.

COM 2 Same as **COM2**.

COM 2 port Same as **COM2**.

com port Abbreviation of **communications port**.

com program Abbreviation of **communications program**.

COM1 The logical name given to the first serial port of a computer. Subsequent ports are COM2, COM3, and so on. Also called **COM1 port**.

COM1 port Same as **COM1**.

COM2 The logical name given to the second serial port of a computer. Also called **COM2 port**.

COM2 port Same as **COM2**.

coma 1. In a CRT, a form of distortion in which a spot moving away from the center of the screen forms a comet shape. May be due, for instance, to a misalignment of the focusing of the electron beams. 2. In an optical system, a form of aberration in which a point source of light forms a comet shape. Also known as **coma aberration**.

coma aberration Same as **coma (2)**.

coma lobes In an antenna, such as a dish antenna, a side lobe that occurs in the radiation or response pattern when only the reflector is tilted back and forth. Used, for instance, to eliminate the need to rotate the entire antenna.

comb filter A filter whose frequency response resembles the teeth of a comb, reflecting its alternately passing and rejecting narrow bands. May be used, for instance, in a TV receiver to separate the luminance and chrominance signals from the composite video signal.

combination microphone A microphone incorporating two or more microphones. Each microphone may be of the same or of a different type.

combination tone A third tone heard when two tones of different frequencies are listened to simultaneously.

combinational circuit A logic circuit whose output at any given moment depends only on its input values at that same instant. Such a circuit has no storage capability. Also called **combinatorial circuit**, or **combinational logic circuit**.

combinational logic A logic circuit or element whose output at any given moment depends only on its input values at that same instant. This contrasts with **sequential logic**, in which the output at any given moment depends, in part, on previous states. Also called **combinatorial logic**.

combinational logic circuit Same as **combinational circuit**.

combinational logic element A logic element whose output at any given moment depends only on its input values at that same instant.

combinatorial circuit Same as **combinational circuit**.

combinatorial logic Same as **combinational logic**.

combined head A magnetic tape or disk head which both reads and writes to the magnetic medium. For instance, such a head in the disk drive of a computer. Also called **read/write head**.

combiner A device that receives multiple separate inputs and unites them to form a common output. Used, for instance, to combine RF signals.

combiner circuit 1. A circuit that receives multiple separate inputs and unites them to form a common output. 2. A circuit that combines the luminance and chrominance signals with the synchronizing signals in a TV camera.

combo box In a GUI, a text-entry field which also has selectable default values. An example is a drop-down menu utilized for font size selection where a user can select from stated sizes or type in that desired.

comm port Abbreviation of **communications port**.

comm program Abbreviation of **communications program**.

command 1. One or more signals which actuate a device. For instance, a remote control sending RF signals to control a device. 2. An instruction that causes a computer action. For instance, pressing a function key to obtain the programmed result.

command buffer A segment of computer memory utilized to temporarily store commands. A command buffer is useful for various functions, such as repeating and editing commands.

command-driven Computer programs which accept commands in the form of special letters, words, or phrases. Although these commands must be learned, such programs are more flexible than **menu-driven programs**, in which menus are used to give commands.

command-driven program A program that is **command-driven**.

command-driven system An operating system that is **command-driven**.

command interpreter A program, which is part of the operating system of a computer, that accepts a given number of commands for the performance of programmed tasks. Such a program may, for example, be used to load an application. Also called **command processor**.

command language A computer language which uses a given number of commands with which it communicates with an operating system or other program.

command line On a computer display screen, the line where a command is typed.

command mode A mode of operation in which a computer or peripheral, such as a modem, accepts commands.

command processor Same as **command interpreter**.

command prompt The location on a command line where input may be entered. Such a prompt may blink to indicate that the computer awaits a command.

command queuing A feature which allows a computer to accept multiple commands as they are entered, then execute them in the order received or in a prescribed sequence.

command set 1. The group of commands a given device, piece of equipment, program, or system uses. 2. A device used to send and/or receive commands.

command shell The user interface for a **command interpreter**. Also called **shell (2)**.

command.com Same as **COMMAND.COM**.

COMMAND.COM The command interpreter for certain operating systems.

comment In computer programming, a statement, or text, that is embedded for any purpose except execution. For instance, a clarifying note detailing the rationale behind a given instruction. Also called **remark**.

commerce server Same as **commercial server**.

commercial carrier Same as **common carrier**.

commercial mobile radio service A business providing a mobile service, such as cellular telephony, connected to the public switched telephone network. Its abbreviation is **CMRS**.

commercial server A network server that handles business transactions such as online purchases made with credit cards, or inventory management. Used extensively over the Internet. Also called **commerce server**, or **merchant server**.

commercial software Programs, such as packaged software or those contained in application suites, which are sold to the general public.

committed information rate In a frame relay network, the rate at which information is guaranteed to be transmitted, averaged over a given time interval. If the network has additional bandwidth available, a user may exceed this amount. Usually expressed as a multiple of bits per second, such as 1.544 Mbps. Its abbreviation is **CIR**.

common 1. Shared by two or more entities, such as circuits, devices, components, systems, users, and so on. As seen, for instance, in a common-source connection. 2. Grounded. As seen, for instance, in a common ground.

common-access method An ANSI standard pertaining to the interface between SCSI peripherals and SCSI host adapters. Its acronym is **CAM**.

common-anode amplifier Same as **cathode follower**.

common-base amplifier A bipolar junction transistor amplifier whose base electrode, which is usually grounded, is common to both the input and output circuits. Also called **grounded-base amplifier**.

common-base circuit For a bipolar junction transistor, a mode of operation in which the base electrode, which is usually grounded, is common to both the input and output circuits. Also called **common-base connection**, or **grounded-base connection**.

common-base connection Same as **common-base circuit**.

common battery 1. A battery shared by two or more circuits, devices, components, pieces of equipment, or systems. 2. A battery which supplies all the DC power of a complete telephone system unit. It is usually located at a central office.

common-battery office A central office which has a **common battery (2)**.

common business-oriented language Same as **COBOL**.

common carrier A telecommunications entity, such as a telephone company, that provides services to the general public. Common carriers are usually regulated by the appropriate authorities, such as governmental agencies. Also called **commercial carrier**, **carrier (3)**, **communications common carrier**, **common communications carrier**, or **public carrier**.

common carrier company A company which is a **common carrier**.

common-cathode amplifier An electron-tube amplifier in which the cathode is at ground potential at the operating frequency. The input is applied between the control grid and ground, and the output is taken from between the plate and ground. It is a widely-used amplifier circuit. Also called **grounded-cathode amplifier**.

common-channel interference Interference resulting from two stations using the same channel.

common-channel interoffice signaling In telecommunications, a technique in which control signals are transmitted over a separate channel, freeing the rest of the channels for the transmission of voice, data, video, and so on. Its abbreviation is **CCIS**.

common-channel signaling In telecommunications, a technique in which signaling information is transmitted separately from user data, using a channel shared by the entire network. Its abbreviation is **CCS**.

common-collector amplifier A bipolar junction transistor amplifier in which the collector electrode, which is usually grounded, is common to both the input and output circuits. Also called **emitter follower**.

common-collector circuit Same as **common-collector connection**.

common-collector connection For a bipolar junction transistor, a mode of operation in which the collector electrode, which is usually grounded, is common to both the input and output circuits. Also called **common-collector circuit**, or **grounded-collector circuit**.

common communications carrier Same as **common carrier**.

Common Desktop Environment A GUI for open systems provided by the Open Group. Its abbreviation is **CDE**.

common-drain amplifier A field-effect transistor amplifier with a common-drain connection, in which the input signal is applied between the gate and drain, and the output signal is taken from between the source and drain. Also called **source follower**.

common-drain circuit Same as **common-drain connection**.

common-drain connection For a field-effect transistor, a mode of operation in which the drain electrode, which is usually grounded, is common to both the input and output circuits. Also called **common-drain circuit**, or **grounded-drain circuit**.

common-emitter amplifier A bipolar junction transistor amplifier in which the emitter electrode, which is usually grounded, is common to both the input and output circuits. Also called **grounded-emitter amplifier**.

common-emitter circuit Same as **common-emitter connection**.

common-emitter connection For a bipolar junction transistor, a mode of operation in which the emitter electrode, which is usually grounded, is common to both the input and output circuits. Also called **common-emitter circuit**, or **grounded-emitter circuit**.

common-gate amplifier A field-effect transistor amplifier in which the gate electrode, which is usually grounded, is common to both the input and output circuits. Also called **grounded-gate amplifier**.

common-gate circuit Same as **common-gate connection**.

common-gate connection For a field-effect transistor, a mode of operation in which the gate electrode, which is usually grounded, is common to both the input and output circuits. Also called **common-gate circuit**, or **grounded-gate connection**.

Common Gateway Interface Its abbreviation is **CGI**. In a World Wide Web server, a specification for the transfer of data between information servers and other applications, such as databases, of the server. Such exchanges, for instance, provide for dynamic Web pages to be tailored to the responses of an Internet user filling out an online order form, or making a search engine query.

Common Gateway Interface program Same as **Common Gateway Interface script**. Its abbreviation is **CGI program**.

Common Gateway Interface script Its abbreviation is **CGI script**. An application which is executed by a World Wide Web server in response to input from an Internet user accessing certain elements of a Web page, such as those contained in an order form. It is the most common way for Web servers to interact dynamically with Internet users. A CGI script, for instance, interprets the received information and transfers it to another program on the server such as a database, and then uses this information to update the Web page tailored to the user's needs. Also called **Common Gateway Interface program**.

common-grid amplifier An electron-tube amplifier in which the control grid is at ground potential at the operating frequency. Its input is applied between the cathode and ground, and the output is taken from between the plate and ground. Also called **grounded-grid amplifier**.

common ground In a circuit, a shared ground.

common-impedance coupling Coupling occurring between two or more circuits through inductance or capacitance.

Common Information Model An object-oriented model that provides a system to describe and share management information, while being application-independent and system-independent. Its abbreviation is **CIM**.

Common Intermediate Format Its abbreviation is **CIF**. A video format which supports both NTSC and PAL signals, and which is used in videoconferencing. Each CIF format is defined by its resolution, the original CIF standard specifying 288 lines and 352 pixels per line, at 30 frames per second. 16CIF, for instance, supports a resolution of 1408 x 1152, at 30 frames per second.

Common Internet File System An Internet file-sharing protocol. Its abbreviation is **CIFS**.

common language A language which is understood by multiple computers and any devices concurrently being used with them. Useful, for instance, in networks.

common LISP Abbreviation of **common list** processor. A standardized version of the LISP programming language. It is frequently used in academic and research settings, and for artificial intelligence applications.

common logarithm A logarithm whose base is 10. Also called **Briggs logarithm**.

Common Management Information Protocol A protocol used with **Common Management Information Services**. Its acronym is **CMIP**.

Common Management Information Services A standard used in network management. It defines functions and mes-

sages used to monitor and control a network. Its acronym is **CMIS**.

common mode Pertaining to signals which are identical in phase and in amplitude to both inputs of a differential device, such as a differential amplifier.

common-mode error A difference in voltage at the output terminals of an operational amplifier, caused by a common-mode voltage at the input.

common-mode rejection The ability of a differential amplifier to reject **common-mode signals**, while responding to signals which are not in phase. Its abbreviation is **CMR**. Also called **in-phase rejection**.

common-mode rejection ratio For a differential amplifier, the ratio of the differential mode gain, to the common mode gain. This is the extent to which the amplifier rejects common-mode signals. Its abbreviation is **CMRR**.

common-mode signal **1.** For a differential device, such as a differential amplifier, a signal applied equally to both inputs. Also called **in-phase signal** (1). **2.** The algebraic average of two signals applied to both ends of a balanced circuit, such as that of a differential amplifier. Also called **in-phase signal** (2).

common-mode voltage For a differential device, such as a differential amplifier, a voltage that appears equally at both inputs. Its abbreviation is **CMV**.

Common Object Request Broker Architecture Same as **CORBA**.

common return A return conductor that is common to two or more circuits.

common-source amplifier A field-effect transistor amplifier in which the source electrode, which is usually grounded, is common to both the input and output circuits. Also called **grounded-source amplifier**.

common-source circuit Same as **common-source connection**.

common-source connection For a field-effect transistor, a mode of operation in which the source electrode, which is usually grounded, is common to both the input and output circuits. Also called **common-source circuit**, or **grounded-source circuit**.

communication Also called **communications** (2), or **telecommunication**. **1.** The transmission of information between two or more points or entities. This includes the equipment, modes, mechanisms, and media used for this purpose. **2.** The information conveyed via **communication** (1). **3.** The use of electrical, electronic, electromagnetic, optical, or acoustic means to transmit information between two or more points. Also, the conveyed information.

communication channel Same as **communications channel**.

communication circuit Same as **communications circuit**.

communication common carrier Same as **communications common carrier**.

communication controller Same as **communications controller**.

communication equipment Same as **communications equipment**.

communication line Same as **communications line**.

communication link Same as **communications link**.

communication network Same as **communications network**.

communication port Same as **communications port**.

communication program Same as **communications program**.

communication protocol Same as **communications protocol**.

communication receiver Same as **communications receiver**.

communication satellite Same as **communications satellite**.

communication security Same as **communications security**.

communication server Same as **communications server**.

communication software Same as **communications software**.

communication speed Same as **communications speed**.

communication system Same as **communications system**.

communications 1. The science dealing with the transmission of information between two or more points or entities. This includes the modes, mechanisms, and media used for this purpose, and all efforts to advance this field. 2. Same as **communication**.

communications channel Same as **circuit (3)**.

communications circuit Same as **circuit (3)**.

communications common carrier Same as **common carrier**.

communications company A company or other entity which offers **communication (1)** services. Also called **telecommunication company**.

communications controller A peripheral which serves to control one or more communications lines linked to a computer. Communications controllers handle tasks such as sending, receiving, coding, decoding, and error-checking, thus freeing resources of the computer it is connected to. Such devices are generally nonprogrammable, but those that can be programmed are called **front-end processors**.

communications equipment The equipment necessary to transmit information between two or more points or entities. Such equipment may consist of a single a radio transmitter, it could be a few telephones, or can be the hardware, usually along with any necessary software, of a LAN or a WAN.

communications line Same as **circuit (3)**.

communications link Also called **information link**, or **data link**. 1. A connection which enables the transmission of information between two points or entities. 2. The resources which facilitate a **communications link (1)**.

communications network A system of computers, transmission channels, and related resources which are interconnected to exchange information. A communications network may be comparatively small, in which case it can be a LAN, or relatively large, in which case it could be a WAN. A LAN may be confined, for instance to a single building, while a WAN may cover an entire country. The communications channels in a network may be temporary or permanent. Also called **computer network (1)**, **network (1)**, or **telecommunications network**.

communications port An interface which serves to connect a peripheral device to a computer. Such a device may be a modem, a mouse, a monitor, another computer, and so on. Its abbreviation is **COM**, **com port**, or **comm port**.

communications program A program that manages the transmission of data between computers. Such programs usually work with a modem, and include file transfer protocols and terminal emulation. Its abbreviation is **com program**. Also called **communications software**.

communications protocol A standard, or a set of rules, which must be agreed upon in order for two or more devices, such as modems, to exchange information effectively. Such conventions must include considerations such as how to initiate and terminate a transmission, the transmission speed, and whether the transmission will be synchronous or asynchronous. In addition, these may include error-detection techniques, encryption, and so on. Examples of communications protocols include Point-to-Point Protocol, and Zmodem. Also called **protocol**.

communications receiver A device intended to receive radio-frequency communications, such as weather transmissions.

communications satellite An artificial satellite which relays, and usually amplifies, radio-wave signals between terrestrial communications stations, terrestrial communications stations and other communications satellites, or between other communications satellites. Such satellites provide high-capacity communications links, offer a very wide coverage area, and may transmit TV, telephone, and data signals, among many others. Its abbreviation is **comsat**. Also called **radio relay satellite**, **relay satellite**, or **repeater satellite**.

communications security The measures taken to help prevent unauthorized access, interception, or alteration of communicated information, such as that in data transmissions. Passwords and encryption are two common precautions. Its abbreviation is **COMSEC**. Also called **telecommunications security**.

communications server A computer that utilizes network-emulation software to connect asynchronous devices to a LAN or WAN. It manages communications, and takes care of protocol conversions. Also called **network-access server**, **remote-access server**, or **access server**.

communications software Same as **communications program**.

communications speed The rate at which data, or any form of information, is transmitted over a communications line. Such a speed may be expressed, for example, in some multiple of bits per second. Also called **transmission speed**.

communications system The equipment, modes, mechanisms, and media used to transmit data, voice, video, or other forms of information from one point to another. Also called **telecommunications system**.

Communicator A popular Web browser.

community-antenna television Same as **cable TV**.

community-antenna TV Same as **cable TV**.

commutating capacitor A capacitor which reverses the current in a silicon-controlled rectifier, so that it goes into cut-off condition.

commutation 1. The transfer of current from one path of a circuit to another path within the same circuit. Also, such a transfer between elements of the same circuit. 2. The reversals of current through the windings of an armature, to provide DC at the brushes. 3. In communications, the repeated, usually cyclical, sampling of multiple quantities, for multiplexing over a single channel.

commutator 1. A device, such as a switch, used to reverse current. 2. In a DC motor or generator, the section that causes the direction of the electric current to be reversed in the armature windings. 3. In a DC motor or generator, the section which maintains electrical continuity between the rotor and the stator. 4. In communications, a circuit or device used for **commutation (3)**.

commutator switch A device used to execute repetitive series of switching operations. Such a switch is usually rotary, and may be electrical or mechanical. Used, for instance, for **commutation (3)**. Also called **sampling switch**, or **scanning switch**.

compact disc Same as **CD (1)**. Also spelled **compact disk**.

compact disc audio Same as **CD audio**.

compact disc burner Same as **CD burner**.

compact disc caddy Same as **CD caddy**.

compact disc carousel player Same as **CD carousel player**.

compact disc changer Same as **CD changer**.

compact disc-DA Same as **CD-DA**.

compact disc digital audio Same as **CD audio**.

compact disc drive Same as **CD drive**.

compact disc-E Same as **CD-E**.

compact disc-erasable Same as **CD-erasable**.

compact disc-extra Same as **CD-extra**.

compact disc-G Same as **CD-G**.

compact disc-I Same as **CD-I**.

compact disc-interactive Same as **CD-interactive**.

compact disc interactive system Same as **CD-interactive system**.

compact disc jukebox Same as **CD jukebox**.

compact disc player Same as **CD player**.

compact disc-plus Same as **CD-plus**.

compact disc-R Same as **CD-R**.

compact disc read-only memory Same as **CD-ROM**.

compact disc read-only memory burner Same as **CD-ROM burner**.

compact disc read-only memory changer Same as **CD-ROM changer**.

compact disc read-only memory drive Same as **CD-ROM drive**.

compact disc read-only memory extended architecture Same as **CD-ROM extended architecture**.

compact disc read-only memory jukebox Same as **CD-ROM jukebox**.

compact disc read-only memory player Same as **CD-ROM player**.

compact disc read-only memory reader Same as **CD-ROM reader**.

compact disc read-only memory recorder Same as **CD-ROM recorder**.

compact disc read-only memory server Same as **CD-ROM server**.

compact disc read-only memory tower Same as **CD-ROM tower**.

compact disc read-only memory writer Same as **CD-ROM writer**.

compact disc read-only memory XA Same as **CD-ROM XA**.

compact disc-recordable Same as **CD-recordable**.

compact disc recorder Same as **CD recorder**.

compact disc-rewritable Same as **CD-rewritable**.

compact disc-ROM Same as **CD-ROM**.

compact disc-ROM drive Same as **CD-ROM drive**.

compact disc-RW Same as **CD-RW**.

compact disc system Same as **CD system**.

compact disc-text Same as **CD-text**.

compact disc tower Same as **CD tower**.

compact disc-universal data format Same as **CD-UDF**.

compact disc video Same as **CDV**.

compact disc-WO Same as **CD-WO**.

compact disc write once Same as **CD write once**.

compact disk Same as **compact disc**.

Compact Flash Abbreviation of **CompactFlash card**.

Compact Flash card Same as **CompactFlash card**.

compact Peripheral Component Interconnect Same as **compactPCI**.

CompactFlash Abbreviation of **CompactFlash card**.

CompactFlash card A flash card which is approximately 43 x 36 mm, has a 50-pin interface, and provides capacities which can exceed 1 GB. Its abbreviation is **CompactFlash**, or **CF card**.

compaction Same as **compression (2)**.

compactPCI Abbreviation of **compact** Peripheral Component Interconnect. A PCI bus specification intended to better address uses within industrial environments. Its features include a more rugged package, high reliability, modularity, and the ability to perform hot swaps.

compander Acronym for **compressor-expander**.

companding A process in which a compressor limits the output signal when recording or transmitting, and an expander increases the input signal when reproducing or receiving. Used, for example, to improve the signal-to-noise ratio.

compandor Same as **compander**.

comparator 1. A circuit, device, or instrument which compares quantities. Used, for example, to compare a variable with a standard measure or a desired value, or for comparing two quantities with each other. 2. A circuit or device, such as a differential amplifier, whose output depends on the comparison of its inputs.

compare The examining of two or more items or quantities, so as to ascertain if they are equivalent or not. If they are not, a determination can be made, for instance, as to their relative magnitude. In computers, compare operations are used, for instance, when analyzing, sorting, or deciding which action to take after a given operation. Also called **compare operation**.

compare operation Same as **compare**.

comparison An operation in which two or more items or quantities are compared.

comparison bridge A bridge circuit which generates an error signal when the output voltage deviates from a reference voltage. It is similar to a four-arm bridge.

compass 1. An instrument which indicates a horizontal reference direction. Such an instrument usually consists of one or more magnetic needles which freely turn on a pivot, and which point to magnetic north. 2. Any instrument which indicates direction. Such a device need not be magnetic.

compatibility 1. The ability of two or more computer systems or components to work together seamlessly, without modification. This also extends to programs and shared files. For instance, a printer that works essentially flawlessly with a given computer, or a program that works equally well across multiple platforms. 2. The extent to which **compatibility** (1) is achieved. 3. The ability of two or more components, devices, pieces of equipment, or systems to work properly together, without modification. 4. The extent to which **compatibility** (3) is achieved.

compatibility mode A mode of operation which makes hardware or software compatible with other hardware or software that would otherwise be incompatible. For instance, a program which is optimized to run within one operating system, but is able to be used in compatibility mode with another.

compatible Characterized by **compatibility** (1), or **compatibility** (3).

compatible color television Same as **compatible color TV system**.

compatible color television system Same as **compatible color TV system**.

compatible color TV Same as **compatible color TV system**.

compatible color TV system Abbreviation of **compatible color television system**. A color TV system whose transmitted color images can be received and viewed by monochrome TV receivers which have not been modified. Also called **compatible color TV**.

compatible components Two or more components which work properly together without modification.

compatible devices Two or more devices which work properly together without modification.

compatible mode A mode in which hardware or software can be used so that it works properly with otherwise incompatible hardware or software.

compatible systems Two or more systems which work properly together without modification.

compensated amplifier A wideband amplifier whose frequency range is increased through the proper selection of circuit components and characteristics.

compensated semiconductor A semiconductor with two types of impurities or imperfections, in which the electrical effects of one type of impurity or imperfection partially cancels the other. For instance, a donor impurity partly annulling the electrical effects of an acceptor impurity.

compensated volume control In an audio system, a volume control which incorporates a circuit that boosts low frequencies when listening at low-volume settings. This compensates for the lower auditory response humans have under these circumstances, thus making the sound more natural. Such a control may also boost high frequencies. Also called **loudness control** (1).

compensating capacitor Also called **compensation capacitor**. 1. In a radio direction finder, a variable capacitor used to improve the accuracy of direction indication. Also called **balancing capacitor**. 2. A capacitor which is utilized to compensate for other components in a circuit. For instance, a temperature-compensating capacitor.

compensating filter A selective filter which is utilized to compensate for a deficiency, irregularity, or otherwise undesirable quantity. Also called **compensation filter**.

compensating leads An additional pair of leads, which are used alongside the working leads of an instrument, to compensate for environmental effects such as changes in temperature. Used, for instance, in a resistance thermometer. Also called **compensation leads**.

compensation The offsetting, counterbalancing, neutralizing, or stabilizing of a component, circuit, device, piece of equipment, or system. Compensation may be used, for instance, to make up for system deficiencies, environmental complications, or for adjusting equipment to meet specific needs.

compensation capacitor Same as **compensating capacitor**.

compensation filter Same as **compensating filter**.

compensation leads Same as **compensating leads**.

compensation signal A signal recorded on a magnetic tape track, which enables electrically correcting errors in tape speed during playback.

compensator A component, circuit, device, or piece of equipment which serves to offset, counterbalance, neutralize, or stabilize. Used, for instance, to make up for system deficiencies, environmental complications, or for adjustments made to meet specific needs.

competitive local exchange carrier A communications entity which offers local telephone service. Such an entity may offer other services, such as long-distance calling, Internet access, and so on. Its abbreviation is **CLEC**.

compilation Also called **compiling**. 1. The process of taking the source code of a program written in a high-level language and translating it into machine language. 2. The process of taking a set of high-level language statements and translating them into a lower-level representation.

compilation error 1. An error occurring during **compilation**. 2. An error detected during **compilation**.

compilation time Same as **compile time**.

compile 1. To take the source code of a program written in a high-level language, and translate it into machine language using a **compiler** (1). 2. To take a set of high-level language statements, and translate them into a lower-level representation using a **compiler** (2).

compile time The time it takes to **compile**. Also called **compilation time**.

compile-time error An error that occurs while a program is being compiled, as opposed to a **runtime error**, which occurs while a program is being executed.

compiled language A computer programming language in which all the code is translated into machine language before being executed. This contrasts with an **interpreted language**, in which each statement is translated then executed, followed by the next statement, and so on. LISP is a programming language that has both compiler and interpreter versions.

compiler Also called **compiler program**, or **compiling program**. 1. A computer program which takes the source code of a program written in a high-level language and translates it into machine language. When using a compiler, all the code is translated before any program instructions are executed, while an **interpreter** translates and executes each statement or instruction before moving on to the next. 2. A computer program which takes a set of high-level language statements and translates them into a lower-level representation.

compiler program Same as **compiler**.

compiling Same as **compilation**.

compiling error 1. An error occurring during **compilation**. 2. An error detected during **compilation**.

compiling program Same as **compiler**.

complement The numerical result obtained when a number is subtracted from the radix, which is the number of digits used in a numbering system. For instance, the complement of 6 in the decimal number system is 4: $(10 - 6) = 4$. The complement of a number in the binary number system is the other: 1 is the complement of 0, and 0 is the complement of 1. Used in computers, for instance, to represent negative numbers. Also called **true complement**, or **radix complement**.

complement number system A system of handling numbers in which arithmetic operations are performed on the complements of numbers. Used, for instance, to handle negative numbers in a simpler manner.

complementary 1. Mutually completing. 2. Compensating for mutual deficiencies. 3. In semiconductors, having components of opposite polarities working together. For instance, incorporating **pnp** and **npn** transistors on the same substrate.

complementary colors Two colors, which when combined in the appropriate proportions, yield an achromatic color. For instance, red and green.

complementary metal-oxide semiconductor Same as **CMOS**.

complementary metal-oxide semiconductor-based Same as **CMOS-based**.

complementary metal-oxide semiconductor battery Same as **CMOS battery**.

complementary metal-oxide semiconductor chip Same as **CMOS chip**.

complementary metal-oxide semiconductor device Same as **CMOS device**.

complementary metal-oxide semiconductor memory Same as **CMOS memory**.

complementary metal-oxide semiconductor RAM Same as **CMOS RAM**.

complementary metal-oxide semiconductor random-access memory Same as **CMOS RAM**.

complementary metal-oxide semiconductor setup Same as **CMOS setup**.

complementary metal-oxide semiconductor technology Same as **CMOS technology**.

complementary network services Additional services offered to clients of telephone companies, such as call waiting, call forwarding, or voice mail. Its abbreviation is **CNS**.

complementary operation In Boolean algebra, an operation which negates another given operation. For example, a NOR operation negates an OR operation. Thus, NOR and OR are said to be complementary.

complementary operator In Boolean algebra, the operator which identifies an operation which negates another.

complementary symmetry Same as **complementary-symmetry circuit**.

complementary-symmetry circuit A circuit which incorporates transistors of opposite polarities, such as **npn** and **pnp**, in an arrangement that achieves push-pull operation. Also called **complementary symmetry**.

complementary transistors Transistors of opposite polarities, such as **npn** and **pnp**, used in the same circuit or device. Used, for instance, to achieve push-pull operation.

complementary wave In a transmission line, such as a coaxial cable, an electromagnetic wave that arises as a result of reflection. This may occur, for instance, at each end of the line.

complete carry In computers, a carry technique in which the carries that result from an operation are propagated to other digit positions. This contrasts with a **partial carry**, where carries are temporarily stored.

complete circuit Same as **closed circuit (1)**.

complete failure A failure which causes the immediate and entire loss of the proper function of the affected component, circuit, device, piece of equipment, or system.

completed call A telephone call that is connected to its destination. A completed call is registered as soon as the connection is made, not when the connection is terminated. Also called **completed call attempt**, or **effective call**.

completed call attempt Same as **completed call**.

complex impedance The expressing of an electrical impedance incorporating both its real number and imaginary number components. Its formula is: $Z = R + jX$, in which Z is the impedance, R is the resistance, X is the reactance, and j is the square root of -1. That is, $j^2 = -1$. Z is expressed in ohms.

complex instruction set computer A computer whose CPU supports instructions of various types and sizes, including numerous address modes. The execution of each instruction generally requires multiple clock cycles. This contrasts with a **reduced-instruction set computer** whose CPU utilizes fewer and simpler instructions. Its acronym is **CISC**.

complex instruction set computing Computing utilizing a **complex instruction set computer**. Its acronym is **CISC**.

complex notation The expressing of a quantity incorporating both its real number and imaginary number components. For instance, acoustic impedance is comprised of acoustic resistance, the real number component, and acoustic reactance, the imaginary number component.

complex number A number in the form of $a + bi$, in which a and b are real numbers, and i is the square root of -1. That is, $i^2 = -1$. In electronics, the symbol utilized to express the square root of -1 is usually j. When $a = 0$, the resulting expression is an **imaginary number**.

complex plane A plane with two perpendicular axes upon which complex numbers are represented. The horizontal axis represents the real number component, while the vertical axis represents the imaginary number component.

complex quantity A quantity that has both real number and imaginary number components. Complex quantities are expressed using complex notation.

complex signal A signal with multiple components.

complex tone A tone consisting of two or more pure sinusoidal frequencies.

complex variable A variable that has both real number and imaginary number components.

complex wave A wave which multiple sine-wave components, each of a different frequency.

complex waveform The shape of a **complex wave**.

compliance The displacement of a mechanical system, such as the cone suspension of a speaker, when a force is applied. It is the mechanical and acoustical equivalent of capacitance.

compliance voltage The output voltage of a constant-current source.

component 1. A constituent part, especially of an organized whole or system. 2. Same as **circuit component**. 3. One of the parts which constitute a magnitude. For example, the real or imaginary number components of impedance.

component density 1. The number of components per unit area. 2. The number of components per unit volume.

component digital Same as **component digital video**.

component digital video A digital signal in which the video information is carried by separate signals, as in the luminance and chrominance signals being separate components. Examples include RGB and YCbCr. Its abbreviation is **component digital**. Also called **digital component video**.

component engineering The application of engineering principles to the fabrication, analysis, uses, testing, compatibility, improvement, and the like, of circuit components.

component failure A failure of a component within a system. May be due to any of various factors, including misuse, wearing-out, or as a consequence of the failure of another system component.

component failure analysis 1. The study of the causes that lead to **component failures**. 2. The calculation of a **component failure rate (2)**.

component failure rate 1. The number of component failures per unit measure, such as time or cycles. 2. The number of expected component failures per unit measure, such as time or cycles.

component layout The physical arrangement of the components of a system, such as those of a circuit.

component lead A conductor, usually a wire, by which circuit components are connected to other components, devices, equipment, systems, points, or materials.

Component Object Model In computers, a binary standard which enables objects to interact with other objects, whether they are written in the same programming language or another. Its acronym is **COM**.

component response 1. A change in the behavior, operation, or function of a component as a consequence of a change in its external or internal environment. For example, a change in its output resulting from a change in its input. 2. A quantified **component response (1)**, such as a frequency response.

component signal Same as **component video**.

component software Software components, such as program routines, which are designed to be combined with other such components to form larger programs. This allows, for example, for applications to be quickly customized by using available constituents whose function is known. Also called **componentware**.

component video In TV transmission and reception, the carrying of video information in separate signals, which are then combined before viewing. For instance, sending the red, green, blue, blanking, and synchronizing signals separately. Also called **component video signal**, or **component signal**.

component video signal Same as **component video**.

componentware Same as **component software**.

Composer A popular Web authoring tool.

composite cable A cable which combines conductors of different types and/or gauges. For example, one incorporating both optical and metallic signal-carrying components.

composite circuit Same as **composited circuit**.

composite color signal Same as **composite video signal**.

composite color sync A TV signal which includes all the synchronizing signals, such as the color burst, necessary for the proper function of a color TV receiver.

composite conductor An electrical conductor composed of two or more strands of different materials. For instance, a conductor with strands of copper and strands of aluminum.

composite digital Same as **composite digital video**.

composite digital video A digital video signal combining the color picture signal, plus all blanking and synchronizing signals. Examples include digitally encoded PAL and NTSC. Its abbreviation is **composite digital**. Also called **digital composite video**.

composite filter A filter which incorporates two or more types of filters connected in cascade.

composite picture signal Same as **composite video signal**.

composite plate The electrodeposition of two or more metals, in successive layers.

composite signal Same as **composite video signal**.

composite video Same as **composite video signal**.

composite video signal A TV signal combining the color picture signal, plus all blanking and synchronizing signals. For example, an NTSC video signal, or an PAL video signal. Also called **composite color signal**, **composite picture signal**, **composite signal**, or **composite video**.

composite wave filter Two or more wave filters connected in cascade.

composited circuit Also called **composite circuit**. **1.** A communications circuit which is used simultaneously for telephony and signaling. **2.** A communications circuit which is used simultaneously for telephony and DC telegraphy.

composition resistor A resistor composed of finely-divided particles, usually carbon, which are mixed with a ceramic binder, formed into a desired shape, then has leads attached to each end. An example is a carbon resistor.

compound **1.** Consisting of two or more components. **2.** Same as **chemical compound**.

compound generator A DC generator having both series and shunt windings.

compound modulation A series of modulation processes in which the modulated wave from one stage becomes the modulating wave for the next. Also called **multiple modulation**.

compound motor A DC motor having both series and shunt windings.

compound semiconductor A semiconductor material consisting of two or more elements. For instance, gallium and arsenic, or boron and phosphorus. Such semiconductors, for instance, may emit radiation in the visible range, and can be used for semiconductor lasers.

compound winding A winding which has both series and shunt windings.

compress To reduce data or information in size or volume. For instance, to compress a video signal or a message sent through a network, or to compact a file. This helps maximize the use of storage space, facilitates faster transfer, and also reduces transmission bandwidth.

compressed-air loudspeaker Same as **compressed-air speaker**.

compressed-air speaker A speaker which uses a stream of compressed air to enhance its reproduction of sound. Also called **compressed-air loudspeaker**.

compressed audio Audio data which is encoded so that it occupies less room. This helps maximize the use of storage space, facilitates faster transfer, and also reduces transmission bandwidth.

compressed data Data which is encoded so that it occupies less room. This helps maximize the use of storage space, facilitates faster transfer, and also reduces transmission bandwidth.

compressed digital video Digital video which is compressed to facilitate faster transfer, while also occupying less resources of the communications channel. Its abbreviation is **CDV**.

compressed file A file whose contents are encoded so as to occupy less room. This helps maximize the use of storage space, facilitates faster transfer, and also reduces transmission bandwidth.

Compressed Serial Line Internet Protocol Same as **Compressed SLIP**.

Compressed SLIP Abbreviation of **Compressed Serial Line Internet Protocol**. A version of SLIP which compresses certain information to reduce overhead, thus making it faster than SLIP. Its acronym is **CSLIP**.

compressed video Video data which is encoded so that it occupies less room. This helps maximize the use of storage space, facilitates faster transfer, and also reduces transmission bandwidth.

compression **1.** A reduction in size or volume. **2.** The encoding of data or information so that it occupies less space and/or bandwidth. There are many algorithms used for compression, and depending on the information being encoded, space savings can range from under 10% to over 99%. Compression can also be achieved by increasing packing density. Also called **compaction**, **data compression**, or **information compression**. **3.** The variation in the gain of a signal, so that the lower magnitude levels have a higher effective gain than the higher magnitude levels. This results in a more balanced output, where lower-level signals have a better signal-to-noise ratio, while higher levels signal do not overload.

compression program Same as **compression utility**.

compression ratio **1.** A ratio which describes the proportion of space saved after data compression. For example, a reduction of 50% in the occupied space may be expressed a 2:1 compression ratio. **2.** The ratio of the gain of a device at a reference signal level, to the gain at a given higher signal level. Usually expressed in decibels. **3.** The ratio of the gain of a device at a given lower signal level, to the gain at a given higher signal level. Usually expressed in decibels.

compression utility A utility, such as WinZip or PKZIP, utilized to compress files. Also called **compression program**.

compression wave A wave which is propagated by means of the compression of a fluid. For instance, sound waves propagating through air. Also called **compressional wave**.

compressional wave Same as **compression wave**.

compressor **1.** A circuit or device which reduces the range of its input signal magnitudes. **2.** A program or routine which compresses data.

compressor-decompressor Hardware and/or software which compresses and decompresses information such as audio or video signals. Its acronym is **codec**.

compressor driver A horn speaker in which the energy radiated from its diaphragm is channeled through a small opening, providing higher-pressure sound waves, which then interact with the lower-pressure area of the horn before being radiated out the mouth of the horn.

compressor-expander Its acronym is **compander**. **1.** A device incorporating a compressor and an expander. **2.** A system in which a compressor limits the output signal when recording or transmitting, and an expander increases the input signal when reproducing or receiving. Used, for example, to improve the signal-to-noise ratio.

Compton-Debye effect Same as **Compton effect**.

Compton effect An increase in the wavelengths of X-rays and gamma rays when they have been scattered by electrons present in matter. After a collision, the scattered radiation has decreased in energy, while the electron has increased in energy. Also called **Compton scattering, Compton process,** or **Compton-Debye effect**.

Compton electron An electron whose energy has increased, as a result of the **Compton effect**. Also called **Compton recoil electron**.

Compton process Same as **Compton effect**.

Compton recoil electron Same as **Compton electron**.

Compton scattering Same as **Compton effect**.

Compton shift The increase in the wavelengths of X-rays and gamma rays as a result of the **Compton effect**.

COMPUSEC Acronym for **computer security**.

computation **1.** The use of a computer. **2.** The action, or manner, of computing. **3.** A computed amount.

computation-bound Same as **CPU-bound**.

compute **1.** To perform a calculation. **2.** To use a computer.

compute-bound Same as **CPU-bound**.

computed axial tomography Same as **computed tomography**. Its acronym is **CAT**.

computed axial tomography scan Same as **computed tomography**. Its abbreviation is **CAT scan**.

computed axial tomography scanner A device which takes a **computed tomography**. Its abbreviation is **CAT scanner**.

computed radiography Radiography performed with the assistance of a computer. An example is computed tomography.

computed tomography A medical diagnostic technique in which a series of X-rays are taken at different angles around a specific part of a body. A computer combines the images of the multiple scans, and produces cross-sectional or three-dimensional pictures of internal organs. It is relatively safe, painless, quick, and very sensitive. Its abbreviation is **CT**. There are several variations, including PET and SPECT. Also called by several other names, including **computed axial tomography scan, computed tomography scan, computer-aided tomography, computerized axial tomography,** and **CAT scan**.

computed tomography scan Same as **computed tomography**. Its abbreviation is **CT scan**.

computed tomography scanner Same as **CT scanner**.

computer A device which accepts an input, processes this information, then presents an output. A computer is composed of hardware and software. The hardware includes all the physical equipment, such as keyboard, mouse, monitor, memory, storage mediums, cables, and the key computing component, which is the CPU. The software, in the form of programs, tells the hardware how to process information. Programs include operating systems, network software, and applications such as word processors. Most computers are digital, although there are other types, such as analog and hybrid. A computer may be contained on a single chip. Also called **computing machine**.

computer-aided design The use of computer hardware and software to design models and products. High-resolution monitors are combined with powerful CPUs and sophisti-

cated software, to enable the design of anything from ICs to airplanes. Such systems allow users to view and manipulate objects in two or three dimensions, using outlines or solid shapes, while keeping track of and updating any changes which affect other components of the object, among many other available functions. Its acronym is **CAD**.

computer-aided design and drafting Similar to **computer-aided design**, but with additional features for drafting, such as dimensioning. Its acronym is **CADD**.

computer-aided design/computer-aided manufacturing Its acronym is **CAD/CAM**. The utilization of computers and specialized machines to design and manufacture a given product. The CAD portion is used for designing, and its output is the set of instructions that tell the CAM part what to do in order to provide the desired result.

computer-aided diagnosis The use of computers to assist in the diagnosis of medical conditions. Such assistance may include correlating symptoms and test results with possible causes, or scanning X-rays and identifying suspicious areas. Its acronym is **CAD**. Also called **computer-assisted diagnosis**.

computer-aided dispatch system A dispatch system, such as that used by emergency services, which utilizes fixed and mobile computers to make better use of the available personnel and vehicle resources. Its acronym is **CADS**.

computer-aided education Same as **computer-aided instruction**. Its abbreviation is **CAE**.

computer-aided engineering The use of computers to analyze engineering designs. It may be used, for instance, for electromagnetic, structural, or circuit analysis. Its abbreviation is **CAE**.

computer-aided graphics Same as **computer graphics (1)**.

computer-aided instruction The incorporation of computers into educational processes. Such learning may take place at an educational institution, a library, a workplace, or the home. Computer resources may help instructors at any educational level to better convey information, and allows students to proceed at a learning rate they are comfortable with. These methods usually make extensive use of graphics, and often allow one-click access to the Internet for yet more information. Its abbreviation is **CAI**. Also called **computer-aided education, computer-aided learning, computer-aided teaching, computer-assisted instruction, computer-assisted learning,** or **computer-assisted teaching**.

computer-aided learning Its abbreviation is **CAL**. Same as **computer-aided instruction**.

computer-aided manufacturing The utilization of computers to automate manufacturing systems. Used, for instance, to control an assembly line in which each station is occupied by industrial robots. Its acronym is **CAM**. Also called **computer-assisted manufacturing**.

computer-aided prototyping The use of computers to help prepare prototypes of products. A computer may be used, for instance, to scan a three-dimensional drawing, process this information, and then send the control signals which form the prototype. An example of computer-aided prototyping is to provide an efficient means of determining and evaluating production requirements before entering the manufacturing phase of a product.

computer-aided prototyping system A system utilized for **computer-aided prototyping**.

computer-aided software engineering The use of computers to assist in the automation of all phases of software design, development, documentation, and maintenance. Its acronym is **CASE**. Also called **computer-aided systems engineering, computer-assisted software engineering,** or **computer-assisted systems engineering**.

computer-aided software engineering technology A software technology which enables **computer-aided software engineering**. Its abbreviation is **CASE technology**.

computer-aided software engineering tools The resources available to software engineers utilizing **computer-aided software engineering** technology.   Such tools include automatic coding and intelligent agents. Its abbreviation is **CASE tools**.

computer-aided systems engineering Same as **computer-aided software engineering**. Its acronym is **CASE**.

computer-aided teaching Same as **computer-aided instruction**. Its acronym is **CAT**.

computer-aided testing Same as **computer-assisted testing**. Its acronym is **CAT**.

computer-aided tomography Same as **computed tomography**. Its acronym is **CAT**.

computer-aided training Same as **computer-assisted training**. Its acronym is **CAT**.

computer-aided translation The use of computers to assist in the translation of one or more natural languages into another or others, and vice versa. Such programs may incorporate features such as pop-up glossaries for specific areas of expertise, the automatic matching of text already translated, and so on. Also called **computer-assisted translation**. Its acronym is **CAT**.

computer animation The use of computers to create and/or modify animated graphics.

computer architecture The design of a computer and its components. This includes for which purposes it can be used, along with its memory and processing requirements and capabilities. This in turn affects variables such as what software may be used with it, and which peripherals are compatible. Also called **architecture (2)**, or **computer system architecture**.

computer-assisted design Same as **computer-aided design**. Its acronym is **CAD**.

computer-assisted diagnosis Same as **computer-aided diagnosis**. Its acronym is **CAD**.

computer-assisted education Same as **computer-aided instruction**. Its acronym is **CAE**.

computer-assisted instruction Same as **computer-aided instruction**. Its acronym is **CAI**.

computer-assisted learning Same as **computer-aided instruction**. Its acronym is **CAL**.

computer-assisted manufacturing Same as **computer-aided manufacturing**. Its acronym is **CAM**.

computer-assisted retrieval The use of computers to assist in the location and/or retrieval of materials stored on diverse media, such as paper, microfilm, and the such. The computer maintains a listing or index, from which the desired materials are selected. Its acronym is **CAR**.

computer-assisted software engineering Same as **computer-aided software engineering**. Its acronym is **CASE**.

computer-assisted systems engineering Same as **computer-aided software engineering**. Its acronym is **CASE**.

computer-assisted teaching Same as **computer-assisted instruction**. Its acronym is **CAT**.

computer-assisted testing The use of computers for performing tests. May be used, for instance, to evaluate a solid model prepared by a CAD program. Its acronym is **CAT**. Also called **computer-aided testing**.

computer-assisted tomography Same as **computed tomography**. Its acronym is **CAT**.

computer-assisted training The use of computers for training and education. Such learning may take place at a workplace, a library, an educational institution, or the home. These methods usually make extensive use of graphics, and may cover practically any educational or occupational discipline. Its acronym is **CAT**. Also called **computer-aided training**, or **computer-based training**.

computer-assisted translation Same as **computer-aided translation**. Its acronym is **CAT**.

computer-automated software engineering Same as **computer-aided software engineering**. Its acronym is **CASE**.

computer-automated systems engineering Same as **computer-aided software engineering**. Its acronym is **CASE**.

Computer Automated Virtual Environment Same as **Computer Automatic Virtual Environment**. Its acronym is **CAVE**.

Computer Automatic Virtual Environment A virtual reality system where real-time images are projected on four or more surfaces, at least three of which must be walls. The use of special glasses gives everything surrounding the user a three-dimensional feel. These glasses also serve to track the path of the user's vision, so that the virtual environment can be changed appropriately. Its acronym is **CAVE**. Also called **Computer Automated Virtual Environment**.

computer axial tomography Same as **computed tomography**. Its acronym is **CAT**.

computer-based education Same as **computer-based instruction**. Its abbreviation is **CBE**.

computer-based instruction The use of computers as the basis for lessons. Such learning may take place at an educational institution, a library, a workplace, or the home. Computer resources help at any educational level to better convey information, and allows students to proceed at a learning rate they are comfortable with. These methods usually make extensive use of graphics, and often allow one-click access to the Internet for yet more information. Its abbreviation is **CBI**. Also called **computer-based education**, **computer-based teaching**, or **computer-based learning**.

computer-based learning Same as **computer-based instruction**. Its abbreviation is **CBL**.

computer-based reference Reference materials, such as CD-ROMs, which require a computer to be accessed. Its abbreviation is **CBR**.

computer-based teaching Same as **computer-based instruction**. Its abbreviation is **CBT**.

computer-based training Same as **computer-assisted training**. Its abbreviation is **CBT**.

computer bus A set of conductors which serve as a channel which provides parallel data transfer from one part of a computer to another. An example is the local bus, which is the pathway between the CPU, the memory, and high-speed peripherals. A bus is subdivided into two parts: the address bus, which identifies specific locations of stored data, and the data bus, which transfers the data. The width of the bus determines the amount of data which can be transferred at a time. A 256-bit bus, for example, can transfer 256 bits at a time. The clock speed determines how fast these transfers take place. Also called **bus (1)**.

computer center A centralized location where computers, data libraries, and associated equipment are housed. Large entities may have many mainframes at such a location. May be used, for instance, for centralized processing. Also called **datacenter**, **data-processing center**.

computer chip A chip, especially the CPU, used in a computer.

computer code **1.** Same as **code (2)**. **2.** A system of symbols or values, accompanied by rules used to represent instructions and information, so as to convert them into a form that is suitable for computer processing. For instance, machine code, source code, or ASCII code.

computer conferencing The use of computers and communications channels to enable two or more people, or groups, to interact. Such conferencing may involve any combination of written, verbal, or visual communication. In addition, other resources may be shared, such as files. Examples include chats and videoconferencing. Also called **computer meeting**.

computer connector A device which serves to connect computers to peripherals, other computers, systems, or networks. Examples include DB connectors and BNC connectors.

computer control The use of a computer or microprocessor to control a component, circuit, device, piece of equipment, machine, mechanism, or system.

computer-controlled A component, circuit, device, piece of equipment, machine, mechanism, or system which is controlled by a computer or microprocessor.

computer-controlled device A device which is controlled by a computer or microprocessor. For instance, a switch.

computer-controlled equipment Equipment which is controlled by a computer or microprocessor. For example, laboratory apparatuses.

computer-controlled machine A machine which is controlled by a computer or microprocessor. For instance, an automated teller machine.

computer-controlled manufacturing Manufacturing which is controlled by a computer or microprocessor. For instance, computer control of an assembly line.

computer-controlled process A process which is controlled by a computer or microprocessor. For instance, the weaving of optical fibers.

computer-controlled robot A robot which is controlled by a computer or microprocessor. Almost all robots are computer-controlled. The computer which controls a robot is called controller.

computer-controlled system A system which is controlled by a computer or microprocessor. For instance, a communications network.

computer data Information which is represented in a manner suitable for computer use.

computer display **1.** A device utilized to display the images produced by a computer. For a computer to generate such a display, it requires a display adapter. Such a display incorporates a viewing screen and its housing. Also called **monitor (1)**, **display (1)**, or **display monitor**. **2.** The visual output of a **computer display (1)**.

computer effects The use of computers and computer graphics to enhance, modify, add, or remove sights and/or sounds in movies, TV, and other multimedia presentations. Such effects are extensively used, and some of the techniques employed include motion capture and tweening. Also called **computer special effects**, or **digital effects**.

computer emergency response team Its acronym is **CERT**. An organization which studies security threats, such as viruses and denial-of-service attacks, to the Internet and WANs. CERT, among other services, offers consultations, and publishes security alerts.

computer-enhanced image An image which has been enhanced with the help of a computer. For this, an image must first be digitized, and then it may be modified in many ways, including adjustment of colors, shades, or scaling. Such techniques have applications in many fields, including art, medicine, and meteorology.

computer-enhanced instruction The use of computers to enhance any education process. The use of computers at educational institutions and libraries, or at home, can help instructors at any educational level to better convey information, and allows students to proceed at a learning rate they are comfortable with. Also called **computer-enhanced learning**.

computer-enhanced learning Same as **computer-enhanced instruction**.

computer environment A specific hardware and software configuration. Many programs, for instance, will only func-

tion properly in a given computer environment. Also called **environment (2)**.

computer file A collection of information which is stored as a unit. Computer files may be retrieved, modified, stored, deleted, or transferred. Each type of file requires the appropriate software for the proper handling of its contents. There are many file types, including data files, program files, system files, and multimedia files. Also called **file**.

computer game A computer program in which one or more users interact with the computer and/or other players, for amusement or competition. Such programs may be of varying complexity, and some feature faithfully realistic simulations of complicated activities such as aircraft piloting. Also called **game**, or **video game (1)**.

computer graphics Its abbreviation is **CG**. **1.** The use of computers to store, manipulate and display images, such as pictures, animation, and video. Computers handle graphics either as vector graphics or as bitmapped graphics, and these are used in programs such as those for computer-aided design, desktop publishing, painting, presentations, or animation. A graphics monitor is required to display the images, and almost all computer monitors are of this type. Also called **computer-aided graphics**, or **graphics (3)**. **2.** Any visible computer output, except for alphanumeric characters. Examples, include icons, pictures, drawings, and graphs. Also called **graphics (4)**.

Computer Graphics Interface A device-independent software standard pertaining to the display and printing of computer graphics. Its abbreviation is **CGI**.

Computer Graphics Metafile A standard file format for storing and communicating graphical information. Its abbreviation is **CGM**.

computer hardware In a computer system, all the physical objects and equipment, such as the CPU, keyboard, mouse, monitor, memory, storage mediums, cables, connectors, cards, and so on. This contrasts with **software**, which consists of programs which tell the hardware what to do. A computer system is composed of both hardware and software. Also called **hardware (1)**.

computer-input microfilm Microfilm information scanned and converted into code suitable for computer use. Its abbreviation is **CIM**.

computer instruction **1.** A command or statement in a computer program or routine. Also called **instruction (1)**. **2.** A computer instruction in machine code. Such an instruction can be directly executed by a processor. Also called **machine instruction**, or **instruction (2)**. **3.** The use of computers in teaching.

computer-integrated manufacturing The use of computers in the automation of tasks pertaining to manufacturing. Its abbreviation is **CIM**.

computer interface **1.** A device which serves to connect a computer to peripheral devices or a network. Such interfaces include SCSI, RS-232, and network adapters. Also called **interface (2)**. **2.** Any interface used within a computer, such as those between programs, between hardware and software, between hardware and a user, or between software and a user. Also called **interface (3)**.

computer keyboard A computer input device which incorporates alphanumeric keys, function keys, control keys, cursor keys, and additional keys, such as enter, backspace, and insert. A 101-key keyboard is commonly used. Also called **keyboard (2)**.

computer language A defined set of characters, symbols, and rules utilized together for communication with a computer, or between computers. The term also applies to communications between computers and peripherals, or between peripherals. There are various examples, including assembly languages, machine languages, command languages, and programming languages. Also called **language (1)**.

computer literacy The sufficient understanding of computers and associated devices, such as printers, to be able to use them to accomplish certain tasks. For instance, a person with computer literacy should be able to effectively use certain applications. It is not necessary for them to command the most advanced functions.

computer logic Also called **logic**. **1.** The functions performed by a computer which involve operations such as mathematical computations and true/false comparisons. Such logic is usually based on Boolean algebra. **2.** The circuits in a computer which enable the performance of logic functions or operations, such as AND, OR, and NOT. These include gates and flip-flops. Also, the manner in which these circuits are arranged. Also called **machine logic (2)**. **3.** The totality of the circuitry contained in a computer. Also called **machine logic (3)**.

computer-managed instruction The use of computers to manage or assist with lessons and in the evaluation a learner's performance. Such learning and assessment may take place at an educational institution, a library, a workplace, or the home. Its abbreviation is **CMI**. Also called **computer-managed learning**.

computer-managed learning Its abbreviation is **CML**. Same as **computer-managed instruction**.

computer meeting Same as **computer conferencing**.

computer memory The locations within a computer that serve for temporarily holding and accessing data in a machine-readable format. Memory chips are used for this purpose, and most are allocated for RAM, or main memory. Memory is usually quantified in multiples of bytes. For example, a computer with 1 gigabyte of RAM can hold approximately 1 billion bytes, or characters, of information, and this is the total temporary workspace this computer has available. Other forms of memory in a computer include ROM, PROM, and EPROM. Although **memory** and **storage** are sometimes used synonymously, storage refers to a more permanent form of holding and accessing data, using magnetic or optical media, such as disks and tapes. Also called **memory**, or **system memory (1)**.

computer mouse A computer pointing device which when moved upon a surface also moves the cursor on the display screen, and which usually has two or more buttons used for selecting, performing functions, or the like. Some mice also have a scroll wheel, which further facilitates navigating through documents, Web pages, and so on. Most mice are either mechanical or optical, and some are cordless. A computer mouse looks more or less like its rodent counterpart, in that it has a compact body and a longish tail, while the moving and clicking of the former is analogous to the scurrying and nose-twitching of the latter, hence its name. Also called **mouse**.

computer music The use of a computer to create, help create, process, or reproduce music. A Musical Instrument Digital Interface is used to enable a computer to communicate with a properly equipped musical instrument or synthesizer.

computer network **1.** Same as **communications network**. **2.** A group of computers which exchange information, complete with the necessary links between them.

computer networking The use of a **computer network** to exchange information.

computer numerical control A form of automatic control of tools and machines. For instance, the instructions of a computer program controlling the position, movements, and speed of a drill in a manufacturing process. Its abbreviation is **CNC**. Also called **computerized numerical control**, or **numerical control**.

computer-on-a-chip A complete computer that is contained on a single chip. Such a chip must have a CPU, memory, a clock, and input/output circuits. These chips may be used in countless items, including automobiles, toys, appliances, clocks, and so on. Also called **microcontroller**, **microcomputer (2)**, or **one-chip computer**.

computer operation A specific action carried out by a computer, such as a compare operation.

computer-oriented language A low-level language, such as an assembly or machine language, which is designed for use only with a specific computer line. Also called **machine-oriented language**.

computer output on microfilm Same as **computer output to microfilm**.

Computer Output to Laser Disc The storing of computer information, such as data, on an optical medium, such as a CD-ROM. Used, for instance, to archive high volumes of information. Its abbreviation is **COLD**.

computer output to microfilm Computer data recorded on microfilm, or microfilm which records computer data. Its abbreviation is **COM**. Also called **computer output on microfilm**.

computer performance evaluation The measuring and evaluating of the performance of computer hardware and software, to determine the efficiency of an existing system for its designated uses, or the suitability of a prospective system. Such an evaluation, for instance, may quantify the ratio of system availability to unavailability, appropriateness of the utilized applications, and so on. Its abbreviation is **CPE**.

computer platform Same as **computing platform**.

computer power Same as **computing power**.

computer program A set of instructions which when translated and executed cause a computer to perform specific operations. Computer programs are written in programming languages, which may be classified as high-level or low-level. High-level languages, such as COBOL, C++, and Java, require a compiler or interpreter to translate statements into machine language. Low-level languages, such as an assembly language, are much closer to machine language, but still require an assembler for conversion into machine language. The only language a computer understands is machine language. Although **program** and **software** are used mostly synonymously, software is a bit more of a general term, as it is used to differentiate from the physical equipment of a computer, the hardware, from the instructions which tell a computer what to do, the software. An application is both a program and software, but when using the term software it usually refers to multiple application programs, or application programs in general, in which case they are called application software. Also called **program (1)**.

computer programming The creating of computer programs. The key components in this process include a full understanding of which tasks the final program will be meant to perform, a command of the programming language being used to write the program, the development of the program logic to suitably address problems to be solved, and the testing and debugging of the program. Also called **programming (2)**.

computer-readable Information that is in a form that can serve as computer input. This includes information in binary form, such as files stored on magnetic disks, and data obtainable through optical-character recognition. Also called **machine-readable**.

computer science The study of the theoretical and practical aspects of computers. These include computer architecture, software engineering, and their integration. Many aspects of computer science require comprehensive knowledge of other realms of expertise, such as electronics and information theory.

computer screen The viewing screen of a computer display. Such a display may be a CRT display, a liquid-crystal display, a plasma display, and so on.

computer security Its acronym is **COMPUSEC**. **1.** The measures taken to protect a computer, the information contained within it, or that which it exchanges with other computers. Protection from eavesdropping or unauthorized access may be afforded, for instance, by passwords and encryption. Other modes of protection include biometric techniques, such as retinal scans, which determine if access is granted to a system. **2.** A security system which incorporates computers. A computer combined with a camera may be used, for example, to monitor the ingress and egress of motor vehicles through a property.

computer simulation The use of a computer to imitate an object or process. Sophisticated software, combined with accurate input devices, enable a computer to respond mathematically to factors such as changing conditions, as if it were the object or process itself. Such simulations may be used to represent or emulate almost anything, including weather conditions or biological processes, and may be utilized to test new theories. Also called **simulation (2)**.

computer software Also called **software**. A set of instructions or programs, which when translated and executed cause a computer to perform specific operations. Software is usually classified into two general categories, which are system software, and application software. System software includes control programs, such as the operating system, while application software refers to programs which perform specific tasks, such as a word processor. Although **program** and **software** are used mostly synonymously, software is a bit more of a general term, as it is used to differentiate from the physical equipment of a computer, the hardware, from the instructions which tell a computer what to do, the software. An application is both a program and software, but when using the term software it usually refers to multiple application programs, or application programs in general, in which case they are called application software.

computer special effects Same as **computer effects**.

computer storage **1.** A device which provides for the holding and accessing of data for an indefinite period. Such information is kept in a machine-readable format, and should be maintained intact in the event of an interruption in system power. The main types of storage media are magnetic, such as tapes and hard disks, and optical, which include holograms and DVDs. Storage is usually quantified in multiples of bytes. For example, a computer hard drive with a 256 gigabyte capacity can hold approximately 256 billion bytes, or characters, of information. Although **storage** and **memory** are often used synonymously, memory refers to a more temporary form of holding and accessing data, such as that in RAM. Also called **storage (1)**. **2.** The act or process of placing data in a **computer storage device**. Also called **storage (2)**.

computer storage device A device which serves for **computer storage (1)**. Also called **storage device**.

computer-supported collaborative learning Computer-aided instruction in which the participants engage in collaborative efforts. Such learning usually makes extensive use of groupware. Its abbreviation is **CSCL**.

computer-supported collaborative work The use of groupware, telepresence, and other computer-based tools to enable people to work together and enhance their efforts. Its abbreviation is **CSCW**.

computer system The complete complement of components required for a computer to function. These include the CPU, keyboard, mouse, monitor, memory, storage mediums, cables, and so on, which comprise the hardware of the computer itself, plus any necessary peripheral devices. In addition, a computer system incorporates the operating system. Also called **system (6)**.

computer system architecture Same as **computer architecture**.

computer telephone integration Same as **computer telephony integration**. Its abbreviation is **CTI**.

computer telephony Same as **computer telephony integration**. Its abbreviation is **CT**.

computer telephony integration The integration of computer and telephony technologies. There are numerous applications, including interactive voice response, automated call distribution, voice mail, and caller identification where the stored information associated with that number appears on a computer screen. Its abbreviation is **CTI**. Also called **computer telephone integration**, or **computer telephony**.

computer telephony integration application An application, such as interactive voice response, used in **computer telephony integration**.

computer terminal **1.** A computer input/output device which incorporates a video adapter, monitor, keyboard, and usually a mouse. Used in networks. Also called **console (3)**, **console terminal**, **terminal (1)**, **network terminal (1)**, or **station (6)**. When such a terminal has no processing capability it is called **dumb terminal**, while a terminal that incorporates a CPU and memory does have processing capability, and is called **intelligent terminal**. **2.** A personal computer or workstation which is linked to a network. Also called **terminal (2)**, **network terminal (2)**, or **station (7)**.

computer tomography Same as **computed tomography**. Its abbreviation is **CT**.

computer tomography scan Same as **computed tomography**. Its abbreviation is **CT scan**.

computer tomography scanner Same as **CT scanner**.

computer TV Also called **PC/TV**, **TV/PC**, or **TV computer**. **1.** A PC, such as that with a TV card, enabled for TV viewing. **2.** A TV with built-in computer capabilities, such as Internet access.

computer utility A program that performs tasks pertaining to the management of computer resources. For instance, such programs may perform diagnostic functions, or handle the management of files. Also called **utility**.

computer virus A computer program or programming code which replicates by seeking out other programs onto which to copy itself. It usually passes from computer to computer through the sharing of infected files, and goes unnoticed until it attacks, unless an antivirus program that recognizes the virus detects it first. The effects of a virus may be as slight as a friendly message appearing once on a screen, it might cause a system crash, or it may reach the extreme of destroying a hard drive. Also called **virus**.

computer vision Simulation of vision by the use of a computer along with sophisticated programs which interpret data from its optical sensors. Used, for example, in robotics.

computer voice A simulation of a human voice generated by a computer. May be used, for instance, in robotics, or to assist a person with reduced vision which is using a computer.

computer word The fundamental unit of storage for a given computer architecture. It represents the maximum number of bits that can be held in its registers and be processed at one time. A word for computers with a 32-bit data bus is 32 bits, or 4 bytes. A word for computers with a 256-bit data bus is 256 bits, or 32 bytes, and so on. Also called **word**, or **machine word**.

computerize **1.** To use a computer to assist with, or control one or more actions which would otherwise be done without one. For instance, to perform calculations, or automate manufacturing. **2.** To equip with one or more computers. **3.** To store in a computer or to convert information into a format a computer can use.

computerized axial tomography Same as **computed tomography**. Its acronym is **CAT**.

computerized axial tomography scan Same as **computed tomography**. Its acronym is **CAT scan**.

computerized axial tomography scanner Same as **CT scanner**. Its acronym is **CAT scanner**.

computerized numerical control Same as **computer numerical control**.

computerized tomography Same as **computed tomography**. Its abbreviation is **CT**.

computerized tomography scan Same as **computed tomography**. Its abbreviation is **CT scan**.

computerized tomography scanner Same as **CT scanner**.

computing machine Same as **computer**.

computing platform A specific hardware and software configuration, including the operating system. When a program or hardware device will only function properly with a particular platform, it is called **platform-dependent**, while those which can work across multiple platforms are called **platform-independent**. Also called **computer platform**, or **platform**.

computing power A measure of the processing power of a computer. May be quantified, for instance, in terms of the floating-point operations per second, or FLOPS, it can perform. For example, a given supercomputer may be able to exceed a speed of 35 teraflops. Also called **computer power**.

comsat Abbreviation of **communications satellite**.

COMSEC Abbreviation of **communications security**.

concatenate To link together, as in a chain. May apply, for instance, to the linking of transmission channels end-to-end, or to connecting pieces of optical-fiber. Also called **catenate**.

concatenation A linking together, as in a chain. Also called **catenation**.

concave Having a surface, side, or boundary that curves inward, such as the inner surface of a sphere. This contrasts with **convex**, which has a surface or boundary that curves outward.

concave lens A lens which is thinner at the center than the edges. Such a lens causes parallel rays of light passing through it to bend away from one another. This contrasts with a **convex lens**, which is thicker at the center than the edges

concavo-convex 1. Concave on one side or surface and convex on the other. Also called **convexo-concave** (1). 2. Having a curvature greater on the concave side than on the convex side.

concavo-convex lens 1. A lens with one concave surface and one convex surface. Also called **convexo-concave lens** (1). 2. A lens with one concave surface and one convex surface, and whose concave surface has a greater curvature.

concentration cell A cell in which the electrodes are each immersed in a solution with a different concentration of the same salt of a metal. The electromotive force produced will depend on the metal and the concentrations of the solutions.

concentrator 1. A device, such as a multiplexer, which combines multiple communications channels or signals. 2. A device which provides a communications path that links many lower-speed channels to one or more higher-speed channels. A benefit of this arrangement is that it accommodates multiple-user connectivity with a reduced amount of circuit use. 3. A device which utilizes optical elements to increase the amount of light, especially sunlight, incident upon a photovoltaic cell or panel.

concentric cable Same as **coaxial cable**.

concentric line Same as **coaxial cable**.

concentric windings Two or more windings, in which the core is surrounded by one winding, which is enclosed by the next winding, and so on. Used, for instance, in a transformer.

conceptual schema A detailed model of the overall structure of a database. This model does not factor in considerations such as how the information is stored or accessed. Also called **logical schema**.

concrete class In object-oriented programming, a class within which objects may be created. This contrasts with an **abstract class**, in which no objects may be created.

concurrency 1. The simultaneous, or nearly simultaneous, accessing of the CPU by multiple computer operations. 2. The ability of a computer to properly handle multiple tasks. 3. The proper handling of multiple tasks by a computer.

concurrent 1. Occurring or functioning at the same time, or in a parallel manner. 2. Pertaining to **concurrent execution**.

concurrent execution Also called **concurrent operation**, **concurrent processing**, or **parallel execution**. 1. The simultaneous execution of multiple computer operations by the CPU. Since microprocessors can work so quickly, it seems simultaneous, even though each operation is usually executed in sequence. 2. The simultaneous execution of multiple computer operations by multiple CPUs. Such processors are usually linked by high-speed channels. Also called **multiprocessing**.

concurrent operation Same as **concurrent execution**.

concurrent processing Same as **concurrent execution**.

concurrent program execution The simultaneous running of multiple computer programs.

condenser 1. A device which changes a gas into a liquid or a solid. 2. A system of mirrors and/or lenses which collects light from a source, then directs it onto a surface. Used, for instance, in a microscope. 3. A nearly obsolete term for **capacitor**.

condenser antenna Same as **capacitor antenna**.

condenser loudspeaker Same as **capacitor speaker**.

condenser microphone Same as **capacitor microphone**.

condenser speaker Same as **capacitor speaker**.

condition code A series of bits that identify a given status, which depends on previous operations. May be used, for instance, to communicate the current state of a device.

conditional Subject to a given circumstance or state. For instance, a conditional branch.

conditional branch In a computer program, an instruction that switches the CPU to another location in memory when a given condition is met. For example, if the result of a calculation has a specific result. Also called **conditional jump**, or **conditional transfer**.

conditional jump Same as **conditional branch**.

conditional stability For a given component, circuit, device, piece of equipment, or system, stability which is obtained only when a given parameter, such as the amplitude of its input signal, is within specified limits. Also called **limited stability**.

conditional statement In a computer program, a statement that is executed when a given condition is met.

conditional transfer Same as **conditional branch**.

conditioned circuit A telephone circuit which is modified so as to carry a digital data signal, instead of an analog voice signal. Such circuits feature noise-filtering components.

conditioning 1. The modifying of devices or equipment, so as to make them compatible with other devices or equipment. Used, for instance, to match certain performance parameters. 2. In telephony, the modifying of twisted pair wire so it can carry a digital data signal. Also called **line conditioning** (1). 3. The modification of a communications line so as to improve performance. Also called **line conditioning** (2).

conductance A measure of the ability of a component, circuit, device, or system to conduct electricity. In a circuit with no reactance it is the reciprocal of resistance. It is the real part of admittance, its symbol is **G**, and is expressed in siemens. Also known as **electrical conductance**.

conducted interference Interference, such as RF noise, which is propagated by means of communications or power lines.

conducting direction For an electrical or electronic component or device, the direction in which there is lesser resistance to the flow of current. Used, for instance, in the context of semiconductors. This contrasts with **inverse direction**, where the converse is true. Also called **forward direction**.

conduction 1. The transmission of energy through a medium without the medium itself moving as a whole. Electrical, acoustic, or heat energy may be propagated in this manner. For instance, the transmission of electrical energy through a metal entailing the migration of electrons through the conductor. This contrasts with **convection**, where the medium itself is moved, and with **radiation**, where waves or particles are emitted. 2. The flow of electrical charge through a medium. This may entail, for instance the movement of electrons through a metal conductor, or the migration of ions in a gas. Also called **electrical conduction**.

conduction band A partially filled energy band in which electrons can move freely under the influence of an electric field, thus producing a net transport of electric charge.

conduction current Electric current which flows through a body by means of electrons present in a **conduction band**.

conduction electrons Electrons present in a **conduction band**. Also called **valence electrons, outer-shell electrons,** or **peripheral electrons**.

conduction field An energy field surrounding a conductor through which an electric current is flowing.

conductive Pertaining to, or capable of **conduction**.

conductive coating A coating which serves to facilitate conduction. For instance, an antistatic coating applied on a TV screen to prevent dust from adhering to the surface.

conductive coupling A direct connection between circuits, utilizing a wire, resistor, or another means of physical contact that passes both AC and DC. Also called **direct coupling**.

conductivity The ease with which an electric current can flow through a body. It is the reciprocal of **resistivity**. Its symbol is σ, and it is expressed in siemens per meter. Also called **electrical conductivity**, or **specific conductance**.

conductivity meter A device which measures and indicates **conductivity**.

conductivity modulation In a semiconductor, a variation in conductivity through the variation of charge-carrier concentration.

conductivity-modulation transistor A transistor whose active properties are achieved through **conductivity modulation**.

conductor 1. A medium suitable for the conduction of electrical, acoustic, heat, or other form of energy. 2. A medium which allows electric current to flow easily. Such a medium may be a metal wire, a dissolved electrolyte, or an ionized gas, among others. Among the elements, silver, copper, and gold are the best electric conductors. Also known as **electric conductor**.

conductor skin effect The tendency of AC to flow near the surface of a conductor. This effect becomes more pronounced as frequency rises, so losses due to resistance become greater with frequency increases. A conductor made from many separately-insulated wire strands which are woven together or braided in a manner that each strand regularly assumes each of the possible positions within the overall cross-section of the conductor minimizes this effect, providing for lower losses at radio frequencies. Also called **skin effect**.

conduit A pipe or tube through which cables or wires are run. A conduit may be made of metal, plastic, or another material which affords protection and/or shielding.

cone 1. A geometric shape with a circular or other closed-plane base, whose sides taper off along a diagonal line to a common vertex. 2. Anything resembling a **cone (1)**. 3. The cone-shaped diaphragm of a **cone speaker**.

cone antenna A wideband antenna whose radiating element is in the shape of a **cone (1)**. Also called **conical antenna**.

cone loudspeaker Same as **cone speaker**.

cone of protection A cone-shaped zone below a lightning rod, which provides a region with a highly reduced probability of a lightning strike. This cone extends 45° from the tip of the rod, which is the vertex of the cone. Such a cone may be used to protect stationary or moving objects.

cone of silence A cone-shaped zone above a radio beacon transmitter, where the intensity of the signal is greatly reduced. This zone is in the form of an inverted cone extending from the antenna, which is the vertex of said cone.

cone resonance Resonance of a **cone speaker**. Such resonance occurs in specific frequency intervals, and produces excessive radiation from the cone. The proper design of the speaker and its enclosure help minimize this undesired vibration.

cone-resonance frequency The frequency at which **cone resonance** is greatest.

cone speaker A speaker utilizing one or more cone-shaped vibrating elements. Such cones may be made of paper, fiber, plastic, metal, or another material. The larger a cone, the better its output of lower frequencies, and the smaller a cone, the better its output of higher frequencies. Frequently, multiple cones are used in such speakers, for a more balanced reproduction of sound. Also called **cone loudspeaker**.

conference call A telephone call in which three or more separate lines are connected at once.

conferencing The holding or participating in a meeting utilizing computer and/or telecommunications technology. For instance, teleconferencing, videoconferencing, or data conferencing. Such conferencing usually includes participants in different locations.

confidence coefficient Same as **confidence level**.

confidence factor Same as **confidence level**.

confidence interval An interval or range of values within which the theoretical probability of an event occurring may be specified. For instance, the probability of a failure occurring within a given time period. The wider the interval, the greater the confidence that the event will occur.

confidence level The probability that a **confidence interval** will contain a given event. May be expressed, for instance, as a percentage. Also called **confidence coefficient**, or **confidence factor**.

confidence limits The upper and lower limits of a **confidence interval**.

confidentiality, integrity, and availability For a given entity, such as a business, procedures implemented to help insure that information is shared only with the intended individuals or entities, that the data transmitted, received, processed, or otherwise handled is complete and accurate, and that the pertinent systems and means are available when needed for delivery, reception, processing, storage and the like.

config.sys Same as **CONFIG.SYS**.

CONFIG.SYS A configuration file in some operating systems.

configuration The arrangement of the components of a circuit, device, piece of equipment, system, or network. For instance, the hardware and/or software setup a computer may have, or the arrangement of circuit components so as to achieve a desired function. In computers, specific products or functions may require certain system configuration minimums, such as RAM exceeding a particular value, or the presence of a given adapter card.

configuration diagram A diagram in which labeled blocks, usually squares and/or rectangles, are used to represent the components of a system which are linked via connecting arrows and/or lines that depict their functional relationships. Such diagrams may show a circuit schematic, hardware and software interconnections, and so on. Also called **block diagram**.

configuration file A file that contains information about a program, file, computer system, or user.

configuration management 1. A system for configuring networks or large projects. 2. A system for monitoring and managing configurations of networks or large projects.

configure To set the **configuration** of a circuit, device, piece of equipment, system, or network.

conformal antenna An antenna, such as that consisting of an amorphous metal, which conforms to the shape of a given surface or area.

conformal coating A thin and non-conductive coating applied to a circuit or device to provide protection from moisture or other environmental concerns. Such a coating, for instance, may be a plastic which is sprayed on.

conic section A geometric figure produced by the intersection of a plane and a cone. Examples include circles, ellipses, parabolas, and hyperbolas.

conical Pertaining to, or in the shape of a **cone** (1). For instance, a conical horn.

conical antenna Same as **cone antenna**

conical horn A horn in the shape of a cone. Horns of such shape may be used in speakers, antennas, and other devices.

conical horn antenna A wideband antenna whose radiating element is a **conical horn**.

conical monopole antenna A wideband antenna consisting of a biconical antenna in which the lower cone is replaced by a ground plane.

conical scanning In radars, scanning in which the major lobe is in the form of a cone. The antenna is at the vertex of the cone.

conjugate branches Two branches of an electric network in which a change in electromotive force in one does not produce a change in electromotive force in the other. Also called **conjugate conductors**.

conjugate bridge A bridge, which when compared to a conventional bridge, has the positions of the detector circuit and supply circuit exchanged.

conjugate conductors Same as **conjugate branches**.

conjugate impedances Two impedances in which the resistive components are equal, and whose reactive components are equal in magnitude, but opposite in sign.

conjugate particles A pair of particles consisting of a particle and its antiparticle. For instance, a proton and an antiproton.

conjunction A logical operation which is true only if all of its elements are true. For example, if three out of three input bits have a value of 1, then the output is 1. Any other combination yields an output of 0. For such functions, a 1 is considered as a true, or high, value, and 0 is a false, or low, value. Also called **AND**, or **intersection** (2).

connect To join together. For example, to connect by means of a communications circuit, or to connect circuit components in parallel.

connect charge The amount that must be paid when there is a fee associated with a given **connect time** (1), or **connect time** (2). Such a fee may be hourly, monthly, by speed of connection, and so on. Also called **connection charge**.

connect time Also called **connection time**. 1. The time during which a user is logged-on to a computer and/or network. 2. The time that elapses during a telephone call. 3. The time it takes to establish a connection.

connected Joined together.

connection 1. The act of joining together, or connecting. 2. The manner in which components are joined together. For instance, a cascade connection. 3. The point at which components are joined together. 4. The state of being joined together.

connection charge Same as **connect charge**.

connection diagram A diagram in which electrical connections between components, devices, equipment, or systems, are detailed.

connection-oriented A mode of communication in which a direct and dedicated connection between two nodes must be established before transmission and reception can occur. Once the transmission is concluded, the connection is terminated. ATM and regular telephone connections are of this nature. This contrasts with **connectionless**.

connection-oriented network protocol A network protocol which in which a direct and dedicated connection between two nodes must be established before transmission and reception can occur. Once the transmission is concluded the connection is terminated. Its abbreviation is **CONP**.

connection time Same as **connect time**.

connectionless A mode of communication in which a direct connection between two nodes is not necessary before transmission. Each packet contains the source and destination addresses, and is routed through the network until it reaches the specified destination. LANs are connectionless. This contrasts with **connection-oriented**.

connectionless network protocol A network protocol in which a direct connection between two nodes is not necessary before transmission. Its abbreviation is **CLNP**.

connectionless network service A network service in which a direct connection between two nodes is not necessary before transmission. Its abbreviation is **CLNS**.

connectivity 1. The ability to establish a connection. 2. The ability to maintain a connection. 3. The state of being connected.

connector 1. That which serves to **connect**. 2. A device or object which serves to join two or more conductors. Examples include terminals and binding posts. Also called **electric connector** (1). 3. A device or object which serves to couple cables with other cables, equipment, or systems. Examples include plugs, jacks, and adapters. Also called **electric connector** (2). 4. A line which connects symbols in a flow chart.

connector insertion tool A tool which assists in the proper insertion of contacts into a connector, connectors into circuit boards, and so on. Also called **insertion tool**.

CONP Abbreviation of **connection-oriented network protocol**.

conservation of charge A law that states that the total charge in an isolated system never changes. This means, for instance, that if a positive charge is created in an isolated system, that an equal negative charge will also appear in the system. Also called **law of charge conservation**, or **charge conservation**.

conservation of energy Also called **energy conservation**. 1. A law which states that energy cannot be created nor destroyed in an isolated system, but that it can be changed from one form to another. For instance, kinetic energy being transformed into potential energy, while the total energy in the system remains constant. Also called **law of energy conservation**. 2. The judicious use of available energy resources.

conservation of mass A law which states that mass, or matter, cannot be cannot be created nor destroyed in an isolated system. For instance, the mass remains constant when a substance changes form a solid to a gas in an isolated system. This law does not always hold true when dealing with subatomic particles. Also called **conservation of matter, law of mass conservation**, or **mass conservation**.

conservation of matter Same as **conservation of mass**.

console 1. A control panel for electronic equipment, such as that at a TV station or radar station. Also known as **control desk**. 2. A terminal utilized to control a computer and monitor its status. Also called **master console**. 3. Same as **computer terminal** (1). 4. A display screen which presents the output of a computer. 5. A cabinet, that houses a device such as a radio or TV, which is designed to stand on the floor.

console switch A switch in a **console** (2) which allows a user to control a computer while it executes instructions.

console terminal Same as **computer terminal** (1).

constant Its abbreviation is **K**. 1. A quantity, value, item, or state which remains the same during a given process. For instance, a fixed value in a computer program, or an attenuation constant. 2. A value, quantity, item, or state which remains the same always. For instance, the rest mass of an electron.

constant-amplitude recording Sound recording in which all the frequencies of equal intensity are recorded with equal amplitude. In this manner, the recorded amplitude is independent of the frequency.

constant angular velocity In magnetic and optical storage, the rotating of a disk in a drive at a steady rate, regardless of what area of the disk is being accessed. Since the tracks on the disk are longer as the outer edge is approached, more information is available there. This contrasts with **constant linear velocity**, in which the disk rotates at a variable rate. Its abbreviation is **CAV**.

constant bit rate An ATM minimum continuous bit rate that is guaranteed to be maintained, for reliable time-sensitive transmissions, such as those including voice and video. Its abbreviation is **CBR**.

constant-conductance network A network having at least one input impedance which is a positive constant. Also called **constant-resistance network**.

constant current A current which is essentially constant, despite variations in variables such as load resistance, line voltage, or temperature, so long as they are within a prescribed range.

constant-current generator A generator whose output current is essentially constant, despite variations of the load resistance within a prescribed range.

constant-current modulation A system of amplitude modulation, in which the output circuits of the signal amplifier and the modulator are connected to a constant-current supply by means of a shared choke coil. Also called **choke-coupled modulation**, or **Heising modulation**.

constant-current power source Same as **constant-current source**.

constant-current power supply Same as **constant-current source**.

constant-current source A power supply whose output current is essentially constant, despite variations in variables such as load resistance, line voltage, and temperature, so long as they are within a prescribed range. Also called **constant-current supply, constant-current power supply, constant-current power source, current-regulated supply**, or **current-regulated power supply**.

constant-current supply Same as **constant-current source**.

constant-current transformer A transformer, which when supplied by an essentially constant voltage source, maintains an essentially constant current in its secondary circuit, despite variations of the load resistance within a prescribed range.

constant-k filter A filter whose product of the series and shunt element impedances is a constant that is independent of frequency.

constant-k network A ladder network whose product of the series and shunt element impedances is a constant that is independent of frequency.

constant linear velocity In magnetic and optical storage, the rotating of a disk in a drive at a varying rate, depending on the area of the disk which is being accessed. Since the tracks on the disk are longer as the outer edge is approached, the disk must be rotated more slowly when accessing there, to ensure a constant data read rate. This contrasts with **constant angular velocity**, in which the disk rotates at a steady rate. Its abbreviation is **CLV**.

constant-resistance network Same as **constant-conductance network**.

constant-speed motor A motor whose speed is essentially constant, despite variations of the load within a prescribed range.

constant voltage A voltage which is essentially constant, despite variations in variables such as load resistance, line voltage, or temperature, so long as they are within a prescribed range.

constant-voltage charge A battery charge in which the voltage at the terminals is essentially constant.

constant-voltage power supply Same as **constant-voltage source**.

constant-voltage source A power supply whose output voltage is essentially constant, despite variations in variables such as load resistance, line voltage, and temperature, so long as they are within a prescribed range. Also called **constant-voltage supply, constant-voltage power supply**, or **voltage-regulated power supply**.

constant-voltage supply Same as **constant-voltage source**.

constant-voltage transformer A transformer that delivers an essentially constant voltage when supplied with a primary voltage that is within a certain range.

constantan An alloy containing between 40 to 45% nickel, and between 55 to 60% copper. Used, for instance, in thermocouples and resistors.

constructive interference The interaction of two or more waves resulting an increase in amplitude. This contrasts with **destructive interference**, where there is a reduction in amplitude, or cancellation.

consumer appliance A powered device intended for home use, which is designed to help do work, perform tasks, or to assist in providing comfort or convenience. These usually use electrical energy, and include dryers, lamps, refrigerators, and microwave ovens. Precision electronics, such as high-fidelity audio components and computers are sometimes not included in this group.

consumer electronics Electronic products intended for purchase and use by consumers. Such products include TVs, computers, cellular phones, DVD players, high-fidelity audio components, and smart appliances. When referring specifically to products to be used in a home, also called **domestic electronics**.

contact 1. The coming together, touching, union, or junction of surfaces or objects. 2. The junction of two conductors, so that current may flow. Also called **electric contact** (1). 3. A part or device which serves to open or close an electric circuit. Such a part or device may or may not act with another part or device. For instance, a blade, metal strip, button, switch, or relay. Also called **electric contact (2)**. **4.** The establishment of a communication. **5.** In radars, the initial detection of a scanned object.

contact arc An arc that forms between electric contacts when a circuit is opened.

contact arcing The formation of a **contact arc**.

contact area The surface area shared by electric contacts when a circuit is closed.

contact blade A flat moving conductor used as an electric contact.

contact bounce An undesired condition in which there is a spontaneous opening, or the intermittent opening or closing of contacts, when such contacts are moved to the open or closed position. Also called **bounce (2)**.

contact button A button used as an electric contact.

contact chatter The continuous, rapid, and undesired opening and closing of electric contacts. May be caused, for instance, by contact bounce. Also called **chatter (2)**.

contact clip A clip which grasps a **contact blade**, to close a circuit.

contact electromotive force Same as **contact emf**.

contact emf A potential difference that is developed between contacts that are made of two dissimilar materials. This potential difference may be a few tenths of a volt. Also known as **contact electromotive force**, **contact potential difference**, **contact potential**, or **Volta effect**.

contact erosion In electric contacts, the gradual loss of material due to sparks, arcs, and the such. The addition of cadmium to an alloy used in contacts may help reduce such erosion. Also called **electrical erosion**.

contact follow The additional distance that contacts travel together, after their initial contact. Also called **contact overtravel**.

contact force **1.** The force holding contacts together. **2.** The force a moving contact exerts on a stationary contact.

contact gap The maximum distance between contacts, when in the open position. Also called **contact separation**.

Contact Image Sensor A image sensor with limited resolution utilized in scanners which are smaller and less expensive than those utilizing CCDs. Its abbreviation is **CIS**.

contact material A material that can serve as an electric contact. Such materials may be metals such as copper, silver, gold, or palladium, or an alloy based on these or other suitable metals. Contact materials should have high thermal and electric conductivities, minimum sticking tendencies, and high resistance to corrosion, among other characteristics.

contact microphone A microphone, such as a throat microphone, which picks up vibrations by being in direct contact with the body producing them. Such microphones may be used, for instance, when extraneous noise is a problem.

contact noise A disturbance, due to fluctuating resistance, produced at the point where two metals or a metal and a semiconductor are joined.

contact overtravel Same as **contact follow**.

contact piston A piston, which as it slides, maintains contact with the inner walls of a waveguide at all times. Also called **contact plunger**.

contact plunger Same as **contact piston**.

contact potential Same as **contact emf**.

contact potential difference Same as **contact emf**.

contact pressure The pressure holding contacts together.

contact protection The use of a **contact protector**.

contact protector A component, device, or system utilized to suppress the arc that forms between electric contacts when a

circuit is opened. A capacitor, for instance, may be used for this purpose.

contact rectifier A rectifier which utilizes one or more metal disks coated with a semiconductor layer. This layer may consist of selenium, copper oxide, or another suitable semiconductor. The rectification occurs as a result of the greater conductivity across the contact in one direction than the other. Also called **semiconductor rectifier (1)**, **metallic-disk rectifier**, **dry rectifier**, or **dry-disk rectifier**.

contact resistance The resistance between two contacts when a circuit is closed. This resistance is very small, usually a fraction of an ohm.

contact sensor A device which detects objects through physical contact with them. These sensors may be used in robots, for example, to determine the location, identity, and orientation of parts to be assembled.

contact separation Same as **contact gap**.

contact spark A spark that forms between electric contacts when a circuit is opened.

contact sparking The formation of a **contact spark**.

contact wire **1.** A thin and flexible wire utilized to make electric contact. Used, for instance, to make electric contact on the surface of a semiconductor. **2.** A wire utilized to make electric contact.

contactless smart card A smart card, such as a proximity card, which does not have to come into contact with the device reading it.

contactless smart card reader A reader, such as a proximity card reader, utilized to detect data contained in **contactless smart cards**.

contactor **1.** A heavy-duty relay utilized to control and/or switch electric circuits, such as power circuits. Also called **power relay (2)**. **2.** In a control system, a device utilized to connect and disconnect a component, circuit, device, piece of equipment, mechanism, or system from a source of power.

contaminant An undesired material or substance that can adversely affect the properties of a component, circuit, device, piece of equipment, system, material, or medium.

content Information, especially that which is available online, which may be any combination of text, audio, video, files, or the like.

content-addressable memory Computer memory which is accessed by content rather than by address. It is often used in memory-management units, and comparison logic is included with each bit of storage. Its acronym is **CAM**. Also known as **content-addressable storage**, **active memory**, **active storage**, **associative storage**, or **associative memory**.

content-addressable storage Same as **content-addressable memory**.

content aggregator An entity that gathers Information from multiple Web sites and provides said content to others. Such content, for instance, can help keep users more time at a the collector's site, might be offered as a premium service, or may be sold to content providers.

content filter A program or utility which seeks to detect advertising and other bothersome or undesirable content before its loaded onto a Web page being accessed. Such a program, for instance, can filter Web page content, protect privacy, prevent pop-up ads from appearing, avert banner ads, eliminate certain JavaScript, stop animated GIFs, turn off ActiveX, disable Web bugs, and so on. Also called **Web filter**, **Web content filter**, **Internet filter**, or **blocking software**.

content provider An entity that provides information content for the Internet, or for software-based products, such as CD-ROMs. Such information may be any combination of text, audio, video, and so on. Examples include those who provide Web page or online encyclopedia content.

content server A server that stores online **content**.

contention **1.** The condition that results when two or more devices simultaneously attempt to use a single resource. For instance, when two or more stations attempt to transmit simultaneously over the same channel. **2.** A method for the allocation of resources, in which two or more devices compete for the same resource. For instance, two or more stations attempting to transmit simultaneously over the same channel, and the outcome being determined through contention resolution.

contention resolution The rules which determine the order in which a resource will be used when two or more devices attempt to access it simultaneously.

context sensitive A computer program feature that varies according to the specific task a user is attempting to accomplish. For instance, a program may provide different tips on the solving of a given problem, depending on the stage that is currently being worked on.

context-sensitive help Help which varies according to the task a user is working on.

context switching The act of changing from one active computer application to another. For instance, the switching from a word-processing program to a Web browser, with both programs running before and after the switch. Also called **task switching**.

contextual menu A menu that only offers choices, items, or the like, which are applicable to the location on the screen clicked upon. Such menus are often accessed by right clicking at the desired point, and vary by application, task, and so on.

contiguous **1.** Touching in some manner, such as sharing an edge or boundary. **2.** Adjacent or nearby. For instance, data in a group of contiguous storage blocks.

contingency plan The measures taken to minimize the deleterious effects of a possible disaster. These include data backup to a remote site, and making the necessary arrangements for key hardware to be available during such an event. Also called **disaster recovery plan (2)**.

continuity **1.** The condition or characteristic of being continuous or unbroken. **2.** Same as **circuit continuity**.

continuity checker Same as **continuity tester**.

continuity test A test to determine whether a circuit has a continuous and complete path for the flow of current. If there isn't a continuous and complete path, such a test may be used to determine where the circuit is broken.

continuity tester A device which performs **continuity tests**. Also called **continuity checker**.

continuity testing The performance of **continuity tests**.

continuous Uninterrupted in time, function, effect, or sequence.

continuous carrier **1.** An uninterrupted carrier signal. **2.** A carrier signal which is not interrupted while transmitting information.

continuous circuit Same as **closed circuit (1)**.

continuous control A method of automatic control in which one or more quantities are monitored continuously, and any necessary adjustments are made in order to maintain operation within the specified parameters. An automatic controller may be used for this purpose.

continuous current **1.** A unidirectional current which has an essentially constant average value. A continuous current may fluctuate, pulse, spike, and so on, but its polarity does not change. Sources of continuous current include batteries, photovoltaic cells, and DC generators. Also called **DC (1)**. **2.** A steady current, as opposed to that which is varied, or started and stopped at intervals.

continuous duty Operation of a device, piece of equipment, or system, at an essentially constant load for an extended, usually indefinite, period of time.

continuous-duty rating A rating which defines the maximum load that a device, piece of equipment, or system can carry for an indefinite period, without exceeding a given increase in temperature.

continuous loading Loading in which the added inductance is provided by a continuous wrapping of a magnetic material around each conductor of the transmission line.

continuous operating temperature The maximum temperature at which a component, device, piece of equipment, or system can continuously operate without harmful effects, such as overheating or an excessive reduction of useful life.

continuous operation A mode of operation in which a component, circuit, device, piece of equipment, or system is operational at all times. This contrasts, for instance, with pulsed or intermittent operation.

continuous power The maximum power an amplifier, such as an audio-frequency amplifier, can deliver for a stated time interval. May be expressed in watts RMS.

continuous rating A rating which defines the maximum load that a device, piece of equipment, or system can carry for an indefinite period without harmful effects, such as exceeding a given increase in temperature. This contrasts with a **short-time rating**, which is that which can be carried for periods not exceeding a given interval.

continuous spectrum A spectrum of frequencies which is uninterrupted, as opposed to being broken up into lines and/or bands.

continuous tone An image, such as a photograph, which is not composed of a dot pattern, as is the case with halftones. A dye-sublimation printer, for instance, may be able to approach such an output.

continuous-transmission frequency-modulated sonar A sonar system in which a continuously transmitted frequency-modulated signal is varied in a sawtooth fashion, and which determines the range of objects by the difference between the emitted and received frequencies at any given moment. Its abbreviation is **CTFM sonar**.

continuous variable A variable that can assume any value within a continuous and infinitely variable range. For instance, the amplitude of an analog signal.

continuous wave Its abbreviation is **CW**. **1.** An electromagnetic wave whose frequency, phase, and amplitude are constant. Also known as **Type A wave**. **2.** One of multiple waves which are transmitted continuously, as opposed to intermittently or pulsed.

continuous-wave Doppler radar Same as **continuous-wave radar**.

continuous-wave laser A laser whose production of coherent light is continuous rather than pulsed. Used, for instance, in some forms of laser welding. Its abbreviation is **CW laser**.

continuous-wave radar A radar system in which RF energy is transmitted continuously. A small proportion of this energy is reflected back by the scanned object, and is received by a separate antenna. Moving targets are detected using the Doppler effect. This contrasts with a **pulsed radar**, in which RF energy is emitted in pulses. Its abbreviation is **CW radar**. Also called **continuous-wave Doppler radar**.

contrast **1.** On a display screen, such as that of a computer or TV, the difference in brightness between areas that are lighter and darker. Also called **screen contrast (1)**. **2.** On a display screen, such as that of a computer or TV, the difference in brightness between the lightest and darkest areas. Also called **screen contrast (2)**. **3.** In optical character recognition, differences in qualities such as color or shading, which help distinguish a character from the background.

contrast control A circuit, device, or system which controls the **contrast** (2) of reproduced images.

contrast medium A substance, usually a dye, which is introduced into an organ or other internal body structure which is to analyzed via X-rays, MRI, or other similar medical diagnostic procedure, so as to enhance resolution.

contrast range The brightness interval between the lightest and darkest areas on a display screen, such as that of a computer or TV.

contrast ratio 1. For a display screen, such as that of a computer or TV, the ratio of the brightness of the lightest areas, to that of the darkest areas. **2.** For a display screen, such as that of a computer or TV, the ratio of the maximum luminance, to the minimum luminance.

control 1. A circuit, device, component, piece of equipment, signal, mechanism, or system, or a combination of these, that operates, regulates, or manages. For instance, a control panel. **2.** The operating, regulating, or managing effect a circuit, device, component, piece of equipment, signal, or system has. **3.** The means by which a control system maintains the desired output. **4.** A single factor or variable which is varied in two or more experiments in which the remaining factors and variables are held constant. This is done to better determine the influence of the factor or variable in question. **5.** An object appearing on a computer screen, such as a push-button or scroll bar, which helps perform an action in a program. **6.** Same as **control key**.

control accuracy In a control system, the level of correspondence between the controlled value and the ideal or specified value.

control action In a control system, an action taken to maintain the desired output.

control agent In a control system, an agent which controls a variable.

control block A block of computer memory that contains information used for control purposes.

control board Same as **control panel** (1).

Control-Break Same as **Ctrl-Break**.

control bus The conductors which carry control information between the CPU and other devices within a computer. For instance, interrupt request signals are sent over these conductors.

control center A location, device, console, terminal, or station which operates, regulates, or manages devices, equipment, or systems.

control channel A channel, such as a communications channel, that transmits control information.

control character 1. A character utilized to control a device, computer, piece of equipment, or system. **2.** Within the ASCII character set, a **control character** (1). Such a character has an assigned numeric value, representing an ASCII code, and controls a function such as backspace. **3.** A character which is typed in conjunction with the control key, such as control-A. The specific function for any given key combination will depend on which program is running.

control circuit 1. A circuit utilized to control a device, piece of equipment, system, or process. **2.** In a computer, a circuit which responds to instructions, such as those of a control program.

control+click To press a computer mouse button while holding down the control key. Used, for instance, to add or remove an item to or from an already populated selection. In this context it is better suited for non-consecutive items, while a **shift+click** is simpler for consecutive items. Its abbreviation is **Ctrl+click**.

control+clicking To select and deselect utilizing **control+clicks**.

control code 1. A code utilized to control a device, computer, piece of equipment, or system. **2.** In computers, a code which controls an action. Such codes may be in the form of control characters, and are utilized to control programs, peripheral devices, and the like.

control component A component utilized to control a device, piece of equipment, system, or process.

control computer A computer utilized in a control system. Used, for instance, to monitor selected parameters and send signals which maintain the desired output.

control counter Same as **control register**.

control data Computer data utilized to control data, programs, or hardware devices. Also called **control information** (1).

control desk Same as **console** (1).

control device A device which controls a given mechanism, piece of equipment, function, process, or system. An example is an infrared remote control for electronic equipment.

control diagram Also called **control flow diagram, flowchart,** or **flow diagram. 1.** A diagram which uses a set of standard symbols to represent the sequence of operations of a system. **2.** A diagram which uses a set of standard symbols to represent the sequence of operations of a computer program or system. Such a chart may show, for instance, the flow of data or the steps of a subroutine.

control electrode An electrode whose input is used to regulate the current of one or more other electrodes. For instance, the gate electrode in a field-effect transistor.

control element An element utilized to control a device, piece of equipment, system, or process.

control field In a computer record, a field which contains control information, such as the type of packet being transmitted.

control flow diagram Same as **control diagram**.

control function A function which controls a given piece of equipment, process, or system. For example, functions of a computer operating system.

control grid The **control electrode** in a vacuum tube. It is usually placed between the cathode and the anode.

control-grid bias In a vacuum tube, the average DC voltage applied between the control grid and the cathode.

control information 1. Same as **control data. 2.** Any information utilized for control purposes.

control instruction A computer instruction utilized to control data, programs, or hardware devices. For instance, an instruction pertaining to the operation of a peripheral. Also called **control statement** (2).

control key A modifier key included on computer keyboards that is used in combination with other keys to generate a function. The specific function for any given key combination will depend on which program is running. Also called **control** (6). Its abbreviation is **Ctrl,** or **Ctrl key**.

control knob A knob utilized to control or adjust the settings of a device or piece of equipment.

control language A set of language statements utilized to control programs or hardware devices. For example, a printer control language.

control logic The sequence of steps that hardware or software follow, to perform control functions.

control mark A control character or code which indicates a subdivision in a magnetic tape file. Also known as **tape mark** (2). Its abbreviation is **CM**.

control panel 1. A panel in which there are multiple indicators and devices, such as switches and dials, which enable a user to monitor and control a system. Used, for instance, to control an aircraft. Also called **control board,** or **panel** (1). **2.** In a computer, a utility program which enables a user to set many system parameters, such as keyboard and mouse characteristics, monitor resolution, and printer settings. Also called **control panel program**.

control panel program Same as **control panel (2)**.

control parallel A computer architecture in which multiple processors simultaneously and independently execute different instructions on different sets of data. Also called **multiple instruction stream-multiple data stream**.

control point In an automatic control system, the target value towards which the system makes adjustments. In the case of a thermostat, for instance, it would be a given temperature.

control processor A processor used in a control system.

control program A program which controls the operations of a computer, performing tasks such as managing system resources. An operating system is an example of such a program.

control register In a CPU, a register that contains the address of the location in memory that is to be accessed by the next instruction. May also refer to the address of the current instruction. Also called by various other names, including **control counter**, **current-instruction register**, **program counter**, **program register**, **instruction register**, **instruction counter**, and **sequence register**.

control rod A material utilized to control the reactivity of a nuclear reactor by absorbing neutrons. Examples include gadolinium, boron, and europium.

control room A room which houses the necessary devices and equipment to monitor and control a facility such as a TV recording studio or a nuclear power plant.

control sequence The order in which computer instructions are executed. For instance, the sequence followed while performing a given task.

control signal 1. A signal utilized to control a device or process. In a computer, for instance, such a signal may be an interrupt request. 2. In telecommunications, a signal that transmits control information. For example, a customer picks up a telephone receiver, hears a dial tone, dials a sequence of digits, and then gets a busy signal. All the tones heard are control signals.

control statement 1. A computer statement which controls the flow of execution of a program. For instance, an IF-THEN statement. 2. Same as **control instruction**.

control station Within a communications network, the station that manages all operations, such as the orderly flow of traffic.

control system A system utilized to maintain one or more output quantities within specified parameters. In a closed-loop control system, a feedback signal is incorporated for this purpose, while in an open-loop control system there is no such feedback. The components of a control system may be electrical, mechanical, thermal, and so on.

control total A total, composed of several numbers taken from a file, which is calculated before, during, and after processing. The numbers utilized to calculate the total do not necessarily have to be taken from numeric data. Control totals are used to verify the accuracy of processed data, or to help ensure that transmitted messages have not been tampered with. At all stages the calculated totals must match, otherwise there is an error. Also called **hash total**.

control track A track on a recordable magnetic medium, such as a tape or a disk, containing control signals such as tape playback speed.

control transformer A transformer utilized to supply a control device.

control unit 1. In a computer, circuitry that performs control functions such as sending control signals, interpreting program instructions, handling peripherals, or managing access to memory locations. 2. A unit which controls a given mechanism, piece of equipment, function, process, or system.

control winding A winding that carries a current that controls the output of a machine.

control word A computer word which stores information used for a control function.

controlled-avalanche device A semiconductor device with precisely defined avalanche voltage characteristics. Such devices can absorb repeated momentary power surges without damage.

controlled-avalanche diode A semiconductor diode with precisely defined avalanche voltage characteristics. Such diodes can absorb repeated momentary power surges without damage, and can be used, for instance, for surge suppression.

controlled-carrier modulation A type of amplitude modulation in which the amplitude of the carrier wave is varied according to the percentage of modulation, providing for an essentially constant modulation factor. Also called **floating-carrier modulation**, or **variable-carrier modulation**.

controlled environment An enclosure, such as a room, in which measures are taken to provide an environment that meets certain requirements, such as maintaining a specified level of temperature and/or humidity, guarding against static electricity or electromagnetic radiation, or isolating from dust. Such environments may be used, for instance, for testing, or to protect sensitive electronic equipment.

controlled-path robot A robot whose movements are dictated by a **controlled-path system**.

controlled-path system A computer control system in which a path of movement is numerically described. Used, for instance, in robotics.

controlled rectifier A rectifier, such as a silicon-controlled rectifier, whose output current may be regulated.

controller 1. A circuit board or device which controls the way peripheral devices access the computer, and vice versa. It is usually contained on a single chip. Examples include disk controllers, graphics controllers, and video controllers. Also called **peripheral controller**, or **host adapter**. 2. A signal, circuit, device, or system which controls any given mechanism, function, process, or piece of equipment. An example is an infrared remote control for electronic equipment. 3. A circuit, mechanism, device, or system, which monitors one or more variables, and automatically makes the necessary adjustments in order to maintain operation within the specified parameters. Also known as **automatic controller**. 4. The computer and programs which control a robot. Also called **controller system**, or **robot controller**.

controller card A circuit board which controls the way peripheral devices access the computer, and vice versa. Examples include disk controllers, graphics controllers, and SCSI controllers.

controller system 1. A system which monitors one or more variables, and automatically makes the necessary adjustments, in order to maintain operation within the specified parameters. 2. Same as **controller (4)**.

convection The transmission of energy or matter through a medium, which is itself moved. For instance, in convection cooling, the air transferring the heat moves along with the heat. This contrasts with **conduction**, where the medium itself is not moved as a whole, and with **radiation**, where waves or particles are emitted.

convection cooling A process by which an object transfers heat to the surrounding air. The heated air is less dense, hence moving upward so that cooler air is then available for further cooling. Used, for instance, to cool components which generate heat, such as transistors.

convection current **1.** A current of air which provides **convection cooling**. **2.** The rate at which charges in an electron stream are transported across a given surface. Also called **convective current**.

convective current Same as **convection current**.

convective discharge An electric discharge in which a stream of charged particles moves away from a body with a high voltage, such as a Van de Graaff generator. Such a discharge may be visible or invisible. Also called **electric wind**.

convenience outlet A power line termination whose socket serves to supply electric power to devices or equipment whose plug is inserted into it. Convenience outlets are usually mounted in a wall, although they may be found elsewhere, such as a floor, or in the back of another electrical device, such as an amplifier. Also called by various other names, including **convenience receptacle**, **outlet**, **electric outlet**, **receptacle**, **power outlet**, and **power receptacle**.

convenience receptacle Same as **convenience outlet**.

conventional current The view of current in an electric circuit as a flow of positive charges. This is the opposite direction of electron flow, which is how current actually flows in a circuit.

convergence **1.** To tend towards an intersecting point. For example, where technologies such as computers and communications meet. **2.** In a multibeam electron tube, such as a TV picture tube with a three-color gun, the intersection of all the electron beams at given point.

convergence coil One of the coils used to help ensure the correct convergence of the electron beams in a TV picture tube with a three-color gun.

convergence control In a three-gun TV picture tube, a control that varies the potential of the convergence electrode. This control, utilizing a variable resistor, adjusts the convergence.

convergence electrode In a multibeam electron tube, such as a TV picture tube with a three-color gun, an electrode whose electric field converges all the electron beams.

convergence magnet One or more magnets utilized to converge two or more electron beams. Used, for instance, in three-gun TV picture tubes. Also called **beam magnet**.

convergence surface In a multibeam electron tube, such as a TV picture tube with a three-color gun, the surface generated by the intersection point of two or more electron beams.

convergent beam A stream of radiated energy which tends towards an intersection point. Also called **converging beam**.

convergent lens Same as **converging lens**.

converging beam Same as **convergent beam**.

converging lens A lens which causes parallel rays of light passing through it to bend toward one another. Such lenses are thicker at the center than the edges. This contrasts with a **diverging lens**, which bends parallel rays of light passing through it away from one another. Also called **convergent lens**, **convex lens**, or **positive lens**.

conversational Same as **conversational mode**.

conversational language A computer language which can be used in **conversational mode**.

conversational mode In computers, a mode of operation in which there is a series of exchanges between a user and a computer, consisting of user commands and computer actions. Also called **conversational**.

conversion **1.** The process of changing something from one use, form, state, or function, to another. For instance, the changing of electrical energy into sound energy. **2.** The process of changing DC into AC, or AC into DC. **3.** The process of changing the frequency of a signal. **4.** The process of changing computer input, data, files, media, pro-grams, hardware, or systems, from one form to another. For example, the conversion of a file from one format to another.

conversion efficiency The efficiency with which a converter performs its function, usually expressed as a ratio of an output magnitude to an input magnitude. For example, the ratio of the electrical energy produced by a photovoltaic cell, to the incident solar energy.

conversion equipment Equipment which converts computer data or files from one medium to another. For example, from a hard disk to a tape drive.

conversion gain ratio For a frequency changer or mixer, the ratio of the output signal power to the input signal power.

conversion program A computer program which makes the appropriate changes to data, files, or programs, so that they may be used with a different computer, software, or network architecture.

conversion rate In analog-to-digital conversions, the frequency with which samples of a variable are taken. For instance, that used in the conversion of analog music into digital form. Usually expressed as samples or cycles per unit time, as in a 192 kHz sampling rate. Sampling must be made at or above the Nyquist rate to prevent aliasing, and for any given sample size, the higher the sampling rate the more accurate the conversion. Also called **sampling rate**.

conversion transducer A transducer whose output frequency is equal to the sum or difference of the input frequency and the local oscillator frequency. Also called **converter** (**4**), or **heterodyne conversion transducer**.

convert **1.** To change something from one use, form, state, or function to another. For instance, to change electrical energy into sound energy. For instance, to change electrical energy into sound energy **2.** To change DC into AC, or AC into DC. **3.** To change the frequency of a signal. **4.** To change computer input, data, files, media, programs, hardware, or systems, from one form to another. For example, to convert a file from one format to another.

converter Also spelled **convertor**. **1.** That which changes something from one use, form, state, or function to another. For instance, a transducer which converts electrical energy into sound energy. **2.** A device which converts DC into AC, or AC into DC. For example, a chopper. **3.** A device which changes the frequency of a signal. For instance, a frequency changer. **4.** Same as **conversion transducer**. **5.** A device which changes computer input, data, files, media, programs, hardware, or systems, from one form to another. For example, an analog-to-digital converter, or a binary decoder. **6.** A device, that connects to a TV, which decodes the incoming signal or otherwise enables viewing of certain channels or programming. Such a device is usually in the form of a box, and may also connect to another device, such as a VCR. Also called **cable box** (**1**). **7.** A device, usually in the form of a box, which serves to tune a TV to all available channels. Also called **cable box** (**2**).

converter amplifier Same as **chopper amplifier**.

converter box Same as **cable box**.

converter tube In a conversion transducer, an electron tube which combines the functions of a mixer and a local oscillator.

convertor Same as **converter**.

convex Having a surface, side, or boundary that curves outward, such as the outer surface of a sphere. This contrasts with **concave**, which has a surface or boundary that curves inward.

convex lens A lens which is thicker at the center than the edges. Such a lens causes parallel rays of light passing through it to bend toward one another. This contrasts with **concave lens**, which is thinner at the center than the edges.

Also called **convergent lens**, **converging lens**, or **positive lens**.

convexo-concave 1. Same as **concavo-convex (1)**. 2. Having a curvature greater on the convex side than on the concave side.

convexo-concave lens 1. Same as **concavo-convex lens (1)**. 2. A lens with one convex surface and one concave surface, and whose convex surface has a greater curvature.

conveyor A system or apparatus which is utilized to transport articles through a facility. Such articles are moved along their path with a continuous motion.

cookie A block of data prepared by a Web server which is sent to a Web browser for storage, and which remains ready to be sent back when needed. A user initially provides key information, such as that required for an online purchase, and this is stored in a cookie file. When a user returns to the Web site of this online retailer, the cookie is sent back to the server, enabling the display of Web pages that are customized to include information such as the mailing address, viewing preferences, or the content of a recent order. Also called **browser cookie**, **Web cookie**, or **Internet cookie**.

cookie file A file containing one or more **cookies**.

coolant A fluid, such as air, water, or oil, which serves to remove heat from a device, component, piece of equipment, machine, or system.

Coolidge tube A high-vacuum X-ray tube in which electrons are produced by a heated cathode. Also called **Coolidge X-ray tube**.

Coolidge X-ray tube Same as **Coolidge tube**.

cooling fin A projecting plate or vane utilized for cooling a component, device, or piece of equipment. It provides an additional surface area for the dissipation of heat. Also called **fin**.

Cooper pair Within the Bardeen-Cooper-Schrieffer theory, a pairing of electrons in a superconductor, due to electron-phonon interactions. It is proposed that such pairing enables electrons to move through a lattice without colliding with it in a disruptive manner.

cooperative multitasking A mode of multitasking in which the foreground task allows background tasks access to the CPU at given times, such as when it is idle. Since the program in the foreground controls access to the CPU, it may monopolize its resources. This contrasts with **preemptive multitasking**, in which all tasks take turns at having the attention of the CPU. Also called **non-preemptive multitasking**.

cooperative processing The use of two or more computers to work on the same task. May be used, for instance, to best allocate the available resources.

coordinate axis Each of the lines which intersect at a common point called the origin, and which are used within a **coordinate system**. When representing three planes, for instance, each axis may be designated by the letters **x**, **y**, and **z**, respectively.

coordinate system A system, such as the Cartesian coordinate system, for representing points in space, whose locations are described by sets of numbers called **coordinates**.

Coordinated Universal Time An internationally agreed time standard based on time kept by atomic clocks, and which represents the local time at the 0° meridian, which passes through Greenwich, England. When utilizing Coordinated Universal Time, each location on the planet has the same time, which is expressed utilizing a 24 hour clock, with a Z frequently appended. For example, 2100Z indicates 9 PM. Since the earth's rotation is gradually slowing, an extra second is added approximately once a year. It is based on International Atomic Time, and its abbreviation is **UTC**. Also called **Universal Time Coordinated**, **Universal Coordinated Time**, **Universal Time**, **World Time**, **Zulu Time**, or **Z time**.

coordinates Each of the sets of numbers which locate a point in space within a coordinate system. If there are three planes, for instance, then a point is located by specifying its position in reference to each of the three **coordinate axes**.

coplanar Lying, occurring, or placed in the same plane. For instance, an antenna array arranged in this manner.

coplanar waveguide A planar microwave transmission line consisting of a central signal line situated between two ground planes, and which is separated from them by a specified gap. The central line and ground planes are affixed to the surface of a dielectric substrate of a given thickness. Its abbreviation is **CPW**.

copper A reddish metallic element whose atomic number is 29. It is lustrous, malleable, diamagnetic, ductile, and an excellent conductor of heat. Among the elements, only silver surpasses it as a conductor of electricity. It has a very high melting point, is not too chemically reactive, and has about 25 known isotopes, of which 2 are stable. Its applications in electronics are many, and include its use in electric wiring, electroplating, and in many useful alloys such as brass. Its chemical symbol is **Cu**.

copper brazing Brazing utilizing an alloy of copper as the filler metal.

copper-brazing alloy An alloy of copper used for brazing.

copper bromide A black powder, or crystals, whose chemical formula is $CuBr_2$. Used as a battery electrolyte. Also called **cupric bromide**.

copper chip A microchip that uses copper, instead of aluminum, for connections. The superior conductive qualities of copper improve the performance of the processor.

copper chloride A yellow to brown powder whose chemical formula is $CuCl_2$. Used in electroplating. Also called **cupric chloride**.

copper-clad A metal which is bonded with, or surrounded by, copper. For example, an aluminum wire surrounded by a copper layer which is bonded to it.

copper-clad aluminum Aluminum which is bonded with, or surrounded by, copper. The aluminum may be in rods, sheets, wire, and so on, depending on the use it will be given.

copper-clad steel Steel which is bonded with, or surrounded by, copper. The steel may be in rods, sheets, wire, and so on, depending on the use it will be given.

copper-clad wire Wire which is surrounded by a layer of copper. The core of such a wire may be made of steel, aluminum, or another metal.

copper coil A coil, such as that used in a transformer, wound with copper wire.

copper conductor 1. A conductor of electricity composed of, or containing copper. For example, copper wire, or copper-clad wire. 2. A conductor of electricity, heat, or another form of energy, which is composed of, or incorporates copper.

copper cyanide A green powder whose chemical formula is $Cu(CN)_2$. Used in electroplating. Also called **cupric cyanide**.

Copper Distributed Data Interface A version of Fiber Distributed Data Interface which utilizes copper cabling instead of fiber-optic cabling. Its abbreviation is **CDDI**.

copper-doped A semiconductor material, such as germanium, which is doped with copper, to vary its conductive properties.

copper doping The use of copper for doping a semiconductor material such as silicon, to vary its conductive properties.

copper ferrocyanide A reddish-brown powder used for lowering the electrical resistance of soil. Also called **cupric ferrocyanide**.

copper fluoride Blue crystals used in ceramics.

copper glycinate Blue crystals used in photometry and in electroplating baths.

copper hydroxide A blue powder whose chemical formula is $Cu(OH)_2$. Used in battery electrodes. Also called **cupric hydroxide**.

copper loss Same as **copper losses**.

copper losses A power loss in copper wires, cables, or windings, due to the resistance of copper conductors. Also known as **copper loss**, or I^2R **losses**.

copper monoxide Same as **cupric oxide**.

copper nitrate Blue crystals used in electroplating baths. Also called **cupric nitrate**.

copper oxide black Same as **cupric oxide**.

copper oxide red Same as **cuprous oxide**.

copper plating The electrolytic deposition of copper onto a substrate. This may be done, for instance, to provide greater electrical conductivity, or resistance to corrosion.

copper selenide Blue-black to green-black crystals whose chemical formula is $CuSe$. Used in semiconductors and solar cells. Also called **cupric selenide**.

copper suboxide Same as **cuprous oxide**.

copper sulfate A chemical compound whose formula is $CuSO_4$. Used in batteries and electroplating, and as a desiccant. Also called **cupric sulfate**.

copper tungstate A light green powder whose chemical formula is $CuWO_4$. Used in semiconductors. Also called **cupric tungstate**.

copper winding A winding, such as that used in a transformer, wound with copper wire.

copper wire A wire consisting of copper. It is usually utilized to conduct electricity, or for wired communications. Copper wire may be rolled, extruded, or drawn, has a uniform cross section, and is available in many diameters.

coprocessor Also spelled **co-processor**. In a computer, a secondary CPU which performs specialized functions that assist the main CPU. For instance, even though a CPU can perform floating-point calculations, a math coprocessor can do so better, and at the same time free the main CPU to address other tasks. Another type of coprocessor is a graphics coprocessor. A CPU may have a coprocessor built into it.

copy 1. A duplicate of an original. For instance, information, such as text or graphics, may be copied from one document to another, to the computer's memory, or to a printer. Similarly, files or directories may be copied from one medium to another, one computer to another, and so on. Copies of digitized information are usually 100% accurate, or very nearly so. 2. To read data from a source, leaving it unchanged, and writing it to another location, so that there is an additional identical version of the original.

copy and paste In computers, to **copy** something from one location, and then place it in another. For instance, to duplicate content such as text and/or images in one document and place it elsewhere in the same document, or in another. Another example is to copy several files from one directory and placing duplicates in another. Its abbreviation is **copy & paste**.

copy protection Measures taken to help avoid the unauthorized copying of software. These include the requirement to enter a registration number when installing a program, or the use of a dongle. Also called **software protection**.

copy & paste Abbreviation of **copy and paste**.

CORBA Acronym for **C**ommon **O**bject **R**equest **B**roker **A**rchitecture. Within an object-oriented environment, an architecture which enables objects, or pieces of programs, to communicate with other objects, regardless of the programming language used, or the platform they are run on. Also called **CORBA architecture**.

CORBA architecture Same as **CORBA**.

Corbino disk A variable-resistance device consisting of a semiconductor disk exhibiting the **Corbino effect**. The intensity of the magnetic field varies the flow of charge carriers.

Corbino effect The generation of an electric current around the circumference of a metallic disk carrying a radial current, when this disk is placed perpendicularly in a magnetic field.

cord A flexible cable, containing one or more insulated conductors, which is generally equipped with connecting terminals such as plugs. For instance, that used with a telephone, mouse, or headphones.

cordless A device, component, or piece of equipment which does not have a **cord**, especially when there might ordinarily be one. For instance, a cordless telephone, a cordless mouse, or a cordless drill. When referring to the transmission of information, such as voice, data, or control signals, also known as **wireless**.

cordless device A device which is **cordless**.

cordless headphones Headphones which do not have a cord. Such headphones receive signals from its base unit which is plugged into an audio amplifier, or similar device. Also called **wireless headphones**.

cordless keyboard A computer keyboard which does not have a cord. Such a keyboard may use infrared or radio-frequency waves to communicate with the computer. Also called **wireless keyboard**.

cordless microphone A microphone which does not have a cord. It has its own power source, and transmits via infrared or radio-frequency signals. Also called **wireless microphone**.

cordless mouse A computer mouse which does not have a cord. Such a mouse may use infrared or radio-frequency waves to communicate with the computer. Also called **wireless mouse**.

cordless phone Same as **cordless telephone**.

cordless switchboard A telephone switchboard in which manually operated keys, as opposed to cords, are used to establish connections.

cordless telephone A telephone which does not have a cord between the base unit and the handset, and which communicates via low-powered radio-frequency signals. In this context, the main difference between a cordless telephone and a **cellular telephone** is that the base unit of the former plugs directly into a land telephone network, while the latter is linked to it via microwaves. Its abbreviation is **cordless phone**. Also called **wireless phone**.

core 1. The central part of something. 2. A form around which a coil or winding is wound. For example, a ferromagnetic core. 3. A magnetic material, such as iron, around which a magnetic coil is wound. Such a core may be used in a device such as an electromagnet or transformer, and increases the inductance of the coil. Also called **magnetic core** (1). 4. Same as **core memory** (1). 5. The center conductor of a coaxial cable, or the bundled optical fibers encased by a protective sheath in a fiber-optic cable. Also called **fiber core**, or **optical fiber core**.

core dump To display, print, copy, or transfer the content of the main memory of a computer. Such a dump may be performed after a process which ends abnormally, for instance, to help pinpoint the source of a problem. A core dump may also be generated automatically. Also called **memory dump**, **dump** (2), or **storage dump** (2).

core frequency Same as **core speed**.

core iron A special iron used for **cores (2)** or **cores (3)**.

core loss The energy dissipated by a ferromagnetic core, such as that found in a transformer or inductor. May be due, for instance, to eddy currents and hysteresis loss. Also called **iron loss**.

core memory Also called **core storage**. **1.** A magnetic material, such as iron oxide, used as a bistable element, multiple units of which are configured in an array which may store data. It is a seldom-used form of non-volatile computer memory. Also called **core (4)**, **magnetic core memory**, or **magnetic core (2)**. **2.** An obsolescent term for the **RAM**, or main memory, of a computer.

core speed The clock rate of a CPU. Also called **core frequency**, **CPU core speed**, **CPU core frequency**, or **processor core speed**.

core storage Same as **core memory**.

core transformer A transformer with a ferromagnetic core, such as ferrite, as opposed to a nonmagnetic core.

CorelDRAW A popular drawing, image editing, and presentation program.

coresident A program which, once loaded, is present in memory at all times, along with one or more other programs which are also present in memory. A notepad is an example of such a program. Also spelled **co-resident**.

corner **1.** The point at which lines, edges, sides, or surfaces meet to form an angle, and the area immediately surrounding this vertex. **2.** An abrupt change in the longitudinal axis of a waveguide. When such a change is 90°, it is called an **elbow bend**.

corner effect For a device, such as a bandstop filter, with one or more sharp bends in a response curve, a rounding in the curve at these points. Ideally, the cutoff frequency should have a rectangular shape at this point in such a curve.

corner reflector **1.** A surface consisting of two or three flat reflecting surfaces which form a corner. This serves to better reflect electromagnetic waves back to their origin. A corner reflector, for instance, enhances the return of a radar signal back to a transmitter. **2.** An antenna incorporating one or more **corner reflectors (1)**. Also called **corner-reflector antenna**.

corner-reflector antenna Same as **corner reflector (2)**.

corona Same as **corona discharge**.

corona discharge An intermittent, luminous, and often audible electrical discharge that originates from a conductor when its potential exceeds a given value, but is not high enough to form a spark. This occurs especially in pointed objects when the electric field near their surfaces surpasses a given amount. For example, an aircraft traveling through an electrical storm may develop such discharges from an antenna, which in turn produces precipitation static. Also called **corona effect**, **corona**, **corposant**, **brush discharge**, or **St. Elmo's fire**.

corona effect Same as **corona discharge**.

corona failure A failure, due to corona discharge, in high-voltage conductors such as terminals. It is usually caused by the degradation of the affected body.

corona loss Energy loss due to corona discharge.

corona resistance The ability of an insulating material to withstand given levels of field-intensified ionization without breakdown occurring.

corona shield A shield placed near a high-potential point, to prevent corona discharge. It does so by redistributing the electrostatic lines of force.

corona start voltage Same as **corona starting voltage**.

corona starting voltage The voltage difference necessary for corona discharge to occur. Also called **corona start voltage**.

corona voltmeter A voltmeter in which the peak voltage is measured and indicated when corona discharge initiates at a known electrode spacing.

corona wire In a laser printer, a wire through which a high voltage is passed to charge the sheet to be printed, so that the oppositely charged toner will be pulled off the drum and onto it.

coroutine A routine, which once loaded, is present in memory at all times. Such a routine may be executed with other routines which are also present in memory at all times. Also spelled **co-routine**.

Corporation for Research and Educational Networking Same as **CREN**.

corposant Same as **corona discharge**.

correction An amount that is factored into a calculation or measurement in order to obtain a more accurate result, or the true value. May be used, for example, to compensate for the readings of an instrument.

correction factor A numerical factor used for **correction**. Such a factor, for instance, may be added, subtracted, or multiplied.

corrective action **1.** An action meant to rectify an undesired condition. **2.** In a control system, an action taken to maintain the desired output.

corrective maintenance **1.** Maintenance performed to rectify an undesired condition. For instance, that carried out after a failure. Also called **remedial maintenance**. **2.** Maintenance performed avoid an undesired condition.

corrective network An electric network inserted into a circuit to enhance certain characteristics, such as its impedance properties. Also called **shaping network**.

correlation A relationship in the change of two variables. The relationship may be causal, parallel, reciprocal, and so on. For example, a comparison of the magnitudes of two signals which vary over time.

correlation detection A method of detection in which a signal is compared with a reference signal which is usually internally generated. Also called **cross-correlation detection**.

correlation detector A device which compares a measured signal with a reference signal, such as a standard.

correlation distance In the propagation of waves by means of tropospheric scatter, the minimum distance between antennas that will give rise to independent fading of the signals received by each antenna.

correlation tracking Tracking using a **correlation tracking system**.

correlation tracking system A tracking system in which multiple signals obtained from a scanned object are compared. The comparison of the signals provides the phase difference between them, which in turn determines the trajectory of the object.

corrosion The gradual disintegration of a metal or alloy as a consequence of chemical reactions with its surrounding environment, such as the oxidation of iron due to the moisture present in air. Corrosion is accelerated by the presence of acids and/or bases.

corrosion protection Measures, such as the application of protective coatings, taken to minimize or inhibit corrosion.

corrosion-resistant Metals or alloys, such as copper or stainless steel, which are comparatively unaffected by corrosion. A material which is susceptible to corrosion may be covered, for example, with a protective paint, so as to make it corrosion-resistant.

corrosive   1. Having the capability to cause **corrosion**. 2. Causing **corrosion**.

corrupted   Computer programs or data which have been damaged. For example, a corrupted file.

corrupted data   Data, such as that within a corrupted file, which has been damaged.

corrupted file   A file that has been damaged by a hardware malfunction, a software failure, a virus, or another cause. Such files are generally unreadable, and data recovery techniques must be employed if the contained information is to be retrieved.

corruption   The damaging of computer programs or data. Causes include hardware malfunctions, software failures, power outages, and viruses.

corundum   A crystalline form of aluminum oxide, whose chemical formula is Al_2O_3. Its applications in electronics include its use as an insulator and as an abrasive.

COS   Abbreviation of **class of service**.

cosine   For a right triangle, the ratio of the length of the side adjacent to an acute angle, to the length of the hypotenuse. This contrasts with **sine**, which is the ratio of the length of the side opposite to an acute angle, to the length of the hypotenuse.

cosine emission law   A law which states that the energy emitted in any direction by a radiating surface is proportional to the cosine of the angle formed between the direction of emission and a perpendicular line extending from the same surface. Also called **cosine law**, or **Lambert's cosine law**.

cosine law   Same as **cosine emission law**.

cosmic background radiation   Same as **cosmic microwave background**. Its abbreviation is **CBR**.

cosmic microwave background   A uniform bath of radiation which is believed to permeate all of space and to have originated with the big bang. The spectrum of this radiation corresponds to that of a blackbody at about 2.73 K, peaking in the microwave region. It is a source of cosmic noise. Its abbreviation is **CMB**. Also called **cosmic microwave radiation**, **cosmic background radiation**, **cosmic microwave background radiation**, or **microwave background**.

cosmic microwave background radiation   Same as **cosmic microwave background**. Its abbreviation is **CMBR**.

cosmic microwave radiation   Same as **cosmic microwave background**. Its abbreviation is **CMR**.

cosmic noise   Also called **cosmic radio noise, Jansky noise,** or **sky noise**. 1. Radio-frequency noise caused by sources outside the earth's atmosphere, such as the cosmic microwave background. 2. Radio-frequency noise that originates outside the earth's atmosphere.

cosmic radiation   Same as **cosmic rays**.

cosmic radio noise   Same as **cosmic noise**.

cosmic radio waves   Radio-frequency waves originating from sources outside the earth's atmosphere, such as the cosmic microwave background.

cosmic rays   High energy particles which travel through space at nearly the speed of light, and which consist mostly of protons and alpha particles. Cosmic rays have many sources, including solar flares, and impinge upon the earth from every direction. Also called **cosmic radiation**, or **primary cosmic rays**.

Cotton-Moulton effect   The ability of some pure liquids, under the influence of a magnetic field, to doubly refract light. For this phenomenon to occur, the magnetic field must be transverse to the light beam.

Cottrell precipitator   A device which uses the **Cottrell process** to precipitate dust.

Cottrell process   The removal of dusts present in gases through their exposure to charged wires or grids. The gas is forced to flow through a chamber containing wires or grids that are maintained at a high voltage, which ionizes the dust particles and makes them migrate to the chamber walls.

coul   Abbreviation of **coulomb**.

coulomb   The SI unit of electric charge. One coulomb is the charge that passes through a given cross section of a conductor with a constant current flow of one ampere, for an interval of one second. It is equal to the charge contained in approximately 6.2415×10^{18} electrons. Its abbreviation is **coul**, and its symbol is **C**.

Coulomb attraction   The electrostatic force of attraction between two oppositely charged particles, according to **Coulomb's law**. Also called **electrostatic attraction**.

Coulomb force   The electrostatic attractive or repulsive force between two charged particles according to **Coulomb's law**. Also called **electrostatic force**.

Coulomb interactions   The interactions between charged particles according to **Coulomb's law**. Also called **electrostatic interactions**.

coulomb meter   Same as **coulometer**.

Coulomb repulsion   The electrostatic force of repulsion between two like charged particles, according to **Coulomb's law**. Also called **electrostatic repulsion**.

Coulomb's law   A law which states that the force between two charged particles is directly proportional to the product of their magnitudes, and inversely proportional to the square of their distance. This force occurs along a straight line between the charges, and is a repulsive force if the charges are like, or an attractive force if they are opposite. Also called **law of electrostatic attraction**.

coulombmeter   Same as **coulometer**.

coulometer   An instrument which measures electric charge, and expresses it in coulombs. It may consist, for instance, of an electrolytic cell in which the mass of a given substance liberated from a solution is correlated to coulombs of flowing current. Also called **coulomb meter, coulombmeter,** or **voltameter**.

count   1. A number reached by counting events, such as pulses. 2. The action of counting events, such as pulses. 3. A single event registered by a radiation counter. 4. A single event, especially when registered by a counter.

counter   1. A circuit, device, register, mechanism, or system used for counting. For example, a counter may store and/or indicate a given total, such as a number of pulses, Web-site hits, changes in state, and so on. 2. A circuit which generates an output after counting a specified number of pulses. Also called **counter circuit**, or **counting circuit**. 3. An instrument which registers and counts ionizing radiation, such as alpha rays, given off by radioactive entities. For example, a Geiger counter or a scintillation counter. Also called **radiation counter, ionization counter,** or **particle counter** (1). 4. In computer programming, a variable that keeps track of a running count.

counter circuit   Same as **counter** (2).

counter electromotive force   Same as **counter emf**.

counter emf   Abbreviation of **counter** electromotive force. The voltage developed in an inductive circuit through which AC flows. The polarity of this voltage is at all times the opposite of that of the applied voltage. Also known as **back electromotive force, back emf, back voltage,** or **reverse voltage** (1).

counter-rotating   The arrangement in which two signal paths, one in each direction, exist in a ring topology.

counter tube   1. An electron tube which produces an output pulse after counting a specified number of pulses. Also called **counting tube** (1). When one output occurs for every ten input pulses it is called **decade counter tube** (1). 2. An electron tube with one input electrode and multiple output electrodes, and in which with each successive input pulse the conduction is transferred in sequence to the next output electrode. Also called **counting tube (2)**. When there are ten output electrodes it is called **decade counter tube (2)**. **3.** An electron tube which registers ionizing radiation, such as alpha rays, and produces an output electric pulse which can be counted. For example, a Geiger-Müller tube. Also called **radiation counter tube**.

counterclockwise In the opposite direction as the moving hands of a clock when viewed from the front. Its abbreviation is **ccw**.

counterclockwise-polarized wave An elliptically polarized transverse electromagnetic wave in which the rotation of the electric field vector is towards the left when viewed along the direction of propagation. Also called **left-hand polarized wave**.

counterelectromotive force Same as **counter emf**.

counterpoise One or more wires mounted close to the ground, but insulated from it, to form a low-impedance and high-capacitance path to said ground. This provides a radio-frequency ground for an antenna. Also called **antenna counterpoise**.

counting circuit Same as **counter (2)**.

counting tube 1. Same as **counter tube (1)**. **2.** Same as **counter tube (2)**.

country code 1. A code used for dialing a country other than that within which a telephone call is originated. Such a code is usually followed by a city code, and then the specific telephone number. Some regions require dialing the country code even when the call is originated in the same country. **2.** Over the Internet, a two-character abbreviation which identifies a country. It appears at the end of a Uniform Resource Locator, or address, as in: **xxxx@zzz.com.qq**, where the country code is **qq**. A given country code in an address does not necessarily mean that the host to that address is physically there. Also called **Internet country code**.

country-specific Pertaining to hardware or software that utilize conventions that incorporate the specific needs of a given country. For instance, a keyboard that has special characters.

couple 1. To join, link, or allow the transfer of energy. For instance, to join circuits. **2.** That which has been joined, linked, or connected in a manner which allows the transfer of energy. For example, coupled circuits. **3.** To place two dissimilar metals in contact with each other. **4.** Two dissimilar metals which have been placed in contact with each other. For instance, a thermocouple.

coupled antennas Two or more antennas which are electromagnetically coupled to each other.

coupled circuits Two or more electric circuits which are coupled. Such coupling may be capacitive, inductive, conductive, and so on.

coupled oscillators Two or more oscillator circuits which are coupled. Such coupling may be, for instance, capacitive or inductive.

coupler 1. A device or means which serves to join, link, or allow the transfer of energy. **2.** A device or means which allows coupling between circuits. Such coupling may be capacitive, inductive, conductive, and so on. **3.** A device or means which allows energy to be transferred between waveguides.

coupling 1. Interaction resulting from joining, linking, or the transferring of energy. **2.** Interaction between circuits in which energy is transferred. The resulting coupling may be capacitive, inductive, conductive, optical, and so on. **3.** A device which serves to join, link, or allow the transfer of energy. For example, a waveguide coupling.

coupling aperture An opening which allows the transfer of energy into or out of a waveguide or cavity resonator. Also called **coupling hole**, or **coupling slot**.

coupling capacitor A capacitor which blocks DC and low-frequency AC, while allowing higher-frequency AC to pass freely. Also called **blocking capacitor**, or **isolation capacitor**.

coupling coefficient Also known as **coupling constant**, **coupling factor**, **coupling ratio**, or **coefficient of coupling**. **1.** A measure of the strength of interaction between two systems, such as particles. **2.** A numerical value which expresses the degree of coupling between two systems, especially circuits. This value is always greater than 0 and less than 1, or 0% and 100% respectively. A coefficient of coupling of 0 would represent no coupling, and 1 would be perfect coupling.

coupling constant Same as **coupling coefficient**.

coupling factor Same as **coupling coefficient**.

coupling hole Same as **coupling aperture**.

coupling loop A small loop of wire inserted into a waveguide or cavity resonator, which allows energy to be transferred in or out. Also called **loop (5)**.

coupling loss Loss occurring during capacitive, inductive, conductive, or optical coupling. Also called **coupling losses**.

coupling losses Same as **coupling loss**.

coupling network An electric network which allows energy to be transferred between circuits.

coupling probe A probe which is inserted into a waveguide or cavity resonator, which allows energy to be transferred in or out. Such a probe is usually a pin or a wire.

coupling ratio Same as **coupling coefficient**.

coupling slot Same as **coupling aperture**.

courseware Software supplementing, or comprising, any given material being learned. Used, for instance, in computer-aided instruction. Also called **educational software**.

courtesy copy Same as **carbon copy**.

covalent bond A chemical bond in which two atoms within a molecule share one or more pairs of electrons. Such a bond may be between atoms of the same element, as in O_2 (molecular oxygen), or between different elements, as in **GaAs** (gallium arsenide). Also called **electron-pair bond**.

coverage 1. The geographical area within which a given transmitter provides effective service. For instance, the zone served by a cellular telephone system, the region within which reception of TV or radio broadcasts is adequate, or the zone a radar can effectively scan. Also called **coverage area**, or **service area**. **2.** The service provided within a coverage area.

coverage area Same as **coverage (1)**.

cp Symbol for **candlepower**.

cP Symbol for **centipoise**.

CPA Abbreviation of **Chirped Pulse Amplification**.

CPA laser Abbreviation of **Chirped Pulse Amplification laser**.

CPE 1. Abbreviation of **customer premises equipment**. **2.** Abbreviation of **computer performance evaluation**.

CPGA Abbreviation of **ceramic pin grid array**.

cpi Abbreviation of **characters per inch**.

CPI Abbreviation of **characters per inch**.

CPM Abbreviation of **critical path method**.

cpm Abbreviation of **cycles per minute**.

cps 1. Abbreviation of **characters per second**. **2.** Abbreviation of **cycles per second**.

CPU Abbreviation of central-processing unit. **1.** The portion of a computer which has the necessary circuits to interpret and execute instructions, and to control all other parts of the computer. The CPU consists of the control unit and the arithmetic-logic unit, both usually contained on a single chip, in which case it is also called **microprocessor**. Also called **chip (4)**, **processor (1)**, or **central processor (1)**. **2.** That contained in a **CPU (1)**, plus memory, buffers, and related components. Also called **processor (2)**, or **central processor (2)**. **3.** The enclosure which houses all the components a computer utilizes, with the exception of the external peripherals, such as monitor, mouse, keyboard, printer, and so on. Also called **system unit**.

CPU-bound Abbreviation of central-processing unit **bound**. A circumstance in which a CPU is overloaded with calculations, which impairs its ability to process. May occur, for example, while recalculating a spreadsheet. Also called **compute-bound**, **computation-bound**, or **process-bound**.

CPU cache Abbreviation of central-processing unit **cache**. Same as **cache (1)**.

CPU chip Abbreviation of central-processing unit **chip**. A **CPU (1)** contained on a single chip. It incorporates the control unit and the arithmetic-logic unit. In order to have the minimum necessary components for computer function, memory and a power supply must be added. Also called **microprocessor**.

CPU cooler Abbreviation of central-processing unit **cooler**. A device which helps maintain a CPU running at a cooler temperature, which increases performance and reliability. Examples of such devices include heat sinks, fans, or refrigeration systems. Also called **chip cooler**.

CPU core frequency Same as **core speed**.

CPU core speed Same as **core speed**.

CPU cycle Abbreviation of central-processing unit **cycle**. Same as **clock cycle**.

CPU speed Abbreviation of central-processing unit **speed**. A measure of the relative processing power of a computer. A 4 GHz computer, for example, processes data twice as fast as a 2 GHz computer. Other factors, such as cache also influence the overall speed of a computer. Usually expressed in multiples of hertz, such as megahertz or gigahertz. Also called **chip speed**.

CPU time Abbreviation of central-processing unit **time**. **1.** The time a CPU is controlled by process, program, device, or the like. **2.** The time occupied by a CPU to perform a given task.

CPW Abbreviation of **coplanar waveguide**.

CQFP Abbreviation of **ceramic quad flat-pack**.

Cr **1.** Chemical symbol for **chromium**. **2.** Within the YCbCr color model, one of the two color-difference signals. The other is **Cb (2)**.

crack To break into a computer system by circumventing or otherwise defeating the protective measures of said system. For example, to guess or otherwise acquire a password to gain access.

cracker A person who breaks into a computer system by circumventing or otherwise defeating the protective measures of said system. Such a person may seek to use resources without paying, to do harm such as destroying files or stealing credit card numbers, or simply to break in and then leave.

cradle **1.** A base unit that a PDA plugs into, so that the user can recharge the batteries and/or communicate with another computer. **2.** A base unit that a cordless telephone or similar device plugs into for recharging.

cramming The unethical practice of changing or adding features or services provided to a telephone customer which has not requested them. Unlike **slamming**, cramming is performed by the telephone company a customer selected, and may involve billing for voice mail, calling cards, caller-ID, and so on, which have not been ordered.

crash **1.** A complete and unexpected program or system halt. Caused by a program deficiency or a hardware failure. Also called **abnormal termination**, **abnormal end**, or **bomb**. **2.** A hard disk failure in which the read/write head collides with the surface of a disk. This may be caused by a foreign object such as dust, or a misalignment. Data is destroyed, and generally both the read/write head and the platter must be replaced. Also called **head crash**.

crash recovery The ability of a computer system to resume operation following a **crash (1)**. This may be accomplished automatically, although there may be partial or extensive data loss.

crawler A program which searches Web sites and organizes the located information. Also called **Web crawler**, **spider (1)**, or **bot (2)**.

crazing A pattern of fine cracks that forms on the surface or within materials such as ceramics and plastics. May be produced, for instance, by mechanical stresses such as repeated heating and cooling. Also, the process via which said cracks are formed.

CRC Abbreviation of **cyclic redundancy check**, or **cyclical redundancy check**.

creep A slow change in a property or value. For example, the deformation of a material under stress or through the passage of time.

creepage Electrical conduction along the surface of a dielectric material.

creeping featurism The addition of features to an existing software product, just for the effect of having more features. This usually leads to an unnecessary toll on system resources, such as RAM and disk space. Also called **feature creep**.

CREN Abbreviation of **C**orporation for **R**esearch and **E**ducational **N**etworking. An organization that runs an academic and research computer network that merged BITNET and CSNET.

crest Also called **crest value**, **peak**, or **peak value**. **1.** The maximum instantaneous value of a voltage, current, signal, or other quantity. **2.** The maximum instantaneous absolute value of the displacement from a reference position, such as zero, for a voltage, current, signal, or other quantity. When referring to a wave or other periodic phenomenon, also called **amplitude**. **2.** The **crest (1)**, or **crest (2)** for a given time interval.

crest factor In a periodically-varying function, such as that of AC, the ratio of the peak amplitude to the RMS amplitude. Also known as **amplitude factor** or **peak factor**.

crest value Same as **crest**.

crest voltage The maximum absolute value of the displacement from a reference position, such as zero, for a voltage. Also called **peak voltage**, or **maximum voltage (1)**.

crest voltmeter An AC voltmeter which indicates the crest value of an applied voltage. Also called **peak voltmeter**.

crimp connection A connection in which the joined conductors are pressed, pinched, or twisted together, as opposed to soldered. Also called **solderless connection**.

crimp contact An electric contact in which the connected conductors are pressed, pinched, or twisted together, as opposed to soldered. A specialized tool may be used for this. Also called **solderless contact**.

crimp terminal A terminal which is joined via a **crimp connection**. Also called **solderless terminal**.

crippled version A computer software or hardware version in which certain key features are unavailable. May be used, for instance, for trial periods. Also called **crippleware**.

crippleware Same as **crippled version**.

critical angle The minimum angle, with respect to the vertical, that will completely reflect an incident electromagnetic wave back into the medium from which it came. If the angle is lesser than this minimum, part or all of the radiation will penetrate or be absorbed by the medium which would have otherwise reflected it. For instance, the minimum angle necessary for radio-frequency waves to be totally reflected by the ionosphere.

critical characteristic A property or process which directly affects the safety functions of a device, piece of equipment, or system. These must be observed at all times to help avoid hazardous conditions for those operating said devices, equipment, or systems.

critical coupling The degree of coupling between two radio-frequency resonant circuits which results in the maximum transfer of energy between them, when both are tuned to the same frequency. Also called **optimum coupling**.

critical current For a given temperature, and in the absence of an external magnetic field, the maximum current a superconductive material can withstand while still maintaining its superconductivity. Also called **superconductor critical current**.

critical damping The degree of damping that provides the fastest transient response without overshoot or oscillation. In an indicator with a needle, for instance, it is the amount of damping that allows the needle to proceed as quickly as possible to a new reading, while not overshooting it or oscillating around it. Also called **optimum damping**.

critical dimension In a waveguide, the dimension of the cross section, which in turn determines its cutoff frequency. Also called **waveguide critical dimension**.

critical field 1. Same as **critical magnetic field**. 2. For a given anode voltage in a magnetron, the smallest theoretical value of magnetic flux density that would prevent an electron emitted from the cathode at zero velocity from reaching the anode. Also called **cutoff field**.

critical flicker frequency 1. For a given luminance, the frequency of light fluctuation that evokes a sensation of flicker. Above this frequency, the perceived flicker disappears. The perception of flickering is also influenced by the retinal field-of-view, and color. Also called **flicker frequency** (2), or **flicker fusion rate**. 2. For a given luminance, the frequency of light fluctuation that evokes a sensation of flicker half the time.

critical frequency 1. The frequency below which a vertically propagated radio wave will be reflected by the ionosphere. That is, any frequency above this will penetrate the ionosphere. This frequency varies by ionospheric layer and time of day. Also called **penetration frequency**. 2. Same as **cutoff** (2).

critical magnetic field For a given temperature, the maximum magnetic field a superconductive material can withstand while still maintaining its superconductivity, so long as the critical current is not exceeded. Also called **critical field** (1), or **superconductor critical field**.

critical path method A technique, used in project management, in which a path connecting the key tasks is drawn. Any delays in this path would ultimately delay the overall project, while tasks that are not critical have a given slack time incorporated. Its abbreviation is **CPM**.

critical potential The minimum potential at which the current flowing through a device increases rapidly.

critical temperature In the absence of an external magnetic field, the maximum temperature a superconductive material can withstand while still maintaining its superconductivity, so long as the critical current is not exceeded. Also called **superconductor critical temperature**.

critical voltage Same as **cutoff voltage** (2).

critical wavelength The wavelength corresponding to the **critical frequency**.

CRO Abbreviation of **cathode-ray oscilloscope**.

crocodile clip A spring-loaded clip utilized to make temporary electric connections. It is long and narrow and has jagged teeth, making for secure yet easy connections at the desired test points. It resembles the head of an crocodile and is also known as **alligator clip**.

Crookes dark space Same as **cathode dark space**.

Crookes radiometer A device consisting of an evacuated bulb enclosing four vanes that are suspended, or otherwise allowed to move with the least possible friction. The vanes are arranged like a turnstile and are black on one side, and white or polished on the other. The incident radiant energy, such as that from sunlight, is converted into motion, and the greater the intensity, the faster the motion. Also called **radiometer** (2).

Crookes tube An early version of a CRT, used to study properties of streams of electrons.

cross 1. A point where wires, lines, conductors, or the like, make contact. Also, to make such contact, or the state of being in such contact. 2. That which combines characteristics of two or more different components, circuits, devices, systems, materials, and so on.

cross antenna An antenna consisting of two horizontal radiators which intersect at right angles. Such an antenna is fed at the intersection, and the radiators may or may not be of equal length.

cross-assembler An assembler that converts assembly language for one hardware platform into machine language for another hardware platform.

cross-band Same as **cross-band communications**.

cross-band communications Two-way communications in which a given frequency or band is used in one direction, while another frequency or band is used in the other. Each direction may have different propagation characteristics. This arrangement may be utilized, for example, in a communications satellite. Also spelled **crossband communications**. Also called **cross-band**, or **cross-band operation**.

cross-band operation Same as **cross-band communications**.

cross-bar switch Same as **crossbar switch**.

cross bearings The intersection point of two or more bearings, which in turn provides a fix.

cross-channel communications Same as **cross-channel operation**.

cross-channel operation Two-way communications in which a given frequency or band is used in one direction, while another frequency or band is used in the other. Each direction has the same propagation characteristics. Also called **cross-channel communications**.

cross-check To verify by means of comparison with another source, or through a different mode for obtaining the same result. For instance, to check a calculation via two means of calculation. Also spelled **crosscheck**.

cross-color Same as **cross-color interference**.

cross-color distortion Same as **cross-color interference**.

cross-color interference Spurious signals resulting from interference between the luminance information and the chrominance information in a TV receiver. May manifest itself, for instance, as rainbow patterns appearing on striped images. Also called **cross-color distortion**, **cross-color**, or **crosstalk** (2).

cross-compiler A compiler that runs on a computer with one hardware platform, and generates machine language to be used with another hardware platform.

cross-connect In communications, a circuit path which can be simply connected and reconnected to establish and change routes. Used, for instance, in a digital cross-connect system. Also called **cross-connection**.

cross-connection Same as **cross-connect**.

cross correlation   The comparison at every point of the change of two variables, to determine their correspondence. Used especially to compare two signals.

cross-correlation detection Same as **correlation detection**.

cross-coupling   1. Unwanted coupling between two communication circuits or components. **2.** Unwanted coupling between two circuits or components.

cross-field amplifier Same as **crossed-field amplifier**.

cross-field device Same as **crossed-field device**.

cross-field interaction   The interaction between perpendicular magnetic and electric fields and an electron beam.

cross-field microwave tube Same as **crossed-field microwave tube**.

cross flux   For a given produced magnetic flux, the perpendicular component.

cross-interleaved Reed-Solomon code   A Reed-Solomon code utilized for compact discs. Its abbreviation is **CIRC**.

cross modulation   1. Modulation of a carrier by a signal other than that intended. For instance, the modulation of one carrier being imposed on another carrier. **2.** Modulation of a carrier by a signal other than that intended, in addition to the intended signal. A manifestation may be the hearing of the desired program in the foreground, while hearing an undesired program in the background.

cross-neutralization   In a push-pull amplifier, a method of neutralization in which capacitors are used for feedback

cross-over cable Same as **crossover cable**.

cross-platform   Computer software or hardware which is designed to work properly with more than one type of platform. An interpreter version of LISP is platform-independent, as is Java. This contrasts with **platform-dependent**, in which software or hardware is designed to work properly only with a specific platform. Also spelled **crossplatform**. Also called **platform neutral**, **platform-independent**, **architecture-independent**, or **multiplatform**.

cross-platform compatibility   Characterized by being **cross-platform**.

cross-point Same as **crosspoint**.

cross-point switch Same as **crosspoint switch**.

cross-polarized waves   Waves whose electric lines of force are at right angles to each other.

cross section   1. A section formed by cutting an object through an axis, usually at a right angle to said axis. For instance, such a section in a wire or a waveguide. **2.** A graphical representation of a **cross section (1)**.

cross-sectional area   The area occupied by a **cross section**. Used mainly in the context of conductors or wires, and may refer to single or multiple conductors.

cross-talk Same as **crosstalk**.

cross-talk coupling Same as **crosstalk coupling**.

cross-talk coupling loss Same as **crosstalk coupling loss**.

cross-talk level Same as **crosstalk level**.

cross-talk loss Same as **crosstalk loss**.

cross-talk unit Same as **crosstalk unit**.

crossband Same as **cross-band communications**.

crossband communications Same as **cross-band communications**.

crossband operation Same as **cross-band communications**.

crossbar switch   A set of vertical and horizontal switches arranged in a matrix, and which serves to connect any verti-

cal point with any horizontal point. Such a switch may be contained on a single chip, and may be used, for instance, for switching in a communications network. Also spelled **cross-bar switch**. Also called **crosspoint switch**.

crosscheck Same as **cross-check**.

crossed-field amplifier   A beam-type microwave amplifier in which the electron beam is influenced by perpendicular magnetic and electric fields. Such an amplifier features high gain and phase stability throughout a wide bandwidth. Also called **cross-field amplifier**.

crossed-field device   A device in which an electron beam is influenced by perpendicular magnetic and electric fields. Utilized to generate microwave radiation. Also called **cross-field device**.

crossed-field microwave tube   A microwave tube in which the electron beam is influenced by perpendicular magnetic and electric fields. Also called **cross-field microwave tube**.

crosshatch   A pattern consisting of evenly spaced intersecting parallel lines. Used, for instance, to fill in areas of computer graphics. Also called **crosshatch pattern**.

crosshatch generator   A signal generator that produces a test pattern in the form of a **crosshatch**.

crosshatch pattern Same as **crosshatch**.

crossover   1. In a circuit, a point at which two properly insulated conductors cross paths. **2.** In a circuit diagram, a point at which the lines representing two conductors, which are not connected, cross paths. **3.** Same as **crossover network**.

crossover cable   A special network connector cable within which two or more conductors are reversed from one end to the other. Such cables may be connected directly to certain Ethernet adapters, and may be used, for instance, to connect two computers for fast and easy file transfers. Also spelled **cross-over cable**.

crossover distortion   In a push-pull amplifier, a form of distortion that occurs when the two matched devices, such as transistors, are not in correct phase with each other.

crossover frequency   The frequency, within a crossover network, at which the power supplied to the two output circuits is equal. For instance, in a two-way speaker, the frequency at which the woofer and tweeter receive the same power. Above this frequency the tweeter gets more power, while the woofer gets more power below it.

crossover network   A filter circuit in a speaker which divides the input audio frequencies and sends the corresponding bands of frequencies to the designated speaker units, such as woofers, midranges, tweeters, super-tweeters, and so on. Also called **crossover (3)**, **dividing network**, **frequency dividing network**, **speaker dividing network**, or **loudspeaker dividing network**.

crossplatform Same as **cross-platform**.

crosspoint   A single intersection point within a **crossbar switch**. Also spelled **cross-point**.

crosspoint switch Same as **crossbar switch**.

crosstalk   Also spelled **cross-talk**. **1.** Interference resulting from the unwanted transfer of energy between channels, circuits, or signal paths. Crosstalk may occur, for instance, between telephone lines, audio channels in an amplifier, between TV audio and video signals, or between antennas. **2.** Same as **cross-color interference**. **3.** The usually unwanted transfer of a recorded signal between sections of a magnetic medium such as a tape. This may occur as a result of the affected sections being in close proximity, as occurs with overlapping segments of a reel of tape. Also called **magnetic printing (1)**, **magnetic transfer**, or **print-through**.

crosstalk coupling   Unwanted coupling occurring between two communication circuits or components, under specified conditions. It is usually measured at specific points of each circuit, and is expressed in decibels. Also spelled **cross-talk**

coupling. Also called **crosstalk loss**, or **crosstalk coupling loss**.

crosstalk coupling loss Same as **crosstalk coupling**. Also spelled **cross-talk coupling loss**.

crosstalk level The level of **crosstalk (1)**. Usually expressed in decibels above a given reference value. Also spelled **cross-talk level**.

crosstalk loss Same as **crosstalk coupling**. Also spelled **cross-talk loss**.

crosstalk unit A unit for quantifying **crosstalk (1)**. Also spelled **cross-talk unit**.

crowbar **1.** A circuit or device which protects against overvoltage, by rapidly placing a low-resistance shunt across the terminals where a given voltage is exceeded. Also called **overvoltage crowbar (1)**. **2.** The low-resistance shunt utilized in a **crowbar (1)**. Also called **overvoltage crowbar (2)**. **3.** The action of a **crowbar (1)**.

crowbar circuit A circuit which serves as a **crowbar (1)**. Also called **overvoltage crowbar circuit**.

CRT Abbreviation of **cathode-ray tube**. An evacuated electron tube which converts electrical signals into a visible form. In it, one or more electron guns produce one or more electron beams which are deflected and focused, en route to striking a phosphor surface on the back of the viewing screen. This surface glows, producing the visible images on the front of the screen. Used extensively as the display unit in computer monitors, TVs, oscilloscopes, automated teller machines, and so on. Also, such a tube and its housing.

CRT computer monitor Abbreviation of **cathode-ray tube computer monitor**. A computer monitor which utilizes a **CRT** to produce the displayed images.

CRT display Abbreviation of **cathode-ray tube display**. **1.** The image produced by a **CRT**. **2.** Same as **CRT screen**. **3.** The displaying of an image on a **CRT**.

CRT display device Abbreviation of **cathode-ray tube display device**. A display device which utilizes a **CRT** to produce the displayed images. Also called **CRT terminal**.

CRT monitor Abbreviation of **cathode-ray tube monitor**. A monitor which utilizes a **CRT** to produce the displayed images.

CRT screen Abbreviation of **cathode-ray tube screen**. The viewing screen of a **CRT**. Also called **CRT display (2)**.

CRT terminal Abbreviation of **cathode-ray tube terminal**. Same as **CRT display device**.

crunch **1.** To process information, especially that requiring a large amount of computations. **2.** To compress data.

cryoelectronic Pertaining to **cryoelectronics**.

cryoelectronics Abbreviation of **cryo**genic **electronics**. Its acronym is **cryotronics**. **1.** The study of electronic components, circuits, devices, equipment, and systems designed for use at cryogenic temperatures. Applications include superconductors, computers, and lasers. **2.** Components, circuits, devices, equipment, and systems utilized at cryogenic temperatures.

cryogen A substance that is utilized to obtain cryogenic temperatures. An example is a cryogenic fluid.

cryogenic Pertaining to **cryogenics**.

cryogenic conductor A material, usually a metal or an alloy, which is capable of exhibiting superconductor properties when cooled to cryogenic temperatures. Also called **super-conductor (3)**.

cryogenic device **1.** A device that exhibits special properties, such as superconductivity, at cryogenic temperatures. **2.** A device operated at cryogenic temperatures.

cryogenic electronics Same as **cryoelectronics**.

cryogenic fluid A substance that remains as a liquid or gas at cryogenic temperatures. For instance, helium remains a

fluid to about 1 K, and only then if also under high pressure. If such a substance is a liquid it is called **cryogenic liquid**, and if a gas, it is called **cryogenic gas**.

cryogenic gas A substance that remains as a gas at cryogenic temperatures.

cryogenic gyroscope A gyroscope in which a central rotating disk of superconducting niobium spins while in levitation at cryogenic temperatures. Also called **superconductor gyroscope**.

cryogenic laser A laser which is operated at cryogenic temperatures.

cryogenic liquid A substance that remains as a liquid at cryogenic temperatures.

cryogenic motor An electric motor designed to be used at cryogenic temperatures.

cryogenic spectroscopy Spectroscopy in which the analyzed substance is kept at a cryogenic temperature.

cryogenic temperature A temperature within a **cryogenic temperature range**.

cryogenic temperature range An interval of very low temperatures, whose range includes absolute zero. For instance, any temperature that falls between 0 K and 50 K.

cryogenics The study of very low temperatures, phenomena occurring at these temperatures, and techniques for achieving and maintaining such temperatures. Its applications in electronics are many, including its use in superconductors, lasers, and particle accelerators.

cryosar A semiconductor device used for switching at cryogenic temperatures. Used, for instance, for high-speed computer memory.

cryotron A superconductor device used for switching at cryogenic temperatures. In it, a magnetic field created by a control current determines whether the device operates as a superconductor or as a normal conductor. Used, for instance, for computer storage.

cryotronics Same as **cryoelectronics**.

cryptanalysis Also called **cryptoanalysis**. **1.** The analysis of a cryptographic system, so as to ascertain its function. **2.** The analysis of encrypted information, so as to ascertain its data content.

crypto Abbreviation of **cryptography**.

cryptoanalysis Same as **cryptanalysis**.

cryptographic Pertaining to **cryptography**.

cryptographic algorithm A sequence of defined steps utilized to implement a cryptographic function. For instance, a cryptographic hash algorithm.

cryptographic equipment Equipment which performs one or more **cryptographic functions**.

cryptographic function Any function used in **cryptography**. Such functions include encryption of data, and authentication of digital signatures.

cryptographic hash Same as **cryptographic hash function**.

cryptographic hash algorithm Same as **cryptographic hash function**.

cryptographic hash function An algorithm, utilized in cryptography, that compresses a variable-sized input into a fixed-sized output. Such an algorithm is one-way, meaning that it is extraordinarily difficult to derive the input value from the output value. Used, for instance, in the creation of digital signatures. Also called **cryptographic hash algorithm**, or **cryptographic hash**.

cryptographic key A mathematical value used to control a cryptographic process such as encryption, authentication, or decryption.

cryptographic logic A sequence of defined steps utilized to perform a cryptographic function.

cryptography The science of coding information so that only the intended recipients can understand it. All aspects of the security of communications are investigated, such as the issue of message interception. Its abbreviation is **crypto**.

cryptologic Pertaining to **cryptology**.

cryptology The science that deals with encrypted, hidden, or disguised communications. It includes the techniques utilized for securing the information, and those for deciphering and/or finding it.

cryptosystem A process, system, or method of **cryptography**. For instance, public-key cryptography.

crystal 1. A homogeneous solid material with geometrically arranged outer surfaces, and a symmetrical internal structure. Crystals may be composed of atoms, molecules, or ions, and each has specific properties, such as index of refraction, hardness, and a unique crystal lattice. The applications of crystals in electronics are many, including their use in lasers, semiconductors, and in miniaturized components. 2. A specific crystal fragment, such as quartz crystal, a piezoelectric crystal, or semiconductor crystal.

crystal activity A measure of the amplitude of the vibrations, under given conditions, of a piezoelectric crystal.

crystal calibrator A crystal oscillator utilized as a frequency standard.

crystal cartridge 1. A phonographic pickup whose transducer is a piezoelectric crystal, which converts the movements of the stylus into the corresponding audio-frequency voltage output. Also called **crystal pickup**, or **piezoelectric pickup**. 2. A microphone whose transducer is a piezoelectric crystal, which converts audio-frequency vibrations into the corresponding audio-frequency voltage output. Also called **crystal microphone**, or **piezoelectric microphone**.

crystal clock A clock which incorporates a quartz-crystal oscillator, whose natural oscillation frequency determines the accuracy of the timepiece. Such a clock can have an analog or digital display, and may have an error of less than 0.1 second per year. Also called **quartz crystal clock**, or **quartz clock**.

crystal control The use of a piezoelectric crystal, usually quartz, to control the frequency of an oscillator. When the crystal is quartz, also called **quartz control**.

crystal-controlled oscillator Same as **crystal oscillator**.

crystal-controlled transmitter An RF transmitter whose carrier frequency is controlled by a piezoelectric crystal oscillator. When the piezoelectric crystal is quartz, it is also called **quartz-controlled transmitter**.

crystal current The current that flows through a crystal. For instance, the RF current that flows through a piezoelectric crystal in a crystal oscillator.

crystal cut A section of a piezoelectric crystal accurately cut along certain axes, to determine characteristics such as its natural vibration frequency. Also serves to classify crystals so cut. For example, an AT or BT cut crystal. Also called **cut (2)**.

crystal defect A flaw in the geometric arrangement of the contained atoms, molecules, or ions of a crystal. Also called **crystal imperfection**, or **lattice defect**.

crystal detector A crystal which converts an AC signal into a pulsating DC. It can be used, for instance, to rectify a modulated RF signal to obtain an audio or video signal.

crystal diffraction The diffraction by a crystal of incident beams, such as that of X-rays or electrons. Used for instance, in a crystal spectrometer.

crystal-diffraction spectrometer Same as **crystal spectrometer**.

crystal diode Also called **crystal rectifier**, or **semiconductor diode**. 1. A semiconductor device with two terminals. 2. A semiconductor device with two terminals, which is utilized for rectification.

crystal earphones Earphones whose transducer is a piezoelectric crystal. Also called **piezoelectric earphones**.

crystal filter A filter circuit utilizing one or more piezoelectric crystals. Such filters feature high selectivity and a good shape factor, and may be used, for instance, in intermediate-frequency amplifiers. Also called **piezoelectric filter**.

crystal growth furnace A furnace designed to provide the proper conditions for growing particular kinds of crystals, such as large single crystals.

crystal headphones Headphones whose transducer is a piezoelectric crystal. Also called **piezoelectric headphones**.

crystal holder A device which serves to support, protect, house, and provide connections for a piezoelectric crystal, so that its properties can be best incorporated into any given device.

crystal imperfection Same as **crystal defect**.

crystal laser A laser, such as a ruby laser, which uses a crystal to generate the coherent beam of light.

crystal lattice The geometric arrangement in space of the atoms, molecules, or ions of a crystal. The pattern of such a lattice is regular and three-dimensional. Also called **space lattice**, or **lattice (2)**.

crystal-lattice filter A filter circuit utilizing an arrangement of piezoelectric crystals to obtain the desired characteristics.

crystal loudspeaker Same as **crystal speaker**.

crystal microphone Same as **crystal cartridge (2)**.

crystal mixer A mixer that uses the nonlinear characteristic of a semiconductor diode crystal to mix two input signals. Used, for instance, in radars.

crystal oscillator An oscillator circuit which utilizes a piezoelectric crystal, usually quartz, to control the oscillation frequency. Such oscillators feature a highly accurate and stable output, especially when in a temperature-controlled environment. Its abbreviation is **XO**. Also called **crystal-controlled oscillator**, or **piezoelectric oscillator**.

crystal oven A chamber whose internal temperature is carefully controlled, so as to provide a piezoelectric crystal with an operating environment which minimizes frequency drift.

crystal pickup Same as **crystal cartridge (1)**.

crystal plate A piezoelectric crystal that has been cut, etched, coated, and otherwise fully prepared to be mounted on its crystal holder. Also called **piezoelectric plate**.

crystal pulling A technique used for cultivating crystals, in which a growing crystal is slowly withdrawn from a molten solution of the crystal. Also called **pulling (3)**.

crystal radio Same as **crystal set**.

crystal receiver Same as **crystal set**.

crystal rectifier Same as **crystal diode**.

crystal resonator A resonant circuit which utilizes a quartz piezoelectric crystal to control the resonance frequency. Such resonators may be used, for instance, to control the frequency of an oscillator, and feature a highly accurate and stable output, especially when in a temperature-controlled environment. Also called **quartz-crystal resonator**, **quartz resonator**, or **piezoelectric resonator**.

crystal sensor A sensor whose detecting element is a piezoelectric crystal, especially quartz. An example is that used in a quartz thermometer. Also called **piezoelectric sensor**.

crystal set A simple radio receiver which incorporates a crystal detector or diode, a tuned circuit, and earphones or a low-power speaker. Such a receiver has no amplifier stages. Also called **crystal receiver**, or **crystal radio**.

crystal speaker A speaker whose transducer is a piezoelectric crystal, which converts its audio-frequency voltage input into the corresponding audio-frequency vibrations. Mostly used for reproduction of high frequencies. Also called **crystal loudspeaker**, or **piezoelectric speaker**.

crystal spectrometer An instrument used to analyze crystal structure by using X-rays. In it, a beam of collimated X-rays strikes the crystal, and a detector measures the angles and intensities of the reflected beam. Applying Bragg's Law to these measurements yields details about the crystal structure. Used, for instance, for X-ray diffraction studies that determine the structure of crystalline DNA. **DNA** is the abbreviation of **d**eoxyribo**n**ucleic **a**cid, and in its strands are contained the chemical basis of heredity. Also known as **crystal-diffraction spectrometer**, **Bragg spectrometer**, or **ionization spectrometer**.

crystal tester A device utilized for testing crystals, such as quartz, used in electronics. For instance, it may verify if a crystal oscillates, its frequency, its temperature coefficient, and so on.

crystal transducer A transducer whose sensitive element is a piezoelectric crystal, especially quartz. Piezoelectric crystals, when subjected to mechanical energy, generate electrical energy, and vice versa. Used, for example, in microphones, pickups, and speakers. Also called **piezoelectric transducer**.

crystal unit A piezoelectric crystal mounted on its crystal holder.

crystal video receiver A video receiver which incorporates a crystal detector that converts the incoming microwave signal into a video and/or audio signal. Used, for instance, in radars.

crystal whisker A single-crystal filament made of a metal, such as cobalt, or a refractory material, such as aluminum oxide. Such crystals feature very high tensile strength and temperature resistance. Also called **whisker (1)**.

crystalline Pertaining to, composed of, or incorporating one or more crystals.

crystallography The science that deals with crystal formation, structure, and phenomena.

Cs Chemical symbol for **cesium**.

CSA Abbreviation of **client/server architecture**.

CSCL Abbreviation of **computer-supported collaborative learning**.

CSCW Abbreviation of **computer-supported collaborative work**.

CSD Abbreviation of **circuit-switched data**.

CSLIP Acronym for **Compressed SLIP**.

CSMA Abbreviation of **carrier sense multiple access**.

CSMA/CA Abbreviation of **carrier sense multiple access with collision avoidance**.

CSMA/CD Abbreviation of **carrier sense multiple access with collision detection**.

CSNET An academic and research computer network which was merged with BITNET to form a network run by CREN. It was originally an abbreviation of Computer and Science Network.

CSP Abbreviation of **chip-scale package**.

CSS Abbreviation of **cascading style sheets**.

CSTN Abbreviation of **color super-twist nematic**.

CSTN display Abbreviation of **color super-twist nematic display**.

CSU Abbreviation of **channel service unit**.

CSU/DSU Abbreviation of **channel service unit/data service unit**.

CSV Abbreviation of **circuit-switched voice**.

CSWR Abbreviation of **current standing-wave ratio**.

CT **1.** Abbreviation of **computed tomography**, **computerized tomography**, or **computer tomography**. **2.** Abbreviation of **center tap**. **3.** Abbreviation of **computer telephony**.

CT-cut crystal A piezoelectric crystal cut used for vibration frequencies below 1 MHz.

CT scan Same as **computed tomography**.

CT scanner An instrument which takes a **computed tomography**. It is an abbreviation of **c**omputed **t**omography **scanner**, **c**omputerized **t**omography **scanner**, or **c**omputer **t**omography **scanner**.

CTD Abbreviation of **charge-transfer device**.

CTE Abbreviation of **coefficient of thermal expansion**.

CTFM sonar Abbreviation of **continuous-transmission frequency-modulated sonar**.

CTFT Abbreviation of **color thin-film transistor**.

CTFT display Abbreviation of **color thin-film transistor display**.

CTI Abbreviation of **computer telephony integration**, or **computer telephone integration**.

Ctrl Abbreviation of **control key**.

CTRL Abbreviation of **control key**.

Ctrl-Alt-Del The simultaneously holding down of the **Ctrl**, **Alt**, and de**l**ete keys on a computer keyboard. Used, for instance, to reboot some systems.

Ctrl-Break Abbreviation of **Control-Break**. The pressing of the break key while holding down the control key on a computer keyboard. This key combination may be used, for instance, to stop a given operation in progress.

CTRL-BREAK Same as **Ctrl-Break**.

Ctrl key Abbreviation of **control key**.

CTRL+ALT+DEL Same as **Ctrl-Alt-Del**.

Ctrl+click Abbreviation of **control+click**.

CTS **1.** Abbreviation of **clear to send**. **2.** Abbreviation of **carpal-tunnel syndrome**.

Cu Chemical symbol for **copper**.

cubical antenna An antenna whose radiating elements are arranged so as to form a cube.

cubical quad antenna An essentially omnidirectional antenna whose elements are in the shape of four-sided loops. Two elements are usually utilized, one driven and the other parasitic. Also called **quad antenna**.

cue circuit A one-way communications circuit utilized to transmit control information. May be used for radio and TV programming.

cumulative backup A data backup technique in which the only files copied are those that have been modified since the last backup, and in which all versions are saved, not just the last one, as is the case of a **differential backup**. This also contrasts with a **full backup**, in which all files are copied, including those that have not been modified since the previous backup. Also called **incremental backup**.

cumulative ionization Also called **cascade (2)**, **avalanche**, **Townsend ionization**, **Townsend avalanche**, or **ion avalanche**. When the only produced charged particles are electrons, it is called **electron avalanche**. **1.** A cumulative ionization process in which charged particles are accelerated by an electric field and collide with neutral particles, creating additional charged particles. These additional particles collide with others, so as to create an avalanche effect. **2.** In semiconductors, the cumulative generation of free charge carriers in an avalanche breakdown.

cup core A core which encloses a coil, providing a magnetic shield.

cupric bromide Same as **copper bromide**.

cupric chloride Same as **copper chloride**.

cupric cyanide Same as **copper cyanide**.

cupric ferrocyanide Same as **copper ferrocyanide**.

cupric hydroxide Same as **copper hydroxide**.

cupric nitrate Same as **copper nitrate**.

cupric oxide A brownish black powder whose chemical formula is **CuO**. It is used in batteries, electrodes, and in electroplating. Also called **copper monoxide**, or **copper oxide black**.

cupric selenide Same as **copper selenide**.

cupric sulfate Same as **copper sulfate**.

cupric tungstate Same as **copper tungstate**.

cuprous cyanide A chemical compound whose formula is **CuCN**. It is used in electroplating.

cuprous oxide Reddish-brown crystals whose chemical formula is Cu_2O. Used in electroplating, ceramics, rectifiers, and photocells. Also called **copper oxide red**, or **copper suboxide**.

cuprous selenide Dark blue to black crystals whose chemical formula is Cu_2Se. Used in semiconductors.

cuprous sulfide A blue to black powder whose chemical formula is Cu_2S. Used in solar cells, rectifiers, and electrodes.

curie A unit of radioactivity defined as exactly 3.7×10^{10} atomic disintegrations per second, which is the approximate decay rate of one gram of pure radium. Its abbreviation is **Ci**.

Curie law A law stating that the susceptibility of certain paramagnetic materials is inversely proportional to its absolute temperature. It doesn't always hold true, especially for liquids or solids.

Curie point Same as **Curie temperature**.

Curie temperature The temperature at which the ferromagnetic properties of a material become paramagnetic. At or above this temperature, the thermal energy in the material is too great to exhibit ferromagnetism. Also called **Curie point**, **ferromagnetic Curie temperature**, or **magnetic transition temperature**.

Curie-Weiss law A variation of the **Curie law** which takes into account the mutual interactions between particles, especially in a liquid or solid. Only holds true at temperatures above the **Curie temperature**.

curium A synthetic element whose atomic number is 96. It is a reactive silvery-white metal, and has about 20 known isotopes, all of which are unstable. It is used, for instance, for compact thermionic or thermoelectric power generation, especially for use in remote areas such as outer space. Its chemical symbol is **Cm**.

current Its symbol is I, and it is expressed in amperes. Also called **electric current**. **1.** A flow of an electric charge through a conductor. Electrons, electron holes, or ions may transport an electric charge. **2.** The rate of flow of electric charge through a conductor.

current amplification Also called **current gain**. **1.** For an amplifying device such as a transistor, electron tube, or photomultiplier tube, the ratio of the output current to the input current. Also called **current ratio**. **2.** The production of an output current which is greater than the input current.

current amplifier A device, such as a transistor or electron tube, whose output current is greater than its input current.

current antinode For a medium having standing waves, such as a transmission line or antenna, the point at which there is a maximum of current. Also called **current loop**.

current attenuation **1.** For an amplifying device such as a transistor, electron tube, or photomultiplier tube, the ratio of the input current to the output current. **2.** The production of an output current which is lesser than the input current.

current balance An ammeter which operates by having a current pass through two nearby coils connected in series. One of the coils is fixed, while the other is attached to the arm of a balance. The attractive force between the coils is counterweighed by the force of gravity acting on a known weight on the arm of the balance, which in turn indicates the current strength on its scale. Also called **Ampère balance**, or **Kelvin balance**.

current calibrator A source whose steady current level serves as a basis for calibrating instruments.

current-carrying capacity The maximum amount of current that a conductor can handle safely.

current coefficient **1.** A coefficient depicting a current change resulting from a variation in another electrical parameter, such as voltage or resistance. **2.** A coefficient depicting a variation in another electrical parameter, such as voltage or resistance, resulting from a current change.

current consumption Same as **current drain**.

current-controlled current source A dependent source whose level of output current depends on its input current. An example is a current amplifier. Its abbreviation is **CCCS**.

current-controlled device A device, such as a switch, whose function is controlled by an input current.

current-controlled switch A switch, such as a semiconductor device, whose switching action is determined by an input current.

current-controlled voltage source A dependent source whose level of output voltage depends on its input current. An example is a transresistance amplifier. Its abbreviation is **CCVS**.

current crest Same as **current peak**.

current density The current flowing per unit of cross-sectional area of a conductor. Its SI unit is amperes per square meter. Its symbol is J. Also called **electric current density**.

current detector A component, circuit, or device which indicates the presence of a current.

current divider A device which serves to deliver a given proportion of the total current to one or more circuits or branches.

current drain The current a circuit or load draws from a power source. Also called **current consumption**, or **drain (1)**.

current feed The feeding of an antenna by connecting its transmission line at a **current antinode**.

current feedback Feedback in which a proportion of the current output to the load is fed back to the input circuit in series.

current flow The flow of an electric charge through a conducting medium. Electrons, electron holes, or ions may transport an electric charge.

current gain Same as **current amplification**.

current generator **1.** A device which generates DC. For instance, a rotating electric machine which converts mechanical power into DC power. **2.** A device which generates AC. For example, a rotating electric machine which converts mechanical power into AC power.

current hogging The undesired condition in which one out of multiple components operated in parallel draws more current than it should. This can lead to a malfunction or failure of the component.

current-instruction register Same as **control register**.

current intensity The magnitude of a current flow. Also called **current strength**, or **current magnitude**. When expressed in amperes it is called **amperage**.

current lag Within a circuit, a change in current which lags behind a change in voltage. For example, in an inductive circuit the current lags behind an applied voltage. This contrasts with **current lead**.

current lead Within a circuit, a change in current which leads a change in voltage. For instance, in a capacitive circuit the current leads an applied voltage. This contrasts with **current lag**.

current leakage Also called **leakage current**. **1.** Current which flows through unwanted paths of a circuit, such as from the output to the input when not intended. **2.** Current which flows between electrodes of an electron tube by any route except the interelectrode space. **3.** Current which flows along the surface or through the body of a dielectric or insulator. **4.** DC which flows through the dielectric of a capacitor. **5.** AC which flows through a rectifier without being rectified. **6.** Current which flows through a component, circuit, or device which is in the off state. Also called **off-state leakage current**.

current limit **1.** The maximum output current a **current limiter** allows. **2.** The maximum current a device can handle without damage.

current limiter A device which limits its output current to a given maximum value, regardless of the applied voltage. Used, for instance, to protect equipment from surges. It may or may not provide protection for lesser fluctuations in current or voltage, which may also be destructive. Also called **current-limiting device**.

current limiting The limiting of an output current to a given maximum value, through the use of a **current limiter**.

current-limiting device Same as **current limiter**.

current-limiting fuse A fuse which interrupts a current that exceeds a given value. Used to protect components and equipment.

current-limiting resistor A resistor that limits the flow of a current to a given maximum value. Used to protect components and equipment. Also called **limiting resistor**.

current loop Same as **current antinode**.

current magnitude Same as **current intensity**.

current maximum Same as **current peak**.

current meter A device which measures and indicates current. For instance, an ammeter or a galvanometer.

current mirror A circuit in which the current on one side is forced to be a replica of that of the other side. Accomplished, for instance, by the bases and emitters of two bipolar junction transistors being connected together.

current-mode logic A logic circuit in which the transistors operate in an unsaturated mode, which provides faster switching. Its abbreviation is **CML**.

current node For a medium having standing waves, such as a transmission line or antenna, the point at which there is a minimum of current, or zero current.

current noise Also called **excess noise**. **1.** Electrical noise produced by current flowing through an electrical component, especially a resistor. **2.** Electrical noise produced by current flowing through a semiconductor material.

current peak Also called **current maximum**, **current peak value**, or **current crest**. **1.** The maximum value of a current. **2.** The maximum value of the displacement from a reference position, such as zero, for a current. **3.** The **current peak** (1), or **current peak** (2) for a given time interval.

current peak value Same as **current peak**.

current rating **1.** The maximum amount of current that a conductor or device can handle safely. **2.** The maximum amount of continuous current that a conductor or device can handle safely, or within a prescribed operating temperature range.

current ratio Same as **current amplification** (1).

current-regulated power supply Same as **constant-current source**.

current-regulated supply Same as **constant-current source**.

current regulation The maintenance of the current flowing through a circuit essentially constant utilizing a **current regulator**.

current regulator A device which maintains the current flowing through a circuit essentially constant, despite variations in variables such as load resistance, line voltage, and temperature, so long as they are within a prescribed range.

current relay A relay which is actuated at a specific current value, as opposed to a given voltage or power value.

current saturation In an electron tube, the condition in which the anode current can not be further increased, regardless of any additional voltage applied to it, since essentially all available electrons are already being drawn to said anode. Also called **plate saturation**, **anode saturation**, **voltage saturation**, or **saturation** (3).

current source A source from which current flows. For example, a power outlet.

current standing-wave ratio For a given transmission line, such as a coaxial cable or waveguide, the ratio of the maximum current to the minimum current. It is a measure of the impedance matching of the line. This ratio is equal to 1 when there is complete impedance matching, in which case the maximum possible RF power reaches the load, such as an antenna. Its abbreviation is **CSWR**. Also called **standing-wave ratio** (2).

current strength Same as **current intensity**.

current surge A sudden and momentary increase in current. May be caused, for instance, by lightning, or faults in circuits. If protective measures are not employed, such a surge may bring about a failure, or significant damage. Also called **surge current**, or **transient current**.

current transformer A transformer utilized for increasing or decreasing current. For example, an arrangement in which the primary winding of such a transformer is connected in series with the main circuit, with the secondary winding to a measuring instrument, can help avoid exposing said instrument to a current whose magnitude is too great.

current-voltage characteristic For a component, circuit, or device, a curve plotting current as a function of voltage. Also called **current-voltage curve**.

current-voltage curve Same as **current-voltage characteristic**.

cursor **1.** On a display, especially that of a computer screen, an indicator such as a blinking underline or a solid rectangle which indicates the location where a keystroke will appear on screen. Also called **insertion point** (1). **2.** On a computer screen, an indicator, such as a small hand, arrow, or I-beam, that moves as the mouse is moved. It serves to select text, menus, and the area of the screen where the next text input or other action will occur, by clicking the mouse there. Also called **pointer** (1), **mouse pointer**, or **mouse cursor**. **3.** A **cursor** (2) indicated by a pointing device, such as a trackball, other than a mouse. **4.** The pointing device of a digitizing tablet. It is similar to a mouse, but is much more accurate, because its location is determined by touching an active surface with an absolute reference. Also called **pen** (1), **puck**, or **stylus** (1).

cursor control The control of the movement or placement of a **cursor** (2) or **cursor** (3). A cursor may be moved or placed in various manners, including the use of keyboard keys, pointing devices such as a mouse, or a system in which a camera detects head motions.

cursor control device A device used for **cursor control**.

cursor key A computer keyboard key which moves the pointer or cursor displayed on the screen. These include the arrows, home, and page down keys. These keys may also be used in association with a control key for more specialized movements.

curtain array An antenna array consisting of a series of wire elements, usually half-wave dipoles, held vertically between two suspension cables.

curve 1. A smooth and continuous line whose points all lie along a single dimension. This definition includes straight lines, and may represent, for instance, the graph of an equation or function. Also, that which has this shape. 2. A smooth and continuous line whose points all lie along a single dimension, and which has no parts that are straight. This definition does not include straight lines.

curve tracer An instrument which charts a voltage or current as a function of another voltage or current, for a given voltage or current. Used, for instance, to chart a characteristic curve of a component.

custom That which is made to order. For instance, custom software, or a custom IC.

custom IC Abbreviation of **custom** integrated circuit. An IC which is designed for a particular function, application, or for given operating conditions. For instance, such an IC may be required for use in an unduly harsh working environment.

custom integrated circuit Same as **custom IC**.

custom-shaped CD An optical disc, such as a CD, CD-ROM, or DVD, that is in any shape other than round. An example is a CD business card.

custom software Software prepared especially to tailor to specific needs or circumstances. This contrasts with **packaged software**, which is sold to the general public. Most programs do, however, offer the ability to customize certain features. Also called **customized software**.

customer premises equipment Communications and/or network equipment situated at a user's location. Examples include regular telephones, DSL splitters, terminal equipment, voice ports, and decoder boxes. Its abbreviation is **CPE**.

customizability The extent to which something can be made or altered to order. For example, the flexibility a computer program has for better addressing different tasks.

customize To make or alter to order. For example, to modify a computer program to better address a given task.

customized software Same as **custom software**.

cut 1. To remove part of something, such as a portion of a document. 2. Same as **crystal cut**.

cut and paste In computers, to remove something from one location, to then place it in another. For instance, to remove content such as text and/or images from one document and place it elsewhere in the same document, or in another. Another example is to remove several files from a directory and place them in another. Its abbreviation is **cut & paste**.

cut-off To shut off, discontinue, separate, isolate, or block.

cut-off attenuator Same as **cutoff attenuator**.

cut-off bias Same as **cutoff bias**.

cut-off current Same as **collector-cutoff current**.

cut-off field Same as **critical field**.

cut-off frequency Same as **cutoff frequency**.

cut-off limiting Same as **cutoff limiting**.

cut-off point Same as **cutoff point**.

cut-off voltage Same as **cutoff voltage**.

cut-off wavelength Same as **cutoff wavelength**.

cut-out Same as **cutout**.

cut & paste Same as **cut and paste**.

Cutler feed An antenna feed system in which RF energy is transferred from a resonant cavity at the end of a waveguide to the reflector.

cutoff 1. The action of shutting off, discontinuing, separating, isolating, or blocking. Also, such an instance. 2. For a device such as a filter, amplifier, waveguide, or transmission line, the frequency at which the attenuation increases rapidly. For example, the frequency above which the output of an electron tube is no longer useful. Also called **cutoff frequency** (1), **critical frequency** (1), or **frequency cutoff** (1). 3. The upper and lower frequencies at which the gain of an amplifier, such as a transistor, drops 3 decibels, or another reference amount. Also called **cutoff frequency** (2), or **frequency cutoff** (2). 4. For a given transmission mode in a waveguide, the frequency below which a traveling wave can not be maintained. That is, the waveguide functions efficiently only above this frequency. Also called **cutoff frequency** (3), **waveguide cutoff**, or **frequency cutoff** (3). 5. The point at which the current flowing through an active device, such as a transistor, is stopped by the control electrode.

cutoff attenuator A non-dissipative attenuator consisting of an adjustable length of waveguide used below its cutoff frequency. Also spelled **cut-off attenuator**.

cutoff bias For an electron tube or transistor, the value of control-electrode bias that stops the flow of output current. Also spelled **cut-off bias**. Also called **cutoff point** (2).

cutoff current Same as **collector-cutoff current**. Also spelled **cut-off current**.

cutoff field Same as **critical field** (2). Also spelled **cut-off field**.

cutoff frequency Also spelled **cut-off frequency**. 1. Same as **cutoff** (2). 2. Same as **cutoff** (3). 3. Same as **cutoff** (4).

cutoff limiting In an electron tube, the limiting of the output voltage by driving the control grid voltage beyond cutoff. Also spelled **cut-off limiting**.

cutoff point Also spelled **cut-off point**. 1. For a device such as a filter, amplifier, waveguide, or transmission line, the frequency at which the attenuation increases rapidly, or at which output is at a given level, such as 70.7% of the maximum. 2. Same as **cutoff bias**.

cutoff voltage Also spelled **cut-off voltage**. 1. For an electron tube or transistor, the value of control-electrode voltage that stops the flow of output current. 2. For a given steady magnetic flux density in a magnetron, the highest theoretical value of anode voltage that would prevent an electron emitted from the cathode at zero velocity from reaching the anode. Also called **critical voltage**. 3. For a cell or battery, the voltage at which discharge is complete. Also called **end-of-discharge voltage**.

cutoff wavelength The wavelength corresponding to the **cutoff frequency**. Also spelled **cut-off wavelength**.

cutout A device which under certain circumstances disconnects a component or circuit, in order to protect equipment. An example is a circuit breaker. Its function is usually automatic. Also spelled **cut-out**.

cutout box A box which houses a set of cutouts or fuses. May be found, for instance, in a dwelling or vehicle.

CV Abbreviation of **computer vision**.

CVD Abbreviation of **chemical-vapor deposition**.

cw Abbreviation of **clockwise**.

CW 1. Abbreviation of **carrier wave**. 2. Abbreviation of **continuous wave**.

CW laser Abbreviation of **continuous-wave laser**.

CW radar Abbreviation of **continuous-wave radar**.

CWIS Abbreviation of **campus-wide information system**.

cyan A greenish blue color. It is one of the primary colors used in certain color processes, such as printing.

cyan-magenta-yellow A color model used for printing in which any color is formed through the appropriate mixture of cyan, magenta, and yellow inks. White means no mixture of colors, and black is produced by adding 100% of each of the colors. Its abbreviation is **CMY**

cyan-magenta-yellow-black A color model similar to **cyan-magenta-yellow**, with the difference that black is produced by a separate ink, instead of adding 100% of each of the other colors. This model provides for sharper black printed output. Its abbreviation is **CMYK**.

cyber- A prefix which serves to produce derivative words pertaining to computers and/or online concepts. For instance, cyberspace. Originally an abbreviation of the word **cyber**netic.

cyber café A public establishment which may serve beverages and/or food, but whose primary service is the offering of one or more terminals providing access to the Internet. Also spelled **cybercafé**. Also called **Internet café**.

cyber warfare Same as **cyberwar**.

cybercafé Same as **cyber café**.

cybercash Also known by various other terms, including **cybermoney, electronic cash, e-money,** and **digital cash. 1.** Any form of money used over the Internet. Used for many purposes, such as effecting online purchases or making credit card payments. Security is a key concern, and considerable measures, such as cryptography, are taken to avoid circumstances such as losses of funds, misplacement of funds, stealing of funds, or the re-use of funds. **2.** A form of currency that is digitally encoded, and that can be converted to legal tender if desired. It may be coded, for instance, through the use of an encrypted serial number. **Legal tender** is a lawfully valid currency that must be accepted for payment of debts, the purchase of goods and/or services, or for other exchanges for value. Circulating paper money is another form of legal tender.

cybercrime Crimes, such as theft or fraud, committed in **cyberspace**.

CyberMall A Web site shared by, or linking to, multiple businesses. For example, a shopper may place products from different merchants in a single shopping cart, for convenient payment and delivery.

cybermoney Same as **cybercash**.

cybernetic 1. Pertaining to **cybernetics. 2.** Pertaining to computers and/or online concepts.

cybernetic organism Same as **cyborg**.

cybernetics The analysis of control and communications processes and systems in living organisms and machines, and how they relate to each other. Cybernetics may be used, for instance, to design a machine capable of duplicating the way a living organism adapts to its environment.

cybersex The use of **cyberspace** as a medium for matters pertaining to eroticism.

cybersitter Software that blocks access to Web sites and/or specific content deemed inappropriate or otherwise undesirable. When utilized to prevent children from such access, also called **Net Nanny**.

cyberspace 1. The universe of interconnected computers. It includes the entire range of information and entertainment resources available through computer networks, especially the Internet. A person browsing the World Wide Web is also navigating through cyberspace. **2.** A worldwide network of interconnected autonomous networks, which is utilized for commerce, education, research, entertainment, and the obtaining of or exchange of information on virtually anything of human interest. Cyberspace currently encompasses hundreds of millions of computers and users, and

spans nearly 200 countries. Among other services, users can exchange email, browse the World Wide Web, purchase and sell goods, and so on. An Internet service provider, for instance, may be used for access. Also called **Internet, information highway (2), information superhighway (2),** or **Infobahn (2)**.

cyberwar Abbreviation of **cyber war**fare. The use of **cyberspace** for information warfare.

cyborg Acronym for **cyb**ernetic **org**anism. A human who has one or more mental and/or physical processes enhanced and/or controlled by devices which may be mechanical, electrical, electromechanical, optical, and so on.

cycle 1. One complete sequence of changes of a periodically repeated phenomenon. For instance, an AC cycle. **2.** One complete sequence of operations performed as a unit. For example, a machine cycle within a computer processor. **3.** One complete sequence of events occurring as a unit, such as sunspot cycles.

cycle count 1. The count a **cycle counter** indicates at any given moment. **2.** A tally of a given number of cycles.

cycle counter A circuit or device which counts **cycles**.

cycle index The number of times a computer has carried out a given cycle.

cycle-index counter 1. A device which counts the number of times a computer has carried out a given cycle. **2.** A variable which indicates the number of times a computer has carried out a given cycle.

cycle life The number of charges and discharges a storage battery can sustain while maintaining adequate performance.

cycle reset The resetting a cycle counter to zero, its initial value, or to a given desired value.

cycle stealing A technique in which a peripheral control unit, such as a direct memory access device, uses the local bus for transfers during cycles that the CPU is not using it.

cycle time The time interval that elapses during a **cycle**.

cycle timer A device, such as a switch, which opens or closes a circuit according to a cycle or schedule.

cycles per minute The frequency per minute that a periodic phenomenon has a complete set of values, or cycles. Its abbreviation is **cpm**.

cycles per second The frequency per second that a periodic phenomenon, such as an electromagnetic wave, has a complete set of values, or cycles. The term currently used for this concept is **hertz**. Its abbreviation is **cps**.

cyclic redundancy check A technique used for error detection while transmitting digital data. In it, the sending computer performs a complex calculation based on the data, and attaches the resulting value. The receiving computer performs the same calculation, and the result is compared to the original value. If the values do not match, an error has occurred. It provides a greater degree of error detection than a checksum. Its abbreviation is **CRC**. Also called **cyclic redundancy code**, or **cyclical redundancy check**.

cyclic redundancy code Same as **cyclic redundancy check**.

cyclical redundancy check Same as **cyclic redundancy check**.

cycling 1. The oscillation of a controlled variable near the desired value, or between two values. Also called **oscillation (3). 2.** To run through one or more cycles.

cyclotron A circular accelerator in which charged particles are given repeated energy boosts by a constant-frequency alternating electric field, while a static magnetic field maintains the particles in their proper spiraling path. The hollow metal box within which the particles are accelerated is divided into two, with each of the sections being called a **dee**. Once the particles reach a given velocity, they spiral out onto the desired target.

cyclotron emission   Same as **cyclotron radiation**.

cyclotron frequency   **1.** The angular frequency of the orbit of a charged particle in a **cyclotron**. **2.** The frequency at which the electric field in a **cyclotron** is alternated so that the particles are further accelerated. **3.** The angular frequency of the orbit of a particle orbiting along an axis which is perpendicular to a uniform magnetic field. Also called **gyrofrequency**.

cyclotron radiation   Also called **cyclotron emission**. **1.** The radiation emitted by particles orbiting in a **cyclotron**. **2.** The radiation emitted by particles orbiting in a magnetic field.

cylinder   **1.** A uniform body with a circular section and straight sides. Such a body may be hollow, partly filled, or solid. Also, the shape described, and that which resembles this shape. **2.** The set of all tracks that have the same track number on each disk surface. In a multiplatter hard drive, there would be two such tracks on each surface, with all tracks being vertically above or below each other. Since an access arm remains stationary until all the tracks in a cylinder are either read or written, access time can be reduced by storing related data in cylinders.

cylindrical   In the form of, or resembling a **cylinder (1)**.

cylindrical antenna   An antenna whose radiating elements are hollow cylinders.

cylindrical capacitor   A capacitor in which two concentric metal cylinders, of equal length, enclose a dielectric. Also called **coaxial capacitor**.

cylindrical wave   A wave whose equiphase surfaces form a series of coaxial cylinders.

cylindrical waveguide   A waveguide in cylindrical form.

cylindrical winding   A coil winding in which one or more layers are helically wound.

Czochralski method   A technique used for cultivating large single crystals. Using this method, a seed crystal is dipped into a molten solution of the crystal, and is slowly withdrawn while being rotated. Used, for instance, to grow semiconductor crystals such as gallium arsenide. Also called **Czochralski process, Czochralski technique**, or **liquid-encapsulated Czochralski**.

Czochralski process   Same as **Czochralski method**.

Czochralski technique   Same as **Czochralski method**.

D

d Abbreviation of **deci-**.

D 1. Symbol for **electric displacement, displacement (2)**, and **electric flux density**. 2. Symbol for **deuterium**. 3. Symbol for **dissipation factor**. 4. Symbol for **drain**.

d/a Abbreviation of **digital to analog**.

D/A Abbreviation of **digital to analog**.

d/a converter Abbreviation of **digital-to-analog converter**.

D/A converter Abbreviation of **digital-to-analog converter**.

D-AMPS Abbreviation of **digital AMPS**.

D'Arsonval galvanometer A DC galvanometer in which a small wire coil is suspended between the poles of a permanent magnet. The current that passes through the coil produces a magnetic field which reacts with that of the magnet, and the resulting movement of the coil is correlated to a calibrated scale. Such galvanometers feature high sensitivity, high damping, and low resistance.

d.c. Same as **DC**.

D channel Abbreviation of **delta channel**, or **data channel**. In ISDN communications, a channel that serves to transmit control information such as that associated with packet-switching, in addition to user-related data such as the identity of the calling party. Its bandwidth is usually 16 Kbps, but can be greater. It works alongside a **B channel (2)**.

d-cinema Same as **digital cinema**.

D display In radars, an oscilloscopic display which plots the bearing of the scanned object in the horizontal plane versus its angles of elevation, which appear in the vertical plane. The blips, which represent the scanned objects, extend vertically to provide a rough estimate of the distance. Also called **D scope, D scanner, D scan**, or **D indicator**.

D flip-flop Abbreviation of **delayed flip-flop**.

D indicator Same as **D display**.

D-layer A layer of the ionosphere that usually only occurs during daylight hours, and which is contained in the D-region. It extends approximately from 60 to 100 kilometers above the surface of the earth, is the lowest ionospheric layer, and is situated below the E-layer. It reflects low-frequency radio waves, while absorbing higher-frequency radio waves.

D-RAM Abbreviation of **dynamic RAM**.

D-region The region of the ionosphere in which the D-layer forms. This region usually only occurs during daylight hours, and extends approximately from 60 to 100 kilometers above the surface of the earth. It is the lowest ionospheric region, is situated below the E-region, and its greatest ionization occurs at midday.

d/s converter Abbreviation of **digital-to-synchro converter**.

D scan Same as **D display**.

D scanner Same as **D display**.

D scope Same as **D display**.

D1 Abbreviation of **D1 Digital Video**.

D1 Digital Video A component digital video format adhering to the ITU-R BT.601 standard, and which utilizes a 19 mm tape. This format provides outstanding quality. Its abbreviation is **D1**.

D2 Abbreviation of **D2 Digital Video**.

D2 Digital Video A composite digital video format for PAL or NTSC signals which utilizes a 19 mm tape. This format provides very high quality, and is not compatible with D1. Its abbreviation is **D2**.

D3 Abbreviation of **D3 Digital Video**.

D3 Digital Video A composite digital video format similar to **D2 Digital Video**, but which utilizes a 0.5 inch digital tape. Its abbreviation is **D3**.

D5 Abbreviation of **D5 Digital Video**.

D5 Digital Video A component digital video format adhering to the ITU-R BT.601 standard, and which utilizes a 0.5 inch tape. Its abbreviation is **D5**.

D9 Abbreviation of **D9 Digital Video**.

D9 Digital Video A component digital video format utilizing a high-density metal particle 0.5 inch tape. Some D9 machines can play back analog S-VHS tapes. Its abbreviation is **D9**. Also called **Digital-S**.

da Abbreviation of **deca-**, or **deka-**.

DAB 1. Abbreviation of **digital audio broadcast**. 2. Abbreviation of **digital audio broadcasting**.

DAC Abbreviation of **digital-to-analog converter**.

DACS Abbreviation of **digital access cross-connect system**.

DAD Abbreviation of **digital audio disc**.

DAE Abbreviation of **digital audio extraction**.

daemon Acronym for **d**isk **a**nd **e**xecution **mon**itor. A program which runs in the background and automatically performs administrative functions when needed, or at specified times. Examples of daemons include email handlers and print spoolers. Also spelled **demon**.

DAGC Abbreviation of **delayed automatic gain control**.

daisy chain 1. A computer hardware configuration in which devices are connected in series, that is, one after the other. The last device in the chain has a terminator, to avoid signals from reflecting back into the line. 2. To connect in a **daisy chain (1)** configuration. Also called **daisy chaining**.

daisy chaining Same as **daisy chain (2)**.

daisy wheel A disk with a given number of raised characters, which is spun so that a hammer may strike each character to be printed. Used in daisy-wheel printers.

daisy-wheel printer A printer that uses a spinning **daisy wheel** to print each character. Such printers are now obsolete. Also called **wheel printer**.

dalton A unit of mass measurement for atoms and molecules. The unit has been arbitrarily defined as 1/12 the mass of an atom of carbon-12, the most common isotope of carbon. Each unit is equal to approximately 1.6605×10^{-24} grams. Also known as **atomic mass unit**.

DAMA Abbreviation of **demand-assigned multiple access**.

damage Any harm which impairs the operation or properties of a component, circuit, device, piece of equipment, system, material, or the like. Damage can be immediately manifested, or may become evident some time later. For example, blasting may damage speakers, or an electrostatic discharge may damage a semiconductor component.

damp Same as **dampen**.

damped A signal, wave, oscillation, or vibration which is having, or has had, its amplitude gradually decreased.

damped meter A meter with a degree of damping which minimizes or prevents overshoot or oscillation.

damped oscillation Oscillation that dies away from an initial maximum amplitude, with each successive oscillation diminishing until the amplitude becomes zero. Also called **damped vibration**.

damped vibration Same as **damped oscillation**.

damped wave A wave whose amplitude dies away from an initial maximum, diminishing over time until its amplitude becomes zero.

dampen To gradually decrease the amplitude of a signal, wave, oscillation, or vibration. For example, the use of acoustic absorption to dissipate incident acoustic waves. Also called **damp**.

damper 1. A diode utilized to limit unwanted oscillations. Also called **damper tube** (1), **damper diode** (1), **damping tube** (1), or **damping diode** (1). 2. In a TV receiver, a diode used to limit unwanted oscillations in the horizontal deflection circuit. Also called **damper tube** (2), **damper diode** (2), **damping tube** (2), or **damping diode** (2). 3. A device which provides **damping** (1), or **damping** (3).

damper diode 1. Same as **damper** (1). 1. Same as **damper** (2).

damper tube 1. Same as **damper** (1). 1. Same as **damper** (2).

damper winding A winding which serves to dampen any oscillations in a synchronous motor. Also called **amortisseur winding**.

damping 1. A gradual reduction or limiting in the amplitude of a vibrating motion, such as an oscillation. Such decreases are due, for instance, to energy losses to friction. In measuring instruments, damping may be introduced to minimize oscillation or overshoot. 2. The extent of the reduction in amplitude **damping** (1) provides. 3. The reduction of the magnitude of vibration or resonance of a surface or material, in order to reduce or eliminate sound waves. Also called **acoustic damping**, or **soundproofing**.

damping coefficient Same as **damping factor** (1).

damping diode 1. Same as **damper** (1). 1. Same as **damper** (2).

damping factor 1. For an underdamped vibrational system, the ratio of the amplitude of a given vibration, to that of the next vibration. For example, the ratio of the amplitude of a damped oscillation to the succeeding one. Also called **damping coefficient**, or **damping ratio**. 2. The natural logarithm of the **damping factor** (1). Also called **decrement** (3), or **logarithmic decrement**. 3. The ratio of the nominal impedance of a speaker, to the total impedance driving it. A higher damping ratio stabilizes the driven speaker, and improves its performance.

damping magnet A permanent magnet whose field is utilized to dampen the movement of a moving conductor, such as a disk.

damping material A material, such as glass wool, which is placed within the enclosure of a speaker to absorb sound and reduce out-of-phase reflections.

damping ratio Same as **damping factor** (1).

damping resistor A resistor that provides **damping** (1). For instance, such a resistor used in a galvanometer to minimize oscillation or overshoot.

damping tube 1. Same as **damper** (1). 1. Same as **damper** (2).

Daniell cell A primary cell consisting of a zinc-mercury anode in dilute sulfuric acid, and a copper cathode in a copper sulfate solution, each solution separated by a porous partition. It has a voltage of approximately 1.08.

DAP Abbreviation of **Directory Access Protocol**.

DAP protocol Same as **Directory Access Protocol**.

daraf A unit of elastance, which is the reciprocal of capacitance. **Daraf** is **farad** spelled backwards, as elastance expressed in darafs is equal to 1 divided by capacitance expressed in farads.

dark conduction Conduction of current that occurs in a photosensitive material when it is not illuminated. An example is residual conduction in a photoconductive cell after the source of illumination is removed.

dark current Current that flows through a photosensitive material when it is off, or not illuminated. An example is a residual current flowing through a photoconductive cell after the source of illumination is removed. Also called **electrode dark current**.

dark discharge An electric discharge, such as that which may occur in a gas, that does not emit light.

dark fiber Fiber-optic cable that is installed but that has no electronics, such as transmitters or regenerators, connected to it. Such cables may be leased or sold to customers that add the required components. This contrasts with **lit fiber**, which does not handle communications traffic.

dark resistance The resistance of a photosensitive material in total darkness.

dark space 1. A dark zone within a glow-discharge tube. For instance, an Aston dark space, or a Crookes dark space. 2. In a glow-discharge tube, a narrow dark space near the cathode, between the cathode glow and the negative glow. Also called **Crookes dark space, cathode dark space**, or **Hittorf dark space**.

dark spot A dark area, that may appear in a received TV image, which is caused by clouds of electrons forming within a TV camera tube.

dark-trace tube A type of CRT whose screen is specially coated, so that it darkens when impacted by electrons. The displayed signals on such a screen are presented as dark traces or blips against a bright background. Also called **skiatron**.

Darlington amplifier An amplifier consisting of a **Darlington pair**. Such an amplifier provides high input impedance and high current gain.

Darlington pair A compound connection of two bipolar junction transistors whose collectors are connected, and in which the emitter of the input transistor is connected to the base of the output transistor. Both transistors are usually mounted on the same transistor housing. Also called **double emitter follower**.

DARS Abbreviation of **d**igital **a**udio **r**adio **s**ervice. A broadcast radio service in which content is digitally encoded. Such broadcasts may or may not be transmitted via satellite.

DAS 1. Abbreviation of **data-acquisition system**. 2. Abbreviation of **dual-attachment station**.

DASD Acronym for **direct-access storage device**.

dashpot A device utilized for damping and controlling motion. It consists of a cylinder with a contained piston, whose motion is damped by a contained liquid or gas.

DAT 1. Abbreviation of **digital audio tape**. 2. Abbreviation of **dynamic address translator**.

data 1. Factual information, such as that derived from observation, experimentation, or calculation, especially that which is used for processing and/or further analysis. 2. Information which is represented in a manner suitable for digital transmission or computer use. 3. Originally the plural form of the word **datum**, although **data** is commonly used in either the plural or singular form.

data abstraction 1. In object-oriented programming, the abstraction of the underlying structure of data. 2. In object-oriented programming, the abstraction of the items contained in a data type.

data acquisition 1. The gathering of data, especially that which is to be used for processing and/or analysis. The collecting of such data may be manual or automatic. 2. The gathering of data which is, or is to be represented in a manner suitable for digital transmission or computer processing.

data-acquisition computer A computer used for **data acquisition**. Such a computer may have analog and/or digital inputs from signals generated by measuring instruments.

data-acquisition system A system, such as a computer, used for **data acquisition**. Its abbreviation is **DAS**.

data analysis 1. The careful study of data from various perspectives, in order to gain a fuller understanding of it. 2. The results of **data analysis** (1).

data attribute A characteristic, such as data type, of a block of data.

data automation The use of automatic processes, equipment, or systems for the purpose of collecting, processing, storing, transmitting, and presenting of data. Pertains especially to the use of computers and peripheral devices, such as GPS and bar-code systems, used for these tasks. Its abbreviation is **datamation**.

data availability The offered breadth, speed, ease of use, and uptime of a source of data.

data bank 1. A depository of data. Such a location may consist, for instance, of a collection of computer tapes, and may be accessed by one or more computers. 2. The data contained in a **data bank** (1).

data base Same as **database**.

data-base engine Same as **database engine**.

data-base machine Same as **database machine**.

data-base management system Same as **DBMS**.

data-base program Same as **database program**.

data-base server Same as **database server**.

data bits In data communications, a block of bits which represent a single character. Such a block usually consists of 8 bits, though it can be as few as 5.

data buffer A segment of computer memory utilized to temporarily store data that awaits transfer. Used, for instance, to compensate for differences in operating speeds

data buffering The use of a segment of computer memory to temporarily store data that awaits transfer. This is useful, for instance, when a computer and a peripheral operate at different speeds.

data bus A channel which provides data transfer from one part of a computer to another. The width of the bus determines the amount of data which can be transferred at a time. A 256-bit bus, for example, can transfer 256 bits at a time.

data-bus connector Its abbreviation is **DB connector**. Any of various types of connectors which are used in communications and computer devices. The number following the DB indicates the number of pins, and such connectors can range from 9 pins to over 50. A DB-25 connector, for instance, is frequently used to interface a computer parallel port.

data cable A cable which serves to transfer data from one device or location to another. Examples include fiber-optic cables, and shielded twisted pair wires.

data capacity The maximum amount of data which can be stored by a medium such as an optical disk.

data capture The collection of data, especially via automatic or semi-automatic means. Such data capture may occur through the use of RFID, bar-code, or voice-recognition systems, to name a few.

data carrier 1. A computer storage medium, such as removable magnetic disk or tape, which serves to transport computer data. 2. A carrier wave intended to be modulated by a data signal.

data cartridge A removable computer storage device which contains a magnetic or optical recordable medium. For instance, a tape cartridge.

data center Same as **datacenter**.

data channel 1. Same as **D channel**. 2. A communications channel used for data transmission. Also called **information channel**. 3. A path along which input/output data is transmitted. Also called **I/O channel**.

data circuit 1. A communications circuit used for data transmission. 2. A communications circuit used for digital data transmission. 3. A communications circuit specially suited for digital data transmission.

data circuit-terminating equipment Same as **data communications equipment**. Its abbreviation is **DCE**.

data code 1. A code, consisting of one or more characters, utilized to represent data in a computer. For example, a bar code. 2. A code, consisting of one or more characters, utilized to abbreviate data in a computer. For instance, a code representing a given product.

data collection 1. The gathering of source data and/or documents. For instance, that collected by an RFID or bar-code reader. 2. The process of organizing gathered data.

data communications Its acronym is **datacom**. 1. The transmission of data between two or more points or entities. 2. The transmission of digital data between two or more points or entities 3. The transmission of digital data between two or more computers.

data communications equipment Its abbreviation is **DCE**. A device which enables, maintains, and terminates connections within a communications system such as a network. A DCE is usually a modem, and may also provide signal conversion between data terminal equipment and a communications channel. Also called **data circuit-terminating equipment**.

data communications network A network, such as a LAN, used for data communications.

data compaction Same as **data compression**.

data compression The encoding of data so that it occupies less space and/or bandwidth. There are many algorithms used for compression, and depending on the information being encoded, space savings can range from under 10% to over 99%. Data compression can also be achieved by increasing packing density. Also called **data compaction**, **compression (2)**, **digital compression**, or **information compression**.

data conferencing The holding or participating in a meeting in which data is shared simultaneously between two or more locations. Aside from file sharing, participants may collaborate through the use of other resources, such as whiteboards. Also called **desktop conferencing**.

data control The automatic or manual control of the access, processing, and use of data. For instance, the tracking of who has accessed or changed any given records, or the scheduling of availability of specific resources.

data conversion The changing of computer data from one form to another. For example, from analog to digital, from that contained on a CD-ROM to a floppy, or from one file format to another.

data converter A circuit or device which performs **data conversions**. For example, an analog-to-digital converter, or a binary decoder.

data corruption The damaging of computer data. Causes include hardware malfunctions, software failures, power outages, and viruses.

data declaration Same as **data definition**.

data definition Computer program statements which define the attributes of data or of a database. Such statements specify characteristics like field size, data type, or the organization of data within a database. Also called **data declaration**.

data definition language A computer language used to define the attributes of data or of a database. It specifies characteristics such as field size, data type, or the organization of data within a database. Its abbreviation is **DDL**. Also called **data description language**.

data description language Same as **data definition language**. Its abbreviation is **DDL**.

data dictionary A file or database which details the contents of one or more databases. It contains information such as file names and specifications, field types and length, and which programs access which data. A data dictionary may update automatically or manually.

data directory A list, table, or index which describes data stored in a computer or computer application, such as a database. Such a directory includes information such as the name of the data, and its attributes.

data display A device, such as a computer monitor, which serves to display data.

data distribution The dissemination of data from a central point, such as a hub in a centralized network.

data element 1. A logical definition of the content of a data field. For example, the colors of a product. Within this data element, a **data item** would be a specific color, such as cyan. Also called **data item** (2). 2. A single unit of computer data, which is stored in a field. Also called **data item** (1).

data encryption The encoding of data so that only the intended recipients can understand it. It is an extremely efficient method to achieve data security, and a code, or key, is used to convert the information back to its original form. Public-key encryption and secret-key encryption are the two common methods.

data encryption key A series of binary digits or characters that are incorporated into data to encrypt or decrypt it. Its abbreviation is **DEK**.

Data Encryption Standard Its abbreviation is **DES**. A widely employed data encryption method. It is a block-cipher algorithm which utilizes a 56-bit key, and each encrypted block is 64 bits in size. There are variations which are even more secure, such as Triple DES.

data entry The process of entering data into a computer. For instance, entering information into a computerized database. Data may be entered automatically or manually, and may be accomplished in various ways, including keyboard entry, voice recognition, or directly from another computer.

data-entry equipment Equipment utilized for **data entry**.

data-entry program A computer program which accepts input data, such as that from a keyboard, and stores it appropriately. Such a program may also perform other functions, such as error-checking, or automatic updating. Also called **input program**.

data-entry terminal A computer terminal used for **data entry**.

data error An error in the input, output, communicating, processing, or storage of data.

data expansion The restoring of data to its original space and/or bandwidth, following data compression.

data-fax modem A data modem that can also handle fax protocols, thus incorporating fax capabilities. Documents that are sent by such a device must be in an electronic form, such as a disk file. Received documents are also stored in electronic form. Also called **fax-data modem**.

data field A space where computer data is stored, such as a cell in a spreadsheet. Most data fields have their associated attributes. For example, it may have a numeric attribute if it contains numbers. A collection of data fields is called a **data record**. Also called **field** (3).

data file A file containing information such as text, as opposed to a program file which contains executable content, such as instructions.

data flow The path that data follows, from its source to its destination. For instance, the route a message follows from its origin to destination, including all the intermediate junctions. Another example is the movement of data that is taken from a storage device, its being processed, and then being sent to an output device such as a monitor. Also spelled **dataflow**.

data flow diagram A diagram in which labeled blocks, usually circles, squares and/or rectangles, are used to represent the flow of data. The blocks are linked with arrows and/or lines, which indicate the direction of the path followed. Also spelled **dataflow diagram**.

data format The structure and properties which an application program applies to data. Computer graphics, for instance, can be in any of various formats, including bit-mapped, JPEG, TIFF, and so on.

data formatting The application of a **data format**.

data frame A block of data of a specific size, or of a maximum size, transmitted in a network. Also called **frame** (3).

data glove A computer input device that is placed over the hand, like a glove. It contains sensors which convert movements into data. Such gloves may also provide tactile feedback, and are used, for instance, in virtual reality applications.

data hiding 1. Any technique or instance of concealing data. These include, for example, the use of steganography, digital watermarks, transparent GIFs, and so on. 2. In object-oriented technology, the use and result of encapsulation.

data independence The separation of data from the programs that use it. In this manner, when data is changed, the applications which manage it are not affected. The greater the data independence, the more generally accessible the data.

data integrity The preservation of the completeness and accuracy of data that is transmitted and received, stored and retrieved, or manipulated in any manner. Also called **integrity**.

Data Interchange Format A standard file format applicable to spreadsheets and databases, for enhanced interchangeability between programs. Its abbreviation is **DIF**.

data item 1. Same as **data element** (2). 2. Same as **data element** (1).

data library 1. Within a datacenter, a cataloged collection of the content of storage media such as disks and tapes. 2. A directory, list, or index of files available from a server.

data line A circuit, or line, through which data is transmitted. For instance, such a line between nodes of a network, or within a data bus. Generally refers to digital data.

data link Also called **communications link**, or **information link**. 1. A connection which enables the transmission of information between two points or entities. 2. The resources which facilitate a **data link** (1).

data link connection identifier In frame relay, a field which identifies a logical connection that is multiplexed into the physical channel. Its abbreviation is **DLCI**.

data link escape Same as **data link escape character**.

data link escape character In data transmission, a control character which indicates that one or more of the characters to follow provides control information. Its abbreviation is **DLE**. Also called **data link escape**.

data-link layer Within the OSI Reference Model for the implementation of communications protocols, the second lowest level, located directly above the physical layer. This layer takes care of functions such as the coding, addressing, and transmission of information between nodes. IEEE 802 standards divide the data-link layer into the Logical Link Control sublayer, and the Media Access Control Sublayer. Also called **layer 2**.

data-link layer address For a device connected to a communications network, an address which uniquely identifies each physical connection. An end device, for instance, has a single data-link layer address, while routers have multiple addresses. Also called **link layer address**.

data management   The handling of data at all stages and levels, including input, coding, storage, processing, retrieval, and output. This includes the functions that the hardware, software, and users perform. For instance, a user can determine which data is to be acquired, an application program can manipulate and organize this information, while the operating system enables the hardware to process and store the data properly.

data management program   A program, such as that used within a DBMS, that performs **data management** functions.

data management system   Same as **DBMS**. Its abbreviation is **DMS**.

data manipulation   Any of the various manners in which data is processed or organized, including sorting, editing, or merging.

data manipulation language   In a DBMS, a language utilized to locate, retrieve, store, update, and delete data from a database. An example is a query language. Its abbreviation is **DML**.

data mart   A smaller version of a **data warehouse**. It may be utilized, for instance, to store information for one of many departments. Also spelled **datamart**.

data medium   Any physical material or medium which serves to store or otherwise contain data. For instance, optical discs, magnetic tapes and disks, microfilm, and paper. Also called **data-storage medium**, **medium (3)**, or **storage medium**.

data migration   1. The process of moving data from one storage source or device, such as a database or tape, to another. 2. The automatic transfer of information between storage media with different priorities. For example, the movement of data from a lower speed medium, such as an optical disk, to a hard disk when needed. This allows more data to be stored than if all was in the higher speed, and hence more expensive, medium. Also called **hierarchical storage management**.

data mining   The process of analyzing data to identify relationships and patterns which may be useful. Such mining may be done manually or automatically, through the use of specialized programs. Also called **mining**.

data mirroring   1. The duplication of stored data at a remote location, on another drive, another medium, or the like. Used, for instance, for backing-up, disaster recovery, or as a security measure. Also called **mirroring (2)**. 2. The maintaining of identical copies of data and files at multiple network sites or servers. For example, an entity which receives many download requests from around the world may have many diversely located mirror sites to facilitate access. Also called **mirroring (3)**.

data model   The structure or design, including the organization, data formats, and interrelations, of a database. Also, a graphical representation of such a model.

data modem   A modem that does not support fax protocols, thus does not have fax capabilities.

data module   A sealed unit which encloses one or more magnetic disks and their associated components, such as access arms.

data name   A name used to designate a data field, data element, or data item.

data network   A system of computers, transmission channels, and related resources which are interconnected for the exchange data. It is a type of communications network.

data organization   The manner in which data is arranged in a database or storage medium. For instance, the sequential storage of data.

Data Over Cable Service Interface Specification   Same as **DOCSIS**.

data packet   Also called **packet**. 1. A block of data transmitted between one location and another within a communica-

tions network. Also called **information packet**. 2. A block of data of a specific size or of a maximum size, transmitted in a packet-switching network. In addition to the payload, a packet contains information such as the source and destination addresses. When used in the context of TCP/IP networks, also called **datagram**.

data parallel   A computer architecture in which multiple processors simultaneously and independently execute the same instruction set on different sets of data. Also called **single instruction stream-multiple data stream**.

data port   A jack or socket used for data communications, especially one to which a modem can be connected to. Also spelled **dataport**.

data processing   Its abbreviation is **DP**. Also called **automatic data processing**, **electronic data processing**, or **information processing**. 1. The processing of information by a computer. 2. The processing of information within a specific application. For example, the manipulation of large amounts of numeric data in a program.

data-processing center   Same as **datacenter**.

data-processing equipment   Equipment used for **data processing**. Such equipment includes computers and any peripherals used in this process. Also called **data-processing system (1)**, or **electronic data-processing equipment**.

data-processing system   Also called **electronic data-processing system**. 1. Same as **data-processing equipment**. 2. The resources and procedures utilized for **data processing**.

data processor   A computer used for **data processing**.

data projector   A device which projects the visual output of a computer onto a remote screen. Such a device may also accept inputs for DVDs, HDTVs, S-Videos, and so on. Examples include DLP projectors and LCD projectors.

data protection   The safeguarding of data against loss, damage, unwanted modification, or unauthorized access. Such safeguards may be administrative, physical, or technical. Also called **data security**, **protection (2)**, or **information security**.

data rate   Same as **data transfer rate**.

data receiver   A point or device, within a communications network, that receives data.

data reception   The reception of data over a communications network.

data record   A group of related fields, each containing information. For instance, a group of fields, each containing one of the following items: a name, a corresponding address, and a contact number. A collection of data fields form a data record, and a collection of data records form a file. Also called **record (2)**. When used in the context of relational databases, also called **tuple**.

data recording   1. The act of registering, retaining, or saving data. 2. Data which has been registered, retained, or saved.

data recovery   1. The salvaging of data stored on a medium, such as a disk or a tape, which has been damaged. Such damage may be a result of a hardware malfunction, a software failure, a power outage, a virus, or physical abuse such as the computer falling from a significant height. 2. The techniques employed to achieve **data recovery (1)**.

data-recovery system   Any system, such as that which includes preparing, storing, and accessing backup disks or tapes, utilized for **data recovery (1)**. Also called **recovery system**.

data reduction   The conversion of data into a more useful form. For instance, through ordering or summarizing.

data redundancy   The replication of a database, or of any portion of it. This may be done, for instance, when a backup is desired.

data register   A high-speed storage area within the CPU, which is utilized to hold data for specific purposes. Before data may be processed, it must be represented in a register. There are various types, including control registers, index registers, arithmetic registers, and shift registers. Also called **register (1)**.

data repository   A location or system, such as a data mart or data warehouse, which stores and organizes the data of an entity, such as a corporation. Also, the data so held. Also called **repository**.

data representation   The manner in which data is represented. This includes its structure, and the digits, symbols, or characters used to describe it.

data retrieval   The process of accessing specific data located in memory or a storage device. Also called **retrieval**.

data scrubbing   The process of fixing or eliminating data which is incomplete, incorrect, duplicated, or the like. Data scrubbing may be necessary, for instance, when database fields are left blank, information is entered poorly, or when multiple databases are improperly combined. Also called **scrubbing**.

data security   Same as **data protection**.

data segment   **1.** An area of computer memory or storage which is allocated for the placement of data used by a program. **2.** An area of computer memory or storage in which data used by a program is located.

data service unit   A device which converts data transmitted or received through an external digital circuit, such as a T1, into the proper format. In addition, the DSU has other functions, such as controlling the flow of data. It is frequently used in conjunction with a channel service unit. Its abbreviation is **DSU**. Also known as **digital service unit**.

data service unit/channel service unit   A device which combines a data service unit and a channel service unit. The data service unit converts data transmitted or received through an external digital circuit, such as a T1, into the proper format, while the channel service unit terminates the external digital circuit at the premises of a customer. Its abbreviation is **DSU/CSU**. Also called **digital service unit/channel service unit**.

data set   A collection of related data or data files. Also spelled **dataset**.

data set ready   A signal, indicating readiness for operation, that a modem sends to a computer. Used in serial communications. Its abbreviation is **DSR**.

data sharing   The ability of two or more programs, nodes, terminals, users, or the like, to access the same data, such as that contained in files. This may occur consecutively or simultaneously.

data sheet   One or more sheets which provide detailed information pertaining to a product. Such information may include specifications, warnings, applications, possible configurations, and so on.

data signal   A data-containing signal transmitted over a communications channel. Such a signal usually consists of a series of bits, and may contain information and/or control data.

data-signaling rate   The rate at which data is transmitted through a given point along a transmission line or communications circuit. May be expressed, for instance, in multiples of bits or bytes, such as gigabits, per second. Its abbreviation is **DSR**.

data sink   A device or component which accepts and stores arriving data signals.

data source   A device, component, apparatus, system, or medium which originates or sends a data signal, or from which data is taken. Also called **source (4)**.

data station   A remote computer terminal used for sending and/or receiving data from a central computer.

data storage   **1.** A device or medium, such as a disk drive or a tape, where data is stored on a permanent or semi-permanent basis. **2.** The act of storing data in a device or medium, such as a disk drive or tape, on a permanent or semi-permanent basis.

data-storage media   The physical materials or mediums which serve to store data on a permanent or semi-permanent basis. These include magnetic disks and tapes, DVDs, holograms, bubble memory, and so on. Also called **storage media (2)**, or **media (2)**.

data-storage medium   Same as **data medium**.

data store   A device, medium, or location, such as a hard disk, optical disc, or data repository, that serves for data storage.

data stream   The continuous flow of data over a communications channel. Also called **stream (2)**.

data striping   Same as **disk striping**.

data structure   The manner in which data is organized. Examples include fields, files, lists, trees, and tables. Data structures simplify the location and processing of data.

data switch   A device which routes data to one of multiple possible locations. For example, such a switch may be used when selecting one of several printers available to a single computer, or when multiple computers use a single printer. Data switches may be automatic or manual.

data system   A system used for one or more processes involving the handling of data. These processes may include the collection, processing, storage, retrieval, or transmission of data, and the devices, programs, channels, and equipment used for these purposes. Also called **information system**.

data tablet   Same as **digitizing tablet**.

data terminal   A device utilized for the input and/or output of data. Examples include keyboards, monitors, and computer terminals.

data terminal equipment   Equipment which serves as the source and/or destination point for transmitted data. It converts user information into signals appropriate for transmission and/or converts a transmitted signal into user information. Data terminal equipment, which usually consists of a computer or a terminal, is generally connected to a communications channel through data communications equipment. Its abbreviation is **DTE**.

data terminal ready   A signal, indicating readiness for operation, that a computer sends to a modem. Used in serial communications. Its abbreviation is **DTR**.

data throughput   The rate at which data is transmitted from one location to another. This may refer, for instance, to an internal pathway of a computer, or to a communications channel. Such a speed is usually expressed in some multiple of bits per second. Also called **throughput (1)**.

data traffic   The exchange of data through communications channels, systems, or networks. Bandwidth describes the traffic capacity of a given channel, system, or network.

data transfer   The movement or copying, without alteration, of data from one location to another. This may be, for instance, across a data bus, a peripheral bus, or between nodes of a network, to name a few. Also called **transfer (2)**.

data transfer rate   The speed at which bits or bytes are transmitted over a communications line or bus. Usually measured in some multiple of bits or bytes per second. Also called **data rate**, **transfer rate**, **bit transfer rate**, **bit rate**, or **line rate**.

data transmission   The transmission of data over a communications network. Its abbreviation is **DT**.

data transmission equipment   Equipment used for **data transmission**.

data transmitter   A point or device, within a communications network, which transmits data.

data type A manner of classifying data. The specification of a data type dictates the way a program handles the data. For instance, most applications can not calculate text. Other examples of data types include integers, real numbers, Boolean, and floating point. Also called **type (4)**.

data unit A group characters which is treated as a unit.

data validation The process of determining if data is accurate, complete, or compliant with specified criteria.

data visualization The showing of computer data in the form of graphics. For example, the presenting of data using 3D models as opposed to text or formulas. Since graphical data is analyzed by the brain at a more intuitive level than text data, greater amounts of information can be more simply displayed, speeding up the navigation through and assimilation of the information being viewed. Also called **information visualization**, or **visualization (1)**.

data warehouse An extensive database which combines all the databases of an entity, such as a large corporation. Some organizations store terabytes of information, and data warehouses provide a means to organize the data, and make decisions based on the information as a whole.

data word A group of bits or bytes treated as a unit of information, as opposed, for instance, to being considered as a control code. Also called **information word**.

database An interrelated collection of information in the form of records, fields, and files. The contents of a database are entered, processed, and accessed by a database program, and the information held may be of any kind, including text, graphics, sounds, and so on. Also spelled **data base**.

database engine Within a DBMS, the software that serves to store and retrieve data. Also spelled **data-base engine**.

database machine One or more computers or devices which are dedicated to database functions. Such devices may use parallel computing to reduce storage and retrieval times by factors of over 100, and are linked to the front-end processor via a high-speed channel. Also spelled **data-base machine**.

database management software Same as **DBMS**.

database management system Same as **DBMS**.

database program An application program designed for the entry, storage, manipulation, organization, security, integrity, and retrieval of data. The most common example of a database program is a DBMS. Also spelled **data-base program**.

database server A computer, within a LAN, that is dedicated to database management functions. It incorporates the databases and the DBMS. Also spelled **data-base server**.

database system Same as **DBMS**.

datacenter A centralized location where computers, data libraries, and associated equipment are housed. Large entities may have many mainframes at such a location. May be used, for instance, for centralized processing. Also spelled **data center**. Also called **data-processing center**, or **computer center**.

datacom Acronym for **data communications**.

dataflow Same as **data flow**.

dataflow diagram Same as **data flow diagram**.

datagram A block of data of a specific size, or of a maximum size, transmitted in a TCP/IP network. In addition to the payload, a datagram contains information such as the source and destination addresses. Also called **IP datagram**.

datamart Same as **data mart**.

datamation Abbreviation of **data automation**.

dataport Same as **data port**.

datascope **1.** Software that extracts data from multiple sources and organizes it for analysis, what-if scenarios, and the like. **2.** A device that can detect bit patterns traveling

along a transmission line. Such a device may also be able to intercept such patterns traveling through air.

dataset Same as **data set**.

datum Although the plural form of **datum** is **data**, **data** is commonly used in either the plural or singular forms. **1.** A single item of factual information, such as that derived from observation, experimentation, or calculation, especially that which is used for processing and/or further analysis. **2.** A single item of information which is represented in a manner suitable for digital transmission or computer use.

daughter board Same as **daughterboard**.

daughterboard A printed-circuit board which plugs directly into another, especially the motherboard. Although similar to an expansion board, a daughterboard accesses the motherboard directly, and not through an expansion bus. Also spelled **daughter board**. Also called **daughtercard**.

daughtercard Same as **daughterboard**.

DAVC Abbreviation of **delayed automatic volume control**.

daylight lamp A lamp whose emitted light has a spectral distribution similar to that of daylight.

dB Abbreviation of **decibel**. **1.** A dimensionless logarithmic unit utilized to express the ratio of two powers or intensities, such as voltages, currents, or sound intensities. The number of decibels is equal to 10 times the common logarithm of the ratio of the measured powers, or when comparing voltages, currents, or analogous acoustic quantities, it is 20 times the common logarithm. It is equal to 0.1 bels, or approximately 0.1151 nepers. **2.** In acoustics, the sound pressure ratio that determines the relative intensity of sounds, against a reference pressure. Such a pressure is usually defined as 20 micropascals, which is assigned a value of 0 decibels. Anything above this 0 decibels threshold is considered to be audible. Before pain sets in, a person with good hearing can listen at 120 decibels, or 20 pascals, which is a trillion times louder than 0 decibels.

Db Chemical symbol for **dubnium**.

DB-9 A **data-bus connector** with 9 pins.

DB-15 A **data-bus connector** with 15 pins.

DB-25 A **data-bus connector** with 25 pins.

DB-37 A **data-bus connector** with 37 pins.

DB-50 A **data-bus connector** with 50 pins.

DB connector Abbreviation of **data-bus connector**.

dB meter Abbreviation of **decibel meter**. An instrument which indicates, in decibels, the power level of a signal in reference to another, or to an arbitrary reference level. Its scale is calibrated logarithmically. Also called **decibel-level meter**.

dB scale Abbreviation of **decibel scale**. A measurement scale based on **dBs**. An example is that utilized when describing dB SPL.

dB sound pressure level Same as **dB SPL**.

dB SPL Abbreviation of **decibel sound pressure level**. A measurement of sound pressure, expressed in decibels, with respect to the threshold of hearing. The threshold of hearing is usually defined as 20 micropascals, which is assigned a value of 0 decibels. For example, leaves gently rustling produce a sound level of approximately 15 dB SPL, a whisper is about 30 dB SPL, the dial tone of a telephone is more or less 80 dB SPL, and an approaching subway train is somewhere around 110 dB SPL. Naturally, each of these approximations may vary quite a bit. For instance, the type of train, its approaching speed, and station acoustics are some of the factors which affect a dB SPL reading in the case of a subway. For a person with good hearing, pain begins somewhere around 120 dB SPL, and there is immediate damage to hearing above 150 dB SPL. Also called **sound pressure level (1)**.

DB2 A popular relational database.

dBa Abbreviation of **adjusted decibels**, or **decibels adjusted**. A measure of noise above a given reference level. The established reference noise power level is usually -85 decibels above 1 milliwatt, or -85 dBm.

DBA Abbreviation of **dynamic bandwidth allocation**.

dBASE A popular relational database management system, format, or language.

dBf Abbreviation of **decibels referred to 1 femtowatt**. The ratio, expressed in decibels, of a given power level to a reference power level of 1 femtowatt. Used, for instance, to express power ratios in transmission systems.

dBi The ratio, expressed in decibels, of the gain of a directional antenna in the direction of maximum radiation, as compared to that of an isotropic antenna.

dBk Abbreviation of **decibels referred to 1 kilowatt**. The ratio, expressed in decibels, of a given power level to a reference power level of 1 kilowatt. Used, for instance, to express power ratios in transmission systems.

dBm Abbreviation of **decibels referred to 1 milliwatt**. The ratio, expressed in decibels, of a given power level to a reference power level of 1 milliwatt. Used, for instance, to express power ratios in transmission systems.

dBm0 A power level in **dBm**, measured at zero transmission level.

DBMS Abbreviation of **d**atabase **m**anagement **s**ystem. Software which serves as the interface between a user and a database. A DBMS manages tasks such as the storage, retrieval, organization, security, and integrity of data. For example, a DBMS may maintain data integrity by insuring that no more than one user updates the same record at the same time. Also called **database system**, **data management system**, or **database management software**.

DBR laser Abbreviation of **distributed Bragg reflector laser**.

dBrn Abbreviation of **decibels above reference noise**. The ratio of a noise level to a lower reference noise level. The reference noise level is usually considered to be -85 decibels above 1 milliwatt, or -85 dBm. Utilized to measure the effect of a noise frequency or band of noise frequencies.

DBS Abbreviation of **direct broadcast satellite**, **direct broadcasting satellite**, or **digital broadcast satellite**.

dBSPL Same as **dB SPL**.

dBV Abbreviation of **decibels referred to 1 volt**. The ratio, expressed in decibels, of a given voltage to a reference level of 1 volt. Used, for instance, to express voltage ratios in transmission systems.

dBW Abbreviation of **decibels referred to 1 watt**. The ratio, expressed in decibels, of a given power level to a reference power level of 1 watt. Used, for instance, to express power ratios in transmission systems.

dBx Abbreviation of **decibels above reference coupling**. The level, expressed in decibels, of coupling between two circuits in relation to a lower reference value of coupling. May be used, for instance, to measure crosstalk coupling.

dc Same as **DC**.

DC Abbreviation of **d**irect **c**urrent. **1.** A unidirectional current which has an essentially constant average value. DC may fluctuate, pulse, spike, and so on, but its polarity does not change. Sources of DC include batteries, photovoltaic cells, and DC generators. Also called **continuous current** (1). **2.** Pertaining to, or utilizing **DC** (1).

DC–AC converter A circuit or device that changes DC voltage to AC voltage. Frequently, the output voltage is much higher than that of the input. Used, for instance, in uninterruptible power supplies. Also called **DC–AC inverter**, **DC inverter**, **inverter** (1), or **static inverter**.

DC–AC inverter Same as **DC–AC converter**.

DC–AC power converter A circuit or device that changes DC power to AC power.

DC amplifier **1.** An amplifier which increases DC voltages. **2.** Abbreviation of **direct-coupled amplifier**.

DC block A circuit or device which blocks DC, while permitting radio frequencies to pass. Such a circuit or device usually employs a capacitance.

DC circuit A circuit which carries DC.

DC component **1.** In a wave having both AC and DC components, the component whose direction does not change. It is the average value of a signal, around which the alternating component varies. **2.** The average value of a signal. In TV, for instance, it is the average luminance of the transmitted TV picture.

DC converter Same as **DC–DC converter**.

DC coupled amplifier Same as **direct-coupled amplifier**.

DC coupling The coupling of circuits or devices directly, or through a resistor, thus allowing for the passing of DC and AC. Used, for example, in a direct-coupled amplifier. Also called **direct coupling**.

DC–DC converter A circuit or device which converts one DC voltage into a DC voltage of another value. This may involve, for instance, a conversion to AC, followed by a change in the value of the AC, and then a conversion back to DC. Also called **DC converter**.

DC dump **1.** The complete removal of DC from a system or a component. **2.** The complete removal of DC from a computer system or component, thereby losing any content in volatile memory.

DC equipment Electrical equipment which can only use a DC power supply.

DC erase Same as **DC erasing**.

DC erasing Magnetic tape erasing employing DC. Also called **DC erase**.

DC generator **1.** A machine which converts mechanical power into DC power. Such machines are usually rotary. Also called **DC source** (1). **2.** Any device, such as a battery, which generates DC power.

DC inverter Same as **DC–AC converter**.

DC leakage **1.** An unwanted flow of DC. **2.** DC which flows through the dielectric of a capacitor.

DC motor A motor which converts DC into mechanical power.

DC noise Noise present during the playback of a magnetic tape, which is due to a large direct current in the recording head at the time of recording.

DC offset A DC bias voltage that is added to the input of a circuit or amplifier. Also called **DC offset voltage**.

DC offset voltage Same as **DC offset**.

DC overcurrent relay A relay which is actuated when the current in a DC circuit exceeds a specified value. Used, for instance, to protect mass-transit system components such as third rails and trolley wires.

DC permanent-magnet motor A permanent magnet motor controlled by DC control signals. The speed-torque curve of this type of motor is essentially linear. Such motors may or may not be brushless.

DC picture transmission The transmission of the DC component of a TV picture signal. This component corresponds to the average illumination of a displayed scene. Also called **DC transmission** (2).

DC plate current The DC in the plate of an electron tube.

DC plate resistance The DC plate voltage divided by the DC plate current. Used in vacuum-tube computations.

DC plate voltage The DC voltage at the plate of an electron tube.

DC power 1. The power in a DC circuit or system. Its unit is the watt. It is equal to the potential difference, expressed in volts, multiplied by the current, expressed in amperes. **2.** The power available from a DC power supply.

DC power source Same as **DC power supply**.

DC power supply A power supply whose output is DC. Examples include batteries, photovoltaic cells, and DC generators. Also known as **DC power source**, or **DC source (2)**.

DC reinsertion Same as **DC restoration**.

DC relay A relay which is actuated by DC.

DC resistance The resistance to DC offered by a circuit, device, or material. Such a resistance obeys Ohm's law. Also called **ohmic resistance**.

DC restoration Also called **DC reinsertion**, or **clamping**. **1.** The restoration of the DC component to a wave or signal. **2.** The adding of a DC component to a wave or signal which lacks such a component. **3.** The establishing of a reference DC level in a signal or device.

DC restorer Also called **clamping circuit, clamper,** or **reinserter**. **1.** A circuit or device which restores the DC component of a wave or signal. **2.** A circuit or device which adds a DC component to a wave or signal which lacks such a component. **3.** A circuit or device which establishes a reference DC level in a signal or device.

DC short A coaxial fitting which provides a path of low resistance between the center and outer conductors for DC. Radio frequencies present are unaffected.

DC signaling A method of communication in which DC is used for signaling.

DC source 1. Same as **DC generator (1)**. **2.** Same as **DC power supply**.

DC-to-AC converter Same as **DC–AC converter**.

DC-to-AC inverter Same as **DC–AC converter**.

DC-to-AC power converter DC–AC power converter.

DC-to-DC converter Same as **DC–DC converter**.

DC transmission 1. The transmission of DC power from one point to another. For instance, from a generating station to a receiving station. **2.** Same as **DC picture transmission**.

DC voltage A voltage that produces a DC. Such a voltage is unidirectional and has an essentially constant average value. Its abbreviation is **dcv**. Also called **direct voltage**.

DC working voltage The maximum continuous DC voltage at which a component, such as a capacitor, may be safely operated. Its abbreviation is **dcwv**.

DCA Abbreviation of **direct chip attach**.

DCC 1. Abbreviation of **digital cross-connect system**. **2.** Abbreviation of **direct cable connection**.

DCE 1. Abbreviation of **data communications equipment**, or **data circuit-terminating equipment**. **2.** Abbreviation of **distributed computing environment**.

DCF Abbreviation of **dispersion-compensating filter**.

DCM Abbreviation of **dispersion-compensating module**.

DCS Abbreviation of **digital cross-connect system**.

DCTL Abbreviation of **direct-coupled transistor logic**.

dcv Abbreviation of **DC voltage**.

dcwv Abbreviation of **DC working voltage**.

DD Abbreviation of **double-density**.

DDA Abbreviation of **digital differential analyzer**.

DDB Abbreviation of **device-dependent bitmap**.

DDBMS Abbreviation of **distributed DBMS**.

DDC 1. Abbreviation of **direct digital control**. **2.** Abbreviation of **Display Data Channel**.

DDD Abbreviation of **direct distance dialing**.

DDL Abbreviation of **data definition language**, or **data description language**.

DDP Abbreviation of **distributed data processing**.

DDR DRAM Same as **DDR SDRAM**.

DDR RAM Same as **DDR SDRAM**.

DDR SDRAM Abbreviation of **Double-Data Rate Synchronous DRAM**. A type of SDRAM which effectively doubles the data transfer rate by enabling data transfers on both the leading and trailing edges of clock cycles. Also called **DDR RAM**, or **DDR DRAM**.

DDS 1. Abbreviation of **digital data service**. **2.** Abbreviation of **digital data storage**.

de Broglie wave The wave associated with a particle, such as an electron, in motion. Also called **matter wave**.

de facto standard A standard adopted through continued use and acceptance, as opposed to being issued or endorsed by a standards entity. Examples include PCL, PDF, and RSA. This contrasts with **de jure standard**.

de jure standard A standard which is issued or endorsed by a standards entity, such as the IEEE or NIST. This contrasts with **de facto standard**.

de-wetting Same as **dewetting**.

deac Abbreviation of **deaccentuator**.

deaccentuator Same as **deemphasis network**. Its abbreviation is **deac**.

deactivate 1. To stop a sequence or process. **2.** To stop, block, or disrupt the operation of a circuit, device, piece of equipment, or system. **3.** To cause to be inactive or ineffective.

dead 1. Describing a component, circuit, or device which is disconnected from a source of voltage. Also known as **cold (1)**. **2.** Describing a circuit or component which is at ground potential. Also known as **cold (2)**. **3.** Not in use, or out of operation. **4.** No longer in use, or no longer operating.

dead band 1. For a device or system, such as an amplifier or control system, a range of values within which the input signal may be varied without affecting the output. For example, an interval of values within which a measuring instrument will not respond. Also called **dead zone**, or **neutral zone**. **2.** A radio-frequency band which is not in use. **3.** A band within the tuning range of a receiver in which there is no reception due to poor sensitivity or improper design of the circuitry.

dead-beat Same as **deadbeat**.

dead-beat instrument Same as **deadbeat instrument**.

dead-beat meter Same as **deadbeat meter**.

dead chamber Same as **dead room**.

dead circuit Also called **neutral circuit**. **1.** A circuit which is disconnected from a source of voltage. **2.** A circuit which is at ground potential.

dead code Routines within a computer program that, although present, are never accessed. For example, all calls to it may have been removed, or may never have been inserted in the first place. Dead code is an example of software rot.

dead enclosure Same as **dead room**.

dead end 1. The portion of a tapped coil through which no current flows. **2.** In an enclosure, such as a sound studio, a wall or area where there is the greatest amount of sound absorption. This contrasts with a **live end (2)**, where sound reflection is maximized.

dead-front board Same as **dead-front panel**.

dead-front panel A panel in which the live terminations and components are not exposed to the front, which is the operating side of the equipment. Such a panel is usually grounded. Also called **dead-front board**, or **dead-front switchboard**.

dead-front switchboard A switchboard with a **dead-front panel**.

dead key A key whose pressing has no immediate effect, but which does affect the following keystroke. For instance, when a keyboard is so programmed, a particular key may be pressed, followed by a letter o, to obtain an accented o, or ó.

dead line 1. A line through which a current or signal is not passing. 2. A line which is not operational due to failure.

dead link A link, such as that between two Web pages, that is no longer valid. This usually generates the frequently seen error 404.

dead room A room whose walls, ceiling, and floor are lined with sound-absorbing materials which minimize or eliminate all sound reflections. Used, for example, to test microphones or sound-level meters. Also called **dead chamber**, **dead enclosure**, **anechoic chamber (1)**, **free-field room**, or **echoless room**.

dead spot 1. In radio communications, an area within the normal range of a transmitter where radio reception is poor or non-existent. May be caused, for instance, by large objects in the vicinity of the receiver. Also called **blind spot**. 2. In a receiver, a section within its tuning range in which reception is poor or non-existent. May be due to poor sensitivity or improper design of the circuitry.

dead time Also spelled **deadtime**. 1. Same as **downtime**. 2. The time interval that elapses between a change in an input, and the response to said signal. Used especially in the context of control systems. 3. The time interval after a signal or event, during which a device, piece of equipment, or system is unable to respond to additional signals or events. During this time, for instance, a radiation counter can not register ionizing radiation. Also called **insensitive time**, or **recovery time (3)**. 4. A time interval intentionally inserted between events. Used, for instance, to prevent the overlap of certain events.

dead zone Same as **dead band (1)**.

deadbeat The coming to rest without oscillating or overshooting. Also, a device or instrument which exhibits this characteristic. For instance, a pointer in an analog meter which proceeds to the next position without wavering. Such devices and instruments incorporate damping. Also spelled **dead-beat**.

deadbeat instrument An instrument with sufficient damping to proceed from one reading to the next without oscillating or overshooting. Also spelled **dead-beat instrument**.

deadbeat meter A meter with sufficient damping to proceed from one reading to the next without oscillating or overshooting. Also spelled **dead-beat meter**.

deadlock A situation that occurs when each of two processes is waiting for the other to complete, before being able to continue. Although operating systems should have ways to solve such situations, it is not always the case.

deadtime Same as **dead time**.

deallocate 1. To undo a previous designation or distribution of equipment or resources. 2. To undo the previous reservation of computer resources for specific purposes.

debounced switch A switch that incorporates **debouncing**.

debouncing Circuitry or measures utilized to prevent the recognition of multiple signals due to contact bounce. Used, for instance, to help insure that a single contact is registered for each computer keystroke.

debug 1. To locate and correct persistent errors or defects in computer software or hardware, especially those which cause malfunctions. 2. To locate and correct defects present in a circuit, device, component, or the like, which cause malfunctions. 3. To remove electronic devices which have been installed for the purpose of surreptitiously listening to and/or recording conversations.

debugger 1. A computer program utilized to **debug (1)** other programs. Such programs, for example, enable a programmer to stop at breakpoints, for the careful evaluation of errors and/or defects. Also called **debugger program**, or **debugging program**. 2. That which is utilized to **debug**.

debugger program Same as **debugger (1)**.

debugging 1. The locating and correcting of persistent errors or defects in computer software or hardware, especially those which cause malfunctions. 2. The locating and correcting of defects present in a circuit, device, component, or the like, which cause malfunctions. 3. The removal of electronic devices which have been installed for the purpose of surreptitiously listening to and/or recording conversations

debugging program Same as **debugger (1)**.

debugging routine A set of computer instructions used for **debugging (1)**. An example is a checking routine.

debunching In a velocity-modulated tube, such as a klystron, the spreading of an electron beam. This is due to their mutual repulsion.

debye A unit of electric dipole moment. One debye is equal to approximately 3.34×10^{-30} coulomb-meters.

Debye length In a plasma, the maximum distance at which a charged particle will be influenced by the electric field of another particle of the opposite charge. It usually refers to the interactions between an electron and any given positive ion. Also called **Debye screening length**, **Debye screening radius**, **Debye shielding distance**, **Debye shielding length**, or **plasma length**.

Debye screening length Same as **Debye length**.

Debye screening radius Same as **Debye length**.

Debye shielding distance Same as **Debye length**.

Debye shielding length Same as **Debye length**.

deca- A metric prefix denoting 10. For example, a decameter, which is 10 meters. Its abbreviation is **da**. Also spelled **deka-**.

decade 1. An interval between two quantities or values, in which the ratio of one to the other is 10 to 1. An example is found in a decade box, where each section has a value equal to 10 times that of the previous section. 2. A series, or group, of ten. An example is found in a **decade counter (2)**, whose display has ten values per section.

decade box A box or other enclosure which houses multiple sections of components, such as precision resistors or capacitors, whose values vary in multiples of ten from section to section. Each section has 10 components of equal value, which provide 10 equal steps. One section may range from 1-9, the next from 10-99, the following from 100-999, and so on. A user employs selector switches to choose the appropriate value from each section to obtain the desired final result.

decade capacitance box Same as **decade capacitor (1)**.

decade capacitor 1. A **decade box** in which multiple sets of capacitors enable the selection of any given capacitance within its range. Also called **decade capacitance box**. 2. A capacitor whose value may be varied in ten equal steps, such as 100 picofarads, 200 picofarads, 300 picofarads, and so on.

decade counter 1. Same as **decade scaler**. 2. A counter whose display consists of one or more sections whose values range from **0** to **9**.

decade counter tube Also called **decade counting tube**. 1. An electron tube which produces an output pulse after counting ten input pulses. 2. An electron tube with one input electrode and ten output electrodes, and in which with each successive input pulse the conduction is transferred in sequence to the next output electrode.

decade counting tube Same as **decade counter tube**.

decade resistance box Same as **decade resistor (1)**.

decade resistor **1.** A **decade box** in which multiple sets of resistors enable the selection of any given resistance within its range. Also called **decade resistance box**. **2.** A resistor whose value may be varied in ten equal steps, such as 100 ohms, 200 ohms, 300 ohms, and so on.

decade scaler A counter which produces an output pulse for every ten input pulses. Also called **decade counter (1)**, or **scale-of-ten circuit**.

decametric waves Electromagnetic waves whose wavelength is within the range of 10 to 100 meters, corresponding to frequencies of between 30 to 3 MHz, respectively. This represents the high-frequency band.

decay **1.** A gradual decrease in the magnitude of a quantity, such as current, voltage, luminescence, or magnetic flux. **2.** The spontaneous disintegration of unstable atomic nuclei, such as those of uranium or curium, mainly through the emission of alpha, beta, or gamma rays. Also called **disintegration (2)**, or **radioactivity (1)**. **3.** A specific instance of **decay (2)**, such as that occurring when an atom of uranium-238 emits an alpha particle to form an atom of thorium-234. Also called **disintegration (3)**, or **radioactivity (2)**. **4.** The reduction over time of the radioactivity of a substance, due to **decay (2)**. Also called **disintegration (4)**, or **radioactivity (3)**.

decay chain A series of nuclides which undergo series decay, which is the transformation of one nuclide into another nuclide through radioactive decay, followed by further transformations until a stable, or nonradioactive, nuclide results. Also called by various other names, including **decay family**, **decay series**, **disintegration chain**, **disintegration series**, **radioactive decay series**, **radioactive series**, or **radioactive chain**.

decay characteristic For a CRT screen, such as that of a cathode-ray oscilloscope, a curve which expresses the relationship between the emitted radiant power and time, after the excitation is removed. Also called **persistence characteristic**.

decay energy The energy released during **decay (3)**. Also called **disintegration energy**.

decay family Same as **decay chain**.

decay rate A quantification of the rate at which decay occurs. For example, the rate at which sound is extinguished in a reverberation chamber. Also called **rate of decay**.

decay series Same as **decay chain**.

decay time **1.** The time that elapses for a given amount of decay to occur. For example, the time it takes a static charge to be reduced to a given percentage of its peak charge. Also called **fall time (1)**, or **storage time (4)**. **2.** The time required for the amplitude of a pulse to decrease from a given percentage of its peak amplitude, such as 90%, to another, such as 10%. Also called **fall time (2)**.

Decca Abbreviation of **Decca** Navigator. A phase-difference hyperbolic navigation system which utilizes a controlling station and two or more secondary stations for the determination of the position of a vessel.

Decca Navigator Same as **Decca**.

decelerating electrode In an electron-beam tube such as a klystron, a charged electrode which creates an electric field which serves to reduce the velocity of the electrons in the beam. Also called **retarding electrode**.

deceleration Slowing down, or the rate at which velocity decreases with respect to time. Also called **negative acceleration**.

deceleration time Also known as **stop time**. **1.** The time required for a moving storage medium to stop after a reading or writing operation. For example, the time it takes a tape reel to stop spinning before or after a read operation. Ap-

plies mostly to magnetic media, but may also include optical media such as CD-ROMs. **2.** The time it takes a moving part of a storage medium to stop after a reading or writing operation. Such a part may be, for instance, an access arm or a read/write head.

decentralized data processing Same as **decentralized processing**.

decentralized network A communications network in which resources, such as processing power and control, are shared by multiple nodes.

decentralized processing Processing facilities which include multiple locations. This contrasts with **centralized processing**, where all processing is performed at a single location. Also called **decentralized data processing**.

deci- A metric prefix denoting one tenth, or 0.1. Its abbreviation is **d**.

decibel Same as **dB**.

decibel-level meter Same as **dB meter**.

decibel meter Same as **dB meter**.

decibel scale Same as **dB scale**.

decibel sound pressure level Same as **dB SPL**.

decibels above 1 femtowatt The ratio, expressed in decibels, of a given power level to a reference power level of 1 femtowatt, when the given power level is greater than 1 femtowatt. Used, for instance, to measure power ratios in transmission systems.

decibels above 1 kilowatt The ratio, expressed in decibels, of a given power level to a reference power level of 1 kilowatt, when the given power level is greater than 1 kilowatt. Used, for instance, to measure power ratios in transmission systems.

decibels above 1 milliwatt The ratio, expressed in decibels, of a given power level to a reference power level of 1 milliwatt, when the given power level is greater than 1 milliwatt. Used, for instance, to measure power ratios in transmission systems.

decibels above 1 watt The ratio, expressed in decibels, of a given power level to a reference power level of 1 watt, when the given power level is greater than 1 watt. Used, for instance, to measure power ratios in transmission systems.

decibels above reference coupling. Same as **dBx**.

decibels above reference noise. Same as **dBrn**.

decibels adjusted Same as **dBa**.

decibels below 1 femtowatt The ratio, expressed in decibels, of a given power level to a reference power level of 1 femtowatt, when the given power level is less than 1 femtowatt. Used, for instance, to measure power ratios in transmission systems.

decibels below 1 kilowatt The ratio, expressed in decibels, of a given power level to a reference power level of 1 kilowatt, when the given power level is less than 1 kilowatt. Used, for instance, to measure power ratios in transmission systems.

decibels below 1 milliwatt The ratio, expressed in decibels, of a given power level to a reference power level of 1 milliwatt, when the given power level is less than 1 milliwatt. Used, for instance, to measure power ratios in transmission systems.

decibels below 1 watt The ratio, expressed in decibels, of a given power level to a reference power level of 1 watt, when the given power level is less than 1 watt. Used, for instance, to measure power ratios in transmission systems.

decibels referred to 1 femtowatt Same as **dBf**.

decibels referred to 1 kilowatt Same as **dBk**.

decibels referred to 1 milliwatt Same as **dBm**.

decibels referred to 1 volt Same as **dBV**.

decibels referred to 1 watt Same as **dBW**.

decimal 1. Based on, or consisting of ten parts, components, or possibilities. 2. Same as **decimal number system**. 3. Same as **decimal number**.

decimal attenuator An attenuator which reduces the amplitude of a signal in decimal steps. That is, in successive powers of 10.

decimal code A code composed of elements which can have one of 10 states or values, namely 0 to 9. The decimal number system is an example of such a code.

decimal-coded digit A decimal digit expressed by a given pattern of zeroes and ones.

decimal digit A digit whose value may be 0, 1, 2, 3, 4, 5, 6, 7, 8, or 9.

decimal notation The utilization of decimal digits to express numerical values.

decimal number A number expressed utilizing the **decimal number system**. Also called **decimal (3)**.

decimal number system A numbering system whose base is 10, meaning that there are 10 digits in this system, specifically 0 to 9. All digits to the left of the decimal point represent successive positive powers of 10, while those to the right of the decimal point are successive negative powers of 10. For example, $456.789 = (4 \times 10^2) + (5 \times 10^1) + (6 \times 10^0) + (7 \times 10^{-1}) + (8 \times 10^{-2}) + (9 \times 10^{-3})$. Also called **decimal (2)**, or **decimal system (2)**.

decimal point The point, within a decimal number, which indicates what power of ten a digit is.

decimal system 1. A system whose components can have one of 10 states or values. 2. Same as **decimal number system**.

decimal-to-binary conversion The conversion of a decimal representation into a binary representation. For example, converting a decimal, or base 10, number into a binary, or base 2, number.

decimeter A unit of distance equal to 0.1 meter. Its abbreviation is **dm**.

decimetric waves Electromagnetic waves whose wavelength is within the range of 0.1 to 1 meter, corresponding to frequencies of between 3000 to 300 MHz, respectively. This represents the ultra-high frequency band.

decipher Same as **decode (1)**.

decision In computers, the determination of a choice or course of action to be taken, based on the particular circumstances and available alternatives. For instance, a computer may compare one set of data to another, to effectuate such a decision. Also called **logical decision (1)**.

decision box A symbol, usually in the form of a diamond, which indicates a point within a flowchart where decision must be made. The result of the decision determines which branch will then be followed.

decision circuit Same as **decision element**.

decision element A circuit which performs logic operations, such as AND or NOT on its input, and whose output is a signal which indicates the result of the operation. Also called **decision circuit**.

decision support software Software utilized within a **decision support system**.

decision support system An analysis and decision-making system which incorporates programs and databases designed to work as a unit. One of the key features of such a system is the ability to help analyze various possible outcomes of any number of decisions, before any concrete actions take place. Its abbreviation is **DSS**.

decision table A **decision tree** in table form.

decision tree A diagram in which the possible outcomes of a problem, with sequential alternatives, are represented as branches. As each decision is made, two or more branches,

representing the next choices, become available. This continues until a conclusion is reached.

deck Abbreviation of tape **deck**. 1. An audio system component which records and plays back magnetic tapes. 2. The tape-transport mechanism of a **deck (1)**.

declination For a given geographic location, the angle representing the difference between magnetic north and true north. Also called **magnetic declination**.

declination angle The angle below a reference plane, such as the horizon, of a descending line. Also called **angle of declination**.

declinometer An instrument used for measuring **declination**.

decode 1. To unscramble information, such as data, with the use of a key. For instance, to convert ciphertext into plaintext. Also called **decipher**, or **decrypt**. 2. To convert encoded data back into its original form. For instance, to convert text back to its original form after it has been expressed in ASCII code.

decoder 1. A circuit, device, or program which converts that which has been encoded back into its original form. 2. In a TV receiver, a circuit or device which transforms the color-difference signals into separate red, green, and blue signals. Also called **color decoder**, or **matrix (4)**. 3. A circuit or device which produces one or more outputs, based on the combination of its input signals. 4. A circuit which responds to a given coded signal, while rejecting all others. 5. Same as **decoder box**.

decoder box A device, that connects to a TV, which decodes an incoming signal or otherwise enables viewing of certain channels, programming, or content. Such a device may serve to permit viewing of satellite or cable programming, HDTV signals, close-captions, and so on. A decoder box may also connect to another device, such as a VCR. Also called **decoder (5)**.

decoder circuit A circuit which serves as a **decoder**. Also called **decoding circuit**.

decoder/demultiplexer A circuit or device which accepts multiple inputs, and which provides one of several mutually exclusive outputs. The output will depend on the combination of input signals.

decoding The actions a **decoder** performs.

decoding circuit Same as **decoder circuit**.

decollimate To convert a beam which is parallel, or nearly parallel, into one which is convergent or divergent.

decollimation The conversion of a beam which is parallel, or nearly parallel, into one which is convergent or divergent, through processes such as diffraction or scattering. Used, for instance, in optical communications.

decommutation The extraction of one or more signals from a composite signal formed through commutation.

decommutator A device, such as a multiplexer, which extracts commutated signals.

decompiler A computer program which attempts to convert machine language back into a high-level source language. It is not always possible to make a completely accurate translation, as some machine code does not have an exact high-level source language equivalent.

decompress The undoing of the effects of compression. That is, to restore data to its original size and/or bandwidth. Also called **uncompress**.

deconvolution The processing of an image or signal to exclude undesired components, such as noise, or to restore it to its original form. Such techniques usually involve the use of Fourier transforms.

decoupling The substantial reduction, or elimination of coupling. Decoupling may be utilized, for instance, between the stages of an amplifier to avoid interstage coupling.

decoupling capacitor A capacitor utilized in a circuit for the substantial reduction, or elimination of coupling. Said especially of such a capacitor used in a decoupling network.

decoupling filter A filter intended to substantially reduce, or eliminate coupling. Such a filter may be used, for instance, between amplifier stages to avoid interstage coupling.

decoupling network A combination of components, such as resistors and capacitors, used between two or more circuits to avoid interstage coupling.

decrement 1. The process of gradually decreasing a quantity or variable. 2. The amount or value lost through the process of gradually decreasing a quantity or variable. 3. Same as **damping factor** (2).

decrypt Same as **decode** (1).

decryption The process of unscrambling information, such as data, with the use of a key.

decryption key A series of binary digits or characters that are utilized to unscramble data which has been encrypted.

DECT Acronym for **Digital-Enhanced Cordless Telephone**, or **Digitally-Enhanced Cordless Telephone**.

dedicated Designated for a specific use or function. For example, a communications circuit for use exclusively by specified users or for a particular function, such as data communication.

dedicated channel A communications channel designated for use exclusively by specified users or for a particular function, such as data communication.

dedicated circuit Same as **dedicated line** (1).

dedicated connection Same as **dedicated line** (1).

dedicated line 1. A permanent communications channel between two or more locations. Such a service includes private switching arrangements and a defined transmission path, and provides an exclusive high-speed connection which is available at all times. T1 and T3 lines are examples. Also called **dedicated circuit**, **dedicated connection**, **private line**, **leased circuit**, or **leased line**. 2. A telephone line which is used for a single purpose, such as for connection to a fax machine.

dedicated server A single computer whose sole function is that of serving a network.

dedicated short-range communication Its abbreviation is **DSRC**. Also called **RFID**. 1. A technology in which RF signals are emitted by an object in response to an interrogator, for purposes of identification, collection of information, or the like. A typical arrangement involves a reader which emits signals which a transponder, or tag, in the object responds to. The reader, which processes the obtained information, and tag need not be in nearly direct contact, as is required, for instance, in a bar-code system. 2. A system using **dedicated short-range communication** (1). Such a system may be utilized, for example, to identify store items being purchased or placed into inventory, to drive through a toll station having the appropriate amount automatically deducted from a smart card attached to the windshield, for the location, complete with proper placement instructions, of parts within a robotic assembly line, or for tracking the movements of people being spied on by attaching such devices on clothing, objects being carried, or via the surgical insertion or embedding of such transceivers. Higher-frequencies, greater transmission power, and enhanced antenna designs, for instance, may be utilized to extend the range of such systems.

dee Either of the two hollow accelerating electrodes in a cyclotron. Originally, such an electrode resembled a letter D, although now it may be in another shape, such as that of a horseshoe.

deemphasis The process of restoring an audio-frequency signal to its original form before reproduction. This is done

to offset the preemphasis of higher frequencies occurring prior to transmission or during recording. May be used, for instance, in frequency-modulated receivers. The utilization of preemphasis followed by deemphasis may help improve the overall signal-to-noise ratio and reduce distortion, among other benefits. Also called **post-emphasis**, or **post-equalization**.

deemphasis circuit Same as **deemphasis network**.

deemphasis network In an FM receiver, a resistance-capacitance filter that restores an audio-frequency signal to its original form. This is done to offset the preemphasis of higher frequencies occurring prior to transmission. Also called **deemphasis circuit**, **deaccentuator**, or **post-emphasis network**.

deenergize To disconnect a component, circuit, device, piece of equipment, or system from a source of power.

deenergized A component, circuit, device, piece of equipment, or system which is disconnected from a source of power.

deep cycle Pertaining to a rechargeable battery which can be repeatedly discharged to a large proportion of its overall capacity, without damage.

deep discharge The discharge of a cell or battery below a given proportion of its overall capacity. For instance, discharging to below 20% of its full charge.

deep level transient spectroscopy A spectroscopy technique utilized for the determination of various parameters of traps in semiconductors. Such parameters include energy levels, concentrations, capture cross-sections, and activation energies. Its abbreviation is **DLTS**.

deep linking Linking to a Web page other than the homepage of a given Web site.

deep ultraviolet lithography Same as **deep ultraviolet photolithography**. Its abbreviation is **deep UV lithography**.

deep ultraviolet photolithography Photolithography in which the resist is exposed to deep ultraviolet light, whose shorter wavelength provides enhanced resolution. The wavelength of such light is close to that of X-rays. Its abbreviation is **deep UV photolithography**. Also called **deep ultraviolet lithography**.

deep UV lithography Same as **deep ultraviolet photolithography**.

deep UV photolithography Same as **deep ultraviolet photolithography**.

deep Web Web content which is not found using search engines. Such content may be located, for instance, by using a given Web site's search feature. Also called **invisible web**.

default A setting of a component, device, piece of equipment, or system which is automatically selected if a user chooses no alternative.

default drive A computer disk drive which is automatically selected if a user chooses no alternative. For instance, a computer will boot up using a specific drive if none other is selected.

default password A password which is set by a manufacturer or provider of a device, piece of equipment, system, software item, or the like. Such passwords are usually extremely easy to guess, or are otherwise known by those seeking to take advantage of individuals or entities that don't bother to change them.

defect 1. A lack of something which is necessary for full utility and/or performance. 2. An imperfection or irregularity which impairs utility and/or performance. 3. A flaw in a crystal lattice. 4. A hole, as opposed to an electron, in a semiconductor crystal.

defect conduction Electric conduction via holes in the valence band of a semiconductor material.

deferred processing The processing of data after a given amount is received, or after a specified time interval. For instance, the processing of data which is stored in blocks. This contrasts with **direct processing**, in which data is processed as it is received.

definition 1. Clarity or distinctness in a received and/or reproduced image. May refer, for instance, to the color reproduction and detail of a televised image. 2. Clarity or distinctness in a received and/or reproduced sound. May refer, for instance, to the fidelity of an audio system. 3. Clarity or distinctness of the text and/or graphics on a printed sheet. May refer, for instance, to the intelligibility of a received fax transmission.

deflect To bend or shift from a straight or forward-moving course. Said, for instance, of a stream, beam, or flow.

deflecting coil Same as **deflection coil**.

deflecting electrode Same as **deflection electrode**.

deflecting plate Same as **deflection electrode**.

deflecting voltage Same as **deflection voltage**.

deflecting yoke Same as **deflection yoke**.

deflection 1. A movement away from a center or zero position. Said, for instance, of the pointer in a measuring instrument. 2. In a CRT, a movement of the electron beam away from a straight-line path. This deflection may be caused by electric or magnetic fields. 3. A movement away from a straight-line path.

deflection angle In CRTs, the maximum angle that an electron beam is diverted during deflection. For example, a TV viewing screen may have a 110° deflection, resulting in greater contrast, detail, and clarity. Also called **angle of deflection**.

deflection coefficient Same as **deflection factor** (2).

deflection coil A coil within a **deflection yoke**. Also called **deflecting coil**.

deflection defocusing In a CRT, defocusing which increases as deflection is increased. As the electron beam is directed further away from the center of the screen, the luminous spot becomes larger and more elliptical, thus becoming less focused.

deflection electrode In a CRT, either of two electrodes utilized for deflecting the electron beam. The electric field which surrounds a pair of such electrodes induces the desired deflection. Also called **deflecting electrode, deflecting plate**, or **deflection plate**.

deflection factor 1. The reciprocal of **deflection sensitivity**. 2. The reciprocal of **instrument sensitivity**, which is the minimum change in a measured quantity that produces an observable change in the indication or output of a measuring instrument. Also called **deflection coefficient**.

deflection plate Same as **deflection electrode**.

deflection sensitivity In a CRT, the displacement of the electron beam per unit change in the deflecting field. May be expressed, for instance, in inches per volt, or inches per ampere. It is the reciprocal of **deflection factor**.

deflection voltage In a CRT, the voltage applied between the **deflection electrodes**. Also called **deflecting voltage**.

deflection yoke In a CRT, a system of coils utilized for magnetic deflection of the electron beam. One possible arrangement consists of two sets of coils, one for horizontal deflection, and the other for vertical deflection. Also called **deflecting yoke, scanning yoke**, or **yoke** (2).

deflector That which serves to bend or shift from a straight or forward-moving course. Examples include deflection electrodes in CRTs, or attachments which direct the acoustic output of a tweeter.

defocus To cause to deviate from accurate focus. Also, the result of this act. For example, to cause a beam of radiation or particles to fail to converge properly on a target surface.

defocusing The causing to deviate from accurate focus, or the result of this.

deformation The alteration of shape, or the shape resulting from such an alteration. For example, the mechanical deformation of a piezoelectric crystal producing a voltage.

deformation potential An electric potential caused by a deformation in a crystal lattice. For instance, that created by acoustic waves traveling through certain crystals.

defrag To rearrange the files located on a computer disk drive, so that they are contiguous. As files are deleted, modified or created on a drive, any available space may be used, which usually leads to fragmentation. The more contiguous the files are, the less the drive heads have to travel, thus access time is optimized. Also called **defragment**.

defrag program Same as **defrag utility**.

defrag utility A program or utility utilized to **defrag**. Also called **defrag program, defragmentation utility, defragmentation program**, or **defragger**.

defragger Same as **defrag utility**.

defragment Same as **defrag**.

defragmentation The process of rearranging the files located on a computer disk drive, so that they are contiguous.

defragmentation program Same as **defrag utility**.

defragmentation utility Same as **defrag utility**.

deg Abbreviation of **degree**.

degas 1. To remove gas from. 2. To remove all gases from. For instance, to remove all gases from an electron tube through heating.

degassing 1. The process of removing gas. 2. The process of removing all gases. For instance, the removal of all gases from an electron tube through heating.

degauss To remove magnetism from a device, piece of equipment, or recordable magnetic medium. For example, to remove unwanted magnetism from tape deck heads. Also called **demagnetize**.

degausser Also known as **demagnetizer**. 1. A circuit or device which removes magnetism. For instance, a circuit or device which demagnetizes a color TV picture tube or a recordable magnetic medium. 2. A device which utilizes a strong magnetic field to erase all information from a recordable magnetic medium, such as a reel of magnetic tape or a floppy disk. In the case of a reel of magnetic tape, for instance, the use of such a device avoids the need for having to run the entire tape past the erase head. Also called **bulk eraser**.

degaussing The process of removing magnetism from a device, piece of equipment, or recordable magnetic medium. For instance, degaussing a diskette before recording on it improves its performance and reliability. Also known as **demagnetizing**, or **demagnetization**.

degaussing circuit A circuit which automatically demagnetizes a color TV picture tube. This process usually takes place while the TV is warming up for use.

degaussing coil A coil which removes magnetism from a device or piece of equipment, such as a CRT monitor. Such a coil usually operates automatically.

degeneracy In a resonator, the condition where two or more modes have a single resonant frequency.

degenerate amplifier Same as **degenerate parametric amplifier**.

degenerate modes In a resonator, one of multiple modes which have a single resonant frequency.

degenerate parametric amplifier A parametric amplifier whose AC power source frequency is twice that of its input signal. Thus, its idler frequency is equal to the frequency of its input signal. Such amplifiers are characterized by low noise figures. Also called **degenerate amplifier**.

degenerate semiconductor A semiconductor whose number of electrons in the conduction band approaches that of a metal. This gives it certain metallic properties, such as good conductivity.

degeneration In an amplifier or system, feedback which is 180° out-of-phase with the input signal, thus opposing it. While this results in a decrease in gain, it also serves reduce distortion and noise, and to stabilize amplification. Also called **degenerative feedback, negative feedback, inverse feedback, reverse feedback**, or **stabilized feedback**.

degenerative feedback Same as **degeneration**.

degradation 1. A gradual deterioration which leads to a lower level of performance, function, quality, condition, intelligibility, or the like. 2. The operation of a computer system at a reduced level of performance. May be due, for instance, to a malfunctioning component, or memory limitations.

degradation failure A failure occurring after the gradual deterioration of a component, device, program, or system. This contrasts with a **catastrophic failure**, where failure is sudden, unexpected, and complete.

degree Its symbol is °. Its abbreviation is **deg**. 1. A unit of temperature interval, such as a kelvin, or a degree Celsius. 2. A standard unit of angular measure, which is equal to 1/360 of a complete revolution of a circle. It is also equal to approximately 0.0174533 radian.

degree absolute A degree within an absolute temperature scale. Since the term degree absolute has been replaced by the term **kelvin**, the latter should be used to refer to this concept.

degree Celsius Its symbol is °C. A unit of temperature interval used in the Celsius scale. This scale is based on the freezing and boiling points of water under certain conditions, assigning values of 0 °C and 100 °C respectively. An increment of 1 degree Celsius equals an increment of 1 kelvin. Formerly called **degree centigrade**.

degree centigrade Its symbol is °C. A unit of temperature interval used in the centigrade scale. Since the term centigrade has been replaced by the term Celsius, this unit of temperature interval is now called **degree Celsius**.

degree Fahrenheit Its symbol is °F. A unit of temperature interval used in the Fahrenheit scale. This scale is based on the freezing and boiling points of water under certain conditions, having an assigned a value of 32 °F and 212 °F, respectively. This temperature scale is rarely used outside of the United States.

degree Kelvin Its symbol is °K. Former name for the term **kelvin**, which is the fundamental SI unit of temperature.

degree Rankine Its symbol is °R. A unit of temperature interval used in the Rankine scale. This scale is the same as the Fahrenheit scale, except that a temperature of 0 °R equals absolute zero. When comparing to the Kelvin temperature scale, 1 °R equals exactly 5/9 kelvin.

degrees of freedom Its abbreviation is **DOF**. 1. The manners in which a system or component can move in space. 2. The manners in which a robotic manipulator or its components can move. A robotic manipulator with six degrees of freedom would have three in the arm and three in the wrist: vertical movement, radial traverse, rotational traverse, wrist pitch, wrist roll, and wrist yaw. Also called **robot degrees of freedom**.

deionization 1. The conversion of an ionic species into a neutral one. 2. The removal of ions from a solution.

deionization potential In a gas-filled tube, the potential at which the ionization of the contained gas ceases, and conduction stops. Also called **extinction potential**.

deionization time In a gas-filled tube, the time required for an ionized gas to become neutral, once the ionizing voltage is removed.

deionization voltage In a gas-filled tube, the voltage at which the ionization of the contained gas ceases, and conduction stops. Also called **extinction voltage**.

DEK Abbreviation of **data encryption key**.

deka- An alternate spelling of **deca-**. Its abbreviation is **da**.

Del Abbreviation of **delete key**.

Del key Abbreviation of **delete key**.

delay Also called **delay time**, or **time delay**. 1. The time interval between two events. 2. The time interval an event is postponed. Also, the act of postponing. 3. The time interval between the sending or emitting of a signal, and its reception or detection. 4. The time interval necessary for a signal to pass through a component, circuit, device, piece of equipment, system, or medium. This includes any additional time the signal is retarded, as with the use of a delay line. 5. The time interval between the instants at which a given point of a wave passes through two specified points of a transmission medium. 6. The time interval between the powering on of a component, circuit, device, piece of equipment, or system, and its starting to operate.

delay circuit A circuit whose output signal occurs a specified time interval after the input signal is received. For example, a sonic delay line. Also called **time-delay circuit**.

delay coincidence circuit 1. A circuit with two or more inputs, which delivers an output pulse only if all inputs receive a pulse within a given time interval. 2. A circuit with two or more inputs, which delivers an output pulse only if a predetermined combination of inputs receive a pulse within a given time interval.

delay counter A counter which introduces a specified delay.

delay distortion Distortion of a signal as it passes through a transmission medium in which different frequencies travel at slightly different speeds. This causes a change in the waveform because the rate of change of phase shift is not constant over the transmitted frequency range. Also called **time-delay distortion, envelope delay distortion, phase distortion**, or **phase-delay distortion**.

delay equalization The corrections a **delay equalizer** makes. Also called **phase correction**, or **phase-delay equalization**.

delay equalizer A corrective network which serves to compensate for the effects of **delay distortion**. It may do so, for instance, by introducing the necessary delays at the appropriate frequencies to offset said distortion. Also called **phase corrector**, or **phase-delay equalizer**.

delay line A device or medium which delays a wave or signal for a specific length of time. May be used, for instance, in surround-sound reproduction to delay one audio signal with respect to another. Also called **artificial delay line**.

delay-line memory A form of computer storage in which a delay line is used to store information. Also called **delay-line storage, circulating memory**, or **circulating storage**.

delay-line register A computer register which utilizes a delay line. Also called **circulating register**.

delay-line storage Same as **delay-line memory**.

delay multivibrator A monostable multivibrator which produces an output pulse at a specified time after receiving a trigger input pulse.

delay relay A relay which introduces a specified delay between the time it is energized and the time it opens or closes. Also called **time-delay relay**.

delay switch A switch that opens, closes, or changes the state of a circuit after a specified delay. Also called **time-delay switch**.

delay time Same as **delay**.

delay timer A timer which introduces a specified delay. May be used, for instance, in a switch.

delay unit A device which introduces a specified delay. May be used, for instance, to delay a signal.

delayed AGC Same as **delayed automatic gain control**.

delayed automatic gain control An automatic gain control circuit which is activated only when the input signal exceeds a predetermined magnitude. This allows for maximum amplification of weaker signals. This contrasts with **instantaneous automatic gain control**, in which the circuit is activated immediately when the amplitude of the input signal changes. Also called **delayed automatic volume control**, **biased automatic gain control**, or **quiet automatic volume control**. Its abbreviation is **DAGC**, or **delayed AGC**.

delayed automatic volume control Same as **delayed automatic gain control**. Its abbreviation is **DAVC**, or **delayed AVC**.

delayed AVC Same as **delayed automatic volume control**.

delayed break The opening of an electric circuit after a specified delay. This contrasts with a **fast break**, where there is no delay. Also called **slow break**.

delayed contact A contact which opens or closes an electric circuit after a specified delay.

delayed flip-flop A flip-flop whose input is delayed one cycle, then appears at the output. That is, any given output is a function of the previous pulse, not the current one. Its abbreviation is **D flip-flop**.

delayed make The closing of an electric circuit after a specified delay. This contrasts with a **fast make**, where there is no delay. Also called **slow make**.

delayed plan position indicator A plan position indicator in which there is a **delayed sweep**. Its abbreviation is **delayed PPI**.

delayed PPI Abbreviation of **delayed plan position indicator**.

delayed repeater satellite A communications satellite which receives and stores information, and then retransmits it after a specified delay.

delayed sweep 1. In a radar or oscilloscope, a sweep which is delayed for a specified time after the initiating pulse. May be used, for instance, to allow a sweep to be initiated anywhere along the X-axis baseline. 2. The technique employed to obtain a **delayed sweep** (1).

delete 1. To remove, erase, or eliminate. Said, for instance, of a computer file, or of a signal. 2. Abbreviation of **delete key**.

delete key On most computer keyboards, a key which when pressed serves to erase any selected content, and when held down or repeatedly pressed, to erase content after the cursor. Its abbreviation is **Del, Del key**, or **delete** (2).

delimiter One or more characters which serve to separate a character string, such as a group of words. For example, a tab character may be used to separate fields of data.

deliquescence The gradual conversion of a solid into a liquid, through the attraction and absorption of atmospheric moisture. Such a substance may be used, for instance, to help keep electronic equipment dry. An example is potassium carbonate.

deliquescent Possessing the quality of **deliquescence**.

deliquescent substance A substance, such as potassium carbonate powder, which gradually becomes a liquid as it attracts and absorbs atmospheric moisture.

Dellinger effect A rapid and complete loss of electromagnetic skywave signals, due to substantially increased ionization in the ionosphere. This ionization is caused by increased solar noise, which itself is caused by solar storms. It may last from a few minutes to several hours. Also called **Dellinger fadeout, Dellinger fading, radio fadeout**, or **fade-out** (2).

Dellinger fadeout Same as **Dellinger effect**.

Dellinger fading Same as **Dellinger effect**.

delta Its symbol is Δ. 1. The fourth letter of the Greek alphabet, and as such may designate the fourth constituent of a series. 2. A net change in the value of a variable. 3. Having the shape of its symbol, or that of a triangle. Said, for instance, of a circuit.

delta channel Same as **D channel**.

delta circuit A combination of three circuit elements, such as resistors, connected in series, and arranged in the form of a triangle, similar to the shape of the Greek letter Δ (delta). It is a type of mesh circuit. Also called **delta connection**.

delta connection Same as **delta circuit**.

delta-matched antenna A single-wire half-wave antenna whose impedance is matched to the characteristic impedance of the transmission line. The two leads of the transmission line are connected to the radiator, forming a Y shape, and since the antenna is not split, this also resembles a Greek letter Δ (delta). Also called **Y-antenna, Y-matched antenna, wye antenna**, or **wye-matched antenna**.

delta matching transformer In a delta-matched antenna the Y, or delta-shaped matching section.

delta modulation Same as **differential pulse-code modulation**. Its abbreviation is **DM**.

delta network An electric network consisting of three branches connected in series, and arranged in the form of a triangle, similar to the shape of the Greek letter Δ (delta). It is a type of mesh network.

delta pulse-code modulation Same as **differential pulse-code modulation**.

delta ray A particle, especially an electron, which causes secondary ionization after being set into motion through a collision with another charged particle, such as an alpha particle.

delta-sigma modulation A form of **delta modulation** in which the integral of the input signal is encoded. This reduces quantization distortion.

delta transformer A three-phase electrical transformer whose three winding ends are connected in a manner which resembles a triangle, or a Greek letter Δ (delta).

demagnetization Same as **degaussing**.

demagnetize Same as **degauss**.

demagnetizer Same as **degausser**.

demagnetizing Same as **degaussing**.

demagnetizing force A magnetizing force whose direction of application reduces the residual induction of a previously magnetized material.

demand 1. The rate at which electric power is consumed. 2. The average rate, over a given interval, at which electric power is consumed. 3. The maximum **demand** (1) of a system. 4. Same as **demand factor** (1). 5. Same as **demand factor** (2).

demand-assigned multiple access In communications, a technique which allocates the available bandwidth, based on demand. For example, each time a given user utilizes the resources of a satellite, a portion of the available pool is occupied. Upon disconnection, this bandwidth is returned to the pool. Its abbreviation is **DAMA**.

demand factor 1. The ratio of the maximum electric power demand by a user or system, to the average power consumed over a given interval. Also called **demand** (4). 2. The ratio of the maximum electric power demand of a system at a given instant, or over a given interval, to the total connected load of the system. Also called **demand** (5).

demand interval A given period of time during which electric power demand is measured. May consist, for instance, of a 15, 30, or 60 minute interval.

demand meter A meter utilized to measure, indicate, and/or record **demand (1)**, **demand (2)**, or **demand (3)**.

demand paging In virtual memory systems, the copying pages of data from a storage device, such as a disk, to the main memory only when needed.

demand processing The processing of data as soon as it becomes available, as opposed to storing it for later processing.

demand pull Technology which is created based on customer needs, as opposed to **technology push**, which is based on the ideas and/or capabilities of the developing entity.

demand-side management Activities intended to lessen the overall demand for electric power by consumers. These include conservation, the use of appliances with increased efficiency, and incentives to distribute the load more evenly, such as reduced rates during nonpeak hours. Its abbreviation is **DSM**.

Dember effect In a semiconductor, the production of a potential difference between two surfaces or regions when one of them is illuminated. Also called **photodiffusion effect**.

demo Abbreviation of **demo**nstrator. **1.** A product which is available for review and close scrutiny, so that another similar or identical product may be obtained brand new. **2.** Same as **demonstration program**.

demo program Same as **demonstration program**.

demo software Same as **demonstration program**.

demodulate To reverse the effects of modulation, thereby restoring the original modulating signal. For instance, a data signal can thus be recovered by extracting it from the carrier signal.

demodulation The process of reversing the effects of modulation, thereby restoring the original modulating signal. Demodulation is utilized for recovering the signal of interest, such as music, images, or data. Also called **detection (2)**, or **signal demodulation**.

demodulator A circuit or device which extracts the original modulating signal from the carrier signal. In this manner the desired signal, such as that containing data or music, is recovered. Also called **detector (2)**.

demodulator-modulator Also known as **modem**, or **modulator-demodulator (1)**. **1.** A device which enables a computer to transmit and receive information over telephone lines. For transmission, the computer modulates a carrier signal so that it contains digital information in a form that can be transmitted over analog lines. It does the converse for reception. Most modems feature data compression and error-correction. **2.** Any signal-conversion device which combines the functions of a demodulator and modulator.

demon Same as **daemon**.

demonstration program A computer software version in which certain key features are unavailable, and that is generally available free of charge. May be used, for instance, for trial periods. Such a program may have an automated sequence highlighting its features. Its abbreviation is **demo program**. Also called **demonstration software**, **demonstration program**, **demo (2)**, **demonstrator**, or **demoware**.

demonstration software Same as **demonstration program**.

demonstrator Same as **demonstration program**.

demoware Abbreviation of **demonstration program**.

demultiplex To reverse the effects of a multiplexer. That is, to separate each of the multiple signals which were combined for transmission over a single channel. Its abbreviation is **demux**.

demultiplexer A circuit or device which reverses the effects of a multiplexer. That is, it separates each of the multiple signals which were combined for transmission over a single channel. Also spelled **demultiplexor**. Its abbreviation is **demux**. Also called **inverse multiplexer (1)**.

demultiplexing The process of reversing the effects of a multiplexer. That is, the separation of each of the multiple signals which were combined for transmission over a single channel. Also called **inverse multiplexing (1)**.

demultiplexing circuit A circuit which serves as a **demultiplexer**.

demultiplexor Same as **demultiplexer**.

demux 1. Abbreviation of **demultiplexer**. **2.** Abbreviation of **demultiplex**.

dendrite A treelike crystalline structure. Said, for instance, of a semiconductor crystal with such an appearance.

dendritic Resembling a **dendrite**.

dendritic web technique A technique for the formation of dendritic crystals. Used, for instance, to make certain polycrystalline materials incorporating silicon.

denial-of-service Same as **denial-of-service attack**.

denial-of-service attack An assault on a network, intended to disrupt or block access to it. For instance, it may consist of an inordinate amount of false messages, which impedes legitimate messages from accessing the servers. Such an attack is not meant to extract information, rather it seeks to do its damage by overwhelming a network. Specific examples include smurf attacks, and SYN floods. Its abbreviation is **DoS attack**, or **denial-of-service**.

dense binary code A binary coding system in which all the possible bit combinations are used. For instance, if there are 8 bits, or 64 bit combinations, all 64 are used.

dense wave division multiplexing Same as **dense wavelength division multiplexing**. Its abbreviation is **DWDM**, or **dense WDM**.

dense wavelength division multiplexing Its abbreviation is **DWDM**, or **dense WDM**. A data transmission technology in which multiple optical signals are multiplexed onto a single optical fiber. Each of the signals has a different wavelength, and with close spectral spacing, can increase the capacity of a single fiber to over 1 Tbps. In addition, DWDM-based networks can carry different types of traffic at different speeds. For instance, ATM and SONET simultaneously. Also called **dense wave division multiplexing**, or **wavelength division multiplexing**.

dense WDM Abbreviation of **dense wavelength division multiplexing**, or **dense wave division multiplexing**.

densimeter Same as **densitometer**.

densitometer Also called **density meter**, **density indicator**, or **densimeter**. **1.** An instrument utilized for measuring and indicating the density of a substance. **2.** An instrument utilized for measuring and indicating the optical density of a substance. Also called **optical density meter**.

density 1. The mass of a given substance or body per unit volume. **2.** The concentration of a given entity per unit of a given totality, such as volume. For example, the logarithm of the ratio of incident light to transmitted light, the number of bits per unit area, the number of charge carriers in a semiconductor material per unit volume, or the number of logic gates per unit area of the surface of an IC.

density indicator Same as **densitometer**.

density meter Same as **densitometer**.

density modulation The varying of the density of an electron beam over time.

departure angle The angle at which electromagnetic radiation departs from a transmitting antenna or the surface of the earth. The angle is measured based on a reference plane, such as the horizon. Also called **angle of departure**, or **angle of radiation**.

dependence The circumstance of relying on, or being controlled, by another circuit, device, process, system, or force. For example, the performance of a computer program depending on the speed and memory of the computer running it.

dependent source A source of electrical energy which depends on a current or a voltage elsewhere in a given circuit. There are four types, which are voltage-controlled current sources, current-controlled current sources, voltage-controlled voltage sources, and current-controlled voltage sources.

dependent variable A variable whose changes are determined by another variable, called **independent variable (1)**.

depletion 1. The reduction or exhaustion of something, especially a resource. 2. In a semiconductor, a reduction in the charge-carrier density in the region of a pn junction.

depletion layer In a semiconductor, the region of a pn junction that is free, or depleted, of charge carriers. Also called **depletion region**, or **space-charge layer**. Two widely used, yet deprecated, terms for this concept are **Barrier layer**, and **barrier (2)**.

depletion-layer capacitance In a semiconductor, the capacitance across the region of a pn junction that is depleted of charge carriers. Also known as **junction capacitance**, **barrier-layer capacitance**, or **barrier capacitance**.

depletion mode A mode of operation of a field-effect transistor, in which current flows at zero gate voltage. As the gate voltage is increased, current decreases. This contrasts with **enhancement mode**, where the converse is true.

depletion-mode FET Abbreviation of **depletion-mode** field-effect transistor. A field-effect transistor which operates in **depletion mode**. Also called **depletion-mode transistor**.

depletion-mode field-effect transistor Same as **depletion-mode FET**.

depletion-mode transistor Same as **depletion-mode FET**.

depletion region Same as **depletion layer**.

depolarization 1. To remove, decrease, or prevent polarization in a primary cell. 2. The process of reflecting light so that it is no longer polarized.

depolarizer 1. A substance which retards, reduces, impedes, or reverses polarization in a primary cell. 2. An optical device which reflects light so that it is no longer polarized.

deposit The concentration of matter at a specific location, or on a given surface. For instance, the electrochemical deposition of two or more metallic elements on a substrate.

deposition 1. The formation of a **deposit**. 2. A process, such as electrodeposition, which leads to the formation of a deposit.

depth finder Same as **depth sounder**.

depth of discharge Its abbreviation is **DOD**. The proportion of energy that has been removed from a storage battery or cell. Usually expressed as a percentage. For example, 80% DOD means that 80% of the ampere-hours of a fully charged storage battery or cell have been depleted.

depth of modulation Also called **modulation factor**, **modulation index**, **modulation coefficient**, or **index of modulation**. 1. A measure of the degree of modulation, usually involving the ratio of a peak variation to a steady non-peak value. The result may be multiplied by 100 to obtain a percentage. 2. In frequency modulation, the ratio of the peak variation of the carrier-wave frequency, to the frequency of the modulating signal. The result may be multiplied by 100 to obtain a percentage. 3. In amplitude modulation, the peak amplitude variation of the composite wave, to the unmodulated carrier amplitude. The result may be multiplied by 100 to obtain a percentage. 4. In amplitude modulation, the ratio of half the difference between the maximum amplitude and minimum amplitude, to the average amplitude. The result may be multiplied by 100 to obtain a percentage.

depth sounder A device which determines and indicates the depth of a body of water, by measuring the time required for transmitted sonic or ultrasonic waves to be reflected from a surface, such as the sea bottom. Also called **depth finder**, **fathometer**, **echo depth sounder**, or **sonic-depth finder**.

dequeue To remove from a queue, that is, to remove from a list of tasks a computer is waiting to perform, or from a data structure where elements are removed in the same order they were originally inserted. **Enqueue** is to place in a queue.

derate To reduce the rating of a device, such as its current rating, to provide an additional safety and/or reliability margin when operating under unusual or extreme conditions, such as elevated temperatures.

derating The reduction of the rating of a device, such as its current rating, to provide an additional safety and/or reliability margin when operating under unusual or extreme conditions, such as elevated temperatures.

derating factor The extent to which a device is derated to compensate for operation under unusual or extreme conditions.

derivative action In a control system, a corrective action whose speed is dictated by the rate of change of the controlled variable. Also called **rate action**.

derivative control An automatic control process whose rate of corrective action is dictated by the rate of change of the error signal.

derived center channel A channel obtained by electrically mixing the left and right channels of a stereophonic system. May be used, for instance, in a surround-sound audio system.

derived class In object-oriented programming, a class derived from a base class via inheritance. Also called **subclass**.

derived units Units which are derived from one or more of the fundamental units for measurement of physical quantities. Fundamental, or base, units include the ampere and the second, while derived units include the coulomb and the farad.

DES Abbreviation of **Data Encryption Standard**.

described video Same as **descriptive video**.

descriptive video In TV programming, the providing of a concise audible description of what is being displayed at a given moment. It is intended to accurately convey information such as who and what is present in each scene, which character is speaking at a specified time, key body language employed, and other visually significant components of what is presented. Useful, for instance, for those with reduced vision. The term may also apply to other video programming so offered. Also called **described video**.

descriptor A word or string which identifies or describes data, a file, or an area of memory.

desensitization The reduction of sensitivity. For instance, a reduction in the sensitivity of a radio receiver to one channel when there is a strong signal of another nearby. Desensitization may be intentionally induced, as with an automatic gain control circuit.

desensitize To reduce sensitivity. For instance, that of a receiver.

deserialize To convert a signal consisting of a serial stream of bits into parallel streams of bits. A parallel stream consists of one more bytes, while a serial stream is one bit after the other.

desiccant A substance that serves to absorb moisture. Copper sulfate, for instance, is effective in this capacity.

design 1. A detailed plan, usually in graphical form, specifying the layout of the components incorporated in a device or process, and how they interrelate with each other. Also, the preparation of such a plan. **2.** The specifying of the electrical elements, and the manner in which they interconnect with each other, in the formation of a circuit which performs a desired function. Also called **circuit design**.

design automation The use of computers to help in the automation of the designing of circuits, components, devices, equipment, and systems.

design compatibility The extent to which components, devices, or pieces of equipment are compatible with each other. This includes, for instance, electromagnetic compatibility.

design optimization The utilization of computers to optimize the design of a product. Computer-aided design is employed, and it is used for example, to enhance the design of a vehicle so as to minimize the effects of air friction upon it while traveling. Also called **automatic design optimization**.

designation **1.** A distinguishing name or sign. **2.** The act of indicating, marking, or identifying.

DeskJet A popular line of inkjet printers.

desktop A representation of a desktop, displayed on a computer monitor, in which a user can manipulate icons and organize tasks as if it were a real desktop. Such icons may include programs, files, shortcuts to network connections, and so on.

desktop accessory A program which performs a function similar to an object which might ordinarily appear on a real desktop, such as a calculator or calendar. Also called **desktop application**.

desktop application Same as **desktop accessory**.

desktop computer A computer specifically designed to be used on top of a desk. This contrasts, for instance, with handheld computers or mainframes.

desktop conferencing Same as **data conferencing**.

desktop management interface Its abbreviation is **DMI**. Within a network, a bi-directional system with which a central computer can ascertain and manage the resources and status of satellite computers and peripherals. A memory-resident agent runs in the background, and performs actions such as indicating readiness for a given task, or reporting a malfunction. In addition, DMI is platform independent.

desktop organizer A program, utility, or function that organizes a desktop, so as to enhance functionality and simplify finding what is most likely to be used. Especially helpful in handheld devices, such as PDAs.

desktop phone Abbreviation of **desktop tele**phone. A telephone which rests on a surface, such as a desk or table, as opposed, for instance, to a wall phone.

desktop publishing The use of a computer, specialized programs, and a high-quality printer to produce an output suitable for commercial printing. Its abbreviation is **DTP**.

desktop telephone Same as **desktop phone**.

desolder To separate leads or surfaces that were previously joined using solder, usually by melting said solder. Also called **unsolder**.

despun antenna On a rotating communications satellite, an antenna which is rotated in the opposite direction, so as to maintain constant the portion of the earth's surface over which it has coverage.

destination The location to which data, a signal, message, file, or the such, is sent, transferred, or directed to.

destination address The address to which data, a signal, message, file, or the such, is sent, transferred, or directed to.

destination computer Also called **object computer**, or **target computer**. **1.** A computer to which a transmission is sent. **2.** A computer into which a program is loaded, or to which data is transferred. **3.** A computer to which compiled, assembled, or translated source code is sent.

destination directory A directory to which data, files, or folders are sent. The directory of origin is the **source directory**. Also called **target directory**.

destination disk A disk to which data, files, or folders are sent. The disk of origin is the **source disk**. Also called **target disk**.

destination drive A disk drive containing a disk to which data, files, or folders are sent. The drive of origin is the **source drive**. Also called **target drive**.

destination file A file to which data is sent. The directory of origin is the **source file (1)**. Also called **target file**.

Destriau effect The emission of light by certain phosphors embedded in an insulating material which is subjected to an alternating electric field. Also called **electroluminescence (2)**.

destructive breakdown The failure of a semiconductor device, such as a transistor, due to the electrical breakdown of its depletion layer.

destructive interference The interaction of two or more waves resulting a reduction in amplitude, or cancellation. This contrasts with **constructive interference**, where there is an increase in amplitude.

destructive read The circumstance in which the reading of data also causes said data to be destroyed. Also called **destructive readout**.

destructive readout Same as **destructive read**. Its abbreviation is **DRO**.

destructive test A test in which that being evaluated is damaged or destroyed. This may consist, for instance, of a prolonged endurance test that ends when that being tested fails.

destructive testing The performance of **destructive tests**.

detection **1.** The process of sensing a signal or other phenomenon. For example, the detection of radioactivity, electromagnetic waves, or of a change in a quantity being monitored. **2.** Same as **demodulation**.

detectivity For a photodetector, such as a photodiode, the gain above noise.

detector **1.** An instrument or device which serves for **detection (1)**. **2.** Same as **demodulator**.

detector bias A voltage which is applied to a detector, to establish a reference level for its operation.

detector circuit **1.** A circuit which extracts the original modulating signal from the carrier signal. **2.** A circuit which serves for **detection (1)**.

detent A mechanical stop, such as a catch, utilized to position and hold one mechanical part in a given position relative to another. Used, for instance, in some rotary switches.

detune **1.** To adjust the frequency of a tuned circuit so that it is different from that of the incoming signal. **2.** To adjust the frequency of a tuned circuit so that it is different from its resonance frequency.

detuning **1.** The process of adjusting the frequency of a tuned circuit so that it is different from that of the incoming signal. **2.** The process of adjusting the frequency of a tuned circuit so that it is different from its resonance frequency.

deuterium An isotope of hydrogen whose nucleus contains a neutron and a proton. It is a highly explosive and flammable gas. It is not radioactive, and is used, for instance, to bombard atomic nuclei. Its symbol is **D**. Also called **heavy hydrogen (1)**.

deuterium oxide A chemical compound whose formula is D_2O. It is water in which the hydrogens are in the deuterium isotopic form. Used, for instance, as a neutron moderator in nuclear reactors. Also called **heavy water (1)**.

deuteron The nucleus of a deuterium atom. It consists of a proton and neutron, has a positive charge, and is used, for example, for nuclear bombardment. Also called **deutron**, or **deuton**.

deuton Same as **deuteron**.

deutron Same as **deuteron**.

Deutsches Industrie Norm A German organization which sets industry standards. For example, this organization has established specifications for many connectors used in various types of electronic products, including computers and video equipment. Its acronym is **DIN**.

development machine A computer that is utilized specifically for the creation of new software, debugging, or for other experimental purposes. It is kept separate from the main systems of an entity to avoid problems such as different versions of programs creating conflicts.

deviation **1.** Divergence from an expected value or result. For example, the difference between an observed value and a theoretical one. **2.** In frequency-modulation, the maximum difference between the instantaneous frequency of the modulated wave, and the frequency of the unmodulated carrier. Also called **frequency deviation** (2). **3.** In optics, the angle between the ray incident on an object and the ray emerging from it. This angle can be due to refraction, reflection, or diffraction. Also called **deviation angle**, or **angle of deviation**.

deviation angle Same as **deviation** (3).

deviation ratio In frequency-modulation, the ratio of the maximum frequency deviation to the maximum modulating frequency.

device **1.** A physical unit or mechanism which performs a specific function or serves for a particular purpose. A device may be electrical, mechanical, electromechanical, and so on, and examples include, resistors, cable connectors, transistors, ICs, disk drives, and computers. **2.** A hardware component or subsystem in a computer, such as a disk drive, keyboard, or port. **3.** An electronic unit which incorporates one or more active components. For instance, a transistor, or an IC. **4.** An active component, such as a transistor or an operational amplifier, which can not be subdivided without disabling its function.

device address An address in a computer's memory which identifies a **device** (2). Each address must be unique to avoid conflicts.

device control character A character, such as an ASCII character, utilized to control a computer device.

device dependence **1.** The circumstance where a program can only run when used with certain hardware. This contrasts with **device independence** (1). **2.** A lack of **device independence** (2), or **device independence** (3). Also, the extent to which such device independence is lacking.

device-dependent bitmap A graphics format that is tied to a specific application or output device. This means, for example, that it will not necessarily be displayed properly through another application, or be printed exactly the same as it was in the original application. This contrasts with **device-independent bitmap**. Its abbreviation is **DDB**.

device driver A program or hardware component which enables a computer operating system to communicate with, and control peripheral devices, such as keyboards, disk drives, or printers. Each hardware device has its own specialized commands, and when a new device is added to the computer its driver must be installed for it to function properly. Also called **driver** (2).

device independence **1.** The circumstance where a program can run equally well with most any hardware. This contrasts with **device dependence** (1). **2.** The ability of any given computing device or peripheral, such as a desktop computer, PDA, Internet appliance, properly equipped TV or cell phone, in-car computer, or mouse, to work properly with any particular operating system or computer application. **3.** The ability of any given operating system or computer application to be able to work properly with any particular computing device or peripheral, such as a desktop computer, PDA, Internet appliance, properly equipped TV or cell phone, in-car computer, or mouse.

device-independent bitmap Its abbreviation is **DIB**. A graphics format that is not tied to a specific application or output device. This helps ensure, for example, that it will be properly displayed through another application, or be printed exactly the same as it was in the original application. This contrasts with **device-dependent bitmap**.

device manager An operating system utility which enables viewing and changing the settings of a computer's devices. A device manager serves, for instance, to change interrupts and communications parameters.

device name In computers, a name given to a **device** (2) to facilitate identification by a given program or software system. An example is the designation given to a port, or that of a logical drive. Also called **logical device**.

device response **1.** A change in the behavior, operation, or function of a device as a consequence of a change in its external or internal environment. For example, a change in its output resulting from a change in its input. **2.** A quantified **device response** (1), such as a frequency response.

device under test A specific circuit or device whose function and/or performance is being assessed. Usually used in the context of ICs. Its abbreviation is **DUT**.

dewetting A condition in which previously applied solder recedes from a surface. Due, for instance, to variations in temperature and the presence of contaminants over extended time periods. Also spelled **de-wetting**.

DF Abbreviation of **direction finder**.

DF antenna Abbreviation of **direction-finder antenna**.

DF antenna system Abbreviation of **direction-finder antenna system**.

DFB laser Abbreviation of **distributed feedback laser**.

DFP port Abbreviation of **digital flat panel port**.

DGPS Abbreviation of **differential GPS**.

DHCP Abbreviation of **Dynamic Host Configuration Protocol**.

DHCP network Abbreviation of **Dynamic Host Configuration Protocol network**.

DHCP server Abbreviation of **Dynamic Host Configuration Protocol server**.

DHTML Abbreviation of **Dynamic HTML**.

diac Abbreviation of **diode alternating-current switch**. A bidirectional avalanche diode with two terminals and three layers, which conducts symmetrically. It allows current to flow when the breakover voltage is exceeded in either direction. Also called **trigger diode**, or **three-layer diode**.

diagnosis The process of determining the causes of a given malfunction. There are various diagnostic resources, such as tests, to help in this task.

diagnostic program **1.** A computer program which serves to test hardware, such as memory and disks. **2.** A computer program which serves to locate and identify problems with software, hardware, or data.

diagnostic routines Also called **diagnostics**, or **diagnostic subroutines**. **1.** Computer program routines which serve to test hardware, such as memory and disks. Such tests are usually performed when a computer is booted. **2.** Computer program routines which serve to locate and identify problems with software, hardware, or data.

diagnostic subroutines Same as **diagnostic routines**.

diagnostic tests 1. Tests utilized to verify that a component, device, piece of equipment, or system is functioning properly. **2.** Tests utilized to pinpoint the causes of a given malfunction.

diagnostics Same as **diagnostic routines**.

diagram 1. A graphical representation designed to show the components of a circuit, device, piece of equipment, or system, and the relationship of each component with the others. Examples include block diagrams and phase diagrams. **2.** A graphical representation, such as a chart or a graph, showing the relationship between two or more variables. **3.** A graphical representation of the electrical elements in a circuit, and the way each is interconnected with each other. Each circuit element is represented by a symbol, while lines represent the wiring. Also called **circuit diagram, schematic diagram, wiring diagram, schematic circuit diagram**, or **wiring schematic**.

dial 1. A face with a graduated scale upon which a pointer or other indicator specifies the level a varying quantity is at. For example, a thermometer dial. Such a dial may also be virtual, such as that displayed on a computer monitor. **2.** A face with a graduated scale upon which a pointer or other indicator specifies the level to which a variable control is set. For instance, a radio dial. Such a dial may also be virtual, such as that displayed on some audio receivers. **3.** A device on a telephone which generates signals which are utilized to place calls. Also called **telephone dial**. **4.** The action of using a **dial (3)** to place a telephone call.

dial-back security In a network accessed by modem, a security measure in which the called server authenticates the user, hangs up, and dials the user back. This helps ensure that the user is accessing the network from an authorized location.

dial cable A cable which moves a pointer along a graduated scale on a **dial (1)**, or **dial (2)**. For instance, such a cable may be attached to a rotating knob.

dial cord A **dial cable** whose composition is not that of a metal.

dial knob A knob which is turned to change the indicated position of a **dial (1)**, **dial (2)**, or **dial (3)**.

dial lamp Same as **dial light**.

dial light A light utilized to illuminate a **dial (1)**, **dial (2)**, or **dial (3)**. Also called **dial lamp**, or **pilot light (2)**.

dial pulse In a rotary telephone, one or more pulses which interrupt a steady direct current, to indicate a dialed digit. For example, to dial a 4, 4 such pulses are sent.

dial pulsing The dialing of telephone number utilizing **dial pulses**. Also called **loop pulsing**.

dial signaling Signaling in which the rotary dial or keypad of a telephone generates pulse trains which are sent to a switching center.

dial telephone A telephone incorporating a dial, as opposed to another dialing mechanism, such as a key pad or one that is voice activated. Also called **dial telephone set**.

dial telephone set Same as **dial telephone**.

dial telephone system A telephone system in which connections between customers are made automatically through dialing.

dial tone A tone heard through a calling device, such as a telephone or modem, indicating that the system is ready for dialing from this line. Also called **dialing tone**.

dial-up Same as **dial-up connection**. Also spelled **dialup**.

dial-up access The use of a **dial-up connection** for communications. Used, for instance, by a computer modem to access a network. Also spelled **dialup access**.

dial-up connection Pertaining to a temporary connection over telephone lines using a public switched network, as opposed to the use of a private network, such as a dedicated circuit.

Ordinary home telephones and dial-up modems use this service. Also spelled **dialup connection**. Also called **dial-up**.

dial-up line A line which enables a **dial-up connection**. Examples include ordinary home lines and ISDN. Also spelled **dialup line**.

dial-up modem A modem through which a **dial-up connection** may be established. Also spelled **dialup modem**.

dial-up network A network, such as the Internet, which is accessed via a **dial-up connection**. Also spelled **dialup network**. Its abbreviation is **DUN**.

dial-up networking The use of a **dial-up network**. Also spelled **dialup networking**.

dial-up service A service, such as that offered by an Internet service provider, which is accessed via a **dial-up connection**. Also spelled **dialup service**.

dialed number identification service A service, offered by telephone companies, which sends and displays the number the calling party has dialed. Serves, for instance, to automatically route incoming calls to specialized representatives within a multifaceted enterprise which utilizes many toll-free numbers. Its abbreviation is **DNIS**

dialer A device which automatically dials telephone numbers. Such numbers may be manually or automatically programmed into memory, and may be dialed when pressing a short code, or at a pre-programmed time. Also called **automatic dialer**, or **telephone dialer**.

dialing The use of a device, such as a telephone or modem, to dial and establish connections.

dialing tone Same as **dial tone**.

dialog box On a computer display, a box that appears to request input, or to provide information. As soon as the input is given, or the message is acknowledged, pressing "OK" or its equivalent closes the box. Such a box may appear, for instance, to request a username and password, or to inform of an error.

dialup Same as **dial-up**.

dialup access Same as **dial-up access**.

dialup connection Same as **dial-up connection**.

dialup line Same as **dial-up line**.

dialup modem Same as **dial-up modem**.

dialup network Same as **dial-up network**.

dialup networking Same as **dial-up networking**.

dialup service Same as **dial-up service**.

diamagnet A material, such as copper, which is **diamagnetic**.

diamagnetic Pertaining to materials whose relative magnetic permeability is less than unity, with the relative magnetic permeability of a vacuum having an assigned value of 1. Such materials are repelled by a magnet, and as such assume a position at right angles to magnetic lines of force. Examples include copper, silver, and bismuth.

diamagnetic material A material, such as copper, whose relative magnetic permeability is less than unity.

diamagnetism The characteristic of certain materials of having a relative magnetic permeability of less than unity.

diamond An allotropic form of the chemical element carbon, occurring as crystals. It is the hardest substance known, and as such has an assigned value of 10 on the Mohs scale. It is a precious stone, has extremely high thermal conductivity, a high refractive index, and may be transparent, translucent, or colored. Its applications include its use in cutting and grinding tools, optical windows, lasers, and semiconductors.

diamond antenna A wideband directional antenna whose long-wire radiators form the sides of a rhombus, and which is fed at one apex. Such antennas feature high power gain over a wide range of frequencies. Also called **rhombic antenna**.

diamond-like carbon An artificially-prepared form of carbon, with a structure similar to that of diamond. Although its properties are not quite as defined as that of diamond, it has many applications in electronics, including its use in capacitors, in optical windows for bar-code readers, in disk-drive heads, and in semiconductors.

diamond needle Same as **diamond stylus**.

diamond stylus A phonograph pickup whose tip is a ground diamond. Also called **diamond needle**.

diaphragm 1. A vibrating membrane in a speaker. It produces a sound-wave output in response to its electric input. Cones and domes are examples of commonly utilized diaphragms in speakers. Also called **speaker diaphragm**, or **loudspeaker diaphragm**. 2. A vibrating membrane in a microphone. It produces an electric output in response to its sound-wave input. Also called **microphone diaphragm**. 3. A pressure-sensitive element in some sensors, such as certain barometers. 4. A membrane, disk, adjustable opening, or other device or mechanism which serves to restrict or regulate the passage of a waves and/or matter, such as light waves or electrolytes.

diathermy The use of high-frequency radio waves for therapeutic purposes. Used, for example, to generate localized heat in body tissues.

diatomic Occurring as, pertaining to, or consisting of two atoms. For example, molecular oxygen, whose chemical formula is O_2.

DIB 1. Abbreviation of **device-independent bitmap**. 2. Abbreviation of **Directory Information Base**.

dibit Any of the four possible combinations of two consecutive bits, which are 00, 01, 10, and 11. Used, for instance, in differential phase-shift keying when four states are represented.

diborane A colorless and reactive gas whose chemical formula is B_2H_6. It is used as a semiconductor dopant.

dice Plural form of **die** (1), or **die** (2).

dichotomizing search A search technique in which the desired item is compared to a list to determine in which half it is located, the other half being discarded. Then, within this half, a similar search is performed. This process of pinpointing the half that contains the item is continued successively until the search is complete. Also known as **binary search**.

dichotomy 1. Characterized by branching into two divisions, such as a routine into two subroutines. 2. Characterized by continued branching into two divisions. For example, two branches, each forming two, for a total of four, then those four forming eight, and so on.

dichroic mirror A mirror which reflects light of given frequencies, while transmitting the rest. Used, for instance, in color TV cameras.

dichroism 1. A property of some crystals, in which different colors are exhibited when viewed from different axes. 2. A property of some substances or materials, in which different colors are exhibited when viewed through different thicknesses. Occurs, for instance, in some solutions.

dichromate cell An electrolytic cell with a carbon electrode immersed in an approximately 12% potassium dichromate solution, and a zinc electrode immersed in an approximately 9% sulfuric acid solution.

dicing The cutting of a semiconductor material into **dice**.

DID Abbreviation of **direct inward dial**, or **direct inward dialing**.

die 1. A piece of semiconductor material or dielectric upon which one or more electrical components may be mounted, etched, or formed. Used, for instance, for fabrication of repetitive units in semiconductor manufacturing. Its plural form is **dice**. Also called **chip** (2), or **semiconductor die**

(1). 2. A piece of semiconductor material or dielectric upon which one or more electrical components has been mounted, etched, or formed. Its plural form is **dice**. Also called **chip** (3), or **semiconductor die** (2). 3. To cease functioning. 4. A tool, device, or mold which serves to impart a desired shape and/or finish to a molten or semisolid material. It may consist, for instance, of a mold into which a molten metal is forced.

die attach The attachment of a **die** (1) to a substrate. An epoxy resin, for instance, may be used for such bonding. Also called **die attachment**, or **die bonding**.

die attachment Same as **die attach**.

die bonding Same as **die attach**.

dielectric A material which is an electric insulator. An electric field may be applied through such a material with little or no energy loss. Dielectrics are used, for instance, to separate the plates in a capacitor, or between the conductors of a cable. Examples of dielectrics include air, oil, glass, paper, rubber, and various plastics and ceramics. A vacuum is a perfect dielectric. Also known as **dielectric material**.

dielectric absorption A property of certain dielectrics, in which electric polarization is retained after removal of the electric field. Such an effect can last up to several years, and it is the basis for electrets. Also called **dielectric hysteresis**, or **electric hysteresis** (1).

dielectric amplifier An amplifier circuit whose active component is a ferroelectric capacitor whose capacitance varies as a function of the applied voltage. Also called **capacitor amplifier**.

dielectric antenna An antenna whose radiating elements are partially or completely composed of a dielectric, such as ceramic.

dielectric breakdown An abrupt increase in the electric current flowing through a dielectric, when the applied voltage exceeds a given critical value. Such a breakdown is usually destructive. Also called **electric breakdown** (2).

dielectric breakdown voltage For a given dielectric, the voltage at which **dielectric breakdown** occurs. Also called **electric breakdown voltage**.

dielectric constant For a given material, the property which determines the electrostatic energy that can be stored per unit volume for a unit potential gradient. This is equivalent to the ratio of the capacitance of a capacitor using the material in question as a dielectric, to the capacitance of a capacitor using a vacuum as a dielectric. The dielectric constant of a vacuum is defined as 1, and for most calculations air is also considered to have this value. Also known as **relative dielectric constant**, **permittivity**, **relative permittivity**, **specific inductive capacity**, or **inductivity**.

dielectric current The current flowing at any given instant through the surface of a dielectric which is in a changing electric field.

dielectric displacement A vector quantity pertaining to the charge displaced by an electric field within a given medium. It is equal to the electric field strength multiplied by the permittivity of the medium, and is usually expressed in coulombs per square meter. Also known by various other names, including **dielectric strain**, **dielectric stress**, **displacement** (2), **electric displacement**, **electric displacement density**, and **electric field induction**.

dielectric fatigue A property of some dielectrics, which have been subjected to a sustained continuous voltage, where there is a reduction in the resistance to breakdown.

dielectric flux density Same as **dielectric displacement**.

dielectric heating The heating of a dielectric, such as a plastic, by applying a high-frequency electric field to it. This is usually accomplished by placing the material between the electrodes of a two-electrode capacitor, and applying an alternating electric field. Dielectric losses produce the heat.

dielectric hysteresis Same as **dielectric absorption**.

dielectric isolation The use of a dielectric to isolate IC regions from each other. Used, for instance, in high-performance analog ICs.

dielectric lens A lens composed of a dielectric. Such a lens refracts radio waves, and is usually used at microwave frequencies.

dielectric loss The energy converted into heat during **dielectric heating**.

dielectric loss angle The difference resulting when the **dielectric phase angle** is subtracted from 90°. Also called **loss angle**.

dielectric material Same as **dielectric**.

dielectric phase angle The angular phase difference between the sinusoidal voltage applied to a dielectric, and the resultant AC with the same period as the applied voltage.

dielectric polarization A phenomenon observed in dielectrics, in which the electrons in each atom are displaced in the direction opposite to that of an applied electric field, while the nucleus of each atom is displaced in the direction of said field. Also called **polarization (5)**, or **electric polarization**.

dielectric power factor The cosine of the **dielectric phase angle**.

dielectric resonator The resonator element in a **dielectric resonator oscillator**. Its abbreviation is **DR**.

dielectric resonator oscillator A microwave oscillator that utilizes a ceramic dielectric with a high Q factor as the resonator element. Such an oscillator can resonate in various modes, and is temperature-stable. Its abbreviation is **DRO**.

dielectric rigidity Same as **dielectric strength**.

dielectric rod antenna A surface-wave antenna whose radiating element is a tapered dielectric rod, and whose radiation pattern is in the direction of said rod.

dielectric strain Same as **dielectric displacement**.

dielectric strength The maximum voltage a dielectric can withstand before an abrupt increase in the electric current flowing through it. May be expressed, for instance, in volts per millimeter. Also called **dielectric rigidity**, or **electric strength**.

dielectric stress Same as **dielectric displacement**.

dielectric susceptibility A dimensionless quantity which measures the ease with which a dielectric may be polarized. It is the ratio of the dielectric polarization, to the product of the electric intensity multiplied by the permittivity of a vacuum. Also called **electric susceptibility**.

dielectric test **1.** A test in which a voltage exceeding the breakdown voltage is applied for a specified time. Such a test is usually destructive, and is utilized to determine the margin of safety of an insulating material against breakdown. **2.** Any test which determines dielectric properties, such as dielectric strength.

dielectric waveguide A waveguide consisting of a solid dielectric, such as glass, surrounded by another dielectric, such as air. An example is an optical fiber.

dielectric wedge A piece of a dielectric, in the form of a wedge, which is inserted in a waveguide to match the impedance with that of another waveguide.

dielectric wire A **dielectric waveguide** utilized for transmitting ultra-high frequency radio waves short distances between parts of a circuit.

dielectric withstand voltage Same as **dielectric withstanding voltage**.

dielectric withstanding voltage The maximum voltage which can be applied to a dielectric without adverse effects, such as dielectric breakdown. Also called **dielectric withstand voltage**, or **withstanding voltage**.

DIF Abbreviation of **Data Interchange Format**.

difference **1.** The degree or amount by which two values vary. **2.** The degree or amount by which two consecutive values vary. **3.** The amount remaining after one quantity is subtracted from another quantity. **4.** In relational databases, an operation utilized to create a new table based on the dissimilarities present in two existing tables.

difference amplifier Same as **differential amplifier**.

difference channel In an audio amplifier, a channel whose output is the difference between the left and right channels. Used, for instance, in surround-sound systems. This contrasts with a **sum channel**, where the output is the sum of the left and right channels.

difference detector A device whose output is the difference between its input signals. For instance, its output may be the difference in amplitudes of its two input waveforms.

Difference Engine An automatic calculator designed in the early 1800s which was steam powered and incorporated rods and rotating wheels.

difference frequency A frequency which is the difference of two other frequencies.

difference signal A signal which is the difference of two other signals. For instance, that of a difference channel.

differential **1.** Pertaining to a difference, such as that between signals, directions, or other variables. **2.** Pertaining to a circuit, device, or system whose operation depends on the difference between two quantities. Examples include differential amplifiers and control systems.

differential amplifier An amplifier whose output signal is a function of the difference between its two input signals. For instance, the instantaneous output voltage of a differential amplifier being proportional to the instantaneous difference between its input voltages. Ideally, such an amplifier would have a zero output when both its inputs are identical. Also called **difference amplifier**, or **differential-input amplifier (1)**.

differential analyzer An analog computer used primarily for solving differential equations.

differential backup A data backup technique in which the only files copied are those that have been modified since the last backup. This contrasts with a **full backup**, in which all files are copied, including those that have not been modified since the previous backup. An **incremental backup** is similar to a differential backup, except that all versions are saved, not just the last one.

differential capacitor A dual variable capacitor incorporating two stator plates and one rotor plate. It is arranged in a manner so that when the rotor is turned, the capacitance of one section increases while that of the other decreases.

differential comparator A circuit which utilizes differential amplifiers to compare an input voltage to a reference voltage, and whose output depends on the difference between these values.

differential cooling Uneven cooling of an object, surface, mixture, or the like.

differential delay For a given frequency band, the difference between the maximum and minimum frequency delays.

differential discriminator A discriminator circuit which only passes pulses whose amplitudes fall within a prescribed interval. Neither of the limits of the interval may have a value of zero.

differential encoding A communications format in which each bit is split into two, providing a self-synchronizing data stream. A 1-bit is transmitted with a high voltage in the first period and a low voltage in the second, with a 0-bit being the converse. One advantage of this encoding is that the difference between a 0-bit and no signal is distinguished, as there is no transition in the latter case. A disadvantage is that it occupies twice the bandwidth. Also called **Manchester code**, or **Manchester encoding**.

differential gain control An automatic control circuit which modifies the gain of a receiver at set intervals, so as to maintain the output signal levels within a desired amplitude, in response to varying input signals. To do so, it reduces sensitivity to strong signals, while increasing it for weak ones. Such modifications are usually anticipated, as when a loran receiver adjusts its gain between successive pulses from fixed transmitters. Also called **sensitivity-time control**, **gain-sensitivity control**, or **gain-time control**.

differential galvanometer A galvanometer in which two currents are passed in opposite directions, through identical coils. The currents neutralize each other, so there is a zero reading when both currents are of equal magnitude.

differential gap In a control system, the interval between the limits of a regulated variable. For example, the thermostat of a room heater set at 20 °C might turn on at 18 °C and shut off at 22 °C, thus having a differential gap of 4 °C.

differential global positioning system Same as **differential GPS**.

differential GPS Abbreviation of **differential g**lobal positioning **s**ystem. In a GPS system, a technique which utilizes data from a known fixed location to compensate for inaccurate signals arriving from an unknown station. Its own abbreviation is **DGPS**.

differential heating Uneven heating of an object, surface, mixture, or the like.

differential input In an amplifier with two inputs, a circuit which rejects identical inputs, and amplifies any differences between said inputs. Also called **differential-input circuit**.

differential-input amplifier 1. Same as **differential amplifier**. 2. A **differential-input amplifier** (1) whose input signals are each measured with respect to a common ground.

differential-input capacitance In a differential amplifier, the capacitance between its two input terminals.

differential-input circuit Same as **differential input**.

differential-input impedance In a differential amplifier, the impedance between its two input terminals.

differential-input measurement In a differential amplifier, a measurement made between its two input terminals. In such a measurement, each input serves as the reference for the other, as opposed to ground. Also called **floating-input measurement**.

differential-input resistance In a differential amplifier, the resistance between its two input terminals.

differential-input voltage 1. In a differential amplifier, the maximum voltage that can be applied across its two input terminals, without damage. 2. In a differential amplifier, the voltage difference between its two input terminals. Also called **differential voltage**, or **differential-mode input**.

differential instrument A measuring instrument which has two identical circuits or coils, through which two currents are passed in opposite directions. The differential between these currents imparts motion to the indicator. An example is a differential galvanometer.

differential keying Keying in which the oscillator is turned on a few milliseconds before the amplifier, and turned off a few milliseconds after, to avoid chirp.

differential Manchester encoding A communications format similar to **differential encoding**, with the difference that a 1-bit is indicated by the first half of the signal being the same as that of the previous bit signal, while the first half of a 0-bit is the opposite of the previous bit signal. Since in either case there is always a transition, the absence of a transition indicates that no signal has been sent. Also called **Manchester differential encoding**.

differential-mode input Same as **differential-input voltage** (2).

differential-mode signal In a balanced three-terminal circuit, a signal applied between the two ungrounded terminals.

differential modulation Modulation in which the selection of the significant condition for any signal element, such as a voltage, depends on the choice made for the previous signal element. An example is delta modulation.

differential PCM Abbreviation of **differential pulse-code modulation**.

differential permeability The rate of change of the induction in relation to the magnetizing force.

differential phase In a color TV signal, a variation in phase of the color subcarrier as the level of the luminance signal is varied from blanking to white.

differential phase-shift keying A phase-shift keying technique in which the phase of the carrier is varied in a manner which provides for more than two possible states, and in which the phase information conveyed in a given interval is relative to the interval preceding it. Its abbreviation is **DPSK**.

differential pressure A difference in pressure between two points or regions, such as those of a differential-pressure transducer.

differential-pressure transducer A transducer whose output is dependent the difference in magnitude of two sensed pressures.

differential pulse-code modulation A pulse-modulation technique used for conversion of an analog signal into a digital signal. In it, each segment of the continuous analog signal is sampled, and on each occasion, one bit is generated. If there is a relative increase in amplitude from segment to segment, a 1 state is produced, a relative decrease produces a 0 state, and when there is no change in relative amplitude there is no change in state. Operates well in the presence of noise. Its abbreviation is **DPCM**, or **differential PCM**. Also called **delta pulse-code modulation**, or **delta modulation**.

differential pulse-height discriminator A circuit which produces a fixed output pulse, but only when the input pulse lies between pre-established limits of amplitude. Also called **pulse-height selector**, **pulse-amplitude selector**, or **amplitude selector**.

differential receiver Same as **differential synchro** (1).

differential relay A relay which responds to a difference in two voltages or currents. For example, such a relay may have two coils and only respond when the respective currents of said coils vary beyond a specified amount.

differential SCSI A version of SCSI which allows cable lengths to be extended to about 25 meters, as opposed to the usual 6 meters.

differential stage A symmetrical amplifier stage in which there is a zero output when both its inputs are identical. When there is a difference between the inputs, such as when only one of them has provides a signal, its output is a function of said difference.

differential synchro 1. A synchro receiver whose output is a mechanical angle obtained by subtracting one electrical angle from another. Such a device utilizes a damper to limit oscillation or spinning. Also called **differential receiver**, or **synchro differential receiver**. 2. A synchro transmitter whose output is an electrical angle obtained by the sum of an electrical angle and a mechanical angle. Also called **differential transmitter**, or **synchro differential transmitter**.

differential thermometer A thermometer utilizing two strips of dissimilar metals bonded together, each with different coefficients of thermal expansion. In this manner, as temperature varies, it will bend in one direction or the other. A calibrated scale may be coupled for temperature readings. Also called **bimetallic thermometer**.

differential transducer A transducer whose output is proportional to the difference between its two inputs.

differential transformer A transformer that is utilized to couple two or more signal sources to a single transmission line.

differential transformer transducer A transformer which produces an AC output which is proportional to the displacement of the core relative to the windings. Also called **linear variable differential transformer**.

differential transmitter Same as **differential synchro (2)**.

differential voltage Same as **differential-input voltage**.

differential voltmeter A voltmeter which only measures the difference between a known and an unknown voltage.

differential winding A coil winding whose magnetic field opposes that of another nearby coil winding, which may or may not be part of the same coil.

differentiating circuit A circuit or electric network whose output is proportional to the differential, with respect to time, of the input signal. For example, a circuit whose output voltage is proportional to the differential of the input voltage with respect to time. The output waveform of such a circuit is the time derivative of the input waveform. Also called **differentiating network**, **differentiator (1)**, or **differentiator circuit**.

differentiating network Same as **differentiating circuit**.

differentiator 1. Same as **differentiating circuit**. 2. A circuit or device whose output voltage is proportional to the differential of the input voltage with respect to time.

differentiator circuit Same as **differentiating circuit**.

Diffie-Hellman A public-key encryption exchange algorithm which results in a shared secret key at both ends.

diffracted radiation Radiation whose path has been altered by **diffraction**, as opposed to reflection or refraction.

diffracted wave A wave whose path has been altered by **diffraction**, as opposed to reflection or refraction.

diffraction The deviation of waves, and particles showing wavelike properties, around obstructions and edges. These include electromagnetic radiation, such as light waves or radio-frequency signals, very small and rapidly moving particles, such as electrons or neutrons, and sound. Diffraction is a consequence of interference, and is most noticeable when the wavelength of the disturbance is comparable to the diameter of the obstacle. Diffraction provides, for instance, for sound to be heard around the corner of an object, or for light to be bent when incident upon the edge of an object.

diffraction grating A surface with many fine parallel lines, grooves, or slits, which are extremely close together, and which serve to disperse light so as to produce a spectrum. Such a surface may be made of metal or glass, and may have over 15,000 lines, grooves, or slits, per centimeter. Also called **grating (2)**.

diffraction instrument Same as **diffractometer**.

diffraction pattern 1. The interference pattern resulting from the superimposition of diffracted waves. 2. A recording of a **diffraction pattern (1)** on a screen or a plate.

diffraction propagation The propagation of electromagnetic waves around obstructions and edges, due to **diffraction**.

diffraction spectrum A spectrum produced by means of a **diffraction grating**. Such a spectrum consists of parallel bands that vary in brightness.

diffractometer An instrument which measures the intensity of diffracted radiation, such as that of electrons, neutrons, or light. Utilized, for instance, in the study of the structure of matter. Also called **diffraction instrument**.

diffractometry The science that deals with the determination of crystal structure through the study of the diffraction of radiation penetrating its structure.

diffuse sound Sound whose energy is evenly dispersed within a given space or enclosure.

diffused-alloy transistor Same as **drift transistor**.

diffused device A semiconductor device, such as a transistor, which is manufactured through the process of **diffusion (6)**. Also called **diffused semiconductor device**.

diffused junction A semiconductor junction formed through the process of **diffusion (6)**.

diffused-junction rectifier A rectifier with a **diffused junction**.

diffused-junction transistor A transistor whose emitters and collectors are formed through the process of **diffusion (6)**. Also called **diffused transistor**.

diffused metal-oxide semiconductor A semiconductor device, such as a transistor, which is manufactured utilizing two successive diffusion processes, one for each region of different conductivity. Such semiconductors have a very precise channel length, feature high gain and speed, and are used, for instance, for microwave applications. Also called **double-diffused metal-oxide semiconductor**, or **double-diffused MOS**. Its abbreviation is **DMOS**, or **diffused MOS**.

diffused MOS Same as **diffused metal-oxide semiconductor**.

diffused resistor A semiconductor resistor formed through the process of **diffusion (6)**.

diffused semiconductor device Same as **diffused device**.

diffused transistor Same as **diffused-junction transistor**.

diffuser A device utilized to disperse light or sound. Used, for instance, to guide the sound emanating from a horn speaker. Also spelled **diffusor**.

diffusion 1. The spontaneous process of one material spreading through and mixing with another. Gases diffuse most readily, followed by liquids, while occurring least in solids. 2. The spontaneous process of a material, such as a gas, dispersing from a region of higher density to that of lower density. 3. The passage of atoms through a crystal lattice. 4. The passage of light through a translucent material. 5. The scattering of light when reflected off an irregular surface. 6. A process employed in the manufacturing of semiconductor devices, in which the desired impurity atoms, usually contained in a gas, are placed in a precisely heated enclosure with the semiconductor material. In this manner, the impurity atoms, or dopants, are distributed uniformly within the crystal lattice of the semiconductor material. Various devices, with a range of properties, may be formed in this manner. Also called **diffusion process**. 7. The movement of charge carriers through a semiconductor.

diffusion bonding A process for joining metals, through the use of heat and pressure, which involves the movement of atoms from one metal to another, through diffusion. Used, for instance, in the manufacturing of semiconductor devices.

diffusion capacitance The capacitance of a forward-biased semiconductor junction.

diffusion constant In a homogeneous semiconductor, the ratio of the density of the diffusion current, to the gradient of the charge-carrier concentration.

diffusion current Current resulting from the movement of charges by diffusion. Occurs, for instance, in electrolytic solutions or semiconductor devices.

diffusion furnace A heated enclosure where **diffusion (6)** takes place. The higher the temperature of the furnace, the faster the dopant penetrates the semiconductor material. Temperatures of operation typically fall within the 700 °C to 1400 °C range.

diffusion length 1. Once a particle is formed, the average distance it travels before being absorbed. 2. Once a charge carrier is generated, the average distance it travels before recombining.

diffusion process Same as **diffusion** (6).

diffusion pump A vacuum pump utilized to create a high vacuum in a given enclosure, such as a vacuum tube. Such a pump may employ a stream of rapidly moving heavy atoms or molecules to drag out most remaining gas molecules, and is usually used in conjunction with a mechanical pump.

diffusion transistor A transistor in which the current flows as a result of **diffusion** (7).

diffusor Same as **diffuser**.

digicam Abbreviation of **digital camera**.

digicash Abbreviation of **digital cash**.

digit A symbol used to represent a unit within a numbering system. In the decimal system, the digits are 0 through 9, while in the binary system they are 0 and 1. For example, the number 2004 has four digits. The number of possible digits for any given numbering system is equal to the radix, or base, of the particular system.

digit period Within a series of pulses, the time interval that elapses between successive pulses.

digital In electronics, **digital** and **analog** can be differentiated as follows: a digital quantity has a range that is divided into a finite number of discrete steps, while an analog quantity has a continuous and infinitely variable range of values. **1.** Pertaining to representation by, and calculation with, **digits**. **2.** Pertaining to information, such as data, represented in the form of discrete units called digits, especially that expressed in binary digits. **3.** Pertaining to that which is computer related, or which involves information in binary form. This includes storing, processing, recording, reproducing, and so on. **4.** Pertaining to a readout or display which indicates in digits.

digital access cross-connect system Same as **digital cross-connect system**. Its abbreviation is **DACS**.

digital advanced mobile phone system Same as **digital AMPS**.

digital advanced television Same as **digital TV**. Its abbreviation is **DTV**.

digital advanced television system Same as **digital TV** (1).

digital advanced TV Same as **digital TV**.

digital ammeter An ammeter whose indications are shown on a digital display, such as an LCD.

digital AMPS Abbreviation of **digital** **a**dvanced **m**obile **p**hone **s**ystem. Its own abbreviation is **D-AMPS**. The digital equivalent of **AMPS**. Through the use of time-division multiple access, AMPS can be upgraded to D-AMPS. D-AMPS, among other benefits, triples the number of available channels.

digital answering machine An answering machine that uses memory chips to store greetings and messages, as opposed to cassettes. Such devices are generally more reliable than those that use cassettes, and have additional features such as instant access to specific messages.

digital audio Audio which is in digital code. Examples include that in digital audio discs, and digital audio tape.

digital audio broadcast Its abbreviation is **DAB**. **1.** A technology which digitally encodes audio for broadcasting. This provides for more channels and better sound quality. **2.** A broadcast utilizing **digital audio broadcast** (1). Also called **digital audio broadcasting**.

digital audio broadcasting Same as **digital audio broadcast** (2). Its abbreviation is **DAB**.

digital audio disc A CD containing music, speech, or other audio. A digital audio disc can usually hold up to 74 minutes of recorded sound. Also spelled **digital audio disk**. Also called **audio CD**, or **music CD**.

digital audio disk Same as **digital audio disc**.

digital audio effects The use of computers to enhance, modify, add, or remove sounds in songs, movies, TV, and other multimedia presentations.

digital audio extraction The ability of a computer to digitally record, onto its hard drive, one or more tracks of a CD. Once a computer contains the content of a CD, the audio format may be changed, for instance, to MP3. A specially-equipped drive is necessary. Its abbreviation is **DAE**.

digital audio input An audio input signal represented by discrete states which correspond to a fixed number of digits. For instance, such an input may consist of two voltage levels, such as high and low, which represent the numbers 0 and 1. An example the input a CD disc provides a CD player.

digital audio output An audio output signal represented by discrete states which correspond to a fixed number of digits. For instance, such an output may consist of two voltage levels, such as high and low, which represent the numbers 0 and 1. An example is the output of a CD player.

digital audio player **1.** A device or piece of equipment which plays audio that is digitally encoded, such as that on a CD, or an MP3 file. **2.** An application which enables a user to access and play digital audio. For example, a media player.

digital audio radio Same as **DARS**.

digital audio radio service Same as **DARS**.

digital audio tape Its abbreviation is **DAT**. Also called **digital tape**. **1.** A helical-scan magnetic storage medium used for recording and reproduction of digitally encoded audio, usually utilizing a sampling rate of up to 48 kHz. **2.** A helical-scan magnetic storage medium used for the recording and storage of digitally encoded data. Such tapes may store from 2 to 24 gigabytes of data, and can transfer at rates of over 2 Mps.

digital barometer A barometer which indicates the measured values of pressure by means of a digital display. Such a device may indicate, in a simple and clear manner, readings expressed in kilopascals, pascals, millimeters of mercury, torr, bars, and so on.

digital broadcast satellite Same as **direct broadcast satellite**. Its abbreviation is **DBS**.

digital broadcasting satellite Same as **direct broadcast satellite**.

digital business Same as **digital commerce**.

digital cable television Same as **digital cable TV**.

digital cable TV Abbreviation of **digital cable television**. Cable TV in which compressed digitally encoded signals are transmitted. This provides for reduced noise and interference, and for many more stations to be transmitted via the same bandwidth. Digital cable TV is usually conveyed via coaxial cable, but may also be transmitted through air. Its own abbreviation is **digital CATV**.

digital camera A camera which records and stores images in digital form. An example is a CCD camera. Its abbreviation is **digicam**.

digital capacitance meter A capacitance meter which indicates the measured values of capacitance by means of a digital display.

digital cash Its abbreviation is **digicash**. Also known by various other terms, including **digital money**, **electronic cash**, and **e-money**. **1.** Any form of money used over the Internet. Used for many purposes, such as effecting online purchases or making credit card payments. Security is a key concern, and considerable measures, such as cryptography, are taken to avoid circumstances such as losses of funds, misplacement of funds, stealing of funds, or the re-use of funds. **2.** A form of currency that is digitally encoded, and that can be converted to legal tender if desired. It may be coded, for instance, through the use of an encrypted serial number. **Legal tender** is a lawfully valid currency that must be accepted for payment of debts, the purchase of goods and/or services, or for other exchanges for value. Circulating paper money is another form of legal tender.

digital CATV Abbreviation of **digital cable TV**.

digital cell phone Same as **digital cellular phone**.

digital cellphone Same as **digital cellular phone**.

digital cellular phone A cellular phone utilizing digital communications technology, such as CDMA, GSM, or TDMA. Also called **digital cellular telephone**, or **digital cell phone**.

digital cellular telephone Same as **digital cellular phone**.

digital certificate An attachment to an electronic transmission which serves to provide an authentication or assurance as to the security, identity, or validity of a message or file. A digital certificate is usually issued by a certificate authority, and serves as a reliable identification for a person, corporation, or other entity from which content is received. Also called **digital ID**, or **digital identification**.

digital channel Same as **digital circuit (2)**.

digital chart A chart which is in digital form, as opposed to paper form. Digital charts are suitable for computer processing, and facilitate use with other electronic devices, such as Electronic Chart Display and Information Systems. Also called **electronic chart**.

digital cinema The use of digital formats and/or equipment in any aspect of movie making, distribution, or presentation. For example, a movie may be prepared using computer effects and digital cameras, sent via satellite to a movie theater, and viewed using a DLP projector accompanied by digital surround sound. Its abbreviation is **d-cinema**. Also called **electronic cinema**.

digital circuit 1. A circuit which may assume one of two stables states. For instance, on or off, high voltage or low voltage, and so on. It operates like a switch, and can perform logic functions. Used extensively in computers and other logic-based devices and equipment. Also called **binary circuit**. 2. A communications channel through which digital information is transmitted bi-directionally. Also called **digital line**, or **digital channel**.

digital clock 1. A device or circuit which provides a steady stream of timed pulses. Used, for instance, as the internal timing device of a digital computer, upon which all its operations depend. 2. An instrument or device which indicates time via a digital display.

digital code Code expressed in terms of the binary digits 0 and 1, which is necessary for processing by a digital computer.

digital commerce The conducting of business utilizing electronic communications, especially computers connected to a network. Digital commerce includes the purchasing and selling of goods and services, electronic data interchange, the use of e-money, and business-related videoconferencing. It may occur over the Internet, a LAN, or any other venue for electronic communication. Also called **digital business**, **ecommerce**, or **ebusiness**.

digital communications The transmission of information, such as text, voice, or video, in binary form. An example is that sent over a T-carrier, or a digital cell phone. An analog signal may be transmitted over a digital cell line, but it first must be encoded in binary form.

digital comparator A circuit whose output depends on the comparison of its two digital input values. Used, for instance, to determine which input is greater in magnitude.

digital component video A digital signal in which the video information is carried by separate signals, as in the luminance and chrominance signals being separate components.

Examples include RGB and YCbCr. Also called **component digital video**.

digital composite video A digital video signal combining the color picture signal, plus all blanking and synchronizing signals. Examples include digitally encoded PAL and NTSC. Also called **composite digital video**.

digital compressed video Video which is encoded so as to occupy less space and/or bandwidth. Two of the benefits afforded by this are the ability to send more information using the same bandwidth, and improved definition. Also called **digitally-compressed video**.

digital compression Same as **data compression**.

digital computer A computer whose input, processing, storing, and output is in binary form. Almost all computers are of this type. This contrasts with an **analog computer**, which processes infinitely variable signals, such as a voltage.

digital computer system The complete complement of components required for a **digital computer** to function. These include the CPU, keyboard, mouse, monitor, memory, storage mediums, cables, and so on, which comprise the hardware of the computer itself, plus any necessary peripheral devices. In addition, a digital computer system incorporates the operating system.

digital content Digital information, especially that which is available online, which may be any combination of audio, video, text, files, or the like.

digital convergence The integration of computers, communications, and consumer electronics. For example, a TV accessing a network computer to program the thermostat of an air-conditioner hour-by-hour, or downloading a limited-release foreign movie to a DVD+RW disc via satellite.

digital converter A circuit or device that transforms a digital input to another form, such an analog output.

digital counter 1. A counter which tallies discrete events. 2. A counter whose indications are shown on a digital display, such as an LCD.

digital cross-connect system A digital switching system in which circuit paths are connected and re-routed utilizing software, eliminating the need for manual changes in the interconnections. Such a system is used in local, long-distance, and broadband communications, and may accommodate T1, T3, or SONET circuits, among others. Its abbreviation is **DCS**, or **DCC**. Also called **digital access cross-connect system**.

digital data Information presented in binary form, which is necessary for a digital computer to be able process it.

digital data service A dedicated communications line for digital circuits such as T1 or T3. To connect to such a service, a device such as a channel service unit/data service unit must be used. Its abbreviation is **DDS**.

digital data storage Information stored in a digital format, such that on a computer hard disk. Its abbreviation is **DDS**.

digital data transmission The sending and/or receiving of digitally coded information.

digital delay circuit Same as a **digital delay line**.

digital delay line Also called **digital delay circuit**, or **digital delayer**. 1. A circuit or device which introduces a time delay in a digital signal, such as digitally encoded audio. One mechanism employed is to store the signal for the desired delay interval. 2. A circuit or device which converts and analog signal into a digital signal, introduces a time delay, and then converts the signal back into analog form.

digital delayer Same as a **digital delay line**.

digital device 1. A device which may assume on of two stable states. 2. A device whose input and/or output is in digital form.

digital differential analyzer A digital computer which performs integration. Such a computer may have multiple parallel computing elements. Its abbreviation is **DDA**.

digital display Also called **digital display device**. **1.** A display whose output is in the form of digits, or numbers. Also called **digital readout**, or **numeric display**. **2.** A video display which accepts digital signals from a computer.

digital display device Same as **digital display**.

digital divider A circuit or device whose output is the quotient of two or more digital inputs.

digital domain All that is in digital form. Most everything pertaining to computers and data is in digital form. A digital-to-analog converter transforms a digital input into an analog output, and an analog-to-digital converter does the converse.

digital effects The use of computers and computer graphics to enhance, modify, add, or remove sights and/or sounds in movies, TV, and other multimedia presentations. Such effects are extensively used, and some of the techniques employed include motion capture and tweening. Also called **computer special effects**, or **computer effects**.

digital electronics Electronic devices and instruments that deal with signals having discrete number of values, such as two. The resulting data may be displayed, for instance, on a digital meter.

Digital-Enhanced Cordless Telephone A cordless telephone incorporating digital enhancements, such as higher speech quality, greater security, and compatibility between handsets and bases of different types and/or manufacturers. Its acronym is **DECT**. Also called **Digitally-Enhanced Cordless Telephone**.

digital envelope An encryption method in which both public and secret keys are used. The message is encoded with secret-key encryption, and the key to decode the message is encrypted using a public key. Also refers to a message sent using this encryption method.

digital film Any storage unit, such as a flash card, utilized to hold images taken by digital cameras.

digital filter A filter which performs actions on digital inputs, as opposed to analog inputs. For example, such a filter may receive its input from an analog-to-digital converter, process the digital signal, and deliver its output to a digital-to-analog converter for reconversion.

digital flat panel port An interface specially designed for a flat panel display to be connected to it. Since there is no need for analog-to-digital or digital-to-analog conversions, the images load faster, and are crisper. Its abbreviation is **DFP port**.

digital form Also called **digital format**. **1.** Information presented in terms of the binary digits 0 and 1, which is necessary for processing by a digital computer. **2.** Information from an analog source which has been assigned a discrete numeric value.

digital format Same as **digital form**.

digital frequency meter A frequency meter whose output is presented via a digital display, such as an LCD.

digital frequency modulation A form of frequency modulation in which the modulating wave shifts the output frequency between fixed values. Switching between these values provides the keying. May be used, for instance, by modems transmitting over telephone lines, by representing the 0 with a lower frequency, and a 1 with a higher frequency. Also called **frequency-shift keying**, or **frequency-shift modulation**.

digital hearing aid A hearing aid that utilizes a chip to process and amplify sounds. Such devices, in comparison to traditional analog hearing aids, produce much less distortion, have higher fidelity, have advanced sound processing, can

be programmed to enhance only the needed frequency intervals, and so on, all in a much smaller space.

digital home A residence with devices and networks such as smart appliances and home networks, which automate everyday household things, such as use of heating and/or cooling, appliances, entertainment devices, security systems, and so on.

digital IC Abbreviation of **digital integrated circuit**. An IC which processes signals which have only two states, and that is used for functions such as switching. This contrasts with a **linear IC**, which is used primarily for analog operations.

digital ID Same as **digital certificate**.

digital identification Same as **digital certificate**.

digital information **1.** Information presented in terms of the binary digits 0 and 1, which is necessary for processing by a digital computer. **2.** Information presented in digital form.

digital ink The coating, such as a flexible array of transistors, of a sheet of **digital paper**. Also called **electronic ink**.

digital input An input signal represented by discrete states which correspond to a fixed number of digits. For example, such an input may consist of two voltage levels, such as high and low, which represent the numbers 0 and 1. An example is the input from a DVD player.

digital instrument An instrument which indicates the measured values by means of a digital display. For instance, a digital multimeter. Also called **digital measuring instrument**.

digital integrated circuit Same as **digital IC**.

digital integrator A device which performs integrations a provides a digital output.

digital light processing The technology employed to control a **digital micromirror device**. Its abbreviation is **DLP**.

digital light-processing projector A projector which utilizes **digital light processing**. Its abbreviation is **DLP projector**.

digital light processor A processor which controls a **digital micromirror device**. Its abbreviation is **DLP**.

digital line Same as **digital circuit (2)**.

digital line protection **1.** Protection of the data transmitted over a digital line. **2.** Protection of equipment, such as modems, connected to digital lines. Digital lines have a higher voltage than analog lines.

digital logic Logic based on Boolean algebra, which is an algebraic system in which variables can have only one of two values, such as 0 or 1, or true or false, and whose primary operations are AND, OR, and NOT, analogous to the way ADD, SUBTRACT, MULTIPLY, and DIVIDE are the primary operations of arithmetic. Digital circuits are based on digital logic, and digital computers are based on digital circuits. Also called **Boolean logic**.

digital logic circuit A circuit which carries out a logic function or operation. Such a circuit has multiple inputs and one output. Its output will depend on specified input conditions, such as a given combination of states. The three basic logic circuits are the AND, OR, and NOT gates. Also called **logic circuit**.

digital loop carrier The equipment, including the lines, which enable the digital multiplexing of two analog lines onto a single twisted copper pair. Used, for example, to provide a single telephone line with two available dial tones. Best suited for voice applications. Its abbreviation is **DLC**.

digital loopback A feature in transmission equipment which enables a terminal to redirect outgoing signals back to itself. Used, for instance, to verify circuit continuity.

digital measuring instrument Same as **digital instrument**.

digital micromirror device A chip fitted with arrays of micromirrors which are used to produce sharp images on a screen utilized in a room with ordinary lighting. Such a chip may have millions of movable mirrors, each representing a pixel on the screen, with each mirror tilting one way for a lit pixel, and the other way for an unlit pixel. Such devices are controlled by a digital light processor, and may be used, or instance, in projection-screen TVs, or in presentation projectors. Its abbreviation is **DMD**.

digital microwave A microwave system through which digital signals are sent. Such a system is available for transmission utilizing DS3, T3, STS-1, and other similar circuits.

digital microwave radio The use of microwave communication links to transmit digital signals, such as voice, video, and data. Also a device used for such communications. Its abbreviation is **DMR**.

digital modem A modem, such as an ISDN modem, which transmits digital information without digital-to-analog, or analog-to-digital conversions. A digital modem sends and receives data over digital lines, while an **analog modem** uses analog lines.

digital modulation The modulation of a carrier wave as a function of two or more discrete states of the modulating signal. Examples include amplitude shift keying, and pulse code modulation.

digital money Same as **digital cash**.

digital monitor A video monitor which only accepts a digital input. The visual output of such a display is analog.

digital multimeter A multimeter whose indications are shown on a digital display, such as an LCD. Aside from measuring voltage, resistance, and current, some also offer additional functions, such as performing continuity tests, or testing diodes. Such meters are generally more sensitive and accurate than their analog counterparts. Its abbreviation is **DMM**.

digital multiplexer A device which combines multiple digital signals into a single bit stream. Also spelled **digital multiplexor**.

digital multiplexing The combination of multiple digital signals into a single bit stream, utilizing a **digital multiplexer**. Digital multiplexing and demultiplexing has several advantages in comparison to analog multiplexing and demultiplexing, including lesser degradation, the ability to use digital compression, and better suitability for the transmission of computer data.

digital multiplexor Same as **digital multiplexer**.

digital multiplier A circuit or device whose output is the product of two or more digital inputs.

digital ohmmeter An ohmmeter whose indications are shown on a digital display, such as an LCD. Its abbreviation is **DOM**.

digital optical input A digital input utilizing a fiber-optic cable, as opposed, for instance, to a coax cable.

digital optical output A digital output utilizing a fiber-optic cable, as opposed, for instance, to a coax cable.

digital oscilloscope An oscilloscope that incorporates an analog-to-digital converter, for transforming its analog input into a digital output. Such an oscilloscope enables permanent signal storage, and extensive waveform analysis and processing. Also called **digital storage oscilloscope**, or **storage oscilloscope** (1).

digital output An output signal represented by discrete states which correspond to a fixed number of digits. For instance, such an output may consist of two voltage levels, such as high and low, which represent the numbers 0 and 1. An example is the output of a DVD player.

digital PABX Same as **digital PBX**.

digital panel meter Its abbreviation is **DPM**. **1.** A panel meter whose output is in the form of digits. **2.** A panel meter which incorporates an analog-to-digital converter which enables it to present an output in the form of digits.

digital paper A storage medium analogous to regular paper, except that it can be written upon or refreshed numerous, often millions, of times via electronic means. Such paper may be made of a plastic coated with the proper electronics, and can be used, for instance, in ebooks, newspapers that update their content via satellite, portable signs, or the like. Digital paper may or may not be fully flexible, and/or offer full color images. Also called **electronic paper**.

digital PBX A PBX that utilizes digital signals. Used, for instance, with digital communications circuits such as T3 or DS3. Also called **digital PABX**.

digital phase shifter A phase shifter whose control pulse is digital.

digital phone Abbreviation of **digital telephone** (1).

digital photography The use of digital cameras and computers to obtain, process, and reproduce photographs.

digital photometer A photometer whose indications are shown on a digital display, such as an LCD.

digital postcard A digital greeting card sent over the Internet. There are many applications and Web sites which can be utilized to prepare such cards, and such greetings may incorporate text, still images, video, animation, audio, and so on, and can be chosen for virtually any occasion or purpose. Also called **electronic card**, or **virtual card**.

digital power meter A power meter whose indications are shown on a digital display, such as an LCD.

digital private network signaling system A signaling standard utilized within or between digital PBXs. Its abbreviation is **DPNSS**.

digital radar A radar system in which computers process the received signals, in contrast to an **analog radar**, which does not incorporate such processing. Digital radars are less susceptible to interference due to atmospheric conditions, and are able to filter undesired components, such as cars and trucks, which might be confused with desired objects, such as aircraft.

digital radio Radio programming or technology, such as a digital audio broadcast, that is digitally encoded.

digital readout Same as **digital display** (1).

digital recording The recording of text, graphics, audio, or video, in which the information is encoded in binary form onto the medium. Examples include digital recordings on discs and tapes.

digital remote control **1.** A remote control which emits a digital signal. Such devices are usually programmable. **2.** A transmitter which controls a digital device, such as a DVD, from a distance.

digital representation Representation of information, such as signals and data, using discrete values. For instance, that which is binary coded.

Digital-S Same as **D9 Digital Video**.

digital satellite radio Radio programming arriving via satellite signals. Also, a technology enabling such a service. Also called **satellite radio**.

digital satellite system A satellite system utilized to transmit digital content. Such content usually includes TV programming, and may also incorporate radio, Internet, and so on. Its signal is usually received by 18 inch antennas, which feed a converter box, which in turn provides the input to other devices, such as TVs, audio systems, or computers. Its abbreviation is **DSS**.

digital selective calling A system for communicating with one or more ships by means of a digitally encoded radio signal. Its abbreviation is **DSC**.

digital service Same as **digital signal** (2).

digital service unit Same as **data service unit**. Its abbreviation is **DSU**.

digital service unit/channel service unit Same as **data service unit/channel service unit**. Its abbreviation is **DSU/CSU**.

digital signal **1.** A signal which is in the form of discrete states which correspond to a fixed number of digits. For example, such a signal may consist of two voltage levels, such as high and low, which represent the numbers 0 and 1. This contrasts with **analog signals**, which can vary in a continuous and infinitely variable manner within any give interval. **2.** A classification system describing the transmission rates and other characteristics of digital communications circuits. For instance, digital signal level 2, or DS2, corresponds to a data signaling rate of 6.312 Mbs. Its abbreviation is **DS**. Also called **digital service**.

digital signal level 0 Its abbreviation is **DS0**. A digital transmission format with a signaling rate of 64 Kbps, which is equivalent to a capacity of 1 voice channel. All DS formats with a higher capacity are based on multiples of DS0.

digital signal level 1 Its abbreviation is **DS1**. A digital transmission format with a signaling rate of 1.544 Mbps, which is equivalent to a capacity of 24 voice channels. It is equivalent to a 24 DS0 channels, or a **T1** circuit.

digital signal level 1C Its abbreviation is **DS1C**. A digital transmission format with a signaling rate of 3.152 Mbps, which is equivalent to a capacity of 48 voice channels, 2 DS1 channels, or a **T1C** circuit.

digital signal level 2 Its abbreviation is **DS2**. A digital transmission format with a signaling rate of 6.312 Mbps, which is equivalent to 4 DS1 channels, 96 voice channels, or a **T2** circuit.

digital signal level 3 Its abbreviation is **DS3**. A digital transmission format with a signaling rate of 44.736 Mbps, which is equivalent to 28 DS1 channels, 672 voice channels, or a **T3** circuit.

digital signal level 4 Its abbreviation is **DS4**. A digital transmission format with a signaling rate of 274.176 Mbps, which is equivalent to 168 DS1 channels, 4032 voice channels, or a **T4** circuit.

digital signal level 5 Its abbreviation is **DS5**. A digital transmission format with a signaling rate of 400.352 Mbps, which is equivalent to 240 DS1 channels, 5760 voice channels, or a **T5** circuit.

digital-signal processing The processing of analog information which has been converted into digital form. For example, sound or images which go through an analog-to-digital conversion, and are thus better suited for manipulation and analysis. Applications include its use telecommunications, and in music processing. Its abbreviation is **DSP**.

digital-signal processing chip Same as **digital-signal processor**.

digital-signal processor Its abbreviation is **DSP**. Also called **digital-signal processing chip**. **1.** A chip designed specifically for **digital-signal processing**. Such chips are especially fast. **2.** A chip designed specifically for the processing of **digital signals** (1). Such chips are especially fast.

digital signature An attachment to an electronic transmission which uniquely identifies and authenticates the sender of a message, file, or other information such as a credit card number. To be effective, a digital signature must not be forged, and measures such as public-key cryptography are employed to insure its integrity. Also called **electronic signature**, or **signature** (2).

Digital Signature Standard A NIST standard used for authentication of **digital signatures**. Its abbreviation is **DSS**.

digital simultaneous voice and data A digital technology which enables compressed voice and data signals to be transmitted simultaneously over a single analog telephone line. Its abbreviation is **DSVD**.

digital sound Sound which is digitally recorded, manipulated, and reproduced. Digital sound has many advantages, including almost unlimited editing possibilities, such as layering and speed-variation for each millisecond of audio, the ability to be dubbed as many times as desired without signal degradation, and extended dynamic range and minimal distortion during playback.

digital speech **1.** Speech which has been encoded digitally. **2.** Speech which is generated by a computer. The sounds are produced either by linking phonemes together, or by drawing from a database containing recorded words. Current technology provides for such speech to be almost completely natural. Its applications include assistance for those with reduced vision, or for retrieving email over any telephone. Also called **digital speech synthesis, speech synthesis**, or **voice synthesis**.

digital speech synthesis Same as **digital speech** (2).

digital storage oscilloscope Same as **digital oscilloscope**. Its abbreviation is **DSO**.

digital subscriber line Same as **DSL**.

digital subscriber line access multiplexer Same as **DSLAM**.

digital subscriber loop Same as **DSL**.

digital switching **1.** Switching of digital signals in a communications network. **2.** The use of digital equipment for switching in a communications network.

digital tape Same as **digital audio tape**.

digital telephone **1.** A telephone enabled for two-way digital voice communications. Such a telephone may have enhanced features, such as improved speech quality, greater security, and multiple programming options. When used over analog lines, a digital-to-analog conversion is made prior to transmission, while an analog-to-digital conversion is made after reception. Its abbreviation is **digital phone**. **2.** A telephone system equipped for digital communication between **digital telephones** (1). Also called **digital telephone system**, or **digital telephony**.

digital telephone system Same as **digital telephone** (2).

digital telephony Same as **digital telephone** (2).

digital television Same as **digital TV**.

digital television converter Same as **digital TV converter**.

digital television receiver Same as **digital TV receiver**.

digital television system Same as **digital TV** (1).

Digital Theater Sound Same as **DTS Surround Sound**. Its abbreviation is **DTS**.

digital thermometer A thermometer which indicates the measured values of temperature by means of a digital display. Such a device indicates accurate temperature readings simply and clearly.

digital-to-analog conversion The conversion of a digital input into an analog output.

digital-to-analog converter A circuit or device that transforms a digital input to an analog output. For instance, a device whose input is a sequence of binary digits, and whose output is an analog voltage waveform. Its abbreviation is **DAC**, or **d/a converter**.

digital-to-synchro converter A circuit or device that transforms a digital input into a synchro output signal. Its abbreviation is **DSC**, or **d/s converter**.

digital transmission The transmission of digital signals. For example, data sent as a bit stream.

digital transmission system A system through which digital signals are transmitted and received. Analog signals may be sent and received with the use of an analog-to-digital converter prior to transmission, and a digital-to-analog converter after reception.

digital TV Abbreviation of **digital** television. Its own abbreviation is **DTV**. Also called **digital advanced TV**. **1.** A TV system which employs digital encoding in the transmission and reception of TV signals. Before transmission, signals are digitally compressed, to incorporate picture and sound of superior quality. This information is decompressed by the receiving device before viewing. Direct broadcast satellite systems, for instance, use this technology. Also called **digital TV system**. **2.** A device, such as a TV receiver, which is compatible with **digital TV (1)**. An example of such a device is an HDTV receiver.

digital TV converter A device which converts a digital TV signal. For example, such a device may serve to convert from one digital standard to another, to decode programming for those authorized to receive it, or to decompress a signal before viewing.

digital TV receiver A TV receiver which is compatible with **digital TV (1)**.

digital TV system Same as **digital TV (1)**.

digital versatile disc Same as **DVD**.

digital versatile disc player Same as **DVD player**.

digital video Video which is digitally encoded. Examples include that in DVDs or CD-ROMs. Its abbreviation is **DV**.

digital video broadcasting Its abbreviation is **DVB**. International digital broadcasting standards which utilize existing communications infrastructures such as satellite and cable. DVB has is open system, enabling greater flexibility in covering a wider array of services, and in simplifying the incorporation of new technologies.

digital video camera A video camera which records and stores images in digital form. An example is a CCD video camera.

digital video compression The encoding of video data so that it occupies less room. This helps maximize the use of storage space, facilitates faster transfer, and also reduces transmission bandwidth. Also called **video compression**.

digital video disc Same as **DVD**.

digital video disc player Same as **DVD player**.

digital video effects The use of computers to enhance, modify, add, or remove images in movies, TV, and other multimedia presentations. Such effects are extensively used, and some of the techniques employed include motion capture and tweening. Its abbreviation is **DVE**. Also called **video effects**.

digital volt-ohm-milliammeter A volt-ohm-milliammeter whose indications are shown on a digital display, such as an LCD. Its abbreviation is **DVOM**.

digital voltmeter A voltmeter whose indications are shown on a digital display, such as an LCD. Its abbreviation is **DVM**.

digital wallet Encryption software which serves to provide the virtual equivalent of a wallet. In it, may be contained digital cash, credit card information, shipping details, and a digital certificate for authentication of the wallet holder. Both merchants and customers benefit from the added security, expedience, and convenience. Also called **virtual wallet**, or **electronic wallet**.

digital watch A watch which indicates time via a digital display. Such watches usually use an LCD display, and may be programmable.

digital watermark Patterns of bits which are incorporated into digital intellectual property, such as voice, video, and/or text, to identify its source. The bits contain information such as the copyright holder, or its intended area of distribution. While ordinary watermarks are usually intended to be visible, digital watermarks are designed to be imperceptible. Such patterns, for instance, must be encoded into a CD or DVD without affecting sounds or images. Also called **watermark**.

digital wattmeter A wattmeter whose indications are shown on a digital display, such as an LCD.

digitally-compressed video Same as **digital compressed video**

digitally encoded That which is in **digital code**.

Digitally-Enhanced Cordless Telephone Same as **Digital-Enhanced Cordless Telephone**. Its acronym is **DECT**.

digitization The conversion of analog information into digital form, which is necessary for a digital computer to process it.

digitize To convert analog information into digital form, which is necessary for a digital computer to process it. A device such as an analog-to-digital converter is used.

digitized An analog input that has been converted into digital form, which is necessary for processing by a digital computer. An example is satellite programming.

digitized image An image which has been converted into digital form, which is necessary for processing by a digital computer. An example is a bitmap. Also called **electric image**.

digitized information Information which has been converted into digital form, which is necessary for processing by a digital computer. This information can be from any analog source, such as a voice or a voltage.

digitized music Music which is converted from an analog source, such as a voice and a guitar, into digital form, such as that found on a CD. The larger the sample of the analog source, and the higher the sampling rate, the more accurate the conversion.

digitized picture A picture which has been converted into digital form, which is necessary for processing by a digital computer. An example is a bitmap.

digitized sound Sound, such as a voice, which has been converted into digital form, such as that found on a CD. The larger the sample of the analog source, and the higher the sampling rate, the more accurate the conversion.

digitized speech Speech which has been converted into digital form, such as that found on a CD. The larger the sample of the analog source, and the higher the sampling rate, the more accurate the conversion.

digitized voice Voice which has been converted into digital form, such as that found on a CD. The larger the sample of the analog source, and the higher the sampling rate, the more accurate the conversion.

digitizer **1.** A device which converts an analog input into digital form, which is necessary for processing by a digital computer. Also called **analog-to-digital converter**. **2.** Same as **digitizing tablet**.

digitizer tablet Same as **digitizing tablet**.

digitizing pad Same as **digitizing tablet**.

digitizing tablet An input device which serves for sketching or tracing images directly into a computer. It consists of an electronic pressure-sensitive tablet which senses the position of the stylus which makes contact with it. The stylus, or puck, is similar to a mouse, but is much more accurate, because its location is determined by touching the active surface of the tablet with an absolute reference. Used, for instance, in computer-aided design. Also called **digitizer tablet**, **digitizer (2)**, **digitizing pad**, **data tablet**, **graphics tablet**, **touch tablet**, or **tablet**.

digitron A display in which all the digits appear in a straight line, as found in calculators or cash registers. Also called **digitron display**.

digitron display Same as **digitron**.

Dijkstra's Algorithm An algorithm utilized for calculating network routing, using comprehensive routing tables. Also called **Shortest-Path First Algorithm**.

dimension **1.** A measure in a given direction, such as length or width. **2.** The number of coordinates required to specify a location in space. Also, the range of such a coordinate. **3.** A fundamental unit, such as ampere, or a derived unit, such as coulomb. Also, the magnitude of such a unit. **4.** Within a series of ordered data elements, a statement which defines a parameter such as its height or width.

dimensional analysis A technique for the verification of an equation through the analysis of the units of measurement, or dimensions, of the variables it contains. To be verified as correct, the units on one side of the equation must be shown to be identical to those on the other.

dimensionless number A number with no associated unit of measure. Examples include ratios of two numbers utilizing the same unit, and logarithms

dimensionless quantity A quantity with no associated unit of measure. Examples include ratios of two numbers utilizing the same unit, and logarithms.

DIMM Acronym for **d**ual **i**nline **m**emory **m**odule. A small circuit board which holds computer memory chips. A DIMM is similar to a **SIMM**, but the former has 64-bit path, while the latter has a 32-bit path. Also called **DIMM module**.

DIMM memory RAM in the form of **DIMMs**.

DIMM module Same as **DIMM**.

DIMM socket A socket on a motherboard into which **DIMMs** are plugged-in.

dimmed On a computer display, a function, operation, or option whose letters or icon is lighter, grayed, or fuzzy, indicating that it is unavailable. For example, the cut function should be dimmed when no content is selected. Also known as **grayed out**.

dimmer A circuit or device which serves to control the intensity of illumination of a light source, such as an incandescent lamp. There are various types, including those employing silicon-controlled rectifiers or rheostats. Also called **light dimmer**.

DIN Acronym for *Deutsches Industrie Norm*.

DIN connector A connector conforming to **DIN** standards. Such connectors are used in various types of electronic products, including computers and video equipment.

DIN jack Same as **DIN socket**.

DIN plug A connector which plugs into a **DIN socket**. A DIN plug has multiple protruding pins.

DIN socket A receptacle for use with a **DIN plug**. Also called **DIN jack**.

diode A two-terminal active electronic device consisting of an anode and a cathode, and which passes current only in one direction. Most diodes are semiconductor devices, and generally consist of a pn junction. Diodes have various applications, including their use in rectifiers, filters, sensors, modulators, and amplifiers. Examples include LEDs, Gunn diodes, Read diodes, and point-contact diodes.

diode AC switch Same as **diac**.

diode alternating-current switch Same as **diac**.

diode amplifier A microwave amplifier that uses a special diode, such as an IMPATT diode, in a tuned cavity. Such amplifiers feature high gain, low noise, high reliability and depending on the diode used, have center frequencies from about 1 GHz to over 300 GHz.

diode array A group diodes contained in a single device. Used, for instance, in some lasers. Also called **diode assembly**.

diode assembly Same as **diode array**.

diode bias A DC voltage applied to a diode to establish a reference level for its operation.

diode bridge A circuit consisting of four diodes in a bridge configuration. Generally utilized as a rectifier whose output polarity is constant, regardless of the input polarity.

diode characteristic For a diode, a curve which describes the relationship between current and voltage.

diode clamp Same as **diode clamping circuit**.

diode clamping circuit A clamping circuit which uses a diode to establish a reference DC level. Also called **diode clamp**.

diode clipper A clipper circuit which utilizes diodes. May be used, for instance, to limit both positive and negative voltage swings in relation to a reference voltage. Also called **diode clipping circuit**, or **diode limiter**.

diode clipping circuit Same as **diode clipper**.

diode-connected transistor A transistor in which two terminals are connected together, to provide diode function. Used, for instance, in a current mirror.

diode current The current flowing through a diode. Current flows when a forward bias is applied. A reverse bias produces a minimal leakage current, until breakdown occurs.

diode demodulator Same as **diode detector**.

diode detector A demodulator circuit which incorporates one or more diodes, and which provides a rectified output whose average value is proportional to that of the original modulation. Also called **diode demodulator**, or **envelope detector**.

diode drop Same as **diode forward voltage**.

diode forward voltage The voltage across the electrodes of a diode carrying current in the forward direction, that is, from anode to cathode. Also called **diode drop**, **diode voltage**, or **forward voltage drop (2)**.

diode gate A logic circuit, such as an AND gate, which utilizes diodes as switching elements.

diode laser A laser in which a forward-biased pn junction diode is used to convert its DC input into a coherent light output. Used, for instance, as a light-pulse generator for transmission of information over fiber-optic lines. An example is a gallium arsenide laser. Also called **diode light source**, **semiconductor laser**, **laser diode**, **injection laser**, **injected laser**, or **injection laser diode**.

diode light source Same as **diode laser**.

diode limiter Same as **diode clipper**.

diode logic Logic circuitry which utilizes diodes as switching elements. Examples include AND and OR circuits.

diode matrix A two-dimensional array of diodes utilized to determine the operation of a circuit. Used, for instance, for code conversion.

diode mixer A circuit which employs a diode to mix frequencies. It makes use of the nonlinear characteristics of diodes, and may be directly incorporated into an RF transmission line.

diode modulator A modulator which employs diodes to combine a modulating signal with a carrier signal.

diode oscillator An oscillator which makes use of the negative resistance or breakdown characteristics of certain diodes, such as Gunn diodes. Used, for instance, at microwave frequencies.

diode pack A device which incorporates two or more diodes, which form a single unit.

diode peak detector A circuit which incorporates a diode for the detection of signal peaks.

diode-pentode A vacuum tube incorporating both a diode and a pentode within the same envelope.

diode probe A probe incorporating one or more diodes. Used, for instance, for demodulation.

diode rectification The use of a diode for conversion of AC into DC.

diode rectifier A circuit or device incorporating one or more diodes, for conversion of AC into DC.

diode resistor A resistor which incorporates one or more diodes.

diode switch A diode in which forward or reverse biasing voltages are applied to the anode, so as to pass or block current, and thus provide switching action. Used, for instance, in microwave applications.

diode-transistor logic A logic circuit in which diodes and transistors are utilized to perform logic functions. For example, the input signal in such a circuit may enter through a diode, while the output is taken from the collector of an inverting transistor. Its abbreviation is **DTL**.

diode-triode A vacuum tube incorporating both a diode and a triode within the same envelope.

diode tube An electron tube with two electrodes, specifically an anode and a cathode. In such a tube the cathode may also be called filament and the anode may be called plate. Also called **tube diode**.

diode voltage Same as **diode forward voltage**.

diode voltage regulator A voltage regulator which utilizes the constant-voltage characteristic and well-defined reverse-breakdown voltages of Zener diodes. Also called **Zener diode voltage regulator**.

DIP 1. Acronym for **dual in-line package**. 2. Abbreviation of **document image processing**.

dip 1. To immerse briefly in a liquid, so as to coat or saturate, as in dip coating. 2. An angle or slant in reference to a horizontal plane. 3. For a given location on the surface of the earth, the angle between the horizontal plane and the direction of the lines of force of the earth's magnetic field. Dip is 0° at the magnetic equator and 90° at each of the magnetic poles. Also called **magnetic dip**, **magnetic inclination**, or **inclination (2)**.

dip coating The application of a layer by immersion in a liquid or an amorphous solid. For example, the application of a plastic protective layer by dipping a component into molten plastic, followed by cooling or heating to solidify. Used, for instance, to apply a liquid photoresist. Also called **dipping**.

dip equator A great line along the surface of the earth that connects all points at which the magnetic dip is 0°. This contrasts with the **geomagnetic equator**, which is a great circle which at each point is 90° from the geomagnetic poles. Also called **magnetic equator**, or **aclinic line**.

DIP package Abbreviation of **dual in-line package**.

dip plating Plating accomplished by immersion of the surface to be plated in a solution where deposition occurs through chemical means. It is a form of electroless plating. Also called **immersion plating**.

dip pole Either of the two locations on the surface of the planet earth where the magnetic dip is 90°. At such a place, the magnetic meridians converge. In the northern hemisphere such spot is called the **north magnetic pole**, while that of the southern hemisphere is the **south magnetic pole**. The earth's magnetic poles do not coincide with its geographic poles, and the latter's position vary over time. Also called **magnetic pole (1)**, or **pole (3)**.

DIP switch One of several small rocker switches incorporated into a **dual in-line package**.

diplex operation The concurrent use of a common element, such as a single antenna or carrier wave, for the transmission or reception of two simultaneous and independent signals.

diplex reception The concurrent use of a common element, such as a single antenna or carrier wave, for the reception of two simultaneous and independent signals.

diplex transmission The concurrent use of a common element, such as a single antenna or carrier wave, for the transmission of two simultaneous and independent signals.

diplexer A device which allows two transmitters to use the same antenna simultaneously or alternately. Used, for example, in TV broadcasting to transmit the audio and video content via the same antenna. Also called **antenna diplexer**.

dipolar Pertaining to a **dipole**.

dipole 1. Two opposite charges of equal magnitude separated by a very small distance. Also, a body, such as a molecule, containing such charges. Also called **doublet (1)**, **electric dipole**, or **electric doublet**. 2. Two equal magnetic poles with opposite polarity separated by a very small distance. Also called **doublet (2)**, **magnetic dipole**, or **magnetic doublet**. 3. Same as **dipole antenna**.

dipole antenna A center-fed antenna usually consisting of a straight radiator half a wavelength long, which is split at its electrical center. Its maximum intensity of radiation is perpendicular to its axis, and such antennas are used mostly for radio-frequencies below 3 GHz. Also called **dipole (3)**, **dipole radiator**, **doublet antenna**, or **half-wave dipole**.

dipole array An antenna incorporating multiple **dipole radiators**.

dipole moment 1. For a **dipole (1)**, the product of the magnitude of one of the charges, and the distance between the centers of the charges. May be expressed in coulomb-meters or abcoulomb-centimeters. Also called **electric dipole moment**, or **electric moment**. 2. For a magnet placed in a magnetic field, the maximum torque experienced by the magnet divided by the magnitude of the magnetic field acting on said magnet. May be expressed in joules/tesla. Also called **magnetic dipole moment**, or **magnetic moment**.

dipole radiator Same as **dipole antenna**.

dipping Same as **dip coating**.

dipulse A form of binary code transmission in which a 1 is represented by a pulse, and a 0 by the absence of a pulse.

direct access The ability to access information directly, as opposed to sequentially. Computer memory and disk drives, for instance, provide direct access. Also called **immediate access (3)**.

direct-access file A file within which individual items can be accessed in any order. Also called **random-access file**.

direct-access storage device Its acronym is **DASD**. A computer storage device whose content can be accessed directly, as opposed to sequentially. For example, a disk drive is a DASD, while a tape drive is not.

direct address A specific memory location. It is an address from which relative addresses may be derived. Also called **absolute address**, **actual address (1)**, **machine address**, or **real address (1)**.

direct addressing An address mode utilizing **direct addresses**.

direct-band-gap semiconductor A semiconductor in which the minimum energy in the conduction band coincides with the maximum energy in the valence band, so transitions between these bands only require a change in energy, and none in momentum. This enables optical transitions between the two bands, so these semiconductors may be used, for instance, in lasers and LEDs. An example is gallium arsenide. Also called **direct-gap semiconductor**.

direct broadcast by satellite Same as **direct broadcast satellite (1)**.

direct broadcast satellite Its abbreviation is **DBS**. Also called **direct broadcasting satellite**, or **digital broadcast satellite**. 1. A technology in which compressed digital signals, containing broadcasts, are transmitted from a satellite with a geostationary orbit, and received by small dish antennas. Also called **direct broadcast by satellite.** **2.** An artificial satellite used for such transmissions.

direct broadcast satellite radio A service in which digitally encoded radio broadcasts are transmitted from satellites or repeater stations. Also called **satellite digital audio radio service.**

direct broadcasting satellite Same as **direct broadcast satellite.**

direct cable connection A feature which enables two computers to exchange data via a direct cable, as opposed to an active interface device such as a modem. Also, such a connection. Its abbreviation is **DCC.**

direct capacitance The capacitance between two conductors.

direct chip attach A configuration in which a chip is mounted directly on a PCB. Also, a technology or technique utilized for such mounting. Its abbreviation is **DCA.** Also called **chip-on-board.**

direct control The automatic control of one device by another. For instance, a computer controlling a peripheral device.

direct-conversion receiver An RF receiver in which the frequency of the local oscillator signal is the same as the carrier frequency of the incoming signal. It is a type of heterodyne receiver, but there is no intermediate frequency.

direct-coupled amplifier An amplifier in which multiple stages are coupled directly, or through a resistor. Used mostly for amplifying DC signals. Its abbreviation is **DC amplifier (2).**

direct-coupled transistor logic Circuit logic in which transistors are directly connected to each other. Its abbreviation is **DCTL.**

direct coupling Same as **DC coupling.**

direct current Same as **DC.**

direct-current amplifier Same as **DC amplifier.**

direct-current block Same as **DC block.**

direct-current circuit Same as **DC circuit.**

direct-current component Same as **DC component.**

direct-current converter Same as **DC converter.**

direct-current coupling Same as **DC coupling.**

direct-current dump Same as **DC dump.**

direct-current equipment Same as **DC equipment.**

direct-current erase Same as **DC erase.**

direct-current erasing Same as **DC erasing.**

direct-current generator Same as **DC generator.**

direct-current inverter Same as **DC inverter.**

direct-current leakage Same as **DC leakage.**

direct-current motor Same as **DC motor.**

direct-current noise Same as **DC noise.**

direct-current offset Same as **DC offset.**

direct-current offset voltage Same as **DC offset voltage.**

direct-current overcurrent relay Same as **DC overcurrent relay.**

direct-current permanent-magnet motor Same as **DC permanent-magnet motor.**

direct-current picture transmission Same as **DC picture transmission.**

direct-current plate current Same as **DC plate current.**

direct-current plate resistance Same as **DC plate resistance.**

direct-current plate voltage Same as **DC plate voltage.**

direct-current power Same as **DC power.**

direct-current power source Same as **DC power source.**

direct-current power supply Same as **DC power supply.**

direct-current reinsertion Same as **DC restoration.**

direct-current relay Same as **DC relay.**

direct-current resistance Same as **DC resistance.**

direct-current restoration Same as **DC restoration.**

direct-current restorer Same as **DC restorer.**

direct-current short Same as **DC short.**

direct-current signaling Same as **DC signaling.**

direct-current source Same as **DC source.**

direct-current-to-alternating-current converter Same as **DC–AC converter.**

direct-current-to-alternating-current inverter Same as **DC–AC converter.**

direct-current-to-alternating-current power converter Same as **DC–AC power converter.**

direct-current-to-direct-current converter Same as **DC–DC converter.**

direct-current transmission Same as is **DC transmission.**

direct-current voltage Same as **DC voltage.**

direct-current working voltage Same as **DC working voltage.**

direct-current–alternating-current converter Same as **DC–AC converter.**

direct-current–alternating-current inverter Same as **DC–AC converter.**

direct-current–alternating-current power converter Same as **DC–AC power converter.**

direct-current–direct-current converter Same as **DC–DC converter.**

direct digital control The use a digital computer for direct control of a device, piece of equipment, system, or process. Its abbreviation is **DDC.** Also called **direct numerical control.**

direct distance dialing A service offered by telephone companies, which enables the dialing of numbers outside the local calling area without the assistance of an operator. Its abbreviation is **DDD.**

direct drive A system in which that which is driven is directly connected to its driving source. For instance, a phonographic turntable's platter being directly connected to the motor. Also called **direct-drive system.**

direct-drive actuator A robotic actuator which is directly linked to the element it is driving, such as an arm.

direct-drive arm A robotic arm whose joints are directly linked to its driving source, such as a torque motor.

direct-drive robot A robot whose driven elements, such as arms, are directly linked to their driving source, such as a torque motor. Such a configuration, for instance, saves space and helps minimize friction and elasticity problems, but makes control tasks more difficult.

direct-drive system Same as **direct drive.**

direct-drive torque motor A torque motor which is directly linked to its load. May be used, for instance in direct-drive robots.

direct electromotive force Same as **direct emf.**

direct emf Abbreviation of **direct** electromotive force. A unidirectional emf whose changes in value are negligible.

direct-gap semiconductor Same as **direct-band-gap semiconductor.**

direct ground A connection to an earth ground.

direct instruction A computer instruction which contains the direct address of its operand.

direct inward dial Same as **direct inward dialing.**

direct inward dialing A service feature which allows incoming calls to reach specific PBX extensions without human intervention. Its abbreviation is **DID**. Also called **direct inward dial**.

direct light Light arriving form a luminous source, such as the sun or a lamp, as opposed to reflected light.

direct measurement The determination of a value without utilizing an intervening process. For instance, the measurement of a capacitance using a capacitance meter, as opposed to a capacitance bridge. This contrasts with an **indirect measurement**, in which the converse is true.

direct memory access The transfer of data from one location in a computer to another, without CPU intervention. For example, data transferred between memory and a peripheral. Such transfers are much faster than if the CPU is used. Its abbreviation is **DMA**.

direct memory access channel A channel through which **direct memory access** transfers take place. Its abbreviation is **DMA channel**.

direct memory access controller A device which controls the flow of data along a **direct memory access channel**. Its abbreviation is **DMA controller**.

direct numerical control Same as **direct digital control**. Its abbreviation is **DNC**.

direct outward dial Same as **direct outward dialing**.

direct outward dialing A PBX feature which allows outgoing calls to be dialed without first entering one or more digits. For example, an outgoing call may be placed without first pressing 9 to obtain a second dial tone. Its abbreviation is **DOD**. Also called **direct outward dial**.

direct path The path of a wave which has not been reflected.

direct pickup 1. A real-time broadcast, especially of TV programming. 2. Reception of a signal via an undesired path, such as that from an antenna, while also receiving a desired cable signal. One possible manifestation is ghosting.

direct piezoelectric effect The generation of an electric charge by subjecting certain crystals or ceramics to mechanical strain. Conversely, **indirect piezoelectric effect** is the mechanical deformation of certain crystals or ceramics when exposed to an electric field. Both are complementary manifestations of the piezoelectric effect. Also called **direct piezoelectricity**.

direct piezoelectricity Same as **direct piezoelectric effect**.

direct processing The processing of data as it is received by a computer. This contrasts with **deferred processing**, in which data is processed after a given amount is received, or after a specified time interval.

direct radiator Same as **direct radiator speaker**.

direct radiator loudspeaker Same as **direct radiator speaker**.

direct radiator speaker Also called **direct radiator**, or **direct radiator loudspeaker**. 1. A speaker whose radiator element acts directly on the surrounding air. This contrasts, for instance, with a speaker whose radiator is coupled with a horn. 2. A speaker in which the sound projects from the front of its enclosure. This contrasts, for instance, with a speaker whose sound emanates from a side of its enclosure.

Direct Rambus DRAM Same as **DRDRAM**.

direct ray A ray which has not been reflected, refracted, or scattered.

direct read after write In the recording of optical disks, such as DVD+RWs, an error-checking technique in which accuracy is verified immediately after it has been written. This contrasts with **direct read during write**. Its abbreviation is **DRAW**.

direct read during write In the recording of optical disks, such as DVD+RWs, an error-checking technique in which accuracy is verified during its recording. Writing and verification occur during the same disk rotation. This contrasts with **direct read after write**. Its abbreviation is **DRDW**.

direct-reading instrument An instrument which measures values though direct actuation by the measured phenomenon.

direct-reading meter A meter which measures and indicates values though direct actuation by the measured phenomenon. For instance, an ammeter which is part of the circuit whose current is being measured.

direct recording 1. A recording in which the produced record is derived directly from the received signals without any intervening processing. 2. The technique employed in **direct recording** (1).

direct-recording instrument An instrument which makes **direct recordings**.

direct reflection Reflection of electromagnetic or acoustic waves in which there is no diffusion or scattering. This may occur, for instance, when reflecting light off a mirror. Also called **specular reflection**, or **regular reflection**.

direct scanning A scanning technique in which the entire scene is illuminated at all times, while each picture element is viewed singly by the TV camera.

direct-sequence spread spectrum A form of spread-spectrum modulation in which the information-carrying bit stream is combined with a pseudorandom bit stream at a higher bit rate, and this combined signal modulates the carrier. Although this spreads the modulated carrier over a much greater bandwidth, it provides greater security, and in many cases when some bits are lost, the ability to recover the data without retransmission. Its abbreviation is **DSSS**.

direct sound Same as **direct sound wave**.

direct sound wave A sound wave which has not been reflected. Most or all the sound waves in an anechoic chamber are of this nature. Also called **direct sound**, or **direct wave** (2).

Direct Stream Digital An encoding format with a sampling rate of approximately 2.82 MHz, utilized for converting analog audio into a digital signal. Direct Stream Digital provides a frequency response that extends from 0 Hz to over 100 kHz, a dynamic range of over 120 dB, and information for multiple independent channels. Used, for instance, in SACDs. Its abbreviation is **DSD**.

direct-view storage tube A CRT whose high-intensity display is the result of secondary emissions of electrons. Such a display remains bright for controllable periods of time, and may be used, for instance, in radars. The displayed image is formed by a writing gun and intensified by a flooding gun. Its abbreviation is **DVST**. Also called **storage tube**.

direct voltage Same as **DC voltage**.

direct wave 1. A wave that is not reflected. For instance, an RF wave which travels from a transmitting antenna to a receiving antenna without being reflected by the ionosphere. 2. Same as **direct sound wave**.

direct-writing galvanometer A galvanometer which directly records on paper its electrical input signal, a pen being mechanically linked to its moving coil.

direct-writing recorder A device which directly records on paper its electrical input signal, a pen being mechanically linked to its moving element.

direction finder A radio receiver which incorporates a highly directional antenna, such as a loop antenna, to determine the direction from which a radio signal arrives. Its abbreviation is **DF**. Also called **radio direction finder**.

direction-finder antenna An antenna suitable for use with a **direction finder**. Such an antenna may be rotated electrically or mechanically. Its abbreviation is **DF antenna**.

direction-finder antenna system Its abbreviation is **DF antenna system**. A **direction-finder antenna**, plus the complete complement of components required for its proper function. 2. An assembly of two or more DF antennas with the proper characteristics, spacing and orientation so as to maximize the intensity of reception in the selected direction.

direction finding The use of a **direction finder** to determine the direction from which a radio signal arrives. Also called **radio direction finding**.

direction-finding station A station with utilizes **direction finders** to determine the direction from which radio signals arrive. Such a station may be located, for instance, along a coastline. Also called **radio direction-finding station**.

direction of lay The lateral direction in which a conductor, or group of conductors, is wound around a cable core, or around another layer of conductors. The direction of lay may be left-hand or right-hand. If the conductors form counterclockwise spirals when traveling away from an observer viewing along the longitudinal axis of a cable, then it is left-hand. The converse is true for right-hand. Also called **lay**.

direction of polarization In a polarized wave, the direction of the electric vector. Also called **polarization direction**.

direction of propagation In a homogeneous isotropic medium, the direction of energy flow at any given point. For instance, the direction of propagation in a uniform waveguide is considered to be along its axis. Also called **propagation direction**.

direction rectifier A rectifier whose DC output voltage magnitude and polarity vary proportionally to the magnitude and relative polarity of an AC selsyn error voltage.

directional 1. Pertaining to, or indicating a direction in space. 2. Capable of sending or receiving signals in only one direction. 3. Capable of sending or receiving signals substantially better in a given direction. 4. Functioning exclusively, or substantially better, in a given direction, or in specific directions.

directional antenna An antenna which radiates or receives significantly more efficiently in a given direction, or in specific directions. Examples include loop antennas, rhombic antennas, and end-fire antennas. Also called **directive antenna**.

directional antenna array An assembly of antenna elements with the proper dimensions, characteristics, and spacing, so as to maximize the intensity of radiation in one or more specified directions. Also called **directional array**.

directional array Same as **directional antenna array**.

directional beam A unidirectional stream of radiated energy, such as radio waves, concentrated in a given direction. For example, that radiated by a directional antenna.

directional characteristic The variation of the properties of a transducer, with respect to direction. For instance, a cardioid microphone has a pickup pattern that is shaped more or less like a heart.

directional coupler Also called **waveguide directional coupler**. 1. A device which couples a primary waveguide system with a secondary waveguide system, so that the energy is transferred to the waves traveling in a specified direction, and not to those traveling in the other direction. Also called **directive feed**. 2. A waveguide device which extracts a small amount of the energy flowing in one direction, while ignoring that flowing in the other. Used, for instance, to monitor power output.

directional diagram Same as **directivity diagram**.

directional filter In a carrier system, a filter which separates the frequencies to be transmitted in one direction from those to be transmitted in the other. May consist, for example, of a bandpass filter. Also known as **directional separation filter**.

directional gain Same as **directivity index**.

directional homing Homing in which a craft maintains as straight a course as possible while approaching the beacon.

directional horn Same as **directional horn antenna**.

directional horn antenna A microwave antenna in the shape of a horn. Also called **directional horn**.

directional hydrophone A hydrophone which is significantly more sensitive in a given direction, or in specific directions.

directional lobe Also called **lobe**. 1. Within an antenna pattern, a three-dimensional section in which radiation and/or reception is increased. Also called **antenna lobe**, and when referring specifically to radiation, known as **radiation lobe**. 2. Within a directional pattern of a transducer, a three-dimensional section in which radiation and/or reception is increased. For instance, such a lobe in the radiation pattern of a speaker.

directional microphone A microphone which is significantly more sensitive in a given direction, or in specific directions. For instance, a cardioid microphone.

directional pattern Same as **directivity pattern**.

directional phase changer Same as **directional phase shifter**.

directional phase shifter A phase shifter in which the introduced shift is different for each direction of transmission. Also called **directional phase changer**.

directional power relay A relay which is actuated when a specified power level is attained in a given direction.

directional relay A relay which responds to the direction of the current, voltage, power, or the like, energizing it. Also called **polarized relay (2)**.

directional response pattern Same as **directivity pattern**.

directional separation filter Same as **directional filter**.

directional transducer A transducer whose properties vary with respect to direction. Examples include directional antennas and directional microphones.

directive An instruction to an assembler or compiler, which is processed at the time of assembly or compilation.

directive antenna Same as **directional antenna**.

directive feed Same as **directional coupler (1)**.

directive gain For an antenna, it is equal to 4π times the ratio of the radiation intensity in a given direction, to the total power radiated by said antenna. It is generally used in the context of directional antennas, and is usually expressed in decibels.

directivity 1. The property of some antennas which causes them to radiate or receive significantly more efficiently in a given direction, or in specific directions. 2. The varying of the properties of a transducer, with respect to direction. For example, a cardioid microphone having a pickup pattern that is shaped more or less like a heart. 3. The value of the **directive gain**, when it is at its maximum.

directivity diagram A diagram representing how the intensity of a quantity varies as a function of direction and distance. Commonly refers to the transmitting or receiving effectiveness of an antenna, and can be described in any plane, although the horizontal and/or vertical planes are generally used. Also called **directional diagram**, **radiation diagram**, or **field diagram**.

directivity factor 1. A quantification of the **directivity (1)**, or **directivity (2)** of an antenna or transducer. 2. For a sound-producing transducer, and at a given frequency, the ratio of the square of the radiated sound intensity at a remote point on the principal axis of said transducer, to the average sound intensity for all directions at the same distance of said point. 3. For a sound-receiving transducer, and at a given frequency, the ratio of the square of the voltage produced in response to the incident acoustic waves parallel to the principal axis of said transducer, to the mean-square voltage that would be produced in a completely diffused sound field of the same frequency and mean-square sound pressure.

directivity index A quantity, expressed in decibels, which measures the directional properties of a transducer. It is equal to 10 times the base-10 logarithm of the **directivity factor (2)**, or **directivity factor (3)**. Also known as **directional gain**.

directivity pattern A pattern representing how the intensity of a quantity varies as a function of direction and distance. Commonly refers to the transmitting or receiving effectiveness of an antenna, and can be measured in any plane, although the horizontal and/or vertical planes are generally used. Also called **directional pattern**, **directional response pattern**, **radiation pattern**, or **field pattern**.

directly grounded Connected to an earth ground.

directly-heated cathode A cathode, within a thermionic tube, which is directly connected to a source of current. Such a cathode, usually in the form of a filament, becomes heated as current passes through it, and emits electrons when sufficiently heated. Also called **filament (2)**, **filament cathode**, **filamentary cathode**, or **filament-type cathode**.

director 1. A parasitic element placed in front of a driven element, so as to increase the gain of the antenna in the direction of the major lobe. Also called **director element**. **2.** In an automatic switching system, a device which translates dialed digits into the signals used to switch the call.

Director A popular authoring tool for interactive multimedia content.

director element Same as **director (1)**.

directory 1. A special kind of file which is used for indexing and organizing other files. It provides information such as the name, type, and size for each listed file. Directories that branch into other directories are called root directories, while those located within a root directory are called subdirectories. In GUIs, the term **folder** is preferred. Also called **file directory**. **2.** In a network, an index listing network resources, such as devices available, or email addresses.

Directory Access Protocol A protocol used for gaining access to X.500 directories. Its abbreviation is **DAP**.

Directory Information Base A database of names and resources in an X.500 system. Also called **white pages (2)**.

directory path A route to a file in which each directory and subdirectory along the way is listed. For example, Documents\Reminders\Today.xyz. Also called **pathname**.

Directory Server Agent Same as **Directory System Agent**.

Directory Service Agent Same as **Directory System Agent**.

Directory Services Markup Language Same as **DSML**.

Directory System Agent An X.500 server program which looks up user addresses in a **Directory Information Base**. Such information requests are made using a Directory User Agent. Its abbreviation is **DSA**. Also called **Directory Service Agent**, or **Directory Server Agent**.

directory tree A graphic representation of a series of directories and subdirectories. Subdirectories, or subfolders, branch out from the root directories, or root folders. Also called **folder tree**.

Directory User Agent An X.500 program that sends requests to a Directory System Agent. Its abbreviation is **DUA**.

DirectX A set of application binary interfaces and application program interfaces utilized by multimedia programs.

dirty power An AC power source which contains noise, voltage fluctuations, and other components which can harm sensitive electronic components or equipment.

disable To turn off, or to prevent from operating. May refer to a component, circuit, device, piece of equipment, function, program, or system.

disabled Turned off, or prevented from operating. May refer to a component, circuit, device, piece of equipment, function, program, or system.

disassembler A program which converts machine language into assembly language. May be used, for instance, in debugging.

disaster recovery Same as **disaster recovery plan (1)**.

disaster recovery plan 1. The plan or procedures employed for recovering from a disaster, such as a fire. The key factors to consider are usually the recovery of data, and the reestablishment of the ability to operate. Such procedures usually include accessing backed-up data, and utilizing key hardware which has been previously earmarked. Also called **disaster recovery**. **2.** The measures taken to minimize the deleterious effects of a possible disaster. These include data backup to a remote site, and making the necessary arrangements for key hardware to be available during such an event. Also called **contingency plan**.

disc Also spelled **disk**. The main distinction between **disc** and **disk**, in the context of computers, is that the former usually refers to optical media, such as DVDs, while the latter usually refers to all others, such as floppy disks and RAM disks. **1.** A digital optical storage medium, usually 12 centimeters in diameter, whose contained information is encoded in microscopic pits on its metallic surface, which is protected by a plastic layer. To recover the recorded content, a laser is focused on the metallic surface of the disc, and the reflected light is modulated by the code on said disc. Since there is no contact between the pickup and the recorded surface, wear is minimized, while the protective layer helps avoid reading errors due to dust or minor marks on the surface of the disc. Any form of data may be recorded onto a disc, and formats include DVDs, DVD+RWs, CDs, and CD-ROMs. Also called **optical disc**. **2.** Same as **disk (1)**. **3.** Same as **disk (3)**.

Disc-at-Once A method of writing to an optical disk, such as a DVD or CD, in which all data is transferred continuously without interruptions. This contrasts, for instance, with **Session-at-Once**, where there are multiple recording sessions. This also contrasts with **Track-at-Once**, where each track is recorded independently. Also spelled **Disk-at-Once**.

disc capacitor Same as **disk capacitor**.

disc coil Same as **disk coil**.

disc file Same as **disk file**.

disc storage Same as **disk storage**.

disc thermistor Same as **disk thermistor**.

disc winding Same as **disk winding**.

discharge 1. The liberation or escape of stored or accumulated energy, especially electric charge. **2.** The diminishing or removal of an electric charge stored or accumulated in a body, such as a battery or capacitor, which is being depleted. Also called **electric discharge (1)**. **3.** The passage of an electric current through a medium, such as a space or air. A spark is an example of such a discharge. Also called **electric discharge (2)**. **4.** The conversion of electrical energy into chemical energy within a battery or cell.

discharge current The current passing or flowing during **discharge**. Also called **discharging current**.

discharge key Same as **discharge switch**.

discharge lamp A lamp which produces light when the gas it contains, usually at a low pressure, becomes ionized by an electric discharge passed through it. An example is a fluorescent lamp. Also known as **electric-discharge lamp**, **gas-discharge lamp**, or **vapor lamp**.

discharge potential The minimum potential at which a gas is ionized in a discharge tube or discharge lamp.

discharge switch A switch that quickly changes a capacitor from a charging source, to a load through which is discharges. Also called **discharge key**.

discharge tube A tube containing a low-pressure gas through which a current passes when it is ionized by a sufficient voltage applied between the electrodes of said tube. Also known as **electric-discharge tube**, or **gas-discharge tube**.

discharge voltage　The minimum voltage at which a gas is ionized in a discharge tube or discharge lamp.

discharger　**1.** A device which short-circuits capacitors, thereby discharging them. **2.** A device used on aircraft to reduce precipitation static. Such devices usually extend from the trailing edges of surfaces of an aircraft, and allow static electricity to discharge into the air. Also called **static discharger**.

discharging current　Same as **discharge current**.

disclosure triangle　In a GUI, a small triangle, usually pointing down or to the right, which when clicked upon provides choices, more information, or the like. Used, for instance, to select font sizes in a word-processing application.

discone antenna　A wideband antenna similar to a biconical antenna, except that one of the cones is replaced by a flat circular disk which is usually parallel to the surface of the earth. The center conductor of a coaxial line terminates at the center of the disk, while the outer conductor of the line terminates at the vertex of the cone.

disconnect　**1.** To terminate a connection or link. For example, to end a telephone connection by hanging up. **2.** To open a circuit by a separating a point where parts of it are joined together. For example, to do so by separating leads. **3.** Same as **disconnect switch**.

disconnect signal　**1.** In a telephone network, a signal which indicates the termination of a telephone connection. **2.** In a telephone network, a signal which terminates a telephone connection.

disconnect supervision　In a central office or PBX, the ability to detect when the remote end of a call has been disconnected.

disconnect switch　Also called **disconnecting switch, disconnect (3)**, or **disconnector**. **1.** A switch utilized for opening or closing a circuit or connection. **2.** A switch used for closing, opening, isolating, or rerouting a connection, after the circuit has been interrupted in another manner. **3.** A switch utilized to quickly open a circuit in case of an overload.

disconnecting switch　Same as **disconnect switch**.

disconnector　Same as **disconnect switch**.

discontinuity　**1.** The condition or characteristic of not being continuous. **2.** The condition in which a circuit does not have a continuous and complete path for the flow of current. **3.** An abrupt change in a medium or a body. For instance, a point at which the impedance of a transmission line rapidly changes. **4.** In a waveguide, an abrupt change in shape which results in reflections. Also called **waveguide discontinuity**.

discontinuous circuit　A circuit which is broken, or which otherwise does not have a complete and uninterrupted path for the flow of current. Also called **open circuit (1)**, or **incomplete circuit**.

discrete　**1.** Pertaining to, or consisting of separate entities which can be distinguished from each other. For example, data contained in the form of bits. **2.** Having a finite number of possible values. For instance, 0 and 1.

discrete circuit　A circuit consisting of **discrete components**.

discrete component　An individual circuit component which is self-contained and has a single identifiable function. Examples include resistors, capacitors, and transistors. Also called **discrete part**, or **discrete element**.

discrete device　A device which is self-contained and has a single identifiable function. Examples include transistors, LEDs, generators, and photocells.

discrete element　Same as **discrete component**.

discrete Fourier transform　A Fourier transform utilizing a discrete or sampled signal, rather than an integral. A fast Fourier transform allows a computer to calculate a discrete Fourier transform very quickly, usually in real-time.

discrete multitone　Its abbreviation is **DMT**. A DSL modulation technique in which the total bandwidth is distributed over a large number of subchannels, which maximizes data transfer. The available frequency interval is usually divided into 256 quadrature amplitude modulated subchannels, each with a 4.3125 kHz bandwidth. Any subchannel which is can not transmit data is not used, while the rest are optimized for modulation. Also called **discrete multitone modulation**.

discrete multitone modulation　Its abbreviation is **DMT modulation**. Same as **discrete multitone**.

discrete part　Same as **discrete component**.

discrete sampling　Sampling in which the time length of each individual sample is extended, to help avoid adversely affecting the frequency response of the channel.

discrimination　**1.** The ability to make fine distinctions. For example, to be able to distinguish between two almost identical values. **2.** The ability to convert a frequency-modulated or phase-modulated signal into an amplitude-modulated signal. **3.** In a tuned circuit, the extent to which unwanted signals are rejected. For example, to be able to tune to a specific frequency within a band, while rejecting all others.

discriminator　**1.** A circuit or device which makes fine distinctions, such as those between two almost identical values. **2.** A circuit or device which converts frequency-modulated or phase-modulated signals into an amplitude-modulated signals.

discriminator circuit　**1.** A circuit which makes fine distinctions, such as those between two almost identical values. **2.** A circuit which converts frequency-modulated or phase-modulated signals into an amplitude-modulated signals.

discriminator transformer　A transformer specifically designed for use in a **discriminator circuit**.

dish　Also called **dish reflector, parabolic reflector, paraboloid reflector**, or **paraboloidal reflector**. **1.** A concave reflecting surface with a paraboloid shape, which concentrates waves into a parallel beam. Used, for instance, in dish antennas. **2.** An antenna incorporating a **dish (1)**. Such an antenna features very high gain, usually operates at microwave frequencies, and has various applications, including its use in satellite communications and in radar. The driven element of a dish antenna is located at the focal point of its dish reflector. Also called **dish antenna, parabolic antenna, paraboloid antenna**, or **paraboloidal antenna**.

dish antenna　Same as **dish (2)**.

dish reflector　Same as **dish**.

disintegration　**1.** To break into smaller entities, or into constituent components. **2.** Same as **decay (2)**. **3.** Same as **decay (3)**. **4.** Same as **decay (4)**.

disintegration chain　Same as **decay chain**.

disintegration energy　Same as **decay energy**.

disintegration family　Same as **decay chain**.

disintegration series　Same as **decay chain**.

disk　Also spelled **disc**. The main distinction between **disc** and **disk**, in the context of computers, is that the former usually refers to optical media, such as DVDs, while the latter usually refers to all others, such as floppy disks and RAM disks. **1.** A digital magnetic storage medium, which is usually in the form of a rotating round plate. Information is encoded by altering the magnetic polarity of minute portions on the surface of such a disk. Disks may be recorded, erased, and rerecorded many times, and the most common formats are hard disks, floppy disks, and removable hard disks. Also called **magnetic disk**. **2.** Same as **disc (1)**. **3.** A thin, circular object, such as a disk capacitor. Also called **disc (3)**.

disk access Access to any computer storage medium for the purposes of storage and/or retrieval. For example, accessing a hard drive.

disk access time All instances apply to computer hard disks, CD-ROMs, DVDs, and other computer disks. **1.** The time that elapses between a request for data from a disk, and its delivery. **2.** The time required to store information onto a disk once the instruction is given. **3.** The time that elapses before a read/write head is positioned over the desired track.

disk and execution monitor Same as **daemon**.

disk array Two or more disks or disk drives utilized as a single unit, to provide additional capacity, access speed, and reliability.

Disk-at-Once Same as **Disc-at-Once**.

disk buffer A segment of computer memory used to temporarily store information that awaits storage onto a disk. Since writing data onto a disk much slower than holding it in memory, a program may uses a disk buffer, and then copy its contents to the disk under certain circumstances, such as when said buffer is filled.

disk cache A form of cache which uses a section of main memory to store recently accessed data from a disk, which can dramatically speed up applications by avoiding the much slower disk accesses.

disk capacitor A fixed capacitor in the shape of a disk. Such a capacitor usually consists of dielectric material with metal-film plates on its surfaces. Also spelled **disc capacitor**.

disk cartridge A removable computer storage device which contains a magnetic or optical disk. For instance, a floppy disk. Also called **cartridge (2)**.

disk cloning The copying of the complete contents of one hard disk to another. Used, for instance, for backup, or when installing a new system.

disk coil A coil that is in the form of a disk. Its winding is flat, as opposed to cylindrical. Also spelled **disc coil**. Also called **disk winding**, or **pancake coil**.

disk compression The compression of the content of a disk, to increase its capacity.

disk controller A chip, along with the associated circuitry, which enables a computer to control a disk drive. A disk controller handles tasks such as positioning the read/write heads, and the transfer of data. It can be incorporated into the drive, built into the motherboard, or may be in the form of an expansion card. Each interface, such as SCSI or IDE, requires a different type of controller.

disk crash A complete and unexpected failure of a disk drive. For example, a hard disk failure in which the read/write head collides with the surface of a disk.

disk drive A computer peripheral which reads data from, and writes data to, a disk. It is an electromechanical device that incorporates a motor which spins one or more disks very rapidly, has one or more heads to read/write data, plus the related connectors, wires, and so on. Each type of disk has its own type of drive, and these include hard disk drives, floppy disk drives, and DVD drives. Also known as **disk unit**.

disk dump To display, print, copy, or transfer the content of a **disk (1)**, with little or no formatting.

disk duplexing A variation of **disk mirroring**, in which separate controllers are used for each disk.

disk failure The condition in which a disk or disk drive no longer performs the function it was intended to, or is not able to do so at a level that equals or exceeds the established minimums. May be due, for instance, to a head crash.

disk farm A large number of hard disks stored at a given location.

disk file A collection of related instructions or data that is treated as a single unit, and which is stored on a magnetic disk. Also spelled **disc file**.

disk image An exact duplicate of the complete contents of an entire magnetic or optical disk.

disk interface The hardware which is utilized to connect a disk drive to a computer, including the plugs and wires. There are several commonly used interfaces, such as SCSI or IDE, and each requires a different type of disk controller.

disk memory A deprecated term for **disk storage**.

disk mirroring A technique in which data is written simultaneously to two or more disks. Used, for instance, for continued access to the content in the case of the failure of one of the disks, or for data recovery. When separate controllers are used for each disk, it is also called **disk duplexing**.

disk operating system The software which runs all the software and hardware of a computer. It is the first program the computer loads when powered on, remains memory-resident, and continuously controls and allocates all resources. Without it, for example, a computer can not recognize input, such as that from a keyboard, it can not process, as the operating system controls the use of the CPU, nor can it provide an output, as it manages all peripherals including the monitor. Its acronym is **DOS**. Also called **executive**, or **supervisor**.

Disk Operating System A common **disk operating system**. Its acronym is **DOS**.

disk pack A stack of hard disks which can be inserted or removed, along with its protective enclosure, from a computer.

disk partition A logical subdivision within a hard disk. Each partition has its own designation.

disk recording **1.** The recording of signals, such as music, onto a phonographic disk. **2.** A phonographic disk so recorded. **3.** The recording of information onto a disk, such as a hard disk or a DVD.

disk-seal tube An ultra-high frequency electron tube with parallel disk-shaped electrodes, which features low interelectrode capacitance and a high-power output. Also called **lighthouse tube**, or **megatron**.

disk server In a communications network, a server which functions as a remote storage medium for a workstation.

disk storage Computer data stored on a disk, such as a hard disk or a floppy disk. Sometimes referred to as **disk memory**, although this latter term is deprecated. Also spelled **disc storage**.

disk striping A technique in which data, such as that contained in a file, is spread over multiple disk drives, so that one disk may be transferring data while the next is locating the following segment. RAID storage, for instance, uses this technique to improve performance. Also called **data striping**, or **striping**.

disk thermistor A thermistor in the shape of a disk. Also spelled **disc thermistor**.

disk unit Same as **disk drive**.

disk winding Same as **disk coil**.

diskette A digital magnetic storage medium which is in the form of a flexible rotating plastic plate. Information is encoded by altering the magnetic polarity of minute portions on the coated surface of such a disk. Diskettes are much slower than hard disks, generally hold 1.44 megabytes of data, and their usual format is 3.5 inches, with a rigid case. Also called **floppy disk**, or **flexible disk**.

diskless workstation A workstation without disk drives. All programs and files are accessed from a file server.

dislocation **1.** A discontinuity in a crystal lattice. A crystal with dislocations may have its electrical properties altered.

The larger the produced crystal, the more likely it is to have dislocations. **2.** A plastic deformation of a metal.

dislocation density   The concentration of **dislocations (1)** within a crystal. Such a concentration depends on various factors, including the size of the crystal and the manner in which it is prepared.

dispatcher   Within a computer operating system, a program which coordinates the use of the resources of the system. For example, it initiates and terminates each task. Also called **scheduler**.

dispersion   **1.** The process or occurrence of scattering or separating into components. **2.** The process by which radiation is separated into components with different wavelengths, energies, speeds, or other characteristics. Also, an instance of such a separation. For example, a prism disperses white light into its component colors, as each travels through it at a different velocity. **3.** The scattering of microwaves by striking an obstruction. **4.** The uniform distribution of sound waves emanating from an acoustic transducer. **5.** The scattering of fine particles within another body. **6.** In spectroscopy, a measure of resolving power. **7. Dispersion (2)** occurring in an optical waveguide, such as an optical fiber or cable. An example is multimode distortion.

dispersion-compensating fiber   An optical fiber that has the opposite dispersion of another fiber along a transmission system, and which compensates for dispersion caused by the latter fiber. Its abbreviation is **DCF**.

dispersion-compensating module   A module that compensates for dispersion caused by an optical fiber along a transmission system. Its abbreviation is **DCM**.

dispersion management   In a fiber-optic system, measures taken, such as the use of dispersion compensating fibers or modules, to minimize or otherwise handle dispersion.

dispersion medium   A medium within which another substance is dispersed. For example, colloidal graphite.

dispersion-shifted fiber   An optical fiber that has nearly zero dispersion at a given wavelength, such as 1550 or 1300 nm. Its abbreviation is **DSF**.

dispersive lens   A lens which causes parallel rays of light passing through it to bend away from one another. Such lenses are thinner at the center than the edges. This contrasts with a **converging lens**, which bends parallel rays of light passing through it toward one another. Also called **divergent lens**, **concave lens**, or **negative lens**.

dispersive line   A delay line which delays each frequency of a wave or signal by a different length of time.

dispersive medium   A medium within which the phase velocity of electromagnetic waves varies as a function of the frequency. A plasma is an example of such a medium. In a **non-dispersive medium**, the phase velocity of electromagnetic waves is independent of frequency.

displacement   **1.** A change in position, or the distance from a given position, especially a starting position. Also, the magnitude of such a change. **2.** Same as **dielectric displacement**. **3.** The distance a memory address is from a base address. Also, the value added to the base address to obtain a second address. Also called **offset (4)**.

displacement current   AC arising in the presence of a time-varying electric field. This current is in addition to ordinary conduction AC. Displacement currents are necessary for the propagation of electromagnetic radiation through outer space. Also called **Maxwell's displacement current**.

displacement transducer   A transducer in which the movement of an object, such as an armature, varies its rate of conversion from one form of energy to another.

display   **1.** A device utilized to display the images generated by a computer, TV, oscilloscope, radar, or other similar device with a visual output. It incorporates the viewing screen and its housing. A specific example is a CRT. Also called **display device**, **display monitor**, **monitor (1)**, or **video monitor (1)**. **2.** The viewing screen of a **display (1)**. Such a display may be a CRT display, a liquid-crystal display, a plasma display, and so on. Also called **display screen**, or **screen (1)**. **3.** The visual output shown on a **display (1)**. For a computer to generate such a display, it requires a display adapter. Also called **readout (1)**.

display adapter   A computer board that enables it to display images. When a computer generates images, this board converts them into the electronic signals that serve as the input to the display device, such as the monitor. The display adapter determines factors such as the maximum resolution and number of colors that can be displayed. An appropriate monitor must be utilized to best benefit from the capabilities of such an adapter. Display adapters usually incorporate their own memory, thus not occupying the RAM of the computer while preparing images. Also known by various other names, including **display board**, **display card**, **display controller**, **graphics adapter**, **video adapter**, **video board**, **video card**, and **video controller**.

display background   On a screen or display, the area that is not being specifically interacted with or observed. This contrasts with **display foreground**. Also called **background (4)**.

display board   Same as **display adapter**.

display card   Same as **display adapter**.

display console   A computer display device, such as a CRT, used for accessing data which is stored or being processed. An accompanying keyboard and/or light pen may be used for input.

display controller   Same as **display adapter**.

display cycle   The complete sequence of operations that must occur prior to an image being shown on a screen or display.

Display Data Channel   A VESA standard for the exchange of information between a monitor and a display adapter. For instance, a monitor can inform the graphics subsystem of the computer about its capabilities, such as maximum resolution. Its abbreviation is **DDC**. Also called **VESA DDC**.

display device   Same as **display (1)**.

display dimmer   A circuit which varies the intensity in the illumination of a meter or other display. Seen, for instance, in home entertainment components which are less distracting to a viewer watching programming under conditions of reduced surrounding light.

display element   A component of a displayed image, such as a cube, text, or the foreground.

display flicker   Brief and irregular fluctuations in an image presented by a **display (1)**. This may occur, for instance, when the refresh rate is too low. Also called **screen flicker**, or **monitor flicker**.

display foreground   On a screen or display, the area that is being specifically interacted with or observed. This contrasts with **display background**. Also called **foreground (3)**.

display image   The combination of the various **display elements** displayed on a screen at any given moment.

display loss   The ratio of the minimum input signal that a receiver can detect, to the minimum output signal that can be detected by a user viewing the display of said receiver. May refer, for instance, of a radar receiver and its operator. Also called **visibility factor**.

display mode   A specific configuration of the settings of a computer monitor. For instance, the resolution, the number of colors that may be displayed, and whether it displays graphics and/or text. Also called **screen mode**, **monitor mode**, or **video mode**.

display monitor   Same as **display (1)**.

display port A port which serves to plug in a **display** (1). Also called **monitor port**, or **video port**.

Display Power Management Signaling A VESA standard pertaining to energy-conservation signaling for monitors. Its abbreviation is **DPMS**. Also called **VESA DPMS**.

display primaries Colors which are combined in an additive mixture to yield a full range of colors to be displayed by a color TV receiver. Red, green, and blue are most commonly used as the additive primary colors. Also called **receiver primaries**.

display resolution Also called **screen resolution**, or **monitor resolution**. **1.** The degree of sharpness and clarity of the displayed image of a monitor. May be expressed, for instance, in terms of the number of pixels displayed, and dot pitch. **2.** The resolution a monitor is set at. A common resolution setting is 800x600, meaning that 800 dots, or pixels, are displayed on each of the 600 lines of the display. Frequently, the number of colors is incorporated into this specification, as in 800x600x256.

display screen Same as **display** (2).

display terminal A computer input/output device which incorporates a monitor, a video adapter, and a keyboard. It also usually includes a mouse. May be used, for instance, in networks. Also called **display unit** (3), or **video display terminal**.

display tube A CRT that serves as a **display** (1).

display unit **1.** A device which provides a visual output. Examples include computer monitors, oscilloscope displays, printers, and meters. **2.** Same as **display** (1). **3.** Same as **display terminal**.

display window In a panoramic display, the width of a section of the presented frequency spectrum. Usually expressed in MHz.

disposable **1.** Designed to be used once, then discarded. **2.** Designed to be more inexpensive to discard when a failure occurs, rather than be repaired.

disruptive discharge A sudden increase of current flow through an insulator, dielectric, or other material separating circuits, when said material fails under electrostatic stress.

dissector tube A camera tube in which the picture to be transmitted is focused upon a photosensitive surface from which electrons are emitted proportionally to the light of each part of the picture. These electrons are focused, in sequence, onto a controller electrode, which produces the output video signal. Also called **image dissector**, or **Farnsworth image dissector**.

dissipation Also called **energy dissipation**. **1.** The loss of another form of energy through its conversion into heat energy. For instance, a circuit suffering electrical energy losses due to resistance. Also called **heat dissipation** (2), **heat loss** (2), or **thermal loss** (2). **2.** The quantified energy loss in **dissipation** (1). **3.** A loss of energy. **4.** A wasteful expenditure of energy.

dissipation constant **1.** The power required to raise the temperature of the body of a thermistor 1 °C in relation to the temperature of its surrounding air. Usually expressed in mW. **2.** A measure of the rate at which an electrically-charged object or particle loses its charge to its surrounding air.

dissipation factor Its symbol is **D**. **1.** The tangent of the dielectric loss angle. **2.** The reciprocal of Q, which is a figure of merit for an energy-storing device. The higher the dissipation factor, the shorter the time for an energy-storing device to discharge.

dissipation line For a rhombic transmitting antenna, a length of resistive transmission line utilized to dissipate power. May consist, for instance, of two parallel lengths of stainless-steel wire terminated by a large noninductive resistor.

dissipator **1.** A device which serves for **dissipation** (1). **2.** A device which serves to draw heat away from another device. For example, a heat sink which helps dissipate heat that a chip generates.

dissociation **1.** To separate two or more entities which were united. Also, such an instance. **2.** The breaking apart of a molecule into simpler constituents. For instance, in electrolytic dissociation, a molecule is broken up into 2 or more ions.

dissymmetrical network Same as **dissymmetrical transducer**.

dissymmetrical transducer A transducer whose input and output image impedances are unequal. Also called **dissymmetrical network**.

distance education Same as **distance learning**.

distance learning Learning in which one or more of the following are true: the instructor or instructors are geographically separated from the student or students, one or more of the students is separated from the rest of the students, or one or more learning resources is available from a remote location. Such learning may incorporate computers, TV, satellite communications, and the Internet. Also called **distance education**, or **teleteaching**.

distance mark On a radar screen, a mark which indicates the distance of a scanned object to the radar receiver. Also called **distance marker**, or **range mark**.

distance marker Same as **distance mark**.

distance-measuring equipment Radio navigation equipment which aids in the determination of the distance of an aircraft to a ground-based transponder. An aircraft sends an interrogation signal which triggers a transponder to emit a return signal. The time elapsed between the emission and reception of the signals is used to calculate the distance. Its abbreviation is **DME**.

distance reception The reception of transmissions from distant radio stations. Certain layers of the ionosphere, for instance, enable such reception. Its abbreviation is **DX**.

distance relay A relay which removes power when a fault occurs within a specified distance from it.

distance resolution Also called **range resolution**, **range discrimination**, **radar resolution**, or **target discrimination**. **1.** The minimum distance between two scanned objects that may be distinguished by a radar set. **2.** The minimum distance between two scanned objects that provides two separate and recognizable indications on a radar screen.

Distance Vector Multicast Routing Protocol A routing protocol which provides connectionless delivery of datagrams to multiple hosts over large networks, especially the Internet. Its abbreviation is **DVMRP**.

distant miking The placement of a microphone at a distance of approximately one meter, or further, from a sound source. Used, for instance, when reflected sounds wish to be incorporated into the signal.

distinctive ringing A service offered by telephone companies, or a PBX feature, which enables one or more phones in the same location to ring differently. For example, a home telephone may have multiple telephone numbers assigned to it, each ringing in a different manner. Or, various extensions in an office may each have its own characteristic ringing pattern.

distortion **1.** The failure of a circuit, device, component, or system to accurately reproduce at its output the essential characteristics of its input. **2.** Any undesired changes in the waveform of a signal passing through a circuit, device, or transmission medium. Also, the extent of such changes. Examples include amplitude, harmonic, intermodulation, and phase distortion. Also called **waveform distortion**. **3.** Any undesired changes occurring in a reproduced image, as compared to the original image. An example is optical distortion. Also called **image distortion. 4.** Any undesired changes in the waveform of a wave, especially an acoustic wave.

distortion-free Same as **distortionless.**

distortion-limited operation The condition in which the distortion of a system limits its performance, as opposed to bandwidth or attenuation.

distortion meter 1. A device which measures **distortion. 2.** An instrument which measures the harmonic content of a wave. Such a device, for instance, may display the harmonic content of an audio-frequency wave. Also called **harmonic distortion meter (2).**

distortion tolerance The maximum amount of distortion a signal can have, and still be useful.

distortionless Without **distortion.** Also called **distortion-free.**

distress frequency A radio frequency designated for emergency use only. For instance, 500 kHz may be used for marine vessels, or aircraft traveling over the sea.

distress signal A brief transmission sent by a craft or vehicle, such as a ship or an airplane, indicating a situation of grave and/or imminent danger, and thereby requesting immediate assistance. One such signal that is internationally recognized is the French expression *m'aider*, which means "come help me!" The English phonetic equivalent of this term is mayday.

distribute 1. To divide or spread throughout a single entity, or among many. **2.** To divide or separate into groups or categories.

distributed 1. Divided or spread throughout a single entity, or among many. For example, distributed computing. **2.** Evenly divided or spread throughout a single entity, or among many. **3.** Evenly spread out within a single entity, as opposed to being lumped at or around a single point. For example, distributed capacitance.

distributed amplifier A wideband multistage amplifier whose active devices, such as transistors or tubes, are distributed along parallel delay lines. The more active devices present, the greater the gain.

distributed Bragg reflector laser Same as **distributed feedback laser.** Its abbreviation is **DBR laser.**

distributed capacitance A capacitance which is not concentrated within a capacitor. Examples include the capacitance between the turns in a coil, or between adjacent conductors of a circuit. This contrasts with **lumped capacitance,** which is concentrated within a capacitor. Also called **self-capacitance, stray capacitance,** or **wiring capacitance.**

distributed circuit A circuit in which properties, such as resistance or capacitance, are distributed throughout. This contrasts with a **lumped circuit,** where such properties are located in discrete components such as resistors or capacitors.

distributed component A circuit parameter, such as resistance or inductance, which is distributed throughout a circuit or along the entire length of a transmission line. For instance, series resistance in a two-wire transmission line, or the distributed resistance of a wire coil. This contrasts with a **lumped component,** which is concentrated within discrete components such as resistors or capacitors. Also called **distributed constant, distributed parameter,** or **stray component.**

distributed computing Same as **distributed processing.**

distributed computing environment A set of standards that facilitate the development of programs that function across multiple platforms. Its abbreviation is **DCE.**

distributed constant Same as **distributed component.**

distributed data processing Same as **distributed processing.** Its abbreviation is **DDP.**

distributed database A database distributed over multiple nodes of a network. This arrangement usually enhances performance, and the combined databases are controlled as one.

distributed database management system Same as **distributed DBMS.**

distributed DBMS Abbreviation of **distributed database management system.** Software which serves as the interface between a user and a distributed database. Such programs manage tasks such as the storage, retrieval, organization, security, and integrity of data. Its own abbreviation is **DDBMS.**

distributed feedback laser A type of semiconductor laser which employs feedback to make certain modes in the resonator oscillate more than others. To do so, the grating spacing is arranged so that the feedback is distributed in both directions, which creates a condition that can approach single-mode oscillation. Its abbreviation is **DFB laser.** Also called **distributed Bragg reflector laser.**

distributed file system A file management system which serves to locate and access files distributed over multiple nodes of a network, or across multiple networks.

distributed impedance An impedance which is evenly distributed throughout a circuit. This contrasts with **lumped impedance,** which is concentrated within a circuit element. Also called **stray impedance.**

distributed inductance An inductance which is evenly distributed throughout a circuit. This contrasts with **lumped inductance,** which is concentrated within a circuit element, such as a coil. Also called **stray inductance.**

distributed intelligence Same as **distributed processing.**

distributed-memory multiprocessing A multiprocessing computer architecture in which multiple CPUs each have their own memory and I/O channels, and which lends itself to massively parallel processing, as opposed to symmetric multiprocessing. Also called **loosely-coupled multiprocessing.**

distributed-mode speaker A speaker which incorporates, or consists of, a **distributed-mode speaker panel.**

distributed-mode speaker panel An acoustic radiator consisting of a flat panel a few millimeters in thickness, which vibrates over its entire surface without pistonic motion. Such panels are very lightweight, disperse sound evenly in all directions at all reproduced frequencies, and have a fairly linear frequency response. Also called **flat-panel speaker (1),** or **flat-panel distributed-mode speaker (1).**

distributed-mode speaker technology The technology employed in speakers incorporating a **distributed-mode speaker panel.** Its abbreviation is **DML technology.**

distributed network 1. An electric network consisting of **distributed components.** Also called **distributed-parameter network. 2.** A communications network in which resources, such as processors and switching equipment, are distributed throughout multiple locations, as opposed to one. **3.** A communications network in which there are multiple routings between nodes. **4.** A computer network in which functions such as processing and storage are handled by multiple nodes, as opposed to a single computer.

distributed parameter Same as **distributed component.**

distributed-parameter network Same as **distributed network (1).**

distributed processing The use of multiple linked processors or computers to process data. Ideally, each processor or computer handles different tasks, and all are connected in a manner that allows the unhindered sharing of resources and information. Also called **distributed data processing, dis-**

tributed processing system, **distributed intelligence**, or **distributed computing**.

distributed processing system Same as **distributed processing**.

distributed-queue dual-bus A broadband protocol implemented on fiber optic. It features a dual bus, each carrying data in both directions, and incorporates a queuing system which maintains transmission order. It supports data, voice, and video traffic. Used, for instance, in metropolitan area networks. Its abbreviation is **DQDB**.

distributed resistance A resistance which is evenly distributed throughout a circuit. It is usually undesired and energy-wasting. This contrasts with **lumped resistance**, which is as concentrated within a circuit element, such as a resistor. Also called **stray resistance**.

distributed sensing The combining of data collected by multiple sensors, so as to derive a synergistic result. Used, for instance, in robotics, and in medical diagnosis. Also called **multisensor data fusion**.

distributed transaction processing The use of multiple nodes of a network for transaction processing. Its abbreviation is **DTP**.

distributed virtual environment A virtual environment in which multiple users interact, as seen, for instance, in tele-immersion, or in MUDs. Its abbreviation is **DVE**.

distributing cable Same as **distribution cable**.

distributing center Same as **distribution center**.

distribution 1. The act of allocating or spreading something. 2. The result of **distribution** (1). 3. A function describing the frequency with which a variable assumes a given value. It may be expressed as a graph of the number of occurrences versus the possible values. Also called **frequency distribution**.

distribution amplifier 1. An amplifier with one input and multiple outputs. 2. A low-impedance RF amplifier with multiple outputs, which serves to feed radio and/or TV signals to several receivers. 3. A low-impedance audio-frequency power amplifier with multiple outputs to feed multiple speakers or receivers.

distribution cable Also called **distributing cable**. 1. A cable which extends from a feeder cable to a specific served area. 2. A cable which extends from a **distribution amplifier**.

distribution center Also called **distributing center**. 1. A point from which a signal is routed to other points. 2. In an AC power system, a location at which generating, conversion, and control equipment is located. The power is distributed from this point.

distribution control In a CRT, such as that of a TV, a control which adjusts the scanning speed of the trace interval, to correct geometric distortion. Also called **linearity control**.

distribution frame In communications, a location where wire and/or cable terminations are located, so that the internal lines of a structure can be linked to external lines. Such a site may be located, for instance, adjacent to a central office. Also called **main distribution frame**.

distribution switchboard 1. A switchboard which serves to route signals to other points. Such a switchboard may also serve as an intermediate point, beyond which signals are routed to other points. 2. A power switchboard which serves to route electric power to points of use. Such a switchboard, for instance, may be located in a building.

distribution system The physical system that delivers electrical energy from a primary source to the specific customers. Such a system includes high-voltage lines, poles, step-down transformers, and voltage dividers. Also called **electric distribution system**.

distribution transformer A step-down transformer used in a **distribution system**.

distributor A circuit, device, or system which allocates or divides something among many devices, paths, and so on. For example, a rotary device which distributes current, or a device which accepts a data stream and distributes it among several lines.

disturbance 1. An undesired interference, interruption, or variation in a condition, quantity, or process. For instance, noise which interferes with the transmission or reception of a signal. 2. In a control system, an unwanted control signal.

dither 1. Same as **dithering** (1). 2. A vibratory force with controlled amplitude and frequency, which is continuously applied to a device which is servomotor driven, so as to avoid it sticking in the null position. Also called **buzz** (2). 3. A vibratory force or oscillation of small amplitude. 4. Same as **dithering** (3).

dithering 1. In graphics, a technique employed to simulate additional colors, or shades of gray, when displaying or printing images. For instance, by varying patterns of white and black dots, variations of gray are produced. Also called **dither** (1). 2. The incorporation of a small perturbation or a little noise, for instance, to minimize the effects of minor nonlinearities. Used, for example, in measuring instruments. Also called **dither** (2).

divalent Pertaining to an element, such as magnesium or oxygen, which has a valence of two.

divergence 1. A moving away from a center, or from a course. 2. In a CRT, the spreading of an electron beam due to the mutual repulsion of the electrons. 3. The spreading of a collimated beam.

divergence angle Also called **angle of divergence**. 1. In CRTs, the angle encompassed by the spread of the electron beam as it travels from the cathode to the screen. 2. The angle encompassed by the spread of a light beam as it travels from a collimating source.

divergence loss The loss of sound energy, due to the spreading of acoustic waves.

divergent beam A stream of radiated energy which tends away from an intersection point. Also called **diverging beam**.

divergent lens Same as **diverging lens**.

diverging beam Same as **divergent beam**.

diverging lens A lens which causes parallel rays of light passing through it to bend away from one another. Such lenses are thinner at the center than the edges. This contrasts with a **converging lens**, which bends parallel rays of light passing through it toward one another. Also called **divergent lens**, **concave lens**, or **negative lens**.

diversity 1. Composed of two or more different elements, qualities, manifestations, paths, methods, and so on. 2. Same as **diversity system**.

diversity communications system Same as **diversity system**.

diversity gain 1. Signal gain through the use of **diversity reception**. 2. Signal gain through the use of **diversity transmission**.

diversity radar A radar system which utilizes two or more sets of transmitters and receivers, each set operating at a different frequency. All sets share a common antenna and display. Also called **frequency-diversity radar**.

diversity receiver A receiver used with **diversity reception**. Also called **dual-diversity receiver**.

diversity reception A mode of reception in which two or more arriving RF signals, each with the same information, are combined, so as to minimize the effects of fading. Alternatively, the stronger signal can be used. There are various examples, including space-diversity reception, in which there are two or more receiving antennas in separate locations, and frequency-diversity reception, where two or more transmissions at different frequencies are received. Also called **dual-diversity reception**.

diversity system A reception or transmission system which employs **diversity reception** or **diversity transmission**, respectively. Also called **diversity (2)**, **diversity communications system**, **dual-diversity system**, or **dual-diversity**.

diversity transmission The converse of **diversity reception**. For example, instead of having two or more receiving antennas in separate locations, there can be two or more physically separate transmitting antennas. Also called **dual-diversity transmission**.

diversity transmitter A transmitter used with **diversity transmission**. Also called **dual-diversity transmitter**.

divide overflow An error which occurs as a result of dividing by zero, or by a number so small that the result exceeds the capability of a program or computer. Also called **overflow error**.

divided-carrier modulation Modulation in which the carrier is divided into two identical components which are 90° out-of-phase. Each component is modulated by a different signal, and the two signals are added together.

divider A circuit or device which divides a number of pulses, cycles, signals, or other input quantity, by a fixed number, such as 2 or 10. For example, counter circuits, voltage dividers, or frequency dividers.

divider probe A test probe that divides a voltage by a given factor, such as 10 or 1000, so that a reading falls within the range of the measuring instrument.

dividing network A filter circuit in a speakers which divides the input audio frequencies and sends the corresponding bands of frequencies to the designated speakers units, such as woofers, midranges, tweeters, super-tweeters, and so on. Also called **crossover network**, **crossover (3)**, **frequency dividing network**, **speaker dividing network**, or **loudspeaker dividing network**.

division 1. The process of separating into two or more parts. Also, the result of such a separation. For instance, the division of multiple communications channels. 2. The process of finding out how many times a quantity fits into another. Also, the result of such a computation.

DLC Abbreviation of **digital loop carrier**.

DLCI Abbreviation of **data link connection identifier**.

DLE Abbreviation of **data link escape character**.

DLL Abbreviation of **dynamic-link library**, **dynamically-linked library**, or **dynamic-linked library**.

DLP 1. Abbreviation of **digital light processing**. 2. Abbreviation of **digital light processor**.

DLP projector Abbreviation of **digital light-processing projector**.

DLTS Abbreviation of **deep level transient spectroscopy**.

dm Abbreviation of **decimeter**.

DM 1. Abbreviation of **delta modulation**. 2. Abbreviation of **document management**.

DMA Abbreviation of **direct memory access**.

DMA/33 A version of **Ultra ATA** with transfer rated of up to 33 MBps.

DMA/66 A version of **Ultra ATA** with transfer rated of up to 66 MBps.

DMA/100 A version of **Ultra ATA** with transfer rated of up to 100 MBps

DMA/133 A version of **Ultra ATA** with transfer rated of up to 133 MBps.

DMA channel Abbreviation of **direct memory access channel**.

DMA controller Abbreviation of **direct memory access controller**.

DMD Abbreviation of **digital micromirror device**.

DME Abbreviation of **distance-measuring equipment**.

DMI Abbreviation of **desktop management interface**.

DML Abbreviation of **data manipulation language**.

DMM Abbreviation of **digital multimeter**.

DMOS Abbreviation of **diffused metal-oxide semiconductor**.

DMR Abbreviation of **digital microwave radio**.

DMS 1. Abbreviation of **data management system**. 2. Abbreviation of **document management system**.

DMT Abbreviation of **discrete multitone**.

DMT modulation Same as **discrete multitone**.

DNC Abbreviation of **direct numerical control**.

DNIS Abbreviation of **dialed number identification service**.

DNS Abbreviation of **Domain Name System**, or **Domain Name Service**. Over the Internet, a system which translates a domain name, such as *www.yipeeee.com*, into an IP address, such as 91.2.133.206. The DNS maintains a database which correlates domain names with the appropriate IP addresses. A domain name may have more than one IP address associated with it, and an IP address may also have more than one domain name associated with it. Although an IP address or a domain name may be entered in the address field of a browser, the latter is more often used as it is generally easier to remember.

DNS error An error 404 resulting from the inability to convert a domain name into an IP address.

DNS query Abbreviation of **Domain Name System query**, or **Domain Name Service query**. A request for a conversion from a domain name to an IP address.

DNS server Abbreviation of **Domain Name System server**, or **Domain Name Service server**. A computer which handles **DNS queries**. It is a type of name server.

do-nothing instruction Same as **dummy instruction**.

dock 1. Same as **docking station**. 2. To plug into a **docking station**.

docking station A base unit that a portable computer plugs into, so that the user can expand its functionalities to that of a desktop computer. Such a base has connections for a monitor, a full-sized keyboard, a printer, AC power, and so on. Also called **dock (1)**.

DOCSIS Acronym for **D**ata **O**ver **C**able **S**ervice **I**nterface **S**pecification. A set of standards for the bi-directional transfer of data utilizing a cable TV network and a cable modem. Data rates are usually between 27 and 36 Mbps downstream, and up to 10 Mbps upstream.

document 1. A file created by a computer application, especially that of a word processor. 2. To explain a program or process. Also, the recorded explanation, be it on paper or in another form.

document-centric Pertaining to an operating system in which a document is the starting point of a task. For example, a document is retrieved, and then the appropriate application is invoked automatically. This contrasts with **application-centric**, where the application is the starting point of a task.

document image processing The storing and retrieving of information in the form of electronic images of paper documents. For example, a series of documents may be scanned, and then stored in the form of bitmapped images. Similar to microfilm, signatures, drawings, hand-written comments, and other markings are left intact, though it does occupy comparatively more storage space. Its abbreviation is **DIP**.

document management The use of computer software, such as database programs, and hardware, such as scanners, to create and manage the documents of an organization. Such documents may be paper-based, or electronic. Its abbreviation is **DM**. Also called **electronic document management**, or **enterprise document management**.

document management system A system which facilitates **document management**. It performs various functions, such as providing search engines, converting between formats, providing access through networks, and determining if a given user can update or only read documents. Its abbreviation is **DMS**. Also called **electronic document management system**.

Document Object Model A specification for how objects, such as images, links, and text, are represented on a Web page. It also establishes what attributes are associated with each object, and how said attributes can be modified. Its abbreviation is **DOM**.

document reader A device which scans printed matter and converts it into a computer-readable format. Such a device may employ, for instance, optical character recognition.

documentation The information that accompanies software, hardware, or equipment. It may detail set-up procedures, instructions for use, optimization, maintenance, troubleshooting, and so on. When the printed documentation for a given product is not available, it can frequently be accessed over the Internet.

DOD 1. Abbreviation of **depth of discharge**. 2. Abbreviation of **direct outward dialing**.

DOF Abbreviation of **degrees of freedom**.

doghouse A weather-resistant enclosure which houses antenna-tuning equipment, and which is usually located at the base of an antenna tower.

Doherty amplifier A linear RF power amplifier incorporating two sections whose inputs and outputs are connected by quarter-wave networks. It is designed so that only one section operates for all input-signal voltages up to one-half maximum amplitude. Beyond this, the other section comes into operation, until maximum amplitude. In this manner, both sections function optimally, thus minimizing distortion at maximum input amplitude.

Dolby 1. A system for the reduction of noise in audio recordings. One variation involves providing additional gain for certain frequencies during recording, and reducing said gain during playback. May be used, for instance, in tape decks. Also called **Dolby noise reduction**, **Dolby noise reduction system**, or **Dolby system**. 2. A multiple-channel sound format which is digitally encoded, and which provides up to six channels of surround sound. Also called **Dolby surround sound**, **Dolby surround**, **Dolby digital surround**, **Dolby digital**.

Dolby AC-3 The coding utilized in the **Dolby** (2) format. It is an abbreviation of **Dolby** Audio Coding-3. Also called **AC-3**.

Dolby digital Same as **Dolby** (2).

Dolby digital surround Same as **Dolby** (2).

Dolby noise reduction Same as **Dolby** (1).

Dolby noise reduction system Same as **Dolby** (1).

Dolby Pro Logic A surround sound technology which delivers four channels from a two-channel source. The provided channels are left, center, right, and a limited-bandwidth surround channel.

Dolby Pro Logic II An enhancement of **Dolby Pro Logic**, in which 5.1 surround sound is simulated using a four channel surround signal.

Dolby surround Same as **Dolby** (2).

Dolby surround sound Same as **Dolby** (2).

Dolby system Same as **Dolby** (1).

dolly An industrial platform provided with wheels or another form of movement, utilized to transport and position electronic equipment such as TV cameras.

DOM 1. Abbreviation of **Document Object Model**. 2. Abbreviation of **digital ohmmeter**.

domain 1. A realm of expertise, knowledge, or activity, such as electronics. 2. Within a ferromagnetic substance, a region in which the atomic or ionic magnetic moments are aligned in the same direction. Such a domain may be used, for instance, to store a bit as a 1 or a 0 on a magnetic tape or disk. Also called **magnetic domain**, or **ferromagnetic domain**. 3. Within a ferroelectric substance, a region whose electric moments are aligned in the same direction. 4. In database management, all the possible values for a given attribute. 5. On the Internet, one of the one of the main registration categories, such as .com, .net, .org, .edu, .gov, or country designators, such as. de for Germany. It consists of no less than two letters, and appears at the end of a domain name, as in the *.com* portion of *www.yipeeee.com*. Also called **top-level domain**. 6. In a network, a collection of resources which are administered as a unit.

domain name On the Internet, a name which identifies one or more IP addresses. For instance, in the following URL, *http://www.yipeeee.com/whoo.html*, the domain name is the *www.yipeeee.com* portion. Sometimes, only the *yipeeee.com* portion is considered to be the domain name. A domain name always ends in a **domain** (5). When a domain name is entered into the address field of a browser, it is converted into the appropriate IP address, after which the appropriate content is transferred to the computer that requested it. Also called **domain name address**, or **Internet domain name**.

domain name address Same as **domain name**.

Domain Name Server query Same as **DNS query**.

Domain Name Service Same as **DNS**.

Domain Name Service server Same as **DNS server**.

Domain Name System Same as **DNS**.

Domain Name System query. Same as **DNS query**.

Domain Name System server Same as **DNS server**.

dome tweeter A tweeter possessing a small, convex-shaped diaphragm. Such diaphragms are usually made of a metal, such as titanium or aluminum, but may also consist of another material, such as silk. Dome tweeters feature a mostly linear response, which frequently extends beyond the range of human hearing.

domestic electronics Electronic products intended for purchase and use by consumers. Such products include TVs, computers, DVD players, high-fidelity audio components, smart appliances, and other such products for use in the home.

dominant mode In a waveguide, the mode of propagation that has the lowest cutoff frequency. Also called **fundamental mode** (1), or **principal mode**.

dominant wave In a waveguide, the electromagnetic wave that has the lowest cutoff frequency.

dominant wavelength A single wavelength of light, which when combined in the proper proportion with white light, and at the correct intensity, matches a given color sample.

dongle A software copy-protection device which usually plugs into a parallel port. It allows backup copies to be made, but aims to prevent unauthorized copying to other computers. An attached dongle does not affect the use of the port for other purposes. Also called **hardware key**.

donor 1. A species, such as an atom, molecule, or ion, which donates one or more electrons to another species. The accepting species is called **acceptor** (2). Also called **electron donor**. 2. An electron-donating impurity which is introduced into a crystalline semiconductor. The donating of electrons makes for an n-type semiconductor. Also called **donor impurity**, or **donor material**.

donor atom **1.** In semiconductors, an atom which donates electrons. For example, phosphorus or arsenic. **2.** An atom which donates electrons.

donor impurity Same as **donor** (2).

donor level In a semiconductor material, an intermediate energy level close to the conduction level. At absolute zero, such a level is filled with electrons, and at higher temperatures said electrons can acquire sufficient energy to pass to the conduction band.

donor material Same as **donor** (2).

donut magnet A permanent magnet in the shape of a doughnut. Used, for instance, in particle accelerators or speakers. Also spelled **doughnut magnet**. Also called **ring magnet**.

doorknob capacitor A high-voltage fixed capacitor whose shape resembles that of a doorknob.

doorknob tube An ultra-high frequency vacuum tube whose shape resembles that of a doorknob.

doorway page A Web page created with given characteristics, such as multiple appearances of specific keywords, so as to rank high on a particular search engine. Also called **bridge page, jump page, entry page**, or **gateway page**.

dopant An impurity which is introduced into a semiconductor material. Such an impurity may be an acceptor impurity, which makes for a p-type semiconductor, or it may be a donor impurity, which makes for an n-type semiconductor. Acceptor impurities include gallium and aluminum, while donor impurities include phosphorus and arsenic. In either case, a dopant increases the conductivity of the semiconductor. Also called **doping agent, impurity** (2), or **semiconductor dopant**.

dope To add a **dopant** to a semiconductor material. The more dopant added to a semiconductor during its manufacturing process, the greater its conductivity.

doped junction A semiconductor junction produced by **doping**.

doped semiconductor A semiconductor material with **doping**, which determines its electrical characteristics, such as concentration of charge carriers. Also called **extrinsic semiconductor**, or **impurity semiconductor**.

doping The adding of a **dopant** to a semiconductor material. Such impurities may be added during manufacturing through any of various processes, including diffusion. Controlled amounts of specific dopants are added to achieve the desired electrical characteristics. Doping is used, for example, in the manufacturing of transistors and diodes. Also called **semiconductor doping**.

doping agent Same as **dopant**.

doping compensation **1.** In a semiconductor material, the addition of a donor impurity to partially compensate for the effects of a previously introduced acceptor impurity. **2.** In a semiconductor material, the addition of an acceptor impurity to partially compensate for the effects of a previously introduced donor impurity.

doping level The level of doping necessary to obtain the desired electrical characteristics.

Doppler broadening In spectroscopy, the widening of a spectral line when the radiating nuclei, atoms, or molecules do not having the same velocities. Since their velocities vary, so does their Doppler shift, accounting for the frequency spreading.

Doppler displacement Same as **Doppler shift** (1).

Doppler effect A change in the observed frequency of a wave, when there is relative motion between the source and the observer. As the distance between the source and the observer decreases, the observed frequency increases, thus the wavelength decreases. In the same manner, as the distance between the source and the observer increases, the observed frequency decreases, thus the wavelength increases. This effect occurs in sound and electromagnetic waves, and an example is the manner in which the pitch of a blowing horn of a car is perceived as higher as the car approaches, and then becomes lower as it moves away. Another example is the way in which the light of a star, as observed from the Earth, shifts toward the violet end of the spectrum, or higher frequencies, when it is approaching. The opposite is true for stars that are receding from the Earth, that is, its light shifts towards the red end of the spectrum. Also called **Doppler shift** (2), or **Doppler principle**.

Doppler-free laser spectroscopy Same as **Doppler-free spectroscopy**.

Doppler-free spectroscopy A spectroscopic technique which employs laser beams to eliminate **Doppler broadening**. This technique offers various benefits, including the measurement of the natural linewidth, and the ability to detect underlying fine structure such as Zeeman splitting. Also called **Doppler-free laser spectroscopy**.

Doppler frequency Same as **Doppler shift** (1).

Doppler navigation The utilization of devices employing the **Doppler effect** as navigation aids.

Doppler principle Same as **Doppler effect**.

Doppler radar A radar which employs the **Doppler effect** to determine the velocity of a scanned object. The Doppler shift of the signal returned by the scanned object indicates its velocity with high accuracy, or if it is stationary. Also called **Doppler system** (1), or **Doppler radar system**.

Doppler radar system Same as **Doppler radar**.

Doppler range Same as **doran**.

Doppler ranging Same as **doran**.

Doppler shift **1.** The extent of the change in the observed frequency of a wave, due to the **Doppler effect**. Also called **Doppler frequency**, or **Doppler displacement**. **2.** Same as **Doppler effect**.

Doppler sonar A sonar system which utilizes the **Doppler effect** for measurements.

Doppler system **1.** Same as **Doppler radar**. **2.** A device, instrument, or system, which utilizes the **Doppler effect** for measurements.

Doppler tracking A tracking system which utilizes the **Doppler effect** for measurements. Used, for instance in Doppler radar.

Doppler ultrasonography An ultrasonography technique with utilizes the **Doppler effect** for measurements. It is highly accurate and non-invasive.

Doppler VOR A VOR system which utilizes the **Doppler effect** for improved accuracy.

doran Acronym for **Do**ppler **ran**ging, or **Do**ppler **ran**ge. A continuous-wave ranging system which employs the **Doppler effect**. It utilizes phase comparison of multiple modulation frequencies on the carrier wave to obtain highly accurate range rata pertaining to scanned objects.

DOS **1.** Acronym for **disk operating system**. **2.** Acronym for **Disk Operating System**.

DoS Same as **denial of service attack**.

DoS attack Abbreviation of **denial of service attack**.

dosage Same as **dose**.

dosage meter Same as **dosimeter**.

dose The amount of ionizing radiation incident upon, or absorbed by, a given mass, volume, or body. May be expressed in roentgens, rems, rads, or reps. Also called **dosage**, or **radiation dose**.

dose meter Same as **dosimeter**. Also spelled **dosemeter**.

dosemeter Same as **dosimeter**. Also spelled **dose meter**.

dosimeter A device which measures **doses**. Such a device, for instance, may be worn by personnel which may be exposed to ionizing radiation. Also called **dosage meter**, or **dosemeter**.

dosimetry Techniques employed in the measurement of **doses**.

dot **1.** On the screen of a CRT, such as that of a TV, a small spot of red, green, or blue phosphor. **2.** A small spot, usually a circle, which is an element within a matrix. Utilized, for instance, to form images on a computer screen or in printed matter. **3.** A small piece of metal alloyed to a semiconductor wafer to form a junction. Also called **button (2)**. **4.** A period. For example, *www.yipeeee.com* may also be read as **yipeeee dot com**. **5.** In a schematic diagram, a symbol representing a junction.

dot addressable A mode of operation in which each **dot (2)** may be programmed.

dot com On the Internet, a top-level domain name suffix. The **com** is an abbreviation of **com**mercial, although the entity employing this suffix need not be commercial. Also called **.com**.

dot-com company A business that does all its selling over the Internet.

dot con An Internet-based fraud or swindle, such as that offering a product or service which does not exist, or that is nothing like that stated. Frequently, such schemes play on common human weaknesses, such as the desire to get rich quickly, to get something for nothing, or thinking that a given business arrangement is favorable at the expense of others.

dot cycle In communications with two alternating signaling conditions, a cycle representing a single alternation.

dot edu On the Internet, a top-level domain name suffix. The **edu** is an abbreviation of **edu**cational, and the entity employing this suffix should be an educational institution such as a college or graduate school. Also called **.edu**.

dot generator In a TV receiver, a signal generator utilized to adjust beam convergence on the picture tube, by producing a pattern of evenly spaced dots on the screen. When the convergence is not correct, the dots appear in groups of three, one each for red, green, and blue. When convergence is correct, the three dots come together to form a single white dot. Alternatively, the dots may be squares.

dot matrix A pattern of dots within a grid. May be used, for example, to form alphanumeric characters.

dot-matrix display A display whose shown characters are formed in a **dot matrix**. Some LCD displays are of this type.

dot-matrix printer A printer which creates alphanumeric characters and images by striking pins against an ink ribbon. There is a matrix of pins, and each pin makes a single dot. The more pins, the higher the resolution of the printed output. Also called **matrix printer**.

dot net On the Internet, a top-level domain name suffix. The **net** is an abbreviation of **net**work, although the entity employing this suffix need not be a network provider. Also called **.net**.

dot org On the Internet, a top-level domain name suffix. The **org** is an abbreviation of **org**anization, although the entity employing this suffix need not be such an entity. Also called **.org**.

dot pattern The pattern a **dot generator** produces on the screen of a TV receiver.

dot pitch On a display screen, the distance between phosphor dots of like color. For example, such an interval between adjacent green dots. Usually expressed in millimeters, and the lower the number, the crisper the image. Also called **phosphor pitch**.

dots per inch Its abbreviation is **dpi**, or **DPI**. A measure of the resolution of displayed or printed images. Generally

used for quantifying the resolution of laser printers. A 2400 dpi printer has 2400 dots per linear inch, and 5.76 million dots per square inch.

double **1.** Consisting of two, or divided into two, not necessarily equal components or parts. **2.** Twice in number or magnitude.

double armature An armature with two separate windings and commutators, but only a single core.

double-balanced mixer A combination of two single-balanced mixers. The benefits of this configuration include minimization of conversion loss, and greater suppression of even-order harmonics.

double-base diode A three-terminal transistor incorporating an emitter and two bases, and having a single pn junction. Utilized mostly for switching, and may be employed, for instance, in relaxation oscillators. Also called **unijunction transistor**.

double-base junction transistor A junction transistor incorporating two base terminals, with the second base connection serving as a fourth electrode. Also known as **tetrode junction transistor**.

double-beam cathode-ray tube Same as **double-beam CRT**.

double-beam CRT Abbreviation of **double-beam** cathode-ray tube. A CRT incorporating two independent electron beams. These beams may be produced by using two electron guns, or by splitting the beam of a single gun. Each beam forms its own trace, and the two traces may overlap. Also called **dual-gun CRT**, or **dual-beam CRT**.

double-beam oscilloscope Same as **dual-trace oscilloscope**.

double-break contacts Within a set of three contacts, a member which is normally closed, and which when opened breaks contact with the other two simultaneously.

double-break switch A switch, which when opened, interrupts a circuit at two points.

double bridge A special bridge circuit whose arrangement minimizes the effects of contact resistance, enabling the accurate measurement of very low resistances. Some such devices can measure resistances as low as few microhms. Also called **Kelvin bridge**, **Kelvin double bridge**, or **Thomson bridge**.

double buffering The use of two buffers to temporarily store information that awaits transfer to or from an input/output device. By using two buffers, one may be emptied while the other is filled, enhancing transfer speed. Also called **ping-pong buffer**.

double-button carbon microphone A carbon microphone with a button on each side of the diaphragm, which provides more gain and less distortion. Also called **double-button microphone**.

double-button microphone Same as **double-button carbon microphone**.

double-byte characters A version of ASCII which uses two bytes per character, as opposed to one. This expands the available characters from 256 to 65,536. This is sufficient to accommodate most of the languages in the world. Also called **Unicode**.

double-channel duplex A method for simultaneous two-way communications utilizing two channels, each transmitting in one direction.

double-channel simplex A method for non-simultaneous two-way communications utilizing two channels, each transmitting in one direction. It is a deprecated term.

double-click To press and release a computer mouse button, or its equivalent, twice in rapid succession, without moving said mouse. If the maximum time interval for both clicks is exceeded, the computer will interpret the input as two independent clicks. This time interval may be adjusted by the user, and double-clicking is used, for instance, to select and open a document in one step.

double-conversion receiver   Same as **dual-conversion receiver**.

double-conversion superheterodyne receiver   Same as **double-conversion receiver**.

double-current generator   A generator which supplies both AC and DC from a single armature winding.

double-density   A disk format indicating that the storage capacity is double that of the standard amount. For instance, if a given disk format holds 1 GB, then a double-density version would have a capacity of 2GB. Its abbreviation is **DD**.

double-density disk   A disk with a **double-density** format.

double-diffused metal-oxide semiconductor   Same as **diffused metal-oxide semiconductor**.

double-diffused MOS   Same as **diffused metal-oxide semiconductor**.

double-diffused transistor   A transistor in which two pn junctions are formed utilizing diffusion of both p- and n-type impurities contained in a gas.

double-diffusion   In the manufacturing of semiconductor devices, the utilization of two diffusion processes, one for each region of different conductivity.

double diode   An electronic component, such as an electron tube, incorporating two diodes. Also called **dual diode**, **duodiode**, or **twin diode**.

double-diode limiter   A limiter circuit, incorporating two diodes, which clips positive and/or negative signals beyond a given amplitude, from a series of positive and negative pulses.

double-doped transistor   A transistor formed by the successive addition of p- and n-type impurities to the melt during crystal growth.

double emitter follower   Same as **Darlington pair**.

double image   On the screen of a TV or radar receiver, an undesired condition in which duplicate images appear. May be due, for instance, to the signal arriving over two paths of different length. Also called **ghosting**, or **ghost (1)**.

double insulation   Additional insulation applied to a conductor. May be employed, for example, to reduce the risk of electric shock in the event of a failure of the insulation already present.

double-level metal   In ICs, a process for contact interconnections in which two vertical layers of metal are separated by an insulating layer. This enables the use of a smaller die size, as compared to a **single-level metal** process. Also called **double-level metal process**.

double-level metal process   Same as **double-level metal**.

double limiter   A limiter circuit which has two or more stages arranged in a manner that the output of one serves as the input for the next. Also called **cascade limiter**.

double-make contacts   Within a set of three contacts, a member which is normally open, and which when closed makes contact with the other two simultaneously.

double-make switch   A switch, which when closed, connects a circuit at two points.

double moding   In a magnetron, the abrupt and random change from one frequency to another.

double modulation   Modulation in which a subcarrier is modulated with the signal containing the desired information, and the resulting modulated subcarrier is then utilized to modulate a second carrier with a higher-frequency.

double-pole double-throw relay   A six-terminal relay arrangement which serves to connect one pair of contacts with either of two available pairs of contacts. Its abbreviation is **DPDT relay**. Also called **double-pole relay (2)**.

double-pole double-throw switch   A six-terminal switch arrangement which serves to connect one pair of contacts with either of two available pairs of contacts. Its abbreviation is **DPDT switch**. Also called **double-pole switch (2)**.

double-pole relay   **1.** Same as **double-pole single-throw relay**. **2.** Same as **double-pole double-throw relay**.

double-pole single-throw relay   A four-terminal relay arrangement which serves to simultaneously make or break two separate circuits, or both sides of the same circuit. Its abbreviation is **DPST relay**. Also called **double-pole relay (1)**.

double-pole single-throw switch   A four-terminal switch arrangement which serves to simultaneously make or break two separate circuits, or both sides of the same circuit. Its abbreviation is **DPST switch**. Also called **double-pole switch (1)**.

double-pole switch   **1.** Same as **double-pole single-throw switch**. **2.** Same as **double-pole double-throw switch**.

double-precision   The use of two computer words to store a single number, providing enhanced precision, and a greater range of magnitudes which can be represented. This contrasts with **single-precision**, in which a single computer word is utilized. For instance, if a single-precision number requires 64 bits, a double-precision number will utilize 128 bits.

double-precision number   A number represented utilizing **double-precision**.

double refraction   The separation of light into two components, each traveling at a different velocity, utilizing a birefringent material. Also called **birefringence**.

double screen   A CRT incorporating an additional layer or screen having a different color and longer persistence.

double shield   A cable or enclosure which incorporates two independent electromagnetic shields.

double-sideband   An amplitude-modulation signal in which the carrier, and both sidebands resulting from the modulation of the carrier, are present. Its abbreviation is **DSB**.

double-sideband amplitude modulation   Same as **double-sideband modulation**.

double-sideband modulation   Amplitude-modulation in which the carrier, and both sidebands resulting from the modulation of the carrier, are present. Its abbreviation is **DSB modulation**. Also called **double-sideband amplitude modulation**.

double-sideband reduced-carrier transmission   **Double-sideband transmission** in which the RF energy of the carrier is reduced. Its abbreviation is **DSBRC transmission**.

double-sideband suppressed-carrier transmission   **Double-sideband transmission** in which the RF energy of the carrier is suppressed. Its abbreviation is **DSBSC transmission**

double-sideband transmission   A form of amplitude-modulated transmission in which the carrier, and both sidebands resulting from the modulation of the carrier, are transmitted. Its abbreviation is **DSB transmission**.

double-sided disk   A disk, such as a floppy or DVD, which can store data on both surfaces.

double speed   An optical disk drive whose spinning speed is double that of the standard rate, or of the previous generation. It is a relative term, as what may currently be double speed may soon be much slower than the standard rate.

double-spot tuning   In superheterodyne reception, the tuning of the signal of a given station at two different local oscillation frequency values.

double-stream amplifier   A traveling-wave microwave amplifier in which the gain is produced by the interaction of two electron beams, each with a different average velocity.

double-stub tuner An impedance-matching tuner incorporating two stubs which are connected in parallel with a transmission line. The stubs are usually ⅜ wavelength apart.

double super-twist nematic A technology used in **double super-twist nematic displays**. Its abbreviation is **DSTN**.

double super-twist nematic display A twisted nematic display in which two separate liquid-crystal display plates are combined to form a single panel, providing greater contrast. Its abbreviation is **DSTN display**.

double superheterodyne receiver Same as **dual-conversion receiver**.

double-throw circuit breaker A circuit breaker which closes by making contact with either of the two available sets of contacts.

double-throw switch A switch that has two operating conditions, each of which connects one of the two available sets of contacts.

double-track recorder A magnetic-tape recorder whose recording head covers half of the width of the tape, which enables the recording of parallel tracks. These can be used either for stereo recording/reproduction along the entire tape, or for monophonic recording with double the reproduction time. Also called **dual-track tape recorder**, or **half-track tape recorder**.

double triode An electronic component, such as an electron tube, incorporating two triodes. Also called **duotriode**, or **dual triode**.

double-tuned amplifier An amplifier which has two different resonance frequencies. This provides a wider bandwidth than that obtained with a **single-tuned amplifier**.

double-tuned circuit A tuned circuit with two nearby resonance frequencies.

double-tuned detector A frequency-modulation discriminator which incorporates two tuned circuits. The resonance frequency of one of the circuits is tuned slightly above the center frequency, and the other is tuned an equal amount below it.

double-V antenna A wideband antenna consisting of two dipole elements folded in the shape of a letter V which are mounted along the same plane. It is used, for instance, for the reception of FM broadcasts. Also spelled **double-vee antenna**. Also called **fan antenna**.

double-vee antenna Same as **double-V antenna**.

double-winding synchronous generator A synchronous generator with two similar windings which are in phase with each other, but which are not connected electrically.

double word A computer word which is twice the usual size. For instance, if the word size for a given microprocessor is 32 bits, then a double word consist of 64 bits.

doubler 1. A circuit or device whose output frequency is twice that of its input frequency. May consist, for instance, of a stage whose resonant output circuit is tuned to the second harmonic of the input frequency. Also called **frequency doubler**. 2. A rectifier circuit whose output DC voltage is about twice the peak value of its input AC voltage. Such a circuit separately rectifies each half-cycle, then adds the rectified voltages. Also called **voltage doubler**.

doublet 1. Same as **dipole** (1). 2. Same as **dipole** (2). 3. A pair of electrons shared by two atoms, which forms a nonpolar bond.

doublet antenna Same as **dipole antenna**.

doubling 1. To make twice as great in amount or magnitude. 2. In a speaker, a form of distortion resulting from the disproportionate reproduction of secondary harmonics.

doubly-refracting Causing **double refraction**.

doughnut magnet Same as **donut magnet**.

down Said of a device, piece of equipment, or system which is not operational. Causes include failures, routine maintenance, repairs, and power outages.

down-conversion The mixing of a signal with a local oscillator, so that the frequency of the output signal is lower than that of the input signal. This contrasts with **up-conversion**, where the output signal has a higher frequency than the input signal. Also spelled **downconversion**.

down-converter A converter whose output frequency is lower than its input frequency. This contrasts with an **up-converter**, whose output frequency is higher than that of the input. Also spelled **downconverter**.

down counter A counter that only counts downwards, as opposed to an **up counter**, which only counts upwards, or an **up-down counter** which does both.

down-lead A transmission line that connects an antenna to a receiver or transmitter. Also called **lead-in**.

down time Same as **downtime**.

downconversion Same as **down-conversion**.

downconverter Same as **down-converter**.

downlink 1. The downward radio-communications path originating from a communications satellite, or other airborne transmitter, to the earth. This contrasts with an **uplink** (1), which is the converse. 2. The establishing, or use, of a **downlink** (1).

downlink frequency The frequency, or band of frequencies, of a **downlink signal**.

downlink signal The signal sent in a **downlink** (1). Such a signal usually occupies a given band of frequencies.

download 1. To receive a data, usually in the form of a file, from a remote computer in a network. For instance, a person navigating the Internet may download a file containing the latest version of a media player. Also, the data or file transferred. This contrasts with an **upload**, where information is sent to a remote computer in a network. 2. To load a font into a printer.

download protocol A protocol utilized for downloading files over a TCP/IP network, such as the Internet. In incorporates functions such as conversions between character codes.

downsampling The reduction of the sampling rate of a sampled signal. For instance, reducing the sampling rate of an audio sample from 44.1 kHz to 22.05 kHz.

downstream In communications, the direction of the flow of information from a content provider to a customer. For example, an ADSL connection may feature a transmission rate of over 10 Mbps when a file is downloaded. Generally utilized in the context of data transfer, but may also refer to other signals from providers, such as cable TV. This contrasts with **upstream**, where the flow of information is in the other direction.

downtime The time during which a device, piece of equipment, or system is not operational. Causes include failures, routine maintenance, repairs, and power outages. Also spelled **down time**. Also called **dead time** (1), or **off time** (2).

downward compatibility The state of being **downward compatible**.

downward compatible Software that is compatible with earlier versions of it, or hardware that is compatible with earlier models of it. Were it not for downward compatibility, upgrading any software or hardware would be unduly complex and expensive. Also known as **backward compatible**.

downward modulation Modulation in which the amplitude of the unmodulated carrier is always greater than the amplitude of the modulated wave.

DP Abbreviation of **data processing**.

DPCM Abbreviation of **differential pulse-code modulation**.

DPDT relay Abbreviation of **double-pole double-throw relay**.

DPDT switch Abbreviation of **double-pole double-throw switch**.

dpi Abbreviation of **dots per inch**.

DPI Abbreviation of **dots per inch**.

DPM Abbreviation of **digital panel meter**.

DPMS Abbreviation of **Display Power Management Signaling**.

DPNSS Abbreviation of **digital private network signaling system**.

DPSK Abbreviation of **differential phase-shift keying**.

DPST relay Abbreviation of **double-pole single-throw relay**.

DPST switch Abbreviation of **double-pole single-throw switch**.

DQDB Abbreviation of **distributed-queue dual-bus**.

DR Abbreviation of **dielectric resonator**.

draft mode Same as **draft quality** (2).

draft quality 1. Printed characters of minimal to adequate quality, usually well below that of letter quality. 2. A printing mode providing a **draft quality** (1) output. Also called **draft mode**.

drag 1. That which retards motion or action. For instance, a retarding force acting on a body. 2. In a GUI, to move an object from one location of a display screen to another. Text, images, files, and the such may be dragged by using a pointing device such as a mouse or stylus, or through keyboard commands. 3. To use a mouse to select text or objects. For instance, to click and hold down a mouse button to select a block of text.

drag-and-drop To utilize a mouse, or other pointing device, to perform operations by dragging objects on the display from one location to another. For instance, a user may drag-and-drop a document icon onto a printer icon for printing. Its abbreviation is **drag & drop**.

drag-cup motor A type of induction motor which features quick start/stops, rapid reversals, and high speed.

drag magnet In motor-type meters, a magnet which provides drag to limit the speed of the rotor or disk. The braking action is due to the effects of eddy currents. Also called **retarding magnet**.

drag & drop Same as **drag-and-drop**.

drain 1. The current, power, or energy drawn from a source. Also, the load or process that draws this current, power, or energy. 2. To draw current, power, or energy until a source is depleted. 3. In a field-effect transistor, the region into which the majority carriers flow from the source. The output of such a transistor is usually taken from the drain, which is analogous to the collector of a bipolar transistor, or the anode of an electron tube. Also, the electrode attached to this region. Its symbol is **D**. Also known as **drain region**, or **drain electrode**.

drain electrode Same as **drain** (3).

drain region Same as **drain** (3).

DRAM Abbreviation of **dynamic RAM**.

draw 1. To shape or elongate wire or metal by hammering, pulling through a die, or the like. 2. To shape or elongate wire or metal by pulling through one or more dies.

DRAW Abbreviation of **direct read after write**.

drawing In wire manufacturing, the pulling of a metal through one or more dies to reduce its diameter to the desired value. Also called **wire drawing**.

drawing program An application which enables the creation of drawings in the form of vector graphics, so picture objects may be created and manipulated as independent ob-

jects. In a **paint program**, drawings are created as bitmaps, which does not lend itself to such handling of illustrations.

DRDRAM Abbreviation of **D**irect **R**ambus **DRAM**. A RAM technology enabling transfers exceeding 10^9 bytes per second, and which supports pipelining. Also, RAM utilizing this technology.

DRDW Abbreviation of **direct read during write**.

Dreamcast A popular gaming system.

dress The arrangement of the connecting wires in a circuit so as to minimize or prevent undesired coupling and feedback.

dribbleware Software which is released in increments, or that has frequent updates and fixes. Users with such programs can usually update their software by downloading the necessary files from the appropriate Web site.

drift 1. A gradual and undesired change in a characteristic or a set adjustment of a component, circuit, device, instrument, piece of equipment, or system. Examples include frequency drift, thermal drift, or changes in power output. Also called **long-term drift** (1). 2. In a semiconductor material, the movement of charge carriers under the influence of an electric field.

drift compensation Measures taken to compensate for **drift** (1).

drift current The movement of charge carriers in a conductor or semiconductor, under the influence of an electric field.

drift field An electric field inherently present within a **drift transistor**.

drift-field transistor Same as **drift transistor**.

drift mobility The average drift velocity of charge carriers in a semiconductor material, per unit electric field. Also called **carrier mobility**, or **mobility**. When referring specifically to electrons, also called **electron mobility**.

drift space 1. In an electron tube, a region which is essentially free of applied electric or magnetic fields. 2. In a velocity-modulated vacuum tube, such as a klystron, a region where bunching occurs.

drift speed The average speed at which electrons or ions move through a medium.

drift transistor A transistor with an internal drift field which reduces the transit time of charge carriers and improves high-frequency response. Also known as **drift-field transistor**, or **diffused-alloy transistor**.

drift velocity The average velocity of charge carriers in a **drift current**.

drip loop A small loop formed in a transmission line, at a point where it enters a structure, to avoid moisture from entering along the surface of said line.

drive 1. The signal, such as a voltage or current, that causes the function of a component, circuit, device, piece of equipment, system, process, or mechanism. Also called **drive signal**, **driving signal** (1), or **excitation** (1). 2. The action of providing **drive** (1). Examples include the application of signal power to a transmitting antenna, a voltage to an oscillating crystal, or a signal voltage to the control electrode of an electron tube. Also called **excitation** (2). 3. In computers, an electromechanical device which reads and/or writes data to a storage medium, such as a disk or a tape. For example, a disk drive. 4. A mechanical device, such as that in a tape deck, which moves a magnetic tape past the heads, for recording, reproducing, or erasing. Such a drive also fast-forwards and rewinds tapes. Also called **tape drive** (1), **tape transport**, or **magnetic tape drive** (1).

drive array Two or more disk drives utilized as a single unit, to provide additional capacity, reduced access speed, and enhanced reliability.

drive bay An opening or shelf in a computer cabinet where additional disk drives, such as DVDs or hard disks, may be installed.

drive circuit A circuit which provides **drive** (1), especially to a motor.

drive control 1. A device which controls **drive** (1). 2. In a TV receiver, a potentiometer which adjusts the output of the horizontal oscillator. Also called **horizontal drive control**.

drive current A current which causes the function of a component, circuit, device, piece of equipment, system, process, or mechanism.

drive signal Same as **drive** (1).

drive tray A tray upon which a storage medium, such as an optical disc, is laid for insertion into its drive.

drive voltage A voltage which causes the function of a component, circuit, device, piece of equipment, system, process, or mechanism.

driven antenna An antenna which receives energy directly from the transmission line. This contrasts with a **parasitic antenna**, which is driven indirectly, that is, through radiation from a driven antenna. Also called **primary radiator**.

driven array An antenna incorporating multiple driven elements, all fed by a common source.

driven element An antenna element which is driven directly by the transmission line. This contrasts with a **parasitic element**, which is driven indirectly, that is, through radiation from a driven element. Also called **active element** (3).

driver 1. A circuit or device which provides and/or controls **drive** (1). 2. Same as **device driver**. 3. In an amplifier, the stage that precedes the output stage. Also called **driver stage**. 4. A transducer which converts electrical energy into sound energy, which is then radiated outward to the surrounding medium. Also called **speaker**, or **loudspeaker**. 5. Within a speaker system, a specialized speaker unit, such as a woofer or tweeter.

driver element Within an antenna incorporating multiple elements, a **driven element**.

driver stage Same as **driver** (3).

driver transformer A transformer which provides **drive** (1). Utilized, for instance, between stages of an amplifier.

driving-point admittance The reciprocal of **driving-point impedance**.

driving-point impedance The complex ratio, measured at the input terminals of a network or transducer, of the applied alternating voltage, to the resulting AC between said terminals.

driving signal 1. Same as **drive** (1). 2. In a TV transmitting system, a timing signal utilized to synchronize the horizontal or vertical scanning of an image.

DRO 1. Abbreviation of **dielectric resonator oscillator**. 2. Abbreviation of **digital readout**. 3. Abbreviation of **destructive readout**.

drone cone A cone speaker which is not driven. It vibrates in response to a driven speaker in the same enclosure. Utilized to reinforce bass frequencies. Also called **passive radiator**.

droop In a pulse train, a decrease in mean pulse amplitude, after maximum amplitude has been attained. Usually expressed as a percentage of maximum amplitude. Also called **pulse droop** (2).

drop 1. Same as **drop-out** (1). 2. Same as **drop cable**. 3. The voltage difference between any two points of a circuit or conductor, due to the flow of current. Also called **voltage drop** (1), or **potential drop** (2). 4. The voltage difference between the terminals of a circuit element, due to the flow of current. Also called **voltage drop** (2), or **potential drop** (3).

drop bar A safety device which grounds a high-voltage capacitor upon opening the door of a protective enclosure.

drop cable An aboveground or underground line which connects a terminal of a distribution cable to a subscriber's premises. Such a line may carry telephone service, cable TV, a link to a network, electric power, and so on. It may consist, for example, of a telephone wire extending from a pole to the home of a subscriber. Also called **drop** (2), **drop wire**, or **service wire**.

drop-down menu In a GUI, a menu which is always available in the active window, but whose contents are revealed only when a user selects said menu with a mouse or keyboard command. Once selected, the menu choices are presented in a vertical array below the menu title. Also spelled **drop-down menu**. Also called **pull-down menu**, or **pop-down menu**.

drop-in The erroneous gaining of bits during a read/write operation.

drop-out Same as **dropout**.

drop repeater In a multichannel communications system, a repeater which incorporates the necessary equipment for the local termination of multiple channels.

drop wire Same as **drop cable**.

dropdown menu Same as **drop-down menu**.

dropout Also spelled **drop-out**. 1. To lose or delete part of a transmitted signal, whether this occurs intentionally or not. Also called **drop** (1). 2. The erroneous loss of bits during a read/write operation. 3. A deterioration or loss of signal when a magnetic tape is dirty or has lost sections of its magnetic coating. In a video tape, for instance, this is manifested as streaks appearing on the screen during playback.

dropout current The maximum current at which a device, such as a relay or circuit-breaker, will release to its deenergized position. Also called **release current**.

dropout value The maximum actuating quantity, such as a current or voltage, at which a device, such as a relay or circuit-breaker, will release to its deenergized position. Also called **release value**.

dropout voltage The maximum voltage at which a device, such as a relay or circuit-breaker, will release to its deenergized position. Also called **release voltage**.

dropping resistor A resistor used in series with a load, which reduces the voltage applied to said load. The voltage reduction is equal to the voltage drop across the terminals of the resistor. Also called **series dropping resistor**.

dross Contaminants, usually oxides, which form on the surface of molten metals or alloys, especially solders.

drum 1. Something whose shape resembles that of a drum. 2. An early type of data-storage device, which consisted of a cylinder coated with a magnetic material, and whose tracks each had their own read/write head. Also called **drum storage**, or **magnetic drum**. 3. A rotating drum used in electrostatic printing, such as that employed by a laser printer to retain a charge.

drum recorder A recorder whose reproduced information is printed on a sheet which is wrapped around a rotating drum. Used, for instance, in some fax devices.

drum scanner A scanner in which the original to be scanned is mounted within a drum which is spun very rapidly during scanning, and which incorporates devices such as microscope lenses and photomultiplier tubes. The resolution attained when utilizing drum scanners is particularly high.

drum storage Same as **drum** (2).

drum switch A switch whose contacts, pins, surfaces, or the like, are placed on the periphery of a revolving drum. Used, for instance, as a motor controller.

drum winding 1. A winding in which the coils are on the outside of a cylindrical surface. 2. A winding in which the coils are on the inside of a cylindrical surface.

dry battery A battery composed of **dry cells**.

dry cell 1. A primary cell in which the positive electrode is carbon, the negative electrode is zinc, and the electrolyte is ammonium chloride. Also called **Leclanché cell**. **2.** A primary cell whose electrolyte is immobilized. It may be, for instance, a paste or a gel. This contrasts with a **wet cell**, whose electrolyte is a liquid

dry circuit A circuit whose maximum voltages and currents are so low, that no arcing occurs between contacts. Under these conditions, the dust particles that accumulate on said contacts are not burned off, leaving an insulating film that may eventually prevent the closing of the circuit.

dry contact An electric contact which does not open or close a circuit.

dry copper pair A pair of wires similar to those utilized for a POTS line, but which have no services, such as a dial tone, connected. Such a line was originally used for intruder alarm monitoring circuits, but can be used for digital communications, such as DSL. Also called **dry wire**, or **dry pair**.

dry-disk rectifier A rectifier which utilizes one or more metal disks coated with a semiconductor layer. This layer may consist of selenium, copper oxide, or another suitable semiconductor. The rectification occurs as a result of the greater conductivity across the contact in one direction than the other. Also called **dry rectifier**, **dry-plate rectifier**, **metallic-disk rectifier**, **semiconductor rectifier (1)**, or **contact rectifier**.

dry electrolytic capacitor An electrolytic capacitor whose electrolyte is in the form of a paste. This contrasts with a **wet electrolytic capacitor**, whose electrolyte is a liquid. A dry electrolytic capacitor has the advantage that it can be used in any position without leaking electrolyte.

dry etching Etching without the use of a solution. It is more precise than **wet etching**, but also more complex. An example is plasma etching.

dry flashover voltage For a clean and dry insulating material, the minimum voltage at which flashover occurs in the surrounding air.

dry joint A faulty soldering connection. May occur, for instance, when the solder has not been sufficiently heated.

dry pair Same as **dry copper pair**.

dry-plate rectifier Same as **dry-disk rectifier**.

dry rectifier Same as **dry-disk rectifier**.

dry-reed relay A relay whose contacts are mounted on magnetic reeds, and which is hermetically sealed in a glass tube. Such a relay is designed to be actuated by an external magnetic field. The contacts are not wetted by a pool of mercury, as is the case with a **mercury-wetted reed relay**. Also called **reed relay**, or **magnetic reed relay**.

dry-reed switch A switch whose contacts are mounted on magnetic reeds, and which is hermetically sealed in a glass tube. Such a switch is designed to be actuated by an external magnetic field. The contacts are not wetted by a pool of mercury, as is the case with a **mercury-wetted reed switch**. Also called **reed switch**, or **magnetic reed switch**.

dry run A practice session or procedure meant to test a device, piece of equipment, or program. For example, executing a computer program on paper, with the objective verifying operation and tracking down bugs.

dry wire Same as **dry copper pair**.

DS Abbreviation of **digital signal**, or **digital service**.

DS0 Abbreviation of **digital signal level 0**.

DS1 Abbreviation of **digital signal level 1**.

DS1C Abbreviation of **digital signal level 1C**.

DS2 Abbreviation of **digital signal level 2**.

DS3 Abbreviation of **digital signal level 3**.

DS4 Abbreviation of **digital signal level 4**.

DS5 Abbreviation of **digital signal level 5**.

DSA Abbreviation of **Directory System Agent**.

DSB Abbreviation of **double-sideband**.

DSB modulation Abbreviation of **double-sideband modulation**.

DSB transmission Abbreviation of **double-sideband transmission**.

DSBRC transmission Abbreviation of **double-sideband reduced-carrier transmission**.

DSBSC transmission Abbreviation of **double-sideband suppressed-carrier transmission**.

DSC **1.** Abbreviation of **digital selective calling**. **2.** Abbreviation of **digital-to-synchro converter**.

DSD Abbreviation of **Direct Stream Digital**.

DSF Abbreviation of **dispersion-shifted fiber**.

DSL Abbreviation of **d**igital **s**ubscriber **l**ine, or **d**igital **s**ubscriber loop. Digital communications technology and equipment which provides high-speed connections over ordinary copper lines. Most versions of this technology enable the same phone line to be split into two, one for analog voice and the other for DSL data. Upstream and downstream data transfer rates can exceed 100 Mbps. There are various versions, including ADSL, DSL Lite, HDSL, and VDSL. Also, such a service.

DSL access multiplexer A device in a central office which combines and separates multiple **DSL lines**.

DSL Light Same as **DSL Lite**.

DSL line A line providing **DSL service**.

DSL Lite A version of **DSL** which is simpler to implement, but is limited to lower speeds, such as 1.5 Mbps downstream, or less. Also spelled **DSL Light**.

DSL modem A modem utilized for **DSL** communications.

DSL network A communications network which utilizes **DSL** technology.

DSL service A service enabled by **DSL** technology and equipment.

DSL splitter A filter which separates the voice and data signals of a telephone line via which **DSL** service is accessed. The voice frequencies are in the lower frequency range, and the data is in the higher interval. Not all DSL services require a POTS splitter. Also called **splitter (3)**, or **POTS splitter**.

DSLAM Abbreviation of **d**igital **s**ubscriber **l**ine **a**ccess **m**ultiplexer. A device in a central office which combines and separates multiple **DSL lines**.

DSM Abbreviation of **demand-side management**.

DSML Abbreviation of **D**irectory **S**ervices **M**arkup **L**anguage. An XML specification for defining the content of directories.

DSO Abbreviation of **digital storage oscilloscope**.

DSP **1.** Abbreviation of **digital-signal processing**. **2.** Abbreviation of **digital-signal processor**.

DSR **1.** Abbreviation of **data set ready**. **2.** Abbreviation of **Dynamic Spatial Reconstructor**. **3.** Abbreviation of **data-signaling rate**.

DSRC Abbreviation of **dedicated short-range communication**.

DSS **1.** Abbreviation of **digital satellite system**. **2.** Abbreviation of **Digital Signature Standard**. **3.** Abbreviation of **decision support system**.

DSSS Abbreviation of **direct-sequence spread spectrum**.

DSTN Abbreviation of **double super-twist nematic**, or **dual-scan STN**.

DSTN display Abbreviation of **double super-twist nematic display**, or **dual-scan STN display**.

DSU Abbreviation of **data service unit**, or **digital service unit**.

DSU/CSU Abbreviation of **data service unit/channel service unit**, or **digital service unit/channel service unit**.

DSVD Abbreviation of **digital simultaneous voice and data**.

DT Abbreviation of **data transmission**.

DT-cut crystal A piezoelectric crystal cut used for vibration frequencies below 500 kHz.

DTE Abbreviation of **data terminal equipment**.

DTL Abbreviation of **diode-transistor logic**.

DTMF Abbreviation of **dual-tone multifrequency**.

DTMF cut-through Abbreviation of **dual-tone multifrequency cut-through**.

DTMF keypad Abbreviation of **dual-tone multifrequency keypad**.

DTMF signaling Abbreviation of **dual-tone multifrequency signaling**.

DTMF tone Abbreviation of **dual-tone multifrequency tone**.

DTP 1. Abbreviation of **distributed transaction processing**. **2.** Abbreviation of **desktop publishing**.

DTR Abbreviation of **data terminal ready**.

DTS Abbreviation of **DTS Surround Sound**, or **Digital Theater Sound**.

DTS Digital Surround Same as **DTS Surround Sound**.

DTS Digital Surround Sound Same as **DTS Surround Sound**.

DTS Sound Abbreviation of **DTS Surround Sound**.

DTS Surround Abbreviation of **DTS Surround Sound**.

DTS Surround Sound A 5.1 surround sound format that is digitally encoded, and which utilizes less compression than other similar systems, such as Dolby digital, for an exceptionally realistic listening experience. The data rate of such a system is 1.44 Mbps, as compared, for instance, with the 384 Kbps or 448 Kbps of Dolby digital. Its abbreviation is **DTS, DTS Surround**, or **DTS Sound**. Also called **DTS Digital Surround, DTS Digital Surround Sound**, or **Digital Theater Sound**.

DTV Abbreviation of **digital TV**.

DTV converter Abbreviation of **digital TV converter**.

DTV system Abbreviation of **digital TV system**.

DTW Abbreviation of **dynamic time warping**.

DUA Abbreviation of **Directory User Agent**.

dual 1. Consisting of two parts or elements. For example, a dual beam. **2.** Two components in the same housing or enclosure. For instance, a dual diode.

dual-attachment station Within a dual-ring FDDI network, a node which has a connection to both the rings. Its abbreviation is **DAS**.

dual beam Two independent, concentrated, and essentially unidirectional streams of radiated energy, such as radio waves, or particles, such as electrons.

dual-beam cathode-ray tube Same as **double-beam CRT**.

dual-beam CRT Same as **double-beam CRT**.

dual-beam oscilloscope Same as **dual-trace oscilloscope**.

dual boot A computer configuration which allows a user to select either of two different operating systems to boot the computer. Although this term usually refers to a choice between two operating systems, it may also refer to instances where there are more than two possible selections.

dual capacitor Two capacitors in the same housing.

dual-channel amplifier An amplifier with two independent channels, usually designated A and B, or left and right. Such a device may incorporate two identical amplifiers on the same chassis. Used, for instance, as a stereo high-

fidelity amplifier. Its abbreviation is **2-channel amplifier**. Also called **two-channel amplifier**.

dual-channel controller A circuit or device which controls two signal paths.

dual-cone loudspeaker Same as **dual-cone loudspeaker**.

dual-cone speaker A speaker enclosure which incorporates two cone drivers. If it is a two-way speaker, one cone usually handles the lower frequencies, while the other handles the rest. Also called **dual-cone loudspeaker**.

dual control A security measure in which two people must provide passwords and/or objects, such as keys, to pass a particular authorization, access, or identification requirement.

dual-conversion receiver A superheterodyne receiver in which there are successive frequency conversions utilizing two local oscillators, and thus having two intermediate frequencies. The first intermediate frequency is higher, for adequate image rejection, while the lower second intermediate frequency provides high selectivity and gain. Also called **dual-conversion superheterodyne receiver, double-conversion receiver, double-conversion superheterodyne receiver, double superheterodyne receiver**, or **triple-detection receiver**.

dual-conversion superheterodyne receiver Same as **dual-conversion receiver**.

dual counter-rotating ring A communications network topology in which there are two rings, each with data traveling in the opposite direction. For example, one ring may serve as the primary ring, providing data transmission during normal operation, while the other, or secondary ring, may provide backup.

dual diode Same as **double diode**.

dual-diversity Same as **diversity system**.

dual-diversity receiver Same as **diversity receiver**.

dual-diversity reception Same as **diversity reception**.

dual-diversity system Same as **diversity system**.

dual-diversity transmission Same as **diversity transmission**.

dual-diversity transmitter Same as **diversity transmitter**.

dual-gate FET Abbreviation of **dual-gate field-effect transistor**. A field-effect transistor with two gate electrodes.

dual-gate field-effect transistor Same as **dual-gate FET**.

dual-gate metal-oxide semiconductor field-effect transistor Same as **dual-gate MOSFET**.

dual-gate MOSFET Abbreviation of **dual-gate metal-oxide semiconductor field-effect transistor**. A metal-oxide semiconductor field-effect transistor with two gate electrodes.

dual-gun cathode-ray tube Same as **double-beam CRT**.

dual-gun CRT Same as **double-beam CRT**.

dual homed 1. Within an FDDI network, a node which is connected to both the primary and secondary rings. **2.** Connected to two networks.

dual in-line memory module Same as **DIMM**.

dual in-line package A rectangular IC package in which the connecting pins are arranged in two parallel lines on opposite sides of the package. Such a package is usually made of plastic or ceramic, and its pins protrude downward from its longer sides. Used, for instance, for computer RAM, or for cache memory. Its acronym is **DIP**. Also spelled **dual inline package**.

dual inline memory module Same as **DIMM**.

dual inline package Same as **dual in-line package**.

dual laser Also called **dual-wavelength laser**. **1.** A laser which utilizes two distinct wavelengths. Such a device may be designed for use by a dentist who employs one wavelength to drill teeth, and the other to work on soft tissue, such as the gums. Another example is its use in some DVD players, to enable reading optical discs with other formats, such as CDs. **2.** A gas laser which simultaneously produces two distinct wavelengths from a single laser beam.

dual meter A meter which enables the monitoring of two aspects of an electric circuit, such as voltage and current. Also called **twin meter (1)**.

dual-mode control A method of control which incorporates two distinct modes of operation. Utilized, for instance, in robotics.

dual modulation The simultaneous modulation of a carrier or subcarrier with two different types of modulation. In each, separate information is conveyed.

dual-output power supply A power supply which provides two outputs. For instance, one may be DC, and the other AC.

dual potentiometer An assembly of two potentiometers. Usually consists of a pair of ganged potentiometers.

dual processors The incorporation of two CPUs in a single computer, to improve performance. One may be utilized, for instance, to perform floating-point operations.

dual-ring topology In an FDDI network, a topology in which there are two rings, each with data traveling in the opposite direction. One ring is the primary ring, which serves for data transmission during normal operation, while the other, or secondary ring, provides backup. Such an arrangement provides for superior reliability and robustness.

dual-scan display Same as **dual-scan STN display**.

dual-scan LCD 1. Same as **dual-scan STN**. **2.** Same as **dual-scan STN display**.

dual-scan STN Abbreviation of **dual-scan** super-twisted nematic. A technology used in **dual-scan STN displays**. Its own abbreviation is **DSTN**. Also called **dual-scan LCD (1)**.

dual-scan STN display Abbreviation of **dual-scan** super-twisted nematic **display**. An STN display with improved brightness and contrast. Its own abbreviation is **DSTN display**. Also called **dual-scan display**, or **dual-scan LCD (2)**.

dual-scan super-twisted nematic Same as **dual-scan STN**.

dual-scan super-twisted nematic display Same as **dual-scan STN display**.

dual-tone multifrequency Its abbreviation is **DTMF**. Also called **dual-tone multifrequency signaling**, or **multifrequency signaling**. **1.** A telephone signaling system which utilizes tones which are a mixture of two frequencies. Each mixture identifies any of the 12 dial keys, 0-9, *, and #, as found on most touch-tone phones. The tone heard when a number 5 is pressed on such a phone, for instance, is a combination of two such frequencies, specifically 770 Hz and 1336 Hz. **2.** A communications method utilizing **dual-tone multifrequency (1)**.

dual-tone multifrequency cut-through The capability of an interactive voice-response system to receive DTMF tones during playback. Such systems, for instance, will accept an extension number any time it is pressed, instead of requiring the caller to wait for the end of the announcement. Its abbreviation is **DTMF cut-through**.

dual-tone multifrequency keypad A telephone keypad utilizing **dual-tone multifrequency (1)**. Its abbreviation is **DTMF keypad**.

dual-tone multifrequency signaling Same as **dual-tone multifrequency**. Its abbreviation is **DTMF signaling**.

dual-tone multifrequency tone A tone generated via **dual-tone multifrequency (1)**. Its abbreviation is **DTMF tone**.

dual trace In a CRT, such as that in an oscilloscope, the incorporation of two independent electron beams, each with its own trace. These beams may be produced by using two electron guns, or by splitting the beam of a single gun.

dual-trace oscilloscope An oscilloscope incorporating a **dual trace**. May be used, for instance, to monitor two phenomena on the same screen. Also called **dual-beam oscilloscope**, or **double-beam oscilloscope**.

dual-track tape recorder Same as **double-track tape recorder**.

dual triode Same as **double triode**.

dual-wavelength laser Same as **dual laser**.

duality 1. The quality or state of being **dual (1)**. **2.** Same as **duality principle**.

duality principle A principle pertaining to the dual or analogous nature of certain entities or properties. For instance, the wave-particle duality of light, in which it has both wave-like properties and particle-like properties. Another example is the analogous nature of certain electric circuits, such as a transistor circuit and an electron tube circuit. Also called **duality (2)**, or **principle of duality**.

dub Also called **re-record**. **1.** For a given audio and/or video recording, to add or replace sounds, or one or more sound tracks. For example, in a movie, to add sound effects or replace dialogue with that of another language. **2.** To transfer of all or part of one recording to another. **3.** To mix two or more sound sources into a single recording. **4.** To record on one device what it being reproduced through another. **5.** That which has been added, replaced, transferred, mixed, or copied through **dubbing**.

dubbing Also called **re-recording**. **1.** For a given audio and/or video recording, the adding or replacing of sounds, or of one or more sound tracks. For example, in a movie, the adding of sound effects or the replacement of dialogue with that of another language. **2.** The transfer of all or part of one recording to another. **3.** The mixing of two or more sound sources into a single recording. **4.** The recording on one device what it being reproduced through another.

dubnium A synthetic radioactive chemical element whose atomic number is 105, and which has about ten identified isotopes. The techniques employed to produce this element are utilized to help discover new elements with higher atomic numbers. Its chemical symbol is **Db**.

duct 1. A pipe, tube, or channel through which a substance or waves are conveyed or propagated. When referring to a pipe or tube with precise dimensions, through which microwave energy is transmitted, also called **waveguide (2)**. **2.** A pipe, tube, or channel through which lines, cables, or wires are run. **3.** Within the troposphere, a layer which, depending on the temperature and humidity conditions, may act as a waveguide for the transmission of radio waves for extended distances. Also called **atmospheric duct, tropospheric duct**, or **wave duct (2)**. **4.** A narrow layer that forms under unusual conditions in the atmosphere or ocean, and which serves to propagate radio waves or sound waves. Also called **wave duct (3)**. **5.** Same as **ducted port**.

ducted port A carefully-dimensioned opening in a speaker enclosure. It is employed for various purposes, but is mainly used to increase and extend the reproduction of low frequencies. Also called **duct (5), port (6), speaker port**, or **vent (3)**.

ductile Easily shaped, hammered into a thin strip or foil, or drawn into wire. Used in the context of metals or alloys, such as gold, silver, or brass.

ducting 1. To convey or propagate through a **duct (1)**. **2.** To enclose in a **duct (2)**. **3.** The occurrence of a **duct (3)**. Also called **tropospheric ducting**.

dull emitter In an electron tube, a cathode which operates at a lower temperature than usual. This is usually achieved by coating said cathode with a metal-oxide compound, such as thorium oxide, in order to enhance its electron emissions.

dumb network A network with little or no processing of the transmitted signals. This contrasts with an **intelligent net-**

work, which has processing power, so that functions such as diagnostics can be incorporated.

dumb terminal A computer input/output device which incorporates a video adapter, monitor, keyboard, and usually a mouse. Such a terminal has no processing capability, and is used in networks. This contrasts with an **intelligent terminal**, which does have processing capability.

dummy 1. A non-operational component, device, or piece of equipment which can be connected or installed as if it were a working component, device, or piece of equipment. Used, for instance, for testing, or to fulfill a requirement. 2. In computers, a character, variable, instruction, address, or the like, which serves as a placeholder. It does nothing more than reserve a space until the intended item replaces it, or to fulfill a requirement.

dummy antenna A device which simulates the electrical characteristics of an antenna, but which dissipates its input as heat instead of radiating RF energy. It provides a nonradiating load for a transmitter, and is used primarily for testing and adjusting. Also called **artificial antenna**.

dummy component A non-operational component which can be connected or installed as if it were a working component. Used, for instance, for testing, or to fulfill a requirement.

dummy device A non-operational device which can be connected or installed as if it were a working device. Used, for instance, for testing, or to fulfill a requirement.

dummy instruction A computer instruction which produces no action. It simply serves to cause the processor to utilize a clock cycle, or to proceed to the next instruction. Used, for instance, to complete a very long instruction word. Also called **do-nothing instruction**, or **no-operation instruction**.

dummy load A device which simulates a transmission line load, such as that of a speaker or antenna. Used primarily for testing and adjusting. Also called **artificial load**.

dump Also called **storage dump**. 1. To display, print, copy, or transfer the content of a computer memory or storage device, with little or no formatting. 2. To display, print, copy, or transfer the content of the main memory of a computer. Such a dump may be performed after a process which ends abnormally, for instance, to help pinpoint the source of a problem. A dump may also be generated automatically. Also called **memory dump**, or **core dump**.

dump power Generated electric power which exceeds that which can be utilized, conserved, or stored.

dumping The occurrence of a **dump**.

dumping resistor 1. A resistor utilized to discharge a capacitor to help insure the safety of personnel. 2. A resistor connected across a voltage source to drain off the charge remaining in capacitors when the power is turned off, while also improving voltage regulation. Also called **bleeder resistor**.

DUN Abbreviation of **dial-up network**.

duodiode Same as **double diode**.

duotriode Same as **double triode**.

duplex 1. Consisting of two parts or elements, especially those which work together. 2. Two-way communication which occurs simultaneously in both directions. Also called **duplex communication** (1), **full-duplex**, or **two-way communication** (2). 3. Two-way communication in which communication can only occur one direction at a time. Also called **duplex communication** (2), **half-duplex**, or **two-way communication** (3).

duplex cable 1. A cable consisting of a twisted-pair of wires. 2. A fiber-optic cable containing two optical fibers.

duplex channel 1. A communications channel utilized for **duplex** (2). Also known as **full-duplex channel**. 2. A communications channel utilized for **duplex** (3). Also known as **half-duplex channel**.

duplex circuit 1. A communications circuit enabled for **duplex** (2). Also known as **full-duplex circuit**. 2. A communications circuit enabled for **duplex** (3). Also known as **half-duplex circuit**.

duplex communication 1. Same as **duplex** (2). 2. Same as **duplex** (3).

duplex operation Also called **duplexing** (1). 1. The operation of a communications channel utilized for **duplex** (2). 2. The operation of a communications channel utilized for **duplex** (3).

duplex system 1. A communications system utilized for **duplex** (2). Also called **full-duplex system**. 2. A communications system utilized for **duplex** (3). Also called **half-duplex system**.

duplex transmission 1. Two-way communication in which there are simultaneous transmissions in both directions. Also called **full-duplex transmission**. 2. Two-way communication in which there are transmissions in only one direction at a time. Also called **half-duplex transmission**.

duplexed system Two systems, such as computers, which work identically. For instance, one may be used if the other fails.

duplexer A device which enables a single antenna to be used to transmit and receive simultaneously or alternately. Used, for example, in radar. Also called **antenna duplexer**.

duplexing 1. Same as **duplex operation**. 2. The use of a **duplexing assembly**.

duplexing assembly A switch that enables a single antenna to be used to transmit and receive simultaneously or alternately. This switch can be manual or automatic, and may be in the form of a circuit, tube, or device. Used, for example, in radar. Also called **transmit-receive switch**.

duplication check In computers, the verification of a result by computing it twice, each time utilizing a different method.

duration control In a sensitivity-time control circuit, the controlling of the duration of the reduced gain.

Duron A popular family of CPU chips.

Dushman equation A fundamental equation for describing the thermionic emission of electrons from the surface of a metal which is heated to high temperatures. Also called **Richardson-Dushman equation**.

dust chamber A dust-collecting chamber in a **dust precipitator**. Also called **dust collector**.

dust collector Same as **dust chamber**.

dust core A magnetic core made from powdered material. The particles are mixed with an appropriate binder, and the core is formed through applied pressure. When said powdered material is ferrite, also called **ferrite core**, or **powdered-iron core**.

dust-ignition-proof motor A motor whose housing is dustproof, thus eliminating the possibility of dust igniting inside the motor.

dust precipitator A device which removes dusts present in gases, such as air, by precipitating them in a chamber containing high-voltage wires or grids. The gas is forced through the chamber, and the wires or grids ionize the dust particles, which in turn makes them migrate to the walls of said chamber. Also called **electrostatic precipitator**, **precipitator**, **electronic air cleaner**, or **electronic precipitator**.

dust-proof Also spelled **dustproof**. Also called **dust-tight**. 1. Impervious to dust. 2. Protected in a manner which does not allow dust to enter an enclosure.

dust-tight Same as **dust-proof**.

dustproof Same as **dust-proof**.

DUT Abbreviation of **device under test**.

duty **1.** The work or operation of a device or piece of equipment. Also, an interval of such work or operation. **2.** The work or operation of a device or piece of equipment, under given conditions. Also, an interval of such work or operation.

duty cycle **1.** For a device or piece of equipment used intermittently, the cycle of starting, operating, and stopping. Also, the time interval that elapses during such a cycle. **2.** For a device or piece of equipment used intermittently, the ratio of its operating time to its rest time, or to total time. Also called **duty ratio** (1), or **duty factor** (1). **3.** The ratio of pulse duration to pulse period, expressed as a percentage. When expressed as a decimal, also called **duty factor** (2). **4.** The ratio of average pulse power to peak pulse power. Also called **duty ratio** (2).

duty factor **1.** Same as **duty cycle** (2). **2. Duty cycle** (3) expressed in decimal form.

duty ratio **1.** Same as **duty cycle** (2). **2.** Same as **duty cycle** (4).

DV Abbreviation of **digital video**.

DVB Abbreviation of **digital video broadcasting**.

DVD Abbreviation of **d**igital **v**ersatile **d**isc or **d**igital **v**ideo **d**isc, but **DVD** also serves as the proper name for such discs. A digital optical storage medium, usually 12 centimeters in diameter, whose contained information is encoded in microscopic pits on its metallic surface, which is protected by a plastic layer. Since there is no contact between the pickup and the recorded surface, wear is minimized, and the protective layer helps avoid reading errors due to dust or minor marks on the surface of the disc. To recover the recorded content, a laser is focused on the metallic surface of the disc, and the reflected light is modulated by the code on said disc. DVDs can be used to store any type of digitally encoded information, and are used especially for movies and/or computer multimedia content, and can hold up to 17 GB. DVD players can usually also play optical discs with other formats, such as CDs and Laserdiscs.

DVD audio A **DVD** which stores music, voices, and/or other sounds. It is superior to CDs in many ways, including its multiple-channel surround capabilities, higher sampling rate, and advanced storage and playback capabilities and features. This helps provide sound reproduction with exceptional characteristics, including an extremely high signal-to-noise ratio, and a very extended dynamic range.

DVD audio disc A disc which contains **DVD audio**.

DVD audio player A device which can read the information stored on a **DVD audio disc**.

DVD audio/video player A **DVD audio player** which can also play optical discs with other formats, such as DVDs and CDs.

DVD burner Same as **DVD recorder**.

DVD carousel player A **DVD player** which utilizes a carousel. A DVD carousel is usually utilized when a few discs are available, so a DVD jukebox is not likely to utilize this method to hold discs awaiting to be swapped, so as to save space. Also called **carousel DVD player**.

DVD changer A **DVD player** that holds more than one DVD disc. Discs may be swapped, but only one can be read at once.

DVD disc A disc intended for use with a **DVD drive** or **DVD player**. Such a disc is used especially for movies and/or computer multimedia content, and can hold up to 17 GB.

DVD drive Same as a **DVD player**. The term **DVD drive** more often refers to such a device used in a computer system, as opposed to an entertainment system, where the term **DVD player** is preferred.

DVD-E Abbreviation of **DVD**-erasable. Same as **DVD**-**rewritable**.

DVD-erasable Same as **DVD-rewritable**.

DVD jukebox A **DVD changer** that can hold many, up to several thousand, DVDs. Discs may be swapped, but only one can be read at once. This contrasts with a **DVD tower**, which can access multiple discs simultaneously.

DVD player A device which can read the information stored on a DVD. To recover the recorded content, a laser is focused on the metallic surface of the disc, and the reflected light is modulated by the code on said disc. DVD players can usually also play optical discs with other formats, such as CDs and Laserdiscs. Also called **DVD system**, **DVD reader**, or **DVD drive**.

DVD-R Abbreviation of **DVD**-recordable. A **DVD** technology that allows such a disc to be recorded. A DVD recorder is used for this purpose, and such discs can only be written on once. A DVD player is used to recover the recorded information.

DVD-R disc A disc intended for use with a **DVD recorder**. Such a disc can store up to 17 GB.

DVD-RAM A **DVD-rewritable** format which can be utilized for reliably rewriting many thousands of times, and whose discs are usually housed in a cartridge.

DVD-read-only memory Same as **DVD-ROM**.

DVD reader Same as **DVD player**.

DVD-recordable Same as **DVD-R**.

DVD-recordable disc A disc intended for use with a **DVD recorder**. Such a disc can store up to 17 GB.

DVD recorder A device used to record **DVDs**. A laser records the desired content by forming the equivalent of pits on the recordable surface by altering the reflectivity of a dye layer. When such a disc is placed in a DVD player, its laser pickup reads the patterns as if they had been permanently stamped into its metallic surface. Also called **DVD burner**, or **DVD writer**.

DVD-rewritable Its abbreviation is **DVD-RW**. A **DVD** technology which allows discs to be rewritten many times. To record, a high-intensity laser followed by a medium-intensity laser anneals the crystals in the recording layer, changing its reflectivity. A low-intensity laser in the DVD player reads the recorded information. Also called **DVD-E**, or **rewritable DVD** (1).

DVD rewritable disc Its abbreviation is **DVD-RW disc**. A **DVD disc** which can be rewritten many times. Also called **rewritable DVD** (2).

DVD-ripper Software that copies data from a **DVD** to a computer hard drive. Also called **ripper** (1).

DVD-ROM Abbreviation of **DVD**-**r**ead-only **m**emory. A **DVD** format which provides higher storage capacity than CD-ROMs, and which is specifically designed to store computer multimedia content.

DVD-ROM disc A disc intended for use with a **DVD-ROM player**. Such a disc usually holds computer multimedia content.

DVD-ROM drive Same as **DVD-ROM player**.

DVD-ROM player A device which can read the information stored on a **DVD-ROM**. It can also play optical discs with other formats, such as DVDs and CDs. Also called **DVD-ROM drive**.

DVD-RW Abbreviation of **DVD-rewritable**.

DVD-RW disc Abbreviation of **DVD rewritable disc**.

DVD+RW A **DVD** technology which allows discs to be rewritten many times. It is similar to **DVD-RW**, but has a larger capacity.

DVD server A **DVD tower** intended for network use.

DVD system Same as **DVD player**.

DVD tower An enclosure containing multiple DVD drives. All the drives are accessible at all times. This contrasts with a **DVD jukebox**, which can only access one disc at a time. Used, for instance, in networks.

DVD video A **DVD** format used for storing full-length movies and other video.

DVD writer Same as **DVD recorder**.

DVE 1. Abbreviation of **digital video effects**. 2. Abbreviation of **distributed virtual environment**.

DVM Abbreviation of **digital voltmeter**.

DVMRP Abbreviation of **Distance Vector Multicast Routing Protocol**.

DVOM Abbreviation of **digital volt-ohm-milliammeter**.

Dvorak Keyboard A keyboard format which is substantially more ergonomically sound than the standard QWERTY format. The QWERTY keyboard was originally designed to slow down a typist as much as possible, as the typewriters then would jam when even modest typing speeds were attained. A Dvorak Keyboard places the keys of each character in a manner which allows for greatly reduced finger travel, in addition to promoting the use of all fingers of both hands when typing, among other benefits.

DVST Abbreviation of **direct-view storage tube**.

DWDM Abbreviation of **dense wavelength division multiplexing**, or **dense wave division multiplexing**.

dwell A delay in a process or an action. For instance, a programmed pause in the work cycle of a robot. Also called **dwell time**.

dwell time Same as **dwell**.

DX 1. Abbreviation of **distance reception**. 2. Abbreviation of **duplex**.

Dy Chemical symbol for **dysprosium**.

dye laser A laser whose active medium is a dye solution. Depending in the dye used, such lasers can operate at wavelengths which include visible, infrared, or ultraviolet light. An example of such a dye is rhodamine B, which has a peak emission wavelength of 610 nanometers.

dye-polymer recording A recording technique utilized with rewritable optical discs. Minute bumps are formed in the dyed layer when recording, which are then read by a laser during reproduction.

dye-sublimation printer A printer in which the dyes are vaporized and then solidified on the page, providing a smoothly mixed color dot. Such printers do not utilize dithering, and can provide photograph-quality images.

dyn Abbreviation of **dyne**.

dynamic Also called as **dynamical**. 1. Pertaining to bodies in motion. 2. Characterized by continuous activity or change. 3. Dependent on changing conditions or parameters. 4. An event or process that occurs during the execution of a program. 5. In computers, an action or process which occurs when needed, as occurs in dynamic allocation.

dynamic address translation The conversion of a relative memory address into an absolute address, during program execution.

dynamic address translator A computer circuit that converts a relative memory address into an absolute address, during program execution.

dynamic allocation 1. The allocation of computer resources during program execution. 2. Same as **dynamic memory allocation**.

dynamic bandwidth allocation Its abbreviation is **DBA**. Bandwidth allocation in which a variable number of users may utilize the full available bandwidth. DBA quickly reapportions bandwidth based on the traffic conditions at any given moment. Used, for instance, in ATM.

dynamic binding The linking of software routines or objects during program execution. This contrasts with **static bind-**

ing, where the linking occurs during program compilation. Also called **late binding**.

dynamic braking 1. A technique for quickly stopping an electric motor. To brake, the motor is disconnected from its power source, thereby becoming a generator. Then, a connected resistor dissipates the energy of rotation into heat. Also called **resistance braking**. 2. **Dynamic braking** (1) in which the generated energy is stored, instead of being dissipated into heat. Used, for instance, in specially-equipped trains or cars. Also called **regenerative braking**.

dynamic caching The allocation of cache based on the available memory, as opposed to that allotted to the application running.

dynamic characteristic The relationship between the instantaneous values of two varying quantities, such as electrode current and electrode voltage, under specified load circumstances. Also called **load characteristic**.

dynamic check Same as **dynamic test**.

dynamic convergence In a picture tube with a three-color gun, the process by which the three electron beams are made to converge on a specified surface during scanning.

dynamic debugging Debugging performed during program execution.

dynamic dump A dump occurring during program execution.

dynamic earphone An earphone consisting of a small or miniature **dynamic speaker**. A pair may be used for stereophonic listening.

dynamic electric field An electric field whose strength varies continuously.

dynamic electricity Electricity produced by electric charges in motion, such as electrons or holes moving through a conductor. This contrasts with **static electricity**, which is that produced by electric charges at rest.

dynamic equilibrium For a given system, a state in which multiple processes vary, while the overall system remains unchanged. The various processes work simultaneously to maintain the system balanced.

dynamic error An error in the varying output of a transducer, as a result of restrictions imposed by the dynamic response of said transducer.

dynamic focus Same as **dynamic focusing**.

dynamic focusing In a color picture tube, the varying of the voltage applied to the focus electrode, to compensate for electron-beam defocusing on the flat surface of the screen. Also called **dynamic focus**.

dynamic headphone A headphone consisting of a small **dynamic speaker**. A pair may be used for stereophonic listening. Also called **dynamic headset**.

dynamic headset Same as **dynamic headphone**.

Dynamic Host Configuration Protocol In a TCP/IP network, a protocol for assigning dynamic IP addresses. This eliminates the need to manually assign IP addresses each time a computer is added to the network. Frequently used with dial-up connections. Its abbreviation is **DHCP**.

Dynamic Host Configuration Protocol network A network utilizing a **Dynamic Host Configuration Protocol**. Its abbreviation is **DHCP network**.

Dynamic Host Configuration Protocol server A server in a **Dynamic Host Configuration Protocol network**. Its abbreviation is **DHCP server**.

Dynamic HTML Abbreviation of **Dynamic HyperText Markup Language**. A technology that enables the content of Web pages to be updated as influenced by user actions, without requiring constant downloads from the Web server. Such a page, for example, could provide varied content for each specific user, depending on factors like the geographic location, or information previously gathered, such as pages recently viewed. It is intended to provide better-tailored content while reducing wait time and resources occupied by the server. Its own abbreviation is **DHTML**.

Dynamic Hypertext Markup Language Same as **Dynamic HTML**.

dynamic impedance The impedance of a device during its operation.

dynamic Internet-Protocol address Same as **dynamic IP address**.

dynamic IP address Abbreviation of **dynamic** Internet-Protocol **address**. An IP address that changes each time a user logs onto a TCP/IP network. Such an IP address may also vary during the same session. This contrasts with a **static IP address**, which is the same each time a user logs on.

dynamic key A encryption technique in which the key changes value each time it is used. In this manner, an intercepted key would not be useful for any future occasions. Also called **dynamic-key encryption**.

dynamic-key encryption Same as **dynamic key**.

dynamic link library A collection of routines and programs which can be called upon as needed by another program the computer may be running. Since DLLs are not loaded into RAM along with the main program, there is a saving in memory resources. Its abbreviation is **DLL**. Also called **dynamically-linked library**, or **dynamic-linked library**.

dynamic-linked library Same as **dynamic link library**. Its abbreviation is **DLL**.

dynamic loudspeaker Same as **dynamic speaker**.

dynamic magnetic field A magnetic field whose strength varies continuously.

dynamic memory A type of computer memory that utilizes series of charges in capacitors to store information. Since these capacitors quickly lose their charge, this form of memory requires constant refreshing in order to avoid the loss of data. Also called **dynamic storage**.

dynamic memory allocation The allocation of computer memory during program execution. Also called **dynamic allocation (2)**.

dynamic microphone A type of microphone in which incident sound waves strike a diaphragm, which is connected to a moving coil. The coil moves within a constant magnetic field, which in turn produces the output signal current. Also called **moving-coil microphone**, or **electrodynamic microphone**.

dynamic noise suppressor A filter circuit which suppresses noise by adjusting its bandpass according to the signal strength. At low signal levels, when noise is most noticeable, the low-frequency and high-frequency bands are reduced. As the signal strength is increased, the action of the filter is reduced accordingly. This helps maintain the signal-to-noise ratio at adequate levels.

dynamic output impedance The output impedance of a device during its operation.

dynamic pickup A phonograph pickup in which the vibrations of the stylus move a coil which is in a continuous magnetic field, which in turn produces the output signal current. Also called **dynamic reproducer, moving-coil cartridge**, or **moving-coil pickup**.

dynamic plate resistance In vacuum tubes, the opposition the plate circuit offers to a small increase in plate voltage. Measured in ohms. Also called **AC plate resistance**.

dynamic RAM Abbreviation of **dynamic** random-access memory. A type of RAM that utilizes series of charges in capacitors to store information. Since these capacitors quickly lose their charge, this type of memory requires constant refreshing in order to avoid the loss of data. The are

various types, including **EDO DRAM** and **SDRAM**. Its abbreviation is **DRAM**.

dynamic random-access memory Same as **dynamic RAM**.

dynamic range 1. For a system or device, the ratio of the maximum producible level of a given parameter, such as signal strength or power, to the minimum detectable level of said parameter. Usually expressed in decibels. 2. For a system or device, the ratio of the maximum producible level of a given parameter, such as signal strength or power, to the minimum detectable level of said parameter, while still maintaining a useful output. Usually expressed in decibels. 3. For an audio component, such as an amplifier or a speaker, the ratio of the maximum reproducible loudness, to the minimum detectable loudness, without excessive noise or distortion. Usually expressed in decibels. 4. For an audio component, such as an amplifier or a speaker, the difference in decibels between the maximum reproducible loudness and the minimum detectable loudness, without excessive noise or distortion.

dynamic reproducer Same as **dynamic pickup**.

dynamic resistance The resistance of a device during its operation.

dynamic routing Routing that adjusts automatically to varying conditions within a communications network, such as malfunctions which arise or increased traffic. For instance, if a circuit fails, it sends the intended data through another route.

dynamic run Same as **dynamic test**.

Dynamic Spatial Reconstructor An ultrafast type of CT scanner. It is capable, for instance, of generating dynamic three-dimensional images of a beating heart. Its abbreviation is **DSR**.

dynamic speaker A type of speaker in which an audio-frequency signal current is sent through a voice coil, which is attached to a diaphragm. The interaction with a constant magnetic field provides the coil with a piston-like movement, which is mechanically transferred to the diaphragm, which in turn generates the sound waves that emanate from the speaker. Also called **dynamic loudspeaker, moving-coil speaker**, or **electrodynamic speaker**.

dynamic SQL SQL in which queries are formulated or modified at runtime.

dynamic stability The ability of a system or object to return to a previously established steady motion, after being perturbed. Used, for instance, to refer to the ability of a walking robot to maintain its balance while moving.

dynamic storage Same as **dynamic memory**.

dynamic test A performance test made during the operation of a component, circuit, device, piece of equipment, or system. Also called **dynamic run**, or **dynamic check**.

dynamic time warping A signal-matching method in which the timescale of a signal representation is modified. Used, for example, in acoustics. Its abbreviation is **DTW**.

dynamic viscosity A measure of the resistance a flowing fluid has to any change in shape. Also called **absolute viscosity**.

dynamic Web page A Web page which varies its content, as influenced by user actions.

dynamical Same as **dynamic**.

dynamically-linked library Same as **dynamic link library**. Its abbreviation is **DLL**.

dynamics The branch of science which deals with motion, and the forces, such as energy and momentum, which affect said motion.

dynamo Abbreviation of **dynamo**electric machine. A machine which converts mechanical energy into electrical energy. Such a machine is usually rotary, and may, for instance, utilize magnetic flux to convert angular displacement into electricity. Also called **generator (2)**, or **electric generator (1)**.

dynamoelectric Pertaining to the conversion of mechanical energy into electrical energy, and vice versa.

dynamoelectric machine Same as **dynamo**.

dynamometer **1.** A device which measures and indicates an output magnitude of a motor, such as power or current. **2.** An instrument which measures and indicates an output electric magnitude of a motor, such as power or current, based on the force between a moving coil and a fixed coil. **3.** A device or instrument which measures and indicates mechanical power.

dynamotor **1.** A rotating electric machine which is utilized to convert one DC voltage to a DC voltage of another value. It is a combination of an electric motor and DC generator, and usually has two armatures and a single field magnet. **2.** A rotating electric machine which is utilized to convert a DC voltage to an AC voltage, or vice versa. It usually has a single armature with two or more windings, and incorporates a commutator for DC operation, and slip rings for AC operation. Also called **rotary converter**, or **synchronous inverter**.

dynatron A type of electron tube which displays a negative-resistance characteristic. It is usually a tetrode, and oscillates at ultra-high and microwave frequencies.

dynatron oscillator An oscillator which utilizes a **dynatron**.

dyne The unit of force in the cgs system. It is equal to the force that accelerates a mass of one gram at the rate of one centimeter per second, per second. It is equal to 10^{-5} newton. Its abbreviation is **dyn**.

dyne-centimeter A unit expressing the work done by a force of one dyne acting through a distance of 1 centimeter. One dyne-centimeter is equal to 1 erg, or 0.1 microjoule. Its abbreviation is **dyne-cm**.

dyne-cm Abbreviation of **dyne-centimeter**.

dyne/cm² Abbreviation of **dynes per square centimeter**.

dynes per square centimeter A unit of pressure equal to 0.1 pascals, or about 1 microbar. Its abbreviation is **dyne/cm²**.

dynistor A semiconductor diode which maintains conduction after the forward voltage is reduced. It is utilized for switching.

dynode In certain electron tubes, especially photomultiplier tubes, one of multiple electrodes which produce secondary emissions of electrons. Each successive dynode in the series produces greater and greater amounts of electrons. Depending on the total number of dynodes, the electron pulse may be millions of times larger than it was at the beginning of the tube. Used, for instance, for detection and measurement of minute flashes of light, and in TV camera tubes. Also called **electron mirror**.

dysprosium A soft silvery metallic chemical element whose atomic number is 66. It has nearly 30 known isotopes, several of which are stable, and is highly magnetic. Its applications in electronics include its use in magnetic alloys, semiconductors, and discharge lamps. Its chemical symbol is **Dy**.

dysprosium oxide A white powder whose chemical formula is **Dy_2O_3**. Used in semiconductors.

dysprosium phosphide A chemical compound used in semiconductors.

E

e- A prefix which serves to produce derivative words pertaining to computer, and/or online concepts. For instance, email, or e-cash. It is an abbreviation of electronic.

e 1. A transcendental number equal to approximately 2.71828, and which is the base for natural logarithms. Also called **Napierian base**. 2. Symbol for **voltage**. 3. Symbol for **emitter**.

e 1. Symbol for **elementary charge**. 2. Symbol for **electron charge**.

E 1. Symbol for **emitter**. 2. Symbol for **exa-**.

E 1. Symbol for **voltage**. 2. Symbol for **electric field strength**, or **electric field vector**. 3. Symbol for **energy**. 4. Symbol for **electromotive force**.

E-7-1 A common modem setting in which there is even parity, there are 7 data bits, and 1 stop bit. Also called **7-E-1**.

E and M lead signaling Same as **E and M signaling**.

E and M leads The two leads utilized in **E and M signaling**.

E and M signaling The use of two leads for communication between a trunk circuit and a signaling circuit. The **E** lead transmits from the signaling circuit to the trunk circuit, and the **M** lead transmits from the trunk circuit to the signaling circuit. E and M originally stood for ear and mouth, respectively. Also called **E and M lead signaling**.

e-banking Abbreviation of **electronic banking**.

E beam Abbreviation of **electron beam**.

E-beam lithography Abbreviation of **electron-beam lithography**.

E bend A smooth waveguide bend in which the direction of the axis remains in a plane parallel to the direction of polarization. The smoothness of such a bend helps prevent undesired reflections. Also called **E-plane bend**.

e-billing Abbreviation of **electronic billing**.

e-bomb Abbreviation of **email bomb**.

e-book Abbreviation of **electronic book**.

e-business Abbreviation of **electronic business**. Same as **ecommerce**.

e-card Abbreviation of **electronic card**.

e-cash Abbreviation of **electronic cash**.

e-check Abbreviation of **electronic check**.

e-cinema Abbreviation of electronic **cinema**. The use of digital formats and/or equipment in any aspect of movie making, distribution, or presentation. For example, a movie may be prepared using computer effects and digital cameras, sent via satellite to a movie theater, and viewed using a DLP projector accompanied by digital surround sound. Also called **digital cinema**.

e-comm Abbreviation of **ecommerce**.

e-commerce Abbreviation of **ecommerce**.

e-communications Abbreviation of **electronic communications**.

e-content Abbreviation of **electronic content**.

E core A transformer core in the shape of a capital **E**, whose coils are wound around one or more of its three horizontal bars.

e-credit Abbreviation of electronic **credit**. The use of credit over a computer network, especially the Internet. It usually consists of credit card transactions, but may also include other forms of credit. Security is a key concern, and considerable measures, such as cryptography, are taken to avoid circumstances such as the fraudulent use of the credit estab-

lished for another. **Credit** is the granting of a given time period to pay for a purchase or a loan.

e-currency Same as **ecash**.

E display In radars, an oscilloscopic display in which the scanned objects appear as intensity-modulated blips. The range of scanned objects is indicated by the horizontal coordinate, while the vertical coordinate provides their elevation. Also called **E scope**, **E scanner**, **E scan**, or **E indicator**.

E field Abbreviation or **electric field**.

e-form Abbreviation of **electronic form**.

e-game Abbreviation of **electronic game**.

E indicator Same as **E display**.

e-ink Abbreviation of **electronic ink**.

E-layer A layer of the ionosphere whose effect is felt mostly during daylight hours, and which is contained in the E-region. It extends approximately from 90 to 130 kilometers above the surface of the earth, and is between the D and F layers. It reflects high to very high frequency radio waves. Also called **Kennelly-Heaviside layer**, or **Heaviside layer**.

e.m.f Abbreviation of **electromotive force**.

E.M.F Abbreviation of **electromotive force**.

e-mail Same as **email**.

E-mail Same as **email**.

e-mail address Same as **email address**.

e-mail attachment Same as **email attachment**.

e-mail bomb Same as **email bomb**.

e-mail bot Same as **email robot**.

e-mail client Same as **email client**.

e-mail filter Same as **email filter**.

e-mail folder Same as **email folder**.

e-mail forwarding Same as **email forwarding**.

e-mail header Same as **email header**.

e-mail program Same as **email program**.

e-mail robot Same as **email robot**.

e-mail server Same as **email server**.

e-mail software Same as **email program**.

e-mail virus Same as **email virus**.

e-money Same as **ecash**.

E notation Floating-point or scientific notation in which an **E** substitutes the radix. For example, 314,000,000 being expressed as 3.14E8, instead of 3.14×10^8.

e-paper Abbreviation of **electronic paper**.

e-payment Abbreviation of **electronic payment**.

E plane The plane of an antenna containing the electric field vector. The magnetic field vector is contained in the **H plane**.

E-plane bend Same as **E bend**.

E-ray Abbreviation of **extraordinary ray**.

E-region The region in the ionosphere within which the E-layer forms.

E scan Same as **E display**.

E scanner Same as **E display**.

E scope Same as **E display**.

e-services Services, especially those that are business related, available via the Internet. Also called **Internet services**.

e-signature Abbreviation of **electronic signature**.

e-support Same as **esupport**.

e-tailer A retailer doing business over the Internet.

e-text Abbreviation of **electronic text**.

E transformer A transformer with an **E core**, and which has its primary winding in the center bar, with the secondary windings in the outer bars.

E vector Same as **electric vector**.

e-wallet Abbreviation of **electronic wallet**.

E wave An electromagnetic wave whose magnetic field vector is at all points perpendicular to the direction of propagation. Usually used in the context of waveguides. Also called **TM wave**, or **transverse magnetic wave**.

e-zine Abbreviation of **electronic magazine**.

E & M lead signaling Same as **E and M signaling**.

E & M leads Same as **E and M leads**.

E & M signaling Same as **E and M signaling**.

E0 A European digital transmission standard, system, or line similar to **DS0**. It has a signaling rate of 64 Kbps, which is equivalent to a capacity of 1 voice channel. Most countries use the European standard.

E1 A European digital transmission standard, system, or line similar to **DS1**, or **T1**, but with greater capacity. It has a signaling rate of 2.048 Mbps, which is equivalent to a capacity of 30 voice channels. Most countries use the European standard.

E1 line A line with a signaling rate of 2.048 Mbps, which is equivalent to DS1, providing a capacity of 30 voice channels, each with 64 Kbps.

E1/PRI Abbreviation is **E1/Primary Rate Interface**. An ISDN line with 30 B channels, each with a speed of 64 Kbps, and one 64 Kbps D channel. This is equivalent to an E1 line. It is similar to a T1/PRI line, except that the latter has 23 B channels and one D channel, so it is equivalent to a T1 line. Multiple B channels can be combined for the desired or needed bandwidth. Also called **Primary Rate Interface**, **PRI/E1**, or **ISDN-PRI**.

E2 A European digital transmission standard, system, or line similar to **DS2**, or **T2**, but with greater capacity. It has a signaling rate of 8.448 Mbps, which is equivalent to a capacity of 120 voice channels. Most countries use the European standard.

E²PROM Same as **EEPROM**.

E3 A European digital transmission standard, system, or line similar to **DS3**, or **T3**, but with lesser capacity. It has a signaling rate of 34.368 Mbps, which is equivalent to a capacity of 480 voice channels. Most countries use the European standard.

E4 A European digital transmission standard, system, or line similar to **DS4**, or **T4**, but with lesser capacity. It has a signaling rate of 139.264 Mbps, which is equivalent to a capacity of 1,920 voice channels. Most countries use the European standard.

E5 A European digital transmission standard, system, or line similar to **DS5**, or **T5**, but with greater capacity. It has a signaling rate of 565.148 Mbps, which is equivalent to a capacity of 7,680 voice channels. Most countries use the European standard.

E911 Abbreviation of **enhanced 911**.

EA Abbreviation of **extended addressing**.

EAPROM Same as **EEPROM**. Abbreviation of **electrically alterable programmable read-only memory**.

ear candy Sounds, such as purportedly trendy music, whose primary purpose is to gratuitously please a user aurally, without providing anything useful. Often utilized as a marketing scheme, and may be heard, for instance, at certain Web sites. Ear candy can be considered especially obtrusive when not expected.

earcon An icon which has an associated sound. Useful, for instance, for those with reduced vision.

early adopter A person who chooses to purchase products whose technology is comparatively new and unproven. Such people usually have an added need for such technology, and often have to overcome barriers such as elevated cost, high incidence of malfunctions, and an elongated learning curve.

early binding The linking of software routines or objects during program compilation. This contrasts with **late binding**, where the linking occurs during program execution. Also called **static binding**.

early-failure period A period, immediately following manufacturing or installation, in which devices or equipment fail at an especially high rate. This may be due, for instance, to manufacturing defects, improper handling, or incorrect installation.

early-warning radar A radar utilized to immediately detect any aircraft which have entered a monitored area.

EAROM Same as **EEPROM**. Abbreviation of **electrically alterable read-only memory**.

earphone 1. A small or miniature speaker which is intended to be placed within the outer ear, or immediately surrounding it. It is utilized for private, portable, or enhanced listening. A pair may be used for stereophonic listening. Also called **headphone (1)**. **2.** An **earphone (1)** specifically designed for use with a hearing aid amplifier. Also called **headphone (2)**. **3.** An **earphone (1)** located in the handset of a telephone, which enables listening. Also called **receiver (4)**, or **telephone receiver (1)**.

earphone coupler A device, used for testing earphones, which is designed to duplicate characteristics of the human ear, such as acoustic impedance, frequency response, and threshold of hearing.

earphone jack A jack into which earphones may be plugged in. Seen, for instance, in audio components, computers, DVDs, and so on.

earphones A device which provides an **earphone (1)** for each ear. Also called **headphones**.

earpiece An **earphone (1)** which is placed within the outer ear.

earth 1. The planet earth. Also, the surface of said planet, especially the land portion. **2.** Same as **electrical ground (1)**. **3.** Same as **electrical ground (2)**. **4.** Same as **electrical ground (3)**. **5.** Same as **electrical ground (4)**.

earth absorption The loss of RF energy due to absorption by the earth. Also called **ground absorption**.

earth connection A connection to an **electrical ground (1)**, or **electrical ground (2)**. Also called **ground connection**.

earth current Also called **ground current**. **1.** A current flowing from a circuit to an **electrical ground (1)**, or **electrical ground (2)**. **2.** A current flowing from an **electrical ground (1)**, or **electrical ground (2)** to a circuit. **3.** A current flowing through the earth. It may be due for instance, to the earth's magnetic field, or solar activity. Also called **telluric current**.

earth electrode One or more conductors which are buried in the ground, and which serve to provide an earth connection. Also called **ground electrode**. When such an electrode consists of a large copper plate, also called **earth plate**.

earth fault Also called **ground fault**. **1.** Any unintentional path to earth of a conductor. May be due, for instance, to a failure of insulation between a conductor and a ground. **2.** A loss of an earth connection.

earth-fault protection Protection against an **earth fault** condition.

earth ground A conducting path to the earth. For instance, a wire which connects to a large copper plate buried in moist soil.

earth inductor An instrument which measures the strength of the earth's magnetic field by means of the electromotive force generated in a rotating coil immersed in said field. The coil delivers an output voltage which is proportional to the intensity of the field. Also called **generating magnetometer (2)**.

earth inductor compass A compass whose indications depend on the current induced in a coil made to revolve in the earth's magnetic field. Also called **induction compass**.

earth-moon-earth Same as **earth-moon-earth communications**.

earth-moon-earth communications Radio communications in which signals are bounced off the moon. Usually utilized for UHF or VHF communications. Its abbreviation is **EME**. Also called **earth-moon-earth**, or **moonbounce communications**.

earth plate A large copper plate buried in the ground, preferably in moist soil, and which serves as an **earth electrode**. Also called **ground plate**.

earth potential A potential arbitrarily considered to be zero. It may be the earth, or a large conducting body whose electric potential is also considered to be zero. Also called **zero potential**, or **ground potential**.

earth satellite A satellite, such as a communications satellite, which orbits around the earth.

earth station A land-based station utilized for transmitting and/or receiving satellite communications. It incorporates devices such as a low-noise amplifier, a down-converter, a multiplexer, and an antenna. An earth station antenna may vary in size from a few centimeters to several meters. Also called **satellite earth station**.

earth wax A yellow to dark brown naturally occurring wax consisting chiefly of hydrocarbons, and which is used as an electrical insulator. Also called **ozokerite**.

earth's magnetic field The magnetic field of the planet earth. The lines of flux of said field extend from the north magnetic pole to the south magnetic pole, which do not coincide exactly with the geographic north and south poles. The field approximately resembles that of a bar magnet, and though its strength varies, it is approximately 50 microtesla at the earth's surface. Also called **geomagnetic field**.

earthed Connected to an **electrical ground (1)**, or **electrical ground (2)**.

earthing Also called **grounding**. **1.** The connection to an earth, or to a conductor connected to an earth. The term earthing refers to an intentional path to earth, while **earth fault** describes an unintentional path to earth. **2.** The process of connecting to an earth, or to a conductor connected to an earth.

EAS Abbreviation of **Emergency Alert System**.

east-west effect An effect where the earth's magnetic field deflects incoming cosmic rays. Although cosmic rays impinge upon the earth from every direction, this effect causes a greater amount to arrive from the west than the east.

easter egg Within a computer application, a hidden message, function, feature, or the like. These may only be found through an obscure sequence of keystrokes, and may contain anything from an animated greeting to a different and complete working program.

eavesdrop To secretly listen in on private conversations. The term includes telephone calls and real-time chats, in addition to furtive viewing of written communications, such as faxes and emails, among others.

Eb Abbreviation of **exabit**.

EB Abbreviation of **exabyte**.

ebanking Abbreviation of **electronic banking**.

EBCDIC Abbreviation of Extended Binary Coded Decimal Interchange Code. A binary code computers use to express text and control characters. It is similar to ASCII, and uses 8 bits to represent 256 characters.

EBIC Abbreviation of **electron-beam induced current**.

ebilling Abbreviation of **electronic billing**.

Ebit Abbreviation of **exabit**.

EBM Abbreviation of **electron-beam machining**.

ebonite A hard rubber utilized as a dielectric.

ebook Abbreviation of **electronic book**.

EBS **1.** Abbreviation of **Emergency Broadcast System**. **2.** Abbreviation of **electronic brainstorming**.

ebusiness Same as **e-commerce**. Abbreviation of **electronic business**.

ebXML Abbreviation of **Electronic Business XML**. An XML specification for effecting business-related exchanges of data, such as transactions and messaging.

Ebyte Abbreviation of **exabyte**.

ecard Abbreviation of **electronic card**.

ecash Abbreviation of **electronic cash**.

ECC **1.** Abbreviation of **error checking and correction**. **2.** Abbreviation of **error-correcting code**, or **error-correction coding**. **3.** Abbreviation of **elliptic curve cryptography**.

ECC memory Abbreviation of error-correcting code **memory**. A type of computer memory which incorporates circuitry which detects and corrects errors as data enters and leaves said memory. Also called **ECC RAM**.

ECC RAM Same as **ECC memory**. Abbreviation of **error-correcting code random-access memory**.

eccentric **1.** Not centered. **2.** Not having the same center. **3.** Deviating from a circular path. For instance, having an elliptical path. **4.** Situated away from a center.

eccentricity The quality or state of being **eccentric**.

Eccles-Jordan circuit A two-stage multivibrator circuit with two stable states. In one of the stable states, the first stage conducts while the other is cut off, while in the other state the reverse is true. When the appropriate input signal is applied, the circuit flips from one state to the other. It remains in this state until the next appropriate signal arrives, at which time it flops back. Used extensively in computers and counting devices. Also known as **Eccles-Jordan multivibrator, flip-flop circuit, bistable multivibrator, bistable multivibrator, bistable flip-flop**, or **trigger circuit (2)**.

Eccles-Jordan multivibrator Same as **Eccles-Jordan circuit**.

ECCM Abbreviation of **electronic countercountermeasures**.

ECDIS Abbreviation of **Electronic Chart Display and Information System**.

ECG Also known as **EKG**. **1.** Abbreviation of **electrocardiogram**. **2.** Abbreviation of **electrocardiograph**.

echeck Abbreviation of **electronic check**.

echelon One within a hierarchy. For example, there may be multiple levels which pertain to the degree of accuracy of instrument calibration, and an echelon would be one such level.

echelon grating A specialized diffraction grating consisting of multiple parallel plates of equal thickness, each arranged slightly beyond the next, so as to resemble stairs. Such a grating features a very high resolution power.

echo **1.** A wave or signal which returns after being reflected. Such a wave or signal must have sufficient magnitude and delay so as to distinguish it from a direct wave or signal. An example is the repetition of a sound after being reflected off a hard surface. **2.** The portion of a transmitted radar signal which a scanned object returns to the radar receiver. Also, the visual indication on a radar screen, such as a blip, of such a reflection. Also called **radar echo**, or **return (3)**.

3. On the screen of a TV or radar receiver, an undesired condition in which duplicate images appear. May be due, for instance, to the simultaneous displaying of a direct signal and a reflected signal. Also called **ghosting**. **4.** A repetition or reflection of a signal within a communications line. For instance, such an echo resulting from impedance mismatches along a telephone line. **5.** A signal, other than the original signal, which is transmitted back to a sender. Used, for example, to test network connections. Also, the transmission of such a signal.

echo area The proportion of the overall area of a scanned object which reflects the same amount of a radar signal that would be reflected by the entire area of said object. Also called **radar cross section, target cross section**.

echo attenuation In a transmission line, the ratio of the transmitted power to the reflected power. Usually expressed in decibels.

echo box A resonant cavity that simulates a radar echo, by storing the energy of a transmitted pulse, and retransmitting it gradually to the radar receiver. Used for testing and tuning. Also known as **phantom target**.

echo cancellation The suppression or elimination of unwanted echoes within a communications line, such as that utilized by a telephone or modem. Such echoes may result, for instance, from impedance mismatches along such a line. Also called **acoustic echo cancellation**.

echo canceller A circuit, device, mechanism, or technique utilized for **echo cancellation**.

echo chamber A reverberant enclosure whose characteristics provide for controlled sound reflections. Used, for instance, to add echo effects during a broadcast.

echo check In communications, an error-checking technique in which the receiving end retransmits the data back to the source. Both sets of data are compared, and if they do not match, an error has occurred. Also called **loop check**.

echo depth sounder Same as **echo sounder**.

echo eliminator Same as **echo suppressor**.

echo intensity On a radar display, the brightness of an **echo** (2). For an intensity-modulated display, the greater the reflected energy, the greater the brightness.

echo location Same as **echo sounding**. Also spelled **echolocation**.

echo matching In an **echo-splitting radar**, the positioning of the antenna to find the direction from which the pulse indications are equal.

echo pulse In radars, the pulse of RF energy that is received after being reflected off a scanned object.

echo ranging Same as **echo sounding**.

echo-ranging sonar Same as **echo-sounding sonar**.

echo signal In radars, the energy reflected from a scanned object. Also called **target signal**.

echo sounder A device which determines and indicates the depth of a body of water, by measuring the time required for transmitted sonic or ultrasonic waves to be reflected from a surface, such as the sea bottom. Also called **echo depth sounder, depth sounder, fathometer**, or **sonic-depth finder**.

echo sounding The use of an **echo sounder** to determine the range and bearing of objects and surfaces. Also spelled **echosounding**. Also called **echo ranging**, or **echo location**.

echo-sounding sonar A sonar device utilizing **echo sounding**. Also called **echo-ranging sonar**.

echo-splitting radar A radar which splits an echo in two, to produce two indications on the radar screen. When the height of the two echo indications is the same, the bearing of a scanned object can be read from a calibrated scale.

echo suppression The use of an **echo suppressor**.

echo suppressor **1.** A circuit or device which desensitizes a navigational instrument after a pulse is received, so as to reject delayed pulses. **2.** Within a communications line, such as that utilized by a telephone or modem, a circuit or device which suppresses or eliminates unwanted echoes.

echo wave A wave which returns after being reflected. Such a wave must have sufficient magnitude and delay so as to distinguish it from a direct wave.

echocardiogram **1.** A displayed or recorded image of an **echocardiography**. **2.** Same as **echocardiography**.

echocardiograph An instrument utilized to produce **echocardiograms (1)**.

echocardiography A painless and noninvasive medical diagnostic technique which utilizes ultrasonic waves to produce images of the internal structures and movements of the heart. A piezoelectric transducer is placed against the chest, and sends an ultrasound beam which is reflected off the various internal structures. The instrument utilized for this procedure is an **echocardiograph**, and the recorded images are called **echocardiograms (1)**. Also called **echocardiogram (2)**.

echoencephalogram **1.** A displayed or recorded image of an **echoencephalography**. **2.** Same as **echoencephalography**.

echoencephalograph An instrument utilized to produce **echoencephalograms (1)**.

echoencephalography A painless and noninvasive medical diagnostic technique which utilizes ultrasonic waves to produce images of the internal structures of the brain. A piezoelectric transducer is placed against the scalp, and sends an ultrasound beam which is reflected off the various internal structures. The instrument utilized for this procedure is an **echoencephalograph**, and the recorded images are called **echoencephalograms (1)**. Also called **echoencephalogram (2)**.

echogram **1.** A graphical representation of multiple **echo soundings**. **2.** Same as **echosonogram**.

echograph An instrument utilized to produce **echograms**.

echoless chamber Same as **echoless room**.

echoless enclosure Same as **echoless room**.

echoless room A room whose walls, ceiling, and floor are lined with sound-absorbing materials which minimize or eliminate all sound reflections. Used, for example, to test microphones or sound-level meters. Also called **echoless chamber, echoless enclosure, anechoic chamber (1), free-field room**, or **dead room**.

echolocation Same as **echo location**.

echoplex A technique for verifying data integrity, in which characters are returned to the sending station for confirmation. The characters are usually displayed on a screen, or printed, to check for accuracy.

echosonogram A displayed or recorded image of a medical diagnostic technique utilizing ultrasonic waves. For instance, an echocardiogram or an echoencephalogram. Also called **echogram (2)**.

echosounding Same as **echo sounding**.

ecinema Same as **e-cinema**.

ECL Abbreviation of **emitter-coupled logic**.

ECM **1.** Abbreviation of **electronic countermeasures**. **2.** Abbreviation of **electrochemical machining**.

ECO Abbreviation of **electron-coupled oscillator**.

ecommerce Abbreviation of electronic **commerce**. The conducting of business utilizing electronic communications, especially computers connected to a network. Electronic commerce includes the purchasing and selling of goods and services, electronic data interchange, the use of e-money, and business-related videoconferencing. It may occur over the Internet, a LAN, or any other venue for electronic communication. Also spelled **e-commerce**. Also called **ebusiness, e-business**, or **digital commerce**.

ecommunications Abbreviation of **electronic communications**.

econtent Abbreviation of **electronic content**.

ECP Abbreviation of **Extended Capabilities Port**, or **Enhanced Capabilities Port**.

ECR **1.** Abbreviation of **electron cyclotron resonance**. **2.** Abbreviation of **electronic cash register**.

ECR ion source Abbreviation of **electron cyclotron resonance ion source**.

ECR plasma Abbreviation of **electron cyclotron resonance plasma**.

ECR source Abbreviation of **electron cyclotron resonance source**.

ecredit Same as **e-credit**.

ECTL Abbreviation of **emitter-coupled transistor logic**.

EDA Abbreviation of **Electronic Design Automation**.

EDAC Abbreviation of **error detection and correction**.

EDC Abbreviation of **error detection and correction**.

EDD Abbreviation of **envelope delay distortion**.

eddy current A circulating current induced in a conductor that is moved through a magnetic field, or which is subjected to a varying magnetic field. Eddy currents are generally undesired, and energy is lost usually in the form of heat. Laminations may be used in magnetic cores to minimize such losses of energy. Also called **Foucault current**.

eddy-current damping Damping provided by **eddy currents** which retard the motion of a moving conductor.

eddy-current heating Heating produced by **eddy currents**. Usually used for materials with high electrical conductivity, such as metals. Eddy-current heating features accuracy, fast heating cycle times, and maximum repeatability, among other benefits. Used, for instance, for soldering, annealing, brazing, metal-to-glass bonding, and shrink fitting. Also called **induction heating**.

eddy-current loss The loss of energy due to **eddy currents**. Such losses are usually in the form of heat. Also called **eddy-current losses**.

eddy-current losses Same as **eddy-current loss**.

EDFA Abbreviation of **erbium-doped fiber amplifier**.

edge **1.** The line at which two surfaces or regions intersect, or where an object or area begins or ends. **2.** The portion of a pulse waveform which first increases or decreases in amplitude. When increasing, it is a leading edge, and when decreasing it is a trailing edge.

edge connector A set of wide contacts which protrude from a circuit board, and which serve to be inserted into another circuit board. Used, for instance, to plug an expansion board into an expansion slot. Also called **edgeboard connector**, or **card-edge connector**.

edge detection In artificial intelligence, the determining of the borders of an object using a detection system incorporating optical sensors and complex programs. Edge detection is a key component in differentiating between objects and all that surrounds them.

edge detector A device or system utilized for **edge detection**.

edge device A device which serves to convert LAN frames into ATM cells, and vice versa.

edge effect An effect observed near the outer edges of the parallel metal plates of a capacitor, in which there is an outward bulging of the lines of force. This effect must be taken into consideration in order to correctly compute the capacitance. Alternatively, a guard ring may be used to eliminate this effect. Also called **fringing (1)**.

edge router A device that routes data packets between LANs and ATM networks. Also called **boundary router**.

edge triggered A device actuated by **edge triggering**.

edge triggering The triggering of a device, such as a flip flop, depending on whether the edge of a clock pulse is rising or falling. For instance, a device can be set to sample data only when the edge of its synchronizing clock pulse is rising, in which case it is positive-edge triggered. If it were to sample data only when the edge of its synchronizing clock pulse is falling, it would then be negative-edge triggered.

edgeboard connector Same as **edge connector**.

edging In a color TV receiver, extraneous colors occurring at the boundaries between zones with different colors, or along the edges of objects. Also called **color edging**, or **fringing (2)**.

EDI Abbreviation of Electronic Data Interchange. The use of computer networks, like the Internet, to conduct business transactions such as buying, selling, sending confirmations, billing, and collecting. EDI follows certain standards, such as X12 or EDIFACT, for the exchange of this information, simplifying and expediting the various processes.

EDIF Acronym for **electronic design interchange format**.

EDIFACT Abbreviation of Electronic Data Interchange for Administration Commerce and Transport. An ISO standard pertaining to **EDI**.

Edison base An internally-threaded lamp base whose diameter is approximately 1 inch. It is commonly used as a base for incandescent bulbs.

Edison battery A storage battery in which the positive electrode is nickel oxide, the negative electrode is iron, and the electrolyte is a solution of approximately 20% potassium hydroxide. Such batteries are rugged, durable, and reliable, but do not recharge very efficiently. Also known as **Edison storage battery**, or **nickel-iron battery**.

Edison cell One of multiple cells comprising an **Edison battery**.

Edison effect The emission of electrons from heated objects, such as a heated electrical conductor. An example is the emission of electrons by a cathode in an electron tube. Also known as **thermionic emission (1)**, or **Richardson effect**.

Edison plug A plug, with two flat blades, which serves to be inserted into a power outlet. In various countries it is the standard plug utilized to connect household devices such as lamps and vacuum cleaners.

Edison storage battery Same as **Edison battery**.

edit To make a change, such as a deletion or a rearrangement, in existing data.

edit decision list In a video editor, a file containing a list of edit choices, such as the desired sequence of frames.

edit mode A mode of operation in which a computer program accepts changes in existing data or files.

editor A computer program which creates text files, or modifies existing text files. A word processor is a more powerful editor program with greater flexibility than most editors. Also called **text editor**.

EDL Abbreviation of **edit decision list**.

EDM **1.** Abbreviation of **electronic document management**, or **enterprise document management**. **2.** Abbreviation of **electrical-discharge machining**, or **electron-discharge machining**.

EDMS Abbreviation of **electronic document management system**, or **enterprise document management system**.

EDO DRAM Abbreviation of extended data out dynamic random-access memory, or extended data output dynamic random-access memory. A type of DRAM whose speed is enhanced by its capability to access the next block of memory while the data from the previous block is being sent to the CPU. Conventional DRAM can only access one block at a time, so this overlap is not possible. Also called **EDO RAM**.

EDO RAM   Same as **EDO DRAM**. Abbreviation of **extended data out random-access memory**, or **extended data output random-access memory**.

EDP   Abbreviation of **electronic data processing**.

EDPE   Abbreviation of **electronic data processing equipment**.

EDRAM   Abbreviation of enhanced **DRAM**. A type of DRAM which incorporates a small amount of SRAM used as a cache. Also called **CDRAM**.

EDS   Abbreviation of **energy dispersive spectroscopy**.

EDSAC   Acronym for **Electronic Delay Storage Automatic Calculator**.

.edu   On the Internet, a top-level domain name suffix. The **edu** is an abbreviation of **edu**cational, and the entity employing this suffix should be an educational institution such as a college or graduate school. Also called **dot edu**.

educational software   Software supplementing, or comprising, any given material being learned. Used, for instance, in computer-aided instruction. Also called **courseware**.

edutainment   Acronym for **edu**cational enter**tainment**. Multimedia content, such as that found over the Internet or in DVDs, which is intended to be both educational and entertaining. An interactive encyclopedia is an example.

EEG   **1.** Abbreviation of **electroencephalogram**. **2.** Abbreviation of **electroencephalography**.

EEMS   Abbreviation of **Enhanced Expanded Memory Specification**.

EEPROM   Abbreviation of electrically erasable programmable read-only memory. A programmable read-only memory chip which is erasable by an electrical signal. As with other types of programmable read-only memory chips, it retains its content without power, but is slower than RAM. Also called **E²PROM**, **EAPROM**, or **EAROM**.

effective   **1.** Producing the desired or expected result or effect. **2.** The actual value or result, as opposed to the ideal or calculated value or result. **3.** The real value or result, for practical purposes.

effective address   The final address a computer program uses, after all modifications have been made.

effective ampere   The mean AC ampere which is equivalent to a steady ampere of DC. One RMS ampere of AC is as effective in producing heat through a resistor as 1 ampere of DC.

effective antenna length   The electrical length of an antenna, as opposed to its physical length.

effective area   For an antenna oriented in a given direction, the ratio of the available power available at the terminals to the power per unit area of the incident wave from that direction.

effective bandwidth   **1.** The actual bandwidth available for data transmission through a communications channel, system, or network. **2.** For a given input signal, the bandwidth an ideal bandpass filter with a rectangular response would pass, for the same amount of energy an actual bandpass filter would pass.

effective call   A telephone call that is connected to its destination. An effective call is registered as soon as the connection is made, not when the connection is terminated. Also called **completed call**.

effective capacitance   The actual capacitance between any two points of an electric circuit.

effective current   The value of AC that is as effective in producing heat through a resistor as a corresponding DC. For a sinusoidal AC, the effective current is equal to 0.707 of the peak value of the current. Also called **RMS current**.

effective cutoff frequency   For a device operating between given impedances, the frequency at which the insertion loss exceeds, by a specified amount, the loss at a given reference frequency.

effective earth radius   A value utilized as the radius of the earth, as opposed to its geometrical radius. This is used to correct for atmospheric refraction when the index of refraction changes linearly with height. Under standard refraction conditions, the effective earth radius is 4/3 that of its actual radius. Utilized, for instance, for calculating the range of an antenna. Also called **effective radius of the earth**.

effective field intensity   A measure of the field strength of a transmitting antenna.

effective height   **1.** The vertical height above the ground level at which the center of radiation of a transmitting antenna is located. **2.** The vertical height above the ground level at which the center of reception of a receiving antenna is located.

effective instruction   The instruction a computer program uses, after all modifications have been made.

effective isotropically radiated power   For a given direction, the product of the antenna input power and the antenna power gain, relative to an isotropic radiator. Its abbreviation is **EIRP**.

effective power   Also called **RMS power**. **1.** A value obtained by squaring multiple instantaneous power measurements, averaging these over a given time interval, and taking the square root of this average. **2.** For a periodic quantity, a value obtained by squaring multiple instantaneous power measurements, averaging these over the time of a complete cycle, and taking the square root of this average. In the specific case of sinusoidal AC, the RMS power value is equal to 0.707 of the peak power value. When expressing power in audio applications, other parameters should be specified, as in 100 watts RMS per channel into 8 ohms, with 0.01% total harmonic distortion and intermodulation distortion at rated power, from 10 Hz to 60 kHz, +/– 0.5 dB.

effective radiated power   Its abbreviation is **ERP**. **1.** For a given direction, the product of the antenna input power and the antenna power gain. Usually expressed in kilowatts. **2.** In the direction of maximum gain, the product of the antenna input power and the antenna power gain. Usually expressed in kilowatts.

effective radius of the earth   Same as **effective earth radius**.

effective resistance   The combined resistance offered by a device in a high-frequency AC circuit. It includes, among others, DC resistance, and resistance due to dielectric and eddy-current losses. Also known as **high-frequency resistance**, **RF resistance**, or **AC resistance**.

effective sound pressure   The value obtained when squaring multiple instantaneous sound-pressure level measurements at a given point, averaging these over the time of a complete cycle, and taking the square root of this average. Usually expressed in pascals, though also expressed in other units, such as dyne/cm², or microbars. Also called **RMS sound pressure**, **sound pressure (3)**, or **acoustic pressure**.

effective thermal resistance   For a semiconductor junction under conditions of thermal equilibrium, the temperature rise above a given external reference, per unit power dissipation.

effective value   Also called **RMS value**. **1.** A value obtained by squaring multiple instantaneous measurements, averaging these over a given time interval, and taking the square root of this average. **2.** For a periodic quantity, a value obtained by squaring multiple instantaneous measurements, averaging these over the time of a complete cycle, and taking the square root of this average. In the specific case of sinusoidal AC, the effective current value is equal to 0.707 of the peak value of the current.

effective voltage The value of an AC voltage that is equally effective as a corresponding DC voltage. For a sinusoidal AC voltage, the effective voltage is equal to 0.707 of the peak value of the voltage. Also called **RMS voltage**.

effectively grounded Grounded by means of an extremely low-impedance connection.

effector In robotics, a device which moves a component such, as a manipulator, in response to signals from the controller. Such a device may be electric, hydraulic, mechanical, or pneumatic. Also called **actuator (3)**, or **robotic actuator**.

efficiency **1.** The ratio of the useful output of a device or system, to its total input. Also called **output efficiency (1)**. **2.** The ratio of the useful power or energy output of a device or system, to its total power or energy input. Also called **output efficiency (2)**, or **power efficiency (1)**. **3.** The proportion of the audio-frequency electrical power input to a speaker which is converted into acoustic energy. Efficiency may be measured, for instance, by driving a speaker with a power input of 1 watt, and measuring its output, in decibels, at a distance of 1 meter. Also called **speaker efficiency**, **loudspeaker efficiency**, **sensitivity (4)**, or **power efficiency (2)**.

EFM Abbreviation of **eight-to-fourteen modulation**.

eform Abbreviation of **electronic form**.

EFT Abbreviation of **electronic funds transfer**.

EGA Abbreviation of **Enhanced Graphics Adapter**.

egame Abbreviation of **electronic game**.

EGP Abbreviation of **Exterior Gateway Protocol**.

EGPWS Abbreviation of **enhanced ground proximity warning system**.

EHF Abbreviation of **extremely high frequency**.

EHT Abbreviation of **extra-high tension**.

EHV Abbreviation of **extra-high voltage**.

EHz Abbreviation of **exahertz**.

Ei- Abbreviation of **exbi-**.

EIA Abbreviation of **Electronics Industries Association**.

EIA-232 Same as **EIA/TIA-232-E**.

EIA-232-E Same as **EIA/TIA-232-E**.

EIA-422 Same as **EIA/TIA-422**.

EIA-423 Same as **EIA/TIA-423**.

EIA-449 Same as **EIA/TIA-449**.

EIA-485 Same as **EIA/TIA-485**.

EIA-530 Same as **EIA/TIA-530**.

EIA-568 Same as **EIA/TIA-568**.

EIA-569 Same as **EIA/TIA-569**.

EIA-606 Same as **EIA/TIA-606**.

EIA-607 Same as **EIA/TIA-607**.

EIA/TIA-232 Same as **EIA/TIA-232-E**.

EIA/TIA-232-E A standard interface for serial communications between a serial device, such as a modem or a mouse, and a computer. Although this is the current term for this interface, it is still better known by its former name, **RS-232C**. Also called **EIA-232**, **EIA/TIA-232**, **EIA-232-E**, or **TIA/EIA-232-E**.

EIA/TIA-422 A standard interface for serial communications similar to **EIA/TIA-232-E**, but providing higher speed and transmission distances, and support for multipoint connections. Also called **EIA-422**, **RS-422**, or **TIA/EIA-422**.

EIA/TIA-423 A standard interface for serial communications similar to **EIA/TIA-232-E**, but providing higher speed and transmission distances. EIA/TIA-423 only supports point-to-point connections. Also called **EIA-423**, **RS-423**, or **TIA/EIA-423**.

EIA/TIA-449 A standard defining the pins for both **EIA/TIA-422** and **EIA/TIA-423** serial communications. Also called **EIA-449**, **RS-449**, or **TIA/EIA-449**.

EIA/TIA-485 A standard interface for serial communications similar to **EIA/TIA-422**, but supporting more nodes per line. Also called **EIA-485**, **RS-485**, or **TIA/EIA-485**.

EIA/TIA-530 A standard defining the pins for both **EIA/TIA-422** and **EIA/TIA-423** serial communications when a DB-25 connector is utilized. Also called **EIA-530**, **RS-530**, or **TIA/EIA-530**.

EIA/TIA-568 A standard detailing specifications for telecommunications cabling systems in structures such as commercial buildings. Also called **EIA-568**, or **TIA/EIA-568**.

EIA/TIA-569 A standard detailing specifications for placement, as in raceways or plenums, of telecommunications cabling systems in structures such as commercial buildings. Also called **EIA-569**, or **TIA/EIA-569**.

EIA/TIA-606 A standard detailing specifications for the telecommunications infrastructure of commercial buildings. Also called **EIA-606**, or **TIA/EIA-606**.

EIA/TIA-607 A standard detailing specifications for telecommunications grounding and bonding in structures such as commercial buildings. Also called **EIA-607**, or **TIA/EIA-607**.

EiB Abbreviation of **exbibyte**.

EIDE Abbreviation of **Enhanced IDE**, or **Enhanced Integrated Device Electronics**.

eight-to-fourteen modulation A process which converts 8-bit computer magnetic media code into 14-bit optical disc code. Used, for instance, when transferring information from a computer hard drive to a CD-RW disc. Its abbreviation is **EFM**.

EIGRP Abbreviation of **Enhanced Interior Gateway Routing Protocol**.

eink Abbreviation of **electronic ink**.

Einstein-de Haas effect The observed rotation in a freely hanging ferromagnetic body when there is a change in its magnetization.

Einstein mass-energy relation A fundamental formula for the interconversion of mass and energy. It is $E = mc^2$, where E is energy, m is mass, and c is the speed of light in a vacuum. Also called **mass-energy equation**.

einsteinium A synthetic chemical element whose atomic number is 99. It has close to 20 known isotopes, all of which are unstable. Used in tracer studies. Its chemical symbol is **Es**.

Einthoven galvanometer Same as **Einthoven string galvanometer**.

Einthoven string galvanometer A galvanometer which measures current by utilizing a conducting thread or fiber which is stretched between the poles of a magnet or electromagnet. The thread or fiber is deflected proportionally to the current passing through it, and is viewed through a microscope. Also called **Einthoven galvanometer**, or **string galvanometer**.

EIR Abbreviation of **excess information rate**.

EIRP Abbreviation of **effective isotropically radiated power**.

EIS Abbreviation of **Executive Information System**.

EISA Acronym for Extended Industry Standard Architecture. A PC bus architecture providing a 32-bit path. ISA cards may be plugged into EISA slots.

EIT Abbreviation of **extreme-ultraviolet imaging telescope**.

ejected electron An electron emitted from a surface or an atom. For instance, those emitted by a cathode in an electron tube, or Auger electrons. Also called **emitted electron**.

EKG Also known as **ECG**. **1.** Abbreviation of **electrocardiogram**. **2.** Abbreviation of **electrocardiograph**.

EL Abbreviation of **electroluminescence**.

EL display Abbreviation of **electroluminescent display**.

elapsed time **1.** The time that passes during a given sequence, process, or operation. **2.** The time interval that apparently elapses during a given sequence, process, or operation. The actual time that a given sequence or process requires may be less. For example, the elapsed time may be the sum of the actual time and any insensitive time.

elapsed-time meter An instrument that measures and displays **elapsed time (1)**. Also called **running-time meter**.

elastance The reciprocal of capacitance. Usually expressed in darafs. Its symbol is **S**.

elastic medium A medium, such as air or water, which can change in shape as a result of a deforming force, and which returns to its original shape when said force is removed.

elastic recoil detection Its abbreviation is **ERD**. Same as **elastic recoil detection analysis**.

elastic recoil detection analysis An ion-beam technique for the determination of light elements on the surface of a sample material. In it, the sample is irradiated with an ion beam consisting of particles which are heavier than the target atoms that wish to be analyzed. The lighter target atoms are then scattered forward and detected. It is especially effective in hydrogen analysis. Its abbreviation is **ERDA**. Also called **elastic recoil detection**.

elastic wave **1.** A wave propagated through an **elastic medium**. **2.** A traveling wave which propagates sound through an elastic medium. The wave is produced by vibrations that are a result of acoustic energy. Also known as **sound wave**, or **acoustic wave**.

elasticity The capability of a material or body to recover its dimensions, after a stress causing deformation is removed. Also called **resilience (2)**.

elasticity modulus For a given material or body, the ratio of stress applied, to the deformation exhibited. Also known as **modulus of elasticity**.

elastomer A substance or material whose elasticity is similar to that of natural rubber. Such a substance or material may be natural or synthetic, and may be used, for instance, in flexible circuits, or as packing in the housing of scintillation counters.

elastomeric Having the elastic properties of an **elastomer**.

elbow Same as **elbow bend**.

elbow bend A curved bend, usually of 90º, in a waveguide. Also called **elbow**, or **waveguide elbow (1)**.

electr- Same as **electro-**.

electra A continuous-wave radio-navigation system that utilizes multiple radio beacons to establish a given number of equisignal zones.

electret A dielectric material which is permanently electrified, and which has oppositely charged extremities. Electrets may be prepared, for instance, by heating a suitable material and letting it cool in a strong electric field. Appropriate materials include certain waxes, plastics, or ceramics. Its behavior in an electric field is similar to that of a permanent magnet in a magnetic field.

electret microphone A microphone whose transducer is an **electret**.

electric Containing, carrying, producing, actuated by, based on, or arising from electricity. For example, electric current or an electric motor. This contrasts with **electrical**, which is all that pertains to and is correlated with electricity, but not having its physical properties or characteristics. Nonetheless, even in technical usage both terms are mostly used interchangeably.

electric alarm **1.** An electrical device which serves to warn by means of a signal, such as sound or light. **2.** A security system whose source of power is electricity.

electric appliance An electronic device designed to help do work, perform tasks, or to assist in providing comfort or convenience. These include dryers, lamps, refrigerators, and microwave ovens. Precision electronics, such as high-fidelity components and computers are sometimes not included in this group. Also called **electrical appliance**.

electric arc A highly luminous and sustained discharge of electricity between two conductors separated by a gas. It is characterized by a high current density and a low voltage drop. Used, for example, in welding and arc lamps. This contrasts with a **spark**, which has a short duration. Also called **electrical arc**, or **arc (1)**.

electric-arc lamp A gas-filled electric lamp whose bright light is produced by a sustained discharge of electricity between its two electrodes, which ionizes the gas. Also known as **arc lamp**.

electric-arc welding Welding utilizing a sustained discharge of electricity between two conductors as the source of heat. Also known as **arc welding**.

electric attraction Same as **electrical attraction**.

electric axis Same as **electrical axis**.

electric battery A source of DC power that incorporates two or more **electric cells**. Also called **electrical battery**, or **battery**.

electric bell A hollow metallic device with a hammer which strikes it to produce a ringing sound. These are electrically actuated, and serve to warn, as in an alarm, or to inform, as in an older telephone.

electric breakdown Also called **electrical breakdown**. **1.** A substantial and abrupt increase in electric current produced by a small increase in voltage. Such an increase may or may not be destructive. Also called **breakdown (4)**. **2.** An abrupt increase in the electric current flowing through a dielectric, when the applied voltage exceeds a given critical value. Such a breakdown is usually destructive. Also called **dielectric breakdown**.

electric breakdown voltage For a given dielectric, the voltage at which **electric breakdown (2)** occurs. Also called **electrical breakdown voltage**, or **dielectric breakdown voltage**.

electric bridge An electric circuit, network, or instrument that has four or more branches, or arms, each with a component such as a resistor or capacitor. It also has a current source and a null detector, and when such a bridge is balanced, its output is zero. An electric bridge allows for an unknown component replacing one of the branches to be accurately measured, as it can be compared to known values. Its characteristic shape is that of a diamond. Also called **electrical bridge**, or **bridge**.

electric cable One or more electric conductors bundled together and encased by a protective sheath. Such cables are usually made with copper or aluminum, or an alloy based on either of them. Used, for instance, for the delivery of electricity to industrial, commercial, or residential locations, or to power electric locomotives. Also called **electric line**.

electric car A car which is powered by an electric motor which utilizes one or more electric batteries.

electric cell A single unit which converts chemical, thermal, nuclear, or solar energy into electrical energy. For example, a solar cell, or a voltaic cell. Also called **electrical cell**, **energy cell**, **cell (1)**.

electric characteristics Same as **electrical characteristics**.

electric charge Also called **electrical charge**, or **electrostatic charge**. **1.** A basic property of subatomic particles, atoms, molecules, and the antimatter counterparts of each of these.

An electric charge may be negative, positive, or there may be no charge. The net charge of a body is the sum of the charges of all its constituents. For instance, a cation has a net deficiency of electrons, so its charge is positive. Each electric charge, whether positive or negative, is equal to 1.6022×10^{-19} coulomb, the charge of an electron, or a whole number multiple of it. Also called **charge** (1). **2.** The electrical energy stored in, or on the surface of an insulated body such as a storage battery or capacitor. Usually expressed in coulombs. Also called **charge** (2).

electric circuit Also called **electrical circuit**. **1.** One or more conducting paths which serve to interconnect electrical elements, in order to perform a desired function such as amplification or filtering. Also known as **circuit** (1). **2.** One or more complete paths through which electrons may circulate. Also known as **circuit** (2).

electric circuit theory The mathematic analysis of electric circuits, and the relationships between the components such circuits contain. Also called **electrical circuit theory**, or **circuit theory**.

electric clock A clock which is driven by electric power. Such a clock may be driven, for instance, by an AC motor.

electric coil One or more conductors wound in a series of turns, so as to introduce inductance, or to produce a magnetic field in an electric circuit. An electric coil may be wound, for example, around a ferromagnetic core, an insulating support, or around air. Used, for instance, in transformers, electromagnets, solenoids, motors, speakers, and so on. Also called **electrical coil, coil, inductance** (3), **inductor, inductance coil**, or **magnetic coil** (1).

electric component **1.** Same as **electrical element**. **2.** In an electromagnetic wave, the electric constituent. It is perpendicular to the magnetic component. Also called **electrostatic component**.

electric conductance Same as **electrical conductance**.

electric conduction Same as **electrical conduction**.

electric conductivity Same as **electrical conductivity**.

electric conductor A medium which allows electric current to flow easily. Such a medium may be a metal wire, a dissolved electrolyte, or an ionized gas, among others. Among the elements, silver, copper, and gold are the best electric conductors. Also known as **electrical conductor, conductor** (2).

electric connection Same as **electrical connection**.

electric connector **1.** A device or object which serves to join two or more electric conductors. Examples include terminals and binding posts. Also called **connector** (2). **2.** A device or object which serves to couple cables with other cables, equipment, or systems. Examples include plugs, jacks, and adapters. Also called **connector** (3).

electric constant The permittivity of a vacuum. It is a constant equal to approximately 8.85419×10^{-12} farad per meter. Its symbol is ϵ_0. Also called **permittivity of free space**.

electric contact Also called **electrical contact**. **1.** The junction of two electric conductors, so that current may flow. Also called **contact** (2). **2.** A part or device which serves to open or close an electric circuit. Such a part or device may or may not act with another part or device. For instance, a button, metal strip, switch, or relay. Also called **contact** (3).

electric control The control of a device, piece of equipment, system, or process through the use of electrical components such as switches, relays, or rheostats. This contrasts with **electronic control**, which utilizes electronic components such as transistors or electron tubes. Also called **electrical control**.

electric controller **1.** A device which regulates the electricity delivered to a device, piece of equipment, or system. **2.** An automatic controller that utilizes **electric control**.

electric coupling Same as **electrical coupling**.

electric current Its symbol is I, and it is expressed in amperes. Also called **electrical current**, or **current**. **1.** A flow of an electric charge through a conductor. Electrons, electron holes, or ions may transport an electric charge. **2.** The rate of flow of electric charge through a conductor.

electric current density The current flowing per unit of cross-sectional area of a conductor. Its SI unit is amperes per square meter. Its symbol is J. Also called **electrical current density**, or **current density**.

electric device Same as **electrical device**.

electric dipole Two opposite charges of equal magnitude separated by a very small distance. Also, a body, such as a molecule, containing such charges. Also called **electric doublet, dipole** (1), or **doublet** (1).

electric dipole moment For an electric dipole, the product of the magnitude of one of the charges, and the distance between the centers of the charges. May be expressed in coulomb-meters or abcoulomb-centimeters. Also called **electric moment**, or **dipole moment** (1).

electric discharge Also called **electrical discharge**. **1.** The diminishing or removal of an electric charge stored or accumulated in a body, such as a battery or capacitor, which is being depleted. Also called **discharge** (2). **2.** The passage of an electric current through a medium, such as a space or air. A spark is an example of such a discharge. Also called **discharge** (3).

electric-discharge lamp A lamp which produces light when the gas it contains, usually at a low pressure, becomes ionized by an electric discharge passed through it. An example is a fluorescent lamp. Also known as **discharge lamp, gas-discharge lamp**, or **vapor lamp**.

electric-discharge machining Same as **electrical-discharge machining**. Its abbreviation is **EDM**.

electric-discharge tube A tube containing a low-pressure gas through which a current passes when it is ionized by a sufficient voltage applied between the electrodes of said tube. Also known as **electrical-discharge tube, discharge tube**, or **gas-discharge tube**.

electric displacement A vector quantity pertaining to the charge displaced by an electric field within a given medium. It is equal to the electric field strength multiplied by the permittivity of the medium, and is usually expressed in coulombs per square meter. Also called by various other names, including **electric flux density, electric displacement density, electric field induction, electric induction, electric stress, displacement** (2), **dielectric displacement**, and **dielectric stress**.

electric displacement density Same as **electric displacement**.

electric distribution system The physical system that delivers electrical energy from a primary source to the specific customers. Such a system includes wires, poles, step-down transformers, and voltage dividers. Also called **electrical distribution system**, or **distribution system**.

electric disturbance Same as **electrical disturbance**.

electric doublet Same as **electric dipole**.

electric eel A type of fish which is capable of delivering electric discharges which may serve to shock prey, communicate with other eels, or aid in navigation. An electric eel can deliver a discharge of around 650 volts. Despite its name, an electric eel is a type of fish.

electric element Same as **electrical element**.

electric energy Also called **electrical energy**. **1.** Same as **electricity** (3). **2.** Same as **electricity** (4). **3.** Same as **electricity** (5).

electric engine An engine whose source of energy is electricity. An electric railroad, for instance, utilizes such an engine.

electric equipment Same as **electrical equipment**.

electric eye A photocell utilized as a detector, such as that used in automatic door openers or intrusion alarms.

electric fence A fence through which one or more electrified wires run. Such a fence may deliver an electric shock intermittently, continuously, or upon contact. Also called **electrical fence**.

electric field 1. The effect which a charged particle or body exerts on charged particles or bodies situated in the medium surrounding it. For instance, if a negatively charged particle is placed within the electric field of a positively charged particle, there will be an attractive force, while there will be a repulsive force if the charges are like. The electric field surrounding a charged particle or body is represented as electric lines of force. **2.** Same as **electric field strength**. **3.** In an electromagnetic field, the electric component. The electric field is perpendicular to the magnetic field.

electric field induction Same as **electric displacement**.

electric field intensity Same as **electric field strength**.

electric field strength Within an electric field, a vector representing the force per unit charge acting upon a given point. Usually expressed in volts per meter, which is equivalent to newtons per coulomb. Its symbol is E. Also called **electric field intensity, electric field (2), electric intensity, electric force, electric vector (2)**, or **electric field vector**.

electric field vector Same as **electric field strength**.

electric filter Same as **electrical filter**.

electric fish A type of fish, such as an electric eel, which is capable of delivering electric discharges which may serve to shock prey, communicate with other fish, or aid in navigation. An electric eel, for instance, can deliver a discharge of around 650 volts.

electric flux Its symbol is ψ. **1.** The amount of electric charge displaced across a given area of a dielectric. It is expressed in coulombs. **2.** For a given region, the electric lines of force. The closer together the lines, the greater the strength of the electric field. Also called **flux (2)**.

electric flux density Same as **electric displacement**.

electric flux lines Same as **electric lines of force**.

electric force Same as **electric field strength**.

electric force lines Same as **electric lines of force**.

electric forming The application of electric energy to a device, such as a semiconductor, to permanently alter its electrical characteristics. Also called **electrical forming**.

electric furnace A furnace in which electricity provides the heat. May consist of a stream of air which is blown past resistance heating coils. Also called **electrical furnace**.

electric fuse A safety device which contains a section of conductor, such as a wire, which melts when the current passing through it exceeds a specified amount, called the fuse rating. When the conductor melts, the circuit is opened, thus preventing damage to the circuit, device, equipment, or installation it is meant to protect. Also called **electrical fuse, fuse**, or **cutout (2)**.

electric generating station Same as **electric power plant (1)**.

electric generator Also called **electrical generator**. **1.** A device which converts mechanical energy into electrical energy. Such a device is usually rotary, and may, for instance, utilize magnetic flux to convert angular displacement into electricity. Also called **dynamo**, or **generator (2)**. **2.** A device which converts another form of energy into electrical energy. Examples include dynamos, batteries, and oscillators. Also called **generator (3)**.

electric glow A glow caused by an electrical discharge. For instance, a glow discharge. Also called **electrical glow**.

electric guitar A guitar which incorporates one or more contact microphones, which convert the acoustic vibrations produced by the strings into an electrical signal suitable for amplification.

electric heating 1. The use of electricity to produce heat. **2.** The conversion of electric energy into heat energy. Examples include dielectric heating and induction heating.

electric hygrometer A hygrometer whose sensitive element has an electrical characteristic, such as conductance, vary proportionally to the level of humidity of the air surrounding it. Also called **electrical hygrometer**.

electric hysteresis 1. A property of certain dielectrics, in which electric polarization is retained after removal of the electric field. Such an effect can last up to several years, and it is the basis for electrets. Also called **electric hysteresis, dielectric absorption**, or **dielectric hysteresis**. In the case of a ferroelectric material, also called **ferroelectric hysteresis**. **2.** Internal friction occurring in a dielectric as a result of exposure to variable electric field. This friction may cause enough heat to cause failure of the dielectric.

electric image 1. An array of electrical charges which represent an actual object. It is the electric counterpart of a real object that is nearby. Also called **image (2)**. **2.** An image which has been converted into digital form, which is necessary for processing by a digital computer. An example is a bitmap. Also called **digitized image**.

electric impedance Same as **electrical impedance**.

electric induction Same as **electric displacement**.

electric inertia Same as **electrical inertia**.

electric installation Same as **electric power plant**.

electric instrument Same as **electrical instrument**.

electric insulation Same as **electrical insulation**.

electric insulator Same as **electrical insulator**.

electric intensity Same as **electric field strength**.

electric interference Same as **electrical interference**.

electric interlock Same as **electrical interlock**.

electric lamp A lamp whose source of power is electricity. Such a lamp may be incandescent, fluorescent, arc, glow, and so on.

electric light The light produced by an **electric lamp**.

electric line Same as **electric cable**.

electric lines of flux Same as **electric lines of force**.

electric lines of force Within an electric field, imaginary lines whose tangent at any given point represent the direction of said electric field at that point. Electric lines of force are utilized to represent the electric field surrounding a charged particle or body, and the closer together the lines, the greater the strength of the said field. Also called **electric lines of flux, electric flux lines, electric force lines, electrostatic flux, lines of force (1)**, or **force lines (1)**.

electric load Same as **electrical load**.

electric locomotive A locomotive which utilizes an electric engine. Such a locomotive may be powered by overhead lines, or an additional rail in the ground. It incorporates a transformer and a power regulator. Among the advantages of an electric locomotive are its lighter weight, greater speed, increased power, improved efficiency, and reduced contamination of the surrounding air.

electric machine 1. A machine which generates electricity. For instance, a dynamo. **2.** A machine whose source of power is electricity.

electric main Same as **electric power transmission line**.

electric means Same as **electrical means**.

electric meter Also called **electrical meter**. **1.** An instrument which measures and indicates electrical power consumption for a given, or combined, time interval. Also called **electric power meter**, or **power meter**. When indicating in units of kilowatt-hours, also called **kilowatt-hour meter**. **2.** An instrument which measures and indicates one or more electrical quantities, such as resistance, current, or voltage. Examples include multimeters, and ammeters.

electric moment Same as **electric dipole moment**.

electric monopole An electric charge distribution that is concentrated around a point, or which is spherically symmetrical. For example, that of an isolated single charge.

electric motor A device or machine which converts electrical energy into mechanical energy, usually torque. Its source of power may be DC or AC. An example is an AC induction motor, such as that utilized in many household appliances. Also called **electrical motor**, or **motor (1)**.

electric network One or more electric circuits incorporating two or more interconnected electrical elements or components, such as resistors, capacitors, coils or generators. There are various ways to classify electric networks. For instance, if a network incorporates active devices, such as amplifiers, it is an active network, while a passive network does not. A bilateral network is one which functions equally well in both directions, while a unilateral network does not. There are many examples of specific networks, and these include bridge, crossover, decoupling, and resistance-capacitance networks. Also called **electrical network**, or **network (2)**.

electric noise Same as **electrical noise**.

electric organ **1.** An organ in certain fish, such as electric eels, which is able to produce an electric discharge. Such an organ consists of multiple electroplaques. **2.** Same as **electronic organ**.

electric oscillation A periodic variation in an electrical quantity. For example, the alternation, at regular intervals, of the direction of AC.

electric outlet Same as **electrical outlet**.

electric piano Same as **electronic piano**.

electric point charge An electric charge considered to occupy a single point in space which has neither area nor volume. Also called **point charge**.

electric polarity Same as **electrical polarity**.

electric polarization A phenomenon observed in dielectrics, in which the electrons in each atom are displaced in the direction opposite to that of an applied electric field, while the nucleus of each atom is displaced in the direction of said field. Also called **electrical polarization, polarization (5)**, or **dielectric polarization**.

electric potential The work required to bring a unit charge from a reference point to a specific point within an electric field. The reference point is usually considered to be an infinite distance from the specific point, whose potential is considered to be zero. When 1 joule is required to bring 1 coulomb of charge, the potential is equal to 1 volt. Its symbol is V, and it is expressed in volts. Also called **electrostatic potential**, or **potential**.

electric potential difference Also called **potential difference**. The difference in electric potential between two points, especially those of a circuit. It is the work required to move a unit charge between these points. Its abbreviation is **PD**, and its symbol is U, or ΔV. When expressed in volts, as it usually is, also called **voltage**.

electric potential energy The energy possessed by an electric charge, depending on its position within an electric field. The energy stored in a capacitor is an example of electric potential energy. When electric potential energy is expressed per unit charge, it is the same as **electric potential**. Also called **electrical potential energy**.

electric power Also called **electrical power**. **1.** The rate at which electric energy is utilized. This may be in its conversion into another form of energy, such as heat, or for doing work. In a DC circuit or system, it is equal to the potential difference, expressed in volts, multiplied by the current, expressed in amperes. In an AC circuit or system, **true power** is the rate at which work is performed or energy is transferred, **apparent power** is the product of the RMS current and the RMS voltage, while **reactive power** is the power in an AC circuit which cannot perform work. Its symbol is P, and it is expressed in watts, or some multiple of watts, such as megawatts. Also called **power (1)**. **2.** Same as **electricity (4)**. **2.** Same as **electricity (5)**.

electric power generation The production of electricity, especially in quantities adequate for residential, commercial, or industrial use. Also called **power generation**.

electric power meter Same as **electric meter (1)**.

electric power outlet Same as **electrical outlet**.

electric power plant Also called **power plant**. **1.** A location which combines the structures, devices, and equipment necessary to produce electricity in quantities adequate for residential, commercial, or industrial use. An electric power plant converts another form of energy, such as hydroelectric, nuclear, or solar, into electrical energy. Also called by various other names, including **electric power station, electric generating station, electric installation, electrical power plant, power station, power generating station**, and **generating station**. **2.** A unit which converts another form of energy, such as hydroelectric, nuclear, or solar, into electrical energy.

electric power station Same as **electric power plant (1)**.

electric power system Same as **electrical system (1)**.

electric power transmission The transmission of generated electricity, especially for residential, commercial, or industrial use. Also called **power transmission**.

electric power transmission line An electric cable utilized for **electric power transmission**. Also called **electric main, electric transmission line**, or **power transmission line**.

electric precipitator Same as **electrostatic precipitator**.

electric pressure A difference in potential which causes current to flow through a circuit or conductor. Also called **electrical pressure**. When referring to a source of electrical energy, also called **electromotive force (1)**.

electric probe **1.** A wire or rod inserted into an electric field, for detection, measurement, or sampling. An example is a waveguide probe. **2.** A probe which serves to discharge an electric current through its tip. May be used, for instance, in certain surgical procedures.

electric propulsion **1.** A form of propulsion in which the propellant consists of charged particles being accelerated by electric and/or magnetic fields. An example is a magnetohydrodynamic generator. Usually used in aircraft or spacecraft. **2.** A form of propulsion involving electricity. For instance, that provided by an electric engine.

electric quantity Same as **electrical quantity**.

electric railroad A railroad which utilizes an electric engine. Such a railroad may be powered by overhead lines, or an additional rail in the ground. It incorporates a transformer and a power regulator. Among the advantages of an electric railroad are its lighter weight, greater speed, increased power, improved efficiency, and reduced contamination of the surrounding air.

electric receptacle Same as **electrical outlet**.

electric recording **1.** The recording of music by electrical means, as opposed to mechanical. Examples include optical recordings, such as that of CDs, and magnetic recordings, such as that of DATs. May also include the recording of images, such as those on DVDs and videocassettes. **2.** Musical recording of electric instruments, such as guitars and pianos, as opposed to nonelectric versions of said instruments. **3.** The recording on a given medium, such as paper, by the passage of an electric current through it. A stylus carrying a current, for instance, may be utilized. Also called **electrosensitive recording**.

electric repulsion Same as **electrical repulsion**.

electric reset Same as **electrical reset**.

electric resistance Same as **electrical resistance**.

electric resistivity Same as **electrical resistivity**.

electric scanning The change in the direction of a radar beam through variations in the phase and/or amplitude of the currents fed to the driven elements of its antenna array. Also called **electrical scanning**.

electric screen **1.** Same as **electrostatic shielding (1)**. **2.** Same as **electrostatic shielding (2)**.

electric screening Same as **electrostatic shielding**.

electric shield **1.** Same as **electrostatic shielding (1)**. **2.** Same as **electrostatic shielding (2)**.

electric shielding Same as **electrostatic shielding**.

electric shock The effect of a passage of current through living tissue. The effects of a shock can range from mild tingling to death, depending on the strength of the current, the points where it enters and leaves the body, whether the heart is along its path, if the skin is wet, and so on. Under many circumstances, a current of 0.1 ampere for 1 second may be fatal. Electric shocks may also be utilized for therapeutic reasons, such as restoring the normal rhythm to a heart which is twitching uncontrollably. Also called **electrical shock**, or **shock (1)**.

electric signal Same as **electrical signal**.

electric socket Same as **electrical outlet**.

electric spark A momentary luminous discharge of electricity between two conductors separated by a gas, which is frequently accompanied by a crackling noise. Used, for instance, for ignition of fuel, or for machining. This contrasts with an **arc (1)**, which has a more sustained duration. Also called **spark**, **spark discharge**, or **sparkover**.

electric starter A device utilized to start an electric motor, and to accelerate it to its operational speed. Such a starter may be powered, for example, by a battery. Also called **starter (2)**.

electric strength The maximum voltage a dielectric can withstand before an abrupt increase in the electric current flowing through it. May be expressed, for instance, in volts per millimeter. Also called **dielectric strength**, or **dielectric rigidity**.

electric stress Same as **electric displacement**.

electric susceptibility A dimensionless quantity which measures the ease with which a dielectric may be polarized. It is the ratio of the dielectric polarization, to the product of the electric intensity multiplied by the permittivity of a vacuum. Also called **dielectric susceptibility**.

electric switch A device which serves for opening, closing, or changing connections in electric circuits. An electric switch may be manual or automatic. There are various types, including mechanical, such as circuit-breakers, and semiconductor, such as transistors. Also called **switch (1)**.

electric switchboard One or more panels which incorporate the switches, circuit-breakers, fuses, and the like, utilized to monitor and operate electric equipment. Also called **switchboard (2)**.

electric system Same as **electrical system**.

electric transducer Same as **electrical transducer**.

electric transient A momentary current or voltage that occurs when the steady-state condition of a circuit has been disturbed. An example is a high-voltage spike.

electric transmission line Same as **electric power transmission line**.

electric tuning Same as **electronic tuning**.

electric vector **1.** In an electromagnetic field, the vector representing the electric field. The electric vector is perpendicular to the magnetic, or H, vector. Also called **E vector**. **2.** Same as **electric field strength**.

electric vehicle A vehicle, such as a car, trolley, or train, which is propelled by electric power. In the case of a car, for instance, a fuel cell may provide the electricity which powers the motor, while an overhead cable may provide the electricity for a train. Its abbreviation is **EV**.

electric-wave filter Same as **electrical filter**.

electric waves Same as **electromagnetic waves**.

electric wind An electric discharge in which a stream of charged particles moves away from a body with a high voltage, such as a Van de Graaff generator. Such a discharge may be visible or invisible. Also called **convective discharge**.

electric wire An electric conductor composed of a single metallic strand, thread, or rod which is flexible and which usually has a circular cross section. Such a wire may or may not have insulation.

electric wiring Also called **wiring**. **1.** The system of wires and/or conductors that connect electrical components, circuits, and devices together. For instance, the wires in a piece of electrical equipment, or the interconnections between components of an IC. **2.** The process of installing or manufacturing **electric wiring (1)**.

electrical All that pertains to and is correlated with electricity, but not having its physical properties or characteristics. For example, electrical symbols or electrical units. This contrasts with **electric**, which is that containing, carrying, or producing electricity, or which is actuated by, based on, or arising from electricity. Nonetheless, even in technical usage both terms are mostly interchangeably.

electrical angle **1.** An angle which specifies a given instant within a cycle of an alternating quantity, such as AC. Such an angle is usually expressed in electrical degrees, and a full cycle has 360°. **2.** An angle which specifies a given phase difference between two alternating quantities. Expressed in electrical degrees.

electrical appliance Same as **electric appliance**.

electrical arc Same as **electric arc**.

electrical attraction An attractive force between two oppositely charged particles or bodies. For instance, that between a cation and an anion. Also called **electric attraction**.

electrical axis The axis along which electrical polarization occurs when a mechanical stress is applied to a piezoelectric crystal. Also called **electric axis**.

electrical bandspread In a radio receiver, the use of a capacitor for improving selectivity, by spreading the bandwidth over a greater electrical range.

electrical battery Same as **electric battery**.

electrical breakdown Same as **electric breakdown**.

electrical breakdown voltage Same as **electric breakdown voltage**.

electrical bridge Same as **electric bridge**.

electrical cell Same as **electric cell**.

electrical center The point at which at which an adjustable component, such as a variable resistor, is divided into two equal electrical values. Such a point does not necessarily occur at the physical center.

electrical characteristics Measurable features, such as resistance, capacitance, and conductivity, which help describe the electrical properties of components, circuits, devices, and so on. Also called **electrical properties**, or **electric characteristics**.

electrical charge Same as **electric charge**.

electrical circuit Same as **electric circuit**.

electrical circuit theory Same as **electric circuit theory**.

electrical coil Same as **electric coil**.

electrical component Same as **electrical element**.

electrical conductance A measure of the ability of a component, circuit, device, or system to conduct electricity. In a circuit with no reactance it is the reciprocal of resistance. It is the real part of admittance, its symbol is **G**, and is expressed in siemens. Also called **electric conductance**, or **conductance**.

electrical conduction The flow of electrical charge through a medium. This may entail, for instance the movement of electrons through a metal conductor, or the migration of ions in a gas. Also called **electric conduction**, or **conduction (2)**.

electrical conductivity The ease with which an electric current can flow through a body. It is the reciprocal of **electrical resistivity**, its symbol is σ, and it is expressed in siemens per meter. Also called **electric conductivity**, or **conductivity**.

electrical conductor Same as **electric conductor**.

electrical connection Also called **electric connection**. **1.** The act of joining together, or connecting two points or electrical components in a circuit. **2.** The manner in which electrical components are joined together. **3.** The point at which two points or electrical components in a circuit are joined together. **4.** The state of two points or electrical components being joined together.

electrical contact Same as **electric contact**.

electrical control Same as **electric control**.

electrical coupling Coupling in which electrical energy is transferred between components or circuits. For example, electrostatic coupling, or direct coupling. Also called **electric coupling**.

electrical current Same as **electric current**.

electrical current density Same as **electric current density**.

electrical degree A unit equal to 1/360 of a complete cycle of an alternating quantity, such as AC.

electrical device 1. A physical unit, such as a resistor, connector, or transistor, which performs a specific electrical function, or serves for a particular electrical use. Also called **electric device**. When such a device can operate on an applied electrical signal, as in amplifying, rectifying, or switching, also called **electronic device (1)**. **2.** Any device requiring electricity to operate.

electrical discharge Same as **electric discharge**.

electrical-discharge machining A method of machining metals, in which minute bits from its surface are vaporized through the use of a series of carefully controlled sparks between an electrode and the metal being worked on, under precise conditions. Used, for instance, for removing metal in hard to reach places, for machining complex patterns, or to work with metals which are otherwise difficult to machine. Its abbreviation is **EDM**. Also called **electric-discharge machining**, or **electron-discharge machining**.

electrical-discharge tube Same as **electric-discharge tube**.

electrical displacement Same as **electric displacement**.

electrical distance The distance between two points in a given medium, expressed in terms of the duration of travel of an electromagnetic wave through an equivalent distance in free space. A unit utilized to express electrical distance is the light-microsecond, which is equal to approximately 300 meters.

electrical distribution system Same as **electric distribution system**.

electrical disturbance A disturbance, such as a random-frequency current or voltage, which is electrical in nature. An example is hum. Also called **electric disturbance**.

electrical element Any of the electrical components that are a part of a circuit. These may include resistors, capacitors, transistors, generators, electron tubes, and so on. Each component has terminals which allow it to be connected to the conducting path. Also called **electrical component**, **electric element**, **electric component (1)**, **element (2)**, **circuit component**, **circuit element**, or **component (2)**.

electrical energy Same as **electric energy**.

electrical engineering The branch of engineering which deals with all aspects of electricity, including its generation, transmission, distribution, and use. It also pertains to the design, construction, operation, and optimization of electric components, circuits, devices, equipment, and systems, to make the best use of electricity.

electrical equipment Equipment which utilizes electricity as its source of power. Their power source may be AC and/or DC. Also, other components and mediums involved in the use of electricity, such as connectors, insulators, and wires. Also called **electric equipment**.

electrical erosion In electric contacts, the gradual loss of material due to sparks, arcs, and the such. The addition of cadmium to an alloy used in contacts may help reduce such erosion. Also called **electric erosion**, or **contact erosion**.

electrical fault In a component or circuit, a defect such as a short circuit, an open circuit, or an unintentional ground. Also, a failure caused by such a defect. Also called **fault (2)**.

electrical fence Same as **electric fence**.

electrical field Same as **electric field**.

electrical field strength Same as **electric field strength**.

electrical filter An electric circuit or device which selectively transmits or rejects signals in one or more intervals of frequencies. The transmitted intervals are called passbands, and the rejected intervals are called stopbands. When a filter incorporates active components, such as transistors, it is an active filter, if not, it is passive. A capacitor, for instance, may serve as a passive filter, because it blocks DC. Filters may be classified as falling within one of the following four categories: low-pass, high-pass, bandpass, and bandstop. There are many examples of filters, including comb, ripple, Butterworth, and loop filters. Also called **electric filter**, **electrical-wave filter**, or **filter (2)**.

electrical forming Same as **electric forming**.

electrical furnace Same as **electric furnace**.

electrical fuse Same as **electric fuse**.

electrical generator Same as **electric generator**.

electrical glow Same as **electric glow**.

electrical ground 1. The earth, which is arbitrarily considered to have an electric potential of zero. Also, a conducting path to the earth. For instance, a path which leads to a large copper plate buried in moist soil. Also, a conducting object, such as a wire, leading to the earth. Also called **earth (2)**, or **ground (2)**. **2.** A large conducting body whose electric potential is arbitrarily considered to be zero. Also, a conducting path to such a body. Also, a conducting object, such as a wire, leading to such a conducting body. Also called **earth (3)**, or **ground (3)**. **3.** Within a circuit, a point which is at zero potential with respect to an **electrical ground (1)**, or **electrical ground (2)**. Also called **earth (4)**, or **ground (4)**. **4.** To connect to an **electrical ground (1)**, **electrical ground (2)**, or **electrical ground (3)**. A path to an electrical ground may be intentional or accidental. Also called **earth (5)**, or **ground (5)**.

electrical hygrometer Same as **electric hygrometer**.

electrical hysteresis Same as **electric hysteresis**.

electrical impedance Its symbol is Z, and it is expressed in ohms. The total opposition a circuit or device offers to the flow of AC. It is a complex quantity, whose real number component is resistance, and whose imaginary number component is reactance. Its formula is: $Z = R + jX$, in which Z is the impedance, R is the resistance, X is the reactance, and j is the square root of -1. That is, $j^2 = -1$. Also called **electric impedance**, or **impedance**.

electrical induction Same as **electric displacement**.

electrical inertia The property of a circuit or conductor which opposes any change to the current flowing through it. That is, it opposes an applied current, and also opposes changes to an already established current. Also called **electrical inertia**, or **inductance (1)**.

electrical installation Same as **electric power plant**.

electrical instrument A device which measures one or more electrical quantities, such as voltages or a currents. For instance, an ammeter, or an electroencephalograph. Also called **electric instrument**.

electrical insulation Also called **electric insulation**. **1.** The use of an **electrical insulator**. **2.** The placing or application of an **electrical insulator**. **3.** Having the isolation, protection, and/or support that an **electrical insulator** provides. **4.** Same as **electrical insulator**.

electrical insulator A material that has a sufficiently high resistance to the passage of electric current, so that current flow through it is minimal or negligible. Used, for instance, to isolate, protect, and support circuits and conductors, and to prevent the loss of current. Examples include rubber, plastic, ceramic, and glass. Also, an object or device made of such a material. Also called **electrical insulation (4)**, **electric insulator**, or **insulator (2)**.

electrical intensity Same as **electric intensity**.

electrical interference In a communications system, any energy which diminishes the ability to receive a desired signal, or that impairs its quality. Sources include electromagnetic noise, undesired signals, parasitic oscillations, and atmospheric conditions. Also called **electric interference**, **interference (2)**, or **radio interference**.

electrical interlock A circuit, device, or mechanism which prevents a piece of equipment from operating under certain potentially hazardous conditions. For instance, a switch that shuts off power when a protective door or panel is opened or removed. Also called **electric interlock**, or **interlock (1)**.

electrical length **1.** The length of a transmission medium, such as a cable or waveguide, expressed in wavelengths, radians, or degrees. **2.** The effective length of an antenna element, usually expressed in wavelengths. The electrical length of an antenna may be greater or lesser than its physical length. For instance, inductive or capacitive elements may be used to change the electrical length of an antenna.

electrical load The electrical energy that is consumed by a component, circuit, device, piece of equipment, or system that is connected to a source of electric power, in order to perform its functions. Also called **electric load**.

electrical means The utilization of electrical components, circuits, devices, equipment, or systems, to perform any given function.

electrical measurement **1.** The measurement of any electrical quantity, such as resistance or current. Expressed in electrical units, such as ohms or amperes. **2.** A specific instance of an **electrical measurement (1)**.

electrical meter Same as **electric meter**.

electrical motor Same as **electric motor**.

electrical network Same as **electric network**.

electrical noise Also called **electric noise**. **1.** Any unwanted electrical signals which produce undesirable effects. Such effects may be imperceptible, they may affect performance, or may even cause irreparable damage. May be produced, for instance, by unwanted voltages or currents. **2.** Any noise which is electrical in origin.

electrical outlet A power line termination whose socket serves to supply electric power to devices or equipment whose plug is inserted into it. Electric outlets are usually mounted in a wall, although they may be found elsewhere, such as a floor, or in the back of another electrical device, such as an amplifier. Also called by various other names, including **electrical power outlet**, **electrical socket**, **electric outlet**, **convenience outlet**, **outlet**, **receptacle**, **power outlet**, and **power receptacle**.

electrical overstress A transient or steady-state condition, in which the voltage and/or current exceeds the ratings and/or capabilities of a device. An example is an excess of electrostatic discharges. Such overstress does not necessarily manifest its caused damage immediately. Its abbreviation is **EOS**.

electrical polarity Also called **electric polarity**. **1.** The property of having an excess or deficiency of electrons. An excess of electrons produces a negative polarity, and a deficiency of electrons produces a positive polarity. This determines the direction of the flow of current, as electrons move from a point with an excess of electrons towards a point where there is a deficiency of electrons. Also called **polarity (2)**. **2.** The characteristic of having two opposite charges, positive and negative, within the same body or system. A battery, for instance, has two terminals, each with opposite polarity, which are the positive terminal and the negative terminal. Also called **polarity (3)**.

electrical polarization Same as **electric polarization**.

electrical porcelain An electrical insulation material consisting of, or incorporating, porcelain.

electrical potential energy Same as **electric potential energy**.

electrical power Same as **electric power**.

electrical power outlet Same as **electrical outlet**.

electrical power plant Same as **electric power plant**.

electrical pressure Same as **electric pressure**.

electrical properties Same as **electrical characteristics**.

electrical quantity Also called **electric quantity**. **1.** The magnitude of an electric charge. Usually expressed in coulombs. **2.** The magnitude of an electrical characteristic, such as resistance or capacitance. **3.** Same as **electrical unit**.

electrical receptacle Same as **electrical outlet**.

electrical repulsion A repulsive force between two similarly charged particles or bodies. For instance, that between two electrons. Also called **electric repulsion**.

electrical reset The resetting of a relay or switch by electrical means. Used, for instance, when remote manual resetting is necessary. Also called **electric reset**.

electrical resistance The opposition a material offers to the flow of current, with the concomitant conversion of electrical energy into heat energy. Resistance may be intentionally introduced into a circuit, for instance, for current regulation or for voltage regulation. In an AC circuit, resistance is the real number component of electrical impedance. Expressed in ohms, or fractions/multiples of ohms. Also called **electric resistance**, or **resistance (1)**.

electrical resistivity Its symbol is **r**, or ρ. A measure of the inherent ability of a material to resist the flow of current. It is the reciprocal of **electrical conductivity**, and depending on their resistivities, materials can be classified as insulators, semiconductors, or conductors. The lower the resistivity, the better conductor a material is. Its formula is: $r = RA/L$, where **r** is resistivity in ohm-meters, R is the resistance in ohms, A is the cross-sectional area of the material in square meters, and L is its length in meters. Also called **electric resistivity**, **resistivity**, or **specific resistance**.

electrical scanning Same as **electric scanning**.

electrical screen **1.** Same as **electrostatic shielding** (1). **2.** Same as **electrostatic shielding** (2).

electrical screening Same as **electrostatic shielding**.

electrical shield **1.** Same as **electrostatic shielding** (1). **2.** Same as **electrostatic shielding** (2).

electrical shielding Same as **electrostatic shielding**.

electrical shock Same as **electric shock**.

electrical signal Electrical events, such as variations in voltage, current, frequency, phase, or duration, which serve to communicate information between two points. An analog electrical signal can vary in a continuous and infinitely variable manner, while digital electrical signals have a finite number of discrete steps within any given interval. Also called **electric signal**.

electrical signaling Signaling utilizing **electrical signals**.

electrical socket Same as **electrical outlet**.

electrical surge A sudden and momentary increase in current or voltage. May be caused, for instance, by lightning, or faults in circuits. If protective measures are not employed, such a surge may bring about a failure or significant damage. Also called **surge** (1).

electrical susceptibility Same as **electric susceptibility**.

electrical symbol A graphical symbol utilized to represent an electrical component or device, such as a resistor, diode, transistor, filter, or antenna. Used, for instance, in circuit diagrams. Also called **graphical symbol**.

electrical system Also called **electric system**. **1.** The network incorporating the wiring, devices, and equipment necessary to transmit and distribute electricity. Also called **electric power system**. **2.** The appropriate arrangement of electrical components and devices, so as to perform a given function within a piece of equipment.

electrical tape A tape consisting of, incorporating, or impregnated with an insulating material. Depending on the specific needs, such tapes, aside from electrical insulation, may offer abrasion resistance, mechanical strength, resistance to heat, or protection against harsh weather. Also called **insulating tape**.

electrical technology Same as **electrotechnology**.

electrical transducer Also called **electric transducer**. **1.** A device which converts a non-electrical signal or form of energy, such as acoustic energy, into an electrical signal or electrical energy. An example is a microphone. **2.** A device which converts an electrical signal or form of energy into a non-electrical signal or form of energy, such as acoustic energy. An example is a speaker. **3.** A transducer which converts one electrical signal into another. For instance, a device whose input is a direct current, and whose output is a proportional alternating current.

electrical unit A unit for expressing an electrical magnitude or measurement. For instance, amperes, coulombs, farads, ohms, volts, or watts. Also called **electrical quantity** (3).

electrical-wave filter Same as **electrical filter**.

electrical wavelength For an electromagnetic wave, the spatial distance between adjacent points of equal phase along the direction of propagation.

electrical zero **1.** The indicated output of a device measuring an electrical quantity, when there is no signal present. Also, such a zero reading after calibration. **2.** An indicated output of zero for a device measuring an electrical quantity, when a given input value, such as a specific voltage, must be applied to obtain said indication of zero. **3.** An electric potential which is arbitrarily considered to be zero, such as that of the ground. **4.** For a rotating device, such as a synchro, the reference point from which angles are measured.

electrically alterable programmable read-only memory Same as **EEPROM**.

electrically alterable read-only memory Same as **EEPROM**.

electrically connected Connected through an electrical conductive path, such as a wire or a resistor, as opposed to an inductive path.

electrically erasable programmable read-only memory Same as **EEPROM**.

electrically erasable programmable ROM Same as **EEPROM**.

electrically erasable PROM Same as **EEPROM**.

electrically operated Requiring electricity to operate.

electricity **1.** The phenomena associated with and arising from moving or stationary electric charges. Electric charge is a basic property of subatomic particles, atoms, molecules, and the antimatter counterparts of each of these, and may be negative or positive. Electricity produced by electric charges in motion is called dynamic electricity, while static electricity is that produced by electric charges at rest. Electricity is one of the fundamental forms of energy, and can produce thermal, radiant, magnetic, and chemical changes. **2.** The science that deals with **electricity** (1). **3.** Energy arising from **electricity** (1). Also, the flow or transfer of this energy. Also called **electric energy** (1). **4.** The power available from an **electrical outlet**. Also called **electric energy** (2), or **electric power** (2). **5.** Electrical energy generated through conversions from other forms of energy, such as chemical, thermal, nuclear, or solar. Also called **electric energy** (3), or **electric power** (3).

electrification **1.** The process and effect of applying an electric charge to a body. For example, to provide a locomotive with electrical energy. **2.** The process and effect of generating an electric charge in a body. For instance, the use of friction to accumulate a static charge in amber. **3.** The generation, distribution, and utilization of electrical energy.

electro- A prefix used in words pertaining to electricity, electronics, and the like. For example, electrical, electronic, electrolytic, electrostatic, etc. A variant form is **electr-**.

electro-oculogram A displayed or recorded image of an **electro-oculography**.

electro-oculography The utilization of electrodes applied to the skin adjacent to the eyes to detect differences in electric potential produced by eye movements. Its abbreviation is **EOG**.

electro-optic Pertaining to **electro-optics**. Also spelled **electro-optical**.

electro-optic device A device, such as a bar-code reader, which incorporates both electronics and optics. Also spelled **electro-optical device**.

electro-optic effect Also spelled **electro-optical effect**. **1.** For a given material, a change in refractive index under the influence of an electric field. **2.** An optical effect produced by an electric field applied to a material through which light passes. An example is the electro-optic Kerr effect.

electro-optic Kerr effect For certain materials, such as transparent dielectrics, double refraction produced by an electric field. Also spelled **electro-optical Kerr effect**. Also called **Kerr effect** (1), or **Kerr electro-optical effect**.

electro-optic material A material which displays an **electro-optic effect**. Such a material, which may be a liquid crystal, can be utilized, for instance, to convert an electrical signal into an optical signal. Also spelled **electro-optical material**.

electro-optic modulator A light modulator which makes use of an **electro-optic effect**. Such a device may modulate the amplitude, direction, frequency, or phase of a light beam. Also spelled **electro-optical modulator**.

electro-optical Same as **electro-optic**.

electro-optical device Same as **electro-optic device**.

electro-optical effect Same as **electro-optic effect**.

electro-optical Kerr effect Same as **electro-optic Kerr effect**.

electro-optical material Same as **electro-optic material**.

electro-optical modulator Same as **electro-optic modulator**.

electro-optics The study and application of the physical and optical properties of electron beams under the influence of electric and/or magnetic fields, in a manner analogous to the behavior of light beams passing through a refractive medium such as a lens. Electro-optics also deals with devices and phenomena incorporating both electronics and optics, as is the case with photoelectricity, lasers, and fiber-optics. It is sometimes utilized as a synonym for **optoelectronics**, but this usage is incorrect. Also called **electron optics**.

electro-osmosis Same as **electroosmosis**.

electroacoustic Pertaining to devices whose operation involves both electric and acoustic energy. Speakers and microphones are examples.

electroacoustic effect The development of a DC voltage in a crystalline or metallic material caused by acoustic waves traveling parallel to its surface. Also known as **acoustoelectric effect**.

electroacoustic transducer 1. A device which transforms acoustic energy into electric energy. For example, a microphone. 2. A device which transforms electric energy into acoustic energy. For example, a speaker.

electroacoustics The science that deals with the transformation of acoustic energy into a corresponding electric energy, and vice versa. Speakers, microphones, and amplifiers are examples of devices used in this field.

electroanalysis 1. The use of electrolysis for chemical analysis. 2. The use of electrochemical methods for chemical analysis.

electrobiology The science pertaining to electrical phenomena in living organisms. This includes, for instance, the study of bioelectricity.

electrocardiogram A graphical recording of the variations in the electrical activity of the heart. An **electrocardiograph** is used to detect and record such activity. Its abbreviation is **ECG**, or **EKG**. Also called **cardiogram**.

electrocardiograph An instrument which detects and records variations in the electrical activity of the heart. An **electrocardiogram** is the graphical recording of this activity. Its abbreviation is **ECG**, or **EKG**. Also called **cardiograph**.

electrocardiography The technique employed to obtain an **electrocardiogram**. Electrocardiography utilizes electrodes applied to various parts of the body, which detect the electrical signals produced by the heart. Also called **cardiography**.

electrocautery The use of an electrically-heated probe to destroy tissue. For instance, one or more platinum wires may be heated to a very high temperature by a direct current or an alternating current.

electrochemical Pertaining to **electrochemistry**.

electrochemical cell A cell which consists of two electrodes immersed in an electrolyte, and in which a chemical reaction occurs that either generates or consumes electrical current. When a current is generated it is a voltaic or galvanic cell, and when current is consumed it is an electrolytic cell.

electrochemical deposition The use of electrolysis for the formation of a metallic layer on a substrate. May be used, for instance, to electroplate.

electrochemical equivalent For an electrolytic solution through which one coulomb of current is passed, the mass, in grams, of any chemical substance produced or consumed.

It may refer, for instance, to the mass of a metal deposited during electroplating.

electrochemical machining The process of removing a metal or alloy from a material, through electrolytic action. The shape of the cathode determines the final shape of the material. It is the reverse of electroplating. Used, for instance, to remove excess metal from an otherwise finished piece. Its abbreviation is **ECM**. Also called **electrolytic machining**.

electrochemical power Electric energy derived from the conversion of chemical energy. For example, the power obtained from a galvanic cell.

electrochemical process 1. A chemical process which gives rise to an electric current. This occurs, for instance, in a galvanic cell. 2. A process in which an electric current gives rise to a chemical process. This occurs, for instance, in electrolysis.

electrochemical series Also called **electromotive series**. 1. A series of elements listed in the order of their relative tendency to lose electrons. The greater the tendency to lose electrons, the higher on the list. Highly reactive metals such as lithium, potassium, and calcium are at the top of such a list, while inert metals such as silver, platinum, and gold are at the bottom of it. Hydrogen is arbitrarily set as the zero point. 2. An **electrochemical series** (1) listing only metals.

electrochemical transducer 1. A transducer that converts an electrical quantity into a chemical change. 2. A transducer that converts a chemical change into an electrical quantity.

electrochemiluminescence Also called **electrogenerated luminescence**. 1. The emission of light as a consequence of a transformation of chemical energy into electric energy. 2. The emission of light as a consequence of a transformation of electric energy into chemical energy.

electrochemistry The science that pertains to the electrical effects of chemical phenomena, and to the chemical effects of electrical phenomena. Also, all that pertains to the transformation of chemical energy into electrical energy, and vice versa. Myriad processes involve electrochemistry, including electrolysis, chemical bonding, electrometallurgy, bioelectricity, and so on.

electrochromic display A passive solid-state display which incorporates a material whose light-transmitting and light-reflecting characteristics are controlled by an electric field.

electrocoagulation The use of a high-frequency electric current to destroy tissue, or to coagulate blood.

electrocochleography A technique for the measurement of electrical activity in the inner ear resulting from sound stimuli.

electrocute To produce death by means of an **electric shock**. May also refer to a severe injury resulting from an electric shock.

electrocution An **electric shock** producing death. May also refer to an electric shock resulting in a severe injury.

electrode A conductor which serves to deliver, receive, collect, emit, deflect, or control electric charge carriers. An electrode provides the path for current entering or leaving a medium such as a dielectric, electrolytic solution, semiconductor, gas, or vacuum. In an electron tube or electrolytic cell, for instance, the anode is the positive electrode, while the cathode is the negative electrode. It is often a metal, and may consist, for instance, of a wire, grid, plate, or region.

electrode admittance A measure of the ease with which AC flows through an electrode. It is the reciprocal of **electrode impedance**.

electrode capacitance The capacitance between an electrode and a reference, such as ground or all other electrodes connected together.

electrode characteristic A curve or graph representing the relationship between the electrode current and the electrode voltage.

electrode conductance The ability of an electrode to conduct electricity. It is the reciprocal of **electrode resistance**.

electrode current The current flowing through an electrode. For instance, the cathode current, or the base current.

electrode dark current Current that flows through a photosensitive material when it is off, or not illuminated. An example is a residual current flowing through a photoconductive cell after the source of illumination is removed. Also called **dark current**.

electrode dissipation Power lost by an electrode, in the form of heat.

electrode drop A decrease in voltage due to **electrode resistance**.

electrode impedance A measure of the impedance encountered by AC which flows through an electrode. Electrode impedance is comprised of **electrode resistance**, the real number component, and **electrode reactance**, the imaginary number component. It is the reciprocal of **electrode admittance**.

electrode potential Also called **electrode voltage**, or **electropotential**. **1.** The potential developed by an electrode material which is in equilibrium with an electrolytic solution of its own ions, as compared to that of a standard hydrogen electrode. **2.** The potential between an electrode and a reference, such as ground, or another electrode.

electrode reactance The imaginary number component of **electrode impedance**.

electrode resistance The opposition an electrode offers to the flow of current. It is the reciprocal of **electrode conductance**, and is also the real number component of **electrode impedance**.

electrode voltage Same as **electrode potential**.

electrodeless discharge A luminous discharge, in a tube or lamp without electrodes, produced by a high-frequency electric field.

electrodeless discharge lamp A discharge lamp without electrodes. Its light is produced by a high-frequency electric field. Such lamps have a longer life than discharge lamps with electrodes.

electrodeposit Also called **electrodeposition (3)**. **1.** An electrical process, such as electrolysis or electrophoresis, which is utilized to deposit a substance on a given surface. **2.** The deposit formed through an **electrodeposit (1)**.

electrodeposition Abbreviation of **electrolytic deposition**. **1.** An electrical process, such as electrolysis or electrophoresis, which is utilized to deposit a material onto an electrode. The deposited material is usually a metal or alloy. **2.** The deposit formed through **electrodeposition (1)**. **3.** Same as **electrodeposit**.

electrodermal response A change in the electrical resistance of the skin in response to emotional stimuli such as anxiety or fear. Such variations are detected by a galvanometer. Also called **galvanic skin response**.

electrodiagnosis The use of electrical and electronic devices to aid in the diagnosis of medical conditions. Such devices include electrocardiographs, and those utilized for magnetic resonance imaging.

electrodialysis The use of an electric current to pull charged particles through selective membranes, thus separating them from the solution that contained them. Used, for instance, to purify water.

electrodynamic **1.** Pertaining to charge carriers in motion, and their associated phenomena. **2.** Pertaining to the relationship between electric, magnetic, and mechanical phenomena.

electrodynamic instrument Same as **electrodynamometer**.

electrodynamic loudspeaker Same as **electrodynamic speaker**.

electrodynamic microphone A type of microphone in which incident sound waves strike a diaphragm, which is connected to a moving coil. The coil moves within a constant magnetic field, which in turn produces the output signal current. Also called **dynamic microphone**, or **moving-coil microphone**.

electrodynamic speaker A type of speaker in which an audio-frequency signal current is sent through a voice coil, which is attached to a diaphragm. The interaction with a constant magnetic field provides the coil with a piston-like movement, which is mechanically transferred to the diaphragm, which in turn generates the sound waves that emanate from the speaker. Also called **electrodynamic loudspeaker**, **dynamic speaker**, or **moving-coil speaker**.

electrodynamics **1.** The branch of science dealing with phenomena associated with charge carriers in motion. For instance, their interaction with accompanying magnetic fields. **2.** The branch of science dealing with the relationships between electric, magnetic, and mechanical phenomena.

electrodynamometer An instrument whose function depends on the interaction between the magnetic fields produced by the current in one or more moving coils, and the current in one or more fixed coils. For example, two stationary coils may be connected in series with a moving coil, which is attached to an indicating needle. The resulting torque can be used to measure electric current, voltage, or power. Also called **electrodynamic instrument**.

electroencephalogram Its abbreviation is **EEG**. **1.** A displayed or recorded image of an **electroencephalography**. **2.** Same as **electroencephalography**.

electroencephalograph An instrument utilized to produce **electroencephalograms (1)**.

electroencephalography A painless and noninvasive medical diagnostic technique in which pairs of electrodes are attached to various parts of the scalp, to record the electrical activity of the brain. The instrument utilized for this procedure is an **electroencephalograph**, and the recorded images are called **electroencephalograms (1)**. Its abbreviation is **EEG**. Also called **electroencephalogram (2)**.

electroform Abbreviation of **electroforming**.

electroforming Its abbreviation is **electroform**. **1.** The electrodeposition of a metal onto a substrate, especially when giving it a desired form or pattern. Used, for instance, to create intricate waveguide patterns, or in the process of creating master CDs. **2.** The use of electricity to alter the characteristics of a component or device, such as a capacitor or semiconductor device. Used, for instance, to condition a pn junction. Electroforming usually takes place during the manufacturing process. Also called **forming (2)**.

electrogalvanized A metal, especially iron or steel, upon which zinc has been electroplated.

electrogalvanizing The electroplating of zinc onto another metal, especially iron or steel.

electrogenerated luminescence Same as **electrochemiluminescence**.

electrogram **1.** A graphical recording of the variations in the electrical activity of the heart, as sensed from within the heart tissue. This contrasts with an **electrocardiogram**, which senses the electrical activity of the heart from the surface of the skin. **2.** A graphical recording of the variations in the electrical activity of a given point in the atmosphere. Such a recording is usually automatic.

electrograph **1.** A graphical representation created through the action of an electric current on a material such as paper. **2.** A graphical representation created through the action of a stylus, such as a pen, which is electrically controlled. **3.** A device, such as a fax, which can send and/or receive text and graphics.

electrographic recording The producing of graphical representations through gaseous discharges between two or more electrodes. The electrostatic field forms the images, which can be recorded on an insulating material.

electrography The formation of electrostatic images, which are recorded on an insulating material. An example is electrographic recording.

electrojet A narrow belt, within the ionosphere, of intense electric current. It may be centered over the magnetic poles, or over the magnetic equator. In the polar regions, electrojets may participate in auroral events.

electrokinetics **1.** The study of moving charged particles, especially within an electric field. **2.** The study of materials moving within electric fields.

electroless deposition A deposition process which does not involve the use of electricity. For instance, that utilized in electroless plating.

electroless plating A plating process, such as immersion plating, which does not involve the use of electricity.

electroluminescence Its abbreviation is **EL**. **1.** The emission of light by certain phosphors when subjected to an applied electric field. **2.** The emission of light by certain phosphors embedded in an insulating material which is subjected to an alternating electric field. Also called **Destriau effect**. **3.** The emission of light due to an electric discharge.

electroluminescent Characterized by **electroluminescence**.

electroluminescent cell Same as **electroluminescent panel**.

electroluminescent display A display which incorporates one or more **electroluminescent panels**. It is a type of flat-panel display, and may be used, for instance, in portable computers. Its abbreviation is **EL display**.

electroluminescent lamp A lamp which incorporates one or more **electroluminescent panels**.

electroluminescent panel A light source which consists of an appropriate phosphor layer sandwiched between dielectric layers, which in turn are surrounded by electrode plates, the front electrode plate being transparent. When a suitable voltage is applied to the electrodes, the phosphor layer emits light. Used, for instance, in electroluminescent displays. Also called **electroluminescent cell**.

electrolysis The effecting of a chemical change in a substance present in an electrolyte through which a current is passed. The change is usually a decomposition into simpler constituents, and the substance may be in gaseous, liquid, molten, or solid form, and may or may not be fully dissolved in the electrolyte, which is a liquid or molten salt. The electrodes between which the current is passed are the anode, towards which the negative constituents travel, and the cathode, towards which the positive constituents travel. Electrolysis is used, for instance, for electroplating, electrorefining, and in electrolytic cells.

electrolyte A substance which dissociates to a certain extent in a medium through which an electric current passes. Electrolytes conduct electricity, and the most common examples are salts dissolved in a solution in which the solvent is water.

electrolytic Pertaining to **electrolysis**. For instance, that which is produced by, or which produces electrolysis.

electrolytic capacitor A capacitor in which the dielectric is a layer of a metal oxide deposited through electrolysis. The oxide serving as dielectric is usually one of aluminum or tantalum, and is formed on aluminum or tantalum foil, or on sintered slugs. The aluminum or tantalum metal serves as one of the plates, while the other is a liquid or paste electrolyte which saturates a paper or gauze. Electrolytic capacitors provide high volume-to-capacitance ratios, but suffer from high leakage currents.

electrolytic cell A cell, such as a galvanic cell, consisting of two or more electrodes separated by an electrolyte, and in which electrochemical reactions occur. Such a cell may also be utilized to induce electrochemical reactions which would not occur spontaneously.

electrolytic conduction The flow of current due to charged particles moving within an electrolyte in which electrodes are placed, and to which a voltage is applied. Negative particles, or anions, travel towards the anode, and positive particles, or cations, towards the cathode.

electrolytic conductivity The ease with which an electric current can flow through an electrolyte.

electrolytic corrosion Corrosion occurring as a result of an electrolytic process.

electrolytic deposition Same as **electrodeposition**.

electrolytic dissociation Within an electrolyte, the breaking apart of a molecule into simpler constituents, each of which is an ion. For instance, sodium chloride dissociating into sodium cations and chloride anions.

electrolytic machining Same as **electrochemical machining**.

electrolytic plating Same as **electroplating**.

electrolytic polarization In an electrolytic cell, the tendency of the products of electrolysis to recombine, thus diminishing the performance of said cell.

electrolytic polishing Same as **electropolishing**.

electrolytic potential In an electrolytic cell, the difference in potential between an electrode and its immediately surrounding electrolyte.

electrolytic process Any process, such as electrolytic conduction or electrolytic dissociation, involving electrolysis.

electrolytic protection The reduction or prevention of corrosion of a metallic material by making it the cathode in a conductive medium that has a sacrificial anode. This conductive medium may occur naturally, or can be applied externally. Used, for instance, in certain pipelines, bridges, and other metal structures requiring corrosion protection over an extended period of time. Also called **cathodic protection**.

electrolytic recording The recording of text and/or images incorporating an electrolytic process. For example, an electric current may be passed through a stylus which records images onto an electrolyte-impregnated paper.

electrolytic rectifier A rectifier incorporating two electrodes, each of a different material, immersed in an electrolyte. The selection of the appropriate electrodes and electrolyte creates a polarizing film on one of the electrodes, which effectively allows current to flow only in one direction. An example of such an arrangement would be an aluminum electrode along with a lead electrode in a sodium bicarbonate solution.

electrolytic refining Same as **electrorefining**.

electrolytic separation The use of electrolysis to separate metals from a solution. The metals are deposited on an appropriate electrode by varying the applied potential.

electrolytic solution A liquid containing a solvent, frequently water, in which an ionic solute is present, and through which electricity is conducted. The solute, which may be a salt, dissociates and may be separated from the solution by being deposited on an electrode.

electrolytic switch A switch whose terminals are immersed in an electrolytic solution.

electrolyze To effect a chemical change, especially decomposition, via electrolysis.

electromachining The use of electric energy to alter the shape of an object. Seen, for instance, in electrical-discharge machining.

electromagnet A magnet which becomes magnetized, or whose magnetization becomes significantly stronger, when an electric current flows through it. Such a magnet consists of a core, usually ferromagnetic, around which a coil is wound. The ferromagnetic core is frequently soft iron, while the coil is an insulated wire. The magnetic field is produced or multiplied when a current flows through the coil.

electromagnetic 1. Pertaining to both electricity and magnetism, their interactions, and the phenomena resulting from such interactions. 2. Pertaining to the interactions between electric and magnetic fields, and the phenomena resulting from such interactions.

electromagnetic attraction The attraction a pole of an electromagnet has for a pole with opposite polarity of another electromagnet.

electromagnetic brake A friction brake controlled by an electromagnet. A **friction brake** is one in which friction provides the stopping resistance. Also called **magnetic brake**.

electromagnetic communications Communications employing **electromagnetic waves**. Most such forms of communications utilize radio-frequencies, such as microwaves.

electromagnetic compatibility Its abbreviation is **EMC**. 1. The ability two or more devices, pieces of equipment, or systems, have to function in the same electromagnetic environment, without the function or performance of any of the devices, pieces of equipment, or systems being adversely affected by any electromagnetic interference created by any of them. Often, there is electromagnetic interference present, but it may not be sufficient to affect other devices. 2. The extent to which **electromagnetic compatibility** (1) exists. 3. The extent to which a device, piece of equipment, or system does not create electromagnetic interference which may deteriorate the function or performance of another device, piece of equipment, or system.

electromagnetic component In an electromagnetic wave, the magnetic constituent. It is perpendicular to the electric component. Also called **electromagnetic component**, or **magnetic component**.

electromagnetic constant Also called **speed of light**. 1. The speed at which electromagnetic radiation propagates through a vacuum. It is the same as the speed of light, which itself is visible electromagnetic radiation. This speed is a physical constant currently defined as 2.99792458×10^8 meters per second. Electromagnetic waves of all frequencies travel at the same speed in a vacuum. Its symbol is c. 2. The speed at which electromagnetic radiation propagates through a given medium. It is generally less than its speed in a vacuum.

electromagnetic coupling Coupling between circuits or devices which share the same electromagnetic field.

electromagnetic current The movement of charged particles, especially in the atmosphere, which give rise to electric and magnetic fields.

electromagnetic deflection 1. The use of a magnetic field to deflect an electron beam. 2. In a CRT, the use of a magnetic field to deflect an electron beam. A deflection yoke may be used for this purpose.

electromagnetic disturbance 1. An electromagnetic phenomenon which causes a degradation in performance, a malfunction, or the failure of a device, piece of equipment, or system. 2. An electromagnetic phenomenon which is superimposed on the intended signal.

electromagnetic energy Same as **electromagnetic radiation**.

electromagnetic environment The distribution of electromagnetic fields surrounding a given location. Such an environment is considered, for instance, when evaluating electromagnetic compatibility. Its abbreviation is **EME**.

electromagnetic field The field associated with **electromagnetic waves**. It is a combination of electric and magnetic fields which are at right angles to each other, and to the direction of motion.

electromagnetic flux The flux associated with an **electromagnetic field**.

electromagnetic focusing In a TV picture tube, the use of a magnetic field to focus the electron beam. This may be accomplished by varying the DC flowing through one or more focusing coils mounted on the neck of the tube. Also called **magnetic focusing**.

electromagnetic force The force responsible for all electromagnetic interactions, such as those which hold atoms and molecules together. It acts on anything with an electric charge, its force carrier is the photon, and its range is infinite. Electromagnetic force is created by the interaction of electric and magnetic fields, and can be attractive or repulsive, as in causing particles with like charges to repel one another and those with opposing charges to attract each other. It is one of the **four fundamental forces of nature**, the others being the gravitational force, the weak force, and the strong force.

electromagnetic forming The use of an intense transient magnetic field to form a metal workpiece into a desired shape. A large capacitor bank is charged, then instantaneously discharged to an induction coil, creating the magnetic field. Also called **magnetic forming**, or **magnetic pulse forming**.

electromagnetic frequency spectrum The **electromagnetic spectrum** in terms of frequency. This encompasses frequencies from just above 0 Hz to beyond 10^{24} Hz, corresponding to wavelengths of over 10^8 meters, to less than 10^{-16} meters, respectively. Also called **frequency spectrum (1)**.

electromagnetic horn A horn-shaped antenna structure which radiates highly directional radio waves, especially microwaves. An example is a horn antenna.

electromagnetic induction The generation of an electromotive force in a circuit or conductor caused by a change in the magnetic flux through said same circuit or conductor. Also called **induction (2)**.

electromagnetic inertia The characteristic lag of the current in a circuit in reference to a voltage which is applied or removed.

electromagnetic interaction Interactions due to **electromagnetic forces**.

electromagnetic interference An electromagnetic disturbance which brings about a degradation in performance, a malfunction, or the failure of a device, piece of equipment, or system. Electromagnetic interference can be produced by countless sources, such as microchips, lightning, power lines, and radio-frequency signals. Devices such as shields and filters are utilized to minimize or prevent its effects. Its abbreviation is **EMI**. When occurring in radio-frequencies, also called **RF interference**.

electromagnetic lens An electron lens which utilizes an electromagnetic field for focusing. Used, for instance, in high-quality image tubes.

electromagnetic mirror A surface or region, such as a building or an ionospheric layer, which reflects electromagnetic waves.

electromagnetic momentum The momentum associated with electromagnetic radiation.

electromagnetic noise Electromagnetic radiation producing noise in a circuit, device, piece of equipment, or system.

electromagnetic oscillograph An oscillograph whose mechanical recording mechanism is controlled by a moving-coil galvanometer.

electromagnetic phenomena Phenomena pertaining to both electricity and magnetism, their interactions, and that resulting from such interactions. Also, phenomena pertaining to the interactions between electric and magnetic fields, and that resulting from such interactions. Examples include electromagnetic attraction, electromagnetic induction, and electromagnetic waves.

electromagnetic propulsion A form of propulsion in which the propellant consists of charged particles being accelerated by an electromagnetic field, as seen, for instance, in a plasma engine.

electromagnetic pulse A momentary high-intensity burst of electromagnetic energy, such as that produced by a nuclear detonation, or lightning. This energy is distributed along a wide interval of frequencies. Such a pulse is capable of severely disrupting or destroying electrical and electronic devices, such as computers and communications equipment. Its abbreviation is **EMP**.

electromagnetic radiation The energy associated with **electromagnetic waves**. This energy oscillates between its electric and magnetic components. Electromagnetic radiation has both wavelike and particle-like properties. Also called **electromagnetic energy**, or **radiation (2)**.

electromagnetic relay A relay which is actuated through the action of a current passing though an electromagnet.

electromagnetic repulsion The repulsion a pole of an electromagnetic has for a pole with the same polarity of another electromagnet.

electromagnetic screen 1. Same as **electromagnetic shielding (1)**. **2.** Same as **electromagnetic shielding (2)**.

electromagnetic screening Same as **electromagnetic shielding**.

electromagnetic shield 1. Same as **electromagnetic shielding (1)**. **2.** Same as **electromagnetic shielding (2)**.

electromagnetic shielding Also called **electromagnetic screening**. **1.** A material or enclosure which blocks the effects of external electric and/or magnetic fields. It may consist, for instance, of a magnetic metal screen, or a grounded enclosure. Used, for example, to help minimize or prevent the effects of external electromagnetic interference. Also called **electromagnetic screen (1)**, or **electromagnetic shield (1)**. **2.** An **electromagnetic shielding (1)** which serves to confine electric and/or magnetic fields within an enclosure. Used, for instance, to help minimize or prevent electromagnetic interference which may affect other devices or systems. Also called **electromagnetic screen (2)**, or **electromagnetic shield (2)**. **3.** The use of an **electromagnetic shielding (1)**, or **electromagnetic shielding (2)**.

electromagnetic shock wave A high-intensity electromagnetic wave arising when multiple waves of varying intensities and velocities coincide in a non-dispersive medium. Used, for instance, to disintegrate kidney stones.

electromagnetic spectrum The range of frequencies of **electromagnetic radiation**. This encompasses frequencies from just above 0 Hz to beyond 10^{24} Hz, corresponding to wavelengths of over 10^8 meters, to less than 10^{-16} meters, respectively. These include, in order of ascending frequency: subsonic frequencies, audio frequencies, radio frequencies, infrared light, visible light, ultraviolet light, X-rays, gamma rays, and cosmic rays. These intervals have been arbitrarily established, may be labeled with alternate names associated with specific applications, and may have subdivisions. Also called **spectrum (5)**.

electromagnetic susceptibility The ease with which a component, circuit, device, piece of equipment, or system may suffer a degradation in performance, a malfunction, or a failure, due to the influence of electromagnetic energy. Also called **electromagnetic vulnerability**.

electromagnetic switch A switch that is actuated through the action of a current passing though an electromagnet.

electromagnetic system of units A system of electromagnetic units within the cgs system of units. Units in this system have an **ab-** prefix, for instance, abvolt or abtesla. The cgs measuring system is obsolescent. Its abbreviation is **emu**. Also called **electromagnetic units**, or **cgs electromagnetic system**.

electromagnetic theory of light The theory which states that light is an electromagnetic wave.

electromagnetic transducer A transducer which converts mechanical energy into electromagnetic radiation, and vice versa.

electromagnetic unit A unit, such as abamapere, within the **electromagnetic system of units**.

electromagnetic units Same as **electromagnetic system of units**.

electromagnetic vulnerability Same as **electromagnetic susceptibility**. Its abbreviation is **EMV**.

electromagnetic wavelength spectrum The **electromagnetic spectrum** in terms of wavelength. This encompasses wavelengths from 10^{-16} meters to over 10^8 meters, corresponding to frequencies of 10^{24} Hz to nearly 0 Hz, respectively.

electromagnetic waves Waves produced by the oscillation or acceleration of an electric charge. Such waves consist of sinusoidal electric and magnetic fields which are at right angles to each other, and to the direction of motion. Electromagnetic waves propagate through a vacuum at the speed of light, as it does through the atmosphere under most conditions. Such waves encompass frequencies from just above 0 Hz to beyond 10^{24} Hz, corresponding to wavelengths of over 10^8 meters, to less than 10^{-16} meters, respectively. Also called **electric waves**, and when occurring within a specified interval of radio-frequencies, called **radio waves** or **Hertzian waves**.

electromagnetics The science that deals with **electromagnetism**. Its abbreviation is **EM**.

electromagnetism 1. The science that studies electromagnetic phenomena. **2.** Magnetic phenomena produced by moving electric charge carriers. Also, electric phenomena produced by varying magnetic fields.

electromechanical 1. A mechanical component, device, piece of equipment, system, or process, which is electrically, electronically, or electromagnetically actuated, operated, or controlled. **2.** An electric, electronic, or electromagnetic component, device, piece of equipment, system, or process, which is mechanically actuated, operated, or controlled. **3.** Pertaining to a combination of electrical and mechanical forces.

electromechanical circuit A circuit incorporating one or more electromechanical components.

electromechanical component A mechanical component which is electrically, electronically, or electromagnetically actuated, operated, or controlled. Also, an electric, electronic, or electromagnetic component which is mechanically actuated, operated, or controlled.

electromechanical device A mechanical device which is electrically, electronically, or electromagnetically actuated, operated, or controlled. Also, an electric, electronic, or electromagnetic device which is mechanically actuated, operated, or controlled. For example, a touch-tone telephone.

electromechanical equipment Mechanical equipment which is electrically, electronically, or electromagnetically actuated, operated, or controlled. Also, electric, electronic, or electromagnetic equipment which is mechanically actuated, operated, or controlled.

electromechanical process A process which involves a combination of electrical and mechanical forces.

electromechanical relay A relay whose moving contact is actuated under the influence of a magnetic field which is produced by a coil through which an electric current is passed. When the coil is energized, it attracts the moving contact towards a stationary contact.

electromechanical switch A switch whose moving contact is actuated under the influence of a magnetic field which is produced by a coil through which an electric current is passed. When the coil is energized, it attracts the moving contact towards a stationary contact.

electromechanical system A system which incorporates a combination of electrical and mechanical components, forces, or the like.

electromechanical transducer 1. A device which converts electrical energy or signals into mechanical energy or signals. For instance, an electric motor. 2. A device which converts mechanical energy or signals into electrical energy or signals. For example, an electric generator.

electromechanics The technology dealing with electromechanical components, devices, equipment, systems, or processes.

electromedical equipment Electrical and electronic equipment, such as electrocardiographs or pacemakers, which serve for the prevention, diagnosis, care, or treatment of disease.

electrometallurgy The separation and processing of metals through electrical or electrochemical processes, such as electrolysis.

electrometer An instrument which measures potential differences drawing very little or no current from the source. Such an instrument may also measure low currents by passing said currents through a high resistance. An electrometer may have an input impedance of greater than 10^{15} ohms.

electrometer amplifier An amplifier whose characteristics, such as sensitivity and precision, enable it to measure very small currents. Some such devices can measure currents in the attoampere range.

electromigration The movement of atoms or ions in a metal conductor, so that they become displaced. It is due to high current densities, and is intensified by heat. Electromigration can eventually open a circuit, or make an unintended electrical connection. Seen for instance, in ICs.

electromotion Motion produced by an electric current. For example, the torque produced by an electric motor.

electromotive force Its abbreviation is **emf**. Expressed in volts. 1. The electric pressure that a source of electrical energy provides, which causes current to flow through a circuit or conductor. Its symbol is E. 2. The potential difference between two electrodes, each of a different composition, immersed in the same electrolyte.

electromotive series Same as **electrochemical series**.

electromyogram A graphical recording of the variations in the electrical activity generated in **electromyography**. Its abbreviation is **EMG**.

electromyograph The instrument utilized to obtain a **electromyogram**.

electromyography A technique for recording the electrical activity generated in muscles that are at rest or active. Also used for recording electrical activity in muscles produced in response to electrical stimulation. Its abbreviation is **EMG**.

electron A negatively charged stable subatomic particle. Electrons may be present as a constituent of an atom, or in a free state. When in the orbitals surrounding the nuclei of atoms, they determine many of its properties, such as chemical bonding. Each chemical element has a unique number of electrons, which is equal to the number of protons in its nucleus. Any inequality of this number results in an ion. As electrons flow through a conductor, an electric current is produced. The charge of an electron is equal to approximately 1.6022×10^{-19} coulomb, and its mass is approxi-

mately equal to 9.1094×10^{-31} kg. The positron is the antiparticle equivalent of the electron.

electron accelerator A device, such as a betatron, which accelerates electrons to increase their energy. A betatron can achieve energies in the GeV range.

electron acceptor A species, such as an atom, molecule, or ion, which accepts one or more electrons from an another species. The donating species is called **electron donor**. Also called **acceptor (2)**.

electron affinity 1. The energy released when an electron is added to a neutral atom, producing a negative ion. 2. The energy required to remove an electron from a negative ion, producing a neutral atom.

electron avalanche A cumulative process in which electrons are accelerated by an electric field and collide with neutral particles, liberating additional electrons. These additional electrons collide with other particles, and so on, so as to create an avalanche effect.

electron beam A narrow stream of electrons traveling in the same direction, with approximately the same speed, and emitted from the same source. Electric and/or magnetic fields are utilized to influence such a beam. Electron beams are used, for example, in CRTs, klystrons, lasers, and for machining or drilling. Its abbreviation is **E beam**.

electron-beam drilling The use of a precise electron beam to drill tiny holes in materials, such as semiconductors. The intense heat of the beam vaporizes the desired locations, which are removed in the vacuum where the drilling occurs.

electron-beam focusing The focusing of an electron beam in a CRT. A focusing electrode may be utilized for this purpose.

electron-beam generator 1. A source of electrons, such as a cathode in an electron tube. 2. A device, such as a klystron, in which extremely high radio frequencies are generated by modulating the velocity of an electron beam.

electron-beam gun Same as **electron gun**.

electron-beam induced current A technique utilized to isolate physical failure sites in semiconductors. It utilizes the electron beam of a scanning electron microscope to generate electron-hole pairs. Anomalies, such as junction defects or subsurface damage, affect the detected current. The detected current is amplified, and displayed on a monitor synchronized with the scan of the electron beam. Its abbreviation is **EBIC**.

electron-beam lithography A lithographic technique in which the resist is exposed by an electron beam, as opposed to light. This provides much greater precision. Used, for instance, to create patterns directly on semiconductor chips. Such patterns may be less than 50 nanometers wide. Its abbreviation is **E-beam lithography**.

electron-beam machining The use of a precise electron beam for machining. The intense heat of the beam vaporizes the desired locations, which are removed in the vacuum where the machining occurs. Used, for instance, for removing material in hard to reach places, for machining complex patterns, or to work with materials which are otherwise difficult to machine. Its abbreviation is **EBM**.

electron-beam magnetometer A device which measures the strength of a magnetic field, through its effect on the direction and intensity of an electron beam which is passed through it.

electron-beam parametric amplifier A parametric amplifier in which a modulated electron beam supplies a variable reactance. Also called **beam parametric amplifier**.

electron-beam recorder A device which utilizes an electron beam to record signals or data onto a substrate, such as film, metal, or plastic. The process occurs in a vacuum. Used, for instance, for CD or DVD mastering, or to transfer computer data onto microfilm.

electron-beam recording 1. A recording produced by an **electron-beam recorder**. 2. The process of producing an **electron-beam recording** (1).

electron-beam tube An electron tube in which one or more electron beams are produced and controlled. Examples include CRTs and klystrons.

electron-beam welding The use of a precise electron beam for welding. The intense heat of the beam joins the desired locations. Electron-beam welding may be utilized in a vacuum, at a low pressure, or at atmospheric pressure.

electron binding energy The minimum energy necessary to remove an electron from an atom or molecule.

electron bunch A group of electrons that form a cluster. Also called **bunch** (2).

electron bunching In a velocity-modulated tube, such as a klystron, the forming of electron bunches. Also called **bunching**.

electron capture A form of radioactive decay in which an electron from an inner shell is absorbed by the nucleus, combining with a proton, thereby reducing the atomic number of the atom by one. A neutron is formed, with the concomitant emission of a neutrino.

electron charge The charge carried by a single electron, which is equal to approximately -1.6022 x 10^{-19} coulomb. It is a fundamental physical constant, and all subatomic particles have an electric charge equal to this value, or a whole-number multiple of it. A single proton has the same magnitude charge, but opposite sign. Its symbol is e.

electron cloud 1. A region where electrons are aggregated, or where they tend to aggregate. For example, a region near a cathode in an electron tube where electrons which do not travel instantaneously to the anode tend to collect. 2. A region around the nucleus of an atom where electrons are predicted to be.

electron configuration The arrangement of an atom's electrons, according to their energy levels. The higher the energy level, the higher the energy of the contained electron.

electron-coupled oscillator A vacuum-tube oscillator which utilizes a screen grid that is connected in a manner that the input and the output of the tube are coupled by the electron beam emitted by the cathode. The screen grid is used as the anode, and the output is taken from the plate. Such an oscillator has reduced susceptibility to the effects of the load. Its abbreviation is **ECO**.

electron coupling The coupling of two circuits within a electron tube, through a shared electron stream. The electron stream between the electrodes of one circuit transfers energy to the electrodes of the other circuit.

electron cyclotron resonance The absorption of microwave energy by electrons which are located in a static magnetic field along with a perpendicular dynamic electric field which varies at a frequency equal to the angular speed of said electrons. This produces higher energy electrons, which increases the ionization of the gases in the chamber, which in turn produces a high-density plasma. Its abbreviation is **ECR**.

electron cyclotron resonance ion source The use of **electron cyclotron resonance** to generate multiply-charged heavy ions. Its abbreviation is **ECR ion source**. Also called **electron cyclotron resonance source**.

electron cyclotron resonance plasma The high-density plasma **electron cyclotron resonance** produces. Its abbreviation is **ECR plasma**.

electron cyclotron resonance source Same as **electron cyclotron resonance ion source**. Its abbreviation is **ECR source**.

electron density 1. The number of electrons per unit volume. 2. The number of free electrons per unit volume.

electron device A device whose main means of conduction is the passage of electrons through a vacuum, gas, or semiconductor. Examples include electron tubes and transistors.

electron diffraction The diffraction of an electron stream when passing through a medium, especially a crystal. The manner in which the electrons scatter is the diffraction pattern. Used, for instance, to study crystal structures.

electron-diffraction analysis The use of **electron diffraction** to analyze structure, especially that of a crystal.

electron-discharge machining Same as **electrical-discharge machining**. Its abbreviation is **EDM**.

electron donor A species, such as an atom, molecule, or ion, which donates one or more electrons to another species. The accepting species is called **electron acceptor**. Also called **donor** (1).

electron drift Same as **electron flow**.

electron emission 1. The freeing of electrons by the surface of a body into the surrounding space. This may be caused by heat, light, an electric field, electron impact, and so on. 2. The freeing of electrons by the surface of an electrode into the surrounding space.

electron emitter 1. A surface from which **electron emission** occurs. 2. An electrode surface from which **electron emission** occurs.

electron flow The movement of electrons in a conductor or semiconductor, under the influence of an electric field. Also called **electron drift**.

electron fluence The passing of electrons through a given area, such as a square centimeter. May be expressed, for instance, in electrons per second.

electron gas A concentration of free electrons within a given medium, such as an electron tube or conductor.

electron gun Also called **electron-beam gun**, or **gun**. 1. A device which produces, and usually controls, an electron beam. Used, for instance, in CRTs, klystrons, electron microscopes, lasers, and for machining or drilling. 2. An **electron gun** (1) in a CRT. It incorporates a cathode which emits the electrons, a control grid, and accelerating and focusing electrodes.

electron hole In a semiconductor, an electron vacancy that is created when an electron jumps the gap from the valence band to the conduction band. Such a hole is mobile and acts as if it were a positively-charged particle, and thus can serve as a charge carrier. Also called **hole**, or **mobile hole**.

electron-hole pair When an **electron hole** is created, a pair consisting of the hole itself, and the electron that jumped to the conduction band.

electron-hole recombination The process by which an electron which had jumped to the conduction band, to create an **electron hole**, returns to the valence band. This results in the electron hole recombining with said electron. This is accompanied by a release of energy, such as radiation. Also called **recombination** (1).

electron image Same as **electronic image**.

electron imaging Same as **electronic imaging**.

electron injection 1. The injection of electrons from one solid material into another, as might occur, for instance, in the transfer of electron charge carriers in semiconductors. 2. The injecting of a beam of electrons into a large electron accelerator, such as a betatron.

electron injector A device utilized for **electron injection** (2).

electron lens A device utilized to focus an electron beam, analogous to the manner in which an optical lens focuses a light beam. Electric and/or magnetic fields may be used for this purpose. Used, for instance, in CRTs, or in electron microscopes. Also called **lens** (2).

electron linear accelerator An electron accelerator in which the particles travel in a straight line, as opposed to having a circular or spiral path. Currently, electrons can be accelerated in this manner to energies in the TeV range. Used, for instance, to study the nature of matter.

electron magnetic resonance Same as **electron spin resonance**. Its abbreviation is **EMR**.

electron mass The mass of an electron at rest. It is a fundamental physical constant equal to approximately 9.10939 x 10^{-31} kg. Its symbol is m_e. Also called **electron rest mass**.

electron microprobe An X-ray device which emits a beam of electrons which is finely focused on a minute point on the surface of the specimen to be studied. The phenomena resulting from the electron bombardment, such as the characteristic X-rays emitted, and the secondary and backscattered electrons, are then analyzed. It is nondestructive, making it suitable for research in many areas, including crystallography, metallurgy, chemistry, and medicine. Also called **electron probe**.

electron microprobe analysis The use of an **electron microprobe** for analysis of a specimen. Also called **electron probe microanalysis**.

electron microscope A microscope which utilizes a beam of electrons focused by electron lenses to observe and record submicroscopic samples. There are various types, including transmission electron microscopes, which pass an electron beam through a thin sample, and scanning electron microscopes in which the electron beam is scanned over the surface of the specimen. Transmission electron microscopes can achieve resolutions of better than 1 angstrom. For reference, the smallest atom, that of hydrogen, has an approximate width of 1 angstrom. Its abbreviation is **EM**.

electron mirror In certain electron tubes, especially photomultiplier tubes, one of multiple electrodes which produce secondary emissions of electrons. Each successive electron mirror, or dynode, in the series produces greater and greater amounts of electrons. Depending on the total number of dynodes, the electron pulse may be millions of times larger than it was at the beginning of the tube. Used, for instance, for detection and measurement of minute flashes of light, and in TV camera tubes. Also called **dynode**.

electron mobility The average drift velocity of electrons in a semiconductor material, per unit electric field.

electron motion The movement of electrons through any given medium, such as a conductor, semiconductor, gas, or vacuum.

electron multiplier An electron tube in which secondary electron emissions produce current amplification. In it, a cathode releases primary electrons, which are then reflected by multiple electron mirrors, or dynodes, each producing more and more electrons. Depending on the number of dynodes, the amplification factor may be in the millions. Also called **electron-multiplier tube (1)**, **multiplier (2)**, or **secondary electron multiplier**.

electron-multiplier phototube A phototube incorporating one or more electron mirrors, or dynodes. Electrons emitted from its photocathode are reflected by the dynodes, each producing more and more electrons. Depending on the number of dynodes, the amplification factor may be in the millions. Also called **electron-multiplier tube (2)**, **multiplier phototube**, **photomultiplier tube**, or **photomultiplier**.

electron-multiplier tube 1. Same as **electron multiplier**. 2. Same as **electron-multiplier phototube**.

electron-optical Pertaining to **electro-optics**.

electron optics Same as **electro-optics**.

electron pair A pair of valence electrons which are shared by two adjacent atoms. Each pair of such electrons forms a bond between said atoms. Also called **pair (4)**.

electron-pair bond A chemical bond in which two atoms within a molecule share one or more pairs of electrons.

Such a bond may be between atoms of the same element, as in O_2 (molecular oxygen), or between different elements, as in **GaAs** (gallium arsenide). Also called **covalent bond**.

electron paramagnetic resonance Same as **electron spin resonance**. Its abbreviation is **EPR**.

electron paramagnetic resonance spectroscopy Same as **electron spin resonance spectroscopy**. Its abbreviation is **EPR spectroscopy**.

electron-positron pair An electron and a positron created simultaneously when a photon is in a high-intensity electric field, such as that of the nucleus of an atom.

electron-positron storage ring A vacuum chamber, with bending and focusing magnets, in which electrons and positrons traveling in opposite directions collide. The results of such collisions, such as annihilation and its related phenomena, are then analyzed.

electron probe Same **electron microprobe**.

electron probe microanalysis Same as **electron microprobe analysis**.

electron projection lithography A projection lithography technique in which an electron beam, as opposed to a light beam, is utilized. Resolutions in the nanometer range are attainable.

electron radius A physical constant equal to approximately 2.8179 x 10^{-13} cm. Its abbreviation is r_e. Also called **classical electron radius**.

electron rest energy The energy of an electron at rest, according to the mass-energy equation. This value is approximately equal to 0.511 MeV. Also called **electron self-energy**.

electron rest mass Same as **electron mass**.

electron scanning The scanning of an electron beam across a given surface, usually in a given pattern at regular intervals. The beam may be moved, for instance, utilizing a magnetic field. Used, for example, in CRTs, scanning electron microscopes, or in electron-beam machining.

electron self-energy Same as **electron rest energy**.

electron shell Any of the several orbits around the nucleus of an atom, in which electrons may be found. Each successive orbit further from the nucleus has greater energy. Each shell may only contain a specific number of electrons, and those in the same shell have the same energy. Electrons in the outer shell produce a net transport of electric charge under the influence of an electric field. Also called **shell (1)**.

electron spectroscopy Any analytical technique which examines the energy of electrons expelled from the surface of a solid which has been irradiated with particles such as electrons, photons, or ions. An example is Auger electron spectroscopy.

electron spectrum The plot, or other visual output, produced when utilizing **electron spectroscopy**. For example, that expressing the number of electrons detected per second, as a function of their energy.

electron spin The rotation of an electron about its own axis, which contributes to the angular momentum of said electron. This motion is independent of the electron orbiting around a nucleus.

electron spin resonance The resonant absorption of microwave radiation by a paramagnetic substance, which has at least one unpaired electron, in the presence of a strong magnetic field. Its abbreviation is **ESR**. Also called **electron paramagnetic resonance, electron magnetic resonance**, or **paramagnetic resonance**.

electron spin resonance spectrometer An instrument utilized for **electron spin resonance spectroscopy**. Also called **electron spin resonance spectrometer**, or **paramagnetic resonance spectrometer**.

electron spin resonance spectroscopy An instrumental analysis technique in which the microwave radiation absorbed in **electron spin resonance** is measured. Its abbreviation is **ESR spectroscopy**. Also called **electron paramagnetic resonance spectroscopy**, or **paramagnetic resonance spectroscopy**.

electron stream Electrons traveling in the same direction, with approximately the same speed, and emitted from the same source. An **electron beam** is a narrow electron stream.

electron synchrotron A electron accelerator in which the path of the particles is circular. Such an accelerator may employ alternating gradients to focus the electrons. Some electron synchrotrons can generate energies in excess of 10 GeV.

electron telescope A telescope which produces an enlarged electron image of a distant object. To do so, an infrared image of the object is focused on a photocathode, followed by enlargement by electron lenses, with the resulting image being displayed on a fluorescent screen.

electron trajectory The path an electron takes in an **electron transfer**.

electron transfer The passage of an electron from one entity to another. For instance, from one atom to another, from a cathode to an anode, from an electron gun to the phosphor surface on the back of the viewing screen of a CRT, and so on.

electron trap An entity or mechanism which serves to capture electrons, especially those that are mobile. For instance, an acceptor impurity in a crystalline semiconductor.

electron tube An active device consisting of a hermetically-sealed envelope within which electrons are conducted between electrodes. The cathode is the source of electrons, the positive electrode to which they travel is the anode or plate, while other electrodes that may be present include control grids and screen grids. Electron tubes may or may not contain a gas, and its presence and concentration affects the characteristics of said tubes. Such tubes have many applications, including their use in amplification, modulation, rectification, and oscillation. There are many examples, including CRTs, phototubes, pentodes, mercury-vapor tubes, and so on. When such a tube is evacuated to a degree that any residual gas present does not affect its electrical characteristics, it is called **vacuum tube**. Also known as **tube (1)**, or **valve**.

electron-tube amplifier An amplifier which incorporates one or more **electron tubes**.

electron unit A unit expressed in terms of a quantity associated with an electron. For instance, a unit of charge equivalent to the charge of an electron, or a unit of mass equivalent to that of an electron.

electron velocity The distance an electron travels per unit time. For instance, electrons in a linear accelerator may travel at nearly the speed of light.

electron volt Same as **electronvolt**.

electron wave The wave associated with an electron in motion.

electronegative 1. Possessing negative electric charge or polarity. 2. Tending to draw electrons to itself. For instance, fluorine and chlorine are electronegative chemical elements.

electronic 1. Pertaining to **electrons**. 2. Pertaining to **electronics**. 3. Pertaining to components and devices which conduct electrons, or other charge carriers such as electron holes or ions, through a vacuum, gas, or semiconductor. These components and devices perform functions such as amplification, rectification, and switching, and include electron tubes and transistors. Also, equipment and systems, such as computers, microwave ovens, and cellular tele-

phones systems, which incorporate such components. 4. Pertaining to computer, and/or online concepts. For instance, electronic mail, or electronic cash.

electronic absorption spectroscopy An instrumental analysis technique in which the electrons of atoms, molecules, or ions become excited as they absorb electromagnetic radiation. The frequencies and their corresponding absorptions are then charted in an **electronic absorption spectrum**.

electronic absorption spectrum The graphical representation produced when utilizing **electronic absorption spectroscopy**.

electronic air cleaner Same as **electrostatic precipitator**.

electronic altimeter 1. An absolute altimeter which bounces radio waves from the terrain below to determine altitude. Also called **radar altimeter**, or **radio altimeter**. 2. An altimeter which utilizes electronic devices to determine altitude.

electronic balance A device which indicates the weight of a sample, based on the force its presence on the weighing pan exerts on a moving coil in a magnetic field. The current needed to balance the sample is precisely measured to derive its weight.

electronic banking Banking in which transactions, such as viewing statements or effecting payments, are performed via a communications network, such as the Internet. The term also includes other operations, such as transfers between banks of funds related to the clearing of checks. Its abbreviation is **e-banking**.

electronic BBS Abbreviation of **electronic b**ulletin **b**oard **s**ystem. A computer system that serves as a message center that is accessed by other computers through a modem. These usually focus on specific interests, supply information, and generally provide a forum to exchange thoughts, files, programs, and email. It is also called **electronic bulletin board**, or **bulletin board system**.

electronic billing Billing for goods and/or services via the Internet. Its abbreviation is **e-billing**, or **ebilling**.

electronic book Its abbreviation is **e-book**. 1. A handheld device which is specially adapted for displaying electronic versions of books. Depending on the specific device, viewed books may be downloaded, or accessed from a disk. 2. A book whose content is in electronic form. Among the benefits of such a format is the ability to search for specific passages quickly and easily. Many e-books also include multimedia content. An electronic book may be accessed by means of a computer, such as a desktop or a handheld, or through an **electronic book (1)**.

electronic brainstorming The use of computers and networks to generate ideas, usually in a collaborative setting such as that facilitated by conferencing and groupware. Its abbreviation is **EBS**.

electronic bug 1. A defect present in a circuit, device, component, or the like, which causes malfunctions. Also called **bug (2)**. 2. An electronic device utilized to surreptitiously listen to and/or record conversations. Also called **electronic listening device**, or **bug (3)**. 3. A semiautomatic telegraph key for producing correctly spaced dots and dashes. Also called **bug (4)**.

electronic bulletin board Same as **electronic BBS**.

electronic bulletin board system Same as **electronic BBS**.

electronic business Same as **ecommerce**.

Electronic Business XML Same as **ebXML**.

electronic calculator An electronic device which performs arithmetic operations. These can only accept numerical input, and usually have liquid-crystal displays or light-emitting diodes for displaying the results of calculations. Almost all calculators can store numbers in memory, many provide additional functions such as logarithms and square-root calculations, and some are programmable. The latter usually incorporate a CPU, and may accept plug-in cards. The only moving parts of such a device are its keys.

electronic card A digital greeting card sent over the Internet. There are many applications and Web sites which can be utilized to prepare such cards, and such greetings may incorporate text, still images, video, animation, audio, and so on, and can be chosen for virtually any occasion or purpose. Its abbreviation is **e-card**. Also called **electronic postcard**, **digital postcard**, or **virtual card**.

electronic cash Its abbreviation is **e-cash**. Also known by various other terms, including **electronic currency**, **electronic money**, **digital money**, or **digital cash**. **1.** Any form of money used over the Internet. Used for many purposes, such as effecting online purchases or making credit card payments. Security is a key concern, and considerable measures, such as cryptography, are taken to avoid circumstances such as losses of funds, misplacement of funds, stealing of funds, or the re-use of funds. **2.** A form of currency that is digitally encoded, and that can be converted to legal tender if desired. It may be coded, for instance, through the use of an encrypted serial number. **Legal tender** is a lawfully valid currency that must be accepted for payment of debts, the purchase of goods and/or services, or for other exchanges for value. Circulating paper money is another form of legal tender.

electronic cash register A cash register which contains electronic components, such as bar-code or RFID readers. Such devices usually incorporate point-of-sale features, such as inventory updating and verification of credit cards. Its abbreviation is **ECR**.

electronic chart A chart which is in digital form, as opposed to paper form. Electronic charts are suitable for computer processing, and facilitate use with other electronic devices, such as Electronic Chart Display and Information Systems. Also called **digital chart**.

Electronic Chart Display and Information System A navigation system which displays data, such as positional information, in electronic form. It makes use of electronic charts, among other resources. Its abbreviation is **ECDIS**.

electronic chart reader A scanning device which converts the information contained in an analog chart into digital form. Used for instance, to present digital displays of navigational or medical charts.

electronic check The electronic equivalent of a check. Used, for instance over the Internet, where the checking account of the check writer is debited, without the use of a paper check. Its abbreviation is **e-check**.

electronic cinema Same as **e-cinema**.

electronic circuit A circuit which contains one or more active components, such as diodes, transistors, ICs, or operational amplifiers.

electronic clock A clock whose accuracy is regulated by electronic components, such as crystal oscillators, as opposed to mechanical devices such as springs. Examples include quartz clocks and cesium clocks.

electronic commerce Same as **ecommerce**.

electronic communications Communications utilizing electronic circuits, components, devices, equipment, and/or systems. Most communications are of this type, and include telephone conversations, faxes, email, real-time chats, SMSs, videoconferencing, and the like, utilizing satellites, LANs, the Internet, and so on. Also, the technology enabling such communications. Its abbreviation is **e-communications**.

electronic commutator A commutator that utilizes active components, such as transistors or electron tubes, as opposed to mechanical switches. This provides greater speed, reliability, and durability, in addition to minimizing noise.

electronic component Same as **electronic element**.

electronic computer Nearly always synonymous with **computer**, as very few computers are not electronic. A pneumatic computer, for instance, has no has no electronic circuits.

electronic conduction The flow of electricity as a result of the movement of electrons.

electronic content Information that is available online, and which may be any combination of audio, video, files, text, or the like. Its abbreviation is **e-content**, or **econtent**.

electronic control The control of a device, piece of equipment, system or process through the use of electronic components, such as transistors or electron tubes. This contrasts with **electric control**, which utilizes electrical components such as switches, relays, or rheostats. Also called **electronic regulation**.

electronic controller An automatic controller that utilizes **electronic control**.

electronic counter An electronic circuit or device used as a counter. For example, it may serve to store and/or indicate a given total, or to generate an output after counting a specified number of pulses. Such counters have no moving parts, providing greater speed, reliability, and durability.

electronic counter-countermeasures Measures taken to minimize the effects of **electronic countermeasures**. Its abbreviation is **ECCM**.

electronic countermeasures Measures taken to reduce the effectiveness of devices, such as radars, which utilize any given portion of the electromagnetic spectrum. An example is jamming. Its abbreviation is **ECM**.

electronic coupling Coupling occurring in association with electronic components or electrons. An example is electron coupling.

electronic credit Same as **e-credit**.

electronic currency Same as **electronic cash**.

Electronic Data Interchange. Same as **EDI**.

electronic data processing Its abbreviation is **EDP**. Also called **data processing**, **information processing**, or **automatic data processing**. **1.** The processing of information by a computer. **2.** The processing of information within a specific application. For example, the manipulation of large amounts of numeric data in a program.

electronic data-processing equipment Equipment used for **electronic data processing**. Such equipment includes computers and any peripherals used in this process. Its abbreviation is **EDPE**. Also called **electronic data-processing system** (1), or **data-processing equipment**.

electronic data-processing system Also called **data-processing system**. **1.** Same as **electronic data-processing equipment**. **2.** The resources and procedures utilized for **electronic data processing**.

Electronic Delay Storage Automatic Calculator A computer completed in 1949 which used mercury delay lines and stored binary digits. It performed approximately 650 instructions per second, and its operating system consisted of 31 words of ROM. Its acronym is **EDSAC**.

Electronic Design Automation The use of sophisticated software to design and simulate the function of electronic circuits and systems. For instance, that of chips, or of telecommunications systems. Its abbreviation is **EDA**.

electronic design interchange format A standard format utilized for the exchange of graphics data, such as that used to describe circuits. Used, for instance, in computer-aided design. Its acronym is **EDIF**.

electronic device **1.** A physical unit, such as a transistor or electron tube, which can operate on an applied electrical signal, as in amplifying, rectifying, or switching. **2.** A device which incorporates one or more **electronic devices (1)**. There are countless examples, including computers, cellular telephones, and audio and video recording and reproducing equipment.

electronic differential analyzer A differential analyzer which uses electronic components.

electronic display A device, incorporating electronic components such as LEDs, utilized to display the images generated by a computer, TV, oscilloscope, radar, or other similar device with a visual output.

electronic document management The use of computer software, such as database programs, and hardware, such as scanners, to create and manage the documents of an organization. Such documents may be paper-based or electronic. Its abbreviation is **EDM**. Also called **enterprise document management**, or **document management**.

electronic document management system A system which facilitates **electronic document management**. It performs various functions, such as providing search engines, converting between formats, providing access through networks, and determining if a given user can update or only read documents. Its abbreviation is **EDMS**. Also called **enterprise document management system**, or **document management system**.

electronic eavesdropping 1. The use of electronic devices, such as bugs, to **eavesdrop**. 2. To eavesdrop **electronic communications**.

electronic efficiency The ratio of the power delivered by an electron stream, to the power supplied to said stream. Utilized as a measure of the effectiveness of an electron stream for the transmission of power.

electronic element A circuit component, such as a transistor or electron tube, which can operate on an applied electrical signal, as in amplifying, rectifying, or switching. **Electrical element** is a more general concept, and also includes any other electrical components that are a part of a circuit, such as resistors, capacitors, and generators. Also called **electronic component**.

electronic engineering The branch of engineering which deals with the design, construction, operation, and optimization of electronic circuits, components, devices, equipment, and systems.

electronic flash 1. A repeatable and artificially produced burst of bright light. Such a flash is usually generated by applying a high voltage to an electrode of a tube containing an inert gas such as xenon. The gas becomes ionized, which permits it to rapidly discharge the energy previously stored in a capacitor. Used, for instance, in photography. Also called **photoflash**, **flash (2)**, or **strobe (3)**. 2. Same as **electronic flash tube**.

electronic flash lamp Same as **electronic flash tube**.

electronic flash tube A tube which produces an **electronic flash (1)**. Also called **electronic flash (2)**, **electronic flash lamp**, **flash tube**, **strobe (2)**, or **strobe light**.

electronic form An online document with blank spaces in which information is entered. May be used, for instance, by an online seller requesting billing and shipping information from a client. When utilized for business transactions, such forms are usually secured with encryption. Its abbreviation is **e-form**.

electronic funds transfer The transfer of money from one account to another, via electronic means such as computer networks. Such transactions should be secure and instantaneous, and usually occur automatically. Examples include electronic transfers between banks, and online payments of bills. Its abbreviation is **EFT**.

electronic game Its abbreviation is **e-game**, or **egame**. 1. A computer game which is played on its own portable, usually

palm-sized, device. 2. A computer game played over the Internet. Also called **Internet game**.

electronic guitar Same as **electric guitar**.

electronic heating The use of a radio-frequency power source, such as an oscillator, to provide heat. Also called **RF heating**, or **high-frequency heating**.

electronic ignition Ignition which is controlled by electronic devices, such as semiconductors.

electronic image An image formed through **electronic imaging**. Also called **electron image**.

electronic imaging The process of converting an image into an electronic equivalent, such as a bitmap, suitable for computer processing, storing, transmitting, or displaying. A charge-coupled device, for instance, may be used to capture such an image. Also called **electron imaging**, or **imaging (2)**.

electronic information processing The use of electronic devices and equipment, such as computers, to process data. This data may be in any form, including text, audio, images, and so on.

electronic ink The coating, such as a flexible array of transistors, of a sheet of **electronic paper**. Its abbreviation is **eink** or **e-ink**. Also called **digital ink**.

electronic instrument 1. An instrument, such as a multimeter or absolute altimeter, which incorporates one or more electronic devices. 2. Same as **electronic musical instrument**.

electronic interference 1. Any electrical or electromagnetic interference which surrounds an electronic device, and which diminishes it performance, such as its ability to receive or process a desired signal, or that causes a malfunction or failure. 2. Any interference an electronic device produces.

electronic inverter An inverter which utilizes one or more electronic components to perform its function.

electronic jamming The intentional use of electromagnetic radiation, especially radio-frequency signals, to disrupt communications. Also called **jamming**.

electronic journal Same as **electronic log**.

electronic line scanning The use of electronic means to move a scanning spot along a scanning line.

electronic listening device Same as **electronic bug**.

electronic lock 1. A lock which can only be opened by entering a specific sequence of signals. Such a lock may be used, for instance, with a key about the size of a credit card which incorporates a magnetic stripe that is swiped to provide the proper signals enabling access or authorization. Electronic locks are easy to reprogram, and usually provide the ability to monitor and log usage. Such locks may be used, for instance, for entrances, or to allow the use of equipment. 2. A lock incorporating one or more electronic devices.

electronic locking The use of an **electronic lock**.

electronic log A record of computer and/or network activity. Used, for example, to find the origin of problems, monitor usage, recover data, or to identify unauthorized access. Also called **electronic journal**, **journal**, or **log (3)**.

electronic magazine A magazine, or similar publication, which is distributed or otherwise made available in digital form. Its abbreviation is **e-zine**.

electronic mail 1. Same as **email (1)**. 2. Same as **email (2)**.

electronic means The utilization of electronic components, circuits, devices, equipment, and/or systems, to perform any given function. There are countless examples, including recording of multimedia content onto a DVD, filing of an income tax return through the Internet, or conversing using a cellular phone.

electronic meeting room A location specially equipped with computers, peripherals, and other electronic devices which facilitate and enhance gatherings of people for work, collaboration on a project involving a common interest, or the like. One or more participants may assist to conferences remotely, via a network such as the Internet.

electronic memory Computer memory, such as semiconductor memory, with no moving parts. Currently, almost all memory is of this type.

electronic messaging Any form of sending, receiving, and exchanging messages, such as emails, instant messages, MMSs, or SMSs, utilizing a communications network.

electronic micrometer A micrometer which utilizes electronic devices, such as semiconductors or electron-tube circuits, for enhanced sensitivity and accuracy.

electronic money Same as **electronic cash**.

electronic motor control The use of an electronic circuit to control the speed of a DC motor which is fed by an AC power line. Also called **motor control (1)**.

electronic multimeter A multimeter which utilizes electronic devices, such as semiconductors or electron-tube circuits, for enhanced sensitivity and accuracy.

electronic music Music which is created and/or enhanced by electronic means. There are many examples, including the use of electronic components to amplify or process sounds, or the use of computers to synthesize music.

electronic music distribution The distribution of music via electronic means, such as a communications network. Such music is in a digital format.

electronic musical instrument Also called **electronic instrument (2)**. **1.** A musical instrument, such as an electronic organ or an electric guitar, which amplifies, synthesizes, or enhances sounds. **2.** A musical instrument, such as an organ or guitar, which is emulated by electronic means. An example is the use of a computer to synthesize sounds corresponding to a given instrument.

electronic navigation The use of electronic devices, instruments, or systems, to navigate or aid in navigation. For instance, the use of radars or GPS systems.

electronic organ An electronic device which synthesizes sounds which emulate those of an organ which utilizes pipes. An electronic organ usually provides many enhancements, such as the ability to interface with a computer. Also called **electric organ (2)**.

electronic organizer A usually handheld device, such as a PDA, which stores information and performs functions which help organize an individual. Such a device may include a telephone, email, and address book, an appointment scheduler, a notepad, and so on. Also called **organizer (1)**.

electronic packaging The process of assembling, interconnecting, embedding, or mounting electronic components, circuits, or devices, so that they can perform their functions while sealed or otherwise protected. An example is a dual in-line package for a chip. Also called **packaging**.

electronic paper A storage medium analogous to regular paper, except that it can be written upon or refreshed numerous, often millions, of times via electronic means. Such paper may be made of a plastic coated with the proper electronics, and can be used, for instance, in ebooks, newspapers that update their content via satellite, portable signs, or the like. Electronic paper may or may not be fully flexible, and/or offer full color images. Its abbreviation is **epaper**. Also called **digital paper**.

electronic payment Payment for goods and/or services via the Internet. Its abbreviation is **e-payment**, or **epayment**. Also called **Internet payment**, or **Web payment**.

Electronic Performance Support System A system which provides help, tips, and other information which assist a user in better performing a given task, while utilizing an electronic device such as a computer. For instance, a user may

be working with a given application, encounter an unknown feature, ask a question out loud, and then be presented with a multimedia explanation of its use. Its abbreviation is **EPSS**.

electronic photometer A photometer which incorporates an electronic device, such as a phototube or phototransistor, for measuring the intensity of light. Also called **photoelectric photometer**.

electronic piano Also called **electric piano**. **1.** An electronic device which synthesizes sounds which emulate those of a piano which utilizes vibrating wire strings. An electric piano usually provides many enhancements, such as the ability to interface with a computer. **2.** A piano whose sound is enhanced and/or amplified electronically.

electronic point-of-sale A point-of-sale with electronic equipment, such as bar-code, RFID, and/or magnetic stripe readers, and a terminal, utilized for pricing and recording transactions, for obtaining purchase authorizations, and the like. Its abbreviation is **EPOS**. Also called **point of sale (3)**.

electronic postcard Same as **electronic card**.

electronic postmark A digital certificate providing an authenticated time and date stamp.

electronic power supply A source of power for electronic components, circuits, devices, equipment, or systems. Such sources include batteries and power packs. Also called **power supply (2)**, or **supply (3)**.

electronic precipitator Same as **electronic air cleaner**.

electronic publishing The providing of information in electronic form, such as in DVDs, or over networks, as opposed to being published in paper form. Such publishing has various advantages, including interactivity, and the ability to keep fully current.

electronic recording The use of electronic means to produce recordings. For example, the recording of sounds onto a voice chip, or those produced by a cathode-ray oscillograph.

electronic rectifier A rectifier which incorporates, or consists of, an electronic component or device. An example is a semiconductor diode.

electronic regulation Same as **electronic control**.

electronic regulator Same as **electronic voltage regulator**.

electronic relay **1.** A relay with no moving parts. **2.** A relay, such as a solid-state relay, which incorporates electronic components or devices.

electronic scanning The use of electronic means to scan, as opposed to mechanical means, such as moving antennas or mirrors. Used, for instance, in radars.

electronic security **1.** The incorporation of electronic components and devices, such as transistors and video monitors, into a security system. **2.** The protection of computer data, communications signals, and networks. Passwords and firewalls are examples of protective measures utilized.

electronic serial number A number which uniquely identifies each cellular phone. Each time a call is made, this number is transmitted to the mobile telephone switching office for verification. Its abbreviation is **ESN**.

electronic signature An attachment to an electronic transmission which uniquely identifies and authenticates the sender of a message, file, or other information such as a credit card number. To be effective, an electronic signature must not be forged, and measures such as public-key cryptography are employed to insure its integrity. Its abbreviation is **e-signature**, or **esignature**. Also called **email signature (2)**, **digital signature**, or **signature (2)**.

electronic software distribution The distribution of software via electronic means. For instance, the use of a network to issue an upgrade to an application. Its abbreviation is **ESD**.

electronic spectroscopy An instrumental analysis technique in which the absorption or emission of radiation associated with electron transitions between energy levels in atoms, molecules, or ions is measured. This occurs mostly in the visible and ultraviolet range. The frequencies and their corresponding absorptions or emissions are then charted in an **electronic spectrum**.

electronic spectrum The graphical representation produced when utilizing **electronic spectroscopy**.

electronic spreadsheet A computer application which manipulates a matrix of interrelated cells, often arranged in rows and columns. Each cell can have labels, numerical values, formulas, or functions. Used primarily for calculations such as budgets, and for what-if scenarios. Also called **spreadsheet (2)**.

electronic stethoscope A stethoscope which incorporates electronic components such as amplifiers and filters, which enable internal body sounds, such as a heart murmurs or lung wheezes, to be heard more clearly and loudly. Such stethoscopes can usually be connected to a computer, enabling storing, further analysis, and transmission of the recorded sounds. Also called **stethoscope (2)**.

electronic support Same as **esupport**.

electronic surge protector 1. A surge protector incorporating electronic components or devices. **2.** A surge protector utilized to safeguard electronic components, devices, equipment, or systems.

electronic switch 1. An electronic component or device, such as a transistor, utilized as a switch. Electronic switches feature high speed and reliability. Used, for instance, in computers. **2.** A device which permits the simultaneous observation of two signals on an oscilloscope with a single electron gun. This is accomplished via a rapid switching rate. **3.** A switch with no moving parts.

electronic switching 1. The use of an **electronic switch**. **2.** The use of electronic switches for telephone switching. For instance, a central office may have a computer which controls all switching.

electronic switching system Its abbreviation is **ESS**. **1.** A switching system consisting exclusively of **electronic switches**. **2.** A telephone switching system which utilizes a computer for all switching. For instance, that used in call routing, and for other functions, such as automatic caller identification and call waiting.

electronic tablet An input device which serves for sketching or tracing images directly into a computer. It consists of an electronic pressure-sensitive tablet which senses the position of the stylus which makes contact with it. The stylus, or puck, is similar to a mouse, but is much more accurate, because its location is determined by touching the active surface of the tablet with an absolute reference. Used, for instance, in computer-aided design. Also called **digitizing tablet, data tablet, tablet**, or **touch tablet**.

electronic technology Same as **electrotechnology**.

electronic temperature control The controlling of temperature utilizing electronic components, such as electronic thermometers.

electronic text A text, such as a book, whose content is in electronic form. An electronic text may be available, for instance, over the Internet, or be contained in an optical disc. Among the benefits of such a format is the ability to search for specific passages quickly and easily, seamless inclusion of multimedia content, and the ability to easily update subject matter. Its abbreviation is **e-text**.

electronic thermometer A thermometer which utilizes an electronic component or device, such as a thermistor, to measure temperature.

electronic timer A timer whose accuracy is regulated by electronic components or devices. For instance, a crystal-controlled timer.

electronic tube Same as **electron tube**.

electronic tuning Tuning utilizing an electronic component, such as a varactor, as opposed to a mechanical device. Also called **electric tuning**.

electronic typewriter A typewriter incorporating electronic components and devices, such as chips. These provide for added functions, such as memory, the ability to connect to a printer or computer, and increased reliability.

electronic voltage regulator 1. A voltage regulator which incorporates electronic components or devices. **2.** A regulator which supplies a steady voltage to electronic components or devices.

electronic voltmeter A voltmeter which utilizes electronic components and devices, such as amplifiers and rectifiers, for enhanced sensitivity and accuracy.

electronic wallet Encryption software which serves to provide the virtual equivalent of a wallet. In it, may be contained digital cash, credit card information, shipping details, and a digital certificate for authentication of the wallet holder. Both merchants and customers benefit from the added security, expedience, and convenience. Its abbreviation is **e-wallet**, or **ewallet**. Also called **virtual wallet**, or **digital wallet**.

electronic warfare The use of electronic devices and/or electromagnetic radiation for military purposes. Also, military efforts to deny adversaries the use of such resources.

electronic watch A watch whose accuracy is regulated by electronic components, such as crystal oscillators, as opposed to mechanical devices such as springs. A quartz watch is an example.

electronics The science that deals with the study, design, use, effects, and improvement of components and devices which conduct electrons, or other charge carriers such as electron holes or ions, through a vacuum, gas, or semiconductor. These components and devices perform functions such as amplification, rectification, and switching, and include electron tubes and transistors. Electronics also includes the equipment and systems which incorporate said components and devices. The applications of electronics are countless, and its technology is the basis of computers, cellular telephony, TV receivers, microwave ovens, sound and video recording and reproducing equipment, and so on.

Electronics Industries Association An organization which sets standards for electronic components and products. Its abbreviation is **EIA**.

electronvolt The energy acquired by an electron which passes through a potential difference of 1 volt. It is equal to approximately 1.6022×10^{-19} joule. Its symbol is **eV**. Electronvolt is also a unit of work. Also spelled **electron volt**.

electroosmosis The movement through a diaphragm, while under the influence of an electric field, of a liquid in which charged particles are suspended. Also spelled **electro-osmosis**.

electrophilic Pertaining to a species, such as an atom, molecule, or ion, which is attracted to electrons or regions of high electron density. Such species are electron acceptors. This contrasts with **nucleophilic**, in which a species is attracted to nuclei or regions of low electron density.

electrophoresis The migration of charged particles suspended in a medium under the influence of an electric field produced by electrodes immersed in the medium. When particles migrate towards the cathode it is called **cataphoresis**, while in **anaphoresis** they migrate towards the anode.

electrophoretic Pertaining to **electrophoresis**.

electrophoretic deposition The formation of a deposit through the use of **electrophoresis**. Used, for instance, to form a desired coat on an electrode.

electrophorus A device utilized to produce electric charges through induction. It consists of a dielectric disk made of resin or hard rubber, and a metal plate with an insulating handle. The disk is charged by rubbing it with fur, and the plate is placed on said disk, which charges said plate by induction. Then, the plate is momentarily grounded, removing negative charge, which leaves the plate with an induced positive charge.

electrophotographic Pertaining to, or utilizing **electrophotography**.

electrophotographic printer A printer using an **electrophotographic process**.

electrophotographic process A process, such as electrophotography, through which images are formed utilizing electricity and light.

electrophotography A method of reproducing images utilizing electricity and light. In it, a photoconductive surface, usually a drum, is positively charged with static electricity, then exposed to an optical image of that which is to be reproduced, forming a corresponding electrostatic pattern. A fine powder called toner, which has been negatively charged is spread over this surface, adhering only to the charged areas of the drum. The toner is then fused to the paper, which has been positively charged, forming the final image. Used, for instance, in laser printers and copy machines. Also called **xerography**.

electrophysiology The branch of science which deals with electrical phenomena associated with physiological processes. Electrophysiology is used, for instance, to study and measure the electrical activity of the brain, heart, nervous system, and so on.

electroplate The formation of a metal deposit on another surface, usually a different metal, via electrolysis. Also called **plate (4)**.

electroplating The process of depositing a metal onto another surface, usually a different metal, via electrolysis. Such a deposit may be thin or thick, and may be used, for instance, to provide coatings which are protective, decorative, or which have any given electrical properties. Also called **electrolytic plating**, or **plating (1)**.

electropneumatic A device which incorporates both electronic and pneumatic components. A robotic actuator, for instance, may be electropneumatic.

electropneumatic actuator A robotic actuator incorporates both electronic and pneumatic components. Such actuators are especially adapted for tasks involving delicate contact with objects.

electropolishing The polishing of a metallic material by making it the anode in an electrolytic cell. Used, for instance, when mechanical polishing is impractical or undesirable. Also called **electrolytic polishing**.

electropositive 1. Possessing positive electric charge or polarity. **2.** Tending to give up electrons. For example, cesium and potassium are electropositive chemical elements.

electropotential Same as **electrode potential**.

electrorefining The utilization of electrolysis to purify a metal. For instance, the use of an electrolytic cell in which the impure metal is the anode, and having it plated at the cathode. Also called **electrolytic refining**.

electroretinogram A graphical recording of the variations in the electrical activity of the retina, when stimulated by light. The **retina** is a light-sensitive membrane within the eye, and is linked to the brain via the optic nerve.

electroretinography The technique utilized to obtain an **electroretinogram**.

electroscope An instrument which detects electric charges by means of electrostatic forces. In a gold-leaf electroscope, for instance, two gold leaves are suspended side by side from a conducting rod which is held by an insulated support, and placed in a grounded enclosure, such as a glass jar. When a charge is applied to a plate to which the rod is connected, the leaves separate due to their mutual repulsion.

electrosensitive recording Same as **electric recording (3)**.

electrostatic Pertaining to electric charges at rest. That is, concerning static electricity. Such charges do not flow along a conducting path, and examples include that produced by friction or induction, or the electric charge stored in a capacitor.

electrostatic accelerator Same as **electrostatic generator**.

electrostatic actuator A device that applies an electrostatic force to the diaphragm of a microphone, for calibration purposes.

electrostatic air cleaner Same as **electrostatic precipitator**.

electrostatic analyzer An instrument that selects particles depending on their **electrostatic deflection**.

electrostatic attraction The electrostatic force of attraction between two oppositely charged particles, according to the **electrostatic attraction law**. Also called **Coulomb attraction**.

electrostatic attraction law A law which states that the force between two charged particles is directly proportional to the product of their magnitudes, and inversely proportional to the square of their distance. This force occurs along a straight line between the charges, and is a repulsive force if the charges are like, or an attractive force if they are opposite. Also called **Coulomb's law**, or **law of electrostatic attraction**.

electrostatic charge Same as **electric charge**.

electrostatic component Same as **electric component (2)**.

electrostatic copier A copier utilizing an **electrostatic printing** process.

electrostatic coupling The coupling of two or more circuits by means of a capacitance. Also called **capacitive coupling**.

electrostatic deflection The bending of an electron beam via the influence of an electrostatic field. In a CRT, for instance, two electrodes, called deflection electrodes, can be used to deflect an electron beam in the desired manner.

electrostatic discharge A quick and spontaneous discharge of static electricity. Such a discharge is usually occurs as a result of two bodies at different electrostatic potentials approaching each other. Without electrostatic discharge protection, such a discharge from a hand, for instance, could damage or destroy a chip. Its abbreviation is **ESD**.

electrostatic discharge damage Damage due to an **electrostatic discharge**. The damage caused to a component, circuit, or device may be immediately manifested, or be evident some time later. Its abbreviation is **ESD damage**.

electrostatic discharge protected area An area where **electrostatic discharge protection** measures are taken. Its abbreviation is **ESD protected area**.

electrostatic discharge protection Measures taken, such as the use of air ionizers, the proper apparel, and special floor mats, to protect against **electrostatic discharges**. Its abbreviation is **ESD protection**.

electrostatic discharge protective packaging Packaging which helps protect against **electrostatic discharge** damage. For instance, a static-dissipative bag. Its abbreviation is **ESD protective packaging**.

electrostatic discharge sensitive Displaying **electrostatic discharge sensitivity**. Its abbreviation is **ESD sensitive**.

electrostatic discharge sensitivity The ease with which a component, circuit, or device may malfunction, be damaged, or be destroyed, due to **electrostatic discharges**. Its abbreviation is **ESD sensitivity**. Also called **electrostatic discharge susceptibility**, or **electrostatic discharge vulnerability**.

electrostatic discharge susceptibility Same as **electrostatic discharge sensitivity**. Its abbreviation is **ESD susceptibility**.

electrostatic discharge vulnerability Same as **electrostatic discharge sensitivity**. Its abbreviation is **ESD vulnerability**.

electrostatic discharge wrist strap Same as **ESD wrist strap**.

electrostatic driver Within a speaker system, a specialized electrostatic speaker, such as a tweeter.

electrostatic energy The energy contained in electric charges at rest. For instance, that contained in a capacitor. It is a form of potential energy.

electrostatic field The electric field associated with electric charges at rest. The intensity of such a field remains constant.

electrostatic field intensity The strength of an **electrostatic field**.

electrostatic field meter A device which measures **electrostatic field intensity**. Also spelled **electrostatic fieldmeter**.

electrostatic fieldmeter Same as **electrostatic field meter**.

electrostatic flux Same as **electric lines of force**.

electrostatic focus Same as **electrostatic focusing**.

electrostatic focusing In a CRT, the use of an electrostatic field to focus an electron beam. For instance, the voltage applied to each of two electrodes, called deflection electrodes, may be varied to focus said beam. Also called **electrostatic focus**.

electrostatic force The electrostatic attractive or repulsive force between two charged particles, according to the **electrostatic attraction law**. Also called **Coulomb force**.

electrostatic generator A device utilized to generate electrostatic charges, such as those produced by induction or friction. Belt generators can generate potentials of several million volts, while the electric potential in lightning can be as great as 100 million volts. Also called **electrostatic accelerator**, **electrostatic machine**, or **static machine**.

electrostatic headphones A pair of headphones, each consisting of a small **electrostatic speaker**.

electrostatic induction The modification of the distribution of electric charge on an body as influenced by another nearby body which is charged. Positive charges will induce negative charges, and vice versa. Also called **induction (3)**.

electrostatic interactions The interactions between charged particles, according to the **electrostatic attraction law**. Also called **Coulomb interactions**.

electrostatic lens An electron lens utilizing electrostatic fields to focus an electron beam.

electrostatic loudspeaker Same as **electrostatic speaker**.

electrostatic machine Same as **electrostatic generator**.

electrostatic memory Same as **electrostatic storage**.

electrostatic microphone A microphone in which a flexible metal diaphragm and a rigid plate form a two-plate air capacitor. As the incident sound waves make the flexible plate vibrate, the capacitance is varied, and this causes variations in the electric current, which in turn can be converted in an audio-frequency signal. Also called **condenser microphone**, or **capacitor microphone**.

electrostatic plotter A plotter utilizing an **electrostatic printing** process.

electrostatic potential Same as **electric potential**.

electrostatic precipitation The use of an **electrostatic precipitator** for the removal of dusts in gases.

electrostatic precipitator A device which removes dusts present in gases, such as air, by precipitating them in a chamber containing high-voltage wires or grids. The gas is forced through the chamber, and the wires or grids ionize the dust particles, which in turn makes them migrate to the walls of said chamber. Also called **electrostatic air cleaner**, **electric precipitator**, **electronic air cleaner**, **precipitator**, or **dust precipitator**.

electrostatic printer A printer utilizing an **electrostatic printing** process.

electrostatic printing A method of reproducing images which utilizes electrostatic charges. For instance, a photoconductive surface, such as drum, is positively charged with static electricity, then exposed to an optical image of that which is to be reproduced, forming a corresponding electrostatic pattern. A fine powder called toner, which has been negatively charged, is spread over this surface, adhering only to the charged areas of the drum. The toner is then fused to the paper, which has been positively charged, forming the final image. Used, for instance, in black and white or color laser printers and copy machines.

electrostatic process 1. Any process involving electrostatic charges. 2. A process for image reproduction involving electrostatic charges.

electrostatic recording A method of recording which utilizes electrostatic charges. For instance, electrostatic printing.

electrostatic repulsion The electrostatic force of repulsion between two similarly charged particles, according to the **electrostatic attraction law**. Also called **Coulomb repulsion**.

electrostatic screen 1. Same as **electrostatic shielding (1)**. 2. Same as **electrostatic shielding (2)**.

electrostatic screening Same as **electrostatic shielding**.

electrostatic separator A device which utilizes a strong electrostatic field to separate finely pulverized materials.

electrostatic shield 1. Same as **electrostatic shielding (1)**. 2. Same as **electrostatic shielding (2)**.

electrostatic shielding Also called **electrostatic screening**, **electric shielding**, **electrical shielding**, **shielding (1)**, or **screening (1)**. 1. A material or enclosure which blocks the effects of an electric field, while allowing free passage to magnetic fields. It may consist, for instance, of a wire mesh or screen which provides a low-resistance path to ground. Utilized, for example, to prevent interaction between circuits, or to enhance the directivity of an antenna. Another example is a metallized plastic bag which helps protect circuit boards against electrostatic discharges during storage and handling, en route to being installed. Such a bag does not have to be grounded. Also called **electrostatic screen (1)**, **electrostatic shield (1)**, **electric screen (1)**, or **electrical screen (1)**. 2. An **electrostatic shielding (1)** which serves to confine an electric field within an enclosure. Also called **electrostatic screen (2)**, **electrostatic shield (2)**, **electric screen (2)**, or **electrical screen (2)**. 3. The use of an **electrostatic shielding (1)**, or **electrostatic shielding (2)**.

electrostatic speaker A speaker in which a flexible metal diaphragm and a rigid plate form a two-plate air capacitor. The mechanical vibration of its thin sound-producing diaphragm is produced by electrostatic fields produced between it and a rigid plate, each with a high audio voltage applied to them. Also called **electrostatic loudspeaker**, **capacitive speaker**, **capacitor speaker**, or **condenser speaker**.

electrostatic storage The storage of data utilizing electrostatic charges. Used, for instance, in storage tubes. Also called **electrostatic memory**.

electrostatic stress The stress an electrostatic field exerts on surrounding bodies. For instance, the force a voltage in a conductor exerts on its surrounding insulation. If this force exceeds a given amount, breakdown occurs.

electrostatic system of units A system of electrostatic units within the cgs system of units. Units in this system have a **stat-** prefix, for instance, statcoulomb or statvolt. The cgs measuring system is obsolescent. Its abbreviation is **esu**. Also called **electrostatic units**, or **cgs electrostatic system**.

electrostatic transducer A transducer consisting of two capacitor plates, one fixed and the other flexible or movable, which converts changes in applied voltage into movement. Also called **capacitive transducer**.

electrostatic tweeter A tweeter consisting of a small electrostatic speaker. Such tweeters feature a mostly linear response which can extend well beyond the range of human hearing.

electrostatic units Same as **electrostatic system of units**.

electrostatic voltmeter A voltmeter in which the voltage to be measured is applied between two plates, one fixed, the other movable and attached to a spring. The resulting electrostatic force displaces the moving plate to an extent which is proportional to the applied voltage, and an attached pointer serves to indicate the voltage on a calibrated scale.

electrostatic wave In a plasma, wave motion whose restoring forces are principally electrostatic.

electrostatics The study of all that pertains to electric charges at rest, including their associated forces, fields, and applications.

electrostimulation The use of electricity to stimulate tissue, such as nerve, muscle, or bone.

electrostriction The mechanical strain that occurs in some materials, such as dielectrics, when placed in an electric field. If a material, such as certain ceramics, also produces an electric charge when under mechanical strain, then it possesses the property of **piezoelectricity**, not electrostriction. Also called **electrostrictive effect**, or **electrostrictive strain** (1).

electrostrictive Possessing, or pertaining to, the property of **electrostriction**.

electrostrictive ceramic A ceramic, such as lead magnesium niobate, which has the property of **electrostriction**.

electrostrictive effect Same as **electrostriction**.

electrostrictive strain 1. Same as **electrostriction**. 2. The extent of the mechanical strain occurring in **electrostriction**.

electrosurgery The use of electricity for surgical procedures. For instance, the application of an electric current to cut tissue, or to stop bleeding.

electrosynthesis A synthesis reaction which is produced by electricity. A **synthesis** occurs when a complex chemical compound is obtained from simpler constituents.

electrotechnology Technology pertaining to electrical and/or electronic products and applications. It is an abbreviation of **electronic technology**, or **electrical technology**.

electrotherapy The use of electricity for medical therapy. Used, for instance, for treatment of certain muscular or skeletal disorders.

electrothermal Also called **electrothermic**. 1. Pertaining to electricity and heat. 2. The conversion of electricity into heat, as occurs, for instance, as a result of a resistance.

electrothermal process Any process, such as the production of an arc, in which an electric current produces heat.

electrothermal propulsion A form of propulsion in which the propellant is heated electrically, as accomplished, for instance, by an electric arc. Usually used in spacecraft.

electrothermic Same as **electrothermal**.

electrotonic Pertaining to **electrotonus**.

electrotonus For nerve or muscle, a change in a characteristic such as conductivity or excitability, when an electric current is passed through it. A steady current is usually used.

electrovalence Also spelled **electrovalency**. 1. The valence an atom acquires when exchanging electrons with another, in the formation of an **electrovalent bond**. 2. The number of electrons gained or lost in attaining **electrovalence** (1).

electrovalency Same as **electrovalence**.

electrovalent Pertaining to **electrovalence**.

electrovalent bond A chemical bond formed by the electrostatic attraction of the oppositely-charged ions which form the resulting compound. In such a bond, each constituent either gains or loses one or more electrons. For instance, in **NaCl** (table salt), the sodium transfers one electron to the chlorine. In the process, the sodium becomes a positively-charged ion, and the chlorine a negatively-charged one. Also called **ionic bond**.

electroweak force A force which unifies the electromagnetic and the weak forces, two of the **four fundamental forces of nature**.

electrowinning The use of an electrochemical process, such as electrolysis, to separate metals from a solution. Used, for instance, to extract metals from their molten ores.

elegant Characterized by simplicity, precision, effectiveness, and the ingenious utilization of available resources. May refer, for instance, to a computer program that is easy to use and performs its designated tasks accurately and efficiently, while occupying the least possible memory and processing resources of the computer it is running on.

element 1. A fundamental, distinct, and usually irreducible constituent of a given entity, process, or concept. For example, any of the various functional parts of an electron tube. 2. Any of the electrical components that are a part of a circuit. These may include resistors, capacitors, transistors, generators, electron tubes, and so on. Each element has terminals which allow it to be connected to the conducting path. Also called **electrical element**, **circuit element**, **circuit component**, or **component** (2). 3. A substance which can not be subdivided into smaller units by chemical means. There are over 110 known chemical elements, each with unique properties, and they comprise all matter above the atomic level. The smallest particle that retains all the properties of an element is an atom. All neutral atoms of a given chemical element have the same number of protons and electrons, and if an element has isotopes, the difference between each is the number of neutrons in its nucleus. Also called **chemical element**. 4. Within an antenna array, one of multiple radiators which contribute to its overall transmission and/or reception characteristics. Such an element may be driven, or parasitic. Also called **antenna element**. 5. In computers, the smallest subunit within a larger set. For instance, picture elements or data elements.

element 110 A synthetic radioactive chemical element whose atomic number is 110, and which currently does not have a single internationally recognized name.

element 111 A synthetic radioactive chemical element whose atomic number is 111, and which currently does not have a single internationally recognized name.

element 112 A synthetic radioactive chemical element whose atomic number is 112, and which currently does not have a single internationally recognized name.

element 113 A synthetic radioactive chemical element whose atomic number is 113, and which currently does not have a single internationally recognized name.

element 114 A synthetic radioactive chemical element whose atomic number is 114, and which currently does not have a single internationally recognized name.

element 115 A synthetic radioactive chemical element whose atomic number is 115, and which currently does not have a single internationally recognized name.

element 116 A synthetic radioactive chemical element whose atomic number is 116, and which currently does not have a single internationally recognized name.

element 117 A synthetic radioactive chemical element whose atomic number is 117, and which currently does not have a single internationally recognized name.

element 118 A synthetic radioactive chemical element whose atomic number is 118, and which currently does not have a single internationally recognized name.

element spacing In an antenna, the spacing between **elements (4)**. Proper spacing is necessary, for instance, to maximize the intensity of radiation in the desired directions.

elemental area Also called **picture element, pixel**, or **scanning spot**. **1.** Within an image, such as that of a computer display, TV, or fax, the smallest unit which can be manipulated. In a CRT, such as a color TV or computer monitor, for instance, it is a trio of color phosphor dots representing a single point. **2.** In a TV picture or fax, the portion of the scanning line being explored at any given instant.

elementary charge The charge carried by a single electron or proton. It is a fundamental physical constant, and is equal to approximately 1.6022×10^{-19} coulomb. All subatomic particles have an electric charge equal to this value, or a multiple of it. Also called **unit electric charge**.

elementary particle Any of the fundamental constituents of all matter in the universe. Elementary particles include protons, neutrons, electrons, photons, and quarks. Categories of such particles include leptons and hadrons. An electron, for instance, is a type of lepton. When an elementary particle is defined as having no internal structure, particles such as protons and neutrons are excluded, as these are composed of quarks. The four fundamental forces of nature are carried by elementary particles. For example, photons carry the electromagnetic force. Also called **fundamental particle**, or **particle (2)**.

elevated duct An atmospheric duct occurring at high altitudes. Such a duct mostly affects the transmission of very-high and ultra-high frequencies.

elevation Same as **elevation angle**.

elevation angle The angle above a reference plane, such as the horizon, of an ascending line. For example, the effective radiated power of a given antenna array may vary as a function of the elevation angle. Also called **elevation**.

elevation-angle error Same as **elevation error**.

elevation error In radars, an error in the calculation of an **elevation angle**. May be due, for instance, to atmospheric conditions. Also called **elevation-angle error**.

elevation indicator An instrument which displays an **elevation angle**.

elevator Within a scroll bar, the box that can be moved in the desired direction, to navigate within a file or window. For instance, by utilizing a mouse to slide the elevator downwards within a vertical scroll bar, a user advances further ahead in a displayed document. Also called **thumb**, or **scroll box**.

ELF Abbreviation of **extremely low frequency**.

eliminator **1.** That which replaces an undesired component or device, or renders it unnecessary. Also, that which diminishes or removes and undesired quantity or signal. **2.** A device that may be utilized to provide DC energy from an AC source. This makes the utilization of a battery unnecessary. Also called **battery eliminator**.

ellipse A geometric shape consisting of a closed curve in the shape of an elongated circle. A circle is also a type of ellipse.

ellipsis A symbol consisting of three dots, **...**, indicating that there is more to follow. For instance, when viewing a pull-down menu, an ellipsis next to a command or operation indicates that there are further choices available within it.

elliptic Same as **elliptical**.

elliptic curve A curve in the shape of an **ellipse**.

elliptic curve cryptography A public-key cryptography method which utilizes points on an elliptic curve to derive a powerful public key. It provides faster calculation times using smaller key sizes, and is especially well suited for devices such as PDAs, cellular phones, and smart cards. Its abbreviation is **ECC**. Also spelled **elliptical curve cryptography**.

elliptic filter Same as **elliptical filter**.

elliptic orbit Same as **elliptical orbit**.

elliptic polarization Same as **elliptical polarization**.

elliptical Pertaining to, or in the shape of an **ellipse**. Also spelled **elliptic**.

elliptical curve cryptography Same as **elliptic curve cryptography**.

elliptical filter A filter similar to a Chebyshev filter, but providing higher selectivity and reduced insertion loss. Used, for instance, in satellite communications. Also spelled **elliptic filter**.

elliptical orbit An orbit which is in the shape of an **ellipse**. The orbits of electrons around a nucleus, or those of satellites, are elliptical. Also spelled **elliptic orbit**.

elliptical polarization Polarization of an electromagnetic wave in which the electric field vector describes an ellipse in a plane perpendicular to the direction of propagation. Also spelled **elliptic polarization**.

elliptical waveguide A waveguide whose cross section is in the form of a **ellipse**.

elliptically-polarized light A light wave with **elliptical polarization**.

elliptically-polarized wave A wave with **elliptical polarization**.

ellipticity In a waveguide, the ratio of the major axis to the minor axis of the polarization ellipse. Also called **axial ratio**.

elongation An extension, or increase in length. Used, for instance, to refer to an extension of the envelope of a signal.

EM **1.** Abbreviation of **electron microscope**. **2.** Abbreviation of **electromagnetics**.

email Abbreviation of **electronic mail**. Also spelled **e-mail**. Also called **mail**. **1.** The transmission of messages, with or without attachments, over a communications network such as the Internet. The first part of an email is the header, which contains information such as the email address of the sender, the time and date sent, and the subject. This is followed by the body, which usually consists of the text that the sender wishes to communicate. There may also be one or more attachments, which are appended files. There are standard protocols, such as SMTP, which define factors such as message format, while standard mail servers, such as IMAP, provide storage of messages in virtual mailboxes until users retrieve them. Email can be accessed by properly equipped devices such as computers, PDAs, cellular phones, and the like, which have a connection to the Internet, and provides a simple and reliable means of sending messages practically instantaneously. **2.** One or more **email (1)** messages. **3.** The sending of **email (1)**.

email address A sequence of characters which uniquely identify an email account. The standard format is user name@domain name. In the case of **xxxx@zzz.com.qq**, *xxxx* is the user name, and *zzz.com.qq* is the domain name. Also called **address (5)**, **Internet address (3)**, or **mail address**.

email attachment A file that is appended to an email. Such a file may be a document, an image, a video, a program, and so on. Certain types of attachments are encoded, thus requiring the recipient to have the appropriate email software to decode them. Also called **enclosure (3)**, **file attachment**, **attachment**, or **mail attachment**.

email autoresponder A feature or program which automatically sends a previously prepared email in response to a received email. Used, for instance, to inform that an email has been received and will be responded to at a later time. Also called **autoresponder**, or **mail autoresponder**.

email bomb The sending of an excessive amount of unwanted email to the same recipient. For instance, the generation of many identical emails which are sent repeatedly to the same email address. It also applies to arranging for a given recipient to receive an inordinate amount or email, for instance, by being placed in numerous mailing lists. In addition to any inconvenience this causes a user, it also wastes available bandwidth, and other network resources. Also, the undesired emails themselves. This contrasts with **spam**, which is unsolicited email directed towards multiple, often millions, of users. Its abbreviation is **e-bomb**. Also called **mail bomb**.

email bot Same as **email robot**.

email client A program which resides in a computer, enabling it to have access to the mail servers of a network. It enables the sending, receiving, and organizing of email. Also called **mail client**.

email filter In an email program, a feature which allows for the automatic sorting of incoming email. For instance, a user can set filters to delete unwanted email from specific spammers, or place emails from a specific individual or entity in its own folder. Also called **mail filter**.

email folder A directory which is used for organizing and storing emails. All email programs are equipped with folders, such as those for incoming mail, sent mail, and trash, and usually allow users to create others, such as those for friends or colleagues, to which incoming mail can be directed. Also called **mail folder**.

email forwarding The resending of an email that has already reached a recipient's address. Once received, an email may be forwarded manually or automatically to one or more email addresses. When an email is manually forwarded, its content may be changed and/or files may be attached. Also called **forwarding (3)**, or **mail forwarding**.

email header The first part of an email, which contains information such as the email address of the sender, that of the recipient or recipients, the time and date sent, the subject, mail protocols utilized, the IP address of the sender, and so on. Also called **mail header**.

email program A program providing the necessary tools to read, prepare, and send and retrieve emails to and from the mail servers of a network. Such a program usually provides features such as word-processing functions, an address book, the capability to insert hyperlinks, filtering, and the ability to retrieve mail from other email accounts. Also called **email software**, or **mail program**.

email robot A program which sends automated responses to emails, or which performs other functions such as removals from mailing lists. Its abbreviation is **email bot**. Also spelled **e-mail robot**. Also called **mail robot**.

email server Also called **mail server**. **1.** A computer which serves to store and/or forward email. **2.** The software which enables an **email server (1)** to perform its functions.

email signature Also called **signature**. **1.** Content, such as a text, image, file, or combination of these, which is appended to an email or message. Examples include favorite quotes, contact information, vCards, or disclaimers. A user may select whether such a signature is included automatically when sending. **2.** Same as **electronic signature**.

email software Same as **email program**.

email virus A computer virus sent through email. Such a virus is usually in the form of an attachment, which if left unopened prevents the harm it may bring. Also called **mail virus**.

emanation To come out of, or be sent from a source. For instance, the light emanating from a lamp.

embed To incorporate in a surrounding mass or enclosure. For example, the embedding of electronic components or devices, so that they can perform their functions while sealed or otherwise protected. Also, to include within a larger entity. For instance, to incorporate of a chip into an automobile, toy, appliance, clock, or the like. Another example is to incorporate a software routine into a program.

embedded That which has been incorporated into a surrounding mass or enclosure. For example, an embedded electronic component or device. Also, that which has been included within a larger entity.

embedded application A computer application which is **embedded**. It may refer, for instance, to an application incorporated into an embedded computer.

embedded command A computer command which is **embedded** within a file, such as that containing text or graphics. Used, for example, for page layout parameters.

embedded component A component which is **embedded**.

embedded computer A computer which is **embedded**. Frequently refers to a microcontroller with the necessary programming to perform one or more tasks. Also called **embedded system**.

embedded controller A controller which is **embedded**. May refer, for instance, to a chip which controls the way peripheral accesses a computer.

embedded help Onscreen help which is provided without a user requesting it. Seen, for instance, in kiosks, where helpful text and/or images guide a user through the various desired steps.

embedded link Within a Web site, a link which directs to a relevant area elsewhere within the same Web site.

embedded program A computer program which is **embedded**. It may refer, for instance to a program incorporated into an embedded computer, or that in ROM.

embedded software One or more **embedded programs**.

Embedded SQL SQL statements which are embedded within another programming language, such as C++.

embedded system Same as **embedded computer**.

embedded Web server A device which has embedded software and/or hardware which performs Web server functions. Examples include robots, instruments, or appliances which incorporate a chip facilitating connection to the Internet.

EMC Abbreviation of **electromagnetic compatibility**.

EME **1.** Abbreviation of **electromagnetic environment**. **2.** Abbreviation of **earth-moon-earth**.

EME communications Abbreviation of **earth-moon-earth communications**.

Emergency Alert System A system which interrupts TV and radio broadcasts to disseminate emergency information and instructions. It incorporates digital and automation technology better than the **Emergency Broadcast System**. The Emergency Alert System has replaced the Emergency Broadcast System. Its abbreviation is **EAS**.

Emergency Broadcast System A system similar to the **Emergency Alert System**, but with less incorporation of digital and automation technology. The Emergency Alert System has replaced the Emergency Broadcast System. Its abbreviation is **EBS**.

emergency channel A channel reserved and/or utilized for **emergency communications**.

emergency communications Radio-frequency communications pertaining to emergencies.

emergency equipment Any equipment and ancillary devices which may be utilized during an emergency. This may include, for instance, backup power sources or alternate communication networks.

emergency light A light which is put into service during a power outage, or when the regular source of lighting has failed. Such lights may be used, for instance, to help locate exits from buildings.

emergency locator beacon A device which sends a radio-frequency beacon signal in case of an emergency. Such a device is usually portable and battery-operated, and may function for up to several days to allow others to home-in after being alerted. Also called **locator beacon (2)**.

emergency position indicating radio beacon A handheld battery-operated maritime device which serves to send a radio-frequency beacon signal in case of an emergency. Its abbreviation is **EPIRB**.

emergency power The electricity available from, or provided by, an **emergency power supply**.

emergency power supply A power supply which is put in use, usually automatically, when the primary power supply is interrupted or otherwise unavailable.

emergency radio channel A radio channel reserved and/or utilized for **emergency communications**.

emf or **EMF** Abbreviation of **electromotive force**.

EMG 1. Abbreviation of **electromyography**. 2. Abbreviation of **electromyogram**.

EMI Abbreviation of **electromagnetic interference**.

emission The process of giving off, or sending out. For example, the emission of electrons by a cathode in an electron tube, the electromagnetic waves radiated by an antenna, or the emission of alpha, beta, or gamma rays by unstable atomic nuclei.

emission bandwidth 1. For a given RF transmitter, the bandwidth utilized. 2. For a given RF transmitter, the authorized bandwidth.

emission delay In a radio navigation system such as loran, the time difference between a transmission by the control station and that of a secondary station.

emission electron microscope An electron microscope in which emitted electrons are projected onto a fluorescent screen.

emission frequency The frequency of a transmitted or radiated signal. For instance, that of an antenna.

emission line 1. In spectroscopy, a characteristic frequency of electromagnetic radiation which is emitted by atoms, corresponding to a transition of an electron to a lower energy level. It appears as a bright spectral line. 2. Any extremely narrow range of emitted wavelengths within the electromagnetic spectrum.

emission security Measures taken, such as cryptography, to help prevent unauthorized access, interception, or alteration of emitted information.

emission spectrometer An instrument which measures the wavelengths of the radiation emitted by chemical elements which are vaporized by an electrical discharge, such as an arc or spark. The light produced is directed through a slit in the spectrometer onto a diffraction grating, and is dispersed to form a spectrum. Utilized, for instance, to measure the concentration of each element in a given sample.

emission spectroscopy The use of **emission spectrometers** to analyze samples containing chemical elements.

emission spectrum The spectrum produced by an **emission spectrometer**.

emissive power Also called **emittance**, or **radiating power**. 1. For a given area of a radiating surface, the total energy radiated per unit time. 2. For a given area of a radiating surface, the total energy radiated per unit time, at a given temperature.

emissivity For a given surface, the ratio of the radiation emitted per unit area, to that which a blackbody would radiate at the same temperature. Its symbol is ϵ.

emit To give off, or send out. For example, for a cathode in an electron tube to emit electrons, an antenna to radiate electromagnetic waves, or unstable atomic nuclei to emit alpha, beta, or gamma rays.

emittance Same as **emissive power**.

emitted electron Same as **ejected electron**.

emitter 1. That which **emits**. 2. In a bipolar transistor, the region from which minority carriers are injected into the base. Also, the electrode attached to this region. Its symbol is **E**. Also called **emitter region**, or **emitter electrode**.

emitter-base junction In a bipolar transistor, the junction between the emitter region and the base region.

emitter bias The voltage applied to the emitter electrode of a bipolar transistor to set its operating point.

emitter-coupled logic A logic circuit design in which pairs of bipolar transistors are coupled by their emitters. Such transistors feature extremely fast switching speeds, and are used in high-performance processors. Its abbreviation is **ECL**. Also called **emitter-coupled transistor logic**.

emitter-coupled transistor logic Same as **emitter-coupled logic**. Its abbreviation is **ECTL**.

emitter current In a bipolar transistor, the current flowing through the emitter electrode.

emitter electrode Same as **emitter (2)**.

emitter follower A transistor amplifier with a common-collector connection, in which the input signal is applied to the base and the output signal is taken from the emitter. Also called **common-collector amplifier**.

emitter junction Either of the pn junctions in a bipolar transistor.

emitter potential The potential at the emitter electrode of a bipolar transistor

emitter region Same as **emitter (2)**.

emitter resistance In a bipolar transistor, the resistance of the emitter electrode.

emitter voltage The voltage at the emitter electrode of a bipolar transistor.

EMM Abbreviation of **Expanded Memory Manager**.

emoney Same as **electronic cash**. Abbreviation of **electronic money**.

emoticon Abbreviation of **emotion icon**. A simple icon usually produced with punctuation marks. They are typically viewed as being sideways. For example, :-) is meant to express a smile, :-D is a bigger smile, and so on for a number of faces or emotions. Also called **smiley**, although emoticons do not always represent smiles.

emotion icon Same as **emoticon**.

EMP Abbreviation of **electromagnetic pulse**.

emphasis Also called **accentuation**. 1. Highlighting of a specific frequency or band of frequencies through amplification. 2. The process of selectively amplifying higher frequencies of an audio-frequency signal before transmission, or during recording. The utilization of emphasis followed by deemphasis may help improve the overall signal-to-noise ratio and reduce distortion, among other benefits. Also called **preemphasis**.

emphasizer **1.** An electric network which serves to provide **emphasis** (**1**). **2.** An electric network which serves to provide **emphasis** (**2**). Also called **preemphasis network**.

empiric Same as **empirical**.

empirical Based on, or depending upon observation, experience, or experimentation, as opposed to theory. Also spelled **empiric**.

empty band **1.** A band which is not in use. **2.** A band which is not occupied. **3.** An energy band which is not occupied by any electrons.

EMR Abbreviation of **electron magnetic resonance**.

EMS **1.** Abbreviation of **electromagnetic susceptibility**. **2.** Abbreviation of **Expanded Memory Specification**. **3.** Abbreviation of **event-management system**.

emu Abbreviation of **electromagnetic system of units**.

emulation **1.** The use of an **emulator**. **2.** The function of an **emulator**.

emulation mode A hardware and/or software operation mode in which another device and/or software is being emulated. For instance, a printer which emulates another printer, so as to be compatible with existing software.

emulator Computer hardware and/or software which is designed to work exactly like another. Seen, for instance, in printer emulation, so that a printer made by one manufacturer is able to work with the software intended for a printer of a different manufacturer. Another example is the design of a computer to run on software intended for another.

EMV Abbreviation of **electromagnetic vulnerability**. Same as **electromagnetic susceptibility**.

enable To turn on, or to put into operation. May refer to a component, device, circuit, piece of equipment, function, program, or system.

enable pulse An pulse that turns on, puts into operation, or prepares for a subsequent action. May refer to a component, circuit, device, piece of equipment, function, program, or system.

enabled Turned on, or operating. May refer to a component, circuit, device, piece of equipment, function, program, or system.

enamel A glass-like coating utilized to provide a hard and/or glossy finish. May be used, for instance, for insulation and/or corrosion protection.

enameled wire Wire which has a baked-on enamel film, which provides insulation. May be used, for instance, in coils.

encapsulant That which serves to **encapsulate** (**2**). For instance, a wax, a plastic, or a ceramic.

encapsulate **1.** To embed or incorporate into something else. **2.** To encase or embed in a surrounding mass or enclosure which houses and protects. For instance, an electronic component may be encapsulated in plastic, to insulate and protect from moisture. Such a plastic would usually be in a molten state when applied to the component, then let solidify for a snug fit.

Encapsulated PostScript A file format for importing and exporting Postscript files. Its abbreviation is **EPS**.

encapsulation **1.** The process of applying an **encapsulant**. May be accomplished, for example, by dipping a component or device in molten glass, then letting cool. **2.** In object-oriented programming, the hiding of the implementation details of an object. The services the object provides are defined and accessible, but their internal workings are not. Also called **information hiding** (**2**). **3.** A technique which enables a network to send data utilizing one protocol, through another network using different protocol. It does so by encapsulating packets using one network protocol within packets being transmitted through the other network. Also called **tunneling** (**2**).

encipher Same as **encode** (**2**).

enclosure **1.** That which serves to house something else. For instance, a cabinet which holds an apparatus. **2.** A cabinet designed to house one or more speaker units. Two common designs are acoustic reflex and acoustic suspension. Within a speaker system, a specialized speaker such as a woofer or tweeter may have its own enclosure, helping enhance performance. Also called **speaker enclosure**, or **loudspeaker enclosure**. **3.** Same as **email attachment**.

encode **1.** To express information utilizing a code. **2.** To scramble information, such as data, in a manner which only those with a key can decipher. Usually used for security purposes. Also called **encrypt**, **encipher**, **code** (**4**), or **scramble** (**1**). **3.** To write a set of computer instructions.

encoded Also called **coded**. **1.** Information, such as data, which is in the form of a code. Also called **encrypted** (**1**). **2.** Information, such as data, which has been scrambled in some manner. Also called **encrypted** (**2**). **3.** Programs or program instructions which have been written.

encoded data Data which has been **encoded**. May be used, for instance, where privacy or security is a concern. Also called **coded data**, or **scrambled data**.

encoded signal A signal which has been **encoded**. May be used, for instance, where privacy or security is a concern. Also called **coded signal**, or **scrambled signal**.

encoded speech Speech which has been **encoded** (**2**), so that it can only be understood with a receiver with the proper circuits and settings. May be used, for instance, where eavesdropping is a concern. Also called **coded speech**, or **scrambled speech**.

encoder **1.** A circuit, device, piece of equipment, program, system, or method utilized to **encode**. For instance, computer hardware and/or software used for such a purpose. **2.** In a TV transmitter, a circuit or device which transforms the separate red, green, and blue camera signals into color-difference signals, and combines these with the chrominance subcarrier. Also called **color encoder**, or **matrix** (**3**). **3.** An electromechanical device, such as a shaft-position encoder, which converts the rotations of a shaft into pulses. **4.** A device which converts an analog quantity into a digital signal. **5.** A device or piece of equipment which prints characters in a certain font and places them in specific locations, so as to facilitate being read by optical character recognition devices.

encoding That processes performed by an **encoder**.

encrypt Same as **encode** (**2**).

encrypted **1.** Same as **encoded** (**1**). **2.** Same as **encoded** (**2**).

encryption The coding of information so that only the intended recipients can understand it. It is an extremely efficient method to achieve data security, and a code, or key, is used to convert the information back to its original form. Public-key encryption and secret-key encryption are the two most common types.

encryption algorithm A set of mathematic formulas utilized to scramble information, such as data, in a manner which only those with a key can decipher.

encryption key A series of binary digits or characters that are incorporated into data to encrypt it.

end bell **1.** In a rotating motor, the part of the housing which supports the bearing and guards the rotating parts. Also called **end bracket**, or **end shield**. **2.** A cable clamp which is affixed to the back of a plug or receptacle.

end bracket Same as **end bell** (**1**).

end cell In a storage battery, a cell which may be connected with the others, so as to adjust the overall voltage.

end device Same as **end instrument**.

end effect In an antenna, an effect due to capacitance at the ends of the radiators. Capacitive coupling with the surrounding environment, such as nearby objects and the humidity of the air, produces this effect.

end-effector A device, tool, or gripping mechanism attached to the wrist of a robot arm. It can be, for example, a drill, a sensor, or a gripper with the ability to move with six degrees of freedom. Also called **hand**, or **robot end-effector**.

end-fed antenna An antenna with an **end feed**.

end feed A feed point located at the end of an antenna radiator. Such a feed is utilized, for instance, in a Zepp antenna.

end-fire antenna Same as **end-fire array**.

end-fire array A linear antenna array whose direction of maximum radiation is along the axis of said array. Such an array may be unidirectional or bidirectional. Also called **end-fire antenna**.

end instrument Also called **end device**. **1.** A device or instrument connected to a terminal of a communications circuit. For example, a telephone or a computer modem. **2.** A device connected to a terminal of a communications circuit, which converts a physical quantity, such as a temperature or pressure, into an electrical quantity, such as a current or voltage. Used, for instance, in telemetering.

end item A finished product which is ready for sale or delivery. For example, a computer which has had all its components properly assembled and burned-in.

end key On most computer keyboards, a key which when pressed serves to move the cursor to the end of a line, the end of a file, the bottom of a screen, and so on.

end mark A symbol or code which represents the end of a unit of information, such as a file or document.

end-of-charge voltage For a cell or battery, the voltage when charging is completed.

end-of-data mark A symbol or code which indicates the end of the last data unit, such as a record, within a data-storage medium.

end-of-discharge voltage For a cell or battery, the voltage when discharge is complete. Also called **end point voltage** (**2**), or **cutoff voltage** (**3**).

end-of-field mark A symbol or code which indicates the end of a field. Used, for instance, with variable-length fields.

end-of-file A code which indicates the ending location of a computer file. Such a code may consist, for instance, of a sequence of characters, or of a special character. Its abbreviation is **EOF**. Also called **end-of-file code**.

end-of-file code Same as **end-of-file**. Its abbreviation is **EOF code**.

end-of-file indicator Its abbreviation is **EOF indicator**. **1.** Same as **end-of-file mark**. **2.** A device indicating that the ending location of a computer file has been reached.

end-of-file label Same as **end-of-file mark**. Its abbreviation is **EOF label**.

end-of-file mark An indicator, such as a code, which denotes the ending location of a computer file. Its abbreviation is **EOF mark**. Also called **end-of-file indicator** (**1**), or **end-of-file label**.

end-of life **1.** The steps leading to a product no longer being manufactured, nor having any form of support, such as repairs or technical help. **2.** A product which has completed all **end-of-life** (**1**) steps. **3.** The end of the **expected life** of a product.

end-of-message A symbol or code indicating the end of a message. For instance, the letters EOM appearing may serve as such an indication. Its abbreviation is **EOM**.

end-of-record A symbol or code indicating the end of a record. For instance, a word in a given format. Its abbreviation is **EOR**.

end-of-tape mark A control character or code which indicates the end of a magnetic tape. It may also be in the form

of a reflective segment. Used, for instance, to signal a change to another tape. Also called **end-of-tape marker**, or **tape mark** (**1**).

end-of-tape marker Same as **end-of-tape mark**.

end-of-text A control character indicating the end of a text file, or of text. Usually used in data transmission. Its abbreviation is **ETX**.

end-of-transmission A control character indicating the end of a transmission. In addition, such a character may serve for other functions, such as releasing a circuit, or disconnecting a terminal after transmission. Its abbreviation is **EOT**.

end office A central office utilized for the interconnection of customer lines and trunks. Its abbreviation is **EO**.

end point Also spelled **endpoint**. **1.** The point at which a process, operation, interval, path, or sequence ends. For instance, a terminal of a communications channel, the end of a series of computations, the maximum or minimum reading of an indicating instrument, or either of the ends the slider can reach in a potentiometer. **2.** In robotics, the point where a path of motion ends, such as the limit of vertical movement of a robotic manipulator.

end point voltage Also called **voltage end point**. **1.** For a cell or battery, the voltage below which any device or piece of equipment connected to it will not operate, or below which operation is inadvisable. **2.** Same as **end-of-discharge voltage**.

end-scale value For an indicating instrument, the electrical quantity corresponding to an indication at the end of the measuring scale.

end setting For an adjustable device, the setting at either of the available limits. For instance, the minimum and maximum settings of the volume control of an audio-frequency amplifier.

end shield Same as **end bell** (**1**).

end-to-end delivery In communications, delivery of data between a source and destination point, followed by acknowledgment of receipt.

end use The use given to an **end item**.

end user The user of an **end item**.

End-User License Agreement A contract between a software manufacturer or provider, and one or more end users. It pertains to matters such as use, distribution, and resale. Most purchased software has such an agreement. Its acronym is **EULA**. Also called **License Agreement**.

endless loop A loop which is repeated indefinitely. Such a loop usually requires user intervention to stop it. An endless loop may be due to a bug, or may be intentional, as in the case of a certain demos. Also called **infinite loop**.

endothermic Pertaining to a reaction or process in which heat is absorbed from the surroundings. This contrasts with **exothermic**, where heat is liberated to the surroundings.

endothermic process A process in which heat is absorbed from the surroundings.

endothermic reaction A reaction in which heat is absorbed from the surroundings.

endpoint Same as **end point**.

energize **1.** To connect electrically to a source of voltage. **2.** To connect to a source of energy. **3.** To provide a signal that makes a component, circuit, device, piece of equipment, or system operational.

energized **1.** Anything connected electrically to a source of voltage. Also known as **live** (**1**), **alive**, or **hot** (**1**). **2.** Anything connected to a source or energy.

energized circuit A circuit that is connected electrically to a source of voltage. Also known as **live circuit**, or **alive circuit**.

energy The capacity to do work. Energy cannot be created nor destroyed in an isolated system, but it can be changed from one form to another. Forms of energy include electrical, chemical, acoustic, atomic, and solar. Usually expressed in joules, but other units may be utilized, such as ergs, calories, or watt-hours.

energy band In a crystalline solid, such as a semiconductor, the range of energy levels within which the electrons must be located. Also called **Bloch band**.

energy-band diagram A graphical representation showing **energy bands**.

energy beam An intense, concentrated, and essentially unidirectional stream of radiated energy, such as a light beam, laser beam, electron beam, or neutron beam. Used, for instance, to cut, shape, drill, or otherwise process workpieces such as metals or ceramics.

energy cell Same as **electric cell**.

energy conservation Also called **conservation of energy**. **1.** A law which states that energy cannot be created nor destroyed in an isolated system, but that it can be changed from one form to another. For instance, kinetic energy being transformed into potential energy, while the total energy in the system remains constant. Also called **energy conservation law**, or **law of energy conservation**. **2.** The judicious use of available energy resources.

energy conservation law Same as **energy conservation (1)**.

energy conversion The transformation of energy from one form to another. For example, the conversion of potential energy into kinetic energy, or chemical energy to electrical energy. Also called **energy transformation**.

energy-conversion device A component or device which serves to convert one form of energy to another. For instance, a photovoltaic cell, which converts radiant energy, such as light, into electrical energy.

energy density **1.** The amount of energy per unit volume. **2.** For an energy-producing device, such as an electric cell, the amount of energy produced per unit volume.

energy diagram Same as **energy-level diagram**.

energy dispersive spectroscopy A technique which generates X-ray spectrums of samples which are bombarded by high-energy electron beams. Each element in the sample produces characteristic X-rays, which can be used for identification. The relative intensities of the X-ray peaks determine the relative concentrations of the component elements. Energy dispersive spectroscopy is usually used in conjunction with scanning electron microscopy, and provides high sensitivity for detection of elements whose atomic number is 6 or higher. Below this number, the detected electrons are likelier to be due to the Auger effect. Its abbreviation is **EDS**.

energy dissipation Also called **dissipation**. **1.** The loss of another form of energy through its conversion into heat energy. For instance, a circuit suffering electrical energy losses due to resistance. **2.** The quantified energy loss in **energy dissipation (1)**. **3.** A loss of energy. **4.** A wasteful expenditure of energy.

energy flux The rate at which energy flows into, through, or from a surface or material which is perpendicular to the flow of said energy.

energy gap **1.** An interval of energies which lies between two permitted energy bands. **2.** An interval of energies which lies between the top valence band and the bottom conduction band.

energy level For a quantized system, one of the allowed energy values which may be adopted. Electrons in such systems can only change from one specific level to another, and in the process absorb or emit energy.

energy-level diagram A graphical representation showing **energy levels**. Also called **energy diagram**.

energy of a charge Electric charge as measured by the following equation: $E = QV/2$, where E is the energy, Q is the charge, and V is the electric potential. Energy may be expressed, for instance, in ergs, charge in statcoulombs, and potential in statvolts.

energy-product curve For a permanent magnetic material, a curve which plots the products of the values of magnetic induction and magnetic field strength, for each point of its demagnetization curve.

energy source A source of energy, such as sunlight or wind, or that arising from reactions involving atomic nuclei or chemicals.

Energy Star In computers, a designation indicating that components, such as monitors, use less than a given maximum of power when inactive. Currently, such a maximum is 30 watts for a monitor, and 60 watts for the complete system including the monitor.

energy storage The accumulation of energy, especially electrical energy. A capacitor, for instance, stores electric charge between conductors which are separated by a dielectric material, when a potential difference exists between said conductors.

energy-storage capacitor A capacitor specially suited for repeated rapid discharging with high energy densities. Used, for instance, in a flash tube.

energy transformation Same as **energy conversion**.

engine **1.** A specialized processor or program that is used for specific, and usually repetitive functions. Examples include database engines, Web search engines, and graphics engines. **2.** A machine or device, such as a heat engine, which converts energy into work or movement.

enhanced 911 A 911 service which incorporates additional call features, such as automatic number identification, selective routing, alternate routing, and detailed call records. It may help, for instance when a caller is unable to specify the location dialed from. **911** is the emergency assistance telephone number in the United States, and a few other countries. Its abbreviation is **E911**.

enhanced BIOS A BIOS incorporating features to better accommodate current computing needs, such as support for larger hard drives.

Enhanced Capabilities Port Same as **Extended Capabilities Port**. Its abbreviation is **ECP**.

enhanced CD Abbreviation of **enhanced compact disc**. A compact disc, or CD, format which contains both audio and data, separated into two sessions. The first session contains up to 98 audio tracks, while the second session has one data track. Such discs may be played in audio CD players or a computer CD drive. An audio CD player will ignore the second session, while a computer CD drive can read both sessions. Also called **enhanced CD-ROM, CD-extra**, or **CD-plus**.

enhanced CD-ROM Same as **enhanced CD**.

enhanced compact disc Same as **enhanced CD**.

enhanced DRAM Same as **EDRAM**.

enhanced dynamic RAM Same as **EDRAM**.

enhanced dynamic random-access memory Same as **EDRAM**.

Enhanced Expanded Memory Specification In some older computer systems, an improved version of the Expanded Memory Specification. Enhancements included allowing entire programs to run in expanded memory. Its abbreviation is **EEMS**.

Enhanced Graphics Adapter An older computer video display standard, which was superseded by VGA. Its abbreviation is **EGA**.

enhanced ground proximity warning system An aircraft ground proximity warning system which incorporates enhancements such as the use of a GPS system, access to a global terrain database, and an extended warning interval when on a collision course with the ground. Its abbreviation is **EGPWS**.

Enhanced IDE   Abbreviation of **Enhanced Integrated Device Electronics**. An enhanced version of the IDE interface. It supports both the ATA-2 and ATAPI interfaces, providing for significantly faster transfer rates when using hard disks, optical discs, or tape drives. In addition, it supports hard disks with much larger capacities. Its own abbreviation is **EIDE**. Also called **Fast ATA**, **ATA-2**, or **Fast IDE**.

Enhanced Integrated Device Electronics   Same as **Enhanced IDE**.

Enhanced Interior Gateway Routing Protocol   A network routing protocol superseding **Interior Gateway Routing Protocol**. It provides improved operating efficiency, and enhanced convergence properties.   Its abbreviation is **EIGRP**.

enhanced keyboard   A computer keyboard usually equipped with 101 or 102 keys, including 12 function keys along the top, and other features, such as additional control and alternate keys. Currently, it is the most prevalent keyboard.

enhanced parallel port   A parallel port standard which supports high-speed, bi-directional communication, multiple devices including printers and external drives, and an extended cable, among other features. It is complaint with the IEEE 1284 standard. Its abbreviation is **EPP**.

enhanced reality   An environment or setting which combines virtual and real images and objects. For example, virtual images may be superimposed upon real objects. Also called **augmented reality**, or **mixed reality**.

enhanced serial port   A serial port with enhanced features, such as speeds of up to 921.6 Kbps, and utilization of UART circuits. Its abbreviation is **ESP**.

enhanced service provider   An entity which provides additional telephone services, such as caller ID, voice messaging, or data services. Its abbreviation is **ESP**.

enhanced small device interface   A standard interface for connecting hard disks to a computer, which allows for high-speed data transfer, with speeds of over 10 Mbps. It has been superseded by other interfaces, such as SCSI. Its abbreviation is **ESDI**.

Enhanced Specialized Mobile Radio   An improved version of **Specialized Mobile Radio**, providing extended coverage through the use of a network of repeaters, along with data services, among others benefits. Its abbreviation is **ESMR**.

enhancement   **1.** A change which provides improvement. For instance, a newer version of computer hardware or software which is faster, provides additional functions, or is more reliable. **2.** Modification which increases suitability for a particular purpose. For instance, the use of image processing to refine images, or coating a cathode for enhanced electron emission.

enhancement mode   A mode of operation of a field-effect transistor, in which an increase in gate voltage increases the current. When the gate voltage is zero, no current flows. This contrasts with **depletion mode**, where the converse is true.

enhancement-mode FET   Abbreviation of **enhancement-mode field-effect transistor**. A field-effect transistor which operates in **enhancement mode**. Also called **enhancement-mode transistor**.

enhancement-mode field-effect transistor   Same as **enhancement-mode FET**.

enhancement-mode transistor   Same as **enhancement-mode field-effect transistor**.

ENIAC   Possibly the first electronic computer, completed near the beginning of 1946. It weighed over 30 tons, had around 18,000 vacuum tubes, occupied about 1,800 square feet, and consumed in excess of 150 kilowatts of power. Its name is an acronym for Electronic Numerical Integrator and Calculator.

ENQ   Abbreviation of **enquiry character**.

enqueue   To place in a queue, that is, to place in a list of tasks a computer is waiting to perform, or into a data structure where elements are removed in the same order they were originally inserted. **Dequeue** is to remove from a queue.

enquiry character   In communications, a control character sent by a transmitting station, requesting a response from the receiving station.   The response may include information such as availability. Its abbreviation is **ENQ**.

ensemble   A group of components which work together to perform a single function, to produce a single effect, or to describe a process or system.

enter key   On a keyboard, a key which when pressed serves to inform the computer that the user wishes the entered information to be processed. For instance, once the information requested by prompt is given, the enter key may be pressed. It may also have program-specific functions, such as its use within a word-processing application to signal the end of a paragraph. Its symbol is often ↵. Also called **return key**.

enterprise computing   All that is related to the use of computers by an enterprise, such as a large organization.

enterprise document management   Same as **electronic document management**. Its abbreviation is **EDM**.

enterprise document management system   Same as **electronic document management system**. Its abbreviation is **EDMS**.

enterprise network   A network serving the needs of an organization. Frequently refers to a large entity which integrates various types of networks, works across multiple platforms, and is geographically dispersed.

enterprise networking   The utilization of an **enterprise network**.

Enterprise Resource Planning   An information-management system designed to serve the needs of an enterprise. It seeks to seamlessly integrate all components of the business, such as personnel, manufacturing, sales, accounting, and follow-up. Its abbreviation is **ERP**.

enterprise storage   All the media and devices utilized to meet the storage needs of an enterprise. Such storage, for instance, should include the data necessary for disaster recovery.

entity   **1.** Anything existing as a particular and discrete unit. **2.** In computers, a group of items or elements that can be treated as a unit. For instance, a data record.

entity bean   A JavaBean that is of a more permanent nature than a **session bean**.

entity-relationship model   A model that divides a database into two parts, the entities, and the relationships between them. For instance, a database may consist a group of patients and a number of physicians in a hospital, the entities, and the relationships may include which patients are attended by which doctors. Its abbreviation is **ER model**.

entrance cable   A cable serving to convey electricity, telephone service, or the like, from an outside transmission line into a structure, such as a building or home. Also called **service entrance cable**.

entropy   **1.** A measure of the disorder of a system, such as the universe. The more order a system has, the less its entropy. The greater the entropy in a system, the lesser its energy available to do work. **2.** In communications, a measure of the information contained in a signal, or of transmission efficiency.

entry   **1.** The entering of information, such as that into a computer.   Such information may be entered manually or automatically. **2.** A specific **entry (1)**, such as a name or number.

entry-level The lowest or simplest level within a hierarchy. For instance, a corporation may offer several computers models in the same line, the simplest being the entry-level system.

entry page A Web page created with given characteristics, such as multiple appearances of specific keywords, so as to rank high on a particular search engine. Also called **doorway page**, **jump page**, **bridge page**, or **gateway page**.

entry point Within a subroutine or program, the first instruction.

envelope 1. A glass housing which encloses the elements of electric lamps, electron tubes, and similar devices. An envelope may or may not be evacuated. Also called **bulb (1)**. **2.** A curve whose points pass through the peaks of a graph, such as that showing the waveform of an amplitude-modulated carrier. **3.** That which serves to encompass or surround, such as the outer sheath of a cable. **4.** The bounds within which an entity or system can be operated. For instance, the range of motion of a robotic end-effector. **5.** In communications, a group of bits, or other information-carrying entity, treated as a unit.

envelope delay 1. The time interval required for the envelope of a wave to pass from one point to another in a transmission system. It is due to different frequencies traveling at slightly different speeds. Often refers to the envelope of a modulated signal. Also called **group delay**. **2.** Same as **envelope delay distortion**.

envelope delay distortion Distortion of a signal as it passes through a transmission medium in which different frequencies travel at slightly different speeds. This causes a change in the waveform because the rate of change of phase shift is not constant over the transmitted frequency range. Its abbreviation is **EDD**. Also called **envelope delay (2)**, **delay distortion**, **time-delay distortion**, **phase distortion**, or **phase-delay distortion**.

envelope demodulator Same as **envelope detector**.

envelope detector A demodulator circuit which incorporates one or more diodes, and which provides a rectified output whose average value is proportional to that of the original modulation. Also called **envelope demodulator**, or **diode detector**.

environment 1. Same as **environmental conditions**. **2.** In computers, a specific hardware and software configuration. Many programs, for instance, will only function properly in a given environment. Also called **computer environment**.

environmental Pertaining to a given **environment**.

environmental chamber An enclosure, such as a room, in which measures are taken to provide an environment that meets certain requirements, such as maintaining a specified level of temperature and/or humidity, guarding against static electricity or electromagnetic radiation, or isolating from dust, and which is utilized for testing. Also called **environmental test chamber**.

environmental conditions The conditions that surround a given area, especially in reference to parameters which may influence the functioning of devices, equipment, or the readings of instruments. These conditions include temperature, pressure, humidity, noise, and light. Also called **environment (1)**, or **ambient conditions**.

environmental control unit A unit or device which enables utilizing or accessing one or more devices or apparatuses via voice, or another form of remote control. Used, for instance, by persons with reduced mobility to switch lights on and off, place telephone calls, control appliances, and so on.

environmental factors Any factors, such as temperature, pressure, humidity, dust, or noise, which affect the overall **environmental conditions**.

environmental light Light which is present in any given environment, especially within which tests or observations are taking place. Also called **room light**, **background light**, or **ambient light**.

environmental noise The noise present in any given environment, especially that within which tests or observations are taking place. Such noise includes nearby and distant sources. Also called **background noise (1)**, **room noise**, **site noise**, **ambient noise**, or **local noise**.

environmental requirements Specifications of **environmental conditions** which must be observed for the proper and/or optimal operation of components, devices, equipment, or systems.

Environmental Stress Screening A type of accelerated life test in which devices are subjected to one or more expected environmental stresses. Such tests are performed within design parameters, and are intended to expose latent flaws. Its abbreviation is **ESS**.

environmental test A test of a component, circuit, device, piece of equipment, or system, which is conducted in a real environment, or a lab environment simulating such conditions.

environmental test chamber Same as **environmental chamber**.

environmental testing The conducting of **environmental tests**.

environmentally sealed That which is isolated from environmental factors, such as moisture, dust, noise, or light, which might impair proper functioning.

EO Abbreviation of **end office**.

EOF Abbreviation of **end-of-file**.

EOF code Abbreviation of **end-of-file code**.

EOF indicator Abbreviation of **end-of-file indicator**.

EOF label Abbreviation of **end-of-file label**.

EOF mark Abbreviation of **end-of-file mark**.

EOG Abbreviation of **electro-oculography**.

EOM Abbreviation of **end-of-message**.

EOR Abbreviation of **end-of-record**.

EOS Abbreviation of **electrical overstress**.

EOT Abbreviation of **end-of-transmission**.

EP Abbreviation of **extended play**.

epaper Abbreviation of **electronic paper**.

epayment Abbreviation of **electronic payment**.

epilayer Abbreviation of **epitaxial layer**.

EPIRB Abbreviation of **emergency position indicating radio beacon**.

epitaxial Pertaining to **epitaxy**.

epitaxial deposition The forming of an **epitaxial layer**.

epitaxial device A semiconductor device with one or more **epitaxial layers**.

epitaxial film A crystal film, especially of a semiconductor material, formed through **epitaxy**.

epitaxial growth Same as **epitaxy**.

epitaxial growth process Same as **epitaxy**.

epitaxial layer A crystal layer, especially of a semiconductor material, formed through **epitaxy**. Its abbreviation is **epilayer**.

epitaxial process Same as **epitaxy**.

epitaxial semiconductor A semiconductor material with one or more **epitaxial layers**.

epitaxial transistor A transistor with one or more **epitaxial layers**.

epitaxy The controlled and oriented growth of a thin single-crystal layer upon the surface of another single crystal, with the deposited layer having the same crystalline orientation as its substrate. The grown crystal and the substrate may or may not be the same single crystal. This method is used extensively in the preparation of semiconductor materials. Common techniques include molecular-beam epitaxy, vapor-phase epitaxy, and liquid-phase epitaxy. Also called **epitaxial growth**, **epitaxial growth process**, or **epitaxial process**.

EPOS Abbreviation of **electronic point-of-sale**.

epoxy glass An **epoxy resin** reinforced with glass fibers. May serve, for instance, in printed-circuit boards as the base material upon which circuits are etched, and to which electrical components are attached. Also called **glass-epoxy resin**.

epoxy resin A class of polymers that are thermosetting, and which feature toughness, chemical resistance, and low expansion. Used, for instance, as an adhesive, or for protective coatings.

EPP Abbreviation of **enhanced parallel port**.

EPR Abbreviation of **electron paramagnetic resonance**.

EPR spectroscopy Abbreviation of **electron paramagnetic resonance spectroscopy**.

EPROM Abbreviation of erasable programmable read-only memory. A programmable ROM chip which is erasable by exposure to ultraviolet light. Unlike PROM, which can be programmed once, EPROMs can be reprogrammed.

EPS Abbreviation of **Encapsulated PostScript**.

EPSS Abbreviation of **Electronic Performance Support System**.

EQ 1. Abbreviation of **equalizer**. 2. Abbreviation of **equalization**.

equal Having the same quantity, value, force, effect, function, or use as another.

equal-energy source A source of energy, such as electromagnetic radiation, in which all frequencies are emitted at equal energy levels throughout the spectrum of said source. An example is equal-energy white.

equal-energy white The light produced by a source which radiates with equal energy for each wavelength of the visible spectrum. The produced light is white, and may be used as a reference.

equal-loudness contours Same as **equal-loudness curves**.

equal-loudness curves A series of curves which plot the sound intensities of pure tones, in decibels, versus frequency, in hertz, for all audible frequencies, so that all frequencies are perceived at the same intensity. The hearing sensitivity of people varies by frequency, and this sensitivity also varies depending on the sound-pressure level. Also called **equal-loudness contours**, **Fletcher-Munson curves**, or **loudness curves**.

equal-loudness level For a given listener, the perception of equal loudness when comparing a sound to a reference sound, such as a 1 kHz tone at 0 decibels. Also called **loudness level**.

equality The quality or state of being **equal**.

equalization The use of an **equalizer**. Also, the resulting response when using an equalizer. Its abbreviation is **EQ**. Also called **frequency-response equalization**.

equalization circuit Same as **equalizer**.

equalization curve A curve depicting the frequency response a circuit, device, piece of equipment, or system should have after equalization. For instance, it may be a curve portraying the equalization necessary to restore an audio-frequency signal to its original form, after preemphasis. Another example is a curve showing the equalization necessary to obtain a frequency response which takes into account the auditory response humans have.

equalization network Same as **equalizer**.

equalizer An electric network whose attenuation or gain varies as a function of frequency, and which is used to modify the frequency response of a circuit, device, piece of equipment, or system. It may counteract distortion, compensate for deficiencies, or shape a frequency response to fit the needs of a given situation. A parametric equalizer in an audio amplifier, for instance, can adjust various intervals within the audible range, to help provide the desired sound experience. Its abbreviation is **EQ**. Also called **equalization circuit**, **equalization network**, **equalizer network**, **equalizer circuit**, **equalizing network**, or **equalizing circuit**.

equalizer circuit Same as **equalizer**.

equalizer network Same as **equalizer**.

equalizing circuit Same as **equalizer**.

equalizing current A current which flows between two compound generators connected in parallel, to balance their output.

equalizing network Same as **equalizer**.

equalizing pulses In a TV signal, pulses, occurring at twice the rate of the line frequency, which are transmitted during blanking intervals immediately before and after the vertical sync pulses. Such pulses improve the precision of scanning-line interlacing.

equation A mathematical statement in which two expressions, one on each side of an equal sign (=), are shown to be equivalent.

equation solver A computer especially suited for solving specific types of complex equations.

equator The primary great circle of the planet earth, or that of another celestial body, which at all points is 90° from the geographic poles.

equatorial electrojet An electrojet centered over, or near, the magnetic equator.

equatorial orbit An orbit, such as that of a communications satellite, in the plane of the earth's equator.

equiphase surface A surface in a wave in which the field vectors all have the same phase.

equiphase zone A region in space in which the phase difference between two signals is negligible.

equipment 1. An assembly of articles, components and other physical resources that are utilized for specific functions or activities. For example, electrical equipment, or data-processing equipment. 2. The providing of **equipment** (1).

equipment chain A chain of units which work together to perform a given function. If any of the components fail, then the overall function is not performed.

equipment failure The condition in which a piece of equipment no longer performs the function it was intended to, or is not able to do so at a level that equals or exceeds established minimums.

equipment ground A ground connection to equipment. Such a ground, for example, may be made between the earth and a non-conducting metal surface of said equipment.

equipment life The lifetime, or average lifetime, of equipment. Refers to the time interval within which the equipment is able to perform its functions adequately.

equipment response 1. A change in the behavior, operation, or function of a piece of equipment as a consequence of a change in its external or internal environment. For example, a change in its output resulting from a change in its input. 2. A quantified **equipment response** (1), such as a frequency response.

equipment test A test performed on equipment. There are various types, including accelerated, bench, and field tests.

equipment testing The performance of **equipment tests**.

equipotential Having the same electric potential at all points.

equipotential cathode A cathode, within a thermionic tube, which is electrically insulated from the heating element. It may consist, for instance, of a filament surrounded by a sleeve with an electron-emitting coating. Such cathodes have the same potential along their entire surface. Also called **unipotential cathode**, **indirectly-heated cathode**, or **heater-type cathode**.

equipotential line An imaginary line having the same electric potential at all points.

equipotential surface A surface having the same electric potential at all points.

equiripple A frequency response which is characterized by equal ripples in the passband and/or stopband. A Chebyshev filter, for instance, has such a response.

equiripple filter A filter with **equiripples** in the passband and/or stopband.

equisignal Pertaining to two signals with equal intensity.

equisignal zone A region in space in which the difference in amplitude of two radio-frequency signals is indistinguishable.

equivalence The property or condition of being **equivalent**. Also spelled **equivalency**.

equivalency Same as **equivalence**.

equivalent That which is identical, or virtually identical, in value, function, or effect.

equivalent absorption Also called **equivalent sound absorption**. **1.** The rate of sound absorption by a given surface. The sabin is the usual unit of expression. **2.** The area of a perfectly absorbing surface, such as an open window, which absorbs the same amount of sound energy as another given surface, such as a wall, under the same conditions. Also called **equivalent absorption area**.

equivalent absorption area Same as **equivalent absorption** (2).

equivalent binary digits The number of binary digits required to express numbers with a different base. For instance, it takes approximately 3.33 times as many binary digits to express an equivalent number of decimal digits.

equivalent capacitance A single concentrated capacitance which would have the same effect as the total capacitance distributed throughout a circuit.

equivalent circuit A combination of circuit elements which, under given conditions, have electrical characteristics equivalent to that of another, generally more complex circuit or device. Used, for instance, in theoretical analysis.

equivalent conductance For an electrolyte, the conductivity divided by the gram-equivalents of solute per cubic centimeter of solvent. Also called **equivalent conductivity**.

equivalent conductivity Same as **equivalent conductance**.

equivalent height The apparent height of an ionized atmospheric layer, as determined by the time interval that elapses between the transmission of a radio-frequency signal and the return of its ionospheric echo. The signal is assumed to be traveling at the speed of light. Also called **virtual height**.

equivalent impedance A single concentrated impedance which would have the same effect as the total impedance distributed throughout a circuit.

equivalent inductance A single concentrated inductance which would have the same effect as the total inductance distributed throughout a circuit.

equivalent network An electric network which, under given conditions, has electrical characteristics equivalent to that of another, generally more complex network. Used, for instance, in theoretical analysis.

equivalent noise input For a photosensitive device under specified conditions, the amount of incident light necessary to produce a given signal-to-noise ratio.

equivalent noise resistance For a given frequency, and under specified conditions, a measure of the spectral density of a noise-voltage generator. Expressed in ohms.

equivalent noise temperature The absolute temperature at which an ideal resistor would have an equal amount of noise as a given electrical element, with the same resistance, at room temperature.

equivalent reactance A single concentrated reactance which would have the same effect as the total reactance distributed throughout a circuit

equivalent resistance A single concentrated resistance which would have the same effect as the total resistance distributed throughout a circuit.

equivalent series resistance The resistance in series with an ideal capacitor which duplicates the performance characteristics of a real capacitor.

equivalent sound absorption Same as **equivalent absorption**.

equivalent weight The mass of a chemical element or compound which, in a reaction, combines with or displaces one mole of hydrogen. For instance, one mole of chlorine combines with one mole of hydrogen to form hydrochloric acid. In this example, the equivalent weight of chlorine is approximately 35.45 grams, as that is the weight of one mole of chlorine. In the case of an element, the equivalent weight is equal to its gram-atomic weight divided by its valence. In the case of a molecule, it is the gram-molecular weight divided by the effective valence. Also called **gram-equivalent**.

Er Chemical symbol for **erbium**.

ER model Abbreviation of **entity-relationship model**.

erasable Capable of being erased. Said for, instance, of a storage medium.

erasable programmable read-only memory Same as **EPROM**.

erasable programmable ROM Same as **EPROM**.

erasable PROM Same as **EPROM**.

erasable storage A storage medium, such as a hard disk, or a DVD-RW, whose stored content may be erased, then rewritten.

erase To remove, delete, or eliminate. Said, for instance, of the deletion of files from a hard disk, or the removal of a signal. A specific example is the removal of the information stored on a magnetic tape by utilizing an AC erase head.

erase current **1.** An AC, generally of a high-frequency, that flows through an erase head, and which erases the content of magnetic tapes. **2.** A DC utilized to erase the content of magnetic tapes.

erase head A device which serves to delete material recorded on a magnetic storage medium, such as a magnetic tape. For instance, an AC erase head employs AC to create an alternating magnetic field which erases the tape. Also called **erasing head**, or **magnetic head (3)**.

erase oscillator An oscillator which provides a high-frequency signal utilized to erase the content of magnetic tapes.

erase speed The rate at which at an erase operation takes place. Utilized, for instance, in the context of chips being repeatedly written and erased. Also called **erasing speed**.

eraser A device, such as a bulk eraser, that serves to delete material recorded on a magnetic storage medium, such as a floppy disk.

erasing head Same as **erase head**.

erasing speed Same as **erase speed**.

erasure The process of removing, deleting, or eliminating. Said, for instance, of the deletion of material recorded on a magnetic storage medium, such as a magnetic tape, or of the removal of a signal.

erbium A soft and malleable metallic chemical element with atomic number 68. It has nearly 30 known isotopes, of which several are stable. Used, for instance, in lasers, semiconductors, nuclear reactors, and in certain ceramics. Its chemical symbol is **Er**.

erbium-doped fiber amplifier Its abbreviation is **EDFA**. A broadband optical amplifier consisting of glass optical fibers doped with erbium. An EFDA increases the optical signal itself, as opposed to having to convert said signal into an electrical signal, which is amplified, then reconverted into light. EFDAs feature low insertion loss, high fiber-to-fiber gain, and no polarization sensitivity, but are limited by the buildup of spontaneous-emission noise.

erbium laser A laser, such as a YAG laser, whose crystal contains erbium impurities. Used, for instance, in surgery.

erbium oxide A pink powder whose chemical formula is Er_2O_3. Used in lasers, semiconductors, special glasses, and as a phosphor activator.

ERD Abbreviation of **elastic recoil detection**.

ERDA Abbreviation of **elastic recoil detection analysis**.

erg The unit of work in the cgs system. It is equal to the work done by a force of one dyne acting through a distance of 1 centimeter. It is also equal to 0.1 microjoule.

ergonomic 1. Pertaining to **ergonomics**. 2. Designed and created according to the principles of ergonomics.

ergonomic keyboard A computer keyboard that conforms to the principles of **ergonomics**. For instance, one that can provides for proper wrist support, and for the fingers to type in a natural position.

ergonomics The science dealing with the design and creation of devices, equipment, and environments which factor in the needs of its human users. That which is ergonomic, for instance, seeks to maximize comfort, posture, safety, and ease of use. Also called **human-factors engineering**.

erlang A time unit used for measuring telephone communications traffic. One erlang equals one call hour, 36 centum call seconds, or 3600 call seconds.

ERP Abbreviation of **effective radiated power**. 2. Abbreviation of **Enterprise Resource Planning**.

error 1. A deviation from a true, expected, or specified result. For example, a discrepancy between a calculated, observed, or measured value, and the theoretically correct one. Another example is the presence of a difference between transmitted and received data. 2. The magnitude or degree of an **error** (1). 3. In computers, an inability to carry out an operation, or an incorrect result, as a result of a software and/or hardware deficiency, or mistakes by a user.

Error 404 An error message displayed by a Web browser when it cannot locate a given Web page, file, or URL, as occurs, for instance, when a page no longer exists, the server is not available, or an address has been incorrectly typed. Also called **404 Error, 404 Not Found**, or **HTTP Error 404**.

error amplifier An amplifier, such as a differential amplifier, which produces an error signal when a sensed output diverges from a reference voltage.

error checking The detection of errors, such as those that may occur in the transmission of data. Any of various methods may be used for this task, such as performing a cyclic redundancy check. Also called **error control**, or **error detection**.

error checking and correction Same as **error detection and correction**. Its abbreviation is **ECC**.

error condition A state or circumstance in which an error is occurring, or has occurred.

error control Same as **error checking**.

error-correcting code Same as **error-correction coding**.

error-correcting code memory Same as **ECC memory**.

error-correcting code random-access memory Same as **ECC memory**.

error correction The correcting of errors, usually through automatic means. Ideally, all errors would be detected, then corrected, without interrupting any ongoing processes.

error-correction coding A coding technique in which sent data is automatically checked for errors upon reception. If any errors are detected, they are automatically corrected. This method has a maximum of detectable errors, and once reached, is no longer reliable in this capacity. Used, for instance, by modems, and in RAIDs. Also called **error-correcting code**.

error-correction mechanism A mechanism or technique, such as error correction coding, which serves to correct errors.

error corrector A device or mechanism which corrects errors.

error current In an automatic control system, a current that is proportional to the difference between the actual operating quantity and a quantity which would be within the specified parameters. Such a current is utilized for automatic correction.

error detection Same as **error checking**.

error detection and correction Also called **error checking and correction**. Its abbreviation is **EDC**, or **EDAC**. 1. A method for detecting errors in data which is transmitted or stored, followed by corrective measures. For instance, error-correction coding may be used for this purpose. 2. Any method of detecting errors which is followed by corrective measures.

error detection mechanism A mechanism or technique, such as feedback or parity checking, which serves to detect errors.

error detector A device or mechanism which detects errors.

error handling The manner in which a routine, program, system, or user responds to encountered errors. For instance, a programming language which inform of errors, and helps in dealing with them.

error interrupt A halt in the execution of a computer program, due to an error.

error log A record of computer and/or network error events. Used, for example, to find the origin of problems.

error message A message from a computer program or operating system indicating that an error has occurred. Such a message usually indicates that the error must be corrected before continuing. Some errors simply require an acknowledgement for correction, while others require stronger measures, such as rebooting.

error range A range of values within which an input will cause an error. For instance, a division by zero, or by a number so small that the result exceeds the capability of a program or computer.

error rate The frequency with which errors occur. In communications, it is a measure of data transmission integrity. It may be quantified, for instance, as a given number of erroneous bits per million transmitted bits.

error ratio The total number of errors, divided by the total number of units of data transmitted, received, transferred, or processed.

error recovery The process of successfully dealing with an error, ideally without interrupting any ongoing operation. If an error causes a halt, then error recovery involves restoring the system to the state it was in before said error occurred.

error routine A software routine that is utilized to address errors that occur. Such a routine may, for instance, generate a detailed report and/or automatically implement corrective measures.

error signal 1. A signal indicating an error. 2. A magnitude which is proportional to the difference between a desired or reference value, and the actual or measured value. In radars, for instance, it may refer to the difference between an actual position and a desired position. When said magnitude is a voltage, also called **error voltage** (1). 3. In an automatic control system, a signal whose magnitude is proportional to the difference between the actual operating quantity and a quantity which would be within the specified parameters. Such a signal is utilized for automatic correction. When said magnitude is a voltage, also called **error voltage** (2).

error voltage 1. A voltage which is proportional to the difference between a desired or reference value, and the actual or measured value. In radars, for instance, may refer to the difference between an actual position and a desired position. 2. In an automatic control system, a voltage that is proportional to the difference between the actual operating quantity and a quantity which would be within the specified parameters. Such a voltage is utilized for automatic correction.

Es Chemical symbol for **einsteinium**.

ES Abbreviation of **expert system**.

Esaki diode A pn-junction diode which is highly doped on each side of the junction. It has a negative resistance at a low voltage in the forward-bias direction, due to the tunnel effect. May be used, for instance, in oscillator or amplifier circuits whose frequencies run well into the microwave range. Also called **tunnel diode**.

ESC 1. Abbreviation of **ESC key**. 2. Abbreviation of **escape character**.

ESC character Abbreviation of **escape character**.

ESC key Abbreviation of **escape key**. On computer keyboards, a key whose function depends on the software running, but which usually aborts an operation that is in progress, such as program execution, data entry, or a Web page loading. It may also serve to send an escape character. Its own abbreviation is **ESC**.

escape character A control character utilized to indicate the beginning of a sequence of characters which send an instruction to a program, or to a peripheral such as a printer. Its abbreviation is **ESC character**, or **ESC**.

escape code A control code utilized to indicate the beginning of a sequence of characters which send an instruction to a program, or to a peripheral such as a printer.

escape key Same as **ESC key**.

escape sequence A series of characters whose first character is an **escape character**.

escape velocity The minimum speed a particle or body must attain in order to escape the gravitational field of a celestial body, such as a planet or a star. For the planet earth, this velocity is approximately 11.2 km/sec.

escrowed encryption An encryption technique in which a third party, especially a governmental body, has the decryption keys. Used, for instance, by governments to decrypt messages utilizing said technique.

escutcheon A decorative and/or protective plate or flange which frames openings or spaces surrounding dials, knobs, indicators, or other panel-mounted components. Also called **escutcheon plate**.

escutcheon plate Same as **escutcheon**.

ESD 1. Abbreviation of **electrostatic discharge**. 2. Abbreviation of **electronic software distribution**.

ESD damage Abbreviation of **electrostatic discharge damage**.

ESD protected area Abbreviation of **electrostatic discharge protected area**.

ESD protection Abbreviation of **electrostatic discharge protection**.

ESD protective packaging Abbreviation of **electrostatic discharge protective packaging**.

ESD sensitive Abbreviation of **electrostatic discharge sensitive**.

ESD sensitivity Abbreviation of **electrostatic discharge sensitivity**.

ESD susceptibility Abbreviation of **electrostatic discharge susceptibility**.

ESD vulnerability Abbreviation of **electrostatic discharge vulnerability**.

ESD wrist strap Abbreviation of electrostatic discharge **wrist strap**. A grounding strip which is typically wrapped around a wrist on one side, and attached to the chassis of the device or piece of equipment being worked on the other, so as to properly channel any static electricity a person may carry or generate. Used, for instance, to prevent a user from harming sensitive computer components when performing tasks such as the installation additional RAM. Also called **ESD wristband**, **wrist strap**, or **antistatic wrist strap**.

ESD wristband Same as **ESD wrist strap**.

ESDI Abbreviation of **enhanced small device interface**.

eservices Same as **e-services**.

ESF Abbreviation of **extended superframe**.

ESF format Abbreviation of **extended superframe format**.

esignature Abbreviation of **electronic signature**.

ESMR Abbreviation of **Enhanced Specialized Mobile Radio**.

ESN Abbreviation of **electronic serial number**.

ESP 1. Abbreviation of **enhanced serial port**. 2. Abbreviation of **enhanced service provider**.

ESR Abbreviation of **electron spin resonance**.

ESR spectroscopy Abbreviation of **electron spin resonance spectroscopy**.

ESS 1. Abbreviation of **electronic switching system**. 2. Abbreviation of **Environmental Stress Screening**.

esu Abbreviation of **electrostatic system of units**.

esupport Abbreviation of electronic **support**. Technical, product, or service support provided via the Internet. An entity could also provide such help via telephone, fax, regular mail, and/or in person, although there is a marked tendency towards esupport whenever practical. Also called **Internet support**.

etailer Same as **e-tailer**.

etalon An optical device consisting of two parallel plates of silvered glass with a gap between them. Used, for instance, in a laser to filter out unwanted modes.

etch 1. To produce patterns on a hard material, such as glass or metal, by selectively removing portions of its surface via a suitable means, such as the use of a plasma, an acid, or a laser beam. Also, to subject a material to such a process. Used, for instance, to create etched circuits. 2. That which is utilized to **etch** (1). Usually refers to a chemical substance, such as an acid, but may also be a plasma, a laser, an ion beam, and so on. Also called **etchant**.

etch rate The speed at which **etching** (1) occurs. Usually used in the context of wet etching.

etchant Same as **etch** (2).

etched circuit A printed circuit formed through **etching** (1). Also called **etched printed circuit**.

etched pattern A pattern, such as that on a semiconductor wafer, formed through **etching** (1).

etched printed circuit Same as **etched circuit**.

etching 1. The production of patterns on a hard material, such as glass or metal, by selectively removing portions of its surface via a suitable means, such as the use of a chemical reaction or a beam. Also, the patterns so produced. Etching techniques include plasma, ion-beam, and laser etching, and those which utilize a chemical agent such as an acid. **2.** A plate, circuit, or other object prepared through **etching (1)**.

etching process   Any process utilized for **etching (1)**.

etching technique   Any technique utilized for **etching (1)**.

etext   Abbreviation of **electronic text**.

Ethernet   A widely-used high-speed LAN defined by the IEEE 802.3 standard. Ethernet can use a bus or star topology, utilizes CSMA/CD, and transmits data in variable-length frames of up to 1,518 bytes. There are various versions, the most common utilizing coax cables, while others use twisted-pair wiring, or fiber-optic cable. Depending on the version, Ethernet can support data transfer rates from 10 Mbps to over 100 Gbps. Also called **Ethernet network**.

Ethernet adapter   Same as **Ethernet card**.

Ethernet address   The physical address of a specific **Ethernet card**. It is a 48-bit number, usually expressed as 12 hexadecimal digits.

Ethernet card   The hardware, usually in the form of an adapter card, which is installed in a computer, and to which the cabling linking to an **Ethernet** is connected. An Ethernet card may also be built into the motherboard. Also called **Ethernet adapter**.

Ethernet network   Same as **Ethernet**.

Ethernet switch   A switch that connects nodes or segments in an **Ethernet network**. Switched Ethernet provides the full bandwidth available to each pair of connected nodes.

EtherTalk   Software employed to use AppleTalk over Ethernet networks.

ETSI   Abbreviation of **European Telecommunications Standards Institute**.

Ettinghausen effect   A phenomenon observed when a strip of a metal conductor is placed perpendicular to a magnetic field. When said strip is longitudinally carrying a current, each end of the strip has a small temperature differential.

Ettinghausen-Nernst effect   A phenomenon observed when a strip of a metal conductor is placed perpendicular to a magnetic field. When said strip is conducting heat longitudinally, a small voltage differential appears at opposite ends of the strip. Also called **Nernst effect**.

ETX   Abbreviation of **end-of-text**.

Eu   Chemical symbol for **europium**.

eudiometer   A device which utilizes electricity to measure the volume of gases. May be used, for instance, to quantify the volume of a gas produced at electrodes during electrolysis.

Eudora   A popular email program.

EULA   Acronym for **End-User License Agreement**.

eureka   In the rebecca-eureka system, the ground transponder beacon.

Euro connector   Same as **Euroconnector**.

Euroconnector   A 21-pin connector utilized for connections between audio and video devices such as TVs and VCRs, and which is commonly utilized in Europe. The devices have recessed openings, while the cable has matching pins at each end. Also spelled **Euro connector**. Also called **SCART**.

European Telecommunications Standards Institute   A European telecommunications organization whose members include manufacturers, service providers, research organizations, and users. It has members in over 50 countries, and strives to promote worldwide standards. Its abbreviation is **ETSI**.

europium   A steel-gray reactive metallic chemical element with atomic number 63. It is soft and malleable, and has nearly 30 known isotopes, of which 2 are stable. Used, for

instance, as a phosphor activator, and as a neutron absorber in nuclear reactors. Its chemical symbol is **Eu**.

europium oxide   A pink powder whose chemical formula is Eu_2O_3. Used in phosphors, and in nuclear reactor control rods.

eutectic   **1.** Pertaining mixtures of two or more substances which have the lowest possible melting point, such as eutectic mixtures, or eutectic alloys. **2.** Same as **eutectic mixture**.

eutectic alloy   An alloy whose constituents form a product with the lowest possible melting point. That is, the proportion of each element in it is such, so that any other overall combination of these elements would have a higher melting point. Such alloys usually make a rapid transition from solid to liquid at its eutectic temperature. Used, for instance, in solders, contacts, and semiconductors.

eutectic attach   The use of an eutectic bond to attach a chip to a substrate. Also called **eutectic bonding**.

eutectic bond   **1.** A bond formed by heating two or more materials in a joint, so as to form an eutectic mixture. To facilitate such bonding, pressure, vibrations, or rubbing may also be used, in addition to the heat. **2.** A bond formed utilizing an eutectic mixture.

eutectic bonding   **1.** The formation of an **eutectic bond**. **2.** Same as **eutectic attach**.

eutectic mixture   A mixture of two or more substances which has the lowest possible melting point. That is, the proportion of each component is such, so that any other overall combination of such substances would have a higher melting point. Low-melting point alloys are usually eutectic. Used, for instance, in solders, contacts, and semiconductors. Also called **eutectic (2)**.

eutectic solder   A solder whose constituents form a product with the lowest possible melting point. That is, the proportion of each element in it is such, so that any other overall combination of these elements would have a higher melting point. For example, a tin-lead solder with 63% tin and 37% lead is an eutectic solder.

eutectic temperature   The melting point of an **eutectic mixture**, or of an **eutectic alloy**.

EV   Abbreviation of **electric vehicle**.

eV   Symbol for **electronvolt**.

evacuate   **1.** To partially or completely remove the contents from an enclosure containing any substance, such as a gas or a liquid. Used, for instance, to refer to an electron tube whose contained gases are being removed. Also called **exhaust (1)**. **2.** The process utilized to **evacuate (1)**. For instance, the use of a vacuum pump to remove all gases from an electron tube.

evacuation   The partial or complete removal the contents from an enclosure containing any substance. For example, the removal of gases or liquids. Also called **exhaustion (1)**.

evanescent field   A field whose amplitude decays over distance, without an accompanying phase shift. In fiber optics, for instance, an evanescent field may be utilized for coupling to another fiber.

evanescent wave   A wave whose amplitude decays over distance. This occurs, for instance, in an optical fiber, when internal reflections take place at the fiber core-cladding interface.

evanescent wave spectroscopy   An instrumental analysis technique which analyzes the absorption of **evanescent waves**.

evaporate   **1.** To convert a liquid into a vapor. **2.** To form a deposit, such as that of a metal film, on a substrate, through selective condensation, vaporization, or sublimation.

evaporation   **1.** The conversion of a liquid into a vapor. **2.** The forming of a deposit, such as that of a metal film, on a substrate, through selective condensation, vaporization, or sublimation.

even field In interlaced scanning, the pattern created by tracing the even-numbered lines. The odd-numbered lines create the **odd field**, and together these compose a displayed frame.

even harmonic A harmonic whose frequency is an even-numbered multiple of the fundamental frequency. For instance, if the fundamental frequency is 200 Hz, then 400 Hz and 2000 Hz are each even harmonics. This contrasts with **odd harmonics**, which are odd-numbered multiples of the fundamental frequency.

even line In a TV, an even-numbered active line. Also called **even-numbered line**.

even-numbered line Same as **even line**.

even parity 1. The presence of an even number of ones or zeroes within a group of bits. 2. An error-detection procedure which verifies if there is **even parity** (1). For each byte transmitted, an additional bit is added, which can be a 0 or 1, so that all bytes are even. At the receiving location, the parity is checked, to be sure it is even. If it is, the transmission has passed this test of data integrity. Also used for data storage. Also called **even parity check**.

even parity check Same as **even parity** (2).

event 1. An action or occurrence. It is similar to an **incident** (2), but an incident may imply a less significant happening than an event. 2. In computers, an **event** (1) which has significance to a task or program. For instance, interrupts, mouse movements, or menu selections.

event counter A circuit, device, register, mechanism, or system used for counting events, such as pulses, or changes in state.

event-driven 1. Pertaining to programs which wait for user actions, such as mouse clicks and key pressings, responds to them, and then returns to waiting for further events. This contrasts with **procedure-driven**, in which a predetermined sequence of steps must be followed. 2. A specific program which is **event-driven** (1). Also called **event-driven program**.

event-driven processing In a computer system, processing of actions which is **event-driven** (1). Interrupts are utilized for all connected devices to request resources. This contrasts with **autopolling**, where there is the periodic detection of the status of all connected devices, to address any needs for resources.

event-driven program Same as **event-driven** (2).

event-management system Software which monitors the activity of a given network. It can provide real-time observation of servers, nodes, and other network devices, and keeps track of events such as successful log-ons, unsuccessful log-ons, time logged-on, resources accessed, and so on. Used, for instance, in the analysis of the use of system resources, to gauge network performance, and for security. Its abbreviation is **EMS**.

ewallet Abbreviation of **electronic wallet**.

exa- A metric prefix representing 10^{18}. For example, exahertz. Its symbol is **E**.

exabit 2^{60} bits, or approximately 1.153×10^{18} bits. Often it is rounded to 1.0×10^{18}. Its abbreviation is **Eb**, or **Ebit**.

exabyte 2^{60} bytes, or approximately 1.153×10^{18} bytes. Often it is rounded to 1.0×10^{18}. Its abbreviation is **EB**, or **Ebyte**.

exahertz 10^{18} Hz. This is within the X-ray region of the electromagnetic spectrum. Its abbreviation is **EHz**.

exalted-carrier reception A form of radio reception in which the carrier is separated from the sidebands, filtered and amplified, and then recombined with the sidebands at an increased level. This helps counteract selective fading.

exbi- A binary prefix meaning 2^{60}, or 1,152,921,504,606,846,976. For example, an exbibyte is equal to 2^{60} bytes. This prefix is utilized to refer to only binary quantities, such bits and bytes Its abbreviation is **Ei-**.

exbibyte 2^{60}, or 1,152,921,504,606,846,976 bytes. Its abbreviation is **EiB**.

Excel A popular spreadsheet program.

except gate A logic gate which provides an output pulse when at least one of its input terminals provides a pulse, and at the same time at least one of its input terminals does not.

exception A condition, usually an error, which diverts the attention of the processor. It is similar to an **interrupt**, the difference being that an exception usually indicates the presence of an error.

exception handling The manner in which **exceptions** are handled by a program. Ideally, most exceptions can be addressed without the program aborting.

exception report A report detailing **exceptions**.

excess 3 code Same as **excess three code**.

excess conduction In a semiconductor material, electrical conduction by excess electrons.

excess electron In a semiconductor material, an electron which is added by a donor impurity. Such electrons are available for conduction.

excess information rate In a frame relay circuit, the rate at which information is transmitted in excess of the committed information rate. Used, for instance, to support burst transmissions. Its abbreviation is **EIR**.

excess minority carrier In a semiconductor material, a minority charge carrier which is in excess of the equilibrium amount.

excess noise Also called **current noise**. 1. Electrical noise produced by current flowing through an electrical component, especially a resistor. 2. Electrical noise produced by current flowing through a semiconductor material.

excess rate A rate of transmission which exceeds a given rate, such as that which is guaranteed.

excess three code A code which represents numbers as the four-bit binary equivalent of a decimal digit plus three. For instance, the number **0** is represented by 0011, which is 3 (decimal 0 + 3). Its abbreviation is **excess 3 code**.

exchange 1. A structure which houses one or more telephone switching systems. At this location, customer lines terminate and are interconnected with each other, in addition to being connected to trunks, which may also terminate there. A typical exchange handles about 10,000 subscribers, each with the same area code plus first three digits of the 10 digit telephone numbers. Also called **central office**, **local central office**, **telephone central office**, or **telephone exchange**. 2. The act or process of substituting a thing or function for another. 3. In computers, to replace the contents of one location with that of the other, and vice versa.

exchange line A telephone line which connects a subscriber to an **exchange** (1).

exchange sort A sorting technique in which adjacent pairs of items within a list are sequentially compared and ordered, until the entire list has been evaluated. This process is then repeated as many times as required to obtain the correct order. Also called **bubble sort**, or **sifting sort**.

excimer laser A gas laser which in which two atoms, one from a reactive gas, such as chlorine or fluorine, the other from an inert gas, such as argon or krypton, form a metasable bond which is electrically excited. The transitions between energy levels generate laser emissions. It is a type of pulsed laser.

excitation 1. The signal, such as a voltage or current, that causes the function of a component, circuit, device, piece of equipment, system, process, or mechanism. Also called **excitation signal**, or **drive (1)**. **2.** The action of providing **excitation (1)**. Examples include the application of signal power to a transmitting antenna, a voltage to an oscillating crystal, or a signal voltage to the control electrode of an electron tube. Also called **drive (2)**. **3.** The process by which a particle, such as an electron or atom, moves from a lower state, usually the ground state, to a state with a higher energy level.

excitation current A current which provides **excitation (1)**.

excitation energy The energy required for **excitation**.

excitation signal Same as **excitation (1)**.

excitation voltage A voltage which provides **excitation (1)**.

excited atom An atom which contains at least one **excited electron**.

excited electron **1.** An electron which is in an energy state higher than the ground state. **2.** An electron which is in an energy state with a higher level of energy than another given state, such as the ground state.

excited state The condition of an atom, molecule, or radical containing at least one **excited electron**.

exciter Also spelled **excitor**. **1.** A component, circuit, or device which provides **excitation (1)**. **2.** An oscillator which generates the carrier frequency of a transmitter. **3.** A part of an antenna which is driven directly by the transmission line. **4.** A small auxiliary generator which supplies field current for an AC generator. **5.** Same as **exciter lamp (1)**. **6.** A probe or loop which projects into a resonant cavity or waveguide.

exciter lamp **1.** A high-intensity incandescent lamp utilized to focus on the optical soundtrack of a motion picture film. Also called **exciter (5)**. **2.** An incandescent lamp utilized to illuminate what is being scanned by a fax or similar device.

exciting current **1.** The current which flows through an **exciter (4)**. Also called **magnetizing current (2)**. **2.** A small current which flows through the primary winding of a transformer to which no load is connected. Also called **magnetizing current (3)**.

exciton **1.** In a semiconductor, a bound electron-hole pair. **2.** In a semiconductor or dielectric, an electron-hole pair which is a mobile concentration of energy, and which does not does not participate in electric conduction.

excitor Same as **exciter**.

excitron A single-anode mercury-pool tube which sustains a continuous cathode spot.

exclusion principle A principle that states that no two fermions can occupy the same quantum state. In the case of electrons, for instance, this means that two electrons in the same atom can not have the same values of all quantum numbers. This leads to there only being able to be two electrons, each with opposite spin, in the same atomic orbital. Also called **Pauli exclusion principle**, or **Pauli principle**.

exclusive NOR A logical operation which is true if all of its elements are the same. For example, if all of its multiple inputs have a value of 0, then the output is 1. And, if all of its inputs have a value of 1, then the output is 1. Any other combination yields an output of 0. For such functions, a 1 is considered as a true, or high, value, and 0 is a false, or low, value. Its abbreviation is **XNOR**. Also called **exclusive NOR operation**.

exclusive NOR circuit Same as **exclusive NOR gate**.

exclusive NOR gate A circuit which has two or more inputs, and whose output is high only if all its inputs are the same. Its abbreviation is **XNOR gate**. Also called **exclusive NOR circuit**.

exclusive NOR operation Same as **exclusive NOR**.

exclusive OR A logical operation which is false if all of its elements are the same. For example, if all of its multiple inputs have a value of 0, then the output is 0. And, if all of its inputs have a value of 1, then the output is 0. Any other combination yields an output of 1. For such functions, a 1 is considered as a true, or high, value, and 0 is a false, or low, value. Its abbreviation is **XOR**. Also called **exclusive OR operation**.

exclusive OR circuit Same as **exclusive OR gate**.

exclusive OR gate A circuit which has two or more inputs, and whose output is low if all of its inputs are the same. Its abbreviation is **XOR gate**. Also called **exclusive OR circuit**.

exclusive OR operation Same as **exclusive OR**.

excursion **1.** For a movable or oscillatory body, a movement from a rest or average position, especially outward and back. For instance, the piston-like movement of the cone of a speaker. Also, the distance traveled in such a movement. **2.** The maximum possible interval or movement in an **excursion (1)**.

executable **1.** That which can be executed. **2.** Same as **executable program**.

executable file A file which is in a format the computer can directly execute. Usually refers to a program.

executable program A computer program which can be run. Such a programs is translated into machine code, enabling it to be executed by the computer it is run on. Also called **executable (2)**.

execute **1.** To carry out, or put into effect, especially according to a given set of requirements. **2.** To run a computer program. Also, to carry out other computer processes, such as commands.

execution **1.** The action or process of carrying out, or putting into effect, especially according to a given set of requirements. **2.** The running of a computer program. Also, the carrying out of other computer processes, such as commands.

execution time **1.** In computers, the time required for the execution of a single instruction or operation. Also, the time required to complete a series of instructions or operations. **2.** The time required for **execution**.

executive The software which runs all the software and hardware of a computer. It is the first program the computer loads when powered on, remains memory-resident, and continuously controls and allocates all resources. Without it, for example, a computer can not recognize input, such as that from a keyboard, it can not process, as the executive controls the use of the CPU, nor can it provide an output, as it manages all peripherals including the monitor. Also called **operating system**, or **supervisor**. When disk-based, also called **disk operating system**.

Executive Information System An information system which provides access to data pertaining to one or more areas of a given business. It should be able to provide the desired data from internal and external sources using a straightforward format, with the presented information preferably in graphical form. Its abbreviation is **EIS**.

exhaust **1.** Same as **evacuate (1)**. **2.** To consume entirely. For instance, to completely drain a battery.

exhaustion **1.** Same as **evacuation**. **2.** The process or action of consuming entirely. For instance, the complete draining of a battery.

exit **1.** In computers, to get out of an operation, routine, or program. For instance, to exit a calling routine and return control to the program that initiated the call. **2.** A point at which an **exit (1)** may, or does take place.

exit point Within a subroutine or program, the last instruction.

exosphere The outermost region of a planet's atmosphere. In the case of the earth, it begins at approximately 500 kilometers above the surface, and within it are contained the Van Allen radiation belts.

exothermic Pertaining to a reaction or process in which heat is liberated to the surroundings. This contrasts with **endothermic**, where heat is absorbed from the surroundings.

exothermic process A process in which heat is liberated to the surroundings.

exothermic reaction A reaction in which heat is liberated to the surroundings.

expand To increase the extent, size, volume, quantity, number, or scope. For instance, to increase the memory or storage of a computer, to restore the dynamic range of a compressed signal, or to widen an electron beam.

expandable Capable of being expanded. For instance, a computer which provides expansion slots.

expanded memory Computer memory whose use is defined by the **Expanded Memory Specification**.

Expanded Memory Manager A memory manager implementing the **Expanded Memory Specification**. Its abbreviation is **EMM**.

Expanded Memory Specification In some older computer systems, a technique for the utilization of more than 1 megabyte of memory. Its abbreviation is **EMS**.

expanded sweep In an oscilloscope, the acceleration of the deflection of the electron beam during a given interval of the sweep.

expander 1. That which serves to **expand**. 2. A circuit or device which automatically increases the volume range of an audio-frequency signal. It works, essentially, to reverse the effect of volume compression, thus restoring the volume range of the original program. Also called **volume expander**. 3. A circuit or device whose output has a greater amplitude range than its input.

expansion 1. An increase in extent, size, volume, quantity, number, or scope. Also, the extent of such an increase. 2. The restoration of the space and/or bandwidth of data after compression. 3. The increase of the volume range of an audio-frequency signal, to reverse the effect of volume compression. 4. The restoration of the gain of a signal, so that so that the higher magnitude levels have a higher effective gain than the lower magnitude levels, reversing the effects of compression. 5. In communications, a circuit with more outputs than inputs.

expansion board Same as **expansion card**.

expansion bus A computer bus to which **expansion cards** connect. When such a card is properly installed in an expansion slot, it is connected to said bus. Examples include PCI and EISA buses.

expansion card A circuit board which is plugged into the bus or an expansion slot of a computer, in order to add functions or resources. Common examples include graphics accelerators, sound cards, and Ethernet adapters. Also called by various other terms, including **expansion board, accessory card, add-on board, adapter card, card** (1), **adapter** (3), and **add-on**.

expansion slot A slot within a computer, into which **expansion cards** are plugged in. When such a card is properly installed, it is connected to the expansion bus. Also called **slot** (3).

expected life The period of time during which a component, circuit, device, piece of equipment, system, apparatus, material, or other item is expected to operate, function, or otherwise be of use. This will depend on various factors, including the environmental conditions surrounding said use. Also called **useful life**.

expendable 1. A component, device, piece of equipment, or system which is designed to be replaced once performance

drops below a given threshold, as opposed to being repaired. 2. That which is naturally consumed during operation.

experiment One or more procedures which are undertaken under controlled conditions, and which serve for research, testing, proving or disproving a hypothesis, or for demonstration.

experimental 1. Pertaining to, or based on **experiments**. 2. Pertaining to that which is still under investigation, evolving, or otherwise requiring further testing and/or research.

experimental conditions Controlled conditions, such as those in a laboratory, utilized for research, tests, proving or disproving hypotheses, or for demonstrations.

experimental data Data collected through experimentation.

experimental design A design which is still under investigation, evolving, or otherwise requiring further testing and/or research.

experimental device A device which is still under investigation, evolving, or otherwise requiring further testing and/or research.

experimental equipment Equipment which is still under investigation, evolving, or otherwise requiring further testing and/or research.

experimental station In communications, a station which utilizes radio-frequency waves for experimentation. Used, for instance, in the development of new techniques and equipment.

experimental system A system which is still under investigation, evolving, or otherwise requiring further testing and/or research.

experimental technology Technology which is under investigation, evolving, or otherwise requiring further testing and/or research

experimentation The conducing of **experiments**.

expert system Its abbreviation is **ES**. 1. An application of artificial intelligence which incorporates a knowledge base and an inference engine. It assists users in problem solving within its knowledge base, and is meant be the equivalent of a human expert, though a well-designed expert system can amply exceed such capabilities. Some such systems incorporate new knowledge with problem-solving experience. Used, for instance, to assist in medical diagnosis, weather projections, and so on. 2. A computer system incorporating an **expert system** (1).

expiration date The date after which a program, component, code, or data is not available, ceases to function, or is no longer valid. For instance, after this date trail software may no longer work, or a cookie or key may expire.

exploded view An illustration, such as a diagram, which shows the components of a structure slightly separated from each other, but drawn in relation to each other and to the whole.

exploring coil A small coil which is inserted into a magnetic field to measure its field strength, or any variations in it. Such a coil is connected to an indicating instrument, such as a ballistic galvanometer or a fluxmeter, and may or may not incorporate an amplifier. An exploring coil may also be utilized to examine the magnetic flux distribution of a magnetic field. Also called **search coil, flip coil, magnetic test coil, magnetic probe**, or **pickup coil**.

explosion A reaction, such as that of a chemical or nuclear nature, in which a large amount of energy is released in a very short time interval, and which is usually accompanied by high temperatures and the release of gases.

explosion-proof Also spelled **explosionproof**. 1. A component, circuit, device, piece of equipment, or system designed so that none of its processes or ensuing results, such as sparking or heating, causes any nearby material to explode. For instance, a device that does not emit sparks, which could in turn make a surrounding gas explode. **2.** Able to withstand a nearby explosion and continue functioning within specified parameters. **3.** That which is contained in an **explosion-proof enclosure** (2).

explosion-proof enclosure 1. An enclosure which is designed so that none of the processes or ensuing results of its contained components, such as sparking or heating, causes any nearby material to explode. **2.** An enclosure which hermetically seals the inside from the outside. In this manner, contained components could emit sparks or radiate heat, but these would not be in contact with the surrounding environment. **3.** An enclosure which is designed to withstand an internal explosion, without provoking any nearby material to explode. **4.** An enclosure which is able to withstand a nearby explosion and continue functioning within specified parameters.

explosion-proof equipment Equipment which is **explosion-proof**.

explosion-proof lighting Lighting which is **explosion-proof**.

explosion-proof motor A motor which is **explosion-proof**.

explosion-proof switch A switch that is **explosion-proof**.

explosionproof Same as **explosion-proof**.

exponent A number indicating the power to which another number, the base, is raised. For example, in 2^8, the 8 is the exponent, and the 2 is the base. Also called **power** (3).

exponential 1. Pertaining to, or involving **exponents**. **2.** Pertaining to, or expressed by an **exponential function**. **3.** Same as **exponential function**.

exponential amplifier An amplifier whose output signal is an **exponential function** of its input signal.

exponential change An increase or decrease in a quantity, such as radiation or charge, whose rate of change follows an **exponential function**.

exponential curve A curve or graph representing an **exponential function**.

exponential decay A decrease in a quantity, such as radiation or charge, whose rate of change follows an **exponential function**. Also called **exponential decrease**.

exponential decrease Same as **exponential decay**.

exponential function Also called **exponential** (3). Such functions are frequently utilized to quantify physical phenomena. **1.** A function which varies as the power of another quantity. Such a function may be expressed, for instance, as $y = a^x$, where a is a constant. **2.** The function $y = e^x$, where e is the base of natural logarithms. An alternate expression for this function is: $y = \exp(x)$ where x is the number to which e is raised. It is the inverse function of the **natural logarithm function**.

exponential growth An increase in a quantity, such as radiation or charge, whose rate of change follows an **exponential function**. Also called **exponential increase**, or **logarithmic growth**.

exponential horn A horn speaker whose cross-sectional area flares out exponentially as the axial distance increases. Usually utilized to reproduce high frequencies.

exponential increase Same as **exponential growth**.

exponential notation A numeric format in which each number is represented by two numbers, and in which the decimal point is not in a fixed location. The first number, the mantissa, specifies the significant digits, while the second number, or exponent, specifies its magnitude. For example, 314,000,000 may be expressed as 3.14×10^8. Although any number, with any radix, may be represented in this manner, it is usually used only for very small or very large numbers. Also called **scientific notation**, or **floating-point notation**. When an E is used instead of a 10, or the radix utilized, it is also called **E notation**.

exponential transmission line A transmission line whose characteristic impedance varies exponentially as a function of its electric length.

exponential waveform A waveform whose rate of change follows an **exponential function**.

export To format, move, or save a file or data in a manner that it can be fully utilized by another application, or within another environment. For example, a given word processor application saving a file in a format which can be used by a given spreadsheet program. This contrasts with **import**, in which the conversion takes place after the data has been received in foreign format.

export filter A filter utilized to **export**.

exposed 1. Unshielded, not insulated, or otherwise unprotected or vulnerable. **2.** Open to view, or not concealed.

exposure 1. The condition or state of being **exposed**. For instance, to light, radiation, or a high voltage. **2.** An act or circumstance which results in being exposed, or exposing something else. **3.** The quantity or extent of **exposure** (1).

exposure meter A device which measures exposure to light or radiation.

exposure time The time a body or material is exposed to light or radiation.

expression 1. In computer programming, a segment of program code which when executed returns a value. Such an expression may include constants, variables, operators, operands, functions, or the like. A Boolean expression is an example. **2.** In mathematics, a combination of numbers, variables, and operations. Also called **arithmetic expression**.

extended addressing The use of additional bits for addressing. May be used, for instance, when added addressing information is required for direct access to memory beyond a given limit. Its abbreviation is **EA**.

extended ASCII An ASCII character set with additional assigned codes for foreign languages, graphics, and other special characters. Extended ASCII is an 8-bit code, providing a total of 256 characters, as compared to the standard 7-bit code providing 128. Extended ASCII numeric values span from 128 to 255, and the specific characters assigned to each number will depend on the systems, programs, and fonts used. Also called **extended ASCII character set**, **extended ASCII set**, or **8-bit ASCII**.

extended ASCII character A character within an **extended ASCII character set**.

extended ASCII character set Same as **extended ASCII**.

extended ASCII code The standard code utilized to express **extended ASCII** text and control characters.

extended ASCII set Same as **extended ASCII**.

Extended Binary Coded Decimal Interchange Code Same as **EBCDIC**.

Extended Capabilities Port A parallel port standard supporting high-speed bi-directional communications between a computer and a peripheral. It is compatible with a Centronics parallel port, but is much faster, and allows for a longer cable to be used. Its abbreviation is **ECP**. Also called **Enhanced Capabilities Port**.

extended data out dynamic random-access memory Same as **EDO DRAM**.

extended data out random-access memory Same as **EDO DRAM**.

extended data output dynamic random-access memory Same as **EDO DRAM**.

extended data output random-access memory Same as **EDO DRAM**.

Extended Graphics Array An enhanced VGA standard supporting resolutions up to 1024 x 768. Its abbreviation is **XGA**.

Extended Industry Standard Architecture Same as **EISA**.

extended memory In certain computer systems, memory beyond 1 megabyte.

extended play Playback time which is longer than ordinarily available. Also refers to storage capacity which is greater than usual. In the case of videocassette recorders, it also refers to recording time which is longer than ordinarily available. Its abbreviation is **EP**.

extended-range communication Communication by means of radio waves that are propagated well beyond line-of-sight distances. This is usually due to scattering by the ionosphere or troposphere. Also called **over-the-horizon communication**, **forward-scatter communication**, **scatter communication**, or **beyond-the-horizon communication**.

extended-range propagation Propagation of radio waves well beyond line-of-sight distances. This is usually due to scattering by the ionosphere or troposphere. Also called **over-the-horizon propagation**, **forward-scatter propagation**, **scatter propagation**, or **beyond-the-horizon propagation**.

extended-range transmission Transmission of radio waves well beyond line-of-sight distances. This is usually due to scattering by the ionosphere or troposphere. Also called **over-the-horizon transmission**, **forward-scatter transmission**, **scatter transmission**, or **beyond-the-horizon transmission**.

extended superframe A T-carrier framing format or standard with enhanced features, such as less-frequent synchronization, and real-time monitoring of the line. It encompasses 24 DS1 frames, while a **superframe** assembles 12. Its abbreviation is **ESF**. Also called **extended superframe format**.

extended superframe format Same as **extended superframe**. Its abbreviation is **ESF format**.

Extended TACACS Abbreviation of **Extended T**erminal **A**ccess **C**ontroller **A**ccess **C**ontrol **S**ystem. An enhanced version of **TACACS**, which provides additional support for auditing and accounting. Its abbreviation is **XTACACS**.

Extended Terminal Access Controller Access Control System Same as **Extended TACACS**.

Extended VGA Abbreviation of **Extended V**ideo **G**raphics **A**rray. Any of various graphics standards which provide for higher resolution than VGA. SVGA resolutions range from 800 x 600 pixels, to 1600 x 1200 pixels, with support for over 16.7 million colors. Also called **SVGA**.

Extended Video Graphics Array Same as **Extended VGA**.

extensible 1. That which can be extended or enhanced. For example, a computer system which provides expansion slots, enabling it to incorporate additional peripherals. 2. That which can be extended or protruded. For instance, a panel which can project out, facilitating its use.

Extensible HTML Abbreviation of **Extensible H**ypertext **M**arkup **L**anguage. A markup language that combines HTML and XML, and which features greater portability and ease of extension and enhancement. Its own abbreviation is **XHTML**.

Extensible Hypertext Markup Language Same as **Extensible HTML**.

Extensible Markup Language A specification for the format of documents and data to be used on the Web. It is a scaled-down version of SGML, and seeks to retain the comparative simplicity of HTML, yet offer greater flexibility in areas such as organization and presentation, while being fully compatible with both. Its abbreviation is **XML**.

Extensible Stylesheet Language Within **Extensible Markup Language**, a standard defining stylesheets. Its abbreviation is **XSL**.

Extensible Stylesheet Language Transformations A language utilized for converting XML documents into other XML documents with different structures. Its abbreviation is **XSLT**.

extension 1. An enlargement in scope, or in length of time. For instance, a plug-in which adds functionality to an application. 2. A set of characters appearing at the end of a filename, indicating the file type. For example, in the filename *notepen.exe*, the *.exe* portion is the extension, and in this case specifies an executable program. Also called **filename extension**. 3. A telephone, modem, or similar device connected to a main line, such as a PBX.

extensometer An instrument which measures small variations in the dimensions of a solid. For instance, it may measure deformation due to stress. An example is a laser extensometer.

extent In a direct-access storage device, such as a hard disk, a contiguous block reserved for a specific data set or program.

Exterior Gateway Protocol A routing protocol used between autonomous systems on the Internet. Its abbreviation is **EGP**.

external bus A bus between a CPU and peripherals. A PCI bus is an example. An **internal bus** runs between a CPU and memory.

external cache A memory cache which utilizes a memory bank between the CPU and main memory. This contrasts with **internal cache**, which is built right into the CPU. Internal cache is faster, but smaller than external cache, while external cache is still faster than main memory. Also called **level 2 cache**, or **L2 cache**.

external capacitor A capacitor which is externally connected to an oscillator, to vary its frequency.

external circuit A circuit, or part of a circuit, which is outside of a main circuit, or any other circuit in question. For instance, that which is externally connected to a battery, including its load.

external device A device which is not part of a central system, but which is utilized with it. For instance, a peripheral, such as a mouse or printer, used with a computer system.

external drive A drive, such as a disk drive or tape drive, that is not located within the system unit of a computer, as opposed to an **internal drive**, which is.

external electric field 1. An electric field other than that being considered. 2. An electric field affecting a given particle or body situated in the medium surrounding said field.

external error An error caused by an **external device**.

external feedback Feedback provided by an **external circuit**.

external field An **external electric field**, or an **external magnetic field**.

external interrupt In a computer, an interrupt generated by an **external device**. This contrasts with an **internal interrupt**, which is generated by the CPU.

external magnetic field 1. A magnetic field other than that being considered. 2. A magnetic field affecting a given particle or body situated in the medium surrounding said field.

external memory A deprecated term for **external storage**.

external modem A modem which is self-contained, and which usually connects to a computer via a cable to a serial port. This contrasts with an **internal modem**, which is plugged into an expansion slot.

external photoelectric effect The ejection of electrons from a surface through the absorption of sufficient incident electromagnetic radiation, such as infrared, visible, or ultraviolet radiation. Said surface is usually a solid, but may be a liquid. Also called **photoemission**.

external power supply A power supply which is physically separated from the device, piece of equipment, or system it is powering. For instance, a robot requiring multiple energy sources, each with different voltages, may be powered by such a supply, thus eliminating the need to convert any voltages within the robot itself.

external Q In a microwave tube, the reciprocal of the difference between the reciprocal of the unloaded Q value and the reciprocal of the loaded Q value.

external reference Within a computer program or routine, a call to a separate program or routine.

external sensor A sensor which is not physically a part of the device or apparatus sampling a given environment. For instance, a remote sensing device which transmits information to a control system.

external storage Also called **external memory**, though this latter term is deprecated. **1.** Computer storage which is removable. For instance, a tape cartridge. **2.** Computer storage which is physically separate from the computer accessing it.

external triggering **1.** The use of external pulses as the triggering signals of an oscilloscope. **2.** The use of external pulses as triggering signals.

external viewer A separate computer application which is called upon to interpret and view files and data which could otherwise not be seen by the current application. Also refers to multimedia files. For instance, a Web browser plug-in utilized to display video in a format that is not currently supported. Also called **viewer**.

extinction coefficient. Also called **absorption ratio, absorption factor, absorptive power**, or **absorptivity**. **1.** The proportion of certain wavelengths or energy retained by a medium to which another has been exposed. **2.** The proportion of certain wavelengths or energy retained by a medium to which another has been exposed, per unit of area or thickness.

extinction potential In a gas-filled tube, the potential at which the ionization of the contained gas ceases, and conduction stops. Also called **deionization potential**.

extinction voltage In a gas-filled tube, the voltage at which the ionization of the contained gas ceases, and conduction stops. Also called **deionization voltage**.

extra-high tension Same as **extra-high voltage**. Its abbreviation is **EHT**.

extra-high voltage A voltage greater than a given amount, such as 240,000 volts or 345,000 volts. Usually used in the context of electric power transmission. Its abbreviation is **EHV**. Also called **extra-high tension**.

extract **1.** To remove via a physical or chemical process. For instance, to utilize electrowinning to extract a metal from its molten ore. **2.** To remove one or more components from a complex signal. For example, in color TV, to extract the chrominance components from the chrominance signal. **3.** In computers, to remove items from a larger group, such as components of expressions, or to remove bits or characters from words. **4.** To decompress data, files, or the like.

extract instruction A computer instruction to remove components from expressions to form new ones, or to remove bits or characters from words to form new ones.

extractor That which serves to **extract**.

extraneous **1.** Coming from the outside, and often undesired. For instance, extraneous interference. **2.** Not forming an indispensable part, or unrelated to the central or intended function. Also called **extrinsic (2)**.

extraneous emission An emission from a transmitter which contains undesired components. For example, an output other than the intended carrier and sidebands.

extraneous response The undesired response of a receiver or recorder, due to an undesired signal, or the combination of desired and undesired signals.

extraneous signal A transmitted or received signal other than that intended. Such a signal may cause interference.

extranet Portions of an intranet which are available to authorized outsiders. For instance, a business which allows existing customers to use the Internet to access certain areas, such as those designated for product information and ordering.

extraordinary component Same as **extraordinary wave**.

extraordinary ray One of the two rays into which a ray of light incident upon a doubly-refracting material is split, the other being the **ordinary ray**. Both rays are plane-polarized perpendicular to each other, and the extraordinary ray passes through the material, while the ordinary ray is completely absorbed. Its abbreviation is **E-ray**.

extraordinary wave One of the two components into which a radio wave entering the ionosphere is split, under the combined influence of the earth's magnetic field and atmospheric ionization, the other being the **ordinary wave**. The electric vector of the extraordinary wave rotates in the sense opposite of that of the ordinary wave. Its abbreviation is **X-wave**. Also called **extraordinary component, extraordinary-wave component**.

extraordinary-wave component Same as **extraordinary wave**.

extrapolation A technique for obtaining an approximation of an unknown value, based on values already known. May be used, for instance, to estimate the value of a variable which is outside of, or between known ranges.

extraterrestrial radiation Electromagnetic radiation from sources not related to the earth. The chief component is solar radiation.

extreme The highest or lowest value, or that furthest from a center or reference value. Used, for instance, to describe the highest or lowest indications of a measuring instrument.

extreme ultraviolet Its abbreviation is **extreme UV**. **1.** Pertaining to **extreme-ultraviolet radiation**. **2.** Same as **extreme-ultraviolet radiation**.

extreme-ultraviolet imaging telescope An instrument which detects emitted and reflected ultraviolet radiation outside the earth's atmosphere, to obtain images of interstellar matter and celestial bodies such as a planets or stars, in addition to other information, such as that pertaining to their compositions and densities. Its abbreviation is **EIT**. Also called **ultraviolet imaging telescope**.

extreme-ultraviolet radiation Electromagnetic radiation in the ultraviolet region which is nearest to X-rays. It corresponds to wavelengths of approximately 4 to 200 nm, though the defined interval varies. Also called **extreme ultraviolet (2)**, or **vacuum ultraviolet**.

extreme UV Abbreviation of **extreme ultraviolet**.

extremely-hard vacuum Same as **extremely-high vacuum**.

extremely high frequency A range of radio frequencies spanning from 30 to 300 GHz, corresponding to wavelengths of 1 cm to 1 mm, respectively. Also, pertaining to this interval of frequencies. Its abbreviation is **EHF**.

extremely-high vacuum An environment or system whose pressure is below 10^{-12} torr. One torr equals 1 millimeter of mercury, or approximately 133.3 pascals. Also called **extremely-hard vacuum**.

extremely low frequency Waves whose frequencies are usually defined to be between 3 and 30 Hz, corresponding to wavelengths of between 100,000 and 10,000 kilometers, respectively. Its abbreviation is **ELF**.

extrinsic **1.** Coming from the outside, and often incorporated for a specific purpose or effect. For instance, extrinsic properties. **2.** Same as **extraneous (2)**.

extrinsic conductivity In a pure semiconductor material, the conductivity provided by imperfections in the crystal and intentionally-introduced impurities, as opposed to **intrinsic conductivity**, which is that inherently present.

extrinsic photoconductivity For a given material, photoconductivity due to the addition of impurities, or to external causes.

extrinsic properties **1.** Properties of a component, circuit, device, piece of equipment, or system which are **extrinsic**. **2.** Electrical characteristics of a semiconductor material which are determined by imperfections in the crystal and intentionally-introduced impurities. This contrasts with **intrinsic properties**, which are those present in the pure crystal.

extrinsic semiconductor A semiconductor material whose intentionally-introduced impurities, called dopants, determine its electrical characteristics, such as concentration of charge carriers. This contrasts with an **intrinsic semiconductor**, which has no dopants added. Also called **extrinsic semiconductor material**.

extrinsic semiconductor material Same as **extrinsic semiconductor**.

extrusion The give a desired shape and/or finish to a molten or semisolid material, by forcing it through a die. Utilized, for instance, to form rods or tubes.

eye candy Images, such as flashy graphics, whose primary purpose is to gratuitously please a user visually, without providing anything useful, such as functionality or information. Often utilized as a marketing scheme, and may be seen, for instance, in certain Web sites.

eye-in-hand system A robot which incorporates a close-range camera in an end-effector. Such an arrangement is useful in ensuring that a gripper properly identifies and grasps objects, or may navigate collision-free through tight and/or unknown environments.

ezine Abbreviation of **electronic magazine**.

F

f Symbol for **femto-**.

f Symbol for **frequency**.

F 1. Symbol for **farad**. 2. Symbol for **fluorine**.

F 1. Symbol for **magnetomotive force**. 2. Symbol for **Faraday constant**. 3. Symbol for **force**.

F connector A common coaxial cable connector which requires screw-on attachment. Used, for instance, to connect antennas, TVs, and VCRs.

F display In radars, a rectangular display where the scanned object appears as a centralized blip when the antenna is aimed directly at it. Horizontal and vertical displacement of the blip away from the center of the display indicates the azimuth and elevation errors respectively. Also called **F scan**, **F scanner**, **F scope**, or **F indicator**.

F indicator Same as **F display**.

F keys Abbreviation of **function keys**.

F layer Within the ionosphere, the combined F_1 and F_2 layers. During the day it consists of both, while at night it consists only of the F_2 layer.

F region Within the ionosphere, the region where the F_1 and F_2 layers form.

F scan Same as **F display**.

F scanner Same as **F display**.

F scope Same as **F display**.

F/V converter Abbreviation of **frequency-to-voltage converter**.

F- Symbol for the negative terminal of an A battery.

F+ Symbol for the positive terminal of an A battery.

F_1 layer Within the ionosphere, a layer that spans an altitude of about 150 to 300 kilometers, and which exists only during daytime. It is contained in the F region, and is useful for medium-distance propagation of medium to high-frequency radio waves during the hours of daylight.

F_2 layer The highest layer in the ionosphere, spanning an altitude of about 200 to 400 kilometers. Although this layer is influenced by factors such as time of day, season, and sunspot activity, it exists at all times, and at night it is the only component of the F region. It is useful for long-distance propagation of high-frequency radio waves day or night. Also known as **Appleton layer**.

fA Abbreviation of **femtoampere**.

Fabry-Perot cavity An optical resonator utilizing two parallel silvered surfaces. Used, for instance, as a laser resonator.

Fabry-Perot interferometer A high-resolution interferometer in which two closely-spaced silvered surfaces repeatedly reflect the incoming light waves, before ultimately transmitting them. Used, for instance, as a laser resonator, and in spectroscopy.

Fabry-Perot laser A laser with a **Fabry-Perot cavity**. The wavelength of the laser is determined by the distance between the reflecting surfaces within the cavity. An example is a standard diode laser.

face 1. A plane surface that bounds a geometric solid. For instance, such a smooth and flat surface of a crystal. 2. The front surface of something. For example, the viewing surface of a dial. 3. Same as **faceplate (1)**.

face plate Same as **faceplate**.

face recognition A biometric technique in which characteristics of a person's face are analyzed for identification purposes. Factors such as the shape of the head and blood flow under the skin may be taken into account.

faceplate Also spelled **face plate**. 1. A transparent or semi-transparent glass surface on the front of a CRT, through which images are seen. Also called **face (3)**. 2. A plate serving to protect a component, device, fixture, or piece of equipment.

facilities 1. A structure, such as a central office, serving a specified purpose. 2. The full complement of resources available to help perform the tasks necessary to reach a given objective.

facsimile Same as **fax**.

facsimile communications Same as **fax communications**.

facsimile machine Same as **fax machine**.

facsimile polling Same as **fax polling**.

facsimile receiver Same as **fax receiver**.

facsimile recorder Same as **fax recorder**.

facsimile signal Same as **fax signal**.

facsimile system Same as **fax system**.

facsimile transceiver Same as **fax transceiver**.

facsimile transmission Same as **fax transmission**.

facsimile transmitter Same as **fax transmitter**.

fade 1. To gradually lose strength, or to slowly disappear. Said for instance, of a signal, image, or sound which progressively diminishes. 2. Same as **fading (2)**.

fade-in 1. The gradual appearance of images and/or sounds, such as might occur at the beginning of a movie or song. 2. A gradual increase in the strength of a signal.

fade margin In a radio system, an allowance made so that a signal can fade up to a given amount, while still maintaining overall performance at an acceptable level. For example, an RF signal may be attenuated by a given number of decibels, yet sustain a signal-to-noise ratio above a specified minimum. Also called **fading margin**.

fade-out Also spelled **fadeout**. 1. A gradual disappearance of images and/or sounds, such as might occur at the end of a movie or song. 2. A rapid and complete loss of electromagnetic skywave signals, due to substantially increased ionization in the ionosphere. This ionization is caused by increased solar noise, which itself is caused by solar storms. It may last from a few minutes to several hours. Also called **radio fadeout**, **Dellinger fadeout**, or **Dellinger effect**. 3. Same as **fading (2)**.

fadeout Same as **fade-out**.

fader 1. A circuit or device which enables gradually changing the strength of a signal. Used, for example, for fade-ins and fade-outs. 2. A circuit or device which enables gradually changing the strength of multiple signals. For instance, a parametric equalizer can increase or decrease the levels of various intervals within the audible range, to help provide the desired sound experience.

fading 1. A gradual loss of strength, or a slow disappearance. Said, for instance, of a received signal. 2. A periodic reduction, or a gradual change in the strength of a signal, due to fluctuations in the transmission medium. For example, a radio signal may vary in this manner due to atmospheric conditions, or to movements of a mobile receiver. Also called **fade (2)**, or **fade-out (3)**. 3. A feature of certain digital music players, such as CDs, in which there is a **fade-out (1)** followed by a **fade-in (1)** when changing tracks or locations on the recorded medium. This provides for a smoother transition which is easier on the listener.

fading margin Same as **fade margin**.

Fahnestock clip A flat spring-like terminal utilized to easily make temporary electrical connections.

Fahrenheit Pertaining to, or expressed by the **Fahrenheit temperature scale**.

Fahrenheit degree Its symbol is °F. The unit of temperature interval used in the **Fahrenheit temperature scale**.

Fahrenheit scale Same as **Fahrenheit temperature scale**.

Fahrenheit temperature scale A temperature scale based on the freezing and boiling points of water under certain conditions, with each having an assigned a value of 32 and 212 degrees, respectively. Its unit of temperature increment is the degree Fahrenheit, or °F. Absolute zero is approximately -459.67 °F. This temperature scale is rarely used outside of the United States. Also called **Fahrenheit scale**.

fail-safe Also spelled **failsafe**. **1.** A device, piece of equipment, or system which continues to function properly after the failure of one or more components. **2.** A device, piece of equipment, or system which continues to function safely after the failure of one or more components. **3.** A component, circuit, device, piece of equipment, or system whose failure causes no significant damage. **4.** A component, circuit, device, piece of equipment, or system designed to withstand most any circumstance without failing.

fail-safe system A system which is **fail-safe**.

fail-soft system A system whose function diminishes after the failure of one or more components. Such a system, for instance, may provide a few minutes of reduced function, so as to lessen the harm a failure might cause, or to give the opportunity for the problem to be resolved.

failback In a network consisting of two or more servers, the restoration of failed resources to the primary server, after **failover**.

failover In a network consisting of two or more servers, the immediate relocation of failed resources to a backup server, thus minimizing interruptions of service.

failsafe Same as **fail-safe**.

failure The condition in which a component, circuit, device, piece of equipment, system, or material no longer performs the function it was intended to, or is not able to do so at a level that equals or exceeds established minimums. Failures may be due to any of various factors, including inherent defects, misuse, wearing-out, or as a consequence of the failure of another component, circuit, device, piece of equipment, or system. Failure may be gradual or sudden, and may be partial, intermittent, or complete.

failure alarm A circuit or device which serves to warn of a **failure**, by means of a signal such as a light. Also, such a warning.

failure analysis **1.** The study of the causes that led to a **failure**. **2.** The calculation of a **failure rate (2)**.

failure criteria The rules which determine whether a failure has occurred or not. For instance, if the performance of a device no longer meets certain minimums, it may be considered to have failed.

failure detection The sensing of a failure in a component, circuit, device, piece of equipment, or system. Failure detection may also serve to indicate the failure mode.

failure detector A device or mechanism which serves to detect a **failure**, and to produce a failure alarm.

failure mechanism Same as **failure mode**.

failure mode The manner in which a failure occurs, factoring in the causes and the surrounding conditions at that time. Also called **failure mechanism**.

failure rate Also called **failure rate level (1)**. **1.** The number of failures per unit measure, such as time or cycles. **2.** The number of expected failures per unit measure, such as time or cycles.

failure rate level **1.** Same as **failure rate**. **2.** The maximum tolerable **failure rate**.

failure ratio **1.** The proportion of failures per unit measure, such as time or cycles. **2.** The proportion of expected failures per unit measure, such as time or cycles.

failure recovery The ability of a component, circuit, device, piece of equipment, or system to reestablish an acceptable operating state following a failure.

fall time **1.** The time that elapses for a given amount of decay to occur. For example, the time it takes a static charge to be reduced to a given percentage of its peak charge. Also called **decay time (1)**, or **storage time (4)**. **2.** The time required for the amplitude of a pulse to decrease from a given percentage of its peak amplitude, such as 90%, to another, such as 10%. Also called **pulse fall time**, or **decay time (2)**. **Rise time (2)** is the converse. **3.** The time required for the output of a circuit to change from a high level, or 1, to a low level, or 0.

false alarm **1.** An alarm signal which is generated without an alarm condition existing. Possible causes include environmental factors, malfunctions, or vandalism. A false alarm may also be intentional, as is the case during testing. **2.** Same as **false echo**.

false drop Within a data search, an item or result which is irrelevant or undesired. For example, a user utilizing a search engine to find information on speaker woofers, and retrieving information on dogs. Also called **false retrieval**.

false echo In radars, an erroneous detection of a scanned object. Also called **false alarm (2)**.

false retrieval Same as **false drop**.

family A group of things which are related. For instance, a line of computers all using the same CPU, or a group of chemical elements with similar properties.

fan **1.** A mechanical device, usually equipped with rotating blades, which is utilized to produce a current of air. Used, for instance, to cool components, or to help purify air. **2.** The volume of space energized by the beam of electromagnetic energy produced by a radar transmitter.

fan antenna A wideband antenna consisting of two dipole elements folded in the shape of a letter V which are mounted along the same plane. It is used, for instance, for the reception of FM broadcasts. Also called **double-V antenna**.

fan beam A radio beam, such as that produced by a radar transmitter, having an elliptical cross-section whose ratio of major to minor axes exceeds 3 to 1.

fan-in **1.** The maximum number of inputs that a logic circuit can accept. **2.** The maximum number of inputs that a circuit can accept. **3.** Multiple inputs that a circuit accepts.

fan-out **1.** The maximum number of outputs which can be fed by a logic circuit. **2.** The maximum number of outputs which can be fed by a circuit. **3.** Multiple outputs that a circuit feeds.

FAQ Abbreviation of Frequently Asked Questions. **1.** A collection of commonly-asked questions pertaining to any given topic or group of topics, followed by their answers. There are FAQs for most any product and/or service, and these are widely available over the Internet. **2.** A document or file containing **FAQ (1)**. Also called **FAQ file**.

FAQ file Abbreviation of Frequently Asked Questions file. Same as **FAQ (2)**.

far-end crosstalk Crosstalk which is propagated between circuits in which the interfering signal travels in the same direction as the desired signal. The terminal where far-end crosstalk is observed is usually distant from the point where it originates. This contrasts with **near-end crosstalk**, where each signal travels in the opposite direction, and is usually observed near or at its origination point. Its abbreviation is **FEXT**.

far field **1.** Same as **far-field region**. **2.** The electromagnetic field further than a given distance from a source of electromagnetic radiation. **3.** The sound field further than a certain distance to a sound source. May be, for instance, that extending beyond two wavelengths.

far-field diffraction pattern The diffraction pattern of a light source which is observed at an effectively infinite distance from the object diffracting it. The light source may be, for instance, the output of an optical fiber, or that of an LED. Also called **Fraunhofer diffraction pattern**.

far-field region The region in which the transmitted energy from an antenna behaves as if it were emanating from a point source in the vicinity of said antenna. Such a region usually begins at a distance of $2D^2/\lambda$ from the antenna, where D is the maximum overall dimension of said antenna, and λ is the wavelength considered. The region of space closer to the antenna than the far field is the **near-field region**. Also called **far field** (1), **far region**, **far zone**, **Fraunhofer region**, or **radiation zone**.

far infrared Its abbreviation is **far IR**. **1.** Pertaining to **far-infrared radiation**. **2.** Same as **far-infrared radiation**.

far-infrared radiation Electromagnetic radiation in the infrared region which is furthest from visible light. It corresponds to wavelengths of approximately 10 to 1000 micrometers, though the defined interval varies. Also called **far infrared** (2), or **long-wave infrared**.

far IR Abbreviation of **far infrared**.

far region Same as **far-field region**.

far ultraviolet Its abbreviation is **far UV**. **1.** Pertaining to **far-ultraviolet radiation**. **2.** Same as **far-ultraviolet radiation**.

far-ultraviolet radiation Electromagnetic radiation which is between near ultraviolet and extreme ultraviolet. It corresponds to wavelengths of approximately 200 to 300 nm, though the defined interval varies. Also called **far ultraviolet** (2), or **long-wave ultraviolet**.

far UV Abbreviation of far ultraviolet.

far zone Same as **far-field region**.

farad The SI unit of capacitance. It is equivalent to the capacitance of a capacitor in which a charge of 1 coulomb increases the potential difference between its plates by 1 volt. In practicality, this unit is so large that it is more common to use microfarads or picofarads. Its symbol is **F**.

faraday Same as **Faraday constant**.

Faraday cage An enclosure which blocks the effects of an electric field, while allowing free passage to magnetic fields. A Faraday cage usually consists of a network of parallel wires which provides a low-resistance path to ground. Such a network, for instance, may be incorporated into special plastic bags, providing safer storage and handling of circuit boards en route to being installed. Such a bag does not have to be grounded. Also called **Faraday shield**, **Faraday shielding** (1), **Faraday screen**, **Faraday electrostatic shield**, or **shielded room**.

Faraday constant Its symbol is F. A physical constant equal to approximately 9.64803×10^4 coulombs per mole. Its formula is: $F = N_A \cdot e$, where N_A is Avogadro's number, and e is the electron charge. It is equal to the electric charge carried by a mole of electrons, or by a mole of singly-charged ions. Also called **faraday**.

Faraday dark space In a glow-discharge tube, a dark space between the negative glow and the positive column.

Faraday effect The rotation of the plane of polarization of an electromagnetic wave as it passes through certain materials, when under the influence of a magnetic field parallel to the direction of propagation of said wave. Also called **Kundt effect**, or **magnetic rotation**.

Faraday electrostatic shield Same as **Faraday cage**.

Faraday rotation The rotation of the plane of polarization of an electromagnetic wave as a result of the **Faraday effect**.

Faraday screen Same as **Faraday cage**.

Faraday shield Same as **Faraday cage**.

Faraday shielding **1.** Same as **Faraday cage**. **2.** The use of a **Faraday cage**.

Faraday's law **1.** Either of **Faraday's laws of electrolysis**. **2.** Same as **Faraday's law of electromagnetic induction**.

Faraday's law of electromagnetic induction A law which states that the electromotive force induced in a conductor is proportional to the rate of change of magnetic flux through it. Also called **Faraday's law of induction**, or **law of induction**.

Faraday's law of induction Same as **Faraday's law of electromagnetic induction**.

Faraday's laws of electrolysis **1.** A law stating that during electrolysis, the amount of a substance deposited onto an electrode or dissolved into solution is directly proportional to the current passed. **2.** A law stating that the amount of a substance deposited onto an electrode or dissolved into solution is proportional to the equivalent weight of said substance.

faradic current An intermittent and nonsymmetrical AC, such as that produced by the secondary winding of an induction coil.

Farnsworth dissector tube Same as **Farnsworth image dissector**.

Farnsworth image dissector A camera tube in which the picture to be transmitted is focused upon a photosensitive surface from which electrons are emitted proportionally to the light of each part of the picture. These electrons are focused, in sequence, onto a controller electrode, which produces the output video signal. Also called **Farnsworth dissector tube**, **image dissector**, or **dissector tube**.

fast-access storage For a given computer system, the storage from which data can be retrieved most rapidly.

Fast ATA An enhanced version of the IDE interface. It supports both the ATA-2 and ATAPI interfaces, providing for significantly faster transfer rates when using hard disks, optical discs, or tape drives. In addition, it supports hard disks with much larger capacities. Also called **Fast IDE, ATA-2**, or **EIDE**.

fast-blow fuse A fuse which opens a circuit at a lower amperage than a normal-blow fuse would. Used, for instance, to protect especially delicate components, circuits, stages, devices, units, equipment, or systems.

fast break The opening of an electric circuit without a delay. This contrasts with a **delayed break**, where there is a specified delay.

fast busy signal A busy signal that indicates that the central office either did not understand the dialed digits, or that a part of the network is too busy to process the call. This contrasts with a **slow busy signal**, which indicates that the dialed number is currently in use.

fast charge The accelerated charging of a rechargeable battery, such as a lithium-ion battery.

fast charger A battery charger used for **fast charges**. Such chargers usually maintain a trickle charge once the full charge is reached.

fast Ethernet An Ethernet standard supporting data transfer rates of up to 100 Mbps. Depending on the specific configuration, it may utilize two or four twisted-pair copper wires, or fiber-optic cables. Also called **100Base-T**.

fast-forward Its abbreviation is **FF**. **1.** To advance a tape rapidly. For instance, than in a VHS cassette or digital audio tape cartridge. The term may also be used to refer to moving forward rapidly within the programming of an optical disc, such as a DVD. **2.** The mechanism utilized to **fast-forward** (1). **3.** A command, button, or function which **fast-forwards** (1).

fast-forward button A button which activates a **fast-forward** (1) function.

fast Fourier transform A set of algorithms utilized to rapidly compute a discrete Fourier transform, usually in real-time. Used, for instance, to analyze complex signals. Its abbreviation is **FFT**.

Fast IDE Same as **Fast ATA**.

fast infrared port A wireless port which utilizes infrared signals to send and/or receive data. Such ports usually require line-of-sight transmission and reception. May be used, for instance, between computers, between computers and peripherals, or between peripherals. Its abbreviation is **FIR port**. Also called **infrared port**.

fast make The closing of an electric circuit without a delay. This contrasts with a **delayed make**, where there is a specified delay.

fast packet Pertaining to a packet-switching technique, protocol, network, or technology, that increases throughput by reducing overhead. Examples include frame relay, cell relay, ATM, and BISDN. Also called **fast packet switching**.

fast packet switching Same as **fast packet**.

fast page mode memory Same as **fast page mode RAM**. Its abbreviation is **FPM memory**.

fast page mode RAM A form of dynamic RAM which enables faster access, as a row of memory bits only needs to be selected once for all columns to be accessed within said row. Other forms of dynamic RAM, such as SDRAM, have superseded fast page mode RAM. Its abbreviation is **FPM RAM**. Also called **fast page mode memory, page mode memory**, or **page mode RAM**.

fast-recovery diode A diode with an ultrahigh operational speed. Such diodes feature minimal carrier storage.

Fast SCSI A SCSI interface which utilizes an 8-bit bus, and supports data rates of up to 10 MBps. It uses a 50-pin cable.

fast-time constant A circuit which responds quickly to any changes in its input. Such circuits emphasize signals of short duration. In radars, for instance, these may be utilized to discriminate against low-frequency components of clutter. Also called **fast-time constant circuit**.

fast-time constant circuit Same as **fast-time constant**.

Fast Wide SCSI A SCSI interface which utilizes a 16-bit bus, and supports data rates of up to 20 MBps. It uses a 68-pin cable.

FAT Abbreviation of **file allocation table**.

fat client Within a client/server architecture, a client which performs most or all of the data processing, thus requiring more powerful computers at the front end. This contrasts with a **thin client**, where the client performs little processing. This also contrasts with **fat server**.

fat server Within a client/server architecture, a server which performs most or all of the data processing. This contrasts with a **thin server**, where the server performs little processing. This also contrasts with **fat client**.

fatal error An error which causes a complete and unexpected program or system halt. Usually caused by a program deficiency or a hardware failure. In most cases, the computer locks up, and any unsaved changes to files are lost.

fatal exception An error condition which causes a complete and unexpected program or system halt. Also, a message from the operating system indicating such a condition. Also called **fatal exception error**.

fatal exception error Same as **fatal exception**.

father file For a given file, such as that being updated, the copy or version saved before it. The current version is called **son file**. Also called **parent file**.

fathom **1.** A unit of distance defined to be 6 feet, or 1.8288 meters. **2.** To determine the depth of a body of water.

fathometer A device which determines and indicates the depth of a body of water, by measuring the time required for transmitted sonic or ultrasonic waves to be reflected from a surface, such as the sea bottom. Also called **echo sounder, depth sounder**, or **sonic-depth finder**.

fatigue **1.** The weakening, deterioration, or failure of a material such as a metal, when subjected to a repeated and/or cyclical stress. **2.** A reduction in the performance of a material over time. For instance, the gradual reduction in efficiency of a light-sensitive material.

fatigue failure Failure of a material, such as a metal, when subjected to a repeated and/or cyclical stress.

fatigue life The number of stress cycles a material, such as a metal, can withstand before failure.

fatigue limit The maximum stress which a material, such as a metal, can be subjected to an unlimited number of cycles without failure.

fault **1.** A defect in a component, circuit, device, piece of equipment, or system, which impairs operation significantly or that causes a failure. Also, a failure caused by such a defect. **2.** In a component or circuit, a defect such as a short circuit, an open circuit, or an unintentional ground. Also, a failure caused by such a defect. Also called **electrical fault**. **3.** A defective region in a component, circuit, or device. **4.** An error or defect in computer software or hardware. Said especially of one which causes a failure. When such a fault is persistent, also called **bug** (1).

fault condition A state or circumstance in which a **fault** has occurred.

fault containment **1.** Mechanisms and techniques employed to minimize the harm caused by a **fault**. **2.** Any harm prevented through the use of **fault containment** (1).

fault current A current flowing in a circuit due to an abnormal condition such as a leakage current, a surge, or a **fault** (2).

fault detection **1.** Mechanisms and techniques utilized to determine the occurrence of a **fault**. Examples include monitoring and periodic testing. **2.** The finding of faults utilizing **fault detection** (1).

fault diagnosis **1.** The process of determining the causes of a given **fault**. **2.** The process of locating a given fault, and determining its causes.

fault finder A device or instrument, such as a test set, which serves to locate **faults**. For instance, any such fault within a communications system.

fault finding The use of a **fault finder** to locate faults. Also called **fault localization**.

fault indication An indication, such as a signal, that informs of a **fault condition**. For example, a panel light blinking, or a message.

fault isolation The use of a **fault finder** to pinpoint the location of a given fault.

fault latency The time that elapses between a **fault** occurring, and its first manifestation.

fault localization Same as **fault finding**.

fault masking The concealment of one **fault** by another existing fault.

fault recovery **1.** Mechanisms and techniques employed to reestablish the desired level of operation of a component, circuit, device, piece of equipment, or system after a **fault** has occurred. **2.** The reestablishment of the desired level of operation of a component, circuit, device, piece of equipment, or system through the use of **fault recovery (1)**.

fault resilience The capability of a component, circuit, device, piece of equipment, or system to recover quickly from a **fault condition**.

fault resilient Characterized by **fault resilience**.

fault signal A signal that informs of a **fault condition**.

fault simulation The introduction of **faults** into a component, circuit, device, piece of equipment, or system, to evaluate its effects and determine how best to identify such faults when not occurring under controlled conditions. Computers are often utilized for such simulations.

fault-tolerance The ability of a component, circuit, device, piece of equipment, or system to continue to function at or above a certain level, regardless of catastrophic events or failures, such as power interruptions, failed components, and/or environmental extremes. For instance, a computer system or network which maintains optimum operation and suffers no data loss under such conditions. Backup power supplies, redundant components, and operating system resilience help in this example.

fault-tolerant system A system characterized by **fault-tolerance**.

favorites The bookmarks that have been saved by a Web browser.

FAX Same as **fax**.

fax Abbreviation of **facsimile**. Also called **telefax**, or **telefacsimile**. **1.** A method of transmitting a printed page between locations, via a telecommunications system. A telephone line is usually used. The document is scanned at the sending location, encoded, and transmitted by the sending device. The receiving device decodes the signal, then prints a copy of the original document, which may include text, graphics, and so on. Computers may also serve to transmit and/or receive such documents, without the need for hard copy at either end. **2.** Same as **fax machine**. **3.** One or more printed pages sent or received via **fax (1)**. **4.** To send or receive a **fax (3)**.

fax-back Same as **fax-on-demand**. Also spelled **faxback**.

fax board Same as **fax card**.

fax broadcast A feature enabling the sending of a single fax simultaneously to multiple recipients. Also called **broadcast fax**.

fax card An expansion card providing fax capabilities. An internal fax-data modem is an example. Also called **fax board**.

fax communications Communications via **fax (1)**.

fax-data modem A data modem that can also handle fax protocols, thus incorporating fax capabilities. Documents that are sent by such a device must be in an electronic form, such as a disk file. Received documents are also stored in electronic form. Also called **fax modem (2)**, or **data-fax modem**.

fax machine A device or piece of equipment utilized to send and/or receive a **fax (3)**. It incorporates a scanner, modem, and printer. Such devices usually include copying functions, and frequently can be used in conjunction with a computer, where they may add scanning abilities. A computer with a fax modem or fax-data modem may also serve to send and/or receive faxes. Also called **fax (2)**, **fax system (2)**, or **facsimile machine**.

fax modem 1. A modem incorporated into a **fax machine**. **2.** Same as **fax-data modem**

fax-on-demand An automated system for the request and reception of faxes. A request is made by telephone, the caller chooses the desired fax or faxes via voice menus, and then the selections are faxed to the calling number, or that provided. Also called **fax-back**.

fax polling A feature that allows a fax machine to call another, and request the remote machine to transmit documents. Also called **facsimile polling**, or **polling (3)**.

fax receiver A device which decodes a fax signal, then prints a copy of the original document transmitted via said signal. Also called **facsimile receiver**.

fax recorder A device which prints a copy of the original document sent by a fax transmitter. Also called **facsimile recorder**.

fax signal An encoded signal sent by a fax transmitter via a telecommunications system, which is then received and decoded by a fax receiver. Also called **facsimile signal**.

fax system 1. Two or more devices which when used together perform the function of a **fax machine**. For instance, a computer equipped with a scanner and fax modem can send and receive faxes. Also called **facsimile system**. **2.** Same as **fax machine**. **3.** A system for communication via **fax (1)**.

fax transceiver A device or piece of equipment which sends and receives fax signals. A full-duplex fax transceiver can send and receive simultaneously, while a half-duplex unit can only do one or the other at any given moment. Also called **facsimile transceiver**.

fax transmission The transmission of one or more printed pages via **fax (1)**. Also called **facsimile transmission**.

fax transmitter A device which scans and encodes a document, then transmits a fax signal via a telecommunications system. Also called **facsimile transmitter**.

faxback Same as **fax-back**.

fc Abbreviation of **foot-candle**.

fC Abbreviation of **femtocoulomb**.

FC-AL Abbreviation of **Fibre-Channel Arbitrated Loop**.

FC connector A rugged fiber-optic connector with a round threaded plug and socket.

FCAL Abbreviation of **Fibre-Channel Arbitrated Loop**.

FCB Abbreviation of **file control block**.

FCC Abbreviation of **Federal Communications Commission**.

FCIF Abbreviation of **Full Common Intermediate Format**.

FCS Abbreviation of **Frame Check Sequence**.

FDDI Abbreviation of **Fiber Distributed Data Interface**.

FDDI 2 Abbreviation of **Fiber Distributed Data Interface 2**.

FDDI II Abbreviation of **Fiber Distributed Data Interface II**.

FDM 1. Abbreviation of **frequency-division multiplexing**. **2.** Abbreviation of **frequency-division multiplex**.

FDMA Abbreviation of **frequency-division multiple access**.

FDTD Abbreviation of **Finite-Difference Time Domain**.

FDX Abbreviation of **full-duplex**.

FDX operation Abbreviation of **full-duplex operation**.

Fe Chemical symbol for **iron**.

FEA 1. Abbreviation of **field-emitter array**. **2.** Abbreviation of **Finite Element Analysis**.

feasibility study An analysis of an enterprise or project which will determine whether it is worth attempting. Factors such as the necessary technology, completion time, net cost, and expected benefits are taken into account. Used, for instance, when considering developing a computer application, a chip, and so on.

feature A characteristic which is meant to be useful, unique, or otherwise favorable. For instance, safety, convenience, or electrical characteristics which are especially suitable for a given task or function.

feature button A button on a telephone or console which is utilized to access a specific feature. Examples include buttons for features such as mute, speakerphone, speed dial, and flash.

feature code A sequence of digits which is dialed to activate a telephone feature or service. For instance, in many areas, dialing *69 activates call return.

feature creep The addition of features to an existing software product, just for the effect of having more features. This usually leads to an unnecessary toll on system resources, such as RAM and disk space. Also called **creeping featurism**.

feature phone A telephone which has additional features, such as caller ID, call forwarding, speed dial, and transferring of calls. Also called **feature telephone**.

feature telephone Same as **feature phone**.

FEC Abbreviation of **forward error correction**.

FECN Abbreviation of **Forward Explicit Congestion Notification**.

FED Abbreviation of **field-emission display**.

Federal Communications Commission A government agency which regulates interstate communications in the United States. Its abbreviation is **FCC**.

federated database A collection of autonomous, heterogeneous, and independent databases, whose integrated access enables collaboration between individuals and entities within a given realm of knowledge. In such an arrangement, there must be the proper balance between the ability to share data, and the preservation of the integrity of the participating databases. Also called **federated database system**.

federated database system Same as **federated database**.

feed 1. To provide power or a signal to a circuit, device, piece of equipment, or system. 2. The location at which power or a signal enters a circuit, device, piece of equipment, or system. 3. The power or signal which enters a circuit, device, piece of equipment, or system. 4. Same as **feeder** (1). 5. The transmission line, such as a coaxial cable, which carries signals between a transmitter and an antenna. Also called **feeder** (2), **feed line**, **antenna transmission line**, or **antenna feed**. 6. To provide a computer with data, or a medium which contains data. For example, to enter data, or to insert a diskette into a drive.

feed-forward control Same as **feedforward control**.

feed horn Same as **feedhorn**.

feed line Same as **feed** (5).

feed point Also spelled **feedpoint**. 1. The point at which a **feed** is received. 2. The point in an antenna where the **feed** (5) is received. Also called **antenna feed point**.

feed point impedance The impedance at a **feed point** (2). Also called **antenna feed point impedance**.

feedback 1. The return of a fraction of the output of a circuit, device, piece of equipment, or system, to the input of the same circuit, device, piece of equipment, or system. Feedback may be intentional, as in a closed-loop control system, or unintentional, as in howling. When the feedback is in phase with the original input signal, it is positive feedback. When the feedback is 180 degrees out-of-phase the original input signal, it is negative feedback. 2. The output returned to the input in **feedback** (1). 3. The mechanism employed to provide **feedback** (1). 4. The use of **feedback** (1) in a control system. Also, the output returned to the input in such a system.

feedback amplifier 1. An amplifier which employs positive and/or negative feedback to modify its performance. Such feedback may serve, for instance, to alter its frequency response. 2. A circuit or device which increases the amplitude of a feedback signal.

feedback attenuation 1. The reduction of the amplitude of a feedback signal. 2. The amount by which the amplitude of a feedback signal is reduced.

feedback circuit 1. A circuit which provides **feedback** (1). 2. A circuit which returns a fraction of its output to its input.

feedback compensation The incorporation of a compensator into a feedback path.

feedback control 1. The use of feedback to control a circuit, device, piece of equipment, system, or process. Utilized, for instance, in a feedback control system. 2. The control of the amount of feedback. This may be accomplished, for instance, with a self-regulating circuit.

feedback control loop In a control system, a closed path with two branches, one direct and the other providing feedback, with one or more mixing points arranged in a manner that an output quantity is maintained within specified parameters. Also called **feedback loop** (2).

feedback control system A control system in which a feedback signal and a reference input are compared in order to control the output quantity, so as to maintain operation within the specified parameters. For example, a thermostat may provide the feedback signal in a heating or refrigerating system. This contrasts with an **open-loop control system**, which has no feedback signal. Also called **closed-loop control system**.

feedback current For a circuit, device, piece if equipment, or system, the output current returned to the input in **feedback** (1).

feedback device 1. A device which provides feedback. 2. A device utilized to sense the orientation of a moving part of a robot, such as an arm or end-effector, and whose information is routed back to the controller.

feedback factor For a circuit, device, piece if equipment, or system, the portion of the output signal that is fed back to its input.

feedback loop 1. A circuit or loop which provides **feedback** (1). 2. Same as **feedback control loop**.

feedback oscillator An oscillator which incorporates an amplifier, and utilizes **feedback** (1). The oscillation frequency is determined by the amplifier components and the feedback.

feedback path The path, such as a circuit, through which feedback arrives. Such a path may or may not be intentional.

feedback regulator 1. A regulator which utilizes feedback. 2. A circuit or device which controls the amount of feedback.

feedback sensor 1. A sensor which provides feedback while it is sampling. 2. In robotics, a mechanism by which information obtained by the sensing devices is routed back to the controller. This can be used, for example, by a robot traversing irregular terrain. Also called **robot feedback sensor**.

feedback signal For a circuit, device, piece if equipment, or system, the output signal returned to the input in **feedback** (1).

feedback transfer function The transfer function of a **feedback path**.

feedback voltage For a circuit, device, piece if equipment, or system, the output voltage returned to the input in **feedback** (1).

feedback winding A winding in a magnetic amplifier, through which a feedback current is introduced.

feeder 1. A conductor or line which provides power or a signal to a circuit, device, piece of equipment, or system. Also called **feed** (4). 2. Same as **feed** (5). 3. A line which carries electric power from one point to another, especially when originating at a generating station or substation.

feeder cable 1. A cable, such as a coaxial cable, which provides power or a signal to a circuit, device, piece of equipment, or system. 2. In a cable TV system, the cable that leads from the headend to a distribution point. A distribution cable then leads to the individual subscribers. Also called **trunk cable** (1). 3. A cable running between a central office and a distribution point. A distribution cable then leads to the local stations and subscribers. Also called **trunk cable (2)**.

feeder line A line which provides power or a signal to a circuit, device, piece of equipment, or system.

feeder loss The reduction of the intensity of a signal traveling through a **feeder line**.

feedforward control A control method in which the correcting signal is applied before the output of a process. This is accomplished by detecting changes at the process input, and applying the appropriate signal. Also spelled **feed-forward control**.

feedhorn In a satellite receiving antenna, a component or device which collects the signal reflected from the convex reflector. This signal is channeled to the LNB. Also spelled **feed horn**.

feedpoint Same as **feed point**.

feedpoint impedance Same as **feed point impedance**.

feedthrough A conductor which makes electrical contact between patterns on both sides of a printed-circuit board or chassis. Also spelled **feedthru**. Also called **interface connection**.

feedthrough capacitor A capacitor that serves as a feedthrough terminal which provides a desired level of capacitance. Used, for instance, as a bypass capacitor in ultra-high frequency circuits.

feedthrough insulator An insulator which serves to pass a center conductor through a physical barrier such as a panel, chassis, or wall.

feedthrough terminal A terminal which serves to pass a center conductor through a physical barrier such as a panel, chassis, or wall.

feedthru Same as **feedthrough**.

feet Plural of **foot**. Its abbreviation is **ft**.

feet per second A unit of velocity which is usually utilized to measure the speed of sound. For example, in dry air at 0° Celsius, and at one atmosphere pressure, sound travels at approximately 1,088 feet per second, which is equal to approximately 331.6 meters per second. Its abbreviation is **fps**, **ft/s**, or **ft/sec**.

feeware Software that is sold. The term is utilized to contrast with **freeware**, which is not.

FEL Abbreviation of **free-electron laser**.

FEM 1. Abbreviation of **field-emission microscope**. 2. Abbreviation of **field-emission microscopy**.

female adapter An adapter with one or more recessed openings, into which another adapter with matching pins or prongs is inserted.

female connector A connector with one or more recessed openings, into which another connector with matching pins or prongs is inserted.

female jack A jack with one or more recessed openings, into which a plug with matching pins or prongs is inserted.

female plug A plug with one or more recessed openings, into which a jack with matching pins or prongs is inserted. Although a plug often refers to a male connector, it is not necessarily so.

femto- A metric prefix representing 10^{-15}. For instance, 1 femtoampere is equal to 10^{-15} ampere. Its abbreviation is **f**.

femtoampere 10^{-15} ampere. It is a unit of measurement of electric current. Its abbreviation is **fA**.

femtocoulomb 10^{-15} coulomb. It is a unit of electric charge. Its abbreviation is **fC**.

femtofarad 10^{-15} farad. It is a unit of capacitance. Its abbreviation is **fF**.

femtohenry 10^{-15} henry. It is a unit of inductance. Its abbreviation is **fH**.

femtometer 10^{-15} meter. It is a unit of measurement of length. Its abbreviation is **fm**.

femtosecond 10^{-15} second. It is a unit of time measurement. Its abbreviation is **fs**, or **fsec**.

femtosiemens 10^{-15} siemens. It is a unit of conductance. Its abbreviation is **fS**.

femtovolt 10^{-15} volt. It is a unit of measurement of potential difference. Its abbreviation is **fV**.

femtowatt 10^{-15} watt. It is a unit of measurement of power. Its abbreviation is **fW**.

FEP Abbreviation of **front-end processor**.

FEP resin Abbreviation of **fluorinated ethylene propylene resin**.

Fermat's principle A principle stating that an electromagnetic wave, such as a light ray, travels between points taking the path that requires the least time. This path may or may not be a straight line. Also called **least-time principle**.

fermi A unit of length equal to 10^{-15} meters, which is the same as 1 femtometer.

Fermi-Dirac distribution function A function describing the probability of an energy level being occupied by a fermion, such as an electron, under conditions of thermal equilibrium.

Fermi-Dirac statistics A system of quantum statistics applied to **fermions**.

Fermi energy Same as **Fermi level**.

Fermi level The energy level at which the **Fermi-Dirac distribution function** has a value of ½. Also called **Fermi energy**.

fermion A particle which has a fractional spin. A fermion can not exist in the same quantum state with another fermion. Examples include electrons, protons, and neutrons. Fermions obey the Pauli exclusion principle.

fermium A synthetic chemical element whose atomic number is 100. It has nearly 20 known isotopes, all of which are unstable. Used in tracer studies. Its chemical symbol is **Fm**.

ferric chloride A brown or black solid whose chemical formula is $FeCl_3$. It is used for etching printed circuits, ceramics, metals, and so on.

ferric oxide A dark red solid whose chemical formula is Fe_2O_3. Its applications in electronics include its use in semiconductors, resistors, and as a coating on magnetic tapes and disks.

ferrimagnetic Displaying **ferrimagnetism**.

ferrimagnetic material A material, such as magnetite, which displays **ferrimagnetism**.

ferrimagnetism A phenomenon observed in some materials, such as ferrites, in which the microscopic magnetic moments are aligned in an antiparallel manner, but are not of equal magnitude. Below a certain temperature, called the Néel temperature, such a material has magnetic properties similar to ferromagnetic materials, while above this temperature it becomes paramagnetic.

ferristor A type of saturable reactor which operates at a high frequency.

ferrite Any of various crystalline or powdered materials whose composition includes ferric oxide and one or more other metals or metal oxides, such as nickel or manganese oxide. Ferrites are often used in pressed and sintered form, and feature very low eddy-current losses at high frequencies. Its many applications in electronics include its use in magnetic cores for inductors and transformers, in semiconductors, in permanent magnets, dielectrics, antennas, and in components utilized in communications, radars, control systems, and so on.

ferrite antenna Same as **ferrite-rod antenna**.

ferrite bead A bead made from **ferrite**. Such beads are often made from powdered ferrite which is pressed and sintered. Used, for instance, as an RF choke, or for magnetic storage.

ferrite choke An RF choke consisting of a ferrite bead or core. For example, such a choke may be slipped around a coaxial cable.

ferrite core A magnetic core made from powdered **ferrite**. The particles are mixed with an appropriate binder, and the core is formed through applied pressure. Used, for instance, as an RF choke, or for magnetic storage. Also called **dust core**, or **powdered-iron core**.

ferrite-core memory A form of magnetic memory consisting of a matrix of ferrite cores linked by wires. It is a seldom-used form of non-volatile computer memory.

ferrite isolator A device, consisting of a ferrite rod, which permits electromagnetic energy to pass freely in one direction, while absorbing that arriving from the opposite direction. Used, for instance, in waveguides.

ferrite loopstick Same as **ferrite-rod antenna**.

ferrite phase shifter A device, consisting of a ferrite rod, which is utilized to alter the phase of the signal passing through it. Used, for instance, in waveguides.

ferrite rod A rod made of **ferrite**. Such rods are often made from powdered ferrite which is pressed and sintered.

ferrite-rod antenna A small receiving antenna consisting of a coil wound around a ferrite rod. Such antennas feature high sensitivity at low and medium frequencies, and are used, for instance, in place of a loop antenna in a radio receiver. Also called **ferrite antenna, ferrite loopstick, ferrod, loopstick antenna**.

ferrite switch A device, composed of ferrite, which when activated blocks the energy flowing through a waveguide. It does so by rotating the electric field vector by 90° when energized.

ferrod Same as **ferrite-rod antenna**.

ferroelectric 1. Pertaining to, or producing **ferroelectricty**. 2. Same as **ferroelectric material**.

ferroelectric capacitor A capacitor in which the dielectric exhibits **ferroelectricty**.

ferroelectric crystal A crystal of a **ferroelectric material**.

ferroelectric hysteresis A property of certain ferroelectric materials, in which electric polarization is retained after removal of the electric field. Also called **electric hysteresis**.

ferroelectric material A dielectric material which exhibits **ferroelectricty**. Examples of such materials include barium titanate, and Rochelle salt. Also called **ferroelectric (2)**.

ferroelectric RAM Same as **FRAM**.

ferroelectricty A property of certain dielectric materials, in which electrical polarization occurs spontaneously. That is, the electric dipoles of the material line up spontaneously due to their mutual interaction. When induced by an electric field, the polarization remains until disturbed. It is analogous to the spontaneous magnetic polarization occurring in **ferromagnetism**.

ferrofluid A dense solution in which microscopic or ultramicroscopic magnetic particles are suspended. Used, for instance, in high-frequency speakers for lubrication, damping, and cooling.

ferromagnet A material, such as a ferromagnetic metal or alloy, which exhibits **ferromagnetism**.

ferromagnetic Displaying **ferromagnetism**.

ferromagnetic alloy An alloy which exhibits **ferromagnetism**. Examples include steel, permalloy, and alnico.

ferromagnetic Curie temperature The temperature at which the ferromagnetic properties of a material become paramagnetic. At or above this temperature, the thermal energy in the material is too great to exhibit ferromagnetism. Also

called **Curie temperature, Curie point**, or **magnetic transition temperature**.

ferromagnetic domain Within a ferromagnetic material, a region in which the atomic or ionic magnetic moments are aligned in the same direction. Such a domain may be used, for instance, to store a bit as a 1 or a 0 on a magnetic tape or disk. Also called **magnetic domain**, or **domain (2)**.

ferromagnetic material A metal or alloy which exhibits **ferromagnetism**. Examples include metals such as iron, nickel, or cobalt, and alloys such as steel, permalloy, or alnico.

ferromagnetic metal A metal which exhibits **ferromagnetism**. Examples include iron, nickel, and cobalt.

ferromagnetic resonance A resonant condition in which the permeability of a ferromagnetic material reaches a peak value at a given microwave frequency. Used, for instance, to measure magnetic properties via microwave absorption. Its abbreviation is **FMR**.

ferromagnetism A property of certain metals and alloys, in which magnetic polarization occurs spontaneously. That is, the magnetic dipoles of the metal or alloy line up spontaneously due to their mutual interaction. Such materials have a very high relative permeability, which varies with the magnetizing force. Above the Curie temperature, the ferromagnetic properties of a material become paramagnetic. It is analogous to the spontaneous electric polarization occurring in **ferroelectricity**.

ferrosoferric oxide A reddish or bluish-black powder whose chemical formula is Fe_3O_4. It is used in magnetic-tape coatings, magnetic inks, and in ferrites. Also called **iron oxide, iron oxide black, black iron oxide**, or **magnetic iron oxide**.

ferrule A mechanical fixture which helps align, hold, or give structural support. For instance, such fixtures at each end of a cartridge fuse, which enable plug-in connection. A ferrule is usually made of metal, ceramic, or plastic.

FET Acronym for field-effect transistor. A semiconductor device whose flow of charge carriers, either electrons or holes, is controlled by an external electric field. Such transistors have three electrodes, which are the source, drain, and gate. The external field controls or modulates the gate, which in turn controls the current flowing from the source to the drain. The electrical path between the source and drain is called channel. Field-effect transistors are one of the two main classes of transistors, the other being bipolar junction transistors. Used extensively in computer chips.

FET voltmeter Abbreviation of field-effect transistor **voltmeter**. A voltmeter which incorporates a **FET**, for enhanced sensitivity and accuracy.

FET VOM Abbreviation of field-effect transistor volt-ohm milliammeter. A volt-ohm milliammeter which incorporates a **FET**, for enhanced sensitivity and accuracy.

fetch To locate an instruction in computer memory and load it into a CPU register. Once an instruction is fetched, it can then be executed. Also called **instruction fetch**.

fetch cycle The time required for a fetch. This interval is measured in clock cycles.

fetch-execute cycle The sequence of steps required to implement computer instructions. An instruction is fetched, then executed, then the next fetch occurs, and so on.

FEXT Abbreviation of **far-end crosstalk**.

fF Abbreviation of **femtofarad**.

FF 1. Abbreviation of **fast-forward**. 2. Abbreviation of **flip-flop**. 3. Abbreviation of **form feed**.

FFC Same as **flat cable**. Abbreviation of **flexible-flat cable**, or **flat-flexible cable**.

FFT Abbreviation of **fast Fourier transform**.

fH Abbreviation of **femtohenry**.

FHSS Abbreviation of **frequency-hopping spread spectrum**.

fiber 1. A slender threadlike object, or a section of material so shaped. For example, a conducting thread, or a fiber of glass. **2.** A thin filament of a transparent material, such as glass or plastic, which is capable of transmitting light signals through successive internal reflections. In order for a fiber to guide light, the proper relationship between the refractive index of the core and its surrounding cladding must be maintained. Also called **optical fiber**, or **light guide**.

fiber Bragg grating An optical-fiber section in which the refraction index of the core provides for the reflection of specific wavelengths. Used, for instance, for filtering, or in wavelength division multiplexing. Also called **Bragg grating**.

fiber bundle A concentrated collection of parallel **optical fibers**. Such a bundle may be flexible or rigid. Also called **fiber-optic bundle**.

Fiber Channel Same as **Fibre Channel**.

Fiber-Channel Arbitrated Loop Same as **Fibre-Channel Arbitrated Loop**.

fiber core The bundled optical fibers encased by a protective sheath in a fiber-optic cable. Also called **optical fiber core**.

Fiber Distributed Data Interface Its abbreviation is **FDDI**. An ANSI standard for fiber-optic transmission of digital data. FDDI networks are token-passing, support data rates of up to several hundred Mbps, and usually have dual counter-rotating logical rings. Used, for instance, as backbones for WANs.

Fiber Distributed Data Interface 2 Same as **Fiber Distributed Data Interface II**.

Fiber Distributed Data Interface II An enhanced version of **Fiber Distributed Data Interface**. It incorporates, for instance, specifications for the transmission of real-time voice and video, in addition to data. Its abbreviation is **FDDI II**.

fiber laser A laser incorporated into an optical fiber.

fiber-optic Also spelled **fiberoptic**. **1.** Pertaining to, or consisting of **fibers (2)**. **2.** Pertaining to, or employing **fiber optics**.

fiber-optic bundle Same as **fiber bundle**.

fiber-optic cable One or more optical fibers bundled together and encased by a protective sheath. Generally used for telecommunications. Such cables usually consist of three layers, which are the core, its surrounding cladding, and the protective jacket. Also called **optical cable, optical fiber cable, cable (2)**, or **light cable**.

fiber-optic channel Same as **fiber-optic circuit**.

fiber-optic circuit A communications circuit whose medium of transmission consists of **fibers (2)**. Also called **fiber-optic channel**.

fiber-optic communications Communications utilizing **fiber-optics**. When compared to communications utilizing metal cables and wires, fiber-optic communication offers the following advantages: much greater bandwidth, lower susceptibility to interference, lower weight and volume, diminished attenuation, greater safety, and increased security, among others.

fiber-optic connector A device which serves to couple fiber-optic cables with other fiber-optic cables, devices, equipment, or systems. Examples include FC connectors, SC connectors, and ST connectors.

Fiber-Optic Inter Repeater Link An older standard for fiber-optic Ethernet, based on the 802.3 IEEE standard. Its abbreviation is **FOIRL**.

fiber-optic link A communications link whose medium of transmission consists of **fibers (2)**.

fiber-optic network A network across which **fiber-optic communications** occur.

fiber-optic probe A probe consisting of a single fiber, or of a small fiber bundle. Used, for instance, for measurements in hard to reach locations, or where other probes, such as those that are electric, may be hazardous.

fiber-optic sensor A sensor consisting of a single fiber, or of a small fiber bundle. Used, for instance, for measurements in hard to reach locations, or where other sensors, such as those that are electric, may be hazardous. Such sensors may be utilized to measure current, voltage, temperature, strain, and so on.

fiber-optic system A communications system utilizing **fiber-optic technology**.

fiber-optic technology Any technology based on, or incorporating **fiber-optics**.

fiber-optics Also spelled **fiberoptics**. **1.** A technology or technique using **fibers (2)** to transmit light from one location to another. **2.** A communications technology or technique using light signals for the transmission of information through **fibers (2)**.

fiber to the curb A communications network in which optical fibers run from a central office until a point close to the locations of the subscribers. From that point to the premises, copper wires are utilized. This contrasts with **fiber to the home**. Its abbreviation is **FTTC**.

fiber to the home A communications network in which optical fibers run from a central office all the way to the locations of the subscribers. In such a network, no copper wires are utilized. This contrasts with **fiber to the curb**. Its abbreviation is **FTTH**.

fiberoptic Same as **fiber-optic**.

fiberoptics Same as **fiber-optics**.

fiberscope An optical instrument consisting of a fiber bundle with a lens through which light passes at one end, and a second lens for viewing on the other. Used, for instance, to observe objects and areas that are not accessible for direct viewing.

Fibre Channel A serial fiber-optic data transmission technology with transfer rates that can range to up to several Gbps. It supports existing technologies such as SCSI, and is used for peripherals, especially storage devices, which have very high bandwidth needs. Also spelled **Fiber Channel**.

Fibre Channel Arbitrated Loop A ring topology utilized in Fibre Channel which supports connection of up to 126 nodes to the loop. Its abbreviation is **FC-AL**. Also spelled **Fiber Channel Arbitrated Loop**. Also called **Arbitrated Loop**.

fibre-optic Same as **fiber-optic**.

fidelity The degree to which the output of a circuit, device, piece of equipment, or system reproduces the essential characteristics of its input.

field 1. A region of space within which a physical force exerts its influence. Examples include electric fields, magnetic fields, and gravitational fields. An electric field, for instance, is one in which a charged particle or body exerts a force on charged particles or bodies situated in the medium surrounding it. Such fields may be represented by lines of force. **2.** In TV, one of two equal parts into which a frame is divided. For example, either of the two scans that are interlaced to make up a frame. **3.** A space where computer data is stored, such as a cell in a spreadsheet. Most fields have their associated attributes. For example, it may have a numeric attribute if it contains numbers. A collection of fields is called a record. Also called **data field. 4.** A setting in which normal operating conditions occur, as opposed, for instance, to a laboratory.

field circuit breaker A circuit breaker which controls the field excitation of a motor.

field coil In an electromagnetic device or machine, a coil of wire which, when current flows through it, is utilized to produce a magnetic field of constant strength. Such coils are used, for instance, in electric motors, electric generators, and electrodynamic speakers. Also called **field winding**.

field conditions The environment a **field** (4) setting provides.

field desorption The application of a strong electric field to a highly curved surface, so as to strip electrons, atoms, or molecules from said surface. Used, for instance, in microscopy or spectroscopy.

field diagram A diagram representing how the intensity of a quantity varies as a function of direction and distance. Commonly refers to the transmitting or receiving effectiveness of an antenna, and can be described in any plane, although the horizontal and/or vertical planes are generally used. Also called **radiation diagram**, **directivity diagram**, or **directional diagram**.

field direction The direction in which a **field** (1) exerts its influence.

field discharge An electric discharge, such as a spark, which occurs across a gap, due to a high potential difference or a sudden change in current.

field effect In a semiconductor device, such as transistor, the control of the flow of charge carriers by an external electric field.

field-effect device A semiconductor device, such as transistor, in which the flow of charge carriers is controlled by an external electric field.

field-effect transistor Same as **FET**.

field-effect transistor volt-ohm milliammeter Same as **FET VOM**.

field-effect transistor voltmeter Same as **FET voltmeter**.

field emission The emission of electrons from a surface, through the application of a strong electric field. Such a surface may be that of a solid or a liquid. Used, for instance, in field-emission microscopy. Also called **cold emission**.

field-emission display A flat-panel display utilizing a **field-emitter array**. Such displays feature very high sensitivity and resolution, compactness, and low power consumption. Its abbreviation is **FED**.

field-emission microscope An instrument which utilizes field emissions from the sharply rounded point of a solid, usually a metal. The emitted electrons are then directed towards a fluorescent or phosphorescent screen, forming an image of the surface of the object. Magnifications of greater than 1 million times are achievable in this manner. Its abbreviation is **FEM**.

field-emission microscopy The use of a **field-emission microscope** to view and analyze the structure and properties of solid surfaces. Its abbreviation is **FEM**.

field-emitter A surface which undergoes **field emission**.

field-emitter array An array of micro-miniature **field emitters**. Used, for instance, in flat-panel display technologies. Its abbreviation is **FEA**.

field-enhanced emission The application of a strong electric field to enhance the emission of electrons from a surface.

field excitation The application of a voltage to the field coils of an electromagnetic device or machine, such as an electric generator or an electrodynamic speaker, so as to establish a steady magnetic field.

field frequency The number of fields transmitted per second in a TV system. It is the product of the frame frequency and the number of fields contained per frame. In the United States, this number is usually 60. Also called **field repetition rate**.

field intensity Same as **field strength**.

field-intensity meter Same as **field-strength meter**.

field-ion microscope A microscope similar to a field-emission microscope, but in which the sharply rounded point is surrounded by helium gas. This induces field ionization of the helium atoms, which are directed towards the screen. This variation has much greater resolving power, and magnifications of greater than 10 million can be attained. Its abbreviation is **FIM**. Also called **ion microscope**.

field-ion microscopy The use of a **field-ion microscope** to view and analyze the structure and properties of solid surfaces at an atomic level. Its abbreviation is **FIM**.

field ionization The ionization of atoms or molecules through the application of a strong electric field.

field length The length of a **field** (3). A field length may be fixed or variable, and is usually expressed in bits or bytes.

field magnet 1. A magnet or electromagnet which provides a magnetic field in an electromagnetic device or machine, such as an electric motor or electric generator. 2. A **field magnet** (1) utilized in a speaker.

field of force Same as **force field**.

field pattern A pattern representing how the intensity of a quantity varies as a function of direction and distance. Commonly refers to the transmitting or receiving effectiveness of an antenna, and can be measured in any plane, although the horizontal and/or vertical planes are generally used. Also called **radiation pattern**, or **directivity pattern**.

field period In a TV system, the time required to transmit one field. It is the reciprocal of the **field frequency**. In the United States, this number is usually 1/60.

field pickup In recording or broadcasting, such as that for radio or TV, the detection of sounds and/or scenes outside a given station or studio. Also called **remote pickup**.

field pole A structure, composed of a magnetic material, upon which a field coil is wound or mounted.

field programmable gate array Its abbreviation is **FPGA**. A gate array which instead of being programmed at a factory, is done so by an end user. FPGAs may usually be programmed thousands of times, and are used, for instance, for making prototypes of chip designs. Also called **programmable gate array**.

field programmable logic array Its abbreviation is **FPLA**. A logic array which instead of being programmed at a factory, is done so by an end user, usually during installation. FPLAs can only be programmed once. Also called **programmable logic array**.

field repetition rate Same as **field frequency**.

field rheostat In an electric motor or generator, a rheostat which controls the flow of current through the field coil.

field separator Within a data record, a character which serves to mark the boundary between fields.

field strength Also called **field intensity**. 1. The strength of a **field** (1). 2. The strength of an electric, magnetic, or electromagnetic field. For instance, the strength of an electric field is the magnitude of the electric field vector, and is usually expressed in volts per meter.

field-strength meter Also called **field-intensity meter**. 1. An instrument which measures **field strength**. 2. An instrument which measures the field strength of a radio transmitter. A specially designed and calibrated radio receiver may be used for this purpose.

field test An evaluation which is performed under actual operating conditions. This contrasts with a **bench test**, which is performed in a test laboratory setting.

field testing The performance of **field tests**.

field winding Same as **field coil**.

FIF Acronym for **Fractal Image Format**.

FIFO Acronym for first-in, first-out. A method of handling data in which the first item stored is the first item to be retrieved. In this manner, that which has been held the longest is that which is used first. Most printer queues use this scheme. This contrasts with **LIFO**, where the next item to be used is that which most recently arrived.

fifth-generation computer A computer utilizing very large-scale integration and ultra-large scale integration. This computer generation started approximately in the mid 1990s.

fifth-generation language A computer programming language that has more advanced natural language processing than fourth-generation languages, and which makes use of expert systems, inference engines, and the like. Its abbreviation is **5GL**.

fifth normal form In the normalization of a relational database, the fifth stage utilized to convert complex data structures into simpler relations. A database must first complete the fourth normal form before proceeding to the fifth. Its abbreviation is **5NF**.

fifth-order filter A filter with ten components, such as five inductors and five capacitors, and which provides a 30 dB rolloff. Used, for instance, as a crossover network.

figure of merit A performance characteristic or rating which determines the suitability of a component, circuit, or device for a given application. For instance, the noise figure for a circuit, the bit-error rate of a demodulator, or the gain-bandwidth product of an amplifier.

filament 1. A threadlike fiber or structure, such as a thin wire. **2.** A cathode, within a thermionic tube, which is directly connected to a source of current. Such a cathode, usually in the form of a wire or ribbon, becomes heated as current passes through it, and emits electrons when sufficiently heated. Also called **filament cathode, filamentary cathode, filament-type cathode,** or **directly-heated cathode**. **3.** A **filament (1)**, within an incandescent lamp, which emits light when heated. Such a filament is often made of tungsten or carbon.

filament battery A battery which supplies current to a **filament (2)**. Also known as **A battery**.

filament cathode Same as **filament (2)**.

filament circuit The circuit through which the **filament current** flows.

filament current The current which is supplied to a **filament (2)**.

filament emission The emission of electrons by a heated **filament (2)**.

filament power supply The source power of utilized to heat a **filament (2)**. Also called **filament supply**.

filament saturation For a thermionic electron tube at a given anode voltage, the condition in which the anodic current can not be further increased with increases in cathode temperature. This is due to a space charge near the cathode. Also called **temperature saturation**, or **saturation (7)**.

filament supply Same as **filament power supply**.

filament transformer A transformer utilized to provide a **filament current**.

filament-type cathode Same as **filament (2)**.

filament voltage The voltage applied to a **filament (2)**.

filament winding Within a power transformer, the winding providing the source of power utilized to heat a **filament (2)**.

filamentary cathode Same as **filament (2)**.

file A collection of information which is stored as a unit. Files may be retrieved, modified, stored, deleted, or transferred. Each type of file requires the appropriate software for the proper handling of its contents. There are many file types, including data files, program files, system files, and multimedia files. Also called **computer file**.

file allocation table Its abbreviation is **FAT**. In some operating systems, a table that is utilized to locate files stored on a disk. Since data may be stored in non-contiguous areas of a disk, a FAT is needed to string together all the pieces of a file. A FAT also keeps track of available disk space, and notes bad sectors to avoid their use.

file and record locking A technique employed to help maintain data integrity in a multiuser environment. When a user accesses any file or record, all other users are blocked from doing so. Once the accessed file or record is updated, it again becomes freely available.

file association The linking of a given type of file to a specific application, usually by using a file extension, such as **.xyz**. It is utilized to alert an operating system to the need to start the necessary application for the desired file.

file attachment A file that is appended to an email. Such a file may be a document, an image, a video, a program, and so on. Certain types of attachments are encoded, thus requiring the recipient to have the appropriate email software to decode it. Also called **email attachment, enclosure (3)**, or **attachment**.

file attribute A characteristic of a file that imposes restrictions on its availability or use. For instance, a file may be read-only, or hidden. Also called **attribute (3)**.

file compression The encoding of a file so that it occupies less space and/or bandwidth. There are many algorithms used for compression, and depending on the information being encoded, space savings can range from under 10% to over 99%.

file control block A block of computer memory that contains information used for purposes of controlling a file. For example, it may hold its location on a disk. Its abbreviation is **FCB**.

file conversion The process of changing a computer file from one form to another. For instance, to convert a file from a given word processor format to ASCII.

file directory A special kind of file which is used for indexing and organizing other files. It provides information such as the name, type, and size for each listed file. File directories that branch into other directories are called root directories, while those located within a root directory are called subdirectories. In GUIs, the term **folder** is preferred. Also called **directory**.

file download To receive one or more files from a remote computer in a network. This contrasts with a **file upload**, where files are sent to a remote computer in a network.

file exporting The conversion of a file or data in one format, into another format required for use by another application or environment, using the appropriate filters. In **file importing** the conversion occurs after files or data are accepted, while in file exporting the conversion occurs before said files or data are moved or saved.

file extension Same as **filename extension**.

file folder A special kind of file which is used for indexing and organizing other files. It provides information such as the name, type, and size for each listed file. File folders that contain other folders are called root folders, while those located within a root folder are called subfolders. In non-graphical user interfaces, the term **file directory** is used. Also called **folder**.

file format The coding of a file which defines what content it has, how it is organized, how it is to be encoded, displayed, and so on. There are many such formats, including those associated with word-processing applications, graphics, and databases.

file fragmentation Same as **fragmentation (1)**.

file gap On a storage medium, such as a magnetic tape or disk drive, an unused physical space between files.

file handle A temporary designation an operating system assigns to an opened file during any given session.

file header An information unit found at the beginning of a file, such as that stored on a tape or disk, which serves to identify it and describe the contained information.

file identification A code, name, label, or the such which serves to identify a file. Also, The assigning of such an identification.

file importing The conversion of a file or data in one format, into another format required for use by the receiving application or environment, using the appropriate filters. In file importing the conversion occurs after files or data are accepted, while in **file exporting** the conversion occurs before said files or data are moved or saved.

file label A record appearing at the beginning of a file, containing information such as its name, size, type, and so on. Also called **header label**.

file layout The manner in which the content of a file is distributed. It consists of the arrangement of the records it contains. Also, a description of such a layout.

file locking 1. A technique employed to help maintain data integrity in a multiuser environment. When a user accesses any file, all other users are blocked from doing so. Once the accessed file is updated, it again becomes freely available. 2. The enabling of a security feature, such as a password, to block access to a file.

file maintenance 1. The updating of master files, based on changes in the data contained in transaction files. 2. Any activity which helps maintain files current and/or optimally accessible. Such activities include the addition, modification, or deletion of contained information, copying or archiving files, or the defragmentation of a disk.

file management system A program, or a part of a program, which performs functions such as the organization, copying, renaming, transferring, viewing, and retrieving of files. An operating system incorporates a file management system, but there are also specialized programs which provide more flexibility and ease of use. Also called **file manager**.

file manager Same as **file management system**.

file name Same as **filename**.

file processing The use of a file. This includes, saving, updating, and manipulating information within it.

file protection 1. Measures taken to prevent the accidental erasure of files, or parts of files. Such protection may be logical or physical. Logical protection, for instance, may involve designating a file as read-only, while physical protection might entail the moving of a tab. 2. Same as **file security**.

file recovery Procedures utilized to restore files, or parts of files, which have been lost due to accidental erasure, software and/or hardware failure, misplacement, and so on. Although there are utility programs that can help in this capacity, the best restoration method is to access a backup copy.

file recovery program Same as **file recovery software**.

file recovery software A program or utility employed for **file recovery**. Also called **file recovery utility**, or **file recovery program**.

file recovery utility Same as **file recovery software**.

file retrieval The accessing of a file, or a part of a file, from a storage location.

file search 1. To look for specific content within a file. 2. To look for a specific file within a file server, database, directory, and so on.

file security The safeguarding of files against loss, damage, unwanted modification, or unauthorized access. Such safeguards may be administrative, physical, or technical. Also called **file protection** (2).

file server In a communications network, a server which manages files and serves as a remote storage location for a workstation. For example, a file server helps insure that multiple requests for files are handled in an orderly manner.

file sharing The accessing of files by multiple users in a network. This may or not occur simultaneously. The files are available from a central computer, such as a file server, which takes care of tasks such as file locking.

file sharing protocol A communications protocol which governs the sharing of files over a network.

file size The length of a file, usually expressed in bytes or a multiple of bytes such as megabytes.

file spec Same as **full path**. Abbreviation of **file specification**.

file specification Same as **full path**.

file system Within a operating system, the manner in which files are named, renamed, organized, copied, transferred, stored, and retrieved.

file transfer The movement, without alteration, of a file from one location to another. This may be, for instance, from one computer to another within a network, or between file directories.

File Transfer, Access, and Management A communications protocol intended to facilitate the access, management, and transfer of files between different computer environments. Its abbreviation is **FTAM**.

file transfer program A program designed to simply, quickly, and safely transfer files between computers, such as those linked via a parallel or USB cable. Its abbreviation is **FTP**.

file transfer protocol A protocol utilized for transferring files over a TCP/IP network, such as the Internet. In incorporates functions such as conversions between character codes, and accepts commands, such as directory listing. Its abbreviation is **FTP**.

file type The kind of information contained in a file, and the format in which it is stored. For instance, an operating system will use this information to select the appropriate application to open when a given file is selected. The file type is usually indicated by a filename extension.

file upload To send one or more files to a remote computer in a network. This contrasts with a **file download**, where files are received from a remote computer in a network.

file viewer A separate application which is called upon to interpret and view files, including those with multimedia content, which could otherwise not be seen by the current application. For instance, a Web browser plug-in utilized to display a video file that is in a format not currently supported.

FileMaker A popular database program.

filename One or more alphanumeric characters utilized to designate a file. Each operating system has restrictions on filenames, such as length, or the inability to use certain characters and/or symbols. Many operating systems support extensions at the end of a filename, which indicate the file type. Also spelled **file name**.

filename extension A set of characters appearing at the end of a filename, indicating the file type. For example, in the filename *notepen.exe*, the **.exe** portion is the extension, and in this case specifies an executable program. Also called **file extension**, or **extension** (2).

filespec Abbreviation of **file specification**. Same as **full path**.

fill 1. That which serves to occupy a space or enclosure. For example, the fill of a gas tube, or the fill between spaces in hard-solder. Also to add such a substance to a space or enclosure. 2. To occupy a virtual space. For instance, the filling of a polygon on a computer screen with a given pattern or color, or to place numbers, data, or formulas in a spreadsheet cell.

filled band   **1.** A band which is in use or occupied. **2.** An energy band which is occupied by electrons. That is, each orbital has its maximum of two electrons.

filler   That which serves to fill.

film   A thin, or very thin layer or coating. A film may be thinner than one femtometer. Examples include acetate films, anode films, epitaxial films, and thin films.

film capacitor   A capacitor in which layers of a metal foil, usually aluminum, are alternated with layers of a dielectric, usually a plastic, to form a roll. Such capacitors are well suited, for instance, for use at high temperatures and high frequencies. When the dielectric is a plastic, also called **plastic-film capacitor**.

film chain   A device which is utilized to transfer film, such as slides or that of motion pictures, to video. It incorporates a film or slide projector, a multiplexer, and a TV camera. Used, for instance, to televise a motion picture on film. Also called **telecine**.

film IC   Abbreviation of **film i**ntegrated **c**ircuit. An IC whose elements are films formed upon an insulating substrate.

film integrated circuit   Same as **film IC**.

film recorder   **1.** A device which takes images stored in a computer file or displayed on a monitor, and captures them on photographic film, such as a slide. **2.** A device whose input is an electronic signal, and whose output is photographic film.

film resistor   A resistor whose resistive material, such as a carbon, is a thin layer deposited on an insulating substrate, such as ceramic. Used, for instance, in low power applications.

FILO   Acronym for first-in, last-out. A method of handling data in which the first item stored is the last item to be retrieved. This is the same as **LIFO**.

filter   **1.** That which serves to selectively allow matter and/or energy to pass, remain, be blocked, or get absorbed. **2.** An electric circuit or device which selectively transmits or rejects signals in one or more intervals of frequencies. The transmitted intervals are called passbands, and the rejected intervals are called stopbands. When a filter incorporates active components, such as transistors, it is an active filter, if not, it is passive. A capacitor, for instance, may serve as a passive filter, because it blocks DC. Filters may be classified as falling within one of the following four categories: low-pass, high-pass, bandpass, and bandstop. There are many examples of filters, including comb, ripple, Butterworth, and loop filters. Also called **electrical filter**, or **electrical-wave filter**. **3.** A device which blocks or absorbs sounds of certain frequencies while leaving others unaffected. Also called **acoustic filter**, or **sound filter**. **4.** An element or device, such as a disk or plate of plastic or glass, which selectively blocks or absorbs one or more intervals of frequencies of electromagnetic radiation, such as light. The optical properties of the element or device determine which frequencies pass, and which are blocked or absorbed. Also called **optical filter**, or **radiation filter (2)**. **5.** In computers, a program, function, or process which transforms data from one format to another. For instance, such a filter may convert a document from the format of one word processing application to that of another. **6.** In computers, a program, function, or process which selectively passes or separates data or items. For example, an email filter.

filter attenuation   The loss in the power of a signal when passing through a filter, as function of frequency, and expressed in decibels.

filter attenuation band   Same as **filter stopband**.

filter bank   A set of filters used together. May be used, for instance, to increase frequency selectivity, or to create multiple bands.

filter capacitor   **1.** A capacitor utilized within a filter. Used, for instance, in a power supply to reduce ripple. **2.** A capacitor utilized as a filter. Capacitors block direct currents and provide a low-reactance path for alternating currents.

filter choke   A coil wound around an iron core, which serves to pass DC while opposing the flow of alternating or pulsating current.

filter crystal   A piezoelectric crystal, such as quartz, utilized in a crystal filter.

filter cutoff   The frequency, or interval of frequencies, between a passband and a stopband of a filter.

filter discrimination   A measure of the selectivity of a filter.

filter pass band   Same as **filter passband**.

filter passband   A continuous range of frequencies which is passed or transmitted by a filter. That is, there is little or no attenuation of a signal in this interval. A filter may have multiple passbands. Also spelled **filter pass band**. Also called **filter transmission band**, or **passband (2)**.

filter response   The behavior of a filter, usually expressed as a function of amplitude versus frequency. Amplitude may be expressed, for instance, as a percentage of its input, or in decibels, while frequency is expressed in hertz.

filter-response curve   A curve depicting a **filter response**.

filter stop band   Same as **filter stopband**.

filter stopband   A continuous range of frequencies which is highly attenuated or rejected by a filter. A filter may have multiple stopbands. Also spelled **filter stop band**. Also called **filter attenuation band**, or **stopband**.

filter transmission band   Same as **filter passband**.

FIM   **1.** Abbreviation of **field-ion microscope**. **2. field-ion microscopy**.

fin   A projecting plate or vane utilized for cooling a component, device, or piece of equipment. It provides an additional surface area for the dissipation of heat. Also called **cooling fin**.

final amplifier   The final amplifier in a cascade or multistage amplifier. For instance, the stage of a transmitter that feeds an antenna. Also called **output amplifier**.

final stage   For a multistage circuit, device, or piece of equipment, such as a cascade or multistage amplifier, the stage delivering the output to the load. Also called **output stage**.

financial software   Programs which help individuals and entities with matters pertaining to finance. Such software may help in areas such as budgeting, asset allocation, investments, cash flow, insurance, education funding, retirement planning, estates, and so on.

find   **1.** To examine or explore a selected text, document, or file, looking for specific character strings, special characters, commands, or the like. Also called **search (3)**. Also, a specific instance of such a find. **2.** To examine or explore a document, file, database, disk, the Internet, or the like, looking for specific data or items. Also called **search (2)**. Also, a specific instance of such a find.

find and replace   To perform a **find (1)** and replace the located content with other content. For example, to find instances where the word *delay* appears within a document, and to put the phrase *time delay* in its place for one or more occurrences. Its abbreviation is **find & replace**. Also called **search and replace**.

find & replace   Abbreviation of **find and replace**.

finder   **1.** In a telephone switching system, a switch or group of relays which determines the circuit or route that a call will take through it. **2.** A device which serves to display an image corresponding to that which is being focused upon by a camera, such as a TV camera. Such a device may be optical and/or electronic. Also called **viewfinder**.

fine adjustment   The adjustment of a setting in relatively small and/or precise increments. This contrasts with a **coarse ad-**

justment, in which the changes are relatively large and/or more approximate.

fine control   An adjustable component which makes **fine adjustments**.

fine-motion planning   In robotics, motion planning involving activities which occur within comparatively small spaces. For example, positioning an end-effector for precise grasping within a comparatively small space. This contrasts with **gross-motion planning**, which involves movement within comparatively large free spaces.

finger   An Internet utility which enables a user to find information on another user. For example, if a user provides the email address of another user which has provided a profile to the system, information such as the name, telephone number, and an indication if that user is currently logged on may be obtained.

fingerprint reader   A device, such as a scanner and a linked computer, which reads and identifies fingerprints. To obtain a match, it compares the scanned images with those that are stored, utilizing fingerprint recognition. Used, for instance, for identification and security purposes.

fingerprint recognition   A technology utilized to identify fingerprints. For instance, a user may place a finger on a glass plate, enabling a biometric system utilizing a charge-coupled device camera along with a computer accessing a fingerprint database to confirm the identity of the user corresponding to the fingerprint. Also called **fingerprint recognition technology**.

fingerprint recognition technology   Same as **fingerprint recognition**.

fingertip controls   Keys that enter information by merely touching, as opposed to pressing them down. Seen, for instance, on microwave ovens, as they are easier to clean.

finish   An outer coating or final treatment for a surface. A finish may be applied to protect, impart desired properties, or to decorate.

finish lead   The lead attached to the last, or outer, turn of a coil. This contrasts with a **start lead**, which is connected to the first turn of a coil. Also called **outside lead**.

finishing rate   The rate at which a battery charges as it approaches the end of the charge cycle. This rate is usually slower than that leading to this stage.

Finite-Difference Time Domain   A computational method for modeling electromagnetic wave interactions, such as wave propagation. Its abbreviation is **FDTD**.

Finite Element Analysis   A method utilized to analyze stresses and structural integrity of objects, by breaking down an overall physical structure into a finite number of discrete elements. For instance, a CAD program can be utilized to study these elements, and display different scenarios under varying thermal and dynamic conditions for the overall structure. Its abbreviation is **FEA**.

Finite-Impulse Response Filter   A filter whose output is a weighed sum of the current and past inputs. It is one of the two primary types of filters utilized in digital signal processing, the other being **Infinite-Impulse Response filters**. Finite-Impulse Response Filters tend to be more linear, but less efficient than Infinite-Impulse Response filters. Its abbreviation is **FIR filter**. Also called **transversal filter**.

finite-state automaton   Same as **finite-state machine**.

finite-state machine   A machine which can be completely described by a finite set of defined states. Such a machine must be in one of these states at any given moment, and there is a set of conditions which determine when it moves from one state to another. Also, a design model which can be so described. Used, for instance, to design specialized digital systems. Its abbreviation is **FSM**. Also called **finite-state automaton**, or **state machine**.

FIR filter   Abbreviation of **Finite-Impulse Response Filter**.

FIR port   Abbreviation of **fast infrared port**.

fire alarm   **1.** An electrical and/or mechanical device which serves to warn of a fire by means of a signal, such as sound and/or light. Also, such a warning. **2.** A device, such as a button or telephone, which serves to alert the proper entities of a fire.

fire detector   A device which senses the presence of a fire. Such a device may detect heat, for instance, and be set to activate a fire alarm and/or a sprinkler system when the temperature exceeds a given limit. Also called **fire sensor**.

fire-proof   Same as **fireproof**.

fire sensor   Same as **fire detector**.

Firebird   A popular relational DBMS.

fireproof   Also spelled **fire-proof**. **1.** Impervious, or highly resistant, to fire or its accompanying effects, such as excessive heat. **2.** Protected in a manner which does not allow fire to enter an enclosure.

firewall   A system which is utilized to help protect a network against threats originating through other networks, such as unauthorized access. A firewall, for example, enables an organization with an intranet to allow its members access to the Internet, while preventing outsiders from accessing its own private resources. A firewall may also control which outside resources are available to what users. A firewall may be implemented through the use of hardware and/or software, and all communications entering or leaving the internal network of the organization are examined to determine if they may proceed. Application gateways, packet filters, and proxy servers are among the available techniques which help afford such protection. A firewall may also help protect a network against internal threats, such as snooping. Also called **Internet firewall**.

FireWire   A serial interface which implements the IEEE 1394 standard. Such an interface features transfer speeds ranging from 100 Mbps to over 2 Gbps, and allows the connection of up to 63 devices. It also supports hot-swapping, and isochronous data transfer. Used, for instance, for real-time transfers of video. Also called **High-Performance Serial Bus**.

firing   **1.** The excitation of a circuit or device via a brief pulse. **2.** The use of a pulse to excite a magnetron or TR tube. **3.** In a gas-discharge tube, the ionization which initiates the flow of current. **4.** In a saturable reactor, the transition from a non-saturated state to a saturated state. **5.** The heating of certain nonmetallic mineral substances, such as silicon dioxide or clay, so as to form a ceramic. Also, other similar bondings using heat.

firing circuit   A circuit which delivers a pulse or signal which initiates **firing**.

firmware   Software, in the form of programs and/or routines, which is stored in ROM, or a form of programmable ROM such as EEPROM. Firmware remains intact with or without power. Used, for instance, to store startup routines.

first detector   Same as **frequency converter (2)**.

first Fresnel zone   For transmitted electromagnetic radiation, the first of a series of concentric areas in space which surround a path between a transmitter and receiver. Its cross-section is circular, and encompasses all paths which are a half-wavelength longer than the line-of-sight path. Utilized, for instance, to describe the first concentric ellipsoid between transmitting and receiving microwave antennas.

first-generation computer   A computer utilizing vacuum-tube technology. Such computers were built from about the mid 1940s to the mid 1950s.

first-generation language   A machine language. Each type of CPU has its own machine language.

first-generation wireless Developed in the late 1970s and early 1980s, provides for analog cellular communications such as those using AMPS. Its abbreviation is **1G**.

first harmonic Same as **fundamental frequency (1)**.

first-in, first-out Same as **FIFO**.

first-in, last-out Same as **FILO**.

first law of thermodynamics A law stating that the total energy in an isolated system remains constant. For a non-isolated system, any added energy, such as heat, must either increase the internal energy of said system, or perform work on its surrounding environment. It is an application of the law of energy conservation.

first-level cache A memory cache that is built right into the CPU. This contrasts with **second-level cache**, which utilizes a memory bank between the CPU and main memory. First-level cache is faster, but smaller than second-level cache, while second-level cache is still faster than main memory. Also called **level 1 cache**, **L1 cache**, **internal cache**, **primary cache**, or **on-chip cache**.

first normal form In the normalization of a relational database, the first stage utilized to convert complex data structures into simpler relations. Its abbreviation is **1NF**.

first-order filter A filter with two components, such as an inductor and a capacitor, and which provides a 6 dB rolloff. Used, for instance, as a crossover network.

first-surface mirror Same as **front-surface mirror**.

fish paper Same as **fishpaper**.

fish tape A flexible metal strip which is utilized to pull wires and cables through walls, conduits, raceways, and the like.

fishbone antenna A directional antenna with an array of elements along each side of a transmission line, arranged in pairs along the same plane, resembling the skeleton of an elongated fish.

fishpaper A vulcanized-fiber paper which is used for electrical insulation. It features great strength, flexibility, and puncture resistance. Used, for instance, to insulate between a transformer core and its windings. Also spelled **fish paper**.

fitting A component or device which joins two parts together, serving a primarily mechanical function, as opposed to an electrical one. An example is a locknut.

fix **1.** In navigation, the determination of an accurate position through the intersection point of two or more bearings. This position is determined without reference to a former position. **2.** In navigation, the determination an accurate position through the use of any available resources, such as radar, loran, ultrasound, landmarks, GPS, and so on. **3.** To make stationary or stable. **4.** To give a final form to. **5.** To repair.

fixed attenuator A circuit or device which reduces the amplitude of a signal, ideally without introducing distortion, by a fixed amount. When such an attenuator consists of a network of fixed resistors, also called **pad (1)**.

fixed bias A bias value that is constant, such as that obtained from a fixed external source. In an electron tube, for instance, such a bias could be obtained from a battery.

fixed capacitor A capacitor with a single capacitance, which cannot be adjusted. This contrasts with a **variable capacitor**, whose capacitance can be varied.

fixed component An electrical component, such as a capacitor or resistor, which has a single value which cannot be adjusted. This contrasts with a **variable component**, whose value can be varied.

fixed contact In a relay or switch, an electrical contact which does not move. Used in conjunction with a **movable contact**. Also called **stationary contact**.

fixed disk **1.** A computer hard disk drive which can not be removed. Most hard disk drives are of this type. **2.** A platter, within a hard disk drive, which can not be removed.

fixed field Same as **fixed-length field**.

fixed-frequency Consisting of, pertaining to, or utilizing a specific frequency. For instance, a single-frequency laser, or a single-frequency receiver. Also called **single-frequency**.

fixed-frequency monitor A monitor which can only accept a video signal within a single frequency range. For example, one that accepts VGA signals, but none other. This contrasts with a **multi-frequency monitor**, which supports a fixed number of frequency ranges, and with a **multiscan monitor**, which responds to signals at any frequency within a wide interval.

fixed-frequency oscillator An oscillator which operates at a single frequency, which cannot be adjusted. This contrasts with a **variable-frequency oscillator**, whose frequency can be varied. Also called **single-frequency oscillator**.

fixed inductor An inductor with a single inductance, which cannot be adjusted. This contrasts with a **variable inductor**, whose inductance can be varied.

fixed Internet-Protocol address Same as **fixed IP address**.

fixed IP address Abbreviation of **fixed** Internet-Protocol address. An IP address which is the same each time a user logs onto a TCP/IP network. Also, such an address corresponding to a server. This contrasts with a **dynamic IP address**, in which a different IP address is assigned each time a user logs on. Also called **static IP address**.

fixed-length field A data field whose size can not be adjusted. It always occupies the same number of bits or bytes, regardless of the amount of data stored. This contrasts with a **variable-length field**, whose size is determined by the amount of data stored. Also called **fixed field**.

fixed-length record A data record composed of **fixed-length fields**. This contrasts with a **variable-length record**, which is composed of variable-length fields.

fixed memory **1.** Computer memory which cannot be altered. An example is ROM. **2.** Computer memory which can only be altered mechanically.

fixed orbit The orbit a **fixed satellite** follows. Also called **synchronous orbit**.

fixed output **1.** An output, such as that of a CD player, whose level can not be changed. **2.** An output whose level is kept at the same level.

fixed-point Pertaining to **fixed-point notation**.

fixed-point arithmetic Arithmetic performed utilizing **fixed-point notation**.

fixed-point calculations Same as **fixed-point operations**.

fixed-point notation A numeric format in which the decimal point is always in a fixed location. This contrasts with **floating-point notation**, in which the decimal point is not in a fixed location. Computer operations on fixed-point numbers usually require less time than that required for floating-point numbers. Also called **fixed-point representation**.

fixed-point number A number expressed in **fixed-point notation**

fixed-point operations Calculations performed utilizing **fixed-point arithmetic**. Also called **fixed-point calculations**.

fixed-point representation Same as **fixed-point notation**.

fixed-point system A numeric system utilizing **fixed-point notation**

fixed resistor A resistor with a single resistance, which cannot be adjusted. This contrasts with a **variable resistor**, whose resistance can be varied.

fixed satellite Also called **synchronous satellite**. **1.** An artificial satellite whose orbit is synchronized with that of the earth, so that it remains fixed over the same spot above the planet at all times. Such a satellite is usually located at an altitude of approximately 35,900 kilometers, and its orbit is in the plane of the equator. As few as three of these satellites can provide worldwide coverage. Used, for instance, for communications and broadcasting. Also called **geostationary satellite**, or **geosynchronous satellite**. **2.** A satellite which remains fixed in relation to a celestial body such as the earth, moon, or sun.

fixed-satellite service   Radiocommunication service between **fixed stations**, utilizing one or more satellites. Also, communications links between said satellites. Its abbreviation is **FSS**.

fixed service   Radiocommunication service between **fixed stations**.

fixed station   A station whose location does not change. For instance, a land station. This contrasts with a **mobile station**, whose location is not fixed.

fixed storage   **1.** A computer storage device, such as a hard disk drive, which can not be removed. **2.** Stored data which can not be altered. Also called **read-only storage**.

fixture   **1.** A mechanical component or device which serves to join, align, or hold two or more parts. For example, a cable connector, or a waveguide gasket. **2.** An item, such as a socket or light, which is permanently attached to a structure, such as a wall or a building.

fl   Same as **fL**.

fL   Abbreviation of **footlambert**.

FL   Same as **fL**.

flag   **1.** Something utilized to notify, remind, identify, comment, or warn. **2.** In computers, a marker, code, or sequence of bits which serves as a **flag (1)**. For example, such a flag indicating the end of a process, or an error condition. **3.** In communications, a marker, code, or sequence of bits which serves as a **flag (1)**. For example, such a flag indicating the end of a transmission, or notifying of an upcoming control code.

flag register   A register whose content indicates aspects of the internal status of a CPU, or a part of a CPU such as an arithmetic-logic unit. Also called **status register**.

flame   **1.** A hot and luminous gaseous body undergoing combustion. **2.** A communication or message, such as an email or a newsgroup posting, which is insulting or otherwise abusive in nature. Also, to send such a message.

flame alarm   An electrical and/or mechanical device which serves to notify or warn of a flame by means of a signal, such as sound and/or light. Also, such a warning.

flame bait   Same as **flamebait**.

flame detector   A device which senses the presence of a flame. Such a device, for instance, may indicate the rate at which fuel is undergoing combustion, and send a signal to a control system regulating said combustion.

flame-proof   Same as **flameproof**.

flame war   **1.** A heated exchange, such as that between the participants of an online newsgroup, intended to insult, degrade, or enrage others. **2.** A heated exchange in a forum such as an online newsgroup.

flamebait   The sending or posting of one or more messages which are insulting or otherwise abusive in nature, so as to elicit a strong reaction or lead to a **flame war (1)**. Also spelled **flame bait**.

flameproof   Also spelled **flame-proof**. **1.** Impervious, or highly resistant, to a flame. **2.** Protected in a manner which does not allow a flame to enter an enclosure.

flammable   That which is easily ignited or burned. Also called **inflammable**.

flange   A protruding rim or edge which guides and/or imparts strength when attaching one object to another. For instance, such a projection which helps insure the snug fit of a plug inserted in a jack.

flanging   The mixing of a signal with a time-delayed copy of the same signal, thus forming a new signal. Used, for instance, to create sound effects.

flap attenuator   A waveguide attenuator in which a sheet or plate of dissipative material is placed through a non-radiating slot. As the sheet or plate is moved into and out of the slot, a variable amount of loss is introduced. Also called **vane attenuator**, **rotary-vane attenuator**, or **guillotine attenuator**.

flapping   In networks, a circumstance in which routers repeatedly advertise the availability or unavailability of a given route. May be caused, for instance, by faulty circuit connections or heavy traffic.

flare   **1.** To open or spread outward. An example is the manner in which the cross-section of a horn-shaped radiating device increases from one end of the tube to the other. **2.** On an oscilloscopic screen, such as that of a radar, an excessively bright area. Such an area is usually enlarged and distorted. **3.** A brief and intense light or flame. **4.** A bright eruption which develops near the surface of the sun. These are of tremendously high temperature, take a few minutes to reach full strength, and subside over the next hour or so. High-energy particles and rays are explosively released, and when they reach the earth may cause radio interference, magnetic storms, or auroras. Also called **solar flare**.

flare angle   The progressive change in the cross-section of a waveguide, horn antenna, or horn speaker.

flash   **1.** A sudden burst of bright light. **2.** A repeatable and artificially produced burst of bright light. Such a flash is usually generated by applying a high voltage to an electrode of a tube containing an inert gas such as xenon. The gas becomes ionized, which permits it to rapidly discharge the energy previously stored in a capacitor. Used, for instance, in photography. Also called **photoflash**, **strobe (3)**, or **electronic flash**. **3.** To occur suddenly, or to proceed rapidly. **4.** To press a telephone switchhook briefly to access a function such as call-waiting, or three-way calling. Also, a button so labeled. Also, to press a button so labeled. Also called **hook flash**. **5.** To write onto **flash memory**.

Flash   A technology utilized for the creation of Web multimedia content, such as animation and streaming graphics. Also, the software utilized for such creations, and the file format. Content is usually displayed by having the appropriate browser plug-ins.

flash arc   In a thermionic tube, a current surge occurring between electrodes which may destroy said tube.

flash BIOS   A computer system's BIOS held in **flash memory (1)**.

flash bulb   A bulb which produces a burst of bright light. It contains a combustible material, such as magnesium, which is ignited by a small voltage. Used, for instance, in photography. A flash bulb may only be used once, while a **flash tube** may be utilized many times.

flash card   Flash memory contained in an expansion card. Used, for instance, for storage in portable or handheld computers such as notebooks and PDAs, MP3 players, or in digital cameras. Also called **flash-memory card**.

flash disk   A storage disk composed of flash memory chips.

flash EEPROM   Flash memory utilized as a programmable ROM chip. It is similar to other programmable ROM chips, such as EEPROMs, but do not permit bit by bit writing and erasing. Also called **flash EPROM**.

flash EPROM   Same as **flash EEPROM**.

flash lamp   Same as **flash tube**. Also spelled **flashlamp**.

flash memory   **1.** A memory chip that retains its content without power, and which is written and erased in blocks of a fixed size. It is similar to programmable ROM chips, such as EEPROMs, but do not permit bit by bit writing and erasing. Used, for instance, for storage in portable or handheld computers such as notebooks and PDAs, in cellular phones, MP3 players, or to store the BIOS of a system. Also called **flash-memory chip**. **2.** The content stored in **flash memory** (**1**).

flash-memory card   Same as **flash card**.

flash-memory chip   Same as **flash memory** (**1**).

flash-over   Same as **flashover**.

flash-over voltage   Same as **flashover voltage**.

flash plating   The electrodeposition of a very thin layer. Used, for instance, to provide a base for a subsequent layer of plating.

flash RAM   Flash memory utilized as RAM.

flash ROM   Flash memory utilized as ROM.

flash test   A method of testing insulation, consisting in momentarily applying a voltage that is much higher than the working voltage.

flash tube   The tube utilized to produce a **flash** (**2**). Also spelled **flashtube**. Also called **flash lamp**, or **electronic flash tube**.

flashback voltage   The peak inverse voltage required for ionization in a gas-discharge tube.

flasher   A circuit or device which serves to **flash** (**2**) one or more lights. Also called **light flasher**.

flashlamp   Same as **flash tube**.

flashover   A disruptive discharge in the form of an arc or spark between conductors, such as electrodes. It is generally caused by high voltages. Also spelled **flash-over**.

flashover voltage   Also spelled **flash-over voltage**. **1.** The voltage at which **flashover** occurs. **2.** The voltage at which a disruptive discharge occurs between conductors, such as electrodes, separated by an insulating material. This discharge may occur around or along the surface of said insulator. Also called **sparkover voltage** (**2**).

FlashPix   A file format providing multiple-resolution storage of images. Used, for instance, to provide the appropriate resolution of graphics displayed by Web pages, based on the connection speed.

flashtube   Same as **flash tube**.

flat   **1.** Lacking curvature. Said, for instance, of a surface, or of a frequency response. **2.** Completely discharged. Said of a battery.

flat address space   An address space in which memory addresses start at zero, and work their way up to the maximum address. Also called **linear address space**.

flat cable   A cable in which the conductors are arranged along the same plane, and laminated or molded into a flat flexible ribbon. Used, for instance, to connect components in computers, LCD displays, DVD players, copiers, and so on. Also called **flat-conductor cable**, **flat-flexible cable**, **flexible-flat cable**, **ribbon cable**, or **tape cable**.

flat-conductor cable   **1.** Same as **flat cable**. **2.** A **flat cable** with flat conductors.

flat display   A video display which is comparatively thin. At present this represents a depth of about 50 mm or less, but this number is being reduced over time. Liquid-crystal, plasma, and electroluminescent displays are examples. Such displays usually feature high resolution, high contrast, minimal image distortion, and a light weight. Used, for instance, in computers and TVs. Also called **flat monitor**, **flat-panel display**, **flat-panel monitor**, or **panel display**.

flat fading   Fading in which all frequency components are equally affected at all times.

flat file   A data file whose contained records are of a single type, and which have no structured relationships between them.

flat-flexible cable   Its abbreviation is **FFC**. Same as **flat cable**.

flat frequency response   Also called **flat response**, or **uniform frequency response**. **1.** For a component, circuit, device, piece of equipment, or system whose output is an interval of frequencies, a response in which all frequencies have equal amplitude. **2.** For a component, circuit, device, piece of equipment, or system whose output is an interval of frequencies, a response in which all frequencies are within a specified interval of amplitudes. For example, a high-fidelity amplifier may have a frequency response of 10 Hz to 60 kHz, +/- 0.5 decibels, and be considered to be flat.

flat line   A radio-frequency transmission line, or an interval of one, whose standing-wave ratio is 1:1, or very nearly so. Also called **flat transmission line**.

flat monitor   Same as **flat display**.

flat pack   A flat rectangular or square IC package with leads projecting horizontally from two or four of its sides. An example is a quad flat pack. Also spelled **flatpack**.

flat-panel display   Same as **flat display**. Its abbreviation is **FPD**.

flat-panel distributed-mode speaker   Same as **flat-panel speaker**.

flat-panel loudspeaker   Same as **flat-panel speaker**.

flat-panel monitor   Same as **flat display**.

flat-panel speaker   **1.** An acoustic radiator consisting of a flat panel a few millimeters in thickness, which vibrates over its entire surface without pistonic motion. Such panels are very lightweight, disperse sound evenly in all directions at all reproduced frequencies, and have a fairly linear frequency response. Also called **flat-panel loudspeaker**, **flat-panel distributed-mode speaker**, or **distributed-mode speaker panel**. **2.** A speaker which is especially thin and usually lightweight. Used, for instance, as computer speakers placed on a desk, or as wall-mounted speakers in a home theater system. Also called **panel speaker**.

flat-panel speaker technology   The technology employed in a **flat-panel speaker**.

flat-panel television   Same as **flat-panel TV**.

flat-panel TV   A TV with a **flat display**.

flat response   Same as **flat frequency response**.

flat screen   The viewing screen of a **flat display**.

flat television screen   The viewing screen of a **flat-panel TV**.

flat top   **1.** The uniform interval of frequencies within a **flat-top response**. **2.** The horizontal part of an antenna. **3.** Same as **flat-top antenna**.

flat-top antenna   An antenna with two or more radiating elements which are parallel to each other and to the ground, each fed at or near their midpoint. Also called **flat top** (**3**).

flat-top response   A response characteristic which shows an interval of frequencies which is transmitted uniformly. Outside of this range the frequencies are highly attenuated, providing for an overall graph with a mostly flat top. Also called **bandpass response**.

flat transmission line   Same as **flat line**.

flat TV screen   The viewing screen of a **flat-panel TV**.

flatbed scanner   An optical scanner which provides a flat surface upon which that to be scanned is placed. This contrasts, for instance, with a handheld scanner.

flatpack   Same as **flat pack**.

flaw   An imperfection, defect, irregularity, or discontinuity. For instance, imperfections in a crystal lattice.

Fleming's left-hand rule   Also called **Fleming's rule** (**1**), or **left-hand rule**. If conventional current is used instead of electron flow, each of these become **right-hand rules**, as conventional current travels in the opposite direction. **1.** A rule stating that if the thumb of a left hand is oriented along the same axis as the flow of electrons through a conducting wire, that the fingers of this hand will curl along the same direction as the magnetic field produced by the wire. **2.** For a conducting wire moving through a magnetic field, a rule stating that if the middle finger, index finger, and thumb of a left hand are extended at right angles to each other, that the middle finger will indicate the direction of the flow of electrons, the index finger the direction of the magnetic field, and the thumb will indicate the direction of the movement of the wire.

Fleming's right-hand rule Also called **Fleming's rule (2)**, **hand rule**, or **right-hand rule**. If electron flow is used instead of conventional current, each of these become **left-hand rules**, as electron flow is in the opposite direction. **1.** A rule stating that if the thumb of a right hand is oriented along the same axis as the current flow through a conducting wire, that the fingers of this hand will curl along the same direction as the magnetic field produced by the wire. **2.** For a conducting wire moving through a magnetic field, a rule stating that if the middle finger, index finger, and thumb of a left hand are extended at right angles to each other, that the middle finger will indicate the current flow, the index finger the direction of the magnetic field, and the thumb will indicate the direction of the movement of the wire. This rule also applies if the conducting wire is substituted by an electron beam.

Fleming's rule Also known as **hand rule**. **1.** Same as **Fleming's left-hand rule**. **2.** Same as **Fleming's right-hand rule**.

Fletcher-Munson curves A series of curves which plot the sound intensities of pure tones, in decibels, versus frequency, in hertz, for all audible frequencies, so that all frequencies are perceived at the same intensity. The hearing sensitivity of people varies by frequency, and this sensitivity also varies depending on the sound-pressure level. Also called **equal-loudness curves**, or **loudness curves**.

Fletcher-Munson effect The non-linear sensitivity, as a function of frequency, that humans have to sound. Generally speaking, sensitivity drops off as sounds approach either extreme of the audible-frequency range. A healthy person with good hearing can usually detect frequencies ranging from about 20 Hz to about 20 kHz.

flex circuit Abbreviation of **flexible circuit**.

flexible circuit A printed circuit mounted on a flexible substrate, allowing a given amount of bending, twisting and/or stretching without damage. Flex circuits may be mounted, for instance, on a thin plastic or polyester film, might have circuitry on both sides of the substrate, and can be single or multi-layer. Its abbreviation is **flex circuit**.

flexible connector An electric connector which permits a given amount of expansion, contraction, or movement of the connected parts. Well suited, for instance, in environments where the temperature varies considerably.

flexible coupling **1.** A coupling which allows a given amount of movement and/or up to a specified amount of misalignment, while still functioning properly. **2.** A **flexible coupling** (1) linking waveguides.

flexible disc Same as **floppy disk**. Same as **flexible disk**.

flexible disk Same as **floppy disk**. Also spelled **flexible disc**.

flexible-flat cable Same as **flat cable**. Its abbreviation is **FFC**.

flexible manufacturing Manufacturing utilizing a **flexible manufacturing system**.

flexible manufacturing system A manufacturing system which integrates automation, computing, and often robotics, to be able to be reprogrammed for varied tasks with a minimum of difficulty. For example, such a system may enable a production line to be quickly rebalanced according to demand. Its abbreviation is **FMS**.

flexible resistor A wire-wound resistor whose wire is rolled around a flexible insulating core.

flexible shaft A shaft, such as that of an adjustable component, made of a flexible material, or of flexible segments.

flexible waveguide A waveguide which can withstand a given amount of bending, twisting and/or stretching without significantly affecting its electrical properties.

flicker **1.** A brief variation or fluctuation, especially of light. **2.** Brief and irregular fluctuations in a displayed image. Seen, for instance in CRTs used in TVs or computer monitors. When such fluctuations occur as a result of unwanted fluctuations in the chrominance and/or luminance signal, it is called **color flicker**.

flicker effect **1.** Erratic variations in the output current of an electron tube, due to an irregularly coated cathode or to random changes in cathode emissions. **2.** The sensation of flicker evoked by a frequency of light fluctuation at or below the **flicker fusion rate**.

flicker frequency **1.** The frequency at which an image on a screen flashes on and off. For images to be perceived as continuous, its frequency must be greater than the **flicker fusion rate**. **2.** Same as **flicker fusion rate**.

flicker fusion rate For a given luminance, the frequency of light fluctuation that evokes a sensation of flicker. Above this frequency, the perceived flicker disappears. The perception of flickering is also influenced by the retinal field-of-view, and color. Also called **flicker frequency (2)**, or **critical flicker frequency**.

flicker noise Noise due to the **flicker effect** (1), and whose amplitude exceeds that of shot noise. Flicker noise may also be produced by semiconductors, as a result of random variations in their electrical activity.

flickering Brief variations or fluctuations, especially of light.

flight control **1.** The utilization of electrical, electronic, optical, magnetic, and mechanical components, devices, equipment, and systems, to monitor and control a flying vehicle. For instance, such a system which monitors and controls surfaces which maneuver and stabilize the craft. **2.** The monitoring and managing of the flow of aircraft in the air or on the ground.

flight-control system A system utilized for **flight control**.

flight telerobotic servicer A remotely controlled robotic device utilized to assemble, maintain, and/or repair satellites or space vehicles.

flight testing The testing, under actual flying conditions, of electrical or electronic components, devices, equipment, or systems which are to be used in flying vehicles.

flint glass A highly-refractive lead-containing glass which is used in electrical equipment, vacuum tubes, and as lenses.

flip chip A surface-mount chip with terminals on one side, and which is flipped face-down for bonding to the substrate or package.

flip chip bonding A bonding method in which a chip is flipped face-down for bonding to the substrate or package. Ball bonding, for instance, may be used to connect the terminals electrically.

flip coil A small coil which is inserted into a magnetic field to measure its field strength, or any variations in it. Such a coil is connected to an indicating instrument, such as a ballistic galvanometer or a fluxmeter, and may or may not incorporate an amplifier. A flip coil may also be utilized to examine the magnetic flux distribution of a magnetic field. Also called **search coil**, **pickup coil**, **magnetic test coil**, or **magnetic probe**.

flip-flop A two-stage multivibrator circuit with two stable states. In one of the stable states, the first stage conducts while the other is cut off, while in the other state the reverse is true. When the appropriate input signal is applied, the circuit flips from one state to the other. It remains in this state until the next appropriate signal arrives, at which time it flops back. Used extensively in computers and counting devices. Its abbreviation is **FF**. Also known as **flip-flop circuit, bistable multivibrator, Eccles-Jordan circuit, bistable flip-flop**, or **trigger circuit (2)**.

flip-flop circuit Same as **flip-flop**.

FLL Abbreviation of **frequency lock loop**.

float 1. To apply a **float charge**. 2. To leave isolated from a ground connection. 3. To leave isolated from a voltage supply.

float charge Also called **floating charge**. 1. The application of a continuous charge to a storage battery, so as to maintain it as close as possible to its fully charged condition. 2. The application of a slow and continuous charge to a battery, so as to compensate for internal losses and small discharges. Also called **trickle charge (1)**.

float switch A switch that is actuated when the surface of a liquid reaches a given level.

float voltage The voltage necessary to maintain a **float charge**.

floating 1. The applying of a **float charge**. 2. The condition of being isolated from a ground connection. Also, a component, circuit, or device so isolated. 3. The condition of being isolated from a voltage supply. Also, a component, circuit, or device so isolated. 3. Not having a fixed position, location, or setting.

floating battery A storage battery receiving a **float charge (1)**. Such a battery is connected in parallel with another source of power, and may be used, for instance, as a backup when the main power source fails or is interrupted.

floating-carrier modulation A type of amplitude modulation in which the amplitude of the carrier wave is varied according to the percentage of modulation, providing for an essentially constant modulation factor. Also called **controlled-carrier modulation**, or **variable-carrier modulation**.

floating charge Same as **float charge**.

floating control A type of control system in which there is a fixed relationship between the error signal and the speed of the control element.

floating input An input circuit which is isolated from a ground connection.

floating junction In a semiconductor, a junction through which no net current flows.

floating neutral A conductor whose voltage to ground varies, depending on the conditions of the circuit.

floating output An output circuit which is isolated from a ground connection.

floating-point Pertaining to **floating-point notation**.

floating-point accelerator Hardware, such as a circuit board or a chip, which speeds up the performance of **floating-point operations**. When it is a chip, also called **floating-point processor**. Its abbreviation is **FPA**.

floating-point arithmetic Arithmetic performed utilizing **floating-point notation**. Used, for instance, by a floating-point processor.

floating-point calculations Same as **floating-point operations**.

floating-point co-processor Same as **floating-point processor**.

floating-point coprocessor Same as **floating-point processor**.

floating-point notation A numeric format in which each number is represented by two numbers, and in which the decimal point is not in a fixed location. The first number, the mantissa, specifies the significant digits, while the second number, or exponent, specifies its magnitude. For ex-

ample, 314,000,000 may be expressed as 3.14×10^8. Although any number, with any radix, may be represented in this manner, it is usually used only for very small or very large numbers. Also called **floating-point representation, scientific notation**, or **exponential notation**. When an E is used instead of a 10, or the radix utilized, it is also called **E notation**.

floating-point number A number expressed in **floating-point notation**.

floating-point operations Calculations performed utilizing **floating-point arithmetic**. For example, an addition or multiplication of two floating-point numbers. Also called **floating-point calculations**.

floating-point operations per second Same as **FLOPS**.

floating-point processor An arithmetic-logic unit designed to perform **floating-point operations**. A CPU may have such a processor incorporated, or a second chip may be utilized. Floating-point processors are especially useful when a computer performs sophisticated calculations, such as those frequently needed in CAD, spreadsheet, and scientific programs. Also called **floating-point coprocessor, floating-point unit, math coprocessor**, or **numeric processor**.

floating-point register A computer register designed to store **floating-point numbers**.

floating-point representation Same as **floating-point notation**.

floating-point system A numeric system utilizing **floating-point notation**.

floating-point unit Its abbreviation is **FPU**. Same as **floating-point processor**.

floating zone In a zone-refining technique, the molten zone in which the impurities are dissolved. Also called **molten zone**.

flood attack A denial-of-service attack in which numerous synchronization packets are maliciously sent to a host or server, which responds to said packets, but which then does not receive the final confirmation from the requesting node. While the host or server waits and eventually times out, legitimate connections desired by other users may not be granted, and enough unanswered requests may lead to an overload or crash. Also called **SYN flood**.

flood gun Same as **flooding gun**.

flooding gun In a storage tube, an electron gun whose emitted electrons uniformly cover the entire screen, intensifying the image produced by the writing gun. Also called **flood gun**.

floor mat A floor covering with properties appropriate for the environment it is used in. For instance, in a computer center, such a mat might be used to reduce or eliminate the buildup of electrostatic charges created by walking or shifting. Such mats are usually grounded.

floor outlet A power line termination, located on a floor, whose socket serves to supply electric power to devices or equipment whose plug is inserted into it.

floppy Same as **floppy disk**.

floppy disc Same as **floppy disk**.

floppy disk A digital magnetic storage medium which is in the form of a rotating plastic plate which is flexible. Information is encoded by altering the magnetic polarity of minute portions on the coated surface of such a disk. Floppy disks are much slower than hard disks, generally hold 1.44 megabytes of data, and their usual format is 3.5 inches, with a rigid case. Also spelled **floppy disc**. Also called **floppy, flexible disk**, or **diskette**.

floppy-disk controller A chip, along with the associated circuitry, which enables a computer to control a floppy-disk drive. A floppy-disk controller handles tasks such as the positioning of read/write heads, and the transfer of data.

floppy-disk drive A disk drive which reads data from, and writes data to, a **floppy disk**. Also called **floppy drive**.

floppy drive Same as **floppy-disk drive**.

floppy optical Same as **floptical**.

flops Same as **FLOPS**.

FLOPS Abbreviation of **fl**oating-point **o**perations **p**er **s**econd. The number of floating-point operations performed per second. It is a commonly used benchmark to measure the speed of a processor. Such speeds are usually indicated in multiples of FLOPS, such as gigaflops, or teraflops.

floptical Acronym for **flop**py op**tical**. A disk or disk drive which incorporates a combination of magnetic and optical technologies, to provide more storage capacity and speed than ordinary floppy disks.

flow **1.** A smooth and uninterrupted motion, progress, or sequence. For example, the movement of a fluid through a duct. **2.** The movement of electric charges. For instance, the flow of electrons through a conductor. **3.** The movement of information through a system. Also, the sequence in which operations are performed. For example, the movement from point to point within a flowchart.

flow chart Same as **flowchart**.

flow control A system which regulates a flow through a given channel, such as a communications line, a conduction path, or a duct. For instance, it may entail a process which times transmitted signals in a manner that faster devices can communicate with slower devices without data loss.

flow diagram Same as **flowchart**.

flow direction The direction, usually indicated by arrows, in which operations proceed in a **flowchart**.

flow line In a **flowchart**, a line or arrow indicating the direction in which operations proceed. Also spelled **flowline**.

flow meter Same as **flowmeter**.

flow soldering A automated method of soldering electronic components to circuit boards, in which molten solder is pumped from a reservoir through a spout to form a wave. The board is passed through the wave via an inclined conveyor. This technique minimizes the heating of the board. Also called **wave soldering**.

flowchart Also spelled **flow chart**. Also called **flow diagram**, or **control diagram**. **1.** A diagram which uses a set of standard symbols to represent the sequence of operations of a system. **2.** A diagram which uses a set of standard symbols to represent the sequence of operations of a computer program or system. Such a chart may show, for instance, the flow of data, or the steps of a subroutine.

flowline Same as **flow line**.

flowmeter An instrument which measures and indicates the flow of a fluid. Also spelled **flow meter**. Also called **fluid meter**, **fluid-flow meter**, or **velocimeter (2)**.

fluctuating current A DC whose value varies in an irregular manner.

fluctuation noise Noise which fluctuates in a random manner, and which usually is the aggregate of a large number of overlapping transient disturbances. The occurrence and magnitude of such noise can not be predicted. Examples include electrical noise and cosmic noise. Also called **random noise**.

fluence The passing of particles, such as electrons or photons, through a given area, such as a square centimeter. May be expressed, for instance, in particles per second. Also called **particle fluence**.

fluid A form of matter which has no defined shape, and whose atoms or molecules move freely past one another. A fluid is usually a gas or a liquid, and tends to assume the shape of its container.

fluid computer A digital computer which utilizes logic elements powered by a fluid, such as air. Such a computer has no electronic circuits, nor any moving parts.

fluid damping The use of a fluid to reduce or limit the amplitude of a mechanical motion, such as a vibration. An example is the use of a ferrofluid for damping the motion of a high-frequency speaker. Also called **viscous damping**.

fluid-flow meter Same as **flowmeter**.

fluid-level indicator An electronic device, instrument, or system which indicates the level of a fluid.

fluid logic Logic operations performed using interactions between fluids. Such operations do not utilize electricity, nor require any moving parts. Seen, for instance, in a fluid computer.

fluid meter Same as **flowmeter**.

fluidic **1.** Of, pertaining to, or characteristic of a **fluid**. Also, in the form of a **fluid**. **2.** Pertaining to, controlled by, or employing **fluidics**.

fluidics A technology which employs fluids in motion to carry out functions that would otherwise be performed by electrical and electronic circuits. Such functions include those involved in control, amplification, processing, logic, and sensing. Devices employing this technology have no moving parts, are characterized by high reliability, ease of maintenance, and extremely high resistance to electromagnetic interference. Used, for instance, where electronic circuits can not be used, such as explosive environments.

fluorescence The emission of electromagnetic radiation, such as light, by a body which has been excited by another form of energy, such as electron bombardment. Fluorescence is a form of luminescence whose persistence is less than about 10^{-8} second, while **phosphorescence** persists longer. That is, within 10^{-8} second or sooner from the excitation ceasing, so does fluorescence. Used, for instance, in spectroscopy, CRTs, and in lighting.

fluorescence microscope A light microscope which utilizes fluorescent light to analyze specimens. Well suited for the study of living cells without damaging samples. Also called **fluorescent microscope**.

fluorescence microscopy The use of a **fluorescence microscope** for the analysis of specimens. Also called **fluorescent microscopy**.

fluorescent lamp A gas-discharge lamp whose light is produced through **fluorescence**. Such a lamp usually consists of a tube coated with a fluorescent substance, and contains mercury vapor. A current passing through the tube excites the vapor, whose ultraviolet emissions excite the coating, which in turn emits visible light. Also called **fluorescent tube**.

fluorescent material A material which emits electromagnetic radiation, especially light, when irradiated with another form of energy, such as X-rays or an electron beam.

fluorescent microscope Same as **fluorescence microscope**.

fluorescent microscopy Same as **fluorescence microscopy**.

fluorescent screen A screen coated with a **fluorescent substance**. Used, for instance, in CRTs, fluoroscopes, or scintillation counters.

fluorescent substance A substance which emits electromagnetic radiation, especially light, when irradiated with another form of energy, such as X-rays or an electron beam. Examples include cesium chloride, and cadmium sulfide.

fluorescent tube Same as **fluorescent lamp**.

fluorinated ethylene propylene resin A thermoplastic material with excellent electrical insulating properties, in addition to resistance to heat and chemicals. Used, for instance, for wire and cable insulation. Its abbreviation is **FEP resin**.

fluorine A chemical element whose atomic number is 9. It is a pale yellow diatomic gas which is poisonous and highly corrosive. It is the most electronegative element, and among the nonmetallic elements is the most reactive. It reacts spontaneously and vigorously with most elements, thus it is never found uncombined in nature. It has about 10 known isotopes, of which one is stable. Fluorine has various applications, including its use in ceramics, glass etching, low-friction materials, and lasers. It is a halogen, and its chemical symbol is **F**.

fluorite Crystals whose color occurrence is varied, and whose chemical formula is CaF_2. Used in ceramics, electrodes, arc welders, phosphors, and in optical equipment.

fluorocarbons Chemical compounds which incorporate both carbon and fluorine atoms. Fluorocarbons are chemically inert, highly heat-resistant, nonflammable, and nontoxic. Used, for instance, as refrigerants, dielectrics, electrical insulators, lubricants, and for nonstick surfaces.

fluorography The photographic recording of an image formed on a fluorescent screen. Also called **photofluorography**.

fluoroscope An instrument in which an X-ray tube is utilized in conjunction with a fluorescent screen. X-ray images of objects placed between them may be directly viewed, without the need to take and develop X-ray photographs. In medicine, the fluorescent screen has been replaced by an image intensifier.

fluoroscopic Pertaining to, or utilizing a **fluoroscope**.

fluoroscopy The use of a **fluoroscope** examination and analysis.

fluorspar Same as **fluorite**.

flush 1. To clear all or part of a temporary storage area, such as a memory buffer. Also, to move the data contained in such a storage area to a more permanent medium, such as a hard disk. 2. Aligned with a given plane or surface.

flutter 1. Rapid fluctuations in a signal or quantity, such as amplitude or frequency. 2. In sound reproduction, a form of distortion manifesting itself as rapid fluctuations of pitch. It is caused by equipment speed variations occurring during recording, dubbing, or reproduction. When occurring at low frequencies, known as **wow**. 3. The rapid fluctuation in the strength of a received signal. May be caused, for instance, by improper tuning, or antenna movements.

flux 1. For a given region, a measure of the strength of a field. For example, that of electric and/or magnetic fields. Flux is usually represented by lines of force. 2. For a given region, the magnetic lines of force. The closer together the lines, the greater the strength of the magnetic field. Also called **magnetic flux (2)**. 3. For a given region, the electric lines of force. The closer together the lines, the greater the strength of the electric field. Also called **electric flux (2)**. 4. The rate of flow of a given quantity across a given area. In the case of light, for instance, the number of photons striking a given surface area perpendicular to its path, per unit time. 5. A material which better prepares surfaces for brazing, soldering, or welding. It may do so, for instance, by removing oxides. Also called **solder flux**.

flux density 1. A manner of describing the strength of a field. It may be expressed, for instance, as the concentration of lines or force, or the rate of flow of a quantity passing perpendicularly through a given area. 2. A quantitative measure of the strength of a magnetic field present within a magnetic medium. It is the number of lines of flux per cross-sectional area of a magnetic circuit, and its SI unit is the tesla. Also called **magnetic flux density**, **magnetic induction (1)**, or **magnetic displacement**. 3. A vector quantity pertaining to the charge displaced by an electric field within a given medium. It is equal to the electric field strength multiplied by the permittivity of the medium. It is usually expressed in coulombs per square meter. Also called by various other names, including **electric flux density**, **electric induction**, or **electric displacement**.

flux-gate Same as **fluxgate**.

flux-gate compass Same as **fluxgate compass**.

flux-gate magnetometer Same as **fluxgate magnetometer**.

flux guide A magnetic material which is utilized to guide **flux (2)** in the desired direction. Used, for instance, in induction heating.

flux leakage Magnetic flux which does not pass through a useful or intended part of a magnetic circuit. Also called **leakage flux**.

flux lines Same as **force lines**.

flux linkage The passage of magnetic flux from one circuit component to another. Also called **magnetic flux linkage**, or **linkage (3)**.

flux meter Same as **fluxmeter**.

flux residue Any excess **flux (5)** which becomes a contaminant near the brazed, soldered, or welded surface. Also called **solder flux residue**.

flux reversal A change in the polarity of a minute magnetic particle on the surface of a magnetic storage medium, such as a tape or a disk. Such a transition changes a 0 to a 1, or vice versa, enabling the recording of data. Also called **flux transition**.

flux transition Same as **flux reversal**.

fluxgate A detector whose output signal amplitude and phase is proportional to the magnitude and direction of an external magnetic field acting along its axis. Used, for instance, to indicate the direction of the earth's magnetic field. Also spelled **flux-gate**.

fluxgate compass A compass incorporating a **fluxgate**. Also spelled **flux-gate compass**.

fluxgate magnetometer A magnetometer incorporating a **fluxgate**. The degree of saturation the magnetic core of the fluxgate determines the strength of the measured magnetic field. Also spelled **flux-gate magnetometer**.

fluxmeter An instrument which measures **flux (2)**. Such an instrument typically incorporates a wire coil which is moved across the flux lines of the magnetic field being measured, and usually provides readings in gauss, teslas, webers, or maxwells. Also spelled **flux meter**.

fly-by-wire Same as **fly-by-wire system**.

fly-by-wire aircraft An aircraft utilizing a **fly-by-wire system**.

fly-by-wire control Control of an aircraft utilizing a **fly-by-wire system**.

fly-by-wire system The use of computers to partially or completely control an airplane in flight. The computer determines how best to control surface movements, the rate of fuel consumption, and so on. A pilot may give instructions to the computer, which in turn sends electrical signals, via wires or cables, which actuate the appropriate control surfaces, so there is no mechanical linkage involved. If desired, the computer can handle all processes automatically, including landings. Since a computer can update all setting thousands of times a second, the use of this system should result in a safer, smoother, and more efficient flight. Also called **fly-by-wire**.

fly-eye lens A lens consisting of numerous small lenses. Such a lens may consist, for instance, of a fiber-optic bundle. Use, for example, in a stereoscopic TV.

flyback 1. In a CRT, the return of the electron beam to its starting point after a sweep or trace. Also called **retrace**, or **kickback (2)**. 2. Same as **flyback time**. 3. Within a sawtooth waveform, the interval in which there is a rapid rise or fall. Ideally, such an increase or decrease would be instantaneous. **4.** A rapid decrease of a current or voltage, which until that instant had been rising.

flyback power supply A power supply that utilizes the **flyback (1)** interval to generate a high-voltage DC utilized by the second anode of a CRT. Also called **kickback power supply**.

flyback time The time interval that elapses during **flyback (1)**. Also called **flyback (2)**, or **retrace time**.

flyback transformer A transformer utilized with a flyback power supply. In a TV receiver, also called **horizontal-output transformer**.

flying erase head In certain video devices, such as VCRs and camcorders, an erase head which can assume multiple positions, as opposed to being fixed. Especially useful when editing.

flying head A head, such as a read/write head, which can assume multiple positions, as opposed to being fixed.

flying spot The small and intensely bright spot utilized in a **flying-spot scanner**.

flying-spot scanner A device which utilizes a small and intensely bright spot which scans an object, scene, screen, film, or the like, and transforms the original image into a series of electrical signals by focusing it onto a photoelectric cell. The scanning may be mechanical or electrical. Used, for instance, for TV transmission, telecine transfer, or character recognition. Also called **scanner (4)**, **light-spot scanner**, or **optical scanner (2)**.

flying-spot scanning The use of a **flying-spot scanner**.

flywheel A disc, usually of a comparatively large mass, which when rotating tends to retain its angular velocity. Used, for instance, to help a tape deck maintain a constant speed, or in flywheel batteries.

flywheel battery A battery that incorporates a flywheel for energy storage. Such a battery may utilize a solar panel to power a motor that converts the generated electric current into mechanical energy by spinning the flywheel. The speed of the flywheel increases as it accumulates energy, and said speed decreases as a load consumes power. Such batteries are very efficient and durable.

flywheel effect In an oscillator, the continuation of oscillations during the intervals between the brief pulses of excitation energy. It may occur, for instance, thorough inductance-capacitance circuit interaction.

fm Abbreviation of **femtometer**.

Fm Chemical symbol for **fermium**.

FM Abbreviation of frequency **m**odulation. A form of modulation in which the instantaneous frequency of a sine-wave carrier is varied above and below the center frequency by a modulating signal. The center frequency is that of the unmodulated carrier, and its variation is proportional to the instantaneous value of the modulating signal. It is a form of angle modulation, and is used, for instance, for broadcasting. This contrasts with **AM**, where the amplitude of the carrier wave and its sidebands are modulated, or varied, but its frequency is kept constant, or with **PM**, where the phase angle of the carrier wave is varied.

FM/AM radio Abbreviation of frequency-modulation/amplitude-modulation **radio**. Same as **FM/AM receiver**.

FM/AM radio receiver Abbreviation of frequency-modulation/amplitude-modulation **radio receiver**. Same as **FM/AM receiver**.

FM/AM receiver Abbreviation of frequency-modulation/amplitude-modulation **receiver**. A radio receiver that can receive either frequency-modulated or amplitude-modulated signals.

FM/AM transmitter Abbreviation of frequency-modulation/amplitude-modulation **transmitter**. A radio transmitter that can send either frequency-modulated or amplitude-modulated signals.

FM broadcast band Abbreviation of frequency-modulation **broadcast band**. A band of frequencies allocated to FM broadcasting stations. In the United States, the standard FM broadcast band is 88 to 108 MHz.

FM broadcasting Abbreviation of frequency-modulation **broadcasting**. The broadcasting of **FM signals**.

FM detector Abbreviation of frequency-modulation **detector**. A circuit or device which extracts a signal which is frequency-modulated.

FM noise Abbreviation of frequency-modulation **noise**. Modulation of the frequency of a carrier wave by signals other than those intended to modulate it.

FM radar Abbreviation of frequency-modulated **radar**. A radar in which a frequency-modulated wave is transmitted. The echo beats with the transmitted wave, and the magnitude of the beat frequency allows for the range of the scanned object to be determined.

FM radio Abbreviation of frequency-modulation **radio**. Same as **FM receiver**.

FM radio receiver Abbreviation of frequency-modulation **radio receiver**. Same as **FM receiver**.

FM receiver Abbreviation of frequency-modulation **receiver**. A radio receiver that can receive FM signals only. In the United States, the standard FM broadcast band is 88 to 108 MHz.

FM signal Abbreviation of frequency-modulated **signal**. A signal whose carrier wave is frequency-modulated.

FM stereo Abbreviation of frequency-modulation **stereo**. Also called **stereo FM**. **1.** The broadcasting of FM signals with two or more tracks. **2.** The transmission and reception of **FM stereo (1)** broadcasts.

FM synthesis Abbreviation of frequency-modulation **synthesis**. The use of FM to synthesize signals. Utilized, for instance, as a technique for approximately simulating the sounds of musical instruments.

FM transmission Abbreviation of frequency-modulated **transmission**. A transmission whose carrier wave is frequency-modulated.

FM transmitter Abbreviation of frequency-modulated **transmitter**. A transmitter whose carrier wave is frequency-modulated.

FM tuner Abbreviation of frequency-modulation **tuner**. A radio tuner that can receive FM signals only. In the United States, the standard FM broadcast band is 88 to 108 MHz.

FM wave Abbreviation of frequency-modulated **wave**. A wave whose frequency varies proportionally to the modulating signal.

FMR Abbreviation of **ferromagnetic resonance**.

FMS Abbreviation of **flexible manufacturing system**.

Fn keys Abbreviation of **function keys**.

FN keys Abbreviation of **function keys**.

focal distance Same as **focal length**.

focal length Also called **focal distance**. **1.** For a lens, the distance between its optical center and its principal focus. **2.** For a mirror, the distance between its surface and its principal focus.

focal point Same as **focus (3)**.

focal spot The spot where the intensity of an electron beam or laser beam is greatest.

focus 1. To make light rays, a beam of radiation, energy, or particles converge. For instance, to make an electron beam converge properly on the screen of a CRT. Also called **focusing (1)**. **2.** To make an adjustment which improves the sharpness of an image. Also, to adjust until the sharpest possible image is obtained. For example, to adjust a camera lens to achieve this effect. Also called **focusing (2)**. **3.** In an optical system, such as a lens or a mirror, a point towards which an incident bundle of parallel rays of light converge. Also called **focal point**. **4.** In a geometric figure, such as an ellipse or hyperbola, a point which helps determine a conic section.

focus coil Same as **focusing coil**.

focus control In a CRT, such as that in a TV receiver, a device which adjusts the size of the luminous spot on the screen so as to obtain the sharpest possible image. A **focusing coil (1)**, for instance, may be used for this purpose.

focusing 1. Same as **focus (1)**. **2.** Same as **focus (2)**.

focusing anode In a CRT, an anode utilized as a **focusing electrode**.

focusing coil Also called **focus coil**. **1.** A coil which produces a magnetic field utilized to focus an electron beam. **2.** A **focusing coil (1)** used in a CRT. The focusing is accomplished by varying the DC flowing through such a coil mounted on the neck of the tube.

focusing electrode 1. An electrode utilized to focus an electron beam. As the potential applied to said electrode is varied, so does the cross-sectional area of the electron beam, which in turn determines its focusing. **2.** A **focusing electrode (1)** used in a CRT.

focusing magnet A magnet whose magnetic field is utilized to focus an electron beam. Used, for instance, in a CRT.

foil A thin sheet of a metal. It may be defined as no thicker than a given amount, such as 0.15 mm. An example is the aluminum or tantalum foil utilized in electrolytic capacitors.

foil capacitor A capacitor which utilizes a metal foil, such as that of aluminum or tantalum, as one of the plates. An example is an electrolytic capacitor.

foil pattern A pattern of thin sheets of metal which serve as conductors in a circuit, as opposed to wires. Utilized, for instance, in printed circuits. A foil pattern may be produced, for instance, via etching or plating.

FOIRL Abbreviation of **Fiber-Optic Inter Repeater Link**.

foldback current limiting In a power supply, an overload protection method in which the output current decreases as overload increases. The output reaches a minimum under short-circuit conditions. Foldback current limiting minimizes internal power dissipation under such conditions.

folded dipole An antenna consisting of two parallel half-wave dipole antennas connected at their ends. It is center-fed, its feedpoint impedance is approximately 300 ohms, and it is used, for instance, for TV or FM reception. Also called **folded dipole antenna**.

folded dipole antenna Same as **folded dipole**.

folded horn An acoustic horn which curls or folds, so as to accommodate a longer length into a more compact enclosure.

folder Same as **file folder**.

folder tree A graphic representation of a series of folders and subfolders. Subfolders, or subdirectories, branch out from the root folders, or root directories. Also called **directory tree**.

folding frequency A frequency which is equal to half the Nyquist rate. That is, it is equal to half of the minimum sampling frequency necessary to prevent aliasing. Also called **Nyquist frequency**.

foldover 1. In analog-to-digital conversion, a form of distortion which arises when the sampling frequency is less than twice the highest frequency component. It would occur, for instance, when a 50 kHz waveform is sampled at any frequency below 100 kHz. An anti-aliasing filter may be used to solve this problem. Also known as **aliasing (1)**. **2.** In a TV picture, a form of distortion in which a white line appears along one of the borders of the screen, obscuring part of the image.

font A set of characters which have the same combination of factors such as typeface, style, and size. Fonts are usually stored either as bitmapped fonts, or as scalable fonts.

font card Same as **font cartridge**.

font cartridge A unit which may be plugged-into an available slot in a printer, and which has one or more stored fonts. This expands the selection of fonts available from said printer. Also called **font card**.

font family A set of fonts which, for a given typeface, provides a full complement of styles, sizes, and so on. For instance, a font family may provide italic and/or boldface variations for sizes ranging from 1 point to 960 points.

foot A unit of length equal to 0.3048 meter. Its abbreviation is **ft**.

foot-candle An obsolescent unit of illuminance or illumination, defined as the illuminance incident upon a surface at a distance of 1 foot from a source whose intensity is 1 standard candle. It is equal to approximately 10.764 lux, the latter being the SI unit. Its abbreviation is **fc**. Also spelled **footcandle**.

foot-lambert Same as **footlambert**.

foot per second A unit of velocity which is usually utilized to measure the speed of sound. One foot per second is equal to 0.3048 meter per second. Its abbreviation is **fps**, **ft/s**, or **ft/sec**.

foot-pound Its abbreviation is **ft-lb**. **1.** A unit of work or energy equal to approximately 1.35582 joule. **2.** A unit of torque equal to approximately 1.35582 newton meter. Also called **pound-foot**.

foot switch A switch that is actuated by the pressure of a foot. Used, for instance, when it is necessary for the hands to be occupied with other tasks at the same instant another device needs to be activated.

footcandle Same as **foot-candle**.

footlambert A unit of luminance equal to approximately 3.42626 candelas per square meter. Its symbol is **fL**. Also spelled **foot-lambert**.

footprint The area or region covered by something. For example, the coverage of a fixed satellite, or the surface area a monitor occupies on a desk.

forbidden band Same as **forbidden energy band**.

forbidden energy band In a crystalline solid, such as a semiconductor, the range of energy levels within which electrons may not be located. This contrasts with **allowed energy bands**, which are energy levels within which electrons may be located. Also called **forbidden band**.

force 1. An influence on a body that can modify its movement, or deform it. For example, if such an influence is applied to a free body, said body would be accelerated. Through the application of a force, work is done. Its symbol is F, and its SI unit is the newton. **2.** Any of the **four fundamental forces of nature**. **3.** In computers, to manually intervene in a process, to effect action that would not ordinarily occur.

force feedback In robotics, a mechanism by which information obtained by the force sensors is routed back to the controller. This is useful, for example, for optimal control of an end-effector.

force field A region in space within which a physical force, such as an electric charge, acts. Also called **field of force**.

force lines Also called **flux lines**, **lines of force**, or **lines of flux**. **1.** Within a field where a force acts, imaginary lines whose tangent at any given point represent the direction of said field at that point. The closer together the lines, the greater the strength of the field. May be used, for instance, in the context of electric, magnetic, or gravitational fields. **2.** Within an electric field, imaginary lines whose tangent at any given point represent the direction of said electric field at that point. Electric force lines are utilized to represent the electric force field surrounding a charged particle or body, and the closer together the lines, the greater the strength of the said field. Also called **electric force lines, electric lines of force, electric lines of flux, electric flux lines**, or **electrostatic flux**. **3.** Within a magnetic field, imaginary lines whose tangent at any given point represent the direction of said magnetic field at that point. Magnetic force lines are utilized to represent the magnetic field surrounding a magnetic body or entity, such as a permanent magnet or a medium carrying a current. The closer together the lines, the greater the strength of the magnetic field. Also called **magnetic force lines, magnetic lines of force, magnetic lines of flux**, or **magnetic flux lines**.

force sensor Also called **robot force sensor**. **1.** In robotics, a sensor which measures the forces exerted by a component of the manipulator. **2.** In robotics, a sensor which measures the forces exerted upon a component of the manipulator.

forced oscillation **1.** Oscillation of a mechanical system which occurs only when the external driving force is present. Also called **forced vibration**. This contrasts with **free oscillation** (1), which occurs without an external force. **2.** In a circuit, oscillation which is maintained through the continuous application of the excitation. This contrasts with **free oscillation** (2), which continues after the excitation is removed.

forced vibration Same as **forced oscillation** (1).

foreground **1.** On a computer screen that has more than one window open, the window which is currently being interacted with. All others are **background** (3) windows. **2.** On a screen or display, the area that is being specifically interacted with or observed. Also called **display foreground**. **3.** On a screen or display, the area where characters and/or graphics are displayed. **4.** In computers, processing and other functions that interrupt those of less priority when needed.

foreground job **1.** In computers, a task which interrupts those of less priority when needed. This contrasts with **background job** (1). **2.** In computers, the task or process that is controlling the system, and is currently accepting input. Also called **foreground task**, or **foreground process**. This contrasts with **background job** (2).

foreground process Same as **foreground job** (2).

foreground task Same as **foreground job** (2).

foreign exchange service A telephone network service in which calls placed to a given central office are forwarded automatically to another. Useful, for example, when an entity changes locations within the same area code and wishes to retain the same phone number.

foreign format A file format other than the default format an application saves files in. The latter is known as its **native format**. Programs usually have filters to export and import data to and from applications using a different native format.

fork oscillator An oscillator whose frequency is determined by a tuning fork. Also called **tuning fork oscillator**.

fork resonator A resonator whose frequency is determined by a tuning fork. Also called **tuning fork resonator**.

form **1.** The shape of something, such as a crystal. **2.** A shape that is desired, such as that attained via electrochemical machining. **3.** A shape or object around which a coil or winding is wound. **4.** In computers, a document or screen display which has blank fields where information is placed.

form factor **1.** For a periodic quantity such as AC, the ratio of the RMS value to the half-period average value. The form factor for a pure sine wave is approximately 1.111. **2.** A

function which takes into account the ratio of the diameter to length of a coil when computing its inductance. Also called **shape factor** (3). **3.** In computers, the size and shape of a given component, such as a circuit board or disk drive.

form feed A command which makes a printer advance to a new page, or one page forward from the current location on a page. Its abbreviation is **FF**.

format **1.** The general size, shape, and organization of something. **2.** A specific manner in which something is structured, presented, or utilized. For instance, an audio or video format. **3.** In computers, a specific manner of arranging, accessing, processing, saving, displaying, or printing. Examples include data formats, file formats, address formats, record formats, text formats, and so on. Also, to assign such properties. **4.** To prepare a computer disk for utilization. In doing so, the operating system performs tasks such as testing the disk, marking any bad sectors, and organizing the available space into specialized compartments along with addressing, for later access and manipulation of data. Also, to prepare a disc or tape for utilization.

formatted capacity In a disk, disc, or tape, the storage capacity after formatting. Data, such as control information, reduces the **unformatted capacity** to this value.

formatting **1.** To provide a **format** (3). **2.** The process of preparing a computer disk for utilization. Formatting includes tasks such as testing the disk, marking any bad sectors, and organizing the available space into specialized compartments along with addressing, for later access and manipulation of data. Also, the preparation of a disc or tape for utilization.

forming The use of electricity to alter the characteristics of a component or device, such as a capacitor or semiconductor device. Used, for instance, to condition a pn junction. Forming usually takes place during the manufacturing process. Also called **electroforming** (2).

formula **1.** A mathematical expression utilized to relate quantities and solve problems. For example, the mass-energy equation is expressed as: $E = mc^2$, where E is energy, **m** is mass, and c is the speed of light in a vacuum. **2.** A combination of chemical symbols and numbers which describes the chemical composition of a substance. For instance, the chemical formula of boron trichloride is BCl_3, indicating that for each boron atom there are three chlorine atoms. Also called **chemical formula** (2).

FORTH An interactive and extensible computer programming language used in artificial intelligence applications. It was originally an abbreviation of **fourth**.

FORTRAN The first high-level programming language, initially developed in the 1950s, and still popular in various fields, especially those of a scientific and mathematical nature. It is an acronym for **for**mula **tran**slator.

forward To send on to a new destination. For example, to forward data, or a telephone call.

forward-backward counter A counter which can count both upwards and downwards. It has, for instance, both an adding input and a subtracting input, thus giving it the ability to count in both directions. Also called **bi-directional counter**, or **up-down counter**.

forward bias Also called **forward voltage**. **1.** A voltage applied in the **forward direction** of a semiconductor device. **2.** A voltage applied in the forward direction of a component or device.

forward breakover voltage The minimum voltage that must be applied to an electrode, or terminal, for conduction in the **forward direction** to occur.

forward chaining In artificial intelligence, a method of reasoning or problem solving that starts with facts and a set of conditions, and works its way toward a solution. This contrasts with **backward chaining**, which starts with a solution, then works its way backwards to a set of conditions and corresponding facts which would bring about this result. Also known as **forward reasoning**.

forward channel In asymmetrical communications, the channel that flows downstream. It is the faster of the two channels, the other being the **backward channel**.

forward compatibility The state of being **forward compatible**.

forward compatible Software that is designed to be compatible with later versions of software or hardware, or hardware that is designed to be compatible with later versions of hardware or software. For instance, a program that was able to run on a given CPU generation also being able to work with a later generation. Forward compatibility is important in being able to upgrade components, programs, or systems in a simple and efficient manner. Also called **upward compatible**.

forward conduction Conduction through device in the **forward direction**.

forward current A current which flows when a **forward bias** is applied.

forward direction For an electrical or electronic component or device, the direction in which there is lesser resistance to the flow of current. Used, for instance, in the context of semiconductors. This contrasts with **reverse direction**, where the converse is true. Also called **conducting direction**.

forward error correction In communications, an error-correction technique in which redundant bits are added before transmission. These bits are utilized to correct detected errors, in an attempt to make retransmission unnecessary. Its abbreviation is **FEC**.

Forward Explicit Congestion Notification In frame relay, a bit which serves to notify of detected congestion in the destination direction of a given packet. Utilized to help control the flow of data. When the congestion is from the source direction of a packet, a **Backward Explicit Congestion Notification** bit is set. Its abbreviation is **FECN**.

forward path In a feedback control loop, the transmission path between the loop actuating signal and the loop output signal.

forward power The power a transmitter delivers to an antenna. Usually expressed in watts.

forward propagation ionospheric scatter Same as **forward scatter (3)**.

forward reasoning Same as **forward chaining**.

forward recovery The reconstruction of a file to a later version. An earlier version and an activity log are used for this purpose. This contrasts with a **backward recovery**, where an earlier version is reconstructed using a later version.

forward recovery time The time that elapses between a forward bias being applied, and the forward current reaching a specified value.

forward resistance The resistance offered to the flow of current in the **forward direction**.

forward scatter Also called **forward scattering**. **1.** Scattering of waves or particles through angles less than 90° with respect to the initial direction of radiation. **2.** In radio-wave propagation, the scattering of waves away from the transmitter. **3.** In radio-wave propagation, the scattering of waves away from the transmitter after being reflected off the ionosphere. Also called **forward propagation ionospheric scatter**, or **ionospheric scatter**.

forward-scatter communication Communication by means of radio waves that are propagated well beyond line-of-sight distances. This is usually due to scattering by the ionosphere or troposphere. Also called **over-the-horizon communication**, **beyond-the-horizon communication**, **scatter communication**, or **extended-range communication**.

forward-scatter propagation Propagation of radio waves well beyond line-of-sight distances. This is usually due to scattering by the ionosphere or troposphere. Also called **over-the-horizon propagation**, **beyond-the-horizon propagation**, **scatter propagation**, or **extended-range propagation**.

forward-scatter transmission Transmission of radio waves well beyond line-of-sight distances. This is usually due to scattering by the ionosphere or troposphere. Also called **over-the-horizon transmission**, **beyond-the-horizon transmission**, **scatter transmission**, or **extended-range transmission**.

forward scattering Same as **forward scatter**.

forward slash In most computer operating systems, a character which serves to denote division, as seen, for instance, in: 22/7. This character also serves to separate elements within an internet address, as seen, for example in: *http://www.yipeeee.com/whoo/yaah.html*. This character has a place on the keyboard, and is shown here between brackets: [/]. Also known as **slash**, or **virgule**.

forward transfer function In a feedback control loop, the transfer function between the loop actuating signal and the loop output signal.

forward voltage Same as **forward bias**.

forward voltage drop **1.** The voltage across a semiconductor junction carrying current in the **forward direction**. **2.** The voltage across the electrodes of a diode carrying current in the forward direction, that is, from anode to cathode. Also called **diode forward voltage**, **diode voltage**, or **diode drop**.

forward wave In a traveling-wave tube, a wave whose group velocity is in the same direction as the electron beam.

forwarding **1.** The sending on to a new destination. For example, the forwarding of data, or of telephone calls. **2.** A service offered by telephone companies, which enables calls to automatically be rerouted from one telephone line to another. Calls may be sent to an alternate location, an answering service, or the like. Some systems have more tailored offerings, such as selective call forwarding for programmed numbers, call forwarding only when the line is busy, or forwarding if there is no answer after a specified number of rings. Also known as **call forwarding**. **3.** The resending of an email that has already reached a recipient's address. Once received, an email may be forwarded manually or automatically to one or more email addresses. When an email is manually forwarded, its content may be changed and/or files may be attached. Also called **email forwarding**.

Foster-Seeley detector Same as **Foster-Seeley discriminator**.

Foster-Seeley discriminator A discriminator whose output signal is proportional to the phase-difference between its input signals. It is utilized for the reception of FM signals. Also called **Foster-Seeley detector**, or **phase-shift discriminator (1)**.

Foucault current A circulating current induced in a conductor that is moved through a magnetic field, or which is subjected to a varying magnetic field. Foucault currents are generally undesired, and energy is lost usually in the form of heat. Laminations may be used in magnetic cores to minimize such losses of energy. Also called **eddy current**.

four-channel sound Its abbreviation is **4-channel sound**. **1.** Same as **four-channel sound system**. **2.** The sound produced by a **four-channel sound system**.

four-channel sound system A sound system in which there are four channels, each feeding its own speaker. It is meant to be an enhancement of two-channel stereo. When recording, the signals of the four channels are encoded into two channels, and when reproducing, the signals are decoded so as to provide the four original channels. Its abbreviation is **4-channel sound system**. Also called **four-channel sound (1)**, **four-channel stereo**, or **quadraphonic sound system**.

four-channel stereo Same as **four-channel sound system**. Its abbreviation is **4-channel stereo**.

four fundamental forces of nature The four forces upon which all forces in nature are based. They are, in order of increasing strength: the **gravitational force**, the **weak force**, the **electromagnetic force**, and the **strong force**.

four-layer device A semiconductor device with four layers of semiconductor material. It consists of alternating layers of p-type and n-type material, thus having three pn junctions. An example is a silicon-controlled rectifier.

four-layer diode A semiconductor diode with four layers of semiconductor material, thus having three pn junctions. It has two terminals, one connected to each of the outer layers. Also called **Shockley diode**.

four-layer laser A laser operating with four energy levels, so that the lower laser level is an excited state, instead of the ground state. This provides for greatly reduced pumping requirements.

four-layer transistor A transistor with four layers of semiconductor material, thus having three pn junctions. It has three terminals. An example is a thyristor.

four-point probe A method of measuring sheet resistance, in which two pairs of precisely spaced probes contact the surface to be evaluated. One pair is utilized to pass a current through the specimen, while the other is used to measure the potential difference developed. The sheet resistance is then calculated from these current and potential difference values. Used, for instance, with semiconductor and superconductor materials. Its abbreviation is **4-point probe**.

four-pole double-throw A switch or relay contact with twelve terminals, and which serves to simultaneously connect two pairs of terminals to either of two other pairs of terminals. Its abbreviation is **4-pole double-throw**.

four-terminal network An electric network with four terminals, two for input and two for output. The input terminals are paired to form the input port, while the output port is formed by the output terminals. Examples include O-networks and H-networks. Its abbreviation is **4-terminal network**. Also called **two-port network**, **quadripole**, or **two-terminal pair network**.

four-track recording The recording of four sound tracks on a **four-track tape**. Its abbreviation is **4-track recording**.

four-track tape A ¼ inch tape upon which four sound tracks are recorded. Such a tape is usually bidirectional, having two tracks in one direction for one stereo signal, and another two tracks in the reverse direction for another. Alternatively, there may be four monaural tracks. Its abbreviation is **4-track tape**.

four-way loudspeaker Same as **four-way speaker**. Its abbreviation is **4-way loudspeaker**.

four-way speaker A speaker which utilizes four individual drivers for each of four intervals of frequencies. Such a speaker usually has a woofer, a mid-woofer, a midrange, and a tweeter. A four-way speaker may have more than one driver dedicated to each covered band. For example, there may be two tweeters. Its abbreviation is **4-way speaker**. Also called **four-way system**, or **four-way loudspeaker**.

four-way system Same as **four-way speaker**. Its abbreviation is **4-way system**.

four-wire circuit Its abbreviation is **4-wire circuit**. **1.** A circuit consisting of two pairs of conductors arranged so that there is simultaneous two-way communication between two points. Each pair of wires, or conductors, forms a single path. **2.** A two-way communications circuit utilizing two

paths, each carrying a signal in one direction. It does not necessarily have to have four wires, as is the case with an optical-fiber link.

four-wire system Its abbreviation is **4-wire system**. **1.** An AC circuit utilizing four wires. **2.** A circuit, such as a communications circuit, utilizing four wires.

four-wire wye A three-phase three-wire circuit incorporating a fourth wire which is a neutral conductor. Its abbreviation is **4-wire wye**. Also called **three-phase four-wire**.

Fourier analysis The analysis of complex waves utilizing **Fourier series** and **Fourier transforms**.

Fourier series The expression of a periodic phenomenon in terms of an infinite series of sine and cosine functions. Utilized, for instance, to evaluate complex waves.

Fourier transform A mathematic expression which converts from the time domain to the frequency domain, and vice versa. Used, for instance, for processing complex signals, and in spectral analysis.

fourth-generation computer A computer utilizing large-scale integration. This computer generation started approximately in the mid 1970s.

fourth-generation language A computer programming language which is intended to be closer to human languages than high-level programming languages such as Pascal, BASIC, Java, or C++. Its abbreviation is **4GL**.

fourth-generation wireless Developed in the 2000s, provides for digital cellular communications which in addition to the capabilities of **third-generation wireless** incorporates technologies such as adaptive processing for clearer reception and transmission, smart antennas which optimize reception and radiation patterns, and bandwidths in the Gbps range. Its abbreviation is **4G**.

fourth normal form In the normalization of a relational database, the fourth stage utilized to convert complex data structures into simpler relations. A database must first complete the third normal form before proceeding to the fourth. Its abbreviation is **4NF**.

fourth-order filter A filter with eight components, such as four inductors and four capacitors, and which provides a 24 dB rolloff. Used, for instance, as a crossover network.

FoxPro A popular database development program.

fp Abbreviation of **freezing point**.

FPA Abbreviation of **floating-point accelerator**.

FPD Abbreviation of **flat-panel display**.

FPGA Abbreviation of **field programmable gate array**.

FPLA Abbreviation of **field programmable logic array**.

FPM memory Abbreviation of **fast page mode memory**.

FPM RAM Abbreviation of **fast page mode RAM**.

fps **1.** Abbreviation of **foot per second**, or **feet per second**. **2.** Abbreviation of **frames per second**.

FPU Abbreviation of **floating-point unit**.

FQDN Abbreviation of **Fully-Qualified Domain Name**.

Fr Chemical symbol for **francium**.

fractal A geometric shape whose structure can be magnified or subdivided, with the resulting shapes being approximately the same as the whole. Used, for instance, in computer graphics to represent natural structures such as mountains, coastlines, clouds, or forests.

Fractal Image Format Its acronym is **FIF**. **1.** A file format utilized to compress **fractals**. The repetitiveness of such shapes can lead to compression ratios of several thousand to one. **2.** A file format which utilizes fractals to compress images.

fractional T1 A communications service in which a fraction of a T1 digital line is utilized. A T1 line is usually subdivided into 24 individual 64 Kbps channels, so any number from 1 to 23 may be used. Its abbreviation is **FT1**.

fractional T3 A communications service in which a fraction of a T3 digital line is utilized. A T3 line may be subdivided in various manners, including multiples of 1.544 Mbps, or 3 Mbps. Its abbreviation is **FT3**.

FRAD Abbreviation of **frame relay access device**, or **frame relay assembler/disassembler**.

fragmentation 1. The storage of files in non-contiguous areas of a disk. This occurs, for instance, when a file is repeatedly enlarged and saved, and the additional space required is not necessarily contiguous to that already utilized by the file. File fragmentation results in slower disk accessing, as there is more head movement. A defragmentation program can rearrange the files located on a computer disk drive, so that they are contiguous. Also called **file fragmentation**. 2. In a communications network, the breaking up of packets into smaller sized units of data. Used, for instance, to accommodate communications protocols whose maximum packet size is smaller. Also called **packet fragmentation**.

FRAM Abbreviation of ferroelectric **RAM**. A form of RAM which retains its content without power. It can be accessed approximately as quickly as other forms of RAM, such as DRAM, has very low power needs, but its storage density is lower. FRAM utilizes ferroelectric crystals to store data. Used, for instance, in wireless devices such as cellular phones and PDAs.

frame 1. The information presented in a single complete picture at any given moment. For instance, a complete picture displayed on a TV, computer monitor, or that of a single complete picture within a motion-picture film. Also, such single images contained in a videotape, or within an animated graphics sequence. 2. Within a pulse train, a single cycle. 3. A block of data of a specific size, or of a maximum size, transmitted in a communications network. Also called **data frame** (3). 4. In fax communications, the rectangular area of copy that a given fax system can handle. 5. A Web browser feature that divides a given window into separate segments which can be independently manipulated. 6. In a speaker, a structure that supports the cone suspension and magnet assemblies. It is usually made of plastic or metal. Also called **basket**. 7. A structure which provides shape and/or support.

frame antenna A directional antenna consisting of a wire coil with one or more turns. Used, for instance, in radio direction finders and portable radio receivers. Also called **loop antenna**, or **coil antenna**.

frame buffer A buffer which stores the data contained in a single **frame** (1) of a computer monitor.

Frame Check Sequence Additional bits added to a **frame** (3) for use in error detection. For instance, in a frame relay frame, it consists of 16 bits. Its abbreviation is **FCS**.

frame frequency Same as **frame rate**.

frame grabber A device, such as that incorporated into a graphics card, that enables a computer to capture frames from video sources such as TVs, VCRs, or digital cameras.

frame period The time interval representing the reciprocal of the **frame rate**. If the frame rate is 30, then the frame period is 1/30 second.

frame rate Also called **frame frequency, frames per second**, or **picture frequency**. 1. The number of frames displayed per second on a monitor or screen. This number should be at least 24 to simulate full motion. 2. Within a TV system, the number of times per second a frame is scanned. In the United States, this number is 30.

frame relay A packet-switching protocol utilized in WANs, and for connections between LANs situated far from each other. It transmits variable-length packets, and supports transmission rates that range from 56 Kbps up to 44.736

Mbps, the signaling rate of a T3 or DS3 circuit. Frame relay utilizes error checking instead of error correction, relying upon the dependable channels utilized.

frame relay access device A device which connects to a frame relay network. At the sending end it assembles data into packets suitable for transmission, and on the receiving end it disassembles them to back into the original data. Its abbreviation is **FRAD**. Also called **frame relay assembler/disassembler**.

frame relay assembler/disassembler Same as **frame relay access device**.

frame relay network A network utilizing a **frame relay** protocol.

frame relay service The communications services provided by a **frame relay network**.

frame switch A switch that connects nodes or segments in a LAN. A frame switch enables all nodes to have the entire bandwidth available, up to 100 Mbps for each in the case of fast Ethernet, as opposed to all nodes having to share the bandwidth, as would be the case when a hub is used. Also called **LAN switch** (1).

frame synchronization In digital communications, the alignment of a **frame** (3). This is usually accomplished through the insertion of a bit in a specific location, for unmistakable identification. Also, the method utilized to obtain such synchronization. Also called **framing** (4).

FrameMaker A popular desktop publishing program.

frames per second Same as **frame rate**. Its abbreviation is **FPS**.

framework In object-oriented programming, the set of classes that provide the structure for building an application. Also called **application framework** (2).

framing 1. The adjusting of a picture, such as that appearing on a monitor or screen, so that it is in the desired position. 2. The synchronization of the vertical component of a video signal, such as that of TV. 3. In fax communications, the adjustment of the picture to a desired position in the direction of line progression. 4. Same as **frame synchronization**.

framing bit 1. A bit utilized for **frame synchronization**. 2. In asynchronous communications, a start bit or a stop bit.

framing control A control which enables **framing** (1), **framing** (2), or **framing** (3).

francium A radioactive metallic chemical element with atomic number 87. It is very rare, and is the most electropositive element. It is also the most reactive of the naturally occurring elements, and has about 30 known isotopes, all unstable. Used in laser trapping studies. Its chemical symbol is **Fr**.

Franklin antenna A vertical antenna consisting of multiple in-phase elements, each on top of the other. Such an antenna is base fed, and several wavelengths high.

Fraunhofer diffraction pattern Same as **far-field diffraction**.

Fraunhofer region Same as **far-field region**.

free-air resonance The frequency at which a speaker will naturally resonate when driven outside an enclosure.

free carrier An electron or hole which is free to transport a charge through a semiconductor.

free charge In a conductor, a charge which is not held by the inductive effect of neighboring charges.

free electron An electron which is not bound to a specific atom. Such an electron can move freely through space or matter under the influence of an electric field, and can produce a net transport of electric charge.

free-electron laser A powerful and efficient laser which generates coherent radiation through the interaction of an electron beam and a periodic transverse magnetic field. The field makes the electrons, moving at nearly the speed of light, oscillate and thus emit radiation. Such a laser may be adjusted to deliver varying power levels through a wide interval of wavelengths, including those encompassing the infrared, visible light, ultraviolet, and X-ray regions. Its abbreviation is **FEL**.

free field **1.** A field that is not affected by any other fields. **2.** A field upon which the effects of other fields are negligible.

free-field chamber Same as **free-field room**.

free-field room A room whose walls, ceiling, and floor are lined with sound-absorbing materials which minimize or eliminate all sound reflections. Used, for example, to test microphones or sound-level meters. Also called **free-field chamber**, **anechoic chamber** (1), **dead room**, or **echoless room**.

free-form database A database which allows a user to enter information without having to follow a given structure or form. Such a database usually simplifies document management.

free-form text Text which does not have a given structure, nor which is entered in any specific format. It may simply consist of a string of words that a user enters as if speaking naturally. Also called **free text**.

free impedance For a transducer, the input impedance when the load impedance is zero. Also called **normal impedance**.

free magnetic pole A magnetic pole that is sufficiently distant or isolated from its opposing pole that the latter has no effect on it.

free net Same as **free network**.

free network A radio network within which stations are free to communicate amongst themselves without first obtaining permission from the network control station. Its abbreviation is **free net**.

free oscillation **1.** Oscillation of a mechanical system which occurs without an external driving force. Also called **free vibration**. This contrasts with **forced oscillation** (2), which only occurs when an external force is present. **2.** In a circuit, oscillation which continues after the excitation is removed. This contrasts with **forced oscillation** (2), in which excitation must be continuously applied.

free radical An atom or molecule which has one or more unpaired electrons. Free radicals are usually highly reactive and unstable. Also called **radical** (1).

free-running frequency The frequency at which a normally synchronized oscillator operates when its synchronizing signal is removed.

free-running multivibrator A multivibrator which alternates between two states, depending on the parameters established for such oscillation. This is done autonomously, that is, without an external trigger. Also called **astable multivibrator**.

free software **1.** Computer software which is free to be copied, modified, and redistributed, though not necessarily free of charge. **2.** **Free software** (1) which is free of charge.

Free Software Foundation An organization that develops, provides, and promotes free software. Its abbreviation is **FSF**.

free space **1.** A region of space devoid of all matter, and in which there are no gravitational or electromagnetic fields. In free space, the speed of light is constant, and at its maximum theoretical value. Also called **space** (3). **2.** A region in which the radiation pattern of an antenna is not affected by any surrounding objects, such as the earth, trees, buildings, and so on. **3.** In computers, an area which can hold data, but which is not occupied at a given moment.

free-space attenuation Same as **free-space loss**.

free-space loss Also called **free-space attenuation**, or **free-space losses**. **1.** The signal attenuation resulting when an antenna radiates through **free space** (2). Such losses may be due, for instance, to beam divergence. **2.** For a radiating antenna, the theoretical signal attenuation resulting when all variables factors are disregarded.

free-space losses Same as **free-space loss**.

free-space pattern The radiation pattern of an antenna radiating through **free space** (2). Also called **free-space radiation pattern**.

free-space propagation The propagation of electromagnetic radiation through **free space** (1), or **free space** (2).

free-space radiation pattern Same as **free-space pattern**.

free text Same as **free-form text**. Also spelled **freetext**.

free vibration Same as **free oscillation** (1).

FreeBSD A popular open-source UNIX operating system which runs on multiple platforms.

FreeHand A popular drawing program.

freetext Same as **free-form text**. Also spelled **free text**.

freeware Computer software which is free of charge, but which may have restrictions pertaining to copying, modifying, and redistribution.

freezing point Its abbreviation is **fp**. Also called **freezing temperature**. **1.** The temperature at which a liquid and a solid of the same substance are at equilibrium with each other. Each pure substance has a specific freezing point, for a given surrounding pressure. For example, at a pressure of 1 atmosphere, the freezing point of water is 0 °C. Also called **melting point** (1). **2.** The temperature at which a liquid substance being cooled begins to solidify.

freezing temperature Same as **freezing point**.

freq Abbreviation of **frequency**.

frequency For a periodic phenomenon, the number of complete cycles per unit of time. When the unit of time is one second, the unit of frequency is the **hertz**. That is, 1 complete cycle per second is 1 Hz. Various multiples of hertz are frequently encountered in electronics, including kilohertz, megahertz, gigahertz, and so on. Examples of periodic phenomena include electromagnetic waves, sound waves, and AC. It is the reciprocal of **period** (2). Its symbol is f, or ν, and its abbreviation is **freq**.

frequency agility The ability of a radar transmitter to quickly shift its operating frequency on a continuous basis. Such changes are automatic, and may follow a programmed algorithm. Used, for instance, to make jamming more difficult.

frequency allocation Also called **frequency assignment**. **1.** The assigning of a segment of the radio-frequency spectrum for authorized uses. **2.** The assignment of radio frequencies by a licensing authority.

frequency analyzer An instrument which measures the intensity of the various component frequencies of a complex waveform or oscillation. Used, for instance, to identify the sources of vibrations.

frequency assignment Same as **frequency allocation**.

frequency band Also called **frequency bandwidth**. **1.** In communications, a specific interval of frequencies utilized for a given purpose, such as radio-channel broadcasting. Also called **band** (1). **2.** A specific interval of frequencies between two limiting frequencies. For example, extremely high frequencies span from 30 GHz to 300 GHz. Also called **band** (2). **3.** Any given interval of frequencies. Also called **band** (3).

frequency bandwidth Same as **frequency band**.

frequency bridge A bridge circuit utilized to measure unknown frequencies. An example is a Wien bridge.

frequency calibration The use of a **frequency calibrator**.

frequency calibrator An instrument with an extremely precise and stable frequency source, which can be used as a frequency standard. Such a device usually incorporates a crystal oscillator, and serves to calibrate frequency counters, radio-frequency receivers, tachometers, and so on.

frequency changer Same as **frequency converter**.

frequency channel A frequency, or band of frequencies, assigned to a particular carrier or for a specific purpose. For example, the band of radio frequencies assigned to a TV broadcast station. Also called **channel (3)**.

frequency characteristic Same as **frequency-response characteristic**.

frequency comparator 1. A circuit or device which detects the difference between two frequencies. A common technique is to mix the two frequencies to obtain a beat frequency. 2. An instrument, such as an oscilloscope or zero-beat indicator, which measures and indicates the difference between two frequencies. Such an instrument incorporates a **frequency comparator (1)**.

frequency compensation The modification of the frequency response of a circuit or device, to tailor it to a specific need. For example, the stabilizing of a circuit to provide a flat response through a given interval of frequencies, or the variable boosting of low frequencies in automatic bass compensation.

frequency control 1. A control which can be adjusted to set or vary the frequency or frequency response of a component, circuit, device, or instrument, such as a receiver, transmitter, or oscillator. 2. A control which is utilized to lock an electronic component onto a chosen frequency. Used, for instance, in TV receivers, radio receivers, and superheterodyne receivers. If automatic, also called **automatic frequency control (1)**. 3. A circuit that maintains the frequency of an oscillator within precise limits. Also called **automatic frequency control (2)**.

frequency conversion 1. The use of a **frequency converter**. 2. The output frequency of a frequency converter.

frequency converter Also called **frequency changer**, or **frequency translator**. 1. A circuit or device whose input is at one frequency, and whose output is at another frequency. 2. In a superheterodyne receiver, the stage at which the incoming modulated radio-frequency signal is combined with the local oscillator, to produce a modulated intermediate frequency. Also called **first detector**, **mixer (2)**, or **mixer stage**. 3. A circuit or device which converts AC at one frequency to AC of another frequency.

frequency counter An instrument which determines the frequency of a periodic wave, by counting the number of cycles or pulses during a given time interval. Such an instrument usually utilizes a crystal as a frequency reference.

frequency cutoff 1. For a device such as a filter, amplifier, waveguide, or transmission line, the frequency at which the attenuation increases rapidly. For example, the frequency above which the output of an electron tube is no longer useful. Also called **cutoff (2)**, **cutoff frequency (1)**, or **critical frequency**. 3. The upper and lower frequencies at which the gain of an amplifier, such as a transistor, drops 3 decibels, or another reference amount. Also called **cutoff (3)**, or **cutoff frequency (2)**. 4. For a given transmission mode in a waveguide, the frequency below which a traveling wave can not be maintained. That is, the waveguide functions efficiently only above this frequency. Also called **cutoff (4)**, **cutoff frequency (3)**, or **waveguide cutoff**.

frequency deviation 1. A divergence in a frequency from an expected or prescribed value. Also, the extent of such a divergence. 2. In frequency-modulation, the maximum difference between the instantaneous frequency of the modulated wave, and the frequency of the unmodulated carrier. Also called **deviation (2)**.

frequency-deviation meter A device which measures and indicates **frequency deviation**. Also called **frequency monitor**.

frequency discriminator 1. A circuit or device which makes fine distinctions between frequencies, such as those between two almost identical values. 2. A circuit whose output voltage is proportional to the differences of its input signals from a given frequency. For example, a circuit which converts frequency-modulated signals into amplitude-modulated signals. 3. A circuit or device which only responds to a given frequency, or frequency interval.

frequency distortion In an amplifier, distortion that occurs when some frequencies are amplified more than others. Also known as **amplitude distortion (1)**, **attenuation distortion**, or **amplitude-frequency distortion**.

frequency distribution A function describing the frequency with which a variable assumes a given value. It may be expressed as a graph of the number of occurrences versus the possible values. Also called **distribution (3)**.

frequency diversity 1. Same as **frequency-diversity reception**. 2. Same as **frequency-diversity transmission**.

frequency-diversity gain Signal gain through the use of **frequency-diversity reception**.

frequency-diversity radar A radar system which utilizes two or more sets of transmitters and receivers, each set operating at a different frequency. All sets share a common antenna and display. Also called **diversity radar**.

frequency-diversity receiver A receiver used with **frequency-diversity reception**.

frequency-diversity reception A form of diversity reception in which two signals, each containing the same information but sent at different frequencies, are received, then combined. Since fading is generally not the same for two frequencies, the effects of fading are thus minimized. Alternatively, the stronger signal can be used. Also called **frequency diversity (1)**.

frequency-diversity system A reception or transmission system which employs **frequency-diversity reception**, or **frequency-diversity transmission**, respectively.

frequency-diversity transmission Simultaneous transmission of radio-frequency signals used for **frequency-diversity reception**. Also called **frequency diversity (2)**.

frequency-diversity transmitter A transmitter used with **frequency-diversity transmission**.

frequency divider 1. A circuit or device whose output signal frequency is an exact integer submultiple of its input signal frequency. For example, such a circuit could convert a 10 MHz signal into a 50 Hz signal with the appropriate combination of counter circuits. This contrasts with **frequency multiplier**. 2. A circuit or device which separates a frequency spectrum into specific intervals. An example is a frequency dividing network.

frequency dividing network A filter circuit in a speaker which divides the input audio frequencies and sends the corresponding bands of frequencies to the designated speaker units, such as woofers, midranges, tweeters, super-tweeters, and so on. Also called **crossover (3)**, **crossover network**, **dividing network**, **speaker dividing network**, or **loudspeaker dividing network**.

frequency-division multiple access A multiplexing method which divides the available frequency spectrum into 30 kHz channels. This technology is used, for instance, in analog cellular telephone networks to provide a 30 kHz channel for each call. Its abbreviation is **FDMA**.

frequency-division multiplex The transmitting of multiple signals over a single transmission path utilizing **frequency-division multiplexing**. Its abbreviation is **FDM**.

frequency-division multiplexing Its abbreviation is **FDM**. A technique in which multiple signals are transmitted simultaneously over a single transmission path by assigning each signal a separate portion of the total available frequency spectrum. For instance, a transmitter of a cable TV signal may utilize FDM to transmit many channels over the same coaxial cable. Also called **frequency multiplexing**.

frequency domain The representation of a signal as a function of frequency, as opposed to time. This contrasts with **time domain**, where the converse is true. A spectrum analyzer displays signals in the frequency domain. A Fourier transform is utilized to convert from the frequency domain to the time domain, and vice versa.

frequency doubler A circuit or device whose output frequency is twice that of its input frequency. May consist, for instance, of a stage whose resonant output circuit is tuned to the second harmonic of the input frequency. Also called **doubler (1)**.

frequency drift A gradual and undesired change in the frequency of a signal, resulting in its departure from a set or intended value. This may occur, for example, in an oscillator or transmitter, and may be due, for instance, to temperature changes or aging of components. Also called **frequency shift (1)**.

frequency guard band A narrow interval of frequencies between channels, which is left vacant to help prevent adjacent-channel interference. Also called **guard band**.

frequency hopping The switching, at given intervals, of the carrier frequency of a radio transmitter. Such changes may be dictated, for instance, by a given algorithm. Used, for example, for security purposes.

frequency-hopping spread spectrum A form of spread-spectrum modulation in which the frequency of the carrier hops among multiple frequencies at a rate determined by a specific code or algorithm. For proper communication, the intended receiver must hop in synchronism with the transmitter. When synchronized, the signal is as clear as if it were on a single logical channel. Otherwise, it appears as indecipherable. Its abbreviation is **FHSS**.

frequency indicator An instrument or display which indicates a frequency, or which enables a frequency to be verified.

frequency interval Same as **frequency range**.

frequency lock Also called **frequency lock loop**. **1.** The use of a feedback loop to maintain a circuit or device at a given frequency. **2.** The use of a feedback loop to maintain an oscillator at a given frequency.

frequency lock loop Same as **frequency lock**. Its abbreviation is **FLL**.

frequency meter **1.** An instrument, such as a cavity frequency meter or frequency counter, utilized to measure the frequency of a signal. Such an instrument is usually graduated in hertz, or multiples of hertz, such as kilohertz or gigahertz. A frequency meter can usually be utilized as a wavemeter, and vice versa, as either quantity can be obtained knowing the other, according to the following formula: $\lambda = c/f$, where λ is wavelength, c is the speed of light in a vacuum, and f is the frequency. **2.** An instrument utilized to measure the frequency of an alternating current.

frequency mixer A circuit, device, or stage which combines two different input frequencies to produce a third output frequency. Examples include frequency converters and beat-frequency oscillators.

frequency-modulated Pertaining to, of having undergone FM.

frequency-modulated radar Same as **FM radar**.

frequency-modulated signal Same as **FM signal**.

frequency-modulated transmission Same as **FM transmission**.

frequency-modulated transmitter Same as **FM transmitter**.

frequency-modulated wave Same as **FM wave**.

frequency modulation Same as **FM**.

frequency-modulation/amplitude-modulation radio Same as **FM/AM radio**.

frequency-modulation/amplitude-modulation radio receiver Same as **FM/AM radio receiver**.

frequency-modulation/amplitude-modulation receiver Same as **FM/AM receiver**.

frequency-modulation/amplitude-modulation transmitter Same as **FM/AM transmitter**.

frequency-modulation broadcast band Same as **FM broadcast band**.

frequency-modulation broadcasting Same as **FM broadcasting**.

frequency-modulation detector Same as **FM detector**.

frequency-modulation noise Same as **FM noise**.

frequency-modulation radio Same as **FM radio**.

frequency-modulation radio receiver Same as **FM radio receiver**.

frequency-modulation receiver Same as **FM receiver**.

frequency-modulation stereo Same as **FM stereo**.

frequency-modulation synthesis Same as **FM synthesis**.

frequency-modulation tuner Same as **FM tuner**.

frequency modulator A circuit or device which effects **FM**.

frequency monitor Same as **frequency-deviation meter**.

frequency multiplexing Same as **frequency-division multiplexing**.

frequency multiplier A circuit or device whose output signal frequency is an exact integer multiple of its input signal frequency. For example, such a circuit could convert a 10 MHz signal into a 30 MHz signal with the use of a frequency tripler. This contrasts with **frequency divider**. Also called **multiplier (3)**.

frequency offset **1.** The difference between a given frequency and a reference frequency. **2.** The difference between a given frequency and a desired frequency. **3.** In a transceiver, the difference between the transmitter frequency and the receiver frequency. **4.** In communications, the difference between the frequency received and the frequency transmitted.

frequency overlap A given interval of frequencies which is shared. For example, that shared by two communications channels.

frequency pulling A variation in the frequency of a circuit, especially that of an oscillator, caused by an external circuit, device, or factor, such as a variation in load impedance. Also called **pulling (1)**.

frequency pushing A variation in the frequency of a circuit, especially that of an oscillator, caused by a change in the supply voltage or current.

frequency quadrupler A circuit or device whose output frequency is four times that of its input frequency. Also called **quadrupler (1)**.

frequency quintupler A circuit or device whose output frequency is five times that of its input frequency. Also called **quintupler (1)**.

frequency range **1.** A continuous interval of frequencies between two limiting frequencies. For instance, all frequencies between 20 Hz and 20 kHz. Also called **frequency interval**, or **frequency span**. **2.** A continuous interval of frequencies within which a component, circuit, device, piece of equipment, or system operates. Said, for instance, of filters, receivers, transmitters, speakers, antennas, and so on.

frequency regulator 1. A circuit or device which maintains another circuit or device at a given frequency. 2. A circuit or device which maintains an AC generator at a given frequency.

frequency relay A relay whose operation depends on the signal frequency. For example, it may be actuated at, above, or below a given frequency.

frequency response 1. A measure of the behavior of a component, circuit, device, piece of equipment, or system, as a function of its input signal frequencies. For example, it may refer to the efficiency of the amplification of a circuit or device as a function of frequency. Also known as **amplitude-frequency response**, **sine-wave response**, or **sine-wave frequency response**. 2. Same as **frequency-response curve**.

frequency-response characteristic The gain or loss of a circuit, device, piece of equipment, or system, as a function of the input signal frequencies. Also called **frequency characteristic**, or **response characteristic**.

frequency-response curve A graph representing the **frequency response** (1) of a component, circuit, device, piece of equipment, or system. Also called **frequency response** (2).

frequency-response equalization The use of an equalizer to modify the frequency response of a circuit, device, piece of equipment, or system. Such equalization may be utilized, for instance, to counteract distortion, compensate for deficiencies, or shape a frequency response to fit the needs for a given situation. Also called **equalization**.

frequency-response measurement The measurement of the **frequency response** (1) of a circuit, device, piece of equipment, or system.

frequency-response run One or more tests performed to obtain the **frequency-response characteristic** of a circuit, device, piece of equipment, or system.

frequency scaling To shift from one frequency, or interval of frequencies, to another.

frequency scanning The scanning through a given interval of frequencies. Such scanning may be automatic or manual, fixed or variable, in discrete steps or throughout a continuous interval, and so on. May be used, for instance, to locate a desired reception or transmission frequency, to find and utilize a clear frequency, or in spectroscopy.

frequency selectivity The degree to which a circuit or device can differentiate between signals of different frequencies, especially when seeking a given frequency.

frequency separation The spacing of frequencies. Used, for instance, to help minimize adjacent-channel interference.

frequency separator In a TV, a circuit which separates the horizontal sync pulses from the vertical sync pulses.

frequency shift 1. Same as **frequency drift**. 2. A change in the frequency of a circuit, device, or piece of equipment, such as a transmitter or oscillator.

frequency-shift keying A form of frequency modulation in which the modulating wave shifts the output frequency between fixed values. Switching between these values provides the keying. May be used, for instance, by modems transmitting over telephone lines, by representing the 0 with a lower frequency, and a 1 with a higher frequency. Its abbreviation is **FSK**. Also called **frequency-shift modulation**, or **digital frequency modulation**.

frequency-shift modulation Same as **frequency-shift keying**.

frequency span Same as **frequency range**.

frequency spectrum 1. The electromagnetic spectrum in terms of frequency. This encompasses frequencies from just above 0 Hz to beyond 10^{24} Hz, corresponding to wavelengths of over 10^8 meters, to less than 10^{-16} meters, respectively. Also called **electromagnetic frequency spectrum**. 2. An interval of frequencies within the **frequency spectrum** (1). For example, the sound, radio, or light spectrum.

frequency splitting In a magnetron, the rapid alternation from one operation mode to another, resulting in a loss of power at the desired frequency.

frequency stability 1. The degree to which a frequency remains constant. 2. The degree to which a frequency remains constant when factors such as voltage, current, and/or temperature vary. 3. The extent to which a component, circuit, device, piece of equipment, or system displays **frequency stability** (2). May be expressed, for instance, in hertz or a multiple of hertz, or as a percent deviation from the desired frequency.

frequency stabilization 1. The maintenance a frequency within precise limits. 2. The maintenance of the frequency of an oscillator within precise limits.

frequency standard A signal source, such as an extremely precise and stable crystal oscillator, whose frequency remains constant, and which can be utilized for calibrating other signal sources. An example is a cesium frequency standard. Also called **primary frequency standard** (1).

frequency swing 1. The peak difference between the maximum and minimum instantaneous frequencies around a center frequency. 2. In frequency modulation, the peak difference between the maximum and minimum instantaneous frequencies around the carrier frequency.

frequency synthesizer A circuit or device which generates precise frequency signals. Such a device usually utilizes one or more crystal oscillators, and can generate equally-spaced frequencies within a given band through the use of frequency multipliers, dividers, mixers, and so on. Also called **synthesizer** (2).

frequency temperature coefficient A numerical value that indicates the relationship between a change in the frequency of a material, component, or device, as a function of temperature. Also called **temperature coefficient of frequency**.

frequency-to-voltage converter A circuit or device whose analog output voltage is proportional to the frequency of its input signal. Its input may be provided, for instance, by a tachometer. This contrasts with a **voltage-to-frequency converter**, which converts a voltage input into a frequency output. Its abbreviation is **F/V converter**.

frequency tolerance The maximum allowable departure of a frequency from an assigned or desired value. For example, the extent to which the carrier frequency of a transmitter may deviate from its assigned frequency. Frequency tolerance may be expressed in hertz or multiples of hertz, as a percentage deviation, or in parts per million.

frequency translation 1. The moving of signals occupying one frequency band to another frequency band within the frequency spectrum. This is done in a manner that the proportionality of the frequency separation between the signals is held constant. 2. The use of a **frequency converter** (1).

frequency translator Same as **frequency converter**.

frequency tripler A circuit or device whose output frequency is triple that of its input frequency. May consist, for instance, of a stage whose resonant output circuit is tuned to the third harmonic of the input frequency. Also called **tripler** (1).

Frequently Asked Questions Same as **FAQ**.

Frequently Asked Questions file Same as **FAQ file**.

Fresnel lens A thin lens with stepped concentric rings enabling it to have the optical properties of lenses which are much thicker and heavier. Used, for instance, in overhead projectors and lighthouses.

Fresnel region The region between an antenna and the **Fraunhofer region**. That is, the region extending from the antenna up to a distance of $2D^2/\lambda$ from the antenna, where D is the maximum overall dimension of said antenna, and λ is the wavelength considered.

Fresnel zone For transmitted electromagnetic radiation, one of a series of concentric areas in space which surround the path between a transmitter and receiver. The first Fresnel zone encompasses all paths ½ wavelength longer than the line-of-sight path, the second 1 wavelength longer than the line-of-sight path, the third 1½ wavelengths longer, and so on. Odd-numbered zones are lobes, while even numbered zones are nulls. Utilized, for instance, to describe the concentric ellipsoids between transmitting and receiving microwave antennas.

friction For two bodies in contact, the resistance one offers to the movement of the other relative to it. Friction may occur between solids, liquids, or gases. Friction between solids may have any of several effects, including the production of heat, wear, or electricity. With fluids, the greater the viscosity, the greater the resistance to the movement of another body through it.

friction feed A method or mechanism for moving papers, or other appropriate mediums, through a printer by pulling or pushing it while pinched between rollers. This contrasts with a **tractor feed**, in which the medium being printed upon has holes which match the pins of rotating wheels on each side of the printer.

frictional electricity The generation of electrostatic charges, or static electricity, through friction. For example, if a dielectric such as glass or amber is rubbed with a dissimilar dielectric such as silk or wool, and each is separated, such charges are produced. Also called **triboelectricity**.

fringe area For a given transmitter, such as that of radio or TV, the area just beyond that where there is consistently satisfactory reception. Proper reception in the fringe area requires, for instance, the use of a high-gain antenna.

fringe howl In a radio-frequency receiver, a howl heard when a circuit is on the verge of oscillation.

fringing 1. An effect observed near the outer edges of the parallel metal plates of a capacitor, in which there is an outward bulging of the lines of force. This effect must be taken into consideration in order to correctly compute the capacitance. Alternatively, a guard ring may be used to eliminate this effect. Also called **edge effect**. 2. In a color TV receiver, extraneous colors occurring at the boundaries between zones with different colors, or along the edges of objects. Also called **color edging**, or **edging**.

frit A combination of ground or powdered glass and other substances, which is utilized for making glazes or enamels.

frit seal A tight seal utilizing a **frit**. Used, for instance, to seal a ceramic package.

front contact In a relay or switch, a contact pair that is open in the resting state, and which is closed when energized. The contact pair consists of a stationary contact and a movable contact. A front contact provides a complete circuit when energized. This contrasts with a **back contact**, in which the contact pair is closed in the resting state. Also called **normally-open contact**, or **make contact**.

front end 1. In a client/server environment, the client part. The server part is the **back end**. 2. In computers, that which provides a user interface. For instance, in an application it may be a GUI where data is entered and seen, as opposed to the processing tasks that are not readily apparent to a user. 3. The first amplification stage of a receiver, such as a radio or TV receiver. May also refer to the first stages.

front-end processor Its abbreviation is **FEP**. 1. A programmable peripheral which serves to control one or more communications lines linked to a computer. Front-end proces-

sors handle tasks such as sending, receiving, coding, decoding, and error-checking, which frees resources of the computer it is connected to. 2. A computer which controls one or more computers which perform specialized functions such as database access or high-speed graphics processing. The controlled computers are called **back-end processors**.

front inputs Audiovisual inputs that are located on the front of a component such as a TV or VCR, as opposed to the back. Useful, for instance, when temporarily connecting a device.

front panel The panel at the front of certain electrical equipment which serves to set, switch, adjust, and monitor the various available settings and functions. For example, an audio receiver's front panel provides a means of setting and viewing the volume level and input signal source, and may also provide choices such as stereo or surround sound, equalization levels, and so on.

front porch In a composite video signal, the interval of the signal between the leading edge of a horizontal blanking pulse and the leading edge of the corresponding sync pulse.

front projection The projection, onto a non-translucent screen, of images sent from a separate unit, and in which the projector and any viewers are on the same side of said screen. Used, for instance, in movie theaters, and front-projection TVs. This contrasts with **rear projection**, where the images are viewed on a translucent screen, and the projector and viewers are on opposite sides of said screen.

front-projection television Same as **front-projection TV**.

front-projection TV Abbreviation of **front-projection** television. A TV with **front projection**.

front-side bus A microprocessor bus that connects the memory and peripherals. It generally runs at a lower speed than that of the CPU, while the **backside bus**, which connects the CPU to a level 2 cache, usually runs at the same speed as the CPU. Also spelled **frontside bus**. Its abbreviation is **FSB**.

front-surface mirror A mirror whose reflective coating is on its front surface. In this manner, there is no refraction, and image distortion is minimized. Used, for instance, in instruments such as microscopes and lasers. This contrasts with **rear-surface mirror**, where the reflective coating is on its rear surface. Also called **first-surface mirror**, or **surface-coated mirror**.

front-to-back ratio 1. For a directional antenna, speaker, or microphone, the ratio of the signal strength in the most favored direction, to that in the opposite direction. Expressed in decibels. 2. For a device, such as a rectifier, the ratio of the magnitude of an electrical characteristic, such as current or resistance, in one direction to that of the opposite direction.

front wave An acoustic wave which is radiated from the front of a speaker. A baffle is used as an acoustic seal which prevents the back waves, which are radiated from the back of a speaker, from canceling the front waves.

FrontPage A popular program for Web site creation and management.

frontside bus Same as **front-side bus**.

fs Abbreviation of **femtosecond**.

fS Abbreviation of **femtosiemens**.

FS Abbreviation of **full scale**.

FSB Abbreviation of **front-side bus**.

FSD Abbreviation of **full-scale deflection**.

fsec Abbreviation of **femtosecond**.

FSF Abbreviation of **Free Software Foundation**.

FSK Abbreviation of **frequency-shift keying**.

FSM Abbreviation of **finite-state machine**.

FSN Abbreviation of **full-service network**.

FSR Abbreviation of **full-scale range**.

FSS Abbreviation of **fixed-satellite service**.

ft Abbreviation of **foot**, or **feet**.

FT-1 Abbreviation of **fractional T1**.

FT-3 Abbreviation of **fractional T3**.

ft-lb Abbreviation of **foot-pound**.

ft/s Abbreviation of **foot per second**, or **feet per second**.

ft/sec Abbreviation of **foot per second**, or **feet per second**.

FT1 Abbreviation of **fractional T1**.

FT3 Abbreviation of **fractional T3**.

FTAM Abbreviation of **File Transfer, Access, and Management**.

FTP 1. Abbreviation of **file transfer protocol**. 2. Abbreviation of **file transfer program**.

FTTC Abbreviation of **fiber to the curb**.

FTTH Abbreviation of **fiber to the home**.

fuel cell A cell which converts chemical energy into electrical energy, and whose reactants are continuously replenished by external reservoirs. A phosphoric acid fuel cell, for instance, utilizes hydrogen and oxygen as the reactants, has carbon electrodes, a phosphoric acid electrolyte, and a platinum catalyst. The products of such a cell are electricity, heat, and water.

fuel cell stack Two or more **fuel cells** that are utilized together.

full adder A logic circuit which accepts three input bits, and whose output is a sum and a carry bit. Two of the input bits are for adding, and the third is a carry bit from another digit position. Also called **three-input adder**.

full backup A data backup technique in which all files are copied, including those that have not been modified since the last backup. This contrasts with an **incremental backup**, in which the only files copied are those that have been modified since the last backup. Also called **archival backup**.

full charge The maximum charge that can be stored in a battery or cell under specified conditions. Also called **full-rated capacity**.

Full CIF Abbreviation of **Full Common Intermediate Format**.

Full Common Intermediate Format A video format which supports both NTSC and PAL signals, and which is used in videoconferencing. Each CIF format is defined by its resolution, the original CIF standard specifying 288 lines and 352 pixels per line, at 30 frames per second. Its abbreviation is **Full CIF**, or **FCIF**.

full-duplex Two-way communication which occurs simultaneously in both directions. This contrasts with **half-duplex**, in which communication can only occur in one direction at a time. Its abbreviation is **FDX**. Also called **full-duplex communication**, **duplex (2)**, or **two-way communication (2)**.

full-duplex channel A communications channel utilized for **full-duplex**. Also called **duplex channel (1)**.

full-duplex circuit A communications circuit enabled for **full-duplex**. Also called **duplex circuit (1)**.

full-duplex communication Same as **full-duplex**.

Full-Duplex Ethernet A switched Ethernet mode of operation in which two stations can simultaneously exchange data using the full available bandwidth, effectively doubling said bandwidth.

full-duplex operation The operation of a communications channel utilized for **full-duplex**. Also called **duplex operation (1)**.

full-duplex system A communications system utilized for **full-duplex**. Also called **duplex system (1)**.

full-duplex transmission Two-way communication in which there are simultaneous transmissions in both directions. Also called **duplex transmission (1)**.

full-featured Devices, equipment, or systems which provide a complete array of capabilities and functions, usually emphasizing the most current, advanced, and powerful features within their category.

full load The maximum load a circuit, device, piece of equipment, or system can carry under specified conditions. Any additional load constitutes an overload.

full-load current The output current from a circuit, device, piece of equipment, or system when carrying a **full load**.

full-load voltage The output voltage of a circuit, device, piece of equipment, or system when carrying a **full load**.

full-motion video 1. Video whose frames are displayed at a rate of at least 24 frames per second, to simulate full motion. 2. Digital video whose frames are displayed at a rate of 30 frames per second.

full path A file pathname that starts with the drive letter, leads through the directories, and ends with a full file name. For example: *C:\Documents\Reminders\today.xyz*. Also known as **full pathname**, **file specification**, **filespec**, **absolute path**, **access path**, or **search path**.

full pathname Same as **full path**.

full radiator An ideal body which would absorb all the radiant energy incident upon it. It would also be a perfect emitter of radiant energy, whose distribution of energy depended solely on its absolute temperature. Also called **blackbody**, **ideal radiator**, or **perfect radiator**.

full-range loudspeaker Same as **full-range speaker**.

full-range speaker A single speaker driver which attempts to reproduce the entire range of audio frequencies. Such a range runs from about 20 Hz to about 20 kHz. Also called **full-range loudspeaker**.

full-rated capacity A value or quantity establishing the maximum amount that can be contained, accommodated, or handled. For instance, the charge that can be withdrawn from a fully charged battery under specified conditions, the maximum current that can be handled by an electric circuit, or the largest number of communication channels that a communications circuit can handle simultaneously. Also called **rated capacity**, or **maximum capacity**.

full scale Its abbreviation is **FS**. 1. The maximum reading or value that a scale can indicate. In an analog readout, for instance, it is usually the reading furthest to the right. 2. Same as **full-scale range**.

full-scale deflection For an analog readout, the maximum reading or value that can be indicated. It is usually the reading furthest to the right. Its abbreviation is **FSD**. Also called **full-scale reading**, or **full-scale value**.

full-scale range The complete range of readings or values a scale can indicate. In an analog readout, for instance, this usually extends from the furthest reading to the left, through the furthest reading to the right. Its abbreviation is **FSR**. Also called **full scale (2)**.

full-scale reading Same as **full-scale deflection**.

full-scale sensitivity The input required to attain a **full-scale (1)** indication.

full-scale value Same as **full-scale deflection**.

full-service network A telecommunications network which provides most or all of the prevailing communications offerings, such as Internet, TV, telephone, and interactive services derived from these, such as a shopping and requests for pay-per-view programming. Its abbreviation is **FSN**.

full-text search A search within one or more documents, records, fields, or the like, in which the entire available text is explored, as opposed to just an index, a set of keywords, or a summary.

full-track head   A tape recorder head utilized for **full-track recording**.

full-track recording   Magnetic tape recording in which nearly the entire width of the tape is utilized for a single track.

full version   A version of a software or hardware product which has the full complement of functions and features, as opposed to a **light version**, which does not.

full-wave antenna   An antenna radiator whose electrical length is equal to one wavelength of the signal being transmitted or received. Also called **full-wavelength antenna**.

full-wave bridge   Same as **full-wave bridge rectifier**.

full-wave bridge rectifier   A rectifier circuit which utilizes four diode or tube rectifiers, arranged in the shape of a diamond, and which provides full-wave rectification. Also called **full-wave bridge**.

full-wave control   A form of phase control which acts on both the positive and the negative half-cycles of an AC cycle.

full-wave rectification   Rectification in which there is a DC output during both the positive and the negative half-cycles of an applied AC input.

full-wave rectifier   A rectifier which rectifies both the positive and the negative half-cycles of an applied AC input, to produce its DC output. An example is a bridge rectifier. Also called **full-wave rectifier circuit**.

full-wave rectifier circuit   Same as **full-wave rectifier**.

full-wavelength antenna   Same as **full-wave antenna**.

fullerene   An allotropic form of carbon arranged in a manner that forms a hollow sphere, cylinder, or the like. These may range from as few as 20 carbon atoms, to several hundred, always in specific multiples of atoms. The most common and stable found to date is buckminsterfullerene, with 60 atoms. Fullerenes are extremely strong and lightweight, are capable of enclosing other atoms, and have applications in various areas, including electronics, optics, superconduction, and medicine.

fully populated   That which has been as completely occupied as possible. For example, a motherboard whose expansion slots have all been filled.

Fully-Qualified Domain Name   An Internet domain name which includes a host name in addition to a domain name. For example, *www.yipeeee.com* is such a name, with the **www** portion denoting the host, and the *yipeeee.com* portion denoting the domain name. Each Fully-Qualified Domain Name must be unique, and have the information necessary for conversion to an IP address. Its abbreviation is **FQDN**.

function   **1.** That which a component, circuit, device, piece of equipment, or system is intended to do. Also, that which it does do. **2.** A mathematical expression in which a quantity depends one or more varying quantities. For example, the frequency response of an amplifier measures its behavior as a function of its input signal frequencies. **3.** Within a computer program, a small group of instructions which perform a given task. Also called **subroutine** (2), **routine** (1), or **procedure** (2). **4.** A **function** (3) which returns a value.

function call   A program statement or instruction which invokes a **function** (3).

function generator   **1.** A signal generator which can produce any of various selectable waveforms, such as those of sine, square, and sawtooth waves, over a wide range of frequencies. Also called **waveform generator**. **2.** A circuit or device within an analog computer whose output signal corresponds to the value of a given function. As its input variables vary, so does its output.

function keys   On a computer keyboard, a set keys, often 10 or 12, whose functions vary depending on the software running and/or the definitions a user has given them. Such keys

may be used, for instance, to access functions or to automate a sequence of actions. Its abbreviation is **F keys**, or **Fn keys**.

function library   A collection of **functions** (3) which can be called upon as needed.

function overloading   The use of the same name for more than one function. The compiler automatically determines which to use in a given context.

function switch   In a device or instrument with multiple functions, a switch utilized to select the desired one. For example, it may be used to select the current meter function of a multimeter.

functional block   An combination of components and materials, such as those incorporated in a crystal oscillator, which perform a given function within a circuit.

functional design   A specification of the functions and objectives of a system, and how its components are interrelated.

functional diagram   A **functional design** in graphic form.

functional programming   Computer programming utilizing a **functional programming language**.

functional programming language   A programming language in which the definition and application of functions is emphasized. Used, for instance, in parallel computing applications.

functional specification   **1.** A detailed description of the design of a product or system, along with its characteristics, purpose, manner of implementation, and intended capabilities. It is used, for instance, to help factor in what will be needed to develop a product or system. **2.** A **functional specification** of an information system.

functional testing   Also called **black-box testing**, or **closed-box testing**. **1.** Any tests utilizing the assistance of a black box. **2.** Tests of software or hardware utilizing the assistance of a black box.

functional unit   One or more devices, entities, or structures which work as a unit for the accomplishment of specified tasks. Examples include amplifier stages and disk drives.

fundamental   Same as **fundamental frequency** (1).

fundamental component   Same as **fundamental frequency** (1).

fundamental frequency   **1.** For a complex signal, wave, or vibration which is periodic, the sinusoidal component having the lowest frequency. All integral multiples are called harmonics, the second harmonic being twice the fundamental frequency, the third harmonic being three times the fundamental frequency, and so on. Any even-numbered multiple of the fundamental frequency is called an even harmonic, while odd-numbered multiples are called odd harmonics. Also called **fundamental, fundamental component**, or **first harmonic**. **2.** For an oscillating system, the lowest natural resonance frequency.

fundamental mode   **1.** In a waveguide, the mode of propagation that has the lowest cutoff frequency. Also called **dominant mode**, or **principal mode**. **2.** In a vibrational system, the mode of vibration with the lowest frequency. Also called **fundamental mode of vibration**.

fundamental mode of vibration   Same as **fundamental mode** (2).

fundamental particle   Any of the fundamental constituents of all matter in the universe. Fundamental particles include protons, neutrons, electrons, photons, and quarks. Categories of such particles include leptons and hadrons. An electron, for instance, is a type of lepton. When a fundamental particle is defined as having no internal structure, particles such as protons and neutrons are excluded, as these are composed of quarks. The four fundamental forces of nature are carried by elementary particles. For example, photons carry the electromagnetic force. Also called **elementary particle**, or **particle** (2).

fundamental suppression In a complex wave, the suppression of the **fundamental frequency** (1). Used, for instance, to measure total harmonic distortion.

fundamental tone For a complex tone, the component with the lowest pitch.

fundamental units Internationally accepted fundamental units for the measurement of physical quantities. All other units and internationally accepted systems of units are based on these fundamental, or absolute, units. Currently, the base, or fundamental, SI units are: the **second**, the **kilogram**, the **meter**, the **ampere**, the **Kelvin**, the **mole**, and the **candela**. Also called **base units**.

fundamental wavelength The wavelength corresponding to a **fundamental frequency**.

fuse A safety device which contains a section of conductor, such as a wire, which melts when the current passing through it exceeds a specified amount, called the fuse rating. When the conductor melts, the circuit is opened, thus preventing damage to the circuit, device, equipment, or installation it is meant to protect. Also called **electric fuse**, or **cutout** (2).

fuse alarm A circuit or device which serves to notify of a blown fuse by means of a signal, such as a light. Also, such a notification.

fuse box A box which houses a set of fuses. May be found, for instance, in a dwelling or vehicle. Also spelled **fusebox**.

fuse current rating Same as **fuse rating**.

fuse rating The maximum current that can flow through a fuse for a specified time interval before melting. Also called **fuse current rating**.

fuse wire A wire whose characteristics, such as a low melting point, make it suitable for use in fuses.

fusebox Same as **fuse box**.

fused junction A junction formed by alloying one or more impurity metals to a semiconductor substrate. Depending on which impurity metal is alloyed, p or n regions will be formed, which in turn compose the junctions. Also known as **alloy junction**.

fused-junction diode A junction diode whose p and n regions are formed by alloying one or more impurity metals to the semiconductor. Also known as **alloy-junction diode**.

fused-junction transistor A junction transistor whose p and n regions are formed by alloying one or more impurity metals to the semiconductor. Also known as **alloy-junction transistor**.

fusible link A wire which performs the function of a fuse in a circuit. Such a link may be intentionally melted or broken, as is the case when utilized in PROM chips.

fusible resistor A low-value resistor used as a fuse. It is usually utilized to protect a circuit, device, piece of equipment, or system when power is first applied. Also called **resistor fuse**.

fusing current The level of current at which a given wire will melt. Also called **wire fusing current**.

fusion splice A fiber-optic splicing method in which the two ends to be joined are cut, aligned, melted, and joined together. This method provides less losses than a **mechanical splice**.

future-proof That which will never become obsolete. It is extremely unlikely that any product is or will be future-proof, and the term is often used as part of a sales pitch.

fuzzy logic A form of logic which evaluates degrees of truthfulness, as opposed to simply categorizing as completely true or completely false. Instead of only using 0 and 1 as the logic values, it consists of variables that can assume any value from 0 to 1. Fuzzy logic is well adapted for dealing with uncertain and/or incomplete data, and for problems with multiple solutions. Used, for instance, in artificial intelligence applications.

fuzzy search A data search employing **fuzzy logic** to provide matches for words that are partially spelled, misspelled, or similar to other words.

fuzzy set A set in which the function that determines membership is not limited to a strict yes or no, or 0 or 1. Instead, said function can assume any value ranging from 0 to 1.

fV Abbreviation of **femtovolt**.

fW Abbreviation of **femtowatt**.

G

g 1. Symbol for **gram**. **2.** Symbol for **grav**.

G 1. Abbreviation of **giga-**. **2.** Symbol for **gravitational force**. **3.** Symbol for **gauss**. **4.** Symbol for **gate**. **5.** Symbol for **generator**.

G **1.** Symbol for **conductance**. **2.** Symbol for **gravitational constant**.

g-cal Abbreviation of **gram-calorie**.

G display In radars, a rectangular display where the scanned object appears as a centralized blip when the antenna is aimed directly at it. As the distance to the scanned object diminishes, the blip appears to grow wings. As said distance continues to decrease, the width of the wings increases. Horizontal and vertical displacement of the blip away from the center of the display indicate the azimuth and elevation errors respectively. Also called **G scan**, **G scanner**, **G scope**, or **G indicator**.

G indicator Same as **G display**.

G line A single wire, with a circular cross-section, coated with a dielectric. Utilized to transmit microwave energy.

G scan Same as **G display**.

G scanner Same as **G display**.

G scope Same as **G display**.

G-Y signal In a color TV receiver, a color-difference signal representing the difference between the green signal and the luminance signal. Thus, adding this signal to the luminance yields the green primary signal. **G** is green, and **Y** is luminance.

G.703 An ITU standard for digital interfaces.

G.711 An ITU standard for encoding audio at 64 Kbps, using Pulse Code Modulation.

G.721 An ITU standard for encoding audio at 32 Kbps, using Pulse Code Modulation.

G.722 An ITU standard for encoding audio at 64Kbps, using Pulse Code Modulation.

G.723 An ITU standard for encoding audio at up to 40Kbps, using Pulse Code Modulation.

G.726 An ITU standard for encoding audio at up to 40Kbps, using Pulse Code Modulation. It supersedes **G.721**, and **G.723**.

G.728 An ITU standard for encoding audio at 16 Kbps, using Code-Excited Linear Predictive.

G.729 An ITU standard for encoding audio at 8 Kbps, using Code-Excited Linear Predictive.

G.804 An ITU standard for DSL.

G.992.2 Same as **G.lite**.

G.lite An ITU standard for DSL. It does not require a splitter to be installed at the customer location. Also known as **G.992.2**.

G3 fax Abbreviation of **Group 3 fax**, or **Group 3 facsimile**.

G4 fax Abbreviation of **Group 4 fax**, or **Group 4 facsimile**.

Ga Chemical symbol for **gallium**.

GaAs FET Abbreviation of **gallium arsenide field-effect transistor**. A field-effect transistor in which the semiconductor material is gallium arsenide. Used, for instance, for high-frequency amplification.

gadget A usually small electrical and/or mechanical device with a given function.

gadolinium A silvery-white metallic chemical element whose atomic number is 64. It is highly magnetic, ferromagnetic, has the highest neutron absorption cross-section of any element, and is superconductive. It has over two-dozen known isotopes, of which several are stable. Its applications include its use in microwave communications, lasers, as a phosphor activator in TV picture tubes, and in control rods. Its chemical symbol is **Gd**.

gadolinium gallium garnet A variety of garnet utilized as a substrate in lasers. Its abbreviation is **GGG**.

gadolinium oxide A whitish powder whose chemical formula is Gd_2O_3. Used as a phosphor activator, in lasers, in ceramic dielectrics, and in microwave communications.

gage Same as **gauge**.

gage pressure Same as **gauge pressure**.

gain Its symbol **A**. **1.** The increase in current, voltage, or power provided by a component, circuit, device, or system. For instance, the ratio of the output power to the input power of a power amplifier. Usually expressed in decibels. **2.** The effectiveness, generally expressed in decibels, of a directional antenna as compared to a given standard, such as a dipole or isotropic antenna.

gain-bandwidth product A figure of merit for amplifiers, especially operational amplifiers. It may be expressed in various manners, such as the product of the gain of an amplifier and its bandwidth. Its abbreviation is **GBP**.

gain coefficient Same as **gain ratio**.

gain constant Same as **gain ratio**.

gain control 1. A circuit or device which adjusts the gain of a component, circuit, device, or system. **2.** The regulating effect a **gain control** (1) has.

gain factor Same as **gain ratio**.

gain margin The level of gain increase that would produce instability in a system. For example, in a feedback control system it would be the increase in gain that would cause oscillation.

gain ratio A ratio indicating the gain a component, circuit, device, or system provides. In the case of a power amplifier, for instance, it would be expressed as the ratio of the output power to the input power. Also called **gain factor**, **gain coefficient**, or **gain constant**.

gain reduction A reduction in the gain of an amplifier at certain frequencies. This may be due, for instance, to its frequency-response characteristic.

gain region In a semiconductor, the area where gain occurs.

gain-sensitivity control Same as **gain-time control**.

gain stability The extent to which the gain of a component, circuit, device, or system remains steady under varying conditions.

gain-time control An automatic control circuit which modifies the gain of a receiver at set intervals, so as to maintain the output signal levels within a desired amplitude, in response to varying input signals. To do so, it reduces sensitivity to strong signals, while increasing it for weak ones. Such modifications are usually anticipated, as when a loran receiver adjusts its gain between successive pulses from fixed transmitters. Also called **gain-sensitivity control**, **differential gain control**, or **sensitivity-time control**.

galactic noise Same as **galactic radio noise**.

galactic radiation Radiation, originating within the Milky Way Galaxy, which travels through space at nearly the speed of light, and which consists mostly of protons and alpha particles.

galactic radio noise Radio-frequency noise originating from the Milky Way Galaxy. Also called **galactic noise**.

galactic radio waves Radio-frequency waves originating from the Milky Way Galaxy.

galena A lustrous blue mineral usually consisting of lead sulfide, but which may also contain other metals such as silver or copper. At a time galena was commonly utilized in crystal sets.

Galileo A European satellite system which competes with, and complements the **GPS System**.

gallium A chemical element whose atomic number is 31. It is a silvery-white solid metal which, upon heating slightly becomes a liquid with a striking silver color. It possesses a very wide temperature range as a liquid, is not particularly chemically reactive, and has about 25 known isotopes, of which 2 are stable. Its applications in electronics include its use as a semiconductor dopant, in lasers, light-emitting diodes, photocells, and in low-melting point alloys. Its chemical symbol is **Ga**.

gallium aluminum arsenide A semiconductor used, for instance, in light-emitting diodes.

gallium antimonide A chemical compound whose formula is GaSb. It is utilized in semiconductor devices.

gallium arsenide Dark gray crystals whose chemical formula is GaAs. It is a semiconductor material featuring high electron mobility, low noise, and exhibiting the Gunn effect. It has various applications, including its use in transistors, light-emitting diodes, ICs, oscillators, and lasers.

gallium arsenide diode A diode in which the semiconductor material is gallium arsenide. Used, for instance, in light-emitting diodes.

gallium arsenide FET Same as **GaAs FET**.

gallium arsenide field-effect transistor Same as **GaAs FET**.

gallium arsenide laser A semiconductor laser in which gallium arsenide is the lasing element. Such lasers emit light in the infrared region.

gallium arsenide phosphide A semiconductor used, for instance, in light-emitting diodes.

gallium indium arsenide A semiconductor used, for instance, in light-emitting diodes.

gallium indium phosphide A semiconductor used, for instance, in light-emitting diodes.

gallium nitride A chemical compound whose formula is GaN. It is utilized in semiconductor devices.

gallium phosphide Amber colored crystals whose chemical formula is GaP. It is utilized in semiconductor devices.

Galton whistle A short pipe which can be utilized to produce high-frequency audible and ultrasonic sound waves.

galvanic Pertaining to the generation or flow of electricity, especially DC, as a result of chemical action. Also called **voltaic**.

galvanic battery A battery composed of **galvanic cells**.

galvanic cell A primary cell which converts chemical energy into electrical energy, and which consists of two electrodes immersed in an electrolyte. Each electrode is of a different metal, and DC is generated. It is a type of electrolytic cell. Also called **voltaic cell**.

galvanic corrosion The electrochemical corrosion of one of the electrodes of a galvanic cell. The electrode which is corroded is that of the less noble metal.

galvanic couple Within a galvanic cell, two dissimilar conductors which generate a potential difference when immersed in the same electrolyte. Such conductors are usually metals, such as silver and zinc. Also called **voltaic couple**.

galvanic current 1. The DC produced by a galvanic cell. 2. An essentially steady DC.

galvanic pile An early battery consisting of a series of alternated disks of dissimilar metals, usually zinc and copper, each separated by paper or cloth soaked in an electrolyte. Also called **voltaic pile**, or **pile (1)**.

galvanic series A series of metals, listed in order of their relative ease of oxidation. The greater the tendency to oxidize, the higher on the list. Less noble metals, such as magnesium and sodium, are at the top of such a list, while noble metals, such as platinum and gold, are at the bottom of it. Alternatively, the more noble metals may appear higher on the list, with the less noble metals lower, when the list is in order of their relative resistance to oxidation.

galvanic skin response A change in the electrical resistance of the skin in response to emotional stimuli, such as anxiety or fear. Such variations are detected by a galvanometer. Its abbreviation is **GSR**. Also called **electrodermal response**.

galvanism 1. A potential difference arising through chemical action. This may occur in the mouth, for instance, with dental fillings of dissimilar metals serving as electrodes, and saliva as the electrolyte. 2. Any utilization of electricity for medical treatment.

galvanization The process of depositing zinc onto another metal, usually iron or steel, to help reduce corrosion. Two processes utilized are dip coating and electroplating. Also called **galvanizing**.

galvanize To deposit zinc onto another metal, usually iron or steel, to help reduce corrosion.

galvanized steel Steel which has been coated with zinc, via galvanization.

galvanizing Same as **galvanization**.

galvanomagnetic effect Any of various electrical or thermal phenomena observed when a strip or bar of a conductor, or of certain semiconductors, is placed perpendicular to a magnetic field. Examples include the Hall effect, the Nernst effect, and the Ettinghausen effect.

galvanometer An instrument utilized for measuring and indicating small currents. A moving-coil galvanometer, for instance, utilizes a small wire coil which is suspended or pivoted in a fixed magnetic field. The current that passes through the coil produces a magnetic field which interacts with the fixed field, and the resulting movement of the coil is correlated to a calibrated scale. Depending on the needs, there are specialized galvanometers. For instance, a ballistic galvanometer is designed to respond to rapid current pulses.

galvanometer constant A constant by which a galvanometer reading is multiplied, in order to obtain the current in the desired units.

galvanometer shunt A resistor connected in parallel with a galvanometer to reduce its sensitivity, and thus increase its range.

galvanoscope An instrument utilized to detect the presence of a small current, and to indicate its direction.

game A computer program in which one or more users interact with the computer and/or other players, for amusement or competition. Such programs may be of varying complexity, and some feature faithfully realistic simulations of complicated activities such as aircraft piloting. Also called **computer game**, or **video game (1)**.

game port An I/O port to which a device such as a joystick is connected. Utilized with computer games.

GameBoy A popular gaming system.

GameCube A popular gaming system.

gamma Its symbol is γ. 1. The third letter of the Greek alphabet, and as such may designate the third constituent of a series. 2. A unit of magnetic flux density equal to 10^{-9} tesla.

gamma camera An instrument which incorporates very sensitive radiation detectors which gather data utilized to produce images of the distribution of radionuclides in a body being scanned. Used, for instance, in computed tomography techniques such as PET and SPECT.

gamma correction Adjustments made to the brightness of displayed images to compensate for tonal irregularities. May refer, for instance, to images displayed on a CRT computer monitor.

gamma decay Radioactive decay in which **gamma rays** are emitted.

gamma emission The emission of **gamma rays** by a radioactive element.

gamma emitter A radioactive element that emits **gamma rays**. Also called **gamma radiator**.

gamma ferric oxide A form of ferric oxide commonly utilized as a coating on magnetic tapes.

gamma irradiation Exposure to **gamma rays**.

gamma match An adapter or device utilized to match an unbalanced antenna transmission line, such as a coaxial cable, and a balanced antenna element, such as a half-wave radiator.

gamma radiation Same as **gamma rays**.

gamma radiator Same as **gamma emitter**.

gamma radiography Radiography utilizing **gamma rays**.

gamma ray A single high-energy photon within **gamma rays**.

gamma-ray detector An instrument which detects **gamma rays**.

gamma-ray source A given quantity of a **gamma emitter**. Used, for instance, in gamma radiography. Also called **gamma source**.

gamma-ray spectrometer An instrument utilized for **gamma-ray spectroscopy**.

gamma-ray spectrometry Same as **gamma-ray spectroscopy**.

gamma-ray spectroscopy An analytical technique which measures the spectrum of energies and relative intensities of gamma rays. It is utilized, for instance, to determine the concentrations of gamma-emitting chemical elements in a given sample. Also called **gamma-ray spectrometry**.

gamma-ray spectrum The plot, or other visual output, produced when utilizing **gamma-ray spectroscopy**.

gamma rays Very high-frequency electromagnetic radiation that is produced during nuclear transitions or reactions. Such rays consist of high-energy photons, are unaffected by magnetic fields, and readily penetrate living tissue, producing serious harm when protective shielding is not utilized. Gamma rays are emitted, for instance, by various radioactive elements, and are one of the products when cosmic rays strike particles in the atmosphere. Such rays have wavelengths of between approximately 10^{-11} and 10^{-13} meter, corresponding to frequencies of approximately 10^{19} and 10^{21} Hz, respectively. Also called **gamma radiation**.

gamma source Same as **gamma-ray source**.

GAN Acronym for **Global-Area Network**.

gang To mechanically connect two electrical components, devices, or controls, so that they can be operated and varied simultaneously, usually using a single knob.

gang capacitor Two or more variable capacitors which are mounted in a manner that enables them to be varied simultaneously. Also called **ganged capacitor**.

gang switch Same as **ganged switches**.

gang tuning Same as **ganged tuning**.

gang tuning capacitor A **gang capacitor** utilized for tuning. Used, for instance, in a radio receiver. Also called **ganged tuning capacitor**.

gang volume control Two or more volume controls which are simultaneously varied by a single knob. Also called **ganged volume control**.

ganged Two or more electrical components, devices, or controls which are connected in a manner that enables them to be operated and varied simultaneously.

ganged capacitor Same as **gang capacitor**.

ganged potentiometer Two or more potentiometers which are mounted in a manner that enables them to be varied simultaneously.

ganged switches Two or more switches which are connected in a manner that enables them to be simultaneously controlled. Also called **gang switch**.

ganged tuning Simultaneous tuning of two or more circuits utilizing a single, usually mechanical, control. Used, for instance, in a radio receiver. Also called **gang tuning**.

ganged tuning capacitor Same as **gang tuning capacitor**.

ganged volume control Same as **gang volume control**.

ganging The connection of two or more electrical components, devices, or controls in a manner that enables them to be operated and varied simultaneously.

Gannt chart In project management, a diagram in which tasks to be performed are charted as a function of time. For example, the horizontal axis may be depict time, while the planned activities are shown as bars in the vertical axis.

gantry robot A robot that has a number of degrees of freedom, usually three, as dictated by the rails and platforms which determine its movements. Also called **gantry-type robot**.

gantry-type robot Same as **gantry robot**.

gap 1. A separation, space, or break in continuity. For instance, an interruption in the flow of transmitted data. 2. A narrow space separating components, electrodes, circuits, or devices. For example, that between contacts. 3. A space of a nonmagnetic material, such as air, between magnetic components, circuits, devices, or materials. For instance, the opening between pole pieces in a tape head. Also called **magnetic gap**, **magnetic air-gap**, or **air gap** (2). 4. A region in which coverage, such as that of a transmitting antenna, is absent or below a given minimum. For example, a zone of inadequate coverage by a radar. 5. A space between blocks of data on a magnetic storage device, such as a disk or tape.

gap analysis 1. The study of the difference between what a given technology is capable of, and the needs of users. 2. The study of the difference between where a given technology is, and where it is desired for said technology to advance.

gap depth The depth of the gap between pole pieces in a tape head. It is measured perpendicular to the surface of the head.

gap filling The electrical or mechanical modification of an antenna array, or the utilization of a supplemental array, to eliminate regions where gaps in coverage occur.

gap length The length of the gap between pole pieces in a tape head.

gap loss A loss in a signal due to a gap. For instance, any loss of power when an optical signal is transferred between fibers which are aligned, but not physically touching.

gap width The width of the gap between pole pieces in a tape head. Also called **head gap** (2).

gapless Not having any **gaps**.

garbage 1. Meaningless, unwanted, incorrect, corrupted, or otherwise useless data. Such data can be entered by a user, or may be produced by a computer. Also called **hash** (6). 2. A signal which is unwanted, unintelligible, or otherwise useless.

garbage collection During program execution, the recovery of dynamically allocated blocks of memory no longer in use. Garbage collection is usually done automatically.

garbage collector A routine or program which performs **garbage collection**.

garbage in, garbage-out An axiom stating that if a user enters meaningless, invalid, or incorrect data into a computer, that the output will also be meaningless, invalid, or incorrect. Its acronym is **GIGO**.

garble To alter data, especially that being sent, in a manner which renders it unintelligible or otherwise difficult to understand. Such alterations may be unintentional, as in a bad connection, or intentional, as in scrambling.

garbled message A message which has been intentionally or unintentionally altered in a manner which renders it unintelligible or otherwise difficult to understand.

garnet A group of silicate minerals incorporating two metal ions, such as iron and aluminum, in fixed proportions. Used, for instance, in lasers, microwave devices, and as an abrasive.

gas A state of matter characterized by very low density, comparatively large changes in volume as pressure and temperature vary, and the tendency to assume the shape of its container. Gases move freely past one another, and diffuse readily into other gases. A gas may be a chemical element, such as argon or oxygen, or a compound, such as carbon dioxide. The other physical states in which matter is known to exist are **solid (1)**, **liquid**, **plasma**, and **Bose-Einstein condensates**.

gas amplification In a radiation-counting device, such as a Geiger counter, the ratio of the total ionization to the initial ionization. Also called **gas multiplication**.

gas breakdown In a gas-filled tube, a cumulative ionization process in which charged particles are accelerated by an electric field and collide with neutral particles, creating additional charged particles. These additional particles collide with others, multiplying the amount of ions present, which in turn makes the gas a good conductor.

gas capacitor A capacitor in which a gas serves as the dielectric between its metallic plates. When the gas is air, also called **air capacitor**.

gas cell An electric cell in which the electrodes absorb gas.

gas detector A device which detects and indicates the presence of gases, such as those that are combustible or poisonous.

gas-discharge display 1. A display, usually consisting of a numerical readout, in which each digit segment is illuminated by the glow of an ionized gas. 2. Same as **gas-plasma display**.

gas-discharge lamp A lamp which produces light when the gas it contains, usually at a low pressure, becomes ionized by an electric discharge passed through it. An example is a fluorescent lamp. Also known as **gas-filled lamp (2)**, **discharge lamp**, **electric-discharge lamp**, or **vapor lamp**.

gas-discharge laser Same as **gas laser**.

gas-discharge tube A tube containing a low-pressure gas through which a current passes when it is ionized by a sufficient voltage applied between the electrodes of said tube. Also known as **discharge tube**, or **electric-discharge tube**.

gas doping The adding of a dopant in the gaseous state to a semiconductor material. For example, the desired impurity atoms may be placed in a precisely heated enclosure with the semiconductor material, so as to be distributed uniformly within the crystal lattice of said semiconductor material.

gas electrode An electrode which absorbs gas. An example is an electrode in a gas cell.

gas-filled cable A cable which contains a comparatively inert gas, such as nitrogen, under pressure. This serves, for instance, to provide insulation and protection from moisture.

gas-filled lamp 1. An incandescent lamp whose filament is surrounded by an inert gas, such as argon. 2. Same as **gas-discharge lamp**.

gas-filled rectifier A gas-filled rectifier tube which incorporates a cold-cathode. Also called **cold-cathode rectifier**.

gas-filled tube Same as **gas tube**.

gas-flow alarm An alarm that is activated when a connected circuit or device detects a given gas flow, such as that exceeding a given rate.

gas-flow indicator An instrument which indicates the flow rate of a gas.

gas-flow meter An instrument which measures and indicates the flow rate of a gas. Such a meter may be ultrasonic, electronic, and so on.

gas focusing In an electron tube, the use of an inert gas to focus an electron beam. The gas along the path of the beam becomes positively charged, and this attracts the electrons, further narrowing the beam. Used, for instance, in a CRT. Also called **ionic focusing**.

gas laser A laser in which one or more gases are stimulated into emitting the laser radiation. The pumping mechanism in such lasers is usually an electric discharge, although other methods may be used, such as chemical reactions or gas compression. Examples include helium-neon, argon, and carbon dioxide lasers. Also called **gas-discharge laser**.

gas magnification In a phototube, an increase in current due to the ionization of the gas it contains.

gas maser A maser in which a gas, such as ammonia or hydrogen, is stimulated to amplify microwave radiation.

gas multiplication Same as **gas amplification**.

gas phototube A phototube in which a small amount of a gas is introduced, so as to enhance sensitivity.

gas-plasma display A flat display which utilizes a matrix of small tubes, each containing a gas. Each tube represents a pixel, and when powered, the gas ionizes to produce a glow. Depending on the contained gas, the glow discharge may be red, green, or blue. Used, for instance, for large-screen and digital TV. Also called **gas-discharge display (2)**, **gas-plasma monitor**, or **plasma display**.

gas-plasma monitor Same as **gas-plasma display**.

gas-plasma screen The viewing screen of a **gas-plasma display**.

gas sensor A component or device which detects the presence of a gas at a given concentration or above. For example, at so many parts per million. A gas sensor may respond to only one gas, or to multiple gases.

gas tube An electron tube containing a small amount of a gas, such as mercury vapor or neon. During operation this gas becomes ionized, to substantially increase current flow. Also called **gas-filled tube**.

gaseous Of, pertaining to, containing, composed of, or existing as a gas.

gasket A flexible seal used between parts to maintain contact or prevent the escape of a fluid. For example, that used to form a seal between a speaker cone and its baffle.

gassing 1. The evolution of one or more gasses. For instance, the bubbles that may form on an electrode during electrolysis. 2. The evolution of gas while charging a storage battery. This occurs especially towards the end of a charge cycle. 3. The process of adding a gas.

gassy tube A vacuum tube which is not sufficiently evacuated, and whose residual gas affects its electrical characteristics. Also called **soft tube**.

gastight Not allowing the passage of gases, including air.

gate 1. A device or barrier which regulates the passage of a signal, charge, pulse, fluid, and so on. 2. A circuit or device which carries out a logic function or operation. Such a circuit or device has multiple inputs and one output. Its output will depend on specified input conditions, such as a given combination of states. The three basic gates are the AND, OR, and NOT gates. Also called **gate circuit (1)**, or **logic gate**. 3. A circuit which determines if another circuit works.

Also called **gate circuit (2)**. **4.** In a field-effect transistor, the electrode which controls the flow of current through the channel. Small variations in the voltage applied to the gate result in large variations in the current passing through a field-effect transistor. The gate is analogous to the base of a bipolar junction transistor. Its symbol is **G**. Also called **gate electrode**. **5.** A control electrode in a device such as a field-effect transistor or thyristor.

gate array An IC consisting of an array of **gates (2)**. The specific arrangement of the gates, and hence their function, is determined late in the manufacturing process. This customization for a specific application can save design and manufacturing time, but a significant proportion of the chip may go unused. Also called **logic array**.

gate circuit **1.** Same as **gate (2)**. **2.** Same as **gate (3)**.

gate-controlled A circuit, switch, or device which is controlled by a gate circuit.

gate-controlled switch **1.** Same as **gate turn-off switch**. **2.** A switch controlled by a gate circuit.

gate current The current flowing through a **gate (3)**, **gate (4)**, or **gate (5)**.

gate electrode Same as **gate (4)**.

gate generator A circuit or device which generates **gate pulses**.

gate pulse A pulse which triggers a gate circuit to pass a signal. Used, for instance, to actuate a gate-controlled semiconductor device.

gate signal A signal which triggers a gate circuit to pass a signal. Used, for instance, to actuate a gate-controlled semiconductor device.

gate trigger current The current necessary to trigger a gate circuit to pass a signal.

gate trigger voltage The voltage necessary to trigger a gate circuit to pass a signal.

gate turn-off A device similar a silicon-controlled rectifier, but whose gate structure provides for more flexibility in switching it on or off. Also spelled **gate turnoff**.

gate turn-off switch A device similar a silicon-controlled rectifier, with the difference that a negative pulse applied to its gate can be used to switch it off. Also spelled **gate turn-off switch**. Also called **gate-controlled switch (1)**.

gate turnoff Same as **gate turn-off**.

gate turnoff switch Same as **gate turn-off switch**.

gate voltage The voltage applied a **gate (3)**, **gate (4)**, or **gate (5)**.

gated flip-flop A flip-flop equipped with a mechanism which does not allow both states to be low, or 0, simultaneously.

gated sweep **1.** In radars, a sweep whose start and duration are controlled, so as to minimize unwanted echoes. **2.** A circuit providing a **gated sweep (1)**.

gateway **1.** A device which enables networks to connect by performing the necessary protocol conversions. For example, a gateway could carry out the translation between T1 and E1, between Ethernet and Token Ring, or between messaging protocols. **2.** In a communications network, or multiple interconnected networks, a device or software which determines where packets, messages, or other signals travel to next. A gateway, using resources such as header information, algorithms, and router tables, establishes the best available path from source to destination. Within the OSI Reference Model, a gateway operates at the network layer. Also called **router**, or **network router**.

gateway page A Web page created with given characteristics, such as multiple appearances of specific keywords, so as to rank high on a particular search engine. Also called **doorway page**, **jump page**, **entry page**, or **bridge page**.

gather write A single data write operation in which two or more non-contiguous buffers or locations are used. A **scatter read** is utilized to access such data.

gating **1.** The process of selecting the portions of a signal or wave that exist during selected time intervals, or that fall between specified amplitude limits. **2.** The use of a gate circuit to control a circuit or device. Also, to control said device during selected portions of a cycle.

gating circuit A circuit utilized for **gating**.

gauge Also spelled **gage**. **1.** A device or instrument which serves to measure a given magnitude, or which is utilized for testing. **2.** A scale or standard of measurement. **3.** A scale or standard utilized for sizing wires, tubing, sheets, rods and the such. Used, for instance, to determine the thickness or diameter of such a material. Also, a measurement expressed in such terms. For example, a 14-gauge wire.

gauge pressure Pressure relative to the ambient pressure. This contrasts with **absolute pressure**, which is measured relative to that found in an absolute vacuum. Also spelled **gage pressure**.

gauss The unit of magnetic flux density or magnetic induction in the cgs system. One gauss is equal to 10^{-4} tesla, or 1 maxwell per centimeter squared. Also known as **abtesla**. Its symbol is **G**, or **Gs**.

Gauss' law **1.** A law which states that the electric flux through any closed surface is proportional to the electric charge contained within said surface. Also called **Gauss' law for electricity**. **2.** Same as **Gauss' law for magnetism**.

Gauss' law for electricity Same as **Gauss' law (1)**.

Gauss' law for magnetism A law which states that the net magnetic flux through any closed surface is zero. That is, the amount of magnetic flux that leaves a closed space is the same as that which enters it. Also called **Gauss' law (2)**.

Gaussian channel A communications channel whose only noise present is **Gaussian noise**.

Gaussian curve A curve corresponding to a **Gaussian distribution**. It is in the shape of a symmetrical bell. Also called **bell-shaped curve**, or **normal curve**.

Gaussian distribution For a random variable, a probability distribution which is symmetrical about the mean value, and which continuously diminishes in value until reaching zero at each extreme. It is utilized to determine the probability of the value of the variable falling within a given interval of values, and when graphed has the shape of a symmetrical bell. Various phenomena, such as some natural frequencies, have a Gaussian distribution. Also called **bell-shaped distribution**, or **normal distribution**.

Gaussian error An error which has a **Gaussian distribution**. Random errors usually have such a distribution.

Gaussian filter A filter whose response curve has a **Gaussian distribution**.

Gaussian noise Noise which has a **Gaussian distribution**. The frequency distribution of such noise follows a Gaussian curve.

Gaussian noise generator A noise generator whose output follows a **Gaussian distribution**.

gaussmeter A magnetometer which is calibrated in gauss, or in a multiple of gauss such as kilogauss.

Gb Abbreviation of **gigabit**.

GB Abbreviation of **gigabyte**.

GbE Abbreviation of **Gigabit Ethernet**.

Gbit Abbreviation of **gigabit**.

Gbits/sec Abbreviation of **gigabits per second**.

GBP Abbreviation of **gain-bandwidth product**.

Gbps Abbreviation of **gigabits per second**.

GBps Abbreviation of **gigabytes per second**.

GBq Abbreviation of **gigabecquerel**.

Gbyte Abbreviation of **gigabyte**.

Gbytes/sec Abbreviation of **gigabytes per second**.

GC Abbreviation of **gigacoulomb**.

GCA Abbreviation of **ground-controlled approach**.

Gd Chemical symbol for **gadolinium**.

Ge Chemical symbol for **germanium**.

gee whiz factor An aspect of design, a mechanism, or a feature that demonstrates technological prowess, but which may or may not have a useful purpose.

geek A person who is inordinately oriented towards and/or dedicated to technology, especially that pertaining to computers and networks. The term usually implies a given level of expertise. Also called **nerd**.

Geiger counter An instrument which detects, measures, and indicates ionizing radiation such as alpha particles. It incorporates a **Geiger tube** (1) to detect and quantify the radiation, and its indications may be visual and/or auditory. Also called **Geiger-Müller counter**, **Geiger-Müller tube** (2), **Geiger tube** (2), **Geiger counter tube** (2), or **Geiger-Müller counter tube** (2).

Geiger counter tube 1. Same as **Geiger tube** (1). 2. Same as **Geiger counter**.

Geiger-Mueller counter Same as **Geiger counter**. Its abbreviation is **GM counter**.

Geiger-Mueller tube Its abbreviation is **GM tube**. 1. Same as **Geiger tube** (1). 2. Same as **Geiger counter**.

Geiger-Müller counter Same as **Geiger counter**. Also spelled **Geiger-Mueller counter**. Its abbreviation is **GM counter**.

Geiger-Müller counter tube 1. Same as **Geiger tube** (1). 2. Same as **Geiger counter**.

Geiger-Müller region Same as **Geiger region**.

Geiger-Müller threshold Same as **Geiger threshold**.

Geiger-Müller tube Also spelled **Geiger-Mueller tube**. Its abbreviation is **GM tube**. 1. Same as **Geiger tube** (1). 2. Same as **Geiger counter**.

Geiger region For a Geiger tube, the interval of operating voltages within which the amplitude of the output is the same, regardless of number of primary ions produced by the initial ionizing event. Also called **Geiger-Müller region**.

Geiger threshold The minimum voltage within the **Geiger region**. Also called **Geiger-Müller threshold**.

Geiger tube 1. A tube which detects and quantifies ionizing radiation, such as alpha particles, beta particles, or gamma radiation. Such a tube contains a gas mixture, such as methane and argon in a given proportion, and has a thin wire anode mounted along the axis of a cylindrical metal chamber which serves as the cathode. A potential difference which is slightly less than that required for electric discharge is maintained between the electrodes at all times. In this manner, the gas becomes ionized whenever said tube is exposed to a radioactive substance, however faint it may be, and the resulting current is amplified for output. Also called **Geiger counter tube** (1), **Geiger-Müller counter tube** (1), or **Geiger-Müller tube** (1). 2. Same as **Geiger counter**.

Geissler tube A tube containing a low-pressure gas, and through which a current passes when it is ionized by a sufficient voltage applied between its two electrodes. The color of the glow depends on the gas used. It is an early form of gas-discharge tube, and modified versions are currently utilized in spectroscopy.

gel battery Same as **gel-cell battery**.

gel cell A cell whose electrolyte has a gelatinous consistency.

gel-cell battery A battery consisting of **gel cells**. Such batteries are sealed, rechargeable, and tend to be virtually maintenance-free. Also called **gel battery**, or **gelled battery**.

gelled battery Same as **gel-cell battery**.

gender bender Same as **gender changer**.

gender changer A device which enables a cable connector to be joined with the same gender of cable connector. For instance, the connection of two phono plugs, as opposed to a phono plug being plugged into a phono jack. Also called **gender bender**, or **cable matcher**.

General Inter-ORB Protocol Same as **GIOP**.

General MIDI A MIDI standard which enables computers and musical devices to communicate effectively, and which helps insure that files created on one computer using this standard are compatible with another utilizing it. It specifies, for instance, 128 standard sounds in specific locations. Its abbreviation is **GM**.

General Packet Radio Service A wireless data communications service, standard, or technology in which data packets are transmitted. It is compatible, for instance, with GSM. Its abbreviation is **GPRS**.

general-purpose Designed to be suitable for multiple functions or applications.

general-purpose computer A computer which can perform most computing tasks adequately. These include, for instance, scientific, educational, and business applications. Most personal computers are of this type.

General-Purpose Interface Bus A computer interface based on the IEEE 488 standard. Used mainly to connect data acquisition and control instruments, industrial automation equipment, medical diagnostic devices, and the like. Its abbreviation is **GPIB**. Also called **GPIB bus**, or **IEEE 488 bus**.

general-purpose language A computer language, such as BASIC, C, or Pascal, which can be used for a variety of applications or classes of problems. This contrasts with **special-purpose language**, which is designed for a single application.

general-purpose register In a CPU, a register which be used for multiple purposes. For example, it may be able to serve as an arithmetic register or as an index register. Its abbreviation is **GPR**.

general-purpose relay A relay which is suitable for a wide variety of applications, as opposed to just one.

general-purpose transistor A transistor which is suitable for a wide variety of applications, as opposed to just one.

generate 1. That which a **generator** does. 2. To convert mechanical energy into electrical energy. 3. To convert another form of energy into electrical energy. 4. To generate AC power. 5. To produce an alternating current or voltage of a desired frequency. 6. To utilize a computer program to create other programs.

generating magnetometer 1. An instrument which measures the strength of a magnetic field by means of the electromotive force generated in a rotating coil immersed in said field. The coil delivers an output voltage which is proportional to the intensity of the field. 2. A **generating magnetometer** (1) utilized to measure the strength of the earth's magnetic field. Also called **earth inductor**.

generating station A location which combines the structures, devices, and equipment necessary to produce electricity in quantities adequate for residential, commercial, or industrial use. A generating station converts another form of energy, such as hydroelectric, nuclear, or solar, into electrical energy. Also called **electric power station**, **power station**, **power plant**, **power generating station**, or **electric power plant** (1).

generation 1. The process of producing something. For instance, energy, a signal, a voltage, a current, a result, a set of instructions, and so on 2. The process of converting mechanical energy into electrical energy. 3. The process of converting another form of energy into electrical energy. **4.** The process of generating AC power. **5.** The process of producing an alternating current or voltage of a desired frequency. **6.** The utilization of a computer program to create other programs. **7.** A group whose members are contemporary to each other, from the technological perspective. For instance, fourth-generation computers. **8.** A member within a family of files. For instance, the copy or version saved before a given file is the parent file. The grandparent file is the generation before that. **9.** The number of recording steps away from a master, or from a given copy.

generation rate In a semiconductor, the rate at which electron-hole pairs are generated. This contrasts with a **recombination rate**, which is the rate at which electron-hole recombination occurs.

generator Its symbol is **G**. **1.** That which produces something. That generated may be energy, a signal, a voltage, a current, a result, a set of instructions, and so on. **2.** A machine or device which converts mechanical energy into electrical energy. Such a machine or device is usually rotary, and may, for instance, utilize magnetic flux to convert angular displacement into electricity. Also called **dynamo**, or **electric generator (1)**. **3.** A device which converts another form of energy into electrical energy. Examples include dynamos, batteries, and oscillators. Also called **electric generator (2)**. **4.** A device which generates AC power. Also called **AC generator (2)**. **5.** An electronic device whose output is an alternating current or voltage of a desired frequency. **6.** A computer program utilized to create other programs. An example is an application generator.

generator efficiency For a **generator (2)**, or **generator (3)**, the ratio of the useful power or energy output, to the total power or energy input.

generator end The input end of a transmission line, where a signal originates. The **sink (2)** is end where the signal is received.

generator noise Electrical noise produced by a **generator (2)**, or **generator (3)**.

generic Pertaining to, or descriptive of an entire group or category.

generic top-level domain A top-level domain which is not country-specific. For example, .com, .net, and .org. Its abbreviation is **gTLD**.

GEO Abbreviation of **geostationary earth orbit**, or **geosynchronous earth orbit**.

geocoding **1.** The use of GPS for the determination of the latitude and longitude coordinates of a given address or location. **2.** The use of GPS for the determination of a given address or location using its latitude and longitude coordinates.

geodesy The science which deals mathematically with the size, shape, and gravitational field of the planet earth. Also, the determination of points on the earth's surface through the use of an earth-based coordinate system.

geodetic distance meter Same as **geodimeter**.

geodimeter An optoelectronic device utilized to measure distances. Usually used in surveying. It is an abbreviation of **geo**detic **distance meter**.

Geographic Information System A computer system utilized for capturing, storing, analyzing, retrieving, and displaying spatial data. Used, for instance, to create, access, manipulate, and view maps. May refer to the complete computer system, or just to the software utilized. Its abbreviation is **GIS**. Also spelled **Geographical Information System**.

geographic north pole On the surface of the planet earth, the point within the northern hemisphere where all meridians meet. It is defined as 90° N, and is the northern end of the axis of rotation of the planet. Also called **north pole (1)**, or **north geographic pole**.

geographic pole Either the **geographic north pole** or the **geographic south pole**.

geographic south pole On the surface of the planet earth, the point within the southern hemisphere where all meridians meet. It is defined as 90° S, and is the southern end of the axis of rotation of the planet. Also called **south pole (1)**, or **south geographic pole**.

Geographical Information System Same as **Geographic Information System**.

geomagnetic Pertaining to **geomagnetism**.

geomagnetic equator A great circle which at each point is 90° from the geomagnetic poles. This contrasts with the **magnetic equator**, which is a great line along the surface of the earth which connects all points at which the magnetic dip is 0°.

geomagnetic field The magnetic field of the planet earth. The lines of flux of said field extend from the north magnetic pole to the south magnetic pole, which do not coincide exactly with the geographic north and south poles. The field approximately resembles that of a bar magnet, and though its strength varies, it is approximately 50 microtesla at the earth's surface. Also called **earth's magnetic field**.

geomagnetic noise Radio-frequency interference caused by **geomagnetism**.

geomagnetic pole On the surface of the planet earth, either of the two oppositely located points which mark the intersection of the planet with the extended axis of a powerful bar magnet hypothetically located at its center. The magnet is assumed to have a field approximating that of the earth.

geomagnetic storm A large-scale, often worldwide, disturbance of the **geomagnetic field**. Such storms are characterized by a sudden onset, and their effects may last from a few minutes to several days. Geomagnetic storms are usually caused by solar flares, and affect radio-frequency communications. Also called **magnetic storm**.

geomagnetism The magnetism of the planet earth. Also called **terrestrial magnetism**.

geometric distortion A form of distortion manifested by changes in the shape of an image, as opposed to the colors or sharpness. Examples include barrel distortion and pincushion distortion. Also spelled **geometrical distortion**.

geometric mean For a product of **n** factors, the nth root. For instance, the geometric mean of 6 and 216 (a total of two factors) is 36, as the square root of 1,296 is 36. Also called **mean (2)**.

geometrical distortion Same as **geometric distortion**.

geophone A transducer which converts vibrations in the earth into electrical signals.

geostationary earth orbit Same as **geostationary orbit**. Its abbreviation is **GEO**.

geostationary orbit The orbit a **geostationary satellite** follows. Also called **geostationary satellite orbit**, or **geostationary earth orbit**.

geostationary satellite An artificial satellite whose orbit is synchronized with that of the earth, so that it remains fixed over the same spot above the planet at all times. Such a satellite is usually located at an altitude of approximately 35,900 kilometers, and its orbit is in the plane of the equator. As few as three of these satellites can provide worldwide coverage. Used, for instance, for communications and broadcasting. Also called **geosynchronous satellite**, **synchronous satellite (1)**, or **fixed satellite (1)**.

geostationary satellite orbit Same as **geostationary orbit**.

geosynchronous earth orbit Same as **geosynchronous orbit**. Its abbreviation is **GEO**.

geosynchronous orbit The orbit a **geosynchronous satellite** follows. Also called **geosynchronous earth orbit**.

geosynchronous satellite Same as **geostationary satellite**.

German silver An alloy consisting of about 65% copper, 18% nickel, and 17% zinc, although the composition may vary. There may also be small amounts of other elements present, such as manganese. It has a silvery-white appearance, and is used, for instance, in electroplating. Also called **nickel silver**.

germanium A lustrous gray-white semimetallic chemical element whose atomic number is 32. It is a semiconductor, may be doped with metals such as arsenic or gallium, and is available in purities of better than 1 part per 10^{10}. It has over 25 known isotopes, of which 5 are stable. In addition to its utilization in semiconductor devices such as transistors and diodes, it is used as a phosphor, and in special glasses and lenses used for spectroscopy and microscopy. Its chemical symbol is **Ge**.

germanium diode A diode in which the semiconductor material is germanium.

germanium dioxide A white powder whose chemical formula is GeO_2. It is utilized in semiconductor devices, as a phosphor, and in glass which transmits infrared radiation.

germanium rectifier A rectifier in which the semiconductor material is germanium.

germanium telluride A semiconductor whose chemical formula is **GeTe**.

germanium transistor A transistor in which the semiconductor material is germanium. Used, for instance, in amplifiers and oscillators.

gesture recognition The ability of a computer system to detect and comprehend gestures such as hand and head movements. The term specifically excludes the use of haptic devices or interfaces. Used, for instance, for entering data using sign language.

getter **1.** A substance which binds other materials to its surface, so as to remove them from a given material or environment. For example, the use titanium to remove undesired impurities from semiconductor materials. **2.** A substance, such as barium, calcium, or cesium which is utilized to remove residual gases from vacuum tubes, or to maintain a high vacuum in said tubes.

gettering The use of a **getter**.

GeV Abbreviation of **gigaelectronvolt**.

GFCI Abbreviation of **ground-fault circuit interrupter**.

GFI Abbreviation of **ground-fault interrupter**.

GFLOPS Abbreviation of **gigaflops**. One billion FLOPS, which is the same as one billion floating-point calculations or operations, per second. Usually used as a measure of processor speed.

Gflops Same as **GFLOPS**.

Gg Abbreviation of **gigagram**.

GGG Abbreviation of **gadolinium gallium garnet**.

ghost **1.** Same as **ghosting**. **2.** The fainter of the two overlapping images in **ghosting**. Also called **ghost image**. **3.** On a computer screen, to display the letters or the icon representing a function, operation, or option in a lighter, grayed, or fuzzy manner, to indicate that it is unavailable. For example, the cut function should be grayed out when no content is selected. **4.** A non-existing person which is created so as to have unauthorized access to confidential and/or valuable information, with the objective of thwarting detection.

ghost image Same as **ghost (2)**.

ghost pulse Same as **ghost signal**.

ghost signal Also called **ghost pulse** when used in the context of radars. **1.** A signal resulting in **ghosting**. **2.** A signal which produces the fainter of the two overlapping images in **ghosting**.

ghost site A Web site that is no longer maintained, but which is still available for viewing. It is essentially the same as an **abandoned site**, except the latter generally does not have any statement to the effect.

ghosted Same as **grayed out**.

ghosting On the screen of a TV or radar receiver, an undesired condition in which duplicate or overlapping images appear. May be due, for instance, to the signal arriving over two paths of different length. Also called **ghost (1)**, or **double image**.

GHz Abbreviation of **gigahertz**.

Gi Abbreviation of **gilbert**.

Gi- Abbreviation of **gibi-**.

giant magnetoresistive head A read/write head used in **giant magnetoresistive head technology**.

giant magnetoresistive head technology An enhancement of **magnetoresistive head technology**, providing increased sensitivity to weaker fields, which in turn affords greater storage density.

GiB Abbreviation of **gibibyte**.

gibi- A binary prefix meaning 2^{30}, or 1,073,741,824. For example, a gibibyte is equal to 2^{30}, or 1,073,741,824 bytes. This prefix is utilized to refer to only binary quantities, such bits and bytes. Its abbreviation is **Gi-**.

gibibit 2^{30}, or 1,073,741,824 bits. Its abbreviation is **Gib**.

gibibits per second 2^{30}, or 1,073,741,824 bits, per second. Its abbreviation is **Gibps**.

gibibyte 2^{30}, or 1,073,741,824 bytes. Its abbreviation is **GiB**.

gibibytes per second 2^{30}, or 1,073,741,824 bytes, per second. Its abbreviation is **GiBps**.

Gibps Abbreviation of **gibibits per second**.

GiBps Abbreviation of **gibibytes per second**.

GIF Acronym for Graphics Interchange Format. A graphics file format which supports 256 colors, multiple resolutions, and compression.

Gig-E Abbreviation of **Gigabit Ethernet**,

giga- A metric prefix representing 10^9, which is equal to a billion. For instance, 1 gigahertz is equal to 10^9 Hz. When referring to binary quantities, such bits and bytes, it is equal to 2^{30}, or 1,073,741,824, although this is frequently rounded to a billion. To avoid any confusion, the prefix **gibi-** may be used when dealing with binary quantities. Its abbreviation is **G**.

gigabecquerel A unit of activity equal to 10^9 becquerels. Its abbreviation is **GBq**.

gigabit 2^{30}, or 1,073,741,824 bits, although this is frequently rounded to a billion. To avoid any confusion, the term **gibibit** may be used when referring to this concept. Its abbreviation is **Gb**.

Gigabit Ethernet An Ethernet technology which supports data-transfer rates of up to 1 gigabit per second. It is part of the IEEE 802.3 standard. Its abbreviation is **GbE, GigE**, or **Gig-E**.

gigabits per second 2^{30}, or 1,073,741,824 bits, per second. Usually used as a measure of data-transfer speed. To avoid any confusion, the term **gibibits per second** may be used when referring to this concept. Its abbreviation is **Gbps**.

gigabyte 2^{30}, or 1,073,741,824 bytes, although this is frequently rounded to a billion. To avoid any confusion, the term **gibibyte** may be used when referring to this concept. Its abbreviation is **GB**.

gigabytes per second 2^{30}, or 1,073,741,824 bytes, per second. Usually used as a measure of data-transfer speed. To avoid any confusion, the term **gibibytes per second** may be used when referring to this concept. Its abbreviation is **GBps**.

gigacoulomb A unit of electric charge equal to 10^9 coulombs. Its abbreviation is **GC**.

gigacycle A unit of frequency equal to 10^9 cycles, or 10^9 Hz. The term currently used for this concept is **gigahertz**.

gigaelectronvolt One gigaelectronvolt is equal to one billion electronvolts, the former being the more correct term to use for this value. Its abbreviation is **GeV**.

gigaflops Same as **GFLOPS**.

gigagram A unit of mass equal to 10^6 kilograms, or 10^9 grams. Its abbreviation is **Gg**.

gigahertz A unit of frequency equal to 10^9 Hz. Its abbreviation is **GHz**. Also called **gigacylcle**.

gigajoule A unit of energy or work equal to 10^9 joules. Its abbreviation is **GJ**.

gigameter A unit of distance equal to 10^9 meters. Its abbreviation is **Gm**.

gigaohm A unit of resistance, impedance, or reactance equal to 10^9 ohms. Its abbreviation is **GΩ**.

gigapascal A unit of pressure equal to 10^9 pascals. Its abbreviation is **GPa**.

gigavolt A unit of potential difference equal to 10^9 volts. Its abbreviation is **GV**.

gigawatt A unit of power equal to 10^9 watts. Its abbreviation is **GW**.

gigawatt-hour A unit of energy equal to 10^9 watt-hours. Its abbreviation is **GWh**, or **GWhr**.

GigE Abbreviation of **Gigabit Ethernet**.

GIGO Acronym for **garbage in, garbage-out**.

GII Abbreviation of **Global Information Infrastructure**.

gilbert The unit of magnetomotive force within the cgs system. One gilbert is equal to approximately 0.7958 ampereturn. Its abbreviation is **Gi**.

gimbal 1. A device with two mutual intersecting and perpendicular axes of rotation, thus providing free angular movement in two directions. A gyroscope, for instance, is mounted on a gimbal so that its spinning wheel will maintain the same orientation in space, regardless of how the fixed support is turned. 2. To mount something, such as a gyroscope or compass needle, on a **gimbal** (1).

GIMP Acronym for **G**NU **I**mage **M**anipulation **P**rogram. A popular image editing program which is open-source and free of charge.

GIOP Abbreviation of **G**eneral **I**nter-**ORB** **P**rotocol. Within a CORBA architecture, a protocol utilized for communication between ORBs.

Giorgi system A measuring system in which the fundamental units for expressing distance, mass, time, and electric current are the meter, kilogram, second, and ampere respectively. Also called **meter-kilogram-second-ampere system**.

GIS Abbreviation of **Geographic Information System**.

GJ Abbreviation of **gigajoule**.

glare 1. An intense, and possibly blinding, light. 2. Light which is reflected from a surface, such as a computer monitor, that interferes with proper viewing.

glare filter A filter intended to minimize the ambient light that reaches a computer or TV viewing screen, causing glare. It can take the form of a plastic that is placed in front of the screen, or it can be incorporated into it. Also called **anti-glare filter**, or **ambient-light filter**.

glass A hard, amorphous, and usually brittle and transparent substance composed of silica combined with other substances. To prepare a glass, all materials are fused at a high temperature, then cooled in a manner that prevents crystallization. Flint glass, for instance, contains silica, lead oxide, and potassium carbonate. Depending on the desired proper-

ties, any of various substances may be added. For example, added boron provides more thermal and electrical resistance, barium increases the refractive index, and cerium increases the absorption of infrared rays.

glass bulb The glass housing which encloses the elements of electric lamps, electron tubes, and similar devices. Also called **glass envelope**, or **glass shell**.

glass capacitor A capacitor in which glass serves as the dielectric between its metallic plates. Also called **glass-plate capacitor**.

glass die attach The use of a melted glass to attach a chip to a substrate. Silver particles are usually added to the glass for enhanced thermal and electrical conductivity.

glass electrode A half-cell in which potential differences are measured through a thin membrane of glass which is in contact with a reference solution and an unknown solution. The potential differences detected are due to differences in the concentrations of hydrogen ion, or another ion, between the two solutions. Utilized especially to determine the pH of aqueous solutions. Also called **glass half-cell**.

glass envelope Same as **glass bulb**.

glass-epoxy resin An epoxy resin reinforced with glass fibers. May serve, for instance, in printed-circuit boards as the base material upon which circuits are etched, and to which electrical components are attached. Also called **glass-fiber epoxy**, or **epoxy glass**.

glass fiber A thin filament of glass, usually defined as less than approximately 2.5×10^{-5} m in thickness. Used, for instance, as an insulating material, or as an optical fiber.

glass-fiber epoxy Same as **glass-epoxy resin**.

glass half-cell Same as **glass electrode**.

glass house A term equivalent to **datacenter**, and which makes reference to the large, glass-windowed rooms which housed mainframes decades ago.

glass laser A laser, such as a neodymium glass laser, which uses a glass to generate the coherent beam of light.

glass-metal seal Same as **glass-to-metal seal**.

glass-plate capacitor Same as **glass capacitor**.

glass shell Same as **glass bulb**.

glass-to-metal seal An airtight seal produced by fusing glass and metal. The metal selected must have a temperature coefficient of expansion similar to that of the utilized glass. Used, for instance, in glass capacitors. Also called **glass-metal seal**.

glass tube An tube with a **glass bulb**.

glassivation A passivation technique in which the protected components and surfaces are encapsulated by glass. For instance, molten glass may be deposited upon the components and surfaces, then be allowed to be hardened. Also, such a technique utilizing a dielectric other than glass. Used, for instance, to seal semiconductor devices and chips.

glaze A thin, smooth, and glass-like coating. Used, for instance, to seal and/or protect.

glint 1. A momentary burst of light. Also called **glitter** (1). 2. In radars, a distorted echo due to reflections from different elements of the scanned object, or to rapid movements of said object. Also called **glitter** (2). 3. The use of the effect in **glint** (2) to reduce the effectiveness of a radar.

glitch 1. A minor malfunction or problem, often of a temporary or random nature. 2. A very brief and undesired high-amplitude transient, such as that which may occur when processing a signal. 3. In a CRT, such as that of a TV or oscilloscope, interference which manifests itself as a narrow horizontal bar moving vertically along the picture.

glitter 1. Same as **glint** (1). 2. Same as **glint** (2). 3. A sparkling light.

global **1.** Pertaining to, or encompassing the entire earth. For example, the GPS system. **2.** Pertaining to, or encompassing an entire program, file, database, or the like. For instance, a global search and replace.

Global-Area Network A communications network whose area is essentially unlimited. Such a network may encompass the globe, and an example is the Internet. Its abbreviation is **GAN**.

Global Information Infrastructure A global network encompassing satellites, computers, communications networks, and other devices and equipment involved in the storage, manipulation, transmission, and reception of information in electronic form, including data, voice, and video. Its abbreviation is **GII**.

Global Maritime Distress and Safety System A system utilized for marine communications pertaining to safety. This system serves to automate alerts of distress situations, deliver meteorological warnings, help coordinate search and rescue procedures, and in general improve communications associated with maritime safety. Its abbreviation is **GMDSS**.

Global Navigation Satellite System Its abbreviation is **GNSS**. **1.** Same as GLONASS. **2.** A general term for a global navigation system utilizing satellites, such as **GPS** or **GLONASS**. **3.** A global navigation system incorporating two or more systems such as **GPS** or **GLONASS**.

global network A network, such as the public switched telephone network, encompassing the globe.

Global Orbiting Navigation Satellite System Same as **GLONASS**.

Global Positioning System Same as **GPS**.

Global Positioning System Receiver Same as **GPS Receiver**.

Global Positioning System satellite Same as **GPS satellite**.

Global Positioning System Surveying Same as **GPS surveying**.

global roaming Roaming, such as that provided by satellite telephones, which spans most or all of the globe.

global search and replace Within a complete document or file, the replacement of a given string of data with another. Its abbreviation is **global search & replace**.

global search & replace Abbreviation of **global search and replace**.

Global System for Mobile Communications Same as **GSM**.

global variable Within a computer program, a variable that can be accessed and modified by any module.

Globalstar system A satellite telephony service providing roaming throughout most of the globe, in addition to data services.

GLONASS A Russian satellite system similar to the **GPS System**. It also has 24 satellites, and is optimized for the northern latitudes. It is an acronym of **Global Navigation Satellite System**, or **Global Orbiting Navigation Satellite System**.

gloss **1.** The ratio of the light reflected in one direction, to the light reflected in all directions. **2.** A shine or luster that a surface displays.

glossmeter A device, such as a photometer, which measures **gloss (1)**.

glove box A sealed enclosure with openings through which special gloves pass. Such an enclosure enables handling materials without being directly exposed to them, and are used, for instance, in semiconductor fabrication, or when working with radioactive materials.

glow discharge In an electron tube, a discharge of electricity which ionizes the contained low-pressure gas, resulting in a glow. The color of said glow will depend on the gas present. Sodium vapor, for instance, produces a characteristic candle-yellow color, and may be used for lighting.

glow-discharge lamp A lamp which utilizes **glow discharge** to produce light. Examples include sodium-vapor lamps and mercury-vapor lamps.

glow-discharge tube A tube which produces **glow discharge**. Also called **glow tube**.

glow lamp An electron tube which produces light when the inert gas it contains becomes ionized by an electric discharge passed between its two electrodes. The gas is at a low pressure, and the light is produced by a glow near the negative electrode. An example is a neon lamp.

glow potential A potential sufficient for **glow discharge** to occur. Said potential must be below the sparking potential.

glow switch A switch incorporating two bimetal strips which come in contact when heated by a **glow discharge**. Used in fluorescent light circuits.

glow tube Same as **glow-discharge tube**.

glow voltage A voltage sufficient for **glow discharge** to occur. Said voltage must be below the sparking voltage.

gluon A particle which is the force carrier of the **strong force**. There are eight identified types of gluons, each of which has no mass nor charge.

glyph A figure or symbol which provides a meaning or information without using words. Examples include icons, such as emoticons, and symbols, such as ⌙.

gm A deprecated abbreviation of **gram**. Instead, its symbol, **g**, should be used.

Gm Abbreviation of **gigameter**.

GM Abbreviation of **General MIDI**.

GM counter Abbreviation of **Geiger-Müller counter**, or **Geiger-Mueller counter**.

GM tube Abbreviation of **Geiger-Müller tube**, or **Geiger-Mueller tube**.

GMDSS Abbreviation of **Global Maritime Distress and Safety System**.

GMT Abbreviation of **Greenwich Mean Time**.

gnd Abbreviation of **ground (2)**, and **ground (3)**.

GNSS Abbreviation of **Global Navigation Satellite System**.

GNU Acronym for GNU's Not Unix. A project, developed and promoted by the Free Software Foundation, providing a Unix-like operating system which can be freely copied, modified, and redistributed.

GNU Image Manipulation Program Same as **GIMP**.

GNU unzip Same as **gunzip**.

GNU zip Same as **gzip**.

GNU's Not Unix Same as **GNU**.

Gnutella A popular file-sharing system.

go/no-go test A test of a component, circuit, device, piece of equipment, or system, in which one or more parameters are evaluated, and which results either in acceptance or rejection without further analysis.

GO TO Same as **GOTO**.

GO TO command Same as **GOTO command**.

GO TO statement Same as **GOTO statement**.

Golay code A code utilized in forward error correction. It can correct up to 3 bit errors in a 23 bit sequence.

gold A dense yellow metallic chemical element with atomic number 79. Gold is chemically inactive, extremely corrosion resistant, has a very high melting point, is the most malleable metal, and among the elements is surpassed only by copper and silver as a conductor of electricity. It has over 30 known isotopes, of which one is stable. Gold is important in many areas of electronics, including its use in electric contacts, electrodes, microelectronic circuits, and thin-film components. Its symbol, **Au**, is taken from the Latin word for gold: **au**rum.

gold chloride Same as **gold trichloride**. Dark orange-red crystals whose chemical formula is $AuCl_3$. Used for gold-plating.

gold cyanide A chemical compound utilized as an electrolyte in electroplating.

gold doping The use of gold as a semiconductor dopant which reduces storage time.

gold foil A thin sheet of gold. Such a sheet is considerably thicker than a gold leaf.

gold leaf A very thin sheet of gold. Gold can be hammered into sheets whose thickness is about 25 nanometers.

gold-leaf electroscope An instrument which detects electric charges by means of electrostatic forces. It consists of two gold leaves which are suspended side by side from a conducting rod which is held by an insulated support, and placed in a grounded enclosure, such as a glass jar. When a charge is applied to a plate to which the rod is connected, the leaves separate due to their mutual repulsion. One variation involves having one fixed plate along with a single leaf.

gold plate **1.** The depositing of a layer of gold onto another surface, via electrolysis. Such a deposit may be thin or thick, and may be used, for instance, to provide coatings which are protective, decorative, or which have any given electrical properties. Such a layer may be added to electrical contacts, for instance, to improve their reliability and resistance to corrosion. **2.** The layer of gold deposited in **gold plate** (1).

gold plated A surface which has a **gold plate** (2).

gold plating **1.** The process of depositing of a layer of gold onto another surface, via electrolysis. **2.** The adding of a feature or function with the intention of bragging about it. Said feature or function may or may not address a given need.

goniometer **1.** A device or instrument which measures angles, such as those between the faces of a crystal. **2.** A device or instrument which is utilized to determine the direction of a received radio signal. **3.** A device or instrument which determines the direction of maximum response to a received radio signal. **4.** A device or instrument which determines the direction of maximum radiation of a transmitted radio signal.

gooey A spelling of the pronunciation of **GUI**.

Google A popular search engine and provider of multiple Web services.

Gopher On the Internet, a document-retrieval utility which presents the results of searches in the form of hierarchical menus.

Gopherspace The totality of information on the Internet which can be accessed through **Gopher**.

Gorizont A Russian satellite system, incorporating over 30 satellites, which provides multiple communications services including telephony and radio and TV broadcasts. *Gorizont* means horizon in Russian.

GOS Abbreviation of **grade of service**.

goto Same as **GOTO**.

GOTO Also spelled **GO TO**. **1.** Same as **GOTO statement**. **2.** Same as **GOTO command**.

goto command Same as **GOTO command**.

GOTO command In an application, a command which directs to a specific record, location, or the like. For example, such a command may be utilized in a word processor to go to a specific page. Also spelled **GO TO command**. Also called **GOTO** (2).

goto statement Same as **GOTO statement**.

GOTO statement In a high level language, a programming statement which transfers execution to another part of the program. It is the high-level programming language equivalent of a jump or branch instruction. Also spelled **GO TO statement**. Also called **GOTO** (1).

.gov On the Internet, a top-level domain name suffix. The **gov** is an abbreviation of **gov**ernment. Utilized by United States governmental agencies.

governor A circuit or device which automatically controls a parameter, such as speed or temperature, of a mechanical device such as an engine.

GPa Abbreviation of **gigapascal**.

GPI Abbreviation of **ground-position indicator**.

GPIB Abbreviation of **General-Purpose Interface Bus**.

GPIB bus Same as **General-Purpose Interface Bus**.

GPR Abbreviation of **general-purpose register**.

GPRS Abbreviation of **General Packet Radio Service**.

GPS Abbreviation of Global Positioning System. **1.** A constellation of 24 artificial satellites which orbit at an altitude of approximately 20,000 meters, and which emit signals that are utilized for precise navigation anywhere on the planet earth. Each satellite is equipped with an atomic clock, and continuously sends a time and identification signal. A receiving device obtaining said signals from multiple satellites is thus provided with measurements of its latitude, longitude, and elevation, in addition to the exact time. Depending on the number of signals received, positioning accuracy may be within a meter. This system may also be utilized to keep track of a fleet of cars, a subway train system, and so on. Also called **GPS System**, or **NAVSTAR**. **2.** A device which receives **GPS** (1) signals. Such a receiver may be located in a ship, an aircraft, a motor vehicle, and so on. Also called **GPS Receiver**.

GPS Receiver Abbreviation of Global Positioning System Receiver. Same as **GPS** (2).

GPS satellite Abbreviation of Global Positioning System satellite. One of the 24 satellites which comprise the **GPS System**.

GPS Surveying Abbreviation of Global Positioning System Surveying. The use of **GPS Receivers** for surveying.

GPS System Same as **GPS** (1).

GPWS Abbreviation of **ground proximity warning system**.

grabber **1.** A device utilized to capture graphical data. Used, for instance, for capturing video from a TV and converting it into digital data suitable for computer storage. **2.** A device which captures data of any kind. This includes audio, video, and text. **3.** A screen tool which enables manipulating objects within a window. It is usually represented by a hand icon.

graceful degradation **1.** A degradation during which there is continued operation at a reduced level of performance after one or more components and/or subsystems fail. **2.** Software and/or hardware which undergoes **graceful degradation** (1), as opposed to a complete failure after one or more components and/or subsystems fail.

grade A position, level, or class within a given scale or accepted standard. For instance, when referring to communication bandwidths, voice grade refers to a frequency range which adequately transmits human speech.

grade of service A measurement of the quality of a communications network, based on the availability of circuits when calls are made. Grade of service is usually based on the busiest hour of the day, and may be expressed, for instance, as a percentage of calls blocked.

gradient The rate of change of a physical quantity, such as voltage or pressure, as a function of a given variable, such as distance.

gradient microphone A microphone which responds to small differences in pressure between the front and back of its diaphragm. The diaphragm is exposed in the front and the back, and the output of such a microphone is proportional to this pressure gradient. Also called **pressure gradient microphone**.

grain boundary A boundary between individual crystals within a polycrystalline structure.

-gram A suffix denoting something written, drawn, or recorded. For instance, electrocardiogram, or spectrogram.

gram A unit of mass equal to 10^{-3} kg. It is the also the unit of mass in the obsolescent cgs system. Its deprecated abbreviation is **gm**. Instead, its symbol, **g**, should always be used.

gram atom Same as **gram-atomic weight**.

gram-atomic weight The atomic weight of an element, expressed in grams per mole. For instance, the gram-atomic weight of silicon is approximately 28.0855 grams, as its atomic weight is approximately 28.0855 atomic mass units. Also called **gram atom**.

gram-calorie The heat energy required to raise the temperature of one gram of water from 14.5 °C to 15.5 °C, at a pressure of one atmosphere. This equals approximately 4.1858 joules. Its abbreviation is **g-cal**.

gram equivalent Same as **gram-equivalent weight**.

gram-equivalent weight The mass of a chemical element or compound which, in a reaction, combines with or displaces one mole of hydrogen. For instance, one mole of chlorine combines with one mole of hydrogen to form hydrochloric acid. In this example, the equivalent weight of chlorine is approximately 35.45 grams, as that is the weight of one mole of chlorine. In the case of an element, the gram-equivalent weight is equal to its gram-atomic weight divided by its valence. In the case of a molecule, it is the gram-molecular weight divided by the effective valence. Also called **gram equivalent**, or **equivalent weight**.

gram mole Same as **gram-molecular weight**.

gram-molecular weight The molecular weight of a substance, expressed in grams per mole. For instance, the gram-molecular weight of sodium chloride is approximately 58.44 grams, as the combined molecular weight of sodium and chlorine is approximately 58.44 atomic mass units. Also called **gram molecule, gram mole**, or **mole (2)**.

gram molecule Same as **gram-molecular weight**.

grammar check 1. Same as **grammar checker. 2.** To use a **grammar checker**.

grammar checker Software which locates possible errors in the structuring of sentences, in punctuation, awkward usage, and the like. A grammar checker may suggest changes on the fly, or when requested by a user. Also called **grammar check (1)**.

Grand Unified Theory Its abbreviation is **GUT. 1.** A theory describing a single force which would result from the unification of the weak, the electromagnetic, and the strong forces. **2.** A theory describing a single force which would result from the unification of the four fundamental forces of nature, which are the gravitational force, the weak force, the electromagnetic force, and the strong force. Also called **Unified-Field Theory (1)**.

grandfather/father/son Three generations of files, the oldest version being the grandfather file, followed by the father, and then the son.

grandfather file For a given file, such as that being updated, the copy or version saved before the father file. That is, the current version is the son, the previous version is the father, and the version before that is the grandfather. Also called **grandparent file**.

grandparent file Same as **grandfather file**.

granular carbon Fragments of carbon in various shapes, whose diameter or length across is roughly a few millime-

ters. Used, for instance, in carbon microphones. Also called **carbon granules**.

granularity A measure of the extent to which the components of a system can be broken down. The greater the granularity, the more flexible and customizable a system should be. May refer, for instance, to computer hardware and/or software.

-graph A suffix denoting a device or instrument which serves to draw or record. For example, an electrocardiograph or a seismograph. It may also serve as a suffix representing that which has been drawn or recorded by such a device or instrument.

graph 1. A visual representation of information, or of relationships between quantities or sets of data. Examples of such presentations include curves utilizing a Cartesian coordinate system, pie charts, and bar graphs. Also called **chart (1)**. **2.** Information, or a specific relationship between quantities or sets of data represented by a **graph (1)**.

graphic 1. Pertaining to, or represented by a **graph**. Also spelled **graphical. 2.** Pertaining to that which has been drawn, recorded, or otherwise pictorially represented. Also spelled **graphical. 3.** Pertaining to **graphics**. Also spelled **graphical. 4.** A pictorial representation, especially that utilized to convey information. For instance, that displayed by a computer monitor or X-ray machine.

graphic analysis Same as **graphical analysis**.

graphic display The display of graphical information by a device such as a CRT or LCD. Also spelled **graphical display**.

graphic equalizer An equalizer in which multiple bands of frequencies may be independently adjusted, so that the overall response of an audio system can be tailored to specific needs or preferences. Such an equalizer usually uses sliding controls to modify each band, the center frequencies of which are fixed. When set, the positions of said sliding controls resemble an approximate graph of the resulting frequency response. Also spelled **graphical equalizer**.

graphic information Same as **graphical information**.

graphic instrument Same as **graphical instrument**.

graphic interface Same as **GUI**.

graphic panel A control panel which incorporates varied colors, lights, graphs, and other visual aids to assist a user which monitors and controls a system. Also spelled **graphical panel**.

graphic recorder An instrument or device which records on paper, or another suitable medium, a graph of a varying quantity or that of a relationship between quantities. An oscillograph is an example. Also spelled **graphical recorder**.

graphic representation Same as **graphical representation**.

graphic symbol Same as **graphical symbol**.

graphic user interface Same as **GUI**.

graphical Same as **graphic (1)**, **graphic (2)**, or **graphic (3)**.

graphical analysis The use of graphs and/or graphics to study phenomena, problems, and so on. Also spelled **graphic analysis**.

graphical display Same as **graphic display**.

graphical equalizer Same as **graphic equalizer**.

graphical information Information which is represented by graphs or graphics. Also spelled **graphic information**.

graphical instrument An instrument whose output is presented in the form of a **graph (2)**. Examples include electrocardiographs, and cathode-ray oscillographs. Also spelled **graphic instrument**.

graphical interface Same as **GUI**.

graphical panel Same as **graphic panel**.

graphical recorder Same as **graphic recorder**.

graphical representation The use of graphs or graphics to represent information. Also, the information so presented. Also spelled **graphic representation**.

graphical symbol A pictorial representation utilized to symbolize an electrical component or device, such as a resistor, diode, transistor, filter, or antenna. Used, for instance, in circuit diagrams. Also spelled **graphic symbol**. Also called **electrical symbol**.

graphical user interface Same as **GUI**.

graphics 1. Pictorial representations, especially those utilized to convey information. **2.** In communications, transmission of non-voice information, such as TV or fax images. **3.** The use of computers to store, manipulate and display images, such as pictures, animation, and video. Computers handle graphics either as vector graphics or as bitmapped graphics, and these are used in programs such as those for computer-aided design, desktop publishing, painting, presentations, or animation. A graphics monitor is required to display the images, and almost all computer monitors are of this type. Also called **computer graphics (1)**. **4.** Any visible computer output, except for alphanumeric characters. Examples, include icons, pictures, drawings, and graphs. Also called **computer graphics (2)**.

graphics accelerator A graphics card which incorporates a chip, and which accelerates and enhances the graphics-handling performance of a system. A graphics accelerator significantly speeds up the updating of the images on a screen, which also frees the CPU to take care of other tasks. Especially useful, for instance, for displaying three-dimensional graphics. Also called **graphics accelerator board**, **graphics accelerator card**, **video accelerator**, or **video accelerator card**.

graphics accelerator board Same as **graphics accelerator**.

graphics accelerator card Same as **graphics accelerator**.

graphics adapter Same as **graphics card**.

graphics application An application specifically designed to create graphics. These include paint, design, presentation, animation, and CAD programs.

graphics board Same as **graphics card**.

graphics buffer A buffer, usually consisting of a segment of memory in a graphics card, which holds the data that awaits to be sent to the display. Also called **video buffer**, or **screen buffer**.

graphics card A computer card that enables it to display images. When a computer generates images, this card converts them into the electronic signals that serve as the input to the display device, such as the monitor. The graphics card determines factors such as the maximum resolution and number of colors that can be displayed. An appropriate monitor must be utilized to best benefit from the capabilities of such a card. Graphics cards usually incorporate their own memory, thus not occupying the RAM of the computer for preparing images. Also known by various other names, including **graphics adapter**, **graphics board**, **graphics controller**, **display adapter**, **display card**, **video adapter**, **video card**, and **video display card**.

graphics controller Same as **graphics card**.

graphics controller card Same as **graphics card**.

graphics conversion The process of changing a graphics file from one format to another. For instance, to convert a JPEG image into a TIFF image.

graphics coprocessor A coprocessor utilized to accelerate the displaying of graphics, significantly speeding up the updating of the images on a screen, and freeing the CPU to take care of other tasks. A graphics coprocessor may be incorporated into a graphics accelerator, or may be part of a separate subsystem. Also called **graphics processor (1)**.

graphics data Data in the form of **graphics**.

graphics display A monitor which displays graphics. This contrasts with an **alphanumeric display**, which can display letters and numbers, but not graphics. Also called **graphics monitor**, or **graphics terminal**.

graphics display adapter Same as **graphics card**.

graphics engine 1. Hardware, such as a graphics accelerator or graphics coprocessor, which handles the displaying of graphics, while allowing the CPU to handle other tasks. **2.** Software utilized for the creation of images. An application sends commands to this software, which in turn accesses the appropriate hardware, such as a graphics accelerator.

graphics file A file containing graphics data. These can be stored either as bitmap graphics or as vector graphics. Also called **image file (1)**.

graphics file format A file format utilized to represent graphics. Such formats fit into one of two categories, which are bitmapped and vector. Specific formats include TIFF, GIF, and JPEG.

Graphics Interchange Format Same as **GIF**.

graphics interface Same as **GUI**.

graphics mode A computer screen mode in which graphics can be displayed. This contrasts with **character mode** or **text mode**, where alphanumeric characters can be displayed, but not graphics.

graphics monitor Same as **graphics display**.

graphics plotter A printer which draws images, such as a curves or other graphical representations, by drawing a series of lines usually using ink pens. Used, for instance, in CAD and architecture. Also called **plotter**.

graphics port A computer port to which a graphics display is connected.

graphics primitive Any element which can be utilized as a building block for a drawing. These includes dots, lines, and geometric shapes.

graphics printer A computer printer, such as a laser or inkjet, which can print graphics. This contrasts with a **character printer (2)**, which can print alphanumeric characters, punctuation marks, and some special symbols, but not graphics.

graphics processor 1. Same as **graphics coprocessor**. **2.** A processor utilized to accelerate the displaying of graphics, significantly speeding up the updating of the images on a screen, and freeing the main CPU to take care of other tasks. A graphics processor may be incorporated into a graphics accelerator, or may be separate. Used, for instance, in 3-D applications. Also called **graphics processor unit**.

graphics processor unit Same as **graphics processor**.

graphics program A computer program specifically designed to create graphics. These include paint, design, presentation, animation, and CAD programs.

graphics tablet An input device which serves for sketching or tracing images directly into a computer. It consists of an electronic pressure-sensitive tablet which senses the position of the stylus which makes contact with it. The stylus, or puck, is similar to a mouse, but is much more accurate, because its location is determined by touching the active surface of the tablet with an absolute reference. Used, for instance, in computer-aided design. Also called **digitizing tablet**, **touch tablet**, **tablet**, or **digitizer (2)**.

graphics terminal Same as **graphics display**.

graphics viewer An application designed to display graphics files. Such a program may also offer thumbnail or preview images, and usually supports and converts between multiple file formats. A graphics viewer may also incorporate image editing functions.

graphite An allotropic form of the chemical element carbon. It is a soft black solid which conducts heat and electricity well. Its applications include its use as electrodes, in brushes, in electroplating, as a coating for CRTs, as a moderator in nuclear reactors, and as a lubricant.

graphite anode An anode composed of **graphite**. It may consist, for instance, of a rod.

graphite cathode A cathode composed of **graphite**. It may consist, for instance, of a rod.

graphite electrode An electrode composed of **graphite**. It may consist, for instance, of a rod.

grass A form of radar clutter produced by background electrical noise. It may be manifested on a display screen, for instance, as a pattern resembling grass. Also, any similar interference appearing on any CRT screen. Also called **hash (2)**.

graticule A grid pattern on a screen of a device such as an oscilloscope, which serves to quantify the displayed information. Such a scale may be a part of the inner surface of the screen, drawn on the screen, placed in front of the screen, or generated electronically.

grating **1.** In a waveguide, fine parallel lines which filter certain types of waves. In a circular waveguide, for instance, radial wires serve to block transverse electric waves. Also called **waveguide grating**. **2.** A surface with many fine parallel lines, grooves, or slits, which are extremely close together, and which serve to disperse light so as to produce a spectrum. Such a surface may be made of metal or glass, and may have over 15,000 lines, grooves, or slits, per centimeter. Also called **diffraction grating**.

grav A unit of acceleration equal to 9.80665 m/s^2, which is approximately equal to the gravitational acceleration of the planet earth. Its symbol is **g**.

gravitation Same as **gravitational force (2)**.

gravitation force Same as **gravitational force**.

gravitational Pertaining to **gravitational force**.

gravitational acceleration The acceleration of a body towards the planet earth, due to gravitation. It is equal to approximately 9.807 m/s^2. This value varies slightly, depending on the location and other conditions.

gravitational attraction Same as **gravitational force (2)**.

gravitational constant A physical constant equal to approximately 6.673×10^{-11} Nm^2/kg^2. Its symbol is **G**. Also called **Newton's gravitational constant**.

gravitational field A region of space within which a gravitational force exerts its influence.

gravitational force Also spelled **gravitation force**. **1.** An attractive force which accelerates one body towards another, due to gravitation. The gravitational acceleration of the planet earth is equal to approximately 9.807 m/s^2. Its symbol is **G**. **2.** The force which provides the mutual attraction of all matter in the universe. Its force carrier is postulated to be the graviton. The gravitational force is always attractive, and although it is the weakest of the **four fundamental forces of nature**, its range is theoretically infinite. The other three fundamental forces are the weak force, the electromagnetic force, and the strong force. Also called **gravitation**, or **gravitational attraction**.

gravitational radiation Same as **gravitational wave**.

gravitational wave The hypothetical gravitational analog of an electromagnetic wave. It is postulated that such waves would be emitted at the speed of light from any accelerating mass. Also called **gravity wave**, or **gravitational radiation**.

graviton The postulated force carrier of **gravitation**.

gravity The effect of a **gravitational force**. Pertains especially to that of celestial bodies, such as a planets or stars.

gravity wave Same as **gravitational wave**.

gray The SI unit of radiation dose. It is equal to the dose of one joule absorbed per kilogram of matter. It is equal to 100 rad. Its abbreviation is **Gy**.

gray body A body whose emitted radiant energy has essentially the same spectral energy distribution as that of a blackbody at any given temperature, but whose emissive power is a constant fraction less. Also spelled **graybody**.

Gray code A modified binary code in which consecutive decimal numbers are represented by binary expressions which differ by only one bit. Since only one bit changes at a time, error detection is facilitated.

gray filter An optical filter that decreases the intensity of light without altering its relative spectral distribution. In the case of visible light, it reduces its intensity without changing its color. Also called **neutral-density filter**.

gray hat hacker A hacker somewhere between a black hat hacker and a white hat hacker. A gray hat hacker may, for instance, post security flaws on the Internet.

gray scale **1.** Same as **grayscale**. **2.** A reference scale consisting of shades of gray between black and white.

graybody Same as **gray body**.

grayed Same as **grayed out**.

grayed out On a computer display, a function, operation, or option whose letters or icon is lighter, grayed, or fuzzy, indicating that it is unavailable. For example, the cut function should be grayed out when no content is selected. Also known as **grayed**, **ghosted**, or **dimmed**.

grayscale Image reproduction in black, white, and multiple shades of gray in between. For example, a grayscale scanned photograph. Also called **monochrome (1)**, or **black-and-white**. Also spelled **gray scale (1)**.

grayscale monitor A display screen which reproduces images in black, white, and multiple shades of gray in between.

greek Also called **greeking**. **1.** On a computer screen, the technique of displaying text as lines, dots, or other unintelligible graphics, to provide a preview of the layout of a document. **2.** The text so displayed in **greek (1)**. **3.** To insert a meaningless text into a document, for viewing a given layout. Also, any use of meaningless text to fill a space.

greeking Same as **greek**.

Green Book A document which details the specifications for the CD-I format.

green gun In a three-gun color TV picture tube, the electron gun which directs the beam striking the green phosphor dots.

green monitor A monitor which is designed to save energy and natural resources. For instance, it may consume an especially low amount of power when in standby mode.

green PC Abbreviation of **green personal computer**. A computer which is designed to save energy and natural resources. For instance, it may arrive with a minimum of packing materials, and consume an especially low amount of power when in standby mode. Also refers to peripherals so designed.

green video voltage In a three-gun color TV picture tube, the signal voltage controlling the grid of the green gun. This signal arrives from the green section of a color TV camera.

Greenwich Civil Time An older term for **Greenwich Mean Time**.

Greenwich Mean Time The local time at the 0° meridian, which passes through Greenwich, England. Greenwich Mean Time has been succeeded by Coordinated Universal Time as the basis for the worldwide system of time. Its abbreviation is **GMT**. Formerly called **Greenwich Civil Time**.

Greenwich meridian The meridian passing through Greenwich, England. It is considered to be the prime meridian, and as such is designated longitude 0°.

grid **1.** A framework of parallel bars, lines, dots, or the like. Also, that resembling such a framework. Seen, for instance, in an optical device, an X-ray machine, an electrostatic shield, or another instrument or device where such a grid selectively passes particles, beams, rays, or waves. **2.** A framework of crisscrossed parallel bars, lines, dots, or the like. Also, that resembling such a framework. Used, for instance, to measure, specify, or determine the position of objects or images. **3.** In an electron tube, an electrode which controls the flow of electrons between the cathode and anode. The structure of a grid, such as a mesh or a plate with orifices, determines how the electron beam travels, and hence controls the flow of current in said tube. There may be more than one grid in an electron tube, and any such grid may also serve to control the passage of ions. If there is one grid, it is the same as **control grid**, and when there are multiple grids, the innermost grid is the control grid. **4.** In a storage cell or battery, a metal plate which serves as a conductor, and which also provides support for the active material. **5.** A network of high-voltage transmission lines which link multiple electric power plants. Such a network may run through an entire country. Also called **power grid**.

grid battery A battery which supplies the bias voltage to the control grid of an electron tube. Also called **C battery**.

grid bias In an electron tube, a steady DC voltage applied to the control grid, to establish a reference level for its operation. Also called **grid voltage (1)**, or **C bias**.

grid capacitor A capacitor connected in series with a **grid** (3). Utilized for blocking.

grid-cathode capacitance In an electron tube, the capacitance between a **grid** (3) and the cathode.

grid characteristic In an electron tube, a curve plotting the grid current versus the grid voltage.

grid circuit In an electron tube, the circuit between the control grid and the cathode.

grid computing The employment of unused processing cycles of many, often millions, of computers which are linked via a network such as the Internet, for work on a common task.

grid control In an electron tube, the controlling of the anode current by varying the potential of the control grid with respect to the cathode.

grid current In an electron tube, the current flowing between the control grid and the cathode.

grid-dip meter An instrument utilized to determine the resonance frequency of a tuned circuit. It usually consists of an electron-tube oscillator with a meter incorporated in the grid circuit. As the frequency of said oscillator approaches the resonance frequency of an external tuned circuit, the grid current dips, or drops, which is then indicated by said meter. Also called **grid-dip oscillator**.

grid-dip oscillator Same as **grid-dip meter**.

grid dissipation In an electron tube, the power dissipated by a **grid** (3) in the form of heat.

grid drive In an electron tube, the signal, such as a voltage, which is applied to a **grid** (3). Also called **grid excitation**.

grid emission In an electron tube, the emission of electrons or ions from a **grid** (3).

grid excitation Same as **grid drive**.

grid leak Same as **grid resistor**.

grid-leak resistor Same as **grid resistor**.

grid limiter In an electron tube, a circuit which incorporates a high-value resistor connected in series with a **grid** (3), to limit grid voltages.

grid limiting The use, or effect, of a **grid limiter**.

grid mesh A framework of metal wires forming a mesh. Used, for instance, as an electrostatic shield.

grid metal A metal, such as a lead alloy, utilized as a **grid** (4) in a battery.

grid modulation In an electron-tube amplifying circuit, the varying of the voltage applied to the control grid, to obtain

amplitude modulation. The audio-frequency signal is applied to the control grid along with the carrier.

grid neutralization The shifting of a portion of the grid-to-cathode AC voltage by 180° to neutralize an amplifier.

grid-plate capacitance In an electron tube, the capacitance between a **grid** (3) and the plate.

grid resistor A high-value resistor in a grid circuit. Such a resistor may serve, for instance, to influence the grid bias, or to prevent accumulation of charge on the grid. Also called **grid-leak resistor**, or **grid leak**.

grid return In an electron tube, the circuit path by which the grid current returns to the cathode.

grid stopper In an electron tube, a resistor connected in series with a **grid** (3), to prevent parasitic oscillations. Also called **grid suppressor**.

grid suppressor Same as **grid stopper**.

grid swing The peak variation between the maximum and minimum instantaneous values of a grid-drive signal, such as a voltage.

grid transformer A transformer which supplies an alternating voltage to a grid circuit.

grid voltage 1. Same as **grid bias**. **2.** In an electron tube, the voltage between a **grid** (3) and the cathode.

grill Same as **grille**.

grill cloth Same as **grille cloth**.

grille Also spelled **grill**. **1.** A grating composed of metal, wood, or another material, which serves as a protective barrier and/or for decoration. **2.** A covering across the front of a speaker which provides protection and decorative appeal. Also called **speaker grille**.

grille cloth An acoustically transparent cloth which covers the front of a speaker to keep dust out, provide a measure of protection, and to give decorative appeal. Also spelled **grill cloth**.

gripper In robotics, the device used by the end-effector to grasp objects. There are many types of grippers, including those which use pressure, suction, or magnetization to take hold of objects. Also called **robot gripper**.

grommet A reinforced ring which serves to strengthen and protect an opening. Used, for instance, to reinforce an opening in a chassis through which a cable or conductor is passed. Said grommet may also seal moisture and dirt out, in addition to helping protect the cable or conductor.

gross-motion planning In robotics, motion planning involving movement within comparatively large free spaces. For example, traveling through an environment while avoiding collisions and maintaining balance. This contrasts with **fine-motion planning**, which involves activities such as positioning an end-effector for precise grasping within a comparatively small space.

ground 1. The surface of the planet earth, especially the land portion. **2.** The earth, which is arbitrarily considered to have an electric potential of zero. Also, a conducting path to the earth. For instance, a path which leads to a large copper plate buried in moist soil. Also, a conducting object, such as a wire, leading to the earth. Its abbreviation is **gnd**. Also called **electrical ground (1)**, or **earth (2)**. **3.** A large conducting body whose electric potential is arbitrarily considered to be zero. Also, a conducting path to such a body. Also, a conducting object, such as a wire, leading to such a conducting body. Also called **electrical ground (2)**, or **earth (3)**. Its abbreviation is **gnd**. **4.** Within a circuit, a point which is at zero potential with respect to a **ground (2)**, or **ground (3)**. Also called **electrical ground (3)**, or **earth (4)**. **5.** To connect to a **ground (2)**, **ground (3)**, or **ground (4)**. A path to a ground may be intentional or accidental. Also called **electrical ground (4)**, or **earth (5)**.

ground absorption The loss of RF energy due to absorption by the earth. Also called **earth absorption**.

ground-air communications Same as **ground-to-air communications**.

ground bus A common connection to which components and devices within equipment or systems are connected. This common connection, or bus, is grounded at one or more points. Also called **grounding bus**.

ground cable Also called **grounding cable**. **1.** A cable which is grounded at one or more points, and which serves to ground devices and equipment. **2.** A heavy-duty **ground cable (1)**.

ground circuit A circuit which provides a conducting path to ground. Also called **grounding circuit**.

ground clamp A clamp which attaches a ground wire to a cold-water pipe or to a ground electrode. Also called **grounding clamp**.

ground clutter In radars, unwanted echoes that appear on the display screen due to signal reflections from the surface of the earth. Also called **ground return (2)**, **surface clutter**, **land return**, or **terrain clutter**.

ground-clutter suppression Techniques employed to suppress **ground clutter**.

ground conductivity The ease with which an electric current can flow through the ground. For example, moist soil conducts better than dry soil.

ground conductor A wire, or other conductor, which provides a path to ground. Also called **grounding conductor**.

ground connection A connection to a **ground (2)**, or **ground (3)**. Also called **grounding connection**, or **earth connection**.

ground control The monitoring and management of aircraft from the ground. This may be accomplished, for instance, via air traffic control.

ground-control approach Same as **ground-controlled approach**.

ground-controlled approach The use of **ground control** to assist in a landing approach. Is usually consists of a radar system which provides information which is relayed to aircraft approaching a landing strip. Its abbreviation is **GCA**. Also called **ground-control approach**.

ground current Also called **earth current**. **1.** A current flowing from a circuit to a **ground (2)**, or **ground (3)**. **2.** A current flowing from a **ground (2)** or **ground (3)** to a circuit. **3.** A current flowing through the earth. It may be due for instance, to the earth's magnetic field, or solar activity. Also called **telluric current**.

ground detector A device or instrument which indicates the presence of a ground at a tested point within a circuit. Also called **ground indicator**.

ground effect Any effects, especially those that are undesired, that the earth has on communications. An example is ground clutter.

ground electrode One or more conductors which are buried in the ground, and which serve to provide a ground connection. Also called **grounding electrode**, or **earth electrode**. When such an electrode consists of a large copper plate, also called **ground plate**.

ground fault Also called **earth fault**. **1.** Any unintentional path to ground of a conductor. May be due, for instance, to a failure of insulation between a conductor and a ground. **2.** An unintentional loss of a ground connection.

ground-fault circuit interrupter Its abbreviation is **GFCI**. A special type of circuit breaker which continuously monitors the current flowing in both the hot and neutral wires. Normally, this amount would be equal, but if even if a slight current leaks to ground, be it through the ground wire or through a person being shocked, there will be more current

in one than the other. The GFCI detects this, and shuts off all current flow within about 0.025 seconds, which should be fast enough to prevent any real harm to a person. Some GFCIs detect current differences of less than 0.001 amperes. There are various types of GFCIs, including the receptacle type, which replaces a conventional outlet, and those installed in a circuit-breaker panel. Also called **ground-fault interrupter**.

ground-fault interrupter Same as **ground-fault circuit interrupter**. Its abbreviation is **GFI**.

ground-fault protection Protection against a **ground fault (1)** condition. An example is that provided by a ground-fault circuit interrupter. Also called **ground protection (2)**.

ground handling equipment Same as **ground support equipment**.

ground indicator Same as **ground detector**.

ground level Same as **ground state**.

ground loop A condition in which two or more points in an electrical system are each connected separately to a common ground, and are also connected via other conductors, inducing undesired circulating currents. Such a loop, for instance, can cause susceptibility to electromagnetic interference which may manifest itself as hum or other noise.

ground loop current A current flowing through a **ground loop**.

ground loop noise Noise resulting from a **ground loop**.

ground lug A lug which connects a grounding conductor to a ground electrode. Also called **grounding lug**.

ground mat Also called **grounding mat**. **1.** A grid of conductors buried in the earth, preferably in moist soil, and which serves as a ground electrode. **2.** A floor mat utilized to protect against electrostatic discharges, by dissipating static electricity.

ground noise **1.** Noise produced by improper grounding. **2.** In sound recording or reproduction, noise produced by improper grounding. **3.** Electrical noise which is inherent in a circuit, device, component, piece of equipment, or system. In sound recording or reproduction, for instance, it is the residual noise in the absence of a signal. Also called **background noise (2)**, or **background (2)**.

ground outlet Same as **grounded outlet**.

ground plane **1.** An artificial radio-frequency ground consisting of an arrangement of wires, rods, or other conductors, which is located at the base of a vertical antenna. Such an arrangement may be above or below the ground, and serves to receive and return radiated energy. **2.** A sheet, plate, or other conductive surface at ground potential which serves as a common reference point for circuit returns and electric potentials. Used, for instance, in circuit boards.

ground-plane antenna A vertical antenna which is mounted above a **ground plane (1)**. Such an antenna is omnidirectional, is usually fed by a coaxial line, and is well-suited for use in the high-frequency and very-high frequency ranges.

ground plate A large copper plate buried in the earth, preferably in moist soil, and which serves as a ground electrode. Also called **grounding plate**.

ground-position indicator An airborne computer that continuously indicates the position of the aircraft based on its airspeed, heading, elapsed time, and drift. Its abbreviation is **GPI**.

ground potential A potential arbitrarily considered to be zero. It may be the earth, or a large conducting body whose electric potential is also considered to be zero. Also called **zero potential**, or **earth potential**.

ground protection **1.** The protection afforded by a **ground connection**. **2.** Same as **ground-fault protection**.

ground proximity warning system An airborne computer which actuates an alarm when the distance of the aircraft to the terrain below is less than a given amount. Such a system utilizes a downward-pointing radar, and may activate a recorded voice, flashing light, horn, or other warning signal. Its abbreviation is **GPWS**.

ground-reflected wave A radio wave which is reflected off the earth at least once en route to a point of reception. It is one of the components of a ground wave.

ground reflection The reflection of a radio wave off the earth.

ground resistance The resistance that the earth offers to the flow of current. Various factors affect this, including the composition and moisture the soil at a specific location.

ground return 1. A conductor which serves as the path to return a circuit to ground. 2. Same as **ground clutter**.

ground-return circuit A circuit in which the earth is utilized to complete the conducting path. This contrasts with a **metallic circuit**, in which an earth ground is not part of the closed path. Also called **single-wire circuit**.

ground rod A rod driven into the earth, and to which ground connections may be made. Such a rod is often composed of copper. Also called **grounding rod**.

ground state For a particle, or system of particles, the lowest energy state. Any state above this is an excited state. Also called **ground level**, or **normal state**.

ground support equipment Any ground-based equipment and related devices which serve to assemble, transport, store, test, adjust, repair, maintain, or otherwise support aircraft or spacecraft. Its abbreviation is **GSE**. Also called **ground handling equipment**.

ground switch A manually-operated switch that is utilized to connect a radio antenna to ground during electrical storms, or when not in use. Also called **lightning switch**.

ground system The portion of an antenna that is associated with, and connected to the earth or another conducting surface which is large in comparison to the dimensions of said antenna.

ground-to-air communications All aircraft-to-ground, and ground-to-aircraft communications. Its principal uses include flight coordination and safety. Also called **ground-air communications**, or **air-ground communications**.

ground wave A radio wave which is propagated parallel to the surface of the earth. Such a wave is composed of three components, which are the direct wave, the ground-reflected wave, and the surface wave. Alternatively, it may be considered to be composed of a space wave and a surface wave, as the space wave incorporates both the direct wave and the ground-reflected wave.

ground-wave propagation Propagation of radio waves parallel to the surface of the earth.

ground wire A wire which provides a conductive path to ground. An example is the grounding conductor connected to the third contact of a grounded outlet. Also called **grounding wire**.

grounded Connected to a ground.

grounded-anode amplifier An electron-tube amplifier in which the anode is at ground potential at the operating frequency. The input is applied between the control grid and ground, and the output is taken from between the cathode and ground. Such an amplifier features high input impedance, low output impedance, and a wide frequency response with little phase distortion. Also known as **grounded-plate amplifier, cathode follower**, or **common-anode amplifier**.

grounded-base amplifier A bipolar junction transistor amplifier whose base electrode, which is usually grounded, is common to both the input and output circuits. Also called **common-base amplifier**.

grounded-base circuit Same as **grounded-base connection**.

grounded-base connection For a bipolar junction transistor, a mode of operation in which the base electrode, which is usually grounded, is common to both the input and output circuits. Also called **grounded-base circuit**, or **common-base connection**.

grounded-cathode amplifier An electron-tube amplifier in which the cathode is at ground potential at the operating frequency. The input is applied between the control grid and ground, and the output is taken from between the plate and ground. It is a widely-used amplifier circuit. Also called **common-cathode amplifier**.

grounded-collector circuit Same as **grounded-collector connection**.

grounded-collector connection For a bipolar junction transistor, a mode of operation in which the collector electrode, which is usually grounded, is common to both the input and output circuits. Also called **grounded-collector circuit**, or **common-collector connection**.

grounded-drain circuit Same as **grounded-drain connection**.

grounded-drain connection For a field-effect transistor, a mode of operation in which the drain electrode, which is usually grounded, is common to both the input and output circuits. Also called **grounded-drain circuit**, or **common-drain connection**.

grounded-emitter amplifier A bipolar junction transistor amplifier in which the emitter electrode, which is usually grounded, is common to both the input and output circuits. Also called **common-emitter amplifier**.

grounded-emitter circuit Same as **grounded-emitter connection**.

grounded-emitter connection For a bipolar junction transistor, a mode of operation in which the emitter electrode, which is usually grounded, is common to both the input and output circuits. Also called **grounded-emitter circuit**, or **common-emitter connection**.

grounded-gate amplifier A field-effect transistor amplifier in which the gate electrode, which is usually grounded, is common to both the input and output circuits. Also called **common-gate amplifier**

grounded-gate circuit Same as **grounded-gate connection**.

grounded-gate connection For a field-effect transistor, a mode of operation in which the gate electrode, which is usually grounded, is common to both the input and output circuits. Also called **grounded-gate circuit**, or **common-gate connection**.

grounded-grid amplifier An electron-tube amplifier in which the control grid is at ground potential at the operating frequency. Its input is applied between the cathode and ground, and the output is connected between the plate and ground. Also called **common-grid amplifier**.

grounded outlet An electrical outlet, which in addition to the current-carrying contacts, has a third contact which serves for connection to a grounding conductor. Devices and equipment which are to benefit from this safety feature must have an appropriate three-prong plug which is inserted into this outlet. There are other possible arrangements for such outlets, including the use of lateral grounding contacts which make contact with metallic strips on the side of the plug. Also called by various other names, including **grounding outlet, ground outlet, grounded receptacle, grounding receptacle, grounded socket, safety outlet**, or **three-prong outlet**.

grounded-plate amplifier An electron-tube amplifier in which the anode is at ground potential at the operating frequency. The input is applied between the control grid and ground, and the output is taken from between the cathode and ground. Such an amplifier features high input impedance, low output impedance, and a wide frequency-response with little phase distortion. Also known as **grounded-anode amplifier**, **cathode follower**, or **common-anode amplifier**.

grounded receptacle Same as **grounded outlet**.

grounded socket Same as **grounded outlet**.

grounded-source amplifier A field-effect transistor amplifier in which the source electrode, which is usually grounded, is common to both the input and output circuits. Also called **common-source amplifier**.

grounded-source circuit Same as **grounded-source connection**.

grounded-source connection For a field-effect transistor, a mode of operation in which the source electrode, which is usually grounded, is common to both the input and output circuits. Also called **grounded-source circuit**, or **common-source connection**.

grounded system A network or system of conductors within which at least one is intentionally grounded. Any grounded conductor may have one or more grounding points.

grounding Also called **earthing**. **1.** The connection to a ground, or to a conductor connected to a ground. The term grounding refers to an intentional path to ground, while **ground fault** describes an unintentional path to ground. **2.** The process of connecting to a ground, or to a conductor connected to a ground.

grounding bus Same as **ground bus**.

grounding cable Same as **ground cable**.

grounding circuit Same as **ground circuit**.

grounding clamp Same as **ground clamp**.

grounding conductor Same as **ground conductor**.

grounding connection Same as **ground connection**.

grounding electrode Same as **ground electrode**.

grounding lug Same as **ground lug**.

grounding mat Same as **ground mat**.

grounding outlet Same as **grounded outlet**.

grounding plate Same as **ground plate**.

grounding receptacle Same as **grounded outlet**.

grounding rod Same as **ground rod**.

grounding socket Same as **grounded outlet**.

grounding wire Same as **ground wire**.

group **1.** A collection of things. Such a collection is usually located together, and/or regarded as a unit. For example, a family of chemical elements with similar properties arranged vertically within a periodic table. **2.** In communications, a number of channels treated as a unit. For instance, a collection of associated voice channels in frequency-division multiplexing. **3.** In computers, a collection of elements regarded as a unit. For example, a set of records in a database.

Group 3 facsimile Same as **Group 3 fax**.

Group 3 fax A fax that is compatible with the **Group 3 protocol**. It is the abbreviation of **Group 3 facsimile**. Its own abbreviation is **G3 fax**.

Group 3 protocol An international protocol for sending faxes over regular telephone lines. It supports data compression, and resolutions of up to 203 x 392 dpi.

Group 4 facsimile Same as **Group 4 fax**.

Group 4 fax A fax that is compatible with the **Group 4 protocol**. It is the abbreviation of **Group 4 facsimile**. Its own abbreviation is **G4 fax**.

Group 4 protocol An international protocol for sending faxes over ISDN lines. It is faster than Group 3, supports data compression, and provides resolutions of up to 400 x 400 dpi.

group busy signal Same as **group busy tone**.

group busy tone An audible signal indicating that there are no idle trunks in a group. Also called **group busy signal**, or **trunk busy signal**.

group delay The time interval required for the envelope of a wave to pass from one point to another in a transmission system. It is due to different frequencies traveling at slightly different speeds. Often refers to the envelope of a modulated signal. Also called **envelope delay (1)**.

group mark A symbol or code which indicates the beginning or end of data treated as a unit.

group velocity The velocity at which a wave group travels. It is the velocity of information propagation, and is not necessarily the same as that of energy propagation.

grouped records A set of records, such as that in a database, treated as a unit.

grouping **1.** The formation of a **group**. **2.** In a fax system, a periodic error in the spacing between recorded lines.

grouping factor The number of records in a disk block. It may be calculated by dividing the block length by the average length of the contained blocks. Also called **blocking factor**.

groupware Software designed to facilitate groups of people, often in different locations, to work together on one more projects. Such software includes emailing, scheduling, file transferring, application sharing, conferencing, the use of a whiteboard, and so on. Also called **workgroup software**, or **teamware**.

GroupWise A popular groupware offering.

Grove cell A primary cell consisting of a platinum electrode in nitric acid, and a zinc electrode in sulfuric acid, with a porous partition separating the electrolytes. It has a voltage of approximately 1.91.

growler **1.** A device or instrument utilized to locate short-circuits, and which emits an audible signal upon finding any. Used especially to locate short-circuited coils in an electric generator or motor armature. **2.** A device or instrument, utilized for testing or detection, which emits an audible signal.

grown-diffused junction A junction formed by diffusing impurities into the semiconductor material after a **grown junction** has been formed.

grown-diffused transistor A transistor with a **grown-diffused junction**.

grown junction A pn junction formed while a semiconductor crystal is being grown from a melt.

grown-junction transistor A transistor with **grown junctions**.

Gs Symbol for **gauss**.

GSE Abbreviation of **ground support equipment**.

GSM A digital cellular network technology which usually utilizes time-division multiple access. It is used in Europe and in many other parts of the world, and may operate in the 900 MHz, 1800 MHz, or 1900 MHz frequency bands. Aside from telephony, it supports voice mail, fax, caller ID, email, Internet access, and error-correction, among others. It is the abbreviation of Global System for Mobile Communications.

GSM 1800 A **GSM** network operating in the 1800 MHz band.

GSM 1900 A **GSM** network operating in the 1900 MHz band.

GSM 900 A **GSM** network operating in the 900 MHz band.

GSR Abbreviation of **galvanic skin response**.

gTLD Abbreviation of **generic top-level domain**.

guard band A narrow interval of frequencies between channels, which is left vacant to help prevent adjacent-channel interference. Also called **frequency guard band**.

guard ring 1. A metallic ring which is placed around a charged body, such as a terminal, and which serves to evenly distribute electric charge over the surrounded surface. 2. A **guard ring** (1) placed around one of the parallel plates of a capacitor to eliminate the edge effect.

guard-ring capacitor A capacitor with a **guard ring** (2) placed around one of its parallel plates to eliminate the edge effect.

guard shield A shielding enclosure surrounding all or part of the input circuit of an amplifier.

guard wire A grounded wire which is situated in a manner which allows it to catch and ground a high-voltage overhead transmission line in the event it breaks. Used as a safety precaution to protect persons or objects which could come in contact with such a line under these circumstances.

guarding The placement of low-impedance conductors at specific points in a circuit, so as to divert leakage currents.

Gudden-Pohl effect A light flash occurring when an electric field is applied to an ultraviolet-irradiated phosphor.

guest A user that logs onto a network or system without having registered or otherwise established an account. Guests typically have restricted access and/or privileges. Also called **visitor** (2).

guest account An account with restricted access and/or privileges provided to a **guest**. Also called **visitor account**.

guest password A generic password assigned to a **guest**. Also called **visitor password**.

GUI Acronym for graphical user interface, or graphic user interface. A user interface which utilizes displayed graphics to provide a simpler and more intuitive manner to interact with a computer. A GUI features a desktop, icons, dialog boxes, menus, buttons, and so on, which provide the choices available to a user at any given moment, and a pointing device such as a mouse or a stylus, so that the user may utilize them simply. Also called **graphical interface**, or **graphics interface**.

guidance Any process, mechanism, or system employed to direct the path of a robot or vehicle. The devices and/or equipment utilized for guidance may be onboard, or controlling signals may be sent from a remote site.

guidance computer A computer utilized for **guidance**. Usually refers to an onboard computer.

guidance system A system employed to direct the path of a robot or vehicle. The devices and/or equipment utilized for guidance may be onboard, or controlling signals may be sent from a remote site.

guide 1. A device which serves to direct along a given course. Also, that which serves to indicate, or to regulate operation. 2. A material medium whose physical boundaries confine and direct propagating electromagnetic waves. A guide, for instance, may be a hollow metallic conductor, a coaxial cable, a fiber-optic cable, or an atmospheric duct. Guides enable propagation of electromagnetic waves with very little attenuation. Also called **waveguide** (1). 3. A **guide** (2) consisting of a hollow metal tube, and which is utilized primarily for propagating microwave energy. The cross-section of such a tube may have any of various shapes, the most common being rectangular, circular, and elliptical. The waves are propagated along the longitudinal axis. Also called **waveguide** (2).

guided propagation The propagation of radio waves through an atmospheric duct. Also called **trapping**.

guided tour Online help in which a user is walked through each of the steps necessary to complete a given task.

guided wave A wave whose energy is confined to given boundaries. An example is a wave traveling in a waveguide.

guillotine attenuator A waveguide attenuator in which a sheet or plate of dissipative material is placed through a non-radiating slot. As the sheet or plate is moved into and out of the slot, a variable amount of loss is introduced. Also called **vane attenuator**, **rotary-vane attenuator**, or **flap attenuator**.

guiltware Freeware, shareware, or the like, which has messages meant to make users feel guilty if they don't send money to the developers.

gull-wing lead On a chip package, a lead which extends out and down, so as to resemble a wing of a gull gliding. J-leads occupy less space.

gun Also called **electron gun**, or **electron-beam gun**. 1. A device which produces, and usually controls, an electron beam. Used, for instance, in CRTs, klystrons, electron microscopes, lasers, and for machining or drilling. 2. A **gun** (1) in a CRT. It incorporates a cathode which emits the electrons, a control grid, and accelerating and focusing electrodes.

Gunn diode A diode exhibiting the **Gunn effect**. Utilized, for instance, to produce microwave oscillations.

Gunn effect An effect observed in certain semiconductor crystals, in which oscillations at microwave frequencies are generated when a sufficiently high DC is applied. A small block of gallium arsenide, for instance, exhibits this effect.

Gunn oscillator An oscillator utilizing a **Gunn diode**.

gunzip Abbreviation of GNU **unzip**. A popular decompression utility.

GUT Abbreviation of **Grand Unified Theory**.

gutta-percha A natural rubber-like substance which is utilized for insulation of cables.

GV Abbreviation of **gigavolt**.

GW Abbreviation of **gigawatt**.

GWh Abbreviation of **gigawatt-hour**.

GWhr Abbreviation of **gigawatt-hour**.

Gy Abbreviation of **gray**.

gyrator A device which reverses the phase of a signal propagated in one direction, while causing no phase shift in signals propagated in the opposite direction. A gyrator is usually utilized in waveguides, and may or may not incorporate active devices.

gyro Abbreviation of **gyroscope**.

gyro horizon A **gyroscope** used to indicate the position of an aircraft with respect to a horizontal reference, usually the horizon.

gyrocompass A compass incorporating a **gyroscope**. Such a compass is not affected by magnetic variations, and provides an accurate line of reference for ships, aircraft, or the like. Also called **gyroscopic compass**.

gyrofrequency The angular frequency of the orbit of a particle orbiting along an axis which is perpendicular to a uniform magnetic field. Also called **cyclotron frequency** (3).

gyromagnetic Pertaining to the magnetic properties of rotating charged particles, such as electrons surrounding the nuclei of atoms.

gyromagnetic effect The rotation induced in a body whose magnetization is changed. Similarly, the magnetization induced in a rotating body. It is a weak effect, and contributes slightly to the earth's magnetic field.

gyropilot An automatic pilot incorporating one or more **gyroscopes**.

gyroscope A mechanical device consisting of a heavy spinning wheel mounted on a gimble. The spinning wheel maintains the same orientation in space, regardless of how the base is turned. When mounted on a double gimbal it is free to rotate in three mutually perpendicular axes. Used, for instance, in a gyrocompass. Its abbreviation is **gyro**.

gyroscopic Pertaining to, or incorporating a **gyroscope**.

gyroscopic compass Same as **gyrocompass**.

gzip Abbreviation of GNU **zip**. A popular compression utility.

GΩ Abbreviation of **gigaohm**.

H

h Symbol for **Planck constant**.

h **1.** Abbreviation of **hour**. **2.** Symbol for **hecto-**.

H Symbol for **magnetic field strength, magnetic field intensity**, or **magnetic intensity**.

H **1.** Chemical symbol for **hydrogen**. **2.** Symbol for **henry**.

H antenna An antenna array consisting of two vertically-stacked collinear elements. It features vertical and horizontal directivity. Also called **lazy-H antenna**.

H-bend A smooth waveguide bend in which the direction of the axis remains in a plane perpendicular to the direction of polarization. The smoothness of such a bend helps prevent undesired reflections. Also called **H-plane bend**.

H channel In ISDN communications, a high-speed channel consisting of 6 B channels, for a total of 384 Kbps.

H display In radars, an oscilloscopic display similar to a B display, but which is modified to include the angle of elevation of the scanned object. The scanned object appears as two closely-spaced blips which approximate a bright line, the slope of which is proportional to sine of the angle of elevation. Also called **H scope, H scanner, H scan,** or **H indicator**.

H indicator Same as **H display**.

H-network A five-branch electric network with two input and two output terminals. It has five impedances, two connected in series between an input and output terminal, two connected in series between the other input and output terminals, and the fifth is connected between the junctions of each pair. Such an arrangement resembles a letter H, and is often used in attenuators and filters. Also called **H-pad**.

H-pad Same as **H-network**.

h-parameters Abbreviation of **hybrid parameters**.

H plane The plane of an antenna containing the magnetic-field vector. The electric field vector is contained in the **E plane**.

H-plane bend Same as **H-bend**.

H-plane T junction In a waveguide, a T junction in which the structure changes along the plane of the magnetic field. Also spelled **H-plane tee junction**. Also called **shunt T junction**.

H-plane tee junction Same as **H-plane T junction**.

H scan Same as **H display**.

H scanner Same as **H display**.

H scope Same as **H display**.

h-sync Abbreviation of **horizontal sync,** or **horizontal synchronization**.

H vector In an electromagnetic field, the vector representing the magnetic field. The H vector is perpendicular to the electric, or E, vector. Also called **magnetic vector**.

H wave An electromagnetic wave whose electric field vector is at all points perpendicular to the direction of propagation. Usually used in the context of waveguides. Also called **TE wave,** or **transverse electric wave**.

H.221 An ITU standard for the frame structure used in audiovisual conferencing over a 64 to 1920 Kbps channel.

H.222 An ITU standard for the frame structure used in audiovisual conferencing over a 392 to 1920 Kbps channel.

H.230 An ITU standard specifying how individual frames in audiovisual conferencing are to be multiplexed onto a digital channel.

H.231 An ITU standard specifying how three or more H.320 compliant audiovisual terminals can simultaneously utilize digital channels up to 1920 Kbps.

H.248 An ITU standard pertaining to signaling and session management during multimedia conferencing. Also called **Media Gateway Control Protocol**.

H.261 An ITU standard codec for digital video communications. It provides for dissimilar video codecs to interpret the encoding and compression of a signal, so as to be able to decode and decompress said signal. It supports both CIF and QCIF, and utilizes one or more 64 Kbps ISDN channels, up to about 2 Mbps.

H.263 An ITU standard for video coding at low bit-rate lines. It supports SQCIF, QCIF, CIF, 4CIF, and 16CIF.

H.310 An ITU standard for audiovisual conferencing over ATM and B-ISDN networks. It defines both unidirectional and bidirectional broadband audiovisual terminals.

H.320 An ITU standard for the use of any combination of audio, video, and data over circuit-switched networks, such as ISDN. Used, for example, for videoconferencing which is application independent.

H.321 An ITU standard for adapting H.320 videoconferencing terminals to B-ISDN networks.

H.322 An ITU standard for videoconferencing over LANs, with a guaranteed quality of service.

H.323 An ITU standard for the use of any combination of audio, video, and data over packet-switched networks, such as the Internet. Used, for example, for videoconferencing which is application and platform independent.

H.324 An ITU standard for the use of any combination of audio, video, and data over POTS, using modem connections. Used, for example, for videoconferencing.

hA Abbreviation of **hectoampere**.

Haas effect An acoustic phenomena in which a listener correctly identifies the direction of a sound source that is heard by both ears, but which arrives at each at slightly different times. If the second arriving sound, that heard by the ear farther to the source, is within about 10 to 35 milliseconds of the first arriving sound, it is not taken into account for locating the sound source. If the sounds arrive within less than 10 milliseconds of each other, the source seems to move towards the location of the first arriving sound. When said sounds arrive beyond 35 milliseconds of each other, two distinct sounds are perceived. This effect is taken into account in the reproduction of stereophonic sound. Also called **precedence effect**.

HAAT Abbreviation of **height above average terrain** The height of an antenna above the average level of the terrain within a given radius surrounding its location. Also called **antenna height (2)**.

hack **1.** To write or refine source code or computer programs. To hack usually refers to either doing so in an ingenious and elegant manner, or in a clumsy and inelegant fashion. **2.** Same as **hacker**. **3.** To break into a computer system by circumventing or otherwise defeating the protective measures of said system.

hacker Also called **hack (2)**. A person who seeks detailed knowledge of computer systems and which delves into, or achieves proficiency, in areas such as programming. When the term is used with a negative connotation, it refers to a person which utilizes this capability to illegally break into computer systems by circumventing or otherwise defeating the protective measures of said systems. When used in this manner, also called **cracker**.

hacking **1.** The writing or refining of source code or computer programs, especially in an ingenious and elegant manner. **2.** The breaking into a computer system by circumventing or otherwise defeating the protective measures of said system.

hadron A subatomic particle which experiences the strong interaction, and which is composed of quarks. Since hadrons have internal structure, they are not elementary particles. Protons and neutrons are examples.

hafnium A lustrous silvery metallic chemical element whose atomic number is 72. It is chemically similar to zirconium, is ductile, has good resistance to corrosion, and is a fine neutron absorber. It has around 30 known isotopes, of which 5 are stable. Its applications include its use in nuclear reactor control rods, and as a getter in lamps and electron tubes. Its symbol is **Hf**.

hafnium carbide A chemical compound whose formula is HfC. It is used in nuclear reactor control rods.

hair hygrometer A hygrometer which utilizes a hair, or a bundle of human hairs, as the sensing element. The hair expands and contracts as the moisture present in the surrounding gas, such as air, varies, and the resulting movement is correlated to a calibrated scale.

hair wire A very thin wire, such as a filament or a whisker.

hairline A very thin line. For example, such a line forming part of the grid pattern in a graticule.

hairspring A thin coiled spring, such as that which helps control the movement in certain timepieces.

halation 1. A glow observed on a photosensitive surface as a result of light being reflected by the front and rear faces of said surface. This may be observed, for example, as a blur surrounding a bright spot on the screen of a CRT when light is reflected on the front and rear faces of the glass. Also called **halo** (2). 2. On a photographic image, a ring of illumination which appears around the image of a bright object. Also called **halo** (3).

half-adder A logic circuit which accepts two input bits, and whose output is a sum bit and a carry bit. Unlike a **full adder**, a half-adder does not accept an input carry bit. Also called **two-input adder**.

half-bridge A rectifier circuit similar to a bridge rectifier, but which substitutes two of the rectifiers with resistors.

half-card A plug-in circuit board which is half the length of a standard, or full-length plug-in circuit board. Also called **short card**.

half-cell A single electrode immersed in an electrolyte. An example is a glass electrode utilized to determine the pH of aqueous solutions.

half-cycle 1. One half of a complete sequence of changes of a periodically repeated phenomenon. For instance, one half of an AC cycle. Also, the time that elapses during such an interval. 2. One half of a complete sequence of operations performed as a unit.

half-duplex Two-way communication in which communication can only occur one direction at a time. This contrasts with **full-duplex**, in which communication can occur simultaneously in both directions. Its abbreviation is **HDX**. Also called **half-duplex communication**, **duplex** (3), or **two-way communication** (3).

half-duplex channel A communications channel utilized for **half-duplex**. Also called **duplex channel** (2).

half-duplex circuit A communications circuit enabled for **half-duplex**. Also called **duplex circuit** (3).

half-duplex communication Same as **half-duplex**.

half-duplex operation The operation of a communications channel utilized for **half-duplex**. Also called **duplex operation** (2).

half-duplex system A communications system utilized for **half-duplex**. Also called **duplex system** (2).

half-duplex transmission Two-way communication in which there are transmissions in only one direction at a time. Also called **duplex transmission** (2).

half-inch tape A magnetic tape, such as that of a VHS cassette or of D3 Digital Video, which is 0.5 inch, or 1.27 cm, across. Its abbreviation is **½ inch tape**.

half-life The time interval that elapses for a given radioactive substance to lose half of its radioactive intensity. That is, it is the time required for half the atoms of a given sample of identical nuclides to undergo decay. The half-life of ^{50}V (an isotope of vanadium), for instance, is approximately 1.4 x 10^{17} years, while that for ^6Be (an isotope of beryllium) is approximately 5.9 x 10^{-21} seconds. Also called **radioactive half-life**.

half-line One of multiple lines diverging from a single point. For example, radii from a circle. Also called **ray** (4).

half-power frequency In a response curve, the frequency at which a **half-power point** occurs.

half-power point In a response curve, or a radiation pattern of an antenna, one of the two points that correspond to a power intensity of half the maximum value. In the case of a bandpass filter, for example, the half-power points are the frequencies at which the power of the filter drops 3 decibels, or 70.7%, in relation to the power output at the center.

half-power width In the radiation pattern of a directional antenna, and within the plane containing the direction of the maximum of a lobe, the angle between the two directions in which the radiation intensity is one half of the maximum value of said lobe.

half-step An interval between two sounds, in which the ratio of the higher to the lower frequency is the twelfth root of two, or approximately 1.0595 to 1. Also called **halftone** (2), or **semitone**.

half tap A bridging circuit or device placed across conductors in a manner that does not affect their continuity.

half tone Same as **halftone**.

half-track head A tape head utilized in a **half-track tape recorder**.

half-track recorder Same as **half-track tape recorder**.

half-track tape A magnetic tape utilized in a **half-track tape recorder**.

half-track tape recorder A magnetic-tape recorder whose recording head covers half of the width of the tape, which enables the recording of parallel tracks. These can be used either for stereo recording/reproduction along the entire tape, or for monophonic recording with double the reproduction time. Also called **half-track recorder**, **double-track recorder**, or **dual-track tape reorder**.

half-wave 1. Pertaining to, or composed of one half of one complete wave cycle. 2. Abbreviation of **half-wavelength**. The electrical length equal to one half of one complete wave cycle.

half-wave antenna An antenna radiator whose electrical length is equal to one-half of the wavelength of the signal being transmitted or received. Also called **half-wavelength antenna**, or **half-wave radiator**.

half-wave dipole A center-fed antenna usually consisting of a straight radiator half a wavelength long, which is split at its electrical center. Its maximum intensity of radiation is perpendicular to its axis, and such antennas are used mostly for radio-frequencies below 3 GHz. Also called **dipole antenna**, or **doublet antenna**.

half-wave line Same as **half-wave transmission line**.

half-wave radiator Same as **half-wave antenna**.

half-wave rectification Rectification provided by a **half-wave rectifier**.

half-wave rectifier A rectifier which provides a DC output only during alternate half-cycles of its input AC. Either the positive half-cycles or the negative half-cycles are rectified, so that the delivered current always has the same polarity.

half-wave rectifier circuit A circuit which provides **half-wave rectification**.

half-wave transmission line An antenna transmission line whose electrical length is equal to one-half of the wavelength of the transmitted or received signal. Also called **half-wave line**.

half-wavelength Same as **half-wave (2)**.

half-wavelength antenna Same as **half-wave antenna**.

half-word For a given computer architecture, half of a computer word.

halftone Also spelled **half tone**. **1.** The process of converting an image with continuous tones, such as a photograph, into an image with appropriately-spaced dots which simulate the original. When digitally printing a grayscale reproduction of such an image, the greater the number of equally-sized dots per unit area, the darker that particular segment is. Another procedure involves photographing the original through a special screen to obtain the halftone. **2.** Same as **half-step**.

halide A compound containing a halogen and at least one other element. There are innumerable examples, including potassium chloride, and ammonium chloride.

halide lamp A gas-filled lamp containing a small proportion of one or more **halides**. Metal-halide lamps are examples.

Hall coefficient A constant of proportionality utilized when performing calculations associated with the **Hall effect**. When calculating the strength of the electric field, E_H, the following formula may be utilized: $E_H = R_H JB$, where R_H is the Hall coefficient, J is the current density, and B is the magnetic flux density. The sign of the Hall coefficient changes with the sign of the charge carrier, so it serves to indicate the sign of the majority carrier. Also called **Hall constant**.

Hall constant Same as **Hall coefficient**.

Hall effect A phenomenon observed when strip or bar of a conductor, or of certain semiconductors, is placed perpendicular to a magnetic field. When said strip or bar is longitudinally carrying a current, an electric field is developed across it. The electric field is perpendicular to both the flow of current and the magnetic field.

Hall-effect gaussmeter A gaussmeter which employs the **Hall effect** to measure the strength of a magnetic field.

Hall-effect generator Same as **Hall generator**.

Hall-effect magnetometer A magnetometer which employs the **Hall effect** to measure the strength of a magnetic field.

Hall-effect switch A switch that employs the **Hall effect**. Such a switch is actuated when in proximity to a magnetic field.

Hall field The electric field developed across a conductor or semiconductor due to the **Hall effect**.

Hall generator A generator which makes use of the **Hall effect** to produce an output voltage proportional to the intensity of a magnetic field. Also called **Hall-effect generator**.

Hall mobility A measure of the mobility of charge carriers in a conductor or semiconductor. It may be quantified utilizing the following formula: $\mu_H = R_H \sigma$, where μ_H is the Hall mobility, R_H is the Hall coefficient, and σ is conductivity.

Hall probe A device which employs the **Hall effect** to measure magnetic flux density.

Hall voltage The voltage developed across a conductor or semiconductor due to the **Hall effect**.

halo **1.** On the screen of a CRT, such as that of a TV, an unwanted light or dark zone which surrounds an image, especially bright spots. **2.** Same as **halation (1)**. **3.** Same as **halation (2)**.

halo antenna A dipole antenna whose elements have been bent into a circle, and which is utilized at very-high frequencies.

halogen Any of the elements in group VII of the periodic table. These are: fluorine, chlorine, bromine, iodine, and astatine.

halogen lamp **1.** A gas-filled lamp containing a small proportion of one or more **halogens**. **2.** A **halogen lamp (1)** whose contained gas is iodine, or that of another halogen or mixture of halogens, and whose filament is composed of tungsten. The gas in these high-efficiency lamps reacts with the tungsten metal which evaporates, recycling the particles back onto the filament surface. Also called **tungsten-halogen lamp**.

halt **1.** Same as **halt instruction**. **2.** A break resulting from a **halt (1)**.

halt instruction In a computer program or routine, an instruction which causes a break in the execution. Also called **halt (1)**, or **stop instruction**.

ham A licensed amateur radio operator.

ham radio Also called **amateur radio**. **1.** A radio receiver and/or transmitter utilized by authorized amateurs, and which is operated in the amateur band. **2.** The operation of a **ham radio (1)**, especially as a hobby.

hammer A part of an apparatus which strikes another. For example, the hammer of a bell, or that of a daisy wheel.

Hamming code In data transmissions, an error-detection code which identifies single-bit and double-bit errors. For each seven bits sent, four have information, and three are check bits. It corrects one-bit errors automatically.

Hamming distance The number of digit positions which are different between two binary words of the same length. For example, the Hamming distance between 1110001 and 1110000 is one, as the last digit is different. Also called **signal distance**.

hand A device, tool, or gripping mechanism attached to the wrist of a robot arm. It can be, for example, a drill, a sensor, or a gripper with the ability to move with six degrees of freedom. Also called **end-effector**, or **robot end-effector**.

hand capacitance A capacitance between a circuit or electronic device and a human body, or a part of it, especially a hand. This may cause interference, noise, or a shift in tuning. Also called **body capacitance**.

hand generator An electric generator which is operated by manually turning a crank. It is used, for instance, in emergency radio transmitters and/or receivers.

hand-held computer Same as **handheld computer**.

hand-held device Same as **handheld device**.

hand-held PC Same as **handheld PC**.

hand-held personal computer Same as **handheld PC**. Its abbreviation is **HPC**.

hand-held scanner Same as **handheld scanner**.

hand-off Also spelled **handoff**. **1.** In cellular telephony, the transfer between cells of a call in progress. This occurs as a user moves to an area where another cell provides better coverage, and should not interrupt the call in any manner. **2.** The act of transferring control of a telephone call.

hand-operated Operated by, or requiring the use of hands. For example, a hand generator, or a hand receiver. Also called **manual (2)**.

hand receiver A listening device, such as an earphone or telephone receiver, which is held to the ear by hand.

hand rule Also called **Fleming's right-hand rule**, or **right-hand rule**. **1.** A rule stating that if the thumb of a right hand is oriented along the same axis as the current flow through a conducting wire, that the fingers of this hand will curl along the same direction as the magnetic field produced by the wire. **2.** For a conducting wire moving through a magnetic field, a rule stating that if the middle finger, index finger, and thumb of a left hand are extended at right angles to each other, that the middle finger will indicate the current flow, the index finger the direction of the magnetic field, and the thumb will indicate the direction of the movement of the wire. This rule also applies if the conducting wire is substituted by an electron beam.

hand set Same as **handset**.

hand shake Same as **handshake**.

hand shaking Same as **handshaking**.

hand wiring The wiring of a circuit by hand, as opposed, for instance, to that accomplished via a printed circuit.

handheld computer A computer which is small enough to hold in a hand or place in a pocket. Such computers generally have a specialized operating system, and provide an interface to communicate with other computers, such as desktop systems. There are many types, including models which are industry-specific, such as those utilized by transport companies to keep track of shipments, those that provide only email functions, or PDAs. Also spelled **hand-held computer**. Also called **palmtop**, or **pocket computer**.

handheld device A device, such as a PDA or handheld PC, which is small enough to hold in a hand or place in a pocket.

Handheld Device Markup Language Same as **HDML**.

handheld PC A **handheld computer** whose functionality more closely approximates that of larger computer, such as a notebook or a desktop system. It is an abbreviation of **handheld** personal computer. Also spelled **hand-held PC**. Its own abbreviation is **HPC**.

handheld personal computer Same as **handheld PC**. Its abbreviation is **HPC**.

handheld scanner An optical scanner which is guided, by hand, across the surface that is being scanned. This contrasts, for instance, with a **flatbed scanner**, which provides a flat surface for scanning. Also spelled **hand-held scanner**.

handle **1.** In computers, one of several small squares surrounding a graphic, and which serves to size it. **2.** A temporary designation which identifies a given memory block, file, object, or the like.

handler A programming routine which conducts a specific task, such as error handling.

handoff Same as **hand-off**.

hands-free A device, especially that utilized for voice communications, which can be operated with little or no use of the hands. An example is a hands-free phone.

hands-free computer A computer whose components are all worn by a user. For example, a head-mounted device which incorporates a head-mounted display, a microphone and earphones for the input and output of data, and all other needed components such as the CPU, RAM, and disk drive. A wearable computer is well-suited, for instance, for the real-time analysis of data by scientists working under field conditions, rural doctors, people conducting business on the go, or to assist those with special needs, such as individuals with reduced vision or motor function. Also called **wearable computer**.

hands-free computing The use of **hands-free computers**.

hands-free phone A phone, such as a speakerphone, which can be operated with little or no use of the hands. Also called **hands-free telephone**.

hands-free telephone Same as **hands-free phone**.

handset A part of a telephone incorporating a microphone and an earphone, and which is designed to be held by the hand while talking and listening. A handset frequently has a dial for placing calls. Also spelled **hand set**. Also called **telephone handset**.

handset cord The cord running from the base of a telephone to the handset.

handshake The coordination signals exchanged in **handshaking**. Also spelled **hand shake**. Also called **hardware handshake**.

handshaking The exchange of signals between communications devices, such as modems, so as to establish contact, agree on a communications protocol, and establish and maintain synchronization while sending and receiving. Also spelled **hand shaking**.

handwriting recognition The technology utilized, or the ability for a computer to recognize hand written characters, words, symbols, signatures, and the like. It may accept direct input, as is the case with PDAs, or that which is scanned after being written on paper or another suitable medium. Also, specific instances of the use of this technology.

hang To have a **hangup**.

hang-up Same as **hangup**.

hangover **1.** An unwanted prolongation in the vibration in a low-frequency speaker after the driving signal has stopped. May be due, for instance, to insufficient damping. Also called **ringing (3)**. **2.** An unwanted prolongation in the decay of a signal. In fax, for example, this may manifest itself as a tail that forms on the lines in the recorded copy. Also called **tailing**.

hangup Also spelled **hang-up**. Also called **unexpected halt**. **1.** A unexpected stoppage in the operation of a computer program or system. It may be due to most any cause, such as a crash, an error condition, or the awaiting of an input that does not arrive. **2.** An unexpected halt or delay in the operation of a device, piece of equipment, or system.

haptic Pertaining to, or based on the sense of touch. Also called **tactile**.

haptic device A device, such as a mouse or a data glove, which utilizes the sense of touch to provide input.

haptic interface An interface, such as a keyboard, mouse, or light pen, which communicates information through touching and some sort of feedback. For instance, a mouse is moved laterally, and this movement is seen by the user on the display, or a keyboard key is pressed and there is an opposing force felt. Haptic interfaces are also employed in robotics and in virtual reality applications.

hard **1.** Resistant to scratching, denting, bending, penetration, stretching, and/or wear. **2.** Having high energy, such as gamma rays or X-rays. **3.** Having the characteristic of readily penetrating matter. Said of radiation, such as that of gamma rays or X-rays. **4.** That which has a permanent or fixed character to it, such as a hard drive or a hard error.

hard boot The starting of a computer by turning on its power. Many program failures are resolved through turning off a computer then turning it back on again. Also called **cold boot**, or **startup (2)**.

hard card An expansion card, incorporating a hard disk drive and its controller, which is plugged into an expansion slot.

hard-coded Hardware or software tasks or features which are incorporated in a manner which cannot be changed. Hard-coded operations are faster, but less flexible, than those that are not hard-coded.

hard copy A printed, or otherwise permanent output of a device such as a printer, or that which incorporates a printing function, such as a fax or oscillograph. When used in the context of computers, this contrasts with **soft copy**, which is data displayed on a screen, or stored. Also spelled **hardcopy**. Also called **printout**.

hard disk A magnetic storage medium used in computers, which consists of one or more rigid platters which rotate at very high speeds. Each of these platters, which are usually made of aluminum, is coated with a material which enables information to be encoded by altering the magnetic polarity of minute portions of the surface of each side of said platters, using read/write heads. Hard-drives are generally not removable, and also incorporate associated devices and wiring, such as a motor, access arms, connectors, and so on. Its abbreviation is **HD**. Also called **hard disk drive (1)**, **hard drive (1)**, or **rigid disk**.

hard disk configuration The configuration of a specific **hard disk type**.

hard disk controller A chip, along with the associated circuitry, which enables a computer to control a hard disk drive. A hard disk controller handles tasks such as positioning the read/write heads and the transfer of data. It can be incorporated into the drive, built into the motherboard, or may be in the form of an expansion card. Each interface, such as SCSI or IDE, requires a different type of controller.

hard disk crash Same as **head crash**.

hard disk drive Its abbreviation is **HDD**. **1.** Same as **hard disk**. **2.** Same as **hard drive**.

hard disk failure The condition in which a hard disk or hard disk drive no longer performs the function it was intended to, or is not able to do so at a level that equals or exceeds established minimums. May be due, for instance, to a head crash.

hard disk interface The hardware which is used to connect a hard disk drive to a computer. These include the plugs and wires. There are several commonly used interfaces, such as SCSI and IDE, and each requires a different type of hard disk controller.

hard disk type A number which specifies the key features of a hard disk, such as capacity, platters, cylinders, and read/write heads. This number may be used, for instance, by the CMOS memory of a computer.

hard-drawn copper wire Copper wire which is not annealed after being drawn. Such wire has increased tensile strength.

hard-drawn wire Wire, such as that of copper or aluminum, which is not annealed after being drawn. Such wire has increased tensile strength in comparison to **annealed wire**.

hard drive Its abbreviation is **HD**. Also called **hard disk drive (2)**. **1.** Same as **hard disk**. **2.** An electromechanical device which reads and/or writes data to a **hard disk**.

hard error A permanent error, such as that which recurs on successive attempts to read data from a disk.

hard failure **1.** A failure from which there is no recovery. The solution may require the replacement of one or more components, or of the device, piece of equipment, or system in question. **2.** Same as **hardware failure**.

hard glass A silicate glass constituted by 5% or more boric oxide. It features a high resistance to corrosion and temperature, and is used, for instance, in light bulbs. Also called **borosilicate glass**.

hard link A directory entry which contains the path leading to the same file or directory. This contrasts with a **symbolic link**, which references a different file or directory. Symbolic links allow for multiple names in different file systems, while hard links do not.

hard macro Also called **macrocell**. **1.** A gate array which with added features, such as storage elements and path controls, which enhance its functionality. **2.** A gate array with a logic function specified, and in which the interconnections between its cells are defined.

hard magnetic material A material, such as alnico alloy, with a high magnetic coercive force. Such a material is suitable for making permanent magnets.

hard radiation High-energy radiation which readily penetrates matter. For example, gamma rays consist of high-energy photons which easily penetrate living tissue, producing serious harm when protective shielding is not utilized.

hard-sectored disk A disk with **hard-sectoring**.

hard-sectoring In magnetic disk storage, sectoring using a physical marking, such as a hole. This contrasts with **soft sectoring**, in which data marks the sector boundaries.

hard solder **1.** An alloy, such as brass, or one formed by copper and nickel, used for **hard-solder**. Also called **brazing alloy**. **2.** A solder that melts at a comparatively high temperature. For example, an alloy of silver, copper, and zinc. This contrasts with a **soft solder**, which melts at a comparatively low temperature. **3.** Same as **hard-soldering**.

hard-soldering The use of a nonferrous metal or alloy to fill the spaces between two metals which are to be joined at a high temperature. The filler metal must have a lower melting point than the metals to be joined. It is similar to soldering, but the term hard-soldering is used when the temperature exceeds an arbitrary value such as 425 °C. More commonly known as **brazing**. Also called **hard solder (3)**.

hard vacuum Same as **high vacuum**.

hard-wire Also spelled **hardwire**. **1.** To connect electrical components and devices directly with wires or cables, as opposed to utilizing intervening switches or radio links. Also called **wire (8)**. **2.** To use wire or cables, as opposed to radio, to transmit information.

hard-wire telemetry Telemetry in which signals are sent via wires and/or cables, as opposed to being sent over radio. Also spelled **hardwire telemetry**. Also called **wire-link telemetry**.

hard-wired Also spelled **hardwired**. **1.** Circuitry utilizing wires or cables to permanently interconnect components and devices. To alter a circuit which is hard-wired, it might be necessary, for instance, to unsolder and resolder. **2.** In computers, functions and other operations which are permanently built into the system, as opposed those which can be changed later through programming. Circuits or devices such as logic gates may be used for this purpose, and physical changes are required to alter such connections.

hard-wired circuit A circuit which is **hard-wired**. Also spelled **hardwired circuit**.

hard-wiring The process of interconnecting components and devices which are to be **hard-wired**. Also, the interconnections themselves. Also spelled **hardwiring**.

hard X-rays X-rays whose wavelengths are closer to the gamma-rays region, as opposed to those nearer to the ultraviolet region. The shorter the wavelength, the greater the energy, which in turn makes these rays penetrate matter more readily than **soft X-rays**.

hardcopy Same as **hard copy**.

hardening The process of making hard or harder. This may be accomplished, for instance by treating with heat and/or through the addition of a suitable material. Also, to impart such hardness.

hardness **1.** The relative resistance a material has to scratching, denting, bending, penetration, stretching, and/or wear. **2.** The relative resistance a material has to scratching, denting, bending, penetration, and/or wear, as determined by a standard test. For example, the relative resistance a material has to scratching, as indicated by the Mohs hardness scale. **3.** The relative readiness with which radiation, especially X-rays, penetrates matter. The shorter the wavelength, the greater the hardness, and the more penetrating the rays.

hardness test A standard test, such as a Brinell hardness test, utilized to determine the hardness of a material.

hardness tester An instrument or device utilized to perform **hardness tests**.

hardware Its abbreviation is **HW**. **1.** In a computer system, all the physical objects and equipment, such as the CPU, keyboard, mouse, monitor, memory, storage mediums, cables, connectors, cards, and so on. This contrasts with **soft-**

ware, which consists of programs which tell the hardware what to do. A computer system is composed of both hardware and software. Also called **computer hardware**. **2.** The physical equipment utilized for a given function, as opposed to ideas or designs detailed on paper. **3.** A collection of tangible devices, such as an assortment of components, devices, and equipment.

hardware address Also called **physical address**. **1.** The address that corresponds to a specific piece of hardware. For instance, an Ethernet address. **2.** When a virtual address is used in computer memory, the address that corresponds to a specific hardware memory location. Memory management takes care of the conversions between the two.

hardware check **1.** The verification or testing of the hardware of a computer system. Such checks are usually done automatically when the system is powered on. Also called **machine check**. **2.** The verification or testing of **hardware (2)**.

hardware-dependent Computer software or hardware which is designed to work properly only with a specific architecture or configuration. Assembly language, for instance, is hardware-dependent. This contrasts with **hardware-independent**, in which software or hardware can run on a variety of architectures or configurations. Also called **machine-dependent**.

hardware description language A computer language utilized for the designing, modeling, and documentation of hardware designs. This language may be used, for example, to simulate the function of a given chip design, and to help write the necessary specifications. Verilog and VHDL are two common languages used for this purpose. Its abbreviation is **HDL**.

hardware diagnostic A computer program utilized to perform **hardware tests**.

hardware failure A failure of a computer hardware component or device. The solution may require the replacement of the component or device in question.

hardware handshake Same as **handshake**.

hardware-independent Computer software or hardware which is designed to work properly with a variety of architectures or configurations. An interpreter version of LISP, for instance, is hardware-independent. This contrasts with **hardware-dependent**, in which software or hardware is designed to work properly only with a specific architecture or configuration. Also called **machine-independent**.

hardware interface **1.** An interface between components, devices, equipment, or systems. **2.** An interface between computer hardware components. For example, between the CPU and a peripheral. **3.** A standardized interface between computer hardware components. For instance, a SCSI or USB interface for the connection of computer peripherals.

hardware interrupt An interrupt generated by a hardware component.

hardware key A software copy-protection device which usually plugs into a parallel port. It allows backup copies to be made, but aims to prevent unauthorized copying to other computers. An attached hardware key does not affect the use of the port for other purposes. Also called **dongle**.

hardware lock A hardware-based lock, as opposed to one based on software. An example is a hardware key. Also called **physical lock**.

hardware monitor A circuit or device which oversees the performance of one or more hardware components. Such a monitor, for instance, may be used to keep track of the voltage and temperature of a CPU, and of the status of its cooling fan.

hardware platform The CPU, or CPU family on which a given operating system or software runs. Each hardware platform has a unique machine language.

hardware profile A set of specific configuration settings for hardware components, especially peripherals and drivers. For example, a portable computer may have two hardware profiles, and the appropriate one can be selected depending on whether it is plugged-into its docking station or not.

hardware recovery Within a disaster recovery plan, the aspect involving the reestablishment of the proper function of hardware components. To facilitate a prompt recovery with a minimum of disruption, it is best to have redundant and/or spare components, devices, equipment, and/or systems.

hardware requirements The minimum hardware availability and configuration a system must have in order for a given program or device to work properly. For example, in order to utilize a given peripheral, a computer needing to have a minimum amount of available memory and disk space, a DVD drive, an available port, and so on. Also called **system requirements**.

hardware test A test performed to confirm the proper function of a hardware component within a computer system, or to ascertain the nature of a hardware error or failure.

hardware testing The performance of **hardware tests**.

hardwire Same as **hard-wire**.

hardwire telemetry Same as **hard-wire telemetry**.

hardwired Same as **hard-wired**.

hardwired circuit Same as **hard-wired circuit**.

hardwiring Same as **hard-wiring**.

harmonic For a complex signal, wave, sound, or vibration which is periodic, a component whose frequency is a whole-number multiple of the fundamental frequency. The fundamental frequency is also called fundamental component, or first harmonic. The frequency of the second harmonic is twice that of the first harmonic, the third harmonic is three times the first harmonic, and so on. Any even-numbered multiple of the first harmonic is called an even harmonic, while odd-numbered multiples are called odd harmonics. **Overtones** are the same as harmonics, except that the first overtone is the same as the second harmonic, the second overtone is the same as the third harmonic, and so on. Also called **harmonic component**.

harmonic analysis The analysis of the harmonic content of a complex wave or periodic function. For example, Fourier series and Fourier transforms may be utilized to analyze complex waves.

harmonic analyzer A device or instrument which measures or analyzes the harmonic content of a complex wave or periodic function. Such an instrument or device may be electronic, electrical, or mechanical. Also called **harmonic wave analyzer**.

harmonic antenna An antenna whose electrical length is a whole-number multiple of the half-wavelength of the signal being transmitted or received. Also called **long-wire antenna (2)**.

harmonic attenuation The attenuation of one or more harmonic components in a complex wave. A low-pass filter, for instance, may be utilized for this purpose. Also called **harmonic reduction**.

harmonic attenuator A circuit or device utilized to obtain **harmonic attenuation**. Also called **harmonic reducer**.

harmonic component Same as **harmonic**.

harmonic content For a complex signal, wave, sound, or vibration which is periodic, all harmonics remaining after the fundamental frequency is removed.

harmonic detector A device or instrument which detects, measures, or responds to a specific harmonic.

harmonic distortion **1.** A form of distortion in which harmonics which were not present in an input signal are present in the output signal. **2.** A form of distortion in which harmonics which were not present in a sinusoidal input signal are present in the output signal. This results in the alteration of the original waveform. The generation of such undesired harmonics is a result of non-linearity of the component, circuit, device, piece of equipment, system, or transmission line in question. **3.** A measured amount of **harmonic distortion (2)**. It is usually expressed as the ratio of the amplitude of a given harmonic to the amplitude of the fundamental frequency, multiplied by 100 to obtain a percentage. If the squared amplitudes of each of these ratios of distorting harmonics are added, the square root is taken from the result, and this number is then multiplied by 100 to obtain a percentage, it is called **total harmonic distortion**. This value is frequently used as an expression of the performance of audio components, especially amplifiers. **4.** A form of distortion in which there is a disproportionate reproduction of harmonics.

harmonic distortion meter **1.** A device which measures **harmonic distortion**. **2.** An instrument which measures the harmonic content of a wave. Such a device, for instance, may display the harmonic content of an audio-frequency wave. Also called **distortion meter (2)**.

harmonic elimination Same as **harmonic suppression**.

harmonic eliminator Same as **harmonic suppressor**.

harmonic filter **1.** A filter which only allows one or more specific harmonic components in a complex input wave to pass. **2.** A filter which prevents one or more specific harmonic components in a complex input wave from passing.

harmonic frequency The frequency of a **harmonic**.

harmonic generator **1.** A signal generator which produces harmonics of a fundamental frequency. **2.** A signal generator which produces strong harmonics of a fundamental frequency.

harmonic interference Any interference resulting from unwanted harmonics. For instance, harmonics of a carrier frequency during radio broadcasting.

harmonic motion Motion whose path is symmetrical about a given equilibrium position, such as that displayed by a swinging pendulum. Harmonic motion is a sinusoidal function of time. Also called **simple harmonic motion**.

harmonic oscillator **1.** An oscillator which produces harmonics of a fundamental frequency. **2.** A crystal oscillator which produces harmonics of the natural frequency of vibration its piezoelectric crystal. **3.** A physical system, such as a pendulum, which exhibits **harmonic motion**. Also called **simple harmonic oscillator**. **4.** For an oscillating system, an equilibrium-restoring force which is proportional to the displacement from the equilibrium position. This contrasts with an **anharmonic oscillator**, in which the forces restoring the equilibrium position are not proportional to the displacement from it.

harmonic producer **1.** A circuit or device which produces harmonics of a fundamental frequency. **2.** An oscillator which utilizes a tuning fork to set a fundamental frequency, and which can produce harmonics of said frequency.

harmonic reducer Same as **harmonic attenuator**.

harmonic reduction Same as **harmonic attenuation**.

harmonic resonance Resonance occurring at a harmonic frequency, which increases any harmonic distortion present.

harmonic ringing A form of selective ringing in which multiple currents that are harmonics of a fundamental frequency are utilized to select the desired ringer. Each given harmonic corresponds to a specific ringer. Also called **harmonic selective ringing**.

harmonic selective ringing Same as **harmonic ringing**.

harmonic suppression The elimination of one or more harmonic components in a complex wave. A low-pass filter, for instance, may be utilized for this purpose. Also called **harmonic elimination**.

harmonic suppressor A circuit or device utilized to obtain **harmonic suppression**. Also called **harmonic eliminator**.

harmonic wave A wave generated by **harmonic motion**.

harmonic wave analyzer Same as **harmonic analyzer**.

harness A bundle of wires or cables which is tied or otherwise attached together so as to be handled, installed, or removed as a unit.

harp antenna An omnidirectional antenna with multiple parallel vertical radiators which are connected at their bottom ends to a horizontal supporting wire.

hartley A unit of information which is equal to approximately 3.322 bits.

Hartley oscillator A variable-frequency oscillator in which the feedback is provided by a tap in the coil of the tank circuit, and in which the cathode of an electron tube or the source or emitter of a transistor is connected to said tap.

Harvard architecture A computer architecture in which instructions and data addresses are stored in different regions and are accessed through separate buses. This enables the system to locate instructions at the same time data is being read and written, which accelerates the rate of computation. This contrasts with **von Neumann architecture**, where instructions and data addresses are stored in the same regions and are accessed through a single bus.

hash **1.** Electrical noise such as that caused by arcing, the contacts of a vibrator, or the brushes of a motor or generator. **2.** A form of radar clutter produced by background electrical noise. It may be manifested on a display screen, for instance, as a pattern resembling grass. Also, any similar interference appearing on a CRT screen. Also called **grass**. **3.** Electrical noise in a radio receiver, such as hissing. **4.** Same as **hash value**. **5.** To calculate a **hash total**, or to produce a **hash value**. **6.** Meaningless, unwanted, incorrect, corrupted, or otherwise useless data. Such data can be entered by a user, or may be produced by a computer. Also called **garbage (1)**.

hash function A mathematical function utilized to convert a variable-sized, and often large, amount of input text into a more compact fixed-sized output, which is called **hash value**. Hash functions are used, for instance, in cryptography and in the creation of digital signatures.

hash total A total, composed of several numbers taken from a file, which is calculated before, during, and after processing. It is calculated by adding the binary value of each character in the block. The numbers used to calculate the total do not necessarily have to be taken from numeric data. Hash totals are used to verify the accuracy of processed data, or to help ensure that transmitted messages have not been tampered with. At all stages the calculated totals must match, otherwise there is an error. Also called **control total**.

hash value A result derived from a **hash function**.

hashing The computation of a **hash total** or a **hash value**.

hassium A synthetic radioactive chemical element whose atomic number is 108. It has about five identified isotopes, all unstable. The techniques employed to produce this element are utilized to help discover new elements with higher atomic numbers. Its chemical symbol is **Hs**.

Hay bridge A four-arm bridge utilized to calculate an unknown inductance, and whose balance is frequency-dependent. A Hay bridge is usually used to measure large inductances.

Hayes-compatible Denoting a modem which is controlled by the AT command set.

hazard A component, circuit, device, piece of equipment, system, material, location, procedure, or circumstance which is a possible source of danger. Also called **safety hazard**.

hazardous location A location, such as that where toxic and/or explosive gases are present, where personnel or others may be in danger.

hazbot A remote-controlled robot designed to work under hazardous circumstances, such as nuclear reactor repair, fire fighting, or space exploration.

HBT Abbreviation of **heterojunction bipolar transistor**.

HC Abbreviation of **Hefner candle**.

HCI 1. Abbreviation of **human-computer interaction**. 2. Abbreviation of **human-computer interface**.

HCMOS Abbreviation of high-speed complementary metal-oxide semiconductor, or high-density complementary metal-oxide semiconductor. A CMOS technology with enhanced speed and performance. Used, for instance, in smart cards.

HD 1. Abbreviation of **high-density**. 2. Abbreviation of **high-definition**. 3. Abbreviation of **hard drive**. 4. Abbreviation of **hard disk**.

HDA Abbreviation of **head drive assembly**.

HDD Abbreviation of **hard disk drive**.

HDF Abbreviation of **Hierarchical Data Format**.

HDL Abbreviation of **hardware description language**.

HDLC Abbreviation of **High-Level Data Link Control**.

HDML Abbreviation of Handheld Device Markup Language. A version of HTML tailored to wireless devices such as properly equipped cell phones and PDAs.

HDSL Abbreviation of high bit-rate digital subscriber line, high-speed digital subscriber line, or high bit-rate **DSL**. A version of DSL utilizing two cable pairs, and which supports up to T1 signaling rates in both directions. There is an equal amount of bandwidth in each direction. For instance, the transmission rate may be 1.544 Mbps upstream, and 1.544 Mbps downstream. E1 is supported with the use of 3 cable pairs.

HDSL2 Abbreviation of high bit-rate digital subscriber line-2, or high bit-rate **DSL-2**. An enhanced version of **HDSL** which utilizes 1 cable pair, and provides other benefits, such as a more reliable connection.

HDSS Abbreviation of **holographic data storage system**.

HDT Abbreviation of **host digital terminal**.

HDTV Abbreviation of high-definition **TV**, or high-definition television. 1. A standard for the transmission and reception of TV signals with increased resolution and sound quality in comparison to standards such as NTSC or PAL. Such a standard provides for the transmission of a signal providing over 1,000 active lines, a 16:9 aspect ratio, and digital-quality sound. Also called **HDTV standard**. 2. A device, such as a TV receiver, which is compatible with **HDTV** (1). Such a receiver provides, among other benefits, an image equal to or greater in resolution to that of a movie theater, a wider screen which more closely matches the human field of vision, and a multiple-channel sound format which is digitally encoded. Also called **HDTV receiver**.

HDTV receiver Abbreviation of high-definition **TV receiver**. Same as **HDTV** (2).

HDTV standard Abbreviation of high-definition **TV standard**. Same as **HDTV** (1).

HDTV system Abbreviation of high-definition **TV system**. A TV system which employs an **HDTV standard**.

HDX Abbreviation of **half-duplex**.

HDX operation Abbreviation of **half-duplex operation**.

He Chemical symbol for **helium**.

He-Ne laser Same as **helium-neon laser**.

head 1. A device which reads, records, or erases signals on a storage medium such as a magnetic tape or a disk drive. Examples include recording heads, read/write heads, and erase heads. 2. In computers, the top or start of something, such as a program, document, page, or the like.

head alignment The alignment of a **head** (1) to maintain the proper spatial relationship with the magnetic medium being read, recorded, or erased. In a tape deck, for instance, it is the adjustment of a playback or record head to achieve the proper 90° alignment with the tape path.

head amplifier An amplifier utilized at the source of an audio or video signal to amplify a signal which would otherwise be too weak for the main amplifier. For example, it may consist of a preamplifier incorporated into a microphone.

head crash A hard disk failure in which the read/write head collides with the surface of a disk. This may be caused by a foreign object such as dust, or a misalignment. Data is destroyed, and generally both the read/write head and the platter must be replaced. Also called **hard disk crash**, or **crash** (2).

head degausser Same as **head demagnetizer**.

head demagnetizer A circuit or device which removes residual magnetism from a tape head. Also called **head degausser**.

head drive assembly For a disk drive, a sealed assembly which includes the read/write heads, access arms, platters, and associated components. Its abbreviation is **HDA**.

head end Same as **headend**.

head gap 1. The space between a read/write head and a computer magnetic storage medium, such as a disk or tape. 2. The width of the gap between pole pieces in a tape head. Also called **gap width**.

head-mounted device A device, such as a head-mounted display, which is worn on the head. Its abbreviation is **HMD**.

head-mounted display Its abbreviation is **HMD**. A computer monitor, or other display, which is worn on the head. An HMD may display to one or both eyes, and may be in the form of goggles, a helmet, or the like. When displaying for one eye, for instance, it allows a user to view it without looking away from the surrounding environment, which is useful in many technical and scientific applications. When displaying to both eyes, it is especially suited for immersion in virtual reality environments. An HMD often has other associated devices working alongside, such as a microphone and earphones.

head-related transfer function Its abbreviation is **HRTF**. A transfer function that takes into account factors involved in sound reaching the ears of a listener, for sound localization and the determination of the arriving spectrum of sound. Variables such as the dimensions of the ear canal, the exterior part of the ear, head, and torso, and how these modify the spectrum of the sound heard are combined with the azimuth, elevation, range, and frequency of the sound source, the slightly different time at which a sound arrives at each ear, and so on. Each listener has a unique HRTF.

head room Same as **headroom**.

head stack In a tape recorder, two or more heads mounted as a unit and utilized for multi-track recording. Also called **head stack assembly**.

head stack assembly Same as **head stack**.

head-stick Same as **headstick**.

head-to-tape contact The physical contact a tape head makes with a magnetic tape being recorded or reproduced. The proper contact helps optimize recording or reproduction, and helps protect the tape from physical damage such as wrinkling. Also called **tape-to-head contact**.

headend 1. In a TV system, such as satellite TV or cable TV, the location at which signals are received, then processed, and from which they are distributed to the individual subscribers. Also spelled **head end**. Also called **cable headend**. 2. Another such similar location from which a communications entity, such as an ISP, distributes information to individual subscribers.

header **1.** That which appears at the start of something, and which serves to identify, describe, route, control, and so on. **2.** In a file, a region which has information such as the name, size, and date created. **3.** In communications, the first part of a message, packet, or the like. A header may contain information such as the originating and destination stations, the length, protocol utilized, and so on. **4.** In word processing, a text which appears at the top of one or more pages. Such text may appear only on a give page, specified pages, even pages, odd pages, or on every page. Also called **page header**. **5.** A mounting plate, usually made of glass, through which one or more insulated terminals or leads originating from a hermetically sealed component, device, or piece of equipment are passed. Utilized, for instance, to expose the leads of a transistor, relay, tube, or the like. Such a header also provides support for said terminals or leads.

header label A record appearing at the beginning of a file, containing information such as its name, size, type, and so on. Also called **file label**.

header record The first within a sequence of records.

heading The horizontal direction in which a craft is pointing, expressed as an angular distance from a reference direction, such as true north or magnetic north. A heading is usually measured clockwise as an angle from 0° to 360° from the reference direction.

headphone Its abbreviation is **phone**. **1.** A small or miniature speaker which is intended to be placed within the outer ear, or immediately surrounding it. It is utilized for private, portable, or enhanced listening. A pair may be used for stereophonic listening. Also called **headset (1)**, or **earphone (1)**. **2.** A **headphone (1)** specifically designed for use with a hearing aid amplifier. Also called **earphone (2)**.

headphone amplifier An audio amplifier intended for driving **headphones**.

headphones A device which provides a **headphone** for each ear. Also called **headset (2)**, or **earphones**.

headroom A measure of the difference between the typical operating level of a device, piece of equipment, or system, and the maximum operating level before clipping or overload occurs. For example, an audio amplifier which nominally operates at +5 dB, and which clips at +25 dB, has 20 dB of headroom. This enables the amplifier to adequately handle peaks. In the case of tape recording, headroom refers to the level above the 0 point on its VU meter at which tape overload occurs. For example, if there in no tape overload until +6 dB, then there are 6 dB of headroom. Also spelled **head room**.

headset **1.** Same as **headphones**. **2.** Same as **headphone (1)**.

headstick A computer input device consisting of a rod-shaped object which is attached to the head. Useful, for instance, to users with reduced motor function when entering commands. Also spelled **head-stick**.

heap An area of memory reserved for data created during execution. It is utilized as temporary storage by a running program. A heap is utilized when the size and order of the data structures can not be determined until the execution of a program.

hearing aid A sound amplifying device designed for those with reduced hearing. Those with conductive or sensorineural hearing loss can be helped by the amplification and/or clarification of sounds. Hearing aids have a microphone, amplifier, speaker, and volume control. The main types are: air conduction, bone conduction, canal aids, postauricular, body aids, and cochlear implants.

hearing-aid battery A battery which is designed for use in a **hearing aid**. Alkaline, or zinc-air batteries, for instance, are appropriate.

hearing loss **1.** A partial or complete loss of the sense of hearing. **2.** The amount of **hearing loss (1)** an individual has, expressed in decibels. An audiogram serves to indicate levels of hearing loss at tested frequencies, while an audiometer is utilized to generate sounds at these frequencies.

heart pacemaker Same as **heart pacer**.

heart pacer An electrical device which delivers small impulses to the heart under specified conditions, and at a predetermined rate, so as to promote its pumping blood at a regular pace. Such a device may or may not be implanted. Also called **heart pacemaker**, **cardiac pacemaker**, **pacemaker**, or **pacer**.

heat The non-mechanical form of energy which is transferred between bodies or regions at different temperatures which are in contact with each other. Heat is transferred from a hotter body or region to a colder body or region. Such regions may also be within the same body. Heat can be transferred via conduction, convection, or radiation. The SI unit for heat is the joule, and other units often utilized are the calorie, and British thermal unit. Also called **heat energy**, or **thermal energy**.

heat absorption The absorption of heat by a body or system from its surroundings, as occurs in an endothermic process. Also called **thermal absorption**.

heat aging The aging of a substance, material, component, device, piece of equipment, or system, by subjecting to comparatively elevated temperatures over time. Also called **thermal aging**.

heat capacity The amount of heat necessary, under specified conditions, to raise the temperature of a body or system one degree Celsius. Also called **thermal capacity**.

heat coil A protective device consisting of a small coil which grounds or opens a circuit when it is heated beyond a specified temperature. This serves to limit the value the current may reach, as heat is produced in the coil via the Joule effect. Also called **thermal coil**.

heat conduction The flow of internal energy from a region or body of higher temperature to a region or body of lower temperature. Also called **thermal conduction**.

heat conductivity Also called **thermal conductivity**. **1.** The ability of a material, body, surface, or medium to conduct heat. **2.** The rate at which **thermal conductivity (1)** occurs. Among the elements, silver is the best conductor of heat, followed by copper, then gold. The thermal conductivity of diamond, an allotropic form of carbon, is significantly higher than that of silver.

heat detector Also called **thermal detector**. **1.** A resistive element which measures electromagnetic radiation by absorbing it and converting it into heat. The increase in its temperature is used to measure the radiant energy. Also called **heat sensor (1)**, or **bolometer**. **2.** An instrument or device, such as a **heat detector (1)**, which senses heat. Also called **heat sensor (2)**. **3.** An instrument or device, such as a thermocouple, which functions or is actuated when exposed to heat. **4.** An instrument or device, such as a thermometer, which quantifies heat.

heat dissipation Same as **heat loss**.

heat endurance The interval of **heat aging** that a material can withstand before failure, or a given degree of degradation.

heat energy Same as **heat**.

heat engine A device or system which converts heat energy into work.

heat exchanger A device or system which transfers heat from one body or medium to another. In doing so, heat is transferred from a hotter body or medium to a colder body or medium. A heat sink is an example.

heat flow Same as **heat transfer**.

heat gain An increase in the amount of heat in a body or region. This may result as a consequence of solar radiation, the heating effect, a fan blowing hotter air into a region, and so on. Heat is transferred from a hotter body or medium to a colder body or medium.

heat loss Also called **heat dissipation**, or **thermal loss**. **1.** A decrease in the amount of heat in a body or region. This may be accomplished, for instance, through the use of heat sinks or fans. Heat is transferred from a hotter body or medium to a colder body or medium. **2.** The loss of another form of energy through its conversion into heat energy. For instance, losses due the heating effect. Also called **dissipation** (1).

heat of evaporation Same as **heat of vaporization**.

heat of fusion The amount of heat necessary, with the pressure and temperature held constant, to convert a given amount of a solid into its liquid form. Also called **latent heat of fusion**.

heat of reaction The amount of heat released or absorbed by a chemical reaction, under specified conditions. In an endothermic reaction heat is absorbed from the surroundings, while in an exothermic reaction heat is liberated to the surroundings.

heat of vaporization The amount of heat necessary, with the pressure and temperature held constant, to convert a given amount of a liquid into its vapor form. Also called **heat of evaporation**, or **latent heat of vaporization**.

heat radiation The process by which heat is radiated by matter. This energy is in the form of electromagnetic waves, and is a result of the temperature of said matter. While such waves may encompass the entire electromagnetic spectrum, a high proportion of this radiation occurs in the infrared region. Also, the heat so radiated. Also called **thermal radiation**, or **radiant heat**.

heat radiator Also called **thermal radiator**. **1.** Matter which emits **heat radiation**. **2.** An apparatus, such as a heat sink, which serves to radiate heat.

heat relay Also called **thermal relay**. **1.** A relay which is actuated as a consequence of the heat produced by a current passing through it. **2.** A relay which is actuated by a given temperature or temperature variation.

heat release The liberation of heat from a body or system to its surroundings, as occurs in an exothermic process.

heat resistance Also called **thermal resistance**. **1.** The extent to which a material or body can withstand the application of heat. **2.** The extent to which a material or body prevents heat from flowing through it.

heat resistant Characterized by **heat resistance**.

heat-seal The use of heat and/or heat-sensitive materials to provide a tight or hermetic seal.

heat-sensitive A material or substance whose characteristics or properties are changed by heat. The properties or characteristics of such a material or substance may be enhanced or degraded, or it may change in size, and so on. An example is heat-sensitive paper.

heat-sensitive paper Paper, such as that used in some faxes or electrocardiographs, which form letters and/or images when exposed to the appropriate heat.

heat sensor **1.** Same as **heat detector** (1). **2.** Same as **heat detector** (2).

heat shield **1.** A barrier which serves to protect against external heat. It may consist, for instance, of a metal or ceramic sheet. **2.** A material, object, or device, such as a heat sink, which serves to remove heat from a piece of equipment, enclosure, or system.

heat shock Any disturbance or other change in a component, circuit, device, piece of equipment, system, material, mechanism, or process, due to a sudden and significant temperature change. Also called **thermal shock**, or **temperature shock**.

heat-shrink plastic A plastic material which shrinks in diameter when heated. Such plastics may be available in any

of various forms, including tape, tubes, strips, bands, sleeves, and so on. Used, for instance, in heat-shrink tubing. Also called **heat-shrinkable plastic**, or **shrink plastic**.

heat-shrink sleeve A sleeve made of a plastic material which shrinks in diameter when heated, providing a snug fit around the cables on each side of a splice, or wherever it is used. Also called **heat-shrinkable sleeve**, or **shrink sleeve**.

heat-shrink tubing Tubing made of a plastic material which shrinks in diameter when heated, providing a snug fit. Such tubing may shrink to about half of its original diameter, and when cooled does not expand. Used, for instance, to protect cable splices from dust and/or moisture. Also called **heat-shrinkable tubing**, or **shrink tubing**.

heat-shrinkable plastic Same as **heat-shrink plastic**.

heat-shrinkable sleeve Same as **heat-shrink sleeve**.

heat-shrinkable tubing Same as **heat-shrink tubing**.

heat sink Also spelled **heatsink**. **1.** A material or object which helps dissipate unwanted heat from components, circuits, devices, equipment, enclosures, or systems, enabling them to continue working within a safe temperature range. It does so by absorbing heat and conducting it away to a surface from which it is dissipated into its surroundings. Heat sinks are used, for instance, to protect power transistors. Also called **sink** (3). **2.** Any material, such as a solid or a fluid, which serves to protect components, circuits, devices, equipment, enclosures, or systems, by removing heat. Also called **sink** (4).

heat sink cooling The use of a **heat sink** to cool a component, circuit, device, piece of equipment, enclosure, or system. Also spelled **heatsink cooling**.

heat switch Also called **thermal switch**. **1.** A switch that is actuated as a consequence of the heat produced by a current passing through it. **2.** A switch that is actuated by a given temperature or temperature variation.

heat transfer The movement of heat between bodies or regions via conduction, convection, or radiation. Heat is transferred from a hotter body or region to a colder body or region. Also called **heat flow**.

heater **1.** A component, device, apparatus, piece of equipment, or system which serves to provide heat. **2.** A resistor utilized as a source of heat, through the conversion of electrical energy. **3.** An element which supplies heat to a **heater-type cathode**.

heater-type cathode A cathode, within a thermionic tube, which is electrically insulated from the heating element. It may consist, for instance, of a filament surrounded by a sleeve with an electron-emitting coating. Such cathodes have the same potential along their entire surface. Also called **equipotential cathode**, **unipotential cathode**, or **indirectly-heated cathode**.

heating effect The heat produced when an electric current passes through a material, due to resistance. A certain portion of the electrical energy passing through a conductor is converted into heat energy due to the resistance of said conductor. Also called **heating effect of a current**, or **Joule effect** (1).

heating effect of a current Same as **heating effect**.

heating element A **heater** (2) utilized in an appliance, such as a stove or electric-heating device. Such an element incorporates the resistor and any related insulation, such as that provided by a ceramic material.

heatsink Same as **heat sink**.

heatsink cooling Same as **heat sink cooling**.

Heaviside bridge A bridge utilized for measuring mutual inductances.

Heaviside-Kennelly layer Same as **Heaviside layer**.

Heaviside layer A layer of the ionosphere whose effect is felt mostly during daylight hours, and which is contained in the

E-region. It extends approximately from 90 to 130 kilometers above the surface of the earth, and is between the D and F layers. It reflects high to very high frequency radio waves. Also called **Heaviside-Kennelly layer**, **E-layer**, or **Kennelly-Heaviside layer**.

heavy-duty Designed and/or built to provide reliable operation while withstanding continued use under harsh circumstances, such as demanding environmental conditions.

heavy hydrogen 1. An isotope of hydrogen whose nucleus contains a neutron and a proton. It is a highly explosive and flammable gas. It is not radioactive, and is used, for instance, to bombard atomic nuclei. Also called **deuterium**. 2. An isotope of hydrogen whose nucleus contains two neutrons and a proton. It is radioactive, and is used, for instance, in cold-cathode tubes, in tracer studies, and to bombard atomic nuclei. Also called **tritium**.

heavy-ion linear accelerator A linear accelerator designed specifically to accelerate heavy ions. Utilized, for instance, to create transuranic chemical elements, to study nuclear reactions, and in spectroscopy. Its acronym is **HILAC**.

heavy load A load which is a comparatively large fraction of the full or rated load.

heavy metal A metal which has a specific gravity greater than a given amount, such as 5. Examples include iron, gold, osmium, and iridium, each of which has a specific gravity of approximately 7.8, 19.3, 22.5, and 22.6, respectively.

heavy water 1. A chemical compound whose formula is D_2O. It is water in which the hydrogens are in the deuterium isotopic form. Used, for instance, as a neutron moderator in nuclear reactors. Also called **deuterium oxide**. 2. A chemical compound whose formula is T_2O. It is water in which the hydrogens are in the tritium isotopic form. Also called **tritium oxide**.

hecto- A metric prefix representing 100. For example, 1 hectometer is equal to 100 meters. Its symbol is **h**.

hectoampere A unit of current equal to 100 amperes. Its abbreviation is **hA**.

hectogram A unit of mass equal to 0.1 kilogram, or 100 grams. Its abbreviation is **hg**.

hectoliter A unit of volume equal to 100 liters, or 0.1 cubic meter. Its abbreviation is **hl**.

hectometer A unit of distance equal to 100 meters. Its abbreviation is **hm**.

hectometric waves Electromagnetic waves whose wavelengths are between 1 and 10 hectometers, corresponding to frequencies of between 3000 and 300 kHz, respectively. This interval represents the medium-frequency band.

hectopascal A unit of pressure equal to 0.1 kilopascal, or 100 pascals. Its abbreviation is **hPa**.

hectowatt A unit of power equal to 100 watts. Its abbreviation is **hW**.

Hefner candle An older unit of luminous intensity, equal to approximately 0.9 candela. Its abbreviation is **HC**, or **HK**. Also called **Hefnerkerze**.

Hefnerkerze Same as **Hefner candle**.

height The vertical distance from a given level, such as the base, to another given level, such as the top.

height above average terrain The height of an antenna above the average level of the terrain within a given radius surrounding its location. Its abbreviation is **HAAT**. Also called **antenna height (2)**.

height control In a TV, a control which adjusts the vertical dimension of the picture.

height effect In a loop antenna, an unbalancing of the directional properties due to interference from nearby objects. This can cause it to act partially as a small vertical antenna. Electrostatic shielding can help minimize this effect. Also called **antenna effect**.

height finder Same as **height-finder radar**.

height-finder radar A radar which determines the height of a scanned object. Also called **height-finding radar**, or **height finder**.

height-finding radar Same as **height-finder radar**.

Heisenberg principle Same as **Heisenberg uncertainty principle**.

Heisenberg uncertainty principle A principle stating that for a given particle it is not possible to simultaneously know with absolute accuracy both its location and momentum. No matter how accurate the measurements are, the mere observing or measuring said particle will interfere with it in an unpredictable fashion. This principle also applies to other observations or measurements. Also called **Heisenberg principle**, **uncertainty principle**, or **indeterminacy principle**.

Heising modulation A system of amplitude modulation, in which the output circuits of the signal amplifier and the modulator are connected to a constant-current supply by means of a shared choke coil. Also called **constant-current modulation**, or **choke-coupled modulation**.

helical 1. Having the shape of a helix. 2. Having the approximate shape of a helix.

helical antenna An antenna whose radiator is in the shape of a helix, and whose axis is perpendicular to a reflecting plane. Such an antenna produces a narrow beam of circularly-polarized waves which can rotate in a clockwise or counterclockwise direction. Usually used at ultra-high and microwave frequencies, such as those utilized by satellites. Also called **helix antenna**, or **spiral antenna**.

helical line A transmission line whose inner conductor is helical. Also called **helical transmission line**, or **spiral line**.

helical potentiometer A precision potentiometer in which multiple turns of the control knob are necessary to move the contact arm from one end of the helically-wound resistance element to the other. Also called **multiturn potentiometer**.

helical recording Recording of a magnetic tape utilizing **helical scanning (3)**.

helical resonator A cavity resonator whose inner conductor is helical.

helical scan Same as **helical scanning**.

helical scanning Also called **helical scan**, or **spiral scanning**. 1. Any scanning which follows the shape of a helix, or approximately so. 2. In radars, scanning in which the pattern of the beam approximates the shape of a helix. 3. In a video tape device, such as a VCR, recording and/or playback in which the heads contact the tape following a diagonal path which forms a helix. The tape is wrapped at an angle around a rapidly rotating drum upon which the video heads are mounted. 4. In fax, scanning which follows the shape of a helix.

helical transmission line Same as **helical line**.

helionics 1. The science of transforming solar energy into electrical energy. 2. A technique utilized to transform solar energy into electrical energy.

heliostat An instrument or device which incorporates a mirror which is automatically moved, so as to always reflect sunlight. A heliostat is usually clock-driven, and may be used, for instance, to maintain solar panels pointing in the optimum direction.

helitron A type of variable oscillator utilized at ultrahigh and microwave frequencies.

helium A chemical element whose atomic number is 2. It is a colorless non-combustible noble gas, and its boiling point and melting point are the lowest of any known substance, the latter occurring at less than one degree of absolute zero. It is the second most abundant element in the known universe, and has 8 known isotopes, of which 2 are stable. Its applications include its use in lasers, chromatography, luminous signs, fuel pressurization, as an inert atmosphere for growing crystals such as those utilized in semiconductors, as a coolant in nuclear reactors, as an inert gas in welding, and in many cryogenic and superconductivity applications and investigations. Its chemical symbol is **He**.

helium-cadmium laser A gas laser utilizing helium and vaporized cadmium, and which produces coherent light in the blue, violet, and ultraviolet ranges. Used, for instance, in spectroscopy.

helium-neon laser A laser in which the active medium is a mixture helium and neon gases. Such a laser usually emits coherent red light, and may be used in bar-code scanners, for alignment, measurement, printing, or as a pointer. Its abbreviation is **HeNe laser**.

helix 1. In the shape of a spiral, or consisting of a spiral. **2.** A coil of wire in the shape of a spiral.

helix antenna Same as **helical antenna**.

Helmholtz coils A pair of coaxially mounted coils with the same diameter and an equal number of turns, and which are separated by the radius of either winding. Such an arrangement is utilized to create a uniform magnetic field in the space between said coils.

Helmholtz double layer In an electrolytic solution, a layer of ions on the surface of an electrode, which itself is surrounded by another layer of oppositely-charged ions from said solution.

Helmholtz resonator A hollow, rigid-walled enclosure utilized as an acoustic resonator. The enclosure has a small opening, and its geometry determines the resonant frequency. An example is a bottle with a small opening which is exposed to air.

help Documentation available on-screen to users of devices such as computers and PDAs, which consists of instructions, hints, and other forms of assistance when using a given program. A program may have a manual readily available on disk which is accessed by mouse or keyboard commands, or it may be available through a Web site. When the help varies according to the task a user is working on, it is context sensitive.

help key On most computer keyboards, a key which when pressed, requests **help**. Often, the F1 key serves for this purpose.

helper application An application which adds functionality to another program which is already running. Browser plug-ins are examples.

hemimorphic Pertaining to, or having asymmetrical axial ends. Said, for instance, of a crystal.

HEMT Acronym for **high-electron mobility transistor**.

HeNe laser Same as **helium-neon laser**.

henry The SI unit of inductance. It is equal to the inductance of a closed circuit in which an electromotive force of one volt is produced by a current changing uniformly at a rate of one ampere per second. Its plural form is either henries or henrys. Its symbol is **H**.

heptode An electron tube with seven electrodes. It usually has an anode, a cathode, and five grids. The innermost grid is a control grid.

hermaphroditic adapter 1. An adapter whose pattern of recessed openings and prongs are identical on each of the connecting surfaces. **2.** An adapter which is neither female nor male, but which can be connected with another similar adapter.

hermaphroditic connector 1. A connector whose pattern of recessed openings and prongs are identical on each of the connecting surfaces. **2.** A connector which is neither female nor male, but which can be connected with another similar connector.

hermetic Completely sealed. Said, for instance, of an enclosure which does not let gases, light, water, contaminants, or the like, enter or escape.

hermetic seal A seal which is **hermetic**.

hermetically sealed Within an enclosure which has a **hermetic seal**.

hermetically-sealed relay A relay with a permanent **hermetic seal**.

hermetically-sealed switch A switch with a permanent **hermetic seal**.

herringbone pattern In a TV picture, interference occurring in a pattern resembling the skeleton of an elongated fish. The lines of such a pattern may wiggle back and forth.

hertz The SI unit of frequency. It is equal to one complete cycle of a periodic phenomenon, such as an electromagnetic wave, per second. Frequencies are expressed in multiples of hertz, such as kilohertz, terahertz, and so on. It is equivalent to **cycles per second**, which is an obsolescent term. Hertz contrasts with **becquerel**, which measures phenomena that occur in an unpredictable manner, such as radioactive disintegrations. Its abbreviation is **Hz**.

Hertz antenna An ungrounded half-wave antenna. Such an antenna is usually installed some distance above the ground, and may be positioned horizontally or vertically. Its feed point may be connected to the radiating element at the center, at either end, or at any point in between. Also called **Hertzian antenna**.

Hertz effect The ionization of a gas exposed to ultraviolet radiation. Also called **Hertzian effect**.

Hertz oscillator An ultrahigh radio-frequency generating device incorporating two metal plates between which there is a spark gap. These act as a capacitor between whose plates oscillatory discharges may occur. Also called **Hertzian oscillator**.

Hertz vector A single vector representing the electromagnetic field of a radio wave. Also called **Hertzian vector**.

Hertz waves Same as **Hertzian waves**.

Hertzian antenna Same as **Hertz antenna**.

Hertzian effect Same as **Hertz effect**.

Hertzian oscillator Same as **Hertz oscillator**.

Hertzian vector Same as **Hertz vector**.

Hertzian waves Electromagnetic waves whose frequencies are within the radio spectrum. Also called **Hertz waves**, or **radio waves**.

heterochromatic 1. Consisting of, or characterized by different colors. **2.** Consisting of, or characterized by different wavelengths or frequencies.

heterodyne 1. To mix two different frequencies in a nonlinear device, to produce two other frequencies. One frequency is the sum of the frequencies, while the other is their difference. **2.** To mix two different frequencies in a nonlinear device. Also, the output of such a mixture, which is called heterodyne frequency.

heterodyne conversion transducer A transducer whose output frequency is equal to the sum or difference of the input frequency and the local oscillator frequency. Also called **conversion transducer**, or **converter (4)**.

heterodyne detection The use of a **heterodyne detector** to obtain an audible heterodyne frequency from an unmodulated RF carrier wave.

heterodyne detector A detector in which an unmodulated RF carrier wave is combined with the signal of a local oscillator, to obtain an audible heterodyne frequency.

heterodyne frequency For two different frequencies which are mixed, either the sum or the difference of the frequencies. Also called **beat frequency**.

heterodyne frequency meter A frequency meter which combines an unknown frequency with a known local oscillator frequency, so as to produce a zero beat. One of the harmonics of the known and/or unknown frequency may also be used. When calibrated in wavelengths, also called **heterodyne wavemeter**.

heterodyne interference A high-pitched heterodyne frequency which is generated in a receiver when two signals of slightly different frequencies are mixed. Also called **heterodyne whistle**.

heterodyne oscillator An oscillator which produces a desired signal frequency within the audible range, by mixing two oscillations at different frequencies. The signal frequency produced is the heterodyne frequency. Also known as **beat-frequency oscillator**.

heterodyne receiver A radio receiver utilizing **heterodyne reception**.

heterodyne reception Radio reception in which the incoming signal is combined with an internally generated signal of a different frequency, to produce an audible beat frequency. Also known as **beat reception**.

heterodyne repeater 1. A repeater in which received signals are converted to an intermediate frequency before retransmission. 2. A repeater in which received signals are converted to an intermediate frequency, amplified, and converted to another frequency band for retransmission.

heterodyne wavemeter A **heterodyne frequency meter** calibrated in wavelengths.

heterodyne whistle Same as **heterodyne interference**.

heterodyning Also called **beating**. 1. Pulsations in amplitude resulting from the mixing of two signals of slightly different frequencies. 2. The production of frequencies which are the sum or difference of two signals of different frequencies which are mixed.

heterogeneous Consisting of, or characterized by having dissimilar elements, constituents, or qualities. For example, heterogeneous radiation. This contrasts with **homogeneous**, which consists of, or is characterized by having similar elements, constituents, or qualities.

heterogeneous environment A computer system in which hardware and/or software from two or more manufacturers is utilized. This contrasts with a **homogeneous environment**, in which all hardware and software is from the same manufacturer.

heterogeneous network This contrasts with **homogeneous network**. 1. A communications network in which not all computers have the same or similar architectures. 2. A communications network in which more than one protocol is used.

heterogeneous radiation This contrasts with **homogeneous radiation**. 1. Electromagnetic radiation composed of different frequencies or wavelengths. 2. Electromagnetic radiation composed of different energies. 3. Radiation composed of different particles.

heterojunction 1. A junction between layers of dissimilar semiconductor materials. 2. A junction between layers of a **heterostructure**.

heterojunction bipolar transistor A bipolar junction transistor with a **heterojunction**, which maximizes the injection of charge carriers from emitter to base. Its abbreviation is **HBT**.

heterojunction FET Same as **high electron mobility transistor**.

heterojunction field effect transistor Same as **high electron mobility transistor**.

heterostructure A structure consisting of two or more layers of semiconductor materials, each with different bandgaps, and whose crystal structure is similar. For instance, a layer

of aluminum gallium arsenide on a layer of gallium arsenide. Used, for instance, in high-performance transistors, and lasers. Also called **semiconductor heterostructure**.

heuristic An approach to problem solving which involves trial-and-error, and learning from experience. By employing directed procedures, the range of possible solutions for any given problem is continually reduced until the solution or answer is found. Any of various methods may be utilized, as opposed to a fixed sequence. Also called **heuristic approach**, or **heuristic method**.

heuristic algorithm An algorithm utilizing a **heuristic approach** to problem solving.

heuristic approach Same as **heuristic**.

heuristic knowledge Knowledge obtained through **heuristic methods**. Used, for instance, by computers which learn from their own mistakes.

heuristic method Same as **heuristic**.

heuristic program A computer program which utilizes a **heuristic approach** to problem solving.

heuristic search A search, such as that for information or for a solution to a problem, which utilizes a **heuristic approach**.

heuristics The study and application of **heuristic approaches** to problem solving.

Heusler alloy An alloy composed of metals which in their pure state are not ferromagnetic. Such an alloy may consist, for instance, of approximately 50% copper, 25% manganese, and 25% tin. Alternatively, the copper may be substituted with silver, and the tin with an element such as aluminum or bismuth.

hex Abbreviation of **hexadecimal**.

hex notation Abbreviation of **hexadecimal notation**.

hex number Abbreviation of **hexadecimal number**.

hex number system Abbreviation of **hexadecimal number system**.

hex representation Abbreviation of **hexadecimal representation**.

hex system Abbreviation of **hexadecimal system**.

hexachloroethane A chemical compound whose formula is C_2Cl_6. Utilized in plasma etching.

hexadecimal Pertaining to the **hexadecimal number system**, or to **hexadecimal notation**. Its abbreviation is **hex**.

hexadecimal notation A notation based on the **hexadecimal number system**. Its abbreviation is **hex notation**. Also called **hexadecimal representation**.

hexadecimal number A number expressed in **hexadecimal notation**. Its abbreviation is **hex number**.

hexadecimal number system A numbering system whose base is 16. The first ten digits utilize the decimal digits 0-9, and the remaining six usually utilize the letters A-F. For example, decimal 10 is the same as hexadecimal A, decimal 20 is hexadecimal 14, and decimal 26 is hexadecimal 1A. Hexadecimal notation is more succinct than binary notation for representing numbers such as machine addresses. For instance, an Ethernet address is specified by 12 hexadecimal digits instead of a 48-bit number. Hexadecimal numbers usually have either a prefix or a suffix to identify them. For example 4D4h has an **h** suffix appended to the hexadecimal number **4D4**. Its abbreviation is **hex number system**. Also called **hexadecimal system**.

hexadecimal representation Same as **hexadecimal notation**. Its abbreviation is **hex representation**.

hexadecimal system Same as **hexadecimal number system**. Its abbreviation is **hex system**.

hexafluoroethane A chemical compound whose formula is C_2F_6. Utilized in plasma etching.

hexagon A closed geometric figure bounded by six straight lines.

Hf Chemical symbol for **hafnium**.

HF Abbreviation of **high-frequency**.

HF band Abbreviation of **high-frequency band**.

HF boost Abbreviation of **high-frequency boost**.

HF compensation Abbreviation of **high-frequency compensation**.

HF heating Abbreviation of **high-frequency heating**.

HF resistance Abbreviation of **high-frequency resistance**.

HF welding Same as **high-frequency welding**.

HFC Abbreviation of **hybrid fiber-coaxial**.

HFC network Abbreviation of **hybrid fiber-coaxial network**.

HFDF Abbreviation of **high-frequency direction finder**.

HFE Abbreviation of **human-factors engineering**.

HFS Abbreviation of **hierarchical file system**.

hg Abbreviation of **hectogram**.

Hg Chemical symbol for **mercury**.

hi Abbreviation of **high**.

hi-8 A video recording and playback system which utilizes an 8 mm tape format, and which provides over 400 horizontal lines of resolution. It is comparable in quality to S-VHS.

hi-fi Abbreviation of **high-fidelity**.

hi-fi amplifier Abbreviation of **high-fidelity amplifier**.

hi-fi receiver Abbreviation of **high-fidelity receiver**.

hi-fi sound Abbreviation of **high-fidelity sound**.

hi-fi stereo Abbreviation of **high-fidelity stereo**.

hi-fi system Abbreviation of **high-fidelity system**.

hi-pot The high potential applied in a **high-potential test**. Also spelled **hipot**.

hi-pot test Abbreviation of **high-potential test**. Also spelled **hipot test**.

hi-res Abbreviation of **high resolution**.

hi tech Abbreviation of **high technology**.

HIC Abbreviation of **hybrid IC**.

HID Abbreviation of **human interface device**.

hidden file A file which is not normally visible to users. A file may be designated as hidden by the operating system, or through the use of an appropriate utility program. A file is usually hidden to prevent corruption, such as that of key operating system files, or to make unauthorized access more difficult.

hidden Markov model A method for inferring a hidden state in a system where the hidden state produces a sequence of observable events. Used extensively in voice recognition software. Its abbreviation is **HMM**.

hierarchical Pertaining to, organized as, or characteristic of a **hierarchy**.

hierarchical artificial neural network Same as **hierarchical neural network**.

Hierarchical Data Format Its abbreviation is **HDF**. A file format utilized for storing and transferring scientific data, especially that of a graphic and numeric nature. HDF is freely available, works across most computer platforms, and also includes tools for writing, analyzing, and viewing data, in addition to supporting other data models.

hierarchical data model A data model utilized in a **hierarchical database**.

hierarchical database A database whose organization is based on a hierarchy, and which utilizes a tree data structure to link records. Each record has one predecessor or owner. Data accessing starts at the top of the hierarchy and works

its way down. A hierarchical database is not well-suited for determining relations between records.

hierarchical file system A file organization system whose arrangement is based on a hierarchy, and which utilizes a tree structure. Such a system may use folders and/or directories. When using folders, for instance, there are root folders, and paths leading through subfolders. Most operating systems use this system. Its abbreviation is **HFS**.

hierarchical menu A menu, such as that displayed on a computer screen, which is subdivided into submenus, and so on.

hierarchical model Same as **hierarchical data model**.

hierarchical network A network which is based on a hierarchy. For instance, it may consist of a large network which is divided into multiple smaller networks, which in turn are subdivided into even smaller networks.

hierarchical neural network A neural network based on a hierarchy. Such a network incorporates multiple stages of subnetworks, in which processing occurs from one stage to the next. Also called **hierarchical artificial neural network**.

hierarchical routing A routing system which is based on a hierarchy. For instance, a large network may be divided into multiple smaller networks, which in turn may be subdivided into even smaller networks. The highest level within the hierarchy would handle routing to the next level, which in turn would take care of the routing to the level below it, and so on.

hierarchical storage management The automatic transfer of information between storage media with different priorities. For example, the movement of data from a lower speed medium, such as an optical disk, to a hard disk when needed. This allows more data to be stored than if all was in the higher speed, and hence more expensive, medium. Its abbreviation is **HSM**. Also called **data migration (2)**.

hierarchy A group or series which is ranked or classified. The usual structure is from highest to lowest, with each successive level below being subordinate to that above it.

high Its abbreviation is **hi**. **1.** Having a great degree of a given magnitude, quantity, or characteristic. Also, having a greater degree of a magnitude, quantity, or characteristic relative to something else. For example, high energy, high voltage, high impedance, and so on. **2.** On the upper end of a given interval or spectrum. Also, that part of an interval or spectrum which is greater than another. For instance, the frequencies which a high-pass filter transmits, as opposed to those it blocks. **3.** Having a great elevation, or a great elevation relative to something else. **4.** In a binary operation, a 1, which also corresponds to on, as opposed to **low**, which corresponds to 0, or off.

high band **1.** In TV, a band which extends from 174 to 216 MHz, corresponding to the VHF channels of 7 to 13. **2.** In communications utilizing one band of frequencies for transmission one direction, and another band for the other, the upper band. The other band, using lower frequencies, is called **low band (2)**. **3.** In communications utilizing multiple frequency bands, the highest band, or a band higher than another given band.

high bit-rate digital subscriber line Same as **HDSL**.

high bit-rate digital subscriber line-2 Same as **HDSL2**.

high bit-rate DSL Same as **HDSL**.

high bit-rate DSL-2 Same as **HDSL2**.

high boost Same as **high-frequency boost**.

high-capacity diskette Same as **high-capacity floppy disk**.

high-capacity floppy Same as **high-capacity floppy disk**.

high-capacity floppy disk A floppy disk whose capacity vastly exceeds a current standard. For example, a floppy featuring a storage capacity of greater than 1 GB when the norm is 1.44 MB. Also called **high-capacity diskette**, or **high-capacity floppy**.

high-capacity floppy drive A disk drive which reads data from, and writes data to, a **high-capacity floppy disk**.

high color Color in which the bit depth is 16, which allows for up to 65,536 possible colors to be displayed. Also called **16-bit color**.

high contrast A level of contrast which amply exceeds a given standard or typical level. For example, a flat display usually features such a level of contrast. In optical character recognition, high contrast improves the distinguishing of a character from the background.

high current **1.** A current level which is high in magnitude as compared to another. **2.** A current level which is greater than usual for a given device, process, application, or operation.

high-definition A level of definition which amply exceeds a given standard or typical level. It may refer, for instance, to color reproduction and detail of a televised image which is highly faithful to the original scene, or to printing which is completely realistic. Its abbreviation is **HD**.

high-definition television Same as **HDTV**.

high-definition television receiver Same as **HDTV receiver**.

high-definition television standard Same as **HDTV standard**.

high-definition television system Same as **HDTV system**.

high-definition TV Same as **HDTV**.

high-definition TV receiver Same as **HDTV receiver**.

high-definition TV standard Same as **HDTV standard**.

high-definition TV system Same as **HDTV system**.

high-density A storage format in which a greater number of bits are stored per unit area than a standard amount. Also, a medium, such as a magnetic disk, with such a storage density. Its abbreviation is **HD**.

high-density complementary metal-oxide semiconductor Same as **HCMOS**.

high-density disk A magnetic disk characterized by **high-density**.

high-density metal-oxide semiconductor Same as **HMOS**.

high-density packaging The packaging of a comparatively large amount of electronic components in a comparatively small space. In chips, large-scale integration, very large-scale integration, and ultra-large scale integration are examples.

high electron mobility transistor A field-effect transistor with a heterojunction, which provides for higher electron mobility. Its acronym is **HEMT**. Also called **heterojunction field effect transistor**, or **modulation-doped field effect transistor**.

high-energy beam A beam consisting of **high-energy particles**.

high-energy electron An electron whose energy exceeds a given amount, such as 1 GeV. An electron linear accelerator, for instance, can accelerate electrons to energies in the TeV range.

high-energy electron beam A beam consisting of **high-energy electrons**.

high-energy neutron A neutron whose energy exceeds a given amount, such as 1 GeV. Such particles may be used, for instance, to collide with other particles to study the nature of matter.

high-energy particle A particle, such as an electron or proton, whose energy exceeds a given amount, such as 1 GeV. Such particles may acquire these energy levels through acceleration by devices such as circular or linear accelerators.

high-energy particle accelerator A device, such as a linac or cyclotron, in which charged particles are greatly accelerated

to achieve high energies. Also called **particle accelerator**, or **accelerator (4)**.

high-energy physics A branch of physics which deals with the properties, interactions, and applications of subatomic particles, especially when they have high energies. Also called **particle physics**.

high-energy proton A proton whose energy exceeds a given amount, such as 1 GeV. Such particles may be used, for instance, to collide with other particles to study the nature of matter.

high-fidelity Indicative of an audio system which reproduces sound faithfully to its input signal. Such a system should feature low distortion, a high signal-to-noise ratio, an extensive dynamic range, and a wide frequency response. Also, a single component within a system, such as a receiver or DVD, with such characteristics. Also, the sound reproduced by such a system. Its abbreviation is **hi-fi**.

high-fidelity amplifier An audio amplifier characterized by **high-fidelity**. Its abbreviation is **hi-fi amplifier**.

high-fidelity receiver An audio receiver characterized by **high-fidelity**. Its abbreviation is **hi-fi receiver**.

high-fidelity sound The sound reproduced by a **high-fidelity** system. Its abbreviation is **hi-fi sound**.

high-fidelity stereo A stereo audio system characterized by **high-fidelity**. Its abbreviation is **hi-fi stereo**.

high-fidelity system An audio system characterized by **high-fidelity**. Its abbreviation is **hi-fi system**.

high filter Also called **low-pass filter**. **1.** A filter which blocks all frequencies above a given cutoff frequency. Thus, all frequencies below this value can pass. An example is that used in an audio amplifier to reduce hiss. **2.** A filter which only passes AC below a given cutoff frequency.

high frequency A range of radio frequencies spanning from 3 to 30 MHz. These correspond to wavelengths of 100 to 10 meters, respectively, which are decametric waves. Also, pertaining to this interval of frequencies. Its abbreviation is **HF**. Also called **high-frequency band**.

high-frequency alternator An alternator which produces a radio-frequency alternating current.

high-frequency band Same as **high frequency**. Its abbreviation is **HF band**.

high-frequency bias In magnetic tape recording, a high-frequency sinusoidal current that is applied to the audio signal to be recorded, in order to optimize performance during playback. This helps improve linearity and extend dynamic range.

high-frequency boost In audio recording and/or reproduction, an increase in the amplification of high frequencies. Its abbreviation is **HF boost**. Also called **high-frequency compensation**, or **high boost**.

high-frequency compensation Same as **high-frequency boost**. Its abbreviation is **HF compensation**.

high-frequency crystal oscillator A **high-frequency oscillator** that is crystal-controlled.

high-frequency direction finder A direction finder which operates in the **high-frequency band**. Its abbreviation is **HFDF**.

high-frequency heating The use of a radio-frequency power source, such as an oscillator, to provide heat. Its abbreviation is **HF heating**. Also called **radio-frequency heating**, or **electronic heating**.

high-frequency loudspeaker Same as **high-frequency speaker**.

high-frequency oscillator An oscillator, such as a crystal oscillator, which generates a radio-frequency output. Such an oscillator may be used, for instance, for many aspects of frequency control in radio communications and high-speed data transmission.

high-frequency resistance The combined resistance offered by a device in a high-frequency AC circuit. It includes, among others, DC resistance, and resistance due to dielectric and eddy-current losses. Its abbreviation is **HF resistance**. Also known as **AC resistance**, **RF resistance**, or **effective resistance**.

high-frequency speaker A small speaker designed to reproduce frequencies above a given threshold, such as 2000 or 5000 Hz. Depending on the design and components, a high-frequency speaker may accurately reproduce frequencies well beyond the limit of human hearing. Such a speaker unit is usually utilized with others, such as woofers and mid-ranges, for reproduction across the full audio spectrum. Also called **high-frequency loudspeaker**, or **tweeter**.

high-frequency welding Welding in which radio-frequency energy is the source of heat. Used, for instance, to join thermoplastic materials such as PVC. Its abbreviation is **HF welding**. Also called **RF welding**.

high impedance 1. An impedance level which is high in magnitude as compared to another. 2. An impedance level which is greater than usual for a given device, process, application, or operation.

high-impedance voltmeter A voltmeter with a comparatively high input impedance. The greater the input impedance, the lesser the disturbance of the circuit tested. Some voltmeters have an input impedance of greater than 10^{15} ohms.

high-level In digital logic, a level within the more positive of the two ranges utilized to represent binary variables or states. A high level corresponds to a 1, or true, value. Also called **high-level logic**, **high logic level**, or **logic high**.

high-level computer language Same as **high-level language**.

High-Level Data Link Control A bit-oriented ISO communications protocol utilized at the data-link layer. Its abbreviation is **HDLC**.

high-level format A format which resets file-allocation tables so that the operating system sees the disk as empty. Such a format does not destroy the data on a disk, while a **low-level format** does.

high-level input current The current flowing into an input when a given **high-level voltage** is applied.

high-level input voltage In digital logic, an input voltage level which is within the more positive of the two ranges utilized to represent the binary variables.

high-level language A programming language that more closely resembles human language than machine code. Such languages are designed to address specific classes of problems, and are essentially hardware-independent. Although high-level languages are easier to write and maintain, they must be translated into machine language by a compiler or interpreter. Examples include Fortran, C, LISP, and Pascal. Its abbreviation is **HLL**. Also called **high-level programming language**, **high-order language**, or **high-level computer language**.

high-level logic Same as **high-level**.

high-level modulation 1. Modulation produced at a point in a system where the power level is roughly equal to that of the total power output of the system. 2. Modulation produced in the plate circuit of the last stage of a radio transmitter.

high-level output current The current flowing into an output, which under specified input conditions, establishes a **high level** at said output.

high-level output voltage A voltage at an output terminal, which under specified input conditions, establishes a **high level** at said output.

high-level programming language Same as **high-level language**.

high-level signal A signal whose amplitude is above a given level, or which is otherwise comparatively high.

high-level voltage In digital logic, a voltage level which is within the more positive of the two ranges utilized to represent binary variables or states.

high logic level Same as **high-level**.

high-memory area The first 64K of extended memory. It usually spans from 1024 to 1088K. Its abbreviation is **HMA**.

high-mu tube A tube with a high amplification factor.

high-order Within a sequence of digits, a location with comparatively higher significance. In a number this is usually to the left, as are **3** and **2** in 321.098.

high-order digit A digit in a **high-order** location.

high-order language Same as **high-level language**.

high-pass filter A filter which passes, with little or no attenuation, all frequencies above a non-zero lower cutoff. Also spelled **highpass filter**. Also called **low filter**.

high-performance A level of performance which amply exceeds a given standard or typical level. An example is high-performance computing.

high-performance computing A level of computing performance which amply exceeds a given standard or typical level. For example, complex research performed with supercomputers is high-performance computing.

High-Performance Parallel Interface A standard for high-speed parallel data transfer, with rates exceeding a given amount, such as 1.6 Gbps, 64 bits at a time. Used, for instance, in high-speed LANs. Its abbreviation is **HIPPI**.

High-Performance Serial Bus A serial interface which implements the IEEE 1394 standard. Such an interface features transfer speeds ranging from 100 Mbps to over 2 Gbps, and allows the connection of up to 63 devices. It also supports hot-swapping, and isochronous data transfer. Used, for instance, for real-time transfers of video. Also called **FireWire**.

high-persistence phosphor A phosphor surface, such as that on the back of the viewing screen of a CRT, which glows for a comparatively long time. Used, for instance, in storage tubes.

high-potential test A test in which a voltage that exceeds the rated voltage is applied to a conductor for a given time, to determine the breakdown voltage of any insulating materials. Used, for instance, to test new devices or equipment, or as a maintenance test for older equipment. Its abbreviation is **hi-pot test**. Also called **high-voltage test** (1).

high-potential testing The performance of **high-potential tests**. Also called **high-potting**.

high-potting Same as **high-potential testing**.

high power Its abbreviation is **HP**. 1. A power level which is high in magnitude as compared to another. 2. A power level which is greater than usual for a given device, process, application, or operation.

high-power amplifier Its abbreviation is **HPA**. 1. In satellite communications, a powerful amplifier which amplifies uplink signals. 2. An amplifier, such as a MOSFET or klystron, which is especially powerful for a given purpose.

high-power rectifier 1. A rectifier designed to function with high currents. 2. A rectifier rated for function with high currents.

high-pressure mercury lamp A mercury vapor lamp in which the partial pressure of the mercury vapor is of the order of 10^5 pascals during operation. Such lamps produce a very intense light. Also called **high-pressure mercury vapor lamp**.

high-pressure mercury vapor lamp Same as **high-pressure mercury lamp**.

high-pressure sodium lamp A sodium vapor lamp in which the partial pressure of the sodium vapor is of the order of 10^4

pascals during operation. Such lamps produce a very intense light. Also called **high-pressure sodium vapor lamp**.

high-pressure sodium vapor lamp Same as **high-pressure sodium lamp**.

high Q For a component, resonant circuit, tuned circuit, periodic device, or energy-storing device, a comparatively high value of Q (2). For instance, for a component or circuit, it could represent a high value for the ratio of reactance to resistance. Also called **high Q factor**.

high Q factor Same as **high Q**.

high resolution A level of displayed resolution that amply exceeds a given standard or typical level. A laser printer featuring high-resolution, for instance, may have a 2400 x 2400 dpi resolution, which can faithfully reproduce intricate detail to help images appear more realistic. A high-resolution monitor, for instance, may have a resolution of well over 1000 x 1000 pixels. Also, a technology utilized to obtain and present such a resolution. As technology progresses, what is considered high resolution is continually improved. Its abbreviation is **hi-res**.

high-resolution display Same as **high-resolution monitor**.

high-resolution electron microscope An electron microscope which incorporates enhancements such as a narrower and more stable electron beam, to attain resolutions which can be better than 1.5 angstroms. Its abbreviation is **HREM**.

high-resolution electron microscopy The use of a **high-resolution electron microscope** for tasks such as the study of atomic structure. Its abbreviation is **HREM**.

high-resolution monitor A monitor, such as a flat display, featuring **high resolution**. Also called **high-resolution display**.

high-resolution radar A radar whose resolution enables it to distinguish between two comparatively close scanned objects. Also, a radar display which enables a user to visually distinguish between two such objects.

high-resolution screen The viewing screen of a **high-resolution monitor**.

high-speed carry In parallel addition, any technique or procedure utilized to accelerate the processing of carries.

High-Speed Circuit-Switched Data An enhancement to GSM which increases data rates to 57.6 Kbps from the usual 9.6 Kbps. Its abbreviation is **HSCSD**.

high-speed complementary metal-oxide semiconductor Same as **HCMOS**.

high-speed digital subscriber line Same as **HDSL**.

high-speed line A communications line, such as a DS3 or T3, via which data is transmitted at high speeds.

high-speed oscilloscope An oscilloscope which can display very high-speed pulses. Some such oscilloscopes can be utilized to study variations in electrical phenomena occurring in the femtosecond range.

high-speed printer A computer printer whose speed significantly exceeds a given prevailing standard or rate.

high-speed relay A relay with an especially fast make or break interval.

high-speed serial interface A serial interface supporting up to 52 Mbps, and which is used for connecting LANs to LANs, or LANs to WANs over high-speed lines. Its abbreviation is **HSSI**.

high-speed switch A switch with an especially fast make or break interval.

high-speed USB An enhanced version of **USB**, providing for data transfer rates of up to 480 Mbps. High-speed USB is fully compatible with USB. Also called **USB 2.0**.

high tech Abbreviation of **high technology**.

high technology A form of technology involving, depending upon, or incorporating highly advanced or specialized components, devices, equipment, systems, methods, or theories. Its abbreviation is **high tech**, or **hi tech**.

high-temperature insulator A material whose electrical and thermal properties enable it to be used as an insulator at elevated temperatures. Certain ceramics, for instance, are suitable for such use.

high-temperature superconductivity Superconductivity occurring at or above a given temperature, such as 77, 90, or 125 K. Its abbreviation is **HTS**.

high-temperature superconductor A material which is exhibits superconduction at or above a given temperature, such as 77, 90, or 125 K. An example is YBCO, which is a ceramic superconductor. Its abbreviation is **HTS**.

high tension Same as **high voltage**. Its abbreviation is **HT**.

high-tension line Same as **high-voltage line**.

high vacuum Also called **hard vacuum**. **1.** An environment or system whose pressure is below 10^{-3} torr, but above 10^{-6} torr. One torr equals 1 millimeter of mercury, or approximately 133.3 pascals. **2.** In an electron tube, a pressure low enough to prevent ionization of any gas present.

high voltage Its abbreviation is **HV**. Also called **high tension**. **1.** A voltage that exceeds a given amount, such as 115, 230, or 500 kilovolts. Such voltages are usually used for carrying electricity over power transmission lines. **2.** A voltage level which is greater than usual for a given device, process, application, or operation.

high-voltage insulation Insulation that is suitable for use at **high voltages**.

high-voltage insulator A material whose electrical properties enable it to be used as an insulator at **high voltages**. Certain ceramics, for instance, are suitable for such use.

high-voltage line A power line carrying a **high voltage**.

high-voltage test **1.** Same as **high-potential test**. **2.** Any test performed utilizing a **high voltage**.

high-voltage testing The performance of **high-voltage tests**.

highlight **1.** On a computer screen, an area which is displayed in a manner that it stands out. For example, a black text with a white background may be selected with the mouse and be converted into white-on-black. This text may then be deleted, copied, moved, have its font changed, and so on. **2.** In a display screen, such as that of a TV, an especially bright area within the overall picture.

highpass filter Same as **high-pass filter**.

HILAC Acronym of **heavy-ion linear accelerator**.

HiperLAN A set of WLAN standards similar to IEEE 802.11.

HIPERLAN Same as **HiperLAN**.

hipot Same as **hi-pot**.

hipot test Abbreviation of **high-potential test**. Also spelled **hi-pot test**.

HIPPI Abbreviation of **High-Performance Parallel Interface**.

hiss A form of random audio-frequency noise which manifests itself as a sharp sibilant sound similar to a sustained **s**. When occurring during magnetic tape playback, it may be caused, for instance, by circuit noise and/or residual magnetism on said tape.

histogram A chart employing bars to communicate information. Such bars are usually vertical, and their height indicates a given value, such as a percentage or frequency of occurrence within a given category, interval, or the like, which is plotted along the horizontal axis. Alternatively, the bars may be horizontal, with the independent variable plotted along the vertical axis.

history A list or record of previous user inputs or actions while utilizing a given program. Such inputs or actions may include keystrokes, commands, menus passed, and the like. A Web browser, for instance, usually has a history folder where recently visited Web pages are listed. A user can usually double-click on an entry within this list to return to any given page.

hit **1.** A successful retrieval of data from a cache, instead of main memory or a disk. Also called **cache hit**. **2.** An instance of a Web page, or any component of it, being accessed by a user over the Internet. **3.** An instance of words, data, files, URLs, or the like matching desired criteria. For instance, if an Internet search engine returns 1,500 documents based on a given combination of keywords, then there were 1,500 hits. **4.** A momentary electrical disturbance in a transmission line, such as that which may be caused by a lightning stroke.

hit rate **1.** The proportion of data accesses that are fulfilled by a cache, instead of main memory or a disk. It is the primary way to measure the effectiveness of a cache. Also called **cache hit rate**. **2.** The frequency a Web page, or any component of it, is accessed by a user over the Internet. For instance, if a user first visits a Web page, then clicks on a link within it, it may count as two hits. Also, if a Web page with 25 graphics is accessed, then 26 hits may be registered, one for accessing the page, and one for each time graphics information is retrieved from the Web server.

Hittorf dark space In a glow-discharge tube, a narrow dark space near the cathode, between the cathode glow and the negative glow. Also called **Crookes dark space**, **cathode dark space**, or **dark space (2)**.

HK Abbreviation of **Hefner candle**, or **Hefnerkerze**.

hl Abbreviation of **hectoliter**.

HLL Abbreviation of **high-level language**.

hm Abbreviation of **hectometer**.

HMA Abbreviation of **high-memory area**.

HMD Abbreviation of **head-mounted display**, or **head-mounted device**.

HMM Abbreviation of **hidden Markov model**.

HMOS Abbreviation of **h**igh-density **m**etal-**o**xide **s**emiconductor. A MOS technology with a greater concentration of transistors than ordinary MOS chips.

Ho Chemical symbol for **holmium**.

hodoscope An instrument which detects and records the path of ionizing particles. It may consist, for instance, of an array of closely-spaced ion counters. Used, for instance, to help study particles in accelerators.

hog horn Same as **horn antenna**

hog horn antenna Same as **horn antenna**.

hold **1.** To keep from moving or progressing without stopping operation. Also, to maintain at a given value. For example, to maintain a field at a specified intensity. **2.** To retain information in a storage device. For instance, to hold data in a computer's RAM. **3.** To interrupt a phone call without disconnecting it. For example, to put a call on hold. **4.** Same as **hold time (1)**. **5.** In a charge-storage tube, the use of electron bombardment to maintain storage elements at equilibrium potentials. **6.** Same as **hold control**.

hold control In a TV receiver, a control that is varied to prevent vertical or horizontal rolling of the picture, by adjusting the frequency of the vertical or horizontal oscillator, respectively. Also called **hold (6)**.

hold mode **1.** An operational mode in which a given movement, progress, value, output, status, or the like, is maintained steady. **2.** In an analog computer, an operational mode in which integration is halted, and all variables are held at their value at that moment.

hold queue A queue where calls, messages, data, tasks, or the like are placed while awaiting further action such as connection, transmission, processing, execution, and so on.

hold recall A feature which rings the same telephone extension or line which placed a given call on hold, after a specified time interval elapses.

hold reminder A feature which provides an alert, such as a tone, to indicate that a given telephone call is still on hold.

hold time Also called **holding time**. **1.** The time interval a phone call makes use of a trunk or channel. This interval may begin, for instance, as soon as the first digit is dialed, and end when the idle state is restored. Also called **hold (4)**. **2.** In resistance welding, the time allotted for a weld to harden. **3.** After a timing pulse, such as a clock pulse, the time interval during which data input to a device must remain stable in order to insure that said data is correct. Used in the context of computing and digital communications.

hold-up time The length of time during which a power supply can operate within given specifications, such as at a stated voltage, following an interruption of its input AC power. Also spelled **holdup time**.

holding beam In an electrostatic cathode-ray storage tube, the electron beam which regenerates the charges which are retained on the surface of the dielectric.

holding circuit A circuit which maintains enough current in a **holding coil** to keep a relay in its actuated position after the triggering current is removed. Also called **locking circuit**.

holding coil A supplementary coil utilized to hold a relay in its actuated position after the triggering current is removed.

holding current The minimum current necessary to maintain a switching device, such as a relay, in its actuated position after the triggering current is removed.

holding gun An electron gun which produces a **holding beam**.

holding time Same as **hold time**.

holdup time Same as **hold-up time**.

hole In a semiconductor material, an electron vacancy that is created when an electron jumps the gap from the valence band to the conduction band. Such a hole is mobile and acts as if it were a positively-charged particle, and thus can serve as a charge carrier. Also called **electron hole**, or **mobile hole**.

hole concentration In a semiconductor material, the number of holes per unit volume. May be expressed for instance, as the number of holes per cubic centimeter. Also called **hole density**.

hole conduction Conduction of electricity via **holes**. Under the influence of an electric field, as holes are formed, electrons occupy them, thus producing other holes. The effective movement of such holes is equivalent to a positive charge moving in the same direction.

hole current A flow of an electric charge through a conductor, as transported by **holes**.

hole density Same as **hole concentration**.

hole-electron pair When a **hole** is created, a pair consisting of the hole itself, and the electron that jumped to the conduction band.

hole injection The introduction of mobile **holes** into a semiconductor material via the application of an electric charge. This may be accomplished through the use of a hole injector.

hole injector A whisker, electrode, layer, or device via which **hole injection** occurs. For example, in an n-type semiconductor this may be accomplished by applying a voltage to a point contact on its surface.

hole mobility The average drift velocity of **holes** in a semiconductor material, per unit electric field.

hole storage The accumulation of **holes** in a semiconductor device, such as a transistor.

hole trap Within a semiconductor crystal, an impurity or lattice defect which cancels **holes** by releasing electrons which fill them.

hollow cathode In a glow-discharge or gas-discharge lamp or tube, a cathode which is not solid throughout and which is closed at one end. Essentially all of the radiation it emits is from the cathode glow.

hollow-cathode lamp A lamp incorporating a **hollow cathode**. Used, for instance, as the radiation source in atomic absorption spectroscopy.

hollow conductor A conductor which is not solid throughout. An example is a waveguide.

hollow core A core, such as a magnetic core, which is not solid throughout. Such a core may have a center hole for mounting.

holmium A lustrous silvery metallic chemical element whose atomic number is 67. It is comparatively soft and malleable, has unusual magnetic properties, and has nearly 30 known isotopes, of which one is stable. Its applications include its use as a getter, in spectroscopy, and in tracer studies. Its chemical symbol is **Ho**.

hologram A three-dimensional image which is recorded on a suitable medium such as a high-resolution photographic plate. The recording medium stores an interference pattern which is formed by a coherent beam of light reflected by a reference mirror, and by light scattered by the object being photographed. A laser usually serves as the source of the coherent light. When suitably illuminated, it produces a three-dimensional image in space. A hologram may be incorporated into a credit card, for instance, as a measure to help prevent fraud. Acoustic or radio waves may also be utilized to create a hologram.

holographic Pertaining to **holograms** or **holography**.

holographic data storage Same as **holographic storage**.

holographic data storage system A system utilized for **holographic storage**. Its abbreviation is **HDSS**.

holographic image The three-dimensional image in space a **hologram** produces.

holographic memory Same as **holographic storage**.

holographic storage The use of holograms to store information in a small optical cylinder, cube, or the like. Using lasers to store computer-generated data in three dimensions provides significantly superior storage densities and access speeds, as compared to magnetic disks. Also called **holographic data storage**, or **holographic memory**.

holography A method of recording and reproducing three-dimensional images of objects, through the creation of **holograms**.

home 1. To navigate towards a given location, or approach a scanned object through **homing** (1). 2. To return to a starting position, as might occur in the case of a stepping relay. 3. A starting location, such as the upper left hand of a computer screen, or a home page. 4. Same as **home key**.

home appliance A powered home device designed to help do work, perform tasks, or to assist in providing comfort or convenience. These usually use electrical energy, and include dryers, lamps, refrigerators, and microwave ovens. Precision electronics, such as high-fidelity components and computers are sometimes not included in this group. Also called **household appliance**.

home button An on-screen icon which when clicked with a mouse, or other pointing device, returns to a starting location such as a home page.

home computer A computer which is used, or designed for use, at a home.

home key On most computer keyboards, a key which when pressed serves to move the cursor to the start of a line, the start of a file, the top of a screen, and so on. Also called **home** (4).

home network A computer network used within a residence, and which controls devices and equipment such as appliances, domestic electronics, security systems, and so on.

home page A Web page that serves as a starting point for navigation. A home page may be that which is first loaded when a Web browser accesses the Internet. It may also be the first or main page within a group of pages, such as those of a business or individual. Also called **start page**, or **welcome page**.

Home Phone Network Alliance Same as **HomePNA**.

Home Phoneline Network Alliance Same as **HomePNA**.

Home PNA Same as **HomePNA**.

Home Radio Frequency Same as **HomeRF**.

Home RF Same as **HomeRF**.

home theater The combining of audio and video equipment to provide a listening and viewing experience approaching that of a fine movie theater. For instance, a DVD movie may be viewed on an HDTV, with surround sound provided by a high-fidelity audio system with speakers which faithfully reproduce their input signals.

homecam A Webcam designed for, or utilized in a home.

HomePNA Abbreviation of **Home Phoneline Network Alliance**, or **Home Phone Network Alliance**. A technology or standard utilizing existing telephone wires in a home to provide a home network or LAN, without affecting ordinary telephone functions such as voice telephony and fax. Its own abbreviation is **HPNA**.

HomeRF Abbreviation of **Home Radio Frequency**. A home network or LAN utilizing the Shared Wireless Access Protocol.

homing 1. The following of a path of waves, such as acoustic or radio, towards their source or point of reflection. Homing may be used, for instance, as an aid in navigation, or to approach a scanned object, and may be automatic or manual, or active or passive. In marine or aeronautical navigation, a beacon is frequently used to emit the guiding or orienting signal. 2. The returning to a starting position, as might occur in the case of a stepping relay.

homing antenna A directional antenna which serves for **homing** (1) when radio waves are being followed.

homing beacon In navigation, a beacon providing the guiding or orienting signal for **homing** (1). Also called **locator beacon** (1).

homing device 1. A transmitting device, such as a beacon, which provides a signal which is used for **homing** (1). 2. A receiving device mounted on a moving object, such as an airplane, which follows the path provided by a **homing device** (1). Also, any receiving device which is guided towards the origin or point of reflection of a signal. 3. A control device which automatically starts its motion or rotation in the desired direction, as opposed to doing so from a fixed point such as an extreme.

homing station A station which has a transmitting device, such as a beacon, which provides a signal which is used for **homing** (1).

homing system A system combining transmitting and receiving **homing devices**.

homodyne receiver A radio receiver utilizing **homodyne reception**.

homodyne reception Radio reception in which the incoming signal is combined with an internally-generated signal of the same frequency. Also called **zero-beat reception**.

homogeneous Consisting of, or characterized by having similar elements, constituents, or qualities. For example, a solution incorporates two or more components to form a homogeneous mixture. This contrasts with **heterogeneous**, which consists of, or is characterized by having dissimilar elements, constituents, or qualities.

homogeneous environment A computer system in which all hardware and software is from the same manufacturer. This contrasts with a **heterogeneous environment**, in which hardware and/or software from two or more manufacturers is utilized.

homogeneous network This contrasts with **heterogeneous network**. **1.** A communications network in which all computers have the same or similar architectures. **2.** A communications network in which only one protocol is used.

homogeneous radiation This contrasts with **heterogeneous radiation**. **1.** Electromagnetic radiation composed of similar frequencies or wavelengths. **2.** Electromagnetic radiation composed of similar energies. **3.** Radiation composed of similar particles.

homojunction A junction between layers of the same semiconductor material, but with each having different properties. It may consist, for instance, of a junction between layers when one is p-type and the other is n-type.

homologation The conformity of a component, device, piece of equipment, system, or specification to one or more standards, such as those established internationally.

homopolar **1.** Pertaining to, or having one polarity. **2.** Having electrical symmetry. **3.** Having an equal distribution of charges.

homopolar generator A DC generator in which the poles are of the same polarity with respect to the armature, thus making a commutator unnecessary. Also called **homopolar machine**, **acyclic machine**, or **unipolar machine**.

homopolar machine Same as **homopolar generator**.

honey pot Same as **honeypot**.

honeycomb coil A coil whose successive windings are crisscrossed. This provides a low distributed capacitance. Also called **lattice winding**.

honeypot A computer system set up for the purpose of attracting crackers and hackers, so as to evaluate their techniques, discovery security vulnerabilities, and so on. Also spelled **honey pot**.

hood **1.** An enclosure which allows a person to work with hazardous materials and/or components, with reduced exposure. **2.** A shield placed over the screen of a CRT to help eliminate ambient light.

hook flash To press a telephone hookswitch briefly to access a function such as call-waiting, or three-way calling. Also, a button so labeled. Also, to press a button so labeled. Also spelled **hookflash**. Also called **flash (4)**.

hook switch Same as **hookswitch**.

hook up **1.** Same as **hookup**. **2.** To assemble a **hookup (1)**.

hook up wire Same as **hookup wire**.

hookflash Same as **hook flash**.

hookswitch On a telephone set, a switch that is depressed when the handset is placed on it. When this switch is depressed, the telephone is not in use. Such a switch can also be located on a handset. A hookswitch may also be used for functions such as call waiting or conference calling. Also spelled **hook switch**. Also called **switchhook**.

hookup Also spelled **hook up**. **1.** A specific configuration of components, circuits, devices, or the like, which operate together for a given purpose. **2.** A diagram of a **hookup (1)**.

hookup wire An insulated and usually tinned copper wire utilized to wire circuits and hookups. Such wire is appropriate for low-current and low-voltage applications. Also spelled **hook up wire**.

hop **1.** An excursion of a radio wave from the earth to the ionosphere and back to the earth. There may be multiple such hops en route between a transmitter and a receiver, and these may be used for long-distance radio communications.

2. In a communications network, a segment between two nodes.

hop count The number of **hops (2)** between a source and destination. Used, for instance, as a routing metric.

horizon **1.** As seen from a given location, the junction at which the earth, or the sea, appears to meet the sky. Also called **apparent horizon**, or **visible horizon**. **2.** A **horizon (1)** as seen from an antenna transmitting site. Such a horizon is the furthest distance that a direct wave may travel. It is the line that is formed by the points at which the direct waves from a radio transmitter become tangential to the surface of the earth. Due to atmospheric refraction, this horizon extends beyond the apparent horizon. Also called **radio horizon**.

horizon sensor A device, such as that which detects the thermal discontinuity between the earth or sea and the sky, which provides a stable vertical reference. Used, for instance, in satellites or spacecrafts to assist in maintaining the proper attitude.

horizontal **1.** Pertaining to, related to, or situated at or near a **horizon (1)**. **2.** Parallel to, or situated along the plane of a **horizon (1)** or a baseline. Also, operating in a plane parallel to a horizon or a base line. **3.** At a right angle to a vertical line.

horizontal amplifier A circuit or device which amplifies the signals which produce a horizontal deflection in an instrument such as an oscilloscope. Also called **X-amplifier**.

horizontal antenna An antenna, such as a dipole antenna, which consists of one or more radiators situated along a horizontal plane.

horizontal axis An axis which is considered to be parallel to, or situated along a horizontal line, the horizon, or the like. In a Cartesian coordinate system, for instance, it is usually considered to be the x-axis.

horizontal beamwidth The angle between the points at which the intensity of an electromagnetic beam is at half of its maximum value, as measured in the horizontal plane.

horizontal blanking In a device such as a CRT, the cutting-off of the electron beam during **horizontal retrace**. This prevents an extraneous line from appearing on the screen during this period. Also called **horizontal retrace blanking**.

horizontal blanking interval The interval within which **horizontal blanking** occurs.

horizontal blanking pulse A pulse which produces **horizontal blanking**. This pulse occurs between the active horizontal lines.

horizontal centering The use of a **horizontal centering control**.

horizontal centering control In a CRT, a control utilized to center the image horizontally on the screen. Also called **horizontal positioning control**.

horizontal convergence control In a color TV tube, a control which varies the voltage of the horizontal dynamic convergence.

horizontal deflection In a CRT, any horizontal deflection of the electron beam.

horizontal deflection coils In a CRT, a pair of coils which serve to deflect the electron beam horizontally.

horizontal deflection electrodes In a CRT, a pair of electrodes which serve to deflect the electron beam horizontally.

horizontal deflection oscillator In a TV receiver, an oscillator which produces signals controlling horizontal sweeps. Also called **horizontal oscillator**.

horizontal deflection plates In a CRT, a pair of plates which serve to deflect the electron beam horizontally.

horizontal dipole A dipole antenna which consists of one or more radiators situated along a horizontal plane. Also called **horizontal dipole antenna**.

horizontal dipole antenna Same as **horizontal dipole**.

horizontal directivity The directivity of an antenna or transducer along the horizontal plane.

horizontal drive control In a TV receiver, a potentiometer which adjusts the output of the horizontal oscillator. Also called **drive control (2)**.

horizontal dynamic convergence In a color TV tube with a three-color gun, the convergence of the three electron beams on a specified surface during horizontal scanning.

horizontal field strength The field strength of an antenna measured in the horizontal plane.

horizontal flyback Same as **horizontal retrace**.

horizontal frequency Also called **horizontal line frequency**. **1.** In a CRT, the number of horizontal sweeps or traces per second. Also called **line frequency (1)**. **2.** The number of horizontal sweeps per second in a given TV system such as NTSC or PAL. Also called **line frequency (2)**.

horizontal hold Same as **horizontal-hold control**.

horizontal-hold control A control which adjusts the frequency of the horizontal deflection oscillator in a TV receiver, to maintain the displayed picture horizontally steady. Also called **horizontal hold**.

horizontal hum bars Comparatively broad horizontal bars which extend over an entire displayed TV picture. These bars may be caused, for instance, by low-frequency electromagnetic fields, are alternately black and white, and may or may not move up and down on the screen. Also called **hum bars**, or **venetian-blind effect (1)**.

horizontal line In a CRT, the line formed by a complete horizontal movement of the electron beam. Also called **line (4)**.

horizontal line frequency Same as **horizontal frequency**.

horizontal linearity control In a CRT, a control which allows the horizontal adjustment of that which is displayed, to reduce or eliminate distortion due to improper image linearity.

horizontal oscillator Same as **horizontal deflection oscillator**.

horizontal output stage In a TV receiver, an output amplifier following the horizontal deflection oscillator.

horizontal output transformer In a TV receiver, a transformer utilized with a flyback power supply. It provides the horizontal deflection voltage, the voltage for the second anode, and the filament voltage for the high-voltage rectifier.

horizontal polarization **1.** Polarization of an electromagnetic wave in which the electric field vector is horizontal. Thus, the magnetic field vector is vertical. **2.** Radio-frequency transmissions utilizing waves with **horizontal polarization (1)**. Under such circumstances, transmitting antenna elements are usually in the horizontal plane, and receiving antenna elements are usually most sensitive in this plane.

horizontal positioning control Same as **horizontal centering control**.

horizontal pulse Same as **horizontal sync pulse**.

horizontal radiator An antenna radiator which is situated along a horizontal plane.

horizontal refresh rate Same as **horizontal scanning frequency**.

horizontal resolution In an image, such as that of a TV or fax, the number of picture elements in the horizontal direction of scanning or recording. May be expressed, for instance, in pixels, or in dots per inch.

horizontal retrace In a CRT, the return of the electron beam to its starting point after a horizontal sweep or trace. Also called **horizontal flyback**, or **line flyback**.

horizontal retrace blanking Same as **horizontal blanking**.

horizontal retrace period The interval during which a horizontal retrace occurs.

horizontal scan Same as **horizontal scanning**.

horizontal scan frequency Same as **horizontal scanning frequency**.

horizontal scan rate Same as **horizontal scanning frequency**.

horizontal scanning Its abbreviation is **horizontal scan**. **1.** In a CRT, the horizontal movement of the electron beam. Also called **horizontal sweep (1)**. **2.** In radars, the rotation of the antenna in the horizontal direction. This rotation may or may not encompass the complete horizon.

horizontal scanning frequency In a CRT, the number of horizontal lines scanned by the electron beam per second. Its abbreviation is **horizontal scan frequency**. Also called **horizontal scanning rate**, **horizontal scan rate**, or **horizontal refresh rate**.

horizontal scanning rate Same as **horizontal scanning rate**.

horizontal scrolling Scrolling to the left or to the right within a computer document or program, while **vertical scrolling** is up or down.

horizontal signal A signal, such as a horizontal blanking pulse, which determines a horizontal parameter.

horizontal sweep **1.** Same as **horizontal scanning (1)**. **2.** In a CRT, to deflect the electron beam so that it has a horizontal movement.

horizontal sweep circuit A circuit producing **horizontal sweeps**.

horizontal sync The scanning synchronization provided by a **horizontal sync pulse**. It is an abbreviation of **horizontal synchronization**. Its own abbreviation is **hsync**.

horizontal sync pulse The pulse which synchronizes the line-by-line scanning of a TV receiver with that of a TV transmitter. When synchronized, the relative position of the electron beam in the picture tube is the same as that of the scanning beam in the camera tube. In a TV system, such a pulse is transmitted at then end of each line, and triggers horizontal retracing and horizontal blanking. It is an abbreviation of **horizontal synchronization pulse**, or **horizontal synchronizing pulse**. Also called **horizontal pulse, horizontal synchronization signal, horizontal synchronizing signal, horizontal synchronization pulse, horizontal synchronizing pulse, horizontal sync signal**, or **line sync pulse**.

horizontal sync signal Same as **horizontal sync pulse**,

horizontal synchronization Same as **horizontal sync**. Its abbreviation is **hsync**.

horizontal synchronization pulse Same as **horizontal sync pulse**.

horizontal synchronization signal Same as **horizontal sync pulse**.

horizontal synchronizing pulse Same as **horizontal sync pulse**.

horizontal synchronizing signal Same as **horizontal sync pulse**.

horizontally-polarized radiation Electromagnetic radiation in which the electric field vector is horizontal. Thus, the magnetic field vector is vertical.

horizontally-polarized wave An electromagnetic wave in which the electric field vector is horizontal. Thus, the magnetic field vector is vertical.

horn **1.** A radiating device, such as an antenna or speaker, whose cross section increases as it approaches the exit opening, which serves to direct and intensify any waves emanating from it. **2.** A tube with a cross section that increases as it approaches the exit opening, which serves to direct and intensify any sound waves emanating from it. Also called **acoustic horn**. **3.** Same as **horn antenna**.

horn antenna A microwave antenna whose radiating element is in the shape of a horn. It is fed at its apex, and the electromagnetic radiation is emitted from its open end, which is wider. The cross section utilized may be rectangular, square, circular, and so on. Also called **horn (1)**, **horn radiator**, **hog horn**, **hog horn antenna**, or **microwave horn**.

horn feed A **horn antenna** utilized to feed a parabolic reflector.

horn loading The use of a horn-shaped structure within a speaker to improve its efficiency.

horn loudspeaker Same as **horn speaker**.

horn mouth The output end of a **horn (1)**. Its cross section is larger than that of the **horn throat**.

horn radiator Same as **horn antenna**.

horn speaker A speaker in which the air excited by the diaphragm is fed to a **horn (1)**. Although the sound quality of such speakers varies by their specific design, they tend to have high efficiency. Also called **horn loudspeaker**.

horn throat The input end of a **horn (1)**. Its cross section is smaller than that of the **horn throat**.

horsepower A unit of power equal to approximately 745.6999 watts. Its abbreviation is **hp**.

horseshoe magnet A magnet whose shape is similar to that of a horseshoe, or a letter U. These are usually permanent magnets, and such a shape is utilized when proximity of the poles is desired.

host **1.** Same as **host computer**. **2.** A business which provides server space for entities such as individuals or corporations. When an entity does not have its own servers, yet desires a presence on the Internet through one or more Web pages, such a host provides the space. Hosting services may be fee-based, depending on the specific needs, such as the memory and bandwidth required, whether it is for commercial use, and so on. Also called **Web host**. **3.** To provide the services of a **host computer (1)**. Also called **hosting (1)**. **4.** To provide the services of a **host (2)**. Also called **hosting (2)**, or **Web hosting**.

host adapter A circuit board or device which controls the way peripheral devices access the computer, and vice versa. It is usually contained on a single chip. Examples include disk controllers, graphics controllers, and video controllers. Also called **controller (1)**, or **peripheral controller**.

host address The physical address of a **host computer**. On the Internet, it would be the same as its IP address.

host computer Also called **host (1)**. **1.** Within a network, a computer that provides users with services such as access to other computers and/or databases, and which may also perform control functions. Over the Internet, for instance, a host computer may be accessed by a user from a remote location who seeks access to information, email services, and so on. **2.** Any computer connected to a network, such as a TCP/IP network.

host digital terminal Its abbreviation is **HDT**. A central office terminal which connects digital telephony switches to hybrid fiber coaxial distribution equipment. An HDT can interface directly with SONET nodes or voice-switching equipment. Such a terminal electronically organizes and manages signal traffic, and serves as the interconnection point to local switching.

host ID Abbreviation of **host id**entification. Within an IP address, the bits which identify a specific **host computer**. For instance, if in the IP address 151.201.4.111, **151.201** identify the network, then **4.111** are the host ID. The network ID can encompass the first, first and second, or first, second, and third sets of numbers separated by the periods. The rest correspond to the host ID. Also spelled **hostID**.

host identification Same as **host ID**.

host name Same as **hostname**.

hostid Same as **host ID**.

hostID Same as **host ID**.

hosting **1.** Same as **host (3)**. **2.** Same as a **host (4)**.

hostname A name which identifies a specific **host computer (1)**. Within the following Internet domain names, *www.yipeee.com*, and *help.yipeeee.com*, the host names are *www* and *help*, respectively. Also spelled **host name**.

hot **1.** Anything connected electrically to a source of voltage. Also known as **energized (1)**, **live (1)**, or **alive**. **2.** Not electrically grounded. **3.** Possessing a high level of energy or radioactivity.

hot-air soldering Soldering which utilizes a narrow jet of air whose temperature is carefully controlled. Used, for instance, for soldering individual joints on PCBs.

hot backup A backup performed while an application, such as a database, is using the data being duplicated. This contrasts with a **cold backup**, in which the data is not in use. Also called **online backup**.

hot-bar soldering Soldering in which a heated bar simultaneously solders multiple leads on a PCB.

hot boot Same as **hot restart**.

hot carrier In a semiconductor, a charge carrier whose energy level is greater than that of the majority carriers usually encountered in such a material.

hot-carrier diode A diode with a metal-semiconductor junction, as opposed to the usual pn junction. Hot carriers are injected from the semiconductor layer into the metal layer, so that there are virtually no minority carriers injected or stored, which provides for a very low forward voltage drop and a very fast switching speed. Such diodes have multiple uses, including several in RF applications, logic circuits, photodiodes, and as rectifiers. Also called **Schottky diode**, **Schottky barrier diode**, or **metal-semiconductor diode**.

hot cathode A cathode, such as that in an electron tube, that emits electrons as a consequence of its being heated. Such a cathode may be directly or indirectly heated. This contrasts with a **cold cathode**, which does not require an applied heat to function. Also called **thermionic cathode**.

hot-cathode ionization gauge An ionization gauge in which ions are produced through collisions with electrons emitted from a **hot cathode**. These ions are then attracted by a negatively charged electrode. Also called **hot-filament ionization gauge**.

hot-cathode tube An electron tube which incorporates a **hot cathode**. Also called **thermionic tube**.

hot desking The use of a given group of desks which are shared by all employees. No one employee has a fixed desk, and there are usually many more employees than desks. Used, for instance, to cater to organizations which have multiple shifts.

hot docking The capability of plugging-in a portable computer into its docking station, while said computer is running.

hot electron In a semiconductor, an electron whose energy level is significantly greater than that than that required for thermal equilibrium with the crystal lattice. This may be caused, for instance, by very high electric fields.

hot-electron injection The injection of **hot electrons** to program semiconductor memory, such as EEPROMs.

hot-filament ionization gauge Same as **hot-cathode ionization gauge**.

hot fix Same as **hotfix**.

hot hole In a semiconductor, a hole whose energy level is significantly greater than that than that required for thermal equilibrium with the crystal lattice. This may be caused, for instance, by very high electric fields.

hot insertion The insertion of a component, device, or piece of equipment into a system which is in operation at that moment.

hot junction **1.** In a two-junction thermocouple circuit, the heated junction. **2.** In a thermocouple circuit, the junction which senses the temperature of a material or object being measured. Also called **measuring junction**.

hot key Same as **hotkey**.

hot line Same as **hotline**.

hot link **1.** A link between two applications via which any changes made in one are immediately reflected in the other. Used, for instance, to update a database from a word processor, or to share real-time data between remote computers. **2.** To establish a **hot link** (1).

hot plug Same as **hot swap**.

hot pluggable Same as **hot swappable**.

hot plugging Same as **hot swapping**.

hot-potato routing In a communications network, routing in which messages or packets are kept moving until reaching its destination. This contrasts, for instance, with **store-and-forward**, where a complete message is received and stored before being passed on to the next node en route to its destination.

hot restart The restarting of a computer without powering down, and in which standby components or systems have all the information pertaining to the main system at the moment of a failure, enabling nearly immediate restarting.

hot spare In a RAID, a hard drive which is ready and automatically placed into use in the event another drive fails or malfunctions.

hot spot **1.** A point, or a comparatively small region, within a surface, component, circuit, device, or system, where there is a higher temperature than that of its surrounding area. For example, such a spot on an electrode. **2.** A point, or a comparatively small region within a surface, component, circuit, device, or system, where there is a greater amount or intensity of given quantity than that of its surrounding area. For example, an especially bright spot on the screen of a CRT. **3.** A point, such as a network node, at which a comparatively greater amount of activity occurs. **4.** On a screen pointer, the specific location which at which an action will be effected. For instance, the tip of a screen pointer in the shape of an arrow. Also called **insertion point** (1). **5.** An area of an image, a section of text, or the like, which when clicked activates a function. For example, such a spot may be a hyperlink.

hot standby An approach utilized to help keep a system, such as a communications network, operating continuously by having components, devices, equipment, and/or other systems ready to take over instantaneously when the main system fails.

hot swap To remove and/or add **hot swappable** components, devices, or equipment. Also called **hot plug**.

hot swappable A component, device, or piece of equipment which can be removed from or added to a system which is in operation at that moment. Usually utilized to refer to the ability to remove and/or add components or peripherals to a computer system which is running at the time, and whose operating system automatically recognizes any changes. USBs, for instance, support hot swapping. Also called **hot pluggable**.

hot swapping The performance of a **hot swap**. Also called **hot plugging**.

hot wire **1.** A wire which is connected electrically to a source of voltage. Also called **live wire** (1). **2.** A wire which is not electrically grounded. **3.** A wire which expands when

heated and contracts when cooled, such as that used in a hot-wire relay or a hot-wire ammeter.

hot-wire ammeter An ammeter which incorporates a stretched **hot wire** (3) whose length at any given moment depends on the current passing through it. Its variations in length deflect an attached pointer, which is correlated to a calibrated scale. Such an ammeter serves to measure both direct and alternating currents. Also called **thermal ammeter**.

hot-wire anemometer An anemometer which measures and indicates wind or air flow speeds by the cooling effect said flow has on a stretched electrically-heated wire. As the intensity of the airflow varies, so does the resistance of the wire. These changes are then correlated to a calibrated scale. For proper readings, the probe must be at a right angle to the flow. Also called **thermal anemometer**.

hot-wire instrument An instrument, such as a hot-wire ammeter, which incorporates a **hot wire** (3) to measure a desired quantity.

hot-wire meter A meter, such as a hot-wire ammeter, which incorporates a **hot wire** (3) to measure a desired quantity.

hot-wire relay A relay which incorporates a stretched **hot wire** (3). As the current flowing through the wire varies its length, the contacts are opened or closed.

hot-wire sensor A sensor, such as a hot-wire anemometer, which incorporates a stretched **hot wire** (3).

HotBot A popular search engine.

hotfix Also spelled **hot fix**. **1.** A repair or maintenance procedure performed while a device, piece of equipment, or system is operating. **2.** A patch, service pack, or the like, intended to remedy one or more bugs or other known problems.

HotJava A Web browser specifically designed to support Java.

hotkey A computer key or sequence of keystrokes which executes the same command, regardless of which programs are running. Hotkeys may be user-defined. Also spelled **hot key**.

hotline A point-to-point communications link which is immediately established when an end device, such as a telephone, goes off-hook. This connection is automatic, thus not requiring additional user actions such as dialing. Also spelled **hot line**.

hour A unit of time equal to 3600 seconds. Its abbreviation is **h**, or **hr**.

household appliance Same as **home appliance**.

housekeeping **1.** Any instructions or routines which are automatically performed by a given program to help provide for itself a more optimum computing environment. This involves performing tasks such as garbage collection, setting counters to their starting values, the creation of files, and so on. **2.** Any **housekeeping** (1) procedures performed manually, such as cache clearing more often than a program ordinarily would.

housing That which serves to enclose, support, and protect. For example, such a housing utilized to contain an instrument.

howl **1.** A phenomenon that occurs when the sound waves produced by speakers interact with an input transducer such as a microphone or phonographic cartridge. If the feedback exceeds a certain amount, any of various undesired effects may occur, such as howling, whistling, motorboating, or excessive cone movement. Also, the sound heard as a consequence of this. Also known as **acoustic feedback**, **acoustic regeneration**, **acoustic howl**, or **sound feedback**. **2.** Same as **howling** (2).

howler An instrument or device which emits an audible signal to inform of a given condition, level, or the like. For instance, a device which emits such a tone when a short-circuit is located.

howling A sustained, undesired, and generally disagreeable audio-frequency tone usually caused by acoustic or electric feedback. Also called **howl (2)**.

hp Abbreviation of **horsepower**.

HP Abbreviation of **high power**.

hPa Abbreviation of **hectopascal**.

HPA Abbreviation of **high-power amplifier**.

HPC Abbreviation of **handheld PC**, or **handheld personal computer**.

HPNA Abbreviation of **HomePNA**.

hr Abbreviation of **hour**.

HREM Abbreviation of **high-resolution electron microscope**, or **high-resolution electron microscopy**.

HRTF Abbreviation of **head-related transfer function**.

Hs Chemical symbol for **hassium**.

HSCSD Abbreviation of **High-Speed Circuit-Switched Data**.

HSM Abbreviation of **hierarchical storage management**.

HSSI Abbreviation of **high-speed serial interface**.

hsync Abbreviation of **horizontal sync**, or **horizontal synchronization**.

HT Abbreviation of **high tension**.

HTML Abbreviation of HyperText Markup Language. A language utilized to create documents on the World Wide Web. HTML utilizes tags and attributes which determine factors such as page layout, graphics, fonts, and hypermedia for linking between Web pages, or sections of pages.

HTML code Abbreviation of HyperText Markup Language **code**. Programming statements and instructions written in **HTML**. Also called **HTML source**, or **HTML source code**.

HTML document Abbreviation of HyperText Markup Language **document**. A document prepared utilizing **HTML**.

HTML editor Abbreviation of HyperText Markup Language **editor**. An editor utilized for creating Web pages utilizing **HTML**. An HMTL editor may enable a user to format Web pages using menus and buttons, in addition to utilizing commands, and display the page being edited the same way it will be displayed on the World Wide Web.

HTML page Abbreviation of HyperText Markup Language **page**. A Web page prepared utilizing **HTML**.

HTML source Abbreviation of HyperText Markup Language **source**. Same as **HTML code**.

HTML source code Abbreviation of HyperText Markup Language **source code**. Same as **HTML code**.

HTML tag Abbreviation of HyperText Markup Language **tag**. Codes used for formatting, hyperlinking, and the like, when utilizing **HTML**.

HTS Abbreviation of **high-temperature superconductivity**, or **high-temperature superconductor**.

http Same as **HTTP**.

HTTP Abbreviation of HyperText Transport Protocol. The communications protocol usually utilized between servers and clients on the World Wide Web. When a Web server transfers the information in an HTML document is to a user's Web browser, HTTP is usually used. For instance, in the following URL, *http://www.yipeeee.com*, the *http* indicates that this protocol is being used. Most Web browsers default to HTTP, so *www.yipeeee.com* would be the same as *http://www.yipeeee.com*.

HTTP 404 Not Found Same as **HTTP Error 404**.

HTTP Error 404 An error message displayed by a Web browser when it cannot locate a given Web page, file, or URL, as occurs, for instance, when a page no longer exists,

the server is not available, or an address has been incorrectly typed. Also called **HTTP 404 Not Found, Error 404, 404 Error,** or **404 Not Found**.

HTTP-NG Abbreviation of HyperText Transport Protocol-Next Generation. An improved version of HTTP, which incorporates enhancements such as increased security, layering, and modularity.

HTTP proxy Abbreviation of HyperText Transport Protocol **proxy**. Same as **HTTP proxy server**.

HTTP proxy server Abbreviation of HyperText Transport Protocol **proxy server**. A proxy server that forwards client requests over the internet utilizing **HTTP**. Also called **HTTP proxy**.

HTTP-Secure Same as **HTTPS**.

HTTP Server Abbreviation of HyperText Transport Protocol Server. A server which stores and retrieves HTML documents, along with other resources, utilizing **HTTP**.

https Same as **HTTPS**.

HTTPS Abbreviation of HyperText Transport Protocol-Secure. A modification of HTTP which is utilized for transmitting data securely between servers and clients on the World Wide Web. A URL will appear with **https** instead of **http** when this protocol is utilized, as seen in: *https://www.yipeeee.com*.

hub 1. The center part of something. For example, the center part of a fan, wheel, tape reel, or disk. 2. A central unit which provides connectivity between two or more devices in a communications network. A hub may be active or passive, but does not provide switching or routing. Used, for instance, in a star network.

hub-and-spoke network A communications network configuration in which all nodes are connected to a central node or hub. Also, a network with such a layout. Also called **star network (1)**.

hue A property of color whose visual sensation corresponds mostly to the wavelength of the light. Hue can refer to any color within the visual spectrum, and is identified by words such as purple, green, blue, orange, yellow, and so on. Achromatic colors, such as black, gray, and white, do not exhibit hue.

hue control In a color TV receiver, a control which varies the phase of the chrominance signals with respect to the burst signal. This changes the hue of the televised images. Also called **phase control (2)**.

Huffman coding A data compression method which is based on the frequency in which the contained elements appear. Characters, symbols, or other elements which have a lower probability of appearing are encoded utilizing more bits, while elements with a higher probability of appearing are encoded utilizing fewer bits. In this way, the shortest bit sequence corresponds to the most frequently occurring element.

hum An electrical disturbance which is generally caused by the frequency of the AC supply, or by one of its harmonics. This may occur, for instance, when the output of a power supply is not properly filtered. In an audio-frequency system, hum may manifest itself as a sustained low-frequency audible disturbance. In a TV receiver, such a disturbance may result in horizontal hum bars.

hum bars Same as **horizontal hum bars**.

hum bucking Same as **humbucking**.

hum bucking coil Same as **humbucking coil**.

hum interference Any disturbance resulting from **hum**.

hum modulation Undesired modulation of a signal, due to **hum interference**.

human-computer interaction The use of one or more **human-computer interfaces**. Its abbreviation is **HCI**. Also called **user-computer interaction**.

human-computer interface An interface between a user and a computer system. This may involve, for instance, the use of keyboards, microphones, cameras, pointing devices, and so on. Its abbreviation is **HCI**. Also called **user-computer interface**.

human engineering Same as **human-factors engineering**.

human error Any malfunction, failure, or other undesired condition attributable to human limitations, such as those leading to the improper installation, operation, or maintenance of a component, circuit, device, piece of equipment, or system.

human-factors engineering Also called **ergonomics**, or **human engineering**. The science dealing with the design and creation of devices, equipment, and environments which factor in the needs of its human users. That which is ergonomic, for instance, seeks to maximize comfort, posture, safety, and ease of use. Its abbreviation is **HFE**.

human interface The interface between a user and a device, piece of equipment, machine, or system. When interacting with a computer system, for instance, user interfaces include keyboards, microphones, cameras, and so on. Also called **user interface**.

human interface device A device providing an interface between a user and another device, piece of equipment, machine, or system. For example, a human-computer interface. Its abbreviation is **HID**. Also called **user interface device**.

human-machine interface The interface, such as controls, between a user and a machine. Also called **user-machine interface**.

humanoid robot A robot made to physically resemble a person. It may incorporate sensors to detect visual, tactile, and auditory stimuli. Most robots utilized in science and industry do not resemble people. Also called **android**.

humbucking The reduction or elimination of hum through the use, for instance, of a **humbucking coil**. Also spelled **hum bucking**.

humbucking coil In a dynamic speaker, an additional coil wound around the field coil, and connected in series opposition with the voice coil, to cancel any effects due to hum. Also spelled **hum bucking coil**.

humidistat An instrument or device which measures and controls relative humidity. It may consist, for instance, of a switch that is actuated when the surrounding humidity reaches a given level. A humidistat may also indicate readings of relative humidity. Also called **hygrostat**.

humidity A measure of the water vapor content in air. Two common manners of expressing this are absolute humidity and relative humidity.

humidity-controlled environment An enclosure, such as a room, in which measures are taken to provide an environment that maintains a specified humidity level. These enclosures may also provide a stable temperature level, or protect against dust, and so on. Such environments may be used, for instance, to protect sensitive electronic equipment.

humidity detector A device which detects and indicates the presence of atmospheric humidity.

humidity meter Same as **hygrometer**.

humidity sensor A sensor which detects and measures atmospheric humidity. In it, an electrical quantity such as resistance or capacitance varies along with the surrounding humidity, and an output voltage or current corresponding to these fluctuations is produced.

hunting A condition in which there are oscillations or other fluctuations about a given desired value, as may occur in an automatic control system when the controlled variable swings back and forth around the intended value. Hunting may be due to any of various factors, such as overcompensation, or inherent instabilities in a given system.

Huygens' principle A principle stating that each point in a wavefront may itself be a source of secondary waves, called Huygens' wavelets, which propagate in all directions. The wave as whole propagates due to this combined effect. This principle, for instance, helps explain diffraction.

Huygens' wavelets Secondary waves occurring according to **Huygens' principle**.

HV Abbreviation of **high voltage**.

hW Abbreviation of **hectowatt**.

HW Abbreviation of **hardware**.

hybrid **1.** Indicative of something that results from a combination of two or more dissimilar components working together. For example, a hybrid circuit. **2.** Same as **hybrid junction**.

hybrid cable A cable which combines two different types of media. For example, a cable with optical fibers and twisted wires.

hybrid circuit **1.** A circuit incorporating two or more different types of components, each with similar functions, working together. For example, a circuit utilizing both transistors and tubes. **2.** Same as **hybrid IC**. **3.** In telecommunications, a circuit which matches a two-wire circuit to a four-wire circuit. Also called **balancing network (3)**.

hybrid coil Same as **hybrid transformer**.

hybrid computer A computer which processes both analog and digital data.

hybrid fiber-coax Abbreviation of **hybrid fiber coaxial**.

hybrid fiber-coax network Same as **hybrid fiber-coaxial network**.

hybrid fiber-coaxial A communications network, or a part of one, which incorporates both optical fiber and coaxial cable. An example is a hybrid fiber coaxial network. Its abbreviation is **HFC**.

hybrid fiber-coaxial network A communications network combining fiber-optic and coaxial cables, and which carries broadband content. A cable TV provider, for instance, might use the fiber-optic portion as a high-speed backbone, and the coax portion to run from the headend to the individual subscribers. Its abbreviation is **HFC network**. Also called **hybrid fiber-coax**.

hybrid IC Abbreviation of **hybrid** integrated circuit. Its own abbreviation is **HIC**. Also called **hybrid circuit (2)**. **1.** An IC composed of parts created utilizing two or more formation techniques, such as thin-film, monolithic, or diffusion. **2.** A circuit which combines one or more ICs with one or more discrete components on the same substrate.

hybrid integrated circuit Same as **hybrid IC**.

hybrid junction A waveguide, resistor, or transformer circuit or device which has four branches which are arranged in a manner that the input signal is equally divided between the adjacent branches. The signal does not reach the fourth, or opposite, branch. Also called **hybrid (2)**, **hybrid tee**, **bridge hybrid**, or **magic tee**.

hybrid microcircuit A microcircuit composed of parts created utilizing two or more formation techniques, such as thin-film, monolithic, or diffusion.

hybrid network A network utilizing two or more topologies. For instance, that combining star and ring topologies.

hybrid parameters Four transistor equivalent-circuit specifications which describe its performance under specified conditions. Its abbreviation is **h-parameters**.

hybrid ring A **hybrid junction** molded into the shape of a ring. Used, for instance, as a high-power duplexer. Also called **hybrid ring junction**, **ring junction**, or **rat race**.

hybrid ring junction Same as **hybrid ring**.

hybrid SACD An SACD player that reproduces SACD discs in addition to standard CDs. A hybrid SACD disc has two layers, with the CD layer located beneath the SACD layer. The SACD layer is transparent to a CD laser, thus can be played in a standard CD player, while a SACD player does not pick up the CD layer, providing for mutual compatibility. Also called **SACD hybrid**.

hybrid set A **hybrid junction** incorporating two or more transformers. Also called **transformer hybrid**.

hybrid-T Same as **hybrid junction**. Also spelled **hybrid-tee**.

hybrid tee Same as **hybrid junction**. Also spelled **hybrid-T**.

hybrid transformer Also called **hybrid coil**. **1.** A **hybrid junction** incorporating a single transformer. **2.** A special kind of transformer used in wire-transmission systems to separate the two directions of transmission. Used, for instance, to convert a four-wire telephone circuit into a two-wire telephone circuit.

hydroacoustic Pertaining to **hydroacoustics**.

hydroacoustics The study of sound energy and phenomena under water surfaces. It may deal, for instance, with the propagation of sound under water. Although this figure varies depending on the specific conditions, sound travels at approximately 1500 m/s in ocean water. An application of hydroacoustics is sonar. Also called **underwater acoustics**.

hydrodynamic Pertaining to **hydrodynamics**.

hydrodynamics The study of fluids in motion, and the interactions between such a fluid and its boundaries.

hydroelectric Pertaining to, generating, or derived from **hydroelectricity** or **hydroelectric power**.

hydroelectric generating station Same as **hydroelectric power station**.

hydroelectric generator A generator which converts the energy contained in flowing or falling water into electrical energy. Also called **hydropower generator**.

hydroelectric power Electrical power generated through the conversion of energy contained in flowing or falling water. For example, a waterfall can be utilized to drive a water turbine coupled to a generator. Also called **hydroelectricity**, or **hydropower**.

hydroelectric power plant Same as **hydroelectric power station**.

hydroelectric power station A location which combines the structures, devices, and equipment necessary to produce **hydroelectric power** in quantities adequate for residential, commercial, or industrial use. Also called **hydroelectric generating station, hydroelectric power plant, hydropower plant**, or **hydropower station**.

hydroelectricity Same as **hydroelectric power**.

hydrogen A chemical element whose atomic number is 1. It is the lightest element, and is a colorless diatomic gas. Hydrogen has 3 isotopes, each with its own name: protium, deuterium, and tritium, with tritium being radioactive. Hydrogen is present in countless chemical compounds, and is the most abundant element in the known universe. It has various applications, including its use as a fuel, in welding, low-pressure research, low-temperature research, and in the production of high-purity metals.

hydrogen atom A single atomic particle of **hydrogen**.

hydrogen bromide A colorless, corrosive, and poisonous gas whose chemical formula is HBr. It is used in plasma etching.

hydrogen chloride A colorless, corrosive, and poisonous gas whose chemical formula is HCl. It is used in plasma etching, and in lasers.

hydrogen electrode Also called **hydrogen half cell**. **1.** A standard reference electrode usually consisting of a solution containing hydrogen atoms at a unit concentration, into which a platinum electrode is immersed, and over which hydrogen gas is passed at a specific pressure. This electrode

has an arbitrarily assigned potential of zero, and other electrode potentials are compared to it. Also called **standard hydrogen electrode**. **2.** An electrode consisting of a solution containing hydrogen atoms, into which an electrode of a noble metal is immersed, and over which hydrogen gas is passed. The noble metal usually used is platinum.

hydrogen equivalent **1.** For an acid, the number of replaceable hydrogen atoms. **2.** For a base, the number of replaceable hydroxyl groups.

hydrogen fluoride A colorless, corrosive, and poisonous gas whose chemical formula is HF. It is used in plasma etching, and in lasers.

hydrogen half cell Same as **hydrogen electrode**.

hydrogen ion A hydrogen atom which has been stripped of its electron. In the case of the protium isotope, which is by far the most common, this means leaving just a proton. Its symbol is H^+.

hydrogen-ion concentration A measure of the acidity or alkalinity of an aqueous solution. It is the amount of hydrogen ion, or H^+, per unit volume in an aqueous solution at a given moment. pH is related to hydrogen-ion concentration by the following formula: $pH = \log_{10}(1/H^+)$.

hydrogen lamp A discharge lamp in which ionized hydrogen gas produces light. Used, for instance, in spectroscopy.

hydrogen laser A gas laser in which hydrogen is utilized to produce a beam of coherent light.

hydrogen maser A gas maser in which hydrogen is stimulated to amplify microwave radiation. The output of such a maser is characterized by great stability and may be used, for instance, as a frequency standard.

hydrogen-oxygen fuel cell A fuel cell in which hydrogen is the fuel and oxygen is the oxidizing agent. The products of such a cell are usually electricity, heat, and water.

hydrogen peroxide A colorless liquid whose chemical formula is H_2O_2. It is used in electroplating and etching.

hydrogen phosphide A colorless, poisonous, and spontaneously flammable gas whose chemical formula is H_3P. It is used in semiconductors. Also called **phosphine**.

hydrogen selenide A colorless and poisonous gas whose chemical formula is H_2Se. It is used in semiconductors.

hydrogen thyratron A thyratron containing hydrogen gas, as opposed to mercury gas, which reduces the effects which changes in the surrounding temperature might have.

hydrolysis A chemical reaction in which water reacts with another substance, and in which the latter is broken into two or more constituents. For example, sodium chloride (NaCl) dissolving in water to produce Na^+ and Cl^- ions.

hydrolytic Pertaining to **hydrolysis**.

hydromagnetic Pertaining to, or utilizing **hydromagnetics**. Also called **magnetohydrodynamics**.

hydromagnetic generator A device which utilizes electrically conducting fluids interacting with a magnetic field to generate electrical energy. For instance, such a device may extract kinetic energy from a jet of plasma to produce electricity. Also called **magnetohydrodynamic generator**.

hydromagnetics The study of the interactions between an electrically conducting fluid and magnetic fields. Plasmas, liquid metals, and ionized gases are examples of such fluids. Hydromagnetics has applications in various fields, including plasma confinement, liquid-metal cooling of nuclear reactors, and in astronomy. Also called **magnetohydrodynamics**, or **magneto-fluid dynamics**. When dealing specifically with electrically conducting gases, also called **magnetogasdynamics**, and when dealing with plasmas, also called **magnetoplasmadynamics**.

hydrometer An instrument which indicates the specific gravity, density, or a similar characteristic of a liquid. A hydrometer is usually simply placed in the liquid in question, and the depth it sinks is to correlated to its vertical scale.

hydronium ion An ion whose chemical formula is H_3O^+. It is a molecule of water to which a proton is added.

hydrophone A transducer which converts underwater sound energy into corresponding electrical signals. Used, for instance, in sonar. Also called **underwater microphone**.

hydropower Same as **hydroelectric power**. Also called **water power**.

hydropower generating station Same as **hydroelectric generating station**.

hydropower generator Same as **hydroelectric generator**.

hydropower plant Same as **hydroelectric power plant**.

hydropower station Same as **hydroelectric power station**.

hydroxide ion A negatively charged ion whose chemical formula is OH^-.

hydroxyl group A chemical unit whose formula is OH. Such a unit is part of many compounds, such as phenol.

hygristor A resistor whose resistance varies proportionally to the humidity surrounding it. Used, for instance, to measure relative humidity.

hygrogram The graphic record made by a **hygrometer**.

hygrograph A **hygrometer** which graphically records its measurements.

hygrometer An instrument for measuring humidity in the atmosphere, especially relative humidity. Examples include electric hygrometers, and hair hygrometers. Also called **humidity meter**.

hygrometric 1. Pertaining to relative humidity and its measurement. 2. Pertaining to devices and instruments which measure and/or indicate relative humidity, or whose function depends on it.

hygroscopic 1. Pertaining to a substance or material which absorbs or attracts moisture. 2. Pertaining to a substance or material whose properties change in the presence of moisture.

hygrostat Same as **humidistat**.

hygrothermograph In instrument which measures temperature and relative humidity, and records them graphically on the same chart. Used, for instance, to monitor these variables in a museum or library.

hyperacoustic zone A region in the atmosphere, at an altitude of approximately 100 to 160 kilometers, where the distance between air molecules is roughly the same as that of the wavelengths of sound, resulting in sound transmission with reduced amplitude. At altitudes above this zone, sound is no longer transmitted.

hyperbola A two-part conic section in which a plane intersects with a double cone. The plane is parallel to the axis of the double cone, and all points of its two branches have a constant difference in distance from two fixed points, called focuses.

hyperbolic Pertaining to, or in the shape of a **hyperbola**.

hyperbolic logarithm A logarithm whose base is e, or approximately 2.71828. Also called **natural logarithm**, or **Napierian logarithm**.

hyperbolic navigation Navigation assisted by a **hyperbolic navigation system**. Also called **hyperbolic radionavigation**.

hyperbolic navigation system A navigation system, such as loran or Decca, in which a ship or aircraft receives synchronized signals from at least two known locations at any given moment. The charting of the time delay with reference to each transmitter yields a hyperbola whose focuses are these transmitting stations. A second pair of stations can be used to draw another hyperbola, and the position of a ship or aircraft is the point on the map where the two curves intersect. Also called **hyperbolic radionavigation system**.

hyperbolic radionavigation Same as **hyperbolic navigation**.

hyperbolic radionavigation system Same as **hyperbolic navigation system**.

HyperCard Authoring software providing a set of tools for developing multimedia presentations, applications, and the like.

hypercardioid microphone A directional microphone whose lateral attenuation is greater than that of an ordinary cardioid microphone, but which has less rear attenuation. Also called **supercardioid microphone**.

hypercube 1. A network architecture with 2^n nodes. For example, a communications network with 2^2 nodes has 4 interconnected nodes. All interconnections are made so as to minimize data travel. Also called **hypercube network**. 2. A parallel processing architecture with 2^n computers. For example, that utilizing 2^4, or 16 interconnected computers. All interconnections are made so as to minimize data travel.

hypercube network Same as **hypercube (1)**.

hyperfrequency waves Electromagnetic waves whose wavelengths range from one centimeter to one meter, corresponding to frequencies of between 30 and 0.3 GHz, respectively. This interval is contained in the microwave range.

hyperlink An element within an electronic document, such as that displayed on a computer monitor, which links to other elements in different locations in the same or different documents. A hyperlink may be displayed as text, an icon, an image, or the like, and is usually accessed by clicking on it with a mouse or its equivalent. When viewing a Web page, for instance, such text, or hypertext, often appears underlined. Also, to create such an element. Also called **link (3)**.

hypermedia The use of multimedia tools to interconnect between sources of information. Hypermedia is one of the principal ways to navigate the Internet. For example, when accessing a Web page, when the mouse cursor forms a hand over certain portions of text or graphics, it is usually indicate the presence of a **hyperlink**. When referring specifically to text, known as **hypertext**.

hyperon A subatomic particle which is heavier than a nucleon.

hypersonic 1. Pertaining to velocities equal to, or greater than, five times the speed of sound through a given medium. For example, if a jet's velocity exceeds five times that of the speed of sound in air, then it is in hypersonic flight. 2. Pertaining to vibrations at or above a given frequency, such as 500 MHz or 1 GHz.

hypersonic speed A speed which is **hypersonic (1)**.

hypersonics 1. The science dealing with velocities equal to, or greater than, five times the speed of sound through a given medium. 2. The science dealing with the production, study, and utilization of vibrations at or above a given frequency, such as 500 MHz or 1 GHz.

hyperspace The universe of interconnected **hypermedia**. It includes the entire range of information and entertainment resources available through the links so created, especially when used over the Internet.

HyperTalk The object-oriented scripting language utilized in **HyperCard**.

hypertext Text which is connected through **hyperlinks**. When viewing a Web page, for example, hypertext usually appears underlined. Hypertext serves, for instance, to make it very simple to navigate between multiple sources when accessing information on a given topic, by simply clicking on the hyperlinked words and/or phrases of interest.

HyperText Markup Language Same as **HTML**.

HyperText Markup Language code Same as **HTML code**.

HyperText Markup Language document Same as **HTML document**.

HyperText Markup Language editor Same as **HTML editor**.

HyperText Markup Language page Same as **HTML page**.

HyperText Markup Language source Same as **HTML source**.

HyperText Markup Language source code Same as **HTML code**.

HyperText Markup Language tag Same as **HTML tag**.

HyperText Transport Protocol Same as **HTTP**.

HyperText Transport Protocol-Next Generation Same as **HTTP-NG**.

HyperText Transport Protocol proxy Same as **HTTP proxy**.

HyperText Transport Protocol proxy server Same as **HTTP proxy server**.

HyperText Transport Protocol-Secure Same as **HTTPS**.

HyperText Transport Protocol Server Same as **HTTP Server**.

hypervelocity A velocity which exceeds a given amount, such as approximately 1000, 1500, or 2500 meters per second.

hypotenuse In a right triangle, the side opposite to the right angle.

hypothesis A theory, proposition, or concept which can be experimentally evaluated so as to be proved or disproved.

hypsometer An instrument or device which utilizes the boiling point of a liquid, such as water, to calibrate thermometers. A hypsometer may also be used to determine height above sea level, as the boiling point of a liquid depends on atmospheric pressure, and the latter varies according to altitude.

hysteresigraph Same as **hysteresisgraph**.

hysteresis 1. A phenomenon in which an induced or observed effect remains after the inducing cause is removed. In this way, two physical quantities are related in a manner that depends on whether one is increasing or decreasing with respect to the other. There are various examples, including magnetic hysteresis, electric hysteresis, and thermal hysteresis. 2. A phenomenon in which the changes in the magnetization induced in a ferromagnetic material lag behind the changes in the magnetizing force. This phenomenon is observed below the Curie point. Also called **magnetic hysteresis**. 3. Any phenomenon in which there is a lag between the cause and the induced or observed effect. For instance, the output of an instrument depending on whether the input resulted from an increase or a decrease from the previous value. Another example is the manner in which a thermostat utilizes the right amount of hysteresis to maintain the desired temperature. Too little hysteresis would result in constant switching between cycles, and too much would allow too large a temperature interval between cycles. 4. In an os-

cillator, an effect in which a given value of a functional parameter may result in multiple values of output power and/or frequency.

hysteresis coefficient A constant, unique to each material, which is utilized when calculating **hysteresis losses**. Also called **hysteresis constant, hysteresis factor**, or **Steinmetz coefficient**.

hysteresis constant Same as **hysteresis coefficient**.

hysteresis curve Same as **hysteresis loop**.

hysteresis distortion Any form of distortion due to **hysteresis**, especially magnetic hysteresis. May be manifested, for instance, in speakers.

hysteresis error For a measuring and/or indicating instrument, a difference in output signals depending on whether the measured variable is increasing or decreasing. Also, the extent of this difference. Also, the maximum observed difference.

hysteresis factor Same as **hysteresis coefficient**.

hysteresis heating Heating resulting from **hysteresis losses**.

hysteresis loop Also called **hysteresis curve**. 1. For a given **hysteretic material**, a closed curve that represents the effects of hysteresis. Rectangular coordinates are usually used, and an example is the curve obtained by plotting magnetization versus magnetic field strength, through a complete magnetizing cycle. 2. Any closed curve representing the effects of hysteresis, be it electric, thermal, and so on.

hysteresis loss Same as **hysteresis losses**.

hysteresis losses Also called **hysteresis loss, hysteretic losses**, or **hysteretic loss**. 1. Power losses, usually in the form of heat, due to **hysteresis**. 2. The energy dissipated, usually in the form of heat, in each complete magnetizing cycle of a **hysteretic material**. It is equal to the area enclosed by the hysteresis loop.

hysteresis meter An instrument which measures and indicates **hysteresis** or **hysteresis losses**.

hysteresis motor A small synchronous motor which utilizes hysteresis and eddy-current losses for starting. Such motors do not require DC excitation, and are used, for instance, in constant-speed applications. An example is that used in a tape deck.

hysteresisgraph An instrument which automatically measures and draws **hysteresis loops**. Also spelled **hysteresigraph**.

hysteretic Pertaining to, arising from, or exhibiting **hysteresis**.

hysteretic loss Same as **hysteresis losses**.

hysteretic losses Same as **hysteresis losses**.

hysteretic material A material, such as a as a ferromagnetic material, exhibiting **hysteresis**.

Hz Abbreviation of **hertz**.

i A mathematical unit equal to the square root of -1. That is, $i^2 = -1$. It is used in complex numbers. In electronics, the symbol utilized to express the square root of -1 is usually *j*.

I **1.** Chemical symbol for **iodine**. **2.** Abbreviation of **input**.

I Symbol for **electric current**, or **current**.

I-beam On a computer screen, a cursor shaped similarly to a capital letter **I**. It is seen, for instance, when a mouse is moved over an area where text is to be entered. Also called **I-beam pointer**.

I-beam pointer Same as **I-beam**.

I-CASE Abbreviation of integrated computer-aided software engineering. The combining of computer-aided software engineering tools in a software package.

I channel In the NTSC color TV system, the channel utilized to transmit cyan-orange color information. It is approximately 1.5 MHz wide. The **Q channel** transmits green-magenta color information.

I display In radars, a display where the scanned object appears as a complete circle when the antenna is aimed directly at it. The radius of the circle is proportional to the distance to the scanned object. When the antenna is not exactly pointed at the scanned object, the circle reduces to a segment of a circle. Also called **I scan**, **I scanner**, **I scope**, or **I indicator**.

I indicator Same as **I display**.

I/O Abbreviation of **input/output**. **1.** The transfer of data between a CPU and a peripheral. Such peripherals include those that serve mostly for input to the CPU, such as keyboards, pointing devices, digitizing tablets, scanners, and microphones, and peripherals which serve mostly for output, including monitors, printers, speakers, and so on. Disk drives, for instance, may engage in frequent transfers in both directions. For any input process there is a corresponding output process, and vice versa. For instance, the output of a CPU is always the input of a peripheral. **2.** The transfer of data between a computer or computer-controlled system, and a peripheral. **3.** Pertaining to any device, piece of equipment, system, method, or media utilized for **I/O** (1), or **I/O** (2). **4.** Pertaining to both the input and output of a component, circuit, device, piece of equipment, or system.

I/O address Abbreviation of input/output **address**. An address which identifies a peripheral. If an I/O address is shared by two devices, conflicts will occur.

I/O area Abbreviation of input/output **area**. Same as **I/O buffer**.

I/O bound Abbreviation of input/output **bound**. Same as **I/O limited**.

I/O buffer Abbreviation of input/output **buffer**. An area of computer memory reserved for **I/O data**. Also called **I/O area**, or **peripheral buffer**.

I/O bus Abbreviation of input/output **bus**. A set of conductors which serve as one or more **I/O channels**. An example is a local bus connecting the CPU and high-speed peripherals.

I/O card Abbreviation of input/output **card**. A circuit board which is plugged into an **I/O bus**, enabling the connection of peripherals to a computer.

I/O channel Abbreviation of input/output **channel**. A path along which **I/O data** is transmitted. Also called **data channel** (3).

I/O control Abbreviation of input/output **control**. The control of one or more **I/O channels**. The tasks pertaining to this are performed by an I/O controller.

I/O controller Abbreviation of input/output **controller**. A chip, circuit board, or device which controls one or more **I/O channels**. A disk controller, for example, incorporates these functions.

I/O data Abbreviation of input/output **data**. Data transferred to the CPU from a peripheral, or vice versa. Also, data which awaits to be so transferred.

I/O device Abbreviation of input/output **device**. Any peripheral which transfers data to a CPU and/or receives data from a CPU. Such devices include keyboards, pointing devices, digitizing tablets, scanners, disk drives, microphones, monitors, printers, and speakers. Also, any external device which otherwise assists, enhances, or depends on computer function. Also called **I/O unit**, or **peripheral** (1).

I/O handler Abbreviation of input/output **handler**. A hardware device or program which controls of one or more **I/O channels**. An operating system, for instance, usually incorporates such a program.

I/O instruction Abbreviation of input/output **instruction**. An instruction resulting in the transfer of **I/O data**.

I/O intensive Abbreviation of input/output **intensive**. An application which requires large amounts of data transfers between the CPU and peripherals. If the speed at which data is exchanged with peripherals is not sufficient, processing becomes I/O limited.

I/O interface Abbreviation of input/output **interface**. An interface, such as a port or expansion slot, which serves for the transfer of data between a CPU and a peripheral.

I/O interrupt Abbreviation of input/output **interrupt**. An interrupt generated by an **I/O device**. Used, for instance, to request I/O data.

I/O limited Abbreviation of input/output **limited**. A state in which the speed of processing exceeds the rate at which input to and output from the CPU can occur. Thus, the computer can proceed only as fast as the **I/O** (1) allows. Also called **I/O bound**.

I/O operation Abbreviation of input/output **operation**. Any operations, such as instructions, transfers, or error-checking, involving **I/O** (1). Also called **I/O process**.

I/O port Abbreviation of input/output **port**. A port, such as a USB port, which serves for the transfer of data between a CPU and a peripheral. Also called **input port** (1), **output port** (1), or **port** (4).

I/O process Abbreviation of input/output **process**. Same as **I/O operation**.

I/O processor Abbreviation of input/output **processor**. A processor which is dedicated to I/O operations, thus freeing resources of the CPU. Such a processor is especially useful, for instance, when dealing with I/O intensive applications.

I/O statement Abbreviation of input/output **statement**. A program statement requesting an **I/O transfer**.

I/O switching Abbreviation of input/output **switching**. A technique utilized in the allocation of **I/O channels**, in which available channels are distributed among the devices which are exchanging data with the CPU. In this manner, for example, a given peripheral may utilize more than one channel.

I/O time Abbreviation of input/output **time**. The time required for one or more **I/O transfers**.

I/O transfer Abbreviation of input/output **transfer**. A data transfer between a CPU and a peripheral, or to or from an I/O buffer.

I/O unit Abbreviation of input/output **unit**. Same as **I/O device**.

I scan Same as **I display**.

I scanner Same as **I display**.

I scope Same as **I display**.

I signal Abbreviation of in-phase **signal**. In the NTSC system, one of the two color-difference signals. The other is the **Q signal**.

I-type semiconductor Same as **intrinsic semiconductor**.

I2 Abbreviation of **Internet2**.

I²L Abbreviation of **integrated-injection logic**.

I2O Abbreviation of **intelligent input/output**, or **intelligent I/O**. An I/O interface which incorporates an auxiliary processor whose architecture is independent of the device driver and host computer operating system. This provides for more efficient transfers of data, and since most tasks are performed independently, it allows the CPU to tend to other processing needs.

I²O Same as **I2O**.

I²R loss Same as **I²R losses**.

I²R losses A power loss in copper wires, cables, or windings, due to the resistance of copper conductors. Also known as **I²R loss**, or **copper losses**.

IAB Abbreviation of **Internet Architecture Board**.

IAD Abbreviation of **Integrated Access Device**.

IAGC Abbreviation of **instantaneous automatic gain control**.

IANA Abbreviation of **Internet Assigned Numbers Authority**.

IAP Abbreviation of **Internet access provider**.

IAVC Abbreviation of **instantaneous automatic volume control**.

IBA Abbreviation of **ion-beam analysis**.

IBG Abbreviation of **inter-block gap**.

iBook A popular line of notebook computers.

IC Abbreviation of **integrated circuit**. A small piece of semiconductor material upon which miniature electronic circuit components, such as transistors or resistors, are placed. An IC may have hundreds of millions of such components. ICs may be classified by the number of electronic components contained. For example, an IC with 1,000,000 or more components is labeled as having ultra-large scale integration. Another way to classify ICs is by construction, as in monolithic IC, thin-film IC, or hybrid IC. Also called **chip (1)**, **microchip**, or **microcircuit**.

IC card Abbreviation of **integrated circuit card**. Also called **smart card**, or **chip card**. **1.** A card which incorporates one or more chips, including a processor and memory. Such cards are usually the size of a credit card, and are used in conjunction with IC card readers which communicate with central computers. Such cards may be used to verify identity, keep digital cash, store medical records, and so on. IC cards are secure, and update the contained information each time it is used. **2.** In portable computers, a **chip card (1)** utilized to add memory, or as a peripheral device such as a modem.

IC card reader Abbreviation of **integrated circuit card reader**. A device into which **IC cards** are inserted, to access and update their contained information. Also called **smart card reader**, or **chip card reader**.

IC housing Same as **IC package**.

IC memory Abbreviation of **integrated circuit memory**. A chip which consists of memory cells and the associated circuits necessary, such as those for addressing, to provide a computer, or other device with memory or storage. Examples include RAM, SRAM, ROM, EEPROM, and flash-memory chips. Also called **memory chip**, **memory IC**, or **chip memory**.

IC package Abbreviation of **integrated circuit package**. The housing, usually made of ceramic or plastic, surrounding an IC. The leads or metallic surfaces of the IC package are used for connection via plugging-in or soldering. IC packages are usually mounted onto printed-circuit boards. Also called **IC housing**, **chip package**, or **package (3)**.

IC tester Abbreviation of **integrated circuit tester**. An instrument which serves to evaluate the function of ICs. Such an instrument, for instance, may be operated automatically for inspection during manufacturing, or manually for research and development purposes. Also called **chip tester**.

IC testing Abbreviation of **integrated circuit testing**. Tests performed on ICs, such as those utilizing an **IC tester**.

ICANN Abbreviation of **Internet Corporation for Assigned Names and Numbers**.

ICE 1. Acronym for **in-circuit emulator**. **2.** Abbreviation of **Information and Content Exchange**.

ice load Also called **ice loading**. **1.** The maximum amount of ice that can safely accumulate on an antenna, or on an overhead wire in a power supply system. May be expressed, for instance in centimeters or kilograms. **2.** The additional stress the presence of ice has on an antenna or an overhead wire in a power supply system.

ice loading Same as **ice load**.

ice proof 1. Impervious to ice. **2.** Impervious to ice up to a given thickness or weight.

Iceland spar A form of calcite utilized for polarizing light.

ICMP Abbreviation of **Internet Control Message Protocol**.

icon On a computer screen, a small displayed image which serves to represent something else, such as a file, program, disk drive, function, and so on. Icons are used in GUIs, and are usually accessed, moved, or otherwise manipulated by using a pointing device such as a mouse.

iconize In a GUI, to hide an application window currently displayed on the screen. The program is not terminated, and is then represented by an icon. Also called **minimize**.

iconoscope An early form of TV camera tube which employs a high-velocity electron beam to scan a photoemissive mosaic screen which stores electrical charge patterns corresponding to the image focused upon said mosaic. Also called **storage camera**.

ICQ A popular instant messaging program. It is an abbreviation of **I seek you**.

ICR Abbreviation of **intelligent character recognition**.

ICRA Abbreviation of **Internet Content Rating Association**.

ICS Abbreviation of **intercom system**.

ICW Abbreviation of **interrupted continuous wave**.

ID 1. Abbreviation of **identification**. **2.** Abbreviation of **inside diameter**.

IDE 1. Abbreviation of **Integrated Drive Electronics**, **Integrated Drive Electronics interface**, or **Intelligent Drive Electronics**. A widely utilized disks and tape drive interface in which the controller is integrated into the disk, disc, or tape drive itself, which simplifies connection. There are enhanced versions, such as Fast IDE. Also called **IDE interface**. The American National Standards Institute formally calls it **advanced technology attachment**, or **AT attachment**. **2.** Abbreviation of **integrated development environment**.

IDE controller The controller built into an **IDE drive**.

IDE drive A disk, disc, or tape drive utilizing an **IDE interface**.

IDE hard drive A hard drive utilizing an **IDE interface**.

IDE interface Same as **IDE (1)**.

IDE-RAID The use of multiple **IDE hard drives** in a RAID. Also called **ATA-RAID**.

IDEA Abbreviation of **International Data Encryption Algorithm**.

ideal Pertaining to a component, circuit, device, piece of equipment, system, process, operation, result, state, substance, or entity which corresponds to a theoretical perfection, or that conforms completely to a theory pertaining to a theoretical perfection. For example, an ideal radiator, or an engine which converts 100% of its input power or energy into useful power or energy, such as work. Also called **perfect (1)**.

ideal capacitor A capacitor with characteristics that correspond to a theoretical perfection. Such a capacitor, for instance, would hold a charge forever, or until discharged. Also called **perfect capacitor**.

ideal component A component with characteristics that correspond to a theoretical perfection. An example is an ideal capacitor. Also called **perfect component**.

ideal coupling A degree of coupling between two systems, especially circuits, corresponding to a theoretical perfection. That is, the coefficient of coupling would be 1, or 100%. Also called **perfect coupling**, or **unit coupling**.

ideal crystal A crystal whose contained atoms, molecules, or ions have a geometric arrangement which is free from lattice defects. Also called **perfect crystal**.

ideal device A device with characteristics that correspond to a theoretical perfection. An example is an engine which converts 100% of its input power or energy into useful power or energy, such as work. Also called **perfect device**.

ideal dielectric A dielectric through which an electric field may be applied with no energy losses. A vacuum is an example, and its dielectric constant is defined as 1. Also called **perfect dielectric**.

ideal inductor An inductor with characteristics that correspond to a theoretical perfection. Such an inductor, for instance, would have zero losses. Also called **perfect inductor**.

ideal radiator An ideal body which would be a perfect emitter of radiant energy, its distribution of energy depending solely on its absolute temperature. It would also absorb all the radiant energy incident upon it. Also called **blackbody**, **perfect radiator**, or **full radiator**.

ideal transducer A transducer with characteristics that correspond to a theoretical perfection. Such an transducer, for instance, would have zero losses. Also called **perfect transducer**.

ideal transformer A transformer with characteristics that correspond to a theoretical perfection. Such a transformer, for instance, would have a coupling coefficient of 1. Also called **perfect transformer**.

identification A process or technique which enables the recognition of an individual, entity, process, and so on. Examples include digital identification, caller identification, and face recognition. When referring to the evidence provided for identification, its abbreviation is **ID**.

identification beacon A guiding, orienting, or warning beacon which serves to unequivocally identify a specific geographic location.

identifier In computers, one or more characters which serve to identify a variable, data element, record, file, program, storage location, or the like.

identity authentication In computers and communications, the process of verifying the authenticity of an individual or entity. Measures such as passwords, digital signatures, and voice recognition are employed. Also called **identity validation**, or **identity confirmation**.

identity confirmation Same as **identity authentication**.

identity validation Same as **identity authentication**.

idiochromatic A crystal or mineral whose color is a result of its pure chemical composition, as opposed to those which are **allochromatic**, whose color is due to the minute presence of an impurity.

idiochromatic mineral A mineral whose color is a result of its pure chemical composition.

idiot-proof An unfriendly term which refers to that which is especially user-friendly.

idle **1.** Not in use and/or not busy. Said, for instance, of a communications circuit. **2.** To function at a at given operational minimum, such as a motor running without a load. **3.** To be functioning under normal operating conditions, but not having an applied signal at the time. Said, for instance, of an amplifier stage during an interval in which there is no input signal. Also called **quiescent (2)**.

idle channel **1.** A communications channel not in use. **2.** An assigned communications channel not in use at a given moment.

idle channel noise The noise present in a communications channel in the absence of a signal.

idle circuit **1.** A communications circuit not in use. **2.** An assigned communications circuit not in use at a given moment.

idle component In an AC circuit, the component of the current, voltage, or power which does not add power. These are, specifically, the idle current, idle voltage, or idle power. Also called **reactive component**, **wattless component**, or **quadrature component**.

idle current The component of an alternating current which is in quadrature with the voltage. Such a component does not add power. Also called **reactive current**, **wattless current**, or **quadrature current**.

idle period A period during which a device, piece of equipment, or system is operational and available, but is not in use.

idle power The power in an AC circuit which cannot perform work. It is calculated by the following formula: $P = I \cdot V \cdot \sin\theta$, where P is the power in watts, I is the current in amperes, V is the voltage in volts, and $\sin\theta$ is the sine of the angular phase difference between the current and the voltage. Also called **reactive power**, **wattless power**, **volt-amperes reactive**, **reactive volt-amperes**, or **quadrature power**.

idle state Also called **resting state**, or **quiescent state**. **1.** A state in which a component, circuit, device, piece of equipment, or system is functioning under normal operating conditions, but does not have an applied signal at a given time. Also called **standby (1)**. **2.** A state in which a device, piece of equipment, or system is operational and available, but is not in use. Also called **standby (2)**.

idle time A time interval during which a device, piece of equipment, or system is operational and available, but is not in use.

idle voltage The voltage component which is in quadrature with the current of an AC circuit. Such a component does not add power. Also called **reactive voltage**, **wattless voltage**, or **quadrature voltage**.

idler frequency In a parametric amplifier, the sum or difference between the signal frequency and the pump frequency.

idler wheel In sound recording or reproducing equipment, a rubber wheel utilized to transfer power through friction. For example, in a tape deck transport, a wheel which traps the tape between itself and the capstan, so that the latter can move the tape at a constant speed.

idling **1.** To be in an **idle (2)** state. **2.** To be in an **idle (3)** state.

idling current 1. The current flowing through a device which is **idling** (1). **2.** The current flowing through a device which is **idling** (2). Also called **quiescent current**.

idling power 1. The power consumed by a device which is **idling** (1). **2.** The power consumed by a device which is **idling** (2). Also called **quiescent power**.

idling voltage 1. The voltage required by a device which is **idling** (1). **2.** The voltage required by a device which is **idling** (2). Also called **quiescent voltage**.

IDN Abbreviation of **integrated digital network**.

IDS Abbreviation of **intrusion detection system**.

IDSL Abbreviation of ISDN digital subscriber line. A digital subscriber line utilizing ISDN facilities. It has upload speeds up to 144 Kbps, or 128 Kbps with compression, using a single pair of copper wires.

IDT Abbreviation of **interdigital transducer**.

IDTV Abbreviation of **improved definition TV**.

IE 1. Abbreviation of **information engineering**. **2.** Abbreviation of **Internet Explorer**.

IEC Abbreviation of **International Electrotechnical Commission**.

IEEE Abbreviation of Institute of Electrical and Electronics Engineers. An organization whose functions include developing and setting electrical, electronics, communications, and computers standards. The IEEE has several hundred thousand members in over 150 countries, and publishes a considerable portion of the world's literature in areas such as electronics and computing.

IEEE 488 bus A computer interface used mainly to connect data acquisition and control instruments, industrial automation equipment, medical diagnostic devices, and the like. Also called **General-Purpose Interface Bus**.

IEEE 488 standard A standard defining the **IEEE 488 bus**.

IEEE 802 IEEE standards pertaining to LANs and MANs. These deal with the data-link layer and part of the physical layer. It divides the data-link layer into the Logical Link Control sublayer, and the Media Access Control sublayer. There are many IEEE 802.x standards, some of which appear in the following entries.

IEEE 802.1 IEEE standards defining high-level network interfaces, including network architecture and management.

IEEE 802.2 IEEE standards defining the Logical Link Control interface between the data-link layer and network layers.

IEEE 802.3 IEEE standards defining CSMA/CD.

IEEE 802.3z IEEE standards defining Gigabit Ethernet.

IEEE 802.4 IEEE standards pertaining to token bus networks.

IEEE 802.5 IEEE standards defining token ring access methods.

IEEE 802.6 IEEE standards pertaining to MANs.

IEEE 802.7 An IEEE technical advisory group pertaining to broadband.

IEEE 802.8 An IEEE technical advisory group pertaining to fiber optics.

IEEE 802.9 IEEE standards pertaining to integrated data and voice networks.

IEEE 802.10 IEEE standards pertaining to network security.

IEEE 802.11 IEEE standards pertaining to wireless networks, such as WLANs.

IEEE 1284 An IEEE standard for high-speed bi-directional parallel ports. It provides for multiple devices including printers, scanners, and external drives to be connected to a parallel port, allows for an extended cable, has multiple modes including ECP and nibble, and is backwards compatible with other standards, such as that of the Centronics parallel interface. Also called **IEEE 1284 standard**.

IEEE 1284 cable Same as **IEEE 1284 printer cable**.

IEEE 1284 compliant A parallel port which is complaint with **IEEE 1284**. Also, a cable used with such a port.

IEEE 1284 printer cable A printer cable used with parallel ports complaint with **IEEE 1284**.

IEEE 1284 standard Same as **IEEE 1284**.

IEEE 1394 A serial interface standard which provides for transfer speeds ranging from 100 Mbps to over 2 Gbps, and allows the connection of up to 63 devices. It also supports hot-swapping, and isochronous data transfer. Used, for example, for real-time transfers of video. FireWire, for instance, implements this standard. Also called **IEEE 1394 standard**.

IEEE 1394 serial bus A serial bus conforming to the **IEEE 1394 standard**.

IEEE 1394 standard Same as **IEEE 1394**.

IEEE Radar Band Designation Any of the 11 frequency intervals established by the IEEE for radio communications. These are, in order of increasing frequency, the L, S, C, X, Ku, K, Ka, V, W, mm, and μmm bands.

IEEE standard Any of the standards published by the IEEE.

IESG Abbreviation of **Internet Engineering Steering Group**.

IETF Abbreviation of **Internet Engineering Task Force**.

IF Abbreviation of **intermediate frequency**.

IF amplifier Abbreviation of **intermediate-frequency amplifier**.

IF interference Abbreviation of **intermediate-frequency interference**.

IF rejection Abbreviation of **intermediate-frequency rejection**.

IF signal Abbreviation of **intermediate-frequency signal**.

IF stage Abbreviation of **intermediate-frequency stage**.

IF-THEN A programming language statement in which **IF** a specified condition is met, **THEN** an action takes place.

IF-THEN-ELSE A programming language statement in which **IF** a specified condition is met, **THEN** an action takes place, or **ELSE** an alternate action takes place.

IF transformer Abbreviation of **intermediate-frequency transformer**.

IFF Abbreviation of **Interchange File Format**.

IFIP Abbreviation of **International Federation of Information Processing**.

IGBT Abbreviation of insulated-gate bipolar transistor. A semiconductor device incorporating MOS gate control, and bipolar current flow mechanisms. Such a device combines high current, high voltage, and a high input impedance. Seen, for instance, in power electronics applications, such as those utilizing power-conversion devices.

IGES Abbreviation of **Initial Graphics Exchange Specification**.

IGFET Abbreviation of insulated-gate field-effect transistor. A field-effect transistor whose gate is insulated from the channel by a thin layer, usually consisting of a metal oxide. All MOSFETs are also IGFETs, and the terms are usually used synonymously.

IGMP Abbreviation of **Internet Group Management Protocol**, and **Internet Group Membership Protocol**.

ignite To raise of the temperature of a substance, such as the fuel mixture in an automobile, to its **ignition point**.

igniter Also spelled **ignitor**. **1.** That which **ignites**. For example, a spark which gives a given substance the minimum temperature necessary for combustion to begin. **2.** In a switching tube, an element which initiates and maintains a discharge. **3.** Same as **igniter electrode**.

igniter electrode An electrode partially immersed in a mercury-pool cathode, and which is utilized to initiate conduction at a given point within each cycle. Also spelled **ignitor electrode**. Also called **igniter (3)**.

ignition 1. The raising of the temperature of a substance, such as the fuel mixture in an automobile, to its **ignition point**. 2. The means utilized to produce **ignition (1)**. This may be accomplished, for instance, electrically or through friction. 3. A switch or other mechanism which activates an **ignition system**. 4. Same as **ignition system**.

ignition coil Within an automotive ignition system, an open-core transformer which converts the low DC voltage from the battery into a voltage that is sufficient to produce the ignition of the fuel mixture.

ignition interference Radio-frequency interference whose source is an **ignition system**.

ignition noise Noise whose source is an **ignition system**. Such noise is usually of very short duration, but may be at a high level.

ignition point For a given substance, the minimum temperature at which combustion begins, and is subsequently maintained. Also called **ignition temperature**.

ignition potential Same as **ignition voltage**.

ignition system In an internal combustion engine, such as that of an automobile or gasoline-driven generator, a system which provides the sparks which ignite the fuel mixture. An example is a capacitive-discharge ignition. Also called **ignition (4)**.

ignition temperature Same as **ignition point**.

ignition voltage The minimum voltage necessary to produce **ignition (1)**, as occurs, for instance, in an ignition system. Also called **ignition potential**.

ignitor Same as **igniter**.

ignitor electrode Same as **igniter electrode**.

ignitron A mercury-pool rectifier utilizing an **igniter electrode** to initiate the arcing between the anode and the cathode at the desired point within each cycle. Used, for instance, for resistance welding.

ignore character 1. A character whose presence indicates that no action is to take place. 2. A character whose presence indicates that a specific action must not take place.

IGP Abbreviation of **Interior Gateway Protocol**.

IGRP Abbreviation of **Interior Gateway Routing Protocol**.

IHV Abbreviation of **independent hardware vendor**.

IIL Abbreviation of **integrated-injection logic**.

IIOP Abbreviation of **Internet Inter-ORB Protocol**. Within a CORBA architecture, a protocol utilized for communication between ORBs over the Internet.

IIR filter Abbreviation of **Infinite-Impulse Response Filter**.

ILD 1. Abbreviation of **interlayer dielectric**. 2. Abbreviation of **injection laser diode**.

ILEC Abbreviation of **incumbent local exchange carrier**.

ILF Abbreviation of **infralow frequency**.

ill-behaved Said of a program which does not function as expected within a given setting, or of that which does not adhere closely to a given standard. For example, an application that bypasses the operating system for certain operations. This contrasts with **well-behaved**, in which the converse is true.

illegal In computing, not acceptable, not understood, not performable, or leading to an error.

illegal character In computing, a character which is not valid or authorized according to specified criteria. For instance, a given alphabetical character which is not a member of a particular language set would be illegal in said set.

illegal operation A computer operation which is not acceptable, not understood, not performable, or which will lead to an error.

illuminance The luminous flux density per unit area of a surface. It is the total amount of visible light incident upon a surface from all directions above said surface. Its SI unit is the lux, although foot-candles are sometimes used. Also called **luminous flux density**.

illuminant 1. A source of light which is defined by its spectral power distribution. 2. A standard source of light. 3. A source of light.

illuminant C An **illuminant (2)** which approximately matches daylight. It is used as a reference white in color TV.

illuminated button Same as **illuminated pushbutton**.

illuminated push-button Same as **illuminated pushbutton**.

illuminated pushbutton Also spelled **illuminated push-button**. Also called **illuminated button**, or **lighted pushbutton**. 1. A push-button which is backlit, or otherwise provided with a source of illumination. 2. A pushbutton which becomes illuminated when pressed or activated.

illuminated remote Abbreviation of **illuminated remote control**. A remote control which is backlit, enabling ease of use in dark conditions. Also called **lighted remote**.

illuminated remote control Same as **illuminated remote**.

illuminated switch Also called **lighted switch**. 1. A switch that is backlit, or otherwise provided with a source of illumination. 2. A switch that becomes illuminated when pressed or activated.

illumination 1. The providing or application of light. This is a qualitative term, while **illuminance** quantifies specifically. 2. A deprecated term for **illuminance**. 3. For an antenna, the geometric distribution of power that reaches the various parts of a reflector dish.

illumination control 1. A photoelectric control which regulates the turning on and off of lighting, according to surrounding light conditions. Such a control may be used, for instance, to provide street lighting when needed. 2. A control which adjusts the intensity of lighting. Such a control may be used, for instance, in a light microscope.

illumination meter Same as **illuminometer**.

illuminator A device which serves to produce, concentrate, or reflect light.

illuminometer A photometric instrument which serves to measure the illumination incident upon a surface. Such an instrument is usually photoelectric. Also called **illumination meter**.

Illustrator A popular full-featured drawing program.

ILS Abbreviation of **instrument landing system**.

IM 1. Abbreviation of **intermodulation**. 2. Abbreviation of **instant messaging**.

IM distortion Abbreviation of **intermodulation distortion**.

IM frequency Abbreviation of **intermodulation frequency**.

IM noise Abbreviation of **intermodulation noise**.

IM products Abbreviation of **intermodulation products**.

iMac A popular line of personal computers.

image 1. A representation, reproduction or counterpart of an object or entity, such as an optical or electrical reproduction or counterpart. These include photographs, graphics, X-ray photographs, holograms, and so on. 2. An array of electrical charges which represent an actual object. It is the electric counterpart of a real object that is nearby. Also called **electric image (2)**. 3. That which is displayed by a TV, computer monitor, oscilloscope, or other display device. Also, that reproduced by a fax, or other similar system. 4. A duplicate or display of the partial or complete contents of a computer storage device, memory, program, file, or the like. An example is a system image. 5. One of two groups of sidebands produced through modulation.

image admittance The reciprocal of **image impedance**.

image antenna A hypothetical or virtual counterpart of an antenna, which is considered to extend below the ground, or ground plane, the same distance that the actual antenna is above it. An image antenna is completely symmetrical with the antenna it is a mirror image of, and is useful, for instance, for calculating the electromagnetic fields emanating from the latter.

image compression The encoding of an image so that it occupies less space and/or bandwidth. There are various standards used for such compression, including JPEG, GIF, and TIFF. MPEG is one of the formats used for video compression.

image converter 1. A device, medium, or process which converts an image from one form or format into another. For example, an optoelectronic device which converts an X-ray image into a visible image, or a program which converts a TIFF image into a JPEG image. 2. Same as **image tube**.

image converter tube Same as **image tube**.

image digitization The conversion of an analog image into digital form, which is necessary for a digital computer to process it.

image digitizer A device which converts an image into digital form, such as a bitmap, which is necessary for digital computer processing.

image dissector A camera tube in which the picture to be transmitted is focused upon a photosensitive surface from which electrons are emitted proportionally to the light of each part of the picture. These electrons are focused, in sequence, onto a controller electrode, which produces the output video signal. Also called **image dissector tube**, **dissector tube**, or **Farnsworth image dissector**.

image dissector tube Same as **image dissector**.

image distortion Any undesired changes occurring in a reproduced image, as compared to the original image. An example is optical distortion. Also called **distortion** (3).

image editing 1. Changes made to an image for any given purpose. This may be accomplished, for instance, through the use of image processing. 2. The use of an **image editor** to make changes in images.

image editing program Same as **image editor**.

image editor An application which is utilized to modify bit-mapped images, such as scanned photographs. Such programs usually have filters that enable them to read and convert many graphics formats, and have other features for enhancing and otherwise altering such images. Also called **image editing program**.

image effect The effect that the reflection of waves off the ground has on the overall electromagnetic radiation of an antenna.

image enhancement Modifications made to images to improve their quality, or their suitability for a particular purpose. An example is the use of image processing to refine images.

image file 1. A file containing graphics data. These can be stored either as bitmap graphics or as vector graphics. Also called **graphics file**. 2. A file containing a duplicate of the partial or complete contents of computer memory, or of a computer storage device, program, file, or the like.

image frequency In superheterodyne radio reception, an unwanted input frequency which results in an output at the intermediate frequency. An image frequency differs from the desired input frequency by twice the value of the intermediate frequency.

image iconoscope A TV camera tube similar to an iconoscope, but in which the image is projected onto a photocathode which emits photoelectrons which are focused onto a target to form the charge image. The sensitivity of such a tube is much greater than that of an iconoscope.

image impedances For an electric network or transducer, impedances which when connected to the input or the output will make the impedances in both directions equal.

image intensification The use of an **image intensifier**.

image intensifier An electronic device which increases the brightness of images, and which can operate at very low levels of light. For example, the image whose brightness will be intensified is first focused onto a photocathode, from which electrons are ejected. These electrons are then accelerated and focused onto a phosphorescent screen, whose displayed image may be thousands of times brighter than the original. Used, for instance, to obtain clear X-ray images with markedly reduced exposure. Also called **light amplifier**.

image-intensifier tube An electron tube utilized as an **image intensifier**.

image interference In a superheterodyne radio receiver, interference due to the presence of an **image frequency**.

image isocon A camera tube similar to an **image orthicon**, but with much greater sensitivity.

image map On the World Wide Web, a single image which has multiple hyperlinks embedded. For example, the image may be a map of a country, and there may be additional information available for each region when clicked. Also, such an image which displays different information when the cursor is placed above different parts of it. Also spelled **imagemap**.

image orthicon A camera tube in which the light from a scene is focused upon a photocathode from which electrons are ejected with an intensity proportional to the intensity of the light in the various locations of said scene. These electrons are focused upon a target, which produces secondary emissions of electrons, in a manner proportional to the electron density of the photocathode, forming an electronic image. The opposite side of this target is scanned by an electron beam, and its interaction with the target forms an intensity-modulated beam which is then multiplied to produce the video output. Utilized in TV broadcasting.

image processing The input, analysis, manipulation, storage, and output of images. For example, pictorial information obtained by scanning is followed by an analog-to-digital conversion. Then, this data is modified, color enhancements may be made, for instance, and finally the image is stored, printed, or displayed. Such processing helps make images more suitable for a desired purpose, and are used, for example, in satellite weather maps, or by robotic image input devices.

image ratio The ratio of the amplitude of the image frequency signal input, to the amplitude of the desired input signal, for identical outputs.

image recognition The processing of the data of an image input device, such as a scanner, for the purpose of comparison to images or components of images stored in computer memory. May be used, for example, to identify fingerprints, or to interpret bar codes. Also known as **pattern recognition**.

image rejection In a superheterodyne radio receiver, the extent to which an **image frequency** is rejected.

image response The extent to which a superheterodyne radio receiver responds to an **image frequency**.

image sensor A sensor composed of light-sensitive cells whose output is a numeric representation of intensity and color. Applications include its use in digital cameras, robotics, and in satellite imaging.

image tube An electron tube which converts an image from one part of the electromagnetic spectrum into another. For example, a tube whose photosensitive surface converts an infrared image into a visible image which is projected onto a fluorescent screen. Also called **image converter (2)**, or **image converter tube**.

image viewer A separate computer application which is called upon to interpret and view images which could otherwise not be seen by the current application. For instance, a Web browser plug-in utilized to display images in a format that is not currently supported.

imagemap Same as **image map**.

imagesetter A typesetting device which takes a computer input, such as that from disk, and produces a high-resolution output on paper, film, or other medium. Such devices usually utilize lasers, and may have resolutions in excess of 3600 dpi. Also called **phototypesetter**.

imaginary number A number in the form of *bi*, in which *b* is a real number, and *i* is the square root of -1. That is, $i^2 = -1$. In electronics, the symbol utilized to express the square root of -1 is usually *j*. A **complex number** has the form of *a* + *bi*, in which *a* and *b* are real numbers, and *i* is the square root of -1. Also called **imaginary quantity**.

imaginary quantity Same as **imaginary number**.

imaging 1. The process of forming an image of an object or medium. 2. The process of converting an image into an electronic equivalent, such as a bitmap, suitable for computer processing, storing, transmitting, or displaying. A charge-coupled device, for instance, may be used to capture such an image. Also called **electronic imaging**.

IMAP Abbreviation of **Internet Message Access Protocol**.

IMD Abbreviation of **intermodulation distortion**.

immediate access 1. Access to memory or storage which is virtually instantaneous. 2. Pertaining to a storage device whose access time is negligible in relation to other storage devices. 3. The ability to access information directly, as opposed to sequentially. Computer memory and disk drives, for instance, provide such access. Also called **direct access**.

immediate access storage Computer storage characterized by **immediate access**.

immediate operand An operand which is contained in a machine instruction. Said instruction contains the value of the operand, a opposed to an address.

immersion plating Plating accomplished by immersion of the surface to be plated in a solution where deposition occurs through chemical means. It is a form of electroless plating. Also called **dip plating**.

immitance A term which encompasses both **im**pedance and ad**mittance**, which are reciprocal quantities. Used, for instance, in the context of certain measuring instruments.

impact avalanche transit time diode Same as **IMPATT diode**.

impact excitation The moving of a particle, such as an electron or atom, from a lower state to a state with a higher energy level, through a collision.

impact ionization The ionization of a particle, such as an atom or molecule, through a collision. This may occur, for instance, in a semiconductor, resulting in an increase in charge carriers.

impact ionization avalanche transit time diode Same as **IMPATT diode**.

impact noise 1. Noise created by impacts, such as those produced by hammering or pneumatic drills. 2. Noise whose amplitude is great in relation to its duration.

impact-noise analyzer An instrument which accurately determines the sound pressure level of **impact noises**. Ordinary sound-pressure meters do not indicate the true effect of such sounds. Used, for instance, to monitor occupational hearing hazards.

impact printer A printer in which the printing process includes mechanically impacting the paper, or other medium, upon which the characters and symbols are being recorded. Examples include daisy-wheel and line-matrix printers.

impact resistance The extent to which a material or body can withstand high-intensity shocks or shock waves. Also called **impact strength**.

impact strength Same as **impact resistance**.

IMPATT amplifier An amplifier incorporating an **IMPATT diode**.

IMPATT diode A pn-junction diode which generates microwave power by utilizing avalanche breakdown. When the junction is reverse-biased into avalanche breakdown, it exhibits negative resistance in the microwave range. Used, for instance, as an amplifier or oscillator. It is an abbreviation of **imp**act **a**valanche **t**ransit **t**ime **diode**, or **imp**act ionization avalanche transit time **diode**.

IMPATT oscillator An oscillator incorporating an **IMPATT diode**.

impedance Its symbol is *Z*. The total opposition a circuit or device offers to the flow of AC. It is a complex quantity, whose real number component is resistance, and whose imaginary number component is reactance. Its formula is: $Z = R + jX$, in which *Z* is the impedance, *R* is the resistance, *X* is the reactance, and *j* is the square root of -1. That is, $j^2 = -1$. *Z* is expressed in ohms. Also called **electrical impedance**.

impedance arm A path within a circuit or network which contains one or more impedances. Also called **impedance leg**, or **impedance branch**.

impedance branch Same as **impedance arm**.

impedance bridge A bridge which is utilized to accurately measure impedances, by comparison to known values. It does so by determining the resistive and reactive components of any impedance in question.

impedance characteristic A graph of the impedance of a circuit, as a function of frequency.

impedance coil An inductor used in a circuit to present a relatively high impedance to frequencies beyond a given value. It impedes the flow of AC, while allowing DC to pass freely. Also called **choke coil**, or **choke (1)**.

impedance compensation The compensation of an impedance, so as to attain **impedance matching**. Also called **impedance correction**.

impedance converter Same as **impedance transformer**.

impedance correction Same as **impedance compensation**.

impedance coupling The coupling of two or more circuits by means of an **impedance**.

impedance drop In an AC circuit, a voltage drop due to **impedance**.

impedance leg Same as **impedance arm**.

impedance match Same as **impedance matching**.

impedance matching The matching of impedances between two circuits, or between a signal source and a load, when said impedances are not equal. This helps insure that the maximum possible power is transferred from source to load, while minimizing distortion. Also called **impedance match**.

impedance-matching network An electric network utilized for **impedance matching**.

impedance-matching transformer A transformer utilized for **impedance matching**. Used, for instance, to compensate for the impedance drop when connecting several speakers to the same audio amplifier in parallel. Also called **impedance transformer (2)**.

impedance meter An instrument which measures and indicates **impedances**. Such an instrument usually provides a digital readout. Its abbreviation is *Z* **meter**.

impedance mismatch A condition in which the impedances between two circuits, or between a signal source and a load, are not equal, thus reducing the transfer of power. Also called **mismatched impedances**.

impedance parameters Within an electric network, circuit parameters pertaining to **impedance**. Its abbreviation is **Z-parameters**.

impedance ratio The ratio of two impedances, such as those that are mismatched.

impedance transformer 1. A transformer utilized to convert one impedance to another. **2.** Same as **impedance-matching transformer**.

impedance triangle A right triangle whose perpendicular sides are proportional to the resistance and reactance of an AC circuit, and whose hypotenuse is proportional to the impedance of said circuit.

impedor A circuit component which has impedance.

imperative statement In a computer program, a statement that is executed without the need to meet any given condition. Also called **unconditional statement**.

imperfect crystal A crystal which has flaws in the geometric arrangement of its contained atoms, molecules, or ions. A **perfect crystal** has no such defects.

imperfection A lattice defect in an **imperfect crystal**.

implantation damage The displacement of host atoms when accelerated ions strike them during ion implantation. Such displacements can be corrected by annealing at a high temperature, such as 700 or 800 ºC.

implantation energy The energy of the implanted ions in a target semiconductor layer when utilizing ion implantation. The energy of the ions determines the depth to which they penetrate.

implanted device 1. An electronic device, such as certain pacemakers, which is surgically inserted or embedded. **2.** A device which is inserted or embedded in another.

implement To put into effect, or to carry out something.

implementation The process of putting into effect, or the carrying out of something. Seen, for instance, when putting a given communications protocol into practical use.

implosion An explosion inward, as might occur when the glass envelope of a vacuum tube shatters, and the surrounding atmosphere rushes in to fill the vacuum, carrying glass fragments with it.

import To accept a file or data in a foreign format, and then convert it to that required for use by the receiving application or environment. The accepting program or system must have the appropriate filters. This contrasts with **export**, where the conversion takes place before the file or data is moved or saved.

import filter A filter utilized to **import**.

impregnate To fill throughout, permeate, or saturate one substance with another. For instance, to use diffusion to uniformly distribute a gaseous dopant within the crystal lattice of a semiconductor material.

impregnated cathode A cathode, such as that of tungsten, which is impregnated with another material, such as barium, to impart the desired properties.

impregnated tape Tape which is impregnated with a given substance to impart the desired properties. Examples include tapes impregnated with magnetizable particles for recording, or those impregnated with certain resins to provide resistance to high temperatures.

impregnation The process of filling throughout, permeating, or saturating one substance with another. For example, the filling of the spaces of an electrical component with an insulating material.

impressed voltage The voltage applied to a component, circuit, or device.

improved definition television Same as **improved definition TV**.

improved definition TV Abbreviation of **improved definition television**. A standard, such as HDTV, for the transmission and reception of TV signals in which the resolution is superior in comparison to standards such as NTSC or PAL. Other factors, such as sound quality, are also usually enhanced. Its abbreviation is **IDTV**. Sometimes used as a synonym for **HDTV**.

impulse 1. A unidirectional surge of current of very short duration. It quickly rises to a maximum, then drops to zero in a similar fashion. Such a surge may be produced, for instance, when switching equipment on or off. Also called **impulse current**, or **pulse current**. **2.** A unidirectional surge of voltage of very short duration. It quickly rises to a maximum, then drops to zero in a similar fashion. Such a surge may be produced, for example, by a lightning stroke. Also called **impulse voltage**, or **pulse voltage**. **3.** A **pulse** of very short duration, although both terms are often used synonymously.

impulse breakdown voltage An **impulse** (2) whose magnitude equals or exceeds that necessary for breakdown.

impulse current Same as **impulse** (1).

impulse excitation The production of oscillation in a circuit via a signal of very short duration. The time interval of the resulting oscillation is much greater than the duration of the exciting impulse. Also called **shock excitation**.

impulse flashover voltage An **impulse** (2) whose magnitude equals or exceeds that necessary for flashover.

impulse frequency The frequency with which **impulses** occur or are generated.

impulse generator A circuit or device which produces **impulses**. One or more capacitors may be used, for instance, for storing and then releasing energy in very short pulses. Also called **surge generator**.

impulse noise 1. Noise produced by **impulses**. The spark-produced voltages in an internal combustion engine may generate such noise. **2.** Noise whose duration is very short. Such noise is generally characterized by large amplitude, and may occur as a series of disturbances which can be random or periodic in nature.

impulse radar A pulsed radar utilizing extremely short high-powered pulses. Such pulses may be in the picosecond range.

impulse ratio 1. The ratio of impulse breakdown voltage to the peak value of the breakdown voltage. **2.** The ratio of impulse flashover voltage to the peak value of the flashover voltage. **3.** The ratio of impulse sparkover voltage to the peak value of the sparkover voltage.

impulse relay 1. A relay which distinguishes between different types of pulses, and is actuated by one type. For instance, a relay which only operates on strong pulses. **2.** A relay which upon receiving a short pulse stores the energy necessary to complete its function.

impulse sparkover voltage An **impulse** (2) whose magnitude equals or exceeds that necessary for sparkover.

impulse test Any test involving **impulses**. For instance, the applying of an impulse voltage to test a dielectric.

impulse transmission In communications, the use of **impulses** for signaling. The impulses denote the transitions in the signals.

impulse voltage Same as **impulse** (2).

impurity 1. A substance whose presence within another reduces or eliminates the purity of the latter. The presence of an impurity in a crystal can change its properties significantly, as seen in the use of dopants in semiconductors, or in the photoconductivity of an allochromatic crystal. **2.** An **impurity** (1) which is introduced into a semiconductor material. Such an impurity may be an acceptor impurity, which makes for a p-type semiconductor, or it may be a donor impurity, which makes for an n-type semiconductor. Acceptor impurities include gallium and aluminum, while donor impurities include phosphorus and arsenic. In either case, such an impurity increases the conductivity of the semiconductor. Also called **dopant**, or **semiconductor dopant**.

impurity band In a semiconductor, an energy band arising from the presence of **impurities (2)**.

impurity concentration In a semiconductor material, the number of **impurities (2)** present, per unit volume. Also called **impurity density**.

impurity density Same as **impurity concentration**.

impurity level 1. In a semiconductor, an energy level arising from the presence of **impurities (2)**. 2. The extent to which **impurities (2)** are present in a semiconductor.

impurity scattering In a semiconductor, scattering of electrons off **impurities (2)**.

impurity semiconductor A semiconductor material with introduced **impurities (2)**, which determine its electrical characteristics, such as concentration of charge carriers. Also called **extrinsic semiconductor**, or **doped semiconductor**.

in 1. Abbreviation of **input**. 2. Abbreviation of **inch**.

In Chemical symbol for **indium**.

in-band signaling 1. In a telephone circuit, the transmission of control signals, such as dial tones and busy signals, via the same channel utilized for voice transmission. This contrasts with **out-of-band signaling**, in which a separate channel is used. 2. In data transmission, signaling in which the least significant bit is taken from the information stream and utilized for control signals and other overhead. Sometimes used over T1 or E1 lines. Also called **robbed-bit signaling**, or **bit robbing**.

in-circuit emulator A chip which emulates a given processor, and which serves for testing and debugging logic circuits. Its acronym is **ICE**.

in-circuit test A test performed on components without removing them from the circuit they are a part of.

in-circuit tester An instrument which tests components without removing them from the circuit they are a part of.

in-lb Abbreviation of **inch-pound**.

in-line code Same as **inline code**.

in-line graphics Same as **inline graphics**.

in-line image Same as **inline image**.

in-line tuning The tuning of all stages, such as those of an amplifier, to the same frequency. Also spelled **inline tuning**.

in parallel The connection of the components within a circuit in a manner that there are multiple paths among which the current is divided, while all the components have the same applied voltage. An example is the connection of components across each other. This contrasts with **in series**, where components are connected end-to-end, and in which there is a single path for the current. Also called **in shunt**, or **parallel connection**.

in-phase A state in which two or more periodic quantities having the same frequency and waveshape pass through corresponding values, such as maximas and minimas, at the same instant at all times. In an **out-of-phase** state, the periodic quantities do not pass through corresponding values at the same instant at all times.

in-phase component 1. Same as **in-phase current**. 2. Same as **in-phase voltage**.

in-phase current Within an AC circuit, the component of the current that is **in phase** with the voltage. Also known as **in-phase component (1)**, **watt current**, **active current**, or **resistive current**.

in-phase rejection The ability of a differential amplifier to reject **in-phase signals**, while responding to signals which are not in phase. Also called **common-mode rejection**.

in-phase signal 1. For a differential device such as an amplifier, a signal applied equally to both inputs. Also called **common-mode signal (1)**. 2. The algebraic average of two signals applied to both ends of a balanced circuit, such as that of a differential amplifier. Also called **common-mode signal (2)**. 3. Same as **I signal**.

in-phase voltage Within an AC circuit, the component of the voltage that is **in phase** with the current. Also known as **in-phase component (2)**, **resistive voltage**, or **active voltage**.

in quadrature Said of two periodic quantities having the same frequency and waveshape which are out-of-phase by 90°, or $\pi/2$ radians.

in/s Abbreviation of **inches per second**.

in series The end-to-end connection of the components within a circuit. There is a single path for the current, which flows through all the components in sequence, while the voltage is divided among the components. This contrasts with **in parallel**, where the connection is side-by-side, or otherwise in a manner where the current is divided among the components. Also called **series connection**.

in shunt Same as **in parallel**.

In-WATS Same as **Inward Wide-Area Telephone Service**.

inaccuracy The quality or condition of being **inaccurate**.

inaccurate Incorrect, in error, or otherwise deviating from a real value. Said, for instance, of a measuring instrument.

inactive 1. Not ready for use, as in an unavailable communications channel. 2. Not currently functioning, as in a satellite which is not transmitting a signal at a given moment. 3. Not energized, as in a circuit that is not powered at the moment. 4. Chemically inert.

inactive application In a computer that has more than one application running, any application other than the one which is currently being interacted with.

inactive arm A component within a transducer that does not vary its electrical characteristics in response to the input of said transducer. Also called **inactive leg**.

inactive computer In a setting with two or more computers, any which is not currently processing.

inactive leg Same as **inactive arm**.

inactive lines In a TV, the lines that do not transmit the visible picture. These lines are above and below the visible picture. Also called **blanking lines**.

inactive window On a computer screen that has more than one window open, any window which is not currently being interacted with. There may be most any number of inactive windows, but there can only be one **active window**, which is the one which will be affected by such actions as keyboard input.

inbox In an email application, the default mailbox which to which incoming messages are directed. The default mailbox for outgoing messages is the **outbox**.

incandescence The emission of visible light by a heated object, such as the filament of an incandescent lamp. Also, the light emitted.

incandescent An object which when heated sufficiently emits visible light. Also, that which incorporates such an object.

incandescent lamp A lamp which incorporates a filament, often made of tungsten or carbon, which emits brilliant visible light when heated by an electric current. Such a lamp is what is commonly referred to as a **light bulb**.

inch A unit of distance equal to exactly 2.54 centimeters, or 0.0254 meters. Its abbreviation is **in**.

inch-pound A unit of work or energy equal to approximately 0.11299 joule. Its abbreviation is **in-lb**.

inches per second A unit of velocity equal to exactly 2.54 centimeters per second, or 0.0254 meters per second. Its abbreviation is **in/s**, or **ips**.

inching The rapidly opening and closing of a circuit, such as that controlling a motor or electromagnet, to produce small movements. Also called **jogging**.

incidence angle The angle formed between the line of an incident ray or wave and a perpendicular line arising from the point of incidence. Also called **angle of incidence**.

incidence plane The plane containing light, sound, particles, waves, or the like, which is reflected off a surface, object, or region. Also called **plane of incidence**.

incident 1. That which strikes, or otherwise falls upon a surface or body. For example, light which is falls upon a photosensitive material, or sound waves striking a diaphragm. **2.** An action or occurrence. It is similar to an **event** (1), but an event may imply a more significant happening than an incident. **3.** In communications and computers, an **incident** (2) with possible adverse consequences, such as a security breach.

incident light Light which falls upon a surface or body.

incident particle A particle which strikes a surface or body.

incident power The power reaching a given point, such as an impedance mismatch in a transmission line, or an air-glass interface in an optical fiber.

incident radiation Radiation which strikes a surface or body.

incident ray A ray which strikes a surface or body.

incident sound Sound which strikes a surface or body.

incident wave 1. A wave which reaches a given point, such as a medium with different propagation characteristics. **2.** A wave which strikes a surface or body.

incidental AM Amplitude modulation of a signal when frequency or phase modulation is desired. It is an abbreviation of **incidental** amplitude modulation. Also called **incidental modulation** (2), or **residual AM**.

incidental amplitude modulation Same as **incidental AM**.

incidental FM Frequency modulation of a signal when amplitude or phase modulation is desired. It is an abbreviation of **incidental** frequency modulation. Also called **incidental modulation** (3), or **residual FM**.

incidental frequency modulation Same as **incidental FM**.

incidental modulation Also called **residual modulation**. **1.** In an RF signal, noise produced by variations of a carrier in the absence of any intended modulation. Also known as **carrier noise**. **2.** Same as **incidental AM**. **3.** Same as **incidental FM**. **4.** Same as **incidental PM**.

incidental phase modulation Same as **incidental PM**.

incidental PM Phase modulation of a signal when frequency or amplitude modulation is desired. It is an abbreviation of **incidental** phase modulation. Also called **incidental modulation** (4), or **residual PM**.

inclination 1. The angle between a given plane and a reference plane. For instance, that between a geocentric orbit and the equator. **2.** For a given location on the surface of the earth, the angle between the horizontal plane and the direction of the lines of force of the earth's magnetic field. Inclination is 0° at the magnetic equator and 90° at each of the magnetic poles. Also called **magnetic dip**, **magnetic inclination**, or **dip** (3). **3.** The action of bending or tilting. **4.** A surface with a slant. Also, the slant itself.

inclinometer An instrument which measures and indicates **inclination** (1), or **inclination** (2).

inclusive OR A logical operation which is true if any of its elements is true. For example, if one or more of its multiple inputs have a value of 1, then the output is 1. Its output is 0 only when all of its inputs have a value of 0. For such functions, a 1 is considered as a true, or high, value, and 0 is a false, or low, value. Although inclusive OR is the same as **OR**, the former is used to distinguish from **exclusive OR**. Also called **inclusive OR operation**.

inclusive OR circuit Same as **inclusive OR gate**. Also called **OR circuit**.

inclusive OR gate A circuit which has two or more inputs, and whose output is high if any of its inputs is high. Also called **inclusive OR circuit**, or **OR gate**.

inclusive OR operation Same as **inclusive OR**.

incoherence Also spelled **incoherency**. **1.** A lack of a definite phase relationship between two or more waves. Two waves, for instance, are in not phase if their crests and troughs do not meet at the same point at the same time. Also called **non-coherence** (1). **2.** Any lack of temporal or phase correlation. Also called **non-coherence** (1). **3.** A lack of clarity or intelligibility, as may occur in a voice transmission.

incoherency Same as **incoherence**.

incoherent Characterized by **incoherence**.

incoherent bundle An assembly of optical fibers in which the relative spatial coordinates of the individual fibers are not the same at both ends. Such a bundle serves to transmit light, but not images. Also called **incoherent fiber bundle**, or **non-coherent bundle**.

incoherent detector A detector whose output signal amplitude does not depend on the phase of its input signal. Also called **non-coherent detector**.

incoherent fiber bundle Same as **incoherent bundle**.

incoherent light Light in which there is not a definite phase relationship between any given points of the waves composing it. Such light may contain waves of different wavelengths. There are many sources of incoherent light, including sunlight, incandescent lamps, and light-emitting diodes. Also called **non-coherent light**.

incoherent light source A source, such as a light-emitting diode, of **incoherent light**. Also called **non-coherent light source**.

incoherent pulses Pulses which are not in a fixed-phase relationship with each other. Also called **non-coherent pulses**.

incoherent radiation Radiation in which there is not a definite phase relationship between any given points of a cross section of the beam. Also called **non-coherent radiation**.

incoherent scattering Scattering in which there is not a definite phase relationship between the incident and dispersed waves or particles. Also called **non-coherent scattering**.

incoherent waves Two or more waves of a single frequency which do not have a definite phase relationship. Also called **non-coherent waves**.

incoming line A line entering a device, piece of equipment, facility, or the like.

incoming mail server A server, using a protocol such as POP or IMAP, which enables the retrieval and storage of emails. Usually utilized in conjunction with an **outgoing mail server**, which uses a protocol such as SMTP.

incomplete circuit A circuit which is broken, or which otherwise does not have a complete and uninterrupted path for the flow of current. Also called **open circuit** (1), or **discontinuous circuit**.

incorporate To unite into a single body or entity, or to combine with something else to constitute a whole.

incorporated United into a single body or entity, or combined into something else to constitute a whole.

increment 1. The process of gradually increasing a quantity or variable. Also, the increase in the quantity or variable. **2.** A small variation in a quantity or variable. **3.** A specific interval representing a variation in a quantity or variable. For example, a single click in either direction of a volume control with fixed steps.

incremental backup A data backup technique in which the only files copied are those that have been modified since the last backup, and in which all versions are saved, not just the last one, as is the case of a **differential backup**. This also contrasts with a **full backup**, in which all files are copied, including those that have not been modified since the previous backup. Also called **cumulative backup**.

incremental computer A computer, such as a differential analyzer, which processes variations in a quantity or variable.

incremental permeability The measured permeability of a material when a small alternating magnetic field is superimposed on a large static field.

incremental representation In an **incremental computer**, the representation of a variable as a function of a change in its value, as opposed to the value of the variable itself.

incremental search A search in which searching begins as soon as the first character is typed, as opposed to waiting for a command such as enter. Seen, for instance, when accessing the help index of an application, where results starting with the typed string are displayed as each letter is entered.

incremental sensitivity For an instrument or device, the smallest detectable change in the value of a quantity or variable.

incumbent local exchange carrier A communications entity which offers local telephone service. Such an entity may also offer other services, such as long-distance calling, Internet access, and so on. The term usually implies that said entity has been in this business for a long time. Its abbreviation is **ILEC**.

incursion An **intrusion** in which damage is done.

independent events Two events for which the occurrence of one in no way alters the probability of the other occurring.

independent failure 1. A failure which is not caused by another component in the same system. 2. A failure which does not affect any other component in the same system.

independent hardware vendor An organization which manufactures computer components and/or peripherals, as opposed to complete systems. Such an organization may, for instance, make circuit boards or disk drives. Its abbreviation is **IHV**.

independent operation 1. Said of a component, circuit, device, piece of equipment, or system which can or does function independently of any other given component, circuit, device, piece of equipment, or system. 2. Said of component, circuit, device, piece of equipment, or system with its own power source.

independent software vendor An organization, which is not owned or controlled by a hardware manufacturer, that develops software. Its abbreviation is **ISV**.

independent variable 1. A variable which determines the change in another variable, called **dependent variable**. 2. A variable whose value does not depend on the changes in another variable.

indeterminacy principle A principle stating that for a given particle it is not possible to simultaneously know with absolute accuracy both its location and momentum. No matter how accurate the measurements are, the mere observing or measuring said particle will interfere with it in an unpredictable fashion. This principle also applies to other observations or measurements. Also called **Heisenberg uncertainty principle**, or **uncertainty principle**.

index It plural form is **indices**, or **indexes**. 1. A list or table which serves to guide or facilitate reference. Also, to create such a list or table. 2. In computers, a list or table together with references or keys which serve to identify and locate the contents of disks, directories, files, records, or the like. Also, to create such a list or table. Also, to locate informa-

tion through the use of the references or keys of said index. 3. An **index** (2) used in a database. Also, to create such an index. Also, to locate information through the use of the references or keys of said index. 4. A number or symbol which usually appears as a subscript or superscript in a mathematical expression, and which indicates an operation to be performed, a specific element within a sequence, and so on. An example would be the use of such a number as a superscript to indicate an exponent. 5. A usually dimensionless quantity indicating a magnitude of a physical quantity or a ratio between two, such as an index of refraction.

index counter In a magnetic tape device, such as a tape deck or VCR, a counter which indicates how much tape has traveled in a given direction, and which serves, for instance, to easily locate a given portion. Also called **tape counter**.

index error In an instrument, an error in indicated values due to a misalignment of the pointer and/or the calibrated scale from which readings are taken. Such an instrument, for instance, would have either a positive or negative reading when there is no input. Depending on the direction of the erroneous reading, adding or subtracting a fixed value is usually an adequate corrective measure.

index hole A hole punched into a floppy disk which serves to indicate the beginning of the first data sectors of said disk.

index mark A signal, code, hole, notch, or other mark which indicates the beginning point of each track on a computer disk.

index of modulation Also called **modulation index**, or **modulation factor**. 1. A measure of the degree of modulation, usually involving the ratio of a peak variation to a steady non-peak value. The result may be multiplied by 100 to obtain a percentage. 2. In frequency modulation, the ratio of the peak variation of the carrier-wave frequency, to the frequency of the modulating signal. The result may be multiplied by 100 to obtain a percentage. 3. In amplitude modulation, the peak amplitude variation of the composite wave, to the unmodulated carrier amplitude. The result may be multiplied by 100 to obtain a percentage. 4. In amplitude modulation, the ratio of half the difference between the maximum amplitude and minimum amplitude, to the average amplitude. The result may be multiplied by 100 to obtain a percentage.

index of refraction Also called **refractive index**. Its symbol is **n**. 1. The ratio of the phase velocity of a wave in free space, to the phase velocity of the same wave in a given medium. Also called **absolute index of refraction** (1). 2. The ratio of the phase velocity of light in free space, to the phase velocity of light in a given medium. Also called **absolute index of refraction** (2). 3. The ratio of the phase velocity of a wave in a given medium, to the phase velocity of the same wave in another medium.

index register A computer register whose stored value can be added to a given address to form an effective address.

indexed address An address which has been modified by an **index register**.

indexed sequential access method A method for accessing records that are sequentially stored, in which an index provides direct access to key fields, providing faster access to the data. Its abbreviation is **ISAM**.

indexing To create or utilize an **index** (1), **index** (2), or **index** (3).

indicating device A device, such as a meter or counter, which indicates the instantaneous, average, peak, or effective value of a magnitude or quantity being measured. Also, a device, such as an indicator lamp, which visually conveys information. Also called **indicator device**.

indicating fuse Same as **indicator fuse**.

indicating instrument An instrument which visually indicates the instantaneous, average, peak, or effective value of a magnitude or quantity being measured. There are many examples, including ammeters, peak voltmeters, digital ohmmeters, and so on. Also called **indicator instrument**.

indicating lamp Same as **indicator lamp**.

indicating light Same as **indicator light**.

indicating scale Same as **indicator scale**.

indicating tube Same as **indicator tube**.

indicator   **1.** A component, circuit, device, instrument, piece of equipment, machine, or mechanism which conveys information pertaining to a magnitude or quantity. Said especially when the information is communicated visually. There many examples, including meters, indicator lamps, gauges, monitors, and so on. **2.** A needle, dial, or other object which serves to convey information in a meter, gauge, or similar device or instrument. **3.** A device, instrument, or piece of equipment which displays the information obtained by another device, instrument, or piece of equipment. For instance, a radar display.

indicator device Same as **indicating device**.

indicator fuse A fuse, such as that with a color-coded flag, which immediately and automatically indicates when it has blown. Also called **indicating fuse**.

indicator instrument Same as **indicating instrument**.

indicator lamp A lamp whose on or off state serves to inform about a given condition in a circuit or device, such as a malfunction. Also called **indicating lamp**.

indicator lamp panel A series of **indicator lamps** which serve to monitor multiple conditions.

indicator light A light whose on or off state serves to inform about a given condition in a circuit or device, such as a malfunction. Also called **indicating light**.

indicator light panel A series of **indicator lights** which serve to monitor multiple conditions.

indicator needle A needle which serves to convey information in a meter, gauge, or similar device or instrument. A specific example is that used in a D'Arsonval galvanometer. Also called **needle (2)**, or **pointer needle**.

indicator scale A scale, such as that on the face of an instrument, which enables quick and easy determination of a measured value. Also called **indicating scale**.

indicator tube An electron-beam tube which conveys information as a function of the size and/or shape of its beam. For instance, a variation in the cross-section of its beam could follow the tuning of a receiver. Also called **indicating tube**.

indices A plural form of **index**.

indirect address An address which points to the storage location of another address. The indicated address may also be another indirect address, or the address of the desired operand. Also called **multilevel address**.

indirect addressing An address mode utilizing **indirect addresses**. Also called **multilevel addressing**.

indirect-band-gap semiconductor A semiconductor in which the minimum energy in the conduction band does not coincide with the maximum energy in the valence band, and in which transitions between these bands require a change in momentum. Thus, optical transitions between the two bands are forbidden. An example is silicon. Also called **indirect-gap semiconductor**.

indirect control The control of a device by another, but requiring human intervention, as opposed to **direct control**, in which said control is automatic.

indirect-gap semiconductor Same as **indirect-band-gap semiconductor**.

indirect instruction A computer instruction containing **indirect addresses**.

indirect light Reflected light, as opposed to that arriving directly form a luminous source, such as the sun or a lamp.

indirect measurement The determination of a value utilizing an intervening process. For instance, the measurement of a capacitance using a capacitance bridge as opposed to a capacitance meter. This contrasts with a **direct measurement**, in which the converse is true.

indirect path The path of a wave which has been reflected, or otherwise does not travel in a straight path.

indirect piezoelectric effect Same as **inverse piezoelectric effect**.

indirect piezoelectricity Same as **inverse piezoelectric effect**.

indirect ray A ray which has been reflected, refracted, or scattered.

indirect-reading meter A meter whose indications are made through **indirect measurements**.

indirect scanning A scanning technique in which a spot of light scans a scene sequentially, and is then reflected to photoelectric cells or phototubes.

indirect wave   **1.** A wave which arrives via any indirect path. For instance, a reflected wave. **2.** A radio-frequency wave which travels via ionospheric propagation.

indirectly-heated cathode A cathode, within a thermionic tube, which is electrically insulated from the heating element. It may consist, for instance, of a filament surrounded by a sleeve with an electron-emitting coating. Such cathodes have the same potential along their entire surface. Also called **equipotential cathode**, **unipotential cathode**, or **heater-type cathode**.

indium A lustrous silvery-white metallic chemical element whose atomic number is 49. It is soft, malleable, ductile, and when bent emits a characteristic metallic cry. It has over 30 known isotopes, of which one is stable. Indium has many applications, including its uses in semiconductor devices, low-melting point alloys, including some which are liquid at room temperature, in photoconductors, in nuclear reactor control rods, in bearings and gaskets that work effectively at cryogenic temperatures, and it can be plated onto another metal to form corrosion-resistant mirrors. Its chemical symbol is **In**.

indium antimonide A crystalline solid whose chemical formula is **InSb**. Used in semiconductors, infrared detectors, and devices employing the Hall effect.

indium arsenide A crystalline solid whose chemical formula is **InAs**. Used in semiconductors and injection lasers.

indium oxide A yellowish white powder whose chemical formula is In_2O_3. Used in special glasses.

indium phosphide A brittle metallic mass whose chemical formula **InP**. Used in semiconductors, injection lasers, and solar cells.

indium selenide Black crystals whose chemical formula is **InSe**. Used in semiconductors and solar cells.

indoor   **1.** Situated in the interior of a building. **2.** Designed for use in the interior of a building, or another enclosure which protects from surrounding weather.

indoor antenna Also called **inside antenna**. **1.** An antenna situated in the interior of a building. **2.** An antenna designed for use in the interior of a building.

indoor apparatus A apparatus designed for use in the interior of a building, or another enclosure which protects it from surrounding weather.

indoor component A component designed for use in the interior of an enclosure which protects it from surrounding weather.

indoor device A device designed for use in the interior of a building, or another enclosure which protects it from surrounding weather.

indoor equipment Equipment designed for use in the interior of a building, or another enclosure which protects it from surrounding weather.

indoor transformer A transformer designed for use in the interior of a building, or another enclosure which protects it from surrounding weather.

induce 1. To produce a given condition or effect by **induction**. 2. To cause or stimulate the occurrence of something.

induced charge A charge produced on a body through **induction (3)**. Also called **induction charge**, or **inductive charge**.

induced current A current that flows through a conductor as a consequence of an applied time-varying electromagnetic field. The current is induced by the variation in the number of lines of magnetic flux intersecting with said conductor. Also called **induction current**, or **inductive current**.

induced electromotive force An electromotive force developed in a circuit or conductor through the variation in the number of lines of magnetic flux intersecting with said circuit or conductor. This occurs by moving a conductor through a magnetic field, or by changing the magnetic field which crosses a conductor. Its abbreviation is **induced emf**.

induced emf Abbreviation of **induced electromotive force**.

induced failure 1. A failure in a component, circuit, device, piece of equipment, or system, due operation beyond its rating. 2. A failure in a component, circuit, device, piece of equipment, or system, due to an external factor.

induced noise Noise in a component, circuit, device, piece of equipment, or system, resulting from **induction (1)**, or **induction (2)**.

induced potential Same as **induced voltage**.

induced voltage A voltage developed in a circuit or conductor through the variation in the number of lines of magnetic flux intersecting with said circuit or conductor. Also called **induced potential**.

inductance Its symbol is L. 1. The property of a circuit or conductor which opposes any change to the current flowing through it. That is, it opposes an applied current, and also opposes changes to an already established current. Inductance makes changes in current lag behind changes in voltage. When quantifying inductance, some multiple of henrys, such as millihenrys or microhenrys is used. Also called **electrical inertia**. 2. The property of a circuit in which a varying current induces an emf in itself or in a neighboring circuit. When the emf is induced in the same circuit it is called **self-induction**, and when the emf is induced in another circuit it is called **mutual induction**. The symbol for self-induction is L, and for mutual induction it is M. When quantifying inductance, some multiple of henrys, such as millihenrys or microhenrys is used. 3. Same as **inductor**.

inductance bridge A bridge utilized to measure an unknown inductance. Hay bridges and Maxwell bridges are examples.

inductance-capacitance Its abbreviation is LC. 1. A combination of inductance and capacitance in a circuit. 2. A circuit or device with a combination of inductance and capacitance, such as that provided by coils and capacitors. For example, certain types of filters. 3. Pertaining to an **inductance-capacitance (1)** or **inductance-capacitance (2)**, or that incorporating or measuring this combination, such as a meter.

inductance-capacitance circuit A circuit composed of, or incorporating inductances and capacitances. Its abbreviation is LC **circuit**.

inductance-capacitance filter A filter composed of, or incorporating inductances and capacitances. Its abbreviation is LC **filter**.

inductance-capacitance meter A meter which measures and indicates inductances and capacitances. Its abbreviation is LC **meter**.

inductance-capacitance network An electric network composed of, or incorporating inductances and capacitances. Its abbreviation is LC **network**.

inductance-capacitance oscillator An oscillator whose frequency is determined by inductances and capacitances. Its abbreviation is LC **oscillator**.

inductance-capacitance-resistance Its abbreviation is LCR. Also called **resistance-capacitance-inductance**. 1. A combination of inductance, capacitance, and resistance in a circuit. 2. A circuit or device with a combination of inductance, capacitance, and resistance, such as that provided by coils, capacitors, and resistors. For example, certain tuned circuits. 3. Pertaining to an **inductance-capacitance-resistance (1)** or **inductance-capacitance-resistance (2)**, or that incorporating or measuring this combination, such as a meter.

inductance-capacitance-resistance meter A meter which measures and indicates inductances, capacitances, and resistances. Its abbreviation is LCR **meter**.

inductance coil Same as **inductor**.

inductance loss Same as **induction loss**.

inductance standard An inductor whose inductance value is highly accurate and stable, and which serves, for instance, as a calibration standard. Also called **standard inductor**.

inductance transducer Same as **inductive transducer**.

induction 1. The generation or modification of electric fields, magnetic fields, voltages, or currents, through the influence of nearby entities. 2. The generation of an electromotive force in a circuit or conductor caused by a change in the magnetic flux through said same circuit or conductor. Also called **electromagnetic induction**. 3. The modification of the distribution of electric charge on an body as influenced by another nearby body which is charged. Positive charges will induce a negative charges, and vice versa. Also called **electrostatic induction**.

induction annealing Annealing in which the heat is generated by an **induced current**.

induction brazing Brazing in which the heat is generated by an **induced current**.

induction charge Same as **induced charge**.

induction coil A transformer which changes a DC into a high-voltage AC. Its primary winding is connected to a DC source which is periodically interrupted, and has comparatively few turns of a thicker wire. This primary winding induces a high voltage in its secondary winding, which is concentrically wound over the primary, and has many turns of a finer wire. An example is a Ruhmkorff induction coil.

induction compass A compass whose indications depend on the current induced in a coil made to revolve in the earth's magnetic field. Also called **earth inductor compass**.

induction current Same as **induced current**.

induction field 1. The portion of an electromagnetic field that is associated with the AC component, and which is responsible for any self or mutual inductance. 2. For a radiator, such as a transmitting antenna, the portion of the electromagnetic field which behaves as if permanently associated with said radiator. This contrasts with the **radiation field**, which is the portion that is propagated outward as electromagnetic waves.

induction furnace An electric furnace in which an **induced current** is passed through the material being heated.

induction generator An asynchronous machine utilized as an AC generator. Used, for example, for induction heating.

induction hardening Hardening in which the necessary heat is generated through **induction heating**.

induction heater A device utilized for **induction heating**. Such a device incorporates an RF generator, and a work coil which induces RF currents in the workpiece it surrounds. Lower frequencies are more effective for deep heat penetration, while higher frequencies are more appropriate for smaller parts or shallow heat penetration.

induction heating Heating in which an **induced current** is passed through the workpiece. The resulting eddy currents produce the heat. Usually used for materials with high electrical conductivity, such as metals. Induction heating features accuracy, fast heating cycle times, and maximum repeatability, among other benefits. Used, for instance, for soldering, annealing, brazing, metal-to-glass bonding, and shrink fitting. Also called **inductive heating**, or **eddy-current heating**.

induction instrument An instrument whose function is based on the interaction between fixed windings through which a current flows, which establishes a magnetic flux, and the current this flux induces in a movable conductor.

induction loss In a conductor, the loss of RF energy through **inductive coupling**. Also called **inductive loss**, or **inductance loss**.

induction loudspeaker Same as **induction speaker**.

induction meter A meter whose function is based on the interaction between fixed windings through which a current flows, which establishes a magnetic flux, and the current this flux induces in a movable conductor.

induction motor An AC motor in which the electric current flowing through the rotor is induced by the AC flowing in its stator. The power source is connected only to the stator. Also called **AC induction motor**.

induction soldering Soldering in which the heat is generated by an **induced current**.

induction speaker A speaker in which an induced audio-frequency current flowing through the diaphragm interacts with a steady magnetic field. This results in the movement that produces sound. Also called **induction loudspeaker**.

induction transducer Same as **inductive transducer**.

induction voltage regulator A regulating transformer in which the primary winding is connected in parallel and the secondary in series with the circuit, and in which the position of the coils relative to each other adjusts the voltage and/or phase relation of said circuit. Its abbreviation is **IVR**.

induction welding Welding in which the heat is generated by an **induced current**.

inductive Pertaining to, causing, utilizing, or arising from **inductance** or **induction**.

inductive charge Same as **induced charge**.

inductive circuit 1. A circuit in which there is **inductance** (1), or **inductance** (2). **2.** A circuit in which the inductive capacitance is greater than the capacitive reactance.

inductive coordination The design, location, and operation of power-supply systems and communications systems, so as to minimize inductive interference.

inductive coupling 1. The coupling of two circuits or devices by means of an **inductance** (2). **2.** The coupling of two circuits by means of a mutual inductance provided by a transformer. In RF circuits, for instance, such coupling minimizes the capacitance between stages, which improves performance. Also called **transformer coupling**.

inductive current Same as **induced current**.

inductive feedback 1. Feedback through an inductor, or via inductive coupling. **2.** Feedback in an amplifier through an inductor, or via inductive coupling.

inductive heating Same as **induction heating**.

inductive interference 1. In a communications system, interference arising from induced voltages, such as those which may be produced by power lines. **2.** Interference arising from induction.

inductive kick In an inductive circuit or coil, a voltage surge produced when the current flowing through it is cut off. The rapidly collapsing magnetic field causes this surge. Also called **inductive surge**, or **kickback** (1).

inductive load A load which is predominantly inductive, so that at the terminals the alternating current lags the alternating voltage. Also called **lagging load**.

inductive loss Same as **induction loss**.

inductive neutralization A method of neutralizing an amplifier, in which capacitance cancels inductance in the feedback circuit. Also called **shunt neutralization**, or **coil neutralization**.

inductive reactance The opposition to the flow of AC due to the inductance of a component or circuit. Its symbol is X_L, it is measured in ohms, and its formula is: $X_L = 2\pi f L$, where π is pi, f is frequency in hertz, and L is inductance in henries.

inductive surge Same as **inductive kick**.

inductive switching The opening and closing of **inductive circuits**. In a semiconductor device, for instance, protective measures must be taken against inductive kick to assure longevity of the switching device.

inductive transducer A transducer in which a variation of a quantity in a given form produces a variation in inductance, which in turn produces a proportional conversion to a quantity in another form. Also called **induction transducer**, or **inductance transducer**.

inductive tuning Radio tuning accomplished by varying an inductance. This may be brought about, for instance, by moving a magnetic core relative to a coil.

inductivity For a given material, the property which determines the electrostatic energy that can be stored per unit volume for a unit potential gradient. This is equivalent to the ratio of the capacitance of a capacitor using the material in question as a dielectric, to the capacitance of a capacitor using a vacuum as a dielectric. Also known as **dielectric constant**, **relative dielectric constant**, **permittivity**, **relative permittivity**, or **specific inductive capacity**.

inductometer A fixed or variable inductor whose inductance is known, and which can be utilized for measuring unknown inductances.

inductor One or more conductors wound in a series of turns, so as to introduce inductance, or to produce a magnetic field in an electric circuit. An inductor may be wound, for example, around a ferromagnetic core. Used, for instance, in transformers, electromagnets, solenoids, motors, speakers, and so on. Also known as **coil**, **inductance** (3), **inductance coil**, **electric coil**, or **magnetic coil** (1).

industrial electronics The branch of electronics that deals with industrial applications, such as computation, communication, automation, instrumentation, control, robotics, simulation, manufacturing, and so on.

industrial robot A robot constructed and programmed to perform industrial tasks such as assembling, drilling, or probing. They are usually reprogrammable, and can perform complex and repetitive tasks reliably for extended periods of time.

industrial robotics The science that deals with the design, programming, construction, operation, and maintenance of robots destined to industrial use.

industrial strength Designed and/or built to provide reliable operation while withstanding continued use under very harsh circumstances, such as extreme environmental conditions.

industrial television Same as **industrial TV**.

industrial TV Abbreviation of **industrial** television. Television utilized, or specifically designed for, industrial applications. For example, closed-circuit TV used for monitoring manufacturing or hazardous areas, or for security surveillance. Its own abbreviation is **ITV**.

Industrial, Scientific, and Medical band Same as **ISM band**.

Industry Standard Architecture Same as **ISA**.

Industry Standard Architecture bus Same as **ISA bus**.

inert Having little or no chemical reactivity with other elements. For instance, platinum and gold are inert elements.

inert gas Any of six chemical elements which are all gases and highly unreactive. They are all in the same group of the periodic table, and are helium, neon, argon, krypton, xenon, and radon. Also called **noble gas**.

inert gas shielded-arc welding Shielded-arc welding in which an inert gas, such as argon or helium, provides the protective atmosphere.

inertance The acoustic equivalent of electrical inductance. Also called **acoustic mass, acoustic inertance, acoustic inertia,** or **acoustic inductance**.

inertia The tendency of a body to resist a change in its momentum. For instance, a body at rest will remain so unless acted upon by an external force.

inertia switch A switch that is actuated when it senses a change in momentum exceeding a given magnitude.

infect To contaminate a computer with a virus, bacterium, worm, Trojan horse or other such malicious piece of code or program.

infection The condition in which a computer virus, bacterium, worm, Trojan horse or other such malicious piece of code or program, has contaminated a file, piece of code, or program. Also, the process by which this contamination occurs.

inference engine Within an expert system or other knowledge-based system, the component which applies principles of reasoning to draw conclusions from the information available in the knowledge base.

InfiniBand An architecture or specification for the movement of data along I/O buses which supports up to 64,000 addressable devices and transfer rates in excess of 10 Gbps, while providing simplicity of operation, security, and reliability.

infinite 1. Having no limits or boundaries. 2. That which exceeds any other arbitrarily set large value. 3. That which is so large that it can not be measured.

infinite attenuation Attenuation which is sufficient to reduce the amplitude of an output signal to zero.

infinite baffle An enclosure or mounting in which there is no path of air from the front to the back of the speaker diaphragm. This may be accomplished, for instance, by mounting a speaker in a wall.

Infinite-Impulse Response filter A filter whose output is a weighed sum of the current inputs and past outputs, thus incorporating feedback. It is one of the two primary types of filters utilized in digital signal processing, the other being **Finite-Impulse Response filters**. Infinite-Impulse Response filters tend to be more efficient, but less linear than Finite-Impulse Response Filters. Its abbreviation is **IIR filter**.

infinite line A transmission line having the characteristics corresponding an ordinary line which is infinitely long. Also called **infinite transmission line**.

infinite loop A loop which is repeated indefinitely. Such a loop usually requires user intervention to stop it. An infinite loop may be due to a bug, or may be intentional, as in the case of a certain demos. Also called **endless loop**.

infinite series A series in which there is an infinite number of added quantities. An example is a Fourier series.

infinite transmission line Same as **infinite line**.

infinitesimal 1. A quantity that approaches zero as a limit. 2. A quantity, or change in quantity, that is extremely small. 3. A quantity so small that it can not be measured.

infinity Its symbol is ∞. 1. That which has no limits or boundaries. 2. A number or quantity which exceeds any other arbitrarily set large value. 3. A number or quantity so large that it can not be measured. 4. A number that exceeds that maximum value that can be stored in a register or memory location.

infix notation A manner of forming mathematical or logical expressions, in which the operators appear between the operands. For instance, $(A + B) \times C$. This contrasts with **prefix notation** where such an expression would appear as *+ABC, and with **postfix notation**, where such an expression would appear as ABC+*.

inflammable That which is easily ignited or burned. Also called **flammable**.

infobahn Abbreviation of **info**rmation Auto**bahn**, the latter being the German word for superhighway. Same as **information highway**.

infomediary Acronym for **info**rmation inter**mediary**. An entity which specializes in information retrieval, and which serves as an intermediary between those who want any given information, and those who can provide it. Some Web sites, for instance, function in this capacity.

informatics Same as **information science**.

information 1. Knowledge which has been acquired through research, experience, instruction, or the like. 2. A collection of facts pertaining to a given topic or area. 3. Data which has been processed, stored, transmitted, or otherwise been given additional meaning within a given setting. Nonetheless, **data** and information are usually used synonymously.

Information and Content Exchange A protocol based on XML which enables a Web site to collect information from other Web sites and display it. Used, for instance, for ecommerce. Its acronym is **ICE**.

information anxiety Any anxiety resulting from **information overload**.

information appliance 1. A device, intended for home use, which is connected to a network, so as to gather and/or distribute information. An example is an Internet appliance. Said especially of those which have an easy-to-use interface. Also called **smart appliance** (1). 2. A home appliance, such as a central air conditioner, which provides a user interface such as a touch screen, for simple programming. Also called **smart appliance** (2).

information bits Bits utilized to represent information, as opposed to those utilized for control purposes.

information center A location where computers, associated equipment, and workspaces are located. Such a center can serve for research, training, and other endeavors, and its resources may be used at the location itself, or remotely.

information channel A communications channel used for data transmission. Also called **data channel** (2).

information compression The encoding of information so that it occupies less space and/or bandwidth. There are many algorithms used for compression, and depending on the information being encoded, space savings can range from under 10% to over 99%. Information compression can also be achieved by increasing packing density. Also called **data compression**, or **compression** (2).

information content Information presented in any form, such as text, audio, video, or any combination of these. This content is obtainable, for instance, via satellite, through the Internet, or software-based products, such as CD-ROMs.

information digitization The conversion of information into digital form, which is necessary for digital computer processing. This information can be from any analog source, such as a voice or a voltage.

information encryption The coding of information so that only the intended recipients can understand it. It is an extremely efficient method to achieve data security, and a code, or key, is used to convert the information back to its original form. Public-key encryption and secret-key encryption are the two most common types.

information engineering The application of a set of techniques for the planning and development of an information system within an enterprise. Its abbreviation is **IE**.

information feedback system **1.** A communications system which utilizes feedback to verify the accuracy of the transmitted data. **2.** A control system which utilizes feedback to assist in monitoring an output quantity.

information hiding **1.** The hiding of the inner workings of a device, program, process, or system. A user simply employs such a device, program, process, or system without needing to be concerned with the complexities involved in their implementation. **2.** In object-oriented programming, the hiding of the implementation details of an object. The services the object provides are defined and accessible, but their internal workings are not. Also called **encapsulation (3)**.

information highway Also called **information superhighway**, or **Infobahn**. **1.** A worldwide network including the Internet, satellite, telephone, and cable communications, and so on. **2.** Same as **Internet**.

information link Also called **communications link**, or **data link**. **1.** A connection which enables the transmission of information between two points or entities. **2.** The resources which facilitate an **information link (1)**.

information management The tasks involved in the planning, safeguarding, use, and distribution of the information resources of an organization.

information overload An excess of information, such as that available via the Internet, TV, radio, and newspapers, which can lead a person to a state of desperation. This condition may result, for instance, from the frustration of knowing that it is not physically possible to assimilate all the information that is desired, the not being able to understand a given proportion of that one tries to digest, or the seemingly infinite barrage of advertisements, programming, and other content vying for attention.

information packet A block of data transmitted between one location and another within a communications network. Also called **data packet (1)**, or **packet (1)**.

information processing Also called **data processing, automatic data processing**, or **electronic data processing**. **1.** The processing of information by a computer. **2.** The processing of information within a specific application. For example, the manipulation of large amounts of numeric data in a program.

information protection Same as **information security**.

information resources Within an organization, information, such as that used for decision making, which is utilized as a resource which helps further its objectives.

information resources management The management of **information resources**. Its abbreviation is **IRM**.

information retrieval The process of locating information for subsequent use and/or analysis. Also, the methods employed for such retrieval. Its abbreviation is **IR**.

information science The science dealing with the gathering, processing, manipulating, storing, and retrieving of recorded knowledge. Also called **informatics**.

information security The safeguarding of data against loss, damage, unwanted modification, or unauthorized access. Such safeguards may be administrative, physical, or techni-

cal. Its acronym is **infosec**. Also called **information protection, data protection, data security**, or **protection (2)**.

information separator A control character which separates data units, such as the fields within a record.

information service Any source of information, especially that accessed through a network such as the Internet.

information services Within an organization, the department which handles data processing. Its abbreviation is **IS**.

information storage The act of storing information in a device or medium, such as a disk drive or tape, on a permanent or semi-permanent basis.

information superhighway Same as **information highway**.

information system A system used for one or more processes involving the handling of data. These processes may include the collection, processing, storage, retrieval, or transmission of data, and the devices, programs, channels, and equipment used for these purposes. Also called **data system**.

information technology The field dealing with the gathering, processing, manipulating, organizing, storing, securing, retrieving, presenting, distributing, and sharing of information, through the use of computers, communications, and related technologies. Its abbreviation is **IT**.

information theory A theory of communications dealing with the analysis of the processes utilized for the encoding, transmission, and reception of information. It helps determine, for instance, the reliability of a communications medium which has a given capacity and noise level, among other variables, through which information is transmitted at a specified rate.

information visualization The showing of computer data in the form of graphics. For example, the presenting of data using 3D models as opposed to text or formulas. Since graphical data is analyzed by the brain at a more intuitive level than text data, greater amounts of information can be more simply displayed, speeding up the navigation through and assimilation of the information being viewed. Also called **data visualization**, or **visualization (1)**.

information warehouse Within an organization, the combined data resources.

information warfare The disruption of the ability of an entity, such as a government, to utilize their computers, so as to create safety, infrastructure, and/or economic havoc.

information word A group of bits or bytes treated as a unit of information, as opposed, for instance, to being considered as a control code. Also called **data word**.

infosec Acronym for **information security**.

infotainment Abbreviation of **info**rmation and enter**tainment**. A given amount of purportedly useful and/or instructional material within an entertainment-based frame. Frequently, any available information is diluted or otherwise diminished by the entertainment component.

infra- A prefix denoting that which is below, or lower than. For example, infrared, or infrasonic.

infralow frequency A range of frequencies spanning from 300 to 3000 Hz, corresponding to wavelengths of 1000 km to 100 km, respectively. Its abbreviation is **ILF**. Also called **ultra low frequency**.

infrared Its abbreviation is **IR**. **1.** Within the electromagnetic spectrum, the portion extending from the lower limit of visible light, to the shortest microwaves. This encompasses from approximately 750 nanometers to approximately 1.0 millimeter, which is the same as 0.75 to 1000 micrometers. This interval itself is subdivided into the near, middle, and far infrared regions. Also called **infrared region**, or **infrared range**. **2.** Pertaining to, generating, detecting, utilizing, or sensitive to **infrared radiation**.

infrared absorption The absorption of infrared radiation. Such absorption, for instance, produces increased molecular vibrational and rotational activity, which can be detected spectroscopically. Its abbreviation is **IR absorption**.

infrared absorption spectroscopy A spectroscopic technique which measures the absorption of infrared radiation by a sample. Used, for instance, for the qualitative analysis of the structure of organic compounds. Its abbreviation is **IR absorption spectroscopy**.

infrared absorption spectrum A display or graph obtained through **infrared absorption spectroscopy**.

infrared astronomy The study of infrared radiation emitted from objects in the universe. Useful, for instance for gathering information from regions which are hidden from optical telescopes due to gas or dust. Its abbreviation is **IR astronomy**.

infrared band Same as **infrared spectrum** (1). Its abbreviation is **IR band**.

infrared beacon 1. A beacon which emits infrared signals which serve to assist marine or aeronautical navigation by guiding, orienting, or warning. 2. A beacon which emits infrared signals which serve as a navigational aid for a robot.

infrared communications Communications utilizing infrared radiation to convey information. Used, for instance, by infrared ports, or by infrared remote controls. Its abbreviation is **IR communications**.

Infrared Data Association Same as **IrDA**.

infrared detection Its abbreviation is **IR detection**. 1. The sensing of infrared radiation utilizing an infrared detector. 2. The use of infrared radiation to sense objects or phenomena.

infrared detector A device or instrument, such as a photocell, thermocouple, bolometer, or radiometer, which detects infrared radiation. Its abbreviation is **IR detector**. Also called **infrared sensor**.

infrared diode Same as **infrared-emitting diode**. Its abbreviation is or **IR diode**.

infrared emission The emission of infrared radiation by a body or medium. Sources include excited atoms or molecules, infrared-emitting diodes, and celestial bodies such as a planets or stars. Its abbreviation is **IR emission**.

infrared-emitting diode A diode which emits infrared radiation when a current is passed through it. Its abbreviation is **IRED**, or **IR emitting diode**. Also called **infrared diode**.

infrared energy Same as **infrared radiation**. Its abbreviation is **IR energy**.

infrared filter A filter which transmits infrared radiation, while absorbing other electromagnetic radiation. Used, for instance, in astronomy and in digital photography. Its abbreviation is **IR filter**.

infrared homing Homing in which infrared radiation is utilized follow the path of a scanned object. Its abbreviation is **IR homing**.

infrared image converter A device which converts infrared images into visible images. For example, an electron tube whose photosensitive surface converts an infrared image into a visible image which is projected onto a fluorescent screen.

infrared image tube An electron tube utilized as an **infrared image converter**.

infrared imaging Imaging in which the source of light is infrared radiation. Used, for instance, in infrared photography or in infrared microscopy. An example would be the detection of fires in areas where visibility is reduced, including hot spots below a surface. Its abbreviation is **IR imaging**.

infrared lamp An incandescent lamp whose emitted energy is mostly, or completely, in the infrared region. Used, for instance, to assist when viewing under conditions of reduced visible light. Its abbreviation is **IR lamp**.

infrared laser A laser which emits coherent infrared radiation. An example is a laser diode used in a CD player or a CD-ROM drive. Its abbreviation is **IR laser**.

infrared light Same as **infrared radiation**. Its abbreviation is **IR light**.

infrared microscope A microscope which utilizes infrared radiation to assist viewing which is unavailable or difficult in the visible range. Used, for instance, in semiconductor and medical research. Its abbreviation is **IR microscope**.

infrared microscopy The use of an **infrared microscope** for viewing and analyzing specimens. Its abbreviation is **IR microscopy**.

infrared motion detector A sensor which detects motion via changes in the infrared energy in a given area or enclosure. Used, for instance, in robotics, or for surveillance. Its abbreviation is **IR motion detector**.

infrared mouse A mouse which utilizes infrared signals to communicate with the computer, as opposed to a cord. Its abbreviation is **IR mouse**.

infrared photoconductor A photoconductor which transmits electricity when exposed to infrared radiation. Its abbreviation is **IR photoconductor**.

infrared photography Photography in which infrared radiation is utilized for illumination, and in which film that is sensitive to said light is utilized. Used, for instance, for photography under conditions of reduced visible light. Its abbreviation is **IR photography**.

infrared port A wireless port which utilizes infrared signals to send and/or receive data. Such ports usually require line-of-sight transmission and reception. May be used, for instance, between computers, between computers and peripherals, or between peripherals. Its abbreviation is **IR port**. Also called **fast infrared port**.

infrared radiation Electromagnetic radiation in the **infrared region**. Any material or entity at a temperature above absolute zero emits infrared radiation. Such radiation may be detected, for instance, by photoelectric cells, thermocouples, and bolometers. Its abbreviation is **IR**, or **IR radiation**. Also called **infrared light, infrared rays**, or **infrared energy**.

infrared radiometer A radiometer which detects and measures infrared radiation. Used, for instance, in space exploration. Its abbreviation is **IR radiometer**.

infrared range Same as **infrared** (1). Its abbreviation is **IR range**.

infrared rays Same as **infrared radiation**. Its abbreviation is **IR rays**.

infrared receiver A receiver which accepts infrared signals carrying information.

infrared reflow Same as **infrared reflow soldering**.

infrared reflow soldering A reflow soldering technique in which the source of heat is infrared radiation. Its abbreviation is **IR reflow**, or **infrared reflow**.

infrared region Same as **infrared** (1). Its abbreviation is **IR region**.

infrared remote control A remote control which utilizes infrared signals to send and/or receive signals between the transmitter and the controlled device or piece of equipment. Infrared remote controls usually require line-of-sight transmission and reception, and are frequently utilized to control consumer electronics, such as TVs, DVDs, high-fidelity audio components, and smart appliances. Its abbreviation is **IR remote control**.

infrared scanner A scanner which responds to infrared radiation. Used, for instance, for exploring areas where visibility is reduced. Its abbreviation is **IR scanner**.

infrared scanning Scanning using an **infrared scanner**. Its abbreviation is **IR scanning**

infrared sensor Same as **infrared detector**. Its abbreviation is **IR sensor**.

infrared signal A signal, such as that used by an infrared port or infrared remote control, in which infrared radiation is utilized to convey information. Its abbreviation is **IR signal**.

infrared soldering Soldering in which the heat is generated by infrared radiation. Used, for instance, for soldering semiconductor chips. Its abbreviation is **IR soldering**.

infrared spectrometer A spectrometer which detects and measures radiant intensities in the infrared region. Used, for instance, for the analysis of organic compounds in a sample. Its abbreviation is **IR spectrometer**.

infrared spectrometry The science and utilization of **infrared spectrometers** for analysis. Its abbreviation is **IR spectrometry**.

infrared spectrophotometer A spectrophotometer operating in the infrared region. Usually utilized to identify organic compounds by detecting molecular vibrational and rotational activity. Its abbreviation is **IR spectrophotometer**.

infrared spectrophotometry The science and utilization of **infrared spectrophotometers** for analysis. Its abbreviation is **IR spectrophotometry**.

infrared spectroscopy An analytical technique in which the wavelengths and corresponding intensities of infrared radiation absorbed by a sample are analyzed. An example is infrared absorption spectroscopy. Such techniques are well suited, for instance, for identification of organic compounds. A light source commonly utilized is a tungsten lamp, and the displayed or graphed output is called **infrared spectrum (2)**. Its abbreviation is **IR spectroscopy**.

infrared spectrum Its abbreviation is **IR spectrum**. **1.** The interval of wavelengths encompassing the **infrared region**. Also called **infrared band**. **2.** A display or graph obtained through **infrared spectroscopy**.

infrared thermometer An instrument or device which measures the emitted infrared radiation of a body, to determine its temperature. Used, for instance, when contact with the body is unadvisable or unduly difficult. Its abbreviation is **IR thermometer**.

infrared transmitter A transmitter which sends infrared signals which carry information.

infrared waves Electromagnetic waves whose wavelengths are in the **infrared region**. Its abbreviation is **IR waves**.

infrared welding Welding in which the heat is generated by infrared radiation. Used, for instance, for welding thermoplastics. Its abbreviation is **IR welding**.

infrasonic Pertaining to, generating, sensitive to, or utilizing **infrasonic frequencies**. Also called **subsonic**.

infrasonic frequency A frequency below the range that humans can hear. That is, below about 20 Hz. Also called **subsonic frequency**.

infrasonic phenomena Phenomena occurring within, or pertaining to **infrasonic frequencies**. Also called **subsonic phenomena**.

infrasonics The science dealing with the study and applications of infrasonic phenomena.

infrasound An acoustic-type disturbance whose frequency is below the range that humans can hear. That is, below about 20 Hz.

inharmonic distortion Distortion occurring at an **inharmonic frequency**.

inharmonic frequency A frequency which is not a whole-number multiple of a fundamental frequency.

inherent Same as **intrinsic**.

inherent error Same as **intrinsic error**.

inherent interference Same as **intrinsic interference**.

inherent noise Same as **intrinsic noise**.

inherent properties Same as **intrinsic properties**.

inherent regulation Same as **intrinsic regulation**.

inheritance In object-oriented programming, the ability of a class to confer properties to a class derived from it.

inherited error In a given sequence, an error which is carried forward from a previous step.

inhibit To stop, block, restrain, or impede from occurring.

inhibit input An input which stops or impedes an event from occurring. For example, an input which disables a gate circuit in a computer. Also called **inhibiting input**.

inhibit pulse A pulse which stops or impedes an event from occurring. For example, a pulse which serves as an inhibiting signal. Also called **inhibiting pulse**.

inhibit signal A signal which stops or impedes an event from occurring. For example, a signal which disables a gate circuit in a computer. Also called **inhibiting signal**.

inhibiting input Same as **inhibit input**.

inhibiting pulse Same as **inhibit pulse**.

inhibiting signal Same as **inhibit signal**.

inhibition The act or process of stopping, blocking, restraining, or impeding from occurring. Also, that which stops, blocks, restrains, or impedes.

inhibitor That which **inhibits**. For example, a substance which impedes, retards, or stops a reaction.

initial charge The charge given to a new battery before placing it in service, or when it is first installed.

initial failure The first instance in which a component, circuit, device, piece of equipment, or system in use fails.

Initial Graphics Exchange Specification An ANSI graphics file format especially suited for describing CAD-created models. Its abbreviation is **IGES**.

initial instructions Computer instructions which aid in the loading of programs into memory.

initial ionizing event In an instrument or device which detects and quantifies ionizing radiation, such as a Geiger counter, the ionizing event which starts the chain of events leading to a count. Also called **primary ionizing event**.

initial permeability The permeability of a material with zero initial magnetization, at small flux densities, such as those under 10 gauss.

initial program load Its abbreviation is **IPL**. **1.** To start up or reset a computer. During this process the computer accesses instructions from its ROM chip, performs self-checks, loads the operating system, and prepares for use by an operator. Such a process may be initiated by turning on the power, pressing a button or switch, by hitting a specific key sequence, or through a program or routine that gives this command. Also called by various other names, including **bootstrap (1)**, **booting**, **boot (1)**, **booting up**, and **bootstrapping (1)**. **2.** To load the operating system into memory during an **initial program load (1)**.

initial surge **1.** A current surge occurring when power is first applied to a circuit, device, piece of equipment, or system. A manifestation is the momentary dimming of the lights in a house when an air conditioner starts. **2.** Same as **initial surge voltage**.

initial surge voltage A voltage surge occurring when electrical power is restored after a power failure. Such a surge may damage connected equipment. Also called **initial surge (2)**.

initial time delay At a given location, the time difference between the arrival of a direct sound wave, and that of the first reflected sound wave.

initial voltage The voltage of a battery at the beginning of a discharge.

initialization 1. In a computer, the process of setting variables, counters, addresses, and the like to their starting value, such as zero. This may occur, for instance, at the start of a routine. **2.** The process of preparing a storage medium, such as a floppy or hard disk, for use. This may include locating bad sectors, indexing, and so on. **3.** The powering up of a computer. **4.** The process of preparing a device or piece of equipment for operation.

initialization string A sequence of events which prepare a peripheral, such as a printer or modem, for proper operation.

initialize 1. In a computer, to set all variables, counters, addresses, and the like to their starting value, such as zero. This may occur, for instance, at the start of a routine. **2.** To prepare a storage medium, such as a floppy or hard disk, for use. This may include locating bad sectors, indexing, and so on. **3.** To power up a computer. **4.** To prepare a device or piece of equipment for operation.

initiate To begin, or otherwise set an operation or process in motion. For instance, to start a transmission.

initiation The act of beginning, or otherwise setting an operation or process in motion. Also an instance of such an act.

inject 1. To apply a signal to a circuit or device. **2.** To introduce charge carriers into a semiconductor.

injected laser Same as **injection laser diode**.

injection 1. The application of a signal to a circuit or device. Also, the process utilized. Also called **signal injection**. **2.** The introduction of charge carriers into a semiconductor. Also, the process utilized. Also called **carrier injection**.

injection efficiency A measure of the efficiency of a pn junction when a forward bias is applied, expressed as the ratio of the current carried by injected current carriers to the total current across said junction.

injection electroluminescence In a semiconductor, light emission resulting from electron-hole recombination as a result of minority carrier injection.

injection laser Same as **injection laser diode**.

injection laser diode A laser in which a forward-biased pn junction diode is used to convert its DC input into a coherent light output. Used, for instance, as a light-pulse generator for transmission of information over fiber-optic lines. Also called **injection laser**, **injected laser**, **diode laser**, **laser diode**, or **semiconductor laser**.

injector That which serves to **inject**.

injector laser Same as **injection laser diode**.

ink jet cartridge Same as **inkjet cartridge**.

ink jet printer Same as **inkjet printer**.

inkjet cartridge A cartridge containing the ink utilized in an **inkjet printer**. For instance, such a printer may utilize four separate cartridges, one each for cyan, magenta, yellow, and black, for better efficiency and quality, as compared to a single cartridge. Also spelled **ink jet cartridge**.

inkjet printer A non-impact printer which sprays small ink droplets onto the paper or other medium being printed upon. Such printers are characterized by comparatively quiet operation and the ability to print colors, but usually have lower resolution than laser printers and may smudge on certain types of paper. Also spelled **ink jet printer**.

inline code Instructions from a low-level computer programming language, such as assembly language, which are embedded in the source code of a high-level computer programming language, such as C. Also spelled **in-line code**.

inline graphics Graphics which are embedded within an HTML document, or the body of a Web page. Such graphics are usually loaded automatically by the Web browser. Also spelled **in-line graphics**.

inline image An image which is embedded within an HTML document, or the body of a Web page. Such images are usually loaded automatically by the Web browser. Also spelled **in-line image**.

inline tuning Same as **in-line tuning**.

inner conductor In a concentric two-conductor line, such as a coaxial cable, the conductor which is on the inside. This conductor usually carries the signal, while the **outer conductor** serves as a shield. A dielectric separates the two conductors. Also called **inside conductor**, or **center conductor**.

inoculate To store information about a file, piece of code, or program, for later verification of the integrity of said file, piece of code, or program. For instance, a checksum may be performed each time a program is run, and if the checksums do not match, there may be a virus, or other infection, present.

inoculation The process of storing information about a file, piece of code, or program, for later verification of the integrity of said file, piece of code, or program.

input Its abbreviation is **I**, or **in**. **1.** That which is applied or otherwise entered into a system, so as to obtain an output or other result. For instance, an applied power to accomplish work. **2.** The energy, voltage, current, or other signal applied to a component, circuit, device, piece of equipment, system, or process. Also, the application of such a signal. For example, a voltage input provided to an electronic device. **3.** The information which is entered into a computer for processing, storing, displaying, or otherwise handling. Also, the entering of such information. Such information may be entered manually or automatically, and this input may be provided by a user, a scanner, a storage device, and so on. Also called **input data (1)**. **4.** The terminals of a component, circuit, device, or piece of equipment to which an **input (2)** is delivered. Also called **input terminals**.

input admittance For a circuit, device, or transmission line, the admittance as seen from the input terminals. It is the reciprocal of **input impedance**.

input area Within the memory or storage of a computer, a segment reserved for the input of data. Also called **input block (1)**, or **input section**.

input bias current 1. A current which is applied to a component or device to establish a reference level for its operation. For instance, the current applied to the base-emitter junction of a transistor to set its operating point. Also called **bias current**. **2.** The current flowing into the inputs of a circuit or device.

input block 1. Same as **input area**. **2.** Same as **input buffer**. **3.** A block of data entered into a computer.

input-bound Same as **input-limited**.

input buffer A segment of computer memory utilized to temporarily store information that awaits transfer or processing. Used, for instance, to compensate for differences in operating speeds. Also called **input block (2)**, **buffer (1)**, or **buffer memory**.

input capacitance For a circuit or device, the capacitance as seen from the input terminals.

input channel A channel via which an input is conveyed.

input circuit A circuit connected to the input electrodes or terminals of a component or device. For example, an exterior circuit connected to an input electrode of an electron tube.

input conductance For a circuit or device, the conductance as seen from the input terminals.

input current 1. The current flowing through an **input circuit**. **2.** The current flowing through an input circuit under specified conditions.

input data 1. Same as **input** (3). 2. Data that awaits to be input.

input device Also called **input unit**, or **input equipment** (2). 1. A peripheral, such as a keyboard, mouse, or disk drive, which provides an input for a computer. 2. Any device which provides an input for a component, circuit, device, piece of equipment, or system.

input electrode An electrode, such as that of an electron tube, to which an **input** (2) is applied.

input equipment 1. Any equipment, such as data-entry equipment or bar-code scanners, utilized to provide an input for a computer. 2. Same as **input device**.

input impedance For a circuit, device, or transmission line, the impedance as seen from the input terminals. It is the reciprocal of **input admittance**.

input-limited A computing process which can proceed only as fast as the input. This contrasts with **processor-limited**, in which the limiting factor is the speed of the CPU. Also called **input-bound**.

input offset current For an operational amplifier, the difference between the two currents that must be supplied to the input terminals so that the output indication is zero.

input offset voltage For an operational amplifier, the voltage that must be applied between the two input terminals so that the output indication is zero volts.

input/output Same as **I/O**.

input/output address Same as **I/O address**.

input/output area Same as **I/O area**.

input/output bound Same as **I/O bound**.

input/output buffer Same as **I/O buffer**.

input/output bus Same as **I/O bus**.

input/output card Same as **I/O card**.

input/output channel Same as **I/O channel**.

input/output control Same as **I/O control**.

input/output controller Same as **I/O controller**.

input/output data Same as **I/O data**.

input/output device Same as **I/O device**.

input/output handler Same as **I/O handler**.

input/output instruction Same as **I/O instruction**.

input/output intensive Same as **I/O intensive**.

input/output interface Same as **I/O interface**.

input/output interrupt Same as **I/O interrupt**.

input/output limited Same as **I/O limited**.

input/output port Same as **I/O port**.

input/output process Same as **I/O process**.

input/output processor Same as **I/O processor**.

input/output statement Same as **I/O statement**.

input/output switching Same as **I/O switching**.

input/output time Same as **I/O time**.

input/output unit Same as **I/O unit**.

input overcurrent An input current exceeding a specified or rated current.

input overcurrent protection Protection, such as fuses, utilized against **input overcurrents**. Used, for instance, to protect chips. Also called **input protection** (1).

input overvoltage An input voltage exceeding a specified or rated voltage.

input overvoltage protection Protection against **input overvoltages**. Used, for instance, to protect chips. Also called **input protection** (2).

input port 1. Same as **I/O port**. 2. A port, such as a keyboard port, which serves for the input of data to a computer.

input power 1. The power presented to the input terminals of a component, circuit, device, piece of equipment, or system. Also called **power input** (1). 2. The power consumed by a circuit or device.

input program A computer program which accepts input data, such as that from a keyboard, and stores it appropriately. Such a program may also perform other functions, such as error-checking, or automatic updating. Also called **data entry program**.

input protection 1. Same as **input overcurrent protection**. 2. Same as **input overvoltage protection**.

input register A computer register which receives input at one speed, and delivers it to the CPU at another, usually greater, speed.

input resistance For a circuit or device, the resistance as seen from the input terminals.

input resonator In a velocity-modulated tube, such as a klystron, the input cavity resonator, where electron velocities are modulated so that they form a bunch. Also called **buncher**, or **buncher cavity**.

input section Same as **input area**.

input sensitivity 1. The input signal level which will drive a device, such as an amplifier, to its full rated output, or to a stated output. 2. The minimum input signal level which will actuate a device, such as a switch.

input sensitivity control A control which serves to vary the **input sensitivity** of a device.

input signal An external signal, such as a voltage or current, which drives a circuit, device, piece of equipment, system, process, or mechanism. Also called **signal input**.

input signal level Also called **signal input level**. 1. The magnitude of an **input signal** which produces a stated output. 2. The magnitude of an input signal.

input storage Within the storage of a computer, a segment reserved for the input of data.

input stream In a computer, a series of control statements associated with the performance of a given task.

input terminals Same as **input** (4).

input transformer A transformer which delivers energy to an **input circuit**. Such a transformer usually also matches impedances.

input transient A transient occurring at the inputs of a circuit or device.

input unit Same as **input device**.

input voltage 1. The voltage across an **input circuit**. 2. The voltage across an input circuit under specified conditions.

input-voltage range 1. The interval of **input voltages** which a circuit, device, piece of equipment, or system can accept, so as to function within given parameters. 2. The interval of input voltages which a circuit, device, piece of equipment, or system can accept without overloading, malfunctioning, failing, or the like.

inquiry In computers or communications, any request for information.

inrush Same as **inrush current**.

inrush current Also called **inrush**. 1. The peak instantaneous current drawn by a circuit or device when first turned on. 2. The peak instantaneous current drawn by a power supply when first turned on.

inrush current limiting A circuit design which limits the value of an **inrush current**. Also, the limiting so provided.

Ins key Abbreviation of **insert key**.

insect robot Multiple robots which work under the supervision of a single controller. Depending on the task, up to several thousand insect robots can work in unison. This contrasts with **autonomous robots**, which each have their own controller.

insensitive time The time interval after a signal or event, during which a device, piece of equipment, or system is unable to respond to additional signals or events. During this time, for instance, a radiation counter can not register ionizing radiation. Also called **dead time (3)**, or **recovery time (3)**.

insert **1.** To place, or to put into or between. **2.** That which is placed, or to put into or between. Also, that which serves to be placed, or to be put into or between. For example, removable components of a die. **3.** To introduce data between existing data items. **4.** Same as **insert key**.

insert earphone An earphone which is partially inserted in the outer ear, and which is used especially for tests.

insert key On computer keyboards, a key whose function is to toggle between insert mode and overwrite mode, although it may also have different functions depending on the application running. Its abbreviation is **Ins key**. Also called **insert (4)**.

insert mode In a computer application, a mode of entering data in which any characters typed on the keyboard are placed between characters already present. This contrasts with **overwrite mode**, in which any newly typed characters replace those already occupying the same area.

insertion force The force required to properly insert plugs, contacts, terminals, adapters, or the like, into the corresponding connector.

insertion gain The gain in power, amplitude, current, or the like, resulting from the insertion of a component or network between portions of a system. For instance, it may be the increase in power resulting when an amplifier is placed between the source and the load. Usually expressed in decibels.

insertion head A apparatus utilized to insert components in a printed-circuit board.

insertion loss The reduction in power, amplitude, current, or the like, resulting from the insertion of a component or network between portions of a system. For instance, it may be the decrease in power resulting when a filter or attenuator is placed between the source and the load. Usually expressed in decibels.

insertion phase shift The phase shift resulting from the insertion of a component or network between portions of a system.

insertion point **1.** On a display, especially that of a computer screen, an indicator such as a blinking underline or a solid rectangle which indicates the location where a keystroke will appear on screen. Also called **cursor (1)**. **2.** On a screen pointer, the specific location which at which an action will be effected. For instance, the tip of a screen pointer in the shape of an arrow. Also called **hot spot (4)**.

insertion tool A tool which assists in the proper insertion of contacts into a connector, connectors into circuit boards, and so on. Also called **connector insertion tool**.

inside antenna Same as **indoor antenna**.

inside diameter For an object with two concentric diameters, such as a pipe, the diameter corresponding to the inner circumference. The **outside diameter** corresponds to the outer circumference. Its abbreviation is **ID**.

inside lead The lead attached to the inner, or first, turn of a coil. This contrasts with an **outside lead**, which is connected to the last turn of a coil. Also called **start lead**.

inside wiring The wiring that is on the customer side of a telephone network interface.

inst **1.** Abbreviation of **instant**. **2.** Abbreviation of **instrument**.

instability The quality of being erratic, fluctuating, or otherwise unstable. For instance, a control system which persis-

tently oscillates between output states in an unwanted manner, due to excessive feedback. Another example is a circuit component whose electrical parameters vary in an unpredictable way.

install To connect or otherwise set in place, so as to prepare for use. For example, to install a computer application by inserting an optical disc and following the prompts until completion.

install program Same as **installation program**.

installation The act or process of connecting or otherwise setting in place, so as to prepare for use. Also, the state of being installed.

installation program A program which installs another, such as an application. It takes care of tasks such as verifying the configuration of the system and creating the appropriate folders. Such a program usually only requires a user to insert a disc and follow the prompts. An installation program may also be used to install hardware. Also called **install program**, or **setup program**.

installation requirements Same as **installation specifications**.

installation specifications Requirements which must be followed so that a component, device, piece of equipment, or system to be installed functions properly and/or safely. Also called **installation requirements**.

installation time The time required for a proper **installation**.

installer A program or device which serves to **install**.

InstallShield A popular program utilized for installing and uninstalling software.

instance **1.** An occurrence of a given event. **2.** A given step within a series or process. **3.** In object-oriented programming, an individual object within a given class. For instance, a triangle, square, pentagon, or hexagon within a class titled geometric shapes.

instance variable In object-oriented programming, a variable which contains information pertaining to a specific instance of the class.

instant Its abbreviation is **inst**. **1.** An extremely brief, or imperceptibly short interval of time. **2.** A specified moment in time. **3.** Without any delay.

instant messaging **1.** A communications service in which users are able to send other users real-time text messages. Such a service usually has a feature where users, after obtaining authorization from other selected users, can see when they are logged-on, in addition to being able to exchange text messages or files, engage in voice or video conferencing, and so on. Also called **instant messenger**, or **instant messaging service**. **2.** The utilization of an **instant messaging service**.

instant messaging service Same as **instant messaging (1)**.

instant messenger Same as **instant messaging (1)**.

instantaneous **1.** Occurring, acting, or completed extremely quickly, or within an imperceptibly short interval of time. **2.** Occurring, acting, or present at a specified moment in time. **3.** Occurring or acting without any delay.

instantaneous AGC Abbreviation of **instantaneous automatic gain control**.

instantaneous amplitude The amplitude of a signal at a specified moment in time.

instantaneous automatic gain control An automatic gain control circuit which is activated immediately when the amplitude of the input signal changes. This contrasts with **delayed automatic gain control**, in which the circuit is activated only when the input signal exceeds a predetermined magnitude. Its abbreviation is **IAGC**, or **instantaneous AGC**. Also called **instantaneous automatic volume control**.

instantaneous automatic volume control Same as **instantaneous automatic gain control**. Its abbreviation is **IAVC**.

instantaneous companding Companding in which the variations are made in response to the instantaneous value of a signal.

instantaneous condition The condition of a dynamic system at a specified moment in time.

instantaneous contacts Contacts, such as those of a timer, which are actuated the instant the driving signal is applied.

instantaneous current The value of a varying current, such as AC, at a particular instant within a cycle. Also called **instantaneous current value**.

instantaneous current value Same as **instantaneous current**.

instantaneous effect Any effect resulting from instantaneous changes in parameters such as amplitude, power, frequency, or impedance. An example is a failure which may occur due to a current surge.

instantaneous frequency 1. The frequency of a signal at a specified moment in time. 2. The time rate of change of a phase angle of a wave divided by 2π.

instantaneous magnitude 1. The magnitude of a varying quantity, such as a current or voltage, at a specified moment in time. 2. The magnitude of a signal at a specified moment in time.

instantaneous power The rate at which power is delivered to a load at a specified moment in time. Also called **instantaneous power output**.

instantaneous power output Same as **instantaneous power**.

instantaneous relay A relay which is actuated the instant the driving signal is applied.

instantaneous sample An individual measurement obtained during **instantaneous sampling**.

instantaneous sampling Sampling in which instantaneous values of a signal or wave are measured.

instantaneous sound pressure The sound pressure at a given point, at a specified moment in time. Used, for instance, to monitor impact noises.

instantaneous switch A switch that is actuated the instant the driving signal is applied.

instantaneous value The value of a varying quantity, such as current or voltage, at a specified moment in time.

instantaneous voltage The value of a varying voltage, such as alternating voltage, at a particular instant within a cycle. Also called **instantaneous voltage value**.

instantaneous voltage value Same as **instantaneous voltage**.

instantiate In object-oriented programming, the creation of an **instance (3)**.

Institute of Electrical and Electronics Engineers Same as **IEEE**.

instruction 1. A command or statement in a computer program or routine. Also called **computer instruction (1)**. 2. A computer instruction in machine code. Such an instruction can be directly executed by a processor. Also called **machine instruction**, or **computer instruction (2)**.

instruction address An address indicating the location of a computer instruction.

instruction address register 1. A register which holds the address of the instruction which is next to be executed while running a program. 2. Same as **instruction register**.

instruction code A coded value or bit string within a machine instruction which specifies the operation to be performed by a processor. The operation may be a branch, add, copy, and so on. Also called **operation code**, or **opcode**.

instruction counter Same as **instruction register**.

instruction cycle The time interval during which an instruction is fetched from memory, decoded, and executed.

instruction fetch To locate an instruction in computer memory and load it into a CPU register. Once an instruction is fetched, it can then be executed. Also called **fetch**.

instruction format The components and layout of an instruction.

instruction mix The different types of instructions contained in a program. For example, I/O instructions, or control instructions.

instruction modification A change in an instruction that results in a different operation being performed when the same instruction is executed the next time.

instruction pointer Same as **instruction register**.

instruction register In a CPU, a register that contains the address of the location in memory that is to be accessed by the next instruction. May also refer to the address of the current instruction. Also called by various other terms, including **instruction counter, instruction address register, instruction pointer, control register, program counter, program register**, and **sequence register**.

instruction set The complete set of machine instructions that a CPU can recognize and execute.

instruction time 1. The time required for an instruction to be fetched from memory. It is the first part of an instruction cycle. 2. The time required to execute an instruction.

instruction word A computer word containing an instruction.

instrument Its abbreviation is **inst**. 1. A device utilized to directly or indirectly measure, indicate, and/or monitor the value of an observed and/or controlled quantity. Such an instrument may also record these variations. There are many examples, including altimeters, ammeters, bridges, circuit analyzers, compasses, digital multimeters, frequency meters, oscilloscopes, and spectrometers. Also called **measurement instrument**. 2. A device which enables the playing or production of music. For instance, an electric guitar or piano. 3. That which is dependent on one or more **instruments (1)**. For example, instrument flying.

instrument accuracy The extent to which a value indicted by an instrument approximates the real value.

instrument amplifier Same as **instrumentation amplifier**.

instrument approach A landing approach utilizing an **instrument approach system**.

instrument approach system A radio navigation system which provides an aircraft with the information necessary for a safe approach. This includes indications of lateral, longitudinal, and vertical guidance during descent from a given altitude, until reaching a point where a landing can be completed.

instrument damping The reduction or limiting of the amplitude of movement of the indicator of an instrument, such as a galvanometer or volume meter, to minimize oscillation or overshoot. An instrument whose damping is sufficient for it to proceed from one reading to the next without oscillating or overshooting is called deadbeat instrument.

instrument error Same as **instrumental error**.

instrument flight Same as **instrument flying**.

instrument flying The flying of an aircraft relying solely on the instrumentation and communications. This usually necessary when visibility is inadequate. Also called **instrument flight**, or **blind flying**.

instrument housing A housing which encloses, supports, and protects an instrument.

instrument lamp Same as **instrument light**.

instrument landing The landing of an aircraft relying solely on instrumentation and communications. This usually necessary when visibility is inadequate. Also called **blind landing**.

instrument landing system A radio navigation system which provides aircraft with the information necessary for a safe approach and landing. This includes indications of distance from the optimum point for landing, and lateral, longitudinal, and vertical guidance through landing. Such a system incorporates multiple land transmitting stations, whose signals are received by the appropriate instrumentation on the aircraft. Used, for instance, when an instrument landing is necessary. Its abbreviation is **ILS**.

instrument light A light utilized to illuminate an **instrument** (1), or its face. Also called **instrument lamp**.

instrument multiplier 1. A precision resistor utilized in series, to extend the voltage range of a voltmeter. 2. A precision resistor utilized in series, to extend the voltage range of an instrument. Also called **voltage-range multiplier (2)**, or **voltage multiplier (2)**.

instrument panel A panel upon which instruments and/or their indicators are mounted. Used, for instance, to monitor and control a system. Also called **panel (2)**.

instrument range The interval within which an instrument operates. Such an interval may consist of a range of frequencies, voltages, currents, resistances, and so on.

instrument relay A sensitive relay whose moving contact is attached to a pointer, and which is utilized to indicate values along the scale of a meter. Also called **meter relay**.

instrument resistor A precision resistor utilized in certain instruments. For example, such a resistor may be utilized in series to extend the voltage range of an instrument.

instrument sensitivity 1. The minimum change in a measured quantity that produces an observable change in the indication or output of a measuring instrument. 2. The minimum change in an observed quantity that a measuring instrument can detect.

instrument shunt A resistor which is connected in parallel, to extend the current range of an instrument. For example, such a shunt utilized to extend the range of an ammeter. In addition, such a resistor helps provide protection against current surges. Also called **shunt (4)**.

instrument system A system incorporating one or more instruments which serve a common purpose. Such an arrangement may assist, for instance, in monitoring and controlling flights, communications, or manufacturing.

instrument takeoff The takeoff of an aircraft relying on instrumentation and communications. Such a takeoff may be utilized to help keep an aircraft clear of obstructions until reaching a given height and distance.

instrument transformer A transformer whose secondary is connected to an instrument, and whose output is a precise fraction or multiple of its input current or voltage, while preserving the proper phase relationship. A current transformer is used to increase or decrease current, while a voltage transformer is utilized to increase or decrease voltage, thus extending the range of an instrument.

instrumental Of, pertaining to, or performed by one or more **instruments**.

instrumental error An error in a measurement and/or indication of an instrument. Such an error may result, for instance, from incorrect use, improper calibration, or to said instrument being damaged. Also, the magnitude of such error. Also called **instrument error**.

instrumentation 1. The use or application of **instruments**. 2. The science dealing with the research, development, manufacturing, and use of instruments. 3. The instruments utilized for a given purpose. 4. The instruments available for a given purpose.

instrumentation amplifier A precision amplifier whose input signal is a voltage, and whose output is a linearly scaled ver-

sion of this signal. Such amplifiers are characterized by high input impedance and high common-mode rejection. Instrumentation amplifiers are especially suited for use in electronic instruments, and may serve, for instance, to increase their sensitivity. Also called **instrument amplifier**.

insulant Same as **insulator**.

insulate Also called **isolate (1)**. 1. To prevent the conduction of electrical, heat, or sound energy into or out of a material or body through the use of an **insulator (1)**. 2. To prevent the passage of electric current through a material or body through the use of an **insulator (2)**. For example, to surround a conductor with an insulating material to isolate it from another.

insulated To be protected, surrounded, or otherwise isolated by an **insulator**.

insulated cable A cable which is surrounded by one or more layers of an **insulator (2)**. This helps prevent shocks, current leakages, short circuits, and the like.

insulated conductor A conductor which is surrounded by one or more layers of an **insulator (2)**. This helps prevent shocks, current leakages, short circuits, and the like.

insulated enclosure An enclosure which is protected, surrounded, or otherwise isolated by an **insulator**.

insulated-gate bipolar transistor Same as **IGBT**.

insulated-gate field-effect transistor Same as **IGFET**.

insulated wire A wire which is surrounded by one or more layers of an **insulator (2)**. This helps prevent shocks, current leakages, short circuits, and the like.

insulating barrier A device, structure, or material which serves as an **insulator**. For instance, an insulating material serving as a barrier to electricity.

insulating layer A layer of an **insulator**. Also called **insulation layer**.

insulating material Same as **insulator**.

insulating oil A quality oil utilized as an **insulator (2)**. An example is transformer oil.

insulating tape A tape consisting of, incorporating, or impregnated with an insulating material. Depending on the specific needs, such tapes, aside from electrical insulation, may offer abrasion resistance, mechanical strength, resistance to heat, or protection against harsh weather. Also called **insulation tape**, or **electrical tape**.

insulation 1. The use of an **insulator**. 2. The placing or application of an insulator. 3. Having the isolation, protection, and/or support that an insulator provides. 4. Same as **insulator**.

insulation breakdown A disruptive electrical discharge through an **insulator (2)** separating circuits. An example is a flashover.

insulation layer Same as **insulating layer**.

insulation rating A measurement, such as dielectric strength or thermal resistance, that helps rate insulators.

insulation resistance 1. The electrical resistance between two conductors separated by an **insulator (2)**. 2. The level of resistance an **insulator (2)** provides to the passage of electric current. Usually expressed in multiples of ohms, such as megohms, for a given interval of voltages.

insulation system The combined insulating materials utilized to provide protection to devices, equipment, facilities, and systems.

insulation tape Same as **insulating tape**.

insulator Also called **insulating material**, **insulant**, or **insulation (4)**. 1. A material which prevents the conduction of electrical, heat, or sound energy. 2. A material that has a sufficiently high resistance to the passage of electric current, so that current flow through it is minimal or negligible. Used, for instance, to isolate, protect, and support circuits and conductors, and to prevent the loss of current. Examples include rubber, plastic, ceramic, and glass. Also, an object or device made of such a material. Also called **electrical insulator**.

insulator arc-over A disruptive discharge in the form of an arc over the surface of an insulator.

insured burst The greatest burst of data above the **insured rate** that an ATM network will temporarily allow.

insured rate The bandwidth an ATM network guarantees under normal network conditions.

integer The set of numbers including zero, and all positive and negative whole numbers. Also called **integral number**.

integer arithmetic Arithmetic in which fractions are ignored. For example, $11/2 = 5$, and not 5.5. It is used for some functions in computers, while floating-point arithmetic is used for others.

integral 1. That which is essential for completeness. Also, that which is complete. 2. Pertaining to, expressed as, or able to be expressed in terms of integers. 3. For a function, the area under a curve above the x-axis.

integral action In a control system, a corrective action whose speed is dictated by the magnitude and duration of any change of the controlled variable. Also called **reset action (5)**.

integral control An automatic control process whose rate of corrective action is dictated by the magnitude and duration of any change of the controlled variable.

integral number Same as **integer**.

integrate 1. To make into a whole by uniting all the component parts. 2. To incorporate into a larger entity. 3. To perform integration.

Integrated Access Device A type of customer-premises equipment enabling broadband access to multiple services, such as analog telephones, PBXs, and data services such as fax and Internet access. Its abbreviation is **IAD**.

integrated amplifier An audio-frequency amplifier which incorporates both a pre-amplifier and a power amplifier in the same chassis. When an integrated amplifier also includes a tuner, it is called **receiver (2)**.

integrated capacitor A capacitor incorporated into an IC. An example is a junction capacitor.

integrated circuit Same as **IC**.

integrated-circuit card Same as **IC card**.

integrated-circuit card reader Same as **IC card reader**.

integrated-circuit memory Same as **IC memory**.

integrated-circuit package Same as **IC package**.

integrated-circuit tester Same as **IC tester**.

integrated-circuit testing Same as **IC testing**.

integrated component A component incorporated into an IC. An example is an integrated resistor.

integrated computer-aided software engineering Same as **I-CASE**.

integrated development environment A software development environment which is integrated into a single application. For instance, an application which coordinates source code and resources, and handles compiling, linking, and the like. Its abbreviation is **IDE (2)**.

Integrated Device Electronics Same as **IDE (1)**.

integrated digital network A communications network utilizing both digital transmission and digital switching. Its abbreviation is **IDN**.

Integrated Drive Electronics Same as **IDE (1)**.

Integrated Drive Electronics interface Same as **IDE (1)**.

integrated electronics The branch of electronics pertaining to the design, manufacturing, optimization, packaging, and utilization of integrated electronic components, such as ICs.

integrated-injection logic IC logic utilizing bipolar transistor gates, and which is characterized by very high packing density, high speed, and low power consumption. Used, for instance, for logic arrays, and various analog and digital applications. Its abbreviation is **IIL**, or **I²L**. Also called **merged transistor logic**.

integrated optical circuit One or more circuits which incorporate, or consist of, solid-state optical components. Examples include photodetectors, and thin-film optical waveguides. Its abbreviation is **IOC**.

integrated programming environment An integrated collection of software resources which are utilized in the development of programs. Its abbreviation is **IPE**.

integrated resistor A resistor incorporated into an IC. An example is a diffused resistor.

Integrated Services Digital Network Same as **ISDN**.

Integrated Services Digital Network line Same as **ISDN line**.

Integrated Services Digital Network modem Same as **ISDN modem**.

Integrated Services Digital Network terminal adapter Same as **ISDN terminal adapter**.

integrated software A software package which seamlessly combines many of the functions provided by separate applications. For instance, that combining a word processor, spreadsheet, and database programs. When using such a package, coordination of information, resources, and tasks is simplified. Also called **integrated software package**.

integrated software package Same as **integrated software**.

integrating amplifier An amplifier whose output is proportional to the integral, with respect to time, of the input signal.

integrating circuit A circuit or electric network whose output is proportional to the integral, with respect to time, of the input signal. For example, a circuit whose output voltage is proportional to the integral of the input voltage with respect to time. The output waveform of such a circuit is the time integral of the input waveform. Also called **integrating network**, **integrator (2)**, or **integrator circuit**.

integrating meter A meter whose indications are the total over time of a measured quantity, such as current. An example is an ampere-hour meter.

integrating network Same as **integrating circuit**.

integrating photometer A photometer whose indications are the total over time of the luminous intensity of a source.

integrating wattmeter A wattmeter whose indications are the total electric power consumed over time.

integration 1. The process of combining constituents to form a whole. Also, a specific instance of such a process, and the state of completeness resulting. 2. The process of placing multiple transistors on a chip. 3. The finding of the value of an **integral (3)**.

integrator 1. That which serves to combine constituents to form a whole. 2. Same as **integrating circuit**. 3. A device which performs **integration (3)**.

integrator circuit Same as **integrating circuit**.

integrity The preservation of the completeness and accuracy of data that is transmitted and received, stored and retrieved, or manipulated in any manner. Also called **data integrity**.

intelligence 1. The ability of a computer to process information. A dumb terminal, for instance, has no processing capability, and thus lacks intelligence. 2. The ability of a device or system to solve problems autonomously, especially when previous problem-solving is effectively applied to future problem solving. 3. The ability of a device or system, such as a robot, to appropriately respond to stimuli. 4. The audio, video, text, or other information conveyed in a signal.

For example, the music utilized to modulate a carrier in an FM signal. Also called **signal intelligence**.

intelligence signal Within a transmission, the signal containing the **intelligence** (4), as opposed, for instance, to overhead or a carrier.

intelligent Characterized by, or possessing **intelligence**.

intelligent agent 1. A computer program that helps automate repetitive tasks, or those that are scheduled. An example is its use in a hand-held personal scheduler. Also called **agent** (2). 2. A computer program that serves to automate the locating and delivering of information over the Internet. Used in Web browsers. Also called **agent** (3).

intelligent cable A cable which incorporates a microprocessor to analyze the signals passing through it, as opposed to just serving as a transfer medium. Also called **smart cable**.

intelligent character recognition Character recognition in which the system can be configured to recognize most any hard to decipher hand-printed or machine-printed characters. Its abbreviation is **ICR**.

intelligent controller A peripheral controller with its own microprocessor.

intelligent database A database which facilitates and enhances the handling of data, by performing some of the processing tasks usually performed by other applications. Such a database may follow intuitively logical rules for searching, provide data validation, and so on.

Intelligent Drive Electronics Same as **IDE**.

intelligent hub Also called **smart hub**. 1. A network hub which performs processing functions such as network management. 2. A network hub which can be electronically reconfigured, be it locally or remotely.

intelligent I/O Same as **I2O**.

intelligent input/output Same as **I2O**.

intelligent network A network with processing power, so that functions such as diagnostics can be incorporated. This contrasts with a **dumb network**, in which there is little or no processing of the transmitted signals.

intelligent robot A robot that has enough sensory input and decision making capability to be able to perform tasks without human intervention.

intelligent terminal A network terminal which incorporates a CPU and memory, so that it has processing capability. This contrasts with a **dumb terminal**, which has no processing capability.

Intelligent Transportation System A transportation network, such as a highway or rail system, which incorporates technologies which enhance safety, reduce delays, save fuel, and so on. Its abbreviation is **ITS**.

intelligibility The extent to which a received signal is **intelligible**. Also called **signal intelligibility**.

intelligibility test A test performed to determine the **intelligibility** of transmitted or reproduced sound.

intelligibility testing The performance of **intelligibility tests**.

intelligible Capable of being understood. Usually used in the context of voice communications. The more intelligible, the clearer the reception of a voice transmission.

intelligible crosstalk Crosstalk which is capable of being comprehended. This contrasts with **unintelligible crosstalk**, which is incomprehensible. Intelligible crosstalk can be more distracting, as it is similar to the information that wishes to be understood.

INTELSAT An international organization which provides commercial satellite services. It is an abbreviation of **Inter**national **Tele**communications **Sat**ellite Organization.

intense 1. Characterized by a given feature which is manifested to an extensive degree. For example, the very intense light of a high-pressure mercury lamp. 2. An extreme of a given characteristic. For instance, intense radiation.

intensify To make more **intense** (1). For instance, to increase the brightness of a screen, such as that of a TV or computer monitor.

intensifying screen A screen coated or impregnated with a substance which fluoresces when exposed to X-rays. This greatly enhances the sensitivity of detection, which serves, for instance, to reduce the exposure of a sample or patient to radiation.

intensity The amount, extent, degree, or strength of a quantity, such as amplitude, magnetization, current, power per unit area, concentration, loudness, or radiation.

intensity control On a monitor, such as that of a computer or TV, a control which adjusts the brightness. Also called **brightness control**.

intensity level 1. In acoustics, a measure of the intensity of a sound relative to a reference intensity. For instance, it may be stated as ten times the common logarithm of the ratio of a given sound in a specified direction, to a reference sound intensity of 1 picowatt per square meter. Expressed in decibels. Also called **sound intensity level**. 2. A measure of the intensity of a physical quantity relative to a reference intensity. For instance, it may be stated as the common logarithm of the ratio of one intensity, to a reference intensity. Usually expressed in decibels.

intensity modulation 1. A form of modulation in which the intensity of a beam is varied proportionally to the intensity of the signal source. Used, for instance, in fiber optics as a mode of transmission in which an analog source modulates the light source. 2. In a CRT, modulation of the electron beam intensity according to the intensity of the signal source. Also called **Z-axis modulation**, or **Z-modulation**.

intensive 1. Of, pertaining to, or characterized by **intensity**. 2. Characterized by, possessing, or requiring a high degree of something. For example, I/O intensive.

inter-block gap Same as **inter-record gap**. Its abbreviation is **IBG**. Also spelled **interblock gap**.

inter-LATA Same as **interLATA**.

inter-layer dielectric Same as **interlayer dielectric**. Its abbreviation is **ILD**.

inter office Same as **interoffice**.

inter-record gap On a storage medium, such as a magnetic tape or disk drive, the unused physical space between consecutive blocks of recorded data. Also spelled **interrecord gap**. Its abbreviation is **IRG**. Also called **inter-block gap**, **block gap**, or **record gap**.

interaction The act or state of two or more entities influencing each other. For example, the electromagnetic force created by the interaction of electric and magnetic fields, the attractive force which accelerates one body towards another due to gravitation, or the modulation resulting from the interaction of wave components as they are transmitted through a nonlinear device.

interaction space A space where fields interact. For instance, a space between a plate and a cathode in a magnetron, where the electric and magnetic fields interact to exert forces upon electrons.

interactive 1. Involving the interaction between a user and a device, piece of equipment, or system. For instance, the exchanges involved in interactive TV, or those of interactive voice responses. 2. Pertaining to a computer program or system in which the user and computer interact, the user providing input, and the computer responding each time in some manner. For example, a user clicking on a hyperlink with a mouse, and the appropriate Web page being displayed, or a user entering commands, which in turn cause computer actions.

interactive computer graphics Same as **interactive graphics**.

interactive display A display, such as that of a computer, TV, or instrument, which allows a user or viewer to provide or select an input, and obtain an appropriate response. For instance, a touch screen in the lobby of an office building which enables a visitor to find the desired location by navigating through multiple screens.

interactive graphics Computer graphics in which a user can modify parameters of an image such as size, colors, design, and content, using a keyboard, digitizing tablet, voice commands, pointing devices, and so on. Used, for instance, in CAD programs and computer games. Also called **interactive computer graphics**.

interactive mode A mode of operation in which a user interacts with a device, piece of equipment, or system. For instance, operating a computer in conversational mode, or using a TV for Internet access, as opposed to simply watching regular programming.

interactive procedure A procedure within an **interactive session**.

interactive processing Processing, such as transaction processing, which requires interaction with a user.

interactive program A computer program which accepts user input, such as that from a keyboard, pointing device, microphone, or the like, and responds with actions such as altering what is displayed on the screen. Most computer programs are of this type. Some, such as compilers and batch programs, however, are not interactive.

interactive session A computer session in which a series of procedures are executed with concurrent user intervention.

interactive television Same as **interactive TV**.

interactive terminal A computer or data terminal which is **interactive**.

interactive TV Abbreviation of **interactive** television. A technology which enables a viewer to interact with a TV set in more ways than just changing channels, adjusting the volume, and the like. Typical interactive TV applications include selecting pay-per-view programming, banking or shopping from home, Internet access, and game playing. Specially-equipped TVs and/or receiver boxes, along with input devices such as keyboards or remote controls, are necessary to employ this technology. Its abbreviation is **ITV**.

interactive video Video, such as that whose source is a DVD or a Web site, which enables a user to interact, and that can be used, for instance, for learning or entertainment.

interactive voice response An automated system in which a telephone caller interacts with a voice menu. The caller usually uses the keypad and/or voice to access information such as airline or movie schedules, account balances and recent transactions, and so on. Depending on the caller input, the appropriate data is accessed from one or more databases, and a synthesized voice provides the desired information. Such a system may also be used for fax-on-demand requests. The key difference between an interactive voice response system and a **voice user interface** is that the former responds only specific words or short phrases, while the latter accepts continuous speech and handles an extensive vocabulary. Its abbreviation is **IVR**. Also called **interactive voice response system**.

interactive voice response system Same as **interactive voice response**. Its abbreviation is **IVR system**, or **IVRS**.

interactive voice response technology The technology utilized in **interactive voice response systems**. Such technology incorporates computers, telephony, and speech recognition.

InterBase A popular relational DBMS.

interbase current The current flowing between the two bases of a unijunction transistor.

interbase resistance The resistance between the two bases of a unijunction transistor.

interblock gap Same as **inter-record gap**. Its abbreviation is **IBG**. Also spelled **inter-block gap**.

intercarrier sound system In a TV receiver, the use of the same intermediate frequency for the sound and video signals. Also called **intercarrier system**.

intercarrier system Same as **intercarrier sound system**.

Interchange File Format A generic file format usually utilized to store audio or video. Its abbreviation is **IFF**.

interchangeability 1. The quality of being **interchangeable**. 2. The extent to which two components, devices, pieces of equipment, or systems are interchangeable. 3. The ability to exchange two interchangeable components, devices, pieces of equipment, or systems. Also, the ease with which this can be done.

interchangeable A component, device, piece of equipment, or system which can be replaced with another, and work identically. For instance, two modules, each with a different design, but with the same specifications and interface characteristics.

interchannel crosstalk Unwanted coupling occurring between two or more communication channels.

interchannel interference Interference between channels, such as adjacent-channel interference.

intercom Abbreviation of **intercom system**, or **intercommunication system**.

intercom system A system utilized to communicate between two or more points within an aircraft, ship, or any given premises, such as a building, house, or park, and in which each station has a microphone and a speaker which enable two-way conversations. Each location employs a speaker which also serves as a microphone, or, an ordinary telephone may be used, in which case it is an **interphone system**. There may also be cameras which provide images from specified locations, for added security. It is an abbreviation of **intercom**munication **system**, and its own abbreviations are **intercom**, or **ICS**.

intercommunication system Same as **intercom system**. Its abbreviation is **intercom**.

interconnect To connect with one another, or to be so connected.

interconnection 1. The manner in which a conducting path is provided between components, circuits, devices, equipment, systems, or materials, so as to form a circuit, or connected circuits. Also, the establishing of such a circuit. 2. The facilities which enable the transfer of electricity between two or more entities. Used, for instance to connect power-generating systems, so as to share the burden during peak loads.

interconnection diagram A diagram depicting the manner in which a conducting path is provided between components, circuits, devices, equipment, systems, or materials, so as to form a circuit, or connected circuits.

interdigital capacitor Same as **interdigitated capacitor**.

interdigital transducer A device which utilizes two comb-shaped metallic patterns on a piezoelectric substrate to electrically convert a microwave voltage into surface acoustic waves, or vice versa. Its abbreviation is **IDT**.

interdigitated capacitor A capacitor in which adjacent conductors forming comb-shaped patterns are separated by a dielectric, to store electric charge. Used in some types of ICs. Also called **interdigital capacitor**.

interelectrode capacitance In an electron tube, capacitance between any given electrodes. When said capacitance is between the plate and the control grid, also called **plate-grid capacitance**.

interexchange carrier Its abbreviation is **IXC**. 1. A telecommunications company which provides long-distance calling services. 2. A telecommunications company which provides service between subscribers served by different central offices.

interface **1.** The point or points where two entities meet. For instance, the connection between devices exchanging information, the boundary between two phases of matter such as that between an electrode and an electrolyte, or between semiconductor regions. **2.** A device which serves to connect a computer to peripheral devices or a network. Such interfaces include SCSI, RS-232, and network adapters. Also called **computer interface (2)**. **3.** Any **interface (1)** used in association with a computer, such as those between programs, between hardware and software, between hardware and a user, or between software and a user. Also called **computer interface (3)**.

interface adapter Same as **interface card**.

interface card A circuit board which is plugged into the bus or an expansion slot of a computer, in order to enable the exchange of data with a peripheral or a network. An example is a network adapter. Also called **interface adapter**.

interface connection A conductor which makes electrical contact between patterns on both sides of a printed-circuit board or chassis. Also called **feedthrough**.

interface routine A program routine, such as a that contained in a device driver, which assists a computer in communicating with a peripheral, or with another system.

interference **1.** Any energy which diminishes the quality of a desired signal, or the performance of a component, circuit, device, piece of equipment, or system. **2.** In a communications system, any energy which diminishes the ability to receive a desired signal, or that impairs its quality. Sources include electromagnetic noise, undesired signals, parasitic oscillations, or atmospheric conditions. Also called **electric interference**, or **radio interference**. **3.** The variations in amplitude occurring in the wave resulting from the superimposition of waves from two or more coherent sources whose phase difference varies. As the phase difference approaches and reaches 0° there is constructive interference, resulting in an increase in amplitude, and as the phase difference approaches and reaches 180°, there is destructive interference, resulting in a reduction or cancellation in amplitude. Interference can occur in waves residing along any part of the electromagnetic spectrum. Also, the additive process by which this phenomenon occurs. Also called **wave interference (2)**. **4.** The mutual effect two or more superimposed waves or vibrations have on each other. Also called **wave interference (1)**.

interference analyzer An instrument that detects sources of interference and/or which serves to analyze its content. Used, for instance, to detect adjacent-channel interference, alternate-channel interference, co-channel interference, and the like.

interference attenuation Same as **interference suppression**.

interference attenuator A circuit or device which reduces the amplitude of interference.

interference control The monitoring of a given interval of frequencies, for the detection of possible interference.

interference eliminator A circuit or device which greatly reduces, or eliminates, the amplitude of interference. An example is an interference filter.

interference fading The fading of a radio signal due to different waves traveling by slightly different paths arriving at the receiver.

interference filter A filter which greatly reduces, or eliminates, the amplitude of interference. Such a filter, for instance, may attenuate interference signals entering a receiver through a power line.

interference pattern The characteristic pattern of amplitude variations which is created by the interaction of two or more superimposed waves from coherent sources whose phase difference varies. This results in a spatial distribution of particle energies, particle densities, pressure, energy flux, or the like. For example, such a pattern which is produced by passing visible light through narrow slits. A hologram stores an interference pattern which is formed by a coherent beam of light reflected by a reference mirror, and by light scattered by the object being photographed.

interference reduction Same as **interference suppression**.

interference rejection The use of components or devices, such as interference filters, to reject interfering signals.

interference signal Also called **interfering signal**. **1.** Any extraneous signal which diminishes the quality of a desired signal, or the performance of a circuit, component, device, piece of equipment, or system. **2.** In a communications system, any extraneous signal which diminishes the ability to receive a desired signal, or that impairs its quality.

interference suppression The reduction or elimination of the amplitude of interference. This may be accomplished, for instance, using an interference filter, or in the case of a received signal, by employing an antenna with the appropriate design. Also called **interference reduction**, or **interference attenuation**.

interference threshold The minimum signal-to-noise ratio needed to help insure error-free communications.

interfering signal Same as **interference signal**.

interferogram The displayed or graphic record made by an **interferometer**.

interferometer An instrument in which a beam of electromagnetic radiation is split, and subsequently recombined after traveling along paths of different length. This results in the production of interference patterns which can be utilized for various purposes, including the extremely accurate measurement of distances and thicknesses, or indices of refraction. In addition, such an instrument can be used for testing optical elements, and to locate and study electromagnetic radiation wherever it occurs, be it in space or at a subatomic level.

interferometry The design, study, and utilization of **interferometers** to measure and analyze interference patterns of electromagnetic waves.

interframe **1.** Anything occurring between multiple frames of video. **2.** Same as **interframe coding**.

interframe coding Video compression in which temporal redundancy is factored in. This form of compression takes advantage of the similarities between successive frames, so only the differences between them are coded, providing for higher compression ratios. This contrasts with **intraframe coding**, in which temporal redundancy is not factored in. Also called **interframe compression**, or **interframe (2)**.

interframe compression Same as **interframe coding**.

interim storage A segment of computer memory or storage which is utilized to temporarily store information that awaits transfer or processing. A buffer is an example. Also called **temporary storage**.

Interior Gateway Protocol A protocol utilized for exchanging routing information within an autonomous system, such as a LAN. Its abbreviation is **IGP**.

Interior Gateway Routing Protocol A network routing protocol superseding Routing Information Protocol. It utilizes multiple criteria, such as reliability, the link's speed, and packet size, to determine the best path for data, and uses less bandwidth. Usually used in autonomous systems, such as LANs. Interior Gateway Routing Protocol itself has an improved version, which is **Enhanced Interior Gateway Routing Protocol**. Its abbreviation is **IGRP**.

interlace scanning Same as **interlaced scanning**.

interlaced Pertaining to, utilizing, or incorporating **interlaced scanning**.

interlaced GIF A GIF image that starts as a fuzzy outline and progresses until fully focused. Seen, for instance, when using a dial-up connection. Also called **progressive GIF**.

interlaced scan Same as **interlaced scanning**.

interlaced scanning A scanning system, such as that used in TVs and some computer monitors, in which the electron beam traces all odd-numbered lines followed by the tracing of all even-numbered lines. The pattern created by tracing the even-numbered lines is called the even field, while the pattern created by tracing the odd-numbered lines is called the odd field. A complete displayed frame consists of both of these interlaced fields. For a given refresh rate, image flicker is less apparent than that of a non-interlaced display. This contrasts with **non-interlaced scanning**, in which the electron beam traces all lines sequentially from top to bottom. Also called **interlaced scan**, **interlacing**, **interlace scanning**, or **line interlace**.

interlacing Same as **interlaced scanning**.

interLATA Between different **LATAs**. A call made from one LATA to another is a long-distance call. Also spelled **inter-LATA**.

interlayer dielectric In an IC or PCB, an insulating medium placed between layers. Such a dielectric helps increase packing density, while reducing interference such as crosstalk. Also spelled **inter-layer dielectric**. Its abbreviation is **ILD**.

interleave To position memory structures in a manner which improves performance. For example, if a data file spans more than one sector on a hard disk, these sectors may be staggered so that all sectors can be read in a single revolution, as opposed to multiple turns. This speeds up access time.

interleaving The positioning of memory structures in a manner which improves performance. An example is memory interleaving.

interlock **1.** A circuit, device, or mechanism which prevents a piece of equipment from operating under certain potentially hazardous conditions. For instance, a switch that shuts off power when a protective door or panel is opened or removed. Also called **electrical interlock**. **2.** A circuit, device, or mechanism which prevents operating devices or ongoing processes from interfering with each other. For example, to prevent a computer routine from running until a given task is completed. **3.** To connect in a manner in which the motion or operation of one or more parts affects one or more other parts.

interlock circuit **1.** A circuit which serves as an **interlock** (1). **2.** A circuit which serves as an **interlock** (2).

interlock device **1.** A device which serves as an **interlock** (1). **2.** A device which serves as an **interlock** (2).

interlock mechanism **1.** A mechanism which serves as an **interlock** (1). **2.** A mechanism which serves as an **interlock** (2).

interlock relay **1.** A relay which serves as an **interlock** (1). **2.** A relay which serves as an **interlock** (2).

interlock switch **1.** A switch that serves as an **interlock** (1). **2.** A switch that serves as an **interlock** (2).

intermediate frequency Its abbreviation is **IF**. **1.** In a superheterodyne receiver, the frequency resulting from the mixing of the received signal with that generated by a local oscillator. The use of one or more intermediate frequencies improves the performance of such a receiver. **2.** A frequency resulting from the mixing of a received signal with an internally generated signal. For example, the audible beat frequency produced in a heterodyne receiver. The use of one or more intermediate frequencies improves the performance of such a receiver.

intermediate-frequency amplifier In a receiver, a stage which amplifies an **intermediate frequency**. Its abbreviation is **IF amplifier**.

intermediate-frequency interference In a receiver, interference from signals at an **intermediate frequency**. Its abbreviation is **IF interference**.

intermediate-frequency rejection The ability of a receiver to reject a received signal at an **intermediate frequency**. Its abbreviation is **IF rejection**.

intermediate-frequency signal Its abbreviation is **IF signal**. **1.** In a superheterodyne receiver, the signal resulting from the mixing of the received signal with that generated by a local oscillator. **2.** In a receiver, the signal resulting from the mixing of a received signal with an internally generated signal.

intermediate-frequency stage In a receiver, a functional unit, such as an intermediate-frequency amplifier, which processes an **intermediate frequency**. Its abbreviation is **IF stage**.

intermediate-frequency transformer A transformer used with an **intermediate-frequency amplifier**. Its abbreviation is **IF transformer**.

intermediate language A computer language that is between a source language and a target language. For example, a language between a low-level language and a high-level language.

intermediate result Any result, other than the final one, in a multiple-step computer operation.

intermediate state Any state between two defined states.

intermediate storage Computer storage for information and values which await further processing.

Intermediate System-to-Intermediate System An OSI link-state protocol. Its abbreviation is **IS-IS**.

intermetallic compound A chemical compound whose constituents are elements which are metals and/or metalloids. There are many such compounds utilized in semiconductors, including gallium arsenide and gallium antimonide.

intermittent Occurring at intervals. For example, that which occurs occasionally, in a non-continuous manner, in a random or unpredictable manner, or at selected times.

intermittent condition A condition occurring at intervals. For instance, an intermittent fault, or contact bounce.

intermittent current **1.** A current which flows at intervals. **2.** A unidirectional current which flows at intervals.

intermittent discharges Discharges occurring at intervals. For instance, corona discharges.

intermittent duty The work or operation of a device, piece of equipment, or system, at intervals. This may consist of alternate periods such as those of load and no-load; load and rest; load, no-load, and rest, and so on.

intermittent-duty rating A rating based on the **intermittent duty** of a device, piece of equipment, or system, as opposed to continuous duty. Also called **intermittent rating**.

intermittent error An error occurring at intervals.

intermittent failure **1.** A failure occurring at intervals, as opposed, for instance, to a complete failure. **2.** A failure occurring at intervals, and which has the same cause on each occasion.

intermittent fault **1.** A fault occurring at intervals. **2.** A fault occurring at intervals, and which has the same cause and/or location on each occasion.

intermittent light Light which flashes on an off at intervals. For example, the light utilized in a pulsed laser.

intermittent malfunction **1.** A malfunction occurring at intervals. **2.** A malfunction occurring at intervals, and which has the same cause on each occasion.

intermittent operation Operation of a component, device, piece of equipment, or system, at intervals. For example, intermittent duty.

intermittent rating Same as **intermittent-duty rating**.

intermittent reception Reception which is interrupted, or markedly diminished at intervals. For instance, reception in a fringe area.

intermittent service Service occurring at intervals. For example, intermittent duty.

intermittent signal A signal occurring at intervals. For instance, that emitted by a beacon.

intermittent tone A tone occurring at intervals. For example, that of a busy signal.

intermittent transmission A transmission occurring at intervals. For instance, that occurring in asynchronous transmission.

intermodulation Its abbreviation is **IM**. **1.** Modulation resulting from the interaction of wave components as they are transmitted through a nonlinear element, device, or system. This produces unwanted frequency components which are equal to the sums and differences of the component frequencies of the original complex wave. **2.** In a nonlinear system, a signal produced through the interaction of an undesired signal with the desired signal. The parasitic output frequencies correspond to the sums and differences of the component frequencies of the inputs. **3.** Same as **intermodulation distortion**.

intermodulation distortion Its abbreviation is **IM distortion**, or **IMD**. Also called **intermodulation (3)**. **1.** Distortion resulting from **intermodulation (1)**, or **intermodulation (2)**. Frequently utilized to evaluate audio components. **2.** Distortion resulting from the interaction of two input signals.

intermodulation frequency Its abbreviation is **IM frequency**. For a nonlinear element, device, or system, frequency components arising from **intermodulation (1)**, or **intermodulation (2)**.

intermodulation interference Its abbreviation is **IM interference**. **1.** Interference resulting from **intermodulation (1)**, or **intermodulation (2)**. **2.** In a superheterodyne receiver, interference resulting from the reception of the signals of two stations whose difference in frequency equals the intermediate frequency of said receiver. This type of interference is manifested when selectivity is too low.

intermodulation noise Its abbreviation is **IM noise**. Noise resulting from **intermodulation interference**.

intermodulation products Its abbreviation is **IM products**. Undesired products resulting from **intermodulation**, such as intermodulation frequencies or intermodulation noise.

internal arithmetic Computations performed by the arithmetic-logic unit of a computer.

internal bus A bus between a CPU and memory. An **external bus** connects the CPU with peripherals.

internal cache A memory cache that is built right into the CPU. This contrasts with **external cache**, which utilizes a memory bank between the CPU and main memory. Internal cache is faster, but smaller than external cache, while external cache is still faster than main memory. Also called **level 1 cache**, **L1 cache**, **on-chip cache**, **primary cache**, or **first-level cache**.

internal drive A drive, such as a disk drive or tape drive, located within the system unit of a computer, as opposed to an **external drive**, which is not.

internal font A font which is permanently stored into a printer's memory. Also called **built-in font**, or **resident font**.

internal impedance **1.** The opposition exhibited by a component, circuit, or device to the flow of AC. **2.** The opposition within a source of electrical energy, such as a cell or generator, to the flow of AC.

internal interrupt In a computer, an interrupt generated by the CPU. This contrasts with an **external interrupt**, which is generated by an external device.

internal memory The RAM of a computer. Also called **internal storage (1)**.

internal modem A modem which resides on an expansion card, and which is plugged into a computer expansion slot. This contrasts with an **external modem**, which is self-contained and usually connects to a computer via a cable.

internal noise The noise inherently present in a component, circuit, device, piece of equipment, or system, as opposed to that arriving from external sources.

internal photoelectric effect In a semiconductor material, the absorption of sufficient photons to move electrons from the valence band to the conduction band, thereby increasing the conductivity of said material.

internal photoionization A process where an excited electron is transferred to a lower energy level without emitting radiation. This excess energy is used instead to expel an electron of the same atom. This may occur in any element except hydrogen and helium. Also called **autoionization**, or **Auger effect**.

internal resistance **1.** The opposition exhibited by a component, circuit, or device to the flow of current. **2.** The opposition within a source of electrical energy, such as a cell or generator, to the flow of current.

internal sort A sort performed utilizing computer memory such as RAM, as opposed to computer storage such as that on a hard disk.

internal storage **1.** Same as **internal memory**. **2.** In a computer, any storage device, such as a hard drive, where data can be held for an indefinite period.

internal thermal shutdown In a semiconductor device, such as a transistor, the junction temperature at which shutdown occurs, to prevent damage from overheating.

internal triggering **1.** The use of internal pulses as the triggering signals of an oscilloscope. **2.** The use of internal pulses as triggering signals.

international ampere An older unit of current equal to approximately 0.9998 ampere.

International Atomic Time An internationally agreed time standard based on time kept by multiple atomic clocks. Through the use of such clocks and satellites, nations on the planet are able to synchronize their time standards. It serves as the basis for Coordinated Universal Time. Its abbreviation is **TAI**.

International Bureau of Weights and Measures English for *Bureau International des Poids et Mesures*. An international organization dedicated to the uniformity of measurements, with emphasis on their adhering to the International System of units (SI).

international call sign A call sign assigned by the International Telecommunications Union, which serves to identify a broadcast radio station. The first two characters, and in some cases just the first character, identify the nationality of the station. Also spelled **international callsign**.

international callsign Same as **international call sign**.

international candle A former name of **candela**, which is the fundamental SI unit of luminous intensity.

international coulomb An older unit of electric charge equal to approximately 0.9998 coulomb.

International Data Encryption Algorithm A symmetric encryption algorithm utilizing 64-bit blocks, and a 128-bit key. Its abbreviation is **IDEA**.

International Electrotechnical Commission An international organization which prepares and publishes standards pertaining to electrical and electronic devices and products. Its abbreviation is **IEC**.

international farad An older unit of capacitance equal to approximately 0.9995 farad.

International Federation of Information Processing An organization which coordinates the efforts of multiple national societies dealing with information processing. Its abbreviation is **IFIP**.

international henry An older unit of inductance equal to approximately 1.0005 henry.

international joule An older unit of work or energy equal to approximately 1.0002 joule.

international ohm An older unit of resistance, impedance, or reactance equal to approximately 1.0005 ohm.

International Organization for Standardization A multinational body which sets international standards, such as those pertaining to communications, with the goal of facilitating economic and technological activity between countries. This entity is very often incorrectly referred to as **International Standards Organization**. Its abbreviation **ISO**, is also a word prefix meaning equal or uniform.

International Standards Organization An incorrect, but very commonly used name for **International Organization for Standardization**. Its abbreviation is **ISO**.

International System of Units Its abbreviation is **SI**, taken from its French name *Système International d'Unités*. A system used for measurement of physical quantities using internationally accepted fundamental units which are defined in an absolute manner. Currently, the base, or fundamental, SI units are: the **second**, for time; the **kilogram**, for mass; the **meter**, for distance; the **ampere**, for electric current; the **Kelvin**, for temperature; the **mole**, for amount of substance; and the **candela**, for luminous intensity. In addition, there are other units defined algebraically in terms of these base units. These include hertz, joule, watt, coulomb, volt, and ohm. Furthermore, the SI allows the use of specific non-approved units, such as electronvolt, bel, and angstrom.

International Telecommunications Satellite Organization Same as **INTELSAT**.

International Telecommunications Union Same as **ITU**.

International Telecommunications Union-Radiocommunication Sector Same as **ITU-R**.

International Telecommunications Union-Telecommunications Standardization Sector Same as **ITU-T**.

international volt An older unit of electric potential equal to approximately 1.0003 volt.

international watt An older unit of power equal to approximately 1.0002 watt.

internet A communications network composed of two or more networks or subnetworks. An internet may consist, for instance of two interconnected LANs. It is an abbreviation of **internet**work. This contrasts with the next entry, **Internet**, in which the term is capitalized.

Internet A worldwide network of interconnected autonomous networks, which is utilized for commerce, education, research, entertainment, and the obtaining of or exchange of information on virtually anything of human interest. The Internet currently encompasses hundreds of millions of computers and users, and spans nearly 200 countries. Among other services, users can exchange email, browse the World Wide Web, purchase and sell goods, and so on. An Internet service provider, for instance, may be used for access. Its abbreviation is **net**. Also called **information highway** (2), **information superhighway** (2), **cyberspace**

(2), or **Infobahn** (2). This contrasts with the previous entry, **internet**, in which the term is not capitalized.

Internet 2 A high-speed network connecting member institutions, and which is intended for academic and research-oriented use. Its abbreviation is **I2**. Also called **Internet II**, or **next-generation Internet** (2).

Internet access The ability to access the Internet. Also, manner in which this is accomplished, such as utilizing the services of an Internet service provider. Also called **Web access**.

Internet access device Same as **Internet appliance**.

Internet access provider Same as **Internet service provider**. Its abbreviation is **IAP**.

Internet address Its abbreviation is **net address**. 1. Same as **IP address**. 2. Same as **Internet domain name**. 3. A sequence of characters which uniquely identify an email account. The standard format is user name@domain name. In the case of **xxxx@zzz.com.qq**, *xxxx* is the user name, and *zzz.com.qq* is the domain name. Also called **email address**, or **address** (5).

Internet advertising Advertisements, such as banner, skyscraper, and pop-up ads, appearing over the Internet. Also, delivery of such ads. Also called **Web advertising**, or **online advertising**.

Internet appliance Any device, such as a properly equipped computer, PDA, TV, or cellular phone, which provides access to the Internet. Also, such a device specifically designed for this purpose. Its abbreviation is **net appliance**. Also called **Internet access device**, or **smart appliance** (3).

Internet Architecture Board An organization which helps oversee various matters pertaining to the Internet, and which collaborates closely with the Internet Engineering Task Force. Its abbreviation **IAB**.

Internet Assigned Numbers Authority The organization originally responsible for assigning IP addresses. Its current functions, such as coordinating with the Internet Engineering Task Force, are now taken care of by the Internet Corporation for Assigned Names and Numbers. Its abbreviation **IANA**.

Internet auction An auction that takes place over the Internet. Countless Web sites offer innumerable items which people or entities wish to sell to the highest bidders. Also called **online auction**.

Internet back bone Same as **Internet backbone**.

Internet backbone A superfast network spanning the globe, linking national Internet service providers at speeds that can exceed 100 Gbps. Local Internet service providers connect to regional Internet service providers, which in turn connect to this backbone, to be a part of the Internet. Also spelled **Internet back bone**. Also called **backbone** (3).

Internet banking Banking in which transactions, such as viewing statements or effecting payments, are performed via the Internet.

Internet-based education Same as **Internet-based instruction**.

Internet-based instruction Learning in which the Internet is used extensively or exclusively for lessons and/or resources. It is usually a part of a computer-based instruction program. Also called **Internet-based education**, or **Internet-based teaching**.

Internet-based learning Learning in which the Internet is used extensively or exclusively for instruction and/or resources. It may be part of a distance learning program.

Internet-based teaching Same as **Internet-based instruction**.

Internet-based training Training in which the Internet is used extensively or exclusively for instruction and/or resources.

Internet bot A program which performs tasks, especially repetitive ones, over the **Internet**. For example, a spider, which searches Websites and organizes the located information. Also known as **Internet robot**, or **bot (1)**.

Internet box **1.** A network computer used exclusively for access to the Internet, or designed for such use. Also called **Internet computer (1)**, **Internet PC (1)**, or **Web box (1)**. **2.** A set-top box, such as that used for Internet TV, which provides an interface to the Internet.

Internet broadcast A specific program offered during **Internet broadcasting**. Also, to transmit such content. Its abbreviation is **net broadcast**, or **netcast**. Also called **Webcast**.

Internet broadcasting Its abbreviation is **net broadcasting**. Also called **Webcasting**. **1.** The use of the Internet for transmission of programming intended for public or general reception. For example, a radio broadcaster may also provide a signal through the Internet, in addition to sending a signal which is received by antennas. **2.** The use of the Internet for transmission of programming intended for individuals and/or entities which have paid a fee or that are otherwise entitled to such content.

Internet browser A program which enables a computer user to browse, and perform other functions, such as downloading files and exchanging email, through the World Wide Web. This program locates and displays pages from the Web, and provides multimedia content. Most of the time there is interactive content on any given page. To have access to certain features, such as streaming video, specific plug-ins may be necessary, in addition to any hardware requirements. Its abbreviation is **net browser (1)**. Also called **Web browser (2)**, or **browser (2)**.

Internet business Same as **Internet commerce**.

Internet cable Internet service provided through a cable TV system. Such a service may plug into a cable modem, or into the TV, utilizing a specially-equipped cable box. Transmission speeds can be up to several Mbps downstream, so it is considered to be broadband. Also called **cable Internet**, or **cable-modem Internet**.

Internet cache An area of computer storage where the most recently downloaded Web pages may be temporarily placed, so that when one of these pages is returned to, it loads faster. Also called **browser cache**, **Web cache**, or **cache (2)**.

Internet café A public establishment which may serve beverages and/or food, but whose primary service is the offering of one or more terminals providing access to the Internet. Its abbreviation is **net café**. Also called **cyber café**.

Internet commerce The conducting of business utilizing the Internet for electronic transactions and communications. Internet commerce includes the purchasing and selling of goods and services, electronic data interchange, the use of e-money, and business-related videoconferencing. Also called **Internet business**.

Internet computer **1.** Same as **Internet box (1)**. **2.** A computer with the necessary hardware and software to connect to the Internet. Also called **Web computer (2)**.

Internet conferencing The holding of conferences, or participation in conferencing, such as teleconferencing, videoconferencing, or data conferencing, via the Internet. Also called **Web conferencing**.

Internet content Information which is accessed via the Internet, and which may consist of any combination of audio, video, files, data, or the like. Also called **Web content (1)**.

Internet content filter Same as **Internet filter**.

Internet content provider An entity that provides information content for the Internet. Such information may be any combination of audio, video, and data, and so on. Examples include the providers of content of Web pages, or of an online encyclopedia.

Internet Content Rating Association An international organization which establishes and promotes ratings for Internet content. The content labels conform to the PICS standard. Its abbreviation is **ICRA**.

Internet Control Message Protocol A TCP/IP protocol utilized for sending error and control messages, such as those indicating that a given IP address is non-existent, or otherwise unavailable. Its abbreviation is **ICMP**.

Internet cookie A block of data prepared by a Web server which is sent to a Web browser for storage, and which remains ready to be sent back when needed. A user initially provides key information, such as that required for an online purchase, and this is stored in a cookie file. When a user returns to the Web site of this online retailer, the cookie is sent back to the server, enabling the display of Web pages that are customized to include information such as the mailing address, viewing preferences, or the content of a recent order. Also called **cookie**, **browser cookie**, or **Web cookie**.

Internet Corporation for Assigned Names and Numbers An organization which oversees many Internet technical administrative functions, such as assignment of IP addresses, and management of protocol parameters. Its abbreviation is **ICANN**.

Internet country code Over the Internet, a two-character abbreviation which identifies a country. It appears at the end of a Uniform Resource Locator, or address, as in: **xxxx@zzz.com.qq**, where the country code is **qq**. A given country code in an address does not necessarily mean that the host to that address is physically there. Also called **country code (2)**.

Internet domain name On the Internet, a name which identifies one or more IP addresses. For instance, in the following URL, *http://www.yipeeee.com/whoo.html*, the Internet domain name is the *www.yipeeee.com* portion. Sometimes, only the *yipeeee.com* portion is considered to be the domain name. A domain name always ends in a **domain (5)**. When a domain name is entered into the address field of a browser, it is converted into the appropriate IP address, after which the appropriate content is transferred to the computer that requested it. Also called **domain name**.

Internet e-mail Same as **Internet email**.

Internet email Also called **Internet mail** **1.** Email which is accessed via a Web browser, as opposed to using an email client. Internet email offers benefits such as access to correspondence from virtually any location in the world, and can usually be utilized to download messages held on POP servers. Also called **Web-based email**. **2.** Any email sent and/or received via the Internet.

Internet Engineering Steering Group Within the Internet Society, a group whose responsibilities include overseeing the Internet Engineering Task Force. Its abbreviation is **IESG**.

Internet Engineering Task Force An organization whose responsibilities include establishing standards for the Internet, helping it work smoothly, and aiding in its evolution. It is managed by the Internet Engineering Steering Group, and has various working groups which deal with areas such as protocols and architectures. Its abbreviation is **IETF**.

Internet etiquette Network etiquette when communicating over the Internet.

Internet Exchange A facility where Internet service providers interconnect. Such a location handles an enormous amount of data traffic, determines how it is routed, and represents a key constituent of the Internet backbone. Also called **Network Access Point**, or **Metropolitan Area Exchange**.

Internet Explorer A popular Web browser. Its abbreviation is **IE**.

Internet facsimile Same as **Internet fax**.

Internet fax The use of the Internet for computer-to-computer, computer-to-fax, or fax-to-fax transmissions of documents. The Internet is used as the linking network, as opposed to a telephone system. It is an abbreviation of **Internet facsimile**.

Internet filter A program or utility which seeks to detect advertising and other bothersome or undesirable content before its loaded onto a Web page being accessed. Such a program, for instance, can filter Web page content, protect privacy, prevent pop-up ads from appearing, avert banner ads, eliminate certain JavaScript, stop animated GIFs, turn off ActiveX, disable Web bugs, and so on. Its abbreviation is **net filter**. Also called **Internet content filter**, **Web blocker**, **Web filter**, **Web content filter**, **content filter**, or **blocking software**.

Internet firewall A system which is utilized to help protect a network against threats originating through other networks, such as unauthorized access. A firewall, for example, enables an organization with an intranet to allow its members access to the Internet, while preventing outsiders from accessing its own private resources. A firewall may also control which outside resources are available to what users. A firewall may be implemented through the use of hardware and/or software, and all communications entering or leaving the internal network of the organization are examined to determine if they may proceed. Application gateways, packet filters, and proxy servers are among the available techniques which help afford such protection. A firewall may also help protect a network against internal threats, such as snooping. Also called **firewall**.

Internet game A computer game played over the Internet. Also called **electronic game** (2).

Internet gateway A device which connects the Internet to another network, such as a LAN, by performing the necessary protocol conversions. For instance, such a gateway could carry out the conversions between messaging protocols.

Internet Group Management Protocol A protocol utilized by IP hosts to report their multicast group membership to adjacent multicast routers. Its abbreviation is **IGMP**. Also called **Internet Group Membership Protocol**.

Internet Group Membership Protocol Same as **Internet Group Management Protocol**. Its abbreviation is **IGMP**.

Internet II Same as **Internet2**.

Internet Inter-ORB Protocol Same as **IIOP**.

Internet keyboard A computer keyboard with additional buttons or keys which simplify and enhance Internet usage. For instance, there may be buttons for one-press access to the home page, email program, preferred search engines, media players, and so on.

Internet mail Same as **Internet email**.

Internet Message Access Protocol A protocol for accessing email or other messages that are stored on a mail server. It has various advantages in comparison to Post Office Protocol, including the capability to access multiple mail servers. It also has the ability to retrieve, store, and manipulate messages for the same email account from more than one user computer, such as those at home and work, without having to transfer said messages back and forth between these computers. Its abbreviation is **IMAP**.

Internet monitoring Also called **Web monitoring**. **1.** Monitoring performed via the Internet, as in the use of a Webcam to observe a given area or environment. **2.** Monitoring performed on Internet usage. For example, the analysis of Internet traffic by time and/or region, or the examination of which users visit which Web sites.

Internet payment Payment for goods and/or services via the Internet. Also called **electronic payment**, or **Web payment**.

Internet PC Abbreviation of **Internet personal computer**. Its own abbreviation is **net PC** (2). Also called **Web PC**. **1.** Same

as **Internet box** (1). **2.** A personal computer with the necessary hardware and software to connect to the Internet.

Internet personal computer Same as **Internet PC**.

Internet phone Same as **Internet telephone**.

Internet portal A Web site, such as *www.google.com*, which serves as a starting point to most any activity on the Internet. Web portals usually provide news, email, search engines, directories, chats, shopping, local interest topics, and so on. Also called **Web portal**, or **portal**.

Internet Printing Protocol An IETF-promoted protocol or standard for printing over the Internet. Its abbreviation is **IPP**.

Internet Protocol Same as **IP**.

Internet Protocol address Same as **IP address**.

Internet Protocol datagram Same as **IP datagram**.

Internet Protocol fragment Same as **IP fragment**.

Internet Protocol header Same as **IP header**.

Internet Protocol multicasting Same as **IP multicasting**.

Internet Protocol network Same as **IP network**.

Internet Protocol next generation Same as **IPng**.

Internet Protocol number Same as **IP address**. Its abbreviation is **IP number**.

Internet Protocol packet Same as **IP packet**.

Internet Protocol security Same as **IPSec**.

Internet Protocol spoofing Same as **IP spoofing**.

Internet Protocol telephony Same as **IP telephony**.

Internet Protocol tunneling Same as **IP tunneling**.

Internet Protocol version 6 Same as **IPv6**.

Internet radio The transmission of radio signals over the Internet. Also called **Web radio**.

Internet Relay Chat A service enabling the use of the Internet for real-time chats. Its abbreviation is **IRC**.

Internet Research Task Force An organization composed of research groups which work on matters concerning Internet protocols, applications, and future technologies. Its abbreviation is **IRTF**.

Internet roaming The accessing of the ISP a given user usually utilizes through another ISP, as may be done, for instance, when traveling.

Internet robot A program which performs tasks, especially repetitive ones, over the **Internet**. For example, a spider, which searches Websites and organizes the located information. Also known as **Internet bot**, or **bot** (1).

Internet security Precautions, such as the utilization of encryption and passwords, taken to safeguard data residing in, or transmitted through the Internet, against loss, damage, unwanted modification, or unauthorized access.

Internet server A server dedicated to providing access to Internet content and services, such as email and Web browsing.

Internet service provider An entity which provides access to the Internet. Customers usually pay a monthly fee for this service, although it usually available for free unless a high-speed connection is desired. Users get a software package to access and browse the World Wide Web, one or more email accounts, and Web pages to have a presence on the Internet. Other offerings may include Web site building and hosting services. Users can connect via a dial-up service or asymmetrical digital subscriber line, among others. Its abbreviation is **ISP**. Also called **Internet access provider**, **access provider**, **service provider**, or **online service provider** (1).

Internet services Services, especially those that are business related, available via the Internet. Also called **e-services**.

Internet Society An international organization which encourages the utilization and advancement of the Internet through means such as training, and the promotion of standards. It supports entities such as the Internet Engineering Task Force, and the Internet Architecture Board. Its abbreviation is **ISOC**.

Internet Software Consortium An organization which helps develop and maintain useful open source over the Internet, and which promotes standards and protocols, such as Dynamic Host Configuration Protocol. Its abbreviation is **ISC**.

Internet support Technical, product, or service support provided via the Internet. An entity could also provide such help via telephone, fax, regular mail, and/or in person, although there is a marked tendency towards Internet support whenever practical. Also called **esupport**.

Internet surfing Exploring, researching, or otherwise spending time navigating from one Web site to the next. Its abbreviation is **net surfing**. Also called **Web surfing**, or **surfing (2)**.

Internet telephone Its abbreviation is **Internet phone**, or **net telephone**. Also called **Web telephone**. **1.** Same as **Internet telephony**. **2.** Software used for **Internet telephony**.

Internet telephony The use of the Internet for computer-to-computer, computer-to-telephone, or telephone-to-telephone calls. When making computer-to-computer calls, both systems must have, in addition to Internet access, a microphone, speakers or headphones, and compatible software. A computer-to-telephone call does not require the receiving party to have anything more than a phone line, while telephone-to-telephone calls use the Internet as the linking network. Also called **Internet telephone (1)**, **net telephony**, or **Web telephony**.

Internet television Same as **Internet TV**.

Internet terminal A network terminal utilized to access the Internet. Also called **Web terminal**.

Internet TV Abbreviation of **Internet** television. Its own abbreviation is **net TV**. Also called **Web TV**. **1.** The use of a TV for Internet access. This may be obtained utilizing Internet-ready TVs, or set-top boxes. **2.** The transmission of TV signals over the Internet.

Internet Usage Policy A set of rules which establish what can and can not be done when accessing the Internet in an office, at a school, or the like.

Internet utility Any utility used with the Internet. Examples include ping, finger, and Archie.

Internet2 Same as **Internet 2**.

internetwork Same as **internet**.

InterNIC An organization that provides Internet domain name registrations services.

interoffice Transmitted, communicated, or occurring between central offices. For instance, a telephone call that originates in one central office, and terminates in another. Also spelled **inter office**.

interoffice trunk A communications channel between central offices.

interoperability **1.** The ability of components, programs, units, or systems to function or communicate with each other. **2.** The extent to which components, programs, units, or systems are able to function or communicate with each other.

interoperable Components, programs, units, or systems which are able to function or communicate with each other.

interphone A privately-owned system utilized to communicate between two or more points within an aircraft, ship, or any given premises, such as a building, house, or park, and in which each location has a telephone, enabling two-way conversations. There may also be cameras which provide images from specified locations, for added security. It is an abbreviation of **interphone** system.

interphone system Same as **interphone**.

interpolate To utilize known values to estimate an unknown value.

interpolation The utilization of known values to estimate an unknown value. For instance, known points which are graphed can be used in this manner to approximate unknown values between them.

interpret To analyze, translate, and execute each computer statement or instruction of a high-level program before handling the next statement or instruction.

interpreted language A computer programming language in which each statement is translated then executed, followed by the next statement, and so on. This contrasts with a **compiled language**, in which all the code is translated into machine language before being executed. LISP is a programming language that has both interpreter and compiler versions.

interpreter A computer program which analyzes, translates, and executes each computer statement or instruction of a high-level program written in an **interpreted language**. This is done in sequence, while a **compiler** translates all the code before any program instructions are executed.

interpretive language A computer language that can not be directly executed by a CPU. Such a language must be interpreted or compiled prior to execution. Also called **pseudocode (2)**.

Interprocess Communication Within a computer or network, the capability of one process to exchange data with another. This may involve a single computer multitasking, or data-sharing among multiple computers linked through a network. Such exchanges may be manual or automatic. Its abbreviation is **IPC**.

interrecord gap Same as **inter-record gap**.

interrogate To send an **interrogation signal**.

interrogation Same as **interrogation signal**.

interrogation signal Also called **interrogation**. **1.** The signal emitted by an **interrogator**. **2.** A signal which triggers a response from receiver.

interrogator **1.** A device which generates signals which trigger the response of a transponder. Used, for instance, in distance-measuring equipment, such as that utilized by an aircraft or ship. Also called **interrogator-transmitter**. **2.** Same as **interrogator-responder**.

interrogator-responder A device which generates signals which trigger the response of a transponder, and which then receives, interprets, and displays said reply. The transponder transmits at a different frequency. Such a device may be used, for instance, to interrogate a radar beacon. Also called **interrogator (2)**, or **interrogator-responsor**.

interrogator-responsor Same as **interrogator-responder**.

interrogator-transmitter Same as **interrogator (1)**.

interrupt A signal to a CPU that enables it to address multiple tasks in an orderly manner. Interrupts are asynchronous events which divert the attention of the processor, and can be generated by hardware or software components. There is an established priority hierarchy in case more than one interrupt arrives simultaneously. Also called **interrupt signal**.

interrupt-driven Same as **interrupt-driven processing**.

interrupt-driven processing In a computer system, processing of actions in which interrupts are utilized for all connected devices to request CPU resources. This contrasts with **autopolling (1)**, where there is the periodic detection of the status of all connected devices, to address any needs for resources. Also called **interrupt-driven**.

interrupt handler A routine which is executed in response to a given interrupt. An interrupt handler may deal, for instance, with input arriving from a peripheral, or the updating of the system clock.

interrupt latency The time interval that elapses between an interrupt being generated, and it being serviced.

interrupt mask A setting or bit which determines whether an interrupt will be serviced. Non-maskable interrupts can not be disabled, while maskable interrupts can be.

interrupt priority 1. The priority hierarchy established in case more than one interrupt arrives simultaneously. If multiple interrupts arrive at once, attention is given only to that with the highest priority. **2.** The level a given interrupt has within an **interrupt priority (1)**.

interrupt request Its abbreviation is **IRQ**. **1.** An interrupt generated by a hardware component. **2.** The hardware line via which an **interrupt request (1)** is sent. Also called **interrupt request line**.

interrupt request line Same as **interrupt request (2)**. Its abbreviation is **IRQ line**.

Interrupt Service Routine A routine which performs the actions required in response to an interrupt.

interrupt signal Same as **interrupt**.

interrupt table Same as **interrupt vector table**.

interrupt vector A memory location containing the address of an **interrupt handler**.

interrupt vector table A list enumerating **interrupt vectors**. Also called **interrupt table**.

interrupted continuous wave A continuous wave which is interrupted at regular intervals. For example, such a wave may be interrupted at an audio-frequency rate so as to provide keying. Its abbreviation is **ICW**.

interrupted current A current which is varied, or started and stopped at intervals, as opposed to a steady current.

interrupter Also spelled **interruptor**. **1.** A device, such as a chopper, which periodically or intermittently interrupts a direct current. **2.** A device, such as a switch, which periodically or intermittently opens a circuit.

interrupter circuit A circuit which serves as an **interrupter**.

interrupting capacity A rating based of the amount of current that a protective device, such as a fuse or circuit breaker, can safely interrupt. Also called **interrupting rating**.

interrupting rating Same as **interrupting capacity**.

interruptor Same as **interrupter**.

Intersecting Storage Rings A device in which beams of protons stored in two interlaced rings are made to collide where said rings meet. The beams are stored in the rings after acceleration in proton synchrotrons, and energies in excess of 60 GeV can be produced. Its abbreviation is **ISR**.

intersection 1. The point where two things meet. It may refer to the meeting of circuits, planes, beams, bearings, and so on. **2.** A logical operation which is true only if all of its elements are true. For example, if three out of three input bits have a value of 1, then the output is 1. Any other combination yields an output of 0. For such functions, a 1 is considered as a true, or high, value, and 0 is a false, or low, value. Also called **AND**, or **conjunction**.

interstage Between stages, such as those of a multistage circuit or device. For example, between stages of a cascade amplifier.

interstage coupling Coupling between stages, such as those of a multistage circuit or device. For example, direct coupling between stages of a cascade amplifier.

interstage transformer A transformer which serves to couple stages of a multistage circuit or device.

interstice A small or narrow space between component, parts, or objects.

interstitial ad An Internet ad that occupies most or all of the viewable browser window, and which frequently forces the user to view and listen to at least part before being allowed to proceed to wherever was intended. Such advertisements usually feature gimmicks such as flashy graphics and sounds.

intersymbol interference In digital communications, a type of interference resulting from the overlap of individual digital pulses, in which case the receiver may not be able to distinguish between changes of state, which in turn can lead to an increased error rate. The closer the signal bandwidth is to the overall channel bandwidth, the greater the possibility of the manifestation of this form of interference.

intersystem communications Communications between two or more computer systems, be it via a direct link of CPUs, shared I/O channels, or the like.

intersystem electromagnetic compatibility 1. The ability two or more systems have to function in the same electromagnetic environment, without the function or performance of any of the systems being adversely affected by any electromagnetic interference created by any of them. Often, there is electromagnetic interference present, but it may not be sufficient to affect other devices. **2.** The extent to which **intersystem electromagnetic compatibility (1)** exists.

interval 1. A space or separation between two points, locations, objects, or the like. For example, the distance between peaks of successive wave cycles. **2.** A given range of values. For instance, an interval of frequencies or temperatures. Also, a specific range of values, such as the readings an instrument can display. **3.** The time that elapses between the occurrence of two specified events, states, or instants. For example, the regular intervals at which AC voltage alternates, the irregular intervals at which a circuit with a loose connection might open and close, or the time within which a pulse rises from 0 to its maximum amplitude.

interval timer Also called **timer**. **1.** An instrument or device utilized to measure time intervals, such as those occurring between events, or those required to complete given operations. **2.** A **timer (1)** that controls the duration of a pulse, operation, event, process, or the like. **3.** A **timer (1)** utilized to power, or otherwise activate a circuit, device, piece of equipment, system, mechanism, process, or the like. **4.** A **timer (1)** utilized for synchronization purposes.

intervalometer An instrument which enables actuation of a device or piece of equipment at precise intervals.

intra-LATA Same as **intraLATA**.

intra office Same as **intraoffice**.

intraframe 1. Anything occurring within a single frame of video. **2.** Same as **intraframe coding**.

intraframe coding Video compression in which temporal redundancy is not factored in. This form of compression does not take advantage of any similarities between successive frames, resulting in lower compression ratios. However, individual frames can be better accessed and manipulated. This contrasts with **interframe coding**, in which temporal redundancy is factored in. Also called **intraframe compression**, or **intraframe (2)**.

intraframe compression Same as **intraframe coding**.

intraLATA Within the same **LATA**. A call made within the same LATA is not a long-distance call. Also spelled **intra-LATA**.

intranet A private network utilizing the same, or similar protocols as those used over the Internet. Such a network may consist, for instance, of the Web site which serves the internal needs of an organization. An intranet is usually protected by a firewall, which enables an organization to allow its members access to the Internet, while its own private resources are protected from outsiders.

intranet server A server dedicated to serving an **intranet**.

intraoffice Transmitted, communicated, or occurring within the same central office. For instance, a telephone call to a neighbor. Also spelled **intra office**.

intrinsic Belonging to, or pertaining to the essential nature of something. Also, arising from the inside of a body or entity, or existing naturally. For example, intrinsic properties. Also called **inherent**.

intrinsic carrier density Same as **intrinsic concentration**.

intrinsic characteristics Same as **intrinsic properties**.

intrinsic coercive force The magnetic field which must be applied to reduce to zero the intrinsic induction of a material.

intrinsic coercivity The maximum value attainable for **intrinsic coercive force**, which is obtained after the material has been fully magnetized.

intrinsic concentration In a semiconductor material, the number of minority charge-carriers exceeding the equilibrium amount. Also called **intrinsic density**, or **intrinsic carrier density**.

intrinsic conductivity The conductivity inherently present in a pure semiconductor material, as opposed to **extrinsic conductivity**, which is that provided by imperfections in the crystal and intentionally-introduced impurities.

intrinsic density Same as **intrinsic concentration**.

intrinsic error An error which arises from the essential nature of something, or that which is built-in or otherwise incorporated. For instance, an error in the design of a computer program which leads to errors. Also called **inherent error**.

intrinsic flux The product of the **intrinsic induction** and the cross-sectional area of a magnetic material.

intrinsic flux density Same as **intrinsic induction**.

intrinsic induction For a given point in a magnetic material or medium, the additional magnetic induction present above that which would be present if said point were in a vacuum, under the influence of a magnetizing force of equal strength. Also called **intrinsic flux density**.

intrinsic interference Same as **intrinsic noise**.

intrinsic layer A layer of semiconductor material whose properties are essentially the same as the intrinsic properties of the bulk of said material.

intrinsic mobility The electron mobility in a pure semiconductor material, as opposed to that provided by imperfections in the crystal, and intentionally-introduced impurities.

intrinsic noise In a component, circuit, device, or piece of equipment, noise which is inherently present, such as that arising from random fluctuations in a current, or thermal agitation. Also called **intrinsic interference**, or **inherent noise**.

intrinsic photoconductivity For a given material, photoconductivity present without the presence of any external causes, or the addition of impurities.

intrinsic properties Also called **intrinsic characteristics**, or **inherent properties**. **1.** Properties of a component, circuit, device, piece of equipment, or system which are **intrinsic**. **2.** Electrical characteristics of a semiconductor material which are inherently present in the pure crystal. This contrasts with **extrinsic properties**, which are those determined by imperfections in the crystal and intentionally-introduced impurities.

intrinsic Q The value of the Q factor for a circuit or device without an external load. Also called **unloaded Q**.

intrinsic region A region of a semiconductor material with little or no doping.

intrinsic regulation Regulation in which the system which is perturbed recovers its equilibrium without the need for a compensating control element. Also called **inherent regulation**.

intrinsic semiconductor A semiconductor material with no dopants. Its electrical characteristics, such as concentration of charge carriers, depend only on the pure crystal. This contrasts with an **extrinsic semiconductor**, which has dopants added. Also called **intrinsic semiconductor material, I-type semiconductor**, or **undoped semiconductor**.

intrinsic semiconductor material Same as **intrinsic semiconductor**.

intrusion An **incident (3)** in which an individual or entity breaks into a computer system or network.

intrusion alarm **1.** An alarm system which is utilized for detecting and informing of the unauthorized presence in, and/or around a protected area. Such a system may consist, for instance, of a sensor, such as a motion detector, a processing unit, such as the alarm control panel, and an output device, such as a siren. Also called **intrusion alarm system**. **2.** Within an **intrusion alarm system (1)**, an electrical and/or mechanical device which serves to warn by means of a signal, such as sound and/or light.

intrusion alarm system Same as **intrusion alarm (1)**.

intrusion detection **1.** The use of an **intrusion alarm (1)** to detect and inform of the unauthorized presence in, and/or around a protected area. **2.** Any technique utilized to detect breaches into computer systems or networks.

intrusion detection system A system, such as an intrusion alarm or that incorporating specialized software for monitoring network activity, utilized for **intrusion detection**. Its abbreviation is **IDS**.

intrusion detector Same as **intrusion sensor**.

intrusion sensor A sensor utilized in an **intrusion alarm system**. Such a sensor may detect, for instance, ultrasonic, infrared, or microwave energy. Also called **intrusion detector**.

invalid In computers, unrecognizable or in error.

invar An alloy of iron and nickel which usually has about 36% nickel, 63.8% iron, and 0.2% carbon. It strong, ductile, and features an extremely low coefficient of thermal expansion. Used, for instance, in precision instruments. Also called **invar alloy**.

invar alloy Same as **invar**.

inverse **1.** That which is the opposite or reverse of something else. For example, a reverse current, or an inverse proportionality. **2.** The opposite operation or function of another. For instance, an inverse logarithm.

inverse bias A bias voltage applied in the proper polarity to a diode or semiconductor junction to cause little or no current to flow. Also called **back bias (2)**, or **reverse bias**.

inverse Chebyshev filter A filter similar to a Chebyshev filter, but with a better group delay characteristic.

inverse current In a device such as a rectifier or semiconductor, current that flows in the opposite direction of that which is normal. Also called **back current**, or **reverse current**.

inverse direction For an electrical or electronic component or device, the direction in which there is greater resistance to the flow of current. Used, for instance, in the context of semiconductors. This contrasts with **conducting direction**, where the converse is true. Also called **reverse direction**.

inverse feedback In an amplifier or system, feedback which is 180° out-of-phase with the input signal, thus opposing it. While this results in a decrease in gain, it also serves reduce distortion and noise, and to stabilize amplification. Also called **negative feedback, reverse feedback, stabilized feedback**, or **degeneration**.

inverse Fourier transform A mathematic expression which converts from the frequency domain to the time domain.

inverse logarithm The number which is the result of a logarithmic operation of another. For example, the logarithm of 1,000 is 3, so the antilogarithm of 3 is 1,000. Also called **antilogarithm**.

inverse multiplexer Also spelled **inverse multiplexor**. **1.** A circuit or device which reverses the effects of a multiplexer.

That is, it separates each of the multiple signals which were combined for transmission over a single channel. Also called **demultiplexer**. **2.** A circuit or device which combines multiple low-speed transmissions into a single high-speed transmission, and vice versa. Also called **multiplexer (2)**.

inverse multiplexing **1.** The process of reversing the effects of a multiplexer. That is, the separation of each of the multiple signals which were combined for transmission over a single channel. Also called **demultiplexing**. **2.** The process of combining multiple low-speed transmissions into a single high-speed transmission, and vice versa. Also called **multiplexing (2)**.

inverse multiplexor Same as **inverse multiplexer**.

inverse-parallel circuit Same as **inverse-parallel connection**.

inverse-parallel connection Two tubes or semiconductor devices connected in parallel, but opposite directions, to control AC without rectification. For instance, the connection of the anode of one diode to the cathode of another, and vice versa. Also called **inverse-parallel circuit, back-to-back connection**, or **back-to-back circuit**.

inverse peak voltage For an electrical or electronic component or device, the maximum instantaneous value of the voltage in the direction in which there is the greater resistance to the flow of current. For example, the maximum reverse-biased voltage a semiconductor can safely handle before avalanche breakdown will occur. Also called **peak inverse voltage**, or **reverse peak voltage**.

inverse piezoelectric effect The mechanical deformation of certain crystals or ceramics when exposed to an electric field. Conversely, **direct piezoelectric effect** is the generation of an electric charge by subjecting certain crystals or ceramics to mechanical strain. Both are complementary manifestations of the piezoelectric effect. Also called **inverse piezoelectricity, indirect piezoelectricity**, or **reverse piezoelectric effect**.

inverse piezoelectricity Same as **inverse piezoelectric effect**.

inverse-square law A law stating that a given physical quantity will vary as the inverse square of the distance from the source. Most forms of energy that are radiated from a point source obey this law. These include electromagnetic, thermal, and nuclear energy. For instance, if the distance to a light source is halved, then the intensity of the received radiation will be four times greater. Also called **law of inverse squares**.

inverse-square law of radiation A law stating that the intensity of radiation emitted from a point source will vary as the inverse square of the distance from said source. It is an example of the inverse-square law.

inverse video On a computer display, the exchanging of the light and dark areas of the screen. That which is normally light appears as dark, and vice versa. When highlighting, for instance, the selected area is displayed as inverse video. Also called **reverse video**.

inverse voltage **1.** For an electrical or electronic component or device, the voltage in the direction in which there is the greater resistance to the flow of current. Also called **reverse voltage (3)**. **2.** The voltage across a rectifier during the half-cycle when current does not flow.

inversion **1.** A reversal of a relationship, position, order, state, phase, or the like. Also, the act or process of effectuating such a reversal. **2.** The operation of forming the inverse of a quantity, operation, element, function, or the like. **3.** The condition of being inside out. **4.** The formation of an image which is rotated 180° about the axis of the same or another image. **5.** A form of speech scrambling in which the frequencies are inverted. **6.** A reversal of the usual decrease or increase with altitude of a given atmospheric property, such as temperature. **7.** The formation of an **inversion layer (2)**.

inversion layer **1.** The atmospheric layer through which an **inversion (6)** occurs. **2.** The formation of a surface layer whose polarity is opposite to that of the bulk of a semiconductor material. This is usually produced as a result of an applied electric field, and is utilized, for instance, to form the channel of an IGFET.

invert To effectuate or subject to an **inversion**.

inverted-cone antenna A wideband antenna consisting of a series of wires which have a common feed point, and whose overall shape is that of a cone with its vertex on the bottom. The wires are attached to supporting poles.

inverted file In data management, an access method in which multiple attributes of the data contained in a file are indexed. These attributes can be used in any combination to identify any records containing them. In this manner, a file can be accessed through an indexed attribute value, as opposed to having to find the attribute value of a specific record, making searches faster.

inverted-L antenna An antenna consisting of a horizontal wire or radiator with a vertical feeder or lead-in wire connected to one end, and whose shape resembles an inverted letter **L**. Used, for instance, for shortwave reception. Also called **L antenna**.

inverted list An index or list containing the attributes of data when indexing **inverted files**.

inverter Also called **inverter circuit**. **1.** A circuit or device that changes a DC voltage to an AC voltage. Frequently, the output voltage is much higher than that of the input. Used, for instance, in uninterruptible power supplies. Also called **DC–AC converter**, or **static inverter**. **2.** A circuit which has a single input, and whose output is high when the input is low, and vice versa. Also called **NOT gate**. **3.** A circuit which converts a positive signal into a negative one and/or vice versa. **4.** A circuit which reverses the polarity of a signal. For instance, an inverting amplifier.

inverter circuit Same as **inverter**.

inverting amplifier An amplifier which introduces a 180° phase shift during the process of amplification. That is, the amplified output signal has the opposite polarity of the input signal. This contrasts with a **non-inverting amplifier**, whose output is in phase with its input.

inverting connection A connection to an **inverting input**. An example is a connection to an input of a differential amplifier or operational amplifier which introduces such a phase shift during the process of amplification.

inverting input An input of a circuit or device whose output has the opposite polarity of said input. For instance, that of an inverting amplifier. Also called **inverting terminal**.

inverting terminal Same as **inverting input**.

inverting transistor A transistor utilized as an inverter for a digital or analog signal.

invisible GIF A GIF which shows the background through all or part of an image. Used, for instance, for smooth blending of graphics on Web pages, or for spying on users utilizing Web bugs. Also called **clear GIF**, or **transparent GIF**.

invisible Web Web content which is not found using search engines. Such content may be located, for instance, by using a given Web site's search feature. Also called **deep web**.

invoke In computers, to call, start, or activate a command, routine, program, function, process, or the like.

Inward Wide-Area Telephone Service A WATS service that is configured only for receiving incoming calls. Such calls, for instance may be originated from a specific region only, or nationwide. This contrasts with an **Outward Wide-Area**

Telephone Service, where only outgoing calls may be placed. Its acronym is **INWATS**.

INWATS Acronym for **Inward Wide-Area Telephone Service**.

IO Same as **I/O**.

IOC Abbreviation of **integrated optical circuit**.

iodine A chemical element whose atomic number is 53. It is a bluish-black solid with a metallic luster, which readily sublimates at ordinary temperatures into a corrosive violet vapor. It is the least reactive of the halides, yet still reacts with many elements, and has over 30 know isotopes, of which one is stable. Its applications include its use in semiconductors, as a contrast medium for X-rays, in lithium batteries, and in photography. Its symbol is **I**.

ion An atom, molecule, or radical with a charge. A negatively charged ion has an excess of electrons relative to those needed for neutrality, and is called **anion**. A positively charged ion has a deficiency of electrons, and is called **cation**. When in solution, ions subjected to an electric potential travel towards, and collect at electrodes. Anions towards the anode, and cations towards the cathode. An ion usually has markedly different properties when compared to its neutral equivalent. For instance, sodium metal is highly caustic to tissue, and chlorine is an irritating and poisonous gas. Yet, sodium chloride, or table salt, is composed of sodium ions and chloride ions, and can be safely ingested.

ion avalanche Also called **avalanche, avalanche effect, cascade (2), cumulative ionization, Townsend ionization,** or **Townsend avalanche**. When the only produced charged particles are electrons, it is called **electron avalanche**. **1.** A cumulative ionization process in which charged particles are accelerated by an electric field and collide with neutral particles, creating additional charged particles. These additional particles collide with others, so as to create an avalanche effect. **2.** In semiconductors, the cumulative generation of free charge carriers in an avalanche breakdown.

ion beam A narrow stream of ions traveling in the same direction, with approximately the same speed, and emitted from the same source. Such beams are used, for example, for ion-beam etching.

ion-beam analysis The study of the surface contours of a material, or the composition of a sample, utilizing a high-energy ion beam. Used, for instance, to analyze the surface of thin-film ICs. The advantages of ion-beam analysis include its generally being non-destructive, its requiring little or no preparation of the sample, and its ability to study very small sections at a time. Its abbreviation is **IBA**.

ion-beam etching An etching technique in which an ion beam is utilized to produce the desired patterns.

ion-beam lithography A lithographic technique similar to electron-beam lithography, but in which the resist is exposed by high-energy ions, instead of electrons. Ion-beam lithography usually offers higher resolution, due to reduced scattering.

ion-beam milling Same as **ion milling**.

ion burn Same as **ion spot**.

ion chamber An instrument for the detection and measurement of radiation. It does so by measuring the electric current that flows when radiation ionizes the gas in a chamber, which makes said gas a conductor of electricity. Also called **ionization chamber**.

ion concentration Also called **ion density, ionization concentration**, or **ionization density**. **1.** The number of ions per unit volume. For instance, the hydrogen-ion concentration is a measure of the acidity or alkalinity of an aqueous solution. **2.** The number of ions per unit volume of a gas. For example, the number of ions per unit volume of air.

ion density Same as **ion concentration**.

ion gun Same as **ion source**.

ion implantation A widely-used technique for introducing dopants into a semiconductor material, in which the impurity atoms are ionized and accelerated to a high energy and injected into the target semiconductor layer. The depth to which the ions penetrate is determined by their energy, and the number of ions implanted can be determined by the current flow through the target, providing for extremely precise placement.

ion implanter A device utilized for **ion implantation**.

ion microprobe A spectroscopic technique in which an ion beam is utilized to sputter ions off the surfaces of samples. These secondary ions produced are then accelerated into a mass spectrometer for separation according to their mass-to-charge ratio. Used, for instance, to detect the elemental composition of the surfaces of geologic samples. Also called **secondary-ion mass spectroscopy**.

ion microscope A microscope similar to a field-emission microscope, but in which the sharply rounded point is surrounded by helium gas. This induces field ionization of the helium atoms, which are directed towards the screen. This variation has much greater resolving power, and magnifications of greater than 10 million can be attained. Also called **field-ion microscope**.

ion migration The migration of ions through a medium, under the influence of an electric field produced by electrodes placed or immersed in said medium.

ion milling An etching technique in which an ion beam is utilized to produce the desired patterns. Also called **ion-beam milling**.

ion potential Same as **ionization potential (1)**.

ion projection lithography A projection lithography technique in which an ion beam, as opposed to a light beam, is utilized. Resolutions in the nanometer range are attainable.

ion source A device which produces, and usually controls, an **ion beam**. Used, for instance, in ion implantation. Also called **ion gun**.

ion spot A spot on the screen of a CRT which is somewhat darker than the area immediately surrounding it, due to bombardment by negative ions. An ion trap is usually utilized to prevent this. Also called **ion burn**.

ion trap **1.** A device or arrangement used to prevent an ion spot from forming on the screen of a CRT. It usually utilizes a magnetic field to bend the electron beam, which then passes through a tiny aperture while the heavier ions are trapped inside the electron gun. Also called **beam bender**. **2.** A device that enables ions to be trapped for a comparatively long period, providing for extremely accurate observation. Used, for instance, in atomic oscillators.

ion-trap magnet One or more magnets placed around the neck of a picture tube, to serve as an **ion trap (1)**.

ionic bond A chemical bond formed by the electrostatic attraction of the oppositely-charged ions which form the resulting compound. In such a bond, each constituent either gains or loses one or more electrons. For instance, in **NaCl** (table salt), the sodium transfers one electron to the chlorine. In the process, the sodium becomes a positively-charged ion, and the chlorine a negatively-charged one. Also called **electrovalent bond**.

ionic conduction **1.** The conduction of a charge through the movement of ions. **2.** In a crystal, the conduction of a charge through the movement of ions within its lattice.

ionic crystal A crystal composed of a lattice of positively-charged and negatively-charged ions which are held together by their electrostatic interaction.

ionic current The current produced through **ionic conduction**.

ionic focusing In an electron tube, the use of an inert gas to focus an electron beam. The gas along the path of the beam becomes positively charged, and this attracts the electrons, further narrowing the beam. Used, for instance, in a CRT. Also called **gas focusing**.

ionic semiconductor A semiconductor in which ions, as opposed to electrons or holes, serve as the primary charge carrier.

ionic solid A solid in which positively-charged and negatively-charged ions are held together by their electrostatic interaction.

ionic spectrum An emission spectrum produced by the effect that a spark between electrodes has on an ion. Also called **spark spectrum**.

ionization The process of converting a neutral atom or molecule into an ion, through the addition or removal of electrons. Salts, for instance, become ionized when dissolved in an appropriate solvent, which increases the conductivity of said solvent. For example, sodium chloride dissociates into sodium and chloride ions when dissolved in water.

ionization chamber An instrument for the detection and measurement of radiation. It does so by measuring the electric current that flows when radiation ionizes the gas in a chamber, which makes said gas a conductor of electricity. Also called **ion chamber**.

ionization concentration Same as **ion concentration**.

ionization counter An instrument which registers and counts ionizing radiation, such as alpha rays, given off by radioactive entities. For example, a Geiger counter or a scintillation counter. Also called **ionization detector**, **radiation counter**, **counter (3)**, or **particle counter (1)**.

ionization current A current flowing as a result of **ionization**. For example, it may be that flowing through a solution via dissolved electrolytes, or that in a gas tube when its contained gas becomes ionized.

ionization density Same as **ion concentration**.

ionization detector Same as **ionization counter**.

ionization energy Same as **ionization potential (1)**.

ionization gage Same as **ionization gauge**.

ionization gauge A vacuum gauge incorporating a vacuum tube whose output current is proportional to the pressure of the gas within it. All gas atoms and molecules present are ionized, producing the detected current. Such instruments are utilized to measure extremely low pressures, including those below 10^{-10} torr. Also spelled **ionization gage**. Also called **ionization vacuum gauge**.

ionization potential Also called **ionization voltage**. **1.** The minimum energy per unit charge necessary to remove an electron, in its ground state, from an atom. Usually expressed in volts. Also called **ion potential**, or **ionization energy**. **2.** In a gas tube, the plate voltage at which conduction begins.

ionization smoke detector A smoke detector which ionizes the air in its sensing chamber, to produce an ionization current. When smoke, or other detected particles suspended in air, enter the ionization area, the electrical conductance varies, thus activating an alarm when a given level is reached.

ionization spectrometer An instrument used to analyze crystal structure by using X-rays. In it, a beam of collimated X-rays strikes the crystal, and a detector measures the angles and intensities of the reflected beam. Applying Bragg's Law to these measurements yields details about the crystal structure. Used, for instance, for X-ray diffraction studies that determine the structure of crystalline DNA. **DNA** is the abbreviation of **deoxyribonucleic acid**, and in its strands are contained the chemical basis of heredity. Also known as **Bragg spectrometer**, **crystal spectrometer**, or **crystal-diffraction spectrometer**.

ionization time In a gas tube, the time interval that elapses between the application of the ionizing energy, and the ionization of the gas.

ionization vacuum gauge Same as **ionization gauge**.

ionization voltage Same as **ionization potential**.

ionized atom An atom with a positive or a negative charge.

ionized gas A gas consisting of ionized atoms and/or ionized molecules.

ionized layer Within the ionosphere, a layer with enhanced ionization. Such ionization is caused by cosmic rays.

ionized molecule A molecule with a net positive or negative charge.

ionizing event Any physical process which results in the converting of neutral atoms or molecules into ions. Exposure to ionizing radiation, for instance, may cause this.

ionizing radiation Radiation whose energy is sufficient to cause ionization of the medium through which it passes. For example, photons, electrons, protons, neutrons, or alpha particles with the necessary energy to ionize directly or indirectly. Gamma rays, for instance, readily penetrate living tissue, producing serious harm when protective shielding is not utilized.

ionosphere The part of the earth's atmosphere in which ions and free electrons exist in comparatively high quantities. Its base is at an altitude of approximately 60 km, and it extends to beyond 400 km, with a more intense concentration between roughly 100 and 300 km altitude. The ionosphere affects the propagation of radio waves, by absorbing, refracting, and reflecting them. Certain regions, for instance, reflect waves in a manner that enables long-distance radio communications. The ionosphere is considered to be subdivided into regions, which are labeled D, E, and F. The D region contains the D layer, the E region the E layer, and the F region contains the F_1 and F_2 layers.

ionospheric correction A correction made to electromagnetic measurements, such as those between satellites and ground stations, to compensate for ionospheric errors.

ionospheric disturbance Any significant disturbance in the ionosphere, such as an ionospheric storm. Such a disturbance, for instance, changes the concentrations of ions and free electrons, and varies the virtual height of one or more layers.

ionospheric error The variations in the path of ionospheric waves, due to the varying composition of the ionosphere. Ionospheric corrections are utilized to compensate for this.

ionospheric layer A layer within the **ionosphere**. These are the D, E, F_1, and F_2 layers.

ionospheric propagation The propagation of radio waves with the assistance of one or more reflections off the ionosphere.

ionospheric region A region within the **ionosphere**. These are the D, E, and F regions.

ionospheric scatter In radio-wave propagation, the scattering of waves away from the transmitter after being reflected off the ionosphere. Also called **ionospheric scattering**, or **forward scatter**.

ionospheric-scatter communication Communication by means of radio waves that are propagated well beyond line-of-sight distances, due to scattering by the ionosphere. Also called **over-the-horizon communication**, **scatter communication**, or **forward-scatter communication**.

ionospheric-scatter propagation Propagation of radio waves well beyond line-of-sight distances, due to scattering by the ionosphere. Also called **over-the-horizon propagation**, **scatter propagation**, or **forward-scatter propagation**.

ionospheric scattering Same as **ionospheric scatter**.

ionospheric storm A significant disturbance within one or more regions of the ionosphere, which results in changes in the concentrations of ions and free electrons, and in variations in the virtual height of one or more layers. Such storms are usually caused by solar flares, and affect RF communications.

ionospheric wave A radio wave whose propagation has been assisted by reflecting off the ionosphere. It is a type of atmospheric radio wave.

IP Abbreviation of **Internet Protocol**. Within the TCP/IP protocol, the protocol running at the level which corresponds approximately to the network layer within the OSI model. IP organizes the data to be transmitted into IP datagrams, routes said datagrams to from the source to the destination, and then reassembles them, with any error packets being lost. In most networks, IP is combined with TCP, so that a virtual connection is established between the source and destination, helping ensure that datagrams will be properly delivered. Also called **IP protocol**.

IP address Abbreviation of Internet Protocol **address**. A 32-bit number which uniquely identifies any computer or device connected to a TCP/IP network. The number is written as four numbers separated by periods, with each number having a possible value which ranges between 0 and 255. Within an IP address, the network ID can encompass the first, first and second, or first, second, and third sets of numbers separated by the periods. The rest correspond to the host ID. For instance, if in the IP address 151.201.4.111, **151.201** identify the network, then **4.111** are the host ID. There are five classes of IP address, class A through class E, which determine the maximum number of hosts and networks which can be accommodated. Also called **Internet address (1)**, or **IP number**.

IP datagram Abbreviation of Internet Protocol **datagram**. A block of data of a specific size, or of a maximum size, transmitted in a TCP/IP network. In addition to the payload, a datagram contains information such as the source and destination addresses. Also called **IP packet**, or **datagram**.

IP fragment Abbreviation of Internet Protocol **fragment**. A component of an **IP datagram**. It may consist, for instance, of the payload or the overhead.

IP header Abbreviation of Internet Protocol **header**. The overhead within an **IP datagram**. Such a header includes control and error-checking information, such as identification, length, protocol number, and source and destination addresses.

IP multicast Abbreviation of Internet Protocol **multicast**. The simultaneous transmission of data to multiple selected destinations within a TCP/IP network. When utilizing IP multicasting, there are no separate connections between the source and destination. There is one source, which sends one set of IP datagrams, and copies of said datagrams are only made where paths diverge at a router. Used, for instance, to transmit corporate messages to employees. Also called **IP multicasting**.

IP multicasting Same as **IP multicast**.

IP network Abbreviation of Internet Protocol **network**. A communications network utilizing the TCP/IP protocol. An example is the Internet.

IP next generation Same as **IPv6**.

IP number Abbreviation of Internet Protocol **number**. Same as **IP address**.

IP packet Abbreviation of Internet Protocol **packet**. Same as **IP datagram**.

IP protocol Same as **IP**.

IP security Abbreviation of Internet Protocol **security**. Same as **IPSec**.

IP spoofing Abbreviation of Internet Protocol **spoofing**. The use of a false IP address to maliciously appears as trusted host or another user.

IP telephony Abbreviation of Internet Protocol **telephony**. The use of IP-based packet-switched connections to transmit voice or other audio over an IP network such as the Internet. IP telephony may also be utilized for fax or video transmission. Also called **packet telephony**.

IP tunneling Abbreviation of Internet Protocol **tunneling**. A technique which enables a TCP/IP network to send data utilizing a different protocol. It does so by encapsulating packets, using the other network protocol, within packets being transmitted through the TCP/IP network.

IP version 6 Same as **IPv6**.

IPC Abbreviation of **Interprocess Communication**.

IPE Abbreviation of **integrated programming environment**.

IPL Abbreviation of **initial program load**.

IPng Same as **IPv6**. Abbreviation of **Internet Protocol next generation**.

IPP Abbreviation of **Internet Printing Protocol**.

ips Abbreviation of **inches per second**.

IPSec Abbreviation of **IP** security. A suite of protocols developed by the Internet Engineering Task Force, for secure communications over the Internet.

IPv6 Abbreviation of Internet Protocol version 6, or **IP version 6**. An IP protocol set forth by the Internet Engineering Task Force, which includes improvements such as increasing the IP address to a 128-bit number, which would provide for a vastly larger amount of available addresses. Also called **IPng**.

Ir Chemical symbol for **iridium**.

IR 1. Abbreviation of **infrared**. 2. Abbreviation of **infrared radiation**. 3. Abbreviation of **information retrieval**.

IR absorption Abbreviation of **infrared absorption**.

IR absorption spectroscopy Abbreviation of **infrared absorption spectroscopy**.

IR astronomy Abbreviation of **infrared astronomy**.

IR band Abbreviation of **infrared band**.

IR communications Abbreviation of **infrared communications**.

IR detection Abbreviation of **infrared detection**.

IR detector Abbreviation of **infrared detector**.

IR diode Abbreviation of **infrared diode**.

IR drop The voltage drop developed across a resistance through which a current flows. This obeys Ohm's law, which is $E = IR$, where E is voltage, I is current, and R is resistance. Also called **resistance drop**.

IR emission Abbreviation of **infrared emission**.

IR emitting diode Abbreviation of **infrared-emitting diode**.

IR energy Abbreviation of **infrared energy**.

IR filter Abbreviation of **infrared filter**.

IR homing Abbreviation of **infrared homing**.

IR imaging Abbreviation of **infrared imaging**.

IR lamp Abbreviation of **infrared lamp**.

IR laser Abbreviation of **infrared laser**.

IR light Abbreviation of **infrared light**.

IR microscope Abbreviation of **infrared microscope**.

IR microscopy Abbreviation of **infrared microscopy**.

IR motion detector Abbreviation of **infrared motion detector**.

IR mouse Abbreviation of **infrared mouse**.

IR photoconductor Abbreviation of **infrared photoconductor**.

IR photography Abbreviation of **infrared photography**.

IR port Abbreviation of **infrared port**.

IR radiation Abbreviation of **infrared radiation**.

IR radiometer Abbreviation of **infrared radiometer**.

IR range Abbreviation of **infrared range**.

IR rays Abbreviation of **infrared rays**.

IR reflow Abbreviation of **infrared reflow**.

IR reflow soldering Abbreviation of **infrared reflow soldering**.

IR region Abbreviation of **infrared region**.

IR remote control Abbreviation of **infrared remote control**.

IR scanner Abbreviation of **infrared scanner**.

IR scanning Abbreviation of **infrared scanning**.

IR sensor Abbreviation of **infrared sensor**.

IR signal Abbreviation of **infrared signal**.

IR soldering Abbreviation of **infrared soldering**.

IR spectrometer Abbreviation of **infrared spectrometer**.

IR spectrometry Abbreviation of **infrared spectrometry**.

IR spectrophotometer Abbreviation of **infrared spectrophotometer**.

IR spectrophotometry Abbreviation of **infrared spectrophotometry**.

IR spectroscopy Abbreviation of **infrared spectroscopy**.

IR spectrum Abbreviation of **infrared spectrum**.

IR thermometer Abbreviation of **infrared thermometer**.

IR welding Abbreviation of **infrared welding**.

IRC Abbreviation of **Internet Relay Chat**.

IrDA Abbreviation of Infrared Data Association. An organization which creates and promotes standards for computer and communications devices utilizing infrared signals to convey information.

IrDA port A port which complies with **IrDA** specifications.

IrDA protocol A protocol which complies with **IrDA** specifications.

IrDA standard A standard created and promoted by **IrDA**.

IRED Abbreviation of **infrared-emitting diode**.

IRG Abbreviation of **inter-record gap**.

iridescence A rainbow-like exhibition of colors created by interference and reflections of light from the surfaces of a thin film, or a thinly-layered material. Some crystals, for instance, display this phenomenon.

iridescent Characterized by **iridescence**.

iridium A chemical element whose atomic number is 77. It is a silver-white and very hard metal whose density is approximately 22.65, which makes it, along with osmium, one of the two most dense elements. In addition, it is the most corrosion-resistant element, and has over 30 know isotopes, of which 2 are stable. Its applications include its use high-temperature apparatuses, in electric contacts and electrodes, and in thermocouples. Its chemical symbol is **Ir**.

iris 1. A diaphragm in a waveguide which introduces an inductance or capacitance. Such a diaphragm may consist of a conductive plate with an aperture, and serves for impedance matching. Also called **waveguide iris**. 2. Same as **iris diaphragm**.

iris diaphragm A mechanical device which serves to vary the diameter of an aperture, such as that of a camera, thereby controlling the amount of light entering. Also called **iris (2)**.

IRIX A popular UNIX-based operating system.

IRM Abbreviation of **information resources management**.

iron A silvery-white ductile and malleable metallic chemical element whose atomic number is 26. It is the second most abundant metallic element in the earth's crust after aluminum, is chemically reactive, is the only metal than can be tempered, and it has over 20 known isotopes, of which 4 are stable. It has various applications in electronics, including its use in magnets, magnetic circuits, and in specialized alloys. Its chemical symbol, **Fe**, is taken from the Latin word for iron: ferrum.

iron-constantan thermocouple A thermocouple in which a wire or strip of iron is bonded with a wire or strip of constantan. Such a thermocouple may be used, for instance, for temperature measurements between 0 °C and about 750 °C, accurate to about 0.1 °C.

iron core 1. A core, such as that of a transformer or armature, which is made of solid or laminated iron. 2. A core, such as that of a transformer or armature, which is made of a solid or laminated ferromagnetic material.

iron-core choke A choke which is wound around an **iron core**.

iron-core coil A coil which is wound around an **iron core**.

iron-core transformer A transformer which is wound around an **iron core**.

iron loss The energy dissipated by an iron core, such as that found in a transformer or inductor. May be due, for instance, to eddy currents and hysteresis loss. Also called **core loss**.

iron oxide A reddish or bluish-black powder whose chemical formula is Fe_3O_4. It is used in magnetic-tape coatings, magnetic inks, and in ferrites. Also called **iron oxide black**, **ferrosoferric oxide**, **black iron oxide**, or **magnetic iron oxide**.

iron oxide black Same as **iron oxide**.

iron oxide tape A magnetic recording tape utilizing iron oxide particles. Such a tape provides barely adequate performance. Also called **Type I tape**.

IRQ Abbreviation of **interrupt request**.

IRQ line Abbreviation of **interrupt request line**.

irradiance The amount of radiant flux incident per unit area of a surface. Also called **radiant flux density**.

irradiate 1. To expose to radiation, such as light or X-rays. 2. To expose to ionizing radiation, such as gamma rays, ultraviolet rays, or X-rays.

irradiation 1. Exposure to radiation, such as light or X-rays. 2. Exposure to ionizing radiation, such as gamma rays, ultraviolet rays, or X-rays. Such exposure may be unwanted, such as that received when working with radioactive materials, or may be desired, as is the case when used therapeutically or in the purification of foods.

irrational number Any real number which cannot be expressed in the form of **a/b**, where **a** is an integer, and **b** is a non-zero integer.

irregular 1. Having an uneven rate of occurrence or duration. 2. Not uniform or symmetrical. 3. Deviating from a given type or standard.

irregularity The quality or state of being **irregular**. Also, that which is irregular.

IRTF Abbreviation of **Internet Research Task Force**.

IS Abbreviation of **information services**.

IS-IS Abbreviation of **Intermediate System-to-Intermediate System**.

ISA Same as **ISA bus**. Abbreviation of **Industry Standard Architecture**.

ISA bus Abbreviation of Industry Standard Architecture **bus**. A personal computer bus architecture that may be used for slower peripherals. Formerly called **AT bus**.

ISAM Abbreviation of **indexed sequential access method**.

ISC Abbreviation of **Internet Software Consortium**.

ISDN Abbreviation of Integrated Services Digital Network. A communications network utilizing ordinary copper telephone wires for the digital transmission of data, voice, and video. An ISDN line typically consists of two 64 kilobits per second B channels, and one 16 kilobits per second D channel, and is called Basic Rate Interface. This allows, for instance, the use of one B channel for voice, and the other for data. There are higher speed ISDN services, including Primary Rate Interface, with signaling rates of over 1.5 Mbps. Broadband ISDN, which uses fiber-optic cables, is capable of transmission speeds exceeding 155 Mbps.

ISDN adapter Same as **ISDN terminal adapter**.

ISDN-Basic Rate Interface Same as **ISDN-BRI**.

ISDN-BRI Abbreviation of **ISDN-B**asic Rate Interface. An ISDN line with two 64 kilobits per second B channels, and one 16 kilobits per second D channel. With the proper configuration, one B channel may be used for conversing on the telephone, and the other for a computer modem. Otherwise, the two B channels can be combined for total available bandwidth of 128 kilobits per second. Also called **Basic Rate Interface**.

ISDN digital subscriber line Same as **IDSL**.

ISDN line Abbreviation of Integrated Services Digital Network **line**. A communications line within an **ISDN**.

ISDN modem Abbreviation of Integrated Services Digital Network **modem**. Same as **ISDN terminal adapter**.

ISDN-PRI Abbreviation of **ISDN-P**rimary Rate Interface. Its own abbreviation is **PRI**. An ISDN line with 23 B channels, each with a speed of 64 Kbps, and one 64 Kbps D channel. This is equivalent to a T1 or DS1 line. In Europe, ISDN-PRI has 30 B channels and one D channel, so it is equivalent to an E1 line. Multiple B channels can be combined for the desired or needed bandwidth. Also called **Primary Rate Interface**, or **T1/PRI**.

ISDN-Primary Rate Interface Same as **ISDN-PRI**.

ISDN TA Abbreviation of **ISDN terminal adapter**.

ISDN terminal adapter Abbreviation of Integrated Services Digital Network **terminal adapter**. A device which interfaces an ISDN line with a user's equipment, such as a computer. Such an adapter may be connected, for example, to a serial port. Its own abbreviation is **ISDN TA**. Also called **ISDN modem**, **ISDN adapter**, or **terminal adapter**.

ISIS Abbreviation of **Intermediate System-to-Intermediate System**.

ISM band Abbreviation of Industrial, Scientific, and Medical **band**. A frequency band that can be used without obtaining a license. One such band centers around 2.4 GHz, and is used, for instance, for Bluetooth.

iso- A prefix meaning equal or uniform. For example, isoelectric, or isotope.

ISO Abbreviation of **International Organization for Standardization**, or **International Standards Organization**.

ISO-9000 A collection of **ISO** standards defining a quality assurance program, and which takes into account factors such as design, development, production, installation, testing, and service.

ISO-9241 An **ISO** standard pertaining to ergonomic requirements for video display terminals.

ISO-9660 An **ISO** standard for defining CD-ROM files.

ISO-OSI Model Abbreviation of **International Organization for Standardization-Open Systems Interconnection Model**. Same as **ISO-OSI Reference Model**.

ISO-OSI Protocol Abbreviation of **International Organization for Standardization-Open Systems Interconnection Protocol**. Same as **ISO-OSI Reference Model**.

ISO-OSI Reference Model Abbreviation of **International Organization for Standardization-Open Systems Interconnection Reference Model**. An ISO standard which defines a seven-layer hierarchical structure for the implementation of communications protocols, with the goal of standardization of current networks along with component hardware and software, and providing guidance for the crea-

tion of future such networks and components. Each higher layer requests and depends on services from the lower layer adjacent to it. The lowest layer is the physical layer, or layer 1, followed by the data-link layer, or layer 2, then comes the network layer, followed by the transport layer, then comes the session layer, and then the presentation layer, and the highest is the application layer, or layer 7. Also known by various other names, including **ISO-OSI Model**, **ISO-OSI Protocol**, **OSI Reference Model**, **Open Systems Interconnection Reference Model**, and **Open Systems Interconnection Model**.

isobar 1. A line or curve representing points along which pressure is the same, such as that which might appear on a weather map. 2. A nuclide which has the same atomic mass as another, but whose atomic number is different. For example, carbon-14 and nitrogen-14 are isobars of each other. This contrasts with an **isotope**, which has the same atomic number, but a different atomic mass.

ISOC Abbreviation of **Internet Society**.

isochromatic Having the same chromaticity or color.

isochromatic lines Lines having the same chromaticity or color.

isochronal Same as **isochronous**.

isochronous Characterized by having a fixed frequency or period, or an equal duration. Also called **isochronal**.

isodynamic line A line representing points along which the earth's magnetic field has the same strength.

isoelectric 1. Having an equal electric potential. That is, having a potential difference of zero. 2. Having a constant electric potential.

isoelectronic Pertaining to atoms which have the same number and configuration of electrons, but not necessarily the same atomic mass.

isofootcandle line Same as **isolux line**.

isolate 1. Same as **insulate**. 2. To disconnect from a power supply. 3. To set apart so that no interaction occurs. 4. To separate one or more components which are combined with others.

isolated 1. To be protected or surrounded by an **insulator**. 2. To have been disconnected from a power supply. 3. To have been set apart, so that no interaction occurs. 4. To be separated, after having been combined with other components.

isolated input 1. An input which is disconnected from a power supply. 2. An input which is insulated and otherwise prevented from interactions. 3. An input which is not grounded.

isolating amplifier Same as **isolation amplifier**.

isolating capacitor Same as **isolation capacitor**.

isolating circuit Same as **isolation circuit**.

isolating diode Same as **isolation diode**.

isolating network Same as **isolation network**.

isolating resistor Same as **isolation resistor**.

isolating switch Same as **isolation switch**.

isolating transformer Same as **isolation transformer**.

isolation 1. The process of protecting with, or surrounding by an **insulator** (2). Also, the condition of being so isolated. 2. The process of disconnecting from a power supply. Also, the condition of being so isolated. 3. The process of setting apart, so that no interaction occurs. Also, the condition of being so isolated. 4. The process of separation, after having been combined with other components. Also, the condition of being so isolated.

isolation amplifier An amplifier which isolates successive stages to help prevent undesirable interactions between them. Also called **isolating amplifier**, **buffer amplifier**, or **buffer** (3).

isolation capacitor A capacitor which blocks DC and low-frequency AC, while allowing higher-frequency AC to pass freely. Also called **isolating capacitor, blocking capacitor,** or **coupling capacitor.**

isolation circuit A circuit which serves to **isolate.** Also called **isolating circuit.**

isolation diode A diode which allows signals to pass in one direction, while blocking those in the other direction. Also called **isolating diode.**

isolation network An electric network which serves to isolate. Also called **isolating network.**

isolation resistor A high-value resistor which allows signals to pass in one direction, while reducing impedance coupling. Also called **isolating resistor.**

isolation switch A switch that disconnects a circuit from a power supply. Also called **isolating switch.**

isolation transformer A transformer which separates sections of circuits, devices, equipment, or systems from direct connections with others, or from the power source, so as to reduce or eliminate undesired interactions. Such a transformer usually has a 1:1 turns ratio, and is used, for instance, to protect against harm to sensitive components or devices, to minimize shock hazard, or to comply with electrical requirements. Also called **isolating transformer.**

isolation voltage The maximum voltage that can be applied to an isolated circuit without breakdown occurring.

isolator That which serves to **isolate.**

isolux line A line representing points along which the illuminance is the same. Also called **isofootcandle line.**

isomagnetic line A line representing points along which the magnetic force has the same intensity.

isomer 1. One of two or more chemical substances which have the same chemical formula, but whose configuration of constituent atoms is different. Two isomeric substances may have very different physical and chemical properties. 2. One of two or more atomic nuclei which having the same atomic number and mass number, but with different energies and radioactive properties.

isothermal A process or region which is maintained at a constant temperature.

isotope An atom which has the same atomic number as another, but a different atomic mass. For example, protium, deuterium, and tritium are the three isotopes of hydrogen. This contrasts with an **isobar (2)**, which has the same atomic mass, but whose atomic number is different.

isotropic Having one or more physical properties which are identical in all directions. For instance, the directivity of an isotropic antenna. This contrasts with **anisotropic,** in which such properties vary depending on the axis along which they are measured.

isotropic antenna A hypothetical antenna which radiates and/or receives equally in all directions. Such an antenna is utilized, for example, as a reference for comparison with real antennas. Also called **isotropic radiator,** or **unipole.**

isotropic medium A medium in which one or more physical properties are identical in all directions.

isotropic radiation Radiation which is the same in all directions. For example, that of an isotropic antenna.

isotropic radiator Same as **isotropic antenna.**

isotropic scattering The scattering of electromagnetic radiation, such as light, equally in all directions.

isotropy The manifestation of **isotropic** properties.

ISP Abbreviation of **Internet service provider.**

ISR 1. Abbreviation of **Interrupt Service Routine.** 2. Abbreviation of **Intersecting Storage Rings.**

ISV Abbreviation of **independent software vendor.**

IT Abbreviation of **information technology.**

Itanium A popular family of CPU chips.

item Within a group, collection, or series, a single article, unit or entity.

iterate To repeat a sequence, process, cycle, or the like.

iteration 1. The act or process of repeating a sequence, process, cycle, or the like. Also, such an instance. For example, the process of repeating a set of instructions. 2. Same as **iterative method.**

iterative impedance For a four-terminal network, the impedance which when connected to one pair of terminals will cause that of the other pair of terminals to match it.

iterative method A problem-solving method in which a sequence of operations is repeated so as to successively approximate a solution. An example is a computer utilizing this method to solve complex equations. Also called **iteration (2).**

ITS Abbreviation of **Intelligent Transportation System.**

ITU Abbreviation of International Telecommunications Union. An international organization which works to standardize communications networks and services. The ITU monitors and studies world telecommunications, and makes recommendations for standardization.

ITU-R Abbreviation of International Telecommunications Union-Radiocommunication Sector. The section of the ITU which develops radio communications standards.

ITU-R BT.601 An ITU standard specifying the format of digital component video. It digitizes both PAL and NTSC signals. Formerly known as the **CCIR 601** standard.

ITU-T Abbreviation of International Telecommunications Union-Telecommunications Standardization Sector. The section of the ITU which develops telecommunications standards.

ITU-TSS Same as **ITU-T.** Abbreviation of International Telecommunications Union-Telecommunications Standardization Sector.

ITV 1. Abbreviation of **interactive TV.** 2. Abbreviation of **industrial TV.**

IVR 1. Abbreviation of **interactive voice response.** 2. Abbreviation of **induction voltage regulator.**

IVR system Abbreviation of **interactive voice response system.**

IVR technology Abbreviation of **interactive voice response technology.**

IVRS Abbreviation of **interactive voice response system.**

IXC Abbreviation of **interexchange carrier.**

J

j A mathematical unit equal to the square root of -1. That is, $j^2 = -1$. It is used in complex numbers. *j* is usually used in electronics, while *i* is usually used in other fields.

J 1. Symbol for **joule**. 2. A high-level programming language especially suited for mathematical applications.

J Symbol for **current density**, or **electric current density**.

J antenna Same as **J-pole antenna**.

J display In radars, a display where there is a circular time base, and in which scanned objects appear as peaks around the displayed circumference. The distance between the peaks is proportional to the range of scanned objects. Also called **J scan, J scanner, J scope,** or **J indicator**.

J indicator Same as **J display**.

J-K flip-flop Same as **JK flip-flop**.

J-lead On a chip package, a lead which extends down and under the housing, so as to resemble a capital letter **J**. Such leads occupy less room than **gull-wing leads**.

J particle An unusually long-lived meson. Also called **psi particle**.

J-pole antenna A vertical half-wave antenna fed at one end by quarter-wave parallel-wire section, and whose shape resembles a capital letter **J**. Also called **J antenna**.

J scan Same as **J display**.

J scanner Same as **J display**.

J scope Same as **J display**.

J2EE Abbreviation of Java 2 Platform, Enterprise Edition. A Java platform oriented towards enterprise applications.

J2ME Abbreviation of Java 2 Platform, Micro Edition. A Java platform oriented towards mobile wireless devices, such as properly equipped cell phones and PDAs.

jabber 1. In a communications network, a condition or instance where a faulty device continuously transmits a random, or otherwise useless signal. For instance, the non-stop broadcasting of the availability of a device, while said availability is known. 2. A data packet which exceeds the maximum allowable size for a given protocol.

jack An electric connector with one or more recessed openings, into which a **plug** with matching pins or prongs is inserted. When said plug is inserted, the circuit is closed. This allows for an easy and rapid connection which is secure. Widely used in communications, computer, and entertainment devices, equipment, and systems. Also called **jack connector**.

jack box A box which houses and enables simple connections for multiple **jacks**.

jack connector Same as **jack**.

jack panel A panel which houses and enables simple connections for multiple **jacks**.

jack screw Same as **jackscrew**.

jacket The outermost layer of a component, connector, cable, floppy disk, or the like, which provides protection and structural support. A jacket may also server to provide additional insulation, even when surrounding an insulator.

jackscrew A screw which is affixed to a two-piece connector, and which serves to draw each half together, or to separate them. Also spelled **jack screw**.

jaggies The jagged appearance that curved or diagonal lines in computer graphics have when there is not enough resolution to show images realistically. Also called **stair-stepping**, or **aliasing** (2).

jammer A device or piece of equipment, such as a radio transmitter, utilized for **jamming**.

jamming The intentional use of electromagnetic radiation, especially radio-frequency signals, to disrupt communications. Also called **electronic jamming**.

Jansky noise Also called **cosmic noise**, or **sky noise**. 1. Radio-frequency noise caused by sources outside the earth's atmosphere, such as the cosmic microwave background. 2. Radio-frequency noise that originates outside the earth's atmosphere.

jar 1. The container holding the electrolyte and plates in a storage cell. 2. An older unit of capacitance equal to approximately 1.11×10^{-9} farad, which is the approximate capacitance of a Leyden jar.

Java A robust object-oriented programming language especially suited for Internet and intranet applications. It is similar to C++, but is simplified, smaller, and easier to use. In addition, Java provides its own memory management, is secure, and essentially architecture-independent.

Java 2 Platform, Enterprise Edition Same as **J2EE**.

Java 2 Platform, Micro Edition Same as **J2ME**.

Java applet An applet written in Java. Such applets are usually interpreted by a Web browser, and can provide video and/or audio effects, perform calculations, or generally provide interactivity to Web pages. Most Java applications are short enough to be considered applets. Also called **Java application** (1).

Java application 1. Same as **Java applet**. 2. Any application written in Java.

Java bytecode The compiled form into which a Java program is converted, to be executed by a Java Virtual Machine.

Java Card A card, such as a smart card, enabled to run Java applications.

Java chip A chip optimized for execution of Java applications, and which executes bytecode natively. Such a chip can be incorporated into computers, handheld devices, cellular telephones, and so on.

Java compiler A compiler which transforms Java source code into machine language.

Java-compliant browser A Web browser which supports Java. Most do.

Java operating system Same as **JavaOS**.

Java sandbox An area within computer memory or storage in which Java-based applications are confined to operate.

Java servlet A Java applet that runs on a server.

Java source code Source code written in Java.

Java Virtual Machine A virtual machine which processes Java bytecode. That is, it provides the environment within which Java applications run. Most Web browsers are equipped with a Java Virtual Machine, and such machines are essentially architecture-independent. Its abbreviation is **JVM**.

JavaBean A reusable Java application, applet, or application component. JavaBeans can be combined to create applications which are essentially architecture-independent. Also, the application interface utilized.

JavaOS Abbreviation of **Java** operating system. An operating system optimized for execution of Java applications. It incorporates a Java Virtual Machine, and may used, for instance, as the sole operating system in network computers and Internet appliances.

JavaScript A script language which is based in part on Java, and which is embedded into HTML documents. JavaScript is easier to use, but less powerful than Java. It is used, for instance, to modify and improve the interactive content of a Web page.

JBOD Abbreviation of **j**ust **a** **b**unch **of** **d**isks, or **j**ust **a** **b**unch **of** **d**rives. Multiple hard disks which are all accessible, but not arranged in any type of RAID configuration, so there is no maximization of performance or fault-tolerance.

JCL Abbreviation of **job control language**.

jewel Same as **jewel bearing**.

jewel bearing A bearing made of a natural or artificial precious or semiprecious stone, such as diamond, sapphire, corundum, or synthetic ruby. Such bearings are used in instruments such as gyros, or fine mechanical watches or clocks. Also called **jewel**.

JFET Abbreviation of **j**unction **f**ield-**e**ffect **t**ransistor. A field-effect transistor whose gate is insulated from the channel by a reverse-biased pn junction.

JFIF Abbreviation of **JPEG File Interchange Format**.

jig A base or device which helps to position, guide, or hold parts being worked on, or the tools utilized for said work.

Jini A technology which endeavors to simplify the connection and maximize the sharing of devices over a network. Such a network does not have to be planned or previously installed, and all connected devices are available to all users. Since each device or service provides the necessary interface, compatibility is maximized. Also called **Jini technology**.

Jini network A network utilizing **Jini technology**.

Jini technology Same as **Jini**.

JIT Abbreviation of **just-in-time**.

JIT compiler Abbreviation of **just-in-time compiler**.

jitter 1. Small and rapid fluctuations in a phenomenon, as might occur, for example, in image or sound reproduction, which are due to factors such as variations in the power supply, mechanical vibrations, or interference. Jitter may manifest itself, for instance, as variations in amplitude, frequency, phase, or time. 2. In a CRT, rapid and irregular fluctuations which make the image appear to shake, and which are usually due to errors in synchronization. 3. In fax, distortion of the received image, due to errors in synchronization.

JK flip-flop A versatile flip-flop whose inputs are designated as **J** and **K**, along with a clock input. Such a flip-flop has no undefined states, no race condition, and is edge-triggered. If both the J and K inputs are at logic 1 when the clock pulse arrives, it toggles between states. If the J and K inputs are at logic 0 when the clock pulse arrives, there is no change. If J is at logic 1 and K at logic 0, a clock pulse drives the output to 1, and if J is at logic 0 and K at logic 1, a clock pulse drives the output to 0. Its abbreviation is **JKFF**.

JKFF Abbreviation of **JK flip-flop**.

job A unit of computer work consisting of multiple steps. A job, for instance, may encompass the use of various programs for several processes, which as a whole are treated as a unit.

job control language A command language used in some mini and mainframe computer systems. Its abbreviation is **JCL**.

job processing Computer operations in which a series of **jobs** are handled, usually without concurrent user intervention.

job queue A list of **jobs** awaiting to be handled. Such a queue may consist, for instance, of a series of programs awaiting execution.

job step A single step within **job**.

job stream A sequence of steps followed to complete a **job**

jogging The rapidly opening and closing of a circuit, such as that controlling a motor or electromagnet, to produce small movements. Also called **inching**.

Johnson counter A counter similar to a ring counter, in which the complement output of the last stage is connected

to the input of the first stage, which provides **2n** states for **n** stages. Also called **twisted-ring counter**.

Johnson noise Broadband noise in a conductor, component, or circuit, due to the thermal agitation of electrons. Such noise increases proportionally to the absolute temperature of the material in question. Also called **thermal noise**.

join In relational databases, the forming of a new relation from two or more relations having one or more common attributes. For example, the joining of two tables through a common field.

joint 1. A connection between two elements, objects, or materials. Also, the location at which such a connection occurs. For instance, the connection point of two conductive paths. 2. In robotics, a point, such as that within an arm, where sections are linked and a given degree of freedom exists.

Joint Photographic Experts Group Same as **JPEG**.

Jones plug A type of polarized connector with multiple contacts.

Josephson effect The flow of current, in the form of Cooper pairs, across a **Josephson junction**. Also called **Josephson tunneling**.

Josephson junction Two superconductors separated by a very thin insulating barrier. Cooper pairs will tunnel through such a barrier. A Josephson junction can be used, for instance, for ultrafast switching functions.

Josephson memory Computer memory utilizing **Josephson junctions**. Such memory is extremely fast.

Josephson tunneling Same as **Josephson effect**.

joule The SI unit of work or energy. It defined as the work accomplished by a force of one newton in moving an object one meter in the direction said force is applied. One joule equals 10^7 ergs, approximately 0.23885 calories. Its symbol is J.

Joule effect 1. The heat produced when an electric current passes through a material, due to resistance. A certain portion of the electrical energy passing through a conductor is converted into heat energy due to the resistance of said conductor. Also called **heating effect**. 2. The mechanical deformation of a material, especially that which is ferromagnetic, when placed in a magnetic field. For example, if such a material is placed in a direction parallel to an applied magnetic field, its length will change. The intensity of the field will determine the extent of mechanical deformation. Also called **magnetostriction**.

Joule heat Heat produced as a consequence of the **Joule effect** (1).

Joule's law For a circuit with a constant resistance, the heat energy generated per unit time equals the resistance multiplied by the square of the current.

journal A record of computer and/or network activity. Used, for example, to find the origin of problems, monitor usage, recover data, or to identify unauthorized access. Also called **electronic journal**, **electronic log**, or **log** (3).

journaling The keeping of a **journal**.

joystick A pointing device consisting of a base with an attached lever which moves in any direction. As the joystick is moved, so does a given object on a screen, and such motion continues in a given direction until the lever is returned to its upright position. A joystick usually has one or more buttons, which provide additional functions. Used, for instance, in video games, or in CAD applications.

JPEG Abbreviation of Joint Photographic Experts Group. 1. An image-compression standard which provides for lossy compression of color and grayscale images. Compression ratios of 20:1, for instance, provide little noticeable loss, and if greater losses are acceptable, ratios of greater than 100:1 can be obtained. 2. Same as **JPEG image**.

JPEG File Interchange Format A file format widely utilized for storing and sharing JPEG images. Its abbreviation is **JFIF**.

JPEG image An image compressed using **JPEG (1)**. Also called **JPEG (2)**.

jughead A utility employed to make searches in Gopherspace.

jukebox 1. An optical disc changer which can hold many, up to several thousand DVDs, CD-ROMs, or the like. Discs may be swapped, but only one can be read at once. Also called **optical jukebox (1)**. 2. A device similar to a **jukebox (1)**, but which stores and enables access to tape cartridges.

jump A computer instruction that switches the CPU to another location in memory. Also called **jump instruction**, or **branch (4)**.

jump instruction Same as **jump**.

jump page A Web page created with given characteristics, such as multiple appearances of specific keywords, so as to rank high on a particular search engine. Also called **doorway page**, **bridge page**, **entry page**, or **gateway page**.

jumper 1. A length of a conductor, such as a wire or a cable, which is utilized to bypass a portion of a circuit, or to attach a device, such as a meter, during testing or troubleshooting. 2. On a circuit board, a small plug or wire whose presence or absence determines whether a given circuit is open or closed, which in turn changes an aspect of the hardware configuration, such as the selection of an interrupt request.

jumper cable A cable which serves as a **jumper (1)**. Such a cable is usually equipped with spring-loaded clamps or clips, and may be used, for instance, to transfer a charge between car batteries, or to join telecommunications circuits.

jumper wire A wire which serves as a **jumper (1)**. Such a wire is usually equipped with a spring-loaded clip, and is used for testing or troubleshooting purposes.

junction 1. A place where things join or meet. Also, the act of joining or meeting. 2. A point in a circuit where two or more conductors meet. For instance, a connection point between sections of a transmission line. 3. A layer or boundary which serves as the interface between semiconductor regions with different properties. For example, a pn junction. Also called **semiconductor junction**. Also, a similar interface between a semiconductor and a metal. 4. A layer or boundary which serves as the interface between materials, or regions within a single material, with different physical properties. For instance, such a junction in a thermocouple. 5. A fitting which serves to join one waveguide section with another. Also called **waveguide junction**.

junction box A box which houses and enables simple connections for multiple junctions, such as those made with wires or cables.

junction capacitance In a semiconductor, the capacitance across the region of a pn junction that is depleted of charge carriers. Also known as **depletion-layer capacitance**, **barrier-layer capacitance**, or **barrier capacitance**.

junction capacitor In a semiconductor, a capacitor which utilizes the capacitance of a reverse-biased pn junction.

junction diode A semiconductor diode consisting of a pn junction, and which is utilized to pass current essentially in only one direction. Such diodes are used, for instance, in solar cells, and in laser diodes. Also called **junction rectifier**, or **pn-junction diode**.

junction field-effect transistor Same as **JFET**.

junction isolation The utilization of a **junction (3)** to isolate components in an IC, as opposed for instance, to the use of oxide isolation.

junction laser A laser in which a forward-biased pn junction diode is used to convert its DC input into a coherent light output. Used, for instance, as a light-pulse generator for transmission of information over fiber-optic lines. Also called **laser diode**, **semiconductor laser**, **injection laser**, or **injection laser diode**.

junction loss Any loss occurring at a junction, such as that at a connection point between sections of a transmission line, or at a waveguide junction.

junction point 1. The point at which a branch, such as that of a circuit or electric network, originates. Also known as **branch point**, or **node (2)**. 2. The point at which a **junction** occurs.

junction rectifier Same as **junction diode**.

junction transistor A transistor with one or more junctions, such as pn junctions. These include bipolar junction transistors, pnpn transistors, and transistors utilizing Josephson junctions.

junk email Email which is sent to one or more recipients who have not requested or authorized it. Junk email is usually synonymous with **spam**. Also called **junk mail**.

junk fax A fax which is sent to one or more recipients who have not requested or authorized it.

junk mail Same as **junk email**.

just-in-time In computers, performed on the fly, or just before a program is run. Its abbreviation is **JIT**.

just-in-time compiler A compiler which translates just before a program is run. Its abbreviation is **JIT compiler**.

justification 1. The altering of the bit rate of a transmission so that it can be received at a different rate, without introducing errors. Also called **pulse stuffing**, or **bit stuffing (2)**. 2. The displaying or printing of text in a manner that all lines in a paragraph, except the last line, are aligned evenly between the left and right margins.

justify 1. To shift the contents of a register in a manner that the most or least significant character is at a specified position, such as the end of said register. 2. To display or print text in a manner that all lines in a paragraph, except the last line, are aligned evenly between the left and right margins.

jute A natural fiber which is formed into rope-like strands which are used, for instance, to fill the spaces of a cable to give it a round cross-section, and to cushion and protect.

jute yarn The rope-like strands formed with **jute**.

JVM Abbreviation of **Java Virtual Machine**.

K

k Abbreviation of **kilo-**.

k Symbol for **Boltzmann constant**.

K **1.** Symbol for **kelvin**. **2.** Symbol for **cathode**, or **negative electrode** (1). **3.** Chemical symbol for **potassium**. **4.** Abbreviation of **kilo-**. **5.** In computing, the number 1024, or 2^{10}. **6.** Symbol for **constant**.

K Band **1.** In communications, a band of radio frequencies extending from 18.00 to 27.00 GHz, as established by the IEEE. This corresponds to wavelengths of approximately 1.7 to 1.1 cm, respectively. **2.** In communications, a band of radio frequencies extending from 10.9 to 36.0 GHz. This corresponds to wavelengths of approximately 27.5 mm to 8.3 mm, respectively.

K display In radars, a modified A display, in which the scanned object appears as two vertical deflections in the horizontal plane, instead of one. When the radar antenna is pointed directly at the scanned object, both blips are of the same height, while any difference in their heights indicates the direction and magnitude of the error. Also called **K scan**, **K scanner**, **K scope**, or **K indicator**.

K electron An electron located in the K shell of an atom.

K indicator Same as **K display**.

K scan Same as **K display**.

K scanner Same as **K display**.

K scope Same as **K display**.

K shell The innermost electron shell of an atom. It is the closest to the nucleus.

kA Abbreviation of **kiloampere**.

Ka Band **1.** In communications, a band of radio frequencies extending from 27.00 to 40.00 GHz, as established by the IEEE. This corresponds to wavelengths of approximately 11 mm to 7.5 mm, respectively. **2.** In communications, a band of radio frequencies extending from 33 to 36 GHz. This corresponds to wavelengths of approximately 9.1 mm to 8.3 mm, respectively.

Karnaugh map A special arrangement of a truth table, which serves to simplify a Boolean expression. It is a tabular equivalent of a Venn diagram. Such a map serves, for instance, to help design the simplest logic circuit for a given function.

kb **1.** Abbreviation of **kilobit**. It is a less utilized, but more proper abbreviation than **Kb**. **2.** Abbreviation of **kilobaud**.

kB Abbreviation of **kilobyte**. It is a less utilized, but more proper abbreviation than **KB**.

Kb **1.** Abbreviation of **kilobit**. **2.** Abbreviation of **kilobaud**.

KB Abbreviation of **kilobyte**.

kb/s Abbreviation of **kilobits per second**.

kB/s Abbreviation of **kilobytes per second**.

kbaud Abbreviation of **kilobaud**.

Kbaud Abbreviation of **kilobaud**.

Kbit Abbreviation of **kilobit**.

kbps Abbreviation of **kilobits per second**.

Kbps Abbreviation of **kilobits per second**.

kBps Abbreviation of **kilobytes per second**.

KBps Abbreviation of **kilobytes per second**.

kBq Abbreviation of **kilobecquerel**.

Kbyte Abbreviation of **kilobyte**.

kc Abbreviation of **kilocycle**.

kC Abbreviation of **kilocoulomb**.

kcal Abbreviation of **kilocalorie**, or **kilogram-calorie**.

kCi Abbreviation of **kilocurie**.

KDP Abbreviation of **potassium dihydrogen phosphate**.

KE Abbreviation of **kinetic energy**.

keep-alive circuit In a transmit-receive tube, a circuit which maintains residual ionization, enabling full ionization to occur more rapidly.

keep-alive message A packet or message sent between devices over a communications network link during idle periods, to verify that the virtual and physical connection between them is operational.

keeper A bar, usually made of soft iron, utilized to join the magnetic poles of a magnet which is not in use. The keeper maintains the magnetic flux circulating continuously through the magnet, which delays demagnetization considerably. Used especially with horseshoe magnets. Also called **magnet keeper**.

Kell factor A factor which is calculated to help determine the practically attainable vertical resolution of a displayed image, such as that of a TV signal. It describes how blurry the lines on a TV appear to human eyes.

kelvin The fundamental SI unit of thermodynamic temperature. It is also the temperature interval used in the Kelvin temperature scale, so 0 K equals absolute zero. An increment of one kelvin equals an increment of one degree Celsius. Please note that the degree symbol is omitted when utilizing kelvins. Also note that the proper way to refer to this concept is **kelvin**, and not **degree Kelvin**, as the latter is its former name. Its symbol is **K**.

Kelvin balance An ammeter which operates by having a current pass through two nearby coils connected in series. One of the coils is fixed, while the other is attached to the arm of a balance. The attractive force between the coils is counterweighed by the force of gravity acting on a known weight on the arm of the balance, which in turn indicates the current strength on its scale. Also called **Ampère balance**, or **current balance**.

Kelvin bridge A special bridge circuit whose arrangement minimizes the effects of contact resistance, enabling the accurate measurement of very low resistances. Some such devices can measure resistances of a few microhms. Also called **Kelvin double bridge**, **double bridge**, or **Thomson bridge**.

Kelvin contacts In a measuring instrument, two sets of contacts which are used at each test point, one carrying the test signal, and the other being connected to said instrument. This enables lead resistance to be eliminated when measurements are taken.

Kelvin double bridge Same as **Kelvin bridge**.

Kelvin effect Also called **Thomson effect**. **1.** When a current flows through a metal whose ends are at different temperatures, heat is absorbed or generated, depending on the metal. **2.** When there is a temperature differential between the ends of a conductor, an electromotive force is developed, the direction of which depends on the metal.

Kelvin scale Same as **Kelvin temperature scale**.

Kelvin temperature scale A temperature scale whose value of zero degrees equals absolute zero, and whose unit of temperature increment is the **kelvin**. Also called **Kelvin scale**.

Kelvin-Varley voltage divider A device which incorporates a cascade of resistors which accurately divide voltages. It minimizes the use of components, and may be used as a standard.

Kendall effect Distortion in a fax record, such as spurious patterns, caused by unwanted modulation products arising from the transmission of a carrier signal.

Kennelly-Heaviside layer A layer of the ionosphere whose effect is felt mostly during daylight hours, and which is contained in the E-region. It extends approximately from 90 to 130 kilometers above the surface of the earth, and is between the D and F layers. It reflects high to very high frequency radio waves. Also called **E-layer**, or **Heaviside layer**.

keraunograph 1. An instrument utilized to detect distant thunderstorms. 2. A pattern made by lightning upon an object which has been struck.

Kerberos An authentication system used in communications networks. The authenticity of a user is established at logon, and a unique key is assigned which identifies such a user throughout the session.

Kermit A communications protocol utilized for asynchronous communications. It is used especially for transferring files over dial-up connections, and is characterized by comparatively slower speeds, but higher accuracy when used over noisy lines.

kernel 1. The fundamental, core, or central part of something. 2. The part of a computer's operating system which provides the most fundamental functions, such as managing memory, loading programs, and allocating system resources.

Kerr cell A glass cell containing a dielectric, usually nitrobenzene, which is utilized to demonstrate the **Kerr effect** (1). The application of an electric field using the electrodes of the cell make the nitrobenzene exhibit double refraction, and the intensity of polarized light passing through the cell can be modulated by varying the field.

Kerr effect 1. For certain materials, such as transparent dielectrics, double refraction produced by an electric field. Also called **Kerr electro-optic effect**, or **electro-optic Kerr effect**. 2. For plane-polarized light reflected from a highly-polished pole of a strong magnet, the rotation of the plane of polarization. The rotation is proportional to the strength of the magnetic field. Also called **Kerr magneto-optic effect**, or **magneto-optic effect** (2).

Kerr electro-optic effect Same as **Kerr effect** (1).

Kerr electro-optical effect Same as **Kerr effect** (1).

Kerr magneto-optic effect Same as **Kerr effect** (2).

Kerr magneto-optical effect Same as **Kerr effect** (2).

keV Abbreviation of **kiloelectronvolt**.

Kevlar A strong synthetic fiber used, for instance, to provide cables with support and protection.

key 1. A crucial or indispensable element. 2. A device part, such as a wedge or pin, which is inserted into a slot or groove for locking, securing, tightening, guiding, or adjusting. 3. A button, lever, or handle which is depressed or pressed, to open or close a circuit, to actuate a mechanism, or the like. Also called **switching key**. 4. A button, such as that on a keyboard or keypad, which is depressed to provide an input or an action. 5. In databases, an identifier for one or more records or files. 6. In computer and communications security, a variable-length bit string which is utilized for encryption and decryption.

key click 1. A sound heard when pressing a key. Used, for instance, to provide feedback when entering data on a computer keyboard. 2. An unwanted transient sound that may be heard when opening or closing a radiotelegraph key.

key command Same as **keyboard shortcut**.

key entry Same as **keyboard entry**.

key escrow The entrusting of a **key** (6) to a purported trusted third party.

key field Within a file, collection of records, or a database, a field which identifies a specific record. The value of the key field in each record must be different.

key in To enter computer data through a keyboard.

key length Also called **key space**. 1. The length, in bits, of a **key** (6). 2. The total number of possible combinations arising from a given **key length** (1). The greater the key length, the greater the protection against a brute-force attack. A 64-bit encryption key has over 1.8×10^{19} possible combinations, a 128-bit key over 3.4×10^{38}, and a 256-bit key has over 1.1×10^{77} combinations.

key logger Same as **keystroke logger**.

key maintenance The generation, storage, secure transmission, and use of a **key** (6).

key pad Same as **keypad**.

key pulse A telephone signaling system in which keys are pressed instead of utilizing a rotary dial.

key pulsing The utilization of **key pulses** for signaling.

key punch Same as **keypunch**.

key recovery 1. The obtaining of a **key** (6) which has been lost. 2. The obtaining of a **key** (6) to have unauthorized access to that which has been encrypted.

key repeat A keyboard feature which enables a key that is held down longer than a given interval to repeat. The time necessary for repeating, and the repeating speed can usually be set. Also called **keyboard repeat**, **typematic**, or **auto-repeat**.

key space Same as **key length**.

key station A station, especially within a communications network, which performs fundamental functions such as traffic management. An example is a control station.

key switch A switch that is operated by a pressing key. For instance, a switch that is actuated by a keyboard entry at a data-entry terminal. Also spelled **keyswitch**.

key telephone system A private telephone system that has a separate button for each outside line. To access a line, as in answering a call or placing one, the appropriate key must be pressed. Its abbreviation is **KTS**.

key-to-disk device A device which accepts keyboard entry and stores it directly on magnetic disks. Also called **key-to-disk unit**.

key-to-disk system A system which accepts keyboard entry and stores it directly on magnetic disks.

key-to-disk unit Same as **key-to-disk device**.

key-to-tape device A device which accepts keyboard entry and stores it directly on magnetic tape. Also called **key-to-tape unit**.

key-to-tape system A system which accepts keyboard entry and stores it directly on magnetic tape.

key-to-tape unit Same as **key-to-tape device**.

key way Same as **keyway**.

key word Same as **keyword**.

keyboard 1. A set of keys which serve to provide input, or cause an action. For example, the keyboard used with a computer or digital multimeter. 2. A computer input device which incorporates alphanumeric keys, function keys, control keys, cursor keys, and additional keys, such as enter, backspace, and insert. A 101-key keyboard is commonly used. Also called **computer keyboard**.

keyboard accelerator Same as **keyboard shortcut**.

keyboard buffer A small segment of computer memory utilized to temporarily store keystrokes that await processing. In the event a typist exceeds the capability of this buffer, the computer usually emits a warning sound. Also called **type-ahead buffer**.

keyboard command Same as **keyboard shortcut**.

keyboard connector Any connector utilized to connect a cable leading from a keyboard to a computer. A 5 or 6 pin DIN connector is common.

keyboard controller Also called **keyboard processor**. **1.** A chip or circuit which detects keystrokes, and transmits the appropriate data. Such a controller is usually installed in a keyboard. **2.** A keyboard which transmits MIDI data.

keyboard entry The use of a keyboard for entering information. Used, especially in the context of data entry utilizing a **keyboard (2)**. Also called **key entry**.

keyboard interrupt An interrupt generated by pressing a key on a computer keyboard.

keyboard layout The physical arrangement of the keys on a keyboard. Although a good number of computer keyboards have 101 keys, these can be arranged in a more ergonomic manner, or in a way that facilitates users with special needs to enter data.

keyboard lockout An feature which prevents the transmission of data from a keyboard when another station using the same circuit is transmitting.

keyboard port The port to which a keyboard is connected. A 5 or 6 pin DIN connector is commonly used.

keyboard processor Same as **keyboard controller**.

keyboard repeat Same as **key repeat**.

keyboard shortcut A key, such as F3, or combination of keys, such as ALT+SHIFT+T, which when pressed executes an action or series of actions within a program or operating system. Such shortcuts can save time or automate sequences. Also called **keyboard command**, **key command**, **keyboard accelerator**, **shortcut key**, or **application shortcut key**.

keyer **1.** A device utilized for automatic **keying**. **2.** A circuit which modulates by keying.

keying **1.** The generation of information-containing signals through the modulation between two discrete values or states of a steady signal or carrier. Examples include frequency-shift keying, amplitude shift keying, and differential phase-shift keying. **2.** The generation of information-containing signals through the interruption of a steady signal or carrier. **3.** The entering of information utilizing a keyboard.

keying frequency In fax, the maximum number of occurrences per second of a black-line signal during scanning.

keypad Also spelled **key pad**. **1.** A supplementary or separate grid of keys having number, function, or cursor keys, which simplify or accelerate keyboard entry. **2.** Any small grid of keys, such as that of a calculator or telephone.

keypunch A device which punches holes in cards or tape. Also spelled **key punch**.

keystone distortion In a TV, a form of distortion in which the image appears as a trapezoid instead of a rectangle. That is, either the top of the screen is wider than the bottom, or vice versa. Also called **keystoning**.

keystoning Same as **keystone distortion**.

keystroke logger A software or hardware program or device which records all keystrokes entered on a keyboard. Used, for instance, to surreptitiously monitor all usage of a terminal. Also called **key logger**.

keystroke monitoring The tracking of keystrokes using means such as keystroke loggers.

keyswitch Same as **key switch**.

keyway The slot or groove a **key (2)** is inserted into. Also spelled **key way**.

keyword **1.** A word which is specified by a user, to help narrow down a search within records, documents, or files. The more significant the word or words selected, the likelier the search will be useful. **2.** In a record, document, or file, a word which is used for indexing, or for otherwise facilitating retrieval of the desired information. **3.** A word, such as a command, which has a special meaning in a program, pro-

gramming language, or other setting, and whose use is restricted or not allowed. Also called **reserved word**.

keyword search A search through records, documents, or files, utilizing one or more **keywords**.

kg Abbreviation of **kilogram**.

kG Abbreviation of **kilogauss**.

kg-m Abbreviation of **kilogram-meter**.

kgcal Abbreviation of **kilogram-calorie**.

kgm Abbreviation of **kilogram-meter**.

kGs Abbreviation of **kilogauss**.

kGy Abbreviation of **kilogray**.

kHz Abbreviation of **kilohertz**.

KHz Abbreviation of **kilohertz**.

Ki- Abbreviation of **kibi-**.

Kib Abbreviation of **kibibit**.

KiB Abbreviation of **kibibyte**.

kibi- A binary prefix meaning 2^{10}, or 1,024. For example, a kibibyte is equal to 2^{10}, or 1,024 bytes. This prefix is utilized to refer to only binary quantities, such bits and bytes. Its abbreviation is **Ki-**.

kibibit 2^{10}, or 1,024 bits. Its abbreviation is **Kib**.

kibibits per second 2^{10}, or 1,024 bits, per second. Its abbreviation is **Kibps**.

kibibyte 2^{10}, or 1,024 bytes. Its abbreviation is **KiB**.

kibibytes per second 2^{10}, or 1,024 bytes, per second. Its abbreviation is **KiBps**.

Kibps Abbreviation of **kibibits per second**.

KiBps Abbreviation of **kibibytes per second**.

kick A sudden movement or jolt, such as might occur during contact bounce.

kickback **1.** In an inductive circuit or coil, a voltage surge produced when the current flowing through it is cut off. The rapidly collapsing magnetic field causes this surge. Also called **inductive kick**, or **inductive surge**. **2.** In a CRT, the return of the electron beam to its starting point after a sweep or trace. Also, the time interval that elapses during this return. Also called **retrace**, or **flyback (2)**.

kickback power supply A power supply that utilizes the **kickback (2)** interval to generate a high-voltage DC utilized by the second anode of a CRT. Also called **flyback power supply**.

Kikuchi lines A pattern of lines observed when an electron beam strikes a solid crystal and electrons are inelastically scattered, providing structural information of said crystal.

killer app Abbreviation of **killer application**. A computer application that is just dynamite.

killer application Same as **killer app**.

killer circuit **1.** A circuit which serves to disable one or more functions, or to shut down a system. For instance, a killer stage, or a muting circuit. **2.** A circuit which produces blanking in a radar receiver.

killer stage In a color TV receiver, a circuit which disables the chrominance circuits when a black-and-white signal is being received. For instance, such a circuit may seek a color burst signal, and if there is none, color decoding is shut off. Also called **color killer**, or **color-killer circuit**.

kilo- A metric prefix representing 10^3, which is equal to a thousand. For instance, 1 kilohertz is equal to 10^3 Hz. When referring to binary quantities, such as bits and bytes, it is equal to 2^{10}, or 1,024, although this is frequently rounded to 1,000. To avoid any confusion, the prefix **kibi-** may be used when dealing with binary quantities. Its proper abbreviation is **k**, although **K** is often used.

kilo Same as **kilogram**.

kilo-oersted A unit of magnetic field strength equal to 10^3, or 1,000 oersteds. Its abbreviation is **kOe**.

kiloampere A unit of current equal to 10^3, or 1,000 amperes. Its abbreviation is **kA**.

kilobaud A unit of data-transmission speed equal to 10^3, or 1,000 baud. Its abbreviation is **kb, Kb, kbaud,** or **Kbaud**.

kilobecquerel A unit of activity equal to 10^3, or 1,000 becquerels. Its abbreviation is **kBq**.

kilobit 2^{10}, or 1,024 bits, although this is frequently rounded to 1,000. To avoid any confusion, the term **kibibit** may be used when referring to this concept. Its abbreviation is **Kb, kb,** or **Kbit**.

kilobits per second 2^{10}, or 1,024 bits, per second. Usually used as a measure of data-transfer speed. To avoid any confusion, the term **kibibits per second** may be used when referring to this concept. Its abbreviation is **Kbps,** or **kbps**.

kilobyte 2^{10}, or 1,024 bytes, although this is frequently rounded to 1,000. To avoid any confusion, the term **kibibyte** may be used when referring to this concept. Its abbreviation is **KB, kB,** or **Kbyte**.

kilobytes per second 2^{10}, or 1,024 bytes, per second. Usually used as a measure of data-transfer speed. To avoid any confusion, the term **kibibytes per second** may be used when referring to this concept. Its abbreviation is **KBps,** or **kBps**.

kilocalorie A unit of heat energy equal to 10^3, or 1,000 calories, or 1 Calorie. Also called **kilogram-calorie**, although kilocalorie is the more proper term. Its abbreviation is **kcal**.

kilocoulomb A unit of electric charge equal to 10^3, or 1,000 coulombs. Its abbreviation is **kC**.

kilocurie A unit of radioactivity equal to 10^3, or 1,000 curies. Its abbreviation is **kCi**.

kilocycle A unit of frequency equal to 10^3, or 1,000 cycles, or 10^3 or 1,000 Hz. The term currently used for this concept is **kilohertz**. Its abbreviation is **kc**.

kiloelectronvolt A unit of energy or work equal to 10^3, or 1,000 electronvolts. Its abbreviation is **keV**.

kilogauss A unit of magnetic flux density, or magnetic induction, equal to 10^3, or 1,000 gauss. Its symbol is **kGs,** or **kG**.

kilogram The SI unit of mass. It is defined as the mass of a platinum-iridium bar kept by the *Bureau International des Poids et Mesures*. It is equal to 10^3, or 1,000 grams. It is the only remaining SI base unit that cannot be determined to a given degree of accuracy by a properly equipped laboratory. Its abbreviation is **kg**. Also called **kilo**.

kilogram-calorie Same as **kilocalorie**. Its abbreviation is **kcal,** or **kgcal**.

kilogram-meter A unit of work or energy equal to approximately 9.8067 joules. Its abbreviation is **kg-m**.

kilogray A unit of radiation dose equal to 10^3, or 1,000 grays. Its abbreviation is **kGy**.

kilohertz A unit of frequency equal to 10^3, or 1,000 Hz. Its proper abbreviation is **kHz**, although **KHz** is frequently utilized. Also called **kilocylcle**.

kilohm A unit of resistance, impedance, or reactance equal to 10^3, or 1,000 ohms. Its abbreviation is **kΩ**. Also spelled **kiloohm**.

kilojoule A unit of energy or work equal to 10^3, or 1,000 joules. Its abbreviation is **kJ**.

kilolumen A unit of luminous flux equal to 10^3, or 1,000 lumens. Its abbreviation is **klm**.

kilomega- An obsolescent and deprecated prefix representing 10^9. The proper prefix for representing 10^9 is **giga-**. Its abbreviation is **kM**.

kilometer A unit of distance equal to 10^3, or 1,000 meters. Its abbreviation is **km**.

kilometers per second A unit of speed expressing the kilometers traveled per second. Its abbreviation is **km/s, km/sec**.

kilometric waves Electromagnetic waves whose wavelength is within the range of 1 to 10 kilometers, corresponding to frequencies of between 300 to 30 kHz, respectively. This represents the low-frequency band.

kiloohm Same as **kilohm**.

kilopascal A unit of pressure equal to 10^3, or 1,000 pascals. Its abbreviation is **kPa**.

kilovar A unit of reactive power equal to 10^3, or 1,000 vars. Its abbreviation is **kvar**. Also called **kilovolt-ampere reactive**.

kilovar-hour A unit of reactive electric energy equal to 10^3, or 1,000 var-hours. Its abbreviation is **kvarh**.

kilovolt A unit of potential difference equal to 10^3, or 1,000 volts. Its abbreviation is **kV**.

kilovolt-ampere A unit of apparent power or true power equal to 10^3, or 1,000 volt-amperes. Its abbreviation is **kVA**.

kilovolt-ampere reactive Same as **kilovar**.

kilovoltmeter A voltmeter whose indications are expressed in kilovolts.

kilowatt A unit of power equal to 10^3, or 1,000 watts. Its abbreviation is **kW**.

kilowatt-hour A unit of energy equal to 10^3, or 1,000 watt-hours. Its abbreviation is **kWh,** or **kWhr**.

kilowatt-hour meter An instrument which measures and indicates, in kilowatt-hours, electrical power consumption for a given, or combined, time interval. Also called **electric meter (1)**.

kinematograph A motion picture camera or projector. Also called **cinematograph**.

kinescope A CRT used in a TV, for viewing the received images. Also called **cathode-ray television tube**, or **picture tube**.

kinescope recording A recording made directly from the images of a TV picture tube. Such a recording was formerly made, for instance, for later re-transmission, or for projection on a motion-picture screen.

kinetic energy The energy a body possesses as a consequence of its motion. Above absolute zero all particles have energy, so even a body which appears to be at rest will have kinetic energy. Its abbreviation is **KE**.

kiosk An unattended free-standing structure incorporating a keyboard and/or touch screen which serves as an interface for those who wish to obtain information, services, and to a limited extent, goods. Examples include ATMs, interactive displays in lobbies of office buildings which enable visitors to find a desired location by navigating through multiple screens, and booths outside movie theaters which dispense tickets which are prepaid, or purchased at that moment.

Kirchhoff's current law A law stating that at a given instant, the sum of the currents flowing into a given point in a circuit or electric network is the same as the sum of the currents flowing out of said point. This is true for any number of branches intersecting at this point. Also called **Kirchhoff's law (1)**, or **Kirchhoff's first law**.

Kirchhoff's first law Same as **Kirchhoff's current law**.

Kirchhoff's law 1. Same as **Kirchhoff's current law**. **2.** Same as **Kirchhoff's voltage law**. **3.** Same as **Kirchhoff's radiation law**.

Kirchhoff's radiation law A law stating that, for a given temperature and wavelength, the ratio of the emissivity to the absorptivity is the same for all bodies, and is equal to the emissivity of a blackbody at said temperature and wavelength. That is, if a body is a good absorber of a given wavelength at a given temperature, then it will also be a good emitter of said wavelength at that same temperature. And, if a body is at poor absorber of a given wavelength at a given temperature, then it will also be a poor emitter of said wavelength at that same temperature. Also called **Kirchhoff's law (3)**.

Kirchhoff's second law Same as **Kirchhoff's voltage law**.

Kirchhoff's voltage law A law stating that at a given instant, the sum of the of all the voltage drops in a closed circuit or electric network is equal to the sum of all the electromotive forces in said circuit or network. Also called **Kirchhoff's law (2)**, or **Kirchhoff's second law**.

kit A set of components or articles which are assembled to work together, or are used for a given purpose.

kJ Abbreviation of **kilojoule**.

klm Abbreviation of **kilolumen**.

kludge 1. A device, piece of equipment, program, or system which is constituted of poorly matched elements, or that otherwise works clumsily or inelegantly. 2. An inelegant solution to a problem.

klystron A linear-beam electron tube which utilizes velocity modulation to amplify microwaves, or to generate microwave oscillations. A two-cavity klystron, for instance, incorporates an electron gun, a buncher cavity, and a catcher cavity. In the buncher cavity, the electron beam is modulated by high-frequency radio waves, bunching occurs in the drift space between the cavities, and the energy in the bunches is transferred to the catcher cavity, causing it to resonate. The output is taken from the catcher cavity. A reflex klystron has one cavity, and there are also klystrons with three or more cavities.

klystron amplifier A **klystron** utilized as a microwave amplifier.

klystron generator Same as **klystron oscillator**.

klystron oscillator A **klystron** utilized as a microwave oscillator. Also called **klystron generator**.

kM Abbreviation of **kilomega**.

km Abbreviation of **kilometer**.

km/s Abbreviation of **kilometers per second**.

km/sec Abbreviation of **kilometers per second**.

kn Abbreviation of **knot**.

knee Anything resembling a human knee, such as a sudden bend in a characteristic curve.

knee current Within a characteristic curve, such a that of a Zener diode, the current at which there is a sudden change between two relatively straight portions.

knee voltage Within a characteristic curve, such a that of a Zener diode, the voltage at which there is a sudden change between two relatively straight portions.

knife-edge diffraction Diffraction over the edges of isolated obstacles, as might occur in the propagation of radio-frequency waves.

knife switch A switch in which the moving element consists of one or more hinged or removable blades. Typically, the blade is slid between two stationary contact clips, which grasp it securely to close the circuit.

knob A protuberance, such as a rounded piece of plastic, which serves to adjust, control, actuate, tighten, or loosen. A knob may, for instance, have a pointer incorporated, to set a desired value within a graduated scale, as seen in a rotary switch.

knockout A perforated portion of a wall, box, or enclosure which can be pushed-out, or otherwise removed. Used, for instance, to provide an opening for a wire or cable during an installation.

knot A unit of velocity equal to 1.852 km/hr, or about 0.51444 m/s. Its abbreviation is **kn**, or **kt**.

knowledge acquisition The process of gathering the information necessary to build a **knowledge base**.

knowledge base A database which contains an abundance of information pertaining to human experience and problem-solving expertise in a given domain or field. Used, for instance, in a knowledge-based system.

knowledge-based system A system which utilizes a **knowledge base** for problem solving in a given domain or field. An expert system is the most common example.

knowledge domain A domain or field covered in a **knowledge base**.

knowledge engineering 1. The discipline concerned with acquiring the information necessary to build a **knowledge base**. 2. The discipline concerned with building and utilizing **knowledge-based systems**.

knowledge representation The manner in which information to be utilized in a **knowledge base** is encoded. Also, the process of encoding such information.

kOe Abbreviation of **kilo-oersted**.

kovar An alloy composed mostly of iron, nickel, and cobalt, whose coefficient of thermal expansion is similar to that of borosilicate glass, making it suitable for glass-to-metal seals. A typical composition may be 53.8% iron, 29% nickel, 17% cobalt, and 0.2% manganese.

kPa Abbreviation of **kilopascal**.

Kr Chemical symbol for **krypton**.

kraft paper A relatively heavy paper which is used, for instance, as an insulator, or as a dielectric in paper capacitors. When used in capacitors, it is usually impregnated with oil or wax.

krypton A colorless noble gas whose atomic number is 18. It has over 35 known isotopes, of which 6 are stable. It applications in electronics include its use in lasers, as a gas filler in bulbs and lamps, and in high-speed photography. Its chemical symbol is **Kr**.

kt Abbreviation of **knot**.

KTS Abbreviation of **key telephone system**.

Ku Band 1. In communications, a band of radio frequencies extending from 12.00 to 18.00 GHz, as established by the IEEE. This corresponds to wavelengths of approximately 2.5 to 1.7 cm, respectively. 2. In communications, a band of radio frequencies extending from about 11 GHz to 13 GHz, although the defined interval varies.

Kundt effect The rotation of the plane of polarization of an electromagnetic wave as it passes through certain materials, when under the influence of a magnetic field parallel to the direction of propagation of said wave. Also called **Faraday effect**, or **magnetic rotation**.

Kundt tube A tube partially filled with dust or powder, which is utilized to measure the speed of sound. The dust or powder, such as that of cork, forms patterns based on sound vibrations, revealing nodes and antinodes, and these provide the length of the standing waves generated in the tube. With the frequency of the acoustic disturbance known, the obtained wavelength can then be utilized to calculate the speed of sound.

kV Abbreviation of **kilovolt**.

kVA Abbreviation of **kilovolt-ampere**.

kvar Abbreviation of **kilovar**.

kvarh Abbreviation of **kilovar-hour**.

kW Abbreviation of **kilowatt**.

kWh Abbreviation of **kilowatt-hour**.

kWhr Abbreviation of **kilowatt-hour**.

kΩ Abbreviation of **kiloohm**.

L

l **1.** Symbol for **length**. **2.** Symbol for **liter**.

L **1.** Symbol for **inductance**. **2.** Symbol for **self-induction**. **3.** Symbol for **Avogadro's number**. **4.** Symbol for **radiance**.

L **1.** Symbol for **lambert**. **2.** Symbol for **liter**.

L antenna An antenna consisting of a horizontal wire or radiator with a vertical feeder or lead-in wire connected to one end, and whose shape resembles an inverted letter **L**. Used, for instance, for shortwave reception. Also called **inverted-L antenna**.

L Band **1.** In communications, a band of radio frequencies extending from 1.00 to 2.00 GHz, as established by the IEEE. This corresponds to wavelengths of approximately 30 to 15 cm, respectively. **2.** In communications, a band of radio frequencies extending from 390 MHz to 1550 MHz. This corresponds to wavelengths of approximately 76.9 to 19.4 cm, respectively.

L carrier A telephone carrier occupying a band of radio frequencies extending from approximately 60 kHz to over 8 MHz.

L-carrier system A telephone system utilizing an **L carrier**.

L display In radars, a display in which the scanned object appears as two horizontal blips, one extending to the right of a central vertical time base, the other to the left. When the radar antenna is pointed directly at the scanned object, both blips are of equal amplitude, while any difference in their relative amplitudes indicate the pointing error. The range is indicated by the position along the baseline. Also called **L scan**, **L scanner**, **L scope**, or **L indicator**.

L electron An electron located in the L shell of an atom.

L indicator Same as **L display**.

L-network An electric network with two impedance branches connected in series, with the junction and free end of one branch connected to one pair of terminals, and the free ends of both branches connected to the other pair. Such an arrangement resembles an inverted letter **L**, and is often used for filters and attenuators.

L-pad An attenuator consisting of a two-branch network which resembles an inverted letter **L**, and which presents essentially the same impedance throughout its entire range. Usually used as a volume control.

L − R In stereo audio, the signal resulting from the difference of the left and right channels. Used, for instance in FM stereo transmission and reception. The term is an abbreviation of **left minus right**.

L + R In stereo audio, the signal resulting from the sum of the left and right channels. Used, for instance in FM stereo transmission and reception. The term is an abbreviation of **left plus right**.

L scan Same as **L display**.

L scanner Same as **L display**.

L scope Same as **L display**.

L-section An electric circuit or network consisting of two or more components which are connected in a manner which resembles an inverted letter **L**. An example is a filter consisting of two or more components arranged in such a way.

L shell The second electron shell of an atom. It is the second closest shell to the nucleus.

L1 cache A memory cache that is built right into the CPU. This contrasts with an **L2 cache**, which utilizes a memory bank between the CPU and main memory. L1 cache is faster, but smaller than L2 cache, while L2 cache is still faster than the main memory. Also called **level 1 cache**, **in-**ternal cache, on-chip cache, first-level cache, or **primary cache**.

L2 cache A memory cache which utilizes a memory bank between the CPU and main memory. This contrasts with an **L1 cache**, which is built right into the CPU. L1 cache is faster, but smaller than L2 cache, while L2 cache is still faster than the main memory. Also called **level 2 cache**, **external cache**, or **second-level cache**.

L2TP Abbreviation of Layer 2 Tunneling Protocol. An IETF-sponsored protocol for the creation of virtual private networks.

L3 cache In computer systems with L1 and L2 cache built into the CPU, a type of memory cache which resides in the motherboard between CPU and main memory. Therefore, L3 cache is the equivalent of L2 cache in systems which have only the L1 cache built into the CPU. Also called **level 3 cache**, or **third-level cache**.

La **1.** Chemical symbol for **lanthanum**. **2.** Symbol for **lambert**.

lab Abbreviation of **laboratory**.

label An item, name, image, or symbol which serves to identify something. For instance, a name or code which is assigned to a computer file, data element, program, subroutine, instruction, variable, storage location, or storage medium. Another example is an adhesive paper which describes the content of the tape or disk it is adhered to. Also, to identify by using such a label.

laboratory A location, such as a room or building, which provides the equipment and conditions necessary for experimentation, research, testing, or for the optimal operation of components, devices, equipment, or systems. A laboratory may be utilized to simulate most any given surroundings. Its abbreviation is **lab**.

laboratory conditions The controlled environment that a **laboratory** may provide. Such conditions, for instance, may contrast field conditions.

laboratory equipment **1.** Equipment designed specifically for use in a **laboratory**, or which is otherwise **laboratory-grade**. **2.** Equipment that can only be utilized in a **laboratory**.

laboratory-grade A component, device, piece of equipment, substance, or material which is suitable for use in a **laboratory**. The term usually denotes a quality standard beyond a given level.

laboratory power supply A power supply which provides for the energy needs of laboratory equipment. Such a supply is usually adjustable, equipped with meters indicating the levels of current and voltage, and may also have a backup source of electricity.

labyrinth enclosure A speaker enclosure that has absorbent panels that form a labyrinth. This helps prevent cabinet resonance and enhances bass response. Also called **acoustic labyrinth**.

labyrinth loudspeaker Same as **labyrinth speaker**.

labyrinth speaker A speaker with a **labyrinth enclosure**. Also called **labyrinth loudspeaker**.

ladar Acronym for **laser radar**, or **Laser Detection and Ranging**.

ladder attenuator An attenuator consisting of a series of symmetrical sections which provide for the input and output impedances to remain constant as the attenuation varies. It is a type of ladder network.

ladder filter A **ladder network** utilized as a filter.

ladder network An electric network composed of multiple H, L, T, and/or π networks connected in cascade. Usually used as filters, such as those with sharp cutoff frequencies, or as analog-to-digital converters. Also called **series-shunt network**.

LADS Acronym for **local-area data service**. A digital communications service, such as DSL, which utilizes a dedicated dry copper pair.

lag 1. The time interval that elapses between events which are considered together. Also, any such difference in phase, such as that between waves. For instance, a lag angle. **2.** The state or condition of **lag** (1) occurring. Also, that which lags. For example, a lagging current. **3.** To fall behind in general. Also, to stay behind. **4.** The time interval that elapses between the transmission of a signal and its reception. **5.** The time interval that elapses between a given action and its intended effect. For instance, the difference between a corrective action, and the change in the desired output. **6.** In a camera tube, the persistence of the electric image after a scene change.

lag angle For two periodic quantities with the same frequency, the lag in phase of one respective to the other. Expressed in radians or degrees. Also called **angle of lag**, or **phase delay** (1).

lagging current An alternating current which lags the alternating voltage producing it. In a purely inductive circuit this lag is 90°. This contrasts with a **leading current**, where the current leads the voltage.

lagging load A load which is predominantly inductive, so that at the terminals the alternating current lags the alternating voltage. This contrasts with a **leading load**, where the current leads the voltage. Also called **inductive load**.

Lalande cell A wet cell with a zinc anode, a cupric oxide cathode, and whose electrolyte is an aqueous solution of sodium hydroxide.

Lamb wave An electromagnetic wave which travels along the surface of a solid whose thickness is approximately equal to the wavelength of said wave.

lambda (λ) **1.** Symbol for **wavelength**. **2.** Symbol for **linear charge density**, or **charge density** (3).

lambda switching A technology utilized to switch individual wavelengths into separate paths when routing multiple data streams to their destination along an optical fiber.

lambda wave An electromagnetic wave which travels along the surface of a solid. Used, for instance, in wave soldering.

lambert A CGS unit of luminance. It is equal to approximately 3183.099 candelas per square meter, or $1/\pi$ candela per square centimeter, or approximately 929.03 footlamberts. Its symbol is **L**, **La**, or **Lb**.

Lambert's cosine law A law which states that the energy emitted in any direction by a radiating surface is proportional to the cosine of the angle formed between the direction of emission and a perpendicular line extending from the same surface. Also called **Lambert's law**, or **cosine emission law**.

Lambert's law Same as **Lambert's cosine law**.

laminated 1. Composed of thin layers bonded together. An example is a laminated core. **2.** A component, device, or surface which has a thin protective coating.

laminated contact A contact, such as that of a switch, composed of multiple laminations whose layers independently make contact with a conducting surface.

laminated core A core composed of thin layers of a material bonded together. An example is a laminated iron core.

laminated iron core A core composed of thin layers of iron, or another ferromagnetic material, bonded together. Such laminations are insulated from each other, which helps reduce eddy currents.

laminated material 1. A material consisting of thin layers bonded together. **2.** A material which has a thin protective coating.

laminated metal 1. A metal consisting of thin layers bonded together. **2.** A metal which has a thin protective coating.

laminated paper 1. Paper which incorporates thin layers of materials bonded together. An example is paper and foil alternated, as may be seen in a paper capacitor. **2.** Paper which has a thin protective coating, such as wax.

laminated surface 1. A surface consisting of thin layers bonded together. **2.** A surface which has a thin protective coating.

lamination 1. The act or process of arranging in thin layers and bonding. **2.** A single thin layer in **lamination** (1). **3.** A thin protective coating for a component, device, or surface. Such a layer may also provide structural support.

lamp 1. A device which produces visible light. Said especially of devices which convert electricity into light. There are many examples, including arc lamps, fluorescent lamps, electroluminescent lamps, gas-filled lamps, glow lamps, and incandescent lamps. **2.** A device which produces light or heat. Examples include infrared lamps, and ultraviolet lamps.

lamp bank A series of lamps which are used together. For example, such a bank may be connected in parallel or in series and be utilized as a resistance load in tests.

lamp black A grayish-black amorphous form of carbon prepared by partially combusting liquid hydrocarbons, and which is utilized in resistors and carbon brushes. Also spelled **lampblack**.

lamp cord A flexible cord containing two insulated wires, and which is utilized to connect lamps and appliances to outlets. Such a cord may consist, for instance, of twisted or parallel 16-gauge copper wire.

lamp dimmer A circuit or device which serves to control the intensity of illumination of a lamp. There are various types, including those employing silicon-controlled rectifiers or rheostats.

lamp driver A circuit or device utilized to drive a lamp.

lamp extractor A tool or device utilized to insert or remove lamps which are small or otherwise difficult to handle, and/or whose socket or holder is hard to work with.

lamp holder Same as **lampholder**.

lamp house Same as **lamphouse**.

lamp jack Same as **lampholder**.

lamp receptacle Same as **lampholder**.

lamp socket Same as **lampholder**.

lampblack Same as **lamp black**.

lampholder A device which serves to connect a lamp to a circuit, in addition to providing mechanical support. Also spelled **lamp holder**. Also called **lamp socket**, **lamp receptacle**, or **lamp jack**.

lampholder adapter An adapter which converts the thread size of a lampholder, so it can accept a lamp of different size thread. Also spelled **lamp holder adapter**.

lamphouse Also spelled **lamp house**. **1.** The part of a film projector which houses the lamp. It incorporates a very bright lamp, often a xenon arc lamp, and a reflector, such as a parabolic mirror. **2.** The part of a microscope, or other instrument or device which requires a light source, which houses the lamp and a reflector.

LAN Acronym for **local-area network**. A computer network which is limited to a small geographical area, usually ranging from a single room through a cluster of office buildings. A LAN consists of a group of nodes, each comprised by a computer or peripheral, which exchange information with each other. In addition to sharing data resources, users can communicate with each other, usually through emails or chats, and share peripherals such as printers. LAN connections may be physical, as with cables, or wireless, as with microwaves or infrared waves. There are various LAN access methods, including Ethernet, Gigabit Ethernet, and token ring. Common topologies include bus, ring, and star.

LAN adapter Same as **LAN card**.

LAN analyzer Software and/or hardware which monitors the activity of a LAN, and which troubleshoots and performs tests such as the simulation of error conditions, so as to help it work smoothly.

LAN card A circuit board which is plugged into the bus or an expansion slot of a computer, and to which the cabling linking to a **LAN** is connected. Also called **LAN adapter**.

LAN Emulation Protocols and other functions which allow Ethernet and token ring networks to be linked utilizing an ATM backbone. At the sending end, Ethernet and token ring packets are converted into ATM cells, and at the receiving end these are converted back to their original form. Its acronym is **LANE**.

LAN fax The use of a LAN for the sending and receiving of faxes among nodes.

LAN gateway A device which performs the protocol conversions necessary to enable a LAN to connect to another network which utilizes different protocols.

LAN hardware The physical equipment, such as repeaters, routers, gateways, switches, and cables, utilized in a **LAN**.

LAN manager Software and/or hardware which monitors and controls a LAN, including its configuration, allocation of resources, and security.

LAN segment A portion of a LAN which is utilized by a given group, such as a department. A bridge is usually utilized to connect these segments.

LAN server Within a LAN, a computer whose hardware and/or software resources are shared by other computers. Such servers, among other functions, control access and manage network resources. There are various types of LAN servers, including file servers, mail servers, application servers, and print servers.

LAN switch 1. A switch, as opposed to a hub, that connects nodes or segments in a LAN. A LAN switch enables all nodes to have the entire bandwidth available, up to 100 Mbps for each in the case of fast Ethernet, for instance, as opposed to all nodes having to share the bandwidth, as would be the case when a hub is used. Also called **frame switch**. 2. A device in a LAN that selects a path or circuit via which data will flow to its next destination. Such a switch usually involves a simpler and faster mechanism than a router, but may also have router functions.

land 1. The solid portion of the surface of the planet earth. 2. On an optical disc, such as a CD or DVD, a non-indented portion, as opposed to an indented portion which is called **pit** (1). The laser beam is reflected off the lands, while being scattered or absorbed by the pits. 3. A surface between grooves, such as those of a diffraction grating or phonograph record. 4. On a printed-circuit board, the printed conductive portion to which components are connected. It may consist, for instance, of the enlarged areas where component leads are soldered. Also called **pad (2)**, or **terminal area**.

Land camera A camera which utilizes a single-step development process to provide a finished positive print. Also called **Polaroid camera**.

land effect A change in direction of a radio wave as it crosses a coastline. It usually only affects ground waves. Also called **coastal bending**, or **shoreline effect**.

Land Grid Array A chip package similar to Ball Grid Array, but which has plated pads for lead connections, instead of tiny balls of solder. Its abbreviation is **LGA**.

land line Same as **landline**.

land-mobile service A mobile radio service between base stations and land-mobile stations, or between land-mobile stations.

land-mobile station Within a land-mobile service, a station which can move within given geographical limits.

land pattern On a PCB, a specific configuration or pattern of lands (4). Also called **PCB land pattern**.

land return In radars, unwanted echoes that appear on the display screen due to signal reflections from the surface of the earth. Also called **ground clutter**, **surface return**, or **terrain clutter**.

land station Within a mobile radio service, such as a maritime mobile service, a station whose location is fixed.

landing beacon A beacon producing a **landing beam**.

landing beam At an airport, a highly directional radio beam emitted to guide an aircraft making an instrument landing.

landing zone On a hard disk, a special section which is utilized to safely park a read/write head when the drive is powered down. This section does not hold data, and usually has a special texture.

landline A communications line consisting of wires or cables, as opposed to a wireless link. A standard home or business telephone uses a landline. Also spelled **land line**. Also called **line (3)**.

landscape 1. Same as **landscape mode**. 2. In computers, pertaining to something whose width is greater than its height, such as a landscape display.

landscape display A computer display whose width is greater than its height. Also, the positioning of a monitor in this fashion when both portrait and landscape orientations are available. Also called **landscape monitor**.

landscape format Same as **landscape mode**.

landscape mode A manner of printing, or presenting information on a screen, in which the width of a page is greater than its height. This contrasts with **portrait mode**, in which the height is greater than the width. Also called **landscape (1)**, **landscape orientation**, or **landscape format**.

landscape monitor Same as **landscape display**.

landscape orientation Same as **landscape mode**.

LANE Acronym for **LAN Emulation**.

Langmuir-Child equation Same as **Langmuir-Child law**.

Langmuir-Child law For a thermionic diode, an equation which yields the cathode current in a space-charge-limited-current state. Also known as **Langmuir-Child equation**, **Langmuir law**, or **Child-Langmuir law**.

Langmuir dark space In a glow-discharge tube, a dark space surrounding a negatively charged probe inserted into the positive column.

Langmuir law Same as **Langmuir-Child law**.

Langmuir probe A probe inserted into a plasma for measurements such as electron density or temperature.

language 1. A defined set of characters, symbols, and rules utilized together for communication with a computer, or between computers. The term also applies to communications between computers and peripherals, or between peripherals. There are various examples, including assembly languages, machine languages, command languages, and programming languages. Also called **computer language**. 2. Any system, along with its rules, utilized for communication.

language processor 1. Same as **language translation program**. 2. A hardware component which performs **language translation**.

language translation The functions a **language translation program** performs.

language translation program Also called **language translator**, or **language processor (1)**. **1.** A program which translates between computer languages such as assembly languages, machine languages, command languages, or programming languages. For instance, a program which translates a higher-level language into a lower-level language, or vice versa. **2.** A computer program which translates to or from a language. For example, a program which translates a natural language into a computer language, or vice versa. **3.** A computer program which translates one natural language into another, or others, and vice versa.

language translator Same as **language translation program**.

lanthanides A series of elements, generally considered to range from atomic number 57 (lanthanum) or 58 (cerium), through 71 (lutetium), inclusive. They have related chemical properties, including high reactivity. Such elements are not rare, and frequently occur together. Also called **rare-earth elements**.

lanthanum A silvery-white, malleable, and ductile chemical element whose atomic number is 57. It is chemically active, when in powder form it ignites spontaneously, and has over 30 known isotopes, of which one is stable. Its applications include its use in oxide form to improve the properties of optical glass, as a phosphor, in superconductors, in semiconductors, and in electronic devices such as vacuum tubes. Its chemical symbol is **La**.

lanthanum fluoride A white powder whose chemical formula is **LaF$_3$**. It is used as a phosphor, and in lasers.

lanthanum oxide A white powder whose chemical formula is **La$_2$O$_3$**. It is used in optical glasses, as a phosphor, in ceramics, and in electrodes.

lap 1. A disk, wheel, or slab utilized to polish, smooth, or reduce the thickness of materials such as crystals or metals. **2.** The abrasive material adhered to the surface of a **lap (1)**. Such a material is usually a solid or a slurry. **3.** To polish, smooth, or reduce a thickness utilizing a **lap (1)**.

LAP-B Same as **LAPB**.

LAP-D Same as **LAPD**.

lap joint A joint made by fastening or bonding overlapping ends.

lap winding An armature winding in which all turns under a given pair of poles are completed before proceeding to the next pair.

LAPB Abbreviation of Link Access Procedure Balanced, or Link Access Protocol Balanced. A data-link layer protocol utilized in X.25 networks.

LAPD Abbreviation of Link Access Procedure-**D** channel, or Link Access Protocol-**D** channel. A data-link layer protocol utilized in an ISDN D channel. It is derived from **LAPB**.

lapel microphone 1. A small microphone which is clipped to a lapel, collar, or the like, so as to be close a person's mouth. Such a microphone is usually wireless. **2.** Same as **Lavalier microphone**.

Laplace transform A mathematical technique which can be utilized to solve a wide variety of problems, such as those in circuit analysis, by transforming differential equations into simpler algebraic equations.

Laplace's law A law which describes the mathematical relationship between electric currents and magnetic fields. Also known as **Ampère's law**.

LapLink A popular file transfer program.

lapping The process of polishing, smoothing, or reducing a thickness utilizing a **lap (1)**. Used, for instance, for reducing the thickness of a wafer by removing controlled amounts from its surface.

laptop Same as **laptop computer**.

laptop computer A portable computer that is AC and/or battery powered, and whose dimensions and weight make it suitable for use during travel. Such computers feature similar computing power, a smaller flat screen, and a keyboard with a more compressed layout than desktop models. In addition, most have a pointing device incorporated, and can be utilized with a docking station. Since the weight and dimensions of such computers has steadily dropped, some models to currently around one kilogram, the term is mostly synonymous with **notebook computer**. Also called **laptop**.

large calorie A unit of heat energy equal to one kilocalorie, or approximately 4.186 kilojoules. Also called **Calorie**.

Large Hadron Collider A particle accelerator which collides protons and ions to produce energies in the TeV range. It is located in the CERN laboratory, and is used, for instance, to study the nature of matter. Its abbreviation is **LHC**.

large-scale integration In the classification of ICs, the inclusion of between 3,000 and 100,000 electronic components, such as transistors, on a single chip. This definition required much fewer components only a few years ago, when just 100 or more electronic components qualified for this level of integration. Its abbreviation is **LSI**.

large-signal This contrasts with **small-signal**. **1.** A signal whose amplitude is comparatively large. **2.** A signal whose amplitude is sufficiently large so that nonlinearities in the component, circuit, device, or system in question are detectable and/or significant. When analyzing the response at such signal levels, nonlinear components can not be ignored.

large-signal analysis The analysis of a component, circuit, device, or system operating at **large-signal** amplitudes.

large-signal bandwidth The useful bandwidth of a device operating at **large-signal** levels.

large-signal characteristics The characteristics of a component, circuit, device, or system operating at **large-signal** amplitudes.

large-signal current amplification Same as **large-signal current gain**.

large-signal current gain The current gain of an amplifier at **large-signal** amplitudes. Also called **large-signal current amplification**.

large-signal diode A diode designed to operate at **large-signal** levels.

large-signal equivalent circuit A circuit which is equivalent to another circuit or device operating at **large-signal** amplitudes.

large-signal gain The gain of an amplifier at **large-signal** amplitudes.

large-signal operation Operation of a component, circuit, device, or system at **large-signal** amplitudes.

large-signal parameters Parameters which are characteristic of a component, circuit, device, or system operating at **large-signal** amplitudes.

large-signal performance The performance of a component, circuit, device, or system at **large-signal** amplitudes.

large-signal transistor A transistor designed to operate at **large-signal** levels.

Larmor orbit The spiral orbit followed by a charged particle in a uniform magnetic field.

laryngophone A microphone which is worn around the throat of the user, and which picks up the vibrations of the larynx directly. Ideally, such a microphone transmits the speaker's voice only, and no background noise. Used, for instance, by emergency personnel, or in airplane cockpits. Also called **throat microphone**.

LASCR Abbreviation of **light-activated silicon-controlled rectifier**.

lase To emit light by the action of a **laser**.

laser Abbreviation of light amplification by stimulated emission of radiation. **1.** A device whose output is a coherent, monochromatic, and essentially non-divergent beam of light in the visible, infrared, ultraviolet, or X-ray region of the electromagnetic spectrum. The input power of a laser is utilized to excite the electrons of atoms, molecules, or ions to higher energy levels, and these particles are made to radiate in phase within a cavity with mirrors at the ends. Any of various materials or mediums may be stimulated through electrical, chemical, or other means into emitting laser radiation, including crystals, dyes, gases, or liquids. The coherent light produced may be continuous, or it may occur in pulses. Lasers have many applications, including their uses in communications, surgery, cutting, drilling, heating, holography, distance measurement, printing, spectroscopy, and disc storage and retrieval. Also called **optical maser**. **2.** Same as **laser beam**.

laser altimeter An instrument which measures and indicates altitude above a reference level, such as a given terrain, through the utilization laser technology.

laser amplifier A laser utilized to amplify light, such as the output of another laser.

laser beam The coherent, monochromatic, and essentially non-divergent beam emitted by a laser. Also called **laser (2)**.

laser-beam communications Same as **laser communications**.

laser-beam printer Same as **laser printer**.

laser cavity A mirrored resonant cavity utilized in a laser. An example is a Fabry-Perot cavity. Also called **laser resonator**.

laser communication The use of lasers to transmit information between two or more points. Lasers are used, for instance, in fiber-optic communications. Also called **laser-beam communications**.

laser cooling The use of a laser to cool particles to extremely low velocities. For example, atomic clocks which utilize lasers to cool atoms to nearly absolute zero should be able to achieve an error of less than one second per billion years in the not too distant future.

laser cutting The use of a laser beam for precise cutting.

laser deposition The use of a laser beam for the precise concentration of matter at a specific location, or on a given surface.

Laser Detection and Ranging Same as **laser radar**. Its acronym is **ladar**.

laser diode A laser in which a forward-biased pn junction diode is used to convert its DC input into a coherent light output. Used, for instance, as a light-pulse generator for transmission of information over fiber-optic lines. Also called **laser diode light source, diode laser, semiconductor laser, injection laser, injected laser,** or **injection laser diode**.

laser diode light source Same as **laser diode**.

laser disc Same as **Laserdisc**.

laser disk Same as **Laserdisc**.

laser Doppler radar Same as **laser radar**. Its acronym is **lopplar**.

laser Doppler velocimetry A method for measuring the velocity of particles or objects, in which the velocity is calculated based on the Doppler shift of laser light scattered by said particles or objects. Used, for instance, to measure the velocity of particles within flowing fluids. Its abbreviation is **LDV**.

laser drilling The use of a laser beam for precise drilling.

laser dye A dye, such as rhodamine B, utilized in a dye laser.

laser extensometer An instrument which utilizes a high-speed laser to measures small variations in the dimensions

of a solid. Such an instrument does not require contact with the specimen, and some can be used at very high temperatures and may be able to measure in the sub-micrometer range.

laser gyro Abbreviation of **laser gyro**scope. A gyroscope in which two laser beams travel in opposite directions over a ring-shaped path utilizing multiple angled mirrors. During any movement of the aircraft, or other vehicle it is mounted on, the angular rotation rate is measured by determining the frequency shift in the beams, thus providing the angular position. Such a gyro is very rugged, and has no moving parts. Also called **ring laser**, or **ring laser gyro**.

laser gyroscope Same as **laser gyro**.

laser heating The use of a laser beam for precise heating.

laser interferometer An interferometer whose source of light is a laser. The resolution of such interferometers can be in the picometer range.

laser light The coherent, monochromatic, and essentially non-divergent light emitted by a laser.

laser memory Computer memory that utilizes a laser for storage and retrieval.

laser printer A printer in which a laser beam is utilized to reproduce images. In it, a photoconductive surface, usually a drum, is positively charged with static electricity, then exposed to an optical image of that which is to be reproduced, forming a corresponding electrostatic pattern. A laser beam, along with a rotating mirror and lenses, is utilized to draw the desired image on the drum. A fine powder, called toner, which has been negatively charged is spread over this surface, adhering only to the charged areas of the drum. The toner is then fused to the paper, which has been positively charged, forming the final image. Laser printers are comparatively quiet, reproduce graphics excellently, and can have resolutions exceeding 2,400 dpi. Most copy machines also utilize this technology. Also called **laser-beam printer**.

laser pulse A short burst of **laser light**.

laser pumping In a laser, the excitation of electrons, molecules, or ions to higher energy levels, so as to initiate and sustain lasing action. Common methods utilized are the application of light, an electrical discharge, or through the effect of a chemical reaction. Also called **pumping (1)**.

laser radar A radar system in which the beam of a laser is used to track objects. Its acronym is **ladar**. Also known as **Laser Detection and Ranging, laser Doppler radar,** or **colidar**.

laser Raman spectroscopy A laser spectroscopic technique that utilizes the Raman effect to obtain information on the vibrational or rotational energy of molecules. Used, for instance, in conjunction with IR spectroscopy to reveal additional detail pertaining to molecular vibrations, as the mechanism for Raman scattering is different from that of IR absorption. Widely used for the determination of quantitative and qualitative aspects of molecular samples, such as structure. Also called **Raman spectroscopy**.

laser rangefinder A rangefinder which utilizes a laser to determine the precise distance to objects.

laser ranging The use of a laser to determine the precise distance to objects. For instance, the time required for a laser beam to be reflected off an object can be utilized to calculated its distance.

laser resonator Same as **laser cavity**.

laser retroreflector In a laser, a device which serves to reflect light back in the direction it came from, or a path parallel to this direction. Its abbreviation is **LR**, or **LRR**. Also called **retroreflector (2)**.

laser retroreflector array Multiple **laser retroreflectors** utilized for tracking objects, such as satellites, or for performing measurements, testing, or the like. Its abbreviation is **LRA**.

laser scissors A laser trap utilized to cut very small objects, such as living cells or the internal structures of cells. Also called **optical scissors**.

laser soldering The use of a laser beam for precise soldering.

laser spectroscopy Any spectroscopic technique in which a laser beam is utilized. For example, a laser may be used as a source of light, as in laser Raman spectroscopy, or to enhance performance, as in Doppler-free spectroscopy.

laser spectrum Within the electromagnetic spectrum, the complete range of laser frequencies. These include the visible, infrared, ultraviolet, and X-ray regions.

laser storage Same as **laser storage medium**.

laser storage medium An optical storage medium, such as a DVD or CD, in which a laser is utilized to read and/or write information. Also called **laser storage**.

laser surgery The use of a laser beam for surgical purposes. Lasers can be utilized to assist in various procedures, including the removal of tumors or kidney stones, or to help reshape the cornea of an eye to correct refractive errors.

laser switch A switch that is actuated by, or otherwise incorporates, a laser. Such a device may provide switching action in the picosecond range, and may be used, for instance, in photonic transistors.

laser technology Technology pertaining to laser devices and applications.

laser threshold The minimum pumping energy necessary to initiate lasing.

laser tracking The use of a laser to track a scanned object.

laser trap A device utilized for **laser trapping**. Also called **optical trap**.

laser trapping The use of a laser to cool a particle to nearly absolute zero, and to then restrict it to an extremely small area. Laser trapping is useful, for instance, for research into the nature of atomic interaction, and improvements in atomic clocks and high-resolution spectroscopy. Also called **optical trapping**.

laser tweezers A laser trap utilized to manipulate very small objects, such as living cells or the internal structures of cells. Also called **optical tweezers**.

laser velocimeter An instrument which incorporates a laser to measure the velocity of particles or objects. Used, for instance, to measure the velocity of particles within flowing fluids.

laser welding The use of a laser beam for precise welding.

Laserdisc Also called **videodisc**. **1.** An optical disc format typically utilized for audio and video, such as movies and music. This format usually provides two analog and two digital audio channels, and its horizontal resolution is generally 425 lines. Also, a disc encoded in this format. Such discs are usually 12 or 8 inches in diameter, and a laserdisc player is required to read them. Its abbreviation is **LD**. Also spelled **laser disc**, **laser disk**, or **laserdisk**. **2.** Any optical disc format or disc.

LaserDisc Same as **Laserdisc**.

laserdisc player A device which can read the information stored on a **Laserdisc**. To recover the recorded content, a laser is focused on the reflecting surface of the disc, and the reflected light is modulated by the code on said disc.

laserdisk Same as **Laserdisc**.

LaserJet A popular line of laser printers.

lasing The emission of light by the action of a **laser**.

last call return Same as **last number redial**.

last-in, first out Same as **LIFO**.

last-in, last out Same as **LILO**.

last mile The link between a service provider's network and end user or customer's location. Usually used in the context of communications. For example, it may be the connection between a telephone company and a client's home, or the link between an Internet service provider and an end user's system. May also refer to a service entrance cable.

last number redial A service offered by telephone companies and/or phones equipped with this feature, which enables a subscriber to dial the last incoming call. A sequence of digits is usually dialed to activate this feature. Frequently, this service continues to dial the desired number if it is initially busy, and alerts when it is available. Some incoming calls may not allow for last number redialing. Also known as **last call return**, or **call return**.

lat Abbreviation of **latitude**.

LATA Acronym for Local-Access Transport Area. A geographic region which is defined to determine which areas called are considered long distance. A call made from one LATA to another is a long-distance call. A single LATA may include multiple central offices, and more than one area code.

latch **1.** That which serves to fasten or hold in place. For example, a device which holds a movable contact in place. **2.** A circuit which reverses its state or logic condition when an input is applied, and which maintains said state or logic condition until the next input, at which point it reverses again.

latch relay Same as **latching relay**.

latch-up Also spelled **latchup**. **1.** In an IC, a condition in which an unwanted self-sustaining high-conductivity path is formed between various pn junctions, resulting in a degradation in performance. Latch-ups may be caused, for instance, by electrostatic discharges, ionizing radiation, or a high supply voltage. **2.** The failure of a component or circuit to return to a previous or desired mode or state after a given stimulus. For example, the changing of a circuit to an unwanted mode by the application of an improper voltage, and a failure to revert to the desired mode after said voltage is removed.

latching current In a thyristor, the minimum current necessary to maintain conduction after the triggering current is removed.

latching relay A relay which maintains its actuated position after the triggering current is removed. It may retain its locked position electrically, magnetically, manually, or through the use of a holding coil. The contacts of such a relay can then be reset manually, electrically, or magnetically. Also called **locking relay**, or **latch relay**.

latchup Same as **latch-up**.

late binding The linking of software routines or objects during program execution. This contrasts with **static binding**, where the linking occurs during program compilation. Also called **dynamic binding**.

late contacts Electric contacts which open or close after other contacts have done so.

latency **1.** The time interval that elapses between the request or triggering of an action, and the obtaining of the desired result, or completion. **2.** The time interval that elapses between initiating a request for data and the transfer of said data. **3.** The time interval required for a signal or data to travel from one communications network node to another. For instance, such latency occurs when a packet is stored, analyzed, and forwarded. Also called **network latency**.

latent failure A failure of a component, circuit, device, piece of equipment, or system, which is not yet manifested, but whose triggering event has already occurred. For instance, an electrical overstress may occur, followed by the failure of a device some time later.

latent heat of evaporation Same as **latent heat of vaporization**.

latent heat of fusion The amount of heat necessary, with the pressure and temperature held constant, to convert a given amount of a solid into its liquid form. Also called **heat of fusion**.

latent heat of vaporization The amount of heat necessary, with the pressure and temperature held constant, to convert a given amount of a liquid into its vapor form. Also called **latent heat of evaporation**, or **heat of vaporization**.

latent image **1.** An image which is not yet visible, but which will become so when developed. **2.** An image which is electrostatically formed or stored, but which is not yet visible.

lateral diffusion In the manufacturing of semiconductor devices, diffusion of the dopant atoms parallel to the surface of the semiconductor material.

lateral transistor A bipolar junction transistor in which current across the base flows parallel to the wafer surface, as opposed to flowing perpendicularly.

laterally-diffused metal-oxide semiconductor Same as **LDMOS**.

laterally-diffused metal-oxide silicon Same as **LDMOS**.

latitude Imaginary circles which are drawn parallel to the equator, and whose circumferences diminish as they approach each pole. All points along the equator have a latitude of 0°, while parallels of latitude above the equator work their way towards 90 °N, and those below the equator work their way towards 90 °S. These circles cross the meridians of **longitude** at right angles, and together serve to define locations on the planet. Its abbreviation is **lat**.

latitude effect Any effect which varies depending on the **latitude** where it occurs or is observed. For example, the increase in cosmic radiation as the latitude increases, due to the geomagnetic field.

lattice **1.** A regular and periodic arrangement of points, particles, components, conductors, lines, or objects throughout a given area or space. For example, a crystal lattice, or a lattice network. **2.** The geometric arrangement in space of the atoms, molecules, or ions of a crystal. The pattern of such a lattice is regular and three-dimensional. Also called **crystal lattice**, or **space lattice**.

lattice coil A coil whose successive windings are crisscrossed. This provides a low distributed capacitance. Also called **honeycomb winding**.

lattice defect In a crystal, a flaw in the geometric arrangement of its contained atoms, molecules, or ions. Also called **lattice imperfection**, or **crystal defect**.

lattice filter A **lattice network** utilized as a filter.

lattice imperfection Same as **lattice defect**.

lattice network A network consisting of four branches and four terminals arranged in a mesh. It has two input terminals, which are nonadjacent, while the other two serve as output terminals. An example is a bridge network.

lattice scattering In a crystal, the scattering of electrons through collisions with vibrating atoms, molecules, or ions within the structure of the lattice. Such scattering may affect the mobility of charge carriers.

lattice wave A wave which propagates through a crystal lattice.

launch **1.** To propel into motion, or to initiate. For example, to load and run a computer program. **2.** To transfer energy from a wire or cable to a waveguide.

launching **1.** The act or process of propelling into motion, or initiating. **2.** The process of transferring energy from a wire or cable to a waveguide. A loop, for instance, may be used for this purpose.

Lauritsen electroscope An electroscope in which the sensitive element is a metallized quartz fiber. Gold is usually plated on the fiber, and such electroscopes are very sensitive. Also called **quartz-fiber electroscope**.

lavalier microphone Same as **lavaliere microphone**.

lavaliere microphone A small microphone which is usually worn on a cord around the neck of a user. Also spelled **lavalier microphone**. Also called **lapel microphone (2)**.

law A statement which describes a phenomenon, or a relationship between phenomena, under specified conditions and which has been proven or is assumed to hold true. For example, the law of energy conservation, or the law of charge conservation.

law of charge conservation A law that states that the total charge in an isolated system never changes. This means, for instance, that if a positive charge is created in an isolated system, that an equal negative charge will also appear in the system. Also called **conservation of charge**, or **charge conservation**.

law of charges Same as **law of electric charges**.

law of electric charges A law that states that there is an attractive force between opposite charges, and a repulsive force between like charges. It is part of the law of electrostatic attraction. Also called **law of charges**.

law of electromagnetic induction A law which states that the electromotive force induced in a conductor is proportional to the rate of change of magnetic flux through it. Also called **law of induction**, or **Faraday's law of electromagnetic induction**.

law of electrostatic attraction A law which states that the force between two charged particles is directly proportional to the product of their magnitudes, and inversely proportional to the square of their distance. This force occurs along a straight line between the charges, and is a repulsive force if the charges are like, or an attractive force if they are opposite. Also called **Coulomb's law**, or **electrostatic attraction law**.

law of energy conservation A law which states that energy cannot be created nor destroyed in an isolated system, but that it can be changed from one form to another. For instance, kinetic energy being transformed into potential energy, while the total energy in the system remains constant. Also called **conservation of energy (1)**, or **energy conservation (1)**.

law of gravitation A law that states that for any two particles or bodies, the force of attraction is directly proportional to the product of their masses, and inversely proportional to the square of the distance between them. This force occurs along a straight line between the particles or bodies, and applies to all matter, including electrons and stars. Also called **Newton's law of gravitation**.

law of induction Same as **law of electromagnetic induction**.

law of inverse squares A law stating that a given physical quantity will vary as the inverse square of the distance from the source. Most forms of energy that are radiated from a point source obey this law. These include electromagnetic, thermal, and nuclear energy. For instance, if the distance to a light source is halved, then the intensity of the received radiation will be four times greater. Also called **inverse-square law**.

law of magnetism A law that states that there is an attractive force between opposite poles, and a repulsive force between like poles.

law of mass conservation A law which states that mass, or matter, cannot be cannot be created nor destroyed in an isolated system. For instance, the mass remains constant when a substance changes form a solid to a gas in an isolated system. This law does not always hold true when dealing with subatomic particles. Also called **law of matter conservation**, **conservation of mass**, or **mass conservation**.

law of matter conservation Same as **law of mass conservation**.

law of reflection A law stating that a wave or ray, such as sound or light, which strikes a reflecting surface at a given angle, is reflected at an equal angle in the opposite direction with respect to a perpendicular line arising from the point of incidence. The striking ray or wave is called incident ray or wave, respectively, and is in the same plane as the reflected ray or wave. Also called **reflection law**.

LAWN Acronym for **local-area wireless network**.

lawnmower In radars, a type of preamplifier which is especially useful for reducing clutter produced by background electrical noise.

lawrencium A synthetic chemical element whose atomic number is 103. It has about 10 known isotopes, all of which are unstable, and is an alpha emitter. The techniques employed to produce this element are utilized to help discover new elements with higher atomic numbers. Its chemical symbol is **Lr**.

lay The lateral direction in which a conductor, or group of conductors, is wound around a cable core, or around another layer of conductors. The direction of lay may be left-hand or right-hand. If the conductors form counterclockwise spirals when traveling away from an observer viewing along the longitudinal axis of a cable, then it is left-hand. The converse is true for right-hand. Also called **direction of lay**.

layer **1.** A defined thickness which is part of a material, or which surrounds it. For example, an atmospheric or ionospheric layer, a layer in a semiconductor, or that produced through plating. **2.** In communications and computers, a defined set of capabilities, functions, and protocols which pertain to a given stratum within a hierarchical arrangement. For instance, any of the seven layers within OSI Reference Model.

layer 1 Within the OSI Reference Model for the implementation of communications protocols, the first, or lowest, of the seven layers. This layer deals with the hardware aspects of sending and receiving data in the network. Layer 1 provides the specifications for the transmission medium, such as fiber-optic cable or coax, for the adapter cards, connectors and their pinouts, and so on. At the sending end this layer accepts the bit stream from layer 2, places it in the physical medium, and at the receiving end extracts the bit stream and returns it to layer 2. Also called **physical layer**.

layer 2 Within the OSI Reference Model for the implementation of communications protocols, the second lowest level, located directly above layer 1. This layer takes care of functions such as the coding, addressing, and transmission of information between nodes. Also called **data-link layer**.

Layer 2 Tunneling Protocol Same as **L2TP**.

layer 3 Within the OSI Reference Model for the implementation of communications protocols, the third lowest level, located directly above layer 2. This layer deals with tasks pertaining to establishing the route the data will take, and ensuring that said data is delivered. It provides specifications for addressing, the creation of virtual circuits, switching, routing, and the like. Also called **network layer**.

layer 4 Within the OSI Reference Model for the implementation of communications protocols, the fourth level, located below layer 5 and above layer 3. This layer helps ensure that data transfers between nodes are complete and error free, handling tasks such as flow control, and error detection and correction. Also called **transport layer**.

layer 5 Within the OSI Reference Model for the implementation of communications protocols, the third highest level, located directly above layer 4. This layer takes care of functions such as establishing, managing, and terminating all exchanges of data. Also called **session layer**.

layer 6 Within the OSI Reference Model for the implementation of communications protocols, the second highest level, located directly above layer 5. This layer is responsible for the selection and use of a common syntax for representing information, and involves tasks such as encryption, decryption, compression, data conversion, text formatting, and text display. Also called **presentation layer**.

layer 7 Within the OSI Reference Model for the implementation of communications protocols, the highest level, located directly above layer 6. This layer takes care of the exchange of information between applications, including tasks such as file transfers, email, and access to a remote computer. Also called **application layer**.

layer winding A coil-winding method in which each layer is wound over the layer within it, all layers being separated by an insulating layer.

layered architecture An architecture, such as that of a communications or neural network, which is hierarchically organized into **layers (2)**.

layered network A network which is hierarchically organized into **layers (2)**.

layout The arrangement of components or devices in an object or system. For example, the layout of the components on a printed circuit. Also, a graphical representation of such an arrangement. Also, the process of arranging said components or devices.

lazy evaluation A programming method in which certain actions only take place when necessary. For instance, a given calculation is performed only when demanded as an input for another calculation.

lazy-H antenna An antenna array consisting of two vertically-stacked collinear elements. It features vertical and horizontal directivity. Also called **H antenna**.

lb Abbreviation of **pound**.

Lb Symbol for **lambert**.

lb ft Abbreviation of **pound-foot**.

LBA **1.** Abbreviation of **logical block addressing**. **2.** Abbreviation of **link budget analysis**.

lbf Abbreviation of **pound force**.

lbf ft Abbreviation of **pound-foot**.

lbf/ft² Abbreviation of **pounds per square foot**.

lbf/in² Abbreviation of **pounds per square inch**.

LC Abbreviation of **liquid crystal**.

LC Abbreviation of **inductance-capacitance**. **1.** A combination of inductance and capacitance in a circuit. **2.** A circuit or device with a combination of inductance and capacitance, such as that provided by coils and capacitors. For example, certain types of filters. **3.** Pertaining to an *LC* (1), or that incorporating such a combination.

LC **circuit** Abbreviation of **inductance-capacitance circuit**. A circuit composed of, or incorporating inductances and capacitances.

LC connector A fiber-optical cable connector that is similar, but smaller, than an SC connector. An SC connector has a 2.5 mm ferrule, while an LC connector has a 1.25 mm ferrule.

LC **filter** Abbreviation of **inductance-capacitance filter**. A filter composed of, or incorporating inductances and capacitances.

LC **meter** Abbreviation of **inductance-capacitance meter**. A meter which measures and indicates inductances and capacitances.

LC **network** Abbreviation of **inductance-capacitance network**. An electric network composed of, or incorporating inductances and capacitances.

LC oscillator Abbreviation of **inductance-capacitance oscillator**. An oscillator whose frequency is determined by inductances and capacitances.

LCC 1. Abbreviation of **leadless chip carrier**. 2. Abbreviation of **leaded chip carrier**.

LCCC 1. Abbreviation of **leadless ceramic chip carrier**. 2. Abbreviation of **leaded ceramic chip carrier**.

LCD Abbreviation of liquid-crystal display. A display incorporating **liquid crystals** sandwiched between glass plates which have a transparent conductive coating, or electrodes, which serve to vary the electric field surrounding said crystals, thus changing their optical properties. LCDs can be reflective or transmissive, and are used in a variety of flat-display devices, such as those of portable computers, calculators, or watches. Improved viewing characteristics may be seen, for instance, in active-matrix displays and nematic-crystal displays. Also called **LCD display** (1).

LCD display 1. Same as **LCD**. 2. Same as **LCD monitor**.

LCD monitor Abbreviation of liquid-crystal display **monitor**. A computer monitor incorporating an **LCD**. Also called **LCD display** (2).

LCD panel Abbreviation of liquid-crystal display **panel**. A data projector that presents the visual output of a computer through an **LCD**. It may incorporate its own light source, or be used in conjunction with an overhead projector for display on a remote screen.

LCD printer Abbreviation of liquid-crystal display **printer**. A type of printer similar to a laser printer, but which utilizes a bright lamp instead of a laser to create an image on the drum or other photoconductive surface.

LCD projector Abbreviation of liquid-crystal display **projector**. A projector which incorporates an **LCD panel** and a light source. Such a projector is usually portable, and has a much more powerful output than an LCD panel without the light source. An LCD projector may also incorporate other features, such as a cordless mouse for simplified operation.

LCD screen Abbreviation of liquid-crystal display **screen**. The viewing screen of an **LCD display**.

LCI Abbreviation of **logical channel identifier**.

LCN Abbreviation of **logical channel number**.

LCP Abbreviation of **Link Control Protocol**.

LCR Abbreviation of **least cost routing**.

LCR Abbreviation of **inductance-capacitance-resistance**. 1. A combination of inductance, capacitance, and resistance in a circuit. 2. A circuit or device with a combination of inductance, capacitance, and resistance, such as that provided by coils and capacitors. For example, certain tuned circuits. 3. Pertaining to an *LCR* (1), or that incorporating or measuring such a combination, such as a meter.

LCR meter Abbreviation of **inductance-capacitance-resistance meter**. A meter which measures and indicates inductances, capacitances, and resistances.

LD Abbreviation of **Laserdisc**.

LDAP Abbreviation of **Lightweight Directory Access Protocol**.

LDC Abbreviation of **leaded chip carrier**.

LDCC Abbreviation of **leaded ceramic chip carrier**.

LDE Abbreviation of **long-delayed echo**.

LDM Abbreviation of **limited-distance modem**.

LDMOS Abbreviation of laterally-diffused metal-oxide semiconductor, or laterally-diffused metal-oxide silicon. A modified metal-oxide semiconductor whose characteristics include added flexibility, enhanced performance, higher efficiency, and superior reliability. Used, for instance, in high-power RF applications, such as wireless communications infrastructure systems.

LDMOS transistor A transistor incorporating **LDMOS** technology.

LDR Abbreviation of **light-dependent resistor**.

LDV Abbreviation of **laser Doppler velocimetry**.

lead 1. A conductor, usually a wire, by which circuit elements or points are connected to components, devices, equipment, systems, points, or materials. Also called **lead wire**. 2. The state or condition of being ahead, or preceding, in time or phase. Also, that which leads. Also, the extent of such a lead. For example a lead angle, or a leading current. 3. A dense and soft grayish silvery-blue metallic chemical element whose atomic number is 82. It is poisonous, malleable, ductile, and corrosion resistant. Lead is a poor electrical conductor, and has about 35 known isotopes, of which 4 are stable. It has many applications including its use in storage batteries, in cable coverings, in solder and fusible alloys, in ceramics, certain glasses, and as a radiation shield. Its chemical symbol, **Pb**, is from the Latin word for lead: **plumbum**.

lead-acid battery A storage battery in which the cathode is lead plated with an oxide, such as lead oxide, the anode is lead, and the electrolyte is dilute sulfuric acid. Each cell produces about 2.1 volts, and a typical car battery, for instance, has 6 such cells for a total output of about 12.6 volts. Also called **lead storage battery**, or **lead battery**.

lead-acid cell A single cell in a **lead-acid battery**. Also called **lead cell**.

lead angle For two periodic quantities with the same frequency, the lead in phase of one respective to the other. Expressed in radians or degrees. Also called **angle of lead**.

lead battery Same as **lead-acid battery**.

lead borate A white powder used in certain ceramic or glass conductive coatings.

lead cell Same as **lead-acid cell**.

lead coplanarity The arrangement of component leads along the same plane.

lead dioxide A brown powder whose chemical formula is PbO_2. It is used in storage batteries and in battery electrodes.

lead fluoride Colorless crystals whose chemical formula is PbF_2. It is used in lasers and optics.

lead-in A transmission line that connects an antenna to a receiver or transmitter. Also called **lead-in wire** (2), or **down-lead**.

lead-in insulator An insulator in the form of a tube through which a **lead-in wire** is passed. Also called **lead-in tube**.

lead-in tube Same as **lead-in insulator**.

lead-in wire 1. A wire by which power is introduced to a structure, such as a building. 2. Same as **lead-in**. 3. A wire by which an electrode is connected to an external circuit.

lead lanthanum zirconate titanate A ferroelectric ceramic which exhibits the piezoelectric effect, and whose properties are similar to **lead zirconate titanate**. Its abbreviation is **PLZT**. The first letter of its abbreviation is taken from **Pb**, the chemical symbol for lead.

lead magnesium niobate An electrostrictive ceramic utilized in ultrasonic imaging devices and sonar systems.

lead oxide A red powder whose chemical formula is Pb_3O_4. It is used in storage batteries and in ceramics. Also called **lead tetroxide**.

lead plating The plating of component leads with a material such as tin, gold, or an eutectic alloy of tin and lead.

lead selenide Gray colored crystals whose chemical formula is $PbSe$. It is used in infrared detectors and in semiconductors.

lead silicate A red powder whose chemical formula is $PbSiO_3$. It is used in ceramics.

lead spacing The distance between adjacent component leads.

lead storage battery Same as **lead-acid battery**.

lead sulfate   White crystals whose chemical formula is $PbSO_4$. It is used in storage batteries.

lead sulfide   Silver colored crystals or a black powder whose chemical formula is PbS. It is used in ceramics, infrared detectors, and in semiconductors.   Lead sulfide is found chiefly in the mineral galena.

lead telluride   Silver-gray crystals whose chemical formula is $PbTe$. It is used in photodetectors, and in semiconductors.

lead tetroxide   Same as **lead oxide**.

lead wire   Same as **lead** (1).

lead zirconate titanate   A ferroelectric ceramic which exhibits the piezoelectric effect.   Used, for instance, as transducers, electro-optic devices such as electro-optic modulators, and as actuators in active control systems.   Its abbreviation is **PZT**.   The first letter of its abbreviation is taken from **Pb**, the chemical symbol for lead.

leaded ceramic chip carrier   Same as **leaded chip carrier**. Its abbreviation is **LDCC**, or **LCCC**.

leaded chip carrier   A ceramic package which has leads consisting of metal prongs or wires, as opposed to metallic contacts which are flush with the package or recessed, as is the case with **leadless chip carriers**.   Its abbreviation is **LDC**, or **LCC**.   Also called **leaded ceramic chip carrier**.

leader   **1.** A section of nonmagnetic tape, usually plastic, which is affixed to the beginning of a length of recording tape, such as a reel. A section attached at the end of a tape is called a **trailer** (1).   Also called **leader tape**.   **2.** In computers, a record which precedes and describes other records. **3.** An initial discharge which sets the path for a subsequent lightning stroke. The leader ionizes the air along its trajectory, and such a discharge may occur between a cloud and the ground, between clouds, or within the same cloud. Also called **leader stroke**.

leader stroke   Same as **leader** (3).

leader tape   Same as **leader** (1).

leadframe   **1.** In a packaged chip, a metal frame which provides the leads for electrical connections.   **2.** In a packaged chip, a metal frame upon which leads may be mounted.

leading current   An alternating current which leads the alternating voltage producing it.   In a purely capacitive circuit this lead is 90°.   This contrasts with a **lagging current**, where the voltage leads the current.

leading edge   **1.** The portion of a pulse waveform which first increases in amplitude.   That is, the rising portion.   The portion which first decreases is the **trailing edge** (1).   **2.** The initial portion of a pulse or signal.   The latter part is called **trailing edge** (2).

leading load   A load in which current leads voltage.   The capacitive reactance exceeds the inductive reactance in such a load.   This contrasts with a **lagging load**, where the voltage leads the current.   Also called **capacitive load**.

leading zero   A zero preceding the leftmost nonzero integer of a number.   The presence or absence of leading zeroes does not change the value of a number, so zeroes preceding the leftmost nonzero integer, but which appear after a decimal point, are not leading zeroes.   Leading zeroes may be used, for instance, to occupy blank spaces in a data field.

leading zero blanking   Same as **leading zero suppression**.

leading zero suppression   The blanking of any **leading zeroes**.   Used, for instance, in some digital meters.   Also called **leading zero blanking**.

leadless ceramic chip carrier   Same as **leadless chip carrier**. Its abbreviation is **LCCC**.

leadless chip carrier   A ceramic package which is hermetically sealed, and which instead of leads consisting of metal prongs or wires, has metallic contacts called castellations, which are flush with the package or recessed.   The contacts are usually on all four sides of the package, although they may be located elsewhere.   Its abbreviation is **LCC**.   Also called **leadless ceramic chip carrier**, or **ceramic leadless chip carrier**.

leaf   **1.** In a hierarchical structure, such as a tree, a node which has no descendants.   In a hierarchical file system, for example, a file is a leaf.   Also called **leaf node**, or **terminal node**. **2.** A very thin sheet of metal.   A leaf is thinner than a **foil**. A gold leaf may have a thickness of about 100 nanometers.

leaf electroscope   An electroscope, especially a gold-leaf electroscope, in which one or two suspended metallic leaves serve to detect electric charges by means of electrostatic forces.

leaf node   Same as **leaf** (1).

leak   **1.** A flaw or fault which allows the undesired escape of something, such as a fluid or energy.   Also, the location of such a flaw or fault.   Also, that which escapes.   May also refer to such a flaw or fault which allows the undesired entry of something.   **2.** The flow of current through unwanted paths of a circuit, along the surface or through the body of a dielectric or insulator, or that which is otherwise undesired or lost.   Also, the current which flows in this manner, or is lost.

leak proof   Constructed and/or arranged in a manner which does not allow **leakage**.   Said, for instance, of an enclosure or apparatus.   Also spelled **leakproof**.   Also called **leak tight**.

leak tight   Same as **leak proof**.   Also spelled **leaktight**.

leakage   **1.** The usually undesired escape or entry of something, such as a fluid or energy, due to a **leak** (1).   **2.** The undesired flow of **leakage current**.

leakage current   Also called **current leakage**.   **1.** Current which flows through unwanted paths of a circuit, such as from the output to the input when not intended.   **2.** Current which flows between electrodes of an electron tube by any route except the interelectrode space.   **3.** Current which flows along the surface or through the body of a dielectric or insulator.   **4.** Direct current which flows through the dielectric of a capacitor.   **5.** Alternating current which flows through a rectifier without being rectified.   **6.** Current which flows through a component, circuit, or device which is in the off state.   Also called **off-state leakage current**.

leakage detection   The use of a **leakage detector**.

leakage detector   An instrument or device which detects **leakage**.   Such a detector may also quantify said leakage. Also called **leakage indicator**.

leakage flux   Magnetic flux which does not pass through a useful or intended part of a magnetic circuit.   Also called **flux leakage**.

leakage indicator   Same as **leakage detector**.

leakage inductance   Self-inductance, especially in a transformer, due to **leakage flux**.

leakage path   The path followed by a **leakage current**.   Also called **sneak path**.

leakage power   Power which is dissipated in an undesired manner.

leakage radiation   Radiation which escapes in an undesired manner.   For example, electromagnetic radiation escaping through an unintended part of an antenna system, or X-rays emanating in an unwanted direction from an X-ray tube.

leakage reactance   Inductive reactance, especially in a transformer, due to **leakage flux**.

leakage resistance   Resistance along the path which **leakage current** flows.   In a capacitor, for instance, this resistance is usually very high.

leakproof   Same as **leak proof**.

leaktight   Same as **leak proof**.   Also spelled **leak tight**.

leaky 1. That which has **leakage**. For example, a leaky waveguide. 2. That which has substantially more leakage than would ordinarily be expected. For instance, a leaky capacitor.

leaky capacitor A capacitor with excessive leakage current, or whose dielectric is flawed.

leaky-wave antenna A wideband antenna with a narrow beam which is used in the microwave range.

leaky waveguide 1. A waveguide with a slot which allows electromagnetic radiation to escape. 2. A waveguide from which electromagnetic radiation escapes in an undesired manner.

leapfrog test In computers, a diagnostic routine which copies itself throughout an available storage medium.

learning bridge In networks, a bridge that can learn the locations of users or systems and deliver messages accordingly. Also known as **transparent bridge**, or **adaptive bridge**.

learning curve The time and processes involved in acquiring a new skill, becoming familiar with a given subject, or the like. For example, the time and effort required to learn to utilize a computer application proficiently. Also, a graph representing such progress.

learning machines Devices and systems that have the ability to apply knowledge that has been previously programmed and recorded, in order to analyze and better face new situations.

learning portal A Web portal that provides multiple education and/or training resources which have been collected from several Web sites.

learning robot A robot that has the ability to apply knowledge that has been previously programmed and recorded, in order to analyze and better face new situations.

leased circuit Same as **leased line**.

leased line A permanent communications channel between two or more locations. Such a service includes private switching arrangements and a defined transmission path, and provides an exclusive high-speed connection which is available at all times. T1 and T3 lines are examples. Also called **leased circuit, dedicated line (1), dedicated circuit, dedicated connection**, or **private line**.

leased-line modem A modem utilizing a **leased line**.

least cost routing A method for automatically selecting the least costly route for a telephone call. For example, specialized software may utilize a table of area codes to determine which fax server out of many should be utilized for broadcasting. Its abbreviation is **LCR**. Also called **automatic route selection**.

least-significant bit Within a sequence of bits, that with the lowest weight or place value. It is usually the bit furthest to the right. Its abbreviation is **LSB**.

least-significant byte Within a sequence of bytes, that with the lowest weight or place value. It is usually the byte furthest to the right. Its abbreviation is **LSB**.

least-significant character Within a sequence of characters, that with the lowest weight or place value. It is usually the character furthest to the right. Its abbreviation is **LSC**.

least-significant digit Within a sequence of digits, that with the lowest weight or place value. It is usually the digit furthest to the right. Its abbreviation is **LSD**.

least-time principle A principle stating that an electromagnetic wave, such as a light ray, travels between points taking the path that requires the least time. This path may or may not be a straight line. Also called **Fermat's principle**.

LEC 1. Abbreviation of **liquid-encapsulated Czochralski**. 2. Abbreviation of **local exchange carrier**.

Lecher lines Same as **Lecher wires**.

Lecher wires Two long insulated wires which are separated from each other by a short distance, with a shorting bar which slides along them. One side of the wires is connected to a source of radio-frequency energy, and the location of the bar is manually varied so as to determine the positions of standing-wave nodes. The distance between said nodes can be utilized to calculate wavelengths. Also called **Lecher lines**.

Leclanché cell A dry primary cell in which the positive electrode is carbon, the negative electrode is zinc, and the electrolyte is ammonium chloride. Its voltage output is about 1.5 volts. Also called **dry cell (1)**.

LED Abbreviation of light-emitting **d**iode. A diode consisting of a pn junction which emits light when forward-biased. The light produced is incoherent, and its wavelength depends on the semiconductor material and doping. Such light may be red, another visible color, or it may be located in the infrared region. LEDs are frequently utilized in alphanumeric readouts and for indicating the on/off status of a function, device, or piece of equipment.

LED display Abbreviation of light-emitting **d**iode **display**. A display incorporating one or more **LEDs**.

LED printer Abbreviation of light-emitting **d**iode **printer**. A type of printer similar to a laser printer, but which utilizes a matrix of LEDs, instead of a laser, to create an image on the drum or other photoconductive surface.

LEED Abbreviation of **low-energy electron diffraction**.

left channel 1. In a stereo audio system, the channel intended for speaker output situated to the left of a listener. Also called **left stereo channel**. 2. In a multiple-channel or surround sound audio system, a channel intended for speaker output situated to the left of a listener. This may be, for instance, either from the left front speaker, or the left rear speaker.

left click To press and release the left button on a computer mouse, or its equivalent. A computer may be programmed to exchange the signals sent by the left and right mouse buttons.

left front speaker In a multiple-channel or surround sound audio system, the speaker whose output is intended to be to the front and left of a listener. Also called **left speaker (1)**.

left-hand lay One or more conductors which form counterclockwise spirals when traveling away from an observer viewing along the longitudinal axis of a cable. Such conductors may be wound around a cable core, or around another layer of conductors. This contrasts with **right-hand lay**, where the converse is true.

left-hand polarized wave An elliptically polarized transverse electromagnetic wave in which the rotation of the electric field vector is towards the left when viewed along the direction of propagation. Also called **counterclockwise-polarized wave**.

left-hand rule Also called **Fleming's left-hand rule**. If conventional current is used instead of electron flow, each of these become **right-hand rules**, as conventional current travels in the opposite direction. 1. A rule stating that if the thumb of a left hand is oriented along the same axis as the flow of electrons through a conducting wire, that the fingers of this hand will curl along the same direction as the magnetic field produced by the wire. 2. For a conducting wire moving through a magnetic field, a rule stating that if the middle finger, index finger, and thumb of a left hand are extended at right angles to each other, that the middle finger will indicate the direction of the flow of electrons, the index finger the direction of the magnetic field, and the thumb will indicate the direction of the movement of the wire.

left minus right Same as **L – R**.

left plus right Same as **L + R**.

left rear speaker In a multiple-channel or surround sound audio system, the speaker whose output is intended to be to the rear and left of a listener.

left shift **1.** In computers, an operation which moves or displaces digits, bit values, or the like, to the left. In a **right shift** (1), the converse is true. **2.** On a computer with two shift keys, the shift key on the left of the keyboard. Some keyboard shortcuts which utilize a combination of keys, for instance, make the distinction between this and the **right shift** (2).

left speaker **1.** Same as **left front speaker**. **2.** The left speaker of a two-channel audio system, such as a that of a home stereo or TV.

left stereo channel Same as **left channel** (1).

leg A path within a circuit or network. Also called **arm** (3), or **branch** (2).

legacy Hardware or software which has been in use sufficient time to be rendered obsolescent or obsolete.

legacy application An application which has been in use sufficient time to be rendered obsolescent or obsolete.

legacy hardware Hardware which has been in use sufficient time to be rendered obsolescent or obsolete.

legacy software Software which has been in use sufficient time to be rendered obsolescent or obsolete.

legacy system A computer system which has been in use sufficient time to be rendered obsolescent or obsolete.

length Its symbol is l. **1.** A distance or dimension which has been measured or calculated. Most properly expressed in meters or fractions/multiples of meters such as millimeters or kilometers, although there are various other terms utilized for specific applications, such as angstroms or parsecs. **2.** The greater or greatest **length** (1) or extent of something. **3.** A time interval, especially between specific events. For instance, the time required for a complete cycle. **4.** The linear storage space occupied by a data structure, block, sector, file, or the like. Usually expressed in bits or bytes.

lens **1.** One or more objects, usually transparent and curved, which serve to refract and sometimes reflect light rays passing through. A lens causes parallel light beams to converge or diverge, is usually made of glass or plastic, and is utilized to form an image. Also called **optical lens**. **2.** A device utilized to focus an electron beam, analogous to the manner in which an optical lens focuses a light beam. Electric and/or magnetic fields may be used for this purpose. Used, for instance, in CRTs, or in electron microscopes. Also called **electron lens**. **3.** A system of disks or other obstacles that refract sound waves similarly to the way an optical lens refracts light. Also called **acoustic lens**, or **sound lens**. **4.** A device which utilizes a magnetic field to focus a particle beam, analogous to the manner in which an optical lens focuses a light beam. Such a lens may consist, for instance, of an array of electromagnets. Also called **magnetic lens**. **5.** One or more objects, structures, or devices utilized to refract, diffract, or otherwise focus or shape any other form of electromagnetic radiation. An example is an antenna lens.

lens antenna An antenna that uses an arrangement of metal and/or dielectric objects as a lens to focus the radio waves. Also called **antenna lens**.

Lenz's law A law which states that the direction of an induced current is that in which the magnetic field it produces opposes the original change in magnetic flux which induced said current.

LEO Acronym for **low earth orbit**.

LEOS Acronym of **low earth orbit satellite**.

LEP Abbreviation of **light-emitting polymer**.

LEP display Abbreviation of **light-emitting polymer display**.

lepton A class of elementary particles which have no strong interactions. Leptons are the lightest class of particles having a non-zero rest mass, and include electrons, muons, neutrinos, and their respective antiparticles.

let-through current The current that flows after a protective device, such as a circuit breaker, has been triggered.

letter bomb An email whose content is intended to harm the recipient's computer, or reduce or eliminate available resources. For example, such an email may contain code which freezes the recipient's computer, an attachment with a worm or virus, or may have a size that overflows the mailbox.

letter quality Printed characters which are dark, crisp, and well-defined. Also, a printing mode with such an output. Most laser printers can print in this mode. Its abbreviation is **LQ**.

letterbox A widescreen image which is presented on a TV with a 4:3 aspect ratio. Since the widescreen image has an aspect ratio of 16:9 or greater, in order to see the entire available width, there will be a black band at the top and bottom of the screen. Also called **letterbox format**.

letterbox format Same as **letterbox**.

level **1.** A relative position, degree, or intensity. For instance, a high-level programming language, a high voltage level, a sound-pressure level, or the intensity of an electrical signal relative to a reference level. **2.** A given amplitude of a phenomenon. For example, a given voltage, resistance, or volume level. **3.** That which maintains itself at a steady value. For instance, a signal of fixed amplitude. **4.** An amplitude which serves as a reference level. For example, a pressure of 20 micropascals, which is assigned a value of 0 decibels, and which provides a reference for determining the relative intensity of sounds. **5.** To place or set on the same level as something else. For instance, to equalize. **6.** Parallel to a horizontal surface, such as the ground.

level 1 cache Same as **L1 cache**.

level 2 cache Same as **L2 cache**.

level 3 cache Same as **L3 cache**.

level above threshold For a given individual, the level, expressed in dB, of a sound relative to the individual's hearing threshold for that sound. For example, a noise bandwidth presented at 30 dB sensation level. Also called **sensation level**.

level compensation The use and/or effect of a **level compensator**.

level compensator A control, mechanism, device, or system which minimizes the amplitude variations of an input or output signal. An example is an automatic level control.

level control **1.** A mechanism, device, or system which controls the amplitude a signal. An example is a volume control. **2.** The adjustments in amplitude obtained using a **level control** (1). **3.** A mechanism, device, or system which controls a **level** (2).

level conversion The use and/or effect of a **level converter**. Also called **level translation**, or **level shifting**.

level converter Also called **level translator**, or **level shifter**. **1.** A circuit, device, mechanism, or system which changes one signal level to another. For example, that which converts a voltage or current level of an input signal. **2.** In a logic circuit or device, the changing of the logic level.

level indicator **1.** An meter which indicates the amplitude a signal. For instance, a volume indicator. **2.** An instrument or device which indicates a **level** (2).

level measurement **1.** The determination of the amplitude a signal. **2.** The determination of a **level** (2).

level shifter Same as **level converter**.

level shifting Same as **level conversion**.

level translation Same as **level conversion**.

level translator Same as **level converter**.

lever switch A switch that is operated by moving a projecting handle or bar.

Leyden jar An early type of capacitor usually utilized for demonstrations. It may consist, for instance, of a cylindrical glass jar which is covered on the inside and outside with aluminum foil. A brass rod extends through the jar's insulating lid, with a knob on the outer end of said rod, and a chain which makes contact with the inner foil surface affixed to the other end. A voltage is then applied to the knob, making the jar a capacitor in which the glass is the dielectric, and the foil conductors are the plates.

LF Abbreviation of **low frequency**.

LFDF Abbreviation of **low-frequency direction finder**.

LFO Abbreviation of **low-frequency oscillator**.

LFSR Same as **linear feedback shift register**.

LGA Abbreviation of **Land Grid Array**.

LHA Same as **LHARC**.

LHARC A freeware data-compression utility. Also called **LHA**.

LHC Abbreviation of **Large Hadron Collider**.

Li Chemical symbol for **lithium**.

Li-ion battery Abbreviation of **lithium-ion battery**.

Li-ion cell Abbreviation of **lithium-ion cell**.

library A collection of routines, programs, files, or storage media. The content of such a collection is usually related in some manner. For example, a data library may consist of a list of files available from a server or storage medium, while a class library is composed of a set of routines and programs that programmers can use to write object-oriented programs.

library routine A routine which is part of a **library**.

License Agreement A contract between a software manufacturer or provider, and one or more end users. It pertains to matters such as use, distribution, and resale. Most purchased software has such an agreement. Also called **End-User License Agreement**.

lidar Acronym for **li**ght **d**etection **a**nd **r**anging. A system similar to radar, but which utilizes a beam of light, such as infrared, instead of a beam of radio-frequency energy, such as microwaves, to obtain measurements of direction, range, altitude, and speed of a scanned object.

lie detector A device which monitors and records changes in physiological variables such as respiration, blood pressure, and perspiration, while a subject is questioned. Such a device purportedly can detect when a person is lying. Also called **polygraph**.

LIF Abbreviation of **low insertion force**. Same as **LIF socket**.

LIF socket Abbreviation of **low insertion force socket**. A socket designed for the use of a minimum of force for insertion or removal of components, thus helping minimize any mechanical stress. Commonly utilized for the insertion and removal of chips.

life Also called **lifetime**. **1.** The time interval during which something exists or functions. **2.** The time interval during which something exists or functions in a given manner. For instance, the operating life, useful life, or shelf life of a device, or the time that elapses between the creation of an electron-hole pair, and its recombination.

life cycle The interval during which a component, device, piece of equipment, system, material, or the like, is useful. Once this period elapses, it becomes increasing less advisable to invest additional resources in maintenance, repairs, replacements, and improvements.

life test A test in which a component, circuit, device, piece of equipment, system, object, or material is operated under, or

subjected to, conditions which are equivalent to a lifetime of use. Such a test, for instance, may be for the actual duration of such a life, or it may be accelerated.

lifetime Same as **life**.

LIFO Acronym for **l**ast-**i**n, **f**irst-**o**ut. A method of handling data in which the last item stored is the first item to be retrieved. In this manner, that which has been held the least time is that which is used first. This contrasts with **FIFO**, where the first item stored is the first item to be retrieved.

lift-off A process utilized for the definition of a pattern on the surface of a semiconductor wafer. In it, the wafer is first covered with a resist which is utilized to produce the desired pattern. A metal, such as gold, is then applied onto the resist and substrate. The resist is then dissolved, in the process lifting-off the metal on it, so that only the desired pattern remains on the substrate. Used, for instance, in cases where etching is difficult or otherwise inadvisable. Also spelled **liftoff**.

liftoff Same as **lift-off**.

light **1.** Electromagnetic radiation whose wavelength enables it to be detected by an unaided human eye. The interval of wavelengths so detectable spans from approximately 750 nanometers (red) to approximately 400 nanometers (violet). Light is currently defined as traveling at 2.99792458×10^8 meters per second. Also called **light radiation**, or **visible radiation**. **2.** Electromagnetic radiation of any wavelength, such as infrared, ultraviolet, or X-rays. **3.** A source of **light** (1), such as a the sun, or a lamp. **4.** The illumination provided by a **light** (3). **5.** Same as **light version**.

light-activated SCR Same as **light-activated silicon-controlled rectifier**.

light-activated silicon-controlled rectifier A silicon-controlled rectifier which has a light-sensitive region which enables incident light to control its switching action. Such a device may also be controlled by a gate. Its abbreviation is **LASCR**, or **light-activated SCR**. Also called **photo SCR**, or **photothyristor**.

light-activated switch A switch in which the presence of light, or the intensity of incident light such as that of a modulated light beam, determines the opening or closing of a circuit. A light-activated switch incorporates a photoelectric device, such as a photocell. An example is a light-activated silicon-controlled rectifier. Also called **light-operated switch**, or **photoelectric switch**.

light adaptation The process by which an eye or a photodetector adapts to changes in illumination.

light amplification by stimulated emission of radiation Same as **laser**.

light amplifier An electronic device which increases the brightness of images, and which can operate at very low levels of light. For example, the image whose brightness will be intensified is first focused onto a photocathode, from which electrons are ejected. These electrons are then accelerated and focused onto a phosphorescent screen, whose displayed image may be thousands of times brighter than the original. Used, for instance, to obtain clear X-ray images with markedly reduced exposure. Also called **image intensifier**.

light beam **1.** A concentrated and essentially unidirectional stream of **light** (1). **2.** A concentrated and essentially unidirectional stream of **light** (2).

light bulb A lamp which incorporates a filament, often made of tungsten or carbon, which emits brilliant visible light when heated by an electric current. Also called **incandescent lamp**.

light cable One or more optical fibers bundled together and encased by a protective sheath. Generally used for telecommunications. Such cables usually consist of three layers, which are the core, its surrounding cladding, and the protective jacket. Also called **optical cable, fiber-optic cable, optical fiber cable,** or **cable (2).**

light chopper A device which periodically interrupts a light beam. This may consist, for instance, of a rotating disk with slots cut at regular intervals, so that such a device can be utilized to obtain optical pulses from a continuous light source.

light communications The use of optical means to transmit information between two or more points. Also, the conveyed information. For instance, communications utilizing a modulated laser beam.

light current An electric current which is produced by incident light. Such a current may occur, for instance, in a photoelectric device.

light-dependent resistor A photocell whose electrical resistance varies as a function of the intensity of light incident upon it. Its abbreviation is **LDR.** Also called **photoconductive cell, photoresistor,** or **photoresistive cell.**

light detection and ranging Same as **lidar.**

light detector Same as **light sensor.**

light dimmer A circuit or device which serves to control the intensity of illumination of a light source, such as an incandescent lamp. There are various types, including those employing silicon-controlled rectifiers or rheostats. Also called **dimmer.**

light-emitting diode Same as **LED.**

light-emitting diode display Same as **LED display.**

light-emitting polymer A polymer, such as polyfluorene, which emits light when electrically stimulated. Used, for instance, for displays such as those of computers, cell phones, or instrument panels, and in electronic paper. Its abbreviation is **LEP.**

light-emitting polymer display A display, such as that of a computer or cell phone, incorporating **light-emitting polymers.** Its abbreviation is **LEP display.**

light energy Radiant energy in the visible spectrum. Also called **luminous energy (2).**

light exposure 1. The condition or state of being exposed to light. **3.** A specific instance of **light exposure (1). 3.** The quantity or extent of **light exposure (1).**

light flasher A circuit or device which provides repeatable and artificially produced bursts of bright light. Also called **flasher.**

light flux Same as **luminous flux.**

light guide A thin filament of a transparent material, such as glass or plastic, which is capable of transmitting light signals through successive internal reflections. In order for such a fiber to guide light, the proper relationship between the refractive index of the core and its surrounding cladding must be maintained. Also spelled **lightguide.** Also called **optical fiber,** or **fiber (2).**

light hood A shield or other object utilized to prevent light from entering or escaping a given area or enclosure. For example, a shield placed over the screen of a CRT to help eliminate ambient light.

light intensity Same as **luminous intensity.**

Light Intensity Modulation Direct Overwrite A magneto-optic technology which combines a two-pass erase-write sequence into a single pass, thereby improving speed. Its acronym is **LIMDOW.**

light load A load which is a comparatively small fraction of the full or rated load.

light meter An instrument or device which measures and indicates the intensity of **light (1),** or **light (2).** An example is an exposure meter.

light microscope A microscope which utilizes reflected light and magnifying lenses to produce enlarged images of objects being viewed. Such microscopes usually use visible light, but may also use electromagnetic radiation close to visible light, such as near infrared. Also called **optical microscope.**

light microscopy The use of a **light microscope** to view and study samples.

light microsecond The distance which light, or electromagnetic waves of any frequency, travels through free space in one microsecond. This is equal to 299.792458 meters.

light modulation The modification of the intensity of a light beam proportionally to a characteristic present in another wave or signal. Also, the result of such a process. Usually utilized to convey meaningful information. Also called **optical modulation.**

light modulator 1. A device, such as an electro-optic modulator, which serves to modulate a light beam. Also called **optical modulator. 2.** A device incorporating a light source and a **light modulator (1)** to reproduce an optical soundtrack of a motion picture.

light-negative Pertaining to a photoconductor which undergoes a decrease in electrical conductivity when exposed to light. Also called **photonegative (1).**

light-operated switch Same as **light-activated switch.**

light pen A handheld computer input device in the shape of a pen, which is utilized to directly change the information presented on a monitor. The tip of a light pen is sensitive to light, and objects on the screen can be drawn, selected, moved, and the like, by pressing it onto the outer surface of the display screen, or by pressing a clip or button on the pen when pointed where desired. Also spelled **lightpen.**

light pipe A conduit made of a transparent material, such as glass or plastic, which is capable of channeling light from one end to the other through successive internal reflections. Such a pipe may be flexible or rigid, and an optical fiber is an example.

light-positive Pertaining to a photoconductor which undergoes an increase in electrical conductivity when exposed to light. Also called **photopositive (1).**

light pulse A short burst of light. Certain lasers, for instance, can generate pulses nearly in the zeptosecond range.

light quantum Also called **photon.** A particle with no mass nor charge, and which composes light and other forms of electromagnetic radiation. Gamma rays and X-rays are examples of higher-energy photons, while infrared rays and radio waves are examples of lower-energy forms. Its energy is equal to hf, where h is the Planck constant, and f is the frequency of the radiation in hertz. Photons are required to explain phenomena such as the photoelectric effect, in which light has particle-like properties. The photon is the carrier of the electromagnetic force, which is the force responsible for all electromagnetic interactions, such as those which hold atoms and molecules together.

light radiation Same as **light (1).**

light ray 1. A **light beam** whose cross section is very small. **2.** A line which represents the path along which light is propagated from a source.

light relay A relay in which the intensity of incident light, such as that of a modulated light beam, determines its opening or closing of a circuit. A light relay incorporates a photoelectric device, such as a photocell. Also called **photoelectric relay.**

light-sensitive Stimulated, actuated, or otherwise responsive to light. That which is light-sensitive has one or more physical and/or chemical properties vary in response to light. Also called **photosensitive.**

light-sensitive detector Same as **light sensor.**

light-sensitive diode A semiconductor diode which converts light into an electrical signal, such as a photocurrent. The reverse current of such a diode is regulated by the intensity of the incident light. Used, for instance, in fiber-optic communications to convert optical power into electrical power. An avalanche photodiode is an example. Also called **photodiode**.

light-sensitive material A material which is **light-sensitive**. Such a material, for instance, may exhibit photoconduction, photoionization, or photoemission. Also called **photosensitive material**.

light-sensitive transistor A bipolar transistor whose conduction is regulated by the intensity of the light incident upon its base region. Such a device is more sensitive than a light-sensitive diode, in addition to providing gain. Also called **phototransistor**.

light sensor Also called **light detector, light-sensitive detector, photodetector,** or **photosensor**. **1.** A device which detects and responds to light. **2.** A transducer whose input is light, and whose output is a corresponding electrical signal. Light sensors are usually semiconductor devices, such as photodiodes.

light source **1.** Any natural or artificial source of light. These include the sun, lamps, and light-emitting diodes. **2.** A lamp, or other source, utilized in devices and instruments such as microscopes and movie projectors, which provides the necessary illumination for proper function.

light spectrum The range of frequencies, within the electromagnetic spectrum, whose wavelengths enable detection by an unaided human eye. The interval of wavelengths so detectable spans from approximately 750 nanometers (red) to approximately 400 nanometers (violet).

light-spot scanner A device which utilizes a small and intensely bright spot which scans an object, scene, screen, film, or the like, and transforms the original image into a series of electrical signals by focusing it onto a photoelectric cell. The scanning may be mechanical or electrical. Used, for instance, for TV transmission, telecine transfer, or character recognition. Also called **scanner** (4), **flying-spot scanner**, or **optical scanner** (2).

light valve A device which incorporates an optical element whose light transmission is varied proportionally to an externally applied electric and/or magnetic variable, such as a voltage, electron beam, or magnetic field. Used, for instance, in a light-valve projector.

light-valve projector A projector utilizing a **light valve**, and which can have a very high light output. Used, for instance, in movie theaters.

light version A version of a software or hardware product which has less features and functionality than the full version. For example, ADSL lite, or an abbreviated offering of an application. Also called **light** (5), or **lite version**.

light wave A wave propagating **light** (1) or **light** (2).

light year The distance light travels through a vacuum in one year. It is unit of distance utilized in astronomy, and is equal to about 9.4605 x 10^{15} meters, or about 0.306591 parsec. Its abbreviation is **ly**.

lighted button Same as **lighted pushbutton**.

lighted push-button Same as **lighted pushbutton**.

lighted pushbutton Also spelled **lighted push-button**. Also called **lighted button**, or **illuminated pushbutton**. **1.** A push-button which is backlit, or otherwise provided with a source of light. **2.** A push-button which becomes lighted when pressed or activated.

lighted remote Abbreviation of **lighted remote** control. A remote control which is backlit, enabling ease of use in dark conditions. Also called **illuminated remote**.

lighted remote control Same as **lighted remote**.

lighted switch Also called **illuminated switch**. **1.** A switch that is backlit, or otherwise provided with a source of light. **2.** A switch that becomes lighted when pressed or activated.

lightguide Same as **light guide**.

lighthouse tube An ultra-high frequency electron tube with parallel disk-shaped electrodes, which features low interelectrode capacitance and a high-power output. Also called **disk-seal tube**, or **megatron**.

lightness **1.** The quality or state of being illuminated. **2.** The visual perception of more or less light reflected by an object. **3.** The quality or state of having a comparatively small mass or weight.

lightning A luminous electrical discharge between a cloud and the ground, between clouds, or within the same cloud, due to potential differences which become strong enough to break down the insulation that the intervening air provides. A lightning discharge may consist of multiple strokes, each of which is preceded by a leader. The current in a typical discharge is about 10,000 to 20,000 amperes, with peaks which can exceed 100,000 amperes. The burst of heat which accompanies a lightning discharge forces the surrounding air to expand explosively, producing shock waves heard as thunder. Also called **lightning discharge**.

lightning arrester Also called **arrester**. **1.** A device which provides a low-resistance path to ground for lightning discharges, thus protecting equipment at risk. **2.** A device which protects equipment by providing a low-resistance path to ground for voltages that exceed a determined amount.

lightning conductor **1.** A system consisting of a lightning rod, a conductor capable of withstanding the current of a lightning discharge, and a ground rod. A lightning conductor does not prevent the occurrence of a lightning discharge, but when properly designed helps ensure that the energy released is safely channeled into the earth. Also called **lightning rod**. **2.** The conductor utilized in a **lightning conductor** (1).

lightning detector An instrument or device, such as a keraunograph, which is utilized to detect lightning in remote locations.

lightning discharge Same as **lightning**.

lightning flash The intensely bright light accompanying a **lightning discharge**.

lightning generator A generator that produces surges similar to **lightning surges**. Usually utilized for testing components, devices, and materials such as insulators.

lightning protection The protection of structures, electrical devices and systems, people, and any other objects or areas which may be vulnerable to the harmful effects of a **lightning discharge**. A **lightning conductor** (1), for instance, may provide such protection.

lightning protection system A system, such as a **lightning conductor** (1), which provides **lightning protection**.

lightning protector A device or system which provides **lightning protection**.

lightning rod **1.** A rod whose pointed shape, placement on top of a structure or area, and conductivity enable it to attract lightning discharges and provide a cone-shaped zone, called a cone of protection, below which there is a highly reduced probability of a lightning strike. A lightning rod is usually metallic, although another good conductor may be utilized. Also called **rod** (2). **2.** Same as **lightning conductor** (1).

lightning strike A direct hit upon an object or structure by a **lightning discharge**.

lightning stroke One of multiple electrical discharges occurring during a single **lightning discharge**.

lightning surge A momentary electrical disturbance, such as a power surge, caused by **lightning**.

lightning switch A manually-operated switch that is utilized to connect a radio antenna to ground during electrical storms, or when not in use. Also called **ground switch**

lightpen Same as **light pen**.

lightwave A wave of **light** (1) or **light** (2).

lightwave system A system which serves to transmit information by means of light signals. A fiber-optic system is an example.

Lightweight Directory Access Protocol A protocol utilized for accessing, storing, and retrieving information such as names, emails, and phone numbers which comprise an online directory. This protocol supports TCP/IP, and is based on the X.500 standard, but is simpler. Its abbreviation is **LDAP**.

LILO Acronym for last-in, last-out. A method of handling data in which the last item stored is the last item to be retrieved. This is the same as **FIFO**.

LIMDOW Acronym for **Light Intensity Modulation Direct Overwrite**.

limen The point, level, or value of a quantity which must be exceeded to have a given effect, result, or response, such as detection, activation, or operation. For instance, the threshold of hearing, the minimum voltage or current necessary to activate a circuit, or the frequency beyond which a speaker does not reproduce sound. Also called **threshold**.

limit 1. The point, line, value, amplitude, magnitude, or the like, beyond which something cannot proceed. For example, a current limit, or a fatigue limit. 2. The greatest or least value of a defined interval or range. For instance, the maximum or minimum value a meter can display, or the boundaries of a frequency band. 3. To confine within a given interval or range. For instance, to limit both positive and negative voltage swings.

limit switch A switch that is actuated when a given limit is reached. For instance, a switch that cuts off power to a device when a moving part moves past a given point of its travel.

limited-distance modem Same as **line driver**. Its abbreviation is **LDM**.

limited signal A signal which is maintained, or which occurs, within a given interval.

limited stability For a given component, circuit, device, piece of equipment, or system, stability which is obtained only when a given parameter, such as the amplitude of its input signal, is within specified limits. Also called **conditional stability**.

limiter A circuit or device which limits the amplitude of its output signal to a predetermined maximum, regardless of the variations of its input. Used, for example, for preventing component, equipment, or media overloads. Also called by various other names, including **limiter circuit, limiting circuit, peak limiter, peak clipper, automatic peak limiter, clipper circuit, clipper**, and **amplitude limiter**.

limiter circuit Same as **limiter**.

limiting The action and effect of a **limiter**. Also called **clipping** (1).

limiting amplifier An amplifier which limits the amplitude of its output signal to a predetermined maximum, regardless of the variations of its input.

limiting circuit Same as **limiter**.

limiting resistor A resistor that limits the flow of a current to a given maximum value. Used to protect components and equipment. Also called **current-limiting resistor**.

limiting resolution A subjective measure of the resolution of a displayed image. It may be defined in various manners, including the maximum number black and white vertical lines that can be seen on a test chart.

linac Acronym of **linear accelerator**.

line 1. A straight or curved length which is considered to have a width of zero. Also, that which resembles this. 2. A

physical medium, such as a wire, cable, or waveguide, which serves to transmit or otherwise convey signals, data, electricity, or electromagnetic radiation between points. Examples include communication lines, power lines, and antenna transmission lines. Also called **transmission line** (1). 3. Same as **landline**. 4. In a CRT, the line formed by a complete horizontal movement of the electron beam. Also called **horizontal line**. 5. In a CRT, the path followed by the electron beam as it moves across the screen. Also called **trace** (2). 6. In fax, one horizontal scanning element. 7. One of multiple **force lines**.

line adapter A device, such as a modem, which modifies a signal so as to make it suitable for transmission over a given medium.

line amplifier An amplifier which boosts the signal of a line utilized for purposes such as TV broadcasting or telephone communication. Such an amplifier may be located at almost any point of such a line.

line analyzer A device which monitors a communications line to help make sure that it is functioning properly, and when this is not the case, to help remedy this.

line balance Also called **transmission line balance**. 1. The extent to which the electrical characteristics of two conductors in a transmission line are similar. Also, the degree of electrical similarity between a conductor and ground. The greater the balance, the lesser the extraneous disturbances, such as crosstalk and hum. 2. A device utilized to help achieve **line balance** (1). An example is a line-balance converter.

line-balance converter A device utilized to match the impedances or transmission characteristics of an unbalanced coaxial transmission line or system, and a balanced two-wire line or system, while minimizing attenuation. Used, for instance, for matching a coaxial cable to a twisted pair. Also called **balun, balanced-to-unbalanced transformer, balanced converter**, or **bazooka**.

line bonding Also called **channel bonding**. 1. The combining of two telephone lines to form a single communications channel in which the transmission speed is doubled. Two modems are required in order to use this technology. Two 56 kilobits per second modems bonded in this manner provide a 112 kilobits per second connection. Also called **modem bonding**. 2. The combining of two communications lines to form a single communications channel in which the transmission speed is doubled.

line circuit A circuit which serves as an interface between a line and a telephone switching system.

line code A code utilized for the transmission of data over a communications line. For example, a code which converts information from digital to analog form for transmission over a telephone line.

line communications Communications utilizing a physical path, such as a wire, cable, or waveguide. For example, communication utilizing a landline.

line concentrator A device which combines multiple communications channels. This may be accomplished, for instance, through multiplexing.

line conditioner 1. A circuit or device utilized for **line conditioning**. 2. A device which improves the quality of the power taken from an AC line. Such a device usually compensates for brownouts and surges, and may also provide noise filtering and a given time interval of backup power during a blackout.

line conditioning 1. In telephony, the modifying of twisted pair wire so it can carry a digital data signal. Also called **conditioning** (2). 2. The modification of a communications line so as to improve performance. Also called **conditioning** (3).

line cord A flexible cord containing two insulated wires, and which terminates in a two-pronged plug. It is utilized to connect an appliance or other electrical devices to a power source, especially an outlet. Such a cord may also have a third conductor, along with a third prong, for a safety connection to ground. Also called **power cord**.

line coupling A component, circuit, or device which serves to join, link, or transfer energy between lines.

line current The current flowing through a **line (2)** at any given moment, or under specific circumstances. For instance, the average current of a landline telephone which is off-hook is about 0.02 to 0.05 amps.

line driver A device which is intended to optimize the use of local data communications facilities, such as those within a building or with a maximum radius of a few kilometers. Such a device, for instance, can condition a digital signal transmitted by an RS232 interface so that it can be reliably transmitted up to several kilometers, instead of the standard 50 to several hundred feet. Also called **limited-distance modem**, or **short-haul modem**.

line drop Also called **line-voltage drop**. **1.** The voltage drop along a communications line or power line. **2.** The voltage drop along a communications line or power line, due to the impedance of said line.

line equalizer An equalizer incorporated into a transmission line. Used, for example, to counteract distortion, compensate for deficiencies, or shape a frequency response to fit the requirements of a given transmission medium. Also called **transmission line equalizer**.

line fault In a line, a defect such as an abrupt change in impedance, a short circuit, an open circuit, or an unintentional ground. Also, a failure caused by such a defect.

line feed A computer control character which advances a cursor or a print head to the same position on the next line. Also, a printer button or command that achieves this result. Also, such a movement. Also spelled **linefeed**.

line filter **1.** A filter inserted into a communications line, so as to remove undesired signal components, or to otherwise selectively transmit or reject signals in one or more intervals of frequencies. For example, a line filter must be installed at all jack points of telephone devices in the premises where an ADSL line is present, to prevent high-frequency components from causing interference when talking or exchanging faxes. The jack point connected to the ADSL modem must not have such a filter. **2.** A filter which prevents the passage of noise signals in a power line, thereby improving the operation of the devices connected to said line. It does so by attenuating radio-frequency noise, while permitting the 50 or 60 Hz current to pass. Also called **power-line filter**.

line finder Same as **linefinder**.

line flyback In a CRT, the return of the electron beam to its starting point after a horizontal sweep or trace. Also called **horizontal flyback**, or **horizontal retrace**.

line frequency **1.** In a CRT, the number of horizontal sweeps or traces per second. Also called **horizontal frequency (1)**. **2.** The number of horizontal sweeps per second in a given TV system such as NTSC or PAL. Also called **horizontal frequency (2)**. **3.** The AC frequency, such as 50 or 60 Hz, of a power line. Also called **power-line frequency**.

line hydrophone A directional hydrophone in the form of a straight line. Such devices are usually used in arrays numbering from a few to over two hundred.

line impedance The impedance a transmission line presents between its terminals. Also called **transmission line impedance**.

line impedance stabilization network An electric network that is inserted in series with an AC power supply, and which establishes a defined source impedance for the measurement of disturbances. Used, for instance, to monitor electromagnetic compatibility and interference for a given range of frequencies. Its abbreviation is **LISN**.

line in Abbreviation of **line input**.

line input In a device or piece of equipment, one or more connectors or terminals into which an input signal is delivered. For example, the jacks in an audio receiver which accept the output of a DVD. Also, the signal delivered. Its abbreviation is **line in**.

line interlace Same as **line interlaced**.

line interlaced A scanning system, such as that used in TVs and some computer monitors, in which the electron beam traces all odd-numbered lines followed by the tracing of all even-numbered lines. The pattern created by tracing the even-numbered lines is called the even field, while the pattern created by tracing the odd-numbered lines is called the odd field. A complete displayed frame consists of both of these interlaced fields. For a given refresh rate, image flicker is less apparent than that of a non-interlaced display. Also called **line interlace**, **interlaced scanning**, or **interlacing**.

line leakage **1.** Same as **line leakage current**. **2.** A flaw or fault in a line which allows for **line leakage current**.

line leakage current Any leakage current present in a communications or power line. For example, a current which flows through an insulator which separates conductors. Also called **line leakage (1)**.

line level The level of a signal at a given point, such as that of a transmission line. For instance, the output level of an audio component, expressed in decibels referred to 1 volt.

line load The load a communications or power line is carrying, as a fraction or percentage of its maximum capacity.

line loop Same as **loop (7)**.

line-loop resistance Same as **loop resistance (3)**.

line loss A loss, such as that of energy, in a communications, power, or antenna transmission line. Also, the magnitude of any such losses. Also called **line losses**.

line losses Same as **line loss**.

line-matching transformer A transformer which serves for matching impedances between sections of a transmission line, between balanced and unbalanced lines, between a line and a terminal device, or between devices.

line-matrix printer A printer which creates alphanumeric characters and images by striking pins against an ink ribbon, and whose matrix of pins is approximately the width of the pages being printed. Such printers are limited to low resolutions, but are fast and rugged.

line microphone A highly directional microphone in the form of a straight line. Such devices may be used alone, or as part of an array. Also called **shotgun microphone**.

line noise Noise generated in a transmission line, in addition to that present in the applied signal. Such noise may be due, for instance, to poor connections, or power-line current or voltage fluctuations. Also called **transmission line noise**.

line of flux One of multiple **lines of flux**.

line of force One of multiple **lines of force**.

line of position In navigation, a line of bearing to a reference or other known point, along which a vessel considered to be located. Such a line may be determined for instance, through the use of radar, loran, ultrasound, landmarks, GPS, and so on. More than one line of position may be utilized for a fix. Its abbreviation is **LOP**. Also called **position line**.

line of propagation The line along which electromagnetic radiation, such as light or microwave radiation, travels.

line of sight Its abbreviation is **LOS**. **1.** A direct and unobstructed path between a transmitting device and a receiving or reflecting device. For instance, such a path between a transmitting antenna and a receiving antenna, or between a radar signal and a scanned object. **2.** A straight line between an eye or lens and that which is observed.

line-of-sight communications Communications in which there is a direct and unobstructed path between a transmitting device and a receiving or reflecting device. Microwaves, for instance, are propagated in this manner, and repeaters can be utilized to extend their range.

line-of-sight distance The distance, free of obstructions, between a transmitting antenna and the horizon. Due to atmospheric refraction, a radio horizon extends beyond an apparent horizon.

line-of-sight path The path followed by a signal in **line-of-sight communications**.

line-of-sight propagation The propagation of radio waves in which there is a direct and unobstructed path between a transmitting device and a receiving device.

line-of-sight transmission The transmission of radio waves in which there is a direct and unobstructed path between a transmitting device and a receiving device. Line-of-sight transmission is required, for instance, when utilizing microwave or infrared signals to convey information.

line out Abbreviation of **line output**.

line output In a device or piece of equipment, one or more connectors or terminals from which an output signal is delivered. For example, the jacks in a DVD which provide a signal which serves as an input for an audio receiver or TV. Also, the signal delivered. Its abbreviation is **line out**.

line plug A two or three-pronged plug which terminates a **line cord**.

line-powered Telephone equipment which is powered solely by the telephone line. Examples include standard home telephones, and many caller-ID display devices.

line printer A printer which prints a single line at a time. An example is a line-matrix printer.

line radiation Electromagnetic radiation produced by power lines. Such radiation may cause interference and other deleterious effects.

line rate The speed at which bits or bytes are transmitted over a communications line or bus. Usually measured in some multiple of bits or bytes per second. Also called **line speed**, **transfer rate**, **data transfer rate**, or **bit transfer rate**.

line regulation **1.** The regulation of the output voltage of a power line, or of the output voltage or current of a power supply. **2.** The maximum steady-state variation in the output voltage or current of a power supply, for a given change in its input line voltage.

line regulator A device which provides **line regulation**.

line seizure In a property-alarm system, the cutting off of regular telephone service while the monitoring station is dialed. This usually occurs when an alarm condition is triggered, and service is restored once such a call is completed.

line spectrum In spectroscopy, discrete spectral lines which represent intervals of wavelengths which are absorbed or emitted by atoms. Molecules produce a **band spectrum**.

line speed Same as **line rate**.

line stretcher A waveguide section whose length can be adjusted so as to provide impedance matching.

line surge A sudden and momentary increase in the current or voltage carried by a line. May be caused, for instance, by lightning, or faults in circuits. If protective measures are not employed, it may cause a failure, or significant damage.

line switching Also known as **circuit switching**. **1.** The temporary linking of two communications terminals in order to establish a communications channel. This circuit must be established before communication can occur, and remains in exclusive use by these terminals until the connection is terminated. Used, for instance, by telephone companies to en-

able dial-up voice conversations. Such a connection is made at a switching center. **2.** The temporary linking of two data terminals, in a manner that the connection remains in exclusive use by these terminals until the connection is terminated.

line sync pulse The pulse which synchronizes the line-by-line scanning of a TV receiver with that of a TV transmitter. When synchronized, the relative position of the electron beam in the picture tube is the same as that of the scanning beam in the camera tube. In a TV system, such a pulse is transmitted at then end of each line, and triggers horizontal retracing and horizontal blanking. It is an abbreviation of **line synchronization pulse**, or **line synchronizing pulse**. Also called **line synchronization signal**, **line synchronizing signal**, or **horizontal sync pulse**.

line synchronization pulse Same as **line sync pulse**.

line synchronization signal Same as **line sync pulse**.

line synchronizing pulse Same as **line sync pulse**.

line synchronizing signal Same as **line sync pulse**.

line termination The point at which a **line (2)** terminates. Also, a device within which such a termination occurs. For example, a power outlet, or a channel service unit. Also called **termination**, or **transmission line termination**.

line transformer A transformer utilized to match transmission line impedances, or for line balance, isolation, or connection to equipment or additional circuits. Also called **transmission line transformer**.

line trap A bandstop filter utilized to minimize transmission line attenuation and losses. Used, for instance, in antenna, power, and telephone systems.

line triggering **1.** The use of power-line pulses as the triggering signals of an oscilloscope. **2.** The use of power-line pulses as triggering signals.

line turnaround In half-duplex communications, the reversal of the transmission direction. In this manner, the sender becomes the receiver, and vice-versa. Also, the time required for such a reversal.

line voltage Also called **power-line voltage**. **1.** The AC voltage supplied by an AC power line. This may vary from country to country, and consists of two nominal voltages. The lower number is primarily for lighting and small appliances, while the larger number is for heating and large appliances. In Canada and the United States, for example, the nominal voltages are approximately 115/230. Also called **AC line voltage**. **2.** The voltage supplied by a power line, especially at the point of utilization.

line-voltage drop Same as **line drop**.

line-voltage regulation The stabilization of a voltage, such as that delivered to a load, through the use of a **line-voltage regulator**.

line-voltage regulator A regulator which maintains an essentially constant output voltage, despite variations in its input voltage, such as that supplied by a power line.

linear **1.** Of, pertaining to, or in the shape of a line. For instance, a system whose components are arranged in a line. **2.** Describing, described by, or following the course of a line. For example, the movement of a robotic arm along the length of a line. **3.** A relationship in which one value is directly proportional to another. For instance, an electronic device whose output is directly proportional to its input. When such a response is graphed, the result is a straight line, which may or may not be sloped. Also, pertaining to such a relationship. **4.** Having only one dimension, or measured only in terms of length.

linear absorption coefficient For given material or medium, the absorption coefficient per unit length.

linear accelerator Its acronym is **linac**. **1.** An accelerator in which the particles travel in a straight line, as opposed to having a circular or spiraling path, as is the case with a **circular accelerator**. Currently, charged particles can be accelerated in this manner to energies in the TeV range. Used, for instance, to study the nature of matter. **2.** A **linear accelerator** (1) in which the accelerated particles are electrons.

linear address space An address space in which memory addresses start at zero, and work their way up to the maximum address. Also called **flat address space**.

linear amplifier An amplifier whose output is directly proportional to its input, over a given range of values. The performance of such an amplifier is linear, and when graphed, its response results in a straight line. Ideally, a linear amplifier does not alter the essential characteristics of its input signal, and a high-fidelity amplifier is an example.

linear antenna array Same as **linear array**.

linear array A directional antenna consisting of a series of elements arranged end to end, horizontally or vertically. The elements are usually of the same type. Also known as **linear antenna array**, or **collinear antenna array**.

linear beam A beam, such as an electron beam in a klystron, in which the particles travel in a straight line.

linear block code A type of forward error correction in which additional bits are added for error correction, as well as error detection.

linear bus A bus network whose nodes are arranged along a straight line. Also called **linear network (2)**.

linear charge density The electric charge per unit length of a line or curve. Usually expressed in coulombs per meter. Its symbol is λ. Also called **charge density (3)**.

linear circuit A circuit whose output is directly proportional to its input, over a given range of values. The performance of such a circuit is linear, and when graphed, its response results in a straight line.

linear component Same as **linear element**.

linear control 1. Same as **linearity control**. **2.** A potentiometer or rheostat whose variations in resistance are uniformly distributed along the entire resistance element.

linear control system A control system in which the output quantities are linearly proportional to its input.

linear detector A detector in which the output quantity or signal is linearly proportional to its input. For instance, a detector whose output voltage varies proportionally along with any changes in a quantity being monitored.

linear device A device in which the output is linearly proportional to the input, over a given range of values.

linear differential transformer A transducer which converts a mechanical motion into a proportional electrical signal or magnitude. For example, that whose output is proportional to the displacement of a core.

linear distortion Distortion in which the waveform of the input signal is altered, but in which there are no harmonics introduced. **Nonlinear distortion (2)**, also introduces harmonics.

linear electric motor Same as **linear motor**.

linear electron accelerator Same as **linear accelerator (2)**.

linear element A circuit element whose output is linearly proportional to its input, over a given range of values. Also called **linear component**.

linear equation An equation in which no variables are raised to a power. When such an equation is graphed, the result is a straight line.

linear expression A mathematical expression in which no variables are raised to a power.

linear feedback Feedback in which the output quantity or signal, such as a corrective action, is linearly proportional to its input.

linear feedback shift register A circuit or mechanism which produces a pseudorandom sequence of bits. Used, for instance, in encryption, error-detection, and CDMA systems. Its abbreviation is **LFSR**.

linear function A function in which no variables are raised to a power. When such a function is graphed, the result is a straight line.

linear IC Abbreviation of **linear integrated circuit**. An IC which is used for analog operations, such as amplification and oscillation. This contrasts with a **digital IC**, which is used for processing signals which have only two states.

linear induction motor An induction motor which provides linear motion, as opposed to rotary motion. It is a type of linear motor.

linear integrated circuit Same as **linear IC**.

linear meter 1. A meter whose indicated value is directly proportional to the quantity being measured. **2.** An analog meter in which the deflection of the pointer is directly proportional to the quantity being measured.

linear modulation Modulation in which the characteristic of a carrier wave being modified is proportional to the amplitude of the information-bearing signal.

linear motor An electric motor which provides linear motion, as opposed to rotary motion. Used, for instance, for propelling rail cars which are magnetically levitated. Also called **linear electric motor**.

linear network 1. An electric network whose output is directly proportional to its input, over a given range of values. The performance of such a network is linear, and when graphed, its response results in a straight line. **2.** Same as **linear bus**.

linear oscillator An oscillator whose output is directly proportional to its input. For instance, an oscillator whose AC output is proportional to its DC input.

linear phase response Variations in phase response in which the changes in phase are linearly proportional to the frequency.

linear polarization Polarization of an electromagnetic wave in which the electric field vector is always situated along a straight line.

linear potentiometer A potentiometer with **linear control**.

linear power amplifier A power amplifier whose output is directly proportional to its input, over a given range of values. The performance of such an amplifier is linear, and when graphed, its response results in a straight line.

Linear Prediction A method utilized for predicting the output waveform of a linear system, and which is often utilized as a speech-compression algorithm that provides high quality audio while utilizing very little bandwidth. Used, for example, for transmission of voice over packet-switching networks. Its abbreviation is **LP**. Also called **Linear Predictive**, **Linear Predictive Coding**, or **Linear Prediction Coding**.

Linear Prediction Coding Same as **Linear Prediction**. Its abbreviation is **LPC**.

Linear Predictive Same as **Linear Prediction**.

Linear Predictive Coding Same as **Linear Prediction**. Its abbreviation is **LPC**.

linear programming A mathematical technique which deals finding the maximum and minimum values of linear expressions incorporating multiple variables, each of which is subject to specified constraints. Used, for instance, to help determine the optimum allocation of available resources.

linear response A response, such as that of a component, circuit, or device, which when graphed results in a straight line. Such a response is considered over a given range of values, and an example is the frequency response of a linear amplifier.

linear scale A scale in which the distance between each indicated value corresponds to an equal magnitude, along the entire length of said scale. This contrasts, for instance, with a logarithmic scale. Also called **uniform scale**.

linear scan Also called **linear scanning**. **1.** In radars, a beam or scan which moves at a constant angular velocity. **2.** In a CRT, such as that of a radar display or TV, the movement of the trace with a constant linear or angular speed.

linear scanning Same as **linear scan**.

linear search A search in which the data in question is examined sequentially from start to finish. Also called **sequential search**.

linear sweep In a CRT, the movement of the electron beam across the screen at a constant speed. This contrasts, for instance, with an expanded sweep.

linear system **1.** A system whose output or response is directly proportional to its input, over a given range of values. The performance of such a system is linear, and when graphed, its response results in a straight line. **2.** A system whose output or response is linear for every part of said system. **3.** A system whose output is the sum of said system's response to each of its multiple inputs considered separately.

linear taper A taper, such as that of a potentiometer, in which the difference in resistance for a given rotation through a given angle is the same along the entire range of the shaft. This contrasts with a **nonlinear taper**, in which the resistance variations are not proportional to shaft rotation.

linear time base In a CRT, such as that utilized in radars, a time base which provides for a **linear sweep**.

linear track A track which is located at a videotapes edge, and which is utilized for mono audio or control purposes. Such tracks are read by a stationary head, as opposed to video and hi-fi audio tracks, which are read by rotary heads.

linear transducer A transducer whose output is directly proportional to its input, over a given range of values.

linear variable differential transformer A transformer which produces an AC output which is proportional to the displacement of the core relative to the windings. Its abbreviation is **LVDT**.

linear video editing The manipulation of analog video. When dealing with video in digital form, **nonlinear video editing** is possible, greatly simplifying processes such as rearranging the order of scenes, copying selections, skipping unwanted parts, and undoing previous actions.

linearity 1. The characteristic or condition of being **linear**. **2.** The extent to which **linearity** (1) exists. **3.** In TV, the extent to which the presented image is free of geometric distortion.

linearity control In a CRT, such as that of a TV, a control which adjusts the scanning speed of the trace interval, to correct geometric distortion. Also called **linear control** (1), or **distribution control**.

linearity error Any deviation from **linearity** (1). Also, the extent to which this deviation occurs.

linearly-graded junction A semiconductor junction in which the concentration of impurities varies linearly, as opposed to changing abruptly.

linearly-polarized light 1. A light wave in which the electric field vector is always situated along a straight line. **2.** A light wave with plane polarization. Also called **plane-polarized light**.

linearly polarized light A light wave with **linear polarization**.

linearly-polarized wave 1. An electromagnetic wave in which the electric field vector is always situated along a straight line. **2.** A wave with plane polarization. Also called **plane-polarized wave**.

linearly polarized wave A wave with **linear polarization**.

linefeed Same as **line feed**.

linefinder A switching device that automatically locates an idle line. An example is a device which locates an unused telephone circuit leading to the desired destination. Also spelled **line finder**.

lines of code A measure of the length of a computer program. The quality and functionality of a program is not necessarily correlated to this number.

lines of flux Same as **lines of force**.

lines of force Also called **lines of flux, force lines**, or **flux lines**. **1.** Within a field where a force acts, imaginary lines whose tangent at any given point represent the direction of said field at that point. The closer together the lines, the greater the strength of the field. May be used, for instance, in the context of electric, magnetic, or gravitational fields. **2.** Within an electric field, imaginary lines whose tangent at any given point represent the direction of said electric field at that point. Electric lines of force are utilized to represent the electric field surrounding a charged particle or body, and the closer together the lines, the greater the strength of the said field. Also called **electric lines of force, electric lines of flux, electric flux lines, electric force lines**, or **electrostatic flux**. **3.** Within a magnetic field, imaginary lines whose tangent at any given point represent the direction of said magnetic field at that point. Magnetic lines of force are utilized to represent the magnetic field surrounding a magnetic body or entity, such as a permanent magnet or a medium carrying a current. The closer together the lines, the greater the strength of the magnetic field. Also called **magnetic lines of force, magnetic lines of flux, magnetic flux lines**, or **magnetic force lines**.

lines of resolution In a video system, especially that utilizing TVs for display, the number of active horizontal lines per frame. For example, the NTSC standard provides for approximately 485 lines of resolution, the PAL system supports approximately 576, standard VHS offers about 250 lines per inch, S-VHS up to about 400, and HDTV has over 1,000 active lines.

lines per minute A measure of the speed of a line printer. Its abbreviation is **lpm**.

link 1. A unit within a series of units which are interconnected. For example, a component connecting moving parts of a machine. Also, to use such units for interconnecting. **2.** A communications line or channel connecting two points or entities, and via which information may be transmitted. Also, the creation of such a link. Also, the resources which facilitate such a link. For instance, a network component connecting two nodes. **3.** An element within an electronic document, such as that displayed on a computer monitor, which connects to other elements in different locations in the same or different documents. A link may be displayed as text, an icon, an image, or the like, and is usually accessed by clicking on it with a mouse or its equivalent. When viewing a Web page, for instance, such text, or hypertext, often appears underlined. Also, to create such an element. Also called **hyperlink**. **4.** In computer programming, an instruction or address which transfers control to a subroutine or another program. **5.** In data management, to interconnect data elements through the use of a pointer.

Link Access Procedure Balanced Same as **LAPB**.

Link Access Procedure-D channel Same as **LAPD**.

Link Access Protocol Balanced Same as **LAPB**.

Link Access Protocol-D channel Same as **LAPD**.

link budget analysis A tool utilized to evaluate the performance and reliability of a communications network, or to determine if a given link is feasible. It takes into effect factors such as power requirements, antenna gain, atmospheric conditions, and transmission path losses. Its abbreviation is **LBA**.

link circuit A circuit utilized for coupling purposes, such as a link coupling.

Link Control Protocol A protocol utilized to ascertain whether a communications link is acceptable. It does so by verifying factors such as packet size limits, authentication of peers, and proper encapsulation. Its abbreviation is **LCP**.

link coupling An inductive coupling utilized to link circuits. An example is a closed loop consisting of two coils, each with a few turns, utilized to inductively couple two circuits. Used, for instance, to connect test instruments to circuits without significant effects on the circuit being tested.

link fuse A fuse consisting of a length of bare fuse wire.

link layer address For a device connected to a communications network, an address which uniquely identifies each physical connection. An end device, for instance, has a single data-link layer address, while routers have multiple addresses. Also called **data-link layer address**.

link margin In link budget analysis, the additional margin provided to help insure reliable communications. It may consist, for instance, of additional power available to compensate for unexpected losses.

link rot The process by which links on a Web page become outdated. This is usually due to links changing location, or being removed.

link-state advertisement A broadcast packet in a communications network utilizing a link-state routing protocol. Its abbreviation is **LSA**. Also called **link-state packet**.

link-state packet Same as **link-state advertisement**. Its abbreviation is **LSP**.

link-state protocol Same as **link state routing protocol**.

link-state routing protocol A routing protocol, such as OSPF, which enables routers to exchange information, so as to choose the best path for data. The determination of the best path is based on factors such as network congestion, the number of hops, and the speed of the communications links. Also called **link state protocol**.

linkage 1. The act or process of creating a link. 2. The quality or state of being linked, and the manner in which such links are created. 3. The passage of magnetic flux from one circuit component to another. Also called **magnetic flux linkage**, or **flux linkage**.

linkage editor A program that provides links between object modules to form other executable programs. The writing of individual object modules makes programming simpler and more flexible. A linkage editor may have other functions, such as converting symbolic addresses into real addresses, or the creation of libraries. Also called **linker**.

linked list In computer data management, a collection of items arranged in a manner that each item contains the address of the next item in sequence. Also called **chained list**.

linker Same as **linkage editor**.

Linpack 1. A collection of FORTRAN subroutines used, for instance, for solving linear equations. 2. A benchmark test utilized to measure FLOPS.

LINUX An open-source version of UNIX which runs on various platforms.

lip microphone A type of close-talking microphone intended to be used near, or touching, the lips of a person.

Lippmann electrometer An electrometer which utilizes a capillary tube containing dilute sulfuric acid and mercury. The displacement within the tube of the interface between the sulfuric acid and the mercury is proportional to the current being measured. Also called **capillary electrometer**.

LIPS Acronym for **logical inferences per second**.

liquid A state of matter which is approximately intermediate between a gas and a solid. Liquids have a definite volume, like a solid, yet have the tendency to assume the shape of its container, like a gas. A liquid will only occupy the proportion of its container which equals the volume of the liquid, and will seek the lowest level. That is, it will not expand to occupy all the available space, as a gas would. The atoms or molecules composing a liquid have enough thermal energy to move freely, but also have significant intermolecular attraction. Liquids are very hard to compress, and when cooled sufficiently are converted into a solids, while when heated sufficiently into gases. The other physical states in which matter is known to exist are **solid** (1), **gas**, **plasma**, and **Bose-Einstein condensates**.

liquid cooling The use of a liquid, such as water or oil, to cool sealed components, circuits, devices, equipment, or systems.

liquid crystal A material whose properties are intermediate between those of a crystal and a liquid. The constituent particles tend to arrange themselves with much greater order than an ordinary liquid, but not enough to become solid crystals. When subjected to variations in a surrounding electric field, its optical properties change. For example, a small applied electrical signal may darken such a crystal, enabling it to be differentiated from a light-colored background, making them suitable for displays. Certain liquid crystals vary, depending on the temperature, the color of the light they reflect, and can be used as thermometers. Its abbreviation is **LC**.

liquid-crystal display Same as **LCD**.

liquid-crystal display monitor Same as **LCD monitor**.

liquid-crystal display panel Same as **LCD panel**.

liquid-crystal display printer Same as **LCD printer**.

liquid-crystal display projector Same as **LCD projector**.

liquid-crystal display screen Same as **LCD screen**.

liquid-encapsulated Czochralski A technique used for cultivating large single crystals. Using this method, a seed crystal is dipped into a molten solution of the crystal, and is slowly withdrawn while being rotated. Used, for instance, to grow semiconductor crystals such as gallium arsenide. Its abbreviation is **LEC**. Also called **Czochralski method**.

liquid-flow alarm An alarm which is activated when a connected circuit or device detects a given liquid flow, such as that which exceeds a specified rate.

liquid-flow indicator An instrument which indicates the flow rate of a liquid.

liquid-flow meter An instrument which measures and indicates the flow rate of a liquid. Such a meter may be ultrasonic, electronic, and so on.

liquid laser A laser whose active medium is a liquid, and which usually consists of molecules or ions in solution. An example is a dye laser.

liquid-level alarm An alarm which is activated when a connected circuit or device detects a given level of a liquid.

liquid-level gauge An instrument which measures the level of a liquid. Such a gauge may be ultrasonic, electronic, and so on.

liquid-level indicator An instrument which indicates the level of a liquid.

liquid-level meter An instrument which measures and indicates the level of a liquid. Such a meter may be ultrasonic, electronic, and so on.

liquid-phase epitaxy A technique for growing epitaxial layers, in which the material to be deposited upon the substrate crystal is in molten form. Used, for instance, in the preparation of LEDs. Its abbreviation is **LPE**.

liquid rheostat A rheostat whose resistance is varied by raising, lowering, or otherwise moving electrodes within a conductive liquid. When the liquid is an aqueous solution, also called **water rheostat**.

liquidus temperature The temperature above which a substance is completely liquid, while the **solidus temperature**

is that below which a substance is completely solid. Used especially in the context of alloys which do not have a single melting point.

LISN Abbreviation of **line impedance stabilization network**.

LISP Abbreviation of **list p**rocessing. A high-level programming language which is frequently used in academic and research settings, and for artificial intelligence applications.

Lissajous figure A symmetrical pattern which is formed when two harmonically-related quantities are superimposed at right angles to each other. A common instance is the pattern which appears on the screen of an oscilloscope when signals are applied simultaneously to the horizontal and vertical deflection plates. For example, if the horizontal signal has twice the frequency of the vertical signal, an upright figure eight will be displayed. Each frequency combination produces its own unique pattern, and such figures are utilized, for instance, for testing and adjustments. Also called **Lissajous pattern**.

Lissajous pattern Same as **Lissajous figure**.

list **1.** A series or number of items, objects, words, or the like, which appear together. Also, to make such a list. Also, to enter into such a list. For example, a print queue, or the adding of a file to a print queue. **2.** A data structure with multiple elements organized in a given order. The items in a list may also be lists themselves.

list box In a GUI, an on-screen display which provides a user various choices which are enclosed within a box. When a desired selection may also be typed in, also called **combo box**.

list processing The manipulation of data in the form of **lists**. This includes accessing, processing, adding, and deleting such data.

list-processing language A computer programming language especially suited for the manipulation of data in the form of **lists**. An example is LISP.

listening angle In the reproduction of stereo audio, the angle that is formed between the speakers and a listener. For example, if both speakers are side by side and separated by the same distance as that to the location of a listener, the listening angle is 60°.

listing A printout of text. Said especially when referring to a printed copy of source code.

lit fiber Fiber-optic cable that is installed, has all electronics such as transmitters and regenerators connected to it, and which carries communications traffic. This contrasts with **dark fiber**, which has no electronics connected to it, and can not handle traffic.

lite version Same as **light version**.

liter A metric unit of volume equal to 0.001 cubic meter, which is same as the volume of a cube whose sides each measure 0.1 meter. Its symbol is l or L.

literal In computer programming, a value which remains unchanged during program execution. The value of a **constant** is also fixed, but a constant is a descriptive name assigned to a value, while a literal is an explicit appearance of a value, such as the number 86. A literal may be a number, character, or string.

lithium A very soft, silvery-white metallic chemical element whose atomic number is 3. It is the least dense of the metals, the least reactive of the alkali metals, the solid element with the highest specific heat, and it has about 8 known isotopes, of which 2 are stable. Its applications include its use in batteries, as a getter and scavenger, in high-strength alloys with low density, and in mediums which transfer heat. Its chemical symbol is Li.

lithium battery A battery composed of **lithium cells**.

lithium carbonate A white powder whose chemical formula is Li_2CO_3. It is used in electrodes and in ceramics.

lithium cell A cell in which the anode is lithium metal, a lithium ion, or a material incorporating lithium. The electrolyte may be any of various substances, such as thionyl chloride or iodine, and said electrolyte may be in a liquid or solid state. Such cells may be primary or secondary, the latter usually being lithium-ion cells. Lithium cells feature high energy density, high capacity, long shelf life, a flat discharge characteristic, and a wide operational temperature. They are used, for instance, in portable computers, computer standby clocks, cameras, cell phones, PDAs, pacemakers, medical instruments, and emergency lighting, among many others.

lithium chloride White crystals whose chemical formula is $LiCl$. It is used in batteries and in solders.

lithium fluoride White crystals or powder whose chemical formula is LiF. It is used in solders, ceramics, mediums which transfer heat, and in spectroscopic instruments.

lithium hydride White crystals or powder whose chemical formula is LiH. It is used in batteries, and in organometallic vapor-phase epitaxy.

lithium hydroxide Colorless crystals whose chemical formula is $LiOH$. It is used in batteries and in photography.

lithium-ion battery A battery composed of **lithium-ion cells**. Its abbreviation is **Li-ion battery**.

lithium-ion cell A secondary **lithium cell** which has a lithium-ion releasing anode, a lithium-ion accepting cathode, and an electrolyte or separator-electrolyte. During charge and discharge cycles lithium ions are shuttled between the anode and the cathode. Its abbreviation is **Li-ion cell**.

lithium manganese dioxide A chemical compound whose structure is $LiMnO_2$. Used in lithium-ion cells.

lithium metal battery A battery composed of **lithium metal cells**.

lithium metal cell A **lithium cell** which has a lithium metal anode.

lithium niobate Crystals whose chemical formula is $LiNbO_3$. Used in lasers, semiconductors, and in infrared detectors.

lithium oxide A white powder whose chemical formula is Li_2O. Used in ceramics and special glasses.

lithium polymer A rechargeable lithium battery technology in which the anode is separated from the cathode by a polymer impregnated with the electrolyte. This enables such a battery to be in virtually any desired shape.

lithium polymer battery A battery incorporating **lithium polymer** technology.

lithium tantalate Crystals whose chemical formula is $LiTaO_3$. Used in communications and optics.

lithographic Pertaining to, or utilizing **lithography**.

lithography A technique utilized to transfer patterns or images from one material or medium to another. In it, an exposing source, such as light or an electron beam, irradiates the resist, with a mask determining the desired patterns. Common techniques include photolithography, electron-beam lithography, ion-beam lithography, and X-ray lithography. Used, for instance, in the manufacturing of chips and thin-film components.

little endian The ordering of a sequence of bytes placing the least significant byte first. This contrasts with **big-endian**, where the most significant byte is placed first.

Litz wire Abbreviation of **Litz**endraht **wire**. A conductor made from many separately-insulated wire strands which are woven together or braided in a manner that each strand regularly assumes each of the possible positions within the overall cross-section of the conductor. This arrangement minimizes skin effect, and provides for low losses at radio frequencies.

Litzendraht wire Same as **Litz wire**.

live 1. Anything connected electrically to a source of voltage. Also known as **energized** (1), **alive**, or **hot** (1). 2. A broadcast occurring at the actual time events take place. Also called **live broadcast**, or **real-time broadcast**. 3. In acoustics, a room that is reverberant.

live broadcast Same as **live** (2).

live chassis A chassis, such as that of a TV or computer, which is connected electrically to a source of voltage.

live circuit A circuit that is connected electrically to a source of voltage. Also known as **energized circuit**, or **alive circuit**.

live conductor 1. A conductor which is connected electrically to a source of voltage. 2. A conductor through which an electric current is flowing.

live end In an enclosure, such as a sound studio, a wall or area where there is the greatest amount of sound reflection. This contrasts with a **dead end** (2), where sound absorption is maximized.

live room A room in which there is significant reverberation, as opposed to a **dead room**, where sound-absorbing materials are utilized to minimize or eliminate all sound reflections.

live wire 1. A wire which is connected electrically to a source of voltage. Also called **hot wire** (1). 2. A wire through which an electric current is flowing.

LiveMotion Popular software utilized to create dynamic and interactive content, especially for the Web.

LLC Abbreviation of **Logical Link Control**.

LLC layer Abbreviation of **Logical Link Control Layer**.

LLC Sublayer Abbreviation of **Logical Link Control Sublayer**.

lm Symbol for **lumen**.

lm-h Abbreviation of **lumen-hour**.

lm-hr Abbreviation of **lumen-hour**.

lm/m² Abbreviation of **lumens per square meter**.

lm-s Abbreviation of **lumen-second**.

lm-sec Abbreviation of **lumen-second**.

lm/W Abbreviation of **lumens per watt**.

LMDS Abbreviation of **local multipoint distribution service**.

LMI Abbreviation of **Local Management Interface**.

ln Symbol for **natural logarithm**

LNA Abbreviation of **low-noise amplifier**.

LNB Abbreviation of **low-noise block down-converter**.

LNP Abbreviation of **local number portability**.

LO Abbreviation of **local oscillator**.

lo Abbreviation of **low**.

lo-res Abbreviation of **low-resolution**.

load 1. The power consumed by a component, device, piece of equipment, machine, or system while performing its functions. This power may be electrical, mechanical, nuclear, wind, and so on. Also, any component, device, piece of equipment, machine, or system consuming this power. Also called **output load** (1). 2. Any component, circuit, device, piece of equipment, or system which consumes, dissipates, radiates, or otherwise utilizes power, especially electricity. There are countless examples, including resistors, amplifiers, TVs, speakers, antennas, lamps, and appliances. Also, the power so consumed. Also called **output load** (2). 3. The electrical power drawn from a source of electricity, such as a generator or power line. Also called **output load** (3). 4. A circuit or device which receives the useful signal output

from a signal source such as an amplifier or oscillator. Also called **output load** (4). 5. To utilize inductors and/or capacitors to increase the electrical length of an antenna, or otherwise alter its characteristics. 6. In dielectric and induction heating, the object or material being heated. 7. To insert a computer storage medium, such as a disc or tape, into a drive or other device utilized to read and/or write to it. 8. To transfer data to or from a computer storage medium, such as a disk or tape. For example, to load a program into memory for execution. Also, to transfer data to or from a database. 9. To place data in a computer register. 10. To insert a disk, cassette, reel, cartridge, drum, or other object composed of, or containing a recordable medium into a device utilized for recording and/or reproduction. 11. In a communications network, the amount of traffic at a given moment.

load-and-go The capability of a program or routine to be executed immediately after loading. Its abbreviation is **load & go**.

load balance Also called **load division**. 1. The even distribution of a load. For example, to equally distribute a work load among a set of parallel processors. 2. To evenly distribute a load among multiple power sources.

load capacitance The capacitance presented by a given load, or as seen from a given point such as the input terminals.

load capacity The maximum load that can be handled safely, without failure, or within a given level or performance. For instance, the maximum number of messages that can be simultaneously exchanged over a given transmission medium, or the greatest weight a robot can manipulate.

load cell A piezoelectric crystal utilized as a strain gauge. The greater the force applied, the greater the potential difference across the crystal.

load characteristic The relationship between the instantaneous values of two varying quantities, such as electrode current and electrode voltage, under specified load circumstances. Also called **dynamic characteristic**.

load circuit The complete circuit via which a load is connected to its power source.

load coil 1. In an induction heater, an AC carrying coil which induces RF currents in the workpiece being heated. Also called **work coil**. 2. Same as **loading coil**.

load current 1. The current drawn by a load. It is the current flowing through a load circuit. 2. The current required by a load for operation.

load division Same as **load balance**.

load factor 1. For a given time interval, the ratio of the average load to the peak load. 2. The ratio of a given load to the maximum load.

load impedance The impedance presented by a given load, or as seen from a given point such as the input terminals.

load leads The conductors which connect a power source to a load.

load line A line which is drawn through a series of characteristic curves to illustrate the effect a given load will have on the relationship between two variables. For instance, such a line drawn through a series of transistor characteristic curves, to show the relationship between the input current and voltage as the load resistance varies.

load management The reduction in overall demand for available energy resources by transferring power needs to off-peak periods. This may entail consumers adjusting usage patterns, or load shifting.

load matching The varying of the impedance of a load circuit, so as to match that of a source. This helps insure that the maximum possible power is transferred from source to load.

load-matching network An electric network utilized for **load matching**.

load module A program or routine which is suitable for loading and execution. A linkage editor is usually utilized to obtain a load module.

load power The power consumed or dissipated by a load.

load regulation **1.** The change in output voltage or current when load conditions are varied. It may be expressed, for instance, as a percentage of the output voltage or current. **2.** The maximum change in output voltage or current when load conditions are varied. It may be expressed, for instance, as a percentage of the output voltage or current. **3.** Same as **load stabilization**.

load regulator Same as **load stabilizer**.

load resistance The resistance presented by a given load, or as seen from a given point such as the input terminals.

load sharing The sharing of processing, or other tasks, among multiple computers.

load shedding In an electric power system, the turning off or disconnection of loads, such as those of lesser priority, to prevent the entire network from failing.

load shifting In an electric power system, the transferring of loads from peak periods to off-peak periods. Usually used as a load management technique.

load stabilization The maintaining of a load voltage or current at a given level, or within a specified interval. Also called **load regulation (3)**.

load stabilizer A circuit or device which provides **load stabilization**. Also called **load regulator**.

load transfer switch In an electric power system, a switch utilized to select a given load circuit out of two or more that are available.

load voltage The voltage across a given load.

load & go Same as **load-and-go**.

loaded antenna An antenna whose characteristics, such as electrical length or resonant frequency, have been altered through the use of inductive and/or capacitive elements, such as loading coils or loading disks.

loaded line A transmission line whose characteristics have been altered through the use of inductive and/or capacitive elements, such as loading coils. Utilized, for instance, for impedance matching. Also called **loaded transmission line**.

loaded Q The value of the Q factor for a circuit or device with an external load. Also called **working Q**.

loaded transmission line Same as **loaded line**.

loader A utility which copies a computer program from a storage device into memory, for execution. Such a utility is usually part of the operating system, can load executable code from an internal or external storage device, and may be a routine or a program.

loader program Same as **loading program**.

loader routine Same as **loading routine**.

loading **1.** The alteration of characteristics of an antenna, such as increasing its electrical length, through the utilization of inductive and/or capacitive elements. Also called **antenna loading**. **2.** The alteration of the characteristics of a transmission line through the use of inductive and/or capacitive elements. Utilized, for instance, to match impedances. **3.** The alteration of the acoustic characteristics of a speaker through the utilization of materials, baffles, or vents. Also called **acoustic loading**. **4.** The insertion of a disk, cassette, reel, cartridge, drum, or other object composed of, or containing a recordable medium into a device utilized for recording and/or reproduction. **5.** The transferring of data to or from a computer storage medium, or the placement of data into a computer register.

loading coil Also called **load coil (2)**. A coil which is inserted into a transmission line or circuit to alter its character-istics. For example, a coil inserted in series with an antenna to increase its electrical length. When loading coils are inserted in series with a telephone line to reduce the effects of line capacitance, each is called a **Pupin coil**.

loading disk A conductive disk, usually metallic, which is placed atop a vertical antenna to increase its electrical length.

loading program A program which serves as a **loader**. Also called **loader program**.

loading routine A routine which serves as a **loader**. Also called **loader routine**.

loadstone Same as **lodestone**.

lobe Also called **directional lobe**. **1.** Within an antenna pattern, a three-dimensional section in which radiation and/or reception is increased. Also called **antenna lobe**, and when referring specifically to radiation also known as **radiation lobe**. **2.** Within a directional pattern of a transducer, a three-dimensional section in which radiation and/or reception is increased. For instance, such a lobe in the radiation pattern of a speaker.

lobe switching In radio-direction finding, mechanically or electrically shifting the orientation of an antenna to determine the direction which provides the greatest accuracy in the determination of the bearing of an object. Also called **beam-lobe switching**, or **beam switching**.

lobing **1.** A pattern of radiation and/or reception maxima and minima which results from reflections off surfaces surrounding an antenna. Some reflections strengthen the main beam, while others weaken it. **2.** In a speaker with two or more drivers whose reproduced frequency intervals overlap, the resulting maxima and minima in a radiation pattern.

local **1.** In a communications network, the resources, such as data and devices, at a given node where a user is present, as opposed to those located elsewhere, which are **remote**. **2.** Anything in a given place, area, enclosure, or environment, as opposed to that outside. For example, local noise, a local call, or a LAN.

Local-Access Transport Area Same as **LATA**.

local action In a cell or battery, chemical reactions occurring which result in the battery slowly discharging even when not connected to an external circuit. This is usually caused by impurities which create potential differences between different portions of an electrode. This produces self-discharge, a process which over months or years may fully deplete an unused cell or battery.

local-area data service A digital communications service, such as DSL, which utilizes a dedicated dry copper pair. Its acronym is **LADS**.

local-area network Same as **LAN**.

local-area wireless network A LAN whose nodes communicate via radio-frequency waves, such as microwaves, or infrared waves. Useful, for instance, in settings where multiple nodes, such as computers, are constantly in motion. Its acronym is **LAWN**. Also called **wireless LAN**.

local battery In wire telephony, a battery which is located at the premises of a customer, as opposed, for instance, to being located at a central office.

local bridge In communications, a bridge which interconnects networks within the same geographical area.

local bus The pathway between the CPU, computer memory, and high-speed peripherals.

local bypass A telephone connection made within the same local calling area, but which does not utilize a local telephone company. Used, for instance, by businesses linking adjacent buildings telephonically.

local call Any call in which no additional charges to the calling or called party are incurred. For example, a call whose origination and destination locations are both served by the same local central office. Also called **local telephone call**.

local calling area The geographical area within which **local calls** are made. Also called **local service area**.

local central office A structure which houses one or more telephone switching systems. At this location, customer lines terminate and are interconnected with each other, in addition to being connected to trunks, which may also terminate there. A typical central office handles about 10,000 subscribers, each with the same area code plus first three digits of the 10 digit telephone numbers. Also called **local exchange, central office, exchange** (1), **telephone central office**, or **telephone exchange**.

local channel 1. The channel between the premises of a customer and a local central office. 2. A broadcast channel which serves a limited area. Such channels are usually limited to a given power level, such as 250 or 1000 watts. Also called **local station**.

local control Control of a circuit, device, piece of equipment, mechanism, or system, at the site where it is located, as opposed to remote control.

local drive In a communications network, a disk drive located at a given user's node, as opposed to a **remote drive**, which is that of another computer, or a **network drive**, which is accessed by multiple nodes.

local exchange Same as **local central office**.

local exchange carrier A communications entity which offers local telephone service, and which may also offer other services, such as long-distance calling, Internet access, and so on. These are usually either competitive local exchange carriers, or incumbent local exchange carriers. Its abbreviation is **LEC**. Also called **local telephone company**, or **local telephone provider**.

local feedback In a multistage amplifier, feedback which is applied individually to each stage. Also called **multiple-loop feedback** (2).

local line Same as **loop** (7).

local loop Same as **loop** (7).

Local Management Interface A series of enhancements to the frame relay protocol. These include global addressing, keep-alive messaging, and continuous status reporting. Its abbreviation is **LMI**.

local memory The memory of a specific CPU, peripheral, program, or function. For example, when using a soft font, said font may be copied from a hard disk to the local memory of a printer.

local multipoint distribution Same as **local multipoint distribution service**.

local multipoint distribution service A wireless broadband service which is capable of providing telephone services, high-speed Internet, and interactive multimedia applications, with data-transfer rates than can exceed 2 Gbps. It operates in the 28 GHz band, offers line-of-sight coverage to about 5 kilometers, and may be used, for instance, where broadband service would otherwise be unavailable for physical or economical reasons. Repeater stations can be used to extend the range, and overlapping cells can help minimize losses due to obstacles. Its abbreviation is **LMDS**. Also called **local multipoint distribution**.

local noise The noise present in any given environment, especially that within which tests or observations are taking place. Such noise includes nearby and distant sources. Also called **background noise** (1), **room noise, site noise, ambient noise**, or **environmental noise**.

local number portability The process via which a subscriber can retain the same telephone number, regardless of the local service provider chosen. Also, the movement of said number between locations. Its abbreviation is **LNP**.

local oscillator Its abbreviation is **LO**. 1. In a superheterodyne receiver, an oscillator whose output is combined with the incoming modulated RF signal, to produce an intermediate frequency. 2. Any oscillator incorporated into a device or piece of equipment, such as a transmitter or receiver.

local program 1. A broadcast by the station which originates it, as opposed to a regional or national program. 2. A program broadcast by a **local channel** (2).

local reception 1. The reception of a **local program**. 2. Reception at the **local side** (1).

local resource In a communications network, a resource, such as a disk drive, located at a given user's node. This contrasts with a **remote resource**, which is that of another computer.

local service area Same as **local calling area**.

local service provider A telephone company which provides an end user with the ability to place and receive **local calls**. Its abbreviation is **LSP**.

local side 1. In communications, the side closer to a given terminal or node. For example, if a terminal is transmitting, the local side is the source, while the destination terminal is the **remote side**. 2. In communications, the portion of a device or piece of equipment which is connected to internal facilities, such as those within a given station.

local station Same as **local channel** (2).

local tandem Same as **local tandem switch**.

local tandem switch A local central office switch that connects calls to and from different central offices. Also called **local tandem**.

local telephone call Same as **local call**.

local telephone company Same as **local exchange carrier**.

local telephone provider Same as **local exchange carrier**.

local traffic Communications traffic which originates and terminates within the same **local calling area**.

local transmission 1. The transmission of a **local program**. 2. Transmission from the **local side** (1).

local trunk A communications channel between **local central offices**.

local variable A variable which only applies to a given computer program, routine, or process.

localhost The computer a given user is currently working on. Localhost enables a user to connect to his or her own system from the same system, and is defined as IP 127.0.0.1 regardless the IP number of a given computer.

localization The process of customizing software for specific regions, countries, or cultures. Localization includes support for local character sets, and translations of all materials to be displayed, such as help menus, messages, commands, and so on. In addition, other user interface factors, such as the placement of icons, or even whether icons are appropriate, should be taken into account.

localizer 1. A beacon which emits radio signals which provide lateral guidance to an aircraft approaching for a landing. It helps the aircraft align itself along the center of a runway, and is part of an instrument landing system. 2. Same as **localizer beam**.

localizer beam The signals emitted by a **localizer** (1). Also called **localizer** (2).

location 1. Any place where data may be stored. 2. A number which indicates the specific place where data is stored within the memory of a computer, peripheral, or disk. Also called **address** (1).

location counter Same as **location register**.

location register In a CPU, a register that contains the address of the location in memory that is to be accessed by the next instruction. May also refer to the address of the current instruction. Also called by various other names, including location counter, control counter, program counter, program register, instruction register, and **instruction counter**.

locator beacon 1. In navigation, a beacon providing the guiding or orienting signal for homing. Also called **homing beacon**. 2. A device which sends a radio-frequency beacon signal in case of an emergency. Such a device is usually portable and battery-operated, and may function for up to several days to allow others to home-in after being alerted. Also called **emergency locator beacon**.

lock 1. To secure, hold, stabilize, or set. 2. To deny access, prevent operation, or preclude any changes or deletion. Also, a setting, device, or feature which puts such restrictions into effect, or which helps implement them.

lock code A code which prevents the use of a device, such as a cellular phone. Such a code may also be utilized to restrict functions, such as the ability to place certain calls, or to reprogram selected features.

lock-in A state in which two oscillating systems are synchronized at the same frequency. Also, the process by which this synchronicity is achieved. For example, the shifts of one or both systems, followed by their being locked in place.

lock-in amplifier A circuit or device utilized to detect and amplify weak input signals whose frequency is matched by an internally-generated frequency, such as that of an internal oscillator. Used, for instance, when signal-to-noise ratios are extremely low. Also called **lock-in detector**.

lock-in detector Same as **lock-in amplifier**.

lock-in range Also called **lock range**. 1. The frequency range within which a closed-loop feedback system can lock onto a signal. 2. The frequency range within which an oscillator can be synchronized by an external signal.

lock-nut Same as **locknut**.

lock-on 1. In radars, the instant the tracking of a scanned object begins. Also, to seek this state. 2. The instant at which a previously varying frequency is fixed. For instance, that required for controlled oscillation. Also, to seek this frequency. 3. The instant at which a previously varying quantity or magnitude is fixed. Also, to seek this state.

lock-out Same as **lockout**.

lock-out circuit Same as **lockout circuit**.

lock range Same as **lock-in range**.

lock-up Same as **lockup**.

lock-up relay A latching relay which electrically or magnetically maintains its actuated position after the triggering current is removed. Also spelled **lockup relay**.

locked oscillator An oscillator whose frequency is synchronized with that of an external signal. Used, for instance, as a frequency divider.

locking circuit A circuit which maintains enough current in a holding coil to keep a relay in its actuated position after the triggering current is removed. Also called **holding circuit**.

locking relay Same as **latching relay**.

locknut Also spelled **lock-nut**. 1. A nut which is utilized to help secure something else, especially another nut. 2. A nut which is self-locking.

lockout Also spelled **lock-out**. 1. The prevention of access, or the use of a given function or resource, to one or more circuits, devices, or processes. Used especially in the context of one circuit, device, or process excluding all others to prevent conflicts or other unwanted conditions. 2. In a telephone system, the prevention of one or more senders or receivers from communicating. This may occur, for instance, as a result of excessive line noise, or to prevent interference. 3. In a telephone system, the disconnection of a line which is not functioning properly. 4. In computers, the allowing of only one device or process to have access to a given re-

source. For example, the prohibiting of reading data which is being updated at that moment. 5. The preventing of a user or entity to gain access to a computer system or network. A lockout may occur, for instance, when an incorrect password is entered more than the allowable number of times.

lockout circuit A circuit which performs **lockout** functions. Also spelled **lock-out circuit**.

lockup A state which a device, piece of equipment, or system does not respond to user input. A hanging computer is an example. Also spelled **lock-up**.

lockup relay Same as **lock-up relay**.

loctal base Same as **loktal base**.

loctal socket Same as **loktal socket**.

loctal tube Same as **loktal tube**.

lodestone A variety of the mineral **magnetite** which exhibits polarity and is a natural magnet. Magnetite and lodestone are often used synonymously. Also spelled **loadstone**.

LOFAR Abbreviation of **Low-Frequency Acquisition and Ranging**. A submarine detection and ranging system.

log 1. Abbreviation of **logarithm**. 2. A record kept of the activity or performance of a device, piece of equipment, or system. Also, to record such activity or performance. 3. A record of computer and/or network activity. Used, for example, to find the origin of problems, monitor usage, recover data, or to identify unauthorized access. Also called **electronic log**, **electronic journal**, or **journal**.

log-in Same as **log-on**. Also spelled **login**.

log-off Same as **log-out**. Also spelled **logoff**.

log-on To initiate a session. Also, the process of initiating a session. This may require, for instance, entering a username and the corresponding password. Also spelled **logon**. Also called **log-in**, or **sign-on**.

log-out To end a session. Also, the process of terminating a session. One or more commands may be required for proper termination. Also spelled **logout**. Also called **log-off**, or **sign-out**.

log-periodic antenna A unidirectional broadband antenna in which the length and spacing of the elements increase logarithmically from one end to the other. The input impedance of such an antenna is essentially constant over a wide range of frequencies. When a log-periodic antenna consists of a series of dipoles arranged along a transmission line, it is also called **log-periodic dipole array**.

log-periodic dipole array A **log-periodic antenna** consisting of a series of dipoles arranged along a transmission line.

log taper Abbreviation of **logarithmic taper**. A taper, such as that of a potentiometer, in which the difference in resistance for a given rotation through a given angle varies logarithmically along the entire range of the shaft. An example is that of a volume control. It is a type of nonlinear taper, and contrasts with a **linear taper**.

logamp Abbreviation of **logarithmic amplifier**.

logarithm The power to which a base, such as e or 10, must be raised, to produce a given number. For example, $\log_e y = n$, where e is the base, y is the power, and n is the resulting number. A logarithm has two portions, the characteristic, which is an integer, and the mantissa, which is a decimal. Its abbreviation is **log**.

logarithmic amplifier An amplifier whose output signal magnitude is a logarithmic function of its input signal magnitude. Used, for instance, for signal compression. Its abbreviation is **logamp**.

logarithmic decrement For an underdamped vibrational system, the natural logarithm of the ratio of the amplitude of a given vibration to that of the next vibration. For example, that of the ratio of the amplitude of a damped oscillation to the succeeding one. Also called **damping factor** (2), or **decrement** (3).

logarithmic growth An increase in a quantity, such as radiation or charge, whose rate of change follows an exponential function. Also called **exponential growth**.

logarithmic response A response, such as that of a component, circuit, or device, which is proportional to the logarithm of an input signal or other stimulus. The human response to loudness is logarithmic, so as to accommodate the wide range of sound pressure levels that can be detected.

logarithmic scale A scale in which the graduations are spaced logarithmically, so that distance between each indicated value corresponds to an equal ratio of increase, along the entire length of said scale. A dB meter, for instance, utilizes a logarithmic scale to indicate values.

logarithmic taper Same as **log taper**.

log$_e$ Symbol for **natural logarithm**.

logic Also called **computer logic**. **1.** The functions performed by a computer which involve operations such as mathematical computations and true/false comparisons. Such logic is usually based on Boolean algebra. **2.** The circuits in a computer which enable the performance of logic functions or operations, such as AND, OR, and NOT. These include gates and flip-flops. Also, the manner in which these circuits are arranged. Also called **machine logic (2)**. **3.** The totality of the circuitry contained in a computer. Also called **machine logic (3)**.

logic 0 Same as **logic low**.

logic 0 input Same as **logic low input**.

logic 1 Same as **logic high**.

logic 1 input Same as **logic high input**.

logic add Same as **logical sum**.

logic analysis The use of a **logic analyzer** for debugging programs and evaluating system performance.

logic analyzer An instrument which performs various functions which assist in the maintenance, testing, and troubleshooting of logic circuits. A logic analyzer, for instance, may monitor hundreds of signals simultaneously to help debug computer hardware or software, and incorporates a display for viewing the various signals being examined.

logic array An IC consisting of an array of logic gates. The specific arrangement of the gates, and hence their function, is determined late in the manufacturing process. This customization for a specific application can save design and manufacturing time, but a significant proportion of the chip may go unused. Also called **gate array**.

logic board A circuit board, such as a motherboard, incorporating logic circuits.

logic bomb A logic error or program routine which causes a complete and unexpected program or system halt. A logic bomb is usually triggered by a specific event, and can result in significant or complete losses of data.

logic chip A chip incorporating logic circuits, as opposed, for instance to memory chips which only store data.

logic circuit A circuit which carries out a logic function or operation. Such a circuit has multiple inputs and one output. Its output will depend on specified input conditions, such as a given combination of states. The three basic logic circuits are the AND, OR, and NOT gates. Also called **digital logic circuit**.

logic controller A chip-based control system often used for industrial applications. It utilizes data links to communicate with other process control components, devices, and equipment, and is utilized for tasks such as complex data manipulation, timing processes, sequencing, and machine control. Also called **programmable logic controller**.

logic design A functional design of a digital computer, or other system utilizing logic elements.

logic diagram A diagram representing logic elements, and the way each is interconnected with each other. Also called **logical diagram**.

logic element A component, circuit, or device which carries out a logic function or operation. It is the smallest entity that can be represented by a logical operator. Gates and/or flip-flops are examples.

logic error An error in a program due to faulty code. Such an error may or may not make a program or system crash. Also called **logical error**.

logic function A function which utilizes logical operators. Examples include the function of AND, OR, XOR, or NOT circuits. Also called **logical function**.

logic gate A circuit or device which carries out a logic function or operation. Such a circuit or device has multiple inputs and one output. Its output will depend on specified input conditions, such as a given combination of states. The three basic logic gates are the AND, OR, and NOT gates. Also called **gate (2)**.

logic high In digital logic, a level within the more positive of the two ranges utilized to represent binary variables or states. A logic high level corresponds to a 1, or true, value. Also called **logic one**, **logic 1**, **logical one**, **one state**, **logic high state**, or **high level**.

logic high input An input, such as that of a flip-flop or logic gate, corresponding to a **logic high state**. Also called **logic one input**, **logic 1 input**, or **one input**.

logic high state Same as **logic high**.

logic input level In digital logic, a voltage or current level which determines the **logic level**.

logic instruction An instruction which executes a **logical operation**.

logic level In digital logic, either of two non-overlapping ranges utilized to represent binary variables or states. The less positive of the two ranges is logic low, or low level, and the more positive of the two ranges is logic high, or high level.

logic low In digital logic, a level within the less positive of the two ranges utilized to represent binary variables or states. A logic low level corresponds to a 0, or false, value. Also called **logic zero**, **logic 0**, **logical zero**, **logic low state**, **low-level**, or **zero state**.

logic low input An input, such as that of a flip-flop or logic gate, corresponding to a **logic low state**. Also called **logic zero input**, **logic 0 input**, or **zero input**.

logic low state Same as **logic low**.

logic network An electric network which carries out a logic function or operation.

logic one Same as **logic high**.

logic one input Same as **logic high input**.

logic operation Same as **logical operation**.

logic operator Same as **logical operator**.

logic probe A probe utilized to determine the logic level of test points within a logic circuit. Usually used for inspection and troubleshooting. Such a probe is useful, for instance, for detecting high-speed transient pulses which could not be seen on an oscilloscope.

logic programming A programming style based on symbolic logic, rules, and relationships between objects. Used, for instance, in artificial intelligence applications, and an example of such a language is Prolog.

logic pulse A transient signal which changes the logic state of a logic circuit.

logic pulser A probe utilized to pulse, or change the logic state, of a logic circuit. Such an instrument may be used in conjunction with a logic probe, to trace a pulse throughout a circuit being tested. Usually used for inspection and troubleshooting.

logic shift Same as **logical shift**.

logic state Either of the two **logic levels** a binary variable can assume.

logic sum Same as **logical sum**.

logic swing In a logic circuit, the difference in voltage between the logic high and logic low states.

logic symbol A graphical symbol utilized to represent a logic element.

logic unit The part of the CPU which performs the arithmetic and logic operations. Some processors contain multiple logic units. Also called **arithmetic and logic unit**, or **arithmetic unit**.

logic value Same as **logical value**.

logic zero Same as **logic low**.

logic zero input Same as **logic low input**.

logical 1. Pertaining to **logic**. 2. The manner in which data, processes, and systems are organized from the user's perspective, as opposed to **physical (2)**, which is the way software and hardware really function. For example, a user may view the content of a file as a series of consecutive paragraphs, while it may actually consist of many data elements stored in multiple locations on one or more disks. 3. In computers, all that pertains to software, as opposed to **physical (1)**, which deals with hardware. 4. Pertaining to Boolean logic.

logical add Same as **logical sum**.

logical address A virtual address assigned within a computer network. Examples include IP addresses and URLs.

logical block addressing A technique utilized to enable a computer BIOS to address IDE or SCSI hard disks larger than 528 megabytes. Its abbreviation is **LBA**.

logical channel identifier Same as **logical channel number**. Its abbreviation is **LCI**.

logical channel number In X.25, the number assigned to a virtual circuit. This number is attached to all packets during a call, differentiating them from other packets in other calls. Up to 4095 total LCNs in both directions are supported by X.25. Its abbreviation is **LCN**. Also called **logical channel identifier**, or **virtual circuit number**.

logical decision 1. In computers, the determination of a choice or course of action to be taken, based on the particular circumstances and available alternatives. For instance, a computer may compare one set of data to another, to effectuate such a decision. Also called **decision**. 2. A decision whose outcome can have only one of two values, such as 0 or 1, or true or false. Also called **Boolean decision**.

logical device In computers, a name given to a device to facilitate identification by a given program or software system. An example is the designation given to a port, or that of a logical drive. Also called **logical device name**, or **device name**.

logical device name Same as **logical device**.

logical diagram Same as **logic diagram**.

logical drive A name given to a drive to facilitate identification by a given program or software system. This contrasts with a **physical drive**, which is the actual hardware. For instance, logical drives C: and D: may be a single partitioned physical drive. Also called **virtual drive**.

logical error Same as **logic error**.

logical expression An expression utilizing logical operators and yielding a logical value. Also called **Boolean expression**.

logical file The manner in which a file is organized from the user's perspective, as opposed to a **physical file**, which is the way it truly is distributed. For example, a user may view the content of a file as a series of consecutive paragraphs, while it may actually consist of many data elements stored in multiple locations on one or more disks.

logical function Same as **logic function**.

logical inferences per second A unit utilized to gauge the speed of artificial intelligence hardware and/or software. Its acronym is **LIPS**.

Logical Link Control One of the two sublayers of the datalink layer, as defined by IEEE 802. It is the closer of the two to the network layer, and handles tasks such as flow control and framing. The other sublayer is the **Media Access Control Sublayer**. Its abbreviation is **LLC**. Also called **Logical Link Control Sublayer**, or **Logical Link Control Layer**.

Logical Link Control Layer Same as **Logical Link Control**. Its abbreviation is **LLC Layer**.

Logical Link Control Sublayer Same as **Logical Link Control**. Its abbreviation is **LLC Sublayer**.

logical lock A software-based lock, as opposed to one based on hardware. An example is a password. Also called **software lock**.

logical network The manner in which a network is organized from the user's perspective, as opposed to the true physical topology, which is known as **physical network**. For example, a user may visualize a bus network as a series of computers arranged along a straight line, while it may truly be a common cable with many twists, turns, and loops.

logical one Same as **logic high**.

logical operation An operation which a **logical operator** performs. Also called **logic operation**, or **operation (4)**.

logical operator An operator within Boolean algebra. Such an operator manipulates logical values, and common examples are AND, OR, NOT, NAND, and XOR. Also called **logic operator**, or **Boolean operator**.

logical record The manner in which a record is organized from the user's perspective, as opposed to a **physical record**, which is the way it truly is distributed. For example, a user may view a record as a single unit, while it may actually consist of many data fields in a database. Also called **record layout (1)**.

logical schema A detailed model of the overall structure of a database. This model does not factor in considerations such as how the information is stored or accessed. Also called **conceptual schema**.

logical shift A shift in which bits are moved to the right or to the left, with zeroes replacing the displaced bits. Also called **logic shift**.

logical sum A logical operation which is true if any of its elements is true. For example, if one or more of its multiple inputs have a value of 1, then the output is 1. Its output is 0 only when all of its inputs have a value of 0. For such functions, a 1 is considered as a true, or high, value, and 0 is a false, or low, value. Also called **logical add**, **logic sum**, **logic add**, or **OR**.

logical value A value which must be one of two possibilities, such as true or false, or 0 or 1. Also called **logic value**, or **Boolean value**.

logical zero Same as **logic low**.

login Same as **log-on**. Also spelled **log-in**.

LOGO A high-level programming language similar to LISP, which used especially for teaching programming and creating graphics.

logoff Same as **log-out**. Also spelled **log-off**.

logon Same as **log-in**. Also spelled **log-on**.

logout Same as **log-out**.

loktal base A vacuum tube base with 8 pins which is secured in a loktal socket. Also spelled **loctal base**.

loktal socket A vacuum tube socket with 8 pins which is secures in a loktal base. Also spelled **local socket**.

loktal tube A vacuum tube whose 8 short pins are securely locked into its corresponding socket. Also spelled **loctal tube**.

LOL In emails, chats, and other online activities, an acronym for laughing out loud, which generally expresses a favorable reaction to a humorous manifestation. When LOL is deemed to be an insufficient reaction, **ROFL** may be used instead.

London penetration depth The minimal distance a magnetic field can penetrate a superconductor without terminating the superconductivity.

long Abbreviation of **longitude**.

long card A standard-sized plug-in circuit board. The term long card is used to differentiate from a **short card**.

long-delayed echo In radio communications, an echo whose delay is significantly longer than expected. Its abbreviation is **LDE**.

long-distance call A telephone call which is originated from, or placed to a destination that is not contained within the same local calling area. A long-distance call does not necessarily imply that there is an additional charge to the caller, as seen in many cellular telephone plans. Also called **toll call**.

long-distance communications Also called **long-haul communications**. **1.** Telephonic communications in which the destination is not contained within the same local calling area as the point of origination. **2.** Communications beyond a given range, such as that possible for ground-wave propagation.

long-distance reception The reception of radio waves beyond a given range, such as that possible for ground-wave propagation.

long-distance transmission The transmission of radio waves beyond a given range, such as that possible for ground-wave propagation.

long-haul Of, pertaining to, or operational over comparatively long distances, while **short-haul** relates to relatively short distances. An example is long-distance reception.

long-haul communications Same as **long-distance communications**.

long-line effect The jumping back and forth by an oscillator between a desired oscillation frequency and a nearby undesired one. This phenomenon is sometimes observed when an oscillator is coupled to a long transmission line.

long-persistence screen A CRT screen whose presented image persists for a given time interval after the electron beam producing it has passed. Used, for instance, in plan position indicators.

long-range navigation Same as **loran**.

long-range radar A radar whose range of detection extends beyond a given distance, such as 500 or 1000 kilometers. Its abbreviation is **LRR**.

long skip In ionospheric propagation, a reflection of comparatively long distance. Such a reflection may be several thousand kilometers, while a **short skip** is usually less than 1000 kilometers.

long-tail pair Same as **long-tailed pair**.

long-tailed pair Two transistors connected with their emitters or sources coupled, with a resistor or a third transistor connected to said emitters or sources. Used, for instance, as a differential amplifier. Also called **long-tail pair**.

long-term Involving or lasting a comparatively long time interval.

long-term drift **1.** A gradual and undesired change in a characteristic or a set adjustment of a device, component, instrument, piece of equipment, or system. Examples include frequency drift, thermal drift, or changes in power output. Also called **drift** (1). **2.** Drift which occurs over a comparatively long time interval, as opposed to **short-term drift**, which occurs more rapidly.

long-term effect An effect, such as long-term instability, lasting a comparatively long time interval.

long-term instability The instability of a component, circuit, device, piece of equipment, system, setting, or value, over a comparatively long time interval.

long-term stability The stability of a component, circuit, device, piece of equipment, system, setting, or value, over a comparatively long time interval.

long throw The excursion of a **long throw woofer**.

long throw woofer A woofer whose suspension allows it to move through an extended excursion, which helps provide improved response along with reduced distortion.

long-wave infrared Electromagnetic radiation in the infrared region which is furthest from visible light. It corresponds to wavelengths of approximately 10 to 1000 micrometers, though the defined interval varies. Also called **far-infrared radiation**.

long-wave ultraviolet Electromagnetic radiation which is between near ultraviolet and extreme ultraviolet. It corresponds to wavelengths of approximately 200 to 300 nm, though the defined interval varies. Also called **far-ultraviolet radiation**.

long waves Electromagnetic waves whose wavelength is within the range of 1 to 10 kilometers, corresponding to frequencies of between 300 to 30 kHz, respectively. This represents the low-frequency band.

long-wire antenna **1.** A linear antenna whose electrical length is at least one wavelength of the signal being transmitted or received, and which is a whole-number multiple of a half-wavelength. It is a directional antenna, and is often several wavelengths long; the longer the radiating element, the greater the gain. **2.** An antenna whose electrical length is a whole-number multiple of the half-wavelength of the signal being transmitted or received. Also called **harmonic antenna**.

longitude Imaginary half-circles that are drawn parallel to the prime meridian, or another reference meridian, and which serve to measure angular displacement from such a meridian. The prime meridian has an angle of 0°, while those to the east work their way towards 180 °E, and those to the west work their way towards 180 °W. These half-circles cross the parallels of **latitude** at right angles, and together serve to define locations on the planet. Its abbreviation is **long**.

longitude effect Any effect which varies depending on the **longitude** where it occurs or is observed.

longitudinal current A current which flows in the same direction along two parallel wires. Another conductor, such as a ground, serves as the return path.

longitudinal magnetization **1.** Magnetization of a material or medium in a manner in which the magnetic flux is essentially parallel to the long axis of the influencing field. In **transverse magnetization** (1) the converse is true. **2.** In magnetic recording, magnetization of a material or medium essentially in the same direction as the material or medium is traveling. In **transverse magnetization** (1) the converse is true.

longitudinal parity The parity of the extra bit added when using **longitudinal redundancy check** for the detection of errors.

longitudinal recording Recording of a magnetic recording medium in which **longitudinal magnetization** (2) is utilized.

longitudinal redundancy check An error-detection technique in which a parity bit is generated for each block of characters on a longitudinal track. This technique enables the detection of single bit errors, and is used, for instance, in some magnetic recording systems. Its abbreviation is **LRC**.

longitudinal wave A wave in which the particles in the medium move parallel to the direction of propagation of the wave. Sound waves are an example. This contrasts with a **transverse wave**, where the particles in the medium move perpendicular to the direction of propagation.

lookup A data search in which a specific value is searched within a predefined table of values, such as a matrix. Also, the function, such as that in a spreadsheet application, which performs such a search. Also, the value so obtained. Also called **table lookup**.

lookup table A table utilized for **lookup** searches. Its abbreviation is **LUT**.

loop 1. A closed signal path. For example, a continuous circuit, or a feedback control loop. 2. A group or combination of branches or components which form a closed current path in an electric network. Also called **mesh** (1). 3. In a standing-wave system, a point of maximum amplitude or displacement. For example, a maximum of voltage or current. Also called **antinode** (1). 4. In a vibrating system, a point of maximum amplitude or displacement. Also called **antinode** (2). 5. A small loop of wire inserted into a waveguide or cavity resonator, which allows energy to be transferred in or out. Also called **coupling loop**. 6. Same as **loop antenna**. 7. A pair of wires, or its equivalent, extending from a telephone central office to the premises of a customer. Also called **local loop**, **local line**, **line loop**, **subscriber loop**, or **subscriber line**. 8. A single turn within a coil. 9. In computers, a set of statements or instructions which are executed repeatedly while a given condition exists. 10. In computers, a set of statements or instructions which are executed a given number of times. 11. In computers, to execute a set of statements or instructions repeatedly, or a given number of times. 12. A tape or reel whose ends have been spliced together, so as to provide uninterrupted play. Also called **tape loop**. 13. A closed curve on a graph, such as a hysteresis loop.

loop actuating signal In a feedback control loop, the signal resulting from the mixing of the input signal and the feedback signal.

loop antenna A directional antenna consisting of a wire coil with one or more turns. Used, for instance, in radio direction finders and portable radio receivers. Also called **loop** (6), **coil antenna**, or **frame antenna**.

loop back Same as **loopback**.

loop check In communications, an error-checking technique in which the receiving end retransmits the data back to the source. Both sets of data are compared, and if they do not match, an error has occurred. Also called **loop checking**, or **echo check**.

loop checking Same as **loop check**.

loop coupling 1. The use of a **loop** (5) to transfer energy in or out of a waveguide. 2. The use of a **loop** (1) to transfer energy in or out of a circuit, device, system, or medium.

loop current The current that flows in a **loop** (7) in an off-hook condition. This current typically has a value of about 0.02 to 0.05 amps.

loop extender A device which enables operation of an unusually long **loop** (7). Such a loop may extend to more than 3.5 kilometers from a local central office.

loop feedback signal In a feedback control loop, the signal fed back into the input circuit.

loop gain The gain of a signal passing through a **loop** (1).

loop input signal In a feedback control loop, the signal which is applied externally.

loop output signal The output signal of a feedback control loop.

loop pulsing The dialing of a telephone number utilizing pulses which interrupt a steady DC, to indicate a dialed digit. For example, to dial a 4, 4 such pulses are sent. Used in rotary telephones. Also called **dial pulsing**.

loop resistance 1. The resistance of the conductors around a **loop** (1). 2. The resistance of the conductors around a **loop** (2). 3. The resistance of the conductors around a **loop** (7). Also called **line-loop resistance**.

Loop-Start Line A line which senses which kind of signaling is required for a phone, or other device using such a line, to operate. Such a line recognizes an off-hook condition, and treats it as request for service. A Loop-Start Line extends from a local central office, and when shared by one or more phones it is a typical landline, such as that of a home, and when shared by multiple devices connected to a PBX or key telephone system, it is a trunk.

Loop-Start Trunk A **Loop-Start Line** shared by multiple devices connected to a PBX or key telephone system.

loop test A test, utilized to locate faults such as discontinuities in a circuit or deficiencies in insulation, in which a conductor is incorporated into a closed loop.

loopback A diagnostic test in which a transmitted signal is returned to its source for error analysis. The signal may pass through the entire network, a given segment, test points, and so on. Utilized, for instance, to troubleshot communications networks. Also spelled **loop back**. Also called **loopback test**.

loopback adapter An adapter which is plugged into a computer port to perform **loopback tests**. Also called **loopback plug**.

loopback plug Same as **loopback adapter**.

loopback signal The diagnostic signal transmitted in a **loopback**.

loopback test Same as **loopback**.

loophole A mechanism which enables defeating computer and/or communications security measures. Such a mechanism may consist of a failure to take one or more factors into consideration, or may be intentional, as is the case of a back door.

looping plug A two-prong plug which serves to loop or patch adjacent jack circuits.

loopstick antenna A small receiving antenna consisting of a coil wound around a ferrite rod. Such antennas feature high sensitivity at low and medium frequencies, and are used, for instance, in place of a loop antenna in a radio receiver. Also called **ferrite-rod antenna**, **ferrite loopstick**, or **ferrod**.

loose coupling Also called **weak coupling**. 1. In a transformer, the arrangement of the primary and secondary windings in a manner which provides for a low degree of energy transfer. 2. In a transformer, the arrangement of the primary and secondary windings in a manner which provides minimum transfer. 3. A level of coupling which is below critical coupling. Also called **undercoupling**.

loosely-coupled multiprocessing A multiprocessing computer architecture in which multiple CPUs each have their own memory and I/O channels, and which lends itself to massively parallel processing, as opposed to symmetric multiprocessing. Also called **distributed-memory multiprocessing**.

LOP Abbreviation of **line of position**.

lopplar Same as **laser radar**. It is the acronym for **laser Doppler radar**.

loran Acronym for **long-range navigation**. A time-difference hyperbolic navigation system which utilizes a controlling station and two or more secondary stations, all spaced far from each other, which transmit synchronized signals which aid in the determination of the position of a vessel.

loran C A version of loran which operates around the 100 kHz band.

loran chain A group of loran stations incorporating a controlling station and two or more secondary stations.

loran D A version of loran which radiates less radio-frequency energy than loran C.

loran fix The determination of an accurate position of a vessel through the use of loran.

loran station In loran, a controlling station or secondary station.

Lorentz force The force exerted on a charged particle moving though an electric and/or magnetic field. This force is described by the following equation: $F = q(E + vB)$, where F is the force, q is the charge, E is the electric field strength, v is the particle velocity, and B is the magnetic flux density.

Lorentz force equation The equation which describes the **Lorentz force**.

LOS Abbreviation of **line of sight**.

loss Also called **losses**. **1.** Energy or power which is dissipated without performing useful work. For example, heat losses. Usually expressed in watts. **2.** The reduction of the strength of a signal which has traveled from one point to another within a communications network. Usually expressed in decibels. **3.** A measure of the extent of the harm to a computer system after an event such as a crash.

loss angle The difference resulting when the dielectric phase angle is subtracted from 90°. Also called **dielectric loss angle**.

loss factor **1.** For a dielectric, the product of the dielectric constant and the power factor. This is a measure of rate at which a dielectric becomes heated, and varies by frequency. Also called **loss index**. **2.** For a given time interval, the ratio of the average power loss to the peak-load power loss.

loss index Same as **loss factor (1)**.

loss modulation A method of amplitude modulation which employs a variable-impedance device which is coupled to the output circuit of the transmitter, to absorb power from the carrier wave. Also called **absorption modulation**.

loss tangent For a given material, the ratio of the imaginary number component of the dielectric constant to the real part.

losses Same as **loss**.

lossless Having no losses. For example, lossless data transmission, or a lossless line.

lossless compression A compression technique in which all data, upon being decompressed, is identical to its original state. This form of compression must be used, for instance, when compressing programs or databases. This contrasts with **lossy compression**.

lossless line A theoretical transmission line which suffers no losses.

lossy **1.** That which suffers losses. **2.** That which suffers more losses than would ordinarily be expected. **3.** That which suffers excessive losses.

lossy compression A compression technique in which a given proportion of the data is lost. Images and sound, for instance, are suitable for such a technique, as redundant or otherwise unnecessary information can be eliminated without noticeable losses in resolution. Lossy compression can be utilized to reduce file sizes to far less than 1% of their original size. This contrasts with **lossless compression**.

lossy line Also called **lossy transmission line**. **1.** A transmission line which suffers more losses than would ordinarily be expected. This may be intentional or unintentional. **2.** A transmission line which suffers excessive losses.

lossy transmission line Same as **lossy line**.

lost call A call that is not completed. This is usually a blocked call, an abandoned call, or an improperly dialed call.

lost cluster On a disk, a cluster which is in use, but is not being utilized to store accessible data. Such a cluster may result, for instance, from program crashes.

lost packet A data packet which is sent, but which does not arrive at the destination.

loudness **1.** The amplitude of a sound, as perceived by a listener. It is a subjective measure of sound intensity, and as such varies from person to person. Loudness is influenced by the absolute amplitude of the sound, its frequency, duration, and to a lesser extent other factors. Also called **volume (1)**. **2.** Same as **loudness control (1)**.

loudness contour In an audio component or system, the boosting of low and high frequencies at low listening levels to compensates for the lower auditory response humans have under these circumstances, thus making the sound more natural. Also, a curve illustrating the sound pressure levels at all frequencies when said compensation is used.

loudness control **1.** In an audio system, a volume control which incorporates a circuit that boosts low frequencies when listening at low-volume settings. This compensates for the lower auditory response humans have under these circumstances, thus making the sound more natural. Such a control may also boost high frequencies. Also called **loudness (2)**, or **compensated volume control**. **2.** A control, such as that of a TV or stereo amplifier, which varies **loudness (1)**. Also called **volume control**.

loudness curves A series of curves which plot the sound intensities of pure tones, in decibels, versus frequency, in hertz, for all audible frequencies, so that all frequencies are perceived at the same intensity. The hearing sensitivity of people varies by frequency, and this sensitivity also varies depending on the sound-pressure level. Also called **Fletcher-Munson curves**, or **equal-loudness curves**.

loudness level The **loudness (1)** perceived by a given listener when comparing a sound to a reference sound, such as a 1 kHz tone at 0 decibels. Also called **equal-loudness level** when the sound is deemed to be at the same intensity.

loudspeaker A transducer which converts electrical energy into sound energy, which is then radiated outward to the surrounding medium. The diaphragm, which is the vibrating membrane of a loudspeaker, may be composed of any of various materials, including paper, fiber, plastic, silk, metal, or it may be a piezoelectric crystal. A loudspeaker may incorporate multiple specialized drivers, including subwoofers, woofers, midranges, tweeters, and supertweeters, to better address various frequency ranges. Loudspeakers provide sound for devices such as TVs, audio systems, computers, DVDs, telephones, and so on. Also called **speaker**, or **driver (4)**.

loudspeaker baffle Also called **speaker baffle**, or **baffle**. **1.** A partition in a loudspeaker enclosure which is used to reduce or eliminate the interaction between the acoustic waves generated by the front of the loudspeaker and those from the back. This is especially important in low-frequency sound reproduction. **2.** The panel on which a loudspeaker is mounted. **3.** A panel used to inhibit the propagation of sound waves. Used, for instance in theaters as part of a sophisticated sound-reproduction system.

loudspeaker crossover network A filter circuit in a loudspeaker which divides the input audio frequencies and sends the corresponding bands of frequencies to the designated loudspeaker units, such as woofers, midranges, tweeters, super-tweeters, and so on. Also called **loudspeaker dividing**

network, **speaker crossover network**, **crossover (3)**, **dividing network**, or **frequency dividing network**.

loudspeaker diaphragm A vibrating membrane in a loudspeaker. It produces a sound-wave output in response to its electric input. Cones and domes are examples of commonly utilized diaphragms in speakers. Also called **speaker diaphragm**, or **diaphragm (1)**.

loudspeaker dividing network Same as **loudspeaker crossover network**.

loudspeaker efficiency The proportion of the audiofrequency electrical power input to a loudspeaker which is converted into acoustic energy. Efficiency may be measured, for instance, by driving a loudspeaker with a power input of 1 watt, and measuring its output, in decibels, at a distance of 1 meter. Also called **speaker efficiency**, **efficiency (3)**, or **sensitivity (4)**.

loudspeaker enclosure A cabinet designed to house one or more loudspeaker units. Two common designs are acoustic reflex and acoustic suspension. Within a loudspeaker system, a specialized loudspeaker such as a woofer or tweeter may have its own enclosure, helping enhance performance. Also called **speaker enclosure**, or **enclosure (2)**.

loudspeaker grill Same as **loudspeaker grille**.

loudspeaker grille A covering across the front of a loudspeaker which provides protection and decorative appeal. Also spelled **loudspeaker grill**. Also called **speaker grille**, or **grille (2)**.

loudspeaker impedance The rated impedance of the voice coil of a loudspeaker. Most speakers have a nominal value assigned, although said impedance varies, depending on the frequency being reproduced. Common values are 4, 8, and 16 ohms, and the driving amplifier must be matched to avoid diminished performance, failure, or damage. Also called **speaker impedance**.

loudspeaker microphone A loudspeaker which also serves as a microphone. Used, for instance, in an intercom system, where a button may be used to switch between functions. Also called **speaker microphone**.

loudspeaker port A carefully-dimensioned opening in a loudspeaker enclosure. It is employed for various purposes, but is mainly used to increase and extend the reproduction of low frequencies. Also called **speaker port**, **port (5)**, **duct (5)**, **ducted port**, or **vent (3)**.

loudspeaker system Also called **speaker system**. **1.** A loudspeaker enclosure which incorporates multiple specialized loudspeakers, such as a woofer, midrange, and tweeter. **2.** Multiple loudspeakers which are designed to work together in providing a desired sound experience. For example, a five loudspeaker arrangement providing surround sound.

loudspeaker voice coil In a dynamic speaker, a coil which is connected to the diaphragm, and through which an audiofrequency signal current is sent to move said coil in a piston-like manner, thus producing a sound-wave output. Also called **speaker voice coil**, or **voice coil (1)**.

louver A slanted opening, or an opening consisting of multiple parallel slats. Used, for instance, to help dissipate heat, or to protect tweeters while allowing sound to pass freely.

low Its abbreviation is **lo**. **1.** Having a small degree of a given magnitude, quantity, or characteristic. Also, having a smaller degree of a magnitude, quantity, or characteristic relative to something else. For example, low voltage, low impedance, or the like. **2.** On the lower end of a given interval or spectrum. Also, that part of an interval or spectrum which is lesser than another. For instance, the frequencies which a high-pass filter blocks, as opposed to those it transmits. **3.** Having a small elevation, or a smaller elevation than something else. **4.** In a binary operation, a 0, which also corresponds to off, as opposed to **high**, which corresponds to 1, or on.

low band **1.** In TV, a band which extends from 54 to 88 MHz, corresponding to the VHF channels of 2 through 6. **2.** In communications utilizing one band of frequencies for transmission one direction, and another band for the other, the lower band. The other band, using higher frequencies, is called **high band**. **3.** In communications utilizing multiple frequency bands, the lowest band, or a band lower than another given band.

low battery A battery whose voltage has fallen below a specified level, and which thus requires recharging or replacement, depending on whether it is rechargeable or not. When the voltage of a battery is low, its performance is impaired, and any components, devices, equipment, or systems being powered may fail or stop operating.

low-battery alarm Same as **low-battery indicator (1)**.

low-battery alert Same as **low-battery indicator (1)**.

low-battery indicator **1.** An alarm, such as an intermittent beep or flashing light, which serves to warn of a **low battery**. Used, for instance, in a smoke detector or cellular phone. Also called **low-battery alert**, or **low-battery alarm**. **2.** A component, circuit, device, instrument, piece of equipment, readout, meter, or mechanism which serves to inform of a low battery.

low-battery warning A signal, such as that which is visible or audible, which warns of a **low battery**.

low-capacitance probe A test probe with an especially low capacitance. Used, for instance, with an oscilloscope to help avoid distorting waveforms.

low current **1.** A current level which is low in magnitude as compared to another. **2.** A current level which is smaller than usual for a given device, process, application, or operation.

low-density A storage format in which a lesser number of bits are stored per unit area than a standard amount. Also, a medium, such as a magnetic disk, with such a storage density.

low earth orbit The orbit a **low earth orbit satellite** follows. Its acronym is **LEO**.

low earth orbit satellite An artificial satellite whose orbit ranges from approximately 500 to 2000 kilometers over the surface if the earth, although the defined value may vary. The time of orbit is much faster than that of a geostationary satellite, and a constellation of low earth orbit satellites are necessary to provide worldwide coverage, as they do not maintain a fixed spot above the planet at all times. Its lower orbit provides for shorter propagation delays, and reduced power requirements. This makes them suitable, for instance, for handheld devices used for paging, Internet access, and real-time conferencing. Its acronym is **LEOS**.

low-energy electron diffraction An electron diffraction technique in which a sample is bombarded with a beam of low-energy electrons. Utilized, for instance, for the study of crystal surfaces. Its abbreviation is **LEED**.

low filter A filter which passes, with little or no attenuation, all frequencies above a non-zero lower cutoff. An example is that used in an audio amplifier to reduce rumble. Also called **high-pass filter**.

low frequency A range of radio frequencies spanning from 30 to 300 kHz. These correspond to wavelengths of 10 km to 1 km, respectively, which are kilometric waves. Also, pertaining to this interval of frequencies. Its abbreviation is **LF**. Also called **low-frequency band**.

Low-Frequency Acquisition and Ranging Same as **LOFAR**.

low-frequency band Same as **low frequency**.

low-frequency compensation **1.** Circuitry which boosts low frequencies to compensate for any deficiencies. An example is automatic bass compensation. **2.** The increase in low frequencies resulting from the use of **low-frequency compensation (1)**.

low-frequency direction finder A direction finder which operates in the **low-frequency band**. Its abbreviation is **LFDF**.

low-frequency loudspeaker Same as **low-frequency speaker**.

low-frequency oscillator An oscillator which can generate frequencies ranging from 20 Hz all the way down to a small fraction of 1 Hz. Used, for instance, to modulate a signal.

low-frequency speaker Also called **low-frequency loudspeaker**, or **woofer**. A large speaker designed to reproduce frequencies below a given threshold, such as 1000 or 300 Hz. Depending on the design and components, a woofer may accurately reproduce frequencies below the limit of human hearing. Such a speaker unit is usually utilized with others, such as midranges and tweeters, for reproduction across the full audio spectrum.

low impedance 1. An impedance level which is low in magnitude as compared to another. **2.** An impedance level which is smaller than usual for a given device, process, application, or operation.

low insertion force Same as **LIF socket**. Its abbreviation is **LIF**.

low insertion force socket Same as **LIF socket**.

low-level Same as **logic low**.

low-level access For software, access to hardware with a minimum of intervening levels, such as translation. A computer's BIOS operates just above the hardware level.

low-level computer language Same as **low-level language**.

low-level contact resistance An extremely small resistance between two contacts when a circuit is closed.

low-level contacts Contacts which are intended for use with low values of current and/or voltage.

low-level format A format which prepares a disk for a specific type of disk controller, and performs tasks such as sector identification. Such a format will destroy all data on a disk, while a **high-level format** only resets file-allocation tables so that the operating system sees the disk as empty. In the case of a hard disk, a low-level format is usually performed by the manufacturer. Also called **physical format**.

low-level input current The current flowing into an input when a given **low-level voltage** is applied.

low-level input voltage In digital logic, an input voltage level which is within the less positive of the two ranges utilized to represent binary variables.

low-level language A computer programming language that more closely resembles machine code than human language. Such languages are architecture-dependent, and an example is assembly language. Also called **low-level programming language**, **low-order language**, or **low-level computer language**.

low-level logic Same as **logic low**.

low-level modulation Modulation produced in a section of a system whose power level is low in relation to the total power output of said system.

low-level output current The current flowing into an output, which under specified input conditions, establishes a **low-level** at said output.

low-level output voltage A voltage at an output terminal, which under specified input conditions, establishes a **low-level** at said output.

low-level programming language Same as **low-level language**.

low-level radio-frequency energy A level of radio-frequency energy below a given threshold, such as 1 microwatt per square centimeter. Used, for instance, in evaluating health and safety matters pertaining to devices which emit electromagnetic radiation. Its abbreviation is **low-level RF energy**.

low-level RF energy Same as **low-level radio-frequency energy**.

low-level signal A signal whose amplitude is below a given level, or which is otherwise comparatively low.

low-level voltage In digital logic, a voltage level which is within the less positive of the two ranges utilized to represent binary variables or states.

low logic level Same as **low-level**.

low-loss dielectric An insulating material which has low dielectric losses, especially at high frequencies.

low-loss line Same as **low-loss transmission line**.

low-loss transmission line A transmission line in which the energy dissipated per unit length is comparatively low. Also called **low-loss line**.

low-noise A component, circuit, device, piece of equipment, system, material, or setting which generates an especially low noise level.

low-noise amplifier An amplifier which contributes an especially small amount of noise to the desired signal. Used, for instance, to amplify the weak satellite signal which is reflected by a satellite dish. Its abbreviation is **LNA**.

low-noise block down-converter A single unit incorporating a low-noise microwave amplifier and a block down-converter. Usually utilized to amplify and down-convert a range of frequencies, for use by a satellite receiver. Its abbreviation is **LNB**.

low-noise microwave amplifier A **low-noise amplifier** utilized to amplify microwaves.

low-order Within a sequence of digits, a location with comparatively lower significance. In a number this is usually to the right, as are **9** and **8** in 321.098.

low-order digit A digit in a **low-order** location.

low-order language Same as **low-level language**.

low-pass filter Also spelled **lowpass filter**. Also called **high filter**. **1.** A filter which blocks all frequencies above a given cutoff frequency. Thus, all frequencies below this value can pass. An example is an anti-aliasing filter. **2.** A filter which only passes AC below a given cutoff frequency.

low power 1. A power level which is low in magnitude as compared to another. **2.** A power level which is smaller than usual for a given device, process, application, or operation.

low-pressure chemical-vapor deposition A chemical-vapor deposition technique in which gaseous reactants at a low pressure surround the substrate. The solid products of said reactants are then deposited onto the substrate. Its abbreviation is **LPCVD**, or **low-pressure CVD**.

low-pressure CVD Same as **low-pressure chemical-vapor deposition**.

low-profile quad flat-pack An extra-thin square mount chip which provides leads on all four sides, affording a high lead count in a small area. Such chips generally also feature a smaller footprint than other quad flat-packs. Its abbreviation is **LQFP**. Also spelled **low-profile quad flatpack**. Also called **low-profile quad flat-package**.

low-profile quad flat-package Same as **low-profile quad flat-pack**.

low-profile quad flatpack Same as **low-profile quad flat-pack**.

low Q For a component, resonant circuit, tuned circuit, periodic device, or energy-storing device, a comparatively low value of Q. For instance, for a component or circuit, it could represent a low value for the ratio of reactance to resistance. Also called **low Q factor**.

low Q factor Same as **low Q**.

low resolution A level of displayed resolution that is below a given standard or typical level. For example, the coarse images displayed by a lower-quality monitor, or that of a low-resolution printer setting. Its abbreviation is **lo-res**.

low tension Same as **low voltage**.

low vacuum An environment or system whose pressure is below 760 torr, but above 1 torr. The lower threshold may be slightly higher, such as 30 torr. One torr equals 1 millimeter of mercury, or approximately 133.3 pascals. Also called **rough vacuum**.

low voltage Also called **low tension**. **1.** A voltage level which is small in magnitude in relation to another. **2.** A voltage level which is smaller than usual for a given device, process, application, or operation.

lower sideband In double-sideband amplitude modulation, the band of frequencies below the carrier frequency. This contrasts with an **upper sideband**, which incorporates the band above the carrier frequency. Its abbreviation is **LSB**.

lowest usable frequency The lowest frequency that can be employed for communications between locations utilizing ionospheric propagation. This frequency will vary according to the angle of incidence and time of day. The highest usable frequency is called **maximum usable frequency**. Its abbreviation is **LUF**.

lowpass filter Same as **low-pass filter**.

LP Abbreviation of **Linear Prediction**.

LPC Abbreviation of **Linear Predictive Coding**, or **Linear Prediction Coding**.

LPCVD Abbreviation of **low-pressure chemical-vapor deposition**.

LPE Abbreviation of **liquid-phase epitaxy**.

lpm Abbreviation of **lines per minute**.

LPT The logical name given to the first parallel port of a computer. A system's primary printer usually utilizes this port. Subsequent parallel ports, if available, are designated LPT2, LPT3, and so on. Originally, LPT was an abbreviation of line printer terminal. Also called **LPT1**, or **LPT port**.

LPT port Same as **LPT**.

LPT1 Same as **LPT**.

LPT2 The logical name given to the second parallel port of a computer. Also called **LPT2 port**.

LPT2 port Same as **LPT2**.

LPT3 The logical name given to the third parallel port of a computer. Also called **LPT3 port**.

LPT3 port Same as **LPT3**.

LQ Abbreviation of **letter quality**.

LQFP Abbreviation of **low-profile quad flat-pack**.

LR Abbreviation of **laser retroreflector**.

Lr Chemical symbol for **lawrencium**.

LRA Abbreviation of **laser retroreflector array**.

LRC Abbreviation of **longitudinal redundancy check**.

LRR **1.** Abbreviation of **long-range radar**. **2.** Abbreviation of **laser retroreflector**.

LSA Abbreviation of **link-state advertisement**.

LSB **1.** Abbreviation of **least-significant bit**. **2.** Abbreviation of **least-significant byte**. **3.** Abbreviation of **lower sideband**.

LSC Abbreviation of **least-significant character**.

LSD Abbreviation of **least-significant digit**.

LSI Abbreviation of **large-scale integration**.

LSP **1.** Abbreviation of **local service provider**. **2.** Abbreviation of **link-state packet**.

Lu Chemical symbol for **lutetium**.

Luddite A person who is afraid of, loathes, and/or is against technology. The term may apply, for instance, to those who believe that technology brings about more harm than good, or that people rely on it to a preposterous degree.

LUF Abbreviation of **lowest usable frequency**.

lug **1.** A fitting or projection which serves to connect a wire by wrapping or through soldering. Also called **terminal lug**. **2.** A fitting or projection which serves to support or grasp.

Lukasiewicz notation A manner of forming mathematical or logical expressions in which the operators appear before the operands. For example, $(A + B)$ x C, which is in **infix notation**, would appear as *+ABC in Lukasiewicz notation. Also called **Polish notation (2)**, or **prefix notation**.

lumen The SI unit of luminous flux. It is equal to the luminous flux emitted within 1 steradian by a point source whose luminous intensity is 1 candela. Its symbol is **lm**.

lumen-hour A unit of light equal to 1 lumen radiated or received during 1 hour. Its abbreviation is **lm-h**, or **lm-hr**.

lumen-second A unit of light equal to 1 lumen radiated or received during 1 second. Its abbreviation is **lm-s**, or **lm-sec**.

lumens per square meter A unit of illuminance equal to a **lux**. Its abbreviation is **lm/m^2**.

lumens per watt Its abbreviation is **lm/W**. **1.** A unit of luminosity factor. **2.** A unit of luminous efficacy, defined as the light produced, in lumens, per watt of power consumed.

luminaire A self-contained unit incorporating one or more lamps, one or more reflectors, housing, cables, and so on required for a lighting system. Used, for instance, in photography.

luminance **1.** The total light emitted or reflected from a surface in a given direction, per unit of projected area. The measurement must be made from the same direction the light is emitted from. May be expressed in foot-lamberts. The nearly obsolete term **brightness (3)** is also used for this concept. **2.** Same as **luminance signal**.

luminance carrier In color TV, the carrier frequency utilized to convey luminance information.

luminance channel In color TV, the channel carrying the luminance signal. This channel may also carry other signals.

luminance flicker Brief and irregular fluctuations in a displayed image, occurring as a result of unwanted fluctuations in the luminance signal. Seen, for instance in CRTs used in TVs or computer monitors.

luminance meter An instrument which measures and indicates luminance.

luminance signal In color TV, the video signal which contains the brightness information. It is composed of 30% red, 59% green, and 11% blue. A black and white TV only displays luminance signals. In a color TV, the sidebands of the chrominance carrier are added to this signal to convey the color information. Also called **luminance (2)**, or **Y signal**.

luminescence The emission of electromagnetic radiation, especially visible light, unaccompanied by high temperatures, such as those present in incandescence. Examples include bioluminescence, chemiluminescence, photoluminescence, and electroluminescence. When luminescence persists for less than 10^{-8} second, it is called **fluorescence**, while **phosphorescence** persists longer. Also, the radiation emitted.

luminescent **1.** Capable of exhibiting the phenomenon of **luminescence**. **2.** Pertaining to **luminescence**.

luminescent material A material which exhibits the phenomenon of **luminescence**. Also called **phosphor**.

luminescent screen **1.** In a CRT, such as that of a TV, a screen which exhibits luminescence when bombarded by an electron beam. **2.** Any screen which exhibits luminescence when bombarded by particles.

luminosity 1. The condition or characteristic of being **luminous**. 2. Same as **luminosity factor**. 3. The subjective perception of the quantity of light emitted by a surface.

luminosity factor For a given wavelength, the ratio of the luminous flux to the radiant flux. Expressed in lumens per watt. Also called **luminosity (2)**.

luminous Emitting or reflecting light.

luminous efficacy For a light source, the ratio of the total luminous flux to the total radiant flux. Expressed in lumens per watt. Also called **luminous efficiency**.

luminous efficiency Same as **luminous efficacy**.

luminous emittance For a given light source, the luminous flux emitted per unit area. Expressed, for instance, in lumens per square meter.

luminous energy 1. The total, over time, of radiant energy in the visible spectrum. This concept may refer to light produced by a source, or that which is received by a surface. Usually expressed in lumen-hours, or lumen-seconds. 2. Radiant energy in the visible spectrum. Also called **light energy**.

luminous exitance For a given surface, the sum of the luminous emittance, plus any reflected radiation, plus any radiation transmitted through said surface. Expressed, for instance, in lumens per square meter.

luminous flux The **luminous energy (2)** per unit time. This concept may refer to light produced by a source, or that which is received by a surface. Its SI unit is the lumen. Also called **light flux**.

luminous flux density The total amount of visible light incident upon a surface, from all directions above said surface, per unit area. Its SI unit is the lux, although foot-candles are sometimes used. Also called **illuminance**.

luminous intensity Its SI unit is the candela. For a point source, the luminous flux, in lumens, per unit solid angle, in steradians, when measured in a specified direction relative to the emitting source. One lumen per steradian equals a source intensity of 1 candela. Also known as **light intensity**.

luminous sensitivity For a transducer, such as a phototube, which converts light into a current, the ratio of the output current to the luminous flux incident upon it.

lumped Concentrated within a single point or entity, as opposed to being evenly spread out. For example, lumped capacitance.

lumped capacitance A capacitance which is concentrated within a capacitor. This contrasts with **distributed capacitance**, an example of which is the capacitance between adjacent conductors of a circuit.

lumped circuit A circuit in which properties, such as resistance or capacitance, are located in discrete components such as resistors or capacitors. This contrasts with a **distributed circuit**, where such properties are distributed throughout.

lumped component A circuit parameter, such as resistance or inductance, which is concentrated within discrete components such as resistors or capacitors. This contrasts with a **distributed component**, where such parameters are distrib-

uted throughout a circuit or along the entire length of a transmission line. Also called **lumped constant**, or **lumped parameter**.

lumped constant Same as **lumped component**.

lumped impedance An impedance which is concentrated within a circuit element. This contrasts with **distributed impedance**, which is evenly distributed throughout a circuit.

lumped inductance An inductance which is concentrated within a circuit element, such as a coil. This contrasts with **distributed inductance**, which is evenly distributed throughout a circuit.

lumped network An electric network consisting of **lumped components**. Also called **lumped-parameter network**.

lumped parameter Same as **lumped component**.

lumped-parameter network Same as **lumped network**.

lumped resistance A resistance which is concentrated within a circuit element, such as a resistor. This contrasts with **distributed resistance**, which is evenly distributed throughout a circuit.

Luneberg lens Same as **Luneburg lens**.

Luneburg lens A lens which focuses energy at ultrahigh frequencies, and which consists of a dielectric sphere whose index of refraction varies as a function of the distance from the center. Also spelled **Luneberg lens**.

lurk To view messages and postings other users enter or place in settings such as chat rooms, without posting, replying, or the like.

LUT Abbreviation of **lookup table**.

lutetium A silvery-white metallic chemical element whose atomic number is 71. It is soft, ductile, and has around 35 known isotopes, of which one is stable. Its applications include its use in semiconductors and in nuclear technology. Its chemical symbol is **Lu**.

lux The SI unit of illuminance. It is also a unit of illumination. It is equal to the illumination of 1 lumen per square meter, which is also equal to approximately 0.0929 foot-candles, or 0.0001 phot. Its symbol is **lx**.

Luxemburg effect The cross modulation of two radio signals which pass through the same region of the ionosphere. This effect may result in the broadcast of a more powerful station being heard at the frequency of a weaker station.

luxmeter An illuminometer whose readings are in luxes.

LVDT Abbreviation of **linear variable differential transformer**.

lx Symbol for **lux**.

ly Abbreviation of **light year**.

Lycos A popular Web portal and search engine.

Lynx A popular text-based Web browser.

LZW Same as **LZW compression**.

LZW compression Abbreviation of Lempel-Zif-Welch **compression**. A data-compression technique which makes use of recurring byte sequences to convert input character strings into output code strings, utilizing pointers to refer to recurring data. Used, for instance, for GIF compression. Also called **LZW**.

M

m **1.** Abbreviation of **milli-**. **2.** Abbreviation of **meter**. **3.** Symbol for **mass**.

m- A prefix which serves to produce derivative words pertaining to mobile communications concepts. For instance, m-business, or m-mail. It is an abbreviation of **mobile**.

M **1.** Abbreviation of **mega-**. **2.** Abbreviation of **megohm**. **3.** Same as MUMPS.

M **1.** Abbreviation of **mutual induction**. **2.** Abbreviation of **mutual inductance**.

m-banking Abbreviation of mobile **banking**. Electronic banking utilizing mobile devices such as cell phones and PDAs.

m-business Abbreviation of mobile **business**. Electronic business effected utilizing mobile devices such as cell phones and PDAs. Also called **m-commerce**.

m-commerce Abbreviation of mobile **commerce**. Same as **m-business**.

M display In radars, an oscilloscopic display similar to an A display, but which incorporates a pedestal which is moved along the baseline until it coincides with the horizontal position of the blip, so as to indicate the range of the scanned object. Also called **M scope**, **M scanner**, **M scan**, or **M indicator**.

M electron An electron located in the **M shell** of an atom.

M indicator Same as **M display**.

m.m.f. Abbreviation of **magnetomotive force**.

m-mail Abbreviation of mobile **mail**. SMSs, email, and the like, sent between mobile devices such as cell phones and PDAs.

m/s Abbreviation of **meters per second**.

M scan Same as **M display**.

M scanner Same as **M display**.

M scope Same as **M display**.

m/sec Abbreviation of **meters per second**.

M shell The third electron shell of an atom. It is the third closest shell to the nucleus.

mA Abbreviation of **milliampere**.

MA **1.** Abbreviation of **magnetic amplifier**. **2.** Abbreviation of **megampere**.

MAC **1.** Abbreviation of **Media Access Control**. **2.** Abbreviation of **Message Authentication Code**. **3.** Abbreviation of **multiplier-accumulator**.

MAC address Abbreviation of **Media Access Control address**.

MAC Layer Abbreviation of **Media Access Control Layer**.

Mac OS A popular operating system.

MAC Sublayer Abbreviation of **Media Access Control Sublayer**.

Mach number A measure of velocity relative to the speed of sound for a given medium. For instance, Mach 2.0 is equal to twice the speed of sound for a given medium. In dry air, for example, at 0° Celsius, and at one atmosphere pressure, sound travels at approximately 331.6 meters per second, so mach 2.0 in this case would be about 1,194 km/hr. Usually utilized to express the speed of an aircraft relative to the speed of sound in air.

machine **1.** A mechanical, electric, electronic, electromechanical, mechanically-actuated electromagnetic apparatus, or the like, which performs a specific function, such as utilizing mechanical energy to perform work. Although the term usually implies carrying out mechanical work, or making it easier, machine can be more generally defined as that which performs or assists in the performance of most any task. **2.** A **computer**. Also, a specific computer, such as that in a network. **3.** Any hardware used for a given task, such as a business machine or automated teller machine.

machine address A specific memory location. It is an address from which relative addresses may be derived. Also called **absolute address**, **real address (1)**, **actual address (1)**, or **direct address**.

machine check The verification or testing of the hardware of a computer system. Such checks are usually done automatically when the system is powered on. Also called **hardware check (1)**.

machine code Same as **machine language**.

machine cycle **1.** A cycle consisting of the four-step sequence a CPU follows when executing a machine language instruction. These are fetch, decode, execute, and store, although the store step may be considered a part of the execute step. **2.** The minimum time interval necessary for a CPU to carry out the operation it performs fastest.

machine-dependent Computer software or hardware which is designed to work properly only with a specific architecture or configuration. Assembly language, for instance, is machine-dependent. This contrasts with **machine-independent**, in which software or hardware can run on a variety of architectures or configurations. Also called **hardware-dependent**.

machine error In computers, an inability to carry out an operation, or an incorrect result, as a result of a hardware deficiency.

machine identification **1.** The identification of a specific machine, such as a network computer. **2.** The use of a machine to identify something, such as printed characters.

machine-independent Computer software or hardware which is designed to work properly with a variety of architectures or configurations. An interpreter version of LISP is machine-independent. This contrasts with **machine-dependent**, in which software or hardware is designed to work properly only with a specific architecture or configuration. Also called **hardware-independent**.

machine instruction A computer instruction in machine code. Such an instruction can be directly executed by a processor. Also called or **instruction (2)**, or **computer instruction (2)**.

machine language The only programming language a computer can understand. All programming languages above this, the lowest level language, must be assembled, compiled, or interpreted, to become the binary-coded machine instructions that are executed by the CPU. Each CPU, or CPU family, has its own machine language. Also called **machine code**.

machine learning The ability of devices and systems to apply knowledge that has been previously programmed and recorded, in order to analyze and better face new situations.

machine logic **1.** The ability of a computer, or a machine incorporating one or more computers, to emulate logic skills, such as reasoning, problem-solving, and appropriate decision making. **2.** The circuits in a computer which enable the performance of logic functions or operations, such as AND, OR, and NOT. These include gates and flip-flops. Also, the manner in which these circuits are arranged. Also called **logic (1)**, or **computer logic (1)**. **3.** The totality of the circuitry contained in a computer. Also called **logic (2)**, or **computer logic (2)**.

machine-oriented language A low-level language, such as an assembly or machine language, which is designed for use only with a specific computer line. Also called **computer-oriented language**.

machine-readable Information that is in a form that can serve as computer input. This includes information in binary form, such as files stored on magnetic disks, and data obtainable through optical-character recognition. Also called **computer-readable**.

machine translation The use of a computer program to translate from one natural language to another, or others, and vice versa. Also called **mechanical translation**.

machine vision 1. The use of hardware and software to emulate functions of an eye. A charge-coupled device, for instance, may be used for the input of images, with the obtained information passed on to a processor. Machine vision is not necessarily limited to the visible spectrum, and may also include interpretation of three-dimensional surfaces. Used, for instance, in optical character recognition, medical imaging systems, and in many areas of robotics. 2. The gathering of visual information using **machine vision (1)**.

machine word The fundamental unit of storage for a given computer architecture. It represents the maximum number of bits that can be held in its registers and be processed at one time. A word for computers with a 32-bit data bus is 32 bits, or 4 bytes. A word for computers with a 256-bit data bus is 256 bits, or 32 bytes, and so on. Also called **word**, or **computer word**.

machining The process of forming workpieces into a desired shape and size by lapping, drilling, polishing, cutting, grinding, vaporizing, and so on.

Macintosh A popular line of personal computers featuring a graphical user interface.

macro 1. A series of computer keystrokes, functions, and/or commands that are recorded for later use. A macro may be saved as a keyboard shortcut such as CTRL-M, and can serve, for instance, to simplify what would otherwise be a tedious process, or to save time by carrying out a lengthy sequence of operations without concurrent user intervention. Also, the name, symbol, or shortcut utilized to access this sequence. Also called **macroinstruction (2)**. 2. A name or subroutine which defines a series of instructions to be carried out when a programming language is assembled or compiled.

macro assembler An assembler that can define and utilize macros.

macro call 1. In computer programming, a statement or instruction which invokes a macro. 2. Same as **macroinstruction**.

macro cell Same as **macrocell**.

macro expansion The process of converting the name, symbol, or shortcut utilized to represent a macro into its defined sequence. Also called **macro substitution**, or **macro replacement**.

macro generator Same as **macro recorder**.

macro language 1. A programming language designed for the defining of macros. 2. A programming language that makes extensive use of macros. 3. The set of macros that a given macro processor can utilize.

macro processor A program or routine which performs macro expansion.

macro recorder A program or routine which generates and stores **macros (1)**. Also called **macro generator**.

macro replacement Same as **macro expansion**.

macro substitution Same as **macro expansion**.

macro virus A virus encoded as a macro, or which is incorporated into a macro. Such viruses can be embedded into a document file, for instance, and infect the computer each time said file is opened.

macrocell Also spelled **macro cell**. Also called **hard macro**. 1. A gate array which with added features, such as storage elements and path controls, which enhance its functionality. 2. A gate array with a logic function specified, and in which the interconnections between its cells are defined.

macroinstruction Also called **macro call (2)**. 1. An instruction which defines or invokes a macro. 2. Same as **macro (1)**.

macroprogramming The use of macros for writing computer programs.

macroscopic 1. Large enough to be perceived by an unaided eye. 2. Pertaining to that which is perceived by an unaided eye.

macrosonics The study and use of sound waves with very large amplitudes.

MAE Abbreviation of **Metropolitan Area Exchange**.

mag Abbreviation of **magnetic**.

MAG Abbreviation of **maximum available gain**.

magamp Abbreviation of **magnetic amplifier**.

magazine A compartment utilized for storage or for supplying a material to be used. For example, the compartment in a camera where a film cartridge or disk is placed.

magenta A purplish red color. It is one of the primary colors used in certain color processes, such as printing.

magic cookie 1. A block of data prepared by a Web server, which is sent to a Web browser for storage, and which remains ready to be sent back when needed. A user initially provides key information, such as that required for an online purchase, and this is stored in a cookie file. When a user returns to the Web site of this online retailer, the cookie is sent back to the server, enabling Web pages that are customized to include information such as the mailing address, viewing preferences, or the content of a recent order. Also called **cookie, Internet cookie, browser cookie**, or **Web cookie**. 2. In UNIX, a data object passed between routines or programs which facilitate interaction.

magic smoke A smoke which is released when an IC is burned, and which is superstitiously believed to enable said chips to function, as they no longer do once this smoke is liberated. This does not, however, explain ICs which fail the smoke test and thus never worked in the first place.

magic T Same as **magic tee**.

magic tee A waveguide, resistor, or transformer circuit or device which has four branches which are arranged in a manner that the input signal is equally divided between the adjacent branches. The signal does not reach the fourth, or opposite, branch. Also spelled **magic T**. Also called **hybrid junction, hybrid tee**, or **bridge hybrid**.

maglev Abbreviation of **magnetic levitation**, or **magnetically levitated**.

magnesium A silvery-white metallic chemical element with atomic number 12. It has about 15 known isotopes, of which 3 are stable. Magnesium is chemically active, hard, and when finely divided easily ignitable upon heating in air, producing a striking white flash which can be used in flash bulbs or pyrotechnics. Other applications include its use in light alloys, batteries, optical mirrors, and precision instruments. Its chemical symbol is **Mg**.

magnesium boride A compound whose chemical formula is MgB_2. Used in superconductors. Also called **magnesium diboride**.

magnesium diboride Same as **magnesium boride**.

magnesium fluoride A white powder, or crystals whose chemical formula is MgF_2. Used in lasers, in infrared and ultraviolet optical elements, and in ceramics.

magnesium oxide A white powder whose chemical formula is MgO. Used in tough glasses, as an electrical insulator, in semiconductors, superconductors, and in lasers.

magnesium silicide Bluish crystals whose chemical formula is Mg_2Si. Used in ceramics and in semiconductors.

magnesium tungstate White crystals whose chemical formula is $MgWO_4$. Used in X-ray screens.

magnet An object or body which produces an appreciable magnetic field around itself. That is, it possesses and exhibits the property of **magnetism**. A magnet thus attracts other magnetic materials, such as iron, nickel, or steel, and will attract or repel other magnets, depending on their mutual orientation of north and south poles. A magnet is usually composed of a ferromagnetic or ferrimagnetic material, and may be permanent or temporary.

magnet brake Same as **magnetic brake**.

magnet charger A device which applies a strong magnetic field to a magnetic material, so as to establish or reestablish magnetic saturation. Such a device, for instance, may use a bank of capacitors to provide a brief and intense current to an electromagnet which provides the magnetic field.

magnet coil A **magnetic coil** wound around the core of an electromagnet.

magnet keeper A bar, usually made of soft iron, utilized to join the magnetic poles of a magnet which is not in use. The keeper maintains the magnetic flux circulating continuously through the magnet, which delays demagnetization considerably. Used especially with horseshoe magnets. Also called **keeper**.

magnet meter An instrument, such as a fluxmeter, utilized to measure the magnetic flux of a magnet. Also called **magnet tester**.

magnet motor A motor which utilizes one or more permanent magnets to produce torque, or another driving force. Also called **permanent-magnet motor**.

magnet steel A steel whose retentivity is especially high. Used, for instance, to make permanent magnets.

magnet tester Same as **magnet meter**.

magnet wire A wire that is suitable for use in coils or windings, such as those of electromagnets or transformers. Such a wire is often composed of copper or aluminum, and is electrically insulated. Also called **winding wire**.

magnetic Its abbreviation is **mag**. **1.** Pertaining to, possessing, or arising from **magnetism**. **2.** Arising from, or pertaining to a **magnet**. **3.** That which is capable of being magnetized, or that can be attracted by a magnet. For instance, a ferromagnetic material. **4.** That which utilizes magnetism. For example, magnetic tape, or a magnetic pickup head.

magnetic air-gap Same as **magnetic gap**.

magnetic amplifier An electromagnetic device which utilizes one or more saturable reactors to obtain amplification. Depending on the design, a magnetic amplifier can achieve a very large power gain, and may be used, for instance, in servo systems requiring sizeable amounts of power to move heavy loads. Its abbreviation is **magamp**, or **MA**. Also called **transductor**.

magnetic analysis **1.** The utilization of magnetic fields for the separation of entities into its constituent parts. An example is mass spectrometry. **2.** The utilization of magnetic fields for the careful study of entities. An example is nuclear magnetic resonance.

magnetic anisotropy The variation of magnetic properties depending on the axis along which they are manifested or measured. This occurs, for instance, in certain magnetic alloys.

magnetic attraction The attraction a pole of a magnet has for a pole with opposite polarity of another magnet.

magnetic axis A straight line which joins the magnetic poles of a magnet.

magnetic azimuth The azimuth of an object utilizing magnetic north as the reference line.

magnetic balance A device or instrument which determines the attraction or repulsion between magnetic poles. It may consist, for instance, of a magnet which is brought near another magnet that is attached to the arm of a balance. The magnitude of the counterweight indicates the force. A magnetic balance may also be used to calculate the strength of magnetic fields.

magnetic bearing The horizontal direction of an object or point, measured clockwise with respect to magnetic north. Usually expressed in degrees.

magnetic bias **1.** A steady magnetic field which is applied to a magnetic circuit, to establish a reference level for its operation. **2.** In magnetic tape recording, a current that is applied to the audio signal to be recorded, in order to optimize performance during playback. This current varies depending on the tape type. Also called **magnetic biasing (2)**, **bias (3)**, or **bias current (2)**.

magnetic biasing **1.** In magnetic tape recording, the superimposition of another magnetic field in addition to the signal being recorded, to help condition the recording medium. **2.** Same as **magnetic bias (2)**.

magnetic bit A bit whose value of 0 or 1 is determined by the magnetic polarity of a material, such as that which is magnetoresistive. A surrounding magnetic field is utilized to set the polarity of said bits, and permanent magnets assist in retaining the stored information in case power is cut off. Used, for instance, in MRAM.

magnetic blow-out Same as **magnetic blowout**.

magnetic blowout Also spelled **magnetic blow-out**. **1.** A device, such as an electromagnet, which produces a powerful magnetic field which extinguishes an arc by lengthening or deflecting it. Used, for instance, as a circuit-breaker. **2.** The extinguishing of an arc through the use of a **magnetic blowout (1)**.

magnetic bottle A device utilizing a magnetic field to confine a plasma within given boundaries. Used, for instance, in controlled nuclear fusion experiments. Also, the magnetic field so used.

magnetic brake A friction brake controlled by an electromagnet. A **friction brake** is one in which friction provides the stopping resistance. Also called **electromagnetic brake**, or **magnet brake**.

magnetic braking The use of a **magnetic brake** for braking.

magnetic bubble memory A type of non-volatile computer memory that utilizes materials, such as crystals, to retain a magnetic polarity. This polarity determines whether ones or zeroes are being stored, and with the assistance of permanent magnets, also serves to retain the stored information in case power is cut off. Used primarily for rugged computing circumstances. Also called **magnetic bubble storage**, or **bubble memory**.

magnetic bubble storage Same as **magnetic bubble memory**.

magnetic card A card which has a portion, such as a magnetic stripe, upon which data can be stored through the selective magnetization of the surface. Examples include credit and debit cards, driver's licenses, telephone calling cards, and so on. The encoded information in such a card is usually accessed using a magnetic card reader. Also called **magnetic stripe card**.

magnetic card reader A device which reads information which has been encoded onto a **magnetic card**. Also called **card reader**.

magnetic cartridge A phonographic pickup that converts the movements of the stylus into the corresponding audio-frequency voltage output by varying the reluctance of an internal magnetic circuit. Also called **magnetic pickup (1)**, or **variable-reluctance pickup**.

magnetic character A character printed with **magnetic ink**.

magnetic character reader A device which scans magnetic characters, and utilizes magnetic character recognition to identify them.

magnetic character recognition The ability of a computer, or other device, to identify scanned magnetic characters, and to convert this information into a digital code suitable for further processing. An example is a magnetic-ink character recognition system utilized to identify bank checks.

magnetic circuit A closed path described by magnetic lines of force. A magnetic circuit is analogous to an electric circuit in that the magnetomotive force corresponds to the electromotive force, reluctance to resistance, and magnetic flux to electric current.

magnetic circuit breaker A circuit breaker which utilizes a magnetic field for opening a circuit under specified conditions. Such devices are especially useful, for instance, for accurate and reliable circuit protection when used over a wide temperature interval, which can be a problem with thermal circuit breakers.

magnetic coercive force The magnetic field which must be applied to a previously magnetized ferromagnetic material, in order to reduce to zero the residual magnetism of said material. Usually expressed in oersteds. Also called **coercive force**.

magnetic coercivity For a given ferromagnetic material, the ease or difficulty in it being magnetized or demagnetized. A material exhibiting low coercivity is easy to magnetize and demagnetize, while a material with high coercivity is hard to magnetize and demagnetize. Also called **coercivity**.

magnetic coil One or more conductors wound in a series of turns, so as to introduce inductance, or to produce a magnetic field in an electric circuit. A magnetic coil may be wound, for example, around a ferromagnetic core. Used, for instance, in transformers, electromagnets, solenoids, motors, loudspeakers, and so on. Also known as **magnet coil, coil, inductor, inductance (3), inductance coil,** or **electric coil**.

magnetic compass A magnetic instrument which indicates a horizontal reference direction. Such an instrument usually consists of one or more magnetic needles which freely turn on a pivot and point to magnetic north.

magnetic component In an electromagnetic wave, the magnetic constituent. It is perpendicular to the electric component. Also called **electromagnetic component**.

magnetic conductivity Same as **magnetic permeability**.

magnetic confinement The confinement of a plasma using magnetic fields. A magnetic bottle may be used for this purpose.

magnetic constant A physical constant equal to $4\pi \cdot 10^{-7}$ henry per meter, or approximately 1.256637×10^{-6} henry per meter. This number represents the absolute permeability of free space. The relative permeability of free space is 1. Its symbol is μ_0. Also called **permeability of free space**.

magnetic contactor A contactor which is actuated via electromagnetic means.

magnetic control The control of a device, piece of equipment, system, or process through the use of electromagnetic components, such as magnetic contactors.

magnetic controller An automatic controller which utilizes **magnetic control**.

magnetic core 1. A magnetic material, such as iron, around which a magnetic coil is wound. Such a core may be used in a device such as an electromagnet or transformer, and increases the inductance of the coil. Also called **core (3)**. **2.** Same as **magnetic core memory**.

magnetic core memory A magnetic material, such as iron oxide, used as a bistable element, multiple units of which are configured in an array which may store data. It is a seldom-used form of non-volatile computer memory. Also called **magnetic core (2), core (4), core storage (1),** or **core memory(1)**.

magnetic coupling Coupling in which magnetic energy is transferred between components or circuits. An example is electromagnetic coupling.

magnetic cycle 1. The variations in the magnetic flux of a material or medium as the strength of an influencing magnetic field varies through a complete cycle. For example, the changes in the magnetization of a body as a magnetizing AC fluctuates. **2.** A cyclical variation in which magnetic properties vary in a regular manner. For example, the 22 year solar magnetic cycle.

magnetic damping The use of one or more magnetic fields for damping. An example would be the use of electromagnetic induction to limit overshoot of the pointer of an instrument.

magnetic declination For a given geographic location, the angle representing the difference between magnetic north and true north. Also called **declination**.

magnetic deflection The use of one or more magnetic fields for deflecting particles or objects. An example is electrostatic deflection.

magnetic density For a given material, medium, or region, the magnetic lines of force per unit area of cross-section.

magnetic detector A device which detects magnetic materials. Such a detector can be used in robotics, for example, to locate materials of a specific composition. Also called **magnetic sensor**.

magnetic dip For a given location on the surface of the earth, the angle between the horizontal plane and the direction of the lines of force of the earth's magnetic field. Magnetic dip is 0° at the magnetic equator and 90° at each of the magnetic poles. Also called **magnetic inclination, inclination (2),** or **dip (3)**.

magnetic dipole Two equal magnetic poles with opposite polarity separated by a very small distance. Two equal magnetic poles with opposite polarity separated by a very small distance. Also called **magnetic doublet, dipole (2),** or **doublet (3)**.

magnetic dipole moment For a magnet placed in a magnetic field, the maximum torque experienced by the magnet divided by the magnitude of the magnetic field acting on said magnet. May be expressed in joules/tesla. Also called **magnetic dipole moment,** or **dipole moment (2)**.

magnetic disc Same as **magnetic disk**.

magnetic disk A digital magnetic storage medium, which is usually in the form of a rotating round plate. Information is encoded by altering the magnetic polarity of minute portions on the surface of such a disk. Magnetic disks may be recorded, erased, and rerecorded many times, and the most common formats are hard disks, floppy disks, and removable hard disks. Also spelled **magnetic disc**. Also called **disk (1)**.

magnetic displacement Same as **magnetic flux density**.

magnetic domain Within a ferromagnetic substance, a region in which the atomic or ionic magnetic moments are aligned in the same direction. Such a domain may be used, for instance, to store a bit as a 1 or a 0 on a magnetic tape or disk. Also called **ferromagnetic domain,** or **domain (2)**.

magnetic doublet Two equal magnetic poles with opposite polarity separated by a very small distance. Also called **magnetic dipole, dipole (2),** or **doublet (3)**.

magnetic drum An early type of data-storage device, which consisted of a cylinder coated with a magnetic material, and whose tracks each had their own read/write head. Also called **drum storage,** or **drum (2)**.

magnetic earphone An earphone consisting of a small or miniature **magnetic speaker**. A pair may be used for stereophonic listening.

magnetic energy 1. The energy of a magnetic field. 2. The energy necessary to establish a magnetic field.

magnetic equator A great line along the surface of the earth that connects all points at which the magnetic dip is 0°. This contrasts with the **geomagnetic equator**, which is a great circle which at each point is 90° from the geomagnetic poles. Also called **dip equator**, or **aclinic line**.

magnetic field 1. The region in space surrounding a magnetic body or entity, such as a permanent magnet or a conductor carrying a current, where an appreciable magnetic force is present. Such a field is represented by magnetic lines of force. The planet earth has a magnetic field, and its lines of force are considered to extend from the north magnetic pole to the south magnetic pole. 2. Same as **magnetic field strength**. 3. In an electromagnetic field, the magnetic field. The magnetic field is perpendicular to the electric field.

magnetic field intensity Same as **magnetic field strength**. Its symbol is H.

magnetic field strength The intensity of a magnetic field at a given point. It may be quantified, for instance, as the force a magnetic field exerts on a unit magnetic pole placed at a given point in space which is in the direction of the line of force for said point. It is a vector quantity, and is usually expressed in amperes per meter, or in oersteds. Its symbol is H. Also called **magnetic field intensity**, **magnetic intensity**, **magnetic force**, **magnetizing force**, or **magnetic field** (2).

magnetic flip-flop A flip-flop in which magnetic properties determine the two stable states. For example, that utilizing a magnet that flips the orientation of its magnetic field at a given temperature, and flips back at another.

magnetic flow meter Same as **magnetic flowmeter**.

magnetic flowmeter An instrument which measures the magnetic flux of a conductive fluid to determine its flow rate. Also spelled **magnetic flow meter**.

magnetic fluid A fluid, such as an oil, with suspended particles of iron or another magnetic material. Such a fluid can be controlled by a magnetic field.

magnetic flux 1. The product of the magnetic flux density, in teslas, and the area, in square meters, of the surface through which the lines of force flow perpendicularly. Expressed in webers. 2. For a given region, the magnetic lines of force. The closer together the lines, the greater the strength of the magnetic field. Also called **flux** (3).

magnetic flux density A quantitative measure of the strength of a magnetic field present within a magnetic medium. It is the number of lines of force per cross-sectional area of a magnetic circuit, and its SI unit is the tesla. Its symbol is B. Also called **magnetic induction** (1), or **magnetic displacement**.

magnetic flux lines Same as **magnetic lines of force**.

magnetic flux linkage The passage of magnetic flux from one circuit component to another. Also called **flux linkage**, or **linkage** (3).

magnetic focusing In a TV picture tube, the use of a magnetic field to focus the electron beam. This may be accomplished by varying the DC flowing through one or more focusing coils mounted on the neck of the tube. Also called **electromagnetic focusing**.

magnetic force Same as **magnetic field strength**.

magnetic force lines Same as **magnetic lines of force**.

magnetic force microscope A scanning probe microscope in which the minute tip is coated with a magnetic material, so that the magnetic fields immediately above the surface of a sample can be imaged with the assistance of a computer. Its abbreviation is **MFM**.

magnetic force microscopy The use of a **magnetic force microscope** to view and analyze specimens. Its abbreviation is **MFM**.

magnetic forming The use of an intense transient magnetic field to form a metal workpiece into a desired shape. A large capacitor bank is charged, then instantaneously discharged to an induction coil, creating the magnetic field. Also called **magnetic pulse forming**, or **electromagnetic forming**.

magnetic friction Same as **magnetic hysteresis**.

magnetic gap A space of a nonmagnetic material, such as air, between magnetic components, circuits, devices, or materials. For instance, the opening between pole pieces in a tape head. Also called **magnetic air-gap**, **gap** (3), or **air gap** (2).

magnetic head Also called **magnetic tape head**. 1. A transducer which converts electrical signals into magnetic signals which can be recorded or stored on magnetic recording media, such as tapes or disks. Also called **magnetic recording head**, **recording head**, or **write head**. 2. A transducer which converts magnetic signals recorded or stored on magnetic recording media, such as tapes or disks, into output electrical signals. When used in the context of tape which has audio and/or video, the signals are usually said to be reproduced, while in the context of data, the signals are usually said to be read. Also called **magnetic read head**, **magnetic reading head**, **read head**, **reading head**, **playback head**, **play head**, or **reproduce head**. 3. A device which serves to delete material recorded on a magnetic storage medium, such as a magnetic tape or disk. For instance, an AC erase head employs AC to create an alternating magnetic field which erases the tape. Also called **erase head**. 4. A transducer which combines the functions of a **magnetic head** (1) and a **magnetic head** (2). Also called **magnetic read/write head**, or **read/write head** (2).

magnetic heading The horizontal direction in which a craft is pointing, expressed as an angular distance from magnetic north.

magnetic hysteresis A phenomenon in which the changes in the magnetization induced in a ferromagnetic material lag behind the changes in the magnetizing force. This phenomenon is observed below the Curie point. Also called **magnetic friction**, or **hysteresis** (2).

magnetic inclination Same as **magnetic dip**.

magnetic induction 1. Same as **magnetic flux density**. 2. The inducing of magnetization in magnetic material, such as iron, when placed in a magnetic field generated, for instance, by an electromagnet.

magnetic ink An ink incorporating particles which can be magnetized, so that the printed characters can be read by magnetic-ink character recognition systems. Magnetic ink can also be read by an unaided eye.

Magnetic-Ink Character Recognition A character recognition system in which characters printed in magnetic ink are scanned and converted into digital code suitable for further processing. In addition to the special ink, said characters have a specified font, to help ensure proper translation. A common example is the system used by financial institutions to sense the information printed at the bottom of checks, for proper handling. Its abbreviation is **MICR**.

magnetic instability 1. A fluctuating, erratic, or otherwise unstable magnetic field. 2. A fluctuating, erratic, or deteriorating magnetic recording medium. For example, a VHS cassette's reduction in image quality due to aging.

magnetic intensity Same as **magnetic field strength**. Its symbol is H.

magnetic iron oxide A reddish or bluish-black powder whose chemical formula is Fe_3O_4. It is used in magnetic-tape coatings, magnetic inks, and in ferrites. Also called **iron oxide, iron oxide black, ferrosoferric oxide**, or **black iron oxide**.

magnetic leakage The usually undesired dispersion, escape, or entry of magnetic flux.

magnetic lens A device which utilizes a magnetic field to focus a particle beam, analogous to the manner in which an optical lens focuses a light beam. Such a lens may consist, for instance, of an array of electromagnets. Also called **lens (4)**.

magnetic levitation Its abbreviation is **maglev**. **1.** The use of magnetic forces to lift, propel, and direct a vehicle or craft, with no physical contact, above its guiding track. One such technology employs a linear motor consisting of electromagnets, on both the guiding track and the rail cars propelled along it, which are energized and deenergized in a manner which provides either acceleration or deceleration. **2.** The use of a magnetic field to induce levitation. **3.** The repelling of a magnetic field to induce levitation.

magnetic lines of flux Same as **magnetic lines of force**.

magnetic lines of force Within a magnetic field, imaginary lines whose tangent at any given point represent the direction of said magnetic field at that point. Magnetic lines of force are utilized to represent the magnetic field surrounding a magnetic body or entity, such as a permanent magnet or a medium carrying a current. The closer together the lines, the greater the strength of the magnetic field. Also called **magnetic lines of flux, magnetic flux lines, magnetic force lines, lines of force (2)**, or **force lines (2)**.

magnetic loudspeaker Same as **magnetic speaker**.

magnetic material A material, especially that which is ferromagnetic, which exhibits magnetic properties naturally.

magnetic media Plural form of **magnetic medium**.

magnetic medium A medium, such as a magnetic disk, tape, or drum, with a magnetizable layer suitable for recording.

magnetic memory Same as **magnetic storage (1)**.

magnetic meridian A horizontal line which at any point is oriented in the direction of the horizontal component of the earth's magnetic field at said point. The needle of a compass without deviation will align itself along such a line. The planet's magnetic meridians converge in two places, which are the north magnetic pole and the south magnetic pole.

magnetic microphone A microphone, such as a dynamic microphone, in which the diaphragm is connected to a coil which moves within a magnetic field to produce the output signal current.

magnetic moment For a magnet placed in a magnetic field, it is equal to the maximum torque experienced by the magnet, divided by the magnitude of the magnetic field acting on said magnet. May be expressed in joules/tesla. Also called **magnetic moment**, or **dipole moment (2)**.

magnetic monopole A hypothetical particle possessing a single isolated magnetic charge analogous to an electric monopole's electric charge. Also called **monopole (2)**.

magnetic needle A needle in a **magnetic compass**. Also called **needle (4)**.

magnetic north For a magnetic compass without deviation, the direction in which the north end of a needle points.

magnetic north pole The location in the northern hemisphere where the magnetic dip is 90°. The corresponding spot in the southern hemisphere is the **south magnetic pole**. At each of these places, the magnetic meridians converge. The north magnetic pole does not coincide with the north geographic pole, and the position of both magnetic poles vary over time. Also called **north magnetic pole, north pole (2)**, or **positive pole (2)**.

magnetic oxide An oxide of iron, such as magnetic iron oxide or ferric oxide, with magnetic properties.

magnetic permeability Its symbol is μ. A quantitative indicator of the response of a material or medium to a magnetic field. It is expressed by the following formula: $\mu = B/H$, where B is the magnetic flux density of the material or medium, and H is the strength of the surrounding magnetic field. Relative permeability is obtained using the following formula: $\mu_r = \mu/\mu_0$, where μ_r is relative permeability, and μ_0 is the permeability of free space. The absolute permeability of free space is a physical constant equal to approximately 1.256637×10^{-6} henry per meter, and its relative permeability is equal to 1, as seen by the this formula. Materials whose relative permeability is less than 1 are diamagnetic, those slightly above 1 are paramagnetic, and those substantially above 1 are ferromagnetic. Also known as **magnetic conductivity**, or **permeability (1)**.

magnetic permeance The ease with which a material or magnetic circuit allows a magnetic field to pass through it. It is expressed as the ratio of the magnetic flux to the magnetomotive force, and is analogous to electrical conductance. It is the reciprocal of **magnetic reluctance**. Its symbol is P, or Λ. Also called **permeance**.

magnetic pickup Same as **magnetic cartridge**.

magnetic pickup coil Same as **magnetic probe**.

magnetic polarity In a magnet or magnetic body, the possession and manifestation of two regions where magnetic intensity is at a maximum, each of which is opposite in nature. One region is labeled as the north, or positive, pole and the other as the south, or negative, pole. Also called **polarity (4)**.

magnetic polarization The lining up of magnetic dipoles in a material. In a ferromagnetic material magnetic polarization occurs spontaneously due to the mutual attraction of said dipoles. Also called **polarization (6)**.

magnetic pole 1. Either of the two locations on the surface of the planet earth where the magnetic dip is 90°. At such a place, the magnetic meridians converge. In the northern hemisphere such spot is called the **north magnetic pole**, while that of the southern hemisphere is the **south magnetic pole**. The earth's magnetic poles do not coincide with its geographic poles, and the latter's position vary over time. Also called **dip pole**, or **pole (3)**. **2.** In a magnet or magnetic body, either of the two regions where magnetic intensity is at a maximum. The magnetic lines of force converge in such areas, which are labeled north pole and south pole. If each of such regions is considered to occupy a single point in space, it could be considered to be analogous to an electric point charge. Also called **pole (4)**.

magnetic potential difference Same as **magnetomotive force**.

magnetic pressure Same as **magnetomotive force**.

magnetic printing 1. The usually unwanted transfer of a recorded signal between sections of a magnetic medium such as a tape. This may occur as a result of the affected sections being in close proximity, as occurs with overlapping segments of a reel of tape. Also called **magnetic transfer, print-through**, or **crosstalk (3)**. **2.** Any printing in which **magnetic ink** is utilized.

magnetic probe A small coil which is inserted into a magnetic field to measure its field strength, or any variations in it. Such a coil is connected to an indicating instrument, such as a ballistic galvanometer or a fluxmeter, and may or may not incorporate an amplifier. A magnetic probe may also be utilized to examine the magnetic flux distribution of a magnetic field. Also called **magnetic test coil, magnetic pickup coil, search coil, flip coil, pickup coil**, or **probe coil**.

magnetic pulse forming The use of an intense transient magnetic field to form a metal workpiece into a desired shape.

A large capacitor bank is charged, then instantaneously discharged to an induction coil, creating the magnetic field. Also called **magnetic forming**, or **electromagnetic forming**.

magnetic quantity 1. The magnitude of a magnetic characteristic, such as susceptibility or coercive force. **2.** Same as **magnetic unit**.

magnetic RAM Same as **MRAM**.

magnetic random-access memory Same as **MRAM**.

magnetic read head Same as **magnetic head (2)**.

magnetic read/write head Same as **magnetic head (4)**.

magnetic reading head Same as **magnetic head (2)**.

magnetic recorder A device utilized for **magnetic recording**.

magnetic recording The utilization of a variable magnetic field to store a variable electrical signal on a magnetic storage medium such as a tape, disk, drum, or wire. For example, sound and/or images which wish to be recorded may be picked up by transducers such as microphones and/or cameras, where the audio and/or visual information is converted into an electrical signal. This signal is fed to a magnetic head, which then converts it into a magnetic signal which is recorded or stored on a suitable medium such as tape with an iron oxide coating.

magnetic recording head Same as **magnetic head (1)**.

magnetic recording medium A medium, such as a tape or disk, suitable for **magnetic recording**.

magnetic reed A reed which is made to vibrate or otherwise move via magnetic means.

magnetic-reed relay A relay whose contacts are mounted on magnetic reeds, and which is hermetically sealed in a glass tube. Such a relay is designed to be actuated by an external magnetic field. The contacts are not wetted by a pool of mercury, as is the case with a **mercury-wetted reed relay**. Also called **dry-reed relay**, or **reed relay**.

magnetic-reed switch A switch whose contacts are mounted on magnetic reeds, and which is hermetically sealed in a glass tube. Such a switch is designed to be actuated by an external magnetic field. The contacts are not wetted by a pool of mercury, as is the case with a **mercury-wetted reed switch**. Also called **dry-reed switch**, or **reed switch**.

magnetic relay A relay, such as a magnetic-reed or electromagnetic relay, which is actuated by a magnetic field. Such a field may be produced externally or internally.

magnetic reluctance The opposition a material or magnetic circuit presents to the passage of magnetic flux. It is expressed as the ratio of the magnetomotive force to the magnetic flux, and is analogous to electrical resistance. Its symbol is R, and it is the reciprocal of **magnetic permeance**. Also called **magnetic resistance**, or **reluctance**.

magnetic reluctivity The reciprocal of **magnetic permeability**. Also called **reluctivity**.

magnetic remanence Any magnetism which remains in a material or medium after a magnetizing force is removed. For example, that still present after a strong external magnetic field is removed. The magnetic remanence of a magnetic tape, for instance, enables it to be able to record and store information. Also called **remanence**.

magnetic repulsion The repulsion a pole of a magnet has for a pole with the same polarity of another magnet.

magnetic resistance Same as **magnetic reluctance**.

magnetic resonance Resonance manifested by certain atoms changing their magnetic spin as they are exposed to a varying electromagnetic field. These atoms absorb energy at specific resonant frequencies while releasing energy. Also known as **spin resonance**.

magnetic resonance imaging The use of magnetic resonance to create cross-sectional images of internal body structures.

In it, a patient is placed in an enclosure surrounded by powerful electromagnets which magnetically align the nuclei of certain atoms in the body. Then, radio waves are directed at these nuclei, usually the proton in hydrogen, and withdrawn, so that when the nuclei revert to their original state they emit radio signals which a computer utilizes to produce the desired images. It is a very sensitive and noninvasive procedure utilized to help study and diagnose a broad range of medical conditions, including cancer, heart disease, and musculoskeletal disorders. Its abbreviations are **MRI**, **MR imaging**, and **MRI imaging**. Also called **nuclear magnetic resonance imaging**.

magnetic retentivity Also called **retentivity**. **1.** The ability of a material to retain magnetism after a magnetizing force is removed. **2.** The degree to which magnetism remains in a material or medium after a magnetizing force is removed. **3.** A specific value of **retentivity (2)**.

magnetic rigidity The product of the magnetic intensity perpendicular to the path of a particle and the radius of the curvature of said particle's motion. It is a measure of the momentum of a particle.

magnetic rotation The rotation of the plane of polarization of an electromagnetic wave as it passes through certain materials, when under the influence of a magnetic field parallel to the direction of propagation of said wave. Also called **Faraday effect**, or **Kundt effect**.

magnetic saturation For a magnetic material, the condition in which an increase in magnetizing force will not produce an increase in magnetization. Also called **saturation (2)**.

magnetic sensor A device which detects magnetic materials. Used in robotics, for example, to locate materials of a specific composition. Also called **magnetic detector**.

magnetic shield Same as **magnetic shielding (1)**.

magnetic shielding 1. A material or enclosure utilized to isolate a magnetic field from its surroundings, or vice versa. Materials with high magnetic permeability are used for this purpose. Also called **magnetic shield**, or **shielding (2)**. **2.** The use of **magnetic shielding (1)**.

magnetic shunt A piece of a ferromagnetic material, such as soft iron, which is utilized to divert a portion of the magnetic flux of a magnet or electromagnet. Used, for instance, to extend the range of a measuring instrument.

magnetic south For a magnetic compass without deviation, the direction opposite to that of the indicated **magnetic north**.

magnetic south pole The location in the southern hemisphere where the magnetic dip is 90°. The corresponding spot in the northern hemisphere is the **north magnetic pole**. At each of these places, the magnetic meridians converge. The south magnetic pole does not coincide with the south geographic pole, and the position of both magnetic poles vary over time. Also called **south magnetic pole, south pole (2)**, or **negative pole (2)**.

magnetic speaker A speaker, such as a dynamic loudspeaker, in which the diaphragm vibrates as influenced by a surrounding magnetic field which varies proportionally to the audio signal. Also called **magnetic loudspeaker**.

magnetic storage 1. A storage medium in which data can be encoded magnetically. Examples include magnetic disks, tapes, and drums. Also called **magnetic memory**. **2.** The saving of data utilizing **magnetic storage (1)**.

magnetic storm A large-scale, often worldwide, disturbance of the geomagnetic field. Such storms are characterized by a sudden onset, and their effects may last from a few minutes to several days. Magnetic storms are usually caused by solar flares, and affect radio-frequency communications. Also called **geomagnetic storm**.

magnetic strip Same as **magnetic stripe**.

magnetic strip reader Same as **magnetic stripe reader**.

magnetic stripe A stripe, such as that on a magnetic card, upon which data can be stored through the selective magnetization of its surface. The encoded information of such a stripe is accessed using a magnetic stripe reader. Also called **magnetic strip**.

magnetic stripe card Same as **magnetic card**.

magnetic stripe reader A device which reads information which has been encoded onto a **magnetic stripe**. Also called **magnetic strip reader**.

magnetic susceptibility A measure of the response of a material to a magnetizing force. It may be expressed, for instance, as the ratio of the magnetization of a material to the magnetizing field. Also called **susceptibility**.

magnetic susceptibility meter An instrument which measures **magnetic susceptibility**. Also called **susceptometer**.

magnetic switch A switch that is actuated by a magnetic field. An example is a magnetic reed switch. Also called **magnetically operated switch**.

magnetic tape A tape utilized for **magnetic recording**. Such a tape usually consists of a strip of plastic, paper, or metal which is coated or impregnated with a magnetizable material such as iron oxide particles. Magnetic tape may be utilized to make analog or digital recordings of data, audio, and/or video. Also called **tape** (2), or **recording tape** (1).

magnetic tape drive 1. A mechanical device, such as that in a tape deck, which moves a magnetic tape past the heads, for recording, reproducing, or erasing. Such a drive also fast-forwards and rewinds tapes. Also called **magnetic tape transport**, **tape drive** (1), or **drive** (4). 2. In computers, a **magnetic tape drive** (1) used for reading or writing data. Also, a peripheral which incorporates such a drive. Also called **magnetic tape unit**, or **tape drive** (2).

magnetic tape head Same as **magnetic head**.

magnetic tape library A collection of magnetic tapes which are utilized to store data. Such tapes are usually contained in cartridges which are accessed for storage and retrieval through the use of one or more magnetic tape drives. A tape library may be used, for instance, by a server which utilizes it for archiving. Also called **tape library**.

magnetic tape player 1. Same as **magnetic tape reader**. 2. Same as **magnetic tape recorder** (2).

magnetic tape reader A device which reads information which has been encoded onto **magnetic tape**. Also called **magnetic tape player** (1).

magnetic tape recorder 1. A device utilized for **magnetic tape recording**. 2. A **magnetic tape recorder** (1) which also reproduces the content of tapes which have been recorded upon. Also called **magnetic tape player** (2).

magnetic tape recording Magnetic recording in which the medium utilized is **magnetic tape**.

magnetic tape transport Same as **magnetic tape drive** (1).

magnetic tape unit Same as **magnetic tape drive** (2).

magnetic test coil Same as **magnetic probe**.

magnetic thin film A very thin film, usually less than 10^{-6} meter, of a magnetic material deposited upon a substrate. In a hard disk, for instance, such a film may cover an aluminum platter, providing a medium suitable for data storage. Also called **thin magnetic film**.

magnetic transducer 1. A device which utilizes a magnet, coil, or magnetic field to convert one form of energy into another. 2. A device which utilizes a magnet, coil, or magnetic field to convert mechanical energy into electrical energy.

magnetic transfer Same as **magnetic printing** (1).

magnetic transition temperature The temperature at which the ferromagnetic properties of a material become paramagnetic. At or above this temperature, the thermal energy in the material is too great to exhibit ferromagnetism. Also called **Curie temperature**, **Curie point**, or **ferromagnetic Curie temperature**.

magnetic unit A unit for expressing a magnetic magnitude or measurement. For instance, tesla, weber, or ampere-turn. Also called **magnetic quantity** (2).

magnetic vector Also called **H vector**. In an electromagnetic field, the vector representing the magnetic field. The magnetic vector is perpendicular to the electric, or E, vector.

magnetic viscosity In a conducting fluid, the damping of the motion of charged particles perpendicular to the flux lines of an external magnetic field. It is analogous to ordinary viscosity, but occurs without mechanical forces.

magnetic wire 1. A wire, such as that in a coil, utilized to create a magnetic field. 2. A wire consisting of a magnetic material which can be recorded upon. May also refer to a wire which is covered with such a magnetically recordable material.

magnetically actuated That which is actuated by a magnetic field. For example, a magnetic switch.

magnetically levitated An object, such as a rail car, which is lifted, propelled, and guided via **magnetic levitation**. Its abbreviation is **maglev**.

magnetically operated That which is operated by a magnetic field. For example, a magnetically levitated vehicle.

magnetically operated switch Same as **magnetic switch**.

magnetics The branch of science that deals with the study, use, and effects of **magnetism**.

magnetism The phenomena arising from, and associated with magnets and magnetic fields. Magnetism occurs due the motion of electric charges, be they through a conductor, or within an atom. Since an electron has a charge and orbits around a nucleus, it produces a minute magnetic field. The magnetic moment of an atom is the vector sum of the magnetic moments of all its contained electrons, and the macroscopic magnetic properties observed in a material are due to the magnetic moments of the atoms, molecules, and ions that constitute said material. The main types of magnetic behavior are ferromagnetism, paramagnetism, ferrimagnetism, antiferromagnetism, and diamagnetism.

magnetite A black mineral consisting primarily of iron oxide, but which can have other metals, such as magnesium or manganese, replacing one or more iron atoms. **Lodestone**, a common variety of this mineral, exhibits polarity and is a natural magnet. The two terms are often used synonymously.

magnetization 1. The process by which a material is magnetized. 2. The extent to which a material has been magnetized. 3. The attribute of being magnetized.

magnetization curve For a magnetic material, a graphical representation showing the relationship between magnetic flux density, or B, in the vertical axis, and magnetizing force, or H, in the horizontal axis. Also called **B-H curve**.

magnetize To impart magnetization to a material.

magnetizer A device or material which imparts magnetization. Used, for instance, to make permanent magnets.

magnetizing current 1. A current which establishes or maintains a magnetic field. 2. The current which flows through an small auxiliary generator which supplies field current for an AC generator. Also called **exciting current** (1). 3. A small current which flows through the primary winding of a transformer to which no load is connected. Also called **exciting current** (2).

magnetizing force Same as **magnetic field strength**.

magneto Abbreviation of **magneto**electric generator. An electric generator which utilizes a permanent magnet to provide the magnetic flux. Used, for instance, in ignition systems. Also called **permanent-magnet generator**.

magneto-electronics Same as **magnetoelectronics**.

magneto-fluid dynamics Same as **magnetohydrodynamics**. Also spelled **magnetofluiddynamics**.

magneto-fluid mechanics Same as **magnetohydrodynamics**.

magneto-hydrodynamics Same as **magnetohydrodynamics**.

magneto-optic Pertaining to **magneto-optics**. Also spelled **magnetooptic**, or **magneto-optical**. Its abbreviation is **MO**.

magneto-optic disc Same as **magneto-optical disc**.

magneto-optic effect Also spelled **magneto-optical effect**. **1.** A phenomenon, such as the Faraday effect, in which an electromagnetic wave interacts with a magnetic field. **2.** For plane-polarized light reflected from a highly-polished pole of a strong magnet, the rotation of the plane of polarization. The rotation is proportional to the strength of the magnetic field. Also called **magneto-optical Kerr effect**, **Kerr magneto-optical effect**, or **Kerr effect (2)**.

magneto-optic Kerr effect Same as **magneto-optic effect (2)**.

magneto-optic recording Same as **magneto-optical recording**.

magneto-optical Same as **magneto-optic**. Its abbreviation is **MO**.

magneto-optical disc A disc utilized for **magneto-optical recording**. Also spelled **magneto-optic disc**, or **magneto-optical disk**. Its abbreviation is **MO disc**.

magneto-optical disk Same as **magneto-optical disc**. Its abbreviation is **MO disk**.

magneto-optical effect Same as **magneto-optic effect**.

magneto-optical Kerr effect Same as **magneto-optic effect (2)**.

magneto-optical recording A high-capacity disc technology which utilizes both optics and magnetism to write, read, and rewrite data. A laser is first utilized to heat the recordable portions on the disc to the magnetic transition temperature, and then the bits are aligned magnetically. A laser is used for subsequent reading, and such discs are usually available either in a 3.5 inch or 5.25 inch diameter format. Also spelled **magneto-optic recording**.

magneto-optical storage Storage on discs utilizing **magneto-optical recording**.

magneto-optics The study of the effects magnetic fields have on light. An example is the magneto-optic effect.

magnetocaloric effect A phenomenon in which certain magnetic materials heat up when placed in a magnetic field, and cool down again when removed from said field. Seen, for instance, in some gadolinium and germanium alloys.

magnetocardiogram A graphical recording of the variations in intensity of the magnetic field of the heart. A **magnetocardiograph** is used to detect and record such activity. Its abbreviation is **MCG**.

magnetocardiograph An instrument which detects and records variations in the intensity of the magnetic field of the heart. A **magnetocardiogram** is a graphical recording of this activity.

magnetocardiography The technique employed to obtain a **magnetocardiogram**. Its abbreviation is **MCG**.

magnetoconductance A variation in the conductance of a material as a function of the strength of an applied magnetic field.

magnetoconductor A material whose conductance varies according to the strength of an applied magnetic field. An example of such a material is magnesium boride.

magnetoelectric Pertaining to the generation of electromotive forces via magnetic techniques, as may occur, for instance, in a dynamo.

magnetoelectric effect The variation of an electric phenomenon in a given material when subjected to a magnetic field. For instance, the induction of electric polarization in a crystal by means of a magnetic field.

magnetoelectric generator Same as **magneto**.

magnetoelectricity The generation of electromotive forces via magnetic techniques.

magnetoelectronics The use of electron spins for data storage in computing devices, as seen, for instance, in MRAM. Also spelled **magneto-electronics**. Also called **spin electronics**.

magnetofluiddynamics Same as **magnetohydrodynamics**.

magnetogasdynamic Pertaining to, or utilizing **magnetogasdynamics**.

magnetogasdynamics **Magnetohydrodynamics** dealing specifically with electrically conducting gases.

magnetograph **1.** An instrument which detects and records a varying magnetic quantity, especially a magnetic field. **2.** An instrument which detects and records a variations in the earth's magnetic field. **3.** An instrument which detects and records variations in the magnetic field of a celestial body, such as the sun.

magnetohydrodynamic Pertaining to, or utilizing **magnetohydrodynamics**. Also called **hydromagnetic**. Its abbreviation is **MHD**.

magnetohydrodynamic generator A device which utilizes electrically conducting fluids interacting with a magnetic field to generate electrical energy. For instance, such a device may extract kinetic energy from a jet of plasma to produce electricity. Its abbreviation is **MHD generator**. Also called **magnetohydrodynamic power generator**, or **hydromagnetic generator**.

magnetohydrodynamic power generation The production of electrical energy utilizing a **magnetohydrodynamic generator**. Its abbreviation is **MHD generation**.

magnetohydrodynamic power generator Same as **magnetohydrodynamic generator**.

magnetohydrodynamics Its abbreviation is **MHD**. The study of the interactions between an electrically conducting fluid and magnetic fields. Plasmas, liquid metals, and ionized gases are examples of such fluids. MHD has applications in various fields, including plasma confinement, liquid-metal cooling of nuclear reactors, and in astronomy. Also spelled **magneto-hydrodynamics**. Also called **magnetofluiddynamics**, **magneto-fluid mechanics**, or **hydromagnetics**. When dealing specifically with electrically conducting gases, also called **magnetogasdynamics**, and when dealing with plasmas, also called **magnetoplasmadynamics**.

magnetometer An instrument utilized for measuring the magnitude of a magnetic field. Such an instrument may also indicate the direction of said field.

magnetomotive force The magnetic pressure that a source of magnetism provides. It is analogous to the electromotive force. It is commonly produced by a current flowing through a coil of wire, said force being proportional to the ampere-turns. Another source of magnetomotive force is a magnetic body. Its abbreviation is **mmf**, its symbol is F, and it is usually expressed in ampere-turns, though gilberts are sometimes used. Also called **magnetic potential difference**, or **magnetic pressure**.

magneton **1.** A physical constant equal to approximately 9.2740×10^{-24} joule/tesla. It is used as a unit of magnetic moment. Its symbol is μ_B. Also called **Bohr magneton**. **2.** A physical constant equal to approximately 5.0508×10^{-27} joule/tesla. It is used as a unit of magnetic moment. Its symbol is μ_N. Also called **nuclear magneton**.

magnetooptic Same as **magneto-optic**. Its abbreviation is **MO**.

magnetooptical Same as **magneto-optic**. Its abbreviation is **MO**.

magnetooptics Same as **magneto-optics**.

magnetopause The region in space where the magnetosphere meets the solar wind. That is, it occurs where the earth's magnetic field balances the pressure of the solar wind.

magnetoplasma dynamics Same as **magnetoplasmadynamics**.

magnetoplasmadynamic Pertaining to, or utilizing **magnetoplasmadynamics**.

magnetoplasmadynamics **Magnetohydrodynamics** dealing specifically with electrically conducting plasmas. Also spelled **magnetoplasma dynamics**.

magnetoresistance The variation of the resistance of a material or resistor as a function of the strength of an applied magnetic field.

magnetoresistive Pertaining to, or exhibiting, **magnetoresistance**. Its abbreviation is **MR**.

magnetoresistive head A read/write head used in **magnetoresistive head technology**. Its abbreviation is **MR head**.

magnetoresistive head technology A disk drive head technology which enables higher storage densities than older methods. It makes use of magnetoresistive materials, which respond better to the weaker magnetic fields of individual bits as storage densities increase. In addition, magnetoresistive heads have separate read/write elements, with each optimized for its function. Its abbreviation is **MR head technology**.

magnetoresistive material Same as **magnetoresistor**.

magnetoresistive RAM Same as **MRAM**.

magnetoresistive random-access memory Same as **MRAM**.

magnetoresistor A material or resistor whose resistance varies according to the strength of an applied magnetic field. An example of such a material is indium antimonide. Also called **magnetoresistive material**.

magnetosphere The region in space where the earth's magnetic field exerts a significant influence. It starts within the ionosphere, and extends to about 10 or so earth radii, although this value may be much higher on the side opposite the sun. The magnetosphere is distorted into a teardrop shape by the solar wind. The term may also refer to such a region of another celestial body.

magnetostatic Pertaining to magnetic properties which are independent of the movement of magnetic fields.

magnetostatic field A magnetic field which does not move nor change direction. Such a field may be produced, for instance, by a single stationary magnetic pole, or by a constant current flowing through a stationary conductor.

magnetostriction The mechanical deformation of a material, especially that which is ferromagnetic, when placed in a magnetic field. For example, if such a material is placed in a direction parallel to an applied magnetic field, its length will change. The intensity of the field will determine the extent of mechanical deformation. Also called **Joule effect (2)**.

magnetostrictive Pertaining to, utilizing, or arising from **magnetostriction**.

magnetostrictive delay line A delay line made of a **magnetostrictive material**.

magnetostrictive material A material, such as nickel, which exhibits **magnetostriction**.

magnetostrictive transducer A transducer in which utilizes the phenomenon of **magnetostriction** to convert a current into a vibration, and vice versa. Used, for instance, in sonar.

magnetostrictor A device which utilizes the phenomenon of **magnetostriction** to convert electric oscillations into mechanical oscillations.

magnetron A microwave tube utilized to generate oscillations in the microwave range of the electromagnetic spectrum. A magnetron usually incorporates a heated cathode, consisting of a hollow cylinder, which is surrounded concentrically by a cylindrical anode with multiple resonant cavities, all of which is contained within a vacuum enclosure. The path of electrons leaving the cathode is influenced by surrounding orthogonal electric and magnetic fields, the latter produced by one or more permanent magnets or electromagnets. In this manner, the electron path to the anode is not straight, but is rather an expanding spiraling orbit, with electrons eventually reaching the anode. The geometry and size of the cavities determine the frequency of the oscillations, and output may be taken, for instance, through a coupled waveguide or a loop to a coaxial line. A magnetron may be used in a pulsed mode, as occurs in radar applications, or its output may be continuous, as is the case when utilized for heating in microwave ovens. Also called **magnetron oscillator**.

magnetron effect In a thermionic tube, the deflection of the electrons emitted by the cathode from a linear path into a curved path due to the presence of a magnetic field, such as that induced by the current in the heater of said cathode.

magnetron oscillator Same as **magnetron**.

magnitude 1. A measure of the extent, influence, size, or significance of something. 2. A numerical value, such as an amplitude. 3. A physical property, such as temperature, that can be quantified. 4. A logarithmic measure of the absolute or relative luminance of a celestial body, such as the sun or moon. The smaller the number, the greater the luminance.

mAh Abbreviation of **milliampere-hour**.

mail Also called **electronic mail**, or **email**. 1. The transmission of messages, with or without attachments, over a communications network such as the Internet. The first part of a mail is the header, which contains information such as the email address of the sender, the time and date sent, and the subject. This is followed by the body, which usually consists of the text that the sender wishes to communicate. There may also be one or more attachments, which are appended files. There are standard protocols, such as SMTP, which define factors such as message format, while standard mail servers, such as IMAP, provide storage of messages in virtual mailboxes until users retrieve them. Email can be accessed by properly equipped devices such as computers, PDAs, cellular phones, and the like, which have a connection to the Internet, and provides a simple and reliable means of sending messages practically instantaneously. 2. One or more **mail (1)** messages. 3. The sending of **mail (1)**.

mail address A sequence of characters which uniquely identify an email account. The standard format is user name@domain name. In the case of **xxxx@zzz.com.qq**, *xxxx* is the user name, and *zzz.com.qq* is the domain name. Also called **email address**, **address (5)**, or **Internet address (3)**.

mail attachment A file that is appended to a **mail (1)**. Such a file may be a document, an image, a video, a program, and so on. Certain types of attachments are encoded, thus requiring the recipient to have the appropriate email software to decode them. Also called **email attachment**, **enclosure (3)**, **file attachment**, or **attachment**.

mail autoresponder A feature or program which automatically sends a previously prepared email in response to a received email. Used, for instance, to inform that an email has been received and will be responded to at a later time. Also called **autoresponder**, or **email autoresponder**.

mail bomb Same as **mailbomb**.

mail bot Same as **mail robot**.

mail box Same as **mailbox**.

mail client A program which resides in a computer, enabling it to have access to the mail servers of a network. It enables the sending, receiving, and organizing of email. Also called **email client**.

mail filter In an email program, a feature which allows for the automatic sorting of incoming email. For instance, a user can set filters to delete unwanted email from specific spammers, or place emails from a specific individual or entity in its own folder. Also called **email filter**.

mail folder A directory which is used for organizing and storing emails. All email programs are equipped with folders, such as those for incoming mail, sent mail, and trash, and usually allow users to create others, such as those for friends or colleagues, to which incoming mail can be directed. Also called **email folder**.

mail forwarding The resending of an email that has already reached a recipient's address. Once received, an email may be forwarded manually or automatically to one or more email addresses. When an email is manually forwarded, its content may be changed and/or files may be attached. Also called **forwarding (3)**, or **email forwarding**.

mail header The first part of an email, which contains information such as the email address of the sender, that of the recipient or recipients, the time and date sent, the subject, mail protocols utilized, the IP address of the sender, and so on. Also called **email header**.

mail merge A feature, usually incorporated into word processing applications, which simplifies the generation and printing of form letters.

mail program A program providing the necessary tools to read, prepare, and send and retrieve emails to and from the mail servers of a network. Such a program usually provides features such as word-processing functions, an address book, the capability to insert hyperlinks, filtering, and the ability to retrieve mail from other email accounts. Also called **mail software**, or **email program**.

mail protocol A protocol utilized for mail functions such as accessing and routing messages. Simple Mail Transfer Protocol and Internet Message Access Protocol are examples.

mail robot A program which sends automated responses to emails, or which performs other functions, such as removals from mailing lists. Its abbreviation is **mail bot**. Also called **email robot**.

mail server Also called **email server**. **1.** A computer which serves to store and/or forward email. **2.** The software which enables a **mail server (1)** to perform its functions.

mail signature Also called **signature**, or **email signature**. **1.** Content, such as a text, image, file, or combination of these, which is appended to an email or message. Examples include favorite quotes, contact information, vCards, or disclaimers. A user may select whether such a signature is included automatically when sending. **2.** An attachment to an electronic transmission which uniquely identifies and authenticates the sender of a message, file, or other information such as a credit card number. To be effective, a digital signature must not be forged, and measures such as public-key cryptography are employed to insure its integrity. Also called **digital signature**.

mail software Same as **mail program**.

mail virus A computer virus sent through email. Such a virus is usually in the form of an attachment, which if left unopened prevents the harm it may bring. Also called **email virus**.

mailbomb Also spelled **mail bomb**. Also called **email bomb**. The sending of an excessive amount of unwanted email to the same recipient. For instance, the generation of many identical emails which are sent repeatedly to the same email address. It also applies to arranging for a given recipient to receive an inordinate amount or email, for instance, by being placed in numerous mailing lists. In addition to any inconvenience this causes a user, it also wastes available bandwidth, and other network resources. Also, the undesired emails themselves. This contrasts with **spam**, which is unsolicited email directed towards multiple, often millions, of users.

mailbot Same as **mail robot**.

mailbox An area within a storage device in which email is placed and stored. Such storage may be at the site of the computer, or at a remote location as is the case with Web-based mail. Also spelled **mail box**.

mailer-daemon A daemon utilized for performing functions associated with the handling of email, such as sending error messages or transporting messages between hosts.

mailing list A group, sometimes massive, of email addresses identified by a name, subject, or the like. Any desired mail can be sent automatically to the members of such a list.

mailing list manager A program which further automates the sending of messages to mailing list recipients. Such a program handles tasks such as accepting posted messages for subsequent sending to all members of a list, and adding or removing subscribers from a mailing list. Its abbreviation is **MLM**.

mailto A hyperlink, which when clicked upon, launches the default email application and pops up an email composition window which has been preaddressed. Used, for instance, to enable a visitor to a Web page to be able to send an email as simply as possible.

main amplifier An amplifier designed to provide an output power which is much greater than the input power. Such an amplifier is utilized to deliver power to a load, such as a loudspeaker. When there are multiple amplifiers, or amplifier stages, the power amplifier is the final, or output amplifier or stage. For example, it is the stage of a transmitter that feeds an antenna. Also called **power amplifier**.

main bang In radars, the transmitted pulse.

main beam Same as **main lobe**.

main distribution frame In communications, a location where wire and/or cable terminations are located, so that the internal lines of a structure can be linked to external lines. Such a site may be located, for instance, adjacent to a central office. Its abbreviation is **MDF**. Also called **distribution frame**.

main line Same as **main loop**.

main lobe Within an antenna pattern, the lobe indicating the direction in which radiation and/or reception is at a maximum. All other lobes are **minor lobes**. Also called **major lobe**, **main beam**, or **primary lobe**.

main loop The key loop in a computer program. It is repeated after each event to help insure that the program performs its tasks properly. Also called **main line**.

main loudspeakers Same as **main speakers**.

main memory Better known as **RAM**. One or more memory chips that can be read and written by the CPU and other hardware, and which serve as the primary workspace of a computer. Instructions and data are loaded here for subsequent execution or processing, and its storage locations may be accessed in any desired order. The amount of main memory a computer has determines factors such as how many programs may be optimally run simultaneously. Also called **main storage**, **random-access memory**, **read/write memory**, **read/write RAM**, **primary memory**, **primary storage**, or **core memory (2)**.

main processor 1. In a computer with two or more CPUs, the main CPU. It is this CPU that controls all others. **2.** Same as **CPU (1)**.

main program 1. The first executed module of a program. Also called **main routine (1)**. **2.** The part of a program which is returned to after a subroutine or branch. Also called **main routine (2)**. **3.** Of a series of related programs, that of the greatest priority or importance.

main routine **1.** Same as **main program (1)**. **1.** Same as **main program (2)**.

main speakers In a surround-sound audio system, the left and right speakers, as opposed to the center or satellite speakers. Also called **main loudspeakers**.

main storage Same as **main memory**.

mainboard Same as **motherboard**.

mainframe **1.** An especially powerful and expensive computer which is capable of quickly processing very large amounts of data, and of serving many, even thousands, of users simultaneously. Although a **supercomputer** can handle a single task or application much more rapidly, a mainframe can usually handle many more tasks and applications at the same time. Used, for instance, by large corporations or governments. Also called **mainframe computer**. **2.** In a computer, the CPU, any other processors such as coprocessors, and main memory. **3.** The enclosure which houses all the components a computer utilizes, with the exception of the external peripherals, such as monitor, mouse, keyboard, printer, and so on. Also called **CPU (3)**.

mainframe computer Same as **mainframe (1)**.

mains A network of lines which serve for the distribution of electric power. Mains may refer to the lines that supply an entire nation or region, or those that provide power in a given structure such as a building or house. For instance, when a device or piece of equipment is plugged into an outlet, it is accessing the electricity carried by the mains. Also called **power mains**.

mains hum An electrical disturbance caused by the frequency of the **mains**, or by one of its harmonics.

maintenance All activities which are performed to help insure the proper operation of components, circuits, devices, equipment, and systems. Maintenance is usually either preventive or corrective, and may includes tests, adjustments, cleaning, and replacements. In the case of software, for instance, maintenance may involve updating or debugging. Also called **upkeep**.

major lobe Same as **main lobe**.

Majordomo A popular mailing list program.

majority carrier In a semiconductor material, the type of charge carrier which constitutes more than half of the total charge carriers. In a p-type semiconductor, for instance, mobile holes are the majority carrier. This contrasts with **minority carrier**, which constitutes less than half.

majority element Same as **majority logic**.

majority gate Same as **majority logic**.

majority logic A logic circuit with multiple inputs and a single output, and whose output is high when a majority of its inputs are high. Also called **majority gate**, or **majority element**.

make **1.** To close a circuit, or a closed circuit. Closing a circuit enables the flow of current. **2.** In a switch or relay, to close or produce contact. **3.** To produce, bring about, form, cause to exist, attain, or the like.

make-before-break contacts Contacts which make a new connection prior to breaking the previous one.

make-before-break switch A switch that makes a new connection prior to breaking the previous one. For instance, the moving contact of such a switch may be wider than the distance between fixed contacts. Also called **shorting switch**, or **short-circuiting switch**.

make contact In a relay or switch, a contact pair that is open in the resting state, and which is closed when energized. The contact pair consists of a stationary contact and a movable contact. A make contact provides a complete circuit

when energized. This contrasts with a **break contact**, in which the contact pair is closed in the resting state. Also called **normally-open contact**, or **front contact**.

make time The time it takes for a switch or relay to close a connection.

male adapter An adapter with one or more pins or prongs, which is inserted into another adapter with matching recessed openings.

male connector A connector with one or more pins or prongs, which is inserted into another connector with matching recessed openings.

male jack A jack with one or more pins or prongs, which is inserted into a plug with matching recessed openings. Although a jack often refers to a female connector, it is not necessarily so.

male plug A plug with one or more pins or prongs, which is inserted into a jack with matching recessed openings.

malfunction **1.** To fail to function. **2.** To fail to function properly.

malfunction detection The sensing of a malfunction. Malfunction detection may also serve to indicate the cause of the malfunction.

malicious code Program code whose purpose is to produce a harmful, unauthorized, or otherwise unwanted effect. Examples include scripts, macros, and HTML code.

malicious logic Computer logic whose purpose is to produce a harmful, unauthorized, or otherwise unwanted effect. Examples include logic bombs, worms, and viruses.

malicious software Same as **malware**.

malleable Easily shaped, as in being able to be hammered into a thin strip or foil. Used especially in the context of metals or alloys, such as gold, silver, or an alloy combining these elements.

malware Acronym for **mal**icious soft**ware**. Software whose purpose is to produce a harmful, unauthorized, or otherwise unwanted effect. Examples include viruses and Trojan horses.

MAN Acronym for **Metropolitan Area Network**.

man-made interference Also called **artificial interference**. Any interference, such as electromagnetic interference, resulting from human creations, actions, or presence. This includes that from oscillators, transmitters, ignition systems, motors, switches, voltage regulators, among many others, and that arising from the use of such devices. This contrasts with **natural interference**, such as atmospherics.

man-made noise Also called **artificial noise**. Any noise, such as radio-frequency noise, resulting from human creations, actions, or presence. This contrasts with **natural noise**, such as cosmic noise.

managed hub A hub which supports network management functions. An **unmanaged hub** does not. It is a type of intelligent hub.

managed service provider Same as **management service provider**. Its abbreviation is **MSP**.

Management Information Base A collection of network management data which can be accessed and maintained using a protocol such as Simple Network Management Protocol or Common Management Information Protocol. Its abbreviation is **MIB**.

management information services Within an organization, the department which handles data processing, especially from the perspective of assisting managerial decisions. Its abbreviation is **MIS**.

management information system An information system utilized for integrating the data needs of an organization, so as to facilitate managerial decisions. Its abbreviation is **MIS**.

management service provider A business that proves information technology services, such as hosting of applications and data sharing, which can be accessed remotely. Its abbreviation is **MSP**. Also called **managed service provider**.

Manchester code Same as **Manchester encoding**.

Manchester coding Same as **Manchester encoding**.

Manchester differential encoding A communications format similar to **Manchester encoding** with the difference that a 1-bit is indicated by the first half of the signal being the same as that of the previous bit signal, while the first half of a 0-bit is the opposite of the previous bit signal. Since in either case there is always a transition, the absence of a transition indicates that no signal has been sent. Also called **differential Manchester encoding**.

Manchester encoding A communications format in which each bit is split into two, providing a self-synchronizing data stream. A 1-bit is transmitted with a high voltage in the first period and a low voltage in the second, with a 0-bit being the converse. One advantage of this encoding is that the difference between a 0-bit and no signal is distinguished, as there is no transition in the latter case. A disadvantage is that it occupies twice the bandwidth. Also called **Manchester code**, **Manchester coding**, or **differential encoding**.

Manchester Mark I A late 1940's computer which utilized a Williams tube and magnetic drums for storage.

manganese A pinkish-gray, lustrous, and brittle chemical element with atomic number 25. It has four allotropic forms, is reactive, and has about 20 isotopes, of which one is stable. It is used in many alloys, including those which are ferromagnetic, and in batteries. Its chemical symbol is **Mn**.

manganese chloride Reddish crystals whose chemical formula is $MnCl_2$. Used in batteries.

manganese dioxide Black crystals or powder whose chemical formula is MnO_2. Used in batteries.

manganese oxide A brownish-black powder whose chemical formula is Mn_3O_4. Used in batteries.

manganin An alloy usually composed of 84% copper, 12% manganese, and 4% nickel, although their relative proportions may vary a little. This alloy features a low temperature coefficient and low contact potential. Used, for instance, in wires for precision wire-wound resistors.

manganin wire A wire composed of **manganin** alloy.

manipulator A robotic mechanism which comprises an arm, a wrist, and an end-effector. It has sets of segments, links, and powered joints, which give it the ability to perform tasks, and to move with up to six degrees of freedom. Also called **robot manipulator**.

manometer An instrument or device utilized to measure the pressure of fluids, especially gases and vapors. A barometer is an example.

mantissa 1. The decimal part of a logarithm. The integer portion is the **characteristic (3)**. For example, if a logarithm of a given number equals 1.2345, the mantissa is .2345, while 1 is the **characteristic**. 2. In a floating-point number, the first part, which specifies the significant digits, and which usually has a value equal to or greater than 1, but less than 10. For example, in 1.234 x 10^{56}, **1.234**. The power to which the base is raised, **56**, is the **characteristic (2)**.

manual 1. Operated or performed by hand, or any other mechanism or procedure which is not automatic. For example, manual tuning, or manual input. 2. Operated by, or requiring the use of hands. For example, a hand generator, or a hand receiver. Also called **hand-operated**. 3. A reference document, book, file, or the like which contains information on the uses and applications of a given device, piece of equipment, system, and so on.

manual input The entering of data into a device, especially a computer, by hand.

manual tuning Tuning performed by hand, or any other mechanism or procedure which is not automatic. Manual tuning may or may not be used with automatic frequency control.

manufacturing automation protocol A set of communications standards used in manufacturing networking and automation processes. Its acronym is **MAP**.

map 1. A representation of an area, region, structure, or the like. For instance, a weather, navigational, or memory map. 2. To prepare a **map (1)**. 3. To represent, associate, or transfer objects. For example, to map a program on a disk into memory, or to map an image in memory onto a display screen.

MAP Acronym for **manufacturing automation protocol**.

MAPI Acronym for **Messaging Application Programming Interface**.

MAR Acronym for **memory address register**.

marching ants A **marquee** in which the dots of the framing line move or flash.

Marconi antenna A vertical quarter-wave antenna whose lower end is grounded.

margin 1. A space separating components, circuits, devices, materials, or objects. 2. An amount that exceeds that which is usually necessary. For example, a margin of safety.

margin of safety An allowance for a level of operation which is more stressful than usual, such as those which may be encountered during brief intervals and/or unforeseeable working conditions.

marginal check Same as **marginal test**.

marginal relay A relay in which the difference between the driving current and the maximum current that may be applied without actuation is minimal.

marginal test A test during which normal or nominal operating conditions are exceeded, to locate potential faults and/or determine margins of safety. Also called **marginal check**.

maritime mobile service A mobile radio service for use between ships and coast stations, between ships, and between coast stations. Its abbreviation is **MMS**.

mark 1. A visible, magnetic, or otherwise detectable trace, impression, symbol, or property, which serves to distinguish, indicate, or identify. Also called **marker (1)**. 2. In a storage medium, such as magnetic tape, a blip, symbol, notch, or other device utilized for identification, timing, or the like. For example, a control character or code which indicates a subdivision in a file, or the end of a record. Also called **marker (2)**. 3. To identify a location, block of text, or other unit of information utilizing a **mark (2)**. 4. In the transmission of information, a high state, or binary 1, as opposed to a **space (11)** which indicates a low state, or binary 0. 5. To identify something for location or guiding purposes. 6. In optical sensing, a line, circle, or other tracing which is recognized by an optical sensor. Also called **optical mark**.

mark detection Same as **mark reading**.

mark reading The recognition of a **mark (6)** by an optical sensor. Used, for instance, to automatically gather information from forms that use checkmarks, filled-in circles, and the like. Also called **mark recognition**, **mark sensing**, **mark scanning**, **mark detection**, or **optical mark reading**.

mark recognition Same as **mark reading**.

mark scanning Same as **mark reading**.

mark sensing Same as **mark reading**.

mark-space ratio Also called **mark-to-space ratio**. 1. In the transmission of information, the ratio of high states to low states. 2. In a pulse or square-wave signal, the ratio of the pulse duration to the interval between successive pulses. This ratio equals one in a perfect square wave.

mark-to-space ratio Same as **mark-space ratio**.

marker 1. Same as **mark** (1). 2. Same as **mark** (2). 3. That which serves to make a **mark** (1) or **mark** (2). 4. On the screen of an oscilloscope, a pip which serves to identify a reference frequency.

marker beacon 1. A beacon which radiates a specific pattern vertically, and which serves to provide positional information to aircraft. Used, for instance, in an instrument landing system. 2. A low-powered beacon which assists marine or aeronautical navigation.

marker frequency A frequency which serves to distinguish, indicate, or identify. For example, that which indicates the upper end of a frequency band.

marker generator 1. A radio-frequency generator, such as an oscillator, which generates **markers** (4). 2. A radio-frequency generator, such as an oscillator, which generates reference pulses, such as those of a specific frequency, duration, or amplitude.

Markov model A manner of representing the associations between data elements utilizing probability. Used extensively in voice recognition software.

markup language A language, such as HTML or XML, utilized for transforming unformatted text into structured documents by inserting hyperlinks, tags, and other display and formatting instructions.

marquee In computer graphics, a dotted line that frames a selected object, such as a picture. When the dots of said line move or flash, also called **marching ants**.

Marx generator A device which charges multiple capacitors in parallel, then discharges them in series, usually using spark gaps. Each discharge produces a high-voltage pulse.

maser Abbreviation of **m**icrowave **a**mplification by **s**timulated **e**mission of **r**adiation. A device whose operation is similar to that of a laser, and which is utilized to amplify or generate coherent microwave radiation. Examples include gas and solid-state masers. Used, for instance, in communications, radio astronomy, radars, and as time and frequency standards.

mask 1. An object, stencil, or other device which is applied or placed upon a surface, so as to permit the selective passing of particles, beams, rays, substances, and so on, to form any desired patterns. 2. The use of a **mask** (1) to selectively shield portions of semiconductor wafers, or other materials, during manufacturing. Used, for instance, in lithography. 3. In a picture tube with a three-color gun, a grill with round holes that is placed behind the screen to make sure that each color beam strikes the correct phosphor dot on said screen. It insures, for instance, that the electron beam intended for the red phosphor dots only hits those. Also called **aperture mask**, or **shadow mask**. 4. To obscure a signal or sound with a stronger one. 5. A pattern of bits or characters which determines whether another set of bits or characters will be selected, transmitted, changed, or discarded. 6. A frame which serves to conceal the edges of a CRT.

mask bit A bit which determines if a corresponding bit will be selected, transmitted, changed, or discarded.

maskable interrupt An interrupt which can be disabled by another interrupt. Such an interrupt may occur, for instance, when a there is a serious problem, or if given task or program needs the undivided attention of the CPU. A **non-maskable interrupt** is one which can not be disabled in this manner.

masking 1. The use of a **mask**. 2. The amount by which the threshold of hearing a sound is increased due to the presence of another, obscuring sound. The level of masking is usually expressed in decibels. Also called **masking effect** (1), **audio masking**, or **aural masking**. 3. The manner in which a signal, property, or phenomenon is obscured by another.

Also, the extent to which this occurs. Also called **masking effect** (2).

masking effect 1. Same as **masking** (2). 2. Same as **masking** (3).

masking sound An obscuring sound whose presence raises the threshold of hearing of another, desired sound. The level of masking is usually expressed in decibels.

masonite A hard board made from pressed wood fibers. Used, for instance, as a panel upon which electrical components may be mounted.

masquerade To attempt to deceive and/or harm by appearing as someone or something else. For example, to send email with the *from* field using the name of another person or entity, or the manner in which a Trojan horse can appear to be a harmless program.

mass 1. The quantity of matter in a body or medium. The mass of a body makes it resist acceleration, and gives it gravitational attraction. Its SI unit is the kilogram, and its symbol is **m**. The **weight** (1) of an object varies depending on the gravitational force exerted upon it, while its mass does not. 2. A given body of matter. 3. A large or very large amount. 4. The principal part of something.

mass absorption coefficient For given material or medium, the linear absorption coefficient divided by the density of said material or medium.

mass conservation A law which states that mass, or matter, cannot be bment be created nor destroyed in an isolated system. For instance, the mass remains constant when a substance changes form a solid to a gas in an isolated system. This law does not always hold true when dealing with sub-atomic particles. Also called **matter conservation**, **conservation of mass**, or **law of mass conservation**.

mass-energy equation A fundamental formula for the interconversion of mass and energy. It is $E = mc^2$, where E is energy, **m** is mass, and c is the speed of light in a vacuum. Also called **Einstein mass-energy relation**.

mass memory Same as **mass storage** (1).

mass number The number of protons and neutrons in the nucleus of an atom. For example, the mass number of the most common isotope of carbon is 12, as it has 6 protons and 6 neutrons. Its symbol is **A**. Also called **nucleon number**.

mass spectrograph A **mass spectrometer** in which the detector is a photographic plate. Also called **mass spectroscope** (2).

mass spectrometer An instrument which identifies ions based on their charge-to-mass ratio. In it, the sample to be analyzed is vaporized, placed in a vacuum, ionized by an electron beam, accelerated by an electric field, then deflected into a curved path by a magnetic field. The amount of deflection of any given ion will depend on its charge-to-mass ratio, so each different species is separated according to its mass. A detector records the distribution of each of the masses, each producing its characteristic peaks. Widely utilized to analyze elements and compounds. Also called **mass spectroscope** (1). Its abbreviation is **MS**.

mass spectrometry The use of a **mass spectrometer** for analysis. Its abbreviation is **MS**.

mass spectroscope 1. Same as **mass spectrometer**. 2. Same as **mass spectrograph**.

mass spectroscopy The use of a **mass spectrometer** to obtain atomic and molecular spectrums. Its abbreviation is **MS**.

mass spectrum The display, plot, or other visual output produced when utilizing **mass spectrometry**.

mass storage 1. An external storage medium, such as a disc or tape, which holds a large amount of data, especially when compared to that which can be placed in the computer's main memory. Also called **mass memory**. **2.** The saving of data to **mass storage** (1).

massively parallel Same as **massively parallel processing**.

massively parallel processing A multiprocessing computer architecture in which multiple CPUs run parallel to each other during the execution of a single program or task. Each CPU has its own copy of the operating system, its own memory, its own copy of the application being run, and its share of the data. There can be most any number of CPUs working simultaneously. Also, processing in this manner. This contrasts with **symmetric multiprocessing**, where all the CPUs share the same memory, and the same copy of the operating system, application, and data being worked on. This also contrasts with **asymmetric multiprocessing**, in which multiple CPUs are each assigned a specific task, with all CPUs controlled by a master processor. Also called **massively parallel**. Its abbreviation is **MPP**.

massively parallel processor A computer system that can perform **massively parallel processing**.

mast A vertical rod which serves as an antenna, or as an antenna support. Also called **antenna mast**.

master **1.** That which controls something else. For example, a loran master station or a master switch. **2.** A reference standard, such as the frequency generated by an atomic clock. **3.** An original from which copies can be made. For instance, a precise workpiece utilized to make others.

master-antenna television Same as **master-antenna TV**.

master-antenna television system Same as **master-antenna TV**.

master-antenna TV Abbreviation of **master-antenna** television. An antenna, cable, or satellite system which receives TV programming, then distributes it to multiple units such as those in a house, hotel, or apartment building. Such a system may also provide signal amplification, and might also offer radio broadcasts. Its own abbreviation is **MATV**. Also called **master-antenna television system**.

master-antenna TV system Same as **master-antenna TV**.

Master Boot Record The data and instructions contained in the first sector of the first hard disk. During the startup process this record indicates the location of the operating system so that it can be loaded into memory. Its abbreviation is **MBR**. Also called **Master Boot Sector**.

Master Boot Sector Same as **Master Boot Record**. Its abbreviation is **MBS**.

master clock **1.** In a computer, a circuit or device which provides a steady stream of timed pulses. This clock serves as the internal timing reference of a digital computer, upon which all its operations depend. Also known as **main clock**, or **system clock**. **2.** A clock which generates pulses which control other clocks, or help maintain their accuracy. **3.** A clock which generates pulses which control other devices, equipment, or systems.

master console A terminal utilized to control a computer and monitor its status. Also called **console** (2).

master control **1.** A console, terminal, station, or location which operates, regulates, or manages devices, equipment, or systems. **2.** A **master control** (1) utilized in a broadcasting station.

master control program **1.** The program, such as the operating system, which runs a computer. **2.** The program, such as the operating system, which runs a computer at any given moment.

master data **1.** Data contained in a **master file**. **2.** Data which is indispensable for the use of other data.

master file **1.** A file containing the key information utilized for any given computer application or task. For example, a file incorporating the information necessary to carry out business, such as products offered, customer information,

and so on. **2.** A file which remains essentially unchanged for extended periods of time, or which has specific information updated at given intervals.

master gain control **1.** A gain control which regulates the gain of an entire system. **2.** A gain control which simultaneously regulates the gain of two or more channels.

master key In computer and communications security, the top-level key within a hierarchy of keys.

master oscillator An oscillator which is utilized to establish the operational frequency of a component, device, or piece of equipment, such as an amplifier or transmitter. Such oscillators usually generate an extremely precise and stable frequency, and may also be used for testing. Its abbreviation is **MO**.

master oscillator-power amplifier A signal generator in which the oscillator is followed by one or more stages of power amplification, which helps prevent undesirable interactions with the output signal. Seen, for example, in a transmitter whose oscillator is isolated from the antenna by a power amplifier. Its abbreviation is **MOPA**.

master record A record contained in a **master file**.

master relay A relay which controls other relays.

master station A station which controls other stations. Seen, for instance in loran.

master switch A switch that controls other switches.

master tape **1.** A tape from which other tapes may be copied. **2.** A tape containing **master files**.

master volume control **1.** A volume control which regulates the volume of an entire system. **2.** A volume control which simultaneously regulates the volume of two or more channels.

match **1.** One which is exactly like another, or which is its counterpart. Also, to make such a pair. **2.** That which is compatible, or which otherwise works properly with something else. **3.** To use compatible components, devices, equipment, or systems together. For example, to match a plug with its corresponding jack. **4.** To attain an equality in a given property or value. For instance, to match impedances. **5.** In computers, a comparison made to determine similarities and/or differences.

matched components Also called **matching components**. **1.** Components which can function properly together without prior modifications. **2.** Components which function properly together, whether adapting modifications have been made or not. **3.** Components which have been specifically designed to work together.

matched devices Also called **matching devices**. **1.** Devices which can function properly together without prior modifications. **2.** Devices which function properly together, whether adapting modifications have been made or not. **3.** Devices which have been specifically designed to work together.

matched filter **1.** A filter which matches the impedance of both its signal source and the output load. **2.** A filter which only responds to signals coded in a specific manner, or whose waveform provides a specified match.

matched impedance An impedance which is matched between two circuits, or between a signal source and a load. This helps insure that the maximum possible power is transferred from source to load, while minimizing distortion. Also called **matching impedance**.

matched line A transmission line whose characteristic impedance is the same as its terminations. In this manner, there is complete absorption of the incident power, and there are no wave reflections back to the source. Also called **matched transmission line**, or **matching line**.

matched load A load whose impedance value is that which results in the maximum possible transfer of power. Also called **matching load**, or **optimum load**.

matched pair A pair, such as left and right loudspeakers, which can function properly together, or which have been specifically designed to work together. Also called **matching pair**.

matched termination A termination which provides for the complete absorption of incident power. Such a termination eliminates all wave reflections back to the source. Also called **matching termination**.

matched transmission line Same as **matched line**.

matching components Same as **matched components**.

matching devices Same as **matched devices**.

matching impedance Same as **matched impedance**.

matching line Same as **matched line**.

matching load Same as **matched load**.

matching network An electric network, such as an inductance-capacitance network, which is utilized to match impedances. Also called **matching pad**.

matching pad Same as **matching network**.

matching pair Same as **matched pair**.

matching termination Same as **matched termination**.

matching transformer A transformer utilized to match impedances.

Materials Requirements Planning An information management system designed to estimate an organization's needs for supplies, inventory, and other materials. For example, such a system may keep track of the components necessary to manufacture any given products, and optimally indicate the dates and amounts each is needed. Its abbreviation is **MRP**.

math co-processor Same as **math coprocessor**.

math coprocessor An arithmetic-logic unit designed to perform floating-point operations. A CPU may have such a processor incorporated, or a second chip may be utilized. Math coprocessors are especially useful when a computer performs sophisticated calculations, such as those frequently needed in CAD, spreadsheet, and scientific programs. Also spelled **math co-processor**. Also called **floating-point processor**, **floating-point coprocessor**, **floating-point unit**, or **numeric processor**.

Mathcad Popular software for expressing, handling, presenting, and applying mathematical concepts.

Mathematica Popular software for expressing, handling, presenting, and applying mathematical concepts.

mathematical logic The use of a system of symbols to represent quantities, operations, and relationships. An example is Boolean logic. Also called **symbolic logic**.

mathematical model A mathematical representation of a concept, process, device, system, program, or the like.

mathematical subroutine A subroutine which performs a mathematical function such as raising to a power or calculating a logarithm.

matrices A plural form of **matrix**.

matrix Its plural form is **matrices**, and less commonly matrixes. **1.** A two-dimensional array utilized to organize, interrelate, or perform operations. For instance, such an array in which numbers are subjected to algebraic operations, a logic network in such a configuration, or the arrangement of the chemical elements in a periodic table. **2.** In computers, a number of elements, such as numbers, dots, or spreadsheet cells, which are arranged in rows and columns, and upon which operations may be performed. **3.** In a TV transmitter, a circuit or device which transforms the separate red, green, and blue camera signals into color-difference signals, and combines these with the chrominance subcarrier. Also called **color encoder**, or **encoder (2)**. **4.** In a TV receiver, a circuit or device which transforms the color-difference sig-

nals into separate red, green, and blue signals. Also called **color decoder**, or **decoder (2)**.

matrix printer A printer which creates alphanumeric characters and images by striking pins against an ink ribbon. There is a matrix of pins, and each pin makes a single dot. The more pins, the higher the resolution of the printed output. Also called **dot-matrix printer**.

matter Anything which has a mass and occupies space. This ranges from subatomic particles through celestial bodies. Matter is understood to exist in five states, which are solid, liquid, gas, plasma, and Bose-Einstein condensates, although there may be others.

matter conservation Same as **mass conservation**.

matter wave The wave associated with a particle, such as an electron, in motion. Also called **de Broglie wave**.

Matteuchi effect The change in magnetization of a material, such as a ferromagnetic rod, subjected to torsional stress.

mattress array A directional antenna consisting of an array of stacked dipoles in front of a flat shared reflector. Each dipole is spaced from ¼ to ¾ of a wavelength apart. Also called **billboard array**, **billboard antenna**, or **bedspring array**.

MATV Abbreviation of **master-antenna TV**.

max Abbreviation of **maximum**.

maxima A plural form of **maximum**.

maxima and minima The highest values, or maxima, and lowest values, or minima, a variable, function, curve, or the like attains. May be used, for instance, to describe the fluctuations of a voltage, or variations in the intensity of a received signal.

maximally flat characteristic A response characteristic, such as that of a filter, which has minimal passband ripple. Also called **Butterworth characteristic**.

maximally flat response A response, such as that of an amplifier, which has minimal passband ripple. Also called **Butterworth response**.

maximize In a GUI, to expand a window, icon, or other object to its largest available size. For example, to enlarge a window so that it occupies the full display screen.

maximize button A virtual button on a computer screen, which when clicked, serves to **maximize** a given window, icon, or other object.

maximum Its plural form is **maxima**, and less commonly maximums. Its abbreviation is **max**. **1.** The greatest possible value, quantity, or degree. **2.** The greatest value, quantity, or degree attained or recorded. **3.** The greatest value a variable, function, curve, or the like, may attain. **4.** The **maximum (3)** for a given interval.

maximum amplitude excursion Also called or **peak-to-peak amplitude**, **peak-to-peak**, or **amplitude excursion**. **1.** For a waveform of an alternating quantity, such as that of AC, the difference between the maximum positive peak and the maximum negative peak. It is the maximum combined range in the amplitude of a signal or other observed quantity. **2.** The difference between the maximum positive peak and the maximum negative peak of any varying quantity. It is the maximum combined range in the amplitude of a signal or other observed quantity.

maximum available gain The gain provided by a circuit or device whose input and output impedances are optimally matched to source and load. Its abbreviation is **MAG**.

maximum burst The bandwidth an ATM network can provide under any circumstances. It is a combination of the insured rate plus any available excess rate.

maximum capacity A value or quantity establishing the maximum amount that can be contained, accommodated, or handled. For instance, the charge that can be withdrawn from a fully charged battery under specified conditions, the maximum current that can be handled by an electric circuit, or the largest number of communication channels that a communications circuit can handle simultaneously. Also called **rated capacity**, or **full-rated capacity** (1).

maximum cell loss ratio For a given ATM Quality-of-Service, the greatest ratio of lost cells to total transmitted cells. Its abbreviation is **MCLR**.

maximum current 1. The maximum absolute value of the displacement from a reference position, such as zero, for a current. Also called **peak current** (1). 2. The **maximum current** (1) that can be applied continuously under specified conditions without harming a component, circuit, device, piece of equipment, or system. Also called **rated current** (2).

maximum demand Also called **maximum power demand**. 1. The highest rate at which electric power is consumed. 2. The highest average rate at which electric power is consumed. 3. The **maximum demand** (1) or **maximum demand** (2) for a given period of time.

maximum operating current The maximum current permissible for the proper operation of a component, circuit, device, piece of equipment, or system.

maximum operating power 1. The maximum power consumed by a component, circuit, device, piece of equipment, or system which is properly operating. 2. The maximum power produced or emitted by a component, circuit, device, piece of equipment, or system which is properly operating.

maximum operating voltage The maximum voltage permissible for the proper operation of a component, circuit, device, piece of equipment, or system.

maximum output Also called **rated output**. 1. The peak output value, such as that of a current or voltage, of a component, circuit, device, piece of equipment, or system. 2. The **maximum output** (1) for a given period of time. 3. The **maximum output** (1) that can be handled safely.

maximum power 1. The peak value of power utilized. 2. The peak value of power utilized for a given period of time. 3. The highest power level a component, circuit, device, piece of equipment, or system can operate at safely.

maximum power demand Same as **maximum demand**.

maximum power output 1. The peak value of output power. 2. The peak value of output power for a given period of time. 3. The highest power output level a component, circuit, device, piece of equipment, or system can handle safely. Also called **rated power output**. 4. The highest permissible power output, such as the maximum transmitting power a regulating authority allows a radio station.

maximum power rating The maximum power level that a component, circuit, device, piece of equipment, or system can produce, consume, dissipate, or otherwise safely handle, for brief periods. Also called **peak power rating**.

maximum power transfer The greatest amount of power that can be transferred from source to load, or between circuits. This occurs when impedances are matched.

maximum power transfer theorem A theorem stating that the greatest possible amount of power is transferred from source to load when they have conjugate impedances. Also called **power transfer theorem**.

maximum rating 1. A rating which defines the highest value, such as a current or voltage, a component, circuit, device, piece of equipment, or system can operate at safely. 2. A rating which defines the highest value, such as a current or voltage, a component, circuit, device, piece of equipment, or system can operate at safely for a given time interval.

maximum recording level For an analog or digital tape, the highest input-signal amplitude which can be recorded without exceeding a specified level of distortion. This may be indicated, for instance, by a VU reading of 0.

maximum signal level 1. The highest input or output signal level permissible. For instance, a maximum recording level. 2. In an amplitude-modulated fax system, the amplitude level corresponding to black or white, whichever is assigned this maximum level.

maximum sound pressure 1. The maximum sound pressure level which a microphone can accept, or that a speaker can deliver, without clipping or exceeding a given level of distortion. Usually expressed in decibels. 2. The **maximum sound pressure** (1) for a given time interval.

maximum transmission unit Its abbreviation is **MTU**. For a given network, the largest packet size or frame that can be transmitted. It is usually expressed in bytes, and any value beyond this must be divided into smaller packets or frames. For example, most Ethernet networks have an MTU of 1500 bytes.

maximum undistorted output 1. The maximum output a circuit, device, or system can deliver without clipping or exceeding a given level of distortion. Used especially in the context of audio components, such as amplifiers or speakers. 2. The **maximum undistorted output** (1) for a given time interval.

maximum usable frequency The highest frequency that can be employed for communications between locations utilizing ionospheric propagation. This frequency will vary according to the angle of incidence and time of day. The minimum usable frequency is called **lowest usable frequency**. Its abbreviation is **MUF**.

maximum voltage 1. The maximum absolute value of the displacement from a reference position, such as zero, for a voltage. Also called **crest voltage**, or **peak voltage**. 2. The **maximum voltage** (1) that can be applied under given conditions without harming a device, instrument, piece of equipment, or system.

maximum wattage The **maximum power** expressed in watts.

maxwell The unit of magnetic flux in the cgs electromagnetic system. There is 10^{-8} weber in a maxwell. Its symbol is **Mx**. Also known as **abweber**.

Maxwell bridge A four-arm AC bridge utilized for measuring an inductance in terms of capacitance and resistance. Alternatively, such a bridge may be used to measure a capacitance in terms of inductance and resistance. Also called **Maxwell-Wien bridge**.

Maxwell's displacement current An AC arising in the presence of a time-varying electric field. This current is in addition to ordinary conduction AC. Displacement currents are necessary for the propagation of electromagnetic radiation through space. Also called **displacement current**.

Maxwell's equations Four equations which serve to describe classical electromagnetism. These equations are based on Gauss' law for electricity, Gauss' law for magnetism, Faraday's law of electromagnetic induction, and Ampère's law.

Maxwell's law A law stating that each part of a movable circuit will experience a force that makes it move in a manner which encloses the maximum possible magnetic flux. Also called **Maxwell's rule**.

Maxwell's rule Same as **Maxwell's law**.

Maxwell-Wien bridge Same as **Maxwell bridge**.

mayday An internationally recognized distress signal utilized by a craft or vehicle in grave and/or imminent danger. It is the English phonetic equivalent of the French expression *m'aider*, which means "come help me!"

mb Abbreviation of **millibar**.

Mb 1. Abbreviation of **megabit**. 2. Abbreviation of **megabar**.

MB Abbreviation of **megabyte**.

Mb/s Abbreviation of **megabits per second**.

MB/s Abbreviation of **megabytes per second**.

Mb/sec Abbreviation of **megabits per second**.

MB/sec Abbreviation of **megabytes per second**.

MBE Abbreviation of **molecular-beam epitaxy**.

MBGA Abbreviation of **Micro Ball Grid Array**.

Mbit Abbreviation of **megabit**.

Mbits/s Abbreviation of **megabits per second**.

Mbits/sec Abbreviation of **megabits per second**.

Mbone Abbreviation of multicast back**bone**. Two or more Internet servers which are specially configured to support IP multicasting, and which work together for the transmission of real-time audio and video.

Mbps Abbreviation of **megabits per second**.

MBps Abbreviation of **megabytes per second**.

MBq Abbreviation of **megabecquerel**.

MBR 1. Abbreviation of **Master Boot Record**. 2. Abbreviation of **memory buffer register**. 3. Abbreviation of **memory-based reasoning**.

MBS Abbreviation of **Master Boot Sector**.

Mbyte Abbreviation of **megabyte**.

Mbytes/s Abbreviation of **megabytes per second**.

Mbytes/sec Abbreviation of **megabytes per second**.

mc Abbreviation of **micro-**.

mC Abbreviation of **millicoulomb**.

MC Abbreviation of **megacoulomb**.

Mc Abbreviation of **megacycle**.

mcd Abbreviation of **millicandela**.

MCG 1. Abbreviation of **magnetocardiography**. 2. Abbreviation of **magnetocardiogram**.

mCi Abbreviation of **millicurie**.

MCi Abbreviation of **megacurie**.

McLeod Gage Same as **McLeod Gauge**.

McLeod Gauge A device which measures the pressure of a gas by measuring its volume at an unknown pressure, compressing it, and then measuring its volume at this higher reference pressure. The original pressure can then be calculated. Useful for measuring pressures down to about 10^{-6} torr. Also spelled **McLeod Gage**.

MCLR Abbreviation of **maximum cell loss ratio**.

MCM Abbreviation of **multichip module**.

MCP 1. Abbreviation of **microchannel plate**. 2. Abbreviation of **multi-chip package**.

MCR Abbreviation of **Minimum Cell Rate**.

MCT Abbreviation of **mercury cadmium telluride**.

MCU 1. Abbreviation of **microcontroller unit**, or **microcontroller**. 2. Abbreviation of **Multipoint Control Unit**.

MCW Abbreviation of **modulated continuous wave**.

MD Abbreviation of **minidisc**, or **minidisk**.

Md Chemical symbol for **mendelevium**.

MDDB Abbreviation of **multidimensional database**.

MDF Abbreviation of **main distribution frame**.

MDI 1. Abbreviation of **Medium Dependent Interface**. 2. Abbreviation of **Multiple Document Interface**.

MDI port Abbreviation of **Medium Dependent Interface Port**.

MDI-X Abbreviation of **Medium Dependent Interface Crossover**.

MDI-X port Abbreviation of **Medium Dependent Interface Crossover Port**.

MDR Abbreviation of **memory data register**.

MDRAM Abbreviation of Multibank **DRAM**, or Multibank Dynamic **RAM**. A type of RAM in which the total available memory is divided into individually-accessible banks, such as those having 32 KB each. This can increase performance, for instance, by facilitating memory interleaving.

MDS 1. Abbreviation of **minimum detectable signal**, or **minimum discernable signal**. 2. Abbreviation of **multipoint distribution service**.

m_e Symbol for **electron mass**, or **electron rest mass**.

ME 1. Abbreviation of **molecular electronics**. 2. A common operating system.

mean 1. The value obtained by first adding together a set of quantities, and then dividing by the number of quantities in the set. Also called **average**, **average value** (1), or **arithmetic mean**. 2. For a product of **n** factors, the nth root. For instance, the geometric mean of 6 and 216 (a total of two factors) is 36, as the square root of 1,296 is 36. Also called **geometric mean**.

mean current The average value of a current flowing through a circuit or device. Also called **average current**.

mean free path 1. In acoustics, the average distance traveled by sound waves between successive reflections. 2. The average distance traveled by a particle between collisions, such as those with other particles, atoms in a crystal lattice, neutrons in a moderator, and so on.

mean free time The average time that elapses between successive collisions of a particle, such as those with other particles, atoms in a crystal lattice, neutrons in a moderator, and so on.

mean life The average lifetime of a particle, substance, device, or unit. Usually refers to atoms, their components, or elementary particles. For instance, the muon, which is an unstable elementary particle, has a mean life of approximately 2.2 microseconds. Also called **average life**.

mean time before failure The average time that elapses before a component, circuit, device, piece of equipment, or system fails for the first time. It is usually expressed in hours, and is a measure of reliability. Its abbreviation is **MTBF**. Also called **mean time to failure**.

mean time between failures The average time that elapses between successive failures of a component, circuit, device, piece of equipment, or system. It is usually expressed in hours, and is a measure of reliability. Its abbreviation is **MTBF**.

mean time to failure Same as **mean time before failure**. Its abbreviation is **MTTF**.

mean time to repair Its abbreviation is **MTTR**. 1. The average time necessary to repair a component, circuit, device, piece of equipment, or system. 2. The average time that elapses between successive needs to repair a component, circuit, device, piece of equipment, or system.

measurand A physical quantity, variable, or property which is measured during observation, analysis, testing, or the like. A measurand may be a voltage, current, temperature, frequency, and so on.

measured rate service Same as **measured service**.

measured service Also called **measured rate service**. 1. Any service for which charges are made according to elapsed connection time. For example, telephone charges based on the number of minutes a connection is maintained. 2. Any service for which charges are made according to elapsed time, frequency of use, speed of connection, distance, or another quantifiable parameter. Also, to charge based on a combination of such factors. For instance, a telephone calling rate based on the distance to the called destination and the duration of the call.

measurement 1. The act or process of ascertaining a value, dimension, magnitude, capacity, or the like. Such a measurement is made against a known standard, and when obtained utilizing only internationally accepted base units, such as time measured in seconds, it is an absolute measurement. **2.** A figure obtained via a **measurement (1)**.

measurement device A physical unit or mechanism which enables ascertaining a value, dimension, magnitude, capacity, or the like. An example is a measurement instrument.

measurement error **1.** A **measurement (2)** in which there is a deviation from a true, expected, or specified result. For example, one where there is discrepancy between a measured value and the theoretically correct one. **2.** The magnitude of a **measurement error (1)**.

measurement instrument A device utilized to directly or indirectly measure, indicate, and/or monitor the value of an observed and/or controlled quantity. Such an instrument may also record these variations. There are many examples, including altimeters, ammeters, bridges, circuit analyzers, compasses, digital multimeters, frequency meters, oscilloscopes, and spectrometers. Also called **measuring instrument**, or **instrument (1)**.

measurement range For a **measurement instrument**, the portion of the total response range within which a given level of accuracy is equaled or exceeded. Also called **measuring range**.

measuring instrument Same as **measurement instrument**.

measuring junction In a thermocouple circuit, the junction which senses the temperature of a material or object being measured. Also called **hot junction (2)**.

measuring range Same as **measurement range**.

mebi- A binary prefix meaning 2^{20}, or 1,048,576. For example, a mebibyte is equal to 2^{20}, or 1,048,576 bytes. This prefix is utilized to refer to only binary quantities, such bits and bytes Its abbreviation is **Mi-**.

mebibit 2^{20}, or 1,048,576 bits. Its abbreviation is **Mib**.

mebibits per second 2^{20}, or 1,048,576 bits, per second. Its abbreviation is **Mibps**.

mebibyte 2^{20}, or 1,048,576 bytes. Its abbreviation is **MiB**.

mebibytes per second 2^{20}, or 1,048,576 bytes, per second. Its abbreviation is **MiBps**.

mechanical **1.** Related to, produced by, or consisting of tangible objects or forces. For example, a switch actuated by centrifugal forces, the deformation of a crystal or ceramic due to the piezoelectric effect, or the access arm of a hard disk drive. **2.** Pertaining to, or arising from the use of machines or tools.

mechanical arm **1.** A mechanical device or element that serves to position objects. **2.** In robotics, a set of links and powered joints which comprises all parts of the manipulator, except for the wrist and the end-effector.

mechanical damping The gradual reduction or limiting in the amplitude of a vibrating motion, such as an oscillation, through the absorption of mechanical energy.

mechanical energy Energy resulting from losses in kinetic energy which do not result in gains in potential energy. Mechanical energy may be generated, for example, by a blowing wind, a changing tide, or by a rotary machine. A dynamo, for instance, converts mechanical energy into electrical energy.

mechanical filter A filter consisting of various pieces of metal, such as disks or rods, which serve to mechanically separate waves of different frequencies. Such a filter is the equivalent of an electric filter, and is used, for instance, in superheterodyne receivers characterized by high selectivity. Also known as **mechanical wave filter**.

mechanical joint **1.** A connection between two elements, objects, or materials, which uses a mechanical device such as a clamp. **2.** The use of a mechanical device, such as a clip, to join electrical conductors, as opposed to soldering.

mechanical mouse A computer mouse in which the mechanical movements of an incorporated ball are converted into cursor or pointer movements on the screen. Such a mouse has a ball underneath that can roll in any direction along two planes, and its movements are converted by sensors into the appropriate signals. An **optical mouse** utilizes light to track mouse movements over a special grid.

mechanical rectifier A rectifier which utilizes a mechanical action to convert AC into DC. For example, that utilizing a commutator actuated by a synchronous motor to pass only the positive or negative half-cycles.

mechanical scanning The use of mechanical means, such as a moving antenna or mirror, or a rotating disk, to scan, as opposed to electronic means.

mechanical splice A fiber-optic splicing method in which a fixture is used to align and join the ends of the fibers. When such splicing is used, a special gel with a matching refractive index may be used to fill the gaps. Even so, this method provides more losses than a **fusion splice**, in which the ends to be joined are cut, aligned, melted, and joined together.

mechanical strain A lengthening, contraction, torsion, or other deformation resulting from an external stress or force. Also called **strain**.

mechanical switch **1.** A switch that is actuated by using a mechanical device such as a lever or button. **2.** A switch with moving parts.

mechanical translation Same as **machine translation**.

mechanical wave filter Same as **mechanical filter**.

mechanics The study of bodies and systems in relation to the matter and forces which influence them.

mechatronics Abbreviation of **mecha**nics and electr**onics**. A science that deals with the study, creation, adaptation, and improvement of electromechanical devices. Robotics is an application of this field.

media **1.** A plural form of **medium**. **2.** The physical materials or mediums which serve to store data on a permanent or semi-permanent basis. These include magnetic disks and tapes, DVDs, holograms, bubble memory, and so on. Also called **storage media (2)**, or **data-storage media**. **3.** The physical materials which serve to store data on any basis. These include magnetic disks and tapes, optical media such as CD-ROMs, computer displays, cards, and paper.

Media Access Control One of the two sublayers of the data-link layer, as defined by IEEE 802. It is the closer of the two to the physical layer, and handles tasks such as packet addressing and error control. The other sublayer is the **Logical Link Control Sublayer**. Its abbreviation is **MAC**. Also called **Media Access Control Layer**, or **Media Access Control Sublayer**.

Media Access Control address In a network, especially a LAN, a hardware address which uniquely identifies each physical connection. In an Ethernet network, it is the same as the Ethernet address. Its abbreviation is **MAC address**.

Media Access Control Layer Same as **Media Access Control**. Its abbreviation is **MAC Layer**.

Media Access Control Sublayer Same as **Media Access Control**. Its abbreviation is **MAC Sublayer**.

media conversion The changing of computer data from one medium to another. For example, from that on a hard disk to that on DVD+RW.

media failure **1.** The condition in which a computer storage medium, such as a hard disk or DVD, no longer performs the function it was intended to, or is not able to do so at a level that equals or exceeds established minimums. May be due, for instance, to a hard disk crash, or a defect in a recorded surface. **2.** The condition in which a computer is not able to read and/or write to a storage medium.

media gateway A device which performs the necessary protocol conversions to link packet-based traffic, such as that of the Internet, to a public switched telephone network.

Media Gateway Control Protocol An ITU standard pertaining to signaling and session management during multimedia conferencing. Its abbreviation is **MGCP**. Also called **H.248**.

media gateway controller A device which performs the tasks necessary, such as signal analysis, to control a **media gateway**.

media gateway controller protocol A protocol utilized for the exchange of information via a **media gateway**.

Media Interface Connector Its abbreviation is **MIC**. Also called **MIC connector**, or **Medium Interface Connector**. **1.** A standard connector for FDDI cables. **2.** A fiber-optic connector.

media player An application which enables a user to access audio, video, or multimedia content, such as music or videos available on CDs, DVDs, or over the Internet.

media processor A processor that is optimized for reproduction of streaming media such as music or movies.

media stream A flow of uninterrupted audio, video, or both over a communications network such as a LAN or the Internet.

median **1.** The middle value within a distribution. For instance, in the sequence 4, 21, 22, 77, 86, the median is 22, as it is in the middle. This contrasts with **mean (1)**, or average value, which in this example would be 42. **2.** Situated in the middle, or essentially in the middle. **3.** Pertaining to that situated in the middle, or essentially in the middle.

medical electronics A branch of electronics which deals with devices, instruments, apparatuses, equipment, and systems with medical applications such as research, examination, diagnosis, care, and treatment. This field also pertains to the theory, design, use, and improvement of such devices, instruments, apparatuses, equipment, and systems.

medical robot A robot which performs one or more medically-related tasks. There are countless applications, including focusing and keeping a camera steady during a procedure, or assisting in the performance of surgery too precise or complex for human hands.

medium **1.** The substance or entity through which something is transmitted, conveyed, carried, or the like. For example, a vacuum, a fluid, a plasma, or a solid. **2.** A surrounding environment within which materials and entities exist, and within which phenomena, such as that of a physical or chemical nature, take place. For example, a vacuum, a fluid, a plasma, or a solid. **3.** Any physical material or medium which serves to store or otherwise contain data. For instance, optical disks, magnetic tapes and disks, microfilm, and paper. Also called **data medium**, or **storage medium**. **4.** Situated or occurring between two values or degrees. For instance, occurring at or near the middle of an interval.

Medium Dependent Interface Same as **Medium Dependent Interface Port**.

Medium Dependent Interface Crossover Same as **Medium Dependent Interface Crossover Port**.

Medium Dependent Interface Crossover Port In a communications network, especially Ethernet, a port which crosses the receiving and transmitting conductors, enabling connections to hubs or switches using a crossover cable. Its abbreviation is **MDI-X port**. Also called **Medium Dependent Interface Crossover**.

Medium Dependent Interface Port In a communications network, especially Ethernet, a port which allows connections to hubs or switches without a crossover cable. Its abbreviation is **MDI port**. Also called **Medium Dependent Interface**, or **uplink port**.

medium earth orbit The orbit a **medium earth orbit satellite** follows. Its acronym is **MEO**.

medium earth orbit satellite An artificial satellite whose orbit ranges from approximately 2,000 to 20,000 kilometers over the surface if the earth, although the defined value may vary considerably. Such satellites are not stationary from a fixed point above the planet, and their orbital paths may be thought of as lying between that of LEOSs and GEOSs. Its acronym is **MEOS**.

medium frequency A range of radio frequencies spanning from 300 to 3000 kHz. These correspond to wavelengths of 1000 to 100 meters, respectively, which are hectometric waves. Also, pertaining to this interval of frequencies. Also called **medium-frequency band**. Its abbreviation is **MF**.

medium-frequency band Same as **medium band**.

Medium Interface Connector Same as **Media Interface Connector**. Its abbreviation is **MIC**.

medium of propagation Also called **propagation medium**. **1.** The medium through which a wave or other phenomenon propagates. This can be the atmosphere, a vacuum, a waveguide, a crystal, water, and so on. **2.** The **medium of propagation (1)** through which an electromagnetic wave travels.

medium-scale integration In the classification of ICs, the inclusion of between 100 and 3,000 electronic components, such as transistors, on a single chip. This definition required much fewer components only a few years ago, when just 10 or more electronic components qualified for this level of integration. Its abbreviation is **MSI**.

medium tension Same as **medium voltage**.

medium vacuum An environment or system whose pressure is below 1 torr, but above 10^{-3} torr. The upper threshold may be slightly higher, such as 30 torr. One torr equals 1 millimeter of mercury, or approximately 133.3 pascals.

medium voltage A voltage which equals or exceeds a given amount, but which is equal to or less than another value. For example, a voltage between 2000 and 15000 volts, between 220 and 5000 volts, between 5000 and 10000 volts, and so on. Its abbreviation is **MV**. Also called **medium tension**.

medium waves Electromagnetic waves whose frequency is within the range of 300 to 3000 kHz, which corresponds to wavelengths of 1000 to 100 meters, respectively. Its abbreviation is **MW**.

meg- Same as **mega-**, but used only when before a vowel, as in megohm instead of megaohm.

mega- A metric prefix representing 10^6, which is equal to a million. For instance, 1 megahertz is equal to 10^6 Hz, or 1,000,000 Hz. When referring to binary quantities, such as bits and bytes, it is equal to 2^{20}, or 1,048,576, although this is frequently rounded to 1,000,000. To avoid any confusion, the prefix **mebi-** may be used when dealing with binary quantities. Its abbreviation is **M**.

mega-electronvolt A unit of energy or work equal to 10^6, or 1,000,000 electronvolts. Also spelled **megaelectronvolt**. Its abbreviation is **MeV**. Also called **million electron volt**.

megabar A unit of pressure equal to 10^6, or 1,000,000 bars. Its abbreviation is **Mb**.

megabecquerel A unit of activity equal to 10^6, or 1,000,000 becquerels. Its abbreviation is **MBq**.

megabit 2^{20}, or 1,048,576 bits, although this is frequently rounded to a million. To avoid any confusion, the term **mebibit** may be used when referring to this concept. Its abbreviation is **Mb**, or **Mbit**.

megabits per second 2^{20}, or 1,048,576 bits, per second. Usually used as a measure of data-transfer speed. To avoid any confusion, the term **mebibits per second** may be used when referring to this concept. Its abbreviation is **Mbps**.

megabyte 2^{20}, or 1,048,576 bytes, although this is frequently rounded to a million. To avoid any confusion, the term **mebibyte** may be used when referring to this concept. Its abbreviation is **MB**, or **Mbyte**.

megabytes per second 2^{20}, or 1,048,576 bytes, per second. Usually used as a measure of data-transfer speed. To avoid any confusion, the term **mebibytes per second** may be used when referring to this concept. Its abbreviation is **MBps**.

megacoulomb A unit of electric charge equal to 10^6, or 1,000,000 coulombs. Its abbreviation is **MC**.

megacurie A unit of radioactivity equal to 10^6, or 1,000,000 curies. Its abbreviation is **MCi**.

megacycle A unit of frequency equal to 10^6, or 1,000,000 cycles, or 10^6 or 1,000,000 Hz. The term currently used for this concept is **megahertz**. Its abbreviation is **Mc**.

megaelectronvolt Same as **mega-electron volt**.

megaflops Same as **MFLOPS**.

megagauss A unit of magnetic flux density, or magnetic induction, equal to 10^6, or 1,000,000 gauss. Its symbol is **MGs**, or **MG**.

megagram A unit of mass equal to 10^6, or 1,000,000 grams, or 1,000 kilograms, or one tonne. Its abbreviation is **Mg**.

megahertz A unit of frequency equal to 10^6, or 1,000,000 Hz. Its abbreviation is **MHz**. Also called **megacycle**.

megajoule A unit of energy or work equal to 10^6, or 1,000,000 joules. Its abbreviation is **MJ**.

megalumen A unit of luminous flux equal to 10^6, or 1,000,000 lumens. Its abbreviation is **Mlm**.

megameter A unit of distance equal to 10^6, or 1,000,000 meters. Its abbreviation is **Mm**.

megampere A unit of current equal to 10^6, or 1,000,000 amperes. Its abbreviation is **MA**.

megaohm Same as **megohm**.

megapascal A unit of pressure equal to 10^6, or 1,000,000 pascals. Its abbreviation is **MPa**.

megaphone A handheld device, usually in the shape of a funnel, which is used to amplify and direct sounds, usually a voice. Such a unit incorporates a microphone, amplifier, and a speaker, and may have additional features such as an alarm siren signal.

megapixel 1. 1,048,576 pixels, although this amount is usually rounded to 1,000,000 pixels. Generally used to describe the resolution of a graphics device, such as a monitor or digital camera. An example is a screen which is has 1,024 horizontal pixels by 1,024 vertical pixels. 2. Over a million pixels. Usually used to describe the resolution of a graphics device, such as a monitor or digital camera.

megapixel display A display, such as that of a monitor or digital camera, with at least one million displayed pixels.

megatron An ultra-high frequency electron tube with parallel disk-shaped electrodes, which features low interelectrode capacitance and a high-power output. Also called **disk-seal tube**, or **lighthouse tube**.

megavar A unit of reactive electric power equal to 10^6, or 1,000,000 vars. Its abbreviation is **Mvar**. Also called **megavolt-ampere reactive**.

megavar-hour A unit of reactive electric energy equal to 10^6, or 1,000,000 var-hours. Its abbreviation is **Mvarh**.

megavolt A unit of potential difference equal to 10^6, or 1,000,000 volts. Its abbreviation is **MV**.

megavolt-ampere A unit of apparent power or true power equal to 10^6, or 1,000,000 volt-amperes. Its abbreviation is **MVA**.

megavolt-ampere reactive Same as **megavar**.

megavoltmeter A voltmeter whose indications are expressed in megavolts.

megawatt A unit of power equal to 10^6, or 1,000,000 watts. Its abbreviation is **MW**.

megawatt-hour A unit of energy equal to 10^6, or 1,000,000 watt-hours. Its abbreviation is **MWh**, or **MWhr**.

megger 1. A portable instrument which is powered by a hand-driven generator, and that is utilized to measure resistance values. It has a wide useful range, provides readings in megohms, and is used, for instance, for measuring insulation resistance, or for continuity tests. 2. Same as **megohmmeter**.

megohm A unit of resistance, impedance, or reactance equal to 10^6, or 1,000,000 ohms. Its proper abbreviation is **MΩ**, although **M** is also used. Also spelled **megaohm**.

megohmmeter An ohmmeter whose readings are in megohms. Also called **megger** (2).

Meissner effect For a material in a superconducting state, the active exclusion of an external magnetic field from penetrating its interior. Under such conditions a superconducting material reflects a magnetic field, and this property may be used to induce magnetic levitation. If a magnetic field of sufficient strength is applied, it penetrates the interior of the material, and the superconductivity is terminated.

meitnerium A synthetic radioactive chemical element whose atomic number is 108, and which has about three identified isotopes. The techniques employed to produce this element are utilized to help discover new elements with higher atomic numbers. Its chemical symbol is **Mt**.

mel A subjective unit of relative pitch. A pure tone whose frequency is 1000 Hz at 40 dB above a given listener's hearing threshold is defined to be 1000 mels. A pitch judged by the same listener as twice as high is 2000 mels, half as high as 500 mels, and so on.

mel frequency cepstral coefficient Its abbreviation is **MFCC**. A technique, often utilized in voice-recognition systems, which takes into account the subjective manner in which each person's auditory system perceives sound, by a using **mel**-based frequency scale. MFCC may also be used, for instance, to better model music signals.

meltdown 1. A complete halt of a computer system or network. May be caused, for instance, by a program deficiency, a hardware failure, or network congestion. 2. The complete halt of a communications network. The term usually entails a condition of excess traffic, although such an occurrence may also be caused by hardware or software. Also called **network meltdown**.

melting point Its abbreviation is **mp**. Also called **melting temperature**. 1. The temperature at which a liquid and a solid of the same substance are at equilibrium with each other. Each pure substance has a specific melting point, for a given surrounding pressure. For example, at a pressure of 1 atmosphere, the melting point of water is 0 °C. Also called **freezing point** (1). 2. The temperature at which a solid substance being heated begins to liquefy.

melting range For a mixture, such as an alloy, the temperature interval spanning from the solidus temperature to the liquidus temperature.

melting temperature Same as **melting point**.

membrane keyboard A computer keyboard in which pressure-sensitive keys are pressed through a plastic or rubber sheet which covers all keys. Usually used in environments where dust, grease, or other contaminants might otherwise enter and harm the inner keyboard components.

membrane keypad A keypad in which pressure-sensitive keys are pressed through a plastic or rubber sheet which covers all keys. Often used in environments where dust, grease, or other contaminants might otherwise enter and harm the inner keypad components.

memory The locations within a computer that serve for temporarily holding and accessing data in a machine-readable format. Memory chips are used for this purpose, and most are allocated for RAM, or main memory. Memory is usually quantified in multiples of bytes. For example, a computer with 1 gigabyte of RAM can hold approximately 1 billion bytes, or characters, of information, and this is the total temporary workspace this computer has available. Other forms of memory in a computer include ROM, PROM, and EPROM. Although **memory** and **storage** are sometimes used synonymously, storage refers to a more permanent form of holding and accessing data, using magnetic or optical media, such as disks and tapes. Also called **computer memory**, or **system memory (1)**.

memory access speed Same as **memory access time**.

memory access time The time that elapses between a request for data and its delivery. For example, the time necessary to exchange information between the CPU and a hard disk. Also called **memory access speed**, or **access time (1)**.

memory address A number which indicates the specific place where data is stored within the memory of a computer.

memory address register Its abbreviation is **MAR**. **1.** The register that contains the addresses of the location in memory currently being accessed. **2.** A register that stores an address. Also called **address register (2)**.

memory address space The amount of memory a CPU can access. It is determined by the size of the address bus. Also called **address space (1)**.

memory allocation The reserving of computer memory for specific purposes. For instance, a running program sets aside a given amount of RAM for certain functions.

memory bank Any given physical location, such as a memory module, that holds data in memory.

memory-based reasoning In a database, a technique utilized for classifying records by comparing them to records that have already been classified. Its abbreviation is **MBR**.

memory board A circuit board containing computer memory. Such a board incorporates one or more memory chips.

memory buffer register A register utilized to temporarily store information that awaits to be written to, or read from, memory. Also called **memory data register**. Its abbreviation is **MBR**.

memory cache Also called **cache**, or **cache memory**. **1.** In computer memory management, a specialized high-speed storage subsystem. Cache works by storing recently and/or frequently accessed information, where it can be accessed much faster than from where it was obtained. A level 1 cache is built right into the CPU, while a level 2 cache utilizes a memory bank between the CPU and main memory. Level 1 cache is faster, but smaller than level 2 cache, while level 2 cache is still faster than main memory. Disk cache uses a section of main memory to store recently accessed data from a disk, and can dramatically speed up applications by avoiding the much slower disk accesses. One way to measure the effectiveness of a cache is the hit rate, which indicates which proportion of data accesses are fulfilled by it, instead of main memory or a disk. **2.** An area of computer memory where the most recently downloaded Web pages may be temporarily stored, so that when one of these pages is returned to, it loads faster. Also called **browser cache**.

memory capacity **1.** The amount of data a computer can store in memory. Usually expressed in multiples of bytes, such as gigabytes or terabytes. **2.** The amount of data that can be placed in a storage medium. Usually expressed in multiples of bytes, such as gigabytes or terabytes. Also called **storage capacity (1)**.

memory card **1.** An expansion card containing computer memory. Examples include flash cards, IC cards, and RAM cards. **2.** A card or module containing RAM chips. These are utilized to expand the RAM available to a computer. A memory card may also incorporate a battery to keep the memory cells charged. Also called **RAM card**.

memory cartridge A plug-in cartridge containing computer memory. Such a cartridge incorporates one or more memory chips, and may be seen, for instance, in printers and portable or handheld computing devices such as notebooks or PDAs.

memory cell A component or circuit which stores one bit of data. It may consist, for instance, of a single transistor.

memory chip A chip which consists of memory cells and the associated circuits necessary, such as those for addressing, to provide a computer, or other device with memory or storage. Examples include RAM, SRAM, ROM, EEPROM, and flash-memory chips. Also called **memory IC**, **IC memory**, or **chip memory**.

memory compression The encoding of data so that it occupies less memory space.

memory cycle An operation in which data is retrieved from memory, or placed in memory, or retrieved, modified, and placed in memory, and so on.

memory cycle time The time required for the completion of a full **memory cycle**.

memory data register Same as **memory buffer register**. Its abbreviation is **MDR**.

memory dialing A telephone feature which allows placing a call by accessing a previously stored number. Such a feature is used, for instance, by pressing a button, entering a short key sequence, by pressing a key for longer than a specified time, and so on. Also called **speed dialing**.

memory dump To display, print, copy, or transfer the content of the main memory of a computer. Such a dump may be performed after a process which ends abnormally, for instance, to help pinpoint the source of a problem. A memory dump may also be generated automatically. Also called **core dump**, **dump (2)**, or **storage dump (2)**.

memory effect A property of some batteries, especially nickel-cadmium, in which battery life is gradually diminished when recharged before becoming completely discharged. This effect can be minimized by periodically completely draining said batteries.

memory IC Same as **memory chip**. Abbreviation of **memory integrated circuit**.

memory integrated circuit Same as **memory chip**. Its abbreviation is **memory IC**.

memory interleaving The use of multiple banks of memory to increase the speed at which data is read and written. For example, a CPU may stagger memory read/writes between two banks of RAM, and perform a write without having to wait for the read to be completed.

memory leak The failure to recover memory which has been allocated to an application, routine, or process which no longer needs it. Unless such memory is properly deallocated, all available memory will eventually be drained. Also, the resulting condition.

memory location **1.** Any place in memory where data may be stored. **2.** A specific place where data is stored within the memory of a computer, peripheral, or disk.

memory management The techniques and operations utilized to help optimize the use of a computer's memory. These include freeing memory when no longer used by a program, handling virtual memory, paging, and swapping, and are usually performed automatically by the operating system. Also called **memory management system**.

memory management system Same as **memory management**.

memory management unit A hardware component that performs **memory management**, especially that related to the use of virtual memory. It is usually incorporated into the CPU. Its abbreviation is **MMU**.

memory manager A program or hardware component which performs **memory management**.

memory map A listing, diagram, or other representation of the locations of instructions and data in a computer's memory.

memory mapped I/O An I/O system in which each I/O location is assigned a memory location as if it were memory.

memory module A module containing RAM chips. These are utilized to expand the RAM available to a computer.

memory protection 1. A mechanism, such as that provided by the operating system, which prevents one program from writing into an area of memory that is being used by another program. 2. A mechanism, such as that provided by the operating system or a hardware device, which prevents modification of any given memory content.

memory register A register in the memory of computer, peripheral, or disk, such as a memory address register.

memory resident Located in memory at all times. Said of programs which are never swapped or otherwise removed from main memory. Certain portions of the operating system, and applications that require immediate access are memory resident. Some programs, such as virus-scanning applications, may be marked as memory resident by a user. Also called **memory-resident program**, **RAM resident**, or **resident program**.

memory-resident program Same as **memory resident**.

Memory Stick A proprietary compact flash card.

MEMS Acronym for **microelectromechanical systems**.

mendelevium A synthetic chemical element whose atomic number is 101. There are over a dozen known isotopes, all of which are unstable. The techniques employed to produce this element are utilized to help discover new elements with higher atomic numbers. Its chemical symbol is **Md**.

menu 1. A list of commands, options, or functions which are available to a computer user. Such menus are common in GUIs, and come in various forms, including pull-down, and may be accessed using a mouse or other pointing device, or the keyboard. A logically organized menu can help users maximize the usefulness of any given application. 2. A list of commands, options, or functions which are available to a user of a given device, piece of equipment, or system.

menu bar A usually horizontal bar from which menu commands, options, or functions can be selected. Also spelled **menubar**.

menu command A command accessed through a **menu**.

menu-driven Computer programs in which menus are utilized to give commands. Such programs are more user-friendly, although less flexible than **command-driven programs**, in which commands are given in the form of special letters, words, or phrases.

menu-driven interface An interface that is **menu-driven**. Seen for instance, in kiosks.

menu-driven program A program that is **menu-driven**.

menu-driven system An operating system that is **menu-driven**.

menu function A function accessed through a **menu**.

menu item Any of the choices, functions, options, or the like, offered by a **menu**. When any such alternative is lighter, grayed, or fuzzy, it indicates that it is unavailable. For example, the cut function should be grayed out when no content is selected.

menubar Same as **menu bar**.

MEO Acronym for **medium earth orbit**.

MEOS Acronym for **medium earth orbit satellite**.

merchant server A network server that handles business transactions such as online purchases made with credit cards, or inventory management. Used extensively over the Internet. Also called **commercial server**.

mercuric chloride Toxic white crystals or powder whose chemical formula is $HgCl_2$. Used in batteries and photography.

mercuric iodide Toxic red crystals whose chemical formula is HgI_2. Used in spectroscopy.

mercuric oxide A toxic bright orange-red powder whose chemical formula is HgO. Used in batteries and ceramics.

mercuric sulfate A toxic white crystalline powder whose chemical formula is $HgSO_4$. Used in batteries.

mercurous sulfate A toxic white to yellow crystalline powder whose chemical formula is Hg_2SO_4. Used in batteries.

mercury A dense metallic chemical element whose atomic number is 80. It is a poisonous silvery liquid, is a very good electrical conductor, and has extremely high surface tension. Mercury has over 30 known isotopes, of which several are stable. Its applications are many, including its use in vapor lamps, arc lamps, switches, amalgams, as neutron absorbers in nuclear plants, and in thermometers and other instruments such as barometers and vacuum pumps. Its chemical symbol, **Hg**, is taken from the Latin word for liquid silver: hydragyrum. Also called **quicksilver**.

mercury arc A highly luminous and sustained discharge of electricity through ionized mercury vapor. Such an arc gives off a brilliant blue-green glow, and emits ultraviolet radiation.

mercury-arc lamp An arc lamp whose light is produced by a **mercury arc**.

mercury-arc rectifier A rectifier tube in which the conducting medium is mercury vapor. Also called **mercury-vapor rectifier**, or **mercury rectifier**.

mercury battery A battery composed of **mercury cells**.

mercury cadmium telluride An alloy whose structure is $HgCdTe$, and which is used in semiconductors and in infrared spectroscopy. Its abbreviation is **MCT**.

mercury cell A primary cell in which the anode is zinc, the cathode is mercuric oxide, and the electrolyte is a solution of potassium hydroxide. It produces an essentially constant output of 1.35V throughout its useful life, and is usually in the form of a small flat disk. It is used in hearing aids, watches, and other small devices, but is being replaced by other cells due to the toxicity of mercury. Also called **zinc-mercuric-oxide cell**.

mercury delay line A delay line in which mercury is the medium of transmission. Used, for instance, to delay acoustic waves.

mercury-displacement relay Same as **mercury relay**.

mercury memory An older form of computer storage utilizing a **mercury delay line**. Also called **mercury storage**.

mercury-pool cathode In a gas tube, a cathode consisting of a pool of mercury. Also called **pool cathode** (1).

mercury rectifier Same as **mercury-arc rectifier**.

mercury relay A relay whose contacts are connected via a moving pool of mercury. For instance, one contact is a fixed piece of metal, while the other is a pool of mercury that is displaced by a plunger assembly, thus coming in contact with the stationary electrode when actuated. Also called **mercury-displacement relay**.

mercury storage Same as **mercury memory**.

mercury switch A switch in which the contacts are sealed in a capsule containing a pool of mercury. When the tube is tilted in one direction, the mercury comes in contact with both electrodes, thus closing the circuit. When tilted the other way, the contacts are exposed to open said circuit. Also called **mercury tilt switch**.

mercury tilt switch Same as **mercury switch**.

mercury tube A gas tube in which the gas ionized during operation is mercury vapor. Also called **mercury-vapor tube**.

mercury-vapor lamp A discharge lamp in which the light is produced by ionizing mercury vapor. Such a lamp provides brilliant blue-green illumination, and emits ultraviolet radiation.

mercury-vapor rectifier Same as **mercury-arc rectifier**.

mercury-vapor tube Same as **mercury tube**.

mercury-wetted reed relay A reed relay whose reed contacts are sealed with a pool of mercury. Capillary action keeps the contacts covered with a mercury film.

mercury-wetted reed switch A reed switch whose reed contacts are sealed with a pool of mercury. Capillary action keeps the contacts covered with a mercury film.

merge To combine or unite. For instance, to combine lists or data.

merged transistor logic IC logic utilizing bipolar transistor gates, and which is characterized by very high packing density, high speed, and low power consumption. Used, for instance, for logic arrays, and various analog and digital applications. Also called **integrated-injection logic**.

meridian 1. An imaginary great circle along the surface of the planet which passes through both geographic poles. All points along a given meridian have the same longitude, with the prime meridian passing through longitude $0°$. Also, a great half-circle joining both geographic poles. **2.** An imaginary great circle along the surface of a celestial body which passes through both its geographic poles, or that joins them.

mesa Within an electronic device, such as a transistor, a raised area which is flat. Such an area serves, for instance, to isolate regions of a semiconductor wafer, and may be produced by etching away all areas surrounding the zone where the mesa is desired.

mesa device A semiconductor device which incorporates a **mesa**.

mesa diode A diode which incorporates a **mesa**.

mesa transistor A transistor which incorporates a **mesa**.

MESFET Acronym for **m**etal **s**emiconductor **f**ield-**e**ffect **t**ransistor. A field-effect transistor which utilizes a metal-semiconductor diode as a gate. It is a unipolar transistor, and the mobility of the charge carriers is higher in relation to MOSFETs. Used, for instance, as a power amplifier in microwave applications.

mesh 1. A group or combination of branches or components which form a closed current path in an electric network. Also called **loop (2)**. **2.** An arrangement of interlocking components or links that form a closed figure, structure, or grid.

mesh circuit A combination of multiple circuit elements, such as resistors, connected in series, and arranged in the form of a polygon. When consisting of three circuit elements arranged in the form a triangle, it is called **delta circuit**. Also called **mesh connection**.

mesh connection Same as **mesh circuit**.

mesh network A communications network in which each node can be reached by at least two pathways.

meson A hadron composed of a quark and antiquark pair, and which may have a charge of +1, -1, or 0.

mesosphere An atmospheric layer extending from about 50 to 80 kilometers above the surface of the planet. It is above the stratosphere and below the thermosphere.

message Its abbreviation is **msg**. **1.** A given amount of information which is transferred from one entity, such as a device, user, application, or system, to another. Also, the information contained in such a communication, which may be text, control characters, commands, images, audio, and so on. Also, to send such a message. **2.** A **message (1)** consisting of an email sent over a communications network such as the Internet. **3.** In programming, a **message (1)** utilized to request or call an action, operation, function, object, or the like.

message authentication In computers and communications, the process of verifying the authenticity of a message. Measures such as passwords and digital signatures may be employed.

Message Authentication Code A string of bits or characters, or a value, computed utilizing both text and a secret key, which is attached to a message in order to authenticate it.

message digest A condensed representation of a message which is used, for instance, to create digital signatures. Such a representation is usually generated via a secure hash algorithm.

message handling The tasks taken care of by an email or other messaging service. These include submission, transfer, switching, storage, and reception functions.

message handling system Same as **messaging system**. Its abbreviation is **MHS**.

message header The first part of a message. In an email, for example, it might contain information such as the email address of the sender, that of the recipient or recipients, the time and date sent, the subject, mail protocols utilized, the IP address of the sender, and so on.

message on hold A prerecorded message which is played when a caller is on hold. Such messages may consist, for instance, of advertisements, a voice menu, or queue status updates. When broadcast programming is provided, the concept is usually called **music on hold**.

message queue A list of messages awaiting transmission or reception. Also, a storage location for such messages.

Message Security Protocol A protocol for the secure handling of messages sent over the Internet. It incorporates security features such a encryption and authentication. Its abbreviation is **MSP**.

message storage 1. The automatic storage of messages, such as emails, which await delivery. **2.** The storage of messages, such as emails, which have already been received. **3.** A location where messages may be held. Also, the act of storing in such a location.

message switch A computer which is used for **message switching**.

message switching In a communications network, the routing of messages through one or more intermediate points, where the complete message is received and stored, before being forwarded towards its destination. This contrasts with **packet switching**, in which messages are broken into packets before being transmitted, routed, forwarded, and so on, to their destination.

message switching system One or more computers which serve for **message switching**.

message transfer agent Within a messaging system, the functional unit which transmits and forwards messages to users or groups of users. Its abbreviation is **MTA**.

message waiting indicator An indicator, such as a flashing light, stutter dial tone, or screen icon which informs that one or more stored messages await.

messaging 1. The sending, receiving, and exchanging of messages. **2.** The sending, receiving, and exchanging of messages with a comparatively small maximum length. For instance, SMS and chat messages.

Messaging API Same as **Messaging Application Programming Interface**.

messaging application An application, such as an email or fax program, which enables **messaging**.

Messaging Application Programming Interface An interface which enables different email applications across different platforms to work together. Its acronym is **MAPI**. Its abbreviation is **Messaging API**.

messaging client An application which enables messaging through the use of one or more remote servers.

messaging gateway In a communications network, a computer or program that links messaging systems by performing the necessary protocol conversions. Used, for instance, between message transfer agents.

messaging middleware Software which performs the necessary tasks, such as format conversions, to enable applications to send and receive messages amongst themselves. Also called **messaging-oriented middleware**.

messaging-oriented middleware Same as **messaging middleware**. Its abbreviation is **MOM**.

messaging protocol A communications protocol, such as SMTP or X.400, used for messaging.

messaging system A group of functional units which work together to provide messaging. For instance, in an email system, the combination of a user agent, message transfer agent, and message storage. Also called **message handling system**.

meta character Same as **metacharacter**.

meta data Same as **metadata**.

meta description Any description of the content of a Web page conveyed by a **meta tag**.

meta element 1. Specific information about a Web page, such as keywords or content, conveyed by a **meta tag**. 2. Same as **meta tag**.

meta file Same as **metafile**.

meta information The information about a Web page conveyed by a **meta tag**. Also spelled **metainformation**.

meta keywords Any keywords, conveyed by a **meta tag**, which would assist in locating a given Web page.

meta-language Same as **metalanguage**.

meta refresh A **meta tag** which redirects to another Web page. For example, if the URL of a given entity changes, such a tag will automatically redirect to the new location.

meta site Same as **metasite**.

meta tag In a language such as HTML or XML, a tag which provides information about a Web page. For example, such a tag may specify keywords, describe content, redirect to another page, or detail the frequency of updates. Meta tags are usually included in the header of Web pages, and any given header may contain multiple such tags. Also spelled **metatag**. Also called **meta element (2)**.

metacharacter A character conveying information about other characters. For example, a wildcard character. A search on a disk specifying *.exe, for instance, would return all files whose extension is exe, as the asterisk is a wildcard, and as such can represent one or more characters. Also spelled **meta character**.

MetaCrawler A popular search engine.

metadata Data conveying information about data. For instance, it may describe its format, name, size, mode of acquisition, field types and so on. A specific example is a data dictionary. Also spelled **meta data**.

metadyne A rotating magnetic amplifier in which a small increase in input power results in a large increase in output power. Also called **amplidyne**.

metafile Also spelled **meta file**. 1. A graphics format which combines elements of bitmap and vector graphics to describe the content of an image. 2. A file that provides information about other files. 3. A file that contains other files.

MetaFrame Popular software facilitating the use of client/server architectures across multiple platforms.

metainformation Same as **meta information**.

metal A chemical element which is usually a hard, malleable, and dense crystalline solid which is lustrous, and a good conductor of heat and electricity. However, not all metals have all these properties. Sodium and potassium, for instance, are soft, mercury is a liquid at ordinary pressures and temperatures, beryllium and bismuth are brittle, and so on. Most naturally occurring elements are metals, and include gold, silver, copper, iron, and cadmium. The properties of each metal may be markedly different than those of others, as seen, for instance in the extremely high reactivity of francium versus the chemical inactivity of gold. Alloys, such as brass and bronze, which exhibit these characteristics are also considered to be metals.

metal-air battery A battery in which the positive electrode is depolarized by atmospheric oxygen. An example is a zinc-air battery. Also known as **air-depolarized battery**.

metal-ceramic A mixture of ceramic and metallic components to obtain a material which combines properties of both. For example, an electric element may combine the dielectric characteristics of a ceramic such as porcelain, with the conductive qualities of a metal such as chromium. Used, for instance, in film resistors. Also called **cerametal, ceramal, ceramet**, or **cermet**.

metal detector An electronic device which serves to detect the presence of metal objects such as wires or pipes in walls, pieces of metal in food products, and so on. Such a device, for instance, may emit electromagnetic waves which interact with any conductive objects encountered, with changes in said field indicating the presence of metals. Also called **metal locator**.

metal-film capacitor Same as **metallized film capacitor**.

metal-film resistor A resistor in which the resistive element is a film of a metal, metal oxide, or alloy, which is deposited on an insulating substrate such as a ceramic. These resistors have excellent operational characteristics, including low noise and minimal drift. Also called **metallized resistor**.

metal-halide lamp A halide lamp in which the contained gas is a mixture of a metallic vapor, such as mercury, and a halide of a metal, such as thallium iodide. Such lamps are very powerful, and are used, for instance, in portable projectors and to illuminate roadways.

metal-insulator-metal capacitor Same as **MIM capacitor**.

metal-insulator semiconductor A semiconductor in which there is an extremely thin layer between the metal contacts and the substrate. Used, for instance, in microstrip devices. Its acronym is **MIS**.

metal locator Same as **metal detector**.

metal-nitride-oxide semiconductor Same as **MNOS**.

metal-organic vapor-phase epitaxy A chemical vapor deposition technique in which at least one organometallic compound is used as a reactant. Its acronym is **MOVPE**. Also called **organometallic vapor-phase epitaxy**.

metal-oxide resistor A **metal-film resistor** in which the resistive element is a film of a metal oxide.

metal-oxide semiconductor Same as **MOS**.

metal-oxide semiconductor capacitor Same as **MOS capacitor**.

metal-oxide semiconductor controlled thyristor Same as **MOS-controlled thyristor**.

metal-oxide semiconductor field-effect transistor Same as **MOSFET**.

metal-oxide semiconductor IC Same as **MOS IC**.

metal-oxide semiconductor integrated circuit Same as **MOS IC**.

metal-oxide semiconductor/large-scale integration Same as **MOS/LSI**.

metal-oxide semiconductor random-access memory Same as **MOS RAM**.

metal-oxide semiconductor technology Same as **MOS technology**.

metal-oxide semiconductor transistor Same as **MOSFET**.

metal-oxide semiconductor/very large-scale integration Same as **MOS/VLSI**.

metal-oxide varistor Its abbreviation is **MOV**. A rugged varistor in which the resistive element is a metal oxide, such as zinc oxide. MOVs are usually used as surge suppressors with reaction times in the nanosecond range.

metal-semiconductor contact A contact between a metal and the semiconductor substrate it is deposited on. It may be either an ohmic contact or a Schottky contact. Also called **semiconductor-metal contact**.

metal-semiconductor diode A diode with a metal-semiconductor junction, as opposed to the usual pn junction. Hot carriers are injected from the semiconductor layer into the metal layer, so that there are virtually no minority carriers injected or stored, which provides for a very low forward voltage drop and a very fast switching speed. Such diodes have multiple uses, including several in RF applications, logic circuits, photodiodes, and as rectifiers. Also called **Schottky diode, Schottky barrier diode, semiconductor-metal diode**, or **hot-carrier diode**.

metal semiconductor field-effect transistor Same as **MESFET**.

metal-semiconductor junction A layer or boundary which serves as the interface between a semiconductor and a metal. Also called **semiconductor-metal junction**.

metal tape A magnetic recording tape in which metal particles, as opposed to oxides of metals, are used. Its performance is superior to chrome, or Type II, tapes. Also called **Type IV tape**.

metalanguage A computer language utilized to describe another computer language. Such a language may also be used to describe itself. Also spelled **meta-language**.

metallic 1. Consisting of, or containing a metal. For example, a metallic chemical element such as aluminum or copper. 2. Of, pertaining to, associated with, or having the characteristics of a metal. For instance, a metallic luster.

metallic bond A chemical bond serving to hold together the atoms in a metal or metallic alloy. In such a bond, the atoms share their valence electrons with neighboring atoms to form a sea of electrons that moves through the lattice, which helps account for the good electrical and thermal conductivity exhibited by metals.

metallic bonding The formation or occurrence of **metallic bonds**.

metallic circuit A circuit, such as a two-wire telephone line, in which the earth is not utilized to complete the conducting path. This contrasts with a **ground-return circuit**, in which an earth ground is part of the closed path.

metallic-disk rectifier A rectifier which utilizes one or more metal disks coated with a semiconductor layer. This layer may consist of selenium, copper oxide, or another suitable semiconductor. The rectification occurs as a result of the greater conductivity across the contact in one direction than the other. Also called **metallic rectifier, dry rectifier, dry-disk rectifier, semiconductor rectifier** (1), or **contact rectifier**.

metallic rectifier Same as **metallic-disk rectifier**.

metallic tape A tape consisting of, or incorporating a metal. For instance, such a tape utilized in an intrusion alarm system.

metallization To deposit a thin metal coating on a substrate. For example, such a coating on a semiconductor, ceramic, or glass, to form the desired conductive paths and/or interconnections, or to provide a protective layer, and so on. Electroplating and chemical-vapor deposition are two common techniques utilized.

metallize To cover, plate, or impregnate a material with a **metal**.

metallized capacitor Same as **metallized film capacitor**.

metallized film A thin, or very thin, layer of metal deposited upon or otherwise coating a substrate.

metallized film capacitor A capacitor in which one or both plates consist of a metal film deposited on a dielectric. Also called **metallized capacitor**, or **metal-film capacitor**.

metallized paper capacitor A **metallized film capacitor** with a paper dielectric.

metallized polycarbonate capacitor A **metallized film capacitor** with a polycarbonate dielectric.

metallized resistor Same as **metal-film resistor**.

metallizing The covering, plating, or impregnating a material with a **metal**.

metalloid A chemical element which has some properties similar to metals, other properties similar to nonmetals, and/or properties which lie somewhere in between. For example, arsenic, germanium, and tellurium. Such elements are usually semiconductors. Also called **semimetal**.

metasearch engine A search engine that searches for and presents results from other search engines.

metasite A Web site that describes or searches other Web sites. For instance, a site which serves as a directory to multiple sites dealing with the same topic. Also spelled **meta site**.

metastable Describing a precarious state of stability which can easily be changed into another, more stable state. For example, a coin that is balanced on its edge.

metatag Same as **meta tag**.

meteor 1. A momentary and bright streak or trail of light which appears in the sky when a meteoroid penetrates, and is subsequently burned, by the atmosphere. 2. Same as **meteoroid**.

meteoric Pertaining to, or arising from, a **meteor** or **meteoroid**.

meteoric scatter 1. The scattering of radio waves back to the surface of the planet by the trails of meteors. 2. Radio wave propagation utilizing **meteoric scatter** (1).

meteorite A **meteoroid** which reaches the surface of the planet without being completely consumed.

meteorogram A displayed or recorded output of a **meteorograph**.

meteorograph An instrument which measures and records multiple meteorological phenomena, such as temperature, pressure, humidity, or wind speed.

meteoroid A small solid mass that orbits, or otherwise moves through space. Also called **meteor** (2).

meteorological Pertaining to **meteorology**.

meteorology The science that deals with the atmosphere, and phenomena related to it, especially weather and its effects.

meter 1. The SI unit of distance. It is defined as the distance that makes the speed of light in a vacuum exactly 299,792,458 meters per second. Thus, a meter is the distance traveled by light in a vacuum during a time interval of 1/299792458 second. Its abbreviation is **m**. Also spelled

metre. **2.** An instrument or device which is utilized to measure, and usually indicate, a physical quantity or other value. There are numerous examples, including ammeters, absorptiometers, barometers, chronometers, dB meters, galvanometers, and voltmeters.

meter accuracy The extent to which a value indicted by a meter approximates the real value.

meter-candle A unit of illuminance or illumination equivalent to a lux. That is, one meter-candle is equal to one lux.

meter-kilogram-second-ampere system A measuring system in which the fundamental units for expressing distance, mass, time, and electric current are the meter, kilogram, second, and ampere respectively. Its abbreviation is **MKSA system.** Also called **Giorgi system.**

meter-kilogram-second system A measuring system in which the fundamental units for expressing distance, mass, and time are the meter, kilogram, and second. Its abbreviation is **MKS system.**

meter multiplier The factor by which a given reading must be multiplied to obtain an equivalent in the desired units. For instance, a disk in a kilowatt-hour meter may be calibrated in watts per revolution, so a meter multiplier can be utilized to obtain kilowatt-hours.

meter rating 1. The maximum error of a meter when operated under specified conditions. **2.** The maximum reading, at or above a given level of accuracy, that a given meter can indicate.

meter reading 1. The amount indicated by a meter. **2.** To verify, usually visually, the amount indicated by a meter.

meter relay A sensitive relay whose moving contact is attached to a pointer, and which is used to indicate values along the scale of a meter. Also called **instrument relay.**

meter resistance The internal resistance of a meter, as measured at its terminals at a given temperature. In a moving-coil galvanometer, for instance, it is the resistance of the coil.

meter sensitivity 1. The minimum change in a measured quantity that produces an observable change in the indication of a meter. **2.** The minimum change in an observed quantity that a meter can detect.

meter shunt A resistor which is connected in parallel, to extend the current range of a meter. For example, such a shunt utilized to extend the range of an ammeter. In addition, such a resistor helps provide protection against current surges.

meters per second A unit of speed expressing the meters traveled per second. It is equal to 0.001 kilometers per second. Its abbreviation is **m/s,** or **m/sec.**

method 1. Any manner, means, or procedure utilized to accomplish something. For instance, an encryption method. **2.** A specific manner, means, or procedure utilized to accomplish something. For example, public-key encryption. **3.** In object-oriented programming, that which is performed or executed by an object when it receives a message.

methodology The body of methods, principles, procedures, and the like, utilized in a given discipline. Methodology implies a disciplined and objective approach, although the term is frequently utilized to refer to most any approach, meticulous or not.

metre An alternate spelling of **meter (1).**

metric system A measuring system based on the decimal system. The meter-kilogram-second system is an example, and the International System of Units is based on the meter-kilogram-second system.

metric ton A unit of mass equal to 1,000 kilograms, or 1 megagram. Its symbol is **t,** or **MT.** Also called **tonne.**

metric waves Electromagnetic waves whose wavelengths are between 1 and 10 meters, corresponding to frequencies of between 300 and 30 MHz, respectively. This interval represents the very high frequency band.

metrology 1. The science that deals with all aspects of measurement, including how to accurately obtain them, and the definition and standardization of the units utilized to express them. Currently, scientific metrology is based on the International System of Units. **2.** A specific system of measurement, such as the meter-kilogram-second system.

Metropolitan Area Exchange A facility where Internet service providers interconnect. Such a location handles an enormous amount of data traffic, determines how it is routed, and represents a key constituent of the Internet backbone. Its abbreviation is **MAE.** Also called **Network Access Point,** or **Internet Exchange.**

Metropolitan Area Network A high-speed computer network that provides services to a geographical area larger than a LAN, but usually smaller than a WAN. Such a network may cover, for instance, a single urban area. Its abbreviation is **MAN.**

MeV Abbreviation of **megaelectronvolt.**

mezzanine card An adapter card that plugs into another adapter card, especially when the latter is plugged into a PCI bus.

mF Abbreviation of **millifarad.**

MF Abbreviation of **medium frequency.**

MFCC Abbreviation of **mel frequency cepstral coefficient.**

MFD 1. Abbreviation of **mode field diameter.** **2.** Abbreviation of **multifunction device.**

Mflops Same as **MFLOPS.**

MFLOPS Abbreviation of **megaflops,** or **million floating-point operations per second.** One million floating-point calculations or operations, per second. Usually used as a measure of processor speed.

MFM 1. Abbreviation of **magnetic force microscope.** **2.** Abbreviation of **magnetic force microscopy.**

MFP 1. Abbreviation of **multifunction printer.** **2.** Abbreviation of **multifunction peripheral.**

mg Abbreviation of **milligram.**

mG Abbreviation of **milligauss.**

Mg 1. Chemical symbol for **magnesium.** **2.** Abbreviation of **megagram.**

MG Abbreviation of **megagauss.**

MGCP Abbreviation of **Media Gateway Control Protocol.**

mGs Abbreviation of **milligauss.**

MGs Abbreviation of **megagauss.**

mH Abbreviation of **millihenry.**

MHD 1. Abbreviation of **magnetohydrodynamics.** **2.** Abbreviation of **magnetohydrodynamic**

MHD generation Abbreviation of **magnetohydrodynamic power generation.**

MHD generator Abbreviation of **magnetohydrodynamic generator.**

mho An older term for a unit equivalent to **siemens.**

MHS Abbreviation of **message handling system.**

MHz Abbreviation of **megahertz.**

Mi Abbreviation of **mile.**

Mi- Abbreviation of **mebi-.**

Mib Abbreviation of **mebibit.**

MiB Abbreviation of **mebibyte.**

Mibps Abbreviation of **mebibits per second.**

MiBps Abbreviation of **mebibytes per second.**

MIC 1. Abbreviation of **microwave IC.** **2.** Abbreviation of **monolithic IC.** **3.** Abbreviation of **Media Interface Connector,** or **Medium Interface Connector.**

mic Abbreviation of **microphone**.

MIC connector Same as **Media Interface Connector**.

mic preamplifier Abbreviation of **microphone preamplifier**.

mica Any of various soft silicate minerals which cleave into thin, flexible, and elastic sheets. Used, for instance, as electrical insulators, in electron tubes, and as a dielectric in capacitors.

mica capacitor A capacitor in which **mica** is the dielectric. Such capacitors feature a high Q factor, and good frequency stability.

Michelson interferometer An interferometer which allows accurate measurements of electromagnetic radiation by varying the path length of one of the two beams. Once the incident beam is split, a fixed mirror and a moving mirror are utilized to produce the interference patterns.

MICR Abbreviation of **Magnetic-Ink Character Recognition**.

micro Abbreviation of **microcomputer (1)**.

micro- A metric prefix representing 10^{-6}, or one millionth. For instance, 1 microampere is equal to 10^{-6}, or one millionth of an ampere. Its abbreviation is μ, or less commonly, **mc**.

micro-ammeter Same as **microammeter**.

micro-ampere Same as **microampere**.

micro-balance Same as **microbalance**.

Micro Ball Grid Array A surface-mount chip package similar to Ball Grid Array, but which uses even smaller balls of solder at each contact for increased component density. Its abbreviation is **MBGA**.

micro-beam Same as **microbeam**.

micro-browser Same as **microbrowser**.

micro-candela Same as **microcandela**.

micro-channel plate Same as **microchannel plate**.

micro-circuit Same as **microcircuit**,

micro-circuitry Same as **microcircuitry**.

micro-component Same as **microelectronic component**. Also spelled **microcomponent**.

micro-computer Same as **microcomputer**.

micro-controller Same as **microcontroller**.

micro-controller unit Same as **microcontroller unit**.

micro-coulomb Same as **microcoulomb**.

micro-crystal Same as **microcrystal**.

micro-curie Same as **microcurie**.

micro-densitometer Same as **microdensitometer**.

micro-display Same as **microdisplay**.

micro-electrode Same as **microelectrode**.

micro-electromechanical systems Same as **microelectromechanical systems**.

micro-electronic Same as **microelectronic**.

micro-electronic circuit Same as **microelectronic circuit**.

micro-electronic component Same as **microelectronic component**.

micro-electronic device Same as **microelectronic device**.

micro-electronics Same as **microelectronics**.

micro-element Same as **microelement**.

micro-etching Same as **microetching**.

micro-farad Same as **microfarad**.

micro-floppy Same as **microfloppy**.

micro-gauss Same as **microgauss**.

micro-henry Same as **microhenry**.

micro-image Same as **microimage**.

micro-instruction Same as **microinstruction**.

micro-joule Same as **microjoule**.

micro kernel Same as **microkernel**.

micro-liter Same as **microliter**.

micro-lithography Same as **microlithography**.

micro-lumen Same as **microlumen**.

micro-lux Same as **microlux**.

micro-machining Same as **micromachining**.

micro-manipulator Same as **micromanipulator**.

micro-miniature Same as **microminiature**.

micro-miniature circuit Same as **microminiature circuit**.

micro-miniature component Same as **microminiature component**.

micro-miniature device Same as **microminiature device**.

micro-module Same as **micromodule**.

micro-newton Same as **micronewton**.

micro-ohm Same as **microhm**. Also spelled **microohm**.

micro-ohmmeter An ohmmeter designed to perform accurate measurements of extremely low resistances. Such an instrument may be able to measure a resistance of a single micro-ohm, but can be utilized for indications in the nanohm range when used with a programmable current source and a nanovoltmeter. Also spelled **microhmmeter**.

micro-optics 1. Optical materials, objects, components, and devices, such as lenses, prisms, and beamsplitters, which are very small or extremely small. **2.** The design, construction, and applications of **micro-optics (1)**.

micro-payment Same as **micropayment**.

micro-positioner Same as **micropositioner**.

micro-power Same as **micropower**.

micro-processor Same as **microprocessor**.

micro-processor unit Same as **microprocessor unit**.

micro-program Same as **microprogram**.

micro-programmable Same as **microprogrammable**.

micro-programming Same as **microprogramming**.

micro-rad Same as **microrad**.

micro-radian Same as **microradian**.

micro-radiography Same as **microradiography**.

micro-rem Same as **microrem**.

micro-roentgen Same as **microroentgen**.

micro-second Same as **microsecond**.

micro-sensor Same as **microsensor**.

micro-siemens Same as **microsiemens**.

micro-spectrophotometer Same as **microspectrophotometer**.

micro-spectroscope Same as **microspectroscope**.

micro-strip Same as **microstrip**.

micro-tesla Same as **micro-tesla**.

micro-volt Same as **microvolt**.

micro-watt Same as **microwatt**.

microammeter A current meter graduated in microamperes. Also spelled **micro-ammeter**.

microampere 10^{-6} ampere. It is a unit of measurement of electric current. Its abbreviation is μA. Also spelled **micro-ampere**.

microampere-hour 10^{-6} ampere-hour. It is a unit of measurement of quantity of electricity. Its abbreviation is μAh, or μA-h.

microbalance A sensitive balance used for measuring light weights, usually well below a gram. Some are able to provide readings in the picogram range. Also spelled **micro-balance**.

microbar 10^{-6} bar. It is a unit of pressure. Its abbreviation is μb.

microbeam A beam, such as that consisting of electrons, photons, or ions, which is extremely thin. Used, for instance, in electron or field-ion microscopy. Also spelled **micro-beam**.

microbeam analysis The use of a **microbeam** to analyze the structure and surfaces of very small samples.

microbrowser A Web browser designed for use by devices with small displays, such as properly equipped cell phones or PDAs. Also spelled **micro-browser**.

microcandela 10^{-6} candela. It is a unit of luminous intensity. Its abbreviation is μcd. Also spelled **micro-candela**.

microchannel plate A compact high-voltage electron multiplier which incorporates a very large number of capillary channels, each consisting of a minute hollow glass tube. Incident photons generate photoelectrons which in turn produce a cascade of secondary electrons. The thousands, or millions, of side-by-side channels enable such plates to be used, for instance, as powerful image intensifiers. Its abbreviation is **MCP**. Also spelled **micro-channel plate**.

microchip A small piece of semiconductor material upon which miniature electronic circuit components, such as transistors or resistors, are placed. A microchip, for instance, may have hundreds of millions of transistors. There are various types, including memory chips, and logic chips. An entire computer may be held on a single microchip with the appropriate components, and such chips may be used in countless items, such as automobiles, toys, appliances, clocks, and so on. Also called **microcircuit (1)**, **chip (1)**, or **IC**.

microcircuit Also spelled **micro-circuit**. **1.** Same as **microchip**. **2.** A miniature electronic circuit, such as that found in a microchip. Microcircuits are composed of microelectronic components. Also called **microelectronic circuit**, or **microminiature circuit**.

microcircuitry Also spelled **micro-circuitry**. **1.** The complete circuits which compose a **microcircuit**. **2.** The physical arrangement of the electrical elements that are part of a **microcircuit**.

microcode A sequence of **microinstructions**, such as that used in microprogramming. Such code is usually hardwired.

microcomponent Same as **microelectronic component**. Also spelled **micro-component**.

microcomputer Also spelled **micro-computer**. **1.** A digital computer utilizing a single microprocessor for all its processing needs. The term is usually used synonymously with **personal computer**. Its abbreviation is **micro**. **2.** Same as **microcontroller**.

microcontroller A complete computer that is contained on a single chip. Such a chip must have a CPU, memory, a clock, and input/output circuits. These chips may be used in countless items, including automobiles, toys, appliances, clocks, and so on. Its abbreviation is **MCU**. Also spelled **micro-controller**. Also called **microcontroller unit**, **microcomputer (2)**, **computer-on-a-chip**, or **one-chip computer**.

microcontroller unit Same as **microcontroller**. Its abbreviation is **MCU**. Also spelled **micro-controller unit**.

microcoulomb 10^{-6} coulomb. It is a unit of electric charge. Its abbreviation is μC. Also spelled **micro-coulomb**.

microcrystal A crystal too small to be seen by the unaided eye. Also spelled **micro-crystal**.

microcurie 10^{-6} curie. It is a unit of radioactivity. Its abbreviation is μCi. Also spelled **micro-curie**.

microdensitometer An optical densitometer whose high sensitivity enables the detection of very small areas, or of very faint spectrum lines. Also spelled **micro-densitometer**.

microdisplay A very small display which features high resolution. For example, a megapixel display which is one centimeter square. Used, for instance, in head-mounted devices. Also spelled **micro-display**.

microelectrode An electrode of very small dimensions. Such an electrode may be made of wire, glass, or another material, and be used, for instance, to electrically simulate individual cells, or record the activity of single neurons. Also spelled **micro-electrode**.

microelectromechanical devices Same as **microelectromechanical systems**.

microelectromechanical machines Same as **microelectromechanical systems**.

microelectromechanical systems Minute machines, whose sizes are usually in the micrometer or nanometer range. Such machines combine electrical and mechanical elements, and may be built, for instance, via photolithography. An example is a miniature accelerometer utilized to trigger air bags in cars. Its acronym is **MEMS**. Also spelled **microelectromechanical systems**. Also called **microelectromechanical machines**, or **microelectromechanical devices**.

microelectronic Pertaining to **microelectronics**. Also spelled **micro-electronic**.

microelectronic circuit Same as **microcircuit (2)**. Also spelled **micro-electronic circuit**.

microelectronic component A miniature component, such as a transistor, diode, capacitor, resistor, transformer, or the like. Also spelled **micro-electronic component**. Also called **microcomponent**, **microelement**, or **microminiature component**.

microelectronic device A device, such as a microchip, consisting of **microelectronic components**. Also spelled **micro-electronic device**. Also called **microminiature device**.

microelectronics The technology that deals with the design, construction, and applications of miniature electronic circuits, and the minute components that are a part of said circuits. An example of this technology is found in microchips. Also spelled **micro-electronics**.

microelement Same as **microelectronic component**. Also spelled **micro-element**.

microetching Etching of very small materials, such as those under observation through a microscope. Also spelled **micro-etching**.

microfarad 10^{-6} farad. It is a unit of capacitance. Its abbreviation is μF. Also spelled **micro-farad**.

microfarad meter A direct-reading meter, for measuring the capacitance of circuits or capacitors, that indicates results in microfarads.

microfiche A flat card of film, usually 4 x 6 inches or 3 x 5 inches, upon which photographically-reduced images of printed sheets, or other information, is stored. A microfiche reader is necessary for viewing.

microfiche reader A machine which optically enlarges and displays images stored on **microfiche**. Such an apparatus may also be able to print the observed sheet.

microfilm A roll of photographic film, usually 35 mm or 16 mm, upon which photographically-reduced images of printed sheets, or other information, is stored. A microfilm reader is necessary for viewing.

microfilm reader A machine which optically enlarges and displays images stored on **microfilm**. Such an apparatus may also be able to print the observed sheet.

microfloppy A digital magnetic storage medium, which is in the form of a rotating plastic plate which is flexible. Information is encoded by altering the magnetic polarity of minute portions of the coated surface of such a disk. Microfloppies are much slower than hard disks, generally hold 1.44

megabytes of data, and their usual format is 3.5 inches, with a rigid case. Also spelled **micro-floppy**.

microform   A medium, such as microfilm or microfiche, which contains photographically-reduced images of printed sheets.

microgauss 10^{-6} gauss. It is a unit of magnetic flux density, or magnetic induction. Its abbreviation is μG, or μGs. Also spelled **micro-gauss**.

microgram 10^{-6} gram. It is a unit of mass. Its abbreviation is μg.

micrographics   The techniques utilized to photographically reduce images onto mediums such as microfilm or microfiche. Also, the technology pertaining to the recording, handling, and use of such images.

microhenry 10^{-6} henry. It is a unit of inductance. Its abbreviation is μH. Also spelled **micro-henry**.

microhm 10^{-6} ohm. It is a unit of resistance, impedance, or reactance. Its abbreviation is $\mu\Omega$. Also spelled **microohm**, or **micro-ohm**.

microhmmeter   Same as **micro-ohmmeter**.

microimage   An image contained in a **microform**. Also spelled **micro-image**.

microinstruction   One of the fundamental operations necessary to carry out a single machine instruction. Microinstructions are even more basic than machine instructions, and a sequence of them may be necessary to perform a single machine instruction. An example is an instruction which moves a bit from one location to another. Also spelled **micro-instruction**.

microjoule 10^{-6} joule. It is a unit of energy or work. Its abbreviation is μJ. Also spelled **micro-joule**.

microkernel   Within a kernel, the hardware-dependent components. By changing the modules layered above a microkernel, one operating system can appear to be most any other, as it provides a platform-independent interface to the rest of the operating system. Also spelled **micro kernel**.

microlambert 10^{-6} lambert. It is a unit of luminance. Its abbreviation is μL.

microliter 10^{-6} liter. It is a unit of volume. Its abbreviation is μl, or μL. Also spelled **micro-liter**.

microlithography   A lithographic technique utilized for transferring patterns or images in the micrometer or nanometer range. Also spelled **micro-lithography**.

microlumen 10^{-6} lumen. It is a unit of luminous flux. Its abbreviation is μlm. Also spelled **micro-lumen**.

microlux 10^{-6} lux. It is a unit of illuminance, or illumination. Its abbreviation is μlx. Also spelled **micro-lux**.

micromachining   Also spelled **micro-machining**. **1.** The manufacturing of minute components and devices, such as microelectronic components. **2.** Machining, such as polishing or cutting, performed on extremely small materials or surfaces.

micromanipulator   A device utilized for manipulating objects at a microscopic level. Also spelled **micro-manipulator**.

micrometer   **1.** 10^{-6} meter. It is a unit of measurement of length. Its abbreviation is μm. Also called **micron**. **2.** An instrument utilized for measuring small, or very small, thicknesses, diameters, or the like.

micromicro-   An obsolescent prefix which is equivalent to **pico-**.

micromicrofarad 10^{-12} farad. It is an obsolescent unit of capacitance equivalent to **picofarad**.

micromicron 10^{-12} meter. It is an obsolescent unit of distance equivalent to **picometer**.

microminiature   Extremely small. Usually utilized in the context of electronics to refer to microminiature components, circuits, or devices. Also spelled **micro-miniature**.

microminiature circuit   Same as **microcircuit (2)**. Also spelled **micro-miniature circuit**.

microminiature component   Same as **microelectronic component**. Also spelled **micro-miniature component**.

microminiature device   Same as **microelectronic device**. Also spelled **micro-miniature device**.

micromodule   A microcircuit which is encapsulated so as to form a very small block, module, card, or the like, which is usually plugged-in. A micromodule may also consist of a single microelement. Also spelled **micro-module**.

micromol   Same as **micromole**.

micromole 10^{-6} mol. It is a unit of amount of substance. Its abbreviation is μmol. Also called **micromole**.

micron   Same as **micrometer (1)**. Its symbol is μ.

micronewton 10^{-6} newton. It is a unit of force. Its abbreviation is μN. Also spelled **micro-newton**.

microoersted 10^{-6} oersted. It is a unit of magnetic field strength. Its abbreviation is μOe.

microohm   Same as **microhm**. Also spelled **micro-ohm**.

microoptics   Same as **micro-optics**.

micropayment   A monetary transaction smaller than a given value, such as a fraction of a cent. Used, for instance, for charging for accessing a given Web page. Also spelled **micro-payment**.

microphone   A transducer which converts sound energy into electrical signals. Most microphones incorporate a diaphragm which vibrates proportionally to the variations in sound pressure surrounding it. These fluctuations are then converted into a corresponding electrical signal such as a current or voltage. There are various types, including dynamic, capacitor, and ribbon microphones. In addition, the pickup pattern of such a transducer can be tailored to specific needs, as seen in cardioid, bi-directional, and omnidirectional microphones. Microphones are used in many devices, such as telephones, hearing aids, computers, and as a signal source for broadcasting, among others. It performs the reverse function of a speaker, and some microphones can also be used as speakers, as is the case with many intercom systems. Its abbreviation is **mike**, or **mic**.

microphone amplifier   Same as **microphone preamplifier**.

microphone boom   A movable mechanical support utilized to place a microphone in a desired location. Used, for example, to suspend a microphone above a scene being filmed or broadcast.

microphone button   In a carbon microphone, the container that holds the carbon granules. Also called **button (3)**, or **carbon button**.

microphone cable   A cable with special shielding that is used to connect a microphone to an amplifier, mixer, or other device.

microphone diaphragm   A vibrating membrane in a microphone. It produces an electric output in response to its sound-wave input. Also called **diaphragm (2)**.

microphone input   In a device or piece of equipment, such as an amplifier or mixer, one or more connectors or terminals via which the signal of a microphone is delivered. Also, the signal delivered.

microphone mixer   A device which combines two or more microphone inputs. Each input usually has its own level adjustment.

microphone output   The output signal a microphone provides. This signal is delivered to a microphone input.

microphone pickup pattern   The directivity pattern for a given microphone. Common patterns include bi-directional, cardioid, and unidirectional.

microphone preamplifier   An amplifier that boosts the low signal level of a microphone to a level suitable as an input for the main amplifier. Such a preamplifier is usually low-noise, provides high-gain, and may be built into the microphone stand, or be contained within the microphone itself. Some microphone preamplifiers also have a volume control. Its abbreviation is **mic preamp**. Also called **microphone amplifier**.

microphone transformer A transformer utilized to match the impedance of a microphone to an amplifier it may be connected to.

microphone windscreen A covering, such as that of foam, placed on a microphone to minimize the sounds a wind would otherwise produce. Also called **windscreen**.

microphonics Unwanted noise in a component, circuit, device, piece of equipment, or system, due to mechanical vibrations by any contained moving elements, materials, or parts. In a microphone, for instance, such noise can be generated by the protective sheath rubbing against the insulator enclosing the conductors as the cable is flexed. Another example is rain or hail hitting an LNB and causing minor disturbances in the televised image.

microphot 10^{-6} phot. It is a unit of illumination. Its abbreviation is μph.

microphotometer A photometer which capable of measuring the luminance of a very small area, such as that being studied under a microscope.

microphysics The branch of physics dealing with matter and systems at the atomic, subatomic, and elementary particle levels.

micropositioner An instrument utilized to position very small objects. Used, for instance, to move objects viewed under an electron microscope. Also spelled **micro-positioner**.

micropower A very low level of power. Used, for instance, to power a microelectronic component, or a microcircuit. Also spelled **micro-power**.

microprobe An instrument, such as an electron microprobe, designed for probing extremely small materials or areas. Another example is a microprobe with an exceptionally small exploring head utilized for testing microcircuits.

microprocessor A CPU contained on a single chip. It incorporates the control unit and the arithmetic-logic unit. In order to have the minimum necessary components for computer function, memory and a power supply must be added. Also spelled **micro-processor**. Also called **microprocessor unit**, **CPU (1)**, **CPU chip**, or **processor (1)**.

microprocessor unit Same as **microprocessor**. Also spelled **micro-processor unit**.

microprogram A program written in **microcode**. Also spelled **micro-program**.

microprogrammable A computer whose **microcode** can be altered by a user. Also spelled **micro-programmable**.

microprogramming The creation of **microprograms**. Also spelled **micro-programming**.

microrad 10^{-6} rad. It is a unit of radiation dose. Its abbreviation is μrad. Also spelled **micro-rad**.

microradian 10^{-6} radian. It is a unit of angular measure. Its abbreviation is μrad. Also spelled **micro-radian**.

microradiography The radiography of very small areas or objects. The resulting radiographs are then amplified to the extent necessary for viewing. Also spelled **micro-radiography**.

microrem 10^{-6} rem. It is a unit of radiation dose. Its abbreviation is μrem. Also spelled **micro-rem**.

microroentgen 10^{-6} roentgen. It is a unit of the ability of radiation to ionize. Its abbreviation is μR. Also spelled **micro-roentgen**.

microscopic 1. Too small to be perceived by an unaided eye. 2. Pertaining to that which is too small to be perceived by an unaided eye. 3. Of, pertaining to, or involving the use of a microscope.

microsecond 10^{-6} second. It is a unit of time measurement. Its abbreviation is μs, or μsec. Also spelled **micro-second**.

microsensor Also spelled **micro-sensor**. 1. A sensor which detects extremely small samples. 2. A sensor which detects extremely small magnitudes.

microsiemens 10^{-6} siemens. It is a unit of conductance. Its abbreviation is μS. Also spelled **micro-siemens**.

microspectrophotometer A spectrophotometer which can study extremely small samples. Such an instrument may incorporate a CCD detector for rigorous analysis of samples smaller than one micron. Also spelled **micro-spectrophotometer**.

microspectroscope A spectroscope utilized to produce and study spectral phenomena of extremely small samples. Such an instrument is used in conjunction with a microscope. Also spelled **micro-spectroscope**.

MicroStation A popular CAD program.

microstrip Also spelled **micro-strip**. 1. Same as **microstrip line**. 2. A thin metallic conductor, usually in the form of a strip, which is mounted on a dielectric substrate.

microstrip antenna An antenna consisting of thin metallic conductor, usually a strip, which is mounted on a dielectric substrate. Such an antenna may be fed, for instance, by a microstrip line.

microstrip component A component, such as a microstrip filter, which incorporates one or more **microstrips**.

microstrip device A device, such as a microstrip antenna, which incorporates one or more **microstrips**.

microstrip filter A filter which incorporates a **microstrip (2)**. Usually used in microwave applications.

microstrip line A microwave transmission line consisting of two thin parallel strips, lines, or plates, which are mounted on opposite sides of the same dielectric substrate. Typically, the substrate is flat, and mounted on a ground plane. Also called **microstrip transmission line**, **microstrip (1)**, or **stripline (2)**.

microstrip transmission line Same as **microstrip line**.

microtesla 10^{-6} tesla. It is a unit of magnetic flux density. Its abbreviation is μT. Also spelled **micro-tesla**.

microvolt 10^{-6} volt. It is a unit of measurement of potential difference. Its abbreviation is μV. Also spelled **micro-volt**.

microvoltmeter A voltmeter designed to provide accurate indications of voltage values in the microvolt range.

microvolts per meter A measure of the intensity of the signal of a radio transmitter. It may be calculated, for instance, by dividing the intensity in microvolts at the receiving antenna, by the height of the receiving antenna. It is also used as a measure of the electromagnetic radiation emitted by microwave ovens, computer displays, celestial bodies, and so on. Its abbreviation is μV/m.

microwatt 10^{-6} watt. It is a unit of measurement of power. Its abbreviation is μW. Also spelled **micro-watt**.

microwatt-hour 10^{-6} watt-hour. It is a unit of measurement of energy or work. Its abbreviation is μW-h, or μW-hr.

microwattmeter A wattmeter designed to provide accurate indications of power values in the microwatt range.

microwave 1. Pertaining to, or arising from **microwaves**, and the interval within the electromagnetic spectrum associated with them. 2. One of multiple **microwaves**. 3. Same as **microwave oven**.

microwave amplification by stimulated emission of radiation Same as **maser**.

microwave amplifier A device, such as a klystron, which amplifies **microwaves**.

microwave antenna An antenna designed for use in the **microwave range (1)**. Such an antenna is usually a dish antenna or a horn antenna.

microwave background A uniform bath of radiation which is believed to permeate all of space and to have originated with the big bang. The spectrum of this radiation corresponds to that of a blackbody at about 2.73 K, peaking in the microwave region. It is a source of cosmic noise. Also called **cosmic microwave background, cosmic background radiation, cosmic microwave background radiation,** or **cosmic microwave radiation**.

microwave band 1. Same as **microwave spectrum**. **2.** Any of the regions **microwaves** are subdivided into, such as the L, S, C, X, and K bands. Also called **microwave region (2)**.

microwave beacon A beacon which utilizes signals whose frequencies are in the microwave range.

microwave cavity An enclosure which is able to maintain an oscillating electromagnetic field when suitably excited. The geometry of the cavity determines the resonant frequency, which is in the microwave range. Used, for instance, in klystrons, or magnetrons, or as a filter. Also known by various other names, including **microwave resonance cavity, cavity (2), cavity resonator, resonant chamber, resonating cavity, waveguide resonator,** and **rhumbatron**.

microwave circuit 1. A circuit especially suited for operation in the microwave range. **2.** A circuit which operates in the microwave range.

microwave circulator A multiport waveguide junction in which microwave energy entering one of its ports is transmitted to an adjacent port, and on to the next, in a predetermined rotation. Also called **circulator**.

microwave communications Communications which utilize microwaves to convey information. For instance, satellite and cellular communications.

microwave detector A device which senses microwave radiation.

microwave diathermy The use of microwaves for therapeutic purposes. Used, for example, to generate localized heat in body tissues.

microwave diode 1. A diode especially suited for operation in the microwave range. **2.** A diode which operates in the microwave range.

microwave dish A dish antenna designed for use in the **microwave range (1)**.

microwave energy Same as **microwave radiation**.

microwave filter A microwave cavity designed to selectively transmit specified microwave frequencies while rejecting or absorbing other intervals of frequencies.

microwave frequency A frequency within the **microwave region**.

microwave generator A device, such as a klystron or magnetron, which generates microwaves. Also called **microwave oscillator**.

microwave heating The use of microwaves for heating, as is the case for instance, in a microwave oven.

microwave horn A microwave antenna whose radiating element is in the shape of a horn. It is fed at its apex, and the electromagnetic radiation is emitted from its open end, which is wider. The cross section utilized may be rectangular, square, circular, and so on. Also called **horn antenna, horn (1),** or **horn radiator**.

microwave IC Abbreviation of **microwave integrated circuit**. An IC designed to work at microwave frequencies. Such a chip usually utilizes hybrid IC technology, and an example is an MMIC. Its own abbreviation is **MIC**.

microwave integrated circuit Same as **microwave IC**.

microwave interferometer An interferometer which measures microwaves traveling through a given medium. Used, for instance, to measure plasma densities.

microwave landing system An instrument landing system which utilizes microwave signals to provide aircraft with the information necessary for a safe approach and landing. Such a system makes use of the high directivity of such waves, for more precise indications. Its abbreviation is **MLS**.

microwave lens A lens, such as that utilizing a collimator or waveguide section, which focuses microwaves.

microwave link Same as **microwave radio relay**.

microwave link system Same as **microwave radio relay system**.

microwave motion detector A sensor which detects motion via changes in the reflection of microwaves within a given area or enclosure. Used, for instance, for surveillance.

microwave oscillator Same as **microwave generator**.

microwave oven A kitchen appliance which utilizes **microwaves** to heat food. A microwave oven incorporates a magnetron utilized in a continuous-output mode, which radiates microwave energy throughout the oven cavity. This power, which can exceed 1000 watts, causes molecules in the food, especially those of water, to vibrate, thereby producing the heat necessary for thawing and/or cooking. Also called **microwave (3),** or **microwave range (3)**.

microwave plumbing The sections and devices, such as elbows and joints, utilized to channel microwave energy. Also called **plumbing (1)**.

microwave radiation Electromagnetic radiation in the **microwave region**. Also called **microwave energy**.

microwave radio relay A radio repeater which is utilized for receiving, amplifying, and transmitting microwaves. Such links are usually placed about 80 kilometers apart and require line-of-sight propagation, but can be spaced much further when factors such as special path conditions are taken into account. Used extensively for cellular communications and wireless networks. Also, the relaying of such waves. Also called **microwave relay, microwave link,** or **microwave repeater**.

microwave radio relay system A series of **microwave radio relays** utilized for microwave communications. Also called **microwave relay system,** or **microwave link system**.

microwave radiometer A device which detects and measures microwave radiation.

microwave range 1. Same as **microwaves**. **2.** The distance emitted microwaves can, or do, travel. **3.** Same as **microwave oven**.

microwave region 1. Same as **microwaves**. **2.** Same as **microwave band**.

microwave relay Same as **microwave radio relay**.

microwave relay system Same as **microwave radio relay system**.

microwave repeater Same as **microwave radio relay**.

microwave resonance cavity Same as **microwave cavity**.

microwave spectrometer A spectrometer which detects and measures wavelengths or indices of refraction in the microwave region.

microwave spectrum The interval of frequencies or wavelengths encompassing the **microwave region**. Also called **microwave band (1)**.

microwave system Any system, such as a communication system, which utilizes microwaves.

microwave transistor 1. A transistor specifically designed to work optimally at microwave frequencies. **2.** A transistor that operates at microwave frequencies.

microwave tube A tube, such as a klystron or magnetron, which amplifies and/or generates microwaves.

microwaves Electromagnetic waves whose wavelengths are between 0.3 and 30 centimeters, corresponding to frequencies of between 100 and 1 GHz, respectively, although the defined interval can vary considerably. For instance, microwaves may also be defined as encompassing wavelengths from 0.1 to 100 centimeters, or between 3 centimeters and 30 centimeters, and so on. Microwaves are further subdivided into bands of frequencies, such as the L, S, C, X, and K bands, and are used extensively in satellite and cellular communications, in radars, and for cooking, among other applications. Also called **microwave region** (1), or **microwave range** (1).

mid-band Also spelled **midband**. 1. Same as **mid-frequency** (1). 2. An interval of frequencies near the middle of a given frequency band. 3. A band between higher and lower frequency bands.

mid-bass An interval of frequencies which partially overlaps those of a woofer and a midrange. Also spelled **midbass**.

mid-frequency Also spelled **midfrequency**. 1. The frequency which equally divides a band of frequencies. Also called **mid-band** (1), or **center frequency** (2). 2. The middle interval of a range or spectrum of frequencies, such as the audible spectrum. Also called **midrange** (3).

mid-frequency range Same as **midrange**.

mid infrared Abbreviation of **middle infrared**.

mid IR Abbreviation of **middle infrared**.

mid-woofer Also spelled **midwoofer**. 1. A speaker designed to reproduce frequencies which partially overlap those of a woofer and a midrange. Seen, for instance, in four-way speakers. 2. A speaker designed to reproduce low and midrange frequencies. Seen, for instance, in two-way speakers.

midband Same as **mid-band**.

midbass Same as **mid-bass**.

middle infrared Its abbreviation is **middle IR**, **mid infrared**, or **mid IR**. 1. Pertaining to **middle-infrared radiation**. 2. Same as **middle-infrared radiation**.

middle-infrared radiation Electromagnetic radiation in the infrared region which is between near IR and far IR. It corresponds to wavelengths of approximately 3 to 10 micrometers, though the defined interval varies. Also called **middle infrared** (2).

middle IR Abbreviation of **middle infrared**.

middleware A software layer between two different types of software. For instance, middleware can serve as the bridge between an application and the operating system, between a client program and a database, or between a network operating system and a DBMS. Middleware takes care of the necessary conversions and translations so that data can be seamlessly exchanged.

midfrequency Same as **mid-frequency**.

MIDI Acronym for **M**usical **I**nstrument **D**igital **I**nterface. A standardized interface, along with its communications protocols, which enables the interconnection of computers, synthesizers, musical instruments, and other electronic music devices. MIDI encodes various aspects of sounds which are recorded or exchanged, such as the pitch, length, delay, and amplitude of each note, as opposed to containing actual music or other sounds. MIDI allows recorded information to be manipulated in many ways, such as combining diverse instruments to create new sounds, or the changing of the key of a song through a simple command. Other features include translating what is played on an instrument into sheet music. Also called **MIDI interface**.

MIDI device A device, such as a computer, synthesizer, or musical instrument, with one or more **MIDI ports**.

MIDI file A file consisting of **MIDI messages**. A MIDI file does not contain music or any other kind of sound. However, it has the information that defines what an instrument or device, such as a synthesizer, needs to reproduce the composed audio.

MIDI interface Same as **MIDI**.

MIDI messages Packets of MIDI information that define aspects of sound, such as what notes are being played, at what amplitude, for what duration, and effects such as reverb, modulation, and sustain, in addition to control data. MIDI messages do not contain music or any other kind of sound. However, they do contain information that defines what has been performed or otherwise composed or programmed.

MIDI port A port utilizing a **MIDI interface**. A MIDI port may be an IN, for input, OUT, for output, or THRU, which mirrors the signal arriving at the IN port. The THRU port serves for daisy-chaining, and any device without such a port must be placed at the end of such a chain.

MIDI sequencer A program or device which enables a user to compose, edit, and play back MIDI files.

midlet A Java application which conforms to **MIDP** specifications.

MIDP Abbreviation of **M**obile **I**nformation **D**evice **P**rofile. A specification for Java applications running on mobile devices such as properly equipped cell phones and PDAs.

midrange Also called **mid-frequency range**. 1. A speaker designed to reproduce frequencies in the middle part of the audible spectrum. This interval extends from about 300 Hz to somewhere in the range of 2000 to 5000 Hz, although these limits can vary. Such a speaker is responsible for the interval of frequencies humans are most sensitive to, and is usually utilized with other speaker units, such as woofers and tweeters, for reproduction across the full audio spectrum. Also called **midrange driver**, **midrange speaker**, or **midrange loudspeaker**. 2. The middle part of the audible spectrum. 3. Same as **mid-frequency** (1).

midrange computer 1. Same as **minicomputer**. 2. A computer whose size, power, and functionality are similar to a minicomputer, with the difference that a midrange computer is only utilized in multiple-user environments.

midrange driver Same as **midrange** (1).

midrange loudspeaker Same as **midrange** (1).

midrange speaker Same as **midrange** (1).

midwoofer Same as **mid-woofer**.

migration 1. The movement of particles, such as atoms or ions, through a material or medium such as a solution or semiconductor. This may be due, for instance, to diffusion or an external electric field. Specific examples include electromigration, and ion migration. 2. To adapt or change a given hardware or software product, platform, or technology to another. Also, the processes involved in doing so.

mike Abbreviation of **microphone**.

mil 1. A unit of distance equal to exactly 0.001 inch, or 2.54×10^{-5} meter. Used, for instance, to specify diameters of wires or thicknesses of sheets.

mile A unit of distance equal to exactly 1609.344 meters. Its abbreviation is **mi**.

milli- A metric prefix representing 10^{-3}, or 0.001. For instance, 1 milliampere is equal to 10^{-3} ampere. Its abbreviation is **m**.

milliammeter A current meter graduated in milliamperes.

milliamp Same as **milliampere**.

milliampere 10^{-3}, or 0.001 ampere. It is a unit of measurement of electric current. Its abbreviation is **mA**, or **milliamp**.

milliampere-hour 10^{-3}, or 0.001 ampere-hour. It is a unit of measurement of quantity of electricity. Its abbreviation is **mAh**, or **mA-h**.

millibar 10^{-3}, or 0.001 bar. It is a unit of pressure. Its abbreviation is **mb**.

millicandela 10^{-3}, or 0.001 candela. It is a unit of luminous intensity. Its abbreviation is **mcd**.

millicoulomb 10^{-3}, or 0.001 coulomb. It is a unit of electric charge. Its abbreviation is **mC**.

millicurie 10^{-3}, or 0.001 curie. It is a unit of radioactivity. Its abbreviation is **mCi**.

millifarad 10^{-3}, or 0.001 farad. It is a unit of capacitance. Its abbreviation is **mF**.

milligauss 10^{-3}, or 0.001 gauss. It is a unit of magnetic flux density, or magnetic induction. Its abbreviation is **mG**, or **mGs**.

milligaussmeter A gaussmeter graduated in milligauss.

milligram 10^{-3}, or 0.001 gram. It is a unit of mass. Its abbreviation is **mg**.

millihenry 10^{-3}, or 0.001 henry. It is a unit of inductance. Its abbreviation is **mH**.

millijoule 10^{-3}, or 0.001 joule. It is a unit of energy or work. Its abbreviation is **mJ**.

millilambert 10^{-3}, or 0.001 lambert. It is a unit of luminance. Its abbreviation is **mL**.

milliliter 10^{-3}, or 0.001 liter. It is a unit of volume. Its abbreviation is **ml**, or **mL**.

millilumen 10^{-3}, or 0.001 lumen. It is a unit of luminous flux. Its abbreviation is **mlm**.

millilux 10^{-3}, or 0.001 lux. It is a unit of illuminance, or illumination. Its abbreviation is **mlx**.

millimeter 10^{-3}, or 0.001 meter. It is a unit of measurement of length. Its abbreviation is **mm**.

millimeter waves Electromagnetic waves whose wavelengths are between 1 and 10 millimeters, corresponding to frequencies of between 300 and 30 GHz, respectively. This interval represents the extremely high frequency band. Also called **millimetric waves**.

millimeters of mercury A unit of pressure equal to approximately 133.3 pascals. Its abbreviation is **mmHg**, or **mm Hg**.

millimetric waves Same as **millimeter waves**.

millimicron An obsolescent unit of distance equal to 10^{-9} meters, or one nanometer. Its abbreviation is **mμ**.

millimol Same as **millimole**.

millimole 10^{-3}, or 0.001 mol. It is a unit of amount of substance. Its abbreviation is **mmol**. Also called **millimol**.

millinewton 10^{-3}, or 0.001 newton. It is a unit of force. Its abbreviation is **mN**.

millioersted 10^{-3}, or 0.001 oersted. It is a unit of magnetic field strength. Its abbreviation is **mOe**.

milliohm 10^{-3}, or 0.001 ohm. It is a unit of resistance, impedance, or reactance. Its abbreviation is **mΩ**.

milliohmmeter An ohmmeter designed to provide accurate indications of resistance values in the milliohm range.

million electron volt Same as **mega-electron volt**.

million floating-point operations per second Same as **MFLOPS**.

million instructions per second Same as **MIPS**.

million operations per second Same as **MFLOPS**.

million samples per second Same as **MSPS**.

milliphot 10^{-3}, or 0.001 phot. It is a unit of illumination. Its abbreviation is **mph**.

millirad 10^{-3}, or 0.001 rad. It is a unit of radiation dose. Its abbreviation is **mrad**.

milliradian 10^{-3}, or 0.001 radian. It is a unit of angular measurement. Its abbreviation is **mrad**.

millirem 10^{-3}, or 0.001 rem. It is a unit of radiation dose. Its abbreviation is **mrem**.

milliroentgen 10^{-3}, or 0.001 roentgen. It is a unit of the ability of radiation to ionize. Its abbreviation is **mR**.

millisecond 10^{-3}, or 0.001 second. It is a unit of time measurement. Its abbreviation is **ms**, or **msec**, the former being more proper.

millisiemens 10^{-3}, or 0.001 siemens. It is a unit of conductance. Its abbreviation is **mS**.

millitesla 10^{-3}, or 0.001 tesla. It is a unit of magnetic flux density. Its abbreviation is **mT**.

millivolt 10^{-3}, or 0.001 volt. It is a unit of measurement of potential difference. Its abbreviation is **mV**.

millivoltmeter A voltmeter designed to provide accurate indications of voltage values in the millivolt range.

millivolts per meter A measure of the intensity of the signal of a radio transmitter. It may be calculated, for instance, by dividing the intensity in millivolts at the receiving antenna by the height of said receiving antenna. Its abbreviation is **mV/m**.

milliwatt 10^{-3}, or 0.001 watt. It is a unit of measurement of power. Its abbreviation is **mW**.

milliwatt-hour 10^{-3}, or 0.001 watt-hour. It is a unit of measurement of energy or work. Its abbreviation is **mW-h**, or **mW-hr**.

milliwattmeter A wattmeter designed to provide accurate indications of power values in the milliwatt range.

MIM capacitor Abbreviation of **metal-insulator-metal capacitor**. A thin-film capacitor in which a metal layer is deposited on a substrate, followed by an insulator layer, and then a by a second metal layer. Used in ICs.

MIMD Abbreviation of **multiple instruction stream-multiple data stream**.

MIME Acronym for Multipurpose Internet Mail Extensions. A protocol utilized to transfer non-text files such as audio or video via email, without said files being converted into ASCII. Also called **MIME protocol**.

MIME compliant Said of a system which supports **MIME types**.

MIME protocol Same as **MIME**.

MIME type One of various encoding modes utilized to send files using the **MIME protocol**. MIME types describe the content of the files, and both the sending and receiving machines must be MIME complaint for encoding and subsequent decoding of the audio, graphics, PDF, or otherwise formatted files.

min 1. Abbreviation of **minimum**. **2.** Abbreviation of **minute (1)**.

mineral A naturally occurring solid homogeneous substance with a definite chemical composition. Minerals usually have a crystalline structure and are inorganic, although there are some that are organic. They are characterized by features such as hardness, luster, cleavage, color, and density. Examples include garnet, magnetite, cobaltite, galena, and mica.

mineral oil A light hydrocarbon oil which is usually a distillate of petroleum, and which is used as a dielectric.

mini Abbreviation of **minicomputer**.

mini CD An optical disc, such as a CD, CD-ROM, or DVD, whose diameter is smaller than the standard 12 cm. Such a CD may be round, in the shape of a business card, or most any other shape.

mini CD-ROM A CD-ROM whose diameter is smaller than the standard 12 cm. Such a CD-ROM may be round, in the shape of a business card, or most any other shape.

mini connector Also spelled **miniconnector**. **1.** A 3.5 millimeter connector, usually either a plug or a jack, frequently used in audio and multimedia devices. Examples include those with three contacts for plugging headphones into personal music players, or a fiber-optic plug for a DVD. **2.** A connector which is comparatively small.

mini-disc Same as **minidisc**.

mini-disk Same as **minidisk**.

mini DVD A DVD whose diameter is smaller than the standard 12 cm. Such a DVD may be round, in the shape of a business card, or most any other shape.

mini floppy An older name for the obsolete 5.25 inch floppy disk.

mini jack A **mini connector** which is a 3.5 millimeter jack. It is utilized with its matching plug. Also spelled **minijack**.

mini-notebook A computer which is lighter and smaller than a notebook computer, but larger than a handheld computer. Such a computer usually weighs less than one kilogram, and generally features computing power similar to a desktop model. Also called **mini-notebook computer**, or **subnotebook computer**.

mini-notebook computer Same as **mini-notebook**.

Mini-PCI A PCI bus, card, standard, or technology utilized in portable computers such as notebooks. The cards utilized are smaller, but the interface is the same.

mini plug A **mini connector** which is a 3.5 millimeter plug. It is utilized with its matching jack. Also spelled **miniplug**.

Mini SQL Same as **mSQL**.

mini-tower Same as **minitower**.

miniature That which is very small, or particularly small for its class. For example, a miniature tube.

miniature battery A small or very small battery, such as that used in hearing aids, watches, or medically-implanted devices.

miniature card A card, such as an expansion card or flash card, which is particularly small for its class.

miniature cell A small or very small cell, such as that used in a miniature battery. A miniature cell is usually referred to as a **miniature battery**, even though the latter consists of two or more cells.

miniature connector **1.** A connector which is smaller than a given size. **2.** A connector which is smaller than the usual size for a given use, application, or device.

miniature jack **1.** A jack that is smaller than a given size. **2.** A jack that is smaller than the usual size for a given use, application, or device.

miniature plug **1.** A plug which is smaller than a given size. **2.** A plug which is smaller than the usual size for a given use, application, or device.

miniature tube A small electron or gas tube. Such tubes often have 7 or 9 pins.

miniaturization The process of designing and manufacturing components, circuits, devices, equipment, and systems on a reduced scale. Also, the ongoing process of making that which has already been miniaturized even smaller.

minicomputer A computer whose size, power, and functionality reside somewhere between a microcomputer and a mainframe. For instance, it may be a high-performance workstation, or a multiprocessor system which supports several dozen users. Its abbreviation is **mini**. Also called **midrange computer** (1).

miniconnector Same as **mini connector**.

minidisc Its abbreviation is **MD**. Also spelled **mini-disc**. **1.** A disc whose diameter is smaller than a more common counterpart. For instance, an optical disc with a 2.5 inch diameter instead of 4.75 inches. **2.** Same as **minidisk**.

minidisk Its abbreviation is **MD**. Also spelled **mini-disk**. **1.** A disk whose diameter is smaller than a more common counterpart. For instance, a few years ago a 3.5 inch floppy was a minidisk in comparison to a 5.25 inch floppy. **2.** Same as **minidisc**.

minijack Same as **mini jack**.

minima A plural form of **minimum**.

minimize In a GUI, to hide an application window currently displayed on the screen. The program is not terminated, and is then represented by an icon. Also called **iconize**.

minimize button A virtual button on a computer screen, which when clicked, serves to **minimize** a given window.

minimum Its plurals form are **minima**, and minimums. Its abbreviation is **min**. **1.** The least possible value, quantity, or degree. **2.** The least value, quantity, or degree attained or recorded. **3.** The least value a variable, function, curve, or the like, may attain. **4.** The **minimum** (3) for a given interval.

Minimum Cell Rate An ATM performance parameter which specifies the minimum guaranteed rate for cell transmission over a given virtual circuit. Its abbreviation is **MCR**.

minimum detectable signal The minimum signal level which produces a given effect, result, or response, such as detection, activation, or operation. For example, the lowest input-signal amplitude which produces a perceptible change in the output of a component, circuit, device, piece of equipment, or system. Also called **minimum discernable signal**, or **threshold signal**. Its abbreviation is **MDS**.

minimum discernable signal Same as **minimum detectable signal**. Its abbreviation is **MDS**.

minimum operating current The minimum current required for the proper operation of a component, circuit, device, piece of equipment, or system.

minimum operating power **1.** The minimum power consumed by a component, circuit, device, piece of equipment, or system which is properly operating. **2.** The minimum power produced or emitted by a component, circuit, device, piece of equipment, or system which is properly operating.

minimum operating voltage The minimum voltage required for the proper operation of a component, circuit, device, piece of equipment, or system.

minimum sampling frequency **1.** The lowest sampling frequency necessary, such as the Nyquist rate, to faithfully reproduce a signal or prevent aliasing. **2.** The lowest sampling frequency required for a given observation, procedure, task, or the like.

minimum shift keying A form of phase-shift keying in which out-of-band interference is diminished by minimizing the changes in phase at each transition. Used, for instance, in differential GPS systems. Its abbreviation is **MSK**.

mining The process of analyzing data to identify relationships and patterns which may be useful. Such mining may be done manually or automatically, through the use of specialized programs. Also called **data mining**.

miniplug Same as **mini plug**.

miniscope A reduced-size optical device or instrument, such as a microscope, telescope, or spectroscope.

minitower A tower, such as that of a computer system or multiple DVD drives, which is more compact than regular-sized towers. Also spelled **mini-tower**.

minor beam Same as **minor lobe**.

minor lobe Within an antenna pattern, any lobe other than the **main lobe**. Also called **minor beam**, **sidelobe**, or **secondary lobe**.

minority carrier In a semiconductor material, the type of charge carrier which constitutes less than half of the total charge carriers. In a p-type semiconductor, for instance, electrons are the minority carrier. This contrasts with **majority carrier**, which constitutes more than half.

minuend A number or quantity that a subtrahend is subtracted from. An **subtrahend** is a newly introduced number or quantity that is subtracted from the minuend, which is already present.

minute 1. A time measure equal to 60 seconds. Its abbreviation is **min. 2.** A unit of angular measure equal to 1/60 degree. Its symbol is '. Also called **arcminute. 3.** Extremely small or insignificant.

MIP mapping A form of texture mapping in which an original high-resolution texture map is scaled multiple times to represent the texture at different distances. Each subsequent map is half the size of the previous one. MIP mapping helps provide for smoother images at various depths, and can be combined with other techniques for added realism or visual appeal. Used, for instance, in CAD and video games. Also spelled **mipmapping**. The MIP portion of the term is taken from the Latin phrase *multum in parvo*, meaning "much in little".

mipmapping Same as **MIP mapping**.

MIPS Acronym for **m**illion **i**nstructions **p**er **s**econd. It generally means one million machine instructions per second, and is an indicator of CPU speed. When evaluating the overall speed of a system, other factors, such as cache, memory speed, and the number of instructions required for given tasks, must also be considered.

mIRC A popular Internet Relay Chat program.

mirror 1. A surface that reflects light. Such a surface is usually smooth, highly polished, and it may consist, for instance, of a thin layer of silver or aluminum on glass. Mirrors have many applications, including their use in lasers, beam splitters, microscopes, TV cameras, and digital micromirror devices. Also called **reflector (2). 2.** To duplicate, replicate, or reflect, as accomplished, for instance, by a current mirror or an electromagnetic mirror.

mirror effect A charged particle traveling into an increasing magnetic field will be reflected back when said field becomes strong enough. This effect is seen, for instance, in a magnetic bottle.

mirror galvanometer A galvanometer in which a small mirror is attached to its coil, or other moving element, so that a light beam can serve as the pointer of its calibrated scale.

mirror image An image in which rights and lefts are reversed, as would appear if viewed reflected by a mirror.

mirror instrument An instrument, such as a mirror galvanometer, whose indications are given by the movements of a beam of light reflected by a mirror.

mirror site One of multiple Web sites which have the same data and files available.

mirroring 1. The creation of a mirror image. **2.** The duplication of stored data at a remote location, on another drive, another medium, or the like. Used, for instance, for backing-up, disaster recovery, or as a security measure. Also called **data mirroring (1). 3.** The maintaining of identical copies of data and files at multiple network sites or servers. For example, an entity which receives many download requests from around the world may have many diversely located mirror sites to facilitate access. Also called **data mirroring (2)**.

MIS 1. Abbreviation of **management information system**. **2.** Abbreviation of **management information services**. **3.** Abbreviation of **metal-insulator semiconductor**.

misalign 1. To not arrange, position, or synchronize a component, circuit, or device in a manner that allows for proper or optimal functioning. For example, to improperly position a read/write head. **2.** To not arrange or position a component, circuit, or device in a manner that allows for proper installa-

tion. For example, to not line up the pins when installing a memory chip.

mismatch 1. To join components, devices, equipment, or systems which are not compatible, or which otherwise do not correspond with each other. For instance, to attempt to match a plug with an incompatible jack. **2.** To not attain an equality in a given property or value. For instance, to not match impedances between two circuits, or between a signal source and a load.

mismatch factor Also called **coefficient of reflection, reflection coefficient, reflection factor, reflectance**, or **reflectivity. 1.** The ratio of the amplitude of a wave reflected from a surface to the amplitude of the same wave incident upon the surface. **2.** The ratio of the current delivered to a load whose impedance is not matched to the source, to the current that would be delivered to the same load if its impedance were fully matched.

mismatched impedances A condition in which the impedances between two circuits, or between a signal source and a load, are not equal, thus reducing the transfer of power. Also called **impedance mismatch**.

mission critical Activities, tasks, and the like which are indispensable for the operation and/or progress of a business or project. For example, if the failure of a given computer application or network connection grinds a business to a halt, it is mission critical.

mix 1. To combine multiple entities into one or more products. For instance, to combine two input signals to provide a single output signal, to combine multiple light beams to form a single beam, to mix two different frequencies in a non-linear device to produce two other frequencies, or to diffuse one material in another. Also, the result of such a combination. **2.** To combine multiple audio input signals to form a composite signal with the desired blend. For example, to combine voices and multiple instruments for a song, or the combination of dialog, music, and effects for a soundtrack. Also, the result of such a combination. Also called **audio mix**, or **sound mix**.

mix down Same as **mixdown**.

mixdown To perform a **mix (2)**. Also spelled **mix down**.

mixed-mode CD A CD incorporating both audio and data. The data is usually located on track one, and the audio, usually in the form of music, occupies the remaining tracks. An example is a CD-extra. Also called **mixed-mode CD-ROM**.

mixed-mode CD-ROM Same as **mixed-mode CD**.

mixed reality An environment or setting which combines virtual and real images and objects. For example, virtual images may be superimposed upon real objects. Also called **augmented reality**, or **enhanced reality**.

mixer 1. A component, circuit, device, piece of equipment, material, medium, or system which serves to **mix (1). 2.** In a superheterodyne receiver, the stage at which the incoming modulated radio-frequency signal is combined with the local oscillator, to produce a modulated intermediate frequency. Also called **mixer stage, first detector**, or **frequency converter (2). 3.** A circuit or device which performs **mixes**. Also called **audio mixer**.

mixer stage Same as **mixer (2)**.

mixing The process of performing a **mix**.

mixing amp Same as **mixing amplifier**.

mixing amplifier An amplifier which also performs **mixes (2)**. Its abbreviation is **mixing amp**.

mixture The result of a **mix**.

mJ Abbreviation of **millijoule**.

MJ Abbreviation of **megajoule**.

MJPEG Abbreviation of **M**otion **J**oint **P**hotographic **E**xperts **G**roup. A video compression method in which every frame consists of an individually-compressed JPEG image.

MKS system Abbreviation of **meter-kilogram-second system**.

MKSA system Abbreviation of **meter-kilogram-second-ampere system**.

ml Abbreviation of **milliliter**.

mL 1. Abbreviation of **milliliter**. 2. Abbreviation of **millilambert**.

mlm Abbreviation of **millilumen**.

Mlm Abbreviation of **megalumen**.

MLM Abbreviation of **mailing list manager**.

MLPPP Abbreviation of **Multilink PPP**.

MLS Abbreviation of **microwave landing system**.

mlx Abbreviation of **millilux**.

mm Abbreviation of **millimeter**.

Mm Abbreviation of **megameter**.

MM Abbreviation of **multimedia messaging**.

mm Band In communications, a band of radio frequencies extending from 110.00 to 300.00 GHz, as established by the IEEE. This corresponds to wavelengths of approximately 2.7 mm to 1.0 mm, respectively.

mm Hg Abbreviation of **millimeters of mercury**.

MMC Same as **MultiMediaCard**.

MMD Abbreviation of **Multichannel Multipoint Distribution**.

MMDS Abbreviation of **Multichannel Multipoint Distribution Service**.

mmf Abbreviation of **magnetomotive force**.

mmHg Abbreviation of **millimeters of mercury**.

MMIC Abbreviation of **monolithic microwave IC**. An IC designed to work at microwave frequencies, and which is built upon a single semiconductor die, such as that composed of gallium arsenide.

mmol Abbreviation of **millimole**, or **millimol**.

MMS 1. Abbreviation of **multimedia messaging service**. 2. Abbreviation of **maritime mobile service**.

MMU Abbreviation of **memory management unit**.

MMX Abbreviation of **multimedia extensions**. Enhancements made to certain CPU chips to accelerate multimedia content.

m_n Symbol for **neutron mass**, or **neutron rest mass**.

mN Abbreviation of **millinewton**.

Mn Chemical symbol for **manganese**.

mnemonic A rhyme, word, formula, or other device utilized as an aid in the memorization of something lengthier or complex. Mnemonics are used extensively in computers, especially in programming using a symbolic language.

mnemonic code Computer code, such as that used in assembly language or a symbolic language, which employs mnemonics as a memory aid. For instance, ADD to represent addition, CLR for clear, or SQR for square root.

MNOS Abbreviation of **metal-nitride-oxide semiconductor**. A semiconductor with two insulating layers, one of which is an oxide, such as silicon oxide, and the other a nitride, such as silicon nitride. Used, for instance, in EEPROMs.

Mo Chemical symbol for **molybdenum**.

MO 1. Abbreviation of **magneto-optical**, or **magneto-optic**. 2. Abbreviation of **master oscillator**.

MO disc Abbreviation of **magneto-optical disc**.

MO disk Abbreviation of **magneto-optical disk**.

mobile 1. That which can be readily moved from one location to another. 2. That which can be readily moved from one location to another while uninterruptedly maintaining proper operation. 3. That which can readily move from one location to another. For example, a mobile robot, or a fluid.

mobile banking Same as **m-banking**.

mobile business Same as **m-business**.

mobile cellular communications A telecommunications system in which mobile, usually portable, telephones are linked to a land telephone network via microwave radio-frequency signals. Individual cell sites provide coverage to a limited area, called a cell, while networks of cell sites provide coverage to large geographical areas. At any given point within the cellular network, the system decides which cell site can provide the best signal to a mobile unit, and as the unit moves, the signal is transferred to the next cell site with the best signal. Also called by various other names, including **mobile cellular telecommunications**, **mobile cellular telephony**, and **cellular communications**.

mobile cellular phone A mobile phone used for **mobile cellular communications**. Also called by various other names, including **mobile cellular radio telephone**, and **cellular telephone** (1).

mobile cellular radio 1. Same as **mobile cellular communications**. 2. Same as **mobile cellular phone**.

mobile cellular radio telephone Same as **mobile cellular phone**.

mobile cellular telecommunications Same as **mobile cellular communications**.

mobile cellular telephone Same as **mobile cellular phone**.

mobile cellular telephone system Same as **mobile cellular communications**.

mobile cellular telephony Same as **mobile cellular communications**.

mobile commerce Same as **m-commerce**.

mobile communications The transmission of information between two or more points or entities, one or more of which is moving or able to move easily. Common examples include cellular communications, mobile services, and Internet access using PDAs. Also, the science dealing with such communications, including the modes, mechanisms, and media used for this purpose, and all efforts to advance this field. Also called **mobile telecommunications**.

mobile computer A computer which can be readily carried from one location to another. This term includes handheld and notebook computers, and may include other devices, such as properly equipped cellular phones.

mobile computing The use of **mobile computers**.

mobile device A device, such as a cell phone or PDA, utilized for mobile communications.

Mobile Information Device Profile Same as **MIDP**.

mobile Internet Internet access utilizing properly equipped **mobile devices**.

Mobile Internet Protocol Same as **Mobile IP**.

Mobile IP Abbreviation of **Mobile Internet Protocol**. A system which enables users to stay connected while moving through networks which use IP addresses different from their home network. Each mobile node is identified by its home IP address, and when in a different network's area, all data is forwarded to the user, care of the remote network. To be able to use this service users need to register with the foreign network, which also must support the same Mobile IP protocol. Also called **Mobile IP Protocol**.

Mobile IP Protocol Same as **Mobile IP**.

mobile mail Same as **m-mail**.

mobile pager A pocket-sized radio receiver or transceiver which serves to receive messages, and in some cases perform various other functions, such as retrieve email. These devices may emit a beep when informing of a new message. Some mobile pagers can also signal by other means, such as vibrations, or a blinking light. Also called **paging device**, **pager** (1), or **beeper** (2).

mobile phone Also called **mobile telephone**. **1.** A portable phone, especially a cellular phone, which allows a user to place and receive calls to or from any point within its coverage area. Cordless phones are usually excluded from this definition. **2.** A cellular telephone installed in a land vehicle, such as a car or truck. Such phones may or may not be removable from the vehicle, and usually have an external antenna. Also called **car phone** (1). **3.** A hand-held cellular telephone which can be placed in a cradle in a land vehicle such as a car or truck. The cradle can serve to charge the phone, and many of them allow the user to engage in conversations without holding the phone. Also called **car phone** (2).

mobile phone service Radiocommunication service for **mobile phones**. Also called **mobile telephone service**.

mobile positioning The use of a system, such as GPS, to keep track of the location of a cell phone, vehicle, fleet of cars, or the like.

mobile positioning system A system, such as GPS, utilized for **mobile positioning**. Its abbreviation is **MPS**.

mobile processor A CPU designed for use in mobile devices such as notebooks, PDAs, or cell phones.

mobile radio service Same as **mobile service**.

mobile receiver A communications receiver which can be readily moved from one location to another while maintaining proper operation.

mobile robot A robot equipped with a movable platform, a rolling mechanism, or mechanical legs, which allow it to move or travel, depending on the tasks it is meant to execute.

mobile-satellite service Radiocommunication service between **mobile stations**, utilizing one or more satellites. Also, communications links between said satellites. Its abbreviation is **MSS**.

mobile service Radiocommunication service between **mobile stations**. Also called **mobile radio service**.

mobile station A station whose location is not fixed. Such a station may be used while in motion, or during stops at any chosen site. A mobile station may be located on a land, water, or air vehicle. This contrasts with a **fixed station**, whose location does not change.

mobile switching center Same as **mobile telephone switching office**. Its abbreviation is **MSC**.

mobile telecommunications Same as **mobile communications**.

mobile telephone Same as **mobile phone**.

mobile telephone service Same as **mobile phone service**.

mobile telephone switching office A structure that houses the computers that monitor and control a mobile cellular communications system. It performs functions such as validating ESNs, establishing connections, tracking calls, arranging handoffs, and gathering billing information. Also called **mobile switching center**. Its abbreviation is **MTSO**.

mobile transmitter A communications transmitter which can be readily moved from one location to another while maintaining proper operation.

mobility The average drift velocity of charge carriers in a semiconductor material, per unit electric field. Also called **carrier mobility**, or **drift mobility**.

mock-up A non-operational model of a device, component, or piece of equipment which can be used, for instance for tests or demonstrations. Also spelled **mockup**.

mockup Same as **mock-up**.

modal bandwidth A measure of the information-carrying capacity of an optical fiber, expressed in units of MHz per kilometer.

modal dispersion A deprecated term for **multimode distortion**.

modal distortion Same as **multimode distortion**.

mode **1.** A manner, procedure, form, method, or way of using, operating, or functioning. **2.** One of multiple available operational states that can be selected from. Also, operation using such a mode. For instance, the use of a transceiver in reception mode. **3.** For a guided electromagnetic wave, one of the possible manners or patterns of oscillation, transmission, or propagation. Such waves can propagate in one of three principal modes, which are as transverse electric waves, transverse magnetic waves, or as transverse electromagnetic waves. Also called **transmission mode** (1). **4.** Same as **mode of propagation**. **5.** A light path through an optical fiber. **6.** Within a group of numbers, series, or set, the value or item which most frequently occurs.

mode changer Same as **mode transducer**.

mode coupling In a waveguide, the transfer of energy between modes.

mode field diameter A measure of the optical power intensity across the end face of a single-mode fiber. Although most of the light is concentrated in the core of such a fiber, some light travels along the inner part of the cladding too. The diameter of the spot of light as it propagates is the mode field diameter. Its abbreviation is **MFD**.

mode filter In a waveguide, a device which lets the energy of one or more modes pass, while others are attenuated or rejected.

mode of propagation Also called **mode** (4). **1.** The mode of propagation of electromagnetic waves in a waveguide. The principal modes are as transverse electric waves, transverse magnetic waves, or as transverse electromagnetic waves. **2.** The mode of propagation of electromagnetic waves. For instance, via a coaxial cable, optical fiber, waveguide, or atmospheric duct.

mode transducer In a waveguide, a device which changes one mode of propagation into another. Also called **mode transformer**, or **mode changer**.

mode transformer Same as **mode transducer**.

model **1.** An object which is built to represent a component, circuit, device, piece of equipment, system, or another object. It can serve, for instance, for illustration, demonstration, or study. **2.** A mathematical, physical, or conceptual representation utilized to help understand a component, circuit, device, piece of equipment, system, mechanism, process, or phenomenon. Such a representation may, for example, help understand a complex circuit or device through one or more analogies. **3.** A specific style or design of a component, circuit, device, piece of equipment, system.

modeling **1.** The creation of a **model**. **2.** The creation of a model with the assistance of a computer.

modem Acronym for **mo**dulator-**dem**odulator. Also known as **demodulator-modulator**. **1.** A device which enables a computer to transmit and receive information over telephone lines. For transmission, a computer modulates a carrier signal so that it contains digital information in a form that can be transmitted over analog lines. It does the opposite for reception. Most modems feature data compression and error-correction. Also called **telephone modem**. **2.** Any signal-conversion device which combines the functions of a demodulator and modulator.

modem bank Same as **modem pool**.

modem bonding The combining of two telephone lines to form a single communications channel in which the transmission speed is doubled. Two modems are required in order to use this technology. For example, two 56 kilobits per second modems bonded in this manner provide a 112 kilobits per second connection. Also called **channel bonding** (1), or **line bonding** (1).

modem eliminator A device, such as a null modem cable, which enables computers to communicate with each other without the use of modems. Such devices usually require the distance between computers to be comparatively short.

modem pool A series of modems which are usually accessed via a single phone number, and which incorporate a switch to route calls to one of any available modems. Such a pool may be provided, for instance, by an ISP. Also called **modem bank**, or **modem server**.

modem port A computer port utilized to connect an external modem via a cable. A serial port is usually used for this.

modem server Same as **modem pool**.

modem standard A standard, such as V.90 or V.92, applying to the exchange of data via modems.

moderator A substance, such as graphite, heavy water, or beryllium, which is utilized in nuclear reactors to slow down neutrons. Also called **neutron moderator**.

MODFET Acronym for **m**odulation-**d**oped **f**ield **e**ffect **t**ransistor. A field-effect transistor with a heterojunction, which provides for higher electron mobility. Also called **heterojunction field effect transistor**, or **high electron mobility transistor**.

modification The altering, adjusting, or otherwise changing of one or more aspects of a component, circuit, device, piece of equipment, system, signal, material, procedure, configuration, function, or the like. Such modifications may be made, for instance, with the intention to improve, adapt, simplify, test, maintain, repair, study, or suit to needs which may vary or are unpredictable. Also, the specific changes made. The term usually implies that major changes are not involved.

modifier key A computer key, such as Alt, Ctrl, or Shift, which is held down while another key is pressed, to give the latter key a different meaning. For example, pressing the alternate key in combination with another key generates a given function, depending on which program is running.

modify To perform one or more **modifications**.

Modula-2 Abbreviation of **Modular Language-2**. A programming language based on Pascal, with enhancements such as support of modular programming.

Modula-3 Abbreviation of **Modular Language-3**. A programming language based on **Modula-2**, with enhancements such as support of object-oriented programming.

modular approach An approach to the design of equipment, systems, programs, and the like, in which independently prepared and self-contained units, or modules, are combined to form the final product. This allows for simultaneously developing and testing the various components, and also helps break down a large and complex system into more manageable parts. Also called **modular principle, modular technique, modular design, building-block approach,** or **building-block design**.

modular connector 1. Same as **modular jack**. 2. Same as **modular plug**.

modular construction The fabrication of circuits, devices, equipment, or systems using a **modular approach**.

modular design Same as **modular approach**.

modular jack A connector with a recessed opening for making contact with multiple conductors, usually for 4, 6, or 8 wires. Such a jack, when used with its matching plug, makes for easy connection and disconnection to and from a communications network. The RJ-11 jack, for instance, is a standard telephone interface found in the walls of homes, and is used to plug in a telephone or modem cord. Also called **modular connector** (1).

modular plug A matching plug which is inserted into a **modular jack**. Also called **modular connector** (2).

modular principle Same as **modular approach**.

modular programming Computer programming utilizing **modules** (1). All units, or modules, must incorporate certain common parameters, such as having compatible interfaces. This facilitates collaborative efforts to expand and improve current product offerings. Object-oriented programming evolved from this approach.

modular robot A robot composed of interconnected **modules**. Used, for instance, in robots whose means of locomotion can be reconfigured.

modular software Programs developed using **modular programming**.

modular technique Same as **modular approach**.

modulate 1. To modify a characteristic of a wave or signal proportionally to a characteristic present in another wave or signal. 2. To modify a characteristic of a carrier wave by an information-bearing signal, as occurs, for instance, in AM, FM, or PM.

modulated amplifier In a transmitter, an amplifying stage in which the modulating signal is introduced.

modulated beam A beam whose intensity is varied so as to convey meaningful information. For instance, a modulated laser, light, or electron beam.

modulated carrier A carrier wave whose amplitude, frequency, or phase has been varied so as to convey meaningful information. Used, for instance, in AM, FM, or PM. Also called **modulated carrier wave**.

modulated carrier wave Same as **modulated carrier**.

modulated continuous wave A carrier wave which is modulated by a continuous audio-frequency tone. Its abbreviation is **MCW**, or **modulated CW**.

modulated CW Same as **modulated continuous wave**.

modulated electron beam An electron beam whose intensity is varied so as to convey meaningful information.

modulated laser beam A laser beam whose intensity is varied so as to convey meaningful information.

modulated light Same as **modulated light beam**.

modulated light beam A light beam whose intensity is varied so as to convey meaningful information. For instance, a light beam modulated by a chopper. Also called **modulated light**.

modulated oscillator An oscillator whose output signal is varied so as to convey meaningful information.

modulated signal A signal whose amplitude, frequency, or phase has been varied so as to convey meaningful information.

modulated stage In a transmitter, the stage in which the carrier wave is modulated by the information-bearing signal.

modulated wave A wave, such as a carrier wave, whose amplitude, frequency, or phase has been varied so as to convey meaningful information.

modulating signal A signal which varies the amplitude, frequency, or phase of a carrier wave so as to convey meaningful information. Also called **modulating wave**.

modulating wave Same as **modulating signal**.

modulation 1. The process of modifying a characteristic of a wave or signal proportionally to a characteristic present in another wave or signal. Also, the result of such a process. Usually utilized to convey meaningful information. 2. The process of modifying a characteristic of a carrier wave by an information-bearing signal, as occurs, for instance, in AM, FM, or PM. Also, the result of such modulation. For example, in FM, the instantaneous frequency of a sine-wave carrier is varied above and below the center frequency by an information-bearing, or modulating, signal. Also called **carrier modulation**, or **RF modulation**.

modulation capability For a transmitter, the maximum modulation attainable while still maintaining distortion below a given level. Usually expressed as a percentage.

modulation code The code utilized to modulate a signal. For instance, that used in pulse-code modulation.

modulation coefficient Same as **modulation factor**.

modulation/demodulation The process of modifying a characteristic of a carrier signal by an information-bearing signal, followed by the process of reversing the effects of this modulation, thereby restoring the original modulating signal. For example, a modem modulates the carrier signal prior to transmission, and upon reception the remote modem demodulates the signal to restore the original information.

modulation depth Same as **modulation factor**.

modulation distortion Distortion in a signal due to modulation or cross modulation.

modulation-doped FET Same as **MODFET**.

modulation-doped field effect transistor Same as **MODFET**.

modulation envelope A curve whose points pass through the peaks of a graph showing the waveform of a modulated carrier.

modulation factor Also called **modulation index, modulation coefficient, index of modulation,** or **depth of modulation**. **1.** A measure of the degree of modulation, usually involving the ratio of a peak variation to a steady non-peak value. The result may be multiplied by 100 to obtain a percentage. **2.** In frequency modulation, the ratio of the peak variation of the carrier-wave frequency, to the frequency of the modulating signal. The result may be multiplied by 100 to obtain a percentage. **3.** In amplitude modulation, the peak amplitude variation of the composite wave, to the unmodulated carrier amplitude. The result may be multiplied by 100 to obtain a percentage. **4.** In amplitude modulation, the ratio of half the difference between the maximum amplitude and minimum amplitude, to the average amplitude. The result may be multiplied by 100 to obtain a percentage.

modulation frequency The frequency of a modulating signal.

modulation index Same as **modulation factor**.

modulation meter An instrument which measures **modulation factor**. Also called **modulometer**.

modulation monitor An instrument, such as a modulation meter, which serves to precisely monitor modulation.

modulation noise Noise in a signal due to modulation.

modulation percentage The **modulation factor** expressed as a percentage. Also called **percent modulation**.

modulation transfer function A mathematical function that describes the capability of an optical device or system, such as a microscope or scanner, to transfer or reproduce various levels of detail from the observed object to the image seen. Its abbreviation is **MTF**.

modulator A circuit or device which effects **modulation**.

modulator-demodulator Same as **modem**.

module **1.** A self-contained, and usually standardized, unit that performs one or more tasks, and which can be incorporated into a complete system. A module has defined performance characteristics, and can be disconnected and removed as a single unit, in addition to being replaced by an equivalent. For example, a circuit board with standardized dimensions and leads. **2.** An independent and self contained unit used within a **modular approach**.

module generation The usually automatic preparation or construction of **modules**.

modulometer Same as **modulation meter**.

modulus A constant, quantity, or value which expresses the extent to which a material or body possesses a physical property. For example, modulus of elasticity.

modulus of elasticity For a given material or body, the ratio of stress applied, to the deformation exhibited. Also known as **elasticity modulus**.

mOe Abbreviation of **millioersted**.

Mohs hardness scale Same as **Mohs scale**.

Mohs scale A hardness scale from 1 to 10, based on ten commonly occurring minerals. It is utilized to determine if one substance can scratch another. Talc is a 1 on the scale, quartz is a 7, corundum a 9, diamond is a 10, to name a few. Any material with a higher number can scratch that with a lower number, so diamond can scratch quartz, but quartz can not scratch diamond. Other materials may have values between the reference minerals. Also called **Mohs hardness scale**.

moire Same as **moiré pattern**. Also spelled **moiré**.

moiré Same as **moiré pattern**. Also spelled **moire**.

moiré pattern An undesirable pattern consisting of wavy lines which at times can be perceived as flickering. It may occur when two repetitive graphic patterns with fine details are superimposed. Such images can appear, for instance, on a computer monitor, a TV screen, or a printed sheet. A specific example is a herringbone pattern. Also spelled **moire pattern**. Also called **moiré**.

moire pattern Same as **moiré pattern**.

moisture-proof Also spelled **moistureproof**. **1.** Impervious, or highly resistant to moisture. **2.** Protected in a manner which does not allow moisture to form within, or to enter an enclosure.

moistureproof Same as **moisture-proof**.

mol Symbol for **mole**.

molar conductance Same as **molar conductivity**.

molar conductivity For a given electrolyte, the electrolytic conductivity divided by its concentration expressed in moles per unit volume. Also called **molar conductance**.

molar solution A solution in which one mole of a given solute is dissolved per liter of solvent. For instance, a 1.0 molar solution of sodium chloride has approximately 58.44 grams of sodium chloride dissolved in 1.0 liter of water.

mold **1.** A cavity which is utilized to give form to a fluid, amorphous, malleable, or plastic substance. Also, that which is given shape using such a cavity. Also, the shape so given. **2.** A frame around which something is shaped. Also, that which is given shape using such a cavity. Also, the shape so given.

molded capacitor A capacitor that is encased in a protective and insulating material, such as plastic, which is molded around it.

molded coil A coil that is encased in a protective and insulating material, such as plastic, which is molded around it. Also called **molded inductor**.

molded component An electrical component, such as a molded capacitor or resistor, that is encased in a protective and insulating material, such as plastic, which is molded around it.

molded inductor Same as **molded coil**.

molded resistor A resistor that is encased in a protective and insulating material, such as plastic, which is molded around it.

mole Its symbol is **mol**. **1.** The SI fundamental unit of amount of substance. Such a substance may be subatomic particles, atoms, molecules, ions, and so on. There are approximately 6.022142×10^{23} elementary entities in a mole. **2.** The molecular weight of a substance, expressed in grams per **mole** (**1**). For instance, a mole of sodium chloride is approximately 58.44 grams, as the combined molecular weight of sodium and chlorine is approximately 58.44 atomic mass units. Also called **gram-molecular weight**.

molectronics Same as **molecular electronics**.

molecular beam A narrow stream of molecules traveling in the same direction, with approximately the same speed, and emitted from the same source. Used, for instance in spectroscopy, and in molecular-beam epitaxy.

molecular-beam epitaxy A technique for growing epitaxial layers, in which the material to be deposited upon the substrate crystal is contained in a molecular or atomic beam. The procedure is carried out under an ultra-high vacuum, and can be used to form very precise and intricate layers as thin as single molecules or atoms. Used, for instance, in the preparation of semiconductors and sophisticated optoelectronic devices. Its abbreviation is **MBE**.

molecular electronics Its abbreviation is **ME (1)**, **moletronics**, or **molectronics**. **1.** The use of molecule-based structures, phenomena, organization, or properties to make electronic components and devices. For example, the use of nanotubes to serve as conducting paths between circuit components, or organic molecules to make electronic components, devices, or chips. **2.** The technology that deals with the design, construction, and applications of electronic circuits and devices whose building blocks are at a molecular level. Such blocks may be designed to self-assemble into circuits and devices, in a manner analogous to the way molecules in cells naturally form structural units.

molecular memory Computer memory in which each memory cell is a single molecule, or very few. In comparison to transistor-based RAM, such as DRAM or SRAM, such memory is vastly smaller and much less expensive, in addition to holding data for a long time without power. Used, for instance, in extremely small computers.

molecular orbital A region of space surrounding the nuclei present in a molecule, within which the probability of an electron being located is greatest. The charge of such an electron is distributed in this space.

molecular weight The ratio of the average mass per molecule or other formula unit, to one-twelfth the mass of a carbon-12 atom, expressed in atomic mass units. It is the same as the sum of the atomic weights of the atoms comprising said molecule. For example, the molecular weight of sodium chloride is 58.44 grams, as the combined atomic weights of sodium and chlorine are approximately 58.44 atomic mass units. Also known as **relative molecular mass**.

molecule The smallest particle of a compound that retains all of its properties. Molecules are composed of two or more atoms, which may or may not be of the same element. For instance, molecular oxygen is O_2, while water is H_2O. Molecules are overall neutral in charge, and if charged would be a type of ion.

moletronics Same as **molecular electronics**.

molten zone In a zone-refining technique, the liquid zone in which the impurities are dissolved. Also called **floating zone**.

molybdenum A silvery-white chemical element with atomic number 42. It is hard, malleable, ductile, has a high melting point, and maintains its strength at elevated temperatures. It has about 25 known isotopes, of which 6 are stable. It has various applications in electronics, including is use as an alloying agent, in electrodes and filaments, and in nuclear energy applications. Its chemical symbol is **Mo**.

molybdenum disilicide A dark gray crystalline powder whose chemical formula is $MoSi_2$. Used in resistors, ceramics, and as a high-temperature coating.

molybdenum pentachloride A green-black crystalline powder whose chemical formula is $MoCl_5$. Used in soldering, and for coatings which resist corrosion and high temperatures.

molybdenum trioxide A white to pale-yellow powder whose chemical formula is MoO_3. Used in alloys and ceramics.

MOM Abbreviation of **messaging-oriented middleware**.

moment of force Also called **torque**. The SI unit of torque is the Newton meter. **1.** A measure of the ability of one or more forces to produce rotation about an axis. **2.** One or more forces producing a turning or twisting motion.

momentary-contact switch A type of switch that returns to its resting state once the actuating force is removed. Such a switch may be normally open or normally closed. An example is a switch that is spring-loaded, and only makes the circuit when a button is held down.

monatomic Occurring as, pertaining to, or consisting of single atoms. For instance, argon, which is a monatomic gas. This contrasts, for instance, with molecular oxygen, which is a diatomic gas. Also spelled **monoatomic**.

monaural Its abbreviation is **mono**. **1.** Pertaining to one ear. **2.** Pertaining to one channel, as in monaural sound. **3.** Same as **monaural sound**. **4.** Same as **monophonic**.

monaural recording A recording utilizing only a single channel or track. There may be multiple inputs or sources, but only one channel or track. Also called **monophonic recording**.

monaural sound Also called **monaural (2)**. **1.** Sound which is recorded utilizing only a single channel or track. Also called **monophonic sound**. **2.** Sound which is heard by only one ear.

Monel metal A corrosion-resistant alloy consisting of nickel, copper, and small quantities of iron, manganese, and sometimes other metals. One composition may consist, for instance, of 67% nickel, 29% copper, 2% iron, and 2% manganese.

monitor **1.** A device utilized to display the images generated by a computer, TV, oscilloscope, radar, camera, or other similar device with a visual output. It incorporates the viewing screen and its housing. A specific example is a CRT. Also called **display (1)**, **display device**, **display monitor**, or **video monitor (1)**. **2.** A **monitor (1)** utilized to present the output of a computer. Also called **video monitor (2)**. **3.** A device or instrument which is utilized to sample, measure, assess, regulate, observe, or control a process or system. Also, to use such a device or instrument, for instance, to assess performance or to make sure an observed value is within the desired interval. **4.** A device, such as a **monitor (1)** or a speaker, which is utilized to keep track of the quality of a recording or broadcast. **5.** Software which supervises a computer program, system, or network, to help insure that it is functioning properly. Also, the use of such software.

monitor amplifier An amplifier which serves to keep track of the quality of a recording, broadcast, or other program. Also called **monitoring amplifier**.

monitor antenna An antenna which serves to keep track of the quality of a broadcast, or other transmission. Such an antenna is usually located near a transmitter. Also called **monitoring antenna**.

monitor flicker Brief and irregular fluctuations in an image presented by a **monitor (1)**. This may occur, for instance, when the refresh rate is too low. Also called **screen flicker**, or **display flicker**.

monitor head In some tape recorders, an additional head which enables monitoring that which is currently being recorded.

monitor loudspeaker Same as **monitor speaker**.

monitor mode A specific configuration of the settings of a computer monitor. For instance, the resolution, the number of colors that may be displayed, and whether it displays graphics and/or text. Also called **display mode**, **screen mode**, or **video mode**.

monitor port A port which serves to plug in a **monitor (2)**. Also called **display port**, or **video port**.

monitor program Same as **monitoring program**.

monitor receiver A receiver which serves to keep track of the quality of a transmission, broadcast, or other program. Also called **monitoring receiver**.

monitor resolution Also called **display resolution**, or **screen resolution**. **1.** The degree of sharpness and distinctness of the displayed image of a monitor. May be expressed, for instance, in terms of the number of pixels displayed, and dot pitch. **2.** The resolution a monitor is set at. A common resolution setting is 800x600, meaning that 800 dots, or pixels, are displayed on each of the 600 lines of the display. Frequently, the number of colors is incorporated into this specification, as in 800x600x256.

monitor software Same as **monitoring software**.

monitor speaker A speaker which is utilized to keep track of the quality of a recording or broadcast. Also called **monitor loudspeaker**.

monitoring The use of a **monitor (3)**, **monitor (4)**, or **monitor (5)**, to regulate, measure, or otherwise keep track of a process or system. For example, the manner in which an automatic pilot maintains a programmed course by monitoring factors such as speed, altitude, and attitude.

monitoring amplifier Same as **monitor amplifier**.

monitoring antenna Same as **monitor antenna**.

monitoring program A program which monitors computer programs, systems, or networks to help insure that they are functioning properly. Also called **monitor program**.

monitoring receiver Same as **monitor receiver**.

monitoring software Software which monitors computer programs, systems, or networks to help insure that they are functioning properly. Also called **monitor software**.

monitoring station A station utilized to regulate, observe, or otherwise keep track of something. For example, a station which monitors those entering or leaving an area under surveillance, or a station where a monitor antenna is located.

monkey chatter Interference of radio or TV reception by other stations using adjacent channels. Such interference may be unilateral or bilateral, and is caused, for instance, by overmodulation. It may be manifested as extraneous confusing, musical, or repetitive sounds being heard along with the desired program. Also known as **adjacent-channel interference**, or **sideband splatter**.

mono **1.** Abbreviation of **monaural**. **2.** Abbreviation of **monophonic**.

monoatomic Same as **monatomic**.

monochromatic **1.** Composed of one color. Also, appearing to have one color. **2.** Of or composed of radiation of a single wavelength, or of an extremely narrow interval of wavelengths. For example, one of the key features of a laser is the monochromatic nature of its output. **3.** Of or composed of particles with the same energy, or an extremely narrow interval of energies.

monochromatic light Light, such as that of a laser or a sodium lamp, which is **monochromatic**.

monochromatic radiation Radiation, such as that of a laser or a sodium lamp, which is **monochromatic**.

monochromator An instrument which isolates narrow portions of a spectrum, so as to produce **monochromatic radiation**. Used, for instance, for analysis or testing, as in spectroscopy.

monochrome **1.** Image reproduction in black, white, and multiple shades of gray in between. For example, that of a monochrome monitor. Also called **black-and-white**, or **grayscale**. **2.** Image reproduction in shades of one color. **3.** Image reproduction with one foreground color and one background color. For example, a green on black monitor.

monochrome display Same as **monochrome monitor**.

monochrome monitor Also called **monochrome display**. **1.** A display which reproduces images in black, white, and shades of gray in between. **2.** A display which reproduces images in shades of one color. **3.** A display which reproduces images with one foreground color and one background color, such as green on black.

monochrome television Same as **monochrome TV**.

monochrome TV Abbreviation of **monochrome television**. TV image reproduction which occurs in black, white, and shades of gray in between. Its abbreviation is **monochrome TV**. Also called **black-and-white TV**.

monoclinic crystal A crystal having three axes, each with a different length. Two of the axes are at right angles to the third, and intersect obliquely with each other.

monolayer **1.** A film or layer one molecule thick. Also called **monomolecular film**. **2.** A film or layer one atom thick.

monolithic **1.** Constructed upon a wafer, die, substrate, or other single mass. **2.** Formed from a single crystal. **3.** Existing as a massive and uniform structure.

monolithic capacitor An **MIM capacitor** used in a **monolithic IC**.

monolithic ceramic capacitor Same as **multilayer ceramic capacitor**.

monolithic IC Abbreviation of **monolithic integrated circuit**. An IC built upon, or formed within, a single semiconductor die. At least one of the components must be formed within, as opposed to above, the semiconductor substrate. Its own abbreviation is **MIC**.

monolithic integrated circuit Same as **monolithic IC**.

monolithic microwave IC Same as **MMIC**.

monolithic microwave integrated circuit Same as **MMIC**.

monomer A molecule which under the influence of heat, pressure, or catalysts, forms repeating units which together compose a polymer.

monometallic Composed of, pertaining to, or containing a single metal.

monomolecular film Same as **monolayer (1)**.

monophase Consisting of a single phase, as opposed to biphase, triphase, or the like.

monophonic Its abbreviation is **mono**. **1.** Pertaining to one channel, as in monophonic sound. Also called **monaural (2)**. **2.** Same as **monophonic sound**. **3.** Same as **monaural**.

monophonic recording Same as **monaural recording**.

monophonic sound Same as **monaural sound**.

monopole **1.** Same as **monopole antenna**. **2.** Same as **magnetic monopole**.

monopole antenna An antenna constructed above a ground plane, and whose radiation pattern is essentially that of a dipole in the half-space above said plane. Also called **monopole (1)**.

monopulse **1.** Same as **monopulse radar**. **2.** A single pulse.

monopulse radar A radar technique in which a single pulse is utilized to obtain directional information concerning the location of a scanned object. Such a radar has a receiving antenna system with two or more partially overlapping lobes. Multiple pulses may be employed when greater accuracy is needed, or for calculating the velocity of a scanned object. Also called **monopulse (1)**.

monoscope A CRT with a test pattern precisely etched into its surface, and which is used to transmit a picture signal which can be used to test and adjust receiving TV equipment. Also called **monotron**.

monostable Having one stable state.

monostable circuit A circuit with one stable state. A trigger signal must be applied to change it to the unstable state, where it remains for a given interval, after which it returns to the stable state.

monostable multivibrator A multivibrator with one stable state and one unstable state. A trigger signal must be applied to change to the unstable state, where it remains for a given interval, after which it returns to the stable state. Also called **one-shot multivibrator, univibrator,** or **single-shot multivibrator.**

monostatic radar A radar system in which a single location is used. Most radars are of this type, and the term is utilized to contrast with **bistatic radar,** a system in which two locations are used.

monotonicity In an analog-to-digital or digital-to analog converter, the condition in which the output increases as the input increases, across the full range of operation.

monotron Same as **monoscope.**

Monte Carlo method A computer-assisted mathematical technique for problem solving, in which repeated calculations or simulations are made using a combination of actual data, when available, and data selected randomly. Especially useful when dealing with problems with too many variables for conventional analytical treatment. May be used, for instance, to numerically plot the trajectories of many particles, and predict the interactions these have with a material through which they travel. The random aspects of this method are loosely analogous to the uncertain results that may be obtained by gambling, and Monte Carlo, Monaco, is a well-known gaming location. Also called **Monte Carlo simulation,** or **Monte Carlo technique.**

Monte Carlo simulation Same as **Monte Carlo method.**

Monte Carlo technique Same as **Monte Carlo method.**

Moog synthesizer One of the first popular synthesizers used for music and other sounds. Such an instrument can simulate a wide assortment of sounds, especially musical instruments, and plays one note at a time.

moon **1.** The only natural satellite of the earth. **2.** A natural satellite of a planet.

moonbounce Same as **moonbounce communications.**

moonbounce communications Radio communications in which signals are bounced off the moon. Usually utilized for UHF or VHF communications. Also called **moonbounce,** or **earth-moon-earth communications.**

Moore's Law A rule-of-thumb stating that the number of transistors that can be fit on a chip doubles every 18 months. The month interval may vary according to the technology at the moment this principle is mentioned.

MOPA Abbreviation of **master oscillator-power amplifier.**

morph To use computer graphics to gradually transform an image, or parts of an image, into another image.

morphing The use of computer graphics to gradually transform an image, or parts of an image, into another image.

Morse code A code utilized for transmitting messages in which each letter and number has a given sequence of dots and dashes, sent as short and long pulses. Alternatively, short and long flashes of light may be employed. Used in telegraphy, and although it is no longer utilized for communications at sea, it is still employed for the identification of aeronautical navigation beacons.

MOS Acronym for **m**etal-**o**xide **s**emiconductor. A unipolar three-layer semiconductor in which there is a thin metal oxide layer between the metal contacts and the substrate. For instance, a silicon dioxide insulating layer may be sandwiched between the metal contacts and a silicon substrate. This arrangement is widely seen, for instance, in MOSFETs. There are various MOS technologies, including CMOS, DMOS, HMOS, LDMOS, NMOS, and PMOS.

MOS capacitor Abbreviation of **m**etal-**o**xide **s**emiconductor **capacitor.** A capacitor in which a metal and a semiconduc-

tor are separated by a thin oxide dielectric, such as silicon oxide. Used, for instance, in chips.

MOS-controlled thyristor Abbreviation of **m**etal-**o**xide **s**emiconductor **controlled thyristor.** A thyristor which utilizes a **MOSFET**-controlled gate. It features very low conduction loss in the on state, high switching speeds, and is used in high power electronic applications.

MOS IC Same as **MOS integrated circuit.**

MOS integrated circuit Abbreviation of **m**etal-**o**xide **s**emiconductor **integrated circuit.** An IC incorporating **MOS-FETs.** Its own abbreviation is **MOS IC.**

MOS/LSI Abbreviation of **m**etal-**o**xide **s**emiconductor/large-scale **i**ntegration. Pertaining to ICs, large-scale integration utilizing **MOS technology.**

MOS RAM Abbreviation of **m**etal-**o**xide **s**emiconductor random-access memory. RAM utilizing **MOS technology.**

MOS technology Abbreviation of **m**etal-**o**xide **s**emiconductor **technology.** A technology employed in the fabrications of metal-oxide semiconductors. There are various MOS technologies, including CMOS, DMOS, HMOS, LDMOS, NMOS, and PMOS.

MOS transistor Acronym for **m**etal-**o**xide **s**emiconductor **transistor.** Same as **MOSFET.**

MOS/VLSI Abbreviation of **m**etal-**o**xide **s**emiconductor/**v**ery large-scale **i**ntegration. Pertaining to ICs, very large-scale integration utilizing MOS technology.

mosaic **1.** A material, usually a thin mica sheet, which on one side is coated with a large number of photosensitive globules, and on the other a photoemissive surface. It is used in TV camera tubes, where the optical image is focused upon the first surface, and the equivalent electrical charge patterns produced on the other are scanned by the electron beam of said tube. Also called **photomosaic.** **2.** The pattern formed by the photosensitive globules of a **mosaic (1). 3.** A pattern formed by setting small pieces on a surface.

Mosaic An early graphics-based Web browser which is still popular.

MOSFET Acronym for **m**etal-**o**xide **s**emiconductor field-effect transistor. A field-effect transistor whose gate is insulated from the channel by a thin layer consisting of a metal oxide, such as silicon dioxide. MOSFETs feature extremely high input impedance and low power consumption, and are used, for instance, in high-gain audio amplifier circuits, and for computer memory. It is a type of **IGFET,** and both terms are usually utilized synonymously, but MOSFET is a much more common term. Also called **MOS transistor.**

MOST Same as **MOS transistor.**

most-significant bit Within a sequence of bits, that with the highest weight or place value. Excluding the sign character, it is usually the bit furthest to the left. Its abbreviation is **MSB.**

most-significant byte Within a sequence of bytes, that with the highest weight or place value. Excluding the sign character, it is usually the byte furthest to the left. Its abbreviation is **MSB.**

most-significant character Within a sequence of characters, that with the highest weight or place value. It is usually the character furthest to the left. Its abbreviation is **MSC.**

most-significant digit Within a sequence of digits, that with the highest weight or place value. Excluding the sign character, it is usually the digit furthest to the left. Its abbreviation is **MSD.**

mote **1.** One of multiple microelectromechanical machines within **smart dust. 2.** One of a collection of microelectromechanical machines which are similar to smart dust, but larger. Used, for instance, for tracking products, monitoring the function of devices, or for spying.

mother crystal A raw piezoelectric crystal, which may occur naturally or be grown artificially.

motherboard The main circuit board of a computer, containing the connectors necessary to attach additional boards. It contains most of the key components of the system, and incorporates the CPU, bus, memory, controllers, expansion slots, and so on. Memory chips may be added to it, and some motherboards allow for the CPU to be replaced. Also called **mainboard, system board,** or **backplane (2).**

Motif 1. A set of graphical user interface standards. **2.** A popular graphical user interface.

motion capture The use of real motion as computer input. For example, the attachment of sensors to a person who is swimming, to better simulate a swimmer in a virtual environment.

motion detector A device which senses movement, or a lack of movement, in a given area, enclosure, device, piece of equipment, or system. Such a device may detect movement through the use of sensors which utilize and/or monitor microwaves, infrared radiation, laser light, vibrations, temperature changes, magnetic fields, and so on. Used, for instance, for intrusion detection, to turn on lights in places where illumination is needed only when someone is present, or to ensure equipment is functioning. Also called **motion sensor.**

Motion Joint Photographic Experts Group Same as **MJPEG.**

Motion-JPEG Same as **MJPEG.**

Motion Picture Experts Group Same as **MPEG.**

motion sensor Same as **motion detector.**

motion video 1. Video whose frames are displayed at a given rate. To be **full-motion video,** said rate must be at least 24 frames per second. **2.** Digital video whose frames are displayed at a given rate. To be full-motion video, said rate is defined to be 30 frames per second.

motor 1. A device or machine which converts electrical energy into mechanical energy, usually torque. Its source of power may be DC or AC. An example is an AC induction motor, such as that utilized in many household appliances. Also called **electric motor. 2.** A device or machine which converts another form of energy into mechanical energy.

motor capacitor 1. Same as **motor-start capacitor. 2.** Same as **motor-run capacitor.**

motor control 1. The use of an electronic circuit to control the speed of a DC motor which is fed by an AC power line. Also called **electronic motor control. 2.** The use of a circuit, device, or mechanism, to control the operation, speed, or other function of a motor.

motor effect The repulsive force between two conductors carrying currents in opposite directions.

motor-generator Same as **motor-generator set.**

motor-generator set An electromechanical device coupling a motor and a generator, which is utilized to change the voltage and/or frequency of a power supply. The usual configuration incorporates a low-voltage motor with a high-voltage generator. Also called **motor-generator.**

motor meter 1. A motor which incorporates a rotor, one or more stators, and a device which makes the speed of the motor proportional to the integral of a measured quantity, such as electrical power. **2.** A meter, such as a power meter, whose moving part is a motor.

motor-run capacitor A capacitor which is connected to the auxiliary winding of a motor, and which is connected in parallel at all times to the main winding of a running motor. Also called **motor capacitor (2).**

motor-speed control A device, such as a triac, or a mechanism, such as feedback, utilized to control the speed of a motor.

motor-start capacitor A capacitor which is connected to the auxiliary winding of a starting motor. When the motor attains a given speed, the auxiliary winding and capacitor are disconnected through the use, for instance, of a centrifugal switch. Also called **motor capacitor (1).**

motorboating A form of acoustic noise due to unwanted oscillations in audio amplifiers, which produces a low-frequency rumble that sounds similar to a motor boat idling.

mount 1. A support or substrate upon which objects may be fixed or assembled. For example, a breadboard. **2.** To fix or assemble on a support or substrate. For instance, to mount a gyroscope on a gimbal.

mounting cord The cord running from the base of a telephone, or similar device, to the phone jack.

mounting flange 1. A flange which serves for, or assists with, mounting. **2.** A rim, which surrounds a speaker, with holes which facilitate mounting on a cabinet, or other housing.

mouse A computer pointing device which when moved upon a surface also moves the cursor on the display screen, and which usually has two or more buttons used for selecting, performing functions, or the like. Some mice also have a scroll wheel, which further facilitates navigating through documents, Web pages, and so on. Most mice are either mechanical or optical, and some are cordless. A computer mouse looks more or less like its rodent counterpart, in that it has a compact body and a longish tail, while the moving and clicking of the former is analogous to the scurrying and nose-twitching of the latter, hence its name. Also called **computer mouse.**

mouse button On a computer mouse, a button which is pressed, or clicked, to select objects, perform functions, or the like. Also called **button (4).**

mouse cursor Same as **mouse pointer.**

mouse pad Same as **mousepad.**

mouse pointer On a computer screen, an indicator, such as a small hand, arrow, or I-beam, that moves as the mouse is moved. It serves to select text, menus, and the area of the screen where the next text input or other action will occur, by clicking the mouse there. Also called **mouse cursor, pointer (1), cursor (2).**

mouse port A port which serves to plug in a **mouse.** Most computers have a dedicated port for the mouse, generally with six pins. Otherwise, a serial port is usually used.

mouse potato A person who spends an inordinate amount of leisure time at the computer, often indulging in snacking, which can help lead to a potato-like physique.

mouse trails A quickly vanishing comet-like tail that moves across a display screen as it follows mouse movements. Activation of this feature helps make such movements more readily apparent when using certain monitors, or to assist people with reduced vision.

MouseKeys A feature that allows the keyboard to be utilized to move the cursor on the display screen as if using a mouse. Useful, for instance, for those with reduced motor function.

mousepad A flat surface upon which a **mouse** is moved. It may consist, for instance, of a rectangular fabric-covered rubber pad which provides a smooth surface that does not create too much friction, yet provides enough traction for the ball underneath. Also spelled **mouse pad.**

mouth 1. The opening of a radiating device such as the horn of a speaker or antenna. The radiated waves emerge from this opening. **2.** For a device such as a horn, the opening with the larger cross section. The **throat** is that with the smaller cross section.

mouth-stick Same as **mouthstick.**

mouthstick A computer input device consisting of a rod-shaped object which is placed in the mouth. Useful, for instance, to users with reduced motor function when entering commands. Also spelled **mouth-stick.**

MOV Abbreviation of **metal-oxide varistor.**

movable contact In a relay or switch, the contact that moves towards or away from a **stationary contact**, to close or open a circuit. Also called **armature contact**.

move 1. The relocation of data, files, programs or the like. **2.** A deprecated synonym for **copy (2)**.

movies-on-demand A service which enables a subscriber to choose from a selection of movies. The term usually refers to a paid digital TV or Internet service whose offered movies are digitally encoded and available for viewing or download at any time, not according to a specific schedule. Although the terms are often used interchangeably, **pay-per-view** is usually a paid cable TV or satellite service, and has a fixed viewing schedule.

moving-coil cartridge A phonograph pickup in which the vibrations of the stylus move a coil which is in a continuous magnetic field, which in turn produces the output signal current. Also called **moving-coil pickup, dynamic reproducer,** or **dynamic pickup**.

moving-coil galvanometer A galvanometer which utilizes a small wire coil which is suspended or pivoted in a fixed magnetic field. The current that passes through the coil produces a magnetic field which reacts with the fixed field, and the resulting movement of the coil is correlated to a calibrated scale. A D'Arsonval galvanometer is an example.

moving-coil instrument An instrument, such as a moving-coil galvanometer, in which the moving element is a small wire coil which is suspended or pivoted in a fixed magnetic field.

moving-coil loudspeaker Same as **moving-coil speaker**.

moving-coil meter A meter, such as a moving-coil galvanometer, in which the moving element is a small wire coil which is suspended or pivoted in a fixed magnetic field.

moving-coil microphone A type of microphone in which incident sound waves strike a diaphragm, which is connected to a moving coil. The coil moves within a constant magnetic field, which in turn produces the output signal current. Also called **dynamic microphone**.

moving-coil pickup Same as **moving-coil cartridge**.

moving-coil speaker A type of loudspeaker in which an audio-frequency signal current is sent through a voice coil, which is attached to a diaphragm. The interaction with a constant magnetic field provides the coil with a piston-like movement, which is mechanically transferred to the diaphragm, which in turn generates the sound waves that emanate from the loudspeaker. Also called **moving-coil loudspeaker, dynamic speaker,** or **electrodynamic loudspeaker**.

moving element Within a device or apparatus with moving parts, an object, part, or section which moves. For instance, an armature, or the needle of a compass or indicating meter.

moving-iron instrument An instrument in which the moving element is a small bar of soft iron which is suspended or pivoted in a fixed magnetic field.

moving-iron meter A meter in which the moving element is a small bar of soft iron which is suspended or pivoted in a fixed magnetic field.

moving-magnet cartridge A phonograph pickup in which a small magnet responds to the vibrations of the stylus as it follows the grooves of a record. The movement of the magnet varies the magnetic field of a fixed-coil structure, which in turn produces the output signal current. Also called **moving-magnet pickup**.

moving-magnet pickup Same as **moving-magnet cartridge**.

Moving Picture Experts Group Same as **MPEG**.

Moving Picture Experts Group Audio Layer 3 Same as **MP3**.

moving robot A robot equipped with a movable platform, a rolling mechanism, or mechanical legs, which allow it to move or travel, depending on the tasks it is meant to execute.

MOVPE Acronym for **metal-organic vapor-phase epitaxy**.

Mozilla A popular open-source Web browser.

m_p Symbol for **proton mass**, or **proton rest mass**.

mp Abbreviation of **melting point**.

MP3 Abbreviation of Moving Picture Experts Group Audio Layer 3. **1.** An audio compression technique which reduces the size of audio files by a factor of 12 or more, without a significant reduction in sound quality. Used extensively to transmit music files over the Internet. **2.** Same as **MP3 file**.

MP3 file An audio file encoded using **MP3 (1)**. Also called **MP3 (2)**.

MP3 player 1. An application which plays **MP3 files**. **2.** A device, such as a palm-sized digital audio player, which plays MP3 files.

MPa Abbreviation of **megapascal**.

MPEG Abbreviation of Moving Picture Experts Group, or Motion Picture Experts Group. **1.** A series of standards for the lossy compression of digital audio and video, and multimedia applications. A system may be equipped with a special board for enhanced playback. **2.** A specific standard within this series, such as MPEG-3.

MPEG-1 An **MPEG** standard that encoded full-motion video designed for use with CD-ROMs with a resolution of 352 x 240 pixels at 30 fps, and CD-quality stereo audio.

MPEG-2 An **MPEG** standard that improved upon MPEG-1, and which provides laserdisc quality video at resolutions up to 1280 x 720 pixels at 30 fps, and CD-quality stereo audio. Used for broadcast TV such as HDTV, and DVDs.

MPEG-3 An **MPEG** standard that was under development, and abandoned when its features were incorporated into **MPEG-2**.

MPEG-4 An **MPEG** standard for the encoding of multimedia applications, especially for use over the Internet. It is similar to MPEG-1, but uses much less bandwidth.

MPEG Audio Layer 3 Same as **MP3**.

MPEG3 Same as **MP3**.

mph Abbreviation of **milliphot**.

MPLS 1. Abbreviation of **Multiprotocol Label Switching**. **2.** Abbreviation of **Multiprotocol Lambda Switching**.

MPOA Abbreviation of Multiprotocol Over ATM. An ATM specification which enables the integration of ATM services with TCP/IP and LAN networks. This, for instance, enables the use of routing protocols from traditional LANs, such as Ethernet, over an ATM backbone.

MPOFR Abbreviation of Multiprotocol Over Frame Relay. An ATM specification which enables the integration of ATM services with frame relay networks such as WANs.

MPP 1. Abbreviation of **massively parallel processing**. **2.** Abbreviation of **massively parallel processor**.

MPPP Abbreviation of **Multilink PPP**.

MPR II A standard establishing limits on the emission of electromagnetic radiation by monitors.

MPS Abbreviation of **mobile positioning system**.

MPU Abbreviation of **microprocessor unit**.

mR Abbreviation of **milliroentgen**.

MR Abbreviation of **magnetoresistive**.

MR head Abbreviation of **magnetoresistive head**.

MR head technology Abbreviation of **magnetoresistive head technology**.

MR Imaging Abbreviation of **magnetic resonance imaging**.

mrad 1. Abbreviation of **milliradian**. **2.** Abbreviation of **millirad**.

MRAM Abbreviation of magnetic **RAM**, or magnetoresistive **RAM**. A type of RAM that stores data using **magnetic bits**.

Such memory is non-volatile, consumes much less power than conventional RAM which depends on the constant replenishment of electrical charges, and eliminates the need for booting, thus enabling computer systems to start up instantaneously.

mrem Abbreviation of **millirem**.

MRI Abbreviation of **magnetic resonance imaging**.

MRI Imaging Abbreviation of **magnetic resonance imaging**.

MRP Abbreviation of **Materials Requirements Planning**.

ms Abbreviation of **millisecond**.

mS Abbreviation of **millisiemens**.

MS 1. Abbreviation of **mass spectrometry**. 2. Abbreviation of **mass spectrometer**. 3. Abbreviation of **mass spectroscopy**.

MS-DOS A common disk operating system.

MSAU Abbreviation of **Multistation Access Unit**.

MSB 1. Abbreviation of **most-significant bit**. 2. Abbreviation of **most-significant byte**.

MSC 1. Abbreviation of **most-significant character**. 2. Abbreviation of **mobile switching center**.

MSD Abbreviation of **most-significant digit**.

msec Abbreviation of **millisecond**.

msg Abbreviation of **message**.

MSI Abbreviation of **medium-scale integration**.

MSK Abbreviation of **minimum shift keying**.

MSP 1. Abbreviation of **management service provider**, or **managed service provider**. 2. Abbreviation of **Message Security Protocol**.

MSPS Abbreviation of million samples per second. The number of samples, in millions, taken per second. It is a unit utilized in data acquisition and digital-signal processing.

mSQL Abbreviation off Mini **SQL**. A popular relational database.

MSS Abbreviation of **mobile-satellite service**.

mT Abbreviation of **millitesla**.

Mt Chemical symbol for **meitnerium**.

MT Abbreviation of **metric ton**.

MT-RJ connector A small fiber-optic connector that has two fibers, and which is approximately the same size as a copper RJ-45 connector.

MTA Abbreviation of **message transfer agent**.

MTBF 1. Abbreviation of **mean time before failure**. 2. Abbreviation of **mean time between failures**.

MTF Abbreviation of **modulation transfer function**.

MTS Abbreviation of **multi-channel TV sound**.

MTSO Abbreviation of **mobile telephone switching office**.

MTTF Abbreviation of **mean time to failure**.

MTTR Abbreviation of **mean time to repair**.

MTU Abbreviation of **maximum transmission unit**.

mu The twelfth letter of the Greek alphabet. Its Greek character, μ, is the symbol for **micro-**, **micron**, and **magnetic permeability** or **permeability** (1).

mu factor In an electron tube, the factor by which the plate voltage increases in proportion to an increase in the grid voltage, where all other voltages and the plate current are held constant. Also known as μ **factor**, or **amplification factor** (2).

mu-law A standard used to convert an analog input, usually voice, into digital form using pulse-code modulation. Currently, it is used only in North America and Japan, while the rest of the world uses the A-law standard. Also known as μ **law**, or **u law**.

mu-metal An alloy consisting of about 75% nickel, 15% iron, 5% copper, 2% chromium, and minor amounts of silicon,

manganese and possibly other elements. It has very high magnetic permeability, and is utilized for magnetic shielding. Also spelled **mumetal**.

MUD Acronym for **mu**lti-**u**ser **d**ungeon, **mu**lti-**u**ser **d**imension, or **mu**lti-**u**ser **d**omain. Interactive multiuser games, and the virtual environment they take place in, played over the Internet. Many MUDs deal with fantasies involving heroics, such as rescues and escapes from dungeons.

MUF Abbreviation of **maximum usable frequency**.

multi-band antenna Same as **multiband antenna**.

multi-band receiver A receiver which can be used to tune to multiple frequency bands. Depending on the unit, a user may be able to select from AM, FM, MW, SW, and TV broadcasts, and so on. Also spelled **multiband receiver**. Also called **universal receiver** (2).

Multi-bank DRAM Same as **MDRAM**.

multi-boot Same as **multiboot**.

multi-cellular horn Same as **multicellular horn**.

multi-channel Pertaining to, having, using, operating, analyzing, characteristic of, or arising from the simultaneous operation of multiple channels. Also spelled **multichannel**.

multi-channel analyzer An instrument which separates its input into multiple channels. When such an input is separated into intervals classified by pulse amplitude or height, also called **pulse-height analyzer**. Also spelled **multichannel analyzer**.

multi-channel loudspeaker Same as **multi-channel speaker**. Also spelled **multichannel loudspeaker**.

multi-channel operation Operation involving multiple channels. For example, the use of a multi-channel transmitter. Also spelled **multichannel operation**.

multi-channel receiver A radio receiver that has multiple channels available, and can use any number of them simultaneously. Also spelled **multichannel receiver**.

multi-channel speaker A speaker which utilizes individual drivers for each of two or more intervals of frequencies. Examples include two-way and three-way speakers. Also spelled **multichannel speaker**. Also called **multi-channel loudspeaker**.

multi-channel stereo A stereophonic system with three or more channels intended to be played simultaneously. For instance, a system whose sound output is provided by left and right stereo speakers plus a subwoofer. Also spelled **multichannel stereo**.

multi-channel television sound Same as **multi-channel TV sound**.

multi-channel transmitter A radio transmitter that has multiple channels available, and can use any number of them simultaneously. Also spelled **multichannel transmitter**.

multi-channel TV sound Abbreviation of **multi-channel television sound**. TV sound which is broadcast over two or more channels. For instance, a transmitter may provide two channels for stereo reception, plus a third channel for an alternate language. Also spelled **multichannel TV sound**. Its own abbreviation is **MTS**.

multi-chip IC Same as **multichip IC**.

multi-chip integrated circuit Same as **multichip IC**.

multi-chip module Same as **multichip module**.

multi-chip package Same as **multichip module**. Also spelled **multichip package**. Its abbreviation is **MCP**.

multi-component remote Abbreviation of **multi-component remote** control. A remote control which is capable of controlling multiple devices, such as TVs, audio amplifiers, DVDs, and so on. Such a device is usually programmable, and is compatible with equipment from multiple manufacturers.

multi-component remote control Same as **multi-component remote**.

multi-computer Same as **multicomputer**.

multi-contact relay A relay with three or more contacts. Also spelled **multicontact relay**.

multi-contact switch A switch with three or more contacts. Also spelled **multicontact switch**.

multi-coupler In radio communications, a device which connects multiple receivers or transmitters to a single antenna, while maintaining impedance matching. Also spelled **multicoupler**.

multi-dimensional database Same as **multidimensional database**.

multi-directional Same as **multidirectional**.

multi-directional microphone Same as **multidirectional microphone**.

multi-drop line Same as **multidrop line**.

multi-element antenna An antenna with multiple radiators which contribute to its overall transmission and/or reception characteristics. Such radiators may be driven or parasitic. Also spelled **multielement antenna**.

multi-frequency Consisting of, pertaining to, or utilizing two or more frequencies or bands. Also spelled **multifrequency**.

multi-frequency monitor A monitor which accepts video signals from a fixed number of frequency ranges. For example, it may accept VGA and Super VGA. This contrasts with a **fixed-frequency monitor**, which only responds to a video signal within a single frequency range, and with a **multiscan monitor**, which responds to signals at any frequency within a wide interval. Also spelled **multifrequency monitor**.

multi-frequency signaling Also spelled **multifrequency signaling**. Also called **dual-tone multifrequency**. **1.** A telephone signaling system which utilizes tones which are a mixture of two frequencies. Each mixture identifies any of the 12 dialed keys, 0-9, *, and #, as found on most touchtone phones. The tone heard when a number 5 is pressed on such a phone, for instance, is a combination of two such frequencies, specifically 770 Hz and 1336 Hz. **2.** A communications method utilizing **multi-frequency signaling (1)**.

multi-function device Same as **multifunction device**.

multi-function meter Same as **multimeter**.

multi-function peripheral Same as **multifunction peripheral**.

multi-function printer Same as **multifunction printer**.

multi-function product Same as **multifunction product**.

multi-handset cordless A cordless telephone device in which a single base unit serves multiple handsets. Each handset only needs to be placed in its cradle when charging. Also spelled **multihandset cordless**.

multi-hop propagation The long-distance propagation of radio waves utilizing multiple reflections between the ionosphere and the earth. Also spelled **multihop propagation**.

multi-hop transmission The long-distance transmission of radio waves utilizing multiple reflections between the ionosphere and the earth. Also spelled **multihop transmission**.

multi-language display A display, such as that of a TV, PDA, or cell phone, which provides the choice of two or more languages. Also spelled **multilanguage display**.

multi-layer Same as **multilayer**.

multi-layer board Same as **multilayer board**.

multi-layer ceramic capacitor Same as **multilayer ceramic capacitor**.

multi-layer circuit Same as **multilayer circuit**.

multi-layer circuit board Same as **multilayer circuit board**.

multi-layer metallization Same as **multilevel metallization**. Also spelled **multilayer metallization**.

multi-layer PCB Same as **multilayer PCB**.

multi-layer printed-circuit board Same as **multilayer PCB**.

multi-level metallization Same as **multilevel metallization**.

multi-mode Same as **multimode**.

multi-mode fiber Same as **multimode fiber**.

multi-mode operation Same as **multimode operation**.

multi-mode optical fiber Same as **multimode optical fiber**.

multi-path Same as **multipath**.

multi-path cancellation Same as **multipath cancellation**.

multi-path delay Same as **multipath delay**.

multi-path fading Same as **multipath fading**.

multi-path interference Same as **multipath interference**.

multi-path reception Same as **multipath reception**.

multi-path transmission Same as **multipath transmission**.

multi-pin connector Same as **multipin connector**.

multi-platform Same as **multiplatform**.

multi-point channel Same as **multipoint channel**.

multi-point circuit Same as **multipoint circuit**.

multi-point conferencing Same as **multipoint conferencing**.

multi-point configuration Same as **multipoint configuration**.

multi-point connection Same as **multipoint connection**.

multi-point distribution service Same as **multipoint distribution service**.

multi-point line Same as **multipoint line**.

multi-point network Same as **multipoint network**.

multi-port repeater Same as **multiport repeater**.

multi-position relay A relay, such as a stepping relay, with multiple contacts. Also spelled **multiposition relay**.

multi-position switch Also spelled **multiposition switch**. Also called **multi-switch (2)**. **1.** A switch, such as a stepping switch, with multiple contacts. **2.** A switch, such as a selectable switch, with multiple operating positions.

multi-processing Same as **multiprocessing**.

multi-processor Same as **multiprocessor**.

multi-programming Same as **multitasking**. Also spelled **multiprogramming**.

Multi-protocol Label Switching Same as **Multiprotocol Label Switching**.

Multi-protocol Lambda Switching Same as **Multiprotocol Lambda Switching**.

multi-protocol router Same as **multiprotocol router**.

multi-purpose instrument An instrument, such as a multimeter, which has two or more functions. Also spelled **multipurpose instrument**.

multi-purpose meter A meter, such as a multimeter, which has two or more functions. Also spelled **multipurpose meter**.

multi-range instrument An instrument which has two or more ranges within which it can measure, indicate, and/or monitor the value of an observed and/or controlled quantity. Also spelled **multirange instrument**.

multi-range meter A meter which has two or more ranges within which it can measure, and usually indicate, a physical quantity or other value. Also spelled **multirange meter**.

multi-scan monitor Same as **multiscan monitor**.

multi-scanning monitor Same as **multiscan monitor**. Also spelled **multiscanning monitor**.

multi-sensor data fusion Same as **multisensor data fusion**.

multi-session CD Same as **multisession CD**.

multi-stable Same as **multistable**.

multi-stage Same as **multistage**.

multi-stage amplifier Same as **multistage amplifier**.

multi-station Also spelled **multistation**. **1.** A communications network having multiple stations that can intercommunicate. **2.** Pertaining to **multi-station (1)**.

multi-switch **1.** A device utilized to amplify and distribute an incoming satellite signal to multiple satellite receivers. **2.** Same as **multi-position switch**.

multi-tasking Same as **multitasking**.

multi-tester Same as **multitester**.

multi-threaded application Same as **multithreaded application**.

multi-threading Same as **multithreading**.

multi-track Same as **multitrack**.

multi-track recording Same as **multitrack recording**.

multi-track recording system Same as **multitrack recording system**.

multi-user Same as **multiuser**.

multi-user dimension Same as **MUD**.

multi-user domain Same as **MUD**.

multi-user dungeon Same as **MUD**.

multi-user system Same as **multiuser system**.

multiband antenna An antenna which is suitable for operation on multiple frequency bands, and which has a single feed line. Also spelled **multi-band antenna**.

multiband receiver Same as **multi-band receiver**.

Multibank DRAM Same as **MDRAM**.

Multibank Dynamic RAM Same as **MDRAM**.

multiboot A computer configuration which allows a user to select any of two or more different operating systems to boot the computer. Also spelled **multi-boot**.

multicast The simultaneous transmission of data to multiple selected destinations within a communications network. An example is an IP multicast. Also called **multicasting**.

multicast address In **multicasting**, a single address that refers to multiple network devices or destinations.

multicast backbone Same as **Mbone**.

Multicast IP address An IP address that corresponds to all hosts and servers in an IP multicast.

multicasting Same as **multicast**.

Multicasting Routing Protocol A routing protocol, such as DVMRP, supporting IP multicasting.

multicellular horn Also spelled **multi-cellular horn**. **1.** A speaker with multiple horn-shaped radiators. **2.** An antenna with multiple horn-shaped radiators.

multichannel Same as **multi-channel**.

multichannel analyzer Same as **multi-channel analyzer**.

multichannel loudspeaker Same as **multi-channel speaker**. Also spelled **multi-channel loudspeaker**.

Multichannel Multipoint Distribution Same as **Multichannel Multipoint Distribution Service**. Its abbreviation is **MMD**.

Multichannel Multipoint Distribution Service A wireless method for sending Cable TV signals using microwaves. It requires line-of-sight transmission and reception, and is used, for instance, to provide service in rural areas where establishing service would otherwise be unavailable for physical or economical reasons. Its abbreviation is **MMDS**. Also called **Multichannel Multipoint Distribution**.

multichannel operation Same as **multi-channel operation**.

multichannel receiver Same as **multi-channel receiver**.

multichannel speaker Same as **multi-channel speaker**.

multichannel stereo Same as **multi-channel stereo**.

multichannel television sound Same as **multi-channel TV sound**.

multichannel transmitter Same as **multi-channel transmitter**.

multichannel TV sound Same as **multi-channel TV sound**.

multichip IC Abbreviation of **multichip** integrated circuit. An IC composed of two or more semiconductor chips which are independently attached to a single substrate or header. Also spelled **multi-chip IC**.

multichip integrated circuit Same as **multichip IC**.

multichip module A package which houses two or more interconnected chips on a common substrate. The connections between them usually consist of multiple layers, each separated by an insulator. Its abbreviation is **MCM**. Also spelled **multi-chip module**. Also called **multichip package**.

multichip package Same as **multichip module**. Also spelled **multi-chip package**. Its abbreviation is **MCP**.

multicomputer Also spelled **multi-computer**. **1.** An architecture in which multiple computers are used, each with its own CPU, memory, and instructions. Used, for instance, for parallel processing. **2.** Consisting of, or incorporating, two or more computers.

multicontact relay Same as **multi-contact relay**.

multicontact switch Same as **multi-contact switch**.

multicoupler Same as **multi-coupler**.

multidimensional database A powerful relational database which processes and accesses information from multiple perspectives. Used, for instance, in Online Analytical Processing. Also spelled **multi-dimensional database**. Its abbreviation is **MDDB**.

multidirectional Also spelled **multi-directional**. Also called **poly-directional**. **1.** Able to operate, or be operated, in two or more directions. **2.** Pertaining to, or utilizing two or more directions.

multidirectional microphone A microphone which is able to switch between multiple pickup patterns. Also spelled **multi-directional microphone**. Also called **poly-directional microphone**.

multidrop line Same as **multipoint line**. Also spelled **multi-drop line**.

multielement antenna Same as **multi-element antenna**.

multifrequency Same as **multi-frequency**.

multifrequency monitor Same as **multi-frequency monitor**.

multifrequency signaling Same as **multi-frequency signaling**.

multifunction device Its abbreviation is **MFD**. Also spelled **multi-function device**. Also called **multifunction product**. **1.** Same as **multifunction peripheral**. **2.** A device, such as a fax machine which also makes copies, which performs multiple functions.

multifunction meter Same as **multimeter**.

multifunction peripheral Its abbreviation is **MFP**. Also spelled **multi-function peripheral**. Also called **multifunction device (1)**. **1.** A computer peripheral which performs multiple functions. For instance, a multifunction printer. **2.** Same as **multifunction printer**.

multifunction printer A computer printer which also performs other functions, such as scanning, copying, and faxing. Also spelled **multi-function printer**. Also called **multifunction peripheral (2)**, or **all-in-one printer**. Its abbreviation is **MFP**.

multifunction product Same as **multifunction device**. Its abbreviation is **MFP**. Also spelled **multi-function product**.

multihandset cordless Same as **multi-handset cordless**.

multihop propagation Same as **multi-hop propagation**.

multihop transmission Same as **multi-hop transmission**.

multilanguage display Same as **multi-language display**.

multilayer Also spelled **multi-layer**. **1.** Pertaining to **multilayer PCBs**, or **multilayer circuits**. **2.** Pertaining to, or consisting of multiple layers.

multilayer board Same as **multilayer PCB**.

multilayer ceramic capacitor An MIM capacitor with multiple alternated metal-ceramic layers. Used in ICs. Also spelled **multi-layer ceramic capacitor**.

multilayer circuit Also spelled **multi-layer circuit**. **1.** Same as **multilayer PCB**. **2.** A circuit with multiple layers of interconnected components.

multilayer circuit board Same as **multilayer PCB**. Also spelled **multi-layer circuit board**.

multilayer metallization Same as **multilevel metallization**. Also spelled **multi-layer metallization**.

multilayer PCB Abbreviation of **multilayer p**rinted-**c**ircuit **b**oard. A printed-circuit board which has two or more layers of circuit patterns which are stacked vertically, each separated by an insulating medium. All layers are laminated together, each having holes which enable interconnection with other layers, helping provide even greater component density. Also spelled **multi-layer PCB**. Also called **multilayer circuit board**, **multilayer board**, or **multilayer circuit (1)**.

multilayer printed-circuit board Same as **multilayer PCB**.

multilevel address An address which points to the storage location of another address. The indicated address may also be another indirect address, or the address of the desired operand. Also called **indirect address**.

multilevel addressing An address mode utilizing **multilevel addresses**. Also called **indirect addressing**.

multilevel metallization Metallization in which there are two or more levels of metal, each separated by an insulator such as silicon dioxide. All incorporated components are interconnected via a highly conductive path. Used, for instance, in monolithic ICs. Also spelled **multi-level metallization**. Also called **multilayer metallization**.

Multilink Point-to-Point Protocol Same as **Multilink PPP**.

Multilink PPP Abbreviation of **Multilink P**oint-to-**P**oint Protocol. An extension of the Point-to-Point Protocol, which enables two or more channels to be linked, with their bandwidths combined. Used, for instance, for channel bonding. Its own abbreviations are **MPPP**, or **MLPPP**.

multimedia **1.** Information which is used or presented in a combination of forms, including full-motion video, animation, graphics, audio, and text. Nearly all computers serve for such presentations and uses, and common examples include watching movies on DVDs, accessing encyclopedic information via the Internet or a CD-ROM, or sending an email with an incorporated voice message. Also called **multimedia content**. **2.** Consisting of, pertaining to, or capable of providing **multimedia (1)**. For instance, a multimedia presentation, or a multimedia computer or cell phone.

multimedia computer **1.** A computer specially designed for the use and presentation of **multimedia**. Such a computer may incorporate a special chip or board, extra video and/or system RAM, and so on. **2.** A computer that can use and present multimedia. Currently, almost all computers can.

multimedia conferencing Conferencing which incorporates **multimedia**.

multimedia content Same as **multimedia (1)**.

multimedia extensions Same as **MMX**.

multimedia messaging The exchange of messages which contain **multimedia content**. Said especially when referring to messaging between devices such as properly equipped cell phones and PDAs. Its abbreviation is **MM**.

multimedia messaging service A service enabling **multimedia messaging**. Its abbreviation is **MMS**.

multimedia monitor A computer monitor which incorporates speakers, and which also may have a microphone and/or digital camera.

multimedia PC Abbreviation of **multimedia** personal computer. **1.** A personal computer specially designed for the use and presentation of **multimedia**. Such a computer may incorporate a special chip or board, extra video and/or system RAM, and so on. **2.** A personal computer that can use and present multimedia. Currently, almost all computers can. **3.** A computer that conforms to certain minimum hardware and software requirements for multimedia use and presentation. Such requirements include having a sound card, an optical drive, and video capabilities exceeding a certain level.

multimedia personal computer Same as **multimedia PC**.

multimedia system A single unit, or multiple components utilized as a unit, which can use and present **multimedia**. Most of the time this refers to a multimedia computer.

MultiMediaCard A flash memory card, usually about the size of a stamp, which is used in devices such as handheld PCs, PDAs, cell phones, digital cameras, and digital music players. Such cards feature high security, speedy data transfers, good storage capacity, and ruggedness. It is similar to an SD card, but is usually smaller. Its abbreviation is **MMC**.

multimeter An instrument which measures and indicates two or more quantities in an electric circuit, such as resistance or voltage. Such devices usually provide an easy manner, such as a switch, to change its function. An example of such an instrument is a volt-ohm-milliammeter. Also called **multifunction meter**, or **circuit analyzer**.

multimode Utilizing, pertaining to, or characteristic of multiple modes. For example, multimode fibers, or multimode distortion.

multimode dispersion A deprecated term for **multimode distortion**.

multimode distortion Within a multimode fiber, dispersion of light resulting from slight differences in the propagation velocity of the different modes. The core used for multimode transmission is generally of a greater diameter than for single mode, and the use of a core with a graded refractive index helps minimize this form of distortion. Also called **modal distortion**. Two deprecated terms for this concept are **multimode dispersion**, and **modal dispersion**.

multimode fiber An optical fiber designed to carry multiple modes simultaneously. Such fibers usually have a larger center core than **single-mode fibers**, and are generally used for comparatively shorter distances. Also spelled **multimode fiber**. Also called **multimode optical fiber**.

multimode operation Also spelled **multi-mode operation**. **1.** For a given circuit, device, piece of equipment, or system, the use of two or more of the available operational states that can be selected from. **2.** For a guided electromagnetic wave, the use of two or more of the possible manners or patterns of oscillation, transmission, or propagation. **3.** The carrying of multiple modes simultaneously via an optical fiber.

multimode optical fiber Same as **multimode fiber**. Also spelled **multi-mode optical fiber**.

multiparty line An arrangement in which the telephones of two or more end users are connected to the same telephone line or local loop. When utilized, a mechanism such as selective ringing may be employed to distinguish for which user an incoming call is meant. Party lines are saddled by privacy and congestion problems. Also called **party line**.

multipath The presence of multiple radio signals arriving simultaneously at the same location, each via a distinct path. The signals usually arrive from the same source, and two paths may result, for instance, if a direct ray and a reflected ray are received at once. Depending on the interaction of the waves, there may be constructive or destructive interference. Also spelled **multi-path**.

multipath cancellation The cancellation of a radio signal due to the presence of multiple signals arriving simultaneously at the same location, each via a different path, when there is enough destructive interference. Also spelled **multi-path cancellation**.

multipath delay The delay between radio signals arriving simultaneously at the same location, each via a different path. Also spelled **multi-path delay**.

multipath fading Fading due to the presence of multiple radio signals arriving simultaneously at the same location, each via a different path, when there is destructive interference. When such interference is complete, there is multi-path cancellation. Also spelled **multi-path fading**.

multipath interference Interference due to the presence of multiple radio signals arriving simultaneously at the same location, each via a different path. Such interference when listening to an FM station may result in static, or in ghosting when watching a TV broadcast. Also spelled **multi-path interference**.

multipath reception The reception of multiple radio signals arriving simultaneously at the same location, each via a different path. This may result, for instance, in multipath interference or multipath fading. Also spelled **multi-path reception**.

multipath transmission The transmission of signals which result in **multipath reception**. Also spelled **multi-path transmission**.

multiphase Consisting of, having, or using two or more phases. For instance, an AC power line with two or more electrical phases, or a relay which responds to the current such a line provides. Also called **polyphase**.

multipin connector A pin connector with two or more pins. Data-bus connectors, for instance range from 9 pins to over 50. Also spelled **multi-pin connector**.

multiplatform Computer software or hardware which is designed to work properly with more than one type of platform. An interpreter version of LISP is platform-independent, as is Java. This contrasts with **platform-dependent**, in which software or hardware is designed to work properly only with a specific platform. Also spelled **multi-platform**. Also called **multiple-platform**, **platform neutral**, **platform-independent**, **architecture-independent**, or **cross-platform**.

multiple access system 1. A communications system which uses a single channel to simultaneously sent multiple signals. Examples include CDMA, FDMA, and TDMA. 2. The ability of a system to receive and transmit data to and from multiple locations. 3. A system in which a single satellite is simultaneously used by two or more earth stations.

multiple crosstalk In communications, the unwanted intermingling of signals between transmission lines. Also called **babble**.

Multiple Document Interface A programming interface utilized to create applications that allow multiple documents to be opened, and worked on, simultaneously. Its abbreviation is **MDI**.

multiple instruction stream-multiple data stream A computer architecture in which multiple processors simultaneously and independently execute different instructions on different sets of data. Its abbreviation is **MIMD**. Also called **control parallel**.

multiple-loop feedback 1. A feedback mechanism in which two or more individual circuits or loops provide feedback. 2. In a multistage amplifier, feedback which is applied individually to each stage. Also called **local feedback**.

multiple modulation A series of modulation processes in which the modulated wave from one stage becomes the

modulating wave for the next. Also called **compound modulation**.

multiple-platform Same as **multiplatform**.

multiple precision Using two or more computer words to represent each number, so as to enhance precision. Also, characterized by such use or precision.

multiple precision arithmetic Arithmetic operations and processes utilizing **multiple precision**.

multiple recipients In communications, the capability to send a message, such as an email or fax, to more than one recipient. Also, the recipients of such messages.

multiple reflections Two or more reflections off surfaces, objects, or regions. For example, repeated reflections of light within a laser cavity. Also called **repeated reflections**.

multiple scattering Scattering of electromagnetic radiation, such as light, or particles, such as electrons, by repeated interactions with dust, objects, ions in a crystal lattice, or the like.

multiple track Same as **multitrack**.

multiple-unit steerable antenna An antenna whose multiple radiating elements can be electrically steered, so as to minimize fading. Its acronym is **MUSA**.

multiple-user system Same as **multiuser system**.

multiple winding A winding incorporating two or more sections that can be connected in parallel, depending on the specific needs.

multiplex 1. To combine two or more signals for transmission over a single channel. Also, the technique utilized. Also, to transmit such a signal. 2. To combine multiple low-speed transmissions into a single high-speed transmission. Also, the technique utilized.

multiplex operation The operation of a communications channel in which there is **multiplexing**.

multiplex stereo The use of a single signal to broadcast the left and right channels of a stereo broadcast. Also, the reception of such a signal.

multiplex system 1. A communications system in which two or more signals are combined for transmission over a single channel. 2. A communications system in which multiple low-speed transmissions are combined into a single high-speed transmission.

multiplexer Its abbreviation is **MUX**. Also spelled **multiplexor**. 1. A circuit or device which simultaneously transmits two or more signals which have been combined into a single channel. An optical multiplexer, for instance, combines multiple signals at different wavelengths. 2. A circuit or device which combines multiple low-speed transmissions into a single high-speed transmission, and vice versa. Also called **inverse multiplexer (2)**.

multiplexing Its abbreviation is **MUXing**. 1. The process of combining two or more signals for transmission over a single channel. Common examples include frequency-division multiplexing, and time-division multiplexing. 2. The process of combining multiple low-speed transmissions into a single high-speed transmission, and vice versa. Also called **inverse multiplexing (2)**.

multiplexor Same as **multiplexer**.

multiplicand A newly introduced number which multiplies another. A **multiplier (6)** is a number that is already present. For instance, in 13 x 3, the multiplier is 13, while the multiplicand is 3.

multiplier 1. A circuit or device which multiplies a number of pulses, cycles, signals, or other input quantity, by a fixed number, such as 2 or 10. For example, electron, frequency, or voltage multipliers. 2. An electron tube in which secondary electron emissions produce current amplification. In it, a cathode releases primary electrons, which are then reflected by multiple electron mirrors, or dynodes, each producing more and more electrons. Depending on the number of dynodes, the amplification factor may be in the millions. Also called **electron multiplier**, or **secondary electron multiplier**. **3.** A circuit or device whose output signal frequency is an exact integer multiple of its input signal frequency. For example, such a circuit could convert a 10 MHz signal into a 30 MHz signal with the use of a frequency tripler. Also called **frequency multiplier**. **4.** A component or circuit whose voltage output is a multiple of its input voltage. For example, a rectifier circuit which delivers a DC output voltage which is a multiple of the peak of its input AC voltage. Also called **voltage multiplier (1)**. **5.** A precision resistor utilized in series, to extend the voltage range of an instrument such as a voltmeter. Also called **multiplier resistor**, or **voltage-range multiplier (2)**. **6.** An already present number by which another is multiplied. A **multiplicand** is a newly introduced number which multiplies the multiplier. For instance, in 13 x 3, the multiplier is 13, while the multiplicand is 3.

multiplier-accumulator Computer circuitry which executes multiplications and additions extremely rapidly, usually within a single clock cycle. Used, for instance, in digital-signal processing. Its abbreviation is **MAC**.

multiplier phototube A phototube incorporating one or more dynodes. Electrons emitted from its photocathode are reflected by the dynodes, each producing more and more electrons. Depending on the number of dynodes, the amplification factor may be in the millions. Also called **multiplier tube**, **electron-multiplier phototube**, **photomultiplier tube**, or **photomultiplier**.

multiplier resistor Same as **multiplier (5)**.

multiplier tube Same as **multiplier phototube**.

multipoint channel Same as **multipoint circuit**. Also spelled **multi-point channel**.

multipoint circuit A circuit or communications channel that interconnects three or more stations or locations. The information transmitted may be accessed by all stations or locations simultaneously. Also spelled **multi-point circuit**. Also called **multipoint channel**.

multipoint conferencing A redundant equivalent for **teleconferencing**. Also spelled **multi-point conferencing**.

multipoint configuration A communications network configuration in which three or more stations or locations are connected to the same line. The usual arrangement provides for a control station and three or more secondary stations. Also spelled **multi-point configuration**.

multipoint connection A connection between three or more stations or locations. The information transmitted may be accessed by all stations or locations simultaneously. Also spelled **multi-point connection**.

Multipoint Control Unit A device which connects three or more terminals engaged in an audiovisual conference. Its abbreviation is **MCU**.

multipoint distribution service A radio service in which microwaves are utilized to broadcast from a single fixed station to multiple fixed stations, with line-of sight coverage. Also spelled **multi-point distribution service**. Its abbreviation is **MDS**.

multipoint line A communications line that interconnects three or more stations or locations. The information transmitted may be accessed by all stations or locations simultaneously. Also spelled **multi-point line**. Also called **multidrop line**.

multipoint network A communications network with a **multipoint configuration**. Also spelled **multi-point network**.

multipolar Pertaining to, or having more than one pair of magnetic poles.

multiport repeater In a communications network, a device which has multiple I/O ports, and which retransmits to all outputs any signal received at any input. Such a device may also perform regenerative functions. Also spelled **multiport repeater**.

multiposition relay Same as **multi-position relay**.

multiposition switch Same as **multi-position switch**.

multiprocessing Also spelled **multi-processing**. Also called **parallel processing, simultaneous processing**, or **concurrent execution (2)**. **1.** The simultaneous execution of multiple computer operations by multiple CPUs in a single computer. Such processors are usually linked by high-speed channels. **2.** The simultaneous execution of multiple computer operations by a single computer, whether it utilizes multiple processors or another mechanism. **3.** The utilization of multiple computers to perform multiple operations simultaneously. **4.** An architecture utilized for **multiprocessing (1)**, **multiprocessing (2)**, or **multiprocessing (3)**. Also called **multiprocessing architecture**.

multiprocessing architecture Same as **multiprocessing (4)**.

multiprocessor Pertaining to, utilizing, or incorporating two or more processors. Seen, for instance, in multiprocessing. Also spelled **multi-processor**.

multiprogramming Same as **multitasking**. Also spelled **multi-programming**.

Multiprotocol Label Switching An IETF specification for accelerating communications traffic flow by labeling each packet with forwarding information, eliminating routing-table lookups. Multiprotocol Label Switching works across various protocols, such as IP and ATM, and supports QoS. Also spelled **Multi-protocol Label Switching**. Its abbreviation is **MPLS**.

Multiprotocol Lambda Switching A protocol utilized to switch individual wavelengths into separate paths when routing multiple data streams to their destination along an optical fiber. Also spelled **Multi-protocol Lambda Switching**. Its abbreviation is **MPLS**.

Multiprotocol Over ATM Same as **MPOA**.

Multiprotocol Over Frame Relay Same as **MPOFR**.

multiprotocol router A router that supports two or more communications protocols. Also spelled **multi-protocol router**.

multipurpose instrument Same as **multi-purpose instrument**.

Multipurpose Internet Mail Extensions Same as **MIME**.

multipurpose meter Same as **multi-purpose meter**.

multirange instrument Same as **multi-range instrument**.

multirange meter Same as **multi-range meter**.

multiscan display Same as **multiscan monitor**.

multiscan monitor A monitor which accepts video signals from frequencies within a wide interval, automatically adapting itself to the signal frequency of the graphics card it is connected to. Such a monitor may support video standards ranging from monochrome to Super VGA. This contrasts with a **fixed-frequency monitor**, which only responds to a video signal within a single frequency range, and with a **multi-frequency monitor**, which supports a fixed number of frequency ranges. Also spelled **multi-scan monitor**. Also called **multiscanning monitor**, or **multiscan display**.

multiscanning monitor Same as **multiscan monitor**. Also spelled **multi-scanning monitor**.

multisensor data fusion The combining of data collected by multiple sensors, so as to derive a synergistic result. Used, for instance, in robotics, and in medical diagnosis. Also spelled **multi-sensor data fusion**. Also called **distributed sensing**.

Multiservice Access Platform A modular, scalable, and high-bandwidth platform which supports multiple communications mediums and protocols. Used, for instance, in central offices or customer premises to simultaneously handle voice and data, using POTS, copper, or optical digital networks.

multisession CD A method of writing to an optical disk, such as a CD-R, in which recording can occur in more than one session. Each session adds overhead, reducing overall capacity. Also, a disc so recorded. This contrasts, for instance, with **Disc-at-Once**, in which all data is transferred continuously without interruptions. Also spelled **multi-session CD**.

multistable Having two or more stable states. Also spelled **multi-stable**.

multistage Having two or more stages. For instance, a multistage amplifier. Also spelled **multi-stage**.

multistage amplifier An amplifier with two or more stages arranged in a manner that the output of one serves as the input for the next, while amplifying at each step. Also spelled **multi-stage amplifier**. Also called **cascade amplifier**.

multistation Same as **multi-station**.

Multistation Access Unit A hub or concentrator that connects computers in a token-ring LAN. Its physical configuration has a star topology, yet its logical configuration has a ring topology, as each message passes through each computer, in a circular manner. One benefit provided by this arrangement is that a failing node can be bypassed without the ring being broken. Its abbreviation is **MSAU**.

multiswitch Same as **multi-switch**.

multitasking The simultaneous execution of two or more programs by a single CPU. Multitasking may be preemptive, in which all tasks take turns at having the attention of the CPU, or cooperative, where the foreground task allows background tasks access to the CPU at given times, such as when it is idle. Either way, it appears to the user as if all programs are executed simultaneously. Currently, most computers are capable of multitasking, and factors such as increased processor speed and additional RAM better equip a system for such use. Also spelled **multi-tasking**. Also called **multiprogramming**.

multitester An instrument, such as a multimeter, which performs two or more test functions. Also spelled **multi-tester**.

multithreaded application Also spelled **multi-threaded application**. **1.** An application which supports **multithreading**. **2.** An application in which multithreading is occurring.

multithreading The simultaneous execution of two or more tasks or processes within a single program. In this context, each task or process is called a **thread**. To occur, multithreading must be supported by the operating system. Also spelled **multi-threading**. Also called **threading**.

multitrack Consisting of, utilizing, or pertaining to two or more tracks, as in multitrack recording. Also spelled **multi-track**. Also called **multiple track**.

multitrack recording Also spelled **multi-track recording**. **1.** A process in which multiple individual sound sources are recorded, each on a single track. All tracks are synchronously locked in time with each other. This provides the flexibility of being able to record and rerecord any given tracks without affecting the rest. **2.** A specific recording made utilizing **multitrack recording** (1).

multitrack recording system A system utilized for **multitrack recording**. Also spelled **multi-track recording system**.

multiturn potentiometer A precision potentiometer in which multiple turns of the control knob are necessary to move the contact arm from one end of the helically-wound resistance element to the other. Also called **helical potentiometer**.

multiuser Pertaining to, consisting of, involving, or able to serve two or more users. For instance, a multiuser system. Also spelled **multi-user**.

multiuser system Also spelled **multi-user system**. Also called **multiple-user system**. **1.** A computer system that can be used by two or more users. Such a computer may be set to load the specific settings of each of its users. **2.** A computer system that can be accessed by two or more users, each of which is located at a different terminal.

multivibrator A relaxation oscillator incorporating two active devices, such as tubes or transistors, which are arranged in a manner that the input of one is derived from the output of the other, and which are coupled in a manner that one is conducting while the other is not. There are various types, including bistable multivibrators, monostable multivibrators, and astable multivibrators.

mumetal Same as **mu-metal**.

MUMPS Abbreviation of **M**assachusetts General Hospital **U**tility **M**ulti**p**rogramming **S**ystem. A high-level programming language used extensively in the health-care business. Its newer name, which still has not spread much, is **M**.

Munsell color system Same as **Munsell system**.

Munsell system A system which enables the precise specification and comparison of colors, based on their hue, chroma, and value. Also called **Munsell color system**.

Muntz metal An alloy containing approximately 60% copper, 39% zinc, and small amounts of lead and iron. It is a type of brass, and corrodes less than copper.

muon A charged lepton whose mass is about 207 times that of an electron.

MUSA Acronym for **multiple-unit steerable antenna**.

muscovite A mineral which is a member of the mica group, and which is used, for instance, for electrical and heat insulation.

music CD A CD containing music, speech, or other audio. A music CD can usually hold up to 74 minutes of recorded sound. Also called **audio CD**, or **digital audio disc**.

music CD player A device which can read the information stored on a **music CD**. To recover the recorded content, a laser is focused on the reflecting surface of the disc, and the reflected light is modulated by the code on the disc. Computer DVD and CD-ROM players can also serve as CD players when a music CD is placed in the drive. Also called **audio CD player**.

music digitization The conversion of an analog music source, such as a voice and a guitar, into digital form, such as that found in a CD. The larger the sample of the analog source, and the higher the sampling rate, the more accurate the conversion.

music on hold Music, or radio or TV broadcast programming which is played when a caller is on hold. The more proper term for prerecorded messages, advertisements, or the like, is **message on hold**.

music power A rating of the output of a power amplifier. It refers to the power output during a peak or other short interval, and any value so obtained can be considered as inflated in comparison to the more accurate and meaningful RMS value.

music synthesizer A device which electronically generates or reproduces sounds that emulate musical instruments and voices, and which can be utilized to compose, play, and record music. A keyboard, real or virtual, is usually used, and two common techniques for generating sounds are FM synthesis and wavetable synthesis. Also called **synthesizer** (3).

Musical Instrument Digital Interface Same as **MIDI**.

mute **1.** A circuit or device which reduces the volume of, or silences sounds. Also, to use such a circuit or device. For example, by pressing the mute button of the remote control of a receiver, the output sound is either significantly reduced or silenced. **2.** A circuit which disables a given sound input or output. For instance, when using a transceiver, the pressing of a button to silence the receiver when transmitting. Also, to use such a circuit or device. **3.** A circuit which automatically disables a given sound input or output. For example, a squelch circuit. Also, the operation of such a circuit.

mute button A button used to activate and deactivate a **mute (1)** or **mute (2)**.

mute control A button, switch, or other control which serves to **mute**.

mute function A function, such as that found in a TV, receiver, or DVD, which serves to **mute**. Also called **muting function**.

mute switch A switch that automatically or manually **mutes**. Also called **muting switch**.

muting **1.** The reducing of the volume of, or silencing of sounds. **2.** The manual or automatic disabling of a given input or output.

muting circuit **1.** A circuit utilized for **muting (1)**. Also called **squelch circuit (2)**. **2.** A circuit utilized for **muting (2)**. Also called **squelch circuit (3)**.

muting function Same as **mute function**.

muting switch Same as **mute switch**.

mutual capacitance Within a cable, the capacitance between two conductors when all the other conductors are connected together and grounded.

mutual conductance Also called **transconductance**. **1.** For an amplifying device, such as a transistor or an electron tube, the ratio of the change in output current to the change in input current. For example, in a tube, it is the ratio of the change in plate current to the change in grid voltage, with the plate voltage held constant. Usually expressed in multiples of siemens or mhos. **2.** For a component, circuit, or device, the ratio of the output current to the input current.

mutual impedance For an AC circuit or network, the ratio of the output voltage to the input current. Also called **transimpedance**.

mutual inductance Its symbol is M. **1.** The extent to which a varying current in a circuit induces an emf in a neighboring circuit. **2.** Same as **mutual induction**.

mutual induction The property of a circuit in which a varying current induces an emf in a neighboring circuit. Its symbol is M, and when quantifying this inductance some multiple of henrys, such as millihenrys or microhenrys, is used. When

the emf is induced in the same circuit it is called **self-induction**. Also called **mutual inductance (2)**.

mutual interference **1.** Interference caused by the interaction of two or more components, channels, circuits, devices, pieces of equipment, or systems, and which affects all entities involved to a given extent. **2. Mutual interference (1)** affecting communications components, channels, circuits, devices, pieces of equipment, or systems.

MUX Abbreviation of **multiplexer**.

muxing Abbreviation of **multiplexing**.

MUXing Abbreviation of **multiplexing**.

MV **1.** Abbreviation of **megavolt**. **2.** Abbreviation of **medium voltage**.

mV Abbreviation of **millivolt**.

mV/m Abbreviation of **millivolts per meter**.

MVA Abbreviation of **megavolt-ampere**.

Mvar Abbreviation of **megavar**.

Mvarh Abbreviation of **megavar-hour**.

MW **1.** Abbreviation of **megawatt**. **2.** Abbreviation of **medium waves**.

mW Abbreviation of **milliwatt**.

mW-h Abbreviation of **milliwatt-hour**.

mW-hr Abbreviation of **milliwatt-hour**.

MWh Abbreviation of **megawatt-hour**.

MWhr Abbreviation of **megawatt-hour**.

Mx Symbol for **maxwell**.

Mycalex A material consisting of a powdered mica bonded to glass, which provides a insulator which can work over a broad temperature range.

mylar A strong, resistant, and thin polyester film utilized as a substrate for magnetic recording media, such as tapes and diskettes. It is also used as an insulator, as a dielectric in capacitors, and in making protective bags and sleeves. Also called **mylar film**.

mylar capacitor A capacitor in which the dielectric is **mylar**.

mylar film Same as **mylar**.

mylar tape Magnetic recording tape in which the substrate is **mylar**.

MySQL A popular open-source relational database.

mμ Abbreviation of **millimicron**.

mΩ Abbreviation of **milliohm**.

MΩ Abbreviation of **megohm**.

N

n 1. Abbreviation of **nano-**. 2. Symbol for **index of refraction**, or **refractive index**.

N 1. Chemical symbol for **nitrogen**. 2. Symbol for **newton**. 3. Abbreviation of **negative**. 4. Symbol for **refractivity**.

N-8-1 A common modem setting in which there is **no** parity bit, there are **8** data bits, and **1** stop bit. Also called **8-N-1**.

N/A Also spelled **NA**. 1. Abbreviation of **not-applicable**. 2. Abbreviation of **not-available**.

N-AMPS Same as **NAMPS**.

n-channel In a field-effect transistor, a conducting channel between the source and the drain formed by an n-type semiconductor.

n-channel FET Abbreviation of **n-channel** field-effect transistor. A field-effect transistor with an n-channel, as opposed to a p-channel.

n-channel field-effect transistor Same as **n-channel FET**.

n-channel JFET Abbreviation of **n-channel** junction field-effect transistor. A JFET with an n-channel, as opposed to a p-channel.

n-channel junction FET Same as **n-channel JFET**.

n-channel junction field-effect transistor Same as **n-channel JFET**.

n-channel metal-oxide semiconductor Same as **NMOS**.

n-channel metal-oxide semiconductor field-effect transistor Same as **NMOSFET**.

n-channel MOS Same as **NMOS**.

n-channel MOSFET Same as **NMOSFET**.

N display In radars, an oscilloscopic display similar to a K display, but which incorporates a pedestal which is moved along the baseline until it coincides with the horizontal position of the blip, so as to indicate the range of the scanned object. Also called **N scope**, **N scanner**, **N scan**, or **N indicator**.

N electron An electron located in the N shell of an atom.

N indicator Same as **N display**.

n-layer Same as **n-region**.

n-region A layer consisting of an **n-type semiconductor**. Also called **n-type layer**, **n-type region**, or **n-layer**.

N scan Same as **N display**.

N scanner Same as **N display**.

N scope Same as **N display**.

N shell The fourth electron shell of an atom. It is the fourth closest shell to the nucleus.

n-type 1. Same as **n-type semiconductor**. 2. Pertaining to an n-type semiconductor, or that which incorporates such a material.

n-type conduction In a semiconductor, conduction via electrons. In **p-type conduction** mobile holes are the charge carriers.

n-type conductivity In a semiconductor, the ease with which **n-type conduction** occurs.

n-type layer Same as **n-region**.

n-type material Same as **n-type semiconductor**.

n-type region Same as **n-region**.

n-type semiconductor A semiconductor material which has the addition of pentavalent dopants, such as arsenic or antimony, which contribute free electrons which increase the conductivity of the intrinsic semiconductor. In this manner, the concentration of conduction-band electrons is much higher, making electrons the majority carrier. This contrasts

with a **p-type semiconductor**, in which holes are the majority carrier. Also called **n-type material**, or **n-type (1)**.

N64 A popular gaming system.

NA 1. Abbreviation of **numerical aperture** 2. Same as **N/A**.

nA Abbreviation of **nanoampere**.

Na Chemical symbol for **sodium**.

N_A Symbol for **Avogadro's number**.

nag screen A screen displayed on a computer monitor which reminds a user to register, or pay, when opening, utilizing, or closing a shareware application. The term also refers to sold products providing such a screen reminding purchasers to register.

nagware A type of shareware which displays a large screen, or other prompt, reminding a user to register, or pay, when opening or closing the application. The term also refers to sold products providing such reminders for registration.

NAK Abbreviation of **negative acknowledge character**.

name resolution The conversion of the name of a user, node, or other resource, to its network address. Used, for instance, in a naming service.

Named Pipes An Interprocess Communication feature supported by various platforms.

namespace A set that determines which names given entities, concepts, objects, devices, or the like may have, so as to provide each with a unique designator, and avoid ambiguities and conflicts. An example is an XML namespace.

naming service In a communications network, a service that ties the name of a user, node, or other resource, to its physical address. For example, such a service converts a user name into an IP number, so as to transmit data.

NAMPS Abbreviation of **n**arrowband **AMPS**. A system similar to AMPS, but which adds digital signaling for added functionality. NAMPS occupies a narrower bandwidth per call in relation to AMPS, thus providing greater calling capacity, but has an increased possibility of interference.

NAND Abbreviation of **Not AND**. A logical operation which is true if any of its elements are false. For example, if all of its multiple inputs have a value of 1, then the output is 0. Any other combination yields an output of 1. For such functions, a 1 is considered as a true, or high, value, and 0 is a false, or low, value. Also called **NAND operation**.

NAND circuit Same as **NAND gate**.

NAND gate A circuit which has two or more inputs, and whose output is high if one or more of these inputs is false. Also called **NAND circuit**.

NAND operation Same as **NAND**.

nanny software Same as **Net Nanny**.

nano- A metric prefix representing 10^{-9}, or one billionth. For instance, 1 nanoampere is equal to 10^{-9}, or one billionth of an ampere. Its abbreviation is **n**.

nano-ohm Same as **nanohm**.

nano-ohmmeter An ohmmeter designed to perform accurate measurements of resistances in the nanohm range.

nanoampere 10^{-9} ampere. It is a unit of measurement of electric current. Its abbreviation is **nA**.

nanobar 10^{-9} bar. It is a unit of pressure. Its abbreviation is **nb**.

nanobot A robot whose dimensions are on a nanometer scale, and which is usually built utilizing nanotechnology.

nanocircuit An electronic circuit whose dimensions are on a nanometer scale.

nanocoulomb 10^{-9} coulomb. It is a unit of electric charge. Its abbreviation is **nC**.

nanocurie 10^{-9} curie. It is a unit of radioactivity. Its abbreviation is **nCi**.

nanoelectronics The technology that deals with the design, construction, and applications of electronic circuits and devices whose overall dimensions or building blocks are on a nanometer scale. Such blocks, for instance, may be designed to self-assemble into circuits and devices, in a manner analogous to the way molecules in cells naturally form structural units.

nanofabrication The fabrication of components, circuits, devices, equipment, systems, mechanisms, and objects whose dimensions are on a nanometer scale. Also, the techniques employed.

nanofarad 10^{-9} farad. It is a unit of capacitance. Its abbreviation is **nF**.

nanogauss 10^{-9} gauss. It is a unit of magnetic flux density, or magnetic induction. Its abbreviation is **nG**, or **nGs**.

nanogram 10^{-9} gram. It is a unit of mass. Its abbreviation is **ng**.

nanohenry 10^{-9} henry. It is a unit of inductance. Its abbreviation is **nH**.

nanohm 10^{-9} ohm. It is a unit of resistance, impedance, or reactance. Its abbreviation is **nΩ**. Also spelled **nanoohm**, or **nano-ohm**.

nanojoule 10^{-9} joule. It is a unit of energy or work. Its abbreviation is **nJ**.

nanolambert 10^{-9} lambert. It is a unit of luminance. Its abbreviation is **nL**.

nanoliter 10^{-9} liter. It is a unit of volume. Its abbreviation is **nl**, or **nL**.

nanolux 10^{-9} lux. It is a unit of illuminance, or illumination. Its abbreviation is **nlx**.

nanomachine **1.** A machine whose dimensions are on a nanometer scale. **2.** A machine whose components are on a nanometer scale.

nanometer 10^{-9} meter. It is a unit of measurement of length. Its abbreviation is **nm**.

nanometer scale Same as **nanoscale**.

nanomol Same as **nanomole**.

nanomole 10^{-9} mol. It is a unit of amount of substance. Its abbreviation is **nmol**. Also called **nanomol**.

nanonewton 10^{-9} newton. It is a unit of force. Its abbreviation is **nN**.

nanoohm Same as **nanohm**.

nanorad 10^{-9} rad. It is a unit of radiation dose. Its abbreviation is **nrad**.

nanoradian 10^{-9} radian. It is a unit of angular measure. Its abbreviation is **nrad**.

nanorem 10^{-9} rem. It is a unit of radiation dose. Its abbreviation is **nrem**.

nanoroentgen 10^{-9} roentgen. It is a unit of the ability of radiation to ionize. Its abbreviation is **nR**.

nanosat Same as **nanosatellite**.

nanosatellite A small artificial satellite which can maintain its position relative to another satellite within a few nanometers or less. Such a satellite, for instance, may weigh less than 1 kilogram. Its abbreviation is **nanosat**.

nanoscale A measurement scale whose dimensions are comparatively near a nanometer. Such an interval may be defined, for instance, as spanning from about 0.01 to 100 nm, from about 0.1 to 10 nm, from 1 to 100 nm, and so on. For reference, the smallest atom, that of hydrogen, has an approximate width of 0.1 nanometer. Also, pertaining to that consisting of, or incorporating such dimensions. Also called **nanometer scale**.

nanoscience The science that deals with the study of structures, materials, properties, and phenomena on a nanometer scale, and the design, use, effects, and improvement of components and devices of such dimensions.

nanosecond 10^{-9} second. It is a unit of time measurement. Its abbreviation is **ns**, or **nsec**.

nanosiemens 10^{-9} siemens. It is a unit of conductance. Its abbreviation is **nS**.

nanotechnology The technology that deals with the design, construction, and applications of components, circuits, devices, equipment, systems, mechanisms, and objects on a nanometer scale. An example is nanoelectronics.

nanotesla 10^{-9} tesla. It is a unit of magnetic flux density. Its abbreviation is **nT**.

nanotransistor A transistor whose dimensions are on a nanometer scale.

nanotube Fullerene that is arranged into a cylindrical shape. It is many times stronger and much lighter than steel, and depending on various factors may be a semiconductor, conductor, or superconductor. Used, for instance, as conducting paths between circuit components on a nanometer scale. Also called **buckytube**.

nanovolt 10^{-9} volt. It is a unit of measurement of potential difference. Its abbreviation is **nV**.

nanovoltmeter A voltmeter designed to provide accurate indications of voltage values in the nanovolt range.

nanovolts per meter 10^{-9} volt/meter. A measure of the intensity of the signal of a radio transmitter. It may be calculated, for instance, by dividing the intensity in nanovolts at the receiving antenna, by the height of said receiving antenna. It is also used as a measure of the electromagnetic radiation emitted by microwave ovens, computer displays, celestial bodies, and so on. Its abbreviation is **nV/m**.

nanowatt 10^{-9} watt. It is a unit of measurement of power. Its abbreviation is **nW**.

NANP Abbreviation of **North America Numbering Plan**.

NAP Abbreviation of **Network Access Point**.

Naperian base Same as **Napierian base**.

Naperian logarithm Same as **natural logarithm**. Also spelled **Napierian logarithm**.

Napierian base A transcendental number equal to approximately 2.71828, and which is the base for natural logarithms. Also spelled **Naperian base**. Also known as **e (1)**.

Napierian logarithm Same as **natural logarithm**. Also spelled **Naperian logarithm**.

Napster A popular site for music in MP3 format.

narrow-band Also spelled **narrowband**. **1.** Operating at, or encompassing a narrow range of frequencies. **2.** Operating at, or encompassing a narrower range of frequencies than is ordinarily available. **3.** Communications at a transmission speed lower than a determined amount. For example, below 1.544 Mbps. **4.** A subdivision of a given frequency band.

narrow-band amplifier An amplifier that operates with an essentially flat frequency response only over a narrow range of frequencies. Also spelled **narrowband amplifier**.

narrow-band AMPS Same as **narrowband AMPS**.

narrow-band antenna An antenna designed to work properly only over a narrow range of frequencies. Also spelled **narrowband antenna**.

narrow-band channel Same as **narrowband channel**.

narrow-band communications Same as **narrowband communications**.

narrow-band FM Same as **narrowband frequency modulation**. Also spelled **narrowband FM**.

narrow-band frequency modulation Same as **narrowband frequency modulation**.

narrow-band Integrated Services Digital Network Same as **narrowband ISDN**. Also spelled **narrowband Integrated Services Digital Network**.

narrow-band interference Same as **narrowband interference**.

narrow-band ISDN Same as **narrowband ISDN**.

narrow-band network Same as **narrowband network**.

narrow-band noise Same as **narrowband noise**.

narrow-band signal Same as **narrowband signal**.

narrow-band transmission Same as **narrowband transmission**.

narrow bandwidth 1. Pertaining to, or encompassing a narrow range of frequencies. 2. Pertaining to, or encompassing a narrower range of frequencies than is ordinarily available. 3. Pertaining to communications at a transmission speed lower than a determined amount. For example, below 1.544 Mbps.

narrowband Same as **narrow-band**.

narrowband amplifier Same as **narrow-band amplifier**.

narrowband AMPS Same as **NAMPS**. Also spelled **narrow-band AMPS**.

narrowband antenna Same as **narrow-band antenna**.

narrowband channel Also spelled **narrow-band channel**. 1. A communication channel that operates over a narrow range of frequencies, or over a narrower range of frequencies than is ordinarily available. 2. A communication channel that can carry only one message at a time, or less than a certain number of messages simultaneously.

narrowband communications Also spelled **narrow-band communications**. 1. Communications utilizing a narrow range of frequencies, or a narrower range of frequencies than is ordinarily available. 2. Communications in which only a single message can be transmitted at a time, or in which less than a certain number of messages can be transmitted simultaneously. 3. Communications at a speed of transmission less than a determined amount. For example, below 1.544 Mbps.

narrowband FM Abbreviation of **narrowband frequency modulation**. Also spelled **narrow-band FM**.

narrowband frequency modulation FM in which the frequency deviation of the carrier wave is below a given maximum. Also spelled **narrow-band frequency modulation**. Its abbreviation is **narrowband FM**, or **NBFM**.

narrowband Integrated Services Digital Network Same as **narrowband ISDN**. Also spelled **narrow-band Integrated Services Digital Network**.

narrowband interference Interference occurring over a narrow range of frequencies. Also spelled **narrow-band interference**.

narrowband ISDN Abbreviation of **narrowband** Integrated Services Digital Network. Any version of ISDN whose transmission speed is below 155 Mbps. For example, Primary Rate Interface. Also spelled **narrow-band ISDN**.

narrowband network A network whose communication channels can carry only one message at a time, or less than a certain number of messages simultaneously. Also spelled **narrow-band network**.

narrowband noise Noise present over a narrow range of frequencies. May refer to electrical noise, thermal noise, radio noise, and so on. Also spelled **narrow-band noise**.

narrowband signal A signal encompassing a narrow range of frequencies. For instance, such a signal sent over a narrowband network. Also spelled **narrow-band signal**.

narrowband transmission A transmission encompassing a narrow range of frequencies. For instance, a voice transmission. Also spelled **narrow-band transmission**.

narrowcast 1. A broadcast to a specified or limited number of recipients. Also, to transmit such signals. 2. In a communications network, the simultaneous transmission of a single message to a specified or limited number of recipients. Also, to transmit such messages.

NAS Abbreviation of **network attached storage**.

NAT Abbreviation of **network address translation**.

National Bureau of Standards Former name of the **National Institute of Standards and Technology**. Its abbreviation is **NBS**.

National Center for Supercomputer Applications Same as **NCSA**.

National Center for Supercomputer Applications server Same as **NCSA server**.

National Center for Supercomputer Applications Telnet Same as **NCSA Telnet**.

national channel A broadcast channel offering **national programs**.

National Electrical Code A United States standard that pertains to the installation and maintenance of electrical wires, cables, fixtures, and related equipment, mainly from the perspective of fire-prevention. Its acronym is **NEC**.

National Electrical Manufacturers Association A United States association that promotes the standardization of electrical components, devices, equipment, and systems. In addition to other functions, it provides analysis of data pertaining to this industry. Its acronym is **NEMA**.

National Institute of Standards and Technology A United States government entity which provides scientific, measurement, and technological standards. Its acronym is **NIST**. Formerly called **National Bureau of Standards**.

national program A broadcast utilizing two or more stations within a large geographic area, such as a country.

national reception The reception of a **national program**.

National Television Standards Committee Same as **NTSC**.

National Television Standards Committee signal Same as **NTSC signal**.

National Television Standards Committee Standard Same as **NTSC Standard**.

national transmission The transmission of a **national program**.

native capacity Also called **raw capacity**. 1. In a disk, disc, or tape, the storage capacity prior to formatting. Data, such as control information, reduces this capacity. Also called **unformatted capacity**. 2. In a disk, disc, or tape, the storage capacity prior to compression. Compression frequently increases this amount 50% or more.

native file format Same as **native format**.

native format The default format an application saves files in. Programs usually have filters to export and import data to and from applications using different native formats. The latter are called **foreign formats**. Also called **native file format**.

natural electricity Any electricity which occurs naturally. For instance, brain waves, nerve impulses, stunning charges by certain fish, or lightning discharges.

natural frequency 1. The frequency at which oscillation of a electrical and/or mechanical system occurs without an external driving force. 2. For a given component, circuit, device, system, or antenna, the lowest resonant frequency. 3. For a given antenna, the lowest resonant frequency without the addition of any inductance or capacitance.

natural interference Any interference which occurs naturally. Atmospherics, for example, are a source of such interference. This contrasts with **artificial interference**, such as that produced through the use of transmitters, ignition systems, motors, and so on.

natural language A language, such as Spanish or German, which evolves naturally over an extended period, as opposed to an **artificial language**, such as machine language, which has been developed for specific needs.

natural language comprehension Same as **natural language processing**.

natural language interface An interface which allows a user to enter information or interact with a computer system via a **natural language**, as occurs, for instance, in a natural language query.

natural language processing The conversion of a **natural language** input into digital form, so that computers may understand and respond to it. Used, for instance, to convert words, phrases, and sentences spoken into a microphone into computer actions, or text which can be displayed, processed, saved, or printed. The term also applies to such processing of a written natural language text which is scanned. Its abbreviation is **NL processing**, or **NLP**. Also called **natural language recognition**, **natural language understanding**, or **natural language comprehension**.

natural language query The use of a question phrased in the manner it would be addressed to another speaker of a **natural language** to obtain information, such as information in a database or help with an application, from a computer. Also called **natural language question**.

natural language question Same as **natural language query**.

natural language recognition Same as **natural language processing**.

natural language support Software and/or hardware which is capable of **natural language processing**, and thus can accept a natural language input.

natural language understanding Same as **natural language processing**.

natural logarithm A logarithm whose base is **e**, or approximately 2.71828. Its symbol is **ln**. Thus, $ln\ x = log_e\ x$. Also called **Napierian logarithm**, or **hyperbolic logarithm**.

natural logarithm function The inverse function of the **exponential function**. That is, if $y = e^x$, then the natural logarithm function is $x = lny$, which is the same as $x = log_e\ y$.

natural magnet A material, such as magnetite, which naturally exhibits the properties of a permanent magnet.

natural noise Any noise, such as radio-frequency noise, which occurs naturally. An example is cosmic noise. This contrasts with **artificial noise**, such as electrical noise.

natural period For a system oscillating at its **natural frequency** (1), the period of an oscillation.

natural radiation Any radiation which occurs naturally. Examples include that present in the radioactive isotopes of naturally-occurring atoms, such as carbon-14, or that arising from cosmic rays.

natural radioactivity The spontaneous disintegration of naturally-occurring unstable atomic nuclei, such as that of uranium-238, mainly through the emission of alpha, beta, or gamma rays.

natural resonance 1. Resonance which occurs in a body or system when no external driving force is present. 2. Resonance in which the period of oscillation of the external driving force is the same as the natural period of a given body or system. Also called **periodic resonance**.

natural resonance frequency Same as **natural resonant frequency**.

natural resonant frequency The frequency at which a body or system resonates without an external driving force. Also called **natural resonance frequency**, **resonant frequency** (2), or **self-resonant frequency**.

natural wavelength 1. The wavelength corresponding to the **natural frequency**. 2. The wavelength corresponding to the **natural resonance frequency**.

nav bar Same as **navigation bar**.

NAVAIDS Abbreviation of **navigational aids**.

navbar Acronym for **navigation bar**.

navigation aid Same as **navigational aid**.

navigation bar Its acronym is **navbar**. 1. Within a Web page, a bar containing a set of icons or buttons which facilitate linking to topics, or otherwise moving around. It is an index or table of contents which usually remains visible while a user navigates different sections of a Web site or related sites. 2. Within a Web browser or Web page, a bar containing a set of icons or buttons which facilitate stopping loading, moving forward or back, linking to topics, going to a home page, printing, and so on.

navigation beacon A beacon which serves as a **navigational aid**. Also spelled **navigational beacon**.

navigation instrument An instrument which serves as a **navigational aid**.

navigation satellite An artificial satellite used in a navigation system. For instance, one of the 24 satellites which comprise the GPS System.

Navigation Satellite Timing and Ranging Same as **NAVSTAR**.

navigation system An electronic system, such as GPS or loran, which serves as a **navigational aid**. Also spelled **navigational system**.

navigational aid An electrical or electronic device, instrument, or system which provides navigational information such as the exact current location, heading, speed, or height, or that otherwise guides, orients, or warns. Examples include beacons, distance-measuring equipment, navigation systems such as loran or electra, and the GPS. In a more general sense, this term would also include items such as charts and landmarks. Also spelled **navigation aid**.

navigational beacon Same as **navigation beacon**.

navigational system Same as **navigation system**.

Navigator A popular Web browser.

NAVSTAR Acronym for **Na**vigation **S**atellite **T**iming **a**nd **R**anging. A constellation of 24 artificial satellites which orbit at an altitude of approximately 20,000 meters, and which emit signals that are utilized for precise navigation anywhere on the planet earth. Each satellite is equipped with an atomic clock, and continuously sends a time and identification signal. A receiving device obtaining said signals from multiple satellites is thus provided with measurements of its latitude, longitude, and elevation, in addition to the exact time. Depending on the number of signals received, positioning accuracy may be within a meter. This system may also be utilized to keep track of a fleet of cars, a subway train system, and so on. Also called **NAVSTAR GPS**, or **GPS** (1).

NAVSTAR GPS Same as **NAVSTAR**.

nb Abbreviation of **nanobar**.

Nb Chemical symbol for **niobium**.

NBFM Same as **narrowband frequency modulation**.

NBS Abbreviation of **National Bureau of Standards**.

nC Abbreviation of **nanocoulomb**.

NC 1. Abbreviation of **numerical control**. 2. Abbreviation of **network computer**. 3. Abbreviation of **normally closed**. 4. Abbreviation of **no circuit**. 5. Abbreviation of **no connection**.

NC contact Abbreviation of **normally-closed contact**.

NCC Abbreviation of **network-centric computing**.

nCi Abbreviation of **nanocurie**.

NCP 1. Abbreviation of **network control program**. 2. Abbreviation of **network control protocol**.

NCSA Abbreviation of National Center for Supercomputer Applications. A center for high-performance computing and networking that provides supercomputer resources, and which develops applications, such as Mosaic.

NCSA server Abbreviation of National Center for Supercomputer Applications **server**. An HTTP server developed by the **NCSA**. It was one of the first tailored to the World Wide Web.

NCSA Telnet Abbreviation of National Center for Supercomputer Applications **Telnet**. A Telenet application developed and distributed by the **NCSA**.

Nd Chemical symbol for **neodymium**.

NDB Abbreviation of **non-directional radio beacon**.

NDMP Abbreviation of **Network Data Management Protocol**.

NDR Same as **non-destructive readout**.

NDRO Same as **non-destructive readout**.

Ne Chemical symbol for **neon**.

near-end crosstalk Crosstalk which is propagated between circuits in which the interfering signal travels in the direction opposite that of the desired signal. The terminal where near-end crosstalk is observed is usually near or at its origination point. This contrasts with **far-end crosstalk**, where each signal travels in the same direction, and is usually observed at a point distant from the where it originates. Its abbreviation is **NEXT**.

near field 1. Same as **near-field region**. 2. The electromagnetic field closer than a given distance from a source of electromagnetic radiation. 3. The sound field close to a sound source. May be, for instance, that within two wavelengths.

near-field recording A data-storage method combining features of magneto-optical recording and magnetic hard disk technologies, to create and utilize extremely narrow bit cells, which in turn provide enhanced capacity and performance. Its abbreviation is **NFR**.

near-field region For a transmitting antenna, the electromagnetic field extending no further than a distance of $2D^2/\lambda$, where D is the maximum overall dimension of said antenna, and λ is the wavelength considered. Beyond this distance, the **far-field region** begins. Also called **near field (1)**, **near region**, or **near zone**

near infrared Its abbreviation is **near IR**. 1. Pertaining to **near-infrared radiation**. 2. Same as **near-infrared radiation**.

near-infrared radiation Electromagnetic radiation in the infrared region which is nearest to visible light. It corresponds to wavelengths of approximately 0.75 to 3 micrometers, though the defined interval varies. Also called **near infrared (2)**.

near-instantaneous companded audio multiplex Same as **NICAM**.

near-instantaneously companded audio multiplex Same as **NICAM**.

near IR Abbreviation of **near infrared**.

near-letter quality Printed characters which are dark, crisp, and well-defined, but not to the extent exhibited in **letter quality**. Also, a printing mode with such an output. Most laser printers can print in this mode. Its abbreviation is **NLQ**.

near online Same as **nearline**.

near region Same as **near-field region**.

near ultraviolet Its abbreviation is **near UV**. 1. Pertaining to **near-ultraviolet radiation**. 2. Same as **near-ultraviolet radiation**.

near-ultraviolet radiation Electromagnetic radiation in the ultraviolet region which is nearest to visible light. It corresponds to wavelengths of approximately 300 to 400 nm, though the defined interval varies. Also called **near ultraviolet**, or **soft ultraviolet**.

near UV Abbreviation of **near ultraviolet**.

Near Video on Demand Cable or satellite programming, especially movies, offered at staggered times. For example, there may be a selection of 30 feature films, each of which starts again every 10 or 15 minutes. Its abbreviation is **NVOD**.

near zone Same as **near-field region**.

nearline Abbreviation of **near online**. Not quite immediately available or accessible. Said, for instance, of data archived in CD-Rs.

NEC Acronym for **National Electrical Code**.

necessary bandwidth For a given type or class of emission, the minimum frequency interval necessary to provide a specified level of quality at a stated transmission speed. Signals outside of this band may be suppressed to help ensure adequate performance.

needle 1. A slender and pointed object, or that resembling such an object. For instance, the shape of the magnetizable particles on the surface a magnetic tape or disk is that of a needle. 2. A **needle (1)** which serves to convey information in a meter, gauge, or similar device or instrument. A specific example is that used in a D'Arsonval galvanometer. Also called **needle pointer**, **indicator needle**, or **pointer needle**. 3. A pointed object which extends from a phonographic pickup, and which serves to follow the undulations of the grooves of a phonographic disc and transmit them as vibrations to said pickup. Such an object is usually made of a metal needle with a diamond or sapphire tip. Also called **stylus (3)**. 4. A needle in a magnetic compass. Also called **magnetic needle**.

needle electrode A slender and pointed electrode, such as that used in electrocautery.

needle pointer Same as **needle (2)**.

needle probe A probe with a slender and pointed tip, such as that used in electrocautery.

Néel temperature For a material exhibiting ferrimagnetism, the temperature below which its magnetic properties are similar to ferromagnetic materials. Above this temperature, such a material becomes paramagnetic.

neg Abbreviation of **negative**.

negation Same as **NOT**.

negative Its abbreviation is **neg**, or **N**. Its symbol is –. 1. Of, or pertaining to, a quantity or value below zero. For instance, a temperature below 0 °C, or a negative number. 2. Of, pertaining to, or the same as the charge or a multiple of the charge of an electron. For example, the minus one charge of a chlorine ion. 3. Of, or pertaining to, a particle, material, or other entity which has an excess of electrons. For instance, an anion. 4. Of, or pertaining to, a quantity, value, entity, phenomenon, or concept which is considered the opposite of an equivalent which would be described as positive. For example, a negative electrode, negative image, negative peak, negative pole, and so on. 5. Same as **negative image**.

negative acceleration Slowing down, or the rate at which velocity decreases with respect to time. Also called **deceleration**.

negative acknowledge character In communications, a message indicating the inability to receive a transmission, or that a transmission has been received incorrectly. This message is sent from the receiving unit or station to the sending unit or station as a means of verification. Its abbreviation is **NAK**.

negative angle An angle which is measured in the clockwise direction from a horizontal axis extending from the origin in a Cartesian coordinate system.

negative bias **1.** A voltage applied to an electrode of an electronic device, such as an electron tube, which makes it negative with respect to another electrode. For example, a voltage that makes the control grid negative with respect to the cathode. **2.** A negative voltage applied to a component or device, especially an electronic device, to establish a reference level for its operation.

negative charge **1.** The charge, or a multiple of the charge, of an electron. Electrons possess a negative charge which is equal in magnitude to the positive charge of a proton. **2.** The charge a material or entity acquires when it has an excess of electrons. For example, if a dielectric such as glass or amber is rubbed with a dissimilar dielectric such as silk or wool, and each is separated, the silk or wool acquires a negative charge, while the glass or amber acquires a positive charge. Also called **negative electrification**, or **negative electricity**.

negative conductor A conductor connected to a **negative terminal**.

negative differential resistance Same as **negative resistance**.

negative edge triggered A device actuated by **negative edge triggering**.

negative edge triggering For a device, such as a flip flop, with edge triggering, actuation when the edge of its clock pulse is falling. For instance, a device which is set to sample data only when the edge of its synchronizing clock pulse is falling. This contrasts with **positive edge triggering**, in which the converse is true.

negative electricity Same as **negative charge (2)**.

negative electrification Same as **negative charge (2)**.

negative electrode Also called **cathode**. **1.** The electrode which is the source of electrons in an electron tube. These electrons travel towards the anode. Its symbol is **K**. **2.** In an electrolytic cell, the electrode towards which positive ions travel. **3.** The electrode where reduction occurs in an electrochemical cell.

negative exponent An exponent which is a negative number, as in 5^{-3}. The resulting value is the reciprocal of the number taken to the positive power. In this example, $5 \times 5 \times 5 = 125$, so 5^{-3} equals $1/125$. Also called **negative power**.

negative feedback In an amplifier or system, feedback which is 180° out-of-phase with the input signal, thus opposing it. While this results in a decrease in gain, it also serves reduce distortion and noise, and to stabilize amplification. Also called **inverse feedback**, **reverse feedback**, **stabilized feedback**, or **degeneration**.

negative-feedback amplifier An amplifier utilizing **negative feedback**.

negative glow In a glow-discharge tube, a luminous region between the cathode dark space and the Faraday dark space.

negative-going Pertaining to a **negative-going pulse** or a **negative-going signal**.

negative-going pulse A pulse whose value or amplitude is increasing in the negative direction.

negative-going signal A signal whose value or amplitude is increasing in the negative direction.

negative grid In an electron tube, a grid which is negative with respect to the cathode.

negative ground A DC electrical system in which the negative terminal is grounded. This is the usual case.

negative grounding The connection to a **negative ground**.

negative half-cycle The half of a complete sequence of changes of a periodically repeated phenomenon during which a minimum is approached. For instance, that of an AC cycle. Also, the time that elapses during such an interval.

negative image An image, such as a photographic image, in which the light and dark areas are reversed in relation to the scene the image is taken from. Also called **negative (5)**, **photonegative (3)**, or **reverse image**.

negative impedance An impedance whose real number component is a **negative resistance**.

negative ion Also called **anion**. A negatively charged ion. Negative ions collect at the anode when subjected to an electric potential while in solution. Chloride ion and hydroxide ion are two common anions.

negative-ion generator A device which generates **negative ions**. Used, for instance, in some electronic air-cleaning devices.

negative lead A conductor connected to a **negative terminal**.

negative lens A lens which causes parallel rays of light passing through it to bend away from one another. Such lenses are thinner at the center than the edges. This contrasts with a **positive lens**, which bends parallel rays of light passing through it toward one another. Also called **diverging lens**, or **concave lens**.

negative line A line, such as a wire, connected to a **negative terminal**.

negative logic Logic in which the high level represents the logic 0 state, and the low level represents the logic 1 state. In **positive logic** the converse is true. Also called **negative true logic**.

negative magnetostriction Magnetostriction in which a material contracts. In **positive magnetostriction** a material expands.

negative modulation **1.** Modulation in which an increase in luminous intensity results in a decrease in carrier modulation or transmission power. In such a case, the carrier signal is greatest when there is no modulation. Used, for instance, in some TV and fax systems. **2.** Same as **negative transmission**.

negative peak Also called **negative peak value**. **1.** The maximum negative value of a voltage, current, or other quantity. **2.** The maximum negative value of the displacement from a reference position, such as zero, for a voltage, current, or other quantity. **2.** The **negative peak (1)** or **negative peak (2)** for a given time interval.

negative peak value Same as **negative peak**.

negative phase sequence In a polyphase system, such as a three-phase AC circuit, the phase sequence opposite that of a **positive phase sequence**. Also called **negative sequence**.

negative photoresist A photoresist that polymerizes or otherwise turns harder when exposed to light. This contrasts with a **positive photoresist**, which depolymerizes or otherwise turns softer when exposed to light.

negative plate In a cell or battery, the plate that is connected to the **negative terminal**. During discharge, electrons flow from this plate, through the external circuit, and into the positive plate. Conversely, in the case of a secondary cell or battery that is charging, the negative plate is connected to the positive terminal.

negative polarity **1.** The property of having an excess of electrons. A deficiency of electrons produces a **positive polarity (1)**. **2.** The polarity of a **negative pole**.

negative pole **1.** In a magnet or magnetic body, one of the two regions where the magnetic lines of force converge, and hence magnetic intensity is at a maximum. The other region is the **positive pole (1)**. Such a pole, when part of a magnet or magnetic body which is freely suspended, will seek a positive pole, as opposite poles attract. Also called **south pole (3)**. **2.** The location in the southern hemisphere where the magnetic dip is 90°. The corresponding spot in the northern hemisphere is the **positive pole (2)**. At each of these places, the magnetic meridians converge. The south magnetic pole does not coincide with the south geographic pole, and the position of both magnetic poles varies over time. Also called **south pole (2)**, **south magnetic pole**, or **magnetic south pole**. **3.** Same as **negative terminal**.

negative potential 1. A potential which is less than a given reference, such as ground. **2.** A potential measured at a negative electrode.

negative power Same as **negative exponent**.

negative proton The antiparticle equivalent of the proton. It has the same mass as the proton, but opposite electric charge and magnetic moment. Also called **antiproton**.

negative resistance For a given component or device, a region within a characteristic curve in which an increase in voltage results in a decrease in current. Negative resistance is observed, for instance, in tunnel diodes, Gunn diodes, and IMPATT diodes. Also called **negative-resistance region**, or **negative differential resistance**.

negative-resistance amplifier An amplifier which exhibits **negative resistance**.

negative-resistance component Same as **negative-resistance element**.

negative-resistance device A device, such as a negative-resistance oscillator, which exhibits **negative resistance**.

negative-resistance diode A diode, such as a tunnel diode, which exhibits **negative resistance**.

negative-resistance element A circuit element, such as a negative-resistance diode, which exhibits **negative resistance**. Also called **negative-resistance component**.

negative-resistance magnetron A magnetron which exhibits **negative resistance**.

negative-resistance oscillator An oscillator which exhibits **negative resistance**.

negative-resistance region Same as **negative resistance**.

negative sequence Same as **negative phase sequence**.

negative temperature coefficient Its abbreviation is **NTC**. **1.** For a given component, device, material, or system, a decrease in a magnitude or property, such as length or resistance, which occurs as a consequence of an increase in temperature. **2.** For a given component, device, or material, a decrease in electrical resistance which occurs as a consequence of an increase in temperature. **3.** A number which quantifies the relationship in **negative temperature coefficient (1)**, or **negative temperature coefficient (2)**.

negative terminal In a source of electrical energy, such as a battery, cell, or generator, the terminal from which electrons flow. The electromotive force provides the electric pressure for electrons to flow from this terminal, through the external circuit, and to the positive terminal. Also called **negative pole (3)**.

negative transmission Transmission of a signal utilizing **negative modulation (1)**. Also called **negative modulation (2)**.

negative true logic Same as **negative logic**.

negative voltage 1. A voltage which is less than a given reference, such as ground. **2.** A voltage measured at a negative electrode.

negatron A negatively-charged electron. The term is utilized to differentiate such a particle from a **positron**, or positively-charged electron.

NEMA Acronym for **National Electrical Manufacturers Association**.

nematic-crystal display A passive-matrix liquid-crystal display in which **nematic liquid crystals** are twisted 90° or more. The glass sheets of the display contain the nematic-crystal material in a manner that they are twisted unless an electric field is applied. When such a field is applied selectively, the affected crystals untwist and become opaque, thus blocking the polarized light passing through, which in turn produces darker pixels. Such displays provide higher contrast and a better viewing angle than ordinary passive LCD screens, which is especially useful when used under highly-lighted circumstances. Used extensively in portable computers. Also known as **twisted nematic display**.

nematic crystals Same as **nematic liquid crystals**.

nematic liquid crystals Twisted and elongated crystals which align themselves longitudinally, but which are otherwise randomly positioned. Nematic liquid crystals have order in one dimension, while **smectic liquid crystals** have order in two. Used, for instance, in nematic-crystal displays. Also called **nematic crystals**, or **twisted nematic liquid crystals**.

neodymium A silver-yellow soft metallic chemical element whose atomic number is 60. It has about 30 known isotopes, of which 5 are stable. Its applications in electronics include its use in lenses and coloring glasses, permanent magnets, and in lasers. Its chemical symbol is **Nd**.

neodymium oxide A blue gray powder whose chemical formula is Nd_2O_3. It is used in ceramic capacitors, coloring glasses, color TV tubes, and in electrodes.

neodymium-YAG laser Abbreviation of **neodymium-yttrium-aluminum-garnet laser**. A YAG laser in which the yttrium-aluminum-garnet crystals are doped with neodymium. Used, for instance, in laser surgery.

neodymium-yttrium-aluminum-garnet laser Same as **neodymium-YAG laser**.

neon A colorless, odorless, and tasteless noble gas whose atomic number is 10. It has over a dozen known isotopes, of which 3 are stable, and ionizes in gas tubes producing a reddish-orange glow. It applications in electronics include its use in lasers, gas tubes and bulbs, as a high-voltage indicator, in Geiger counters, and as a cryogenic refrigerant. Its chemical symbol is **Ne**.

neon bulb Same as **neon lamp**.

neon glow lamp Same as **neon lamp**.

neon indicator An indicator light, such as that utilized to inform about a given condition in a circuit or device, consisting of a **neon lamp**. Also called **neon indicator light**, or **neon indicator lamp**.

neon indicator lamp Same as **neon indicator**.

neon indicator light Same as **neon indicator**.

neon lamp Also called **neon bulb, neon glow lamp, neon tube**, or **neon light (1)**. **1.** A glow lamp in which the inert gas that becomes ionized by an electric discharge is neon. Such lamps consume little power and emit an orange-red light. Used, for instance, to indicate the status of a circuit or device, or for decorative lighting. **2.** A lamp similar to a **neon lamp (1)**, but which uses other gases to obtain other desired colors.

neon light 1. Same as **neon lamp**. **2.** The light produced by a **neon lamp**.

neon tube Same as **neon lamp**.

neoprene A synthetic rubber which is rugged, water-resistant, and especially impervious to oil. Use, for instance, for highly weather-resistant cable jackets. Also called **polychloroprene**.

NEP Abbreviation of **noise equivalent power**.

neper A dimensionless logarithmic unit utilized to express the ratio of two powers or intensities, such as voltages, currents, or signal levels. Nepers are similar to decibels, except that the former is based on natural logarithms, and the latter on common logarithms. One neper is equal to approximately 8.6858 decibels. Its abbreviation is **Np**.

neptunium A radioactive transuranic chemical element whose atomic number is 93, and which has over 20 identified isotopes, all unstable. It has three allotropic forms, and is used in neutron-detection instruments. Its chemical symbol is **Np**.

nerd A person who is inordinately oriented towards and/or dedicated to technology, especially that pertaining to computers and networks. The term usually implies a given level of expertise. Also called **geek**.

Nernst effect A phenomenon observed when a strip of a metal conductor is placed perpendicular to a magnetic field. When said strip is conducting heat longitudinally, a small voltage differential appears at opposite ends of the strip. When this effect occurs in certain crystals, also called **Nernst-Ettinghausen effect**.

Nernst-Ettinghausen effect The **Nernst effect**, as exhibited by certain crystals, such as those that are piezoelectric.

Nernst glower An incandescent lamp in which a rod, usually composed of zirconium oxide combined with rare-earth oxides such as yttrium oxide, is heated to emit a brilliant light. Used, for instance, in infrared spectroscopy. Also called **Nernst lamp**.

Nernst lamp Same as **Nernst glower**.

nest In computers, to incorporate a structure or entity into another similar structure or entity. For example, a loop within another loop, a procedure within another procedure, or a record within another record.

nested interrupt An interrupt which is incorporated into another interrupt.

nested loop A loop which is incorporated into another loop.

nested procedure A procedure which is incorporated into another procedure.

nested record A record which is incorporated into another record.

nested task A task which is incorporated into another task.

nested transaction A transaction which is incorporated into another transaction. Also called **subtransaction**.

nesting The process of incorporating a structure or entity into another similar structure or entity. For example, a loop within another loop, a procedure within another procedure, or a record within another record.

.net On the Internet, a top-level domain name suffix. The **net** is an abbreviation of **net**work, although the entity employing this suffix need not be a network provider. Also called **dot net**.

net 1. Abbreviation of **network**. 2. Abbreviation of **Internet**. 3. A final amount or value that remains after making any necessary adjustments. For example, a net current or a net voltage.

net address Abbreviation of **Internet address**.

net appliance Abbreviation of **Internet appliance**.

net broadcast Abbreviation of **Internet broadcast**.

net broadcasting Abbreviation of **Internet broadcasting**.

net browser Abbreviation of **Internet browser**.

net café Abbreviation of **Internet café**.

net capacitance The resulting capacitance when multiple capacitors are incorporated in a circuit. For example, the net capacitance of two capacitors connected in parallel is the sum of their respective capacitances.

net filter Abbreviation of **Internet filter**.

net gain The overall gain of a component, circuit, device, system, or transmission line, after all attenuation is accounted for. In the case of an amplifier, for instance, it can be calculated as the ratio of the input power to the output power. A ratio of greater than 1 results in a net gain, while a ratio of less than one is a **net loss**.

net loss The overall loss of a component, circuit, device, system, or transmission line, after all attenuation is accounted for.

Net Nanny Software intended to prevent children from viewing or interacting with Web sites whose content, such as pornography, is deemed inappropriate or otherwise undesirable. A Net Nanny may block access to given Web sites and/or specific content. Also called **nanny software**.

Net PC 1. Abbreviation of **Net**work **PC**. A simple and inexpensive PC that is designed specifically for network use, and which is typically configured as a thin client. Net PCs usually do not have floppy or optical drives, nor expansion slots, so as to prevent a user from adding anything that is not available via the server. 2. Abbreviation of **Internet PC**.

net phone Abbreviation of **Internet telephone**.

net power The overall power of a component, circuit, device, system, or transmission line, after all attenuation is accounted for.

net surfing Abbreviation of **Internet surfing**.

net telephone Abbreviation of **Internet telephone**.

net telephony Abbreviation of **Internet telephony**.

net TV Abbreviation of **Internet TV**.

NetBeans An open-source Java-based integrated development environment. Also called **NetBeans IDE**.

NetBeans IDE Same as **NetBeans**.

NetBEUI Abbreviation of **NetBIOS** Enhanced User Interface. An enhanced version of **NetBIOS** working at the transport layer.

NetBIOS Abbreviation of **Net**work **B**asic **I**nput/**O**utput **S**ystem. A network API or protocol working at the session layer within the OSI Reference Model.

NetBIOS Enhanced User Interface Same as **NetBEUI**.

NetBSD A popular open-source UNIX operating system which runs on multiple platforms.

NetBus A Trojan horse which takes advantage of security deficiencies of Windows operating systems to install server functions in a computer to be hacked, and which enables a remote system to have complete control of the targeted computer when it is connected to the network.

NetCam A video camera primarily intended for the transmission of images over the Internet. Such a camera may be utilized, for instance, for videoconferencing, in chat programs with video capabilities, for sending video email, or for monitoring a given area or environment. Also called **Webcam**, or **videocam**.

netcast Abbreviation of **Internet broadcast**.

netiquette Acronym for **net**work **etiquette**. A set of rules of etiquette which should be observed by users of communications networks such as LANs or the Internet. For example, using ALL CAPITALS in an email or chat is equivalent to shouting.

netizen A user of a network, especially the Internet. It is an acronym for **net**work **citizen**.

netlist A description, listing, or computer file detailing the electrical elements in a circuit, and the way each is interconnected with each other. A netlist helps in the transition from a logical circuit design to a physical circuit design.

netmask Abbreviation of **network mask**.

NetMeeting A popular Internet conferencing program.

NetNews The news content within **Usenet**.

NetObjects Fusion A popular Web authoring tool.

NetPC Same as **Net PC**.

NetWare A popular LAN operating system which runs across multiple platforms.

network Its abbreviation is **net**. 1. A system of computers, transmission channels, and related resources which are interconnected to exchange information. A communications network may be comparatively small, in which case it can be a LAN, or relatively large, in which case it could be a WAN. A LAN may be confined, for instance to a single building, while a WAN may cover an entire country. The communications channels in a network may be temporary or permanent. Also called **communications network**, or **telecommunications network**. **2.** One or more electric circuits incorporating two or more interconnected electrical elements or components, such as resistors, capacitors, coils or generators. There are various ways to classify electric networks. For instance, if a network incorporates active devices, such as amplifiers, it is an active network, while a passive network does not. A bilateral network is one which functions equally well in both directions, while a unilateral network does not. There are many examples of specific networks, and these include bridge, crossover, decoupling, and resistance-capacitance networks. Also called **electric network**. **3.** A network of stations connected in a manner that they operate as a group. For instance, a radio network, TV network, radar network, or the like. Also called **chain (2)**. **4.** The entity producing or emitting that which is broadcast throughout a **network (3)**, such as a radio network. **5.** Any interconnected group or structure, especially when complex.

network access Entry or connection to a communications network. For instance, accessing the Internet or an intranet.

Network Access Point A facility where Internet service providers interconnect. Such a location determines handle an enormous amount of data traffic, and determine how it is routed, and represents a key constituent of the Internet backbone. Its abbreviation is **NAP**. Also called **Metropolitan Area Exchange**, or **Internet Exchange**.

network-access server A computer that utilizes network-emulation software to connect asynchronous devices to a LAN or WAN. It manages communications, and takes care of protocol conversions. Also called **access server, remote-access server**, or **communications server**.

network adapter Same as **network card**.

network adapter card Same as **network card**.

network adapter card driver Same as **network card driver**.

network adapter driver Same as **network card driver**.

network address An address, such as an IP number, which uniquely identifies a node within a communications network. Also called **node address**.

network address translation A standard which enables multiple computers in a LAN, or other private network, to share a single IP address. This allows for the better use of a limited number of available IP addresses, and helps protect specific IP addresses in the LAN from outside networks. Its abbreviation is **NAT**.

network administration Same as **network management**.

network administrator The person responsible for administrative tasks, such as supervision of activity, security, upgrades, expansion, and maintenance, of a communications network.

network analysis **1.** The careful study of an electric network to determine its properties, such as the values of its components, and its overall performance, including factors such as reliability. **2.** The careful study of a communications network to determine its properties, such as impedances and transmission power, and its overall performance, including factors such as signal strength under various conditions.

network analyzer **1.** Software and/or hardware which monitors a communications network, and which troubleshoots and performs tests such as the simulation of error conditions, so as to help it work smoothly. Also called **protocol analyzer**. **2.** A device or instrument which performs **network analysis**. Used, for instance, to simulate networks and systems.

network architecture The design of a network and its components. This includes the hardware, software, and protocols. Also called **architecture (3)**.

network attached storage One or more mass-storage units dedicated to sharing files across a communications network. Such units may be connected to a server, or to other network resources. Its abbreviation is **NAS**.

network authentication **1.** A system which is employed to authenticate users, transmissions, applications, or the like, which utilize communications networks. For example, Kerberos serves to authenticate users. **2.** The process of authenticating users, transmissions, applications, or the like, which utilize communications networks.

Network Basic Input/Output System Same as **NetBIOS**.

Network BIOS Same as **NetBIOS**.

network bridge A device which connects networks or segments of networks at the data-link layer, utilizing the same communications protocols. When connecting LANs, however, such bridges are protocol-independent. Also called **bridge (2)**.

network browser **1.** A program which enables a computer user to browse, and perform other functions, such as downloading files and exchanging email, through a network. **2.** Same as **Internet browser**.

network bus A common cable, or wire, that connects all nodes in a communications network. Also called **bus (2)**.

network card An expansion card that enables a computer or other device, such as a printer, to access a network. Such a card may provide a network connection via fiber-optic cables, coaxial cables, infrared radiation, and so on. Also called **network interface card, network adapter**, or **network adapter card**.

network card driver A network device driver which controls the function of a **network card**. Also called **network interface card driver, network adapter driver**, or **network adapter card driver**.

network-centric computing Computing which makes extensive use of, or totally depends on a communications network. Such computing usually consists of a group, often very large, of network computers whose processing, applications, and storage are provided by one or more network servers. The benefits of such an arrangement include the need to only install or upgrade software in one, or a few computers, along with centralized administration. Its abbreviation is **NCC**.

network computer Its abbreviation is **NC**. **1.** Any computer within a communications network. **2.** A computer designed specifically or exclusively for use within a communications network. Such a computer, for instance, may have limited processing power and memory, a reduced operating system, and little or no storage capability. It instead relies on network servers for greater processing power, access to applications, and data storage. When such a computer is used solely for Internet access, it is called **Internet box**.

network constant A value, such as a capacitance, resistance, or inductance, in an electric circuit.

network control program A program which takes care of tasks related to the administration of a communications network, including traffic management, error control, polling, security, and allocation of resources. Its abbreviation is **NCP**.

network control protocol The part of the Point-to-Point Protocol concerned with the network-layer aspects of a data transmission. Its abbreviation is **NCP**.

Network Data Management Protocol An open standard protocol for backing-up networked-based data. Its abbreviation is **NDMP**.

network database 1. Any database utilized within a network. **2.** A database containing information pertaining to a network, such as the addresses of users. **3.** A database whose organization is based on multiple connections which link records. It is more flexible than a hierarchical database, and unlike a relational database, requires an established index.

network device In a communications network, a device or peripheral, such as a printer, which is accessed by multiple nodes.

network device driver A program which enables the operating system of a computer to communicate with, and control, a network card.

network directory A directory of the users, groups, computers, resources, or the like, of a communications network. Such a directory is usually contained in a file, and facilitates, for instance, the communication between nodes and the utilization of resources without knowledge of where or how they are connected.

network drive In a communications network, a disk drive which is accessed by multiple nodes. This contrasts with a **local drive**, which is located at a specific user's node, or a **remote drive**, which is that of a computer other than that of a given user, but is not necessarily used by multiple nodes. Also called **networked drive**.

network etiquette Same as **netiquette**.

network file system Its abbreviation is **NFS**. **1.** A protocol which enables files to be shared within a communications network. Computers utilizing such a protocol access files from a network or remote drive as if from a local drive. **2.** A file system utilized or distributed over a network

network filter A filter circuit used in an electric network. Used, for instance, to pass a given interval of frequencies with little or no attenuation, while rejecting all frequencies outside this range.

network ID Abbreviation of **network id**entification. Within an IP address, the part that is not the host ID. For instance, if in the IP address 151.201.4.111, **151.201** identify the network, then **4.111** are the host ID. The network ID can encompass the first, first and second, or first, second, and third sets of numbers separated by the periods. The rest correspond to the host ID.

network identification Same as **network ID**.

network information center An organization that provides information and support services to users of a communications network. Such an entity may also provide administrative services. An example is InterNIC. Its abbreviation is **NIC**.

network information service A service that provides information to all users of a communications network. Such information may be informative, useful, indispensable, and so on. Its abbreviation is **NIS**.

network interface 1. A device, such as a network card, which serves to connect a computer or other device to a communications network. **2.** Any interface with a network, such as that between a user's equipment and a telephone network.

network interface card Same as **network card**. Its abbreviation is **NIC**.

network interface card driver Same as **network card driver**. Its abbreviation is **NIC driver**.

network interface device A unit which serves as an interface between computers, or other devices such as telephones, and a communications network. Such a unit, for instance, may perform functions such as protocol conversions and buffering. An example is a network card. Its abbreviation is **NID**. Also called **network interface unit**.

network interface unit Same as **network interface device**. Its abbreviation is **NIU**.

network latency The time interval required for a signal or data to travel from one communications network node to another. For instance, such latency occurs when a packet is stored, analyzed, and forwarded. Also called **latency (3)**.

network layer Within the OSI Reference Model for the implementation of communications protocols, the third lowest level, located directly above the data-link layer. This layer deals with tasks pertaining to establishing the route the data will take, and ensuring that said data is delivered. It provides specifications for addressing, the creation of virtual circuits, switching, routing, and the like. Also called **layer 3**.

network management The processes and techniques utilized to monitor and control a communications network, including its configuration, allocation of resources, error management, and security. Also, the day-to-day management of such a network utilizing these processes and techniques. Also called **network administration**.

network manager Software and/or hardware which monitors and controls a network, including its configuration, allocation of resources, error management, and security.

network mask A bit combination which identifies which portion of an IP address corresponds to the network or subnetwork, and blocks out the rest. For example, a network may use the same values in the first three address fields of a four field address such as WWW.XXX.YYY.ZZZ, and block out, or mask, all but the ZZZ portion, since it is the only one that will vary. Its abbreviation is **netmask**. Also called **subnetwork mask**, or **address mask**.

network meltdown The complete halt of a communications network. The term usually entails a condition of excess traffic, although such a occurrence may also be caused by hardware or software. Also called **meltdown (2)**.

network model 1. A specific structure, layout, implementation, or the like, of one or more aspects of a communications network. For instance, the OSI Reference Model. **2.** A specific topography or architecture of a network.

network modem A modem shared by multiple users in a network.

Network News Transfer Protocol A protocol utilized to post, distribute, and retrieve Usenet content. Its abbreviation is **NNTP**.

network node Same as **node (3)**.

Network of Workstations A network of clustered systems interconnected via high-speed channels, and which are utilized for parallel computing. Its acronym is **NOW**.

network operating system An operating system which enables two or more independent computers to act as a single system within a communications network. In the context of server-based networks, such an operating system handles tasks such as centralized administration and security, allocation of resources, error handling, and the orderly handling of multiple simultaneous requests by multiple nodes. An example is NetWare. Its acronym is **NOS**, and its abbreviation is **network OS**.

network operations center A central location within a communications network where tasks such as maintenance and improvements are supervised. Its acronym is **NOC**.

network OS Same as **network operating system**.

network parameters The operating values, such as resistances and capacitances, of the components, such as resistors and capacitors, of a given electric network.

Network PC Same as **Net PC (1)**.

Network Personal Computer Same as **Net PC (1)**.

network printer In a communications network, a printer which is accessed by multiple nodes.

network protocol A protocol utilized by a communications network, or by any of its layers.

network provisioning Also called **provisioning**. **1.** The setting up of a communications network. This may include the allocation of hardware, such as multiplexers and interface units, the selection of the mediums of transmission, such as coaxial cables and optical fibers, and establishing the connections between nodes. Alternatively, it may involve simply programming a computer to make the appropriate arrangements within an established network. **2. Network provisioning** (1) for a specific user or entity.

network-ready Hardware and/or software which is designed for use within a communications network.

network router In a communications network, or multiple interconnected networks, a device or software which determines where packets, messages, or other signals travel to next. A network router, using resources such as header information, algorithms, and router tables, establishes the best available path from its source to destination. Within the OSI Reference Model, a router operates at the network layer. Also called **router**, or **gateway** (2).

network security Measures taken to help prevent the unauthorized access to a communications network, its directories, files, and so on. Also, the prevention of interception or alteration of any data sent across it. The term may also include the protection of network hardware.

network segment A portion of a network which is utilized by a given group, such as a department. A bridge is usually utilized to connect these segments.

network server Also called **server**. **1.** Within a communications network, a computer whose hardware and/or software resources are shared by other computers. Servers, among other functions, control access to the network and manage network resources. There are various types of servers, including application servers, file servers, network access servers, and Web servers. **2.** Within a network with a client/server architecture, a computer and/or program which responds to requests made by clients.

Network Service Provider Its abbreviation is **NSP**. **1.** An entity which provides ISPs access to an Internet backbone. **2.** An entity which offers services, such as access to resources, available from a network.

network software **1.** Programs which are used in a communications network. **2.** Programs which are designed specifically, or exclusively, for use in networks.

network structure The way in which a communications network, a part of a network, or an aspect of a network is structured. For instance, a network architecture or topography, or the manner in which protocols are implemented.

network switch A device in a communications network that selects a path or circuit via which data will flow to its next destination. A switch usually involves a simpler and faster mechanism than a router, but may also have router functions. Also called **switch** (4).

network synthesis The derivation of the components, their values, and their arrangement within an electric circuit, so as to synthesize a circuit which provides a given output signal in response to a given input signal.

network terminal **1.** A computer input/output device which incorporates a video adapter, monitor, keyboard, and usually a mouse. Used in networks. Also called **computer terminal** (1), **terminal** (1), **console** (3), or **station** (6). When such a terminal has no processing capability it is called **dumb terminal**, while a terminal that incorporates a CPU and memory does have processing capability, and is called **intelligent terminal**. **2.** A personal computer or workstation which is linked to a network. Also called **computer terminal** (2), **terminal** (2), or **station** (7).

network terminator 1 Same as **NT1**.

network terminator 2 Same as **NT2**.

Network Time Protocol Its abbreviation is **NTP**. **1.** A protocol utilized to synchronize the clocks of computers and devices connected via the Internet. An extremely accurate source, such as International Atomic Time is usually used. Synchronizations of better than one nanosecond may be achieved among multiple machines, depending on the source, network conditions, and the equipment itself. **2.** A protocol utilized to synchronize the clocks of computers and devices within a communications network, using an extremely accurate source.

network-to-network interface An interface, such as a network bridge, between communications networks. Its abbreviation is **NNI**.

network topology For a communications network, the arrangement of the nodes and their interconnections. Common topologies include bus, ring, and star. Network topologies may be physical or logical. Also called **topology** (1).

network traffic management The measures taken to monitor and control the traffic of a communications network. Used, for example, to avoid situations of excessive congestion. Widely utilized, for instance, in ATM. Also called **traffic management**.

network traffic measurement In the analysis of communications lines or networks, a measure of the maximum traffic that can be handled. For instance, call hours may serve to quantify telephone communications traffic.

networked drive Same as **network drive**.

Neumann's law A quantitative expression of Faraday's law of electromagnetic induction.

neural network **1.** A network composed of many processing modules that are interconnected by elements which are programmable and that store information, so that any number of them may work together. Its goal is to mimic the action of biologic neural networks, displaying skills such as the ability to solve problems, find patterns and learn. Also called **artificial neural network**. **2.** An interconnected system of nerves.

neuristor Acronym for **neuron transistor**.

neuroelectricity Electricity generated by, or present, in the nervous system of an organism.

neuroengineering The science dealing with the study and technological applications of nerves, related structures, and nerve-based entities such as neural networks. An example is the attempt to reverse-engineer the brain to better understand its functions.

neuron transistor Its acronym is **neuristor**. **1.** A semiconductor device which simulates the functions and behaviors of a nerve cell. **2.** A semiconductor device incorporating a nerve cell.

neutral **1.** That which has no electric charge. For example, an atom or molecule with an equal number of protons and electrons. **2.** That which has no voltage. For instance, a circuit at ground potential. **3.** Same as **neutral conductor**. **4.** That which is neither acid nor basic. For instance, a solution whose pH is 7.0. **5.** That which does not depend on a specific architecture or configuration. For example, a platform-independent computer application.

neutral architecture Hardware or software which can work with more than one type of architecture. This provides greater flexibility, but may sacrifice performance. Also called **platform-neutral architecture**.

neutral bus A common connection for multiple neutral conductors. For example, the common connection between the neutral wire from a power company, the neutral conductors in a home, and a ground electrode.

neutral conductor Also called **neutral** (1), **neutral line**, or **neutral wire**. **1.** A conductor connected to one or more neutral points of a circuit. **2.** In a two-wire circuit, the conductor which completes the mains path, and which is grounded at the supply end. **3.** In a polyphase circuit or single-phase three-wire circuit, a conductor whose potential and phase are meant to be intermediate between that of the other conductors.

neutral-density filter An optical filter that decreases the intensity of light without altering its relative spectral distribution. In the case of visible light, it reduces its intensity without changing its color. Also called **gray filter**.

neutral ground A ground connection made to one or more neutral points of a circuit, device, piece of equipment, or system.

neutral line Same as **neutral conductor**.

neutral point **1.** A point in a circuit which is at zero potential. **2.** A point in a circuit which is grounded, and as such is considered to be at zero potential.

neutral wire Same as **neutral conductor**.

neutral zone For a device or system, such as an amplifier or control system, a range of values within which the input signal may be varied without affecting the output. For example, an interval of values within which a measuring instrument will not respond. Also called **dead band (1)**, or **dead zone**.

neutralization **1.** The process of making **neutral**. **2.** The process of counteracting or compensating for something, to remedy an unwanted condition, improve performance, or otherwise meet specific needs. For instance, the use of feedback to avoid undesired oscillation in an amplifier, or the minimization of feedback to avoid howl. **3.** The process of adding a sufficient amount of an acid or base to render a solution neutral.

neutralize **1.** To make **neutral**. **2.** To counteract or compensate for something, in order to remedy an unwanted condition, improve performance, or otherwise meet specific needs. For instance, to use feedback to avoid undesired oscillation in an amplifier, or to minimize feedback to avoid howl. **3.** To add a sufficient amount of an acid or base to render a solution neutral.

neutrino A neutral elementary particle with zero mass, or nearly zero mass. It is a type of lepton, and is produced, for instance during beta decay. Neutrinos travel at the speed of light, and pass through matter with little or no interaction.

neutron An elementary particle that has no electric charge, and whose mass slightly exceeds that of a proton. It is in all atomic nuclei except protium, which is hydrogen-1. It is a baryon, and when part of a nucleus is very stable. When free, a neutron decays in a few minutes into a proton, an electron, and an antineutrino.

neutron absorber A substance or material which is particularly effective in absorbing free neutrons. Boron, gadolinium, europium, hafnium, and mercury are elements used for this task, usually in a nuclear reactor.

neutron capture A process in which a neutron is acquired by an atomic nucleus. For example, a neutron collides with a nucleus, producing an excited state, said state decays emitting gamma radiation, and the neutron remains captured in the nucleus.

neutron fluence The passing of neutrons through a given area, such as a square centimeter. May be expressed, for instance, in neutrons per second.

neutron mass Same as **neutron rest mass**.

neutron moderator A substance, such as graphite, heavy water, or beryllium, which is utilized in nuclear reactors to slow down neutrons. Also called **moderator**.

neutron rest mass The mass of a neutron at rest. It is a fundamental physical constant equal to approximately 1.67493×10^{-27} kg. Its symbol is m_n. Also called **neutron mass**.

neutron separation energy The energy required to remove a neutron from a nucleus.

neutron star A star whose mass greater than the sun, yet whose diameter is just a few kilometers, making it extremely dense. It consists mostly of neutrons, and an example is a pulsar.

new candle A former name of **candela**, which is the fundamental SI unit of luminous intensity.

newbie A person who is new to something, especially computers, the Internet, or the like.

news group Same as **newsgroup**.

news reader Same as **newsreader**.

news server A server that stores, organizes, and distributes **newsgroup** messages.

newsgroup A forum where users, especially over the Internet, post and/or read messages expressing opinions, giving advice, or otherwise providing or deriving information on virtually any topic. Also spelled **news group**.

newsreader Client software that formats and displays information being accessed from a **news server**. Also spelled **news reader**.

newton The SI unit of force. One newton is the force that will accelerate a mass of one kilogram at a rate of one meter per second, per second. It is equal to 10^5 dynes. Its symbol is **N**.

newton meter The SI unit of torque. Its abbreviation is **Nm**.

Newton's gravitational constant A physical constant equal to approximately 6.673×10^{-11} Nm^2/kg^2. Its symbol is G. Also called **gravitational constant**.

Newton's law of gravitation A law that states that for any two particles or bodies, the force of attraction is directly proportional to the product of their masses, and inversely proportional to the square of the distance between them. This force occurs along a straight line between the particles or bodies, and applies to all matter, including electrons and stars. Also called **Newton's law of universal gravitation**, or **law of gravitation**.

Newton's law of universal gravitation Same as **Newton's law of gravitation**.

Newton's laws of motion Three laws pertaining to motions and forces. The first states that a body remains in its state of motion, be it at rest or at a constant velocity, unless an external force acts upon it. The seconds states that for a constant mass, force equals mass times acceleration. The third states that every force or action has an equal and opposite reaction.

newtons per coulomb A unit of electric field strength.

NEXT Abbreviation of **near-end crosstalk**.

next-generation Internet Its abbreviation is **NGI**. **1.** A project intended to provide for the incorporation of new technologies, vastly faster data transfer rates, and enhanced security, in addition to other improvements, relative to the current Internet. Also, any implementation of such an initiative. **2.** A high-speed network connecting member institutions, and which is intended for academic and research-oriented use. Also called **Internet2**.

nF Abbreviation of **nanofarad**.

NF **1.** Abbreviation of **noise figure**. **2.** Abbreviation of **noise factor**.

NFR Abbreviation of **near-field recording**.

NFS Abbreviation of **network file system**.

ng Abbreviation of **nanogram**.

nG Abbreviation of **nanogauss**.

NG Abbreviation of **no good**.

NGI Abbreviation of **next-generation Internet**.

nGs Abbreviation of **nanogauss**.

nH Abbreviation of **nanohenry**.

NI Abbreviation of **non-interlaced**.

Ni Chemical symbol for **nickel**.

Ni-MH battery Same as **nickel-metal hydride battery**.

nibble Half a byte, which is the same as 4 adjacent bits. Also spelled **nybble**.

NIC 1. Abbreviation of **network interface card**. 2. Abbreviation of **network information center**.

NIC driver Abbreviation of **network interface card driver**.

Nicad battery Abbreviation of **nickel cadmium battery**.

Nicad cell Abbreviation of **nickel cadmium cell**.

NICAM Acronym for near-instantaneously companded audio multiplex. A standard for the digital encoding of TV audio, which produces sound equivalent to a CD. Used in some European countries.

NiCd battery Abbreviation of **nickel cadmium battery**.

NiCd cell Abbreviation of **nickel cadmium cell**.

nichrome An alloy usually composed of 60% nickel, 25% iron, and 15% chrome, although it may also have a small amount carbon. It has a high electrical resistance, a substantial tolerance to heat, and good resistance to oxidation. Used, for instance, in resistors and heating elements.

nichrome wire A wire made of **nichrome**. Used, for instance, in resistors and heating elements.

nickel A lustrous silver-white chemical element whose atomic number is 28. It is malleable, hard, ferromagnetic, conducts electricity and heat well, and resists corrosion. It has about 25 known isotopes, of which 5 are stable. Its applications in electronics include its use in alloys, for electroplating, in ceramics, cells and batteries, permanent magnets, and in electrodes. Its chemical symbol is **Ni**.

nickel ammonium sulfate A crystalline chemical compound used in electroplating.

nickel cadmium battery Its abbreviations are **NiCd battery**, and **Nicad battery**. 1. A rechargeable battery consisting of **nickel cadmium cells**. 2. Same as **nickel cadmium cell**.

nickel cadmium cell A rechargeable cell in which the anode is cadmium metal, the cathode is nickel hydroxide, and in which the electrolyte is alkaline, usually potassium hydroxide. Each cell is rated at about 1.2 volts, and batteries made of such cells are widely used portable devices including cell phones, portable computers, and music players. Such cells usually suffer from memory effect. Its abbreviation is **NiCd cell**, or **Nicad cell**. Also called **nickel cadmium battery (2)**.

nickel carbonate Light-green crystals or a brown powder used in electroplating.

nickel carbonyl A colorless liquid whose chemical formula is $Ni(CO)_4$. It is used in electroplating.

nickel chloride A brown or green powder whose chemical formula is $NiCl_2$. It is used in electroplating.

nickel cyanide A light-green powder whose chemical formula is $Ni(CN)_2$. It is used in electroplating.

nickel-iron battery A storage battery in which the positive electrode is nickel oxide, the negative electrode is iron, and the electrolyte is a solution of approximately 20% potassium hydroxide. Such batteries are rugged, durable, and reliable, but do not recharge very efficiently. Also known as **Edison battery**.

nickel-metal hydride battery Its abbreviations are **NiMH battery**. 1. A rechargeable battery consisting of **nickel-metal hydride cells**. 2. Same as **nickel-metal hydride cell**.

nickel-metal hydride cell A cell similar to a **nickel cadmium cell**, except that the anode is a hydrogen-absorbing alloy. The nominal voltage of such a cell is also approximately 1.2V, it provides greater energy density than nickel-

cadmium cells, enhanced performance, no memory effect, and eliminates concerns over cadmium toxicity. Its abbreviations are **NiMH cell**. Also called **nickel-metal hydride battery (2)**.

nickel nitrate Green crystals whose chemical formula is $Ni(NO_3)_2$. It is used in electroplating.

nickel oxide A green powder whose chemical formula is **NiO**. It is used in storage cells.

nickel silver An alloy consisting of about 65% copper, 18% nickel, and 17% zinc, although the composition may vary. There may also be small amounts of other elements present, such as manganese. It has a silvery-white appearance, and is used, for instance, in electroplating. Also called **German silver**.

nickel sulfate Yellow, green, or blue crystals whose chemical formula is $NiSO_4$. It is used in electroplating and in ceramics.

nickname An alternate name utilized to identify oneself, another individual, a group, or an entity. Used, for instance, to identify oneself when utilizing an instant messaging service, or to identify an email addressee.

Nicol prism A prism made by appropriately cementing together two pieces of calcite. A Nicol prism is birefringent, and is used to produce plane-polarized light from unpolarized light. It does so by allowing all extraordinary rays to pass through, while the ordinary rays are lost through total internal reflection.

NID Abbreviation of **network interface device**.

night effect Same as **night error**.

night error Polarization error occurring during twilight or night, as is usually the case. Also called **night effect**.

night-vision binoculars A night-vision device which is handheld, and which incorporates the appropriate lenses for image enlargement.

night-vision device A device, such as night vision goggles, which enable enhanced vision during the nighttime, or under other low-light conditions. A powerful unit may incorporate two parallel image intensifiers, one for each eye, and some devices also include the appropriate lenses for image enlargement.

night-vision goggles A night-vision device which is head-mounted. Such a device may or may not provide image enlargement.

nil Zero, or nothing.

NiMH battery Same as **nickel-metal hydride battery**.

NiMH cell Same as **nickel-metal hydride cell**.

nine's complement Same as **nines complement**.

nines complement The complement when using the decimal number system. The complement is obtained when subtracting a number from one less than the radix. For example, since the radix in the decimal number system is 10, then the nine's complement of 6 is 3: $(10 - 1) - 6 = 3$. Its abbreviation is **9's complement**.

niobium A silvery and lustrous, ductile and malleable metallic chemical element whose atomic number is 41. It has over 25 known isotopes, of which one is stable. Its applications in electronics include its use in superconductor alloys, and as a getter in vacuum tubes. It was formerly known as columbium, a designation still used mostly by metallurgists. Its chemical symbol is **Nb**.

NIS Abbreviation of **network information service**.

NIST Acronym for **National Institute of Standards and Technology**.

nit A unit of luminance equal to one candela per square meter. Its abbreviation is **nt**.

Nitinol A shape-memory alloy consisting nickel and titanium.

nitrogen A colorless, odorless, and tasteless diatomic gas whose atomic number is 7. It comprises approximately 78% by volume of dry air in the atmosphere, and although as an element is rather inactive, numerous of its compounds are very reactive. It has over a dozen known isotopes, of which 2 are stable. Its applications include its use in thermometers, bulbs, as a refrigerant, and in many compounds necessary for life. Its chemical symbol is **N**.

NIU Abbreviation of **network interface unit**.

Nixie tube **1.** A gas tube filled with neon and a small amount of mercury and/or argon, whose ten cathodes are in the shape of numerals. When a voltage is applied to the appropriate pins, the tube glows displaying the desired number. Used, for instance, in calculators. **2.** A Nixie tube utilized to display other information, such as letters, symbols, or the like. Used, for instance, in signs.

nJ Abbreviation of **nanojoule**.

nl Abbreviation of **nanoliter**.

nL **1.** Abbreviation of **nanoliter**. **2.** Abbreviation of **nanolambert**.

NL processing Abbreviation of **natural language processing**.

NLP Abbreviation of **natural language processing**.

NLQ Abbreviation of **near-letter quality**.

NLS Abbreviation of **natural language support**.

nlx Abbreviation of **nanolux**.

nm Abbreviation of **nanometer**.

Nm Abbreviation of **newton meter**.

NMI Abbreviation of **non-maskable interrupt**.

nmol Abbreviation of **nanomole**, or **nanomol**.

NMOS Abbreviation of **n-channel MOS**, which itself is an abbreviation of **n-channel metal-oxide semiconductor**. **1.** A MOS technology in which an n-type channel is used, making electrons the charge carriers. Since electrons move faster than holes, such devices are faster than **PMOS**, in which mobile holes are the charge carriers. Also called **NMOS technology**. **2.** A semiconductor device incorporating the technology described in **NMOS (1)**. Also called **NMOS device**.

NMOS device Same as **NMOS (2)**.

NMOS technology Same as **NMOS (1)**.

NMOS transistor A field-effect transistor utilizing **NMOS (1)**. Also called **NMOSFET**.

NMOSFET Same as **NMOS transistor**.

NMR **1.** Abbreviation of **nuclear magnetic resonance**. **2.** Abbreviation of **normal-mode rejection**.

NMR Imaging Abbreviation of **nuclear magnetic resonance imaging**.

NMR spectrometer Abbreviation of **nuclear magnetic resonance spectrometer**.

NMR spectroscopy Abbreviation of **nuclear magnetic resonance spectroscopy**.

NMRR Abbreviation of **normal-mode rejection ratio**.

NNI Abbreviation of **network-to-network interface**.

NNTP Abbreviation of **Network News Transfer Protocol**.

No Chemical symbol for **nobelium**.

NO Abbreviation of **normally open**.

no circuit The circumstance where there is no trunk available for an attempted call. Its abbreviation is **NC**.

no connection Its abbreviation is **NC**. **1.** In a circuit diagram, indicative that there is no connection between wires, leads, or other conductors. **2.** Indicative that there is no connection for a given pin, such as that of a connector or vacuum tube.

NO contact Abbreviation of **normally-open contact**.

no good Indicative that a given component, circuit, device, piece of equipment, system, material, or the like, is broken, malfunctioning, or otherwise not fit for use. Its abbreviation is **NG**.

no-load Operation under **no-load conditions**, or pertaining to such operation.

no-load conditions The circumstance where a component, circuit, device, piece of equipment, or system is properly set-up and operational, but is not providing an output to a load. For instance, a transformer with no load present.

no-load current The current which flows through a component, circuit, device, piece of equipment, or system, under **no-load conditions**. Also called **open-circuit current**.

no-load loss Losses occurring under **no-load conditions**. For instance, those of a transformer with no load present. Also called **no-load losses**.

no-load losses Same as **no-load loss**.

no-load operation Operation under **no-load conditions**

no-load voltage The voltage level present under **no-load conditions**. For example, the voltage at the output terminals of a power supply when 0% load is applied, or that of a battery in the absence of a charge or discharge current. Also called **open-circuit voltage**.

no-op Abbreviation of **no-operation instruction**.

NO-OP Abbreviation of **no-operation instruction**.

no-op instruction Abbreviation of **no-operation instruction**.

no operation Same as **no-operation instruction**.

no-operation instruction A computer instruction which produces no action. It simply serves to cause the processor to utilize clock cycles, or to proceed to the next instruction. Used, for instance, to complete a very long instruction word. Its abbreviation is **NO-OP**, **NOP**, or **no-op instruction**. Also called **no operation**, **do-nothing instruction**, or **dummy instruction**.

no parity A modem setting in which there is no parity bit.

No. Abbreviation of **number**.

nobelium A synthetic radioactive chemical element whose atomic number is 102, and which has over a dozen identified isotopes, all unstable. The techniques employed to produce this element are utilized to help discover new elements with higher atomic numbers. Its chemical symbol is **No**.

noble **1.** Pertaining to, or having the attributes of a **noble metal**. **1.** Pertaining to, or having the attributes of a **noble gas**.

noble gas Any of six chemical elements which are all gases and highly unreactive. They are all in the same group of the periodic table, and are helium, neon, argon, krypton, xenon, and radon. Also called **inert gas**, or **rare gas**.

noble metal A metal, such as platinum or gold, which is inert and highly resistant to corrosion.

NOC Abbreviation of **network operations center**.

noctovision A TV system which transmits infrared views of objects and mediums, allowing for enhanced viewing of nocturnal scenes.

nodal Pertaining to, consisting of, or located in or around a **node**.

node **1.** A point or location where entities meet or cross paths. For instance, a point where a continuous curve crosses itself. **2.** The point at which a branch, such as that of a circuit or electric network, originates. Also, the point at which a junction occurs. Also known as **branch point**, or **junction point**. **3.** Within a communications network, a device, such as a personal computer, printer, or server, which is connected to, and is able to exchange information with other devices. Also called **network node**. **4.** In a standing-wave system, a point of minimum amplitude or displacement. For example, a minimum, usually zero, of voltage or current.

This contrasts with an **antinode (1)**, where the converse is true. **5.** In a vibrating system, a point of minimum amplitude or displacement. This contrasts with an **antinode (2)**, where the converse is true. **6.** In a data structure, such as a tree structure, a point from which subordinate items originate, or where two or more lines meet. Also, the interconnected items in such a structure.

node address Same as **network address**.

node analysis In the analysis of electric networks, the determination of the voltages of **nodes (2)**.

noise **1.** A usually unwanted electrical disturbance, often random and/or persistent in nature, which affects the quality or usefulness of a signal, or which adversely affects the operation of a device. Such noise may occur naturally, or be artificially created. There are many examples, including AC noise, background noise, amplification noise, atmospheric noise, cosmic noise, and thermal noise. Noise which occurs over a wide range of frequencies, regardless of its origin, is called broadband noise. **2.** Unwanted sound that occurs within the range of frequencies that humans can hear, and which can result in an unpleasant, uneven, distracting, or otherwise undesired listening experience. Examples include hum and hiss. Such noise may also harm equipment. Also called **audio noise, audio-frequency noise**, or **acoustic noise (2)**. **3.** In communications, **noise (1)** which results in transmitted data which is meaningless, superfluous, or whose information content is otherwise impaired. **4.** Any disturbance which adversely affects a signal, or the operation of a device.

noise abatement **1.** Measures taken to reduce the intensity of noises and/or vibrations, and the deleterious impact these have on the surrounding environment. **2.** Reductions in noise resulting from **noise abatement (1)**.

noise analysis The careful study of noise, to determine characteristics such as intensity and frequency components.

noise analyzer An instrument utilized for **noise analysis**.

noise bandwidth The interval of frequencies that any given noise spans.

noise blanker A circuit or device which serves to suppress or cut-off noise. For instance, a circuit which reduces pulse-type noises when a desired signal is present.

noise canceling Same as **noise cancellation**.

noise-canceling microphone **1.** A microphone which is designed to selectively minimize certain noises. For instance, that which is highly directional. Also called **antinoise microphone**. **2.** A microphone whose characteristics require it to be used in close proximity to its sound source. Such microphones are well-suited for use in surroundings with excessive background noise, such as a room full of telephone operators. Also called **close-talking microphone**.

noise cancellation The suppression or elimination of unwanted noises. Used, for instance, to enhance the quality of a signal, or to help control noise pollution from commercial, automotive, and industrial sources. Destructive interference of the acoustic waves, or sound absorbers, for instance, may be employed for this purpose. Also called **noise canceling**.

noise current An electrical current that is unwanted, and causes interference in a component, circuit, device, piece of equipment, or system.

noise diode **1.** A diode whose output is a precise noise. **2.** A diode which suppresses noise.

noise distortion Distortion arising from the presence of noise in a signal.

noise dose meter Same as **noise dosimeter**.

noise dosimeter A device or instrument which measures the cumulative noise levels it is exposed to. Such a device incorporates a microphone, and may also record average levels, peaks, and so on, in addition to providing a computer interface for detailed analysis. Used, for instance, to monitor the noise levels personnel are exposed to. Also called **noise dose meter**.

noise elimination The diminishing or removal of undesired noise signals. Also called **noise reduction (2)**, or **noise suppression (1)**.

noise eliminator A circuit, device, piece of equipment, or system utilized for **noise elimination**. For example, a signal processor which removes all electrical and ignition-type noise from a TV signal arriving via satellite.

noise equivalent power The power of a desired signal that is equivalent to the noise level present. That is, it is the signal level required to obtain a signal-to-noise ratio of unity. For instance, the radiant power that produces a signal-to-noise ratio of unity, for a given bandwidth. Its abbreviation is **NEP**.

noise factor For a specified bandwidth, the ratio of the total noise output to the total noise input for a given component, circuit, device, piece of equipment, or system. It is the additional noise a component, circuit, device, piece of equipment, or system adds. Usually measured at a noise temperature of 290 K. When expressed in decibels, also called **noise figure**, although the terms are often used synonymously.

noise figure The **noise factor** expressed in decibels, although the terms are often used synonymously. Used, for instance, as a figure of merit.

noise filter **1.** A filter which is utilized to reduce or block noise interference. For instance, that used in an audio amplifier to pass the desired frequencies while reducing or blocking noise signals at other frequencies. **2.** A filter which is utilized to reduce or block noise interference from entering or circulating through an AC power line.

noise filtering The use of a **noise filter** to reduce or block noise interference.

noise floor **1.** The level of noise present in a component, circuit, device, piece of equipment, or system, which is powered on, but not passing a signal. **2.** The weakest desired signal that can be detected, received, or reproduced. This signal just barely exceeds the **noise floor (1)**.

noise generator A device or instrument which produces precise or random noise signals. Used, for instance, for testing, adjusting, and troubleshooting. Also called **noise source (2)**.

noise immunity **1.** The extent to which a component, circuit, device, piece of equipment, or system is insensitive to noise. **2.** The extent to which a component, circuit, device, piece of equipment, or system minimizes the deleterious effects of noise.

noise interference Interference arising from the presence of noise in a signal.

noise level **1.** The amplitude, degree, or intensity of a noise signal. Said, for instance, of that present in a circuit, device, or system. **2.** The amplitude, degree, or intensity of all noise signals present. Said, for instance, of that of a given environment or system. **3.** The **noise level (1)**, or **noise level (2)** for a specified interval of frequencies.

noise limiter A circuit or device which limits the maximum amplitude of an input signal, so as to minimize noise. It usually is the same as **automatic noise limiter**. Also called **noise suppressor (1)**, or **noise silencer**.

noise limiting The effect obtained using a **noise limiter**. Also called **noise suppression (2)**.

noise margin In a logic circuit, the maximum allowable variations in logic input levels that will still not adversely affect logic output levels. Used, for instance, to avoid spurious triggering.

noise meter 1. An instrument which measures **noise levels**. 2. An instrument which measures acoustic **noise levels**.

noise pollution Noise which adversely affects its surrounding environment. Sources include automobiles, aircraft, trains, sea vessels, construction equipment, excessively amplified music, gunfire, explosives, and many other human-created devices and human-based activities.

noise power The power level of a noise signal.

noise pulse A momentary high-intensity burst of noise, such as that produced by an ignition system. Also called **noise spike**.

noise quieter A circuit in a radio receiver that suppresses noise when listening to programming or other signals. Also called **quieter**, or **squelch circuit** (1).

noise quieting The noise suppression resulting from the use of a **noise quieter**. Also called **quieting**.

noise reduction 1. A system for the reduction of noise in audio recordings. One variation, for instance, involves providing additional gain for certain frequencies during recording, while reducing said gain during playback. 2. Same as **noise elimination**.

noise reduction system A system, such as Dolby, used for **noise reduction**.

noise signal A signal which has noise present. The noise can be unwanted, as in an interfering signal, or desired, as in a test signal.

noise silencer Same as **noise limiter**.

noise source 1. A source of unwanted **noise**. 2. Same as **noise generator**.

noise spike Same as **noise pulse**.

noise suppression 1. Same as **noise elimination**. 2. Same as **noise limiting**.

noise suppressor 1. Same as **noise limiter**. 2. A circuit or device utilized for **noise suppression**.

noise temperature For a given pair of terminals at a specified frequency, the temperature of a passive system which would have the same noise power as the actual terminals of the electric network under consideration. The lower the noise temperature of a device or system, the better.

noise voltage A voltage that is unwanted, and causes interference in a component, circuit, device, piece of equipment, or system.

noiseless Characterized by, or pertaining to, a lack of **noise**.

nominal Pertaining to, or consisting of, a **nominal value**.

nominal bandwidth 1. For a given communications channel, the assigned band of frequencies, including frequency guard bands. 2. For a given component, circuit, device, piece of equipment, system, or channel, the nominal operational interval of frequencies. In the case of a filter, for instance, it would be the difference between the upper and lower nominal cutoff frequencies.

nominal capacitance A rated or named value, stating the capacitance of a capacitor when used in a given manner. The actual value, or interval of values, may not coincide with this number. Also called **rated capacitance**.

nominal current A rated or named value, stating the current that a component, circuit, device, piece of equipment, system consumes, produces, or safely tolerates, when used in a given manner. For example, it may be a rating of the current drawn expressed in RMS amperes after a device is turned on and warmed up. The actual value, or interval of values, may not coincide with this number. Also called **nominal current rating**, or **rated current** (1).

nominal current rating Same as **nominal current**.

nominal frequency A rated or named value, stating the center frequency of a crystal or crystal unit, such as that of quartz.

The actual value, or interval of values, may not coincide with this number. Also called **rated frequency**.

nominal impedance A rated or named value, stating the impedance of a circuit or device when used in a given manner. The actual value, or interval of values, may not coincide with this number. For instance, the nominal impedance of a speaker may be 8 ohms, but depending on the frequency reproduced may dip below 3, or exceed 50 ohms. Also called **rated impedance**.

nominal power A rated or named value stating the power that a component, circuit, device, piece of equipment, or system can produce, consume, dissipate, or otherwise safely handle, when used in a given manner. For instance, the nominal power of a speaker may indicate the RMS power that can be delivered to a dynamic speaker for extended periods, without harming the voice coil or other components. The actual value, or interval of values, may not coincide with this number. Also called **nominal power rating**, or **rated power** (1).

nominal power rating Same as **nominal power**.

nominal Q A rated or named value, stating the Q, or quality factor, of a component, circuit, or device, when used in a given manner. The actual value, or interval of values, may not coincide with this number.

nominal rating Same as **nominal value**.

nominal resistance A rated or named value, stating the resistance of a resistor when used in a given manner. The actual value, or interval of values, may not coincide with this number. Also called **rated resistance**.

nominal value A rated or named value, or interval of values, describing a property or the behavior of a component, circuit, device, piece of equipment, or system operated under given conditions. Examples include nominal power and nominal impedance. The actual observed value, or interval of values, may not coincide with these numbers. For instance, the voltage supplied by an AC power line in a given country. Also called **nominal rating**, or **rating** (1).

nominal voltage A rated or named value, stating the available, input, or output voltage of a component, circuit, device, piece of equipment, or system, under specified conditions. For instance, the stated AC line voltages for a given locality may be 115/230, where the lower number is primarily for lighting and small appliances, and the larger number is for heating and large appliances. The actual values, or interval of values, may not coincide with these numbers. Also called **nominal voltage rating**, or **rated voltage** (1).

nominal voltage rating Same as **nominal voltage**.

nomogram Same as **nomograph**.

nomograph A chart used to solve numerical equations. Two known variables and an unknown are graphed in a manner that the results of calculations may be obtained by tracing an intersecting line through known values to obtain the desired value. May be used, for instance, to solve many electronics equations. Also called **nomogram**, or **alignment chart**.

non- A prefix meaning not. For example, nonlinear, or nonperiodic.

non-blocking system A telephone system that is configured to make sure all calls are connected. For instance, it may provide more connection paths than telephones. Also spelled **nonblocking system**.

non-bridging In a selectable switch, the action of the movable contact when it is not wide enough to touch two consecutive contacts. In this manner, the circuit is broken during switching. Also spelled **nonbridging**.

non-coherence Also spelled **noncoherence**. 1. A lack of a definite phase relationship between two or more waves. Two waves, for instance, are in not phase if their crests and troughs do not meet at the same point at the same time.

Also called **incoherence (1)**. **2.** Any lack of temporal or phase correlation. Also called **incoherence (2)**.

non-coherent Characterized by **non-coherence**. Also spelled **noncoherent**.

non-coherent bundle An assembly of optical fibers in which the relative spatial coordinates of the individual fibers are not the same at both ends. Such a bundle serves to transmit light, but not images. Also spelled **noncoherent bundle**. Also called **incoherent bundle**.

non-coherent demodulation Same as **non-coherent detection**. Also spelled **noncoherent demodulation**.

non-coherent detection Demodulation which does not depend on the phase of the carrier signal. Also spelled **noncoherent detection**. Also called **non-coherent demodulation**.

non-coherent detector A detector whose output signal amplitude does not depend on the phase of its input signal. Also spelled **noncoherent detector**. Also called **incoherent detector**.

non-coherent light Light in which there is not a definite phase relationship between any given points of the waves composing it. Such light may contain waves of different wavelengths. There are many sources of incoherent light, including sunlight, incandescent lamps, and light-emitting diodes. Also spelled **noncoherent light**. Also called **incoherent light**.

non-coherent light source A source, such as a light-emitting diode, of **non-coherent light**. Also spelled **noncoherent light source**. Also called **incoherent light source**.

non-coherent pulses Pulses which are not in a fixed-phase relationship with each other. Also spelled **noncoherent pulses**. Also called **incoherent pulses**.

non-coherent radiation Radiation in which there is not a definite phase relationship between any given points of a cross section of the beam. Also spelled **noncoherent radiation**. Also called **incoherent radiation**.

non-coherent scattering Scattering in which there is not a definite phase relationship between the incident and dispersed waves or particles. Also spelled **noncoherent scattering**. Also called **incoherent scattering**.

non-coherent waves Two or more waves of a single frequency which do not have a definite phase relationship. Also spelled **noncoherent waves**. Also called **incoherent waves**.

non-conductor Also spelled **nonconductor**. **1.** A medium which is not suitable for the conduction of electrical, acoustic, heat, or other form of energy. **2.** A medium which does not allow electric current to flow easily. Such a medium may be a dielectric such as air, oil, glass, paper, or any of various plastics or ceramics.

non-contact detector Same as **non-contact sensor**. Also spelled **noncontact detector**.

non-contact sensing device Same as **noncontact sensor**.

non-contact sensor A proximity sensing device used to determine the distance to one or more objects, without coming in contact with that which is detected. They are many types, including optical, acoustic, and magnetic. Frequently used in robotics. Also spelled **noncontact sensor**. Also known as **non-contact detector**, **non-contact sensing device**, or **proximity sensor**.

non-contact temperature measurement The determination of the temperature of objects or mediums utilizing an instrument or device which does not come in contact with that being measured. For example, the use of an infrared thermometer to keep track of the temperature of stored food products. Also spelled **noncontact temperature measurement**.

non-contact thermometer An instrument or device, such as an infrared thermometer, utilized for **non-contact temperature measurement**. Also spelled **noncontact thermometer**.

non-corrosive flux A flux, such as that used for brazing, soldering, or welding, which contains no substances that will corrode the materials being joined. Also spelled **noncorrosive flux**.

non-crystalline Not having a crystalline structure. For example, glass. Also spelled **noncrystalline**. Also called **amorphous (2)**.

non-dedicated server A computer which performs both network server and client functions. Also spelled **nondedicated server**.

non-destructive read The circumstance in which the reading of data does not cause said data to be destroyed, as would be the case with a **destructive readout**. Also spelled **nondestructive read**. Also called **non-destructive readout**.

non-destructive readout Same as **non-destructive read**. Its abbreviation is **NDRO**, or **NDR**. Also spelled **nondestructive readout**.

non-destructive test A test in which that being evaluated is not damaged, destroyed, or otherwise impaired. Included, for instance, are tests involving optical, ultrasonic, or magnetic imaging. Also spelled **nondestructive test**.

non-destructive testing The performance of **non-destructive tests**. Also spelled **nondestructive testing**.

non-directional Also spelled **nondirectional** **1.** Not pertaining to, or indicating a specific direction in space. **2.** Sending or receiving signals in any given direction. **3.** Capable of sending or receiving signals with the same efficiency or sensitivity in all directions. Also called **omnidirectional (3)**. **4.** Functioning at essentially the same level of performance in all directions. Also called **omnidirectional (4)**.

non-directional antenna Also spelled **nondirectional antenna**. Also known as **omnidirectional antenna**. **1.** An antenna which radiates or receives with essentially the same efficiency or sensitivity in all directions. **2.** An antenna which radiates or receives with essentially the same efficiency or sensitivity in all directions within a given plane.

non-directional beacon Same as **non-directional radio beacon**. Also spelled **nondirectional beacon**.

non-directional microphone A microphone whose sensitivity is the same in all directions. Also spelled **nondirectional microphone**. Also known as **omnidirectional microphone**, or **astatic microphone**

non-directional radio beacon Also spelled **nondirectional radio beacon**. Its abbreviation is **NDB**. Also called **non-directional beacon**, or **omnidirectional radio beacon**. **1.** A beacon which emits signals in all directions. **2.** A beacon which emits signals in all directions within the horizontal plane.

non-dispersive medium A medium within which the phase velocity of electromagnetic waves is independent of frequency. A vacuum is an example of such a medium. In a **dispersive medium**, the phase velocity of electromagnetic waves varies as a function of the frequency. Also spelled **nondispersive medium**.

non-electric Not incorporating, utilizing, or based on electricity, electrical components, circuits, devices, equipment, or systems. For instance, a switch actuated by centrifugal forces. Also spelled **nonelectric**.

non-electrical Not pertaining to, or correlated with, electricity. For instance, that which is mechanical. Also spelled **nonelectrical**.

non-electronic Also spelled **nonelectronic**. **1.** Not pertaining to electrons. **2.** Not pertaining to electronics. **3.** Not pertaining to components and devices, such as semiconductors, which conduct electrons, or other charge carriers such as ions or mobile holes, through a vacuum, gas, or semiconductor. **4.** Not pertaining to computer, and/or online concepts. For instance, a paper document as opposed to a scanned document.

non-erasable memory Also spelled **nonerasable memory**. **1.** Memory, such as ROM, whose content may be read, but not modified or erased. **2.** Same as **non-erasable storage**.

non-erasable storage Also spelled **nonerasable storage**. **1.** A storage medium, such as a CD-ROM, whose stored content may be read, but not modified or erased. **2.** Same as **non-erasable memory**.

non-flammable That which is not easily ignited or burned. Also spelled **nonflammable**.

non-harmonic For a complex signal, wave, sound, or vibration which is periodic, a component whose frequency is not a whole-number multiple of the fundamental frequency. Also spelled **nonharmonic**. Also called **non-harmonic component**.

non-harmonic component Same as **non-harmonic**.

non-impact printer A printer which does not mechanically impact the paper, or other medium, upon which the characters and symbols are being recorded. Examples include laser, inkjet, and thermal printers. Also spelled **nonimpact printer**.

non-inductive That which has little or no inductance. Also spelled **noninductive**.

non-inductive capacitor A capacitor whose design provides for little or no inductance. Also spelled **noninductive capacitor**.

non-inductive circuit A circuit whose design provides for little or no inductance. Also spelled **noninductive circuit**.

non-inductive load A load with little or no inductance. Also spelled **noninductive load**.

non-inductive resistor A resistor whose design provides for little or no inductance. Also spelled **noninductive resistor**.

non-inductive winding A winding whose design provides for little or no inductance. Also spelled **noninductive winding**.

non-interlaced Pertaining to, utilizing, or incorporating **non-interlaced scanning**. Also spelled **noninterlaced**. Its abbreviation is **NI**.

non-interlaced display A display which utilizes **non-interlaced scanning**. Also spelled **noninterlaced display**.

non-interlaced monitor A monitor which utilizes **non-interlaced scanning**. Also spelled **noninterlaced monitor**.

non-interlaced scan Same as **non-interlaced scanning**. Also spelled **noninterlaced scan**.

non-interlaced scanning A scanning system, such as that of computer monitors, in which the electron beam traces all lines sequentially from top to bottom. This contrasts with **interlaced scanning**, in which the electron beam traces all odd-numbered lines followed by the tracing of all even-numbered lines. Interlaced scanning may produce a fluttering effect when displaying certain graphics, a drawback not exhibited by a non-interlaced monitor. Also spelled **noninterlaced scanning**. Also called **non-interlaced scan, non-interlacing, progressive scanning**, or **sequential scanning**.

non-interlacing Same as **non-interlaced scanning**. Also spelled **noninterlacing**.

non-inverting amplifier An amplifier whose output signal has the same polarity as that of the input signal. This contrasts with an **inverting amplifier**, whose output has the opposite polarity of its input signal. Also spelled **noninverting amplifier**.

non-inverting connection A connection to a **non-inverting input**. An example is a connection to an input of a differen-

tial amplifier or operational amplifier whose output is in phase with its input. Also spelled **noninverting connection**.

non-inverting input An input of a circuit or device whose output has the same polarity as said input. For instance, that of an noninverting amplifier. Also spelled **noninverting input**. Also called **non-inverting terminal**.

non-inverting terminal Same as **non-inverting input**. Also spelled **noninverting terminal**.

non-ionizing radiation Radiation whose energy is not sufficient to cause ionization of the medium through which it passes. Examples include radio waves, infrared, and visible light. Also spelled **nonionizing radiation**.

non-linear Same as **nonlinear**.

non-linear amplifier Same as **nonlinear amplifier**.

non-linear capacitor Same as **nonlinear capacitor**.

non-linear circuit Same as **nonlinear circuit**.

non-linear component Same as **nonlinear component**.

non-linear control system Same as **nonlinear control system**.

non-linear detection Same as **nonlinear detection**.

non-linear detector Same as **nonlinear detector**.

non-linear device Same as **nonlinear device**.

non-linear dielectric Same as **nonlinear dielectric**.

non-linear distortion Same as **nonlinear distortion**.

non-linear element Same as **nonlinear element**.

non-linear feedback Same as **nonlinear feedback**.

non-linear material Same as **nonlinear material**.

non-linear modulation Same as **nonlinear modulation**.

non-linear network Same as **nonlinear network**.

non-linear oscillator Same as **nonlinear oscillator**.

non-linear phase response Same as **nonlinear phase response**.

non-linear resistor Same as **nonlinear resistor**.

non-linear response Same as **nonlinear response**.

non-linear system Same as **nonlinear system**.

non-linear taper Same as **nonlinear taper**.

non-linear transducer Same as **nonlinear transducer**.

non-linear video editing Same as **nonlinear video editing**.

non-linearity Same as **nonlinearity**.

non-linearity error Same as **nonlinearity error**.

non-magnetic Also spelled **nonmagnetic**. **1.** That which is not capable of being magnetized, or which is neither attracted nor repelled by a magnet. For instance, air, glass, or wood. **2.** That which does not utilize magnetism. For instance, a switch actuated by centrifugal forces.

non-maskable interrupt An interrupt which requires the immediate attention of the CPU, even if it is handling another interrupt. Non-maskable interrupts are usually reserved for grave situations, such as hardware parity errors. Such an interrupt can not be disabled by another interrupt, while a **maskable interrupt** can be. Also spelled **non-maskable interrupt**. Its abbreviation is **NMI**.

non-metal Same as **nonmetal**.

non-metallic Same as **nonmetallic**.

non-microphonic A component, circuit, device, piece of equipment, or system, which is designed to minimize or eliminate unwanted noise due to mechanical vibrations by any contained moving elements, materials, or parts. Also spelled **nonmicrophonic**.

non-operational That which is not ready or able to function or operate. For instance, a mock-up. Also spelled **nonoperational**.

non-parametric test A statistical test in which no assumptions are made about the data utilized. For example, it does not assume that the data is taken from a normal distribution. In a **parametric test** such assumptions are made. Also spelled **nonparametric test**.

non-parity memory   Computer memory, such as DRAM, which does not utilize parity checking. Also spelled **non-parity memory**. Also called **non-parity RAM**.

non-parity RAM   Same as **non-parity memory**. Also spelled **nonparity RAM**.

non-periodic signal   Also spelled **nonperiodic signal**. Also called **aperiodic signal**. **1.** A signal which does not occur in a regular and repetitive pattern. **2.** A non-repetitive signal.

non-periodic waveform   Also spelled **nonperiodic waveform**. Also called **aperiodic waveform**. **1.** A waveform which does not occur in a regular and repetitive pattern. **2.** A non-repetitive waveform.

non-planar   Not situated in, consisting of, pertaining to, or utilizing a single plane. For instance, that which is three-dimensional. Also spelled **nonplanar**.

non-planar circuit   A circuit which is three-dimensional. This contrasts with a **planar circuit**, which is two-dimensional. Also spelled **nonplanar circuit**.

non-polar   Same as **nonpolar**.

non-polarized capacitor   A capacitor which can be connected either way in a circuit. The leads of a **polarized capacitor** must each be connected to specific positive and negative terminals. Also spelled **nonpolarized capacitor**.

non-polarized electrolytic capacitor   An electrolytic capacitor which can be connected either way in a circuit. The leads of a **polarized electrolytic capacitor** must each be connected to specific positive and negative terminals. Also spelled **nonpolarized electrolytic capacitor**.

non-preemptive multitasking   A mode of multitasking in which the foreground task allows background tasks access to the CPU at given times, such as when it is idle. Since the program in the foreground controls access to the CPU, it may monopolize its resources. This contrasts with **preemptive multitasking**, in which all tasks take turns at having the attention of the CPU. Also spelled **nonpreemptive multitasking**. Also called **cooperative multitasking**.

non-procedural language   A programming language which does not require specific statements, instructions, or subroutines which need to be executed in a specified sequence. Instead, such a language utilizes a data-processing system to understand what is stated so that it can be acted upon. Such a language emphasizes what needs to be done, as opposed to how it must be carried out, as would be the case with a **procedural language**. Used, for instance, in relational database applications. Also spelled **nonprocedural language**.

non-radiative transition   Same as **nonradiative transition**.

non-reactive   Same as **nonreactive**.

non-real-time variable bit rate   Same as **non-real-time VBR**.

non-real-time VBR   Abbreviation of **non-real-time** variable bit rate. In ATM, variable bit rate service which does not support real-time applications such as audio and/or video conferencing. Its own abbreviation is **nrt-VBR**.

non-rechargeable battery   A battery composed of **non-rechargeable cells**. Also spelled **nonrechargeable battery**. Also called **primary battery**.

non-rechargeable cell   A cell, such as an alkaline or mercury cell, whose manner of discharge does not lend itself to recharging. For instance, the current produced may involve the dissolution of one of its plates. This contrasts with a **rechargeable cell**, which can be recharged multiple times. Also spelled **nonrechargeable cell**. Also called **primary cell**.

non-recurrent   Also spelled **nonrecurrent**. Also called **non-recurring**. **1.** That which does not occur repeatedly. **2.** That which does not repeat itself periodically.

non-recurring   Same as **non-recurrent**. Also spelled **nonrecurring**.

non-resonant   Also spelled **nonresonant**. **1.** Not exhibiting resonance. For instance, a non-resonant line. **2.** Pertaining to that which is **non-resonant (1)**. **3.** That which does not tend to oscillate or vibrate. For example, non-resonant materials may be incorporated into a speaker cabinet.

non-resonant antenna   An antenna that has an approximately constant input impedance over a wide range of frequencies. A diamond antenna is an example. Also spelled **nonresonant antenna**. Also called **aperiodic antenna**, or **untuned antenna**.

non-resonant line   Also spelled **nonresonant line**. Also called **non-resonant transmission line**. **1.** A transmission line which has no reflections, and therefore no standing waves. **2.** A transmission line which does not resonate at the frequency utilized.

non-resonant transmission line   Same as **non-resonant line**. Also spelled **nonresonant transmission line**.

non-return to zero   Its abbreviation is **NRZ**. Also spelled **nonreturn to zero**. Also called **non-return to zero encoding**, **non-return to zero coding**, or **non-return to zero code**. **1.** A data-encoding method in which the signal does not return to a zero state between sent bits. For instance, 1s may be represented by a positive voltage and 0s by a negative voltage, with no zero-voltage condition. Timing is utilized to distinguish each bit from the next one. Used for digital data transmission. NRZ requires half of the occupied bandwidth as compared to Manchester encoding. **2.** A magnetic tape recording technique which utilizes **non-return to zero (1)**.

non-return to zero code   Same as **non-return to zero**. Its abbreviation is **NRZ code**. Also spelled **nonreturn to zero code**.

non-return to zero coding   Same as **non-return to zero**. Its abbreviation is **NRZ coding**. Also spelled **nonreturn to zero coding**.

non-return to zero encoding   Same as **non-return to zero**. Its abbreviation is **NRZ encoding**. Also spelled **nonreturn to zero encoding**.

non-return to zero inverted   Its abbreviation is **NRZI**. Also called **non-return to zero inverted encoding**, **non-return to zero inverted code**, or **non-return to zero inverted coding**. **1.** A data-encoding method in which 0s are represented by a change to the opposite state of the signal, while 1s have no change in state. Alternatively, the 1s may have a change in state while the 0s do not. Used for digital data transmission. **2.** A magnetic tape recording technique which utilizes **non-return to zero inverted (1)**.

non-return to zero inverted code   Same as **non-return to zero inverted**. Its abbreviation is **NRZI code**.

non-return to zero inverted coding   Same as **non-return to zero inverted**. Its abbreviation is **NRZI coding**.

non-return to zero inverted encoding   Same as **non-return to zero inverted**. Its abbreviation is **NRZI encoding**.

non-short-circuiting switch   Same as **non-shorting switch**.

non-shorting switch   A switch that opens a connection prior to making a new connection. For instance, the moving contact of such a switch may be narrower than the distance between fixed contacts. Also spelled **nonshorting switch**. Also called **non-short-circuiting switch**, or **break-before-make switch**.

non-sinusoidal wave   A wave with a **non-sinusoidal waveform**. Also spelled **nonsinusoidal wave**.

non-sinusoidal waveform A waveform, such as that of a square wave, which is not sinusoidal in shape. Also spelled **nonsinusoidal waveform.**

non-synchronous Also spelled **nonsynchronous. 1.** Not synchronous. That is, not occurring at the same time or in equal and fixed intervals. Also called **asynchronous (1). 2.** Not coordinated, especially as regards to time. Also called **asynchronous (2).**

non-text file 1. A file which has no text. **2.** A file whose content is not limited just to text. **3.** A file that is in a format utilizing sequences of 8 bits. Almost all computer files are in this 8-bit format. This contrasts, for instance, with 7-bit ASCII files. Also called **binary file.**

non-trigger voltage The maximum voltage that can be applied to a circuit or device, such as a thyristor, without triggering its operation. Also spelled **nontrigger voltage.**

non-uniform electric field An electric field in which the exerted force does not have the same strength at all surrounding points. Also spelled **nonuniform electric field.**

non-uniform field A field, such as an electric, magnetic, or gravitational field, in which the exerted force does not have the same strength at all surrounding points. Also spelled **nonuniform field.**

non-uniform magnetic field A magnetic field in which the exerted force does not have the same strength at all surrounding points. Also spelled **nonuniform magnetic field.**

non-uniform memory access A parallel-processing computer architecture in which the total available memory is divided into multiple shared segments, as opposed to being in one centralized location. The processor's own memory is accessed the fastest, while the rest of the memory banks are accessed at varying speeds depending on their location. Its abbreviation is **NUMA.** Also spelled **nonuniform memory access.**

non-volatile memory Computer memory which retains its content when the power is interrupted or turned off. Examples include ROM, EEPROM, and flash memory. Also spelled **nonvolatile memory.** Also called **non-volatile storage (2),** or **permanent memory.**

non-volatile RAM Abbreviation of **non-volatile** random-access memory. RAM, such as MRAM, which retains its content when the power is interrupted or turned off. Its own abbreviation is **NVRAM.**

non-volatile random-access memory Same as **non-volatile RAM.**

non-volatile storage Also spelled **nonvolatile storage.** Also called **permanent storage. 1.** Computer storage which retains its content when the power is interrupted or turned off. Almost all computer storage, such as that utilizing disks or tapes, is of this nature. **2.** Same as **non-volatile memory.**

non-zero dispersion-shifted fiber A single-mode optical fiber whose zero-dispersion wavelength is shifted from the value it would otherwise occur at, such as 1550 or 1300 nm. Also spelled **nonzero dispersion-shifted fiber.** Its abbreviation is **NZDSF.**

nonblocking system Same as **non-blocking system.**

nonbridging Same as **non-bridging.**

noncoherence Same as **non-coherence.**

noncoherent Same as **non-coherent.**

noncoherent bundle Same as **non-coherent bundle.**

noncoherent demodulation Same as **non-coherent detection.** Also spelled **non-coherent demodulation.**

noncoherent detection Same as **non-coherent detection.**

noncoherent detector Same as **non-coherent detector.**

noncoherent light Same as **non-coherent light.**

noncoherent light source Same as **non-coherent light source.**

noncoherent pulses Same as **non-coherent pulses.**

noncoherent radiation Same as **non-coherent radiation.**

noncoherent scattering Same as **noncoherent scattering.**

noncoherent waves Same as **non-coherent waves.**

nonconductor Same as **non-conductor.**

noncontact detector Same as **non-contact detector.**

noncontact sensing device Same as **non-contact sensing device.**

noncontact sensor Same as **non-contact sensor.**

noncontact temperature measurement Same as **non-contact temperature measurement.**

noncontact thermometer Same as **non-contact thermometer.**

noncorrosive flux Same as **non-corrosive flux.**

noncrystalline Same as **non-crystalline.**

nondedicated server Same as **non-dedicated server.**

nondestructive read Same as **non-destructive read.**

nondestructive readout Same as **non-destructive readout.**

nondestructive test Same as **non-destructive test.**

nondestructive testing Same as **non-destructive testing.**

nondirectional Same as **non-directional.**

nondirectional antenna Same as **non-directional antenna.**

nondirectional beacon Same as **non-directional radio beacon.** Also spelled **non-directional beacon.**

nondirectional microphone Same as **non-directional microphone.**

nondirectional radio beacon Same as **non-directional radio beacon.**

nondispersive medium Same as **non-dispersive medium.**

nonelectric Same as **non-electric.**

nonelectrical Same as **non-electrical.**

nonelectronic Same as **non-electronic.**

nonerasable memory Same as **non-erasable memory.**

nonerasable storage Same as **non-erasable storage.**

nonflammable Same as **non-flammable.**

nonharmonic Same as **non-harmonic.**

nonharmonic component Same as **non-harmonic component.**

nonimpact printer Same as **non-impact printer.**

noninductive Same as **non-inductive.**

noninductive capacitor Same as **non-inductive capacitor.**

noninductive circuit Same as **non-inductive circuit.**

noninductive load Same as **non-inductive load.**

noninductive resistor Same as **non-inductive resistor.**

noninductive winding Same as **non-inductive winding.**

noninterlaced Same as **non-interlaced.**

noninterlaced display Same as **non-interlaced display.**

noninterlaced monitor Same as **non-interlaced monitor.**

noninterlaced scan Same as **non-interlaced scan.**

noninterlaced scanning Same as **non-interlaced scanning.**

noninterlacing Same as **non-interlacing.**

noninverting amplifier Same as **non-inverting amplifier.**

noninverting connection Same as **non-inverting connection.**

noninverting input Same as **non-inverting input.**

noninverting terminal Same as **non-inverting terminal.**

nonionizing radiation Same as **non-ionizing radiation.**

nonlinear Also spelled **non-linear. 1.** Not in the shape of a straight line. Also, pertaining to that which is not in the shape of a straight line. **2.** A relationship in which one value is not directly proportional to another. For instance, an electronic device whose output is not directly proportional to its input. When such a response is graphed, the result is a curve. Also, pertaining to such a relationship. **3.** Not following the course of a straight line. Also, that which is described by a path which is not a straight line.

nonlinear amplifier An amplifier in which the output signal is not linearly proportional to the input. Also spelled **nonlinear amplifier**.

nonlinear capacitor A capacitor in which the capacitance is not linearly proportional to the applied voltage. Also spelled **non-linear capacitor**.

nonlinear circuit A circuit whose output is not linearly proportional to its input. Also spelled **non-linear circuit**.

nonlinear component Same as **nonlinear element**. Also spelled **non-linear component**.

nonlinear control system A control system in which the output quantities are not directly proportional to its input. Also spelled **non-linear control system**.

nonlinear detection The use of a **nonlinear detector**. Also spelled **non-linear detection**.

nonlinear detector A detector in which the output quantity or signal is not linearly proportional to its input. Also spelled **non-linear detector**.

nonlinear device A device in which the output quantity or signal is not linearly proportional to the input. Also spelled **non-linear device**.

nonlinear dielectric A dielectric in which the dielectric constant is not linearly proportional to the applied voltage. Also spelled **non-linear dielectric**.

nonlinear distortion Also spelled **non-linear distortion**. **1.** Distortion in which the waveform of the input signal is altered, due to a nonlinear response. **2.** Distortion in which the waveform of the input signal is altered, and in which there are harmonics introduced. **Linear distortion** does not introduce harmonics. **3.** In an amplifier, the component of the output signal which alters the essential characteristics of the input signal, due to a nonlinear response. Also called **amplitude distortion (2)**.

nonlinear element A circuit element whose output is not linearly proportional to its input. Also spelled **non-linear element**. Also called **nonlinear component**.

nonlinear feedback Feedback in which the output quantity or signal, such as a corrective action, is not linearly proportional to its input. Also spelled **non-linear feedback**.

nonlinear material A material whose variation in one or more properties is not linearly proportional to an external force. For instance, a magnetic material whose permeability changes nonlinearly relative to a change in magnetomotive force. Also spelled **non-linear material**.

nonlinear modulation Modulation in which the characteristic of a carrier wave being modified is not proportional to the amplitude of the information-bearing signal. Also spelled **non-linear modulation**.

nonlinear network An electric network whose output is not linearly proportional to its input. Also spelled **non-linear network**.

nonlinear oscillator An oscillator whose output is not linearly proportional to its input. Also spelled **non-linear oscillator**.

nonlinear phase response Variations in phase response in which the changes in phase are not linearly proportional to the frequency. Also spelled **non-linear phase response**.

nonlinear resistor A resistor in which the resistance is not linearly proportional to the applied voltage. Also spelled **non-linear resistor**.

nonlinear response A response, such as that of a component, circuit, or device, which when graphed does not result in a straight line. That is, its output is not linearly proportional to its input. Also spelled **non-linear response**.

nonlinear system Also spelled **non-linear system**. **1.** A system whose output or response is not linearly proportional to its input. **2.** A system whose output is the not the sum of said system's response to each of its multiple inputs considered separately.

nonlinear taper A taper, such as that of a potentiometer, in which resistance variations are not proportional to shaft rotation. An example is an audio taper. This contrasts with a **linear taper**, in which the resistance variations for a given rotation through a given angle is the same along the entire range of the shaft. Also spelled **non-linear taper**.

nonlinear transducer A transducer whose output is not linearly proportional to its input. Also spelled **non-linear transducer**.

nonlinear video editing The use of computers to store and manipulate video. By having the images in digital form, tasks such as rearranging the order of scenes, copying selections, skipping unwanted parts, and undoing previous actions are greatly simplified. Also spelled **non-linear video editing**.

nonlinearity Also spelled **non-linearity**. **1.** The quality and/or state of being **nonlinear**. **2.** The degree to which **nonlinearity (1)** exists.

nonlinearity error Any error occurring as a consequence of a nonlinear component, circuit, device, piece of equipment, or system. Also spelled **non-linearity error**.

nonmagnetic Same as **non-magnetic**.

nonmaskable interrupt Same as **non-maskable interrupt**.

nonmetal Also spelled **non-metal**. **1.** A chemical element which usually accepts electrons to form negative ions, whose oxides are acidic, and which tend to be poor conductors of heat and electricity, although some have semiconductor properties. At ordinary temperatures and pressures, a nonmetallic element may be a solid, such as carbon or silicon, a liquid, such as bromine, or a gas, such as oxygen or xenon. Some nonmetals, such as boron and carbon, have crystalline allotropic forms. **2.** Any element or material which is not a metal.

nonmetallic Consisting of, incorporating, pertaining to, or characteristic of a **nonmetal**. Also spelled **non-metallic**.

nonmicrophonic Same as **non-microphonic**.

nonoperational Same as **non-operational**.

nonparametric test Same as **non-parametric test**.

nonparity memory Same as **non-parity memory**.

nonparity RAM Same as **non-parity RAM**.

nonperiodic signal Same as **non-periodic signal**.

nonperiodic waveform Same as **non-periodic waveform**.

nonplanar Same as **non-planar**.

nonplanar circuit Same as **non-planar circuit**.

nonpolar Also spelled **non-polar**. **1.** Having no poles. **2.** Having no electric dipole moment.

nonpolarized capacitor Same as **non-polarized capacitor**.

nonpolarized electrolytic capacitor Same as **non-polarized electrolytic capacitor**.

nonpreemptive multitasking Same as **non-preemptive multitasking**.

nonprocedural language Same as **non-procedural language**.

nonradiative transition An energy transition which does not result in the emission of radiation. For instance, one in which a particle losing energy does not emit a photon. The converse is true in a **radiative transition**. Also spelled **non-radiative transition**.

nonreactive Also spelled **non-reactive**. **1.** Said of a chemical element or compound which reacts very little, or not at all, with other elements or compounds. **2.** Said of a component, circuit, or device, with little or no reactance.

nonrechargeable battery Same as **non-rechargeable battery**.

nonrechargeable cell Same as **non-rechargeable cell**.

nonrecurrent Same as **non-recurrent**.

nonrecurring Same as **non-recurrent**. Also spelled **non-recurring**.

nonresonant Same as **non-resonant**.

nonresonant antenna Same as **non-resonant antenna**.

nonresonant line Same as **non-resonant line**.

nonresonant transmission line Same as **non-resonant line**. Also spelled **non-resonant transmission line**.

nonreturn to zero Same as **non-return to zero**.

nonreturn to zero code Same as **non-return to zero code**.

nonreturn to zero coding Same as **non-return to zero coding**.

nonreturn to zero encoding Same as **non-return to zero encoding**.

nonshorting switch Same as **non-shorting switch**.

nonsinusoidal wave Same as **non-sinusoidal wave**.

nonsinusoidal waveform Same as **non-sinusoidal waveform**.

nonsynchronous Same as **non-synchronous**.

nontrigger voltage Same as **non-trigger voltage**.

nonuniform electric field Same as **non-uniform electric field**.

nonuniform field Same as **non-uniform field**.

nonuniform magnetic field Same as **non-uniform magnetic field**.

nonuniform memory access Same as **non-uniform memory access**.

nonvolatile memory Same as **non-volatile memory**.

nonvolatile storage Same as **non-volatile storage**.

nonzero dispersion-shifted fiber Same as **non-zero dispersion-shifted fiber**.

NOP Abbreviation of **no-operation instruction**.

NOR Abbreviation of **Not OR**. A logical operation which is true only if all of its elements are false. For example, if all of its multiple inputs have a value of 0, then the output is 1. Any other combination yields an output of 0. For such functions, a 1 is considered as a true, or high, value, and 0 is a false, or low, value. Also called **NOR operation**.

NOR circuit Same as **NOR gate**.

NOR gate A circuit which has two or more inputs, and whose output is high only when all inputs are low. Also called **NOR circuit**.

NOR operation Same as **NOR**.

normal **1.** Conforming to that which is standard, common, typical, or expected. **2.** A value or magnitude which is **normal** (1). **3.** Perpendicular to. For example, a line which is at a right angle to a surface. **4.** A chemical solution whose concentration is one equivalent weight of solute per liter of solution. Also called **normal solution**.

normal-blow fuse A fuse other than a slow-blow or a fast-blow fuse. The term is utilized to contrast with these terms.

normal channel A communications channel utilized for ordinary communications. This contrasts, for instance, with an emergency channel.

normal conditions **1.** A combination of environmental conditions which are established for any given setting or purpose. For example, that in which the temperature is 0 °C, the pressure is 1 atmosphere, and there is 0% relative humidity. **2.** Same as **normal temperature and pressure**.

normal curve A curve corresponding to a **normal distribution**. It is in the shape of a symmetrical bell. Also called **bell-shaped curve**, or **Gaussian curve**.

normal distribution For a random variable, a probability distribution which is symmetrical about the mean value, and which continuously diminishes in value until reaching zero at each extreme. It is utilized to determine the probability of the value of the variable falling within a given interval of values, and when graphed has the shape of a symmetrical bell. Various phenomena, such as some natural frequencies, have a normal distribution. Also called **bell-shaped distribution**, or **Gaussian distribution**.

normal electrode An electrode, such as a hydrogen electrode, which serves as a reference standard.

normal form In **normalization** (1), one of the stages utilized to convert complex data structures into simpler relations. The higher the normal form, the more advanced the normalization. For example, the second normal form is more refined than the first, the third more than the second, and so on.

normal impedance For a transducer, the input impedance when the load impedance is zero. Also called **free impedance**.

normal mode A typical manner in which a component, circuit, device, piece of equipment, or system is operated or used. There may be more than one such mode.

normal-mode rejection Its abbreviation is **NMR**. **1.** The ability of an instrument or device, such as an amplifier, to reject noise occurring at the power-line frequency, usually 50 or 60 Hz, or any harmonics of said frequency. **2.** The extent to which **normal-mode rejection** (1) occurs. Also called **normal-mode rejection ratio**.

normal-mode rejection ratio Same as **normal-mode rejection** (2). Its abbreviation is **NMRR**.

normal operating conditions The circumstances, such as ambient conditions, input voltage, and output current, which are required for the proper functioning of a component, circuit, device, piece of equipment, or system. Also called **operating conditions** (1), or **standard operating conditions**.

normal position The resting state or deenergized position of the contacts of a switch or relay. For example, a normally-open switch has a contact pair that is open in the resting state, and closed when energized.

normal pressure A pressure, such as 1 atmosphere, which is used under a given setting, such as that in a laboratory. Also called **standard pressure**.

normal solution Same as **normal** (4).

normal state For a particle, or system of particles, the lowest energy state. Any energy state above this is an excited state. Also called **ground state**.

normal temperature A temperature, such as 0 °C, which is used under a given setting, such as that in a laboratory. Also called **standard temperature**.

normal temperature and pressure Its abbreviation is **NTP**. Also called **normal conditions** (2), **standard temperature and pressure**, or **standard conditions** (2). **1.** A temperature of 0 °C, and a pressure of 1 atmosphere. Used, for instance, as a practical and easily maintained condition for laboratory experimentation. **2.** An established combination of temperature and pressure, such as of 25 °C and 1 atmosphere, utilized for any given setting or purpose.

normalization **1.** In relational database management, the process of converting complex data structures into simpler relations which minimize or eliminate redundancies, inconsistencies, and ambiguities. Normalization involves the creation of multiple tables which provide for the most flexible and stable data structure obtainable, so that future maintenance of said structure is kept to a minimum. **2.** In programming, the process of changing the format of a number so that it fits within a prescribed range. For example, to adjust the mantissa and exponent of a floating-point number so that portions of it are in a specified range.

normalize 1. In relational database management, to organize data in a manner which minimizes redundancies, inconsistencies, and ambiguities. **2.** In programming, to change the format of a number so that it fits within a prescribed range.

normalized admittance The reciprocal of **normalized impedance**.

normalized impedance The ratio of a wave impedance to the characteristic impedance of a transmission line, such as waveguide. Its reciprocal is **normalized admittance**.

normally closed Same as **normally-closed contact**. Its abbreviation is **NC**.

normally-closed contact In a relay or switch, a contact pair that is closed in the resting state, and which is opened when energized. The contact pair consists of a stationary contact and a movable contact. A normally closed contact provides a complete circuit when not energized. This contrasts with a **normally-open contact**, in which the contact pair is open in the resting state. Its abbreviation is **NC contact**. Also called **normally closed**, **break contact**, or **back contact**.

normally-closed relay A relay with **normally-closed contacts**.

normally-closed switch A switch with **normally-closed contacts**.

normally open Same as **normally-open contact**. Its abbreviation is **NO**.

normally-open contact In a relay or switch, a contact pair that is open in the resting state, and which is closed when energized. The contact pair consists of a stationary contact and a movable contact. A normally-open contact provides a complete circuit when energized. This contrasts with a **normally-closed contact**, in which the contact pair is closed in the resting state. Its abbreviation is **NO contact**. Also called **normally open**, **make contact**, or **front contact**.

normally-open relay A relay with **normally-open contacts**.

normally-open switch A switch with **normally-open contacts**.

North America Numbering Plan Telephones in North America usually consist of ten digits, in the following format: NPA-NXX-XXXX. NPA is the area code, NXX is the exchange, and XXXX the line number. N can be any digit between 2 and 9, P can be any digit except 9, and X can be any digit between 0 and 9. Its abbreviation is **NANP**.

North bridge Same as **Northbridge**.

North bridge chip Same as **Northbridge chip**.

North bridge chipset Same as **Northbridge chipset**.

north geographic pole Same as **north pole (1)**.

north magnetic pole The location in the northern hemisphere where the magnetic dip is 90°. The corresponding spot in the southern hemisphere is the **south magnetic pole**. At each of these places, the magnetic meridians converge. The north magnetic pole does not coincide with the north geographic pole, and the position of both magnetic poles vary over time. Also called **north pole (2)**, **magnetic north pole**, or **positive pole (2)**.

north pole 1. On the surface of the planet earth, the point within the northern hemisphere where all meridians meet. It is defined as 90° N, and is the northern end of the axis of rotation of the planet. Also called **north geographic pole**, or **geographic north pole**. **2.** Same as **north magnetic pole**. **3.** In a magnet or magnetic body, one of the two regions where the magnetic lines of force converge, and hence magnetic intensity is at a maximum. The other region is the **south pole (3)**. Such a pole, when part of a magnet which is

freely suspended, will seek a south pole, as opposite poles attract. Also called **positive pole (1)**.

north-seeking pole In a magnet which is freely suspended, the pole which will point in the direction of the **north magnetic pole**.

Northbridge In certain CPU chipset architectures, one or more chips that connect the CPU to the RAM, AGP, PCI buses, and L2 cache, while the **Southbridge** controls the CPUs I/O functions, connecting to the USB and IDE buses, keyboard controller, and the like. Also spelled **North bridge**.

Northbridge chip A **Northbridge** consisting of a single chip. Also spelled **North bridge chip**.

Northbridge chipset A **Northbridge** consisting of a multiple chips. Also spelled **North bridge chipset**.

northern lights An aurora occurring in the northern hemisphere. An aurora is a luminous phenomenon of the upper atmosphere occurring mostly in the high latitudes of both hemispheres. They are caused by the interaction of excited particles from space and particles of the upper atmosphere, and usually affect radio communications. Also called **aurora borealis**.

Norton's theorem In the analysis of electric networks, a theorem stating that combinations of two-terminal AC voltage sources and impedances can be represented by a single current source and a parallel impedance. A DC source would be represented by a single current source and a parallel resistor.

NOS Acronym for **network operating system**.

NOT A logic operator which inverts the input. If its single input is 0, then its output is 1. And, if its input is 1, then its output is 0. For example, a NOT operation converts 10101 into 01010. For such operations, a 1 is considered as a true, or high, value, and 0 is a false, or low, value. Also called **NOT operation**, or **negation**.

Not AND Same as **NAND**.

Not AND circuit Same as **NAND circuit**.

Not AND gate Same as **NAND gate**.

not-applicable Pertaining to something which does not, or can not, apply for a specified use, or within a given context. For example, the Curie law not applying to certain liquids or solids. Its abbreviation is **N/A**, or **NA**.

not-available Pertaining to something which is not available in a given context, or at a specified time. For instance, a device which is not ready for use at the moment. Its abbreviation is **N/A**, or **NA**.

NOT circuit Same as **NOT gate**.

NOT gate A circuit which has a single input, and whose output is high when the input is low, and vice versa. Also called **NOT circuit**, or **inverter (2)**.

NOT operation Same as **NOT**.

NOT-OR Same as **NOR**.

NOT-OR circuit Same as **NOR circuit**.

NOT-OR gate Same as **NOR gate**.

not ready A message indicating that a circuit, device, piece of equipment, or system, is not available or prepared for the desired use or function.

notation A system of characters or symbols utilized to represent numbers, values, or quantities. For example, scientific notation, complex notation, floating-point notation, binary notation, or decimal notation.

notch 1. A V-shaped cut or indentation. Also, to make such a cut or indentation. **2.** A cut, indentation, perforation, or the like which serves to distinguish, indicate, or identify. For example, that utilized to indicate where a record ends. **3.** That which is in the shape of a **notch (1)**, such as the frequency response of a notch filter.

notch filter A type of bandstop filter which attenuates a very narrow band. The dip in the frequency response of such a filter resembles a notch.

note 1. A tone of a specified pitch. 2. A symbol utilized to represent a **note (1)**.

notebook Same as **notebook computer**.

notebook computer A portable computer similar to a laptop computer, but which is generally lighter and smaller. Since the weight and dimensions of laptop computers have steadily dropped, some models to currently around one kilogram, the terms are mostly synonymous. Also called **notebook**.

Notes A popular groupware product.

NOW Acronym for **Network of Workstations**.

noy A subjective unit of noisiness which depends on the noise level perceived by a given person listening to a random noise signal at 40 dB SPL between the frequencies of 910 and 1090 Hz. A sound of 2 noys is twice as noisy as a 1 noy sound, 2 is half as noisy as 4, and so on. It is based on a linear scale.

np Abbreviation of **negative-positive**. It refers to the region where a p-type semiconductor and an n-type semiconductor meet. In a p-type semiconductor, holes, which act as positively-charged particles, are the majority carrier, while in an n-type semiconductor electrons, which are negative particles, are the majority carrier. Also called **pn**.

Np 1. Chemical symbol for **neptunium**. 2. Abbreviation of **neper**.

NP Same as **np**.

np junction The relatively narrow region where a p-type semiconductor and an n-type semiconductor meet. A homojunction is that between layers of the same semiconductor material, each layer having different properties, and a heterojunction is the boundary between layers of different semiconductor materials. Most diodes consist of an np junction, and bipolar junction transistors have two np junctions. Also called **pn junction**.

NPA-NXX-XXXX format The number format used in the North America Numbering Plan.

NPN transistor A bipolar junction transistor with n-type material emitter and collector regions separated by a p-type material base region. A **PNP transistor** has p-type emitter and collector regions separated by an n-type base region.

NPO capacitor A type of capacitor whose capacitance remains essentially the same over a wide temperature interval.

nR Abbreviation of **nanoroentgen**.

nrad 1. Abbreviation of **nanorad**. 2. Abbreviation of **nanoradian**.

nrem Abbreviation of **nanorem**.

nrt-VBR Abbreviation of **non-real-time VBR**.

NRZ Abbreviation of **non-return to zero**.

NRZ code Abbreviation of **non-return to zero code**.

NRZ coding Abbreviation of **non-return to zero coding**.

NRZ encoding Abbreviation of **non-return to zero encoding**.

NRZI Same as **non-return to zero inverted**.

NRZI code Abbreviation of **non-return to zero inverted code**.

NRZI coding Abbreviation of **non-return to zero inverted coding**.

NRZI encoding Abbreviation of **non-return to zero inverted encoding**.

ns Abbreviation of **nanosecond**.

nsec Abbreviation of **nanosecond**.

NSP Abbreviation of **Network Service Provider**.

NT A common operating system.

nt Abbreviation of **nit**.

nT Abbreviation of **nanotesla**.

NT1 Abbreviation of **network terminator 1**. An ISDN line-terminating unit located at a customer location. It accepts a two-wire signal from the communications provider and converts it into a four-wire signal used within the premises of a customer. It may be built into devices or equipment, or may stand alone. A T-interface links an NT1 and an **NT2**.

NT2 Abbreviation of **network terminator 2**. In ISDN, a device, such as a PBX, located at a customer location, and which interfaces, via a T-interface, with an **NT1**.

NTC Abbreviation of **negative temperature coefficient**.

nth Pertaining to an unspecified number. For example, the nth root.

NTP 1. Abbreviation of **Network Time Protocol**. 2. Abbreviation of **normal temperature and pressure**.

NTSC Abbreviation of National Television Standards Committee. 1. The entity which established the **NTSC Standard**. 2. Same as **NTSC Standard**.

NTSC signal Abbreviation of National Television Standards Committee **signal**. A TV signal adhering to the **NTSC Standard**.

NTSC Standard Abbreviation of National Television Standards Committee **Standard**. A standard for the generation, transmission, and reception of TV signals. It provides for a resolution of 525 horizontal lines per frame, with 60 interlaced half-frames per second for a total of 30 frames per second. Of the 525 total lines, approximately 485 are active, while the aspect ratio is 4:3. NTSC is used in most of America and some Asian countries, while the other two main world standards are **PAL** and **SECAM**. All three are mutually incompatible. Also called **NTSC (2)**.

nuclear 1. Of, pertaining to, utilizing, or derived from **nuclei**. 2. Of, pertaining to, utilizing, or derived from **nuclear energy**.

nuclear battery A battery in which radioactive energy is converted into electrical energy. These batteries are ideally suited for uses requiring high energy density, reliability, and longevity. Also known as **atomic battery**.

nuclear bombardment The subjecting of a nucleus to impacts by high-energy particles.

nuclear chain reaction The bombardment of neutrons upon an unstable nucleus, such as that of uranium, which initiates a sequence of nuclear fissions whose products include other elements, a substantial amount of energy, and more neutrons, which further propagate the reaction. Also called **chain reaction (3)**.

nuclear energy Energy released by reactions involving atomic nuclei. There are three types of nuclear energy: (1) the splitting, or fission, of nuclei, (2) the union, or fusion, of nuclei, and (3) the radioactive decay of nuclei. Also known as **atomic energy**.

nuclear fission The splitting of atomic nuclei, with the concomitant release of an enormous amount of energy. Used, for instance, for nuclear power.

nuclear force The force which holds together an atomic nucleus. Its range is extremely short, and its force carrier is the gluon. It is the strongest of the **four fundamental forces of nature**, the others being the gravitational force, the weak force, and the electromagnetic force. Also called **strong force**, or **strong nuclear force**.

nuclear fusion The joining of two or more atomic nuclei to form a larger nucleus, with the concomitant release of a large amount of energy. Used, for instance, for nuclear power, or to synthesize atoms with higher mass numbers.

nuclear magnetic resonance Its abbreviation is **NMR**. 1. Resonance of certain atomic nuclei which are exposed to radio waves in the presence of a magnetic field. The powerful magnetic field makes certain nuclei in molecules, especially those incorporating hydrogen-1 and carbon-13, orient themselves parallel to this external field. The RF energy is pulsed to make the nuclei oscillate between energy states, different frequencies provide varying levels of energy absorption, and such differences provide information about the structure of analyzed molecules. Observed, for instance, in nuclear magnetic resonance imaging. **2.** An analytical technique utilizing **nuclear magnetic resonance (1)** for the determination of the structure and composition samples. Also called **nuclear magnetic resonance spectroscopy**.

nuclear magnetic resonance imaging The use of magnetic resonance to create cross-sectional images of internal body structures. In it, a patient is placed in an enclosure surrounded by powerful electromagnets which magnetically align the nuclei of certain atoms in the body. Then, radio waves are directed at these nuclei, usually the proton in hydrogen, and withdrawn, so that when the nuclei revert to their original state they emit radio signals which a computer utilizes to produce the desired images. It is a very sensitive and noninvasive procedure utilized to help study and diagnose a broad range of medical conditions, including cancer, heart disease, and musculoskeletal disorders. Also called **magnetic resonance imaging**. Its abbreviation is **NMR imaging**.

nuclear magnetic resonance spectrometer An instrument utilized for **nuclear magnetic resonance spectroscopy**. Its abbreviation is **NMR spectrometer**.

nuclear magnetic resonance spectroscopy Same as **nuclear magnetic resonance (2)**. Its abbreviation is **NMR spectroscopy**.

nuclear magneton A physical constant equal to approximately 5.0508×10^{-27} joule/tesla. It is used as a unit of magnetic moment. Its symbol is μ_N. Also called **magneton (2)**.

nuclear medicine The use of radioactive isotopes to help prevent, diagnose, and treat diseases. In prevention and diagnosis, for instance, small amounts of short-lived radioisotopes are introduced into the body. Specific organs, bones, and tissues absorb varying amounts, which can enhance techniques utilized for imaging internal body structures. In the treatment of disease, for example, radioisotopes are used to combat cancerous cells.

nuclear pile A deprecated term for **nuclear reactor**.

nuclear power Also known as **atomic power**. **1.** Power derived by reactions involving atomic nuclei. The three source of such power are fission, fusion, and radioactive decay. **2.** The use of **nuclear power (1)** to generate electricity.

nuclear power plant A location which combines the structures, devices, and equipment necessary to produce nuclear power in quantities adequate for residential, commercial, or industrial use. The key component of such a location is the **nuclear reactor**, whose generated heat can be utilized to produce steam which drives the turbines of an electric generator. Also called **nuclear power station**, or **atomic power plant**.

nuclear power station Same as **nuclear power plant**.

nuclear reaction Any reaction, such as fission or fusion, which changes the structure or composition of an atomic nucleus. Also called **atomic reaction**.

nuclear reactor A device which utilizes controlled nuclear fusion or fission to generate energy. A fission reactor, for instance, has the following key components: a fissionable fuel, such as uranium-235, a moderator, such as heavy water, shielding, such as that provided by thick concrete, control rods, such as those made with boron, and a coolant such as water. The heat produced by the nuclear reactions can be utilized to generate electricity. Also called **atomic reactor**,

or **reactor (2)**. Deprecated terms for this concept include **nuclear pile**, **atomic pile**, and **pile (2)**.

nuclei A plural form of **nucleus**.

nucleon A proton or a neutron.

nucleon number The number of protons and neutrons in the nucleus of an atom. For example, the mass number of the most common isotope of carbon is 12, as it has 6 protons and 6 neutrons. Its symbol is **A**. Also called **mass number**.

nucleonics **1.** The science dealing with the study, characteristics, and phenomena of nucleons or of atomic nuclei. **2.** The science dealing with the study, characteristics, phenomena, and applications of nuclear energy.

nucleophilic Pertaining to a species, such as an ion or polar molecule, which is attracted to nuclei or regions of low electron density. Such species are electron donors. This contrasts with **electrophilic**, in which a species is attracted to electrons or regions of high electron density.

nucleus Its plural forms are **nuclei** and **nucleuses**. **1.** The positively-charged dense central core of an atom. All atomic nuclei, except protium, which is hydrogen-1, consist of protons and neutrons. The nucleus of hydrogen-1 consists of a single proton. In an uncharged atom, the number of electrons is the same as the number of protons. When two or more atoms have the same atomic number, but different atomic masses, they are isotopes of the same element. A nucleus is held together by the nuclear force, and comprises almost all of the mass of an atom. Also called **atomic nucleus**. **2.** The central or key part of something. **3.** The part of a control program, such as an operating system, which is memory resident.

nuclide **1.** Any species of atom that exits for an interval greater than a given minimum, such as one yoctosecond. Nuclides are distinguished by their atomic mass, atomic number, and energy state of the nucleus. For example, carbon-14 is a nuclide of carbon. A nuclide which has the same atomic mass as another, but whose atomic number is different is called an **isobar (2)**. **2.** Any **nuclide (1)** distinguished only by its atomic mass and atomic number.

NUL Same as **null character**.

null **1.** A quantity of zero. Also, that which has zero magnitude, or is immeasurably close to such a value. **2.** An instrument or meter reading of zero. **3.** A position or condition which results in a value or output of zero. For example, when a bridge is balanced, its output is zero. **4.** Within the directivity pattern of an antenna, a point of minimum or zero signal strength. **5.** Same as **null character**.

null balance **1.** A condition or configuration in which there is no current flow. **2.** A condition or configuration in which there is no output.

null character A character code which has no value, but which occupies space. Used, for instance, to reserve a space that may later have a value. Also called **NUL**, or **null (5)**.

null current For a circuit or device, an output current of zero, as is the case when a bridge is balanced.

null detection The reading of a **null detector** when an analyzed circuit is balanced. Also called **null indication**.

null detector An instrument or device which detects and indicates a **null reading**. Also called **null indicator**, or **null meter**.

null frequency A frequency at which a **null reading** is obtained.

null indication Same as **null detection**.

null indicator Same as **null detector**.

null meter Same as **null detector**.

null method A method of measurement in which a reading of zero is at the center of the scale. Indicated values may be greater than or less than the zero reading. One variation is the use of an audible signal for measurement, in which case the zero reading would be silent. Also called **zero method**, or **balanced method**.

null modem   **1.** To use a **null modem cable** to connect two computers. **2.** Same as **null modem cable**.

null modem cable   A special cable which allows two computers to be connected to each other without using modems. For instance, the use of a cable which crosses the inputs and outputs to connect two computers via an RS-232 port. In this way, the signal from the output pins of one computer serves as the input for the other. Used, for example, for exchanging computer files, or for interactive play between users of different computers located in the same room. Such cables usually require the distance between the computers to be comparatively short. Also called **null modem (2)**.

null point   The point at which a **null reading** is obtained. For instance, the point at which a bridge is balanced.

null pointer   **1.** In programming, a pointer which directs to, or specifies a value of zero, or which points to nothing. **2.** In programming, a pointer with a value of zero.

null reading   An indication of an output value, such as a current or a voltage, of zero. Used, especially in the context of bridge circuits.

null string   In programming, a string containing no elements or characters.

null voltage   For a circuit or device, an output voltage of zero, as is the case when a bridge is balanced.

num lock   Abbreviation of **numlock key**.

num lock key   Same as **numlock key**.

NUMA   Abbreviation of **non-uniform memory access**.

number   Its abbreviation is **No. 1.** An arithmetic value utilized to represent a quantity, magnitude, order, rank, or the like. **2.** A symbol utilized to represent a **number (1)**. **3.** A **number (1)** which serves for reference or identification.

number cruncher   A computer, such as a supercomputer, capable of performing enormous amounts of calculations extremely rapidly. Also, an application, such as a statistical program, which is intense in numerical computations.

number crunching   **1.** What a **number cruncher** does. **2.** Pertaining to a **number cruncher**.

number field   A field into which only numbers may be entered. This contrasts, for instance, with a text field.

number system   Same as **numbering system**.

numbering system   A system utilized to name and/or represent numbers. Examples include the binary, decimal, and hexadecimal number systems. Also called **number system**, **numeration system**, or **numeral system**.

numeral system   Same as **numbering system**.

numeration system   Same as **numbering system**.

numeric beeper   Same as **numeric pager**.

numeric co-processor   Same as **numeric coprocessor**.

numeric coprocessor   An arithmetic-logic unit designed to perform floating-point operations. A CPU may have such a processor incorporated, or a second chip may be utilized. Numeric coprocessors are especially useful when a computer performs sophisticated calculations, such as those frequently needed in CAD, spreadsheet, and scientific programs. Also spelled **numeric co-processor**. Also called **floating-point processor**, **floating-point coprocessor**, **floating-point unit**, or **math coprocessor**.

numeric data   Data presented in the form of numbers. Also called **numerical data**.

numeric display   A display whose output is in the form of digits, or numbers. Also called **numerical display**, **numeric readout**, **numerical readout**, **digital display**, or **digital readout**.

numeric keypad   A small grid of keys consisting of numbers, and which usually incorporates other keys frequently utilized with such numbers. For example, such a keypad on a telephone, calculator, or that occupying the right portion of a full computer keyboard. Also called **numerical keypad**.

numeric pager   A pocket-sized radio receiver or transceiver that can only display received messages which consist of up to a given amount of numbers. Also called **numeric beeper**.

numeric readout   Same as **numeric display**.

numerical analysis   **1.** The branch of mathematics dealing with the creation and use of calculations and algorithms to analyze and solve problems. **2.** The application of **numerical analysis (1)** to solve specific problems.

numerical aperture   A measure of the light-gathering capability of an optical system or element, such as a lens or fiber. In the case of a lens or optical fiber, for instance, it can be approximately calculated as the sine of the acceptance angle multiplied by the refractive index of the material or medium in contact with the fiber, which is usually air. Its abbreviation is **NA**.

numerical control   A form of automatic control of tools and machines. For instance, the instructions of a computer program controlling the position, movements, and speed of a drill in a manufacturing process. Its abbreviation is **NC**. Also called **computerized numerical control**, or **computer numerical control**.

numerical data   Same as **numeric data**.

numerical display   Same as **numeric display**.

numerical keypad   Same as **numeric keypad**.

numerical readout   Same as **numeric display**.

numlock key   On computer keyboards, a key whose function is to toggle the numeric keypad between number entry and cursor and page location movement. Also spelled **num lock key**. Its abbreviation is **num lock**.

nutation   A periodic variation, such as a wobble, in a spinning or rotating body, such as a gyroscope.

nutator   An electrical and/or mechanical device utilized to rotate the axis of the beam of an antenna, radar, beacon, or the like.

nV   Abbreviation of **nanovolt**.

nV/m   Abbreviation of **nanovolts per meter**.

NVOD   Abbreviation of **Near Video on Demand**.

NVRAM   Abbreviation of **non-volatile RAM**.

nW   Abbreviation of **nanowatt**.

NXX code   The code, or prefix, which identifies a central office. In the NPA-NXX-XXXX ten-digit telephone number format, it is the NXX portion, where N can be any digit between 2 and 9, and X can be any digit between 0 and 9. A single central office may serve more than one NXX code. Also called **central office prefix**, **central office code**, or **prefix (1)**.

nybble   Same as **nibble**.

nylon   A strong and flexible synthetic thermoplastic that is highly impact-resistant, and which absorbs very little moisture. It has high tensile strength, has electrical insulation properties, and is available in many forms, including sheets and molded plastics.

Nyquist diagram   A diagram which facilitates the prediction of the stability and performance of a closed-loop control system by observing its open-loop behavior.

Nyquist frequency   A frequency which is equal to half the **Nyquist rate**. Also called **folding frequency**.

Nyquist rate   The minimum sampling rate that enables a signal to be faithfully reproduced. Sampling at or above this frequency prevents aliasing.

NZ-DSF Abbreviation of **non-zero dispersion-shifted fiber**.

NZDSF Abbreviation of **non-zero dispersion-shifted fiber**.

nΩ Abbreviation of **nanohm**.

O

O 1. Chemical symbol for **oxygen**. 2. Abbreviation of **output**.

O display In radars, an oscilloscopic display similar to an A display, but which incorporates an adjustable pedestal to indicate the range of the scanned object. Also called **O scope**, **O scanner**, **O scan**, or **O indicator**.

O electron An electron located in the O shell of an atom.

O indicator Same as **O display**.

O-network An electric network with four impedance branches that are connected in series, and which has two input and two output terminals. Two adjacent terminals serve as input, the other two as output.

O-ray Abbreviation of **ordinary ray**. One of the two rays into which a ray of light incident upon a doubly-refracting material is split, the other being the **E-ray**. Both rays are plane-polarized perpendicular to each other, and the ordinary ray is completely absorbed, while the extraordinary ray passes through the material.

O scan Same as **O display**.

O scanner Same as **O display**.

O scope Same as **O display**.

O shell The fifth electron shell of an atom. It is the fifth closest shell to the nucleus.

O-wave Abbreviation of **ordinary wave**.

O-wave component Abbreviation of **ordinary-wave component**.

OA Abbreviation of **optical axis** or **optic axis**.

OATS Abbreviation of **open-area test site**.

object 1. A material or virtual item or thing. For instance, a ball, or a ball displayed on a computer screen. 2. In computers, a single item or entity which can be selected and manipulated. An example of such an item is a ball displayed on a screen while running a graphics program. 3. In object-oriented programming, an item or variable which is treated as a distinct entity, and which incorporates both data and the associated routines necessary to manipulate it. Examples include a spell checker, or a graphics routine utilized to draw a ball. Objects are characterized by modularity, and utilize information hiding. 4. A figure seen through, or imaged by, an optical system. The image so formed may be real or virtual. An object may be considered as a collection of points serving as the source of light rays for such a system.

object-based Based on, or pertaining to, an **object-oriented approach**.

object bus A software bus which serves as the interface via which **objects** (3) are transferred. Typically used within a CORBA architecture.

object code Computer code which has been compiled or translated into executable form. This is performed by an assembler, compiler, or interpreter, which transforms source code to machine code, which is then directly executed by the CPU. Object code is frequently the same as machine code, or is very similar to it.

object computer Also called **target computer**, or **destination computer**. 1. A computer to which a transmission is sent. 2. A computer into which a program is loaded, or to which data is transferred. 3. A computer to which compiled, assembled, or translated source code is sent.

object database Same as **object-oriented database**.

object database management system Same as **object-oriented database management system**. Its abbreviation is **ODBMS**, or **object DBMS**.

object DBMS Abbreviation of **object database management system**.

object ID Same as **object identifier**.

object identifier A pointer, name, or symbol which uniquely identifies an **object** (3). Its abbreviation is **object ID**.

object language 1. The low-level language, such as machine language, into which a source language is converted by an assembler, compiler, or interpreter. The source language is usually a high-level language. Also called **target language**. 2. Same as **object-oriented programming language**.

Object Linking and Embedding A technique utilized for linking and embedding **objects** (3) into documents, which facilitates the sharing of data between applications. When an object is embedded, the document itself contains a copy of the object, and when an object is linked, it points to the application which contains the object. Its acronym is **OLE**.

Object Management Architecture An architecture or application within which **objects** (3) interact via an Object Request Broker, according to an Object Management Group specification. Its abbreviation is **OMA**.

Object Management Group An organization that endorses open standards for object-oriented applications, and which distributes the specifications it prepares for such an environment. Its abbreviation is **OMG**.

object mobility The ability to move an **object** (3) from one system to another without any source code changes.

object model 1. A description or representation that includes the data attributes and functions of an **object** (3). 2. A description or representation of an object-oriented application.

Object Modeling Technique A developmental methodology utilized for the analysis, design, and use of **objects** (3). Its abbreviation is **OMT**.

object module All or part of the output of an assembler, compiler, or interpreter, which is ready for linking with other such output.

object oriented Its abbreviation is **OO**. 1. Same as **object-oriented approach**. 2. Pertaining to, or characteristic of, an **object-oriented program**, **object-oriented approach**, **object-oriented database**, or the like, which supports **objects** (3).

object-oriented analysis An analysis of a problem or system from the perspective of components, or objects, which interact with each other to perform the necessary tasks for completion. Its abbreviation is **OOA**.

object-oriented application An application created utilizing **object-oriented programming**.

object-oriented approach A modular approach to the design and development of programs, in which independently prepared and self-contained units, or **objects** (3), are combined to form the final product. Such an approach emphasizes flexibility and scalability, and generally focuses on the problems which wish to be solved, as opposed to how tasks will be performed. Its abbreviation is **OOA**. Also called **object oriented** (1).

object-oriented database A database specifically designed to work in an object-oriented programming environment. Data of various types may be stored, including text, graphics, sound, and video, and it provides database management system capabilities to **objects** (3) created by object-oriented programming languages. Its abbreviation is **OODB**. Also called **object database**.

object-oriented database management system A database management system that supports the modeling, creation, storage, and retrieval of data as **objects** (3). It combines object-oriented programming and database capabilities. Its abbreviation is **OODBMS**, or **object-oriented DBMS**. Also called **object database management system**.

object-oriented DBMS Abbreviation of **object-oriented database management system**.

object-oriented design The design of a model, program, or system utilizing an object-oriented approach. An object-oriented analysis is often performed before an object-oriented design. Its abbreviation is **OOD**.

object-oriented graphics Images which are represented in the form of equations and graphics primitives. Used, for instance, in CAD and drawing programs. This contrasts with **bitmap graphics**, in which data structures composed of rows and columns of bits serve to represent the information of an image. Object-oriented graphics are more flexible, require less memory, and are manipulated with greater ease than bitmap graphics. Also called **vector graphics**.

object-oriented interface A computer user interface, such as a GUI, which makes use of icons and other on-screen objects, along with pointing devices to access said objects.

object-oriented language Same as **object-oriented programming language**.

object-oriented operating system An operating system designed using an object-oriented approach. Such an operating system, for instance, facilitates the development of object-oriented applications. Its abbreviation is **object-oriented OS**.

object-oriented OS Abbreviation of **object-oriented operating system**.

object-oriented program A program created utilizing **object-oriented programming**.

object-oriented programming A modular programming technique which supports objects, classes, inheritance, encapsulation, abstraction, and polymorphism. Each of these features enables individual software modules to be linked, exchanged, used multiple times, have characteristics transferred, and so on, which greatly enhances the flexibility and functionality of object-oriented programs. Java and C++ are examples of object-oriented programming languages. Its abbreviation is **OOP**.

object-oriented programming language A language, such as Java or C++, utilized for **object-oriented programming**. Its abbreviation is **OOPL**. Also called **object-oriented language**, or **object language (2)**.

object-oriented programming system A programming technique or system which utilizes an **object-oriented approach**. Its acronym is **OOPS**.

object-oriented relational database Same as **object-relational database**.

object-oriented technology The application of an **object-oriented approach** to the development of computer programs. Its abbreviation is **OOT**. Also called **object technology**.

object program Also called **target program**. **1.** The machine-language program output of a compiler or assembler. **2.** A program in a form which can be directly executed by a given computer.

Object Query Language A query language that supports the use of **objects (3)**. Used, for instance, in object-relational databases. Its abbreviation is **OQL**.

object recognition In machine vision, the identification or distinguishing of objects. Object recognition is used in robotics, for instance, for visual guidance, or in the assembly of complex units. Also, any technique or system utilized for such recognition.

object-relational database A database combining features of an object-oriented database and a relational database. Also called **object-oriented relational database**.

object-relational database management system A relational database management system that supports the modeling, creation, storage, linking, and retrieval of data as **objects (3)**. It combines object-oriented programming and relational database managing capabilities. Its abbreviation is **object-relational DBMS**, or **ORDBMS**.

object-relational DBMS Abbreviation of **object-relational database management system**.

Object Request Broker The interface which enables **objects (3)** to communicate with other objects. This usually takes place within a CORBA architecture. Its abbreviation is **ORB**.

object technology Same as **object-oriented technology**.

Objective-C An object-oriented version of the C programming language.

objective lens In an optical system, the first lens through which light rays pass. In a light microscope, for instance, it is the lens nearest to the sample being observed.

observed failure rate The failure rate of a component, circuit, device, piece of equipment, system, or material for a given time interval, based on controlled conditions.

OC **1.** Abbreviation of **open circuit**. **2.** Abbreviation of **open collector**. **3.** Abbreviation of **optical carrier**.

OC-1 Abbreviation of **optical carrier-1**. Within the SONET standard, a transmission rate of 51.8 Mbps. The equivalent electrical signal is called **STS-1**. An OC-1 can carry 1 DS-3 within its payload.

OC-3 Abbreviation of **optical carrier-3**. Within the SONET standard, a transmission rate of 155.52 Mbps, which is 3 times OC-1. The equivalent electrical signal is called **STS-3**. An OC-3 can carry 3 DS-3s within its payload, which is the same as **STM-1**.

OC-9 Abbreviation of **optical carrier-9**. Within the SONET standard, a transmission rate of 466.56 Mbps, which is 9 times OC-1. The equivalent electrical signal is called **STS-9**. An OC-9 can carry 9 DS-3s within its payload, which is the same as **STM-3**.

OC-12 Abbreviation of **optical carrier-12**. Within the SONET standard, a transmission rate of 622.08 Mbps, which is 12 times OC-1. The equivalent electrical signal is called **STS-12**. An OC-12 can carry 12 DS-3s within its payload, which is the same as **STM-4**.

OC-18 Abbreviation of **optical carrier-18**. Within the SONET standard, a transmission rate of 933.12 Mbps, which is 18 times OC-1. The equivalent electrical signal is called **STS-18**. An OC-18 can carry 18 DS-3s within its payload, which is the same as **STM-6**.

OC-24 Abbreviation of **optical carrier-24**. Within the SONET standard, a transmission rate of 1.24416 Gbps, which is 24 times OC-1. The equivalent electrical signal is called **STS-24**. An OC-24 can carry 24 DS-3s within its payload, which is the same as **STM-8**.

OC-36 Abbreviation of **optical carrier-36**. Within the SONET standard, a transmission rate of 1.86624 Gbps, which is 36 times OC-1. The equivalent electrical signal is called **STS-36**. An OC-36 can carry 36 DS-3s within its payload, which is the same as **STM-12**.

OC-48 Abbreviation of **optical carrier-48**. Within the SONET standard, a transmission rate of 2.48832 Gbps, which is 48 times OC-1. The equivalent electrical signal is called **STS-48**. An OC-48 can carry 48 DS-3s within its payload, which is the same as **STM-16**.

OC-96 Abbreviation of **optical carrier-96**. Within the SONET standard, a transmission rate of 4.97664 Gbps, which is 96 times OC-1. The equivalent electrical signal is called **STS-96**. An OC-96 can carry 96 DS-3s within its payload, which is the same as **STM-32**.

OC-192 Abbreviation of **optical carrier-192**. Within the SONET standard, a transmission rate of 9.95328 Gbps, which is 192 times OC-1. The equivalent electrical signal is called **STS-192**. An OC-192 can carry 192 DS-3s within its payload, which is the same as **STM-64**.

OC-768 Abbreviation of **optical carrier-768**. Within the SONET standard, a transmission rate of 39.81312 Gbps, which is 768 times OC-1. The equivalent electrical signal is called **STS-768**. An OC-768 can carry 768 DS-3s within its payload, which is the same as **STM-256**.

OC-3072 Abbreviation of **optical carrier-3072**. Within the SONET standard, a transmission rate of 159.25248 Gbps, which is 3072 times OC-1. The equivalent electrical signal is called **STS-3072**. An OC-3072 can carry 3072 DS-3s within its payload, which is the same as **STM-1024**.

OC1 Same as **OC-1**.

OC3 Same as **OC-3**.

OC9 Same as **OC-9**.

OC12 Same as **OC-12**.

OC18 Same as **OC-18**.

OC24 Same as **OC-24**.

OC36 Same as **OC-36**.

OC48 Same as **OC-48**.

OC96 Same as **OC-96**.

OC192 Same as **OC-192**.

OC768 Same as **OC-768**.

OC3072 Same as **OC-3072**.

occluded gas A gas which is trapped in an enclosure or material. For example, that in an electron tube or within a crystal lattice.

occlusion 1. The process of obstructing, closing, or preventing the passage of. For instance, the blocking of electromagnetic radiation. **2.** That which obstructs, closes, or prevents from passing. For instance, an obstacle which blocks electromagnetic radiation. **3.** That which is obstructed, closed, or prevented from passing. For instance, a gas that is trapped within a crystal lattice.

occupied band A frequency band that is in use, or that is designated for use.

occupied bandwidth 1. An interval of frequencies or a transmission capacity that is being utilized, or that is designated or reserved for utilization. **2.** An interval of frequencies outside of which a given percentage of the total power is emitted. For example, an interval where 0.5% of the total average radiated power is emitted above the upper limit, and 0.5% of the total average radiated power is emitted below the lower limit.

OCR 1. Abbreviation of **optical character recognition**. **2.** Abbreviation of **optical character reader**. **3.** Abbreviation of **optical character reading**.

OCR software Abbreviation of **optical character recognition software**.

oct Abbreviation of **octal**.

octagon A closed geometric figure bounded by eight straight lines.

octal Pertaining to the **octal number system**, or to **octal notation**.

octal base An electron-tube base with eight projecting pins or contacts. Such a base is inserted into an **octal socket**.

octal notation A notation based on the **octal number system**. Also called **octal representation**.

octal number A number expressed in **octal notation**.

octal number system A numbering system whose base is 8, and whose utilized digits are 0-7. For example, decimal 10 is the same as octal 12, decimal 20 is octal 24, and decimal 26 is octal 32. Octal numbers usually have either a prefix or

a suffix to identify them. For example, O32 has a letter **O** prefix to indicate that what follows is octal number **32**. Also called **octal system**.

octal representation Same as **octal notation**.

octal socket A socket into which an **octal base** is inserted.

octal system Same as **octal number system**.

octal tube An electron tube with an **octal base**.

octave 1. An interval of frequencies whose ratio of its upper limit to its lower limit is 2 to 1. For instance, a frequency interval from 1,000 to 2,000 Hz. **2.** An interval of frequencies considered to have eight steps, including the upper and lower limits. Notes one octave apart are perceived by human ears as the same note at different pitches, and not as different notes. The ratio of the upper frequency limit to the lower frequency limit is 2 to 1.

octave band A frequency band in which the ratio of the upper frequency limit to the lower frequency limit is 2 to 1. When making acoustic measurements, ten bands may be utilized, with the following approximate center frequencies: 31.5, 63, 125, 250, 500, 1000, 2000, 4000, 8000, and 16000 Hz. The 1000 Hz octave band, for instance, comprises frequencies ranging from about 707 Hz to 1414 Hz. Also called **octave frequency band**.

octave-band analysis The use of an **octave-band analyzer** to determine sound magnitudes. Used, for instance, to study the noise levels that are present in a given environment.

octave-band analyzer An instrument which analyzes sounds levels an octave band at a time. For example, it may analyze a band from 707 Hz to 1414 Hz, among various selectable ranges. Such a device may also measure one-third octave bands, and provide a computer interface for analysis and storage of gathered data. Also called **octave-band noise analyzer**.

octave-band filter Same as **octave filter**.

octave-band noise analyzer Same as **octave-band analyzer**.

octave filter A bandpass filter whose ratio of upper cutoff frequency to lower cutoff frequency is 2 to 1. Also called **octave-band filter**.

octave frequency band Same as **octave band**.

octet A unit of data consisting of 8 adjacent bits, and representing a single alphabetic character, symbol, or decimal digit. It is the smallest addressable unit of computer storage. Much better known as **byte**.

octode An electron tube with eight electrodes. Such a tube usually has an anode, a cathode, a control electrode, and five grids.

octopus cable A cable with a single connector on one end, and multiple connectors at the other. Used, for instance, to connect multiple peripherals to a computer with less cabling tangles.

OD 1. Abbreviation of **optical density**. **2.** Abbreviation of **outside diameter**.

ODBC Abbreviation of **Open DataBase Connectivity**.

ODBMS Abbreviation of **object database management system**.

odd-even check An error-detection procedure which verifies if there is even or odd parity. For each byte transmitted, an additional bit is added, which can be a 0 or 1, so that all bytes are even or odd. At the receiving location, this parity is checked, to be sure it is the same it was before sending. If it is, the transmission has passed this test of data integrity. Also used for data storage. Also called **parity check**.

odd field In interlaced scanning, the pattern created by tracing the odd-numbered lines. The even-numbered lines create the **even field**, and together these compose a displayed frame.

odd harmonic A harmonic whose frequency is an odd-numbered multiple of the fundamental frequency. For instance, if the fundamental frequency is 200 Hz, then 600 Hz and 1000 Hz are each odd harmonics. This contrasts with **even harmonics**, which are even-numbered multiples of the fundamental frequency.

odd line In a TV, an odd-numbered active line. Also called **odd-numbered line**.

odd-numbered line Same as **odd line**.

odd parity **1.** The presence of an odd number of ones or zeroes within a group of bits. **2.** An error-detection procedure which verifies if there is **odd parity** (1). For each byte transmitted, an additional bit is added, which can be a 0 or 1, so that all bytes are odd. At the receiving location, the parity is checked, to be sure it is odd. If it is, the transmission has passed this test of data integrity. Also used for data storage. Also called **odd parity check**.

odd parity check Same as **odd parity** (2).

ODI Abbreviation of **Open Data-Link Interface**.

odometer A device or instrument which measures distance traveled by a vehicle, robot, or the like. Also, an indicator or readout of such a device or instrument.

ODS Abbreviation of **operational data store**.

Oe Symbol for **oersted**.

OE Abbreviation of **optoelectronics**.

OEB Abbreviation of **Open eBook**.

OEM Abbreviation of **original equipment manufacturer**.

oersted The unit of magnetic field strength within the cgs system. It is equal to approximately 79.5775 ampere-turns per meter. Its symbol is **Oe**.

OES Abbreviation of **optical emission spectroscopy**. A form of atomic emission spectroscopy which measures emissions in the infrared, visible, and ultraviolet ranges.

OFC Abbreviation of **oxygen-free copper**.

off Also called **off state**. **1.** A state in which no current, voltage, or other signal passes, so that a component, circuit, switch, device, piece of equipment, or system is disconnected or not otherwise operating. **2. Off** (1), as opposed to **on** (1), when there are only two such states available to a component, circuit, switch, device, piece of equipment, or system.

off air Same as **off the air**.

off-center-fed antenna An antenna whose transmission line is connected to any point other than the center of its radiators.

off-center feed The connection of the transmission line of an antenna to any point other than the center of its radiators.

off current Same as **off-state leakage current**.

off-delay The time interval that elapses between the removal of the actuating signal and the subsequent disconnection of a circuit or device.

off-hook The condition where a telephone is in use. This state occurs, for instance, when the handset is removed from its cradle, which releases the hookswitch, which in turn occupies an available line. Also, when another device, such as a fax, is in use. This contrasts with **on-hook**, where the telephone or other device is not in use. Also spelled **offhook**. Its abbreviation is **OH**. Also called **off-hook state**, or **off-hook condition**.

off-hook condition Same as **off-hook**. Also spelled **offhook condition**.

off-hook state Same as **off-hook**. Also spelled **offhook state**.

off-isolation The level of isolation from a circuit or device that a switch provides when in the off state. May be expressed in decibels, or as a percentage.

off-line Same as **offline**.

off-line backup Same as **offline backup**.

off-line browser Same as **off-line reader**. Also spelled **off-line browser**.

off-line navigator Same as **off-line reader**. Also spelled **offline navigator**.

off-line power supply A power supply that operates directly from an AC line, and which does not utilize a transformer prior to rectification and filtering.

off-line reader Same as **offline reader**.

off-line storage Same as **offline storage**.

off-peak **1.** Not at a maximum. **2.** Not operating at a maximum. **3.** Not occurring during a period of comparatively high, or maximum demand.

off-peak energy Electrical energy supplied or consumed during periods of comparatively low system demands.

off period **1.** A time interval during which a component, circuit, device, piece of equipment, or system is not on or otherwise operating. **2.** A time interval during which no current passes. **3.** In an electron tube, a time interval during which no current passes. **4.** In on/off operation, the time interval during which an off state occurs.

off position **1.** For a switch or relay, a position in which current, or another signal, does not pass, so that a circuit or device is disconnected or otherwise not operating. **2.** For a component, switch, circuit, device, piece of equipment, or system, the position corresponding to the off state.

off-scale Beyond the interval of indicated values in a meter or indicating instrument. For example, a needle whose location is beyond the maximum marked value of the calibrated scale of an instrument.

off-scale reading For a meter or indicating instrument, a reading which is **off-scale**.

off state Same as **off**.

off-state current Same as **off-state leakage current**.

off-state leakage current Current which flows through a component, circuit, or device which is in the off state. Also called **off-state current**, **off current**, or **leakage current** (6).

off-state voltage The voltage across a component, circuit, or device which is in the off state.

off the air In radio, TV, or another form of broadcasting, not currently transmitting, nor recording future programming. Also called **off air**.

off-the-shelf **1.** That which is available for sale in the state it is in. **2.** That which can be purchased and put to use without making any modifications.

off time **1.** The time during which a component, circuit, device, piece of equipment, or system is in the off state. **2.** The time during which a device, piece of equipment, or system is not operational. Causes include failures, routine maintenance, repairs, and power outages. Also called **downtime**, or **dead time** (1).

offhook Same as **off-hook**. Its abbreviation is **OH**.

offhook condition Same as **off-hook**. Also spelled **off-hook condition**.

offhook state Same as **off-hook**. Also spelled **off-hook state**.

Office A popular application suite.

office application A computer application intended for office use. Examples include word processing, database, spreadsheet, and presentation programs.

office automation The use, and integration of, devices, machines, and systems to perform or automate office tasks which would otherwise be handled manually. Office automation provides humans with additional resources which save time and enhance productivity. There are many examples, including the use of computers for entering data, preparing reports, and storing information, and the use of emails, real-time messages, faxes, and teleconferencing for communications.

office suite An software package incorporating a group of **office applications**.

OfficeJet A popular line of multifunction printers.

offline Also spelled **off-line**. This contrasts with **online**. **1.** Not connected to a computer. Said, for instance, of a peripheral, or of data stored in a removable medium which is not inserted at the moment. **2.** Not installed in a computer, said for example, of hardware or software. **3.** Not connected to a computer or communications network. Said, for instance, of a user, or of a computer system. **4.** Not under the control of a computer. Said, for example, of a manufacturing process. **5.** Connected to, or controlled by, a computer, but not in an operational state. Said, for instance, of a printer which is connected, but not ready to receive data. **6.** Not in an operational mode or state. Said, for example, of a piece of equipment that is in standby.

offline backup A backup performed while the data being duplicated is not in use. This contrasts with an **online backup**, in which an application is using the data. Also spelled **off-line backup**. Also called **cold backup**.

offline browser Same as **offline reader**. Also spelled **off-line browser**.

offline navigator Same as **offline reader**. Also spelled **off-line navigator**.

offline reader An application which downloads email, Web pages, and other online messages and content to a local disk, where they are available for viewing once a network connection, such as that to the Internet, is terminated. This may serve, for instance, to save paid connection time, or free resources for other uses. Also spelled **off-line reader**. Also called **offline browser**, or **offline navigator**.

offline storage Storage of data in a medium which is not immediately accessible by a computer system. For example, discs kept in a data library. Also spelled **off-line storage**.

offload **1.** To transfer data to a peripheral. **2.** To transfer tasks or processing demands from one computer, device, or system, to another.

offset **1.** That which serves to balance, counteract, neutralize, or otherwise compensate for something else. Also, to balance, counteract, neutralize, or otherwise compensate for something else. **2.** An imbalance between two halves or sections of a circuit, such as that of a differential amplifier. Also, the magnitude or value of such an imbalance. **3.** In a control system, the steady-state difference between the theoretical or desired value of the controlled variable, and the actual value. **4.** The distance a memory address is from a base address. Also, the value added to the base address to obtain a second address. Also called **displacement** (3).

offset current For an operational amplifier, the DC that must be applied to the inputs to produce a zero output voltage.

offset QPSK Abbreviation of **offset quadrature phase-shift keying**.

offset quadrature phase-shift keying A form of quadrature phase-shift keying in which an offset is introduced, for improved spectral efficiency. Its abbreviations are **offset QPSK**, or **OQPSK**.

offset voltage For an operational amplifier, the DC input voltage that must be applied to the inputs to produce a zero output voltage.

OFHC Abbreviation of **oxygen-free high-conductivity copper**.

OH Abbreviation of **off-hook**.

ohm The SI unit of resistance, impedance, and reactance. It is defined as the resistance between two points through which a current of one ampere will flow, when there is a potential difference of one volt between said points. Its symbol is Ω (omega).

ohm-centimeter A unit of resistivity equal to 0.01 **ohm-meter**. Its symbol is Ω-cm.

ohm-meter The SI unit of resistivity. It is the resistance of a material or conductor, expressed in ohms, multiplied by its cross sectional area, expressed in square meters, divided by its length in meters. Its symbol is Ω-m.

Ohm's law A law stating that the current the flowing through a DC circuit is directly proportional to the voltage, and inversely proportional to the resistance. It may be expressed as follows: $I = E/R$, where I is the current, E is the voltage, and R is the resistance. For an AC circuit, the relationship would be: $I = E/X$, where I is the current, E is the voltage, and X is the impedance. When a conductor or material obeys this law, it is said to be ohmic. Ohm's law does not apply to all conductors of electricity

ohmage Same as **ohmic value**.

ohmic **1.** A conductor or material which obeys **Ohm's law**. **2.** Pertaining to a conductor or material which obeys Ohm's law. **3.** Pertaining to a region of a characteristic curve which obeys Ohm's law.

ohmic contact A contact between two materials in which an increase in voltage results in an increase in current, as dictated by Ohm's law. A metal-semiconductor contact which exhibits this characteristic is an ohmic contact, while a **Schottky contact** has rectifying properties.

ohmic dissipation Same as **ohmic losses**.

ohmic heating Heating of a material or conductor through which an electric current passes, due to the resistance of said material or conductor.

ohmic loss Same as **ohmic losses**.

ohmic losses Power losses due to **ohmic heating**. Also called **ohmic dissipation**, or **ohmic loss**.

ohmic region For a given device, a region within a characteristic curve in which an increase in voltage results in an increase in current, as dictated by Ohm's law. Usually utilized in the context of negative-resistances devices, to contrast with negative-resistance regions.

ohmic resistance The resistance to DC offered by a circuit, device, or material. Such a resistance obeys Ohm's law. Also called **DC resistance**.

ohmic response A response, such as that of a component, circuit, or device, which obeys Ohm's law.

ohmic value Resistance expressed in **ohms**, or in multiples of ohms, such as megohms. Also called **ohmage**.

ohmmeter An instrument graduated in ohms, or fractions/multiples of ohms, which is utilized to measure and indicate the magnitude of electric resistance of conductors, insulators, and other materials. Such an instrument usually has multiple ranges to choose from, and some may provide accurate readings as low as a single micro-ohm, while others may extend to the teraohm range.

ohms per square A measure of the resistance between parallel edges of a thin-film resistive material, such as a metal or semiconductor. The resistance of a four-sided material is defined as the resistance, in ohms, multiplied by its width, and divided by its length. Since the length and width of a square are equal, the measured resistance value will be the same for any size square. The resistance of such a material could then be stated in ohms, but in this context is expressed in units of ohms per square.

ohms per volt A measure of the sensitivity of a instrument, such as a voltmeter. It refers to the resistance, in ohms, divided by the full-scale voltage value for a given range. The higher the ohms-per-volt rating, the greater the sensitivity of the meter.

ohnosecond In computers and networks, the instant during which a user realizes that the wrong key was pressed, and that a significant mistake was made. For example, pressing "no" when prompted to save a document which has been worked on a while, or clicking "send" during a real-time chat when an unadvisable text was present.

oil capacitor Also called **oil-filled capacitor**, or **oil-immersed capacitor**. 1. A capacitor in which the dielectric consists of oil-impregnated paper. 2. A capacitor in which the dielectric is oil.

oil circuit breaker A circuit breaker in which the contacts are immersed in a special oil which extinguishes an electric arc which may form between said contacts. Its abbreviation is **OCB**. Also called **oil switch (1)**, or **oil-filled circuit breaker**.

oil-cooled transformer Same as **oil-filled transformer**.

oil-cooled unit A unit, such as a transformer, in which a circulating oil is utilized for cooling.

oil dielectric A highly refined oil utilized as a dielectric in components and devices such as capacitors and transformers. Examples include castor oil and mineral oil.

oil-diffusion pump A diffusion pump which utilizes rapidly moving molecules of a special oil to create a high vacuum.

oil-filled cable A cable in which an oil, such as mineral oil, is contained at an approximately steady pressure. The presence of such an oil helps prevent voids in the cable's insulation and helps provide high voltage and power capacity to maximize transmission economy, in addition to being better-suited to self monitoring. Also called **oil-immersed cable**.

oil-filled capacitor Same as **oil capacitor**.

oil-filled circuit breaker Same as **oil circuit breaker**.

oil-filled switch Same as **oil switch**.

oil-filled transformer A transformer in which a circulating oil is utilized for cooling. Also called **oil-cooled transformer**, or **oil-immersed transformer**.

oil-immersed cable Same as **oil-filled cable**.

oil-immersed capacitor Same as **oil capacitor**.

oil-immersed switch Same as **oil switch**.

oil-immersed transformer Same as **oil-filled transformer**.

oil-impregnated A material, such as paper, which is impregnated with oil.

oil-impregnated paper Paper, such as kraft paper, which is impregnated with oil. Used, for instance, in oil capacitors.

oil switch Also called **oil-filled switch**, or **oil-immersed switch**. 1. Same as **oil circuit breaker**. 2. A switch that is immersed in oil. Immersion in oil serves for cooling, and suppresses an arc which may form between the contacts.

oiled paper A insulating paper, such as kraft paper, which is impregnated with oil. Used, for instance, in capacitors, or to help protect from moisture.

OLAP Acronym for **O**nline **A**nalytical **P**rocessing. Decision support software which enables a user to quickly, often in real-time, analyze considerable amounts of data which has been collected then processed from multiple perspectives. Used, for instance, to expedite complex business decisions such as market trends. Also, the technology employed for such processing.

OLAP database Abbreviation of **O**nline **A**nalytical **P**rocessing **database**. A database where the raw information utilized in **OLAP** is collected and processed.

OLAP server Abbreviation of **O**nline **A**nalytical **P**rocessing **server**. A high-capacity server specifically designed for **OLAP**.

OLE Acronym for **Object Linking and Embedding**.

OLED Abbreviation of **Organic Light-Emitting Diode**, or **Organic Light-Emitting Device**.

OLED display Abbreviation of **Organic Light-Emitting Diode Display**, or **Organic Light-Emitting Display**.

OLTP Abbreviation of **online transaction processing**.

OMA Abbreviation **Object Management Architecture**.

omega (Ω) A Greek letter utilized as a symbol for **ohm**.

Omega A phase-difference radio navigation system which utilizes the same signal from multiple fixed transmitters to determine a line-of-position.

OMG Abbreviation of **Object Management Group**.

omni-bearing Same as **omnibearing**.

omni-directional Same as **omnidirectional**.

omni-directional antenna Same as **omnidirectional antenna**.

omni-directional beacon Same as **omnidirectional radio beacon**. Also spelled **omnidirectional beacon**.

omni-directional hydrophone Same as **omnidirectional hydrophone**.

omni-directional loudspeaker Same as **omnidirectional speaker**. Also spelled **omnidirectional loudspeaker**.

omni-directional microphone Same as **omnidirectional microphone**.

omni-directional radio beacon Same as **omnidirectional radio beacon**.

omni-directional radio range Same as **omnidirectional range**. Also spelled **omnidirectional radio range**.

omni-directional range Same as **omnidirectional range**.

omni-directional range station Same as **omnidirectional range station**.

omni-directional signal Same as **omnidirectional signal**.

omni-directional speaker Same as **omnidirectional speaker**.

omni-range Same as **omnirange**.

omni-range station Same as **omnidirectional range station**. Also spelled **omnirange station**.

omnibearing A bearing obtained utilizing an **omnidirectional range** as a reference. Also spelled **omni-bearing**.

omnidirectional Also spelled **omni-directional**. 1. Pertaining to, or indicating all directions in space. 2. Sending or receiving signals in all directions. 3. Capable of sending or receiving signals with the same efficiency or sensitivity in all directions. Also called **non-directional (3)**. 4. Functioning at essentially the same level of performance in all directions. Also called **non-directional (4)**.

omnidirectional antenna Also spelled **omni-directional antenna**. Also called **non-directional antenna**. 1. An antenna which radiates or receives with essentially the same efficiency or sensitivity in all directions. 2. An antenna which radiates or receives with essentially the same efficiency or sensitivity in all directions within a given plane.

omnidirectional beacon Same as **omnidirectional radio beacon**. Also spelled **omni-directional beacon**.

omnidirectional hydrophone A hydrophone whose sensitivity is the same in all directions. Also spelled **omni-directional hydrophone**.

omnidirectional loudspeaker Same as **omnidirectional speaker**. Also spelled **omni-directional loudspeaker**.

omnidirectional microphone A microphone whose sensitivity is the same in all directions. Also spelled **omni-directional microphone**. Also known as **non-directional microphone**, or **astatic microphone**.

omnidirectional radio beacon Also spelled **omni-directional radio beacon**. Also called **omnidirectional beacon**, or **non-directional radio beacon**. 1. A beacon which emits signals in all directions. 2. A beacon which emits signals in all directions within the horizontal plane.

omnidirectional radio range Same as **omnidirectional range**. Also spelled **omni-directional radio range**.

omnidirectional range A radio navigation system in which each of its multiple stations transmit signals which provide bearings in all directions to and from the transmitter. Such a transmitter emits beacon signals of equal strength in all directions along a given plane, which provide directional guidance within a given area. A common example is VOR. Used extensively by pilots and aircraft. Also spelled **omni-directional range**. Its abbreviation is **omnirange**. Also called **omnidirectional radio range**.

omnidirectional range station A land station within an **omnidirectional range** service. Also spelled **omni-directional range station**. Also called **omnirange station**.

omnidirectional signal A signal whose intensity is the same in all directions. Also spelled **omni-directional signal**.

omnidirectional speaker A speaker that produces essentially equal amounts of acoustic energy in all directions. Also spelled **omni-directional speaker**. Also called **omnidirectional loudspeaker**.

omnirange Abbreviation of **omnidirectional range**. Also spelled **omni-range**.

omnirange station Same as **omnidirectional range station**. Also spelled **omni-range station**.

OMR 1. Abbreviation of **optical mark reading**. 2. Abbreviation of **optical mark reader**. 3. Abbreviation of **optical mark recognition**.

OMT Abbreviation of **Object Modeling Technique**.

OMVPE Acronym for **organometallic vapor-phase epitaxy**.

on Also called **on state**. 1. A state in which a current, voltage, or other signal passes, so that a component, circuit, switch, device, piece of equipment, or system is connected or otherwise operating. 2. **On** (1), as opposed to **off** (1), when there are only two such states available to a component, circuit, switch, device, piece of equipment, or system.

on air Same as **on the air**.

on-board computer A computer contained within another device, apparatus, or system. Examples include computers in satellites, car engines, navigation systems, appliances, and toys. Also spelled **onboard computer**.

on-chip cache A memory cache that is built right into the CPU. This contrasts with **external cache**, which utilizes a memory bank between the CPU and main memory. On-chip cache is faster, but smaller than external cache, while external cache is still faster than main memory. Also called **L1 cache**, **level 1 cache**, **first-level cache**, **internal cache**, or **primary cache**.

on-course signal A signal indicating that an aircraft is following a given fixed or desired course. For instance, a continuous tone heard from an AN range.

on current Same as **on-state current**.

on-delay The time interval that elapses between the arrival of the actuating signal and the subsequent connection of a circuit or device.

on-demand 1. Same as **on-demand system**. 2. Pertaining to an **on-demand system**. 3. That which is available only when paid or requested.

on-demand programming 1. TV programming immediately available on upon payment or request. An example is movies-on-demand. 2. Programming immediately available on upon payment or request.

on-demand system A system utilized for on-demand programming, or another service where delivery or access is provided immediately after payment or request. Also called **on-demand** (1).

on-hook The condition where a telephone is in not use. This state occurs, for instance, when the handset is resting on its cradle, which keeps the hookswitch depressed. Also, when another device, such as a fax, is not in use. This contrasts with **off-hook**, where the telephone or other device is in use. Also spelled **onhook**. Also called **on-hook state**, or **on-hook condition**.

on-hook condition Same as **on-hook**. Also spelled **onhook condition**.

on-hook state Same as **on-hook**. Also spelled **onhook state**.

on interval Same as **on period**.

on-line Same as **online**.

On-line Analytical Processing Same as **OLAP**. Also spelled **Online Analytical Processing**.

On-line Analytical Processing database Same as **OLAP database**. Also spelled **Online Analytical Processing database**.

On-line Analytical Processing server Same as **OLAP server**. Also spelled **Online Analytical Processing server**.

on-line backup Same as **online backup**.

on-line community Same as **online community**.

on-line computer system Same as **online computer system**.

on-line content Same as **online content**.

on-line data reduction Same as **online data reduction**.

on-line help Same as **online help**.

on-line information service Same as **online information service**.

on-line service Same as **online service provider**. Also spelled **online service**

on-line service provider Same as **online service provider**.

on-line transaction processing Same as **online transaction processing**.

on/off control 1. A control system where the controlled output quantity or device is either fully on or fully off. The term is utilized to distinguish from systems where there are intermediate settings or levels of operation. 2. Same as **on/off switch**.

on/off keying Amplitude shift keying in which two signal levels are used, one of which is zero. For example, a mark can be made with the energized level, and a space with the zero level. Its abbreviation is **OOK**. Also called **binary amplitude shift keying**.

on/off operation 1. For a switch, component, circuit, device, piece of equipment, or system, a transition from an on state to an off state. 2. For a switch, component, circuit, device, piece of equipment, or system, operation which involves transitions from an on state to an off state. For example, on/off keying.

on/off switch Also called **on/off control** (2). 1. A switch that serves to change between the on state and the off state. That is, the switch either turns something on, or it turns it off. Also called **power switch** (1). 2. A switch that serves to change between the on state and a standby state. Also called **on/standby switch**, or **power switch** (2). 3. A switch that serves to change between states in on/off keying.

on-peak energy Electrical energy supplied or consumed during periods of comparatively high system demands.

on period Also called **on interval**, or **on time**. 1. The time during which a component, circuit, device, piece of equipment, or system is in the **on state**. 2. A time interval during which current passes. 3. In an electron tube, a time interval during which current passes. 4. In on/off operation, the time interval during which an on state occurs.

on position 1. For a switch or relay, a position in which current, or another signal, passes, so that a circuit or device is connected or otherwise operating. 2. For a component, switch, circuit, device, piece of equipment, or system, the position corresponding to the on state.

on resistance Same as **on-state resistance**.

on-screen display Its abbreviation is **OSD**. Also called **OSD display**. **1.** A control, used for monitor or TV settings, which is displayed on the screen. Used, for instance, to adjust brightness, contrast, horizontal centering, and so on. **2.** A display on the screen of a TV, or other projection device, which provides menus, settings, and other options and information. An example is the content showed when utilizing a device with an on-screen menu or on-screen programming.

on-screen menu A menu, such as that of a DVD or VCR, which is shown on the screen of the TV, or other device, it is displayed through. Such a menu serves to select from choices pertaining to audio, language, programming, playback mode, and so on. Such menus are usually text-based, but may incorporate icons to facilitate navigation.

on-screen programming For a device such as a DVD recorder or VCR, the showing of programming options and settings through the TV, or other device, it is displayed through.

on/standby switch Same as **on/off switch** (**2**).

on state Same as **on**.

on-state current The current flowing thorough a component, circuit, or device which is in the on state. Also called **on current**.

on-state resistance The resistance of a component, circuit, or device which is in the on state. Also called **on resistance**.

on-state voltage The voltage across a component, circuit, or device which is in the on state. Also called **on voltage**.

on the air In radio, TV, or another form of broadcasting, currently transmitting programming. Also, currently recording for a future broadcast. Also called **on air**.

on the fly Pertaining to that which occurs concurrently with, and in response to something else. For example, dynamic allocation, an AutoCorrect spell-check feature, or a dynamic Web page.

on time Same as **on period**.

on voltage Same as **on-state voltage**.

onboard computer Same as **on-board computer**.

ondogram The record produced by an **ondograph**.

ondograph An instrument which graphically records oscillatory vibrations, especially an alternating current.

ondoscope A glow-discharge tube which is utilized to detect intense RF fields, without the need for a direct connection. The contained gas ionizes when placed in a field of sufficient intensity.

one-address instruction In computer programming, an instruction which contains a single address part. Also called **single-address instruction**.

one-chip computer A complete computer that is contained on a single chip. Such a chip must have a CPU, memory, a clock, and input/output circuits. Such chips may be used in countless items, including automobiles, toys, appliances, clocks, and so on. Also called **microcontroller, microcomputer** (**2**), or **computer-on-a-chip**.

one condition Same as **one state**.

one input An input, such as that of a flip-flop or logic gate, corresponding to a **one state**. Also called **logic one input**, or **logic high input**.

one output An output, such as that of a flip-flop or logic gate, corresponding to a **one state**. Also called **logic one output**, or **logic high output**.

one-pass compiler A compiler that translates source code or language statements after a single pass.

one-shot Same as **one-shot multivibrator**.

one-shot circuit Same as **one-shot multivibrator**.

one-shot execution Same as **one-shot operation**.

one-shot multivibrator A multivibrator with one stable state and one unstable state. A trigger signal must be applied to change to the unstable state, where it remains for a given interval, after which it returns to the stable state. Also called **one-shot circuit, one-shot, monostable multivibrator, univibrator**, or **single-shot multivibrator**.

one-shot operation A computer operational mode in which each instruction, or part of an instruction, is executed in response to an external signal which is usually controlled manually by a user. Utilized, for instance, for debugging, or for detecting hardware malfunctions. Also called by various other names, including **one-shot execution, single-step operation**, and **step-by-step operation**.

one state In digital logic, a level within the more positive of the two ranges utilized to represent binary variables or states. A high level corresponds to a 1, or true, value. Also called **one condition, logic one, logic 1, logic high**, or **high level**.

one-third octave band A frequency band in which the ratio of the upper frequency limit to the lower frequency limit is equal to one third of an octave. The 1000 Hz one-third octave band, for instance, comprises frequencies ranging from about 891 Hz to 1122 Hz. Also called **one-third octave frequency band**, or **third octave band**.

one-third octave-band analysis The use of a **one-third octave-band analyzer** to determine sound magnitudes. Used, for instance, to study the noise levels that are present in a given environment.

one-third octave-band analyzer An instrument which analyzes sounds levels one-third octave band at a time. For example, it may analyze a band from 891 Hz to 1122 Hz, among various selectable ranges. Such a device may also provide a computer interface for analysis and storage of gathered data. Also called **one-third octave-band noise analyzer**.

one-third octave-band noise analyzer Same as **one-third octave-band analyzer**.

one-third octave frequency band Same as **one-third octave band**.

one-time pad A form of encryption which utilizes a random key the same length as the message. Such a key is only used once. If only the recipient has the key, such communication is totally secure.

one-time password A password which can only be used once. Such passwords may be generated by special software, and help thwart malicious programs which intercept passwords for future use.

one-time programmable That which can only be programmed once. Examples include ROM chips and CD-ROMs. Its abbreviation is **OTP**.

one-time PROM Same as **OTPROM**.

one-tuner picture-in-picture A picture-in-picture feature in which the TV does not have a second tuner dedicated to the second picture. The source of the second picture must then be another input, such as that of a DVD or VHS. This contrasts with a **two-tuner picture-in-picture** TV, where the second picture may also be that of another channel. Its abbreviation is **one-tuner PIP**.

one-tuner PIP Abbreviation of **one-tuner picture-in-picture**.

one-way communication Communication in which one or more locations receive, but do not transmit. Examples include broadcasting, and that utilized by one-way intercom systems. Also called **simplex communication** (**1**).

one-way hash function A hash function which converts a text input into a fixed string of digits. It is termed one-way due to the difficulty in reversing the function, that is, obtaining the original text from the numbers.

one-way intercom An intercom system which only communicates in one direction. Used, for instance, as a baby alarm.

one-way radio Radio communication in which one or more locations receive, but do not transmit. An example is broadcasting. Also called **one-way radio communications**.

one-way radio communications Same as **one-way radio**.

one-way repeater 1. A repeater which retransmits, and usually amplifies, radio signals arriving or traveling in one direction. 2. A repeater which retransmits, and usually amplifies one-way communications signals.

one-way trunk A trunk between two central stations via which traffic may originate from one end. Once the trunk is occupied in this manner, two-way communication may take place.

one's complement The complement when using the binary number system. The complement of a number in the binary system is the other. That is, the complement of 1 is 0, and the complement of 0 is 1. Used in computers, for instance, to represent negative numbers.

ones density In a digital communications circuit, a minimum amount of one bits that must appear per unit bits, in order to maintain synchronization. For example, no more than 15 consecutive zero bits may be transmitted, nor less than 12.5% one bits out of the overall transmission.

onhook Same as **on-hook**.

onhook condition Same as **on-hook**. Also spelled **on-hook condition**.

onhook state Same as **on-hook**. Also spelled **on-hook state**.

online Also spelled **on-line**. This contrasts with **offline**. 1. Connected to a computer. Said, for instance of a peripheral, or of data stored in a removable medium which is inserted at the moment. 2. Installed in a computer, said for example, of hardware or software. 3. Connected to a computer or communications network. Said, for instance, of a user, or of a computer system. 4. Connected to the Internet. Said, for instance, of a user, or of a computer system. 5. Under the control of a computer. Said, for example, of a manufacturing process. 6. Connected to, or controlled by, a computer, and in an operational state. Said, for instance, of a printer which is connected, and ready to receive data. 7. In an operational mode or state. Said, for example, of a piece of equipment that is active. 8. Immediately available or accessible.

online advertising Advertisements, such as banner, sky-scraper, and pop-up ads, appearing over the Internet. Also, delivery of such ads. Also called **Web advertising**, or **Internet advertising**.

Online Analytical Processing Same as **OLAP**. Also spelled **On-line Analytical Processing**.

Online Analytical Processing database Same as **OLAP database**. Also spelled **On-line Analytical Processing database**.

Online Analytical Processing server Same as **OLAP server**. Also spelled **On-line Analytical Processing server**.

online auction An auction that takes place over the Internet. Countless Web sites offer innumerable items which people or entities wish to sell to the highest bidders. Also called **Internet auction**.

online backup A backup performed while an application, such as a database, is using the data being duplicated. This contrasts with an **offline backup**, in which the data is not in use. Also spelled **on-line backup**. Also called **hot backup**.

online community A collection of users of an online service or group. For example, all people who connect to the Internet on a regular basis, or the members of a newsgroup. Also

spelled **on-line community**. Also called **virtual community**.

online computer system A computer which is connected to a communications network. Also spelled **on-line computer system**.

online content Information which is accessed through a network such as the Internet, and which may consist of any combination of audio, video, files, data, or the like. Also spelled **on-line content**.

online data reduction The processing of data the instant it is received by, or entered into, a computer system. Also spelled **on-line data reduction**.

online help Also spelled **on-line help**. 1. Help, such as that needed while using an application, which is immediately available by automatically accessing a disk, disc, or other storage medium. 2. Help which is available through a network connection. For example, that obtained from a Web site.

online information service Any source of information which is accessed through a network such as the Internet. Also spelled **on-line information service**.

online service Same as **online service provider**. Also spelled **on-line service**.

online service provider Its abbreviation is **OSP**. Also spelled **on-line service provider**. Also called **online service**. 1. An entity which provides access to the Internet. Customers usually pay a monthly fee for this service, although it usually available for free unless a high-speed connection is desired. Users get a software package to access and browse the World Wide Web, one or more email accounts, and Web pages to have a presence on the Internet. Other offerings may include Web site building and hosting services. Users can connect via a dial-up service or asymmetrical digital subscriber line, among others. Also called **Internet service provider**, **access provider**, or **service provider**. 2. An entity which offers network, especially Internet, services and content.

online transaction processing The processing of transactions the instant each is received by, or entered into, a computer system. Online transaction processing not only maintains all master files constantly current, it also enables entering or retrieving information without delays in either process. Used, for instance in mail order businesses which have multiple locations which accept Internet, telephone, and fax orders. Also spelled **on-line transaction processing**. Its abbreviation is **OLTP**. Also called **transaction processing**, or **real-time transaction processing**.

OO Abbreviation of **object oriented**.

OOA 1. Abbreviation of **object-oriented approach**. 2. Abbreviation of **object-oriented analysis**.

OOD Abbreviation of **object-oriented design**.

OODB Abbreviation of **object-oriented database**.

OODBMS Abbreviation of **object-oriented database management system**.

OOK Abbreviation of **on/off keying**.

OOP Abbreviation of **object-oriented programming**.

OOPL Abbreviation of **object-oriented programming language**.

OOPS Acronym for **object-oriented programming system**.

OOT Abbreviation of **object-oriented technology**.

op 1. Abbreviation of **operation**. 2. Abbreviation of **operational**. 3. Abbreviation of **operating**.

op amp Abbreviation of **operational amplifier**.

op code Abbreviation of **operation code**.

OPA Abbreviation of **optical parametric amplifier**.

opacimeter An instrument which measures **opacity**.

opacity For a given body, material, or medium, a measure of the inability to transmit radiant energy, especially light, or sound. In the case of light, it may be quantified as the ratio of light flux incident upon a material, to the light flux transmitted by said material. It is the reciprocal of **transmittance**.

opamp Abbreviation of **operational amplifier**.

opaque 1. A body, material, or medium which does not pass radiant energy, such as light, or sound. 2. A body, material, or medium which does not reflect radiant energy, such as light, or sound. 3. A body, material, or medium which does not pass one form of radiant energy, such as ultraviolet light, but passes others, such as visible light. 4. A body, material, or medium which does not pass one type of particle, such as ions, but passes others, such as photons. 5. A body, material, or medium which does not pass any type of particle.

opcode Abbreviation of **operation code**.

open 1. Discontinuous and/or broken. For instance, an open circuit does not have an uninterrupted path for current. A switch in the open position is the same as being off. Also, to make incomplete. 2. To start or initiate. For instance, to open a program or window, or to establish a communications link. 3. Said of a communications line or circuit which is available for use. Also, to make such a line or circuit available. 4. Not forming a self-contained unit or system. 5. Not blocking or otherwise obstructing. Also, to unblock. 6. Having a space or gap, as opposed to not having such a space or gap. Also, to make a space or gap. 7. A file which is being used or accessed at a given moment. Also, to open such a file. 8. Accessible to all, or designed to work across different architectures or with varied products. For example, an open architecture. Also, to make accessible in this manner.

open architecture A device or system whose technical specifications are made public. In computers, for instance, this encourages third-party vendors to develop compatible products. This contrasts with a **closed architecture**, in which technical specifications are not made public.

open-area test site A space utilized to determine the electromagnetic compatibility of components, devices, equipment, and systems, and for other measurements of electromagnetic radiation, such as those required for compliance with given standards. Such a site may be located outside, or within a specially-designed enclosure, and its ground surface usually consists of a material which provides a uniform ground plane. Its acronym is **OATS**

open-back cabinet 1. A speaker cabinet in which there are no sound-reflecting panels in the back. 2. A cabinet in which there is no back panel.

open box Also called **white box**, **structural box**, or **clear box**. 1. A component, device, or unit whose internal structure and function are known in great detail. Usually used in the context of its insertion into a system. 2. In computers, a hardware or software unit whose internal structure and function is known in great detail.

open-box testing Also called **white-box testing**, **structural testing**, or **clear-box testing**. 1. Any tests utilizing the assistance of an **open box** (1). This contrasts with **closed-box testing** (1), in which the inner structure is not known, or is not necessary to be known. 2. Tests of software or hardware utilizing the assistance of an **open box** (2). This contrasts with **closed-box testing** (2), in which the inner structure is not known, or is not necessary to be known.

open-captioned TV programming which has **open-captioning**.

open-captioning Closed-captioning which has been decoded before transmission, or which otherwise does not require a decoder. Some programs, for instance, provide open-captioning, which usually appears as white text in a black box somewhere on the screen. This contrasts with **subtitles**, which usually do not have a black background.

open captions The text and symbols seen when viewing programming with **open-captioning**.

open-circuit Its abbreviation is **OC**. 1. A circuit which is broken, or which otherwise does not have a complete and uninterrupted path for the flow of current. Also called **incomplete circuit**, or **discontinuous circuit**. 2. A communications circuit which is available for use. 3. A circuit utilized under no-load conditions.

open circuit breaker A circuit breaker whose contacts are open, or which otherwise does not allow an uninterrupted path for current.

open-circuit current The current which flows through a component, circuit, device, piece of equipment, or system, under no-load conditions. Also called **no-load current**.

open-circuit impedance In a four-terminal electric network, the impedance at one end when the terminals at the other end are open. When the terminals at the far end are short-circuited, it is called **short-circuit impedance**.

open-circuit jack A jack whose circuit is open until the corresponding plug is inserted.

open-circuit resistance In a four-terminal electric network, the resistance at one end when the terminals at the other end are open. When the terminals at the far end are short-circuited, it is called **short-circuit resistance**.

open-circuit voltage The voltage level present under no-load conditions. For example, the voltage at the output terminals of a power supply when 0% load is applied, or that of a battery in the absence of a charge or discharge current. Also called **no-load voltage**.

open collector In a bipolar transistor, a configuration in which the collector, or output, is left disconnected from the internal circuitry, and is available for external connection. In ICs, external connections are available from the collector of output transistors. Used, for instance, for high-current or high-voltage loads. Its abbreviation is **OC**. Also called **open-collector configuration**.

open-collector configuration Same as **open collector**.

open-collector transistor A transistor or IC with an **open-collector configuration**.

open computing Computing utilizing **open systems**.

open core A magnetic core in which the paths of the magnetic lines of force pass through air. Such a design allows the use of thicker wire and can thus handle more current than that of **closed cores**, but can generate more EMI. Such cores are usually cylindrical in shape.

Open Data-Link Interface An interface which enables a network card to support multiple data-link layer protocols. Its abbreviation is **ODI**.

Open Database Connectivity A database API. Its abbreviation is **ODBC**.

open-delta connection Also called **V connection**. 1. The connection of two transformers so as to form the shape of a letter V, as opposed to the combination of three to form a **delta connection**. 2. The connection of two circuit elements, such as resistors, so as to form the shape of a letter V, as opposed to the combination of three to form a delta connection.

Open eBook An open standard pertaining to electronic books. Also, a format based on this standard. Its abbreviation is **OEB**.

open-ended Characteristic of that which can be expanded, improved upon, or the like. Said, for instance, of a device, piece of equipment, system, process, or program.

open feeder Same as **open-wire line**.

Open Group An organization which promotes open systems and standards.

open line 1. A line which is discontinuous or broken. Such a line does not have an uninterrupted path for current. **2.** A communications line which is available for use. **3.** Same as **open-wire line**.

open loop 1. In computer programming, a loop which repeats itself a specified number of times. A **closed loop (1)** repeats itself indefinitely, unless interrupted or stopped by external intervention. **2.** In a control system, a circuit in which no feedback signal is utilized as a basis for comparison to a reference input. Thus, the output does not have an effect on the input. A **closed loop (1)** does incorporate such feedback.

open-loop control system A control system in which there is no feedback signal to monitor the output. Such control systems are generally less accurate but simpler in design than **closed-loop control systems**, which do have a feedback signal. Also called **open-loop system**.

open-loop gain The gain of an amplifier when it has no feedback signal. When such a loop is present it is called **closed-loop gain**.

open-loop output impedance The impedance of an amplifier when it has a feedback signal. When no such loop is present it is called **closed-loop impedance**.

open-loop output resistance The resistance of an amplifier when it has a feedback signal. When no such loop is present it is called **closed-loop resistance**.

open-loop system Same as **open-loop control system**.

open-loop voltage gain In an amplifier with no feedback signal, the ratio of the output voltage to the input voltage. When a feedback signal is present, it is called **closed-loop voltage gain**.

open magnetic circuit 1. A magnetic circuit in which there is not a complete and uninterrupted path for magnetic flux circulation around a ferromagnetic core. **2.** A magnetic circuit in which there is not a complete and uninterrupted path for magnetic flux circulation.

open plug A plug utilized to hold a jack in the open position. Usually refers to a jack which has normally-closed contacts.

Open Shortest Path First A routing protocol used in TCP/IP networks to determine the shortest path between nodes. Its abbreviation is **OSPF**.

open source Computer program source code which is freely available. Open source helps outside programmers point out and provide fixes for bugs, suggest improvements, or customize an application for the needs of a given entity. Popular software with open source code include Linux and Mozilla.

open standard A set of hardware and/or software specifications which is freely available. Such standards promote interoperability, and help in the proliferation of new technologies, platforms, interfaces, methods, applications, devices, and so on. This contrasts with a **closed standard**, in which technical specifications are not made public.

open subroutine A subroutine that can be included in a routine from wherever said subroutine is stored, while a **closed subroutine** must be placed in a specified location to be accessed by a routine.

open system A system whose technical specifications are made public. In computers, for instance, this encourages third-party vendors to develop compatible products. This contrasts with a **closed system**, in which technical specifications are not made public.

Open Systems Interconnection Same as **OSI Reference Model**. Its abbreviation is **OSI**.

Open Systems Interconnection Model Same as **OSI Reference Model**. Its abbreviation is **OSI Model**.

Open Systems Interconnection Protocol Same as **OSI Reference Model**. Its abbreviation is **OSI Protocol**.

Open Systems Interconnection Reference Model Same as **OSI Reference Model**.

open wire 1. Same as **open-wire line**. **2.** A wire which is bare, and usually uninsulated.

open-wire feeder Same as **open-wire line**.

open-wire line A transmission line in which the conductors are in the form of bare, uninsulated wire. Materials such as ceramic, glass, or plastic are utilized to physically attach this wire to poles or another support, as such lines are always surrounded by air. Short circuits between such conductors are avoided by having the proper spacing. Overhead lines are a common example. Also called **open-wire transmission line, open-wire feeder, open feeder, open line (3)**, or **open wire (1)**.

open-wire transmission line Same as **open-wire line**.

OpenBSD A popular open-source UNIX operating system which runs on multiple platforms.

OpenGL A 2D and 3D graphics language, API, or standard which works across multiple platforms.

OpenView Popular network management software.

OpenVMS A popular operating system.

Opera A popular Web browser.

operand 1. An entity upon which an operation or function is performed. For instance, in $(A + B) \times C$, A, B, and C are operands, while + and x are **operators**. **2.** A symbol representing an **operand (1)**.

operate current A current that actuates or otherwise causes the function of a relay, switch, circuit, device, piece of equipment, system, process, or mechanism.

operate delay 1. Same as **operate time (1)**. **2.** Same as **operate time (2)**.

operate interval Same as **operate time**.

operate lag 1. Same as **operate time (1)**. **2.** Same as **operate time (2)**.

operate time Also known as **operate interval, operating time**, or **operation time (2)**. **1.** The time interval occurring between the application of an operate current, operate voltage, or other signal which actuates or otherwise causes function, and the actual actuation or function. Also called **operate delay (1)**, or **operate lag (1)**. **2.** The **operate time (1)** for a switch or relay. Also called **operate delay (2)**, or **operate lag (2)**. **3.** The time required for an event or other task. **4.** The time a component, circuit, device, piece of equipment, or system is in operation.

operate voltage A voltage that actuates or otherwise causes the function of a relay, switch, circuit, device, piece of equipment, system, process, or mechanism.

operating 1. That which is on or otherwise functioning at any given moment. For example, a device or system which is in the on state. **2.** That which is performing a function on something else at a given moment.

operating ambient temperature Same as **operating temperature range**.

operating angle For an amplifier circuit, the electrical angle of the input signal during which the anode or plate current flows in an electron tube. In the case of a transistor, it is the electrical angle during which the collector or drain current flows. For a Class A amplifier, the operating angle is 360°. That is, in a class A amplifier there is a current output at all times, regardless of the stage of the input signal. For a class AB amplifier the operating angle is greater than 180°, but less than 360°, for a class B amplifier it is 180°, and for a class C amplifier it is less than 180°.

operating bias Also called **operational bias**. **1.** The bias required for normal operation. **2.** The bias required for optimal operation.

operating conditions Also called **operational conditions**. **1.** The circumstances, such as ambient conditions, input voltage, and output current, which are required for the proper functioning of a component, circuit, device, piece of equipment, or system. Also called **standard operating conditions**, or **normal operating conditions**. **2.** The conditions which characterize the operation of a component, circuit, device, piece of equipment, or system.

operating current The current required for the proper operation of a component, circuit, device, piece of equipment, or system. Also called **operational current**.

operating cycle One complete sequence of operations performed as a unit by a component, circuit, device, piece of equipment, or system which is operating. Also called **operation cycle**, or **operational cycle**.

operating frequency Also called **operational frequency**. **1.** The frequency at which a component, circuit, device, piece of equipment, or system operates. For example, that of a radar transmitter. **2.** The **operating frequency (1)** in which a given level of performance is equaled or exceeded.

operating frequency band Same as **operating frequency range**.

operating frequency range Also called **operational frequency range**, **operating frequency band**, or **operating range (2)**. **1.** The frequency interval over which a component, circuit, device, piece of equipment, or system operates. For example, that of an amplifier or radio transmitter. **2.** The **operating frequency range (1)** in which a given level of performance is equaled or exceeded.

operating life Also called **operational life**. **1.** The time interval during which a component, circuit, device, piece of equipment, or system is calculated or expected to operate. **2.** The time interval during which a component, circuit, device, piece of equipment, or system actually operates.

operating mode Also called **operational mode**. **1.** A given manner of operation when more than one is available. For example, presentation in portrait mode as opposed to landscape mode. **2.** A manner of operation in which a component, circuit, device, piece of equipment, system, or mechanism is functional, as opposed to a **non-operational mode**, in which it is not. **3.** A manner of operation in which a component, circuit, device, piece of equipment, system, or mechanism is operating, as opposed to a **stand-by mode**, in which it is ready for immediate activation.

operating overload Any overload which a component, circuit, device, piece of equipment, or system is subjected to during normal operation, and which does not hinder proper function. For instance, a current surge occurring when power is first applied to a device or piece of equipment. Also called **operational overload**.

operating point **1.** Within a family of characteristic curves of an active device such as a transistor or vacuum tube, the point corresponding to the values of current and voltage under specified operating conditions. **2.** Same as **operating position (3)**.

operating position **1.** The designated position or placement for the proper operation of a component, device, piece of equipment, or mechanism. **2.** The position where a moving part must be for the proper operation of a component, device, piece of equipment, or mechanism. **3.** In a switch or relay, the point where the contacts initially touch. Also called **operating point (2)**.

operating power Also called **operational power**. **1.** The power consumed by a component, circuit, device, piece of equipment, or system which is properly operating. **2.** The power produced or emitted by a component, circuit, device, piece of equipment, or system which is properly operating.

operating range The range over which a component, circuit, device, instrument, piece of equipment, or system operates. For example, an operating frequency range, an operating temperature range, an amplitude range, and so on.

operating ratio **1.** The proportion of time a component, circuit, device, piece of equipment, or system is fully operational. May be expressed, for instance, as a percentage. **2.** The ratio of the time a component, circuit, device, piece of equipment, or system is operating, to the time it is available for operation. May be expressed, for instance, as a percentage.

operating reliability Same as **operational reliability**.

operating system The software which runs all the software and hardware of a computer. It is the first program the computer loads when powered on, remains memory-resident, and continuously controls and allocates all resources. Without it, for example, a computer can not recognize input, such as that from a keyboard, it can not process, as the operating system controls the use of the CPU, nor can it provide an output, as it manages all peripherals including the monitor. Its abbreviation is **OS**. Also called **executive**, or **supervisor**. When disk-based, also called **disk operating system**.

operating temperature Same as **operating temperature range**.

operating temperature range Also called **operational temperature**, **operating temperature**, **operating ambient temperature**, or **ambient-temperature range**. **1.** The ambient temperature interval within which an instrument is operated. **2.** The ambient temperature interval within which an instrument is designed to work without malfunctioning or being damaged. **3.** The ambient temperature interval required for an instrument to function optimally. **4.** The **operating temperature range (1)**, **operating temperature range (2)**, or **operating temperature range (3)** for a component, circuit, device, piece of equipment, system, or material.

operating time Same as **operate time**.

operating voltage The voltage required for the proper operation of a component, circuit, device, piece of equipment, or system. Also called **operational voltage**, or **working voltage (3)**.

operation **1.** The act, process, or state of functioning or being functional. Also, the manner in which something functions. **2.** One or more steps within a specific task or process. Also, a complete sequence within a task or process. **3.** The action or actions resulting from a single computer instruction. **4.** A function which a logical operator performs. Also called **logical operation**.

operation code A coded value or bit string within a machine instruction which specifies the operation to be performed by a processor. The operation may be a branch, add, copy, and so on. Its abbreviation is **opcode**, or **op code**. Also called **instruction code**.

operation cycle Same as **operating cycle**.

operation number Within a subroutine or other process which involves multiple operations, a number specifying the position of a given step.

operation register A register which stores an **operation code**.

operation time **1.** The time required for an **operation (2)**, or **operation (3)**. **2.** Same as **operate time**.

operational **1.** That which is ready to function or operate. For instance, a circuit or device which can be put in the on state at any given moment. **2.** Pertaining to, or arising from, one or more operations.

operational amplifier A direct-coupled amplifier which features very high gain, stability, immunity to oscillation, and high input impedance. Such an amplifier has various applications, including computing functions such as integration and differentiation, signal-conditioning functions such as filtering, and in sensors. Its abbreviation is **op amp**.

operational bias Same as **operating bias**.

operational conditions Same as **operating conditions**.

operational current Same as **operating current**.

operational cycle Same as **operating cycle**.

operational data store A database similar to a data warehouse, except that its data is continuously updated, while that of a data warehouse is not. Its abbreviation is **ODS**.

operational frequency Same as **operating frequency**.

operational frequency band Same as **operating frequency band**.

operational frequency range Same as **operating frequency range**.

operational life Same as **operating life**.

operational mode Same as **operating mode**.

operational overload Same as **operating overload**.

operational power Same as **operating power**.

operational reliability The actual reliability exhibited by a component, circuit, device, instrument, piece of equipment, or system which has been operating, as opposed to the calculated or expected reliability. Also called **operating reliability**.

operational temperature Same as **operating temperature range**.

operational transconductance amplifier Its abbreviation is **OTA**. Also called **transconductance amplifier**. **1.** An amplifier which converts its input voltage into an output current proportional to said voltage. **2.** A differential amplifier which converts the difference between its input voltages into an output current proportional to said difference.

operational voltage Same as **operating voltage**.

operations research The application of mathematical and/or scientific techniques to model, solve, and optimize complex decisions, problems, and systems involving people, data, and physical devices such as machines and installations. Used, for instance, in business decision-making. Its abbreviation is **OR**.

operator **1.** An operation or function which is performed on an entity. For instance, in **(A + B) x C**, + and x are operators, while **A, B**, and **C** are **operands**. **2.** A symbol representing an **operator (1)**. **3.** A person who utilizes a device, piece of equipment, machine, or system.

operator overloading In computer programming, the assigning of multiple meanings to the same operator.

operator precedence The hierarchy which defines the sequence in which multiple operators within an expression are applied.

OPO Abbreviation of **optical parametric oscillator**.

opposition **1.** The resistance something offers to something else. For example, the opposition a given material offers to the flow of current or to the passage of magnetic flux. **2.** The circumstance or state where two periodic quantities having the same frequency and waveshape are out-of-phase by 180°. Also called **phase opposition**.

opt-in To explicitly accept, authorize, and/or request some product and/or service. For example, to sign-up for an email newsletter.

opt-out To cancel that which had been previously accepted, authorized, and/or requested. For example, to cancel a subscription to an email newsletter.

Optacon A device which converts visual information into a tactile format. It usually consists of a unit which translates printed text into a vibrating tactile version which can be felt by fingertips. It is an abbreviation of **op**tical-to-**tac**tile con verter.

Opteron A popular family of CPU chips.

optic **1.** Of or pertaining to vision or the eye. **2.** Any component, such as a lens, mirror, or prism, of an optical instrument such as a camera, telescope, microscope, or laser. **3.** Same as **optical**.

optic axis Same as **optical axis**.

optical **1.** Of, pertaining to, or utilizing visible light, vision, or optics. Visible light spans from approximately 750 nanometers (red) to approximately 400 nanometers (violet), although the term optical sometimes includes near infrared and near ultraviolet radiation. **2.** That which assists vision or the utilization of light. For instance, an optical instrument. **3.** Same as **optic**.

optical aberration In optics, the inability of an optical lens to produce a perfect correlation between an object and its resulting image. This can be due to various factors, including the physical properties of the lens. Types of optical aberration include curvature, astigmatism, spherical aberration, and chromatic aberration. Also called **aberration**.

optical activity The capacity of a substance to rotate the plane of polarization of plane-polarized light that passes through it. Chiral molecules have this characteristic.

optical amplifier **1.** An amplifier that increases an optical signal without converting it into an electrical signal which is amplified and reconverted into light. An example is an erbium-doped fiber amplifier. **2.** An amplifier that increases an optical signal.

optical axis Also spelled **optic axis**. Its abbreviation is **OA**. **1.** A line passing through both centers of curvature of a lens. Light travels along this axis with no deviation. Also called **principal axis (1)**. **2.** A line passing through the centers of curvature of a lens or other optical system. Light travels along this axis with no deviation. Also called **principal axis (2)**. **3.** The axis along which light is transmitted through a doubly-refracting crystal or medium without double refraction. Also called **principal axis (3)**.

optical bar code A precise arrangement of parallel vertical bars and spaces of varying widths, encoded with characters such as letters, numbers, or symbols. These bars are read by optical devices called bar-code readers, which may scan the data horizontally and/or vertically, or at any angle. These serve for rapid and error-free entry of information, and are used, for instance in retail stores, libraries, or to help robots identify components in assembly facilities. Also called **bar code**.

optical cable One or more optical fibers bundled together and encased by a protective sheath. Generally used for telecommunications. Such cables usually consist of three layers, which are the core, its surrounding cladding, and the protective jacket. Also called **optical fiber cable**, **optical fiber bundle**, **fiber-optic cable**, **cable (2)**, or **light cable**.

optical cable assembly An optical cable which is fitted with the appropriate connectors, and which is ready for installation.

optical carrier Its abbreviation is **OC**. Within the SONET standard, one of the defined transmission speeds. OC speeds range from 51.8 Mbps, to hundreds of Gbps. OC-1 has a transmission rate of 51.84 Mbps, OC-3 runs at three times OC-1, or 155.52 Mbps, OC-96 at 4.97664 Gbps, and so on. The equivalent electrical signal is called **STS**, so STS-1 corresponds to OC-1.

optical carrier-1 Same as **OC-1**.

optical carrier-3 Same as **OC-3**.

optical carrier-9 Same as **OC-9**.

optical carrier-12 Same as **OC-12**.

optical carrier-18 Same as **OC-18**.

optical carrier-24 Same as **OC-24**.

optical carrier-36 Same as **OC-36**.

optical carrier-48 Same as **OC-48**.

optical carrier-96 Same as **OC-96**.

optical carrier-192 Same as **OC-192**.

optical carrier-768 Same as **OC-768**.

optical carrier-3072 Same as **OC-3072**.

optical center A point on the axis of a lens at which emergent rays are parallel to the incident rays. That is, light passes through this point with no deviation.

optical character reader A device which scans printed or written characters, and utilizes **optical character recognition** to identify them. Its abbreviation is **OCR**.

optical character reading Same as **optical character recognition**. Its abbreviation is **OCR**.

optical character recognition The ability of a computer or other device to identify optically scanned printed or written characters, and to convert this information into a digital code suitable for further processing. Used, for instance, to scan volumes of books which will subsequently become available online. Its abbreviation is **OCR**. Also called **optical character reading**.

optical character recognition software The software a computer or peripheral utilizes to perform **optical character recognition**. Its abbreviation is **OCR software**.

optical communications The use of light for the transmission and reception information between two or more points or entities. Examples include communication utilizing laser beams, infrared radiation, visible light, or ultraviolet radiation, along with the appropriate devices which enable the exchange of such information. Also, the conveyed information.

optical computer 1. A computer designed to utilize light to process or store data. Examples include systems which employ magneto-optical recording of data, or those which make use of optical circuits capable of ultra-fast parallel processing. **2.** A computer which can utilize light to enter, process, store, or present data, be it through its design, modifications performed, or the in incorporation of the necessary peripherals. For instance, a system utilized for recognition and identification of fingerprints, faces, or the patterns of ocular blood vessels.

optical computing The use of an **optical computer**.

optical coupler Same as **optoisolator**.

optical coupling 1. The utilization of a light beam or a light pipe for coupling circuits. **2.** The utilization of an **optoisolator**.

Optical Cross-Connect A cross connect utilized for switching and routing in optical communications. Used, for example, to configure, reconfigure, or restore fiber optic networks quickly, efficiently, and reliably. Usually utilized for long-haul and high-speed signals, such as OC-768. Its abbreviation is **OXC**.

optical density A measure of the opacity of a translucent body, material, or medium. It may be defined as the base-10 logarithm of the reciprocal of its transmittance. Its abbreviation is **OD**.

optical density meter An instrument utilized for measuring and indicating the optical density of a substance. Also called **densitometer (2)**.

optical detector A transducer which converts optical energy into another form of energy. For example, one that generates an output voltage which is proportional to the frequency of the incident light.

optical disc A digital optical storage medium, usually 12 centimeters in diameter, whose contained information is encoded in microscopic pits on its metallic surface, which is protected by a plastic layer. To recover the recorded content, a laser is focused on the metallic surface of the disc, and the reflected light is modulated by the code on said disc. Since there is no contact between the pickup and the recorded surface, wear is minimized, while the protective layer helps avoid reading errors due to dust or minor marks on the surface of the disc. Any form of data may be recorded onto a disc, and formats include DVDs, DVD+RWs, CDs, and CD-ROMs. Also spelled **optical disk**. Also called **disc (1)**.

optical disc drive Same as **optical drive**. Also spelled **optical disk drive**.

optical disc library Same as **optical jukebox**. Also spelled **optical disk library**.

optical disk Same as **optical disc**.

optical disk drive Same as **optical drive**. Also spelled **optical disc drive**.

optical disk library Same as **optical jukebox**. Also spelled **optical disc library**.

optical dispersion The process by which light is separated into components with different wavelengths, speeds, or other characteristics. Also, an instance of such a separation. For example, a prism disperses white light into its component colors, as each travels through it at a different velocity.

optical distance Same as **optical path**.

optical drive A computer drive that accepts **optical discs**. Such a drive reads from, and may also write to such discs, and may be compatible with multiple formats. Also called **optical disc drive**, or **optical disk drive**.

optical element A component within an optical instrument or system which acts upon any light passing through it. Examples include lenses, mirrors, and prisms.

optical emission spectroscopy A form of atomic emission spectroscopy which measures emissions in the infrared, visible, and ultraviolet ranges. Its abbreviation is **OES**.

optical encoder A sensor which measures linear motion, rotary motion, or other positional information by detecting the interruptions of a fixed light beam. Used, for instance, to count revolutions, or in robotics to identify parts.

optical fiber A thin filament of a transparent material, such as glass or plastic, which is capable of transmitting light signals through successive internal reflections. In order for an optical fiber to guide light, the proper relationship between the refractive index of the core and its surrounding cladding must be maintained. Also called **fiber (2)**, or **light guide**.

optical fiber bundle Same as **optical cable**.

optical fiber cable Same as **optical cable**.

optical fiber core The bundled optical fibers encased by a protective sheath in a fiber-optic cable. Also called **fiber core**.

optical filter An element or device, such as a disk or plate of plastic or glass, which selectively blocks or absorbs one or more intervals of frequencies of electromagnetic radiation, such as light. The optical properties of the element or device determines which frequencies pass, and which are blocked or absorbed. Also called **filter (4)**, or **radiation filter (2)**.

optical glass A glass whose composition, homogeneity, and finish render it appropriate for the faithful transmission of light.

optical heterodyning The combining of two laser beams, one an incoming signal and the other locally produced, in a nonlinear device, to produce an AC whose frequency is the difference of the frequencies of the two beams. Used, for instance, in laser Doppler velocimetry.

optical interference Interference occurring in light waves.

optical isolator Same as **optoisolator**.

optical jukebox Also called **optical disc library**, or **optical library**. 1. An optical disc changer which can hold many, up to several thousand DVDs, CD-ROMs, or the like. Discs may be swapped, but only one can be read at once. Also called **jukebox** (1). 2. An **optical jukebox** (1) which can read and/or write from or to multiple discs simultaneously.

optical length Same as **optical path**.

optical lens One or more objects, usually transparent and curved, which serve to refract and sometimes reflect light rays passing through. A lens causes parallel light beams to converge or diverge, is usually made of glass or plastic, and is utilized to form an image. Also called **lens** (1).

optical lever A device or instrument which measures small amounts of rotation. The rotating device or object has a mirror towards which a narrow beam of light is directed. The reflected beam directs a light spot upon a scale, from which readings are taken. The greater the distance between the mirror and the scale, the greater the sensitivity.

optical library Same as **optical jukebox**.

optical link Same as **optoisolator**.

optical lithography A lithographic technique which utilizes a light beam to transfer patterns or images from one material or medium to another. Light serves as the exposing source, and such a beam irradiates the resist, with a mask determining the desired patterns. A common example is photolithography.

optical mark In optical sensing, a line, circle, or other tracing which is recognized by an optical sensor. Also called **mark** (6).

optical mark reader A device which scans forms, or other sheets, and utilizes **optical mark reading** to identify them. Its abbreviation is **OMR**.

optical mark reading The recognition of an **optical mark** by an optical sensor. Used, for instance, to automatically gather information from forms that use checkmarks, filled-in circles, and the like. Its abbreviation is **OMR**. Also called **optical mark recognition**, **optical mark sensing**, **mark reading**, **mark recognition**, **mark sensing**, **mark scanning**, or **mark detection**.

optical mark recognition Same as **optical mark reading**. Its abbreviation is **OMR**.

optical mark sensing Same as **optical mark reading**.

optical maser A term sometimes utilized to refer to a **laser**.

optical media All mediums utilized for **optical storage**.

optical memory 1. The use of an optical system for the writing and reading of data to and from a computer's main memory. An example is holographic memory. 2. Same as **optical storage**.

optical microscope A microscope which utilizes reflected light and magnifying lenses to produce enlarged images of objects being viewed. Such microscopes usually use visible light, but may also use electromagnetic radiation close to visible light, such as near infrared. Also called **light microscope**.

optical modulation The modification of the intensity of a light beam proportionally to a characteristic present in another wave or signal. Also, the result of such a process. Usually utilized to convey meaningful information. Also called **light modulation**.

optical modulator A device, such as an electro-optic modulator, which serves to modulate a light beam. Also called **light modulator** (1).

optical mouse 1. A computer mouse whose movements are converted into electrical signals via optical means. For example, such a mouse may rest upon a reflective grid which detects the movements of a light situated under said mouse. 2. A mouse, such as an infrared mouse, which communi-

cates with a computer using optical signals, as opposed to a cord.

optical network A communications network, such as a fiber-optic network, which utilizes light signals to convey information.

optical parametric amplifier Same as **optical parametric oscillator**. Its abbreviation is **OPA**.

optical parametric oscillator An oscillator which utilizes a parametric amplifier inside a resonant optical cavity for the generation of tunable radiation. Usually used as a powerful source of tunable laser radiation. Its abbreviation is **OPO**. Also called **optical parametric amplifier**, or **parametric oscillator**.

optical path Also called **optical distance**, or **optical length**. 1. The actual distance light travels along a given path. 2. The actual distance light travels through a given substance, multiplied by the refractive index of said substance.

optical phenomena Phenomena pertaining to, and arising from, the generation, transmission, reception, detection, modulation, and deflection of light, especially that in the visible, infrared, and ultraviolet regions.

optical plastic A plastic whose composition, homogeneity, and finish render it appropriate for the faithful transmission of light.

optical power The total optical energy radiated by a source. Usually expressed in watts.

optical prism A transparent optical material or lens, such as that made of glass, with no less than two polished plane surfaces which are at an angle relative to each other so as to reflect or refract light. A prism may be utilized, for instance, to disperse white light into its component colors, to produce plane-polarized light from unpolarized light, or to invert an image. Used, for instance, in cameras and binoculars. Also called **prism**.

optical projection lithography A projection lithography technique in which a light beam, usually that of a laser, is utilized to produce an image of the photomask on the substrate.

optical proximity detector Same as **optical proximity sensor**.

optical proximity sensor A sensor which uses reflected light to sense the distance to one or more objects. Lasers are frequently used. There are many applications in robotics, including floor detection, location of parts, and collision avoidance. Also known as **optical proximity detector**.

optical pulse A short burst of optical energy. Certain lasers, for instance, can generate pulses nearly in the zeptosecond range.

optical pumping In a laser, the utilization of incident light to excite electrons, molecules, or ions to higher energy levels. Also called **photopumping**.

optical pyrometer A device or instrument which determines the temperature of very hot bodies, surfaces, or sources, by comparing its brightness with a standardized source of luminance. The comparison is usually performed visually.

optical radiation Electromagnetic radiation corresponding to the **optical spectrum** (1).

optical reader A device which utilizes optical sensors to detect and interpret characters, marks, bar codes, and the like. Examples include optical character readers, optical mark readers, and bar-code readers. Also called **optical scanner** (3), or **scanner** (5).

optical resolution The resolution of an optical device such as an optical scanner.

optical return loss The ratio of incident optical power to reflected optical power. Usually expressed in dB. Its abbreviation is **ORL**.

optical scanner **1.** An optical device which scans a sheet, film, area, scene, or the like, and converts it into data, an electrical image, or another form which enables the original visual information to be stored, processed, transmitted, or reproduced. Also called **scanner (2)**. **2.** A device which utilizes a small and intensely bright spot which scans an object, scene, screen, film, or the like, and transforms the original image into a series of electrical signals by focusing it onto a photoelectric cell. The scanning may be mechanical or electrical. Used, for instance, for TV transmission, telecine transfer, or character recognition. Also called **scanner (4)**, **light-spot scanner**, or **flying-spot scanner**. **3.** Same as **optical reader**. **4.** A computer peripheral which converts the visual information of a paper, film, or the like, and into an image which is stored as binary data, such as a bitmap. Such an image can then be displayed or printed. When used with optical character recognition software, such images can also be processed, allowing, for instance, the characters of printed documents to be converted into electronic text. Also called **scanner (6)**.

optical scanning The use of an **optical scanner**.

optical scissors An optical trap utilized to cut very small objects, such as living cells or the internal structures of cells. Also called **laser scissors**.

optical signal A light signal carrying information.

optical sound recorder A photoelectric device utilized for **optical sound recording**. Also called **photographic sound recorder**.

optical sound recording A technique in which a light beam modulated by an audio source is utilized to record a sound track on photographic film. Also called **photographic sound recording**.

optical sound reproducer A photoelectric device utilized to reproduce an **optical sound recording**.

optical sound track The sound record produced or reproduced when utilizing **optical sound recording**.

optical spectroscopy Any analytical technique in which the wavelengths and corresponding intensities of optical radiation absorbed or emitted by a sample are analyzed.

optical spectrum **1.** The portion of the electromagnetic spectrum ranging from the far-infrared through the extreme ultraviolet. This interval includes visible light, and spans from approximately 4 nm to 1 mm, though the defined interval varies. **2.** The graphical representation produced when utilizing **optical spectroscopy**.

optical storage The use of an optical system for the storage of data, images, sounds, and so on. Also, the mediums, such as discs, cylinders, plates, or cubes utilized for such storage. Storage using CDs, DVDs, and holograms are examples. Also called **optical memory (2)**.

optical surface A surface, such as that of a lens or mirror, which refracts or reflects light.

optical switch **1.** A switch that utilizes optical means to change between states. A photoswitch is an example. **2.** A device which diverts an optical signal from one path or circuit to another. Used, for instance, in optical networks, or to perform logic operations.

optical system A set of lenses, mirrors, prisms, and other optical devices which are used together to perform optical functions such as refraction, reflection, polarization, absorption, or dispersion of light.

optical tachometer A tachometer which utilizes optical means, as opposed, for instance, to mechanical means, to make speed measurements.

Optical Time Domain Reflectometer An instrument that evaluates the transmission characteristics of an optical fiber by sending a pulse and measuring the time interval that elapses and the intensity of reflected light. Used, for instance, to troubleshoot optical networks. Its abbreviation is **OTDR**.

Optical Time Domain Reflectometry The use of an **Optical Time Domain Reflectometer** to evaluate the transmission characteristics of an optical fiber. abbreviation is **OTDR**.

optical-to-tactile converter Same as **Optacon**.

optical transition A change between atomic or molecular energy states which absorbs or emits light.

optical trap A device utilized for **optical trapping**. Also called **laser trap**.

optical trapping The use of a laser to cool a particle to nearly absolute zero, and to then restrict it to an extremely small area. Optical trapping is useful, for instance, for research into the nature of atomic interaction, and improvements in atomic clocks and high-resolution spectroscopy. Also called **laser trapping**.

optical tweezers An optical trap utilized to manipulate very small objects, such as living cells or the internal structures of cells. Also called **laser tweezers**.

optical waveguide A waveguide which has the ability to direct the flow of radiant energy occurring within the **optical spectrum (1)**. An example is a fiber-optic cable.

optically active Said of a substance which exhibits **optical activity**.

optically-coupled isolator Same as **optoisolator**.

optics **1.** The branch of science which deals with all aspects of light, including its nature, properties, generation, detection, measurement, propagation, transmission, control, and applications. Optics may refer strictly to visible radiation, but more generally pertains to electromagnetic radiation whose wavelengths span from microwaves to X-rays, with emphasis on the infrared, visible, and ultraviolet regions. **2.** The branch of science that deals with vision.

optimal Same as **optimum**.

optimal bunching Same as **optimum bunching**.

optimal conditions Same as **optimum conditions**.

optimal coupling Same as **optimum coupling**.

optimal current Same as **optimum current**.

optimal damping Same as **optimum damping**.

optimal frequency Same as **optimum frequency**.

optimal impedance Same as **optimum impedance**.

optimal load Same as **optimum load**.

optimal performance Same as **optimum performance**.

optimal resistance Same as **optimum resistance**.

optimal value Same as **optimum value**.

optimal voltage Same as **optimum voltage**.

optimal working frequency Same as **optimum working frequency**.

optimization The process of making as functional, effective, or useful as possible. Also, the techniques utilized towards these objectives of improved performance.

optimize To make as functional, effective, or useful as possible. Also, the techniques utilized towards these objectives of improved performance.

optimum The condition, state, or point reached when a component, circuit, device, piece of equipment, system, method, or process is at its most effective, functional, or useful. Also called **optimal**.

optimum bunching In a velocity-modulated tube, such as a klystron, the electron-bunching condition which results in the maximum output. Also called **optimal bunching**.

optimum conditions The operational and/or ambient conditions which provide for optimum performance. Also called **optimal conditions**.

optimum coupling The degree of coupling between two radio-frequency resonant circuits which results in the maximum transfer of energy between them, when both are tuned to the same frequency. Also called **optimal coupling**, or **critical coupling**.

optimum current The value of current which provides for optimum performance. Also called **optimal current**.

optimum damping The degree of damping that provides the fastest transient response without overshoot or oscillation. In an indicator with a needle, for instance, it is the amount of damping that allows the needle to proceed as quickly as possible to a new reading, while not overshooting it or oscillating around it. Also called **optimal damping**, or **critical damping**.

optimum frequency Also called **optimal frequency**. **1.** The operational frequency resulting in optimum performance. **2.** The frequency resulting in optimum transmission and reception.

optimum impedance The value of impedance which provides for optimum performance. Also called **optimal impedance**.

optimum load A load whose impedance value is that which results in the maximum possible transfer of power. Also called **optimal load, matched load,** or **matching load**.

optimum performance Performance of a component, circuit, device, piece of equipment, system, method, or process which is at its most effective, functional, or useful. Also called **optimal performance**.

optimum resistance The value of resistance which provides for optimum performance. Also called **optimal resistance**.

optimum value A value, such as a voltage or current, which provides for optimum performance. Also called **optimal value**.

optimum voltage The value of voltage which provides for optimum performance. Also called **optimal voltage**.

optimum working frequency In the transmission of radio waves via ionospheric reflection, the frequency resulting in optimum transmission and reception. Also called **optimal working frequency**.

options Within a program, choices a user has for given settings. Used, for instance, to determine how an application will handle certain tasks, the manner in which data is displayed, or which of multiple operating systems will be booted. When no preference is chosen for a given option, the default setting is utilized. Also called **preferences**.

options menu A menu, such as a drop-down menu, providing **options**. Also called **preferences menu**.

opto-acoustic Same as **optoacoustic**.

opto-acoustic cell Same as **optoacoustic cell**.

opto-acoustic device Same as **optoacoustic device**.

opto-acoustic modulator Same as **optoacoustic modulator**.

opto-acoustics Same as **optoacoustics**.

opto-electronic device Same as **optoelectronic device**.

opto-electronics Same as **optoelectronics**.

optoacoustic Pertaining to **optoacoustics**. Also spelled **opto-acoustic**.

optoacoustic cell A device which is utilized to modulate, deflect, and focus light waves, through their interactions with acoustic waves. Used, for example, in laser equipment for control of the intensity and position of laser beam radiation. Also spelled **opto-acoustic cell**. Also called **opto-acoustic device, optoacoustic modulator, acousto-optical cell,** or **Bragg cell**.

optoacoustic device Same as **optoacoustic cell**. Also spelled **opto-acoustic device**.

optoacoustic modulator Same as **optoacoustic cell**. Also spelled **opto-acoustic modulator**.

optoacoustics The science that deals with the interactions between acoustic waves and light waves in solid mediums. Acoustic waves can be made to modulate, deflect, and focus light waves. Also spelled **opto-acoustics**. Also called **acousto-optics**.

optocoupler Same as **optoisolator**.

optoelectronic device Also spelled **opto-electronic device**. **1.** A device which converts optical energy into electrical energy. Also, a device which converts an optical signal into an electrical signal. An example is a photodiode. **2.** A device which converts electrical energy into optical energy. Also, a device which converts an electrical signal into an optical signal. An example is an LED. **3.** A device, such as a laser, which combines optics and electronics.

optoelectronic isolator Same as **optoisolator**.

optoelectronics The science that deals with the study of, and applications of electronic components, circuits, devices, and systems which generate, transmit, receive, detect, and modulate light, especially that in the visible, infrared, and ultraviolet regions. There are many applications, including those in communications, lasers, and displays. It is sometimes utilized as a synonym for **electro-optics**, but this usage is incorrect. Also spelled **opto-electronics**. Its abbreviation is **optotronics**, or **OE**.

optoisolator A semiconductor device which allows electrical signals to be transferred between circuits, devices, or systems, while maintaining electrical isolation between them. An optoisolator may consist, for instance, of an LED or IRED for signal transmission, and a photodetector, such as a photodiode, for signal reception. The signal transmitter converts the electrical signal into an optical signal, while the receiver restores the original electrical signal. Used, for instance, in communications, and in control systems. Also called **optical coupler, optical isolator, optocoupler, optically-coupled isolator, optoelectronic isolator, optical link, photocoupler,** or **photoisolator**.

optophone A device which converts visual information into an auditory format. It usually incorporates a photoelectric cell which translates printed text into electrical signals which are converted into sounds of varying pitch.

optotronics Abbreviation of **optoelectronics**.

OQL Abbreviation of **Object Query Language**.

OQPSK Abbreviation of **offset quadrature phase-shift keying**.

OR **1.** A logical operation which is true if any of its elements is true. For example, if one or more of its multiple inputs have a value of 1, then the output is 1. Its output is 0 only when all of its inputs have a value of 0. For such functions, a 1 is considered as a true, or high, value, and 0 is a false, or low, value. Although OR is the same as **inclusive OR**, the latter is used to distinguish from **exclusive OR**. Also called **OR operation**, or **logical sum**. **2.** Abbreviation of **operations research**.

OR circuit Same as **OR gate**. Also called **inclusive OR circuit**.

OR gate A circuit which has two or more inputs, and whose output is high if any of its inputs is high. Also called **OR circuit**, or **inclusive OR gate**.

OR operation Same as **OR** (1).

Orange book A document which details the specifications for certain CD write-once formats.

ORB Abbreviation of **Object Request Broker**.

orbit **1.** A path followed by a particle or body, especially a closed path surrounding an object or location. **2.** A path followed by an electron around a nucleus. **3.** A closed path, usually circular or elliptical, followed by a natural or artificial satellite around the earth or another object. Also, a complete revolution. Such a path is determined by forces such as gravitation. **4.** A closed path, usually circular or elliptical, followed by one celestial body, such as a star, around another celestial body or location. Also, a complete revolution. Such a path is determined by forces such as gravitation.

orbital 1. A region of space surrounding one or more atomic nuclei, in which the probability of an electron being located is greatest. The charge of such an electron is distributed in this space. No more than two electrons may occupy the same orbital. Also called **atomic orbital**. **2.** Pertaining to an **orbit**.

orbital electron An electron in an **orbital** (1). Also called **planetary electron**.

orbital period The time interval that elapses between successive passages of a satellite through the same point of an orbit.

orbital velocity 1. The velocity of a particle or body in a given orbit. **2.** The minimum velocity necessary to maintain a given orbit.

ORDBMS Abbreviation of **object-relational database management system**.

order 1. A logical sequence of events, elements, or things. Also, to place in such a sequence. **2.** A specific manner in which successive events, elements, or things are organized. **3.** A degree of quality or quantity utilized for classification, or within a hierarchy. **4.** Same as **order of magnitude**.

order of magnitude A range of values where each successively higher number is ten times larger, and each successively lower number is ten times smaller. Used extensively in the metric system. For instance, a teravolt is 24 orders of magnitude above a picovolt. Also called **order** (4).

order tone A short tone sent over a trunk which indicates readiness for an order. It is a type of **zip tone**.

ordinary component Same as **ordinary wave**.

ordinary ray One of the two rays into which a ray of light incident upon a doubly-refracting material is split, the other being the **extraordinary ray**. Both rays are plane-polarized perpendicular to each other, and the ordinary ray is completely absorbed, while the extraordinary ray passes through the material. Its abbreviation is **O-ray**.

ordinary wave One of the two components into which a radio wave entering the ionosphere is split, under the combined influence of the earth's magnetic field and atmospheric ionization, the other being the **extraordinary wave**. The ordinary wave has characteristics more closely resembling those expected in the absence of a magnetic field. Its abbreviation is **O-wave**. Also called **ordinary component**, or **ordinary-wave component**.

ordinary-wave component Same as **ordinary wave**. Its abbreviation is **O-wave component**.

ordinate In a two-coordinate system, the vertical coordinate. The **abscissa** is the horizontal coordinate. Its symbol is **y**, or **Y**. Also called **y-coordinate**.

.org On the Internet, a top-level domain name suffix. The **org** is an abbreviation of **org**anization, although the entity employing this suffix need not be such an entity. Also called **dot org**.

organic 1. Pertaining to, associated with, or arising from, living organisms. **2.** Pertaining to carbon and its compounds, reactions, and so on.

organic electricity Electricity, such as brain waves, nerve impulses, or stunning charges by certain fish, occurring in, or arising from living organisms.

Organic LED Abbreviation of **Organic Light-Emitting Diode**.

Organic Light-Emitting Device Same as **Organic Light-Emitting Diode**. Its abbreviation is **OLED**.

Organic Light-Emitting Diode A device that utilizes very thin layers of an **organic semiconductor** applied to a substrate, such as glass, for light emission. Such devices provide sharp contrast, do not require a light source behind the viewing surface, have a wide viewing angle, and consume little power. Used, for instance, in displays such as those of cell phones, digital cameras, audio and video components such as receivers and DVD players, PDAs, and notebook computers. Also called **Organic Light-Emitting Device**. Its abbreviation is **OLED**, or **Organic LED**.

Organic Light-Emitting Diode Display Same as **Organic Light-Emitting Display**. Its abbreviation is **OLED display**.

Organic Light-Emitting Display A display incorporating an **Organic Light-Emitting Diode**. Its abbreviation is **OLED display**. Also called **Organic Light-Emitting Diode Display**.

organic semiconductor A semiconductor material based on, or composed of, organic materials. Used, for instance, in semiconductors which are utilized to form a thin film on a flexible substrate such as plastic. An example is pentacene.

organizer 1. A usually handheld device, such as a PDA, which stores information and performs functions which help organize an individual. Such a device may include a telephone, email, and address book, an appointment scheduler, a notepad, and so on. Also called **electronic organizer**. **2.** An application, such as a personal information manager, which performs the functions of an **organizer** (1).

organometallic compound A compound that has at least one carbon-metal bond.

organometallic vapor-phase epitaxy A chemical vapor deposition technique in which at least one organometallic compound is used as a reactant. Its acronym is **OMVPE**. Also called **metal-organic vapor-phase epitaxy**.

orientation 1. The location or position of something relative to a point, direction, object, or the like. Also, the act of locating or positioning. For instance, the alignment of the magnetic particles on a magnetic tape, or the direction in which a vector, or a set of vectors point in. **2.** The location or position of a directional device. For instance, the direction a directional antenna or microphone is pointed in. **3.** The location or position of something relative to Cartesian axes, the markings of a compass, or the like. **4.** When there is a choice between landscape mode and portrait mode, the selected orientation.

orifice 1. An opening, such as a hole, gap, slit, vent, port, mouth, outlet, or perforation, through which electrons, light, waves, or any form of radiant energy may pass. Also called **aperture** (1). **2.** In a waveguide, resonant cavity, loudspeaker, or other similar enclosure, the opening through which energy is transmitted.

origin 1. In a coordinate system, such as the Cartesian coordinate system, the point where all axes intersect. **2.** The point where something comes into existence, begins, or departs from.

origin server A Web server which provides original Web pages or content, as opposed to a **cache server**, which stores that which has been previously downloaded.

original equipment manufacturer Its abbreviation is **OEM**. **1.** A company which produces hardware. This hardware can be sold by such a company directly, or it may be purchased by other companies which sell it under their own brand name. **2.** A company which purchases components from multiple manufacturers, and assembles or integrates them into products which they offer for sale under their own brand name.

ORL Abbreviation of **optical return loss**.

orphan file A file which no longer serves any purpose, but which remains stored in a computer system. For example, orphan files may remain when an applications is uninstalled.

orthicon A camera tube similar to an image orthicon, except that it utilizes a low-velocity electron beam, making it more suitable for use in conditions of reduced light.

orthoferrite A ferrite incorporating a rare-earth element such as samarium or dysprosium. Such materials exhibit magneto-optic effects such as the Faraday effect or the Kerr magneto-optic effect.

orthogonal Perpendicular to, or at right angles with something else. Also, related to that which is perpendicular or at right angles relative to something else. For example, Cartesian axes are orthogonal to each other.

Os Chemical symbol for **osmium**.

OS Abbreviation of **operating system**.

OS/2 A popular operating system.

OS X A popular and robust UNIX-based operating system.

osc Abbreviation of **oscillator**.

oscillate To periodically vary a physical quantity or object around a set point, a given value, an average value, or between two extreme values. For example, the manner in which AC fluctuates, or the moving position of a pendulum.

oscillating circuit Also called **oscillatory circuit**. **1.** A circuit which produces oscillations. For instance, one whose output is an alternating current. **2.** A circuit which is capable of free oscillation. **3.** A circuit which is capable of oscillating. **4.** A circuit which is oscillating at any given moment.

oscillating crystal **1.** A crystal which produces oscillations. For instance, a quartz piezoelectric crystal. **2.** A crystal which is oscillating at any given moment.

oscillating current A current which periodically increases and decreases its amplitude. An example is AC. Also called **oscillatory current**.

oscillating electric field An electric field which periodically increases and decreases its intensity, or which otherwise varies over time. An oscillating electric field produces an oscillating magnetic field, and vice versa. Also called **oscillatory electric field**.

oscillating field A field, such as an electric or magnetic field, which periodically increases and decreases its intensity, or which otherwise varies over time. Also called **oscillatory field**.

oscillating magnetic field A magnetic field which periodically increases and decreases its intensity, or which otherwise varies over time. An oscillating magnetic field produces an oscillating electric field, and vice versa. Also called **oscillatory magnetic field**.

oscillating surge Same as **oscillatory transient**.

oscillating transient Same as **oscillatory transient**.

oscillating voltage A voltage which periodically increases and decreases its amplitude. An example is an AC voltage. Also called **oscillatory voltage**.

oscillation **1.** The periodic variation of a physical quantity or object around a set point, a given value, an average value, or between two extreme values. For example, AC, the movement of a pendulum, or the states of an astable multivibrator. Oscillation may be desired, as in that of a crystal utilized as a frequency standard, or unwanted, as in hunting. Also called **vibration** (2). **2.** A single variation, movement, or cycle during **oscillation** (1). **3.** The **oscillation** (1) of a controlled variable near the desired value, or between two values. Also called **cycling** (1).

oscillation control **1.** Any measure, such as damping, utilized to control **oscillation**. **2.** A device or mechanism which controls an oscillation frequency. An example is a crystal oscillator.

oscillation frequency The frequency at which **oscillation** occurs. For instance, the frequency at which a crystal oscillates.

oscillator Its abbreviation is **osc**. **1.** That which **oscillates**. **2.** A circuit, device, piece of equipment, object, or system which produces and/or maintains **oscillation** (1). **3.** A circuit or device which produces the **oscillation** (1) of an electrical quantity, such as an alternating or pulsating current or voltage. Examples include crystal oscillators, klystrons, magnetrons, and Colpitts oscillators. Most oscillators utilize positive or negative feedback, and the frequency of their output may be fixed or variable. When referring to a circuit, also called **oscillator circuit**. **4.** A circuit or nonrotating device which converts a DC input into a periodically varying electrical quantity.

oscillator circuit Same as **oscillator** (3).

oscillator coil A coil which is utilized in an oscillator circuit. For example, such a coil utilized in a superheterodyne receiver.

oscillator drift A gradual and undesired change in the frequency of an oscillator.

oscillator frequency The frequency at which oscillation is maintained by an oscillator.

oscillator interference Interference produced by an **oscillator** (1), or **oscillator** (2).

oscillator-mixer A circuit, device, or stage, such as that in a receiver, which combines the functions of an oscillator and mixer.

oscillator-multiplier A circuit or device which combines the functions of an oscillator and frequency multiplier.

oscillator padder In a heterodyne receiver, an adjustable capacitor inserted in series with the oscillator tuning circuit, for calibration at the lower end of the tuning range. Also called **padder**, **padder capacitor**, or **padding capacitor**.

oscillator power supply A power supply which provides for the energy needs of an oscillator. For example, that which provides the high DC necessary for operation of a Gunn diode.

oscillator radiation Electromagnetic radiation emitted by an oscillator. For instance, RF energy emitted by the oscillator of a superheterodyne receiver.

oscillator stabilization A device or mechanism utilized to stabilize an oscillator. For example, a mechanism which compensates for oscillator drift.

oscillator stage An oscillating unit within a device or system.

oscillator synchronization The synchronization of an oscillator with another device or system.

oscillatory circuit Same as **oscillating circuit**.

oscillatory current Same as **oscillating current**.

oscillatory field Same as **oscillating field**.

oscillatory surge Same as **oscillatory transient**.

oscillatory transient A sudden and momentary increase in current or voltage whose instantaneous value changes polarity rapidly. Also called **oscillating transient**, **oscillating surge**, or **oscillatory surge**.

oscillatory voltage Same as **oscillating voltage**.

oscillogram A displayed or recorded image of an **oscilloscope**. It may consist, for instance, of a photograph of a trace displayed by an oscilloscope.

oscillograph An instrument which produces a permanent record of one or more varying electrical quantities as a function of time or another variable, especially when produced by an **oscilloscope**. Such a record, for instance, may be plotted by a pen, or be photographed.

oscillography The use of an **oscilloscope** to observe and study one or more varying electrical quantities.

oscillometer A device utilized to measure and indicate oscillations.

oscilloscope Its abbreviation is **scope (1)**. **1.** An instrument which uses a CRT to produce visible patterns of one or more varying electrical quantities, or nonelectrical quantities, such as acoustic waves, with the assistance of a transducer. An oscilloscope typically displays variations in voltage plotted versus time, and may be used, for instance, to monitor signals such as brain waves. Also called **cathode-ray oscilloscope**. **2.** An **oscilloscope (1)** which uses another type of display, such as an LCD or plasma display.

oscilloscope camera A camera specially adapted and fitted to an **oscilloscope** to produce oscillographs.

oscilloscope tube The CRT incorporated into an **oscilloscope** to produce visible patterns of one or more varying electrical signals.

oscilloscopic Pertaining to an **oscilloscope**.

OSD Abbreviation of **on-screen display**.

OSD display Same as **on-screen display**.

OSI Same as **OSI Reference Model**. Abbreviation of **Open Systems Interconnection**.

OSI Model Same as **OSI Reference Model**. Abbreviation of **Open Systems Interconnection Model**.

OSI Protocol Same as **OSI Reference Model**. Abbreviation of **Open Systems Interconnection Protocol**.

OSI Reference Model Abbreviation of **O**pen **S**ystems **I**nterconnection **Reference Model**. An ISO standard which defines a seven-layer hierarchical structure for the implementation of communications protocols, with the goal of standardization of current networks along with component hardware and software, and providing guidance for the creation of future such networks and components. Each higher layer requests and depends on services from the lower layer adjacent to it. The lowest layer is the physical layer, or layer 1, followed by the data-link layer, or layer 2, then comes the network layer, followed by the transport layer, then comes the session layer, and then the presentation layer, and the highest is the application layer, or layer 7. Its own abbreviation is **OSI-RM**. Also known by various other names, including **OSI Model**, **OSI**, **OSI Protocol**, and **ISO-OSI Reference Model**.

OSI-RM Abbreviation of **OSI Reference Model**.

osmiridium A very hard alloy consisting mostly of osmium and iridium, and which also contains smaller quantities of platinum, rhodium, and ruthenium. It is used, for instance, in electrical contacts.

osmium A chemical element whose atomic number is 76. It is a bluish-white lustrous metal which is very hard, and whose density is approximately 22.61, which makes it, along with iridium, one of the two most dense elements. In addition, it is the most corrosion-resistant element, and has over 35 know isotopes, of which 5 are stable. Its applications include its use in hard alloys, and in instrument bearings. Its chemical symbol is **Os**.

osmosis For two solutions, each with different concentrations and separated by a partially permeable membrane, the process via which solvent molecules pass from one side to the other, without solute molecules passing.

OSP Abbreviation of **online service provider**.

OSPF Abbreviation of **Open Shortest Path First**.

OTA Abbreviation of **operational transconductance amplifier**.

OTDR **1.** Abbreviation of **Optical Time Domain Reflectometer**. **2.** Abbreviation of **Optical Time Domain Reflectometry**.

OTHR Abbreviation of **over-the-horizon radar**.

OTL Abbreviation of **output transformerless**.

OTP Abbreviation of **one-time programmable**.

OTPROM Abbreviation of one-time **PROM**. A PROM chip that can only be programmed once. Most PROM chips are of this type. This contrasts, for instance, with EEPROM, which can be rewritten multiple times.

ounce Its abbreviation is **oz**. **1.** A unit of mass equal to approximately 0.0283495 kg. Also called **avoirdupois ounce**. **2.** A unit of mass equal to approximately 0.0311035 kg. Also called **troy ounce**. **3.** A unit of weight or force equal to the force of gravity on one **ounce (1)**. This value varies by location, and is equal to approximately 0.278014 newtons. Also abbreviated **ozf**, which is more proper than **oz**. Also called **ounce force**.

ounce force Same as **ounce (3)**.

out Abbreviation of **output**.

out-of-band Beyond, or otherwise located outside a given interval of frequencies.

out-of-band signaling In a telephone circuit, the transmission of control signals, such as dial tones and busy signals, utilizing a dedicated channel which is separate from any channels utilized for voice transmission. This contrasts with **in-band signaling (1)**, in which in the same channel is utilized.

out-of-phase A state in which two or more periodic quantities having the same frequency and waveshape do not pass through corresponding values, such as maximas and minimas, at the same instant at all times. In an **in-phase** state, the periodic quantities pass through corresponding values at the same instant at all times.

out-of-phase component **1.** Same as **out-of-phase current**. **2.** Same as **out-of-phase voltage**.

out-of-phase current Within an AC circuit, the component of the current that is **out-of-phase** with the voltage. Also known as **out-of-phase component (1)**.

out-of-phase signal A signal that is **out-of-phase** relative to another.

out-of-phase voltage Within an AC circuit, the component of the voltage that is **out-of-phase** with the current. Also known as **out-of-phase component (2)**.

Out-WATS Abbreviation of **Outward Wide-Area Telephone Service**.

outage **1.** An interruption or complete loss of AC power arriving from the power line. Also called **power outage (1)**, or **power failure (1)**. **2.** An interruption or complete loss of power arriving from any source, such as a power line or generator. Also called **power outage (2)**, or **power failure (2)**. **3.** An interruption or complete loss of a signal, such as that of a radio transmitter.

outbox In an email application, the default mailbox in which outgoing messages are stored. The default mailbox for incoming messages is the **inbox**.

outdiffusion The undesired diffusion of dopant atoms from a material with a higher doping level to that with a lower doping level, when sufficient heat is applied. This may occur, for instance, during the formation of epitaxial layers if the temperature is not properly maintained.

outdoor **1.** Situated in the exterior of a building. **2.** Designed for use in the exterior of a building, or outside of another enclosure which protects from surrounding weather.

outdoor antenna Also called **outside antenna**. **1.** An antenna situated in the exterior of a building. **2.** An antenna designed for use in the exterior of a building.

outdoor apparatus An apparatus designed for use in the exterior of a building, or outside of another enclosure which protects it from surrounding weather.

outdoor component A component designed for use in the exterior of a building, or outside another enclosure which protects it from surrounding weather.

outdoor device A device designed for use in the exterior of a building, or outside another enclosure which protects it from surrounding weather.

outdoor equipment Equipment designed for use in the exterior of a building, or outside another enclosure which protects it from surrounding weather.

outdoor transformer A transformer designed for use in the exterior of a building, or outside another enclosure which protects it from surrounding weather.

outer conductor In a concentric two-conductor line, such as a coaxial cable, the conductor which is on the outside. This conductor serves as a shield, while the **inner conductor** usually carries the signal. A dielectric separates the two conductors. Also called **outside conductor**.

outer-shell electrons Electrons which are present in the outermost shell of an atom. Such electrons can move freely under the influence of an electric field, thus producing a net transport of electric charge. Outer-shell electrons also determine the chemical properties of the atom. Also called **valence electrons**, **peripheral electrons**, or **conduction electrons**.

outer space For a given celestial body, the region of space beyond a specified limit. In the case of the planet earth, it is usually defined as the region beyond the atmosphere

outgassing **1.** The removal or release of any occluded gases from an enclosure, crystal, material, or system. Heating, reductions in pressure, or the use of getters, for instance, may be used to accomplish this task. **2.** The **outgassing** (1) of an electron tube, so as to form a vacuum tube.

outgoing line A line leaving a device, piece of equipment, facility, or the like.

outgoing mail server A server, using a protocol such as SMTP, which accepts sent or forwarded emails. Usually utilized in conjunction with an **incoming mail server**, which uses a protocol such as POP or IMAP.

outlet A power line termination whose socket serves to supply electric power to devices or equipment whose plug is inserted into it. Outlets are usually mounted in a wall, although they may be found elsewhere, such as a floor, or in the back of another electrical device, such as an amplifier. Also called by various other names, including **power outlet**, **convenience outlet**, **electrical outlet**, or **receptacle**.

outlet box A box at which there is a power line termination, and whose sockets serve to supply electric power to devices or equipment whose plugs are inserted into them.

outline font A font in which the basic outlines of each character are stored, and scaled into the appropriate size when printed or displayed. This contrasts with a **bitmapped font**, in which each character has its own stored bitmap.

Outlook A popular email client.

output Its abbreviation is **O**, or **out**. **1.** That which is delivered by a component, circuit, device, piece of equipment, system, or process, in response to an input. For instance, the work accomplished in exchange for an energy input. **2.** The energy, voltage, current, or other signal delivered or produced by a component, circuit, device, piece of equipment, system, or process. For example, a voltage output taken from an electronic device. Also, to deliver or provide such a signal. **3.** The information which is generated by a computer after entering, receiving, processing, storing, or otherwise handling input data. Also, the generation or delivery of such information. A computer output may be, for instance, in the form of displayed images, printed pages, sounds, or data sent to peripherals or other computers linked via a network. Also called **output data** (1). **4.** The terminals of a compo-

nent, circuit, device, or piece of equipment from which an **output** (2) is delivered. Also called **output terminals**.

output admittance For a circuit, device, or transmission line, the admittance as seen from the output terminals. It is the reciprocal of **output impedance**.

output amplifier The final amplifier in a cascade or multistage amplifier. For instance, the stage of a transmitter that feeds an antenna. Also called **final amplifier**.

output area Within the memory or storage of a computer, a segment reserved for the output of data. Also called **output block** (1), or **output section**.

output block **1.** Same as **output area**. **2.** Same as **output buffer**. **3.** A block of data awaiting output from a computer.

output-bound Same as **output-limited**.

output buffer A segment of computer memory utilized to temporarily store information that awaits output. Also called **output block** (2).

output capability Also called **output capacity**. **1.** The maximum output power a component, circuit, stage, device, piece of equipment, or system can produce or withstand without damage or failure. **2.** The maximum output power a component, circuit, stage, device, piece of equipment, or system can produce or withstand while maintaining a given level of performance, or under stated conditions.

output capacitance For a circuit or device, the capacitance as seen from the output terminals.

output capacity Same as **output capability**.

output channel A channel via which an output is conveyed.

output circuit A circuit connected to an output electrode or terminal of a component or device. For example, an exterior circuit connected to the output electrode of an electron tube, or to a load.

output conductance For a circuit or device, the conductance as seen from the output terminals.

output current **1.** The current flowing through an **output circuit**. **2.** The current flowing through an output circuit under specified conditions.

output data **1.** Same as **output** (3). **2.** Data that awaits to be output.

output device Also called **output unit**, or **output equipment** (2). **1.** A peripheral, such as a display, printer, or speaker, which provides an output for a computer. **2.** Any device which provides an output for a component, circuit, device, piece of equipment, or system.

output efficiency **1.** The ratio of the useful output of a device or system, to its total input. Also called **efficiency** (1). **2.** The ratio of the useful power or energy output of a device or system, to its total power or energy input. Also called **output efficiency** (2).

output electrode An electrode, such as that of an electron tube, from which an **output** (2) is taken.

output equipment **1.** Any equipment, such as a data projector, utilized to provide an output for a computer. **2.** Same as **output device**.

output gap A gap or cavity from which an output is taken from a velocity-modulated tube such as a klystron.

output impedance For a circuit, device, or transmission line, the impedance as seen from the output terminals. It is the reciprocal of **output admittance**.

output indicator **1.** A component, circuit, device, instrument, piece of equipment, machine, or mechanism which conveys information pertaining to an output magnitude or quantity. Said especially when the information is communicated visually. There many examples, including meters, indicator lamps, gauges, monitors, and so on. **2.** A needle, dial, or other object which serves to indicate the output of a meter, gauge, or similar device or instrument.

output leakage current For a transistor in the off state, the current into the output terminal at a specified input voltage. Also, such a current flowing through an output transistor of an IC.

output-limited A computing process which can proceed only as fast as the output. Also called **output-bound**.

output limiter A circuit or device which limits the amplitude of its output signal to a predetermined maximum, regardless of the variations of its input. Used, for example, for preventing component, equipment, or media overloads. Also called by various other names, including **limiter**, **limiter circuit**, **limiting circuit**, **peak limiter**, **clipper**, **automatic peak limiter**, and **amplitude limiter**.

output limiting The use and effect of an **output limiter**.

output load 1. The power consumed by a component, device, piece of equipment, machine, or system while performing its functions. This power may be electrical, mechanical, nuclear, wind, and so on. Also, any component, device, piece of equipment, machine, or system consuming this power. Also called **load** (1). 2. Any component, circuit, device, piece of equipment, or system which consumes, dissipates, radiates, or otherwise utilizes power, especially electricity. There are countless examples, including resistors, amplifiers, TVs, speakers, antennas, lamps, and appliances. Also, the power so consumed. Also called **load** (2). 3. The electrical power drawn from a source of electricity, such as a generator or power line. Also called **load** (3). 4. A circuit or device which receives the useful signal output from a signal source such as an amplifier or oscillator. Also called **load** (4).

output load current 1. The current drawn by an **output load**. 2. The current drawn by an output load under specified operating conditions.

output medium A medium, such as a display, speaker, or printer, via a which an output is delivered or presented.

output meter An instrument or device which is utilized to measure, and usually indicate, an output magnitude or value. An example is a VU meter.

output port 1. A port, such as a USB port, which serves for the transfer of data between a CPU and a peripheral. Also called **I/O port**, or **input port** (1). 2. A port, such as a display port, which serves for the output of data from a computer.

output power Also called **power output**, or **amplifier power**. 1. The extent to which an amplifier can amplify an input signal. Also called **power amplification** (3). 2. The delivering of the output of an amplifier to a load, such as a loudspeaker. Expressed in watts Also, the power so delivered.

output regulation For a power or signal source, such as a power supply or amplifier, the regulation of the output current, voltage, power, or other signal.

output regulator A component, circuit, device, system, or mechanism which provides **output regulation**. For instance, a line regulator.

output resistance For a circuit or device, the resistance as seen from the output terminals.

output resonator In a velocity-modulated tube, such as a klystron, the cavity from which the output is taken. It is located where maximum electron bunching occurs. Also called **catcher**.

output section Same as **output area**.

output signal The energy, voltage, current, or other signal delivered or produced by a component, circuit, device, piece of equipment, system, or process. Also, to deliver or provide such a signal. Also called **signal output**.

output signal level Also called **signal output level**. 1. The magnitude of an **output signal**. 2. A specific **output signal level** (1), such as a rating.

output stage For a multistage circuit, device, or piece of equipment, such as an amplifier, the final stage. Also called **final stage**.

output storage Within the storage of a computer, a segment reserved for the output of data.

output stream In a computer, a series of control statements sent to a destination, in association with the performance of a given task.

output terminals Same as **output** (4).

output transformer A transformer which delivers energy from an output circuit. Such a transformer may be used, for instance, to couple an amplifier to a load.

output transformerless A circuit, device, or piece of equipment, such as an amplifier, which has no output transformer. Its abbreviation is **OTL**.

output transient A transient occurring at the outputs of a circuit or device.

output transistor A transistor from which an output is taken. For example, a MOSFET utilized in the final stage of a power amplifier.

output tube A tube from which an output is taken. For example, a tube utilized in the final stage of a power amplifier.

output unit Same as **output device**.

output voltage 1. The voltage across an **output circuit**. 2. The voltage across an output circuit under specified conditions.

output voltage accuracy For a power supply, the extent to which the output voltage matches its nominal value, under specified conditions. May be expressed, for instance, as a percentage.

output winding In a transformer or saturable reactor, the winding delivering the power to the load.

outside antenna Same as **outdoor antenna**.

outside conductor Same as **outer conductor**.

outside diameter For an object with two concentric diameters, such as a pipe, the diameter corresponding to the outer circumference. The **inside diameter** corresponds to the inner circumference. Its abbreviation is **OD**.

outside lead The lead attached to the outer, or last, turn of a coil. This contrasts with an **inside lead**, which is connected to the inner turn of a coil. Also called **finish lead**.

Outward Wide-Area Telephone Service A WATS service that is configured only for placing outgoing calls. Such calls, for instance may be destined to a specific region only, or nationwide. This contrasts with an **Inward Wide-Area Telephone Service**, where only incoming calls may be received. Its acronym is **OUTWATS**.

OUTWATS Acronym for **Outward Wide-Area Telephone Service**.

oven An enclosure which serves for drying, hardening, heating, or baking. A crystal oven is an example.

over-charging Same as **overcharging**.

over-clock Same as **overclock**.

over-coupling Same as **overcoupling**.

over-current Same as **overcurrent**.

over-current circuit breaker Same as **overcurrent circuit breaker**.

over-current protection Same as **overcurrent protection**.

over-current relay Same as **overcurrent relay**.

over-current release Same as **overcurrent release**.

over-damping Same as **overdamping**.

over-drive Same as **overdrive**.

over-dub Same as **overdub**.

over-dubbing Same as **overdubbing**.

over-excite Same as **overdrive**. Also spelled **overexcite**.

over-load Same as **overload**.

over-modulation Same as **overmodulation**.

over-modulation indicator Same as **overmodulation indicator**.

over-potential Same as **overpotential**.

over-sampling Same as **oversampling**.

over-scan Same as **overscan**.

over-scanning Same as **overscan**. Also spelled **overscanning**.

over-shoot Same as **overshoot**.

over-swing Same as **overshoot (2)**. Also spelled **overswing**.

over-temperature Same as **overtemperature**.

over-temperature protection Same as **overtemperature protection**.

over-the-horizon communication Communication by means of radio waves that are propagated well beyond line-of-sight distances. This is usually due to scattering by the ionosphere or troposphere. Also called **beyond-the-horizon communication, forward-scatter communication, scatter communication**, or **extended-range communication**.

over-the-horizon propagation Propagation of radio waves well beyond line-of-sight distances. This is usually due to scattering by the ionosphere or troposphere. Also called **beyond-the-horizon propagation, forward-scatter propagation, scatter propagation**, or **extended-range propagation**.

over-the-horizon radar A radar in which the transmitted and reflected beams are bounced off the earth's ionospheric layer. This may allow detection of scanned objects well beyond line-of-sight distances. Its abbreviation is **OTHR**.

over-the-horizon transmission Transmission of radio waves well beyond line-of-sight distances. This is usually due to scattering by the ionosphere or troposphere. Also called **beyond-the-horizon transmission, forward-scatter transmission, scatter transmission**, or **extended-range transmission**.

over-travel Same as **overtravel**.

over-voltage Same as **overvoltage**.

over-voltage crowbar Same as **overvoltage crowbar**.

over-voltage crowbar circuit Same as **overvoltage crowbar circuit**.

over-voltage protection Same as **overvoltage protection**.

over-voltage protection circuit Same as **overvoltage protection circuit**.

over-voltage relay Same as **overvoltage relay**.

over-voltage release Same as **overvoltage release**.

over-voltage switch Same as **overvoltage switch**.

over-voltage test Same as **overvoltage test**.

overall distortion The total distortion introduced by a given device, piece of equipment, or system, as opposed to that of a given component or stage.

overall efficiency The extent to which power is transferred from source to load under specified conditions. It is usually stated as the ratio of power emitted or supplied by the source, to the power absorbed or consumed by the load.

overall gain The gain of an entire device or system, as opposed to that of a single stage. For example, the total gain of a power amplifier with multiple stages.

overall noise The total noise introduced by a given device, piece of equipment, or system, as opposed to that of a given component or stage.

overall sound-pressure level Same as **overall SPL**.

overall SPL Abbreviation of **overall sound-pressure level**. **1.** The total SPL taking into account multiple sources of sound energy. For example, the SPL produced by a speaker with multiple drivers, or that of a specific location within a city where cars, trucks, buses, trains, construction sites, factories, people, and many other sources contribute the overall noise present. **2.** The SPL for a given interval of frequencies which encompasses other intervals. For instance, the SPL produced by a speaker system, as opposed to that produced by the subwoofer only.

overbunching In a velocity-modulated tube, such as a klystron, bunching which exceeds the optimum level.

overcharging The continued application of a charging current once a storage battery or storage cell is fully charged. This can lead to overheating, impaired function, loss of electrolyte, evolution of excessive gas, and so on. Also spelled **over-charging**.

overclock To run a processor or system faster than it was rated to. For example, running a 4 GHz processor at 4.4 GHz. Although overclocking will speed-up performance, doing so improperly may produce unwanted consequences such as overheating or failure of components. Also spelled **over-clock**.

overcoupling In a resonant circuit, coupling which exceeds critical coupling. Also spelled **over-coupling**.

overcurrent For a component, circuit, device, piece of equipment, or system, a current that exceeds the optimum, operational, or rated current. Such a current may produce a malfunction or failure, and may result, for instance, from a short circuit. Also spelled **over-current**.

overcurrent circuit breaker A circuit breaker which is automatically actuated when a current exceeds a set amount. Also spelled **over-current circuit breaker**.

overcurrent protection The use of components, devices, or mechanisms, such as fuses or circuit breakers, to prevent damage due to overcurrents. Also spelled **over-current protection**.

overcurrent relay A relay which automatically opens a circuit when a current exceeds a set amount. Also spelled **over-current relay**. Also called **overload relay (2)**.

overcurrent release A device or mechanism, such as an overcurrent circuit-breaker, which opens a circuit when the current exceeds a specified maximum. Also spelled **over-current release**.

overdamping Damping which exceeds critical damping. In an indicator with a needle, for instance, overdamping may prevent rapidly changing values to be indicated. Also spelled **over-damping**.

overdrive Also spelled **over-drive**. Also called **overexcite**. **1.** To provide a voltage, current, power, or other input whose signal level exceeds that which a component, circuit, device, piece of equipment, or system is designed to safely handle. **2.** To provide an amplifier with an input signal whose amplitude level exceeds that which can be properly or safely handled.

overdriven amplifier An amplifier, or amplifier stage, whose input signal amplitude exceeds that which it was designed to properly handle. When overdriven, an amplifier goes into saturation and cutoff.

overdub For a given audio and/or video recording, to add or replace sounds, or one or more sound tracks, while listening to previously recorded sounds. Used, for instance, when building a song track-by-track. Also spelled **over-dub**. Also called **overdub recording.**.

overdub recording Same as **overdub**.

overdubbing For a given audio and/or video recording, the adding or replacing of sounds, or of one or more sound tracks, while listening to previously recorded sounds. Also spelled **over-dubbing**.

overexcite Same as **overdrive**. Also spelled **over-excite**.

overflow Also called **arithmetic overflow 1.** The condition in which the result of an operation exceeds the capacity of its designated storage location. For example, an operation whose exponent exceeds the number of bits allotted to it. Also called **overflow condition. 2.** The amount by which the result of an operation exceeds the capacity of its designated storage location.

overflow condition Same as **overflow (1).**

overflow error An error which occurs as a result of dividing by zero, or by a number so small that the result exceeds the capability of a program or computer. Also called **divide overflow.**

overflow indicator A component, device, mechanism, or message which alerts to an **overflow condition.**

overflow record 1. A data record which is too large to fit in the allotted storage space. Such a record, for instance, may be stored in another area, or may be discarded. **2.** A location or record indicating any discarded records, such as those that were too large or not processed in time.

overhaul 1. The careful examination, or disassembling, of devices, equipment, or systems, seeking out any components, units, or mechanisms which require adjustment, repair, and/or replacement. **2.** To make extensive adjustments, repairs, and/or replacements.

overhead 1. Time, space, or procedures which are necessary for a given function or task, but which do not necessarily form a part of it. **2.** In communications, bandwidth occupied by control codes, origin and destination information, error detection and correction coding, encryption, and so on. In ATM, for instance, each cell consists of 53 bytes, 48 of which are payload, and 5 of which are overhead. **3.** In computers, procedures or processing time required for desired tasks, such as running applications, but which do not necessarily form a part of a given task. Housekeeping is an example. **4.** Something located, operating, or originating from above something else.

overhead bits In communications, the bits utilized for **overhead (2).**

overhead line 1. Same as **overhead power line. 2.** Same as **overhead transmission line.**

overhead power line A power line whose conductors are suspended above the ground using supports, such as poles, towers, or other structures such as buildings. Since open-wire lines are usually used, a dielectric such as ceramic or glass is utilized to safely attach such lines to their supports. Also called **overhead line (1).**

overhead projector A projector utilized to reproduce transparencies. Such a projector usually incorporates a Fresnel lens.

overhead transmission line A transmission line, such as that utilized for power or communications, whose conductors are suspended above the ground using supports, such as poles, towers, or other structures such as buildings. If open-wire lines are used, a dielectric such as ceramic or glass is utilized to safely attach such lines to their supports. Also called **overhead line (2).**

overlaid windows On a computer screen, a group of windows arranged in a manner that they overlap one another. The title bars are generally visible, allowing a user to know which windows are open. Also called **cascading windows.**

overlap 1. The performance of all or part of an operation while one or more other operations take place. Also, the time during which this occurs. **2.** To partially cover something, or have an area, region, or segment in common. Also, the area covered or in common.

overlay 1. To cover a surface with a layer or design. Also, that which is placed over, or which otherwise covers something else. For example, a graticule placed over the screen of an oscilloscope. **2.** The superimposition or insertion of one image onto or into another. Used, for instance, to place a computer-generated image onto a filmed image. **3.** The use of the same area of a computer's RAM for multiple segments of the same program. When the size of a program exceeds the memory capacity of a given system, segments that are not needed at a given instant are shuttled out of the main memory, and when needed are shuttled back in, each time overwriting the segment which had been there earlier. **4.** Same as **overlay segment. 5.** An area code whose boundaries overlap with those of another area code. Also called **area code overlay (1). 6.** An area code whose boundaries are the same as those of another area code. That is, both area codes serve the same specific region or zone. Also called **area code overlay (2).**

overlay network One of multiple communications networks which share the same geographic region. For example, a cellular network in the same area as a land-based telephone system.

overlay program 1. A program which may be divided into **overlay segments. 2.** A program which is divided into **overlay segments.**

overlay segment One of multiple segments of a program, which, one at a time, occupies the same area of RAM in **overlay (3).** Also called **overlay (4).**

overload Also spelled **over-load. 1.** A load which exceeds the optimum, operational, or rated load of a component, circuit, device, piece of equipment, machine, or system. Such a load may produce, for instance, distortion, excessive heating, malfunction, or failure. Also, to deliver a signal which leads to this condition. **2.** In a communications network, a level of traffic which exceeds that which can be reliably handled. When such a condition occurs in a telephone network, for instance, calls may be blocked or delayed, the quality of connected calls may be diminished, and so on. **3.** Same as **overloading (2).**

overload capacity The extent to which a component, circuit, device, piece of equipment, machine, or system can withstand an overload condition without permanent damage.

overload condition A state in which an **overload (1)** exists.

overload distortion Distortion resulting from an **overload (1).**

overload indicator A component, device, display, mechanism, or alarm which alerts to an **overload condition.**

overload level 1. The operational level beyond which an **overload (1)** begins. **2.** The maximum level of **overload (1)** which can be tolerated without malfunction, failure, or danger. **3.** The extent to which an **overload (1)** occurs.

overload protection The utilization of **overload protectors.**

overload protector A device or mechanism, such as a circuit breaker or peak limiter, which is utilized to prevent **overloads (1),** or their harmful effects.

overload recovery time The time it takes a component, circuit, device, piece of equipment, or system, to resume ordinary operation after an **overload (1).** Some devices can recover within a nanosecond after being overdriven. Also called **recovery time (2).**

overload relay 1. A relay which is actuated when a load exceeds a given value. Used, for instance, as an overload protector. **2.** Same as **overcurrent relay.**

overload release A device or mechanism, such as a peak limiter, which opens a circuit when the load exceeds a specified maximum.

overload time 1. The duration of an **overload (1). 2.** The maximum time an **overload (1)** can occur without causing permanent damage or creating a safety hazard.

overloaded circuit 1. An electric circuit whose load exceeds that which it can handle safely. Such an overload, for instance, may blow a fuse, or cause permanent damage. **2.** A communications circuit whose traffic or bandwidth exceeds that which it can reliably handle. Such an overload, for example, may lead to blocked calls or transmission errors.

overloading **1.** The condition where a load exceeds the optimum, operational, or rated load of a component, circuit, device, piece of equipment, machine, or system. Also, the delivery of a signal which leads to this condition. **2.** In computer programming, the assigning of multiple meanings to the same variable, operator, procedure, or routine. Also called **overload (3)**.

overmodulation Amplitude modulation of a carrier which exceeds a given maximum, such as 100%. This causes distortion, an example of which is sideband splatter. Also spelled **over-modulation**. Also called **overthrow (3)**.

overmodulation indicator A component, device, display, mechanism, or alarm which alerts to an **overmodulation condition**. Also spelled **over-modulation indicator**.

overpotential Same as **overvoltage**. Also spelled **over-potential**.

overpower relay A relay which is actuated when the power in a circuit exceeds a given threshold.

overpower switch A switch that is actuated when the power in a circuit exceeds a given threshold.

override **1.** To counteract or nullify an operation or mechanism which would have otherwise automatically occurred. Also, such an act of counteracting or nullifying. **2.** To counteract or nullify the influence of an automatic control system. Also, such an act of counteracting or nullifying.

oversampling Sampling above the Nyquist rate, or other minimum sampling frequency, for a given signal. The higher the sampling rate, the more accurate the conversion, and some devices feature rates which are hundreds of times the minimum necessary. For example, if the Nyquist rate for a given signal is 44.1 kHz, then 2X oversampling means that the signal is sampled at 88.2 kHz. In this case 8X oversampling would yield a rate of 352.8 kHz, and so on. Also spelled **over-sampling**.

overscan In a display, such as a CRT, the deflection of the scanning beam beyond the edges of the screen. This may result in the loss of images at the edges of the display. Also spelled **over-scan**. Also called **overscanning**.

overscanning Same as **overscan**. Also spelled **over-scanning**.

overshoot Also spelled **over-shoot**. **1.** To go beyond or exceed that which is necessary, desired, or expected. For example, a momentary excursion of a pulse or parameter outside of set limits, as might occur when turning on a device. Also called **overthrow (1)**. **2.** In an analog indicating device, a momentary excursion of the pointer beyond the value to be indicated. Proper damping minimizes this. Also called **overswing**, or **overthrow (2)**. **3.** The extent to which **overshoot (1)** or **overshoot (2)** occurs.

overstress A transient or steady-state condition in which the ratings and/or capabilities of a device are exceeded. For instance, the voltage or current rating being exceeded. Such overstress does not necessarily manifest the damage it causes immediately.

overswing Same as **overshoot (2)**. Also spelled **over-swing**.

overtemperature A temperature that exceeds that at which a component, circuit, device, piece of equipment, or system can operate safely. Operation at such temperatures may result in overheating, reduced operational life, malfunction, or failure. Also spelled **over-temperature**.

overtemperature protection Devices or mechanisms, such as thermostats, which prevent the temperature of a component, circuit, device, piece of equipment, or system to exceed a specified value. Also spelled **over-temperature protection**.

overthrow **1.** Same as **overshoot (1)**. **2.** Same as **overshoot (2)**. **3.** Same as **overmodulation**.

overtone For a complex tone, sound, signal, wave, or vibration which is periodic, a component whose frequency is a whole-number multiple, greater than one, of the fundamental frequency. Overtones are the same as **harmonics**, except that the first overtone is the same as the second harmonic, the second overtone is the same as the third harmonic, and so on. The fundamental frequency is also called fundamental component, or first harmonic.

overtone content For a complex tone, sound, signal, wave, or vibration which is periodic, all overtones.

overtone crystal A quartz crystal which oscillates at an **overtone frequency**.

overtone crystal oscillator An oscillator which incorporates an **overtone crystal**. Also called **overtone oscillator (1)**.

overtone frequency The frequency of an **overtone**.

overtone oscillator **1.** Same as **overtone crystal oscillator**. **2.** An oscillator which produces overtones.

overtravel For a component, device, piece of equipment, or mechanism, any movement beyond the operating position which does not result in damage. For example, pressing a switch beyond that necessary for the contacts to initially touch, but not far enough to damage the mechanism. Also spelled **over-travel**.

overtype Same as **overwrite mode**.

overtype mode Same as **overwrite mode**.

overvoltage Also spelled **over-voltage**. Also called **overpotential**. **1.** A voltage which exceeds the optimum, operational, or rated value of a component, circuit, device, piece of equipment, machine, or system. Such a voltage may produce, for instance, distortion, excessive heating, malfunction, or failure. **2.** The extent to which an **overvoltage (1)** occurs.

overvoltage crowbar Also spelled **over-voltage crowbar**. **1.** A circuit or device which protects against overvoltage, by rapidly placing a low-resistance shunt across the terminals where a given voltage is exceeded. Also called **crowbar (1)**. **2.** The low-resistance shunt utilized in a **overvoltage crowbar (1)**. Also called **crowbar (2)**.

overvoltage crowbar circuit A circuit which serves as an **overvoltage crowbar (1)**. Also spelled **over-voltage crowbar circuit**. Also called **crowbar circuit**.

overvoltage protection The use of components, devices, or mechanisms, such as an overvoltage crowbar, to prevent damage due to overvoltage. Also spelled **over-voltage protection**. Its abbreviation is **OVP**.

overvoltage protection circuit A circuit, such as an overvoltage crowbar circuit, which serves to protect against overvoltage. Also spelled **over-voltage protection circuit**.

overvoltage relay A relay which is actuated when the voltage in a circuit exceeds a given threshold. Also spelled **overvoltage relay**.

overvoltage release A device or mechanism, such as an overvoltage crowbar, which opens a circuit when the voltage exceeds a specified maximum. Also spelled **over-voltage release**.

overvoltage switch A switch that is actuated when the voltage in a circuit exceeds a given threshold. Also spelled **overvoltage switch**.

overvoltage test A test, such as a high-potential test, in which a voltage exceeding the rated or operational voltage is applied to a conductor, circuit, device, piece of equipment, or system. Used, for instance, to evaluate new devices or equipment. Also spelled **over-voltage test**.

overwrite **1.** To write into an area of storage where data was already present, thereby destroying the data that had occupied said area. This is done, for instance, each time a file of a disk or disc is updated. **2.** Same as **overwrite mode**.

overwrite mode In a computer application, a mode of entering data in which any characters typed on the keyboard replace those already occupying the same area. This contrasts with **insert mode**, in which any newly typed characters are placed between characters already present. Also called **overwrite (2)**, **overtype mode**, **overtype**, or **typeover mode**.

OVP Abbreviation of **overvoltage protection**.

Owen bridge A four-arm AC bridge utilized to measure an inductance and its series resistance.

OXC Abbreviation of **Optical Cross-Connect**.

oxidation The converse is **reduction**. **1.** A chemical reaction in which an atom, molecule, ion, or radical combines with oxygen. For instance, iron combining with oxygen to form iron oxide. **2.** A chemical reaction in which an atom, molecule, ion, or radical loses one or more electrons. For example, a reaction in which iron combines with oxygen, where the iron undergoes a change in valence from 0 to +3.

oxidation-reduction A chemical reaction in which at least one atom, molecule, ion, or radical gains at least one electron, while at least one atom, molecule, ion, or radical loses at least one electron. Oxidations almost always involve a simultaneous reduction, and vice versa. For example, in a reaction where iron combines with oxygen to form iron oxide, the iron is oxidized, as its valence increases from 0 to +3, while the oxygen is reduced, as its valence decreases from 0 to -2. Its abbreviation is **redox**.

oxide-coated cathode A cathode which is coated with a metal-oxide compound, such as thorium oxide, in order to enhance its electron emissions. Also called **coated cathode**.

oxide-coated filament A filament which is coated with a metal-oxide compound to enhance its properties. An oxide-coated cathode is an example. Also called **coated filament**.

oxide film A thin, or very thin layer or coating of an oxide, such as iron oxide. Such a coating may be protective, decorative, or functional. Specific examples include a magnetic oxide layer upon a plastic substrate in magnetic recording tape, a magnetic thin film deposited upon an aluminum substrate in a hard disk, or that of an oxide-coated cathode.

oxide isolation In an IC, the use of an oxide, such as silicon dioxide, to isolate components, as opposed for instance, to the use of junction isolation.

oxide masking In semiconductor manufacturing, the use of an oxide, usually silicon dioxide, to mask portions of wafers.

oxide passivation A passivation technique in which the protected components and surfaces are protected by a layer of an oxide, usually silicon dioxide.

oxygen A chemical element that is a colorless, odorless, and tasteless diatomic gas whose atomic number is 8. It comprises about 21% of the atmosphere, nearly 50% of the earth's crust, and almost 90% of water, making it the most abundant element on the planet. It is very reactive, forming compounds with almost every element, and has over one dozen known isotopes, of which 3 are stable. Its applications include its use in most combustion reactions, such as those occurring during the utilization of most fuels. Its chemical symbol is **O**.

oxygen-free copper Copper which is essentially free of oxides, thus maximizing its conductivity. Used, for instance, to make very-high quality speaker cables, as it minimizes sound losses and coloration. Its abbreviation is **OFC**. Also called **oxygen-free high-conductivity copper**.

oxygen-free copper wire Wire made from **oxygen-free copper**.

oxygen-free high-conductivity copper Same as **oxygen-free copper**. Its abbreviation is **OFHC**.

oz Abbreviation of **ounce**.

ozf Abbreviation of **ounce force**.

ozocerite Same as **ozokerite**.

ozokerite A yellow to dark brown naturally occurring wax consisting chiefly of hydrocarbons, and which is used as an electrical insulator. A refined form is called **ceresin**. Also spelled **ozocerite**. Also called **earth wax**.

ozone A poisonous and highly reactive blue gas whose chemical formula is O_3, making it an allotropic form of oxygen. It may be formed, for instance, when oxygen is exposed to electrical discharges, or in certain photochemical reactions. Ozone has a characteristic odor, which is sometimes detected near a lightning stroke, and although it is toxic, a layer of ozone in the atmosphere helps shields living organisms from excessive ultraviolet radiation from the sun.

P

p Abbreviation of **pico-**.

P **1.** Abbreviation of **peta-**. **2.** Symbol for **pressure**. **3.** Chemical symbol for **phosphorus**. **4.** Symbol for **poise**. **5.** Abbreviation of **positive**. **6.** Abbreviation of **primary winding**.

P **1.** Symbol for **electric power**, or **power (1)**. **2.** Symbol for **permeance**, or **magnetic permeance**.

P Band In communications, a band of radio frequencies extending from about 230 MHz to 1,000 MHz, although the defined interval varies.

p-channel In a field-effect transistor, a conducting channel between the source and the drain formed by a p-type semiconductor.

p-channel FET Abbreviation of **p-channel field-effect transistor**. A field-effect transistor with a p-channel, as opposed to an n-channel. Its own abbreviation is **PFET**.

p-channel field-effect transistor Same as **p-channel FET**.

p-channel JFET Abbreviation of **p-channel junction field-effect transistor**. A JFET with a p-channel, as opposed to an n-channel.

p-channel junction FET Same as **p-channel JFET**.

p-channel junction field-effect transistor Same as **p-channel JFET**.

p-channel metal-oxide semiconductor Same as **PMOS**.

p-channel metal-oxide semiconductor field-effect transistor Same as **PMOSFET**.

p-channel MOS Same as **PMOS**.

p-channel MOSFET Same as **PMOSFET**.

p-code Abbreviation of **pseudocode**.

P display Same as **plan position indicator**.

P electron An electron located in the P shell of an atom.

p-i-n diode Same as **PIN diode**.

p-i-n photodiode Same as **PIN photodiode**.

P indicator Same as **plan position indicator**.

p-layer Same as **p-region**.

p-p Abbreviation of **peak-to-peak**.

p-p amplitude Abbreviation of **peak-to-peak amplitude**.

p-p value Abbreviation of **peak-to-peak value**.

p-p voltage Abbreviation of **peak-to-peak voltage**.

p-region A layer consisting of a **p-type semiconductor**. Also called **p-type region**, **p-type layer**, or **p-layer**.

P scan Same as **plan position indicator**.

P scanner Same as **plan position indicator**.

P scope Same as **plan position indicator**.

P shell The sixth electron shell of an atom. It is the sixth closest shell to the nucleus.

p-static Abbreviation of **precipitation static**.

p-type **1.** Same as **p-type semiconductor**. **2.** Pertaining to a **p-type semiconductor**, or that which incorporates such a material.

p-type conduction In a semiconductor, conduction via mobile holes. In **n-type conduction** electrons are the charge carriers.

p-type conductivity In a semiconductor, the ease with which **p-type conduction** occurs.

p-type layer Same as **p-region**.

p-type material Same as **p-type semiconductor**.

p-type region Same as **p-region**.

p-type semiconductor A semiconductor material which has the addition of trivalent dopants, such as boron or gallium, which contribute mobile holes which increase the conductivity of the intrinsic semiconductor. In this manner, the concentration of holes is much higher, making them the majority carrier. This contrasts with an **n-type semiconductor**, in which electrons are the majority carrier. Also called **p-type material**, or **p-type (1)**.

P2P architecture Abbreviation of **peer-to-peer architecture**.

P2P communications Abbreviation of **peer-to-peer communications**.

P2P computing Abbreviation of **peer-to-peer computing**.

P2P network Abbreviation of **peer-to-peer network**.

pA Abbreviation of **picoampere**.

Pa **1.** Symbol for **pascal**. **2.** Chemical symbol for **protactinium**.

PA **1.** Abbreviation of **power amplifier**. **2.** Abbreviation of **pulse amplifier**.

PA system Abbreviation of **public-address system**.

PABX Abbreviation of **private automatic branch exchange**.

pacemaker An electrical device which delivers small impulses to the heart under specified conditions, and at a predetermined rate, so as to promote its pumping blood at a regular pace. Such a device may or may not be implanted. Also called **pacer**, **cardiac pacemaker**, or **heart pacer**.

pacer Same as **pacemaker**.

pack To convert data into a more compact form making use of characteristics, or the current state, of the data and/or storage medium. For instance, to store two digits per octet, as opposed to one digit per octet. It is a form of compression.

package **1.** The enclosure in which a component, circuit, device, piece of equipment, apparatus, or system, is housed. Also, to assemble, place or otherwise house that which is contained in such an enclosure. **2.** A set of items which are designed to work together for a given purpose. Unlike a **kit**, such a set does not necessarily require assembly. **3.** The housing, usually made of ceramic or plastic, surrounding an IC. The leads or metallic surfaces of the package are used for connection via plugging-in or soldering. Such packages are usually mounted onto printed-circuit boards. Also called **chip package**, or **IC package**. **4.** A combination of application programs which are tailored to a given type of work, and which function especially well together. For example an office suite incorporating word processing, database, spreadsheet, presentation, and communications programs. Also called **software package (1)**, **application suite**, **suite**, or **bundled software**. **5.** Same as **packaged software**.

packaged software An application program which is sold to the general public with all that should be necessary for it to work properly. This contrasts with **custom software**, which is designed to meet specific needs or circumstances. Also called **package (5)**, or **software package (2)**.

packaging The process of assembling, interconnecting, embedding, or mounting electronic components, circuits, or devices, so that they can perform their functions while sealed or otherwise protected. An example is a dual in-line package for a chip. Also called **electronic packaging**.

packaging density Same as **packing density**.

packed data Data which has been converted into a more compact form, by making use of characteristics, or the current state, of the data and/or storage medium. For example, that stored utilizing two digits per octet, as opposed to one digit per octet.

packed decimal A storage mode in which two consecutive decimal digits, each occupying four bits, comprise a single byte. A byte ordinarily represents a single decimal digit.

packet Also called **data packet**. **1.** A block of data transmitted between one location and another within a communications network. Also called **information packet**. **2.** A block of data of a specific size or of a maximum size, such as that transmitted in a packet-switching network. In addition to the payload, a packet contains information such as the source and destination addresses. When used in the context of TCP/IP networks, it is also called **datagram**.

packet assembler/disassembler A device or interface which splits a data stream into discrete packets for transmission, and joins them upon reception. Used, for instance, to enable data-terminal equipment not equipped for packet communications to access a packet-switching network. Its acronym is **PAD**.

packet classification In a communication service with QoS standards, such as ATM, the determination of the level a packet has within the priority hierarchy.

packet communications Communications based on **packet switching**.

packet data network Same as **packet-switching network**.

packet data transmission Same as **packet transmission**.

packet drops Same as **packet loss**.

packet filter A functional unit which performs **packet filtering**.

packet filtering The determination, usually based on IP source or destination address, whether packets will enter or leave a network, such as a LAN. It is a firewall technique, and filtering may also be based on other factors, such as UDP source or destination ports.

packet filtering router A router that performs **packet filtering**.

packet fragmentation In a communications network, the breaking up of packets into smaller sized units of data. Used, for instance, to accommodate communications protocols whose maximum packet size is smaller. Also called **fragmentation (2)**.

packet header The first part of a packet. A packet header contains information such as the originating and destination stations, and is used, for instance, by a router to determine the best path for a given packet to travel en route to its destination address.

Packet Internet Groper Same as **ping (2)**.

packet length The length, in bits, of a **packet (2)**.

packet loss Any losses of **packets (2)** due to excessive network congestion, routing errors, a bad connection, or the like. Also called **packet losses**, or **packet drops**.

packet losses Same as **packet loss**.

packet network Same as **packet-switching network**.

packet over SONET The sending of IP packets directly over SONET networks. SONET offers a lower packet overhead than ATM, providing for higher transmission speeds. Its abbreviation is **POS**.

packet overhead In a packet-switched network, the overhead occupied by each packet. In ATM, for instance, each cell consists of 53 bytes, 48 of which are payload, and 5 of which are overhead.

packet protocol A protocol, such as frame relay, utilized for **packet communications**.

packet radio Communications in which **packets (2)** are transmitted wirelessly. Used, for instance, in a General Packet Radio Service. Also called **packet radio communications**.

packet radio communications Same as **packet radio**.

packet radio network A network, such as that using GSM, utilized for **packet radio communications**.

packet sequence The sequence in which packets are sent in a **packet-switching network**.

packet sniffer A hardware or software device which monitors network traffic, examining each packet, looking for problems such as bottlenecks. A packet sniffer may also be used maliciously, to capture private information which is not properly encrypted.

packet sniffing The use of a **packet sniffer**.

packet-switched network Same as **packet-switching network**.

packet switching In a communications network, the transmission, routing, forwarding, and the like, of messages which are broken into **packets (2)**. Since each contains a destination address, each of the packets of a single message may take different paths, depending on the availability of channels, and may arrive at different times, with each complete message being reassembled at the destination. Also, the technology employed for such communications. This contrasts with **message switching**, in which complete messages are transmitted, routed, forwarded, and so on, to their destination. It also contrasts with **circuit switching**, in which two terminals are linked to establish a communication channel which remains in exclusive use by these terminals until the connection is terminated.

packet-switching network A network based on **packet switching**. Its abbreviation is **PSN**. Also called **packet-switched network, packet network**, or **packet data network**.

packet telephony The use of IP-based packet-switched connections to transmit voice or other audio over an IP network such as the Internet. IP telephony may also be utilized for fax or video transmission. Also called **IP telephony**.

packet transmission The transmission of data in a **packet-switching network**. Also called **packet data transmission**.

packet writing A method of writing to an optical disk, such as a CD-R, in which multiple blocks of data are written incrementally in packets with a fixed or variable length. This contrasts, for instance, with **Disc-at-Once**, in which all data is transferred continuously without interruptions, or **Track-at-Once** in which each track, as opposed to a block, is recorded independently.

packetized voice The transmission of audio over a packet-switched network, as in Internet telephony. An example is voice over IP.

packets per second For a device, such as a router or switch, in a packet-based network, the number of packets forwarded per second. It is a measure of speed, throughput, or performance. Its abbreviation is **PPS**.

packing **1.** The conversion of data into a more compact form making use of characteristics, or the current state, of the data and/or storage medium. For instance, the storing of two digits per octet, as opposed to one digit per octet. It is a form of compression. **2.** The insertion of a material into a cavity or space for a given purpose, such as reinforcement. Also, the material so placed. **3.** The sticking together of granules in a carbon microphone. This reduces sensitivity.

packing density Also called **packaging density**, or **packing factor**. **1.** The number of components per unit, unit area, or unit volume in a device such as an IC. For example, the number of transistors on a chip. **2.** The number of bits or other storage measure per unit length, unit area, or unit volume of a storage device. For instance, the number tracks per inch on a hard disk.

packing factor Same as **packing density**.

pad **1.** A fixed attenuator consisting of a network of fixed resistors. Also called **resistance pad**. **2.** On a printed-circuit board, the printed conductive portion to which components are connected. It may consist, for instance, of the enlarged areas where component leads are soldered. Also called **terminal area**, or **land** (4). **3.** One of multiple metal pads on the surface of a semiconductor device onto which connections may be made. Also called **bonding pad**. **4.** A layer of a material placed over something else to provide cushioning. **5.** The use of **padding characters**, or **padding bits**.

PAD Acronym for **packet assembler/disassembler**.

pad bits Same as **padding bits**.

pad characters Same as **padding** (1).

padder In a heterodyne receiver, an adjustable capacitor inserted in series with the oscillator tuning circuit, for calibration at the lower end of the tuning range. Also called **padder capacitor**, **padding capacitor**, or **oscillator padder**.

padder capacitor Same as **padder** (1).

padding **1.** In computers and communications, characters, such as a blank characters, which are utilized to occupy empty positions of data structures. Also called **padding characters**, or **pad characters**. **2.** Same as **padding bits**.

padding bits Bits which are added to occupy empty positions of data structures. Used, for instance, to avoid restricted bit patterns in a transmission along a communications channel. Also called **pad bits**, or **padding** (2).

padding capacitor Same as **padder** (1).

padding characters Same as **padding** (1).

paddle A computer input device which moves the screen cursor, or another object, along a single axis.

paddle switch A toggle switch whose actuating lever is flattened and comparatively wide, similar to the end of a boat paddle.

page **1.** The text, graphics, and so on which occupy the entire screen of a computer monitor at any given moment. **2.** The text, graphics, and so on which occupy a sheet of paper printed by a computer. **3.** A fixed-length block of data possessing a virtual address that is transferred as a unit. **4.** A program segment possessing a virtual address that is transferred as a unit. **5.** A document within the World Wide Web. Such a document may be written, for instance, in HTML, and may occupy more or less space than a **page** (1). Such a page may provide text, graphics, audio, interactive features, and so on. Also called **Web page**. **6.** A message received via a **pager** (1).

page-description language A programming language, such as PostScript, which is utilized to specify how documents will be printed or displayed. Such a language defines how the elements of each page, such as text and graphics, will appear. This software resides in the output device, usually a printer, thus making such programs machine-independent. Its abbreviation is **PDL**.

page down On a computer keyboard, a key which moves the cursor down a **page** (1), a **page** (2), or another amount depending on the application and/or any modifier key pressed at the same time. Its abbreviation is **PgDn**. Also called **page down key**.

page down key Same as **page down**. Its abbreviation is **PgDn key**.

page fault An interrupt which occurs when an application requests data which is not currently loaded into the RAM of the computer. Such an interrupt may trigger paging of data from virtual memory to real memory.

page fault handler System software which handles **page faults**.

page frame In the RAM of a computer, a segment of memory the size of a **page** (3).

page header In word processing, a text which appears at the top of one or more pages. Such text may appear only on a give page, specified pages, even pages, odd pages, or on every page. Also called **header** (4).

page mode memory Same as **page mode RAM**.

page mode RAM Abbreviation of **page mode random-access memory**. A form of dynamic RAM which enables faster access, as a row of memory bits only needs to be selected once for all columns to be accessed within said row. Other forms of dynamic RAM, such as SDRAM, have superseded page mode RAM. Also called **page mode memory**, **fast page mode memory**, or **fast page mode RAM**.

page mode random-access memory Same as **page mode RAM**.

page orientation When there is a choice between landscape mode and portrait mode for a given page, the selected orientation.

page printer A printer which prints full pages, one at a time, after being stored in memory. An example is a laser printer.

page reader A document reader which scans a full page at a time.

page source The source code of a Web page, such as HTML code. Also called **page source code**, or **Web page source**.

page source code Same as **page source**.

page up On a computer keyboard, a key which moves the cursor up a **page** (1), a **page** (2), or another amount depending on the application and/or any modifier key pressed at the same time. Its abbreviation is **PgUp**. Also called **page up key**.

page up key Same as **page up**. Its abbreviation is **PgUp key**.

page view An instance of a Web page, or any component of it, being viewed by a user over the Internet. A **hit** (2) does not necessarily have to be viewed to be counted.

PageMaker Popular desktop publishing software.

pager **1.** A pocket-sized radio receiver or transceiver which serves to receive messages, and in some cases perform various other functions, such as retrieve email. These devices may emit a beep when informing of a new message. Some pagers can also signal by other means, such as vibrations, or a blinking light. Also called **paging device**, **mobile pager**, or **beeper**. **2.** A public-address system, such as that in an airport or shopping center, which serves convey voice messages. Also called **paging system**. **3.** In a cordless phone, a feature in which pressing a button on the base unit serves to emit a beep heard in the handset. Used, for instance, for people who have difficulty locating the handset due to reduced vision, or for those who are simply forgetful and/or disorganized. Also, the button itself.

pages per minute The number of pages a printer can print, per second. It is a measure of printer speed, although this number will vary considerably depending on the graphics density per page, whether it is the same page being printed over and over again, and so on. Its abbreviation is **PPM**.

pages per second The number of pages a printer can print, per second. It is a measure of printer speed, although this number will vary considerably depending on the graphics density per page, whether it is the same page being printed over and over again, and so on. Its abbreviation is **PPS**.

pagination In text processing, the division of a document into pages, or the numbering of its pages.

paging **1.** The transfer of **pages** (3) or **pages** (4) in and out of memory. The translations between virtual addresses and physical addresses are usually performed by the memory management unit. **2.** The sending of **pages** (6).

paging device Same as **pager** (1).

paging system **1.** A communications service which enables **pagers** (1) to receive alerts and messages, and otherwise exchange information. **2.** Same as **pager (2)**. **3.** The mechanism utilized for **paging (1)**.

paint program An application which enables the creation of drawings in the form of bitmaps. In a **drawing program**, drawings are created utilizing vector graphics, so picture objects may be created and manipulated as independent objects. An **image editor** is an application which has many advanced tools to modify bitmapped images.

Painter A popular paint program.

pair 1. Two of anything. **2.** Two items which correspond to each other. For example, Cooper pairs, electron-hole pairs, or Darlington pairs. **3.** A transmission line consisting of two similar conductors. For example, a two-wire circuit, or a two-wire telephone line. Also called **wire pair. 4.** Two valence electrons which are shared by two adjacent atoms. Each pair of such electrons forms a bond between said atoms. Also called **electron pair**.

paired cable A cable consisting of one or more **pairs (3)**. The conductors of such a cable are usually twisted or braided together, which case they are called **twisted pair**. Used, for instance, for telephone communications. Also called **twin cable**.

pairing 1. The creation or matching of a **pair. 2.** In TV, an image defect in which even and odd scan lines overlap, reducing vertical resolution.

PAL 1. Acronym for phase-alternating line, phase-alternation line, or phase-alternate line. A standard for the generation, transmission, and reception of TV signals. It provides for a resolution of 625 horizontal lines per frame, with 50 interlaced half-frames per second for a total of 25 frames per second. Of the 625 total lines, approximately 576 are active, while the aspect ratio is 4:3. In PAL, the phase of certain color information is inverted from scan line to scan line, so as to minimize phase distortion. PAL is used throughout most of Europe and many countries around the world, such as China and Brazil, while the other two main world standards are **NTSC** and **SECAM**. All three are mutually incompatible. Also called **PAL standard. 2.** Acronym for **programmable array logic**.

PAL signal Abbreviation of phase-alternating line **signal**, phase-alternation line **signal**, or phase-alternate line **signal**. A TV signal adhering to the **PAL Standard**.

PAL standard Abbreviation of phase-alternating line **standard**, phase-alternation line **standard**, or phase-alternate line **standard**. Same as **PAL (1)**.

palladium A silver-white chemical element whose atomic number is 46. It does not tarnish in air, and when finely divided can absorb about 900 times its own volume of hydrogen. It has about 30 known isotopes, of which 6 are stable. Its applications in electronics include its use in electric contacts, relays, switches, lasers, and superconductors. Its chemical symbol is **Pd**.

palletize In robotics, to place parts in different positions within pallets or trays.

palletizing In robotics, the placing of parts in different positions within pallets or trays.

Palm A popular PDA.

palm-top Same as **palmtop**.

palm-top computer Same as **palmtop**. Also spelled **palm-top computer**.

palmtop A computer which is small enough to hold in a hand or place in a pocket. Such computers generally have a specialized operating system, and provide an interface to communicate with other computers, such as desktop systems. There are many types, including models which are industry-specific, such as those utilized by transport companies to keep track of shipments, those that provide only email func-

tions, or PDAs. Also spelled **palm-top**. Also called **palm-top computer, pocket computer**, or **handheld computer**.

palmtop computer Same as **palmtop**. Also spelled **palm-top computer**.

PAM Acronym for **pulse-amplitude modulation**.

pan 1. To perform a **panoramic sweep. 2.** To move a movie or TV camera in a horizontal and/or vertical plane, so as to create a panoramic view. In this context it is an abbreviation of **pan**orama or **pan**oramic.

PAN Abbreviation of **Personal Area Network**.

pan adapter Abbreviation of **panoramic adapter**.

pan-and-scan The conversion of an image source whose aspect ratio is 16:9, to a ratio of 4:3. Used, for instance, to present motion pictures on a TV with a 4:3 aspect ratio. In the process, a significant amount of picture information is lost. Its abbreviation is **pan & scan**.

pan & scan Abbreviation of **pan-and-scan**.

panadapter Abbreviation of **panoramic adapter**.

pancake coil A coil that is in the form of a disk. Its winding is flat, as opposed to cylindrical. Also called **pancake winding**, or **disk coil**.

pancake winding Same as **pancake coil**.

panel Also called **panel board. 1.** A rigid surface, sheet, or board upon which multiple indicators and devices, such as switches and dials are mounted, so as to enable a user to monitor and control a system. Used, for instance, to control an aircraft. Also called **control panel (1). 2.** A surface, sheet, or board upon which instruments and/or their indicators are mounted. Also called **instrument panel. 3.** A rigid surface, sheet, or board which serves to protect, cover, or hide. For example, that on the back of an instrument or device. Such a panel often has connectors, controls, and so on. **4.** A device usually consisting of multiple solar cells which are connected to produce a given power output. Also called **solar panel**.

panel board Same as **panel**.

panel display A video display which is comparatively thin. At present this represents a depth of about 50 mm or less, but this number is being reduced over time. Liquid-crystal, plasma, and electroluminescent displays are examples. Such displays usually feature high resolution, high contrast, minimal image distortion, and a light weight. Used, for instance, in computers and TVs. Also called **flat display, flat monitor, flat-panel display**, or **flat-panel monitor**.

panel indicator An indicator mounted on a **panel (1)** or **panel (2)**.

panel light 1. A indicating light within a **panel (1)** or **panel (2)**. For example, one which blinks to inform of a fault condition. **2.** A light which illuminates the viewing surface of a **panel (1)** or **panel (2)**. Such a light may also be located behind the viewing surface, or a specific indicator.

panel loudspeaker Same as **panel speaker**.

panel meter A meter mounted on a **panel (1)** or **panel (2)**.

panel speaker A speaker which is especially thin and usually lightweight. Used, for instance, as computer speakers placed on a desk, or as wall-mounted speakers in a home theater system. Also called **panel loudspeaker**, or **flat-panel speaker**.

panic button A button, switch, or other such device which serves to alert of imminent danger or of an emergency. For instance, in a home-security system, there may be several such buttons strategically located throughout the house, so that any individual can alert the monitoring station of an emergency.

panning 1. The movement of a radar beam so as to scan a wide area. **2.** The movement of a movie or TV camera in a horizontal and/or vertical plane, so as to create a panoramic view. **3.** In computers, to move the viewing area in a manner similar to **panning** (2).

panoramic adapter Its abbreviation is **pan adapter**. **1.** A camera adapter which facilitates viewing, filming, and photographing panoramic views. **2.** An adapter which widens the viewed frequency interval of a device such as an oscilloscope.

panoramic display Also called **panoramic indicator**. **1.** A display which shows all the signals present in a given frequency interval. **2.** A display which shows all signals present for a full 360° around a given point.

panoramic indicator Same as **panoramic display**.

panoramic potentiometer An audio mixer control that is utilized to position the signal of a channel somewhere between the left and right speakers. Its acronym is **panpot**.

panoramic radar A radar which transmits a wide beam in a given direction, as opposed to scanning.

panoramic receiver **1.** A radio receiver which is tuned to receive a wide interval of frequencies. **2.** A radio receiver which automatically tunes multiple frequency bands for given intervals, so as to cover a wide interval of frequencies.

panoramic sweep In radars, scanning which sequentially covers a wide area.

panoramic view An unbroken view of an entire area which is comparatively wide.

panpot Acronym for **panoramic potentiometer**.

pantography A system utilized to transmit and automatically record radar data at a remote location.

Pantone Matching System A color-matching system used in printing and in graphics programs. Its abbreviation is **PMS**.

PAP Acronym for **Password Authentication Protocol**.

paper capacitor A fixed capacitor in which paper, such as kraft paper, is used as a dielectric. The usual arrangement has the paper, typically impregnated with oil or wax, sandwiched between two metal foils, with all three sheets rolled together.

paper feed A mechanism which moves a paper, or other suitable medium, through a printer or printing mechanism.

paperless office An idealized office which stores all documents in any form other than paper. Documents would all be stored magnetically, optically, and so on.

PAR Abbreviation of **precision approach radar**.

parabola A curve which is formed by a conic section whose distance from a fixed point, called focus, is the same as that to a fixed line, called directrix. The curve obtained when a quadratic equation is graphed is in the shape of a parabola.

parabola generator A signal generator that produces a test pattern in the form of a **parabola**.

parabolic antenna Same as **parabolic reflector** (2).

parabolic dish Same as **parabolic reflector**.

parabolic microphone A microphone which is suspended, attached, or mounted at the focal point of parabolic reflector. Such microphones are highly directional and sensitive

parabolic reflector Also called **paraboloid reflector, paraboloidal reflector, parabolic dish, paraboloid** (2), or **dish**. **1.** A concave reflecting surface with a **paraboloid** (1) shape, which concentrates waves into a parallel beam. Used, for instance, in dish antennas. **2.** An antenna incorporating a **parabolic reflector** (1). Such an antenna features very high gain, usually operates at microwave frequencies, and has various applications, including its use in satellite communications and in radar. The driven element of a dish antenna is located at the focal point of its dish reflector. Also called **parabolic antenna, paraboloid antenna, paraboloidal antenna**, or **dish antenna**.

paraboloid **1.** A solid or surface formed by rotating a **parabola** around its axis of symmetry. **2.** Same as **parabolic reflector**.

paraboloid antenna Same as **parabolic reflector** (2).

paraboloid reflector Same as **parabolic reflector**.

paraboloidal antenna Same as **parabolic reflector** (2).

paraboloidal reflector Same as **parabolic reflector**.

paraffin wax A white wax obtained by refining crude oil. It is used for waterproofing, and as an insulator.

parallax The apparent displacement or change of direction of an object when viewed from different positions. In an indicating instrument for example, different readings may be obtained depending on the position from which the pointer is viewed.

parallax error An error, such as that occurring when reading an instrument, due to **parallax**.

parallax-second Same as **parsec**.

parallel **1.** A state or condition in which there is an equal distance at all points. For example, two infinitely long straight lines which do not intersect, or two curves or surfaces which are equidistant from each other at all points. Also, pertaining to that maintains this relationship. **2.** Characteristic of, or pertaining to a **parallel circuit** or **parallel connection**. **3.** Acted upon, performed, functioning, or occurring simultaneously. For instance, parallel processors working together. **4.** The simultaneous processing, storing, transfer, transmission, reception, or the like, of multiple bits, characters, or data units. This contrasts with **serial** (3), where each is handled one after the other. **5.** Any of the great lines which circle the earth, and which are **parallel** (1) to the plane of the equator.

parallel access Also called **simultaneous access**. **1.** Storage and/or retrieval of data to or from a computer storage medium in which multiple bits, characters, or data units are transferred simultaneously. This contrasts with **serial access** (1), where bits, characters, or data units are transferred sequentially. Also called **parallel storage**. **2.** Reading and/or writing of data from or to computer memory in which multiple bits, characters, or data units are transferred simultaneously. This contrasts with **serial access** (2), where bits, characters, or data units are transferred sequentially.

parallel adder **1.** In computers, a logic circuit which adds all the digit positions simultaneously. **2.** In computers, a logic circuit which adds two or more numbers or quantities simultaneously.

parallel addition The function a **parallel adder** performs.

parallel algorithm An algorithm in which multiple defined steps are carried out simultaneously.

parallel capacitance The total capacitance of **parallel capacitors**. Also called **shunt capacitance**.

parallel capacitors Two or more capacitors connected in parallel. The total capacitance is the sum of each of the individual capacitances, thus, the multiple capacitors act like a single larger capacitor. Also called **shunt capacitors**.

parallel channel **1.** A communications channel in which multiple bits, characters, or data units are transferred simultaneously. Along a **serial channel** these are transferred sequentially. **2.** One of multiple channels utilized in **parallel communications**.

parallel circuit A circuit whose components are connected in parallel. That is, they have a **parallel connection**. Also called **shunt circuit**.

parallel communications Communications in which data is transmitted simultaneously over multiple channels. This contrasts with **serial communications**, in which data is transmitted sequentially over a single channel.

parallel computation Same as **parallel processing**.

parallel computer **1.** A computer that has multiple processors which are utilized to perform **parallel processing** (1). **2.** A computer which is utilized to perform **parallel processing** (2). **3.** One of multiple computers working together in **parallel processing** (3).

parallel computing Same as **parallel processing**.

parallel-conductor line Same as **parallel-wire line**.

parallel-conductor transmission line Same as **parallel-wire line**.

parallel connection The connection of the components within a circuit in a manner that there are multiple paths among which the current is divided, while all the components have the same applied voltage. An example is the connection of components across each other. This contrasts with **series connection**, where components are connected end-to-end, and in which there is a single path for the current. Also called **in parallel**. Also called **shunt connection**.

parallel data interface Same as **parallel interface**.

parallel data transfer Same as **parallel transfer**.

parallel data transmission Same as **parallel transmission**.

parallel database A database which utilizes **parallel processing**. Used, for instance, for data mining.

parallel execution Also called **concurrent execution, concurrent processing**, or **multiprocessing**. **1.** The simultaneous execution of multiple computer operations by the CPU. Since microprocessors can work so quickly, it seems simultaneous, even though each operation is usually executed in sequence. **2.** The simultaneous execution of multiple computer operations by multiple CPUs. Such processors are usually linked by high-speed channels.

parallel-fed That which has a **parallel feed**. Also, pertaining to that which has a parallel feed. Also called **shunt-fed**.

parallel feed The simultaneous application of a DC operating voltage and an AC signal voltage to a component or device, each via a different path. Also called **shunt feed**.

parallel generator An electric generator whose armature and field windings are connected in parallel. Also called **parallel-wound generator**, or **shunt generator**.

parallel I/O Abbreviation of **parallel** input/output. The transfer of data between a CPU and a peripheral in which multiple bits, characters, or data units are transferred simultaneously. This contrasts with **serial I/O**, where bits, characters, or data units are transferred sequentially.

parallel impedance An impedance acting in parallel with another quantity or magnitude, such as another impedance. Also called **shunt impedance**.

parallel inductance An inductance acting in parallel with another quantity or magnitude, such as a capacitance. Also called **shunt inductance**.

parallel inductors Two or more inductors connected in parallel. When said inductors are sufficiently shielded or apart from each other, the magnetic inductance is considered to be negligible. Otherwise, such inductance, or coupling, must be factored in. Also called **shunt inductors**.

parallel input/output Same as **parallel I/O**.

parallel interface An interface, such as a Centronics parallel interface, which transfers multiple bits, characters, or data units simultaneously. Used, for instance, for connecting a computer to a printer. This contrasts with a **serial interface**, in which bits, characters, or data units are transferred sequentially. Also called **parallel data interface**.

parallel lines Lines which lie in the same plane, but which do not intersect.

parallel motor An electric motor whose armature and field windings are connected in parallel. Also called **parallel-wound motor**, or **shunt motor**.

parallel network An electric network whose components are connected in parallel. That is, they have a **parallel connection**. Also called **shunt network**.

parallel operation **1.** Same as **parallel processing**. **2.** The performance of two or more operations simultaneously. **3.** Operation of a circuit or device whose components are connected in parallel. Also called **shunt operation**.

parallel output One of multiple outputs which are delivered simultaneously. Seen, for instance, in parallel transmission. This contrasts with **serial output**, in which each output is delivered in sequence.

parallel-plate capacitor A capacitor incorporating two parallel plates separated by a dielectric. The plates may be movable, in which case the capacitance can be varied as the gap between the plates is adjusted.

parallel-plate waveguide A waveguide in which electromagnetic radiation is propagated between two parallel conducting plates.

parallel port A computer port utilizing a **parallel interface**.

parallel printer A printer that connects to a computer via a **parallel interface**.

parallel processing Also called **parallel computing, parallel computation, parallel operation** (1), **simultaneous processing, multiprocessing**, or **concurrent execution** (2). **1.** The simultaneous execution of multiple computer operations by multiple CPUs in a single computer. Such processors are usually linked by high-speed channels. **2.** The simultaneous execution of multiple computer operations by a single computer, whether it utilizes multiple processors or another mechanism. **3.** The utilization of multiple computers to perform multiple operations simultaneously. **4.** An architecture utilized for **parallel processing** (1), **parallel processing** (2), or **parallel processing** (3). Also called **parallel processing architecture**.

parallel processing architecture Same as **parallel processing** (4).

parallel processors Two or more processors utilized for **parallel processing**. These processors may be located in a single computer, or in multiple computers. Also called **simultaneous processors**.

parallel programming Computer programming in which multiple steps or operations are performed simultaneously. In **serial programming**, steps or operations are performed in sequence.

parallel regulator A regulator, such as a voltage regulator, which is connected in parallel with a device or load. Also called **shunt regulator**.

parallel resistance The total resistance of **parallel resistors**.

parallel resistors Two or more resistors connected in parallel. The total resistance is the sum of the reciprocal of each of the individual resistors, thus, the multiple resistors have a lower total resistance than the lowest valued resistor. Also called **shunt resistors**.

parallel resonance **1.** Resonance in a **parallel-resonant circuit**. Also called **antiresonance** (1). **2.** In a **parallel-resonant circuit**, the frequency at which the parallel impedance is highest. Also called **antiresonance** (2). **3.** In a **parallel-resonant circuit**, the frequency at which the inductive capacitance equals the capacitive reactance. Also called **antiresonance** (3).

parallel-resonant circuit A resonant circuit in which a capacitor and an inductor are connected in parallel with an AC source. Resonance occurs at or near the maximum impedance of the circuit. Also called **antiresonance circuit**, or **tank circuit** (1).

parallel-resonant frequency The resonant frequency of a **parallel-resonant circuit**. Also called **antiresonance frequency**.

parallel-series Also called **parallel-series circuit**. **1.** A circuit which incorporates both parallel connections and serial connections of its components. For instance, a circuit with multiple parallel paths, each with components connected in series. Also, pertaining to such a circuit. **2.** A circuit whose components can be arranged either in a parallel or series manner.

parallel-series arrangement Same as **parallel-series connection**.

parallel-series circuit Same as **parallel-series**.

parallel-series connection The arrangement of a **parallel-series circuit**. Also called **parallel-series arrangement**.

parallel server A server which utilizes **parallel processing**. Used, for instance, in symmetric multiprocessing.

parallel storage Same as **parallel access (1)**.

parallel stream The simultaneous transmission of bits, characters, or data units utilizing multiple lines, fibers, channels, or the like. A **serial stream** provides for the transmission of bits, characters, or data units one at a time in sequence.

parallel-T network An electric network consisting of two T-networks connected in parallel. Also spelled **parallel-tee network**. Also called **twin-T network**, or **twin-tee network**.

parallel-tee network Same as **parallel-T network**.

parallel transfer The simultaneous transfer of bits, characters, or data units utilizing multiple lines, fibers, channels, or the like. For example, the use of eight wires to simultaneously transfer each of the eight bits of a byte. This contrasts with **serial transfer**, in which bits, characters, or data units are sent one at a time in sequence. Also called **parallel data transfer**.

parallel transmission The simultaneous transmission of bits, characters, or data units utilizing multiple lines, fibers, channels, or the like. For example, the use of eight wires to simultaneously transmit each of the eight bits of a byte. This contrasts with **serial transmission**, in which bits, characters, or data units are sent one at a time in sequence. Also called **parallel data transmission**.

parallel winding In an electric motor or generator, an armature winding connected in parallel with the field winding. Also called **shunt winding**.

parallel-wire line A transmission line consisting of two parallel wires or conductors. Also called **parallel-wire transmission line**, **parallel-conductor line**, or **parallel-conductor transmission line**.

parallel-wire transmission line Same as **parallel-wire line**.

parallel-wound generator Same as **parallel generator**.

parallel-wound motor Same as **parallel motor**.

parallelism The quality or state of being parallel, operating in a parallel manner, or maintaining a parallel relationship.

parallelogram A quadrilateral with two pairs of parallel sides. A **rhombus** is a quadrilateral with four equal sides and possessing no right angles, a **rectangle** is a quadrilateral with four right angles and two pairs of sides of equal length, and a **square** is a quadrilateral with four right angles and four equal sides.

parallelogram of forces A parallelogram whose diagonal corresponds to the sum of two forces. Each of the added forces is represented by the sides of said parallelogram.

parallelogram of vectors A parallelogram whose diagonal corresponds to the sum of two vectors. Each of the added vectors is represented by the sides of said parallelogram.

paramagnet A material, such as aluminum, which is **paramagnetic**.

paramagnetic Pertaining to materials whose relative magnetic permeability is slightly greater than unity, with the relative magnetic permeability of a vacuum having an assigned value of 1. Such materials are moderately attracted by a magnet, and aluminum and liquid oxygen are examples.

paramagnetic material A material, such as aluminum, whose relative magnetic permeability is slightly greater than unity.

paramagnetic resonance The resonant absorption of microwave radiation by a paramagnetic substance, which has at least one unpaired electron, in the presence of a strong magnetic field. Also called **electron spin resonance**, **electron paramagnetic resonance**, or **electron magnetic resonance**.

paramagnetic resonance spectrometer An instrument utilized for **paramagnetic resonance spectroscopy**. Also called **electron spin resonance spectrometer**.

paramagnetic resonance spectroscopy An instrumental analysis technique in which the microwave radiation absorbed in **paramagnetic resonance** is measured. Also called **electron spin resonance spectroscopy**, or **electron paramagnetic resonance spectroscopy**.

paramagnetism The characteristic of certain materials of having a relative magnetic permeability which is slightly greater than unity.

parameter **1.** For a given system, one of multiple factors or variables, such as temperature or current, which help determine and quantify its behavior. **2.** A specific value for a circuit component, such as the resistance of a resistor, or the capacitance of a capacitor, in a given circuit configuration. Also called **circuit parameter**. **3.** A quantity which can be assigned an arbitrary value for one set of specific conditions, and another for others. **4.** A value or expression that is passed between functions or programs for processing. This information is used to carry out operations. For instance, in the logarithmic function $\log (x)$, the argument is x. Also called **argument**.

parameter RAM Same as **PRAM**.

parameter random-access memory Same as **PRAM**.

parametric amplifier A microwave amplifier whose main oscillator is tuned to the received frequency while a pumping oscillator of a different frequency varies one or more parameters, such as capacitance or reactance, of the main oscillator. The energy from the pumping action amplifies the applied signal. Such amplifiers feature a very low noise level, and are used, for instance, in satellite communications and radars. Its abbreviation is **paramp**.

parametric converter A parametric device which converts an input signal frequency into an output at a higher or lower frequency. Also called **parametric frequency converter**.

parametric device A device whose function depends on the varying of one or more parameters, such as capacitance, inductance, or reactance.

parametric down-converter A parametric converter whose output frequency is lower than its input frequency. This contrasts with **parametric up-converter**, whose output frequency is higher than that of the input.

parametric equalizer An equalizer similar to a graphic equalizer, but with the added functionality that each of the multiple center frequencies is adjustable. Some parametric equalizers also allow adjustment of the bandwidth of each interval of frequencies.

parametric frequency converter Same as **parametric converter**.

parametric oscillator An oscillator which utilizes a parametric amplifier inside a resonant optical cavity for the generation of tunable laser radiation. Usually used as a powerful source of tunable laser radiation. Also called **optical parametric oscillator**, or **optical parametric amplifier**.

parametric test A statistical test in which there are assumptions made about the data utilized. For example, it may be assumed that the data is taken from a normal distribution. In a **non-parametric test** no such assumptions are made.

parametric up-converter A parametric converter whose output frequency is higher than its input frequency. This contrasts with **parametric down-converter**, whose output frequency is lower than that of the input.

parametron A device which utilizes Josephson junctions to achieve ultra-fast switching.

paramp Abbreviation of **parametric amplifier**.

paraphase amplifier An amplifier with a single input, and whose output consists of two signals which are 180° out-of-phase with each other. Also called **phase splitter (3)**.

parasitic In a circuit, a generally undesired, useless, and energy-wasting current, oscillation, capacitance, resistance, or other parameter or signal.

parasitic antenna An antenna which is driven indirectly, that is, through radiation from a driven antenna. This contrasts with a **driven antenna**, which receives energy directly from the transmission line.

parasitic capacitance In a circuit, a generally undesired, useless, and energy-wasting capacitance. An example is stray capacitance.

parasitic current In a circuit, a generally undesired, useless, and energy-wasting current. An example is an eddy current.

parasitic element An antenna element which is driven indirectly, that is, through radiation from a driven element. This contrasts with a **driven element**, which is driven directly by the transmission line. Also called **parasitic reflector**, **passive element (2)**, or **reflector element**.

parasitic inductance In a circuit, a generally undesired, useless, and energy-wasting inductance. An example is stray inductance.

parasitic oscillation In a circuit, a generally undesired, useless, and energy-wasting oscillation. An example is that occurring in a radio receiver at a frequency other than the operating frequency.

parasitic reflector Same as **parasitic element**.

parasitic resistance In a circuit, a generally undesired, useless, and energy-wasting resistance. An example is distributed resistance.

parasitic signal In a circuit, a generally undesired, useless, and energy-wasting signal. An example is circuit noise.

parasitic suppressor **1.** A component or device which suppresses **parasitic oscillations**. It may consist, for instance of a coil and resistor in parallel. **2.** A component or device which suppresses **parasitics**.

PARD Acronym for **periodic and random deviation**.

parent **1.** A process which initiates another process, which is called **child (1)**. **2.** Data upon which other data is dependent, which is called **child (2)**. **3.** A component or process upon which another component or process is dependent, which is called **child (3)**.

parent-child Same as **parent-child relationship**.

parent-child relationship Used within the context of computers. Also called **parent-child**. **1.** A process, called parent, which initiates another process, called child. **2.** Data, called parent, upon which other data is dependent, called child. **3.** A component or process, called parent, upon which another component or process is dependent, which is called child.

parent file For a given file, such as that being updated, the copy or version saved before it. The current version is called **son file**. Also called **father file**.

parental blocking The programming of a device, such as a TV, satellite receiver, or DVD, so that certain channels or programming are blocked. Used in the context of parents or guardians intending to impede viewing of certain content by youngsters. Also called **parental control**.

parental control Same as **parental blocking**.

parental-control software Software intended to enable parents or guardians to limit or prevent viewing of certain content, such as that available over the Internet, by youngsters.

parenthesis-free notation Same as **Polish notation (1)**.

Pareto Principle A rule of thumb that applies in a variety of settings. For instance, roughly 20% of the features of a given application are used about 80% of the time, or about 80% of the failures or defects of a given product such as a chip are due to the top 20% or so of causes, among countless examples. Also called **80/20 rule**.

parity **1.** The quality or state of being equal, or having an equivalent function or effect. **2.** The presence of an even or odd number of ones or zeroes within a group of bits. For example, for a given group of bits, the total number of zeroes being an even number, thus having even parity. **3.** In data transmission, the sameness of the **parity bit**. **4.** Of two integers, if both are odd, or both are even, they have the same parity.

parity bit In data transmission, an additional bit, 0 or 1, that is added to each byte sent, so that all bytes are have either an even or odd number of ones or zeroes. At the receiving location the additional bit is checked, to be sure it is the same. If it is, the transmission has passed this test of data integrity.

parity check An error-detection procedure which verifies if there is even or odd parity. For each byte transmitted, an additional bit is added, which can be a 0 or 1, so that all bytes are even or odd. At the receiving location, this parity is checked, to be sure it is the same it was before sending. If it is, the transmission has passed this test of data integrity. Also used for data storage. Also called **parity checking**, or **odd-even check**.

parity checking Same as **parity check**.

parity error An error in parity found when performing a **parity check**.

parity memory Computer RAM which utilizes **parity checking**. Also called **parity RAM**.

parity RAM Same as **parity memory**.

park In a disk drive, to safely position the read/write heads, so as to protect them and the data stored on the disks. Most drives do this automatically upon disconnection of power.

parse **1.** To examine closely and break down into components. **2.** In computers, to analyze and separate into components which are more easily processed, converted, or the like. For example, to parse statements into expressions.

parsec Acronym for **par**allax-**sec**ond. A unit of distance utilized in astronomy equal to approximately 3.08572 x 10^{16} meters, or about 3.261671 light years. Its abbreviation is **pc**.

parser A routine or algorithm that performs **parsing**.

Parseval's theorem A theorem which relates RMS values of a composite signal to the RMS values of its components.

parsing **1.** The process of examining closely and breaking down into components. **2.** In computers, the process of analyzing and separating into components which are more easily processed, converted, or the like.

part **1.** Any piece, item, component, portion, or subdivision of a whole. **2.** A **part (1)** which can be individually removed or replaced from a whole. **3.** Any of the electrical elements which compose a circuit. These may include resistors, capacitors, transistors, generators, electron tubes, and so on. Each part has terminals which allow it to be connected to the conducting path. Also called **circuit component, component (2), electrical element**, or **electrical component**.

part detection Same as **part recognition**.

part failure The condition in which a part of a circuit, device, piece of equipment, or system no longer performs the function it was intended to, or is not able to do so at a level that equals or exceeds established minimums.

part identification Same as **part recognition**.

part recognition In robotics, the detection and identification of components or parts. Artificial vision may be used, for instance, to distinguish between parts that are nearly identical. Also called **part identification**, or **part detection**.

partial 1. Not a whole. Also, pertaining to that which is not a whole. **2.** One of the multiple components of a complex tone, such as a harmonic, though it need not be a harmonic.

partial carry In computers, a carry technique in which the carries that result from an operation are temporarily stored. This contrasts with a **complete carry**, where carries are propagated to other digit positions.

partial node In a standing-wave system or vibrating system, a point between the minimum and maximum amplitude or displacement. The amplitude of a partial node lies between that of an antinode and that of a node.

partial pressure The pressure exerted by one of multiple gases which are mixed together.

Partial Response Maximum Likelihood A technique utilizing digital signal processing to differentiate a desired signal from a noise signal. Used, for instance, to increase the storage density of a magnetic disk. Its abbreviation is **PRML**.

particle 1. A very small or infinitesimal portion of matter, such as a molecule, atom, proton, or quark. **2.** Any of the fundamental constituents of all matter in the universe. Elementary particles include protons, neutrons, electrons, photons, and quarks. Categories of such particles include leptons and hadrons. An electron, for instance, is a type of lepton. When an elementary particle is defined as having no internal structure, particles such as protons and neutrons are excluded, as these are composed of quarks. The four fundamental forces of nature are carried by elementary particles. For example, photons carry the electromagnetic force. Also called **elementary particle**. **3.** A comparatively small subdivision of matter, such as magnetic particles or dust.

particle accelerator A device, such as a linac or cyclotron, in which charged particles are greatly accelerated to achieve high energies. Also called **accelerator** (4), or **high-energy particle accelerator**.

particle beam A narrow stream of particles traveling in the same direction, with approximately the same speed, and emitted from the same source. Electron beams and ion beams are examples.

particle bombardment The subjecting of a target to impacts by high-energy particles. Also called **bombardment**.

particle counter Also called **particle detector**. **1.** An instrument which registers and counts ionizing radiation, such as alpha rays, given off by radioactive entities. For example, a Geiger counter or a scintillation counter. Also called **radiation counter**, **counter** (3), or **ionization counter**. **2.** A device which registers and counts particles.

particle detector Same as **particle counter**.

particle fluence The passing of particles, such as electrons or photons, through a given area, such as a square centimeter. May be expressed, for instance, in particles per second. Also called **fluence**.

particle physics A branch of physics which deals with the properties, interactions, and applications of subatomic particles, especially when they have high energies. Also called **high-energy physics**.

particle velocity 1. The distance a particle travels per unit time. For instance, electrons in a linear accelerator may travel at nearly the speed of light. **2.** In a sound field, the velocity of an infinitesimal part of the medium, relative to the medium as a whole. May be expressed, for instance, in meters per second. A velocity microphone responds to particle velocity rather than to sound pressure.

particle-wave duality The dual nature of electromagnetic radiation, in which it has particle-like properties, as seen in the photoelectric effect, and wavelike properties, as evidenced by phenomena such as diffraction and interference. Certain particles such as electrons also exhibit this duality, as they have wavelike properties, as seen, for instance, in electron diffraction. Also called **wave-particle duality**.

partition One of multiple portions created by **partitioning**.

partition boot sector The first sector of a hard disk or a bootable floppy disk. During startup, its contained data and instructions start the process of loading the operating system into memory.

partition noise In an active device, such as a transistor or electron tube, noise resulting from the random distribution of electrons between electrodes.

partition table A table which lists the partitions of a hard disk. Such a table includes the information necessary to access each partition, and indicates which is used for booting.

partitioning 1. The dividing of a hard disk into multiple virtual disks, each of which appear as a separate disk to the operating system. For instance, a user may have drives labeled C:, D:, E:, and F:, but only have a single physical disk. Used, for instance, when setting up disks with different operating systems. **2.** The dividing of computer memory into fixed portions. **3.** The dividing of a block of data into fixed portions

partner robots Same as **personal robots**.

parts per billion A unit of proportion, equal to a given amount of anything per billion of the same thing or something else. Used, for instance, as a measure of concentration or to quantify error. For example, five erroneous bits per billion transmitted bits. Its abbreviation is **ppb**.

parts per million A unit of proportion, equal to a given amount of anything per million of the same thing or something else. Used, for instance, as a measure of concentration or to quantify error. For example, nine atoms of xenon per million of air. Its abbreviation is **ppm**.

parts per thousand A unit of proportion, equal to a given amount of anything per thousand of the same thing or something else. Used, for instance, as a measure of concentration. For example, 35 molecules of salt per thousand of water. Its abbreviation is **ppt**.

party line An arrangement in which the telephones of two or more end users are connected to the same telephone line or local loop. When utilized, a mechanism such as selective ringing may be employed to distinguish for which user an incoming call is meant. Party lines are saddled by privacy and congestion problems. Also called **multiparty line**.

PAS Abbreviation of **photoacoustic spectroscopy**.

pascal The SI unit of pressure. It is equal to the pressure resulting from a force of one Newton applied uniformly to an area of one square meter. That is, $Pa = N/m^2$. It is also equal to approximately 9.869233×10^{-6} atm., 1.0×10^{-5} bar, about 7.500615×10^{-3} mmHg, approximately 7.500617 torr, or about 0.0208854 lbf/ft^2. Its symbol is **Pa**.

Pascal A procedure-oriented high-level programming language which lends itself to structured programming.

Paschen-Back effect In spectroscopy, an effect in which spectral lines split by the Zeeman effect seem to coalesce when the light source is subjected to a strong magnetic field.

Paschen's law A law stating that the breakdown potential that initiates a discharge between two parallel plates in a gas is a function of the product of the gas pressure and the distance between said plates. Other factors, such as surface irregularities, need to be taken into account for more precise measurements.

pass 1. To proceed or be allowed to proceed. For instance, the frequencies which a filter transmits. **2.** To be transmitted or transferred from one point to another, or between entities. For example, the transmission of messages in a communications network, the sending of signals between a CPU and peripheral, or the transfer of control from one device to another. **3.** To proceed from one state or condition to another. For instance, to go from a forward bias to a reverse bias. **4.** In computers, a complete read or write cycle. **5.** A single circuit made by a satellite within its orbit. **6.** An interval during which a satellite is within its telemetry range.

pass band Same as **passband**.

pass band filter Same as **passband filter**.

pass band ripple Same as **passband ripple**.

pass-through A component, circuit, device, unit, piece of equipment, or system, which acts an intermediary. Such an intermediary may act upon a signal, as in effecting a protocol conversion, or it may not as is the case of a USB hub. Also spelled **passthrough**.

passband Also spelled **pass band**. **1.** A frequency interval within which a signal is transmitted or passed. A frequency-sensitive device, such as an amplifier or filter, may have multiple passbands. Also called **transmission band**. **2.** A continuous range of frequencies which is passed or transmitted by a filter. That is, there is little or no attenuation of a signal in this interval. A filter may have multiple passbands. Also called **filter passband**.

passband filter A filter which selectively transmits or passes one or more frequency intervals, each of which is called **passband**. Also spelled **pass band filter**.

passband ripple Slight variations in attenuation occurring within a **passband**. A Butterworth characteristic has minimal passband ripple. Also spelled **pass band ripple**.

passivate To protect junctions, regions, layers, or surfaces of semiconductors through **passivation**.

passivation The protection of junctions, regions, layers, or surfaces of semiconductors by depositing a material which renders them chemically passive or unreactive. Used, for instance, to protect from oxidation, contamination, and other perils of harsh environments. Two common techniques are glassivation and oxide passivation.

passive 1. A component or device which does not operate on an applied electrical signal, as in amplifying, rectifying, or switching, nor is a source of energy. Resistors, capacitors, and transformers are examples. **2.** Chemically inert.

passive absorber A medium or material which dissipates sound energy as heat.

passive antenna 1. An antenna with no active components or devices, such as amplifiers. **2.** An antenna, such as a parasitic antenna, which is not connected to a transmission line.

passive circuit A circuit which does not incorporate any active components or devices. Such a circuit consists exclusively of passive components and devices.

passive communications satellite An artificial satellite which reflects signals between stations. Such a satellite does not provide signal amplification. Also known as **passive comsat**, or **passive satellite**.

passive component An electronic component, such as a capacitor or resistor, that can not operate on an applied electrical signal, as in amplifying, rectifying, or switching, nor is a source of energy. Also called **passive element (1)**.

passive comsat Same as **passive communications satellite**.

passive control system A control system that does not regulate inputs and outputs to maintain or improve operating conditions. For example, one which relies on the natural forces of convection and diffusion to regulate heat.

passive crossover A speaker crossover which does not require power to operate.

passive detection Detection in which the location of the detector is not revealed.

passive device A device, such as a transformer, that does not operate on an applied electrical signal, as in amplifying, rectifying, or switching, nor is a source of energy.

passive display Same as **passive-matrix display**.

passive electric network An electric network with no active components or devices. Also called **passive network**.

passive element 1. Same as **passive component**. **2.** Same as **parasitic element**. **3.** A chemically inert element.

passive equalizer An equalizer that does not require power to operate.

passive filter A filter that does not incorporate active components or devices, such as transistors, to assist in signal passing or rejection.

passive homing Homing in which guidance is provided by a signal or energy transmitted or radiated by a beacon or scanned object. This contrasts with **active homing**, in which a signal or energy is emitted or transmitted from a source and reflected back by a scanned object.

passive hub A central connecting device, such as a computer or router, which only passes signals sent through it, without regenerating them. This contrasts with **active hubs**, which regenerate and retransmit the signals sent through them.

passive infrared detection The detection of infrared radiation emitted by an object, such as a human body. This contrasts with **active infrared detection**, in which infrared energy is directed to, and reflected by, objects being detected. Its abbreviation is **passive IR detection**.

passive infrared detector A device utilized for **passive infrared detection**. Its abbreviation is **passive IR detector**. Also called **passive infrared sensor**.

passive infrared sensor Same as **passive infrared detector**. Its abbreviation is **passive IR sensor**.

passive IR detection Abbreviation of **passive infrared detection**.

passive IR detector Abbreviation of **passive infrared detector**.

passive IR sensor Abbreviation of **passive infrared sensor**.

passive matrix A technology used in **passive-matrix displays**.

passive-matrix display A liquid-crystal display in which the pixels of the screen are powered by a grid of wires. This contrasts with an **active-matrix display**, in which each pixel of the screen is powered separately and continuously by a transistor, which provides for a brighter image with higher contrast. Also called **passive display, passive-matrix LCD, passive-matrix LCD display**, or **passive-matrix screen**.

passive-matrix LCD Same as **passive-matrix display**.

passive-matrix LCD display Same as **passive-matrix display**.

passive-matrix liquid-crystal display Same as **passive-matrix display**.

passive-matrix screen Same as **passive-matrix display**.

passive mixer A mixer, such as that combining input signals or light beams, which does not utilize active components or devices, such as transistors.

passive modulator A modulator which utilizes no active devices such as transistors.

passive network Same as **passive electric network**.

passive optical network A broadband optical network in which an optical splitter not located at a headend, central office, or similar site, is utilized to provide multiple users optical terminations. Its abbreviation is **PON**.

passive radar A radar technique which relies on microwave energy which is ordinarily radiated and reflected from any given object. Such a radar does not transmit microwaves to enhance the detection of objects. This contrasts with **active radar**, in which microwaves are transmitted for the detection of scanned objects, in addition to providing other information, such as distance, direction, and speed.

passive radiator A cone loudspeaker which is not driven. It vibrates in response to a driven loudspeaker in the same enclosure. Utilized to reinforce bass frequencies. Also called **drone cone**.

passive reflector An object surface utilized to reflect electromagnetic radiation, such as microwaves.

passive satellite Same as **passive communications satellite**.

passive solar building design Same as **passive solar design**.

passive solar design A building design which takes into account factors such as the orientation of the structure, the proper sizing of windows and overhangs, and the judicious selection of structural elements, so as to provide lighting, and to help heat and/or cool by taking advantage of the sun's energy and the local climate. Also called **passive solar building design**.

passive sonar Sonar equipment which can only receive ultrasonic sounds, and is thus able to detect bearing only. This contrasts with **active sonar**, which can transmit and receive ultrasonic sounds, enabling detection of the bearing and range of objects.

passive star In a star network, a central connecting device, such as a computer or router, that does not regenerate and retransmit the signals sent through it.

passive substrate A substrate, such as a printed-circuit board, upon which components are mounted. In an **active substrate (1)**, components are formed directly within said substrate.

passive system A radio or radar system capable only of receiving. This contrasts with an **active system**, which can transmit and receive.

passive tracking system 1. A system which does not transmit or re-transmit signals for tracking objects, or which does not require an external energy source. For example, a mechanism which keeps solar panels aligned with the sun without external power. 2. A system which does not exchange signals with any satellite it is tracking. 3. A tracking system which utilizes passive detection.

passive transducer 1. A transducer which does not utilize active components or devices, such as transistors. 2. A transducer without an internal power source.

passive transponder A transponder with no internal power source. The power requirements of such a transponder are provided by the reader and/or interrogator electromagnetic field. An **active transponder** has its own power source.

passthrough Same as **pass-through**.

password A character string which serves for authentication or the fulfillment of other security requirements. Used, for instance, to log on to a system, make system changes, protect files, gain access to networks or specific sections of networks, and so on. Passwords such as those consisting of the actual name of a user, that of a close relative, words commonly found in dictionaries, or those having only a couple letters repeated tend to be easy to crack. Those with non-alphabetic characters, mixed case alphabetic characters, or combinations of these tend to be harder to crack. In addition, passwords should have at least six characters, ought to be frequently changed, and must never be written down.

Password Authentication Protocol Its acronym is **PAP**. An authentication protocol utilized to authenticate usernames and their corresponding passwords. PAP does not usually utilize encryption for transmission, so it is not as safe, for instance, as Challenge Handshake Authentication Protocol.

password field A text-entry field where a password is typed. Such fields usually conceal the typed password, instead displaying bullets, asterisks, or the like, as characters are entered.

password hint Same as **password reminder**.

password protection The use of one or more passwords as a protective measure.

password reminder A hint that jogs the memory of a person who has forgotten a password. An obvious hint with an obvious answer can help a cracker to easily obtain someone else's password. Also called **password hint**.

password sniffing The examination of network traffic in an attempt to maliciously collect passwords.

paste 1. In a battery or cell, an immobilized electrolyte which is in the form of a paste, as opposed, for instance, to a liquid. 2. The placement of text, images, files, or other content in a location other than that where it was copied or cut from.

paste solder A metal solder in the form of a fine powder which is combined with a flux.

PAT Abbreviation of **port address translation**.

patch 1. A temporary connection made between terminals of a **patch panel**. Also, to be connected in this manner. 2. A temporary connection made between terminals, using for instance, an alligator clip. Also, to be connected in this manner. 3. A piece of code added to a program to fix some problem, such as a bug. Also, the modification provided by the insertion of such code. Used, for instance, as a provisional remedy between product releases.

patch bay Same as **patch panel**.

patch board Same as **patch panel**. Also spelled **patchboard**.

patch cable A cable, usually equipped with pins or plugs, utilized for **patching**.

patch cord A cord, equipped with connecting terminals such as plugs, utilized for **patching**. Also spelled **patchcord**.

patch-in To make a connection via a **patch (1)** or **patch (2)**.

patch-out To break a connection made via a **patch (1)** or **patch (2)**.

patch panel A panel upon which the terminals of components or systems are readily accessible for temporary connection. Used, for instance, in communications, computers, and for testing purposes. Also known as **patch board**, **patch bay**, **patching board**, **patching panel**, **patchboard**, or **board (2)**.

patchboard Same as **patch panel**. Also spelled **patch board**.

patchcord Same as **patch cord**.

patching 1. The making of a connection via a **patch (1)** or **patch (2)**. 2. The use and effect of a **patch (3)**.

patching board Same as **patch panel**.

patching panel Same as **patch panel**.

path 1. The route along which something travels. For instance, current, data, beams, crafts, and so on. 2. The route along which a current travels. For example, the course of the flow of current through a circuit. 3. In a printed circuit, the conductive strips, foils, pads, segments, or the like, via which a current flows. 4. The route along which an electromagnetic wave travels, such as that between a transmitter and receiver. 5. In a network, a route between two nodes. There may be more than one route between two specific nodes, and each path consists of one or more branches. 6. Same as **pathname**. 7. In computers, a sequence of instructions followed within a routine. 8. In database management, the route between data or sets of data. For instance, the logical connections between two records.

path attenuation Same as **path loss**.

path determination In a communications network, the determination of the path a packet will travel by. A router or switch takes care of this task, usually aided by a routing algorithm.

path loss The attenuation suffered by a signal traveling along a given trajectory between a transmitter and a receiver. Possible causes include absorption, and free-space losses. Usually expressed in dB. Also called **path attenuation**, or **path losses**.

path losses Same as **path loss**.

path name Same as **pathname**.

pathname A route to a file in which each directory and subdirectory along the way is listed. For example, Documents\Reminders\Today.xyz. Also spelled **path name**. Also called **path (6)**, or **directory path**.

pattern 1. A configuration, arrangement, form, design, figure, graph, or the like, consisting of, or representing, a phenomenon, quantity, or entity which is varied, or that varies in some manner. For instance, a radiation pattern, an interference pattern, a test pattern, a bit pattern, or an etched pattern. 2. A diagram or other model which is followed in creating or doing something. 3. A regular, repetitive, or symmetrical image displayed on a device such as an oscilloscope or TV. For example, a test pattern or a herringbone pattern. 4. A distribution, arrangement, graph, or other representation of the variation in intensity of a quantity as a function of direction and distance. An example is the transmitting or receiving effectiveness of an antenna, and such a pattern can be measured in any plane, although the horizontal and/or vertical planes are generally used. Also called **directivity pattern**, **radiation pattern**, or **field pattern**. 5. The path followed, or meant to be followed, by an aircraft during a landing approach.

pattern recognition The processing of the data of an image input device, such as a scanner, for the purpose of comparison to images or components of images stored in computer memory. May be used, for example, to identify fingerprints, or to interpret bar codes. Also known as **image recognition**.

Pauli exclusion principle A principle that states that no two fermions can occupy the same quantum state. In the case of electrons, for instance, this means that two electrons in the same atom can not have the same values of all quantum numbers. This leads to there only being able to be two electrons, each with opposite spin, in the same atomic orbital. Also called **Pauli principle**, or **exclusion principle**.

Pauli principle Same as **Pauli exclusion principle**.

pause 1. An interruption in execution. A pause is usually not an exit. 2. A key on a computer keyboard which creates a **pause (1)**. Also called **pause key (1)**. 3. An interruption in processing, execution, transmission, communication, or the like. Also, a command which creates such a pause. Also called **break (4)**. 4. A key on a computer keyboard which creates a **pause (3)**. Also called **pause key (2)**, or **break (5)**. 5. A temporary interruption. 6. A slightly delayed response. 7. A preprogrammed break.

pause key 1. Same as **pause (2)**. 2. Same as **pause (4)**.

PAX Abbreviation of **private automatic exchange**.

pay-per-view A paid service which enables a subscriber to choose from a selection of movies shown according to a fixed viewing schedule. A cable TV or satellite subscriber usually chooses the desired movie with a remote control, and is automatically billed. Although the terms are often used interchangeably, **movies-on-demand** is usually a paid digital TV or Internet service, and selections can be viewed or downloaded anytime. Its abbreviation is **PPV**.

pay television Same as **pay TV**.

pay TV Abbreviation of **pay television**. Also called **subscription TV**. 1. A service in which payment is made to be able to view TV content which is otherwise unavailable at a given location. Examples include pay-per-view, and premium channels. A decoder box is usually necessary for such programming. 2. Any TV service which requires payment. These include **pay TV (1)**, basic satellite, premium cable, and so on.

payload In communications, the bandwidth occupied by the information which wishes to be transmitted. The rest, such as control codes, origin and destination information, and error detection and correction coding, is **overhead (2)**. In ATM, for instance, each cell consists of 53 bytes, 48 of which are payload, and 5 of which are overhead.

payware Software which is sold, as opposed to **freeware**, which is not.

pb Abbreviation of **picobar**.

Pb 1. Chemical symbol for **lead**. 2. Abbreviation of **petabit**. 3. Within the YPbPr color model, one of the two color-difference signals. The other is **Pr (2)**.

PB Abbreviation of **petabyte**.

PB SRAM Same as **PBSRAM**.

PBGA Abbreviation of **Plastic Ball Grid Array**.

Pbit Abbreviation of **petabit**.

Pbps Abbreviation of **petabits per second**.

PBps Abbreviation of **petabytes per second**.

PBq Abbreviation of **petabecquerel**.

PBS Abbreviation of **polarization beam splitter**.

PBSRAM Abbreviation of pipeline burst static **RAM**. A type of RAM utilizing both bursts and pipelining, to improve the performance of a processor. Usually used as a form of cache.

PBX Abbreviation of **private branch exchange**.

PBX tie line Same as **PBX tie trunk**. Abbreviation of private branch exchange **tie line**.

PBX tie trunk Abbreviation of private branch exchange **tie trunk**. A telephone line that connects two PBXs. Also called **PBX tie line, tie trunk**, or **tie line**.

Pbyte Abbreviation of **petabyte**.

pc Abbreviation of **parsec**.

pC Abbreviation of **picocoulomb**.

PC 1. Abbreviation of **personal computer**. 2. Abbreviation of **printed circuit**. 3. Abbreviation of **photocell**. 4. Abbreviation of **point contact**. 5. Abbreviation of **program counter**. 5. Abbreviation of **programmable controller**. 6. Abbreviation of **polycarbonate**.

PC board Abbreviation of **printed-circuit board**.

PC card 1. Abbreviation of **PCMCIA card**. 2. Any adapter card that can be used with a personal computer.

PC card slot A slot within a computer, into which a **PC card** is plugged in.

PC-clone Abbreviation of personal computer **clone**. A computer system which does the same thing as another, earlier product. A PC-clone must be 100% compatible with the well-known computer it is based on. Also called **clone PC**.

PC compatible A system, or hardware or software computer product which is 100% compatible with a long-standing well-known PC brand name, but is produced by another brand.

PC memory card A memory card utilized to provide or expand the memory of a PC.

PC/TV Also called **TV/PC, TV computer**, or **computer TV**. 1. A PC, such as that with a TV card, enabled for TV viewing. 2. A TV with built-in computer capabilities, such as Internet access.

PCB 1. Abbreviation of **printed-circuit board**. 2. Abbreviation of **power circuit breaker**.

PCB land pattern On a printed-circuit board, a specific configuration or pattern of **pads (2)**. Also called **land pattern**.

PCBs   Abbreviation of **polychlorinated biphenyls.**

pCi   Abbreviation of **picocurie.**

PCI   Abbreviation of Peripheral Component Interconnect. A local bus standard which allows for high-speed interconnection between a CPU, the computer memory, and high-speed peripherals. PCI features plug-and-play capabilities, the sharing of IRQs, and bus mastering. Also called **PCI bus,** or **PCI local bus.**

PCI bus   Abbreviation of Peripheral Component Interconnect bus. Same as **PCI.**

PCI Industrial Computer Manufacturers Group.   Same as **PICMG.**

PCI local bus   Abbreviation of Peripheral Component Interconnect **local bus.** Same as **PCI.**

PCI mezzanine card   A mezzanine card plugged into a **PCI bus.** Its abbreviation is **PMC.**

PCL   Abbreviation of Printer Control Language. A de facto standard utilized for communication between computers and printers.

PCM   1. Abbreviation of **pulse-code modulation.** 2. Abbreviation of **process-control monitor.**

PCMCIA   Abbreviation of Personal Computer Memory Card International Association. An organization that establishes and promotes standards for the interoperability of adapter cards used in portable computers and other devices.

PCMCIA adapter card   Same as **PCMCIA card.**

PCMCIA card   Abbreviation of Personal Computer Memory Card International Association **card.** Its own abbreviation is **PC Card** (1). Also called **PCMCIA adapter card. 1.** A flat electronic card that is inserted into portable computers, such as handheld computers, to provide or enhance features such as memory, storage, multimedia, or networking, and which conforms to PCMCIA specifications. The typical PCMCIA card is rugged, has a 68-pin interface, and is 85.6 mm long by 54.0 mm wide. The thickness varies, depending on whether it is a Type I, Type II, or Type III card. A PCMCIA card is inserted into a PCMCIA slot. **2.** A **PCMCIA card** (1) designed for and utilized in other devices, such as digital cameras.

PCMCIA card slot   Same as **PCMCIA slot.**

PCMCIA connector   A 68-pin connector used as an interface for a **PCMCIA card.**

PCMCIA device   A device, such as a portable computer or digital camera, which has a slot which accepts a **PCMCIA card.**

PCMCIA slot   The slot, in a portable computer or other PCMCIA device, which accepts a **PCMCIA card.** Also called **PCMCIA card slot,** or **PCMCIA socket.**

PCMCIA socket   Same as **PCMCIA slot.**

PCMCIA specification   Abbreviation of Personal Computer Memory Card International Association **specification.** A set of standards which define factors such as card dimensions, interfaces, and the software utilized to ensure compatibility between PC cards and the devices using them.

PCMCIA Type I card   A **PCMCIA card** which is 3.3 mm thick and is typically used for memory. Also called **PCMCIA Type I PC card,** or **Type I PC card.**

PCMCIA Type I PC card   Same as **PCMCIA Type I card.**

PCMCIA Type II card   A **PCMCIA card** which is 5.0 mm thick and which may be used, for instance, for additional memory, or as a network card. Also called **PCMCIA Type II PC card,** or **Type II PC card.**

PCMCIA Type II PC card   Same as **PCMCIA Type II card.**

PCMCIA Type III card   A **PCMCIA card** which is 10.5 mm thick and which may be used, for instance, for disk storage or as a wireless network adapter. Also called **PCMCIA Type III PC card,** or **Type III PC card.**

PCMCIA Type III PC card   Same as **PCMCIA Type III card.**

PCS   1. Abbreviation of **Personal Communications Services.** 2. Abbreviation of **process control system.**

PCTV   Same as **PC/TV.**

Pd   Chemical symbol for **palladium.**

PD   1. Abbreviation of **potential difference.** 2. Abbreviation of **pulse duration.** 3. Abbreviation of **phase detector.**

PD software   Same as **public-domain software.**

PDA   Abbreviation of personal digital assistant. A handheld computer which stores information and performs functions which help organize an individual. Such a device may include a telephone, email, and address book, an appointment scheduler, a notepad, and so on. PDAs are usually enabled for Internet access, and when so equipped can send and retrieve email, in addition to performing limited browsing functions. Such devices usually utilize a stylus for menu selections, and may or may not have a small keyboard. When no keyboard is present, the stylus may be used to enter characters on its touch-screen display. PDAs have specialized software, and communicate with other computers via a cable or wirelessly.

PDC   Abbreviation of **Primary Domain Controller.**

PDF   1. Abbreviation of Portable Document Format. A de facto standard file format for the preparation, distribution, viewing, manipulating, and printing of documents. PDF always displays files exactly as created, and works across most platforms. **2.** Abbreviation of **printer description file.**

PDIP   Abbreviation of plastic dual in-line package, or plastic **DIP.** A dual in-line package made of plastic, as opposed, for instance, to ceramic.

pdl   Abbreviation of **poundal.**

PDL   Abbreviation of **page-description language.**

PDM   1. Abbreviation of **pulse-duration modulation.** 2. Abbreviation of **Product Data Management.**

PDN   Abbreviation of **public data network.**

PDP   Abbreviation of **plasma display panel.**

PE   Abbreviation of **potential energy.**

peak   Also called **peak value, crest,** or **crest value. 1.** The maximum instantaneous value of a voltage, current, signal, or other quantity. **2.** The maximum instantaneous absolute value of the displacement from a reference position, such as zero, for a voltage, current, signal, or other quantity. When referring to a wave or other periodic phenomenon, also called **amplitude. 3.** The **peak** (1), or **peak** (2) for a given time interval.

peak amplitude   In a wave or other periodic phenomenon, the maximum absolute value of the displacement with respect to zero. Also known as **amplitude peak.**

peak anode current   In an electron tube, the maximum instantaneous value of the anode current. Also known as **peak plate current.**

peak anode voltage   In an electron tube, the maximum instantaneous value of the anode voltage. Also known as **peak plate voltage.**

peak chopper   Same as **peak limiter.**

peak clipper   Same as **peak limiter.**

peak current   The maximum absolute value of the displacement from a reference position, such as zero, for a current. Also called **maximum current** (1).

peak demand   Same as **peak hour demand.**

peak detection   The utilization of a **peak detector.**

peak detector An instrument or device which senses a given signal, and which provides an output corresponding to a peak value, such as a voltage, of said signal.

peak distortion 1. For a given signal, the maximum instantaneous value of distortion, such as amplitude, harmonic, intermodulation, or phase distortion. 2. The **peak distortion** (1) for a given time interval.

peak envelope power For a radio transmitter operating under normal conditions, the average power supplied to the transmission line during one cycle at the peak of the modulation envelope. Utilized, for instance, to determine compliance with RF radiation regulations. Its abbreviation is **PEP**. Also called **peak power (7)**, or **peak power output (2)**.

peak factor In a periodically-varying function, such as that of AC, the ratio of the peak amplitude to the RMS amplitude. Also known as **crest factor** or **amplitude factor**.

peak filter A filter which attenuates bands whose amplitude are greatest within a response curve.

peak forward voltage For an electrical or electronic component or device, the maximum instantaneous value of the voltage in the direction in which there is the least resistance to the flow of current.

peak hour The time interval during which power or communications usage is at a maximum. The duration is not necessarily equal to an hour, and refers more to a general time of day. Also called **peak period**, **busy hour**, or **busy period**.

peak hour demand The average electric power consumed during **peak hours**. Also called **peak demand**.

peak hour traffic The average telecommunications traffic during **peak hours**. Also called **peak traffic**.

peak indication Same as **peak reading**.

peak indicator An instrument or meter, such as a VU meter or peak voltmeter, which provides readings of peaks. When digital, such devices may store peak values in memory, and when analog the indicating needle may fall slowly or may remain at a peak reading until reset or a higher peak is attained. Also called **peak meter**, or **peak-reading meter**.

peak inductor current The maximum instantaneous value of the current flowing through an inductor.

peak intensity The amplitude, magnitude, or other value or quantity corresponding to a **peak**.

peak inverse voltage For an electrical or electronic component or device, the maximum instantaneous value of the voltage in the direction in which there is the greatest resistance to the flow of current. For example, the maximum reverse-biased voltage a semiconductor can safely handle before avalanche breakdown will occur. Its abbreviation is **PIV**. Also called **peak reverse voltage**, **inverse peak voltage**, or **reverse peak voltage**.

peak level The level, such as that of a signal, corresponding to a **peak**.

peak limiter A circuit or device which limits the amplitude of its output signal to a predetermined maximum, regardless of the variations of its input. Used, for example, for preventing component, equipment, or media overloads. Also called by various other names, including **peak clipper**, **peak chopper**, **limiter**, **limiter circuit**, **clipping circuit**, **amplitude limiter**, **automatic peak limiter**, and **clipper**.

peak limiting The use and effect of a **peak limiter**.

peak load Same as **peak power**.

peak meter Same as **peak indicator**.

Peak Music Power Same as **Peak Music Power Output**. Its abbreviation is **PMP**.

Peak Music Power Output A misleading measure of the power handling capacity of audio components, such as amplifiers or speakers, based on a measure of power peaks. This results in significantly overblown figures, which are sometimes utilized to make inferior products seem better. When evaluating and stating power capabilities of such devices, a measure such as RMS power is desirable. Its abbreviation is **PMPO**.

peak output 1. The maximum output that a component, circuit, device, piece of equipment, or system can produce without immediate damage. 2. Same as **peak power (4)**.

peak period Same as **peak hour**.

peak picker A circuit, device, or instrument which detects and indicates the peaks of a given waveform.

peak plate current Same as **peak anode current**.

peak plate voltage Same as **peak anode voltage**.

peak power Also called **peak load**. 1. The maximum amount of power a component, circuit, device, piece of equipment, or system utilizes. 2. The maximum amount of power a component, circuit, device, piece of equipment, or system utilizes for a given period of time. 3. The highest power level a component, circuit, device, piece of equipment, or system can operate at safely. 4. The maximum output that a power supply can produce without immediate damage. Also called **peak output (2)**, or **peak power output (1)**. 5. The maximum amount of power a speaker can handle during peaks without immediate damage. 6. The maximum electric power required of a generating system, such as that needed to meet momentary periods of high demand. 7. Same as **peak envelope power**.

peak power output 1. Same as **peak power (4)**. 2. Same as **peak envelope power**.

peak power rating The maximum power level that a component, circuit, device, piece of equipment, or system can produce, consume, dissipate, or otherwise safely handle, for brief periods. Also called **maximum power rating**.

peak reading The value, quantity, or other reading which is displayed and/or recorded during a **peak**. Also called **peak indication**.

peak-reading meter Same as **peak indicator**.

peak-reading voltmeter Same as **peak voltmeter**.

peak reverse voltage Same as **peak inverse voltage**.

peak signal level The signal level, such as that displayed by a VU meter, during a **peak**.

peak sound pressure Same as **peak SPL**.

peak sound-pressure level Same as **peak SPL**.

peak SPL Abbreviation of **peak sound-pressure level**. The greatest sound pressure level during a peak, such as that produced by a speaker, or as a consequence of impulse noise. Ordinary sound-pressure meters do not indicate the true effect of such sounds, and an impact-noise analyzer should be utilized for accurate readings. Also called **peak sound pressure**.

peak-to-peak Its abbreviation is **p-p**, **pk-pk**, or **pp**. Also called **peak-to-peak amplitude**, **peak-to-peak value**, **maximum amplitude excursion**, or **amplitude excursion**. 1. For a waveform of an alternating quantity, such as that of AC, the difference between the maximum positive peak and the maximum negative peak. It is the maximum combined range in the amplitude of a signal or other observed quantity. 2. The difference between the maximum positive peak and the maximum negative peak of any varying quantity. It is the maximum combined range in the amplitude of a signal or other observed quantity.

peak-to-peak amplitude Same as **peak-to-peak**. Its abbreviation is **p-p amplitude**, **pk-pk amplitude**, or **pp amplitude**.

peak-to-peak value Same as **peak-to-peak**. Its abbreviation is **p-p value**, **pk-pk value**, or **pp value**.

peak-to-peak voltage Its abbreviation is **p-p voltage**, **pk-pk voltage**, or **pp voltage**. 1. For an AC, the difference between the maximum positive peak voltage and the maximum negative peak voltage. **2.** For a varying DC, the difference between the maximum positive peak voltage and the maximum negative peak voltage.

peak traffic Same as **peak hour traffic**.

peak value Same as **peak**.

peak voltage The maximum absolute value of the displacement from a reference position, such as zero, for a voltage. Also called **crest voltage**, or **maximum voltage (1)**.

peak voltmeter An AC voltmeter which indicates the peak value of an applied voltage. Also called **peak-reading voltmeter**, or **crest voltmeter**.

peaking 1. The reaching of one or more peaks within a response curve. **2.** The increasing of the amplitude of a device, instrument, piece of equipment, or system, so as to have a peak output or indication. **3. Peaking (2)** for a given band of frequencies.

peaking circuit 1. A circuit utilized for **peaking (2)**, or **peaking (3)**. **2.** A circuit utilized to modify the response curve of an amplifier, so as to improve its high-frequency response.

peaking coil Also called **peaking inductor**. **1.** A coil utilized for **peaking (2)**, or **peaking (3)**. **2.** A coil utilized to modify the response curve of an amplifier, so as to improve its high-frequency response. Such coils may be placed in series and/or parallel.

peaking inductor Same as **peaking coil**.

peaking transformer A transformer whose core is saturated by a comparatively low current, and whose output consists of sharply-peaked voltage pulses.

pebi- A binary prefix meaning 2^{50}, or 1,125,899,906,842,624. For example, a pebibyte is equal to 2^{50} bytes. This prefix is utilized to refer to only binary quantities, such as bits and bytes Its abbreviation is **Pi-**.

pebibit 2^{50}, or 1,125,899,906,842,624 bits. Its abbreviation is **Pib**.

pebibits per second 2^{50}, or 1,125,899,906,842,624 bits per second. Its abbreviation is **Pibps**.

pebibyte 2^{50}, or 1,125,899,906,842,624 bytes. Its abbreviation is **PiB**.

pebibytes per second 2^{50}, or 1,125,899,906,842,624 bytes per second. Its abbreviation is **PiBps**.

PECL Abbreviation of **positive emitter-coupled logic**.

PECVD Abbreviation of **plasma-enhanced chemical-vapor deposition**.

pedestal 1. A support or base for something, such as an antenna. **2.** Same as **pedestal level**.

pedestal level In a composite picture signal, the level which separates the range containing the picture information from that of the synchronizing information. Also known as **pedestal (2)**, or **blanking level**.

peek To view, read, or display the content of a memory location. Also the command to perform this action. To **poke** is to alter or store this content.

peer In a communications network, a hardware or software device, or other functional unit, which operates at the same layer as another. For example, a component which operates at the same layer within the OSI Reference Model. Also called **peer entity**.

peer entity Same as **peer**.

peer protocol A communication protocol utilized between **peers**.

peer-to-peer architecture A network architecture in which each node has the same capabilities and responsibilities. Since dedicated servers are not utilized, performance under heavy loads may be less than that achievable with a client/server architecture. Its abbreviation is **P2P architecture**.

peer-to-peer communications Communications between **peers**. Its abbreviation is **P2P communications**.

peer-to-peer computing Computing in which the resources of multiple machines interconnected by a network, such as the Internet, are pooled. Seen, for instance, in the utilization of unused CPU cycles for distributed processing. Its abbreviation is **P2P computing**.

peer-to-peer network A network with a **peer-to-peer architecture**. Used, for instance, for groups of users to swap files from each other's hard drives over the Internet. Its abbreviation is **P2P network**.

pel Abbreviation of **pixel**, or **picture element**.

Peltier effect An effect in which a temperature change is produced at the junction of a thermocouple when a current passes through said junction. When the current flows in one direction, the temperature of one of the materials rises, while that of the other material comprising the thermocouple falls. When the current flows in the other direction, the temperature changes in the materials are reversed. It is the converse of the **Seebeck effect**.

PEM Abbreviation of **Privacy Enhanced Mail**.

pen 1. The pointing device of a digitizing tablet. It is similar to a mouse, but is much more accurate, because its location is determined by touching an active surface with an absolute reference. Also called **puck**, **cursor (4)**, or **stylus (1)**. **2.** The pointing device of a PDA, which serves for selection or input. Also called **stylus (2)**.

pen-based computing Same as **pen computing**.

pen computer A computer, such as a small PDA, which utilizes a **pen (2)** as the primary manual input device, as opposed to a keyboard.

pen computing Computing which utilizes a **pen (2)** as the primary manual input device, as opposed to a keyboard. Also called **pen-based computing**.

pen plotter A plotter which utilizes one or more pens to draw. This contrasts, for instance, with a plotter utilizing an electrostatic printing process.

pen recorder A graphic recorder, such as that of an electroencephalograph or seismograph, in which a pen records signal variations on a paper or other chart.

pencil 1. Same as **pencil beam**. **2.** Something in the shape of a pencil, such as the tip of a probe.

pencil beam A beam of radiated energy, such as light, or particles, such as electrons, whose cross-section is in the approximate shape of a circle with a comparatively small diameter. Such a beam is usually made parallel, or nearly parallel, through the use of one or more collimators. Used, for instance, in astronomy and radars. Also called **pencil (1)**.

pencil-beam antenna A directional antenna whose major lobe has a cross-section in the approximate shape of a circle with a comparatively small diameter.

pendant control box Same as **pendant control station**.

pendant control station A control panel, usually consisting of one or more pushbuttons, that is connected to a suspended cable. Used, for example, as an emergency shut-off for high-voltage equipment, or to control an industrial robot. Also called **pendant control box**, or **pendant pushbutton station**.

pendant pushbutton station Same as **pendant control station**.

pendulum A rigid body which freely swings about a fixed point. A simple pendulum may be considered to have a point mass whose path is symmetrical about a given equilibrium position, and whose movement is influenced only by gravity. Used, for instance, to demonstrate simple harmonic motion, or to determine the value of the gravitational acceleration of the planet earth.

pendulum switch A switch that is actuated by being jostled or tilted.

penetrating frequency Same as **penetration frequency**.

penetrating radiation **1.** Radiation that readily penetrates a given material or medium. For example, gamma rays and X-rays penetrate living tissue without difficulty, producing serious harm when protective shielding is not utilized. **2.** Radiation that penetrates a given material or medium under certain conditions. For instance, a vertically propagated radio wave above the critical frequency will penetrate the ionosphere.

penetration An intrusion or otherwise unauthorized entry into a computer system or network. For instance, the cracking of a password using a brute-force attack, followed by the obtaining of key information such as other passwords.

penetration frequency The frequency below which a vertically propagated radio wave will be reflected by the ionosphere. That is, any frequency above this will penetrate the ionosphere. This frequency varies by ionospheric layer and time of day. Also called **penetrating frequency**, or **critical frequency (1)**.

penetration testing The performance of tests which attempt to gain unauthorized access into a computer system or network. Used, for instance, to locate security vulnerabilities.

pentacene A compound whose chemical formula is $C_{22}H_{14}$. It is used as an organic semiconductor.

pentagon A closed geometric figure bounded by five straight lines.

pentavalent Pertaining to an element, such as arsenic or antimony, which has a valence of five. Such elements are used, for instance, as dopants which contribute free electrons which increase the conductivity of an intrinsic semiconductor. Doping with a pentavalent chemical element makes a p-type semiconductor, while **trivalent** elements provide for n-type semiconductors.

Pentium A popular family of CPU chips.

pentode An electron tube with five electrodes. Such a tube usually has an anode, a cathode, a control electrode, and two grids.

PEP Abbreviation of **peak envelope power**.

perceived level The level of a magnitude or phenomenon as subjectively perceived by a given individual. For example, brightness is the perception, as judged by a person, of more or less light emitted from a source.

perceived noise level A noise level as judged by a given listener. For instance, the extent to which a noise source, such as an aircraft, is perceived by an individual as a nuisance.

percent One part per hundred. For example, 20 is ten percent of 200, as it is ten parts per hundred, multiplied by two. Its symbol is %.

percent accuracy The accuracy of a measured or calculated value, expressed as a **percentage**. For example, that of a reading taken by an instrument. Also called **percentage accuracy**.

percent amplitude The amplitude of a wave, or other periodic phenomenon, expressed as a **percentage**. For instance, the amplitude of an output signal. Also called **percentage amplitude**.

percent capacity A proportion of the total capacity of a component, circuit, device, piece of equipment, or system, expressed as a **percentage**. For example, the number of occupied communications circuits at a given moment expressed as a percentage of the total available circuits. Also called **percentage capacity**.

percent distortion The distortion of a component, circuit, device, piece of equipment, or system, expressed as a per-

centage. For example, that of an amplifier. Also called **percentage distortion**.

percent efficiency The efficiency of a component, circuit, device, piece of equipment, or system, expressed as a **percentage**. For instance, that of an antenna or transistor. Also called **percentage efficiency**.

percent error A deviation from a true, expected, or specified result, expressed as a **percentage**. For instance, the magnitude, expressed as a percentage, of an instrumental error. Also called **percentage error**.

percent harmonic distortion The harmonic distortion of a component, circuit, device, piece of equipment, or system, expressed as a **percentage**. For example, that of a transmission line. Also called **percentage harmonic distortion**.

percent modulation The modulation factor expressed as a percentage. Also called **percentage modulation**, or **modulation percentage**.

percent ripple The ratio of the ripple voltage to the nominal or average total voltage, expressed as a **percentage**. For instance, that of a filter. Also called **percentage ripple**, or **ripple percentage**.

percentage A fraction or ratio expressing a **percent**. A fraction with 100 as a denominator represents a percentage. For example, 50/100 is 50%, or 300/100 is 300%. A ratio is multiplied by 100 to obtain a percentage. For instance, a ratio whose value is 0.5 represents 50%.

percentage accuracy Same as **percent accuracy**.

percentage amplitude Same as **percent amplitude**.

percentage capacity Same as **percent capacity**.

percentage distortion Same as **percent distortion**.

percentage efficiency Same as **percent efficiency**.

percentage error Same as **percent error**.

percentage harmonic distortion Same as **percent harmonic distortion**.

percentage modulation Same as **percent modulation**.

percentage ripple Same as **percent ripple**.

perceptron A single-layer artificial neural network with limited learning abilities.

percussive maintenance The attempt to solve a hardware problem by slapping, banging, kicking, or otherwise assaulting the device, piece of equipment, or system in question. Rarely advisable, yet the occasional resetting of a loose connection via this method can justify an otherwise futile act.

perfboard Abbreviation of **perforated board**.

perfect **1.** Pertaining to a component, circuit, device, piece of equipment, system, process, operation, result, state, substance, or entity which corresponds to a theoretical ideal, or that conforms completely to a theory pertaining to a theoretical ideal. For example, a perfect radiator, or an engine which converts 100% of its input power or energy into useful power or energy, such as work. Also called **ideal**. **2.** Having all that is essential to constitute a whole, or that which is complete.

perfect capacitor A capacitor with characteristics that correspond to a theoretical perfection. Such a capacitor, for instance, would hold a charge forever, or until discharged. Also called **ideal capacitor**.

perfect component A component with characteristics that correspond to a theoretical perfection. An example is an ideal capacitor. Also called **ideal component**.

perfect coupling A degree of coupling between two systems, especially circuits, corresponding to a theoretical perfection. That is, the coefficient of coupling would be 1, or 100%. Also called **ideal coupling**, or **unit coupling**.

perfect crystal A crystal whose contained atoms, molecules, or ions have a geometric arrangement which is free from lattice defects. Also called **ideal crystal**.

perfect device A device with characteristics that correspond to a theoretical perfection. An example is an engine which converts 100% of its input power or energy into useful power or energy, such as work. Also called **ideal device**.

perfect dielectric A dielectric through which an electric field may be applied with no energy losses. A vacuum is an example, and its dielectric constant is defined as 1. Also called **ideal dielectric**.

perfect inductor An inductor with characteristics that correspond to a theoretical perfection. Such an inductor, for instance, would have zero losses. Also called **ideal inductor**.

perfect radiator An ideal body which would be a perfect emitter of radiant energy, its distribution of energy depending solely on its absolute temperature. It would also absorb all the radiant energy incident upon it. Also called **blackbody**, **ideal radiator**, or **full radiator**.

perfect transducer A transducer with characteristics that correspond to a theoretical perfection. Such an transducer, for instance, would have zero losses. Also called **ideal transducer**.

perfect transformer A transformer with characteristics that correspond to a theoretical perfection. Such an transducer, for instance, would have a coupling coefficient of 1. Also called **ideal transformer**.

perfect vacuum A hypothetical space which contains no matter, and whose absolute pressure is defined as zero. Empty outer space approaches a perfect vacuum. Also called **absolute vacuum**, or **vacuum (1)**.

perforated board A plastic, resin, or laminated board with rows and columns of equally-spaced holes, upon which electronic devices and components can be mounted and interconnected, for testing, experimenting, and preparation of prototype circuits. Its abbreviation is **perfboard**.

performance 1. The level of functioning of a component, circuit, device, piece of equipment, system, material, or the like. Also, the act of functioning. 2. A comparative measure of the level of operation or function of a component, circuit, device, piece of equipment, system, material, or the like. For instance, high-performance, or small-signal performance.

performance characteristics Measurable features that help describe and distinguish the level of operation of components, circuits, devices, systems, materials, or the like.

performance chart A chart that describes, and helps evaluate, the level of operation of a component, circuit, device, system, material, or the like.

performance curve A curve that describes, and helps evaluate, the level of operation of a component, circuit, device, system, material, or the like. For example, a curve plotting the harmonic distortion of an amplifier as a function of frequency.

performance data Data which describes, and helps evaluate, the level of operation of a component, circuit, device, system, material, or the like. For example, such data gathered under harsh operating conditions.

performance monitor Software and/or hardware which enables the supervision of one or more performance aspects of a process, mechanism, system, or the like. For example, an onboard computer evaluating the operation of a car.

performance test A test that evaluates the level of operation of a component, circuit, device, system, material, or the like. For instance, tests performed keeping a load stable while varying other system parameters, so as to optimize performance.

performance testing The carrying out of **performance tests**.

period 1. An interval of time during which one or more phenomena, conditions, or events occur. For example, a busy period, or an early-failure period. 2. The interval of time that elapses between two consecutive events of a periodically repeated phenomenon. For instance, the duration of a complete cycle. Also, the time interval elapsing between successive occurrences of the same phases of a cyclic event. Its symbol is *T*. 3. A series of elements which form a horizontal row within a periodic table of chemical elements.

periodic 1. Occurring, appearing, or characterized by regular and repetitive intervals or cycles. For instance, AC, the chopping of a signal, or the oscillation of a crystal. 2. Occurring, appearing, or characterized by irregular intervals. For example, the backing up of data occasionally. 3. Characterized or ordered by repeated properties. For instance, the sequencing of elements within a periodic table.

periodic and random deviation For a power supply, the total variation in output factoring in components such as ripple, load, noise, hum, stability, and temperature coefficient. Its acronym is **PARD**.

periodic antenna An antenna whose input impedance varies as a function of the operating frequency. An example is a dipole array.

periodic curve A curve or graph representing a **periodic function**. An example is a sine curve.

periodic duty A form of intermittent duty in which the work or operation of a device, piece of equipment, or system, occurs in regular and repetitive intervals or cycles.

periodic filter A filter whose response reflects its alternately passing and rejecting frequency bands. A comb filter is an example.

periodic function A function, such as a sine function, characterized by regular and repetitive intervals or cycles.

periodic law A law which states that the physical and chemical properties of chemical elements are a periodic function of their atomic number. Thus, when the elements are arranged in order of atomic number, as occurs in a periodic table, elements with similar properties will recur at regular intervals. This enables, for instance, to calculate properties of elements yet to be discovered.

periodic magnitude A magnitude whose variations are characterized by regular and repetitive intervals or cycles.

periodic noise Noise whose variations are characterized by regular and repetitive intervals or cycles.

periodic quantity A quantity whose variations are characterized by regular and repetitive intervals or cycles.

periodic resonance Resonance in which the period of oscillation of the external driving force is the same as the natural period of a given body or system. Also called **natural resonance (2)**.

periodic signal A signal whose variations are characterized by regular and repetitive intervals or cycles.

periodic table A table in which chemical elements are arranged so as to exploit patterns in physical and chemical properties, as provided by periodic law. A typical periodic table is arranged in columns and rows, where the columns represent groups of elements with the same electronic configuration in their outermost shell, which gives them similar properties. As each outer shell is filled, a new row of elements is started, and there is a one-by-one increase in atomic number as the elements progress from left to right.

periodic value A value whose variations are characterized by regular and repetitive intervals or cycles.

periodic wave A wave whose variations are characterized by regular and repetitive intervals or cycles.

periodic waveform 1. A waveform which occurs in a regular and repetitive pattern. 2. A repetitive waveform.

periodicity 1. The quality or state of being **periodic**. 2. The manner in which something is periodic. Also, that which makes something periodic.

peripheral **1.** Any device which transfers data to a CPU and/or receives data from a CPU. Such devices include keyboards, pointing devices, digitizing tablets, scanners, disk drives, microphones, monitors, printers, and speakers. Also, any external device which otherwise assists, enhances, or depends on computer function. Also called **peripheral device**, **peripheral equipment** (1), or **I/O device**. **2.** Anything which is auxiliary in nature. For example, an auxiliary device as opposed to a main device.

peripheral buffer An area of computer memory reserved for data awaiting transfer to the CPU from a peripheral, or vice versa. Also called **I/O buffer**.

peripheral bus A bus, such as a local bus, which provides parallel data transfer between the CPU, the memory, and high-speed peripherals.

peripheral card A circuit board, such as a sound card or graphics card, which is plugged into the bus or an expansion slot of a computer, in order to add and control a peripheral. Also called by various other terms, including **expansion card**, **expansion board**, **accessory card**, **add-on board**, **adapter card**, **card** (1), and **adapter** (3).

Peripheral Component Interconnect Same as **PCI**.

Peripheral Component Interconnect bus Same as **PCI**.

Peripheral Component Interconnect local bus Same as **PCI**.

peripheral controller A circuit board or device which controls the way peripheral devices access the computer, and vice versa. It is usually contained on a single chip. Examples include disk controllers, graphics controllers, and video controllers. Also called **controller** (1), or **host adapter**.

peripheral device Same as **peripheral** (1).

peripheral electrons Electrons which are present in the outermost shell of an atom. Such electrons can move freely under the influence of an electric field, thus producing a net transport of electric charge. Peripheral electrons also determine the chemical properties of the atom. Also called **valence electrons**, **outer-shell electrons**, or **conduction electrons**.

peripheral equipment **1.** Same as **peripheral** (1). **2.** Any equipment which transfers data to a CPU and/or receives data from a CPU, or which otherwise assists, enhances, or depends on computer function. For example, a machine utilized for laser cutting which is controlled by an external computer.

peripheral interface adapter A circuit board which is plugged into the bus or an expansion slot of a computer, in order to enable the exchange of data with a peripheral. An example is a network adapter. Its abbreviation is **PIA**. Also called **peripheral interface card**.

peripheral interface card Same as **peripheral interface adapter**.

Perl Acronym for **P**ractical **E**xtraction **R**eport **L**anguage. A UNIX-based programming language which is often used for system administrative tasks, or for the creation of scripts.

permalloy A ferromagnetic alloy consisting of about 79% nickel, 17% iron, and 4% molybdenum, although the relative proportions may vary. Other elements, such as manganese, may be present is small quantities. It has high permeability, and is used for instance, in tape deck heads, or to shield components from magnetic fields

permanent cookie Same as **persistent cookie**.

permanent magnet A high-retentivity material, such as alnico alloy or certain steels, which can be magnetized sufficiently to retain its magnetism indefinitely. Permanent magnets have applications in various areas of electronics, including their use in sensing devices and speakers. Its abbreviation is **PM**.

permanent-magnet dynamic loudspeaker Same as **permanent-magnet speaker**. Its abbreviation is **PM dynamic loudspeaker**.

permanent-magnet dynamic speaker Same as **permanent-magnet speaker**. Its abbreviation is **PM dynamic speaker**.

permanent-magnet focusing The utilization of one or more permanent magnets to focus an electron beam. Used, for instance, in klystrons.

permanent-magnet generator An electric generator which utilizes a permanent magnet to provide the magnetic flux. Used, for instance, in ignition systems. Its abbreviation is **PMG**. Also called **magneto**.

permanent-magnet loudspeaker Same as **permanent-magnet speaker**. Its abbreviation is **PM loudspeaker**.

permanent-magnet motor A motor which utilizes one or more permanent magnets to produce torque, or another driving force. Also called **magnet motor**.

permanent-magnet speaker A dynamic speaker incorporating a permanent magnet which produces the constant magnetic field. Its abbreviation is **PM speaker**. Also called **permanent-magnet loudspeaker**, **permanent-magnet dynamic speaker**, or **permanent-magnet dynamic loudspeaker**.

permanent-magnet stepper motor A stepper motor in which the rotor is a permanent magnet. Also called **permanent-magnet stepping motor**.

permanent-magnet stepping motor Same as **permanent-magnet stepper motor**.

permanent memory Computer memory which retains its content when the power is interrupted or turned off. Examples include ROM, EEPROM, and flash memory. Also called **non-volatile memory**.

permanent-split capacitor motor A capacitor motor in which the auxiliary winding and capacitor are energized at all times. Also called **capacitor start-run motor**.

permanent storage **1.** Storage, such as that of a ROM chip, which can not be altered once data has been placed there. **2.** Computer storage which retains its content when the power is interrupted or turned off. Almost all computer storage, such as that utilizing disks or tapes, is of this nature. Also called **non-volatile storage** (1). **3.** Computer memory which retains its content when the power is interrupted or turned off. Examples include ROM, EEPROM, and flash memory. Also called **non-volatile storage** (2).

permanent virtual circuit In a communications network, a virtual circuit between two nodes which is established before any actual data transfers takes place, and which is available at all times. Used, for instance, in frame relay. This contrasts with a **switched virtual circuit**, which is established only when needed. Its abbreviation is **PVC**.

permeability **1.** Its symbol is μ. A quantitative indicator of the response of a material or medium to a magnetic field. It is expressed by the following formula: $\mu = B/H$, where B is the magnetic flux density of the material or medium, and H is the strength of the surrounding magnetic field. Relative permeability is obtained using the following formula: $\mu_r = \mu/\mu_0$, where μ_r is relative permeability, and μ_0 is the permeability of free space. The absolute permeability of free space is a physical constant equal to approximately 1.256637×10^{-6} henry per meter, and its relative permeability is equal to 1, as seen by the this formula. Materials whose relative permeability is less than 1 are diamagnetic, those slightly above 1 are paramagnetic, and those substantially above 1 are ferromagnetic. Also known as **magnetic permeability**. **2.** The quality or state of being **permeable**.

permeability of a vacuum Same as **permeability of free space**.

permeability of free space A physical constant equal to $4\pi \cdot 10^{-7}$ henry per meter, or approximately 1.256637×10^{-6} henry per meter. This number represents the absolute permeability of free space. The relative permeability of free space is 1. Its symbol is μ_0. Also called **permeability of a vacuum**, or **magnetic constant**.

permeability tuning The tuning of a resonant circuit by varying the permeability of an inductance-capacitance circuit.

permeable **1.** Said of a material, substance, body, or entity that allows a substance, especially a gas or a liquid, to penetrate or pass through it,. **2.** The extent to which a material, substance, body, or entity allows a substance to penetrate or pass through it.

permeameter An instrument or device which measures the permeability of materials or mediums, especially ferromagnetic materials. Such an instrument or device may also be utilized to measure other magnetic properties of a given specimen

permeance The ease with which a material or magnetic circuit allows a magnetic field to pass through it. It is expressed as the ratio of the magnetic flux to the magnetomotive force, and is analogous to electrical conductance. It is the reciprocal of **reluctance**. Its symbol is P, or Λ. Also called **magnetic permeance**.

Permendur A soft alloy composed of 49% iron, 49% cobalt, and 2% vanadium. It features very high flux density at saturation, low magnetic coercive force, and its properties are maintained at elevated operating temperatures.

perminvar An alloy consisting of 45% nickel, 30% iron, and 25% cobalt, which maintains an essentially constant permeability within a wide interval of magnetic intensities.

permittivity The property of a material which determines the electrostatic energy that can be stored per unit volume for a unit potential gradient. This is equivalent to the ratio of the capacitance of a capacitor using the material in question as a dielectric, to the capacitance of a capacitor using a vacuum as a dielectric. Its symbol is ϵ. Also called **dielectric constant**, **relative permittivity**, **inductivity**, and **specific inductive capacity**.

permittivity of free space The **permittivity** of a vacuum. It is a constant equal to approximately 8.85419×10^{-12} farad per meter. Its symbol is ϵ_0. Also called **electric constant**.

perpendicular magnetization Magnetization of a material or medium in a plane perpendicular to the plane of said material or medium. Multilayer materials may exhibit perpendicular magnetization. Also called **vertical magnetization**.

perpendicular recording Recording of a magnetic recording medium in which **perpendicular magnetization** is utilized. Also called **vertical recording**.

persistence **1.** The length of time that the phosphors on the screen of a CRT glow after the electron beam has passed. **2.** The property or tendency of certain materials, especially phosphors, to glow after the excitation has passed.

persistence characteristic For a CRT screen, such as that of a cathode-ray oscilloscope, a curve which expresses the relationship between the emitted radiant power and time, after the excitation is removed. Also called **decay characteristic**.

persistence of vision A characteristic of vision in which the retina briefly retains an image after it, or its source, has been removed. This interval is usually approximately 0.05 second. Thus, a light that switches on and off at a frequency of greater than 20 cycles per second generally appears as non-flickering. Persistence of vision and the **Phi phenomenon** are exploited to enable a movie or TV presentation to be perceived as consisting of continuous motion, even though they are comprised of frames.

persistent cookie A cookie that is kept indefinitely, or until deleted. This contrasts with a **session cookie**, which is deleted or otherwise disappears when a Web browser is closed, or after a comparatively short time. Also called **permanent cookie**.

persistent current A current, such as that in a superconducting material, which flows undiminished. Also, such a current flowing through a closed path. A persistent current produces a persistent magnetic field, or is induced by one.

persistent data Data which is stored between sessions. This contrasts with **transient data**, which is discarded from session to session.

persistent magnetic field **1.** A magnetic field produced by a **persistent current**. **2.** A magnetic field which produces a persistent current.

persistent object In object-oriented technology, an object that continues to exist beyond the duration of the process that created it.

Personal Area Network A network in which multiple devices are wirelessly interconnected when located within a given radius, such as ten meters, and which is suitable for a single user moving around employing PDAs, desktop computers, printers, and so on. Its abbreviation is **PAN**. Also called **Wireless Personal Area Network**.

personal communications device Any device which is portable and used for communications such a voice calls, email, pages, or faxes. These include cellular phones, pagers, and properly equipped portable computers such as notebooks or PDAs. Also called **personal communicator**.

Personal Communications Services Its abbreviation is **PCS**. A class of communications services which utilize all-digital technology for transmission and reception, and which operates just below the 2 GHz range. PCS technologies include GSM, CDMA, and TDMA.

personal communicator Same as **personal communications device**.

personal computer A usually stand-alone microcomputer which is primarily intended for use by an individual. Its abbreviation is **PC**.

personal computer clone. Same as **PC-clone**.

Personal Computer Memory Card International Association Same as **PCMCIA**.

personal digital assistant Same as **PDA**.

personal equation A factor which is intended to compensate for the errors introduced by a given observer during any observation. For instance, an individual detecting the start of an event will have a delay, usually in the 0.2 to 0.5 second range, before being able to depress a stopwatch button. The personal equation, which can be calculated through the statistical analysis of the reaction times of an individual over many observations and recordings, can then be incorporated to reduce this source of error.

personal identification number **1.** Same as **PIN (2)**. **2.** Same as **PIN (3)**.

personal information manager An application which helps a user organize information such as names, telephone numbers, emails, personal notes, and so on. Its acronym is **PIM**.

personal robots Also called **partner robots**. **1.** Robots, such as seeing-eye robots, intended to help and provide comfort to humans. Other examples include those utilized for nursing care, or as pets. **2.** A robot intended to entertain or perform chores around a home or office. An example is a toy which follows voice commands to perform simple tasks.

perturbation **1.** A change in a body or system. For example, the influence of an external body on the orbit of an artificial satellite. Also, an action or event leading to such a change. **2.** A **perturbation** (1) of small magnitude. Also, an action or event causing such a change.

pervasive computing The presence of interconnected computing devices, such as motes, smart appliances, and wearable computers, which permeate a given environment. Pervasive computing may be used, for instance, to enable immediate access to computing tasks under nearly all circumstances, or for keeping track of innumerable objects and/or persons regardless of where they are. Also called **ubiquitous computing**.

perveance In an electron tube, a magnitude which determines the saturation current which can flow for a given geometry. It is a proportionality constant equal to $J/V^{3/2}$, where J is current density, and V is collector voltage. This value changes as a given tube ages.

PET Abbreviation of **positron emission tomography**.

PET scan Abbreviation of **positron emission tomography scan**.

PET scanner Abbreviation of **positron emission tomography scanner**.

peta- A metric prefix representing 10^{15}, which is equal to a quadrillion. For instance, 1 petahertz is equal to 10^{15} Hz, or 1,000,000,000,000,000 Hz. When referring to binary quantities, such as bits and bytes, it is equal to 2^{50}, or 1,125,899,906,842,624, although this is frequently rounded to 1,000,000,000,000,000. To avoid any confusion, the prefix **pebi-** may be used when dealing with binary quantities. Its abbreviation is **P**.

peta-electronvolt A unit of energy or work equal to 10^{15} electronvolts. Also spelled **petaelectronvolt**. Its abbreviation is **PeV**.

petabecquerel A unit of activity equal to 10^{15} becquerels. Its abbreviation is **PBq**.

petabit 2^{50}, or 1,125,899,906,842,624 bits, although this is frequently rounded to 1,000,000,000,000,000 bits. To avoid any confusion, the term **pebibit** may be used when referring to this concept. Its abbreviation is **Pb**, or **Pbit**.

petabits per second 2^{50}, or 1,125,899,906,842,624 bits, per second. Usually used as a measure of data-transfer speed. Its abbreviation is **Pbps**. To avoid any confusion, the term **pebibits per second** may be used when referring to this concept.

petabyte 2^{50}, or 1,125,899,906,842,624 bytes, although this is frequently rounded to 1,000,000,000,000,000 bytes. To avoid any confusion, the term **pebibyte** may be used when referring to this concept. Its abbreviation is **PB**, or **Pbyte**.

petabytes per second 2^{50}, or 1,125,899,906,842,624 bytes, per second. Usually used as a measure of data-transfer speed. Its abbreviation is **PBps**. To avoid any confusion, the term **pebibytes per second** may be used when referring to this concept.

petacycle A unit of frequency equal to 10^{15} cycles, or 10^{15} Hz. The term currently used for this concept is **petahertz**.

petaelectronvolt Same as **peta-electron volt**.

petaflops Same as **PFLOPS**.

petagram A unit of mass equal to 10^{15} grams. Its abbreviation is **Pg**.

petahertz A unit of frequency equal to 10^{15} Hz. Its abbreviation is **PHz**. Also called **petacycle**.

petajoule A unit of energy or work equal to 10^{15} joules. Its abbreviation is **PJ**.

petameter A unit of distance equal to 10^{15} meters. Its abbreviation is **Pm**.

petaohm Same as **petohm**.

petavolt A unit of potential difference equal to 10^{15} volts. Its abbreviation is **PV**.

petawatt A unit of power equal to 10^{15} watts. Its abbreviation is **PW**.

petohm A unit of resistance, impedance, or reactance equal to 10^{15} ohms. Its abbreviation is **PΩ**. Also spelled **petaohm**.

PeV Abbreviation of **peta-electronvolt**.

pF Abbreviation of **picofarad**.

PF Abbreviation of **power factor**.

PFC 1. Abbreviation of **power-factor correction**. 2. Abbreviation of **power-factor corrector**.

PFET Abbreviation of **p-channel FET**.

Pflops Same as **PFLOPS**.

PFLOPS Abbreviation of **petaflops**. 10^{15} FLOPS, which is the same as 10^{15} floating-point calculations or operations, per second. Usually used as a measure of processor speed.

PFM Abbreviation of **pulse-frequency modulation**.

pg Abbreviation of **picogram**.

Pg Abbreviation of **petagram**.

pG Abbreviation of **picogauss**.

PGA 1. Abbreviation of **programmable gate array**. 2. Abbreviation of **pin grid array**. 3. Abbreviation of **programmable gain amplifier**.

PgDn Abbreviation of **page down**.

PgDn key Abbreviation of **page down key**.

PGP Abbreviation of **Pretty Good Privacy**.

pGs Abbreviation of **picogauss**.

PgUp Abbreviation of **page up**.

PgUp key Abbreviation of **page up key**.

ph Abbreviation of **phot**.

pH 1. A measure of the acidity or alkalinity of a solution. For an aqueous solution it can be expressed as $pH = \log_{10}(1/H^+)$, where H^+ is the hydrogen-ion concentration. A pH of 7 is defined as neutral, below 7 is acidic, and above 7 is basic, or alkaline. It is an abbreviation of potential for hydrogen. Also called **pH value**. 2. Abbreviation of **picohenry**.

pH indicator 1. Same as **pH meter**. 2. An instrument, device, substance, or object, which indicates **pH values**. For instance, a powder or paper which changes color depending on whether a solution is acidic or alkaline.

pH meter An instrument or device, such as that utilizing a glass electrode, which serves to measure and indicate **pH values**. Also called **pH indicator (1)**.

pH scale A scale for expressing **pH (1)** values, which can range from 0 to 14. Since the scale is logarithmic, a pH of 1 is 100 times more acidic than a reading of 3, a pH of 11 is a thousand times more basic than a reading of 8, and so on.

pH value Same as **pH (1)**.

phantom channel In audio reproduction, a virtual channel created by two or more real channels. For instance, the use of left and right speakers to produce a center mono channel which is perceived by the listener as if there were a center speaker.

phantom circuit A third communications circuit, without a wire, derived from two circuits with wires which are appropriately coupled.

phantom signal A signal presented on a CRT, such as that of a radar or TV, whose origin can not be readily ascertained. For instance, that arising from a possible propagation anomaly which can not be pinpointed.

phantom target A resonant cavity that simulates a radar echo, by storing the energy of a transmitted pulse, and retransmitting it gradually to the radar receiver. Used for testing and tuning. Also known as **echo box**.

phase 1. Within a periodic phenomenon or process, a specified stage of progress. For instance, a moon phase. 2. For a given periodic phenomenon, process, or quantity, the portion of a complete cycle that has been completed, as measured from a given reference point. Two or more periodic quantities having the same frequency and waveshape that pass through corresponding values, such as maximas and minimas, at the same instant at all times are **in-phase**. While periodic quantities that do not pass through corresponding values at the same instant at all times are **out-of-phase**. A phase angle expresses the difference between the phases of two such quantities. Phase is usually expressed in degrees or radians, where a complete cycle is 360° or 2π, respectively. Its symbol is ϕ. **3.** In a polyphase system, such as a three-phase AC circuit, one of the circuits. **4.** In chemistry, a homogeneous region that can be observed and separated from another it is present with. For example, ice in water.

phase-alternate line Same as **PAL (1)**.

phase-alternate line signal Same as **PAL signal**.

phase-alternate line standard Same as **PAL standard**.

phase-alternating line Same as **PAL (1)**.

phase-alternating line signal Same as **PAL signal**.

phase-alternating line standard Same as **PAL standard**.

phase-alternation line Same as **PAL (1)**.

phase-alternation line signal Same as **PAL signal**.

phase-alternation line standard Same as **PAL standard**.

phase angle For two periodic quantities with the same frequency, the difference between their respective phases. Expressed in radians or degrees. For example, the phase difference between an AC and the voltage producing it. Also known as **phase difference (1)**, or **angular phase difference**.

phase-angle meter Same as **phase meter**.

phase-angle voltmeter A voltmeter which also indicates the phase angle of a measured voltage.

phase center For a given antenna, the apparent center of signal transmission or reception. This point varies, and depends, for instance, on the signal frequency. Also called **antenna phase center**.

phase change Same as **phase shift**.

phase-change coefficient Same as **phase constant**.

phase comparator Same as **phase detector**.

phase comparison Same as **phase detection**.

phase compensation The compensation of undesired phase differences. For example, the switching of capacitors in and out of a power-distribution network to compensate for variations in the power factor.

phase-compensation circuit An electric circuit utilized for **phase compensation**. Also called **phase-compensation network**.

phase-compensation network Same as **phase-compensation circuit**.

phase compensator A circuit, device, mechanism, or system utilized for **phase compensation**.

phase conductor In a polyphase circuit or system, a conductor other than a neutral conductor.

phase constant The imaginary number component of the propagation constant. The real part is the **attenuation constant**. Usually expressed in radians per unit length. Also called **phase-change coefficient**, or **wavelength constant**.

phase control Also called **phase-shift control 1.** A circuit, device, or mechanism utilized to vary the phase angle of a signal. **2.** In a color TV receiver, a control which varies the phase of the chrominance signals with respect to the burst signal. This changes the hue of the televised images. Also called **hue control**.

phase converter A device which changes the number of phases of an AC source, without changing its frequency. For instance, a rotary converter whose input is 50 Hz single-phase 220VAC, and whose output is 50 Hz three-phase 220VAC.

phase correction The corrections a **phase corrector** makes. Also called **phase-delay equalization**, **phase equalization**, or **delay equalization**.

phase corrector A corrective network which serves to compensate for the effects of **phase distortion**. It may do so, for instance, by introducing the necessary delays at the appropriate frequencies to offset said distortion. Also called **phase-delay equalizer**, **phase equalizer**, or **delay equalizer**.

phase delay Also called **phase lag**, **phase difference (2)**, or **phase displacement**. **1.** For two periodic quantities with the same frequency, the lag in phase of one respective to the other. Expressed in radians or degrees. Also called **angle of lag**, or **lag angle**. **2.** In the transmission of a single-frequency wave from one point to another through a transmission line or system, the time interval between the instants at which a given point of a wave passes through two specified points of said transmission medium.

phase-delay distortion Same as **phase distortion**.

phase-delay equalization Same as **phase correction**.

phase-delay equalizer Same as **phase corrector**.

phase detection The use of a **phase detector**. Also called **phase discrimination**, or **phase comparison**.

phase detector Its abbreviation is **PD**. Also called **phase discriminator**, **phase comparator**, **phase-sensitive detector**, or **phase-shift discriminator (2)**. **1.** A circuit, device, or instrument which detects any phase difference between two signals. **2.** A circuit, device, or instrument which detects any phase difference between two signals, and whose output voltage is proportional to said difference.

phase-detector circuit A circuit utilized for **phase detection**.

phase deviation In phase modulation, the peak difference between the instantaneous phase angle of the modulated wave and that of the unmodulated carrier.

phase difference 1. Same as **phase angle**. **2.** Same as **phase delay**.

phase discrimination Same as **phase detection**.

phase discriminator Same as **phase detector**.

phase displacement Same as **phase delay**.

phase distortion Distortion of a signal as it passes through a transmission medium in which different frequencies travel at slightly different speeds. This causes a change in the waveform because the rate of change of phase shift is not constant over the transmitted frequency range. Also called **phase-delay distortion**, **phase-frequency distortion**, **delay distortion**, **envelope delay distortion**, or **time-delay distortion**.

phase equalization Same as **phase correction**.

phase equalizer Same as **phase corrector**.

phase error Any defect or error caused by an undesired phase difference.

phase-frequency distortion Same as **phase distortion**.

phase inversion Also called **phase inverting**, **phase reversal**, or **phase reverting**. **1.** A change in phase of 180°, or π radians, which is the same as a half-cycle. **2.** A change in phase of 180°, or π radians, in an AC signal.

phase-inversion switch Same as **phase-reversal switch**.

phase inverter A circuit or device which changes the phase of a signal by 180°, or π radians. Used, for instance, to drive one side of a push-pull amplifier.

phase inverting Same as **phase inversion**.

phase jitter Small and rapid fluctuations in **phase (2)**.

phase lag Same as **phase delay**.

phase-lock A technique utilized to make the phase of an oscillator signal follow the phase of a reference signal. The phase angles of the two signals are compared utilizing a phase detector, and the appropriate corrective adjustments are made.

phase-lock loop Same as **phase-locked loop**. Its abbreviation is **PLL**.

phase-locked Pertaining to two signals whose phases are synchronized with each other, as would occur when utilizing a phase-locked loop.

phase-locked loop A circuit which serves to make the phase of an oscillator signal follow the phase of a reference signal. The usual case is that of a local oscillator being locked in phase with an input signal. The phase angles of the two signals are compared utilizing a phase detector, and the appropriate corrective adjustments are made. In this manner, the input signal frequency can be regenerated without any jitter that may have been present. Widely used in frequency control systems, as frequency synthesizers, as demodulators for FM signals, for the synchronization of TV scanning signals, in microwave communications, and so on. Its abbreviation is **PLL**. Also called **phase-lock loop**.

phase-locked oscillator Its abbreviation is **PLO**. **1.** An oscillator whose output signal is kept locked in phase with a reference signal. **2.** An oscillator whose output signal is kept locked in phase with another signal whose frequency is the same.

phase margin The difference between the shift a component, device, or system introduces into the phase of a signal, and a 180° or π shift.

phase meter An instrument which determines **phase angles**. Such a device should be able to provide accurate readings through a comparatively wide range of signal levels, and should also be reasonably immune to distortion and noise. Used, for instance, to monitor a stereo signal to determine mono compatibility. Also spelled **phasemeter**. Also called **phase-angle meter**.

phase modulation A form of modulation in which the information-bearing signal varies the phase angle of the carrier wave. The amount of phase shift is proportional to the amplitude of the modulating signal, but the amplitude of the carrier remains constant at all times. Used, for instance, for data transmission. Its abbreviation is **PM**.

phase modulator A circuit or device which effects **phase modulation**.

phase multiplier A circuit or device which multiplies the frequency of waves whose phase wish to be compared. At higher frequencies the resolution of such assessments is greater.

phase noise Unwanted disturbances in phase, usually of a random nature. Phase jitter is an example.

phase opposition The circumstance or state where two periodic quantities having the same frequency and waveshape are out-of-phase by 180°. Also called **opposition (2)**.

phase quadrature A phase difference of 90°, or $\pi/2$ radians. Two periodic quantities having the same frequency and waveshape which are out-of-phase by 90° are said to be **in quadrature**. Also called **quadrature**.

phase relationship For two periodic quantities with the same frequency, the comparison of their respective phases. Any difference is usually expressed as a phase angle.

phase resolution For a given component, circuit, device, instrument, piece of equipment, or system, the minimum detectable difference in phase.

phase resonance Resonance in which the phase angle between the fundamental components of oscillation and the applied signal is 90°. Also called **velocity resonance**.

phase response A change in the phase of a circuit, device, or system, as a function of frequency. For example, the phase shift introduced by a filter.

phase response curve A graph of the changes in the phase of a circuit, device, or system, as a function of frequency. For instance, that of a filter.

phase reversal Same as **phase inversion**.

phase-reversal keying Binary phase-shift keying in which the phase angles are 180° apart, as is usually the case. Its abbreviation is **PRK**.

phase-reversal switch A switch utilized to invert the phase of a signal. Usually used in stereo audio components or systems to insure that the phase of a loudspeaker is correct, or that desired. Also called **phase-inversion switch**.

phase reverting Same as **phase inversion**.

phase-rotation relay Same as **phase-sequence relay**.

phase-sensitive detector Same as **phase detector**. Its abbreviation is **PSD**.

phase sequence In a polyphase system, such as a three-phase AC circuit, the order in which phases appear.

phase-sequence indicator An instrument or device that indicates the phase sequence in a polyphase system, such as a three-phase AC circuit.

phase-sequence relay In a polyphase system, such as a three-phase AC circuit, a relay whose actuation depends on the order in which the phase voltages reach their maximum positive amplitudes. Also called **phase-rotation relay**.

phase sequencer An instrument or device which indicates the order in which phases appear in a polyphase system.

phase shift Also called **phase change**. **1.** A change in the phase relationship between two signals. For example, the changing of the phase of a signal as it passes through a filter, or that utilized in phase-shift keying. Also, the magnitude of such a change. **2.** A change in the phase relationship between two periodic or alternating quantities. Also, the magnitude of such a change.

phase-shift circuit Same as **phase-shift network**.

phase-shift control Same as **phase control**.

phase-shift detector Same as **phase-shift discriminator (1)**.

phase-shift discriminator **1.** A discriminator whose output signal is proportional to the phase-difference between its input signals. It is utilized for the reception of FM signals. Also called **phase-shift detector**, or **Foster-Seeley discriminator**. **2.** Same as **phase detector**.

phase-shift keying In the transmission of data, a modulation technique in which the phase of the carrier is varied between two or more angles. For instance, to represent binary data only two angles are needed, each 180° apart, and the phase shift could be -90° for a 0, and +90° for a 1. If higher data transfer rates are needed, more phase shifts can be used. For example, four phase angles allow for two bits to be encoded into each signal element, while eight phase angles provide for three bits. When two phase shifts are used, it is called **binary phase-shift keying**, and when four are used it is called **quadrature phase-shift keying**. Its abbreviation is **PSK**.

phase-shift network Also called **phase-shift circuit**. **1.** An electric network which is utilized to vary the phase relationship between two signals. For example, it may produce an output signal whose phase is different from its input signal. **2.** An electric network utilized to vary the phase relation between two alternating currents. For instance, that which changes the phase angles between the components of a two-phase system.

phase-shift oscillator An oscillator which produces a phase inversion between the input and output, or between successive stages. Such an oscillator usually employs a phase-shift network in its feedback path.

phase shifter **1.** A circuit or device utilized to vary the phase relationship between two signals. For example, it may produce an output signal whose phase is different from its input signal. **2.** A circuit or device utilized to vary the phase relationship between two alternating currents. For instance, that which changes the phase angles between the components of a two-phase system. **3.** A device which changes the electrical length of a waveguide. Also called **phaser (3)**.

phase-shifting transformer A transformer which introduces a phase shift between secondary voltages relative to the primary voltages.

phase speed Same as **phase velocity**.

phase splitter Also called **phase-splitting circuit**. **1.** A circuit with a single input, and whose output consists of two signals which are 180° out-of-phase with each other. **2.** A circuit with a single input, and whose output consists of two signals which are out-of-phase with each other. **3.** Same as **paraphase amplifier**.

phase-splitting circuit Same as **phase splitter**.

phase transformer A transformer which converts, utilizes, or produces single-phase or multiple-phase current. An example is a three-phase transformer.

phase velocity The speed at which a wave whose phase is constant propagates through a given medium. It is the velocity at which given points within it, such as crests and troughs, travel. In a vacuum it is equal to the product of the frequency and wavelength. In a dispersive medium the phase velocity of electromagnetic waves varies as a function of the frequency, while in a non-dispersive medium the phase velocity is independent of the frequency. Also called **phase speed**.

phase voltage **1.** In a polyphase system, the voltage of a given phase. **2.** In a polyphase system, the potential difference between individual phases.

phase winding In a polyphase device or machine, such as a generator or motor, one of the individual windings, each of which has a different phase.

phased antenna array Same as **phased array**.

phased antennas Same as **phased array**.

phased array An array of antennas, or antenna elements, which are simultaneously fed, and whose relative phase can be varied electronically. An example is such an array of dipoles. Used, for instance, to produce multiple simultaneous beams, or to integrate signal information received from its spatially separated elements. Also called **phased array antenna**, **phased antenna array**, or **phased antennas**.

phased array antenna Same as **phased array**.

phased array radar A radar system utilizing a **phased array**. Utilized, for instance, for simultaneously tracking many scanned objects. Also called **array radar**.

phasemeter Same as **phase meter**.

phaser **1.** A circuit, device, or mechanism utilized to vary or synchronize the phase of two signals. **2.** In a fax, a device which synchronizes the transmitting unit and the receiving unit, so as to insure that the transmitted images have the same relative positions on the received pages. **3.** Same as **phase shifter (3)**.

phasing **1.** The utilization of a circuit, device, or mechanism to vary or synchronize the phase of two signals. Also, the synchronization so obtained. **2.** In a fax, the use of a device which synchronizes the transmitting unit and the receiving unit, so as to insure that the transmitted images have the same relative positions on the received pages. Also, the synchronization so obtained.

PHEMT Abbreviation of **pseudomorphic high-electron mobility transistor**.

phenol A white or colorless solid whose chemical formula is C_6H_5OH, and which is used, for instance, in thermosetting plastics. Also called **carbolic acid**.

phenol-formaldehyde Same as **phenol-formaldehyde resin**.

phenol-formaldehyde resin A thermosetting synthetic resin used as a dielectric or for making hard plastic objects, such as knobs, utilized in electronics. Also called **phenol-formaldehyde**.

phenolic resin A resin, such as a phenol-formaldehyde resin, in which phenol is polymerized with another substance.

phenomena Within scientific usage, the proper plural form of **phenomenon**.

phenomenon Its proper plural form within scientific usage is **phenomena**, although phenomenons is sometimes used. **1.** An event, circumstance, or characteristic which can be perceived by the senses. For example, an acoustic or optical phenomenon. **2.** An event, circumstance, or characteristic which can be observed and/or recorded utilizing instruments and devices. For instance, electromagnetic phenomena. **3.** An event, circumstance, or characteristic which can be theorized upon, without necessarily being able to be observed and/or recorded. For example, the passing of an interval of time beyond the measuring ability of the most precise instrument.

Phi phenomenon The illusion of movement perceived when presented with the same, or essentially the same, object in different places in rapid succession. This phenomenon, along with **persistence of vision**, enables a movie or TV presentation to be perceived as consisting of continuous motion, even though they are comprised of frames.

phon A unit which takes into account the non-linear response of human hearing in describing loudness levels of sounds or noises. To define a phon scale, a listener is subjected to a reference tone of 1 kHz at a sound pressure of 2.0×10^{-5} pascal, while listening to another sound alternately. The reference tone is increased to match the sound, and the number of decibels the reference tone needs to be augmented is equated to the number of phons. That is, an increase in 20 decibels of the reference tone means that the other sound has an intensity of 20 phons.

phon scale A sound-measurement scale based on **phons**. The phon scale, due to its subjective nature, is not necessarily equal to the decibel scale, which is objective.

phone **1.** Abbreviation of **telephone**. **2.** Abbreviation of **headphone**.

phone amplifier Abbreviation of **telephone amplifier**.

phone answering machine Abbreviation of **telephone answering machine**.

phone communications Abbreviation of **telephone communications**.

phone connector **1.** Same as **phone plug**. **2.** Same as **phone jack**.

phone dialer Abbreviation of **telephone dialer**.

phone feature Abbreviation of **telephone feature**.

phone handset Abbreviation of **telephone handset**.

phone jack Also called **phone connector (2)**, or **phone socket**. **1.** An electric connector with a recessed opening into which a **phone plug (1)** with a matching prong is inserted. **2.** Abbreviation of **telephone jack**.

phone line Abbreviation of **telephone line**.

phone link Abbreviation of **telephone link**.

phone network Abbreviation of **telephone network**.

phone number Abbreviation of **telephone number**.

phone pickup Abbreviation of **telephone pickup**.

phone plug Also called **phone connector (1)**. **1.** A two- or three-conductor plug whose diameter is usually 0.25 or 0.125 inches. It consists of a single rod which has two or three contacts, each separated by an insulator. Two-conductor plugs are used for mono signals, and three-conductor plugs for stereo. Widely utilized to connect audio components, such as headphones and microphones. A phone plug is plugged into a **phone jack (1)**. **2.** Abbreviation of **telephone plug**.

phone receiver Abbreviation of **telephone receiver**.

phone repeater Abbreviation of **telephone repeater**.

phone service Abbreviation of **telephone service**.

phone set Abbreviation of **telephone set**.

phone socket Same as **phone jack**. Abbreviation of **telephone socket**.

phone station Abbreviation of **telephone station**.

phone subscriber Abbreviation of **telephone subscriber**.

phone system Abbreviation of **telephone system**.

phone transmitter Abbreviation of **telephone transmitter**.

phoneme For a given natural language, one speech sound within the complete set. Each of these sounds, such as *p* in *pat* and *b* in *bat*, and help distinguish different words. Used, for instance, in speech recognition and synthetic speech systems.

phonetic alphabet A set of standardized words utilized to identify the letters of a given alphabet. For instance, a collection of easily identifiable words which help ensure that spelling during a voice communication is not misunderstood. In radiotelephony, for example, words such as Delta, for D, and Kilo, for K, are used to distinguish letters.

phonic Of, pertaining to, or arising from sound, especially that of speech.

phono- A prefix used in words pertaining to sound or voice. For instance, phonocardiography, or phonograph.

phono cartridge In a phonograph, an electromechanical transducer whose stylus vibrates as it follows the grooves of a record, and which converts these vibrations into electric signals. There are various types, including moving coil and piezoelectric, and each attaches to a phonograph arm. Also called **phonograph pickup**, **phonograph cartridge**, **phono pickup**, **pickup (2)**, **pickup cartridge**, or **cartridge (4)**.

phono connector Also called **RCA connector**. 1. Same as **phono plug**. 2. Same as **phono jack**.

phono jack An electric connector with a center opening into which a **phono plug** with a matching prong is inserted. Also called **phono connector (2)**, **phono socket**, or **RCA jack**.

phono pickup Same as **phono cartridge**.

phono plug A plug with a prong whose diameter is usually 0.125 inches, and whose outer sleeve is 0.375 inches in diameter. The shell is connected to the outer conductor of the attached coaxial cable, and is often slotted to further facilitate connection and disconnection. The inner prong is connected to the center conductor, which carries the signal. Widely utilized to connect audio and video components, such as DVDs, CDs, TVs, and amplifiers. A phono plug is plugged into a **phono jack**. Also called **phono connector (1)**, or **RCA plug**.

phono socket Same as **phono jack**. Also called **RCA socket**.

phonocardiogram A displayed or recorded image of a **phonocardiography**.

phonocardiograph An instrument utilized to produce **phonocardiogram**.

phonocardiography A medical diagnostic technique utilized to detect and record heart sounds, such as murmurs. The instrument utilized for this procedure is a **phonocardiograph**, and the recorded images are called **phonocardiograms**.

phonocatheter A catheter equipped with a microphone. Such a device is utilized, for instance, to listen to heart sounds with reduced interference from any background noises or patient movements, as may occur with a stethoscope.

phonograph An obsolescent device which reproduces sound which has been recorded in the spiral grooves of a disc called phonograph record. A phonograph incorporates a motor-driven turntable which rotates a platter at a constant speed. The disk is placed on the platter, and a tone arm, which holds the cartridge, travels progressively towards the center of the disk. The stylus of the cartridge follows the

variations in the grooves, and converts them into electrical signals which are fed into an amplifier, preamplifier, receiver, or other such device. Also called **phonograph turntable (1)**, **record player**, or **turntable (1)**.

phonograph cartridge Same as **phono cartridge**.

phonograph disc Same as **phonograph record**.

phonograph needle A pointed object which extends from a phonographic pickup, and which serves to follow the undulations of the grooves of a phonographic disc and transmit them as vibrations to said pickup. Such an object is usually made of a metal needle with a diamond or sapphire tip. Also called **phonograph stylus**, **needle (3)**, or **stylus (3)**.

phonograph pickup Same as **phono cartridge**.

phonograph record A disc, usually composed of vinyl, which has sound recorded as undulations in its spiral grooves. A phonograph is utilized to extract the sound so recorded. Also called **phonographic record**, **phonograph disc**, **phonographic disc**, or **record (4)**.

phonograph stylus Same as **phonograph needle**.

phonograph turntable Also called **turntable**. 1. Same as **phonograph**. 2. The rotating platform which rotates the platter at a constant speed in a phonograph.

phonographic Pertaining to a **phonograph**, or that incorporated in a phonograph, such as a cartridge.

phonographic disc Same as **phonograph record**.

phonographic record Same as **phonograph record**.

phonon The quantum of lattice vibrational energy in a crystalline solid, defined as hf, where h is the Planck constant, and f is the frequency of the vibration. It is analogous to a photon, which is the quantum of electromagnetic radiation.

phosphine A colorless, poisonous, and spontaneously flammable gas whose chemical formula is PH_3. Also called **hydrogen phosphide**.

phosphor A material which is capable of luminescence. Used, for instance, for phosphor coatings. Also called **phosphor material**, **phosphorescent material**, or **luminescent material**.

phosphor bronze A copper-based alloy with anywhere from 1 to 10% tin, and up to 1% phosphorus. It is used in electrical switches and contacts.

phosphor coating A surface, such as that on the inner surface of a CRT, composed of a phosphor material such as cerium or lanthanum. In a CRT, each dot of this material glows when struck by focused electrons delivered by one or more electron guns. Also called **phosphor surface**, **phosphorescent coating**, or **phosphorescent surface**.

phosphor material Same as **phosphor**.

phosphor pitch On a display screen, the distance between phosphor dots of like color. For example, such an interval between adjacent green dots. Usually expressed in millimeters, and the lower the number, the crisper the image. Also called **dot pitch**.

phosphor screen A screen, such as that of a CRT, which has a **phosphor coating**. Also called **phosphorescent screen**.

phosphor surface Same as **phosphor coating**.

phosphorescence The luminescence that remains after the exciting source is removed. For instance, the phosphors on the screens of CRTs glow after the electron beam has passed. Also called **afterglow**.

phosphorescent coating Same as **phosphor coating**.

phosphorescent material Same as **phosphor**.

phosphorescent screen Same as **phosphor screen**.

phosphorescent surface Same as **phosphor coating**.

phosphoric acid A transparent crystalline solid or a colorless liquid whose chemical formula is H_3PO_4. It is used, for instance, as an etchant, in ceramics, and as an electrolyte.

phosphoric acid fuel cell A fuel cell which utilizes hydrogen and oxygen as the reactants, has carbon electrodes, a phosphoric acid electrolyte, and a platinum catalyst. The products of such a cell are electricity, heat, and water.

phosphorous Of, pertaining to, or containing **phosphorus**.

phosphorous oxychloride A colorless fuming liquid whose chemical formula is $POCl_3$. It is used as a semiconductor dopant.

phosphorus A nonmetallic chemical element whose atomic number is 15. It has three allotropic forms, one consists of white or colorless crystals, another of a black solid resembling graphite, and the third is a red to violet powder. White phosphorous is poisonous, ignites spontaneously in air, and is phosphorescent. Red phosphorous is more stable and is used in semiconductors and electroluminescent coatings. Black phosphorous is electrically conductive. Phosphorus has about 20 known isotopes, of which one is stable. Its chemical symbol is **P**.

phot The cgs unit of illumination. It is equal to approximately 10,000 lux. Its abbreviation is **ph**.

photo- A prefix used in words pertaining to light and images, as in photography, photoelectric, photonuclear, and so on.

photo-acoustic effect Same as **photoacoustic effect**.

photo-acoustic spectroscopy Same as **photoacoustic spectroscopy**.

photo-anode Same as **photoanode**.

photo-cathode Same as **photocathode**.

Photo CD A system or format for storing photographs on CDs. Also, a disc with such photos stored.

photo-cell Same as **photocell**.

photo-cell amplifier Same as **photocell amplifier**.

photo-chemical detection Same as **photochemical detection**.

photo-chemical reaction Same as **photochemical reaction**.

photo-conduction Same as **photoconduction**.

photo-conductive Same as **photoconductive**.

photo-conductive cell Same as **photoconductive cell**.

photo-conductive device Same as **photoconductive device**.

photo-conductive material Same as **photoconductive material**.

photo-conductivity Same as **photoconductivity**.

photo-conductor Same as **photoconductor**.

photo-coupler Same as **photocoupler**.

photo-current Same as **photocurrent**.

photo-Darlington Same as **photodarlington**.

photo-decomposition Same as **photodecomposition**.

photo-detector Same as **photodetector**.

photo-device Same as **photodevice**.

photo-diffusion effect Same as **photodiffusion effect**.

photo-diode Same as **photodiode**.

photo-disintegration Same as **photodisintegration**.

photo editing **1.** Changes made to a photo for any given purpose. This may be accomplished, for instance, through the use of image processing. **2.** The use of a **photo editor** to make changes in images.

photo editing program Same as **photo editor**.

photo editor An application which is utilized to digitally modify photographs. Such programs usually have filters that enable them to read and convert many graphics formats, and have other features for enhancing and otherwise altering such images. Also called **photo editing program**.

photo-elastic Same as **photoelastic**.

photo-elastic effect Same as **photoelastic effect**.

photo-elastic material Same as **photoelastic material**.

photo-elasticity Same as **photoelasticity**.

photo-electric Same as **photoelectric**.

photo-electric absorption Same as **photoelectric absorption**.

photo-electric alarm Same as **photoelectric alarm**.

photo-electric cell Same as **photoelectric cell**.

photo-electric color comparator Same as **photoelectric color comparator**.

photo-electric colorimeter Same as **photoelectric colorimeter**.

photo-electric constant Same as **photoelectric constant**.

photo-electric control Same as **photoelectric control**.

photo-electric counter Same as **photoelectric counter**.

photo-electric current Same as **photoelectric current**.

photo-electric detector Same as **photoelectric detector**.

photo-electric device Same as **photoelectric device**.

photo-electric effect Same as **photoelectric effect**.

photo-electric emission Same as **photoelectric emission**.

photo-electric material Same as **photoelectric material**.

photo-electric photometer Same as **photoelectric photometer**.

photo-electric photometry Same as **photoelectric photometry**.

photo-electric process Same as **photoelectric process**.

photo-electric relay Same as **photoelectric relay**.

photo-electric sensor Same as **photoelectric sensor**.

photo-electric smoke detector Same as **photoelectric smoke detector**.

photo-electric switch Same as **photoelectric switch**.

photo-electric threshold Same as **photoelectric threshold**.

photo-electric transducer Same as **photoelectric transducer**.

photo-electric tube Same as **photoelectric tube**.

photo-electric work function Same as **photoelectric work function**.

photo-electricity Same as **photoelectricity**.

photo-electrode Same as **photoelectrode**.

photo-electromotive force Same as **photo-emf**. Also spelled **photoelectromotive force**.

photo-electron Same as **photoelectron**.

photo-electron spectroscopy Same as **photoelectron spectroscopy**.

photo-emf Abbreviation of **photoelectromotive force**. The electromotive force produced via photovoltaic action, such as that of a photocell.

photo-emission Same as **photoemission**.

photo-emissive material Same as **photoemissive material**.

photo-emitter Same as **photoemitter**.

photo-fabrication Same as **photofabrication**.

photo-flash Same as **photoflash**.

photo-fluorography Same as **photofluorography**.

photo-generation Same as **photogeneration**.

photo-ionization Same as **photoionization**.

photo-isolator Same as **photoisolator**.

photo-lithographic process Same as **photolithographic process**.

photo-lithography Same as **photolithography**.

photo-magnetic effect Same as **photomagnetic effect**.

photo-mask Same as **photomask**.

photo-mosaic Same as **photomosaic**.

photo-multiplier Same as **photomultiplier**.

photo-multiplier tube Same as **photomultiplier tube**.

photo-negative Same as **photonegative**.

photo-neutron Same as **photoneutron**.

photo-nuclear reaction Same as **photonuclear reaction**.

photo-phone Same as **photophone**.

photo-positive Same as **photopositive**.

photo printer A device which prints photographs, especially from a digital source such as those that are scanned or taken by a digital camera. Such a printer usually works with an image editor, and its output tends to be highly realistic.

photo-proton Same as **photoproton**.

photo-pumping Same as **photopumping**.

photo-relay Same as **photorelay**.

photo-resist Same as **photoresist**.

photo-resistance Same as **photoresistance**.

photo-resistive cell Same as **photoresistive cell**.

photo-resistor Same as **photoresistor**.

photo scanner A scanner specially designed to convert the images of printed photographs into bitmaps.

photo SCR Abbreviation of **photo** silicon-controlled rectifier. A silicon-controlled rectifier which has a light-sensitive region which enables incident light to control its switching action. Such a device may also be controlled by a gate. Also called **photothyristor**, or **light-activated silicon-controlled rectifier**.

photo-sensitive Same as **photosensitive**.

photo-sensitive device Same as **photosensitive device**.

photo-sensitive material Same as **photosensitive material**.

photo-sensitivity Same as **photosensitivity**.

photo-sensor Same as **photosensor**.

photo-switch Same as **photo-switch**.

photo-thyristor Same as **photo SCR**. Also spelled **photo-thyristor**.

photo-transistor Same as **phototransistor**.

photo-tube Same as **phototube**.

photo-typesetter Same as **phototypesetter**.

photo-voltaic Same as **photovoltaic**.

photo-voltaic array Same as **photovoltaic array**.

photo-voltaic cell Same as **photovoltaic cell**.

photo-voltaic device Same as **photovoltaic device**.

photo-voltaic effect Same as **photovoltaic effect**.

photo-voltaic material Same as **photovoltaic material**.

photo-voltaic module Same as **photovoltaic module**.

photo-voltaic panel Same as **photovoltaic panel**.

photo-voltaic system Same as **photovoltaic system**.

photoacoustic effect For a given material or medium, the generation of an acoustic signal when exposed to modulated light. Certain semiconductors exhibit this effect. Also spelled **photo-acoustic effect**.

photoacoustic spectroscopy A spectroscopic technique which provides an optical spectrum, especially infrared, based on the sounds produced when a sample is irradiated with light. A highly-sensitive microphone is utilized as the detector. Used, for instance, to analyze below the surface of opaque samples. Its abbreviation is **PAS**. Also spelled **photo-acoustic spectroscopy**.

photoanode An anode which emits electrons from its surface when exposed to electromagnetic radiation, especially light. Also spelled **photo-anode**.

photocathode A cathode which emits electrons from its surface when exposed to electromagnetic radiation, especially light. This is due to the **photoelectric effect (1)**. Photocathodes are used in light amplifiers, photomultiplier tubes, image orthicons, and electron telescopes, among others. Also spelled **photo-cathode**.

PhotoCD Same as **Photo CD**.

photocell Abbreviation of **photo**electric **cell**. A device whose electrical characteristics, such as voltage, current, resistance, or capacitance, vary with incident light. Such incident light may be infrared, visible, or ultraviolet, and the variations in the electrical quantities will depend on the frequency of the light, and the specific photosensitive substance being used. Such devices generally incorporate a light-sensitive semiconductor material, and are usually utilized as a transducer which converts light energy into electrical energy. When employed as a detector, such as that used in automatic door openers or intrusion alarms, also called **electric eye**. Its abbreviation is **PC**. Also spelled **photo-cell**.

photocell amplifier A device utilized to amplify the signal of a **photocell**. Also spelled **photo-cell amplifier**.

photochemical detection The detection and measurement of electromagnetic radiation produced through a **photochemical reaction**. Also spelled **photo-chemical detection**.

photochemical reaction A chemical reaction caused, or greatly influenced, by light, such as ultraviolet radiation. Also spelled **photo-chemical reaction**.

photochromic Pertaining to, or arising from, **photochromic materials**.

photochromic glass A glass made of, or incorporating, a **photochromic material**.

photochromic material A material which changes color when exposed to light of a given wavelength, and which returns to its original color once the light source is removed. Said especially of light in the visible, near infrared, or near ultraviolet regions. Such a material, for example, may also change from transparent to a given color, and vice-versa when exposed to the appropriate light. Used, for instance, in optical switches, optical filters, optical modulation devices, lenses, optical recording media, charge-coupled devices, and for energy-efficient windows and display panels.

photochromic plastic A plastic made of, or incorporating, a **photochromic material**.

photoconduction An increase the electrical conduction of a material, due to the absorption of electromagnetic radiation, such as light. In such materials, an increase in the intensity of the incident light results in a decrease in resistance. Seen, for instance, in certain semiconductors. Also spelled **photoconduction**.

photoconductive Pertaining to, or exhibiting **photoconduction**. Also spelled **photo-conductive**.

photoconductive cell A photocell, such as a selenium cell, whose electrical resistance varies as a function of the intensity of light incident upon it. Also spelled **photoconductive cell**. Also called **photoresistor**, **photoresistive cell**, or **light-dependent resistor**.

photoconductive device A device, such as a photoconductive cell, whose electrical resistance varies as a function of the intensity of light incident upon it. Used, for instance, to measure luminous intensity. Also spelled **photo-conductive device**.

photoconductive material Same as **photoconductor**. Also spelled **photo-conductive material**.

photoconductivity A variation in the electrical conductivity of a material as a function of the intensity of incident electromagnetic radiation, such as light. Observed, for instance, in certain semiconductors, where photons provide electrons in the valence band with enough energy to pass into the conduction band. Also spelled **photo-conductivity**. Also called **photoresistance**.

photoconductor A material, such as cadmium sulfide, which exhibits **photoconduction**. Also spelled **photo-conductor**. Also called **photoconductive material**.

photocoupler A semiconductor device which allows electrical signals to be transferred between circuits, devices, or systems, while maintaining electrical isolation between them. A photocoupler may consist, for instance, of an LED or IRED for signal transmission, and a photodetector, such as a photodiode, for signal reception. The signal transmitter converts the electrical signal into an optical signal, while the receiver restores the original electrical signal. Used, for instance, in communications, and in control systems. Also spelled **photo-coupler**. Also called **photoisolator, optical coupler, optical isolator, optocoupler, optically-coupled isolator**, or **optoelectronic isolator**.

photocurrent An electric current which is produced by incident radiant energy, such as light. It is the current that flows through photoelectric devices, and may result, for instance, from the photoelectric effect or the photovoltaic effect. Also spelled **photo-current**.

photodarlington A Darlington amplifier whose input transistor is a phototransistor. Also spelled **photo-Darlington**.

photodecomposition Chemical breakdown through exposure to radiant energy, such as light. Also spelled **photo-decomposition**.

photodetector Also spelled **photo-detector**. Also called **photosensor, photodevice, photoelectric detector, light sensor, light detector**, or **light-sensitive detector**. **1.** A device which detects and responds to light. **2.** A transducer whose input is light, and whose output is a corresponding electrical signal. Light sensors are usually semiconductor devices, such as photodiodes.

photodevice Same as **photodetector**. Also spelled **photo-device**.

photodiffusion effect In a semiconductor, the production of a potential difference between two surfaces or regions when one of them is illuminated. Also spelled **photo-diffusion effect**. Also called **Dember effect**.

photodiode A semiconductor diode which converts light into an electrical signal, such as a photocurrent. The reverse current of such a diode is regulated by the intensity of the incident light. Used, for instance, in fiber-optics to convert optical power into electrical power. An avalanche photodiode is an example. Also spelled **photo-diode**. Also called **light-sensitive diode**.

photodisintegration The disintegration or transformation of an atomic nucleus due to bombardment by high-energy radiation, such as gamma rays. Also spelled **photo-disintegration**.

photoelastic Said of a material which exhibits **photoelasticity**. Also spelled **photo-elastic**.

photoelastic effect Same as **photoelasticity**. Also spelled **photo-elastic effect**.

photoelastic material A material, such as that constituting certain epoxy resins, which exhibits **photoelasticity**. Also spelled **photo-elastic material**.

photoelasticity The change in certain optical properties of a material when subjected to mechanical stress. This occurs, for instance, in certain transparent dielectrics, and can result in birefringence. The varying of the plane of polarization of a light wave passing through such a material, as a function of the applied stress, is another example. Also spelled **photo-elasticity**. Also called **photoelastic effect**.

photoelectric Pertaining to, or arising from phenomena involving the electrical effects of electromagnetic radiation, especially light. Examples include photoconduction and photoemission. Also spelled **photo-electric**.

photoelectric absorption For a given material, the absorption of photons resulting in any of the **photoelectric effects**. Also spelled **photo-electric absorption**.

photoelectric alarm An alarm, such as an intrusion alarm, which utilizes a photoelectric device such as an electric eye. Also spelled **photo-electric alarm**.

photoelectric cell Same as **photocell**. Also spelled **photo-electric cell**.

photoelectric color comparator An instrument which uses one or more photoelectric devices to compare a color with a given standard. Also spelled **photo-electric color comparator**. Also called **color comparator**.

photoelectric colorimeter A colorimeter incorporating one or more **photoelectric devices**. Also spelled **photo-electric colorimeter**.

photoelectric constant The ratio of the Planck constant to the electron charge. Also spelled **photo-electric constant**.

photoelectric control The incorporation of a photoelectric device, such as a photocell, in a control system. Also spelled **photo-electric control**.

photoelectric counter A counting device which incorporates a photoelectric device, such as a photocell. Such a counter may work, for instance, by counting the times a light path is interrupted by objects being counted. Also spelled **photo-electric counter**.

photoelectric current A current resulting from any of the **photoelectric effects**. Also spelled **photo-electric current**.

photoelectric detector Same as **photodetector**. Also spelled **photo-electric detector**.

photoelectric device A device, such as a photocell or a phototube, whose function involves the electrical effects of electromagnetic radiation, especially light. Also spelled **photo-electric device**.

photoelectric effect Also spelled **photo-electric effect**. **1.** The emission of electrons from a material exposed to electromagnetic radiation, especially in the infrared, visible, and ultraviolet regions. The frequency of the radiation must exceed a given minimum, or threshold frequency, for electrons to be emitted. Beyond this value, the energy of the emitted electrons depends on the frequency of the radiation. The greater the intensity of the incident radiation, the greater the number of liberated electrons. **2.** Any observed effect in which electromagnetic radiation, especially light, induces changes in the electrical properties of materials, surfaces, or regions exposed to it. These include the photoconductive effect, the photovoltaic effect, and the photodiffusion effect.

photoelectric emission The emission of electrons due to any of the **photoelectric effects**. Also spelled **photo-electric emission**.

photoelectric material A material, such as cesium or potassium, exhibiting any of the **photoelectric effects**. Also spelled **photo-electric material**.

photoelectric photometer A photometer which incorporates an electronic device, such as a phototube or phototransistor, for measuring the intensity of light. Also spelled **photo-electric photometer**. Also called **electronic photometer**.

photoelectric photometry The science and utilization of **photoelectric photometers** for sensing and analysis. Also spelled **photo-electric photometry**.

photoelectric process A process, such as the emission of electrons through the absorption of photons, due to any of the **photoelectric effects**. Also spelled **photo-electric process**.

photoelectric relay A relay in which the intensity of incident light, such as that of a modulated light beam, determines its opening or closing of a circuit. A photoelectric relay incorporates a photoelectric device, such as a photocell. Also spelled **photo-electric relay**. Also called **photorelay**, or **light relay**.

photoelectric sensor A sensor, such as that used in an electric eye, incorporating a **photoelectric device**. Also spelled **photo-electric sensor**.

photoelectric smoke detector A smoke detector that incorporates a **photoelectric device**. In such a device, for instance, a light source and a photosensitive device may be arranged in a manner that the rays from said source ordinarily do not strike the sensor. When particles of smoke cross the path of the light rays, light scattered by said smoke makes some of the light fall upon the sensor. Alternatively, the light beam may be directed at the sensor, and light scattered by the smoke reduces the intensity of the light reaching the detector. Also spelled **photo-electric smoke detector**. Also called **projected-beam smoke detector**.

photoelectric switch A switch in which the presence of light, or the intensity of incident light such as that of a modulated light beam, determines the opening or closing of a circuit. A photoelectric switch incorporates a photoelectric device, such as a photocell. An example is a photo SCR. Also spelled **photo-electric switch**. Also called **photoswitch**, **light-activated switch**, or **light-operated switch**.

photoelectric threshold For a given material, the minimum frequency of incident radiation leading to the emission of electrons via the **photoelectric effect** (1). Also spelled **photo-electric threshold**.

photoelectric transducer Also spelled **photo-electric transducer**. **1.** A transducer which converts light energy into electrical energy. **2.** A transducer which utilizes any of the **photoelectric effects** to convert light energy into electrical energy.

photoelectric tube Same as **phototube**. Also spelled **photo-electric tube**.

photoelectric work function The minimum energy required by an incident photon to eject an electron from a given material, especially a metal. Usually expressed in electron-volts. Also spelled **photo-electric work function**.

photoelectricity Also spelled **photo-electricity**. **1.** Electricity arising from one of the **photoelectric effects**. **2.** The study and applications of one or more photoelectric effects.

photoelectrode An electrode which emits electrons from its surface when exposed to electromagnetic radiation, especially light. Such an electrode is usually a photocathode. Also spelled **photo-electrode**.

photoelectromotive force Same as **photo-emf**. Also spelled **photo-electromotive force**.

photoelectron Any electron that is emitted from a material or surface which has absorbed sufficient incident electromagnetic radiation, such as infrared, visible, or ultraviolet light. Also spelled **photo-electron**.

photoelectron spectroscopy A spectroscopic technique which studies the electrons ejected from the surface of atoms or molecules via photoionization, to determine the composition and electronic state of a sample material. The two main methods utilized are X-ray photoelectron spectroscopy and ultraviolet photoelectron spectroscopy, each named after the source of exciting radiation. Also spelled **photo-electron spectroscopy**.

photoemission The ejection of electrons from a surface through the absorption of sufficient incident electromagnetic radiation, such as infrared, visible, or ultraviolet light. Said surface is usually a solid, but may be a liquid. Also spelled **photo-emission**. Also called **external photoelectric effect**.

photoemissive material A material, such as cesium, which exhibits **photoemission**. Used, for instance, in photocathodes. Also spelled **photo-emissive material**. Also called **photoemitter**.

photoemitter Same as **photoemissive material**. Also spelled **photo-emitter**.

photofabrication The manufacturing of components, circuits, and devices, such as ICs and printed-circuit boards, utilizing **photolithography**. Also spelled **photo-fabrication**.

photoflash A repeatable and artificially produced burst of bright light. Such a flash is usually generated by applying a high voltage to an electrode of a tube containing an inert gas such as xenon. The gas becomes ionized, which permits it to rapidly discharge the energy previously stored in a capacitor. Used, for instance, in photography. Also spelled **photo-flash**. Also called **electronic flash**, **flash** (2), or **strobe** (3).

photofluorography The photographic recording of an image formed on a fluorescent screen. Also spelled **photo-fluorography**. Also called **fluorography**.

photogeneration In a semiconductor material, the generation of charge carriers due to incident electromagnetic radiation, such as infrared, visible, or ultraviolet light. Also spelled **photo-generation**.

photograph An image or other visual representation which is recorded with a camera, and subsequently reproduced on a suitable surface, such as a flat sheet, or through a display device, such as a computer monitor. Also called **picture** (2).

photographic Of, pertaining to, or utilized in photography or photographs.

photographic sound recorder A photoelectric device utilized for **photographic sound recording**. Also called **optical sound recorder**.

photographic sound recording A technique in which a light beam modulated by an audio source is utilized to record a sound track on photographic film. Also called **optical sound recording**.

photographic sound reproducer A photoelectric device utilized to reproduce a **photographic sound recording**.

photography The techniques and procedures involved in obtaining, processing, and reproducing photographs.

photoionization The removal of one or more electrons from an atom or molecule as a consequence of light absorption. Also spelled **photo-ionization**.

photoisolator Same as **photocoupler**. Also spelled **photo-isolator**.

photolithographic process Same as **photolithography**. Also spelled **photo-lithographic process**.

photolithography A lithographic technique in which precise patterns are created on substrates, such as metals or resins, through the use of photographically-produced masks. Typically, a substrate is coated with a photoresist film, which is dried or hardened, and then exposed through irradiation by light, such as ultraviolet light, shining through the photomask. The unprotected areas are then removed, usually through etching, which leaves the desired patterns. Used, for instance, in the preparation of components, circuits, and devices, such as thin-film components, ICs, and printed-circuit boards. Also spelled **photo-lithography**. Also called **photolithographic process**.

photoluminescence Luminescence resulting from the absorption of light, especially that in the visible, infrared, or ultraviolet ranges.

photolysis Chemical decomposition through exposure to radiant energy, such as light.

photomagnetic effect A change in the magnetic susceptibility of certain materials due to variations of incident light. Also spelled **photo-magnetic effect**.

photomask A mask utilized to selectively shield portions of materials or substrates, such as semiconductor wafers, during photolithography. Also spelled **photo-mask**.

photometer **1.** An instrument utilized to measure and indicate luminous intensity. **2.** An instrument utilized to measure and indicate one or more photometric quantities, such as luminous intensity, illuminance, or color.

photometric Pertaining to the measurement of the properties, such as luminous intensity, of visible light.

photometry The science dealing with the measurement of the properties of visible light. Instruments and devices, such as photometers, are used for this purpose. When also dealing with infrared and ultraviolet light, also called **radiometry** (1).

photomicrography The photographing of small or extremely small objects. This is usually accomplished utilizing a microscope fitted with a camera.

photomosaic Also spelled **photo-mosaic**. 1. A material, usually a thin mica sheet, which on one side has a light-sensitive surface, and on the other a photoemissive surface. It is used in TV camera tubes, where the optical image is focused upon the first surface, and the equivalent electrical charge patterns produced on the other face are scanned by the electron beam of said tube. Also called **mosaic** (1). 2. The pattern formed by the photosensitive globules of a **photomosaic** (1).

photomultiplier Same as **photomultiplier tube**. Also spelled **photo-multiplier**.

photomultiplier tube A phototube incorporating one or more dynodes. Electrons emitted from its photocathode are reflected by the dynodes, each producing more and more electrons. Depending on the number of dynodes, the amplification factor may be in the millions. Also spelled **photomultiplier tube**. Its abbreviation is **PMT**. Also called **photomultiplier**, **electron-multiplier phototube**, or **multiplier phototube**.

photon A particle with no mass nor charge, and which composes light and other forms of electromagnetic radiation. Gamma rays and X-rays are examples of higher-energy photons, while infrared rays and radio waves are examples of lower-energy forms. Its energy is equal to hf, where h is the Planck constant, and f is the frequency of the radiation in Hz. Photons are required to explain phenomena such as the photoelectric effect, in which light has particle-like properties. The photon is the carrier of the electromagnetic force, which is the force responsible for all electromagnetic interactions, such as those which hold atoms and molecules together. Also called **light quantum**.

photon fluence The passing of photons through a given area, such as a square centimeter. May be expressed, for instance, in photons per second.

photon flux The number of photons striking a given surface area perpendicular to the path of a light beam, per unit time. For instance, the number of photons striking the photocathode of a photomultiplier tube, per unit time.

photonegative Also spelled **photo-negative**. 1. Pertaining to a photoconductor which undergoes a decrease in electrical conductivity when exposed to light. Also called **light-negative**. 2. Pertaining to that which undergoes a decrease in a given property when exposed to light, or which otherwise has a negative response to light. 3. An image, such as a photographic image, in which the light and dark areas are reversed in relation to the scene the image is taken from. Also called **negative** (5), **negative image**, or **reverse image**.

photoneutron A neutron released during a **photonuclear reaction**. Also spelled **photo-neutron**.

photonic Pertaining to **photons** or **photonics**.

photonic device A device whose operation is based on **photonics**.

photonic transistor A transistor in which light signals are utilized to perform logic functions, as opposed to electrical signals. The switching power for such a transistor may be in the nanowatt range, with switching action in the picosecond range.

photonics The science dealing with the generation and manipulation of radiant energy, such as light. It includes amplification, modulation, emission, transmission, deflection, detection, and reception. Photonics combines elements from various realms, including optics, electronics, and chemistry, and specific applications include lasers, communications, data storage and retrieval, spectroscopy, printing, and energy generation.

photonuclear reaction A nuclear reaction caused by a colliding photon. Also spelled **photo-nuclear reaction**.

photophone Also spelled **photo-phone**. 1. A telephone that transmits still pictures in addition to voice. This contrasts with a **videophone**, which also transmits full-motion video. 2. An early telephone transmission system utilizing modulated light.

photopositive Also spelled **photo-positive**. 1. Pertaining to a photoconductor which undergoes an increase in electrical conductivity when exposed to light. Also called **light-positive**. 2. Pertaining to that which undergoes an increase in a given property when exposed to light, or which otherwise has a positive response to light. 3. An image, such as a photographic image, in which the light and dark areas are the same as the scene the image is taken from. Also called **positive** (5), or **positive image**.

photoproton A proton released during a **photonuclear reaction**. Also spelled **photo-proton**.

photopumping In a laser, the utilization of incident light to excite electrons, molecules, or ions to higher energy levels. Also spelled **photo-pumping**. Also called **optical pumping**.

photorelay Same as **photoelectric relay**. Also spelled **photo-relay**.

photoresist A photosensitive material which is applied to a substrate during photolithography, to enable a desired pattern to be imaged. A properly exposed and developed photoresist masks portions of the substrate with extremely high precision. A photoresist that polymerizes or otherwise turns harder when exposed to light is a negative photoresist, while a positive photoresist depolymerizes or otherwise turns softer when exposed to light. Also spelled **photoresist**.

photoresistance Same as **photoconductivity**. Also spelled **photo-resistance**.

photoresistive cell Same as **photoconductive cell**. Also spelled **photo-resistive cell**.

photoresistor Same as **photoconductive cell**. Also spelled **photo-resistor**.

photosensitive Stimulated, actuated, or otherwise responsive to light. That which is photosensitive has one or more physical and/or chemical properties vary in response to light. Also spelled **photo-sensitive**. Also called **light-sensitive**.

photosensitive device A device, such as a photodiode, photoswitch, or photomultiplier tube, which is stimulated by, actuated, or otherwise responsive to light. Also spelled **photo-sensitive device**.

photosensitive material A material which is **photosensitive**. Such a material, for instance, may exhibit photoconduction, photoionization, or photoemission. Also spelled **photosensitive material**. Also called **light-sensitive material**.

photosensitivity Also spelled **photo-sensitivity**. 1. The property of being stimulated, actuated, or otherwise responsive to electromagnetic radiation, especially infrared, visible, or ultraviolet light. 2. The property of being stimulated, actuated, or otherwise responsive to visible light.

photosensor Same as **photodetector**. Also spelled **photosensor**.

Photoshop A popular image editor.

photoswitch Same as **photoelectric switch**.

photothyristor Same as **photo SCR**. Also spelled **photo-thyristor**.

phototransistor A bipolar transistor whose conduction is regulated by the intensity of the light incident upon its base region. Such a device is more sensitive than a photodiode, in addition to providing gain. Also spelled **photo-transistor**. Also called **light-sensitive transistor**.

phototube An electron tube incorporating a photosensitive electrode, usually a photocathode, from which electrons are emitted when exposed to electromagnetic radiation, especially light. The output signal of such a tube is proportional to the intensity of the incident light, and the photosensitive electrode may be chosen so as to selectively respond to specific regions of frequencies. Also spelled **photo-tube**. Also called **photoelectric tube**.

phototypesetter A typesetting device which takes a computer input, such as that from disk, and produces a high-resolution output on paper, film, or other medium. Such devices usually utilize lasers, and may have resolutions in excess of 3600 dpi. Also spelled **photo-typesetter**. Also called **imagesetter**.

photovoltaic Pertaining to a material or device which generates an electric current when exposed to radiant energy, especially light. Examples include photovoltaic materials and photovoltaic cells. Its abbreviation is **PV**. Also spelled **photo-voltaic**.

photovoltaic array A group of **photovoltaic cells** which are connected to produce a given power output. An example is a solar panel. Its abbreviation is **PV array**. Also spelled **photo-voltaic array**. Also called **photovoltaic panel**, or **photovoltaic module**.

photovoltaic cell A semiconductor device which generates an electric current when exposed to radiant energy, especially light. These are often made of silicon, selenium, or germanium. Such cells can be used in arrays to power anything from street lights to satellites. Its abbreviation is **PV cell**. Also spelled **photo-voltaic cell**. Also known as **barrier-layer photocell**. When optimized for the conversion of solar energy, also called **solar cell**.

photovoltaic device A device, such as a solar panel, which incorporates one or more **photovoltaic cells**. Its abbreviation is **PV device**. Also spelled **photo-voltaic device**.

photovoltaic effect The production of a voltage difference across the junction between two dissimilar materials, such as a metal and a semiconductor, when light strikes it. It is an example of a photoelectric effect, and is utilized, for instance, in photovoltaic cells to produce a current. Also spelled **photo-voltaic effect**.

photovoltaic material A material, such as silicon or cadmium telluride, which generates a voltage when exposed to light. Its abbreviation is **PV material**. Also spelled **photo-voltaic material**.

photovoltaic module Same as **photovoltaic array**. Its abbreviation is **PV module**. Also spelled **photo-voltaic module**.

photovoltaic panel Same as **photovoltaic array**. Its abbreviation is **PV panel**. Also spelled **photo-voltaic panel**.

photovoltaic system A power system incorporating one or more photovoltaic arrays, along with the necessary structural supports, conductors, storage batteries, inverters, fuses, grounds, and so on. When utilized to convert solar energy, also called **solar system (1)**. Its abbreviation is **PV system**. Also spelled **photo-voltaic system**.

phreak A person who breaks into secure communications systems, especially telephone networks, and utilizes services which would otherwise be paid for. Also, to engage in such activity. For example, to send audio codes using a telephone line to obtain free calls.

physical 1. In computers, all that pertains to hardware, as opposed to **logical (3)**, which deals with software. 2. The manner in which software and hardware really function, as opposed to **logical (2)**, which is the way data, processes, and systems are organized from the user's perspective. For example, a user may view the content of a file as a series of consecutive paragraphs, while it may actually consist of many data elements stored in multiple locations on a disk. 3. Pertaining to matter, energy, the laws of nature, and the fields dealing with them, especially physics.

physical address Also called **hardware address**. 1. The address that corresponds to a specific piece of hardware. For instance, an Ethernet address. 2. When a virtual address is used in computer memory, the address that corresponds to a specific hardware memory location. Memory management takes care of the conversions between the two.

physical characteristics Same as **physical properties**.

physical colocation A physical location within the facilities of a telecommunications company, which interconnects the equipment of a customer or competitor. This has benefits such as reducing the cost of operations, or being able to provide superior service. Used, for instance, by telephone companies or Internet providers. Also called **colocation (1)**.

physical drive The actual hardware comprising a drive, as opposed to a **logical drive**, which is a name given to a drive to facilitate identification by a given program or software system. For instance, logical drives C: and D: may be a single partitioned physical drive.

physical file The manner in which a file is truly organized, as opposed to a **logical file**, which is the way in which a file is distributed from the user's perspective. For example, a user may view the content of a file as a series of consecutive paragraphs, while it may actually consist of many data elements stored in multiple locations on one or more disks.

physical format A format which prepares a disk for a specific type of disk controller, and performs tasks such as sector identification. Such a format will destroy all data on a disk, while a **high-level format** only resets file-allocation tables so that the operating system sees the disk as empty. In the case of a hard disk, a physical format is usually performed by the manufacturer. Also called **low-level format**.

physical layer Within the OSI Reference Model for the implementation of communications protocols, the first, or lowest, of the seven layers. This layer deals with the hardware aspects of sending and receiving data in the network. Layer 1 provides the specifications for the transmission medium, such as fiber-optic cable or coax, for the adapter cards, connectors and their pinouts, and so on. At the sending end this layer accepts the bit stream from the data-link layer, places it in the physical medium, and at the receiving end extracts the bit stream and returns it to the data-link layer. Also called **layer 1**.

physical lock A hardware-based lock, as opposed to one based on software. An example is a dongle. Also called **hardware lock**.

physical memory The RAM chips a computer has, as opposed to **virtual memory**, which is the memory, as it appears to an application, enhanced by the utilization of secondary storage, such as a hard disk. Also called **physical storage (1)**, or **real memory**.

physical network The true physical topology of a network, as opposed to the manner in which a network is organized from the user's perspective, which is known as **logical network**. For example, a user may visualize a bus network as a series of computers arranged along a straight line, while it may truly be a common cable with many twists, turns, and loops.

physical properties Properties of matter which can vary without changes in chemical composition. These include density, melting point, elasticity, and malleability. Also called **physical characteristics**.

physical record The true distribution of a record, as opposed to the manner in which a record is organized from the user's perspective. For example, a user may view a record as a single unit, while it may actually consist of many data fields in a database. Also called **record layout (2)**.

physical security Measures such as locking the enclosure or housing of a device, restricting entry into certain areas, or using tamper-evident materials, which help to physically avoid security breaches.

physical storage Also called **real storage**. **1.** Same as **physical memory**. **2.** Within the virtual memory available to a system, the portion which actually uses the RAM of the computer.

physics The science that deals with the properties and interactions of matter and energy.

PHz Abbreviation of **petahertz**.

Pi- Abbreviation of **pebi-**.

pi An irrational number whose value is approximately 3.14159265358979, although it continues indefinitely in a non-repeating manner. It is used in many calculations. Its symbol is π.

pi filter A filter incorporating a series element and two parallel elements whose configuration resembles the Greek letter pi (π).

pi meson Same as **pion**. Also known as π **meson**.

pi mode A magnetron operation mode in which the phase difference between successive cavities is π. This mode generally produces the highest output.

pi network A four terminal electric network with three impedance branches connected in series. Usually utilized to match antenna impedances. Also known as π**-network**.

PIA Abbreviation of **peripheral interface adapter**.

Pib Abbreviation of **pebibit**.

PiB Abbreviation of **pebibyte**.

Pibps Abbreviation of **pebibits per second**.

PiBps Abbreviation of **pebibytes per second**.

PIC Acronym for **Programmable Interrupt Controller**.

pick-and-place robot A robot which picks up parts in one location and places them in another. Its abbreviation is **pick & place robot**.

pick-off Same as **pickoff**.

pick-up Same as **pickup**.

pick-up arm Same as **pickup arm**.

pick-up cartridge Same as **pickup cartridge**.

pick-up coil Same as **pickup coil**.

pick-up current Same as **pickup current**.

pick-up pattern Same as **pickup pattern**.

pick-up tube Same as **pickup tube**.

pick-up voltage Same as **pickup voltage**.

pick & place robot Same as **pick-and-place robot**.

picket fencing Variations in signal strength resulting from movements of a receiving and/or transmitting antenna. The fluctuations are a result of the same signal arriving via different paths en route from the transmitter to the receiver, and may be observed, for instance, as fluctuations in the FM reception of a moving car.

pickoff A device, such as a **pickup (1)**, which converts a mechanical motion into an electrical signal proportional to said motion. Also spelled **pick-off**.

pickup Also spelled **pick-up**. **1.** An instrument, device, or object which converts a physical magnitude or other measurable or detectable quantity into a corresponding electrical signal. These include devices which translate flashing lights, sounds, the undulations of the grooves of a phonographic disc, the lands and pits of an optical disc, or a scene,

into electrical signals. Microphones, phonograph pickups, and TV cameras are examples. A pickup is also a transducer when it converts one form of energy into another, as is the case of a microphone. **2.** Same as **phono cartridge**. **3.** Same as **pickup tube**. **4.** The minimum magnitude, such as that of a current or voltage, required to actuate a relay. **5.** The sounds and/or scenes which are being detected for recording or broadcast, as for TV or radio.

pickup arm In a phonograph, the arm which holds the cartridge. Also spelled **pick-up arm**. Also called **tone arm**.

pickup cartridge Same as **phono cartridge**. Also spelled **pick-up cartridge**.

pickup coil A small coil which is inserted into a magnetic field to measure its field strength, or any variations in it. Such a coil is connected to an indicating instrument, such as a ballistic galvanometer or a fluxmeter, and may or may not incorporate an amplifier. A pickup coil may also be utilized to examine the magnetic flux distribution of a magnetic field. Also spelled **pick-up coil**. Also called **probe coil**, **search coil**, **flip coil**, **magnetic test coil**, or **magnetic probe**.

pickup current The current necessary to actuate or close a relay or switch. Also spelled **pick-up current**. Also called **pull-in current**.

pickup pattern Also spelled **pick-up pattern**. **1.** A directivity pattern for a device receiving sound or electromagnetic radiation, such as a microphone or antenna. **2.** The **pickup pattern (1)** of a microphone. Common patterns include bidirectional, cardioid, and unidirectional. Also called **microphone pickup pattern**.

pickup tube An electron tube within a TV camera which serves as the transducer that converts the optical image into electric video signals. Also spelled **pick-up tube**. Also known as **pickup (3)**, **TV camera tube**, or **camera tube**.

pickup voltage The voltage necessary to actuate or close a relay or switch. Also spelled **pick-up voltage**. Also called **pull-in voltage**.

PICMG Abbreviation of PCI Industrial Computer Manufacturers Group. An organization which develops and promotes open specifications for PCI, and telecommunications and other computing applications.

pico- A metric prefix representing 10^{-12}, or one trillionth. For instance, 1 picoampere is equal to 10^{-12}, or one trillionth of an ampere. Its abbreviation is **p**.

pico-ohm Same as **picohm**.

picoammeter A current meter graduated in picoamperes.

picoampere 10^{-12} ampere. It is a unit of measurement of electric current. Its abbreviation is **pA**.

picobar 10^{-12} bar. It is a unit of pressure. Its abbreviation is **pb**.

picocoulomb 10^{-12} coulomb. It is a unit of electric charge. Its abbreviation is **pC**.

picocurie 10^{-12} curie. It is a unit of radioactivity. Its abbreviation is **pCi**.

picofarad 10^{-12} farad. It is a unit of capacitance. Its abbreviation is **pF**.

picogauss 10^{-12} gauss. It is a unit of magnetic flux density, or magnetic induction. Its abbreviation is **pG**, or **pGs**.

picogram 10^{-12} gram. It is a unit of mass. Its abbreviation is **pg**.

picohenry 10^{-12} henry. It is a unit of inductance. Its abbreviation is **pH**.

picohm 10^{-12} ohm. It is a unit of resistance, impedance, or reactance. Its abbreviation is **pΩ**. Also spelled **picoohm**, or **pico-ohm**.

picojoule 10^{-12} joule. It is a unit of energy or work. Its abbreviation is **pJ**.

picoliter 10^{-12} liter. It is a unit of volume. Its abbreviation is **pl**, or **pL**.

picometer 10^{-12} meter. It is a unit of measurement of length. Its abbreviation is **pm**.

picomol Same as **picomole**.

picomole 10^{-12} mol. It is a unit of amount of substance. Its abbreviation is **pmol**. Also called **picomol**.

piconewton 10^{-12} newton. It is a unit of force. Its abbreviation is **pN**.

picoohm Same as **picohm**.

picorad 10^{-12} rad. It is a unit of radiation dose. Its abbreviation is **prad**.

picoradian 10^{-12} radian. It is a unit of angular measure. Its abbreviation is **prad**.

picosat Same as **picosatellite**.

picosatellite A satellite similar to a nanosatellite, except that is usually smaller. A picosatellite may weigh less than 100 grams. Its abbreviation is **picosat**.

picosecond 10^{-12} second. It is a unit of time measurement. Its abbreviation is **ps**, or **psec**.

picosiemens 10^{-12} siemens. It is a unit of conductance. Its abbreviation is **pS**.

picotesla 10^{-12} tesla. It is a unit of magnetic flux density. Its abbreviation is **pT**.

picovolt 10^{-12} volt. It is a unit of measurement of potential difference. Its abbreviation is **pV**.

picovoltmeter A voltmeter designed to provide accurate indications of voltage values in the picovolt range.

picowatt 10^{-12} watt. It is a unit of measurement of power. Its abbreviation is **pW**.

PICS Abbreviation of **Platform for Internet Content Selection**.

pictorial diagram Same as **pictorial wiring diagram**.

pictorial wiring diagram A representation of a circuit similar to a wiring diagram, except that each circuit element is drawn or photographed, along with the wiring between said components. Also called **pictorial diagram**, or **picture diagram**.

picture Its abbreviation is **pix**. **1.** An image, or other visual representation, which is photographed or otherwise presented on a flat surface. **2.** Same as **photograph**. **3.** An image, or sequence of images, presented on a display device such as a TV or computer monitor.

picture black **1.** In TV, the peak video signal level corresponding to the darkest part of the presented image. **2.** In fax communications, the amplitude of the signal produced when scanning the maximum darkness of the transmitted material. Also called **black signal**.

picture carrier In a TV, the carrier frequency which is modulated to convey picture information.

picture demodulator Same as **picture detector**.

picture detector In a TV, a circuit or device which demodulates a modulated **picture carrier**, so as to extract picture information. Also called **picture demodulator**.

picture diagram Same as **pictorial wiring diagram**.

picture digitization The conversion of a picture into digital form, which is necessary for a digital computer to process it.

picture digitizer A device which converts a picture into digital form, such as a bitmap, which is necessary for digital computer processing.

picture element Same as **pixel**. Its abbreviation is **pel**.

picture frequency Within a TV system, the number of times per second a frame is scanned. In the United States, this number is 30. Also called **frame frequency**, or **frame rate**.

picture-in-picture A TV feature in which a second picture, occupying a given portion of the viewing screen, may be seen at the same time as the main picture. The source of the second picture may another input, such as that of a DVD or VHS, or in the case of a TV equipped with two-tuner picture-in-picture, another channel. Its abbreviation is **PIP**.

picture information In a TV, the visual content presented via the active lines. An HDTV featuring over 1,000 active lines has several times the picture information of a TV adhering to the NTSC standard which provides for approximately 485 active lines. Also, the signal providing such content.

picture monitor A TV monitor, which is usually located at a studio, utilized to keep track of the quality of a recording or broadcast.

picture noise Any noise displayed on a device, such as a CRT, which displays a picture. An example is grass.

picture reception The reception of a signal containing one or more pictures, as might occur with a TV or a receiving fax.

picture signal In color TV, the signal which contains the picture information.

picture size In TV, the dimensions of the viewable area. May be expressed, for instance, in inches, as in 40" diagonal, or 25" by 35", or as an area such 875 sq in. Also, the viewable area of a picture-in-picture image.

picture transmission The transmission of a signal containing one or more pictures, as might occur with a TV transmitter or a transmitting fax.

picture transmitter Also called **video transmitter**, or **visual transmitter**. **1.** A transmitter, such as a broadcast transmitter, which sends signals containing one or more pictures. Also, the equipment utilized for this purpose. **2.** The equipment utilized by a TV transmitter to send the video signal. This signal is combined with that of the sound transmitter for regular TV viewing.

picture tube A CRT used in a TV, for viewing the received images. Also called **kinescope**, or **cathode-ray television tube**.

picture white **1.** In TV, the peak video signal level corresponding to the lightest part of the presented image. **2.** In fax communications, the amplitude of the signal produced when scanning the lightest portion of the transmitted material. Also called **white signal**.

Pierce crystal oscillator Same as **Pierce oscillator**.

Pierce oscillator An oscillator, similar to a Colpitts oscillator, which incorporates a quartz crystal in the feedback circuit, which provides for high stability. Also called **Pierce crystal oscillator**.

piezodielectric A material whose dielectric constant changes when subjected to mechanical strain.

piezoelectric **1.** Pertaining to, arising from, or exhibiting **piezoelectricity**. **2.** Pertaining to, or arising from a **piezoelectric material**.

piezoelectric ceramic A ceramic, such as lead zirconate titanate, which exhibits **piezoelectricity**.

piezoelectric crystal A crystal, such as quartz, which exhibits **piezoelectricity**.

piezoelectric earphones Earphones whose transducer is a piezoelectric crystal. Also called **crystal earphones**.

piezoelectric effect The generation of a potential difference across opposite faces of a **piezoelectric material** when subjected to mechanical strain.

piezoelectric filter A filter circuit utilizing one or more piezoelectric crystals. Such filters feature high selectivity and a good shape factor, and may be used, for instance, in intermediate-frequency amplifiers. Also called **crystal filter**.

piezoelectric headphones Headphones whose transducer is a piezoelectric crystal. Also called **crystal headphones**.

piezoelectric loudspeaker Same as **piezoelectric speaker**.

piezoelectric material A material, such as quartz or lead zirconate titanate, which exhibits **piezoelectricity**.

piezoelectric microphone A microphone whose transducer is a piezoelectric crystal, and which converts audio-frequency vibrations into the corresponding audio-frequency voltage output. Also called **crystal microphone**, or **crystal cartridge (2)**.

piezoelectric oscillator An oscillator circuit which utilizes a piezoelectric crystal, usually quartz, to control the oscillation frequency. Such oscillators feature a highly accurate and stable output, especially when in a temperature-controlled environment. Also called **crystal-controlled oscillator**, or **crystal oscillator**.

piezoelectric pickup A phonographic pickup whose transducer is a piezoelectric crystal, which converts the movements of the stylus into the corresponding audio-frequency voltage output. Also called **crystal pickup**, or **crystal cartridge (1)**.

piezoelectric plate A piezoelectric crystal that has been cut, etched, coated, and otherwise fully prepared to be mounted on its crystal holder. Also called **crystal plate**.

piezoelectric resonator A resonant circuit which utilizes a quartz piezoelectric crystal to control the resonance frequency. Such resonators may be used, for instance, to control the frequency of an oscillator, and feature a highly accurate and stable output, especially when in a temperature-controlled environment. Also called **quartz-crystal resonator, quartz resonator**, or **crystal resonator**.

piezoelectric sensor A sensor whose detecting element is a piezoelectric crystal, especially quartz. An example is that used in a quartz thermometer. Also called **crystal sensor**.

piezoelectric speaker A speaker whose transducer is a piezoelectric crystal, which converts its audio-frequency voltage input into the corresponding audio-frequency vibrations. Mostly used for reproduction of high frequencies. Also called **piezoelectric loudspeaker**, or **crystal speaker**.

piezoelectric transducer A transducer whose sensitive element is a piezoelectric crystal, especially quartz. Piezoelectric crystals, when subjected to mechanical energy, generate electrical energy, and vice versa. Used, for example, in microphones, pickups, and loudspeakers. Also called **crystal transducer**.

piezoelectric tweeter A **piezoelectric speaker** utilized as a high-frequency driver.

piezoelectricity Electricity generated by subjecting certain crystals or ceramics to mechanical strain. The magnitude and orientation of the deformation will determine the electrical polarization. The most common substance exhibiting this phenomenon is quartz, and some of the applications of piezoelectricity include its use in oscillators, transducers, motors, and in instruments such as clocks and navigational aids. Conversely, when a material capable of generating piezoelectricity is exposed to an electric field, it will undergo mechanical deformation.

piezoresistance In certain materials, a change in resistance when subjected to mechanical strain.

piezoresistant A material whose resistance changes when subjected to mechanical strain.

piggyback control An automatic control system in which the control units are linked in a chain where each unit controls the succeeding one. Also called **cascade control**.

pigtail A connection, usually consisting of a braided wire, in which one end is stationary and the other has a limited range of motion.

pile 1. An early battery consisting of a series of alternated disks of dissimilar metals, usually zinc and copper, each separated by paper or cloth soaked in an electrolyte. Also

called **voltaic pile**, or **galvanic pile**. 2. A deprecated term for **nuclear reactor**. 3. A stack, such as a carbon pile, of elements or objects utilized together.

pillow speaker A usually small and flat speaker designed for use when laying down or otherwise resting the head. Such a speaker is usually placed under a pillow.

pilot A signal transmitted via a **pilot channel**. Also called **pilot signal**.

PILOT Acronym for **P**rogrammed **I**nquiry **L**earning or **T**eaching. A high-level programming language used, for instance, to develop interactive tutorials.

pilot channel In communications, a channel, such as that transmitting a timing signal, usually utilized for control purposes.

pilot circuit A circuit carrying a **pilot signal**.

pilot frequency A frequency transmitted via a **pilot channel**.

pilot indicator Same as **pilot light**.

pilot indicator light Same as **pilot light**.

pilot lamp Same as **pilot light**.

pilot light Also called **pilot lamp, pilot indicator light**, or **pilot indicator**. 1. A small lamp utilized to indicate that a circuit is energized, that a given device is functioning, or the like. 2. A light utilized to illuminate a dial. Also called **dial light**.

pilot model A preliminary model of a component, circuit, device, piece of equipment, or system, utilized for experimentation, debugging, and the like. Such a model is used, for instance, to help define a production process.

pilot production A small production run of a component, circuit, device, piece of equipment, or system, utilized for testing or to refine a production process.

pilot signal Same as **pilot**.

pilot subcarrier A subcarrier utilized as a **pilot signal**.

pilot test A test, such as that performed on a pilot model, utilized to evaluate operation under actual operating conditions.

PIM 1. Abbreviation of **pulse-interval modulation**. 2. Abbreviation of **personal information manager**.

PIN 1. Acronym for **p**ositive-**i**ntrinsic-**n**egative. A semiconductor configuration in which a lightly-doped intrinsic layer is sandwiched between heavily-doped **p** and **n** layers. 2. Acronym for **p**ersonal **i**dentification **n**umber. A usually numeric password used to help authenticate a person wishing to perform banking, purchasing, or other types of transactions. 3. Acronym for **p**ersonal **i**dentification **n**umber. A password utilized to help authenticate a person wishing to activate, deactivate, or otherwise use a security system such as that protecting a home.

pin In a plug-in electrical component such as an IC or tube, a terminal which allows physical connection with compatible sockets in other devices or components. It may also provide structural support. Also known as **prong**, or **base pin**.

pin compatible An IC, tube, or other component or device whose pins match those of another.

pin connector 1. Same as **pin jack**. 2. Same as **pin plug**.

PIN diode A three-layer diode consisting of a lightly-doped intrinsic layer sandwiched between heavily-doped **p** and **n** layers. Used, for instance, as an attenuator, for switching, or as a photodiode. Also spelled **p-i-n diode**.

pin grid array A chip package, usually made of ceramic or plastic, capable of providing up to several hundred pins, all located on its underside. Used, for instance, to package computer chips. The design seeks to minimize the distance signals must travel from the chip to each designated pin. Its abbreviation is **PGA**.

pin-hole Same as **pinhole**.

pin-hole detector Same as **pinhole detector**.

pin jack A receptacle for a **pin plug**. Also called **pin connector** (1).

pin-out Same as **pinout**.

PIN photodiode A **PIN diode** used as a **photodiode**. Also spelled **p-i-n photodiode**.

pin plug A connecting plug consisting of one or more metal prongs whose diameter is roughly that of an ordinary pin. Used, for instance, for connecting video equipment. A **pin plug** is inserted into a **pin jack**. Also called **pin connector** (2).

pin straightener A device utilized to straighten the pins or terminals of components such as electron tubes.

pin switch A switch whose position is selected by pushing, pulling, toggling, or otherwise repositioning a pin.

pin-through hole Same as **plated-through hole**. Its abbreviation is **PTH**.

pinch effect A magnetic attraction occurring between parallel conductors which each carry a large unidirectional current in the same direction. Also, the constriction of a fluid or plasma which is exposed to such currents. This phenomenon can keep an enclosed plasma away from its container walls.

pinch-off The point at which the current flowing through a FET is stopped by the gate. Also, the state of having the current cut off in this manner. Also spelled **pinchoff**.

pinch-off voltage The value of the gate voltage producing **pinch-off**. Also spelled **pinchoff voltage**.

pinch roller In a tape deck or drive, a cylinder or roller, usually made of rubber, which forces the tape against the capstan to provide proper traction. Also called **pressure roller**.

pinchoff Same as **pinch-off**.

pinchoff voltage Same as **pinch-off voltage**.

pincushion Same as **pincushion distortion**.

pincushion correction The utilization of circuitry to minimize or eliminate **pincushion distortion**.

pincushion distortion In TV reception, a form of distortion where all four sides of the displayed image curve inward. It is a type of geometric distortion. Also called **pincushion**, or **pincushioning**.

pincushioning Same as **pincushion distortion**.

PINE A character-based email client for UNIX systems. It is an abbreviation of **P**rogram for **I**nternet **N**ews and **E**mail, or it is an acronym for **P**ine **I**s **N**ot **E**lm, **elm** being the program it replaced.

ping **1.** A sonic or ultrasonic pulse, such as that utilized in sonar. **2.** A utility used to determine whether a given IP address is connected to the Internet at a given moment. A packet is sent to the desired address, and a response is awaited. Used, for instance, for troubleshooting. Also, to employ such a utility. Although ping is often considered to be an acronym for **P**acket **I**nternet **G**roper, its origin is based on the sending of a signal in expectation of a response, as is the case with a **ping** (1).

ping of death A ping that is maliciously utilized to crash the computer of the recipient, or otherwise have harmful effects. It is usually caused by sending a datagram larger than that allowed by a given IP protocol.

ping packet A packet sent to determine whether a given IP address is connected to the Internet at a given moment.

ping pong In a communications network, a reversal in the transmission direction. That is, the transmitter becomes the receiver, and vice versa.

ping-pong buffering The use of two buffers to temporarily store information that awaits transfer to or from an input/output device. By using two buffers, one may be emptied while the other is filled, enhancing transfer speed. Also called **double buffer**.

ping request The sending of a **ping packet**.

pinhole Also spelled **pin-hole**. **1.** A tiny hole or puncture. For instance, the aperture of certain optical devices or systems. **2.** A **pinhole** (1) resulting from a defect or fault. For example, that formed in a metal during electroplating.

pinhole detector A device which detects **pinholes** (2). A typical arrangement involves the use of a precise light source on one side of the material, usually a strip, being evaluated, with a photomultiplier on the other side detecting any incident light aside from the ambient light. Some detectors can sense pinholes of less than 1 micrometer at strip speeds approaching one hundred kilometers per hour. Also spelled **pin-hole detector**.

pink noise Noise whose sound energy is equal or constant per octave band. For example, the sound intensity would be constant throughout the 1000 Hz octave band, which comprises frequencies ranging from about 707 Hz to 1414 Hz, as it would be for the 250 Hz or 4000 Hz octave band, or any other octave band specified. The intensity of such noise is inversely proportional to frequency over the stated range. **White noise** passed through a filter with a 3 dB/octave rolloff results in pink noise.

pink noise generator An instrument or device which generates **pink noise**. Used, for instance, as a reference noise source, or for masking other noises.

pinout A diagram, graphic, table, text, or the like, which describes the location and function of each pin in a multipin connector. Also spelled **pin-out**.

PIO mode Abbreviation of **P**rogrammed **I**nput/**O**utput **Mode**. The I/O standard utilized by IDE drives. Each PIO mode has a given transfer rate.

PIO mode 0 A PIO mode with a transfer rate of up to 3.3 Mbps.

PIO mode 1 A PIO mode with a transfer rate of up to 5.2 Mbps.

PIO mode 2 A PIO mode with a transfer rate of up to 8.3 Mbps.

PIO mode 3 A PIO mode with a transfer rate of up to 11.1 Mbps.

PIO mode 4 A PIO mode with a transfer rate of up to 16.6 Mbps.

PIO mode 5 A PIO mode with a transfer rate of up to 22.2 Mbps.

pion The lightest type of meson, consisting of a quark and antiquark pair, and having a charge of +1, -1, or 0. Charged pions are slightly heavier than neutral pions. Also called **pi meson**.

pip Also called **blip**. **1.** The display of a received pulse on a CRT, especially a radar screen. It may appear, for instance, as a spot of light. **2.** A short pulse that serves as a signal, such as that used in the Morse code. **3.** A small mark or spot on a recordable medium, which is used for tracking purposes.

PIP Abbreviation of **picture-in-picture**.

pipe **1.** A hollow tube through which something passes. For example, a conduit via which cables are run. **2.** A virtual conduit through which one section of memory can pass its output to another section of memory, which uses it as input. Also called **pipeline** (2).

pipeline **1.** A sequence of process segments handled simultaneously in **pipelining**. **2.** Same as **pipe** (2).

pipeline burst cache A form of cache utilizing a **PBSRAM** chip.

pipeline burst SRAM Same as **PBSRAM**.

pipeline burst static RAM Same as **PBSRAM**.

pipeline processing Processing utilizing **pipelining**.

pipeline processor A processor capable of **pipeline processing**.

pipelining 1. The simultaneous performance of multiple program instructions at varying stages, so as to provide parallel processing. For instance, while one instruction is being executed, another is being decoded, while another is being fetched. 2. The use of a **pipe (2)** for **pipelining (1)**.

piracy 1. The unauthorized copying and subsequent use and/or sale of software with copyright protection. When purchasing a program, an individual or entity becomes a licensed user, and can usually make backup copies when possible. Also called **software piracy**. 2. The unauthorized copying and subsequent use and/or sale of material, such as that contained in a book or DVD, with copyright protection.

Pirani gauge A pressure gauge utilizing a hot wire placed in the gas whose pressure is to be measured. The rate of heat loss of the wire, which is connected to a bridge, depends on the rate at which the gas conducts heat away from said wire. The voltage across the wire is maintained constant, and since its resistance varies proportionally to its temperature, changes in resistance enable the measurement of the surrounding pressure. Conversely, the resistance of the wire can be held constant, so that the potential difference across it corresponds to the pressure. Utilized to measure low gas pressures.

piston 1. In a waveguide, a sliding cylinder, usually consisting of a metal or dielectric, which is utilized for tuning. Also called **plunger**, **waveguide plunger**, or **tuning piston**. 2. A cylinder or disk which is fitted closely within a larger cylinder, and which moves against the pressure of a fluid. An example is that in a dashpot.

pistonphone A chamber, equipped with a piston, which is utilized for precise measurements of sound pressure. Used, for instance, to calibrate microphones.

pit 1. On an optical disc, such as a CD or DVD, an indented portion, as opposed to a non-indented portion which is called **land (2)**. The laser beam is reflected off the lands, while being scattered or absorbed by the pits. 2. A small cavity or depression on a surface or material. Such a pit may be unwanted, as in a damaged surface, or desired, as is the case of a **pit (1)**.

pitch 1. The frequency of a given sound. 2. A subjective attribute of a sound, which depends primarily on its frequency, and to a lesser extent on factors such as its intensity, location, and duration. Pitch is perceived to run along a scale ranging from low to high. A fixed-frequency sound may be perceived as changing in pitch, as, for instance, a high-frequency sound whose intensity is steadily increased will appear to be rising in pitch. 3. The distance between similar elements arranged regularly or in a fixed pattern. For example, dot pitch, or the distance between the centers of adjacent turns of a coil. 4. In a robot, up and down movement of a wrist, end-effector, or other such moving part. 5. The up and down movement of a part, object, or device. 6. In computers, the number of printed characters per inch.

PIV Abbreviation of **peak inverse voltage**.

pivot A generally short shaft upon which a related moving part rotates, balances, oscillates, or the like. For example, the pivot which supports the magnetic needle of a compass, or that upon which the wire coil of a galvanometer pivots.

pix Abbreviation of **picture**, or pictures.

PIX firewall A popular family of firewall products.

pixel Abbreviation of **picture (pix) element**. Also called **scanning spot**, or **elemental area**. Its own abbreviation is **pel**. 1. Within an image, such as that of a computer display, TV, or facsimile, the smallest unit which can be manipulated. In a CRT, such as a color TV or computer monitor, for instance, it is a trio of color phosphor dots representing a single point. In computers, also refers to the storage of information used to define a pixel in an image. 2. In a TV picture or facsimile, the portion of the scanning line being explored at any given instant.

pixel aspect ratio The ratio of the width to the height of a single pixel.

pixel depth The number of bits used to define a pixel in an image, as determined by the hardware and software. For instance, a 24-bit video adapter allows for over 16.7 million colors to be displayed. Also called **bit depth**, or **color depth**.

pixel image An image, or representation of an image, composed of or corresponding to pixels on a screen. A bitmap is an example.

pixel map A data representation which describes a **pixel image**. Such a map includes attributes such as color, resolution, bit depth, dimensions, and format.

pixelated That which exhibits **pixelation**.

pixelation The appearance of individual or multiple pixels in a digital image, which gives it a jagged or stair-stepped appearance. Pixelation will appear, for instance, if a bitmapped image is enlarged enough.

pJ Abbreviation of **picojoule**.

PJ Abbreviation of **petajoule**.

pk-pk Abbreviation of **peak-to-peak**.

pk-pk amplitude Abbreviation of **peak-to-peak amplitude**.

pk-pk value Abbreviation of **peak-to-peak value**.

pk-pk voltage Abbreviation of **peak-to-peak voltage**.

PKI Abbreviation of **Public-Key Infrastructure**.

PKUNZIP A software utility which enables the uncompression of files which have been compressed using **PKZIP**.

PKZIP A utility which places multiple files in a single, compressed file format. **PKUNZIP** is utilized to uncompress such files.

pl 1. Abbreviation of **picoliter**. 2. Abbreviation of **poundal**.

pL Abbreviation of **picoliter**.

PL/SQL Abbreviation of Procedural Language/SQL. A procedural language extension of SQL.

PLA Abbreviation of programmable logic array. A logic array which instead of being programmed at a factory, is done so by an end user, usually during installation. PLAs can only be programmed once.

plain old telephone service Same as **POTS**.

plain old telephone service splitter Same as **POTS splitter**.

plain text Same as **plaintext**.

plain vanilla Providing only the bare essentials for programs, equipment, or systems to function. This contrasts with **bells and whistles**, where all the latest features and items may be present.

plaintext Text which is not encrypted, as opposed to **ciphertext**, which has been. It is simply regular text before or after encryption. Also spelled **plain text**. Also called **clear text**.

plan position indicator In radars, an oscilloscopic display in which the radar site is in the center, and the scanned objects appear as blips surrounding it. The distance from the center indicates the range of a scanned object, and the radial angle provides its bearing. Its abbreviation is **PPI**. Also called **P display**, **P scope**, **P scanner**, **P scan**, or **P indicator**.

planar Of, pertaining to, consisting of, or situated along a single plane or the same plane as something else. For instance, a planar transistor or a planar array.

planar array An antenna array in which all the elements lie along the same plane. Such an array may incorporate parasitic elements, and if the array is phased, multiple identical beams may be transmitted simultaneously. Also called **planar-array antenna**.

planar-array antenna Same as **planar array**.

planar circuit A circuit which is two-dimensional. This contrasts with a **non-planar circuit**, which is three-dimensional.

planar device A semiconductor device, such as a diode or transistor, whose terminals all lie along the same plane or face.

planar diode A diode whose terminals lie along the same plane or face. Usually manufactured utilizing a planar process.

planar process In the manufacturing of semiconductor devices, the production of all components along a single plane, as opposed to three-dimensionally. It is a widely-utilized process in which an oxide layer is first grown on the substrate, followed by multiple steps of diffusion and etching to produce the desired junctions, and usually ending with another oxide layer for insulation, stability, and protection.

planar transistor A transistor whose terminals all lie along the same plane or face. Usually manufactured utilizing a planar process.

Planck constant A fundamental physical constant equal to approximately 6.62607×10^{-34} joule-second. It is the ratio of the energy of a photon to its frequency. Its symbol is *h*.

Planck's law A fundamental law stating that electromagnetic energy is composed of discrete quanta of energy, called photons. The energy of a photon, or light quantum, is equal to *hf*, where *h* is the **Planck constant**, and *f* is the frequency of the radiation in hertz.

plane A flat surface considered to extend endlessly in all directions.

plane of incidence The plane containing light, sound, particles, waves, or the like, which is reflected off a surface, object, or region. Also called **incidence plane**.

plane of polarization For an electromagnetic wave, the plane which contains both the electric field vector and the direction of propagation of said wave. Also called **polarization plane**.

plane of reflection The plane containing light, sound, particles, waves, or the like, which are incident upon a surface, object, or region. Also called **reflection plane**.

plane polarization Polarization of an electromagnetic wave in which the electric field vector is always situated in a plane that also contains the direction of propagation of said wave.

plane-polarized light A light wave with **plane polarization**. Also called **linearly-polarized light** (2).

plane-polarized wave A wave with **plane polarization**. Also called **linearly-polarized wave** (2).

planetary electron An electron in an orbital. Also called **orbital electron**.

plant 1. Same as **power generating station**. 2. One or more building or structures utilized to manufacture products.

plant load factor For a given power generator or group of generators, the ratio of the actual energy produced, to the amount that would have been produced at the maximum continuous duty rating. Usually expressed as a percent. Its abbreviation is **PLF**.

plasma A state of matter consisting of a gas that has been completely ionized, and which contains equal numbers of positive and negative particles, such as positive ions and free electrons. A plasma is electrically conductive, and susceptible to magnetic fields. Such a state of matter may be observed, for instance, in stars, interstellar gases, thermonuclear reactions, and in gas-discharge tubes. A gas which has

been sufficiently ionized so as to become conductive and susceptible to magnetic fields is usually considered a plasma as well. The other physical states in which matter is known to exist are **solid** (1), **liquid**, **gas**, and **Bose-Einstein condensates**.

plasma confinement The confinement of a plasma with a region, such as the central part of an enclosure. One technique utilized for this purpose is magnetic confinement. Also called **plasma containment**.

plasma containment Same as **plasma confinement**.

plasma display A flat display which utilizes a matrix of small tubes, each containing a gas. Each tube represents a pixel, and when powered, the gas ionizes to produce a glow. Depending on the contained gas, the glow discharge may be red, green, or blue. Used, for instance, for large-screen and digital TV. Also called **plasma monitor**, **plasma display panel**, or **gas-plasma display**.

plasma display panel Same as **plasma display**. Its abbreviation is **PDP**.

plasma-enhanced chemical-vapor deposition A chemical-vapor deposition technique in which gaseous reactants are placed into a plasma. The makes high temperatures unnecessary, as the plasma provides species which are highly energetic and reactive. Its abbreviation is **PECVD**, or **plasma-enhanced CVD**.

plasma-enhanced CVD Same as **plasma-enhanced chemical-vapor deposition**.

plasma etching A dry-etching technique in which a plasma provides species which are highly energetic and reactive, which chemically etch patterns into the immediately surrounding material.

plasma length In a plasma, the maximum distance at which a charged particle will be influenced by the electric field of another particle of the opposite charge. It usually refers to the interactions between an electron and any given positive ion. Also called **Debye length**, **Debye screening radius**, **Debye shielding distance**, or **Debye shielding length**.

plasma monitor Same as **plasma display**.

plasma screen The viewing screen of a **plasma display**.

plastic A solid material which can flow into a given desired shape through the application of heat and/or pressure. A plastic is ordinarily synthetic, and its chief constituent is usually a polymerized organic molecule, such as polyvinyl chloride. The two main classifications for plastics are thermosetting and thermoplastic. Plastics are electrical insulators, and have many applications in electronics, including their use in components such as capacitors, recording media such as floppy disks and optical disks, optical fibers, lenses, IC packages, in frames, as an encapsulant, and for tubing, among many others.

Plastic Ball Grid Array A surface-mount ceramic chip package that utilizes tiny balls of solder to attach its leads to a printed-circuit board. Plastic Ball Grid Arrays feature low induction, compact size, and a high lead count in a small area. Its abbreviation is **PBGA**.

plastic chip package Same as **plastic package**.

plastic DIP Same as **PDIP**.

plastic dual in-line package Same as **PDIP**.

plastic-film capacitor A capacitor in which layers of a metal foil, usually aluminum, are alternated with layers of a plastic, such as polystyrene, to form a roll. Such capacitors are well suited, for instance, for use at high temperatures and high frequencies.

plastic-leaded chip carrier A square or rectangular plastic chip package having J-leads on all four sides. This occupies less room than comparable packages with gull-wing leads. Its abbreviation is **PLCC**.

plastic-leadless chip carrier A plastic chip package which is hermetically sealed, and which instead of leads consisting of metal prongs or wires, has metallic contacts called castellations which are flush with the package or recessed. The contacts are usually on all four sides of the package, although they may be located elsewhere. Its abbreviation is **PLCC**.

plastic package A chip package made of plastic. Also called **plastic chip package**.

plastic PGA Abbreviation of **plastic pin grid array**.

plastic pin grid array A plastic package capable of providing up to several hundred pins, all located on its underside. Used, for instance, to package computer chips. The design seeks to minimize the distance signals must travel from the chip to each designated pin. Its abbreviation is **PPGA**, or **plastic PGA**.

plastic quad flat-pack A plastic surface-mount chip package in the form of a square, which provides leads on all four sides. Such a package affords a high lead count in a small area. Its abbreviation is **PQFP**. Also spelled **plastic quad flatpack**. Also called **plastic quad flat-package**.

plastic quad flat-package Same as **plastic quad flat-pack**.

plastic quad flatpack Same as **plastic quad flat-pack**.

plate 1. The positive electrode in an electron tube. Electrons emitted by the cathode travel towards it. Also known as **anode** (1). 2. One of the electrodes in a capacitor. Also called **capacitor plate**. 3. One of the electrodes in an electrochemical cell. 4. The formation of a metal deposit on another surface, usually a different metal, via electrolysis. Also called **electroplate**. 5. A thin, or very thin, layer or coat deposited or applied to a metal. Also, to apply such a plate. 6. A light-sensitive sheet, usually of glass or metal, upon which a photographic image may be recorded.

plate battery A battery which supplies the anode, or plate, current in electron tubes. Also known as **anode battery**, or **B battery** (1).

plate capacitance In an electron tube, the capacitance between the plate and another electrode, especially the cathode. Also called **anode capacitance**.

plate characteristic In an electron tube, the variation of the current of the anode relative to the voltage applied to it. Also known as **anode characteristic**.

plate circuit A circuit which includes all of the components connected between the anode and the cathode in an electron tube, including the anode voltage source. Also called **anode circuit**.

plate current In an electron tube, the electron flow from the cathode to the anode. Also known as **anode current**.

plate detection The function of a **plate detector**.

plate detector An electron tube detector in which the anode circuit rectifies the input signals. Also called **anode detector**.

plate dissipation In an electron tube, the power dissipated by the anode in the form of heat. This loss is caused by the anode being bombarded by electrons and anions. Also called **anode dissipation**.

plate efficiency In an electron tube, the ratio between the AC load circuit power and the DC anode input power. Also called **anode efficiency**.

plate-grid capacitance In an electron tube, capacitance between the plate and the control grid.

plate impedance In an electron tube, the total impedance between the anode and the cathode, without taking into account the electron stream. Also called **plate-load impedance, anode-load impedance**, or **anode impedance**.

plate input power In a vacuum tube, the DC power consumed by the anode. Also called **plate power input**, or **anode input power**.

plate-load impedance Same as **plate impedance**.

plate modulation In an electron tube, amplitude modulation obtained by varying the voltage of the anode proportionally to the fluctuations in the modulating wave. Also called **anode modulation**.

plate potential Same as **plate voltage**.

plate power input Same as **plate input power**.

plate power supply In an electron tube, the DC applied to the anode to place it at a high current potential relative to the cathode. Also called **plate supply**, or **anode supply**.

plate pulse modulation In an electron tube, modulation produced by applying external voltage pulses to the anode. Also called **anode pulse modulation**.

plate resistance In an electron tube, the ratio of a minimal change in the anode voltage to a minimal change in the anode current. All other voltages must be held constant. Also called **anode resistance**.

plate saturation In an electron tube, the condition in which the plate current can not be further increased, regardless of any additional voltage applied to it, since essentially all available electrons are already being drawn to said plate. Also called **anode saturation, current saturation, voltage saturation**, or **saturation** (3).

plate spacing In a capacitor, the distance between plates. The capacitance value of a capacitor varies as a function of various variables, including the distance between its plates. A variable capacitor, for instance, may utilize plates which are moved relative to each other.

plate supply Same as **plate power supply**.

plate voltage Also called **plate potential**, or **anode voltage**. 1. In an electron tube, the difference in potential between the anode and the cathode. 2. In an electron tube, the difference in potential between the anode and a specific point of the cathode.

plateau Within the characteristic curve of a component or device, such as a tube or transistor, a region where an increase in one variable, such as current, has little or no effect on another variable, such as voltage. Also, any such flat region within a response curve.

plated-through hole Within a printed-circuit board, a hole made through plating, and which serves to mount and connect components by connecting their pins or leads. Its abbreviation is **PTH**. Also called **pin-through hole**.

platen A cylindrical roller against which a print head strikes in an impact printer, typewriter, or similar device. A paper, or other suitable medium which is being printed upon, is guided and supported by this roller.

platform A specific hardware and software configuration, including the operating system. When a program or hardware device will only function properly with a particular platform, it is called **platform-dependent**, while those which can work across multiple platforms are called **platform-independent**. Also called **computing platform**.

platform-dependent Computer software or hardware which is designed to work properly only with a specific platform. Assembly language, for instance, is platform-dependent. This contrasts with **platform-independent**, in which software or hardware can work with more than one type of platform. Also called **architecture-dependent**.

Platform for Internet Content Selection Its abbreviation is **PICS**. A rating system utilized to classify Web sites. PICS itself provide no ratings, but promotes a uniform template for others to utilize.

platform-independent Computer software or hardware which is designed to work properly with more than one type of platform. An interpreter version of LISP is platform-independent, as is Java. This contrasts with **platform-dependent**, in which software or hardware is designed to work properly only with a specific platform. Also called **platform neutral**, **cross-platform**, **multiplatform**, or **architecture-independent**.

platform neutral Same as **platform-independent**.

platform-neutral architecture Hardware or software which can work with more than one type of architecture. This provides greater flexibility, but may sacrifice performance. Also called **neutral architecture**.

plating 1. The process of depositing a metal onto another surface, usually a different metal, via electrolysis. Such a deposit may be thin or thick, and may be used, for instance, to provide coatings which are protective, decorative, or which have any given electrical properties. Also called **electroplating**. 2. The process of depositing or applying a thin, or very thin, layer or coat upon a metal.

platinotron A magnetron designed for wideband amplification of microwave oscillations.

platinum A dense and lustrous silvery-white metallic chemical element whose atomic number is 78. It is chemically inactive, extremely corrosion resistant, malleable, ductile, and can absorb large amounts of hydrogen. It has around 35 known isotopes, of which 5 are stable. Its applications in electronics include its use in contacts, electrodes, thermocouples, and in electroplating. Its chemical symbol is **Pt**.

platinum-iridium alloy An alloy containing platinum and from 1% to 30% iridium. Used in electrical contacts, fuse wire, thermocouples, and where very high corrosion resistance is essential.

platinum resistance thermometer A resistance thermometer which utilizes a platinum wire. Such a thermometer is highly stable, and may be able to provide very accurate readings throughout a temperature range spanning from below -200 °C, to over 1000 °C. Its abbreviation is **PRT**.

platinum-rhodium alloy An alloy containing platinum and from 2% to 40% rhodium. Used in thermocouples, and in high temperature applications.

platter 1. One of the rigid disks comprising a hard drive. Such platters rotate at very high speeds, and are coated, usually on both sides, with a material which enables information to be encoded by altering the magnetic polarity of minute portions of the surface of each side of said platters, using read/write heads. 2. The rotating disk upon which a phonographic record rests when playing. 3. A surface upon which one or more disks or discs rest when playing or awaiting to be played. For example, a rotating tray upon which the discs rest in a multiple CD or DVD player.

play head Same as **playback head**.

playback The reproduction of a sound and/or video recording, such as that of a magnetic tape or optical disc.

playback head A transducer which converts magnetic signals recorded or stored on magnetic recording media, such as tapes or disks, into output electrical signals. When used in the context of tape which has audio and/or video, the signals are usually said to be reproduced, while in the context of data, the signals are usually said to be read. Also called **play head**, **magnetic read head**, **magnetic reading head**, **read head**, **reading head**, or **reproduce head**.

player 1. A device which plays back, reproduces, or reads that which has been stored in a recording medium such as a magnetic tape or optical disk. Examples include DVD players, CD players, digital audio players, cassette players, and VHS players. 2. A device or application which enables a user to access content available on recorded media, or through a network such as the Internet. An example is a media player.

PlayStation A popular gaming system.

PLC 1. Abbreviation of **programmable logic controller**. 2. Abbreviation of **power-line communication**.

PLCC 1. Abbreviation of **plastic-leaded chip carrier**. 2. Abbreviation of **plastic-leadless chip carrier**.

PLD Abbreviation of **programmable logic device**.

plenum 1. An enclosure whose gas pressure is higher than the surrounding environment. 2. In a structure, such as an office building, a space between real ceilings and drop ceilings, or between walls or floors, which can be used to run wires and cables.

plenum cable A cable whose flammability and smoke characteristics make it safe for running through a plenum, without the need for a conduit.

PLF Abbreviation of **plant load factor**.

PLL Abbreviation of **phase-locked loop**, or **phase-lock loop**.

PLM Abbreviation of **pulse-length modulation**.

PLO Abbreviation of **phase-locked oscillator**.

plot 1. A curve or other graphical representation. Also, to draw, trace, print, or display such a curve or graphical representation. 2. To produce an image, such as a curve or other graphical representation, by drawing a series of lines.

plotter A printer which draws images, such as a curves or other graphical representations, by drawing a series of lines usually using ink pens. Used, for instance, in CAD and architecture. Also called **graphics plotter**.

plug 1. An electric connector with one or more pins or prongs, which is inserted into a **jack** with matching recessed openings. When said plug is inserted, the circuit is closed. This allows for an easy and rapid connection which is secure. Widely used in communications, computer, and entertainment devices, equipment, and systems. Also called **plug connector**. 2. A **plug** (1) used in the context of power line terminations. Such a plug is inserted into a power outlet, socket, or receptacle, so as to access electric power. Also called **power plug**.

plug adapter 1. A fitting utilized to convert a plug in one format to that of another. 2. A fitting utilized to insert a plug into a jack, outlet, socket, or receptacle which would otherwise not match.

plug-and-play Its abbreviation is **PnP**. 1. The ability of a computer system to automatically establish the proper configuration for a peripheral or expansion card which is connected to it. A PnP device, once installed, should be able to be used immediately, without any manual configuration. 2. A peripheral or expansion card, such as a monitor or sound card, which supports **plug-and-play** (1).

plug-board Same as **plugboard**.

plug connector Same as **plug** (1).

plug fuse A fuse which screws into an Edison base.

plug-in Also spelled **plugin**. 1. A program which is used in association with a larger program, to provide the latter with additional functionality. A browser plug-in is an example. Also called **plug-in software**, **plug-in program**, **plug-in application**, or **plug-in component** (2). 2. A component, circuit, circuit board, device, or functional unit which has the proper terminals to be plugged-in to another assembly or unit. For instance, a capacitor, transistor, fuse, meter, or module with pins, prongs, clips, or other terminations which match for easy insertion or removal.

plug in To insert a **plug** into a matching jack, outlet, socket, or receptacle. Also, to insert a plug into an adapter. Also, to insert a matching component, circuit, circuit board, device, or functional unit into another assembly. For instance, to plug an adapter card into a computer.

plug-in application Same as **plug-in** (1). Also spelled **plugin application**.

plug-in capacitor A capacitor which has the proper terminals to allow for easy insertion and removal from a matching socket. Also spelled **plugin capacitor**.

plug-in circuit A circuit, such as a circuit board, which has the proper terminals to be plugged-in to another assembly or unit. Also spelled **plugin circuit**.

plug-in coil A coil which has the proper terminals to allow for easy insertion and removal from a matching socket. Also spelled **plugin coil**.

plug-in component Also spelled **plugin component**. 1. A component, such as a capacitor or resistor, which has the proper terminals to be plugged-in to another assembly or unit. 2. Same as **plug-in** (1).

plug-in device A device, such as a meter, which has the proper terminals to be plugged-in to another assembly or unit. Also spelled **plugin device**.

plug-in fuse A fuse, such as a cartridge fuse, which has the proper terminals to allow for easy insertion and removal from a matching socket. Also spelled **plugin fuse**.

plug-in meter A meter, such as a voltmeter, which has the proper terminals to be easily plugged-into or removed from a circuit. Also spelled **plugin meter**.

plug-in module A module, such as a circuit board, which has the proper terminals to be plugged-in to another assembly or unit. Also spelled **plugin module**.

plug-in program Same as **plug-in** (1). Also spelled **plugin program**.

plug-in resistor A resistor which has the proper terminals to allow for easy insertion and removal from a matching socket. Also spelled **plugin resistor**.

plug-in software Same as **plug-in** (1). Also spelled **plugin software**.

plug-in unit A unit, such as an amplifier, which has the proper terminals to be plugged-in to another assembly or unit. Also spelled **plugin unit**.

plug & play Same as **plug-and-play**.

plugboard A board upon which the terminals of components or systems are readily accessible for temporary connection, especially by inserting or removing plugs from jacks or sockets. Also spelled **plug-board**.

pluggable That which has a **plug**, and which can therefore be plugged in.

plugin Same as **plug-in**.

plugin application Same as **plug-in** (1). Also spelled **plug-in application**.

plugin capacitor Same as **plug-in capacitor**.

plugin circuit Same as **plug-in circuit**.

plugin coil Same as **plug-in coil**.

plugin component Same as **plug-in component**.

plugin device Same as **plug-in device**.

plugin fuse Same as **plug-in fuse**.

plugin meter Same as **plug-in meter**.

plugin module Same as **plug-in module**.

plugin program Same as **plug-in** (1). Also spelled **plug-in program**.

plugin resistor Same as **plug-in resistor**.

plugin software Same as **plug-in** (1). Also spelled **plug-in software**.

plugin unit Same as **plug-in unit**.

plumbicon A TV camera tube similar to a vidicon, but with greater sensitivity.

plumbing 1. The sections and devices, such as elbows and joints, utilized to channel microwave energy. Also called **microwave plumbing**. 2. The sections and devices, such as elbows and tees, utilized to channel energy in a waveguide. Also, the installation of such a system. Also called **waveguide plumbing**.

plunger Same as **piston** (1).

plutonium A silvery-white radioactive transuranic metallic chemical element whose atomic number is 94. It is chemically active, has around 20 identified isotopes, all unstable, and is used in nuclear power reactors, and for power generation in remote areas such as outer space. Its chemical symbol is **Pu**.

PLZT Abbreviation of **lead lanthanum zirconate titanate**. A ferroelectric ceramic which exhibits the piezoelectric effect, and whose properties are similar to **PZT**. The first letter of its abbreviation is taken from **Pb**, the chemical symbol for lead.

pm Abbreviation of **picometer**.

Pm 1. Chemical symbol for **promethium**. 2. Abbreviation of **petameter**.

PM 1. Abbreviation of **phase modulation**. 2. Abbreviation of **pulse modulation**. 3. Abbreviation of **permanent magnet**.

PM dynamic loudspeaker Abbreviation of **permanent-magnet dynamic loudspeaker**.

PM dynamic speaker Abbreviation of **permanent-magnet dynamic speaker**.

PM loudspeaker Abbreviation of **permanent-magnet loudspeaker**.

PM speaker Abbreviation of **permanent-magnet speaker**.

PMBX Abbreviation of **private manual branch exchange**.

PMC Abbreviation of **PCI mezzanine card**.

PMD Abbreviation of **polarization mode dispersion**.

PMG Abbreviation of **permanent-magnet generator**.

PMMA Abbreviation of **polymethylmethacrylate**.

pmol Abbreviation of **picomole**, or **picomol**.

PMOS Abbreviation of **p**-channel **MOS**, which itself is an abbreviation of **p**-channel metal-oxide semiconductor. 1. A MOS technology in which a p-type channel is used, making mobile holes the charge carriers. Since holes move slower than electrons, such devices are slower than **NMOS**, in which electrons are the charge carriers. Also called **PMOS technology**. 2. A semiconductor device incorporating the technology described in **PMOS** (1). Also called **PMOS device**.

PMOS device Same as **PMOS** (2).

PMOS technology Same as **PMOS** (1).

PMOS transistor A field-effect transistor utilizing **PMOS** (1). Also called **PMOSFET**.

PMOSFET Same as **PMOS transistor**.

PMP Abbreviation of **Peak Music Power**.

PMPO Abbreviation of **Peak Music Power Output**.

PMS Abbreviation of **Pantone Matching System**.

PMT Abbreviation of **photomultiplier tube**.

pN Abbreviation of **piconewton**.

PN Same as **pn**.

pn Abbreviation of **positive-negative**. It refers to the region where a p-type semiconductor and an n-type semiconductor meet. In a p-type semiconductor, holes, which act as positively-charged particles, are the majority carrier, while in an n-type semiconductor electrons, which are negative particles, are the majority carrier. Also called **np**.

pn diode Same as **pn-junction diode**.

pn junction The relatively narrow region where a p-type semiconductor and an n-type semiconductor meet. A homojunction is that between layers of the same semiconductor material, each layer having different properties, and a heterojunction is the boundary between layers of different semiconductor materials. Most diodes consist of a pn junction, and bipolar junction transistors have two pn junctions. Also called **np junction**.

pn-junction diode A semiconductor diode consisting of a **pn junction**, and which is utilized to pass current essentially in only one direction. Such diodes are used, for instance, in solar cells, and in laser diodes. Also called **pn diode, junction diode**, or **junction rectifier**.

PN sequence Abbreviation of **pseudonoise sequence**. A spread-spectrum digital code technique in which a pseudo-random sequence of noise signals spreads multiple information signals along a wideband path. The receiver uses the same sequence to decode the signals, providing for increased signal density for a given bandwidth, plus greater security. Used, for instance, in CDMA.

pneumatic **1.** Pertaining to gases, especially air. **2.** That which utilizes air, especially compressed air, to operate. For example, a robotic actuator may incorporate pneumatic components.

pneumatic computer A digital computer which utilizes logic elements powered by a gas, such as air. Such a computer has no electronic circuits, nor any moving parts. It is a type of fluid computer.

PNG Acronym for **Portable Network Graphics**.

PNNI Abbreviation of **Private Network-to-Network Interface**.

PnP Abbreviation of **plug-and-play**.

PNP transistor A bipolar junction transistor with p-type material emitter and collector regions separated by a n-type material base region. An **NPN transistor** has n-type emitter and collector regions separated by an p-type base region.

Po Chemical symbol for **polonium**.

pocket computer A computer which is small enough to hold in a hand or place in a pocket. Such computers generally have a specialized operating system, and provide an interface to communicate with other computers, such as desktop models. There are many types, including those that are industry-specific, such as those utilized by transport companies to keep track of shipments, those that provide only email functions, or PDAs. Also called **palmtop**, or **handheld computer**.

point **1.** Within space, a location or object which has no properties except location. Such a point has no length, width, area, volume, or the like. For instance, a point charge, a point along a line, or the point where two lines intersect. **2.** A location or object whose dimensions are extremely small, or exceedingly small in relation to something else. An example is a point source. **3.** A tip or end which is tapered, sharpened, or otherwise has an extremely small area. For instance, the tip of a cat whisker. **4.** A symbol, usually a dot or comma, which separates the integral part from the fractional part of a number. For example, a decimal point. **5.** A given moment in time, or within an interval of time. For instance, the starting point of a process. **6.** A defined condition, state, degree, or limit. For example, the melting point of a pure substance. **7.** The spot where something is located or occurs. For instance, a network access point, a branch point, an antenna feed point, a breakpoint, or a cutoff point. **8.** A condition, magnitude, or value which is desired or serves as a reference. For example, a control point, or a half-power point. **9.** A unit of angle measure equal to 1/32 of a circle, or 11°15'. Used, for instance, in navigation. **10.** A unit of proportion equal to 0.01 or 1%. **11.** A unit of mass equal to 2 milligrams, or 2.0×10^{-6} kg. Usually used in the context of precious stones. **12.** To indicate with a **pointer**.

point-and-click The ability to move a cursor over a desired location on a computer display, then clicking, as with a mouse or other pointing device, for selection. Used, for instance, to access menu choices or to activate programs. Also, the act of so pointing and clicking. Its abbreviation is **point & click**.

point charge An electric charge considered to occupy a single point in space which has neither area nor volume. Also called **electric point charge**.

point contact Its abbreviation is **PC**. **1.** The touching of two objects at a single point. Also, the point at which such contact is made. **2.** A contact between a pointed metallic wire, such as a cat whisker, and a semiconductor surface.

point-contact diode A diode whose operation depends on a **point contact** (2).

point-contact transistor A transistor whose operation depends on two **point contacts** (2), one serving as an emitter, the other as a collector.

point defect A flaw in the geometric arrangement of one, or few, of the contained atoms, molecules, or ions of a crystal.

point impedance The impedance at a given point, such as an antenna feed point.

point of presence Its abbreviation is **POP**. **1.** The physical location where a long-distance telephone network interfaces with a local exchange carrier. **2.** The physical location where a cellular telephone network interfaces with a local exchange carrier. **3.** An access point where different communications services interface, such as the location where a LAN meets a local exchange carrier.

point of sale Its abbreviation is **POS**. **1.** A location at which purchases are made. **2.** An electronic system incorporating equipment such as bar-code, RFID, and/or magnetic stripe readers, and terminals, for pricing and recording transactions, updating inventories, obtaining purchase authorizations, and the like. Also called **point-of-sale system**. **3.** A specific location which is a part of a **point of sale** (2). For instance, a department store may have several dozen such stations. Also called **electronic point-of-sale**.

point-of-sale keyboard A keyboard utilized at a **point-of-sale terminal**. Such a keyboard usually incorporates magnetic stripe, bar-code, and/or smart card readers. Its abbreviation is **POS keyboard**.

point-of-sale system Same as **point of sale** (2). Its abbreviation is **POS system**.

point-of-sale terminal A terminal, such as an electronic cash register, used in a **point-of-sale system**. Its abbreviation is **POS terminal**.

point source A source, such as that of electromagnetic radiation or sound, whose dimensions are so small in relation to the distance traveled to its detector or receiver, that it can be considered to be just a point.

point-to-multipoint From one point to many, as in point-to-multipoint communication.

point-to-multipoint communication Communication between one point and multiple points. For example, a transmission from a satellite to many receivers.

point-to-point From one point to another, as in point-to-point communication.

point-to-point communication Communication between two precisely defined points, such as relay satellites or fixed stations.

point-to-point configuration Communication between two specified points. For example, a communications circuit with two nodes.

point-to-point connection The connection established in point-to-point communications.

point-to-point motion In robotics, motion which occurs along a limited number of points along a given path, as opposed to an infinitely continuous path.

Point-to-Point Protocol Same as **PPP**.

Point-to-Point Protocol over Ethernet Same as **PPPoE**.

point-to-point transmission The transmission of signals between two precisely defined points, such as relay satellites or fixed stations.

Point-to-Point Tunneling Protocol Same as **PPTP**.

point-to-point wiring On a circuit board, the manual wiring between components, as opposed to printed wiring.

point & click Abbreviation of **point-and-click**.

pointer 1. On a computer screen, an indicator, such as a small hand, arrow, or I-beam, that moves as the mouse is moved. It serves to select text, menus, and the area of the screen where the next text input or other action will occur, by clicking the mouse there. Also called **mouse pointer, cursor** (2), or **mouse cursor**. 2. A **pointer** (1) indicated by a pointing device, such as a trackball, other than a mouse. 3. In a meter, gauge, or similar device or instrument, an object, such as a needle or wire, which serves to convey information. 4. Same as **pointer needle**.

pointer needle A needle which serves to convey information in a meter, gauge, or similar device or instrument. A specific example is that used in a D'Arsonval galvanometer. Also called **pointer** (4), **needle** (2), or **indicator needle**.

pointing device An input device, such as a mouse, puck, or touchpad, which moves the **pointer** (1) or **pointer** (2) on a computer display.

pointing stick A pointing device consisting of a small tip, approximately the size of a pencil eraser, placed near the middle of a keyboard. Seen, for instance, in notebook computers.

poise The cgs unit of dynamic viscosity. One poise equals 0.1 pascal second. Its symbol is **P**.

Poisson distribution In statistics, a probability distribution which indicates the likelihood that one or more specified events will occur within a given time frame.

poke To alter or store the content of a memory location which has been **peeked**. Also the command to perform this action.

POL Abbreviation of **problem-oriented language**.

polar 1. Of, or pertaining to, having, or characterized, by one or more **poles**. 2. Utilizing a pole, such as a magnetic or geographic pole, as a reference, or located near such a pole. 3. Passing over the earth's poles, as in a polar orbit.

polar angle In **polar coordinates**, the angle formed between a straight line joining a specific point to the origin, and the polar axis.

polar axis 1. In **polar coordinates**, the straight line relative to which the polar angle is measured. 2. In a crystal, an axis of symmetry with no plane of symmetry perpendicular to it.

polar coordinates A system of coordinates in which the position of a point within a given plane is determined by its distance from the origin, and the angle, called polar angle, formed between a straight line joining said point to said origin and another line, which is fixed, called the polar axis.

polar diagram A diagram utilizing polar coordinates to show how the intensity of a quantity varies as a function of direction and distance. An example is an antenna diagram.

polar orbit An orbit, such as that of a communications satellite, that passes over, or close to, the earth's poles. Such an orbit provides a different view each time around, as the earth rotates underneath.

polar-orbiting satellite A satellite following a **polar orbit**. Also called **polar satellite**.

polar pattern A pattern utilizing polar coordinates to show how the intensity of a quantity varies as a function of direction and distance. An example is an antenna pattern.

polar relay Same as **polarized relay**.

polar satellite Same as **polar-orbiting satellite**.

polarimeter An instrument which measures the rotation of the plane of polarization of a sample placed within it. Chiral molecules, for instance, rotate the plane of polarization of plane-polarized light, so a polarimeter measures optical activity.

polarimetry The measurement of the rotation of the plane of polarization of radiant energy, such as light, usually through the use of a polarimeter.

polariscope An instrument which combines a polarizer and an analyzer to study the effects a given sample has on polarized light.

polarity 1. The possession or manifestation of two opposite or opposing attributes, magnitudes, or the like. For example, electrical or magnetic polarity. Also, the indication or orientation of a state or condition arising from such polarity, as in positive or negative polarity. 2. The property of having an excess or deficiency of electrons. An excess of electrons produces a negative polarity, and a deficiency of electrons produces a positive polarity. This determines the direction of the flow of current, as electrons move from a point with an excess of electrons towards a point where there is a deficiency of electrons. Also called **electrical polarity** (1). 3. The characteristic of having two opposite charges, positive and negative, within the same body or system. A battery, for instance, has two terminals, each with opposite polarity, which are the positive terminal and the negative terminal. Also called **electrical polarity** (2). 4. In a magnet or magnetic body, the presence and manifestation of two regions where magnetic intensity is at a maximum, each of which is opposite in nature. These regions are labeled north, or positive, pole and south, or negative, pole. Also called **magnetic polarity**.

polarity-reversal switch Same as **polarity switch**.

polarity-reversing switch Same as **polarity switch**.

polarity switch A switch utilized to reverse polarity. Used, for instance in one device to match that of another. Also called **polarity-reversing switch**, or **polarity-reversal switch**.

polarization 1. The production or creation of a state in which there are two opposite or opposing attributes, magnitudes, or the like. For instance, the application of an external electric field so as to separate the electric charges of an object or body, thus forming a positive and a negative pole. 2. A property of an electromagnetic wave which describes the time-varying direction of the electric field vector. Such polarization may be classified, for instance, as circular, elliptical, or linear, each of which describes the shaped traced by the electric field vector as a function of time. Polarization may also be defined in terms of the direction of the electric field with respect to a transmitting antenna or another reference, such as the surface of the earth, in which case it is said to have horizontal or vertical polarization. Also called **wave polarization**. 3. The confining of the vibrations of an electromagnetic wave, such as light, to a single plane, as opposed to the innumerable planes rotating around the vector axis. For instance, the production of plane-polarized light from unpolarized light. 4. The orientation of the radiated and received electric lines of flux of an antenna, in relation to the surface of the earth. Also called **antenna polarization**. 5. A phenomenon observed in dielectrics, in which the electrons in each atom are displaced in the direction opposite to that of an applied electric field, while the nucleus of each atom is displaced in the direction of said field. Also called **dielectric polarization**, or **electric polarization**. 6. The lining up of magnetic dipoles in a material. In a ferromagnetic material magnetic polarization occurs spontaneously due to the mutual attraction of said dipoles. Also called **magnetic polarization**. 7. In certain electric cells, chemical processes which shorten the useful life. For example, bubbles which may form around an electrode, hence increasing the internal resistance of such a cell.

polarization beam splitter An optical or optoelectronic device which splits a beam of light into polarized beams. Used, for instance, in lasers. Its abbreviation is **PBS**.

polarization direction In a polarized wave, the direction of the electric vector. Also called **direction of polarization**.

polarization diversity Same as **polarization-diversity system**.

polarization-diversity gain Signal gain through the use of **polarization-diversity reception**.

polarization-diversity radar A form of radar reception which utilizes two sets of antennas and receivers, each set tuned to the same signal. One antenna is vertically polarized, while the other is horizontally polarized.

polarization-diversity receiver A receiver used with **polarization-diversity reception**.

polarization-diversity reception A mode of reception in which two arriving radio-frequency signals, each with the same information, are received by two antennas, one vertically polarized, the other horizontally polarized. Two receivers are used, and the two signals are combined so as to minimize the effects of fading.

polarization-diversity system A reception system which employs **polarization-diversity reception**. Also called **polarization diversity**.

polarization error An error arising from a change in the polarization of waves between a transmitter and receiver. This may occur, for instance, as a consequence of reflection off the ionosphere. When used in the context of direction-finding, this leads to errors in bearing or course indications, and is also called **night error**, as this effect is manifested mostly at night, especially during twilight, when comparatively rapid changes occur in the ionosphere.

polarization fading Fading in the reception of radio waves due to variations in the polarization of the arriving signal relative to the orientation of the receiving antenna.

polarization mode dispersion In an optical fiber, a form of dispersion resulting from differences in the propagation velocities of orthogonal polarization modes. This results in a spreading of the pulse. Its abbreviation is **PMD**.

polarization modulation A form of modulation in which the polarization of a beam is varied proportionally to the intensity of the signal source. Used, for instance, in optical communications.

polarization plane For an electromagnetic wave, the plane which contains both the electric field vector and the direction of propagation of said wave. Also called **plane of polarization**.

polarized capacitor A capacitor whose leads must each be connected to specific positive and negative terminals. A **non-polarized capacitor** can be connected either way.

polarized component A component, such as a diode or a polarized capacitor, whose leads must each be connected in a specific orientation or manner.

polarized electrolytic capacitor An electrolytic capacitor whose leads must each be connected to specific positive and negative terminals. A **non-polarized electrolytic capacitor** can be connected either way.

polarized electromagnetic radiation Same as **polarized radiation**.

polarized light **1.** Electromagnetic radiation in the visible range whose electric field vector is confined to a single plane. **2.** Electromagnetic radiation in the visible range which is polarized, be it elliptically, circularly, plane-polarized, or the like.

polarized outlet A power outlet into which a **polarized plug** is inserted. Also called **polarized receptacle**.

polarized plug A power plug which can only be inserted into a corresponding outlet in one orientation. Used, for instance, to help insure that the ground of the power line is connected to the ground terminal of a device or piece of equipment using such a plug. A **polarized plug** plugs into a **polarized outlet**.

polarized radiation Also called **polarized electromagnetic radiation**. **1.** Electromagnetic radiation whose electric field vector is confined to a single plane. **2.** Electromagnetic radiation which is polarized, be it elliptically, circularly, plane-polarized, or the like.

polarized receptacle Same as **polarized outlet**.

polarized relay Also called **polar relay**. **1.** A relay which responds to the direction of the current energizing it. **2.** A relay which responds to the direction of the current, voltage, power, or the like, energizing it. Also called **directional relay**.

polarizer That which polarizes, or which changes polarization. For instance, a Nicol prism, which produces plane-polarized light from unpolarized light.

polarizing angle For unpolarized light incident upon a material, the angle at which reflected light has the greatest polarization. At this angle of incidence, the angle between reflected and transmitted rays is 90°. Also called **Brewster angle**.

polarizing filter A filter that polarizes electromagnetic radiation, especially light, passing thorough it. For instance, such a filter may produce plane-polarized light, and might be composed of plastic sheets with birefringent crystals appropriately oriented. Used, for example, in photography.

polarographic analysis Same as **polarography**.

polarography Any of various techniques for quantitative analysis utilizing current-voltage curves generated by low-concentration samples in an electrolytic cell. As a variable potential is applied to such a cell, the relationship of changes in voltage to changes in current are graphed, which can be utilized to identify and study the samples. Also called **polarographic analysis**.

Polaroid camera A camera which utilizes a single-step development process to provide a finished positive print. Also called **Land camera**.

pole **1.** Each of the input or output contacts or terminals of a switch or relay. The term may also refer to independent sets of contacts used to open or close a circuit, each of which would be a pole. **2.** Each of the input or output terminals, electrodes, or lines of a component, circuit, or device. For example, the poles of a battery. **3.** Either of the two locations on the surface of the planet earth where the magnetic dip is 90°. At such a place, the magnetic meridians converge. In the northern hemisphere such spot is called the **north magnetic pole**, while that of the southern hemisphere is the **south magnetic pole**. The earth's magnetic poles do not coincide with its geographic poles, and the latter's position vary over time. Also called **magnetic pole (1)**, or **dip pole**. **4.** In a magnet or magnetic body, either of the two regions where magnetic intensity is at a maximum. The magnetic lines of force converge in such areas, which are labeled north pole and south pole. If each of such regions is considered to occupy a single point in space, it could be considered to be analogous to an electric point charge. Also called **magnetic pole (2)**. **5.** A point of concentration of electric charge, such as each of the two opposite charges of equal magnitude separated by a very small distance in a dipole. **6.** A post which serves to support suspended cables or conductors, such as those of electricity or communications. **7.** Either of the extremities of an axis which runs through a sphere. For example, the north geographic pole of the earth. **8.** The origin in a system utilizing polar coordinates.

pole face In a pole piece, a surface facing an air gap.

pole piece 1. A piece or section of a magnetic material on either end of a magnet or electromagnet. 2. In a magnet or electromagnet, a piece or section of a magnetic material on either end of an air gap.

pole shoe A **pole piece** (1) which helps reduce the reluctance of an air gap.

poling The adjustment of **polarity**. For example, to align electric dipoles utilizing an external field.

Polish notation 1. A manner of forming mathematical or logical expressions, in which the operators appear before or after, but not between the operands. When utilizing Polish notation parenthesis are never used, which is better suited for the way digital computers ordinarily evaluate expressions. Polish notation may be in the form of prefix notation or postfix notation. Also called **parenthesis-free notation**. 2. A manner of forming mathematical or logical expressions in which the operators appear before the operands. For example, $(A + B) \times C$, which is in **infix notation**, would appear as *+ABC in Polish notation. The same expression would appear as ABC+* in **reverse Polish notation**. Also called **prefix notation**, or **Lukasiewicz notation**.

polling Also called **autopolling**. 1. In a system, the periodic detection of the status of all connected devices, to address any needs for resources. This contrasts with an event-driven allocation scheme, where interrupts are sent to request resources. 2. In networking, the sequential determination of which terminals wish to transmit data, to allow doing so in an orderly manner. 3. A feature that allows a facsimile machine to call another, and request the remote machine to transmit documents. Also called **fax polling**.

polling cycle A complete **polling** (1) or **polling** (2) sequence. For example, in one such cycle, each connected terminal is interrogated once. Also called **autopolling cycle**.

polonium A radioactive chemical element whose atomic number is 84. It has over 40 identified isotopes, more than any other known element, yet very rarely occurs naturally. It is used as a nuclear power source, and for compact thermoelectric power generation ideally suited for use in remote areas such as outer space. Its chemical symbol is **Po**.

poly-directional Also spelled **polydirectional**. Also called **multi-directional**. 1. Able to operate, or be operated, in two or more directions. 2. Pertaining to, or utilizing two or more directions.

poly-directional microphone A microphone which is able to switch between multiple pickup patterns. Also spelled **polydirectional microphone**. Also called **multi-directional microphone**.

polycarbonate A clear polymer which is highly impact-resistant, features excellent dimensional stability, and which is used, for instance, in optical lenses, filters, and discs, and in protective eyewear and windows. Its abbreviation is **PC**.

polychlorinated biphenyls A highly toxic group of organic compounds which are stable and heat-resistant. Once widely used in capacitors, transformers, and batteries. Its abbreviation is **PCBs**. Also called **chlorinated biphenyls**, or **chlorobiphenyls**.

polychloroprene A synthetic rubber which is rugged, water-resistant, and especially impervious to oil. Use, for instance, for highly weather-resistant cable jackets. Also called **neoprene**.

polychromatic 1. Composed of more than one color. Also, appearing to have more than one color. 2. Of or composed of radiation of multiple wavelengths, or of a wide interval of wavelengths. For example, white light. 3. Of or composed of particles with different energy levels.

polychromatic radiation Electromagnetic radiation, such as white light, which is **polychromatic**.

polycrystalline Pertaining to, or consisting of, a polycrystalline structure or polycrystalline material.

polycrystalline material 1. A crystalline material in which there is short-range order. The overall crystalline structure is composed of highly-ordered units, with grain boundaries between individual crystals. A **single-crystal material (1)** has no such boundaries, so there is complete uniformity over the entire crystal. The grain boundaries decrease the efficiency of any semiconductor device consisting of a polycrystalline material. Used, for instance, as semiconductor materials which do not have to be of the highest quality. 2. A crystalline material consisting or more than one type of crystal.

polycrystalline silicon Same as **polysilicon**.

polycrystalline structure The crystal structure present in a **polycrystalline material**.

polydirectional Same as **poly-directional**.

polydirectional microphone Same as **poly-directional microphone**.

polyester A strong and resistant synthetic fiber which is used as an insulator and dielectric, and as a substrate for magnetic recording media.

polyester capacitor Same as **polyester film capacitor**.

polyester film A film made of **polyester**. Used, for instance, in polyester film capacitors.

polyester film capacitor A capacitor in which the dielectric is **polyester film**. Also called **polyester capacitor**.

polyethylene A synthetic water-resistant, flexible, and thermoplastic material which is made via the polymerization of ethylene, whose chemical formula is C_2H_4. It is used as an insulator and dielectric, and in containers and pipes.

polygon An enclosed geometric figure bounded by three or more straight lines. A **triangle** is a three-sided polygon, a **quadrilateral** is a four-sided polygon, a **pentagon** has five sides, a **hexagon** six, a **septagon** seven, an **octagon** eight, and so on.

polygraph A device which monitors and records changes in physiological variables such as respiration, blood pressure, and perspiration, while a subject is questioned. Such a device purportedly can detect when a person is lying. Also called **lie detector**.

polyimide A strong, and temperature-resistant resin which is wear and corrosion-resistant, and which is used, for instance, as a dielectric, as a protective film, or as a substrate in printed-circuit boards.

polymer A compound or substance composed of repeating subunits called monomers, which form larger units under the influence of heat, pressure, or catalysts. Polymers are often plastics or resins, and an example is polyethylene.

polymerization The processes involved in forming a polymer out of repeating units of monomers.

polymethylmethacrylate A clear thermoplastic resin which is used, for instance, in optical instruments, lighting fixtures, and as a substrate for optical discs. Its abbreviation is **PMMA**.

polymorphic virus A computer virus which produces different copies of itself each time it infects. Each variation is operational, and it is intended to thwart antivirus software.

polymorphism 1. The occurrence of different forms or shapes. 2. In object-oriented programming, the ability to behave differently, or elicit a different response, depending on the circumstances. For example, a routine may be handled in a given manner depending on the characteristics of an object.

polyphase Consisting of, having, or using two or more phases. For instance, an AC power line with two or more electrical phases, or a relay which responds to the current such a line provides. Also called **multiphase**.

polyphase circuit An electric circuit in which the voltages or currents are out-of-phase with each other by a specified amount. For example, in a three-phase AC circuit, the voltages may each differ in phase by 120°, or one-third of a cycle.

polyphase current A current, such as that flowing through a polyphase circuit, which is out-of-phase with another current by a specified amount.

polyphase meter A meter, such as a wattmeter, utilized in a polyphase circuit.

polyphase motor A motor which utilizes polyphase current.

polyphase power 1. The power utilized by a polyphase circuit or system. 2. The power provided by a polyphase circuit or system.

polyphase system A system, such as a polyphase circuit, which utilizes two or more voltages or currents, each of which is out-of-phase with the others by a specified amount.

polyphase transformer A transformer utilized to supply a polyphase circuit, device, piece of equipment, or system.

polyphase voltage A voltage, such as that supplying a polyphase circuit, which is out-of-phase with another voltage by a specified amount. For example, in a three-phase AC circuit, the voltages may each differ in phase by 120°, or one-third of a cycle.

polyphonic ringtones A feature, especially that of a cell phone, enabling multiple tones or notes to be played simultaneously, resulting in a fuller and more realistic sound experience.

polypropylene A synthetic and tough thermoplastic material which is made via the polymerization of propylene, whose chemical formula is C_3H_6. It is water-resistant, oil-resistant, is used as an insulator and dielectric, and in molded articles such as containers.

polyrod antenna An antenna consisting of a parallel array of rods composed of a dielectric material such as polystyrene. Used for transmitting microwaves.

polysilicon Abbreviation of **poly**crystalline **silicon**. A polycrystalline form of silicon which is used, for instance, in semiconductors, and in LCD displays.

polysilicon LCD Abbreviation of **polysilicon** liquid-crystal display. A panel or screen consisting of three LCD layers, one each for red, green, and blue, and which provides enhanced colors, brightness, and contrast in comparison to traditional LCDs. Used, for instance, in LCD projectors. Also, the technology pertaining to such panels or screens.

polysilicon liquid-crystal display Same as **polysilicon LCD**.

polysilicon TFT Abbreviation of **polysilicon TFT LCD**.

polysilicon TFT LCD A technology enabling active-matrix displays to allow more light to pass through LCDs at higher temperatures. Also, a screen or panel incorporating this technology. It is an abbreviation of **polysilicon thin-film transistor** liquid-crystal display. Its own abbreviation is **polysilicon TFT**.

polystyrene A clear, synthetic, water-resistant, and tough thermoplastic material which is made via the polymerization of styrene, whose chemical formula is C_8H_8. It is used, for instance, as an electrical and thermal insulator, as a dielectric, and in molded articles and packaging.

polystyrene capacitor A capacitor in which polystyrene film serves as the dielectric. Also called **polystyrene film capacitor**.

polystyrene film A film made of **polystyrene**. Used, for instance, in polystyrene capacitors.

polystyrene film capacitor Same as **polystyrene capacitor**.

polysulfone A clear and tough thermoplastic with high heat resistance and excellent dimensional stability, and which is used, for example, as an electrical insulator and dielectric.

polytetrafluoroethylene A chemically-resistant, waterproof, and heat-tolerant fluorocarbon-based resin with an extremely low coefficient of friction. It has a self-lubricating effect which makes it ideally suited for use in bearings and joints. Its other applications include its use for electrical and thermal insulation. Its abbreviation is **PTFE**. Much better known as **Teflon**. Also called **polytetrafluoroethylene resin**, or **tetrafluoroethylene**.

polytetrafluoroethylene resin Same as **polytetrafluoroethylene**.

polyurethane A thermoplastic or thermosetting synthetic polymer which is waterproof, impact resistant, durable, and hard yet flexible. It has many applications, including its use as an electrical and thermal insulator, and for impact protection for components and devices.

polyvinyl chloride A synthetic, water-resistant, and tough thermoplastic material which is made via the polymerization of vinyl chloride, whose chemical formula is C_2H_3Cl. It also resistant to fire and chemicals, and is used, for instance, as an electrical insulator, as a dielectric, and in molded articles and pipes. Its abbreviation is **PVC**.

PON Abbreviation of **passive optical network**.

pool cathode 1. In a gas tube, a cathode consisting of a pool of mercury. Also called **mercury-pool cathode**. 2. A cathode consisting of a pool of a liquid metal, usually mercury.

POP 1. Abbreviation of **point of presence**. 2. Abbreviation of **Post Office Protocol**.

pop To remove the first item from the top of a stack. This contrasts with **push** (1), which is to add an item to the top of a stack.

pop-down menu Same as **pull-down menu**.

POP server Abbreviation of Post Office Protocol **server** A server which handles **Post Office Protocol** tasks, such as storage, and email client requests for downloads.

pop-under ad Similar to a **pop-up ad**, except that it displays in a new window behind the currently open browser window. In this manner, it is seen, for instance, when the browser is closed. Such ads are just as abusive as pop-up ads.

pop-up Same as **pop-up window**. Also spelled **popup**.

pop-up ad An advertisement that appears on a Web page in a new window. Many Web sites generate multiple such ads, and these are often regarded as abusive. Also spelled **popup ad**.

pop-up help Help offered in the form of pop-up windows. For example, context-sensitive help which opens a smaller window providing information when a given topic is selected. Also spelled **popup help**.

pop-up menu A menu which is provided in the form of pop-up windows. When a final choice is selected, the window closes. Also spelled **popup menu**.

pop-up message A message, such as an error message, which appears in the form of a pop-up window. Such messages usually requires a user to acknowledge it by pressing "OK," before being able to continue. Also spelled **popup message**.

pop-up window A window, such as that of a pop-up message, which appears on a computer display under specified circumstances. Also spelled **popup window**. Also called **pop-up**.

popcorn noise Random noise which sounds similar to popping corn. Observed, for instance, in certain audio circuits utilizing semiconductor devices with impurities in the junctions.

populate To place or supply with components, devices, data, members, or the like. For example, to mount components on a circuit board.

population 1. In statistics, an aggregate of items from which samples are taken. 2. The placement or supplying of components, devices, data, members, or the like. For example, the mounting of components on a circuit board.

population inversion For an atomic system, the condition where there are more electrons in a higher energy state than a given lower state, such as the ground state. Population inversion is required to establish and maintain lasing action. Since atomic systems tend to prefer more electrons in lower states, external energy, such as electromagnetic radiation of a suitable frequency, is required to maintain population inversion.

popup Same as **pop-up**.

popup ad Same as **pop-up ad**.

popup help Same as **pop-up help**.

popup menu Same as **pop-up menu**.

popup message Same as **pop-up message**.

popup window Same as **pop-up window**.

porcelain A hard white ceramic which is strong, translucent, and which usually has a hard glaze. Used, for instance, as an insulator, a dielectric, and in capacitors.

porcelain capacitor A capacitor in which a high-grade porcelain serves as the dielectric.

porcelain insulator An electrical insulator composed of, or incorporating, porcelain.

porch In a composite video signal, either the front porch or the back porch.

port 1. A point or opening which provides an input, output, or interface. For example, an I/O port, a speaker port, or an input to an electric network. 2. An input, output, or other point where a signal or energy can be provided to, or taken from, a circuit or system. For instance, a point where a meter may be placed for readings. 3. An input, output, or interface between a computer and a peripheral, another computer, or a network. For instance, a data port. 4. An input, output, or interface between a CPU and a peripheral. For example a USB port, or a serial port. Also called **I/O port**. 5. To modify a program so that it can function properly within a different system or architecture. 6. A carefully-dimensioned opening in a loudspeaker enclosure. It is employed for various purposes, but it is mainly used to increase and extend the reproduction of low frequencies. Also called **speaker port, duct (5), ducted port**, or **vent (3)**. 7. An opening in a waveguide through which energy can be fed to, or taken from. For example, an opening utilized to take measurements. Also called **waveguide port**.

Port 80 A port number commonly used over the Internet for HTTP traffic.

port address Same as **port number**.

port address translation The process of redirecting a given IP address or port to a different IP address or port. It is a type of network address translation. Its abbreviation is **PAT**. Also called **port mapping**.

port aggregation The use of multiple paths between network nodes, so as to increase transmission speed.

port density The number of ports for a given network device. Such ports may be physical or logical.

port expander A device which serves to connect multiple inputs to a single port. An example is a USB hub.

port mapping Same as **port address translation**.

port number In a communications network such as a TCP/IP, a number which identifies the type of port. Port 80, for instance, is commonly used over the Internet for HTTP traffic. Also called **port address, protocol port**, or **protocol port number**.

port replicator A device which has multiple computer ports, such as those utilized for keyboards, monitors, printers, and other peripherals, and which provides a single socket into which a portable computer is plugged in. This saves the time and inconvenience of plugging in all components each time the computer is at this location. A port replicator is similar to a docking station, except that the latter provides additional slots for expansion.

port speed The data transfer rate of a given **port (4)**. For example, a USB 2.0 port supports rates of up to 480 Mbps, FireWire provides up to over 2 Gbps, and a 100-Gigabit Ethernet port supports data transfer rates of up to 100 Gbps.

portable 1. That which can be easily moved from one location to another. For example, a portable computer. 2. Computer software which can be utilized across multiple platforms with little or no modifications.

portable computer A computer, such as a PDA or notebook, that can be easily moved from one location to another.

Portable Document Format Same as **PDF**.

portable language A computer language, such as **C**, which works essentially equally well across multiple platforms.

Portable Network Graphics A format for storing bitmapped images, which features lossless compression, grayscale support, and color correction across multiple platforms. Its acronym is **PNG**.

Portable Operating System Interface for UNIX Same as **POSIX**.

portable station A station which can be moved from one location to another. Such a station may or may not be reasonably carried by a single person.

portal A Web site, such as *www.google.com*, which serves as a starting point to most any activity on the Internet. Web portals usually provide news, email, search engines, directories, chats, shopping, local interest topics, and so on. Also called **Web portal**, or **Internet portal**.

ported cabinet A speaker enclosure which incorporates one or more **ports (5)**. Also called **ported enclosure**, or **vented cabinet**.

ported enclosure Same as **ported cabinet**.

ported loudspeaker Same as **ported speaker**.

ported speaker A speaker with a **ported cabinet**. Also called **ported loudspeaker**.

portrait 1. Same as **portrait mode**. 2. In computers, pertaining to something whose height is greater than its width, such as a portrait display.

portrait display A computer display whose height is greater than its width. Also, the positioning of a monitor in this fashion when both portrait and landscape orientations are available. Also called **portrait monitor**.

portrait format Same as **portrait mode**.

portrait mode A manner of printing, or presenting information on a screen, in which the height of a page is greater than its width. This contrasts with **landscape mode**, in which the width is greater than the height. Also called **portrait (1), portrait orientation**, or **portrait format**.

portrait monitor Same as **portrait display**.

portrait orientation Same as **portrait mode**.

pos 1. Abbreviation of **positive**. 2. Abbreviation of **position**.

POS 1. Abbreviation of **point of sale**. 2. Abbreviation of **packet over SONET**.

POS keyboard Same as **point-of-sale keyboard**.

POS system Abbreviation of **point-of-sale system**.

POS terminal Abbreviation of **point-of-sale terminal**.

position Its abbreviation is **pos**. 1. The specific location of something within space. For example, the location of a magnetic pole, a ship, or where a laser beam is directed. Also, to place in such a location. 2. The location of something relative to something else. For instance, a line of position, or the displacement of the value of a voltage from a reference position. Also, to place or be situated at such a location. **3.** The manner or orientation in which something is placed, arranged, or installed. Also, to place, arrange, or install in this manner. For example, the precise positioning of an object by a mechanical arm, or through the use of a micropositioner. **4.** A **position** (1) as opposed to another such position, or other possible positions. For instance, the on position of a switch, or an operating position. **5.** The **position** (1) a component, circuit, device, piece of equipment, or system is set at, when multiple such positions are available. For example, the sensitivity setting of a measuring instrument.

position control **1.** A component, circuit, device, instrument, piece of equipment, signal, mechanism, or system, or a combination of these, that operates, regulates, or manages the position of something. For example, a position control system. **2.** A control, such as a rotary switch, which determines the position or setting of a component, circuit, device, instrument, piece of equipment, mechanism, or system.

position control system A control system that maintains the position of something within specified limits. Such a system, for instance, may alter variables like velocity and orientation, so as to counteract any disturbances which may arise.

position detection Same as **position sensing**.

position detector Same as **position sensor**.

position fixing In navigation, the process of determining an accurate position through the intersection point of two or more bearings. Such a position is determined without reference to a former position.

position indication A reading from a **position indicator**.

position indicator An instrument or device which indicates a given position. Examples include ground-position indicators, and tape counters.

position line In navigation, a line of bearing to a reference or other known point, along which a vessel considered to be located. Such a line may be determined for instance, through the use of radar, loran, ultrasound, landmarks, GPS, and so on. More than one position line may be utilized for a fix. Also called **line of position**.

position sensing The detection of a position utilizing a **position sensor**. Also called **position detection**.

position sensor A device or instrument which detects the position of one or more objects, and converts these into signals suitable for processing, transmission, control, or the like. Also called **position detector**, or **position transducer**.

position setting A setting established by a control, such as a rotary switch, which determines the position of a component, circuit, device, instrument, piece of equipment, mechanism, or system.

position switch A switch utilized to adjust **position settings**.

position transducer Same as **position sensor**.

positional 3D audio The reproduction of sound that produces as faithfully as possible a three-dimensional sound image. Specialized coding and decoding equipment is required, along with an appropriate loudspeaker array. Also known by various other terms, including **positional 3D sound, positional audio, positional sound, three-dimensional audio, 3D sound**, and **ambisonics**.

positional 3D sound Same as **positional 3D audio**.

positional audio Same as **positional 3D audio**.

positional notation A system of representing numbers in which the position of each digit indicates its value. For example, in the decimal number system, all digits to the left of the decimal point represent successive positive powers of 10, while those to the right of the decimal point are successive negative powers of 10. Also called **positional representation**.

positional representation Same as **positional notation**.

positional sound Same as **positional 3D audio**.

positioning The setting, adjusting, controlling, or otherwise seeking to obtain a specific position.

positive Its abbreviation is **pos**, or **P**. Its symbol is **+**. **1.** Of, or pertaining to, a quantity or value above zero. For instance, a temperature greater than absolute zero, or a positive number. **2.** Of, pertaining to, or the same as the charge or a multiple of the charge of a proton. For example, the plus one charge of a sodium ion. **3.** Of, or pertaining to, a particle, material, or other entity which has a deficiency of electrons. For instance, a cation. **4.** Of, or pertaining to, a quantity, value, entity, phenomenon, or concept which is considered the opposite of an equivalent which would be described as negative. For example, a positive electrode, positive image, positive peak, positive pole, and so on. **5.** Same as **positive image**.

positive angle An angle which is measured in the counterclockwise direction from a horizontal axis extending from the origin in a Cartesian coordinate system.

positive bias **1.** A voltage applied to an electrode of an electronic device, such as an electron tube, which makes it positive with respect to another electrode. For example, a voltage that makes the control grid positive with respect to the cathode. **2.** A positive voltage applied to a component or device, especially an electronic device, to establish a reference level for its operation.

positive charge **1.** The charge, or a multiple of the charge, of a proton. Protons possess a positive charge which is equal in magnitude to the negative charge of an electron. **2.** The charge a material or entity acquires when it has a deficiency of electrons. For example, if a dielectric such as glass or amber is rubbed with a dissimilar dielectric such as silk or wool, and each is separated, glass or amber acquires a positive charge, while the silk or wool acquires a negative charge. Also called **positive electrification**, or **positive electricity**.

positive column In an glow-discharge tube, a luminous glow which occurs between the Faraday dark space and the anode.

positive conductor A conductor connected to a **positive terminal**.

positive distortion In optics, distortion which causes an image to bulge convexly on all sides, similar to the shape of a barrel. Seen, for instance on computer or TV screens. Also called **barrel distortion**.

positive edge triggered A device actuated by **positive edge triggering**.

positive edge triggering For a device, such as a flip flop, with edge triggering, actuation when the edge of its clock pulse is rising. For instance, a device which is set to sample data only when the edge of its synchronizing clock pulse is rising. This contrasts with **negative edge triggering**, in which the converse is true.

positive electricity Same as **positive charge** (2).

positive electrification Same as **positive charge** (2).

positive electrode Also called **anode**. **1.** The electrode toward which electrons emitted by the cathode travel in an electron tube. **2.** In an electrolytic cell, the electrode towards which negative ions travel. **3.** The electrode where oxidation occurs in an electrochemical cell.

positive electron Same as **positron**.

positive emitter-coupled logic A logic circuit design similar to emitter-coupled logic, except that the power supply voltage is positive with respect to a reference, such as ground. Its abbreviation is **PECL**.

positive exponent An exponent which is a positive number, as in 5^3, as opposed to a **negative exponent**, as in 5^{-3}. Also called **positive power**.

positive feedback In an amplifier or system, feedback which is in phase with the input signal, thus reinforcing it. While this results in an increase in gain, an excess can destabilize amplification, and produce distortion and noise. Howl may occur, for instance, if positive feedback exceeds a given threshold. Also called **regenerative feedback**, or **regeneration (2)**.

positive-feedback amplifier An amplifier utilizing **positive feedback**.

positive-going Pertaining to a **positive-going pulse** or a **positive-going signal**.

positive-going pulse A pulse whose value or amplitude is increasing in the positive direction.

positive-going signal A signal whose value or amplitude is increasing in the positive direction.

positive grid In an electron tube, a grid which is positive with respect to the cathode.

positive ground A DC electrical system in which the positive terminal is grounded. Occasionally seen in some vehicles.

positive grounding The connection to a **positive ground**.

positive half-cycle The half of a complete sequence of changes of a periodically repeated phenomenon during which a maximum is approached. For instance, that of an AC cycle. Also, the time that elapses during such an interval.

positive image An image, such as a photographic image, in which the light and dark areas are the same as the scene the image is taken from. Also called **positive (5)**, or **photopositive (3)**.

positive impedance An impedance whose real number component is a **positive resistance**.

positive-intrinsic-negative Same as **PIN**.

positive ion Also called **cation**. A positively charged ion. Positive ions collect at the cathode when subjected to an electric potential while in solution. Sodium ion and potassium ion are two common cations.

positive lead A conductor connected to a **positive terminal**.

positive lens A lens which causes parallel rays of light passing through it to bend toward one another. Such lenses are thicker at the center than the edges. This contrasts with a **negative lens**, which bends parallel rays of light passing through it away from one another. Also called **converging lens**, or **convex lens**.

positive line A line, such as a wire, connected to a **positive terminal**.

positive logic Logic in which the high level represents the logic 1 state, and the low level represents the logic 0 state. In **negative logic** the converse is true.

positive magnetostriction Magnetostriction in which a material expands. In **negative magnetostriction** a material contracts.

positive modulation 1. Modulation in which an increase in luminous intensity results in an increase in carrier modulation or transmission power. Used, for instance, in some TV and fax systems. 2. Same as **positive transmission**.

positive peak Also called **positive peak value**. 1. The maximum positive value of a voltage, current, or other quantity. 2. The maximum positive value of the displacement from a reference position, such as zero, for a voltage, current, or other quantity. 3. The **positive peak (1)**, or **positive peak (2)** for a given time interval.

positive peak value Same as **positive peak**.

positive phase sequence In a polyphase system, such as a three-phase AC circuit, the phase sequence which is in the usual order. In a **negative phase sequence** this order is reversed. Also called **positive sequence**.

positive photoresist A photoresist that depolymerizes or otherwise turns softer when exposed to light. This contrasts with a **negative photoresist**, which polymerizes or otherwise turns harder when exposed to light.

positive plate In a cell or battery, the plate that is connected to the positive terminal. During discharge, electrons flow from the negative plate, through the external circuit, and into the positive plate. Conversely, in the case of a secondary cell or battery that is charging, the positive plate is connected to the negative terminal.

positive polarity 1. The property of having a deficiency of electrons. An excess of electrons produces a **negative polarity (1)**. 2. The polarity of a **positive pole**.

positive pole 1. In a magnet or magnetic body, one of the two regions where the magnetic lines of force converge, and hence magnetic intensity is at a maximum. The other region is the **negative pole (1)**. Such a pole, when part of a magnet which is freely suspended, will seek a negative pole, as opposite poles attract. Also called **north pole (3)**. 2. The location in the northern hemisphere where the magnetic dip is 90°. The corresponding spot in the southern hemisphere is the **negative pole (2)**. At each of these places, the magnetic meridians converge. The north magnetic pole does not coincide with the north geographic pole, and the position of both magnetic poles varies over time. Also called **north pole (2)**, **north magnetic pole**, or **magnetic north pole**. 3. Same as **positive terminal**.

positive potential 1. A potential which is greater than a given reference, such as ground. 2. A potential measured at a positive electrode.

positive power Same as **positive exponent**.

positive resistance For a given device, a region within a characteristic curve in which an increase in voltage results in an increase in current.

positive sequence Same as **positive phase sequence**.

positive temperature coefficient Its abbreviation is **PTC**. 1. For a given component, device, material, or system, an increase in a magnitude or property, such as length or resistance, which occurs as a consequence of an increase in temperature. 2. For a given component, device, or material, an increase in electrical resistance which occurs as a consequence of an increase in temperature. 3. A number which quantifies the relationship in **positive temperature coefficient (1)**, or **positive temperature coefficient (2)**.

positive terminal In a source of electrical energy, such as a battery, cell, or generator, the terminal to which electrons flow. The electromotive force provides the electric pressure for electrons to flow from the negative terminal, through the external circuit, and on to the positive terminal. Also called **positive pole (3)**.

positive transmission Transmission of a signal utilizing **positive modulation (1)**. Also called **positive modulation (2)**.

positive true logic Same as **positive logic**.

positive voltage 1. A voltage which is greater than a given reference, such as ground. 2. A voltage measured at a positive electrode.

positron Also called **positive electron**, or **antielectron**. 1. The antiparticle equivalent of the electron. It has the same mass as the electron, but opposite electric charge and magnetic moment. 2. A positively-charged electron. The term is utilized to differentiate such a particle from a **negatron**, or negatively-charged electron.

positron emission tomography Its abbreviation is **PET**. A medical imaging technique which detects the movement and concentration of an injected radioactive tracer, to monitor and study metabolic activity in a patient. The tracer emits positrons, which collide with electrons in the body and produce gamma radiation which is then detected and quantified. PET is especially useful in examining and diagnosing brain function and capabilities, in addition to tumors and strokes. Also called **positron emission tomography scan**.

positron emission tomography scan Same as **positron emission tomography**. Its abbreviation is **PET scan**.

positron emission tomography scanner An instrument which takes **positron emission tomographies**. Its abbreviation is **PET scanner**.

POSIX Abbreviation of Portable Operating System Interface for UNIX. A set of ISO and IEEE standards aimed at making applications portable between platforms, especially between UNIX systems and others.

post 1. A upright length of a material, such as wood or metal, which serves for support, attachment, connection, or the like. For example, that utilized to join two or more conductors. 2. A terminal, used for making electrical connections, that incorporates a screw which is usually adjusted by hand. Used, for instance, to connect audio speaker cables. Also called **binding post**, **binding screw**, or **screw terminal**. 3. A **post** (1) placed in a waveguide to introduce a reactance, susceptance, or the like. An example is a capacitive post. Also called **waveguide post**. 4. In a communications network, especially the Internet, to submit a message to a newsgroup, online bulletin board, or the like.

POST Acronym for **Power-On Self-Test**.

post-acceleration Also spelled **postacceleration**. Also called **post-deflection acceleration**. 1. In a CRT, acceleration of an electron beam after its deflection. 2. In an electron tube, acceleration of an electron beam after its deflection.

post-deflection acceleration Same as **post-acceleration**. Also spelled **postdeflection acceleration**.

post-emphasis The process of restoring an audio-frequency signal to its original form before reproduction. This is done to offset the preemphasis of higher frequencies occurring prior to transmission or during recording. May be used, for instance, in frequency-modulated receivers. The utilization of preemphasis followed by postemphasis may help improve the overall signal-to-noise ratio and reduce distortion, among other benefits. Also spelled **postemphasis**. Also called **post-equalization**, or **deemphasis**.

post-emphasis network An electric network which serves to provide **post-emphasis**. Also spelled **postemphasis network**. Also called **deemphasis network**, or **deaccentuator**.

post-equalization Same as **post-emphasis**. Also spelled **postequalization**.

post-mortem Same as **postmortem**.

Post Office Protocol A widely-used protocol to store, access, and send email over the Internet. Its abbreviation is **POP**.

Post Office Protocol server Same as **POP server**.

post-processor Same as **postprocessor**.

postacceleration Same as **post-acceleration**.

postdeflection acceleration Same as **post-acceleration**. Also spelled **post-deflection acceleration**.

postemphasis Same as **post-emphasis**.

postemphasis network Same as **post-emphasis network**.

postequalization Same as **post-emphasis**. Also spelled **post-equalization**.

postfix notation A manner of forming mathematical or logical expressions in which the operators appear after the operands. For example, $(A + B) \times C$, which is in **infix notation**, would appear as $ABC+^*$ in postfix notation. Also called **reverse Polish notation**, or **suffix notation**.

postmortem The analysis or study of a recent event, such as a malfunction, crash, or failure. Also spelled **post-mortem**.

postprocessor Hardware or software which performs processing after another device or program has processed data. Also spelled **post-processor**.

PostScript A de facto page-description language standard which is supported by most printing establishments and many printers.

pot 1. Abbreviation of **potentiometer**. 2. Abbreviation of **potential**.

pot core A ferrite magnetic core in which a center post, around which the coil is wound, is surrounded by an enclosure in the general form of a kitchen pot. Such a configuration shields against stray flux.

potassium A soft silvery-white metallic element whose atomic number is 19. It is an alkali metal, extremely reactive, and oxidizes rapidly or ignites in moist air. It has over 20 known isotopes, of which 2 are stable. Its applications in electronics include its use in high-sensitivity magnetometers, optical devices, and many of its compounds are of significance in this field. Its chemical symbol, **K**, is taken from the Latin word for alkali: kalium.

potassium bichromate Same as **potassium dichromate**.

potassium bifluoride Colorless crystals whose chemical formula is KHF_2. Used for etching, and as an electrolyte.

potassium bromide White crystals or powder whose chemical formula is KBr. Used in spectroscopy, and in photography.

potassium carbonate A white powder whose chemical formula is K_2CO_3. Used in optical glasses, TV tubes, and in electroplating.

potassium chloride White crystals or powder whose chemical formula is KCl. Used in electrode cells, and in photography.

potassium cyanide A white powder or mass whose chemical formula is KCN. Used in electroplating.

potassium dichromate Bright orange-red crystals whose chemical formula is $K_2Cr_2O_7$. Used in batteries, electroplating, photolithography, and in ceramics. Also called **potassium bichromate**.

potassium dihydrogen phosphate A ferroelectric substance whose chemical formula is KH_2PO_4. Its abbreviation is **KDP**.

potassium ferricyanide Ruby-red crystals used in etching and electroplating.

potassium ferrocyanide Pale yellow crystals or powder used in etching and electroplating.

potassium fluoride A white powder whose chemical formula is KF. Used in hard soldering, and in etching.

potassium hydroxide A white powder whose chemical formula is KOH. Used in fuel cells, batteries, as an electrolyte, and in electroplating.

potassium iodide A white powder whose chemical formula is KI. Used in spectroscopy and other analytical techniques.

potassium manganate Dark green crystals whose chemical formula is K_2MnO_4. Used in batteries, and in photography.

potassium perchlorate Colorless or white crystals or powder whose chemical formula is $KClO_4$. Used in analytical techniques, and in photography.

potassium permanganate Dark purple crystals whose chemical formula is $KMnO_4$. Used in analytical techniques, and in photography.

potassium silicate A colorless or yellow solid used in ceramics and electrodes.

potassium sodium tartrate Colorless to white crystals displaying piezoelectricity and ferroelectricity. Used in piezoelectric devices, such as crystal microphones, and for silvering mirrors. Also called **Rochelle salt**, or **Seignette salt**.

potential The work required to bring a unit charge from a reference point to a specific point within an electric field. The reference point is usually considered to be an infinite distance from the specific point, whose potential is considered to be zero. When 1 joule is required to bring 1 coulomb of charge, the potential is equal to 1 volt. Its symbol is V, and it is expressed in volts. Its abbreviation is **pot**. Also called **electric potential**, or **electrostatic potential**.

potential barrier Within a field of force, a region in which the potential is such that it opposes the passage of a particle through it, or the movement of a particle subject to said field. Observed, for instance, in junction diodes. Also called **potential hill**.

potential difference The difference in potential between two points, especially those of a circuit. It is the work required to move a unit charge between these points. Its abbreviation is **PD**, and its symbol is U, or ΔV. When expressed in volts, as it usually is, also called **voltage**. Also called **electric potential difference**.

potential divider A component, circuit, or device which is utilized to provide an output which is a desired fraction of the input potential difference. For instance, a resistor, variable resistor, tapped resistor, or combination of resistors with taps, movable contacts, or junctions may be utilized for this purpose. Also called **voltage divider**.

potential drop **1.** The potential difference between any two points of a circuit or conductor. **2.** The potential difference between any two points of a circuit or conductor, due to the flow of current. Also called **voltage drop** (1), or **drop** (3). **3.** The potential difference between the terminals of a circuit element, due to the flow of current. Also called **voltage drop** (2), **drop** (4).

potential energy The energy a body possesses as a consequence of its position, configuration, or state. Potential energy provides a body with the capacity to do work. Examples include the energy stored in a capacitor, or that in a compressed spring. Its abbreviation is **PE**.

potential gradient For a given conductor or dielectric, the voltage per unit length. Also called **voltage gradient**.

potential hill Same as **potential barrier**.

potential transformer A transformer utilized to provide an output voltage different than the input voltage. It is a step-down transformer when the output voltage is lower than the input voltage, or a step-up transformer when the output is higher than the input voltage. Used, for instance, as an instrument transformer. Also called **voltage transformer**.

potentiometer Its abbreviation is **pot**. **1.** A variable resistor incorporating a sliding contact or tap, so as to allow a variable proportion of the total resistance to be included in a circuit. It is usually utilized as a potential divider. **2.** A **potentiometer** (1) used as a volume control. **3.** A device or instrument which incorporates one or more calibrated resistors, and which is utilized to measure unknown potential differences when compared with a precisely known potential difference, such as that of a standard cell.

potentiometry **1.** The use of **potentiometers** to study and measure unknown potential differences. **2.** The study and measurement unknown potential differences, such as those in quantitative electrochemical analysis.

POTS Acronym for **p**lain **o**ld **t**elephone **s**ervice. Analog land-based telephone service with no added features, consisting simply of one or more single-line telephone instruments connected to a public switched telephone network.

POTS splitter Abbreviation of **p**lain **o**ld **t**elephone **s**ervice **splitter**. A filter which separates the voice and data signals of a telephone line via which DSL service is accessed. The voice frequencies are in the lower frequency range, and the data is in the higher interval. Not all DSL services require a POTS splitter. Also called **splitter** (3), or **DSL splitter**.

potted circuit An electric circuit which is embedded, encased, or immersed in a material to protect it from its surrounding environment. For example, a circuit which is immersed in oil or embedded in rubber, and surrounded by a metal case.

potting The process of embedding, encasing, or immersing a circuit in a material to protect it from its surrounding environment.

potting material A material, such as wax, plastic, rubber, or resin, utilized for **potting**.

pound Its abbreviation, **lb**, is taken from the Latin word for pound, libra. **1.** A unit of mass equal to exactly 453.59237 grams. Also called **pound mass**. **2.** A unit of mass equal to approximately 373.242 grams. Used in certain contexts, such as the indication of mass of precious metals or stones. **3.** A unit of weight or force equal to the force of gravity on one **pound** (1). This value varies by location, and is equal to approximately 4.44822 newtons. Also abbreviated **lbf**, which is more proper than **lb**. Also called **pound force**.

pound-foot A unit of torque equal to approximately 1.35582 newton meter. Its abbreviation is **lbf ft**, or **lb ft**. Also called **foot-pound** (2).

pound force Same as **pound** (3).

pound mass Same as **pound** (1).

poundal A unit of force equal to approximately 0.138255 newton. Its abbreviation is **pdl**, or **pl**.

pounds per square foot A unit of pressure equal to approximately 47.88026 pascals. Its abbreviation is $\mathbf{lbf/ft^2}$, or **psf**.

pounds per square inch A unit of pressure equal to approximately 6.894757 kilopascals. Its abbreviation is $\mathbf{lbf/in^2}$, or **psi**.

powdered-iron core A magnetic core made from powdered ferrite. The particles are mixed with an appropriate binder, and the core is formed through applied pressure. Used, for instance, as a radio-frequency choke, or for magnetic storage. Also called **ferrite core**, or **dust core**.

power **1.** The rate at which energy is expended, or at which work is performed. Its symbol is P, its abbreviation is **pwr**, and it is usually expressed in watts, or some multiple of watts, such as megawatts. When dealing specifically with the rate at which electric energy is utilized in a DC circuit or system, it is equal to the potential difference, expressed in volts, multiplied by the current, expressed in amperes. In an AC circuit or system, **true power** is the rate at which work is performed or energy is transferred, **apparent power** is the product of the RMS current and the RMS voltage, while **reactive power** is the power in an AC circuit which cannot perform work. Also called **electric power** (1). **2.** To provide **power** (1). **3.** A number indicating the exponent to which another number, the base, is raised. For example, in 2^8, the 8 is the power, and the 2 is the base. Also called **exponent**. **4.** The product obtained when multiplying a given number by a given **power** (3). For instance, in 2^8, 2 is raised to the 8th power, which results in 256. **5.** The multiplication of a given number or value by another, to express a multiple of a given standard or reference. For instance, 256X oversampling, or a 64x DVD drive. **6.** A measure of the ability of a lens, or optical system, to magnify an image or object.

power adapter A converter that changes the AC available from an electrical outlet into a low-voltage source of DC power. Utilized, for instance, to provide for the energy requirements of portable devices such as notebook computers, in cell phone chargers, and so on. Also called **AC adapter**.

power amplification **1.** For a given component, circuit, device, stage, piece of equipment, or system, the production of an output power which is greater than the input power. **2.** For an amplifying device such as a transistor, electron tube, or photomultiplier tube, the ratio of the output power to the input power. Also called **power gain** (2), or **power ratio**. **3.** Same as **power output** (1).

power amplifier An amplifier designed to provide an output power which is much greater than the input power. Such an amplifier is utilized to deliver power to a load, such as a loudspeaker. When there are multiple amplifiers, or amplifier stages, the power amplifier is the final, or output amplifier or stage. For example, it is the stage of a transmitter that feeds an antenna. Its abbreviation is **PA**. Also called **main amplifier**.

power amplifier stage A power-amplifying unit within a device or system.

power-amplifier tube Same as **power tube**.

power attenuation **1.** For a given component, circuit, device, stage, piece of equipment, or system, the production of an output power which is less than the input power. Also, the extent of such attenuation. **2.** For an amplifying device such as a transistor, electron tube, or photomultiplier tube, the ratio of the input power to the output power. Also called **power loss** (2).

power band-width Same as **power bandwidth**.

power bandwidth For a power amplifier, the frequency interval over which it produces no less than half of its rated power at or below a given level of distortion. Used, for instance, to help evaluate the performance of audio amplifiers. Also spelled **power band-width**.

power blackout A complete loss of AC power arriving from a power line. Also called **blackout**.

power brownout A temporary reduction in the AC power arriving from a power line. It may be due to unusually heavy power demands, a malfunction of the generation or distribution system, or it may be done intentionally by the power company to counter excessive demand. It usually goes unnoticed by power consumers, except when it affects sensitive electronic equipment, such as computers. Also called **brownout**.

power circuit **1.** A circuit that provides the power to a component, device, piece of equipment, machine, or system which consumes power while performing its functions. **2.** A circuit that provides the power to a component, circuit, device, piece of equipment, machine, or system which consumes, dissipates, radiates, or otherwise utilizes power, especially electricity.

power circuit breaker Its abbreviation is **PCB**. **1.** A device which automatically cuts off the power delivered to a component, circuit, device, piece of equipment, or system under specified conditions. **2.** A circuit or switch utilized to apply or remove power from a component, circuit, device, piece of equipment, or system.

power consumption **1.** The power drawn from a source, such as that utilized by a load. Also called **power drain** (1). **2.** The rate at which power is drawn from a source. Usually expressed in watts. Also called **power drain** (3).

power control A circuit, device, mechanism, or system which regulates or manages the power delivered to a component, circuit, device, piece of equipment, or system.

power converter A circuit or device that changes from one AC power source to another of a different voltage, that converts AC power to DC power, or that changes DC power to AC power.

power cord A flexible cord containing two insulated wires, and which terminates in a two-pronged plug. It is utilized to connect an appliance or other electrical devices to a power source, especially an outlet. Such a cord may also have a third conductor, along with a third prong, for a safety connection to ground. Also called **line cord**.

power density Same as **Poynting vector**.

power derating The reduction of the power rating of a device, so as to provide an additional safety and/or reliability margin when operating under unusual or extreme conditions, such as elevated temperatures.

power detector A detector designed to accurately handle signals of great amplitude without damage.

power diode A diode, such as that used in diode lasers, which can operate at an especially high power level.

power dissipation **1.** The loss of power through its conversion into heat energy. **2.** The quantified energy loss in **power dissipation** (1). **3.** Any loss of power. **4.** A wasteful expenditure of power. **5.** The power consumed by a load.

power divider A divider capable of distributing large amounts of power. Used, for instance, to supply multiple antennas with the desired proportion of energy.

power down To turn off a device, piece of equipment, or system.

power drain **1.** Same as **power consumption** (1). **2.** The load or process that draws the power utilized in **power consumption** (1). **3.** Same as **power consumption** (2).

power efficiency **1.** The ratio of the useful power or energy output of a device or system, to its total power or energy input. Also called **efficiency** (2). **2.** The proportion of the audio-frequency electrical power input to a speaker which is converted into acoustic energy. Power efficiency may be measured, for instance, by driving a speaker with a power input of 1 watt, and measuring its output, in decibels, at a distance of 1 meter. Also called **efficiency** (3).

power factor In an AC circuit, the ratio of the true power to the apparent power. If both the voltage and the current are sinusoidal, the power factor is equal to the cosine of the phase angle between them. If both the voltage and the current are in phase, then the power factor is 1. Its abbreviation is **PF**.

power-factor controller A circuit or device which regulates a **power factor**. Also called **power-factor regulator**.

power-factor correction Any technique utilized to increase the power factor of a circuit or supply. For example, the utilization of power-factor correction capacitors. Its abbreviation is **PFC**.

power-factor correction capacitor A capacitor utilized for **power-factor correction**. Such capacitors help, for instance, to compensate for inductive reactance, which helps reduce the phase difference between the voltage and current.

power-factor corrector A component or device utilized for **power-factor correction**. Its abbreviation is **PFC**.

power-factor meter An instrument which measures and indicates **power factors**.

power-factor regulator Same as **power-factor controller**.

power failure Also called **power outage**. **1.** An interruption or complete loss of AC power arriving from the power line. Also called **outage** (1). **2.** An interruption or complete loss of power arriving from any source, such as a power line or generator. Also called **outage** (2).

power fluctuation Irregular variations in power, such as sudden increases or decreases in that available from a power line.

power frequency Same as **power-line frequency**.

power gain **1.** The increase in power provided by a component, circuit, device, piece of equipment, or system. For instance, the ratio of the input power to the output power of a power amplifier. Usually expressed in decibels. **2.** Same as **power amplification** (2).

power generating station A location which combines the structures, devices, and equipment necessary to produce electricity in quantities adequate for residential, commercial, or industrial use. A power generating station converts another form of energy, such as hydroelectric, nuclear, or solar, into electrical energy. Also called by various other names, including **power station**, **power plant** (1), **plant** (1), **electric power station**, **electric power plant** (1), and **generating station**.

power generation The production of electricity, especially in quantities adequate for residential, commercial, or industrial use. Also called **electric power generation**.

power generator A machine or device, such as a solar power generator, which serves for **power generation**.

power grid A network of high-voltage transmission lines which link multiple power generating stations. Such a network may run through an entire country. Also called **grid (5)**.

power-handling ability Same as **power-handling capability**.

power-handling capability Also called **power-handling capacity**, or **power-handling ability**. **1.** The maximum input or output power that a component, circuit, device, stage, piece of equipment, or system can provide or handle without damage. For example, the maximum amount of power that can be provided by an amplifier without clipping, or that can be fed to a speaker without overheating a voice coil. **2.** The **power-handling capability** (1) under given conditions, or for a stated time interval.

power-handling capacity Same as **power-handling capability**.

power IC Abbreviation of **power** integrated circuit. An IC designed to handle high currents and voltages. Such an IC may incorporate multiple power transistors.

power input **1.** The power presented to the input terminals of a component, circuit, device, piece of equipment, or system. Also called **input power** (1). **2.** The power absorbed, utilized, or otherwise dissipated by a component, circuit, device, piece of equipment, system, material, or medium. **3.** The level of power required for proper operation.

power integrated circuit Same as **power IC**.

power level **1.** The magnitude or amount of power being produced, such as that delivered to a load, at any given moment. **2.** The magnitude or amount of power absorbed, utilized, or otherwise dissipated by a component, circuit, device, piece of equipment, system, material, or medium at any given moment. **3.** A specific **power level** (1) or **power level** (2) required for a given level of operation.

power-level indicator A device or instrument, such as a VU meter, which measures and indicates power levels. Also called **power-level meter**.

power-level meter Same as **power-level indicator**.

power line **1.** A line which supplies AC power to customers, and whose source is usually a power generating station. **2.** A line which supplies electric power.

power-line carrier A carrier frequency utilized for communications over power lines. Such a carrier is usually well above the 50 or 60 Hz power-line frequency, and may be used for control purposes, or for the transmission of data.

power-line communications Data communications which utilize power lines for transmission and reception. In such communications, data is transmitted at frequencies well above the 50 or 60 Hz power-line frequency. Its abbreviation is **PLC**.

power-line filter A filter which prevents the passage of noise signals in a power line, thereby improving the operation of the devices connected to said line. It does so by attenuating radio-frequency noise, while permitting the 50 or 60 Hz current to pass. Also called **line filter** (2).

power-line frequency The AC frequency, such as 50 or 60 Hz, of a power line. Also called **power frequency**, or **line frequency** (3).

power-line interference Any interference, such as hum, which is caused by power lines.

power-line monitor A device or instrument which is utilized to sample, measure, assess, regulate, observe, or control a power line.

power-line noise Also called **AC noise**. **1.** Electromagnetic interference occurring in power lines. **2.** Electromagnetic interference, occurring in power lines, which adversely affects the performance of electronic components.

power-line outlet Same as **power outlet**.

power-line voltage Also called **line voltage**. **1.** The AC voltage supplied by an AC power line. This may vary from country to country, and consists of two nominal voltages. The lower number is primarily for lighting and small appliances, while the larger number is for heating and large appliances. In Canada and the United States, for example, the nominal voltages are approximately 115/230. Also called **AC line voltage**. **2.** The voltage supplied by a power line, especially at the point of utilization.

power loss Also called **power losses**. **1.** Any reduction in power. **2.** Power which is dissipated without performing useful work. For example, heat losses. Usually expressed in watts. **3.** A reduction of in the power level of a signal which has traveled from one point to another. Usually expressed in decibels. **4.** Same as **power attenuation** (2).

power losses Same as **power loss**.

Power Mac A popular PC family.

power mains A network of lines which serve for the distribution of electric power. Mains may refer to the lines that supply an entire nation or region, or those that provide power in a given structure such as a building or house. For instance, when a device or piece of equipment is plugged into an outlet, it is accessing the electricity carried by the mains. Also called **mains**.

power management The administration and judicious use of available power resources. Used, for instance, in energy conservation, or to maximize the time a battery can power a portable computer.

power margin The excess power capacity of a component, circuit, device, piece of equipment, or system. Such power, for example, may provide a margin of safety.

power metal-oxide semiconductor field-effect transistor Same as **power MOSFET**.

power meter An instrument which measures and indicates electrical power consumption for a given, or combined, time interval. Also called **electric meter** (1), or **electric power meter**. When indicating in units of kilowatt-hours, also called **kilowatt-hour meter**.

power MOSFET Abbreviation of **power** metal-oxide semiconductor field-effect transistor. A MOSFET capable of handling high currents and voltages. For example, such a transistor might have a current capacity of 100 amps, and a voltage rating of 100 kV.

power network A network utilized for **power transmission**.

power on/off switch Same as **power switch**.

Power-On Self-Test A series of diagnostic tests automatically performed when a computer boots. The tests are stored in the ROM, and include verifying the proper function of the RAM, drives, and so on. Its acronym is **POST**.

power outage Same as **power failure**.

power outlet A power line termination whose socket serves to supply electric power to devices or equipment whose plug is inserted into it. Power outlets are usually mounted in a wall, although they may be found elsewhere, such as a floor, or in the back of another electrical device, such as an amplifier. Also called by various other names, including **power-line outlet**, **power receptacle**, **power socket**, **convenience outlet**, **outlet**, **electrical outlet**, or **receptacle**.

power output Also called **output power**, or **amplifier power**. **1.** The extent to which an amplifier can amplify an input signal. Also called **power amplification** (3). **2.** The delivering of the output of an amplifier to a load, such as a loudspeaker. Expressed in watts Also, the power so delivered.

power-output tube Same as **power tube**.

power pack A usually sealed unit which provides the power required for a device or piece of equipment. Such a unit usually incorporates a transformer to convert the power-line voltage to that needed, and when a specific device or piece of equipment uses DC, a rectifier.

power pentode A power tube with five electrodes.

power plant 1. Same as **power generating station**. 2. A unit which converts another form of energy, such as hydroelectric, nuclear, or solar, into electrical energy.

power plug Same as **plug (2)**.

power rail Same as **power-supply rail**.

power rating A rated or named value stating the power that a component, circuit, device, piece of equipment, or system can produce, consume, dissipate, or otherwise safely handle, when used in a given manner, or under specified conditions. For instance, the power rating of a speaker may indicate the RMS power that can be delivered to a dynamic speaker for extended periods, without harming the voice coil or other components. The peak power rating would specify the maximum power level which can be briefly tolerated without harm.

power ratio Same as **power amplification (1)**.

power receptacle Same as **power outlet**.

power rectifier A rectifier designed for increased power-handling capability, and which is usually used in power supplies.

power relay 1. A relay which is actuated at a given power level. An underpower relay is actuated when the power in a circuit drops below a given threshold, while an overpower relay is actuated when the power in a circuit exceeds a given threshold. 2. A heavy-duty relay utilized to control and/or switch electric circuits, such as power circuits. Also called **contactor (1)**.

power resistor A heavy-duty resistor which is utilized, for instance, in electric power systems, or to dissipate significant current levels.

power semiconductor A semiconductor, such as a transistor or diode, which can operate at especially high power levels without damage.

power socket Same as **power outlet**.

power source Same as **power supply**.

power spectral density For a given bandwidth of electromagnetic radiation consisting of a continuous interval of frequencies, the total power divided by the specified bandwidth. Usually expressed in multiples of watts per hertz, such as mW/Hz, or kW/Hz. Used, for instance, to analyze the power content of complex signals. Its abbreviation is **PSD**. Also called **spectral density**, **spectral power density**, or **spectral energy density**.

power spike An instantaneous power increase of great magnitude, such as that cause by a lightning strike. A spike can arrive via power lines, phone lines, network lines, or the like, and may cause extensive damage if protective measures are not taken.

power station Same as **power generating station**.

power supply Also called **power source**. 1. A source of power or electricity. For example, a power line or battery. Also called **supply (2)**. 2. A source of power for electronic components, circuits, devices, equipment, or systems. Such sources include batteries and power packs. Also called **electronic power supply**, or **supply (3)**.

power-supply circuit A circuit whose output serves as a source of power. Such a circuit, for instance, usually converts AC to DC, and may or may not change the value of the voltage. A voltage doubler, for instance, provides an output DC voltage that is about twice the peak value of its input AC voltage.

power-supply filter A filter utilized to remove all frequencies above or below a given value from the output of a power supply. A capacitor filter is the simplest type of power-supply filter.

power-supply rail A metal rail that passes through a component, device, or piece of equipment, delivering power. Also called **power rail**, or **rail**.

power-supply rejection ratio Its abbreviation is **PSRR**. 1. For a given component, circuit, or device, the ratio of a change in output, to the a disturbance in the input provided by a power supply. For example, it may be the ratio of the change in the output voltage of a circuit, to the change in its power-supply voltage. 2. The **power-supply rejection ratio (1)** for an amplifier.

power-supply sensitivity 1. The minimum change in the input provided by a power supply which produces a change in the output of a given component, circuit, or device. 2. The **power-supply sensitivity (1)** of an amplifier.

power-supply transformer Same as **power transformer**.

power surge Also called **surge power**. 1. A sudden and momentary increase in power. May be caused, for instance, by lightning. If protective measures are not employed, such a surge may bring about a failure or significant damage. 2. A sudden and momentary increase in current or voltage.

power switch Also called **power on/off switch**. 1. A switch that serves to connect and disconnect a component, circuit, device, piece of equipment, or system, from a source of power, such as a power line. 2. A switch that serves to change between the on state and the off state. That is, the switch either turns something on, or it turns it off. Also called **on/off switch (1)**. 3. A switch that serves to change between the on state and a standby state. Also called **on/off switch (2)**.

power switching 1. Switching between an on and off state. 2. Switching between power supplies, especially those at high power levels.

power tetrode A power tube with four electrodes.

power transfer The transfer of power between circuits, devices, equipment, or systems. For example, the transfer between a source and a load, or from a group of power transmission lines to another.

power transfer theorem A theorem stating that the greatest possible amount of power is transferred from source to load when they have conjugate impedances. Also called **maximum power transfer theorem**.

power transformer A transformer generally utilized to couple a power supply to components, circuits, devices, equipment, or systems. The primary winding of such a transformer is usually connected to a power line, and each of the secondary windings provides a different current and/or voltage output. The rating of a power transformer is stated in terms of the maximum voltage and current-delivering capacity of its secondary windings. Alternatively, there may be a single secondary winding with multiple taps. Also called **power-supply transformer**.

power transistor A transistor, such as a power MOSFET, designed for increased power-handling capability without damage, and which is used for power amplification.

power transmission The transmission of generated electricity, especially for residential, commercial, or industrial use. Also called **electric power transmission**.

power transmission line An electric cable utilized for the transmission of generated electricity, especially for residential, commercial, or industrial use. Also called **electric power transmission line**, **electric main**, or **electric transmission line**.

power triode A **power tube** with three electrodes.

power tube An electron tube designed for increased power-handling capability, and which is used for power amplification. Such tubes are usually highly sensitive and efficient. Also called **power-amplifier tube**, or **power-output tube**.

power unit 1. A single unit, such as a power pack, which serves as a power supply. 2. Any unit, such as watt or femtowatt, utilized to describe or quantify power.

power-up 1. To turn on a device, piece of equipment, or system. 2. To bootstrap a computer.

power user 1. A computer user who is thoroughly familiar with various software applications, and which often has considerable knowledge of the hardware aspects of computing. 2. A consumer of electrical power.

power winding In a saturable reactor, the winding delivering the power to the load.

PowerBook A popular family of notebook computers.

PowerMac A popular PC family.

PowerPoint A popular presentation program.

Poynting theorem A theorem, based on Maxwell's equations, which relates losses in energy of an electromagnetic field to work is done on the particles.

Poynting vector For an electromagnetic wave, a vector equal to the product of the electric field strength and the magnetic field strength, each of which is also a vector quantity. A Poynting vector represents the power density of an electromagnetic wave, in addition to pointing in the direction this energy is flowing. Usually expressed in watts per square meter. Also called **power density**.

pp Abbreviation of **peak-to-peak**.

pp amplitude Abbreviation of **peak-to-peak amplitude**.

pp value Abbreviation of **peak-to-peak value**.

pp voltage Abbreviation of **peak-to-peak voltage**.

ppb Abbreviation of **parts per billion**.

PPGA Abbreviation of **plastic pin grid array**.

PPI Abbreviation of **plan position indicator**.

ppm Abbreviation of **parts per million**.

PPM 1. Abbreviation of **pulse-position modulation**. 2. Abbreviation of **pages per minute**.

PPP Abbreviation of **Point-to-Point Protocol**. A widely-used protocol to connect a computer to the Internet via dial-up access. PPP works at the data link layer, can handle synchronous or asynchronous communications, and provides error-checking and secure password verification.

PPPoE Abbreviation of **Point-to-Point Protocol over Ethernet**. A specification, technique, or protocol for running PPP over Ethernet. Used, for instance, to share a single DSL line among multiple nodes.

PPS 1. Abbreviation of **pulses per second**. 2. Abbreviation of **pages per second**. 3. Abbreviation of **packets per second**.

ppt Abbreviation of **parts per thousand**.

PPTP Abbreviation of **Point-to-Point Tunneling Protocol**. A protocol utilized to send secure transmissions over an IP network, such as the Internet. PPTP supports encapsulation of encrypted packets, and is used, for instance, to establish a VPN over the Internet.

PPV Abbreviation of **pay-per-view**.

PQFP Abbreviation of **plastic quad flat-pack**.

Pr 1. Chemical symbol for **praseodymium**. 2. Within the YPbPr color model, one of the two color-difference signals. The other is **Pb (3)**.

Practical Extraction Report Language Same as **Perl**.

prad 1. Abbreviation of **picorad**. 2. Abbreviation of **picoradian**.

PRAM Acronym for **parameter random-access memory**. A battery-powered RAM utilized in some computers to back-up vital system information such as connected devices, and the time.

praseodymium A silvery-yellow metallic chemical element whose atomic number is 59. It is soft, malleable, ductile, and has about 35 known isotopes, of which one is stable. It is used in carbon-arc lamps, phosphors, glasses, and lasers. Its chemical symbol is **Pr**.

pre-alarm signal Same as **pre-alarm warning**.

pre-alarm warning A signal which indicates that an alarm condition is being approached. For instance, an alert when smoke and/or heat levels increase towards the point where an alarm would be set off, so as to enable investigation before resources are committed. Also called **pre-alarm signal**.

pre-amplifier Same as **preamplifier**.

pre-compile Same as **precompile**.

pre-compiler Same as **precompiler**.

pre-distortion Same as **predistortion**.

pre-emphasis Same as **preemphasis**.

pre-emphasis network Same as **preemphasis network**.

pre-emptive multitasking Same as **preemptive multitasking**.

pre-equalization Same as **preequalization**.

pre-fetch Same as **prefetch**.

pre-packaged software Same as **packaged software**. Also spelled **prepackaged software**.

pre-processor Same as **preprocessor**.

pre-recorded tape Same as **prerecorded tape**.

pre-scaler Same as **prescaler**.

pre-selector Same as **preselector**.

pre-set Same as **preset (1)**.

pre-shoot Same as **preshoot**.

pre-store Same as **prestore**.

pre-travel Same as **pretravel**.

preamp Abbreviation of **preamplifier**.

preamplifier An amplifier utilized to boost signal levels, without distorting them, to those sufficient for a power amplifier to handle them effectively. Preamplifiers are often used in processing radio, TV, audio, and microwave signals. Also spelled **pre-amplifier**. Its abbreviation is **preamp**.

precedence effect An acoustic phenomena in which a listener correctly identifies the direction of a sound source that is heard by both ears, but which arrives at each at slightly different times. If the second arriving sound, that heard by the ear farther to the source, is within about 10 to 35 milliseconds of the first arriving sound, it is not taken into account for locating the sound source. If the sounds arrive within less than 10 milliseconds of each other, the source seems to move towards the location of the first arriving sound. When said sounds arrive beyond 35 milliseconds of each other, two distinct sounds are perceived. This effect is taken into account in the reproduction of stereophonic sound. Also called **Haas effect**.

precipitation 1. Any of the forms of water which fall from the atmosphere and reach the surface of the earth, including rain, snow, sleet, and hail. Water in any of these forms which evaporates before reaching the ground may also be considered as precipitation. Also, the amount of such precipitation for a given time interval. 2. The deposition of solid particles from a solution, especially a supersaturated solution. 3. The deposition of dust, smoke, and other particles from air, or another gas.

precipitation attenuation An attenuation in the energy of electromagnetic waves traveling through the atmosphere where precipitation is present, due to scattering and absorption.

precipitation clutter In radars, unwanted echoes that appear on the display screen due to signal reflections from precipitation such rain or hail. Also called **precipitation return**, **precipitation echo**, or **rain return**.

precipitation echo Same as **precipitation clutter**.

precipitation return Same as **precipitation clutter**.

precipitation static Static, such as that produced by rain or snow striking or blocking an antenna, due to precipitation. Precipitation static may also be considered to arise from other atmospheric phenomena such as lightning or dust storms, as might be observed, for instance, in an airplane flying through lightning. Its abbreviation is **p-static**.

precipitator A device which removes dusts present in gases, such as air, by precipitating them in a chamber containing high-voltage wires or grids. The gas is forced through the chamber, and the wires or grids ionize the dust particles, which in turn makes them migrate to the walls of said chamber. Also called **electrostatic precipitator**, **dust precipitator**, **electrostatic air cleaner**, or **electronic precipitator**.

precision 1. A measure of the repeatability of a given result or procedure, such as a measurement. For example, a measurement made several times in which the same, or nearly the same value is obtained, indicates precise measurement. This does not necessarily mean that such measurements specify or approximate a real value, thus a precise measurement may not be accurate. Still, even in technical usage the term is often used synonymously with **accuracy**. 2. The number of significant digits provided in a measurement or other value. The more significant digits, the more precise. 3. Pertaining to components, devices, equipment, and systems which are characterized by exactness, such as those which adhere closely to a given standard.

precision approach radar A ground-based radar which assists flights approaching a landing, and which provides both vertical and horizontal guidance. Its abbreviation is **PAR**.

precision electronic instrument An electronic instrument characterized by accuracy, stability, and reproducibility of performance or results.

precision instrument An instrument characterized by accuracy, stability, and reproducibility of performance or results.

precision potentiometer A potentiometer characterized by accuracy, stability, and reproducibility of values or measurements.

precompile To prepare source code for translation by a compiler. Also spelled **pre-compile**.

precompiler A program that prepares source code for translation by a compiler. Also spelled **pre-compiler**.

predetermined counter A counter which stops and/or provides an alert when a given number is reached.

predication A CPU technique in the which execution of all possible branches is performed in parallel. When the correct result is arrived at, the other branches are dropped. Used in parallel execution.

predistortion Same as **preemphasis**. Also spelled **predistortion**.

preemphasis The process of selectively amplifying higher frequencies of an audio-frequency signal before transmission, or during recording. The utilization of preemphasis followed by postemphasis may help improve the overall signal-to-noise ratio and reduce distortion, among other benefits. Also spelled **pre-emphasis**. Also called **preequalization (1)**, **predistortion**, **emphasis (2)**, or **accentuation (2)**.

preemphasis network An electric network which serves to provide **preemphasis**. Also spelled **pre-emphasis network**. Also called **emphasizer (2)**.

preemptive multitasking A mode of multitasking in which all the tasks have periodic access to the CPU. This contrasts with **non-preemptive multitasking**, in which the foreground task must allow background tasks access to the CPU. Also spelled **pre-emptive multitasking**.

preequalization Also spelled **pre-equalization**. 1. Same as **preemphasis**. 2. Equalization of a signal before transmission or during recording.

preferences Within a program, choices a user has for given settings. Used, for instance, to determine how an application will handle certain tasks, the manner in which data is displayed, or which of multiple operating systems will be booted. When no preference is chosen for a given option, the default setting is utilized. Also called **options**.

preferences menu A menu, such as a drop-down menu, providing **preferences**. Also called **options menu**.

prefetch To locate an instruction in computer memory and load it into a CPU register, before it is requested. Also spelled **pre-fetch**.

prefix 1. The code which identifies a central office. In the NPA-NXX-XXXX ten-digit telephone number format, it is the NXX portion, where N can be any digit between 2 and 9, and X can be any digit between 0 and 9. A single central office may serve more than one prefix. Also called **central office prefix**, **central office code**, or **NXX code**. 2. A word element which is affixed to the front of a word to produce a form with a different or derived meaning. For example, the **tera-** portion of terabyte. 3. A letter, number, symbol, or the like, placed before something else. For instance, the **Q** in Q15.

prefix notation Same as **Polish notation (2)**.

preform 1. That which has been given a shape beforehand. For example, that given to a component, object, or material. Also, to determine the shape of something beforehand. 2. A specific object, such as a component, which has been given a shape beforehand.

premium channel A cable or satellite TV channel whose content may be viewed in exchange for payment. For example, channels which present unedited movies 24 hours a day without commercials.

premium content Information, especially that which is available online, which may be accessed and/or downloaded in exchange for payment. For example, an online newspaper may offer per article charges, fees for access to given sections, and/or subscription packages for one day, one month, one year, or the like.

prepackaged software Same as **packaged software**. Also spelled **pre-packaged software**.

preprocessor Hardware or software which performs processing before another device or program is to process data. Also spelled **pre-processor**.

prerecorded tape A magnetic tape which has been recorded or written upon. Also spelled **pre-recorded tape**. Also called **recorded tape**.

prescaler A scaler that extends the frequency range of a counter by dividing or multiplying an input frequency. For instance, such a device which extends the upper frequency limit by dividing the input frequency by 10, 64, 100, 256, or the like. Also spelled **pre-scaler**.

preselector Also spelled **pre-selector**. 1. A device which selects a setting, mechanism, or the like, before a given process, such as activation or tuning, occurs. 2. A component, circuit, stage, device, or mechanism which is used ahead of a stage or unit within a receiver to provide for maximum sensitivity and selectivity. For instance, a stage placed before the mixer stage in a superheterodyne receiver.

presence In sound reproduction, the impression that the original source is providing the listening experience. The greater the presence, the more natural and faithful the sound.

presence boost An increase in **presence frequencies**.

presence frequencies An interval of audio frequencies, such as 2 kHz to 5 kHz, which when boosted increases the sense of **presence**.

presentation graphics Software utilized mainly to prepare business presentations which incorporate graphics, such as charts and graphs. Also, the output produced by such software.

presentation layer Within the OSI Reference Model for the implementation of communications protocols, the second highest level, located directly above the session layer. This layer deals with tasks pertaining to the selection and use of a common syntax for representing information, and involves tasks such as encryption, decryption, compression, data conversion, text formatting, and text display. Also called **layer 6**.

preset 1. To select a control or setting beforehand. Also spelled **pre-set**. 2. One of multiple settings which can be selected beforehand.

preset counter A counter which performs a function, such as emitting a pulse or performing a reading, each time a given number of inputs is counted.

preshoot In a pulse, a transient blip or distortion preceding a major transition. Also spelled **pre-shoot**.

press-to-talk Same as **push-to-talk**. Its abbreviation is **PTT**.

press-to-talk operation Same as **push-to-talk operation**. Its abbreviation is **PTT operation**.

press-to-talk switch Same as **push-to-talk switch**.

pressure 1. Force per unit area. The SI unit is the pascal, which is equal to the pressure resulting from a force of one Newton applied uniformly to an area of one square meter. Other units include atmospheres, bars, millimeters of mercury, torr, and pounds per square inch. Its symbol is **P**. 2. A specific value or reading of **pressure** (1). 3. The application of a force over a given surface area.

pressure cable A cable in which a fluid under pressure surrounds the conductors, providing insulation and protection. An example is a gas-filled cable.

pressure contact An electric contact requiring pressure to open or close a circuit.

pressure gage Same as **pressure gauge**.

pressure gauge An instrument, such as a barometer, which measures pressures above or below atmospheric pressure. A **vacuum gauge** only measures values below atmospheric pressure. Also spelled **pressure gage**.

pressure gradient microphone A microphone which responds to small differences in pressure between the front and back of its diaphragm. The diaphragm is exposed in the front and the back, and the output of such a microphone is proportional to this pressure gradient. Also called **gradient microphone**.

pressure microphone 1. A microphone in which only one side the diaphragm is exposed to the incident sound pressure. 2. A microphone whose output varies proportionally to the incident sound pressure. Most microphones, including dynamic and capacitor, are of this type.

pressure pad In a tape deck or drive, a pad, usually made of a felt-type material, which forces the tape into close contact with the head.

pressure pickup Same as **pressure transducer**.

pressure roller In a tape deck or drive, a cylinder or roller, usually made of rubber, which forces the tape against the capstan to provide proper traction. Also called **pinch roller**.

pressure-sensitive A component, device, instrument, material, system, or mechanism which responds to pressure, and/or changes in pressure. Examples include pressure microphones and pressure switches.

pressure-sensitive switch Same as **pressure switch**.

pressure sensor 1. A component, device, instrument, material, or system which detects pressure, and/or changes in pressure. A pressure transducer is an example. 2. Same as **pressure transducer**.

pressure switch Also called **pressure-sensitive switch**. 1. A switch that is actuated by the application of pressure. 2. A switch that is actuated by the application of pressure exceeding a given value. 3. A switch that is actuated by a specified variation in pressure. 4. A switch that is actuated when the internal or external pressure of a component, device, or system is above or below a given threshold.

pressure transducer A transducer which converts changes in pressure into a corresponding electrical quantity, such as current, voltage, or resistance. Examples include absolute pressure transducers, and differential-pressure transducers. Also called **pressure pickup**, or **pressure sensor** (2).

pressure welding Welding, such as cold welding, in which the application of pressure is the main or only process involved in joining the surfaces or parts.

pressure zone A zone of high air pressure adjacent to a surface upon which sound waves strike.

pressure zone microphone A microphone which utilizes a flat surface or plate which detects variations in the air pressure adjacent to said surface or plate. Such microphones tend to be very sensitive, and are often placed on flat surfaces such as floors, tables, or walls. Its abbreviation is **PZM**.

pressurization 1. The use of a fluid to maintain the pressure within an enclosure above atmospheric pressure. Used, for instance, in cables, to keep moisture out. 2. The raising of the pressure of a fluid above atmospheric pressure, or another reference pressure.

prestore To store data in a register or computer memory in anticipation to future use in a routine or program. Also spelled **pre-store**.

pretersonics The branch of electronics that deals with acoustic waves at microwave frequencies that travel along the surface or through crystals or metallic objects. Also known as **acoustoelectronics**.

pretravel The distance a contact, button, switch, or the like, moves between the free position and the actuated or operating position. Also spelled **pre-travel**.

Pretty Good Privacy A public-key encryption program which is easy to use and provides rather high security. It is widely utilized to encrypt email, and non-commercial versions are freely distributed. Its abbreviation is **PGP**.

preventive maintenance 1. Maintenance performed to avoid undesired conditions, such as a failures, or to extend operating life. 2. **Preventive maintenance** (1) performed at regular intervals. 3. Maintenance performed to maintain a given level of operation.

preview Same as **print preview**.

PRF Abbreviation of **pulse repetition frequency**.

pri Abbreviation of **primary winding**.

PRI Abbreviation of **Primary Rate Interface**.

PRI/E1 Same as **Primary Rate Interface**.

PRI/T1 Same as **Primary Rate Interface**.

primaries Same as **primary colors**.

primary 1. That which occurs first within a given sequence or process. 2. That which occurs before something else. For instance, a stage which precedes another. 3. That which has the highest rank, degree, or importance. For example, a primary standard. 4. Same as **primary winding**. 5. A high-voltage conductor in a power-distribution system. 6. One of two or more **primary colors**. 7. Of or pertaining to one or more primary colors.

primary battery A battery composed of **primary cells**. Also called **non-rechargeable battery**.

primary beam Same as **primary lobe**.

primary cache A memory cache that is built right into the CPU. This contrasts with **secondary cache**, which utilizes a memory bank between the CPU and main memory. Primary cache is faster, but smaller than secondary cache, while secondary cache is still faster than main memory. Also called **level 1 cache, L1 cache, internal cache, first-level cache,** or **on-chip cache**.

primary cell A cell, such as an alkaline or mercury cell, whose manner of discharge does not lend itself to recharging. For instance, the current produced may involve the dissolution of one of its plates. This contrasts with a **secondary cell**, which can be recharged multiple times. Also called **non-rechargeable cell**.

primary circuit A circuit, especially that in a transformer, which accepts an input power or signal. This power is then inductively transferred to a **secondary circuit**.

primary coil Same as **primary winding**.

primary colors Colors which are combined in an additive mixture to yield a full range of colors. Red, green, and blue are usually used as the additive primary colors. They are called this way because they themselves are not formed by the combination of other colors. Also called **primaries, additive primary colors,** or **color primaries**.

primary cosmic rays High energy particles which travel through space at nearly the speed of light, and which consist mostly of protons and alpha particles. Cosmic rays have many sources, including solar flares, and impinge upon the earth from every direction. Also called **cosmic rays,** or **cosmic radiation**.

primary current The current flowing through a **primary winding**. Also called **transformer primary current**.

Primary Domain Controller In a communications network, the server that maintains user and machine profiles for a domain, and which handles tasks such as authentication of users and systems. Its abbreviation is **PDC**.

primary electron 1. An electron which produces the secondary emission of electrons. In an electron multiplier, for example, a cathode releases primary electrons, which are then reflected by electron mirrors which produce **secondary electrons**. 2. An electron, incident upon a surface or material, which produces secondary electrons.

primary emission The emission of electrons by any another process other than the bombardment by charged particles, such as electrons. For example, the emission of electrons from a heated cathode.

primary failure A failure which arises independently of other components, devices, or units. A **secondary failure** is directly or indirectly caused by the failure of another component, device, or unit.

primary frequency standard 1. A signal source, such as an extremely precise and stable crystal oscillator, whose frequency remains constant, and which can be utilized for calibrating other signal sources. An example is a cesium frequency standard. Also called **frequency standard**. 2. A **primary frequency standard** (1) utilized by national institutions.

primary inductance The inductance measured across a **primary winding** under no-load conditions.

primary instrument An instrument which can be calibrated without utilizing another device or instrument.

primary ionizing event In an instrument or device which detects and quantifies ionizing radiation, such as a Geiger counter, the ionizing event which starts the chain of events leading to a count. Also called **initial ionizing event**.

primary lobe Within an antenna pattern, the lobe indicating the direction in which radiation and/or reception is at a maximum. All other lobes are **secondary lobes**. Also called **primary beam, main lobe, major lobe,** or **main beam**.

primary measuring element The portion of a measuring device which senses or detects a desired phenomenon. For instance, an electrode in an electroencephalograph, or a piezoelectric element in an accelerometer.

primary memory Better known as **RAM**. One or more memory chips that can be read and written by the CPU and other hardware, and which serve as the primary workspace of a computer. Instructions and data are loaded here for subsequent execution or processing, and its storage locations may be accessed in any desired order. The amount of primary memory a computer has determines factors such as how many programs may be optimally run simultaneously. Computer storage, such as disks, discs, and tapes, are considered as **secondary memory**. Also called **primary storage, random-access memory, read/write memory, read/write RAM, main memory, main storage,** or **core memory (2)**.

primary power 1. The power in a primary circuit. 2. The main source of power, as opposed to a backup source of power.

primary radiation Radiation which arrives directly from its source, as opposed to that which is scattered, reflected, or refracted.

primary radiator In an antenna with parasitic elements, a radiator which receives energy directly from a transmission line. A primary radiator drives parasitic elements, which in turn are **secondary radiators**.

Primary Rate Interface Its abbreviation is **PRI**. An ISDN line with 23 B channels, each with a speed of 64 Kbps, and one 64 Kbps D channel. This is equivalent to a T1 or DS1 line. In Europe, PRI has 30 B channels and one D channel, so it is equivalent to an E1 line. Multiple B channels can be combined for the desired or needed bandwidth. Also called **ISDN-PRI**. The T1 equivalent is also called **PRI/T1**, or **T1/PRI**, while the E1 equivalent is called **PRI/E1**, or **E1/PRI**.

primary ring Within a dual-ring topology, the ring which serves for data transmission during normal operation, while the **secondary ring** provides backup.

primary service area A region within which coverage is maintained at or above a given level. For example, the area within which the signal of a broadcast station has minimal interference or fading.

primary standard 1. The definition of fundamental units of measure, such as the base SI units. For example, the time unit **second** is defined as 9,192,631,770 oscillations of cesium atoms at their natural resonance frequency. 2. A reference which has been determined with complete, or nearly complete accuracy, and which serves as a primary basis for measurements. For instance, an atomic frequency standard, which can be used in atomic clocks, which in turn may be used to correct and calibrate other clocks.

primary storage Same as **primary memory**.

primary turns The number of turns in a **primary winding**. The ratio of the number of primary turns to the number of secondary turns yields the turns ratio.

primary voltage The voltage applied to a **primary winding**. Also called **transformer primary voltage**.

primary winding In a transformer, the winding that is directly connected to the input power. One or more **secondary windings** provide the output to the load or loads. For example, a primary winding accepts the input energy an AC line provides, while a secondary winding provides the energy to the load at a different voltage. Its abbreviation is

pri, and its symbol is **P**. Also called **primary (4)**, or **primary coil**.

prime meridian The meridian which has an assigned longitude of 0°. The Greenwich meridian is generally considered as the prime meridian of the planet earth. Also called **zero meridian**.

prime mover **1.** A device, apparatus, or system which converts a natural source of energy, such as sunlight, wind, or flowing water, into mechanical energy or power. Examples include heat engines and turbines. **2.** A **prime mover (1)** within a power plant.

primitive **1.** In computer graphics, a basic element, such as a line, curve, or cone, which can be combined with others for more complex images. **2.** In computer programming, a basic element, such as an operation, which can be combined with others for more sophisticated operations.

principal axis **1.** A line passing through both centers of curvature of a lens. Light travels along this axis with no deviation. Also called **optical axis (1)**. **2.** A line passing through the centers of curvature of a lens or other optical system. Light travels along this axis with no deviation. Also called **optical axis (2)**. **3.** The axis along which light is transmitted through a doubly-refracting crystal or medium without double refraction. Also called **optical axis (3)**. **4.** A reference axis chosen in a manner which provides for maximum symmetry, greatest response, or the like.

principal focus In an optical system, such as a lens or a mirror, a point towards which an incident bundle of parallel rays of light converge. Also called **focus (3)**, or **focal point**.

principal mode In a waveguide, the mode of propagation that has the lowest cutoff frequency. Also called **fundamental mode (1)**, or **dominant mode**.

principal ray Within multiple rays, the ray that passes through the optical center of a lens or optical system.

principle of duality A principle pertaining to the dual or analogous nature of certain entities or properties. For instance, the wave-particle duality of light, in which it has both wave-like properties and particle-like properties. Another example is the analogous nature of certain electric circuits, such as a transistor circuit and an electron tube circuit. Also called **duality (2)**, or **duality principle**.

principle of superposition For a physical system with multiple influences, a theorem stating that the resultant influence is the algebraic or vector sum of the each of the influences applied separately. Also called **superposition principle**.

print **1.** The markings, such as text and graphics, a printer makes when providing its output. Also, a command to produce said output. **2.** Any letters, symbols, or the like presented on paper, plastic, or another medium which allows for reading or other interpretation. Also, to produce said marking. **3.** A design or pattern transferred from one medium to another, as in photolithography. Also, to produce such designs or patterns. **4.** A photographic image transferred from a negative, or via another means. Also, to produce such an image.

print buffer A segment of computer memory utilized to temporarily store data that awaits printing. Used, for instance, to enable a system to pass any documents on to the printer, and let the printer print in the background while the processor tends to other tasks. Also called **printer buffer**.

print engine Same as **printer engine**.

print head Same as **printhead**.

print job One or more documents which are printed as a unit.

print manager A program which manages tasks pertaining to printing, such as handling a job order. Also spelled **printer manager**.

print mode One of multiple user-selectable modes available when printing. For example, a document may be printed at a given resolution, in portrait or landscape mode, and so on.

print-out Same as **printout**.

print preview In an application such as a word processor, a feature which enables viewing a document as it will be printed. Also, the command which provides such a preview. When a program features WYSIWYG, there is no need to use this command, as the displayed images faithfully represent what would become the printed images. Also called **preview**.

print queue Two or more documents awaiting printing. Also, a storage location for such documents. Also called **printer queue**.

print screen A command to send to the printer exactly what is displayed on the monitor at that instant. Also, the key or function which sends this command.

print server Within a computer or communications network, a server which controls multiple printers. A print server may offload printing from a file server, improving performance and printing availability. Also called **printer server**.

print spooler A program which manages printing tasks. Such a program can add, remove, change the order of, or cancel print jobs, in addition to placing documents in a memory or storage location to await printing at the printer's pace. Also called **printer spooler**, or **spooler (2)**.

print-through The usually unwanted transfer of a recorded signal between sections of a magnetic medium such as a tape. This may occur as a result of the affected sections being in close proximity, as occurs with overlapping segments of a reel of tape. Also called **magnetic printing (1)**, **magnetic transfer**, or **crosstalk (3)**.

print to file A command to create a version of a document formatted for printing, but which is saved as a file.

printed circuit Its abbreviation is **PC**. Also called **printed electronic circuit**. **1.** An electronic circuit whose conducting paths are formed by strips, foils, pads, segments, or the like, as opposed to wires, and which is mounted on a suitable substrate such as an insulating board. The board is usually coated with a conductive material, such as copper, and a technique such as photolithography is utilized to deposit a protective layer on portions of the conductors, so as to form a specified pattern. The unprotected metal is then etched away, followed by the addition of the desired circuit components, such as chips, transistors, diodes, resistors, and capacitors. Printed circuits lend themselves to mass-production, and when a fault arises an entire board may be replaced without the need to trace the source of the problem. **2.** Same as **printed-circuit board**.

printed-circuit board A board upon which a **printed circuit (1)** is mounted. Printed-circuit boards may be made, for instance, of a glass-epoxy resin, and can be single-sided, double-sided, or multilayered. Its abbreviation is **PCB**, or **PC board**. Also called **printed circuit (2)**, or **printed-wiring board**.

printed-circuit wiring Same as **printed wiring**.

printed electronic circuit Same as **printed circuit**.

printed wiring The strips, foils, pads, segments, or the like which form the conducting paths in **printed circuits**. Also called **printed-circuit wiring**.

printed-wiring board Same as **printed-circuit board**.

printer **1.** A device which produces a computer output consisting of letters, numbers, symbols, graphics, or the like, presented on paper, plastic, or another medium which allows for reading or other interpretation. There are printers whose output is provided a page, line, or character at a time, and printer types include laser, inkjet, and dot matrix. **2.** Any device or unit which serves as a **printer (1)**. For example, a unit within a cash register or a printing calculator which performs this function.

printer buffer Same as **print buffer**.

printer cable A cable, such as an IEEE 1284 printer cable, utilized to connect a printer to a computer.

Printer Control Language Same as **PCL**.

printer controller A circuit board or device which controls the way a printer accesses the computer, and vice versa. It may be contained in a single chip which is located in the printer or computer.

printer description file A file containing all the information pertinent to a given printer, such as configuration, resident fonts, memory, media supported, and so on. Its abbreviation is **PDF**.

printer driver A program which enables a computer operating system to communicate with, and control a printer. Each printer has its own specialized commands, and when a new printer is added to the computer, its driver must be installed for it to function properly.

printer engine Within a printer, the parts and mechanisms which perform the actual printing on the paper or other suitable medium. Most laser printers, for instance, have easily replaceable self-contained units containing the printer engine. Also spelled **print engine**.

printer file A file in a format which is ready for printing.

printer font 1. A font intended for printing, as opposed to a **screen font**, which is intended for displaying. **2.** A font which is resident in a printer.

printer head Same as **printhead**.

printer manager Same as **print manager**.

printer port A port, such as LPT, used by a computer for connecting a printer.

printer queue Same as **print queue**.

printer server Same as **print server**.

printer spooler Same as **print spooler**.

printhead In a character printer, a component or mechanism which deposits or otherwise transfers ink onto the paper or other suitable medium. Also spelled **print head**. Also called **printer head**.

printing calculator A calculator which incorporates a unit providing a printed output consisting of numbers, symbols, and the like.

printout A printed, or otherwise permanent output of a device such as a printer, or that which incorporates a printing function, such as a fax or oscillograph. When used in the context of computers, this contrasts with **soft copy**, which is data displayed on a screen, or stored. Also spelled **print-out**. Also called **hard copy**.

priority The condition of having precedence over something else. For instance, that which is at a higher level than something else within a hierarchy.

priority indicator Any characters, symbols, or other information which indicates the priority of something. For instance, coding which sends an email as high priority, or a flag representing the reception of such an email.

priority interrupt An interrupt, such as a non-maskable interrupt, which takes precedence over others.

priority processing Processing which follows a hierarchy.

priority queuing Queuing which follows a hierarchy, as opposed, for instance, to FIFO.

prism A transparent optical material or lens, such as that made of glass, with no less than two polished plane surfaces which are at an angle relative to each other so as to reflect or refract light. A prism may be utilized, for instance, to disperse white light into its component colors, to produce plane-polarized light from unpolarized light, or to invert an image. Used, for instance, in cameras and binoculars. Also called **optical prism**.

privacy The state of being concealed from others, especially against unsanctioned intrusion. Used, for instance, in the context of data preparation, transmission, reception, and storage.

privacy code A code utilized to help ensure that only authorized receivers may receive a transmission. A tone, for instance, may be transmitted along with a voice message, and only be heard by receivers with a matching code.

Privacy Enhanced Mail An Internet standard for email which provides security measures such as encryption and digital signatures. Also, email which adheres to this standard. Its abbreviation is **PEM**.

privacy policy A statement, such as that seen on many Web sites, pertaining to measures taken to protect the privacy of visitors, users, subscribers, and the like.

private automatic branch exchange A private branch exchange with automatic, as opposed to manual, switching. Its abbreviation is **PABX**. Also called **private automatic exchange**.

private automatic exchange Same as **private automatic branch exchange**. Its abbreviation is **PAX**.

private branch exchange Its abbreviation is **PBX**. An in-house telephone switching system that makes outside lines available to a given number of extensions in an office, group of offices, building, or the like. PBX features may include call forwarding, internal paging, call-detail recording, conference calling, caller ID, least-cost routing, and reduced-digit dialing for internal calls. A PBX is subscriber-owned, and may also offer data interfaces. When call routing requires no manual intervention, also called **private automatic branch exchange**.

private exchange A subscriber-owned telephone exchange which is usually connected to the public switched network. A private branch exchange is an example. Its abbreviation is **PX**.

private file A file which is not available to the public in general, as opposed to a **public file**, which is.

private folder In a communications network, a folder which is not available to the public in general, as opposed to a **public folder**, which is.

private key In public-key encryption, the key which is utilized for decryption, and which is available only to the intended recipient. The **public key** is available to the public in general, and is utilized to encrypt.

private-key cryptography Same as **private-key encryption**.

private-key encryption An encryption method in which the same key is used for encryption and decryption of messages. The fundamental concern with this method is insuring that the intended recipient indeed is the one to receive the key. Also called **private-key cryptography**, **secret-key encryption**, or **symmetric encryption**.

private line A permanent communications channel between two or more locations. Such a service includes private switching arrangements and a defined transmission path, and provides an exclusive high-speed connection which is available at all times. T1 and T3 lines are examples. Also called **dedicated line (1)**, **dedicated circuit**, **dedicated connection**, **leased circuit**, or **leased line**.

private manual branch exchange A private branch exchange with manual, as opposed to automatic, switching. Its abbreviation is **PMBX**.

Private Network-to-Network Interface Its abbreviation is **PNNI**. A routing protocol that enables different ATM switches to work together in the same network. PNNI can dynamically reroute packets, so any given switch can determine the path to any other switch.

private Web site A Web site within an intranet, or an intranet composed of a Web site.

privileged instruction An instruction which is executed only when a system is running in **privileged mode**. The resetting of registers or the setting of interrupts are examples.

privileged mode An operational state which allows certain operations which are otherwise restricted, such as memory management, to be carried out. The operating system usually operates in this mode. Also called **privileged state**, or **supervisor mode**.

privileged state Same as **privileged mode**.

PRK Abbreviation of **phase-reversal keying**.

PRML Abbreviation of **Partial Response Maximum Likelihood**.

probability 1. The likelihood that something will happen. 2. A statistical calculation to determine the quantitative likelihood that something will happen.

probe 1. A usually slender object or device which is utilized to detect, sample, or measure a phenomenon, by being placed in proximity or in direct contact with an object, material, or area. A probe may consist of an electrode, lead, rod, optic fiber, or the like. Probes can be used to test and quantify any number of phenomena, including voltage, current, resistance, magnetic flux strength, temperature, pressure, electron density in a plasma, logic levels within a logic circuit, and so on. Also, to sample with or otherwise use such a probe. Also called **test probe**. 2. A **probe** (1) which is inserted into a waveguide or cavity resonator, allowing energy to be transferred in or out. Such a probe may be rod, pin, or a wire. Also called **waveguide probe**.

probe coil Same as **pickup coil**.

probe microphone A small or miniature microphone which is utilized to detect sound pressure. Used, for instance, in hard to reach places, or when detection of sound must be obtained with a minimum of effect on the surrounding sound field.

probing The use of a **probe**.

problem-oriented language A computer language which is designed to solve specific problems. FORTRAN, for instance, is especially useful when dealing with problems of a scientific and mathematical nature. Its abbreviation is **POL**.

problem reduction In artificial intelligence, techniques, procedures, and processes utilized to simplify complex problems.

problem solving The ability to assess a given problem or task, and to be able to arrive at an effective solution or manner of completion. In artificial intelligence, problem solving involves proceeding through a series of operations which lead from an initial state to a goal state.

problem space In artificial intelligence, the formulation of a problem into states, such as the initial state, intermediate, and goal states, and the determination of the necessary operations to proceed to completion. Also called **state space**.

procedural language A programming language, such as C or FORTRAN, which requires specific statements, instructions, or subroutines which need to be executed in a specified sequence. Such a language emphasizes how something must be carried out, as opposed to what needs to be done, as would be the case with a **non-procedural language**.

Procedural Language/SQL Same as **PL/SQL**.

procedure 1. A sequence of operations required to perform a given task. 2. Within a computer program, a small group of instructions which perform a given task. Also called **routine** (1), **subroutine** (2), or **function** (3).

procedure call A statement that invokes a specific **procedure** (2).

procedure-driven Pertaining to programs in which users must follow a given sequence of steps, each of which is dictated by said program. This contrasts with **event-driven**, in which programs wait for user actions, such as mouse clicks and key pressings, responds to them, and then returns to waiting for further events. Also called **procedure-oriented**.

2. A specific program which is **procedure-driven** (1). Also called **procedure-driven program**.

procedure-driven program Same as **procedure-driven** (2).

procedure-oriented Same as **procedure-driven**.

process 1. A sequence of steps, events, operations, functions, or the like, which leads to a given result. Also, the following of such a sequence. 2. In computers, handling, manipulating, or otherwise operating on data, as occurs, for instance, when compiling, calculating, converting, generating, displaying, or transmitting. Also, the steps involved in such operations.

process-bound A circumstance in which a CPU is overloaded with calculations, which impairs its ability to process. May occur, for example, while recalculating a spreadsheet. Also called **compute-bound**, or **CPU-bound**.

process control The monitoring and management of any given process.

process-control monitor A circuit, device, mechanism, or system which effects **process control**. For instance, a testing device placed at various locations of a semiconductor wafer to check performance characteristics of the surrounding components. Its abbreviation is **PCM**.

process-control system A system, such as a control system, which monitors and manages a given process, or which maintains a variable as close as possible to a desired value. Its abbreviation is **PCS**.

process technology The processes undertaken to convert blank wafers of silicon, or another semiconductor material, into finished wafers composed of hundreds or thousands of chips.

processing The function and procedures of a **processor**.

processing-limited Same as **processor-limited**.

processor 1. The portion of a computer which has the necessary circuits to interpret and execute instructions, and to control all other parts of the computer. The processor consists of the control unit and the arithmetic-logic unit, both usually contained on a single chip, in which case it is also called **microprocessor**. Also called **CPU** (1), or **central processor** (1). 2. Software or hardware that in any manner processes data. For instance, a word processor, a language translation program, or a **processor** (1). 3. A component, circuit, device, unit, piece of equipment, system, or mechanism which is utilized in the completion of a given **process** (1).

processor core frequency Same as **processor core speed**.

processor core speed The clock rate of a CPU. Also called **processor core frequency**, **core speed**, or **CPU core speed**.

processor-limited A computing process which can proceed only as fast as the CPU speed allows. This contrasts with **input-limited**, in which the limiting factor is the rate at which input is entered. Also called **processing-limited**.

prod A point, tip, prong, or terminal, such as that of a probe, which facilitates testing or otherwise monitoring. Also called **test prod**.

Product Data Management Within an enterprise, an information system which manages, organizes, and controls the data pertaining to a product throughout all its stages, from planning and design, all the way through its useful life. Its abbreviation is **PDM**.

product demodulator Same as **product detector**.

product detector A demodulator whose output is the product of two signals. It may consist, for instance, of the product of the input signal of a radio receiver and the signal of a local oscillator. Also called **product demodulator**.

product modulator A modulator whose output is the product of two signals. It may consist, for instance, of the product of the carrier signal of a radio receiver and the modulating signal.

production model   A model which is in its final form, and is thus ready to serve as the basis for the manufacturing of components, circuits, devices, equipment, or units, which are similar to it identical.

production system   An expert system which applies IF-THEN rules to expressions to produce new expressions with the goal of problem solving. Also called **rule-based system**.

productivity software   Applications, such as word processors, databases, spreadsheets, and presentation graphics, which are designed to enable users to work, and present their work, more efficiently and professionally.

productivity suite   An application suite comprised of multiple **productivity software** offerings bundled together.

program   **1.** A set of instructions which when translated and executed cause a computer to perform specific operations. Computer programs are written in programming languages, which may be classified as high-level or low-level. High-level languages, such as COBOL, C++, and Java, require a compiler or interpreter to translate statements into machine language. Low-level languages, such as an assembly language, are much closer to machine language, but still require an assembler for conversion into machine language. The only language a computer understands is machine language. Although **program** and **software** are used mostly synonymously, software is a bit more of a general term, as it is used to differentiate from the physical equipment of a computer, the hardware, from the instructions which tell a computer what to do, the software. An application is both a program and software, but when using the term software it usually refers to multiple application programs, or application programs in general, in which case they are called application software. Also called **computer program**. **2.** To create a **program** (1). **3.** The content of a given broadcast.

program code   A set of computer program instructions. Also called **programming code**, or **code** (2).

program compatibility   **1.** The ability of two or more computer programs to seamlessly work together, without modification. For instance, a program that works equally well across multiple platforms. **2.** The extent to which **program compatibility** (1) is achieved.

program control   In a control system, the utilization of a program to maintain or vary one or more variables or values within given limits.

program counter   Same as **program register**.

program file   **1.** A file consisting of a computer program. **2.** A file which is a part of a computer program.

program flow chart   Same as **program flowchart**.

program flow diagram   Same as **program flowchart**.

program flowchart   A diagram which uses a set of standard symbols to represent the sequence of operations of a computer program. Such a chart may show, for instance, the steps of a subroutine. Also spelled **program flow chart**. Also called **program flow diagram**.

program generator   A program, such as an application generator, which enables the generation of other programs.

program library   A collection of routines and programs. The content of such a collection is usually related in some manner. For example, a class library is composed of a set of routines and programs that programmers can use to write object-oriented programs.

program listing   A printed copy of the source code of a program.

program logic   The rules, instructions, and algorithms utilized by a program to achieve the desired results.

program maintenance   The revision and updating of computer programs, so as to remedy problems and better adapt to new needs.

program register   In a CPU, a register that contains the address of the location in memory that is to be accessed by the next instruction. May also refer to the address of the current instruction. Also called by various other names, including **program counter**, **control register**, **instruction register**, and **sequence register**.

program signal   In TV, a complex signal containing the sound and picture information.

program specification   A document or file which describes in detail the objectives, structure, and functions of a computer program, so as to enable programming and facilitate maintenance.

program statement   In a high-level language, a syntactic unit which includes one or more operators and one or more operands. A program statement generates one more machine language instructions, which in turn direct a computer to perform one or more actions. Also called **statement**.

program step   Within a computer program, a single instruction. May also refer to a single statement, which can contain multiple instructions. Also called **step** (4).

program version   The numbers, letters, and/or words which identify a specific version of a program. For example, *1.0*, *5.01a*, or *2004b beta build #1111*. Changes in whole numbers usually imply significant modifications, while fractions, letters, or builds often identify versions which correct bugs, provide minor enhancements, or the like. Also called **program version number**, **release version number**, or **version number**.

program version number   Same as **program version**.

programmability   The extent to which something is **programmable**. Also, characterized by being programmable.

programmable   That which is capable of accepting instructions, and subsequently following them. All computers can be programmed, and many other devices, such as appliances and calculators, may also be programmable.

programmable array logic   A logic chip whose OR gates are predefined, and whose AND gates may be programmed. A **programmable logic array** may also have its OR gates programmed. Its acronym is **PAL**.

programmable calculator   An electronic calculator that is **programmable**. Such calculators are usually programmed either by entering sequences of keystrokes, or by inserting a card or disk.

programmable controller   A controller, such as an automatic controller, whose functions and tasks may be programmed. abbreviation is **PC**.

programmable counter   A counter which can be programmed. For example, such a counter may be set to generate one or more outputs based on one or more set pulse counts.

programmable device   A device which accepts instructions, or whose settings can be changed, so as to tailor its function to perform a given task in the desired manner. For instance, a programmable calculator.

programmable function keys   On a computer keyboard, a set keys, often 10 or 12, whose functions may be programmed within a given application. Such keys may be used, for instance, to access functions or to automate a sequence of actions. Programmable function keys may also be defined as including most any on the keyboard, including letters and symbols, as modifier keys can be used to change their function.

programmable gain amplifier   An amplifier whose signal gain can be programmed. For instance, such an amplifier may be set to apply a variable signal gain which depends upon the input voltage. Its abbreviation is **PGA**.

programmable gate array Its abbreviation is **PGA**. A gate array which instead of being programmed at a factory, is done so by an end user. PGAs may usually be programmed thousands of times, and are used, for instance, for making prototypes of chip designs. Also called **field programmable gate array**.

programmable IC Abbreviation of **programmable** integrated circuit. An IC, such as a programmable logic array, which can be programmed by an end user.

programmable integrated circuit Same as **programmable IC**.

Programmable Interrupt Controller A chip or other hardware component which manages interrupts. Its acronym is **PIC**.

programmable logic array Its abbreviation is **PLA**. A logic chip which instead of being programmed at a factory, is done so by an end user, usually during installation. PLAs can only be programmed once. Also called **field programmable logic array**.

programmable logic controller A chip-based control system often used for industrial applications. It utilizes data links to communicate with other process control components, devices, and equipment, and is utilized for tasks such as complex data manipulation, timing processes, sequencing, and machine control. Its abbreviation is **PLC**. Also called **logic controller**.

programmable logic device An IC, such as a programmable logic array, or a PROM, which can be programmed by an end user. Its abbreviation is **PLD**.

programmable power supply A power supply whose output can be programmed. For instance, such a supply may be set to provide a variable output voltage which depends upon the input current or resistance, or on a digital code.

programmable read-only memory Same as **PROM**.

programmable read-only memory chip Same as **PROM**.

programmable robot A robot which can be programmed for a given task, then reprogrammed for others, if desired. Each programmed set of steps may involve any number of complex and repetitive tasks. Also called **reprogrammable robot**.

programmable ROM Same as **PROM**.

programmatic interface The interface between a user and a given programming language or operating system. An application program interface is an example.

programmed halt Same as **programmed stop**.

Programmed Input/Output Mode Same as **PIO mode**.

Programmed Inquiry Learning or Teaching Same as **PILOT**.

programmed stop An instruction causing a break in a computer program or routine. Also called **programmed halt**.

programmer 1. An individual that designs, writes, debugs, and tests programs. **2.** A hardware device that programs chips, such as PROMs.

programming 1. The entering of settings and/or instructions into a programmable device, so that it performs specific tasks. For instance, programming a VCR to record a desired program. **2.** The creating of computer programs. The key components in this process include a full understanding of which tasks the final program will be meant to perform, a command of the programming language being used to write the program, the development of the program logic to suitably address problems to be solved, and the testing and debugging of the program. Also called **computer programming**. **3.** The program offerings of one or more TV or radio broadcasters.

programming code Same as **program code**.

programming interface A set of instructions that determine how a computer application interacts with the operating system. This contrasts with an **application binary interface**, which interacts with the operating system and the hardware. Also called **application program interface**, or **software interface**.

programming language A language which is utilized to create computer programs, through the writing of instructions and statements which comprise source code. Natural languages were formerly excluded from this definition, but currently some words and phrases may be used as well for programming.

progress indicator On a computer screen, a dialog box showing how far along a given process has progressed. Progress may be expressed, for instance, as a percentage completed, as an estimated time to completion, or as the number of items remaining.

progressive download The playing of audio and/or video while a portion of the download is still in progress. It is intermediate between true streaming and performing a complete download before play begins. Also called **pseudostreaming**.

progressive GIF A GIF image that starts as a fuzzy outline and progresses until fully focused. Seen, for instance, when using a dial-up connection. Also called **interlaced GIF**.

progressive JPEG A JPEG image that starts as a fuzzy outline and progresses until fully focused. Seen, for instance, when using a dial-up connection.

progressive scan Same as **progressive scanning**.

progressive scanning A scanning system, such as that computer monitors, in which the electron beam traces all lines sequentially from top to bottom. This contrasts with **interlaced scanning**, in which the electron beam traces all odd-numbered lines followed by the tracing of all even-numbered lines. Interlaced scanning may produce a fluttering effect when displaying certain graphics, a drawback not exhibited by monitors with progressive scanning. Also called **progressive scan**, **non-interlaced scanning**, or **sequential scanning**.

progressive wave A wave which propagates freely through a given medium, such as a transmission line or vacuum. In such a wave the antinodes and nodes progress along the medium at the speed of the wave, while in a **standing wave** the antinodes and nodes do not change position. Also called **traveling wave**.

projected-beam smoke detector Same as **photoelectric smoke detector**.

projection cathode-ray tube Same as **projection CRT**.

projection CRT Abbreviation of **projection** cathode-ray tube. In a **projection TV**, a CRT which produces an intensely bright image which is projected onto the translucent screen. Also called **projection tube**.

projection lithography A photolithographic technique in which an optical system is used to produce an image of the photomask on the substrate. When a light beam, usually that of a laser, is focused through the mask it is called **optical projection lithography**. Particles, such as electrons or ions, may be used to achieve better resolution, as in electron projection lithography, or ion projection lithography.

projection panel A panel, such as an LCD panel, which is used in conjunction with a **projector (1)**.

projection television Same as **projection TV**.

projection tube Same as **projection CRT**.

projection TV Abbreviation of **projection** television A TV in which images sent from a separate unit are projected onto a translucent screen. In a front-projection TV the projector and any viewers are on the same side of said screen, while in a rear-projection TV the projector and any viewers are on opposite sides of the screen.

projector 1. A device which transmits a beam of light, especially that conveying images, onto a surface, such as a translucent screen. An overhead projector is an example. 2. A device or object, such as a horn, which serves to direct and intensify sound.

Prolog Acronym for **pro**gramming **log**ic. A high-level programming language which is frequently used for artificial intelligence applications, such as expert systems.

PROM Acronym for programmable read-only memory. A chip which may be programmed once, usually by a customer. This contrasts with **EEPROM** chips, which can be reprogrammed, and with **ROM** chips, which are programmed when manufactured. Also called **PROM chip**.

PROM burner A device utilized to program **PROMs**. To write code or data onto a PROM chip is referred to as burning, as connections are melted when programming. Also called **PROM programmer**.

PROM chip Same as **PROM**.

PROM programmer Same as **PROM burner**.

promethium A radioactive metallic chemical element whose atomic number is 61. It has about 30 known isotopes, all unstable, and is used in nuclear-powered batteries, semiconductors, and as an x-ray source. Its chemical symbol is **Pm**.

promiscuous mode 1. In a communications network, a mode of operation in which all data packets can be received by all nodes. This can lead to security problems, such as the interception of passwords, email, account names, and so on. 2. In a communications network, a node engaged in packet sniffing.

prompt A symbol or message which is displayed on a monitor, especially that of a computer, indicating that input is awaited, and when provided will be entered there.

prong Same as **pin**.

propagate 1. To spread or extend from a smaller to a larger area. 2. For a wave move through a medium, such as the atmosphere, a vacuum, or a transmission line. For example, an electromagnetic wave traveling via a transmission line, or an acoustic wave traveling through water. 3. To move through a medium.

propagation 1. The process of spreading or extending from a smaller to a larger area. 2. The movement of a wave through a medium, such as the atmosphere, a vacuum, or a transmission line. For example, an electromagnetic wave traveling via a transmission line, or an acoustic wave traveling through water. Also, the process by which this occurs. Also called **wave propagation**, or **wave motion**. 3. Movement through a medium.

propagation coefficient Same as **propagation constant**.

propagation constant A complex quantity whose real number component is the **attenuation constant**, and whose imaginary number component is the **phase constant**. It describes the effect a transmission medium or line has on a wave traveling through or along it. Also called **propagation coefficient**.

propagation delay Also called **propagation time**, or **propagation time delay**. 1. The time interval required for a signal to move from one point to another. 2. The time interval required for a signal to move through a component, circuit, stage, unit, device, system, or medium.

propagation direction In a homogeneous isotropic medium, the direction of energy flow at any given point. For instance, the direction of propagation in a uniform waveguide is considered to be along its axis. Also called **direction of propagation**.

propagation loss Also called **propagation losses**. 1. Any losses in the strength of a signal traveling from one point to another. Such losses may be due to absorption, scattering, dispersion, or the like. 2. **Propagation loss** (1) occurring in a signal traveling from a transmitting antenna to a receiving antenna.

propagation losses Same as **propagation loss**.

propagation medium Also called **medium of propagation**. 1. The medium through which a wave or other phenomenon propagates. This can be the atmosphere, a vacuum, a waveguide, a crystal, water, and so on. 2. The **propagation medium** (1) through which an electromagnetic wave travels.

propagation mode For a guided electromagnetic wave, one of the possible manners of propagation. Such waves can propagate in one of three principal modes, which are as transverse electric waves, transverse magnetic waves, or transverse electromagnetic waves. Also called **waveguide mode**.

propagation path The path, such as that between a transmitter and a receiver, a wave follows.

propagation speed The speed at which a wave propagates through a given medium. It is the speed at which given points within it, such as crests and troughs, travel. In a vacuum it is equal to the product of the frequency and wavelength. In a dispersive medium the propagation speed of electromagnetic waves varies as a function of the frequency, while in a non-dispersive medium the propagation speed is independent of the frequency. Also called **wave speed**, **wave celerity**, or **celerity**.

propagation time Same as **propagation delay**.

propagation time delay Same as **propagation delay**.

propagation velocity The **propagation speed** in a given direction. Also called **wave velocity**.

property 1. An attribute or characteristic that helps describe, distinguish, and define a component, circuit, device, system, material, and so on. 2. A **property** (1) which is possessed by all members of a given class.

proportional 1. Having, establishing, or maintaining a relationship with one or more entities or values. For instance, the force between two charged particles is directly proportional to the product of their magnitudes, and inversely proportional to the square of their distance. 2. A relationship between two or more quantities, in which the variation in one causes a change in the other. For example, a transducer whose input is a DC, and whose output is a proportional AC.

proportional action An action, such as a corrective action in a control system, which is proportional to a given process, state, or variable.

proportional band A frequency band whose values are equally distributed along the stated interval of frequencies. For example, the first 10% of such a band comprises the same interval of hertz as any other equally-sized segment, such as the middle or last 10%.

proportional control Control in which the corrective action is proportional to a given process, state, or variable.

proportional counter 1. A radiation counter which incorporates a **proportional counter tube**. 2. Same as **proportional counter tube** (1).

proportional counter tube 1. A counter tube whose output is proportional to the number of detected ionized particles. Also called **proportional counter** (2). 2. A counter tube whose output is proportional to the number of detected pulses or events.

proprietary Private, only used by a single or few groups, exclusively owned, or otherwise not utilizing a common or standardized format, composition, mechanism, or the like. For instance, proprietary software, or a proprietary file format.

proprietary file format A file format which is used by a single organization or entity, and which is not compatible with other formats.

proprietary hardware Hardware which is used by a single organization or entity, and which is not compatible with other formats.

proprietary protocol A protocol which is used by a single organization or entity, and which is not compatible with other formats.

proprietary software Software which is used by a single organization or entity, and which is not compatible with other formats.

proprioceptor A sensor which receives stimuli from within a body, such as that of a robot, and which responds to changes in position and other movements.

prosodic features Components of speech such as duration, stress, loudness, and rhythm, which impart additional meaning to spoken words and phrases. A worthy voice-recognition system should take these factors into consideration.

prosthesis An artificial device which replaces a missing body part, or that with reduced function. For example, a cochlear implant to improve or restore hearing.

protactinium A shiny silvery radioactive metallic chemical element whose atomic number is 91. It is malleable, ductile, and has about 25 known isotopes, all unstable. It is used in nuclear power reactors. Its chemical symbol is **Pa**. Sometimes spelled **protoactinium**.

protected area 1. An area that is under surveillance, or otherwise secured. Devices and systems such as motion detectors and closed-circuit TV, for instance, may be utilized in such an area. Also called **protected region**, **protected zone**, or **protected location** (2). 2. Same as **protected location** (1).

protected location 1. A computer memory location which is not accessible in some manner. For example, an area which is reserved for CPU use, and which is not available to programs. Also called **protected memory area**, or **protected area** (2). 2. Same as **protected area** (1).

protected memory area Same as **protected location** (1).

protected region Same as **protected area** (1).

protected zone Same as **protected area** (1).

protection 1. Any materials, devices, or measures, such as fuses, coatings, insulators, shields, or ground-fault circuit interrupters, which help prevent components, circuits, devices, equipment, systems, or personnel, from being damaged, or otherwise exposed to harmful conditions. Specific examples include overload protection, electrostatic discharge protection, and ground-fault protection. 2. The safeguarding of data against loss, damage, unwanted modification, or unauthorized access. Such safeguards may be administrative, physical, or technical. Also called **data protection**, or **information security**. 3. The act of providing **protection** (1), or **protection** (2). Also the condition of having such protection.

protective circuit A circuit, such as a ground-fault circuit interrupter, which affords **protection** (1).

protective device 1. A device, such as a circuit breaker, which provides **protection** (1). 2. A device, such as a dongle, which provides **protection** (2).

protective mechanism A mechanism which affords **protection** (1) or **protection** (2).

protective relay A relay which provides **protection** (1).

protective resistor A resistor, such as a bleeder, which affords **protection** (1).

protective switch A switch that provides **protection** (1).

protective system A system which affords **protection** (1) or **protection** (2).

protector That which provides **protection** (1) or **protection** (2).

protium An isotope of hydrogen whose nucleus contains a single proton. Of all the known nuclei of chemical elements, only that of protium has no neutrons. It composes the vast majority of the matter of the known universe.

protoactinium Same as **protactinium**.

protoboard Abbreviation of **prototyping board**.

protocol A standard, or a set of rules which must be agreed upon in order for two or more devices, such as modems, to exchange information effectively. Such conventions must include considerations such as how to initiate and terminate a transmission, the transmission speed, and whether the transmission will be synchronous or asynchronous. In addition, these may include error-detection techniques, encryption, and so on. Examples of communications protocols include Point-to-Point Protocol, and Zmodem. Also called **communications protocol**.

protocol analyzer Software and/or hardware which monitors a communications network, and which troubleshoots and performs tests such as the simulation of error conditions, so as to help it work smoothly. Also called **network analyzer** (1).

protocol conversion The process of translating from one protocol to another, when two devices utilizing different protocols are exchanging data.

protocol port Same as **port number**.

protocol port number Same as **port number**.

protocol stack A group of communications network layers which work together and follow a hierarchy. The OSI Reference Model is an example.

protocol suite The complete set of protocols used by a given network. Such a suite can form, or be a part of a protocol stack.

proton A positively charged extremely stable subatomic particle which is present in all atomic nuclei. Its mass is about 1,840 times that of an electron, and slightly less than that of a neutron. A protium nucleus consists of a single proton. It is the lightest baryon, and its antiparticle is the antiproton. The number of protons in a nucleus is what differentiates all chemical elements. Protons are used, for instance, in particle accelerators.

proton fluence The passing of protons through a given area, such as a square centimeter. May be expressed, for instance, in protons per second.

proton mass Same as **proton rest mass**.

proton microscope A microscope similar to an electron microscope, except that it utilizes a beam of protons.

proton number The number of protons in the nucleus of an atom. It is also the number of electrons that surround the uncharged atom. The atomic number of carbon, for instance, is 6. Also known as **atomic number**.

proton rest energy The energy of a proton at rest, according to the mass-energy equation. This value is approximately equal to 938 MeV. Also called **proton self-energy**.

proton rest mass The mass of a proton at rest. It is a fundamental physical constant equal to approximately 1.67262×10^{-27} kg. Its symbol is m_p. Also called **proton mass**.

proton self-energy Same as **proton rest energy**.

proton separation energy The energy required to remove a proton from a nucleus.

proton synchrotron A synchrotron in which protons are the accelerated particles. In a tevatron, for instance, particle energies in excess of 1 TeV may be attained, with collision energies near 2 TeV.

prototype A usually full-scale original model suitable for evaluation and testing, and which is utilized as a basis for production or future models.

prototyping The creation of **prototypes**.

prototyping board A perforated board or chassis on which electronic devices and components can be mounted and interconnected simply and quickly, for testing, experimenting, and preparation of prototype circuits. Its abbreviation is **protoboard**. Also called **breadboard** (1).

proustite A red mineral whose chemical formula is Ag_3AsS_3. It is soft, relatively dense, and translucent. Used, for instance, in optics research.

provisioning Also called **network provisioning**. **1.** The setting up of a communications network. This may include the allocation of hardware, such as multiplexers and interface units, the selection of the mediums of transmission, such as coaxial cables and optical fibers, and establishing the connections between nodes. Alternatively, it may involve simply programming a computer to make the appropriate arrangements within an established network. **2. Provisioning** (1) for a specific customer.

proximity alarm **1.** An alarm which is activated when the distance between two objects is less than or greater than a given amount. **2.** An alarm incorporating a **proximity sensor**.

proximity card A card, such as a smart card, whose data can be accessed by a **proximity card reader**. Used, for instance, for purchases or identification.

proximity card reader A device which detects data contained by **proximity cards** that are held or otherwise located nearby. A card and reader may communicate, for instance, via RFID technology.

proximity detector Same as **proximity sensor**.

proximity effect **1.** The altering of the distribution of current in a conductor due to the presence of another nearby conductor. **2.** In a microphone, a boost in the low-frequency response when in close proximity to a sound source.

proximity relay A relay which responds to very small variations in capacitance, similar to those produced by a human body, or a part of it, such as a hand. Such a device may be used, for instance, as an intrusion alarm. Also called **capacitance relay**.

proximity sensing **1.** The utilization of a **proximity sensor**. **2.** The function of a **proximity sensor**.

proximity sensor A sensing device utilized to determine the distance to one or more objects, without coming in contact with that which is detected. They are many types, including optical, acoustic, and magnetic. Frequently used in robotics. Also known as **proximity detector**, **non-contact sensor**, or **non-contact detector**.

proximity switch A switch that is actuated when an object or body is in close proximity. Such a switch may incorporate a proximity sensor.

proximity warning A warning, such as an alarm, produced when the distance between two objects is less than or greater than a given amount.

proxy Same as **proxy server**.

proxy cache The caching function of a **proxy server**.

proxy server A server, application, or system that serves as an intermediary between a private network, such as a LAN, and all other networks. Its two main functions are to provide document caching and access control. When caching, a proxy first attempts to access data which it has cached, and if not present there, it fetches it from a remote server where said data resides. When controlling access, it serves as a firewall. Also called **proxy**, or **application level gateway**.

PRR Abbreviation of **pulse repetition rate**.

PRT Abbreviation of **platinum resistance thermometer**.

ps Abbreviation of **picosecond**.

pS Abbreviation of **picosiemens**.

PSD **1.** Abbreviation of **power spectral density**. **2.** Abbreviation of **phase-sensitive detector**.

psec Abbreviation of **picosecond**.

pseudo code Same as **pseudocode**.

pseudo instruction Same as **pseudoinstruction**.

pseudo-random Same as **pseudorandom**.

pseudo-random number Same as **pseudorandom number**.

pseudo-random sequence Same as **pseudorandom sequence**.

pseudo-streaming Same as **progressive download**. Also spelled **pseudostreaming**.

pseudocode Also spelled **pseudo code**. Its abbreviation is **p-code**. **1.** A description, in a natural language, of the steps of a program, routine, or algorithm. It is intended to be detailed and sufficiently clear, so that programmers can understand enough of the design of a program to make changes and improvements. Such a language is not necessarily suitable for computer processing, although there are programs that convert pseudocode into a programming language. **2.** A computer language that can not be directly executed by a CPU. Such a language must be interpreted or compiled prior to execution. Also called **interpretive language**. **3.** An intermediate language which is processed by a virtual machine. The virtual machine converts the instructions into machine instructions most any CPU will understand, making such code essentially architecture-independent. For example, Java bytecode is processed by a Java Virtual Machine. Also called **bytecode**

pseudoinstruction An instruction written in **pseudocode**. Also spelled **pseudo instruction**.

pseudomorphic high-electron mobility transistor A gallium-arsenide transistor with enhanced electron mobility, which is used, for instance, as a low-noise microwave amplifier. Its abbreviation is **PHEMT**.

pseudonoise sequence Same as **PN sequence**.

pseudorandom Generated by a specific process, such as an algorithm, but still having sufficient randomness for most purposes. For a number to be truly random, there could not be a specific pattern or even the objective of seeking such a number. Also spelled **pseudo-random**.

pseudorandom number A number which is **pseudorandom**. Such a number passes one or more tests of randomness, but is not truly random. Also spelled **pseudo-random number**.

pseudorandom sequence A sequence, such as a series of numbers, which is **pseudorandom**. Also spelled **pseudo-random sequence**.

pseudostreaming Same as **progressive download**. Also spelled **pseudo-streaming**.

psf Abbreviation of **pounds per square foot**.

psi Abbreviation of **pounds per square inch**.

psi particle An unusually long-lived meson. Also called **J particle**.

PSK Abbreviation of **phase-shift keying**.

PSN **1.** Abbreviation of **packet-switching network**. **2.** Abbreviation of **public switched network**.

psophometer An instrument that measures and indicates the level of noise in a communications system, such as that utilized for telephony. Such a device usually incorporates a weighting network, whose characteristics depend on the type of circuit being evaluated. The measured results approximate the effects of noise at different frequencies on a human listener.

PSRR Abbreviation of **power-supply rejection ratio**.

PSTN Abbreviation of **public switched telephone network**.

psychoacoustics The science that deals with the manner in which sound is perceived. Factors such as head-related transfer functions, and the number of sound sources and their locations, are taken into account. Used, for instance, in designing surround-sound systems, and in formulating audio compression techniques.

psychogalvanometer A device which measures galvanic skin response.

pT Abbreviation of **picotesla**.

Pt Chemical symbol for **platinum**.

PTC Abbreviation of **positive temperature coefficient**.

PTFE Abbreviation of **polytetrafluoroethylene**.

PTFE resin Abbreviation of **polytetrafluoroethylene resin**.

PTH Abbreviation of **plated-through hole**, or **pin-through hole**.

PTM Abbreviation of **pulse-time modulation**.

PTT Abbreviation of **push-to-talk**, or **press-to-talk**.

PTT operation Abbreviation of **push-to-talk operation**, or **press-to-talk operation**.

PTT switch Abbreviation of **push-to-talk switch**, or **press-to-talk switch**.

Pu Chemical symbol for **plutonium**.

public-address amplifier An amplifier specially designed for use in a **public-address system**. The features of such an amplifier may include multiple microphone inputs, and muting of all other inputs when telephone paging.

public-address system A system, such as that utilized in airports or shopping centers, which serves convey voice messages through a distributed group of speakers powered by one or more amplifiers. Its abbreviation is **PA system**.

public carrier A telecommunications entity, such as a telephone company, that provides services to the general public. Public carriers are usually regulated by the appropriate authorities, such as governmental agencies. Also called **common carrier**, **carrier (3)**, **commercial carrier**, or **communications common carrier**.

public data network A communications network which is established and operated by a telecommunications entity such as a public carrier. Its abbreviation is **PDN**.

public-domain software Software which has been given or otherwise relinquished to the public, and which has no protections such as a copyrights. Its abbreviation is **PD software**.

public file A file available to the public in general, as opposed to a **private file**, which is not.

public folder In a communications network, a folder available to the public in general, as opposed to a **private folder**, which is not.

public key In public-key encryption, the key which is utilized to encrypt, and which is available to the public in general. The **private key** is used for decryption, and is only intended for the recipient.

public-key cryptography Same as **public-key encryption**.

public-key encryption An encryption method which uses two keys for successful encryption and decryption of messages. The first key is used to encrypt and is public, as most anyone can look it up. To decrypt, the second key is necessary. This key is private, as only the recipient knows it. Since it is not necessary to send the decryption key in any message, this method eliminates the vulnerability inherent in its being sent to the recipient. Also called **public-key cryptography**, or **asymmetric encryption**.

Public-Key Infrastructure A system in which entities such as certification authorities issue digital certificates, or their equivalent, in order to authenticate, use, and manage cryptographic keys. Such an infrastructure helps provide a secure environment for communications and transactions using a public network such as the Internet. Its abbreviation is **PKI**.

public switched network Its abbreviation is **PSN**. **1.** Same as **public switched telephone network**. **2.** Any public network providing switching, which includes telephone networks and public data networks.

public switched telephone network Its abbreviation is **PSTN**. The multiple interconnected telephone networks in place worldwide, which enable most any phone to dial most any other phone around the globe. POTS is the basic service a PSTN provides. Such a network may also be used for data services, such as dial-up Internet access. Also called **public telephone network**, **public switched network (1)**, or **switched telephone network**.

public telephone network Same as **public switched telephone network**.

public Web site A Web site which can normally be accessed by any user. The term is utilized to contrast with a **private Web site**.

puck The pointing device of a digitizing tablet. It is similar to a mouse, but is much more accurate, because its location is determined by touching an active surface with an absolute reference. Also called **cursor (4)**, **pen (1)**, or **stylus (1)**.

pull **1.** Same as **pull technology**. **2.** To receive information utilizing **pull technology**.

pull-down menu In a GUI, a menu which is always available in the active window, but whose contents are revealed only when a user selects said menu with a mouse or keyboard command. Once selected, the menu choices are presented in a vertical array below the menu title. Also called **pop-down menu**, or **drop-down menu**.

pull-in current Same as **pickup current**.

pull-in voltage Same as **pickup voltage**.

pull switch A switch that is actuated by pulling. Such a switch, for instance, may use a cord that is pulled in case of an emergency.

pull technology Data distribution, such as that over the Internet, in which users receive information by requesting it. For example, a user requesting the download of a given page by entering its address or by clicking on a desired hyperlink. In **push technology**, material is delivered whether the user wants it or not. Also called **pull (1)**.

pulling **1.** A variation in the frequency of a circuit, especially that of an oscillator, caused by an external circuit, device, or factor, such as a variation in load impedance. Also called **frequency pulling**. **2.** An effect in which a component, circuit, or device induces another to change from a desired value. **3.** A technique used for cultivating crystals, in which a growing crystal is slowly withdrawn from a molten solution of the crystal. Also called **crystal pulling**.

pulsar A neutron star which emits sharp and brief pulses of energy, as opposed to the usual steady stream of radiation. Pulsar pulse rates have been observed ranging from around one millisecond, to approximately 5 seconds.

pulsating current A current which passes through regular cycles, and whose average is not zero. Such a current may result, for instance, from the sum of an alternating current and a direct current. Also called **undulating current**.

pulsating DC Abbreviation of **pulsating direct current**. A DC which passes through regular cycles. While the polarity of such a current never changes, its amplitude does.

pulsating direct current Same as **pulsating DC**.

pulse For a quantity which is normally constant, a well-defined variation which increases from a steady value to a maximum, then back to or near the original value, all in a comparatively short time. Pulses may be desired or unwanted. Applications of intentionally-generated pulses include clock pulses, gate pulses, dial pulses, radar pulses, blanking pulses, and certain light pulses. Undesired pulses include noise pulses, spikes, and certain light pulses. A pulse of very short duration is an **impulse (3)**, although both terms are often used synonymously.

pulse amplifier Its abbreviation is **PA**. **1.** An amplifier specifically designed to amplify **pulses**. **2.** An amplifier which delivers a single output pulse for each input pulse.

pulse amplitude Also called **pulse height**. **1.** For a pulse, the maximum absolute value of the displacement from a reference position, such as zero, or from a steady value which is otherwise maintained. **2.** For a pulse, the instantaneous value of the displacement from a reference position, such as zero, or from a steady value which is otherwise maintained. **3.** For a pulse, the average value of the displacement from a reference position, such as zero, or from a steady value which is otherwise maintained.

pulse-amplitude modulation Pulse modulation in which the amplitude of the pulses in a pulse train, which serve as the carrier, are varied according to the information-bearing signal. Its acronym is **PAM**.

pulse-amplitude selector Same as **pulse-height selector**.

pulse analysis The utilization and function of a **pulse analyzer**.

pulse analyzer An instrument which measures and indicates pulse characteristics, such as amplitude and frequency.

pulse band-width Same as **pulse bandwidth**.

pulse bandwidth The interval of frequencies within which a given percentage of the **pulse amplitude (1)** is maintained. Also spelled **pulse band-width**.

pulse carrier A pulse train utilized as a carrier. Used, for instance, in pulse-amplitude modulation.

pulse characteristics Same as **pulse form**.

pulse code **1.** A code consisting of one or more combinations of pulses. **2.** A pulse train which is modulated to convey information.

pulse-code modulation Modulation in which an analog signal is sampled, the amplitude of each sample is quantized, and then encoded as a digital signal for transmission. In order for the signal to be faithfully reproduced, the sampling rate must exceed the Nyquist rate. At the receiving end a decoder is utilized to recover the original information. Widely used to convert analog signals into digital form. Its abbreviation is **PCM**.

pulse coder A circuit or device which converts a signal into a coded pulse train. Also called **pulse encoder**.

pulse communications Communications utilizing any form of **pulse modulation**.

pulse counter A device detects and counts pulses. A binary counter, for example, produces one output pulse for every two input pulses, while a decade scaler produces a single pulse for every ten input pulses. Other pulse counters tally the number of pulses for a given time interval.

pulse counting The use and function of a **pulse counter**.

pulse current A unidirectional surge of current of very short duration. It quickly rises to a maximum, then drops to zero in a similar fashion. Such a surge may be produced, for instance, when switching equipment on or off. Also called **impulse current**, or **impulse (1)**.

pulse decay time Same as **pulse fall time**.

pulse decoder A circuit or device which converts that which has undergone pulse modulation back into its original form. Also called **pulse detector (1)**.

pulse decoding The function and use of a **pulse decoder**.

pulse detector **1.** Same as **pulse detector**. **2.** A device, such as a pulse counter, which senses pulses. Also called **pulse sensor**.

pulse dialer In telephony, a device which utilized for **pulse dialing**.

pulse dialing Telephone dialing in which one or more pulses interrupt a steady DC, to indicate each dialed digit. For example, when dialing a 4, four such pulses are sent. Also called **rotary dialing**.

pulse discriminator A circuit or device which makes fine distinctions between nearly identical pulses. Used, for instance, to respond only to pulses of a given shape, period, amplitude, or the like.

pulse droop **1.** A detectable downward slope present in an otherwise flat top of a rectangular pulse. **2.** In a pulse train, a decrease in mean pulse amplitude, after maximum amplitude has been attained. Usually expressed as a percentage of maximum amplitude. Also called **droop**.

pulse duration Its abbreviation is **PD**. **1.** For a given pulse, the time interval that elapses between the end of the rise time and the start of the decay time. The pulse duration may also be defined in other manners, such as the time interval between the 50% points of peak pulse amplitude. Also called **width (2)**. Also known as **pulse length**, or **pulse width**, although both are deprecated terms. **2.** The time interval that a pulse is emitted or transmitted. For example, the time during which a radar pulse is transmitted.

pulse-duration modulation A form of modulation in which the duration of pulses are varied to convey meaningful information. The position of the leading and/or trailing edges of each pulse may be changed relative to its unmodulated location. Its abbreviation is **PDM**. Also called **pulse-length modulation**, or **pulse-width modulation**, although both are deprecated terms.

pulse duty factor In a pulse train, the ratio of the pulse duration, to the pulse interval.

pulse encoder Same as **pulse coder**.

pulse excitation Excitation, such as that of a laser, provided in pulses.

pulse fall time The time required for the amplitude of a pulse to decrease from a given percentage of its peak amplitude, such as 90%, to another, such as 10%. **Pulse rise time** is the converse. Also called **pulse decay time**, **fall time (2)**, or **decay time (2)**.

pulse form The characteristics of a **pulse**. These include the shape, such as rectangular, the amplitude, the duration, any pulse droop present, and so on. Also called **pulse shape**, or **pulse characteristics**.

pulse forming The altering of one or more characteristic of a pulse or pulse train. Also called **pulse shaping**.

pulse-forming circuit A circuit which serves for **pulse forming**. Examples include pulse stretchers and pulse regenerators. Also called **pulse-shaping circuit**.

pulse-forming line A transmission line or artificial line which serves for **pulse forming**. Used, for instance, in radars to produce high-intensity pulses. Also called **pulse-shaping line**.

pulse-forming network An electric network which serves for **pulse forming**. Examples include pulse stretchers and pulse regenerators. Also called **pulse-shaping network**.

pulse frequency Same as **pulse repetition frequency**.

pulse-frequency modulation A form of modulation in which the repetition rate of pulses within a train is varied so as to convey meaningful information. Its abbreviation is **PFM**. Also called **pulse-repetition-rate modulation**.

pulse generator A circuit or device which produces **pulses**. Such a circuit or device may produce any number of desired waveforms, with varying amplitudes, duration, and pulse repetition frequency. Also called **pulser**.

pulse group Same as **pulse train**.

pulse height Same as **pulse amplitude**.

pulse-height analyzer An instrument which separates its input into multiple channels based on pulse amplitude or height. It is a type of multi-channel analyzer.

pulse-height discriminator A circuit that acts upon pulses whose amplitude exceed a determined value. Used in detectors. Also known as **amplitude discriminator**.

pulse-height selector A circuit which produces a fixed output pulse, but only when the input pulse lies between pre-established limits of amplitude. Also called **pulse-amplitude selector, amplitude selector**, or **differential pulse-height discriminator**.

pulse interval Within a pulse train, the time interval between successive pulses. Also called **pulse spacing, pulse period**, or **pulse repetition period**.

pulse-interval modulation Same as **pulse-position modulation**. Its abbreviation is **PIM**.

pulse jitter Within a pulse train, slight variations in the intervals between successive pulses.

pulse length A deprecated term for **pulse duration (1)**.

pulse-length modulation A deprecated term for **pulse-duration modulation**. Its abbreviation is **PLM**.

pulse mode **1.** Operation of a component, circuit, device, piece of equipment, or system in which pulses are generated or emitted, as opposed to providing a continuous output. Also called **pulse operation**. **2.** Operation of a laser in **pulse mode (1)**.

pulse modulation Its abbreviation is **PM**. **1.** Any form of modulation in which the amplitude, duration, repetition rate, or the like, of pulses are varied to convey meaningful information. Examples include pulse-code modulation, pulse-duration modulation, and pulse-frequency modulation. **2.** Any form of modulation in which a pulse train serves as the carrier, as in pulse-amplitude modulation.

pulse modulator A circuit or device which effects **pulse modulation**.

pulse narrower A pulse shaper which produces an output pulse whose width is narrower, or whose duration is lesser, than the input pulse.

pulse operation Same as **pulse mode (1)**.

pulse oscillator An oscillator, such as that producing pulsating current, whose output consists of pulses. Also called **pulsed oscillator**.

pulse period Same as **pulse interval**.

pulse-position modulation A form of modulation in which the timing of the pulses within a train is varied so as to convey meaningful information. Also called **pulse-interval modulation**.

pulse radar Same as **pulsed radar**.

pulse rate Same as **pulse repetition frequency**.

pulse regeneration **1.** The process of restoring a pulse to its original form, amplitude, duration, or the like. **2.** The process of restoring a pulse train to its original characteristics, such as timing.

pulse regenerator A circuit or device which effects **pulse regeneration**.

pulse repeater A circuit or device which retransmits pulses. Such a circuit or device may also perform pulse-forming functions, such a pulse stretching.

pulse repetition frequency The number of pulses emitted or occurring per unit time. Its abbreviation is **PRF**. Also called **pulse repetition rate, pulse rate**, or **pulse frequency**.

pulse repetition period Same as **pulse interval**.

pulse repetition rate Same as **pulse repetition frequency**. Its abbreviation is **PRR**.

pulse-repetition-rate modulation Same as **pulse-frequency modulation**.

pulse resolution For a circuit, device, or instrument, the minimum distance between input pulses that can be detected.

pulse rise time The time required for the amplitude of a pulse to increase from a given percentage of its peak amplitude, such as 10%, to another, such as 90%. **Pulse fall time** is the converse. Also called **rise time (2)**.

pulse scaler A scaler that produces an output each time a given number of input pulses are received. For example, a decade scaler produces an output pulse for every ten input pulses.

pulse sensor Same as **pulse detector (2)**.

pulse separation Within a pulse train, the time interval between the trailing edge of a pulse, and the leading edge of the next.

pulse sequence Same as **pulse train**.

pulse shape Same as **pulse form**.

pulse shaper A circuit or device which serves for **pulse forming**.

pulse shaping Same as **pulse forming**.

pulse-shaping circuit Same as **pulse-forming circuit**.

pulse-shaping line Same as **pulse-forming line**.

pulse-shaping network Same as **pulse-forming network**.

pulse spacing Same as **pulse interval**.

pulse stretcher A pulse shaper which produces an output pulse whose duration is greater than the input pulse.

pulse string Same as **pulse train**.

pulse stuffing The altering of the bit rate of a transmission so that it can be received at a different rate, without introducing errors. Also called **justification (1)**, or **bit stuffing (2)**.

pulse test Any test involving **pulses**. For instance, the applying of a pulse voltage to test a dielectric.

pulse-time modulation A form of modulation in which the timing of a given pulse, or pulse train, characteristic is varied so as to convey meaningful information. Examples include pulse-position modulation and pulse-duration modulation. Its abbreviation is **PTM**.

pulse train A series of pulses with similar characteristics, and which follow a regular recurrence pattern. Also called **pulse sequence, pulse string**, or **pulse group**.

pulse transformer A transformer specially designed to transfer pulses, essentially without changing their characteristics.

pulse transmission **1.** Any form of transmission utilizing pulses. Examples include all forms of pulse modulation. **2.** Transmission utilizing a pulse mode, as opposed to a continuous output.

pulse transmitter A transmitter utilized for **pulse transmission**.

pulse voltage A unidirectional surge of voltage of very short duration. It quickly rises to a maximum, then drops to zero in a similar fashion. Such a surge may be produced, for example, by a lightning stroke. Also called **impulse voltage**, or **impulse (2)**.

pulse waveform The shape, such as rectangular or spiked, of a pulse.

pulse width A deprecated term for **pulse duration (1)**.

pulse-width modulation A deprecated term for **pulse-duration modulation**. Its abbreviation is **PWM**.

pulsed Doppler radar A Doppler radar system in which RF energy is transmitted in pulses, as opposed to a **continuous-wave Doppler radar**, in which RF energy is emitted continuously.

pulsed laser A laser whose output consists of short bursts of light. Certain very powerful lasers, for instance, can generate pulses nearly in the zeptosecond range, and may be utilized to study activities in atomic nuclei.

pulsed light Light which is emitted in pulses, such as that produced by a pulsed laser.

pulsed operation A mode of operation in which circuits, devices, equipment, or systems utilize discrete pulses.

pulsed oscillator Same as **pulse oscillator**.

pulsed radar A radar which transmits pulses of RF energy. This contrasts with a **continuous-wave radar**, in which RF energy is emitted continuously. Also called **pulse radar**.

pulser Same as **pulse generator**.

pulses per second A unit of pulse repetition frequency, expressing the number of pulses produced or detected per second. For instance, the rate at which a laser delivers pulses, per second. Its abbreviation is **PPS**.

pulsing circuit A circuit serving as a **pulse generator**.

pump 1. In a parametric amplifier, an oscillator which varies a parameter, such as the capacitance or reactance, of the main oscillator. The resulting energy from this pumping action enables amplification of the input signal. Also called **pump oscillator**. 2. The signal provided by a **pump (1)**. Also called **pump signal**. 3. To apply a **pump signal**. 4. In a laser, to excite electrons, molecules, or ions to higher energy levels, so as to initiate and sustain lasing action. Also, such to provide such excitation in a maser. Also, a device which provides such pumping.

pump frequency In a device such as a laser or parametric amplifier, the frequency at which pumping is provided. Also called **pumping frequency**.

pump laser A powerful laser utilized to drive an optical amplifier, or for other optical signal amplification.

pump oscillator Same as **pump (1)**.

pump signal Same as **pump (2)**.

pump voltage In a parametric amplifier, an alternating voltage provided by a **pump (1)**.

pumping 1. In a laser, the excitation of electrons, molecules, or ions to higher energy levels, so as to initiate and sustain lasing action. Common methods utilized are the application of light, an electrical discharge, or through the effect of a chemical reaction. Also called **laser pumping**. 2. **Pumping (1)** in another device, such as a maser or parametric amplifier.

pumping frequency Same as **pump frequency**.

punch 1. A tool utilized for stamping a design onto a surface, forcing an object or material into or out of a hole, or that pierces or cuts materials or objects. 2. A device which punches holes in **punched cards**.

punch-through In a bipolar junction transistor, a destructive discharge occurring when the collector-base voltage is sufficient for the collector space charge region to spread over the entire base region, coming in contact with the emitter region. Also spelled **punchthrough**.

punch-through voltage The voltage at which **punch-through** occurs. Also spelled **punchthrough voltage**.

punched card A card which holds 80 or 96 columns of data, each representing one character, used by computers with card readers. This is a practically obsolete storage medium. Also called **card (3)**.

punched card reader A device which reads **cards (3)**. Also called **card reader**.

punched card system A computer system in which **cards (3)** are used for input and output. Such systems are practically obsolete. Also called **card system**.

punched tape An obsolete form of data storage consisting of a long strip of paper, or another material, with holes punched.

punchthrough Same as **punch-through**.

punchthrough voltage Same as **punch-through voltage**.

puncture 1. A disruptive electrical discharge through an insulator, dielectric, or other material separating circuits. Also called **breakdown (2)**. 2. A hole or depression made by a pointed object. Also, to make such a hole or depression.

puncture voltage The voltage at which a disruptive electrical discharge occurs through an insulator, dielectric, or other material separating circuits. Also called **breakdown voltage (1)**.

Pupin coil One of multiple loading coils which are inserted in series with a telephone line, to reduce the effects of line capacitance.

pure color A color in which there is no white component present. Such a color has 100% saturation. Also called **saturated color**.

pure sine wave A sine wave with no distortion nor harmonics.

pure sound Same as **pure tone**.

pure tone A tone, such as that produced by striking a tuning fork, whose acoustic energy occurs at a single frequency. Such a tone would appear on an oscilloscope as a sine wave with little or no harmonic content. Also called **pure sound**, or **simple tone**.

pure wave A wave with no distortion nor harmonic content. An example is a pure sine wave.

purge 1. To remove any gases from an enclosure. 2. To remove any impurities from a substance or material. 3. To remove all data from memory, a storage medium, a file, or the like.

purging 1. The removal any gases from an enclosure. 2. The removal of any impurities from a substance or material. 3. The removal of all data from memory, a storage medium, a file, or the like.

purification The removal of impurities or other undesired constituents.

purify To remove impurities or other undesired constituents.

purity 1. The extent to which a substance, material, or the like, is free from impurities. 2. The extent to which a signal, waveform, data or the like, is free from distortion or errors. 3. The extent to which a color is not mixed with other colors. Said especially of a primary color which is not mixed with other primary colors. Also called **color purity (2)**. 4. For a given color, the degree to which all unwanted components are excluded. Such a color only contains the proper proportion of the desired primary colors. Also called **color purity (1)**.

purity adjustment The adjustment of the **purity (4)** of the displayed images of a CRT, using, for instance, a purity magnet.

purity coil A coil mounted around the neck of a CRT to adjust the **purity (4)** of the displayed images.

purity magnet A magnet mounted around the neck of a CRT to adjust the **purity (4)** of the displayed images.

purple plague A purplish compound formed when bonding gold to aluminum. The formation of such a compound may cause the failure of semiconductor devices or microelectronic circuits.

push 1. To add an item to the top of a stack. This contrasts with **pop**, which is to remove the first item from the top of a stack. 2. Same as **push technology**. 3. To send data utilizing **push technology**.

push-button Same as **pushbutton**.

push-button dialing Telephone dialing in which buttons are pressed to select numbers. Also spelled **pushbutton dialing**.

push-button switch A switch that is operated by pressing a pushbutton. Also spelled **pushbutton switch**. Also called **button switch**.

push-button tuner A tuner whose selections are made by pressing a button. Also spelled **pushbutton tuner**.

push-down list A data list in which the last item placed is the first item to be retrieved, as in LIFO. Also spelled **push-down list**.

push-down stack A stack in which the last item placed is the first item to be retrieved, as in LIFO. Also spelled **push-down stack**.

push-down storage Data storage in which the last item placed is the first item to be retrieved, as in LIFO. Also spelled **pushdown storage**.

push-pull 1. Same as **push-pull configuration**. 2. Pertaining to a circuit or device with a **push-pull configuration**.

push-pull amplifier An amplification circuit incorporating two identical signal branches which each operate in phase opposition. The active devices may be two transistors or two tubes, and their outputs each comprise half of the combined amplified output. Such an arrangement provides for cancellation of even harmonics, which reduces distortion. Such amplifiers are frequently utilized for class A, class AB, or class B amplification, such as that used in audio high-fidelity systems.

push-pull arrangement Same as **push-pull configuration**.

push-pull circuit A circuit which incorporates two identical signal branches, each operated in phase opposition. Used, for instance in push-pull amplifiers, or push-pull oscillators.

push-pull configuration A circuit in which there are two input signals, each in phase opposition, provided separately to two identical active devices. The output of these two signal branches is combined for the overall output. Such an arrangement is used, for instance, in push-pull amplifiers. Also called **push-pull (1)**, or **push-pull arrangement**.

push-pull currents For the two conductors of a balanced line, currents flowing which are equal in magnitude, but opposite in phase at all points along said line. Also called **balanced currents**.

push-pull operation Operation utilizing a **push-pull configuration**.

push-pull oscillator An oscillator in which two amplifying devices, such as transistors or tubes, are utilized in phase opposition.

push-pull transformer A transformer utilized in a **push-pull circuit**.

push-pull voltages Voltages which are equal in magnitude but opposite in polarity with respect to ground. Also called **balanced voltages**.

push-push amplifier An amplifier with a **push-push configuration**.

push-push circuit A circuit with a **push-push configuration**.

push-push configuration A circuit in which there are two input signals, each in phase opposition, provided separately to two identical active devices, with the outputs of said devices connected in parallel. Such an arrangement cancels odd-harmonics, including the fundamental frequency, and is used, for instance, in push-push amplifiers.

push switch A switch that is actuated by pushing. Such a switch, for instance, may use a pushbutton that is pressed in case of an emergency.

push technology Data distribution, such as that over the Internet, in which users receive information, usually in the form of advertisements, without requesting it. Examples include spam, or banner ads and pop-up ads appearing when a Web page is loaded. **Pull technology** refers to information obtained by a user who requests it. Also called **push (2)**.

push-to-talk Pertaining to, or requiring **push-to-talk operation**. Also called **push-to-talk**. Its abbreviation is **PTT**.

push-to-talk operation Communication in which the talker must push or press a button for voice transmission. Used, for instance, when both ends are utilizing the same frequency. Its abbreviation is **PTT operation**. Also called **press-to-talk operation**.

push-to-talk switch A switch that must be pushed when talking during **push-to-talk operation**. Its abbreviation is **PTT switch**. Also called **press-to-talk switch**.

push-up list A data list in which the first item placed is the first item to be retrieved, as in FIFO. Also spelled **pushup list**.

push-up stack A stack in which the first item placed is the first item to be retrieved, as in FIFO. Also spelled **pushup stack**.

push-up storage Data storage in which the first item placed is the first item to be retrieved, as in FIFO. Also spelled **pushup storage**.

pushbutton A knob, disk, or the like, which is pressed to activate or operate a circuit, device, component, machine, and so on. For instance, a pushbutton may be pressed to activate an electric circuit. A pushbutton may also be virtual, as seen when a computer mouse presses such a button on-screen by clicking the mouse. Also spelled **push-button**. Also called **button (1)**.

pushbutton dialing Same as **push-button dialing**.

pushbutton switch Same as **push-button switch**.

pushbutton tuner Same as **push-button tuner**.

pushdown list Same as **push-down list**.

pushdown stack Same as **push-down stack**.

pushdown storage Same as **push-down storage**.

pushup list Same as **push-up list**.

pushup stack Same as **push-up stack**.

pushup storage Same as **push-up storage**.

pV Abbreviation of **picovolt**.

PV 1. Abbreviation of **petavolt**. 2. Abbreviation of **photovoltaic**.

PV array Abbreviation of **photovoltaic array**.

PV cell Abbreviation of **photovoltaic cell**.

PV device Abbreviation of **photovoltaic device**.

PV material Abbreviation of **photovoltaic material**.

PV module Abbreviation of **photovoltaic module**.

PV panel Abbreviation of **photovoltaic panel**.

PV system Abbreviation of **photovoltaic system**.

PVC 1. Abbreviation of **polyvinyl chloride**. 2. Abbreviation of **permanent virtual circuit**.

pW Abbreviation of **picowatt**.

PW Abbreviation of **petawatt**.

PWM Abbreviation of **pulse-width modulation**.

pwr Abbreviation of **power**.

PX Abbreviation of **private exchange**.

pyramidal horn antenna A horn antenna in which the sides of the radiating element form a pyramid.

pyranometer 1. An instrument which measures solar radiation. 2. An instrument which measures direct solar radiation, along with any other sky radiation, such as that which is diffused or reflected.

Pyrex A variety of borosilicate glass which is highly resistant to chemicals and heat. Used, for instance, in laboratory glassware, and in telescopes.

pyrheliometer An instrument which measures solar radiation. Such an instrument is usually mounted on a base which tracks the sun, so as to be aimed directly at the sun through-out the day.

pyroelectric crystal A crystal, such as that of quartz or tour-maline, which exhibits **pyroelectricity**.

pyroelectric material A material composed of a **pyroelectric crystal**.

pyroelectricity The development of opposite electrical charges between opposite faces of certain asymmetric crys-tals when heated.

pyromagnetic Pertaining to, or arising from, a combination of heat and magnetism.

pyrometer 1. Any instrument or device, such as an optical pyrometer, which measures and indicates very high tempera-tures. **2.** Any instrument or device which measures and in-dicates temperatures. Examples include electronic ther-mometers, quartz thermometers, infrared thermometers, and bimetallic thermometers. Also called **thermometer**.

Python An interpreted object-oriented programming language which is available for many platforms, and whose source code is freely distributed.

PZM Abbreviation of **pressure zone microphone**.

PZT Abbreviation of **lead zirconate titanate**. A ferroelectric ceramic which exhibits the piezoelectric effect. Used, for instance, as transducers, electro-optic devices such as elec-tro-optic modulators, and as actuators in active control sys-tems. The first letter of its abbreviation is taken from **Pb**, the chemical symbol for lead.

pΩ Abbreviation of **picohm**.

PΩ Abbreviation of **petohm**.

Q

Q **1.** Symbol for **charge (1)**. **2.** Symbol for **Quality factor**. A figure of merit for a component, resonant circuit, tuned circuit, periodic device, or energy-storing device. One measure is the ratio of the reactance to the resistance. In the case of a capacitor, for instance, the higher the *Q*, the longer the time for it to discharge. In the case of an oscillator, for example, the higher the *Q*, the more stable its output.

Q Abbreviation of **quantum**.

Q Band In communications, a band of radio frequencies extending from 36 to 46 GHz. This corresponds to wavelengths of approximately 8.3 mm to 6.5 mm, respectively.

Q channel In the NTSC color TV system, the channel utilized to transmit green-magenta color information. It is approximately 0.5 MHz wide. The **I channel** transmits cyan-orange color information.

Q display In radars, an oscilloscopic display similar to an A display, but which incorporates an adjustable pedestal to indicate the range of the scanned object. Also called **Q scope, Q scanner, Q scan,** or **Q indicator**.

Q factor Abbreviation of **Quality factor**. Same as *Q* **(2)**.

Q indicator Same as **Q display**.

Q meter An instrument utilized to measure *Q* **(2)**. Also called **Quality factor meter**.

Q multiplier A component or device, such as an amplifier or filter, which raises the effective *Q* **(2)** of a circuit or device.

Q point Abbreviation of **quiescent point**.

Q scan Same as **Q display**.

Q scanner Same as **Q display**.

Q scope Same as **Q display**.

Q signal Abbreviation of quadrature **signal**. In the NTSC system, one of the two color-difference signals. The other is the **I signal**.

Q-switched laser A laser utilizing **Q switching**.

Q switching Manipulation of the *Q* **(2)** of a laser, so as to create very powerful pulses, generally in the nanosecond range. The usual method is to keep the *Q* low during population inversion, then increasing the *Q* just as lasing action is about to begin.

QA 1. Abbreviation of **quality assurance**. **2.** Abbreviation of **quality audit**. **3.** Abbreviation of **quality analysis**. **4.** Abbreviation of **quality assessment**.

QAGC Abbreviation of **quiet automatic gain control**.

QAM Abbreviation of **quadrature amplitude modulation**.

QAVC Abbreviation of **quiet automatic volume control**.

QBE Abbreviation of **query by example**.

QC Abbreviation of **quality control**.

QCD Abbreviation of **quantum chromodynamics**.

QCIF Abbreviation of **Quarter Common Intermediate Format**.

QE Abbreviation of **quality enhancement**.

QED Abbreviation of **quantum electrodynamics**.

QFP Abbreviation of **quad flat-pack**.

QFT Abbreviation of **quantum field theory**.

QIC Abbreviation of **quarter-inch cartridge**.

QM Abbreviation of **quality management**.

QoS Abbreviation of quality of service. **1.** A level of service, such as that which is provided at or above a given degree. **2.** In communications, a guaranteed level of performance, including factors such as bandwidth and throughput. ATM, for instance, has QoS standards.

QPS Abbreviation of **queries per second**.

QPSK 1. Abbreviation of **quadrature phase-shift keying**. **2.** Abbreviation of **quaternary phase-shift keying**.

QSO Abbreviation of **quasi-stellar object**.

QSR 1. Abbreviation of **quasi-random signal**. **2.** Abbreviation of **quasi-stellar radio source**.

QSS Abbreviation of **quasi-stellar source**.

qty Abbreviation of **quantity**.

quad 1. A group or set of four which have something in common. For example, the four sides of a quad flat-pack, or four components working together in a circuit or device. **2.** A group of four conductors, such as two twisted pairs. **3.** Abbreviation of **quadraphonic**. **4.** Same as **quad antenna**.

quad antenna An essentially omnidirectional antenna whose elements are in the shape of four-sided loops. Two elements are usually utilized, one driven and the other parasitic. Also called **quad**, or **cubical quad antenna**.

quad flat-pack Its abbreviation is **QFP**. A square surface-mount chip which provides leads on all four sides. It comes in various varieties, such as plastic QPF, and afford a high lead count in a small area. Also spelled **quad flatpack**. Also called **quad flat-package**.

quad flat-package Same as **quad flat-pack**.

quad flatpack Same as **quad flat-pack**.

quad speed Four times the usual or standard speed. For example, a quad speed CD-ROM drive.

quad speed CD-ROM drive A CD-ROM drive that is four times the usual or standard speed. Its abbreviation is **4X CD-ROM drive**.

quadbit A group of four bits used in communications.

quadded cable A cable incorporating one or more **quads (2)**, all of which are housed within the same jacket.

quadrant 1. A circular arc of 90°, occupying one fourth of the circumference of a circle. **2.** The area occupied by a **quadrant (1)**. **3.** Something, such as a part, in the shape of a **quadrant (2)**. **4.** One of four equal areas a plane, region, or space is divided into. **5.** One of the four regions, within the same plane, divided by the axes of a Cartesian coordinate system.

quadrant electrometer An electrometer in which a needle, vane, or blade is suspended in the center of a brass circle divided into quadrants. Each opposite pair of quadrants is at the same potential, and such a device is used for measuring small potential differences.

quadraphonic Its abbreviation is **quad**. **1.** Same as **quadraphonic sound**. **2.** Pertaining to **quadraphonic sound**, or a **quadraphonic sound system**.

quadraphonic sound 1. Same as **quadraphonic sound system**. **2.** The sound produced by a **quadraphonic sound system**.

quadraphonic sound system A sound system in which there are four channels, each feeding its own loudspeaker. It is meant to be an enhancement of two-channel stereo. When recording, the signals of the four channels are encoded into two channels, and when reproducing, the signals are decoded so as to provide the four original channels. Also called **quadraphonic sound (1)**, **quadraphony (1)**, or **four-channel sound system**.

quadraphony 1. Same as **quadraphonic sound system**. **2.** Recorded in, reproducing, utilizing, or pertaining to **quadraphonic sound**.

quadratic equation An equation in which the highest power to which any unknown value is raised is two. An example is

$ax^2 + bx + c = 0$, where a, b, an c are constants, and x is a variable.

quadrature A phase difference of 90°, or $\pi/2$ radians. Two periodic quantities having the same frequency and waveshape which are out-of-phase by 90° are said to be **in quadrature**. Also called **phase quadrature**.

quadrature amplifier An amplifier which in addition to providing gain, introduces a phase difference of 90°, or $\pi/2$ radians.

quadrature amplitude modulation A form of **quadrature modulation** in which the amplitude of each of the two carriers are modulated to convey meaningful information. Its abbreviation is **QAM**.

quadrature component In an AC circuit, the component of the current, voltage, or power which does not add power. These are, specifically, the quadrature current, quadrature voltage, or quadrature power. Also called **idle component, reactive component**, or **wattless component**.

quadrature current The component of an alternating current which is in quadrature with the voltage. Such a component does not add power. Also called **idle current, reactive current**, or **wattless current**.

quadrature modulation Modulation in which there are two carriers, each out-of-phase by 90°, and each modulated by a different signal.

quadrature phase-shift keying A phase-shift keying technique in which four phase angles are utilized to transmit information. The four angles are usually 90° apart each, and two bits can be encoded into each signal element. Its abbreviation is **quadrature PSK**, or **QPSK**. Also called **quaternary phase-shift keying**.

quadrature power The power in an AC circuit which cannot perform work. It is calculated by the following formula: $P = I \cdot V \cdot \sin\theta$, where P is the power in watts, I is the current in amperes, V is the voltage in volts, and $\sin\theta$ is the sine of the angular phase difference between the current and the voltage. Also called **reactive power, idle power, wattless power, volt-amperes reactive**, or **reactive volt-amperes**.

quadrature PSK Abbreviation of **quadrature phase-shift keying**.

quadrature signal Same as **Q signal**.

quadrature voltage The voltage component which is in quadrature with the current of an AC circuit. Such a component does not add power. Also called **reactive voltage, wattless voltage**, or **idle voltage**.

quadrilateral A closed geometric figure bounded by four straight lines. Examples include squares, rectangles, and parallelograms.

quadrillion A number equal to 10^{15}.

quadripartite Consisting of, or divided into four parts. For example, a plane divided into four regions.

quadripole An electric network with four terminals, two for input and two for output. The input terminals are paired to form the input port, while the output port is formed by the output terminals. Examples include O-networks and H-networks. Also called **two-port network, four-terminal network**, or **two-terminal pair network**.

quadrivalent Pertaining to an element, such as carbon or silicon, which has a valence of four. Also called **tetravalent**.

quadruped robot A mobile robot which uses four mechanical legs to allow it to move or travel.

quadruple speed An optical disk drive whose spinning speed is quadruple that of the standard rate, or of the previous generation. It is a relative term, as what may currently be quadruple speed may soon be much slower than the standard rate.

quadrupler 1. A circuit or device whose output frequency is four times that of its input frequency. Also called **frequency quadrupler**. 2. A rectifier circuit whose output DC voltage is about four times the peak value of its input AC voltage. Also called **voltage quadrupler**.

quadrupole 1. A four-pole magnet. Used, for instance, to focus and direct particle beams. 2. An arrangement or system having four electric or magnetic poles. For example, four closely-spaced charges of equal magnitude which are arranged in a manner that they form the vertices of a parallelogram, with opposite vertices each having the same charge.

qualification test One of multiple tests performed in **qualification testing**.

qualification testing Tests performed to ascertain the appropriateness, usefulness, performance, and reliability of a component, circuit, device, piece of equipment, system, or material which is being evaluated for acceptance.

qualitative analysis An analysis whose results are expressed in non-numerical terms. For example, an analysis in which it is determined which chemical species are present in a sample. A **quantitative analysis** would also determine, for instance, the proportions in which each were present.

qualitative test A test, such as a qualitative analysis, whose results are expressed in non-numerical terms.

quality 1. A property or other distinguishing characteristic. For instance, audibility is the quality of a sound pressure level being sufficient to be detected. 2. Within a given group or realm, superiority or excellence. For example, the reproduction of an audio system which faithfully reproduces the essential characteristics of the source music.

quality analysis Any analysis or evaluation of the quality of a component, circuit, device, piece of equipment, product, or system which is being produced. Also called **quality assessment**. Its abbreviation is **QA**.

quality assessment Same as **quality analysis**. Its abbreviation is **QA**.

quality assurance The measures taken to insure that there is **quality control**. Its abbreviation is **QA**.

quality audit A systematic and usually independent examination which determines the level of quality of a given product or process. Such an audit may also verify compliance with established parameters or standards. A quality audit may be followed by recommendations for improvement. Its abbreviation is **QA**.

quality control The system, procedures, mechanisms, and the like, which are utilized to verify that a given component, circuit, device, piece of equipment, product, material, or system which is being produced or modified is at or above established standards and/or specifications. Its abbreviation is **QC**.

quality enhancement Measures taken to improve quality, such as that of a manufactured product. Also, the improvements so made. Its abbreviation is **QE**.

Quality factor Same as Q (2).

Quality factor meter Same as **Q meter**.

quality management The measures taken to insure that quality is maintained at a given level or above. It may also include quality enhancement. Its abbreviation is **QM**.

quality of service Same as **QoS**.

quanta Plural form of **quantum**.

quantitative analysis A test, such as a quantitative analysis, whose results are expressed in numerical terms.

quantitative analysis An analysis whose results are expressed in numerical terms. For example, an analysis in which it is determined which chemical species are present in a sample, and in what proportions. A **qualitative analysis** would only determine, for instance, the species present.

quantity Its abbreviation is **qty**. **1.** A number, amount, or value. **2.** A specified number, amount, or value. **3.** A large number, amount, or value. **4.** For a given component, circuit, device, system, material, or the like, a property which is measurable, countable, or can otherwise be expressed as a **quantity** (1). **5.** An entity which has a value or magnitude, upon which mathematical operations may be performed.

quantization **1.** The division of a quantity or phenomenon, such as a wave, with an infinitely variable range of values into one or more ranges with finite values, each called a quantized value. For example, the conversion of an analog input into a digital output. Since the number of subranges created can not be infinite, there will always be a loss, however minor, of information, called quantization error. **2.** The division of a range of values into labeled subranges. For example, any number between 1 and 30 is **a**, between 31 and 60 is **b**, between 61 and 90 is **c**, and so on. **3.** The description of an interval of values as a discrete number possible values. For example, anything occurring between 12 AM day one and 12 AM day two, being recorded as occurring on day one, regardless of the hour. **4.** The process of applying quantum mechanics or quantum theory to something.

quantization distortion Distortion introduced in the process of **quantization** (1). This distortion is due to the quantization error present. Also called **quantizing distortion**.

quantization error The information which is lost in the process of **quantization** (1). This causes noise and distortion. Also called **quantizing error**.

quantization noise Noise introduced in the process of **quantization** (1). This noise is due to the quantization error present. Also called **quantizing noise**.

quantize **1.** To perform the process of **quantization** (1), or **quantization** (2). **2.** To apply quantum mechanics or quantum theory to something.

quantized **1.** That which has undergone the process of **quantization** (1), or **quantization** (2). **2.** That which has had quantum mechanics or quantum theory applied to it.

quantized system A system in which only certain allowed energy values may be adopted. Electrons in such a system can only change from one specific level to another, and in the process absorb or emit energy.

quantized value One of the finite values derived from a quantity or phenomenon with an infinitely variable range of values, through the process of **quantization** (1). Also called **quantum** (2).

quantizer That which performs the process of **quantization** (1), or **quantization** (2). For example, a circuit or device serving as an analog-to-digital converter.

quantizing The process of **quantization** (1), or **quantization** (2).

quantizing distortion Same as **quantization distortion**.

quantizing error Same as **quantization error**.

quantizing noise Same as **quantization noise**.

quantum Its abbreviation is **Q**. Its plural form is **quanta**. **1.** For a given physical phenomenon, such as electromagnetic radiation, the smallest quantity, such as that of energy, that can exist independently. For such phenomena, any quantity above this can only exist in multiples of this unit. In the case of light, for instance, energy can be absorbed or radiated only in multiples of these discrete packages called photons. **2.** A **quantum** (1) utilized as a unit. For example, the quantum of electromagnetic radiation is the photon, which is also called light quantum. **3.** Same as **quantized value**. **4.** Any given quantity which can be counted or measured.

quantum bit Same as **qubit**.

quantum chromodynamics The area of quantum theory dealing with the relationships between quarks, especially the strong interaction via gluons. Its abbreviation is **QCD**.

quantum computer A computer whose basic unit of computing is the **qubit**. A quantum computer operates on all qubits simultaneously, thus is exponentially faster than conventional computers based, for instance, on the charge of a capacitor in RAM or on the magnetization of macroscopic particles on a hard disk.

quantum computing The use of **quantum computers**.

quantum cryptography Cryptography which makes use of quantum mechanics to code information or create keys which are unbreakable. Used, for instance, in one-time pads. Also called **quantum encryption**.

quantum dot A semiconductor structure forming a three-dimensional **quantum well**. Used, for instance, in semiconductor lasers, and to study the behavior of the electrons of atoms so constrained.

quantum-dot laser A semiconductor laser utilizing a **quantum dot**. Such a laser is highly temperature insensitive, provides an extremely broad gain spectrum, and an exceedingly narrow line width.

quantum efficiency Also called **quantum yield**. **1.** The number of electrons released by a photoemissive surface, such as a photocathode, per photon of incident radiation. **2.** The number of photon-induced reactions, per incident photon. For instance, the ratio number of photons emitted by a surface, to the number of photons absorbed.

quantum electrodynamics The area of quantum theory dealing with electromagnetic interactions between elementary particles, such as electrons and muons, especially exchanges of photons. Its abbreviation is **QED**.

quantum electronics The application of quantum mechanics, especially the energy states of matter, to electronics. Applied, for instance, in masers.

quantum encryption Same as **quantum cryptography**.

quantum field theory The area of quantum theory that deals with the quantum-mechanical interactions between elementary particles and fields. An application is quantum electrodynamics. Its abbreviation is **QFT**.

quantum jump A transition or change in energy whose magnitude is a **quantum** (2). For example, a change in the orbit of an electron in which a quantum is absorbed or emitted. Also called **quantum transition**.

quantum mechanics **1.** The science dealing with the application of quantum theory to the mechanics of elementary and atomic particles and systems. **2.** Same as **quantum theory**.

quantum number A number, with integer or half-integer values, which characterizes a property or state of a particle or system. For example the spin of an electron may be characterized by the quantum numbers $+\frac{1}{2}$, or $-\frac{1}{2}$.

quantum physics The branch of science which utilizes quantum theory to analyze, explain, and predict the physical properties of a system.

quantum state A state in which a particle or system can exist in, according to quantum theory. Such a state is described by quantum numbers.

quantum statistics The application of statistical methods to particles and systems that obey the rules of quantum mechanics. For instance, the distribution of energy levels of the particles of a given system.

quantum system A system which can only be accurately described through the use of quantum physics.

quantum theory The theory according to which energy is emitted or absorbed in discrete units called **quanta**. It describes the behavior of atomic and subatomic particles and systems. According to quantum theory, electromagnetic radiation has both particle-like properties, as seen in the photoelectric effect, and wavelike properties, as evidenced by phenomena such as diffraction and interference. Such radiation is composed of photons, whose energy is equal to hf, where h is the Planck constant, and f is the frequency of the radiation in hertz. Also called **quantum mechanics (2)**.

quantum transition Same as **quantum jump**.

quantum well A semiconductor heterostructure in which an extremely thin semiconductor layer with a narrower bandgap is surrounded by two layers with a wider bandgap. This allows charge carriers to move perpendicular to the direction of the crystal growth, but not in the same direction. A quantum well has one dimension, and is used, for instance, in transistors and semiconductor lasers.

quantum-well laser A semiconductor laser utilizing a **quantum well**. Such a laser is highly-tunable and produces an extremely narrow line width.

quantum wire A semiconductor structure forming a two-dimensional **quantum well**. Used, for instance, in semiconductor lasers, and to carry signals.

quantum yield Same as **quantum efficiency**.

quarantine The moving of a virus-infected file to a location where it less likely to do harm to the system. Such a location may be a folder only accessible by the anti-virus software, where it can be examined, or otherwise held for future use or deletion.

quark An elementary particle, with a fractional charge of ⅓ or ⅔, combinations of which are considered to constitute all hadrons. So far six types of quarks have been identified, and a proton, consists of two up quarks and a down quark. A quark and antiquark pair form a meson. So far, an isolated quark has not been observed.

Quarter Common Intermediate Format Its abbreviation is **QCIF**. A video format which supports both NTSC and PAL signals, and which is used in videoconferencing. QCIF resolution is 144 lines by 176 pixels per line, at 30 frames per second. This is one quarter of the original CIF standard.

quarter-inch cartridge A standard utilized for tape drives used for back-up. Also, the technology utilized for storage on such tapes. Also, a given cartridge used for such back-up. Its abbreviation is **QIC**.

quarter-phase Having, or pertaining to, a phase difference of 90°, or $\pi/2$ radians. Said, for instance, of two voltages or currents which are 90° out-of-phase relative to the other. Also called **two-phase (1)**.

quarter-wave Abbreviation of **quarter-wavelength**. **1.** Pertaining to, or composed of one-quarter of one complete wave cycle. **2.** The electrical length equal to one-quarter of one complete wave cycle. Also, having such a length.

quarter-wave antenna An antenna radiator whose electrical length is equal to one-quarter of the wavelength of the signal being transmitted or received. Also called **quarter-wavelength antenna**, or **quarter-wave radiator**.

quarter-wave dipole A dipole antenna consisting of a radiator whose electrical length is equal to one-quarter of the wavelength of the signal being transmitted or received.

quarter-wave line Same as **quarter-wave transmission line**.

quarter-wave monopole A monopole antenna consisting of a radiator whose electrical length is equal to one-quarter of the wavelength of the signal being transmitted or received.

quarter-wave plate In optics, a plate made of a double-refracting crystal whose density provides for a difference of one-quarter of a cycle between the ordinary and extraordinary rays.

quarter-wave radiator Same as **quarter-wave antenna**.

quarter-wave stub Same as **quarter-wave transmission line**.

quarter-wave transformer Same as **quarter-wave transmission line**.

quarter-wave transmission line An antenna transmission line whose electrical length is equal to one-quarter of the wavelength of the fundamental frequency of the signal to be transmitted. Used, for instance, for impedance matching, and filtering of harmonics. Also called **quarter-wave line**, **quarter-wave transformer**, or **quarter-wave stub**.

quarter-wavelength Same as **quarter-wave**.

quarter-wavelength antenna Same as **quarter-wave antenna**.

quartz A mineral whose chemical formula is SiO_2. It is the most commonly occurring mineral in the earth's crust. It is a crystalline form of silicon dioxide, has a hardness of 7 on the Mohs scale, and when pure it is a colorless and called **rock crystal**, although there are many colored varieties. Quartz has many applications, including its use in oscillators, frequency standards, clocks, piezoelectric devices, electronic components, and in optical instruments.

quartz clock A clock which incorporates a quartz-crystal oscillator, whose natural oscillation frequency determines the accuracy of the timepiece. Such a clock can have an analog or digital display, and may have an error of less than 0.1 second per year. Also called **quartz crystal clock**, or **crystal clock**.

quartz control The use of a quartz piezoelectric crystal to control the frequency of an oscillator.

quartz-controlled transmitter A radio-frequency transmitter whose carrier frequency is controlled by a quartz-crystal oscillator.

quartz crystal A piezoelectric crystal composed of naturally-occurring or artificially grown quartz or silicon dioxide. Used, for instance, in oscillators and piezoelectric devices. Also called **quartz piezoelectric crystal**.

quartz crystal clock Same as **quartz clock**.

quartz-crystal oscillator An oscillator circuit which utilizes a quartz piezoelectric crystal to control the oscillation frequency. Such oscillators feature a highly accurate and stable output, especially when in a temperature-controlled environment. Also called **quartz oscillator**, **crystal-controlled oscillator**, or **crystal oscillator**.

quartz-crystal resonator Same as **quartz resonator**.

quartz cut A section of a quartz crystal accurately cut along certain axes, to determine characteristics such as its natural vibration frequency.

quartz fiber A slender threadlike filament of quartz. Used, for instance, in a quartz-fiber electroscope.

quartz-fiber electroscope An electroscope in which the sensitive element is a metallized quartz fiber. Gold is usually plated on the fiber, and such electroscopes are very sensitive. Also called **Lauritsen electroscope**.

quartz-halogen lamp A gas-filled lamp with a quartz envelope and a small proportion of one or more halogens, such as iodine. Used, for instance, for intense light, as quartz is more heat-resistant than glass.

quartz lamp A mercury vapor lamp with a quartz envelope. Quartz does not absorb the emitted ultraviolet radiation, as ordinary glass would.

quartz lock The use of a quartz-crystal oscillator to regulate the frequency or speed of a circuit, device, piece of equipment, or system. Used, for instance, in receivers, transmitters, clocks, and turntables.

quartz oscillator Same as **quartz-crystal oscillator**.

quartz piezoelectric crystal Same as **quartz crystal**.

quartz plate A quartz crystal that has been cut, etched, coated, and otherwise fully prepared to be mounted on its holder.

quartz resonator A resonant circuit which utilizes a quartz piezoelectric crystal to control the resonance frequency.

Such resonators may be used, for instance, to control the frequency of an oscillator, and feature a highly accurate and stable output, especially when in a temperature-controlled environment. Also called **quartz-crystal resonator, piezo-electric resonator**, or **crystal resonator.**

quartz thermometer A thermometer which incorporates a specially selected and cut quartz crystal. The oscillation frequency of such a crystal will fluctuate along with its surrounding temperature. Such thermometers are extremely accurate and have a wide operational range.

quartz transducer A transducer whose sensitive element is a quartz crystal. Quartz crystals, when subjected to mechanical energy, generate electrical energy, and vice versa. Used, for example, in microphones, pickups, and loudspeakers.

quartz watch A watch that incorporates a quartz-crystal oscillator whose natural oscillation frequency determines the accuracy of the timepiece. Such a watch may have an analog or digital display, and may have an error of less than 0.1 seconds per year.

quasar It is an abbreviation of **quasi**-stellar object, **quasi**-stellar radio source, or **quasi**-stellar source. A class of celestial objects which when viewed through a telescope appear as stars, but which emit the energy of hundreds or thousands of galaxies. They are considered to be the most luminous objects in the known universe. Quasars that are also sources of radio waves are **quasi-stellar radio sources**, or **quasi-stellar sources**.

quasi- A prefix meaning to a certain extent or degree. For example, quasi-random.

quasi-linear system A system whose output or response is nearly linearly proportional to its input, over a given range of values, or for every part of the system.

quasi-optical Possessing properties of, or behaving as if light waves, without being light waves. For instance, certain microwaves propagate as if light waves.

quasi-peak detector A detector which records the peaks of a given signal, with weighting based on the repetition frequency of the peaks. Thus, the more frequent the occurrence of the peaks, the higher the reading. Used, for instance, in tests to determine electromagnetic compatibility.

quasi-random Nearly random. For example, a sequence of numbers in which part is random, and part is not. That which is generated by a specific process, such as an algorithm, would be **pseudorandom**, but that which is quasi-random may arise by other means. Also spelled **quasirandom**.

quasi-random number A number which **quasi-random.**

quasi-random signal A signal which **quasi-random**. Such a signal may consist of a sequence of bits which has portions which are random enough to avoid certain prohibited sequences, such as 14 consecutive zeroes. Also spelled **quasirandom signal**. Its abbreviation is **QSR**.

quasi-sine-wave A wave whose shape approximates that of a sine wave, without being a true sine wave.

quasi-square-wave A wave whose shape approximates that of a square wave, without being a true square wave.

quasi-stellar object Same as **quasar**. Its abbreviation is **QSO**.

quasi-stellar radio source Same as **quasar**. Its abbreviation is **QSR**.

quasi-stellar source Same as **quasar**. Its abbreviation is **QSS**.

quasirandom Same as **quasi-random.**

quasirandom signal Same as **quasi-random signal**.

quaternary Consisting of, incorporating, or appearing in, groups of four.

quaternary alloy An alloy consisting of four elements.

quaternary phase-shift keying Same as **quadrature phase-shift keying**. Its abbreviation is **QPSK**, or **quaternary PSK**.

quaternary PSK Abbreviation of **quaternary phase-shift keying**.

Quattro Pro A popular spreadsheet program.

qubit Abbreviation of **qu**antum **bit**. An individual particle utilized to store information. Particles, which obey the laws of quantum mechanics, can have their spin influenced for storage, and detected for retrieval. A single qubit can be used to represent a one, a zero, or both, with a coefficient representing the probability for each state. Two qubits can represent every two-bit number, which specifically are 00, 01, 10, and 11. Each additional qubit increases the number representation exponentially, at a rate of 2^n, where n is the number of qubits. For example, 50 qubits can represent every number between zero and beyond a quadrillion.

quench **1.** To extinguish or end abruptly. For example, to extinguish an electric arc. **2.** To suppress, or reduce the degree of something. For instance, to suppress oscillation, or reduce the level of phosphorescence of a material. **3.** To cool abruptly. For example, to treat a metal by heating it and immersing it in a cold liquid.

quenched gap A spark gap in which conduction may be stopped abruptly. Also called **quenched spark gap**, or **quenched spark**.

quenched spark Same as **quenched gap**.

quenched spark gap Same as **quenched gap**.

quenching **1.** The process extinguishing or ending abruptly. For example, the extinguishing an electric arc. **2.** The suppression, or reduction in the degree of something. For instance, the suppression of oscillation, or the reduction in the level of phosphorescence of a material. **3.** The process of cooling abruptly. For example, the treating of a metal by heating it and immersing it in a cold liquid.

queries per second A measure of database performance based on the number of queries that can be handled per second. Its abbreviation is **QPS**.

query In databases, a request for information based on specific criteria. This may be accomplished, for instance, by choosing parameters from a menu, through query by example, or via a query language. Also, to make such a request.

query box A dialog box which presents a user with a question, and provides multiple selectable options or a text-entry field for answering.

query by example In database management, a query method in which a user provides an example of what is desired. In it, the system provides a blank record with a space for each field, allowing the selection of which fields and values are desired for a given search. Its abbreviation is **QBE**.

query language A computer language specifically designed for requesting information from a database. Such a language, for instance, may allow a user to express what is desired in the form of a question. An example is SQL. It is a type of data manipulation language.

queue **1.** A list of tasks, in a specified order, that await to be performed by computer. Examples include print queues and message queues. Also, a storage location for such documents, messages, or the like. **2.** A sequence or list which awaits further action in a specified order. For instance, a hold queue. **3.** A data structure in which items are inserted and removed in a specified order, usually FIFO.

quick-blow fuse A fuse which opens a circuit instantaneously. Also called **quick-break fuse**.

quick break The opening of an electric circuit rapidly or instantaneously.

quick-break fuse Same as **quick-blow fuse**.

quick-break switch A switch that opens an electric circuit rapidly or instantaneously.

quick charge **1.** A partial charge of a rechargeable battery, done for a short interval and at a high current. Also called **booster charge**, or **boost charge**. **2.** A complete charge of a rechargeable battery, done for a short interval and at a high current.

quick-disconnect Enabling rapid and easy connection and disconnection.

quick-disconnect connector A connector, such as a phono plug, which allows for rapid and easy connection and disconnection.

quick-disconnect plug A plug, such as a phono plug, which allows for rapid and easy connection and disconnection.

quick make The closing of an electric circuit rapidly or instantaneously.

quick make/quick break A switch or mechanism which opens or closes an electric circuit rapidly or instantaneously. This helps minimize or prevent arcing.

quick-make switch A switch that closes an electric circuit rapidly or instantaneously.

QuickDraw A graphics display system utilized to present images on computer monitors, and for printing.

Quicken A popular program to help manage personal finances.

quicksilver An alternate name for the chemical element **mercury**.

quicksort A sort algorithm which first separates all elements which are greater than the median value from those that are less. Then, the two resulting sides are sorted in the same manner, yielding four lists. This process is continued until the sort is complete.

QuickTime A technology utilized for the creation, modification, and reproduction of multimedia files. Also, an application or media player utilized to prepare and playback such files. Also, a file in this format.

quiescent **1.** To be inactive. **2.** To be functioning under normal operating conditions, but not having an applied signal at the time. Said, for instance, of an amplifier stage during an interval in which there is no input signal. Also called **quiescent operation**, or **idle (3)**.

quiescent current The current flowing through a device which is **quiescent (2)**. Also called **idling current (2)**.

quiescent operation Same as **quiescent (2)**.

quiescent point Within a load line, the point that represents the **quiescent value**. Its abbreviation is **Q point**.

quiescent power The power consumed by a device which is **quiescent (2)**. Also called **idling power (2)**.

quiescent state Also called **resting state**, or **idle state**. **1.** A state in which a component, circuit, device, piece of equipment, or system is functioning under normal operating conditions, but does not have an applied signal at a given time. Also called **standby (1)**. **2.** A state in which a device, piece of equipment, or system is operational and available, but is not in use. Also called **standby (2)**.

quiescent value A voltage or current value for a device which is **quiescent (2)**.

quiescent voltage The voltage required by a device which is **quiescent (2)**. Also called **idling voltage (2)**.

quiet automatic gain control Same as **quiet automatic volume control**. Its abbreviation is **QAGC**.

quiet automatic volume control An automatic gain control circuit which is activated only when the input signal exceeds a predetermined magnitude. This allows for maximum amplification of weaker signals. Its abbreviation is **QAVC**. Also called **quiet automatic gain control, delayed automatic gain control, delayed automatic volume control**, or **biased automatic gain control**.

quiet battery A battery whose design provides for power with a minimum of noise.

quiet tuning Tuning in which sound is suppressed when not optimally tuned to a signal.

quieter A circuit in a radio receiver that suppresses noise when listening to programming or other signals. Also called **quieting circuit, noise quieter**, or **squelch circuit (1)**.

quieting The noise suppression resulting from the use of a **quieter**. Also called **noise quieting**.

quieting circuit Same as **quieter**.

quieting sensitivity In an FM receiver, the minimum input-signal amplitude necessary to provide an output with a signal-to-noise ratio at a given level or above.

quintillion A number equal to 10^{18}.

quintupler **1.** A circuit or device whose output frequency is five times that of its input frequency. Also called **frequency quintupler**. **2.** A rectifier circuit whose output DC voltage is about five times the peak value of its input AC voltage. Also called **voltage quintupler**.

quit To exit a program in an orderly manner, as opposed, for instance, to a crash. Also called **shut down (1)**.

QWERTY keyboard The standard English language keyboard. It is named after the first six letters on the top row containing alphabetical characters. Other languages which utilize similar characters usually use a keyboard with a similar layout.

R

r 1. Symbol for **resistivity**. 2. Symbol for **roentgen**.

R 1. Symbol for **resistance**. 2. Symbol for **reluctance**, or **magnetic reluctance**.

R Symbol for **roentgen**.

R-ADSL Abbreviation of **Rate-Adaptive Digital Subscriber Line**.

R display In radars, an oscilloscopic display similar to an A display, but whose time base near the blip is expanded, providing more accurate range information. Also called **R scope**, **R scanner**, **R scan**, or **R indicator**.

R indicator Same as **R display**.

R-S flip-flop Same as **RS flip-flop**.

R scan Same as **R display**.

R scanner Same as **R display**.

R scope Same as **R display**.

R/W Abbreviation of **read/write**.

R/W cycle Abbreviation of **read/write cycle**.

R/W head Abbreviation of **read/write head**.

R-Y signal In a color TV receiver, a color-difference signal representing the difference between the red signal and the luminance signal. Thus, adding this signal to the luminance yields the red primary signal. R is red, and Y is luminance.

R&D Abbreviation of **research and development**.

Ra Chemical symbol for **radium**.

rabbit-ear antenna Same as **rabbit ears**.

rabbit ears A type of antenna, usually used to receive TV broadcasts when there is no cable or satellite, consisting of two rods which can be positioned as desired. The rods are usually telescopic, and when seeking to optimize reception they must be repositioned in a manner similar to the way rabbits locate sounds by twisting and directing their ears. Also called **rabbit-ear antenna**.

RAC Abbreviation of **remote-access concentrator**.

race condition 1. A condition in which two or more processes or states overlap in a manner that leads to erroneous or otherwise undesired results. For example, a gate receiving inputs from two or more paths without the proper sequencing. 2. A situation in which a final result depends on the order in which two or more processes are executed.

raceway A channel through which cables or wires are run, and which serves to protect them between locations where they are connected to devices and equipment. Also called **cable raceway**.

rack 1. A framework or stand, usually made of metal, which holds and supports panels, devices, or pieces of equipment. Used, for instance, to install multiple components of an audio system. 2. A **rack** (1) with standardized dimensions. Also called **relay rack**. 3. A straight bar which has teeth which can be meshed with a round bar with teeth, as in a rack and pinion.

rack and pinion A device which converts rotary motion into linear motion, and vice versa, in which a round bar with teeth, called pinion, is matched with a **rack** (3). Used, for instance, as an actuator movement mechanism. Its abbreviation is **rack & pinion**.

rack-mountable That which can be **rack-mounted**.

rack-mounted That which is installed, placed upon, or built into a **rack** (1).

rack-mounting The installing, placing upon, or building into a **rack** (1).

rack unit In the context of **rack-mounting**, a unit of vertical mounting space equal to 1.75 in, or 4.445 cm. Its abbreviation is **RU**, or **U**.

rack & pinion Same as **rack and pinion**.

racon Acronym for **radar beacon**.

rad 1. Abbreviation of **radian**. 2. A unit of radiation dose equal to 0.01 gray. It is equal to the dose of 100 ergs absorbed per gram of matter. It is an acronym for radiation absorbed dose. Its abbreviation is **rd**.

RAD Acronym for **Rapid Application Development**.

rad/s Abbreviation of **radian per second**.

rad/sec Abbreviation of **radian per second**.

RAD tool Abbreviation of **Rapid Application Development tool**.

radar Acronym for **ra**dio **d**etection **a**nd **r**anging. 1. A system or technique in which microwave radiation is beamed by a transmitter, and the radiation reflected off objects is detected and analyzed to indicate the presence of said objects. A radar may also be able to provide information on the range, bearing, speed, altitude, and other characteristics of one or more scanned objects. Radars can emit the microwave energy in pulses, or in a continuous manner. In an active radar, which is the norm, microwaves are transmitted to enhance the detection of scanned objects, while a passive radar detects the microwave energy which is ordinarily radiated and reflected from any given object. A Doppler radar employs the Doppler effect to determine the velocity of a scanned object. There may also be multiple sets of transmitters and receivers, as in frequency-diversity radar. Used, for instance, to track and guide aircraft and ships, or to monitor weather conditions. 2. An apparatus utilized to generate, transmit, receive, and display information utilizing **radar** (1). A typical set has a source of microwave power, such as a magnetron, transmitting and receiving antennas, a receiver, and a display, such as a CRT or LCD. Also called **radar system**, **radar unit**, **radar set**, **radar apparatus**, **radar equipment**, or **radar installation**.

radar-absorbing material A material designed to absorb radar energy.

radar altimeter An absolute altimeter which bounces radio waves from the terrain below to determine altitude. Also called **radio altimeter**, or **electronic altimeter** (1).

radar altitude A determination of altitude utilizing a **radar altimeter**. Also called **radio altitude**.

radar antenna An antenna used in **radar** to emit and/or detect microwave energy.

radar apparatus Same as **radar** (2).

radar approach control The use of radar equipment as part of an air traffic control service, as is usually the case. Also called **radar control**, or **terminal radar approach control**.

radar astronomy The use of radar to study astronomical phenomena, such as the location and movements of celestial bodies.

radar backscatter The scattering of microwaves by a scanned object back towards a radar transmitter. Also called **radar scatter** (1).

radar band A frequency band, such as the L band, utilized in radar. Also called **radar frequency**, or **radar frequency band**.

radar beacon A transponder which when interrogated by a radar automatically responds with a signal which appears on the display of the triggering radar. This signal provides information concerning the range and bearing of the interrogating aircraft or ship. Its acronym is **racon**. Also called **radar transponder**, or **radar beacon transponder**.

radar beacon transponder Same as **radar beacon**.

radar beam The beam of microwave energy emitted by a radar. This beam may be directed, for instance, in a circular, helical, or horizontal manner.

radar bearing A bearing obtained through the use of a radar.

radar clutter In radars, unwanted echoes that appear on the display screen. Clutter may be caused by rain, antenna movements, vegetation, and so on. Also known as **clutter**, or **background return**.

radar-clutter suppression Techniques employed to suppress **radar clutter**.

radar contact The appearance of a scanned object on a radar display, such as a PPI. When radar contact is lost, the necessary radar information is no longer being received.

radar control Same as **radar approach control**.

radar coverage The area or region within which a radar can effectively detect scanned objects. Also, the use of a radar within this area or region.

radar cross section The proportion of the overall area of a scanned object which reflects the same amount of a radar signal that would be reflected by the entire area of said object. Its abbreviation is **RCS**. Also called **echo area**, or **target cross section**.

radar detector A device which detects the presence of radar.

radar display Also called **radar presentation**. **1.** Same as **radarscope**. **2.** The information displayed by a **radarscope**.

radar display screen Same as **radarscope**.

radar dome Same as **radome**.

radar echo The portion of a transmitted radar signal which a scanned object returns to the radar receiver. Also, the visual indication on a radar screen, such as a blip, of such a reflection. Also called **radar return**, **return (3)**, **return echo**, or **echo (2)**.

radar energy The microwave energy emitted by a radar or reflected by a scanned object.

radar equation An equation which describes the relationship between the microwave energy emitted and received by a radar, and the radar cross section of a scanned object.

radar equipment Same as **radar (2)**.

radar fix The determination of an accurate position of a vessel or craft through the use of radar.

radar frequency Same as **radar band**.

radar frequency band Same as **radar band**.

radar homing Homing in which radar is utilized follow the path of a scanned object.

radar horizon The angle of elevation at which the energy emitted by a radar becomes tangential to the surface of the earth. The topography of the surface of the earth will affect this value.

radar image An image of an object or region obtained utilizing radar. Used, for instance, to monitor and indicate precipitation and storms in a given area.

radar indicator Same as **radarscope**.

radar installation Same as **radar (2)**.

radar meteorology The use of radar to assist in monitoring, analyzing, and presenting meteorological phenomena.

radar navigation Navigation in which one or more radars provide information such as location, heading, speed, or altitude.

radar net Same as **radar network**.

radar network Two or more radar systems which work together for an expanded coverage area. Used, for instance, in meteorology. Its abbreviation is **radar net**.

radar parameters Measurable characteristics analyzed in **radar (1)**, such as the angle of arrival, frequency, repetition rate, and amplitude of the reflected microwave energy.

radar presentation Same as **radar display**.

radar pulse A pulse of microwave energy emitted by a radar transmitter.

radar range **1.** The distance to a scanned object, as determined by radar. **2.** The maximum distance at which a radar effectively detects scanned objects. It may be defined, for instance, as the distance at which a radar set can detect a given object half the time.

radar receiver Within a radar set, a radio receiver that detects and amplifies the microwave signals picked up by a radar antenna.

radar reflectivity A measure of the extent to which a scanned object reflects a radar beam. Factors such as size, shape, and composition influence this.

radar reflector A surface, such as a corner reflector, which is utilized to enhance the return of a radar signal back to a transmitter.

radar resolution Same as **range resolution**.

radar return Same as **radar echo**.

radar scan **1.** The motion of a radar while searching for scanned objects. This movement may be circular, helical, horizontal, and so on. Also called **radar scanning**. **2.** A full cycle during a **radar scan (1)**. For instance, a movement of the radar beam through 360° during circular scanning.

radar scanning Same as **radar scan (1)**.

radar scatter Also called **radar scattering**. **1.** Same as **radar backscatter**. **2.** The scattering of the microwave energy emitted by a radar.

radar scattering Same as **radar scatter**.

radar scope Same as **radarscope**.

radar screen Same as **radarscope**.

radar set Same as **radar (2)**.

radar shadow An area, such as that behind a mountain, where radar coverage is not available.

radar signal processing Processing of radar signals so as to improve resolution, reduce noise, and the like.

radar station A location at which a radar set operates.

radar system **1.** Same as **radar (2)**. **2.** A **radar (2)** of specific type, such as a Doppler radar.

radar target An object which reflects a sufficient amount of the energy beam emitted by a radar to produce a radar echo. Also called **target (2)**, or **scanned object**.

radar tracking A tracking system which utilizes a radar to follow movements of a scanned object. Also called **radio tracking (2)**.

radar transmitter Within a radar set, the section which produces and emits microwave energy. A radar usually utilizes a magnetron to generate microwave oscillations.

radar transponder Same as **radar beacon**.

radar unit Same as **radar (2)**.

radar wave A microwave which is either emitted by a radar or reflected by a scanned object.

radar wind Same as **rawin**.

radarscope The viewing screen, such as that of a CRT or LCD, utilized in a radar set. Also spelled **radar scope**. Also called **radar screen**, **radar display (1)**, **radar display screen**, or **radar indicator**.

radial **1.** Extending from, or directed towards a center or point. Also, that which spreads or is distributed uniformly from a center or point, or in all directions. **2.** Pertaining to a radius or ray. Also, similar to a radius or ray. **3.** Same as **radial conductor**.

radial conductor One of multiple conductors which extend outward. Also called **radial (3)**.

radial field 1. A field which extends from, or is directed towards a center or point. 2. A field which spreads or is distributed uniformly from a center, or in all directions.

radial ground A ground system, such as that of an antenna, which consists of multiple conductors which extend outward from a central point.

radial lead A conductor extending perpendicular to the axis of a component such as a resistor or capacitor.

radial-lead component A component, such as a capacitor or resistor, with **radial leads**.

radial leaded That which has **radial leads**.

radial traverse In a robotic arm, extension and retraction. This varies the effective length of the arm.

radian A unit of angle measure equal to approximately 57.2957795 degrees. π radians equals 180°. Its abbreviation is **rad**.

radian frequency The frequency of a periodic quantity expressed in radians per second. It is obtained by multiplying the frequency by 2π. Also called **angular frequency, circular frequency**, or **angular velocity**.

radian per second A unit of angular velocity equal to approximately 57.2957795 degrees per second, or 9.5492966 rpm. Its abbreviation is **rad/s**, or **rad/sec**.

radiance For a source of radiant energy, the radiant flux per unit solid angle, per unit of projected area. Usually expressed in watts per steradian per square meter. Its symbol is L.

radiant 1. Emitting radiation, such as light or heat. 2. A point, particle, body, object, or region from which radiation, such as light or heat, is emitted. 3. Emitted in all directions from a given location or object, such as a point source or body. 4. Consisting of, or pertaining to, radiation, such as light or heat.

radiant efficiency For a given source of radiation, the ratio of the radiant flux emitted, to the power consumed.

radiant energy Energy which is emitted, propagated, absorbed, reflected, or received, in the form of **radiation**.

radiant flux The rate at which radiant energy is incident upon, passes through, or is emitted by a surface or body. Its symbol is Φ. Usually expressed in watts. Also called **radiant power**.

radiant flux density The amount of radiant flux incident per unit area of a surface. Also called **irradiance**.

radiant heat The process by which heat is radiated by matter. This energy is in the form of electromagnetic waves, and is a result of the temperature of said matter. While such waves may encompass the entire electromagnetic spectrum, a high proportion of this radiation occurs in the infrared region. Also, the heat so radiated. Also called **heat radiation**, or **thermal radiation**.

radiant intensity For a source of radiation emitting in a given direction, radiant flux per unit solid angle. Usually expressed in watts per steradian.

radiant power Same as **radiant flux**.

radiate 1. To emit in the form of waves or rays. For example, an antenna radiates radio waves, light rays are emitted from a lamp, or heat is radiated from a hot object. 2. To extend or spread from a center or point. 3. To emit light or radiation, or to expose to light or radiation.

radiated emissions 1. Electromagnetic energy radiated from a component, circuit, device, piece of equipment, or system. Such emissions may cause electromagnetic interference in other components, circuits, devices, equipment, or systems. The determination of radiated emissions is a key factor in ascertaining electromagnetic compatibility. 2. **Radiated emissions (1)** for a given frequency interval.

radiated energy Energy which extends or spreads from a center, point, component, circuit, device, piece of equipment, system, object, or region. Also, the quantified energy level, such as that of a transmitting antenna.

radiated field A field which extends or spreads from a center, point, component, circuit, device, piece of equipment, system, object, or region. For instance, that surrounding a conductor through which an electric current is flowing.

radiated interference 1. Interference, such as that caused by radiated noise, which extends or spreads from a center, point, component, circuit, device, piece of equipment, system, object, or region. 2. Interference resulting from the radiation of another component, circuit, device, piece of equipment, system, object, or region.

radiated noise 1. Noise, such as solar radio noise, which extends or spreads from a center, point, component, circuit, device, piece of equipment, system, object, or region. 2. Noise resulting from the radiation of another component, circuit, device, piece of equipment, system, object, or region. For instance, noise occurring in an audio component whose cables are affected by the field of adjacent power lines.

radiated power Power which extends or spreads from a center, point, component, circuit, device, piece of equipment, system, object, or region. Also, the quantified power level, such as that of a transmitting antenna.

radiated signal A signal which extends or spreads from a center, point, component, circuit, device, piece of equipment, system, object, or region. For instance, that emitted by an antenna.

radiated susceptibility The extent to which a component, circuit, device, piece of equipment, or system is susceptible to radiated emissions, radiated interference, or radiated noise.

radiating element An antenna element which radiates electromagnetic energy.

radiating power Also called **emissive power**, or **emittance**. 1. For a given area of a radiating surface, the total energy radiated per unit time. 2. For a given area of a radiating surface, the total energy radiated per unit time, at a given temperature.

radiating surface A surface which radiates electromagnetic energy.

radiation 1. The propagation of energy via electromagnetic waves, particle beams, sound, or the like, through a given medium such as space, the atmosphere, or a solid object. Also, the energy so propagated. Also, the process via which this occurs. 2. The energy associated with electromagnetic waves. This energy oscillates between its electric and magnetic components. Electromagnetic radiation has both wavelike and particle-like properties. Also called **electromagnetic radiation**. 3. A stream of particles, such as electrons, neutrons, protons, or the like. Also, the emission of such particles. Also, the energy propagated via such particles. 4. The emission of particles or waves by a radioactive substance. Also, the particles or waves so emitted. 5. The exposure to **radiation (1)**, **radiation (2)**, **radiation (3)**, or **radiation (4)**.

radiation absorbed dose Same as **rad (2)**.

radiation absorber A surface, material, body, or region which absorbs radiation such as electromagnetic energy, heat, or sound.

radiation angle The angle at which electromagnetic radiation departs from a transmitting antenna or the surface of the earth. The angle is measured based on a reference plane, such as the horizon. Also called **angle of radiation**, or **angle of departure**.

radiation belt 1. A region of charged particles that surround a celestial body possessing magnetic fields. 2. Each of the two regions of ionizing radiation which surround the planet earth, and which interact with its magnetic field. These belts are composed mostly of protons and electrons which originate from the solar wind which become trapped by the earth's magnetic field. These two belts extend from about 700 kilometers to beyond 50,000 kilometers, and their contained particles travel along the planet's lines of force. Also called **Van Allen radiation belt**.

radiation counter An instrument which registers and counts ionizing radiation, such as alpha rays, given off by radioactive entities. For example, a Geiger counter or a scintillation counter. Also called **radiation detector (1)**, **ionization counter**, **counter (3)**, or **particle counter**.

radiation counter tube An electron tube which registers ionizing radiation, such as alpha rays, and produces an output electric pulse which can be counted. For example, a Geiger-Müller tube. Also called **counter tube (3)**.

radiation damage 1. Damage to a semiconductor device, such as a transistor, as a consequence of exposure to ionizing radiation. An antirad material may be used, for instance, to help prevent such damage. 2. Damage to a component, circuit, device, piece of equipment, system, material, or the like, due to exposure to ionizing radiation.

radiation detector 1. Same as **radiation counter**. 2. A component, circuit, device, piece of equipment, system, material, or process which detects radiation.

radiation diagram A diagram representing how the intensity of a quantity varies as a function of direction and distance. Commonly refers to the transmitting or receiving effectiveness of an antenna, and can be described in any plane, although the horizontal and/or vertical planes are generally used. Also called **directional diagram**, **directivity diagram**, or **field diagram**.

radiation dose The amount of ionizing radiation incident upon, or absorbed by, a given mass, volume, or body. May be expressed in roentgens, rems, rads, or reps. Also called **radiological dose (2)**, **dosage**, or **dose**.

radiation efficiency For an antenna, device, piece of equipment, or system, the ratio of the power radiated to the power consumed.

radiation excitation 1. Excitation of particles, atoms, or molecules by ionizing radiation. 2. Excitation of particles, atoms, or molecules by electromagnetic radiation.

radiation exposure 1. The condition or state of not being shielded, or otherwise protected, from radiation, especially ionizing radiation. 2. An act or circumstance which results in being exposed to radiation. 3. The quantity or extent of **radiation exposure (2)**.

radiation field For a radiator, such as a transmitting antenna, the portion of the electromagnetic field which is propagated outward as electromagnetic waves. This contrasts with the **induction field (2)**, which behaves as if permanently associated with said radiator.

radiation filter 1. An element or device, such as a disk or plate of plastic or glass, which selectively blocks or absorbs one or more intervals of frequencies of radiation. 2. An element or device, such as a disk or plate of plastic or glass, which selectively blocks or absorbs one or more intervals of frequencies of electromagnetic radiation, such as light. The optical properties of the element or device determine which frequencies pass, and which are blocked or absorbed. Also called **filter (4)**, or **optical filter**.

radiation hardened That which has undergone **radiation hardening**.

radiation hardening 1. Techniques utilized to enhance the resistance of components, circuits, devices, equipment, systems, and materials to ionizing radiation. Also, the additional resistance so obtained. 2. Techniques utilized to enhance the resistance of components, circuits, devices,

equipment, systems, and materials to electromagnetic radiation. Also, the additional resistance so obtained.

radiation hazard A possible source of danger to components, circuits, devices, equipment, systems, or personnel, resulting from radiation exposure.

radiation intensity 1. The magnitude of a given radiation. 2. For a given antenna, the power radiated per unit solid angle. 3. For a source of radiant energy, the power radiated per unit solid angle.

radiation ionization 1. Ionization of atoms or molecules by ionizing radiation. For instance, the ionization of the mercury vapor present in a gas tube. 2. Ionization of atoms or molecules by electromagnetic radiation.

radiation lobe Within the directional pattern of an antenna, a three-dimensional section in which radiation is increased.

radiation loss Any losses, such as RF or heat energy, due to radiation. Also called **radiation losses**.

radiation losses Same as **radiation loss**.

radiation monitor A component, circuit, device, piece of equipment, system, material, or process utilized to monitor ionizing radiation. Such a monitor may be equipped with a mechanism, such as an alarm, to indicate when exposure exceeds a given threshold.

radiation noise Radiation, especially electromagnetic radiation, producing noise in a circuit, device, piece of equipment, or system.

radiation pattern A pattern representing how the intensity of a quantity varies as a function of direction and distance. Commonly refers to the transmitting or receiving effectiveness of an antenna, and can be measured in any plane, although the horizontal and/or vertical planes are generally used. Also called **directivity pattern**, or **field pattern**.

radiation physics The science that deals with **radiation**, especially ionizing radiation, and the effects it has on matter.

radiation pressure 1. The very slight pressure that electromagnetic radiation exerts upon a surface. 2. The pressure that a sound wave exerts upon a surface.

radiation pyrometer Same as **radiation thermometer**.

radiation resistance For a given antenna, the radiated power divided by the square of the effective current, measured at the feed point.

radiation scattering The scattering of radiation due to interactions or collisions with particles. For example, atmospheric scattering, Bragg scattering, or multiple scattering.

radiation shield 1. A material or enclosure which limits or blocks the effects of radiation, such as ionizing radiation. Also called **radiation shielding (1)**. 2. A **radiation shield (1)** which serves to confine radiation within an enclosure. Used, for instance, to help minimize or prevent electromagnetic interference which may affect other devices or systems. Also called **radiation shielding (2)**.

radiation shielding 1. Same as **radiation shield (1)**. 2. Same as **radiation shield (2)**. 3. The use of a **radiation shield (1)**, or **radiation shield (2)**.

radiation source 1. Any source, such as a nucleus or particle accelerator, of ionizing radiation. 2. Any source of **radiation**.

radiation therapy Same as **radiotherapy**.

radiation thermometer An instrument or device, such as an infrared thermometer, which measures the radiation emitted by a body to determine its temperature. It is a type of noncontact thermometer. Also called **radiation pyrometer**.

radiation zone The region in which the transmitted energy from an antenna behaves as if it were emanating from a point source in the vicinity of said antenna. Such a region usually begins at a distance of $2D^2/\lambda$ from the antenna, where D is the maximum overall dimension of said antenna, and λ is the wavelength considered. The region of space closer to the antenna than the radiation zone is the **near-field region**. Also called **far-field region, far field (1)**, or **Fraunhofer region**.

radiative recombination A recombination in which energy is released in the form of a photon.

radiative transition An energy transition which results in the emission of radiation. For instance, one in a which a particle losing energy emits a photon. The converse is true in a **nonradiative transition**.

radiator 1. Any particle, body, material, or system, which emits **radiation**. 2. The part of an antenna or element which emits radio waves. Also called **antenna radiator**. 3. A device or surface which vibrates to produce sound waves. Examples include headphone diaphragms, speaker cones, and distributed-mode speaker panels. Also called **acoustic radiator**, or **sound radiator**. 4. An enclosure which is sealed, except for a small opening through which radiant energy may enter or escape. The radiation of such an enclosure approximates that of a blackbody. Also called **cavity radiator**.

radical 1. An atom or molecule which has one or more unpaired electrons. Radicals are usually highly reactive and unstable. Also called **free radical**. 2. A group of atoms, such as a hydroxyl group, which reacts as a single unit. 3. Arising from, oriented towards, or pointing to an origin, root, or source.

radio- 1. A prefix denoting concepts pertaining to **radiant energy**, or its uses. For example, radiometer. 2. A prefix denoting **radioactive**, or pertaining to radioactivity. For instance, radionuclide.

radio 1. The transmission and/or reception of information, such as voice, video, data, or control signals, in which there are no connecting wires. Instead, such communication is achieved by means of electromagnetic waves, such as radio waves or infrared waves, or via acoustic waves. Also, the specific content. 2. Same as **radio receiver**. 3. The broadcasting of signals which are intended for reception by a radio receiver. 4. Same as **radio transmitter**. 5. Same as **radio transceiver**. 6. Abbreviation of **radio-frequency**. Same as **RF**.

radio altimeter Same as **radar altimeter**.

radio altitude Same as **radar altitude**.

radio antenna An antenna utilized by a **radio (2)**, **radio (4)**, or **radio (5)**.

radio astronomy The study of celestial bodies by detecting and observing RF energy which is emitted and/or reflected by them. Also spelled **radioastronomy**.

radio beacon A non-directional radio transmitter station that emits signals from which bearing information can be derived by using a radio direction finder. Radio beacons may operate continuously, or in response to interrogation signals. Some also provide range information. Also spelled **radiobeacon**. Also known as **radiophare**, or **aerophare**.

radio beam 1. A concentrated and essentially unidirectional stream of radio waves, such as that emitted by a highly directional antenna. 2. An RF signal transmitted in a given direction, such as that of an antenna, radar, or radio beacon.

radio bearing A bearing obtained utilizing a radio direction finder.

radio blackout A complete loss of a radio signal, or of radio-wave propagation. Also called **blackout (2)**.

radio broadcast An RF transmission intended for public or general reception. Also, to transmit such signals. Refers especially to radio programming.

radio broadcast band A band of frequencies allocated to broadcasting radio stations. For instance, the standard AM broadcast band is 535 to 1705 kHz.

radio broadcast channel Same as **radio broadcasting channel**.

radio broadcast receiver A device designed to receive radio transmissions intended for public or general reception.

radio broadcast reception The receiving of radio transmissions intended for public or general reception.

radio broadcast service Same as **radio broadcasting service**

radio broadcast station Same as **radio broadcasting station**.

radio broadcast transmission A transmission of radio signals intended for reception by the general public.

radio broadcast transmitter A transmitter of radio signals intended for reception by the general public.

radio broadcasting The transmission of RF signals intended for public or general reception. Refers especially to radio programming.

radio broadcasting channel Same as **radio broadcasting station**.

radio broadcasting service A radio-communications service in which radio signals intended for reception by the general public are transmitted or retransmitted. Also called **radio broadcast service**.

radio broadcasting station A station within a **radio broadcasting service**. Also called **radio broadcast station**, **radio broadcasting channel**, or **radio broadcast channel**.

radio button One of multiple buttons presented on a computer display which allow for the selection of a given option. When a selection is made, all other possibilities are automatically excluded. This contrasts with a **check box**, where multiple choices may be selected together.

radio channel 1. A frequency, or band of radio frequencies, assigned to a particular carrier or station for a specific purpose. For example, the band of radio frequencies assigned to a radio broadcasting station. 2. A frequency, or band of radio frequencies, utilized for radio communications. 3. A wireless medium via which information is conveyed between two or more locations. Also called **radio circuit (2)**. 4. Same as **RF channel**.

radio circuit 1. Same as **RF channel**. 2. Same as **radio channel (3)**. 3. A circuit within a radio receiver, radio transmitter, or radio transceiver.

radio clock A clock whose time is set via radio signals. For instance, a clock whose indicated time is synchronized with signals received by a radio station which in turn bases its signals on a national time standard. The time kept by such clocks is usually very accurate.

radio common carrier A common carrier which provides RF communications services, such as paging, to the general public. Such entities are usually regulated by the appropriate authorities, such as governmental agencies. This term usually excludes entities providing landline telephone services. Its abbreviation is **RCC**.

radio communication Also spelled **radiocommunication**. 1. Same as **radio communications (1)**. 2. Same as **radio communications (2)**.

radio communications Also spelled **radiocommunications**. 1. The transmission of information, such as audio, video, data, or control signals, without the use of connecting wires. Instead, such communication is achieved by means of electromagnetic waves, such as radio waves or infrared waves. Also called **radio communication (1)**. 2. The information conveyed in **radio communications (1)**. Also called **radio communication (2)**. 3. The science dealing with **radio communications (1)**.

radio compass A radio receiver that automatically and continuously indicates the direction from which a radio signal arrives. Also called **automatic direction finder**, **automatic radio compass**, or **automatic radio direction finder**.

radio control The use of electromagnetic waves, such as radio waves or infrared waves, for remote control, as opposed to the use of connecting wires. Also spelled **radio-control**.

radio detection **1.** The use of radio signals to detect objects. **2.** The detection of radio signals emitted or reflected by objects.

radio detection and ranging Same as **radar**.

radio detector A device which serves for **radio detection (2)**.

radio device A device, such as an antenna or receiver, which utilizes, operates on, emits, receives, or whose function otherwise involves radio signals. Also called **RF device**.

radio direction finder A radio receiver which incorporates a highly directional antenna, such as a loop antenna, to determine the direction from which a radio signal arrives. Its abbreviation is **RDF**. Also called **direction finder**.

radio direction finding The use of a **radio direction finder** to determine the direction from which a radio signal arrives. Its abbreviation is **RDF**. Also called **direction finding**.

radio direction-finding station Its abbreviation is **RDF station**. A station with utilizes radio direction finders to determine the direction from which radio signals arrive. Such a station may be located, for instance, along a coastline. Also called **direction-finding station**.

radio-electronic Also spelled **radioelectronic**. Pertaining to, characteristic of, or utilizing **radio electronics**.

radio electronics Also spelled **radioelectronics**. **1.** The branch of electronics that deals with **radio communications (1)**. **2.** The branch of electronics that deals with devices and apparatuses, such as radars, which utilize radio waves.

radio element **1.** Same as **radioactive element**. **2.** Same as **radioisotope**.

radio emission The process of giving off, or sending out electromagnetic waves. For example, such radiation by an antenna.

radio engineering The branch of engineering which deals with all aspects of radio communications, including transmission, reception, and the design, manufacturing, and optimization of the equipment utilized.

radio facsimile Same as **radio fax**.

radio fade-out A rapid and complete loss of electromagnetic skywave signals, due to substantially increased ionization in the ionosphere. This ionization is caused by increased solar noise, which itself is caused by solar storms. It may last from a few minutes to several hours. Also spelled **radio fadeout**. Also called **fadeout, Dellinger fadeout,** or **Dellinger effect**.

radio fadeout Same as **radio fade-out**.

radio fax Abbreviation of **radio facsimile**. Fax communications utilizing radio signals.

radio field intensity Same as **radio field strength**.

radio field strength The intensity of an electromagnetic field produced by **radio waves**. Also called **radio field intensity**.

radio fix The determination of an accurate position of a vessel through the use radio signals, such as those detected by radio direction finders.

radio frequency Same as **RF**. Its abbreviation is **radio (6)**. Also spelled **radiofrequency**.

radio-frequency amplification Same as **RF amplification**.

radio-frequency amplifier Same as **RF amplifier**.

radio-frequency antenna Same as **RF antenna**.

radio-frequency band Same as **radio spectrum**. Its abbreviation is **RF band**.

radio-frequency carrier Same as **RF carrier**.

radio-frequency channel Same as **RF channel**.

radio-frequency choke Same as **RF choke**.

radio-frequency circuit Same as **RF circuit**.

radio-frequency coil Same as **RF coil**.

radio-frequency communications Same as **RF communications**.

radio-frequency converter Same as **RF converter**.

radio-frequency current Same as **RF current**.

radio-frequency detector Same as **RF detector**.

radio-frequency device Same as **radio device**. Its abbreviation is **RF device**.

radio-frequency diode Same as **RF diode**.

radio-frequency energy Same as **RF energy**.

radio-frequency field Same as **RF field**.

radio-frequency filter Same as **RF filter**.

radio-frequency generator Same as **RF generator**.

radio-frequency harmonic Same as **RF harmonic**.

radio-frequency heating Same as **RF heating**.

radio-frequency IC Same as **RF IC**.

radio-frequency ID Same as **RFID**.

radio-frequency identification Same as **RFID**.

radio-frequency indicator Same as **RF indicator**.

radio-frequency integrated circuit Same as **RF IC**.

radio-frequency interference Same as **RF interference**.

radio-frequency lamp Same as **RF lamp**.

radio-frequency line Same as **RF line**.

radio-frequency link Same as **RF link**.

radio-frequency modulation Same as **RF modulation**.

radio-frequency noise Same as **radio noise**. Its abbreviation is **RF noise**.

radio-frequency oscillator Same as **RF oscillator**.

radio-frequency power Same as **RF power**.

radio-frequency power amplifier Same as **RF power amplifier**.

radio-frequency power supply Same as **RF power supply**.

radio-frequency probe Same as **RF probe**.

radio-frequency pulse Same as **RF pulse**.

radio-frequency radiation Same as **RF energy**. Its abbreviation is **RF radiation**.

radio-frequency range Same as **RF range**.

radio-frequency reception Same as **RF reception**.

radio-frequency resistance Same as **RF resistance**.

radio-frequency response Same as **RF response**.

radio-frequency sensitivity Same as **RF sensitivity**.

radio-frequency sensor Same as **RF sensor**.

radio-frequency shield Same as **RF shield**.

radio-frequency shielding Same as **RF shielding**.

radio-frequency signal Same as **RF signal**.

radio-frequency signal generator Same as **RF signal generator**.

radio-frequency spectrum Same as **radio spectrum**. Its abbreviation is **RF spectrum**.

radio-frequency splitter Same as **RF splitter**.

radio-frequency stage Same as **RF stage**.

radio-frequency system Same as **RF system**.

radio-frequency transformer Same as **RF transformer**.

radio-frequency transistor Same as **RF transistor**.

radio-frequency transmission Same as **RF transmission**.

radio-frequency transmission line Same as **RF line**. Abbreviation of **RF transmission line**.

radio-frequency waves Same as **RF waves**.

radio-frequency welding Same as **RF welding**.

radio galaxy A galaxy, such as *Cygnus A*, which emits detectable amounts of radio waves.

radio homing The use of radio signals to track an object.

radio horizon A horizon as seen from an antenna transmitting site. Such a horizon is the furthest distance that a direct wave may travel. It is the line that is formed by the points at which the direct waves from a radio transmitter become tangential to the surface of the earth. Due to atmospheric refraction, this horizon extends beyond the apparent horizon. Also called **horizon (2)**.

radio interference In a communications system, any energy which diminishes the ability to receive a desired signal, or that impairs its quality. Sources include electromagnetic noise, undesired signals, parasitic oscillations, and atmospheric conditions. Its abbreviation is **RI**. Also called **electric interference**, or **interference (2)**.

radio interferometer An interferometer which acts on radio waves. Used, for instance, in astronomy.

radio interferometry The design, study, and utilization of **radio interferometers** to measure and analyze interference patterns of radio waves.

radio isotope Same as **radioisotope**.

radio jamming The intentional use of RF signals to disrupt communications.

radio link Also spelled **radiolink**. **1.** A connection which enables the transmission of information between two points or entities, without the use of connecting wires. **2.** The resources which facilitate a **radio link (1)**.

radio local loop The use of RF waves, such as microwaves, for last mile connections between a telephone company's central office and the customers it serves. May be used, for instance, to provide telecommunications access to new communities without having to run wires or cables to each user. Its abbreviation is **RLL**. Also called **wireless local loop**.

radio location Same as **radiolocation**.

radio locator A device, such as radar, utilized for **radio location**.

radio magnetic indicator An instrument which indicates both the bearing and heading of an aircraft. Its abbreviation is **RMI**.

radio modem A modem which transmits data without utilizing connecting wires or cables. Some systems have ranges that exceed 50 kilometers. Used, for instance, where data transmission is not offered over existing phone lines.

radio navigation The utilization of radio-electronic devices and apparatuses, such as radars and radio beacons, as navigational aids. Also spelled **radionavigation**.

radio navigation aid A radio-electronic device, apparatus, or system which provides navigational information such as the exact current location, heading, speed, or height, or that otherwise guides, orients, or warns. Also spelled **radionavigational aid**, or **radionavigation aid**.

radio navigation system A system, such as loran, omnidirectional range, Omega, or instrument approach system, utilized for **radio navigation**. Also spelled **radionavigation system**.

radio navigational aid Same as **radio navigation aid**. Also spelled **radionavigational aid**.

radio net Same as **radio network**.

radio network Its abbreviation is **radio net**. **1.** A group of radio stations that broadcast essentially the same programming. Such broadcasts may or may not utilize the same frequencies, nor do they have to be simultaneous. **2.** A group of radio stations that communicate with each other utilizing a common frequency.

radio noise Electromagnetic noise occurring at radio frequencies. Examples include cosmic noise and geomagnetic noise. Also called **radio-frequency noise**, or **RF noise**.

radio-opaque Same as **radiopaque**.

radio phone Same as **radiotelephone**. Also spelled **radio-phone**.

radio photo Abbreviation of **radio photo**graph. A photo sent via a radio link. Also spelled **radiophoto**.

radio photograph Same as **radio photo**. Also spelled **radio-photograph**.

radio program The content of a given radio broadcast.

radio programming The program offerings of one or more radio broadcasters.

radio pulse Same as **RF pulse**.

radio range **1.** A device, apparatus, unit, facility, or system which emits radio signals which serve as a navigational aid. An example is omnidirectional range. **2.** The distance to a given transmitter of radio waves. For example, the distance between a scanned object and a radar transmitter. Also called **range (10)**.

radio range station A land station within a service providing radio signals which serve as a navigational aid.

radio receiver A circuit, device, piece of equipment, or system which converts radio waves into audio, video, or other information-bearing signals. AM, FM, and AM/FM receivers are examples. Its abbreviation is **radio (2)**. Also called **radio set (1)**, **receiver (3)**, **radio receiving set**, **radio receiver set**, **radio system (2)**, **receiving set**, or **receiver set**.

radio receiver set Same as **radio receiver**.

radio receiving set Same as **radio receiver**.

radio reception The reception of radio waves, such as those emitted by a radio transmitter, or those arriving from space.

radio relay Same as **radio repeater**.

radio-relay link Same as **radio repeater link**.

radio-relay network Same as **radio repeater network**.

radio-relay satellite Same as **relay satellite**.

radio-relay station Same as **radio repeater station**.

radio-relay system Same as **radio repeater system**.

radio repeater A device, location, or system which receives, amplifies, and transmits radio-wave signals between locations such as terrestrial communications stations, between terrestrial communications stations and communications satellites, or between other communications satellites. Such repeaters provide high-capacity communications links, can be combined to offer a very wide coverage area, and may transmit TV, telephone, and data signals, among many others. A one-way repeater works in one direction, and a two-way repeater in both. Also called **radio relay**, **repeater (2)**.

radio repeater link A fixed connection between **radio repeater stations**. Also called **radio-relay link**.

radio repeater network A network linking multiple **radio repeaters**. Also called **radio-relay network**.

radio repeater station A station, such as a satellite earth station, which serves as a **radio repeater**. Also called **radio-relay station**, **repeater station**, **relay station**, **relay transmitter (2)**, **relay receiver (2)**, or **repeater transmitter (1)**.

radio repeater system A system, such as a microwave radio relay system, which serves as a **radio repeater**. Also called **radio-relay system**, or **relay system**.

radio scanner A radio receiver that can scan multiple radio bands. With such a unit, for instance, a user may tap buttons or turn a dial for frequency selection, or the receiver may proceed automatically from one frequency in use to another, between preprogrammed frequencies, and the like. Also called **scanning radio**.

radio screen 1. Same as **RF shielding** (1). 2. Same as **RF shielding** (2).

radio screening Same as **RF shielding**.

radio sender Same as **radio transmitter**.

radio set 1. Same as **radio receiver**. 2. Same as **radio transmitter**. 3. Same as **radio transceiver**.

radio shield Same as **RF shield**.

radio shielding Same as **RF shielding**.

radio signal 1. A signal sent by means of electromagnetic waves, such as radio waves or infrared waves, as opposed utilizing connecting wires. 2. Same as **RF signal**.

radio source A source, such as the cosmic microwave background, of **radio waves**.

radio spectrum A range of frequencies, within the electromagnetic spectrum, which is utilized primarily for communications. Usually defined as spanning from approximately 3 kHz to approximately 300 GHz, which corresponds to wavelengths of 100 km to 1 mm, respectively. The radio spectrum has several arbitrarily established intervals, which in order of ascending frequency are: very low frequency, low frequency, medium frequency, high frequency, very high frequency, ultrahigh frequency, superhigh frequency, and extremely high frequency. There are additional terms utilized for classifying ranges within the radio spectrum. For instance, microwave frequencies, depending on the defined interval, usually overlap with ultrahigh frequencies, superhigh frequencies, and extremely high frequencies. Also called **radio-frequency spectrum**, or **RF band**.

radio star A star, such as the sun, known to be a **radio source**.

radio station 1. A location equipped with one more receivers and/or transmitters, antennas, or the like, which enable the emission and/or reception of radio signals. Two or more such stations comprise a radio network. 2. A **radio station** (1) utilized for broadcasting.

radio system 1. A system, such as radio telephony, utilized for **radio communications**. 2. Same as **radio receiver**.

radio telegram Same as **radiogram** (2). Also spelled **radio-telegram**.

radio telegraph Same as **radiotelegraph**.

radio telegraphy Same as **radiotelegraphy**.

radio telemetry Telemetry in which connecting wires or cables are not utilized. Instead, signals are usually transmitted via electromagnetic waves, such as radio waves or infrared waves.

radio telephone Same as **radiotelephone**.

radio telephone communication Same as **radiotelephone communication**.

radio telephony Also spelled **radiotelephony**. 1. The utilization of modulated radio waves to transmit sounds, such as speech and/or music. Also, a system or technology enabling such communication. 2. Telephonic communications utilizing a radio link, as opposed to connecting wires. Also, a system or technology enabling such communication.

radio telescope An instrument which detects, amplifies, and analyses radio waves originating outside the earth's atmosphere. A radio telescope incorporates one or more parabolic reflectors or other highly directional antennas, very sensitive receivers, and computers which analyze the detected signals, which are then displayed and/or recorded. Since radio wavelengths are much longer than those of visible light, such telescopes tend to be rather large when capable of detecting extremely weak radio sources.

radio teletype A teletype which utilizes radio waves for transmission and/or reception, as opposed to wires. Also spelled **radioteletype**. Its abbreviation is **RTTY**.

radio teletypewriter A teletypewriter which utilizes radio waves for transmission and/or reception, as opposed to wires. Its abbreviation is **RTTY**.

radio therapy Same as **radiotherapy**.

radio tower 1. A tower upon which one or more antennas are mounted. 2. A tower serving as an antenna.

radio tracking 1. A tracking system which utilizes a radio beam to follow the movements of a scanned object. 2. Same as **radar tracking**.

radio transceiver A single unit incorporating both a **radio transmitter** and a **radio receiver**. Its abbreviation is **radio** (5). Also called **radio set** (3). When components or circuits are shared, also called **transceiver** (1).

radio transmission The emission of radio waves from a radio transmitter, or another radio source.

radio transmitter A circuit, device, or system which converts audio, video, or other content into radio waves which are emitted. AM, FM, and AM/FM transmitters are examples. Its abbreviation is **radio** (4). Also called **radio sender**, **radio set** (2), **transmitter** (2), or **sender** (2).

radio-transparent Also spelled **radiotransparent**. This contrasts with **radiopaque**. 1. Allowing the passage of ionizing radiation, such as X-rays. Said, for instance, of bodies, materials, or mediums. 2. Allowing the passage of radiant energy, such as light. Said, for instance, of bodies, materials, or mediums.

radio tube An electron tube, such as a microwave tube, utilized at radio frequencies.

radio wave propagation The movement of a radio wave through a medium, such as the atmosphere, a vacuum, or a transmission line. Also, the process by which this occurs. Also spelled **radiowave propagation**.

radio waves Electromagnetic waves whose frequencies are within the **radio spectrum**. Also spelled **radiowaves**. Also called **RF waves**, or **Hertzian waves**.

radio wind Same as **rawin**.

radio window An interval of radio frequencies which pass through the atmosphere. This interval varies, in general ranging from approximately 5 MHz to 40 GHz, though under certain conditions it may extend beyond 300 GHz. The lower end of the spectrum is limited by ionospheric reflection of waves back into space, while the higher end is limited by absorption by atmospheric constituents such as water and carbon dioxide.

radioactive 1. Pertaining to, characteristic of, or arising from **radioactivity**. 2. That which exhibits or utilizes radioactivity.

radioactive chain Same as **radioactive decay series**.

radioactive chain decay The transformation of a nuclide into another nuclide through radioactive decay, followed by further transformations until a stable, or nonradioactive, nuclide results. In such a series, the first member is called the parent, and the last is the end product. For instance, the chemical element uranium undergoes a series of radioactive decay steps which eventually ends with a stable isotope of lead. Also called **chain disintegration**, **series decay**, **series disintegration**, or **chain decay**.

radioactive dating Same as **radiometric dating**.

radioactive decay Same as **radioactivity**.

radioactive decay series A series of nuclides which undergo **radioactive chain decay**, which is the transformation of one nuclide into another nuclide through radioactive decay, followed by further transformations until a stable, or nonradioactive, nuclide results. Also called **radioactive chain**, **radioactive disintegration series**, **radioactive family**, **radioactive series**, **decay chain**, **disintegration series**, or **disintegration family**.

radioactive disintegration Same as **radioactivity**.

radioactive disintegration series Same as **radioactive decay series**.

radioactive element A chemical element, such as uranium or radon, exhibiting **radioactivity**. Also called **radio element (1)**, or **unstable element**.

radioactive family Same as **radioactive decay series**

radioactive half-life The time interval that elapses for a given radioactive substance to lose half of its radioactive intensity. That is, it is the time required for half the atoms of a given sample of identical nuclides to undergo decay. The half-life of ^{50}V (an isotope of vanadium), for instance, is approximately 1.4×10^{17} years, while that for ^6Be (an isotope of beryllium) is approximately 5.9×10^{-21} seconds. Also called **half-life**.

radioactive isotope Same as **radioisotope**.

radioactive material A material which incorporates one or more **radioactive substances**.

radioactive nuclide Same as **radionuclide**.

radioactive series Same as **radioactive decay series**.

radioactive source A material or substance, such as an alpha, beta, or gamma emitter, which serves as a source of radiation.

radioactive substance A substance, such as a radioactive element, exhibiting **radioactivity**.

radioactive tracer A radioisotope which substitutes a non-radioactive isotope of the same element, enabling detection and monitoring which would otherwise be more difficult or impracticable. Used, for instance, in medical diagnostic techniques such as positron emission tomography. Also called **radiotracer**, or **tracer**.

radioactivity Also called **radioactive decay**, or **radioactive disintegration**. **1.** The spontaneous disintegration of unstable atomic nuclei, such as those of uranium or curium, mainly through the emission of alpha, beta, or gamma rays. Also called **decay (2)**. **3.** A specific instance of **radioactivity (1)**, such as that occurring when an atom of uranium-238 emits an alpha particle to form an atom of thorium-234. Also called **decay (3)**. **4.** The reduction over time of the radioactivity of a substance, due to **radioactivity (1)**. Also called **decay (4)**.

radioastronomy Same as **radio astronomy**.

radiobeacon Same as **radio beacon**.

radiocarbon dating A form of **radiometric dating** in which carbon-14 is the radioisotope utilized.

radiocommunication Same as **radio communication**.

radiocommunications Same as **radio communications**.

radiocontrol Same as **radio control**.

radioelectronic Same as **radio-electronic**.

radioelectronics Same as **radio electronics**.

radiofrequency Same as **RF**. Its abbreviation is **radio (6)**. Also spelled **radio frequency**.

radiogenic Produced by, or pertaining to, **radioactivity**.

radiogenic dating Same as **radiometric dating**.

radiogram **1.** Same as **radiograph**. **2.** Abbreviation of **radio telegram**. A message sent via radiotelegraphy. Also called **wireless telegram**.

radiograph An image produced on a radiosensitive surface, such as a photographic film or plate, when exposed to penetrating radiation, especially X-rays. Typically, the object to be analyzed is placed between the X-ray source and the film or plate, with the relative opacity determining the intensity of the shadows produced. Used, for instance, as a medical diagnostic technique, for ascertaining product defects, or in the quantitative analysis of materials. Also called **radiogram (1)**, **roentgenogram**, **roentgenograph**, **X-ray (3)**, X-ray photograph, X-ray image, shadowgraph, or skiagraph.

radiography The obtaining of **radiographs**. Also, the science and techniques involved in obtaining such images, including reducing the exposure necessary for photographing, improvement of resolution, and the like. Also called **roentgenography**, **X-ray photography**, **skiagraphy**, or **shadowgraphy**.

radioisotope For a given chemical element, an isotope which exhibits **radioactivity**. Radium, for instance, as over 30 identified isotopes, all of which are radioactive. Also spelled **radio isotope**. Also called **radioactive isotope**, **radio element (2)**, or **unstable isotope**.

radioisotopic dating Same as **radiometric dating**.

radiolink Same as **radio link**.

radiolocation The use of radio waves, such as the microwaves emitted by a radar, to determine the range and bearing of objects. Also spelled **radio location**.

radiological **1.** Of, pertaining to, or arising from radiation or radioactive materials. **2.** Of, or pertaining to radiology.

radiological dose **1.** The energy absorbed per unit mass of tissue which is exposed to ionizing radiation. Usually expressed in grays or rads. **2.** Same as **radiation dose**.

radiology **1.** The science that deals with radiation and radioactive materials, and their applications in medical diagnosis and treatment. For example, the use of X-rays in radiography, fluoroscopy, or in the treatment of cancer. **2.** The science that deals with radiation and radioactive materials, and their applications, medical and otherwise.

radiolucency **1.** The characteristic of being transparent to radio waves, especially X-rays. Also, the extent to which such waves pass. **2.** The characteristic of being transparent to X-rays. Also, the extent to which such waves pass.

radiolucent **1.** Transparent to radio waves, especially X-rays. **2.** Transparent to X-rays.

radioluminescence Luminescence induced by radioactivity. It is the emission of light by a radioactive material, unaccompanied by high temperatures.

radioluminescent **1.** Capable of exhibiting the phenomenon of **radioluminescence**. **2.** Pertaining to **radioluminescence**.

radiolysis The effecting of a chemical change in a substance due to radiation. The change is usually a decomposition into simpler constituents.

radiometer **1.** An instrument or device which detects and/or measures radiant energy such as infrared, visible, or ultraviolet light. Examples include Crookes radiometers and infrared radiometers. **2.** A device consisting of an evacuated bulb enclosing four vanes that are suspended, or otherwise allowed to move with the least possible friction. The vanes are arranged like a turnstile and are black on one side, and white or polished on the other. The incident radiant energy, such as that from sunlight, is converted into motion, and the greater the intensity, the faster the motion. Also called **Crookes radiometer**.

radiometric dating An analytical technique in which the approximate age of an object or material is determined by the extent to which a given radioisotope has decayed. For instance, after death, tissue no longer absorbs carbon-14, so its remaining proportion present can estimate when such an organism died. Also called **radioisotopic dating**, **radioactive dating**, or **radiogenic dating**. It is called **radiocarbon dating** when carbon-14 is the radioisotope utilized.

radiometry **1.** The science dealing with the measurement of the properties of infrared, visible, and ultraviolet light. Instruments and devices, such as photometers, are used for this purpose. When only dealing with visible light, it is called **photometry**. **2.** The science dealing with the construction, use, and applications of **radiometers**.

radionavigation Same as **radio navigation**.

radionavigation aid Same as **radio navigation aid**. Also spelled **radio navigational aid**.

radionavigation system Same as **radio navigation system**.

radionavigational aid Same as **radio navigation aid**. Also spelled **radio navigation aid**.

radionuclide Abbreviation of **radio**active **nuclide**. A nuclide, such as carbon-14, which exhibits radioactivity.

radioopaque Same as **radiopaque**.

radiopaque Also spelled **radio-opaque**, or **radioopaque**. This contrasts with **radio-transparent**. **1.** Not allowing the passage of ionizing radiation, such as X-rays. Said, for instance, of bodies, materials, or mediums. **2.** Not allowing the passage of radiant energy, such as light. Said, for instance, of bodies, materials, or mediums.

radiophare Same as **radio beacon**.

radiophone Same as **radiotelephone**. Also spelled **radio phone**.

radiophoto Same as **radio photo**.

radiophotograph Same as **radio photo**. Also spelled **radio photograph**.

radiosensitive **1.** Sensitive to ionizing radiation. May refer, for instance, to tissue or to photographic film. **2.** Sensitive to radiant energy, such as light. May refer, for instance, to tissue or to photographic film.

radiosensitivity The extent to which a body, material, or surface is **radiosensitive**.

radiosity A rendering technique in which the light radiated, absorbed, and reflected by objects and surfaces is carefully calculated, producing lifelike shadows, reflections, and the like. The images presented are more realistic than those obtained via **ray tracing**.

radiosonde A balloon-borne instrument which is utilized to measure and monitor meteorological parameters such as temperature, humidity, pressure, and wind speed. The data such an instrument gathers is sent to one or more monitoring stations via RF signals.

radiotelegram Same as **radiogram (2)**. Also spelled **radio telegram**.

radiotelegraph Also spelled **radio telegraph**. Also called **wireless telegraph**. **1.** The transmission of **radiograms (2)**. **2.** A device utilized to send **radiograms (2)**.

radiotelegraphy Telegraphy in which radio waves are utilized, as opposed to wires. Also spelled **radio telegraphy**. Also called **wireless telegraphy**.

radiotelephone Also spelled **radio telephone**. Its abbreviation is **radiophone**. **1.** A device, such as a transmitter, receiver, or telephone utilized in **radio telephony**. **2.** Pertaining to **radio telephony**.

radiotelephone communication A communication, such as a conversation, utilizing **radio telephony**. Also spelled **radio telephone communication**.

radiotelephony Same as **radio telephony**.

radioteletype Same as **radio teletype**.

radiotherapy Also spelled **radio therapy**. Also called **radiation therapy**. **1.** The use of ionizing radiation, such as X-rays, for therapeutic purposes. **2.** The use of radiant energy, such as infrared radiation, for therapeutic purposes.

radiotracer Same as **radioactive tracer**.

radiotransparent Same as **radio-transparent**.

radiowave propagation Same as **radio wave propagation**.

radiowaves Same as **radio waves**.

radium A brilliant white metallic radioactive chemical element whose atomic number is 88. It has over 30 identified isotopes, all of which are radioactive. Radium emits alpha, beta, and gamma radiation, and neutrons when mixed with beryllium. It is used as a neutron and gamma ray source, in luminous paints, and in the treatment of diseases, such as cancer. Its chemical symbol is **Ra**.

radium bromide White, yellow, or brown crystals whose chemical formula is $RaBr_2$. Used in the treatment of diseases, such as cancer.

radium chloride White, yellow, or brown crystals whose chemical formula is $RaCl_2$. Used in the treatment of diseases, such as cancer.

RADIUS Acronym for **Remote Access Dial-in User Service**.

radix The base, or the number of digits used, in a numbering system. For example, 2 in the binary system or 10 in the decimal system. The radix also serves as the multiplier within its numbering system. In the decimal system, for instance, each single position movement to the right of a digit represents a division by 10, while a movement to the left is a multiplication by 10. This can be seen, for example, in the number 153, where the 1 is in the hundreds position, the 5 is the tens, and the 3 is the units. Also called **radix number**, **base (8)**, or **base number**.

radix complement The numerical result obtained when a number is subtracted from the radix, which is the number of digits used in a numbering system. For instance, the complement of 6 in the decimal number system is 4: $(10 - 6) =$ 4. The complement of a number in the binary number system is the other: 1 is the complement of 0, and 0 is the complement of 1. Used in computers, for instance, to represent negative numbers. Also called **complement**, or **true complement**.

radix number Same as **radix**.

radix point A point, within a number, which indicates the separation between the integer portion and the fractional portion. An example is a decimal point.

radome Abbreviation of **ra**dar **dome**. A structure, such as a dome, which houses and protects radar equipment. Such a structure is usually made of a dielectric, and is transparent to radio waves.

radon A colorless noble gas whose atomic number is 86. It is the densest known gas, and has over 30 known isotopes, all of which are unstable. It is used in the treatment of diseases, such as cancer, and as a neutron source. Its chemical symbol is **Rn**.

RADSL Acronym for **Rate-Adaptive Digital Subscriber Line**.

RAID Abbreviation of **R**edundant **A**rray of **I**ndependent **D**isks, or **R**edundant **A**rray of **I**nexpensive **D**isks. Also called **RAID storage**, or **RAID system**. **1.** A storage unit consisting of multiple hard disks among which data is distributed, for increased performance, fault-tolerance, and data backup. A RAID employs disk striping to improve performance, and may also provide disk mirroring and error correction for fault-tolerance and data recovery. Such a disk array has a single controller and array manager. Used, for instance, in network servers. Also called **RAID array**, or **RAID disk array**. **2.** A method of data storage utilizing a **RAID (1)**.

RAID-0 A **RAID** level which provides disk striping for greater performance, but which does not offer fault-tolerance.

RAID-1 A **RAID** level which provides disk mirroring, for fault-tolerance, but does not offer disk striping.

RAID-2 A **RAID** level which utilizes parity checking and Hamming code for error detection and error-correction coding.

RAID-3 A **RAID** level similar de **RAID-2**, but which utilizes an improved error-detection mechanism and a dedicated disk for error-correction data.

RAID-4 A **RAID** level similar de **RAID-3**, but which stores entire files in each drive, as opposed to byte-sized units.

RAID-5 A **RAID** level with speed approaching that of **RAID-0**, and fault-tolerance similar to **RAID-3**, and which in addition provides one or more dedicated disks for error-correction data.

RAID-6 A **RAID** level similar de **RAID-5**, but which also provides an additional parity-checking scheme for greater fault-tolerance.

RAID-7 A **RAID** level which provides asynchronous I/O transfers, and caching via a high-speed bus.

RAID-10 A **RAID** level which combines features of **RAID-0** and **RAID-1**.

RAID-53 A **RAID** level which combines features of **RAID-0** and **RAID-3**.

RAID array Same as **RAID** (1).

RAID array controller Same as **RAID controller**.

RAID controller A single controller utilized for a **RAID array**. Also called **RAID array controller**.

RAID disk One of multiple hard disks utilizes in a **RAID**.

RAID disk array Same as **RAID** (1).

RAID level A level, such as RAID-5, which serves to distinguish between RAID configurations.

RAID storage Same as **RAID**.

RAID system Same as **RAID**.

RAID technology The technology employed in creating and utilizing **RAIDs**.

rail A metal rail that passes through a component, device, or piece of equipment, delivering power. Also called **power-supply rail**, or **power rail**.

rain attenuation The attenuation of a signal due to rain or atmospheric moisture, such as that contained in clouds. It is the extent to which the amplitude of a signal is reduced when passing through such a medium en route from a transmitter to a receiver, and is usually due to absorption and/or scattering.. Rain attenuation is an important consideration in satellite communications. Also called **rain losses**, or **rain loss**.

rain clutter In radars, unwanted echoes that appear on the display screen due to signal reflections from rain. Also called **rain return**.

rain-clutter suppression Techniques employed to suppress **rain clutter**.

rain loss Same as **rain attenuation**.

rain losses Same as **rain attenuation**.

rain return Same as **rain clutter**.

RAM Acronym for random-access memory. One or more memory chips that can be read and written by the CPU and other hardware, and which serve as the primary workspace of a computer. Instructions and data are loaded here for subsequent execution or processing, and its name is based on the ability to access storage locations in any desired order. The amount of RAM a computer has determines factors such as how many programs may be optimally run simultaneously. Although some other types of memory, such as ROM, may also be accessed randomly, RAM may be volatile, while the others are not. The two main types of RAM are dynamic RAM and static RAM. RAM may also called **read/write memory**, or **read/write RAM**, to distinguish it from ROM, and **main memory**, **main storage**, **primary memory**, or **primary storage** to differentiate it from storage devices such as hard disks. Also called **core memory** (2).

RAM cache Any cache utilizing **RAM**. For example, disk cache uses a section of RAM to store recently accessed data from a disk, which can dramatically speed up applications by avoiding the much slower disk accesses.

RAM card A card or module containing RAM chips. These are utilized to expand the RAM available to a computer. A RAM card may also incorporate a battery to keep the memory cells charged. Also called **memory card** (2).

RAM chip Abbreviation of random-access memory **chip**. A chip which consists of memory cells and the associated circuits necessary, such as those for addressing, to provide a computer with RAM. Examples include DRAM and SRAM.

RAM disk A portion of RAM which has been configured to simulate a disk drive. This speeds up applications considerably, as RAM is much faster than mechanical hard disks for tasks such as storage and retrieval of data. Since RAM is volatile, files must be copied back to hard disks before ending a given session. To avoid loss of data during use of a RAM disk, as might occur during a power failure, a battery backup may be utilized. Also spelled **RAMdisk**. Also called **RAM drive**.

RAM drive Same as **RAM disk**.

RAM refresh To **refresh** (2) the RAM of a computer.

RAM resident Located in RAM at all times. Said of programs which are never swapped or otherwise removed from main memory. Certain portions of the operating system, and applications that require immediate access are memory resident. Some programs, such as virus-scanning applications, may be marked as memory resident by a user. Also called **RAM-resident program**, **memory resident**, or **resident program**.

RAM-resident program Same as **RAM resident**.

Raman effect An effect observed when monochromatic light passes through a transparent medium, and in which some of the scattered light has frequencies above or below that of the incident radiation. This results from inelastic collisions between incident photons and molecules in the medium, which change the vibrational or rotational energy of the latter. This produces additional, fainter, lines in the spectrum. Also called **Raman scattering**.

Raman lidar A form of lidar which utilizes the Raman effect to measure atmospheric parameters such as water vapor or ozone concentration, and temperature. In it, a laser beam is directed into the atmosphere, with scattering intensity being correlated with any measurements being made. Its abbreviation is **RL**. Also called **scattering Raman lidar**.

Raman scattering Same as **Raman effect**.

Raman spectrometer An instrument utilized for **Raman spectroscopy,**

Raman spectroscopy A laser spectroscopic technique that utilizes the **Raman effect** to obtain information on the vibrational or rotational energy of molecules. Used, for instance, in conjunction with IR spectroscopy to reveal additional detail pertaining to molecular vibrations, as the mechanism for Raman scattering is different from that of IR absorption. Widely used for the determination of quantitative and qualitative aspects of molecular samples, such as structure. Also called **laser Raman spectroscopy**.

Raman spectrum The spectrum obtained via **Raman spectroscopy**.

Rambus DRAM Same as **RDRAM**.

Rambus dynamic random-access memory Same as **RDRAM**.

RAMDAC Abbreviation of Random-Access Memory Digital-to-Analog Converter. A controller chip utilized in some video adapters. A RAMDAC incorporates RAM and one or more digital-to-analog converters.

RAMdisk Same as **RAM disk**.

ramp Same as **ramp wave**.

ramp generator A signal generator that produces **ramp waves**.

ramp wave A sawtooth wave in which the amplitude increases linearly to a given value, drops essentially instantaneously to zero, and rises again. Such a wave includes all harmonics, and resembles a series of ramps placed one after the other. Also called **ramp**.

ramp waveform The shape of a **ramp wave**.

random That which has no regular pattern, possessing an equal chance of all possible events, locations, or the like occurring in consecutive instances. Although certain phenomena, such as some natural frequencies, may have a random nature, they can have a regular distribution such as a Gaussian distribution.

random access For a given storage device, such as computer memory or a hard drive, the ability to access information or storage locations in any desired order, with neither the content nor position of subsequent searches depending on those performed previously.

random-access device A storage device which allows for **random access**. Also called **random-access storage device**.

random-access file A file within which individual items can be accessed in any order. Also called **direct-access file**.

random-access memory Same as **RAM**.

random-access memory chip Same as **RAM chip**.

Random-Access Memory Digital-to-Analog Converter Same as **RAMDAC**.

random-access storage Computer storage which allows for **random access**.

random-access storage device Same as **random-access device**.

random digit A single digit within a **random number**.

random error **1.** Error of a **random** nature. Such error may be predictable on a statistical basis. **2.** A specific instance of **random error** (1).

random event An event which is **random** nature. Such an event may be predictable on a statistical basis.

random fading Random reductions in the strength of radio waves as they are propagated through the atmosphere.

random fluctuations Fluctuations, such as those in the strength of a signal, which are **random** in nature.

random interference Interference, such atmospherics, which is **random** in nature.

random noise Noise which fluctuates in a **random** manner, and which usually is the aggregate of a large number of overlapping transient disturbances. The occurrence and magnitude of such noise can not be predicted. Examples include electrical noise and cosmic noise. Also called **fluctuation noise**.

random number **1.** A number in which each successive digit has an equal chance of being any of all possible digits. The frequency distribution of the digits occurring in such a number may be predictable, but the order of individual digits is not. An algorithm can generate numbers which are nearly random, or **pseudorandom**, but not random in every respect. **2.** A number arrived at or obtained completely by chance.

random number generation The utilization of a **random number generator**. Also, the numbers so obtained.

random number generator A circuit, device, program, or the like which generates numbers which pass one or more tests of randomness, but which are not truly random.

random play A feature, such as that of a CD player, which orders the tracks to be played at random.

random process **1.** A process in which there are one or more **random variables**. Also called **stochastic process**. **2.** A process within which each successive event occurs completely by chance.

random signal **1.** A signal which starts and stops, or which appears and disappears, in a **random** manner. **2.** A signal which varies in a random manner.

random variable A variable that can assume any of all the possible values, and in which each occurrence can not be predicted, but whose overall distribution of values may be predictable on a statistical basis.

random variation A variation, such as that in the amplitude of a signal, which is **random** in nature.

random winding A winding whose successive turns do not follow any particular order.

random-wire antenna An antenna whose length has not been calculated for a given purpose, such as operating frequency or radiation pattern. It simply is a length of wire which works adequately for a given need, and is usually end-fed.

randomize To make **random** in nature. For instance, to make one or more variables in an experiment random.

randomness The quality or state of being **random**.

range **1.** The upper and lower limits within which a given fluctuation or value may occur. Also, any interval which has upper and lower limits, such as the continuous interval of frequencies within which a component, circuit, device, piece of equipment, or system operates. **2.** The upper and lower limits within which a given fluctuation or value actually occurs. For instance, a temperature range, or an audibility curve. **3.** A **range** (1) specified for a given parameter, such as that required for proper operation. For example, an amplitude range, an input-voltage range, or a carrier-frequency range. **4.** A **range** (1) which is arbitrarily established. For instance, the infrared range, or a cryogenic temperature range. **5.** The maximum distance within which a given process or level or performance is maintained. For example, a dynamic range, or the maximum useful distance from a transmitting antenna to a receiver. **6.** The distance between two points, objects, or locations. For instance, that between a location from where an object is being observed, and said object. **7.** The maximum distance a part, such as a robotic arm, may move or travel. Also called **range of motion**. **8.** The maximum distance between two particles or entities at which there is a measurable effect. For example, the maximum distance at which two particles may experience an attractive or repulsive force. **9.** The complete **range** (1) of readings or values a scale can indicate, or the upper and lower limits of such a scale. **10.** Same as **radio range** (2).

range-amplitude display In radars, an oscilloscopic display which plots time or distance in the horizontal plane versus the scanned object, which appears in the vertical plane. Also called **A display**, **A scope**, **A indicator**, or **A scanner**.

range-bearing display In radars, an oscilloscopic display which plots azimuth of the scanned object in the horizontal coordinate, versus its range, which appears in the vertical coordinate. The target appears as a bright spot on the display. Also called **B display B scope**, **B indicator**, or **B scanner**.

range discrimination Same as **range resolution**.

range-height indicator In radars, an oscilloscopic display which plots the range of the scanned object in the horizontal coordinate, versus its height, which appears in the vertical coordinate. Its abbreviation is **RHI**.

range mark On a radar screen, a mark which indicates the distance of a scanned object to the radar receiver. Also called **range marker**, or **distance mark**.

range marker Same as **range mark**.

range of motion Same as **range** (7).

range resolution Also called **range discrimination**, **radar resolution**, **target discrimination**, or **distance resolution**. **1.** The minimum distance between two scanned objects that may be distinguished by a radar set. **2.** The minimum distance between two scanned objects that provides two separate and recognizable indications on a radar screen.

range sensing The measurement of a **range** (**7**).

ranging 1. The determination of the **range** (**6**) of an object. **2.** The utilization of a **radio range. 3.** In an instrument with multiple ranges of measurement, switching between intervals. This may be accomplished automatically or manually.

Rankine degree The unit of temperature interval used in the **Rankine temperature scale**. Its symbol is **°R**.

Rankine scale Same as **Rankine temperature scale**.

Rankine temperature scale A temperature scale which is the same as the Fahrenheit scale, except that a temperature of 0 degrees equals absolute zero. Its unit of temperature increment is the degree Rankine, or **°R**. When comparing to the Kelvin temperature scale, 1 °R equals exactly 5/9 kelvin. Zero degrees Rankine equals 0 K, or approximately -459.67 °F. Also called **Rankine scale**.

Rapid Application Development An approach to software development in which prototyping of program segments is utilized. This technique usually employs computer-aided software engineering and user input throughout the various processes. Its acronym is **RAD**.

Rapid Application Development tool Any programming tool utilized for **Rapid Application Development**. Its abbreviation is **RAD tool**.

rapid prototyping The process of generating an actual object directly from its digital representation, usually utilizing CAD/CAM. Used, for instance, in the modeling of concepts. Its abbreviation is **RP**.

rapid quenching Quenching which is performed nearly instantaneously. Quench rates of better than 10^9 K/s may be attained, and such a process may be utilized, for instance, to make crystals, layers, surfaces, or other structures which are far from equilibrium points.

rapid-start lamp A lamp which incorporates a device which provides for immediate illumination upon activation. For instance, a fluorescent lamp whose ballast keeps a low flow of current running through the filaments at all times.

rare earth 1. Same as **rare-earth element. 2.** Pertaining to, or characteristic of, **rare-earth elements**.

rare-earth elements A series of elements, generally considered to range from atomic number 57 (lanthanum) or 58 (cerium), through 71 (lutetium), inclusive. They have related chemical properties, including high reactivity. Such elements are not rare, and frequently occur together. Also called **rare-earth metals, rare earth** (**1**), or **lanthanides**.

rare-earth magnet A permanent magnet, such as that made from samarium cobalt, which incorporates one or more rare-earth elements.

rare-earth metals Same as **rare-earth elements**.

rare gas Any of six chemical elements which are all gases and highly unreactive. They are all in the same group of the periodic table, and are helium, neon, argon, krypton, xenon, and radon. Also called **noble gas**, or **inert gas**.

rarefaction A reduction in the density of a fluid, such as air, caused by the passage of a sound wave. Also, the region where this occurs.

rarefied gas 1. A gas whose pressure is much lower than that of the atmosphere. **2.** A gas at a very low pressure.

RARP Same as **reverse address resolution protocol**.

RAS Abbreviation of **remote-access server**.

raster A pattern of scanning lines which cover the viewing area of a display, such as that of a computer monitor or TV screen. The electron beam follows these lines to produce images. Also called **scanning pattern** (**1**).

raster display A display, such as that of a computer monitor or TV screen, which utilizes a **raster scan** (**1**) to produce images. Also called **raster-scan display**.

raster graphics Images formed by bitmaps. Common formats include JPEG, GIF, and TIFF. This contrasts with **vector graphics**, in which images are represented in the form of equations and graphics primitives. Also called **raster images, bitmapped graphics**, or **bitmap images**.

raster image processing The utilization of a **raster image processor**. Its acronym is **RIP**.

raster image processor A hardware device and/or software product that converts vector graphics into raster graphics. Used, for instance, in printers or plotters. Its acronym is **RIP**.

raster images Same as **raster graphics**.

raster line One of multiple lines within a **raster**.

raster scan Also called **raster scanning. 1.** A technique or technology for generating, recording, or presenting images, in which a beam is swept across a **raster**. In a CRT, for instance, the intensity of the electron beam is varied depending on the color and brightness a given pixel will have within the overall image. **2.** In radars, horizontal or vertical scanning one line at a time to obtain the desired coverage, such as that having a desired shape, or encompassing a given plane.

raster-scan display Same as **raster display**.

raster scanning Same as **raster scan**.

rasterization The process of converting vector graphics into raster graphics, using a **raster image processor**. Also called **vector-to-raster** (**2**).

rasterize To convert vector graphics into raster graphics, using a **raster image processor**. Also called **vector-to-raster** (**1**).

rat race Same as **ring junction**.

rate 1. A quantity or change in quantity measured with respect to time. For example, a speed of 1000 km/s, or a frequency of 1000 Hz. **2.** A quantity or change in quantity measured with respect to the same or another quantity. For instance, a bit error rate. **3.** A quantity measured with respect to a whole, or to a given standard. For instance, a committed information rate. **4.** To establish or specify a rating, such as a power rating. **5.** The cost per unit time, or per other quantifiable measure. For example, the cost per hour or per kilometer.

rate action In a control system, a corrective action whose speed is dictated by the rate of change of the controlled variable. Also called **derivative action**.

rate adaptive Pertaining to a communications technology, such as Rate-Adaptive Digital Subscriber Line, in which the communications line or circuit is tested to determine the most favorable transmission rate.

Rate-Adaptive Digital Subscriber Line A DSL technology in which the modem verifies certain aspects of the line, such as length and quality, and adapts its operating speed to maximize data-transfer speed. Its acronym is **RADSL**. Its abbreviation is **Rate-Adaptive DSL**.

Rate-Adaptive DSL Abbreviation of **Rate-Adaptive Digital Subscriber Line**.

rate control In a control system, the regulation of the rate at which one or more output quantities are varied, so as to keep them within specified parameters.

rate of decay A quantification of the rate at which decay occurs. For example, the rate at which sound is extinguished in a reverberation chamber. Also called **decay rate**.

rated accuracy The accuracy of the measured or calculated values of a device or instrument, when used under given conditions, as expressed by the manufacturer.

rated capacitance A rated or named value, stating the capacitance of a capacitor when used in a given manner. The actual value, or interval of values, may not coincide with this number. Also called **nominal capacitance**.

rated capacity A value or quantity establishing the maximum amount that can be contained, accommodated, or handled. For instance, the charge that can be withdrawn from a fully charged battery under specified conditions, the maximum current that can be handled by an electric circuit, or the largest number of communication channels that a communications circuit can handle simultaneously. Also called **maximum capacity**, or **full-rated capacity (1)**.

rated conditions For a given component, circuit, device, piece of equipment, or system, the operating conditions, such as voltage, current, temperature, and/or humidity, required for a given **rating**.

rated current 1. A rated or named value, stating the current that a component, circuit, device, piece of equipment, system consumes, produces, or safely tolerates, when used in a given manner. For example, it may be a rating of the current drawn expressed in RMS amperes after a device is turned on and warmed up. The actual value, or interval of values, may not coincide with this number. Also called **nominal current**. 2. The maximum current that can be applied continuously under specified conditions without harming a component, circuit, device, piece of equipment, or system. Also called **maximum current (2)**.

rated frequency A rated or named value, stating the center frequency of a crystal or crystal unit, such as that of quartz. The actual value, or interval of values, may not coincide with this number. Also called **nominal frequency**.

rated impedance A rated or named value, stating the impedance of a circuit or device when used in a given manner. The actual value, or interval of values, may not coincide with this number. For instance, the nominal impedance of a speaker may be 8 ohms, but depending on the frequency reproduced may dip below 3, or exceed 50 ohms. Also called **nominal impedance**.

rated load The maximum power that is consumed by a component, circuit, device, piece of equipment, or system that is connected to a source of electricity, while safely performing its functions.

rated output Also called **maximum output**. 1. The peak output value, such as that of a current or voltage, of a component, circuit, device, piece of equipment, or system. 2. The **rated output (1)** for a given period of time. 3. The **rated output (1)** that can be handled safely.

rated power 1. A rated or named value stating the power that a component, circuit, device, piece of equipment, or system can produce, consume, dissipate, or otherwise safely handle, when used in a given manner. For instance, the nominal power of a speaker may indicate the RMS power that can be delivered to a dynamic speaker for extended periods, without harming the voice coil or other components. The actual value, or interval of values, may not coincide with this number. Also called **nominal power**, or **nominal power rating**. 2. The maximum power a component, circuit, device, piece of equipment, or system can handle safely.

rated power output The highest power output level a component, circuit, device, piece of equipment, or system can handle safely. Also called **maximum power output (3)**.

rated resistance A rated or named value, stating the resistance of a resistor when used in a given manner. The actual value, or interval of values, may not coincide with this number. Also called **nominal resistance**.

rated voltage 1. A rated or named value, stating the available, input, or output voltage of a component, circuit, device, piece of equipment, or system, under specified conditions. For instance, the stated AC line voltages for a given locality

may be 115/230, where the lower number is primarily for lighting and small appliances, and the larger number is for heating and large appliances. The actual values, or interval of values, may not coincide with these numbers. Also called **nominal voltage**, or **nominal voltage rating**. 2. The maximum voltage that can be applied continuously under specified conditions without harming a component, circuit, device, piece of equipment, or system.

rated wattage The **rated power** expressed in watts.

rating 1. A rated or named value, or interval of values, describing a property or the behavior of a component, circuit, device, piece of equipment, or system operated under given conditions. For instance, a rated frequency. The actual observed value, or interval of values, may not coincide with these numbers. For instance, the voltage supplied by an AC power line in a given country. Also called **nominal value**, or **nominal rating**. 2. A rated or named value, or interval of values, describing the operational limits of a component, circuit, device, piece of equipment, or system operated under given conditions. For instance, a power rating.

ratio A relationship between two quantities consisting of one being divided by the other. For example, a signal-to-noise ratio, a charge-mass ratio, an aspect ratio, a cell loss ratio, a compression ratio, or gain expressed as a ratio of an output to an input.

ratio arms In a bridge, such as a Wheatstone bridge, two adjacent arms which serve to indicate ratios of resistance.

ratio control The maintaining of two variables at a given ratio, as seen, for instance, in a ratio control system.

ratio control system A control system in which one variable is kept at a preset ratio with another. Used, for instance, when mixing two component materials.

ratio detector An FM detector that utilizes two diodes, and which converts frequency variations in the input signal voltage into proportional variations in output current. Such a detector demodulates signals without limiter stages.

ratio meter An instrument or device which measures and indicates the ratio of two quantities.

ratio of transformation For a given transformer, the ratio of the primary turns to the secondary turns. Also called **transformer ratio**, **transformation ratio**, or **turns ratio**.

rational number Any real number which can be expressed in the form of **a/b**, where **a** is an integer, and **b** is a non-zero integer.

RAW Abbreviation for **read-after-write**.

raw capacity Also called **native capacity**. 1. In a disk, disc, or tape, the storage capacity prior to formatting. Data, such as control information, reduces this capacity. Also called **unformatted capacity**. 2. In a disk, disc, or tape, the storage capacity prior to compression. Compression frequently increases this amount 50% or more.

raw data 1. Data which has not been processed, or that has not been processed to the full extent intended. 2. Data which is not in any given proprietary format. 3. Data which has been gathered but not analyzed or otherwise acted upon.

rawin The use of radar to determine wind speed and direction, usually by tracking a specially-equipped balloon. It is an acronym for **ra**dar **win**d, or **ra**dio **win**d.

rawinsonde A **radiosonde** which is tracked by a radar, or other device which utilizes radio waves, so as to better monitor wind characteristics.

ray 1. A line of radiant energy, such as light, whose cross section is very small. Also, a representation of such a line. For example, the path of a single photon. 2. A particle or beam of particles emitted or radiated from a source. For instance, alpha rays, beta rays, cathode rays, or gamma rays. 3. A particle or beam of particles whose cross section is small or very small. For example, that utilized in electron-beam lithography. **4.** One of multiple lines diverging from a single point. For example, radii from a circle. Also called **half-line**.

ray tracing A rendering technique in which the paths of all rays from light sources are traced. It takes into consideration factors such as absorption, refraction, and reflections by and through objects and surfaces. It is very computing intensive, yet **radiosity** provides images that are more realistic.

Raydist A radionavigation system which utilizes phase comparisons for the determinations of position.

Rayleigh fading For a mobile antenna or receiver, fading due to the presence of multiple radio signals arriving simultaneously, each via a different path, when there is destructive interference. An example is picket fencing.

Rayleigh scattering Scattering of electromagnetic radiation, such as light, by particles which are much smaller than the wavelength of the propagating radiation. For example, the scattering of solar radiation by particles in the atmosphere makes the sky overhead appear bluish, and sunsets as reddish, as light seen above is comparatively scattered, while that of a sunset is comparatively unscattered.

Rb Chemical symbol for **rubidium**.

RBS **1.** Abbreviation of **Rutherford backscattering**. **2.** Abbreviation of **Rutherford backscattering spectrometry**.

RC **1.** Abbreviation of **remote control**. **2.** Abbreviation of **reflection coefficient**.

RC Abbreviation of **resistance-capacitance**.

RC circuit Abbreviation of **resistance-capacitance circuit**.

RC filter Abbreviation of **resistance-capacitance filter**.

RC meter Abbreviation of **resistance-capacitance meter**.

RC network Abbreviation of **resistance-capacitance network**.

RC oscillator Abbreviation of **resistance-capacitance oscillator**.

RCA cable A cable with **RCA plugs** on each end.

RCA connector Also called **phono connector**. **1.** Same as **RCA plug**. **2.** Same as **RCA jack**.

RCA jack Also called **RCA connector (2)**, **RCA socket**, or **phono jack**. An electric connector with a center opening into which a **RCA plug** with a matching prong is inserted.

RCA plug Also called **RCA connector (1)**, or **phono plug**. A plug with a prong whose diameter is usually 0.125 inches, and whose outer sleeve is 0.375 inches in diameter. The shell is connected to the outer conductor of the attached coaxial cable, and is often slotted to further facilitate connection and disconnection. The inner prong is connected to the center conductor, which carries the signal. Widely utilized to connect audio and video components, such as DVDs, CDs, TVs, and amplifiers. An RCA plug is plugged into an **RCA jack**.

RCA socket Same as **RCA jack**. Also called **phono socket**.

RCC Abbreviation of **radio common carrier**.

RCD Abbreviation of **residual current device**.

RCL Abbreviation of **resistance-capacitance-inductance**.

RCP Abbreviation of **remote copy protocol**.

RCR Abbreviation of **reverse-current relay**.

RCS Abbreviation of **radar cross section**.

rd **1.** Abbreviation of **rad (2)**. **2.** Abbreviation of **rutherford**.

Rd Abbreviation of **rutherford**.

RDBMS Same as **relational database**. Abbreviation of **relational database management system**.

RDF **1.** Abbreviation of **radio direction finder**. **2.** Abbreviation of **radio direction finding**. **3.** Abbreviation of **Resource Description Framework**.

RDF station Abbreviation of **radio direction-finding station**.

RDRAM Abbreviation of **Rambus DRAM**, or **Rambus dynamic random-access memory**. A RAM technology enabling transfers exceeding ten billion bytes per second, and which supports pipelining. Also, RAM utilizing this technology.

r_e Symbol for **classical electron radius**.

Re Chemical symbol for **rhenium**.

re-activate Same as **reactivate**.

re-activation Same as **reactivation**.

re-assembly Same as **reassembly**.

re-broadcast Same as **rebroadcast**.

re-format Same as **reformat**.

re-formatting Same as **reformatting**.

re-initialization Same as **reinitialization**.

re-initialize Same as **reinitialize**.

re-initiate Same as **reinitiate**.

re-initiation Same as **reinitiation**.

re-install Same as **reinstall**.

re-load Same as **reload**.

re-modulation Same as **remodulation**.

re-modulator Same as **remodulator**.

re-radiation Also spelled **reradiation**. **1.** Subsequent radiation of RF energy which has been received from an incident wave. The energy which is subsequently radiated may or may not of the same frequency or interval of frequencies. Used, for instance, for capturing a satellite signal and radiating the obtained data to multiple receivers. **2.** Undesired **re-radiation (1)**. This may produce interference. **3.** The radiating of a surface, body, or region, which has already been radiated upon.

re-record Also spelled **rerecord**. Also called **dub**. **1.** For a given audio and/or video recording, to add or replace sounds, or one or more sound tracks. For example, in a movie, to add sound effects or replace dialogue with that of another language. **2.** To transfer of all or part of one recording to another. **3.** To mix two or more sound sources into a single recording. **4.** To record on one device what it being reproduced through another

re-recording Also spelled **rerecording**. Also called **dubbing**. **1.** For a given audio and/or video recording, the adding or replacing of sounds, or of one or more sound tracks. For example, in a movie, the adding of sound effects or the replacement of dialogue with that of another language. **2.** The transfer of all or part of one recording to another. **3.** The mixing of two or more sound sources into a single recording. **4.** The recording on one device what is being reproduced through another.

re-route Same as **redirect**. Also spelled **reroute**.

re-router Same as **redirector**. Also spelled **rerouter**.

re-routing Same as **redirection**. Also spelled **rerouting**.

re-sample Same as **resample**.

re-sampling Same as **resampling**.

re-start Same as **restart**.

re-transmission Same as **retransmission**.

re-transmission system Same as **retransmission system**.

re-transmit Same as **retransmit**.

re-transmitter Same as **retransmitter**.

re-writable Same as **rewritable**.

re-writable disc Same as **rewritable disc**.

re-writable optical disc Same as **rewritable disc**. Also spelled **rewritable optical disc**.

re-write Same as **rewrite**.

reactance The imaginary number component of **impedance**. It is the part of the total impedance which is not due to pure resistance. When reactance is due to a pure capacitance it is called **capacitive reactance**, and when due to a pure inductance it is an **inductive reactance**. Expressed in ohms. Its symbol is X.

reactance modulator A modulator whose reactance changes in response to an input signal. Used, for instance, in FM.

reactivate Also spelled **re-activate**. **1.** To start a sequence or process again. For example, to enter a computer command to initiate a sequence. **2.** To make operational again. For example, to recharge a secondary battery. **3.** To again treat a substance or material to make it more effective for a designated purpose. For example, the treatment of a cathode to improve its thermionic emission once it has dropped below a given threshold.

reactivation Also spelled **re-activation**. **1.** The initiation of a sequence or process again. For example, the entering of a computer command to repeat a sequence. **2.** The process of making operational again. For example, the recharging of a secondary battery. **3.** The process of again treating a substance or material to make it more effective for a designated purpose. For example, the treatment of a cathode to improve its thermionic emission once it has dropped below a given threshold.

reactive **1.** Tending to easily form part of a chemical reaction. For example, fluorine is the one of the most reactive of all chemical elements, and forms compounds with almost every other element, usually quite vigorously. **2.** Tending to respond to an action or stimulus. Also, tending to respond to an action or stimulus more than others within the same category or group. **3.** Pertaining to, causing, utilizing, or arising from **reactance**.

reactive component In an AC circuit, the component of the current, voltage, or power which does not add power. These are, specifically, the reactive current, reactive voltage, or reactive power. Also called **idle component, wattless component**, or **quadrature component**.

reactive current The component of an alternating current which is in quadrature with the voltage. Such a component does not add power. Also called **idle current, wattless current**, or **quadrature current**.

reactive factor The ratio of the reactive power to the apparent power.

reactive ion-beam etching An etching technique similar to **reactive ion etching**, but in which physical reactions, due to the momentum of the particles, are relied upon more than chemical reactions. Its abbreviation is **RIBE**.

reactive ion etching An etching technique in which high-energy ions are utilized to produce the desired patterns. This process relies more on chemical reactions than physical reactions to perform the etching. Its abbreviation is **RIE**.

reactive load A load with an inductive or capacitive reactance, or which is predominantly reactive.

reactive power The power in an AC circuit which cannot perform work. It is calculated by the following formula: $P = I \cdot V \cdot \sin\theta$, where P is the power in watts, I is the current in amperes, V is the voltage in volts, and $\sin\theta$ is the sine of the angular phase difference between the current and the voltage. Also called **reactive volt-amperes, idle power, wattless power, quadrature power**, or **volt-amperes reactive**.

reactive sputtering Sputtering in which a reactive gas combines with the target material to form a deposited compound. For example, the sputtering of silicon in a plasma containing oxygen resulting in the deposition of silicon dioxide.

reactive volt-amperes Same as **reactive power**.

reactive voltage The voltage component which is in quadrature with the current of an AC circuit. Such a component does not add power. Also called **idle voltage, wattless voltage**, or **quadrature voltage**.

reactivity **1.** The extent to which an atom, molecule, ion, radical, or the like, is **reactive**. **2.** The extent to which something responds to an action or stimulus.

reactor **1.** That which reacts with something else, or that responds to an action or stimulus. **2.** A device which utilizes controlled nuclear fusion or fission to generate energy. A fission reactor, for instance, has the following key components: a fissionable fuel, such as uranium-235, a moderator, such as heavy water, shielding, such as that provided by thick concrete, control rods, such as those made with boron, and a coolant such as water. The heat produced by the nuclear reactions can be utilized to generate electricity. Also called **nuclear reactor**, or **atomic reactor**. **3.** A component, circuit, or device which introduces a reactance to a circuit. Capacitors and inductors are examples.

read **1.** To access, copy, or otherwise locate and transfer data from a memory location, storage device, magnetic stripe, or the like. Examples include copying data from a hard disk to RAM, from RAM to the CPU, from a printer to RAM, or from a magnetic card to a magnetic card reader. The converse is a **write (1)**. **2.** Same as **read operation**. **3.** To indicate, register, or otherwise display. For example, a value displayed by a meter. Also, to view such a reading. **4.** To comprehend a received radio message.

read-after-write Its abbreviation is **RAW**. In a tape drive, the verification of data after it is written. Also, the technique utilized for this. This contrasts with **read-while-write (2)**, where only a single pass is utilized.

read command A computer instruction to perform a **read operation**.

read cycle The complete sequence of events required to perform a **read operation**. Also, a single such cycle.

read cycle time The time interval that elapses between the starts of successive **read cycles**.

read error **1.** An attempted read operation which is not successful. Possible causes include corrupted storage locations, program errors, or a malfunction of a read/write head. **2.** A read operation in which there are one or more erroneous bits delivered or transferred.

read head A transducer which converts magnetic signals recorded or stored on magnetic recording media, such as tapes or disks, into output electrical signals. When used in the context of tape which has audio and/or video, the signals are usually said to be reproduced, while in the context of data, the signals are usually said to be read. Also called **reading head, reproducing head, reproduce head, magnetic read head, magnetic reading head**, or **playback head**.

read-in To place data into a storage location.

read notification Same as **read receipt**.

read-only Pertaining to memory, storage, data, files, disks, or the like, whose content can be read, but not changed. For example, ROM, or read-only files.

read-only access Access to memory, storage, data, files, disks, or the like, in which read operations are permitted, but no writing is allowed. Such a file may or may not be able to be deleted.

read-only attribute A file attribute that imposes restrictions on its ability to be changed. Such a file may or may not be able to be deleted.

read-only file A file whose content can be read, but not changed. Such a file may or may not be able to be deleted or copied.

read-only memory Same as **ROM**.

read-only memory basic input/output system Same as **ROM BIOS**.

read-only memory card Same as **ROM card**.

read-only memory cartridge Same as **ROM cartridge**.

read-only memory chip Same as **ROM chip**.

read-only memory emulator Same as **ROM emulator**.

read-only memory socket Same as **ROM socket**.

read-only storage 1. Same as **ROM**. 2. Stored data which can not be altered. Also called **fixed storage**.

read operation The act or operation of performing a **read (1)**. A destructive read causes accessed data to be destroyed, while a non-destructive read does not. Also called **read (2)**.

read-out Same as **readout**.

read-out device Same as **readout device**.

read receipt A confirmation or notification given to a sender that a given email has been read or opened by a recipient. Also called **read notification**.

read the manual Its abbreviation is **RTM**. A suggestion to read the documentation, and/or online information, which is available for a given software or hardware product. This does not necessarily involve a frustration-free experience, which may lead to its more colorful abbreviation, **RTFM**. The reader may deduce what the **F** stands for.

read time The time interval that elapses between the initiation of a **read operation** and the delivery or transfer of the desired information.

read-while-write Its abbreviation is **RWW**. 1. In a memory or storage device, such as flash memory, the ability to perform a read operation in one part of memory or storage, while performing a write operation in another. Also, the technique utilized for this. 2. In a tape drive, the ability to verify data on the same pass as it is written. Also, the technique utilized for this. This contrasts with **read-after-write**, where a second pass is required.

read/write Its abbreviation is **R/W**. 1. Pertaining to that which can simultaneously perform both read operations and write operations, or have them performed upon it. 2. Pertaining to that which can perform both read operations and write operations, or have them performed upon it. Such operations need not occur concurrently.

read/write cycle A complete cycle encompassing a read operation and a write operation. These operations may or may not occur simultaneously. Its abbreviation is **R/W cycle**.

read/write head Its abbreviation is **R/W head**. 1. A magnetic tape or disk head which both reads and writes to the magnetic medium. For instance, such a head in the disk drive of a computer. Also called **combined head**. 2. A transducer which converts electrical signals into magnetic signals which can be recorded or stored on magnetic recording media, such as tapes or disks, and which also converts magnetic signals recorded or stored on said media into output electrical signals, and which can also delete such recorded material. Also called **magnetic head (4)**, or **magnetic read/write head**.

read/write memory Same as **RAM**.

read/write RAM Same as **RAM**.

reader That which is capable of performing read operations, or otherwise obtaining information or other content. For example, a read head, a bar-code reader, a card reader, or a document reader.

reading 1. The performance of a **read (1)**, **read (2)**, or **read (3)** operation. 2. The information conveyed via a **read (3)**. Also, to take such a reading.

reading head Same as **read head**.

readme Abbreviation of **README file**.

README Abbreviation of **README file**.

README file A file containing information which supplements the documentation which accompanies a software or hardware product. Frequently it is a plaintext file, and on occasions may be the only information readily available for a given offering. Its abbreviation is **README**.

readme file Same as **README file**.

readout Also spelled **read-out**. 1. The visual output shown on a display device. 2. The viewing screen on a display device. 3. The visible output of a device such as a computer, meter, calculator, or analytical instrument. This includes, for instance, that which is shown on a display, through a series of LEDs, via an indicating needle, or through a printer.

readout device A device, such as a monitor or analytical instrument, which provides a **readout (1)**, or **readout (3)**. Also spelled **read-out device**.

ready A message indicating that a circuit, device, piece of equipment, or system, is available and prepared for the desired use or function.

real address 1. A specific memory location. It is an address from which relative addresses may be derived. Also called **absolute address**, **actual address (1)**, **machine address**, or **direct address**. 2. In virtual memory, a storage location in RAM, as opposed to that in a secondary storage device, such as a hard disk.

real image In an image-forming optical system, an image in which the beam of light converges towards a given point. In a **virtual image (1)**, the beam of light diverges from a given point, producing an image at the apparent source.

real memory The RAM chips a computer has, as opposed to **virtual memory**, which is the memory, as it appears to an application, enhanced by the utilization of secondary storage, such as a hard disk. Also called **real storage (1)**, or **physical memory**.

real number A number which is either rational or irrational, but not imaginary. Zero is also a real number.

real power In an AC circuit, the rate at which work is performed or energy is transferred, measured in watts. It is the product of the voltage portion of the signal being measured, multiplied by the current that is in phase with that voltage. It is also equal to the difference between the apparent power and the reactive power. Also called **active power**, **actual power (1)**, or **true power**.

real storage 1. Same as **real memory**. 2. Within the virtual memory available to a system, the portion which actually uses the RAM of the computer.

real-time Also spelled **realtime**. 1. That which occurs instantaneously, or so quickly that processing, entering, adaptation, or any other response is least as fast as a triggering event or circumstance. Said especially when the process being responded to is characterized by continuous activity, change, or varying parameters. For instance, a real-time control system, or an event-management system. 2. In computers, that which is processed, or otherwise acted upon, without any delay. Examples include real-time transaction processing and the computation of a discrete Fourier transform. 3. In communications, that which occurs without a discernable delay. For instance, telephonic voice conversations, real-time chats, or real-time broadcasts.

real-time animation Animation which is created and updated in real-time. Such animation allows, for instance, for more realistic virtual environments, such as those encountered in certain multi-user games. Also spelled **realtime animation**.

real-time audio Audio which is transmitted over a communications channel in real-time. Internet telephony is an example. Also spelled **realtime audio**.

real-time audio conferencing Audio conferencing in which participants at more than one site can interact in real-time. Also spelled **realtime audio conferencing**.

real-time broadcast A broadcast occurring at the actual time events take place. Also called **live broadcast**, or **live (2)**. Also spelled **realtime broadcast**.

real-time chat A real-time written and/or verbal conversation between two or more individuals via a network such as the Internet. Also spelled **realtime chat**. Also called **chat**.

real-time clock In a computer, an independent circuit which keeps track of the time of day. A real-time clock is not usually used to generate clock pulses. Its abbreviation is **RTC**. Also spelled **realtime clock**. Also called **time-of-day clock**.

real-time compression Compression of data occurring in real-time. Used, for instance, in real-time conferencing. Also spelled **realtime compression**.

real-time conferencing A conference in which participants at more than one site can interact in real-time. Such a conference may incorporate audio, video, written chat, shared files and applications, and so on. Also spelled **realtime conferencing**. Also called **teleconferencing**.

real-time control system A control system in which controlled events or circumstances are reacted to instantaneously, or with negligible delays. Also spelled **realtime control system**.

real-time data Data which is acquired, processed, transmitted, or otherwise acted upon in real-time. Also spelled **real-time data**.

real-time data system Same as **real-time information system**. Also spelled **realtime data system**.

real-time decompression Decompression of data occurring in real-time. Used, for instance, in real-time conferencing. Also spelled **realtime decompression**.

real-time information system An information system which handles processes in real-time. Used, for instance, to track a mass-transit system. Also spelled **realtime information system**. Also called **real-time data system**.

real-time operating system An operating system which handles all inputs and processes in real-time. Such an operating system should also provide for high fault-tolerance and error and exception-handling capabilities, in addition to providing support for non-real-time applications. Also spelled **real-time operating system**. Its abbreviation is **realtime OS**, or **RTOS**.

real-time operation Also spelled **realtime operation**. **1.** In a computer, an operation occurring in real-time. **2.** The operation of a system in which all processing, events, or the like occur in real-time.

real-time OS Same as **real-time operating system**. Also spelled **realtime OS**.

real-time processing Also spelled **realtime processing**. **1.** In a computer, processing which occurs in real-time. **2.** Same as **real-time transaction processing**.

Real-Time Streaming Protocol Its abbreviation is **RTSP**. A protocol for the real-time transmission of audio, video, or multimedia content. Used, for instance, over a TCP/IP network such as the Internet to broadcast a video presentation. **RTP** is similar to RTSP, with the main difference being that the former only allows one-way communication, while the latter is two-way. Thus, RTSP would permit the use of features such as rewinding or accessing chapters directly, while RTP would not.

real-time system Also spelled **realtime system**. **1.** A computer system that processes in real-time. **2.** A system, such as a real-time control system, which operates or reacts in real-time.

real-time transaction processing The processing of transactions the instant each is received by, or entered into, a computer system. Online transaction processing not only maintains all master files constantly current, it also enables entering or retrieving information without delays in either process. Used, for instance in businesses which have multiple locations which accept Internet, telephone, fax, postal, and point-of-sale orders. Also spelled **realtime transaction processing**. Also called **real-time processing (2)**, **transaction processing**, or **online transaction processing**.

Real-Time Transport Protocol Its abbreviation is **RTP**. A protocol for the real-time transmission of audio, video, or multimedia content. Used, for instance, over a TCP/IP network such as the Internet to broadcast a movie. RTP is similar to **RTSP**, with the main difference being that the former only allows one-way communication, while the latter is two-way. Thus, RTSP would permit the use of features such as rewinding or accessing chapters directly, while RTP would not. Also called **RTP protocol**.

real-time variable bit rate Same as **real-time VBR**.

real-time VBR Abbreviation of **real-time** variable bit rate. In ATM, variable bit rate service which supports real-time applications such as audio and/or video conferencing. Its own abbreviation is **rt-VBR**.

real-time video Full-motion video which is transmitted over a communications channel in real-time. Compression and specialized protocols may be employed to reduce the required bandwidth, which would otherwise be comparatively large. Also spelled **realtime video**.

real-time video conferencing Videoconferencing in which participants at more than one site can interact in real-time. Also spelled **realtime video conferencing**.

RealAudio A popular format for streaming audio.

reality check A simple test performed on software and/or hardware, so as to ascertain whether certain basics are present and operational. For example, hardware may be subjected to a smoke test, or an application may be required to display written text followed by its being printed.

RealMedia A popular streaming media technology which supports audio, video, animations, text, and other media formats.

RealPlayer A browser plug-in that enables playback of RealAudio, RealVideo, and multimedia content in other formats, such as DVDs.

RealServer Server software which enables streaming of RealAudio, RealVideo, and other media formats.

realtime Same as **real-time**.

realtime animation Same as **real-time animation**.

realtime audio Same as **real-time audio**.

realtime audio conferencing Same as **real-time audio conferencing**

realtime broadcast Same as **real-time broadcast**.

realtime chat Same as **real-time chat**.

realtime clock Same as **real-time clock**.

realtime compression Same as **real-time compression**.

realtime conferencing Same as **real-time conferencing**.

realtime data Same as **real-time data**.

realtime data system Same as **real-time information system**. Also spelled **real-time data system**.

realtime decompression Same as **real-time decompression**.

realtime information system Same as **real-time information system**.

realtime operating system Same as **real-time operating system**.

realtime operation Same as **real-time operation**.

realtime OS Same as **real-time OS**.

realtime processing Same as **real-time processing**.

realtime system Same as **real-time system**.

realtime transaction processing Same as **real-time transaction processing**.

realtime video Same as **real-time video**.

realtime video conferencing Same as **real-time video conferencing**.

RealVideo A popular format for streaming video.

Réamur scale Same as **Réamur temperature scale**.

Réamur temperature scale An obsolete temperature scale in which water freezes at 0 degrees and boils at 80 degrees. Also called **Réamur scale**.

rear projection The projection, onto a translucent screen, of images sent from a separate unit, and in which the projector and any viewers are on opposite sides of said screen. Used, for instance, in rear-projection TVs. This contrasts with **front projection**, where the images are viewed on a non-translucent screen, and the projector and viewers are on the same side of said screen.

rear-projection television Same as **rear projection TV**.

rear projection TV Abbreviation of **rear-projection** television. A TV with **rear projection**.

rear-surface mirror A mirror whose reflective coating is on its rear surface, as is the case with a household mirror. This contrasts with **front-surface mirror**, in which the reflective coating is on the front surface. Also called **second-surface mirror**, or **back-surface mirror**.

rear wave An acoustic wave which is radiated by the back of a loudspeaker, which can cancel the waves radiated from the front of the loudspeaker. An infinite baffle completely isolates the back waves from the front waves, while a port helps both waves to be in phase. Also called **back wave**.

reassembly In a packet-switched network, the reconstitution of a packet at the receiving end, after **segmentation** at the transmission end. Also spelled **re-assembly**.

rebecca In the **rebecca-eureka system**, the airborne interrogator.

rebecca-eureka Same as **rebecca-eureka system**.

rebecca-eureka system A radar homing system which utilizes an airborne interrogator, called rebecca, and a ground transponder beacon, called eureka. Also called **rebecca-eureka**.

reboot To reset a computer. During this process the computer accesses instructions from its ROM chip, performs self-checks, loads the operating system, and prepares for use by an operator. A rebooting may be initiated by pressing a button or switch, by hitting a specific key sequence, or through a program or routine that gives this command. It is an abbreviation of **rebooting**.

rebooting Same as **reboot**.

rebroadcast Also spelled **re-broadcast**. **1.** To repeat or reemit a broadcast. **2.** A broadcast that is repeated or relayed by a station other than that emitting the original broadcast.

rebuild Also called **reconstruct**, or **remodel**. **1.** To make repairs that are so extensive that it is equivalent to building again. **2.** To build again.

recall **1.** To restore to a former condition. Also, the act of restoring to a former condition. **2.** In computers, a retrieval of information. Also, the act of retrieving information. **3.** To request that a defective manufactured product be returned for adjustment, repair, or disposal. Also, the process of informing of such a recall, and making the changes. **4.** Same as **redial**.

receipt notification A confirmation or notification given to a sender that a given email has been received or opened by a recipient.

received power **1.** The power an antenna receives from a transmitter or other signal source. **2.** The power a device receives from a transmitter or other signal source.

receiver Its abbreviation is **RX**. **1.** A component, device, piece of equipment, or system which accepts information-bearing signals, and which can extract the meaningful information contained. There are many types of receivers, including those utilized in communications, TV and entertainment, radars, and so on. Signals may arrive from land-based antennas, satellites, remote controls, and so on. **2.** A single audio-frequency component which incorporates a preamplifier, a power amplifier, and a tuner. Such a component usually has multiple inputs for CDs, DVD, TVs, tape decks, and so on, and may have circuitry for specialized sound reproduction, such as Dolby surround sound. **3.** Same as **radio receiver**. **4.** A small loudspeaker located in the handset of a telephone, which enables listening. Also called **earphone (3)**, or **telephone receiver (1)**. **5.** That which is a destination or which otherwise accepts signals, energy, particles, waves, or the like, which move from one point to another.

receiver bandwidth **1.** The interval of frequencies within which a receiver can produce a given proportion of its maximum output. Usually calculated at 50% or 90% of full power. **2.** The interval of frequencies within which the performance of a receiver falls within certain limits.

receiver muting In a transceiver or transmitter-receiver, the reduction or silencing of the receiving input while transmitting. This may occur manually or automatically.

receiver noise **1.** The electrical noise generated by a receiver in the absence of an input signal. **2.** Any noise generated by a receiver.

receiver noise factor For a specified bandwidth, the ratio of the total noise output to the total noise input for a given receiver. It is the additional noise a receiver adds to any signal it accepts. Usually expressed in decibels, in which case it is called **receiver noise figure**.

receiver noise figure The **receiver noise factor** expressed in decibels.

receiver off-hook The condition where a telephone is not in use, yet the handset is removed from its cradle. This may occur, for instance, by not hanging up after a call. There is usually a loud warning tone to indicate that this condition exists. Its abbreviation is **ROH**.

receiver primaries Colors which are combined in an additive mixture to yield a full range of colors to be displayed by a color TV receiver. Red, green, and blue are most commonly used as the additive primary colors. Also called **display primaries**.

receiver selectivity **1.** The degree to which a receiver can differentiate between a desired signal and other signals. **2.** The degree to which a receiver rejects the signals of channels adjacent to that which is desired. Also called **adjacent-channel selectivity**.

receiver sensitivity **1.** For a receiver, the minimum input signal level which will produce a discernable output signal. **2.** For a receiver, the minimum input signal level which will produce an output signal with a signal-to-noise ratio equal to or greater than a given value.

receiver set Same as **radio receiver**.

receiver synchro In a synchro system, the synchro which converts the voltage received from the synchro transmitter into the corresponding angular position of its stator. Also called **synchro receiver**.

receiving antenna An antenna which picks up electromagnetic radiation. There are many types of receiving antennas, including those utilized in communications, TV and entertainment, radars, and so on. A single antenna may be able to be used for both transmission and reception.

receiving device A device, such as a GPS Receiver, a beacon receiver, or a receiving fax, which picks up or receives a signal emitted or sent by a transmitting device utilizing the same system.

receiving equipment Equipment, such as radar sets, which incorporate one or more **receiving devices**.

receiving set Same as **radio receiver**.

receiving station A station, such as an automatic repeater station, or a direction-finding station, which incorporates one or more **receiving devices**.

receiving tube A tube specially designed for use in **radio receivers**.

receptacle A power line termination whose socket serves to supply electric power to devices or equipment whose plug is inserted into it. Receptacles are usually mounted in a wall, although they may be found elsewhere, such as a floor, or in the back of another electrical device, such as an amplifier. Also called **convenience outlet**, **outlet**, **electrical outlet**, **power outlet**, or **power receptacle**.

reception Also called **signal reception**. **1.** The conversion of information-bearing signals, such as those conveyed via electromagnetic waves, into the signals of interest, such as music, images, or data. For example, the use of a receiver to demodulate an FM signal, so as to recover the original music. **2.** The quality or fidelity attained during **reception (1)**.

reception mode In a transceiver, operation in which signals are received. This contrasts with **transmission mode**, in which signals are sent. Switching between the two modes may be automatic or manual.

recharge The replenishing of the energy available to a rechargeable battery or cell. Such a charge may be partial, complete, or floating.

rechargeable **1.** Having the ability to accept **recharges**. **2.** Pertaining to **rechargeable batteries** or **rechargeable cells**.

rechargeable battery A battery composed of **rechargeable cells**. Examples include lithium polymer batteries, lead-acid batteries, and gel-cell batteries. Also called **accumulator battery**, **storage battery**, or **secondary battery**.

rechargeable cell A cell whose available energy can be replenished after discharge, with the process being able to be repeated safely multiple times. Examples include lithium-ion cells, and nickel-metal hydride cells. This contrasts with a **non-rechargeable cell**, whose manner of discharge does not lend itself to recharging. Also called **storage cell (1)**, or **secondary cell**.

recharger That which serves to **recharge**.

reciprocal **1.** A number or quantity which is related to another in that the product of both is 1. For example, the reciprocal of 5 is 1/5, or the reciprocal of R = 1/R. **2.** Interchangeable in one or more manners.

reciprocal ohm An older term for a unit equivalent to **siemens**.

reciprocity The maintaining of a **reciprocal** relationship or state.

reciprocity theorem **1.** Any theorem in which two quantities, variables, components, states, or the like, are interchangeable in one or more manners. **2.** A theorem stating that if a voltage source at one point produces a given current at another point in an electric network, then the source and current may be interchanged.

recirculation air Air that flows from a clean room, through a filtering system, and back into the clean room.

reclosing circuit breaker A circuit breaker which, under given circumstances, automatically closes a circuit after being opened.

reclosing relay A relay which, under given circumstances, automatically closes after being opened.

reclosing switch A switch that, under given circumstances, automatically closes after being opened.

recombination **1.** The process by which an electron which had jumped to the conduction band, to create an electron

hole, returns to the valence band. This results in the electron hole recombining with said electron. This is accompanied by a release of energy, such as radiation. Also called **electron-hole recombination**. **2.** The process by which a particle or object is formed by the combination of charged particles or objects. For example, the formation of a neutral molecule via the combination of a cation and an anion.

recombination current Current flow resulting from **recombination (1)**.

recombination process Any process via which **recombination** occurs.

recombination radiation In semiconductor, radiation emitted as a consequence of **recombination (1)**. Such radiation is responsible for the light output of diode lasers.

recombination rate In a semiconductor, the rate at which **recombination (1)** occurs. This contrasts with a **generation rate**, which is the rate at which electron-hole pairs are generated.

recompile To compile a computer program again, as may be done after changes are incorporated.

reconstruct Same as **rebuild**.

record **1.** To write, store, or otherwise preserve images, sounds, data, signals, or any other form of information, especially when in a permanent form suitable for later reproduction. This includes recording upon magnetic tapes, magnetic disks, optical discs, or wires, or producing papers, photographs, faxes, or the like, with content. **2.** A group of related fields, each containing information. For instance, a group of fields, each containing one of the following items: a name, a corresponding address, and a contact number. A collection of fields form a record, and a collection of records form a file. Also called **data record**. When used in the context of relational databases, also called **tuple**. **3.** A set of data elements treated as a unit. **4.** A disc, usually composed of vinyl, which has sound recorded as variations in its spiral grooves. A phonograph is utilized to extract the sound so recorded. Also called **phonograph record**, **phonographic record**, or **phonograph disc**. **5.** The printed images provided by a graphic recorder, such as an oscillograph.

record block A group of computer records stored and transferred as a unit. Also called **block (4)**.

record blocking The formation of **record blocks**.

record count The number of records in a data structure such as a file.

record deck In dubbing, the deck to which signals are transferred or copied, and whose recorded signals originate from a **source deck (2)**.

record format Same as **record layout**.

record gap On a storage medium, such as a magnetic tape or disk drive, the unused physical space between consecutive blocks of recorded data. Also called **inter-record gap**, **inter-block gap**, or **block gap**.

record head Same as **recording head**.

record layout Also called **record format**, or **record structure**. **1.** The manner in which a record is organized from the user's perspective. For example, a user may view a record as a single unit, while it may actually consist of many data fields in a database. Also called **logical record**. **2.** The true distribution of a record, as opposed to the manner in which a record is organized from the user's perspective. For example, a user may view a record as a single unit, while it may actually consist of many data fields in a database. Also called **physical record**.

record length The length of a **record (2)** or **record (3)**, expressed in bits, bytes, characters, fields, or another unit of data. Also called **record size**.

record locking A technique employed to help maintain data integrity within a multiuser environment. When a user accesses any given record, all other users are blocked from doing so. Once the accessed record is updated, it again becomes freely available.

record mark A symbol or code indicating the start or end of a data record.

record number Within a database, file, or other data structure, a number which uniquely identifies a record.

record player An obsolescent device which reproduces sound which has been recorded in the spiral grooves of a disc called phonograph record. A record player incorporates a motor-driven turntable which rotates a platter at a constant speed. The disk is placed on the platter, and a tone arm, which holds the cartridge, travels progressively towards the center of the disk. The stylus of the cartridge follows the variations in the grooves, and converts them into electrical signals which are fed into an amplifier, preamplifier, receiver, or other such device. Also called **phonograph**, or **turntable** (1).

record separator character A character that serves as a logical boundary between **records** (2) or **records** (3).

record size Same as **record length**.

record structure Same as **record layout**.

recorded tape A magnetic tape which has been recorded or written upon. Also called **prerecorded tape**.

recorder An instrument or device which makes **recordings**. Examples include DVD recorders, tape recorders, fax recorders, and oscillographs. Also called **recording instrument**.

recording 1. The process of writing, storing, or otherwise preserving images, sounds, data, signals, or any other form of information, especially when in a permanent form suitable for later reproduction. This includes recording upon magnetic tapes, magnetic disks, optical discs, or wires, or producing papers, photographs, faxes, or the like, with content. Also, that which has been written, stored, or otherwise preserved. 2. The process of tracing, or otherwise printing the output of a graphic recorder, such as an oscillograph. Also, the recording so produced.

recording channel 1. A channel, or signal path, that has been recorded. 2. A channel, or signal path, designated for or available for recording.

recording density For a given storage medium, such as a magnetic tape or disk, the number of bits, or other storage units, that can be recorded per unit length or area.

recording disc A disc, such as an optical disc, upon which **recordings** (1) are made.

recording disk A disk, such as a hard disk, upon which **recordings** (1) are made.

recording head A transducer which converts electrical signals into magnetic signals which can be recorded or stored on magnetic recording media, such as tapes or disks. Also called **record head**, **magnetic head** (1), or **write head**.

recording instrument Same as **recorder**.

recording level 1. For an analog or digital tape, the amplitude of the input signal which is being recording. 2. For an analog or digital tape, the amplitude of the input signal which can be recording without exceeding a specified level of distortion.

recording medium A medium, such as a recording disk, upon which **recordings** (1) are made.

recording noise Noise introduced during **recording** (1).

recording/playback head A head which performs both recording and reproducing functions. Also called **recording/reproducing head**.

recording/reproducing head Same as **recording/playback head**.

recording spot The spot where recording, such as that onto an optical disk or fax sheet, takes place at a given instant.

recording tape 1. A magnetic tape utilized for **recording** (1). Such a tape usually consists of a strip of plastic, paper, or metal which is coated or impregnated with a magnetizable material such as iron oxide particles. Also called **magnetic tape**, or **tape** (2). 2. Any tape suitable for **recording** (1).

recording thermometer A thermometer which makes a record of the temperatures it measures. Also called **thermograph** (1).

recover 1. To reset or otherwise re-establish a former state or condition, a given operational state, or a given level or value. 2. To **recover** (1) after an error or malfunction.

recoverable error An error, such as that in a computer program, which can be simply remedied without further deleterious effects.

recovery 1. The process of resetting or otherwise re-establishing a former state or condition, a given operational state, or a given level or value. Also, the result of such a process. 2. A **recovery** (1) after an error or malfunction. Also, the result of such a process. 3. The restoration of computer data to the state it was in before an error or failure.

recovery delay Same as **recovery time**.

recovery system Any system, such as that which includes preparing, storing, and accessing backup disks or tapes, utilized for data recovery. Also called **data-recovery system**.

recovery time Also called **recovery delay**. 1. For a component, circuit, device, piece of equipment, system, mechanism, or process, the time interval that must elapse after a given event or condition, before ordinary operation may resume or a specified operational state is attained. For example, the time required for the return of the electron beam to its starting point after a sweep or trace, or the time that elapses between a forward bias being applied, and the forward current reaching a specified value. 2. The time it takes a component, circuit, device, piece of equipment, or system, to resume ordinary operation after an overload. Some devices can recover within a nanosecond after being overdriven. Also called **overload recovery time**. 3. The time interval after a signal or event, during which a device, piece of equipment, or system is unable to respond to additional signals or events. During this time, for instance, a radiation counter can not register ionizing radiation. Also called **dead time** (3), or **insensitive time**.

Recreational Software Advisory Council An organization that established and promoted ratings for Internet content, and which became a part of the **Internet Content Rating Association**. Its abbreviation is **RSAC**.

rectangle A quadrilateral with four right angles and two pairs of sides of equal length. When all four sides are equal, it is a **square**.

rectangular axis Each of the mutually perpendicular lines which intersect at a common point called the origin, and which is used within a **rectangular coordinate system**. When representing three planes, for instance, each axis is designated by the letters **x**, **y**, and **z**, respectively. Also called **Cartesian axis**.

rectangular cavity A cavity resonator in the shape of a **rectangle**.

rectangular-coordinate robot A robot whose movements are along Cartesian, or rectangular, axes. For example, a manipulator may be able to move along the **x**, **y**, and **z** axes. Used in robotic applications such as assembly, gluing, or arc welding. Also called **Cartesian-coordinate robot**.

rectangular coordinate system Also called **Cartesian coordinate system**. A coordinate system in which the locations of points are given in reference to the axes, numbering two or more. Each of these coordinate axes is perpendicular to the others, and all intersect at a common point called the origin. Each coordinate axis, or Cartesian axis, represents a plane called a Cartesian plane. A number of lines, equal to the number of planes, may be drawn from the origin to any given point on a plane or in space, and the exact location of this point can be specified by giving the coordinates with respect to the axes.

rectangular coordinates Each of the sets of numbers which locates a point on a plane or in space, according to a **rectangular coordinate system**. If there are two planes, for instance, then a point is located by specifying the position in reference to each of the axes, one that is usually horizontal and denominated **x**, and one that is usually vertical and denominated **y**. Also called **Cartesian coordinates**.

rectangular pulse A pulse whose amplitude varies between two essentially steady values, with transitions between said values occurring instantaneously. When the pulse duration of each of these fixed values is the same, it is a **square pulse**.

rectangular wave A wave whose amplitude varies between two essentially steady values, with transitions between said values occurring instantaneously. When the pulse duration of each of these fixed values is the same, it is a **square wave**.

rectangular-wave generator A signal generator whose output consists of **rectangular waves**.

rectangular waveguide A waveguide whose cross section is in the form of a **rectangle**.

rectenna Acronym for **rec**tifying an**tenna**. An antenna that converts received microwave energy into a direct current. Such antennas tend to be rather large, and may be used, for instance, for power generation.

rectification 1. The process of converting an AC into a unidirectional current, especially DC. Also, the result of such a process. 2. Any process, such as demodulation, a **rectifier** performs. Also, the result of such a process.

rectified current Electrical current which has undergone **rectification** (1).

rectifier 1. A component, circuit, or device which passes current in one direction only. Although semiconductor rectifiers are prevalent, there are also electrolytic and mechanical devices which perform this function, in addition to rectifier tubes, among others. Rectifiers are utilized mostly to convert an AC into a DC, but may also be used, for instance, to demodulate a radio-frequency signal to obtain an audio or video signal. A full-wave rectifier rectifies both the positive and the negative half-cycles of an applied AC input, while a half-wave rectifier provides a DC output only during alternate half-cycles of its input current. 2. A **rectifier** (1) utilized to convert an AC into a DC.

rectifier circuit A circuit, such as a full-wave bridge rectifier, which provides **rectification** (1).

rectifier diode A diode utilized as a **rectifier**.

rectifier filter A filter, such as a smoothing filter, utilized to help condition the output of a rectifier.

rectifier/filter A unit combining a rectifier and a filter. For example, a module which provides rectification and subsequent smoothing of the output DC.

rectifier stack A rectifier consisting of two or more rectifying elements, such as plates, utilized together.

rectifier transformer A transformer whose secondary supplies a rectifier.

rectifier tube An electron tube utilized as a **rectifier**.

rectifier unit A module which incorporates all the elements necessary for rectification.

rectify To perform a process of **rectification**.

rectifying antenna Same as **rectenna**.

rectilinear 1. Consisting of, bounded by, or characterized by, one or more straight lines. 2. Moving along a path following one or more straight lines.

rectilinear scan Scanning, such as that utilized in a CRT, whose path follows one or more straight lines. Also called **rectilinear scanning**.

rectilinear scanner A device, such as a CRT, which employs a **rectilinear scan**.

rectilinear scanning Same as **rectilinear scan**.

recurrence frequency Same as **repetition rate**.

recurrence rate Same as **repetition rate**.

recurrent Also called **recurring**. 1. That which occurs repeatedly. 2. That which repeats itself periodically.

recurrent circuit A circuit incorporating one or more repeating units. Also called **recurring circuit**.

recurrent error Same as **recurring error**.

recurrent network A network incorporating one or more repeating units. Also called **recurring network**.

recurring Same as **recurrent**.

recurring circuit Same as **recurrent circuit**.

recurring decimal Same as **repeating decimal**.

recurring error An error, such as a bug, which occurs again and again. Also called **recurrent error**, or **repetitive error**.

recurring network Same as **recurrent network**.

recursion A programming process or technique in which a routine or subroutine calls itself. LISP, for instance, utilizes such a technique extensively.

recursive filter A digital filter, such as an Infinite-Impulse Response filter, whose output factors in previous samples of its output.

recursive procedure Same as **recursive subroutine**.

recursive routine A routine which calls itself. Used, for instance, for problem solving.

recursive subroutine A subroutine which calls itself. Used, for instance, for problem solving. Also called **recursive procedure**.

recycle bin An icon which represents a directory or folder where deleted files and folders are stored. A recycle bin allows for files to be restored if deleted by accident or if a user decides for any reason to keep a file. Also, the directory or folder where deleted files are stored. Also called **trash can** (1).

Red Book A document which details the specifications for the CD-DA format.

red-green-blue Same as **RGB**.

red-green-blue display Same as **RGB display**.

red-green-blue monitor Same as **RGB monitor**.

red-green-blue signal Same as **RGB signal**.

red-green-blue system Same as **RGB system**.

red-green-blue video Same as **RGB video**.

red gun In a three-gun color TV picture tube, the electron gun which directs the beam striking the red phosphor dots.

red video voltage In a three-gun color TV picture tube, the signal voltage controlling the grid of the red gun. This signal arrives from the red section of a color TV camera.

redial A feature which allows a device, such as a telephone or modem, to dial again the last entered telephone number, or one that has been programmed into its memory. The device may redial a specific number of times, for a given time interval, or until a connection is established. Also, a button activating said feature. Also called **recall** (4).

redirect To send through an alternate channel or path, or to otherwise divert from the accustomed destination. For example, to send a message to an alternate recipient when the preferred recipient is not available. Also called **reroute**.

redirection The process of sending through an alternate channel or path, or otherwise diverting from the accustomed destination. Also, the result of such a redirection. Also called **rerouting**.

redirector That which serves to **redirect**. For example, a software or hardware device in a communications network which redirects data which is en route from a client to a server, or vice versa. Also called **rerouter**.

redo A command that reverses an **undo** operation. Also, to perform such a reversal.

redox Abbreviation of **oxidation-reduction**.

redraw Same as **refresh** (1).

reduced-instruction set computer Same as **RISC** (1).

reduced-instruction set computing Same as **RISC** (2).

reduction The converse is **oxidation**. **1.** A chemical reaction in which oxygen is removed from an atom, molecule, ion, or radical. For instance, iron oxide losing oxygen to form iron. **2.** A chemical reaction in which an atom, molecule, ion, or radical gains one or more electrons. For example, a reaction in which iron combines with oxygen, where the oxygen undergoes a change in valence from 0 to -2.

redundancy **1.** The quality or state of being **redundant**. **2.** The duplication of components, circuits, devices, equipment, systems, mechanisms, procedures, or the like, to allow for continued operation after a failure, malfunction, error, or other problem. For example, the incorporation of redundant components for fault-tolerance, the use of additional bits in error correction, or the use of a RAID for fault-tolerance and data backup.

redundancy check A technique, such as parity check or cyclic redundancy check, in which a value resulting from a calculation is transmitted with the data, for error detection.

redundant **1.** That which is duplicated or otherwise present as a precaution against failures, malfunctions, errors, or the like. **2.** That which exceeds what is necessary or required.

Redundant Array of Independent Disks Same as **RAID**.

Redundant Array of Inexpensive Disks Same as **RAID**.

redundant bits Bits in addition to those needed to convey the desired information. For instance, in forward error correction redundant bits are included in a transmission to correct detected errors.

redundant code Code which includes information beyond that required to convey what is intended, or which duplicates functions performed elsewhere. Used, for instance, for error correction or fault-tolerance.

redundant data Data which duplicates other data, or which is otherwise present as a precaution against failures, malfunctions, errors, or the like.

redundant operation The operation of a device, piece of equipment, or system, with parallel power supplies, so as to provide continuous power in the event of the failure of one.

reed A thin and flexible blade, leaf, bar, or strip which is made to vibrate or otherwise move. Used, for instance, in reed relays and switches.

reed relay A relay whose contacts are mounted on magnetic reeds, and which is hermetically sealed in a glass tube. Such a relay is designed to be actuated by an external magnetic field. The contacts are not wetted by a pool of mercury, as is the case with a **mercury-wetted reed relay**. Also called **dry-reed relay**, or **magnetic reed relay**.

Reed-Solomon code Any of various error-correction systems in which coding is based on groups of bits, such as bytes, as opposed to individual zeroes and ones. Used extensively in CD technology. Its abbreviation is **RS code**.

reed switch A switch whose contacts are mounted on magnetic reeds, and which is hermetically sealed in a glass tube. Such a switch is designed to be actuated by an external magnetic field. The contacts are not wetted by a pool of mercury, as is the case with **mercury-wetted reed switch**. Also called **dry-reed switch**, or **magnetic reed switch**.

reel **1.** An object, usually with a circular cross-section, around which something is rolled or wound. For example, a reel upon which a magnetic tape, film, wire, cable, or other flexible material is wound. Also called **spool** (1). **2.** That which is wound around a **reel** (1). Also called **spool** (2). **3.** A **reel** (1) including that which is wrapped around it. Also called **spool** (3).

reentrant **1.** Entering again, or pointing inward. For example, a reentrant angle of a ray. **2.** Pertaining to **reentrant code**.

reentrant cavity A cavity resonator with one or more sections directed inward.

reentrant code Program code which can be utilized by two or more programs at the same time. Employed, for instance, in multithreading.

ref Abbreviation of **reference**.

refactoring In object-oriented programming, procedures utilized for code improvement which are implemented in a manner which does not impair the overall functionality of the program.

reference Its abbreviation is **ref**. **1.** That which serves as a basis for comparison with something else. For example, a reference level or position. **2.** A component, circuit, device, or piece of equipment whose specifications lend it to be compared to others in its category. **3.** An act of mentioning something else. **4.** A written work utilized as a source of information.

reference amplifier **1.** An amplifier utilized in a recording studio or other setting in which the quality of a recording is monitored or studied. **2.** An amplifier against which other amplifiers may be compared. Said, for instance, of an amplifier with impeccable specifications which is part of an audio system providing a superior sound experience.

reference antenna An antenna with one or more characteristics, such as radiation pattern, may be used as a basis for comparison with other antennas. Such an antenna may be real or theoretical, and an example is an isotropic antenna.

reference black level In a video signal, the amplitude level that corresponds to the maximum permitted black peaks, that is, to the screen being black. Also called **black level**.

reference burst In a TV receiver, a color-synchronizing signal at the beginning of each scanning line, which establishes a frequency and phase reference for the chrominance signal. Also called **color burst**, **burst** (4), or **color-sync signal**.

reference CD **1.** A CD, such as a CD-ROM, containing reference material, such as an encyclopedia. **2.** Same as **reference CD player**.

reference CD player Also called **reference CD** (2). **1.** A CD player utilized in a recording studio or other setting in which the quality of a recording is monitored or studied. **2.** A CD player against which other CD players may be compared. Said, for instance, of a player with impeccable specifications which is part of an audio system providing a superior sound experience.

reference CD-ROM **1.** A CD-ROM containing reference material, such as an encyclopedia. **2.** Same as **reference CD-ROM player**.

reference CD-ROM player A CD-ROM player, such as that with impeccable specifications, against which other CD-ROM players may be compared. Also called **reference CD-ROM** (2).

reference current A current level whose steady value serves as a basis for comparison or operation.

reference diode A diode with one or more characteristics, such as forward voltage drop, which may be used as a basis for comparison with other diodes.

reference dipole A dipole which serves as a **reference antenna**.

reference direction A direction, such as true north or magnetic north, which serves as a reference for angular measurements.

reference DVD 1. A DVD containing reference material, such as an encyclopedia. **2.** Same as **reference DVD player**.

reference DVD player Also called **reference DVD (2)**. **1.** A DVD player utilized in a recording studio or other setting in which the quality of a recording is monitored or studied. **2.** A DVD player against which other DVD players may be compared. Said, for instance, of a player with impeccable specifications which is part of a home theater system providing a superior visual and aural experience.

reference electrode An electrode which generates a potential against which comparisons may be made. For example, a standard hydrogen electrode, which has an arbitrarily assigned potential of zero.

reference frequency A frequency, such as a carrier or cutoff frequency, against which other frequencies may be compared. A crystal oscillator, for instance, may be utilized to generate such a frequency.

reference input A input, such as that at a given voltage, which serves as a basis for comparison or operation.

reference level 1. A signal with a specified current, voltage, amplitude, frequency, or the like, against which other signals may be compared. For example, in adjusted decibels the established reference noise power level is usually -85 decibels above 1 milliwatt. Also called **reference signal level**. **2.** A level which serves as a basis for comparison. For instance, that of a given terrain, or a baseline.

reference line A line, such as an axis or that along which magnetic north points, which serves as a basis for comparison or measurement.

reference loudspeakers Same as **reference speakers**.

reference noise A stated noise level against which noise or other signals may be compared. An example is decibels above reference noise.

reference output An output, such as that at a given voltage, which serves as a basis for comparison or operation.

reference power A power level whose steady value serves as a basis for comparison or operation.

reference power supply A power supply whose steady output serves as a basis for comparison or operation.

reference pressure A pressure which serves as a basis for comparison or measurement. For example, in acoustics, 20 micropascals has an assigned a value of 0 decibels.

reference receiver 1. A receiver utilized in a recording studio or other setting in which the quality of a recording is monitored or studied. **2.** A receiver against which other receivers may be compared. Said, for instance, of a receiver with impeccable specifications which is part of an audio system providing a superior sound experience.

reference signal level Same as **reference level (1)**.

reference speakers Also called **reference loudspeakers**. **1.** Speakers utilized in a recording studio or other setting in which the quality of a recording is monitored or studied. **2.** Speakers against which other speakers may be compared. Said, for instance, of speakers with impeccable specifications which are part of an audio system providing a superior sound experience.

reference supply A source or current, voltage, or power whose steady output serves as a basis for comparison or operation.

reference system A system, such as an audio system, whose specifications lend it to be compared to others in its category.

reference temperature A temperature, such as absolute zero, which serves as a basis for comparison or measurement.

reference time 1. A specified instant which serves a basis for comparison or operation. For example, that at which a given sequence of events or instructions is initiated. **2.** Time, such as atomic time, which is kept so accurately that other time measuring and indicating devices utilize it as a reference.

reference tone A tone whose stated frequency is utilized as a basis for comparison. For example, a 1 kHz tone utilized as a basis for the phon scale.

reference transistor A transistor with one or more characteristics, such current gain, which may be used as a basis for comparison with other transistors.

reference voltage A voltage level whose steady value serves as a basis for comparison or operation.

reference volume A volume level, such as that producing a 0 reading on a VU meter, which serves as a basis for recording, reproducing, comparing, or measuring.

reference white level In a video signal, the amplitude level that corresponds to the maximum permitted white peaks, that is, to the screen being white. Also called **white level**.

referential integrity A feature which helps insure that data is consistent throughout a relational database. For instance, if a record is deleted from table A, which is linked with table B, then the linked record in table B must also be deleted. Also, the database integrity so obtained.

reflect To bounce off or otherwise throw back from a surface, object, or region. For example, to reflect an image off a mirror, or to bounce sound energy off a wall.

reflectance Also called **reflection factor**, **reflection coefficient**, **reflection ratio**, **reflectivity**, or **mismatch factor**. **1.** The ratio of the amplitude of a wave reflected from a surface to the amplitude of the same wave incident upon the surface. **2.** The ratio of the current delivered to a load whose impedance is not matched to the source, to the current that would be delivered to the same load if its impedance were fully matched.

reflected current Same as **return current (2)**.

reflected impedance In a transformer, the impedance as seen by the primary when the secondary has a specified load impedance.

reflected power 1. The ratio of the power returned to a source, to the transmitted power. It utilized, for instance, as a parameter to indicate or evaluate the extent to which there is impedance matching between two points, connectors, or devices in a transmission line. The lower the reflected power, the better the impedance matching. **2.** Any power that is returned to, or towards, its source. For example, from a load back to a generator.

reflected-power meter An instrument or device which measures and indicates **reflected power**.

reflected ray A ray which has bounced off a given surface, object, region, discontinuity, or the like. Also, the path taken by such a ray after reflection.

reflected resistance In a transformer, the resistance as seen by the primary when the secondary has a resistive load.

reflected wave A wave which has bounced off a given surface, object, region, discontinuity, another wave, or the like. For example, an electromagnetic wave reflected by the ionosphere, that which is bounced off a scanned object back to a radar, or a wave that encounters a discontinuity along a transmission line.

reflecting antenna Same as **reflector antenna**.

reflecting element Same as **reflector element**.

reflecting grating Same as **reflection grating**.

reflecting satellite Same as **reflector satellite**.

reflecting telescope Same as **reflector telescope**.

reflection The act of bouncing off or otherwise throwing back from a surface, object, or region. Also, that which is reflected, such as an image, sound, heat, particles, or waves.

reflection angle The angle formed between the line of a reflected ray or wave and a perpendicular line arising from the point of reflection. Also called **angle of reflection**.

reflection coating Same as **reflective coating**.

reflection coefficient Same as **reflectance**. Its abbreviation is RC.

reflection factor Same as **reflectance**.

reflection grating Also called **reflecting grating**, or **reflector grating**. **1.** In a waveguide, fine parallel lines which serve to reflect certain types of waves. **2.** A surface with many fine parallel lines, grooves, or slits, which are extremely close together and which serve to reflect certain wavelengths of light. Used, for instance, in lasers.

reflection high-energy electron diffraction An electron diffraction technique in which a sample is bombarded with a beam of high-energy electrons which is directed at a very small angle, so as to graze said sample. The electrons are diffracted, and then detected by a phosphor screen on the other side of the electron gun. Utilized, for instance, to study the surface contours of samples, or to monitor the growth of epitaxial films. Its abbreviation is **RHEED**.

reflection law A law stating that a wave or ray, such as sound or light, which strikes a reflecting surface at a given angle, is reflected at an equal angle in the opposite direction with respect to a perpendicular line arising from the point of incidence. The striking ray or wave is called incident ray or wave, respectively, and is in the same plane as the reflected ray or wave. Also called **law of reflection**.

reflection loss Same as **return loss**.

reflection-loss meter Same as **return-loss meter**.

reflection losses Same as **return loss**.

reflection mapping A rendering method in which an image is mapped upon another to simulate light reflections. This technique adds realism to an image without the complex computations of ray tracing.

reflection plane The plane containing light, sound, particles, waves, or the like, which are incident upon a surface, object, or region. Also called **plane of reflection**.

reflection ratio Same as **reflectance**.

reflective coating Also called **reflection coating**. **1.** A coating, such as that used in a mirror, via which reflection of waves or rays is maximized. **2.** A coating which maximizes the reflection of waves, rays, particles, heat, or the like.

reflective display A display which does not have its own light source behind the viewing surface. Such a display works best when there is plenty of ambient light, as too little makes it hard to read or view properly

reflective LCD Abbreviation of **reflective** liquid-crystal display. A liquid-crystal display which does not have its own light source behind the viewing surface. Such a display works best when there is plenty of ambient light, as too little makes it hard to read or view properly. Also called **reflective LCD display**.

reflective LCD display Same as **reflective LCD**.

reflective liquid-crystal display Same as **reflective LCD**.

reflective screen The viewing screen of a **reflective display**.

reflectivity Same as **reflectance**.

reflectometer **1.** An instrument which measures and indicates the **reflectance** (1) of a surface. **2.** A device or instrument which measures and indicates any form of **reflection**.

reflector **1.** An object, material, device, or system utilized to **reflect** images, sounds, heat, particles, waves, or the like. **2.** A surface that reflects light. Such a surface is usually smooth, highly polished, and it may consist, for instance, of a thin layer of silver or aluminum on glass. Reflectors have many applications, including their use in lasers, beam splitters, microscopes, TV cameras, and digital micromirror devices. Also called **mirror** (1). **3.** Same as **reflector electrode**. **4.** A material which is placed around the core of a nuclear reactor to reflect neutrons back to said core, preventing their escape. Also called **tamper** (1).

reflector antenna An antenna which is utilized to radiate towards a reflector, so as to enhance directivity, or radiate in a given pattern. Also called **reflecting antenna**.

reflector electrode In a **reflex klystron**, the electrode which reverses the direction of the electron beam. Also called **reflector** (3), or **repeller**.

reflector element An antenna element which is driven indirectly, that is, through radiation from a driven element. This contrasts with an **active element** (3), which is driven directly by the transmission line. Also called **reflecting element, parasitic element, parasitic reflector**, or **passive element** (2).

reflector grating Same as **reflection grating**.

reflector satellite An artificial satellite whose surface reflects electromagnetic waves. Also called **reflecting satellite**.

reflector telescope A telescope employing mirrors, usually two, to gather and focus light. Also called **reflecting telescope**.

reflector voltage In a **reflex klystron**, the voltage between the reflector electrode and the cathode.

reflex circuit A circuit in which a single active device, such as a tube, is utilized to amplify the same signal more than once.

reflex klystron A klystron with one cavity. In such a klystron, the cavity first serves as a buncher, then the electron beam is reflected back through the cavity by a reflector electrode, with the output taken from this same cavity, which then serves as a catcher.

reflow soldering Soldering in which surfaces are joined via the heating of solder which has been previously deposited at selected points.

reformat Also spelled **re-format**. **1.** To again prepare a computer disk, or other storage medium, for utilization. When formatting again all data on the disk will be lost. **2.** To update, adapt, or otherwise change the format of something.

reformatting Also spelled **re-formatting**. **1.** The process of again preparing a computer disk, or other storage medium, for utilization. When formatting again all data on the disk will be lost. **2.** The process of updating, adapting, or otherwise changing the format of something.

refract To deflect light via **refraction**.

refracted ray The ray that results from the **refraction** of an incident ray. In an optical fiber, for instance, it is a ray that is refracted from the core into the cladding.

refracted wave A wave which emerges from a medium, such as the ionosphere, after **refraction**.

refraction The process via which a wave abruptly changes direction when passing from a medium within which it has one propagation velocity, into another in which it has a different propagation velocity. The ray entering the second medium is the incident ray, while the refracted ray is that resulting after bending. For example, a light ray passing at an angle from air into glass is refracted, as light travels more slowly through glass. Other forms of electromagnetic radiation, such as microwaves or X-rays may also be refracted, in addition to sound waves. Also, the effect refraction has on a wave.

refraction angle The angle formed between the line of a refracted ray or wave and a perpendicular line arising from the point of refraction. Also called **angle of refraction**.

refraction index Same as **refractive index**.

refractive Pertaining to, characteristic of, or arising from **refraction**. For example, a refractive index.

refractive index Also called **refraction index**, or **index of refraction**. Its symbol is **n**. **1.** The ratio of the phase velocity of a wave in free space, to the phase velocity of the same wave in a given medium. Also called **absolute index of refraction (1)**. **2.** The ratio of the phase velocity of light in free space, to the phase velocity of light in a given medium. Also called **absolute index of refraction (2)**. **3.** Same as **relative index of refraction**.

refractivity Its symbol is **N**. A measure of the extent to which a medium will **refract** a ray. May be expressed as $N = (n - 1)$, where **n** is the refractive index of a given medium.

refractometer An instrument or device which measures the **refractive index** of a given material or medium. An example is an instrument which measures the critical angle of a medium.

refractometry The measurement of **refractive indices** of materials or mediums. Also, the techniques employed for such measurements.

refractory Also called **refractory material**. **1.** A material, such as alumina or silicon carbide, with a high melting point. **2.** A material, such as a ceramic, which is difficult to make soft, in addition to withstanding high temperatures.

refractory ceramic A ceramic, such as porcelain, which is difficult to make soft, in addition to withstanding high temperatures.

refractory material Same as **refractory**.

refractory metal A metal, such as tungsten, which has a high melting point, and which is difficult to make soft.

refrangible That which can be refracted.

refresh **1.** To renew the image presented on a display screen, such as that of a computer, by resending the appropriate signals. This is necessary, as the phosphors only glow for a comparatively short time. Also called **redraw (1)**. **2.** To maintain the condition of memory cells or modules, by sending the appropriate pulses. Most forms of RAM, for instance, require refreshing to avoid loss of data. **3.** To restore to a previous condition. **4.** To clear all options, windows, or the like, on a display screen. **5.** To update the contents of fields, records, or the like.

refresh cycle All the processes involved in a **refresh (2)**. This may occur several thousand times per second. Some types of RAM can self-refresh, which reduces power consumption.

refresh rate The number of times per second that a **refresh (1)** or **refresh (2)** occurs. For example, a monitor may have a refresh rate of 200 Hz, while a DRAM chip may refresh 4000 times per second.

refrigerant A substance which serves to cool an object, material, region, component, device, piece of equipment, and so on. At cryogenic temperatures neon may be used as a refrigerant, at low temperatures nitrogen could be utilized, and air may be used at high temperatures, as in its cooling of heat sinks.

refrigerator A compartment, appliance, room, or other enclosure utilized to keep that inside cooler than a given temperature. An adiabatic demagnetization refrigerator, for instance, can cool to temperatures near absolute zero.

refurbish To clean and repair to bring back to the original condition.

regenerate **1.** To again generate. For example, to refresh a signal, or to create anew a state which is improved. **2.** To restore a pulse or pulse train to its original shape. Said for instance of a signal which has been distorted.

regeneration **1.** The process of again generating, or of restoring a pulse or pulse train to its original shape. **2.** Same as **regenerative feedback**.

regenerative amplifier An amplifier circuit which utilizes **regenerative feedback**, for increased gain and sensitivity.

regenerative braking A form of dynamic braking in which the generated energy is stored, instead of being dissipated into heat. Used, for instance, in specially-equipped trains or cars. Also called **dynamic braking (2)**.

regenerative detector A detector circuit which utilizes **regenerative feedback**, for increased gain and sensitivity.

regenerative feedback In an amplifier or system, feedback which is in phase with the input signal, thus reinforcing it. While this results in an increase in gain, an excess can destabilize amplification, and produce distortion and noise. Howl may occur, for instance, if positive feedback exceeds a given threshold. Also called **regeneration (2)**, or **positive feedback**.

regenerative receiver A receiver circuit which utilizes **regenerative feedback**, for increased gain and sensitivity.

regenerative repeater A **repeater** which restores pulses to their original shape, amplifies, and retransmits. Also called **regenerator (2)**.

regenerator **1.** A circuit or device which effects **regeneration (1)**. An example is a pulse regenerator. **2.** Same as **regenerative repeater**.

region **1.** A given area or zone. For example, the D region of the ionosphere, a far-field region, or a p-region of a semiconductor. **2.** A defined interval, such as the IR, microwave, or UV regions. **3.** A contiguous area of an image which is treated as a unit.

regional channel A broadcast channel offering **regional programs**.

regional program A broadcast utilizing two or more stations within a geographic area, as opposed to a local or national program.

regional reception The reception of a **regional program**.

regional transmission The transmission of a **regional program**.

register **1.** A high-speed storage area within the CPU, which is utilized to hold data for specific purposes. Before data may be processed, it must be represented in a register. There are various types, including control registers, index registers, arithmetic registers, and shift registers. Also called **data register**, or **storage register (1)**. **2.** A device which records and/or stores a value or number. Also called **storage register (2)**. **3.** To indicate. For example, for an instrument to register a given value. Also, that which is so indicated. **4.** To align, or be in proper alignment. Said, for instance of superimposed images.

register capacity The interval of values which can be accepted or stored by a **register (1)** or **register (2)**.

register length The number of digits, characters, or the like, which can be accepted or stored by a **register (1)** or **register (2)**.

registered jack Same as **RJ**.

registration **1.** The act or process of recording and/or storing a value or number. **2.** The act or process of indicating a value. **3.** The act or process of aligning.

regression testing The running of the same sequence of tests on a program each time a modification is made. This helps ensure that any new code works properly, and that said code does not adversely affect the code already present.

regular reflection Reflection of electromagnetic or acoustic waves in which there is no diffusion or scattering. This may occur, for instance, when reflecting light off a mirror. Also called **specular reflection**, or **direct reflection**.

regulate **1.** To adjust, manage, or otherwise control something. For example, to regulate a temperature to avoid overheating, or to adjust a mechanism for the proper function of a device. **2.** In a control system, to maintain one or more output quantities within specified parameters. **3.** To maintain a voltage, current, or the like, within specified values.

regulated power supply A power supply that has circuitry to maintain the output voltage constant when the input line or load varies.

regulating system A system, such as a control system, which regulates something else.

regulating transformer A transformer which maintains the voltage of the output essentially constant, despite variations in variables such as load resistance, line voltage, and temperature, so long as they are within a prescribed range. Also called **voltage-regulating transformer**.

regulation **1.** The process of adjusting, managing, or otherwise controlling something. **2.** In a control system, the maintaining of one or more output quantities within specified parameters. **3.** The process of maintaining a voltage, current, amplitude, power level, or the like, within specified values. **4.** For a transformer, power supply, or generator, the maximum difference in voltage between no-load and full-load conditions, operational temperature extremes, line-voltage fluctuations, or the like. Also, the change in output voltage for a given variation in a parameter such as load current.

regulator **1.** A system which monitors one or more variables, and automatically makes the necessary adjustments in order to maintain operation within the specified parameters. Thermostats and voltage regulators are examples. Also known as **automatic regulator**. **2.** A device which serves to **regulate** a quantity. For example, a rheostat.

regulator diode A diode, such as a Zener diode, which serves for voltage regulation.

reinitialization Also spelled **re-initialization**. **1.** In a computer, the process of again setting variables, counters, addresses, or the like to their starting value, such as zero. **2.** The process of again preparing a storage medium, such as a floppy or hard disk, for use. **3.** The again powering up of a computer. **4.** The process of again preparing a device or piece of equipment for operation.

reinitialize Also spelled **re-initialize**. **1.** In a computer, to again set all variables, counters, addresses, or the like to their starting value, such as zero. **2.** To again prepare a storage medium, such as a floppy or hard disk, for use. **3.** To again power up a computer. **4.** To again prepare a device or piece of equipment for operation.

reinitiate To again begin, or otherwise set an operation or process in motion. Also spelled **re-initiate**.

reinitiation The act of again beginning, or otherwise setting an operation or process in motion. Also an instance of such an act. Also spelled **re-initiation**.

reinserter Also called **restorer (2)**, **clamping circuit**, **DC restorer**, or **clamper**. **1.** A circuit or device which restores the DC component of a wave or signal. **2.** A circuit or device which adds a DC component to a wave or signal which lacks such a component. **3.** A circuit or device which establishes a reference DC level in a signal or device.

reinstall To again install. For instance, to reinstall a computer program which has become increasingly problematic over time. Also spelled **re-install**.

reject **1.** To not allow, or otherwise refuse. For instance, to reject access to a network, or to reject all signals except those with a specific coding. **2.** To block or otherwise impede. For instance, to reject other signals which might interfere with that which is desired, or for a filter to block frequencies above a given value.

reject band Same as **rejection band**.

reject circuit Same as **rejector circuit**.

reject filter Same as **rejection filter**.

rejecter Same as **rejector**.

rejecter circuit Same as **rejector circuit**.

rejection **1.** The act or process of not allowing, or otherwise refusing. **2.** The act or process of blocking or otherwise impeding.

rejection band A band of frequencies which is blocked by a **rejection filter**. Also called **reject band**.

rejection circuit Same as **rejector circuit**.

rejection filter A filter, such as a bandstop filter, which highly attenuates all frequencies above or below a given cutoff point, or within two specific cutoff points. All other frequencies pass with little or no attenuation. Also called **reject filter**.

rejector That which **rejects**. For example, a rejector circuit. Also spelled **rejecter**.

rejector circuit A circuit, such as a filter, which highly attenuates all frequencies within two specific cutoff points. All frequencies above and below this interval pass with little or no attenuation. Also spelled **rejecter circuit**. Also called **reject circuit**, or **rejection circuit**.

relation **1.** A connection or association between two or more entities. For instance, the correlation of the changes in a dependent variable to those of an independent variable. **2.** A reference for comparison, location, or the like. For example, a phase relationship, a satellite which maintains a fixed relationship to the earth, or something whose dimensions are small in comparison to something else. **3.** A two-dimensional table within a relational database.

relational **1.** Pertaining to, or arising from, a **relation**. **2.** Pertaining to a relational database or model.

relational algebra The rules and operators utilized to express and manipulate relations, or tables, within a relational database. Common operations include difference, join, and union.

relational data model Same as **relational model**.

relational database A type of database or database management system in which data is stored in multiple tables, whose rows represent records, and whose columns represent fields. The number of columns is fixed, while the number of rows is not, and relational algebra is utilized to manipulate the relationships between such tables. Relational databases are characterized by referential integrity and normalization, and a relational language, such as SQL, is utilized to access and manage such databases. Also called **relational database management system**, or **relational DBMS**.

relational database management system Same as **relational database**. Its abbreviation is **relational DBMS**, or **RDBMS**.

relational DBMS Same as **relational database**. Abbreviation of **relational database management system**.

relational expression An expression with one or more **relational operators**.

relational language A language, such as SQL, which is utilized to query, update, and otherwise access and manage a relational database.

relational model A data model whose structure is based on a set of relations, as seen in a relational database. Also called **relational data model**.

relational operator An operator, such as less than, equal to, or greater than, which serves to compare two or more values.

relational query A query, such as that in SQL, made to a **relational database**.

relational structure A data structure based on a **relational model**.

relative accuracy For a given instrument, the extent to which a measured, calculated, and/or indicated value approximates the real value. Usually expressed as a percentage difference from the real value.

relative address An address which identifies a storage location by means of its displacement from a base address. When a relative address is added to a base address, the result is an absolute address.

relative atomic mass The average mass of all the isotopes of an element, expressed in atomic mass units. It is the combined mass, or weight, of the protons, neutrons, and electrons of an atom. The atomic weight of carbon, for example, is approximately 12.011 atomic mass units, not 12. The reason is that carbon has three naturally occurring isotopes, of which carbon-12 is by far the most common. Also called **atomic weight**.

relative attenuation For a given filter, the reduction in output in its stopband, minus the insertion loss in its passband.

relative bearing In navigation, a bearing in which the reference line is the heading of the craft itself.

relative density The ratio of the density of a given material or medium, to that of a standard material or medium. For example, the ratio of the density of a material to that of water at 4 °C at one atmosphere, or the ratio of the density of a gas to that of dry air at 0 °C at one atmosphere. Also called **specific gravity**.

relative dielectric constant Same as **relative permittivity**.

relative error For a given quantity that is measured or indicated, the ratio of the absolute error to the real value.

relative gain 1. For a given antenna, the ratio of the gain in a given direction, to that a reference antenna such as an isotropic antenna would have. 2. The gain of a component, circuit, device, piece of equipment, or system, in relation to a given standard.

relative humidity For a given temperature, the ratio of the absolute humidity to the maximum humidity attainable without precipitation in the same volume of air. Usually expressed as a percentage. Its abbreviation is **RH**.

relative index of refraction 1. The ratio of the phase velocity of a wave in a given medium, to the phase velocity of the same wave in another medium. The ratio of the phase velocity of wave in free space, to that of another medium is the **absolute index of refraction** (1). 2. The ratio of the phase velocity of light in a given medium, to the phase velocity of light in another medium. The ratio of the phase velocity of light in free space, to that of another medium is the **absolute index of refraction** (2).

relative ionospheric opacity meter Same as **riometer**.

relative level The ratio of a given level, such as that of a signal, to a reference level.

relative luminosity For a given wavelength, the ratio of the luminosity factor to that of a given standard or other reference level.

relative molecular mass The ratio of the average mass per molecule or other formula unit, to one-twelfth the mass of a carbon-12 atom, expressed in atomic mass units. It is the same as the sum of the atomic weights of the atoms comprising said molecule. For example, the molecular weight of sodium chloride is 58.44 grams, as the combined atomic weights of sodium and chlorine are approximately 58.44 atomic mass units. Also known as **molecular weight**.

relative movement Any movement in which a new position is based on a former position. For instance, when a mouse is moved, each new location of the cursor on the screen is based on its former location.

relative path Same as **relative pathname**.

relative pathname A pathname stated in terms of the current directory. For instance, *C:\Documents\Reminders\today.xyz*

or *C:\Documents\Reminders\tomorrow.xyz* would be **full pathnames**, but when working between documents within the Reminders folder, the *C:\Documents\Reminders* portion would be implied, so only *today.xyz* and *tomorrow.xyz*, or just *today* and *tomorrow* could be used. Also called **relative path**.

relative permeability The ratio of the permeability of a given material or medium, to the permeability of free space. It is obtained using the following formula: $\mu_r = \mu/\mu_0$, where μ_r is relative permeability, μ is the permeability of a given material or medium, and μ_0 is the permeability of free space. According to this formula, the relative permeability of free space is 1. Materials whose relative permeability is less than 1 are diamagnetic, those slightly above 1 are paramagnetic, and those substantially above 1 are ferromagnetic.

relative permittivity For a given material, the property which determines the electrostatic energy that can be stored per unit volume for a unit potential gradient. This is equivalent to the ratio of the capacitance of a capacitor using the material in question as a dielectric, to the capacitance of a capacitor using a vacuum as a dielectric. Also called **relative dielectric constant, dielectric constant, permittivity, inductivity**, or **specific inductive capacity**.

relative power Also called **relative power level**. 1. The ratio of a given power level to a reference power level. 2. Same as **relative transmission level**.

relative power level Same as **relative power**.

relative refraction index Same as **relative index of refraction**.

relative refractive index Same as **relative index of refraction**.

relative response For a given transducer, the ratio of the response under stated conditions, to that of a given standard or other reference level.

relative sensitivity For an instrument or device, the ratio of the change detected in an observed quantity or variable, to the actual change in the value of said quantity or variable.

relative signal level The ratio of a given signal level to a reference signal level.

relative transmission level For a given point of a transmission line or system, the ratio of the signal power level to that of a reference point, such as the origin of said line or system. Usually expressed in decibels. Also called **relative power** (2), or **transmission level**.

relative uniform resource locator Same as **relative URL**.

relative URL Abbreviation of **relative uniform resource locator**. Its own abbreviation is **relURL**. A URL whose location is specified relative to another URL, usually the base document where the URL resides. Such a URL omits certain information, such as the domain name and certain directories, often leaving just the document name. An **absolute URL** specifies the full path.

relaxation 1. The return of a system towards or to equilibrium, after an abrupt change. 2. The return of a system to its previous state, after an abrupt change.

relaxation oscillator An oscillator whose output signal, such as a current or voltage, changes abruptly from one level to another. For example, a relaxation oscillator circuit may be arranged so that in each cycle energy is stored in a capacitor, and later discharged at stated time intervals. The output waveform is nonsinusoidal, and may be, for instance, sawtooth or rectangular. Examples include multivibrators and unijunction transistors.

relaxation time 1. The time required for a system to return to equilibrium after an abrupt change. 2. The time required for a system to return to its previous state, after an abrupt change. 3. The time during which an electron can travel in a metal or semiconductor before it is scattered and loses its momentum.

relay **1.** A device in which a variation in a parameter, such as capacitance, current, frequency, or voltage, controls a component, circuit, another device, piece of equipment, system, or mechanism. A common example is a device in which a varying current or voltage controls the contacts which make or break a circuit. Relays may be electrical, electronic, electromechanical, electromagnetic, photoelectric, and so on, and there are numerous types, including armature, capacitance, current, reed, and frequency. In addition, relays may be classified as automatic, bi-directional, bistable, delay, differential, directional, high-speed, latching, and the like. **2.** A **relay** (1) used as a switch. **3.** In communications, a device that repeats, retransmits, or otherwise passes on a signal. Also, to pass on a signal utilizing such a device.

relay amplifier Same as **relay driver**.

relay antenna An antenna utilized in a **radio repeater system**.

relay armature **1.** The moving contact in an electromagnetic relay. **2.** The moving contact a relay.

relay channel A channel utilized as a **relay**. For instance, a communications channel via which a signal is repeated or otherwise passed on.

relay chatter The continuous, rapid, and undesired opening and closing of relay contacts. May be caused, for instance, by contact bounce.

relay circuit A circuit utilized as a **relay**. For instance, a communications circuit via which a signal is repeated or otherwise passed on.

relay contacts A pair of contacts in a relay, usually consisting of one movable contact and one stationary contact. When a relay is actuated, said contacts are closed.

relay control system A control system, such as that employing a bang-bang mechanism, which is regulated by a relay.

relay-controlled A component, circuit, device, piece of equipment, system, or mechanism, which is regulated by a relay. Also called **relay-operated**.

relay driver An amplifier which increases a weak signal to a level sufficient to actuate a relay. Also called **relay amplifier**.

relay magnet In a relay, a magnet which attracts the armature when actuated.

relay-operated Same as **relay-controlled**.

relay rack Same as **rack** (2).

relay receiver **1.** A receiver within a **radio repeater station**. **2.** Same as **radio repeater station**.

relay satellite An artificial satellite which relays, and usually amplifies, radio-wave signals between terrestrial communications stations, terrestrial communications stations and other communications satellites, or between other communications satellites. Such satellites provide high-capacity communications links, offer a very wide coverage area, and may transmit TV, telephone, and data signals, among many others. Also called **radio-relay satellite**, **repeater satellite**, **communications satellite**, or **satellite relay**.

relay set Two or more relays utilized together, and which are usually mounted on the same support.

relay station Same as **radio repeater station**.

relay system Same as **radio repeater system**.

relay transmitter **1.** A transmitter within a **radio repeater station**. **2.** Same as **radio repeater station**.

release **1.** A device, such as a mechanical or electromechanical switch, that opens a circuit. Also, the action of such a device. **2.** The opening of the contacts of a relay or switch. Also, a signal or condition which produces this opening. **3.** To free from an influencing force. For instance, to release a brake. **4.** To allow to escape. For example, to release a gas from an enclosure. **5.** To allow to become available. For

instance, to release a communications circuit. **6.** Same as **release version**.

release current The maximum current at which a device, such as a relay or circuit-breaker, will release to its deenergized position. Also called **dropout current**.

release force In a relay or switch, the amount the actuating force must be reduced for the contacts be released.

release time **1.** The time required for the contacts of a relay or switch to open, once the actuating signal or condition is removed. **2.** The time required for a circuit to open once the actuating signal, such as a current or voltage, is removed. **3.** The time required for something, such as a communications circuit, to become available once that which prevent its use is removed. **4.** The time it takes a complementary action to begin, once the other action is terminated. For example, in a transceiver, the time it takes a receiver to be actuated once the transmitter stops operating.

release value The maximum actuating quantity, such as a current or voltage, at which a device, such as a relay or circuit-breaker, will release to its deenergized position. Also called **dropout value**.

release version A specific version of a given program or software package. For example *Program, release 1.01*. Also called **release** (6), **revision level**, or **version**.

release version number The numbers, letters, and/or words which identify a specific release version of a given program or software package. For example, *1.0*, *5.01a*, or *2004b beta build #1111*. Changes in whole numbers usually imply significant modifications, while fractions, letters, or builds often identify versions which correct bugs, provide minor enhancements, or the like. Also called **version number**, or **program version**.

release voltage The maximum voltage at which a device, such as a relay or circuit-breaker, will release to its deenergized position. Also called **dropout voltage**.

reliability **1.** The extent to which a component, circuit, device, piece of equipment, system, mechanism, object, or the like, is **reliable**. **2.** The characteristic of being reliable.

reliability engineering The area of engineering that deals with the reduction or prevention of failures and malfunctions, and the measures that can be taken to manage and minimize the effects of such occurrences.

reliability test A test which evaluates the extent to which a component, circuit, device, piece of equipment, system, mechanism, object, or the like, is **reliable**.

reliability testing The performance of **reliability tests**.

reliable **1.** That which can be depended upon to provide a level of operation or function at or above a given level, when used under specified conditions. **2.** That which displays consistency. For example, multiple experiments which provide similar results.

reload Also spelled **re-load**. **1.** To obtain an up-to-date version of a given Web page from a server. For example, if a user returns to the browser's home page, what is displayed may be a cached version. By reloading, the most current version of the page is provided. **2.** To again load a program into memory for execution. Done, for instance, after a crash. **3.** To again transfer data to or from a computer storage medium, such as a disk or tape. **4.** To again place data in a computer register. **5.** To again insert a disk, cassette, reel, cartridge, drum, or other object composed of, or containing a recordable medium into a device utilized for recording and/or reproduction.

relocatable code Program code which can be loaded into any part of memory, as opposed to a given location.

reluctance The opposition a material or magnetic circuit presents to the passage of magnetic flux. It is expressed as the ratio of the magnetomotive force to the magnetic flux, and is analogous to electrical resistance. Its symbol is R, and it is the reciprocal of **permeance**. Also called **magnetic reluctance**, or **magnetic resistance**.

reluctance motor A brushless synchronous motor that does not utilize a permanent magnet, and which does not have a coil in its rotor. Such a motor has pointed poles in both the stator and rotor, and the coils of the stator are fed the energizing current, which creates a magnetic pulling force that enables rotation. Reluctance motors are comparatively robust and efficient, and can be used when very high ambient temperatures are present. Also called **switched-reluctance motor**, or **variable-reluctance motor**.

reluctivity The reciprocal of **magnetic permeability**. Also called **magnetic reluctivity**.

relURL Abbreviation of **relative URL**.

rem 1. A unit of radiation dose equivalent equal to 0.01 sievert. It is an acronym for **r**oentgen **e**quivalent **m**an. 2. Abbreviation of **remark**.

rem statement Same as **remark**. Abbreviation of **remark statement**.

remanence Any magnetism which remains in a material or medium after a magnetizing force is removed. For example, that still present after a strong external magnetic field is removed. The magnetic remanence of a magnetic tape, for instance, enables it to be able to record and store information. Also called **magnetic remanence**.

remark In computer programming, a statement or text that is embedded for any purpose except execution. For instance, a clarifying note detailing the rationale behind a given instruction. Also called **remark statement**, or **comment**. Its abbreviation is **rem** (2).

remark statement Same as **remark**. Its abbreviation is **rem statement**.

remedial maintenance Maintenance performed to rectify an undesired condition. For instance, that carried out after a failure. Also called **corrective maintenance** (1).

remodel Same as **rebuild**.

remodulation The conversion from one form of modulation to another. For instance, a conversion of AM to FM, or from single-sideband modulation to double-sideband modulation. Also spelled **re-modulation**.

remodulator A circuit or device which converts one form of modulation into another. For instance, a circuit that converts AM to FM, or from single-sideband modulation to double-sideband modulation. Also spelled **re-modulator**.

remote 1. Located at a point other than that where something occurs, is acted upon, or is situated. For instance, a remote terminal. 2. Located at a comparatively distant point. For example, a remote pickup. 2. Controlled, or otherwise operated upon from a different location. For instance, remote tuning. 3. Same as **remote control**. 4. A broadcast, such as that of radio or TV, originating from a location other than a broadcast studio.

remote access Access to one or more computers via **remote computing**.

remote-access concentrator A **remote-access server** that supports one or more T1 or E1 lines. Its abbreviation is **RAC**.

remote-access program A program utilized for **remote computing**. Also called **remote-control program**.

remote-access server A computer that utilizes network-emulation software to connect asynchronous devices to a LAN or WAN. It manages communications, and takes care of protocol conversions. Its abbreviation is **RAS**. Also called **network-access server, access server**, or **communications server**.

remote-access software Software utilized for **remote computing**. Also called **remote-control software**.

remote administration The administration of a computer or network from a remote location.

remote alarm An alarm situated at a location other than where that being monitored or controlled is located. For instance, an alarm which alerts a fire station of smoke and/or fire from a device in a home which emits signals via a telephone line.

remote amplifier 1. An amplifier which is not located at the source of the signal to be acted upon. Used, for instance, in sensors which are very small. 2. An amplifier which accepts commands from a remote control.

Remote Authentication Dial-in User Service Its acronym is **RADIUS**. A client/server protocol utilized for remote-access servers and central servers to communicate pertaining to tasks such as secure authentication and billing. RADIUS is a de facto standard.

remote boot Abbreviation of **remote boot**ing. The booting of a computer from a remote location, such as a client by a server.

remote booting Same as **remote boot**.

remote call forwarding A telephone company service in which call forwarding features may be programmed from most any phone line other than that which will have calls rerouted.

remote communications 1. A redundant term for **communications**. 2. Communications to and from locations which are considered very distant or hard to reach, such as space.

remote computer In a computer or communications network, any computer other than that of a given user. For instance, a server being accessed by a client. Also called **remote computer system**, or **remote system**.

remote computer system Same as **remote computer**.

remote computing The control of one or more computer systems by another, which is at a different location. Each of these computers has remote-access software installed, and one or all computers may have control abilities. All computers usually have the same information displayed at any given time. Also called **telecomputing**.

remote console A console which serves to control devices, equipment, or processes at a remote location.

remote control Its abbreviation is **RC**. Also called **telecontrol**. 1. The operating, regulating, or managing of a component, circuit, device, apparatus, piece of equipment, system, function, quantity, or the like, from a location other than where that being controlled is situated. To effect remote control, radio, infrared, or acoustic waves may be employed, or connecting wires or cables may be used, among other means. When remote control is wireless, it is also called **radio control**. 2. Same as **remote-control device** (1).

remote-control apparatus 1. An apparatus which serves to control devices, equipment, or systems situated at one or more remote locations. 2. An apparatus which is controlled by devices or mechanisms situated at a remote location. Also called **remote-controlled apparatus**.

remote-control device 1. A device which serves to control other devices, equipment or systems situated at one or more remote locations. Also called **remote control** (2), **remote-control unit** (1), or **remote-control transmitter** (3). 2. A device which is controlled by devices or mechanisms situated at a remote location. Also called **remote-controlled device, remote-control unit** (2), or **remote-controlled unit**.

remote-control equipment 1. Equipment which serves to control devices, other equipment, or systems situated at one or more remote locations. 2. Equipment which is controlled by devices or mechanisms situated at a remote location. Also called **remote-controlled equipment**.

remote-control operation The operation of a component, circuit, device, apparatus, piece of equipment, system, function, quantity, or the like, from a remote location.

remote-control program Same as **remote-access program**.

remote-control receiver 1. Within a **remote-control device** (2), the unit which receives the controlling signals. 2. A receiver, such as that of a satellite, which is controlled from a remote location. Also called **remote-controlled receiver**.

remote-control software Same as **remote-access software**.

remote-control station 1. A station that serves to control devices, equipment, other stations, or systems situated at one or more remote locations. 2. A station that is controlled by devices, mechanisms, or stations situated at a remote location. Also called **remote-controlled station**.

remote-control switch 1. A switch that serves to control devices, equipment, or systems situated at one or more remote locations. 2. A switch that is controlled by devices or mechanisms situated at a remote location. Also called **remote-controlled switch**.

remote-control system 1. A system which serves to control devices, equipment, or other systems situated at one or more remote locations. 2. A system which is controlled by devices or mechanisms situated at a remote location. Also called **remote-controlled system**. 3. A system or mechanism utilized to effect **remote control** (1).

remote-control transmitter 1. Within a **remote-control device** (1), the unit which generates and transmits the controlling signals. 2. A transmitter, such as that of a satellite, which is controlled from a remote location. Also called **remote-controlled transmitter**. 3. Same as **remote-control device**.

remote-control unit 1. Same as **remote-control device** (1). 2. Same as **remote-control device** (2).

remote-controlled Operated, regulated, or managed via **remote control** (1).

remote-controlled device Same as **remote-control device** (2).

remote-controlled equipment Same as **remote-control equipment** (2).

remote-controlled receiver Same as **remote-control receiver** (2).

remote-controlled station Same as **remote-control station** (2).

remote-controlled switch Same as **remote-control switch** (2).

remote-controlled system Same as **remote-control system** (2).

remote-controlled transmitter Same as **remote-control transmitter** (2).

remote-controlled unit Same as **remote-control device** (2).

remote copy protocol A protocol utilized for copying files, which utilizes TCP to help insure reliable delivery of the data. Its abbreviation is **RCP**.

remote data storage Same as **remote storage**.

remote detection Same as **remote sensing**.

remote detector Same as **remote sensor**.

remote drive In a communications network, a disk drive at a node other than that of a given user, as opposed to a **local drive**, which is contained in a user's computer, or a **network drive**, which is accessed by multiple nodes.

remote indicator An indicator which displays the information obtained by a device, instrument, piece of equipment, or sensing element at a remote location. Used, for instance, for probing very small areas, or while taking measurements from locations that are hard to reach.

remote job entry The submission of jobs via transmission to a remote computer. Once each job is completed, the out-

come may or may not be returned to the local computer. Its abbreviation is **RJE**.

remote location 1. Any location other than a given location. For instance, that where a remote terminal is situated. 2. A location other than where that being controlled is located. For example, the location of a remote computer which is being accessed by a local computer.

remote login Same as **rlogin**.

remote maintenance Maintenance which is activated and/or controlled via telecommunications. The site where the maintenance is performed must have the necessary devices and equipment to properly perform any required tasks. Utilized, for instance, for installations where physical access is impracticable, such as satellites. Also called **telemaintenance**.

remote metering The automatic gathering of information from one location while transmitting it to a remote location. To effect remote metering, radio, infrared, or acoustic waves may be employed, or connecting wires or cables may be used, among other means. When remote metering is wireless, it is called **radio telemetry**. Also called **telemetry** (1), or **telemetering**.

remote node Within a communications network, any node other than that where a given user is located, from the perspective of said user. Such a node may be another computer, a server, a printer, or the like.

remote PC Abbreviation of **remote** personal computer. A PC that serves as a **remote computer**.

remote personal computer Same as **remote PC**.

remote pickup In recording or broadcasting, such as that for radio or TV, the detection of sounds and/or scenes outside a given station or studio. Also called **field pickup**.

Remote Procedure Call A protocol that allows a program in a local computer to access a different program in a remote computer, especially a server. Once the remote computer completes the desired task, the outcome is returned to the local computer. Its abbreviation is **RPC**. Also called **Remote Procedure Call Protocol**.

Remote Procedure Call Protocol Same as **Remote Procedure Call**.

Remote Procedure Call Protocol Specification A specification which defines a **Remote Procedure Call** protocol.

remote programming Also called **teleprogramming**. 1. Programming by a device or system situated at a location other than that which is being programmed. 2. Programming utilizing a remote control.

remote reception Reception at the **remote side** (1).

remote resource In a communications network, a resource, such as a disk drive, located at a computer other than that of a given user. This contrasts with a **local resource**, which is that of the user's computer.

remote sensing The sensing of a signal or other phenomenon utilizing a detector, such as an infrared detector, which does not come into physical contact with that being measured or monitored. Also called **remote detection**, or **teledetection**.

remote sensor A device, such as a photocell, utilized for **remote sensing**. Also called **remote detector**.

remote side 1. In communications, the side further to a given terminal or node. For example, if a terminal is transmitting, the remote side is the destination, while the source terminal is the **local side**. 2. In communications, the portion of a device or piece of equipment which is connected to external facilities, such as those outside a given station.

remote storage The storage of data at a **remote location**. Used, for instance, for back-up or as a security measure. Also called **remote data storage**.

remote system Same as **remote computer**.

remote transmission Transmission from the **remote side (1)**.

removable disc Same as **removable disk**.

removable disk A disk, such as a DVD disc or floppy disk, which can be removed from its drive when not in use. Also called **removable disc**.

removable disk drive A disk drive, such as a DVD drive, floppy drive, or hard drive, which can be removed from a computer system when not in use. Also called **removable drive (1)**.

removable drive 1. Same as **removable disk drive**. 2. Same as **removable media drive**. 3. Same as **removable optical drive**.

removable hard drive A hard drive which can be removed from a computer system when not in use.

removable media Media, such as DVD discs, floppy disks, or magnetic tapes, which can be removed from its drive when not in use.

removable media drive A drive, such as a DVD drive, floppy drive, optical drive, or tape drive, which can be removed from a computer system when not in use. Also called **removable drive (2)**.

removable optical drive An optical drive which can be removed from a computer system when not in use. Also called **removable drive (3)**.

removable optical media Optical media, such as CD-ROMs and DVDs, which can be removed from its drive when not in use.

removable storage Data storage provided by **removable media**. Also called **removable data storage**.

REN Abbreviation of **ringer equivalence number**.

render To create more realistic two-dimensional images by incorporating three-dimensional attributes such as absorption, refraction, and reflection by objects and surfaces.

rendering In computer graphics, the process of creating more realistic two-dimensional images by incorporating three-dimensional attributes such as absorption, refraction, and reflection by objects and surfaces. Radiosity and ray tracing are two techniques utilized.

renovate To repair, remodel, or otherwise restore.

rep An obsolescent unit of radiation dose equivalent to 0.93 rad, or 0.0093gray. It is an acronym for roentgen equivalent physical.

repair 1. To restore to a given level of operation after failure, breakage, or other damage. 2. That which has been restored to a given level of operation after failure, breakage, or other damage.

repeat counter A counter which keeps track of repetitive events, such as a given number of loops executed by a program.

repeat key A key, or combination of keys, which repeat a procedure such as a computation or insertion of text. Most every key on a computer keyboard repeats when held down, which is useful, for instance, when a single character requires repeating or multiple paragraph breaks are desired.

repeatability The ability to duplicate readings, operations, results, or the like, in successive instances. For example, for an instrument to consistently provide the same indication of voltage or current under identical circumstances. Also called **reproducibility**.

repeated reflections Two or more reflections off surfaces, objects, or regions. For example, multiple reflections of light within a laser cavity. Also called **multiple reflections**.

repeater 1. A device, location, or system which receives, amplifies, and transmits signals between points or locations. Some repeaters may not perform amplification functions. A one-way repeater works in one direction, and a two-way repeater in both. 2. Same as **radio repeater**.

repeater satellite Same as **relay satellite**.

repeater station Same as **radio repeater station**.

repeater transmitter 1. Same as **radio repeater station**. 2. A transmitter within a radio repeater station.

repeating decimal A decimal with a digit, or sequence of digits, which repeats itself indefinitely. For example, 0.333333333333, and 0.345634563456. Also called **recurring decimal**.

repeating station Same as **radio repeater station**.

repeating transmitter 1. Same as **radio repeater station**. 2. A transmitter within a radio repeater station.

repeller Same as **reflector electrode**.

repetition frequency Same as **repetition rate**.

repetition instruction 1. A computer command to repeat one or more instructions. 2. An instruction to repeat a given operation or process.

repetition rate The rate at which something periodic repeats itself. Also called **recurrence rate, recurrence frequency,** or **repetition frequency**.

repetitive error Same as **recurring error**.

repetitive-motion injury Same as **repetitive-strain injury**. Its abbreviation is **RMI**.

repetitive-strain injury Damage to tissue, muscles, ligaments, tendons, bones, or the like, caused by repeated trauma. For example, even when using an ergonomically sound wrist pad, if a person types at a keyboard long enough, carpal-tunnel syndrome may result. Another example is trigger finger. Its abbreviation is **RSI**. Also called **repetitive-motion injury**.

replace The placing of text, numbers, or other data where other text, numbers, or data previously was. For example, to replace all occurrences of the word *delay* with the phrase *time delay* in a text document. Seen, for instance, in search and replace procedures.

replication 1. The process of making a copy of something, such as a text, file, or disc. 2. **Replication (1)** utilized to keep distributed databases synchronized.

reply 1. Same as **response**. 2. A response, such as an echo, or an email answering another.

report generator An application which allows users to create reports by choosing the information to be included, and presenting it in the desired format. Such programs, for instance, can access a database for material to be inserted into charts or graphs, automate many processes such as the placement of images and page headers and footers, and present the final output in a format, such as PDF, which is portable. Once a presentation is prepared, subsequent presentations can be automatically generated, for instance, by inserting updated data. Also called **report writer**.

report writer Same as **report generator**.

repository A location or system, such as a data mart or data warehouse, which stores and organizes the data of an entity, such as a corporation. Also, the data so held. Also called **data repository**.

reproduce head Same as **read head**.

reproducibility Same as **repeatability**.

reproducing head Same as **read head**.

reproducing system A device, or combination of devices, which serve to reproduce sound within the range of frequencies that humans can hear. Examples include high-fidelity audio systems and public-address systems. Also called **sound-reproduction system**.

reproduction 1. The playing back of that which has been recorded. For instance, the reproduction of sound. 2. The presenting of an original or another source of data or content. For example, the reproduction of data.

reproduction of sound The output of a **reproducing system**. When utilized in the context of music, the audio portion of a home theater, or the like, such reproduction may be monaural, stereophonic, multi-channel, or ambisonic.

reprogrammable That, such as a robot or memory chip, which can be programmed again and again.

reprogrammable robot A robot which can be programmed for a given task, then reprogrammed for others, if desired. Each programmed set of steps may involve any number of complex and repetitive tasks. Also called **programmable robot**.

repulsion The forces that act upon entities, tending to push them apart. Examples include the repulsion between similarly charged particles, or that between like magnetic poles. Also called **repulsive force**.

repulsion motor An AC motor whose brushes are short-circuited.

repulsion-start induction motor A motor that starts as a **repulsion motor**, and which beyond a determined speed runs as an induction motor.

repulsive force Same as **repulsion**.

request to send In communications, a signal some devices send in order to secure permission to send a transmission. The appropriate response from the other station is **clear to send**. Its abbreviation is **RTS**.

reradiation Same as **re-radiation**.

rerecord Same as **re-record**.

rerecording Same as **re-recording**.

reroute Same as **redirect**. Also spelled **re-route**.

rerouter Same as **redirector**. Also spelled **re-router**.

rerouting Same as **redirection**. Also spelled **re-routing**.

rerun Also called **rollback** (3). **1.** In computers, to repeat the execution of a program or job from the beginning. Done, for instance, to rectify an error. **2.** In computers, to repeat the execution of a program or job from any given point.

res Abbreviation of **resolution**.

resample To again sample, or to change a sampling rate. Also spelled **re-sample**.

resampling The process of again sampling, or the changing of a sampling rate. Also spelled **re-sampling**.

research and development The efforts, such as scientific investigation, involved in deriving and applying new knowledge to develop and/or improve upon an existing product. Its abbreviation is **R&D**.

Reservation Protocol Same as **Resource Reservation Protocol**. Its abbreviation is **RSVP**.

reserve **1.** To allocate, hold, or otherwise set aside for a specified purpose or content. For example, a memory area which is reserved for use by a given program. **2.** An additional amount which is factored in, or otherwise available, for a specific purpose. For example, a power reserve.

reserve battery A battery which is ready for use, or that can be prepared for use quickly and simply, which is put into use when needed. For example, a backup battery in a handheld computer.

reserved character A character, such as /, ?, or >, which has a special meaning in a program, programming language, or other setting, and whose use is restricted or not allowed. For example, the email address *where?here/n>w@zzz.com.qq* may not be allowed.

reserved word A word, such as a command, which has a special meaning in a program, programming language, or other setting, and whose use is restricted or not allowed. Also called **keyword** (3).

reset **1.** To set again. For example, to reset a circuit breaker. **2.** To restore a former condition, or to start again. For ex-

ample, to reboot a computer. **3.** Same as **reset button**. **4.** To restore a storage element or storage location to its zero state. Also called **clear** (5). **5.** To change a setting. For example, to reset a clock.

reset action **1.** The action of setting again. **2.** The action of restoring a former condition, or starting again. **3.** The action of restoring a storage element or storage location to its zero state. **4.** The changing of a setting. **5.** In a control system, a corrective action whose speed is dictated by the magnitude and duration of any change of the controlled variable. Also called **integral action**.

reset button Also called **reset** (3). **1.** A button, which when pressed, resets a computer. **2.** A button, which when pressed, resets a component, circuit, device, piece of equipment, system, or mechanism.

reset command A command which resets a component, circuit, device, piece of equipment, system, or mechanism.

reset current A current which resets a component, circuit, device, piece of equipment, system, or mechanism.

reset pulse **1.** A pulse that sets again. **2.** A pulse that restores to a former condition. **3.** A pulse that restores a storage element or storage location to its zero state.

reset-set flip-flop Same as **RS flip-flop**.

reset signal **1.** A signal that sets again. **2.** A signal that restores to a former condition. **3.** A signal that restores a storage element or storage location to its zero state.

reset switch A switch, which when actuated, resets a component, circuit, device, piece of equipment, system, or mechanism.

reset time **1.** The time that elapses after a reset signal or command is provided, to the actual resetting of a component, circuit, device, piece of equipment, system, or mechanism. **2.** The time that elapses before a device which automatic resets does so after opening, failing, or the like. For example, the time required for an automatic circuit breaker to reset.

reset timer A timer which keeps track of a given **reset time** (2).

reset voltage A voltage which resets a component, circuit, device, piece of equipment, system, or mechanism.

resident font A font which is permanently stored into a printer's memory. Also called **internal font**, or **built-in font**.

resident module A self-contained unit, or module, within a computer program which is located in memory at all times.

resident program A program which is located in memory at all times. Such programs are never swapped or otherwise removed from main memory. Certain portions of the operating system, and applications that require immediate access are memory resident. Some programs, such as virus-scanning applications, may be marked as memory resident by a user. Also called **RAM-resident program**, or **memory-resident program**.

residual AM Amplitude modulation of a signal when frequency or phase modulation is desired. It is an abbreviation of **residual** amplitude modulation. Also called **residual modulation** (2), or **incidental AM**.

residual amplitude modulation Same as **residual AM**.

residual charge **1.** In a capacitor, a charge remaining after an initial discharge or a rapid discharge. **2.** In a battery, any charge level remaining prior to recharging. In batteries with memory effect, the lower the residual charge, the better.

residual current A current that continues to flow through a component, circuit, or device, after the power supplied is removed.

residual current device Its abbreviation is **RCD**. A special type of circuit breaker which continuously monitors the current flowing in both the hot and neutral wires. Normally, this amount would be equal, but if even if a slight current leaks to ground, be it through the ground wire or through a person being shocked, there will be more current in one than the other. The RCD detects this, and shuts off all current flow within about 0.025 seconds, which should be fast enough to prevent any real harm to a person. Some RCDs detect current differences of less than 0.001 amperes. There are various types of RCDs, including the receptacle type, which replaces a conventional outlet, and those installed in a circuit-breaker panel. Also called **ground-fault circuit interrupter**.

residual error Also called **undetected error**. **1.** All error remaining after that which can be accounted for is taken into account. **2.** All error remaining after all error-correcting measures have been taken.

residual flux density Same as **remanence**.

residual FM Frequency modulation of a signal when amplitude or phase modulation is desired. It is an abbreviation of **residual** frequency modulation. Also called **residual modulation (3)**, or **incidental FM**.

residual frequency modulation Same as **residual FM**.

residual gas **1.** Any gas remaining in an electron tube after evacuation. A vacuum tube is one in which such gases are evacuated to a degree that its electrical characteristics are not affected. **2.** Any gas remaining in an enclosure after evacuation.

residual induction Same as **remanence**.

residual magnetic flux Same as **remanence**.

residual magnetic induction Same as **remanence**.

residual magnetism Same as **remanence**.

residual modulation Also called **incidental modulation**. **1.** In an RF signal, noise produced by variations of a carrier in the absence of any intended modulation. Also known as **carrier noise**. **2.** Same as **residual AM**. **3.** Same as **residual FM**. **4.** Same as **residual PM**.

residual phase modulation Same as **residual PM**.

residual PM Phase modulation of a signal when frequency or amplitude modulation is desired. It is an abbreviation of **residual** phase modulation. Also called **residual modulation (4)**, or **incidental PM**.

residual resistance The resistance of a material, especially a conductor, which is present regardless of temperature variations.

residual resistivity The resistivity that remains in a material as it is cooled to temperatures near absolute zero. This value is proportional to the impurities, lattice imperfections, grain boundaries, and the like, present in said material.

residual voltage A voltage that is present when a null reading is produced by a component, circuit, or device, such as a bridge.

resilience **1.** The capability a device, unit, piece of equipment, or system has to recover quickly from a fault condition, malfunction, error, or the like. **2.** The capability of a material or body to recover its dimensions, after a stress causing deformation is removed. Also called **elasticity**.

resilient Characterized by **resilience**.

resin Any of a class of amorphous substances, which can range from soft to very hard. Most resins are polymers, and may be naturally occurring, as is the case with amber, or synthetic, as in acrylic, epoxy, and phenolic resins.

resist A material whose chemical properties change when irradiated. Resists are usually utilized as thin films applied to substrates to selectively protect areas from energy beams. Utilized, for instance, to transfer patterns in semiconductor manufacturing techniques such as lithography.

resistance **1.** The opposition a material offers to the flow of current, with the concomitant conversion of electrical energy into heat energy. Resistance may be intentionally introduced into a circuit, for instance, for current regulation or for voltage regulation. In an AC circuit, resistance is the real number component of impedance. Expressed in ohms, or fractions/multiples of ohms. Its symbol is R. Also called **electrical resistance**. **2.** A specific **resistance (1)** value, such as that of a conductor, component, circuit, or device. Also called **resistance value**. **3.** The real number component of acoustic impedance. Expressed in acoustical ohms. Also called **acoustic resistance**. **4.** Same as **resistor**. **5.** The capability or extent to which a material or body can withstand something. For example, abrasion, impact, or heat resistance.

resistance alloy An alloy, such as constantan or nichrome, utilized to make resistors, resistance wires, or the like.

resistance box A box or other enclosure which houses multiple sections of precision resistors, whose values vary in given multiples from section to section. A user employs selector switches, or other mechanisms, to choose the appropriate value from each section to obtain the desired final resistance value. An example is a resistance decade box. Also called **resistor box**.

resistance braking A technique for quickly stopping an electric motor. To brake, the motor is disconnected from its power source, thereby becoming a generator. Then, a connected resistor dissipates the energy of rotation into heat. Also called **resistor braking**, **resistive braking**, or **dynamic braking**.

resistance brazing Brazing in which a high current is made to pass through the desired points, with the resulting heat melting the filler metal and thus joining the parts.

resistance bridge Also called **resistive bridge**. **1.** A bridge, such as a Wheatstone bridge, which incorporates only resistors. **2.** A bridge utilized to measure unknown resistances.

resistance bulb Same as **resistance lamp**.

resistance-capacitance Its abbreviation is RC. **1.** A combination of resistance and capacitance in a circuit. **2.** A circuit or device with a combination of resistance and capacitance, with little or no inductance. **3.** Pertaining to a **resistance-capacitance (1)**, or that incorporating or measuring such a combination, such as a meter.

resistance-capacitance circuit A circuit composed of, or incorporating resistances and capacitances. Such a circuit has little or no inductance. Its abbreviation is RC **circuit**.

resistance-capacitance filter A filter composed of, or incorporating resistances and capacitances. Such a filter has no inductances. Its abbreviation is RC **filter**.

resistance-capacitance-inductance Its abbreviation is RCL. Also called **inductance-capacitance-resistance**. **1.** A combination of resistance, capacitance, and inductance in a circuit. **2.** A circuit or device with a combination of resistance, capacitance, and inductance, such as that provided by coils and capacitors. For example, certain tuned circuits. **3.** Pertaining to a **resistance-capacitance-inductance (1)**, or that incorporating or measuring such a combination, such as a meter.

resistance-capacitance meter A meter which measures and indicates resistances and capacitances. Its abbreviation is RC **meter**.

resistance-capacitance network An electric network composed of, or incorporating resistances and capacitances. Such a network has little or no inductance. Its abbreviation is RC **network**.

resistance-capacitance oscillator An oscillator whose frequency is determined by resistances and capacitances. Its abbreviation is RC **oscillator**.

resistance-coupled amplifier An amplifier whose stages are coupled via **resistors**.

resistance coupling Same as **resistive coupling**.

resistance decade A decade box in which multiple sets of resistors enable the selection of any given resistance within its range. Also called **resistance decade box, resistor decade, resistor decade box, decade resistor (1)**, or **decade resistance box**.

resistance decade box Same as **resistance decade**.

resistance drop The voltage drop developed across a resistance through which a current flows. This obeys Ohm's law, which is $E = IR$, where E is voltage, I is current, and R is resistance. Also called **resistive drop**, or **IR drop**.

resistance element Also called **resistor element, resistive element**, or **resistive component (1)**. **1.** An element, such as a strip or wire of a resistance material, used within a **resistor**. **2.** Same as **resistor**.

resistance lamp An incandescent lamp utilized as a resistor. Used, for instance, to prevent the current flowing through a circuit from exceeding a given limit. Also called **resistance bulb**.

resistance loss Same as **resistive losses**.

resistance losses Same as **resistive losses**.

resistance material A material, such as carbon, certain metals, or resistance alloys, which can serve as a **resistor element**. Also called **resistor material**, or **resistive material**.

resistance meter A device or instrument which serves to measure and indicate resistance values.

resistance pad A fixed attenuator consisting of a network of fixed resistors. Also called **resistor pad, resistive pad**, or **pad (1)**.

resistance soldering Soldering in which a high current is made to pass through the desired points, with the resulting heat melting the solder and thus joining the parts.

resistance spot welding Spot welding in which a high current is made to pass through the desired points, with the resulting heat joining the parts.

resistance standard A resistor whose resistance value is highly accurate and stable, and which serves, for instance, as a calibration standard. Also called **standard resistance**.

resistance strain gage Same as **resistance strain gauge**.

resistance strain gauge A strain gauge which incorporates a strip or wire consisting of a resistance material. As such a material is stretched or compressed, its resistance varies, which in turn allows for the computation of applied strain. Also spelled **resistance strain gage**.

resistance strip A strip of a **resistance material**. Also called **resistor strip**, or **resistive strip**.

resistance temperature coefficient A change in the resistance of a component, device, or material as a function of temperature. A negative temperature coefficient of resistance indicates a decrease in resistance as temperature is raised, while a positive temperature coefficient of resistance indicates an increase in resistance as temperature is raised. Most conductors have a positive temperature coefficient of resistance, while most semiconductors and insulators have a negative temperature coefficient of resistance. Also called **temperature coefficient of resistance**.

resistance temperature detector The sensing element in a **resistance thermometer**. Its abbreviation is **RTD**. Also called **resistance temperature sensor**, or **resistive temperature detector**.

resistance temperature sensor Same as **resistance temperature detector**.

resistance thermometer A thermometer in which the sensing element is a resistance element, usually in the form of a wire or strip. The resistance material utilized in said element has a precisely known correlation between its resistance and increases or decreases in temperature, thus enabling accurate readings. Some resistance thermometers can provide indications ranging from below –200 °C, to above 1,000 °C.

resistance transducer Same as **resistive transducer**.

resistance value Same as **resistance (2)**.

resistance welding Welding in which a high current is made to pass through the desired points, with the resulting heat joining the parts.

resistance wire A wire made of a **resistance material**. Used, for instance, in wire-wound resistors. Also called **resistor wire**, or **resistive wire**.

resistive braking Same as **resistance braking**.

resistive bridge Same as **resistance bridge**.

resistive component **1.** Same as **resistance element**. **2.** In AC, the real number component of the impedance.

resistive coupling The coupling of two circuits or devices by means of a resistor. Also called **resistance coupling**.

resistive current Within an AC circuit, the component of the current that is in phase with the voltage. Also known as **watt current, in-phase current**, or **active current**.

resistive drop Same as **resistance drop**.

resistive element Same as **resistance element**.

resistive feedback The return of part of the energy from the output of a circuit back to its input by means of a common resistance.

resistive load A load in which the current is in phase with the voltage. Such a load consists essentially of a pure resistance.

resistive loss Same as **resistive losses**.

resistive losses Losses in power due to the resistance of the material, such as a conductor, through which current passes. An example is copper loss. Also called **resistive loss, resistance loss**, or **resistance losses**.

resistive material Same as **resistance material**.

resistive pad Same as **resistance pad**.

resistive strip Same as **resistance strip**.

resistive temperature detector Same as **resistance temperature detector**.

resistive transducer A transducer in which a variation in a quantity or signal produces a variation in resistance, which in turn produces a proportional conversion to a quantity or signal in another form. For example, in a resistance thermometer, a change in temperature causes a change in the resistance of the resistive element, which in turn produces a signal that is interpreted for readings. Also called **resistance transducer**.

resistive voltage Within an AC circuit, the component of the voltage that is in phase with the current. Also known as **active voltage**, or **in-phase voltage**.

resistive wire Same as **resistance wire**.

resistivity Its symbol is **r**, or ρ. A measure of the inherent ability of a material to resist the flow of current. It is the reciprocal of **electrical conductivity**, and depending on their resistivities, materials can be classified as insulators, semiconductors, or conductors. The lower the resistivity, the better conductor a material is. Its formula is: $r = RA/L$, where **r** is resistivity in ohm-meters, R is the resistance in ohms, A is the cross-sectional area of the material in square meters, and L is its length in meters. Also called **electric resistivity**, or **specific resistance**.

resistor A component or device with a fixed **resistance (2)**. A resistor opposes the flow of current, with the concomitant conversion of electrical energy into heat energy. There are various examples, including carbon, chip, film, thin-film, and wire-wound resistors. A rheostat, or variable resistor, is one whose resistance may be adjusted, while that of a fixed resistor cannot be varied. Also called **resistance element (2)**, **resistor element (2)**, or **resistance (4)**.

resistor box Same as **resistance box**.

resistor braking Same as **resistance braking**.

resistor color code A system utilizing markings, such as color dots and bands, to indicate resistor information such as values, multipliers, and tolerance.

resistor core The insulated core around which resistance wire is wound, as in the case of a wire-wound resistor.

resistor decade Same as **resistance decade**.

resistor decade box Same as **resistance decade**.

resistor element 1. Same as **resistance element**. 2. Same as **resistor**.

resistor fuse A low-value resistor used as a fuse. It is usually utilized to protect a circuit, device, piece of equipment, or system when power is first applied. Also called **fusible resistor**.

resistor material Same as **resistance material**.

resistor network 1. An electric network composed of multiple resistors. For multiple resistors connected in parallel, for instance, the total resistance is the sum of the reciprocal of each of the individual resistors. 2. An electric network incorporating multiple resistors.

resistor pad Same as **resistance pad**.

resistor strip Same as **resistance strip**.

resistor-transistor logic An early form of semiconductor logic in which resistors and transistors carry out logic functions. Its abbreviation is **RTL**.

resistor wire Same as **resistance wire**.

resistors in parallel Two or more resistors connected in parallel. The total resistance is the sum of the reciprocal of each of the individual resistors, thus, the multiple resistors have a lower total resistance than the lowest valued resistor. Also called **parallel resistors**, or **shunt resistors**.

resistors in series Two or more resistors connected in series. The total resistance is the sum of the values of each the individual resistors. Also called **series resistors**.

resize To change the dimensions of something. For example, that of an image or window displayed on a computer monitor. Also called **size (4)**, or **scale (6)**.

resolution Its abbreviation is **res**. 1. The minimum difference between two values, quantities, or entities, which can be distinguished by, separated, or acted upon, by a detector, sensor, measuring device, or the like. Also, such differentiation. 2. The smallest detail which can be distinguished by, or acted upon, by a detector, sensor, measuring device, or the like. Also, such differentiation. 3. The minimum distance between two objects which can be distinguished. For example, the angular resolution of a telescope, or the range resolution of a radar. Also called **resolving power (1)**. 4. The minimum distance between two separate and recognizable objects presented on a display screen, such as that of a radar. Also called **resolving power (2)**. 5. The extent to which an optical system can reproduce the points, lines, and surfaces of an object as distinct entities in an image. 6. The precision with which a quantity or value can be measured or indicated. 7. The minimum adjustment which is detectable, or which can be made by a given system or mechanism, such as a control system. 8. The process of separating something into its constituent parts. Also, such a separation. For instance, the dispersion of white light into its component colors by a prism. 9. The degree of sharpness, clarity, and fineness of detail which a display screen, scanner, printer, or other such device is capable of presenting, processing, or producing. Monitors, for instance, may express resolution in terms of the number of pixels displayed and dot pitch. Also, the resolution setting of such a device. Also, the fine-

ness present in such a displayed, processed, or printed image.

resolution chart A chart, such as that with groups of parallel lines of varying width, which serves to determine the resolution of a TV or monitor. It is a type of test pattern. Also called **resolution pattern**.

resolution pattern Same as **resolution chart**.

resolving power 1. Same as **resolution (3)**. 2. Same as **resolution (4)**. 3. In spectroscopy, the ability to separate or differentiate between measured parameters of nearly equal magnitude or value. For example, the separation of particles with nearly identical mass by a mass spectrometer, or the differentiation of spectral lines of nearly equal wavelength by a spectroscope.

resonance 1. The condition in which a circuit, body, or system exhibits enhanced vibration or oscillation at a given frequency, called **resonant frequency**. There are many examples, including acoustic, atomic, beam, magnetic, and spin resonance. 2. The condition in which a circuit, body, or system exhibits enhanced vibration or oscillation when acted upon by an external force which is at or near the resonant frequency of said circuit, body, or system. 3. The condition in which the amplitude or intensity of a phenomenon is enhanced at frequencies or energies which are at or near a given value. 4. The condition in which the response of a circuit or system is enhanced when the driving signal is at or near the resonant frequency of said circuit or system. Also, the condition of exhibiting maximum response at said frequency. 5. The condition in which of acoustic waves are intensified at certain frequencies due to **resonance (1)**. Also called **acoustic resonance (1)**. 6. The condition in which an object close to a sound source vibrates along with said source.

resonance absorption Absorption, such as that of rays or particles, which is enhanced at given frequencies.

resonance bridge A four-arm AC bridge that incorporates a series- or parallel-resonant arm, and which is balanced at its resonant frequency. Also called **resonant bridge**.

resonance cavity Same as **resonant cavity**.

resonance characteristic Same as **resonance curve**.

resonance circuit Same as **resonant circuit**.

resonance curve A curve depicting the response of a resonant circuit, or other body or system exhibiting resonance. In the case of a circuit, for instance, such a graph usually shows current or voltage in the vertical axis, and frequency in the horizontal axis. Also called **resonance characteristic**.

resonance filter Same as **resonant filter**.

resonance fluorescence Same as **resonance radiation**.

resonance frequency Same as **resonant frequency**.

resonance luminescence Same as **resonance radiation**.

resonance radiation The emission of electromagnetic radiation, such as light, by a body which has been excited at its resonant frequency by another form of energy, such as electron bombardment. Also called **resonance luminescence**, **resonance fluorescence**, **resonant radiation**, **resonant luminescence**, or **resonant fluorescence**.

resonance width The interval of frequencies within which a resonator can resonate. The narrower the resonance width, the higher the Q of a given resonator.

resonant antenna An antenna whose parameters, such as capacitance and inductance, provide for resonance at its operating frequency. Also called **tuned antenna**.

resonant bridge Same as **resonance bridge**.

resonant cavity An enclosure which is able to maintain an oscillating electromagnetic field when suitably excited. The geometry of the cavity determines the resonant frequency. Used, for instance, in lasers, klystrons, or magnetrons, or as a filter. Also known by various other names, including **resonating cavity**, **resonant chamber**, **resonance cavity**, **rhumbatron**, **cavity (2)**, **cavity resonator**, **microwave cavity**, **microwave resonance cavity**, and **waveguide resonator**.

resonant chamber Same as **resonant cavity**.

resonant circuit A circuit whose parameters provide for resonance at a single frequency. In a parallel-resonant circuit, for instance, a capacitor and an inductor are connected in parallel with an AC source, which provides for resonance at or near the maximum impedance of such a circuit. Also called **resonance circuit**.

resonant dipole A dipole antenna whose parameters, such as capacitance and inductance, provide for resonance at its operating frequency. Also called **resonant dipole antenna**, or **tuned dipole**.

resonant dipole antenna Same as **resonant dipole**.

resonant filter A filter whose action results in the boosting of certain frequencies, inducing resonance. Also called **resonance filter**.

resonant fluorescence Same as **resonance radiation**.

resonant frequency Also called **resonance frequency**. **1.** A frequency at which a circuit, body, system, or phenomena exhibits enhanced vibration, oscillation, response, absorption, amplitude, or the like. For instance, the vibration of a piezoelectric transducer is greatest with the least applied voltage at its resonant frequency. That of a resonant cavity is determined by its geometry, and so on. **2.** The frequency at which a body or system resonates without an external driving force. Also called **natural resonant frequency**, or **self-resonant frequency**.

resonant line Also called **resonant transmission line**. **1.** A transmission line which has reflections, and therefore has standing waves. **2.** A transmission line which resonates at the frequency utilized.

resonant luminescence Same as **resonance radiation**.

resonant radiation Same as **resonance radiation**.

resonant resistance The resistance of a resonant circuit or resonant line at its resonant frequency.

resonant transmission line Same as **resonant line**.

resonate To exhibit or produce **resonance**.

resonating cavity Same as **resonant cavity**.

resonator That which exhibits or produces **resonance**. For example, an acoustic cavity, or a crystal resonator.

resource A means utilized to perform one or more operations, or to accomplish a given task. In computers, for instance, resources include storage devices, processors, and applications.

resource allocation **1.** The designation or distribution of **resources**. **2.** The reserving of computer resources for specific purposes.

Resource Description Framework A framework for describing and processing metadata. Used, for instance, to enhance the capabilities of search engines, or for cataloging Web site content. Its abbreviation is **RDF**.

resource requirements The minimum software and/or hardware a system must have in order for a given program or device to work properly. For example, to run a given application a computer needing to have a minimum amount of available memory and disk space, one of the supported operating systems, an available port, and so on.

Resource Reservation Protocol A communications protocol which allows for the reservation of a given bandwidth, and which confers a high priority to the packets sent during the subsequent transmission. Used, for instance, for multicasting of real-time of audio and/or video programming. Its abbreviation is **RSVP**. Also called **Resource Reservation Setup Protocol**, or **Reservation Protocol**.

Resource Reservation Setup Protocol Same as **Resource Reservation Protocol**. Its abbreviation is **RSVP**.

resource sharing The distribution of **resources**, or their or simultaneous use by multiple users or entities.

resources management The management of **resources**.

responder Also called **responsor**. **1.** A device or system which incorporates both a transmitter and a receiver, and which emits the appropriate signals automatically when the specified trigger signals are received. Used, for instance, in satellite communications, or in radars. Also called **transponder (1)**. **2.** The part of a **responder (1)** which transmits signals when triggered by an **interrogator (1)**. An example is a radar beacon.

response **1.** A change in the behavior, operation, or function of a component, circuit, device, piece of equipment, system, mechanism, process, material, or the like, as a consequence of a change in its external or internal environment. For example, a change in output resulting from a change in input. **2.** A quantified **response (1)**, such as a frequency response. **3.** The automatic emission of the appropriate signals when specified trigger signals are received, as seen, for instance, in a **responder (1)**. Also called **reply**.

response characteristic The gain or loss of a circuit, device, piece of equipment, or system, as a function of its input signal frequencies. Also called **frequency-response characteristic**, or **frequency characteristic**.

response curve A graph representing the **response (1)** of a component, circuit, device, piece of equipment, or system. An example is a frequency-response curve.

response delay Same as **response time**.

response interval Same as **response time**.

response time Also called **response delay**, or **response interval**. **1.** The time interval that elapses between the reception of a trigger signal and the corresponding **response (1)**. **2.** The time interval that elapses between a change in the external or internal environment of a component, circuit, device, piece of equipment, system, mechanism, process, material, or the like, and the corresponding **response (1)**. For example, the time required for an instrument to update its reading when a monitored parameter varies.

responsor Same as **responder**.

rest energy The energy of a particle at rest, according to the mass-energy equation. The rest energy of an electron, for instance, is approximately 0.511 MeV, while that of a proton is approximately 938 MeV. Also called **self-energy**.

rest mass The mass of a particle at rest. The rest mass of an electron, for instance, is equal to approximately 9.10939×10^{-31} kg, while that of a proton is approximately 1.67262×10^{-27} kg.

rest position The position a device, mechanism, or object is in when no signal is applied. Also, the position returned to once a signal is no longer present. Also called **resting position**.

rest state Same as **resting state**.

restart To again start up a device, piece of equipment, unit, or system. For example, a checkpoint/restart. Also spelled **re-start**.

resting position Same as **rest position**.

resting state Also called **rest state**, **idle state**, or **quiescent state**. **1.** A state in which a component, circuit, device, piece of equipment, or system is functioning under normal operating conditions, but does not have an applied signal at a given time. Also called **standby (1)**. **2.** A state in which a device, piece of equipment, or system is operational and available, but is not in use. Also called **standby (2)**.

restore 1. To make repairs that are sufficient to bring back to the original condition and/or functionality. **2.** To reset or otherwise go back to a former condition. **3.** To return computer data to the state it was in before an error or failure.

restorer 1. That which serves to **restore**. **2.** Same as **reinserter**.

restricted function 1. A function which cannot be used by a given application, or under certain circumstances. **2.** A function which is not available to a given user.

retarding electrode In an electron-beam tube such as a klystron, a charged electrode which creates an electric field which serves to reduce the velocity of the electrons in the beam. Also called **decelerating electrode**.

retarding field A field, such as that produced by a retarding electrode, which decelerates electrons.

retarding magnet In motor-type meters, a magnet which provides drag to limit the speed of the rotor or disk. The braking action is due to the effects of eddy-currents. Also called **drag magnet**.

retention period 1. The time interval during which data, records, files, archives, or the like must be stored. **2.** Same as **retention time**.

retention time The time interval during which a storage medium, such as RAM, a hard disk, a smart card, or an optical disc, can reliably retain data. Also called **retention period (2)**, or **storage time (1)**.

retentivity Also called **magnetic retentivity**. **1.** The ability of a material to retain magnetism after a magnetizing force is removed. **2.** The degree to which magnetism remains in a material or medium after a magnetizing force is removed. **3.** A specific value of **retentivity (2)**.

reticle 1. In an optical instrument, a grid or pattern which helps determine a scale or location. **2.** In semiconductor manufacturing, a photomask which identifies the boundaries of components within an IC.

retrace In a CRT, the return of the electron beam to its starting point after a sweep or trace. Also, the time interval that elapses during this return. Also called **return trace**, **return (6)**, **flyback**, or **kickback (2)**.

retrace blanking In a device such as a CRT, the cutting-off of the electron beam during **retrace**. Also called **beam blanking**, or **blanking (2)**.

retrace interval Same as **retrace time**.

retrace line In a CRT, a line traced by the electron beam as it returns to its starting point. Also called **return line (2)**.

retrace period Same as **retrace time**.

retrace time The time interval that elapses during **retrace**. Also called **return time**, **retrace interval**, **return interval**, **retrace period**, or **flyback time**.

retransmission The process of again transmitting a signal or data. For instance, the reemission of a broadcast, or the resending of data. Also spelled **re-transmission**.

retransmission system A system, such as a radio repeater system, which serves for **retransmission**. Also spelled **re-transmission system**.

retransmit To again transmit a signal or data. For example, to reemit a broadcast, or to resend data. Also spelled **re-transmit**.

retransmitter A device, such as a radio repeater, which serves to **retransmit**. Also spelled **re-transmitter**.

retrieval The process of accessing specific information located in memory or a storage device. Also called **data retrieval**.

retrieve To access specific information located in memory or a storage device.

retro-reflector Same as **retroreflector**.

retro-reflector array Same as **retroreflector array**.

retrofit To add, modify, or substitute components, circuits, devices, or modules to an existing product, when such items were not available or otherwise provided when said product was originally built. Also, that which has been upgraded or otherwise modified by such additions or changes.

retrograde orbit The orbit of an artificial satellite whose followed path is in the direction opposite of the rotation of the earth.

retroreflector Also spelled **retro-reflector**. **1.** A device which serves to reflect incident radiation, especially light, back in the direction it came from, or a path parallel to this direction. **2.** A **retroreflector (1)** utilized to reflect light in a laser. Also called **laser retroreflector**.

retroreflector array Multiple **retroreflectors (1)** utilized for tracking objects, such as satellites, or for performing measurements, testing, or the like. Also spelled **retro-reflector array**.

return 1. To go back, bring back, or send back, to an earlier state, place, setting, or the like. Also, the act of going, bringing, or sending back. **2.** To reinstate control of a computer program to the point where an instruction, routine, function, or other operation was called. This is done once such an operation is completed. Also, the reinstating of control. **3.** Same as **radar echo**. **4.** Same as **return key**. **5.** The action resulting from pressing a **return key**. **6.** Same as **retrace**.

return beam A beam which is reflected or otherwise sent back to its source. Seen, for instance, in radars.

return call A service offered by telephone companies and/or phones equipped with this feature, which enables a subscriber to dial the last incoming call. A sequence of digits is usually dialed to activate this feature. Frequently, this service continues to dial the desired number if it is initially busy, and alerts when it is available. Some incoming calls may not allow for call returning. Also known as **call return**, or **last number redial**.

return channel In two-way communications, a channel via which data, messages, or control information is received, while the channel which serves for transmission is the **send channel**.

return circuit A circuit via which current returns to a power station.

return code In computer programming, a code that influences the execution of subsequent instructions or processes.

return conductor A conductor through which current must pass to complete a circuit. Also called **return lead**, **return line (1)**, or **return wire**.

return current 1. Current which flows through a **return conductor**. **2.** Current which returns to its source upon encountering a discontinuity in a transmission line. Also called **reflected current**.

return echo Same as **radar echo**.

return instruction After completion of an instruction, routine, function, or other operation which has been called, the command or statement which reinstates control to the calling program. Also called **return jump**.

return interval Same as **retrace time**.

return jump Same as **return instruction**.

return key On a keyboard, a key which when pressed serves to inform the computer that the user wishes the entered information to be processed. For instance, once the information requested by prompt is given, the enter key may be pressed. It may also have program-specific functions, such as its use within a word-processing application to signal the end of a paragraph. Also called **return (4)**, or **enter key**.

return lead Same as **return conductor**.

return line 1. Same as **return conductor**. **2.** Same as **retrace line**.

return loss Also called **return losses, reflection loss,** or **reflection losses.** **1.** At a discontinuity in a transmission system, or at any point where there are mismatched impedances, the difference between the incident power and the reflected power from said discontinuity or point. Usually expressed watts. **2.** At a discontinuity in a transmission system, or at any point where there are mismatched impedances, the ratio of incident power to the reflected power from said discontinuity or point. Usually expressed in dB, or as a percentage. **3.** Any losses caused by reflections, such as those within a transmission line, or off a surface.

return-loss meter An instrument or device which measures and indicates **return losses.** Also called **reflection-loss meter.**

return losses Same as **return loss.**

return signal **1.** The signal emitted by a **responder.** This signal which triggers this response is an **interrogation signal.** **2.** In communications, a signal returning to its source. **3.** In communications, a signal arriving via a **return channel.**

return statement After completion of an instruction, routine, function, or other operation which has been called, the statement which reinstates control to the calling program.

return time Same as **retrace time.**

return to zero Its abbreviation is **RZ,** or **RTZ.** **1.** A data-encoding method in which the signal returns to a zero state between sent bits. For instance, ones may be represented by a positive voltage and zeroes by a negative voltage, with a zero-voltage condition between successive bits. Used for digital data transmission. Also called **return to zero encoding,** or **return to zero code.** **2.** A magnetic tape recording technique which utilizes **return to zero (1).**

return to zero code Same as **return to zero (1).** Its abbreviation is **RZ code.**

return to zero encoding Same as **return to zero (1).** Its abbreviation is **RZ encoding.**

return trace Same as **retrace.**

return wire Same as **return conductor.**

rev **1.** Abbreviation of **reverse.** **2.** Abbreviation of **revolution.**

reverberation **1.** The persistence of sound in an enclosed space, especially a reverberation chamber, after the source is stopped. **2.** The tailing-off of sound in an enclosure, especially a reverberation chamber. **3.** The condition of being repeatedly reflected. May refer to light, sound, heat, and so on.

reverberation chamber A room in which the walls and other surfaces have a composition and placement specifically designed to reflect sound as much as possible. Used, for instance, for acoustic testing. It is the converse of an **anechoic chamber.** Also called **reverberation room.**

reverberation room Same as **reverberation chamber.**

reverberation time For a given enclosure, especially a reverberation chamber, the time interval that elapses for the sound-pressure level to decay 60 dB after the sound source is stopped. This represents a million-fold reduction in sound pressure.

reverberation unit A device which generates artificial echoes, so as to simulate a **reverberation chamber.**

reverse Its abbreviation is **rev.** **1.** In the opposite position, order, direction, side, or the like, of the usual. For instance, a reverse current, or a reverse image. **2.** In the opposite position, order, direction, side, or the like, of something else. For example, a back panel in relation to a front panel. **3.** Having the opposite phase or polarity as something else. For instance, reverse feedback. **4.** To invert a given direction, order, phase, polarity, action, or the like. For example,

in a stereo audio component, to switch the signals of the left and right channels.

reverse address resolution protocol A protocol for determining the IP address of a specific computer or node within the Internet, when the hardware, or physical, address is known. It is the converse of **address resolution protocol.** Its abbreviation is **reverse ARP,** or **RARP.**

reverse ARP Abbreviation of **reverse address resolution protocol.**

reverse bias A bias voltage applied in the proper polarity to a diode or semiconductor junction to cause little or no current to flow. Also called **back bias (2),** or **inverse bias.**

reverse breakdown voltage The level of reverse bias which will cause a semiconductor junction to break down and conduct in the reverse direction.

reverse channel In asymmetrical communications, the channel that flows upstream. It is the slower of the two channels, the other being the **forward channel.** Also called **backward channel.**

reverse conduction Conduction through a component or device in the **reverse direction.**

reverse current In a device such as a rectifier or semiconductor, current that flows in the opposite direction of that which is normal. Also called **back current,** or **inverse current.**

reverse-current relay A relay that is actuated by **reverse current.** Used, for instance, to protect rectifiers. Its abbreviation is **RCR.**

reverse direction For an electrical or electronic component or device, the direction in which there is greater resistance to the flow of current. Used, for instance, in the context of semiconductors. This contrasts with **forward direction,** where the converse is true. Also called **inverse direction.**

Reverse DNS A request for a conversion from an IP address to a domain name. It is the converse of a **DNS query.**

reverse emission In a vacuum tube, the emission of electrons in the reverse direction, that is, from anode to cathode. Also called **back emission.**

reverse engineer To arrive at a device, system, program, or the like, through the process of **reverse engineering.**

reverse engineering **1.** The process of analyzing an existing device or system, identifying its components, their functions and interrelationships, so as to better understand, duplicate, or improve upon it. **2.** The utilization of **reverse engineering (1)** to see how a computer program works, to duplicate it, or improve upon it. This starts by reversing a program's machine code back to the source code it was written in, and working backwards to the original program.

reverse feedback In an amplifier or system, feedback which is 180° out-of-phase with the input signal, thus opposing it. While this results in a decrease in gain, it also serves reduce distortion and noise, and to stabilize amplification. Also called **inverse feedback, negative feedback, stabilized feedback,** or **degeneration.**

reverse image An image, such as a photographic image, in which the light and dark areas are reversed in relation to the scene the image is taken from. Also called **negative (5),** or **negative image.**

reverse path forwarding The use of the source address, as opposed to the destination address, for the routing of data packets. Used, for instance, for deducing the next hops for broadcast and multicast packets. Its abbreviation is **RPF.**

reverse peak voltage For an electrical or electronic component or device, the maximum instantaneous value of the voltage in the direction in which there is the greatest resistance to the flow of current. For example, the maximum reverse-biased voltage a semiconductor can safely handle before avalanche breakdown will occur. Also called **peak inverse voltage.**

reverse piezoelectric effect The mechanical deformation of certain crystals or ceramics when exposed to an electric field. Conversely, **direct piezoelectric effect** is the generation of an electric charge by subjecting certain crystals or ceramics to mechanical strain. Both are complementary manifestations of the piezoelectric effect. Also called **inverse piezoelectric effect, inverse piezoelectricity,** or **indirect piezoelectricity.**

reverse Polish notation A manner of forming mathematical or logical expressions in which the operators appear after the operands. For example, **(A + B) x C,** which is in **infix notation,** would appear as **ABC+*** in reverse Polish notation. Also called **postfix notation,** or **suffix notation.** Its abbreviation is **RPN.**

reverse recovery time The time interval that elapses after switching between forward bias and reverse bias, and the reverse current reaching saturation.

reverse resistance The resistance offered to the flow of current in the **reverse direction.**

reverse solidus In most computer operating systems, a character which serves to separate folders and directories in file pathnames, as seen, for instance, in: C:\Documents\Reminders\Today.xyz. This character has a place on the keyboard, and is shown here between brackets: [\ ]. Also known as **backslash.**

reverse video On a computer display, the exchanging of the light and dark areas of the screen. That which is normally light appears as dark, and vice versa. When highlighting, for instance, the selected area is displayed as reverse video. Also called **inverse video.**

reverse voltage **1.** The voltage developed in an inductive circuit through which AC flows. The polarity of this voltage is at all times the opposite of that of the applied voltage. Also known as **back electromotive force, back voltage,** or **counter electromotive force.** **2.** A voltage whose polarity is at all times the opposite of another voltage. **3.** For an electrical or electronic component or device, the voltage in the direction in which there is the greater resistance to the flow of current. Also called **inverse voltage (1).**

reversible actions Actions which can be undone, or which otherwise allow to return to a former state. In computer applications, for instance, examples include undo commands and dialog boxes reminding a user that a document will be closed without saving it.

reversible counter A counter which can count both upwards and downwards. It has, for instance, both an adding input and a subtracting input, thus giving it the ability to count in both directions. Also called **bi-directional counter, forward-backward counter,** or **up-down counter.**

reversing motor A motor whose direction of rotation can be reversed. Used, for instance, in certain electromechanical tools such as drills.

reversing switch A switch that reverses a direction, position, order, polarity, phase, or the like. For example, a switch inside of a transceiver which changes between transmission and reception modes, or a polarity switch.

revert **1.** To return to a former condition or state. **2.** To return to a former version of a document. For example, to revert to the latest saved version.

revision level Same as **release version.**

revolution Its abbreviation is **rev.** **1.** Rotational motion about an axis or body. Also, a single complete rotation. For example, 360° represents a complete revolution around a circle. **2.** Orbital motion about a point or body. Also, a single complete rotation. For instance, a revolution of the earth about the sun.

revolutions per minute A unit of angular velocity indicating the number of 360° rotations completed per minute. This is equal to about 0.1047198 radian per second. Its abbreviation is **rpm.**

revolutions per second A unit of angular velocity indicating the number of 360° rotations completed per second. This is equal to about 6.2831853 radians per second. Its abbreviation is **rps.**

revolving field Same as **rotating field.**

revolving magnetic field Same as **rotating magnetic field.**

REW Abbreviation of **rewind.**

rewind Its abbreviation is **REW.** **1.** To rapidly wind a tape in the direction opposite of playback. For instance, that in a VHS cassette or digital audio tape cartridge. The term may also be used to refer to moving backwards rapidly within the content of an optical disc, such as a DVD. **2.** The mechanism utilized to **rewind (1).** **3.** A command or function which **rewinds (1).** **4.** To wind again.

rewind button A button which activates a **rewind (1)** function.

rewritable Also spelled **re-writable.** Its abbreviation is **RW.** **1.** The ability to be erased and written or record upon multiple times. Said, for instance, of optical discs, chip cards, magnetic disks, magnetic tapes, or the like. **2.** A specific medium, such as a DVD disc, which is **rewritable (1).**

rewritable CD Its abbreviation is **RW CD.** **1.** A CD technology which allows discs to be rewritten many times. A laser records the desired content, forming the equivalent of pits on the recordable surface by altering the reflectivity of a dye layer. When such a disc is placed in a player, its laser pickup reads the patterns as if they had been permanently stamped into its metallic surface. Also called **CD-RW,** or **CD-rewritable.** **2.** A CD disc which can be rewritten many times. Also called **CD rewritable disc.**

rewritable disc Also spelled **re-writable disc.** Its abbreviation is **RW disc.** Also called **rewritable optical disc.** **1.** A technology which allows optical discs, such as DVDs, to be rewritten many times. **2.** An optical disc which can be rewritten many times.

rewritable DVD Its abbreviation is **RW DVD.** **1.** A DVD technology which allows discs to be rewritten many times. To record, a high-intensity laser followed by a medium-intensity laser anneals the crystals in the recording layer, changing its reflectivity. A low-intensity laser in the DVD player reads the recorded information. Also called **DVD-rewritable,** or **DVD-E.** **2.** A DVD disc which can be rewritten many times. Also called **DVD rewritable disc.**

rewritable optical disc Same as **rewritable disc.** Also spelled **re-writable optical disc.** Its abbreviation is **RW optical disc.**

rewrite To erase and write or record upon again. Said for instance, of RAM, flash cards, optical discs, hard disks, magnetic tape, and the like. Also spelled **re-write.**

Rf Chemical symbol for **rutherfordium.**

RF Abbreviation of radio frequency. Also called **RF frequency,** or **radio (6).** **1.** Within the electromagnetic spectrum, a frequency or interval of frequencies utilized for communications. Also called **RF range (1).** **2.** A specific interval of RF frequencies, especially the **radio spectrum.** Also called **RF range (2).** **3.** A specific **RF (1),** such as that of a carrier wave or VHF.

RF amplification Abbreviation of radio-frequency **amplification.** The use of an **RF amplifier.** Also, the resulting amplification.

RF amplifier Abbreviation of radio-frequency **amplifier.** A circuit or device utilized to amplify RF signals.

RF antenna Abbreviation of radio-frequency **antenna.** An antenna designed for transmitting and/or receiving radio frequencies. There are many examples, including bowtie, coaxial, dipole, dish, helical, horn, loop, and Yagi antennas.

RF band Abbreviation of radio-frequency **band**. Same as **radio spectrum**.

RF carrier Abbreviation of radio-frequency **carrier**. A carrier wave whose frequency is in the radio spectrum.

RF channel Abbreviation of radio-frequency **channel**. A channel utilized for **RF communications**. Also called **RF circuit**, **RF link**, **radio circuit** (1), or **radio channel** (4).

RF choke Abbreviation of radio-frequency **choke**. A low-inductance coil utilized to block radio-frequency alternating currents, by presenting a relatively high impedance at such frequencies. Its own abbreviation is **RFC**.

RF circuit Abbreviation of radio-frequency **circuit**. Same as **RF channel**.

RF coil Abbreviation of radio-frequency **coil**. A coil utilized for carrying **RF current**.

RF communications Abbreviation of radio-frequency **communications**. Communications in which radio frequencies, such as VLF, EHF, or microwaves, are utilized to convey information.

RF converter Abbreviation of radio-frequency **converter**. A converter whose input and/or output is within the RF range. For example, a converter which changes the frequency of a signal or current from one radio frequency to another.

RF current Abbreviation of radio-frequency **current**. An AC whose frequency is within the radio spectrum. Used, for instance, in communications, for medical diagnostic or therapeutic purposes, or in heating.

RF detector Abbreviation of radio-frequency **detector**. A circuit, device, or apparatus which detects RF signals. Used, for instance, to detect electronic bugs.

RF device Abbreviation of radio-frequency **device**. Same as **radio device**.

RF diode Abbreviation of radio-frequency **diode**. **1.** A diode specifically designed to work optimally at radio frequencies. **2.** A diode that operates at radio frequencies.

RF energy Abbreviation of radio-frequency **energy**. The energy associated with **radio waves**. Also called **RF radiation**.

RF field Abbreviation of radio-frequency **field**. The field associated with **radio waves**. It is a combination of electric and magnetic fields which are at right angles to each other, and to the direction of propagation.

RF filter Abbreviation of radio-frequency **filter**. **1.** A filter which passes a given interval of radio frequencies with little or no attenuation. **2.** A filter which highly attenuates a given interval of radio frequencies.

RF frequency Same as **RF**.

RF generator Abbreviation of radio-frequency **generator**. A generator whose output is at radio frequencies. Used, for instance, for RF heating.

RF harmonic Abbreviation of radio-frequency **harmonic**. A harmonic of a radio frequency, such as a carrier wave.

RF heating Abbreviation of radio-frequency **heating**. The use of a radio-frequency power source, such as an oscillator, to provide heat. Also called **electronic heating**, or **high-frequency heating**.

RF IC Same as **RFIC**.

RF ID Same as **RFID**.

RF identification Same as **RFID**.

RF indicator Abbreviation of radio-frequency **indicator**. An indicator, such as a lamp, which reveals the presence of RF energy. Used, for instance, to verify that a transmitter is transmitting.

RF integrated circuit Same as **RFIC**.

RF interference Abbreviation of radio-frequency **interference**. An electromagnetic disturbance, occurring at radio frequencies, which brings about a degradation in performance, a malfunction, or the failure of a component, circuit, device, piece of equipment, or system. Such interference can be produced by countless sources, such as microchips, lightning, RF amplifiers, ignition systems, geomagnetism, and so on. Devices such as shields and filters are utilized to minimize or prevent its effects. Its own abbreviation is **RFI**.

RF lamp Abbreviation of radio-frequency **lamp**. An electrodeless lamp which is powered by RF energy. Such lamps provide higher efficiency and longer life than comparable incandescent and fluorescent lamps.

RF line Abbreviation of radio-frequency **line**. A transmission line utilized to carry RF signals or energy. An example is an antenna feed. Also called **RF transmission line**.

RF link Abbreviation of radio-frequency **link**. Same as **RF channel**.

RF modulation Abbreviation of radio-frequency **modulation**. The process of modifying a characteristic of an RF carrier wave by an information-bearing signal, as occurs, for instance, in AM, FM, or PM. Also, the result of such modulation. For example, in FM, the instantaneous frequency of a sine-wave carrier is varied above and below the center frequency by an information-bearing, or modulating, signal. Also called **modulation** (2), or **carrier modulation**.

RF noise Abbreviation of radio-frequency **noise**. Same as **radio noise**.

RF oscillator Abbreviation of radio-frequency **oscillator**. An oscillator whose output is at radio frequencies. Klystrons and magnetrons, for instance, produce microwave oscillations.

RF power Abbreviation of radio-frequency **power**. Power whose frequency is within the **radio spectrum**. An oscillator, for instance, serves as a source for such power.

RF power amplifier Abbreviation of radio-frequency **power amplifier**. A power amplifier whose output is at radio frequencies. An example is the output stage of a radio transmitter.

RF power supply Abbreviation of radio-frequency **power supply**. A power supply, such as that used for a transmitting antenna, whose output is **RF power**.

RF probe Abbreviation of radio-frequency **probe**. A probe, such as a waveguide probe, which is utilized to transfer RF energy into or out of an enclosure.

RF pulse Abbreviation of radio-frequency **pulse**. A pulse, such as that utilized in nuclear magnetic resonance, occurring at a radio frequency. Also called **radio pulse**.

RF radiation Abbreviation of radio-frequency **radiation**. Same as **RF energy**.

RF range Abbreviation of radio-frequency **range**. **1.** Same as **RF** (1). **2.** Same as **RF** (2).

RF reception Abbreviation of radio-frequency **reception**. The reception of RF waves, such as those emitted by a radio transmitter, or arriving from space.

RF resistance Abbreviation of radio-frequency **resistance**. The combined resistance offered by a device in a high-frequency AC circuit. It includes, among others, DC resistance, and resistance due to dielectric and eddy-current losses. Also known as **high-frequency resistance**, **AC resistance**, or **effective resistance**.

RF response Abbreviation of radio-frequency **response**. **1.** A response, such as that of a receiver, to RF waves. **2.** A quantified **RF response** (1).

RF screen **1.** Same as **RF shielding** (1). **2.** Same as **RF shielding** (2).

RF screening Same as **RF shielding**.

RF sensitivity Abbreviation of radio-frequency **sensitivity**. **1.** The input RF signal level which will drive a device, such as a receiver, to its full rated output, or to a stated output. **2.** The minimum input RF signal level which will be detected by a device, such as a receiver.

RF sensor Abbreviation of radio-frequency **sensor**. Same as **RF detector**.

RF shield Abbreviation of radio-frequency **shield**. Also called **radio shield**. **1.** Same as **RF shielding** (1). **2.** Same as **RF shielding** (2).

RF shielding Abbreviation of radio-frequency **shielding**. Also called **RF screening**, **radio shielding**, or **radio screening**. **1.** A material or enclosure which blocks the effects of RF fields. Used, for instance, to help minimize or prevent the effects of external electromagnetic interference. Also called **RF shield** (1), **RF screen** (1), or **radio screen** (1). **2.** An **RF shielding** (1) which serves to confine RF fields within an enclosure. Used, for instance, to help minimize or prevent electromagnetic interference which may affect other devices or systems. Also called **RF shield** (2), **RF screen** (2), or **radio screen** (2). **3.** The use of an **RF shielding** (1), or **RF shielding** (2).

RF signal Abbreviation of radio-frequency **signal**. A signal whose frequency is within the **radio spectrum**. Also called **radio signal** (2).

RF signal generator Abbreviation of radio-frequency **signal generator**. A circuit or device that generates **RF signals**. Used, for instance, for calibrating and testing.

RF spectrum Abbreviation of radio-frequency **spectrum**. Same as **radio spectrum**.

RF splitter Abbreviation of radio-frequency **splitter**. A device, such as a coaxial splitter, which divides an RF signal so that it will travel over multiple paths.

RF stage Abbreviation of radio-frequency **stage**. A stage within a radio device, such as a receiver.

RF system Abbreviation of radio-frequency **system**. Any system, such as a communication system, which utilizes radio waves.

RF transformer Abbreviation of radio-frequency **transformer**. A transformer which works at radio frequencies.

RF transistor Abbreviation of radio-frequency **transistor**. **1.** A transistor specifically designed to work optimally at radio frequencies. **2.** A transistor that operates at radio frequencies.

RF transmission Abbreviation of radio-frequency **transmission**. The emission of RF waves from a radio transmitter, or another radio source.

RF transmission line Same as **RF line**.

RF waves Abbreviation of radio-frequency **waves**. Same as **radio waves**.

RF welding Abbreviation of radio-frequency **welding**. Welding in which radio-frequency energy is the source of heat. Used, for instance, to join thermoplastic materials such as PVC. Also called **high-frequency welding**.

RFC Abbreviation of **RF choke**.

RFC 1490 A standard for encapsulating SNA and LAN traffic within a frame relay WAN network.

RFI Same as **RF interference**.

RFIC Abbreviation of radio-frequency integrated circuit. An IC designed to work at radio frequencies. Used, for instance, in low-noise amplifiers, power amplifiers, IF circuits, frequency synthesizers, and so on.

RFID Abbreviation of radio-frequency identification. Also called **dedicated short-range communication**. **1.** A technology in which RF signals are emitted by an object in response to an interrogator, for purposes of identification, collection of information, or the like. A typical arrangement involves a reader which emits signals which a transponder, or tag, in the object responds to. The reader, which proc-

esses the obtained information, and tag need not be in nearly direct contact, as is required, for instance, in a bar-code system. Also called **RFID technology**. **2.** A system using **RFID technology**. Such a system may be utilized, for example, to identify store items being purchased or placed into inventory, to drive through a toll station having the appropriate amount automatically deducted from a smart card attached to the windshield, for the location, complete with proper placement instructions, of parts within a robotic assembly line, or for tracking the movements of people being spied on by attaching such devices on clothing, objects being carried, or via the surgical insertion or embedding of such transceivers. Higher-frequencies, greater transmission power, and enhanced antenna designs, for instance, may be utilized to extend the range of such systems. Also called **RFID system**.

RFID interrogator Same as **RFID reader**.

RFID reader In an **RFID system**, a device which sends an interrogating signal which is responded to by an RFID tag. Also called **RFID interrogator**.

RFID system Same as **RFID** (2).

RFID tag A transponder within an **RFID system**. Such a transponder is usually in the form of an IC, and responds to interrogating signals arriving from an RFID reader. Such a tag, for instance, may be incorporated into a box label, a smart card, or a key chain wand which is waved near a reader to effect payments. Also called **RFID transponder**, or **tag** (4).

RFID technology Same as **RFID** (1).

RFID transponder Same as **RFID tag**.

RG-6 A type of coaxial cable that is rated at 75 ohms, and which is used, for instance, for satellite dish cabling.

RG-8 A type of coaxial cable that is rated at 50 ohms, and which is used, for instance, for cable TV.

RG-11 A type of coaxial cable that is rated at 75 ohms, and which is used, for instance, for long cable runs between an antenna and TV receivers.

RG-58 A type of coaxial cable that is rated at 50 ohms, and which is used, for instance, for LANs.

RG-59 A type of coaxial cable that is rated at 75 ohms, and which is used, for instance, for broadband cable.

RG-62 A type of coaxial cable that is rated at 93 ohms, and which is used, for instance, for LANs.

RGB Abbreviation of red-green-blue. A color model used for displaying in which any color is formed through the appropriate mixture of red, green, and blue.

RGB display Same as **RGB monitor**. Abbreviation of red-green-blue **display**.

RGB monitor Abbreviation of red-green-blue **monitor**. A monitor which utilizes three electron guns, one each for red, green, and blue, to represent all displayed colors. Such a monitor accepts separate red, green, and blue signals. Widely utilized as TV and computer displays. Also called **RGB display**.

RGB signal Abbreviation of red-green-blue **signal**. Each of the three signals which control the red, blue, and green electron guns of an **RGB monitor** respectively. Also, the three signal considered together.

RGB system Abbreviation of red-green-blue **system**. A display system utilizing **RGB signals**.

RGB video Abbreviation of red-green-blue **video**. Video utilizing **RGB signals**.

Rh Chemical symbol for **rhodium**.

RH Abbreviation of **relative humidity**.

RHEED Abbreviation of **reflection high-energy electron diffraction**.

rhenium A silver-white or gray metallic chemical element whose atomic number is 75. Among the elements, its melting point is exceeded only by carbon and tungsten, and its density is greater than all but iridium, osmium, and platinum. It has over 30 known isotopes, of which one is stable. It has various applications in electronics, including its use in filaments, contacts, high-temperature thermocouples, flashbulbs, electroplating, semiconductors, and superconductors. Its chemical symbol is **Re**.

rheostat A resistor whose resistance may be varied by a mechanical device such as a sliding contact. Also known as **adjustable resistor (1)**, or **variable resistor (1)**.

RHI Abbreviation of **range-height indicator**.

rho (*ρ*) **1.** Symbol for **volume charge density**, or **charge density (2)**. **2.** Symbol for **resistivity**.

rhodium A lustrous silvery-white chemical element whose atomic number is 45. It is rare, resists almost every acid, and has nearly 30 known isotopes, of which one is stable. Its applications in electronics include its use in corrosion-resistance contacts, electrodes, thermocouples, high-reflectivity mirrors, optical instruments, and in oxidation-resistant and high-temperature alloys with platinum and iridium. Its chemical symbol is **Rh**.

rhombic antenna A wideband directional antenna whose long-wire radiators form the sides of a rhombus, and which is fed at one apex. Such antennas feature high power gain over a wide range of frequencies. Also called **diamond antenna**.

rhombus A quadrilateral with four equal sides, but having no right angles. A **square** is a quadrilateral with four right angles and four equal sides.

rhumbatron Same as **resonant cavity**.

RI 1. Abbreviation of **radio interference**. **2.** Abbreviation of **ring indicator**.

RIAA curve Abbreviation of Recording Industry Association of America **curve**. A curve representing the equalization curve that helps phonograph records have a flatter response and less distortion when played.

ribbon cable A cable in which the conductors are arranged along the same plane, and laminated or molded into a flat flexible ribbon. Used, for instance, to connect components within a computer. Also called **flat cable**, or **tape cable**.

ribbon microphone A dynamic microphone whose moving element is a thin metal ribbon which is suspended within a magnetic field. It is the most common type of velocity microphone.

RIBE Abbreviation of **reactive ion-beam etching**.

rich content Same as **rich media**.

rich email Email that incorporates **rich media**.

rich media Multimedia content, such as that encountered in Internet advertising, which incorporates richer graphics, audio, and/or video which are intended to induce a user to interact without leaving the page or email where it appears. Attention-getting devices may include Flash or Shockwave presentations. Also called **rich content**.

rich text 1. ASCII text which includes formatting commands such as fonts, and attributes such as bold. **2.** Text which is presented with embellishments such as graphics, audio, and/or video.

Rich Text Format A format for presenting **rich text**. Its abbreviation is **RTF**.

Richardson-Dushman equation A fundamental equation for describing the thermionic emission of electrons from the surface of a metal which is heated to high temperatures. Also called **Richardson equation**, or **Dushman equation**.

Richardson effect The emission of electrons from heated objects, such as a heated electrical conductor. An example is the emission of electrons by a cathode in an electron tube. Also known as **thermionic emission (1)**, or **Edison effect**.

Richardson equation Same as **Richardson-Dushman equation**.

ricinus oil A pale yellow viscous oil used in electrical insulation compounds and in capacitors. Also called **castor oil**.

ride gain To monitor a given level, such as that of a VU meter, while making adjustments, such as those made with a fader.

ridge waveguide A waveguide with one or more longitudinal projections which serve, for instance, for tuning. Used, for example, in diode lasers.

RIE Abbreviation of **reactive ion etching**.

right channel 1. In a stereo audio system, the channel intended for speaker output situated to the right of a listener. Also called **right stereo channel**. **2.** In a multiple-channel or surround sound audio system, a channel intended for speaker output situated to the right of a listener. This may be, for instance, either from the right front speaker, or the right rear speaker.

right click To press and release the right button on a computer mouse, or its equivalent. A computer may be programmed to exchange the signals of the left and right mouse buttons.

right front speaker In a multiple-channel or surround sound audio system, the speaker whose output is intended to be to the front and right of a listener. Also called **right speaker (1)**.

right-hand lay One or more conductors which form clockwise spirals when traveling away from an observer viewing along the longitudinal axis of a cable. Such conductors may be wound around a cable core, or around another layer of conductors. This contrasts with **left-hand lay**, where the converse is true.

right-hand polarized wave An elliptically polarized transverse electromagnetic wave in which the rotation of the electric field vector is towards the right when viewed along the direction of propagation. Also called **clockwise-polarized wave**.

right-hand rule Also called **Fleming's right-hand rule**, or **hand rule**. If electron flow is used instead of conventional current, each of these become **left-hand rules**, as electron flow is in the opposite direction. **1.** A rule stating that if the thumb of a right hand is oriented along the same axis as the current flow through a conducting wire, that the fingers of this hand will curl along the same direction as the magnetic field produced by the wire. **2.** For a conducting wire moving through a magnetic field, a rule stating that if the middle finger, index finger, and thumb of a left hand are extended at right angles to each other, that the middle finger will indicate the current flow, the index finger the direction of the magnetic field, and the thumb will indicate the direction of the movement of the wire. This rule also applies if the conducting wire is substituted by an electron beam.

right rear speaker In a multiple-channel or surround sound audio system, the speaker whose output is intended to be to the rear and right of a listener.

right shift 1. In computers, an operation which moves or displaces digits, bit values, or the like, to the right. In a **left shift (1)**, the converse is true. **2.** On a computer with two shift keys, the shift key on the right of the keyboard. Some keyboard shortcuts which utilize a combination of keys, for instance, make the distinction between this and the **left shift (2)**.

right speaker 1. Same as **right front speaker**. **2.** The right speaker of a two-channel audio system, such as a that of a home stereo or TV.

right stereo channel Same as **right channel (1)**.

rigid disk An occasionally used synonym for **hard disk**.

RIMM The module utilized with **RDRAM** chips. Also called **RIMM module**.

RIMM module Same as **RIMM**.

ring 1. An object or arrangement of objects in the form of a circle whose center is not filled. For example, a ring magnet, or a ring network. **2.** Movement whose path resembles a circle. Also, to form such rings around a given point or location. For instance, the path a signal follows in a ring network. **3.** One of multiple turns, such as those of a coil or helix. **4.** To surround with a circular arrangement. For example, to place a guard ring. **5.** Of the two wires of a POTS line, the more electrically negative. The more electrically positive is the **tip (3)**. Also called **ring lead**, or **ring wire**. **6.** For a telephone or similar device, such as a fax, to emit a sound which indicates that there is an incoming call. **7.** Same as **ring network**.

ring-back Same as **ringback**.

ring-back signal Same as **ringback**. Also spelled **ringback signal**.

ring-back tone Same as **ringback**. Also spelled **ringback tone**.

ring cadence For a telephone receiving an incoming call, the timing pattern of the incoming ring signal. This varies from country to country, or zone to zone, and may consist, for instance, of the ring voltage being applied for two seconds, followed by four seconds off, then back on for two seconds, and so on, until the phone is answered or the calling party hangs up, or a maximum number of rings is reached. If there is distinctive ringing, different phones in the same location may each have their own unique cadence. Also called **ringing cadence**, **ring cycle**, **ring pattern**, **ring interval**, **ringing cycle**, **ringing pattern**, or **ringing interval**.

ring circuit 1. A circuit via which a **ring signal** is sent. Also called **ringing circuit**. **2.** A circuit arranged roughly in the form of a circle, or in which signals travel along a path which forms a closed ring. For example, a household electrical circuit in which all appliances are connected in series, with each end connected to the power supply.

ring configuration The physical arrangement or layout of a **ring network**.

ring counter A counter whose multiple stages are arranged in a circle, and in which only one stage is active at any given moment. Each input pulse advances the counter through each step of a stage, and then on to the next stage, successively.

ring current Same as **ringing current**.

ring cycle Same as **ring cadence**.

ring-down Same as **ringdown**.

ring-down circuit Same as **ringdown circuit**.

ring-down signaling Same as **ringdown**. Also spelled **ring-down signaling**.

ring generator A device which generates a **ring signal**. Also called **ringing generator**.

ring indicator In a modem, an indicator, such as a signal from a specific pin, which alerts to an incoming call. Its abbreviation is **RI**. Also called **ringing indicator**.

ring interval Same as **ring cadence**.

ring junction A hybrid junction molded into the shape of a ring. Used, for instance, as a high-power duplexer. Also called **rat race**, or **hybrid ring**.

ring laser Same as **ring laser gyro**.

ring laser gyro Abbreviation of **ring laser gyro**scope. A gyroscope in which two laser beams travel in opposite directions over a ring-shaped path utilizing multiple angled mirrors. During any movement of the aircraft, or other vehicle it is mounted on, the angular rotation rate is measured by de-

termining the frequency shift in the beams, thus providing the angular position. Such a gyro is very rugged, and has no moving parts. Its own abbreviation is **RLG**. Also called **ring laser**, or **laser gyro**.

ring laser gyroscope Same as **ring laser gyro**.

ring latency In a ring network, such as a token-ring network, the time required for a given packet or signal to travel all the way around the ring.

ring lead Same as **ring (5)**.

ring magnet A permanent magnet in the shape of a doughnut. Used, for instance, in particle accelerators, or loudspeakers. Also called **donut magnet**.

ring main A power mains which is wired as a **ring circuit (2)**.

ring modulator A modulator in which the incorporated diodes are arranged in a manner that forms a closed ring. Used, for instance, as a mixer, demodulator, or balanced modulator.

ring network A communications network in which each node has two branches connected to it, so as to form a continuous loop or ring. As a given message or packet travels around such a ring, the appropriate nodes accept it, otherwise said message or packet is regenerated, enabling a ring network to span considerable distances. Used, for instance in LANs. Also called **ring (7)**.

ring oscillator An oscillator whose components are arranged so as to form a closed ring.

ring pattern Same as **ring cadence**.

ring signal In a telephone or similar device, such as a fax, a signal sent by a telephone company which alerts to an incoming call. In the United States it is typically between 70 and 90 volts at 20 Hz, although it varies from country to country, or from zone to zone. The receiving unit may emit an audible tone and/or flash a light to indicate this condition, among other possibilities, with the ring cadence depending on whether there is distinctive ringing, and so on. Also called **ringing signal**.

ring time Same as **ringing time**.

ring tone A **ring signal** which consists of an audible tone. Also called **ringing tone**.

ring topology The topology of a **ring network**.

ring voltage The voltage of a **ring signal**. The typical value in the United States lies between 70 and 90 volts at 20 Hz, but this amount may vary from country to country, or from zone to zone. Also called **ringing voltage**.

ring winding A winding whose turns are wound around a doughnut-shaped core.

ring wire Same as **ring (5)**.

ringback In telephony, a tone which is heard by a calling party when the receiving party's telephone is ringing or is otherwise being alerted to an incoming call. The sound heard by the calling party may bear no relationship to the sound, or other altering signal, at the receiving end. Also spelled **ring back**. Also called **ringback tone**, or **ringback signal**.

ringback signal Same as **ringback**. Also spelled **ring-back signal**.

ringback tone Same as **ringback**. Also spelled **ring-back tone**.

ringdown Also spelled **ring-down**. Also called **ringdown signaling**. **1.** To manually signal another telephonic device. Used especially in the context of communicating with an operator. **2.** Same as **ringdown circuit**. **3.** To signal another telephonic device by making it ring. This may be accomplished manually or automatically.

ringdown circuit A circuit that connects two telephonic devices, and in which an off-hook condition in one automatically rings the other. There is no dialing in such a circuit. Used, for instance, for emergency communications. Also spelled **ring-down circuit**. Also called **ringdown (2)**, or **automatic ringdown circuit**.

ringdown signaling Same as **ringdown**. Also spelled **ring-down signaling**.

ringer 1. A device or mechanism which serves to **ring (6)**. 2. A signal, device, or mechanism which actuates a **ringer (1)**.

ringer equivalence number Its abbreviation is **REN**. A number which indicates the load a telephone, answering machine, fax, or similar device, places on a telephone ringing circuit. If such a circuit is capable of driving a total of five REN, then five telephones with a REN of one may ring simultaneously, or 4 with a REN of 1.25, and so on. However, if, for instance, three devices each with a REN of 2 were connected to this line, then one or all may fail to ring in the event of an incoming call. A REN may also indicate the total load a given telephone line can drive in terms of the RENs of multiple connected devices.

ringing 1. The transmission of a ring signal to alert to an incoming call. The **ring cadence** is the timing pattern of such a signal. 2. The tendency of a circuit to momentarily oscillate when its input signal suddenly changes. 3. An unwanted prolongation in the vibration in a low-frequency speaker after the driving signal has stopped. May be due, for instance, to insufficient damping. Also called **hangover (1)**. 4. An unwanted prolongation in oscillation, or in the decay of a signal.

ringing cadence Same as **ring cadence**.

ringing circuit Same as **ring circuit (1)**.

ringing current An AC which flows when a **ring signal** is sent. Also called **ring current**.

ringing cycle Same as **ring cadence**.

ringing generator Same as **ring generator**.

ringing indicator Same as **ring indicator**.

ringing interval Same as **ring cadence**.

ringing pattern Same as **ring cadence**.

ringing signal Same as **ring signal**.

ringing time The interval during which **ringing** occurs. Also called **ring time**.

ringing tone Same as **ring tone**.

ringing voltage Same as **ring voltage**.

riometer Abbreviation of relative ionospheric opacity **meter**. An instrument which detects and indicates levels of cosmic noise. A riometer performs its measurements by monitoring the changes in the absorption of electromagnetic energy within the ionosphere.

RIP 1. Acronym for **Routing Information Protocol**. 2. Acronym for **raster image processor**. 3. Acronym for **raster image processing**.

ripper 1. Software that copies data from a DVD to a computer hard drive. Also called **DVD-ripper**. 2. Software that copies data from an audio CD to a computer hard drive. Also called **CD-ripper**.

ripple 1. An AC component present in a DC output, such as that of a rectifier or power supply. This results in slight fluctuations in the instantaneous value of said output. Usually due to incomplete or improper filtering. Also called **ripple component**, or **ripple current (1)**. 2. Slight fluctuations in the instantaneous value of the response of a component, circuit, or device. Passband ripple is an example. Usually expressed in dB. 3. Same as **ripple amplitude**.

ripple amplitude The peak-to-peak value of **ripple (1)** or **ripple (2)**. Used, for instance, as a performance characteristic. Also called **ripple (3)**.

ripple component Same as **ripple (1)**.

ripple counter A binary counter incorporating multiple flip-flops which are cascaded.

ripple current 1. Same as **ripple (1)**. 2. A current whose instantaneous value has slight fluctuations. Such a voltage may be direct or alternating.

ripple factor The ratio of the ripple voltage to the nominal or average total voltage. Also called **ripple ratio**. When expressed as a percentage, also called **ripple percentage**.

ripple filter 1. In rectifiers or DC generators, a filter which reduces the AC component of the output, while passing the DC. Also known as **smoothing filter (1)**. 2. A filter, such as a Butterworth filter, designed to minimize **ripple (2)**.

ripple frequency The frequency at which **ripple** occurs. For example, in a full-wave rectifier this frequency would be 120 Hz if the line frequency is 60 Hz.

ripple percentage The ratio of the ripple voltage to the nominal or average total voltage, expressed as a percentage. For instance, that of a filter. Also called **percent ripple**.

ripple ratio Same as **ripple factor**.

ripple voltage 1. The AC voltage producing **ripple (1)**. 2. A voltage whose instantaneous value has slight fluctuations. Such a voltage may be direct or alternating.

RISC 1. Acronym for reduced-instruction set computer. A computer whose CPU supports relatively few and simple instructions. The execution of each instruction is generally accomplished within a single clock cycle. This contrasts with a **complex-instruction set computer** whose CPU utilizes numerous and complex instructions. 2. Acronym for reduced-instruction set computing. Computing utilizing a **RISC (1)**. Also called **RISC computing**.

RISC computing Same as **RISC (2)**.

RISC CPU Same as **RISC processor**.

RISC processor A processor utilized for **RISC computing**. Also called **RISC CPU**.

rise time 1. The time required for a signal or magnitude to increase from a given level, such as zero, to another level, such the highest attainable value. 2. The time required for the amplitude of a pulse to increase from a given percentage of its peak amplitude, such as 10%, to another, such as 90%. **Fall time (2)** is the converse. Also called **pulse rise time**.

riser A vertical shaft within a structure, such as a building, which serves for running cables, usually from floor to floor.

riser cable A cable, often consisting of several hundred pairs, which is run through **risers**.

riser card An expansion card which allows other cards to be directly plugged into it, and which is generally plugged into a motherboard. Used, for instance, when a computer with no available expansion slots must be expanded.

rising-sun magnetron A magnetron whose anode has several cavity resonators, with larger and smaller cavities alternated, each having a different resonant frequency.

Rivest-Shamir-Adleman Same as **RSA**.

RJ Abbreviation of registered jack. A series of telephone and communications network interfaces, standards, and connectors, such as RJ-11, and RJ-45.

RJ-11 Abbreviation of registered jack-11. A common telephone connector that usually utilizes four wires, but can have up to six available. Although jack generally refers to the connector into which a plug is inserted, RJ-11 may refer to the jack, plug, or both. Also called **RJ-11 connector**.

RJ-11 connector Same as **RJ-11**.

RJ-12 Abbreviation of registered jack-12. A connector similar to **RJ-11**, except that all six wires are used. Although jack generally refers to the connector into which a plug is inserted, RJ-12 may refer to the jack, plug, or both. Used, for instance, to connect two phone lines. Also called **RJ-12 connector**.

RJ-12 connector Same as **RJ-12**.

RJ-14 Abbreviation of registered jack-14. A connector similar to **RJ-11**, except that the four wires are used for two phone lines. A single phone line usually only requires two wires, which are tip and ring. Although jack generally refers to the connector into which a plug is inserted, RJ-14 may refer to the jack, plug, or both. Also called **RJ-14 connector**.

RJ-14 connector Same as **RJ-14**.

RJ-21 Abbreviation of registered jack-21. A 50-pin connector used, for instance, for tip and ring conductors for up to 25 telephone lines. Although jack generally refers to the connector into which a plug is inserted, RJ-21 may refer to the jack, plug, or both. Also called **RJ-21 connector**.

RJ-21 connector Same as **RJ-21**.

RJ-22 Abbreviation of registered jack-22. A connector similar to **RJ-11**, except it is a bit narrower, accommodating only four wires. It is utilized to connect the telephone with the handset. Although jack generally refers to the connector into which a plug is inserted, RJ-22 may refer to the jack, plug, or both. Also called **RJ-22 connector**.

RJ-22 connector Same as **RJ-22**.

RJ-45 Abbreviation of registered jack-45. A connector that holds up to eight wires, and which utilized for the transmission of digital data over ordinary phone lines and LANs. Although jack generally refers to the connector into which a plug is inserted, RJ-45 may refer to the jack, plug, or both. Also called **RJ-45 connector**.

RJ-45 connector Same as **RJ-45**.

RJ-48 Abbreviation of registered jack-48. A connector that holds up to eight wires, and which is utilized for terminations, for instance, in T1 or E1 service. Although jack generally refers to the connector into which a plug is inserted, RJ-48 may refer to the jack, plug, or both. Also called **RJ-48 connector**.

RJ-48 connector Same as **RJ-48**.

RJE Abbreviation of **remote job entry**.

RL Abbreviation of **Raman lidar**.

RLE Abbreviation of **run-length encoding**.

RLG Abbreviation of **ring laser gyro**.

RLL 1. Abbreviation of **run-length limited**. 2. Abbreviation of **radio local loop**.

rlogin Abbreviation of remote **login**. A command that enables a computer to log onto a remote server as if physically located there.

RMI 1. Abbreviation of **radio magnetic indicator**. 2. Abbreviation of **repetitive-motion injury**.

rms Same as **RMS**.

RMS Abbreviation of root mean square. 1. The square root of the average of the squares of a set of related values. 2. The square root of the average of the squares of a set of values.

RMS current Abbreviation of root-mean-square **current**. The value of AC that is as effective in producing heat through a resistor as a corresponding DC. For a sinusoidal AC, the RMS current is equal to 0.707 of the peak value of the current. Also called **effective current**.

RMS current rating Abbreviation of root-mean-square **current rating**. A current-handling rating, such as that of a device, based on an **RMS value**, as opposed, for instance, to an instantaneous value.

RMS meter Abbreviation of root-mean-square **meter**. A meter which provides **RMS values**, as opposed, for instance, to instantaneous measurements.

RMS output Abbreviation of root-mean-square **output**. An output, such as a current or voltage, based on an **RMS value**, as opposed, for instance, to an instantaneous value.

RMS power Abbreviation of root-mean-square **power**. Also called **effective power**. 1. A value obtained by squaring multiple instantaneous power measurements, averaging these over a given time interval, and taking the square root of this average. 2. For a periodic quantity, a value obtained by squaring multiple instantaneous power measurements, averaging these over the time of a complete cycle, and taking the square root of this average. In the specific case of a sinusoidal AC, the RMS power value is equal to 0.707 of the peak power value. When expressing power in audio applications, other parameters should be specified, as in 100 watts RMS per channel into 8 ohms, with 0.01% total harmonic distortion and intermodulation distortion at rated power, from 10 Hz to 60 kHz, +/− 0.5 dB.

RMS power per channel Abbreviation of root-mean-square **power per channel**. The **RMS power** (2) for each channel of an audio power amplifier or receiver.

RMS power rating Abbreviation of root-mean-square **power rating**. A power-handling rating, such as that of a component or system, based on an **RMS value**, as opposed, for instance, to an instantaneous value.

RMS rating Abbreviation of root-mean-square **rating**. A rating, such as that of the current-, power-, or voltage-handling capacity of a component or device, based on an **RMS value**, as opposed, for instance, to an instantaneous value.

RMS reading Abbreviation of root-mean-square **reading**. A reading, such as that of a meter, of an **RMS value**, as opposed, for instance, to an instantaneous value.

RMS sound pressure Abbreviation of root-mean-square **sound pressure**. The value obtained when squaring multiple instantaneous sound-pressure level measurements at a given point, averaging these over the time of a complete cycle, and taking the square root of this average. Usually expressed in pascals, though also expressed in other units, such as dyne/cm^2, or microbars. Also called **effective sound pressure**, **acoustic pressure**, or **sound pressure** (3).

RMS value Abbreviation of root-mean-square **value**. Also called **effective value**. 1. A value obtained by squaring multiple instantaneous measurements, averaging these over a given time interval, and taking the square root of this average. 2. For a periodic quantity, a value obtained by squaring multiple instantaneous measurements, averaging these over the time of a complete cycle, and taking the square root of this average. In the specific case of sinusoidal AC, the RMS current value is equal to 0.707 of the peak value of the current.

RMS voltage Abbreviation of root-mean-square **voltage**. The value of an AC voltage that is equally effective as a corresponding DC voltage. For a sinusoidal AC voltage, the RMS voltage is equal to 0.707 of the peak value of the voltage. Also called **effective voltage**.

RMS voltage rating Abbreviation of root-mean-square **voltage rating**. A voltage-handling rating, such as that of a device, based on an **RMS value**, as opposed, for instance, to an instantaneous value.

Rn Chemical symbol for **radon**.

roam To use a cellular telephone in an area outside of its assigned home or calling area. Availability may vary, but the cost per minute is usually higher unless such coverage is part of the service.

roaming The use of a cellular telephone in an area outside of its assigned home area.

robbed-bit signaling In data transmission, signaling in which the least significant bit is taken from the information stream and utilized for control signals and other overhead. Sometimes used over T1 or E1 lines. Also called **bit robbing**, or **in-band signaling** (2).

robot **1.** A mechanical device with an incorporated computer, that is able to perform programmed tasks and which may or may not be autonomous. Robots can be equipped to detect environmental conditions and to react to them. They have many applications, and are most often used in assembly or manufacturing, being able to work with precision for very extended periods of time. In addition, the complexity of a given task, its repetitiveness, or the conditions under which it is done, can far exceed that which any human could duplicate. Common elements which compose a robot are a controller and manipulators. **2.** A program which performs tasks, especially repetitive ones, over networks such as the Internet. Searching Websites and organizing the located information is an example of a typical task for a bot program called spider. More commonly known as **bot**.

robot actuator In robotics, a device which moves a component, such as a manipulator, in response to signals from the controller. Such a device may be electric, hydraulic, mechanical, or pneumatic. Also called **robotic actuator, actuator (3)**, or **effector**.

robot arm A set of links and powered joints which comprises all parts of the manipulator, except for the wrist and the end-effector. Also called **robotic arm**, or **arm (4)**.

robot color sensing In robotics, the use of image sensors to differentiate between colors. Used, for instance in robots which inspect certain finished products. Also called **robotic color sensing**.

robot computer The computer which control a robot. Also called **robotic computer**, or **robot control system (1)**.

robot control system **1.** Same as **robot computer**. **2.** Same as **robot controller**.

robot controller The computer and programs which control a robot. Also called **robotic controller, robot control system (2)**, or **controller (4)**.

robot degrees of freedom The manners in which a robotic manipulator or its components can move. A robotic manipulator with six degrees of freedom would have three in the arm and three in the wrist: vertical movement, radial traverse, rotational traverse, wrist pitch, wrist roll, and wrist yaw. Also called **robotic degrees of freedom**, or **degrees of freedom (2)**.

robot end-effector A device, tool, or gripping mechanism attached to the wrist of a robot arm. It can be, for example, a drill, a sensor, or a gripper with the ability to move with six degrees of freedom. Also called **robotic end-effector, robot hand, end-effector**, or **hand**.

robot feedback sensor A mechanism by which information obtained by the sensing devices of a robot is routed back to the controller. This can be used, for example, if the robot is traversing irregular terrain. Also known as **robotic feedback sensor**, or **feedback sensor (2)**.

robot force sensor Also called **robotic robot force sensor**, or **force sensor**. **1.** A sensor which measures the forces exerted by a component of the manipulator. **2.** A sensor which measures the forces exerted upon a component of the manipulator.

robot gripper In robotics, the device used by the end-effector to grasp objects. There are many types of grippers, including those which use pressure, suction, or magnetization to take hold of objects. Also called **robotic gripper**, or **gripper**.

robot guidance Any process, mechanism, or system employed to direct the path of a robot. For instance, a robot controller may incorporate input from robotic vision and acoustic proximity sensors in determining the motion of the robot. Also called **robotic guidance**.

robot hand Same as **robot end-effector**.

robot image processing Techniques and procedures used by a robotic controller to process data from its image input device. Also called **robotic image processing**.

robot image recognition The processing of the data of an image input device of a robot for the purpose of comparison to images or components of images stored in its controller. May be used to identify parts to be assembled, or variations in a terrain being traveled upon. Also called **robotic image recognition**, or **robot pattern recognition**.

robot learning The ability of a robot to apply knowledge that has been previously programmed and recorded, in order to analyze and better face new situations.

robot manipulator A robotic mechanism which comprises an arm, a wrist and an end-effector. It has sets of segments, links, and powered joints, which give it the ability to perform tasks, and to move with up to six degrees of freedom. Also known as **robotic manipulator**, or **manipulator**.

robot pattern recognition Same as **robot image recognition**.

robot programming language A language which serves to define instructions for the control of a robot. It is the highest layer of a robotic software system, and provides the interface between the human user and a robot. Also called **robotic programming language**.

robot proximity detector Same as **robot proximity sensor**.

robot proximity sensor A non-contact sensing device used in robots to determine the distance to one or more objects. They are many types, including optical, acoustic, or magnetic. Also called **robot proximity detector**.

robot tactile sensor A robotic device, usually a gripper, that detects objects through physical contact with them. These sensors may be used, for example, to determine the location, identity, and orientation of parts to be assembled. Also called **robotic tactile sensor**.

robot vision Simulation of vision by a robot through the use of sophisticated controller programs which interpret data from its optical sensors. Also called **robotic vision**.

robot voice Robotic simulation of a human voice through the use of controller programs and other components, which include a speaker. Also called **robotic voice**.

robotic Pertaining to **robots** or **robotics**.

robotic actuator Same as **robot actuator**.

robotic arm Same as **robot arm**.

robotic color sensing Same as **robot color sensing**.

robotic computer Same as **robot computer**.

robotic control system Same as **robot control system**.

robotic controller Same as **robot controller**.

robotic degrees of freedom Same as **robot degrees of freedom**.

robotic end-effector Same as **robot end-effector**.

robotic feedback sensor Same as **robot feedback sensor**.

robotic force sensor Same as **robot force sensor**.

robotic gripper Same as **robot gripper**.

robotic guidance Same as **robot guidance**.

robotic hand Same as **robot hand**.

robotic image processing Same as **robot image processing**.

robotic image recognition Same as **robot image recognition**.

robotic manipulator Same as **robot manipulator**.

robotic pattern recognition Same as **robot pattern recognition**.

robotic programming language Same as **robot programming language**.

robotic tactile sensor Same as **robot tactile sensor**.

robotic vision Same as **robot vision**.

robotic voice Same as **robot voice**.

robotics The science which deals with the study, design, construction, development, control, applications, maintenance, and improvement of **robots (1)**.

robust 1. Built and/or assembled in a manner which provides for reliable operation even under harsh conditions. 2. Software which as few or no bugs, and which performs well under most any conditions.

Rochelle salt Colorless to white crystals displaying piezoelectricity and ferroelectricity. It is used in piezoelectric devices, such as crystal microphones, and for silvering mirrors. Also called **potassium sodium tartrate**, or **Seignette salt**.

rock crystal Pure quartz, as opposed to quartz mixed with impurities.

rocker switch A switch that is actuated by tapping or pressing the top or bottom part. Usually used as an on/off switch, or to choose between two settings, as in the case of a DIP switch.

rod 1. A straight bar or piece of a material which usually has a particular function or use. Examples include control, ferrite, ground, and lightning rods. 2. A **rod (1)** whose pointed shape, placement on top of a structure or area, and conductivity enable it to attract lightning discharges and provide a cone-shaped zone, called a cone of protection, below which there is a highly reduced probability of a lightning strike. A lightning rod is usually metallic, although another good conductor may be utilized. Also called **lightning rod**.

rod antenna An antenna, such as a whip antenna, consisting of a metal **rod (1)**.

rod thermistor A thermistor in the shape of a rod.

roentgen A unit of radiation exposure equal to the amount of ionizing radiation which will produce a charge of 2.58×10^{-4} coulombs per kilogram of air. Its symbol is **R**, or **r**. Also spelled **röntgen**.

roentgen equivalent man Same as **rem (1)**.

roentgen equivalent physical Same as **rep**.

roentgen rays Better known as **X-rays**. Electromagnetic radiation whose wavelengths range between approximately 10^{-8} and 10^{-11} meter, which corresponds to frequencies of 10^{16} to 10^{19} Hz respectively, although the defined interval varies. Such rays are located between ultraviolet and gamma radiation. Roentgen rays consist of high-energy photons, and penetrate living tissue and other matter to varying degrees, with less penetrating rays called soft X-rays, and more penetrating rays, whose energy and frequency are higher, known as hard X-rays. X-rays are produced, for instance, when matter is bombarded by electrons with sufficient energy, as occurs in an X-ray tube. X-rays are used as a medical diagnostic technique, for therapy, and in the study of the structure and quantitative analysis of materials.

roentgenogram Same as **radiograph**.

roentgenograph Same as **radiograph**.

roentgenography Same as **radiography**.

ROFL In emails, chats, and other online activities, an abbreviation of rolling on the floor laughing, which is similar to **LOL**, expect that it conveys an even greater reaction. Also spelled **ROTFL**.

ROH Abbreviation of **receiver off-hook**.

role-playing game A computer game, such as a MUD, in which one or more players assume the identity of fictional characters while playing. Its abbreviation is **RPG**.

roll 1. To move along a surface by revolving around an axis. For instance, for a mouse ball to roll over a mouse pad. Also, to cause to move in this manner. 2. To move, or be moved, along a surface using wheels or other devices which **roll (1)**. For example, a robot with wheels traversing a terrain. 3. That which is wrapped around itself. For instance, a roll of film. Also, to wrap around itself. Also, that which is usually rolled around itself, even if it is not wound at a given moment. 4. Movement in a clockwise or counterclockwise manner. For example, the wrist roll of a robot. 5. A slow vertical movement of the entire image presented on a TV display, due to improper vertical synchronization.

roll-off To gradually attenuate the level of a signal, above or below a specified frequency or value. Also, to experience such an attenuation.

rollback 1. In computers, to reverse a step, transaction, or process. Done, for instance, to rectify an error. 2. In computers, to return to a previous state. Used, for instance, for recovery after a malfunction. 3. Same as **rerun**.

roller inductor A variable inductor which is usually utilized to tune antennas.

rolling The slow vertical movement of the entire image presented on a TV display, due to improper vertical synchronization.

rolloff The gradual attenuation in the level of a signal, above or below a specified frequency or value. For instance, the rate at which a filter attenuates a signal beyond a cutoff frequency. Usually expressed in dB.

rollover To move the pointer over a **rollover button** on a computer screen.

rollover button On a computer screen, a button that changes in appearance and/or generates a sound when the pointer is moved over it. For example, an application may highlight an icon accessing a given feature when the mouse pointer is moved over it. In addition, the feature name and/or a brief description may be shown, a tone may be produced, and so on.

ROM Acronym for read-only memory. A chip, cartridge, or other storage device, whose content can be read, but not changed. ROM is usually programmed by the manufacturer, although some types, such as PROM, are user-programmable. Certain varieties of ROM, such as EEPROM, can be erased, but once programmed the content can not be altered. It is a non-volatile form of memory, and is utilized, for instance, to store the BIOS of a computer, fonts in a printer, or to contain video games.

ROM BIOS Abbreviation of read-only memory basic input/output system. 1. The BIOS of a computer system stored in a ROM chip. 2. The BIOS of a computer system stored in a ROM chip, cartridge, or the like.

ROM card Abbreviation of read-only memory **card**. A card which has data or software, such as printer fonts or computer games, stored permanently.

ROM cartridge Abbreviation of read-only memory **cartridge**. A cartridge which has data or software, such as printer fonts or computer games, stored permanently.

ROM chip Abbreviation of read-only memory **chip**. A chip which consists of memory cells and the associated circuits necessary, such as those for addressing, to provide a computer with **ROM**.

ROM emulator Abbreviation of read-only memory **emulator**. A circuit or device which utilizes RAM to simulate the ROM of a computer. Such a circuit or device is plugged into the ROM socket, and is used, for instance, to debug ROM chips.

ROM socket Abbreviation of read-only memory **socket**. A receptacle within a computer or peripheral which accepts a ROM chip or ROM cartridge.

röntgen Same as **roentgen**.

roof mount 1. To fix, assemble, or otherwise place or install on a roof. For example, a roof-mounted antenna. 2. A support or substrate, located on a roof, upon which objects may be fixed or assembled. For instance, a mast located on a roof which serves to support an antenna.

roof-mount antenna Same as **roof-mounted antenna**.

roof-mounted antenna An antenna, such as that used in a car or home, which is fixed or otherwise placed on a roof. Also called **roof-mount antenna**, or **rooftop antenna**.

roof-top antenna Same as **roof-mounted antenna**. Also spelled **rooftop antenna**.

rooftop antenna Same as **roof-mounted antenna**. Also spelled **roof-top antenna**.

room noise The noise present in any given environment, especially that within which tests or observations are taking place. Such noise includes nearby and distant sources. Also called **background noise (1)**, **local noise**, **site noise**, **ambient noise**, or **environmental noise**.

room resonance Any **resonance (5)** present in a room, due to its geometry, composition of surfaces such as walls, and so on.

room temperature Its abbreviation is **RT**. **1.** The temperature in any room or chamber, especially where a device or piece of equipment is functioning, or where tests or observations are taking place. **2.** A specific temperature, or temperature interval, which is considered to be average or appropriate for everyday use or placement of something. For example, 24 °C, or 22 to 26 °C.

room-temperature superconductivity Superconductivity occurring at or above a given temperature, such as 250 or 300 K, which is close to room temperature when compared to cryogenic temperatures.

room-temperature superconductor A material which is exhibits superconductive properties at or above a given temperature, such as 250 or 300 K, which is close to room temperature when compared to cryogenic temperatures.

room tone The specific acoustic characteristics, such as background noise and the composition of surfaces, of a given room or chamber. Room tone helps, for instance, to determine the extent to which components, devices, and systems, such as speakers or sound systems, are suitable for use there.

root **1.** The top level within a hierarchy. For example, a root directory. **2.** A computer user with access to all levels of operation. Such a user, for instance, can perform tasks at the system level. Also called **root user**, or **superuser**.

root account A computer account which provides complete access to all data and systems. Such an account may be used for viewing and manipulating all files, and for performing operating system tasks. Root accounts may be held, for instance, by system administrators. Also called **superuser account**.

root directory A directory which branches into other directories, which in turn are called **subdirectories**. In GUIs, the terms **root folder** and **subfolder** are preferred, respectively.

root domain Within an Internet domain name, the portion preceding the top-level domain category or designator. For example, in *www.yipeeee.com* or *www.yipeeee.qq*, it is the *yipeeee* portion. Multiple root domains may be served by the same IP address. Also called **second-level domain**.

root domain name A **root domain** plus the domain category or designator. For example, in *www.yipeeee.com* it is the *yipeeee.com* portion, while in *www.yipeeee.qq*, it would be the *yipeeee.qq* portion. Multiple root domain names may be served by the same IP address. Also called **second-level domain name**.

root folder A folder which branches into other folders, which in turn are called **subfolders**. In non-graphical user interfaces, the terms **root directory** and **subdirectory** are preferred, respectively.

root mean square Same as **RMS**.

root-mean-square current Same as **RMS current**.

root-mean-square current rating Same as **RMS current rating**.

root-mean-square meter Same as **RMS meter**.

root-mean-square output Same as **RMS output**.

root-mean-square power Same as **RMS power**.

root-mean-square power per channel Same as **RMS power per channel**.

root-mean-square power rating Same as **RMS power rating**.

root-mean-square rating Same as **RMS rating**.

root-mean-square reading Same as **RMS reading**.

root-mean-square sound pressure Same as **RMS sound pressure**.

root-mean-square value Same as **RMS value**.

root-mean-square voltage Same as **RMS voltage**.

root-mean-square voltage rating Same as **RMS voltage rating**.

root password A password utilized to gain access to a **root account**.

root server A server within the **root server system**.

root server system A worldwide system of servers which contain the databases utilized to match all top-level domain names with their corresponding IP addresses. This system is made available to, and utilized by, routers around the globe.

root user Same as **root (2)**.

rosin An amber-colored resin obtained from certain trees, which is used, for instance, in some soldering compounds.

rosin-core solder A solder, which is usually presented in a tubular or cylindrical form, whose core is composed of a rosin which serves as a flux.

ROT13 encryption An encryption technique in which each letter is replaced by the letter which would appear 13 places in front of it in the alphabet. For example, *cat* would become *png*. It is used, for instance, to protect a user from reading a movie spoiler in a review. Such encryption is not intended for secure communications.

rotary amplifier A magnetic amplifier in which a small change in the field input power results in enormous gain. An amplidyne is an example. Also called **rotating amplifier**.

rotary antenna Same as **rotating antenna**.

rotary beam Same as **rotating beam**.

rotary converter A rotating electric machine which is utilized to convert a DC voltage to an AC voltage, or vice versa. It usually has a single armature with two or more windings, and incorporates a commutator for DC operation, and slip rings for AC operation. Also called **dynamotor (2)**, or **synchronous inverter**.

rotary dial A telephone dial which moves in a circular manner as numbers are selected, effecting **rotary dialing**. It is an obsolescent signaling method.

rotary dialing Telephone dialing in which one or more pulses interrupt a steady DC, to indicate each dialed digit. For example, when dialing a 4, four such pulses are sent. Also called **pulse dialing**.

rotary encoder **1.** A electromechanical device which converts the rotations of a disk or shaft into pulses. If the output is binary, such a device may be used as an analog-to-digital converter. **2.** A digital control which emulates the function of a knob utilized to manually adjust settings such as volume or clock adjustments.

rotary gap Same as **rotary spark gap**.

rotary phone Abbreviation of **rotary** telephone. A telephone which incorporates a **rotary dial**.

rotary relay A relay which turns around an axis when opening or closing contacts. Also called **rotary stepping relay**, or **stepping relay**.

rotary selector switch A selector switch that turns around an axis when opening or closing contacts.

rotary spark gap A spark gap which incorporates one or more fixed electrodes and one or more electrodes which rotate on a disk. There is sparking only when electrodes are physically aligned. Also called **rotary gap**.

rotary stepping relay Same as **rotary relay**.

rotary stepping switch Same as **rotary switch**.

rotary switch A switch that turns around an axis when opening or closing contacts. Such a switch is usually adjusted by a knob, and may have detents to hold in a given position. Used, for instance, to choose from multiple settings. Also called **rotary stepping switch**, **rotating switch**, or **stepping switch**.

rotary telephone Same as **rotary phone**.

rotary transformer A rotating machine, such as a dynamotor, which serves as a transformer. Also called **rotating transformer**.

rotary-vane attenuator A waveguide attenuator in which a sheet or plate of dissipative material is placed through a non-radiating slot. As the sheet or plate is moved into and out of the slot, a variable amount of loss is introduced. Also called **vane attenuator**, **flap attenuator**, or **guillotine attenuator**.

rotatable antenna Same as **rotating antenna**.

rotate To turn around an axis or a given point.

rotating amplifier Same as **rotary amplifier**.

rotating antenna An antenna which can rotate along a given axis, so as to maximize signal strength in a given direction. A device such as a nutator is utilized to rotate the axis of such an antenna. Also called **rotary antenna**, or **rotatable antenna**.

rotating beam A beam, such as that transmitted by a rotating antenna, which can be rotated along a given axis. Also called **rotary beam**.

rotating field A field which rotates around a center or point. For instance, that in a multiphase generator which results in rotational motion. Also called **revolving field**.

rotating machine A machine, such as a dynamo or dynamotor, which has one or more components capable of turning around an axis or a given point and/or whose movement is influenced by a rotating field.

rotating magnetic field A magnetic field which rotates around a given axis or point. For example, in an induction motor, the rotor is made to rotate under the influence of the rotating magnetic field in the stator. Also called **revolving magnetic field**.

rotating switch Same as **rotary switch**.

rotating transformer Same as **rotary transformer**.

rotation rate Same as **rotation speed**.

rotation speed For a moving object, the change in direction per unit time, as expressed in radians per second. Also known as **rotation rate**, or **angular speed**.

rotational delay Same as **rotational latency**.

rotational latency In a hard drive, the time interval that elapses before a disk rotates to the location a read/write head will access. Also called **rotational delay**.

rotational traverse In a robotic arm, side to side motion. May be caused by moving the robot itself, or by moving the arm horizontally.

rotator That which **rotates**, or induces rotation.

ROTFL Same as **ROFL**.

rotor In an electrical and/or mechanical component, device, apparatus, or machine, a member which can **rotate**. For instance, the moving part or armature of a motor, the moving plates of a variable capacitor, or the moving contacts of a rotary switch. A stationary part in such a component, device, apparatus, or machine is a **stator**.

rotor blade A blade, such as that in a rotary switch, that can **rotate**.

rotor coil A coil, such as that in a motor, that can **rotate**. Also called **rotor winding**.

rotor contact A contact, such as that in a rotary switch, that can **rotate**.

rotor plate A plate, such as that in a variable capacitor, that can **rotate**.

rotor winding Same as **rotor coil**.

rough vacuum An environment or system whose pressure is below 760 torr, but above one torr. The lower threshold may be slightly higher, such as 30 torr. One torr equals one millimeter of mercury, or approximately 133.3 pascals. Also called **low vacuum**.

round To perform a **rounding** (1) procedure. Also called **round off**.

round number The number resulting from a **rounding** (1) procedure.

round off Same as **round**.

round-off error Same as **rounding error**. Also spelled **roundoff error**.

round-trip time The time interval required for a signal or data to travel from one communications network node to another, and back. For example, the time required for a message to arrive at its destination, plus the time required for the acknowledgement to be received at the source. Its abbreviation is **RTT**.

rounding 1. The dropping of one or more decimal digits from a number, with the rightmost remaining digit being increased if the next number was 5 or higher. For example, 0.44444 rounded to four decimal places would be 0.4444, while 0.66666 rounded to four decimal places would be 0.6667. When **truncating**, digits are simply dropped, with the rightmost remaining digit never changing. 2. The smoothing of the corners of a rectangular wave or curve with sharp transitions. Seen, for instance, in the corner effect.

rounding error Any error in a calculation or processing introduced as a consequence of a **rounding** (1) procedure. Also called **roundoff error**.

roundoff error Same as **rounding error**. Also spelled **round-off error**.

routable protocol A communications protocol, such as TCP/IP, which utilizes network addresses in addition to that of the destination, which allows for packets and messages, for instance, to be routed to multiple networks.

route 1. The path something takes to proceed from one point or location to another. For example, the route a data packet follows from its source to its destination. 2. A path that can be taken for a given purpose. For instance, any of various communications paths such as a primary route and multiple alternate routes. 3. To send something via a given route. For example, to send or forward data packets through a network. 4. To position wires, cables, or other conductors for later use for the transmission of signals.

route selection A process, such as that incorporating routing algorithms, via which a router determines the path a packet or message will take.

route server A server which has access to numerous routes, and which may work at multiple layers within the OSI Reference Model.

router In a communications network, or multiple interconnected networks, a device or software which determines where packets, messages, or other signals travel to next. A router, using resources such as header information, algorithms, and router tables, establishes the best available path from its source to destination. Within the OSI Reference

Model, a router that operates at the network layer. Also called **network router**, or **gateway (2)**.

router page A Web page whose primary purpose is to route users to another Web page. Used, for instance, to display some information while an automatic redirection to a new page takes place.

router protocol Same as **routing protocol**.

router table Same as **routing table**.

routine 1. Within a computer program, a group of instructions which perform a given task. The term is usually utilized synonymously with **subroutine**, **function (3)**, or **procedure (2)**, and in some cases may also refer to a **macro (2)**, **bead (2)**, a **module** within a program, or even a **program**. **2.** A sequence of steps which is followed to perform a given task. Examples include bootstrap routines and check routines.

routing In a communications network, the process of determining the path which packets, messages, or other signals will follow. Examples include dynamic routing, and least cost routing. Also, the sending of packets, messages, or other signals through the selected route.

routing algorithm In a communications network, an algorithm which is utilized for the determination of the path a packet or message will travel by. Such an algorithm usually takes into account multiple routing metrics.

routing domain A set of routers and hosts which operate under the same administrative domain.

Routing Information Protocol A network routing protocol that determines the best route for data based on the hop count between the source and destination. Other factors, such as the line's speed and utilization are ignored. It works best with smaller environments, but because of its limitations has been superseded by **Interior Gateway Routing Protocol**. Its acronym is **RIP**.

routing metric Any of the parameters which are factored into a routing algorithm, or any other method utilized in determining the best possible route for a packet or message. These include hop count, bandwidth, round-trip time, link speed, load, reliability, and the maximum transmission unit.

routing protocol A protocol, such as Enhanced Interior Gateway Routing Protocol, utilized for the implementation of a routing algorithm, or any other method utilized by a router to determine the best possible path for a packet or message. The protocol also takes care of tasks such as the proper communication between all routers involved. Also called **router protocol**.

routing switch In a communications network, a switch that also performs **routing** operations.

routing table In a communications network, a table which lists and keeps track of all possible routes between nodes. Such a table may also store the routing metrics associated with each path. Also called **router table**.

Routing Table Maintenance Protocol A protocol utilized for establishing and maintaining routing tables and related information. Its abbreviation is **RTMP**.

row A horizontal arrangement or series. For instance, a row of pixels or digits. This contrasts with a **column**, which is a vertical arrangement or series.

RP Abbreviation of **rapid prototyping**.

RPC Abbreviation of **Remote Procedure Call**.

RPF Abbreviation of **reverse path forwarding**.

RPG Abbreviation of **role-playing game**.

rpm Abbreviation of **revolutions per minute**.

RPM meter Abbreviation of revolutions per minute meter. A device or instrument which measures and indicates angular velocity, and which is graduated in **revolutions per minute**.

RPN Abbreviation of **reverse Polish notation**.

rps Abbreviation of **revolutions per second**.

RS-170 An EIA NTSC video standard.

RS-232 Same as **RS-232C**.

RS-232 interface Same as **RS-232C**.

RS-232 port Same as **RS-232C port**.

RS-232C Abbreviation of recommended standard **232C**. A standard interface for serial communications between a serial device, such as a modem or a mouse, and a computer. Although the current term for this interface is **EIA/TIA-232-E** or **TIA/EIA-232-E**, it is still better known as **RS-232C**. Also called **RS-232**, **RS-232 interface**, or **RS-232C interface**.

RS-232C interface Same as **RS-232C**.

RS-232C port A serial port utilizing an **RS-232 interface**. Also called **RS-232 port**.

RS-232C standard The standard which defines serial transfers utilizing an **RS-232C interface**.

RS-343 An EIA high-resolution non-broadcast video standard.

RS-366 An EIA standard for dialing commands using an **RS-232C** interface.

RS-422 A standard interface for serial communications similar to **RS-232C**, but providing higher speed and transmission distances, and support for multipoint connections. Also called **EIA/TIA-422**, or **TIA/EIA-422**.

RS-423 A standard interface for serial communications similar to **RS-232C**, but providing higher speed and transmission distances. RS-423 only supports point-to-point connections. Also called **EIA/TIA-423**, or **TIA/EIA-423**.

RS-449 A standard defining the pins for both **RS-422** and **RS-423** serial communications. Also called **EIA/TIA-449**, or **TIA/EIA-449**.

RS-485 A standard interface for serial communications similar to **RS-422**, but supporting more nodes per line. Also called **EIA/TIA-485**, or **TIA/EIA-485**.

RS-530 A standard defining the pins for both **RS-422** and **RS-423** serial communications when a DB-25 connector is utilized. Also called **EIA/TIA-530**, or **TIA/EIA-530**.

RS code Abbreviation of **Reed-Solomon code**.

RS flip-flop Abbreviation of reset-set **flip-flop**. A flip-flop whose inputs are designated as **R** and **S**, along with a clock input. Its outputs are similar to those of a JK flip-flop for any given combination of inputs, except that a race condition can occur in an RS flip-flop when both inputs are at logic 1, thus such a condition should be avoided. Also called **SR flip-flop**.

RS170 Same as **RS-170**.

RS232 Same as **RS-232C**.

RS232 interface Same as **RS-232C**.

RS232 port Same as **RS-232C port**.

RS232C Same as **RS-232C**.

RS232C interface Same as **RS-232C**.

RS232C port Same as **RS-232C port**.

RS232C standard Same as **RS-232C standard**.

RS343 Same as **RS-343**.

RS366 Same as **RS-366**.

RS422 Same as **RS-422**.

RS423 Same as **RS-423**.

RS449 Same as **RS-449**.

RS485 Same as **RS-485**.

RS530 Same as **RS-530**.

RSA Abbreviation of Rivest-Shamir-Adleman. A powerful public-key encryption technology which utilizes an algorithm which takes advantage of the enormous computing power required to factor very large numbers. It is a de facto standard for encrypting data sent over the Internet. Also called **RSA encryption**.

RSA algorithm An algorithm utilized for **RSA encryption**.

RSA chip A chip utilized for encoding and decoding data using **RSA encryption**.

RSA encryption Same as **RSA**.

RSAC Abbreviation of **Recreational Software Advisory Council**.

RSI Abbreviation of **repetitive-strain injury**.

RSVP Abbreviation of **Resource Reservation Protocol**, **Resource Reservation Setup Protocol**, or **Reservation Protocol**.

RT Abbreviation of **room temperature**.

rt-VBR Abbreviation of **real-time VBR**.

RTC Abbreviation of **real-time clock**.

RTD Abbreviation of **resistance temperature detector**.

RTF Abbreviation of **Rich Text Format**.

RTFM Same as **read the manual**.

RTL Abbreviation of **resistor-transistor logic**.

RTM Same as **read the manual**.

RTMP Abbreviation of **Routing Table Maintenance Protocol**.

RTOS Abbreviation of **real-time operating system**.

RTP Abbreviation of **Real-Time Transport Protocol**.

RTP protocol Same as **Real-Time Transport Protocol**.

RTS Abbreviation of **request to send**.

RTSP Abbreviation of **Real-Time Streaming Protocol**.

RTT Abbreviation of **round-trip time**.

RTTY 1. Abbreviation of **radio teletype**. 2. Abbreviation of **radio teletypewriter**.

RTZ Abbreviation of **return to zero**.

Ru Chemical symbol for **ruthenium**.

RU Abbreviation of **rack unit**.

rubber 1. A natural or synthetic polymer which upon vulcanization becomes elastic, tough, and resistant to cold, heat, abrasions, water, and many chemicals. Used, for instance, for insulation, as a protective layer, and as a filler material. 2. **Rubber** (1) which has been vulcanized.

rubber-coated Insulated, protected, or otherwise covered by **rubber**. Also called **rubber-covered**.

rubber-coated cable A cable which is insulated, protected, or otherwise covered by **rubber**. Also called **rubber-covered cable**.

rubber-coated wire A wire which is insulated, protected, or otherwise covered by **rubber**. Also called **rubber-covered wire**.

rubber-covered Same as **rubber-coated**.

rubber-covered cable Same as **rubber-coated cable**.

rubber-covered wire Same as **rubber-coated wire**.

rubber-insulated Insulated by a layer, jacket, or the like, composed of **rubber**.

rubber-insulated cable A cable which is insulated by a **rubber** layer, jacket, or the like.

rubber-insulated wire A wire which is insulated by a **rubber** layer, jacket, or the like.

rubber-insulation Insulation provided by a layer, jacket, or the like, composed of **rubber**.

rubber jacket A jacket, such as that utilized around a cable or connector, composed of **rubber**.

rubidium A soft silver-white metal whose atomic number is 37. It is extremely reactive, forming compounds with most every element, reacts vigorously with water, and may ignite when exposed to air. It is the most electropositive element after cesium, becomes a liquid at temperatures above 39°, and has around 30 known isotopes, of which one is stable. It is used in photocells, as a getter in vacuum tubes, in atomic clocks, and in special glasses. Its chemical symbol is **Rb**.

rubidium atomic clock Same as **rubidium clock**.

rubidium atomic frequency standard Same as **rubidium frequency standard**.

rubidium carbonate A white powder whose chemical formula is Rb_2CO_3. It is used in special glasses and ceramics.

rubidium clock A clock whose accuracy is governed by the natural resonance frequency of rubidium atoms. It is less accurate than a cesium clock, but much less expensive. Also called **rubidium atomic clock**.

rubidium frequency standard A frequency standard based on the natural resonance frequency of rubidium. Used, for instance, in atomic clocks. Also called **rubidium atomic frequency standard**.

rubidium oscillator An oscillator whose accuracy is governed by the natural resonance frequency of rubidium atoms. Used, for instance, as a frequency standard.

rubidium-vapor frequency standard A frequency standard based on the natural resonance frequency of rubidium vapor.

ruby A precious stone consisting of corundum with trace chromium impurity atoms, which give it its transparent red appearance. Used, for instance, in precision instruments such as lasers.

ruby laser A solid-state laser in which a ruby crystal or rod is utilized to generate the coherent beam of light. The first lasers were ruby lasers, and these emit in the red region of the visible spectrum.

Ruhmkorff coil Same as **Ruhmkorff induction coil**.

Ruhmkorff induction coil An induction coil utilized to produce sparks of lengths of up to about 30 centimeters. Also called **Ruhmkorff coil**.

rule-based expert system Same as **rule-based system**.

rule-based system An expert system which applies IF-THEN rules to expressions to produce new expressions with the goal of problem solving. Also called **rule-based expert system**, or **production system**.

rumble 1. An unwanted low-frequency vibration which is audible. For example, the vibrations of a platter and motor of a phonograph turntable. 2. A quantified level of **rumble** (1).

rumble filter A filter which highly attenuates **rumble** (1). Such a filter may, for instance, have a 24 dB/octave rolloff below 50 Hz.

run 1. To execute a computer program. 2. To perform one or more jobs or tasks. Also, the performance of one or more jobs or tasks. 3. To operate, cause to operate, or be operated. Also, an instance of operation, or a period of operation. 4. To lay, suspend, or otherwise place cabling or wiring.

run-length encoding A compression technique in which a series of consecutive characters which are identical are converted into a given code. Its abbreviation is **RLE**.

run-length limited An encoding technique in which groups of bits are coded instead of encoding each bit one at a time. Utilized, for instance, to store data on floppy or hard disks. Its abbreviation is **RLL**.

run time The time during which a computer program is run. Used, for instance, to contrast with **compile time**.

run-time Same as **runtime**.

run-time error Same as **runtime error**.

run-time version Same as **runtime version**.

runaway A condition in which heat generated by a component, circuit, or device causes additional increases in generated heat, which leads to greater heat levels. This is usually due to the heating effect, and if protective measures are not taken usually results in the failure or destruction of the component, circuit, or device. Also called **thermal runaway**.

running-time meter An instrument that measures and displays the time that elapses during a given sequence, process, or operation. Also called **elapsed-time meter**.

runtime That which occurs while a computer program is being executed or run, such as a runtime error. Also spelled **run-time**.

runtime error An error that occurs while a program is being executed, as opposed to a **compile-time error**, which occurs while a program is being compiled. Also spelled **run-time error**.

runtime version Also spelled **run-time version**. **1.** A version of one program that enables another program to be run. Used, for instance, to provide a software version with limited capabilities as compared to the regularly purchased version. **2.** A version of a program which is ready to be run.

rupture **1.** The opening or breaking open of a circuit, insulator, or the like. **2.** The opening resulting from a **rupture (1)**.

rupture capacity A rating based on the level of current a circuit, device, or material can withstand before **rupture (1)**. When used in the context of circuit breakers or fuses, also called **interrupting capacity**.

ruthenium A hard and lustrous silvery-white metallic chemical element whose atomic number is 44. It has over 25 known isotopes, of which 7 are stable. It is utilized as a hardener for platinum, improves the corrosion-resistance of titanium, and is used in contacts, filaments, analytical instruments, and solar cells. Its chemical symbol is **Ru**.

rutherford A unit of radioactivity equal to 10^6 becquerel, or approximately 2.7027027×10^{-5} curie. Its abbreviation is **rd**, or **Rd**.

Rutherford back scattering Same as **Rutherford backscattering**.

Rutherford backscattering Its abbreviation is **RBS**. Also spelled **Rutherford back scattering**. Also called **Rutherford scattering**. **1.** The scattering of ions at angles equal to, or near 180°, by a sample which has been bombarded by high-energy ions. Used, for instance, in Rutherford backscattering spectrometry. **2.** Same as **Rutherford backscattering spectrometry**.

Rutherford backscattering spectrometry An analytical technique in which a high-energy beam of ions, such as He^{+2}, is directed at a sample, with a detector positioned so as to collect particles which are scattered back at or close to 180°. The energies of the collected ions are analyzed to provide extremely detailed information on the composition, structure, thickness, and flaws of samples, in addition to the presence and distribution of any impurities. Its abbreviation is **RBS**. Also called **Rutherford backscattering spectroscopy**, or **Rutherford backscattering (2)**.

Rutherford backscattering spectroscopy Same as **Rutherford backscattering spectrometry**.

Rutherford scattering Same as **Rutherford backscattering**.

rutherfordium A synthetic radioactive chemical element whose atomic number is 104, and which has about a dozen identified isotopes. The techniques employed to produce this element are utilized to help discover new elements with higher atomic numbers. Its chemical symbol is **Rf**.

RW Abbreviation of **rewritable**.

RW CD Abbreviation of **rewritable CD**.

RW disc Abbreviation of **rewritable disc**.

RW DVD Abbreviation of **rewritable DVD**.

RW optical disc Abbreviation of **rewritable optical disc**.

RWW Abbreviation of **read-while-write**.

RX Abbreviation of **receiver**.

RZ Abbreviation of **return to zero**.

RZ code Abbreviation of **return to zero code**.

RZ encoding Abbreviation of **return to zero encoding**.

S

s Abbreviation of **second** (1).

S **1.** Chemical symbol for **sulfur.** **2.** Symbol for **siemens.** **3.** Symbol for **secondary winding.** **4.** Symbol for **elastance.** **5.** Symbol for **source** (2).

S-100 bus An early 100-pin computer bus which is still used in some systems.

S Band **1.** In communications, a band of radio frequencies extending from 2.00 to 4.00 GHz, as established by the IEEE. This corresponds to wavelengths of approximately 15 to 7.5 cm, respectively. **2.** In communications, a band of radio frequencies extending from 1.550 GHz to 5.2 GHz. This corresponds to wavelengths of approximately 19.4 cm to 5.8 cm, respectively.

S-CDMA Abbreviation of synchronous code division multiple access. A version of CDMA utilized for transmission over coaxial transmission lines. S-CDMA is highly noise resistant.

s/d converter Abbreviation of **synchro-to-digital converter.**

S-HTTP Same as **SHTTP.**

S interface In ISDN, an interface which links an NT2 with a terminal adapter, a telephone, or another ISDN device.

S.M.A.R.T. Same as **SMART.**

S-meter **1.** Same as **signal-strength indicator.** **2.** A signal-strength indicator calibrated in **S units** and/or decibels.

S/MIME Abbreviation of Secure Multipurpose Internet Mail Extensions. A version of MIME that provides additional security, including encryption and authentication. S/MIME works across multiple email clients.

S/N Abbreviation of **signal-to-noise ratio.**

S/N ratio Abbreviation of **signal-to-noise ratio.**

s-parameters Abbreviation of **scattering parameters.**

S-R flip-flop Same as **SR flip-flop.**

S-RAM Same as **SRAM.**

S/RF meter In a radio receiver, transmitter, or transceiver, a meter which indicates signal strength in **S units**, as well as another unit, such as watts.

S units Any of various arbitrary units utilized to quantify signal strength, such as that displayed on a signal-strength indicator. An S unit is often equivalent to six decibels.

S-VHS Abbreviation of super-VHS. A VHS technology or system providing for resolution of up to 400 lines per inch, and an improved signal-to-noise ratio as compared to standard VHS. Also, such a recorded or reproduced signal.

S-VHS recorder Same as **S-VHS VCR.**

S-VHS tape Abbreviation of super-VHS tape. A tape recorded with an **S-VHS** signal.

S-VHS VCR A VCR which records and reproduces **S-VHS tapes.** An S-VHS VCR can also record and reproduce standard VHS tapes. Also called **S-VHS recorder.**

S-video Abbreviation of super video. Video transmitted via an **S-video connector.** Also, video in such a format.

S-video cable Abbreviation of super **video cable.** A cable utilized to carry signals between devices with **S-video connectors.**

S-video connector Abbreviation of super **video connector.** A video connector which handles the chrominance and luminance signals separately. Such a connector provides for the transfer of higher quality images between devices such as DVDs and TVs when compared to an RCA connector. A digital optical cable, however, provides a noticeably sharper image than an S-video connector under the same circumstances. An S-video connector usually has 4 or 7 pins.

S-video input An input to a device which accepts **S-video** signals. Also, the connector located on such a device.

S-video output An output from a device which sends **S-video** signals. Also, the connector located on such a device.

sabin A unit equivalent to the sound energy absorption by one square foot of a perfectly absorbing surface, such as an open window.

SACD Abbreviation of Super Audio **CD.** A CD format which offers significantly higher audio quality as compared to standard CDs. SACD uses Direct Stream Digital to provide a sampling rate in the MHz range, a frequency response that extends from 0 Hz to over 100 kHz, and a dynamic range that exceeds 120 dBs. SACD may be in a stereo or multichannel format. Also called **SACD format.**

SACD format Same as **SACD.**

SACD hybrid An **SACD player** that reproduces SACD discs in addition to standard CDs. An SACD hybrid disc has two layers, with the CD layer located beneath the SACD layer. The SACD layer is transparent to a CD laser, thus can be played in a standard CD player, while a SACD player does not pick up the CD layer, providing for mutual compatibility. Also called **hybrid SACD.**

SACD player Abbreviation of Super Audio **CD player.** A device which reproduces discs in an **SACD format.** An SACD player may also reproduce standard CDs, while a **SACD hybrid** plays both.

safe overload An overload, such as an operating overload, which a component, circuit, device, piece of equipment, or system is subjected to, without causing permanent damage or creating a safety hazard.

safety alarm An alarm intended to notify of a location, procedure, or circumstance which is a possible source of danger.

safety-critical system A system whose reliable performance helps insure the health and safety of users or others. For example, a computer system that monitors electronic devices keeping track of the vital signs of hospital patients.

safety cutout A cutout intended to prevent injuries to personnel, and/or damage to equipment.

safety factor An amount by which a given rating, such as a power rating, can be exceeded without causing permanent damage or creating a safety hazard. Also called **safety margin.**

safety fuse A fuse intended to prevent injuries to personnel, and/or damage to equipment.

safety glass **1.** A glass, such as tempered glass, whose composition and design reduces its possibility of breaking or shattering. **2.** A glass, such as tempered glass, designed to minimize or prevent injury if broken or shattered.

safety ground An electrical ground intended to prevent injuries to personnel, and/or damage to equipment

safety hazard A component, circuit, device, piece of equipment, system, material, location, procedure, or circumstance which is a possible source of danger. Also called **hazard.**

safety margin Same as **safety factor.**

safety outlet An electrical outlet, which in addition to the current-carrying contacts, has a third contact which serves for connection to a grounding conductor. Devices and equipment which are to benefit from this safety feature must have an appropriate three-prong plug which is inserted into this outlet. There are other possible arrangements for such outlets, including the use of lateral grounding contacts which make contact with metallic strips on the side of the plug. Also called **grounding outlet, grounding receptacle,** or **three-prong outlet.**

safety relay A relay, such as an interlock relay, intended to prevent injuries to personnel, and/or damage to equipment.

safety signal A signal, such as an alarm signal, intended to notify of a location, procedure, or circumstance which is a possible source of danger.

safety switch A switch, such as an interlock switch, intended to prevent injuries to personnel, and/or damage to equipment.

sag **1.** A brownout lasting no longer than a few seconds. **2.** The amount that an overhead transmission line dips between supports such as poles or towers. The optimum amount will depend on the type and weight of a given cable, the surrounding climate, and so on.

sal ammoniac A crystalline salt used as an electrolyte in dry cells and for electroplating. Its chemical formula is NH_4Cl. Also known as **salmiac,** or **ammonium chloride.**

salami slicing **1.** A scheme in which minimal amounts of money, usually less than 0.1 cent, are taken from multiple, often millions, of accounts or transactions, so as to accumulate a large amount of stolen money while attempting to avoid detection due to the minor individual quantities involved. **2.** To reduce anything favorable, such as the level of quality control and/or service, very gradually in the attempt to escape notice by those affected.

sales automation The use of electronic devices such as computers and telephones to automate tasks related to sales activities such as lists and descriptions of product offerings, inventory maintenance, the taking of client orders, and the like. Used, for instance, for Internet-based sales.

sales force automation The use of electronic devices such as computers, telephones, and PDAs, to automate tasks related to sales force activities such as client and product data, marketing, follow-up, service, and the like. Its abbreviation is **SFA.**

salient pole A pole, such as that used in a generator or motor, which extends from a given structure or part.

salmiac Same as **sal ammoniac.**

salt-spray test An accelerated test in which a salty mist is utilized to test the corrosion resistance of components, circuits, devices, equipment, systems, or materials.

SAM Abbreviation of **sequential-access method.**

samarium A hard silver-white metallic chemical element whose atomic number is 62. It has about 30 known isotopes, of which 4 are stable. It is used in carbon-arc lamps, as a dopant in lasers, as a neutron absorber, and it alloys with cobalt to form extremely powerful permanent magnets. Its chemical symbol is **Sm.**

samarium cobalt A silver-gray metal whose chemical formula is $SmCo_5$, or Sm_2Co_{17}. Utilized as extremely powerful permanent magnets which have high coercivity, corrosion resistance, and excellent thermal stability.

samarium cobalt magnet A powerful permanent magnet made of **samarium cobalt.**

samarium oxide A cream-colored powder whose chemical formula is Sm_2O_3. Utilized as a neutron absorber and in special glasses.

sample **1.** To measure or obtain data in discrete steps or intervals, as opposed to continuously. Said especially in the context of analog-to-digital conversions. **2.** In statistics, a group or subset within a population. Such a sample is meant to be representative of said population. Also, to select such a group or subset. **3.** A given item within a group of similar items. Also, to select such an item. For example, a device which is selected and tested.

sample and hold **1.** To hold a changing value at a given level, for more processing, recording, or measurement.

Used, for instance, in sample-and-hold circuits. **2.** To store a reading, measurement, or value for a given time interval. Seen, for example, in peak indicators.

sample-and-hold amplifier An operational amplifier which holds a changing value, such as a voltage, at a given level, for more accurate processing, recording, or measurement. Used, for instance, with an analog-to-digital converter. Its abbreviation is **SHA.**

sample-and-hold circuit A circuit, such as an analog-to-digital converter, which holds a changing value, such as a voltage, at a given level, for more accurate processing, recording, or measurement.

sample interval Same as **sampling interval.**

sample rate Same as **sampling rate.**

sample size In statistics, the size of a **sample (2).**

sampled-data control system A control system which samples data at given intervals.

sampled-data system A system which samples data at given intervals. In order to avoid aliasing, the sampling rate must equal or exceed the Nyquist rate.

sampler **1.** That which serves to **sample (1).** **2.** That which serves as a **sample (3).**

sampling **1.** The taking of **samples.** **2.** A given **sample (2).**

sampling frequency Same as **sampling rate.**

sampling interval The interval between successive **samples (1).** Also called **sample interval.**

sampling oscilloscope A digital oscilloscope with an extremely high sampling rate which is used, for instance, for measuring jitter present in high-speed communications systems, or for testing chips.

sampling period The reciprocal of the **sampling rate.**

sampling rate In analog-to-digital conversions, the frequency with which samples of a variable are taken. For instance, that used in the conversion of analog music into digital form. Usually expressed as samples or cycles per unit time, as in a 192 kHz sampling rate. Sampling must be made at or above the Nyquist rate to prevent aliasing, and for any given sample size, the higher the sampling rate the more accurate the conversion. Also called **sample rate, sampling frequency,** or **conversion rate.**

sampling switch A device used to execute repetitive series of switching operations. Such a switch is usually rotary, and may be electrical or mechanical. Used, for instance, for sampling of multiple quantities for multiplexing over a single channel. Also called **scanning switch,** or **commutator switch.**

sampling window The time interval within which a **sample (1)** is taken.

samurai A person who is paid to hack into systems for purposes such as ascertainment of security vulnerabilities.

SAN **1.** Acronym for **storage-area network. 2.** Acronym for **system-area network.**

sandbox An area within computer memory or storage in which applications, especially those which are Java-based, are confined to operate. By restricting the ability to access certain resources of the operating system and memory, a given level of protection is provided in case any code is bad, or if a malicious applet is contained.

SAP **1.** Abbreviation of **secondary audio program. 2.** Abbreviation of **service access point. 3.** Abbreviation of **Service Advertising Protocol.**

SAPI Same as **speech API.**

sapphire A precious stone consisting of corundum with trace titanium and cobalt or iron impurity atoms, which give it is transparent blue appearance. A sapphire may also be of most any color except red, as that variety of corundum is known as **ruby**. Used, for instance, in precision instruments such as lasers, and as phonograph styli.

sapphire needle Same as **sapphire stylus**.

sapphire stylus A phonograph pickup whose tip is a ground sapphire. Also called **sapphire needle**.

sapphire substrate A semiconductor substrate composed of sapphire.

SAR 1. Acronym for **segmentation and reassembly**. 2. Abbreviation of **synthetic aperture radar**. 3. Abbreviation of **successive approximation register**.

SAS Abbreviation of **single-attachment station**.

SASD Abbreviation of **sequential-access storage device**.

SASL Abbreviation of **Simple Authentication and Security Layer**.

sat 1. Abbreviation of **satellite**. 2. Abbreviation of **saturation**.

satcom Acronym for **satellite communications**.

satellite Its abbreviation is **sat**. 1. A body that revolves around another. 2. A celestial body that revolves around another. For example, a moon around a planet. 3. An artificial object which orbits a celestial body under the gravitational influence of the latter. Also known as **artificial satellite (1)**. 4. A **satellite (3)** intended or utilized for communications, broadcasting, observation, testing, research, or the like. Also called **artificial satellite (2)**.

satellite antenna 1. An antenna designed for reception of **satellite signals**. Such an antenna, for instance, may have an 18-inch parabolic reflector that collects the microwave energy and focuses it into a feedhorn, with the signal then being channeled to an LNB for amplification and down-conversion to a range of frequencies utilized by a satellite receiver. 2. An antenna, such as a helical antenna, utilized for the transmission of **satellite signals**.

satellite communications The use of **satellites (4)** for relaying, and usually amplifying, radio-wave signals between terrestrial communications stations, terrestrial communications stations and communications satellites, or between communications satellites. Such communications can offer high-capacity, a very wide coverage area, and may transmit TV, telephone, and data signals, among many others. Its acronym is **satcom**.

satellite communications system A system, incorporating both space elements and ground elements, utilized for **satellite communications**.

satellite computer A remote computer which is secondary to a main or host computer.

satellite content Information in any form, such as TV programming, music, or Web pages, which is obtained via satellite.

satellite converter Same as **satellite decoder**.

satellite coverage 1. The geographical area within which a satellite transmitter provides effective service. For example, as few as three synchronous satellites can provide worldwide coverage. Also called **satellite footprint**. 2. The service provided via **satellite coverage (1)**.

satellite decoder Also called **satellite converter**. 1. A decoder which extracts information-bearing signals transmitted via satellite. 2. Same as **satellite decoder box**.

satellite decoder box A decoder box which enables reception of certain channels, programming, or other satellite content. Also called **satellite decoder (2)**.

satellite digital audio radio service Its abbreviation is **SDARS**. A service in which digitally encoded radio broadcasts are transmitted from satellites or repeater stations. Also called **direct broadcast satellite radio**.

satellite dish A satellite receiving antenna which utilizes a parabolic reflector. Also, the reflecting dish itself. Such a dish may range in size from a few centimeters to several meters.

satellite downlink 1. The downward signal path originating from a satellite, such as a communications satellite, to the earth. 2. The establishing, or use, of a **satellite downlink (1)**.

satellite earth station A land-based station utilized for transmitting and/or receiving satellite communications. It incorporates devices such as a low-noise amplifier, a down-converter, a multiplexer, and an antenna. An earth station antenna may vary in size from a few centimeters to several meters. Also called **earth station**.

satellite footprint Same as **satellite coverage (1)**.

satellite Internet Internet access via satellite signals. Also, a service providing such access.

satellite link A communications line or channel between satellites, or which has a satellite at either end.

satellite LNB An LNB whose output signal is fed to a **satellite receiver**.

satellite loudspeakers Same as **satellite speakers**.

satellite master-antenna television Same as **satellite master-antenna TV**.

satellite master-antenna television system Same as **satellite master-antenna TV**.

satellite master-antenna TV Abbreviation of **satellite master-antenna** television. A satellite system which receives TV programming, then distributes it to multiple units such as those in a house, hotel, or apartment building. Such a system may also provide signal amplification, and might also offer radio broadcasts. Its own abbreviation is **SMATV**. Also called **satellite master-antenna TV system**.

satellite master-antenna TV system Same as **satellite master-antenna TV**.

satellite mobile phone Same as **satellite phone**.

satellite mobile telephone Same as **satellite phone**.

satellite modem A device which enables a computer to transmit and receive information over a **satellite link**. The transponder such a modem communicates with may also carry other content, such as TV programming.

satellite network A group of satellites which work together to receive, amplify, relay, store, process, and transmit signals. Such a network may provide worldwide coverage, and an example is the GPS System.

satellite orbit The orbit, such as a synchronous orbit, a **satellite** follows.

satellite phone 1. Same as **satellite telephony**. 2. Same as **satellite phone service**. 3. A mobile or stationary telephone which is utilized for **satellite telephony**. Some units provide both satellite and cellular functionality.

satellite phone service A service based on **satellite telephony**. Also called **satellite telephone service**, or **satellite phone (2)**.

satellite phone system Same as **satellite telephony**.

satellite programming Program offerings, such as those of TV or radio, which are transmitted via satellite.

satellite radio Radio programming arriving via satellite signals. Also, a technology enabling such a service. Also called **digital satellite radio**.

satellite receiver The component of a satellite system that accepts information-bearing signals transmitted via satellite, and extracts the meaningful information contained. A satellite receiver obtains its signal from an LNB or similar device. In addition, it accepts user commands, such as those pertaining to channel selection and menu navigation.

satellite reception The receiving of **satellite signals**.

satellite relay An artificial satellite which relays, and usually amplifies, radio-wave signals between terrestrial communications stations, terrestrial communications stations and other communications satellites, or between other communications satellites. Such satellites provide high-capacity communications links, offer a very wide coverage area, and may transmit TV, telephone, and data signals, among many others. Also called **satellite repeater, communications satellite, relay satellite, radio-relay satellite,** or **repeater satellite.**

satellite repeater Same as **satellite relay.**

satellite service A service in which TV, radio, data, or other signals are transmitted or retransmitted via satellite. For example, a broadcasting-satellite service, satellite Internet, or the Global Positioning Service.

satellite signal A microwave signal which is transmitted by a **satellite (4)**, and which is used for TV reception, telephone communications, or Internet access, among many others. Such a signal is usually compressed.

satellite speakers Speakers, such as those providing the sounds associated with special effects, which are situated around a listener. Such speakers may be for instance, the left rear and right rear speakers. Also called **satellite loud-speakers.**

satellite splitter A device which divides a signal received by a satellite antenna, so as to provide multiple paths. Used, for instance, to enable multiple TVs to access the signal of a single satellite

satellite system A system, such as that utilized to view TV programming, which incorporates satellite signals, and the appropriate hardware, including satellite antennas and receivers. The term may also include devices such as TVs, audio systems, or computers, through which the satellite content is utilized.

satellite telephone Same as **satellite phone (3)**.

satellite telephone service Same as **satellite phone service.**

satellite telephone system Same as **satellite telephony.**

satellite telephony A telephone system utilizing a network of communications satellites, such as LEOS, to provide global, or partially global coverage for voice communications. Used, for instance, to provide coverage where there is no other wireless or land-based service available. Some systems also provide data services. Also, any technology based on, or enabling such a system. Also called **satellite phone (1), satellite phone system,** or **satellite telephone system.**

satellite television Same as **satellite TV.**

satellite transmission The transmission of **satellite signals.**

satellite transmitter A unit, such as that within a communications satellite, which transmits or retransmits satellite signals.

satellite transponder In a communications satellite, a unit which receives a signal from one location, processes and amplifies it, and retransmits it to other locations. A single satellite may have many transponders. Also called **transponder (2).**

satellite TV Abbreviation of **satellite** television. TV programming, such as that of networks, movie channels, or pay-per view, arriving via satellite signals.

satellite uplink 1. The upward signal path originating from an earth-based transmitter towards a satellite, such as a communications satellite. **2.** The establishing, or use, of a **satellite uplink (1)**.

saturable core A core, such as that utilized in a saturable reactor, whose level of **saturation (2)** can be easily varied.

saturable-core reactor Same as **saturable reactor.**

saturable-core transformer Same as **saturable transformer.**

saturable reactor A coil which is wound around a core whose permeability is varied, so as to vary the reactance of said coil. Such a device incorporates a control coil that carries DC, which is utilized to vary the level of **saturation (2)** of the magnetic core. Also called **saturable-core reactor.**

saturable transformer A transformer which regulates its output voltage by varying the level of **saturation (2)** of its magnetic core. Also called **saturable-core transformer.**

saturated color A color in which there is no white component present. Such a color has 100% **saturation (4)**. Also called **pure color.**

saturated logic Logic in which the transistors are in **saturated mode.**

saturated mode The operation of a transistor at or beyond its **saturation point.**

saturated operation Operation of a component, core, circuit, or device, such as a magnetic core or transistor, at or beyond its **saturation point.**

saturated solution A solution which contains the maximum amount of a given substance that it can dissolve in equilibrium. This maximum amount of solute will vary depending on the temperature and pressure.

saturation Its abbreviation is **sat. 1.** The condition where a maximum effect has been attained, so that any additional increase in the causative variable will not achieve any additional results. **2.** For a magnetic material, the condition in which an increase in magnetizing force will not produce an increase in magnetization. Also called **magnetic saturation. 3.** In an electron tube, the condition in which the anode current can not be further increased, regardless of any additional voltage applied to it, since essentially all available electrons are already being drawn to said anode. Also called **plate saturation, current saturation, voltage saturation,** or **anode saturation. 4.** The extent to which a given color lacks a white component. 100% saturation means there is no white present. Also called **color saturation,** or **chroma (3). 5.** A point at which a chemical solution, such as an electrolyte, contains the maximum amount of a substance that it can dissolve in equilibrium. **6.** In magnetic media, the condition where the magnetizable particles are fully aligned. **7.** For a thermionic electron tube at a given anode voltage, the condition in which the anode current can not be further increased with increases in cathode temperature. Also called **temperature saturation. 8.** In a bipolar transistor, the condition where the collector becomes positive with respect to the base. Ordinarily the converse is true. Also called **transistor saturation.**

saturation current For a given component or device, the current flowing at or beyond its **saturation point.** In an electron tube, for example, it is the level at which the anode current is not further increased, regardless of any additional voltage applied to it.

saturation curve A curve, such as a characteristic curve correlating current and voltage, which depicts the manner in which a given parameter approaches and reaches saturation.

saturation flux density Same as **saturation magnetic flux density.**

saturation induction Same as **saturation magnetic flux density.**

saturation magnetic flux density For a magnetic material or medium, the maximum attainable magnetic flux density. Also called **saturation flux density,** or **saturation induction.**

saturation magnetization For a magnetic material, the maximum attainable magnetization.

saturation magnetostriction For a given material, the maximum attainable magnetostriction. Once this value is reached, magnetic deformation will not increase regardless of the applied magnetic field.

saturation point 1. For a component, device, or material, the point at which saturation is attained. For a magnetic material, for instance, an increase in magnetizing force beyond this point will not result in an increase in magnetization. **2.** For a component or device, the point within a characteristic curve correlating the current and voltage at which saturation is attained. Once this point is reached, no increase in voltage will result in an increase in current.

saturation region For a component or device, an operating range at or beyond its **saturation point.**

saturation resistance For a component or device at or beyond its **saturation point,** the ratio of the voltage to the current.

saturation signal A signal which produces the maximum possible response, output, or reading.

saturation state An operational state in which one or more parameters are at or beyond their **saturation point.**

saturation value The value of a given parameter, such as current, voltage, or magnetizing force, required for a component, device, or material to reach its **saturation point.**

saturation voltage For a given component or device, the minimum voltage required to produce **saturation current.**

save 1. To copy or store data from a temporary area, such as RAM, to a permanent medium such as a hard drive or optical disc. Each time a save is performed, the previous version of the same data or file is replaced with the most recent. Many programs can be set to save at fixed intervals, as any work done between saves is usually lost if there is a system malfunction or failure. **2.** A command or menu item which produces a **save** (1). Also, to issue such a command, or select this item from a menu.

save as A command or menu item that allows the current version of data or a file to be renamed, placed in a different directory or format, or the like.

SAW Abbreviation of **surface acoustic wave.**

SAW filter Abbreviation of **surface acoustic wave filter.**

saw-tooth Same as **sawtooth wave.** Also spelled **sawtooth.**

saw-tooth generator Same as **sawtooth generator.**

saw-tooth oscillator Same as **sawtooth oscillator.**

saw-tooth pulse Same as **sawtooth pulse.**

saw-tooth wave Same as **sawtooth wave.**

saw-tooth-wave generator Same as **sawtooth-wave generator.**

saw-tooth waveform Same as **sawtooth waveform.**

sawing A technique for cutting a semiconductor material into dies or chips, in which high-speed precision saws, with diamond-coated blades which can be thinner than one micrometer, are used. Scribing is usually performed prior to sawing to facilitate cutting. Also called **wafer sawing.**

sawtooth Same as **sawtooth wave.** Also spelled **saw-tooth.**

sawtooth generator A circuit, device, or instrument which produces **sawtooth waves.** Also spelled **saw-tooth generator.**

sawtooth oscillator An oscillator which produces **sawtooth waves.** Also spelled **saw-tooth oscillator.**

sawtooth pulse A pulse whose graphed or displayed amplitude resembles the teeth of a saw. Also spelled **saw-tooth pulse.**

sawtooth wave A wave, such as a ramp wave, whose shape resembles the teeth of a saw. Also spelled **saw-tooth wave.** Also called **sawtooth.**

sawtooth-wave generator A signal generator whose output consists of **sawtooth waves.** Also spelled **saw-tooth wave generator.**

sawtooth waveform The shape of a **sawtooth wave.** Also spelled **saw-tooth waveform.**

SAX Abbreviation of **Simple API for XML.** An event-driven API that allows the interpretation of Web documents and data using XML.

sb Symbol for **stilb.**

Sb Chemical symbol for **antimony.**

SB Abbreviation of **sideband.**

SBC Abbreviation of **single-board computer.**

Sc Chemical symbol for **scandium.**

SC Abbreviation of **suppressed carrier.**

SC connector A fiber-optical cable connector that is pushed into place for a secure connection, as opposed, for instance, to being pushed and twisted. Such connectors are available for both single and multimode fibers, and are helpful in locations where there is a high density of connections. An SC connector has a 2.5 mm ferrule.

SCA Abbreviation of **Single Connector Attachment.** A peripheral connector which incorporates pins for both data and power transfer. Such a connector usually has 80 pins, and may be used, for instance, with SCSI devices such as disk drives.

SCADA Acronym for **s**upervisory **c**ontrol **a**nd **d**ata **a**cquisition, or **s**ystem **c**ontrol **a**nd **d**ata **a**cquisition. An application or system utilized for the real-time collection of data from one or more remote locations, with the obtained information being utilized to monitor and manage ongoing processes. Used, for instance, to supervise and control a factory, a transportation system, or a power grid.

scalability The ability to expand or otherwise adapt to changing needs. For instance, the capability of a computer application to maintain proper operation when moved from one operating system to another. Ideally, such an application would maximize its ability to work within each system, depending on the resources offered by said systems. Also, the extent of such scalability.

scalable 1. Characterized by **scalability. 2.** Able to be changed in size, configuration, operating mode, or the like. Said, for instance, of software, hardware, or a system.

Scalable Coherent Interface A standard high-speed data link which is used, for instance, between massively parallel processors. Its abbreviation is **SCI.**

scalable parallel processing Parallel processing in which processors, I/O capacity, and/or users may be added without significant reductions in performance. Its abbreviation is **SPP.**

Scalable Processor Architecture Same as **SPARC.**

Scalable Vector Graphics A vector graphics format used within XML. Used, for instance, to prepare scalable pictures for use in Web documents. Its abbreviation is **SVG.**

scalar 1. Same as **scalar quantity. 2.** A quantity having a single value or consisting of a single item. **3.** A quantity or item in a field.

scalar quantity A quantity, such as mass or temperature, which has a magnitude but not direction. A **vector quantity** has a magnitude and direction. Also called **scalar** (1).

scale 1. A series of marks which serve as a reference when measuring and indicating magnitudes or values. Seen, for example, in meters and indicating instruments. Such a scale may be, for instance, linear or logarithmic. Also, to change from one such scale to another, when more than one is available. **2.** An instrument or device which measures and indicates mass. **3.** A progressive, graduated, or hierarchical ordering or series of items, values, or the like. For example, a temperature scale, or a dB scale. **4.** An established standard for describing, measuring, and stating values. For instance, the Mohs scale. **5.** A qualitative or quantitative indication of relative magnitudes. For instance, nanoscale, or large-scale. **6.** The ratio of the size of a body or region to a model or representation of it. For example, a scale used in lithography. **7.** Same as **size (4)**. **8.** To change the manner in which a quantity is expressed, so as to fit in a different range. For example, to change from kilometers to meters. **9.** A thin piece, slice, or flake. For instance, an iron oxide scale formed on the surface of a piece of iron.

scale divider **1.** A number which divides a measurement or reading, so as to provide the final displayed magnitude or value. Used, for instance, to convert a quantity into a value that is in the units displayed by a given scale. **2.** A number which divides a measurement or reading, so as to change its scale of magnitude. Such a factor is usually a negative power the base of the number system utilized is raised to.

scale division The separation or interval between successive markings of a **scale (1)**.

scale factor **1.** A number which is added, multiplied, or otherwise factored in to a measurement or reading, so as to provide the final displayed magnitude or value. Used, for instance, to convert a quantity into a value that is in the units displayed by a given scale. **2.** A number which divides or multiplies a measurement or reading, so as to change its scale of magnitude. Such a factor is usually an exponent the base of the number system utilized is raised to.

scale-factor error Any error introduced when utilizing a **scale factor**.

scale length The separation or interval between one end of a scale and the other. Also called **scale span (2)**.

scale multiplier **1.** A number which multiplies a measurement or reading, so as to provide the final displayed magnitude or value. Used, for instance, to convert a quantity into a value that is in the units displayed by a given scale. **2.** A number which multiplies a measurement or reading, so as to change its scale of magnitude. Such a factor is usually a positive power the base of the number system utilized is raised to.

scale-of-ten circuit A counter circuit which produces an output pulse for every ten input pulses. Also called **scale-of-ten counter**, **scale-of-ten scaler**, or **decade scaler**.

scale-of-ten counter Same as **scale-of-ten circuit**.

scale-of-ten scaler Same as **scale-of-ten circuit**.

scale-of-two circuit Same as **scale-of-two counter**.

scale-of-two counter Also known as **scale-of-two circuit**, **binary scaler**, or **binary counter**. **1.** A counter circuit which produces one output pulse for every two input pulses. **2.** A counter whose elements may assume one of two stable states.

scale range The difference between the maximum and minimum values of a scale. Also called **scale span (1)**.

scale span **1.** Same as **scale range**. **2.** Same as **scale length**.

scaler **1.** A circuit or device, such as a counter circuit, which generates an output after counting a specified number of pulses. Used, for instance, in radiation counters. **2.** A **scale multiplier** or a **scale divider**.

scaling circuit A circuit, such as a scale-of-ten circuit, which serves as a **scaler**.

scaling factor The number of input pulses a **scaler (1)** counts per output pulse. Also called **scaling ratio**.

scaling ratio Same as **scaling factor**.

scan **1.** To move a detector or a focused beam of electromagnetic radiation, light, sound, or particles, in a fixed pattern over a surface or through a region, to sense, examine, reproduce, transmit, or the like. Seen, for example, in the tracing of the electron beam in a CRT, in the motion of a radar while searching for scanned objects, the helical path of a head in contact with a magnetic tape, or in the conversion of a scene into an electrical image. **2.** A complete **scan (1)**. **3.** Same as **scanning (1)**. **4.** A **scan (1)** performed by an antenna, as in radars. **5.** To search records, files, the Internet, or other sources of information, looking for specific data. **6.** An analytical or medical diagnostic technique, such as computed

tomography, in which one or more **scans (1)** are taken. **7.** To utilize a device, such as a radar or scanner, which performs **scans (1)**. **8.** To examine while seeking something in particular. For instance, to scan communications channels looking for an available circuit.

scan conversion The conversion of one video scan format into another utilizing a **scan converter**.

scan converter A device which changes one video scan format into another. For example, that which converts the video output of a computer into an NTSC TV signal, or from an NTSC signal to a PAL signal, a PAL signal into an HDTV signal, or plan position indicator images into those which can be viewed by a computer monitor or TV. Also called **video scan converter**.

scan-converter tube A CRT utilized as a **scan converter**.

scan frequency Same as **scanning frequency**.

scan head In a scanning device, such as a barcode scanner or fax, the sensing device which is moved or otherwise employed while scanning. Also called **scanning head**.

scan line A horizontal line that is detected, read, displayed, or produced by a scanning device such as an optical scanner or CRT.

scan rate Same as **scanning frequency**.

scandium A silvery-white metal whose atomic weight is 21. It is comparatively soft, reacts quickly with most acids, and has over 15 known isotopes, of which one is stable. It is used in semiconductors. Its chemical symbol is **Sc**.

scanned area The geographical area within which a radar has effective coverage.

scanned object An object which reflects a sufficient amount of the energy beam emitted by a radar to produce a radar echo. Also called **radar target**, or **target (2)**.

scanner **1.** That which **scans**. Examples include optical scanners, magnetic character readers, computerized tomography scanners, plan position indicators, and radio scanners. Also, one that scans. **2.** An optical device which scans a sheet, film, area, scene, or the like, and converts it into data, an electrical image, or another form which enables the original visual information to be stored, processed, transmitted, or reproduced. Also called **optical scanner (1)**. **3.** Same as **scanning device**. **4.** A device which utilizes a small and intensely bright spot which scans an object, scene, screen, film, or the like, and transforms the original image into a series of electrical signals by focusing it onto a photoelectric cell. The scanning may be mechanical or electrical. Used, for instance, for TV transmission, telecine transfer, or character recognition. Also called **optical scanner (2)**, **light-spot scanner**, or **flying-spot scanner**. **5.** A device which utilizes optical sensors to detect and interpret characters, marks, bar codes, and the like. Examples include optical character readers, optical mark readers, and bar-code readers. Also called **optical reader**, or **optical scanner (3)**. **6.** A computer peripheral which converts the visual information of a paper, film, or the like, and into an image which is stored as a bitmap. Such an image can then be displayed or printed. When used with optical character recognition software, such images can also be processed, allowing, for instance, the characters of printed documents to be converted into electronic text. Also called **optical scanner (4)**. **7.** A radio receiver that searches frequencies or bands for signals. Such a receiver may proceed from station to station until manually stopped. When multiple bands may be scanned, also called **scanning radio**. **8.** The part of a scanning device which performs the actual **scanning (1)**.

scanning **1.** The motion of a detector or of a focused beam of electromagnetic radiation, light, sound, or particles in a fixed pattern over a surface or region, to detect, examine, reproduce, transmit, or the like. Such scanning may be circular, helical, horizontal, vertical, rectilinear, spiral, conical, and so on. Also, a specific instance of such scanning. Also called **scan** (3). **2.** The movement of an antenna while **scanning** (1), as in radars. **3.** The searching through records, files, the Internet, or other sources of information, looking for specific data. For instance, data mining. **4.** The performance of an analytical or medical diagnostic technique, such as computed tomography, in which **scanning** (1) takes place. **5.** The process of utilizing a device, such as a radar or optical scanner, which performs **scans** (1). Also, the **scanning** (1) such a device performs. **6.** The process of examining while seeking something in particular. For instance, the scanning of communications channels looking for an available circuit.

scanning antenna An antenna, such as that of a radar, which moves in a **scanning** (1) manner.

scanning beam A beam, such as that of an electron gun in a CRT, which moves in a **scanning** (1) manner.

scanning circuit A circuit which generates **scans** (1). Also called **sweep circuit** (2).

scanning device A device, such as an optical scanner or radar, which performs **scans**. Also called **scanner** (3).

scanning electron microscope An electron microscope in which the electron beam is scanned over the surface of the specimen. Such a microscope produces a three-dimensional view on a film or display screen. This contrasts, for instance, with a **transmission electron microscope**, which passes an electron beam through a thin sample. Its abbreviation is **SEM**.

scanning electron microscopy The use of a **scanning electron microscope** to view and analyze specimens. Its abbreviation is **SEM**.

scanning frequency The frequency at which **scanning** (1) takes place. In a CRT, for instance, it could be the number of horizontal or vertical lines scanned by the electron beam per second. Also called **scanning rate, scan frequency, scan rate, scanning speed,** or **sweep frequency** (2).

scanning head Same as **scan head**.

scanning line 1. A line along which **scanning** (1) occurs, as in a TV camera. **2.** A line produced by **scanning** (1), as in a TV receiver.

scanning pattern 1. A pattern of scanning lines which cover the viewing area of a display, such as that of a computer monitor or TV screen. The electron beam follows these lines to produce images. Also called **raster. 2.** The pattern followed when **scanning** (1).

scanning probe microscope A microscope utilized in **scanning probe microscopy**. Its abbreviation is **SPM**.

scanning probe microscopy Microscopy, such as atomic force microscopy, in which a tiny probe is utilized to scan the surface of a sample at an atomic level. Its abbreviation is **SPM**.

scanning radar A radar, such as an air traffic control radar, which scans a given region.

scanning radio A radio receiver that can scan multiple radio bands. With such a unit, for instance, a user may tap buttons or turn a dial for frequency selection, or the receiver may proceed automatically from one frequency in use to another, between preprogrammed frequencies, and the like. Also called **scanning receiver, scanning radio receiver, sweeping receiver,** or **radio scanner**.

scanning radio receiver Same as **scanning radio**.

scanning rate Same as **scanning frequency**.

scanning receiver Same as **scanning radio**.

scanning sequence The order and/or pattern followed when **scanning** (1).

scanning sonar Sonar in which a given area is scanned side-to-side by the transmitted ultrasonic sounds. Used, for instance, to identify shapes and features of a sea-floor sector. Also called **side-scanning sonar**.

scanning speed Also called **stroke speed. 1.** Same as **scanning frequency. 2.** Same as **spot speed**.

scanning spot Also called **pixel, picture element,** or **elemental area. 1.** Within an image, such as that of a computer display, TV, or facsimile, the smallest unit which can be manipulated. In a CRT, such as a color TV or computer monitor, for instance, it is a trio of color phosphor dots representing a single point. Also called **spot** (3). **2.** In a TV picture or facsimile, the portion of the scanning line being explored at any given instant. Also called **spot** (4).

scanning switch Same as **sampling switch**.

scanning transmission electron microscope An electron microscope which combines the three-dimensional imaging of a scanning electron microscope, and the higher resolution of a transmission electron microscope. Its abbreviation is **STEM**.

scanning transmission electron microscopy The use of **scanning transmission electron microscopes** to study and analyze samples. Its acronym is **STEM**.

scanning tuning Tuning, such as that of a scanning radio, which scans through one or more radio bands.

scanning tunneling microscope An electron microscope which has a minute probe which scans the surface of a sample at an atomic level. The extremely sharp tip, from which a single atom projects, is brought sufficiently close to the surface so as to create a tunneling current. The rapid change in tunneling current as a function of distance enables the atomic resolution. In addition, the scanning produces a three-dimensional view on a film or display screen. Its abbreviation is **STM**.

scanning tunneling microscopy The use of **scanning tunneling microscopes** to study and analyze samples. Its abbreviation is **STM**.

scanning yoke In a CRT, a system of coils utilized for magnetic deflection of the electron beam. One possible arrangement consists of two sets of coils, one for horizontal deflection, and the other for vertical deflection. Also called **deflection yoke,** or **yoke** (2).

SCART A 21-pin connector utilized for connections between audio and video devices such as TVs and VCRs, and which is commonly utilized in Europe. The devices have recessed openings, while the cable has matching pins at each end. It is an abbreviation of **S**yndicat des **C**onstructeurs d'**A**ppareils **R**adiorécepteurs et **T**éléviseurs. Also called **SCART connector,** or **Euroconnector**.

SCART connector Same as **SCART**.

scatter 1. To spread out, deflect, diffuse, or otherwise disperse electromagnetic radiation, particles, or acoustic waves via striking or propagating through a surface, region, or medium. **2.** To cause to **scatter** (1). **3.** The act of **scattering**.

scatter communication Communication by means of radio waves that are propagated well beyond line-of-sight distances. This is usually due to scattering by the ionosphere or troposphere. Also called **over-the-horizon communication, forward-scatter communication, beyond-the-horizon communication,** or **extended-range communication**.

scatter propagation Propagation of radio waves well beyond line-of-sight distances. This is usually due to scattering by the ionosphere or troposphere. Also called **over-the-horizon propagation, forward-scatter propagation, beyond-the-horizon propagation,** or **extended-range propagation**.

scatter read A single data read operation in which two or more non-contiguous buffers or locations are accessed. A **gather write** was previously utilized to store such data.

scatter transmission Transmission of radio waves well beyond line-of-sight distances. This is usually due to scattering by the ionosphere or troposphere. Also called **over-the-horizon transmission**, **forward-scatter transmission**, **beyond-the-horizon transmission**, or **extended-range transmission**.

scattering 1. The process via which electromagnetic radiation, such as light, or particles, such as electrons or neutrons, are spread out, deflected, diffused, or otherwise dispersed after striking or propagating through a surface, region, or medium. Sound waves may also undergo such scattering. Specific examples include X-rays being scattered by the atoms in a crystal lattice, or electromagnetic waves being scattered by the ionosphere. When there is a definite phase relationship between the incident and dispersed waves or particles, there is coherent scattering, otherwise it is non-coherent scattering. **2.** The result of **scattering (1)**.

scattering loss Losses due to **scattering (1)**. Also called **scattering losses**.

scattering losses Same as **scattering loss**.

scattering parameters Network parameters pertaining to reflection and transmission coefficients between the incident and reflected waves of a given network. Its abbreviation is **s-parameters**.

scattering Raman lidar A form of lidar which utilizes the Raman effect to measure atmospheric parameters such as water vapor or ozone concentration, and temperature. In it, a laser beam is directed into the atmosphere, with scattering intensity being correlated with any measurements being made. Its abbreviation is **SRL**. Also called **Raman lidar**.

SCbus A bus for connections between computer telephony components using **SCSA**.

schedule To allocate time periods and/or resources. For example, to program a computer to perform a backup at a given, time, to place a print job in a queue, or to set a switch to open or close a circuit at stated intervals. Also, a list, chart, or other representation of that which is scheduled.

scheduled maintenance Maintenance which is performed on the basis of a given number of elapsed days, hours of operation, cycles completed, or the like. Such maintenance may be programmed to take place at specified hours or on given days of the week to minimize the effect on overall operations.

scheduler Within a computer operating system, a program which coordinates the use of the resources of the system. For example, it initiates and terminates each task. Also called **dispatcher**.

scheduling algorithm An algorithm utilizing for establishing schedules, such as those used by **schedulers**.

schema Its plural forms are **schemata** and **schemas**. **1.** A description of the structure or aspects of a given database or database management system in a language such a database or database management system supports. For instance, a description in SQL code of the overall structure of a database, the attributes of the fields, and so on. **2.** In artificial intelligence and object-oriented computing, a formal expression of an inference rule utilized in applications. A schema provides a general template, such as a class, while specific values, such as objects, are replaced in specific instances. **3.** An outline or model of something.

schemata A plural form of **schema**.

schematic 1. Same as **schematic diagram**. **2.** Pertaining to a **schematic diagram**.

schematic capture The entering of a **schematic diagram** into a computer-aided design or computer-aided engineering tool. Used, for instance, for digital simulations or prototyping.

schematic circuit diagram Same as **schematic diagram**.

schematic diagram A graphical representation of the electrical elements in a circuit, and the way each is interconnected with each other. Each circuit element is represented by a symbol, while lines represent the wiring. Also called **schematic**, **schematic circuit diagram**, **circuit diagram**, **wiring diagram**, **wiring schematic**, or **diagram (3)**.

schematic symbol A symbol utilized in a **schematic diagram**.

Schering bridge A four-arm AC bridge utilizing capacitors and resistors, which is used for measuring capacitances and dissipation factors of capacitors.

Schmitt circuit Same as **Schmitt trigger**.

Schmitt trigger A bistable circuit which converts an input signal of varying waveform into a square-wave output by a switching action which is triggered at a specified point within the positive and negative swing of said input signal. Used, for instance, in binary logic circuits with high noise immunity. Also called **Schmitt trigger circuit**, or **Schmitt circuit**.

Schmitt trigger circuit Same as **Schmitt trigger**.

Schottky barrier A metal-semiconductor contact or junction which has rectifying properties. A metal-semiconductor contact in which an increase in voltage results in an increase in current is an **ohmic contact**. Also called **Schottky contact**.

Schottky barrier diode Same as **Schottky diode**.

Schottky contact Same as **Schottky barrier**.

Schottky diode A diode with a metal-semiconductor junction, as opposed to the usual pn junction. Hot carriers are injected from the semiconductor layer into the metal layer, so that there are virtually no minority carriers injected or stored, which provides for a very low forward voltage drop and a very fast switching speed. Such diodes have multiple uses, including several in radio-frequency applications, logic circuits, photodiodes, and as rectifiers. Also called **Schottky barrier diode**, **semiconductor-metal diode**, **hot-carrier diode**, or **metal-semiconductor diode**.

Schottky effect In a thermionic tube, an increase in electron emission due to a decrease in the thermionic work function. This reduction is induced by an external electric field applied at the surface of the emitter.

Schottky logic IC logic incorporating **Schottky diodes** for enhanced speed.

Schottky noise Same as **shot noise**.

Schottky photodiode A photodiode which incorporates a **Schottky barrier**. The wavelengths of maximum sensitivity to incident light will depend on the metals and semiconductors utilized.

Schottky transistor-transistor logic Same as **Schottky TTL**.

Schottky TTL Abbreviation of **Schottky** transistor-transistor logic. Transistor-transistor logic which incorporates **Schottky diodes**. This provides for faster operating speeds than standard TTL, plus lower power consumption and greater immunity to noise.

Schrödinger equation An equation utilized to obtain the wave function of a particle. It is the key equation in non-relativistic quantum mechanics. Also spelled **Schroedinger equation**.

Schroedinger equation Same as **Schrödinger equation**.

SCI Abbreviation of **Scalable Coherent Interface**.

scientific language A computer language, such as FORTRAN, designed specifically for scientific applications, computations, and problem-solving.

scientific notation A numeric format in which each number is represented by two numbers, and in which the decimal point is not in a fixed location. The first number, the mantissa, specifies the significant digits, while the second number, or exponent, specifies its magnitude. For example, 314,000,000 may be expressed as 3.14×10^8. Although any number, with any radix, may be represented in this manner, it is usually used only for very small or very large numbers. Also called **floating-point notation**, or **exponential notation**. When an E is used instead of a 10, or the radix utilized, it is also called **E notation**.

scientific visualization The use of computers to represent, display, and manipulate scientific data and phenomena, such as the magnetic field around a point or object, or for the simulation of processes within a cell when metabolizing nutrients or when faced with a harmful substance. Enormous processing power must be combined with specifically tailored applications and display devices, and supercomputers are frequently utilized for such work. Since graphical data is analyzed by the brain at a more intuitive level than text data, more complex information can be more simply transmitted, helping provide even deeper insight into that being researched. Also called **visualization** (2).

scintillating crystal Same as **scintillation crystal**.

scintillating material Same as **scintillation material**.

scintillation **1.** A flash or spark. **2.** A flash of visible light produced by a crystal or other material which is exposed to ionizing radiation. Each incident photon produces a flash of light, whose frequency depends on the energy of the incident radiation. **3.** Rapid fluctuations, usually of small amplitude, in a parameter of a radio transmitter or receiver, or in transmission or reception. For example, such variations in the carrier wave of a transmitter. **4.** Rapid fluctuations, usually of small amplitude, observed in one or more parameters of a component, circuit, device, piece of equipment, or system. For instance, such variations in the irradiance levels within the cross-section of a laser beam. **5.** In radars, rapid fluctuations in the apparent location of a scanned object. Such variations oscillate around the mean position of said object. Also called **target glint**, **target scintillation**, or **wander** (2). **6.** Rapid fluctuations in the apparent position, brightness, color, or the like, of distant luminous objects, such as stars, as viewed through the atmosphere. These are due, for instance, to turbulence in the atmosphere.

scintillation counter A radiation counter which incorporates a **scintillation material**. As ionizing particles strike said material, the emitted bursts of light are detected by a photomultiplier tube. The output of said tube is then analyzed to determine the level of activity. Used, for instance, for the detection of high-energy photons such as X-rays or gamma rays. Also called **scintillator** (2), **scintillation detector**, **scintillation meter**, **scintillator detector**, **scintillator counter**, or **scintillometer**.

scintillation crystal A crystal, such as sodium iodide, which emits flashes of visible light when exposed to ionizing radiation. Also called **scintillating crystal**, or **scintillator crystal**.

scintillation detector Same as **scintillation counter**.

scintillation material A material, such as a scintillation crystal, which emits flashes of visible light when exposed to ionizing radiation. Also called **scintillating material**, **scintillator** (1), or **scintillator material**.

scintillation meter Same as **scintillation counter**.

scintillation spectrometer A **scintillation counter** which identifies specific radioactive nuclides according to their energies. The higher the frequency, or energy, of the radiation incident upon its photocathode, the greater the energy of the electrons emitted by said cathode.

scintillator **1.** Same as **scintillation material**. **2.** Same as **scintillation counter**.

scintillator counter Same as **scintillation counter**.

scintillator crystal Same as **scintillation crystal**.

scintillator detector Same as **scintillation counter**.

scintillator material Same as **scintillation material**.

scissoring In computer graphics, the cutting off of a displayed image beyond a given boundary. Also called **clipping** (4).

SCM Abbreviation of **single-chip module**.

SCMS Acronym for Serial Copy Management System. A copy-protection scheme in which a single digital copy is allowed. SCMS may allow multiple analog copies. It is pronounced *scums*.

scope **1.** Abbreviation of **oscilloscope**. **2.** An abbreviation or suffix denoting an instrument utilized for viewing. For example, microscope, radarscope, or telescope. **3.** A range of magnitude, function, motion, capabilities, thoroughness, or the like. For example, the scope of a computer application, or that of a spot check.

SCP Abbreviation of **service control point**. Within a telephone network, a node that links to databases which handle inquiries and other services, such as calling card verifications, 800 numbers, collect calls, and the like.

SCR **1.** Abbreviation of **silicon-controlled rectifier**. A semiconductor device with four layers, p-n-p-n, which has an input terminal, or gate, output terminal, or anode, and a terminal common to both called cathode. Used, for instance, as a solid-state switch that turns large amounts of power on and off very quickly. Also called **thyristor** (2). **2.** Abbreviation of **space-charge region**.

scramble **1.** To alter information, such as data, in a manner which only those with a key can decipher. Usually used for security purposes. Also called **encode** (2), **encrypt**, or **code** (1). **2.** To distort, disorganize, or otherwise alter a signal so that it can only be understood with a receiver with the proper circuits and settings.

scrambled data Data which has been altered in a manner which only allows those with a key to decipher it. May be used, for instance, where privacy or security is a concern. Also called **coded data**, or **encoded data**.

scrambled signal A signal, such as speech, which has been distorted, disorganized, or otherwise altered, so that it can only be understood with a receiver with the proper circuits and settings. May be used, for instance, where eavesdropping is a concern. Also called **coded signal**, or **encoded signal**.

scrambled speech Speech which has been distorted, disorganized, or otherwise altered, so that it can only be understood with a receiver with the proper circuits and settings. May be used, for instance, where eavesdropping is a concern. Also called **coded speech**, or **encoded speech**.

scrambler **1.** A circuit or device utilized for encoding a **scrambled signal**. Such a circuit or device, for instance, may divide the audio-frequency spectrum into bands which are displaced prior to transmission, and placed in their proper order by the receiving device. Also called **speech scrambler**, **speech inverter**, or **voice scrambler**. **2.** A circuit or device utilized to **scramble**.

scratch Abbreviation of **scratchpad**.

scratch disk A disk, or part of a disk, utilized for **scratchpad** purposes.

scratch disk space A part of a disk utilized for **scratchpad** purposes.

scratch file A file utilized for **scratchpad** purposes.

scratch-pad Same as **scratchpad**.

scratch-pad memory Same as **scratchpad memory**.

scratch-pad RAM Same as **scratchpad RAM**.

scratch register A register utilized for **scratchpad** purposes.

scratch space Same as **scratchpad**.

scratchpad A location or section, such as that in memory or on a disk, or a file, where data is held temporarily. Used, for instance, for intermediate storage in a computer, or to record a number in cell phone. Also spelled **scratch-pad**. Its abbreviation is **scratch**. Also called **scratch space**.

scratchpad memory Memory utilized for **scratchpad** purposes. Also spelled **scratch-pad memory**.

scratchpad RAM RAM utilized for **scratchpad** purposes. Also spelled **scratch-pad RAM**.

screen **1.** The viewing area of a display device, such as that of a computer or TV. A screen may be the viewing surface of a CRT display, a liquid-crystal display, a plasma display, or the like. Also called **display (1)**, or **monitor (1)**. **2.** Same as **shielding (1)**. **3.** Same as **shielding (2)**. **4.** A covering, enclosure, material, or device which provides protection from heat, sparks, light, radioactive particles, noise and so on. **5.** Same as **screen grid**.

screen buffer A buffer, usually consisting of a segment of memory in a graphics card, which holds the data that awaits to be sent to the display. Also called **video buffer**, or **graphics buffer**.

screen capture The sending of the content displayed at a given moment by a computer monitor to a file, buffer, or printer. Also called **screenshot**.

screen contrast **1.** On a **screen (1)**, such as that of a computer or TV, the difference in brightness between areas that are lighter and darker. Also called **contrast (1)**. **2.** On a **screen (1)**, such as that of a computer or TV, the difference in brightness between the lightest and darkest areas. Also called **contrast (2)**.

screen dump Printing or storing the full content being displayed at a given moment on a computer screen.

screen flicker Brief and irregular fluctuations in an image presented by a **screen (1)**. This may occur, for instance, when the refresh rate is too low. Also called **display flicker**, or **monitor flicker**.

screen font A font intended for displaying, as opposed to a **printer font (1)**, which is intended for printing.

screen frequency When creating halftone images by photographing through a screen, the density of dots, as expressed in lines per inch or centimeter. A screen with 200 lines per inch is composed of dots that are one half the size of the dots on a screen of 100 lines per inch, so the former has higher resolution. This term is unrelated to concepts such as scanning frequency or picture frequency.

screen grabber **1.** A device utilized to save the graphical data presented on a computer screen at any given moment. **2.** A device utilized to save the graphical data presented on any portion of a computer screen at any given moment.

screen grid In an electron tube, such as a tetrode or pentode, a grid placed between the control grid and the anode. Such a grid is kept at a positive potential, so as to reduce the electrostatic influence the anode has on the control grid. Also called **screen (5)**.

screen-grid tube An electron tube, such as a tetrode or pentode, which incorporates a **screen grid**.

screen illumination The light source that enables a **screen (1)**, such as that of a liquid-crystal display or oscilloscope, to have the proper brightness, contrast, and resolution.

screen material A material, such as a phosphor or a fluorescent substance, utilized to coat a **screen (1)**.

screen mode A specific configuration of the settings of a computer monitor. For instance, the resolution, the number of colors that may be displayed, and whether it displays graphics and/or text. Also called **display mode**, **monitor mode**, or **video mode**.

screen name A user-chosen name utilized for communications with other users in chats, instant messaging, and the like. A screen name may be a user name, a real name, or whatever occurs to a user at a given instant. Also called **username (2)**.

screen overlay **1.** A usually transparent sheet which is mounted or placed before a computer display, and which serves for making choices or the like on a touch screen. Also called **touch screen overlay**. **2.** A screen or sheet which is mounted or placed before a computer display. Used, for instance, to reduce glare, or as a **screen overlay (1)**.

screen phone **1.** An Internet appliance whose display screen is utilized to access the Internet or place telephone calls. **2.** A telephone with a display screen indicating information such as the number dialed, duration of call, caller ID data, information of address book entries, and so on.

screen reader An application that interprets that which is displayed on a computer screen at a given moment, and converts it into speech. Useful, for instance, for those with reduced vision, or in situations when performing tasks in which looking at the monitor is not convenient or feasible.

screen resolution Also called **display resolution**, or **monitor resolution**. **1.** The degree of sharpness and distinctness of the displayed image of a monitor. May be expressed, for instance, in terms of the number of pixels displayed, and dot pitch. **2.** The resolution a monitor is set at. A common resolution setting is 800x600, meaning that 800 dots, or pixels, are displayed on each of the 600 lines of the display. Frequently, the number of colors is incorporated into this specification, as in 800x600x256.

screen room Same as **shielded room**.

screen saver An image, sequence of images, text, or just a blank screen which is presented on a computer monitor after a given time of user inactivity. The images and time before screen saver activation are programmable, and any activity, such as a mouse movement or keystroke, restores the screen to its former image. Currently, most monitors are not susceptible to burn-in, so screen savers usually serve to display decorative or entertaining graphics. A screen saver may be programmed to require a password before restoring the previous screen, which serves as a security measure when desired.

screen scraper Software that provides a GUI to a system which is character-based. Used, for instance, with some older mainframes.

screen scraping The use of a **screen scraper**.

screen shot Same as **screen capture**. Also spelled **screenshot**.

screen size The total size of a display screen, such as that of a TV or computer monitor, usually measured diagonally. In the case of CRTs, for instance, the screen size may be larger than the **viewable image** size, as there is usually a given space on the sides of the useful image. In the case of flat displays, however, stated screen sizes are generally equal to the size of the viewable image.

screen voltage **1.** The voltage a **screen grid** is held at. **2.** The voltage of a **screen grid** at any given moment.

screened pair A transmission line similar to a **shielded pair**, but with a thinner, and usually more flexible, shielding.

screened twisted pair A transmission line similar to a **shielded twisted pair**, but with a thinner, and usually more flexible, shielding. Its abbreviation is **ScTP**.

screening **1.** Same as **shielding (1)**. **2.** Same as **shielding (2)**.

screening room Same as **shielded room**.

screenshot Same as **screen capture**. Also spelled **screen shot**.

screw terminal A terminal, used for making electrical connections, that incorporates a screw which is usually adjusted by hand. Used, for instance, to connect audio loudspeaker cables. Also called **binding post**, **binding screw**, or **post (2)**.

scribe To cut a grid pattern of grooves in a material to facilitate breaking, cutting, or the like.

scribe line The lines or grooves formed when **scribing**.

scriber A tool utilized for **scribing**. Such a tool may, for instance, have a diamond tip.

scribing The process of cutting a grid pattern of grooves in a material to facilitate breaking, cutting, or the like. Used, for instance, when processing a semiconductor wafer into individual chips.

script **1.** A sequence of computer instructions, or a program, which is executed or carried out by another program, as opposed to the CPU. In the context of the Internet, such instructions or programs may also be run by a Web server. For example, certain content encountered when navigating the World Wide Web may be executed by the browser of a user computer, or be processed on a Web server. **2.** Instructions which determine how a multimedia sequence proceeds, depending on user interaction. Used, for instance, to fill out forms or provide additional information at Web sites. **3.** A series of commands, such as a macro, which is executed without concurrent user intervention.

script error An error encountered within a **script**. Such an error may result, for instance, in a Web page being displayed incorrectly, or in execution being stopped.

script kiddie A derogatory term for unsophisticated hackers which utilize readily available techniques, routines, and programs to exploit well-known vulnerabilities. Also spelled **script kiddy**.

script kiddy Same as **script kiddie**.

script language Same as **scripting language**.

script software Software utilized to write **scripts**. Its abbreviation is **scriptware**.

scripting language A programming language, such as JavaScript or Perl, utilized to write **scripts**. Also called **script language**.

scriptlet A reusable **script** fragment, such as that within HTML coding.

scriptware Abbreviation of **script software**.

scroll In a GUI, to move up, down, left, or right within a screen to proceed to different locations within that displayed, or to go to other sections which do not fit on the screen at a given moment. Scrolling may be accomplished, for instance, by pressing arrow keys, or using a mouse to move a scroll bar. Used, for example, to access the content of a Web page exceeding that which can be displayed at any given moment.

scroll arrow An arrow at either end of a **scroll bar**, which is used to scroll in the desired direction.

scroll bar In a GUI, a vertical or horizontal bar located at a screen side or bottom, and which contains a box. This box can be moved with a mouse, or other pointing device, in the desired direction within the file or window it is located. The box is called elevator, as it resembles an elevator in its shaft, which would be the scroll bar.

scroll box Within a scroll bar, the box that can be moved in the desired direction, to navigate within a file or window. For instance, by utilizing a mouse to slide the elevator downwards within a vertical scroll bar, a user advances further ahead in a displayed document. Also called **scroll elevator**, **thumb**, or **elevator**.

scroll elevator Same as **scroll box**.

Scroll Lock On many keyboards, a key intended to toggle between a scrolling mode and non-scrolling mode. Nonetheless, such a key generally does not perform this function, which is of little importance, as it is rarely if ever used.

scroll mouse A mouse equipped with a **scroll wheel**. Also called **wheel mouse**.

scroll wheel A wheel, usually located between the left and right buttons of a computer mouse, which serves to scroll up and down documents, zoom in and out, and the like, depending on the application or function. Such a wheel may also serve as a third mouse button.

scrollable window A window which occupies more space than can be displayed at a given moment, and which can be scrolled up and down and/or left and right.

scrolling In a GUI, the process of moving an image or window which occupies more space than can be displayed at a given moment up and down and/or left and right.

scrubbing The process of fixing or eliminating data which is incomplete, incorrect, duplicated, or the like. Scrubbing may be necessary, for instance, when database fields are left blank, information is entered poorly, or when multiple databases are improperly combined. Also called **data scrubbing**.

SCS Abbreviation of silicon-controlled switch. A semiconductor device similar to an **SCR**, except that all four layers are available to the external circuit, making it a four-terminal switching device.

SCSA Abbreviation of Signal Computing System Architecture. A standard architecture or framework for providing computer telephony services such as call origination and control, conference calling, and voice recognition, plus information services such as fax and email.

SCSI Acronym for Small Computer System Interface. A flexible high-speed parallel interface for the connection of multiple peripherals, such as disk drives, DVDs, scanners, and printers. SCSI supports various platforms, and there are multiple standards, including SCSI-2, and Ultra-3, providing for cable lengths of up to 12 meters, transfer rates up to 640 MBps, and connections for up to 16 devices. Also called **SCSI Interface**.

SCSI-1 A SCSI interface which utilizes an 8-bit bus, allows up to a 6 meter cable, and supports data rates up to 5 MBps. It uses a 25-pin cable.

SCSI-2 A SCSI interface which utilizes an 8-bit bus, allows up to a 6 meter cable, and supports data rates up to 5 MBps. It uses a 50-pin cable.

SCSI adapter Same as **SCSI controller**.

SCSI bus A parallel bus carrying data and signals between SCSI devices and a SCSI controller. SCSI buses are either 8 or 16 bits wide.

SCSI chain Multiple SCSI peripherals which are connected in a daisy chain fashion.

SCSI connector A connector utilized to attach SCSI devices. Such a connector usually has 50 or 68 pins.

SCSI controller A circuit board, chip, or device, along with the associated circuitry, which controls the way SCSI peripheral devices access the computer, and vice versa. Also called **SCSI adapter**, or **SCSI host adapter**.

SCSI device A peripheral device, such as a disk drive, DVD, scanner, or printer, utilizing a **SCSI Interface**. Also called **SCSI peripheral**.

SCSI host adapter Same as **SCSI controller**.

SCSI Interface Same as **SCSI**.

SCSI peripheral Same as **SCSI device**.

SCSI port A computer port which serves to connect peripherals to a SCSI bus.

SCSI termination The last device in a **SCSI chain**. Such a device must have a terminator to avoid signals from reflecting back into the line.

ScTP Abbreviation of **screened twisted pair**.

scuzzy A spelling of the pronunciation of **SCSI**.

SD Abbreviation of **standard deviation**.

SD card Abbreviation of **secure digital card**. A flash memory card, usually about the size of a stamp, which is used in devices such as handheld PCs, PDAs, cell phones, digital cameras, and digital music players. Such cards feature high security, speedy data transfers, good storage capacity, and ruggedness.

SDA Abbreviation of **source data automation**.

SDARS Abbreviation of **satellite digital audio radio service**.

SDC Abbreviation of **synchro-to-digital converter**.

SDH Abbreviation of **Synchronous Digital Hierarchy**.

SDI Abbreviation of **serial digital interface**.

SDIP **1.** Abbreviation of **shrink DIP**. **2.** Abbreviation of **skinny DIP**.

SDK Abbreviation of **software development kit**. A set of routines, utilities, and other software tools utilized to assist in the development of programs. An SDK may be provided, for instance, by a software or hardware vendor, for the enhanced operation of software for a given operating system, computer system, user interface, or the like.

SDL Abbreviation of **Specification and Description Language**.

SDLC Abbreviation of **synchronous data-link control**.

SDR Abbreviation of **signal-to-distortion ratio**.

SDRAM Abbreviation of **synchronous DRAM**. A form of dynamic RAM which is synchronized with the clock speed of the CPU it is running with, to optimize its speed. This synchronization enables SDRAM to have no wait states, and in addition, such memory supports burst modes and interleaving. The speed of such chips is usually expressed in MHz. Also, the technology pertaining to such chips.

SDSL **1.** Abbreviation of **symmetric DSL**. A form of DSL in which the upstream and downstream data rates are the same. SDSL can support data transfer rates exceeding 3 Mbps. **2.** Abbreviation of **single-line DSL**.

SDTS Abbreviation of **Spatial Data Transfer Standard**.

SDTV Abbreviation of Standard Definition **TV**. A digital format which has video and audio quality equal to or exceeding NTSC, but below that of HDTV. There are multiple SDTV standards, each having differences in frames per second and lines of resolution, providing either a 4:3 or 16:9 aspect ratio, while surround sound may or may not be supported.

Se Chemical symbol for **selenium**.

sea clutter In radars, unwanted echoes that appear on the display screen due to signal reflections from the sea, especially waves. Also called **sea return**.

sea-clutter suppression Techniques employed to suppress **sea clutter**.

sea return Same as **sea clutter**.

seaborgium A synthetic radioactive chemical element whose atomic number is 106, and which has nearly 10 identified isotopes. The techniques employed to produce this element are utilized to help discover new elements with higher atomic numbers. Its chemical symbol is **Sg**.

seal **1.** To close securely or hermetically to avoid any leaking in or out, of air, moisture, contaminants, light, radiation, or the like. Also, the hermeticity so obtained. Also, the process of conferring such hermeticity. **2.** A material or device which serves to make a **seal** (1). Also, the hermeticity so conferred. **3.** To affix a label or other binding which when broken evidences tampering. Also, such a label or binding. **4.** To affix a label or other certificate which attests to the authenticity of something. Also, such a label.

sealed cabinet A speaker enclosure, such as an acoustic suspension enclosure, with no vents, as opposed to a **ported cabinet**, which has one or more such openings. Also called **sealed enclosure**.

sealed circuit A circuit that is hermetically or securely closed, so as to prevent air, moisture, contaminants, light, radiation, or the like, from entering.

sealed enclosure Same as **sealed cabinet**.

sealed instrument **1.** An instrument which is hermetically or securely closed, so as to prevent air, moisture, contaminants, light, radiation, or the like, from entering. **2.** A meter which has a **seal** (3), or **seal** (4).

sealed loudspeaker Same as **sealed speaker**.

sealed meter **1.** A meter which is hermetically or securely closed, so as to prevent air, moisture, contaminants, light, radiation, or the like, from entering. **2.** A meter which has a **seal** (3), or **seal** (4).

sealed relay A relay which is hermetically or securely closed, so as to prevent air, moisture, contaminants, light, radiation, or the like, from entering.

sealed speaker A speaker with a **sealed cabinet**. Also called **sealed loudspeaker**.

sealed switch A switch that is hermetically or securely closed, so as to prevent air, moisture, contaminants, light, radiation, or the like, from entering.

sealed tube A tube that is hermetically or securely closed, so as to prevent air, moisture, contaminants, light, radiation, or the like, from entering or leaving.

sealing compound Same as **sealing material**.

sealing material A material, such as a wax or adhesive, which is utilized to provide a **seal** (1). Also called **sealing compound**.

seamless integration The incorporation of a new software and/or hardware item to an existing system, with subsequent proper operation.

search **1.** To thoroughly examine or explore, looking for something. Also called **seek** (2). Also, a specific instance of such a search. **2.** To examine or explore a document, file, database, disk, the Internet, or the like, looking for specific data or items. Also called **seek** (3), or **find** (2). Also, a specific instance of such a search. **3.** To examine or explore a selected text, document, or file, looking for specific character strings, special characters, commands, or the like. Also called **seek** (4), or **find** (1). Also, a specific instance of such a search.

search and replace To perform a **search** (2) and replace the located content with other content. For example, to find instances where the word *delay* appears within a document, and to put the phrase *time delay* in its place for one or more occurrences. Its abbreviation is **search & replace**. Also called **find and replace**.

search coil A small coil which is inserted into a magnetic field to measure its field strength, or any variations in it. Such a coil is connected to an indicating instrument, such as a ballistic galvanometer or a fluxmeter, and may or may not incorporate an amplifier. A search coil may also be utilized to examine the magnetic flux distribution of a magnetic field. Also called **search probe**, **pickup coil**, **magnetic test coil**, or **magnetic probe**.

search criteria The parameters chosen for a given **search**.

search engine **1.** Software which examines or explores the Internet, looking through Web pages, documents, files, images, audio, multimedia, news, message boards, and the like, using selected keywords. A search engine may incorporate, for instance, a program which searches all available or applicable Web sites as selected by a proprietary algorithm, another program which creates an index or catalog of these sites, and another program which compares a given keyword search with all entries in the index and gives the user the results. A typical search can take just a fraction of second. A search engine may be limited to specific areas of interest, such as art or electronics, specific formats, such as mp3 or

JPEG, and so on. Also called **Web search engine. 2.** Software that examines or explores documents, files, databases, disks, the Internet, or the like, looking for specified keywords or other data or items.

search key 1. Abbreviation of **search keyword. 2.** A word, part of a word, phrase, number, or other string which serves as the basis for a search through a document or database.

search keyword A word specified by a user which guides a **search**. Combinations of such keywords help focus or narrow the results of such searches. Its abbreviation is **search key (1)**.

search path A file pathname that starts with the drive letter, leads through the directories, and ends with a full file name. For example: *C:\Documents\Reminders\Today.xyz*. Also known as **absolute path**, **access path**, **full path**, or **filespec**.

search probe Same as **search coil**.

search radar A radar system, such as scanning radar, intended to scan a given region.

search radio Same as **search receiver**.

search receiver A radio receiver, such as a scanning radio, intended to scan multiple radio bands. Also called **search radio**.

search site A Web site, such as *www.google.com*, which provides one or more **search engines (1)**. Also called **search Web site**, or **Web search site**.

search sonar A sonar system, such as scanning sonar, intended to scan a given area.

search string A string which serves as the basis for a search through a document or database. Such a string may consist of a word, part of a word, phrase, number, and so on.

search time The time interval that elapses during the completion of a **search**.

search Web site Same as **search site**.

search & replace Same as **search and replace**.

searchable database A database within which searches may be performed using queries. All databases are of this type, so the term is redundant.

seasonal effects The effects, such as those on ionospheric propagation, due to **seasonal variations**.

seasonal factors Any factors, such as the level of absorption of waves by the ionosphere, which undergo **seasonal variations**.

seasonal variations Any variations in conditions, composition, properties, or the like, which vary by season or in other more or less annual cycles. Ionospheric propagation, for instance, varies by season, due to the relative position of the sun and its effect on the ionization density in the various ionospheric layers.

seat 1. To fix firmly and correctly in place. For example, to seat a chip or circuit board in its socket. **2.** A surface, location, or part upon which something rests.

sec 1. Abbreviation of **second (1)**. **2.** Abbreviation of **secondary winding**.

SECAM Acronym for the French term *Système Électronique Couleur Avec Mémoire*, or *Séquentiel Couleur Avec Mémoire*, although there are various similar variations. A standard for the generation, transmission, and reception of TV signals. It provides for a resolution of 625 horizontal lines per frame, with 50 interlaced half-frames per second for a total of 25 frames per second. Of the 625 total lines, approximately 576 are active, while the aspect ratio is 4:3. SECAM is used in a few European, some Asian, and many African countries, while the other two main world standards are **PAL** and **NTSC**. All three are mutually incompatible. Also called **SECAM standard**.

SECAM signal A TV signal adhering to the **SECAM Standard**.

SECAM standard Same as **SECAM**.

second 1. The fundamental SI unit of time. One second is currently defined as 9,192,631,770 oscillations of a cesium-133 atom at its natural resonance frequency. Its abbreviation is **s**, or **sec. 2.** A unit of angular measure equal to 1/60 arcminute, or 1/3600 degree. Its symbol is **"**. Also called **arcsecond**.

second anode In a CRT, the anode which accelerates the electron beam. Also called **ultor**, or **ultor anode**.

second breakdown In a transistor, a form of destructive breakdown in which local heating causes an increase in charge carriers, which in turn reduces resistivity, which enables higher current density. The increases in current density generate more heat, which reduces resistivity further, and so on. Such a breakdown is nearly instantaneous, and leads to damage or destruction.

second-channel interference Interference of radio or TV reception by other stations using alternate channels. It can be unilateral or bilateral. Also called **alternate-channel interference**.

second detector In a superheterodyne receiver, the stage which demodulates the modulated IF signal. It follows the **first detector**, which modulates the incoming radio-frequency signal to produce said IF signal.

second dial tone A dial tone which is obtained after dialing at least one digit. For example, a second dial tone obtained when using a conference-call feature, or that obtained when accessing an outside line from a business. A second dial tone may or may not have a different pitch.

second-generation computer A computer utilizing discrete components, such as transistors. These computers were built from about the mid 1950s to the mid 1960s.

second-generation language An assembly language. Each type of CPU has its own assembly language.

second-generation wireless Developed in the early 1990s, provides for digital cellular communications such as those using GSM, CDMA, and TDMA. Its abbreviation is **2G**.

second law of thermodynamics A law which states that heat will not be transferred spontaneously from a colder system to a warmer system.

second-level cache A memory cache which utilizes a memory bank between the CPU and main memory. This contrasts with **first-level cache**, which is built right into the CPU. First-level cache is faster, but smaller than second-level cache, while second-level cache is still faster than the main memory. Also called **secondary cache**, **level 2 cache**, or **external cache**.

second-level domain Within an Internet domain name, the portion preceding the top-level domain category or designator. For example, in *www.yipeeee.com* or *www.yipeeee.qq*, it is the *yipeeee* portion. Multiple second-level domains may be served by the same IP address. Its abbreviation is **SLD**. Also called **root domain**.

second-level domain name A **second-level domain** plus the domain category or designator. For example, in *www.yipeeee.com* it is the *yipeeee.com* portion, while in *www.yipeeee.qq*, it would be the *yipeeee.qq* portion. Multiple second-level domain names may be served by the same IP address. Also called **root domain name**.

second normal form In the normalization of a relational database, the second stage utilized to convert complex data structures into simpler relations. A database must have completed the first normal form before proceeding to the second. Its abbreviation is **2NF**.

second-order filter A filter with four components, such as two inductors and two capacitors, and which provides a 12 dB rolloff. Used, for instance, as a crossover network.

second-surface mirror A mirror whose reflective coating is on its rear surface, as is the case with a household mirror. This contrasts with **front-surface mirror**, in which the reflective coating is on the front surface. Also called **back-surface mirror**, or **rear-surface mirror**.

secondary 1. That which occurs after something else, as in a sequence, or which has lesser importance. **2.** That which does not have the highest rank, degree, or importance. For example, a secondary standard. **3.** Same as **secondary winding. 4.** A color which is formed by the mixing of two primary colors. For example, red and green producing yellow. Also called **secondary color. 5.** Of or pertaining to a secondary color.

secondary audio program Its abbreviation is **SAP**. In a TV broadcast, an audio signal which is carried alongside the main audio channel. SAP can provide an alternate language soundtrack, a commentary on that being viewed, teletext, or other information.

secondary battery Same as **storage battery**.

secondary beam Same as **sidelobe**.

secondary cache Same as **second-level cache**.

secondary cell Same as **storage cell (1)**.

secondary channel A supplemental communications channel, which can serve, for instance, as a backup, or for transmission of a secondary audio program. Also called **supplementary channel**, or **auxiliary channel**.

secondary circuit A circuit, especially that in a transformer, which inductively receives power from a **primary circuit**. Such a circuit may serve to provide energy to a load.

secondary coil Same as **secondary winding**.

secondary color Same as **secondary (4)**.

secondary communications channel A supplemental communications channel, which can serve, for instance, as a backup, or for transmission of a secondary audio program. Also called **supplementary communications channel**, or **auxiliary channel**.

secondary cosmic rays The radiation which results from the interaction of primary cosmic rays with atomic nuclei, and particles, such as electrons, in the atmosphere. The resulting high-energy particles, such as muons, are thus produced in the atmosphere, while **primary cosmic rays** originate in space.

secondary current The current flowing through a **secondary winding**. Also called **transformer secondary current**.

secondary device A device which serves in a supplemental or auxiliary capacity, or that is added to enhance the performance of the main device. Also called **supplementary device**, or **auxiliary device**.

secondary electron An electron which is produced as a result of bombardment by **primary electrons**. In an electron multiplier, for example, a cathode releases primary electrons, which are then reflected by electron mirrors to produce secondary electrons.

secondary electron multiplier An electron tube in which secondary electron emissions produce current amplification. In it, a cathode releases primary electrons, which are then reflected by multiple electron mirrors, or dynodes, each producing more and more electrons. Depending on the number of dynodes, the amplification factor may be in the millions. Its abbreviation is **SEM**. Also called **multiplier (2)**, or **electron multiplier**.

secondary emission The emission of electrons from a surface as a result of bombardment by charged particles, such as electrons.

secondary equipment Equipment which serves in a supplemental or auxiliary capacity, or that is added to enhance performance. Also called **supplemental equipment**, or **auxiliary equipment**.

secondary failure A failure caused directly or indirectly by the failure of another component, device, or unit. A **primary failure** arises independently of other components, devices, or units.

secondary frequency standard A frequency standard which does not have the extreme precision or stability which characterize a **primary frequency standard**. A secondary frequency standard is calibrated against a primary frequency standard.

secondary inductance The inductance measured across a **secondary winding** under no-load conditions.

secondary instrument An instrument which can not be calibrated without utilizing another device or instrument. It may be calibrated by a **primary instrument**, or through the use of secondary standards.

secondary-ion mass spectrometer An instrument utilized for **secondary-ion mass spectroscopy**. Its acronym is **SIMS**.

secondary-ion mass spectroscopy A spectroscopic technique in which an ion beam is utilized to sputter ions off the surfaces of samples. These secondary ions produced are then accelerated into a mass spectrometer for separation according to their mass-to-charge ratio. Used, for instance, to detect the elemental composition of the surfaces of geologic samples. Its acronym is **SIMS**. Also called **ion microprobe**.

secondary lobe Same as **sidelobe**.

secondary memory Computer storage, such as disks, discs, and tapes, as opposed to RAM, which is considered **primary memory**. Also called **secondary storage (1)**.

secondary power 1. An alternate or supplemental source of electrical power that may be used when the main power source is interrupted, depleted, or insufficient. Also called **supplemental power**, or **auxiliary power. 2.** The power in a **secondary circuit**.

secondary radiation 1. Radiation, such as that is which is scattered, reflected, or refracted, which does not arrive directly from a source. That which arrives directly from a source is **primary radiation. 2.** Radiation, such as secondary cosmic rays, which is produced by the action of primary radiation, such as cosmic rays.

secondary radiator In an antenna with parasitic elements, a radiator which is driven by a **primary radiator**. A primary radiator receives its energy directly from a transmission line.

secondary ring Within a dual-ring topology, the ring which backs up the **primary ring**, which serves for data transmission during normal operation.

secondary service area A region beyond a primary service area, but which still has coverage. For example, the area within which the signal of a broadcast station is received, but has significant interference or fading.

secondary standard 1. A unit which is a specified multiple of a **primary standard (1)**. For example, a nanosecond, which is 10^{-9} second. **2.** A reference which has been determined with very high accuracy, but which must be calibrated or otherwise verified by a **primary standard (2)**. Such a secondary standard may be adequate for many applications, and might itself be used as a reference for other devices and instruments.

secondary storage 1. Same as **secondary memory. 2.** Any storage which supplements RAM, such as external storage or auxiliary storage.

secondary turns The number of turns in a **secondary winding**. The ratio of the number of primary turns to the number of secondary turns yields the turns ratio.

secondary voltage The voltage produced in a **secondary winding**. Also called **transformer secondary voltage**.

secondary winding In a transformer, the winding that provides the output to the load or loads. There may be multiple secondary windings, while the winding that is directly connected to the input power is the **primary winding**. For example, a primary winding accepts the input energy an AC line provides, while a secondary winding provides the energy to the load at a different voltage. Its abbreviation is **sec**, and its symbol is **S**. Also called **secondary (3)**, or **secondary coil**.

secret-key cryptography Same as **symmetric encryption**.

secret-key encryption Same as **symmetric encryption**.

section 1. A subdivision or cut portion of something. Also, the process or act of subdividing or cutting something. 2. A plane surface formed by cutting through a solid. Also, the shape or area of such a section. For example, a conic section. 3. A stage or other functional unit within a circuit, device, piece of equipment, or system. For instance, an amplifier stage. 4. In radio communications, the transmission span between radio repeaters.

sector 1. The smallest addressable portion of a storage unit such as a hard disk, optical disc, or tape. Each **track** consists of a given number of sectors. 2. A specific **sector (1)** within a storage unit. 3. A part or division of something.

sector display In radars, a display which allows viewing of only a given sector of the total scanned area. Such a sector is usually selectable.

sector interleave On a magnetic or optical disk, to position **sectors (2)** in a manner which improves performance. For example, if a data file spans more than one sector on a hard disk, these sectors may be staggered so that all sectors can be read in a single revolution, as opposed to multiple turns. This speeds up access time. Also called **sector map (1)**.

sector interleaving On a magnetic or optical disk, the positioning of **sectors (2)** in a manner which improves performance.

sector map 1. Same as **sector interleave**. 2. A map or table describing sector interleaving, the location of sectors, or the location of bad sectors on a disk. 3. A table utilized to translate between a table listing sectors and the actual physical sectors. Used, for instance, by an operating system for interleaving.

sector scan In radars, a scan through only a limited angle. Such a sector is usually selectable.

secular variation Slow variations in the magnetic field of the earth. Such changes occur over centuries and are usually considered to be mostly internal in origin, as opposed to rapid fluctuations which tend to be external in origin.

secure 1. Safe from unauthorized use, access, interception, monitoring, or the like. For example, secure communications. Also, to ensure such security. 2. Not presenting a safety hazard. 3. Not likely to fail or malfunction.

secure area An area, such as a sandbox, where files which may be unsafe are opened, so as to minimize the possibility of other parts of the system being adversely affected. Used, for instance, to open executable files which may have viruses.

secure channel A communications channel in which all data and messages are encrypted or otherwise safe from unauthorized use, access, interception, monitoring, or the like. Also called **secure circuit**, **secure link**, or **secure line**.

secure circuit Same as **secure channel**.

secure communications Communications which are safe from unauthorized use, access, interception, monitoring, or the like, using means such as encryption.

secure digital card Same as **SD card**.

secure electronic transaction A transaction that is safe from unauthorized access, interception, monitoring, or the like, using means such as encryption. Used, for instance, for sending financial data. Its abbreviation is **SET**. Also called **secure transaction**.

secure hash algorithm An algorithm that generates a condensed representation of a message, called message digest, and is used, for instance, to create digital signatures. Its abbreviation is **SHA**.

secure hash algorithm-1 An enhanced version of **secure hash algorithm**. Its abbreviation is **SHA1**.

Secure HyperText Transport Protocol Same as **SHTTP**.

secure line Same as **secure channel**.

secure link Same as **secure channel**.

secure message A message, such as that containing sensitive data or which is otherwise confidential, that is protected by encryption.

secure messaging The exchange of **secure messages**.

Secure MIME Same as **S/MIME**.

secure mode Within multiple available modes of operation, one that provides additional or maximum security. HTTPS, for example, is a secure mode within HTTP.

Secure Multipurpose Internet Mail Extensions Same as **S/MIME**.

secure server A server that is safe from unauthorized use, access, monitoring, or the like. Messages, such as those containing credit card information, to and from such servers are encrypted using protocols such as SSL, SHTTP, or TLS.

secure site A Web site which is safe from unauthorized use, access, monitoring, or the like. Messages, such as those containing credit card information, to and from such sites are encrypted using protocols such as SSL, SHTTP, or TLS. Also called **secure Web site**.

Secure Sockets Layer Same as **SSL**.

Secure Sockets Layer encryption Same as **SSL encryption**.

Secure Sockets Layer protocol Same as **SSL protocol**.

secure transaction Same as **secure electronic transaction**.

secure Web server 1. A Web server that is safe from unauthorized use, access, monitoring, or the like. Messages, such as those containing credit card information, to and from such servers are encrypted using protocols such as SSL, SHTTP, or TLS. 2. A Web server which utilizes a firewall or other system to protect an intranet from outsiders.

secure Web site Same as **secure site**.

security 1. The measures and techniques utilized to safeguard against losses, damage, unwanted modifications, or unauthorized access. Used, for instance, in the context of computers and communications. Examples include computer security, data protection, and network security. 2. The act of providing **security (1)**. Also the condition of having such security.

security audit An assessment of the hardware, software, procedures, and records of a computer or system, to determine proper compliance with established security parameters and procedures. It may be followed by recommendations for improvement.

security breach An intrusion, incursion, or other incident in which security, such as that of a computer system or network, is compromised. Also called **security incident**.

security card 1. A card, such as a smart card, which serves to verify identity or for other such security purposes. 2. A card, such as a smart card, whose contained data is encrypted, or otherwise protected from unauthorized access, manipulation, or the like.

security code A code, such as a password or PIN, which is utilized to help safeguard against losses, damage, unwanted modifications, or unauthorized access. For example, such a code which prevents the accidental deletion of data programmed into a wireless phone, or that utilized to activate and deactivate a home security system.

security incident Same as **security breach**.

security protocol A protocol, such as SSL, SHTTP, or TLS, utilized for the authentication and/or encryption of messages. Used, for instance, for secure messaging.

security robot A robot utilized to assist or replace human personnel in hazardous situations, for surveillance, or the like. Such a robot may have a navigating system, an array of sensors, such as those detecting fires or intruders, and may incorporate cameras, infrared, ultraviolet, and microwave detectors, in addition to being able to work in groups to cover larger areas. Also called **sentry robot**.

security system Any system, such as a firewall or that incorporating infrared sensors and video monitors, intended to provide **security**.

security token An object or device, such as a smart card, which serves to authenticate a user in order to gain access to a computer network. Such a card, for instance, may be inserted into a smart card reader, with the holder prompted for a password or PIN for added security. Also called **authentication token**.

Seebeck effect An effect in which an electromotive force is developed when a there is a temperature differential at the junction of a thermocouple. That is, a current will flow across this junction when the temperature of one of the materials is higher than that of the other material comprising said thermocouple. It is the converse of the **Peltier effect**.

Seebeck emf An electromotive force produced by the **Seebeck effect**.

seed crystal A crystal utilized to set the growth pattern of one or more crystals. Used, for instance, in the Czochralski method.

seeing-eye robot A robot intended to assist those with reduced vision. Such a robot, for instance, might locate a crosswalk, detect the present color of a traffic signal, scan for vehicles and other dangers, and indicate that it is safe to cross using a simulated voice.

seek **1.** To position the access arm or mechanism of a direct-access storage device, such as a disk drive or optical drive, to a given track or position. **2.** Same as **search** (1). **3.** Same as **search** (2). **4.** Same as **search** (3).

seek time The time interval that elapses between consecutive **seek** (1) operations. Usually refers to the time required for an access arm to proceed from one track to the next. **Access time** is longer, as it requires more steps, such as those involved in data transfer.

segment Any part into which a whole is divided. For instance, a section, a network segment, an overlay segment, a data segment, or one of the bars utilized to formulate characters in an LCD display. Also, to cut, apportion, or otherwise divide something.

segment mark A signal, character, code, hole, notch, barrier, impression, or other mark which indicates where one **segment** ends, and the other begins.

segmentation In **segmentation and reassembly**, the breaking up of a packet into smaller units at the transmission end.

segmentation and reassembly Its acronym is **SAR**. In a packet-switched network, the breaking up of a packet into smaller units at the transmission end, with the subsequent reassembling at the receiving end. SAR helps speed up transmissions, and in the compliance with packet size restrictions imposed by a given protocol. Used, for instance, in ATM.

segmented address space An address space divided into blocks of a given size, such as 64 KB.

segmented display A display, such as an LCD display, which is divided into **segments**.

Seignette salt Colorless to white crystals displaying piezoelectricity and ferroelectricity. It is used in piezoelectric devices, such as crystal microphones, and for silvering mirrors. Also called **Rochelle salt**, or **potassium sodium tartrate**.

seismogram A graphic record produced by a **seismograph**.

seismograph An instrument or device which detects and records vibrations in the earth. Such a device usually incorporates one or more geophones. Also called **seismometer** (1).

seismometer **1.** Same as **seismograph**. **2.** An instrument or device which detects vibrations in the earth.

seismoscope An instrument which indicates the presence of an earthquake. A **seismograph** also indicates other aspects of earthquakes, such as their intensity.

select all A menu item or command which is utilized to highlight or otherwise choose all that is available within a given window or document.

select none A menu item or command which is utilized to undo a **select all** operation, or to unselect one or more items which are available.

selective That which exhibits or is characterized by **selectivity**.

selective absorption **1.** The absorption of only certain frequencies of electromagnetic or acoustic waves by a surface, object, or region. Also called **absorption** (2). **2.** The absorption of only particles with given energies by a surface, object, or region. Also called **absorption** (3). **3.** The absorption of only certain particles by a surface, object, or region. Also called **absorption** (4).

selective amplifier An amplifier that exhibits or is characterized by **selectivity**.

selective calling The ability of a communications node, station, network, or system, to send a message or other transmission only to certain nodes, stations, or receivers.

selective calling system A communications system utilized for **selective calling**.

selective circuit A circuit that exhibits or is characterized by **selectivity**.

selective component A component that exhibits or is characterized by **selectivity**.

selective diffusion In the manufacturing of semiconductor devices, diffusion of the dopant atoms on selected surfaces of the semiconductor material, as opposed, for instance, to being uniformly distributed throughout.

selective dump A memory dump involving only selected storage locations or areas.

selective epitaxial growth Epitaxy performed on specified portions of the crystal substrate. Also called **selective epitaxy**.

selective epitaxy Same as **selective epitaxial growth**.

selective fading In the propagation of electromagnetic waves, decreases in amplitude in which different frequencies are affected to varying degrees. This contrasts with **amplitude fading**, in which such decreases are uniform throughout all the frequency components of the signal.

selective interference Interference occurring only at a given frequency, or within a narrow frequency band.

selective receiver A receiver that exhibits or is characterized by **selectivity**.

selective reflection **1.** The reflection of only certain frequencies of electromagnetic or acoustic waves off a surface, object, or region. **2.** The reflection of only particles with given energies off a surface, object, or region. **3.** The reflection of only certain particles off a surface, object, or region.

selective ringing The ringing of only one of the available telephone units. Used, for instance in the context of party lines to distinguish for which user an incoming call is meant.

selective scattering **1.** The scattering of only certain frequencies of electromagnetic or acoustic waves off a surface, object, or region. **2.** The scattering of only particles with given energies off a surface, object, or region. **3.** The scattering of only certain particles off a surface, object, or region.

selectivity 1. The ability of a component, circuit, device, piece of equipment, or system, to operate only at a given frequency, or within a given band. Also, the degree to which this selectivity is attained. **2.** The ability of a component, circuit, device, piece of equipment, or system, to distinguish, separate, or otherwise act upon a given frequency or band. Also, the degree to which this selectivity is attained. **3. Selectivity (1)** or **selectivity (2)** manifested by a radio receiver. An example is adjacent-channel selectivity. Also, the degree to which this selectivity is attained.

selectivity control A control which adjusts the **selectivity** of a component, circuit, device, piece of equipment, or system. For instance, that of a receiver.

selector 1. That which serves to choose from multiple positions, options, operational modes, channels, or the like. **2.** Same as **selector switch**.

selector channel A channel which connects high-speed peripherals to the memory of a computer.

selector switch A manually operated multi-position switch. Such a switch is usually adjusted by a knob or handle, and may have detents to hold in a given position. Used, for instance, in devices or instruments with multiple functions, ranges, or modes of operation. Such a switch is usually rotary. Also called **selector (2)**.

selenium A chemical element whose atomic number is 34. It has several allotropic forms, and may be a red amorphous powder, consist of red crystals, or occur in the form of gray metallic crystals, among others. It has over 25 known isotopes, of which 5 are stable. It has many applications in electronics, including its use in photocells, batteries, rectifiers, TV cameras, photography, and semiconductors. Its chemical symbol is **Se**.

selenium cell A photocell in which the light-sensitive material is a layer or disk coated with selenium. Used, for instance, in exposure meters. Also called **selenium photocell**.

selenium photocell Same as **selenium cell**.

selenium rectifier A metallic-disk rectifier in which one side of each metal disk is covered with a selenium layer.

self-absorption Absorption of energy, particles, or waves, by the same surface, body, or medium which emitted said energy, particles, or waves.

self-adapting The ability of a component, circuit, device, piece of equipment, system, material, mechanism, or process to adjust its operation or characteristics depending on changing environmental conditions or other variables.

self-adjusting communications Communications utilizing a system that monitors internal and external parameters automatically and continuously in order to makes adjustments to better operate in its varying environment. Also known as **self-optimizing communications**, or **adaptive communications**.

self-alarm A device which automatically generates an alert under certain alarm conditions. For example, a one-way communicator dialing an emergency services number under specified circumstances. Also called **autoalarm**.

self-aligned gate In the fabrication of ICs incorporating MOSFETs, a process in which the gate electrodes are put in place before ion implantations or diffusions of the source or drain are made. Also, the technology employed in such a process.

self-bias In a transistor or vacuum tube, obtaining of the correct bias utilizing a dropping resistor instead of an external bias voltage. Also called **automatic bias (1)**, or **automatic grid bias**.

self-booting 1. A disk which has the necessary components of an operating system for booting a computer. **2.** That which boots a computer automatically. For example, a command which triggers a boot during the installation of an application.

self-capacitance Same as **stray capacitance**.

self-check A mechanism which automatically monitors the operation of a component, device, or system, seeking to maintain performance within specified parameters. May serve, for instance, to detect malfunctions and to optimize performance.

self-cleaning contacts Electrical contacts which wipe or slide past each other, resulting in a cleaning action when used. Also called **self-wiping contacts**, or **wiping contacts (2)**.

self-clocking A format or transmission in which clock pulses or other synchronizing information is incorporated into the sent signal. Manchester encoding, for instance, provides a self-synchronizing data stream.

self-contained That which has all the parts necessary to operate as an independent or complete unit. Examples include autonomous robots, external modems, and black boxes. Standardized self-contained units may be used, for instance, in a modular approach.

self-contained device A device, such as a self-contained instrument, which is **self-contained**.

self-contained instrument An instrument which requires no accessories nor external power to perform measurements and indications.

self-contained module A module, such as a resident module, which is **self-contained**.

self-contained unit A unit, such as an external modem, which is **self-contained**.

self-diagnostics Same as **self-testing**.

self-diffusion Diffusion occurring within different regions of the same material. For example, the spontaneous movement of an atom to a new site within the same crystal.

self-discharge The loss of energy of a battery which is in storage or otherwise not connected to a load. The battery chemistry and ambient temperature influence this rate considerably. Also called **battery self-discharge**.

self-discharge rate The rate at which **self-discharge** occurs. Battery chemistry and ambient temperature influence this rate considerably.

self-driven Same as **self-excited**.

self-energy The energy of a particle at rest, according to the mass-energy equation. The self-energy of an electron, for instance, is approximately 0.511 MeV, while that of a proton is approximately 938 MeV. Also called **rest energy**.

self-error correction 1. In communications, a system which automatically detects and rectifies errors during transmission or reception. **2.** A system which automatically detects and rectifies errors.

self-excited A component, circuit, device, piece of equipment, system, process, or mechanism which provides its own excitation signal. That is, it does not require an external signal, such as a current or voltage, to operate. An example is a self-excited oscillator. Also called **self-driven**.

self-excited oscillator An oscillator which maintains oscillation without the aid of an external signal.

self-extracting An archive, file, program, or the like, which automatically decompresses when run or accessed. Seen, for instance, when downloading an application.

self-extracting archive An archive which is **self-extracting**.

self-extracting file A file which is **self-extracting**.

self-extracting program A program which is **self-extracting**.

self-focus The ability of an optical system or device to automatically make adjustments which improve the sharpness of an image. Seen, for instance, in digital cameras.

self-healing That which can automatically compensate for a missing component, a malfunction, or otherwise fix itself under most circumstances, so as to maintain proper operation.

self-healing capacitor A capacitor which self-extinguishes arcs. This provides a more gradual failure when dielectric breakdown occurs, resulting in greater reliability.

self-impedance The impedance at the feed point of an antenna when all other terminals are open-circuited.

self-inductance The induction of an emf in a circuit caused by a varying current in the same circuit.

self-induction The property of a circuit in which a varying current induces an emf in the same circuit. Its symbol is L, and when quantifying this inductance some multiple of henrys, such as millihenrys or microhenrys, is used. When the emf is induced in a neighboring circuit it is called **mutual induction**.

Self-Monitoring Analysis and Reporting Technology Same as **SMART**.

self-optimizing communications Same as **self-adjusting communications**.

self-organizing map A map representing the relationships of the components of an artificial neural network. Its abbreviation is **SOM**.

self-powered **1.** A component, device, piece of equipment, or system which does not require external power to operate. Such a component, device, piece of equipment, or system may have its own batteries, generator, or the like. **2.** A component, device, piece of equipment, or system which does not require a separate connection to an external power source to operate. For instance, a USB peripheral, such as an ADSL modem, which draws its power from the USB port or a USB hub, and not from a power outlet.

self-powered device A device which does not require external power to operate. Such a device may have its own batteries, a generator, or the like.

self-powered equipment Equipment which does not require external power to operate. Such equipment may have its own batteries, a generator, or the like.

self-powered system A system which does not require external power to operate. Such a system may have its own batteries, a generator, or the like

self-quenching Quenching which occurs internally or automatically.

self-refresh The ability of some types of RAM to refresh the state of memory cells or modules without intervention on the part of the CPU, which reduces power consumption. Used, for instance, in notebook computers.

self-regulation The ability of a component, circuit, device, piece of equipment, system or mechanism, to monitor its own operation, and automatically make the appropriate changes. Seen, for instance, in control systems.

self-reset The automatic return to a former position or condition after a specified requirement, such as the passing of a given time period, is met. Seen, for instance, in automatic circuit breakers.

self-resetting Said of a component, circuit, device, piece of equipment, system, or mechanism which automatically returns to a former position or condition after a specified requirement, such as the passing of a given time period, is met. Certain circuit breakers, relays, and switches are examples.

self-resonance frequency Same as **self-resonant frequency**.

self-resonant frequency The frequency at which a body or system resonates without an external driving force. Also called **self-resonance frequency**, **resonant frequency (2)**, or **natural resonant frequency**.

self-scattering Scattering of electromagnetic radiation or sound by the same surface, region, or medium which emitted said radiation or sound.

self shutdown Same as **self shutoff**.

self shutoff Also called **self shutdown**, or **automatic shutoff**. **1.** In a computer, the process of automatically saving all data, quitting all applications, and turning off the power. **2.** The process of automatically finishing a process in an orderly manner, and turning off the power of a device.

self-sustained oscillation Oscillation, such as that of a self-excited oscillator, which is maintained without the aid of an external signal.

self-test One of multiple tests performed in **self-testing**.

self-testing Also called **self-diagnostics**. **1.** Tests which enable a component, circuit, device, piece of equipment, system, or mechanism to ascertain automatically and without user intervention if anything is malfunctioning, has failed, or if there is any other type of problem. An indicator or alarm of some type is usually utilized to alert of any unwanted conditions. **2.** A feature which provides for **self-testing (1)**.

self-wiping contacts Same as **self-cleaning contacts**.

selsyn Same as **synchro**.

selsyn generator Same as **synchro transmitter**.

selsyn motor Same as **synchro receiver**.

selsyn receiver Same as **synchro receiver**.

selsyn system Same as **synchro system**.

selsyn transmitter Same as **synchro transmitter**.

SEM **1.** Abbreviation of **scanning electron microscope**. **2.** Abbreviation of **scanning electron microscopy**. **3.** Abbreviation of **secondary electron multiplier**. **4.** Abbreviation of **single-electron memory**.

semantic error An error caused by using the wrong statements in a computer program. A violation of any rules which determine how statements are structured results in a **syntax error**.

semantic net Abbreviation of **semantic network**.

semantic network A representation of knowledge in which nodes represent physical or conceptual objects whose interrelationships are represented by arcs. Such a depiction is usually graphical, in which case it resembles a data flow diagram. Used, for instance, in knowledge-based systems. Its abbreviation is **semantic net**.

semaphore A mechanism, technique, or system used to coordinate or synchronize multiple processes that utilize the same operating system resources. A semaphore helps insure that only one process at a time accesses a given resource, and is used, for instance, to share a common memory space or group of files.

semi-anechoic chamber A room similar to an anechoic chamber, except that the floor is not lined with sound or radio-wave absorbing materials. Also called **semi-anechoic room**, or **semi-anechoic test chamber**.

semi-anechoic room Same as **semi-anechoic chamber**.

semi-anechoic test chamber Same as **semi-anechoic chamber**.

semi-duplex operation Same as **semiduplex operation**.

semiconductor A material, usually a crystal, whose conductivity lies somewhere between that of an electric conductor, such as a metal, and that of an insulator, such as rubber. A semiconductor may be a pure metal, such as silicon or germanium, or a compound such as gallium arsenide or indium phosphide. The electrical characteristics of an intrinsic semiconductor depend solely on a pure crystal, while those of an extrinsic semiconductor rely on dopants which are introduced. In a p-type semiconductor trivalent dopants are added to increase conductivity, while pentavalent dopants are added to enhance the conductivity n-type semiconductors. Mobile electrons, holes, or ions transport charges through semiconductors. Used extensively to make semiconductor devices. Also called **semiconductor material**.

semiconductor circuit A circuit consisting entirely of semiconductor components.

semiconductor circuit board A circuit board consisting entirely of semiconductor components.

semiconductor component A component, such as a semiconductor diode, consisting of or incorporating semiconductor materials.

semiconductor counter Same as **semiconductor detector**.

semiconductor detector A radiation detector which incorporates a semiconductor material. An example is a PIN photodiode. Also called **semiconductor counter**.

semiconductor device A device, such as a diode, transistor, or photocell, incorporating one or more semiconductor materials.

semiconductor die 1. A piece of semiconductor material or dielectric upon which one or more electrical components may be mounted, etched, or formed. Used, for instance, for fabrication of repetitive units in semiconductor manufacturing. Also called **die** (1), or **chip** (2). 2. A piece of semiconductor material or dielectric upon which one or more electrical components has been mounted, etched, or formed. Also called **die** (2), or **chip** (3).

semiconductor diode Also called **crystal diode**, or **crystal rectifier**. 1. A semiconductor device with two terminals. 2. A semiconductor device with two terminals, which is utilized for rectification.

semiconductor dopant An impurity which is introduced into a semiconductor material. Such an impurity may be an acceptor impurity, which makes for a p-type semiconductor, or it may be a donor impurity, which makes for an n-type semiconductor. Acceptor impurities include gallium and aluminum, while donor impurities include phosphorus and arsenic. In either case, a semiconductor dopant increases the conductivity of the semiconductor. Also called **dopant**, or **impurity** (2).

semiconductor doping The adding of a **semiconductor dopant** to a semiconductor material. Such impurities may be added during manufacturing through any of various processes, including diffusion. Controlled amounts of specific dopants are added to achieve the desired characteristics. Doping is used, for example, in the manufacturing of transistors and diodes. Also called **doping**.

semiconductor film Same as **semiconductor thin film**.

semiconductor heterostructure A structure consisting of two or more layers of semiconductor materials, each with different bandgaps, and whose crystal structure is similar. For instance, a layer of aluminum gallium arsenide on a layer of gallium arsenide. Used, for instance, in high-performance transistors, and lasers. Also called **heterostructure**.

semiconductor junction A layer or boundary which serves as the interface between semiconductor regions with different properties. For example, a pn junction. Also called **junction** (3).

semiconductor laser A laser in which a forward-biased pn junction diode is used to convert its DC input into a coherent light output. Used, for instance, as a light-pulse generator for transmission of information over fiber-optic lines. Also called **diode laser**, **laser diode**, **injection laser**, **injected laser**, or **injection laser diode**.

semiconductor material Same as **semiconductor**.

semiconductor memory Computer memory, such as chip memory, in which information is stored in memory cells incorporating semiconductor devices such as transistors or semiconductor capacitors. Also called **semiconductor storage**.

semiconductor-metal contact A contact between a metal and the semiconductor substrate it is deposited on. It may be either an ohmic contact or a Schottky contact. Also called **metal-semiconductor contact**.

semiconductor-metal diode Same as **Schottky diode**.

semiconductor-metal junction A layer or boundary which serves as the interface between a semiconductor and a metal. Also called **metal-semiconductor junction**.

semiconductor rectifier 1. A rectifier which utilizes one or more metal disks coated with a semiconductor layer. This layer may consist of selenium, copper oxide, or another suitable semiconductor. The rectification occurs as a result of the greater conductivity across the contact in one direction than the other. Also called **metallic-disk rectifier**, **dry rectifier**, **dry-disk rectifier**, or **contact rectifier**. 2. A rectifier consisting of, or incorporating, semiconductor materials. An example is a diode rectifier.

semiconductor region An area or layer within a semiconductor material which consists either of an n-type semiconductor or a p-type semiconductor.

semiconductor storage Same as **semiconductor memory**.

semiconductor substrate A body or material composed of a semiconductor material, upon which components, circuits, and/or devices are formed, etched, mounted, or otherwise placed.

semiconductor thin film A very thin layer or coating composed of a semiconductor material. For example, an organic semiconductor may consist of a thin film of a semiconductor on a plastic substrate. Also called **semiconductor film**.

semiconductor transistor An active semiconductor device which can serve as a switch, amplifier, or oscillator, among other possible functions. The two main classes are bipolar junction transistors, and field-effect transistors. Each of these has three regions or electrodes, and charge carriers flow between two of these terminals, while a third controls this flow. Transistors may be used as discrete devices or in components such as ICs, and can be found in countless items, such as computers, robots, TVs, cellular telephones, wireless transmitters, audio amplifiers, cars, and so on. A transistor may have more than three electrodes, as is the case with a tetrode junction transistor. Also called **transistor**.

semiconductor wafer Same as **slice** (2).

semiduplex operation Also spelled **semi-duplex operation**. 1. In communications, operation which is duplex at one end and half-duplex at the other. 2. In communications, operation which is duplex at one end and simplex at the other.

semimetal A chemical element which has some properties similar to metals, other properties similar to nonmetals, and/or properties which lie somewhere in between. For example, arsenic, germanium, and tellurium. Such elements are usually semiconductors. Also called **metalloid**.

semimetallic Consisting of, pertaining to, or characteristic of a **semimetal**.

semitone An interval between two sounds, in which the ratio of the higher to the lower frequency is the twelfth root of two, or approximately 1.0595 to 1. Also called **half-step**, or **halftone** (2).

send channel In two-way communications, a channel via which data, messages, or control information is sent, while the channel which serves for reception is the **return channel**.

sender 1. A component, device, piece of equipment, system, object, or entity which emits, radiates, or otherwise sends signals. Transmitters which send information-bearing signals are used in many areas, including communications, TV and entertainment, radars, and so on. Also called **transmit-**

ter (1). 2. A circuit, device, or system which converts audio, video, or other content into radio waves which are emitted. AM, FM, and AM/FM transmitters are examples. Also called **transmitter (2)**, **radio transmitter**, or **radio sender**.

sensation level For a given individual, the level, expressed in dB, of a sound relative to the individual's hearing threshold for that sound. For example, a noise bandwidth presented at 30 dB sensation level. Its abbreviation is **SL**. Also called **level above threshold**.

sense 1. The ability of a component, device, piece of equipment, system, or entity, to detect or otherwise perceive a given phenomenon, state, or condition. Also, the action of such perception. For example, to detect a direction or polarity, or to be aware of the state of a relay. 2. In computers, to read data. Also, a read operation.

sense amp Abbreviation of **sense amplifier**.

sense amplifier Its abbreviation is **sense amp**. 1. A circuit or device that detects and amplifies very low voltages. 2. A **sense amplifier** (1) whose output is a fixed voltage.

sense antenna An antenna which indicates the direction from which a radio signal arrives. Used, for instance, in a direction finder. Also called **sensing antenna**.

sense circuit Same as **sensing circuit**.

sense resistor A resistor utilized as a **sensor**. Used, for instance, to detect a current or a voltage in a circuit. Also called **sensing resistor**.

sensing antenna Same as **sense antenna**.

sensing circuit A circuit that serves as a **sensor**. For example, a circuit in a pacemaker which detects electrical activity in the heart. Also called **sense circuit**.

sensing element A probe or other part or section of a **sensor** (1), which samples or otherwise detects. Also called **sensor (2)**.

sensing resistor Same as **sense resistor**.

sensitive 1. Able to detect or sense phenomena or stimulation. Also, able to change or otherwise respond to a given phenomenon, action, or circumstance. For example, that which is photosensitive. 2. Able to detect and respond to very small or subtle changes. Said, for instance, of an instrument. 3. Able to distinguish between two or more levels, settings, states, or the like. For example, case sensitive. 4. Harmed or otherwise adversely affected by a given phenomenon, condition, level, or the like. For instance, electrostatic discharge sensitive.

sensitive device 1. A device which responds to a given stimulus or signal. For example, a photosensitive device, or a passband filter. 2. A **sensitive device** (1) that responds to very slight stimuli or signals.

sensitivity 1. The ability to detect or otherwise perceive a given phenomenon, state, condition, or level. Also, the degree of such sensitivity. 2. The magnitude of the response of a component, circuit, device, instrument, piece of equipment, system, or material, to a given stimulus, level, or other input signal. For instance, the ohms-per-volt rating of a voltmeter, or the magnitude of the deflection of an indicating needle in response to a given change in a detected variable. 3. The minimum input signal that produces a given output signal. For example, the minimum input signal level that a specific radio receiver needs to produce a stated output level. 4. The proportion of the audio-frequency electrical power input to a loudspeaker which is converted into acoustic energy. Efficiency may be measured, for instance, by driving a loudspeaker with a power input of 1 watt, and measuring its output, in decibels, at a distance of 1 meter. Also called **speaker efficiency**, **loudspeaker efficiency**, or **efficiency (3)**.

sensitivity adjustment An adjustment made utilizing a **sensitivity control**.

sensitivity control A control, such as that in a receiver, which adjusts **sensitivity**.

sensitivity-time control An automatic control circuit which modifies the gain of a receiver at set intervals, so as to maintain the output signal levels within a desired amplitude, in response to varying input signals. To do so, it reduces sensitivity to strong signals, while increasing it for weak ones. Such modifications are usually anticipated, as when a loran receiver adjusts its gain between successive pulses from fixed transmitters. Its abbreviation is **STC**. Also called **sensitivity-time control circuit**, **differential gain control**, **gain-time control**, or **gain-sensitivity control**.

sensitivity-time control circuit Same as **sensitivity-time control**.

sensitometer An instrument which measures and indicates the photosensitivity of materials.

sensitometry The measurement and techniques involved in the use of **sensitometers**.

sensor 1. A component, circuit, device, instrument, piece of equipment, system, material, or mechanism which serves to **sense** (1), and respond. A sensor may, for instance, detect a given value, or any change in values. There are many examples, including acoustic, pressure, contact, force, motion, proximity, position, light, optical, radiation, infrared, ultraviolet, radio-frequency, magnetic, photoelectric, and time sensors. Most sensors convert a non-electrical phenomenon into a proportional electrical equivalent, and nearly all transducers are sensors of some kind. A sensor may provide an analog or digital output, the latter being used, for instance, for computer applications. Contact sensors detect objects and phenomena through physical contact with an object or medium, while non-contact sensors have no such contact. 2. Same as **sensing element**.

sentry robot Same as **security robot**.

separation 1. The act of dividing something. For instance, a disk partitioning, the removal of a gas from a solid it was trapped in, or the separation of individual pulses within a train. 2. The location or area where a **separation** (1) occurs. 3. Any space or interval resulting from a **separation** (1). 4. In a stereo audio component, the extent to which the information from one channel is absent from the other. Usually expressed in decibels. Also called **stereo separation**, or **channel separation** (2).

separation circuit A circuit which serves to separate signals or components of signals. For example, a circuit separating the chrominance and luminance signals within a video signal.

separation energy The energy required to remove a particle, such as a proton, from a nucleus.

separator That which serves to divide, cut, sort, isolate, or the like. For example, an electrostatic separator, a frequency separator, an amplitude separator, an information separator, or the dielectric material which isolates the conductors of a capacitor.

septa The plural form of **septum**.

septagon A closed geometric figure bounded by seven straight lines

septillion A number equal to 10^{24}.

septum A partition or membrane which divides two cavities. For instance, that which partitions a waveguide. Its plural form is **septa**.

sequence 1. An arrangement of two or more things, entities, or events, in a given order. 2. An arrangement of two or more things, entities, or events, in a successive order. 3. That which is ordered by a **sequence** (1) or **sequence** (2). 4. To arrange in a **sequence** (1) or **sequence** (2).

sequence counter Same as **sequence register**.

sequence register In a CPU, a register that contains the address of the location in memory that is to be accessed by the next instruction. May also refer to the address of the current instruction. Also called by various other names **sequence counter**, **control register**, **program register**, and **instruction register**.

sequence relay A multi-contact relay whose contacts open and close in a given sequence. Also called **sequential relay**.

sequence signal A signal that controls the steps within a given sequence.

sequence switch A multi-contact switch whose contacts open and close in a given sequence. Also called **sequential switch**.

sequence timer A timer that controls the steps within a given sequence. Also called **sequential timer**.

Sequenced Packet Exchange A communications protocol utilized at the transport layer within the OSI Reference Model. Its abbreviation is **SPX**.

sequencer That which serves to **sequence (4)**. For instance, an automatic call sequencer, or a MIDI sequencer.

sequential Consisting of, having, following, or characterized by a **sequence**. For instance, sequential control, or sequential scanning.

sequential access Same as **serial access**.

sequential-access memory Same as **sequential-access storage**.

sequential-access method The retrieval of data in the same sequence it was originally stored. Its abbreviation is **SAM**.

sequential-access storage Computer storage whose content is accessed sequentially, as opposed to directly. Also called **sequential-access memory**.

sequential-access storage device Its acronym is **SASD**. A computer storage device whose content is accessed sequentially, as opposed to directly. For example, a tape drive is a SASD, while a disk drive is a **direct-access storage device**.

sequential circuit A logic circuit whose output at any given moment depends, in part, on previous states. Also called **sequential logic circuit**.

sequential control Operation of a component, circuit, device, piece of equipment, system, mechanism, or process, which is controlled following a specified sequence.

sequential logic A logic circuit or element whose output at any given moment depends, in part, on previous states. This contrasts with **combinational logic**, whose output at any given moment depends only on its input values at that same instant.

sequential logic circuit Same as **sequential circuit**.

sequential logic element A logic element whose output at any given moment depends, in part, on previous states.

sequential memory Same as **serial access (2)**.

sequential programming Same as **serial programming**.

sequential relay Same as **sequence relay**.

sequential scanning A scanning system, such as that computer monitors, in which the electron beam traces all lines sequentially from top to bottom. This contrasts with **interlaced scanning**, in which the electron beam traces all odd-numbered lines followed by the tracing of all even-numbered lines. Interlaced scanning may produce a fluttering effect when displaying certain graphics, a drawback not exhibited by monitors with sequential scanning. Also called **non-interlaced scanning**, or **progressive scanning**.

sequential search A search in which the data in question is examined sequentially from start to finish. Also called **linear search**.

sequential storage Same as **serial access (1)**.

sequential switch Same as **sequence switch**.

sequential timer Same as **sequence timer**.

Séquentiel Couleur Avec Mémoire Same as **SECAM**.

serial 1. Of, pertaining to, occurring as, or resembling, a **series**. 2. Acted upon, performed, or operating sequentially, or one after the other. For example, serial access. 3. The sequential processing, storing, transfer, transmission, reception, or the like, of multiple bits, characters, or data units. This contrasts with **parallel (4)**, where such events and processes occur simultaneously.

serial access Also called **sequential access**. 1. Storage and/or retrieval of data to or from a computer storage medium in which bits, characters, or data units are transferred sequentially, or one after the other. Such transfers must follow a given order, such as one running from start to finish or from top to bottom. This contrasts with **parallel access (1)**, where multiple bits, characters, or data units are transferred simultaneously. Also called **serial storage**, or **sequential storage**. 2. Reading and/or writing of data from or to computer memory in which bits, characters, or data units are transferred sequentially. This contrasts with **parallel access (2)**, where multiple bits, characters, or data units are transferred simultaneously. Also called **serial memory**, or **sequential memory**.

serial adder 1. In computers, a logic circuit which adds pairs of corresponding digit positions sequentially. 2. In computers, a logic circuit which adds two or more numbers or quantities sequentially.

serial addition The function a **serial adder** performs.

serial bus A bus, such as High-Performance Serial Bus or USB, which transmits bits in a serial stream.

serial channel A communications channel in which bits, characters, or data units are transferred sequentially. Along a **parallel channel (1)** these are transferred simultaneously.

serial communications Communications in which data is transmitted sequentially over a single channel. This contrasts with **parallel communications**, in which data is transmitted simultaneously over multiple channels.

serial computer 1. A computer that has a single processor, and which performs **serial processing**. 2. A computer which is utilized to perform serial processing.

Serial Copy Management System Same as **SCMS**.

serial data interface Same as **serial interface**.

serial data transfer Same as **serial transfer**.

serial data transmission Same as **serial transmission**.

serial digital interface A standard for the serial transmission of digital video, such as that used for broadcasting. Also, an interface adhering to this standard. Its abbreviation is **SDI**.

serial I/O Abbreviation of **serial input/output**. The transfer of data between a CPU and a peripheral in which bits, characters, or data units are transferred sequentially. This contrasts with **parallel I/O**, where multiple bits, characters, or data units are transferred simultaneously.

serial infrared A technology for providing a connection between devices supporting the use of infrared signals to convey information. Also, a device utilizing this technology. Its abbreviation is **SIR**.

serial input/output Same as **serial I/O**.

serial interface An interface, such as a USB, which transfers bits, characters, or data units sequentially. Used, for instance, for connecting a computer to a printer. This contrasts with a **parallel interface**, in which bits, characters, or data units are transferred simultaneously. Also called **serial data interface**.

Serial Line Internet Protocol Same as **SLIP**.

Serial Line IP Same as **SLIP**.

serial memory Same as **serial access (2)**.

serial mouse A computer mouse that connects to a **serial port**. This contrasts, for instance, with a bus mouse.

serial number 1. A number or sequence of characters which uniquely identifies an item or product. 2. A number that identifies an item or position within a sequence.

serial operation 1. The performance of two or more operations in sequence. 2. Operation of a circuit or device whose components are connected in series.

serial output One of multiple outputs which are delivered in sequence. Seen, for instance, in serial transmission. This contrasts with **parallel output**, in which all outputs are delivered simultaneously.

serial/parallel Supporting or utilizing both serial and parallel functions or devices. For example, a controller card which supports both serial and parallel interfaces.

serial port A computer port utilizing a **serial interface**.

serial printer 1. A printer which prints a single character at a time. Examples include daisy-wheel printers and dot-matrix printers. Also called **character printer (1)**. 2. A printer that connects to a computer via a serial interface.

serial processing The execution of multiple computer operations one after the other. This contrasts with **parallel processing (2)** in which multiple operations are executed simultaneously.

serial programming Computer programming in which steps or operations are performed in sequence. In **parallel programming**, multiple steps or operations are performed simultaneously. Also called **series programming**, or **sequential programming**.

serial SCSI The use of a SCSI Interface implementing a serial protocol. Since SCSI provides a parallel interface, the appropriate signal conversions must be performed.

serial storage Same as **serial access (1)**.

Serial Storage Architecture Same as **SSA**.

serial stream The transmission of bits, characters, or data units one at a time in sequence. A **parallel stream** provides for the simultaneous transmission of bits, characters, or data units utilizing multiple lines, fibers, channels, or the like.

serial transfer The transfer of bits, characters, or data units one at a time in sequence. This contrasts with **parallel transfer**, in which multiple bits, characters, or data units are transferred simultaneously. Also called **serial data transfer**.

serial transmission The transmission of bits, characters, or data units one at a time in sequence. This contrasts with **parallel transmission**, in which multiple bits, characters, or data units are transmitted simultaneously. Also called **serial data transmission**.

serialize To convert parallel streams of bits into a serial stream of bits. A parallel stream consists of one more bytes, while a serial stream has one bit after the other.

series 1. An arrangement of two or more related things, entities, or events, in a given order. 2. An arrangement of two or more related things, entities, or events, in a successive order. 3. That which is ordered by a **series (1)** or **series (2)**. 4. Characteristic of, or pertaining to a **series circuit** or **series connection**. 5. The sum of a finite or infinite sequence of numbers or terms. 6. A group of related objects or entities which vary or occur in a given sequence. For example, a radioactive decay series.

series-aiding Two or more sources of emf which aid in current flowing in the same direction. In sources which are **series-opposing** the converse is true.

series capacitance The total capacitance of **series capacitors**.

series capacitors Two or more capacitors connected in series. The total capacitance is the reciprocal of the sum of the reciprocal of the value of each of the individual capacitances. Also called **capacitors in series**.

series circuit A circuit whose components are connected in series. That is, they have a **series connection**.

series connection The end-to-end connection of the components within a circuit. There is a single path for the current, which flows through all the components in sequence, while the voltage is divided among the components. This contrasts with **parallel connection**, where the connection is side-by-side, or otherwise in a manner where the current is divided among the components. Also called **in series**.

series decay The transformation of a nuclide into another nuclide through radioactive decay, followed by further transformations until a stable, or nonradioactive, nuclide results. In such a series, the first member is called the parent, and the last is the end product. For instance, the chemical element uranium undergoes a series of radioactive decay steps which eventually ends with a stable isotope of lead. Also called **series disintegration**, **chain decay**, or **radioactive chain decay**.

series disintegration Same as **series decay**.

series dropping resistor A resistor used in series with a load, which reduces the voltage applied to said load. The voltage reduction is equal to the voltage drop across the terminals of the resistor. Also called **dropping resistor**.

series-fed That which has a **series feed**. Also, pertaining to that which has a series feed.

series feed The application, in series, of a DC voltage and an AC voltage to a component or device.

series generator An electric generator whose armature and field windings are connected in series. Also called **series-wound generator**.

series impedance An impedance acting in series with another quantity or magnitude, such as another impedance.

series inductance An inductance acting in series with another quantity or magnitude, such as a capacitance.

series inductors Two or more inductors connected in series. When said inductors are sufficiently shielded or apart from each other, the magnetic inductance is considered to be negligible. Otherwise, such inductance, or coupling, must be factored in.

series motor An electric motor whose armature and field windings are connected in series. Also called **series-wound motor**.

series network An electric network whose components are connected in series. That is, they have a **series connection**.

series operation 1. The performance of two or more operations sequentially. 2. Operation of a circuit or device whose components are connected in series.

series-opposing Two or more sources of emf which oppose current flowing in the same direction. In sources which are **series-aiding** the converse is true.

series-parallel Also called **series-parallel circuit**. 1. A circuit which incorporates both parallel connections and serial connections of its components. For instance, a circuit with multiple parallel paths, each with components connected in series. Also, pertaining to such a circuit. 2. A circuit whose components can be arranged either in a parallel or series manner.

series-parallel arrangement Same as **series-parallel connection**.

series-parallel circuit Same as **series-parallel**.

series-parallel connection The arrangement of a **series-parallel circuit**. Also called **series-parallel arrangement**.

series-parallel switch A switch that allows two or more components to be connected in series or in parallel.

series programming Same as **serial programming**.

series regulator A regulator, such as a voltage regulator, which is connected in series with a device or load.

series resistance The total resistance of **series resistors**.

series resistors Two or more resistors connected in series. The total resistance is the sum of the values of each the individual resistors. Also called **resistors in series**.

series resonance 1. Resonance in a **series-resonant circuit**. 2. In a series-resonant circuit, the frequency at which the series impedance is lowest. 3. In a series-resonant circuit, the frequency at which the inductive capacitance equals the capacitive reactance.

series-resonant circuit A resonant circuit in which a capacitor and an inductor are connected in series with an AC source. Resonance occurs at or near the minimum impedance of the circuit.

series-resonant frequency The resonant frequency of a **series-resonant circuit**.

series-shunt network A network composed of multiple H, L, T, and/or π networks connected in cascade. It may consist, for instance, of a cascaded series of H or L networks. Usually used as filters, or as analog-to-digital converters. Also called **ladder network**.

series transformer A current transformer in which the primary winding is connected in series with the main circuit, and the secondary winding to a measuring instrument.

series-tuned Colpitts oscillator A Colpitts oscillator with a tuning capacitor in series with the resonant tank coil, which provides greater stability. Also called **Clapp oscillator**.

series winding In an electric motor or generator, an armature winding connected in series with the field winding.

series-wound generator Same as **series generator**.

series-wound motor Same as **series motor**.

serrated vertical pulse In TV, a vertical sync pulse which is subdivided into six component pulses, each occurring at twice the horizontal scanning frequency. Also called **serrated vertical sync pulse**.

serrated vertical sync pulse Same as **serrated vertical pulse**.

server Also called **network server**. 1. Within a communications network, a computer whose hardware and/or software resources are shared by other computers. Servers, among other functions, control access to the network and manage network resources. There are various types of servers, including application servers, file servers, network access servers, and Web servers. 2. Within a network with a client/server architecture, a computer and/or program which responds to requests made by clients.

server appliance A self-contained server which incorporates both hardware and software, and which is designed to be installed and maintained with a minimum of effort and support. Such a server is usually plugged into an existing network, with all supported applications preinstalled, and is used, for instance, as a Web server, mail server, or file server. Server appliances are configured and accessed via Web browsers. Also called **appliance server**.

server application 1. An application intended to run on a server. 2. An application running on a server. 3. In a client/server environment, an application residing and/or running on a server instead of a client.

server-based That which resides, occurs, or runs on a server. For example, server-based archiving, or a server-based application. That which is **client-based** resides, occurs, or runs on a client.

server-based application An application which resides and/or runs on a server, as opposed to a **client-based application** which resides and/or runs on a client.

server-based archiving Archiving which occurs at a server, as opposed to **client-based archiving** which takes place at a client.

server-based computing The use of applications stored and/or run on a server. This contrasts with **client-based computing**, in which applications are stored and/or run on a client.

server cluster Same as **server farm**.

server farm Two or more networked servers which are housed at a single location. Used, for instance, to balance the overall load of a network, or for redundant backup of resources or data. Also called **server cluster**.

Server Information Table A table containing the information gathered utilizing a **Service Advertising Protocol**.

Server Message Block A file-sharing protocol utilized, for instance, over the Internet. Its abbreviation is **SMB**. Also called **Server Message Block Protocol**.

Server Message Block Protocol Same as **Server Message Block**.

server not found An error 404 resulting from the server not being available. Also called **cannot find server**.

server operating system Its abbreviation is **SOS (4)**. 1. An operating system which resides on a server. 2. An operating system which runs a server.

server-parsed HTML Same as **SHTML**.

server program Also called **server software**. 1. A program intended to run on a server. 2. A program running on a server. 3. In a client/server environment, a program residing and/or running on a server instead of a client.

server side Any activity, such as processing, occurring at the server within a **client/server architecture**.

server-side include Additional or alternate information, such as the date of the most recent update or the inclusion of selected text, which is incorporated into an HTML file before being sent from the server side to the client side, resulting in an SHTML document. Also, a command to include such information. Its abbreviation is **SSI**.

server-side script A script run on a server, such as a Web server. A **client-side script** is run on a client, such as a Web browser.

server software Same as **server program**.

service 1. Any activity involving installation, supplying, maintenance, or repairs. Such activities are meant to be performed by qualified personnel, and may or may not be a part of a warranty or other agreement to provide said services. 2. An established or otherwise organized system which is meant to supply the needs of the public. For example, a broadcasting, telephone, satellite, or information service. 3. Proper function or operation. For instance, the service life of a device. 4. A level or category within multiple offerings or possibilities. For example, dial-up, cable, or DSL access to the Internet.

service access point Its abbreviation is **SAP**. 1. A junction that interconnects networks, network layers, or network services. 2. A point where a circuit may be accessed.

Service Advertising Protocol In a communications network, a protocol utilized to identify information such as the addresses and services of available servers. A **Server Information Table** contains the data so gathered. Its abbreviation is **SAP**.

service area The geographical area within which a given transmitter provides effective service. For instance, the zone served by a cellular telephone system, the region within which reception of TV or radio broadcasts is adequate, or the zone a radar can effectively scan. Also called **coverage (1)**, or **coverage area**.

service band A frequency band allocated to a given class of communications services.

service cable Same as **service wire**.

service charge 1. Any charge made in exchange for **services**. **2.** A specific charge made for services, such as a per hour rate. **3.** Any charge made in exchange for services which extend beyond that which is ordinary. For instance, an installation charge for a device which normally is not professionally installed.

service code A short telephone number dialed to reach a given service, such as a three digit number utilized to dial emergency services.

service control point Same as **SCP**.

service entrance The location where a **service entrance cable** enters the premises of a customer.

service entrance cable A cable serving to convey electricity, telephone service, or the like, from an outside transmission line into a structure, such as a building or home. Also called **entrance cable**.

Service Level Agreement A contract specifying the level of service a provider is committed to deliver to a client or user. Such an agreement may cover performance minimums pertaining to downtime, bandwidth, and response time, in addition to the availability of technical support, replacement components, and the like. Its abbreviation is **SLA**.

service life 1. The cumulative length of time that a component, circuit, device, piece of equipment, system, or material is used for its intended purpose. **2.** The average length of time that a component, circuit, device, piece of equipment, system, or material can be used for its intended purpose. **3.** The **service life (2)** as stated by a manufacturer.

service line Same as **service wire**.

service maintenance 1. In batteries, the proportion of the full-rated capacity which remains after a given period of time. **2.** Maintenance provided under an agreement to provide services, such as a contract.

service meter A meter which monitors and indicates usage of a given service. A kilowatt-hour meter is an example.

service pack A software update which is intended, for instance, to remedy a bug or other problem, provide drivers enabling support of newer devices, or the like. Service packs are almost always available for download from the Internet.

service pack 1 The first in a series of service packs. Service packs usually serve to enhance the software in question, but the first often addresses glaring security or bug issues which should have been resolved previously. Its abbreviation is **SP-1**.

service pack 2 The second in a series of service packs. All service packs subsequent to the first usually detect which items have been installed by prior packs. Its abbreviation is **SP-2**.

Service Profile Identifier A number which a telephone company assigns to an ISDN terminal B channel, to define the services a user is subscribed to. Such a number usually consists of 14 digits, but may be longer or shorter. Its abbreviation is **SPID**.

service provider An entity which provides access to the Internet. Customers usually pay a monthly fee for this service, although it usually available for free unless a high-speed connection is desired. Users get a software package to access and browse the World Wide Web, one or more email accounts, and Web pages to have a presence on the Internet. Other offerings may include Web site building and hosting services. Users can connect via a dial-up service or asymmetrical digital subscriber line, among others. Also called **Internet service provider**, **access provider**, or **online service provider (1)**.

service switch A switch that serves to connect and disconnect a **service wire**.

service switching point A local exchange in an SS7 network. Its abbreviation is **SSP**.

service test A test which is performed under actual operating conditions, or after a component, circuit, device, piece of equipment, system, or material is installed and in use for a given time.

service wire An aboveground or underground line which connects a terminal of a distribution cable to a subscriber's premises. Such a line may carry telephone service, cable TV, a link to a network, electric power, and so on. It may consist, for example, of a telephone wire extending from a pole to the home of a subscriber. Also called **service line**, **service cable**, **drop wire**, or **drop cable**.

servlet A small program, such as a Java applet, that runs on a server.

servo 1. Abbreviation of **servomotor**. **2.** Abbreviation of **servomechanism**. **3.** Abbreviation of **servosystem**.

servo amplifier An amplifier utilized in a **servomechanism** or **servosystem**. Also spelled **servoamplifier**.

servo driven Same as **servomotor driven**.

servo-driven robot Same as **servo robot**.

servo mechanism Same as **servomechanism**.

servo motor Same as **servomotor**.

servo motor driven Same as **servomotor driven**.

servo robot A robot whose positions and/or movements are driven by one or more **servomechanisms**, as directed by one or more controllers. Also called **servo-driven robot**.

servo system Same as **servosystem**.

servoamplifier Same as **servo amplifier**.

servomechanism An electrical and/or mechanical closed-loop control system which automatically controls the position and/or speed of a device, piece of equipment, transducer, apparatus, system, or mechanism. One or more feedback mechanisms enable monitoring and automatically keeping specified parameters within the desired limits, with servomotors usually being utilized to drive the load. Used, for instance, to control electromechanical systems such as machinery, robots, aircraft, or satellites. Also spelled **servo mechanism**. Its abbreviation is **servo**. Also called **servosystem (1)**.

servomotor A motor which drives the load in a **servomechanism**. Such a motor may be electrical, mechanical, pneumatic, hydraulic, and so on. Also spelled **servo motor**. Its abbreviation is **servo**.

servomotor driven A device, piece of equipment, transducer, apparatus, system, or mechanism that is driven by a **servomotor**. Also spelled **servo motor driven**. Also called **servo driven**.

servosystem Also spelled **servo system**. Its abbreviation is **servo**. **1.** Same as **servomechanism**. **2.** A system incorporating one or more servomechanisms.

session 1. A period during which a user is connected to a computer or communications network. Also, all the activities which take place during such a session. Also called **user session (1)**. **2.** A period during which an individual or entity uses an application or program.

Session-at-Once A method of writing to an optical disk, such as a DVD or CD, in which multiple recording sessions are utilized. Used, for instance, to record a single music CD from multiple sources on different occasions. This contrasts, for instance, with **Disc-at-Once**, in which all data is transferred continuously without interruptions. This also contrasts with **Track-at-Once**, where each track is recorded independently.

session bean A JavaBean that is of a more temporary nature than an **entity bean**.

session cookie A cookie that is deleted or otherwise disappears when a Web browser is closed, or after a comparatively short time. This contrasts with a **persistent cookie**, which is kept indefinitely or until deleted. Also called **temporary cookie**, or **transient cookie**.

session layer Within the OSI Reference Model for the implementation of communications protocols, the third highest level, located directly above the transport layer. This layer takes care of functions such as establishing, managing, and terminating all exchanges of data. Also called **layer 5**.

set 1. To adjust for proper operation. Also, to adjust selecting any of multiple operational modes, positions, or the like. For example, to tune a radio receiver. 2. To adjust based on a given standard. Also, to adjust to a desired setting. For instance, to set a clock based on an atomic standard. 3. To place or put in a secure or stable position or location. Also, to place in a specific location. For example, to install a memory chip. 4. To place in a given position. Also, to place in a given manner. For instance, to set upright. 5. To place in a given state. For example, to set a storage cell to a one, or high, state. 6. To assign a specific value. For instance, to assign a one to a high state. 7. To put in an operational state. For example, to turn on. 8. To become hardened, solidified, or fixed. Also, to be so hardened or set. Said, for instance, of a thermoplastic. 9. Two or more objects or entities which form a group, are within the same category, or otherwise belong together. For example, a character set or a data set. 10. A device, piece of equipment, or system which sends and/or receives radio waves, especially a radio or TV receiver. 11. A component, device, piece of equipment, apparatus, or system which is used as a unit for a given purpose. For example, a telephone set, a radar set, or a pair of headphones. 12. For a celestial body, such as the sun or moon, to descend below the horizon.

SET Abbreviation of **secure electronic transaction**.

set analyzer An instrument or device which analyzes radio sets, radar sets, TVs, and so on.

set noise The noise present in a set, such as a receiver or radar, in the absence of a signal.

set point 1. The point, level, or setting which a control system strives to maintain. 2. The point, level, or setting which a component, circuit, device, piece of equipment, or system is selected to operate at.

set pulse A pulse which sets an operating mode, state, position, or the like.

set-reset flip-flop Same as **SR flip-flop**.

set screw A screw, which when tightened, is intended to hold parts together. For example, that which holds two or more adjustable parts in place.

set-top box A box or similar device, such as a decoder box, which rests on top of, or is otherwise placed near a TV or ancillary device such as a VCR. Also spelled **settop box**. Its abbreviation is **STB**.

set-up Same as **setup**.

set-up program Same as **setup program**.

set-up requirements Same as **setup requirements**.

set-up specifications Same as **setup specifications**.

set-up string Same as **setup string**.

set-up time Same as **setup time**.

setting The position, direction, level, value, scale, or other manner in which something is set.

settling time 1. The time required for a component, circuit, device, instrument, piece of equipment, or system, to settle and remain at a given value once an input signal or pulse is received. 2. The time required for a component, circuit, device, piece of equipment, or system, to stop or otherwise cease functioning once the proper signal is received or removed. 3. The time required for an instrument to provide a

reading, within a given degree of accuracy, when a monitored parameter varies.

settop box Same as **set-top box**.

setup Also spelled **set-up**. 1. To install, mount, adjust, position, or otherwise prepare for use or operation. Also, the processes involved in such a setup. 2. The manner or disposition in which something is **setup** (1). 3. In communications, to establish a physical or logical connection, such as that needed for a telephone call or network access. To **tear down** is to end such a connection.

setup program A program which installs another, such as an application. It takes care of tasks such as verifying the configuration of the system and creating the appropriate folders. Such a program usually only requires a user to insert a disc and follow the prompts. An installation program may also be used to install hardware. Also spelled **set-up program**. Also called **installation program**.

setup requirements Same as **setup specifications**. Also spelled **set-up requirements**.

setup specifications Requirements which must be followed so that a component, device, piece of equipment, or system is properly **setup** (1). Also spelled **set-up specifications**. Also called **setup requirements**.

setup string A set of consecutive bits or characters which prepare a device, such as a peripheral, for use. Also spelled **set-up string**.

setup time The time required for a proper **setup** (1). Also spelled **set-up time**.

SEU Abbreviation of **single-event upset**.

seven-segment display An LCD alphanumeric indicator, such as that seen in watches and calculators, in which each digit consists of seven bars which are selectively turned on and off to represent numbers and/or letters. Its abbreviation is **7-segment display**.

sextillion A number equal to 10^{21}.

SF Abbreviation of **superframe**.

SF format Abbreviation of **superframe format**.

SFA Abbreviation of **sales force automation**.

sferics Same as **spherics**.

SFG Abbreviation of **signal-flow graph**.

SFS Abbreviation of **shared file system**.

Sg Chemical symbol for **seaborgium**.

SGML Abbreviation of Standard Generalized Markup Language. An ISO standard which describes how to specify a markup language or set of tags. Used, for instance, to establish tagging rules which are implemented by HTML, which is an SGML-based language.

SGRAM Abbreviation of synchronous graphics **RAM**. A form of RAM that is synchronized with the clock speed of the CPU it is running with, optimizing its speed, and which is utilized for video memory. SGRAM is usually contained in a graphics card.

SHA 1. Abbreviation of **sample-and-hold amplifier**. 2. Abbreviation of **secure hash algorithm**.

SHA-1 Abbreviation of **secure hash algorithm-1**.

SHA1 Abbreviation of **secure hash algorithm-1**.

shaded-pole induction motor Same as **shaded-pole motor**.

shaded-pole motor A single-phase induction motor with low starting torque which is suitable for low-power applications. Such a motor incorporates a **shading coil** (1) for starting, and is used, for instance, in domestic appliances such as small fans and hair dryers. Also called **shaded-pole induction motor**.

shading 1. Any enhancements made to shadows present in an image or scene, or the creation of shadows for a given visual effect, such as for highlighting, realism, or 3D. Used, for instance, in computer graphics and TV. **2.** In TV, the compensation of spurious signals generated by a camera tube during flyback.

shading coil Also called **shading ring**. **1.** A shorted copper ring or band which is usually utilized to produce a rotating magnetic field in a shaded-pole motor. **2.** A shorted coil surrounding a part of the core of an AC relay to prevent contact chatter.

shading ring Same as **shading coil**.

shadow **1.** An image, shape, or area which is dark due to the interception of light rays by an opaque body or entity. **2.** An area of relative darkness. Also, the dark or darker portions of an image. **3.** A region where radiation, such as light, X-rays, or sound does not reach, due to the interception by a body, entity, or medium through which such radiation can not pass. Also, such a region present on a film, such as that of an X-ray.

shadow area Same as **shadow region**.

shadow attenuation The attenuation of radio waves in a **shadow region**.

shadow effect A reduction in radio reception due to obstructions, as occurs in **shadow regions**. Also called **shadow loss**, or **shadow losses**.

shadow loss Same as **shadow effect**.

shadow losses Same as **shadow effect**.

shadow mask In a picture tube with a three-color gun, a grill with round holes that is placed behind the screen to make sure that each color beam strikes the correct phosphor dot on the screen. It insures, for instance, that the electron beam intended for the red phosphor dots only hits those. Also called **aperture mask**, or **mask (3)**.

shadow memory Same as **shadow RAM**.

shadow RAM The copying of portions of a system's BIOS to a special area of RAM. RAM is typically much faster than BIOS. Also called **shadow memory**, or **shadow ROM**.

shadow region Also called **shadow area**, or **shadow zone**. **1.** In radio communications, a region where reception is significantly weaker, due to obstructions. Also called **blind zone (1)**. **2.** In radars, an area with no echoes, due to obstructions. Also called **blind zone (2)**. **3.** In radio communications, an area within the normal range of a transmitter, where radio reception is poor or non-existent, due to large objects in the vicinity of the receiver.

shadow ROM Same as **shadow RAM**.

shadow zone Same as **shadow region**.

shadowgram Same as **shadowgraph**.

shadowgraph An image produced on a radiosensitive surface, such as a photographic film or plate, when exposed to penetrating radiation, especially X-rays. Typically, the object to be analyzed is placed between the X-ray source and the film or plate, with the relative opacity determining the intensity of the shadows produced. Used, for instance, as a medical diagnostic technique, for ascertaining product defects, or in the quantitative analysis of materials. Also called **shadowgram, skiagraph, skiagram, radiograph, X-ray (3), X-ray photograph, X-ray image, roentgenogram**, or **roentgenograph**.

shadowgraphy The obtaining of **shadowgraphs**. Also, the science and techniques involved in obtaining such images, including reducing the exposure necessary for photographing, improvement of resolution, and the like. Also called **skiagraphy, radiography, roentgenography**, or **X-ray photography**.

shaft **1.** A long and narrow rod or pole. **2.** A ray or beam of light. **3.** A vertical passageway.

shaft-angle encoder Same as **shaft-position encoder**.

shaft-position encoder A device or system which senses the angular position of a shaft and converts this into a proportional magnitude. Used, for instance, as an analog-to-digital converter. Also called **shaft-angle encoder**.

shake table A platform used in conjunction with a device which delivers controlled vibrations. Utilized to test the ability of components, devices, equipment, systems, and materials, to withstand vibrations, shocks, jolts, and the like. Also called **shaking table, shaker**, or **vibration table**.

shake-table test A test performed using a **shake table**. Also called **shake test**.

shake-table testing The performance of **shake-table tests**. Also called **shake testing**, or **shaking-table testing**.

shake test Same as **shake-table test**.

shake testing Same as **shake-table testing**.

shaker Same as **shake table**.

shaking table Same as **shake table**.

shaking-table testing Same as **shake-table testing**.

Shannon-Hartley theorem Same as **Shannon theorem**.

Shannon limit For a given communications channel or line, the best signal-to-noise ratio attainable, according to the **Shannon theorem**.

Shannon theorem A theorem which relates the capacity of a given communications channel or line to its bandwidth and signal-to-noise ratio. Also called **Shannon-Hartley theorem**.

shape coding The coding of a control, such as a button or knob, by a distinctive shape. For example, an on/off button of a TV being round, while the volume control buttons being in the shape of arrowheads. Such coding may also be used in combination with size and color coding.

shape factor **1.** The ratio of two relative bandwidths for a bandpass or a bandstop filter. For example, the ratio of the 60 dB rejection bandwidth, to the 3 dB bandwidth of the passband. **2.** For a low-pass or high-pass filter, the ratio of a stopband frequency to a cutoff frequency. **3.** A function which takes into account the ratio of the diameter to length of a coil when computing its inductance. Also called **form factor (2)**.

shape-memory alloy An alloy that can undergo significant plastic deformation, then be returned to its original shape upon being heated. Used, for instance, in medical devices such as prosthetic limbs or dynamic instruments utilized for minimally invasive surgical procedures, in robotics, in cell phone antennas, and to automatically open and close greenhouse windows. A common example is Nitinol. Its abbreviation is **SMA**.

shaping circuit Same as **shaping network**.

shaping network An electric network inserted into a circuit to enhance certain characteristics, such as its impedance properties. Also called **shaping circuit**, or **corrective network**.

share-level security Security in which a network or network resource, such as a database or printer, requires a password, as opposed to **user-level security**, in which each user must be authenticated before utilizing a network or resource. Share-level security is appropriate, for instance, when all authorized users have the same privileges.

shared-channel interference Also called **cochannel interference**. **1.** Interference arising from two or more signals of the same type being transmitted via the same communications channel. **2.** Interference arising from two or more simultaneous transmissions over the same communications channel.

shared data Data, such as that contained in files, which can be accessed by two or more programs, nodes, terminals, users, or the like. This may occur consecutively or simultaneously. Also, data enabled for such sharing.

shared directory A directory, such as a network directory, which can be accessed by two or more nodes, terminals, users, or the like. This may occur consecutively or simultaneously. Also, a directory enabled for such sharing.

shared drive A disk drive which is utilized by two or more terminals, nodes, users, or the like. This may occur consecutively or simultaneously. Also, a drive enabled for such sharing.

shared Ethernet An Ethernet network in which all nodes must compete for a portion of the total available bandwidth, as opposed to **switched Ethernet**, in which the full bandwidth is available to each pair of connected nodes.

shared file A file which can be accessed by two or more nodes, terminals, users, or the like. This may occur consecutively or simultaneously. Also, a file enabled for such sharing.

shared file system A system enabling or implementing the use of **shared files**. In such a system all participating nodes may see the same set of available files, thus making files transfers unnecessary. Its abbreviation is **SFS**.

shared folder A folder which can be accessed by two or more programs, nodes, terminals, users, or the like. This may occur consecutively or simultaneously. Also, a folder enabled for such sharing.

shared logic A computer whose processing is shared by two or more terminals, nodes, or the like. This may occur consecutively or simultaneously.

shared medium A data medium, such as a disk, or a data transmission medium, such as a cable, which is utilized by two or more terminals, nodes, users, or the like. This may occur consecutively or simultaneously.

shared memory Memory, such as RAM, which can be accessed by two or more resources, programs, systems, terminals, users, or the like. This may occur consecutively or simultaneously.

shared memory architecture A system architecture which allows for **shared memory**. Used, for instance, to enable a video card to use of part of the main memory, as opposed to having its own RAM. Its abbreviation is **SMA**. Also called **unified memory architecture**.

shared printer A printer which can be accessed by two or nodes, terminals, users, or the like. This may occur consecutively or simultaneously. Also, a printer enabled for such sharing.

shared resource A resource, such as specific data or a peripheral, which can be accessed by two or more programs, nodes, terminals, users, or the like. This may occur consecutively or simultaneously. Also, a resource enabled for such sharing.

shared variable A variable which can be accessed by two or more programs, routines, or the like. This may occur consecutively or simultaneously.

shared whiteboard A whiteboard which supports use by two or more users at different locations, as may occur, for instance, during real-time conferencing. Also, a whiteboard so shared.

Shared Wireless Access Protocol A protocol utilized for wireless networking in a home, employing cordless telephone and WLAN technologies. Utilized for short-range telephonic and data communications, such as those between computers and properly equipped devices such as cell phones, PDAs, and smart applicances. Its acronym is **SWAP**.

shareware Software that is distributed at no cost for a stated trial period. Once this time, such as 30 or 90 days, has passed, nag screens may start to appear, or access to the program or given features may become unavailable unless a payment is made. Frequently, after payment and registra-

tion, a user in entitled to technical support, enhanced features, and other benefits.

sharing violation An attempt to access a file that is already in use by another application or process. A sharing violation would not occur if said file is enabled for use by two or more nodes, terminals, users, or the like.

sharp pulse A pulse, such as a spike, of short or very short duration.

sharpness The extent to which a component, circuit, device, piece of equipment, or system is characterized by selectivity or resolution.

sheath 1. A protective covering, case, or enclosure, especially for a blade or blade-like object. 2. A protective outer covering for a cable. Depending on the intended use for the cable, and the operational environment, it may protect against moisture, abrasion, magnetic fields, radiation, and so on. Some cables have more than one sheath. Also called **cable jacket**, or **cable sheath**. 3. In a gas tube, a layer of electrons which surrounds the anode when its current is high. Also called **anode sheath**.

sheet-fed scanner Same as **sheetfed scanner**.

sheet feeder An apparatus or mechanism which feeds sheets into a printer, fax, or similar device.

sheet resistance The resistance of a thin-film resistive material, such as a metal or semiconductor. It is typically measured utilizing a four-point probe, and expressed in ohms per square.

sheetfed scanner An optical scanner which only allows individual sheets to be scanned, as opposed, for instance, to a **flatbed scanner**, which provides a flat surface upon which books, or other thick objects, may be placed. Also spelled **sheet-fed scanner**.

shelf life Also called **storage life**. 1. The maximum length of time that a component, circuit, device, piece of equipment, system, or material can be stored and still retain its performance characteristics. The light, temperature, moisture, electrical fields, magnetic fields, vibrations, and other surrounding conditions may affect this interval considerably. In the case of batteries, for instance, those with nickel cadmium cells can self-discharge in a few weeks, while alkaline batteries may retain most of their charge for several years. 2. A **shelf life** (1) as stated by a manufacturer.

shelfware 1. Software, or software licenses, which are bought but not subsequently utilized. Such programs may wind up on a forgotten shelf, hence the name. 2. Software that is sold through ordinary retail channels. Such programs may be displayed on a shelf, hence the name.

shell 1. Any of the several orbits around the nucleus of an atom, in which electrons may be found. Each successive orbit further from the nucleus has greater energy. Each shell may only contain a specific number of electrons, and those in the same shell have the same energy. Electrons in the outer shell produce a net transport of electric charge under the influence of an electric field. Also called **electron shell**. 2. The user interface for a command interpreter, which is a part of a computer's operating system that accepts a given number of commands for the performance of programmed tasks. Also called **command shell**. 3. A hard outer case, such as that of an audio or video cassette. 4. The outer layer or housing of something, such as the envelope of an electric lamp or electron tube.

shell account An Internet account in which a command line is utilized to give commands, as opposed to the use of a graphical user interface. Lynx, for instance, is a text-based Web browser, and can provide access for such an account.

shell out To temporarily access the operating system while using a given application, as opposed to exiting the application, performing operating system functions, and re-opening the application.

shell script A text file containing a script to be executed or carried out by the **shell (2)** of an operating system, especially that which is UNIX-based.

SHF Abbreviation of **super high frequency**.

shield 1. An enclosure, housing, wall, panel, sheet, projection, or other structure or entity which serves to protect from something. 2. Same as **shielding (1)**. 3. Same as **shielding (2)**. 4. A wall or other housing, usually of concrete and/or lead, built around a nuclear reactor to help prevent radiation from escaping. Also, such a shield placed around any source of radiation. Also called **shielding (3)**.

shield box A box or similar enclosure which serves to **shield** that which is placed inside. Also called **shielding box**.

shield braid A usually flexible sheath that serves to shield the inner wires it encloses. It may consist, for instance, of helically interwoven metal or fiber filaments which also provide structural support, and which may be used as a conductor or for grounding. Also called **shielding braid**.

shield can A can or similar enclosure which serves to **shield** that which is placed inside. Also called **shielding can**.

shield disk A disk, such as that protecting from heat, which serves as a **shield**. Also called **shielding disk**.

shield enclosure Also called **shielded enclosure**. 1. An enclosure, such as a shield box, which serves to **shield** that which is placed inside. 2. Same as **shielded room**.

shield grid Also called **shielding grid**. 1. A grid, such as that in a shielded room, which provides **shielding**. 2. In an electron tube, a grid that protects the control grid.

shield panel A panel, such as a baffle, which serves as a **shield**. Also called **shielding panel**.

shield partition A partition, such as a baffle, which serves as a **shield**. Also called **shielding partition**.

shield plate A plate, such as that in an amplifier, which serves as a **shield**. Also called **shielding plate**.

shield room Same as **shielded room**.

shielded Protected by, or otherwise incorporating a **shield** or **shielding**.

shielded-arc welding Arc welding in which a protective atmosphere or flux is utilized.

shielded cable A cable incorporating one or more conductors which are **shielded**.

shielded enclosure Same as **shielded room**.

shielded line A transmission line, such as a shielded twisted pair or coaxial cable, which is **shielded**. Also called **shielded transmission line**.

shielded pair A transmission line consisting of two similar conductors which are each **shielded**. The protective sheath is usually metallic.

shielded room A room or similar enclosure which blocks the effects of an electric field, while allowing free passage to magnetic fields. A shielded room usually consists of a network of parallel wires which provides a low-resistance path to ground. Used, for instance, for testing emissions and susceptibility of components, circuits, devices, equipment, systems, materials, and the like. Also called **shielded enclosure**, **screen room**, **screening room**, **shield room**, **shielding room**, **shield enclosure**, **Faraday cage**, **Faraday shield**, or **Faraday screen**.

shielded transmission line Same as **shielded line**.

shielded twisted pair A transmission line consisting of two similar conductors which are twisted around each other and **shielded**. The protective sheath is usually metallic, and provides additional protection from interference. Its abbreviation is **STP**.

shielded wire A wire, such as that protected by a shield braid, which is **shielded**.

shielding 1. A material or enclosure which blocks the effects of an electric field, while allowing free passage to magnetic fields. It may consist, for instance, of a wire mesh or screen which provides a low-resistance path to ground. Utilized, for example, to prevent interaction between circuits, or to enhance the directivity of an antenna. Another example is a metallized plastic bag which helps protect circuit boards against electrostatic discharges during storage and handling, en route to being installed. Such a bag does not have to be grounded. Also, a shielding which serves to confine an electric field within an enclosure, or the use of such a shielding. Also called **shield (2)**, **screening (1)**, **screen (2)**, **electrostatic shielding**, or **electric shielding**. 2. A material or enclosure utilized to isolate a magnetic field from its surroundings, or vice versa. Materials with high magnetic permeability are used for this purpose. Also called **shield (3)**, **screening (2)**, **screen (3)**, or **magnetic shielding (1)**. 3. Same as **shield (4)**.

shielding box Same as **shield box**.

shielding braid Same as **shield braid**.

shielding can Same as **shield can**.

shielding disk Same as **shield disk**.

shielding effectiveness The extent to which **shielding (1)** or **shielding (2)** blocks the effect of fields. It may be expressed, for instance, as the ratio of the field strength without the material or enclosure, to that with said material or enclosure.

shielding enclosure Same as **shielded enclosure**.

shielding grid Same as **shield grid**.

shielding panel Same as **shield panel**.

shielding partition Same as **shield partition**.

shielding plate Same as **shield plate**.

shielding room Same as **shielded room**.

shift 1. To move from one place to another, or from one position to another. Also, such a shift. 2. The movement of one or more characters or words by a given number of positions to the left or right within a register or memory location. Such a shift may result in a loss of characters or data. 3. Same as **shift key**.

shift+click To press a computer mouse button while holding down the shift key. Used, for instance, to extend a selection of consecutive items from the location of the cursor to the location where a shift+click is made. In this context it is better suited for consecutive items, while a **control+click** is utilized for non-consecutive items.

shift+clicking To perform **shift+click** operations, such as selections.

shift key A modifier key included on computer keyboards that is frequently used in combination with other keys to produce uppercase characters, although the specific function for any given key combination will depend on which program is running. The key may also be held down while clicking on a mouse, or other pointing device, for alternate functions. Its symbol is often ↑. Also called **shift (3)**.

shift pulse A pulse which initiates a **shift (1)** or **shift (2)**.

shift register A high-speed register within which all bits are moved a fixed number of positions to the left or right at each clock cycle. Used, for instance, for multiplications or divisions, timing, or frequency conversions.

shim A thin object or material which is utilized to fill gaps, level, or otherwise help fit properly. Also, to fill or adjust using shims.

shimming The adjustment of a magnetic field by placing, removing, or relocating shims composed of magnetic materials or coils. Used, for instance, in NMR optimization.

ship-shore communications Same as **ship-to-shore communications**.

ship station   Within a mobile radio service, such as a maritime mobile service, a station located on a ship. Such a station is usually understood to be on a ship that is not permanently moored.

ship-to-ship communications   Communications between ships at sea.

ship-to-shore communications   Communications between ships at sea and land stations. Also called **ship-shore communications**, or **shore-to-ship communications**.

SHM   Abbreviation of **short-haul modem**.

shock   **1.** The effect of a passage of current through living tissue. The effects of a shock can range from mild tingling to death, depending on the strength of the current, the points where it enters and leaves the body, whether the heart is along its path, if the skin is wet, and so on. Under many circumstances, a current of 0.1 ampere for 1 second may be fatal. Electric shocks may also be utilized for therapeutic reasons, such as restoring the normal rhythm to a heart which is twitching uncontrollably. Also called **electric shock**. **2.** A rapid pulse, signal, acceleration, vibration, acceleration, impact, or other such event which is applied to a component, circuit, device, piece of equipment, system, material, or the like. **3.** Any disturbance or other change due to a **shock (2)**.

shock absorber   A material, device, or mechanism which helps absorb or otherwise minimize physical shocks such as jolts or vibrations. Used, for instance, to protect susceptible devices, equipment, or systems.

shock excitation   The production of oscillation in a circuit via a signal of very short duration. The time interval of the resulting oscillation is much greater than the duration of the exciting impulse. Also called **impulse excitation**

shock-excited   That which is actuated via **shock excitation**.

shock hazard   A component, circuit, device, piece of equipment, system, material, location, procedure, or circumstance which could lead to a **shock (1)**.

shock mount   A mount which incorporates shock absorbers to protect against physical shocks and/or unwanted acoustic energy.

shock therapy   Electric shock utilized for therapeutic purposes. Used, for instance, to restore the normal rhythm to a heart which is twitching uncontrollably.

shock wave   A large-amplitude compression wave created by an object moving through a fluid at a speed which exceeds the speed of sound for said fluid, as occurs, for example, when an airplane breaks the sound barrier. Shock waves may also be created by an explosion or other violent disturbance in a fluid, as is the case with lightning. The sound heard as result of the shock waves emanating from an object or medium is called **sonic boom**. Also spelled **shockwave**.

Shockley diode   A semiconductor diode with four layers of semiconductor material, thus having three junctions. It has two terminals, one connected to each of the outer layers. Also called **four-layer diode**.

Shockwave   A technology utilized for the presentation of multimedia over Web pages. Also, a format, plug-in, player, file, or presentation using such a technology.

shockwave   Same as **shock wave**.

shopping cart   **1.** A virtual location where items selected for purchase are held for reference, while effecting shopping over the Internet. Before payment, the items in a shopping cart may be reviewed, replaced, removed, modified, and so on. **2.** A program enabling the use of a **shopping cart (1)**.

shoran   Acronym for **short-range navigation**. Navigation utilizing aids which have a comparatively short range. For instance, that utilizing short-range radar.

shore effect   Same as **shoreline effect**.

shore-to-ship communications   Same as **ship-to-shore communications**.

shoreline effect   A change in direction of a radio wave as it crosses a shoreline. It usually only affects ground waves. Also called **shore effect**, **land effect**, or **coastal bending**.

short   **1.** Abbreviation of **short circuit**. **2.** Abbreviation of **short-circuit**.

short bar   Same as **shorting bar**.

short card   A plug-in circuit board which is half the length of a standard, or full-length plug-in circuit board. Also called **half-card**.

short circuit   A low-resistance connection which is established between two points in a circuit, bypassing any paths with higher resistance. Also, such a connection between the conductors of a transmission line. Such a short may be intentional, but when accidental or otherwise unintentional usually causes damage due to excessive current flow. Its abbreviation is **short**.

short-circuit   To establish or cause a **short circuit** connection. Its abbreviation is **short**.

short-circuit current   **1.** The current flowing under **short circuit** conditions. **2.** The current flowing through a circuit which has no little or no resistance or load. This would occur, for instance, when directly connecting the positive and negative terminals of a cell via a low-resistance conductor.

short-circuit impedance   In a four-terminal electric network, the impedance at one end when the terminals at the other end are short-circuited. When the terminals at the far end are open, it is called **open-circuit impedance**.

short-circuit resistance   In a four-terminal electric network, the resistance at one end when the terminals at the other end short-circuited open. When the terminals at the far end are open, it is called **open-circuit resistance**.

short-circuiting bar   Same as **shorting bar**.

short-circuiting switch   A switch that makes a new connection prior to breaking the previous one. For instance, the moving contact of such a switch may be wider than the distance between fixed contacts. Also called **shorting switch**, or **make-before-break switch**.

short-haul   Of, pertaining to, or limited to comparatively short distances, while **long-haul** relates to relatively long distances. An example is the operational radius of a short-haul modem.

short-haul modem   A device which is intended to optimize the use of local data communications facilities, such as those within a building or with a maximum radius of a few kilometers. Such a device, for instance, can condition a digital signal transmitted by an RS232 interface so that it can be reliably transmitted up to several kilometers, instead of the standard 50 to several hundred feet. Also called **line driver**, or **limited-distance modem**.

short messaging   Same as **SMS messaging**.

short messaging service   Same as **SMS**.

short-range navigation   Same as **shoran**.

short-range radar   A radar whose maximum line-of-sight range is relatively short. The defined interval varies, but may be, for instance, up to 20 kilometers, or up to 200 kilometers.

short skip   In ionospheric propagation, a reflection of comparatively short distance. Such a reflection is usually less than 1000 kilometers, while a **long skip** may be several thousand kilometers.

short-term   Involving or lasting a comparatively short time interval.

short-term drift   Drift which occurs over a comparatively short time interval, as opposed to **long-term drift**, which takes longer.

short-term effect   An effect, such as short-term instability, lasting a comparatively short time interval.

short-term instability The instability of a component, circuit, device, piece of equipment, system, setting, or value, over a comparatively short time interval.

short-term stability The stability of a component, circuit, device, piece of equipment, system, setting, or value, over a comparatively short time interval.

short-time rating A rating which defines the maximum load that a device, piece of equipment, or system can carry for periods not exceeding a given interval, without harmful effects, such as exceeding a given increase in temperature. This contrasts with a **continuous rating**, which is that which can be carried for an indefinite period.

short-wave Same as **shortwave**.

short-wave antenna Same as **shortwave antenna**.

short-wave broadcast Same as **shortwave broadcast**.

short-wave broadcasting Same as **shortwave broadcasting**.

short-wave converter Same as **shortwave converter**.

short-wave radio Same as **shortwave radio**.

short-wave radio receiver Same as **shortwave radio** (1). Also spelled **shortwave radio receiver**.

short-wave receiver Same as **shortwave radio** (1). Also spelled **shortwave receiver**.

short-wave transmitter Same as **shortwave transmitter**.

shortcut A key, key combination, icon, or file which provides quick and simple access to a program, function, file, device, or the like. For instance, double-clicking on an icon which represents a given file will open said file, along with the application utilized to access it.

shortcut key A key, such as F3, or combination of keys, such as ALT+SHIFT+T, which when pressed executes an action or series of actions within a program or operating system. Such shortcuts can save time or automate sequences. Also called **application shortcut key**, or **keyboard shortcut**.

Shortest-Path First Algorithm An algorithm utilized for calculating network routing, using comprehensive routing tables. Its abbreviation is **SPF**, or **SPF Algorithm**. Also called **Dijkstra's Algorithm**.

shorting bar A bar, composed of a low-resistance conductor, which is placed between two points, such as binding posts, which wish to be **short-circuited**. Also called **short bar**, or **short-circuiting bar**.

shorting link A link, such as a shorting bar, utilized to short-circuit.

shorting switch Same as **short-circuiting switch**.

shortwave A wave whose frequency lies between 3 and 30 MHz, corresponding to wavelengths of 100 to 10 meters, respectively. These are decametric waves, and comprise the high frequency band. There may be other defined bands for shortwaves, such as approximately 1.5 to 30 MHz, approximately 1.6 to 30 Mhz, approximately 5.95 to 26.10 Mhz and so on. Also spelled **short-wave**. Its abbreviation is **SW**.

shortwave antenna An antenna specially designed for reception and/or transmission of **shortwaves**. Also spelled **short-wave antenna**.

shortwave broadcast A radio transmission intended for public or general reception utilizing **shortwaves**. Also spelled **short-wave broadcast**.

shortwave broadcasting The transmission of radio signals intended for public or general reception utilizing **shortwaves**. Also spelled **short-wave broadcasting**.

shortwave converter A frequency converter utilized to change a **shortwave** input frequency into an output of a different frequency. Used, for instance, to enable pickup by radio receivers which otherwise could not. Also spelled **short-wave converter**.

shortwave radio Also spelled **short-wave radio**. **1.** A radio receiver that can receive **shortwave** signals. Also called

shortwave radio receiver, or **shortwave receiver**. **2.** The broadcasting and reception of **shortwaves**.

shortwave radio receiver Same as **shortwave radio** (1). Also spelled **short-wave radio receiver**.

shortwave receiver Same as **shortwave radio** (1). Also spelled **short-wave receiver**.

shortwave transmitter A radio transmitter that radiates **shortwaves**. Also spelled **short-wave transmitter**.

shot effect Same as **shot noise**.

shot noise Also called **shot effect**, or **Schottky noise**. **1.** A random form of noise arising from the discrete nature of electrons. Such noise may occur, for instance, as electrons randomly overcome a given potential barrier. In the specific case of a thermionic emitter, shot noise results in fluctuations in particle emissions, while in an electron tube or transistor it causes variations in the flow of electrons between electrodes. **2. Shot noise** (1) observed in junction devices, such as pn diodes or bipolar junction transistors.

shotgun microphone A highly directional microphone in the form of a straight line. Such devices may be used alone, or as part of an array. Also called **line microphone**.

shoulder surfing The use of visual means, such as looking over a person's shoulder or viewing through binoculars, to obtain confidential information such as passwords and entry codes. In the case of ATMs, for instance, simply getting close to the console and cupping the other hand around the keypad when entering digits can be enough to thwart such measures.

shout To use all capital letters when communicating via email, messaging, chat, or the like. It is more respectful and generally nicer, for instance, to emphasize using asterisks, as in *really*, as opposed to REALLY.

shovelware Software, usually consisting of freeware and/or shareware, which is gathered from any existing sources, and put into packages, such as those available on optical discs, with little or no regard for selectivity, usefulness, or cohesion.

shrink DIP Abbreviation of **shrink d**ual **i**n-line **p**ackage. A dual in-line package that is usually longer and wider, and which provides a higher pin count. Its own abbreviation is **SDIP**.

shrink dual in-line package Same as **shrink DIP**.

shrink plastic A plastic material which shrinks in diameter when heated. Such plastics may be available in any of various forms, including tape, tubes, strips, bands, sleeves, and so on. Used, for instance, in heat-shrink tubing. Also called **shrinkable plastic**, or **heat-shrinkable plastic**.

shrink sleeve A sleeve made of a plastic material which shrinks in diameter when heated, providing a snug fit around the cables on each side of a splice, or wherever it is used. Also called **shrinkable sleeve**, or **heat-shrink sleeve**.

shrink tubing Tubing made of a plastic material which shrinks in diameter when heated, providing a snug fit. Such tubing may shrink to about half of its original diameter, and when cooled does not expand. Used, for instance, to protect cable splices from dust and/or moisture. Also called **shrinkable tubing**, or **heat-shrink tubing**.

shrink-wrap Same as **shrink-wrapped software**.

shrink-wrapped Same as **shrink-wrapped software**.

shrink-wrapped software Software that is mass-produced and available for purchase. Such programs may be bought through retailers, such as Internet merchants, and can be shipped in a plastic-wrapped box, or may be downloaded. Also called **shrink-wrap**, or **shrink-wrapped**.

shrinkable plastic Same as **shrink plastic**.

shrinkable sleeve Same as **shrink sleeve**.

shrinkable tubing Same as **shrink tubing**.

SHTML Abbreviation of **server-parsed HTML**. An HTML document or file which incorporates information which is modified by a server before being sent to a Web page being browsed at the moment. Such pages may appear slightly differently each time viewed. SHTML allows the server side to include additional or alternate information, such as the date of the most recent update or the inclusion of selected text, before being sent to the client side.

SHTTP Abbreviation of **Secure HyperText Transport Protocol**. An HTTP protocol utilized for effecting secure transactions, and for the transmission of secure messages, using mechanisms such as encryption and digital signatures.

shunt **1.** Characteristic of, or pertaining to a **shunt circuit** or **shunt connection**. Also, such a circuit or connection. **2.** A low-impedance path which allows a current to flow around one or more circuits or components, as opposed to through it. This path may be intentionally or unintentionally created. Also, to create such a path. Also called **bypass (1)**. **3.** A conductor, component, or other object which creates a **shunt (2)**. **4.** A resistor which is connected in parallel, to extend the current range of an instrument. For example, such a shunt utilized to extend the range of an ammeter. In addition, such a resistor helps provide protection against current surges. Also called **instrument shunt**. **5.** To turn or otherwise move to an alternate course or path. **6.** A ferromagnetic bar or other object, especially that composed of soft iron, utilized to divert magnetic flux. Examples include magnetic shunts and magnet keepers.

shunt capacitance The total capacitance of **shunt capacitors**. Also called **parallel capacitance**.

shunt capacitors Two or more capacitors connected in shunt. The total capacitance is the sum of each of the individual capacitances, thus, the multiple capacitors act like a single larger capacitor. Also called **parallel capacitors**.

shunt circuit A circuit whose components are connected in shunt. That is, they have a **shunt connection**. Also called **parallel circuit**.

shunt connection The connection of the components within a circuit in a manner that there are multiple paths among which the current is divided, while all the components have the same applied voltage. An example is the connection of components across each other. This contrasts with **series connection**, where components are connected end-to-end, and in which there is a single path for the current. Also called **parallel connection**.

shunt-fed That which has a **shunt feed**. Also, pertaining to that which has a shunt feed. Also called **parallel-fed**.

shunt feed The simultaneous application of a DC operating voltage and an AC signal voltage to a component or device, each via a different path. Also called **parallel feed**.

shunt generator An electric generator whose armature and field windings are connected in shunt. Also called **shunt-wound generator**, or **parallel generator**.

shunt impedance An impedance acting in shunt with another quantity or magnitude, such as another impedance. Also called **parallel impedance**.

shunt inductance An inductance acting in shunt with another quantity or magnitude, such as a capacitance. Also called **parallel inductance**.

shunt inductors Two or more inductors connected in shunt. When said inductors are sufficiently shielded or apart from each other, the magnetic inductance is considered to be negligible. Otherwise, such inductance, or coupling, must be factored in. Also called **parallel inductors**.

shunt motor An electric motor whose armature and field windings are connected in shunt. Also called **shunt-wound motor**, or **parallel motor**.

shunt network An electric network whose components are connected in shunt. That is, they have a **shunt connection**. Also called **parallel network**.

shunt neutralization A method of neutralizing an amplifier, in which capacitance cancels inductance in the feedback circuit. Also called **inductive neutralization**, or **coil neutralization**.

shunt operation Operation of a circuit or device whose components are connected in shunt. Also called **parallel operation (3)**.

shunt regulator A regulator, such as a voltage regulator, which is connected in shunt with a device or load. Also called **parallel regulator**.

shunt resistors Two or more resistors connected in shunt. The total resistance is the sum of the reciprocal of each of the individual resistors, thus, the multiple resistors have a lower total resistance than the lowest valued resistor. Also called **parallel resistors**.

shunt T junction In a waveguide, a T junction in which the structure changes along the plane of the magnetic field. Also spelled **shunt tee junction**. Also called **H-plane T junction**.

shunt tee junction Same as **shunt T junction**.

shunt winding In an electric motor or generator, an armature winding connected in shunt with the field winding. Also called **parallel winding**.

shunt-wound generator Same as **shunt generator**.

shunt-wound motor Same as **shunt motor**.

shunting The creation, incorporation, or utilization of a shunt.

shut down **1.** To exit a program in an orderly manner, as opposed, for instance, to a crash. Also called **quit**. **2.** To exit all programs in use, and turn off a computer system. **3.** To properly turn off or stop a component, device, piece of equipment, system, process, or mechanism.

Si Chemical symbol for **silicon**.

SI Abbreviation of *Système International d'Unités*. French for **International System of Units**. A system utilized for measurement of physical quantities using internationally accepted fundamental units which are defined in an absolute manner. Currently, the base, or fundamental, SI units are: the **second**, for time; the **kilogram**, for mass; the **meter**, for distance; the **ampere**, for electric current; the **Kelvin**, for temperature; the **mole**, for amount of substance; and the **candela**, for luminous intensity. In addition, there are other units defined algebraically in terms of these base units. These include hertz, joule, watt, coulomb, volt, and ohm. Furthermore, the SI allows the use of certain non-approved units, such as electronvolt, bel, and angstrom.

SI units Any of the units defined within the **SI** system.

SIC Abbreviation of **specific inductive capacity**, **specific inductive capacitance**, or **specific inductive capacitancy**.

side effect An undesired or indirect consequence caused by an operation, expression, or subroutine. For instance, a change to a variable caused by a function not working directly on it.

side frequency Same as **sideband frequency**.

side lobe Same as **sidelobe**.

side-lobe suppression Same as **sidelobe suppression**.

side-scanning sonar Same as **scanning sonar**.

sideband Its abbreviation is **SB**. **1.** The frequencies above or below a carrier wave. These bands result from the effects of the modulation process, and the frequencies above said carrier's center frequency comprise the upper sideband, while those below are the lower sideband. Sidebands contain the meaningful information which wishes to be conveyed, and when only one of the two sidebands is transmitted it is called single-sideband transmission. During such transmissions, the carrier and the other sideband are usually suppressed. **2.** The wave components of one or both **sidebands (1)**. Also called **sideband components**. **3.** A band of frequencies above or below a center frequency. These are the upper and the lower sidebands, respectively. **4.** Pertaining to a **sideband (1)** or **sideband (3)**.

sideband attenuation The attenuation of one or both **sidebands**.

sideband components Same as **sideband (2)**.

sideband cutting **1.** The attenuation or suppression of either the upper or the lower **sideband**. **2.** The attenuation of both sidebands.

sideband frequency A frequency within a **sideband**. The lower sideband, for instance, incorporates frequencies which are the difference between the carrier frequency and the modulation frequencies. Also called **side frequency**.

sideband interference Interference, such as sideband splatter, caused by one or both **sidebands**.

sideband power The power contained in both **sidebands**.

sideband splash Same as **sideband splatter**.

sideband splatter Interference of radio or TV reception by other stations using adjacent channels. Such interference may be unilateral or bilateral, and is caused, for instance, by overmodulation. It may be manifested as extraneous confusing, musical, or repetitive sounds being heard along with the desired program. Also known as **sideband splash, splatter, monkey chatter** or **adjacent-channel interference**.

sideband suppression The suppression of either the upper or the lower **sideband**.

sideband transmission A radio transmission in which one or both **sidebands** are transmitted.

sidelobe Within an antenna pattern, any lobe other than the main lobe. Also spelled **side lobe**. Also called **secondary lobe, secondary beam, minor lobe**, or **minor beam**.

sidelobe suppression The suppression of any **sidelobes** present. Also spelled **side-lobe suppression**.

sidetone **1.** The reproduction by the receiver of sounds from the same telephone's transmitter. Although a caller's own voice is usually heard over the same telephone, if the amplitude of this signal exceeds a given threshold it is usually unwanted. Also called **telephone sidetone**. **2.** An attenuated portion of transmitted audio which is returned to the sender. This signal may be desired or undesired.

sidetone adjustment The increasing or decreasing of the sidetone level, utilizing a sidetone circuit.

sidetone circuit A circuit which varies the **sidetone level**.

sidetone level The level of **sidetone** present in a signal.

siemens Its symbol is **S**. Formerly called **mho**, or **reciprocal ohm**. **1.** The SI unit of conductance. A conductor or circuit has a conductance of one siemens when it carries one ampere per volt. **2.** A unit of susceptance. **3.** A unit of admittance.

sifting sort A sorting technique in which adjacent pairs of items within a list are sequentially compared and ordered, until the entire list has been evaluated. This process is then repeated as many times as required to obtain the correct order. Also called **bubble sort**, or **exchange sort**.

sigma (σ) **1.** Symbol for **surface charge density**. **2.** Symbol for **conductivity**.

sign **1.** Anything, such as a symbol, action, condition, state, or the like, intended to convey information. **2.** A symbol, such as + or **x**, used in mathematics to denote an operation. **3.** The positive or negative nature of a number. Also, the symbols, + and -, which indicate this nature. **4.** To affix or append a signature, such as that in an email.

sign bit A bit indicating the sign of a number, signal, or other value.

sign-in Same as **sign-on**. Also spelled **signin**.

sign-off Same as **sign-out**. Also spelled **signoff**.

sign-on To initiate a session. Also, the process of initiating a session. This may require, for instance, entering a username and the corresponding password. Also spelled **signon**. Also called **sign-in**, or **log-on**.

sign-out To end a session. Also, the process of terminating a session. One or more commands may be required for proper termination. Also spelled **signout**. Also called **sign-off**, or **log-out**.

signal **1.** Anything, such as a light, sound, location, or movement, which serves to convey information. **2.** A varying quantity, state, or parameter which serves to convey information. For example, a fluctuating amplitude, frequency, or waveform, a series of tones, flashing lights, field strength variations, changes in pulse duration, and so on. Used, for instance, in communications, for sending control signals, to activate processes, to alert to a given condition, and so on. **3.** A fluctuating electrical quantity, such as a current or voltage, which serves as a **signal (2)**. **4.** The sounds, images, data, or other information transmitted and/or received in communications, radio, TV, radar, and so on. **5.** A specific **signal (1)** or **signal (2)**, such as a busy signal, a carrier signal, or a chrominance signal.

signal amplitude The amplitude of a given **signal**. For example, the displacement from a reference position, such as zero, of an amplitude-modulated wave.

signal attenuation The attenuation of a given **signal**. For instance, that of a sideband.

signal averaging A technique utilized to improve signal-to-noise ratios, in which signals are averaged to help extract a greater proportion of the desired information.

signal bandwidth The bandwidth of a given **signal**. For example, that utilized to transmit a TV signal.

signal bell A bell, such as a ringer in an older telephone, which serves to **signal**.

signal bit A bit which serves to **signal**. For instance, a bit which provides a start signal.

signal booster A component, circuit, or device, such as a booster amplifier or preamplifier, which increases, amplifies, or reinforces a signal.

signal cable A cable, such as a coaxial cable, through which a **signal** is transmitted.

signal carrier A signal which is modulated in frequency, amplitude, or phase, in order for it to carry information. For instance, an AM radio transmitter modulates the amplitude of a carrier signal. Also called **carrier signal**.

signal channel **1.** A channel, such as a radio or TV channel, via which a **signal** is sent. **2.** A channel along which a given type of signal is sent, as opposed to another. For instance, one carrying control signals as opposed to a conversation.

signal circuit **1.** A circuit, such as a communications circuit, through which a **signal** is sent. **2.** A circuit through which a given type of signal is sent, as opposed to another. For instance, one carrying control signals as opposed to a conversation.

signal code A code, such as that utilizing varying frequencies, employed to convey meaningful information via a **signal**.

signal component A component within in a complex **signal**.

Signal Computing System Architecture Same as **SCSA**.

signal conditioner A circuit or device which serves for **signal conditioning**.

signal conditioning **1.** The processing of a signal to improve its intelligibility, signal-to-noise ratio, or the like. **2.** The processing or conversion of a signal, so as to make it compatible with a given channel, device, or system. For example, line conditioning. Also called **signal conversion (1)**.

signal conversion **1.** Same as **signal conditioning (2)**. **2.** The conversion of a signal from one form, such as analog, to another, such as digital.

signal-conversion equipment A device or piece of equipment which serves for **signal conversion**.

signal converter A component, circuit, device, piece of equipment, or system, which serves for **signal conversion**.

signal current The electrical current of a **signal**. Such a current, for instance, may be varied to convey meaningful information.

signal degradation The gradual deterioration of the quality or intelligibility of a **signal**. Also, the results of such degradation.

signal delay The delay of a **signal**, as in the time it takes to proceed between nodes in a network. Also called **signal time delay**.

signal demodulation The process of reversing the effects of modulation, thereby restoring the original modulating signal. Demodulation is utilized for recovering the signal of interest, such as music, images, or data. Also called **demodulation**, or **detection (2)**.

signal device A device, such as a laser, utilized to generate or send a **signal**.

signal diode A diode that is utilized to modulate, demodulate, amplify, or otherwise transmit, receive, or process a **signal**.

signal distance The number of digit positions which are different between two binary words of the same length. For example, the signal distance between 1110001 and 1110000 is one, as the last digit is different. Also called **Hamming distance**.

signal distortion Any undesired change in the intelligibility or quality of a **signal**. For example, an undesired change in the waveform of a signal passing through a circuit, device, or transmission medium. Also, the extent of such changes.

signal duration The time interval of a **signal** or signal element, such as a pulse.

signal edge In a transmitted **signal**, the portion of a pulse wave which first increases or decreases in amplitude. When increasing, it is a leading edge, and when decreasing it is a trailing edge.

signal element A discrete component of a **signal**, such as a pulse within a pulse train.

signal encoder A circuit or device utilized for **signal encoding**.

signal encoding The conversion of a **signal** from one form into another. For instance, a color encoder in a TV transmitter.

signal enhancement The use and effect of a **signal enhancer**.

signal enhancer A component, circuit, device, piece of equipment, process, or system, such as a signal conditioner, which improves the intelligibility or quality of a signal.

signal equipment Equipment, such as communications equipment, utilized to send **signals**.

signal-flow diagram Same as **signal-flow graph**.

signal-flow graph A diagram which uses a set of standard symbols to represent the transmission or flow of a signal through an electric network, device, piece of equipment, or system. Each node in such a diagram represents a parameter, and is connected to others nodes via branches. It is a type of flowchart. Its abbreviation is **SFG**. Also called **signal-flow diagram**.

signal forming Same as **signal shaping**.

signal frequency The frequency, such as that of a carrier, of a **signal**.

signal gain The gain, such as that provided by an amplifier or a radio repeater, provided to a **signal**.

signal generation The process of producing a **signal**. For instance, a circuit whose output is an electrical parameter which can be varied to convey meaningful information.

signal generator A component, circuit, device, piece of equipment, system, or process utilized for **signal generation**. Also called **signal source**.

signal ground In a signal circuit or system, a point which is at zero potential or zero signal potential. Such a ground provides a reference upon which all signaling is based.

signal injection The application of a signal to a circuit or device. Also, the process utilized. Also called **injection (1)**.

signal input An external signal, such as a voltage or current, which drives a circuit, device, piece of equipment, system, process, or mechanism. Also called **input signal**.

signal input level Also called **input signal level**. **1.** The magnitude of a **signal input** which produces a stated output. **2.** The magnitude of a **signal input**.

signal integrity The preservation of the intelligibility or other qualities of a signal that is transmitted, received, or processed in any manner.

signal intelligence The audio, video, text, or other information conveyed in a **signal**. For example, the music utilized to modulate a carrier in an FM signal. Also called **intelligence (4)**.

signal intelligibility The extent to which a received signal is **intelligible**. Also called **intelligibility**

signal intensity Same as **signal strength**.

signal interference Any energy which diminishes the quality of a desired **signal**, such as that which lessens its ability to be received, processed, or understood.

signal inversion A reversal, such as that in the order of, or phase, of a **signal**.

signal lamp Same as **signal light**.

signal leakage An undesired escape of a **signal**. For instance, cracked cables leaking RF energy which interferes with nearby components and devices.

signal level **1.** The strength of a signal at a given point or moment, relative to a reference such as a baseline. **2.** A specific **signal level (1)**, as opposed to another. For instance, those used in amplitude shift keying. **3.** Same as **signal strength**.

signal-level indicator Same as **signal-level meter**.

signal-level meter A device or instrument which indicates **signal levels**. Also called **signal-level indicator**.

signal light Also called **signal lamp**. **1.** A light, such as that whose output is modulated, which serves to send signals. **2.** A light, such as that within a display, which provides a signal. For example, a lamp indicating an electrical fault.

signal line A line, such as a communications line, along which a **signal** travels.

signal loss The reduction of the intensity of a signal traveling through a given component, circuit, device, piece of equipment, system, or medium. Examples include line, free-space, gap, junction, feeder, insertion, and radiation losses. Usually expressed in decibels. Also called **signal losses**.

signal losses Same as **signal loss**.

signal message A message, such as an error message, pertaining to a signal or lack of a signal.

signal meter An instrument or device which measures and indicates **signal levels**.

signal mixer A circuit, device, or system, such as a multiplexer, which combines multiple signals.

signal/noise ratio Same as **signal-to-noise ratio**.

signal output The energy, voltage, current, or other signal delivered or produced by a component, circuit, device, piece of equipment, system, or process. Also, to deliver or provide such a signal. Also called **output signal**.

signal output level Also called **output signal level**. **1.** The magnitude of a **signal output**. **2.** A specific **signal output level (1)**, such as a rating.

signal path The path along which a **signal** travels. For example, the route an electromagnetic wave travels between a transmitter and a receiver.

signal peak **1.** The maximum instantaneous value of a signal. **2.** The maximum instantaneous absolute value of the displacement from a reference position, such as zero, for a signal. **3.** The **signal peak** (**1**), or **signal peak** (**2**) for a given time interval.

signal phase The phase, such as that of a chrominance signal with respect to a burst signal, of a **signal**.

signal power The power, usually expressed in dBs or watts, of a **signal**.

signal processing The processing, such as demodulation, amplification, or conversion, of a **signal**.

signal processor A component, circuit, device, piece of equipment, system, or process which performs **signal processing**.

signal pulse **1.** A pulse, such as that within a pulse train, which conveys a signal. **2.** A signal, such as that which starts a device, consisting of a single pulse.

signal quality **1.** A quality, such as intelligibility, of a signal. Also, a combination of such qualities. **2.** A quality, such as brightness, associated with an audio or visual signal.

signal receiver A device, such as a radio receiver, suitable for **signal reception**.

signal reception Also called **reception**. **1.** The conversion of information-bearing signals, such as those conveyed via electromagnetic waves, into the signals of interest, such as music, images, or data. For example, the use of a receiver to demodulate an FM signal, so as to recover the original music. **2.** The quality or fidelity attained during **signal reception** (**1**).

signal rectification The process of converting an AC signal into a DC signal. Also, the result of such a process.

signal regeneration The process of again generating, or of restoring a signal to its original shape or form.

signal regenerator A circuit or device which is utilized for **signal regeneration**.

signal return The return of a signal back to its source. This occurs in radars, for instance, when a scanned object reflects microwave energy.

signal selector A control or switch that allows the selection of a signal reception and/or transmission mode when there are multiple options. Used, for instance, to transmit at a given frequency or within a specified band.

signal separation The separation of one or more components from a complex signal. An example is demultiplexing.

signal separator A circuit, device, or system, utilized for **signal separation**.

signal shaper A circuit or device which is utilized for **signal shaping**.

signal shaping The altering of one or more characteristics of a signal. For instance, changing a characteristic of a pulse train, so as to convey meaningful information. Also called **signal forming**.

signal-shaping circuit A circuit utilized for **signal shaping**. Also called **signal-shaping network**.

signal-shaping network Same as **signal-shaping circuit**.

signal source Same as **signal generator**.

signal speed The speed, such as that of a modem, at which a signal is received and/or transmitted.

signal station A station, such as a radio beacon, which sends and/or receives signals.

signal strength The intensity of a **signal**. For example, the voltage amplitude of a signal, or the output, in watts, of a radio transmitter. Also called **signal intensity**, or **signal level** (**3**).

signal-strength indicator A device or instrument which indicates **signal strength**. Also called **signal-strength meter**, or **S-meter** (**1**).

signal-strength meter Same as **signal-strength indicator**.

signal system A system, such as pulse-code modulation, utilized to code or send signals.

signal time delay Same as **signal delay**.

signal-to-distortion ratio The ratio of the magnitude of a given parameter of the desired signal, to that of the same parameter for any distortion present. It is usually the ratio of their respective amplitudes, and is expressed in decibels. Its abbreviation is **SDR**.

signal-to-interference ratio The ratio of the magnitude of a given parameter of the desired signal, to that of the same parameter for any interference present. It is usually the ratio of their respective amplitudes, and is expressed in decibels. Its abbreviation is **SIR**.

signal-to-noise Same as **signal-to-noise ratio**.

signal-to-noise-and-distortion ratio The ratio of the magnitude of a given parameter of the desired signal, to that of the same parameter for all undesired energy present combined, including noise and distortion. It is usually the ratio of their respective amplitudes, and is expressed in decibels. Its abbreviation is **SINAD**.

signal-to-noise ratio Its abbreviations are **S/N**, **S/N ratio**, **SNR**, **signal-to-noise**, or **signal/noise ratio**. For a given signal, the ratio of the magnitude of a parameter of the useful or desired signal, to that of the same parameter for any noise present. An S/N ratio may be expressed in many ways, such as the ratio of the signal power of the desired signal, to the noise power, expressed in decibels. The ratio of peak voltages is usually utilized for pulse noise, and RMS values for broadband or random noise. An S/N ratio may be stated for a given point in a circuit or transmission medium, for a given bandwidth, and so on. A higher S/N ratio provides for a better-quality signal, in addition to affording greater noise immunity. Digital recordings and transmissions tend to have higher S/N ratios than equivalent analog recordings or transmissions.

signal tracer A component, circuit, device, or instrument which is utilized to follow a signal through a component, circuit, device, or system, such as a receiver or amplifier. Used, for instance, for testing and troubleshooting.

signal tracing The use of a **signal tracer**.

signal transfer point Within a communications network, such as a telephone network, a point or location, such as a central office, where transfers are made from one signaling link to another. Its abbreviation is **STP**.

signal transmission The conversion of information-bearing signals, such as audio, video, or other content, into electromagnetic waves, or another form, which is transmitted in a form suitable for subsequent **signal reception** (**1**).

signal transmitter A device, such as a radio transmitter, which serves for **signal transmission**.

signal unit A unit, such as a group of bits of a given size, which serves to convey information signals. Its abbreviation is **SU**.

signal voltage The voltage of a **signal**. Unless otherwise specified, it is the RMS voltage value of a given signal.

signal wave A wave, such as an electromagnetic wave emitted by an antenna, which conveys a **signal**.

signal wire A wire, such as that in a twisted pair, through which a **signal** is transmitted.

signaling **1.** The transmission or emission of **signals**, using electrical signals, tones, flashing lights, and so on. **2.** A specific method, technique, or technology for **signaling** (**1**), such as binary signaling, bit robbing, carrier signaling, common-channel signaling, or multi-frequency signaling.

signaling rate The rate, such as a data-signaling rate, at which a signal passes through a given point.

Signaling System 7 Same as **SS7**.

signature Also called **email signature**. **1.** Content, such as a text, image, file, or combination of these, which is appended to an email or message. Examples include favorite quotes, contact information, vCards, or disclaimers. A user may select whether such a signature is included automatically when sending. **2.** An attachment to an electronic transmission which uniquely identifies and authenticates the sender of a message, file, or other information such as a credit card number. To be effective, a digital signature must not be forged, and measures such as public-key cryptography are employed to insure its integrity. Also called **electronic signature**, or **digital signature**.

signature block A block of text appended as a **signature (1)**.

signature file A file appended as a **signature (1)**.

significant digits Same as **significant figures**.

significant figures The precision with which a number is expressed, based on the number of digits that are known with a given degree of confidence. For instance, *3.141592* is pi expressed to seven significant digits, with the last digit being rounded. Each of the following has four significant digits: *5,016, 7.348 x 10⁶, 0.003502*, and *142,100*. When performing mathematical calculations, such as addition or division, the precision, in significant digits, of the final answer can be no greater than that of the least accurate figure. Also called **significant digits**.

signin Same as **sign-on**. Also spelled **sign-in**.

signoff Same as **sign-out**. Also spelled **sign-off**.

signon Same as **sign-on**.

signout Same as **sign-out**.

silence suppression The removal of pauses in speech or other silence when recording or transmitting.

silencer A circuit or device, such as a noise limiter, which attenuates noise.

silent zone Same as **skip zone**.

silica Same as **silicon dioxide**.

silica gel A gelatinous form of silica utilized as a drying agent. Used, for instance, to help protect devices and instruments from moisture.

silicide A compound in which silicon is an anion. Examples include cobalt silicide, magnesium silicide, and tungsten silicide.

silicon A nonmetallic chemical element whose atomic number is 14, and which has two allotropic forms. One form is a brown powder, and the other consists of dark gray crystals with a bluish tinge. Silicon is very common in nature, comprising about 25% of the earth's crust by weight, and is usually found in the form of silicon dioxide. It has over 15 known isotopes, of which 3 are stable. It is used extensively in semiconductors such as transistors, diodes, rectifiers, and photocells, as an alloying agent, and in optics. Its chemical symbol is **Si**.

silicon avalanche diode A silicon diode with a very high ratio of reverse-to-forward resistance, until avalanche breakdown occurs. After this, voltage drop is nearly constant and independent of current. May be used, for instance, in surge suppressors with reaction times in the picosecond range.

silicon capacitor A capacitor in which a silicon crystal serves as the dielectric.

silicon carbide Very hard bluish-black crystals whose chemical formula is SiC. It is used in semiconductors, in rugged high-temperature electronics, optics, ceramics, and as an abrasive.

silicon cell A photovoltaic cell in which the light-sensitive material is a silicon crystal.

silicon chip A chip whose substrate or semiconductor material is silicon.

silicon-controlled rectifier Same as **SCR**.

silicon-controlled switch Same as **SCS**.

silicon crystal **1.** Pure silicon in crystalline form, such as that utilized to make chips. **2.** A **silicon crystal (1)** which has been doped to alter its electrical properties.

silicon detector A detector, such as that utilized to detect x-rays, which incorporates a silicon diode.

silicon diode A crystal diode in which silicon is the semiconductor. Used, for instance, as a diode rectifier, or as a photodetector.

silicon dioxide Colorless crystals or an amorphous white powder whose chemical formula is SiO_2. Occurs commonly in nature as quartz, sand, and flint, among others. Used in semiconductors, optical glasses, and ceramics. Also called **silica**, or **silicon oxide**.

silicon gate **1.** A MOS gate composed of polysilicon. **2.** The technology utilized in forming **silicon gates (1)**. Also called **silicon-gate technology**.

silicon-gate technology Same as **silicon gate (2)**.

silicon germanium A heterostructure consisting of silicon and germanium. Such a semiconductor material can be used to make transistors that, in comparison to silicon transistors, provide improved power and operating frequency capabilities, in addition to greater immunity to noise. Used, for instance, in RF devices such as radars, GPS receivers, cell phones, and analog-to-digital converters. Also called **silicon-germanium heterostructure**.

silicon-germanium heterostructure Same as **silicon germanium**.

silicon monoxide A black solid whose chemical formula is SiO. Used as a dielectric, in thin-film and optical protective coatings, in semiconductors, and as an abrasive.

silicon nitride A gray to white powder whose chemical formula is Si_3N_4. Used in semiconductors, refractory coatings, and as an insulator.

silicon-on-insulator An IC fabrication technique or technology in which silicon is epitaxially grown on an insulating substrate, such as sapphire or silicon dioxide. Used, for instance, for fast MOS chips with improved processing performance and lower power consumption. Its abbreviation is **SOI**.

silicon-on-sapphire A **silicon-on-insulator** technique or technology in which the substrate is sapphire. Its abbreviation is **SOS (2)**.

silicon oxide Same as **silicon dioxide**.

silicon oxynitride A chemical substance incorporating silicon, oxygen, and nitrogen, which is used in semiconductors, optical coatings, and as a dielectric.

silicon photocell A photocell in which the light-sensitive material is a silicon crystal.

silicon photovoltaic cell A photovoltaic cell in which the light-sensitive material is a silicon crystal.

silicon rectifier **1.** A diode rectifier in which a silicon alloy junction provides the rectification. **2.** A semiconductor rectifier in which the semiconductor layer is composed of a silicon crystal.

silicon resistor **1.** A resistor incorporating, or composed of, silicon. **2.** A thin-film resistor utilizing silicon as a substrate.

silicon solar cell A solar cell in which the light-sensitive material is a silicon crystal.

silicon solar panel A solar panel composed of multiple **silicon solar cells**.

silicon steel A steel that contains a given proportion of silicon, usually between 1.5 and 5.0% Used, for instance, for transformer cores with low hysteresis.

silicone A polymer incorporating a chain of alternating silicon and oxygen atoms, with radicals attached to silicon atoms. Depending on the degree of polymerization and the attached radicals, a silicone may be a solid or semi-solid, including elastomer, rubber, or resin forms. Silicones are stable over a wide temperature interval, and serve as excellent dielectrics, lubricants, heat-transfer agents, and for vibration damping, among many other uses.

silver A lustrous white metallic chemical element whose atomic number is 47. Silver is soft, extremely ductile and malleable, and is the best conductor of heat and electricity among the elements. It has over 35 known isotopes of which 2 are stable, is not chemically active, and does not oxidize in air, but does tarnish due to reactions with atmospheric sulfur compounds. Its applications in electronics are many, including its use in electric contacts and conductors, electrical components, plating, and batteries. Its chemical symbol, **Ag**, is taken from the Latin word for silver: **argentum**.

silver brazing Brazing utilizing an alloy of silver as the filler metal.

silver-brazing alloy Same as **silver solder** (1).

silver bromide A pale yellow powder whose chemical formula is $AgBr$. It is used photography, and in special glasses.

silver chloride A white powder whose chemical formula is $AgCl$. It is used photography, optics, spectroscopy, and in plating.

silver cyanide A white powder whose chemical formula is $AgCN$. It is used in plating.

silver glass die attach The use of a melted glass containing silver particles to attach a chip to a substrate. The silver particles are added to the glass for enhanced thermal and electrical conductivity.

silver iodide A light yellow powder whose chemical formula is AgI. It is used in photography.

silver-mica capacitor A capacitor in which one or more sides of one or more mica sheets or plates are silvered.

silver nitrate Colorless crystals whose chemical formula is $AgNO_3$. It is used in plating, for silvering mirrors, and in photography.

silver oxide A charcoal-gray powder whose chemical formula is AgO. It is used in batteries.

silver-oxide battery Same as **silver-oxide cell**.

silver-oxide cell A cell with a zinc anode, a silver oxide cathode, and a potassium hydroxide electrolyte. Usually encountered as button cells which feature high energy density, a comparatively flat discharge curve, and long shelf life. Also called **silver-oxide battery**, **silver-zinc cell**, **silver-zinc battery**, or **zinc-silver oxide cell**.

silver plating 1. The process of depositing silver metal onto another surface, usually a different metal, via electrolysis. Such a deposit may be thin or thick, and may be used, for instance, to provide coatings which are protective, decorative, or which have any given electrical properties. 2. The process of depositing or applying a thin, or very thin, layer or coat of silver upon a metal.

silver potassium cyanide White crystals whose chemical formula is $KAg(CN)_2$. Used in plating.

silver solder 1. An alloy composed of silver, copper, and zinc, which is used for brazing. Also called **silver-brazing alloy**. 2. An alloy, containing silver, which is used for soldering.

silver-zinc battery Same as **silver-oxide cell**.

silver-zinc cell Same as **silver-oxide cell**.

SIM Abbreviation of Subscriber Identity Module. Same as **SIM card**.

SIM card Abbreviation of Subscriber Identity Module **card**. A smart card that is inserted into a device such as cellular telephone, and which serves to identify a user, store information such as telephones numbers, and which may also have applications resident. Also called **SIM**.

SIMD Abbreviation of **single instruction stream-multiple data stream**.

SIMM Acronym for single inline memory module. A small circuit board which holds computer memory chips. A SIMM is similar to a **DIMM**, but the former has a 32-bit path, while the latter has a 64-bit path. Also called **SIMM module**.

SIMM memory RAM in the form of **SIMMs**.

SIMM module Same as **SIMM**.

SIMM socket A socket on a motherboard into which **SIMMs** are plugged-in.

Simple API for XML Same as **SAX**.

Simple Authentication and Security Layer A specification for adding an authentication mechanism to connection-based protocols such as IP. An example would be IMAP sending a command to use Kerberos to authenticate a user before email retrieval. Its abbreviation is **SASL**.

simple harmonic motion Motion whose path is symmetrical about a given equilibrium position, such as that displayed by a swinging pendulum. Simple harmonic motion is a sinusoidal function of time, and is also called **harmonic motion**.

simple harmonic oscillator A physical system, such as a pendulum, which exhibits **simple harmonic motion**. Also called **harmonic oscillator** (3).

Simple Mail Transfer Protocol Same as **SMTP**.

Simple Network Time Protocol A version of Network Time Protocol which uses simpler algorithms, and which has a slight reduction in accuracy. Its abbreviation is **SNTP**.

Simple Networking Management Protocol Same as **SNMP**.

simple sound source An ideal source that radiates sound uniformly in all directions, under free-field conditions.

simple tone A tone, such as that produced by striking a tuning fork, whose acoustic energy occurs at a single frequency. Such a tone would appear on an oscilloscope as a sine wave with little or no harmonic content. Also called **pure tone**, or **pure sound**.

simplex 1. Same as **simplex communication** (1). 2. A deprecated term for **half-duplex**.

simplex channel 1. A communications channel which allows transmission in one direction only. 2. A deprecated term for **half-duplex channel**.

simplex circuit 1. A communications circuit which allows transmission in one direction only. 2. A deprecated term for **half-duplex circuit**.

simplex communication 1. Communication in which data, voice, or the like is transmitted in one direction only. One or more locations receive, but cannot transmit. Examples include broadcasting, and that utilized by one-way intercom systems. Also called **simplex** (1), or **one-way communication**. 2. A deprecated term for **half-duplex communication**.

simplex operation 1. The operation of a communications channel utilized for **simplex communication** (1). 2. A deprecated term for **half-duplex operation**.

simplex system 1. A communications system utilized for **simplex communication** (1). 2. A deprecated term for **half-duplex system**.

simplex transmission 1. Transmission of data, voice, or the like, in one direction only. 2. A deprecated term for **half-duplex transmission**.

SIMS 1. Acronym for **secondary-ion mass spectroscopy**. 2. Acronym for **secondary-ion mass spectrometer**.

simulation 1. An imitation, model, or other representation of an object, situation, process, feature, manner of operation, or the like. Such a simulation is intended to faithfully reproduce the real or theoretical objects, situations, processes, and so on, it is based on, and can be used, for instance, for analysis, testing, or training. **2.** The use of a computer to imitate an object or process. Sophisticated software, combined with accurate input devices, enable a computer to respond mathematically to factors such as changing conditions, as if it were the object or process itself. Such simulations may be used to represent or emulate almost anything, including weather conditions or biological processes, and may be utilized to test new theories. Also called **computer simulation**. **3.** A mathematical model which is utilized to represent a physical component, circuit, device, piece of equipment, system, process, or phenomenon. **4.** The use of software and/or hardware which enables programs written for one computer to work properly with another system. **5.** The use of software and/or hardware which enables programs written for one computer to work properly with another system which would have otherwise been incompatible.

Simulation Program with Integrated Circuit Emphasis Software utilized to simulate or model ICs at the transistor level. Its abbreviation is **SPICE**.

simulator 1. A device, computer, apparatus, enclosure, location, or the like which is utilized to prepare and/or experience simulations. **2.** Software and/or hardware which enables programs written for one computer to work properly with another system. **3.** Software and/or hardware which enables programs written for one computer to work properly with another system which would have otherwise been incompatible.

simulcast 1. The simultaneous broadcasting of the same program over two different transmission systems. For example, an AM and an FM station transmitting the same program at the same time, or the concurrent transmission of DTV content along with regular TV content. **2.** The simultaneous broadcasting of the same program over two or more stations utilizing the same transmission system. For example, the same TV content presented at the same time over multiple networks.

simultaneous 1. Occurring, performed, or existing at the same instant. **2.** Occurring, performed, or existing concurrently. **3.** Pertaining to that which is **simultaneous** (1), or **simultaneous** (2).

simultaneous access Also called **parallel access**. **1.** Storage and/or retrieval of data to or from a computer storage medium in which multiple bits, characters, or data units are transferred simultaneously. This contrasts with **serial access** (1), where bits, characters, or data units are transferred sequentially. Also called **simultaneous storage**. **2.** Reading and/or writing of data from or to computer memory in which multiple bits, characters, or data units are transferred simultaneously. This contrasts with **serial access** (2), where bits, characters, or data units are transferred sequentially.

simultaneous processing Also called **parallel processing**, **multiprocessing**, or **concurrent execution** (2). **1.** The simultaneous execution of multiple computer operations by multiple CPUs in a single computer. Such processors are usually linked by high-speed channels. **2.** The simultaneous execution of multiple computer operations by a single computer, whether it utilizes multiple processors or another mechanism. **3.** The utilization of multiple computers to perform multiple operations simultaneously.

simultaneous processors Two or more processors utilized for **simultaneous processing**. These processors may be located in a single computer, or in multiple computers. Also called **parallel processors**.

simultaneous storage Same as **simultaneous access** (1).

simultaneous transmission 1. The simultaneous emission or transmission of multiple signals, waves, or messages, to more than one recipient or location. **2.** Simultaneous transmission by two or more locations or nodes in both directions at the same time, as occurs in duplex transmission.

simultaneous voice and data A technology, such as DSVD, which enables voice and data signals to be transmitted simultaneously over a single analog telephone line. Such signals may or may not be compressed. Its abbreviation is **SVD**.

SINAD Abbreviation of **signal-to-noise-and distortion ratio**.

SINAD ratio Abbreviation of **signal-to-noise-and distortion ratio**.

sine For a right triangle, the ratio of the length of the side opposite to an acute angle, to the length of the hypotenuse. This contrasts with **cosine**, which is the ratio of the length of the side adjacent to an acute angle, to the length of the hypotenuse.

sine curve The graph obtained from the following equation: $y = \sin x$. Such a curve proceeds smoothly from 0 to 1 to 0 to −1 to 0 to 1 and so on. Also called **sinusoid**.

sine law A law which states that the intensity of radiation from a linear source is proportional to the sine of the angle formed between the axis of said source and the specific direction being considered.

sine wave A periodic wave whose amplitude follows the values of a sine curve. Simple harmonic motion is a sinusoidal function of time. Power mains current without harmonic current has this waveform. Also called **sinusoidal wave**.

sine-wave frequency response A measure of the behavior of a component, circuit, device, piece of equipment, or system, as a function of its input signal frequencies. For example, it may refer to the efficiency of the amplification of a circuit or device as a function of frequency. Also known as **sine-wave response**, **frequency response** (1), or **amplitude-frequency response**.

sine-wave response Same as **sine-wave frequency response**.

sine waveform The shape of a **sine wave**.

singing An unwanted self-sustained oscillation, usually resulting from excessive feedback. It may be manifested, for instance, as a continued whistle or howl in an amplified telephone circuit.

single-address instruction In computer programming, an instruction which contains one address part. Also called **one-address instruction**.

single-attachment station Within a dual-ring FDDI network, a node which a connection to one of the rings, usually the primary ring. Its abbreviation is **SAS**.

single-balanced mixer A mixer circuit which suppresses one of its inputs, usually that of the local oscillator. Balance may be achieved, for instance, utilizing a ring configuration of diodes. Used, for example, in superheterodyne receivers. Also called **balanced mixer**.

single-board Pertaining to a **single-board computer**.

single-board computer A complete computer contained on a single printed-circuit board. Such a board must have a CPU, memory, a clock, and input/output circuits. Its abbreviation is **SBC**.

single-chip module A package which houses a single chip, as opposed to a **multichip module**, which houses multiple interconnected chips on a common substrate. Its abbreviation is **SCM**.

Single Connector Attachment Same as **SCA**.

single-crystal Pertaining to, or consisting of, a **single-crystal structure** or **single-crystal material**.

single-crystal material **1.** A crystalline material in which there is long-range order. Such a material has no grain boundaries, so it is completely uniform throughout the entire crystal, regardless of its size. The lack of grain boundaries provides for higher efficiency than a **polycrystalline material** (1), which has said boundaries. Used, for instance, as semiconductor materials of the highest quality. **2.** A crystalline material consisting of one type of crystal.

single-crystal structure The crystal structure present in a **single-crystal material**.

single density A disk format indicating that the storage capacity is the standard amount. If the customary capacity for a given disk format is 1 GB, it would usually be labeled as single density, while a double-density version in this case would have a capacity of 2GB.

single-density disk A disk with a **single density** format.

single-electron memory Computer memory in which the presence, movement, position, or state of an electron determines the difference between logic 0 and logic 1. Its abbreviation is **SEM**.

single-ended Consisting of, or pertaining to a component, circuit, device, transmission line, signal, or system in which one side, or terminal, of the input and/or output is connected to ground. Thus, that which is single-ended is asymmetric with respect to ground. For instance, a single-ended line, or a single-ended circuit. Also called **unbalanced** (1).

single-ended amplifier An amplifier in which the input and/or output are **single-ended**. Also called **unbalanced amplifier**.

single-ended circuit A circuit which is **single-ended**. Also called **unbalanced circuit**.

single-ended input An input in which is **single-ended**. Also called **unbalanced input**.

single-ended line A transmission line, such as a coaxial line, which is **single-ended**. Also called **single-ended transmission line**, or **unbalanced line**.

single-ended output An output which is **single-ended**. Also called **unbalanced output**.

single-ended signal A signal which is **single-ended**. Also called **unbalanced signal**.

single-ended transmission line Same as **single-ended line**. Also called **unbalanced transmission line**.

single-event upset A phenomenon in which radiation incident upon ICs causes errors, such as changed bits, malfunctions, or failures. Its abbreviation is **SEU**.

single-frequency Consisting of, pertaining to, or utilizing a specific frequency. For instance, a single-frequency laser, or a single-frequency receiver. Also called **fixed-frequency**.

single-frequency laser A laser whose output is at a single frequency, or within an extremely narrow interval of frequencies.

single-frequency operation Operation of a component, circuit, device, piece of equipment, or system, at a single frequency.

single-frequency oscillator An oscillator which operates at a single frequency, which cannot be adjusted. This contrasts with **variable-frequency oscillator**, whose frequency can be adjusted. Also called **fixed-frequency oscillator**.

single-frequency receiver A receiver that can only receive a single frequency, as opposed to being tunable throughout a frequency band.

single-frequency signal A signal comprised of a single frequency, or an extremely narrow interval of frequencies.

single-gun Consisting of, utilizing, or pertaining to, one electron gun.

single-gun cathode-ray tube Same as **single-gun CRT**.

single-gun CRT Abbreviation of **single-gun** cathode-ray tube. A CRT with a single electron gun.

single-hop propagation The long-distance propagation of radio waves utilizing a single reflection off the ionosphere.

single-hop transmission The long-distance transmission of radio waves utilizing a single reflection off the ionosphere.

single in-line memory module Same as **SIMM**.

single in-line package A rectangular IC housing in which the connecting pins are arranged along a single side of the package. Such a package is usually made of plastic or ceramic, and its pins protrude downward. Used, for instance, for computer RAM. Its acronym is **SIP**. Also spelled **single inline package**. Also called **single in-line pin package**.

single in-line pin package Same as **single in-line package**. Also spelled **single inline pin package**. Its abbreviation is **SIPP**.

single inline memory module Same as **SIMM**.

single inline package Same as **single in-line package**.

single inline pin package Same as **single in-line package**. Also spelled **single in-line pin package**. Its abbreviation is **SIPP**.

single instruction-multiple data stream Same as **single instruction stream-multiple data stream**.

single instruction-single data stream Same as **single instruction stream-single data stream**.

single instruction stream-multiple data stream A computer architecture in which multiple processors simultaneously and independently execute the same instruction set on different sets of data. Its abbreviation is **SIMD**. Also called **single instruction-multiple data stream**, or **data parallel**.

single instruction stream-single data stream A computer architecture in which multiple processors simultaneously execute the same instruction set sequentially on a stream of data. Its abbreviation is **SISD**. Also called **single instruction-single data stream**.

single-layer coil A coil with one wire layer, which is usually wound helically.

single-layer solenoid A solenoid with one wire layer, which is usually wound helically.

single-level metal In ICs, a process for contact interconnections in which one layer of metal is used. This contrasts with **double-level metal**, where two vertical layers of metal are separated by an insulating layer. Also called **single-level metal process**.

single-level metal process Same as **single-level metal**.

single-line digital subscriber line Same as **single-line DSL**.

single-line DSL Abbreviation of **single-line** digital subscriber line. A DSL technology, such as VDSL, using a single pair of copper wires, as opposed, for instance, to that using two or more pairs, as is the case with HDSL. Its abbreviation is **SDSL** (2).

single-loop feedback A feedback mechanism in which two or more individual circuits or loops provide feedback.

single-mode fiber An optical fiber designed to carry just a single mode. Such fibers usually have a smaller center core than **multimode fibers**, and are generally used for comparatively longer distances. Also spelled **singlemode fiber**. Also called **single-mode optical fiber**.

single-mode optical fiber Same as **single-mode fiber**.

single-phase **1.** Consisting of, utilizing, or pertaining to one AC phase. **2.** Consisting of, utilizing, or pertaining to one phase.

single-phase bridge rectifier A bridge rectifier which works with a single-phase AC input.

single-phase circuit A circuit in which there is a single AC phase, as opposed to a **polyphase circuit**, where there are multiple phases. Such a circuit, for instance, may utilize two wires which differ in phase by 180°, or a half-cycle.

single-phase current A current, such as that flowing through a single-phase circuit, which has a single phase.

single-phase full-wave Pertaining to a circuit or device, such as a full-wave rectifier, which utilizes single-phase current.

single-phase generator A generator of single-phase current.

single-phase half-wave Pertaining to a circuit or device, such as a half-wave rectifier, which utilizes single-phase current.

single-phase meter A meter, such as a wattmeter, utilized in a single-phase circuit.

single-phase motor A motor which utilizes single-phase current.

single-phase power 1. The power utilized by a single-phase circuit or system. 2. The power provided by a single-phase circuit or system.

single-phase rectifier A rectifier which works with a single-phase AC input.

single-phase system A system, such as a single-phase circuit, which has a single AC voltage.

single-phase three-wire A single-phase circuit with three wires. The AC voltage between any two wires are either in phase, or phase opposition with each other.

single-phase transformer A transformer utilized to supply a single-phase circuit, device, piece of equipment, or system.

single-phase two-wire A single-phase circuit with two wires, each of which differs in phase by 180°, or a half-cycle

single-phase voltage A voltage, such as that supplying a single-phase circuit, in which there is a single AC phase, as opposed to a **polyphase voltage**, where there are multiple phases.

single-photon emission computed tomography Its acronym is **SPECT**. A computed tomography medical diagnostic technique which measures the emission of photons of a given energy emitted by radioactive tracers introduced into the body being scanned. The resolution and sensitivity of SPECT is significantly lower than that of PET, but its comparatively affordable cost makes it more generally accessible. Used, for instance, to study lesions and blood flow through the brain. Also called **single-photon emission computerized tomography**, or **single-photon emission computed tomography scan**.

single-photon emission computed tomography scan Same as **single-photon emission computed tomography**. Its abbreviation is **SPECT scan**.

single-photon emission computed tomography scanner An instrument which takes a **single-photon emission computed tomography**. Its abbreviation is **SPECT scanner**.

single-photon emission computed tomography scanning The use of a **single-photon emission computed tomography scanner**. Its abbreviation is **SPECT scanning**.

single-photon emission computerized tomography Same as **single-photon emission computed tomography**.

single-point ground A single ground point in a circuit, stage, or system, which connects all grounds and returns. Such a ground, for instance, helps minimize unwanted coupling. Also called **star ground**.

single-point grounding The use of a single ground point in a circuit, stage, or system, to connect all grounds and returns. Also called **star grounding**.

single-pole double-throw relay A three-terminal relay arrangement in which one contact can connect to either of two other available contacts, but not both. Its abbreviation is **SPDT relay**. Also called **single-pole relay (2)**.

single-pole double-throw switch A three-terminal switch arrangement in which one contact can connect to either of

two other available contacts, but not both. Its abbreviation is **SPDT switch**. Also called **single-pole switch (2)**.

single-pole relay 1. Same as **single-pole single-throw relay**. 2. Same as **single-pole double-throw relay**.

single-pole single-throw relay A two-terminal relay arrangement which serves to open or close a single circuit. Its abbreviation is **SPST relay**. Also called **single-pole relay (1)**.

single-pole single-throw switch A two-terminal switch arrangement which serves to open or close a single circuit. Its abbreviation is **SPST switch**. Also called **single-pole switch (1)**.

single-pole switch 1. Same as **single-pole single-throw switch**. 2. Same as **single-pole double-throw switch**.

single-precision The use of one computer word to store a single number. This contrasts with **double-precision**, in which two computer word are utilized. For instance, if a single-precision number requires 64 bits, a double-precision number will utilize 128 bits, which provides greater accuracy.

single-precision number A number represented utilizing **single-precision**.

single program-multiple data A computer architecture in which multiple processors simultaneously execute the same program, each using a different subset of data. Its abbreviation is **SPMD**.

single scattering Scattering of electromagnetic radiation, such as light, or particles, such as electrons, by a single interaction with a particle, an ion within a crystal lattice, or the like.

single-shot Consisting of, utilizing, or pertaining to operation in which there is a single pulse, action, step, cycle, stable state, or the like.

single-shot execution Same as **single-step operation**.

single-shot multivibrator A multivibrator with one stable state and one unstable state. A trigger signal must be applied to change to the unstable state, where it remains for a given interval, after which it returns to the stable state. Also called **monostable multivibrator**, **one-shot multivibrator**, or **univibrator**.

single-shot operation Same as **single-step operation**.

single-sideband An amplitude-modulation signal in which only one of the sidebands resulting from the modulation of the carrier is present. When the carrier is suppressed, as it usually is, also called **single-sideband suppressed carrier**, or **single-sideband signal**. Its abbreviation is **SSB**.

single-sideband amplitude modulation Same as **single-sideband modulation**.

single-sideband communications Communications in which only a **single-sideband** is utilized. Its abbreviation is **SSB communications**.

single-sideband modulation Amplitude-modulation in which only one of the sidebands resulting from the modulation of the carrier is present. The carrier may or may not be suppressed. Its abbreviation is **SSB modulation**. Also called **single-sideband amplitude modulation**.

single-sideband signal Same as **single-sideband**.

single-sideband suppressed carrier Single-sideband transmission in which the carrier is suppressed. Its abbreviation is **SSSC**, or **SSB-SC**.

single-sideband transmission The transmission of a **single-sideband signal**. Its abbreviation is **SSB transmission**.

single-sided disk A disk, such as a floppy or DVD, which can store data on one surface.

single sign-on In a communications network, a single authentication process which enables a user to access multiple authorized applications, resources, functions, areas, or the like.

The user should not encounter any further authentication prompts until the next session. Its abbreviation is **SSO**.

single-stage Having a single stage. For instance, a single-stage amplifier.

single-stage amplifier An amplifier with a single amplification stage, as opposed to a **multi-stage amplifier**, which has two or more.

single-step Pertaining to **single-step operation**.

single-step execution Same as **single-step operation**.

single-step operation A computer operational mode in which each instruction, or part of an instruction, is executed in response to an external signal which is usually controlled manually by a user. Utilized, for instance, for debugging, or for detecting hardware malfunctions. Also called **single-step execution, step-by-step operation, step-by-step execution, single-shot operation, single-shot execution, step-through operation, step-through execution, one-shot operation**, or **one-shot execution**.

single sweep 1. On an oscilloscopic display, such as that of a radar, a complete sweep. **2.** The ability of an oscilloscope to display and/or record only a single sweep at a time, thus preventing multiple unwanted oscillograms.

single-sweep operation For an oscilloscope, an operational mode in which only a single sweep at a time is displayed and/or recorded, thus preventing multiple unwanted oscillograms.

single-threaded application 1. An application which supports only **single threading**. **2.** An application in which **single threading** is occurring.

single threading The execution of a single task or process within a computer program. This contrasts with **multithreading**, where two or more tasks or processes are simultaneously executed. In this context, each task or process is called a **thread**.

single-throw switch A switch that has two or more poles, but which only serves to open or close a single circuit.

single-track recorder A magnetic-tape recorder whose recording head covers the full width of the tape, or which otherwise permits the recording of only one track. Also called **single-track tape recorder**.

single-track recording Recording utilizing a **single-track recorder**.

single-track tape recorder Same as **single-track recorder**.

single-tuned amplifier An amplifier which has a single resonance frequency. This provides a narrower bandwidth than that obtained with a **double-tuned amplifier**.

single-tuned circuit A tuned circuit with one resonance frequency.

single-turn coil A coil having one turn of wire, or of another conductor.

single-turn potentiometer A potentiometer whose moving contact can travel once, or nearly once, around its shaft or axis.

single-user Pertaining to, consisting of, involving, or able to be used or accessed by only one user. For instance, a single-user system.

single-user system 1. A computer system that can be used by a single user. This contrasts with a **multiuser system (1)**, which can be used by two or more users. **2.** A computer system that can be accessed only by a single user. This contrasts with a **multiuser system (2)**, which can be accessed by two or more users, each located at a different terminal.

single-wire circuit A circuit in which the earth is utilized to complete the conducting path. Also called **ground-return circuit**.

single-wire line A transmission line in which the earth is utilized to complete the conducting path. Also called **single-wire transmission line**.

single-wire transmission line Same as **single-wire line**.

singlemode fiber Same as **single-mode fiber**.

singlemode optical fiber Same as **single-mode fiber**.

sink 1. A point, region, or object where current, power, energy, matter, a signal, or flux is absorbed or terminates. A **source (2)** is where it originates. **2.** The output end of a transmission line, where a signal is received. The **generator end** is that where the signal originates. **3.** A material or object which helps dissipate unwanted heat from components, circuits, devices, equipment, enclosures, or systems, enabling them to continue working within a safe temperature range. It does so by absorbing heat and conducting it away to a surface from which it is dissipated into its surroundings. Heat sinks are used, for instance, to protect power transistors. Also called **heat sink (1)**. **4.** Any material, such as a solid or a fluid, which serves to protect components, circuits, devices, equipment, enclosures, or systems, by removing heat. Also called **heat sink (2)**.

sinter The performance of a **sintering** process. Also, the objects or bodies resulting from such a process.

sintered That which has undergone a **sintering** process. Ferrites, for instance, are often used in sintered form.

sintering The process of forming solid objects or masses from a powdered material which is heated to a temperature below its melting point. The result is a solid, and usually porous, mass. Sintering may be performed, for instance, on metals, glasses, certain oxides, and plastics.

sinusoid Same as **sine curve**.

sinusoidal Pertaining to, in the form of, or resembling a **sine curve**.

sinusoidal wave Same as **sine wave**.

SIP Abbreviation of **single in-line package**.

SIPP Abbreviation of **single in-line pin package**.

SIR 1. Abbreviation of **signal-to-interference ratio**. **2.** Abbreviation of **serial infrared**. **3.** Acronym for **surface insulation resistance**.

SISD Abbreviation of **single instruction stream-single data stream**.

SIT Acronym for **static induction transistor**.

site A given location within the World Wide Web. A site may have a home page which serves as the starting point for multiple pages, or may consist simply of a single Web page. Also called **Web site**.

site license A software license authorizing use within a single location, faculty, campus, or the like. Such a license usually specifies a maximum number of available and/or simultaneous users.

site management The multiple tasks involved in the administration of a Web site, including hosting, monitoring, verification that all contained links are active, maintenance, and so on. Also called **Web site management**.

site map A table, diagram, or other visual representation which summarizes the contents or pages of a **site**. A site map should provide a direct and intuitive presentation which enables a user to go directly to the desired content, when said user does not wish to navigate throughout the site. Such a map is frequently organized hierarchically, and the need for a site map increases proportionally to the size and complexity of the information offered. Also spelled **sitemap**. Also called **Web site map**.

site noise The noise present in any given environment, especially that within which tests or observations are taking place. Such noise includes nearby and distant sources. Also called **background noise (1)**, **local noise**, **room noise**, **ambient noise**, or **environmental noise**.

site search A Web search limited to a specific **site**. Also called **Web site search**.

sitemap Same as **site map**.

six degrees of freedom Six manners in which a system or component, such as a robotic manipulator, can move in space. For instance, a robotic manipulator with six degrees of freedom would have three in the arm and three in the wrist: vertical movement, radial traverse, rotational traverse, wrist pitch, wrist roll, and wrist yaw. Its abbreviation is **6-DOF**.

sixth-generation computer A computer generation which started in the early 2000s and includes progress in areas such as artificial intelligence, quantum computing, and the ability of computers to fully comprehend natural human input such as spoken language and gestures.

sixth-order filter A filter with twelve components, such as six inductors and six capacitors, and which provides a 36 dB rolloff. Used, for instance, as a crossover network.

size **1.** The dimensions, extent, magnitude, proportion, influence, or the like, of something. **2.** A **size (1)** relative to another size. **3.** To order of otherwise classify by **size (1)**. **4.** To change the dimensions of something. For example, that of an image or window displayed on a computer monitor. Also called **scale (6)**, or **resize**.

size coding The coding of a control, such as a button or knob, by different sizing. For example, an on/off button of a TV may be larger, while the volume control buttons are smaller. Such coding may also be used in combination with shape and color coding.

skew **1.** To distort in some manner, especially in reference to a given desired center, location, position, axis, or instant. For example, clock skew, the distortion resulting from a misaligned head, or that arising from improper horizontal sync. **2.** To take or move in a slanted direction. A skewed sheet working its way through a printer, for instance, may result in a jam. **3.** A slanting position, movement, or direction. For example, a line which is neither parallel with, nor intersecting with another.

skiagram Same as **shadowgraph**.

skiagraph Same as **shadowgraph**.

skiagraphy Same as **shadowgraphy**.

skiatron A type of CRT whose screen is specially coated, so that it darkens when impacted by electrons. The displayed signals on such a screen are presented as dark traces or blips against a bright background. Also called **dark-trace tube**.

skin depth The depth to which a current penetrates beneath the surface of a conductor, as determined by the **skin effect**.

skin effect The tendency of AC to flow near the surface of a conductor. This effect becomes more pronounced as frequency rises, so losses due to resistance become greater with frequency increases. A conductor made from many separately-insulated wire strands which are woven together or braided in a manner that each strand regularly assumes each of the possible positions within the overall cross-section of the conductor minimizes this effect, providing for lower losses at radio frequencies. Also called **conductor skin effect**.

skinny DIP Abbreviation of **skinny d**ual **i**n-line **p**ackage. A dual in-line package which is usually shorter and thinner, and which provides a lower pin count. Its own abbreviation is **SDIP**.

skinny dual in-line package Same as **skinny DIP**.

skip **1.** To bypass something while proceeding from one point, position, or location, to another. For example, to ignore one or more instructions within a sequence of instruc-

tions. **2.** In radio-wave propagation, a single reflection off the ionosphere.

skip distance The distance, along the surface of the earth, a wave travels in a single **skip (2)**. A short skip is usually considered to be that of less than 1000 kilometers, while long skips exceed this distance.

skip zone A zone which runs from where ground waves become too weak for reception to the point where a sky wave returns to the surface of the earth. Also called **silent zone**, or **zone of silence (2)**.

sky noise Also called **space noise**, **cosmic noise**, or **Jansky noise**. **1.** Radio-frequency noise caused by sources outside the earth's atmosphere, such as the cosmic microwave background. **2.** Radio-frequency noise that originates outside the earth's atmosphere.

sky wave Also spelled **skywave**. Also called **atmospheric radio wave**. A radio wave whose propagation has been assisted by reflecting off the atmosphere. When it reflects off the ionosphere it is called **ionospheric wave**.

skyscraper ad Abbreviation of **skyscraper ad**vertisement. On the Internet, a form of advertising which appears on Web pages, and which may be clicked upon to obtain more information, usually by being transferred to the home page of an entity. These are often presented as a tall rectangular color graphic measuring, for instance, 600 pixels high by 120 pixels wide, and are placed on either side of the page. A **banner ad** is usually rectangular and located at the top or bottom of a page.

skyscraper advertisement Same as **skyscraper ad**.

skywave Same as **sky wave**.

SL **1.** Abbreviation of **sensation level**. **2.** Abbreviation of **stereolithography**.

SLA Abbreviation of **Service Level Agreement**.

slab A broad, flat, and usually thick piece or object, such as a thick quartz crystal from which blanks will be cut.

slag A program or routine, such as a Trojan horse or a logic bomb, which causes significant damage, such as the corruption of the data contained in a computer.

slamming **1.** The unethical practice of changing the telephone service provided to a customer without permission, through coercion, via a maze of confusing, misleading, and often fraudulent marketing statements, or the like. **2. Slamming (1)** employed to switch the long-distance provider of a customer.

slash In most computer operating systems, a character which serves to denote division, as seen, for instance, in: 22/7. This character also serves to separate elements within an internet address, as seen, for example in: *http://www.yipeeee.com/whoo/yaah.html*. This character has a place on the keyboard, and is shown here between brackets: [/]. Also known as **forward slash**, or **virgule**.

slave That which is controlled by something else. For example, a secondary station within a loran system, or a slave switch.

slave clock A clock which is controlled by a **master clock (2)**.

slave oscillator An oscillator which is controlled by a **master oscillator**.

slave relay A relay which is controlled by a **master relay**.

slave station A station which is controlled by a **master station**.

slave switch A switch that is controlled by a **master switch**.

SLD Abbreviation of **second-level domain**.

sleep **1.** A state in which there is a suspension of activities. For example, an inactive period in a program induced by a code sent by another program. **2.** A state in which a system or terminal is inactive until an input, such as the appropriate code, is received.

sleep mode A mode of operation in which all but the essential components and operations are suspended. Once the sleep mode is terminated, by providing an input such as a mouse movement or keystroke, the computer quickly returns to the exact state it was in before entering this mode. Used, for instance, for power conservation. Also called **suspend mode**.

sleeve **1.** An outer covering, usually in the form of a tube, which serves to protect wires, components, or devices. Such a sleeve, for instance, may provide physical protection and/or electromagnetic shielding. Also called **sleeving**. **2.** The grounded part of a phone plug, phono plug, or similar connector.

sleeve antenna A vertical antenna fed by a coaxial line, in which the inner conductor of the coaxial cable is extended ¼ wavelength, while the outer conductor which formerly enclosed it is folded back by ¼ wavelength. Also called **coaxial antenna**.

sleeving Same as **sleeve (1)**.

slew rate **1.** For a component, circuit, device, piece of equipment, or system, the maximum rate at which an output voltage can swing from one extreme to another across its full dynamic range. May be expressed, for instance, in volts per microsecond. **2.** The **slew rate (1)** of an operational amplifier. **3.** The maximum rate of change of an output. **4.** The maximum rate of change at which a control system can react to a adjustment or change. **5.** The rate of change of an output in response to an input.

SLF Abbreviation of **super low frequency**.

SLIC Acronym for **subscriber line interface circuit**.

slice **1.** A thin and broad piece cut from a larger object, such as a slab. Also, to cut or divide into such a slice. **2.** A **slice (1)** of a semiconductor material which serves as a substrate for fabrication processes such as masking and etching. Also, to create such slices. Also called **wafer**, or **semiconductor wafer**.

slicer A circuit or device which passes only those portions of an input signal which lie between two fixed amplitude boundaries. These boundaries are usually close together. Also called **slicer amplifier**, **clipper-limiter**, or **amplitude gate**.

slicer amplifier Same as **slicer**.

slide To move smoothly over a surface while maintaining continuous contact, as with sliding contacts. Also called **wipe**.

slide-rule dial A dial, such as that of a meter, whose pointer moves along a straight scale that resembles that of a slide rule.

slide switch A switch that is actuated by sliding a button, bar, knob, or other protrusion between two or more fixed positions. Usually utilized as an on/off switch.

slide wire A straight length of resistance wire and a sliding contact which are used together to control resistance levels in potentiometers or bridges.

slide-wire bridge A bridge that incorporates a **slide wire** in one or more of its branches.

slide-wire potentiometer A potentiometer which utilizes a **slide wire** and a sliding contact to control the resistance level.

slider Also called **wiper**. **1.** An arm, brush, blade, strip, surface, or the like, which **slides**. For example, a sliding contact. **2.** A movable contact that maintains continuous contact with a stationary contact.

slideware Software which does not exist, but which is nonetheless hyped in sales presentations such as those using slides. It is a form of **vaporware (1)**.

sliding action The wiping contact made by **sliding contacts**.

sliding contacts Also called **wiping contacts**. **1.** Electrical contacts which slide or wipe past each other while maintaining smooth and continuous contact for added reliability. Such contacts may also be self-cleaning. **2.** Same as **self-cleaning contacts**.

slime In electrolytic refining or plating, an insoluble residue that collects at the bottom of the solution or on the surface of the anode. In the electrolytic refining of copper, for example, this mud usually contains platinum, gold, and silver. Also called **anode mud**, or **anode slime**.

slip **1.** Same as **slip speed**. **2.** In fax, distortion in the received image due to slippage in the mechanism which advances the paper.

SLIP Acronym for Serial Line Internet Protocol. A protocol utilized for the transfer of IP packets over a dial-up connection using a serial port. It is slow, does not provide error-correction, and has been superseded by PPP. Also called **SLIP protocol**.

SLIP protocol Same as **SLIP**.

slip ring A conductive rotating ring which works in conjunction with one or more stationary brushes, in order to maintain a continuous electrical connection. Used, for instance, in an AC generator. Also called **collector ring**.

slip speed In an induction motor, the difference between the synchronous speed and the actual operating speed. Also called **slip (1)**.

SLM Abbreviation of **spatial light modulator**.

slope **1.** A line, plane, surface, orientation, or direction which is inclined. Also, to move in such a manner. **2.** A deviation from a horizontal line, plane, surface, orientation, or direction. Also, the extent of such a deviation. **3.** A measure of the change in a variable in the vertical axis divided by a change in the value of a variable measured along the horizontal axis. For example, a horizontal line has a slope of zero. **4.** A measure of the change in a variable relative to a change in the value of another variable. **5.** In a transmission line, the rate of change of attenuation relative to frequency. Expressed, for instance, in dB per Hz. **6.** Within a straight or nearly straight portion of a characteristic curve, the deviation from such a graphed line.

slope detector A tank circuit tuned slightly to one side of a given desired carrier frequency. It detects to FM signals based on the **slope (6)**.

slope equalization The use and effect of a **slope equalizer**.

slope equalizer A circuit or device utilized to adjust the **slope (5)** of a transmission line.

slope resistance The ratio of the change in a voltage applied to an electrode, to the change in current in said electrode.

slot **1.** A narrow opening, slit, or groove. For instance, an armature slot, a coupling aperture, those in a slot mask, or that via which a disk is inserted into a drive. **2.** To cut or otherwise make a **slot (1)**. **3.** A **slot (1)** within a computer, into which expansion cards are plugged in. Also called **expansion slot**. **4.** To place in a **slot (1)**.

Slot 1 A 242-pin motherboard slot used for certain CPUs.

Slot 2 A 330-pin motherboard slot used for certain CPUs.

Slot A A 242-pin motherboard slot used for certain CPUs.

slot antenna An antenna in which energy is radiated through slots made in the wall of a waveguide, or along a conducting surface. Usually utilized at microwave frequencies. Also called **slot radiator**, or **slotted-waveguide antenna**.

slot antenna array Same as **slot array**.

slot array An array of **slot antennas**. Used, for instance, for creating a given radiation pattern. Also called **slot antenna array**.

slot cell A material or sheet of insulation placed in a slot, such as that of a magnetic core.

slot coupling The coupling of energy between a waveguide and a coaxial cable using narrow apertures.

slot insulation The placement of an insulating material or sheet in slots or spaces in devices or equipment. Also, the insulator so placed.

slot line Same as **slotted line**. Also spelled **slotline**.

slot mask In a picture tube with a three-color gun, a grill with long and narrow apertures that is placed behind the screen to insure that each color beam strikes the correct phosphor dot on the screen. For example, the electron beam intended for the green phosphor dots only hitting those.

slot radiator Same as **slot antenna**.

slotline Same as **slotted line**. Also spelled **slot line**.

slotted line A section in a transmission line, such as a waveguide or coaxial line, in which a lengthwise slot is cut into the outer conductor, with an adjustable probe placed in said slot. Used, for instance, for the determination of load impedance or standing-wave ratios in microwave systems. Also called **slotted section**, **slotline**, or **slotted waveguide**.

slotted section Same as **slotted line**.

slotted waveguide Same as **slotted line**.

slotted-waveguide antenna Same as **slot antenna**.

slow-blow fuse A fuse whose melting conductor can withstand moderate overcurrents for brief time intervals. Such a fuse may be used, for instance, to allow an additional margin for inrush current.

slow break The opening of an electric circuit after a specified delay. This contrasts with a **fast break**, where there is no delay. Also called **delayed break**.

slow busy signal A busy signal that indicates that the dialed number is currently in use. This contrasts with a **fast busy signal**, where the central office either did not understand the dialed digits, or a part of the network is too busy to process the call.

slow charge 1. Charging of a rechargeable battery, such as that done overnight, which takes an especially long time. 2. A full charging of a rechargeable battery, other than a fast charge. Some batteries perform better when not fast-charged.

slow charger A battery charger used for **slow charges**. Such chargers may or may not maintain a trickle charge once the full charge is reached.

slow death Gradual and undesired changes in characteristics, or reductions in performance of a component, circuit, device, piece of equipment, or system.

slow make The closing of an electric circuit after a specified delay. This contrasts with a **fast make**, where there is no delay. Also called **delayed make**.

slow-scan television Same as **slow-scan TV**.

slow-scan TV Abbreviation of **slow-scan** television. A system in which still images are scanned and displayed over a TV or computer monitor. This allows, for instance, for images to be transmitted via a telephone line, whose bandwidth would otherwise be too narrow for regular TV viewing. Using this system, a new image may be presented every few seconds. Its abbreviation is **SSTV**.

slow-wave structure A structure, such as a helical coil surrounding a vacuum tube in a traveling-wave tube, which serves to slow down the path of a beam. Its abbreviation is **SWS**.

SlowKeys 1. Keys on a computer keyboard which are programmed or set to accept an input only after being pressed longer than a given interval. Used, for instance, to avoid unwanted input when pressing keys inadvertently. This contrasts, for instance, with **BounceKeys**, which ignore re-

peated pressings of a given key within a given time interval. 2. A feature which prevents keystrokes from being registered until a given time interval has elapsed while pressing. Useful, for instance, for those with reduced motor function.

SLSI Abbreviation of **super-large-scale integration**.

slug 1. A movable ferrite core which is utilized to vary the inductance of a coil. Also called **tuning slug**. 2. A lump or piece of metal which is to be further processed. 3. A heavy copper ring which is placed on the core of a relay to obtain delayed operation.

slug-tuned coil A coil tuned by a **slug** (1).

slug tuner A tuner which utilizes a **slug-tuned coil**.

slug tuning Tuning utilizing a **slug tuner**.

slurry A diluted mixture of a liquid, especially water, and finely divided particles or substances which are mostly insoluble in said liquid. In semiconductor manufacturing, for instance, a slurry may be seen as a fine particulate deposit on a wafer which is not properly cleaned.

Sm Chemical symbol for **samarium**.

SMA 1. Abbreviation of **surface-mount assembly**. 2. Abbreviation of **shared memory architecture**. 3. Abbreviation of **shape-memory alloy**.

SMA connector A type of fiber-optic connector that is secured using a threaded ring.

Small Computer System Interface Same as **SCSI**.

small office/home office A work environment such as that of a small office or an office located in a home. Many applications and devices are designed for such a setting. Its acronym is **SOHO**.

small-outline DIMM A DIMM which is smaller, has fewer pins, and supports lower-bit transfers. Used, for instance, in portable computers. Its abbreviation is **SO-DIMM**, or **SODIMM**.

small-outline gull-wing Same as **small-outline integrated circuit**. Its abbreviation is **SOG**.

small-outline IC Abbreviation of **small-outline integrated circuit**.

small-outline integrated circuit A small surface-mounted chip package with gull-wing leads. Its abbreviation is **small-outline IC**, or **SOIC**. Also called **small-outline gull-wing**.

small-outline J-lead A small surface-mounted chip package with J-leads. Its abbreviation is **SOJ**.

small-outline package A surface-mounted chip or transistor package with gull-wing leads. Its abbreviation is **SOP**.

small-outline transistor A surface-mounted transistor package with gull-wing leads. Its abbreviation is **SOT**.

small-scale integration In the classification of ICs, the inclusion of 100 or less electronic components, such as transistors, on a single chip. Its abbreviation is **SSI**.

small-signal This contrasts with **large-signal**. 1. A signal whose amplitude is comparatively small. 2. A signal whose amplitude is sufficiently small to be able to consider the operation of the component, circuit, device, or system in question to be linear. When analyzing the response at such signal levels, nonlinear components can usually be ignored.

small-signal analysis The analysis of a component, circuit, device, or system operating at **small-signal** amplitudes.

small-signal bandwidth The useful bandwidth of a device operating at **small-signal** levels.

small-signal characteristics The characteristics of a component, circuit, device, or system operating at **small-signal** amplitudes.

small-signal current amplification Same as **small-signal current gain**.

small-signal current gain The current gain of an amplifier at **small-signal** amplitudes. Also called **small-signal current amplification**.

small-signal diode A diode designed to operate at **small-signal** levels.

small-signal equivalent circuit A circuit which is equivalent to another circuit or device operating at **small-signal** amplitudes.

small-signal gain The gain of an amplifier at **small-signal** amplitudes.

small-signal operation Operation of a component, circuit, device, or system at **small-signal** amplitudes.

small-signal parameters Parameters which are characteristic of a component, circuit, device, or system operating at **small-signal** amplitudes.

small-signal performance The performance of a component, circuit, device, or system at **small-signal** amplitudes.

small-signal transistor A transistor designed to operate at **small-signal** levels.

Smalltalk An object-oriented programming language integrated with an operating system, which is utilized to provide an object-oriented programming environment.

SMART Abbreviation of Self-Monitoring Analysis and Reporting Technology. A standard, technology, or interface utilized to keep track of the state of hard drive, and for warning of possible future failures.

smart Of, pertaining to, consisting of, or arising from a component, circuit, device, piece of equipment, or system, which is highly automated, adapts rapidly to changing environmental or operational conditions, emulates human thought or behavior processes, or the like.

smart appliance 1. A device, intended for home use, which is connected to a network, so as to gather and/or distribute information. An example is an Internet appliance. Said especially of those which have an easy-to-use interface. Also called **information appliance** (1). **2.** A home appliance, such as a central air conditioner, which provides a user interface such as a touch screen, for simple programming. Also called **information appliance** (2). **3.** Any device, such as a properly equipped computer, PDA, TV, or cellular phone, which provides access to the Internet. Also, such a device specifically designed for this purpose. Also called **Internet appliance**, or **Internet access device**.

smart cable A cable which incorporates a microprocessor to analyze the signals passing through it, as opposed to just serving as a transfer medium. Also called **intelligent cable**.

smart card Also called **chip card**, or **IC card**. **1.** A card which incorporates one or more chips, including a processor and memory. Such cards are usually the size of a credit card, and are used in conjunction with smart card readers which communicate with central computers. Such cards may be used to verify identity, keep digital cash, store medical records, and so on. Smart cards are secure, and update the contained information each time it is used. **2.** In portable computers, a **smart card** (1) utilized to add memory, or as a peripheral device such as a modem.

smart card reader A device into which chip **smart cards** are inserted, to access and update their contained information. Also called **chip card reader**, or **IC card reader**.

smart dust A collection of microelectromechanical machines which incorporate sensors, processing capabilities, a power supply, and the ability to intercommunicate wirelessly with each other so as to form a single computing unit. Each smart dust particle is called a **mote** (1), and such dust is light enough to remain suspended in the air of a given area or enclosure. Used, for instance, for weather sensing, verification that groups of devices are functioning properly, or for spying.

smart home Same as **smart house**.

smart house A house, such as that equipped with **smart appliances**, with a given degree of automation. Such a dwelling may enable, for instance, remotely turning on the heat and/or lights, starting the washing machine, rolling the drapes, taking a peek at an indoor camera before entering, and so on. Features may be accessed via a cell phone, PDA, a computer linked to the Internet, and the like. Also called **smart home**.

smart hub Also called **intelligent hub**. **1.** A network hub which performs processing functions such as network management. **2.** A network hub which can be electronically reconfigured, be it locally or remotely.

smart label A label, such as that placed on a box, which stores information such as a serial number. Used, for instance, in an RFID system.

smart material A material, such as a shape-memory alloy, which can respond to stimuli in its environment.

Smart Media Same as **SmartMedia**.

Smart Media card Same as **SmartMedia**. Also spelled **SmartMedia card**.

smart phone Same as **smartphone**.

smart telephone Same as **smartphone**.

smart terminal A network terminal which incorporates a CPU and memory. Such a terminal has processing capability, although to a lesser extent than an **intelligent terminal**.

SmartMedia A compact memory card, such as an SD card, which consists of a flash memory chip. Used, for instance, in handheld PCs, PDAs, cell phones, digital cameras, and digital music players. Such cards usually do not have an incorporated controller, with said controller being contained in the slot such cards are inserted into. Also spelled **Smart Media**. Also called **SmartMedia card**.

SmartMedia card Same as **SmartMedia**. Also spelled **Smart Media card**.

smartphone A telephone which has features and services such as access to email, the Internet, unified messaging, personal information management, voice-recognition capabilities, a USB port, reception of GPS signals, and so on, in addition to traditional telephone functions. Also spelled **smart phone**. Also called **smart telephone**.

SMATV Abbreviation of **satellite master-antenna TV**.

SMATV system Abbreviation of **satellite master-antenna TV system**.

SMB Abbreviation of **Server Message Block**.

SMB Protocol Same as **Server Message Block Protocol**.

SMD Abbreviation of **surface-mount device**.

SMDS Abbreviation of Switched Multimegabit Data Service. A high-speed packet-switched data service that is connectionless. It is usually in the form a public data network, and serves to link LANs, and sometimes WANs.

smear In TV or fax, a form of distortion in which objects appear to be extended horizontally beyond their usual boundaries, with concomitant blurring. Also called **smearing**.

smearing Same as **smear**.

smectic crystals Same as **smectic liquid crystals**.

smectic liquid crystals Liquid crystals where molecules are arranged in layers, and in which there is order within their parallel layer, but much less so between adjacent layers. Smectic liquid crystals have order in two dimensions, while **nematic liquid crystals** have so in one. Also called **smectic crystals**.

SMIL Acronym for **Synchronized Multimedia Integration Language**.

smiley A simple icon usually produced with punctuation marks. They are typically viewed as being sideways. For example, :-) is meant to express a smile, :-D is a bigger smile, and so on for a number of faces or emotions. Smileys do not always represent smiles. Also called **emoticon**.

Smith chart A chart utilized to aid in calculations, such as those pertaining to reflectance and impedances, circuits, transmission lines, waveguides, or antennas.

smoke alarm An electrical and/or optical device which serves to warn of the presence of smoke by means of a signal, such as sound and/or light. Also, such a warning.

smoke detector A component, circuit, or device which senses the presence of smoke in a given enclosure, structure, area, or the like. Such a detector serves as the input for a smoke alarm. Examples include photoelectric smoke detectors, and ionization smoke detectors. Also called **smoke sensor**.

smoke sensor Same as **smoke detector**.

smoke test A test performed on a component, circuit, printed circuit, device, piece of equipment, or system. In it, the item in question is powered on, or otherwise put into operation,, and if any smoke is visible the test is failed. The observation of sparks or other such electrical effects also constitute a failed test.

smooth 1. Having a surface which is free from roughness, projections, irregularities, or the like. 2. To polish or otherwise rid a surface of roughness, projections, irregularities, or the like. 3. To remove voltage and/or current irregularities, such as ripple, present in a signal or generated by a source. 4. To remove irregularities, inconsistencies, differences, or the like. 5. Having an even motion, movement, or trajectory.

smoothing choke A choke utilized to reduce ripple.

smoothing circuit In rectifiers or DC generators, a circuit which reduces the AC component of the output, while passing the DC.

smoothing filter 1. In rectifiers or DC generators, a filter which reduces the AC component of the output, while passing the DC. Also known as **ripple filter** (1). 2. A filter which blocks all frequencies above a given cutoff frequency, before an analog-to-digital conversion. The filter insures that no input signals have a higher frequency than half the digital sampling rate. Also called **low-pass filter** (1), or **anti-aliasing filter**.

SMP 1. Abbreviation of **symmetric multiprocessing**. 2. Abbreviation of **symmetric multiprocessor**. 3. Abbreviation of **switched-mode power supply**.

SMR Abbreviation of **Specialized Mobile Radio**.

SMS Abbreviation of short messaging service. 1. A service which allows for the sending of messages no longer than a given length, such as 160 characters, or a given number of bytes, such as 140. It is usually used in the context of wireless devices such as cell phones and PDAs, and may consist of text or simple graphics. Also called **SMS service**. 2. A specific message sent via an **SMS service**. Also called **SMS message**. 3. A standard enabling such messaging. Also called **SMS standard**. 4. A technology enabling such a service. Also called **SMS technology**.

SMS message Same as **SMS (2)**.

SMS messaging The sending of **SMS messages**. Also called **short messaging**.

SMS service Same as **SMS (1)**.

SMS standard Same as **SMS (3)**.

SMS technology Same as **SMS (4)**.

SMT Abbreviation of **surface-mount technology**.

SMTP Abbreviation of Simple Mail Transfer Protocol. A protocol used for sending, forwarding, and routing emails. It is usually utilized along with an email protocol such as POP or IMAP, for message retrieval and storage. Thus, an email program might use SMTP as the outgoing mail server, and POP or IMAP as the incoming mail server.

SMTP server A server utilized to route emails send using SMTP. A mail server such as POP or IMAP is usually used for reception and storage of such mail.

smurf Same as **smurf attack**.

smurf attack A denial-of-service attack in which a network is inundated with ping packets. In it, a ping request is sent to a broadcast address of a target network, where all nodes receive and immediately respond to said packets. Since each request can be replicated up to 255 times, thousands or millions of such requests per second, along with their responses, can quickly paralyze a network. If IP broadcasting is disabled, such an attack can not be made. Also called **smurfing**, or **smurf**.

smurfing Same as **smurf attack**.

Sn Chemical symbol for **tin**.

SNA Abbreviation of **Systems Network Architecture**.

snail mail Mail handled by a postal service, as opposed to email. Aside from distinguishing between physical and electronic delivery, the term makes reference to the sluggishness of postal services. Also spelled **snailmail**.

snailmail Same as **snail mail**.

snake 1. An object or device which assists in running wires or cables through tight or otherwise difficult places, such as narrow conduits. 2. To run wires or cables using a **snake** (1).

snap-action switch A mechanical switch that rapidly transfers to the opposite position once its actuating button, lever, paddle, or the like, is moved past a comparatively small distance. Also called **snap switch**.

snap-in connector A connector, such as a phone or phono plug, which snaps into place by aligning and applying pressure.

snap-on connector A connector which requires an applied pressure or a twisting action to secure. A BNC is an example.

snap switch Same as **snap-action switch**.

snapshot 1. Same as **snapshot dump**. 2. The data contained in memory, storage locations, a database, or the like, at a given point in time.

snapshot dump A copy of all or part of the data contained in memory, storage locations, a database, or the like, at a given point in time. Such a version may be displayed, printed, or saved. Also called **snapshot (1)**.

sneak current Current which flows via a **sneak path**.

sneak path The path followed by a leakage current, such as that which flows from an output to an input when not intended. Also called **leakage path**.

sneakernet A means of transporting data between computers which are not part of the same network. For example, a user may copy files onto an optical disk, walk to the target computer, and paste said files.

Snell's law Same as **Snell's law of refraction**.

Snell's law of refraction For light undergoing refraction, a law which states that the ratio of the sine of the angle of incidence to the sine of the angle of refraction is a constant, for specific wavelengths of light and particular sets of mediums. Utilized to calculate refraction angles. Also called **Snell's law**.

sniffer A hardware or software device, such as a packet sniffer, which monitors network traffic looking for problems such as bottlenecks. A sniffer may also be used maliciously, to capture private information which is not properly encrypted.

snippet A small piece of programming code. Also called **code snippet**.

SNMP Abbreviation of Simple Networking Management Protocol. A protocol utilized for network management and monitoring. SNMP sends messages to network devices, such as terminals, routers and hubs, which in turn provide information about themselves which they have stored in a Management Information Base.

SNMP-2 An enhanced version of **SNMP** which provides continuous feedback and improved security.

SNMP agent An **SNMP** compliant device. Such a device, when queried, provides information about itself, such as status and activity, which has been stored in a Management Information Base.

SNOBOL Acronym for String Oriented Symbolic Language. A computer programming language used for string processing and pattern matching.

snow On a display screen, such as that of a TV or radar, an unwanted pattern resembling a heavy snow storm. Such a pattern covers the entire viewing area, may be due to random noise, and is seen when a signal is absent or too weak to overcome said noise.

SNR Abbreviation of **signal-to-noise ratio**.

SNTP Abbreviation of **Simple Network Time Protocol**.

snubber An *RC* network utilized to reduce spikes, or the rate at which voltage rises in switching applications.

SO-DIMM Abbreviation of **small-outline DIMM**.

SoC Abbreviation of **system-on-a-chip**.

SOC Abbreviation of **state of charge**.

social engineering 1. The maliciously obtaining of confidential information by conning others into unwittingly providing it. For instance, to call an office impersonating a senior staff member who barks the need a forgotten password immediately. 2. The maliciously obtaining of confidential information without resorting to technical expertise such as hacking. For example, the guessing of obvious passwords, shoulder surfing, or sifting through dumpsters for sheets containing passwords or access codes.

socket 1. A fixture into which other components, such as plugs, connecting terminals, CPUs, or circuit boards are plugged in. 2. A power line termination whose openings serve to supply electric power to devices or equipment whose plug is inserted into it. Also called **electric socket**, or **power outlet**. 3. A software request which connects an application with a service, via a network. Each socket has an associated address, consisting of a port number and network address.

socket 0 A 168-pin motherboard socket used for certain older CPUs.

socket 1 A 169-pin motherboard socket used for certain older CPUs.

socket 2 A 238-pin motherboard socket used for certain older CPUs.

socket 3 A 237-pin motherboard socket used for certain older CPUs.

socket 4 A 273-pin motherboard socket used for certain older CPUs.

socket 5 A 320-pin motherboard socket used for certain older CPUs.

socket 6 A 235-pin motherboard socket used for certain older CPUs.

socket 7 A 321-pin motherboard socket used for certain older CPUs.

socket 8 A 387-pin motherboard socket used for certain older CPUs.

socket 370 A 370-pin motherboard socket used for certain CPUs.

socket 423 A 423-pin motherboard socket used for certain CPUs.

socket 478 A 478-pin motherboard socket used for certain CPUs.

socket 603 A 603-pin motherboard socket used for certain CPUs.

socket A A 462-pin motherboard socket used for certain older CPUs.

SODIMM Abbreviation of **small-outline DIMM**.

sodium A soft silvery-white metallic element whose atomic number is 11. It is an alkali metal, extremely reactive, oxidizes rapidly in air, and reacts vigorously with water to produce sodium hydroxide. It is less dense than water, and has over 15 known isotopes, of which one is stable. Its applications in electronics include its use in sodium lamps, as a heat-transfer agent, in solar power, and as a coolant in nuclear reactors. Its chemical symbol, **Na**, is taken from the Latin word for sodium: **na**trium.

sodium acetate Colorless crystals used in photography and electroplating.

sodium bicarbonate A white crystalline powder or granules whose chemical formula is $NaHCO_3$. Used in plating.

sodium bifluoride A white crystalline powder whose chemical formula is $NaHF_2$. Used in plating and in etching.

sodium bisulfite A white crystalline powder whose chemical formula is $NaHSO_3$. Used in plating.

sodium borate White crystals or crystalline powder whose chemical formula is $B_4Na_2O_7$. Used in soldering and glasses.

sodium bromide A white crystalline powder whose chemical formula is $NaBr$. Used in photography.

sodium chloride Colorless or white crystals or powder whose chemical formula is $NaCl$. It is one of the most common compounds on the planet, and is essential to most of its life forms. It is better known as *common salt*, *table salt*, or *salt*, and its applications include its use in spectroscopy, metallurgy, ceramics, photography, and in nuclear reactors.

sodium chloroplatinate A yellow powder whose chemical formula is Na_2PtCl_6. Used in etching, photography, and microscopy.

sodium fluoride Clear crystals or a white powder whose chemical formula is NaF. Used in electroplating, spectroscopy, and glass manufacturing.

sodium hydroxide A white chemical compound whose formula is $NaOH$. Its applications in electronics include its use in electroplating and etching. Also called **caustic soda**.

sodium hypophosphite White granules or powder whose chemical formula is NaH_2PO_2. Used in plating.

sodium iodide White scintillating crystals whose chemical formula is NaI.

sodium lamp Same as **sodium-vapor lamp**.

sodium nitrate Colorless crystals whose chemical formula is $NaNO_3$. Used in glass manufacturing.

sodium nitrite Yellow crystals, granules, or powder whose chemical formula is $NaNO_2$. Used in photography.

sodium stannate Whitish crystals whose chemical formula is Na_2SnO_3. Used in ceramics, electroplating, and in glass manufacturing.

sodium thiosulfate White crystals or powder whose chemical formula is $Na_2S_2O_3$. Used in photography and plating.

sodium-vapor lamp A discharge lamp in which the light is produced by ionizing sodium vapor. Such a lamp provides a characteristic yellow light, and is used, for instance, for outdoor lighting, such as that of streets. Also called **sodium lamp**.

SOFAR Acronym for **s**ound **f**ixing **a**nd **r**anging. A sonar system which utilizes signals traveling within a **SOFAR channel**.

SOFAR channel Abbreviation of **s**ound **f**ixing **a**nd **r**anging **channel**. Within the ocean, a channel via which acoustic waves travel nearly horizontally due to refraction. Reductions in water temperature result in lower sound speed, while increases in pressure speed up sound. Since sound tends to bend, or refract, towards a region of lower sound speed, a horizontal channel is formed, as the sound waves are refracted down by the temperature reductions, and up again by the pressure increases.

soft 1. Easy to cut, work, or shape. 2. Having energy that is comparatively lower than **hard radiation**, as do soft X-rays. 3. Not readily penetrating matter. Said of radiation, such as soft X-rays. 4. That which is flexible, adaptable, or temporary. For example, soft copy, or a soft error.

soft boot The restarting of a computer without powering down. This may be accomplished for instance, by entering an appropriate key sequence. If a program failure is not resolved through a soft boot, a **hard boot** may help. Also called **warm boot**.

soft copy Computer data displayed on a screen, or stored, as opposed to **hard copy**, which is printed.

soft-drawn wire Wire, such as that of copper or aluminum, which is annealed after being drawn. Such wire has decreased tensile strength in comparison to **hard-drawn wire**. Also called **annealed wire**.

soft error An error from which recovery is possible. For example, an error resulting in the inability to read data, but which is corrected on the following attempt.

soft failure A failure from which recovery is possible. For instance, a failure resulting from operation in a harsh environment, but which is remedied when working under more favorable conditions.

soft handoff In cellular communications such as those utilizing CDMA, the smooth transition from one cell site to another. This is accomplished, for instance, by overlapping the coverage provided by adjacent cell sites and momentarily using signals from both repeaters until the handoff is secure.

soft iron Iron which is easily magnetized and demagnetized. Used, for instance, in armature cores, electromagnets, and magnet keepers.

soft key Same as **softkey**.

soft link Same as **symbolic link**.

soft magnetic material A material, such as regular iron, with a low magnetic coercive force. Such a material is not suitable for making permanent magnets.

soft modem An application that performs modem functions using the system's hardware, such as the CPU, as opposed to chips and other components contained in a modem. Soft modems are more flexible and better adapt to newer technologies, but in exchange utilize more system resources such as available RAM and CPU power.

soft radiation Radiation whose energy is comparatively lower than **hard radiation**, and which thus does not readily penetrate matter.

soft-sectored disk A disk with **soft-sectoring**.

soft sectoring In magnetic disk storage, sectoring in which data marks the sector boundaries. This contrasts with **hard sectoring**, which utilizes a physical marking, such as a hole.

soft solder A solder that melts at a comparatively low temperature. For example, an alloy of lead and tin. This contrasts with a **hard solder**, which melts at a comparatively high temperature.

soft start 1. The use of a component, circuit or device, to limit the peak instantaneous current drawn by a component, cir-

cuit, device, piece of equipment, or system, such as a power amplifier or power supply, when first turned on. 2. A specific instance of a **soft start** (1).

soft-start circuit A circuit utilized to limit the peak instantaneous current drawn by a circuit, device, piece of equipment, or system, when first turned on.

soft tube A vacuum tube which is not sufficiently evacuated, and whose residual gas affects its electrical characteristics. Also called **gassy tube**.

soft ultraviolet Electromagnetic radiation in the ultraviolet region which is nearest to visible light. It corresponds to wavelengths of approximately 300 to 400 nm, though the defined interval varies. Also called **near-ultraviolet radiation**.

soft vacuum 1. A low pressure in an electron tube, but which is still high enough to allow ionization of a present gas. 2. A low pressure in an electron tube, but which is still high enough to exhibit a given property.

soft X-rays X-rays whose wavelengths are closer to the ultraviolet region, as opposed to those nearer to the gamma-ray region. The longer the wavelength, the lesser the energy, which in turn makes these rays penetrate matter less readily than **hard X-rays**.

softkey A key, such as that on a computer keyboard or cell phone, that can serve for different functions or purposes, depending on the application running, mode of operation, or the like. A cell phone, for instance, may use a single key to access menu selections and make entries, with the current function indicated by a keyword on its LCD display. Also spelled **soft key**.

softswitch Abbreviation of **soft**ware **switch**. Software which performs the necessary protocol conversions to link packet-based traffic, such as that of the Internet, to a public switched telephone network.

software A set of instructions or programs, which when translated and executed cause a computer to perform specific operations. Software is usually classified into two general categories, which are system software, and application software. System software includes control programs, such as the operating system, while application software refers to programs which perform specific tasks, such as a word processor. Although **software** and **program** are used mostly synonymously, software is a bit more of a general term, as it is used to differentiate from the physical equipment of a computer, or the **hardware**, from the instructions which tell a computer what to do, or the software. An application is both a program and software, but when using the term software it is usually to refer to multiple application programs, or application programs in general, in which case they are called application software. Also called **computer software**. Its abbreviation is **SW**.

software architecture The design of software and its components. This determines what other software it can interact with, the hardware it can work with, its reliability, and its flexibility and expandability. Also called **architecture** (4).

software bloat The additional consumption of system resources, such as RAM and disk space, resulting from the inclusion of new features to existing software products. Such features may or may not be necessary, and also tend to consume more of the user's time and patience.

software bug A persistent error that creates malfunctions or stoppages of a computer program. These are usually remedied by rewriting part of the software.

software configuration 1. The specific manner in which software is set up for a given system, architecture, or use. 2. The process of setting a given **software configuration** (1).

software conversion The process of changing a program from one form to another. For example, to adapt a program to a different platform.

software developer kit Same as **SDK**.

software development kit Same as **SDK**.

software engineering The discipline dealing with the design, development, implementation, testing, documentation, maintenance, and enhancement of software.

software failure A failure of a computer program. Such a failure may be caused, for instance, by a program deficiency such as a bug, or a virus.

software house A business entity which sells or develops software tailored to specific customer needs. This contrasts, for instance, with a **software vendor**, which sells packaged software.

software interface A set of instructions that determine how a computer application interacts with the operating system. This contrasts with an **application binary interface**, which interacts with the operating system and the hardware. Also called **application program interface**, or **programming interface**.

software interrupt An interrupt generated by a computer program.

software lock A software-based lock as opposed to one based on hardware. An example is a password. Also called **logical lock**.

software package **1.** A combination of application programs which are tailored to a given type of work, and which function especially well together. For example an office suite incorporating word processing, database, spreadsheet, presentation, and communications programs. Also called **suite, software suite, suite of applications, package (4), application suite**, or **bundled software**. **2.** An application program which is sold to the general public with all that should be necessary for it to work properly. This contrasts with **custom software**, which is designed to meet specific needs or circumstances. Also called **packaged software, package (5)**, or **prepackaged software**.

software piracy The unauthorized copying and subsequent use and/or sale of software with copyright protection. When purchasing a program, an individual or entity becomes a licensed user, and can usually make backup copies when possible. Also called **piracy (1)**.

software portability The extent to which computer software can be utilized across multiple platforms with little or no modifications.

software program A redundant term equivalent to **software** or **program**.

software protection Measures taken to help avoid the unauthorized copying of software. These include the requirement to enter a registration number when installing a program, or the use of a dongle. Also called **copy protection**.

software publisher A business entity that develops, markets, and sells software. Such a business may also market and sell software they have not developed.

software release Same as **software version**.

software requirements The software and configuration a system must have in order for a given program or device to work properly. For example, in order to utilize a given application, a computer needing to have one of the supported operating systems, a Web browser, and so on.

software reuse **1.** The ability to use code and routines over and over in the same application or program or those being developed. Software reuse is facilitated for instance, within an object-oriented approach. **2.** A specific instance of **software reuse (1)**.

software rot A program, or parts of a program, which previously worked, but no longer does. Also, parts of a program or routines which although present, are never accessed.

software suite Same as **software package (1)**.

software support **1.** To provide technical support for a given software product. **2.** The support, by a given platform, of a given application or other computer program.

software switch Same as **softswitch**.

software tool A program or utility that assists in the development, analysis, testing, and maintenance of software. Such tools include compilers, decompilers, debuggers, editors, and the like. Also called **tool (2)**.

software version The numbers, letters, and/or words which identify a specific version of a software package. For example, *1.0*, *5.01a*, or *2004b beta build #1111*. Changes in whole numbers usually imply significant modifications, while fractions, letters, or builds often identify versions which correct bugs, provide minor enhancements, or the like. Also called **software version number**, or **software release**.

software version number Same as **software version**.

SOG Abbreviation of **small-outline gull-wing**.

SOH Abbreviation of **start-of-heading**.

SOHO Acronym for **small office/home office**.

SOI Abbreviation of **silicon-on-insulator**.

SOIC Abbreviation of **small-outline integrated circuit**.

SOJ Abbreviation of **small-outline J-lead**.

sol Abbreviation of **solution**.

solar access The availability of direct sunlight, such as that needed for the optimum operation of solar devices.

solar activity The occurrence of events, such as sunspots and solar flares, which change variables such as the energy released or the radio noise produced by the sun.

solar array Same as **solar panel**.

solar battery Same as **solar panel**.

solar battery charger A charger that utilizes a solar panel to charge conventional rechargeable batteries.

solar cell A semiconductor device which generates an electric current when exposed to radiant energy, especially sunlight. These are often made of silicon or selenium. Solar cells can be used in arrays to power anything from street lights to satellites. It is a photovoltaic cell which is optimized for the conversion of solar energy. Also known as **solar generator**, or **sun cell**.

solar-cell array Same as **solar panel**.

solar concentrator A device which utilizes optical elements to increase the amount of sunlight incident upon a solar cell or panel.

solar constant The rate at which solar radiation is received just outside of the earth's atmosphere. It is calculated based on a surface perpendicular to the incident radiation at the earth's mean distance from the sun. This value is not constant, and is equal to approximately 1,370 watts per square meter.

solar cosmic rays Cosmic rays whose source is solar activity, such as solar flares.

solar cycle The periodic variation in the number of sunspots, and in the levels of other solar activity, such as solar flares. Each cycle lasts approximately eleven years. Also called **sunspot cycle**.

solar device **1.** A device, such as a solar panel, which makes use of sunlight. **2.** A device which makes use of radiant energy, especially sunlight.

solar energy **1.** The electromagnetic radiation generated by the sun. **2.** **Solar energy (1)** which is incident upon the planet earth. **3.** **Solar energy (2)** which is converted into other forms of energy, such as electricity. In a single hour, the sun provides more energy than the entire level expended by human activities in a full year.

solar-energy conversion The conversion of **solar energy (2)** into other forms of energy, such as electricity.

solar-energy conversion device A device, such as a solar panel, utilized for **solar-energy conversion**.

solar flare A bright eruption which develops near the surface of the sun. These are of tremendously high temperature, take a few minutes to reach full strength, and subside over the next hour or so. High-energy particles and rays are explosively released, and when they reach the earth may cause radio interference, magnetic storms, or auroras. Also called **flare (4)**.

solar flux A measure of the solar radio noise at a given frequency, such as 2,800 MHz.

solar generator Same as **solar cell**.

solar irradiation Exposure to **solar radiation**.

solar module Same as **solar panel**.

solar noise Same as **solar radio noise**.

solar panel A device usually consisting of multiple **solar cells** which are connected to produce a given power output. Also called **solar array, solar module, solar battery, solar-cell array, sun battery**, or **panel (4)**.

solar power 1. Electric energy derived from the conversion of **solar energy (2)**. For example, the power obtained from a solar panel. **2.** The use of solar energy to generate electricity, or another form of energy.

solar radiation Electromagnetic energy radiated by the sun, which includes a wide interval of wavelengths. The solar radiation arriving on the earth's surface is mostly in the visible light, infrared, and ultraviolet regions.

solar radio emission **Solar radiation** occurring in the RF region.

solar radio noise RF noise produced by the sun. Solar noise increases during solar activity such as flares. Also called **solar noise**.

solar radio waves RF waves originating from the sun.

solar spectrum The distribution of electromagnetic radiation emitted by the sun. Although the solar spectrum essentially covers the entire electromagnetic spectrum, most of the energy is concentrated in an interval whose wavelengths span from approximately 300 nm to 3000 nm.

solar system 1. A power system incorporating one or more solar panels, along with the necessary structural supports, conductors, storage batteries, inverters, fuses, grounds, and so on. **2.** A system consisting of a star, along with the planets, asteroids, and other celestial bodies which orbit around it. **3.** The system consisting of the sun and the planets, asteroids, and other celestial bodies which orbit around it.

solar wind Streams of plasma flowing outward from the sun. The magnetosphere is distorted into a teardrop shape by the solar wind, and the Van Allen radiation belts are composed mostly of protons and electrons which originate from the solar wind which become trapped by the earth's magnetic field. The velocity of such particles is typically around 400 kilometers per second, but during periods of heavy solar activity can exceed 800 kilometers per second.

Solaris A UNIX-based operating system designed for networking, which is characterized by stability and scalability.

solder 1. A metal or alloy that is utilized to join two metals, such as the leads of circuit components to conductive portions of printed circuits. The solder must have a lower melting point than that which is to be joined, as the former is melted, and the latter is not. Solders with comparatively low melting points, such as an eutectic alloy of tin and lead, are called soft solders, while those with comparatively high melting points, such as an alloy of silver, copper, and zinc, are called hard solders. A solder flux may be utilized to help prepare a surface for soldering. **2.** To join surfaces with **solder (1)**.

solder ball A tiny ball of solder used, for instance, in Ball Grid Array chip packages. Also called **solder bump (1)**.

solder bridge The use of solder to form an electric connection between terminals, wires, or pins. Also called **bridging (3)**.

solder bump 1. Same as **solder ball**. **2.** A solder ball utilized in the context of flip chip bonding.

solder die attach The use of solder to attach a chip to a substrate. Used mainly for high-power devices because of its fine thermal and electrical conductivity.

solder flux A material which better prepares surfaces for **soldering**. It may do so, for instance, by removing oxides. Also called **soldering flux**, or **flux (5)**.

solder flux residue Any excess **solder flux** which becomes a contaminant near the soldered surface. Also called **flux residue**.

solder gun Same as **soldering gun**.

solder iron Same as **soldering iron**.

solder mask A material or device which is placed upon a surface so that only areas that wish to be soldered are exposed. Used, for instance, in the preparation of printed-circuit boards.

solder paste A **solder flux** in the form of a paste. Also called **solder paste flux, soldering paste**, or **soldering paste flux**.

solder paste flux Same as **solder paste**.

solder pencil Same as **soldering pencil**.

solder wetting The coating of a surface to be soldered with a smooth adherent film of molten solder. This helps ensure proper heating, the minimization of the transfer of impurities, and the obtaining of a sound joint. Also called **wetting (2)**.

solderability The ability of a surface to be joined to another using **solder (1)**. This may depend, for instance, on how well such a surface lends itself to **solder wetting**.

soldering The use of a **solder (1)** to join surfaces. Also, the process involved.

soldering flux Same as **solder flux**.

soldering gun A **soldering iron** in the form of a portable handheld device equipped with a button or switch allowing for intermittent application of heat. Also called **solder gun**.

soldering iron A device or tool which incorporates a tip that is utilized to apply heat to a **solder (1)**. The application of a thin coat of solder to the tip of a soldering iron, and/or the conductors or surfaces to be soldered, helps ensure proper heating, the minimization of the transfer of impurities, and the obtaining of a sound joint. Also called **solder iron**.

soldering paste Same as **solder paste**.

soldering paste flux Same as **solder paste**.

soldering pencil A **soldering iron** whose shape is roughly similar to that of a pencil. The wattage of such a device is usually comparatively small, such as 35 or 50 watts. Also called **solder pencil**.

solderless Not utilizing or requiring **soldering**. For instance, a solderless contact.

solderless breadboard A breadboard upon which components are mounted and interconnected without employing **soldering**. Connections on such a board are made, for instance, by inserting the leads of components into jacks.

solderless connection A connection in which the joined conductors are pressed, pinched, or twisted together, as opposed to soldered. Also called **crimp connection**.

solderless contact An electric contact in which the connected conductors are pressed, pinched, or twisted together, as opposed to soldered. A specialized tool may be used for this. Also called **crimp contact**.

solderless terminal A terminal which is joined via a **solderless connection**. Also called **crimp terminal**.

solenoid **1.** An electromechanical device which incorporates a coil wound around a movable core, and which produces a magnetic field when a current flows through it, inductively moving said core. The core is usually consists of a ferromagnetic cylindrically-shaped plunger or rod. Used, for instance, to convert electrical energy into the mechanical energy utilized to operate electromechanical devices and equipment. **2.** A coil which has been wound into the shape of a cylinder. When a moveable ferromagnetic core is inserted in its center, the result is a **solenoid** (1). Also called **solenoid coil** (2).

solenoid coil **1.** The coil in a **solenoid** (1). **2.** Same as **solenoid** (2).

solenoid magnet An electromagnet produced by the current flowing through a **solenoid coil**.

solenoid-operated Operated by, or requiring the use of a **solenoid** (1). An electrical door locking device, for instance, may be solenoid-operated.

solenoid plunger A usually ferromagnetic plunger or rod which serves as the armature of a **solenoid** (1).

solenoid switch A switch whose movement is assisted by a **solenoid** (1). Used, for instance, to activate the starter motor of a car.

solenoid valve A valve, such as that controlling the flow of a fluid such as air or water, that is opened and closed by a **solenoid** (1).

solid **1.** A state of matter characterized by having both a definite shape and a definite volume, and which tends to resist forces that tend to change said shape and volume. The atoms, molecules, or ions that compose solids are very close together, and are held firmly in place by intermolecular attraction which exceeds that of liquids. Matter in the solid state has little or no motion, and any movement of its constituent particles is usually confined to vibrations around a fixed position. Solids usually have a high density, and are either crystalline or amorphous. Solids can be distinguished from each other via characteristics such as density, hardness, malleability, and brittleness. The other physical states in which matter is known to exist are **liquid**, **gas**, **plasma**, and **Bose-Einstein condensates**. Also called **solid state**. **2.** A geometric figure with three dimensions.

solid angle A three-dimensional conical surface formed by all the rays that extend from a given point through a closed curve. Usually measured in steradians.

solid circuit A circuit consisting of a single piece of material, usually a semiconductor. The components of such a circuit are formed by etching, deposition, diffusion, alloying, and the like.

solid conductor A conductor composed of a single wire, strip, bar, or the like. The term is used, for instance, to contrast with a stranded or otherwise divided conductor.

solid earth Same as **solid ground**.

solid electrolyte An electrolyte, such as that utilized in a fuel cell, which is in the **solid state**.

solid ground A direct connection to an earth ground. Also called **solid earth**.

solid-ink printer A printer whose ink is in the form of a solid stick or mass. When printing, the appropriate amount of ink is melted and transferred to the sheet being printed, where it rapidly cools and solidifies to form crisp images.

solid model A three-dimensional geometric model which represents both the external shapes and the internal structures of an object. This contrasts, for instance, with a **surface model**, which appears externally as a solid, but which does not have internal structure. A solid model has attributes such as thickness and volume, making it more complex and processing-intensive than a surface model.

solid modeling The creation and representation of **solid models**, using computer-aided design and/or mathematical techniques.

solid state Same as **solid** (1).

solid-state **1.** Consisting of, pertaining to, characteristic of, or arising from matter whose state is **solid** (1). **2.** Based on, utilizing, or consisting of materials which are **solids** (1), especially crystalline semiconductor materials. The function of that which is solid-state usually depends on electrical, magnetic, and/or optical phenomena occurring within solids, such as the movements of charge carriers in response to light. Examples include solid-state components, circuits, and devices. **3.** That which has no moving parts. Used, for instance, in the context of that which formerly required or otherwise utilized moving parts, such as a solid-state switch as opposed to a mechanical switch.

solid-state amplifier An amplifier, such as that consisting of transistors, which is composed entirely of solid-state components.

solid-state battery A battery whose electrolyte and terminals are all in the solid state. Such batteries usually feature very high energy density, excellent charge retention, high thermal stability, an extended temperature operating range, and can be made into most any shape, in addition to being flexible. Used, for instance, in thin-film technology.

solid-state camera A camera, such as a CCD camera, which utilizes solid-state technology to convert optical images into electric charges.

solid-state capacitor A capacitor, such as a solid-tantalum capacitor, whose dielectric is in the solid state.

solid-state circuit A circuit, such as a semiconductor circuit, composed entirely of solid-state components.

solid-state component A circuit component, such as a solid-state capacitor or transistor, which is in the solid state.

solid-state device A device, such as a semiconductor device, which is in the solid state.

solid-state disk A storage disk, such as a flash disk, consisting of memory chips. The access time of such disks is much lower than that of conventional magnetic disks. Its abbreviation is **SSD**.

solid-state disk drive A disk drive which reads data from, and writes data to, a **solid-state disk**. Such a drive is usually very rugged, incorporates its own battery backup, and should be seen by an operating system as identical to a regular magnetic disk drive.

solid-state floppy disk card A memory card which is intended to perform the functions of a floppy disk and drive. Its abbreviation is **SSFDC**.

solid-state image sensor An image sensor, such as a CCD image sensor, which utilizes solid-state technology to convert optical images into electric charges.

solid-state lamp A lamp, such as an LED, whose light-emitting material is in the solid state.

solid-state laser A laser whose active medium is in the solid state. Such a solid is usually a glass or a crystal, and examples include ruby and YAG lasers. Semiconductor lasers are sometimes excluded from this definition.

solid-state memory Computer memory, such as chip memory, which is in the solid state.

solid-state physics The branch of physics which deals with the properties, interactions, and applications of **solids** (1). An area of great importance currently is the theory, operation, and technologies pertaining to semiconductor materials.

solid-state relay Also called **static relay**. **1.** A relay that incorporates only solid-state components. **2.** A relay with no moving parts.

solid-state switch Also called **static switch**. **1.** A switch that incorporates only solid-state components. **2.** A switch with no moving parts.

solid-state technology Any technology, such as that utilizing semiconductors, based on, incorporating, or pertaining to solid-state materials.

solid-tantalum capacitor A tantalum capacitor whose dielectric is tantalum oxide in the solid state.

solid wire A single undivided wire, as opposed to that which is stranded or otherwise divided.

solidus temperature The temperature below which a substance is completely solid, while the **liquidus temperature** is that above which a substance is completely liquid. Used especially in the context of alloys which do not have a single melting point.

soliton An optical pulse that maintains its waveform over very long distances. For example, that of a specially-designed laser pulse traveling long distances along an optical fiber.

solubility The extent to which a solute can dissolve in a solvent, to form a saturated solution.

solute The substance which is dissolved in a **solvent** to form a solution. A solute may be a gas, liquid, or solid.

solution Its abbreviation is **sol**. **1.** A homogeneous material, usually in the liquid phase, in which one substance, the solute, becomes completely dispersed in another, the solvent. For example, an electrolytic solution may consist of salt, the solute, dissolved in water, the solvent. **2.** The state of a solute being dissolved in a solvent to form a **solution** (1). **3.** The answer or approach to a given problem. **3.** The processes utilized to arrive at a **solution** (3).

solution conductivity The ease with which an electric current can flow through a solution, such as an electrolytic solution.

solvent The substance into which a **solute** is dissolved to form a solution. A solvent is usually a liquid, but may also be a solid or gas.

SOM **1.** Abbreviation of **self-organizing map**. **2.** Abbreviation of **start-of-message**.

son file A current version of a file, such as that being updated. The copy or version saved before it is called **father file**. Also called **child file**.

sonar Acronym for **so**und **na**vigation and **r**anging. **1.** A system utilized for the detection and location of underwater surfaces, objects and structures, via the reception of sonic or ultrasonic waves reflected by them. Active sonar equipment transmits and receives sonic or ultrasonic sounds, for detection of the bearing and range of objects, while passive sonar equipment can only receive sonic or ultrasonic sounds, and is thus able to detect bearing only. A radar operates utilizing similar principles, but usually employs microwaves. **2.** An apparatus utilized to generate, transmit, receive, and present information utilizing **sonar** (1). A typical set has a source of acoustic or ultrasonic power, usually an underwater sound projector, hydrophones to detect the reflected sonic or ultrasonic waves, a receiver, one or more speakers, and a display, such as a CRT or LCD. Also called **sonar unit**, **sonar device**, **sonar equipment**, **sonar system**, or **sonar set**.

sonar device Same as **sonar** (2).

sonar equipment Same as **sonar** (2).

sonar navigation Navigation utilizing sound waves to detect and locate objects. Also called **sonic navigation**, or **acoustic navigation**.

sonar receiver In a sonar system, the hydrophones and receiver which detect, amplify, and process the sonic or ultrasonic waves reflected off underwater surfaces.

sonar set Same as **sonar** (2).

sonar system Same as **sonar** (2).

sonar transducer In a sonar system, a device which converts electrical energy into sound energy, and/or vice versa.

sonar transmitter In a sonar system, the source of acoustic or ultrasonic power, which is usually an underwater sound projector, along with the transmitter which generates the signals which drive said projector.

sonar unit Same as **sonar** (2).

sonde A device, such as a radiosonde, which obtains meteorological data by being directly in contact with the atmosphere.

sone A unit utilized to describe subjective loudness. One sone is defined as a pure 1000 Hz tone at 40 decibels above the hearing threshold of a given listener. A sound, as perceived by the same individual, which is twice as loud is two sones, and so on.

SONET Acronym for **S**ynchronous **O**ptical **Net**work. **1.** An ANSI standard for the synchronous transmission of data through optical fibers. SONET provides for transmission rates starting at 51.8 Mbps, called OC-1, through OC-3072 which has a rate of 159.25248 Gbps. SONET provides sophisticated network management including self-healing, a standard interface, and its compatibility with SDH enables the interconnection of digital networks internationally. **2.** A network implementing a **SONET** (1) standard. Also called **SONET network**.

SONET network Same as **SONET** (2).

SONET ring The physical or logical ring architecture utilized when implementing a **SONET network**.

sonic **1.** Of, pertaining to, or arising from **sound**. **2.** Characteristic of, pertaining to, or arising from velocities approaching, equaling, or exceeding the speed of sound. An example is sonic boom.

sonic barrier The significant increase in drag that a body, such as an aircraft, experiences when approaching the speed of sound for a given medium, such as atmospheric air. Also called **sound barrier**.

sonic boom The loud and explosive sound heard when an object or medium exceeds the speed of sound. The sound heard is a result of the shock waves emanating from said object or medium. An aircraft traveling above the speed of sound will produce shock waves, as does the explosive expansion of the air surrounding a lightning discharge. The latter is referred to as **thunder**.

sonic cleaning The use of high-frequency sound waves to clean surfaces which are immersed in an appropriate solvent. Used, for instance, for surfaces that would otherwise be inaccessible for cleansing. The term is usually synonymous with **ultrasonic cleaning**.

sonic delay line A circuit or device that delays the transmission of sound signals. This is accomplished by having the signals pass through an appropriate medium. Also called **acoustic delay line**.

sonic-depth finder A device which determines and indicates the depth of a body of water, by measuring the time required for transmitted sonic or ultrasonic waves to be reflected from a surface, such as the sea bottom. Also called **echo depth sounder**, **depth sounder**, or **fathometer**.

sonic frequency Same as **sound frequency**.

sonic imaging Same as **sound imaging**.

sonic navigation Same as **sonar navigation**.

sonic speed Same as **speed of sound**.

sonic velocity Same as **sound velocity**.

sonobuoy A buoy that incorporates a receiver to detect underwater sounds, and a transmitter to send radio signals to a remote location. A **buoy** is a device which floats above water to mark, indicate, warn, and in this case detect and transmit.

sonogram The images produced while performing an examination utilizing **sonography**. Also called **sonograph**, or **ultrasonogram**.

sonograph Same as **sonogram**.

sonography The use of ultrasonic energy to produce images of structures within the human body. The reflections of the high-frequency sound waves directed at the desired regions are processed and converted into images suitable for observation and analysis. It is noninvasive, and is used, for instance, to monitor a fetus during a pregnancy. Also called **ultrasonography**.

sonoluminescence The emission of light by a liquid which is excited by sound. Such light is usually in the form of extremely short flashes.

SOP 1. Abbreviation of **small-outline package**. **2.** Abbreviation of **sum-of-products**.

sophisticated Very complex, and usually involving recently developed and/or unconventional technology. Said, for instance, of electronics or robots.

sorption Any process, such as absorption or adsorption, in which something is taken up and held. Also called **sorption process**.

sorption process Same as **sorption**.

sort 1. To order, arrange, or group according to category, size, type, function, or the like. Also, such an arrangement or group. **2.** To order, arrange, or group data or data units into a sequence, or to reorder data or data units into a new sequence. Also, such an ordering, arrangement, or grouping. Examples include the sorting of records, text, or files, by date, alphabetically, or any other desired criteria.

sort algorithm One or more mathematical formulas and/or defined steps utilized to order, arrange, or group data or data units into a sequence.

sort field Same as **sort key (1)**.

sort key Also called **sorting key**. **1.** A field which is utilized as the basis for the sequencing of items within a file, set, or the like. Also called **sort field**. **2.** A key utilized as the basis for the sequencing of items within a file, set, or the like.

sorter That, such as a program or algorithm, which serves to **sort**.

sorting key Same as **sort key**.

SOS 1. An internationally recognized distress signal, comprised of the letters S,O, and S in Morse code. This particular combination letters was not selected to mean anything; they were chosen for their simplicity of transmission. In Morse code SOS is three dots, followed by three dashes, and then three dots again. It is rarely used anymore, having been superseded by automated systems, such as those utilizing satellites. **2.** Abbreviation of **silicon-on-sapphire**. **3.** Abbreviation of **systems-on-silicon**. **4.** Abbreviation of **server operating system**.

SOT Abbreviation of **small-outline transistor**.

sound 1. A vibration or other disturbance which travels through an elastic medium, such as air, water, or metal, in the form of longitudinal waves. In dry air, for example, at 0 °C, and at one atmosphere pressure, sound travels at approximately 331.6 meters per second, while in copper sound travels at approximately 3,360 meters per second. Sound waves may be reflected, refracted, scattered, or absorbed, and are also subject to constructive and destructive interference. Characteristics which help distinguish between sounds include frequency, amplitude, and timbre. **2. Sound (1)** whose frequency is within the audible spectrum. Sound is usually heard, but if the frequency is sufficiently low, it may be felt. Sound below 20 Hz is called infrasound, while ultrasound is above 20,000 Hz. Also, the auditory sensation that such sound produces.

sound absorber A material, surface, or medium which absorbs a proportion of the sound energy that strikes it. Also called **acoustic absorber**.

sound absorption Sound energy which is retained or dissipated by a medium which has had sound waves pass though or strike it. Also called **acoustic absorption**.

sound absorption coefficient The proportion of the incident sound energy which is absorbed by a surface or medium. Also called **sound absorption factor**, **sound absorptivity**, **acoustic absorption factor**, **acoustic absorptivity**, or **acoustic absorption coefficient**.

sound absorption factor Same as **sound absorption coefficient**.

sound absorption loss The loss of sound energy due to its retention or dissipation by a medium or surface. This loss in incident acoustic energy may be due, for example, to conversion into heat. Also called **acoustic absorption loss**.

sound absorptivity Same as **sound absorption coefficient**.

sound-activated relay Same as **sound-operated relay**.

sound-activated switch Same as **sound-operated switch**.

sound adapter Same as **sound card**.

sound amplifier Also called **sound-frequency amplifier**, **audio amplifier**, or **audio-frequency amplifier**. **1.** A device which increases the strength of sound waves. **2.** An amplifier of signals that operates within the range of frequencies that humans can hear. **3.** A high-fidelity amplifier that operates within and beyond the range of frequencies humans can hear. For instance, such an amplifier may have a frequency response of 5 Hz to 100 kHz. **4.** A device which intensifies sounds by its shape. A horn is an example.

sound analyzer An instrument which utilizes components such as amplifiers and filters to determine characteristics of sound, such as frequency, amplitude, timbre, and spectral features.

sound band Also called **audio band 1.** The band of frequencies that humans can hear. A healthy person with good hearing can usually detect frequencies ranging from about 20 Hz to about 20 kHz. **2.** A band of frequencies encompassing all or part of the interval of 20 Hz to 20 kHz.

sound barrier Same as **sonic barrier**.

sound bars Unwanted horizontal bars which appear on a TV display, resulting from audio signals interfering with video signals.

Sound Blaster Same as **SoundBlaster**.

sound board Same as **sound card**.

sound buffer In a computer, a buffer which holds the audio data that awaits to be sent to the speakers. Also called **audio buffer**.

sound card An expansion board that enables computers to handle sounds. Audio cards serve to record, playback, and synthesize sounds. They generally provide external jacks for a microphone, a line-in, speakers or headphones, and a MIDI port. Also called **sound board**, **sound adapter**, **audio adapter**, **audio card**, or **audio board**.

sound carrier In TV, the carrier frequency which is modulated to convey sound information. Also called **audio carrier**.

sound chamber A room or other enclosure, such as an anechoic chamber, utilized to perform sound tests, measurements, and experiments. Also called **acoustic chamber**.

sound channel Also called **audio channel**. **1.** A channel which carries a signal within the range of frequencies that humans can hear. **2.** In a system transmitting and/or receiving audio and video signals, the sound channel. **3.** A channel within a high-fidelity component, such as the left channel, or center channel. **4.** In a system where multiple audio signals are available, one of the sound channels, such a version in another language.

sound codec 1. Abbreviation of **sound co**der/**dec**oder. Hardware and/or software which converts analog audio to digital code, and vice versa. 2. Abbreviation of **sound** compressor/**dec**ompressor. Hardware and/or software which compresses and decompresses audio signals.

sound component Also called **audio component**. 1. In a wave or signal, the audio-frequency component. 2. In a multimedia system, a device that processes and/or reproduces audio frequencies.

sound compression The encoding of audio data so that it occupies less room. This helps maximize the use of storage space, and also reduces transmission bandwidth. Also called **audio compression**.

sound-damping Same as **soundproofing**.

sound-deadening Same as **soundproofing**.

sound delay A delay in the transmission of sound signals. Also called **acoustic delay**.

sound detector 1. A device, instrument, transducer, or object which senses sound. Microphones and hydrophones are examples. Also called **sound sensor** (1). 2. A sensor whose mode of detection utilizes sound waves. Also called **sound sensor** (2), or **acoustic sensor**. 3. A detector which extracts a sound signal from a modulated signal, such as that of an FM transmission.

sound digitization The conversion of sound from an analog source, such as a voice, into digital form, such as that found on a compact disc. The larger the sample of the analog source, and the higher the sampling rate, the more accurate the conversion.

sound digitizer A device, such as that incorporating an analog-to-digital converter, which performs **sound digitization**.

sound dispersion The separation of a complex sound wave into its frequency components by passing it through an appropriate medium. Also called **acoustic dispersion**.

sound editor A computer program which serves to synthesize and manipulate sounds, such as those within a soundtrack.

sound energy The energy created by sound waves. It is the additional energy that particles in a medium have because of the presence of sound. Also called **acoustic energy**.

sound feedback A phenomenon that occurs when the sound waves produced by speakers interact with an input transducer such as a microphone or phonographic cartridge. If the feedback exceeds a certain amount, any of various undesired effects may occur, such as howling, whistling, motorboating, or excessive cone movement. Also, the sound heard as a consequence of this. Also known as **acoustic feedback**, **acoustic regeneration**, **acoustic howl**, **audio-frequency feedback**, **audio feedback**, or **howl** (1).

sound field Also called **acoustic field**. 1. A region containing sound waves. 2. A specially-tailored **sound field** (1), such as that produced by a home theater system.

sound film 1. Motion-picture film which includes the sound track on the same strip. 2. A motion picture with sound.

sound filter A device which blocks or absorbs sounds of certain frequencies while leaving others unaffected. Also called **filter** (3), or **acoustic filter**.

sound fixing and ranging Same as **SOFAR**.

sound fixing and ranging channel Same as **SOFAR channel**.

sound frequency A frequency that is within the range that humans can hear. A healthy person with good hearing can usually detect frequencies ranging from about 20 Hz to about 20 kHz. Also called **sonic frequency**, **audible frequency**, **acoustic frequency**, or **audio frequency**

sound-frequency amplifier Same as **sound amplifier**.

sound-frequency distortion Distortion occurring in the range of frequencies that humans can hear, an example of which is total harmonic distortion. Also called **audio-frequency distortion**.

sound-frequency meter An instrument used to measure frequencies in the range that humans can hear. There are several types, including analog and digital, the latter being highly accurate. Also called **audio-frequency meter**.

sound-frequency output Same as **sound output**.

sound-frequency range The spectrum of frequencies detectable by human ears. A healthy person with good hearing can usually detect frequencies ranging from about 20 Hz to about 20 kHz. Also known as **audio-frequency range**.

sound-frequency response 1. The efficiency of the amplification of a circuit, device, or system, as a function of input signals with frequencies within the range that humans can hear. Also called **audio response** (1). 2. A graph in which the output of a circuit, device, or system, is plotted against signals with frequencies within the range that humans can hear. Also called **audio response** (2).

sound-frequency spectrum Same as **sound spectrum**.

sound generator A transducer which converts another form of energy, such as electrical or mechanical, into sound energy. For example, a tuning fork converting mechanical energy into sound energy. Also called **acoustic generator**.

sound grating A series of equally sized rods or slits that are situated a fixed distance apart, which serves as an obstacle to sound waves. This causes the waves to be diffracted according to their wavelengths. Also called **acoustic grating**.

sound imaging The use of sound energy to create images of the internal structure of non-transparent objects. A common example is the use in medicine of ultrasonic waves to provide images of internal organs. Also known as **sonic imaging**, **acoustic imaging**, or **ultrasonic imaging**.

sound intensity The sound power transmitted per unit area, expressed in watts per square meter. Also called **acoustic intensity**.

sound intensity level In acoustics, a measure of the intensity of a sound relative to a reference intensity. For instance, it may be expressed as ten times the common logarithm of the ratio of a given sound in a specified direction, to a reference sound intensity of 1 picowatt per square meter. Expressed in decibels. Also called **sound level** (1), or **intensity level** (1).

sound lens A system of disks or other obstacles that refract sound waves similarly to the way an optical lens refracts light. Also called **lens** (3), or **acoustic lens**.

sound level 1. Same as **sound intensity level**. 2. Same as **sound pressure level**.

sound level meter 1. An instrument which measures and indicates **sound pressure levels**. Also called **sound pressure meter**. 2. An instrument which measures and indicates sound levels. 3. An instrument, such as a VU meter, which measures sound pressure levels in units other than decibels, using a frequency weighting, or the like.

sound mix To combine multiple audio input signals to form a composite signal with the desired blend. For example, to combine voices and multiple instruments for a song, or the combination of dialog, music, and effects for a soundtrack. Also, the result of such a combination. Also called **audio mix**, or **mix** (2).

sound navigation and ranging Same as **sonar**.

sound-operated relay A relay that is actuated by sound. Also called **sound-activated relay**, or **sound relay**.

sound-operated switch A switch that is actuated by sound. Also called **sound-activated switch**, or **sound switch**.

sound oscillator A oscillator which produces electric waves with frequencies within the range that humans can hear.

May be used, for instance, for long-distance surveys of submerged cables. Also called **audio oscillator**.

sound output Also called **sound-frequency output**. **1.** The output of a source of audio frequencies, such as an amplifier or oscillator. For instance, the audio power an amplifier delivers to a speaker. Also called **audio output (1)**. **2.** Any audible output from a device, component, or system. For instance, that produced by a speaker. Also called **audio output (2)**. **3.** In computers, an audible output, such as a chime when a Web page is loaded, or the simulation a of human voice reading a document on screen. Also called **audio output (4)**.

sound peak limiter A circuit or device which limits the amplitude of its audio-frequency output signal to a predetermined maximum, regardless of the variations of its input. Used, for example, for preventing component, equipment, or media overloads. Also known as **audio peak limiter**.

sound phenomena Phenomena pertaining to, and arising from, the generation, production, propagation, transmission, reception, detection, perception, control, and processing of sound. Also called **acoustic phenomena**.

sound power The total sound power radiated by a source. Usually expressed in watts, or ergs per second. Also called **acoustic power**.

sound power level A measure of the magnitude of sound power to a given reference, such as decibels referred to 1 milliwatt.

sound pressure **1.** The pressure that sound waves exert on an area or surface. For example, the minute fluctuations in atmospheric pressure resulting from the passage of a sound wave. **2.** The **sound pressure (1)** for a specific area. May be expressed, for instance, in dynes per square centimeter, or newtons per square meter. **3.** The value obtained when squaring multiple instantaneous sound pressure level measurements at a given point, averaging these over the time of a complete cycle, and taking the square root of this average. Usually expressed in pascals, though also expressed in other units, such as dyne/cm^2, or microbars. Also called **RMS sound pressure**, **effective sound pressure**, or **acoustic pressure**.

sound pressure level Its abbreviation is **SPL**. Also called **sound level (2)**. **1.** A measurement of sound pressure, expressed in decibels, with respect to the threshold of hearing. The threshold of hearing is usually defined as 20 micropascals, which is assigned a value of 0 decibels. For example, leaves gently rustling produce a sound level of approximately 15 dB SPL, a whisper is about 30 dB SPL, the dial tone of a telephone is more or less 80 dB SPL, and an approaching subway train is somewhere around 110 dB SPL. Naturally, each of these approximations may vary quite a bit. For instance, the type of train, its approaching speed, and station acoustics are some of the factors which affect a dB SPL reading in the case of a subway. For a person with good hearing, pain begins somewhere around 120 dB SPL, and there is immediate damage to hearing above 150 dB SPL. Also called **dB SPL**. **2.** A **sound pressure level (1)** measurement or indication in units other than decibels, such as micropascals.

sound pressure meter Same as **sound level meter (1)**.

sound probe A slender or otherwise small microphone utilized to detect, sample, or measure sound phenomena.

sound-proofing Same as **soundproofing**.

sound radiation Sound waves that travel through a medium. Also called **acoustic radiation**.

sound radiator A device or surface which vibrates to produce sound (acoustic) waves. Examples include headphone diaphragms, speaker cones, and distributed-mode speaker panels. Also called **radiator (3)**, or **acoustic radiator**.

sound range The spectrum of frequencies detectable by human ears. A healthy person with good hearing can usually detect frequencies ranging from about 20 Hz to about 20 kHz. Also known as **audio range**.

sound recorder An instrument or device, such as a tape recorder, which makes **sound recordings**. Also called **audio recorder**.

sound recording The process of producing a permanent, or semi-permanent, record of audio which is suitable for later reproduction. This includes recording upon magnetic tapes, magnetic disks, optical discs, phonographs, films, wires, and so on. Also, a specific recording. Also, that which has been recorded. Also called **audio recording**.

sound recording system A system, such as that incorporating microphones, amplifiers, filters, mixers, converters, computers, and digital recorders, utilized for **sound recording**. Also called **audio recording system**.

sound reflection coefficient Same as **sound reflectivity**.

sound reflection factor Same as **sound reflectivity**.

sound reflectivity The ratio of the sound energy reflected from a given surface to that striking the surface. Also called **sound reflection coefficient**, **sound reflection factor**, **acoustic reflectivity**, **acoustic reflection coefficient**, or **acoustic reflection factor**.

sound refraction The bending of sound waves as they travel through a medium which varies in temperature, pressure, or other physical characteristics. These differences make sound travel at different speeds, as is the case when sound passes from warmer water to cooler water. Also called **acoustic refraction**.

sound reinforcement **1.** The intensification of sound using mechanical objects or devices such as horns or resonant chambers. **2.** The intensification or other alterations of sound which is being recorded and/or reproduced, using circuits, devices, equipment, or systems, such as amplifiers, filters, or computers.

sound relay Same as **sound-operated relay**.

sound-reproducing system Same as **sound-reproduction system**.

sound reproduction The output of a **sound-reproduction system**. When utilized in the context of music, home theater, or the like, such reproduction may be monaural, stereophonic, multi-channel, or ambisonic.

sound-reproduction system A device, or combination of devices, which serve to reproduce sound within the range of frequencies that humans can hear. Examples include high-fidelity audio systems and public-address systems. Also called **sound-reproducing system**, or **reproducing system**.

sound response Also called **audio response**. **1.** The efficiency of the amplification of a circuit, device, or system, as a function of input signals with frequencies within the range that humans can hear. **2.** A graph in which the output of a circuit, device, or system, is plotted against signals with frequencies within the range that humans can hear. **3.** An audible sound produced by a component, device, or system. For example, the sound of a Geiger counter, or keyboard feedback for those with reduced hearing.

sound scattering The irregular and unpredictable dispersion of sound waves, due to diffraction, reflection, or refraction. Also called **acoustic scattering**.

sound sensor **1.** Same as **sound detector (1)**. **2.** Same as **sound detector (2)**.

sound shielding A barrier that prevents the penetration of sound waves. Also called **acoustic shielding**.

sound signal A signal that is within the range of frequencies that humans can hear. A healthy person with good hearing can usually detect frequencies ranging from about 20 Hz to about 20 kHz. Also called **audio signal**.

sound spectrogram The recorded or displayed output of a sound spectrograph.

sound spectrograph An instrument that performs time, amplitude, and frequency analyses of audio signals. Such an instrument, for example, details the spectral composition of sounds. Used, for instance, for voice analysis. Its recorded or displayed output is a **sound spectrogram**.

sound spectrum 1. The spectrum of frequencies detectable by human ears. A healthy person with good hearing can usually detect frequencies from about 20 Hz to about 20 kHz. Also known as **audio spectrum, audio range, audio-frequency range**, or **audio-frequency spectrum**. 2. The complete range of sound frequencies. Also known as **acoustic spectrum** (1). 3. An interval of sound frequencies. Also known as **acoustic spectrum** (2). 4. A graph depicting an interval of sound frequencies. Also known as **acoustic spectrum** (3).

sound switch Same as **sound-operated switch**.

sound system A device, or combination of devices, which serve to process and/or reproduce sound within the range of frequencies that humans can hear. Examples include high-fidelity audio systems and public-address systems. Depending on the components, such a system may also record sound. Also called **audio system**.

sound track Same as **soundtrack**.

sound transducer Also called **acoustic transducer**. 1. A device which transforms sound energy into another form of energy. For example, a microphone, which converts sound energy into electrical energy. 2. A device which transforms another form of energy into sound energy. For example, a speaker, which converts electrical energy into sound energy.

sound transmission The movement of sound energy through a medium. For example, sound waves traveling through water. Also called **acoustic transmission**.

sound transmission coefficient The ratio of the sound (acoustic) energy transmitted through a material to the sound energy incident on said material. Also called **acoustic transmission coefficient, acoustic transmissivity**, or **acoustic transmittivity**.

sound transmitter Also called **audio transmitter**, or **aural transmitter**. 1. A transmitter, such as a broadcast transmitter, which sends signals containing audio. Also, the equipment utilized for this purpose. 2. The equipment utilized by a TV transmitter to send the sound signal. This signal is combined with that of the video transmitter for regular TV viewing.

sound trap A circuit which helps prevent a sound signal or channel from interfering with a video signal or channel.

sound treatment The use of absorbers, diffusers, and reflectors to help give a room the desired sound characteristics. Used, for instance, to set up a recording studio. Also called **acoustic treatment**.

sound velocity The **speed of sound** in a given direction. Also called **sonic velocity**, or **velocity of sound**.

sound wave A traveling wave which propagates sound through an elastic medium. The wave is produced by vibrations that are a result of sound energy. Also known as **acoustic wave**, or **elastic wave** (2).

SoundBlaster A popular line of sound cards which has become a de facto standard.

sounding 1. Probing which detects physical properties in nature. Usually refers to detection of underwater environments, or above a given level of the atmosphere. Also, a specific instance of such probing. 2. The act or process of determining the depth of water, as seen, for instance, in echo sounding.

soundproofing The reduction of the magnitude of vibration or resonance of a surface or material, in order to reduce or eliminate sound waves. Also spelled **sound-proofing**. Also called **sound-deadening, sound-damping, acoustic damping**, or **damping** (3).

soundtrack Also spelled **sound track**. 1. A track, such as that located in a strip along a side of a motion-picture film, which contains the audio content. 2. The audio portion of a movie, TV program, or the like.

source Its symbol is **S**. 1. The point or region where something comes into existence, or from which something arrives or is obtained. 2. A point, region, or object where current, power, energy, matter, a signal, or flux is emitted or originates. A **sink** (1) is where it is absorbed or terminates. 3. In a field-effect transistor, the region from which majority carriers flow into the channel, en route to the drain. Also, the electrode attached to this region. Also known as **source region**, or **source electrode**. 4. A device, component, apparatus, system, or medium which originates or sends a data signal, or from which data is taken. Also called **data source**. 5. Same as **source code**.

source code A set of programming instructions and statements that are expressed in a form suitable for input into an assembler, compiler, or translator, which in turn transforms said code into machine code. Source code is usually written in a high-level or assembly language which is understandable by humans, while only machine code can be directly executed by the CPU. Also called **source** (5), or **code** (3).

source computer Also called **source machine**. 1. A computer from which a transmission originates. 2. A computer from which a program is loaded or from which data is transferred. 3. A computer on which source code is compiled, assembled, or translated.

source data The data which serves as the source for a document, file, database, application, or the like. Source data may be taken, for instance, from source documents.

source data automation The use of automatic processes, equipment, or systems for the purpose of collecting, processing, and storing data which is obtained from a source, utilizing devices such as optical scanners, bar-code readers, or magnetic-stripe readers. Its abbreviation is **SDA**.

source deck 1. A tape deck, such as that which is part of an audio system, that serves as the source of a signal. 2. In dubbing, the deck from which the signal to be recorded is taken. The **record deck** is that to which signals are transferred or copied to.

source directory A directory from which data, files, or folders are taken. The receiving directory is the **target directory**.

source disk A disk from which data, files, or folders are taken. The receiving disk is the **target disk**.

source document A document, such as a form, into which data is entered, and whose contained information is subsequently utilized.

source drive A disk drive containing a disk from which data, files, or folders are taken. The receiving drive is the **target drive**.

source electrode Same as **source** (3).

source file 1. A file from which data is taken. The receiving file is the **target file**. 2. A file containing **source code**.

source follower A transistor amplifier with a common-drain connection, in which the input signal is applied between the gate and drain and the output signal is taken from between the source and drain. Also called **source-follower amplifier**, or **common-drain amplifier**.

source-follower amplifier Same as **source follower**.

source impedance The impedance presented by a source to the input terminals of a component, circuit, device, or instrument. The source impedance should be matched with the input impedance, to help ensure maximum energy or signal transfer.

source instruction **1.** An instruction used in a **source language**. **2.** An instruction within **source code** which has been written.

source language A programming language used in a **source program**. A source language is usually a high-level language, such as Fortran, C, LISP, or Pascal.

source machine Same as **source computer**.

source program A program which is suitable for conversion by an assembler, compiler, or interpreter. A source program is written in a source language, and is converted into a low-level **target language**, such as machine language. A source program is the program that a computer programmer writes.

source region Same as **source (3)**.

source route bridge In communications networks, a bridge that is aware of the locations of users or systems prior to message delivery. This contrasts, for instance, with a **transparent bridge**, which can learn the locations of users or systems and deliver messages accordingly. Its abbreviation is **SRB**.

source route bridging A communications protocol in which a sending station or node is aware of the locations of network bridges prior to message delivery. Its abbreviation is **SRB**.

source statement **1.** A statement used in a **source language**. **2.** A statement within **source code** which has been written.

South bridge Same as **Southbridge**.

South bridge chip Same as **Southbridge chip**.

South bridge chipset Same as **Southbridge chipset**.

south geographic pole Same as **south pole (1)**.

south magnetic pole The location in the southern hemisphere where the magnetic dip is 90°. The corresponding spot in the northern hemisphere is the **north magnetic pole**. At each of these places, the magnetic meridians converge. The south magnetic pole does not coincide with the south geographic pole, and the position of both magnetic poles vary over time. Also called **south pole (2)**, **magnetic south pole**, or **negative pole (2)**.

south pole **1.** On the surface of the planet earth, the point within the southern hemisphere where all meridians meet. It is defined as 90° S, and is the southern end of the axis of rotation of the planet. Also called **south geographic pole**, or **geographic south pole**. **2.** Same as **south magnetic pole**. **3.** In a magnet or magnetic body, one of the two regions where the magnetic lines of force converge, and hence magnetic intensity is at a maximum. The other region is the **north pole (3)**. Such a pole, when part of a magnet or magnetic body which is freely suspended, will seek a north pole, as opposite poles attract. Also called **negative pole (1)**.

south-seeking pole In a magnet which is freely suspended, the pole which will point in the direction of the **south magnetic pole**.

Southbridge In certain CPU chipset architectures, one or more chips that control the CPUs I/O functions, connecting to the USB and IDE buses, keyboard controller, and the like, while the **Northbridge** connects the CPU to the RAM, AGP, PCI buses, and L2 cache. Also spelled **South bridge**.

Southbridge chip A **Southbridge** consisting of a single chip. Also spelled **South bridge chip**.

Southbridge chipset A **Southbridge** consisting of a multiple chips. Also spelled **South bridge chipset**.

southern lights An aurora occurring in the southern hemisphere. An aurora is a luminous phenomenon of the upper atmosphere occurring mostly in the high latitudes of both hemispheres. They are caused by the interaction of excited particles from space and particles of the upper atmosphere,

and usually affect radio communications. Also called **aurora astralis**.

SP-1 Abbreviation of **service pack 1**.

SP-2 Abbreviation of **service pack 2**.

sp gr Abbreviation of **specific gravity**.

SP1 Abbreviation of **service pack 1**.

SP2 Abbreviation of **service pack 2**.

space **1.** The expanse where the entire universe exists. **2.** The **space (1)** extending from beyond the earth's atmosphere. **3.** A region of **space (1)** devoid of all matter, and in which there are no gravitational or electromagnetic fields. In free space, the speed of light is constant, and at its maximum theoretical value. Also called **free space (1)**. **4.** An expanse of a surface, object, or medium. **5.** A specific region, or that in the vicinity of an object or another space. For instance, a Faraday dark space. **6.** An area or volume which is not occupied or used. For example, a space between words. **7.** A given interval of time. **8.** A location, interval, or the like, which is available at the moment. For instance, space available in memory. **9.** A location, interval, or the like, which has a specific use or is otherwise designated. An example is scratch disk space. **10.** An area or volume where something specific occurs. For instance, an interaction space. **11.** In the transmission of information, a low state, or binary 0, as opposed to a **mark (4)** which indicates a high state, or binary 1.

space charge **1.** The net electric charge within a given region, due to the presence of charged particles such as electrons, protons, ions, or the like. A space charge may also be created by electron holes in semiconductors. Examples include a cloud of electrons around a cathode or anode in an electron tube, or a space-charge layer in a transistor. **2.** A charge distributed throughout a given region.

space-charge density The space charge per unit volume in a **space-charge region**.

space-charge effect An effect, such as anode fall or cathode drop, due to the presence of a **space charge**.

space-charge field A field produced by a **space charge**.

space-charge layer In a semiconductor, the region of a pn junction that is free, or depleted, of charge carriers. Also called **depletion layer**. Two widely used, yet deprecated, terms for this concept are **Barrier layer**, and **barrier (2)**.

space-charge limited current A current barrier that a **space charge** imposes, as in filament saturation.

space-charge region A region within which there is a **space charge**. Its abbreviation is **SCR (2)**.

space diversity **1.** Same as **space-diversity reception**. **2.** Same as **space-diversity transmission**.

space-diversity gain **1.** Signal gain through the use of **space-diversity reception**. **2.** Signal gain through the use of **space-diversity transmission**.

space-diversity receiver A receiver used for **space-diversity reception**.

space-diversity reception A form of diversity reception in which there are two receiving antennas in separate locations whose signals are combined. Since any fading present is generally not the same for both locations, said effects are minimized. Also called **space diversity (1)**.

space-diversity system A reception or transmission system which employs **space-diversity reception** or **space-diversity transmission**, respectively.

space-diversity transmission The converse of **space-diversity reception**. That is, instead of having two receiving antennas in separate locations, there are two physically separate transmitting antennas. Also called **space diversity (2)**.

space-diversity transmitter A transmitter used for **space-diversity transmission**.

space lattice The geometric arrangement in space of the atoms, molecules, or ions of a crystal. The pattern of such a lattice is regular and three-dimensional. Also called **crystal lattice**, or **lattice (2)**.

space noise Same as **sky noise**.

space wave The component of a ground wave that travels through space between a transmitter and receiver. A space wave includes both direct waves and ground-reflected waves. The **surface wave (1)** component of a ground wave travels along the surface of the earth between a transmitter and receiver.

spaghetti Slender tubing utilized to insulate wires or other conductors.

spaghetti code Source or programming code which is unnecessarily cumbersome and convoluted, and which is often characterized by excessive use of branch instructions such as GOTO. Incessant fixes for bugs which should have been worked out prior to releasing software is a common cause.

spam 1. Unsolicited and generally unwelcome email which is sent to multiple, often millions of users. Common varieties of spam include get-rich quick schemes, chain letters, supposed discount offers, invitations to pornographic Web sites, and appearance-related ruses such as those pertaining to sex, body-part enhancement, weight reduction, and so on. There are various views on the origin of the term, including references to a meat-like product, which like its email counterpart is ubiquitous, yet generally unappetizing and unwanted. One measure to counter spam which usually backfires is clicking on a link offering removal from a list, or sending an email to the effect, as this only serves to inform the sender that this address is indeed currently active, which in turn paves the way for even more spam. **2.** The sending of the same messages or postings to numerous newsgroups or users.

spambot 1. A bot that specializes in gathering email address for use when **spamming**. Such a bot may, for instance, obtain email addresses from Web pages containing the "@" sign. **2.** A bot which sends **spam**.

spamdexing The practice of including words and/or other content in a Web page so as to deceive search engines into listing it at or near the top of search results, when such a position is unwarranted. The term refers to deceptive practices, such as including keywords hundreds of times to move up through weighted relevance, incorporating competitor's product names to lure potential clients from where they meant to go, utilizing the kinds of words that people tend to use as keywords, and so on. Such content may be hidden from users, for instance, by employing a white text on a white background. It is an abbreviation of **spam indexing**.

spamming The sending of **spam**.

span The distance, interval, range, or other measure between two points, extremities, or limits. For instance, the span between the lowest and highest readings of an analog meter, or the dimensions between the ends of a component.

SPARC Acronym for Scalable Processor Architecture. A RISC architecture or microprocessor which supports various operating systems.

spark A momentary luminous discharge of electricity between two conductors separated by a gas, which is frequently accompanied by a crackling noise. Used, for instance, for ignition of fuel, or for machining. This contrasts with an **arc (1)**, which has a more sustained duration. Also called **spark discharge**, **sparkover**, or **electric spark**.

spark arrester A component, circuit, or device which reduces, eliminates, or prevents **sparks**. Also called **spark suppressor**, **spark quencher**, or **spark killer**.

spark coil A small coil utilized to produce **spark discharges**.

spark discharge Same as **spark**.

spark gap 1. The distance between electrodes across which **spark discharges** occur. The insulation, usually provided by air, is self-restoring after each spark. Used, for instance, to protect equipment, or for ignition. Also, an arrangement of such electrodes. **2.** The distance between an electrode and a workpiece across which there is a spark discharge.

spark interference Interference, such as ignition interference, caused by **sparking**.

spark killer Same as **spark arrester**.

spark noise Noise, such as ignition noise, caused by **sparking**.

spark-over Same as **spark**.

spark quencher Same as **spark arrester**.

spark spectrum An emission spectrum produced by the effect that a spark between electrodes has on an ion. Also called **ionic spectrum**.

spark suppressor Same as **spark arrester**.

spark voltage The voltage at which a disruptive electrical discharge occurs in a gas. Also called **sparking voltage**, **sparking potential**, or **breakdown voltage**.

sparking The occurrence of **sparks**.

sparking potential Same as **spark voltage**.

sparking voltage Same as **spark voltage**.

sparkover Same as **spark**.

sparkover voltage 1. The voltage at which **sparkover** occurs. **2.** The voltage at which a disruptive discharge occurs between conductors, such as electrodes, separated by an insulating material. This discharge may occur around or along the surface of said insulator. Also called **flashover voltage (2)**.

spatial data Data which is represented by, or represents, two-dimensional or three-dimensional images, such as maps. A Geographic Information System, for instance, can be utilized for capturing, storing, analyzing, retrieving, and displaying spatial data.

Spatial Data Transfer Standard A format established for the transfer of **spatial data** between dissimilar computer systems. Its abbreviation is **SDTS**.

spatial distribution The arrangement, orientation, occurrence, or dispersion of something within a space or volume. For example, the arrangement of the atoms within a chiral molecule, or the radiation pattern of an antenna.

spatial light modulator An object, device, or transducer which serves to modulate a laser beam so as to input data into a computer system utilizing holographic storage. Such data input can occur at nearly the speed of light. Its abbreviation is **SLM**.

spatial resolution A measure of the smallest spatial feature that can be resolved by an imaging system such as a telescope, microscope, robotic eye, display, or the like. Any object or structure which is smaller will not be detected as a discrete fraction of the overall image. May be expressed in many manners, including line pairs per unit distance.

SPC Abbreviation of **statistical process control**.

SPDT relay Abbreviation of **single-pole double-throw relay**.

SPDT switch Abbreviation of **single-pole double-throw switch**.

speaker A transducer which converts electrical energy into sound energy, which is then radiated outward to the surrounding medium. The diaphragm, which is the vibrating membrane of a speaker, may be composed of any of various materials, including paper, fiber, plastic, silk, metal, or it may be a piezoelectric crystal. A speaker may incorporate multiple specialized drivers, including subwoofers, woofers, midranges, tweeters, and supertweeters, to better address various frequency ranges. Speakers provide sound for devices such as TVs, audio systems, computers, DVDs, telephones, and so on. Also called **loudspeaker**, or **driver** (4).

speaker adaptation 1. A manner in which a speech recognition system adapts to the voice of a given speaker. This usually involves, among other things, the obtaining of repeated samples from said speaker. 2. A specific instance of **speaker adaptation** (1).

speaker authentication Also called **speaker identification**, or **speaker verification**. 1. The use of speech for identity authentication processes. Used, for instance, to obtain access to a given facility. 2. A specific instance of **speaker authentication** (1).

speaker baffle Also called **loudspeaker baffle**, or **baffle**. 1. A partition in a speaker enclosure which is used to reduce or eliminate the interaction between the acoustic waves generated by the front of the speaker and those from the back. This is especially important in low-frequency sound reproduction. 2. The panel on which a speaker is mounted. 3. A panel used to inhibit the propagation of sound waves. Used, for instance in theaters as part of a sophisticated sound-reproduction system.

speaker crossover network A filter circuit in a speaker which divides the input audio frequencies and sends the corresponding bands of frequencies to the designated speaker units, such as woofers, midranges, tweeters, super-tweeters, and so on. Also called **speaker dividing network**, **loudspeaker crossover network**, **crossover** (3), **dividing network**, or **frequency dividing network**.

speaker-dependent A speech-recognition system which must be trained through the input of multiple samples from a given speaker, or which only functions properly with one or more specific speakers.

speaker diaphragm A vibrating membrane in a loudspeaker. It produces a sound-wave output in response to its electric input. Cones and domes are examples of commonly utilized diaphragms in speakers. Also called **loudspeaker diaphragm**, or **diaphragm** (1).

speaker dividing network Same as **speaker crossover network**.

speaker efficiency Same as **sensitivity** (4).

speaker enclosure A cabinet designed to house one or more loudspeaker units. Two common designs are acoustic reflex and acoustic suspension. Within a loudspeaker system, a specialized loudspeaker such as a woofer or tweeter may have its own enclosure, helping enhance performance. Also called **loudspeaker enclosure**, or **enclosure** (2).

speaker grill Same as **speaker grille**.

speaker grille A covering across the front of a loudspeaker which provides protection and decorative appeal. Also spelled **speaker grill**. Also called **loudspeaker grille**, or **grille** (2).

speaker identification Same as **speaker authentication**.

speaker impedance The rated impedance of the voice coil of a loudspeaker. Most speakers have a nominal value assigned, although said impedance varies, depending on the frequency being reproduced. Common values are 4, 8, and 16 ohms, and the driving amplifier must be matched to avoid diminished performance, failure, or damage. Also called **loudspeaker impedance**.

speaker-independent A speech-recognition system which functions properly with speakers which have rarely or never used said system previously.

speaker microphone A loudspeaker which also serves as a microphone. Used, for instance, in an intercom system, where a button may be used to switch between functions. Also called **loudspeaker microphone**.

speaker phone Same as **speakerphone**.

speaker port A carefully-dimensioned opening in a speaker enclosure. It is employed for various purposes, but is mainly used to increase and extend the reproduction of low frequencies. Also called **loudspeaker port**, **port** (6), **duct** (5), **ducted port**, or **vent** (3).

speaker recognition 1. The ability of a device, piece of equipment, computer, or system, to only detect and recognize words spoken by one or more specific speakers. It is a specific type of **speech recognition**, but the latter is meant to understand speech in general. 2. A specific instance of **speaker recognition** (1). 3. The technology employed in **speaker recognition** (1).

speaker-recognition system A system, such as that incorporating the appropriate software and hardware, utilized for **speaker recognition** (1), or **speaker recognition** (2).

speaker system Also called **loudspeaker system**. 1. A loudspeaker enclosure which incorporates multiple specialized loudspeakers, such as a woofer, midrange, and tweeter. 2. Multiple loudspeakers which are designed to work together in providing a desired sound experience. For example, a five loudspeaker arrangement providing surround sound.

speaker telephone Same as **speakerphone**.

speaker verification Same as **speaker authentication**.

speaker voice coil In a dynamic speaker, a coil which is connected to the diaphragm, and through which an audio-frequency signal current is sent to move said coil in a piston-like manner, thus producing a sound-wave output. Also called **loudspeaker voice coil**, or **voice coil** (1).

speakerphone A feature which enables a telephone user to place calls and participate in conversations without the need to hold the handset near the mouth. Utilized, for instance, to allow multiple persons at the same location to use the same telephone unit for a conference call. Also, a telephone equipped with this feature. Also spelled **speaker phone**. Also called **speaker telephone**.

special character A character that is not a letter, digit, or space, such as @, %, <, or +.

special-purpose computer A computer which is utilized to perform a very specialized class of computing tasks, or for solving highly complex problems. For instance, a computer which performs scientific visualization.

special-purpose language A computer language, such as LISP or Prolog, which is designed for a single application or class of problems. This contrasts with a **general-purpose language**, which can be used for a variety of applications.

Specialized Mobile Radio Its abbreviation is **SMR**. A two-way radio communications system in which multiple transceivers are linked by a single repeater. SMR usually operates in the VHF or UHF ranges, and uses a pair of channels, one for reception and the other for transmission. **Enhanced Specialized Mobile Radio** utilizes a network of repeaters for extended coverage, and provides data services, among other improvements.

specific address A specific memory location. It is an address from which relative addresses may be derived. Also called **absolute address**, **real address** (1), **actual address** (1), **direct address**, or **machine address**.

specific conductance The ease with which an electric current can flow through a body. It is the reciprocal of resistivity. Its symbol is σ, and it is expressed in siemens per meter. Also called **conductivity**, or **electrical conductivity**.

specific conductivity The conductance of a material per unit volume. Usually expressed in siemens per cubic centimeter.

specific gravity The ratio of the density of a given material or medium, to that of a standard material or medium. For example, the ratio of the density of a material to that of water at 4 °C at one atmosphere, or the ratio of the density of a gas to that of dry air at 0 °C at one atmosphere. Its abbreviation is **sp gr**. Also called **relative density**.

specific inductive capacitance Same as **specific inductive capacity**. Its abbreviation is **SIC**.

specific inductive capacity For a given material, the property which determines the electrostatic energy that can be stored per unit volume for a unit potential gradient. This is equivalent to the ratio of the capacitance of a capacitor using the material in question as a dielectric, to the capacitance of a capacitor using a vacuum as a dielectric. Its abbreviation is SIC. Also called **specific inductive capacitance, dielectric constant, permittivity, relative permittivity**, or **inductivity**.

specific insulation resistance A measure of the inherent ability of a material to resist the flow of current, per unit volume. It is usually calculated as the resistance between opposite faces of a one centimeter cube, and expressed in ohm-centimeters. Also called **volume resistivity**.

specific ionization The number of ion pairs, both primary and secondary, formed per unit track length of an ionizing particle moving through matter.

specific resistance A measure of the inherent ability of a material to resist the flow of current. It is the reciprocal of **electrical conductivity**, and depending on their resistivities, materials can be classified as insulators, semiconductors, or conductors. The lower the specific resistance, the better conductor a material is. Its formula is: $r = RA/L$, where r is resistivity in ohm-meters, R is the resistance in ohms, A is the cross-sectional area of the material in square meters, and L is its length in meters. Also called **resistivity**, or **electric resistivity**.

specific resistivity The reciprocal of **specific conductivity**.

specification **1.** A detailed statement which describes and defines aspects such as the function, characteristics, design, manner of implementation, and intended capabilities of a component, circuit, device, piece of equipment, system, process, or the like. **2.** The act or process of preparing a **specification** (1). **3.** A **specification** (1) pertaining to hardware, software, or networks. For example, that of a DVD format, or for the transfer of data within a network. **4.** A set of requirements. For instance, the power requirements, operational temperatures, intended capabilities, and compatible connectors for a given device.

Specification and Description Language An object-oriented language utilized to specify event-driven systems in which multiple real-time activities occur simultaneously. Used, for instance, to describe telecommunications, robotics, and satellite systems. Its abbreviation is **SDL**.

SPECmark Abbreviation of **S**ystems **P**erformance **E**valuation Cooperative **Mark**. A computer benchmark utilized, for instance, to measure performance pertaining to floating-point operations.

SPECT Acronym for **single-photon emission computed tomography**, or **single-photon emission computerized tomography**.

SPECT scan Abbreviation of **single-photon emission computed tomography scan**.

SPECT scanner Abbreviation of **single-photon emission computed tomography scanner**.

SPECT scanning Abbreviation of **single-photon emission computed tomography scanning**.

spectra A plural form of **spectrum**.

spectral Of, pertaining to, or derived from a **spectrum**.

spectral analysis The processing and study of **spectral responses**. A spectral analysis provides a frequency-domain representation of a signal, and Fourier transforms may be utilized for conversions to the time domain, and vice versa.

spectral band In spectroscopy, a compact series of spectral lines which represent an interval of wavelengths which are absorbed or emitted by molecules. Atoms produce a **line spectrum**. Also called **band** (4), or **band spectrum**.

spectral characteristic Same as **spectral response**.

spectral density For a given bandwidth of electromagnetic radiation consisting of a continuous interval of frequencies, the total power divided by the specified bandwidth. Usually expressed in multiples of watts per hertz, such as mW/Hz, or kW/Hz. Used, for instance, to analyze the power content of complex signals. Also called **spectral power density, spectral energy density**, or **power spectral density**.

spectral energy density Same as **spectral density**.

spectral line Within a spectrum, a line representing the absorption or emission of electromagnetic radiation at a specific wavelength.

spectral power density Same as **spectral density**.

spectral purity The extent to which a signal is monochromatic.

spectral response The response of a component, circuit, device, piece of equipment, system, transducer, or material, to electromagnetic radiation, as a function of frequency. The greater the sensitivity to a given frequency or band, the greater the response. For instance, the output of a photocell in relation to the wavelength of incident light. Also called **spectral sensitivity**, or **spectral characteristic**.

spectral sensitivity Same as **spectral response**.

spectral width **1.** A measure of the interval encompassed by a spectrum. It may refer, for instance, to the wavelengths emitted by a laser beam. **2.** A **spectral width** (1) under given conditions, such as that at half of the peak power.

spectrogram The recorded output of a **spectrograph**.

spectrograph **1.** A **spectroscope** equipped to produce a photographed or otherwise recorded output. **2.** A device or instrument which enables plotting, photographing, or otherwise recording the output of instruments such as spectrometers or spectrophotometers.

spectrometer **1** An instrument which is utilized to obtain **spectrums**, by measuring wavelengths, energies, masses, ionization potentials, and so on. Examples include mass spectrometers, emission spectrometers, scintillation spectrometers, ionization spectrometers, nuclear magnetic resonance spectrometers, and acoustic spectrometers. **2.** A spectroscope which is equipped with a calibrated scale suitable for measurements of wavelengths, refractive indices, radiant intensities, or the like, pertaining specifically to light.

spectrometry The science and utilization of **spectrometers** for measurement and analysis.

spectrophotometer An instrument which measures the intensity of radiation as a function of wavelength. Such instruments usually analyze visible light, in which case they incorporate a photometer, but may also be designed to assess other intervals of the electromagnetic spectrum, such as the infrared, microwave, or X-ray regions. A spectrophotometer may be utilized to quantify absorption, emission, reflectance, transmittance, absorptance, and so on. Also called **spectroradiometer**.

spectrophotometric analysis Same as **spectrophotometry**.

spectrophotometry The science and utilization of **spectrophotometers** for measurement and analysis. Also called **spectrophotometric analysis**.

spectroradiometer Same as **spectrophotometer**.

spectroscope An instrument which splits radiant energy, especially light, into components of various wavelengths. Such an instrument may, for instance, incorporate a prism, grating, slit, or collimator lens. Used, for instance, for spectral analysis. A spectroscope produces a spectrum suitable for visual observation.

spectroscopy The branch of science dealing with the obtaining and subsequent analysis of **spectra**. There are many examples, including absorption spectroscopy, emission spectroscopy, nuclear magnetic resonance spectroscopy, electron spectroscopy, mass spectroscopy, gamma-ray spectroscopy, laser spectroscopy, and photoacoustic spectroscopy.

spectrum Its plural forms are **spectra** and **spectrums**. **1.** A distribution, display, plot, or other output which is based on an order of increasing or decreasing magnitude of a property, such as mass, energy, wavelength, or ionization potential. For example, a mass spectrum is a distribution, display, plot, or other visual output of the charge-to-mass ratios of ions within a given sample. **2.** A range of energies arranged in order of increasing or decreasing wavelengths or frequencies. For example, an absorption spectrum, an emission spectrum, or a sound spectrum. Also, a distribution, display, plot, or other output depicting such a range. **3.** A **spectrum** (1) or **spectrum** (2) based on a property of light. For example, the energy emitted by a light source as a function of wavelength. Also, a distribution, display, plot, or other output depicting such a range. **4.** A frequency band utilized for a given purpose or which identifies a given region. For example, the infrared, visible, ultraviolet, or radio spectrums. **5.** The range of frequencies of electromagnetic radiation. This encompasses frequencies from just above 0 Hz to beyond 10^{24} Hz, corresponding to wavelengths of over 10^8 meters, to less than 10^{-16} meters, respectively. These include, in order of ascending frequency: subsonic frequencies, audio frequencies, radio frequencies, infrared light, visible light, ultraviolet light, X-rays, gamma rays, and cosmic rays. These intervals have been arbitrarily established, may be labeled with alternate names associated with specific applications, and may have subdivisions. Also called **electromagnetic spectrum**.

spectrum analysis The utilization of a **spectrum analyzer** for the processing and study of signals.

spectrum analyzer An instrument which analyzes the frequency components of a signal, along with their corresponding amplitudes. A spectrum analyzer represents signals in the frequency domain, and its output is usually presented via a display such as an oscilloscope. Used, for instance, to monitor and troubleshoot broadband networks.

specular reflection Reflection of electromagnetic or acoustic waves in which there is no diffusion or scattering. This may occur, for instance, when reflecting light off a mirror. Also called **direct reflection**, or **regular reflection**.

speech The enunciation of words. Speech may be natural, as in humans talking, or artificial, as in digital speech.

speech amplifier An amplifier which is specially designed to amplify **speech frequencies**. Such an amplifier may be utilized, for instance, in a public-address or intercom system. Also called **voice amplifier**.

speech analyzer A circuit, device, piece of equipment, or system that evaluates various aspects of human voice. An example is a sound spectrograph. Used, for instance, in voice-recognition systems. Also called **voice analyzer**.

speech API Abbreviation of **speech** application program interface. An application program interface which facilitates tasks such as speech recognition and speech-to-text and text-to-speech conversions. Its own abbreviation is **SAPI**. Also called **voice API**.

speech application program interface Same as **speech API**.

speech band Same as **speech bandwidth**.

speech bandwidth The interval of frequencies encompassing **speech frequencies**. Also called **speech band**, or **voice bandwidth**.

speech clipper A circuit or device which limits the amplitude of its output signal to a predetermined maximum, regardless of the variations of its input. Used, for example, for pre-venting component, equipment, or media overloads. Also called **voice clipper**.

speech clipping Also called **voice clipping**. **1.** The action and effect of a **speech clipper**. **2.** The unintended limiting of the amplitude of a speech signal, due to the exceeding of the capabilities of an amplifier or circuit.

speech codec Also called **voice codec**. **1.** Abbreviation of **speech co**der/**dec**oder. Hardware and/or software which converts speech to digital code, and vice versa. **2.** Abbreviation of **speech c**ompressor/**dec**ompressor. Hardware and/or software which compresses and decompresses speech signals.

speech compression The encoding of speech so that it occupies less room. This helps maximize the use of storage space, and also reduces transmission bandwidth. There are various speech-compression algorithms, including Code-Excited Linear Prediction. Also called **voice compression**.

speech control The operation or regulation of a component, circuit, device, piece of equipment, system, process, or mechanism, using a voice. For example, the use of voice commands to control navigation aids, temperature settings, audio selections, or seat adjustments in an automobile. Also called **voice control**.

speech-controlled A component, circuit, device, piece of equipment, system, process, or mechanism which is operated or regulated by speech. Also called **speech-operated**, or **voice-controlled**.

speech-controlled computer A computer equipped with the proper interfaces, software, and processing capabilities to be fully and accurately controlled by voice commands. Used, for instance, by a surgical team during an operation. Also called **voice-controlled computer**.

speech digitization The conversion of speech into digital form. This makes storage, processing, and transmission easier and more accurate. Also called **voice digitization**.

speech digitizer A device which converts speech into digital form. The larger the sample of the voice, and the higher the sampling rate, the more accurate the conversion. Also called **voice digitizer**.

speech enhancement The use of **speech processing** to augment the intelligibility of speech. Used, for instance, to help those with reduced hearing. Also called **voice enhancement**.

speech filter A filter which selectively transmits or rejects signals in one or more intervals of speech frequencies. Used, for instance, to improve intelligibility. Also called **voice filter**.

speech frequencies In the audio spectrum, the frequencies within which speech occurs. This interval varies by region, but for communications may be defined as spanning from approximately 200 Hz to 6000 Hz. Components, circuits, devices, transducers, equipment, and systems processing, converting, transmitting, or receiving speech must be able to properly handle this interval. Also called **voice frequencies**.

speech input Speech signals received by a detecting device, such as a microphone. Used, for instance, in the context of recording or speech recognition.

speech intelligibility The extent to which speech is capable of being understood. Usually used in the context of voice communications. The more intelligible, the clearer the reception of a voice transmission. Also called **voice intelligibility**.

speech interface Same as **speech user interface**.

speech inverter Same as **scrambler**.

speech network A network, such as a telephone network, via which speech is transmitted. Also called **voice network**.

speech-operated Same as **speech-controlled**.

speech output Computer output consisting of digital speech. Also called **voice output**.

speech power The power of an audio signal that is within the speech frequencies. Also called **voice power**, or **talk power**.

speech processing Any form of processing performed on speech. These include digitization, amplification, compression, filtering, scrambling, and synthesis. Also called **voice processing**.

speech processor A circuit, device, piece of equipment, or system which serves for **speech processing**. For instance, a speech-recognition system. Also called **voice processor**.

speech recognition Also called **voice recognition**. **1.** The ability of a device, piece of equipment, computer, or system, to detect and recognize spoken words. A speech-recognition system, for instance, must be able to properly identify all phonemes of a given language to properly process speech. A computer so enabled converts spoken words into commands, text, or the like, for activating features, entering text, surfing the Internet, engaging in a voice chat, and so on. Also used, for instance, in an interactive voice response system. This contrasts with **speaker recognition**, which is meant to only detect and recognize words spoken by one or more specific speakers. **2.** A specific instance of **speech recognition (1)**. **3.** The technology employed in **speech recognition (1)**. Also called **speech-recognition technology**.

speech-recognition application An application utilized for **speech recognition**. Used, for instance, for speech-to-text and text-to-speech conversions. Also called **voice-recognition application**.

speech-recognition device A device, such as a handheld computer, which is utilized for **speech recognition (1)**. Also called **voice-recognition device**.

speech-recognition system A system, such as that incorporating the appropriate software and hardware, utilized for **speech recognition (1)**. Also called **voice-recognition system**.

speech-recognition technology Same as **speech recognition (3)**.

speech scrambler Same as **scrambler**.

speech synthesis Speech which is generated by a computer. The sounds are produced either by linking phonemes together, or by drawing from a database containing recorded words. Current technology provides for such speech to be almost completely natural. Its applications include assistance for those with reduced vision, or for retrieving email over any telephone. Also called **digital speech**, or **voice synthesis**.

speech synthesizer A device, piece of equipment, computer, or system, which is utilized for **speech synthesis**. Also called **voice synthesizer**.

speech-to-text The conversion of a speech input into a text. Also called **speech-to-text conversion**, or **voice-to-text**.

speech-to-text conversion Same as **speech-to-text**.

speech user interface Its acronym is **SUI**. A user interface which utilizes speech recognition to enable a system to respond to voice commands and entries. Used, for instance, in computer systems, PDAs, or cell phones. The key difference between a SUI and an **interactive voice response system** is that the former accepts continuous speech and handles an extensive vocabulary, while the latter responds only specific words or short phrases. Also called **speech interface**, or **voice user interface**.

speed 1. The rate at which motion occurs. Also, the magnitude of a velocity. Speed is a scalar quantity, while **velocity** is a vector quantity. Speed is usually measured as the distance traveled per unit time. Angular speed is the change in direction per unit time. **2.** The rate at which something occurs. For example, a communications speed. **3.** One of multiple available rates of operation. For example, extended play. **4.** A specific rate of operation. For example, a CPU speed. **5.** A relative rate of operation. For example, double speed. **6.** A rating which expresses the sensitivity of a given photographic medium, such as film or paper, to light. **7.** A measure of the ability of a lens to accumulate light at a given aperture.

speed buffer A segment of computer memory utilized for **speed buffering**.

speed buffering The use of a data buffer to compensate for different rates of operation, transfer, transmission, or the like. Used, for instance, when a computer and a peripheral operate at different speeds.

speed-control system A control system that maintains the speed of something, such as a motor or rate of operation, within specified limits.

speed dial Same as **speed dialing**.

speed dialing A telephone feature which allows placing a call by accessing a previously stored number. Such a feature is used, for instance, by pressing a button, entering a short key sequence, by pressing a key for longer than a specified time, and so on. Also called **speed dial**, or **memory dialing**.

speed of light Also called **electromagnetic constant**. **1.** The speed at which light propagates through a vacuum. This speed is a physical constant currently defined as 2.99792458 x 10^8 meters per second. Electromagnetic waves of all frequencies travel at the same speed in a vacuum. Its symbol is *c*. **2.** The speed at which light propagates through a given medium. It is generally less than its speed in a vacuum.

speed of sound The speed at which sound (acoustic) waves travel through a specific medium. This is influenced by conditions such as the temperature, pressure, and density of the medium of propagation. In dry air, for example, at 0 ºC, and at one atmosphere pressure, sound travels at approximately 331.6 meters per second, while in copper sound travels at approximately 3,360 meters per second. Also known as **sonic speed**, or **acoustic speed**.

speed of transmission The rate at which data, or any form of information, is transmitted over a communications line. Such a speed may be expressed, for example, in some multiple of bits per second. Also called **communications speed**.

spell check 1. Same as **spell checker**. **2.** To use a **spell checker**.

spell checker Software which locates possible errors in the spelling of words. A spell checker may suggest changes on the fly, or when requested by a user. Such a program does not distinguish when a word is misspelled and results in another correctly spelled word. For example, meaning to say brought (past tense of bring) and spelling bought (past tense of buy). Also called **spell check (1)**, or **spelling checker**.

spelling checker Same as **spell checker**.

SPF Abbreviation of **Shortest-Path First Algorithm**.

SPF Algorithm Abbreviation of **Shortest-Path First Algorithm**.

sphere 1. A solid figure in which all points on its surface are the same distance from the center. Also, that which resembles a sphere, such as the planet earth, or another celestial body. **2.** A three-dimensional surface within which all points are same distance from the center.

sphere gap A spark gap between two **spherical electrodes**.

spherical Having the shape of, pertaining to, characteristic of, or arising from, a **sphere**.

spherical aberration In optics, the inability of a lens, mirror, or reflector to produce a perfect correlation between an object and its resulting image, due to the rays from the edges being focused at a location other than those at the center.

This results from such objects having a spherical shape, and this form of distortion can be avoided by using a parabolic shape.

spherical angle The angle which is formed at the intersection of two arcs of a sphere.

spherical antenna 1. An antenna incorporating a **spherical reflector**. 2. An antenna whose over all shape, or whose radiators, are spherical.

spherical degree A unit of surface for spheres, which is equal to 1/720 of the total surface area.

spherical divergence 1. The propagation or spreading, in a spherical manner, of the energy of an electromagnetic wave. 2. The propagation or spreading, in a spherical manner, of sound waves. Also, decreases in sound energy resulting from this spreading.

spherical electrode An electrode in the shape of a **sphere (1)**.

spherical reflector A concave surface with a **spherical** shape. Seen, for instance, in a dish antenna which uses such a reflector as opposed to a parabolic reflector.

spherical wave A wave in which the wavefronts form concentric spheres.

spherics Interference of radio waves caused by natural atmospheric electrical phenomena, such as lighting. Atmospheric noise is a specific example of such a disturbance. Also spelled **sferics**. Also called **strays, atmospherics, atmospheric interference**, or **atmospheric strays**.

SPICE Abbreviation of **Simulation Program with Integrated Circuit Emphasis**.

SPID Abbreviation of **Service Profile Identifier**.

spider 1. A program which searches Web sites and organizes the located information. Also called **crawler, Web crawler**, or **bot (2)**. 2. In a speaker, a component which holds the voice coil and/or the rear of the diaphragm near the magnet, without letting them touch. A spider is also connected to the basket, for support, and helps, for instance, to return the speaker to its rest position when no signal is present. 3. A device, object, arrangement, mechanism, or movement which is similar in any manner to spiders. For example, a hub with radial spokes.

spike 1. An instantaneous current or voltage increase of great magnitude, such as that cause by a lightning strike or when power is restored after a blackout. A spike can arrive via power lines, phone lines, network lines, or the like, and may cause extensive damage if protective measures are not taken. 2. An very short pulse superimposed on a pulse of longer duration. 3. A **spike (1)** or **spike (2)** as displayed on a oscilloscope or other monitor, a printed plot, such as that of an EEG, and so on.

spike antenna A small antenna with a single element which is suitable for placement on a table, installation on cabinets, or the like.

spike arrester A component, circuit, device, system, mechanism, object, or the like, which prevents or minimizes the effects of **spikes (1)**.

spike noise 1. Noise produced by **spikes**. 2. Noise whose duration is extremely short. Such noise is characterized by large amplitude, and may occur as a series of disturbances which may be random or periodic in nature.

spike suppressor A component, circuit, device, system, mechanism, object, or the like, which prevents or minimizes the effects of **spikes (1)** or **spikes (2)**.

spillover An antenna signal, such as that of a satellite, which extends beyond the intended coverage area.

spin 1. The intrinsic angular momentum of particles, such as electrons, protons, photons, and antiparticles. A particle has spin even when at rest. 2. Rotation of an object or body around an axis or central point. Also, to quickly rotate in this manner. 3. To cause to **spin (2)**.

spin coating The application of a layer by pouring the desired solution or material onto a substrate, and spinning said substrate to expel excess material and provide a uniform thickness. Used, for instance, to apply a liquid photoresist.

spin electronics The use of electron spins for data storage in computing devices, as seen, for instance, in MRAM. Its abbreviation is **spintronics**. Also called **magnetoelectronics**.

spin resonance Resonance manifested by certain atoms changing their magnetic spin as they are exposed to a varying electromagnetic field. These atoms absorb energy at specific resonant frequencies while releasing energy. Also known as **magnetic resonance**.

spindle 1. A part, such as a rod or shaft, which serves as an axis or point around which a revolving motion takes place. An example is a capstan. A spindle may also serve to position and hold things, as is the case with the central shaft which holds a phonographic record in place on the platter. 2. The rotating shaft of a hard drive or optical drive, around which a platter or disc revolves.

spintronics Abbreviation of **spin electronics**.

spiral 1. A curve which winds around a fixed point while continuously increasing or decreasing its distance from said point. 2. That which is in the shape of a **spiral (1)**. 3. A coil of wire in the shape of a **spiral (1)**.

spiral antenna An antenna whose radiator is in the shape of a helix, and whose axis is perpendicular to a reflecting plane. Such an antenna produces a narrow beam of circularly-polarized waves which can rotate in a clockwise or counter-clockwise direction. Usually used at ultra-high and micro-wave frequencies, such as those utilized by satellites. Also called **helical antenna**.

spiral coil A coil, such as a pancake coil, wound in the form of a spiral. Also called **spiral winding**, or **spiral inductance**.

spiral inductance Same as **spiral coil**.

spiral line A transmission line whose inner conductor is spiral. Also called **spiral transmission line**, or **helical line**.

spiral scan Same as **spiral scanning**.

spiral scanning Also called **spiral scan**, or **helical scanning**. 1. Any scanning which follows the shape of a spiral, or approximately so. 2. In radars, scanning in which the pattern of the beam approximates the shape of a spiral. 3. In a video tape device, such as a VCR, recording and/or playback in which the heads contact the tape following a diagonal path which forms a spiral. The tape is wrapped at an angle around a rapidly rotating drum upon which the video heads are mounted. 4. In fax, scanning which follows the shape of a spiral.

spiral transmission line Same as **spiral line**.

spiral winding Same as **spiral coil**.

SPL Abbreviation of **sound pressure level**.

splash page The page that is displayed when certain Web sites are first accessed. It may have a multimedia presentation which leads automatically to the home page, or may give a user specified options, such as the desired language. Also called **splash screen (1)**.

splash screen 1. Same as **splash page**. 2. The first screen seen when certain applications or operating systems are loading or just loaded. Such a screen may provide simple information, such as the name of the program, or specified options, such as the desired application within a suite.

splatter Same as **sideband splatter**.

splice 1. To join two things by their ends. For instance, to join two pieces of film, or two optical fibers at their ends. Also, the act of so joining, and the place where such a joint occurs. 2. To join by overlapping, as occurs for example, when using a shrink sleeve. Also, the act of so joining, and the place where such a joint occurs. 3. To join by intertwining. For instance, to join wires by twisting their ends together. Also, the act of so joining, and the place where such a joint occurs.

splice tray A container or location utilized to organize and protect spliced optical fibers.

splicer That which serves to effect a **splice**.

splicing block A device which facilitates tape splicing by holding the ends in place.

splicing tape A non-magnetic, flexible, and durable adhesive tape specially designed for performing tape splices.

split pair The use of one wire from each of two wire pairs to make a pair. This may be done accidentally or intentionally. Such an arrangement is usually beset by interference.

split-phase motor A motor which has an auxiliary winding connected with the main winding when starting, and which after reaching a specified speed runs utilizing only the main winding. Used, for instance, where neither high cycle rates nor high torques are required, as in the case of small blowers.

split screen The division of the display screen of a computer into two or more parts or windows. Used, for instance, when keeping track of one program while simultaneously working on another.

splitter 1. That which serves to divide or separate something. For example, a beam splitter or a phase splitter. 2. A device which divides the signals received, or those to be sent, of an antenna, so as to provide multiple paths. Used, for instance, to enable multiple TVs to access the signal of a single satellite. Also called **antenna splitter**. 3. A filter which separates the voice and data signals of a telephone line via which DSL service is accessed. The voice frequencies are in the lower frequency range, and the data is in the higher interval. Not all DSL services require a splitter. Also called **POTS splitter**, or **DSL splitter**.

splitterless ADSL An ADSL service in which the splitter is located at the central office, or another location other than the premises of the customer.

splitterless DSL A DSL service in which the splitter is located at the central office, or another location other than the premises of the customer.

SPM 1. Abbreviation of **scanning probe microscopy**. 2. Abbreviation of **scanning probe microscope**.

SPMD Abbreviation of **single program-multiple data**.

spontaneous emission The natural emission particles or waves by a substance, such as that which is radioactive.

spoof To create and use false email return address, IP address, or the like.

spoofing The creation and use of a false email return address, IP address, or the like. Used, for instance, by spammers which wish to hide their identity, elicit confidence, or otherwise deceive a recipient. Spoofing may also be used by individuals, programs, or entities wishing to extract confidential information from unwary users.

spool 1. An object, usually with a circular cross-section, around which something is rolled or wound. For example, a reel upon which a magnetic tape, film, wire, cable, or other flexible material is wound. Also called **reel** (1). 2. That which is wound around a **spool** (1). Also called **reel** (2). 3. A **spool** (1) including that which is wrapped around it. Also called **reel** (3). 4. To place a document or data in a **spooler** (2).

spooler 1. That which serves to wind or roll something around a **spool** (1). 2. A program which manages printing tasks. Such a program can add, remove, change the order of, or cancel print jobs, in addition to placing documents in a memory or storage location to await printing at the printer's pace. Also called **print spooler**.

sporadic E Same as **sporadic E-layer**.

sporadic E-layer An ionospheric layer, within the E-region, which occasionally forms during periods of increased ionization. Also called **sporadic E**.

sporadic-E propagation The propagation of radio waves with the assistance of one or more reflections off a **sporadic E-layer**.

sporadic-E reflections Sharply defined reflections of radio waves at frequencies above the critical frequency in **sporadic-E layers** of the ionosphere. Also called **sporadic reflections**.

sporadic reflections Same as **sporadic-E reflections**.

spot 1. A specified point or comparatively small location or region. For example, a cathode spot, a dead spot, a hot spot, a focal spot, or a recording spot. 2. On a display, such as a CRT, the luminous area created at any given instant by an electron beam. 3. Same as **scanning spot** (1). 4. Same as **scanning spot** (2).

spot beam 1. A satellite signal that is sent to a limited geographic area. 2. An antenna signal that is sent to a limited area.

spot check An occasional and/or random check, inspection, or test which is usually limited in scope.

spot frequency 1. A single frequency, as opposed to a frequency band. 2. A frequency which serves a reference for others.

spot size The dimensions of the region illuminated by an electron beam in a CRT.

spot speed The speed at which a **scanning spot** moves. Also called **scanning speed** (2).

spot welding Welding, especially resistance welding, performed on small areas. The size and shape of the electrodes determine the size and shape of the weld.

SPP Abbreviation of **scalable parallel processing**.

SPQ Abbreviation of **statistical quality control**.

spray coating The application of a layer by spraying a solution onto a substrate, which is then allowed to dry. Used, for instance, to apply a liquid photoresist.

spread 1. To extend, or to extend from a given point or location. Also, to have extended, or to be extended from a given point or location. 2. To distribute over a surface, or within a volume. Also, to have distributed over a surface, or within a volume. 3. To increase the size of a gap. Also, to have the size of a gap increased. 4. The limits within which a value fluctuation may occur. Also, the limits within which a value fluctuation actually occurs.

spread spectrum 1. Any modulation technique in which the bandwidth of the information-bearing signal is intentionally spread over a much wider bandwidth than would otherwise be necessary. This may be done, for instance, for added security or for the ability to recover the data without retransmission. Examples include frequency-hopping spread spectrum, in which the frequency of the carrier hops among multiple frequencies at a rate determined by a specific code or algorithm, and direct-sequence spread spectrum, where the information-carrying bit stream is combined with a pseudorandom bit stream at a higher bit rate, and this combined signal modulates the carrier. Also, the technology utilized to implement any such technique. Also called **spread-spectrum modulation**, or **spread-spectrum technique**. 2. A transmission utilizing a **spread spectrum** (1) technique or technology.

spread-spectrum modulation Same as **spread spectrum** (1).

spread-spectrum technique Same as **spread spectrum** (1).

spread-spectrum transmission A data transmission utilizing a **spread-spectrum technique**.

spreader An object or material, such as a bar or insulator, which helps keep wires or other conductors apart. Used, for instance, to space the conductors of a transmission line separated by air.

spreading resistance For a contact with a small area, such as a point contact, the resistance which does not lie strictly along the path between electrodes. This may occur, for instance, in semiconductor materials whose dimensions are large in relation to such a contact. Used, for example, to evaluate the dopant profile of a semiconductor junction.

spreadsheet 1. A computer application which manipulates a matrix of interrelated cells, often arranged in rows and columns. Each cell can have labels, numerical values, formulas, or functions. Used primarily for calculations such as budgets, and for what-if scenarios. Also called **spreadsheet program**, **spreadsheet application**, or **electronic spreadsheet**. **2.** The matrix of interrelated cells a **spreadsheet (1)** manipulates. Also called **worksheet**.

spreadsheet application Same as **spreadsheet (1)**.

spreadsheet program Same as **spreadsheet (1)**.

spring-loaded Secured through the use of a compressing spring. For example, an alligator clip is spring-loaded.

SPS 1. Abbreviation of **standby power supply**. **2.** Abbreviation of **standby power system**.

SPST relay Abbreviation of **single-pole single-throw relay**.

SPST switch Abbreviation of **single-pole single-throw switch**.

spurious Not proceeding from a true source, not proceeding from an apparent source, not intended, undesired, or false or otherwise not genuine.

spurious emission Also called **spurious radiation**. **1.** For a transmitter, such as that of TV or radio, an emission at a frequency other than the intended frequency, or which is outside the intended bandwidth. This may cause noise or interference. **2.** Any undesired emission of electromagnetic radiation from a component, circuit, device, piece of equipment, or system.

spurious modulation Modulation, such as incidental FM, which is undesired.

spurious oscillation Oscillation, such as that occurring at an unintended frequency, which is undesired.

spurious radiation Same as **spurious emission**.

spurious rejection The ability of a radio receiver to reject **spurious signals**.

spurious response Any response by a radio receiver to a frequency other than that intended. For instance, the reception of one channel while tuned to another.

spurious signal A signal, such as that occurring at an unintended frequency, which is undesired.

sputter 1. To cause particles, such as atoms or ions, to be removed from a surface, as occurs, for example, in an ion microprobe. Also, such an instance. **2.** To coat a surface with particles, such as atoms or ions, in short bursts. Also, such an instance. **3.** To eject or emit particles, signals, sounds, or the like, in short and usually irregular bursts. Also, such an instance.

sputter etching Etching accomplished via **sputtering (1)**.

sputtering 1. The causing of particles, such as atoms or ions, to be removed from a surface, as occurs, for instance, through particle bombardment. **2.** The coating of a surface with particles, such as atoms or ions, in short bursts. **3.** The ejection or emission of particles, signals, sounds, or the like, in short and usually irregular bursts. **4.** In a vacuum tube, the emission of particles from a cathode which is disintegrating as a consequence of bombardment by high-energy ions. Also called **cathode sputtering (1)**. **5.** The use of **sputtering (4)** to deposit a thin film of a metal onto a surface such as glass, metal, or plastic. Also called **cathode sputtering (2)**.

SPX Abbreviation of **Sequenced Packet Exchange**.

spyware Software that tracks and gathers information about where a user navigates within the Web. Spyware may be used, for instance, for statistical purposes, or may be employed to scrutinize unsuspecting Web surfers.

SQL Acronym for Structured Query Language. A standardized language utilized to query, update, and otherwise access and manage a database or relational database. Usually pronounced *seaquill*. Also called **SQL language**.

SQL database A database or relational database that can be queried, updated, or otherwise accessed and managed via **SQL**.

SQL language Same as **SQL**.

SQL query A query structured in **SQL**.

SQL server A DBMS that can be queried, updated, or otherwise accessed and managed via **SQL**.

square A quadrilateral with four right angles and four equal sides. A **rhombus** is a quadrilateral with four equal sides, but having no right angles.

square-law demodulation Same as **square-law detection**.

square-law demodulator Same as **square-law detector**.

square-law detection The use and output of a **square-law detector**. Also called **square-law demodulation**.

square-law detector A detector, such as a diode detector, whose output is proportional to the square of the input voltage. Also called **square-law demodulator**.

square-law response For a component, circuit, device, piece of equipment, or system, an output which is proportional to the square of its input.

square pulse A rectangular pulse in which the pulse duration of each of the fixed values is the same.

square rooter Also called **square-rooting circuit**. **1.** A circuit or device whose output is proportional to the square root of its input. **2.** A circuit or device which calculates square roots.

square-rooting circuit Same as **square rooter**.

square wave A rectangular wave in which the pulse duration of each of the fixed values is the same.

square-wave frequency response Same as **square-wave response**.

square-wave generator A signal generator whose output consists of **square waves**.

square-wave modulation The modulation provided by a **square-wave modulator**. Used, for instance, in AM.

square-wave modulator A modulator whose output consists of **square waves**.

square-wave response A measure of the behavior of a component, circuit, device, piece of equipment, or system, as a function of a square-wave input signal. Also known as **square-wave frequency response**.

square-wave test A test of a component, circuit, device, piece of equipment, or system, utilizing a **square-wave generator**.

squarer Same as **squaring circuit**.

squaring circuit Also called **squarer**. **1.** A circuit whose output is proportional to the square of its input. **2.** A circuit which converts a sine, or other wave or pulse, into a square wave or pulse. **3.** A circuit or device which calculates squares.

squeal An unwanted high-pitched sound emitted by a device or piece of equipment. For example, a loud shrilling sound resulting from howling. Also called **squealing**.

squealing Same as **squeal**.

squegger Same as **squegging oscillator**.

squegging The action of a **squegging oscillator**.

squegging oscillator An oscillator which stops operating for a predetermined time after completing one or more cycles. Its grid bias increases during oscillation until oscillation stops, then decreases until oscillation is reestablished. Also called **squegger**, or **blocking oscillator**.

squelch The use and action of a **squelch circuit**.

squelch circuit 1. A circuit in a radio receiver that suppresses noise when listening to programming or other signals. Also called **quieter, quieting circuit**, or **noise quieter**. 2. A circuit utilized for reducing of the volume of, or silencing sounds. Also called **muting circuit (1)**. 3. A circuit utilized for manually or automatically disabling of a given input or output. Also called **muting circuit (2)**.

SQUID Acronym for **superconducting quantum interference device**.

squint 1. The angle between the axis of a radar antenna and another axis, such as that of the main lobe. 2. The angle between the axes of the main lobes when utilizing beam switching in radars.

squirrel-cage induction motor An induction motor incorporating a **squirrel-cage rotor**. Also called **squirrel-cage motor**.

squirrel-cage motor Same as **squirrel-cage induction motor**.

squirrel-cage rotor A winding that is shorted at both ends by rings, and which is wound around the rotor of an induction motor. Also called **squirrel-cage winding**.

squirrel-cage winding Same as **squirrel-cage rotor**.

sr Symbol for **steradian**.

Sr Chemical symbol for **strontium**.

SR flip-flop Abbreviation of set-reset **flip-flop**. A flip-flop whose inputs are designated as **R** and **S**, along with a clock input. Its outputs are similar to those of a JK flip-flop for any given combination of inputs, except that a race condition can occur in an RS flip-flop when both inputs are at logic 1, thus such a condition should be avoided. Also called **RS flip-flop**.

SR motor Abbreviation of **switched-reluctance motor**.

SRAM Abbreviation of static random-access memory. A type of RAM which does not require constant refreshing. It is faster than **DRAM**, which requires constant refreshing, and is utilized for instance, as memory cache.

SRB 1. Abbreviation of **source route bridging**. 2. Abbreviation of **source route bridge**.

SRD Abbreviation of **step-recovery diode**.

SRL Abbreviation of **scattering Raman lidar**.

SS7 Abbreviation of Signaling System 7. An ITU-defined protocol that is utilized to set-up, manage, and terminate telephone calls. SS7 uses out-of-band signaling, so control signals, such as those required to provide services such as call set-up, call-forwarding, roaming, toll-free dialing, caller ID, and call termination, use a dedicated channel other than that transmitting the voice signals, providing for a more efficient, secure, and flexible arrangement. In addition, SS7 handles other tasks such as the verification that a given number can access given features, in addition to determining the routing of calls. Also called **SS7 protocol**.

SS7 network A telephone network using the **SS7 protocol**.

SS7 protocol Same as **SS7**.

SSA Abbreviation of Serial Storage Architecture. An open standard for high-speed serial transfers to and from peripherals such as disk drives and printers. SSA self-configures, supports hot-swapping, and allows connection of up to 128 devices simultaneously, at transfer speeds up to 160 MBps. Also, a serial interface adhering to this standard.

SSB Abbreviation of **single-sideband**.

SSB communications Abbreviation of **single-sideband communications**.

SSB modulation Abbreviation of **single-sideband modulation**.

SSB-SC Abbreviation of **single-sideband suppressed carrier**.

SSB transmission Abbreviation of **single-sideband transmission**.

SSD Abbreviation of **solid-state disk**.

SSFDC Abbreviation of **solid-state floppy disk card**.

SSG Abbreviation of **standard signal generator**.

SSI 1. Abbreviation of **server-side include**. 2. Abbreviation of **small-scale integration**.

SSL Abbreviation of Secure Sockets Layer. A protocol utilized for the encryption of messages and documents sent over the Internet, helping insure authentication, privacy, and data integrity. SSL is supported by most browsers, and is used, for instance, when transmitting financial account information or when paying for purchases. SSL is now part of **TLS**, which is a more complete protocol, however the two are not mutually compatible. Also called **SSL protocol**.

SSL encryption Abbreviation of Secure Sockets Layer **encryption**. Encryption utilized when sending data during an SSL session. For instance, a credit card number can be encrypted at one side of the connection, the data sent using the SSL protocol, and decrypted at the other end.

SSL protocol Abbreviation of Secure Sockets Layer **protocol**. Same as **SSL**.

SSO Abbreviation of **single sign-on**.

SSP 1. Abbreviation of **service switching point**. 2. Abbreviation of **storage service provider**.

SSSC Abbreviation of **single-sideband suppressed carrier**.

SSTV Abbreviation of **slow-scan TV**.

ST connector An older fiber-optic connector utilizing a bayonet base and socket.

St. Elmo's fire An intermittent, luminous, and often audible electrical discharge that originates from a conductor when its potential exceeds a given value, but is not high enough to form a spark. This occurs especially in pointed objects when the electric field near their surfaces surpasses a given amount. For example, an aircraft traveling through an electrical storm may develop such discharges from an antenna, which in turn produces precipitation static. Also called **corona discharge, corona effect, corona, brush discharge**, or **corposant**.

stability 1. The quality or state of being **stable**. 2. The extent to which **stability (1)** is maintained.

stability factor A measure of the **stability** of a component, circuit, device, piece of equipment, system, process, or material.

stability platform Same as **stable platform (1)**.

stabilization 1. The process of maintaining, causing, or achieving **stability**. 2. The extent to which **stabilization (1)** is maintained, caused or achieved.

stabilized feedback In an amplifier or system, feedback which is 180° out-of-phase with the input signal, thus opposing it. While this results in a decrease in gain, it also serves reduce distortion and noise, and to stabilize amplification. Also called **inverse feedback, reverse feedback, negative feedback**, or **degeneration**.

stabilized platform Same as **stable platform (1)**.

stable 1. Tending to remain in a given position, state, setting, mode, or the like. Also, tending to avoid unwanted variations. For instance, free of unwanted oscillation. 2. Tending to return to a former or desired position, state, setting, mode, or the like, after a displacement or other change occurs. An example is a speaker with proper damping. 3. Tending to maintain a given value, quantity, intensity, characteristic, or the like. Said, for instance, of the output of a control system. **4.** Not tending to easily respond to a change, action, stimulus, or variation. Said, for instance, of a voltage regulator which maintains the same output despite changes in load resistance. **5.** A chemical substance not tending to easily form part of a chemical reaction. **6.** A particle which does not undergo radioactive decay.

stable local oscillator A high-precision local oscillator which is used, for instance, to generate a reference carrier. Its acronym is **STALO**.

stable oscillation **1.** Oscillation which is very or extremely regular, such as that of a quartz crystal. **2.** Same as **steady-state oscillation**.

stable platform **1.** A platform or surface, such as that utilizing gyros, which is intended to maintain a specified orientation in space. Used, for instance, for mounting instruments utilized to study astronomical phenomena. Also called **stabilized platform**, or **stability platform**. **2.** A platform which is especially steady, solid and/or rugged. Used, for instance, to mount a robotic arm. **3.** A software or hardware platform which is not prone to malfunction or failure under difficult or unusual conditions.

stable state A state characterized by **stability**. Said, for instance, of either of the two stable states of a flip-flop or other bistable device.

stack **1.** An orderly pile in which elements are arranged in layers, related groups, or the like. For example, a fuel cell stack, a rectifier stack, a head stack, or a protocol stack. The elements of such a stack may or may not be identical. Also, to arrange in such a stack. **2.** A set of registers, memory locations, or other storage areas or units which are ordered in a manner that provides for the handling of data, interrupts, or the like, in a specified order, such as FIFO or LIFO. To push is to add an item to the top of a stack, and to pop is to remove the first item from the top of a stack.

stack overflow An error condition arising when attempting to push or otherwise add an item to a stack that is full, while a **stack underflow** attempts to pop or otherwise remove an item out of an empty stack.

stack pointer A register that keeps track or identifies the last item entered into a **stack (2)**, or the current location in such a stack.

stack size The size, usually expressed in some multiple of bytes, allocated to a **stack (2)**.

stack underflow An error condition arising when attempting to pop or otherwise remove an item out of an empty stack, while a **stack overflow** attempts to push or otherwise add an item to a stack that is full.

stacked array An antenna array in which elements, or groups of elements, are arranged one above the other. Used, for instance, to provide for greater gain.

stacked-dipole array A **stacked array** consisting of multiple identical dipoles. Also called **stacked dipoles**.

stacked dipoles Same as **stacked-dipole array**.

stacking The preparation or arrangement of a **stack (1)**. For example, the arrangement of a stacked-dipole array.

stage A component or functional unit within a circuit, device, piece of equipment, or system. For example, an amplifier stage, an intermediate-frequency stage, an oscillator stage, or a mixer stage.

stage gain Any gain introduced by a given **stage**.

stage loss Any losses introduced by a given **stage**. Also called **stage losses**.

stage losses Same as **stage loss**.

stagger **1.** To place at alternate sides of a center or given location, time, or the like. For example, occurrence fifteen minutes before and after an hour. **2.** To position or distribute in non-consecutive or non-adjacent locations. For instance, staggered tuning, or memory interleaving. **3.** In fax,

an error resulting from variations in the position of the recording spot along a given line.

stagger tuning Tuning to frequencies which are slightly above or slightly below a given frequency. Used, for instance, to create a wider bandwidth. Also called **staggered tuning**.

staggered tuning Same as **stagger tuning**.

staggering **1.** The placement at alternate sides of a center or given location, time, or the like. **2.** The positioning or distribution in non-consecutive or non-adjacent locations.

stair-stepping The jagged appearance that curved or diagonal lines in computer graphics have when there is not enough resolution to show images realistically. Also called **aliasing (2)**, or **jaggies**.

STALO Acronym for **stable local oscillator**.

stand-alone Also spelled **standalone**. **1.** Self-contained and able to be operated independently of other devices, equipment, or systems. For example, a stand-alone computer. **2.** Pertaining to that which is **stand-alone (1)**.

stand-alone computer A computer which is not a part of a network such as a LAN or WAN. A computer which connects to the Internet when desired, and which otherwise does not form part of a network is also considered to be stand-alone. Also spelled **standalone computer**.

stand-alone device A device, such as an overhead projector, fax, or a PDA, which is self-contained and able to function independently of other devices, equipment, or systems. Also spelled **standalone device**.

stand-alone PC Abbreviation of **stand-alone** personal computer. A personal computer which is not a part of a network such as a LAN or WAN. A personal computer which connects to the Internet when desired, and which otherwise does not form part of a network is also considered to be stand-alone. Also spelled **standalone PC**.

stand-alone personal computer Same as **stand-alone PC**. Also spelled **standalone personal computer**.

stand-by Same as **standby**.

stand-by battery Same as **standby battery**.

stand-by button Same as **standby button**.

stand-by command Same as **standby command**.

stand-by computer Same as **standby computer**.

stand-by condition Same as **standby**. Also spelled **standby condition**.

stand-by current Same as **standby current**.

stand-by device Same as **standby device**.

stand-by equipment Same as **standby equipment**.

stand-by function Same as **standby function**.

stand-by mode Same as **standby (3)**. Also spelled **standby mode**.

stand-by operation Same as **standby operation**.

stand-by power Same as **standby power**.

stand-by power supply Same as **standby power supply**.

stand-by power system Same as **standby power system**.

stand-by processor Same as **standby processor**.

stand-by state Same as **standby**. Also spelled **standby state**.

stand-by switch Same as **standby switch**.

stand-by time Same as **standby time**.

stand-by unit Same as **standby unit**.

standalone Same as **stand-alone**.

standalone computer Same as **stand-alone computer**.

standalone device Same as **stand-alone device**.

standalone PC Same as **stand-alone PC**.

standalone personal computer Same as **stand-alone PC**.

standard Its abbreviation is **std. 1.** An established reference against which comparisons or verifications may be made. A de facto standard is adopted through continued use and acceptance, while a de jure standard is issued or endorsed by a entity which establishes standards. **2.** A component, circuit, device, instrument, piece of equipment, or system, whose parameters or specifications are known precisely, and which serves as a basis for verification, comparison, or the like. For instance, a standard capacitance, or a standard resistance. **3.** A component, circuit, device, instrument, piece of equipment, or system whose output, such as a frequency, voltage, amplitude, level, or the like, is precisely known, and which serves as a basis for comparisons or verifications. For example, an atomic frequency standard. **4.** A specification, such as that for software, hardware, a format, or protocol, which is widely accepted, or assigned to serve as a model or basis. For instance, the OSI Reference Model. **5.** A transmission and reception system which is established for a given communications or broadcast format. Examples include the high-definition, NTSC, and PAL TV standards.

standard ASCII An ASCII character set that uses 7-bit code, and which provides a total of 128 characters. These characters are approximately the same as those found on a common computer keyboard, plus some control characters such as start-of-text and end-of-text. **Extended ASCII**, for instance, has an 8-bit code providing a total 256 characters. Also called **standard ASCII character set**, or **7-bit ASCII**.

standard ASCII character A character within an **standard ASCII character set**.

standard ASCII character set Same as **standard ASCII**.

standard ASCII code The standard code utilized to express **standard ASCII** text and control characters.

standard atmosphere A unit of pressure intended to equal the pressure of the earth's atmosphere at sea level. One standard atmosphere equals 101.325 kilopascals, or 760 torr. Also called **atmosphere** (2).

standard broadcast band A band of frequencies allocated to broadcasting stations. For instance, the standard AM broadcast band is 535 to 1705 kHz. Also called **broadcast band**.

standard candle A former name of **candela**.

standard capacitance A capacitor whose capacitance value is precisely known, and which serves to measure or compare unknown capacitances. May be used, for instance in a Schering bridge. Also called **standard capacitor**, or **capacitance standard**.

standard capacitor Same as **standard capacitance**.

standard cell **1.** A primary cell whose voltage at a given temperature is precisely known, and which serves as a voltage reference standard. An example is a Weston standard cell. **2.** An IC design which incorporates previous designs.

standard conditions **1.** A combination of environmental conditions which are established for any given setting or purpose. For example, that in which the temperature is 0 °C, the pressure is 1 atmosphere, and there is 0% relative humidity. **2.** Same as **standard temperature and pressure**.

Standard Definition television Same as **SDTV**.

Standard Definition TV Same as **SDTV**.

standard deviation A statistical measure of the dispersion or spread around the mean, for a given set, series, or population. It is equal to the square root of the variance. If a data set has as a bell-shaped distribution, for instance, 68.3% of the data will fall on either side of one standard deviation, 95.5% within two standard deviation, 99.7% within three, and so on. Its abbreviation is **SD**. Its symbol is σ.

standard frequency A frequency whose value is known with extreme precision, and which is used as a reference for calibration, comparison, or verification.

standard-frequency oscillator An oscillator, such as an extremely precise and stable crystal oscillator, whose output is a **standard frequency**.

standard-frequency signal A signal at a **standard frequency**, which is utilized, for instance, for calibration and testing of radio equipment.

Standard Generalized Markup Language Same as **SGML**.

standard hydrogen electrode A standard reference electrode usually consisting of a solution containing hydrogen atoms at a unit concentration, into which a platinum electrode is immersed, and over which hydrogen gas is passed at a specific pressure. This electrode has an arbitrarily assigned potential of zero, and other electrode potentials are compared to it. Also called **hydrogen electrode** (1).

standard inductance Same as **standard inductor**.

standard inductor An inductor whose inductance value is highly accurate and stable, and which serves, for instance, as a calibration standard. Also called **standard inductance**, or **inductance standard**.

standard light source A light source, such as that which faithfully duplicates sunlight, which is utilized as a standard.

standard operating conditions The circumstances, such as ambient conditions, input voltage, and output current, which are required for the proper functioning of a component, circuit, device, piece of equipment, or system. Also called **operating conditions** (1), or **normal operating conditions**.

standard parallel port A parallel port, such as a Centronics parallel port, which has been shipping with new computers for several years, as opposed that incorporating a newer technology such as an Extended Capabilities Port.

standard pitch A specific frequency utilized as a standard. For many purposes this frequency is taken as 440 Hz.

standard pressure A pressure, such as 1 atmosphere, which is used under a given setting, such as that in a laboratory. Also called **normal pressure**.

standard resistance A resistor whose resistance value is highly accurate and stable, and which serves, for instance, as a calibration standard. Also called **standard resistor**, or **resistance standard**.

standard resistor Same as **standard resistance**.

standard serial port A serial port, such as an RS-232C port, which has been shipping with new computers for several years, as opposed that incorporating a newer technology such as a USB port.

standard signal generator A circuit, device, instrument, piece of equipment, or object, which generates one or more **standard frequencies**. Used, for instance, for the calibration and testing of FM and AM equipment. Its abbreviation is **SSG**.

standard temperature A temperature, such as 0 °C, which is used under a given setting, such as that in a laboratory. Also called **normal temperature**.

standard temperature and pressure Its abbreviation is **STP**. Also called **standard conditions** (2), or **normal temperature and pressure**. **1.** A temperature of 0 °C, and a pressure of 1 atmosphere. Used, for instance, as a practical and easily maintained condition for laboratory experimentation. **2.** An established combination of temperature and pressure, such as of 25 °C and 1 atmosphere, utilized for any given setting or purpose.

standard time **1.** An hour of day based on a celestial parameter, such as the rotation of the planet. It may be, for instance, any of the established time zones along the surface of the planet. **2.** The average or usual time required for a given task, step, process, operation, or the like.

Standard Wire Gage Same as **Standard Wire Gauge**.

Standard Wire Gauge A standard, such as Brown and Sharp gauge or Birmingham Wire Gauge, utilized for designating the diameter of wires, or the thickness of sheets, tubing, or the like. Also spelled **Standard Wire Gage**.

standardization The creation, promotion, and implementation of **standards**. Also, the adherence to standards.

standardize To create, promote, and implement **standards**. Also, to adhere to standards.

standby Also spelled **stand-by**. Also called **standby state**, or **standby condition**. **1.** A state in which a component, circuit, device, piece of equipment, or system is functioning under normal operating conditions, but does not have an applied signal at a given time. Also called **standby operation** (1), **resting state** (1), **idle state** (1), or **quiescent state** (1). **2.** A state in which a device, piece of equipment, or system is operational and available, but is not in use. Also called **standby operation** (1), **resting state** (2), **idle state** (2), or **quiescent state** (2). **3.** A state in which a component, circuit, device, piece of equipment, or system is not in use, but is ready for immediate activation. Power consumption in the standby state is minimal, while in the off state it should be zero. Also called **standby mode**. **4.** A button, command, condition, or function which places a component, circuit, device, piece of equipment, or system in **standby** (1), **standby** (2), or **standby** (3).

standby battery A battery which is operational, available, and ready for use during emergencies, power failures, or the like. Also spelled **stand-by battery**.

standby button A button which places a component, circuit, device, piece of equipment, or system in **standby** (1), **standby** (2), or **standby** (3). Also spelled **stand-by button**.

standby command A command which places a component, circuit, device, piece of equipment, or system in **standby** (1), **standby** (2), or **standby** (3). Also spelled **stand-by command**.

standby computer A computer which is ready to be used if a main computer fails, requires maintenance, or is otherwise not available for regular service. Also spelled **stand-by computer**.

standby condition Same as **standby**. Also spelled **stand-by condition**.

standby current The current drained by a component, circuit, device, piece of equipment, or system which is not in use, but ready for immediate activation. Power consumption in the standby state is usually minimal. Also spelled **stand-by current**.

standby device A device, such as a backup device, which is ready for use when a similar device fails or is otherwise not available for use. Also spelled **stand-by device**.

standby equipment Equipment, such as backup equipment, which is ready for use when unforeseen circumstances, such as multiple failures in similar equipment, occur. Also spelled **stand-by equipment**.

standby function A function which places a component, circuit, device, piece of equipment, or system in **standby** (1), **standby** (2), or **standby** (3). Also spelled **stand-by function**.

standby mode Same as **standby** (3). Also spelled **stand-by mode**.

standby monitor In a token-passing network, any station except the one with control of the token, which is called **active monitor**. Only one station can have control of the token at any given moment.

standby operation Also spelled **stand-by operation**. **1.** Same as **standby** (1). **2.** Same as **standby** (2).

standby power Also spelled **stand-by power**. **1.** The power drained by a component, circuit, device, piece of equipment, or system which is not in use, but ready for immediate acti-

vation. Power consumption in the standby state is usually minimal. **2.** The power provided by a **standby power supply**.

standby power supply A power supply, such as that provided by a backup power supply or a standby battery, which is ready for use in the case of an event such as a failure of the regular source of power. Also spelled **stand-by power supply**. Its abbreviation is **SPS**.

standby power system A system, such as that incorporating a **standby power supply**, that provides power when needed. Also spelled **stand-by power system**. Its abbreviation is **SPS**.

standby processor A CPU which is ready to be used if the main CPU fails. Also spelled **stand-by processor**.

standby state Same as **standby**. Also spelled **stand-by state**.

standby switch A switch that places a component, circuit, device, piece of equipment, or system in **standby** (1), **standby** (2), or **standby** (3). Also spelled **stand-by switch**.

standby time Also spelled **stand-by time**. **1.** The amount of time a battery can power a portable phone once fully charged. This time is usually calculated for a telephone which is turned on and able to accept calls. **2.** The time a battery can power a device, such as a PDA, once fully charged. This time is usually calculated for a device which is turned on and able to function, but which is not used for any purpose. **3.** The time during which a component, circuit, device, piece of equipment, or system is in standby mode.

standby unit A unit, such as a backup device, which is ready for use when a similar device fails or is otherwise not available for use. Also spelled **stand-by unit**.

standing wave A wave in which the antinodes and nodes do not change position through a given medium, such as a transmission line. Such waves usually occur as a result of the total or partial reflection of said wave by a barrier such as a discontinuity in a transmission line. This contrasts with a **progressive wave**, whose antinodes and nodes progress along the medium at the speed of the wave. Also called **stationary wave**.

standing-wave detector Same as **standing-wave ratio meter**.

standing-wave indicator Same as **standing-wave ratio meter**.

standing-wave loss For a given transmission line, any losses resulting from a **standing-wave ratio** other than 1. Also called **standing-wave losses**.

standing-wave losses Same as **standing-wave loss**.

standing-wave meter Same as **standing-wave ratio meter**.

standing-wave ratio Its abbreviation is **SWR**. **1.** For a given transmission line, such as a coaxial cable or waveguide, the ratio of the maximum voltage to the minimum voltage. It is a measure of the impedance matching of the line. This ratio is equal to 1 when there is complete impedance matching, in which case the maximum possible RF power reaches the load, such as an antenna. Also called **voltage standing-wave ratio**. **2.** For a given transmission line, such as a coaxial cable or waveguide, the ratio of the maximum current to the minimum current. It is a measure of the impedance matching of the line. This ratio is equal to 1 when there is complete impedance matching, in which case the maximum possible RF power reaches the load, such as an antenna. Also called **current standing-wave ratio**.

standing-wave ratio bridge A bridge utilized to measure **standing-wave ratios**. Its abbreviation is **SWR bridge**.

standing-wave ratio detector Same as **standing-wave ratio meter**. Its abbreviation is **SWR detector**.

standing-wave ratio indicator Same as **standing-wave ratio meter**. Its abbreviation is **SWR indicator**.

standing-wave ratio meter A device or instrument that detects standing waves, or which determines standing-wave ratios. Its abbreviation is **SWR meter**. Also called **standing-wave ratio indicator**, **standing-wave ratio detector**, **standing-wave meter**, **standing-wave detector**, or **standing-wave indicator**.

standoff insulator An insulator, such as that in the form of a post, which separates conductors from a surface they are mounted on, such as a chassis.

star 1. A large self-luminous celestial body, usually composed of gases, which derives its energy from nuclear energy within its core. The sun is an example. **2.** That which has a central hub or node and is connected to other things which surround said center. Also that which is similarly configured. For example, a star network or a star ground.

star bus A communications network topology combining elements of both a star and bus layout. The nodes are arranged in stars around the hubs, which are connected together via a common bus.

star circuit Same as **star network (2)**.

star configuration The physical arrangement or layout of a **star network (1)**.

star connection Same as **star network (2)**.

star ground A single ground point in a circuit, stage, or system, which connects all grounds and returns. Such a ground, for instance, helps minimize unwanted coupling. Also called **single-point ground**.

star grounding The use of a single point in a circuit, stage, or system, to connect all grounds and returns. Also called **single-point grounding**.

star network 1. A communications network configuration in which all nodes are connected to a central node or hub. Also, a network with such a layout. Also called **hub-and-spoke network**. **2.** An electric network configuration in which three or more branches, each with one terminal, are connected to a common central point or node. Also called **star circuit**, or **star connection**. When there are three arms, such as impedance branches, also called **wye network**, **Y network**, or **T network**.

star topology The topology of a **star network (1)**.

start bit In asynchronous communications, a bit transmitted before a character or packet.

start element Same as **start signal**.

start lead The lead attached to the first, or inner, turn of a coil. This contrasts with a **finish lead**, which is connected to the last turn of a coil. Also called **inside lead**.

start-of-heading A control character or code which comprises the first character of a heading. Its abbreviation is **SOH**.

start-of-message Its abbreviation is **SOM**. A symbol or code indicating the beginning of a message. For instance, the letters SOM appearing may serve as such an indication.

start page A Web page that serves as a starting point for navigation. A start page may be that which is first loaded when a Web browser accesses the Internet. It may also be the first or main page within a group of pages, such as those of a business or individual. Also called **home page**, or **welcome page**.

start signal In asynchronous communications, a signal, such as a given sequence of bits, transmitted before a character or packet. Also called **start element**.

start/stop communications Communications which are not synchronized by a timing signal. Telephonic communication, in which two or more parties can talk simultaneously, is asynchronous. Otherwise, each person would have to wait a determined time interval before speaking. In computers, asynchronous communication may start or end at any time, with these events being signaled respectively by start and stop bits. Also called **asynchronous communications**.

start/stop data The data conveyed in a **start/stop transmission**.

start/stop data transmission Same as **start/stop transmission**.

start/stop device Also called **asynchronous device**. **1.** A device whose operation does not depend on a timing mechanism, but rather on each preceding event having been completed. **2.** An AC device which does not run at a speed which is synchronized to the frequency of the power operating it.

start/stop input/output A computer's ability to simultaneously have an inflow and outflow of data.

start/stop logic In hardware or software, logic which depends on events concluding, and not on clock-generated synchronization signals.

start/stop mode A mode of operation which does not depend on a timing mechanism, but rather on each preceding event having been completed. Also called **asynchronous mode**.

start/stop operation Operation which does not depend on a timing mechanism, but rather on each preceding event having been completed. Also called **asynchronous operation**.

start/stop protocol A communications protocol which governs a **start/stop transmission**.

start/stop transmission Data transmission that occurs intermittently, rather than in a steady stream. Each character has its own start and stop bits and is sent individually without being synchronized by timing signals. This contrasts with a **synchronous transmission**, which occurs in a steady stream and is synchronized by timing signals. Also called **start/stop data transmission**, **asynchronous data transmission**, or **asynchronous transmission**.

start-up Same as **startup**.

start-up disk Same as **system disk**. Also spelled **startup disk**.

start-up routine Same as **startup routine**.

start-up screen Same as **startup screen**.

start-up sequence Same as **startup routine**. Also spelled **startup sequence**.

starter 1. That which serves to initiate an event, sequence, process, or the like. **2.** A device utilized to start an electric motor, and to accelerate it to its operational speed. Such a starter may be powered, for example, by a battery. Also called **electric starter**. **3.** Same as **starter electrode**.

starter electrode An electrode in a gas tube which is utilized to provide the high voltage necessary to establish glow discharge, arc discharge, or the like. Also called **starter (3)**.

starting current 1. The current necessary to initiate an event, sequence, process, or the like. **2.** The current necessary to start an electric motor. **3.** The current necessary to initiate glow discharge, arc discharge, or the like, in a gas tube.

starting rate The rate at which a battery charges as it starts a charge cycle. This rate is not necessarily the same as that which follows.

starting voltage 1. The voltage necessary to initiate an event, sequence, process, or the like. **2.** The voltage necessary to start an electric motor. **3.** The voltage necessary to initiate glow discharge, arc discharge, or the like, in a gas tube. Also called **striking voltage**.

startup Also spelled **start-up**. **1.** To start up or reset a computer. During this process the computer accesses instructions from its ROM chip, performs self-checks, loads the operating system, and prepares for use by an operator. A startup may be initiated by turning on the power, pressing a button or switch, by hitting a specific key sequence, or through a program or routine that gives this command. Also called by various other names, including **bootstrap**, **booting**, **booting up**, and **initial program load (1)**. **2.** The starting of a computer by turning on its power. Many program failures are resolved through turning off a computer then turning it back on again. Also called **cold boot**, or **hard boot**.

startup disk Same as **system disk**. Also spelled **start-up disk**.

startup routine The automatic steps a computer follows when performing a **startup**. Also spelled **start-up routine**. Also called **startup sequence**, or **bootstrap routine**.

startup screen The screen displayed when first starting-up or loading a computer, application, Web page, or the like. An example is a splash screen. Also spelled **start-up screen**.

startup sequence Same as **startup routine**. Also spelled **start-up sequence**.

stat- A prefix which identifies units conforming to the cgs electrostatic system. Statvolt is an example.

stat mux Abbreviation of **statistical multiplexer**.

statA Abbreviation of **statampere**.

statampere The unit of current within the cgs electrostatic system. There are approximately 3.33564×10^{-10} ampere in a statampere. Its abbreviation is **statA**.

statC Abbreviation of **statcoulomb**.

statcoulomb The unit of electrical quantity within the cgs electrostatic system. There are approximately 3.33564×10^{-10} coulomb in a statcoulomb. Its abbreviation is **statC**.

state 1. A description of the conditions or circumstances of a particle, entity, or system, which encompasses one or more key attributes. Also, such a state. For example, a ground state, an excited state, or a quantum state. 2. The condition in which a component, circuit, device, piece of equipment, system, material, process, or setting is in at a given moment. For example, and idle state, or a logic state. 3. The condition in which a computer program, device, system, setting, sequence, or the like is in at a given moment. 4. Same as **state of matter**.

state machine A machine which can be completely described by a finite set of defined states. Such a machine must be in one of these states at any given moment, and there is a set of conditions which determine when it moves from one state to another. Also, a design model which can be so described. Used, for instance, to design specialized digital systems. Also called **finite-state machine**.

state of charge Its abbreviation is **SOC**. 1. For a battery or cell, the amount of charge remaining, as a percentage of the full charge, at a given moment. 2. For a battery or cell, the amount of charge remaining at a given moment, expressed in ampere-hours, coulombs, volts, or the like.

state of matter Any of the manners in which matter is currently known to exist, based on structure, phase, form, composition, and the like. These are solid, liquid, gas, plasma, and Bose-Einstein condensates. Also called **state** (4).

state-of-the-art Consisting of, or incorporating the latest technology in its field. What the term entails at a given time is usually not the same at a later time. For example, some time back an 8 MHz CPU was considered state-of-the-art, while currently an 8 GHz CPU is more like it. The term is frequently utilized to hype components, devices, equipment, systems, software, materials, and so on.

state space In artificial intelligence, the formulation of a problem into states, such as the initial state, intermediate, and goal states, and the determination of the necessary operations to proceed to completion. Also called **problem space**.

stateful A system, program, or process which keeps track of information concerning configuration, what has occurred previously, and certain other details. Most applications, such as word processors are stateful. That which is **stateless** has no information pertaining to such details.

stateless A system, program, or process which has no information concerning configuration, what has occurred previously, or certain other details. For example, a return to a Web page when a browser has not saved any information concerning such a page, such as cookies or a cache copy. That which is **stateful** does keep track of such information.

statement In a high-level language, a syntactic unit which includes one or more operators and one or more operands. A statement generates one more machine language instructions, which in turn direct a computer to perform one or more actions. Also called **program statement**.

statF Abbreviation of **statfarad**.

statfarad The unit of capacitance in the cgs electrostatic system. There are approximately 1.11265×10^{-12} farad in a statdfarad. Its abbreviation is **statF**.

statH Abbreviation of **stathenry**.

stathenry The unit of inductance in the cgs electrostatic system. There are approximately 8.98758×10^{11} henrys in a stathenry. Its abbreviation is **statH**.

static 1. Having no motion, activity, change, or the like. For example, that which is fixed, set, at rest, or idle. 2. Pertaining to that which is **static** (1). 3. Pertaining to **static electricity**. 4. Unwanted and distracting random noises, such as hissing and crackling, which affect the quality of a received radio signal. Said especially of such noise resulting from spherics. 5. Any disturbance, such as crackling heard from a speaker or specks seen on a TV screen, which is a result of random radio noise.

static allocation 1. The allocation of computer resources before program execution. 2. Same as **static memory allocation**.

static binding The linking of software routines or objects during program compilation. This contrasts with **dynamic binding**, where the linking occurs during program execution. Also called **early binding**.

static characteristic The relationship between the instantaneous values of two varying quantities, while all other related variables are kept constant. For example, collector current versus base voltage, with all other operating voltages held at a fixed value.

static charge An electric charge that does not move, such as that bound to, or possessed by, an object.

static check Same as **static test**.

static convergence In a picture tube with a three-color gun, the process by which the three electron beams are made to converge on a specified surface in the absence of scanning.

static debugging Debugging performed while a program is not executing. An example is that performed at the end of a routine.

static device 1. A device which does not move. 2. A device with no moving parts. 3. A device which does not cycle.

static discharger A device used on aircraft to reduce precipitation static. Such devices usually extend from the trailing edges of surfaces of an aircraft, and allow static electricity to discharge into the air. Also called **discharger** (2).

static dump A dump occurring when a program is not executing. For instance, that performed at the end of a routine.

static electric field An electric field whose strength does not vary continuously.

static electricity Energy in the form of electric charges at rest. Examples include electricity produced by friction or induction, or the electric charge stored in a capacitor. This contrasts with **dynamic electricity**, which is that produced by electric charges in motion. Under certain circumstances, an amount of static electricity which is undetectable by humans may be sufficient to damage a chip.

static eliminator 1. A circuit or device which reduces or eliminates **static** (4), or **static** (5). 2. A device or object which reduces or eliminates **static electricity**.

static friction Same as **stiction**.

Static HTML Abbreviation of **Static HyperText Markup Language**. HTML which once presented on a Web page does not change, and is thus non-interactive. Also, the technology enabling such HTML. This contrasts with **Dynamic HTML**, which can be updated as influenced by user actions.

Static HyperText Markup Language Same as **Static HTML**.

static induction transistor A field-effect transistor with an especially short channel. Its acronym is **SIT**.

static Internet-Protocol address Same as **static IP address**.

static inverter A circuit or device that changes a DC voltage to an AC voltage. Frequently, the output voltage is much higher than that of the input. Used, for instance, in uninterruptible power supplies. Also called **inverter** (1), or **DC–AC converter**.

static IP address Abbreviation of **static** Internet-Protocol **address**. An IP address which is the same each time a user logs onto a TCP/IP network. Also, such an address corresponding to a server. This contrasts with a **dynamic IP address**, in which a different IP address is assigned each time a user logs on. Also called **fixed IP address**.

static machine A device utilized to generate electrostatic charges, such as those produced by induction or friction. Belt generators can generate potentials of several million volts, while the electric potential in lightning can be as great as 100 million volts. Also called **electrostatic generator**, **electrostatic accelerator**, or **electrostatic machine**.

static memory Computer memory that does not require constant refreshing in order to avoid the loss of data. Static memory may be volatile, as in SRAM or non-volatile, as in ROM. Also called **static storage**.

static memory allocation The allocation of computer memory before program execution. Also called **static allocation** (2).

static RAM Same as **SRAM**.

static random-access memory Same as **SRAM**.

static relay Same as **solid-state relay**.

static route A path taken by data when utilizing **static routing**.

static routing Routing which follows a specifically configured path. Such routing does not adapt to varying conditions, such as malfunctions or increased traffic. This contrasts with **dynamic routing**, which automatically makes such adjustments.

static run Same as **static test**.

static SQL SQL in which queries are formulated prior to runtime.

static stability The ability of a system or object to maintain its stability when not in motion or in equilibrium. Used, for instance, to refer to the ability of a walking robot to maintain its balance when stopped mid-stride.

static storage Same as **static memory**.

static switch Same as **solid-state switch**.

static test A performance test made during the operation of a component, circuit, device, piece of equipment, or system, in which the driving signals and environmental conditions are kept steady. Also called **static run**, or **static check**.

statics The branch of mechanics which deals with bodies which are stationary or in equilibrium.

station 1. A place, facility, site, or position where something is placed, installed, or is otherwise located. 2. A **station** (1) where communications equipment is situated. For instance, a receiving station, a transmitting station, a land station, or a

ship station. 3. A **station** (2) within a radiocommunications service in which TV, radio, or other signals intended for reception by the general public are received, transmitted, or retransmitted. Also called **broadcasting station**. 4. A place designated for study, observation, measurement, testing, or the like. For example, a radar station. 5. Any place along a communications network or system where there is a signal input or output. 6. A computer input/output device which incorporates a video adapter, monitor, keyboard, and usually a mouse. Used in networks. Also called **terminal** (1), **computer terminal** (1), **network terminal** (1), or **console** (3). When such a terminal has no processing capability it is called **dumb terminal**, while a terminal that incorporates a CPU and memory does have processing capability, and is called **intelligent terminal**. 7. A personal computer or workstation which is linked to a network. Also called **terminal** (2), **computer terminal** (2), or **network terminal** (2).

station authentication In computers and communications, the process of verifying the authenticity of a given station. Measures such as passwords may be employed.

station authorization The authorization, such as that granted by a licensing authority, which enables a communications station to broadcast legally.

stationary 1. That which does not move. 2. That which is unable to move. 3. That which is able to move, but is not moving at a given moment. 4. Showing no detectable change or variation.

stationary battery 1. A battery that is designed for use in a fixed location. 2. A battery which is not intended to be shook excessively, inverted, or the like, while in use.

stationary blade Same as **stator blade**.

stationary contact In a relay or switch, an electrical contact which does not move. Used in conjunction with a **movable contact**. Also called **stator contact**, or **fixed contact**.

stationary orbit Same as **synchronous orbit**.

stationary plate Same as **stator plate**.

stationary satellite Same as **synchronous satellite**.

stationary wave Same as **standing wave**.

stationary winding Same as **stator coil**.

statistical multiplexer A multiplexer which allocates its bandwidth dynamically, depending on the traffic at any given moment. This is especially useful when there are burst transmissions. Also spelled **statistical multiplexor**. Its abbreviation is **stat mux**.

statistical multiplexor Same as **statistical multiplexer**.

statistical process control The use of statistical methods to assist in, analyze, and improve process control. Its abbreviation is **SPC**.

statistical quality control The use of statistical methods to assist in, analyze, and improve quality control. Its abbreviation is **SQC**.

statmho Same as **statsiemens**.

statohm The unit of resistance, impedance, or reactance in the cgs electrostatic system. There are approximately 8.98758×10^{11} ohms in a statohm. Its abbreviation is **statΩ**.

stator In an electrical and/or mechanical component, device, apparatus, or machine, a member which is **stationary**. For example, the stationary part or windings of a motor, the stationary plates of a variable capacitor, or the fixed contacts of a rotary switch. A rotating part in such a component, device, apparatus, or machine is a **rotor**.

stator blade A blade, such as that in a rotary switch, that does not move. Also called **stationary blade**.

stator coil A coil, such as that in a motor, that does not move. Also called **stator winding**, or **stationary winding**.

stator contact Same as **stationary contact**.

stator plate A plate, such as that in a variable capacitor, that does not move. Also called **stationary plate**.

stator winding Same as **stator coil**.

statS Abbreviation of **statsiemens**.

statsiemens The unit of conductance in the cgs electrostatic system. There are approximately 1.11265×10^{-12} siemens in a statsiemen. Its abbreviation is **statS**. Also called **statmho**.

statT Abbreviation of **stattesla**.

stattesla The unit of magnetic flux density in the cgs electrostatic system. There are approximately 2.99793×10^6 teslas in a stattesla. Its abbreviation is **statT**.

status The state or condition of something at a given moment. For example, the availability of a device connected to a network, or the on/off status of a piece of equipment.

status bar A bar, usually located at the bottom of a window, which provides information pertaining to specific aspects of an application. For example, a browser may indicate via the status bar that a given page has been fully loaded, and whether there is a secure connection or not.

status code A code, such as that sent through a network, indicating the state or condition of a device, system, process, or the like.

status register A register whose content indicates aspects of the internal status of a CPU, or a part of a CPU such as an arithmetic-logic unit. Also called **flag register**.

statV Abbreviation of **statvolt**.

statvolt The unit of potential difference in the cgs electrostatic system. There are approximately 299.793 volts in a statvolt. Its abbreviation is **statV**.

statW Abbreviation of **statwatt**.

statwatt The unit of power in the cgs electrostatic system. There is 10^{-7} watt in a statwatt. Its abbreviation is **statW**.

statWb Abbreviation of **statweber**.

statweber The unit of magnetic flux in the cgs electrostatic system. There are approximately 299.793 webers in a statwebber. Its abbreviation is **statWb**.

statΩ Abbreviation of **statohm**.

STB Abbreviation of **set-top box**.

STC Abbreviation of **sensitivity-time control**.

std Abbreviation of **standard**.

steady state 1. For a given component, circuit, device, piece of equipment, system, mechanism, or body, a state which is attained after transients, fluctuations, and other changes have died down or otherwise disappeared. 2. For a given component, circuit, device, piece of equipment, system, mechanism, or body, a state which does not oscillate, alternate, increase, or otherwise vary.

steady-state current A current which does not vary.

steady-state error In a control system, any error that remains after any changes in the system have died down or otherwise disappeared.

steady-state operation Operation of a component, circuit, device, piece of equipment, system, or mechanism, once transients, fluctuations, and other changes have died down or otherwise disappeared.

steady-state oscillation For an oscillating system, oscillation occurring once transients, fluctuations, and other changes have died down or otherwise disappeared. Also called **stable oscillation (2)**.

steady-state voltage A voltage which does not vary.

stealth virus A virus which effectively conceals its presence. Such a virus, for instance, may alter the data of the disk directory it is located in, so as to hide the additional bytes it adds.

steerable antenna A directional antenna that can be rotated, or which can otherwise change the direction of its major lobe.

Stefan-Boltzmann constant A constant, pertaining to blackbody radiation, whose value is approximately 5.6704×10^{-8} W/m^2K^4. Its symbol is σ.

Stefan-Boltzmann law A law stating that the energy radiated by a blackbody per second per unit area is proportional to the fourth power of its absolute temperature.

steganography The utilization of unused data or locations within a file or data to hide additional information, such as concealed messages. A message hidden in this manner may or may not be encrypted, with its key security feature being its undetectability. For instance, steganography may use redundant bits within a given file format, such as that of an image, to embed the data in an imperceptible manner. When utilized to hold bit patterns which identify the sources of digital intellectual property, such as voice, video, or data, known as **watermark**.

Steinmetz coefficient A constant, unique to each material, which is utilized when calculating hysteresis losses. Also called **hysteresis coefficient**.

STEM 1. Acronym for **scanning transmission electron microscope**. 2. Acronym for **scanning transmission electron microscopy**.

step 1. One within a series, sequence, order, or process. 2. To cause to proceed one **step (1)** at a time, or from one step to another. 3. The interval encompassing a single **step (1)**. 4. Within a computer program, a single instruction. May also refer to a single statement, which can contain multiple instructions. Also called **program step**. 5. To cause to execute a single **step (4)**.

step attenuator A circuit or device which reduces the amplitude of a signal, ideally without introducing distortion, in precise steps.

step-by-step execution Same as **single-step operation**.

step-by-step operation Same as **single-step operation**.

step change The process of proceeding from one **step (1)** to another. Also, the interval between these steps.

step counter A circuit, device, register, mechanism, or system utilized for counting steps, such as those in arithmetic processes.

step-down 1. To decrease in precise steps. Also, pertaining to that which decreases in precise steps. 2. To decrease a voltage. Also, pertaining to that which decreases a voltage, such as a step-down transformer.

step-down ratio For transformer, a turns ratio of greater than one. A step-down transformer has such a ratio.

step-down transformer A transformer whose output voltage is lower than its input voltage. A secondary winding of such a transformer has fewer turns than the primary. Such a transformer may have multiple secondary windings. Used, for instance, to decrease the voltage of electricity as it leaves the transmission system and enters the distribution system. The output voltage of a **step-up transformer** is higher than its input voltage. Also spelled **stepdown transformer**.

step frame The capturing of video images a single frame at a time. Used, for instance, when there is not enough bandwidth or system resources for real-time video.

step function A function which has a value of zero before a given instant, and a fixed non-zero value after said instant. When the non-zero value is one, also called **unit-step function**.

step-index fiber An optical fiber whose core has a uniform refractive index.

step-recovery diode A diode in which a charge is stored when forward biased, and which briefly conducts when it becomes reverse biased, after which it returns to a high-impedance state. Used, for instance, as a harmonic generator. Its abbreviation is **SRD**.

step relay Same as **stepping relay**.

step switch Same as **stepping switch**.

step-through execution Same as **single-step operation**.

step-through operation Same as **single-step operation**.

step-up 1. To increase in precise steps. Also, pertaining to that which increases in precise steps. **2.** To increase a voltage. Also, pertaining to that which increases a voltage, such as a step-up transformer.

step-up ratio For transformer, a turns ratio of less than one. A step-up transformer has such a ratio.

step-up transformer A transformer whose output voltage is higher than its input voltage. A secondary winding of such a transformer has more turns than the primary. Such a transformer may have multiple secondary windings. Used, for instance, to increase the voltage of electricity as it leaves the power generating station and enters the transmission system. The output voltage of a **step-down transformer** is lower than its input voltage. Also spelled **stepup transformer**.

stepdown transformer Same as **step-down transformer**.

stepped leader In lightning, an initial streamer which sets the path for subsequent streamers.

stepper motor A motor whose rotation occurs in discrete steps, as opposed to a continuous motion. Used, for instance, to move an access arm in a disk drive. Also called **stepping motor**.

stepper relay Same as **stepping relay**.

stepper switch Same as **stepping switch**.

stepping motor Same as **stepper motor**.

stepping relay A relay which turns around an axis when opening or closing contacts. Also called **stepper relay, step relay**, or **rotary relay**.

stepping switch A switch that turns around an axis when opening or closing contacts. Such a switch is usually adjusted by a knob, and may have detents to hold in a given position. Used, for instance, to select from multiple settings. Also called **stepper switch, step switch**, or **rotary switch**.

stepup transformer Same as **step-up transformer**.

steradian A unit of measure of solid angles. There are 4π steradians in a sphere, thus one steradian equals about 0.079577 sphere. Its symbol is **sr**.

stereo 1. Abbreviation of **stereophonic**. **1.** Abbreviation of **stereo sound**, or **stereophonic sound**. **3.** Abbreviation of **stereo system**, or **stereophonic system**. **4.** Abbreviation of **stereoscope**.

stereo AM Also called **AM stereo**. **1.** The broadcasting of AM signals with two or more tracks. **2.** The transmission and reception of **stereo AM (1)** broadcasts.

stereo amplifier An audio amplifier utilized for delivering **stereo sound**.

stereo audio system Same as **stereo system**.

stereo broadcast A broadcast utilizing two or more channels or tracks, as in stereo FM or stereo TV.

stereo broadcasting Broadcasting utilizing two or more channels or tracks, as in stereo FM or stereo TV.

stereo effect The reproduction of sound in which there is a slight delay between channels, so as to create a listening experience where sounds appear to arrive from two or more locations. In **monaural sound** the audio seems to come from a single point, while **ambisonics** strives to provide a faithful three-dimensional sound image.

stereo FM Also called **FM stereo**. **1.** The broadcasting of FM signals with two or more tracks. **2.** The transmission and reception of **stereo FM (1)** broadcasts.

stereo receiver An audio receiver utilized for delivering **stereo sound**.

stereo reception The reception of **stereo sound**, as in stereo FM.

stereo recording A recording utilizing two or more channels or tracks. Also called **stereophonic recording**.

stereo separation Same as **separation (4)**.

stereo signal A signal having two or more channels or tracks. For example, that of a stereo broadcast. Also called **stereo sound (3)**.

stereo sound Its abbreviation is **stereo (2)**. Also called **stereophonic sound**. **1.** Sound reproduction in which two channels are utilized. Stereo sound, as opposed to monophonic sound, provides a more natural, three-dimensional sound experience. **2.** Sound reproduction in which two or more channels are utilized. **3.** Same as **stereo signal**.

stereo sound system Same as **stereo system**.

stereo system A sound-reproduction system providing **stereo sound**. Such a system incorporates an amplifier or receiver, two or more speakers, and one or more signal sources such as tuners, DVD players, CD players, TVs, or cable TV or satellite signals. Its abbreviation is **stereo (3)**. Also called **stereo sound system, stereo audio system, stereophonic sound system**, or **stereophonic system**.

stereo television Same as **stereo TV**.

stereo transmission The transmission of **stereo sound**, as in stereo TV.

stereo tuner A tuner that can receive **stereo signals**, such as those of stereo FM or stereo TV.

stereo TV Abbreviation of **stereo** television. **1.** A standard for the transmission and reception of TV signals in which two or more audio channels are utilized. **2.** A device, such as a TV receiver, which is compatible with **stereo TV (1)**.

stereolithography A manufacturing technique or technology in which a photosensitive material, usually a polymer resin, is formed into the desired shape through the use of a laser which cures said material layer by layer. Used, for instance, to create a solid model directly from CAD data in rapid prototyping. Its abbreviation is **SL**.

stereophonic Its abbreviation is **stereo (1)**. **1.** Pertaining to two or more channels, as in stereo sound. **2.** Pertaining to the recording, transmission, and reproduction of sound utilizing two or more channels.

stereophonic recording Same as **stereo recording**.

stereophonic sound Same as **stereo sound**. Its abbreviation is **stereo (2)**.

stereophonic sound system Same as **stereo system**.

stereophonic system Same as **stereo system**. Its abbreviation is **stereo (3)**.

stereoscope An instrument which is utilized to view two overlapped two-dimensional images, so as to create the impression of depth. Each pair of images, such as photographs, is taken at slightly different angles. Its abbreviation is **stereo (4)**.

stereoscopic Pertaining to a **stereoscope**, or to **stereoscopy**.

stereoscopic machine vision In robotics, a method of vision which allows for interpretation of three-dimensional surfaces, by combining the input from each of the two optical sensors. Used, for instance, in obstacle-detection systems in mobile robots. Also called **binocular machine vision**.

stereoscopic television Same as **stereoscopic TV**.

stereoscopic TV Abbreviation of **stereoscopic** television. TV in which two-dimensional images are presented so as to create the impression of depth. One such system uses multiple cameras, with the positions, angles, and the like computed taking into account the manner in which the brain processes visual input, to display images which appear as three-dimensional. Also called **three-dimensional TV**.

stereoscopy The creation and viewing of two-dimensional images with the impression of depth. Pairs of two-dimensional images, such as photographs, are taken at different angles designed for subsequent viewing using **stereoscopes**. Also, the techniques utilized to create and view such images.

stethoscope 1. An instrument or device which is utilized to listen to internal body sounds, such as heart beats. **2.** A stethoscope (1) which incorporates electronic components such as amplifiers and filters, which enable internal body sounds, such as a heart murmurs or lung wheezes, to be heard more clearly and loudly. Such stethoscopes can usually be connected to a computer, enabling storing, further analysis, and transmission of the recorded sounds. Also called **electronic stethoscope**.

sticking The tendency of a contact, flip-flop, rotary switch, or the like, to preferentially remain in one of the available positions, or to resist changing.

StickyKeys A feature which makes modifier keys, such as shift, control, or alt, on a computer keyboard to appear as if still pressed, once released. Used, for instance when pressing two keys simultaneously is difficult for a given user. Also, keys so enabled.

stiction Acronym for **static friction**. Friction which opposes motion between movable parts which are at rest at a given moment.

stilb A cgs unit of luminance. It is equal to one candela per square centimeter. Its symbol is **sb**.

still A single stationary image from a motion picture, or that presented on a TV, computer monitor, or the like. Also called **still frame**.

still frame Same as **still**.

stimulated emission The emission of particles or waves by a substance in response to an external stimulation, such as an incident photon. For example, the electrons of atoms which has been raised to higher energy levels releasing photons when jumping back to lower levels. Seen, for instance, in lasers.

STM 1. Abbreviation of **scanning tunneling microscope**. **2.** Abbreviation of **scanning tunneling microscopy**. **3.** Abbreviation of Synchronous Transport Module. Within the SONET standard, one of the defined transmission speeds. STM speeds range from 155.52 Mbps, to hundreds of Gbps. STM-1 has a transmission rate of 155.52 Mbps, STM-3 runs at three times STM-1, or 466.56 Mbps, STM-64 at 9.95328 Gbps, and so on. The equivalent electrical signal is called **STS**, with STS-3 corresponding to STM-1, STS-12 corresponding to STM-4, STS-192 corresponding to STM-64, and so on.

STM-1 Abbreviation of Synchronous Transport Module-1. Within the SONET standard, a transmission rate of 155.52 Mbps. The equivalent electrical signal is called **STS-3**. STM-1 can carry 3 DS-3s within its payload, which is the same as **OC-3**.

STM-3 Abbreviation of Synchronous Transport Module-3. Within the SONET standard, a transmission rate of 466.56 Mbps. The equivalent electrical signal is called **STS-9**. STM-3 can carry 9 DS-3s within its payload, which is the same as **OC-9**.

STM-4 Abbreviation of Synchronous Transport Module-4. Within the SONET standard, a transmission rate of 622.08 Mbps. The equivalent electrical signal is called **STS-12**. STM-4 can carry 12 DS-3s within its payload, which is the same as **OC-12**.

STM-6 Abbreviation of Synchronous Transport Module-6. Within the SONET standard, a transmission rate of 933.12 Mbps. The equivalent electrical signal is called **STS-18**.

STM-6 can carry 18 DS-3s within its payload, which is the same as **OC-18**.

STM-8 Abbreviation of Synchronous Transport Module-8. Within the SONET standard, a transmission rate of 1.24416 Gbps. The equivalent electrical signal is called **STS-24**. STM-8 can carry 24 DS-3s within its payload, which is the same as **OC-24**.

STM-12 Abbreviation of Synchronous Transport Module-12. Within the SONET standard, a transmission rate of 1.86624 Gbps. The equivalent electrical signal is called **STS-36**. STM-12 can carry 36 DS-3s within its payload, which is the same as **OC-36**.

STM-16 Abbreviation of Synchronous Transport Module-16. Within the SONET standard, a transmission rate of 2.48832 Gbps. The equivalent electrical signal is called **STS-48**. STM-16 can carry 48 DS-3s within its payload, which is the same as **OC-48**.

STM-32 Abbreviation of Synchronous Transport Module-32. Within the SONET standard, a transmission rate of 4.97664 Gbps. The equivalent electrical signal is called **STS-96**. STM-32 can carry 96 DS-3s within its payload, which is the same as **OC-96**.

STM-64 Abbreviation of Synchronous Transport Module-64. Within the SONET standard, a transmission rate of 9.95328 Gbps. The equivalent electrical signal is called **STS-192**. STM-64 can carry 192 DS-3s within its payload, which is the same as **OC-192**.

STM-256 Abbreviation of Synchronous Transport Module-256. Within the SONET standard, a transmission rate of 39.81312 Gbps. The equivalent electrical signal is called **STS-768**. STM-256 can carry 768 DS-3s within its payload, which is the same as **OC-768**.

STM-1024 Abbreviation of Synchronous Transport Module-1024. Within the SONET standard, a transmission rate of 159.25248 Gbps. The equivalent electrical signal is called **STS-3072**. STM-1024 can carry 3072 DS-3s within its payload, which is the same as **OC-3072**.

STN 1. Abbreviation of **switched telephone network**. **2.** Abbreviation of **super-twist nematic**, or **super-twisted nematic**.

STN display Abbreviation of **super-twist nematic display**, or **super-twisted nematic display**.

stochastic 1. In statistics, incorporating one or more random variables. **2.** Based, in part, on chance and/or guesswork.

stochastic process A process in which there are one or more random variables. Also called **random process (1)**.

Stokes' law A law stating that radiation resulting in luminescence has a greater wavelength than the radiation which caused said luminescence.

stop band Same as **stopband**.

stop band filter Same as **stopband filter**.

stop band ripple Same as **stopband ripple**.

stop bit In asynchronous communications, a bit transmitted after a character or packet.

stop code A code which causes a **stop instruction**.

stop element Same as **stop signal**.

stop instruction In a computer program or routine, an instruction which causes a break in the execution. Also called **halt instruction**.

stop signal In asynchronous communications, a signal, such as a given sequence of bits, transmitted after a character or packet. Also called **stop element**.

stop time Also known as **deceleration time**. **1.** The time required for a moving storage medium to stop after a reading or writing operation. For example, the time it takes a tape reel to stop spinning before or after a read operation. Applies mostly to magnetic media, but may also include optical media such as CD-ROMs. **2.** The time it takes a moving part of a storage medium to stop after a reading or writing operation. Such a part may be, for instance, an access arm or a read/write head.

stopband Also spelled **stop band**. **1.** A frequency interval within which a signal is not transmitted or passed. A frequency-sensitive device, such as an amplifier or filter, may have multiple stopbands. **2.** A continuous range of frequencies which is highly attenuated or rejected by a filter. A filter may have multiple stopbands. Also called **filter stopband**.

stopband filter A filter which selectively blocks or rejects one or more frequency intervals, each of which is called **stopband**. Also spelled **stop band filter**.

stopband ripple Slight variations in attenuation or rejection occurring within a **stopband**. Also spelled **stop band ripple**.

stopping potential The negative potential that must be applied to a surface to prevent photoelectric emission of electrons.

storage **1.** A device which provides for the holding and accessing of data for an indefinite period. Such information is kept in a machine-readable format, and should be maintained intact in the event of an interruption in system power. The main types of storage media are magnetic, such as tapes and hard disks, and optical, which include holograms and DVDs. Storage is usually quantified in multiples of bytes. For example, a computer hard drive with a 256 gigabyte capacity can hold approximately 256 billion bytes, or characters, of information. Although **storage** and **memory** are often used synonymously, memory refers to a more temporary form of holding and accessing data, such as that in RAM. Also called **computer storage** (1). **2.** The act or process of placing data in a **storage device** (1). Also called **computer storage** (2). **3.** The charging, retention of charge, or recharging of a cell or capacitor. **4.** The accumulation and retention of a form of energy, such as heat or electricity. **5.** The placing of something in a given location or object for later use. Also, such a location or object.

storage allocation The reserving of computer storage areas for specific purposes.

storage-area network Its acronym is **SAN**. A high-speed network which links different forms of storage devices to one or more data servers. SANs are compatible with fibre channel, SCSI, and other data transmission technologies, and support backup and response, archiving, disk mirroring, data migration, IP storage, and so on. Also called **storage network**.

storage array Multiple disks, such as a RAID, utilized together for storage.

storage battery A battery composed of **storage cells** (1). Examples include lithium polymer batteries, lead-acid batteries, and gel-cell batteries. Also called **secondary battery**, **rechargeable battery**, or **accumulator battery**.

storage camera An early form of TV camera tube which employs a high-velocity electron beam to scan a photoemissive mosaic screen which stores electrical charge patterns corresponding to the image focused upon said mosaic. Also called **iconoscope**.

storage capacity **1.** The amount of data that can be placed in a storage medium. Usually expressed in multiples of bytes, such as gigabytes or terabytes. Also called **memory capacity** (2). **2.** The total amount of charge that can be withdrawn from a fully charged battery or cell under specified conditions. Usually expressed in ampere-hours.

storage cathode-ray tube Same as **storage tube**.

storage cell **1.** A cell whose available energy can be replenished after discharge, with the process being able to be repeated safely multiple times. Examples include lithium-ion

cells, and nickel-metal hydride cells. This contrasts with a **non-rechargeable cell**, whose manner of discharge does not lend itself to recharging. Also called **secondary cell**, or **rechargeable cell**. **2.** The smallest unit of information a digital computer can handle. Usually synonymous with **bit**. **3.** In computers, an elementary unit of storage, such as a binary cell.

storage CRT Same as **storage tube**.

storage density **1.** The compactness of data storage. Expressed, for instance, in gigabytes per square inch. **2.** The compactness of charge storage in a battery. Expressed, for example, in watt-hours per kilogram.

storage device A device which serves for **storage** (1). Also called **computer storage device**.

storage dump Also called **dump**. **1.** To display, print, copy, or transfer the content of a computer memory or storage device, with little or no formatting. **2.** To display, print, copy, or transfer the content of the main memory of a computer. Such a dump may be performed after a process which ends abnormally, for instance, to help pinpoint the source of a problem. A dump may also be generated automatically. Also called **memory dump**, or **core dump**.

storage life Also called **shelf life**.

storage location **1.** Any place in storage where data may be placed. **2.** A specific place where data is placed within the storage of a computer, peripheral, or disk.

storage management The tasks pertaining to the administration of data storage, including selection of mediums, security, backups, and disaster recovery. Hierarchical storage management, for instance, provides for the automatic transfer of information between storage media with different priorities.

storage media **1.** Two or more **storage mediums** considered together. **2.** The physical materials or mediums which serve to store data on a permanent or semi-permanent basis. These include magnetic disks and tapes, DVDs, holograms, bubble memory, and so on. Also called **data-storage media**, or **media** (2).

storage medium Any physical material or medium which serves to store or otherwise contain data. For instance, optical discs, magnetic tapes and disks, microfilm, and paper. Also called **data medium, data-storage medium**, or **medium** (3).

storage network Same as **storage-area network**.

storage oscilloscope **1.** An oscilloscope that incorporates an analog-to-digital converter, for transforming its analog input into a digital output. Such an oscilloscope enables permanent signal storage, and extensive waveform analysis and processing. Also called **digital oscilloscope**. **2.** An analog oscilloscope capable of retaining a displayed image for an extended period of time. Used, for instance, for studying and analyzing the presented images.

storage register **1.** A high-speed storage area within the CPU, which is utilized to hold data for specific purposes. Before data may be processed, it must be represented in a register. There are various types, including control registers, index registers, arithmetic registers, and shift registers. Also called **data register**, or **register** (1). **2.** A device which records and/or stores a value or number. Also called **register** (2).

storage service provider An entity that provides storage space for data, in addition to other services such as security, periodic backup, replication of data, and recovery after a disaster. Its abbreviation is **SSP**.

storage technology Any technology, such as magnetic, optical, or magneto-optic, utilized for **storage** (1).

storage temperature The recommended temperature range for storage of components, circuits, devices, equipment, systems, or materials. This interval is usually greater than the operating temperature range. A proper storage temperature can extend shelf life considerably.

storage time 1. The time interval during which something is stored, or the time required to effect such storage. 2. The time interval during which a storage medium, such as RAM, a hard disk, a smart card, or an optical disc, can reliably retain data. Also called **retention time.** 3. For a semiconductor junction, such as a diode, the time during which current reverses and stays at a constant level, when bias is switched from forward to reverse. 4. The time that elapses for a given amount of decay to occur. For example, the time it takes a static charge to be reduced to a given percentage of its peak charge. Also called **decay time (1),** or **fall time.**

storage tube A CRT whose high-intensity display is the result of secondary emissions of electrons. Such a display remains bright for controllable periods of time, and may be used, for instance, in radars. The displayed image is formed by a writing gun and intensified by a flooding gun. Also called **storage CRT, storage cathode-ray tube,** or **direct-view storage tube.**

store To place, replenish, or retain in **storage.** Also, a location or object which serves for such storage.

store-and-forward A manner of delivering data in which a complete message is received and stored before being passed on to the next node en route to its destination. Its abbreviation is **store & forward.**

store & forward Abbreviation of **store-and-forward.**

stored-value card A smart card that stores a given level of money which can be applied to purchase goods and/or services. For example, a prepaid telephone card, or that issued by a retailer to a customer in exchange for cash given or returned items.

storm In communications, a condition in which an excess of simultaneous traffic overloads a network. An example is a broadcast storm.

STP 1. Abbreviation of **shielded twisted pair.** 2. Abbreviation of **standard temperature and pressure.** 3. Abbreviation of **signal transfer point.**

straight dipole A dipole antenna with a single radiator, which is usually split at its electrical center.

straight-line code Program code without branch instructions such as GOTO.

straight-line path A path, such as that between microwave radio relays, which follows a straight line.

strain A lengthening, contraction, torsion, or other mechanical deformation resulting from an external force. Also called **mechanical strain.**

strain gage Same as **strain gauge.**

strain-gage bridge Same as **strain-gauge bridge.**

strain-gage transducer Same as **strain-gauge transducer.**

strain gauge An instrument or device which measures and indicates the **strain** exhibited by a material or body. It may consist, for instance, of a transducer which varies an electrical property, such as resistance or voltage, proportionally to the magnitude of the detected strain. Examples include resistance strain gauges, and load cells. Also spelled **strain gage.**

strain-gauge bridge The utilization of a bridge arm, such as that in a Wheatstone bridge, as a **strain gauge.** The changes in resistance indicate the magnitude of the strain. Also spelled **strain-gage bridge.**

strain-gauge transducer A transducer, such as that which converts a change in length into a proportional change in resistance, utilized as a **strain gauge.** Also spelled **strain-gage transducer.**

strand 1. One of multiple conductors twisted together in a **stranded wire** or **stranded cable.** 2. One of multiple groups of conductors twisted together in a stranded wire.

stranded cable 1. A cable consisting of multiple non-insulated wires which are twisted together. 2. A cable consisting of multiple groups of stranded wire which are braided together. Each group is usually insulated from the others.

stranded conductor A conductor consisting of one or more groups of **stranded wire.**

stranded copper wire A wire consisting of multiple non-insulated copper wires which are twisted together. Multiple groups of such wires may be braided together.

stranded wire A wire consisting of multiple non-insulated wires which are twisted together. Multiple groups of such wires may be braided together.

stratosphere The layer within the atmosphere that occurs between the troposphere and the mesosphere. It ranges from approximately 10 kilometers to about 50 kilometers.

stray capacitance A capacitance which is not concentrated within a capacitor. It is usually undesired, and examples include the capacitance between the turns in a coil, or between adjacent conductors of a circuit. This contrasts with **lumped capacitance,** which is concentrated within a capacitor. Also called **self-capacitance, distributed capacitance,** or **wiring capacitance.**

stray component A circuit parameter, such as resistance or inductance, which is distributed throughout a circuit or along the entire length of a transmission line. This contrasts with **lumped component,** which is concentrated within discrete components such as resistors or capacitors. For instance, series resistance in a two-wire transmission line, or the distributed resistance of a wire coil. Also called **stray parameter,** or **distributed component.**

stray field Any undesired and usually energy-wasting leakage of a portion of an electric or magnetic field.

stray impedance An impedance which is evenly distributed throughout a circuit. This contrasts with **lumped impedance,** which is concentrated within a circuit element. Also called **distributed impedance.**

stray inductance An inductance which is evenly distributed throughout a circuit. This contrasts with **lumped inductance,** which is concentrated within a circuit element, such as a coil. Also called **distributed inductance.**

stray parameter Same as **stray component.**

stray resistance A resistance which is evenly distributed throughout a circuit. It is usually undesired and energy-wasting. This contrasts with **lumped resistance,** which is as concentrated within a circuit element, such as a resistor. Also called **distributed resistance.**

strays Same as **sferics.**

streaking 1. A form of distortion in which a displayed or printed image, such as that of a TV or fax, has horizontal lines or smudges which are dark or light. In a TV image such streaks may move up and down the screen. 2. A form of distortion in which objects extend horizontally beyond their ordinary boundaries, forming darker or lighter lines or smudges at the edges. Seen, for instance, in TV or fax. In a TV image such streaks may move up and down the screen.

stream 1. A flow of data from a source to a destination, such as that between nodes in a network, or from a disk to RAM. 2. A **stream (1)** which occurs in an uninterrupted manner over a communications channel. Also called **data stream.** 3. The data being transmitted during **streaming.** 4. A steady flow. Also, to produce a steady flow, or to flow in a steady manner.

stream cipher A cipher that operates on a continuous data stream, as opposed to blocks of a fixed size, as is the case with a **block cipher**.

streamer In lightning, a high-ion density stream which sets the path for a subsequent leader stroke.

streaming Transmission of information over a computer network, such as the Internet, in a manner which allows for accessing while the transmission is in progress. Once transmission begins, a given time interval is buffered in the receiving computer, and data may be accessed while the download is being completed. Used, for instance, for streaming video.

streaming audio Audio which is transmitted over a computer network in a manner which allows for listening while the transmission is in progress. Once transmission begins, a given time interval is buffered in the receiving computer, and sound may be accessed while the download is being completed.

streaming data Data, such as streaming audio or video, which allows for accessing while the transmission is in progress.

streaming tape A magnetic tape which records and reads without stops, such as those which could occur at inter-record gaps. Used, for instance, for the bulk transfer of data between a tape and a disk.

streaming video Video which is transmitted over a computer network in a manner which allows for viewing while the transmission is in progress. Once transmission begins, a given time interval is buffered in the receiving computer, and images may be accessed while the download is being completed.

stress **1.** A force per unit area that causes **strain**. Stress is caused by the internal forces of a body which resist the lengthening, contraction, torsion, or other mechanical deformation caused by an external force. Also, to subject to such a force. **2.** To subject to operating conditions which are harsher than usual.

stress test A test in which the usual operating conditions of a component, circuit, device, piece of equipment, system, or material are exceeded.

striking voltage Same as **starting voltage (3)**.

string **1.** A sequence of similar elements, such as bits or characters, which are treated as a unit. **2.** A set of consecutive bits or characters. **3.** A series or succession of circuits, stages, units, devices, equipment, or the like.

string galvanometer A galvanometer which measures current by utilizing a conducting thread or fiber which is stretched between the poles of a magnet or electromagnet. The thread or fiber is deflected proportionally to the current passing through it, and is viewed through a microscope. Also called **Einthoven string galvanometer**.

String Oriented Symbolic Language Same as **SNOBOL**.

string variable **1.** A string that can store any type of information, including text and numbers. **2.** A name or value assigned to a specific alphanumeric sequence.

strip **1.** A long and narrow piece or section. For example, a magnetic stripe. **2.** To remove the insulation or jacket from a wire or cable. **3.** To remove from an object or material. For instance, to strip an electron from a surface.

strip-chart recorder A graphical recorder which creates its written record using a pen which traces upon a moving strip of paper which is moved at a fixed speed. Used, for instance, to continuously plot a given variable, such as temperature or humidity, as time passes.

strip fuse A fuse whose melting conductor is a strip, as opposed for, instance, to a wire.

strip line Same as **stripline**.

strip transmission line Same as **strip line**.

stripe pitch On the display screen of a picture tube with a shadow mask, the distance between phosphor dots of like color between one phosphor stripe and the next. Usually expressed in millimeters, and the lower the number, the crisper the image. Also called **aperture pitch**.

striping A technique in which data, such as that contained in a file, is spread over multiple disk drives, so that one disk may be transferring data while the next is locating the following segment. RAID storage, for instance, uses this technique to improve performance. Also called **disk striping**, or **data striping**.

stripline Also spelled **strip line**. Abbreviation of **strip transmission line**. **1.** A transmission line consisting of one or more thin conductors, usually strips, which are either mounted above, or between conducting surfaces. The conducting surfaces are usually ground planes. An example is a **stripline (2)**. **2.** A microwave transmission line consisting of two thin parallel strips, lines, or plates, which are mounted on opposite sides of the same dielectric substrate. Typically, the substrate is flat, and mounted on a ground plane. Also called **microstrip line**, **microstrip transmission line**, or **microstrip**.

strobe **1.** Abbreviation of **stroboscope**. **2.** Abbreviation of **strobe light**. **3.** A repeatable and artificially produced burst of bright light. Such a flash is usually generated by applying a high voltage to an electrode of a tube containing an inert gas such as xenon. The gas becomes ionized, which permits it to rapidly discharge the energy previously stored in a capacitor. Used, for instance, in photography, or in visual special effects. Also called **electronic flash**, **flash (2)**, or **photoflash**. **4.** To cause a **strobe (3)** to flash. **5.** On a display device such as a radar screen, a spot, such as a blip, which serves as a reference for deriving information such as range, bearing, or elevation.

strobe lamp Same as **strobe light**.

strobe light A lamp utilized to produce a **strobe (3)**. Its abbreviation is **strobe (2)**. Also called **strobe lamp**, **stroboscopic light**, **stroboscopic lamp**, **flash lamp**, or **electronic flash tube**.

strobe pulse **1.** A pulse whose duration is shorter than other pulses, and which is utilized for synchronization, as a reference, or the like. **2.** A pulse generating a **strobe (3)**.

stroboscope A device or instrument which makes vibrating, rotating, or otherwise moving bodies appear stationary. It usually does so by emitting bursts of bright light, or by actuating a shutter, which enables such bodies to be seen intermittently. Used, for instance, for synchronization or observation. Its abbreviation is **strobe (1)**.

stroboscopic Pertaining to a **strobe**, **strobe light**, or **stroboscope**.

stroboscopic lamp Same as **strobe light**.

stroboscopic light Same as **strobe light**.

stroke speed Same as **scanning speed**.

strong coupling **1.** Same as **strong electron-phonon coupling**. **2.** In a transformer, the arrangement of the primary and secondary windings in a manner which provides for a high degree of energy transfer. Also known as **close coupling (1)**, or **tight coupling (1)**. **3.** In a transformer, the arrangement of the primary and secondary windings in a manner which provides maximum energy transfer. Also known as **close coupling (2)**, or **tight coupling (2)**.

strong electron-phonon coupling Within the Bardeen-Cooper-Schrieffer theory, the interaction of electrons and phonons leading to the formation of Cooper pairs. Also called **strong coupling (1)**.

strong encryption A form of encryption whose cryptographic key is so large that it ensures that a brute-force attack is not feasible. A 256-bit key, for instance, has over 1.1×10^{77} possible combinations.

strong force The force which holds together an atomic nucleus. Its range is extremely short, and its force carrier is the gluon. It is the strongest of the **four fundamental forces of nature**, the others being the gravitational force, the weak force, and the electromagnetic force. Also called **strong nuclear force**, or **nuclear force**.

strong nuclear force Same as **strong force**.

strong typing In programming, the strict enforcement of the data type of a given variable. Strong typing detects more errors when compiling than **weak typing**, which allows certain exceptions to data-typing rules.

strontium A soft and pale yellow metallic chemical element whose atomic number is 38. It is an alkaline-earth metal, is extremely reactive, and has around 30 known isotopes, of which 4 are stable. It is used, for instance, as a getter in vacuum tubes, and in nuclear batteries. Its chemical symbol is **Sr**.

strontium carbonate A white powder or solid whose chemical formula is $SrCO_3$. It is a phosphor, and is used, for instance, to coat the glass of CRTs.

strontium chloride White crystals whose chemical formula is $SrCl_2$. Used, for instance, in electron tubes.

strontium chromate A light yellow powder whose chemical formula is $SrCl_2$. Used, for instance, as a coating to prevent corrosion, and in electroplating.

strontium fluoride A white powder whose chemical formula is SrF_2. Used, for instance, in lasers and in optics.

strontium titanate A colorless ceramic whose chemical formula is $SrTiO_3$. Used, for instance, in lasers and in optics.

structural testing Also called **white-box testing**, **clear-box testing**, or **open-box testing**. **1.** Any tests utilizing the assistance of a white box. **2.** Tests of software or hardware utilizing the assistance of a white box.

structure The manner in which something consisting of multiple parts or components is put together or organized. Examples include data structures and network structures.

structured data Data which has a defined structure, such as that in a database.

structured program A computer program written utilizing **structured programming**.

structured programming A programming method in which a hierarchy of modules is utilized, each with a single entry point and a single exit point. This enables the flow of control to be clearly defined and readily recognized, facilitating systematization, modification, debugging, and ease of maintenance.

structured programming language A programming language, such as PASCAL or ADA, which lends itself to, or follows **structured programming**.

Structured Query Language Same as **SQL**.

STS Abbreviation of Synchronous Transfer Signal. The equivalent electrical signal to **optical carrier** transmission speeds. STS-1 corresponds to OC-1, STS-3 to OC-3, and so on.

STS-1 Synchronous Transfer Signal-1. The equivalent electrical signal to **OC-1**, which corresponds to a signaling rate of 51.8 Mbps.

STS-3 Synchronous Transfer Signal-3. The equivalent electrical signal to **OC-3**, which corresponds to a signaling rate of 155.52 Mbps.

STS-9 Synchronous Transfer Signal-9. The equivalent electrical signal to **OC-9**, which corresponds to a signaling rate of 466.56 Mbps.

STS-12 Synchronous Transfer Signal-12. The equivalent electrical signal to **OC-12**, which corresponds to a signaling rate of 622.08 Mbps.

STS-18 Synchronous Transfer Signal-18. The equivalent electrical signal to **OC-18**, which corresponds to a signaling rate of 933.12 Mbps.

STS-24 Synchronous Transfer Signal-24. The equivalent electrical signal to **OC-24**, which corresponds to a signaling rate of 1.24416 Gbps.

STS-36 Synchronous Transfer Signal-36. The equivalent electrical signal to **OC-36**, which corresponds to a signaling rate of 1.86624 Gbps.

STS-48 Synchronous Transfer Signal-48. The equivalent electrical signal to **OC-48**, which corresponds to a signaling rate of 2.48832 Gbps.

STS-96 Synchronous Transfer Signal-96. The equivalent electrical signal to **OC-96**, which corresponds to a signaling rate of 4.97664 Gbps.

STS-192 Synchronous Transfer Signal-192. The equivalent electrical signal to **OC-192**, which corresponds to a signaling rate of 9.95328 Gbps.

STS-768 Synchronous Transfer Signal-768. The equivalent electrical signal to **OC-768**, which corresponds to a signaling rate of 39.81312 Gbps.

STS-3072 Synchronous Transfer Signal-3072. The equivalent electrical signal to **OC-3072**, which corresponds to a signaling rate of 159.25248 Gbps.

stub **1.** A short section of a transmission line, such as a coaxial stub, which is utilized to change the properties of said line. Used, for instance, for impedance matching. **2.** A program routine that substitutes for another while the latter is subsequently accessed or written. A program utilizing Remote Procedure Calls, for instance, may use such a stub while awaiting for the remote computer to complete its task.

stub-matching The use of one or more **stubs** (1) for impedance matching.

stub tuner A tuning device utilizing one or more **stubs** (1).

studio A room, building, or other location specifically designed for producing broadcast programming.

stuffing bits Additional bits inserted into a data stream to avoid the appearance of unintended control sequences, or to conform to a required frame size. These bits are removed at the receiving location to restore the original message.

StuffIt A compression program utilized when copying one or more files to multiple floppy disks.

stutter dial tone **1.** In certain phone systems, an intermittent dial tone which alerts to voice-mail messages that await. **2.** In certain phone systems, an intermittent dial tone which notifies that a code, such as that deactivating call waiting, has been accepted.

style sheet A predefined layout which serves as the basis for a document, such as a monthly report prepared utilizing a word-processing application, or that composing a Web page. A style sheet specifies parameters such as page size, margins, headers, footers, fonts, hyperlinks, and so on. Also called **template** (3).

styli A plural form of **stylus**.

stylus **1.** The pointing device of a digitizing tablet. It is similar to a mouse, but is much more accurate, because its location is determined by touching an active surface with an absolute reference. Also called **cursor** (4), **pen** (1), or **puck**. **2.** The pointing device of a PDA, which serves for selection or input. Also called **pen** (2). **3.** A pointed object which extends from a phonographic pickup, and which serves to follow the undulations of the grooves of a phonographic disc and transmit them as vibrations to said pickup. Such an object is usually made of a metal needle with a diamond or sapphire tip. Also called **needle** (3), or **phonograph needle**.

styrofoam A light, resilient, and water-resistant polystyrene plastic which is used, for instance, for insulation and cushioning devices being transported.

SU Abbreviation of **signal unit**.

sub-allocation Same as **suballocation**.

sub-assembly Same as **subassembly**.

sub-atomic particle Same as **subatomic particle**.

sub-audible frequency Same as **subsonic frequency**. Also spelled **subaudible frequency**.

sub-band Same as **subband**.

sub-carrier Same as **subcarrier**.

sub-carrier frequency Same as **subcarrier frequency**.

sub-chassis Same as **subchassis**.

sub-command Same as **subcommand**.

sub-directory Same as **subdirectory**.

sub-domain Same as **subdomain**.

sub-folder Same as **subfolder**.

sub-frequency Same as **subfrequency**.

sub-harmonic Same as **subharmonic**.

sub-menu Same as **submenu**.

sub-millimeter waves Same as **submillimeter waves**.

sub-miniature connector Same as **subminiature connector**.

sub-miniature device Same as **subminiature device**.

sub-miniature jack Same as **subminiature jack**.

sub-miniature plug Same as **subminiature plug**.

sub-miniature tube Same as **subminiature tube**.

sub-multiple Same as **submultiple**.

sub-nanosecond Same as **subnanosecond**.

sub-network Same as **subnetwork**.

sub-network mask Same as **subnetwork mask**.

sub-notebook Same as **subnotebook computer**. Also spelled **subnotebook**.

sub-notebook computer Same as **subnotebook computer**.

sub-panel Same as **subpanel**.

sub-program Same as **subprogram**.

sub-routine Same as **subroutine**.

sub-schema Same as **subschema**.

sub-sonic Same as **subsonic**.

sub-sonic frequency Same as **subsonic frequency**.

sub-sonic phenomena Same as **subsonic phenomena**.

sub-station Same as **substation**.

sub-string Same as **substring**.

sub-system Same as **subsystem**.

sub-transaction Same as **subtransaction**.

sub-tree Same as **subtree**.

sub-woofer Same as **subwoofer**.

suballocation A resource, piece of equipment, interval of frequencies, or the like, which is part of a larger allocation. Also spelled **sub-allocation**.

subassembly A functional unit which is designed to fit into a larger assembly. Each subassembly is usually manufactured separately. Also, the putting together of such units. Also spelled **sub-assembly**.

subatomic particle Any particle which is smaller than an atom. For instance, an atomic nucleus, or an elementary particle such as an electron or photon. Also spelled **sub-atomic particle**.

subaudible frequency Same as **subsonic frequency**. Also spelled **sub-audible frequency**.

subband An interval of frequencies which form part of a larger band. For example, a portion of the band of frequencies transmitted by a satellite antenna. Also spelled **sub-band**.

subcarrier Also spelled **sub-carrier**. **1.** A carrier wave that is utilized to modulate another carrier. Used, for instance, to carry additional signals such as audio or other information. **2.** In color TV, the carrier whose sidebands are added to the luminance signal, in order to convey the color information. The frequency of this carrier signal is set at 3.579545 MHz. Also called **color subcarrier**, **color carrier**, **chrominance carrier**, or **chrominance subcarrier**.

subcarrier frequency The frequency of a **subcarrier**, such as that of a chrominance-subcarrier. Also spelled **sub-carrier frequency**.

subchassis A supporting frame which is part of a larger chassis. Also spelled **sub-chassis**.

subclass In object-oriented programming, a class derived from a superclass via inheritance. Also called **derived class**.

subcommand A command within a **submenu**. Also spelled **sub-command**.

subdirectory A directory which is located within another directory, which in turn is called **root directory**. In GUIs, the terms **subfolder** and **root folder** are used, respectively. Also spelled **sub-directory**.

subdomain A domain within a second-level domain. For example, in *www.wheee.yipeeee.qq*, the *wheee* is a subdomain of *yipeeee*. Also spelled **sub-domain**.

subfolder A folder which is located within another folder, which in turn is called **root folder**. In non-graphical user interfaces, the terms **subdirectory** and **root directory** are used, respectively. Also spelled **sub-folder**.

subfrequency A frequency, or band of frequencies, which is part of a larger interval. Also spelled **sub-frequency**.

subharmonic Also spelled **sub-harmonic**. **1.** An integral submultiple of a first harmonic. For example, the third sub-harmonic of a 9 MHz harmonic is 3 MHz. **2.** An integral submultiple of a harmonic or other reference frequency.

submarine cable A cable designed for use under water. Such a cable, for instance, may be laid below the seabed between continents, or along coastlines.

submenu A menu that appears after a higher-level menu has been selected. The higher-level menu may itself also be a submenu of another menu. Also spelled **sub-menu**.

submillimeter waves Electromagnetic waves whose wavelengths are between 0.1 and 1 millimeter, corresponding to frequencies of between 3000 and 300 GHz, respectively. Also spelled **sub-millimeter waves**.

subminiature connector Also spelled **sub-miniature connector**. **1.** A connector which is smaller than a given size. **2.** A connector which is smaller than the usual size for a given use, application, or device.

subminiature device Also spelled **sub-miniature device**. **1.** A device that is smaller than a given size. **2.** A device that is smaller than the usual size for a given use, application, or the like.

subminiature jack Also spelled **sub-miniature jack**. **1.** A jack that is smaller than a given size. **2.** A jack that is smaller than the usual size for a given use, application, or device.

subminiature plug Also spelled **sub-miniature plug**. **1.** A plug which is smaller than a given size. **2.** A plug which is smaller than the usual size for a given use, application, or device.

subminiature tube A very small electron or gas tube which is designed for use in subminiature devices. Also spelled **sub-miniature tube**.

submultiple The result obtained when dividing a number by another number. For example, 15 is a submultiple of 45, as the latter has been divided by 3, and 22.05 is a submultiple of 44.1, as the latter has been divided by 2. Also spelled **submultiple**.

subnanosecond A time interval shorter than a nanosecond. Also spelled **sub-nanosecond**.

subnet Abbreviation of **subnetwork**.

subnet mask Abbreviation of **subnetwork mask**.

subnetwork A network within a larger network. For example, the Internet is a worldwide network of interconnected autonomous networks, each of which is a subnetwork. Its abbreviation is **subnet**. Also spelled **sub-network**.

subnetwork mask A bit combination which identifies which portion of an IP address corresponds to the network or subnetwork, and blocks out the rest. For example, a network may use the same values in the first three address fields of a four field address such as WWW.XXX.YYY.ZZZ, and block out, or mask, all but the ZZZ portion, since it is the only one that will vary. Its abbreviation is **subnet mask**. Also spelled **sub-network mask**. Also called **address mask**, or **network mask**.

subnotebook Same as **subnotebook computer**. Also spelled **sub-notebook**.

subnotebook computer A computer which is lighter and smaller than a notebook computer, but larger than a handheld computer. Such a computer usually weighs less than one kilogram, and generally features computing power similar to a desktop model. Also spelled **sub-notebook computer**. Also called **subnotebook**, or **mini-notebook**.

subpanel A panel which forms part of a larger panel. For instance, that which provides additional controls. Also spelled **sub-panel**.

subprogram A part, such as a subroutine or module, of a larger program. For example, a subroutine can be invoked by a program to perform a given task, then return control to the invoking program. Also spelled **sub-program**.

subroutine Within a computer program, a small group of instructions which perform a given task. Also spelled **subroutine**. Also called **routine (1)**, **function (3)**, or **procedure (2)**.

subschema A subset within a **schema**. It may consist, for instance, of a database description from the perspective of a specific application, or the part of a database pertaining to a given department within an enterprise. Also spelled **subschema**.

subscribe 1. To pay or otherwise sign-up for a given service, such as CATV. 2. To authorize an entity to send information. For example to agree to become part of a mailing list.

subscriber A user or other entity which has been authorized to utilize or receive one or more services. For example, a user who has paid for access to the Internet, telephone service, or satellite channels.

Subscriber Identity Module Same as **SIM**.

Subscriber Identity Module card Same as **SIM card**.

subscriber line A pair of wires, or its equivalent, extending from a telephone central office to the premises of a customer. Also called **subscriber loop**, **loop (7)**, **local loop**, **local line**, or **line loop**.

subscriber line interface circuit A circuit, usually in the form of a chip, that converts a two-wire subscriber line signal into a four-wire signal, and vice versa, at a central office, PBX, or the like, and which in addition performs other functions such as line supervision. Its acronym is **SLIC**.

subscriber loop Same as **subscriber line**.

subscription television Same as **subscription TV**.

subscription TV Abbreviation of **subscription** television. Also called **pay TV**. 1. A service in which payment is made to be able to view TV content which is otherwise unavailable at a given location. Examples include pay-per-view,

and premium channels. A decoder box is usually necessary for such programming. 2. Any TV service which requires payment. These include **subscription TV (1)**, basic satellite, premium cable, and so on.

subset A set which is contained within or forms part of another larger set.

subsonic Pertaining to, generating, sensitive to, or utilizing **subsonic frequencies**. Also spelled **sub-sonic**. Also called **infrasonic**.

subsonic frequency A frequency below the range that humans can hear. That is, below about 20 Hz. Also spelled **sub-sonic frequency**. Also called **subaudible frequency**, or **infrasonic frequency**.

subsonic phenomena Phenomena occurring within, or pertaining to **subsonic frequencies**. Also spelled **sub-sonic phenomena**. Also called **infrasonic phenomena**.

substation A station which is a part of a larger station, which supplements a larger station, or which is located between larger stations or between a larger station and another location. For example, a station between a power station and a group of homes. Also spelled **sub-station**.

substitution method 1. A method of measurement in which an unknown quantity is determined via the replacement with known quantities which have the same effect as that which has been replaced. For example, the determination of an unknown resistance by placing a resistor with a known value in its place. 2. The substitution of a component or device with an equivalent one, for testing, troubleshooting, or performing repairs.

substrate 1. The base layer, or other surface upon which something is deposited, etched, attached, or otherwise prepared or fabricated. A substrate also provides physical support and insulation. For example, the base film of a magnetic tape, or the plastic base of a compact disc. 2. A **substrate (1)** utilized in the manufacturing of circuits and microcircuits. Such a substrate may be made, for instance, of ceramic, plastic, glass, or a semiconductor material. An active substrate has active elements, such as transistors, formed directly on it, while passive substrates have components mounted upon them.

substring A subset within, or portion of, a **string**. Also spelled **sub-string**.

subsystem Also spelled **sub-system**. 1. A system which is a part of, or assists, a larger system. Cache memory, for instance, is a specialized high-speed storage subsystem. 2. A system which is subordinate to another system.

subterranean Located or utilized beneath the surface of the earth.

subtitles In TV programming, and similar multimedia presentations such as films presented in theaters, the providing of text, and symbols such as ♫, which are intended to accurately transcribe the dialogue and describe the sounds of that being presented. A decoder is not necessary to show subtitles, as the displayed text and symbols are always superimposed on the viewed image. Useful, for instance, for those with reduced hearing. This contrasts with **closed-captioning**, which requires a decoder to appear on-screen.

subtracter Same as **subtractor**.

subtracter circuit Same as **subtractor**. Also spelled **subtractor circuit**.

subtractive color A color formed by a **subtractive mixture**.

subtractive color process A method of producing colors through a **subtractive mixture**. Also called **subtractive process**, or **subtractive synthesis**.

subtractive color system A system of combining subtractive primary colors in varying proportions to yield a full range of colors.

subtractive mixture A combination of subtractive primary colors utilized to form other colors. For example, a printer may use cyan, magenta, and yellow inks to form a full range of colors.

subtractive primaries Same as **subtractive primary colors**.

subtractive primary colors Colors which are combined in a subtractive mixture to yield a full range of colors. Cyan, magenta, and yellow are usually used as the subtractive primary colors. They are called this way because they are considered to be subtracted from white. Also called **subtractive primaries**.

subtractive process Same as **subtractive color process**.

subtractive synthesis Same as **subtractive color process**.

subtractor Also spelled **subtracter**. Also called **subtractor circuit**. **1.** In computers, a logic circuit which subtracts two or more numbers or quantities. **2.** A circuit in which two or more input signals are combined to yield one output signal, which is proportional to the subtraction of the input signals.

subtractor circuit Same as **subtractor**. Also spelled **subtracter circuit**.

subtrahend A number or quantity that is subtracted from another. It is the newly introduced number or quantity. This contrasts with the **minuend**, which is the number or quantity already present from which a subtrahend is subtracted.

subtransaction A transaction which is incorporated into another transaction. Also spelled **sub-transaction**. Also called **nested transaction**.

subtree A branch point within a tree. For example, a node within a directory tree. Also spelled **sub-tree**.

subwoofer A speaker utilized to reproduce frequencies below a given threshold, such as 100 or 200 Hz. A subwoofer should be able to faithfully reproduce audio down to around 30 Hz, although some extend to below 15 Hz. Used, for instance, in a home theater system. Also spelled **sub-woofer**.

successive approximation register An analog-to-digital converter which incorporates a digital-to-analog converter to determine the output string successively, bit by bit. Its abbreviation is **SAR**.

suffix notation A manner of forming mathematical or logical expressions in which the operators appear after the operands. For example, $(A + B) \times C$, which is in **infix notation**, would appear as $ABC+*$ in suffix notation. Also called **postfix notation**, or **reverse Polish notation**.

SUI Abbreviation of **speech user interface**.

suite Same as **software package (1)**.

suite of applications Same as **software package (1)**.

sulfation In a lead-acid battery or cell, the growth of large lead sulfate crystals on a plate or other internal part, which makes recharging more difficult. It may result, for instance, from self-discharge, or improper charging or maintenance. The lead sulfate crystals occurring in such a battery or cell are usually much smaller.

sulfur A non-metallic chemical element whose atomic number is 16. It is known to exist in four allotropic forms, two of which are yellow and crystalline, while the other two are dark amorphous solids. It has close to 20 known isotopes, of which 4 are stable. It is widely found in nature, and its applications include its use in the preparation of sulfuric acid, and in vulcanization. It is also spelled **sulphur**, and its chemical symbol is **S**.

sulfur chloride A reddish fuming liquid with a penetrating odor, whose chemical formula is S_2Cl_2. Used in vulcanization.

sulfur dioxide A colorless and irritating liquid whose chemical formula is SO_2. It is a major pollutant, and is used in the production of sulfuric acid.

sulfur hexafluoride A colorless gas whose chemical formula is SF_6. It is used in semiconductors, lasers, and as a dielectric.

sulfuric acid A colorless, dense, and strongly corrosive liquid whose chemical formula is H_2SO_4. It is the most widely produced chemical in the world, and is used, for instance, as an electrolyte in batteries, in electroplating, as an etchant, and in metallurgy.

sulphur Same as **sulfur**.

sum channel In an audio amplifier, a channel whose output is the sum of the left and right channels. The resulting signal is monophonic. This contrasts with a **difference channel**, where the output is the difference between the left and right channels.

sum check Abbreviation of **sum**mation **check**. An error-detection technique, such as a checksum, in which digits are added for comparison with a previously obtained value.

sum frequency A frequency which is the result of the sum of two other frequencies.

sum-of-products A logic circuit which expresses the products of terms. The input of such a circuit consists of multiple AND gates, and the output has OR gates. Its abbreviation is **SOP**.

summation check Same as **sum check**.

summation circuit Same as **summing circuit (2)**.

summation network Same as **summing circuit (2)**.

summer **1.** Same as **summing circuit**. **2.** Same as **summing amplifier**.

summing amplifier An amplifier whose output is proportional to the sum of two or more inputs. Also called **summer (2)**.

summing circuit Also called **summer (1)**, **adder**, or **adder circuit**. **1.** In computers, a logic circuit which adds two or more numbers or quantities. **2.** A circuit in which two or more input signals are combined to yield one output signal, which is proportional to the sum of the input signals. Also called **summing network**, **summation circuit**, or **summation network**.

summing network Same as **summing circuit (2)**.

sun battery Same as **solar panel**.

sun cell Same as **solar cell**.

sun lamp **1.** A lamp whose produced light is rich in ultraviolet radiation, and which is used, for instance, for therapeutic or cosmetic purposes. **2.** Same as **sunlight lamp**.

sun relay A relay in which the intensity of incident sunlight determines its opening or closing a circuit. It is a type of photoelectric relay.

sun switch A switch in which the intensity of incident sunlight determines its opening or closing a circuit. It is a type of photoelectric switch.

sunlight lamp A lamp whose produced light is intended to have a spectral distribution that matches as closely as possible that of the sun. Also called **sun lamp (2)**.

SunOS Abbreviation of **Sun o**perating **s**ystem. A component of Solaris.

sunspot A region of the sun in which intense magnetic fields are formed. Such an area is slightly cooler and darker than its surrounding surface. Sunspots affect the ionosphere, and can affect radio transmissions. The average sunspot is larger than the planet earth, and the number of sunspots occurring follows cycles lasting approximately 11 years.

sunspot cycle Same as **solar cycle**.

Super Audio CD Same as **SACD**.

Super Audio CD player Same as **SACD player**.

super heterodyne Same as **superheterodyne**.

super high frequency A range of radio frequencies spanning from 3 to 30 GHz. These correspond to wavelengths of 10 cm to 1 cm, respectively. Also, pertaining to this interval of frequencies. Also spelled **superhigh frequency**. Its abbreviation is **SHF**. Also called **super high frequency band**.

super high frequency band Same as **super high frequency**.

super-large-scale integration Its abbreviation is **SLSI**. **1.** In the classification of ICs, the inclusion of over 1,000,000,000 electronic components, such as transistors, on a single chip. **2.** A term sometimes utilized to refer to **very-large-scale integration**, or to **ultra-large-scale integration**.

super low frequency Waves whose frequencies are usually defined to be between 30 and 300 Hz, corresponding to wavelengths of between 10,000 and 1,000 kilometers, respectively. Its abbreviation is **SLF**.

super-regenerative circuit Same as **superregenerative circuit**.

super-regenerative receiver Same as **superregenerative receiver**.

super-saturated solution Same as **supersaturated solution**.

super-saturation Same as **supersaturation**.

super tweeter An additional tweeter some speakers provide. Used, for instance, when a speaker has a driver that covers the upper-midrange and lower-tweeter intervals, or to extend the frequency range of a speaker beyond the threshold of hearing. Also spelled **supertweeter**.

super-twist display Same as **super-twist nematic display**. Also spelled **supertwist display**.

super-twist nematic A technology used in **super-twist nematic displays**. Also spelled **supertwist nematic**. Its abbreviation is **STN**.

super-twist nematic display A twisted nematic display in which the crystals are twisted from 180° to 270°. This provides greater contrast, and may be used, for instance, in portable computers, PDAs, or cellular telephones. Also spelled **supertwist nematic display**. Its abbreviation is **STN display**.

super-twisted nematic Same as **super-twist nematic**. Also spelled **supertwisted nematic**. Its abbreviation is **STN**.

super-twisted nematic display Same as **super-twist nematic display**. Also spelled **supertwisted nematic display**. Its abbreviation is **STN display**.

super user Same as **superuser**.

super-user account Same as **superuser account**.

super-user password Same as **superuser password**.

Super VGA Same as **SVGA**.

super-VHS Same as **S-VHS**.

super-VHS tape Same as **S-VHS tape**.

super video Same as **S-video**.

super video cable Same as **S-video cable**.

super video connector Same as **S-video connector**.

Super Video Graphics Array Same as **SVGA**.

supercardioid microphone A directional microphone whose lateral attenuation is greater than that of an ordinary cardioid microphone, but which has less rear attenuation. Also called **hypercardioid microphone**.

superclass In object-oriented programming, a class from which from a subclass is derived, via inheritance. Also called **base class**.

supercomputer A computer which is exceedingly fast, tremendously powerful, and particularly expensive. A supercomputer may have thousands of processors, and perform over 10^{15} floating-point calculations or operations per second. Such a computer is utilized for performing highly sophisticated and complex calculations, such as the study of the nature of matter, or the calculation of the highest prime

number at a given time. A **mainframe** can usually handle many more tasks and applications at the same time, but a supercomputer can handle a single task or application considerably more rapidly.

superconducting Possessing, exhibiting, or capable of **superconductivity**.

superconducting alloy An alloy, such as certain niobium alloys, which can exhibit **superconductivity** under the proper conditions.

superconducting cable A cable in which **superconductivity** is attained and maintained through the constant pumping of liquid nitrogen, which keeps the temperature below a given threshold. Used, for instance, for power transmission. Such cables can carry more power, are lighter, and have much lower losses than copper cables. Alternatively, liquid helium, or another extremely cold or cryogenic liquid may be utilized.

superconducting ceramic A ceramic, such as YBCO, which can exhibit **superconductivity** under the proper conditions.

superconducting material Same as **superconductor**.

superconducting quantum interference device An instrument which incorporates one or more Josephson junctions to for the detection of extremely weak fields, currents, or voltages. Such an instrument can be used, for instance, to monitor changes in the electromagnetic field generated by a human body. Its acronym is **SQUID**.

superconducting solenoid A solenoid in which the coil is a superconducting wire, so as to generate a powerful magnetic field.

superconducting solenoid coil The superconducting coil in a **superconducting solenoid**.

superconducting solenoid magnet An extremely powerful electromagnet produced by the current flowing through a **superconducting solenoid coil**.

superconducting thin-film A thin-film of a superconducting material which is incorporated into a device, instrument, piece of equipment, or system. Used, for instance, in highly sensitive detection systems, or in filters utilized at microwave frequencies.

superconducting transition For a given superconductor material, a transition to a superconducting state, under the proper conditions. These include, for instance, the superconductor critical temperature not being exceeded.

superconduction The exhibiting of **superconductivity**. Also, an instance in which superconductivity occurs.

superconductive Possessing, exhibiting, capable of, utilizing, pertaining to, or arising from **superconductivity**.

superconductive properties Same as **superconductivity**.

superconductivity The flow of current with a complete, or nearly complete disappearance of all electrical resistance. Superconductive properties have been observed in many elements, alloys, compounds, and other material materials, such as ceramics, under the proper conditions. Under such circumstances, usually occurring at very low temperatures, such materials become perfect, or nearly perfect, conductors. The Bardeen-Cooper-Schrieffer theory provides an explanation for superconductivity occurring at cryogenic temperatures, but may not hold up for high-temperature, or room-temperature superconductivity. In theory, perfect superconduction can continue indefinitely without decay. When superconductivity occurs at or above a given temperature, such as 77, 90, or 125 K, it is called high-temperature superconductivity, and when above a higher threshold, such as 250 or 300 K, it is called room-temperature superconductivity. Also called **superconductive properties**.

superconductor Also called **superconductor material**, or **superconducting material**. **1.** An element, alloy, compound, or other material which exhibits **superconductivity**.

2. An element, alloy, compound, or other material which exhibits superconductivity below a given temperature. **3.** A material, usually a metal or an alloy, which is capable of exhibiting superconductivity when cooled to cryogenic temperatures. Also called **cryogenic conductor**.

superconductor critical current For a given temperature, and in the absence of an external magnetic field, the maximum current a superconductive material can withstand while still maintaining its superconductivity. Also called **critical current**.

superconductor critical field For a given temperature, the maximum magnetic field a superconductive material can withstand while still maintaining its superconductivity, so long as the critical current is not exceeded. Also called **critical field (1)**, or **critical magnetic field**.

superconductor critical temperature In the absence of an external magnetic field, the maximum temperature a superconductive material can withstand while still maintaining its superconductivity, so long as the critical current is not exceeded. Also called **critical temperature**.

superconductor gyroscope A gyroscope in which a central rotating disk of superconducting niobium spins while in levitation at cryogenic temperatures. Also called **cryogenic gyroscope**.

superconductor material Same as **superconductor**.

superframe A T-carrier framing format or standard with enhanced features, such as less-frequent synchronization, and real-time monitoring of the line. It encompasses 12 DS1 frames, while an **extended superframe** assembles 24. Its abbreviation is **SF**. Also called **superframe format**.

superframe format Same as **superframe**. Its abbreviation is **SF format**.

superhet Abbreviation of **superheterodyne**.

superheterodyne Also spelled **super heterodyne**. Its abbreviation is **superhet**. **1.** In a radio receiver, to mix an incoming signal with a signal generated by an internal oscillator, to provide an intermediate frequency. Since it is easier to process a modulated signal at a single frequency, this intermediate frequency can be amplified, filtered, and the like, and then demodulated to obtain the desired content. In this manner, the performance of such a receiver can be enhanced across its entire reception band. **2.** Same as **superheterodyne receiver**.

superheterodyne circuit In a radio receiver, a circuit that mixes an incoming signal with a signal generated by an internal oscillator, to provide an intermediate frequency.

superheterodyne radio receiver Same as **superheterodyne receiver**.

superheterodyne receiver A radio receiver which incorporates a **superheterodyne circuit**. Also called **superheterodyne radio receiver**, or **superheterodyne (2)**.

superhigh frequency Same as **super high frequency**.

superhigh frequency band Same as **super high frequency**.

superlattice A structure composed of multiple extremely thin layers of semiconductor materials, each of a uniform thickness. Such structures can be designed to have very specific electronic and photonic properties, and are used, for instance, in thermoelectric devices, and for cooling microelectronics such as CPUs.

supermalloy An alloy, containing nickel, iron, and molybdenum, with very high permeability.

supermendur An alloy, containing vanadium, cobalt, and iron, with very high flux density.

superposition The additive combination of signals or phenomena, such as the vector sums of multiple waves to form a complex wave.

superposition principle For a physical system with multiple influences, a theorem stating that the resultant influence is the algebraic or vector sum of the each of the influences applied separately. Also called **principle of superposition**.

superposition theorem For a given point along a linear circuit with multiple sources of current, a theorem stating that the total current is the algebraic sum of each source acting separately.

superregenerative circuit A receiver circuit that utilizes a squegging oscillator to increase regeneration to a given point, so as to provide enhanced sensitivity without excessive feedback. Also spelled **super-regenerative circuit**.

superregenerative receiver A radio receiver incorporating a **superregenerative circuit**. Also spelled **super-regenerative receiver**.

supersaturated solution A solution containing an amount of solute which exceeds that required for saturation. Also spelled **super-saturated solution**.

supersaturation A state in which a solution contains an amount of solute which exceeds that required for saturation. Also spelled **super-saturation**.

superscalar **1.** A CPU that can execute more than one instruction per clock cycle. Currently, most processors are of this type. Also called **superscalar processor**. **2.** An architecture enabling a processor to execute more than one instruction per clock cycle.

superscalar processor Same as **superscalar (1)**.

supersensitive Exhibiting extraordinary sensitivity. For example, an ammeter capable of measuring currents in the attoampere range.

superserver A network server with exceptional capabilities, such as large amounts of cache, ample redundant disk storage, outstanding fault-tolerance, extra high speed, or the like.

supersonic **1.** Pertaining to, or arising from speeds exceeding the **speed of sound**. **2.** Pertaining to, or arising from sounds at **supersonic frequencies**. Also called **ultrasonic (1)**. **3.** Pertaining to, or arising from **supersonics**. Also called **ultrasonic (2)**.

supersonic frequency A sound whose frequency is beyond the range of human hearing. That is, above about 20 kHz. Also called **ultrasonic frequency**.

supersonic sound waves Same as **supersonic waves**.

supersonic waves Sound waves whose frequency is beyond the range of human hearing. That is, above about 20 kHz. Used, for instance, in medical diagnosis and treatment, and in sonar. Also called **supersonic sound waves**, or **ultrasonic waves**.

supersonics The study of the phenomena pertaining to and arising from **supersonic waves**, and their applications. Also called **ultrasonics**.

superstructure A structure built upon a foundation or something else, such as another structure. A semiconductor heterostructure is an example.

supertweeter Same as **super tweeter**.

supertwist display Same as **super-twist nematic display**. Also spelled **super-twist display**.

supertwist nematic Same as **super-twist nematic**. Its abbreviation is **STN**.

supertwist nematic display Same as **super-twist nematic display**. Its abbreviation is **STN display**.

supertwisted nematic Same as **super-twist nematic**. Also spelled **super-twisted nematic**. Its abbreviation is **STN**.

supertwisted nematic display Same as **super-twist nematic display**. Also spelled **super-twisted nematic display**. Its abbreviation is **STN display**.

superuser A computer user with access to all levels of operation. Such a user, for instance, can perform tasks at the system level. Also spelled **super user**. Also called **root** (2).

superuser account A computer account which provides complete access to all data and systems. Such an account may be used for viewing and manipulating all files, and for performing operating system tasks. Root accounts may be held, for instance, by system administrators. Also spelled **super-user account**. Also called **root account**.

superuser password A password utilized to gain access to a **superuser account**. Also spelled **super-user password**.

supervisor The software which runs all the software and hardware of a computer. It is the first program the computer loads when powered on, remains memory-resident, and continuously controls and allocates all resources. Without it, for example, a computer can not recognize input, such as that from a keyboard, it can not process, as the supervisor controls the use of the CPU, nor can it provide an output, as it manages all peripherals including the monitor. Also called **operating system**, or **executive**. When disk-based, also called **disk operating system**.

supervisor mode An operational state which allows certain operations which are otherwise restricted, such as memory management, to be carried out. The operating system usually operates in this mode. Also called **supervisor state**, or **privileged mode**.

supervisor state Same as **supervisor mode**.

supervisory control and data acquisition Same as **SCADA**.

supervisory signal A signal, such as a hook flash or that transmitted by hanging up, which alerts a central office or PBX to a given command, such as switching between calls when using call waiting, or to release the line when ending a call.

supplemental power Same as **secondary power** (1).

supplementary channel Same as **secondary channel**.

supplementary communications channel Same as **secondary communications channel**.

supplementary device Same as **secondary device**.

supplementary equipment Same as **secondary equipment**.

supply 1. To make something available for use, to equip, or to provide sufficiently for. Also, that which is supplied, or is available for supplying. Also, such an act of supplying. **2.** A source of power or electricity. For example, a power line or battery. Also called **power supply** (1). **3.** A source of power for electronic components, circuits, devices, equipment, or systems. Such sources include batteries and power packs. Also called **electronic power supply**, or **power supply** (2).

supply current The current obtained from a **supply** (2) or **supply** (3).

supply frequency The frequency of the AC power obtained from an AC power **supply** (2) or **supply** (3).

supply power The power obtained from a **supply** (2) or **supply** (3).

supply reel In a tape player or system, the reel which holds the tape before it is played or recorded upon. The **take-up reel** winds the tape after reproduction or recording.

supply voltage The voltage obtained from a **supply** (2) or **supply** (3).

support 1. To bear the weight of something, provide a base, or otherwise help provide physical stability. **2.** The ability to function properly with specific software and/or hardware, or to have a given capability. For example, to support multiple operating systems, real-time applications, natural language input, hot-plugging, or a data rate of 100 Gbps. **3.** To provide assistance. For instance, to provide technical support.

suppressed carrier A carrier wave that is suppressed before transmission. The carrier must be restored at the receiving location prior to demodulation. Its abbreviation is **SC**.

suppressed-carrier modulation Modulation, such as that utilized in suppressed-carrier transmission, in which the carrier is suppressed.

suppressed-carrier transmission Also called **carrier suppression**. **1.** A mode of transmission in which the carrier wave is suppressed after carrier modulation, in which case only one or both sidebands are transmitted. The carrier wave is restored at the receiving end for demodulation. **2.** A mode of transmission in which the carrier wave is suppressed when there is no modulating signal.

suppressed-zero instrument An instrument in which the zero reading is below the lower limit of the indicator scale.

suppression The reduction, minimization, or elimination of signals, components of signals, energy, activity, segments, or the like. Examples include noise, interference, harmonic, carrier, sideband, and leading zero suppression, among many others.

suppressor 1. A component, circuit, device, unit, piece of equipment, system, or material which serves to significantly reduce or eliminate unwanted energy, undesired signals, interference signals, excessive currents, noise, echoes, or the like. **2.** Same as **suppressor grid**.

suppressor grid In an electron tube, such as a tetrode or pentode, a grid placed between the screen grid and the plate, so as to minimize secondary emissions from the plate. Also called **suppressor** (2).

surf Abbreviation of **surf** the net, or **surf** the Internet. To navigate from one place to another on a network such as the Internet. When surfing, a user may be looking for something specific, such as research material, news items, a place to exchange impressions, or one may simply follow hyperlinks from Web page to Web page aimlessly.

surf the Internet Same as **surf**.

surf the net Same as **surf**.

surface acoustic wave An acoustic or ultrasonic wave which travels along the surface of a solid, usually a highly-polished piezoelectric substrate. Used, for instance, as delay lines, filters, attenuators, oscillators, or phase shifters. Its abbreviation is **SAW**.

surface acoustic wave filter A filter incorporating a highly-polished piezoelectric substrate upon which acoustic or ultrasonic waves travel. Used, for example, as bandpass and bandstop filters employed in the megahertz and gigahertz ranges. Its abbreviation is **SAW filter**.

surface analyzer An instrument which measures and indicates any irregularities present on a surface. Used, for instance, to detect wafer imperfections, or for monitoring the growth of epitaxial films during the preparation and fabrication of semiconductor materials and devices.

surface barrier A potential barrier formed on the surface of a semiconductor material, due to trapped charge carriers.

surface charge density The electric charge per unit area. Usually expressed in coulombs per square meter. Its symbol is σ. Also called **charge density** (1).

surface clutter Same as **surface return**.

surface-coated mirror A mirror whose reflective coating is on its surface. In this manner, there is no refraction, and image distortion is minimized. Used, for instance, in instruments such as microscopes and lasers. Also called **front-surface mirror**, or **first-surface mirror**.

surface conductivity The reciprocal of **surface resistivity**.

surface density The mass or concentration of a given substance, energy, field, or entity, per unit area of a surface. For example, bit density, component density, charge density, or flux density.

surface effect An effect occurring on or along the surface of a material, as opposed, for instance to a **bulk effect**. An example is the Auger effect.

surface insulation resistance The electrical resistance between two conductors separated by an insulator under specified electrical and environmental conditions. Its acronym is **SIR**.

surface leakage A leakage current which flows along the surface of a dielectric. Also called **surface leakage current**.

surface leakage current Same as **surface leakage**.

surface model A geometric model which provides a representation of an object which appears externally as a solid, but which does not have internal structure, as does a **solid model**. A surface model does not have attributes such as thickness and volume, making it less complex and less processing-intensive than a solid model.

surface modeling The creation and representation of **surface models**, using computer-aided design and/or mathematical techniques.

surface mount 1. A technology for mounting components and devices in which the connections are soldered to a surface, as opposed to having leads which pass through said surface. Since the substrates upon which such components and devices are mounted do not need holes, the component density can be much higher, with the added benefit of lending itself more readily to production via automated processes. Also called **surface-mount technology**. **2.** To install or affix a component or device utilizing **surface-mount technology**. Also, a specific instance of such a surface mount. **3.** Pertaining to that which has a **surface mount (1)**. **4.** To set, install, of fix something upon a surface. For example, a switch mounted on a wall.

surface-mount assembly The assembly of **surface-mount PCBs**. Its abbreviation is **SMA**.

surface-mount capacitor A capacitor which is **surface mounted**.

surface-mount chip A chip, such as a ceramic quad flat-pack, which is **surface mounted**. Also called **surface-mount IC**.

surface-mount chip package An IC package which is **surface mounted**. Ball Grid Arrays, for instance, utilize tiny balls of solder to attach its leads to a printed-circuit board. Also called **surface-mount package**, or **surface-mount IC package**.

surface-mount circuit board Same as **surface-mount PCB**.

surface-mount component A component, such as a capacitor or transistor, which is **surface mounted**.

surface-mount device A device, such as an IC, which is **surface mounted**. Its abbreviation is **SMD**.

surface-mount diode A diode which is **surface mounted**.

surface-mount IC Same as **surface-mount chip**.

surface-mount IC package Same as **surface-mount chip package**.

surface-mount integrated circuit Same as **surface-mount chip**.

surface-mount integrated circuit package Same as **surface-mount chip package**.

surface-mount package Same as **surface-mount chip package**.

surface-mount PCB Abbreviation of **surface-mount** printed-circuit board. A printed-circuit board upon which components are **surface mounted**. Also called **surface-mount circuit board**.

surface-mount printed-circuit board Same as **surface-mount PCB**.

surface-mount resistor A resistor which is **surface mounted**.

surface-mount technology Same as **surface-mount (1)**. Its abbreviation is **SMT**.

surface-mount transistor A transistor which is **surface mounted**.

surface-mounted A component or device which has a **surface mount**.

surface passivation The protection of surfaces of semiconductors by depositing a material which renders them chemically passive or unreactive. Used, for instance, to protect from oxidation, contamination, and other perils of harsh environments. Two common techniques are glassivation and oxide passivation.

surface recombination rate In a semiconductor, the rate at which recombination occurs at the surface. The **volume recombination rate** is that which occurs within the overall body of a semiconductor material.

surface resistivity The resistivity of a material between opposite sides of a unit square of its surface. May be expressed, for instance, in ohms per square.

surface return In radars, unwanted echoes that appear on the display screen due to signal reflections from the surface of the earth. Also called **surface clutter**, **terrain return**, **land return**, or **ground clutter**.

surface wave 1. The component of a ground wave that travels along the surface of the earth between a transmitter and receiver. The **space wave** component travels through space between a transmitter and receiver, and incorporates both direct waves and ground-reflected waves. **2.** A wave, such as a surface acoustic wave or a ground wave, which travels along the surface of a plate, crystal, the ground, an interface between dissimilar materials, and so on.

surfing 1. The navigation from one place to another on a network such as the Internet, for purposes of exploration, research, entertainment, or the like. **2. Surfing (1)** over the Internet. Also called **Web surfing**, or **Internet surfing**.

surge 1. A sudden and momentary increase in current or voltage. May be caused, for instance, by lightning, or faults in circuits. If protective measures are not employed, it may cause a failure, or significant damage. Also called **electrical surge**. **2.** A sudden and momentary increase in a magnitude such as power or current.

surge arrester Same as **surge suppressor**.

surge current A sudden and momentary increase in current. May be caused, for instance, by lightning, or faults in circuits. If protective measures are not employed, such a surge may bring about a failure, or significant damage. Also called **current surge**, or **transient current**.

surge diverter Same as **surge suppressor**.

surge generator A device which produces a unidirectional surge of voltage or current of very short duration. One or more capacitors may be used, for instance, for storing and then releasing energy in very short pulses. Also called **impulse generator**.

surge impedance The impedance of a circuit, which when connected to the output terminals of a uniform transmission line, makes the line appear infinitely long. Under these conditions, the transmission line has no standing waves, and the ratio of voltage to current at any given frequency is the same at any point of the line. Its symbol is Z_0. Also called **characteristic impedance**.

surge power Also called **power surge**. **1.** A sudden and momentary increase in power. May be caused, for instance, by lightning. If protective measures are not employed, such a surge may bring about a failure or significant damage. **2.** A sudden and momentary increase in current or voltage.

surge protection Same as **surge suppression**.

surge protector Same as **surge suppressor**.

surge suppression The use and effect of a **surge suppressor**. Also called **surge protection**.

surge suppressor A component, circuit, or device which reduces, eliminates, or prevents **surges**, especially those occurring through power or communications lines. A surge suppressor should prevent peak AC voltages from exceeding given thresholds, thus preventing failures or significant damage. An avalanche diode used in this capacity can have reaction times in the picosecond range. Also called **surge protector**, **surge arrester**, or **surge diverter**.

surge voltage A sudden and momentary increase in voltage. May be caused, for instance, by lightning, or faults in circuits. If protective measures are not employed, such a surge may bring about a failure or significant damage. Also called **voltage surge**, or **transient voltage**.

surround sound 1. An audio experience in which the listener feels as if surrounded by voices, music, special effects, and other sounds which accompany that which is being viewed on a TV, at a movie theater, or the like. In a home theater system, for instance, there may be six channels, a left front channel, a left rear channel, a center channel, a right front channel, a right rear channel, and that which drives a subwoofer. Surround sound may also be digitally simulated using two channels and two speakers. **2.** A multiple-channel sound format, such as Dolby surround, which is digitally encoded to provide **surround sound** (1). **3.** Said of a component, device, or system which supports or provides **surround sound** (1).

surround-sound system A sound system which provides **surround sound** (1).

surveillance radar A radar utilized to monitor a given region, such as that surrounding an airport.

susceptance The imaginary number component of **admittance**. Expressed in siemens. Its symbol is **B**.

susceptibility A measure of the response of a material to a magnetizing force. It may be expressed, for instance, as the ratio of the magnetization of a material to the magnetizing field. Also called **magnetic susceptibility**.

susceptibility meter Same as **susceptometer**.

susceptometer An instrument which measures **susceptibility**. Also called **susceptibility meter**, or **magnetic susceptibility meter**.

suspend mode Same as **sleep mode**.

suspension 1. Any device or system which serves to cushion objects from vibrations, jolts, shocks, or other movements. A suspension also provides structural support to that which is cushions. **2.** A mixture in which macroscopically visible solid particles are present in a liquid or gas, but do not float nor settle on the bottom of the vessel. Also, the solid particles so suspended. **3.** A device or structure which serves to attach a speaker to the basket. **4.** A fine wire that supports the moving element of a meter, such as that in a moving-coil galvanometer.

sustained data rate Same as **sustained transfer rate**.

sustained data transfer rate Same as **sustained transfer rate**.

sustained oscillation 1. Uninterrupted oscillation which is maintained at or near a resonance frequency. When such oscillation is maintained without the aid of an external signal, it is self-sustained oscillation. **2.** Oscillation which continues beyond a desired interval.

sustained transfer rate Also called **sustained data transfer rate**, or **sustained data rate**. **1.** The data transfer rate that can be maintained for an extended period of time. **2.** The maximum data transfer rate that can be maintained for an extended period of time.

SVC Abbreviation of **switched virtual circuit**.

SVD Abbreviation of **simultaneous voice and data**.

SVG Abbreviation of **Scalable Vector Graphics**.

SVGA Abbreviation of Super Video Graphics Array. Any of various graphics standards which provide for higher resolution than VGA. SVGA resolutions range from 800 x 600 pixels, to 1600 x 1200 pixels, with support for over 16.7 million colors. Also called **Super VGA**, or **Extended VGA**.

SW 1. Abbreviation of **shortwave**. **2.** Abbreviation of **switch** (1), or **switch** (2). **3.** Abbreviation of **software**.

swap 1. To move a program, parts of a program, or data, from a hard drive to RAM and vice versa. By swapping applications and data in and out of the system's memory as needed, the operating system is able to address more memory than is physically available. Used, for instance, when a program is too large for the memory available to a given computer. **2.** To exchange one thing for another, as occurs, for instance, in hot swapping.

SWAP Acronym for **Shared Wireless Access Protocol**.

swap file A file utilized to temporarily save a program, parts of a program, or data when **swapping** (1). The operating system determines the use, size, and location of such a file on a hard disk.

swap in To move a program, parts of a program, or data, from a hard drive to RAM, during a **swap** (1).

swap out To move a program, parts of a program, or data, from RAM to a hard drive, during a **swap** (1).

swapping 1. The transfer of a program, parts of a program, or data, from a hard drive to RAM and vice versa during a **swap** (1). In the context of virtual memory, such moves are also called **paging** (1). **2.** The exchanging of one thing for another, as occurs, for instance, in hot swapping.

sweep 1. The steady motion of an electron beam across the screen of a CRT, usually in the horizontal direction. Also, one complete sweep. **2.** The performance of a **scan** (1), or a **scan** (2). **3.** To search for electronic bugs, usually with the intention of removing and/or disabling them. Also such an instance of searching. **4.** To brush or wipe, as in self-cleaning contacts.

sweep circuit 1. A circuit which generates **sweeps** (1). **2.** Same as **scanning circuit**.

sweep expansion Same as **sweep magnification**.

sweep frequency Also called **sweep rate**. **2.** The frequency at which **sweeps** (1) occur. **2.** Same as **scanning frequency**.

sweep generator Also called **sweep oscillator**, or **sweep-signal generator**. **1.** A circuit, device, or instrument whose output signal sweeps through a given frequency range. Such a signal generator may produce single scans, periodic scans, continuous scans, or the like. The rate at which scans are performed is usually adjustable. **2.** A **sweep generator** (1) utilized for testing, calibration, and/or troubleshooting. The response of any tested device is usually monitored via an oscilloscope. **3.** A circuit which generates **sweeps** (1).

sweep magnification The expansion of a sweep displayed by an oscilloscope, so as to enable closer observation of complex waveforms. Also called **sweep expansion**.

sweep oscillator Same as **sweep generator**.

sweep rate Same as **sweep frequency**.

sweep signal 1. A signal whose frequency is varied through a given frequency range. **2.** A signal generating a **sweep** (1).

sweep-signal generator Same as **sweep generator**.

sweep test A test, such as that revealing a frequency response, which is performed utilizing a **sweep generator**.

sweep time The time interval that elapses during a **sweep** (1).

sweep voltage The voltage of a **sweep signal**.

sweeping receiver Same as **scanning radio**.

SWG Abbreviation of **Standard Wire Gauge**.

swing The total available variation, from the lowest amplitude or other value or quantity, to the highest, or the peak difference between maximum instantaneous absolute values. Examples include carrier swing, frequency swing, and grid swing.

Swing Java components and tools utilized to prepare graphical user interfaces. Also called **Swing components**.

Swing components Same as **Swing**.

swinging choke A choke utilized to perform voltage regulation in power supplies.

switch **1.** A device which serves for opening, closing, or changing connections in electric circuits. A switch may be manual or automatic. There are various types, including mechanical, such as circuit-breakers, and semiconductor, such as transistors. Also called **electric switch**. Its abbreviation is **SW**. **2.** A device that activates or turns on, deactivates or turns off, or changes between given levels or modes of operation. Such a switch may be mechanical, electrical, magnetic, optical, acoustic, gravitational, and so on. Examples include rotary, paddle, light-activated, Hall-effect, inertia, lever, magnetic-reed, mercury, button, centrifugal, and diode switches, among many others. Its abbreviation is **SW**. **3.** To change from one state, connection, mode of operation, or the like, to another, using a **switch (1)** or **switch (2)**. **4.** A device in a communications network that selects a path or circuit via which data will flow to its next destination. A switch usually involves a simpler and faster mechanism than a router, but may also have router functions. Also called **network switch**. **5.** To change from one possibility, alternative, setting, or the like, to another. For example, to switch from landscape to portrait mode. **6.** In programming, a choice of a jump or process from multiple available jumps or processes. Also, a bit or byte which represents, keeps track of, or modifies such a choice.

switch box A box or unit which provides connectivity between multiples devices, pieces of equipment, or systems, by allowing the selection from one or more inputs to provide one or more outputs. For instance, such a box can be utilized to choose which of the multiple printers connected to a network will print a given document, or to determine which units within a building will receive which cable TV programming signals. Also spelled **switchbox**.

switch chatter The continuous, rapid, and undesired opening and closing of switch contacts. May be caused, for instance, by contact bounce.

switch current **1.** The current flowing through a switch at any given moment. **2.** The current necessary to actuate a switch.

switch debouncing In a switch, circuitry or measures utilized to prevent the recognition of multiple signals due to contact bounce.

switch hook Same as **switchhook**.

switch leakage current Current which flows through a switch that is in the off state.

switch voltage The voltage necessary to actuate a switch.

switchboard **1.** A piece of equipment that serves to manually or automatically connect telephone extensions, lines and/or trunks. Examples include private branch exchanges, and telephone exchanges. Also called **telephone switchboard**. **2.** One or more panels which incorporate the switches, circuit-breakers, fuses, and the like, utilized to monitor and operate electric equipment. Also called **electric switchboard**.

switchbox Same as **switch box**.

switched capacitor An active filter incorporating a capacitor and a switch. Banks of such devices are used, for instance, in ICs. Also called **switched-capacitor filter**.

switched-capacitor filter Same as **switched capacitor**.

switched circuit A communications or power circuit which can be turned on and off, activated or deactivated, or which can be made to follow any of multiple paths.

switched Ethernet An Ethernet network that uses switches instead of hubs and repeaters, and which provides the full bandwidth available to each pair of connected nodes. Depending on which version is used, the dedicated bandwidth between the sending and receiving stations can range from 10 Mbps to over 100 Gbps. In **shared Ethernet** all nodes must compete for any available bandwidth.

switched line A communications or power line which can be turned on and off, activated or deactivated, or which can be made to follow any of multiple paths.

switched-mode power supply A power supply whose input is switched very quickly. This helps reduce losses at higher frequencies, and allows for a more compact design than those whose input is 50 or 60 Hz. The switching devices, however, may produce more noise or interference. Its abbreviation is **SMPS**.

Switched Multimegabit Data Service Same as **SMDS**.

switched network A communications or power network which can be turned on and off, activated or deactivated, or which can be made to follow any of multiple paths. An example is a switched telephone network.

switched-reluctance motor A brushless synchronous motor which does not utilize a permanent magnet, and which does not have a coil in its rotor. Such a motor has pointed poles in both the stator and rotor, and the coils of the stator are fed the energizing current, which creates a magnetic pulling force which enables rotation. Reluctance motors are comparatively robust and efficient, and can be used in very high ambient temperatures. Its abbreviation is **SR motor**. Also called **reluctance motor**, or **variable reluctance motor**.

switched service Said of telephone service, such as POTS, which is switched or channeled through a central office. Also called **switched telephone service**.

switched telephone network Its abbreviation is **STN**. The multiple interconnected telephone networks in place worldwide, which enable most any phone to dial most any other phone around the globe. POTS is the basic service a PSTN provides. Such a network may also be used for data services, such as dial-up Internet access. Also called **public switched telephone network**, **public telephone network**, or **public switched network (1)**.

switched telephone service Same as **switched service**.

switched virtual circuit In a communications network, a virtual circuit between two nodes which is established only when needed. Such a connection lasts only as long as a data transfer lasts. This contrasts with a **permanent virtual circuit**, which is established before any actual data transfers takes place. Its abbreviation is **SVC**.

switchgear One or more devices, pieces of equipment, or systems, utilized to perform **switching** functions. Used, for instance, in the context of power transmission.

switchhook On a telephone set, a switch that is depressed when the handset is placed on it. When this switch is depressed, the telephone is not in use. Such a switch can also be located on a handset. A switchhook may also be used for functions such as call waiting or conference calling. Also spelled **switch hook**. Also called **hookswitch**.

switching The opening, closing, connecting, disconnecting, making, or breaking of circuits, or the changing of paths, routes, settings, modes of operation, levels, or the like. Said especially when performed utilizing a **switch**.

switching center A location, such as a central office, which performs **switching** functions.

switching circuit A circuit, such as a Schmitt trigger, which performs **switching** functions.

switching component A component, such as a transistor, which performs **switching** functions. Also called **switching element**.

switching device A device, such as a commutator switch, which performs **switching** functions.

switching diode A diode utilized for **switching**, as opposed, for instance, to rectifying.

switching element Same as **switching component**.

switching equipment Equipment, such as that located at a central office, which performs **switching** functions.

switching frequency 1. The rate at which **switching** occurs. Also called **switching rate**. **2.** The frequency at which a switch is actuated.

switching gate A gate, such as a logic gate, which performs **switching** functions.

switching hub A hub which provides the temporary linking of two communications terminals in order to establish a communications channel. This circuit must be established before communication can occur, and remains in exclusive use by these terminals until the connection is terminated. Used, for instance, by telephone companies to enable dial-up voice conversations, or for the temporary linking of two data terminals via a dedicated channel.

switching key A button, lever, or handle which is depressed or pressed, to open or close a circuit, to actuate a mechanism, or the like. Also called **key (3)**.

switching loss Same as **switching losses**.

switching losses Power losses occurring during **switching**, such as those suffered by a power transistor. Also called **switching loss**.

switching matrix A matrix, such as a crossbar switch, that performs **switching** functions.

switching mechanism The mechanism utilized to perform a given **switching** function.

switching rate Same as **switching frequency (1)**.

switching regulator A **switching circuit** that utilizes a closed loop to regulate the output voltage. Used, for instance, in power supplies.

switching speed The rate at which packets are transported in a packet-switching network.

switching system A system, such as a digital cross-connect system, which performs **switching** functions.

switching time The time necessary for a switch to change a state, open a circuit, reroute, and so on, once the actuating signal or action is provided.

switching transistor A transistor utilized for **switching**, as opposed, for instance, to amplifying.

switching tube A tube, such as an antitransmit/receive tube, which performs **switching** functions.

SWR Abbreviation of **standing-wave ratio**.

SWR bridge Abbreviation of **standing-wave ratio bridge**.

SWR detector Abbreviation of **standing-wave ratio detector**.

SWR indicator Abbreviation of **standing-wave ratio indicator**.

SWR meter Abbreviation of **standing-wave ratio meter**.

SWS Abbreviation of **slow-wave structure**.

symbol 1. That which is utilized to represent something else. **2.** A sign, notation, character, mnemonic, or the like, which represents and identifies a value, entity, relationship, operation, function, concept, component, device, and so on. For example, an electrical, chemical, logic, or schematic symbol, or that indicating a mathematical operation. The meaning of a given symbol may vary from one context to another.

symbol set A defined list of symbols which computer hardware and/or software recognizes as forming a unique group. An example is a character set.

symbol table A data structure where symbols and their associated attributes are stored. Each programming language has its own symbol table.

symbolic address A symbol, or other identifier, which is arbitrarily named by a programmer and represents an address or a location.

symbolic coding The preparation of computer programs and routines using a **symbolic language**, or another system of symbols.

symbolic language A computer programming language in which operations, operands, addresses, results, and the like, are named and represented in symbolic form. All high-level programming languages are symbolic languages.

symbolic link A directory entry which contains the path leading to a different file or directory. This contrasts with a **hard link**, which references the same or directory. Symbolic links allow for multiple names in different file systems, while hard links do not. Its abbreviation is **symlink**. Also called **soft link**.

symbolic logic The use of a system of symbols to represent quantities, operations, and relationships. An example is Boolean logic. Also called **mathematical logic**.

symbolic programming Computer programming utilizing a **symbolic language**, or which is otherwise based on symbols.

symlink Abbreviation of **symbolic link**.

symmetric Same as **symmetrical**.

symmetric circuit Same as **symmetrical circuit**.

symmetric compression A compression technique with which the time and/or processing power required to decompress is approximately the same as that required to originally compress. Used, for instance, in some real-time applications. Also spelled **symmetrical compression**.

symmetric conductivity Same as **symmetrical conductivity**.

symmetric cryptography Same as **symmetric encryption**.

symmetric digital subscriber line Same as **SDSL**.

symmetric DSL Same as **SDSL**.

symmetric encryption An encryption method in which the same key is used for encryption and decryption of messages. The fundamental concern with this method is insuring that the intended recipient indeed is the one to receive the key. Also called **symmetric cryptography**, **secret-key cryptography**, **secret-key encryption**, or **private-key cryptography**.

symmetric multiprocessing A multiprocessing computer architecture in which multiple CPUs, usually in the same cabinet, share the same memory, and the same copy of the operating system, application, and data being worked on, during the execution of a single program or task. As more CPUs are desired or needed, they can be added. Also, processing in this manner. This contrasts with **massively parallel processing**, where each CPU has its own copy of the operating system, its own memory, its own copy of the application being run, and its own share of the data. This also contrasts with **asymmetric multiprocessing**, in which multiple CPUs are each assigned a specific task, with all CPUs controlled by a master processor. Also spelled **symmetrical multiprocessing**. Its abbreviation is **SMP**. Also called **tightly-coupled multiprocessing**.

symmetric multiprocessor A computer system that can perform **symmetric multiprocessing**. Also spelled **symmetrical multiprocessor**. Its abbreviation is **SMP**. Also called **tightly-coupled multiprocessor**.

symmetric wave A wave in which there is symmetry present. For instance, that in which the positive portion is identical to the negative portion. Also spelled **symmetrical wave**.

symmetrical Pertaining to, exhibiting, or arising from **symmetry**. Also spelled **symmetric**.

symmetrical circuit A circuit in which there is **symmetry** present. In a push-pull circuit, for instance, each signal branch is identical, but operated in phase opposition. Another example would be a full-duplex circuit which allows simultaneous communications to occur in both directions at the same transmission speed. Also spelled **symmetric circuit**.

symmetrical compression Same as **symmetric compression**.

symmetrical conductivity Also spelled **symmetric conductivity**. **1.** Conductivity which is the same in both directions. **2.** Conductivity which is the same throughout the cross section of a conductor.

symmetrical digital subscriber line Same as **SDSL**.

symmetrical DSL Same as **SDSL**.

symmetrical multiprocessing Same as **symmetric multiprocessing**.

symmetrical multiprocessor Same as **symmetric multiprocessor**.

symmetrical wave Same as **symmetric wave**.

symmetry A balance, correspondence, proportion, or relationship, such as that observed in forms, magnitudes, or values existing or occurring on opposite sides of a line, axis, or plane. For example, a symmetric wave, symmetrical conductivity, or the arrangement of a push-push circuit.

sympathetic vibration A vibration in one object or system which is induced by another object or system vibrating at the same frequency.

SYN flood A denial-of-service attack in which numerous synchronization packets are maliciously sent to a host or server, which responds to said packets, but which then does not receive the final confirmation from the requesting node. While the host or server waits and eventually times out, legitimate connections desired by other users may not be granted, and enough unanswered requests may lead to an overload or crash. Also called **SYN flood attack**, or **flood attack**.

SYN flood attack Same as **SYN flood**.

sync 1. Abbreviation of **synchronization**. **2.** To **synchronize**.

sync bit Abbreviation of **sync**hronization **bit**. In communications, a bit utilized to establish and maintain synchronization during a data transmission.

sync byte Abbreviation of **sync**hronization **byte**. In communications, a byte utilized to establish and maintain synchronization during a data transmission.

sync character Abbreviation of **sync**hronization **character**. In communications, a character utilized to establish and maintain synchronization during a data transmission.

sync generator A circuit or device which produces synchronizing signals or pulses. For example, that which generates line, horizontal, or vertical sync pulses.

sync pulse Abbreviation of **sync**hronization **pulse**, or **sync**hronizing **pulse**. **1.** A pulse, such as a horizontal sync pulse, which serves to **synchronize**. **2.** A pulse within a **sync signal**.

sync separator In a TV receiver, a circuit that extracts the sync pulses from the composite video signal. Also called **sync stripper**.

sync signal Abbreviation of **sync**hronization **signal**, or **sync**hronizing **signal**. A signal, such as a color burst, that which establishes and maintains synchronous communications, or which otherwise serves to **synchronize**.

sync stripper Same as **sync separator**.

syncDRAM Same as **SDRAM**.

synchro Also called **selsyn**. **1.** An electromechanical device that converts a mechanical angular position into an electrical

signal which is utilized to synchronize the angular position of a remote device. For example, one unit, the synchro transmitter, produces a voltage which is proportional to the angle of its rotor, and another unit, the synchro receiver, matches the position of its rotor to that of the transmitter based on the voltage sent by said transmitter. **2.** Same as **synchro transmitter**. **3.** Same as **synchro receiver**. **4.** Same as **synchro system**. **5.** Abbreviation of **synchronous**.

synchro differential receiver A synchro receiver whose output is a mechanical angle obtained by subtracting one electrical angle from another. Such a device utilizes a damper to limit oscillation or spinning. Also called **differential receiver**, or **differential synchro (1)**.

synchro differential transmitter A synchro transmitter whose output is an electrical angle obtained by the sum of an electrical angle and a mechanical angle. Also called **differential transmitter**, or **differential synchro (2)**.

synchro generator Same as **synchro transmitter**.

synchro motor Same as **synchro receiver**.

synchro receiver In a synchro system, the synchro which converts the voltage received from the synchro transmitter into the corresponding angular position of its stator. Also called **synchro motor, synchro (3), selsyn receiver, selsyn motor**, or **receiver synchro**.

synchro system A system which incorporates one or more synchro transmitters and one or more synchro receivers. Used, for instance, to control the motors of separate audio and video recorders, so as to maintain complete consistency throughout the filming of a given scene. Also called **synchro (4)**, or **selsyn system**.

synchro-to-digital converter A circuit or device that transforms a synchro input signal into a digital output signal. Its abbreviation is **SDC**, or **s/d converter**.

synchro transmitter In a synchro system, the synchro which transmits a voltage corresponding to the position of its rotor. Also called **synchro generator, synchro (2), selsyn transmitter, selsyn generator**, or **transmitter synchro**.

synchrocyclotron A cyclotron in which the frequency of the electric field is adjusted so as to compensate for the increased mass of the particles as their speed approaches that of light. This can result in energies exceeding 1 GeV.

synchroflash 1. A device which synchronizes one or more flashes with the opening of the shutter of a camera. **2.** A flash that is synchronized with the opening of the shutter of a camera.

synchronism The quality or state of being **synchronous**.

synchronization Its abbreviation is **sync (1)**. **1.** The causing to occur, operate, transmit, or move at the same time as something else. For example, the synchronization of the soundtrack and the images of a movie. **2.** The occurring, operating, transmitting, or moving at the same time as something else.

synchronization bit Same as **sync bit**.

synchronization byte Same as **sync byte**.

synchronization character Same as **sync character**.

synchronization error Any error arising from incomplete **synchronization**.

synchronization pulse Same as **sync pulse**.

synchronization signal Same as **sync signal**.

synchronize 1. To cause to occur, operate, transmit, or move at the same time as something else. For example, to synchronize the soundtrack of a movie, so as to match the corresponding images. **2.** To occur, operate, transmit, or move at the same time as something else.

synchronized 1. That which is caused to occur, operate, transmit, or move at the same time as something else. For instance, a movie soundtrack which is synchronized with the corresponding images. **2.** That which occurs, operates, transmits, or moves at the same time as something else.

Synchronized Multimedia Integration Language Its acronym is **SMIL**. A markup language utilized to identify and synchronize multimedia elements presented on Web pages. For example, streaming audio, still images, and movies can be sent separately from different locations and be displayed as a cohesive unit. SMIL may have multiple versions of each presentation, so as to accommodate for different needs, such the bandwidth available to any given user.

synchronizer That which serves to **synchronize**.

synchronizing clock A clock which emits synchronizing pulses or signals.

synchronizing pulse Same as **sync pulse**.

synchronizing signal Same as **sync signal**.

synchronous Its abbreviation is **synchro** (**5**). **1.** Occurring at the same time, having the same rate, or operating in unison with something else. **2.** Having the same frequency and/or phase.

synchronous circuit A circuit which is controlled by clock-generated synchronization signals, as opposed to an **asynchronous circuit**, which is not.

synchronous clock An electric clock which is driven by a **synchronous motor**.

synchronous code division multiple access Same as **S-CDMA**.

synchronous communications Communications which are synchronized by a timing signal. This contrasts with **asynchronous communication**, which may start or end at any time, with these events being signaled respectively by start and stop bits.

synchronous component **1.** A component whose operation depends on a timing mechanism. **2.** An AC component which runs at a speed which is synchronized to the frequency of the power operating it.

synchronous computer A computer in which each event is controlled by a clock-generated synchronization signal. This contrast with an **asynchronous computer**, in which each operation starts as a result of another being completed.

synchronous control Control which depends on a signal generated by a clock, as opposed to **asynchronous control**, which is based on events concluding.

synchronous converter **1.** A device which makes the necessary conversions between **synchronous devices**. **2.** Same as **synchronous inverter**.

synchronous data Data sent via a **synchronous transmission**.

synchronous data-link control A bit-oriented communications protocol utilized for the control of synchronous transmissions over communications links. Its abbreviation is **SDLC**.

synchronous data transmission Same as **synchronous transmission**.

synchronous demodulator Same as **synchronous detector**.

synchronous detector A detector which restores the suppressed carrier in a suppressed-carrier transmission, prior to demodulation. Also called **synchronous demodulator**.

synchronous device **1.** A device whose operation depends on a timing mechanism, as opposed to an **asynchronous device**, in which events occur as each preceding event is completed. **2.** An AC device which runs at a speed which is synchronized to the frequency of the power operating it. This contrasts with an **asynchronous device**, which is not.

Synchronous Digital Hierarchy Its abbreviation is **SDH**. An ITU standard for transmitting digital data over optical networks, at rates ranging from 155.52 Mbps, with higher speeds based on multiples of this number, such as 2.48832

or 39.81312 Gbps. The compatibility between SDH and SONET enables the interconnection of digital networks internationally.

synchronous DRAM Same as **SDRAM**.

synchronous dynamic RAM Same as **SDRAM**.

synchronous dynamic random-access memory Same as **SDRAM**.

synchronous gate **1.** A gate circuit whose function is triggered by clock-synchronized pulses. **2.** A gate circuit whose output is synchronized with its input.

synchronous generator An AC generator whose AC power output is proportional to the speed at which its armature is rotated.

synchronous graphics RAM Same as **SGRAM**.

synchronous graphics random-access memory Same as **SGRAM**.

synchronous inverter A rotating electric machine which is utilized to convert a DC voltage to an AC voltage, or vice versa. It usually has a single armature with two or more windings, and incorporates a commutator for DC operation, and slip rings for AC operation. Also called **synchronous converter** (**2**), **dynamotor** (**2**), or **rotary converter**.

synchronous logic In hardware or software, logic which depends on clock-generated synchronization signals, as opposed to **asynchronous logic**, which depends on events concluding.

synchronous machine An AC machine which runs at a speed which is synchronized to the frequency of the power operating it, as opposed to an **asynchronous machine**, which is not.

synchronous mode A mode of operation which depends on a timing mechanism, as opposed to an **asynchronous mode**, which requires each preceding event to have been completed.

synchronous motor An AC motor which runs at a speed which is synchronized to the frequency of the power operating it, as opposed to an **asynchronous motor**, which is not.

synchronous network A network utilized for **synchronous communications**.

synchronous operation Operation which depends on a timing mechanism, as opposed to **asynchronous operation**, which requires each preceding event to have been completed.

Synchronous Optical Network Same as **SONET**.

synchronous orbit The orbit a **synchronous satellite** follows. Also called **stationary orbit**, or **fixed orbit**.

synchronous protocol A communications protocol that governs a **synchronous transmission**.

synchronous satellite Also called **stationary satellite**, or **fixed satellite**. **1.** An artificial satellite whose orbit is synchronized with that of the earth, so that it remains fixed over the same spot above the planet at all times. Such a satellite is usually located at an altitude of approximately 35,900 kilometers, and its orbit is in the plane of the equator. As few as three of these satellites can provide worldwide coverage. Used, for instance, for communications and broadcasting. Also called **geosynchronous satellite**, or **geostationary satellite**. **2.** A satellite which remains fixed in relation to a celestial body such as the earth, moon, or sun.

synchronous speed **1.** For an AC motor, the rate at which the magnetic field rotates. **2.** The rate at which a synchronous transmission occurs.

Synchronous Transfer Signal Same as **STS**.

synchronous transmission Data transmission that occurs in a steady stream which is synchronized by timing signals, as opposed to an **asynchronous transmission**, which has start and stop bits. Also called **synchronous data transmission**.

Synchronous Transport Module Same as **STM** (**3**).

synchroscope A device or instrument which indicates whether two values, magnitudes, voltages, frequencies, phases, or the like, are synchronized. An oscilloscope, for instance, may indicate the synchronization, or any difference in phases, frequencies, or the like which are being monitored. Used, for instance, to verify that two or more generators are in phase.

synchrotron A particle accelerator similar to a cyclotron, and in which the frequency of the field is adjusted so as to compensate for the increased mass of the particles as their speed approaches that of light. A proton synchrotron accelerates protons, an electron synchrotron accelerates electrons, and there are specialized variations, such as tevatrons, which can generate energies in excess of 1 TeV.

synchrotron radiation The electromagnetic radiation emitted by charged particles which are moving within a magnetic field at speeds approaching that of light. This is observed, for instance, in a synchrotron.

syntax **1.** In a natural language, the rules which describe how words and phrases are combined to form expressions and sentences. Such rules are taken into account, for instance, in voice-recognition programs. **2.** The rules which determine how statements are formed, so as to be understood by a given computer program.

syntax error An error in a computer program arising from the improper use **syntax (2)**.

synthesis The proper combination of elements, components, substances, and so on, to form a given whole. Examples include the synthesis of digital speech, musical sounds, circuits with a given response, or chemical compounds.

synthesis telescope A computer-controlled telescope which utilizes two or more pairs of antennas which sequentially cover sections of the total aperture, in order to gather the information equivalent to that obtained by a much larger single telescope. Also called **aperture synthesis telescope**.

synthesize To unite the proper combination of elements, components, substances, and so on, to form a given whole.

synthesizer **1.** That which serves to **synthesize**. **2.** A circuit or device which generates precise frequency signals. Such a device usually utilizes one or more crystal oscillators, and can generate equally-spaced frequencies within a given band through the use of frequency multipliers, dividers, mixers, and so on. Also called **frequency synthesizer**. **3.** A device which electronically generates or reproduces sounds that emulate musical instruments and voices, and which can be utilized to compose, play, and record music. A keyboard, real or virtual, is usually used, and two common techniques for generating sounds are FM synthesis and wavetable synthesis. Also called **music synthesizer**.

synthetic **1.** Pertaining to, characteristic of, involving, or arising from **synthesis**. **2.** Produced by synthesis, especially that which is artificial. Examples include synthetic radioactive chemical elements, resins, plastics, crystals, fibers, rubbers, polymers, and so on.

synthetic aperture radar An airborne radar that emits microwave energy along a given path, and which utilizes the Doppler effect to process the phase of the returned radar signals. Such a radar effectively synthesizes the equivalent of a very long aperture, and is used, for instance, for ground mapping. Its abbreviation is **SAR**.

synthetic voice Simulation of a human voice through the use of a device which incorporates a loudspeaker and a computer. Used, for example, in robotics, or to assist those with reduced speech abilities.

syntonic Pertaining to two oscillating circuits which have the same resonant frequency.

syntony The circumstance of two oscillating circuits having the same resonant frequency.

sysadmin Acronym for **system administrator**.

sysop Acronym for **system operator**.

SysReq Same as **SysRq**.

SysReq key Same as **SysRq**. Abbreviation of **System Request key**. Also spelled **SysRq key**.

SysRq Abbreviation of **System Request**. On many computer keyboards, a key intended to access specific operating system functions. Such a key is sometimes programmed to perform special functions within a given operating system environment or application. Also spelled **SysReq**. Also called **SysRq key**.

SysRq key Same as **SysRq**. Abbreviation of **System Request key**. Also spelled **SysReq key**.

system **1.** A set of interrelated and/or interdependent components which form a complex whole serving for one or more purposes or functions. There are countless examples, including control, sound, expert, radar, alarm, computer, carrier, biometric, and satellite systems. **2.** A **system (1)** incorporating components which are mechanical, electrical, electronic, magnetic, optical, and so on, or any combination of these. For instance, electromechanical, optoelectronic, or magnetohydrodynamic systems. **3.** A set of objects, entities, characteristics, phenomena, or rules, utilized to describe, classify, organize, compare, or analyze. For example, a system which enables the precise specification and comparison of colors, a scale which allows the classification of materials according to their resistance to scratching or denting, a coordinate system, or a quantum system. **4.** A set of related objects, entities, characteristics, or phenomena, which occur naturally, such as a solar system. **5.** A set of components, including equipment, media, and channels, which incorporate all that is necessary for any form of communication from one or more points to others. For example, telephone, TV, or radio systems, computer networks, and so on. **6.** The complete complement of components required for a computer to function. These include the CPU, keyboard, mouse, monitor, memory, storage mediums, cables, and so on, which comprise the hardware of the computer itself, plus any necessary peripheral devices. In addition, a computer system incorporates the operating system. Also called **computer system**.

system administration The functions performed by a **system administrator**. These include the allocation of resources, security, maintenance, improvements, and the like.

system administrator A person who is responsible for, supervises, and/or manages a computer or communications network. Its acronym is **sysadmin**.

system analysis Same as **systems analysis**.

system analysis and design Same as **systems analysis**.

system-area network A high-speed network which serves to link processors, I/O systems, or servers. Its acronym is **SAN**.

system basic input/output system Same as **system BIOS**.

system BIOS Acronym for **system basic input/output system**. **1.** A set of indispensable software routines that enable a computer to boot itself. It has the code necessary to control all peripherals and perform other functions, such as testing, and is generally stored on a ROM chip so that disk failures do not disable it. Currently, it is usually contained in a flash-memory chip, and is typically copied to the RAM at startup, as RAM is faster. Also called **BIOS**. **2.** The **system BIOS (1)** of a computer which is contained in the motherboard, as opposed to that contained in a peripheral.

system board The main circuit board of a computer, containing the connectors necessary to attach additional boards. It contains most of the key components of the system, and incorporates the CPU, bus, memory, controllers, expansion slots, and so on. Memory chips may be added to it, and some motherboards allow for the CPU to be replaced. Also called **motherboard**, **mainboard**, or **backplane (2)**.

system bus The pathway between the CPU and the computer memory. Also called **front side bus**.

system clock In a computer, a circuit or device which provides a steady stream of timed pulses. This clock serves as the internal timing reference of a digital computer, upon which all its operations depend. Also known as **system timer**, **master clock (1)**, or **main clock**.

system console Within a computer or communications network, a terminal which serves for monitoring and control purposes.

system control and data acquisition Same as **SCADA**.

system design Same as **systems design**.

system development The steps involved in designing, prototyping, testing, implementing, and improving a computer system.

system disk A disk that contains the files necessary to boot a computer. It is usually used when the disk from which the computer ordinarily boots fails. The operating system is included in the contents of a boot disk, and it may be a floppy disk, or an optical disk such as a CD-ROM. Also called **startup disk**, **boot disk**, or **bootable disk**.

system engineering Same as **systems engineering**.

system error An error which results from an operating system malfunction, or which otherwise prevents the operating system from functioning properly. Such an error is usually remedied by rebooting.

system failure **1.** The condition in which a system no longer performs the function it was intended to, or is not able to do so at a level that equals or exceeds established minimums. **2.** A **system failure (1)** which prevents a computer system from functioning properly. Such an error is usually remedied by rebooting.

system file A file within the operating system of a computer.

system flowchart **1.** A flowchart representing the sequence of operations and/or the interrelationships of the components of a system. **2.** A flowchart representing the sequence of operations and/or the flow of data in a computer system.

system folder A folder containing **system files**.

system font A font that is utilized to display text and messages pertaining to the operating system of a computer. A system may have more than one such font, depending on the type of text or message, and may also be user-chosen.

System I/O Former name for **InfiniBand**.

system image The complete contents, including the operating system, applications running, and data, in the memory of a computer system at any given moment.

system life cycle **1.** The interval during which a system is useful. Once this period elapses, it becomes increasing less advisable to invest additional resources in maintenance, repairs, and improvements. **2.** The **system life cycle (1)** of a computer.

system management Same as **systems management**.

system memory **1.** The locations within a computer that serve for temporarily holding and accessing data in a machine-readable format. Memory chips are used for this purpose, and most are allocated for RAM, or main memory. Memory is usually quantified in multiples of bytes. For example, a computer with 1 gigabyte of RAM can hold approximately 1 billion bytes, or characters, of information, and this is the total temporary workspace this computer has available. Other forms of memory in a computer include ROM, PROM, and EPROM. Also called **memory**, or **computer memory**. **2.** System memory (1) utilized by the operating system.

System Network Architecture Same as **Systems Network Architecture**.

system noise Noise which is inherent in a system. Such noise is present in the absence of an input signal.

system of units A system utilized for measurement of physical quantities using units which are precisely defined and internationally accepted. Currently, the International System of Units is the most important, and other units systems in use include the meter-kilogram-second-ampere, and the centimeter-gram-second systems. Also called **units system**.

system-on-a-chip A computer-on-a-chip which also incorporates other components, such as memory control, an operating system required to run applications, peripheral interfaces, and the like. Its abbreviation is **SoC**. Also called **system-on-chip**, **systems-on-silicon**, or **system-on-silicon**.

system-on-chip Same as **system-on-a-chip**.

system-on-silicon Same as **system-on-a-chip**.

system operator Its acronym is **sysop**. **1.** A person who operates, monitors, and/or manages a system, such as an alarm system. **2.** A person who operates, monitors, and/or manages a computer system, a computer network, or a communications network.

system program A program, such as a data management program, which serves to control the manner in which a computer system operates.

system prompt A symbol or message which is displayed on a computer monitor indicating the location where a command may be entered.

system recovery **1.** The process of resetting or otherwise reestablishing a system to a former state or condition, a given operational state, or a given level or value. Also, the result of such a process. **2.** A **system recovery (1)** after an error or malfunction. Also, the result of such a process. **3.** The restoration of the operation of a computer system, including the recuperation of all contained data, to the state it was in before an error or failure.

System Request Same as **SysRq**.

System Request key Same as **SysRq**.

system requirements The minimum hardware availability and configuration a system must have in order for a given program or device to work properly. For example, in order to utilize a given peripheral, a computer needing to have a minimum amount of available memory and disk space, a DVD drive, an available port, and so on. Also called **hardware requirements**.

system resources The means utilized by a computer system to perform its operations, functions, and tasks. These include processors, memory, storage devices, operating system, applications, peripherals, IRQs buses, and so on.

system response **1.** A change in the behavior, operation, or function of a system, as a consequence of a change in its external or internal environment. For example, a change in its output resulting from a change in its input. **2.** A quantified **system response (1)**, such as a frequency response.

system software Software, such as an operating system, which serves to control the manner in which a computer system operates.

system support **1.** To provide assistance, such as technical support, for a **system**. **2.** The ability to function properly with a specific **system (5)** or **system (6)**. For example, a TV which properly displays received PAL signals, or an application that functions in a given computer environment.

system timer Same as **system clock**.

system tray An area, such as that located on the lower right-hand side of the taskbar, which serves to display small icons of certain programs, such as a time/day clock, volume control, status of a modem, and the like. The functions of such applications can easily be accessed by double-clicking or right-clicking on the corresponding icon. Its abbreviation is **systray**.

system unit The enclosure which houses all the components a computer utilizes, with the exception of the external peripherals, such as monitor, mouse, keyboard, printer, and so on. Also called **CPU (3)**.

systematic error Non-random error which is a consequence of a measurement procedure, or the design or properties of a component, circuit, device, piece of equipment, or system. If an experiment is repeated multiple times, systematic error will remain the same, while **random errors** will vary each time.

Système Électronique Couleur Avec Mémoire Same as **SECAM**.

Système International d'Unités Same as **SI**.

systems administrator Same as **system administrator**.

systems analysis Also called **systems analysis and design**. **1.** The careful study of an activity, procedure, technique, problem, or the like, describing aspects such as the concepts, components, their purposes, functions, and the way they interact, with the overall objective of designing and implementing a system that can solve, simplify, or improve upon that being analyzed. **2.** The use of computers for **systems analysis (1)**. **3.** A **systems analysis (1)** performed on a computer system with the goal of improving that which is already present, or for better adapting computers to human activities.

systems analysis and design Same as **systems analysis**.

systems design The design and implementation sections within **systems analysis**.

systems development Same as **system development**.

systems engineering The branch of engineering which is concerned with the study, design, development, uses, and technology of integrated systems, based on **systems analysis**.

systems flowchart Same as **system flowchart**.

systems life cycle Same as **system life cycle**.

systems management The tasks pertaining to the administration of the computers or information technology systems of an enterprise. Systems management includes the obtaining of the necessary hardware and software, allocation of resources, maintenance, upgrading, security, network and Web interfacing, network administration, database management, and so on, depending on the entity involved.

Systems Network Architecture Its abbreviation is **SNA**. A network model or group of protocols utilized for intercommunication between mainframes, terminals, and peripherals. SNA is subdivided into functional layers, in a manner similar to the OSI Reference Model.

systems-on-silicon Same as **system-on-a-chip**. Its abbreviation is **SOS (3)**.

systems program Same as **system program**.

systems software Same as **system software**.

systray Abbreviation of **system tray**.

T

t **1.** Symbol for **tonne**, or **metric ton**. **2.** Symbol for **ton**.

T **1.** Symbol for **temperature**, or **thermodynamic temperature**. **2.** Abbreviation of **tera-**. **3.** Symbol for **tesla**. **4.** Symbol for **tritium**. **5.** Symbol for **ton**. **6.** Shaped in the approximate form of a capital letter **T**, as is the case of a T junction. Also spelled **tee**.

T Symbol for **period (2)**.

T-1 Same as **T1**.

T-1 line Same as **T1 line**.

T-1/PRI Same as **T1/PRI**.

T-1C Same as **T1C**.

T-2 Same as **T2**.

T-3 Same as **T3**.

T-4 Same as **T4**.

T-5 Same as **T5**.

T adapter Same as **T junction**.

T antenna An antenna consisting of one or more horizontal radiators, each with a feed line connected at approximately its center. Also spelled **tee antenna**, or **T-type antenna**.

T-carrier Any of various digital transmission services, formats, or lines with signaling rates currently ranging from 1.544 Mbps for T1, through 400.352 Mbps for T5. Widely utilized for the transmission of multiplexed voice and data signals.

T circuit Same as **T network**.

T connection Same as **T network**.

T-connector A coaxial connector, with one plug and two jacks, which is in the shape of a capital letter **T**. Also, such a connector with two plugs and one jack.

T flip-flop Abbreviation of **toggle flip-flop**. A flip-flop that changes state with each positive or negative pulse. Used, for instance, in counter circuits.

T-interface A four-wire ISDN interface between an NT1 and an NT2.

T junction Also spelled **tee junction**. Also called **T adapter**, or **tee adapter**. **1.** A waveguide junction whose longitudinal axis is intersected by a branch at a right angle, so as to resemble a capital letter **T**. **2.** A point in a circuit where two or more conductors meet at right angles, so as to resemble a capital letter **T**. Also, such a point along a transmission line, such as a coaxial cable.

T network An electric network configuration in which three branches, such as impedance arms, are connected to a common central point or node. Also called **T circuit, T connection, tee circuit, tee network, tee connection, wye network**, or **Y network**.

T-Span **1.** A unit of 24 voice channels, which is the equivalent of a T1 line or circuit. **2.** Same as **T1**.

T-type antenna Same as **T antenna**.

T.4 An ITU standard pertaining to G3 fax page and content formats.

T.6 An ITU standard pertaining to G3 fax coding and control functions.

T.30 An ITU standard pertaining to G3 fax protocols.

T.36 An ITU standard pertaining to G3 fax security.

T.38 An ITU standard pertaining to G3 fax over IP networks.

T.120 An ITU standard for real-time data conferencing and application sharing, including protocols for multimedia conferencing.

T.121 An ITU recommendation for the development and implementation of protocols pertaining to **T.120**.

T.122 An ITU description of the services available to participants transmitting from one or more locations during real-time data conferencing.

T.123 An ITU standard specifying protocol stacks utilized in multimedia conferencing.

T.124 An ITU standard pertaining to the control of real-time conferences, including maintenance of lists of participants, compatible applications, and conference termination.

T.125 An ITU standard pertaining the implementation of **T.122** protocols. It specifies how data is transmitted during real-time conferences.

T.126 An ITU standard establishing protocols pertaining to the handling of images during real-time conferences, including the use of whiteboards.

T.127 An ITU standard establishing protocols pertaining to file transfers during real-time conferences.

T.128 An ITU standard pertaining to application-sharing during real-time conferences.

T1 A digital transmission standard, system, or line with a signaling rate of 1.544 Mbps, which is equivalent to DS1, providing a capacity of 24 voice channels, each with 64 Kbps. Such channels may be combined, and can serve to transmit voice and/or data. T1 is used mostly in the Unites States, Canada, and Japan, and is similar to **E1**, which is used by most other countries. E1 has a signaling rate of 2.048 Mbps. Also called **T-Span (2)**.

T1 line A line with a signaling rate of 1.544 Mbps, which is equivalent to DS1, providing a capacity of 24 voice channels, each with 64 Kbps.

T1/PRI Abbreviation is T1/Primary Rate Interface. An ISDN line with 23 B channels, each with a speed of 64 Kbps, and one 64 Kbps D channel. This is equivalent to a T1 or DS1 line. In Europe, PRI has 30 B channels and one D channel, so it is equivalent to an E1 line. Multiple B channels can be combined for the desired or needed bandwidth. Also called **Primary Rate Interface, PRI/T1**, or **ISDN-PRI**.

T1C A digital transmission standard, system, or line with a signaling rate of 3.152 Mbps, which is equivalent to **DS1C**, which provides a capacity of 48 voice channels.

T2 A digital transmission standard, system, or line with a signaling rate of 6.312 Mbps, which is equivalent to DS2, which provides a capacity of 96 voice channels. E2 is similar, but with greater capacity, is used in most countries, and has a signaling rate of 8.448 Mbps.

T²L Abbreviation of **transistor-transistor logic**.

T3 A digital transmission standard, system, or line with a signaling rate of 44.736 Mbps, which is equivalent to DS3, which provides a capacity of 672 voice channels. E3 is similar, but with lesser capacity, is used in most countries, and has a signaling rate of 34.368 Mbps.

T4 A digital transmission standard, system, or line with a signaling rate of 274.176 Mbps, which is equivalent to DS4, which provides a capacity of 4032 voice channels. E4 is similar, but with lesser capacity, is used in most countries, and has a signaling rate of 139.264 Mbps.

T5 A digital transmission standard, system, or line with a signaling rate of 400.352 Mbps, which provides a capacity of 5760 voice channels. E5 is similar, but with greater capacity, is used in most countries, and has a signaling rate of 565.148 Mbps.

T9 Abbreviation of **T9 text input**.

T9 text input A feature of handheld devices such as cell phones and PDAs, which simplifies entering text by displaying words that may be spelled by a given sequence of letters typed using a telephone keypad. For example, entering 7-8-6 would display *run* or multiple possible matches, from which a user could choose the desired word. Its abbreviation is **T9**.

T568A Within EIA/TIA-568, a standard dealing with four-pair wiring, such as that utilized for Ethernet.

T568B Within EIA/TIA-568, a standard dealing with four-pair wiring, such as that utilized for Ethernet.

Ta Chemical symbol for **tantalum**.

TA Abbreviation of **terminal adapter**.

TAB Acronym for **tape automated bonding**.

tab 1. Same as **tab key**. **2.** To move from one **tab stop** to the next, or through successive tab stops. **3.** In a GUI, one of multiple buttons, segments, or the like, which is selectable by clicking on it with the mouse or other pointing device, when two or more menu choices are available. Such tabs serve, for instance, to subdivide categories of options.

tab character A control character which is utilized to indicate a **tab stop**.

tab key Also called **tab** (1). **1.** A computer keyboard key that moves the cursor from one **tab stop** to the next, or through successive tab stops. **2.** A computer keyboard key which when pressed inserts a tab stop.

tab stop When a tab key is pressed, a point at which the cursor stops, or to which the on-screen highlight advances. In a word-processing document, for instance, there are tab stops at regular intervals along lines, and a user may add more such stops at any desired location. In database or spreadsheet applications, tab stops progress through cells or fields. In a Web page, the tab key advances the on-screen highlight from specific location to specific location on the screen, such as from hyperlink to hyperlink.

table 1. An arrangement of words, numbers, symbols, or the like, in rows and columns, or in another manner which assists in locating information. For example, a periodic table. **2.** Data that is arranged in rows and columns, or in another manner in which each contained item is uniquely identified, thus assisting in locating the contained information. Examples include the data stored in relational databases, that contained in spreadsheets, or in a file allocation table.

table look-up Same as **table lookup**.

table lookup A data search in which a specific value is searched within a predefined table of values, such as a matrix. Also, the function, such as that in a spreadsheet application, which performs such a search. Also, the value so obtained. Also spelled **table look-up**. Also called **lookup**.

tablet An input device which serves for sketching or tracing images directly into a computer. It consists of an electronic pressure-sensitive surface which senses the position of the stylus which makes contact with it. The stylus, or puck, is similar to a mouse, but is much more accurate, because its location is determined by touching the active surface of the tablet with an absolute reference. Used, for instance, in computer-aided design. Also called **touch tablet**, **digitizing tablet**, **graphics tablet**, or **digitizer** (2).

tablet computer A portable computer incorporating a **tablet** as its primary input device. Such a computer may also accept input from a camera, microphone, and/or an attachable keyboard, and nearly its entire top or front surface may consist of the pressure-sensitive surface. Such a computer may have the same functionality as a desktop model. Also called **tablet PC**, or **tablet personal computer**.

tablet PC Same as **tablet computer**. Abbreviation of **tablet personal computer**.

tablet personal computer Same as **tablet computer**.

tabulate 1. To arrange information in the form of a **table**. **2.** To obtain a total, or perform another calculation on a column or row of numbers.

tabulator 1. That which serves to read information from an obsolete data medium, such as punched cards or perforated tape. The results may be calculations, lists, or the like. **2.** That which serves to place into tables, or to calculate values placed in such a table.

TACACS Acronym for Terminal Access Controller Access Control System. An authentication protocol in which an access server communicates with an authentication server to determine whether access to a network will be granted. TACACS has mostly been superseded by TACACS+, XTACACS, or RADIUS,

TACACS+ An enhanced version of **TACACS**, which includes data encryption and a challenge/response system. Despite its name, TACACS+ is not compatible with TACACS, nor XTACACS.

tacan Acronym for **tac**tical **a**ir **n**avigation. An air navigation system providing bearing and range information using UHF signals.

TACAN Same as **tacan**.

tach Abbreviation of **tachometer**.

tachometer A device or instrument which measures and indicates angular velocity. When graduated in revolutions per minute, also called **RPM meter**. Its abbreviation is **tach**.

tachometer generator A generator whose output energy is proportional to the rotational speed of a connected shaft. The frequency of the output energy may also be proportional to said speed.

tachyon A hypothetical particle whose speed would exceed that of light.

tactical air navigation Same as **tacan**.

tactile Pertaining to, or based on the sense of touch. Also called **haptic**.

tactile sensing device Same as **tactile sensor**.

tactile sensor A device that detects objects through physical contact with them. These sensors may be used in robots, for example, to determine the location, identity, and orientation of parts to be assembled. Also called **tactile sensing device**.

TAD Abbreviation of **telephone answering device**.

tag 1. That which serves to mark or otherwise enable monitoring or identification. For example, a radioactive tracer, or an RFID tag. Also, to mark, monitor, or identify with such a tag. **2.** One or more bits or characters which serve to identify data, such as that in a database. Also, the name of such a tag. **3.** A command or code which specifies how a document, or a part of a document, must be formatted. Examples include HTML tags and meta tags. Also, to so mark a document. **4.** A transponder within an RFID system. Such a transponder is usually in the form of an IC, and responds to interrogating signals arriving from an RFID reader. Such a tag, for instance, may be incorporated into a box label, a smart card, or a key chain wand which is waved near a reader to effect payments. Also called **RFID tag**.

tag-based language A language, such as HTML or XML, which incorporates **tags** (3).

tag RAM Abbreviation of **tag r**andom-**a**ccess **m**emory. A special area of RAM which holds addresses, and which determines the amount of memory which can be cached.

tag switching A network switching technology that uses **tags** (2) to store forwarding information.

Tagged Image File Format Same as **TIFF**.

TAI Abbreviation of **International Atomic Time**.

tail 1. The rear, bottom, or lowest part of something. **2.** That which resembles the rear, bottom, or lowest part of something. **3.** A small pulse that follows, and is in the same direction as the main pulse that produced it. **4.** Same as **trailing edge**. **5.** A brief signal that remains after the actuating signal is removed. For example, a brief signal emitted by a transmitter after talking has stopped.

tailing An unwanted prolongation in the decay of a signal. In fax, or example, this may manifest itself as a tail that forms on the lines in the recorded copy. Also called **hangover (2)**.

take-down Same as **takedown**.

take-down time Same as **takedown time**.

take-up reel In a tape player or system, the reel which winds the tape after it is played or recorded upon. The **supply reel** holds the tape before reproduction or recording. Also spelled **takeup reel**.

takedown After an operating cycle, the series of operations required to prepare a component, circuit, device, piece of equipment, system, material, or mechanism for the next cycle. Also spelled **take-down**.

takedown time The time interval required to complete a **takedown**. Also spelled **take-down time**.

takeup reel Same as **take-up reel**.

talk battery The DC voltage, typically 48V, which a central office supplies via the local loop for operation of a telephone. The ring voltage is usually higher.

talk-listen switch In an intercom or similar device, a switch that changes from the transmitting to the receiving mode, and vice versa. When a single speaker is used, it also serves to change to and from its microphone function.

talk power The power of an audio signal that is within the speech frequencies. Also called **speech power**, or **voice power**.

talk time The length of time a cell phone can be utilized for conversations. Usually expressed in minutes or hours. Also spelled **talktime**.

talker A command-driven technology which enables multiple users to interact with each other over the Internet, and which is utilized especially for real-time chatting.

talktime Same as **talk time**.

TAM Abbreviation of **telephone answering machine**.

tamper **1.** A material which is placed around the core of a nuclear reactor to reflect neutrons back to said core, preventing their escape. Also called **reflector (4)**. **2.** To interfere in a harmful or unauthorized manner. For example, to tamper with a delicate mechanism without the proper expertise.

tamper-evident A security feature which makes any **tampering** evident. Used, for instance, to help protect smart cards against fraudulent reprogramming.

tamper-proof Not allowing, or impervious to **tampering**. It is very difficult to attain such a degree of protection.

tamper-resistance **1.** The property or characteristic of affording a comparatively high degree of protection against **tampering**. **2.** The extent to which something provides **tamper-resistance (1)**.

tamper-resistant That which is characterized by **tamper-resistance**.

tamper switch A switch that is activated when something is interfered with in a harmful or unauthorized manner. Such a switch, for instance, may activate an alarm to indicate that a security mechanism is being meddled with.

tampering The process of interfering in a harmful or unauthorized manner. Said, for instance, of the improper use or alteration of a component, circuit, device, piece of equipment, system, material, or mechanism.

tan Abbreviation of **tangent (1)**.

tandem The connection of circuits or devices in a manner that the output of one is the input of the next. Seen, for example, in a cascade amplifier. Also called **cascade (3)**.

tandem office A central office utilized mainly to connect trunks, as opposed to subscriber lines.

tandem office switch Same as **tandem switch**.

tandem processors Two or more processors utilized or connected together. Used, for instance, to provide fault-tolerance, or for multiprocessing.

tandem switch A central office switch utilized mainly to connect trunks, as opposed to subscriber lines. Also called **tandem office switch**.

tangent **1.** For a right triangle, the ratio of the length of the side opposite to an acute angle, to the length of the side adjacent to said acute angle. It is the same as the ratio of the sine to the cosine. Its abbreviation is **tan**. **2.** A line, curve, or plane which touches another curve or surface at a single point, but does not intersect it.

tangent galvanometer A galvanometer whose readings are proportional to the tangent of the angle of deflection of a needle from its position when no current is present.

tangential mode The reflection of sound waves off four of the six surfaces of a room.

tank **1.** Same as **tank circuit**. **2.** An acoustic delay line, such as a mercury delay line, which uses a recirculating fluid as the transmission medium. **3.** A capsule or container which holds a liquid. For example, that holding a pool of mercury in a mercury switch. **4.** A large container or reservoir for storing or holding fluids.

tank circuit Also called **tank (1)**. **1.** A resonant circuit in which a capacitor and an inductor are connected in parallel with an AC source. Resonance occurs at or near the maximum impedance of the circuit. Also called **antiresonance circuit**, or **parallel-resonant circuit**. **2.** A circuit, such as a **tank circuit (1)**, which serves to store energy. **3.** A **tank circuit (2)** which stores energy over a band of frequencies distributed around the resonant frequency.

tantalum A blue-gray hard metallic chemical element whose atomic number is 73. It has over 30 known isotopes, of which one is stable. It is more corrosion-resistant than platinum, it is very ductile, and can be drawn into very thin wires. Its applications in electronics include its use in semiconductors, lasers, analytical instruments, capacitors, vacuum tubes, and in alloys which provide very high corrosion resistance, ductility, and strength. Its chemical symbol is **Ta**.

tantalum capacitor An electrolytic capacitor in which the anode is composed of a tantalum foil, bar, or slug. Also called **tantalum electrolytic capacitor**.

tantalum carbide A crystalline solid nearly as hard as diamond, and whose chemical formula is TaC. It is used in electrodes, and objects requiring extreme resistance to corrosion and high strength, such as certain cutting tools.

tantalum chip capacitor A tantalum capacitor constructed in chip form. These are very small, and are used, for instance, in hybrid ICs.

tantalum electrolytic capacitor Same as **tantalum capacitor**.

tantalum oxide A white crystalline solid whose chemical formula is Ta_2O_5. It is used in lasers, semiconductors, optics, capacitors, and as a dielectric. Also called **tantalum pentoxide**.

tantalum pentoxide Same as **tantalum oxide**.

tap **1.** A connection made to a component, circuit, bus, or the like, at a point other than an end. Also, to make such a connection. **2.** A **tap (1)** utilized to connect a node to a network. **3.** A device that cuts internal threads, such as those of a screw. Also, to make such cuts. **4.** A concealed device which serves to surreptitiously intercept a communication, especially a telephonic conversation. Also called **wiretap (1)**. **5.** To install a **tap (4)**, or make such a connection. Also called **wiretap (2)**. **6.** To monitor and/or record a communication using a **tap (4)**. Also called **wiretap (3)**.

TAP Same as **TAP protocol**. Abbreviation of **Telelocator Alphanumeric Protocol**.

tap changer A device which is utilized to adjust the voltage ratio of a transformer.

TAP protocol A protocol utilized to transmit messages to alphanumeric pages. Its abbreviation is **TAP**.

tap switch A multi-contact switch utilized to connect or adjust one or more **taps (1)**.

tape **1.** A narrow and flexible strip of a material such as plastic, paper, metal, or cloth. For example, electrical tape, magnetic tape, or fish tape. **2.** A stripe or band of a material utilized for **tape recording**. Such a tape usually consists of a strip of plastic, paper, or metal which is coated or impregnated with a magnetizable material such as iron oxide particles. Such a tape may be utilized to make analog or digital recordings of data, audio, and/or video. Also called **magnetic tape**, or **recording tape (1)**. **3.** A length, reel, cassette, cartridge, or the like, of **tape (2)**. **4.** A **tape (2)** or **tape (3)** which has been recorded upon. **5.** The process or act of recording upon a **tape (2)**.

tape-and-reel packaging A format for packaging, transporting, storing, and placing components and devices. The desired components and devices, such as capacitors or chips, are securely adhered to a tape which is wound upon a reel, providing a simple and protective manner of packaging, transporting, and storing. The reels can then be utilized with special equipment which provides for automatic insertion or placement of the parts so held. Its abbreviation is **tape & reel packaging**.

tape archive Its acronym is **tar**. **1.** A utility that is used to store or archive multiple files together. Tar does not compress files, and may also be used for mediums other than tapes. **2.** A command for creating files using **tar (1)**. **3.** To create a combined file using **tar (1)**.

tape automated bonding An automatic method or process for bonding chips and leadframes, in which metal leads are formed onto a tape, such as that of Mylar, to be automatically bonded to pads via the proper alignment and the application of pressure and/or heat. Its acronym is **TAB**.

tape azimuth The angular relationship between the head gap and the tape path.

tape backup The use of a magnetic tape medium, such as a tape cassette, for data backup.

tape cable A cable in which the conductors are arranged along the same plane, and laminated or molded into a flat flexible ribbon. Used, for instance, to connect components in computers, LCD displays, DVD players, copiers, and so on. Also called **flat-conductor cable**, **flat cable**, or **ribbon cable**.

tape cartridge A container, which houses a length of magnetic tape, that can be plugged into the appropriate device in a manner that eliminates the need for directly handling the tape. For instance, a videotape, or a continuous-loop tape cartridge. Also called **tape magazine**.

tape cassette A flat compact case containing magnetic tape and two floating reels, one for supply, the other for take-up. It is designed for easy insertion and removal from devices intended for their use. Utilized, for instance for audio or videotapes, or for data backup. Also called **cassette (1)**.

tape core **1.** The hub or center around which a **tape (1)** or **tape (2)** is wound. **2.** Same as **tape-wound core**.

tape counter In a magnetic tape device, such as a tape deck or VCR, a counter which indicates how much tape has traveled in a given direction, and which serves, for instance, to easily locate a given portion. Also called **index counter**.

tape deck Also called **deck**. **1.** An audio system component which records and plays back magnetic tapes. Also called **tape player (1)**. **2.** The tape-transport mechanism of a **tape deck (1)**.

tape drive **1.** A mechanical device which moves a magnetic tape past the heads, for recording, reproducing, or erasing. Such a drive also fast-forwards and rewinds tapes. Also called **tape transport**, **transport**, **magnetic tape drive (1)**, or **drive (4)**. **2.** In computers, a **tape drive (1)** used for reading or writing data. Also, a peripheral which incorporates such a drive. Also called **tape unit**, or **magnetic tape drive (2)**.

tape dump To display, print, copy, or transfer the content of a tape cartridge, cassette, or the like, with little or no formatting.

tape eraser A device, such as a bulk eraser, that serves to delete material recorded on a **tape (2)**.

tape head A device which reads, records, or erases signals on a magnetic tape.

tape label An item, name, image, or symbol which serves to identify a tape and/or its contents. For instance, a name or code which is assigned to a computer file contained on a tape, or an adhesive paper which describes the content of the tape it is adhered to.

tape library A collection of magnetic tapes which are utilized to store data. Such tapes are usually contained in cartridges which are accessed for storage and retrieval through the use of one or more magnetic tape drives. A tape library may be used, for instance, by a server which utilizes it for archiving. Also called **magnetic tape library**.

tape loop A tape or reel whose ends have been spliced together, so as to provide uninterrupted play. Also called **loop (12)**.

tape magazine Same as **tape cartridge**.

tape mark **1.** A control character or code which indicates the end of a magnetic tape. It may also be in the form of a reflective segment. Used, for instance, to signal a change to another tape. Also called **end-of-tape mark**. **2.** A control character or code which indicates a subdivision in a magnetic tape file. Also known as **control mark**.

tape player **1.** Same as **tape deck (1)**. **2.** An audio system component which plays back magnetic tapes.

tape recorder **1.** A device utilized for **tape recording**. **2.** A **tape recorder (1)**, such as an audio system component, which also reproduces the content of tapes which have been recorded upon.

tape recording The utilization of a variable magnetic field to store a variable electrical signal on a **tape (2)**. For example, sound and/or images which wish to be recorded may be picked up by transducers such as microphones and/or cameras, where the audio and/or visual information is converted into an electrical signal. This signal is fed to a magnetic head, which then converts it into a magnetic signal which is recorded or stored on a suitable medium such as tape with an iron oxide coating.

tape skew The circumstance in which a moving magnetic tape is not properly aligned with the recording and/or playback heads. This can result in lowered S/N ratios, signal leakage between channels, and uneven head wear, among other problems.

tape speed The rate at which a **tape (2)** moves past a head. For example, with multiple other factors held constant, the faster an audio tape moves past the head, the better the performance in areas such as frequency response, headroom, and reduced wow and flutter.

tape splice To join two lengths of tape by their ends. Also, the place where such a joint is made.

tape splicer That which serves to effect **tape splices**.

tape-to-head contact The physical contact a tape head makes with a magnetic tape being recorded or reproduced. The proper contact helps optimize recording or reproduction, and helps protect the tape from physical damage such as wrinkling. Also called **head-to-tape contact**.

tape transmitter A device which transmits in a code determined by the perforations in a punched paper tape.

tape transport Same as **tape drive (1)**.

tape unit Same as **tape drive (2)**.

tape width The width of a **tape (2)**. A typical VHS tape, for instance, has a width of approximately 12.65 mm, while a digital audio tape might have a width of about 3.81 mm.

tape-wound core A magnetic core with a ferromagnetic tape which is wound in a manner that each turn is directly aligned with all previous turns. Also called **tape core (2)**.

tape & reel packaging Abbreviation of **tape-and-reel packaging**.

taper 1. A gradual reduction in the thickness, width, or diameter of an object or section. Also, the overall reduction in thickness, width, or diameter, per unit length. 2. A gradual reduction in intensity or amplitude. Also, the rate of such a reduction per unit time, length, or the like. 3. A uniform rate of change in a property which is proportional to a change in a mechanical position or setting such as rotation or length. For example, the changes in resistance in a potentiometer as a sliding tap is moved, or a change in the cross-section of a waveguide.

tapered waveguide A waveguide whose cross-section **tapers (3)**.

TAPI Acronym for Telephony Application Programming Interface. A programming interface which facilitates the interaction of computers and voice services, especially telephony.

tapped coil A coil, such as that of a variable inductor, whose inductance is varied utilizing one or more **taps (1)**. Also called **tapped inductor**.

tapped component A component, such as a tapped coil or a tapped resistor, incorporating one or more **taps (1)**.

tapped device A device, such as a tapped transformer, incorporating one or more **taps (1)**.

tapped inductor Same as **tapped coil**.

tapped resistor A resistor, such as a rheostat, whose resistance is varied utilizing one or more **taps (1)**.

tapped transformer A transformer, such as a variable transformer, which incorporates one or more **taps (1)**.

tapped winding A winding, such as that of an autotransformer, which incorporates one or more **taps (1)**.

tar Acronym for **tape archive**.

Targa A format utilized for bitmap files.

target 1. Anything which is aimed at or which represents the objective of something. For example, a scanned object, a target electrode, particles subjected to bombardment, a target value in a control system, or a target language. 2. An object which reflects a sufficient amount of the energy beam emitted by a radar to produce a radar echo. Also called **radar target**, or **scanned object**. 3. In a TV camera tube, an electrode, surface, or other structure upon which the image is stored, and which is scanned by an electron beam to produce the output signal. 4. A particle, group of particles, surface, object, material, or the like, which is subjected to impacts by high-energy particles, such as electrons, neutrons, or alpha particles. 5. In an X-ray tube, the electrode which is bombarded, and from which X-rays are emitted. 6. In computers, the file, disk, system, or the like, towards which something is transferred or otherwise sent.

target acquisition 1. The process required for the detection of **targets (2)**. Also, the moment when such a target is detected. 2. A specific instance of **target acquisition (1)**.

target computer Also called **target machine, object computer**, or **destination computer**. 1. A computer to which a transmission is sent. 2. A computer into which a program is loaded, or to which data is transferred. 3. A computer to which compiled, assembled, or translated source code is sent.

target cross section The proportion of the overall area of a scanned object which reflects the same amount of a radar signal that would be reflected by the entire area of said object. Also called **radar cross section**, or **echo area**.

target directory A directory to which data, files, or folders are sent. The directory of origin is the **source directory**. Also called **destination directory**.

target discrimination Also called **target resolution, range resolution, radar resolution**, or **distance resolution**. 1. The minimum distance between two **targets (2)** that may be distinguished by a radar set. 2. The minimum distance between two **targets (2)** that provides two separate and recognizable indications on a radar screen.

target disk A disk to which data, files, or folders are sent. The disk of origin is the **source disk**. Also called **destination disk**.

target drive A disk drive containing a disk to which data, files, or folders are sent. The drive of origin is the **source drive**. Also called **destination drive**.

target electrode An electrode, such as an anode of an X-ray tube, which is subject to particle bombardment.

target file A file to which data is sent. The directory of origin is the **source file (1)**. Also called **destination file**.

target glint In radars, rapid fluctuations in the apparent location of a scanned object. Such variations oscillate around the mean position of said object. Also called **target scintillation, scintillation (5)**, or **wander (2)**.

target language The low-level language, such as machine language, into which a source language is converted by an assembler, compiler, or interpreter. The source language is usually a high-level language. Also called **object language (1)**.

target machine Same as **target computer**.

target program Also called **object program**. 1. The machine-language program output of a compiler or assembler. 2. A program in a form which can be directly executed by a given computer.

target resolution Same as **target discrimination**.

target scintillation Same as **target glint**.

target signal In radars, the energy reflected from a scanned object. Also called **echo signal**.

target surface 1. A surface, such as that of a target electrode, which is subject to particle bombardment. 2. A surface of a **target (2)** from which a wave reflects back to a radar.

tariff A system or structure which establishes how services, such as those provided by telephone or power companies, are charged.

task 1. A piece of work, or a step or process within a given piece of work. 2. In a multitasking environment, an independently running program or subprogram.

task bar Same as **taskbar**.

task environment The location or area, along with its conditions, within which a robot performs any given tasks. This may be a lab, a building, an open terrain, and so on.

task management Within a multitasking environment, the part of an operating system that monitors, controls, and allocates resources to the **tasks (2)** running.

task swapping The act of changing from one **task** (2) to another by copying or saving the data of one program, then loading another program.

task switching The act of changing from one **task** (2) to another. For instance, the switching from a word-processing program to a Web browser, with both programs running before and after the switch. Also called **context switching**.

taskbar A toolbar which displays the applications which are currently running, in addition to providing access to other applications which may have a small icon present. A user simply clicks with the mouse, or its equivalent, to make selections. There may also be other information present, such as the time and date, status of a modem, and so on. A taskbar may be adjusted in size, and is usually located at the bottom of a computer screen, but can be placed elsewhere. Also spelled **task bar**.

taut-band meter A meter, such as certain galvanometers, whose moving element is suspended and kept taut. For instance, such a meter may incorporate a coil whose movements or twisting are correlated to its output readings.

Tb 1. Chemical symbol for **terbium**. 2. Abbreviation of **terabit**.

TB Abbreviation of **terabyte**.

Tbit Abbreviation of **terabit**.

Tbps Abbreviation of **terabits per second**.

TBps Abbreviation of **terabytes per second**.

Tbyte Abbreviation of **terabyte**.

Tc Chemical symbol for **technetium**.

TC Abbreviation of **temperature coefficient**.

TCG Abbreviation of **time-code generator**.

TCM Abbreviation of **Trellis-Coded Modulation**.

TCO Abbreviation of **total cost of ownership**.

TCP Abbreviation of Transmission Control Protocol. Within the **TCP/IP** protocol, the protocol which handles the task of delivering the data. TCP keeps track of the packets that a given message is divided into, to ensure proper routing, and that the entire message is properly reassembled. TCP works at the transport-layer within the OSI model. Also called **TCP protocol**.

TCP/IP Abbreviation of Transmission Control Protocol over Internet Protocol. A set of protocols which enable different types of computer systems to communicate via different types of computer networks. **TCP** works at the transport-layer within the OSI model, while **IP** does so at the network layer. The TCP/IP suite may include other protocols, such as UDP, or RTSP. It is currently the most widely used protocol for delivery of data over networks, including the Internet. Also called **TCP/IP protocol**.

TCP/IP protocol Same as **TCP/IP**.

TCP/IP stack A group of protocols working together within **TCP/IP**. Such a stack, in addition to TCP and IP protocols, may include DHCP, UDP, FTP, RTSP, and so on.

TCP protocol Same as **TCP**.

TCS 1. Abbreviation of **trusted computer system**. 2. Abbreviation of **transmission convergence sublayer**.

TCXO 1. Abbreviation of **temperature-compensated crystal oscillator**. 2. Abbreviation of **temperature-controlled crystal oscillator**.

TD-SCDMA Abbreviation of time-division synchronous code division multiple access. A technology combining TDMA and CDMA in a 3G technology.

TDD 1. Abbreviation of **Telecommunications Device for the Deaf**. 2. Abbreviation of **time-division duplexing**. 3. Abbreviation of **time-division duplex**.

TDES Abbreviation of **Triple DES**.

TDM 1. Abbreviation of **time-division multiplexing**. 2. Abbreviation of **time-division multiplex**. 3. Abbreviation of **time-division multiplexer**.

TDMA Abbreviation of **time-division multiple access**.

TDR 1. Abbreviation of **time-domain reflectometry**. 2. Abbreviation of **time-domain reflectometer**. 3. Abbreviation of **time-delay relay**.

TDS 1. Abbreviation of **time-division switching**. 2. Abbreviation of **time-delay spectrometry**.

Te Chemical symbol for **tellurium**.

TE Abbreviation of **terminal equipment**.

TE mode Abbreviation of **transverse electric mode**.

TE wave Abbreviation of **transverse electric wave**.

TEA laser Abbreviation of **transversely-excited atmosphere laser**.

teach box A device which is utilized to program mechanical motions into computers, such as robot controllers. Used, for instance, for subsequent task execution by programmable robots.

teach pendant A handheld control panel utilized to guide a robot through the steps required to complete a task. It is connected to the robot by means of a suspended cable. All motions are recorded by the robot for future reference.

teamware Software designed to facilitate groups of people, often in different locations, to work together on one more projects. Such software includes emailing, scheduling, file transferring, application sharing, conferencing, the use of a whiteboard, and so on. Also called **groupware**, or **workgroup software**.

tear down In communications, to end a physical or logical connection, such as that utilized for a telephone call or network access. To **setup** (3) is to establish such a connection.

tear-off menu In a GUI, a menu which can be moved, and possibly resized. Used, for instance, for optimal positioning of tools when switching frequently between a tool palette and an image being edited. Also spelled **tearoff menu**.

tearing In TV, a form of distortion in which groups of horizontal lines are displaced in an irregular manner, resembling the ripping of a fabric. It is caused by improper horizontal sync.

tearoff menu Same as **tear-off menu**.

tebi- A binary prefix meaning 2^{40}, or 1,099,511,627,776. For example, a tebibyte is equal to 2^{40}, or 1,099,511,627,776 bytes. This prefix is utilized to refer to only binary quantities, such bits and bytes Its abbreviation is **Ti-**.

tebibyte 2^{40}, or 1,099,511,627,776 bytes. Its abbreviation is **TiB**.

tech support Abbreviation of **technical support**.

technetium A silver-gray radioactive metal whose atomic number is 43. It has close to 30 known isotopes, all unstable, and is used in nuclear medicine, tracer studies, lasers, and superconductors. Its chemical symbol is **Tc**.

technical support The technical help a provider of hardware, software, network services, or the like, is supposed to provide to customers. Its abbreviation is **tech support**.

technobabble Terminology which is specialized to a given field, such as computers or communications, and which is usually not understood by those not sufficiently familiar with said field. Such terms are often utilized improperly to confuse, intimidate, marginalize, or otherwise make a person uncomfortable.

technological Pertaining to, incorporating, affected by, or arising from **technology**.

technology 1. The application of science, engineering, and other areas of expertise to the creation and improvement of that which is utilized for commerce and/or industry. 2. A specific technological method, technique, use, or approach utilized in manufacturing, software development, communications, analysis, miniaturization, and so on. For example, CMOS technology, object-oriented technology, electrotechnology, fiber-optic technology, fingerprint recognition technology, or nanotechnology.

technology push Technology which is created based on the ideas and/or capabilities of the developing entity, as opposed to **demand pull**, which is based on customer needs.

Technology Without An Interesting Name Same as **TWAIN**.

technophobia An excessive fear of technology, or a fear whose degree is unwarranted. Frequently, however, users may be labeled as technophobic when in fact they resist using badly designed operating systems and interfaces which can justifiably lead to a great deal of frustration and apprehension.

tee Same as **T (6)**.

tee adapter Same as **T junction**.

tee antenna Same as **T antenna**.

tee circuit Same as **T network**.

tee connection Same as **T network**.

tee junction Same as **T junction**.

tee network Same as **T network**.

Teflon A chemically-resistant, waterproof, and heat-tolerant fluorocarbon-based resin with an extremely low coefficient of friction. It has a self-lubricating effect which makes it ideally suited for use in bearings and joints. Its other applications include its use for electrical and thermal insulation. Also known as **tetrafluoroethylene, tetrafluoroethylene resin**, or **polytetrafluoroethylene**.

tel 1. Abbreviation of **telephone**. 2. Abbreviation of **telegram**. 3. Abbreviation of **telegraph**.

tel- Same as **tele-**.

telco Abbreviation of **telephone company**.

tele- A prefix used in words pertaining to distance, or from a distance. Examples include words such as telephone, television, telecommunication, and telecontrol. Also spelled **tel-**.

tele-immersion The sharing of a common virtual environment by multiple participants in a teleconference. Used, for instance, by scientists at multiple locations working together while handling objects in a virtual laboratory.

tele-teaching Same as **teleteaching**.

telecamera Same as **TV camera**. Abbreviation of **television camera**.

telecast Same as **TV broadcast**. Abbreviation of **television broadcast**.

telecasting Same as **TV broadcasting**. Abbreviation of **television broadcasting**.

telecine A device which is utilized to transfer film, such as slides or that of motion pictures, to video. It incorporates a film or slide projector, a multiplexer, and a TV camera. Used, for instance, to televise a motion picture on film. Also called **film chain**.

telecine camera A camera used in a **telecine**.

telecom Abbreviation of **telecommunications**.

telecommunication Also called **telecommunications (2)**, or **communication**. 1. The transmission of information between two or more points or entities. This includes the equipment, modes, mechanisms, and media used for this purpose. 2. The information conveyed in **telecommunication (1)**. 3. The use of electrical, electromagnetic, optical, or acoustic means to transmit information between two or more points. Also, the conveyed information.

telecommunication informatics Same as **telematics**.

telecommunications Its abbreviation is **telecom**. 1. The science dealing with the use of electrical, electromagnetic,

optical, or acoustic means to transmit information between two or more points. In addition to the modes, mechanisms, and media used for this purpose, the term also encompasses all efforts to advance this field. 2. Same as **telecommunication**.

telecommunications common carrier A telecommunications company, such as a telephone company, that provides services to the general public. Such entities are usually regulated by the appropriate authorities, such as governmental agencies. Also called **carrier (3)**, **commercial carrier**, **public carrier**, or **common carrier**.

telecommunications company A company or other entity which offers **telecommunication (1)** services. Also called **communications company**.

Telecommunications Device for the Deaf A device or piece of equipment that facilitates telephonic communications for those with reduced hearing and/or speech abilities. Such a device usually has a keyboard for inputting messages, and a display for viewing incoming messages. Its abbreviation is **TDD**. Also known by an alternate term, **Telecommunications Device for the Hearing Impaired**, which reflects a bit more empathy for those with special needs. Also called **text telephone**, or **teletypewriter (2)**.

Telecommunications Device for the Hearing Impaired An alternate, and more empathic term for **Telecommunications Device for the Deaf**.

Telecommunications Industry Association An organization that represents manufacturers and suppliers, and which establishes and promotes telecommunications standards. Its abbreviation is **TIA**.

Telecommunications Management Network An ITU framework pertaining to the compatibility of diverse devices and operating systems utilized across different communications networks. Its abbreviation is **TMN**.

telecommunications network A system of computers, transmission channels, and related resources which are interconnected to exchange information. A communications network may be comparatively small, in which case it can be a LAN, or relatively large, in which case it could be a WAN. A LAN may be confined, for instance, to a single building, while a WAN may cover an entire country. The communications channels in a network may be temporary or permanent. Also called **network (1)**, or **communications network**.

telecommunications satellite An artificial satellite which relays, and usually amplifies, radio-wave signals between terrestrial communications stations, terrestrial communications stations and other communications satellites, or between other communications satellites. Such satellites provide high-capacity communications links, offer a very wide coverage area, and may transmit TV, telephone, and data signals, among many others. Also called **communications satellite**, **radio relay satellite**, **relay satellite**, or **repeater satellite**.

telecommunications security The measures taken to help prevent unauthorized access, interception, or alteration of communicated information, such as that in data transmissions. Passwords and encryption are two common precautions. Also called **communications security**.

telecommunications system The equipment, modes, mechanisms, and media used to transmit data, voice, video, or other forms of information from one point to another. Also called **communications system**.

telecommute To work at one location, such as a home, using a computer or similar device to communicate with and access the resources of another location, such as an office.

telecommuter A person who **telecommutes**. Also called **teleworker**.

telecomputing The control of one or more computer systems by another, which is at a different location. Each of these computers has remote-access software installed, and one or all computers may have control abilities. All computers usually have the same information displayed at any given time. Also called **remote computing**.

teleconferencing A conference in which participants at more than one site can interact in real-time. Such a conference may incorporate audio, video, written chat, shared files and applications, and so on. Also called **real-time conferencing**.

telecontrol Also called **remote control**. **1.** The operating, regulating, or managing of a component, circuit, device, apparatus, piece of equipment, system, function, quantity, or the like, from a location other than where that being controlled is located. To effect remote control, radio, infrared, or acoustic waves may be employed, or connecting wires or cables may be used, among others. When remote control is wireless, it is called **radio control**. **2.** A device which serves for **telecontrol** (1).

telecopy Same as **telefax** (4).

telecopying The use of **telefax** (4).

teledetection The sensing of a signal or other phenomenon utilizing a detector, such as an infrared detector, which does not come into physical contact with that being measured or monitored. Also called **remote sensing**.

telefacsimile Same as **telefax**.

telefax Abbreviation of **telefacsimile**. Also called **fax**, or **facsimile**. **1.** A method of transmitting a printed page between locations, via a telecommunications system. A telephone line is usually used. The document is scanned at the sending location, encoded, and transmitted by the sending device. The receiving device decodes the signal, then prints a copy of the original document, which may include text, graphics, and so on. **2.** A device or piece of equipment utilized to send and/or receive a **telefax** (1). It incorporates a scanner, modem, and printer. **3.** One or more printed pages sent or received via **telefax** (1). **4.** To send or receive a **telefax** (3). Also called **telecopy**.

telegram A message sent via **telegraph** (1). Its abbreviation is **tel**. Also called **telegraph** (3).

telegraph Its abbreviation is **tel**. **1.** An obsolescent communication system in which a signal code, such as Morse code, is utilized to send messages to remote locations. Such a system usually uses connected wires, but may alternatively utilize radio waves for transmission. Also called **telegraph system**. **2.** To transmit a message using a **telegraph** (1). **3.** Same as **telegram**.

telegraph channel The path the transmitted signal follows when transmitting along a **telegraph circuit**.

telegraph circuit A communications link via which **telegrams** are sent. Such a circuit usually uses connected wires, but may alternatively utilize radio waves for transmission.

telegraph code A code, such as Morse code, utilized for communications via a **telegraph system**.

telegraph key A key which is pressed and released to transmit the code utilized in a **telegraph system**.

telegraph system Same as **telegraph** (1).

telegraphy **1.** Communications via **telegraph**. **2.** A system for communications via **telegraph**.

Telelocator Alphanumeric Protocol Same as **TAP protocol**.

telemaintenance Maintenance which is activated and/or controlled via telecommunications. The site where the maintenance is performed must have the necessary devices and equipment to properly perform any required tasks. Utilized, for instance, for installations where physical access is impracticable, such as satellites. Also called **remote maintenance**.

telemarketing Selling or promoting something over the telephone, usually without securing prior authorization from the recipient of the call.

telematics Acronym for **tele**communication infor**matics**. The use of computers along with telecommunications systems. Examples include the utilization of PDAs to send and receive email, finding a location using a GPS receiver in an automobile, or the use of such a system to keep track of a subway train system.

telemedicine The use of telepresence, videoconferencing, or similar communications systems to assist in, collaborate with, and/or perform medical procedures such as diagnosis, therapy, or surgery.

telemeter **1.** A device, instrument, or apparatus utilized for **telemetry** (1). **2.** To utilize a **telemeter** (1), and to transmit readings so obtained.

telemetering Same as **telemetry** (1).

telemetry **1.** The automatic gathering of information from one location while transmitting it to a remote location. To effect telemetry, radio, infrared, or acoustic waves may be employed, or connecting wires or cables may be used, among other means. When telemetry is wireless, it is called **radio telemetry**. Also called **telemetering**, or **remote metering**. **2.** The science dealing with the devices and modes of transmission utilized in **telemetery**(1).

telemetry receiver A device to which information gathered via **telemetry** is sent.

telemetry system A system, including a transmitter and receiver, utilized for **telemetry**.

telemetry transmitter A device which emits or otherwise sends signals detected or otherwise received by a **telemetry receiver**.

teleoperation The direct and continuous control of a **teleoperator**.

teleoperator The human operator of a **telepresence system**.

telephone Its abbreviation is **tel**, or **phone**. **1.** A device which is utilized to send and receive signals via **telephony** (1). Such a set may consist, for instance, of a telephone handset, a switchhook, and the appropriate circuits and devices required for signaling, switching, and so on. Also called **telephone set**. Telephone may also refer to a device, apparatus, or system which incorporated these functions, such as a computer or a fax. **2.** Same as **telephony** (2).

telephone amplifier Its abbreviation is **phone amplifier**. **1.** A small amplifier that increases the volume heard through a telephone receiver. **2.** Same as **telephone repeater**.

telephone answering device Same as **telephone answering machine**. Its abbreviation is **TAD**.

telephone answering machine A device which automatically answers calls, plays a message, and records messages left by callers. Such a device may be incorporated into a telephone, into a similar device, such as a properly equipped fax or computer, and so on. Its abbreviation is **phone answering machine**, or **TAM**. Also called **telephone answering device**, or **answering machine**.

telephone central office Same as **telephone exchange**.

telephone channel A communications channel utilized for **telephony** (1).

telephone circuit A communications circuit utilized for **telephony** (1).

telephone communications Same as **telephony** (1). Its abbreviation is **phone communications**.

telephone company A company or other entity which offers **telephone services**. Its abbreviation is **telco**.

telephone connector **1.** Same as **telephone plug**. **2.** Same as **telephone jack**.

telephone dial A device on a telephone which generates signals which are utilized to place calls. Also called **dial** (3).

telephone dialer A device which automatically dials telephone numbers. Such numbers may be manually or automatically programmed into memory, and may be dialed when pressing a short code, or at a pre-programmed time. Its abbreviation is **phone dialer**. Also called **automatic dialer**, or **dialer**.

telephone exchange A structure which houses one or more telephone switching systems. At this location, customer lines terminate and are interconnected with each other, in addition to being connected to trunks, which may also terminate there. A typical central office handles about 10,000 subscribers, each with the same area code plus first three digits of the 10 digit telephone numbers. Its abbreviation is **CO**. Also called **telephone central office**, **central office**, **exchange** (1), or **local central office**.

telephone feature Its abbreviation is **phone feature**. **1.** A feature, such as speakerphone, mute, or flash, which is available using a **telephone set**. **2.** A feature, such as caller ID, conference calling, or call waiting, which is available using a **telephone service**.

telephone handset A part of a telephone incorporating a microphone and an earphone, and which is designed to be held by the hand while talking and listening. A handset frequently has a dial for placing calls. Its abbreviation is **phone handset**. Also called **telephone receiver** (2), or **handset**.

telephone jack A connector, such as that in a wall, utilized to connect to a telephone system. A **telephone plug** is inserted into a telephone jack. Its abbreviation is **phone jack** (2). Also called **telephone socket**, or **telephone connector** (2).

telephone line A physical medium, such as a wire or cable, which serves to transmit telephone signals between points. Its abbreviation is **phone line**.

telephone link A communications line or channel connecting two points or entities, via which telephone signals may be exchanged. Such a link may use wires or cables, or may be wireless. Its abbreviation is **phone link**.

telephone modem A device which enables a computer to transmit and receive information over telephone lines. For transmission, a computer modulates a carrier signal so that it contains digital information in a form that can be transmitted over analog lines. It does the opposite for reception. Most modems feature data compression and error-correction. Also called **modem** (1).

telephone network A communications network based on **telephony** (1). Its abbreviation is **phone network**.

telephone network interface The interface, such as a network interface device, between a telephone network and the equipment of a customer.

telephone number A number which identifies a given telephone station or subscriber. Its abbreviation is **phone number**.

telephone pickup Its abbreviation is **phone pickup**. **1.** A microphone which enables the monitoring and/or recording of telephone conversations. Used, for instance, for multiple people in a room to listen to a call, or to surreptitiously monitor and/or record a conversation. **2.** A device incorporating a **telephone pickup** (1).

telephone plug A connector, such as an RJ-11 connector, which is utilized to plug a telephone into a **telephone jack**. Its abbreviation is **phone plug**. Also called **telephone connector** (1).

telephone receiver Its abbreviation is **phone receiver**. **1.** A small loudspeaker located in the handset of a telephone, which enables listening. Also called **receiver** (4), or **earphone** (3). **2.** Same as **telephone handset**.

telephone repeater A device, location, or system which receives, amplifies, and transmits telephone signals between points or locations. Its abbreviation is **phone repeater**. Also called **telephone amplifier** (2).

telephone service A communications service based on **telephony** (1). Its abbreviation is **phone service**.

telephone set Same as **telephone** (1). Its abbreviation is **phone set**.

telephone sidetone The reproduction by the receiver of sounds from the same telephone's transmitter. Although a caller's own voice is usually heard over the same telephone, if the amplitude of this signal exceeds a given threshold it is usually unwanted. Also called **sidetone** (1).

telephone socket Same as **telephone jack**. Its abbreviation is **phone socket**.

telephone station A location at which one or more **telephones** (1) are linked to a telephone system. Its abbreviation is **phone station**.

telephone subscriber A user or other entity which has been authorized to use or receive one or more telephone services. Its abbreviation is **phone subscriber**.

telephone switchboard A piece of equipment that serves to manually or automatically connect telephone extensions, lines and/or trunks. Examples include private branch exchanges, and telephone exchanges. Also called **switchboard** (1).

telephone system A communications system based on **telephony** (1). Examples include POTS, cellular phone systems, cable phone systems, and satellite phone systems. Its abbreviation is **phone system**.

telephone transmitter A microphone located in the handset of a telephone, which converts sound energy into electrical signals. Its abbreviation is **phone transmitter**, or **transmitter** (3).

telephony **1.** Communication via a system which converts sounds into electrical signals which are transmitted, received at one or more remote locations, and converted back into sounds. Telephony may be utilized to transmit speech or data, and signals may be sent over wires, optical fibers, or wirelessly, as is the case with cellular and satellite phone systems. At each end of a connection there is a telephone, or a device, such as a computer or PDA, which incorporates these functions. Internet telephony, for instance, uses the Internet to link computer-to-computer, computer-to-telephone, or telephone-to-telephone calls. Also called **telephone communications**. **2.** A communications service, system, or technology based on **telephony** (1). **3.** The branch of science dealing with **telephony** (1).

Telephony API Same as **TAPI**.

Telephony Application Programming Interface Same as **TAPI**.

telephony server A server which integrates communications networks, such as the Internet, and voice services, especially telephony.

Telephony Server API Same as **TSAPI**.

Telephony Server Application Programming Interface Same as **TSAPI**.

Telephony Services API Same as **TSAPI**.

Telephony Services Application Programming Interface Same as **TSAPI**.

telephoto **1.** The sending of photos over a fax system. **2.** The sending of photos over a telecommunications system. **3.** Same as **telephoto lens**.

telephoto lens A lens whose longer focal length and narrower field of view makes objects and scenes appear larger than with a normal lens. Used, for instance, to magnify distant objects. Also called **telephoto** (3).

telepointer In applications such as shared whiteboards, a pointer controlled by a remote user, and which is seen on the monitor of the local user.

telepresence 1. A system which enables a person in one location to perceive conditions, such as those of the surrounding environment, of a remote location, and perform actions at said remote location. Telepresence provides the person, or teleoperator, the necessary sensory input, including that required for visual, auditory, and tactile perception, and incorporates the essential communications and mechanical systems required to be able to manipulate objects and otherwise interact with that present at the remote site. Used, for instance, to perform surgery from a remote location, or to control a robot at another location. Also called **telepresence system**. 2. The feedback received by a **teleoperator**. 3. The actions performed by a **teleoperator** at the remote location.

telepresence system Same as **telepresence (1)**.

teleprinter 1. A terminal which can print signals received from remote locations, but which does not have a keyboard for transmission. 2. Same as **teletypewriter (1)**.

teleprocessing monitor Same as **TP monitor**.

teleprogramming Also called **remote programming**. 1. Programming by a device or system located at a location other than that which is being programmed. 2. Programming utilizing a remote control.

teleran A navigational aid in which images of ground-based radar systems are televised to aircraft, for viewing by the piloting crew. It is an abbreviation of **tele**vision **ra**dar **nav**igation.

telescoping antenna An antenna, usually consisting of a vertical rod, that extends by pulling on its end. Used, for instance, in portable devices.

Telescript A communications-oriented programming language that seeks to allow messaging across most platforms and operating systems, and which utilizes intelligent agents. Also, any technology utilized to develop such a language, or a program created using this programming language.

teleteaching Teaching in which one or more of the following are true: the instructor or instructors are geographically separated from the student or students, one or more of the students is separated from the rest of the students, or one or more learning resources is available from a remote location. Such teaching may incorporate computers, TV, satellite communications, and the Internet. Also spelled **tele teaching**. Also called **distance learning**.

teletext A service provided by TV broadcasters, in which text and simple graphics are presented on the TV screen, employing an otherwise unused portion of the transmitted signal. A PAL or SECAM TV may come equipped with this feature, or a decoder box may be utilized. Used, for instance, to provide programming information, news, or the like.

teletype Same as **teletypewriter (1)**. Its abbreviation is **TTY**.

teletypewriter Its abbreviation is **TTY**. 1. A typewriter or terminal that transmits and receives messages sent to and received from remote locations, usually using coded signals sent, for instance, via telephone circuits. Also called **teletype**, or **teleprinter (2)**. 2. Same as **Telecommunications Device for the Deaf**.

teleview To monitor or otherwise observe utilizing a remote viewing system such as a TV or CCTV.

televise 1. To broadcast via a TV system. 2. To pick up a scene and convert it into a TV signal.

television 1. Same as **TV (1)**. 2. Same as **TV (2)**. 3. Same as **TV (3)**. 4. Same as **TV (4)**.

television band Same as **TV band**.

television board Same as **TV card**.

television broadcast Same as **TV broadcast**.

television broadcast band Same as **TV broadcast band**.

television broadcast channel Same as **TV broadcast channel**.

television broadcast receiver Same as **TV broadcast receiver**.

television broadcast reception Same as **TV broadcast reception**.

television broadcast service Same as **TV broadcast service**.

television broadcast station Same as **TV broadcast station**.

television broadcast transmission Same as **TV broadcast transmission**.

television broadcast transmitter Same as **TV broadcast transmitter**.

television broadcaster Same as **TV broadcaster**.

television broadcasting Same as **TV broadcasting**.

television broadcasting channel Same as **TV broadcasting channel**.

television broadcasting service Same as **TV broadcasting service**.

television broadcasting station Same as **TV station**.

television camera Same as **TV camera**.

television camera tube Same as **TV camera tube**.

television card Same as **TV card**.

television channel Same as **TV station**.

television decoder Same as **TV decoder**.

television decoder box Same as **TV decoder box**.

television interference Same as **TV interference**.

television monitor Same as **TV (2)**.

television program Same as **TV program**.

television programming Same as **TV programming**.

television projector Same as **TV projector**.

television receiver Same as **TV (2)**.

television repeater Same as **TV repeater**.

television screen Same as **TV screen**.

television set Same as **TV (2)**.

television signal Same as **TV signal**.

television standard Same as **TV standard**.

television station Same as **TV station**.

television system Same as **TV (1)**.

television transmitter Same as **TV transmitter**.

television tuner Same as **TV tuner**.

televisor A TV transmitter or broadcaster.

telework The work performed by a **telecommuter**.

teleworker Same as **telecommuter**.

telex 1. An obsolescent communications service using devices such as teletypewriters to exchange messages using telephone lines. 2. A device utilized to send or receive messages via **telex (1)**. Also called **telex terminal**. 3. To send messages via **telex (1)**. 4. A message sent or received via **telex (1)**.

telex terminal Same as **telex (2)**.

telluric current A current flowing through the earth. It may be due for instance, to the earth's magnetic field, or solar activity. Also called **earth current (3)**, or **ground current (3)**.

tellurium A lustrous silvery-white semi-metallic chemical element whose atomic number is 52. It is a semiconductor which exhibits increased conduction with exposure to light, and its conductivity varies depending on its crystal alignment. It has about 35 know isotopes, of which 5 are stable. Its applications include its use in semiconductors, lasers, vulcanization, and batteries. Its chemical symbol is **Te**.

Telnet 1. A protocol that enables a user to log-on to, and enter commands on a remote computer as if at the remote site. Authorization must be granted by the remote, or host, computer before commands may be entered. This contrasts, for example, with FTP, which is utilized to access and download files without logging-on to the remote computer. Also called **Telnet protocol. 2.** A utility or program using **Telnet** (1) to log-on to and enter commands on a host computer.

Telnet protocol Same as **Telnet** (1).

TEM 1. Abbreviation of **transmission electron microscope. 2.** Abbreviation of **transmission electron microscopy**

TEM mode Abbreviation of **transverse electromagnetic mode.**

TEM wave Abbreviation of **transverse electromagnetic wave.**

temp Abbreviation of **temperature.**

temp directory Abbreviation of **temporary directory.**

temp file Abbreviation of **temporary file.**

temp folder Abbreviation of **temporary folder.**

temper To subject a material to sustained heating followed by cooling, each at suitable rates, to change its properties. This process can be performed, for instance, to realign the of atoms in a crystal, or to stabilize certain electrical properties.

temperature Its symbol is **T**, and its abbreviation is **temp. 1.** A quantitative measure of the hotness or coldness of a body or region. Temperature may be expressed using a standard scale, such as the thermodynamic temperature scale, the Kelvin temperature scale, or the Celsius temperature scale. Absolute zero equals 0 K, or approximately -273.15 °C. Heat flows from a body or region which has a higher temperature to that with a lower temperature. **2.** A measure of the average kinetic energy of the particles in a body or region. The higher the temperature, the greater the kinetic energy.

temperature coefficient 1. A numerical value that indicates the relationship between a change in a physical quantity, phenomena, or parameter, such as resistance, current, expansion, reactivity, or accuracy, as a function of temperature. A negative temperature coefficient indicates a decrease in a magnitude or property as temperature is raised, while a positive temperature coefficient indicates an increase in a magnitude or property as temperature is raised. Its abbreviation is **TC. 2.** Same as **temperature coefficient of resistance.**

temperature coefficient of capacitance A numerical value that indicates the relationship between a change in the capacitance of a component, device, material, or area, as a function of temperature. Also called **capacitance temperature coefficient.**

temperature coefficient of frequency A numerical value that indicates the relationship between a change in the frequency of a material, component, or device, as a function of temperature. Also called **frequency temperature coefficient.**

temperature coefficient of resistance A change in the resistance of a component, device, or material as a function of temperature. A negative temperature coefficient of resistance indicates a decrease in resistance as temperature is raised, while a positive temperature coefficient of resistance indicates an increase in resistance as temperature is raised. Most conductors have a positive temperature coefficient of resistance, while most semiconductors and insulators have a negative temperature coefficient of resistance. Also called **temperature coefficient** (2), or **resistance temperature coefficient.**

temperature-compensated crystal oscillator A crystal oscillator whose temperature is automatically maintained within a very narrow interval. This provides enhanced accuracy, reduced power consumption, and eliminates warm-up time. Its abbreviation is **TCXO.** Also called **temperature-controlled crystal oscillator.**

temperature compensation 1. The use of devices or mediums to stabilize the temperature surrounding a component, circuit, device, piece of equipment, system, or material. Utilized, for instance, to minimize sources of error resulting from temperature variations. **2.** A specific allowance made to account for temperature variations beyond stated limits.

temperature control 1. A control utilized to adjust the temperature of a component, circuit, device, piece of equipment, system, or material, or its surrounding area. **2.** Any adjustment made to a temperature, such as that of a given location. **3. Temperature control** (1) or **temperature control** (2) which is performed automatically.

temperature-controlled crystal oscillator Same as **temperature-compensated crystal oscillator.** Its abbreviation is **TCXO.**

temperature-controlled environment An enclosure, such as a room, in which measures are taken to provide an environment that automatically maintains a specified temperature level. These enclosures may also provide a stable humidity level, or protect against dust, and so on. Such environments may be used, for instance, to protect sensitive electronic equipment.

temperature derating The reduction of the rating of a device, such as its current rating, to provide an additional safety and/or reliability margin when operating under unusual or extreme temperature conditions.

temperature error Any error resulting from temperature variations. Temperature compensation helps minimize such error.

temperature gradient The rate of change of temperature as a function of a given variable, especially distance.

temperature interval Same as **temperature range.**

temperature inversion Also called **thermal inversion. 1.** Within the atmosphere, an increase in temperature along with an increase in height. This is opposite the usual pattern. **2.** Any inversion of the usual decrease or increase of a property, as a function of temperature.

temperature meter Same as **thermometer.**

temperature range A specific temperature interval, such as an operating temperature range. Also called **temperature interval.**

temperature saturation For a thermionic electron tube at a given anode voltage, the condition in which the anodic current can not be further increased with increases in cathode temperature. This is due to a space charge near the cathode. Also called **filament saturation,** or **saturation** (7).

temperature scale 1. An established standard for stating temperature values. An absolute temperature scale, such as the Kelvin scale, is based on absolute zero, so its zero degree reading equals absolute zero. Another scale is the Celsius scale, which is based on the freezing and boiling points of water under certain conditions. **2.** The scale on a thermometer which enables reading temperature values expressed in degrees. Also called **thermometer scale.**

temperature-sensitive resistor Same as **thermistor.**

temperature sensor In a device such as a thermometer or thermostat, a probe or other part or section which detects temperature.

temperature shock Same as **thermal shock.**

tempered glass Glass which is subjected to sustained heating followed by cooling, each at suitable rates, to enhance its properties. Such glass flexes more, and resists impact, shock, thermal, and mechanical stresses. If broken, tempered glass fragments into comparatively harmless rounded pellets.

template **1.** A pattern or design usually cut out of a metal or plastic plate or sheet, which is used as a guide for shaping, drawing, carving, or the like. Such patterns or designs may also be available or otherwise prepared via computer applications. Used, for instance, to help in preparing circuit diagrams or flowcharts. **2.** An overlay usually made of plastic or cardboard which is placed over a computer keyboard, or a part of one, which labels or describes the functions of keys within a given application. **3.** A predefined layout which serves as the basis for a document, such as a monthly report prepared utilizing a word-processing application, or that composing a Web page. A style sheet specifies parameters such as page size, margins, headers, footers, fonts, hyperlinks, and so on. Also called **style sheet**.

temporary cookie A cookie that is deleted or otherwise disappears when a Web browser is closed, or after a comparatively short time. This contrasts with a **persistent cookie**, which is kept indefinitely or until deleted. Also called **transient cookie**, or **session cookie**.

temporary directory A directory which holds **temporary files**. Its abbreviation is **temp directory**.

temporary file A file that is used only when needed, after which time it is deleted. For example, such files created during a session in which a given application is in use. Temporary files may be located in memory and/or a disk. Its abbreviation is **temp file**.

temporary folder A folder which holds **temporary files**. Its abbreviation is **temp folder**.

temporary magnet A magnet, such as that made of soft iron, which retains its magnetism for brief periods, as opposed to a **permanent magnet**, which can remain magnetized indefinitely. Temporary magnets are usually characterized by low reluctance and low retentivity.

temporary storage A segment of computer memory or storage which is utilized to temporarily store information that awaits transfer or processing. A buffer is an example. Also called **interim storage**.

ten-turn potentiometer A potentiometer whose moving contact can travel approximately ten times around its shaft or axis. Its abbreviation is **10-turn potentiometer**.

ten's complement Same as **tens complement**.

TENS Acronym for **transcutaneous electrical nerve stimulation**.

tens complement The complement when using the decimal number system. For instance, the complement of six in the decimal system is four: $(10 - 6) = 4$. It is obtained by subtracting a number from the radix, which in the decimal system is 10. Its abbreviation is **10's complement**.

tension **1.** The potential difference or electromotive force between two points, such as those in a circuit, expressed in volts, or in multiples of volts, such as kilovolts or millivolts. Also called **voltage**. **2.** The condition of being stretched or pulled, often tightly. Also, the process or act of so stretching or pulling. **3.** A force creating and/or maintaining a **tension (2)**.

tera- A metric prefix representing 10^{12}, which is equal to a trillion. For instance, 1 terahertz is equal to 10^{12} Hz, or 1,000,000,000,000 Hz. When referring to binary quantities, such as bits and bytes, it is equal to 2^{40}, or 1,099,511,627,776, although this is frequently rounded to a trillion. To avoid any confusion, the prefix **tebi-** may be used when dealing with binary quantities. Its abbreviation is **T**.

tera-electronvolt A unit of energy or work equal to 10^{12} electronvolts. Also spelled **teraelectronvolt**. Its abbreviation is **TeV**.

terabecquerel A unit of activity equal to 10^{12} becquerels. Its abbreviation is **TBq**.

terabit 2^{40}, or 1,099,511,627,776 bits, although this is frequently rounded to a trillion bits. To avoid any confusion, the term **tebibit** may be used when referring to this concept. Its abbreviation is **Tb**, or **Tbit**.

terabits per second 2^{40}, or 1,099,511,627,776 bits, per second. Usually used as a measure of data-transfer speed. Its abbreviation is **Tbps**. To avoid any confusion, the term **tebibits per second** may be used when referring to this concept.

terabyte 2^{40}, or 1,099,511,627,776 bytes, although this is frequently rounded to a trillion bytes. To avoid any confusion, the term **tebibyte** may be used when referring to this concept. Its abbreviation is **TB**, or **Tbyte**.

terabytes per second 2^{40}, or 1,099,511,627,776 bytes, per second. Usually used as a measure of data-transfer speed. Its abbreviation is **TBps**. To avoid any confusion, the term **tebibytes per second** may be used when referring to this concept.

teracycle A unit of frequency equal to 10^{12} cycles, or 10^{12} Hz. The term currently used for this concept is **terahertz**.

teraelectronvolt Same as **tera-electronvolt**.

teraflops Same as **TFLOPS**.

teragram A unit of mass equal to 10^{12} grams. Its abbreviation is **Tg**.

terahertz A unit of frequency equal to 10^{12} Hz. Its abbreviation is **THz**. Also called **teracycle**.

terajoule A unit of energy or work equal to 10^{12} joules. Its abbreviation is **TJ**.

terameter A unit of distance equal to 10^{12} meters. Its abbreviation is **Tm**.

teraohm A unit of resistance, impedance, or reactance equal to 10^{12} ohms. Its abbreviation is **TΩ**.

teraohmmeter An ohmmeter designed to provide accurate indications of resistance values in the teraohm range.

terapascal A unit of pressure equal to 10^{12} pascals. Its abbreviation is **TPa**.

teravolt A unit of potential difference equal to 10^{12} volts. Its abbreviation is **TV (5)**.

terawatt A unit of power equal to 10^{12} watts. Its abbreviation is **TW**.

terawatt-hour A unit of energy equal to 10^{12} watt-hours. Its abbreviation is **TWh**, or **TWhr**.

terbium A soft silver-gray metallic chemical element whose atomic number is 65. It is ductile, malleable, and is a rare-earth element. It has nearly 30 known isotopes, of which one is stable. It is used, for instance, in phosphors, lasers, and as a dopant. Its chemical symbol is **Tb**.

terminal **1.** A computer input/output device which incorporates a video adapter, monitor, keyboard, and usually a mouse. Used in networks. Also called **computer terminal (1)**, **network terminal (1)**, **console (3)**, or **station (6)**. When such a terminal has no processing capability it is called **dumb terminal**, while a terminal that incorporates a CPU and memory does have processing capability, and is called **intelligent terminal**. **2.** A personal computer or workstation which is linked to a network. Also called **computer terminal (2)**, **network terminal (2)**, or **station (7)**. **3.** A device, such as a wire or adapter, which serves to make electrical, network, or other types of connections. Also, the specific point or location where this occurs. **4.** A device, such as a post or a strand of metal, which protrudes from an electrical component to facilitate connection within a circuit. **5.** That which is at an end, or consists of the end of something, such as that of a series, or of a transmission line.

Terminal Access Controller Access Control System Same as **TACACS**.

terminal adapter Abbreviation of ISDN **terminal adapter**. A device which interfaces an ISDN line with a user's equipment, such as a computer. Such an adapter may be connected, for example, to a serial port. Its own abbreviation is **TA**.

terminal area On a printed-circuit board, the printed conductive portion to which components are connected. Said especially of such portions which run along the edge of said board. Also called **terminal pad, pad (2)**, or **land (4)**.

terminal block An insulated block equipped with multiple terminals, which enables the mounting and interconnection of circuits. Used, for instance, to connect multiple telephone pairs.

terminal board An insulated board equipped with multiple terminals, which enables the interconnection of wires or cables. Such terminals may be, for instance, in the form of binding posts or lugs. Also called **terminal strip**.

terminal emulation The ability to make one computer terminal appear as another type of terminal to another. Terminal emulation is used, for instance, to enable a user with a graphical-user interface to log-on to and enter commands into a text-based system such as an older mainframe. In this manner, such a terminal appears to the accessed mainframe as another mainframe.

terminal equipment The equipment at the end of a communications line or channel. For example, ISDN equipment located at the premises of a customer. Its abbreviation is **TE**.

terminal ID Abbreviation of **terminal id**entification. The number or sequence of characters which identifies a specific network terminal, or a component such as a card. Examples include Ethernet addresses and those of network nodes.

terminal identification Same as **terminal ID**.

terminal impedance For a circuit, device, or transmission line, the impedance as seen from the input or output terminals.

terminal lug A fitting or projection which serves to connect a wire by wrapping or through soldering. Also called **lug (1)**.

terminal node In a hierarchical structure, such as a tree, a node which has no descendants. In a hierarchical file system, for example, a file is a leaf. Also called **leaf (1)**, or **leaf node**.

terminal pad Same as **terminal area**.

terminal pair A pair of terminals which are linked. For example, the positive and negative terminals of a speaker.

terminal radar approach control The use of radar equipment as part of an air traffic control service, as is usually the case. Its acronym is **TRACON**. Also called **radar approach control**.

terminal server A computer, controller, or other hardware device which provides terminals or nodes with a common connection point to a network such as a LAN or WAN. A terminal server enables the connection of multiple computer terminals, printers, or the like, to one or more host computers.

terminal session A session during which a terminal, such as a node or a mainframe, is in use.

terminal strip Same as **terminal board**.

terminal voltage The voltage at the output terminals of a component, circuit, or device. An example is a no-load voltage.

terminate 1. To end a transmission line with a **termination**. **2.** To end a program, routine, process, or the like.

terminate and stay resident Same as **TSR**.

terminated line A transmission line with a **termination**, so as to avoid reflections and standing waves. Also called **terminated transmission line**.

terminated transmission line Same as **terminated line**.

termination The point at which a **transmission line (1)** ends. Also, a device within which such a termination occurs. For example, a power outlet, or a channel service unit. Also called or **transmission line termination**, or **line termination**.

terminator 1. A character or code which indicates the end of a string. **2.** A hardware device or component which serves as the last device when multiple devices or nodes are connected in series, so as to avoid signals from reflecting back into the line. Used, for instance, as a SCSI termination, or as a transmission line termination.

Terms Of Service Norms regulating the usage of computer services, especially those of networks such as the Internet. The providers of the service establish what activities are and are not acceptable when using said service. A common example is an Internet service provider prohibiting sending unsolicited email or using an account for certain commercial purposes. Its abbreviation is **TOS**. Also called **acceptable use policy**.

ternary 1. Consisting of, incorporating, or appearing in, groups of three. **2.** Having three possible values, states, or the like. **3.** Having a base of three. Said of a number system.

ternary alloy An alloy consisting of three elements.

terrain-clearance indicator An instrument which measures and indicates altitude above a given terrain by the utilization of radio, radar, laser, sonic, or capacitive technology. It is a type of absolute altimeter.

terrain clutter In radars, unwanted echoes that appear on the display screen due to signal reflections from the surface of the earth. Also called **terrain return, terrain echoes, surface clutter, land return**, or **ground clutter**.

terrain echoes Same as **terrain clutter**.

terrain return Same as **terrain clutter**.

terrestrial magnetism The magnetism of the planet earth. Also called **geomagnetism**.

Terrestrial Microwave A system, method, technology, or service, such as Multichannel Multipoint Distribution Service, which utilizes microwave line-of-sight communications between sending and receiving units located on the ground or on towers, as opposed to a sender and/or receiver antenna being located on a communications satellite. Used, for instance, for telephone, TV, and/or data services. Also called **Terrestrial Microwave Radio**.

Terrestrial Microwave Radio Same as **Terrestrial Microwave**.

tertiary coil Same as **tertiary winding**.

tertiary winding A third winding on a transformer. Also called **tertiary coil**.

tesla The SI unit of magnetic flux density. It is defined as the field strength generating one newton of force per ampere per meter of a conductor. One tesla is equal to one weber per square meter, or 10^4 gauss. Its symbol is **T**.

Tesla coil An air-core transformer which develops a high voltage at high frequencies. In a typical configuration, the primary coil consists of a few turns of large-diameter wire, while the secondary has many turns of a small-diameter wire. A properly selected capacitor is charged to a voltage of several thousand volts, then discharged through the primary, with the process being repeated hundred of times with the assistance of a spark gap. The secondary inductively absorbs energy from the primary, magnifying the voltage further, to levels of hundreds of thousands or millions of volts. Also called **Tesla transformer**.

Tesla transformer Same as **Tesla coil**.

tessellation In computer-assisted modeling, the representation of 3D objects by creating a mosaic formed by small triangles, squares, or the like.

test **1.** One or more procedures utilized to critically ascertain, observe, examine, or otherwise evaluate. Tests may or may not be carried out under controlled conditions, such as those afforded by a laboratory, and may be qualitative or quantitative in nature. Tests may be performed on components, circuits, devices, pieces of equipment, systems, processes, materials, hardware, software and the like, and there are myriad examples, including life, overvoltage, shake, continuity, hardness, pulse, smoke, and stress tests. Tests may be classified as destructive versus non-destructive, static as opposed to dynamic, parametric versus non-parametric, and so on. **2.** To subject to, or perform a **test** (1).

test bed Same as **testbed**.

test board Same as **testboard**.

test card Same as **test pattern**.

test chamber An enclosed section, compartment, space, or room, such as a shielded room or an environmental chamber, utilized for **testing**. Also called **testing chamber**, **test room**, or **test enclosure**.

test chart Same as **test pattern**.

test circuit A circuit, such as an oscillator circuit, utilized for **testing**. Also called **testing circuit**.

test clip A clip, such as an alligator clip, utilized to make temporary electric connections during **testing**.

test data Data specifically chosen or utilized to perform software and/or hardware tests. Such data, for instance, may be used to test a program being developed, or to verify the status of a node within a communications network.

test device A device, such as an artificial antenna or a burst generator, utilized for **testing**. Also called **testing device**.

test enclosure Same as **test chamber**.

test equipment Equipment, such as oscilloscopes or automatic test equipment, utilized for **testing**. Also called **testing equipment**.

test instrument An instrument, such as an interferometer or logic analyzer, utilized for **testing**. Also called **testing instrument**.

test lab Abbreviation of **test laboratory**.

test laboratory A laboratory which provides the equipment and conditions necessary for **testing**. Such a laboratory may be utilized to simulate most any given surroundings. Its abbreviation is **test lab**. Also called **testing laboratory**.

test lead A lead, such as that at the end of a test probe, utilized for **testing**. Also called **testing lead**.

test pattern Also called **test chart**, or **test card**. **1.** A chart consisting of special patterns with lines, geometric figures, and colors, which serve to test and adjust TV receivers. Such an image may be transmitted, for instance, by a TV broadcast station when there is otherwise no programming. **2.** A pattern, such as a Lissajous figure, which serves for testing.

test points Locations on components, along circuits, or on devices, equipment, or systems, utilized for **testing**. Also called **testing points**.

test probe A usually slender object or device which is utilized to detect, sample, or measure a phenomenon, by being placed in proximity or in direct contact with an object, material, or area. A probe may consist of an electrode, lead, rod, optic fiber, or the like. Probes can be used to test and quantify any number of phenomena, including voltage, current, resistance, magnetic flux strength, temperature, pressure, electron density in a plasma, logic levels within a logic circuit, and so on. Also, to sample with or otherwise use such a probe. Also called **testing probe**, or **probe** (1).

test prod A point, tip, prong, or terminal, such as that of a probe, which facilitates testing or otherwise monitoring. Also called **testing prod**, or **prod**.

test program A computer program, such a utility, employed for **testing**. Also called **testing program**.

test room Same as **test chamber**.

test set A combination of instruments, devices, and equipment, such as those located in a laboratory, utilized for **testing**. Also called **testing set**.

test signal A signal, such as an artificial echo or that of a standard signal generator, utilized for **testing**. Also called **testing signal**.

test-signal generator A component, circuit, device, instrument, piece of equipment, or system, that generates **test signals**.

test software Software, such as BIOS, utilized for **testing**. Also called **testing software**.

test tone A tone, such as that sent through a transmission system to evaluate losses, utilized for **testing**.

testbed A location or area where tests are performed. Such a site usually has the proper devices, instruments, equipment, systems, and software required for the desired testing. Also spelled **test bed**.

testboard A board, panel, or the like, which has the necessary components, devices, instruments, equipment, cables, and terminals necessary to perform electrical tests. Also spelled **test board**. Also called **testing board**.

tester An apparatus or human used for testing.

testing The carrying out of **tests**.

testing board Same as **test board**.

testing chamber Same as **test chamber**.

testing circuit Same as **test circuit**.

testing device Same as **test device**.

testing enclosure Same as **test chamber**.

testing equipment Same as **test equipment**.

testing instrument Same as **test instrument**.

testing lab Same as **test laboratory**.

testing laboratory Same as **test laboratory**.

testing lead Same as **test lead**.

testing points Same as **test points**.

testing probe Same as **test probe**.

testing prod Same as **test prod**.

testing program Same as **test program**.

testing room Same as **test chamber**.

testing set Same as **test set**.

testing signal Same as **test signal**.

testing software Same as **test software**.

tethered robot A robot which is controlled by wiring which is physically linked to it by a tethered cable. A robot linked to a teach pendant is an example.

tetrafluoroethylene Same as **Teflon**. Its abbreviation is **TFE**.

tetrafluoroethylene resin Same as **Teflon**.

tetravalent Pertaining to an element, such as carbon or silicon, which has four valence electrons. Also called **quadrivalent**.

tetrode An electron tube with four electrodes. Such a tube usually has an anode, a cathode, a control electrode, and a grid.

tetrode junction transistor A junction transistor incorporating two base terminals, with the second base connection serving as a fourth electrode. Also known as **tetrode transistor**, or **double-base junction transistor**.

tetrode transistor Same as **tetrode junction transistor**.

TeV Abbreviation of **tera-electronvolt**.

tevatron A synchrotron which can produce particle energies in the tera-electronvolt range. Such a device, for instance, can achieve energies in excess of 1 TeV, with collision energies near 2 TeV.

text 1. Data in the form of alphanumeric characters, as opposed for instance, to audio or graphics. 2. Data in the form of words, numbers, or the like, which are intended to convey meaningful information to users, as opposed to for instance, to control codes.

text-based browser Same as **text-based Web browser**.

text-based Web browser A Web browser based on text, as opposed to graphics. Useful, for instance, for persons with reduced vision. A popular text-based browser is Lynx. Also called **text-based browser**.

text box A box, such as dialog box, into which text is entered. Also called **text-entry box**.

text editing The use of a **text editor**.

text editor A computer program which creates text files, or modifies existing text files. A word processor is a more powerful editor program with greater flexibility than most editors. Also called **editor**.

text entry 1. The entering of text, such as that into a computer. Such information may be entered manually or automatically. 2. A specific **text entry** (1), such as a name or number.

text-entry box Same as **text box**.

text field A field into which only text may be entered. This contrasts, for instance, with a number field.

text file A file, such as an ASCII file, which contains only text, as opposed for instance, to a graphics file. Also called **text-only file**.

text message A message consisting of text, or which incorporates simple graphics. An example is an SMS message.

text messaging The sending of **text messages**.

text mode A computer screen mode in which alphanumeric characters can be displayed, but not graphics. Also called **character mode**.

text-only file Same as **text file**.

text telephone Same as **Telecommunications Device for the Deaf**. Its abbreviation is **TT**, or **TTY**.

text-to-speech The conversion of a text into a speech output. Also called **text-to-speech conversion**, **text-to-voice**, or **text-to-voice conversion**.

text-to-speech conversion Same as **text-to-speech**.

text-to-voice Same as **text-to-speech**.

text-to-voice conversion Same as **text-to-speech**.

texture In computer graphics, shading, graininess, reflections, and other attributes which are incorporated into an image to more authentically convey the appearance and feel of a real surface. Also, to add such texture.

texture map A two-dimensional representation of the texture of a given surface, such as that of wood, marble, cloth, or the like. Used in the context of **texture mapping**.

texture mapping In computer graphics, the adding of **texture** to surfaces of three-dimensional images or objects. This is usually accomplished by wrapping a texture map around the desired object. The result is a three-dimensional object whose surfaces have a two-dimensional texture. It would be the equivalent, for instance, of covering a wooden cube with a marble veneer.

texture sensing The ability of a robot end-effector to determine the extent to which a surface is smooth or rough, usually using a laser directed at the surface in question.

TFE Same as **Teflon**. Abbreviation of **tetrafluoroethylene**.

Tflops Same as **TFLOPS**.

TFLOPS Abbreviation of **teraflops**, which itself is an abbreviation of **trillion floating-point operations per second**. 10^{12} FLOPS, which is the same as 10^{12} floating-point calculations or operations, per second. Usually used as a measure of processor speed.

TFT Abbreviation of **thin-film transistor**.

TFT LCD Abbreviation of **thin-film transistor liquid-crystal display**.

TFT LCD display Same as **thin-film transistor liquid-crystal display**.

TFT liquid-crystal display Abbreviation of **thin-film transistor liquid-crystal display**.

TFTP Abbreviation of **Trivial File Transfer Protocol**.

Tg Abbreviation of **teragram**.

Th Chemical symbol for **thorium**.

thallium A lustrous silver-gray metal whose atomic number is 81. It is soft, malleable, when exposed to dry air quickly forms a heavy oxide coating, and when in moist air or water rapidly forms a hydroxide. It has about 35 known isotopes, of which 2 are stable. It is used, for instance, in semiconductors, lasers, cryogenic switches, and optics. Its chemical symbol is **Tl**.

THD Abbreviation of **total harmonic distortion**.

theory 1. A set of statements, concepts, principles, and/or rules whose intention is to explain and/or predict phenomena using logical and scientific arguments. Examples include quantum theory, the Bardeen-Cooper-Schrieffer theory, and the Grand Unified Theory. 2. A branch of science in which a **theory** (1) is applied, as opposed to practice. Examples include circuit theory, and information theory.

thermal Of, pertaining to, generating, or arising from **thermal energy**.

thermal absorption The absorption of heat by a body or system from its surroundings, as occurs in an endothermic process. Also called **heat absorption**.

thermal aging The aging of a substance, material, component, device, piece of equipment, or system, by subjecting to comparatively elevated temperatures over time. Also called **heat aging**.

thermal agitation 1. The random movements of electrons in a conductor, due to the effects of heat. Such movements generate thermal noise. Also called **thermal effect** (1). 2. The random movements of charge carriers in a semiconductor, due to the effects of heat. Such movements generate noise. Also called **thermal effect** (2).

thermal ammeter An ammeter which incorporates a stretched wire whose length at any given moment depends on the current passing through it. This wire expands when heated and contracts when cooled, and its variations in length deflect an attached pointer which is correlated to a calibrated scale. Such an ammeter serves to measure both direct and alternating currents. Also called **hot-wire ammeter**.

thermal anemometer An anemometer which measures and indicates wind or air flow speeds by the cooling effect said flow has on a stretched electrically-heated wire. As the intensity of the airflow varies, so does the resistance of the wire. These changes are then correlated to a calibrated scale. For proper readings, the probe must be at a right angle to the flow. Also called **hot-wire anemometer**.

thermal battery A battery composed of **thermal cells**.

thermal capacity The amount of heat necessary, under specified conditions, to raise the temperature of a body or system one degree Celsius. Also called **heat capacity**.

thermal cell A cell which is activated by raising the temperature above a given threshold. Such a cell, for instance, may have an electrolyte which melts to become activated.

thermal circuit breaker A circuit breaker which automatically opens a circuit when the temperature caused by the flow of current exceeds a set amount.

thermal coil A protective device consisting of a small coil which grounds or opens a circuit when it is heated beyond a specified temperature. This serves to limit the value the current may reach, as heat is produced in the coil via the Joule effect. Also called **heat coil**.

thermal compression bonding Same as **thermocompression bonding**.

thermal conduction The flow of internal energy from a region or body of higher temperature to a region or body of lower temperature. Also called **heat conduction**.

thermal conductivity Also called **heat conductivity**. **1.** The ability of a material, body, surface, or medium to conduct heat. **2.** The rate at which **thermal conductivity (1)** occurs. Among the elements, silver is the best conductor of heat, followed by copper, then gold. The thermal conductivity of diamond, an allotropic form of carbon, is significantly higher than that of silver.

thermal cut-out Same as **thermal cutout**.

thermal cutout A cutout which disconnects a component or circuit when the operating temperature exceeds a given threshold. Its function is usually automatic. Also spelled **thermal cut-out**.

thermal cycling A type of accelerated testing in which the tested components, devices, materials, or the like, are subjected to repeated cycles of heating and cooling.

thermal detector Also called **heat detector**. **1.** A resistive element which measures electromagnetic radiation by absorbing it and converting it into heat. The increase in its temperature is used to measure the radiant energy. Also called **thermal sensor (1)**, or **bolometer**. **2.** An instrument or device, such as a **thermal detector (1)**, which senses heat. Also called **thermal sensor (2)**. **3.** An instrument or device, such as a thermocouple, which functions or is actuated when exposed to heat. **4.** An instrument or device, such as a thermometer, which quantifies heat.

thermal device A device, such as a thermocouple, which detects and/or measures heat, or which functions or is actuated when exposed to heat.

thermal dissipation Same as **thermal loss**.

thermal drift Drift caused by the heat generated by a component, circuit, device, piece of equipment, system, or material during operation or use, or by the surrounding environment.

thermal effect **1.** Same as **thermal agitation (1)**. **2.** Same as **thermal agitation (2)**. **3.** Any effect, such as activation or deterioration, due to the effects of heat.

thermal electromotive force Same as **thermal emf**.

thermal emf Abbreviation of **thermal** electromotive force. An electromotive force arising from a temperature difference at two points along a circuit, across a junction, or within an object, as observed, for instance, in the Seebeck effect.

thermal energy The non-mechanical form of energy which is transferred between bodies or regions at different temperatures which are in contact with each other. Thermal energy is transferred from a hotter body or region to a colder body or region. Such regions may also be within the same body. Thermal energy can be transferred via conduction, convection, or radiation. Also called **heat energy**, or **heat**.

thermal grease A grease-like material which is utilized to enhance the transfer of heat between two surfaces or bodies. Used, for instance, between a heat sink and a chip to fill the tiny air gaps present due to surface roughness.

thermal imager Same as **thermograph (2)**.

thermal imaging Same as **thermography**.

thermal instrument An instrument, such as a hot-wire instrument, which detects and/or measures heat, or which functions or is actuated when exposed to heat.

thermal inversion Same as **temperature inversion**.

thermal ionization Ionization caused by elevated temperatures.

thermal loss Also called **thermal dissipation**, or **heat loss**. **1.** A decrease in the amount of heat in a body or region. This may be accomplished, for instance, through the use of heat sinks or fans. Heat is transferred from a hotter body or medium to a colder body or medium. **2.** The loss of another form of energy through its conversion into heat energy. For instance, losses due the heating effect. Also called **dissipation (1)**.

thermal management The mechanisms employed to keep the temperature of a battery or cell within a given range during charging and discharging.

thermal meter A meter, such as a thermometer, which detects and/or measures heat, or which functions or is actuated when exposed to heat.

thermal noise Broadband noise in a conductor, component, or circuit, due to the thermal agitation of electrons. Such noise increases proportionally to the absolute temperature of the material in question. Also called **Johnson noise**.

thermal photograph A photograph taken utilizing **thermography**.

thermal photography Same as **thermography**.

thermal printer A printer which forms letters and/or images by the selective application of heat to a heat-sensitive paper. Used, for instance, in some faxes or electrocardiographs.

thermal protection The protection afforded by a **thermal protector**.

thermal protector A device, such as a thermal circuit breaker or thermal cutout, which protects a component, circuit, device, piece of equipment, or system, from excessive heat. Used, for instance, to shut down a power supply before it overheats.

thermal radiation The process by which heat is radiated by matter. This energy is in the form of electromagnetic waves, and is a result of the temperature of said matter. While such waves may encompass the entire electromagnetic spectrum, a high proportion of this radiation occurs in the infrared region. Also, the heat so radiated. Also called **heat radiation**, or **radiant heat**.

thermal radiator Also called **heat radiator**. **1.** Matter which emits **thermal radiation**. **2.** An apparatus, such as a heat sink, which serves to radiate heat.

thermal relay Also called **heat relay**. **1.** A relay which is actuated as a consequence of the heat produced by a current passing through it. **2.** A relay which is actuated by a given temperature or temperature variation.

thermal resistance Also called **heat resistance**. **1.** The extent to which a material or body can withstand the application of heat. **2.** The extent to which a material or body prevents heat from flowing through it.

thermal resistor Same as **thermistor**.

thermal runaway A condition in which heat generated by a component, circuit, or device causes additional increases in generated heat, which leads to greater heat levels. This is usually due to the heating effect, and if protective measures are not taken, usually results in the failure or destruction of the component, circuit, or device. Also called **runaway**.

thermal sensor **1.** Same as **thermal detector (1)**. **2.** Same as **thermal detector (2)**.

thermal shock Any disturbance or other change in a component, circuit, device, piece of equipment, system, material, mechanism, or process, due to a sudden and significant temperature change. Also called **temperature shock**, or **heat shock**.

thermal switch Also called **heat switch**. **1.** A switch that is actuated as a consequence of the heat produced by a current passing through it. **2.** A switch that is actuated by a given temperature or temperature variation.

thermal wax printer A non-impact printer that uses heat to melt a wax which is transferred from a ribbon onto an appropriate medium such as paper. Such a wax may be colored or black. Also called **thermal wax-transfer printer**.

thermal wax-transfer printer Same as **thermal wax printer**.

thermally-sensitive resistor Same as **thermistor**.

thermion A particle, such as an electron or ion, emitted by a body heated to an elevated temperature. For example, electrons emitted by a cathode in an electron tube.

thermionic **1.** Pertaining to **thermions**. **2.** Pertaining to electrons emitted by a body at an elevated temperature.

thermionic cathode A cathode, such as that in an electron tube, that emits electrons as a consequence of its being heated. Such a cathode may be directly or indirectly heated. This contrasts with **cold cathode**, which does not require an applied heat to function. Also called **hot cathode**.

thermionic current Electric current caused by the flow of **thermions**, as occurs, for instance, in an electron tube.

thermionic diode A diode tube which incorporates a **thermionic cathode**.

thermionic emission **1.** The emission of electrons from heated objects, such as a heated electrical conductor. An example is the emission of electrons by a cathode in an electron tube. Also known as **Edison effect**, or **Richardson effect**. **2.** The emission of **thermions** from heated objects.

thermionic tube An electron tube which incorporates a **thermionic cathode**. Also called **thermionic valve**, or **hot-cathode tube**.

thermionic valve Same as **thermionic tube**.

thermionic work function The energy necessary to move an electron from the Fermi level to its surrounding medium. It is the energy required to free an electron from a surface such as a hot cathode.

thermionics The study and applications of **thermionic emissions**, and of the devices whose operation is based on such emissions.

thermistor Acronym for **therm**al res**istor**. A resistor which has either a known positive temperature coefficient or a known negative temperature coefficient. Thermistors are usually made of semiconductors, may have very significant temperature coefficients, and are generally in the form of a rod, bead, or disk. Used, for instance, in temperature measuring devices, or to compensate for temperature variations in other components. Also called **temperature-sensitive resistor**, or **thermally-sensitive resistor**.

thermistor bridge A bridge in which one or more arms are replaced by **thermistors**.

thermistor probe A probe incorporating a **thermistor**. Seen, for instance, in a thermometer probe.

thermistor thermometer A thermometer whose temperature-sensing element is a **thermistor**.

thermoammeter Same as **thermocouple ammeter**.

thermocompression bonding Bonding in which pressure, with or without the presence of heat, is utilized to join materials or surfaces, as opposed for instance, to using a current or solder. Also called **thermal compression bonding**.

thermocouple A device incorporating two dissimilar metals which are welded or otherwise joined at their ends, and in which a temperature difference at the junction generates a proportional electromotive force. The converse is also true,

that is, a temperature change is produced at said junction when a current passes through the device. The former phenomenon is explained by the Seebeck effect, and the latter by the Peltier effect. Thermocouples typically use semiconductor metals, such as bismuth and tellurium, and are utilized, for instance, as thermometers, or for the conversion of heat energy into electric energy.

thermocouple ammeter An ammeter incorporating a **thermocouple**. Its abbreviation is **thermoammeter**.

thermocouple bridge A bridge in which one or more arms are replaced by **thermocouples**.

thermocouple converter A **thermoelectric generator** utilizing a thermocouple to convert heat energy into electric energy.

thermocouple instrument An instrument, such as a thermocouple ammeter, which incorporates one or more **thermocouples**.

thermocouple junction The metal-to-metal junction in a **thermocouple**.

thermocouple meter A meter, such as a thermocouple ammeter, which incorporates one or more **thermocouples**.

thermocouple thermometer A thermometer whose temperature-sensing element is a **thermocouple**.

thermocouple vacuum gauge A vacuum gauge that incorporates a **thermocouple**. The thermocouple detects the thermal conduction of any gas present.

thermodynamic temperature A temperature measured using the **thermodynamic temperature scale**. Expressed in kelvins. Its symbol is **T**.

thermodynamic temperature scale A temperature scale which is based on defined points which can be experimentally reproduced, such as the temperature of the triple point of water, which is approximately 273.16 K. Values are expressed in kelvins.

thermodynamics The branch of science that deals with the principles, relationships, and conversions between heat and other forms of energy. There are three fundamental laws that govern thermodynamics. The first states that the total energy in an isolated system remains constant, the second asserts that heat will not be transferred spontaneously from a colder system to a warmer system, and the third affirms that at absolute zero the entropy of a perfect crystalline substance is zero.

thermoelectric Pertaining to, characteristic of, utilizing, or arising from **thermoelectricity**.

thermoelectric converter Same as **thermoelectric generator**.

thermoelectric cooler A device which utilizes the Peltier effect for cooling. The current is made to flow in the direction that cools the desired side of the junction, while the heat on the other side is dissipated into its surroundings. Multiple such modules can be utilized for additional cooling capacity. Also called **thermoelectric module**, or **thermoelectric heat pump**.

thermoelectric cooling The use of one or mode **thermoelectric coolers** for cooling.

thermoelectric device A device, such as a thermoelectric generator, which makes use of, or operates as a result of a **thermoelectric effect**.

thermoelectric effect An effect in which heat is converted into electric energy, or vice versa. Examples include the Seebeck, Peltier, and Kelvin effects.

thermoelectric element **1.** Same as **thermoelectric generator**. **2.** A block of a semiconductor material forming part of a **thermoelectric generator**. Also called **thermoelement**.

thermoelectric generator A device, such as that incorporating a thermocouple, which converts heat energy into electric energy. Used, for instance, to convert sunlight into electricity. Also called **thermoelectric converter, thermal generator, thermoelectric power generator,** or **thermoelectric element (1).**

thermoelectric heat pump Same as **thermoelectric cooler.**

thermoelectric heater A device similar to a **thermoelectric cooler,** except that the current is made to flow in the direction that heats the desired side of the junction.

thermoelectric heating The use of one or mode **thermoelectric heaters** for heating.

thermoelectric junction Same as **thermojunction.**

thermoelectric material A material, such as bismuth telluride, which converts heat energy into electric energy.

thermoelectric module Same as **thermoelectric cooler.**

thermoelectric power generator Same as **thermoelectric generator.**

thermoelectricity The generation of electricity as a consequence of heat flow, as occurs, for instance, in a thermocouple.

thermoelectron An electron emitted by a body heated to an elevated temperature. It is a type of thermion.

thermoelement Same as **thermoelectric element (2).**

thermogram The recorded output of a **thermograph.**

thermograph 1. A **thermometer** which makes a record of the temperatures it measures. Also called **recording thermometer.** 2. A device or apparatus utilized for **thermography.** Also called **thermal imager.**

thermography Acronym for **therm**al photo**graphy.** Any method, such as infrared photography, in which infrared radiation is utilized to form images. Used, for instance, as a medical diagnostic technique. Also called **thermal imaging.**

thermojunction Abbreviation of **thermoelectric junction.** The point or surface where the dissimilar metals make contact within a thermocouple.

thermoluminescence Luminescence produced by a material upon heating, or through continued heating. Observed, for instance, in a material which has been previously exposed to ionizing radiation.

thermomagnetic effect Any phenomena in which a temperature change results in a change in magnetism, or vice versa. An example is the Ettinghausen effect.

thermometer Any instrument or device which measures and indicates **temperatures.** Examples include electronic thermometers, quartz thermometers, infrared thermometers, and bimetallic thermometers. Also called **temperature meter,** or **pyrometer (2).**

thermometer probe In certain thermometers, the portion which is utilized to detect temperatures.

thermometer scale Same as **temperature scale (2).**

thermometry 1. The measurement of **temperatures.** 2. The branch of science dealing with temperature measurement.

thermonuclear reaction A reaction at extremely high temperatures in which lighter atomic nuclei are fused to form heavier nuclei. Enormous amounts of energy are released during such reactions, and plasmas may be observed.

thermopile A device consisting of two or more thermocouples connected in series or in parallel. When connected in series, a thermopile provides an output with a higher voltage, and when in parallel, serve for the conversion of heat energy into electric energy.

thermoplastic 1. A material, such as polystyrene or polyethylene, which becomes softer when heated, and harder when cooled. A thermoplastic material may be able to undergo such cycles indefinitely without damage. Also called **thermoplastic material.** 2. Becoming softer when heated, and harder when cooled.

thermoplastic material Same as **thermoplastic (1).**

thermoplastic resin A resin, such as polymethylmethacrylate, which is a **thermoplastic (1).**

thermoregulator A component, circuit, or device, such as a thermostat, which helps keep an enclosure, medium, material, or the like, within specified temperature limits.

thermosetting Becoming permanently solid or hardened when heated.

thermosetting material A material, such as bakelite or epoxy resin, which upon heating becomes permanently solid or hardened.

thermosetting resin A resin, such as bakelite or epoxy resin, which upon heating becomes permanently solid or hardened.

thermosphere The outermost layer of the earth's atmosphere, lying between the mesosphere and outer space. Temperatures rise with increased altitude in this atmospheric shell.

thermostat A switch that is automatically actuated when a temperature deviates by a specified amount from a target value or interval. A thermostat serves to open or close a circuit which controls a heating and/or cooling mechanism, so as to maintain a temperature within the desired limits. A thermostat often incorporates a bimetallic element which bends in a given direction to turn the circuit on and off, and must have the proper amount of hysteresis to prevent constant switching between cycles or too large a temperature interval between cycles Used, for instance, in heating, cooling, or refrigerating systems, or for overtemperature protection. Also called **thermostatic switch.**

thermostatic switch Same as **thermostat.**

Thévenin's theorem In the analysis of electric networks, a theorem stating that combinations of two-terminal AC voltage sources and impedances can be represented by a single voltage source and a series impedance. A DC source would be represented by a single voltage source and a series resistor. Also spelled **Thevenin's theorem.**

Thevenin's theorem Same as **Thévenin's theorem.**

THF Abbreviation of **tremendously high frequency.**

Thick Ethernet Same as **ThickNet.**

thick film 1. A technology utilized to apply upon a substrate a film which is ten micrometers or greater in thickness, although the defined thickness varies. Thick-film technology is usually employed to form passive circuit components and interconnections, and involves applying a liquid, solid, or paste coating through a screen or mask with the desired patterns upon a ceramic or glass substrate, followed by heating. Used, for instance, in the formation of hybrid ICs. This contrasts with **thin film (1),** which can involve a thicknesses of less than 0.1 micrometer. Also called **thick-film technology.** 2. A film applied utilizing **thick-film technology.**

thick-film capacitor A capacitor formed upon a substrate using **thick-film technology.**

thick-film circuit A circuit incorporating, or consisting exclusively of, **thick-film components.**

thick-film component A component, such as a capacitor or resistor, formed upon a substrate using **thick-film technology.**

thick-film material A material which is deposited upon a substrate using **thick-film technology.**

thick-film resistor A resistor formed upon a substrate using **thick-film technology.**

thick-film technology Same as **thick film (1).**

ThickNet Abbreviation of **Thick** Ethernet. An Ethernet standard using a coaxial cable which is about one centimeter in diameter, as opposed to **ThinNet** whose cabling is about 0.5 centimeter in diameter. Also called **ThickWire,** or **10Base5.**

ThickWire Same as **ThickNet.**

thin client Within a client/server architecture, a client which performs little data processing, allowing the use of less powerful computers at the front end. This contrasts with a **fat client**, where the client performs most or all of the processing. This also contrasts with **thin server**.

Thin Ethernet Same as **ThinNet**.

thin film **1.** A technology utilized to apply upon a substrate a film which is less than ten micrometers in thickness, although the thickness may be less than 0.1 micrometer. Thin-film technology is usually employed to form active or passive circuit components and interconnections, using techniques such as sputtering or chemical-vapor deposition. Used, for instance, in the formation of ICs, to apply magnetic thin films, or in the preparation of organic semiconductors. This contrasts with **thick film (1)**, which can involve thicknesses of greater than 100 micrometers. Also called **thin-film technology**. **2.** A film applied utilizing **thin-film technology**.

thin-film capacitor A capacitor formed upon a substrate using **thin-film technology**.

thin-film circuit A circuit incorporating, or consisting exclusively of, **thin-film components**.

thin-film component A component, such as a transistor or resistor, formed upon a substrate using **thin-film technology**.

thin-film deposition The deposition of a **thin film (1)** upon a substrate, as accomplished, for instance via sputtering or chemical-vapor deposition.

thin-film heads Disk drive heads using minute coils which are deposited onto a thin film, and which convert the variations in magnetic flux of the rotating platter into data. Also called **thin-film inductive heads**.

thin-film inductive heads Same as **thin-film heads**.

thin-film material A material which is deposited upon a substrate using **thin-film technology**.

thin-film photovoltaic cell A photovoltaic cell which is formed upon a substrate using **thin-film technology**. Its abbreviation is **thin-film PV cell**.

thin-film PV cell Same as **thin-film photovoltaic cell**.

thin-film resistor A resistor formed upon a substrate using **thin-film technology**.

thin-film semiconductor A semiconductor formed upon a substrate using **thin-film technology**.

thin-film solar cell A solar cell which is formed upon a substrate using **thin-film technology**.

thin-film technology Same as **thin film (1)**.

thin-film transistor A transistor formed upon a substrate using **thin-film technology**. Used, for instance, in LCD displays. Its abbreviation is **TFT**.

thin-film transistor LCD Abbreviation of **thin-film transistor liquid-crystal display**.

thin-film transistor liquid-crystal display A liquid-crystal display in which each pixel of the screen is powered separately and continuously by a transistor. This provides for a brighter image with higher contrast than a **passive-matrix display**. Its abbreviation is **thin-film transistor LCD**. Also known as **active-matrix display**, **active-matrix thin-film transistor display**, or **active display**.

thin magnetic film A very thin film, usually less than 10^{-6} meter, of a magnetic material deposited upon a substrate. In a hard disk, for instance, such a film may cover an aluminum platter, providing a medium suitable for data storage. Also called **magnetic thin film**.

thin quad flat-pack A square surface-mount chip which provides leads on all four sides, affording a high lead count in a small area. Such chips are generally thinner than other quad flat-packs. Its abbreviation is **TQFP**. Also spelled **thin quad flatpack**. Also called **thin quad flat-package**.

thin quad flat-package Same as **thin quad flat-pack**.

thin quad flatpack Same as **thin quad flat-pack**.

thin server Within a client/server architecture, a server which performs little data processing. This contrasts with a **fat server**, where the server performs most or all of the processing. This also contrasts with **thin client**.

thin small outline package A surface-mounted chip or transistor package similar to a small-outline package, but thinner. Its abbreviation is **TSOP**.

ThinNet Abbreviation of **Thin** Ethernet. An Ethernet standard using a coaxial cable which is about 0.5 centimeter in diameter, as opposed to **ThickNet** whose cabling is about one centimeter in diameter. Also called **ThinWire**, **10Base5**, or **cheapernet**.

ThinWire Same as **ThinNet**.

thionyl chloride A yellow-red suffocating liquid whose chemical formula is $SOCl_2$. Used, for instance, as an electrolyte in lithium cells.

third-generation computer A computer utilizing ICs. This computer generation started approximately in the mid 1960s.

third-generation language A high-level computer programming language, such as Pascal, BASIC, Java, or C++. Its abbreviation is **3GL**.

third-generation wireless Developed in the late 1990s and early 2000s, provides for digital cellular communications which are more flexible and provide a much wider bandwidth than that available with **second-generation wireless**, in addition to seamless global roaming. Its abbreviation is **3G**.

third law of thermodynamics A law stating that at absolute zero, the entropy of a perfect crystalline substance is zero.

third-level cache In computer systems with first-level and second-level cache built into the CPU, a type of cache which resides in the motherboard between CPU and main memory. Thus, L3 cache is the equivalent of L2 cache in systems which have only the L1 cache built into the CPU. Also called **L3 cache**, or **level 3 cache**.

third normal form In the normalization of a relational database, the third stage utilized to convert complex data structures into simpler relations. A database must first complete the second normal form before proceeding to the third. Its abbreviation is **3NF**.

third octave band A frequency band in which the ratio of the upper frequency limit to the lower frequency limit is equal to one third of an octave. The 1000 Hz one-third octave band, for instance, comprises frequencies ranging from about 891 Hz to 1122 Hz. Also called **third octave frequency band**, or **one-third octave band**.

third octave frequency band Same as **third octave band**.

third-order filter A filter with six components, such as three inductors and three capacitors, and which provides an 18 dB rolloff. Used, for instance, as a crossover network.

Thomson bridge A special bridge circuit whose arrangement minimizes the effects of contact resistance, enabling the accurate measurement of very low resistances. Some such devices can measure resistances of a few microhms. Also called **Kelvin bridge**, **Kelvin double bridge**, or **double bridge**.

Thomson effect Also called **Kelvin effect**. **1.** When a current flows through a conductor whose ends are at different temperatures, heat is absorbed or generated, depending on the metal. **2.** When there is a temperature differential between the ends of a conductor, an electromotive force is developed, the direction of which depends on the metal.

Thomson scattering The scattering of electromagnetic radiation by charged particles, especially electrons, according to classical physics.

thoriated-tungsten filament In an electron tube, a tungsten filament with a small amount of thorium added for enhanced electron emission.

thorium A lustrous silver-white radioactive metallic element whose atomic number is 90. It is soft, ductile, and can be extruded, drawn, and welded. It has about 30 known isotopes, all unstable. It is used, for instance, as a nuclear fuel, in lamps, lasers, semiconductors, and in photoelectric devices. Its chemical symbol is **Th**.

thorium dioxide Same as **thorium oxide**.

thorium oxide A heavy white crystalline powder whose chemical formula is ThO_2. Of all known oxides, it is that with the highest melting point. It is used, for instance, in high-temperature ceramics, as a nuclear fuel, and in optics. Also called **thorium dioxide**.

thrashing 1. A state in which computer processes or tasks progress very slowly or not at all. This may occur, for instance, when an operating system is having difficulties in allocating resources for one program, and when accomplished, creates limitations in another program, which starts another such cycle. 2. **Thrashing** (1) due to excessive paging in and out of memory.

thread 1. In computers, a process within another process, usually utilizing the resources of the latter. 2. One of multiple tasks or processes within those being simultaneously executed in **threading**. 3. A chain of messages, postings, or the like, which follows a current event, movie, music genre, or any chosen topic. 4. A thin strand, such as that of a conductor.

threaded connector A connector with helical or spiral grooves or ridges, which requires screwing in to tighten. Used, for instance, at the ends of coaxial cables.

threaded discussion A **thread** (3) maintained or followed by an online forum, such as a newsgroup.

threading The simultaneous execution of two or more tasks or processes, or threads, within a single program. To occur, threading must be supported by the operating system. Also called **multithreading**.

three-channel sound Its abbreviation is **3-channel sound**. 1. Same as **three-channel sound system**. 2. The sound produced by a **three-channel sound system**.

three-channel sound system A sound system in which there are three channels, each feeding its own speaker. The channels so provided are the left, right, and center. Its abbreviation is **3-channel sound system**. Also called **three-channel sound** (1), or three-channel stereo.

three-channel stereo Same as **three-channel sound system**. Its abbreviation is **3-channel stereo**.

three-click rule A principle stating that no more than three mouse clicks should be required to access a given program feature, a desired Web site content, or the like. Although the number may vary from user to user, the requirement of more than three clicks is usually considered sufficient to result in frustration, exiting from the Web site, or the like. Its abbreviation is **3-click rule**.

three-D Abbreviation of **three-dimensional**. Its own abbreviation is **3D**.

three-dimensional Its abbreviation is **three-D**, or **3D**. 1. Of, pertaining to, existing in, or consisting of three dimensions. 2. That which has, or appears to have, the effect of depth.

three-dimensional animation Animation with height, width, and depth. The objects appearing in such animation can be rotated and viewed from different angles, providing different perspectives. Used, for instance, in virtual reality settings. Its abbreviation is **3D animation**.

three-dimensional array In the organization of data, an array whose elements are described or defined using three numbers or dimensions. For instance, a table organized into rows, columns, and layers. Its abbreviation is **3D array**.

three-dimensional audio Same as **three-dimensional sound**. Its abbreviation is **3D audio**.

three-dimensional graphics Graphics with height, width, and depth. Such graphics can be rotated and viewed from different angles, providing different perspectives. Also, hardware, software, or a system supporting such graphics. Its abbreviation is **3D graphics**.

three-dimensional modeling The creation and representation of solid models, or models with height, width, and depth. Its abbreviation is **3D modeling**.

three-dimensional sound The reproduction of sound that produces as faithfully as possible a three-dimensional sound image. Specialized coding and decoding equipment is required, along with an appropriate loudspeaker array. Also known by various other terms, including **three-dimensional audio**, **3D sound**, **positional 3D audio**, and **ambisonics**.

three-dimensional television Same as **three-dimensional TV**.

three-dimensional TV Abbreviation of **three-dimensional television**. TV in which two-dimensional images are presented so as to create the impression of depth. One such system uses multiple cameras, with the positions, angles, and the like computed taking into the manner in which the brain processes visual input, to display images which appear as three-dimensional. Its own abbreviation is **3D-TV**. Also called **stereoscopic TV**.

three-gun Consisting of, utilizing, or pertaining to, three electron guns.

three-gun cathode-ray tube Same as **three-gun CRT**.

three-gun color picture tube Same as **three-gun CRT**.

three-gun color television picture tube Same as **three-gun CRT**.

three-gun color TV picture tube Same as **three-gun CRT**.

three-gun CRT Abbreviation of **three-gun cathode-ray tube**. A CRT with three electron guns, one for each primary color, which are red, green, and blue. Its own abbreviation is **3-gun CRT**. Also called **three-gun color television picture tube**, **three-gun color TV picture tube**, or **three-gun color picture tube**.

three-input adder A logic circuit which accepts three input bits, and whose output is a sum and a carry bit. Two of the input bits are for adding, and the third is a carry bit from another digit position. Its abbreviation is **3-input adder**. Also called **full adder**.

three-layer diode Same as **trigger diode**.

three-phase Its abbreviation is **3-phase**. 1. Consisting of, utilizing, or pertaining to three AC phases. Each phase is usually $2\pi/3$, or 120° apart. 2. Consisting of, utilizing, or pertaining to three phases.

three-phase AC Alternating current with three phases. Each phase usually has a phase difference of $2\pi/3$, or 120° in relation to the other two. Its abbreviation is **3-phase AC**.

three-phase bridge rectifier A bridge rectifier which works with a three-phase AC input. Its abbreviation is **3-phase bridge rectifier**.

three-phase circuit A circuit in which there are three AC phases, as opposed, for instance, to a **single-phase circuit**, where there is one AC phase. Such a circuit, for example, may utilize three wires whose currents differ in phase by 120°. Its abbreviation is **3-phase circuit**.

three-phase current A current, such as that flowing through a three-phase circuit, which has a three phases, each usually differing in phase by 120° in relation to the other two. Its abbreviation is **3-phase current**.

three-phase four-wire   A **three-phase three-wire** circuit incorporating a fourth wire which is a neutral conductor. Its abbreviation is **3-phase 4-wire**. Also called **four-wire wye**.

three-phase full-wave   Pertaining to a circuit or device, such as a full-wave rectifier, which utilizes three-phase current. Its abbreviation is **3-phase full-wave**.

three-phase generator   A generator of three-phase AC. Its abbreviation is **3-phase generator**.

three-phase half-wave   Pertaining to a circuit or device, such as a half-wave rectifier, which utilizes three-phase current. Its abbreviation is **3-phase half-wave**.

three-phase meter   A meter, such as a wattmeter, utilized in a three-phase circuit. Its abbreviation is **3-phase meter**.

three-phase motor   A motor which utilizes three-phase current. Its abbreviation is **3-phase motor**.

three-phase power   Its abbreviation is **3-phase power**. **1.** The power utilized by a three-phase circuit or system. **2.** The power provided by a three-phase circuit or system.

three-phase rectifier   A rectifier which works with a three-phase AC input. Its abbreviation is **3-phase rectifier**.

three-phase system   A system, such as a three-phase circuit, which has a three-phase current or voltage. Its abbreviation is **3-phase system**.

three-phase three-wire   A three-phase circuit with three wires. The phase difference between any two wires is usually $2\pi/3$, or 120°. Its abbreviation is **3-phase 3-wire**.

three-phase transformer   A transformer utilized to supply a three-phase circuit, device, piece of equipment, or system. Its abbreviation is **3-phase transformer**.

three-phase voltage   A voltage, such as that supplying a three-phase circuit, in which there are three AC phases, as opposed, for instance, to a **single-phase voltage**, where there is one AC phase. Its abbreviation is **3-phase voltage**.

three-prong electrical outlet   Same as **three-prong outlet**. Its abbreviation is **3-prong electrical outlet**.

three-prong outlet   An electrical outlet, which in addition to the current-carrying contacts, has a third contact which serves for connection to a grounding conductor. Devices and equipment which are to benefit from this safety feature must have an appropriate three-prong plug which is inserted into this outlet. Its abbreviation is **3-prong outlet**. Also called **three-prong electrical outlet**, **grounding outlet**, **grounding receptacle**, or **safety outlet**.

three-state logic   Same a **tri-state logic**. Its abbreviation is **3-state logic**.

three-terminal regulator   A power IC housed in a standard three-terminal transistor package. Its abbreviation is **3-terminal regulator**.

three-tier client/server   A client/server architecture in which the user interface is structured into three layers or tiers, each of which may have one or more components. Its abbreviation is **3-tier client/server**.

three-way loudspeaker   Same as **three-way speaker**. Its abbreviation is **3-way loudspeaker**.

three-way speaker   A speaker which utilizes three individual drivers for each of three intervals of frequencies. Such a speaker usually has a woofer, a midrange, and a tweeter. A three-way speaker may have more than one driver dedicated to each covered band. For example, there may be two tweeters. Its abbreviation is **3-way speaker**. Also called **three-way loudspeaker**, or **three-way system**.

three-way system   Same as **three-way speaker**. Its abbreviation is **3-way system**.

three-wire system   Its abbreviation is **3-wire system**. **1.** An AC circuit utilizing three wires. **2.** A circuit, such as a communications circuit, utilizing three wires.

threshold   The point, level, or value of a quantity which must be exceeded to have a given effect, result, or response, such as detection, activation, or operation. For instance, the threshold of hearing, the minimum voltage or current necessary to activate a circuit, or the frequency beyond which a loudspeaker does not reproduce sound. Also called **limen**.

threshold current   The minimum current level which produces a given effect, result, or response, such as detection, activation, or operation. For example, the lowest current needed to sustain lasing action in a diode laser.

threshold detector   A sensor, device, or instrument, which detects a **threshold**. For example, that which detects a given minimum level of electromagnetic radiation.

threshold frequency   **1.** The minimum frequency which produces a given effect, result, or response, such as detection, activation, or operation. For example, the lowest frequency of electromagnetic radiation required for electrons to be emitted by a surface as a consequence of the photoelectric effect. **2.** The minimum or maximum frequency which produces a given effect, result, or response, such as detection, activation, or operation. For example, each of the cutoff frequencies of a filter, amplifier, or waveguide.

threshold of audibility   Same as **threshold of hearing**.

threshold of detectability   Same as **threshold of hearing**.

threshold of hearing   The minimum sound pressure level necessary to be detected by human ears. Since hearing acuity varies considerably from human to human, it has been internationally agreed that a value of 20 micropascals is the equivalent of 0 decibels to a person with good hearing. Thus, sound pressure level equal to or greater than 0 decibels is considered audible. Before pain sets in, a person with good hearing can listen at 120 decibels, or 20 pascals, which is a trillion times louder than 0 decibels. Also called **threshold of audibility**, **threshold of detectability**, or **audibility threshold**.

threshold of pain   The minimum sound pressure level sufficient to cause pain in a person with good hearing. This level is usually considered to be 120 decibels, or 20 pascals.

threshold signal   The minimum signal level which produces a given effect, result, or response, such as detection, activation, or operation. For example, the lowest input-signal amplitude which produces a perceptible change in the output of a component, circuit, device, piece of equipment, or system. Also called **minimum detectable signal**, or **minimum discernable signal**.

threshold voltage   The minimum voltage level which produces a given effect, result, or response, such as detection, activation, or operation. For example, the lowest voltage needed to switch a transistor from a blocking state to a conducting state.

throat   For a device such as a horn, the opening with the smaller cross section. The **mouth (2)** is that with the larger cross section.

throat microphone   A microphone which is worn around the throat of the user, and which picks up the vibrations of the larynx directly. Ideally, such a microphone transmits the speaker's voice only, and no background noise. Used, for instance, by emergency personnel, or in airplane cockpits. Also called **laryngophone**.

through-hole   Within a printed-circuit board, a hole made through plating, and which serves to mount and connect components by connecting their pins or leads. Also spelled **thru-hole**. Also called **plated-through hole**.

throughput   **1.** The rate at which data is transmitted from one location to another. This may refer, for instance, to an internal pathway of a computer, or to a communications channel. Such a speed is usually expressed in some multiple of bits per second. Also called **data throughput**. **2.** The effective rate at which a CPU processes data. Throughput takes into account factors such as processing speed, the rate at which data is transferred between the CPU and peripherals, the specific application being utilized, and so on. **3.** The speed or rate at which something is outputted or completed.

thru-hole Same as **through-hole**.

thulium A lustrous silver-white metallic element whose atomic number is 69. It is soft, malleable, and ductile, and has about 35 known isotopes, of which one is stable. It is used, for instance, as an X-ray source, and in arc lamps. Its chemical symbol is **Tm**.

thumb Within a scroll bar, the box that can be moved in the desired direction, to navigate within a file or window. For instance, by utilizing a mouse to slide the elevator downwards within a vertical scroll bar, a user advances further ahead in a displayed document. Also called **elevator**, or **scroll box**.

thumb wheel switch Same as **thumbwheel switch**.

thumbnail A small version of an image, page, or the like. Used, for instance, in groups which facilitate locating and selecting multiple images from a set, or to economize bandwidth by displaying the larger image only when clicked upon by a user.

thumbnail image A small version of an image.

thumbnail page A small version of a page.

thumbwheel switch A multi-position switch incorporating a small knob which is usually turned using a single finger, such as the thumb. For example, four such switches may be placed side-by-side for selection of values or settings ranging from 0000 through 9999 inclusive. Also spelled **thumb wheel switch**.

thump An unwanted low-frequency transient heard through a speaker, especially when powering a connected amplifier without the volume setting at its minimum.

thunder The sonic boom heard as a consequence of the shock waves produced by the explosive expansion of the air surrounding a lightning discharge.

thyratron A hot-cathode gas-filled tube which incorporates one or more control electrodes. The control electrodes assist in the initiation of current flow, and usually do not otherwise affect the operation of the tube. Used, for instance, as a switching device for modulators.

thyristor **1.** A solid-state equivalent of a **thyratron**. Examples includes SCRs, and three-layer diodes. **2.** A semiconductor device with four layers, p-n-p-n, which has an input terminal, or gate, output terminal, or anode, and a terminal common to both called cathode. Used, for instance, as a solid-state switch that turns large amounts of power on and off very quickly. Also called **SCR (1)**.

THz Abbreviation of **terahertz**.

Ti Chemical symbol for **titanium**.

Ti- Abbreviation of **tebi-**.

TIA Abbreviation of **Telecommunications Industry Association**.

TIA-232 Same as **TIA/EIA-232-E**.

TIA-232-E Same as **TIA/EIA-232-E**.

TIA-422 Same as **TIA/EIA-422**.

TIA-423 Same as **TIA/EIA-423**.

TIA-449 Same as **TIA/EIA-449**.

TIA-485 Same as **TIA/EIA-485**.

TIA-530 Same as **TIA/EIA-530**.

TIA-568 Same as **TIA/EIA-568**.

TIA-569 Same as **TIA/EIA-569**.

TIA-606 Same as **TIA/EIA-606**.

TIA-607 Same as **TIA/EIA-607**.

TIA/EIA-232 Same as **TIA/EIA-232-E**.

TIA/EIA-232-E A standard interface for serial communications between a serial device, such as a modem or a mouse, and a computer. Although this is the current term for this interface, it is still better known by its former name, **RS-232C**. Also called **TIA-232**, **TIA/EIA-232**, **TIA-232-E**, or **EIA/TIA-232-E**.

TIA/EIA-422 A standard interface for serial communications similar to **EIA/TIA-232-E**, but providing higher speed and transmission distances, and support for multipoint connections. Also called **TIA-422**, **RS-422**, or **EIA/TIA-422**.

TIA/EIA-423 A standard interface for serial communications similar to **EIA/TIA-232-E**, but providing higher speed and transmission distances. EIA/TIA-423 only supports point-to-point connections. Also called **TIA-423**, **RS-423**, or **EIA/TIA-423**.

TIA/EIA-449 A standard defining the pins for both **EIA/TIA-422** and **EIA/TIA-423** serial communications. Also called **TIA-449**, **RS-449**, or **EIA/TIA-449**.

TIA/EIA-485 A standard interface for serial communications similar to **EIA/TIA-422**, but supporting more nodes per line. Also called **TIA-485**, **RS-485**, or **EIA/TIA-485**.

TIA/EIA-530 A standard defining the pins for both **EIA/TIA-422** and **EIA/TIA-423** serial communications when a DB-25 connector is utilized. Also called **TIA-530**, **RS-530**, or **EIA/TIA-530**.

TIA/EIA-568 A standard detailing specifications for telecommunications cabling systems in structures such as commercial buildings. Also called **TIA-568**, or **EIA/TIA-568**.

TIA/EIA-569 A standard detailing specifications for placement, as in raceways or plenums, of telecommunications cabling systems in structures such as commercial buildings. Also called **TIA-569**, or **EIA/TIA-569**.

TIA/EIA-606 A standard detailing specifications for the telecommunications infrastructure of commercial buildings. Also called **TIA-606**, or **EIA/TIA-606**.

TIA/EIA-607 A standard detailing specifications for telecommunications grounding and bonding in structures such as commercial buildings. Also called **TIA-607**, or **EIA/TIA-607**.

TiB Abbreviation of **tebibyte**.

tick A single clock pulse or clock cycle.

tickler Same as **tickler coil**.

tickler coil A small coil through which energy is inductively transferred between the output of a circuit or device and the input of said circuit or device, so as to provide positive feedback. Used, for instance, in an oscillator. Also called **tickler**.

tie **1.** A wire, cord, clamp, clip, ring, bracket, strap, or other object or device which serves to fasten or secure. **2.** To fasten or secure using a **tie (1)**. Also, to be so fastened or secured. **3.** To bring together two or more wires, cables, bundles, pieces, objects, or the like, and secure or fasten using a **tie (1)**. **4.** A beam, post, or other object which joins parts and provides mechanical support.

tie cable **1.** A cable that joins two distribution points within a network. For example, that connecting a PBX with another PBX or with a central office. **2.** A cable that joins two or more other cables or circuits.

tie line Same as **tie trunk**.

tie point A terminal, with a fitting such as a lug, which serves to join two or more wires, cables, or conductors.

tie trunk A telephone line that connects two PBXs. Also called **tie line**, **PBX tie trunk**, or **PBX tie line**.

Tier 1 **1.** An ISP with a direct connection to the Internet backbone. It is the highest, and fastest, tier. Also called **Tier 1 ISP**. **2.** The highest tier within a hierarchy.

Tier 1 ISP Same as **Tier 1 (1)**.

Tier 2 **1.** An ISP with a connection to a regional service provider, which in turn has a direct connection to the Internet backbone. Also called **Tier 2 ISP**. **2.** The second highest tier within a hierarchy

Tier 2 ISP Same as **Tier 2 (1)**.

Tier 3 **1.** An ISP with a connection to a local Internet provider, which links to a regional service provider, which in turn has a direct connection to the Internet backbone. Also called **Tier 3 ISP**. **2.** The third highest tier within a hierarchy.

Tier 3 ISP Same as **Tier 3 (1)**.

TIFF Acronym for Tagged Image File Format. A format which is widely utilized to store, compress, and share bitmap images. There are multiple versions, with the supported resolution depending on which is utilized. Also called **TIFF format**.

TIFF file A file in the **TIFF** format.

TIFF format Same as **TIFF**.

tight coupling **1.** In a transformer, the arrangement of the primary and secondary windings in a manner which provides for a high degree of energy transfer. Also called **strong coupling (2)**, or **close coupling (1)**. **2.** In a transformer, the arrangement of the primary and secondary windings in a manner which provides maximum energy transfer. Also called **strong coupling (3)**, or **close coupling (2)**. **3.** Pertaining to, involving, or arising from **tightly-coupled multiprocessing**.

tightly-coupled multiprocessing A multiprocessing computer architecture in which multiple CPUs, usually in the same cabinet, share the same memory, and the same copy of the operating system, application, and data being worked on, during the execution of a single program or task. As more CPUs are desired or needed, they can be added. This contrasts with **massively parallel multiprocessing**, where each CPU has its own copy of the operating system, its own memory, its own copy of the application being run, and its own share of the data. Also, processing in this manner. Also called **symmetric multiprocessing**.

tightly-coupled multiprocessor A computer system that can perform **tightly-coupled multiprocessing**. Also called **symmetric multiprocessor**.

tilde A character, ~, that is present on most computer keyboards, and which serves, for instance, as a symbol, abbreviation, or to change an *n* into an *ñ*.

tiled windows On a computer screen, a group of windows arranged in a manner that they do not overlap one another. The borders of each are visible, enabling resizing, and the more windows open, the smaller the space allotted to each.

tilt switch A switch, such as a mercury switch, which opens or closes a circuit when tilted beyond a given angle. Used, for instance, as a motion detector, or as a level indicator.

timbre The characteristics or qualities of a sound which enable distinguishing it from other sounds having the same pitch and loudness. The harmonic composition of a sound, and the intensity and phase relationships between said harmonics affect the timbre of a sound.

time Its symbol is **t**. Its SI unit is the **second**. **1.** A dimension that enables two otherwise identical events occurring at the same point in space to be distinguished. Time orders the sequence of these events. **2.** A measure of the duration of one or more actions, processes, events, or the like. **3.** A measure of the **time (1)** that elapses between actions, processes, events, or the like. **4.** The instant at which a given action, process, event, or the like, occurs. **5.** A specific instant within a day, based on a given parameter or phenomena. For example, standard time is based on a celestial parameter, and atomic time is based on the natural resonance frequencies of atoms.

time base Also spelled **timebase**. **1.** A circuit or device that provides the voltage or signal which deflects an electron beam across the screen of a CRT during each sweep. Such a circuit or device usually utilizes a sweep generator which produces sawtooth waves, thus providing for a linear scan of the spot across the screen. Used in CRT displays such as those of TVs and radars. Also called **time-base generator (1)**. **2.** The voltage or signal provided by a **time base (1)**. **3.** The line formed by the voltage or signal provided by a **time base (1)**.

time-base corrector A circuit or device that corrects timing errors during playback of video signals, such as those of recorded VHS tapes. Also spelled **timebase corrector**.

time-base generator Also spelled **timebase generator**. **1.** Same as **time base (1)**. **2.** A circuit or device, especially a sweep generator, which produces sweep signals utilized for timing or synchronizing.

time-code generator A pulse generator which produces a code which serves to precisely time events, or to provide the exact time. Used, for instance, in GPS. Its abbreviation is **TCG**.

time constant The time required for an electrical quantity, such as a current or a voltage, to rise from zero or another initial value to a given proportion of the final value or total possible change. For example, that required to increase approximately 63.21% $(1 - 1/e)$ of the final steady value in response to an influencing change in a circuit.

time-controlled system A system, such as a digital computer, whose operations are governed by a timing mechanism. Also called **clock-control system**.

time delay Also called **time lag, delay time**, or **delay**. **1.** The time interval between two events. **2.** The time interval an event is postponed. Also, the act of postponing. **3.** The time interval between the sending or emitting of a signal, and its reception or detection. **4.** The time interval necessary for a signal to pass through a circuit, device, or medium. This includes any additional time the signal is retarded, as with the use of a delay line. **5.** The time interval between the instants at which a given point of a wave passes through two specified points of a transmission medium. **6.** The time interval between the powering of a circuit or device, and its starting to operate.

time-delay circuit A circuit whose output signal occurs a specified time interval after the input signal is received. For example, a sonic delay line. Also called **delay circuit**.

time-delay distortion Distortion of a signal as it passes through a transmission medium in which different frequencies travel at slightly different speeds. This causes a change in the waveform because the rate of change of phase shift is not constant over the transmitted frequency range. Also called **delay distortion, envelope delay distortion, phase distortion**, or **phase-delay distortion**.

time-delay fuse A fuse which opens a circuit after a given delay once the current passing through it exceeds a specified amount.

time-delay relay A relay which introduces a specified delay between the time it is energized and the time it opens or closes. Its abbreviation is **TDR**. Also called **delay relay**.

time-delay switch A switch that opens, closes, or changes the state of a circuit after a specified delay. Also called **delay switch**.

time-division duplex The transmitting of two signals over a single transmission path utilizing **time-division duplexing**. Its abbreviation is **TDD**.

time-division duplexing Duplexing in which a single channel is utilized to transmit two signals, each with its allocated time slot. The same frequency may be utilized to transmit and receive simultaneously. Its abbreviation is **TDD**.

time-division multiple access A multiplexing method which divides a high-speed channel into time slots, each assigned to different signals. Each packet or frame within its channel is uniquely timed, increasing throughput while avoiding collisions. This technology is used, for instance, in satellite systems, and in digital cellular telephone networks, such as GSM and digital AMPS. Its abbreviation is **TDMA**.

time-division multiplex The transmitting of multiple signals over a single transmission path utilizing **time-division multiplexing**. Its abbreviation is **TDM**.

time-division multiplexer A circuit or device which combines signals in **time-division multiplexing**. Also spelled **time-division multiplexor**. Its abbreviation is **TDM**.

time-division multiplexing Multiplexing in which a single channel is utilized to transmit multiple signals, each with its allocated time slot. Using this technique, several lower speed signals are multiplexed into a higher speed channel for transmission, and are demultiplexed at receiving locations. Used, for instance, to combine voice or data channels into high-bandwidth circuits such as T1 or E1. Also, the technology enabling such multiplexing. Its abbreviation is **TDM**.

time-division multiplexor Same as **time-division multiplexer**.

time-division switching A **time-division multiplexing** switching method in which bits, or other units of data, are shifted between time slots, such as those of time-division multiplexing frames. Its abbreviation is **TDS**.

time-division synchronous code division multiple access Same as **TD-SCDMA**.

time domain The representation of a signal as a function of time, as opposed to frequency. This contrasts with **frequency domain**, where the converse is true. An oscilloscope displays signals in the time domain. A Fourier transform is utilized to convert from the time domain to the frequency domain, and vice versa.

time-domain reflectometer An instrument, such as an Optical Time Domain Reflectometer, which is utilized to evaluate the characteristics of a transmission line by sending one or more pulses and measuring the time interval that elapses and intensity of the reflected signals. The resulting signals are usually observed using an oscilloscope. Used, for instance, to troubleshoot communications networks. Its abbreviation is **TDR**.

time-domain reflectometry The use of a **time-domain reflectometer** to evaluate the transmission characteristics of a communications line, such as an optical fiber. abbreviation is **TDR**.

time factor Same as **time scale**.

time-interval counter An instrument which measures very short, or extremely short, time intervals, based on counting pulses. Such counters may be able to provide values in the femtosecond range, and are used, for instance, to calibrate frequency standards.

time lag Same as **time delay**.

time modulation A form of modulation, such as pulse-time modulation, in which the timing of a pulse, pulse train, or other signal is varied so as to convey meaningful information.

time-of-day clock In a computer, an independent circuit which keeps track of the time of day. A time-of-day clock is not usually used to generate clock pulses. Also called **real-time clock**.

time out Same as **timeout**.

time scale Also called **time factor**. **1.** An established standard utilized to quantify the passage of time. The fundamental SI unit of time is the **second**. A time scale may be expressed in submultiples or multiples of seconds, such as

nanoseconds or kiloseconds, although intervals longer than one second are more frequently expressed as minutes, hours, days, and so on. **2.** A scale, such as that describing an axis within a coordinate system, in which time is utilized. For instance, the use of time coordinates for the x-axis in a Cartesian coordinate system. **3.** The ratio of one event to another. For example, the ratio of the duration of an event in theory, to that actually observed.

time sharing Also spelled **timesharing**. **1.** The simultaneous use of the same resource by two or more components, circuits, devices, pieces of equipment, systems, processes, users, or the like. **2. Time sharing (1)** in the context of computers and/or communications. For example, the concurrent accessing of a file or other resource by two or more users within a communications network.

time signal **1.** A signal, such as pulse, which serves to time or synchronize operations, mechanisms, or the like. For example, those generated by a CPU to synchronize all system operations. Also called **timing signal (1)**. **2.** A high-precision signal which is broadcast by radio, and which serves to provide the exact time, or to calibrate or adjust timepieces.

time slice **1.** During multitasking, a brief time interval during which a given program is given the attention of the CPU. **2.** During time sharing, a brief time interval which is allotted to a given component, circuit, device, piece of equipment, system, process, user, task, or the like.

time slot A specific interval designated for a specified use or process. For example, the time slots corresponding to the allocation of multiple channels in time-division multiplexing. Also spelled **timeslot**.

time stamp Same as **timestamp**.

time standard A device which produces pulses that serve as time signals, or for the precise measurement of time intervals.

time switch A switch that opens or closes one or more circuits at specified times.

Time-to-Live A value in a header which indicates whether a given packet has exceeded the time or number of hops it is permitted for travel within a network. Bad routing, for instance, may lead to a packet to otherwise go through endless loops. When the value reaches zero, the packet is discarded and an error message is sent to the originating host. Its abbreviation is **TTL**.

time-varying electric field An electric field whose strength fluctuates over time. Such a field may produce a displacement current.

time-varying magnetic field A magnetic field whose strength fluctuates over time. Such a field may produce eddy currents.

time zone On the planet earth, a region throughout which the same standard time is utilized. Each is such zone encompasses 15° longitude in width, providing for 24 zones around the globe, each an hour apart. The standard time maintained in many countries and areas does not coincide with the time zone established by these meridians.

timebase Same as **time base**.

timebase corrector Same as **time-base corrector**.

timebase generator Same as **time-base generator**.

timeout Also spelled **time out**. **1.** A condition in which a predetermined time interval for a given event is exceeded. Once this interval elapses the desired process is stopped, and an error message is displayed. For example, a Web browser will eventually timeout if a requested Web page is provided by a Web server that is down. **2.** A specific instance of a **timeout (1)**. **3.** To exceed the time required to generate a **timeout (1)** condition.

timer Also called **interval timer**. **1.** An instrument or device utilized to measure time intervals, such as those occurring between events, or that required to complete a given operation. **2.** A **timer** (1) that controls the duration of a pulse, operation, event, process, or the like. **3.** A **timer** (1) utilized to power, or otherwise activate a circuit, device, piece of equipment, system, mechanism, process, or the like. **4.** A **timer** (1) utilized for synchronization purposes.

timer interrupt An interrupt generated by an internal timer.

timesharing Same as **time sharing**.

timeshift The postponement of an event or process, such as reception or transmission, to a later time.

timeslot Same as **time slot**.

timestamp Also spelled **time stamp**. **1.** A record of the exact time an event occurs. Used, for instance, to synchronize events across a communications network, or to establish the time an email is sent. **2.** To record a **timestamp** (1).

timing diagram A chart or diagram depicting the relative timing of multiple inputs and outputs of logic circuits.

timing signal **1.** Same as **time signal**. **2.** A signal which is recorded or transmitted with another, and which serves to synchronize the flow of information, or to mark the exact time.

tin **1.** A metallic element whose atomic number is 50. It has two known allotropic forms, the more common is soft, malleable, and silvery-white, the other is brittle gray. Tin can be hammered into an extremely thin foil, and when bent emits a characteristic cry. Tin and some of its alloys have been known since ancient times. It has close to 40 known isotopes, of which 10 are stable. Its applications include its use in plating, corrosion-resistant coatings, solders, wires, and in many alloys with uses in electronics and metallurgy. Its chemical symbol, **Sn**, is taken from the Latin word for tin: stannum. **2.** Same as **tinning**.

tin coating Same as **tin plating**.

tin plating The deposition of tin onto a metal substrate, usually to provide it with greater resistance to corrosion, and sometimes for improvement of appearance. Also called **tin coating**.

tinned wire A wire, usually of copper, which has a thin coat of tin. Used, for instance, help prevent oxidation and corrosion of soldered connections.

tinning The application of a thin coat of solder to the tip of a soldering iron, and/or the conductors or surfaces to be soldered, to help ensure proper heating, to minimize the transfer of impurities, and to obtain a sound joint. Also called **tin** (2).

tinsel cord A cord which has a thin metal strip of foil incorporated for additional strength. Used, for instance, in headphones or microphones, which are subject to frequent flexions.

tinsel wire A single wire within a **tinsel cord**.

tint control In a TV receiver, a circuit, device, or system which adjusts the color balance of reproduced images by varying the relative intensities of the three electron beams.

tip **1.** The end of something that has a projection or which is otherwise pointed. For example, the prong or rod of a phone or phono plug. **2.** An attachment or other object, such as a ferrule or plastic cap, intended to be fitted to the end of something else. **3.** Of the two wires of a POTS line, the more electrically positive. The more electrically negative is the **ring** (5). Also called **tip lead**, or **tip wire**.

Tip and Ring The two wires which comprise a POTS line. The more electrically positive wire is called **tip** (3), and the more electrically negative wire is called **ring** (5). Its abbreviation is **Tip & Ring**.

tip connector A **tip jack** or a **tip plug**.

tip jack A small jack, such as a phone jack, with a single recessed opening, into which a matching **tip plug** is inserted.

tip lead Same as **tip** (3).

tip plug A small plug, such as a phone plug, which is inserted into a matching **tip jack**.

tip wire Same as **tip** (3).

Tip & Ring Same as **Tip and Ring**.

TIPS Acronym for trillion instructions per second. It generally means one trillion machine instructions per second, and is an indicator of CPU speed. When evaluating the overall speed of a system, other factors, such as cache, memory speed, and the number of instructions required for given tasks, must also be considered.

TIR Abbreviation of **total internal reflection**.

titanium A metallic element whose atomic number is 22. It has two known allotropic forms, the more common is a lustrous silver-white metal, and the other is a dark gray powder. Titanium is strong, light, ductile, and when heated malleable. It has about 15 known isotopes, of which 5 are stable. Its applications include its use in electrodes, in corrosion-resistant parts, lasers, semiconductors, X-ray tubes, and as a structural material for vessels such as ships, aircraft, and spacecraft. Its chemical symbol is **Ti**.

titanium boride A very hard corrosion-resistant solid whose chemical formula is TiB_2. Used, for instance, in high-temperature conductors, in resistant coatings, cermetals, and for various metallurgical processes.

titanium carbide A very hard crystalline solid whose chemical formula is TiC. Used, for instance, in electrodes, cermetals, and cutting tools.

titanium dioxide A white powder whose chemical formula is TiO_2. Used, for instance, in resistant coatings, ceramics, lasers, semiconductors, and high-temperature transducers.

titanium hydride A black powder whose chemical formula is TiH_2. Used, for instance in powder metallurgy where it can hold very large amounts of hydrogen, in solders, and as a vacuum-tube getter.

titanium nitride A yellow-brown crystalline solid whose chemical formula is TiN. Used, for instance in cermetals, lasers, and semiconductors.

title bar The bar at the top of a given window. It usually identifies the application, provides a file or page name, and allows moving, resizing, maximizing, minimizing, and closing said window.

TJ Abbreviation of **terajoule**.

Tl Chemical symbol for **thallium**.

TLD Abbreviation of **top-level domain**.

TLF Abbreviation of **tremendously low frequency**.

TLM Abbreviation of **transmission line matrix**, or **transmission line modeling**.

TLS Abbreviation of **transport-layer security**.

Tm **1.** Chemical symbol for **thulium**. **2.** Abbreviation of **terameter**.

TM mode Abbreviation of **transverse magnetic mode**.

TM wave Abbreviation of **transverse magnetic wave**.

TMN Abbreviation of **Telecommunications Management Network**.

tn Symbol for **ton**.

TN Abbreviation of **twisted nematic**.

TN display Abbreviation of **twisted nematic display**.

toggle To switch or alternate between states, positions, or settings. For example, two change the position of a toggle switch, or switch between upper case and lower case using a Caps lock key.

toggle button In a GUI, a button that alternates between two or more states. For example, a button which when clicked upon alternates between bolding and unbolding.

toggle flip-flop Same as **T flip-flop**.

toggle key A key, such as a Caps lock key, that alternates between two or more states.

toggle switch A switch, such as a bat-handle or paddle switch, which has a projecting lever which serves to manually alternate between positions or settings, such as on and off.

token A data frame that is circulated throughout a **token passing-network**. A given number of empty data frames, or tokens, circulate along the nodes, and when a specific node wishes to transmit, it inserts a message. All nodes monitor these circulating tokens, and when the destination node examines it, it accepts the message, resetting the status of said token to available. Thus, this token is first available, becomes occupied, then is freed again. To avoid collisions, a token can only be sent by a node which has possession of it at the moment.

token bus Same as **token bus network**.

token bus network A LAN which utilizes a bus topology, and in which **tokens** are circulated from node to node. In such a network the ring formed is virtual not physical. Also called **token bus**.

token passing The circulating of **tokens** from node to node in a token-passing network.

token-passing network A network, such as a token ring network or a token bus network, in which **tokens** are circulated from node to node.

token ring Same as **token ring network**. It's abbreviation is **TR**.

token ring adapter A network card utilized in a **token-passing network**, especially a token ring network. Also called **token ring card**, **token ring adapter card**, or **token ring network card**.

token ring adapter card Same as **token ring adapter**.

token ring card Same as **token ring adapter**.

token ring LAN Same as **token ring network**.

token ring network A LAN which utilizes a ring topology, and in which **tokens** are circulated from node to node. Also called **token ring**, or **token ring LAN**.

token ring network card Same as **token ring adapter**.

tolerance 1. The maximum permissible deviation, error, or other variation in a property, value, setting, dimension, or the like. For instance, distortion tolerance, frequency tolerance, the greatest allowable error in a reading displayed by a measurement instrument, or the maximum fluctuations in the voltage supplied by an AC power line which do not impair the function of fed devices or equipment. 2. The **tolerance** (1) relative to a given nominal, standard, or desired value.

toll call A telephone call which is originated from, or placed to a destination that is not contained within the same local service area. A long-distance call does not necessarily imply that there is an additional charge to the caller, as seen in many cellular telephone plans. Also called **long-distance call**.

tomography The use of X-rays to view and/or produce detailed images of a given plane below the surface of a material, surface, or body. Used especially as a medical diagnostic technique, and specific forms include computed tomography, positron emission tomography, and single-photon emission computed tomography.

ton A unit of mass equal to 907.18474 kilograms, 0.90718474 tonne, or 9.0718474×10^5 grams. Its symbol is **tn**, **T**, or **t**.

tone 1. An audible sound which usually has a definite frequency, specific characteristics such as harmonics, and a given duration. 2. The timbre of a given instrument, sound, or voice. 3. The characteristics or sensory perceptions elicited by colors, sounds, or the like. For example, the audio characteristics of a room, or the quality of a given color.

tone arm In a phonograph, the arm which holds the cartridge. Also spelled **tonearm**. Also called **pickup arm**.

tone beeper Same as **tone pager**.

tone burst A pulse or other short signal consisting of a single tone, which is utilized for testing, measurement, and/or calibration. Used, for instance, to differentiate a desired signal from spurious signals.

tone-burst generator An instrument or device, such as an oscillator, which generates **tone bursts**.

tone control 1. A circuit or device which is utilized to modify the relative frequency response of a component, device, piece of equipment, or system, especially an audio amplifier. 2. A control, such as a bass control in an audio amplifier, which is utilized to adjust a **tone control** (1). 3. Adjustments made using a **tone control** (1).

tone dialer In telephony, a device, such as a keypad, which is utilized for **tone dialing**.

tone dialing 1. Telephone dialing in which tones or tone pairs are utilized to represent the available keys. Dual-tone multi-frequency, for instance, uses tones which are a mixture of two frequencies, with each combination identifying the 0-9, *, and # keys found on most touch-tone phones. 2. **Tone dialing** (1) using a keypad, as opposed for instance, to the tones generated by a modem. Also called **touch-tone dialing**.

tone generator An instrument or device, such as an oscillator, which generates **tones** (1). Used, for instance, for testing, measurement, and/or calibration.

tone modulation 1. The modulation of an RF carrier at a fixed audio frequency, so as to convey meaningful information. 2. The modulation of a **tone** (1), so as to convey meaningful information.

tone pager A pocket-sized radio receiver or transceiver which serves to receive messages. These devices emit a beep when informing of a new message. Also called **tone beeper**, **paging device**, **pager** (1), or **beeper**.

tone ringer A solid-state device which provides a telephone with an audible ring signal. In the case of cellular telephones and similar devices, a tone ringer may provide tones which can be chosen from an extensive stored menu, or downloaded for even more choices.

tonearm Same as **tone arm**.

toner An ink, usually in powdered form, which is used, for instance, in laser printers and copiers.

toner cartridge A disposable or recyclable cartridge containing **toner**. Such a cartridge usually incorporates other components, such as a drum, providing for greater simplicity and reliability.

tonne A unit of mass equal to 1,000 kilograms, or 1 megagram. Its symbol is **t**. Also called **metric ton**.

tool 1. A device, machine, system, or object which helps facilitate or complete manual or mechanical work. 2. A program or utility that assists in the development, analysis, testing, and maintenance of software. Such tools include compilers, decompilers, debuggers, editors, and the like. Also called **software tool**. 3. One of the functions, features, or the like, available for use via a **toolbar**.

tool palette A **toolbar** with an especially large number of available functions, features, or settings, and which has many icons, samples, and/or controls available at a given moment. Seen, for instance, in paint programs.

toolbar A vertical or horizontal bar which contains icons and/or buttons which can be clicked upon to access functions, features, or the like, within a given application, such as a word processor, Web browser, or image editor. A toolbar can usually be customized, moved, or hidden, and examples include taskbars and tool palettes. Also called **toolbox (2)**.

toolbox 1. A set of routines and/or utilities which assist in developing, using, and/or maintaining applications. Such a kit may, for instance, consist of a toolbar. Also called **toolkit. 2.** Same as **toolbar.**

toolkit Same as **toolbox** (1).

tools menu A menu which allows the selection and setting of program preferences, settings, functions, features, or the like.

top-down approach Same as **top-down reasoning.**

top-down design Same as **top-down reasoning.**

top-down programming A method of programming that first defines higher-level functions, and then breaks these down into simpler functions. This contrasts with **bottom-up programming**, where the simpler functions are defined, and then this base is utilized to proceed to higher levels.

top-down reasoning A method of reasoning or problem solving that first defines higher-level functions or tasks, and then breaks these down into simpler functions or tasks. This contrasts with a **bottom-up approach**, where the simpler functions or tasks are defined, and then this base is utilized to proceed to higher levels. Also known as **top-down approach**, or **top-down design.**

top-level domain On the Internet, one of the one of the main registration categories, such as .com, .net, .org, .edu, .gov, or country designators, such as. de for Germany. It consists of no less than two letters, and appears at the end of a domain name, as in the **.com** portion of *www.yipeeee.com*. Its abbreviation is **TLD**. Also called **domain** (5).

top-loaded antenna A vertical antenna whose electrical length is adjusted by varying an inductance or capacitance at its upper tip. Also called **top-loaded vertical antenna.**

top-loaded vertical antenna Same as **top-loaded antenna.**

top-of-file 1. The beginning of a file. This may be, for instance, a header, the first line, or the first record **2.** A symbol, code, marker, or the like, which indicates a **top-of-file** (1).

topology 1. For a communications network, the arrangement of the nodes and their interconnections. Common topologies include bus, ring, and star. Network topologies may be physical or logical. Also called **network topology. 2.** The manner in which something is physically laid out. For example, the physical layout of the conductors in a circuit.

toroid 1. A surface generated by rotating a closed curve around an axis in the same plane as the curve, but which does not contain nor intersect with said axis. **2.** A body having the shape of a **toroid** (1), which corresponds approximately to that of a doughnut. **3.** Same as **toroidal coil.**

toroid coil Same as **toroidal coil.**

toroid core Same as **toroidal core.**

toroidal coil A **toroid** (2), usually composed of powdered ferrite, which has one or more coils wound around it. Such a coil is able to produce a concentrated magnetic field with minimal flux leakage. Also called **toroid** (3), or **toroid coil.**

toroidal core A core, such as that in a toroidal coil, in the shape of a **toroid** (2). Also called **toroil core.**

torque Also called **moment of force.** The SI unit of torque is the Newton meter. **1.** A measure of the ability of one or more forces to produce rotation about an axis. **2.** One or more forces producing a turning or twisting motion.

torr A unit of pressure equal to about 133.3224 pascals, 1.31579×10^{-3} atm., 1.333224×10^{-3} bar, or 0.9999998 mmHg.

torsion A twisting stress or deformation in an object produced by a **torque**, while a part of said object is held still. Also, the effect of such twisting. Also, the act of so twisting.

torsion wave Same as **torsional wave.**

torsional wave A wave motion in which the vibrations in the propagating medium are rotational movements around the direction of energy transfer. Also called **torsion wave.**

TOS Abbreviation of **Terms Of Service.**

Toslink An optical-fiber cable utilized, for instance, to channel the output of a DVD to a TV, or from a CD to a receiver. Also, a connector on either side of such a cable.

total binding energy The minimum energy required to dissociate a system into its component parts. For example, the energy necessary to dissociate a nucleus into its component protons and neutrons. Also known as **binding energy** (2).

total bypass In telecommunications, avoiding the use of both local and long-distance telephone companies as a transmission pathway. A satellite, for instance, may serve as an alternative route.

total cost of ownership A model which factors in all direct and indirect costs associated with owning, operating, and maintaining a computer. In addition to the initial disbursement for a given system, such costs may include technical training, software and/or hardware upgrades, technical support, administrative fees, the adapting of a given location for the use of such a system, remote storage charges, and so on. The term also applies to the total cost of ownership of other devices, equipment, or systems, such as communications networks. Its abbreviation is **TCO.**

total distortion A sum of all the forms of distortion affecting the waveform of a signal.

total harmonic distortion A measure of the distortion caused by all the harmonics present in an output signal which were not present in a sinusoidal input signal. It may be calculated as follows: the square of the ratio of the amplitude of the first harmonic to the fundamental frequency, and the square of the ratio of the second harmonic to the fundamental frequency, and so on are all added together, and the square root is taken from the result, and this number is then multiplied by 100 to express it as a percentage. This value is frequently used as an expression of the performance of audio components, especially amplifiers. Its abbreviation is **THD.**

total internal reflection The total reflection of electromagnetic radiation occurring when it strikes a boundary or interface with a lower refractive index at an angle of incidence greater than the critical angle. For example, the successive internal reflections of light within an optical fiber. Its abbreviation is **TIR.**

touch pad Same as **touchpad.**

touch screen A computer screen that detects where contact is made upon it. Any of various technologies may be used. For example, such a screen may have an overlay which is coated with a thin electrically conductive grid which detects changes in an electrical parameter, such as resistance or capacitance, when touched. Such screens, may be used, for instance, to make selections and the like on displays at kiosks, in early-childhood learning centers, or on PDAs. Also called **touch-sensitive screen, touch-sensitive display**, or **touch-sensitive monitor.**

touch screen overlay A usually transparent sheet which is mounted or placed before a computer display, and which serves for making choices or the like on a **touch screen.** Also called **screen overlay** (1).

touch-sensitive 1. A technology which enables a surface, device, or system to identify a point where contact is made. Used, for instance, to obtain an input based on locations pressed on a touch screen. **2.** A device or surface, such as a touch screen or touch screen overlay, which detects contact or pressure.

touch-sensitive display Same as **touch screen.**

touch-sensitive monitor Same as **touch screen.**

touch-sensitive screen Same as **touch screen.**

touch sensor A device that detects objects through physical contact with them. These sensors may be used by robots, for example, to determine the location, identity, and orientation of parts to be assembled.

touch tablet Same as **tablet**.

touch-tone Pertaining to **touch-tone dialing**, or a **touch-tone phone**.

touch-tone dialing Same as **tone dialing (2)**.

touch-tone phone Abbreviation of **touch-tone** telephone. A telephone, usually equipped with twelve dialing pushbuttons, whose keypad enables **tone dialing (1)**.

touch-tone telephone Same as **touch-tone phone**.

touchpad A pressure-sensitive pad upon which a finger, or other object, is moved to move the pointer shown on a computer monitor. Such a pad may have specific areas which are tapped to serve as the equivalent of left or right mouse clicks. Other pads allow taps anywhere for such clicking. Also spelled **touch pad**. Also called **trackpad**.

tourmaline One of a group of crystals which contain various elements including boron, aluminum, and silicon, and which exhibit birefringence and piezoelectric properties. Used, for instance, in oscillators, gauges, and for the polarization of light.

tower 1. A structure which is significantly taller than it is wide. For example, an antenna tower. 2. A **tower (1)** which extends from another structure. For example, a radio tower. 3. A **tower (1)** or **tower (2)** which supports an antenna or is used as an antenna radiator. Also called **tower antenna**, or **tower radiator**. 4. In computers and communications, a cabinet or other enclosure which is taller than it is wide. For example, a DVD tower.

tower antenna Same as **tower (3)**.

tower radiator Same as **tower (3)**.

Townsend avalanche Also called **Townsend ionization, avalanche, avalanche effect, cascade (2), cumulative ionization**, or **ion avalanche**. When the only produced charged particles are electrons, it is called **electron avalanche**. 1. A cumulative ionization process in which charged particles are accelerated by an electric field and collide with neutral particles, creating additional charged particles. These additional particles collide with others, so as to create an avalanche effect. 2. In semiconductors, the cumulative generation of free charge carriers in an avalanche breakdown.

Townsend discharge In a gas-discharge tube, a luminous discharge that occurs at voltages that are too low to maintain self-sustained conduction.

Townsend ionization Same as **Townsend avalanche**.

TP Abbreviation of **twisted pair**.

TP monitor Abbreviation of transaction processing **monitor**, or teleprocessing **monitor**. Software which monitors transactions as they are passed through multiple terminals and/or processing stages. A TP monitor helps insure, for instance, that the integrity of each transaction is maintained and in the proper format, and that no transactions are lost.

TPa Abbreviation of **terapascal**.

tpi Abbreviation of **tracks per inch**.

TPI Abbreviation of **tracks per inch**.

TPM Abbreviation of **transactions per minute**.

TPS Abbreviation of **transactions per second**.

TQFP Abbreviation of **thin quad flat-pack**.

TR Abbreviation of **token ring**.

TR switch Abbreviation of **transmit-receive switch**.

TR tube Abbreviation of **transmit-receive tube**.

trace 1. To follow the course of something. For example, to record all computer system activities which lead to a crash, or to trace a telephone call. Also, the record, listing, or other

documentary evidence so obtained. 2. In a CRT, the path followed by the electron beam as it moves across the screen. Also called **line (5)**. 3. The movement or path followed by a scan or sweep. Also, to follow such a movement. 4. The line drawn by a recording instrument such as a spectrograph, oscillograph, or cardiograph. Also, to draw such a line. 5. A comparatively minute quantity of something, such as an element. 6. An amount which is too small to be measured, but which is know to be present.

trace interval The time required for a **trace (2)**, **trace (3)** to be completed.

trace route Same as **traceroute**.

tracer A radioisotope which substitutes a non-radioactive isotope of the same element, enabling the detection and monitoring which would otherwise be more difficult or impracticable. Used, for instance, in medical diagnostic techniques such as positron emission tomography. Also called **radioactive tracer**, or **radiotracer**.

traceroute Also spelled **trace route**. 1. An Internet utility that indicates the route taken by a packet from a client machine to a remote host, specifying the number of hops. 2. A command initiating a **traceroute (1)**.

track 1. A path along which sounds, images, or other signals are recorded. For example, a soundtrack, a control track, or that of one of the multiple audio signals which are mixed for a recording. 2. A specific song or selection from a sound recording, such as that available on a CD. 3. On a data medium, such as a disk or tape, a channel, band, or other path associated with the sequential access of data. On an optical disc, a track usually follows a continuous spiral, hard disk tracks usually form concentric circles, while tracks generally run in a parallel manner on tapes. Each track consists of a given number of **sectors**. 4. One or more grooves, rails, or ridges, often of metal, which serves to provide support and guide movement. For instance, a track along which a heavy apparatus may be moved more easily. 5. The path along which something moves. For instance, the trajectory of a reflected particle. Also, to follow the path along which something moves. For example, to track a scanned object.

Track-at-Once A method of writing to an optical disk, such as a DVD or CD, in which each track is recorded independently. Used, for instance, to record a single music CD from multiple sources. This contrasts, for instance, with **Disc-at-Once**, in which all data is transferred continuously without interruptions. This also contrasts with **Session-at-Once**, where there are multiple recording sessions.

track density A measure, usually tracks per inch, which indicates the density of the tracks present on a disk or tape.

track pitch The spacing between tracks, such as those of an optical disc, hard disk, or magnetic tape.

trackball A pointing device which incorporates a ball which is rotated by a hand or by fingers upon its stationary base. The movements of the ball translate into cursor movements on the display screen, and one or more buttons serve for selecting, performing functions, or the like.

tracking 1. The following of the path of something that moves. For example, the following of the movements of a scanned object by a radar. 2. The following of a changing variable or characteristic, while making corresponding adjustments. For instance, the condition in which simultaneously tuned circuits maintain close resonance for a given tuned frequency. 3. A mechanism that adjusts the angle and speed at which a magnetic tape passes the heads, as occurs, for instance, in a VCR. 4. A mechanism which adjusts the lateral pressure of a phonograph needle as it follows a groove.

tracking force The force a stylus exerts on a phonographic disk.

tracking generator An instrument which can manipulate a signal, so as to change its otherwise linear or sinusoidal waveform, into one that tracks another given shape, such as that of an oscillator which produces square or sawtooth waves, while maintaining the sweeps synchronized.

tracking mode An operational mode in which a component, circuit, device, piece of equipment, system, or mechanism follows the actions, movements, variations, or the like of something else, while making corresponding adjustments.

trackpad Same as **touchpad**.

tracks per inch A measure of the density of the **tracks** (3) on a disk or tape. Some disks, for instance, have hundreds of thousands of tracks per inch of surface. Its abbreviation is **TPI**, or **tpi**.

TRACON Acronym for **terminal radar approach control**.

tractor feed A method or mechanism for moving papers, or other appropriate mediums, through a printer by using pins which are mounted on wheels on each side of the printer. The medium being utilized must have holes which match the pins. This contrasts with a **friction feed**, in which the medium is advanced while pinched between rollers.

traffic **1.** The volume of users, messages, transmissions, or the like present within a communications network at a given moment. This may refer to telephone calls connected, data being downloaded, emails being received, faxes being sent, and so on. **2.** The volume of data being transmitted over a communications network at a given moment. **3.** The volume of users utilizing a communications network at a given moment. **4.** The volume of users accessing a Web site at a given moment. **5.** The **traffic** (1), **traffic** (2), **traffic** (3), or **traffic** (4), for a stated time period, such as a second, minute, hour, day, month, year, and so on.

traffic intensity A measure, such as that taken during peak hours, which indicates the level of usage a communications network is capable of. It may consist, for example, of a determination of the number of connections that can be reliably maintained simultaneously.

traffic management The measures taken to monitor and control the traffic of a communications network. Used, for example, to avoid situations of excessive congestion. Widely utilized, for instance, in ATM. Also called **network traffic management**.

trailer **1.** A section of nonmagnetic tape, usually plastic, which is affixed to the end of a length of recording tape. A section attached at the beginning of a tape is called a **leader** (1). **2.** In communications, the last part of a message, packet, or the like. Such a trailer, for instance, may incorporate error-checking data.

trailer label **1.** A label, such as that consisting of a block, identifying the last record within a file, such as that stored on a tape. **2.** A label identifying the last data unit, or the end of a transmission of data. For example, an end-of-data mark.

trailing antenna An antenna which hangs from the rear of an aircraft. Formerly utilized to extend the range of communications to and from aircraft so equipped.

trailing edge Also called **tail** (3). **1.** The portion of a pulse waveform which first decreases in amplitude. That is, the falling portion. The portion which first increases is the **leading edge** (1). **2.** The latter portion of a pulse or signal. The initial part is called **leading edge** (2).

train A comparatively long line or sequence, such as a pulse train, or a large succession of events, components, or the like.

trans **1.** Abbreviation of **transaction**. **2.** Abbreviation of **transverse**.

transaction Its abbreviation is **trans**. An event or activity in which something is exchanged between two entities, or which elicits a corresponding action. For example, a command which calls for a processing action, an exchange between a user and an interactive system, a change in a record which updates a master file, or an online or point-of-sale purchase.

transaction file A file containing **transaction records**. Also called **change file**.

transaction monitor Same as **TP monitor**.

transaction processing The processing of transactions the instant each is received by, or entered into, a computer system. Online transaction processing not only maintains all master files constantly current, it also enables entering or retrieving information without delays in either process. Used, for instance in mail order businesses which have multiple locations which accept Internet, telephone, and fax orders. Also called **online transaction processing**, or **real-time transaction processing**.

transaction processing monitor Same as **TP monitor**.

transaction record In a computer database, a record which changes information in the corresponding master file. Also called **change record**.

Transaction Tracking System A system that protects data by undoing incomplete transactions due to software and/or hardware failures. When a transaction is undone, data is returned to its previous state. Its abbreviation is **TTS**.

transactions per minute A performance measure which indicates the number of transactions, such as those a database server can perform, per minute. Its abbreviation is **TPM**.

transactions per second A performance measure which indicates the number of transactions, such as those a database server can perform, per second. Its abbreviation is **TPM**.

transadmittance In a circuit or device, the ratio of the output current to the input voltage, when one or both values may have an imaginary number component.

transceiver Acronym for **trans**mitter-**receiver**. **1.** A unit which combines a radio transmitter and a receiver in the same housing, usually sharing certain circuits and components. **2.** A device which combines the functions of transmitter and a receiver. For example, that which enables a computer to both transmit and receive signals to and from a communications network. **3.** A device which both transmits and receives, as opposed to just performing either function. For example, a two-way pager.

transcendental number A number, such as π or e, which is not the root of a polynomial equation with integer coefficients.

transcode To convert from one code or format to another. For example, to convert analog music into digitally encoded music, or to change from one digital music format to another.

transconductance Also called **mutual conductance**. **1.** For an amplifying device, such as a transistor or an electron tube, the ratio of the change in output current to the change in input current. For example, in a tube, it is the ratio of the change in plate current to the change in grid voltage, with the plate voltage held constant. Usually expressed in multiples of siemens or mhos. **2.** For a component, circuit, or device, the ratio of the output current to the input current.

transconductance amplifier Also called **operational transconductance amplifier**. **1.** An amplifier which converts its input voltage into an output current proportional to said voltage. **2.** A differential amplifier which converts the difference between its input voltages into an output current proportional to said difference.

transcribe **1.** To copy or transfer data from one medium or another, performing the necessary conversions, such as those involving formatting, to the new medium. **2.** To record for future use. Usually utilized in the context of programming that is recorded for later broadcasting.

transcriber That which serves to **transcribe**.

transcription 1. The process of copying or transferring data from one medium to another, while performing the necessary conversions, such as those involving formatting, to the new medium. Also, the result of such a process. **2.** The process of recording for future use. Also, the recording so made.

transcutaneous electrical nerve stimulation The use of electrodes placed on the skin for the application of weak electrical discharges intended to reduce pain. Its acronym is **TENS**. Also called **transcutaneous nerve stimulation**.

transcutaneous nerve stimulation Same as **transcutaneous electrical nerve stimulation**.

transducer 1. A component, circuit, or device which converts a non-electrical parameter, magnitude, signal, or phenomenon into a proportional or corresponding electrical equivalent. Also, that which performs the converse transformation. For example, a microphone converts acoustic energy into electrical energy, while a speaker converts electrical energy into acoustic energy. There are many other examples, including piezoelectric, magnetic, pressure, photoelectric, and resistive transducers. Nearly all transducers are sensors of some kind. **2.** A component, circuit, or device which converts an input in one energy form and produces an output in another energy form. For example, a photoelectric cell converts light energy into electrical energy. **3.** A component, circuit, or device which converts a parameter, magnitude, signal, or phenomenon in one form into that in another form. For example, that which converts a movement into a change in pressure, as occurs in a piston.

transducer amplifier 1. An amplifier that increases the output signal of a transducer. **2.** An amplifier that increases the input signal of a transducer.

transducer efficiency For a given transducer, the ratio of the power output in the desired form, to the power input.

transductor An electromagnetic device which utilizes one or more saturable reactors to obtain amplification. Depending on the design, a magnetic amplifier can achieve a very large power gain, and may be used, for instance, in servo systems requiring sizeable amounts of power to move heavy loads. Also called **magnetic amplifier**.

transfer 1. To move or cause to move or pass from one location or entity to another. For example, a charge, power, heat, or funds transfer. Also, such a movement or passing. **2.** The movement or copying, without alteration, of data from one location to another. This may be, for instance, across a data bus, a peripheral bus, or between nodes of a network, to name a few. Also called **data transfer**.

transfer characteristic 1. The relationship between an input magnitude, such as a voltage, to an output magnitude, such as a current. Also, the relation between an output magnitude, such as a voltage, to an input magnitude, such as a current. Usually expressed as a graph or plot. **2.** The relationship of the voltage of one electrode to the current of another, with all other electrodes having a fixed voltage. **3.** The relationship between the illumination incident upon a camera tube and the output signal current, under specified conditions.

transfer check A verification of the accuracy of a **transfer** (2).

transfer function A mathematical function which expresses the relationship between the output of a component, circuit, device, piece of equipment, or system, and its input. For instance, the relationship between the output of a linear control system and its input.

transfer impedance A measure of the shield effectiveness of a transmission line, such as a cable. The lower the transfer impedance, the more effective the shielding.

transfer protocol A protocol utilized for transferring data and/or files over a communications network, such as the Internet.

transfer rate The speed at which bits or bytes are transmitted over a communications line or bus. Usually measured in some multiple of bits or bytes per second. Also called **data rate**, **data transfer rate**, **bit transfer rate**, **bit rate**, or **line rate**.

transfer resistor Same as **transistor**.

transfer time 1. The time that elapses between one set of contacts being opened and another set of contacts being closed. **2.** In an uninterruptible power supply, the time that elapses between the failure of the primary source, such as mains AC, and the transfer of power to the battery or generator. **3.** The time required for data to be transferred from one point to another. For example, that required for transfers between nodes, or for a download to be completed.

transform 1. To change in nature, function, operation, state, format, or the like, from one form or manifestation into another. **2.** To transfer energy using a **transformer**.

transformation 1. The process of changing in nature, function, operation, state, format, or the like, from one form or manifestation into another. Also, having undergone such a transformation. **2.** The transfer of energy using a **transformer**.

transformation ratio Same as **turns ratio**.

transformer A component or device which is utilized to transfer electrical energy from one circuit to another via magnetic or electromagnetic induction. A transformer has no moving parts and consists of two or more coils of wire separately wrapped around a magnetic core. The coils are inductively linked by magnetic lines of force, and do not come into physical contact. A transformer is usually utilized to change electrical energy at one AC voltage to another, with a step-down transformer providing an output voltage lower than its input voltage, and a step-up transformer providing an output voltage higher than its input voltage. The winding that is directly connected to the input power is the primary, while the winding or windings that provide the output to the load or loads are the secondaries. The output of a variable transformer can be adjusted between given values.

transformer-coupled amplifier An amplifier whose stages are coupled via **tranformers**.

transformer coupling The coupling of two circuits by means of a mutual inductance provided by a transformer. In RF circuits, for instance, such coupling minimizes the capacitance between stages, which improves performance. Also called **inductive coupling** (2).

transformer efficiency For a given transformer, the ratio of the output power to the input power. Usually expressed as a percentage.

transformer hybrid A hybrid junction incorporating two or more transformers. Also called **hybrid set**.

transformer input current Same as **transformer primary current**.

transformer input voltage Same as **transformer primary voltage**.

transformer loss Also called **transformer losses**. **1.** Any power which is dissipated by a transformer without performing useful work. **2.** The ratio of the power an ideal transformer would deliver to a load under specified conditions, to the power a given transformer actually delivers to the same load under the same conditions. Usually expressed in decibels.

transformer losses Same as **transformer loss**.

transformer oil A specially-processed insulating oil, such as mineral oil, utilized in transformers.

transformer output current Same as **transformer secondary current**.

transformer output voltage Same as **transformer secondary voltage**.

transformer primary current The current flowing through a primary winding of a **transformer**. Also called **transformer input current**, or **primary current**.

transformer primary voltage The voltage applied to a primary winding of a **transformer**. Also called **transformer input voltage**, or **primary voltage**.

transformer ratio Same as **turns ratio**.

transformer secondary current The current flowing through a secondary winding of a **transformer**. Also called **transformer output current**, or **secondary current**.

transformer secondary voltage The voltage produced in a secondary winding of a **transformer**. Also called **transformer output voltage**, or **secondary voltage**.

transient 1. A very brief, and usually pronounced, condition, circumstance, event, or phenomenon. Examples include spikes, a sudden liberation of energy, or an abrupt increase in the magnitude of a sound. A system is said to be in a steady state once the effects of a transient have passed, and transients may be considered as temporary phenomena occurring between steady states. 2. A voltage or current spike occurring as a result of a temporary or unexpected condition or event, such as a lightning strike or a significant change in an input or driving signal. 3. That whose duration is very brief. For example, an impulse.

transient arrester Same as **transient suppressor**.

transient cookie Same as **temporary cookie**.

transient current A sudden and momentary increase in current. May be caused, for instance, by lightning, or faults in circuits. If protective measures are not employed, such a surge may bring about a failure, or significant damage. Also called **surge current**, or **current surge**.

transient data Data which is discarded between sessions. This contrasts with **persistent data**, which is stored from session to session.

transient disturbance A disturbance, such as a transient overvoltage, which is brief and intense.

transient field A field, such as a magnetic field utilized in magnetic pulse forming, which is brief and intense.

transient overcurrent An overcurrent that is brief and intense, such as that occurring during power switching.

transient overshoot Overshoot of an especially short duration.

transient overvoltage An overvoltage that is brief and intense, such as that occurring during power switching.

transient phenomena Any phenomena, such as those occurring between steady states, which is transient in nature.

transient protector Same as **transient suppressor**.

transient pulse A powerful impulse. Such pulses are detected in logic circuits, for instance, using logic probes.

transient recovery time The recovery time of a component, circuit, device, piece of equipment, system, mechanism, or process, after a transient condition such as an overvoltage.

transient response The speed with which a component, circuit, device, piece of equipment, or system changes its output in response to a sudden and intense change in its input.

transient signal A signal that is brief and intense. An electronic device or instrument may suffer a temporary loss of sensitivity following such a signal.

transient suppression The use and effect of a **transient suppressor**.

transient suppressor A component, circuit, or device, such as a surge suppressor or a clipper, which reduces the effects of,

eliminates, or prevents unwanted **transients**. Also called **transient protector**, or **transient arrester**.

transient voltage A sudden and momentary increase in voltage. May be caused, for instance, by lightning, or faults in circuits. If protective measures are not employed, such a surge may bring about a failure or significant damage. Also called **voltage surge**, or **surge voltage**.

transimpedance For an AC circuit or network, the ratio of the output voltage to the input current. Also called **mutual impedance**.

transimpedance amplifier An amplifier that converts an input current, such as that from a photodiode, into an output voltage.

transistor An active semiconductor device which can serve as a switch, amplifier, or oscillator, among other possible functions. The two main classes are bipolar junction transistors, and field-effect transistors. Each of these has three regions or electrodes, and charge carriers flow between two of these terminals, while a third controls this flow. Transistors may be used as discrete devices or in components such as ICs, and can be found in countless items, such as computers, robots, TVs, cellular telephones, wireless transmitters, audio amplifiers, appliances, cars, toys, and so on. A transistor may have more than three electrodes, as is the case with a tetrode junction transistor. The term was originally an acronym for **trans**fer re**sistor**. Also called **semiconductor transistor**.

transistor amplifier An amplifier which utilizes only transistors as its active components, as opposed, for instance, to those using electron tubes.

transistor computer Same as **transputer**.

transistor oscillator An oscillator which utilizes only transistors as its active components, as opposed, for instance, to those using crystals or tubes.

transistor parameters Parameters, such as small-signal parameters, which describe the behavior or performance of transistors under specified conditions.

transistor radio A radio which incorporates one or more transistors, as opposed to electron tubes. Such radios are usually portable.

transistor saturation In a bipolar transistor, the condition where the collector becomes positive with respect to the base. Ordinarily the converse is true. Also called **saturation (8)**.

transistor tester An instrument which serves to evaluate the function of transistors. Such an instrument, for instance, may be operated automatically for inspection during manufacturing, or manually, for research and development purposes.

transistor-transistor logic Logic similar to diode-transistor logic, except that bipolar transistors are utilized to perform logic functions. Its abbreviation is **TTL**, or **T^{2}L**.

transistor voltmeter A voltmeter, such as an FET voltmeter, which incorporates one or more transistors for enhanced sensitivity and accuracy.

transistorize To incorporate a component, circuit, device, piece of equipment, apparatus, or system with one or more transistors.

transistorized That which has one or more transistors incorporated, as opposed, for instance, to using electron tubes.

transit time 1. The time required for a charge carrier to travel from one point to another. For instance, that necessary for an electron or hole to travel between different points of a semiconductor. 2. The time required for an electron to travel from one electrode to another. For example, that necessary to travel from a cathode to an anode.

transition 1. A passage or change from one state, level, stage, mode, location, or the like, to another. For example, the transition of an electron from one energy level to another. **2.** To make a **transition (1)**.

transition band The interval of frequencies which surround a passband or stopband interface.

transition factor The ratio of the current delivered to a load whose impedance is not matched to the source, to the current that would be delivered to the same load if its impedance were fully matched. Also called **transition ratio**, **mismatch factor (2)**, **reflection factor (2)**, **reflectance (2)**, or **reflectivity (2)**.

transition ratio Same as **transition factor**.

transition region An area, zone, or interval between regions having different properties or characteristics. For instance, the region between n-type and p-type semiconductors, or the part of a spectrum between a passband and a stopband.

transition temperature 1. The temperature below which a material becomes superconductive. **2.** The temperature at which a **transition** occurs. For instance, the temperature at which a substance changes phase, or the temperature at which the ferromagnetic properties of a substance become paramagnetic.

translate 1. To convert one language, such as a computer language, into another. **2.** In computers, to convert from one format, such as that of data or a file, into another. **3.** To convert from one format, function, system, or level, to another. **4.** In computer graphics, to change the position of an image or object on the screen without rotating said image or object.

translation 1. The process of converting one language into another. **2.** In computers, the process of converting from one format, such as that data or a file, into another. **3.** The process of changing from one format, function, system, or level, to another. **4.** In computer graphics, the process of changing the position of an image or object on the screen without rotating said image or object.

translator 1. A device, software, hardware, system, or person which converts one language into another. **2.** In computers, an application which serves to convert one language into another. For example, from a high-level language into machine language, or from a natural language to a programming language. **3.** In computers, that which serves to convert from one format, such as that data or a file, into another. **4.** That which serves to change from one format, function, system, or level, to another. For example, a frequency translator, or a level translator.

translucency The quality or state of being **translucent**.

translucent That which transmits light, but is not completely transparent. The light diffusion present enables objects to be seen, but not clearly. Examples include porcelain and frosted glass. Optical density is a measure of the opacity of a translucent body, material, or medium.

transmission 1. The act or process of conveying a signal or any form of information, such as data, from one location to another, via wires, optical cables, waveguides, electromagnetic waves, acoustic waves, or any other means of communication. **2.** The **transmission (1)** of data. **3.** A specific instance of **transmission (1)** or **transmission (2)**. **4.** The signal, data, or other information conveyed via a **transmission (1)** or **transmission (2)**. **5.** Same as **transmittance**. **6.** The broadcasting of TV signals.

transmission band A frequency interval within which a signal is transmitted or passed. A frequency-sensitive device, such as an amplifier or filter, may have multiple transmission bands. Also called **passband (1)**.

transmission channel A path along which information is transmitted. For instance, a fiber-optic link carrying data between nodes of a network, or a bus between computer devices. Also called **channel (1)**.

transmission coefficient Same as **transmission ratio**.

Transmission Control Protocol Same as **TCP**.

Transmission Control Protocol over Internet Protocol Same as **TCP/IP**.

transmission convergence sublayer In ATM, a sublayer within the physical layer that prepares cells for transmission. Its abbreviation is **TCS**.

transmission electron microscope An electron microscope in which the electron beam is passed through a thin sample. Such microscopes can achieve resolutions of better than 1 angstrom. For reference, the smallest atom, that of hydrogen, has an approximate width of 1 angstrom. This contrasts, for instance, with a **scanning electron microscope**, in which the electron beam is scanned over the surface of the specimen. Its abbreviation is **TEM**.

transmission electron microscopy The use of **transmission electron microscopes** to view and analyze specimens. Its abbreviation is **TEM**.

transmission factor Same as **transmission ratio**.

transmission gain 1. An increase in the strength of a signal during its transmission from one point to another. Usually expressed in decibels. **2. Transmission gain (1)** expressed as the ratio of the power at the second or reception point, to the power at the first or transmission point.

transmission grating 1. A surface with many fine parallel lines, grooves, or slits, which are extremely close together and which serve to pass certain wavelengths of light. Used, for instance, in lasers. **2.** In a waveguide, fine parallel lines which serve to pass certain types of waves.

transmission level For a given point of a transmission line or system, the ratio of the signal power level at a reference point, such as the origin of said line or system. Usually expressed in decibels. Also called **relative transmission level**, or **relative power**.

transmission line 1. A physical medium, such as a wire, cable, or waveguide, which serves to transmit or otherwise convey signals, data, electricity, or electromagnetic radiation between points. Examples include communication lines, power lines, and antenna transmission lines. Also called **line (2)**. **2.** One or more wires, cables, or conductors which serve to supply electric power, especially AC power. **3.** A **transmission line (1)** utilized for communications.

transmission line analyzer A device which monitors a transmission line to help make sure that it is functioning properly, and when this is not the case, to help remedy this.

transmission line balance Also called **line balance**. **1.** The extent to which the electrical characteristics of two conductors in a transmission line are similar. Also, the degree of electrical similarity between a conductor and ground. The greater the balance, the lesser the extraneous disturbances, such as crosstalk and hum. **2.** A device, such as a balun, utilized to help achieve **transmission line balance (1)**.

transmission line equalizer An equalizer incorporated into a transmission line. Used, for example, to counteract distortion, compensate for deficiencies, or shape a frequency response to fit the requirements of a given transmission medium. Also called **line equalizer**.

transmission line impedance The impedance a transmission line presents between its terminals. Also called **line impedance**.

transmission line level The level of a signal at a given point of a transmission line.

transmission line loss A loss, such as that of energy, in a transmission line, such as an antenna feed. Also, the magnitude of any such losses. Also called **transmission line losses**.

transmission line losses Same as **transmission line loss**.

transmission line matrix A time-domain computational method for modeling electromagnetic wave interactions, such as wave propagation. Also called **transmission line modeling**. Its abbreviation is **TLM**.

transmission line modeling Same as **transmission line matrix**. Its abbreviation is **TLM**.

transmission line noise Noise generated in a transmission line, in addition to that present in the applied signal. Such noise may be due, for instance, to poor connections, or power-line current or voltage fluctuations. Also called **line noise**.

transmission line termination Same as **termination**.

transmission line transformer A transformer utilized to match transmission line impedances, or for line balance, isolation, or connection to equipment or additional circuits. Also called **line transformer**.

transmission loss Also called **transmission losses**. **1.** A reduction in the strength of a signal during its transmission from one point to another. Usually expressed in decibels. **2. Transmission loss** (1) expressed as the ratio of the power at the second or reception point, to the power at the first or transmission point.

transmission losses Same as **transmission loss**.

transmission mode **1.** For a guided electromagnetic wave, one of the possible manners or patterns of oscillation, transmission, or propagation. Such waves can propagate in one of three principal modes, which are as transverse electric waves, transverse magnetic waves, or transverse electromagnetic waves. Also called **mode** (3). **2.** In a transceiver, operation in which signals are sent. This contrasts with **reception mode**, in which signals are received. Switching between the two modes may be automatic or manual.

transmission ratio Also called **transmission coefficient**, or **transmission factor**. **1.** The ratio of the flux transmitted by a body, to the flux incident upon the same body. **2.** The ratio of the radiant energy transmitted from a body, to the total radiant energy incident upon the same body.

transmission speed The rate at which data, or any form of information, is transmitted over a communications line. Such a speed may be expressed, for example, in some multiple of bits per second. Also called **communications speed**.

transmissivity Same as **transmittivity**.

transmissometer An instrument or device which measures the level of transmission of light through a medium, such as the atmosphere.

transmit Its abbreviation is **XMIT**, or **XMT**. **1.** To transfer something from one entity to another. For example, to cause a disturbance to propagate through a given medium, or for energy to be transferred from one body to another. **2.** To allow particles and/or energy to pass through a given material or medium. For example, for light to pass through the atmosphere. **3.** To transfer a signal or any form of information, such as data, from one location to another, via wires, optical cables, waveguides, air, or another medium, using electromagnetic waves, acoustic waves, or any other means of communication. **4.** To **transmit** (3) data. **5.** To broadcast TV signals.

transmit-receive switch A switch that enables a single antenna to be used to transmit and receive simultaneously or alternately. This switch can be manual or automatic, and may be in the form of a circuit, tube, or device. Used, for example, in radar. Its abbreviation is **TR switch**. Also called **duplexing assembly**.

transmit-receive tube A switching tube that enables a transmitter and receiver to use the same antenna. Generally used in radar systems. Its abbreviation is **TR tube**.

transmittance For a body or material which is irradiated, the proportion of the incident radiation transmitted. Put another way, **transmittance** = (1 − **absorptance**). Also called **transmittancy**, or **transmission** (5).

transmittancy Same as **transmittance**.

transmitter Its abbreviation is **TX**. **1.** A component, device, piece of equipment, system, object, or entity which emits, radiates, or otherwise sends signals. Transmitters which send information-bearing signals are used in many areas, including communications, TV and entertainment, radars, and so on. Also called **sender** (1). **2.** A circuit, device, or system which converts audio, video, or other content into radio waves which are emitted. AM, FM, and AM/FM transmitters are examples. Also called **radio transmitter**, **radio sender**, or **sender** (2). **3.** Same as **telephone transmitter**. **4.** That which originates or otherwise emits signals, energy, particles, waves, or the like, which move from one point to another.

transmitter muting In a transceiver, the reduction or silencing of the transmitting output unit while receiving. This may occur manually or automatically.

transmitter-receiver Same as **transceiver**.

transmitter-responder Same as **transponder**.

transmitter synchro In a synchro system, the synchro which transmits a voltage corresponding to the position of its rotor. Also called **synchro transmitter**.

transmitting antenna An antenna which emits electromagnetic radiation. There are many types of transmitting antennas, including those utilized in communications, TV and entertainment, radars, and so on. A single antenna may be used for both transmission and reception.

transmitting device A device, such as a transmitter on a GPS satellite, a beacon transmitter, or a transmitting fax, which emits or sends a signal intended for a receiving device utilizing the same system.

transmitting equipment Equipment, such as radar sets, which incorporate one or more **transmitting devices**.

transmitting station A station, such as a radio or homing station, which incorporates one or more **transmitting devices**.

transmittivity The ratio of the radiation transmitted from one medium to another, to the radiation perpendicularly incident to the surface between the two mediums. Also called **transmissivity**.

transmultiplexer A device or piece of equipment that converts multiplexed signals in one form, such as frequency-division multiplex, to another, such as time-division multiplex, and vice versa. Also spelled **transmultiplexor**.

transmultiplexor Same as **transmultiplexer**.

transonic At, near, or approaching the speed of sound. A transonic interval may encompass from Mach 0.9 to 1.1, Mach 0.5 to 1.5, Mach 0.5 to 0.9, and so on.

transonic speed A speed, such as that of an aircraft, which is **transonic**.

transparence Same as **transparency** (1).

transparency **1.** The quality or state of being **transparent**. Also spelled **transparence**. **2.** The extent to which something is transparent. **3.** An object, such as a photographic slide, which is viewed by allowing light to shine through it.

transparent **1.** A body, material, or medium which freely passes radiant energy, such as light, or sound. **2.** A body, material, or medium which freely passes one form of radiant energy, such as visible light, but does not pass others, such as infrared light. **3.** A body, material, or medium which freely passes one type of particle, such as photons, but does not pass others, such as ions. **4.** A body, material, or medium which passes any type of particle. **5.** The passing of a signal through a communications network with no changes, or no detectable changes, as occurs, for instance, in transparent routing. **6.** In computers and communications, a change in hardware and/or software which does not produce any detectable changes, especially those which impair operation.

transparent bridge In networks, a bridge that can learn the locations of users or systems and deliver messages accordingly. Also known as **adaptive bridge**, or **learning bridge**.

transparent cache A cache, such as that of a proxy server, whose retrieval of data is invisible to a user. For example, a user requests a Web page, and if already present on a cache server, the page loads as if sent from the address it is located on the Internet.

transparent GIF A GIF which shows the background through all or part of an image. Used, for instance, for smooth blending of graphics on Web pages, or for spying on users utilizing Web bugs. Also called **clear GIF**, or **invisible GIF**.

transparent routing 1. Routing in which all protocols, IP addresses, and the like, automatically have the necessary conversions made, for a seamless connection end-to-end. 2. Routing in which the sender does not need to know the path to the recipient.

transponder Acronym for **trans**mitter-res**ponder**. 1. A device or system which incorporates both a transmitter and a receiver, and which emits the appropriate trigger signals automatically when the specified trigger signals are received. Used, for instance, in satellite communications, or in radars. Also called **responder** (1). 2. In a communications satellite, a unit which receives a signal from one location, processes and amplifies it, and retransmits it to other locations. A single satellite may have many transponders. Also called **satellite transponder**.

transport Same as **tape drive** (1).

transport layer Within the OSI Reference Model for the implementation of communications protocols, the fourth level, located below the session layer and above the network layer. This layer helps ensure that data transfers between nodes are complete and error free, handling tasks such as flow control, and error detection and correction. Also called **layer 4**.

transport-layer security A protocol, utilized for the encryption of messages and documents sent over the Internet, which incorporates and supersedes SSL. Enhancements include measures taken to ensure that no third parties can intercept or otherwise tamper with messages sent between clients and servers. Even though SSL is now part of TLS, the two are not mutually compatible. Its abbreviation is **TLS**.

transport protocol A communications protocol which handles tasks pertaining to the **transport layer**.

transportable That which can be moved from one location to another. That which is **portable** is generally smaller and more convenient to move and operate.

transportable computer A computer that can be moved from one location to another. A **portable computer** is generally smaller and more convenient to move and operate.

transputer Acronym for **trans**istor com**puter**. A **computer-on-a-chip** utilized in the context of parallel processing. Many transputers, even thousands, may be joined via high-speed links to enable very fast and powerful processing. In other contexts, a computer-on-a-chip is usually employed alone in a device, and only able to work alongside a limited number of other such computers.

transresistance For a component or circuit, the ratio of an output voltage to an input current.

transresistance amplifier An amplifier which converts its input current into an output voltage proportional to said current.

transuranic A chemical element having an atomic number greater than that of uranium, whose atomic number is 92. Also called **transuranium**.

transuranium Same as **transuranic**.

transversal filter A filter whose output is a weighed sum of the current and past inputs. It is one of the two primary types of filters utilized in digital signal processing, the other being **Infinite-Impulse Response filters**. Transversal filters tend to be more linear, but less efficient than Infinite-Impulse Response filters. Also called **Finite-Impulse Response Filter**.

transverse Crossing from side to side, or lying across. Said, for instance, of something that runs across something else, as opposed to running lengthwise. Its abbreviation is **trans**.

transverse electric mode A waveguide mode in which the electric field vector is perpendicular to the direction of propagation. Its abbreviation is **TE mode**.

transverse electric wave An electromagnetic wave whose electric field vector is at all points perpendicular to the direction of propagation. Used, for instance, in the context of waveguides. Its abbreviation is **TE wave**. Also called **H wave**.

transverse electromagnetic mode A waveguide mode in which the electric field and magnetic field vectors are perpendicular to the direction of propagation. Its abbreviation is **TEM mode**.

transverse electromagnetic wave An electromagnetic wave whose magnetic field vector and electric field vectors are at all points perpendicular to the direction of propagation. Used, for instance, in the context of waveguides. Its abbreviation is **TEM wave**.

transverse magnetic mode A waveguide mode in which the magnetic field vector is perpendicular to the direction of propagation. Its abbreviation is **TM mode**.

transverse magnetic wave An electromagnetic wave whose magnetic field vector is at all points perpendicular to the direction of propagation. Used, for instance, in the context of waveguides. Its abbreviation is **TM wave**. Also called **E wave**.

transverse magnetization 1. Magnetization of a material or medium in a manner in which the magnetic flux is essentially perpendicular to the long axis of the influencing field. In **longitudinal magnetization** (1) the converse is true. 2. In magnetic recording, magnetization of a material or medium essentially in the direction perpendicular to that the material or medium is traveling. In **longitudinal magnetization** (2) the converse is true.

transverse wave A wave in which the particles in the medium move perpendicular to the direction of propagation of the wave. Radio and light waves are examples. This contrasts with a **longitudinal wave**, where the particles in the medium move parallel to the direction of propagation.

transversely-excited atmosphere laser A powerful gas laser which utilizes electrical discharges across the optical axis to produce population inversion. The gas may be at a pressure comparable to or greater than one atmosphere. Its abbreviation is **TEA laser**.

trap 1. That which serves to catch, deflect, or otherwise separate energy, particles, waves, signals, signal components, or the like. For example, and ion trap, or a sound trap. 2. An entity or mechanism in a semiconductor which serves to capture electrons, or an impurity or lattice defect which cancels holes. 3. A resonant circuit which is connected in series or in parallel with an antenna receiving system to help suppress unwanted signals at a given frequency. Also called **wave trap**. 4. A mechanism or program which detects specified computer actions, processes, or conditions, such as those resulting in errors.

trap door Same as **trapdoor**.

trapdoor A hidden software or hardware mechanism that allows access to a program, system, or network in a manner which completely eludes all security measures. It can be activated, for instance, by a seemingly random key sequence, and is usually left in place by the software developer. It is intended to allow for functions such as maintenance, and is a significant security risk when known by an unintended user or program. Also spelled **trap door**. Also known as **back door**.

trapezium 1. A quadrilateral with no parallel sides. 2. Same as **trapezoid**.

trapezium distortion Same as **trapezoidal distortion**.

trapezoid A quadrilateral with one pair of parallel sides. A **parallelogram** is a quadrilateral with two pairs of parallel sides. Also called **trapezium (2)**.

trapezoid distortion Same as **trapezoidal distortion**.

trapezoid pattern Same as **trapezoidal pattern**.

trapezoid pulse Same as **trapezoidal pulse**.

trapezoid wave Same as **trapezoidal wave**.

trapezoidal distortion In TV reception, a form of distortion in which the displayed image is in the form of a **trapezoid**. It is a type of geometric distortion. Also spelled **trapezoid distortion**. Also called **trapezium distortion**.

trapezoidal pattern A pattern, such as that displayed on an oscilloscope, in the shape of a **trapezoid**. Also spelled **trapezoid pattern**.

trapezoidal pulse A pulse whose amplitude increases quickly to a given value, then steadily rises to a higher value, then drops quickly to the original value, where it is steady for a given interval, before repeating itself. Such pulses are in the shape of **trapezoids**. Also spelled **trapezoid pulse**.

trapezoidal wave A wave whose amplitude increases quickly to a given value, then steadily rises to a higher value, then drops quickly to the original value, where it is steady for a given interval, before repeating itself. Such waves are in the shape of **trapezoids**. Also spelled **trapezoid wave**.

trapping The propagation of radio waves through an atmospheric duct. Also called **guided propagation**.

trash Same as **trash can**.

trash can Also called **trash**. 1. An icon which represents a directory or folder where deleted files and folders are stored. A trash can allows for files to be restored if deleted by accident or if a user decides for any reason to keep a file. Also, the directory or folder where deleted files are stored. Also called **recycle bin**. 2. A **trash can (1)** which also serves to eject optical discs or floppies by dragging and dropping the icon representing the desired drive.

trashware Software whose design is especially bad, which is generally useless, or which otherwise is suitable for immediate discarding.

Travan A proprietary technology, cartridge, or drive utilized for tape back-up.

traveling wave A wave which propagates freely through a given medium, such as a transmission line or vacuum. In such a wave the antinodes and nodes progress along the medium at the speed of the wave, while in a **standing wave** the antinodes and nodes do not change position. Also called **progressive wave**.

traveling-wave amplifier Same as **traveling-wave tube amplifier**. Its abbreviation is **TWA**.

traveling-wave tube A microwave tube that can be used for amplification or oscillation, and which features low noise, high gain, and a wide bandwidth. In such a tube, an electron beam produced by an electron gun interacts with a microwave signal which travels along a slow-wave structure, so that they move in approximate synchronism. This facilitates the transfer of energy from the electron stream to the microwave signal, thus providing gain. The output is taken from, or near, the collector. A backward-wave oscillator is a type of traveling-wave tube in which the power transfer re-

sults in oscillation. Used, for instance, in satellite and wireless communications, and in radars. Its abbreviation is **TWT**.

traveling-wave tube amplifier An amplifier which incorporates one or more **traveling-wave tubes**. Its abbreviation is **TWTA**. Also called **traveling-wave amplifier**.

treble Within the audio-frequency spectrum, the high end of the detectable frequencies. This interval usually spans from about 2,000 or 5,000 Hz to about 20,000 Hz, although the defined interval varies. Also called **treble frequencies**.

treble boost 1. An amplifier circuit which effectively boosts high frequencies when listening to an audio system at low-volume settings. This compensates for the lower auditory response humans have under these circumstances, thus making the sound more natural. 2. The increase in high frequencies resulting from the use of a **treble boost (1)**.

treble compensation Emphasis of high frequencies utilizing **treble boost**.

treble control In an audio amplifier, a manual tone control which effectively boosts or attenuates high frequencies. It may be adjusted in discrete steps, or through a continuous interval.

treble cut The attenuation of high audio frequencies. May be used, for instance, to reduce or eliminate hiss. Seen especially in audio amplifiers. Also called **treble suppression**.

treble filter A filter which attenuates all frequencies above a given cutoff frequency, such as 5,000 Hz.

treble frequencies Same as **treble**.

treble response The ability of a component, circuit, device, piece of equipment, or system to handle the frequencies in the high end of the audible spectrum. This interval usually spans from about 2,000 or 5,000 Hz to about 20,000 Hz, and may refer, for instance, to the response of speakers, microphones, or amplifiers.

treble roll-off 1. The attenuation of frequencies above a given value, such as 2,000 or 5,000 Hz. 2. The amount by which frequencies above a given value, such 2,000 or 5,000 Hz, are attenuated.

treble suppression Same as **treble cut**.

trebly Descriptive of sound reproduction of a speaker or audio system which overly emphasizes **treble frequencies**.

tree 1. A diagram or other graphic representation which describes, lists, or otherwise represents in branching subunits a hierarchical structure of paths or alternatives. Examples include folder trees and decision trees. 2. A circuit or network with multiple branches, but no meshes.

tree network A communication network which combines a bus network with another configuration, such as that of a star or ring. For instance, it may consist of several star networks linked by a common bus. Seen, for example, in a setting where small groups of nearby computer are joined by wire cables, with all groups linked via optical cables.

tree structure A structure, such as that of data, in which each node has branching subunits which follow a hierarchical structure of paths or alternatives, all the way down until reaching nodes with no branches, called leaves. Examples include binary trees and decision trees.

tree topology The topology of a **tree network**.

Trellis-Coded Modulation A modulation technique used by certain modems, which augments noise immunity and incorporates error detection and correction. Used, for instance, to enhance quadrature amplitude modulation. Its abbreviation is **TCM**.

Trellis coding A form of error detection and correction used in some modems.

tremendously high frequency A range of radio frequencies spanning from 300 to 3000 GHz. These correspond to wavelengths of 1 mm to 100 μm, respectively. Also, pertaining to this interval of frequencies. Its abbreviation is **THF**.

tremendously low frequency Waves whose frequencies are usually defined to be less than 3 Hz, corresponding to wavelengths greater than 100,000 km. Its abbreviation is **TLF**.

TRF amplifier Abbreviation of **tuned-radio frequency amplifier**.

TRF receiver Abbreviation of **tuned-radio frequency receiver**.

tri-state logic Also called **three-state logic**. **1.** Digital logic which can assume any of three states, which are true, false, and undefined. **2.** Digital logic which can provide any of three outputs, which are high, low, and off.

triac Abbreviation of **tri**ode **a**lternating-**c**urrent switch. A gate-controlled thyristor in which current is able to flow in both directions. Used, for instance, for switching current in either direction when properly triggered.

triad In a CRT, a triangular group composed of a single red, green, and blue phosphor dot. If all three electron beams strike the three dots in a triad with equal intensity at the same time, it will appear as white.

triangle A closed geometric figure bounded by three straight lines.

triangular prism A prism with three triangular bases.

triangular pulse A pulse whose amplitude rises along a straight line to a given maximum, then returns quickly to the baseline again following a linear path. Alternatively, such a pulse may rise quickly to a given peak, then drop gradually following a linear path.

triangular wave A wave whose amplitude rises along a straight line to a given maximum, then returns quickly to the baseline again following a linear path. Alternatively, such a wave may rise quickly to a given peak, then drop gradually following a linear path.

triangular-wave generator A signal generator whose output consists of **triangular waves**.

triangulation The location of an unknown point, such as that of a ship or airplane, by forming a triangle in which two known points serve as the other two vertices, and using geometry to determine the location of said unknown point. Used, for instance, in radio navigation.

triatomic Occurring as, pertaining to, or consisting of three atoms. For instance, ozone, which is a triatomic gas. This contrasts, for instance, with molecular oxygen, which is a diatomic gas.

triax Same as **triax cable**.

triax cable A cable similar to a coaxial cable, except that the second conductor is surrounded by a dielectric material, followed by a third conductor, with a protective jacket enveloping the third conductor. Also called **triaxial cable**, or **triax**.

triax connector A fixture which enables connection of a **triax cable** between electrical devices. Also called **triaxial connector**.

triaxial cable Same as **triax cable**.

triaxial connector Same as **triax connector**.

triaxial loudspeaker Same as **triaxial speaker**.

triaxial speaker A speaker system which incorporates three concentric drivers, usually consisting of a tweeter mounted within a midrange, which is itself mounted inside a woofer, so that the sound radiates from a common axis. Alternatively, two smaller speakers may be mounted side-by-side within a larger speaker. Also called **triaxial loudspeaker**.

triboelectric Pertaining to **triboelectricity**.

triboelectricity The generation of electrostatic charges, or static electricity, through friction. For example, if a dielec-

tric such as glass or amber is rubbed with a dissimilar dielectric such as silk or wool, and each is separated, such charges are produced. Also called **frictional electricity**.

triboluminescence Luminescence generated by a mechanical action, especially friction. Seen, for instance, when vigorously rubbing quartz.

trichromatic Of, pertaining to, consisting of, or incorporating three colors. For example, the use of red, green, and blue as primary colors.

trickle charge **1.** The application of a slow and continuous charge to a storage battery, so as to compensate for internal losses and small discharges. Also called **float charge (2)**. **2.** The application of a slow and continuous charge to a storage battery, so as to maintain it as close as possible to its fully charged condition. **3.** The application of a slow and continuous charge to a storage battery.

trickle charger A device providing a **trickle charge**.

tricobalt tetroxide Black or gray crystals whose chemical formula is Co_3O_4. Used in semiconductors.

tricolor picture tube A CRT which produces the color images displayed on color TV receivers and computer monitors. A three-gun color picture tube emits three beams of electrons, which are focused and directed towards color phosphors, which in turn produce the color images on the screen. Also called **color picture tube, color TV picture tube**, or **color kinescope**.

trigatron An electron tube with a spark gap that can be triggered to start and stop conduction. Used, for instance, in radars.

trigger **1.** A pulse or other signal that initiates or activates a component, circuit, device, piece of equipment, system, action, process, or mechanism. Also, to so initiate or activate. **2.** A **trigger (1)** which serves to initiate or activate instantaneously. **3.** A **trigger (1)** which serves to terminate or stop an action, process, or mechanism. **4.** A **trigger (1)** which serves to alternate between two operational or stable states, as occurs in a flip-flop. **5.** In computers, a command or mechanism which initiates an action under specified conditions. For example, that which triggers the execution of an applet.

trigger circuit **1.** A circuit which serves as a **trigger (1)**, **trigger (2)**, **trigger (3)**, or **trigger (4)**. **2.** A two-stage multivibrator circuit with two stable states. In one of the stable states, the first stage conducts while the other is cut off, while in the other state the reverse is true. When the appropriate input signal is applied, the circuit flips from one state to the other. It remains in this state until the next appropriate signal arrives, at which time it flops back. Used extensively in computers and counting devices. Also known as **bistable multivibrator, flip-flop circuit, flip-flop, Eccles-Jordan circuit, bistable flip-flop**. **3.** A circuit whose output changes drastically in response to a minimal change in its input.

trigger current A current which serves as a **trigger (1)**, **trigger (2)**, **trigger (3)**, or **trigger (4)**. Also called **triggering current**.

trigger diode A bi-directional avalanche diode with two terminals and three layers, which conducts symmetrically. It allows current to flow when the breakover voltage is exceeded in either direction. Also called **diac**, or **three-layer diode**.

trigger finger A locking of a finger into a fixed position, or flexion in which motion is jerky. The index finger is a likely candidate, and such a condition can result from incessant mouse clicking during computer and/or network activity. It is a type of repetitive-strain injury.

trigger level The signal level or amplitude necessary for **triggering**. Also called **triggering level**.

trigger pulse A pulse, such as the output of a **trigger circuit** (1), which serves as a **trigger** (1), **trigger** (2), **trigger** (3), or **trigger** (4). Also called **triggering pulse**.

trigger signal A signal, such a pulse, which serves as a **trigger** (1), **trigger** (2), **trigger** (3), or **trigger** (4). Also called **triggering signal**.

trigger voltage A voltage which serves as a **trigger** (1), **trigger** (2), **trigger** (3), or **trigger** (4). Also called **triggering voltage**.

triggering The action of a **trigger** (1), **trigger** (2), **trigger** (3), or **trigger** (4).

triggering current Same as **trigger current**.

triggering level Same as **trigger level**.

triggering pulse Same as **trigger pulse**.

triggering signal Same as **trigger signal**.

triggering voltage Same as **trigger voltage**.

trillion A number equal to 10^{12}.

trillion floating-point operations per second Same as **TFLOPS**.

trillion instructions per second Same as **TIPS**.

trillion operations per second Same as **TFLOPS**.

trim To make fine adjustments in an instrument, setting, level, or the like. Performed, for instance, when calibrating an instrument.

trimmer **1.** In a radio receiver, an adjustable capacitor inserted in parallel with a larger capacitor, and which is utilized for fine-tuning adjustments. Also called **trimmer capacitor** (1), or **trimming capacitor** (1). **2.** An adjustable capacitor utilized for making fine adjustments. Also called **trimmer capacitor** (2), or **trimming capacitor** (2). **3.** A component, such as an adjustable capacitor or resistor, utilized for making fine adjustments.

trimmer capacitor **1.** Same as **trimmer** (1). **2.** Same as **trimmer** (2).

trimmer potentiometer A potentiometer, such as that with an adjustable screw, utilized for making fine adjustments. Such a potentiometer allows for precise placement of the sliding contact or tap, and reduces the likelihood of accidentally changing its position. Also called **trimming potentiometer**.

trimming The use of a **trimmer**.

trimming capacitor **1.** Same as **trimmer** (1). **2.** Same as **trimmer** (2).

trimming potentiometer Same as **trimmer potentiometer**.

Trinitron A type of CRT that uses an aperture grill picture tube.

triode **1.** An electron tube with three electrodes, which are an anode, a cathode, and a control electrode. Also called **triode tube**. **2.** A three-electrode active device, such as a transistor.

triode alternating-current switch Same as **triac**.

triode amplifier An amplifier that uses, or consists of, one or more **triodes** (1).

triode-hexode A triode tube and a hexode working together, or within the same envelope.

triode tube Same as **triode** (1).

trip To release, trigger, activate, or otherwise set something into operation. For example, to trigger a circuit breaker.

trip coil In a circuit breaker, a coil which opens the protected circuit when the current flowing through it exceeds a given threshold. Also called **tripping coil**.

trip device In a circuit breaker, a device, such as a trip coil, which opens the protected circuit under specified conditions. Also called **tripping device**.

triphase Consisting of three phases, as in three-phase current.

Triple DES Abbreviation of **Triple D**ata **E**ncryption **S**tandard. An enhanced version of DES providing even stronger encryption. One version utilizes three keys in succession, as opposed to one. Its own abbreviation is **TDES**, or **3DES**.

triple-detection receiver A superheterodyne receiver in which there are successive frequency conversions utilizing two local oscillators, and thus having two intermediate frequencies. The first intermediate frequency is higher, for adequate image rejection, while the lower second intermediate frequency provides high selectivity and gain. Also called **double-conversion superheterodyne receiver, double-conversion receiver, dual-conversion receiver, dual-conversion superheterodyne receiver,** or **double superheterodyne receiver**.

triple diode An electronic component, such as an electron tube, containing three diodes.

triple point The temperature and pressure at which the gas, liquid, and solid phases of a substance are in equilibrium. The triple point of water, for instance, occurs at approximately 273.16 K (0.01 °C), and a pressure of approximately 611.2 Pa.

triple super-twist nematic A technology used in **triple super-twist nematic displays**.

triple super-twist nematic display A twisted nematic display in which three separate liquid-crystal display plates are combined to form a single panel, providing greater contrast. Its abbreviation is **TSTN display**.

tripler **1.** A circuit or device which serves to triple a value, such as that of a current, frequency, or voltage. **2.** A circuit or device whose output frequency is triple that of its input frequency. May consist, for instance, of a stage whose resonant output circuit is tuned to the third harmonic of the input frequency. Also called **frequency tripler**. **3.** A rectifier circuit whose output DC voltage is about three times the peak value of its input AC voltage. Also called **voltage tripler**.

triplexer A device that allows the use of two radar receivers simultaneously.

tripping The releasing, triggering, activating, or otherwise setting of something into operation. For example, the triggering of a circuit breaker. Also, an instance of such tripping.

tripping coil Same as **trip coil**.

tripping device Same as **trip device**.

tristimulus values The amounts of each of the three primary colors that must be combined in order to match a given color sample.

tritium An isotope of hydrogen whose nucleus contains two neutrons and a proton. It is radioactive, and is used, for instance, in cold-cathode tubes, in tracer studies, and to bombard atomic nuclei. Its symbol is **T**. Also called **heavy hydrogen** (2).

tritium oxide A chemical compound whose formula is T_2O. It is water in which the hydrogens are in the tritium isotopic form. Also called **heavy water** (2).

trivalent Pertaining to an element, such as boron or gallium, which has a valence of three. Such elements are used, for instance, as dopants which contribute mobile holes which increase the conductivity of an intrinsic semiconductor. Doping with a trivalent chemical element makes a p-type semiconductor, while **pentavalent** elements provide for n-type semiconductors.

Trivial File Transfer Protocol A version of FTP with no security features. Its abbreviation is **TFTP**.

Trojan Same as **Trojan horse**.

Trojan horse A program which appears to be useful and/or harmless, but which actually causes serious harm to a computer it is run on. A Trojan house may be utilized, for in-

stance, to collect private information, such as passwords, or to destroy data. Unlike viruses, Trojan horses do not replicate themselves. Also called **Trojan**. A popular misnomer for this term is **Trojan virus**.

Trojan virus A misnomer for **Trojan horse**.

troll **1.** To attempt to induce other network users, especially over the Internet, to respond to intentionally inflammatory messages, postings, emails, or the like. **2.** To browse or surf the Internet. **3.** To be present in a chat room or other similar setting without saying or typing anything. **4.** A user that **trolls (1)**, **trolls (2)**, or **trolls (3)**.

troposcatter Abbreviation of **tropospheric scatter**.

troposphere The innermost layer of the earth's atmosphere, extending from the surface to an altitude of about 10 to 15 kilometers. Most weather-related phenomena occur in this atmospheric shell.

tropospheric bending The deviation of the path of radio waves as they propagate through the **troposphere**.

tropospheric duct Within the troposphere, a layer which, depending on the temperature and humidity conditions, may act as a waveguide for the transmission of radio waves for extended distances. Also called **atmospheric duct**, or **duct (3)**.

tropospheric ducting The occurrence of a **tropospheric duct**.

tropospheric propagation Propagation of radio waves well beyond line-of-sight distances due to **tropospheric scatter**.

tropospheric scatter The scattering of radio waves as they collide with particles and turbulences in the **troposphere**. The scattering increases as the frequency of the waves increases. Its abbreviation is **troposcatter**. Also called **tropospheric scattering**.

tropospheric scattering Same as **tropospheric scatter**.

tropospheric wave A radio wave whose propagation has been assisted by reflecting off the **troposphere**. It is a type of atmospheric radio wave.

troubleshoot To systematically investigate the origin or cause of a malfunction, fault, or problem, in order to remedy it. For instance, to use a logic analyzer to troubleshoot logic circuits, or a network analyzer to troubleshoot a communications network.

troubleshooting The systematic investigation of the origin or cause of a malfunction, fault, or problem, in order to remedy it. Troubleshooting may be performed on components, circuits, devices, equipment, systems, networks, software, hardware, and so on.

troubleshooting test A test, such as that utilizing a protocol analyzer, performed during **troubleshooting**.

troubleshooting testing The performance of **troubleshooting tests**.

trough Also called **trough value**. **1.** The minimum instantaneous value of a voltage, current, signal, or other quantity. **2.** The minimum instantaneous absolute value of the displacement from a reference position, such as zero, for a voltage, current, signal, or other quantity.

trough value Same as **trough**.

troy ounce A unit of mass equal to approximately 0.0311035 kg. Also called **ounce (2)**.

true BASIC A structured-programming version of BASIC.

true color Color in which the bit depth is 24, which allows for over 16.7 million possible colors to be displayed. 24 bits means that there are 8 for each of the primary colors, which are red, green, and blue. Each pixel is composed of a trio of color phosphor dots, one for each primary color. Also called **24-bit color**.

true complement The numerical result obtained when a number is subtracted from the radix, which is the number of dig-

its used in a numbering system. For instance, the complement of 6 in the decimal number system is 4: $(10 - 6) = 4$. The complement of a number in the binary number system is the other: 1 is the complement of 0, and 0 is the complement of 1. Used in computers, for instance, to represent negative numbers. Also called **complement**, or **radix complement**.

true north The direction of the north geographic pole, as opposed to the north magnetic pole.

true power In an AC circuit, the rate at which work is performed or energy is transferred, measured in watts. It is the product of the voltage portion of the signal being measured, multiplied by the current that is in phase with that voltage. It is also equal to the difference between the apparent power and the reactive power. Also called **real power**, **active power**, or **actual power (1)**.

true south The direction of the south geographic pole, as opposed to the south magnetic pole.

TrueType A widely-utilized outline font technology. Also, a font presented or printed using this technology. Also called **TrueType font**.

TrueType font Same as **TrueType**.

truncate To perform a **truncation** procedure.

truncated number The number resulting from a **truncation (1)** procedure.

truncating Same as **truncation**.

truncation Also called **truncating**. **1.** The dropping one or more decimal digits from a number, with the rightmost remaining digit never changing. When **rounding (1)**, the rightmost remaining digit is increased if the next number was 5 or higher. For example, when truncating 0.66666 to four decimal places the result would be 0.6666, while it would be 0.6667 if rounding. **2.** The deletion of leading or trailing characters or digits from a string. For instance, the removal from a field of all text that exceeds a given number of characters. When truncating, there is no concern pertaining to the accuracy of that which remains.

trunk **1.** A physical link, such as a wire, cable, or bus, over which information or power in transferred between two points. **2.** A communications path between two central offices or switching systems. Also called **trunk circuit (1)**. **3.** A communications path between a central office and a PBX or the premises of a customer. Also called **trunk circuit (2)**. **4.** A path via which data travels within a computer. **5.** A path utilized to interconnect two power generating stations or power distribution networks. Also called **trunk main**, or **trunk feeder**.

trunk busy signal An audible signal indicating that there are no idle trunks in a group. Also called **trunk busy tone**, or **group busy tone**.

trunk busy tone Same as **trunk busy signal**.

trunk cable **1.** In a cable TV system, the cable that leads from the headend to a distribution point. A distribution cable then leads to the individual subscribers. Also called **feeder cable (2)**. **2.** A cable running between a central office and a distribution point. A distribution cable then leads to the local stations and subscribers. Also called **feeder cable (3)**.

trunk circuit **1.** Same as **trunk (2)**. **2.** Same as **trunk (3)**.

trunk exchange A central office which interconnects **trunks (2)**.

trunk feeder Same as **trunk (5)**.

trunk group Two or more associated **trunks (2)** or **trunks (3)** which are usually utilized for the same purpose. For example, those utilized only for incoming calls.

trunk main Same as **trunk (5)**.

TRUSTe An initiative which seeks to build the trust of Internet users by promoting the full disclosure of Web site privacy-related matters, such as what data is collected and how it is used.

trusted computer system A computer system that is considered to be highly, or completely, resistant to intrusion, sabotage, or the like. Its abbreviation is **TCS**.

truth table A table or diagram that describes a logic function or operation by indicating the output truth values for all possible combinations of input truth values. Used, for instance, to represent logic functions, and in designing circuits which implement such functions.

truth value In logical operations, the truth or falsity of a proposition. When dealing with such operations, a 1 is considered as a true value, and 0 is a false value.

TSAPI Abbreviation of Telephony Services Application Programming Interface, or Telephony Server Application Programming Interface. A programming interface which facilitates the interaction of servers and voice services, especially telephony.

Tschebyscheff filter A filter characterized by minimal passband ripple and a steep cutoff response. Used, for instance, in audio applications. Also spelled **Chebyshev filter**.

TSOP Abbreviation of **thin small outline package**.

TSR Abbreviation of terminate and stay resident. A utility that once loaded stays in memory until terminated. Used, for instance, to quickly load a calendar program in an older system which is unable to perform multitasking.

TSTN Abbreviation of **triple super-twist nematic**.

TSTN display Abbreviation of **triple super-twist nematic display**.

TT Abbreviation of **text telephone**.

TTL 1. Abbreviation of **Time-to-Live**. 2. Abbreviation of **transistor-transistor logic**.

TTS Abbreviation of **Transaction Tracking System**.

TTY 1. Abbreviation of **teletypewriter**. 2. Abbreviation of **teletype**. 3. Abbreviation of **text telephone**.

tube 1. An active device consisting of a hermetically-sealed envelope within which electrons are conducted between electrodes. The cathode is the source of electrons, the positive electrode to which they travel is the anode or plate, while other electrodes that may be present include control grids and screen grids. Electron tubes may or may not contain a gas, and its presence and concentration affects the characteristics of said tubes. Such tubes have many applications, including their use in amplification, modulation, rectification, and oscillation. There are many examples, including CRTs, phototubes, pentodes, mercury-vapor tubes, and so on. When such a tube is evacuated to a degree that any residual gas present does not affect its electrical characteristics, it is called **vacuum tube**. Also known as **electron tube**, or **valve**. 2. A long and hollow object, usually cylindrical in shape, which serves for the passage of fluids or other materials, for the placement of cables, or to hold substances.

tube amplifier An amplifier which incorporates one or more vacuum tubes. Also called **vacuum-tube amplifier**.

tube diode An electron tube with two electrodes, specifically an anode and a cathode. In such a tube the cathode may also be called filament and the anode may be called plate. Also called **diode tube**.

tube noise 1. The electrical noise generated by a **tube (1)** in the absence of an input signal. 2. Any noise generated by a **tube (1)**.

tube socket A socket into which a **tube (1)** is inserted. Such a socket provides the necessary connections for the tube's terminals, in addition to providing mechanical support.

tube tester An instrument which serves to evaluate the function of **tubes (1)**. Such an instrument may indicate transconductance, noise levels, grid emissions, shorts, and can have sockets for three-, four-, five-, six-, seven-, eight-, or

nine-pin tubes, seven- or nine-pin miniature tubes, and so on. Some also test transistors and diodes.

tunable That which can be **tuned**. Also spelled **tuneable**.

tunable dye laser A dye laser whose output of coherent light can be tuned over a continuous range of frequencies. Its output can be varied by adjusting optical tuning elements and/or changing the dye utilized.

tunable laser A laser, such as a tunable dye laser, whose output of coherent light can be varied over a continuous range of frequencies.

tune 1. To adjust to a desired frequency. For instance, to tune a receiver or an oscillator. Also, such an adjustment. 2. To adjust a component, circuit, body, or system, so as to make it resonate with a given signal, force, or the like. Also, to adjust a resonant frequency. 3. To adjust a component, circuit, device, piece of equipment, or system, so as to attain optimum performance or output.

tuneable Same as **tunable**.

tuned 1. That which has been adjusted to a desired frequency. Said, for instance, of a receiver. 2. A component, circuit, body, or system which has been adjusted so as to make it resonate with a given signal, force, or the like. Also, a resonant frequency which has been adjusted. 3. A component, circuit, device, piece of equipment, or system which has been adjusted so as to attain optimum performance or output.

tuned amplifier 1. An amplifier designed for operation at a single frequency, or within a very narrow band. 2. An amplifier which can be tuned over a given range of frequencies.

tuned antenna An antenna whose parameters, such as capacitance and inductance, provide for resonance at its operating frequency. Also called **resonant antenna**.

tuned cavity An enclosure which is able to maintain an oscillating electromagnetic field when suitably excited. The geometry of the cavity determines the resonant frequency. Used, for instance, in lasers, klystrons, or magnetrons, or as a filter. Also known by various other names, including **cavity (2)**, **cavity resonator**, **resonant chamber**, **resonating cavity**, **microwave cavity**, **microwave resonance cavity**, **waveguide resonator**, and **rhumbatron**.

tuned circuit Also called **tuning circuit**. 1. A resonant circuit whose resonance frequency can be adjusted. For example, a tank circuit with variable components. Used, for instance, in tuned amplifiers. 2. A **tuned circuit (1)** which is set to a specific frequency. 3. A **tuned circuit (1)** utilize for tuning. Used, for instance, in a receiver or oscillator.

tuned dipole A dipole antenna whose parameters, such as capacitance and inductance, provide for resonance at its operating frequency. Also called **tuned dipole antenna**, or **resonant antenna**.

tuned dipole antenna Same as **tuned dipole**.

tuned-radio frequency amplifier Its abbreviation is **TRF amplifier**, or **tuned RF amplifier**. 1. An RF amplifier designed for operation at a single frequency, or within a very narrow band. 2. An RF amplifier which can be tuned over a given range of frequencies.

tuned-radio frequency receiver Its abbreviation is **TRF receiver**, or **tuned RF receiver**. 1. An RF receiver designed for operation at a single frequency, or within a very narrow band. 2. An RF receiver which can be tuned over a given range of frequencies.

tuned RF amplifier Abbreviation of **tuned-radio frequency amplifier**.

tuned RF receiver Abbreviation of **tuned-radio frequency receiver**.

tuned transformer A transformer designed to resonate at the frequency of its input AC voltage.

tuner **1.** A component, circuit, device, or mechanism which serves to select a desired frequency. Used, for instance, in a radio receiver. **2.** A **tuner** (**1**) which is a subunit within a device or piece of equipment such as a receiver in a sound system. **3.** A separate sound system component which is utilized as a **tuner** (**1**) for selecting from radio bands such as AM and FM. A **receiver** (**2**) is sound system component which incorporates such a tuner, along with a preamplifier and a power amplifier. **4.** That which serves to adjust the resonance frequency of a component, circuit, body, or system.

tuner dial Same as **tuning dial**.

tuner range Same as **tuning range**.

tuner sensitivity Same as **tuning sensitivity**.

tungsten A lustrous silver-gray metal whose atomic number is 74. It is very hard, dense, ductile, has great corrosion resistance, and has the highest melting point and lowest vapor pressure of all known metals. At elevated temperatures it also has the highest tensile strength. It has almost 35 known isotopes, of which 3 are stable. Its applications include its use as lamp filaments, in electron tubes, in X-ray tubes, hard parts such as rocket nozzles, and in high-speed cutting devices and tools. Also called **wolfram**, from where it gets its chemical symbol, **W**.

tungsten carbide A very hard gray powder whose chemical formula is **WC**. Used, for instance, instead of diamonds in cutting and grinding tools, and in resistors.

tungsten filament A lamp filament, such as that in a tungsten-halogen lamp, composed of **tungsten**.

tungsten-halogen lamp A halogen lamp whose contained gas is iodine, or that of another halogen or mixture of halogens, and whose filament is composed of tungsten. The gas in these high-efficiency lamps reacts with the tungsten metal which evaporates, recycling the particles back onto the filament surface. Also called **halogen lamp** (**2**).

tungsten silicide A very hard blue-gray solid whose chemical formula is WSi_2. Used, for instance in lasers, semiconductors, resistors, optics, and as a mask.

tuning **1.** The process of adjusting to a desired frequency. For instance, the tuning of a receiver or an oscillator. **2.** The process of adjusting a component, circuit, body, or system, so as to make it resonate with a given signal, force, or the like. Also, the process of adjusting a resonant frequency. **3.** The process of adjusting a component, circuit, device, piece of equipment, or system, so as to attain optimum performance or output.

tuning capacitor A variable capacitor utilized for **tuning**.

tuning circuit Same as **tuned circuit**.

tuning coil A variable inductor utilized for **tuning**. Also called **tuning inductor**.

tuning core Same as **tuning slug**.

tuning dial A dial, such as that on a radio receiver, utilized for **tuning**. Also called **tuner dial**.

tuning diode A diode, such as a varactor, utilized for **tuning**.

tuning fork A steel instrument with a handle and a U-shaped portion that can be made to vibrate. A vibrating tuning fork produces an essentially pure tone which can be maintained for several seconds. Used, for instance, for testing, tuning, and calibration.

tuning fork oscillator An oscillator whose frequency is determined by a **tuning fork**. Also called **fork oscillator**.

tuning fork resonator A resonator whose frequency is determined by a **tuning fork**. Also called **fork resonator**.

tuning indicator **1.** A device, such as a meter or an indicator light, which indicates when a radio receiver is properly tuned to the frequency of a given station. **2.** A device, such as a meter or an indicator light, which indicates the proper

tuning of a component, circuit, device, piece of equipment, or system. Used, for instance, to calibrate antennas, or for locating satellite signals.

tuning inductor Same as **tuning coil**.

tuning meter A meter, such as a digital meter, serving as a **tuning indicator**.

tuning piston In a waveguide, a sliding cylinder, usually consisting of a metal or dielectric, which is utilized for tuning. Also called **piston** (**1**), **plunger**, or **waveguide plunger**.

tuning range The upper and lower limits within which a tuner, such as that in a radio receiver, can select frequencies. Also called **tuner range**.

tuning sensitivity A measure, such as the ratio of the frequency change per unit voltage change, of the sensitivity of a tuner. Also called **tuner sensitivity**.

tuning slug A movable ferrite core which is utilized to vary the inductance of a coil. Also called **tuning core**, or **slug** (**2**).

tuning stub A stub, such as a coaxial stub, which is utilized to tune a transmission line. Used, for instance, for impedance matching.

tuning voltage A voltage, such as that applied to a varactor, utilized for **tuning**.

tunnel **1.** To pass through a potential barrier via the **tunnel effect**. **2.** To send data utilizing **tunneling** (**2**).

tunnel diode A pn-junction diode which is highly doped on each side of the junction. It has a negative resistance at a low voltage in the forward-bias direction, due to the **tunnel effect**. May be used, for instance, in oscillator or amplifier circuits whose frequencies run well into the microwave range. Also called **Esaki diode**.

tunnel-diode amplifier A microwave amplifier incorporating one or more **tunnel diodes**.

tunnel-diode oscillator A microwave oscillator incorporating one or more **tunnel diodes**.

tunnel effect The passage of a particle, such as an electron, through a potential barrier that could not be crossed according to classical mechanics. Quantum mechanics, however, takes into account the wavelike properties of particles, and provides for the existence of a probability of an electron passing through this region. The wave function of such a particle allows for a given probability that the particle will appear on the other side of said barrier. The tunnel effect, for instance, explains alpha decay, and is employed, for example, in tunnel diodes. Also called **tunneling** (**1**).

tunneling **1.** Same as **tunnel effect**. **2.** A technique which enables a network to send data utilizing one protocol, through another network using different protocol. It does so by encapsulating packets using one network protocol within packets being transmitted through the other network. Also called **encapsulation** (**3**).

tunneling current The current created by particles that undergo **tunneling** (**1**).

tuple A group of related fields, each containing information. For instance, a group of fields, each containing one of the following items: a name, a corresponding address, and a contact number. The term is used in the context of relational databases. Also called **record** (**2**).

turbidimeter An instrument that measures and indicates the loss in the intensity of a light beam as it passes through a solution, slurry, or the like.

Turbo C A proprietary C compiler.

Turbo C++ A proprietary C++ compiler.

Turbo Pascal A proprietary Pascal compiler.

turbo pump Abbreviation of **turbomolecular pump**.

turbomolecular pump A vacuum pump that uses a turbine that spins very rapidly to mechanically push gas molecules out of a chamber. A turbomolecular pump can provide ultra-high vacuums, and is usually used in conjunction with a mechanical pump. However, it is not especially effective with very light gases such as hydrogen and helium. Its abbreviation is **turbo pump**.

turbulence A disturbed flow of a fluid, especially air or water, which is random in nature. In the atmosphere, for instance, it can be manifested as abrupt and unpredictable shifts in air speed and direction, and can scatter radio waves.

Turing machine A hypothetical computing machine that served as a model for digital computing, which was developed later. Such a machine would have an infinite amount of storage and would have a finite number of internal states, with its behavior at any given moment determined by the state of the machine and the symbol or character being read at that moment.

Turing test A test in which an intelligent person interrogates both a computer and another person, without knowing which is which, then states which is which based on the answers. If the interrogator is convinced that they are equal, or has no way of distinguishing between them, then the system passes this test of artificial intelligence.

turn A single wind or loop of a wire or conductor, such as that in a coil.

turn-off time The maximum time required for a component, circuit, switch, device, piece of equipment, or system to completely stop operation once the actuating signal or driving power is removed.

turn-on time The maximum time required for a component, circuit, switch, device, piece of equipment, or system to be fully operational once the actuating signal or driving power is provided.

turnaround time **1.** The time necessary for a result to be obtained once all that is needed has been furnished. For example, the time that elapses between the entering of a database query and receiving the results. **2.** In a half-duplex circuit, the time a unit requires to switch from a receiving mode to a transmitting mode, and vice versa. **3.** The time required for a transceiver to switch from a receiving mode to a transmitting mode, and vice versa.

turns ratio For a given transformer, the ratio of the primary turns to the secondary turns. Also called **transformer ratio**, **transformation ratio**, or **ratio of transformation**.

turnstile antenna An antenna with one or more levels of two crossed dipoles, so that each layer has the approximate appearance of a turnstile. Such antennas are well-suited for use in the ultra-high-frequency and very-high frequency ranges.

turntable Also called **phonograph turntable**. **1.** An obsolescent device which reproduces sound which has been recorded in the spiral grooves of a disc called phonograph record. A turntable incorporates a motor-driven **turntable (2)** which rotates a platter at a constant speed. The disk is placed on the platter, and a tone arm, which holds the cartridge, travels progressively towards the center of the disk. The stylus of the cartridge follows the variations in the grooves, and converts them into electrical signals which are fed into an amplifier, preamplifier, receiver, or other such device. Also called **phonograph**, or **record player**. **2.** The rotating platform which rotates the platter at a constant speed in a **turntable (1)**.

turntable rumble **1.** An unwanted low-frequency vibration that is audible, and which is due to the vibrations of a platter and motor of a phonograph turntable. **2.** A quantified level of **turntable rumble (1)**.

turret tuner A tuner with a rotary switch, such as that on some older TVs, with detents which click in position at each selected station.

TV **1.** Abbreviation of **television**. A communications system utilized for picking up visual and auditory information which is transmitted via radio waves, cables, or wires, for reproduction by receivers at remote locations. A TV system may use an analog or digital standard, such as HDTV, NTSC, PAL, or SECAM, for the transmission and reception of broadcast signals. Also called **TV system**. **2.** Abbreviation of **television**. A device which reproduces the images and sounds received through a **TV system**. Also called **TV receiver**, **TV set**, or **TV monitor** (1). **3.** Abbreviation of **television**. The transmission of images and sounds via a **TV system**. **4.** Abbreviation of **television**. The programming and other offerings viewed using a **TV** (2). **5.** Abbreviation of **teravolt**.

TV band Same as **TV broadcast band**. Abbreviation of **television band**.

TV board Same as **TV card**. Abbreviation of **television board**.

TV broadcast Abbreviation of **television broadcast**. A transmission of TV signals intended for public or general reception, as opposed, for instance, to CCTV. Also, to transmit such signals. Its own abbreviation is **telecast**.

TV broadcast band Abbreviation of **television broadcast band**. A band of frequencies allocated to broadcasting TV stations. Examples include the VHF and UHF broadcast bands. Also called **TV band**.

TV broadcast channel Same as **TV station**. Abbreviation of **television broadcast channel**.

TV broadcast receiver Abbreviation of **television broadcast receiver**. A device designed to receive TV transmissions intended for public or general reception.

TV broadcast reception Abbreviation of **television broadcast reception**. The receiving of TV transmissions intended for public or general reception.

TV broadcast service Same as **TV broadcasting service**. Abbreviation of **television broadcast service**.

TV broadcast station Same as **TV station**. Abbreviation of **television broadcast station**.

TV broadcast transmission Abbreviation of **television broadcast transmission**. A transmission of TV signals intended for reception by the general public.

TV broadcast transmitter Abbreviation of **television broadcast transmitter**. A transmitter of TV signals intended for reception by the general public.

TV broadcaster Abbreviation of **television broadcaster**. An entity which transmits **TV broadcasts**.

TV broadcasting Abbreviation of **television broadcasting**. The transmission of TV signals intended for public or general reception, as opposed, for instance, to CCTV. Its own abbreviation is **telecasting**.

TV broadcasting channel Same as **TV station**. Abbreviation of **television broadcasting channel**.

TV broadcasting service Abbreviation of **television broadcasting service**. A radio-communications service in which TV signals intended for reception by the general public are transmitted or retransmitted. Also called **TV broadcast service**.

TV broadcasting station Same as **TV station**.

TV camera Abbreviation of **television camera**. A device that converts images formed by lenses into electric signals. A photosensitive surface in the contained camera tube serves as the transducer which converts the optical image into electric video signals suitable for broadcasting, recording, or the like. Its own abbreviation is **telecamera**. Also known as **camera** (2), or **video camera** (1).

TV camera tube Abbreviation of **television camera tube** An electron tube within a TV camera which serves as the transducer that converts the optical image into electric video signals. Also known as **camera tube**, **pickup tube**, or **pickup (3)**.

TV card Abbreviation of television **card**. An expansion card that enables TV viewing via a computer. Also called **TV board**.

TV channel Same as **TV station**. Abbreviation of television **channel**.

TV computer Same as **TV/PC**.

TV decoder Abbreviation of television **decoder**. **1.** Same as **TV decoder box**. **2.** A device incorporated into a TV which performs the functions of a TV decoder box.

TV decoder box Abbreviation of television **decoder box**. A device, that connects to a TV, which decodes an incoming signal or otherwise enables viewing of certain channels, programming, or content. Such a device may serve to permit viewing of satellite or cable programming, HDTV signals, close-captions, and so on. A decoder box may also connect to another device, such as a VCR. Also called **TV decoder**, or **decoder box**.

TV interference Abbreviation of television **interference**. Any interference which diminishes the ability to receive a desired TV signal, or that impairs the quality of its reproduction. Examples include monkey chatter, alternate-channel interference, cross-color interference, glitches, herringbone patterns, hum, jitter, and ghosting. Its own abbreviation is **TVI**.

TV monitor Abbreviation of television **monitor**. **1.** Same as **TV (2)**. **2.** A monitor utilized at a TV station or other location to keep track of the quality of a recording or broadcast. **3.** A **TV (2)** utilized for closed-circuit viewing, as a computer monitor, or for purposes other than watching TV programming.

TV-out Abbreviation of **TV-out** port. A port, such as that providing an S-video connector, which enables a TV to be used as a computer display. Also called **video-out port**.

TV-out port Same as **TV-out**.

TV/PC Also called **TV computer**, **PC/TV**, or **computer TV**. **1.** A PC, such as that with a TV card, enabled for TV viewing. **2.** A TV with built-in computer capabilities, such as Internet access.

TV program Abbreviation of television **program**. The content of a given TV broadcast.

TV programming Abbreviation of television **programming**. The program offerings of one or more television broadcasters.

TV projector Abbreviation of television **projector**. In a projection TV, the unit or device which transmits the images onto a translucent screen.

TV receiver Same as **TV (2)**. Abbreviation of television **receiver**.

TV repeater Abbreviation of television **repeater**. A device, location, or system which receives, amplifies, and transmits TV signals between points or locations.

TV screen Abbreviation of television **screen**. The viewing area of a TV. Such a screen may be the viewing surface of a CRT display, a liquid-crystal display, a plasma display, or the like.

TV set Same as **TV (2)**. Abbreviation of television **set**.

TV signal Abbreviation of television **signal**. A TV signal adhering to a given **TV standard**.

TV standard Abbreviation of television **standard**. A standard, such as HDTV, NTSC, PAL, or SECAM, for the generation, transmission, and reception of TV signals.

TV station Abbreviation of television **station**. A station within a **TV broadcasting service**. Such a station has an assigned interval of frequencies. Also called **TV channel**,

TV broadcast station, **TV broadcasting station**, **TV broadcasting channel**, or **TV broadcast channel**.

TV system Same as **TV (1)**. Abbreviation of television **system**.

TV transmitter Abbreviation of television **transmitter**. An entity or device which transmits **TV signals**.

TV tuner Abbreviation of television **tuner**. A tuner utilized to select from given TV stations.

TVI Abbreviation of **TV interference**.

TW Abbreviation of **terawatt**.

TWA Abbreviation of **traveling-wave amplifier**.

TWAIN A de-facto programming interface standard for use with image-capturing devices such as scanners and digital cameras. Currently it is considered an acronym for Technology Without An Interesting Name, although the word *twain* was originally meant to convey the difficulties software had in bridging between image-capturing devices and computers.

tweak To make fine adjustments, such as those performed when calibrating or tuning.

tweaking The making of fine adjustments, such as those performed when calibrating or tuning.

tweek A form of atmospheric interference, usually caused by lightning, which when detected by the appropriate radio receivers produces a sound more or less like the crackling heard when stepping across a gravel path.

tween The generation of frames in **tweening**.

tweening In computer graphics, the process of generating frames which are approximately intermediate to two images, so as to smooth the transition between said images. Used, for instance, in animation and digital video effects.

tweeter A small speaker designed to reproduce frequencies above a given threshold, such as 2000 or 5000 Hz. Depending on the design and components, a tweeter may accurately reproduce frequencies well beyond the limit of human hearing. Such a speaker unit is usually utilized with others, such as woofers and midranges, for reproduction across the full audio spectrum. Also called **high-frequency speaker**.

TWh Abbreviation of **terawatt-hour**.

TWhr Abbreviation of **terawatt-hour**.

twin-axial cable Same as **twinaxial cable**.

twin cable A cable consisting of one or more **pairs (3)**. The conductors of such a cable are usually twisted or braided together, which case they are called **twisted pair**. Used, for instance, for telephone communications. Also called **paired cable**.

twin diode An electronic component, such as an electron tube, incorporating two diodes. Also called **double diode**, or **dual diode**.

twin-lead **1.** Abbreviation of **twin-lead cable**. **2.** Abbreviation of **twin-lead line**.

twin-lead cable A cable with two leads side-by-side, as opposed, for instance, to the concentric arrangement of a coaxial cable. Its abbreviation is **twin-lead (1)**.

twin-lead line A transmission line with two leads side-by-side, as opposed, for instance, to the concentric arrangement provided by a coaxial cable. Its abbreviation is **twin-lead (2)**. Also called **twin-lead transmission line**.

twin-lead transmission line Same as **twin-lead line**.

twin meter **1.** A meter which enables the monitoring of two aspects of an electric circuit, such as voltage and current. Also called **dual meter**. **2.** A **twin meter (1)** with separate displays.

twin-T network An electric network consisting of two T-networks connected in parallel. Also spelled **twin-tee net-**

work. Also called **parallel-T network**, or **parallel-tee network**.

twin-tee network Same as **twin-T network**.

twin triode Two triodes working together, or within the same envelope.

twinaxial cable Also spelled **twin-axial cable**. **1.** A coaxial cable with two inner conductors, as opposed to one. **2.** Two coaxial cables encased by the same protective jacket.

twisted nematic A technology used in **twisted nematic displays**. Its abbreviation is **TN**.

twisted nematic display A passive-matrix liquid-crystal display in which nematic liquid crystals are twisted 90° or more. The glass sheets of the display contain the nematic-crystal material in a manner that they are twisted unless an electric field is applied. When such a field is applied selectively, the affected crystals untwist and become opaque, thus blocking the polarized light passing through, which in turn produces darker pixels. Such displays provide higher contrast and a better viewing angle than ordinary passive LCD screens, which is especially useful when used under highly-lighted circumstances. Used extensively in portable computers. Its abbreviation is **TN display**.

twisted nematic liquid crystals Twisted and elongated molecules which are used in **nematic-crystal displays**. Such crystals align themselves longitudinally, but are otherwise randomly positioned. Also called **nematic liquid crystals**, or **nematic crystals**.

twisted pair A transmission line consisting of one or more pairs of similar conductors which are insulated from each other, and twisted or braided together. Used, for instance, for telephone communications. When multiple pairs are used together, they are usually color-coded. Its abbreviation is **TP**. Also called **twisted-pair cable**, or **twisted-pair wire**.

twisted-pair cable Same as **twisted pair**.

twisted-pair wire Same as **twisted pair**.

twisted-ring counter A counter similar to a ring counter, in which the complement output of the last stage is connected to the input of the first stage, which provides **2n** states for **n** stages. Also called **Johnson counter**.

two-binary, one quaternary A line coding technique in which four signal levels are encoded as two bits. Used, for instance, in ISDN. Its abbreviation is **2B1Q**.

two-channel amplifier An amplifier with two independent channels, usually designated A and B, or left and right. Such a device may incorporate two identical amplifiers on the same chassis. Used, for instance, as a stereo high-fidelity amplifier. Its abbreviation is **2-channel amplifier**. Also called **dual-channel amplifier**.

two-dimensional Of, pertaining to, existing in, or consisting of two dimensions, especially length and width. Its abbreviation is **2D**.

two-dimensional array In the organization of data, an array whose elements are described or defined using two numbers or dimensions. An example is a matrix. Its abbreviation is **2D array**.

two-dimensional graphics Graphics with height and width, but no depth. Such graphics are flat. Also, hardware, software, or a system supporting such graphics. Its abbreviation is **2D graphics**.

two-element beam A beam antenna, usually a Yagi antenna, with two elements, one driven and the other parasitic. Its abbreviation is **2-element beam**.

two-input adder A logic circuit which accepts two input bits, and whose output is a sum bit and a carry bit. Unlike a **three-input adder**, a two-input adder does not accept an input carry bit. Its abbreviation is **2-input adder**. Also called **half-adder**.

two-phase Its abbreviation is **2-phase**. **1.** Having, or pertaining to, a phase difference of 90°, or $\pi/2$ radians. Said, for instance, of two voltages or currents which are 90° out-of-phase relative to the other. Also called **quarter-phase**. **2.** Having, or pertaining to, a phase difference of 180°, or π radians. Said, for instance, of two voltages or currents which are 180° out-of-phase relative to the other. **3.** Having two phases.

two-phase AC Alternating current with two phases. Each phase usually has a phase difference of 90° or 180° in relation to the other. Its abbreviation is **2-phase AC**.

two-phase circuit A circuit in which there are two AC phases, each usually differing in phase by 90° or 180°. Its abbreviation is **2-phase circuit**.

two-phase commit A feature which helps ensure that two or more synchronized databases are properly updated. Each change must either be performed successfully on all databases, or said databases are returned to the state they were in prior to the updating transaction. Its abbreviation is **2-phase commit**.

two-phase current A current, such as that flowing through a two-phase circuit, which has a two phases, each usually differing in phase by 90° or 180°. Its abbreviation is **2-phase current**.

two-phase generator A generator of two-phase AC. Its abbreviation is **2-phase generator**.

two-phase motor A motor which utilizes two-phase current. Its abbreviation is **2-phase motor**.

two-phase power Its abbreviation is **2-phase power**. **1.** The power utilized by a two-phase circuit or system. **2.** The power provided by a two-phase circuit or system.

two-phase system A system, such as a two-phase circuit, which has a two-phase current or voltage. Its abbreviation is **2-phase system**.

two-phase transformer A transformer utilized to supply a two-phase circuit, device, piece of equipment, or system. Its abbreviation is **2-phase transformer**.

two-phase voltage A voltage, such as that supplying a two-phase circuit, in which there are two AC phases. Its abbreviation is **2-phase voltage**.

two-port network An electric network with four terminals, two for input and two for output. The input terminals are paired to form the input port, while the output port is formed by the output terminals. Examples include O-networks and H-networks. Its abbreviation is **2-port network**. Also called **two-terminal pair**, **two-terminal pair network**, **four-terminal network**, or **quadripole**.

two-state device A device having two stable states, such as on or off, or conducting or not conducting. An example is a flip-flop. Its abbreviation is **2-state device**.

two-state logic Its abbreviation is **2-state logic**. **1.** Digital logic which can assume either of two states, which are true or false. **2.** Digital logic which can provide either of two outputs, which are high and low, or on and off.

two-terminal pair Same as **two-port network**.

two-terminal pair network Same as **two-port network**.

two-tier client/server A client/server architecture in which the user interface is structured into two layers or tiers, each of which may have one or more components. Its abbreviation is **2-tier client/server**.

two-track recording The recording of two sound tracks on a **two-track tape**. Its abbreviation is **2-track recording**.

two-track tape A magnetic tape upon which two sound tracks are recorded. Such a tape may provide for one stereo signal, or two monaural tracks. Its abbreviation is **2-track tape**.

two-tuner picture-in-picture A picture-in-picture feature in which the TV has a second tuner dedicated to the second picture, enabling the viewing of two channels simultaneously. This contrasts with a **one-tuner picture-in-picture** TV, where the second picture must be that another input, such as a DVD or VHS. Its abbreviation is **two-tuner PIP**, or **2-tuner picture-in-picture**.

two-tuner PIP   Abbreviation of **two-tuner picture-in-picture**. Its own abbreviation is **2-tuner PIP**.

two-way beeper   Same as **two-way pager**. Its abbreviation is **2-way beeper**.

two-way communication   Its abbreviation is **2-way communication**. **1.** Communication in which both locations can send and receive. This may or may not occur simultaneously. **2. Two-way communication** (1) which occurs simultaneously in both directions. Also called **duplex** (2), or **full-duplex**. **3. Two-way communication** (1) in which communication can only occur one direction at a time. Also called **duplex** (3), or **half-duplex**.

two-way intercom   An intercom system which communicates in both directions. Most intercom systems are of this type. Its abbreviation is **2-way intercom**.

two-way loudspeaker   Same as **two-way speaker**. Its abbreviation is **2-way loudspeaker**.

two-way pager   A pocket-sized radio transceiver which serves to send and receive messages, and in some cases perform various other functions, such as retrieve email. These devices may emit a beep when informing of a new message. Some two-way pagers can also signal by other means, such as vibrations, or a blinking light. Its abbreviation is **2-way pager**. Also called **two-way beeper**.

two-way radio   Its abbreviation is **2-way radio**. **1.** Radio communication in which both locations can receive and transmit. Also called **two-way radio communications**. **2.** A device, such as a transceiver, utilized for two-way radio communications.

two-way radio communications   Same as **two-way radio** (1). Its abbreviation is **2-way radio communications**.

two-way repeater   Its abbreviation is **2-way radio repeater**. **1.** A repeater which retransmits, and usually amplifies, radio signals arriving or traveling in two directions. **2.** A repeater which retransmits, and usually amplifies, two-way communications signals.

two-way speaker   A speaker which utilizes two individual drivers for each of two intervals of frequencies. Such a speaker usually has a woofer or a mid-woofer, along with a tweeter. Its abbreviation is **2-way speaker**. Also called **two-way system**, or **two-way loudspeaker**.

two-way system   Same as **two-way speaker**. Its abbreviation is **2-way system**.

two-wire circuit   Same as **two-wire system** (2). Its abbreviation is **2-wire circuit**.

two-wire line   Same as **two-wire system** (2). Its abbreviation is **2-wire line**.

two-wire system   Its abbreviation is **2-wire system**. **1.** An AC circuit utilizing two wires. **2.** A circuit, such as a communications circuit, utilizing two wires. Also called **two-wire circuit**, or **two-wire line**.

two's complement   Same as **twos complement**.

twos complement   Within the binary number system, the result obtained when all the ones and zeroes of a number are interchanged, and 1 is added to the sum. For example, the two's complement of 1011 is 0101, and is calculated as follows: the exchange of ones and zeroes of 1011 yields 0100, to which 1 is added for a result of 0101. Its abbreviation is **2's complement**.

TWT   Abbreviation of **traveling-wave tube**.

TWTA   Abbreviation of **traveling-wave tube amplifier**.

twystron   A microwave tube which is a hybrid combining a klystron and a traveling-wave tube.

TX   Abbreviation of **transmitter**.

TXT file   An ASCII text file format.

type   **1.** Things which have one or more characteristics in common which enable them as a group to be distinguished from other classes. Also, the traits that differentiate such a type. **2.** To classify by **type** (1). **3.** A specific thing which is of a specified **type** (1). **4.** A manner of classifying data. The specification of a data type dictates the way a program handles the data. For instance, most applications can not calculate text. Other examples of data types include integers, real numbers, Boolean, and floating point. Also called **data type**. **5.** To use a typewriter to write or store characters. **6.** To use a computer keyboard for data entry. **7.** Abbreviation of **typeface**.

Type A connector   A USB connector which is rectangular in shape and is plugged into a USB port.

Type A wave   An electromagnetic wave whose frequency, phase, and amplitude are constant. Also known as **continuous wave**.

type-ahead buffer   A small segment of computer memory utilized to temporarily store keystrokes that await processing. In the event a typist exceeds the capability of this buffer, the computer usually emits a warning sound. Also spelled **typeahead buffer**. Also called **keyboard buffer**.

Type B connector   A USB connector which is square in shape and is plugged into a USB peripheral.

type check   A verification, such as that made by a compiler or interpreter, that a given **type** (4) is treated the same throughout a given process.

type checking   The process of verifying that a given **type** (4) is treated the same throughout a given process.

type declaration   In a program, the specific classification a **type** (4) is given.

type family   Abbreviation of **typeface family**.

Type I card   Same as **Type I PC card**.

Type I PC card   A PCMCIA card which is 3.3 mm thick and is typically used for memory. Also called **Type I card**, **Type I PCMCIA card**, or **PCMCIA Type I card**.

Type I PCMCIA card   Same as **Type I PC card**.

Type I tape   A magnetic recording tape utilizing iron oxide particles. Such a tape provides barely adequate performance. Also called **iron oxide tape**.

Type II card   Same as **Type II PC card**.

Type II PC card   A PCMCIA card which is 5.0 mm thick and which may be used, for instance, for additional memory, or as a network card. Also called **Type II card**, **Type II PCMCIA card**, or **PCMCIA Type II card**.

Type II PCMCIA card   Same as **Type II PC card**.

Type II tape   A magnetic recording tape utilizing chromium dioxide particles. Such tape, with the proper deck, provides excellent frequency response and a wide dynamic range. Also known as **chrome tape**, or **chromium tape**.

Type III card   Same as **Type III PC card**.

Type III PC card   A PCMCIA card which is 10.5 mm thick and which may be used, for instance, for disk storage, or as a wireless network adapter. Also called **Type III card**, **Type III PCMCIA card**, or **PCMCIA Type III PC card**.

Type III PCMCIA card   Same as **Type III PC card**.

Type III tape   An obsolete dual-layer tape combining Type I tape and Type II tape.

Type IV tape   A magnetic recording tape in which metal particles, as opposed to oxides of metals, are used. Its performance is superior to Type II tapes. Also called **metal tape**.

typeahead buffer Same as **type-ahead buffer**.

typeface The name given to a specific or characteristic design a given set of printed or displayed characters has. For example, the typeface utilized throughout this dictionary is Times New Roman. Other popular typefaces include Arial and Helvetica. Its abbreviation is **type** (7).

typeface family A group of **typefaces**, with specific variations, as in Times New Roman normal, Times New Roman bold, Times New Roman italic, and Times New Roman bold-italic. Its abbreviation is **type family**.

typematic A keyboard feature which enables a key that is held down longer than a given interval to repeat. The time necessary for repeating, and the repeating speed can usually be set. Also called **key repeat, keyboard repeat,** or **auto-repeat**.

typematic delay The time interval that must pass before a key repeats when using the **typematic** feature.

typematic rate The speed at which a key repeats when using the **typematic** feature.

typeover mode In a computer application, a mode of entering data in which any characters typed on the keyboard replace those already occupying the same area. This contrasts with **insert mode**, in which any newly typed characters are placed between characters already present. Also called **overwrite mode,** or **overtype mode**.

TΩ Abbreviation of **teraohm**.

U

U **1.** Chemical symbol for **uranium**. **2.** Abbreviation of **rack unit**. **3.** Within the **YUV** color model, one of the two color-difference signals. The other is **V (3)**.

U Symbol for **potential difference**.

U interface In ISDN, an interface which links a central office with an NT1.

u-law A standard used to convert an analog input, usually voice, into digital form using pulse-code modulation. Currently, it is used only in North America and Japan, while the rest of the world uses the A-law standard. Also spelled ulaw. Also known as **mu-law**, or *μ* **law**.

UA Abbreviation of **user agent**.

UART Abbreviation of Universal **A**synchronous **R**eceiver **T**ransmitter. A chip or circuit utilized to transmit and receive data asynchronously via a computer serial port. It has a parallel-to-serial converter for data being transmitted from the CPU, and a serial-to-parallel converter for data being received by the CPU. In addition, a UART provides buffering, and handles interrupts generated by the keyboard and mouse, among other functions.

UATA Abbreviation of **Ultra ATA**.

ubiquitous computing The presence of interconnected computing devices, such as motes, smart appliances, and wearable computers, which permeate a given environment. Ubiquitous computing may be used, for instance, to enable immediate access to computing tasks under nearly all circumstances, or for keeping track of innumerable objects and/or persons regardless of where they are. Also called **pervasive computing**.

ubitron A type of high-power traveling-wave tube.

UBR Abbreviation of **Unspecified Bit Rate**.

UCD Abbreviation of **user-centered design**.

UCE Abbreviation of **Unsolicited Commercial Email**.

UCITA Acronym for Uniform Computer Information Transactions Act. A law providing software vendors with powers such as the ability to arbitrarily shut down, from a remote location, an application paid for by a user or entity.

UCT Abbreviation of **Universal Coordinated Time**.

UDDI Abbreviation of **U**niversal **D**escription, **D**iscovery and **I**ntegration. An XML-based registry which enables commercial entities to list themselves on the Internet.

UDF Abbreviation of **Universal Disk Format**, or **Universal Disc Format**.

UDMA Same as **Ultra ATA**. Abbreviation of **Ultra DMA**.

UDP Abbreviation of User Datagram Protocol. A connectionless protocol that is a part of the TCP/IP suite, and which is similar to **TCP**, except that it does not verify that the packets received are in the correct sequence, or even that they have been delivered. Since it provides a less reliable connection than TCP, it is utilized, for instance, in situations where retransmission is impractical, such as real-time audio and video broadcasts.

UDP/IP UDP utilized over IP.

UHF Abbreviation of **ultra high frequency**. A range of radio frequencies spanning from 0.3 to 3 GHz. These correspond to wavelengths of 1 meter to 10 cm, respectively. Also, pertaining to this interval of frequencies. Also called **UHF band**, or **UHF range**.

UHF adapter Same as **UHF connector**. Abbreviation of ultra high frequency **adapter**. Also spelled **UHF adaptor**.

UHF adaptor Same as **UHF connector**. Abbreviation of ultra high frequency **adapter**. Also spelled **UHF adapter**.

UHF antenna Abbreviation of ultra high frequency **antenna**. An antenna that operates in the **UHF** range. Used, for instance, for TV reception.

UHF band Same as **UHF**.

UHF channel Same as **UHF TV channel**. Abbreviation of ultra high frequency **channel**.

UHF connector Abbreviation of ultra high frequency **connector**. Also called **UHF adapter**. **1.** A connector which joins a UHF antenna to a cable, such as that of a TV. **2.** A connector that converts a **UHF connector** (1) into another type of connector, such as BNC.

UHF converter Abbreviation of ultra high frequency **converter**. A circuit or device that converts a UHF signal to a higher or lower frequency. For example, a circuit that converts a UHF TV signal into a VHF signal for use in some older TVs.

UHF loop Abbreviation of ultra high frequency **loop**. A loop antenna that operates in the **UHF** range. Also called **UHF loop antenna**.

UHF loop antenna Same as **UHF loop**. Abbreviation of ultra high frequency **loop antenna**.

UHF range Same as **UHF**. Abbreviation of ultra high frequency **range**.

UHF receiver Abbreviation of ultra high frequency **receiver**. A receiver which responds to **UHF** signals.

UHF signal Abbreviation of ultra high frequency **signal**. An RF signal in the **UHF** range.

UHF television antenna Same as **UHF TV antenna**.

UHF television channel Same as **UHF TV channel**.

UHF transmitter Abbreviation of ultra high frequency **transmitter**. A transmitter that emits **UHF** signals.

UHF tuner Abbreviation of ultra high frequency **tuner**. A tuner, such as that of a TV, utilized for reception of signals in the **UHF** range.

UHF TV antenna Abbreviation of ultra high frequency **TV antenna**. A **UHF antenna** utilized for TV reception.

UHF TV channel Abbreviation of ultra high frequency **TV channel**. A TV channel whose assigned frequency is in the **UHF** band. Also called **UHF channel**.

UHV **1.** Abbreviation of **ultra-high voltage** **2.** Abbreviation of **ultra-high vacuum**.

UI Abbreviation of **user interface**.

UJT Abbreviation of **unijunction transistor**. A three-terminal transistor incorporating an emitter and two bases, and having a single pn junction. Utilized mostly for switching, and may be used, for instance, in relaxation oscillators.

UL Abbreviation of Underwriters Laboratories. An organization that specializes in product testing and safety certifications.

ulaw Same as **u-law**.

ULAW Same as **u-law**.

ULF Abbreviation of **ultra low frequency**.

ULSI Abbreviation of **ultra large-scale integration**.

ultimate sensitivity The greatest sensitivity attainable by an instrument or meter. This depends on both the resolution and the lowest measurement range.

ultimate trip current The current or RMS current at which an overcurrent protection device, such as a circuit breaker, will trip. This is usually specified for given conditions, such as an operating temperature range.

ultor In a CRT, the anode which accelerates the electron beam. Also called **ultor anode, ultor electrode,** or **second anode.**

ultor anode Same as **ultor.**

ultor electrode Same as **ultor.**

ultor voltage A high-voltage DC utilized by an **ultor.** It is provided by a flyback power supply.

ultra- A prefix denoting that which is above, greater than, or exceeding a given interval, range, or scope.

Ultra ATA Abbreviation of **Ultra A**dvanced **T**echnology **A**ttachment. A protocol utilized for the transfer of data between a hard drive and the RAM of a computer. It provides increased data integrity and faster transfer rates than ATA or IDE, and currently features speeds ranging up to 133 MBps. Its own abbreviation is **UATA.** Also called **Ultra DMA,** or **Ultra IDE.**

Ultra ATA/33 A version of **Ultra ATA** with transfer rated of up to 33 MBps. Also called **Ultra DMA/33, ATA/33,** or **DMA/33.**

Ultra ATA/66 A version of **Ultra ATA** with transfer rated of up to 66 MBps. Also called **Ultra DMA/66, ATA/66,** or **DMA/66.**

Ultra ATA/100 A version of **Ultra ATA** with transfer rated of up to 100 MBps. Also called **Ultra DMA/100, ATA/100,** or **DMA/100.**

Ultra ATA/133 A version of **Ultra ATA** with transfer rated of up to 133 MBps. Also called **Ultra DMA/133, ATA/133,** or **DMA/133.**

Ultra Direct Memory Access Same as **Ultra ATA.**

Ultra DMA Same as **Ultra ATA.** Abbreviation of **Ultra Direct Memory Access.** Its own abbreviation is **UDMA.**

Ultra DMA/33 Same as **Ultra ATA/33.**

Ultra DMA/66 Same as **Ultra ATA/66.**

Ultra DMA/100 Same as **Ultra ATA/100.**

Ultra DMA/133 Same as **Ultra ATA/133.**

Ultra Extended Graphics Array Same as **UXGA.**

ultra-fast switch Same as **ultrafast switch.**

ultra-fast switching Same as **ultrafast switching.**

ultra-hard vacuum Same as **ultra-high vacuum.**

ultra high frequency Same as **UHF.**

ultra high frequency adapter Same as **UHF adapter.**

ultra high frequency adaptor Same as **UHF adaptor.**

ultra high frequency antenna Same as **UHF antenna.**

ultra high frequency band Same as **UHF.**

ultra high frequency channel Same as **UHF channel.**

ultra high frequency connector Same as **UHF connector.**

ultra high frequency converter Same as **UHF converter.**

ultra high frequency loop Same as **UHF loop.**

ultra high frequency loop antenna Same as **UHF loop antenna.**

ultra high frequency range Same as **UHF range.**

ultra high frequency receiver Same as **UHF receiver.**

ultra high frequency signal Same as **UHF signal.**

ultra high frequency television antenna Same as **UHF TV antenna.**

ultra high frequency television channel Same as **UHF TV channel.**

ultra high frequency transmitter Same as **UHF transmitter.**

ultra high frequency tuner Same as **UHF tuner.**

ultra high frequency TV antenna Same as **UHF TV antenna.**

ultra high frequency TV channel Same as **UHF TV channel.**

ultra-high vacuum An environment or system whose pressure is below 10^{-9} torr, but above 10^{-12} torr. One torr equals 1 millimeter of mercury, or approximately 133.3 pascals. Also spelled **ultrahigh vacuum.** Its abbreviation is **UHV.**

ultra-high vacuum diffusion A diffusion process, such as that utilized to introduce semiconductor dopants, performed within an ultra-high vacuum. Also spelled **ultrahigh vacuum diffusion.**

ultra-high voltage A voltage that exceeds a given amount, such as 345, 765, or 1,000 kilovolts. Such voltages are usually used for carrying electricity over power transmission lines. Also spelled **ultrahigh voltage.** Its abbreviation is **UHV.**

Ultra IDE Same as **Ultra ATA.**

ultra-large-scale integration In the classification of ICs, the inclusion of over 1,000,000 electronic components, such as transistors, on a single chip. This definition may require many more components in a few years to qualify for this level of integration, as miniaturization progresses. Its abbreviation is **ULSI.**

ultra-low distortion For a given component, circuit, device, equipment, or system category, an especially low percentage of distortion, such as total harmonic distortion. A value that would qualify for this designation will vary by the area in question. For example, an audio amplifier with 0.01% total harmonic distortion and intermodulation distortion at its rated power of 100 watts RMS per channel into 8 ohms, from 10 Hz to 60 kHz, would be considered to feature ultra-low distortion. Also spelled **ultralow distortion.**

ultra low frequency Waves whose frequencies are usually defined to be between 300 and 3000 Hz, corresponding to wavelengths of between 1000 and 100 km, respectively. Its abbreviation is **UHF.** Also spelled **ultralow frequency.** Also called **infralow frequencies.**

ultra-microscope Same as **ultramicroscope.**

ultra-microscopy Same as **ultramicroscopy.**

Ultra SCSI A SCSI interface which utilizes an 8-bit bus and supports data rates of up to 20 MBps. It uses a 50-pin cable.

ultra-short waves Radio waves whose wavelengths are shorter than a given length, such as 10 meters or 0.1 meter. Also spelled **ultrashort waves.**

ultra-sonic Same as **ultrasonic.**

ultra-sonics Same as **ultrasonics.**

ultra-violet Same as **ultraviolet.**

Ultra Wide SCSI Same as **Ultra2 SCSI.**

ultra-wideband radar A radar, such as an impulse radar, which utilizes a very wide range of frequencies. Also spelled **ultrawideband radar.**

Ultra XGA Same as **UXGA.**

Ultra2 SCSI A SCSI interface which utilizes an 8-bit bus and supports data rates of up to 40 MBps. It uses a 50-pin cable. Also called **Ultra Wide SCSI.**

Ultra3 SCSI A SCSI interface which utilizes a 16-bit bus and supports data rates of up to 160 MBps. It uses a 68-pin cable. Also known as **Ultra160,** or **wide Ultra3 SCSI.**

Ultra4 SCSI A SCSI interface which utilizes a 16-bit bus and supports data rates of up to 320 MBps. It uses a 68-pin cable. Also known as **Ultra320.**

Ultra5 SCSI A SCSI interface which utilizes a 16-bit bus and supports data rates of up to 640 MBps. It uses a 68-pin cable. Also known as **Ultra640.**

Ultra160 Same as **Ultra3 SCSI.**

Ultra320 Same as **Ultra4 SCSI.**

Ultra640 Same as **Ultra5 SCSI.**

ultrafast switch A switch that capable of exceedingly fast action. A laser can be utilized for switching in the zeptosecond range. Also spelled **ultra-fast switch.**

ultrafast switching Switching functions performed exceedingly rapidly. A laser can be utilized for switching in the zeptosecond range. Also spelled **ultra-fast switching**.

ultrafiche Abbreviation of **ultra**microfiche. A microfiche format whose reduction ratio is greater than a given amount, such as that providing for 1,000 pages of text per card or film.

ultrahigh frequency Same as **UHF**.

ultrahigh vacuum Same as **ultra-high vacuum**.

ultrahigh vacuum diffusion Same as **ultra-high vacuum diffusion**.

ultrahigh voltage Same as **ultra-high voltage**.

ultralow distortion Same as **ultra-low distortion**.

ultralow frequency Same as **ultra low frequency**.

ultramicrofiche Same as **ultrafiche**.

ultramicroscope Also spelled **ultra-microscope**. **1.** A microscope, such as an atomic force microscope or a transmission electron microscope, utilized for observation and recording of samples smaller than a given size, such as 10 angstroms. **2.** A light microscope whose resolution is especially high. **3.** A microscope whose resolution exceeds that of a light microscope.

ultramicroscopy The use of **ultramicroscopes** to view and analyze specimens. Also spelled **ultra-microscopy**.

ultrashort waves Same as **ultra-short waves**.

ultrasonic Also spelled **ultra-sonic**. **1.** Pertaining to, or arising from sounds at **ultrasonic frequencies**. Also called **supersonic** (2). **2.** Pertaining to, or arising from **ultrasonics**. Also called **supersonic** (3).

ultrasonic beam A concentrated and essentially unidirectional stream of **ultrasonic waves**.

ultrasonic bonding Bonding in which **ultrasonic waves** to unite two or more items which are forced together. Used, for instance when dealing with materials which can not, or should not be heated.

ultrasonic cleaning The use of **ultrasonic waves** to clean surfaces which are immersed in an appropriate solvent. Used, for instance, for surfaces that would otherwise be inaccessible for cleansing.

ultrasonic communication Communication utilizing **ultrasonic waves**. Used, for instance, for underwater communications.

ultrasonic delay line A device or circuit, such as a mercury delay line, that delays the transmission of **ultrasonic signals**.

ultrasonic detector A device, instrument, or material which detects, measures, or responds to **ultrasonic waves**. Such an instrument may incorporate, for instance, a piezoelectric transducer.

ultrasonic diagnosis The use of **ultrasonic waves** to aid in the diagnosis of medical conditions. Used, for instance, in the assessment of musculoskeletal conditions.

ultrasonic drill A drill which bores holes by using **ultrasonic** vibrations. Used, for instance, in dentistry.

ultrasonic energy The energy created by **ultrasonic waves**.

ultrasonic flaw detection The use of an **ultrasonic flaw detector** for the examination of solid materials.

ultrasonic flaw detector A device or instrument which generates and emits **ultrasonic waves** which are utilized to study the cracks, deterioration, or other flaws in a solid material. A specific application is in the detection of metal fatigue.

ultrasonic flow meter Same as **ultrasonic flowmeter**.

ultrasonic flowmeter An instrument which utilizes **ultrasonic waves** to determine the flow rate of a fluid. Such an instrument, for example, may penetrate a pipe with ultrasonic waves, and detect the rate of flow using the Doppler effect. Also spelled **ultrasonic flow meter**.

ultrasonic frequency A sound whose frequency is beyond the range of human hearing. That is, above about 20 kHz. Also called **supersonic frequency**.

ultrasonic generator A circuit or device, such as an oscillator, which produces **ultrasonic waves**.

ultrasonic hologram A three-dimensional image which is recorded via **ultrasonic holography**.

ultrasonic holography A method of holography utilizing **ultrasonic waves**. In it, a coherent beam of ultrasonic energy is directed towards the object whose image is being taken, and the resulting interference pattern is recorded to form an ultrasonic hologram.

ultrasonic imaging The use of **ultrasonic energy** to create images of the internal structure of non-transparent objects. A common example is the use in medicine of ultrasonic waves to provide images of internal organs. Also known as **acoustic imaging**.

ultrasonic inspection The use of **ultrasonic waves** for the examination of solid materials as performed, for instance, by an ultrasonic flaw detector.

ultrasonic level detector A device or instrument which utilizes **ultrasonic waves** to determine the level of a liquid or other material. Such a device, for example, may reflect such waves off a surface when the monitored level drops below a given point. Conversely, such a device may not detect reflected waves when the level exceeds a given point.

ultrasonic light diffraction The diffraction of light through its interaction with an **ultrasonic wave** field.

ultrasonic machining The use of a precise **ultrasonic beam** for machining. Used, for instance, for surfaces which can not, or should not be heated.

ultrasonic probe **1.** A probe which serves to detect **ultrasonic waves**. **2.** A probe which serves to emit ultrasonic waves.

ultrasonic receiver A receiver which accepts **ultrasonic signals** which carry information. Used, for instance, along with an ultrasonic transmitter in a remote-control system.

ultrasonic relay A relay which is actuated by **ultrasonic waves**.

ultrasonic scanner A scanner, such as that utilized for ultrasonic imaging, which employs **ultrasonic waves**.

ultrasonic scanning Scanning, such as that performed during ultrasonic imaging, which utilizes **ultrasonic waves**.

ultrasonic sealing **1.** Same as **ultrasonic soldering** (1). **2.** The use of **ultrasonic soldering** (1) to seal materials, such as thermoplastics.

ultrasonic signal Any signal, such as a sound or vibration, which occurs at **ultrasonic frequencies**.

ultrasonic soldering **1.** Soldering in which **ultrasonic waves** join surfaces, without the generation of heat. Used, for instance, for surfaces which can not, or should not be heated. Also called **ultrasonic sealing** (1). **2.** Soldering which is enhanced by ultrasonic waves. The ultrasonic vibrations help remove oxides and otherwise better prepare a surface for soldering, making a flux unnecessary.

ultrasonic sound Same as **ultrasound** (1).

ultrasonic sound waves Same as **ultrasonic waves**.

ultrasonic sounding Sounding in which **ultrasonic waves** are utilized to detect surfaces and depths.

ultrasonic switch A switch that is actuated by **ultrasonic waves**.

ultrasonic testing The utilization of **ultrasonic waves** for the testing. Used, for instance, for the evaluation of solid materials, or for medical diagnosis.

ultrasonic therapy The use of **ultrasonic waves** for therapeutic purposes. Used, for instance, in physical therapy.

ultrasonic thickness gage Same as **ultrasonic thickness gauge**.

ultrasonic thickness gauge A device or instrument which utilizes **ultrasonic waves** to determine thicknesses of surfaces, layers, sheets, objects, and the like. Such an instrument usually determines distance by measuring the time of travel of an ultrasonic beam. Also spelled **ultrasonic thickness gage**.

ultrasonic transducer A device, such as that incorporating a piezoelectric material, which converts electrical energy into **ultrasonic energy**, or vice versa.

ultrasonic transmitter A transmitter that emits **ultrasonic signals** which carry information. Used, for instance, along with an ultrasonic receiver in a remote-control system.

ultrasonic waves Sound waves whose frequency is beyond the range of human hearing. That is, above about 20 kHz. Used, for instance, in medical diagnosis and treatment, for soldering and welding, and in sonar. Also called **supersonic waves**.

ultrasonic welding 1. Welding in which **ultrasonic waves** join surfaces, without the generation of heat. Used, for instance, for surfaces which can not, or should not be heated. 2. Welding which is enhanced by ultrasonic waves. The ultrasonic vibrations help remove oxides and otherwise better prepare a surface for welding, making a flux unnecessary.

ultrasonic wire bonding Wire bonding in which **ultrasonic waves** are utilized to attach the wires.

ultrasonics The study of the phenomena pertaining to and arising from **ultrasonic waves**, and their applications. Also spelled **ultra-sonics**. Also called **supersonics**.

ultrasonogram The images produced while performing an examination utilizing **ultrasonography**. Also called **ultrasonograph**, or **sonogram**.

ultrasonograph Same as **ultrasonogram**.

ultrasonography The use of ultrasonic energy to produce images of structures within the human body. The reflections of the high-frequency sound waves directed at the desired regions are processed and converted into images suitable for observation and analysis. It is noninvasive, and is used, for instance, to monitor a fetus during a pregnancy. Also called **sonography**.

ultrasound 1. Any sound or vibration which occurs at **ultrasonic frequencies**. Also called **ultrasonic sound**. 2. The utilization of **ultrasonic waves** for medical diagnostic or therapeutic purposes. Used, for instance, in ultrasonography, or for ultrasonic therapy.

UltraSPARC An enhanced version of **SPARC**.

ultraviolet Its abbreviation is **UV**. Also spelled **ultra-violet**. 1. Within the electromagnetic spectrum, the portion extending from the upper limit of visible light, to the longest X-rays. This encompasses from approximately 4 to 400 nanometers. This interval itself is subdivided into the near, far, and extreme ultraviolet regions. Also called **ultraviolet region**, or **ultraviolet range**. 2. Pertaining to, generating, detecting, utilizing, or sensitive to **ultraviolet radiation**.

ultraviolet A Same as **UVA**.

ultraviolet absorption The absorption by a surface, material, body, or medium, of **ultraviolet radiation**. Used, for instance, in ultraviolet absorption spectroscopy. Its abbreviation is **UV absorption**.

ultraviolet absorption spectroscopy A spectroscopic technique which measures the absorption of **ultraviolet radiation** by a sample. Used, for instance, for the determination of free radicals in samples. Its abbreviation is **UV absorption spectroscopy**.

ultraviolet absorption spectrum A display or graph obtained through **ultraviolet absorption spectroscopy**. Its abbreviation is **UV absorption spectrum**.

ultraviolet astronomy The study of **ultraviolet radiation** emitted from objects in the universe. Useful, for instance, for gathering information pertaining to the composition and densities of interstellar matter and celestial bodies. Its abbreviation is **UV astronomy**.

ultraviolet B Same as **UVB**.

ultraviolet band Same as **ultraviolet spectrum** (1). Its abbreviation is **UV band**.

ultraviolet C Same as **UVC**.

ultraviolet detection Its abbreviation is **UV detection**. 1. The sensing of ultraviolet radiation utilizing an **ultraviolet detector**. 2. The use of **ultraviolet radiation** to sense objects or phenomena.

ultraviolet detector A device or instrument which detects **ultraviolet radiation**. Used, for instance, in ultraviolet astronomy. Its abbreviation is **UV detector**. Also called **ultraviolet sensor**.

ultraviolet emission The emission of **ultraviolet radiation** by a body or medium. Sources include interstellar matter, and celestial bodies such as a planets or stars. Its abbreviation is **UV emission**.

ultraviolet energy Same as **ultraviolet radiation**. Its abbreviation is **UV energy**.

ultraviolet filter Its abbreviation is **UV filter**. 1. A filter which transmits **ultraviolet radiation**, while absorbing other electromagnetic radiation. Used, for instance, in ultraviolet astronomy. 2. A filter which absorbs **ultraviolet radiation**, while transmitting other electromagnetic radiation. Used, for instance, to protect from damage which may be caused by UV radiation.

ultraviolet imaging Imaging in which the source of light is emitted and/or reflected **ultraviolet radiation**. Used, for instance, in ultraviolet imaging telescopes. Its abbreviation is **UV imaging**.

ultraviolet imaging telescope An instrument which detects emitted and reflected **ultraviolet radiation** outside the earth's atmosphere, to obtain images of interstellar matter and celestial bodies such as a planets or stars, in addition to other information, such as that pertaining to their compositions and densities. Also called **extreme-ultraviolet imaging telescope**.

ultraviolet lamp A lamp, such as a mercury-vapor lamp, whose emitted energy is mostly, or completely, in the **ultraviolet region**. Used, for instance, to assist viewing in conditions of reduced visible light. Its abbreviation is **UV lamp**.

ultraviolet laser A laser, such as an excimer laser, which emits coherent **ultraviolet radiation**. Used, for instance, in laser surgery. Its abbreviation is **UV laser**.

ultraviolet light Same as **ultraviolet radiation**. Its abbreviation is **UV light**.

ultraviolet photoelectron spectroscopy A photoelectron spectroscopy technique in which the source of exciting radiation is **ultraviolet radiation**. Its abbreviation is **UV photoelectron spectroscopy**, or **UPS**.

ultraviolet radiation Electromagnetic radiation in the **ultraviolet region**. Sources of such radiation include celestial bodies, interstellar matter, and mercury-vapor and mercury-arc lamps. Its abbreviation is **UV radiation**. Also called **ultraviolet light, ultraviolet rays**, or **ultraviolet energy**.

ultraviolet radiometer A radiometer which detects and measures **ultraviolet radiation**. Used, for instance, to measure the UV energy emitted by the sun. Its abbreviation is **UV radiometer**.

ultraviolet range Same as **ultraviolet** (1). Its abbreviation is **UV range**.

ultraviolet rays Same as **ultraviolet radiation**. Its abbreviation is **UV rays**.

ultraviolet region Same as **ultraviolet (1)**. Its abbreviation is **UV region**.

ultraviolet sensor Same as **ultraviolet detector**. Its abbreviation is **UV sensor**.

ultraviolet signal A signal in which **ultraviolet radiation** is present. Its abbreviation is **UV signal**.

ultraviolet spectrometer A spectrometer which detects and measures radiant intensities in the **ultraviolet region**. Used, for instance, in ultraviolet astronomy. Its abbreviation is **UV spectrometer**.

ultraviolet spectrometry The science and utilization of **ultraviolet spectrometers** for analysis. Its abbreviation is **UV spectrometry**.

ultraviolet spectrophotometer A spectrophotometer operating in the **ultraviolet region**. Used, for instance, to identify solids dissolved in samples. Its abbreviation is **UV spectrophotometer**.

ultraviolet spectrophotometry The science and utilization of **ultraviolet spectrophotometers** for analysis. Its abbreviation is **UV spectrophotometry**.

ultraviolet spectroscopy Its abbreviation is **UV spectroscopy**. **1.** An analytical technique in which the wavelengths and corresponding intensities of **ultraviolet radiation** emitted and/or reflected by a surface, material, body, or medium is analyzed. Used, for instance, in ultraviolet astronomy. The displayed or graphed output is called **ultraviolet spectrum (2)**. **2.** An analytical technique in which the wavelengths and corresponding intensities of ultraviolet radiation absorbed by a sample are analyzed. An example is ultraviolet absorption spectroscopy. The displayed or graphed output is called **ultraviolet spectrum (2)**.

ultraviolet spectrum Its abbreviation is **UV spectrum**. **1.** The interval of wavelengths encompassing the **ultraviolet region**. Also called **ultraviolet band**. **2.** A display or graph obtained through **ultraviolet spectroscopy**.

ultraviolet therapy The use of ultraviolet radiation for therapeutic purposes. Used, for instance, for certain skin conditions. Its abbreviation is **UV therapy**.

ultraviolet waves Electromagnetic waves whose wavelengths are in the **ultraviolet region**. Its abbreviation is **UV waves**.

ultrawideband radar Same as **ultra-wideband radar**.

UM Abbreviation of **unified messaging**.

UMA Abbreviation of **unified memory architecture**.

umbilical cord A physical medium, such as a wire, cable, or tube, which serves to control and/or power another device. Used, for instance, with certain robots.

umbra **1.** A dark area, such as that of a shadow, which is blocked from light rays emanating from a given source. Also, the darkest area of such a shadow. **2.** A dark area arising from the shadow cast by a celestial body, such as the sun or moon, during an eclipse. Also, the darkest area of such a shadow.

umbrella antenna An antenna incorporating multiple wires which are projected downward from a central mast, so as to resemble an umbrella without is webbing.

UML Abbreviation of Unified Modeling Language. A notational language utilized for developing complex software, especially in the context of object-oriented technology.

UMS **1.** Abbreviation of **unified messaging system**. **2.** Abbreviation of **unified messaging service**.

UMTS Abbreviation of Universal Mobile Telecommunications System. A 3G broadband wireless phone system or service with features such as global roaming, data transfer rates exceeding 2 Mbps, and support for multimedia content. Also, a technology enabling this service or system.

unary expression An expression with a single operand.

unary operation An operation with a single operand.

unattended operation **1.** Operation which does not require supervision, human or otherwise. It may involve self-acting and/or self-regulating mechanisms, and utilize an autonomous information processing system. **2.** One or more steps or processes which do not require supervision, human or otherwise.

unattenuated **1.** A physical quantity which does not suffer a reduction over time or distance. Said, for instance, of an amplitude or energy. **2.** A physical quantity which has not suffered a reduction over time or distance. Said, for instance, of an amplitude or energy. For example, a signal which passes through a filter with no attenuation, or light traversing a fiber-optic cable with no losses.

unbalance **1.** To disturb the balance or state of equilibrium of something. Also, to be in such an unbalanced state. **2.** To perturb the equalization or symmetry between one or more variables among circuits, components, devices, systems, networks, or ground. Also, to be in such an unbalanced state.

unbalanced **1.** Consisting of, or pertaining to a component, circuit, device, transmission line, signal, or system in which one side, or terminal, of the input and/or output is connected to ground. Thus, that which is unbalanced is asymmetric with respect to ground. For instance, an unbalanced line, or an unbalanced circuit. Also called **single-ended**. **2.** That which is not in a state of balance.

unbalanced amplifier An amplifier in which the input and/or output are **unbalanced (1)**. Also called **single-ended amplifier**.

unbalanced bridge A bridge in which branch values are not adjusted in a manner which provides an output voltage of zero. At this time a response other than zero is obtained from the null detector.

unbalanced circuit A circuit which is **unbalanced (1)**. Also called **single-ended circuit**.

unbalanced input An input in which is **unbalanced (1)**. Also called **single-ended input**.

unbalanced line A transmission line, such as a coaxial line, which is **unbalanced (1)**. Also called **unbalanced transmission line**, or **single-ended line**.

unbalanced output An output which is **unbalanced (1)**. Also called **single-ended output**.

unbalanced signal A signal which is **unbalanced (1)**. Also called **single-ended signal**.

unbalanced transmission line Same as **unbalanced line**.

unbuffered Not utilizing a buffer, as occurs for instance, with real-time exchanges of data.

UNC Abbreviation of Uniform Naming Convention, or Universal Naming Convention. A convention or system utilized for specifying network resources, such as servers, directories, and files. The system uses slashes and backslashes to state pathnames, with an example of a typical format being \\server\directory\file.

uncalibrated **1.** That which has not gone through a calibration process to ensure high accuracy. Said, for instance, or an instrument, a measurement, or an output frequency. **2.** That which has not been calibrated recently.

uncalibrated device A device which is **uncalibrated**.

uncalibrated frequency An output frequency which is **uncalibrated**.

uncalibrated instrument An instrument which is **uncalibrated**.

uncalibrated measurement A measurement which is **uncalibrated**.

uncalibrated meter A meter which is **uncalibrated**.

uncalibrated reading A reading which has been made with a **uncalibrated** instrument.

uncalibrated signal An output signal which is **uncalibrated**.

uncertainty The estimated amount by which a measured or calculated value differs from the true value. Expressed, for instance, in terms of standard deviations.

uncertainty in measurement The **uncertainty** of a given measurement.

uncertainty principle A principle stating that for a given particle it is not possible to simultaneously know with absolute accuracy both its location and momentum. No matter how accurate the measurements are, the mere observing or measuring said particle will interfere with it in an unpredictable fashion. This principle also applies to other observations or measurements. Also called **Heisenberg uncertainty principle**, or **indeterminacy principle**.

uncharged **1.** Having no electric charge. For instance, a neutron has no charge, and molecules have no net charge. **2.** Having no charge stored. Said, for instance, of a depleted battery.

uncompress The undoing of the effects of compression. That is, to restore data to its original size and/or bandwidth. Also called **decompress**.

uncompression program Same as **uncompression utility**.

uncompression utility A utility, such as WinZip or PKUNZIP, utilized to **uncompress** files. Also called **uncompression program**.

unconditional Not subject to a given circumstance or state. For instance, an unconditional branch.

unconditional branch In a computer program, a branch that is taken on all occasions it is encountered. This contrasts with a **conditional branch**, which requires a given condition to be met. Also called **unconditional jump**, or **unconditional transfer**.

unconditional jump Same as **unconditional branch**.

unconditional stability For a given component, circuit, device, piece of equipment, or system, stability which is not dependent on a given parameter, such as the amplitude of its input signal, being within specified limits.

unconditional statement In a computer program, a statement that is executed without the need to meet any given condition. Also called **imperative statement**.

unconditional transfer Same as **unconditional branch**.

undamped oscillation Oscillation that does not die away from an initial maximum amplitude with each successive oscillation. This contrasts with **damped oscillation**, which dies away from an initial maximum amplitude, with each successive oscillation diminishing until the amplitude becomes zero. Also called **undamped vibration**.

undamped vibration Same as **undamped oscillation**.

undamped wave A wave whose amplitude does not die away from an initial maximum over time. This contrasts with a **damped wave**, whose amplitude dies away from an initial maximum, diminishing over time until its amplitude becomes zero.

undelete To restore that which has been previously deleted, such as a text, file, or directory. If a file is in a trash can, it can be easily undeleted with a command. However, after a trash can is emptied, a special utility is necessary to recover such files. Also, a command which causes this function to be performed. Also called **unerase**.

undeliverable That, such as an improperly addressed email, which can not be delivered to the intended recipient.

under-bunching Same as **underbunching**,

under-charging Same as **undercharging**.

under-coupling Same as **undercoupling**.

under-current Same as **undercurrent**.

under-current protection Same as **undercurrent protection**.

under-current relay Same as **undercurrent relay**.

under-damping Same as **underdamping**.

under-load Same as **underload**.

under-loading Same as **underloading**.

under-modulation Same as **undermodulation**.

under-modulation indicator Same as **undermodulation indicator**.

under-potential Same as **undervoltage**. Also spelled **underpotential**.

under-power relay Same as **underpower relay**.

under-power switch Same as **underpower switch**.

under-rate Same as **underrate**.

under-shoot Same as **undershoot**.

under-voltage Same as **undervoltage**.

under-voltage protection Same as **undervoltage protection**.

under-voltage protection circuit Same as **undervoltage protection circuit**.

under-voltage relay Same as **undervoltage relay**.

under-voltage release Same as **undervoltage release**.

under-voltage switch Same as **undervoltage switch**.

underbunching In a velocity-modulated tube, such as a klystron, bunching which is below the optimum level. Also spelled **under-bunching**.

undercharging The applying of an insufficient charge to a storage battery or storage cell. Consistent undercharging of a battery may reduce cell life, lead to one or more cells being completely discharged before others, and produce recharges which are incomplete yet more time consuming. Also spelled **under-charging**.

undercoupling A level of coupling which is below critical coupling. Also spelled **under-coupling**. Also called **loose coupling (3)**, or **weak coupling (3)**.

undercurrent For a component, circuit, device, piece of equipment, or system, a current that is below the optimum, operational, or rated current. Such a current may produce a malfunction or failure. Also spelled **under-current**.

undercurrent protection The use of components, devices, or mechanisms, such as fuses or circuit breakers, to prevent damage due to **undercurrents**. Also spelled **under-current protection**.

undercurrent relay A relay which automatically opens a circuit when a current falls below a set amount. Also spelled **under-current relay**.

underdamping Damping which is below critical damping. Also spelled **under-damping**.

underfill In flip chip bonding, an encapsulant deposited between the chip and substrate to help insure reliable connections.

underflow Also called **underflow condition**, or **arithmetic underflow**. **1.** The condition in which the result of an operation is more than zero, but less than the lowest value that can be expressed. **2.** The amount by which a non-zero result of an operation is less than the lowest value that can be expressed.

underflow condition Same as **underflow (1)**.

underflow error An error resulting from an **underflow (1)**.

underground antenna An antenna which is buried, located in a tunnel, or is otherwise placed or utilized beneath the ground.

underground cable A cable which is buried, located in a tunnel, or is otherwise placed or utilized beneath the ground.

underground line A transmission line, such as a power line, which is buried, located in a tunnel, or is otherwise placed or utilized beneath the ground. Also called **underground transmission line**.

underground power line A power line which is buried, located in a tunnel, or is otherwise placed or utilized beneath the ground.

underground transmission line Same as **underground line**.

underlap In fax, a reproduction defect in which the scanning lines are too close together.

underline cursor On a computer screen, a cursor in the shape of an underline, as opposed to a solid block.

underload Also spelled **under-load**. **1.** A load which is lower than the optimum, operational, or rated load of a component, circuit, device, piece of equipment, machine, or system. If an underload occurs due to a condition such as a load not dissipating or otherwise utilizing the provided power, a malfunction, failure, or overheating may occur. **2.** A load which is lower than the usual load of a component, circuit, device, piece of equipment, machine, or system.

underload condition A state in which an **underload** (1) exists.

underload protection The utilization of **underload protectors**.

underload protector A device or mechanism, such as a circuit breaker, which is utilized to prevent **underloads** (1) or their harmful effects.

underload relay **1.** A relay which is actuated when a load is lower than a given value. Used, for instance, as an underload protector. **2.** Same as **undercurrent relay**.

underload release A device or mechanism, such as a circuit breaker, which opens a circuit when the load falls below a specified minimum.

underloaded circuit An electric circuit whose load is below that which it can handle safely. If such an underload occurs due to a condition such as a load not dissipating or otherwise utilizing the provided power, a malfunction, failure, or overheating may occur.

underloading The condition where a load is lower than the optimum, operational, or rated load of a component, circuit, device, piece of equipment, machine, or system. Also spelled **under-loading**.

undermodulation Amplitude modulation of a carrier which is below that which is necessary for proper transmission or recording. This may lead, for instance, to the desired signal being masked by noise. Also spelled **under-modulation**.

undermodulation indicator A component, device, display, mechanism, or alarm which alerts to an **undermodulation** condition. Also spelled **under-modulation indicator**.

Undernet A popular Internet Relay Chat network.

underpotential Same as **undervoltage**. Also spelled **under-potential**.

underpower relay A relay which is actuated when the power in a circuit drops below a given threshold. Also spelled **under-power relay**.

underpower switch A switch that is actuated when the voltage in a circuit drops below a given threshold. Also spelled **under-power switch**.

underrate To state a rating below the level which a component, circuit, device, piece of equipment, or system can be operated under given conditions. Used, for instance, to provide a margin of security, or to help ensure optimum performance. Also spelled **under-rate**.

underscore A keyboard character, shown here between quotes, "_", which is utilized, for instance, when a space is not allowed or otherwise can not be used. Seen, for example, in email addresses.

undershoot Also spelled **under-shoot**. **1.** To fall short of that which is necessary, desired, or expected. For instance, in an analog indicating device, a pointer excursion which does not reach the value that should be indicated. **2.** In a control system, the amount by which a response falls short of that which is necessary to maintain the controlled parameter at the desired level. **3.** The extent to which **undershoot** (1) or **undershoot** (2) occurs.

undervoltage Also spelled **under-voltage**. Also called **underpotential**. Its abbreviation is **UV**. **1.** A voltage which is below the optimum, operational, or rated value of a component, circuit, device, piece of equipment, machine, or system. Such a voltage may produce, for instance, distortion, a malfunction, or failure. In computers and similar devices, undervoltages can lead to data losses. **2.** An **undervoltage** (1) in a power line which lasts more than a given time interval, such as several seconds.

undervoltage protection The use of circuits, components, devices, or mechanisms, to prevent damage due to **undervoltage**. Also spelled **under-voltage protection**.

undervoltage protection circuit A circuit which serves to protect against **undervoltage**. Also spelled **under-voltage protection circuit**.

undervoltage relay A relay which is actuated when the voltage in a circuit falls below a given threshold. Also spelled **under-voltage relay**.

undervoltage release A circuit, device, or mechanism, such as an undervoltage protection circuit, which opens a circuit or deactivates a device or piece of equipment when the voltage falls below a specified minimum. Also spelled **under-voltage release**.

undervoltage switch A switch that is actuated when the voltage in a circuit falls below a given threshold. Also spelled **under-voltage switch**.

underwater acoustics The study of acoustic energy and phenomena under water surfaces. It may deal with, for instance, the propagation of sound under such conditions. Although this figure varies depending on the specific conditions, sound travels at approximately 1500 m/s in ocean water. An application of underwater acoustics is sonar. Also called **hydroacoustics**.

underwater antenna An antenna that is submerged, located in a tunnel beneath or within a body of water, or which is otherwise placed or utilized beneath water surfaces.

underwater cable A cable that is submerged, located in a tunnel beneath or within a body of water, or which is otherwise placed or utilized beneath water surfaces.

underwater line A transmission line, such as a power line, that is submerged, located in a tunnel beneath or within a body of water, or which is otherwise placed or utilized beneath water surfaces. Also called **underwater transmission line**.

underwater microphone A transducer which converts underwater sound energy into corresponding electrical signals. Used, for instance, in sonar. Also called **hydrophone**.

underwater power line A power line that is submerged, located in a tunnel beneath or within a body of water, or which is otherwise placed or utilized beneath water surfaces.

underwater speaker A speaker whose enclosure enables it to be utilized under water. Used, for instance, in sonar, or for diver safety.

underwater transmission line Same as **underwater line**.

Underwriters Laboratories Same as **UL**.

undetected error Also called **residual error**. **1.** All error remaining after that which can be accounted for is taken into account. **2.** All error remaining after all error-correcting measures have been taken.

undistorted **1.** Lacking any distortion, such as that in the reproduction of sounds or images. **2.** Lacking any meaningful or measurable distortion.

undistorted audio Same as **undistorted sound**.

undistorted image An image which is **undistorted**.

undistorted output An output, such as that of power, which is **undistorted**.

undistorted signal A signal, such as that carrying data, which is **undistorted**.

undistorted sound Sound which is **undistorted**. Also called **undistorted audio**.

undistorted wave A wave without any undesired changes in its waveform.

undo A command that reverses the last action or change, such as a deletion, change in spelling, or image alteration. Most programs allow undoing multiple edits, and some allow undoing all edits of the session in progress. Also, to perform such a reversal.

undock To unplug from a **docking station**.

undoped A semiconductor material which has not had a dopant added.

undoped semiconductor A semiconductor material with no dopants. Its electrical characteristics, such as concentration of charge carriers, depend only on the pure crystal. Also called **intrinsic semiconductor**.

undulating current A current which passes through regular cycles, and whose average is not zero. Such a current may result, for instance, from the sum of an alternating current and a direct current. Also called **pulsating current**.

unerase Same as **undelete**.

unexpected halt Also called **hangup**. **1.** A unexpected stoppage in the operation of a computer program or system. It may be due to most any cause, such as a crash, an error condition, or the awaiting of an input that does not arrive. **2.** An unexpected halt or delay in the operation of a device, piece of equipment, or system.

unfiltered That which has not gone through a filter or any process which selectively allows matter and/or energy to pass, remain, be blocked, or get absorbed.

unformatted capacity In a disk, disc, or tape, the storage capacity prior to formatting. Data, such as control information, reduces this capacity. Also called **raw capacity (1)**, or **native capacity (1)**.

ungrounded Not connected to a ground. Said, for instance, of a component, circuit, object, surface, outlet, or the like.

ungrounded circuit A circuit which is not connected to a ground.

ungrounded device A device which is not connected to a ground.

ungrounded outlet An outlet which does not have a contact which serves for connection to a grounding conductor.

ungrounded system A system which is which no leads are connected to a ground.

ungrounded wire A wire which is not connected to a ground.

unhandled exception An exception which, due to a software deficiency, is not handled properly, resulting in the application responsible being shut down. This may also result in a system halt.

UNI Abbreviation of User Network Interface, or User-to-Network Interface. An interface between a network device, such as an ATM switch, and a given user device.

Unibus A proprietary bus architecture.

unicast In a communications network, the transmission from one node to another. This contrasts, for instance, with **multicast**, in which one node transmits to many.

Unicode A version of ASCII which uses two bytes per character, as opposed to one. This expands the available characters from 256 to 65,536. This is sufficient to accommodate most of the languages in the world. Also called **double-byte characters**.

unidirectional Operating, moving, transferring, flowing, transmitting, receiving, or responding in a single direction. For example, a unidirectional bus, a unidirectional current, or a unidirectional counter.

unidirectional antenna An antenna, such as a directional antenna, whose directivity pattern consists of a single lobe, or of a single principal lobe.

unidirectional antenna array An assembly of antenna elements with the proper dimensions, characteristics, and spacing, so as to have a directivity pattern consisting of a single lobe, or of a single principal lobe. Also called **unidirectional array**.

unidirectional array Same as **unidirectional antenna array**.

unidirectional bus In computers, a bus that carries signals in a single direction.

unidirectional counter A counter which can count either upwards or downwards, but not in both directions.

unidirectional coupler In a waveguide, a device which can sample either incident or reflected waves, but not both. Used, for instance, to measure power.

unidirectional current A current, such as DC, which always flows in the same direction.

unidirectional data bus A data bus that carries data in one direction only.

unidirectional device A device, such as a unidirectional microphone, which exhibits unidirectional properties.

unidirectional loudspeaker Same as **unidirectional speaker**.

unidirectional microphone An microphone, such as a directional microphone, which is sensitive in only one direction, or significantly more sensitive in a given direction.

unidirectional parallel port A parallel port which supports communication only from the computer to the peripheral device connected to said port.

unidirectional pattern The directivity pattern of a unidirectional device, such as a unidirectional antenna or a unidirectional microphone.

unidirectional printing The ability of a computer printer to print in one direction only.

unidirectional pulses Pulses which have a single polarity. For example, those of rectified AC.

unidirectional speaker A speaker that produces acoustic energy in a single direction, such as the front or rear. Also called **unidirectional loudspeaker**.

unidirectional transducer **1.** A transducer which only can measure input from one direction from a reference point, such as zero. **2.** A transducer which can only operate in a single direction. Also called **unilateral transducer**.

Unified-Field Theory Also called **Unified Theory**. **1.** A theory describing a single force which would result from the unification of the four fundamental forces of nature, which are the gravitational force, the weak force, the electromagnetic force, and the strong force. Also called **Grand Unified Theory**. **2.** A theory which describes a single force which would result from the unification of two or more of the four fundamental forces of nature. For example, that describing a single force which would result from the unification of the gravitational force and the electromagnetic force.

unified memory architecture A system architecture which allows for sharing memory. Used, for instance, to enable a video card to use of part of the main memory, as opposed to having its own RAM. Its abbreviation is **UMA**. Also called **shared memory architecture**.

unified messaging Its abbreviation is **UM**. **1.** The ability to access voice messages, SMSs, emails, faxes, and the like, from a single device such as a cell phone, PDA, or desktop computer. Received emails and faxes, for instance, which are accessed via a regular telephone are converted into audio files. **2.** A system providing **unified messaging** (1). Also called **unified messaging system**. **3.** A service providing **unified messaging** (1). Also called **unified messaging service**.

unified messaging service   Same as **unified messaging** (3). Its abbreviation is **UMS**.

unified messaging system   Same as **unified messaging** (2). Its abbreviation is **UMS**.

Unified Modeling Language   Same as **UML**.

Unified Theory   Same as **Unified-Field Theory**.

unifilar   Having, consisting of, or utilizing a single thread, fiber, filament, wire, or the like. For example, a unifilar suspension.

unifilar suspension   In measuring instruments, the suspending of the movable part by a single thread or wire. A **bifilar suspension**, which uses parallel threads or wires, usually provides more stability.

uniform cable   A cable, such as a coaxial cable, with essentially identical electrical properties throughout its length.

uniform circular motion   Circular motion, such as that of a particle, in which the speed remains constant.

Uniform Computer Information Transactions Act   Same as **UCITA**.

uniform electric field   An electric field in which the exerted force has the same strength at all surrounding points.

uniform field   A field, such as an electric, magnetic, or gravitational field, in which the exerted force has the same strength at all surrounding points.

uniform frequency response   Also called **uniform response**, or **flat frequency response**. **1.** For a component, circuit, device, piece of equipment, or system whose output is an interval of frequencies, a response in which all frequencies have equal amplitude. **2.** For a component, circuit, device, piece of equipment, or system whose output is an interval of frequencies, a response in which all frequencies are within a specified interval of amplitudes. For example, a high-fidelity amplifier may have a frequency response of 10 Hz to 60 kHz, +/− 0.5 decibels, and be considered to be flat.

uniform line   A transmission line, such as a coaxial cable, with essentially identical electrical properties throughout its length. Also called **uniform transmission line**.

uniform magnetic field   A magnetic field in which the exerted force has the same strength at all surrounding points.

Uniform Naming Convention   Same as **UNC**.

Uniform Resource Identifier   Same as **URI**.

Uniform Resource Locator   Same as **URL**.

Uniform Resource Name   Same as **URN**.

uniform response   Same as **uniform frequency response**.

uniform scale   A scale in which the distance between each indicated value corresponds to an equal magnitude, along the entire length of said scale. This contrasts, for instance, with a logarithmic scale. Also called **linear scale**.

uniform transmission line   Same as **uniform line**.

uniform waveguide   A waveguide with essentially identical electrical and physical properties throughout its axial length.

UniForum   An association which helps in the development and promotion of open standards, technologies, and systems.

unijunction transistor   A three-terminal transistor incorporating an emitter and two bases, and having a single pn junction. Used mostly for switching, and may be employed, for instance, in relaxation oscillators. Its abbreviation is **UJT**. Also called **double-base diode**.

unilateral transducer   Same as **unidirectional transducer** (2).

uninstall   To remove software or hardware which has been previously installed in a computer. Some applications have

their own uninstall utility, and when this is not the case an uninstall program may be employed. When a program is properly uninstalled not only are its files removed, but system files are restored to their former state.

uninstall program   Same as **uninstaller**.

uninstall utility   Same as **uninstaller**.

uninstaller   A program or utility that serves to **uninstall** other programs. Also called **uninstall program**, **uninstall utility**, **uninstaller program**, or **uninstaller utility**.

uninstaller program   Same as **uninstaller**.

uninstaller utility   Same as **uninstaller**.

unintelligible crosstalk   Crosstalk which is incomprehensible. This contrasts with **intelligible crosstalk**, which is capable of being comprehended. Unintelligible crosstalk is less distracting, as it is usually perceived as miscellaneous noise.

uninterruptible power   The power provided by an **uninterruptible power supply**.

uninterruptible power source   Same as **uninterruptible power supply**. Its abbreviation is **UPS**.

uninterruptible power supply   A device or system which provides continuous power whether the primary source, such as mains AC, fluctuates irregularly or fails. Such a source may consist of one or more batteries, or of a generator. An uninterruptible power supply may provide the power needs of devices or equipment through brownouts and blackouts, and in the case of computers affords time for properly executed shutdowns which help protect against data losses and equipment damage. Also called **uninterruptible power system**, or **uninterruptible power source**. Its abbreviation is **UPS**.

uninterruptible power system   Same as **uninterruptible power supply**. Its abbreviation is **UPS**.

union   **1.** That which serves to join, couple, or bond parts. Also, such a union. Also, the state of being so joined. **2.** The joining or merging of data entities such as rows or files.

unipolar   **1.** Having a single pole or polarity. **2.** Pertaining to transistors in which either electrons or mobile holes serve as charge carriers, but not both.

unipolar input   An input which only accepts signals of a single polarity.

unipolar machine   A DC generator in which the poles are of the same polarity with respect to the armature, thus making a commutator unnecessary. Also called **homopolar generator**, **homopolar machine**, or **acyclic machine**.

unipolar output   An output signal which has only a single polarity.

unipolar transistor   A transistor, such as a FET, which utilizes only one type of charge carrier. Bipolar junction transistors utilize both electrons and mobile holes as charge carriers.

unipole   A hypothetical antenna which radiates and/or receives equally in all directions. Such an antenna is utilized, for example, as a reference for comparison with real antennas. Also called **isotropic antenna**, or **isotropic radiator**.

unipotential cathode   A cathode, within a thermionic tube, which is electrically insulated from the heating element. It may consist, for instance, of a filament surrounded by a sleeve with an electron-emitting coating. Such cathodes have the same potential along their entire surface. Also called **equipotential cathode**, **indirectly-heated cathode**, or **heater-type cathode**.

uniprocessor   Pertaining to, utilizing, or incorporating a single processor, as opposed to **multiprocessor**.

unique user   A specific visitor to a Web site, as distinguished from other such visitors. For example, for a given period a Web site may have 100 hits, all by a single user, or 10 by 10 unique users, and so on. Individual users are usually distinguished by their IP number, but this may not be accurate, as a single IP number may identify multiple users, and a single user may have a dynamic IP address. Also called **unique visitor**, or **visitor (1)**.

unique visitor Same as **unique user**.

unit 1. An item, group, structure, or entity which is regarded as a single entity or whole. 2. A **unit (1)** which is a part of a larger item, group, structure, or entity. For example, a control unit, a data unit, a logic unit, or a crystal unit. 3. A piece, part, or assembly which can independently perform one or more functions. For instance, a radar unit, an amplifier, a power pack, or a rectifier unit. 4. An amount or magnitude of a physical quantity, such as mass or distance, which is utilized as a standard to express such quantities or magnitudes. For example, the defined SI unit for mass is the kilogram, and the meter serves to express distances. Such units are also used in multiples to state values, as in picogram, or kilometer. Also called **unit of measurement**.

unit cell Within a crystalline solid, the smallest repeating unit which has all the characteristics, such as physical properties and symmetry, of the overall structure.

unit charge 1. Same as **unit electric charge**. 2. A unit, such as the coulomb, utilized to express electric charge.

unit coupling A degree of coupling between two systems, especially circuits, corresponding to a theoretical perfection. That is, the coefficient of coupling would be 1, or 100%. Also called **perfect coupling**, or **ideal coupling**.

unit electric charge The charge carried by a single electron or proton. It is a fundamental physical constant, and is equal to approximately 1.6022×10^{-19} coulomb. All subatomic particles have an electric charge equal to this value, or a multiple of it. Also called **unit charge (1)**, or **elementary charge**.

unit function Same as **unit-step function**.

unit length A standard measure, such as the meter, or fractions/multiples of meters, utilized to quantify length or characteristics or properties which depend on length. Used, for instance, in the context of transmission line attenuation, a voltage gradient, or the recording density of a storage medium.

unit magnetic pole Same as **unit pole**.

unit of measurement Same as **unit (4)**.

unit pole A unit which describes the strength of a magnetic pole. In a vacuum, a unit magnetic pole would repel an identical pole at a distance of one centimeter with a force of one dyne. This equals about 1.25664×10^{-7} webers, or approximately 12.5664 maxwells. Also called **unit magnetic pole**.

unit power factor In an AC circuit, the power factor when both the voltage and the current are in phase.

unit-step function A step function whose non-zero value is one. Also called **unit function**.

units system A system utilized for measurement of physical quantities using units which are precisely defined and internationally accepted. Currently, the International System of Units is the most important, and other units systems in use include the meter-kilogram-second-ampere, and the centimeter-gram-second systems. Also called **system of units**.

unity 1. A value of one. Also, a result of one, as in the ratio of two equal values in the same units. 2. The number one. 3. The state or quality of being one.

unity coupling A coupling coefficient of 1, which is that of an ideal transformer.

unity gain A gain ratio of 1. In the case of a power amplifier, it would be a 1:1 ratio of the output power to the input power.

unity gain bandwidth The frequency bandwidth of a component or device, such as an amplifier, at **unity gain**.

unity power factor A power factor of 1. This occurs when both the voltage and the current in an AC circuit are in phase.

UNIVAC An early computer which utilized vacuum tubes and punched cards. It is an acronym for **U**niversal **A**utomatic **C**omputer.

univalent Pertaining to an element, such as sodium or chlorine, which has a valence of one.

universal adapter An adapter that works with various different formats. For example, an AC adaptor that works with multiple types of plugs. Also spelled **universal adaptor**.

universal adaptor Same as **universal adapter**.

Universal Asynchronous Receiver Transmitter Same as **UART**.

universal bridge A bridge which is equipped for measurements of two or more circuit parameters, such as resistance, capacitance, inductance, or frequency.

universal client In a client/server architecture, a client that can access multiple applications, networks, systems, or the like.

universal connector A connector that works with various different formats. For example, a connector that works with multiple types of cables.

Universal Coordinated Time Same as **Universal Time Coordinated**. Its abbreviation is **UCT**.

Universal Description, Discovery and Integration Same as **UDDI**.

Universal Disc Format Same as **Universal Disk Format**.

Universal Disk Format A format which is compatible with multiple optical disc technologies, such as DVD and CD-R. Its abbreviation is **UDF**. Also spelled **Universal Disc Format**.

universal filter A filter that can be utilized in different formats, such as low-pass, high-pass, bandpass, or bandstop.

universal joint A joint that transmits rotary motion between shafts which do not lie along the same line. Used, for instance, in robotics.

Universal Mobile Telecommunications System Same as **UMTS**.

universal motor A motor that runs on AC or DC. Also called **ac/dc motor**.

Universal Naming Convention Same as **UNC**.

Universal Plug-and-Play Its abbreviation is **UPnP**, **Universal PnP**, or **Universal Plug & Play**. 1. A plug-and-play technology that works across multiple platforms, and is compatible with devices such as desktop computers, PDAs, smart appliances, and so on. 2. A peripheral, device, or computer which supports **Universal Plug-and-Play (1)**.

Universal Plug & Play Abbreviation of **Universal Plug-and-Play**.

Universal PnP Abbreviation of **Universal Plug-and-Play**.

universal port replicator A device which connects to a portable computer, such as a notebook, and which provides multiple ports, such as those for a monitor, keyboard, mouse, USB devices, networks, and so on.

universal product code A bar code which is printed on items for sale for identification purposes. Such codes are read by bar-code scanners, and are used, for instance, to tally sales at a point-of-sale, and to maintain inventory. Its abbreviation is **UPC**.

universal receiver 1. A receiver which accepts multiple types of information-bearing signals. 2. A receiver which can be used to tune to multiple frequency bands. Depending on the unit, a user may be able to select from AM, FM, MW, SW, and TV broadcasts, and so on. Also called **multi-band receiver**.

universal remote Abbreviation of **universal remote control**.

universal remote control Its abbreviation is **universal remote**. **1.** A remote control that works with multiple home entertainment components, such as amplifiers, TVs, DVDs, and so on. **2.** A remote control that works with multiple devices and appliances, such as lighting, heating, garage doors, ovens, and so on.

Universal Resource Identifier Same as **URI**.

Universal Resource Locator Same as **URL**.

Universal Resource Name Same as **URN**.

Universal Serial Bus Same as **USB**.

universal shunt A high-resistance shunt utilized to reduce the sensitivity of a measuring instrument, such as a galvanometer. This increases its range. Also called **Ayrton shunt**.

Universal Synchronous Receiver Transmitter Same as **USRT**.

Universal Time Same as **Universal Time Coordinated**. Its abbreviation is **UT**.

Universal Time Coordinated An internationally agreed time standard based on time kept by atomic clocks, and which represents the local time at the 0° meridian, which passes through Greenwich, England. When utilizing Universal Time Coordinated, each location on the planet has the same time, which is expressed utilizing a 24 hour clock, with a Z frequently appended. For example, 2100Z indicates 9 PM. Since the earth's rotation is gradually slowing, an extra second is added approximately once a year. It is based on International Atomic Time, and its abbreviation is **UTC**. Also called **Universal Coordinated Time, Universal Time, Coordinated Universal Time, World Time, Zulu Time**, or **Z time**.

universal transmitter **1.** A transmitter, such as a universal remote control, which works with various different formats. **2.** A transmitter which emits signals in any of multiple frequency bands.

universal Turing machine A Turing machine that can act as any Turing machine.

univibrator A multivibrator with one stable state and one unstable state. A trigger signal must be applied to change to the unstable state, where it remains for a given interval, after which it returns to the stable state. Also called **monostable multivibrator, one-shot multivibrator**, or **single-shot multivibrator**.

UNIX A popular multitasking operating system designed for multiple users, and which is utilized especially by servers and workstations. It is powerful, flexible, can be used across various platforms, and current versions include LINUX and Solaris. Also called **UNIX operating system**.

UNIX operating system Same as **UNIX**.

UNIX server A server utilizing a **UNIX operating system**.

UNIX shell account A shell account which utilizes **UNIX** commands.

unknown host An error message indicating that a specific address, such as an IP address, could not be located for a given host.

unknown recipient An error message indicating that a specific destination address could not be found within an online directory a messaging server is accessing.

unload **1.** To remove a computer storage medium, such as a disc or tape, from a drive or other device utilized to read and/or write from and/or to it. **2.** To transfer data from a computer storage medium, such as a disk or tape, or a database. Also, to remove a program from memory. **3.** To remove a disk, cassette, reel, cartridge, drum, or other object composed of, or containing a recordable medium from a device utilized for recording and/or reproduction. **4.** To remove an output load.

unloaded antenna An antenna whose characteristics, such as electrical length or resonant frequency, have not been altered through the use of inductive and/or capacitive elements, such as loading coils or loading disks.

unloaded line A transmission line whose characteristics have not been altered through the use of inductive and/or capacitive elements, such as loading coils. Also called **unloaded transmission line**.

unloaded Q The value of the Q factor for a circuit or device without an external load. Also called **intrinsic Q**.

unloaded transmission line Same as **loaded line**.

unmanaged hub A hub which does not support network management functions. A **managed hub** does.

unmark To remove or undo a mark which has been previously made. For example, to remove a symbol which identifies data, the end of a record, or a subdivision of a file, or to unselect a block of data or a choice within a menu.

unmatched components **1.** Components which can not function properly together without prior modifications. **2.** Components which have not been specifically designed to work together. **3.** Components which do not have an equality in a given property or value. For instance, those whose impedances are not matched.

unmatched devices **1.** Devices which can not function properly together without prior modifications. **2.** Devices which have not been specifically designed to work together. **3.** Devices which do not have an equality in a given property or value. For instance, those whose impedances are not matched.

unmatched filter A filter which does not match the impedance of both its signal source and the output load.

unmatched impedance An impedance which is unmatched between two circuits, or between a signal source and a load. This contrasts with a **matched impedance**, which helps insure that the maximum possible power is transferred from source to load.

unmatched line A transmission line whose characteristic impedance is not the same as its terminations. Under such circumstances, there is not a complete absorption of the incident power. Also called **unmatched transmission line**.

unmatched load A load whose impedance value is that not that which results in the maximum possible transfer of power.

unmatched pair A pair of components, devices, or the like, which does not function properly together, or which has not been specifically designed to work together.

unmatched termination A termination which does not provide for the complete absorption of incident power. Such a termination will not eliminate wave reflections back to the source.

unmatched transmission line Same as **unmatched line**.

unmodulated **1.** A beam, wave, signal, or the like, which has not undergone modulation. **2.** A beam, wave, signal, or the like, which has not yet undergone modulation. For instance, an unmodulated carrier.

unmodulated beam A beam whose intensity is not varied so as to convey meaningful information. For instance, a unmodulated laser, light, or electron beam.

unmodulated carrier A carrier wave, such as that to be used for AM or FM, whose amplitude, frequency, or phase has not yet been varied so as to convey meaningful information. Also called **unmodulated carrier wave**.

unmodulated carrier wave Same as **unmodulated carrier**.

unmodulated electron beam An electron beam whose intensity is not varied so as to convey meaningful information.

unmodulated laser beam A laser beam whose intensity is not varied so as to convey meaningful information.

unmodulated light Same as **unmodulated light beam**.

unmodulated light beam A light beam whose intensity is not varied so as to convey meaningful information. Also called **unmodulated light**.

unmodulated oscillator An oscillator whose output signal is not varied so as to convey meaningful information.

unmodulated signal A signal whose amplitude, frequency, or phase has not been varied so as to convey meaningful information.

unmodulated wave A wave, such as a carrier wave, whose amplitude, frequency, or phase has not been varied so as to convey meaningful information.

unpack To restore data to its original space and/or bandwidth, following **packing (1)**.

unpacking The restoring of data to its original space and/or bandwidth, following **packing (1)**.

unpolarized electromagnetic radiation Same as **unpolarized radiation**.

unpolarized light Electromagnetic radiation in the visible range whose electric field vector is not confined to a single plane.

unpolarized outlet A power outlet into which an **unpolarized plug** is inserted. Also called **unpolarized receptacle**.

unpolarized plug A power plug which does not have to be inserted into a corresponding outlet in any given orientation, while a **polarized plug** must be. An unpolarized plug plugs into a **unpolarized outlet**.

unpolarized radiation Electromagnetic radiation whose electric field vector is not confined to a single plane. Also called **unpolarized electromagnetic radiation**.

unpolarized receptacle Same as **unpolarized outlet**.

unrecoverable error An error, such as a fatal error, which can be can not be simply remedied without further deleterious effects. At the very least the computer must be re-booted, and in some cases additional measures must be taken.

unreflected ray A ray which has not bounced off a given surface, object, region, discontinuity, or the like.

unreflected wave A wave which has not bounced off a given surface, object, region, discontinuity, another wave, or the like.

unregulated 1. That which is not adjusted, managed, or otherwise controlled. **2.** A voltage, current, or the like, which is not maintained within specified values.

unregulated power supply A power supply that is unregulated, that is, having no circuitry to maintain output voltage constant when the input line or load varies. Also called **brute power supply**.

unreliable protocol A protocol, such as UDP, that does not implement error detection and/or correction measures, and which in exchange provides faster data transfer rates. Such a protocol may be used, for instance, when such measures are not needed, or are addressed by an external form of error detection and/or correction.

unsaturated 1. The condition where a maximum effect has not been attained, thus any additional increase in the causative variable may achieve additional results. **2.** For a magnetic material, the condition in which an increase in magnetizing force will still produce an increase in magnetization. **3.** In an electron tube, the condition in which the anode current can be further increased by applying an additional voltage. **4.** A color which has a white component. White light is 100% unsaturated. **5.** A chemical solution, such as an electrolyte, which does not contain the maximum amount of a substance that it can dissolve in equilibrium.

unscramble To decrypt, decode, or otherwise remove or undo that which has been scrambled.

unshielded Not protected by, or otherwise incorporating a shield, or shielding.

unshielded cable A cable incorporating one or more conductors which are insulated from each other, but which as a group do not have an additional outer sheath which helps protect from external interference. Used, for instance, for electric power transmission, or as speaker cables.

unshielded line A transmission line, such as an unshielded twisted pair or an unshielded cable, in which the individual conductors are insulated from each other, but which as a group do not have an additional outer sheath which helps protect from external interference. Also called **unshielded transmission line**.

unshielded pair A transmission line, such as an unshielded twisted pair, consisting of two similar conductors which are individually insulated from each other, but which as a pair do not have an additional outer sheath which helps protect from external interference.

unshielded transmission line Same as **unshielded line**.

unshielded twisted pair A transmission line consisting of two similar conductors which are individually insulated and twisted around each other, but which as a pair do not have an additional outer sheath which helps protect from external interference. Widely utilized for ordinary telephone wiring. Its abbreviation is **UTP**.

unshielded twisted pair Ethernet Same as **UTP Ethernet**.

unshielded twisted pair wiring Wiring, such as that utilized for ordinary telephone service, consisting of **unshielded twisted pairs**.

unsolder To separate leads or surfaces that were previously joined using solder, usually by melting said solder. Also called **desolder**.

Unsolicited Commercial Email Spam of a commercial nature. Most spam is of this type. Its abbreviation is **UCE**.

Unspecified Bit Rate An ATM service in which any available bandwidth is allocated on a best-effort basis. Used, for instance, for data which is not time-critical.

unstable 1. Not tending to remain in a given position, state, setting, mode, or the like. For instance, unstable oscillation. **2.** Not tending to return to a former or desired position, state, setting, mode, or the like, after a displacement or other change occurs. For example, a speaker with improper damping. **3.** Not tending to maintain a given value, quantity, intensity, characteristic, or the like. **4.** Tending to easily respond to a change, action, stimulus, or variation. **5.** A chemical substance which easily forms part of a chemical reaction. **6.** A particle or isotope which exhibits radioactive decay.

unstable element A chemical element, such as uranium or radon, exhibiting **radioactivity**. Also called **radioactive element**.

unstable isotope For a given chemical element, an isotope which exhibits **radioactivity**. Radium, for instance, as over 30 identified isotopes, all of which are radioactive. Also called **radioisotope**, or **radioactive isotope**.

unstable oscillation Oscillation, such as that produced by an unstable oscillator, which is irregular.

unstable oscillator An oscillator, such as a relaxation oscillator, whose output signal, such as a current or voltage, is irregular.

unstable state A state which is erratic, fluctuating, or otherwise unstable. For instance, that of a control system which persistently oscillates between output states in an unwanted manner, due to excessive feedback.

unstructured data Data which does not have to have a defined structure, such as the text of an email, or an image.

unsubscribe 1. To cancel a given service which had been signed-up for, such as CATV. **2.** To cancel the authorization previously given to an entity to send information. For example, to cancel a subscription to a mailing list.

untar 1. To separate the individual files which were combined using a **tape archive** utility. 2. A utility utilized to **untar** (1).

untuned 1. That which has not been adjusted to a desired frequency. 2. A component, circuit, body, or system which has been not been adjusted so as to make it resonate with a given signal, force, or the like. 3. A component, circuit, device, piece of equipment, or system which has not been adjusted so as to attain optimum performance or output.

untuned antenna An antenna that has an approximately constant input impedance over a wide range of frequencies. A diamond antenna is an example. Also called **non-resonant antenna**, or **aperiodic antenna**.

unzip To uncompress files which have been compressed using a program such as WinZip or PKZIP.

unzipping The uncompression of files which have been compressed using a program such as WinZip or PKZIP.

up A device, piece of equipment, or system which is operational, or ready for use.

up-conversion The mixing of a signal with a local oscillator, so that the frequency of the output signal is higher than that of the input signal. This contrasts with **down-conversion**, where the output signal has a lower frequency than the input signal. Also spelled **upconversion**.

up-convert The performance of an **up-conversion**. Also spelled **upconvert**.

up-converter A converter whose output frequency is higher than its input frequency. This contrasts with a **down-converter**, whose output frequency is lower than that of the input. Also spelled **upconverter**.

up counter A counter that only counts upwards, as opposed to a **down counter**, which only counts downwards, or an **up-down counter** which does both.

up-down counter A counter which can count both upwards and downwards. It has, for instance, both an adding input and a subtracting input, thus giving it the ability to count in both directions. Also called **forward-backward counter**, or **bi-directional counter**.

up-time Same as **uptime**.

UPC Abbreviation of **universal product code**.

upconversion Same as **up-conversion**.

upconvert Same as **up-convert**.

upconverter Same as **up-converter**.

update 1. To make a change in an existing file or data, especially to make it more current. 2. A newer version of an existing program or software package. Each update has a higher version number than the last.

upgrade 1. A newer version of software and/or hardware. For example, an update, or the replacement of a CPU with a newer model. 2. To add power, speed, capacity, or the like, to existing hardware. For example, to increase RAM to 2 GB from 512 MB. 3. To perform or install an **upgrade** (1) or **upgrade** (2).

upkeep All activities which are performed to help insure the proper operation of components, circuits, devices, equipment, and systems. Maintenance is usually either preventive or corrective, and may includes tests, adjustments, cleaning, and replacements. In the case of software, for instance, maintenance may involve updating or debugging. Also called **maintenance**.

uplink 1. The upward radio-communications path originating from an earth-based transmitter towards a communications satellite, or other airborne receiver. This contrasts with a **downlink** (1), which is the converse. 2. The establishing, or use, of an **uplink** (1).

uplink frequency The frequency, or band of frequencies, of an **uplink signal**.

uplink port In a communications network, especially Ethernet, a port which allows connections to hubs or switches without a crossover cable. Also called **Medium Dependent Interface Port**.

uplink power The power of an **uplink signal**.

uplink signal The signal sent in an **uplink** (1). Such a signal usually occupies a given band of frequencies.

uplink station A location from which an **uplink signal** is sent.

upload To send data, usually in the form of a file, to a remote computer in a network. For instance, a person updating a Web page maintained on the Internet may upload a file with new information. Also, the data or file transferred. This contrasts with a **download** (1), where information is sent from a remote computer in a network.

UPnP Abbreviation of **Universal Plug-and-Play**.

upper atmosphere The portion of the atmosphere above the troposphere.

upper sideband In double-sideband amplitude modulation, the band of frequencies above the carrier frequency. This contrasts with a **lower sideband**, which incorporates the band below the carrier frequency. Its abbreviation is **USB**.

UPS 1. Abbreviation of **uninterruptible power supply**, **uninterruptible power system**, or **uninterruptible power source**. 2. Abbreviation of **ultraviolet photoelectron spectroscopy**.

upsampling The increase of the sampling rate of a sampled signal. For instance, increasing the sampling rate of an audio sample from 44.1 kHz to 352.8 kHz.

upstream In communications, the direction of the flow of information from a customer to a content provider. For example, a person navigating the Internet sends information upstream when requesting a Web page. Generally utilized in the context of data transfer, but may also refer to other signals from customers, such as requests for pay-per-view programming. This contrasts with **downstream**, where the flow of information is in the other direction.

uptime The time during which a device, piece of equipment, or system is functioning or is otherwise operational. Also spelled **up-time**.

upward compatibility The state of being **upward compatible**.

upward compatible Software that is designed to be compatible with later versions of software or hardware, or hardware that is designed to be compatible with later versions of hardware or software. For instance, a program that was able to run on a given CPU generation also being able to work with a later generation. Forward compatibility is important in being able to upgrade components, programs, or systems in a simple and efficient manner. Also called **forward compatible**.

uranium A silver-white radioactive metallic element whose atomic number is 92. It is very dense, malleable, and ductile, and is a poor conductor of electricity. It is highly reactive, extremely toxic, and has over 20 known isotopes, all unstable. Its applications include its use as a source of nuclear power, and for conversion into plutonium. Its chemical symbol is U.

uranium dioxide Black crystals whose chemical formula is UO_2. Used as a nuclear fuel. Also called **uranium oxide**.

uranium oxide Same as **uranium dioxide**.

urea-formaldehyde Same as **urea-formaldehyde resin**.

urea-formaldehyde resin A thermosetting synthetic resin used as a dielectric. Used, for instance, to provide insulation in the form of a foam. Also called **urea-formaldehyde**.

URI Abbreviation of **U**niform **R**esource **I**dentifier, or **U**niversal **R**esource **I**dentifier. A character string which identifies any resource, such as a page, image, text, or audio, available over the Internet. An example is a **URL**.

URL Abbreviation of **U**niform **R**esource **L**ocator, or **U**niversal **R**esource **L**ocator. An Internet address which directs a browser to a specific location where an Internet resource, such as a Web page or document, is located. For example in the following URL, *http://www.yipeeee.com/whoo.html*, **http** is the protocol, the ***www.yipeeee.com*** portion is the domain name, and ***whoo.html*** is a document named *whoo* created utilizing HTML.

URN Abbreviation of **U**niform **R**esource **N**ame, or **U**niversal **R**esource **N**ame. A location-independent identifier of any resource available over the Internet. It is a type of **URI**, except that with a URN only the name of the resource needs to be known, regardless of its location. A URN which states an address is a **URL**.

usability 1. The extent to which a program, network, Web site, piece of hardware, system, or the like, is easy to use. **2. Usability** (1) for individuals, such as those with reduced visual and/or motor functions, with special needs.

usable frequency A frequency that can be utilized for communications between locations utilizing ionospheric propagation. The interval of frequencies that can be utilized will vary according to the angle of incidence and time of day.

USB Abbreviation of **U**niversal **S**erial **B**us. **1.** A serial interface which has a maximum data transfer rate of up to 12 Mbps, and which can link up to 127 peripherals. USB also supports hot swapping and Plug-and-Play, and enhanced versions, such as USB 2.0 can increase throughput significantly. A USB also provides a given amount of power to connected peripherals, often eliminating the need for such peripherals to have a separate supply. Most computers are equipped with multiple USB ports. Also called **USB bus**, or **USB interface**. **2.** A standard defining a **USB bus**. **3.** Abbreviation of **upper sideband**.

USB 2.0 An enhanced version of **USB**, providing for data transfer rates of up to 480 Mbps. USB 2.0 is fully compatible with USB. Also called **high-speed USB**.

USB bus Same as **USB** (1).

USB cable A cable utilized to connect a USB peripheral to a **USB port**.

USB camera A digital camera which plugs into a **USB port**.

USB CD-ROM An external CD-ROM which plugs into a **USB port**.

USB connector A connector on either side of a **USB cable**. A Type A connector is rectangular and is plugged into a USB port, and a Type B connector is square and is plugged into a USB peripheral.

USB DVD An external DVD which plugs into a **USB port**.

USB hub A device which provides multiple **USB ports**. A USB hub plugs into a USB port on a computer, and may or may not be self-powered. USB hubs may be daisy chained for accommodating even more peripherals.

USB interface Same as **USB** (1).

USB modem An external modem which plugs into a **USB port**.

USB peripheral A peripheral, such as a printer, scanner, or modem, which plugs into a **USB port**.

USB port A computer port utilizing a **USB interface**. A USB hub has multiple such ports.

USB printer A printer which plugs into a **USB port**.

USB scanner A scanner which plugs into a **USB port**.

useful life The period of time during which a component, circuit, device, piece of equipment, system, apparatus, material, or other item is expected to operate, function, or other-

wise be of use. This will depend on various factors, including the environmental conditions surrounding said use. Also called **expected life**.

useful line In a scanning device such as a fax, the proportion of the scanning line available for image information. Also called **available line** (2).

Usenet Acronym for **User net**work. A bulletin board system with tens of thousands of discussion groups or newsgroups, covering most any imaginable topic, with millions of participants throughout the world.

USENET Same as **Usenet**.

user Anyone who employs something such as a computer system, application, communications network, email, program function, or telephone, among countless examples.

user account An account, which is established by a user or entity, that authorizes access to a system, network, information service, or the like. Each user account usually has an associated username and a password.

user agent Within an email system, a functional unit which allows a user to interact with the messaging system. A user agent enables the creation, submission, and reception of messages. Its abbreviation is **UA**.

user authentication That, such as a password or digital signature, which serves to authenticate a user.

user-centered design Product design which is oriented around the real needs of users, as opposed, for instance, to perceived needs. Such design usually involves users throughout the process. Its abbreviation is **UCD**.

user-computer interaction The use of one or more **user-computer interfaces**. Also called **human-computer interaction**.

user-computer interface An interface between a user and a computer system. This may involve, for instance, the use of keyboards, microphones, cameras, pointing devices, and so on. Also called **human-computer interface**.

User Datagram Protocol Same as **UDP**.

user-defined That which is selected or otherwise defined by a user, as opposed, for instance, to a default setting.

user-defined data type A data type which is specified by a user, as opposed to a default classification.

user-defined function A programmable function which is defined by a user.

user-defined function keys Programmable function keys whose functions are defined by a user.

user-friendly 1. Easy to learn and use. **2.** Designed to be easy to learn and use. **3.** Purported to be easy to learn and use. It is often utilized as a marketing buzzword.

user group A group of people who share a given hardware and/or software product, or a particular area of interest usually pertaining to computers, such as digital effects. Used, for instance, to exchange impressions, or to request advice.

user ID Abbreviation of **user identification**.

user identification That, such as a username or face recognition, which serves to identify a user. Its abbreviation is **user ID**, or **userID**.

user interface The interface between a user and a device, piece of equipment, machine, or system. When interacting with a computer system, for instance, user interfaces include keyboards, microphones, cameras, and so on. Its abbreviation is **UI**. Also called **human interface**.

user interface device A device providing an interface between a user and another device, piece of equipment, machine, or system. For example, a user-computer interface. Also called **human interface device**.

user-level security Security in which a given user must be authenticated before utilizing a network or resource, such as a database or printer, as opposed to **share-level security**, in which just the network or network resource requires a password. User-level security is appropriate, for instance, when authorized users do not necessarily have the same privileges.

user-machine interface The interface, such as controls, between a user and a machine. Also called **human-machine interface**.

user name Same as **username**.

User Network Interface Same as **UNI**.

user profile A set of preferences, settings, authorizations, email addresses, and the like, which is stored for each user of a multiuser system.

user session **1.** A period during which a user is connected to a computer or communications network. Also, all the activities which take place during such a session. Also called **session** (1). **2.** A period during which a user utilizes an application or program.

User-to-Network Interface Same as **UNI**.

userid Abbreviation of **user identification**.

userID Abbreviation of **user identification**.

username Also spelled **user name**. **1.** A name which identifies a specific user accessing a computer, network, or other service. Such a name may be selected by the user, and usually has an associated password. **2.** A user-chosen name utilized for communications with other users in chats, instant messaging, and the like. A screen name may be a user name, a real name, or whatever occurs to a user at a given instant. Also called **screen name**.

USRT Abbreviation of Universal Synchronous Receiver Transmitter. A chip or circuit utilized for the synchronous transmission and reception of data via a computer serial port. It converts bytes, of other data units, into serial bits, and vice versa.

UT Abbreviation of **Universal time**.

UTC Abbreviation of **Universal Time Coordinated**, and **Coordinated Universal Time**.

utility A program that performs tasks pertaining to the management of computer resources. For instance, such programs may perform diagnostic functions, or handle the management of files. Also called or **utility program**, or **computer utility**.

utility box A location or enclosure which is suitable for housing or storing electrical or communications components, circuits, devices, equipment, materials, cables, connections, and the like. Such a box, for instance, may be portable, or might be recessed into a wall or floor.

utility program Same as **utility**.

UTP Abbreviation of **unshielded twisted pair**.

UTP Ethernet Abbreviation of **unshielded twisted pair Ethernet**. Ethernet, such as 1000BaseCX or 100Base-TX, which uses unshielded twisted pairs.

uucoding A program which converts files in an 8-bit format into 7-bit ASCII characters, for transmission over the Internet. Files are uuencoded prior to transmission, and uudecoded upon reception. This allows non-text files, such as images, to be sent. Widely utilized for sending email attachments.

UUcoding Same as **uucoding**.

uudecode **1.** A program which performs **uudecoding**. **2.** To perform **uudecoding**.

uudecoding The decoding of files which have undergone **uuencoding**. This is done upon reception.

uuencode **1.** A program which performs **uuencoding**. **2.** To perform **uuencoding**.

uuencoding The encoding of 8-bit files into 7-bit ASCII characters, for transmission over the Internet. This is done prior to transmission.

UUencoding Same as **uuencoding**.

UV **1.** Abbreviation of **ultraviolet**. **2.** Abbreviation of **undervoltage**.

UV absorption Abbreviation of **ultraviolet absorption**.

UV absorption spectroscopy Abbreviation of **ultraviolet absorption spectroscopy**.

UV absorption spectrum Abbreviation of **ultraviolet absorption spectrum**.

UV astronomy Abbreviation of **ultraviolet astronomy**.

UV band Abbreviation of **ultraviolet band**.

UV detection Abbreviation of **ultraviolet detection**.

UV detector Abbreviation of **ultraviolet detector**.

UV emission Abbreviation of **ultraviolet emission**.

UV energy Abbreviation of **ultraviolet energy**.

UV filter Abbreviation of **ultraviolet filter**.

UV imaging Abbreviation of **ultraviolet imaging**.

UV lamp Abbreviation of **ultraviolet lamp**.

UV laser Abbreviation of **ultraviolet laser**.

UV light Abbreviation of **ultraviolet light**.

UV photoelectron spectroscopy Abbreviation of **ultraviolet photoelectron spectroscopy**.

UV radiation Abbreviation of **ultraviolet radiation**.

UV radiometer Abbreviation of **ultraviolet radiometer**.

UV range Abbreviation of **ultraviolet range**.

UV rays Abbreviation of **ultraviolet rays**.

UV region Abbreviation of **ultraviolet region**.

UV sensor Abbreviation of **ultraviolet sensor**.

UV signal Abbreviation of **ultraviolet signal**.

UV spectrometer Abbreviation of **ultraviolet spectrometer**.

UV spectrometry Abbreviation of **ultraviolet spectrometry**.

UV spectrophotometer Abbreviation of **ultraviolet spectrophotometer**.

UV spectrophotometry Abbreviation of **ultraviolet spectrophotometry**.

UV spectroscopy Abbreviation of **ultraviolet spectroscopy**.

UV spectrum Abbreviation of **ultraviolet spectrum**.

UV therapy Abbreviation of **ultraviolet therapy**.

UV waves Abbreviation of **ultraviolet waves**.

UVA Abbreviation of **ultraviolet A**. Electromagnetic radiation in the ultraviolet region which corresponds to wavelengths ranging from approximately 320 nm to approximately 400 nm, though the defined interval varies.

UVB Abbreviation of **ultraviolet B**. Electromagnetic radiation in the ultraviolet region which corresponds to wavelengths from approximately 290 nm to approximately 320 nm, though the defined interval varies.

UVC Abbreviation of **ultraviolet C**. Electromagnetic radiation in the ultraviolet region which corresponds to wavelengths from approximately 200 nm to approximately 290 nm, though the defined interval varies.

UXGA Abbreviation of **Ultra XGA**, which in turn is an abbreviation of **Ultra** Extended Graphics Array. A graphics standard providing for a screen resolution of 1600 x 1200 pixels.

v Abbreviation of **velocity**.

v Symbol for **voltage**.

V 1. Symbol for **volt**. 2. Chemical symbol for **vanadium**. 3. Within the **YUV** color model, one of the two color-difference signals. The other is **U** (3).

V 1. Symbol for **voltage**. 2. Symbol for **potential**, or **electric potential**. 3. Symbol for **volume**.

V-antenna A center-fed antenna consisting of two radiators arranged so as to form the shape of a letter V. Also spelled **vee antenna**.

V Band 1. In communications, a band of radio frequencies extending from 40.00 to 75.00 GHz, as established by the IEEE. This corresponds to wavelengths of approximately 7.5 mm to 4 mm, respectively. 2. In communications, a band of radio frequencies extending from 46 to 56 GHz. This corresponds to wavelengths of approximately 6.5 mm to 5.4 mm, respectively.

V-chip A chip that is installed in a TV, DVD, VCR, converter box, or similar device, which prevents viewing TV programming based on a rating. The term also applies to other content viewed through TVs, such as motion pictures recorded on DVDs. Although V-chip is used as the proper name for this circuit, it was originally an abbreviation of violence **chip**.

V connection Also spelled **vee connection**. Also called **open-delta connection**. 1. The connection of two transformers so as to form the shape of a letter V, as opposed to the combination of three to form a **delta connection**. 2. The connection of two circuit elements, such as resistors, so as to form the shape of a letter V, as opposed to the combination of three to form a delta connection.

V/F converter Abbreviation of **voltage-to-frequency converter**.

V/m Abbreviation of **volts per meter**.

v-mail Abbreviation of **video mail**, or **video email**.

v-sync Abbreviation of **vertical sync**, or **vertical synchronization**.

V.17 An ITU standard pertaining to fax, with transmission rates of up to 14,400 bps.

V.21 An ITU standard pertaining to full-duplex modems, with transmission rates of up to 300 bps.

V.22 An ITU standard pertaining to full-duplex modems, with transmission rates of up to 1,200 bps.

V.22bis An ITU standard pertaining to full-duplex modems, with transmission rates of up to 2,400 bps.

V.23 An ITU standard pertaining to half-duplex modems, with transmission rates of up to 1,200 bps downstream and up to 75 bps upstream.

V.24 An ITU standard similar to the EIA/TIA-232-E standard.

V.25 An ITU standard pertaining to answering machines.

V.26 An ITU standard pertaining to full-duplex modems, with transmission rates of up to 2,400 bps downstream and up to 75 bps upstream, while using four wire-lines.

V.27 An ITU standard pertaining to full-duplex modems, with transmission rates of up 4,800 bps, while using four wire-lines.

V.27ter An ITU standard pertaining to full-duplex modems, with transmission rates of up 4,800 bps. Used by Group 3 fax.

V.29 An ITU standard pertaining to full-duplex modems, with transmission rates of up 9,600 bps, while using four wire-lines. Used by Group 3 fax on two-wire lines.

V.32 An ITU standard pertaining to full-duplex modems, with transmission rates of up 9,600 bps, using Trellis-Coded Modulation.

V.32bis An ITU standard pertaining to full-duplex modems, with transmission rates of up 14,400 bps, using Trellis-Coded Modulation.

V.35 An ITU standard pertaining to modems combining multiple telephone circuits for higher bandwidth, with transmission rates of up to 64,000 bps.

V.42 An ITU standard pertaining to the utilization of error detection and correction by modems, incorporating the Link Access Protocol.

V.42bis An ITU standard pertaining to modem data compression.

V.44 An ITU standard pertaining to modem data compression, achieving higher compression rates than V.42bis.

V.54 An ITU standard pertaining to the performance of loopback tests by modems.

V.90 An ITU standard pertaining to modems with transmission rates of up 56,000 bps downstream and 33,600 upstream.

V.92 An ITU standard similar to **V.90**, but with enhancements such as faster connection times, and upstream speeds of up to 48,000 bps.

V.120 An ITU standard pertaining to serial communications over ISDN lines.

VA Abbreviation of **volt-ampere**.

vac Abbreviation of **vacuum**.

Vac Abbreviation of **volts AC**.

VAC Abbreviation of **volts AC**.

vacancy 1. A space or location which is empty or otherwise unoccupied. Also, the condition of being so empty or otherwise unoccupied. 2. In a crystal lattice, a defect in which a position is not occupied by a nucleus, atom, molecule, or ion.

vacua A plural form of **vacuum**.

vacuum Its abbreviation is **vac.** 1. A hypothetical space which contains no matter, and whose absolute pressure is defined as zero. Free space approaches such a vacuum. Also called **absolute vacuum**, or **perfect vacuum**. 2. A region of space, an environment, or a system whose pressure is below a given threshold, such as 10^{-12} torr. 3. A region of space, an environment, or a system whose pressure is significantly below atmospheric pressure. A low vacuum has a pressure between 760 and 1 torr, a medium vacuum has a pressure between 1 and 10^{-3} torr, a high vacuum between 10^{-3} and 10^{-6} torr, a very high vacuum between 10^{-6} and 10^{-9} torr, an ultra-high vacuum between 10^{-9} and 10^{-12} torr, and an extremely-high vacuum has a pressure lower than 10^{-12} torr. 4. A region of space, an environment, an enclosure, or a system in which the pressure of any gases, such as air, is sufficiently low so as to not affect any desired or monitored characteristics or parameters. For example, a vacuum tube is an electron tube which is evacuated to a degree that any residual gas present does not affect its electrical characteristics.

vacuum brazing Brazing which takes place at especially low pressures. Used, for instance, for special glass-to-metal or ceramic-to-metal bonding.

vacuum capacitor A capacitor sealed within an evacuated envelope or tube, and in which the created vacuum serves as the dielectric between its metallic plates. Such a capacitor has an elevated breakdown voltage.

vacuum chamber An enclosed section, compartment, space, or room which is held at a high, very high, ultra-high, or extremely high vacuum.

vacuum circuit breaker A circuit breaker sealed within an evacuated envelope or enclosure, and in which the created vacuum separates its contacts. Such circuit breakers usually feature high performance through a wide temperature interval, and should work reliably through tens of thousands of operating cycles.

vacuum degassing The heating of a metal or alloy in a vacuum, so as to remove from them any gases present.

vacuum deposition Deposition which is performed in a vacuum chamber. Used, for instance, in thin-film technology.

vacuum diffusion A diffusion process, such as that utilized to introduce semiconductor dopants, performed within a vacuum, such as an ultra-high vacuum.

vacuum drying The removal of liquids, such as water, from solid materials placed in a vacuum chamber. Used, for instance, for materials which can not, or should not be exposed to heat.

vacuum enclosure An enclosure which is held at a high, very high, ultra-high, or extremely high vacuum.

vacuum envelope An envelope, such as that of a vacuum tube, which is held at a high, very high, ultra-high, or extremely high vacuum.

vacuum evaporation The deposition of thin films on substrates via evaporation within an enclosure held at a high, very high, ultra-high, or extremely high vacuum. Used, for instance, in semiconductor manufacturing.

vacuum fluorescent display An alphanumeric or graphic display in which the emitted light results from a phosphor-coated surface being struck by electrons which are emitted in a vacuum envelope. The presented content is usually blue-green, but can be filtered to show other colors. Used, for instance, for instrument displays, those of audio and/or video components, indicators in automobiles, in calculators displays, and so on. Its abbreviation is **VFD**.

vacuum gage Same as **vacuum gauge**.

vacuum gauge An instrument which measures pressures below atmospheric pressure. Examples include ionization vacuum gauges, thermocouple vacuum gauges, and Pirani gauges. A **pressure gauge** measures values above and/or below atmospheric pressure. Also spelled **vacuum gage**.

vacuum impregnation An impregnation process performed in a vacuum chamber, to help ensure that components and devices are hermetically sealed.

vacuum metallizing The covering, plating, or impregnating a material with a metal, with said process occurring within a vacuum chamber.

vacuum microelectronics A branch of microelectronics in which active semiconductor devices, such as transistors, utilize the operational principles of vacuum tubes. Used, for instance to make solid-state counterparts of tubes which are much smaller, more reliable, require less energy, and which can operate at much higher frequencies.

vacuum phototube A phototube which is evacuated to a degree that any residual gas present does not affect its electrical characteristics.

vacuum plating Plating which is performed in a vacuum chamber. Used, for instance, to create hermetic and long-lasting glass-to-metal seals.

vacuum pump A pump, such as a diffusion pump, utilized to create a low, medium, high, very high, ultra-high, or extremely high vacuum in a given enclosure, such as a vacuum tube.

vacuum relay A relay which is sealed within an evacuated envelope or enclosure, and in which the created vacuum separates its contacts. Such an arrangement minimizes

sparking even when said contacts are in very close proximity.

vacuum seal A seal which maintains a vacuum within a given enclosure, such as a vacuum tube or chamber.

vacuum switch A switch that is sealed within an evacuated envelope or enclosure, and in which the created vacuum separates its contacts. Such an arrangement minimizes sparking even when said contacts are in very close proximity.

vacuum tube An electron tube which is evacuated to a degree that any residual gas present does not affect its electrical characteristics.

vacuum-tube amplifier An amplifier which incorporates one or more **vacuum tubes**. Also called **tube amplifier**.

vacuum-tube circuit A circuit incorporating one or more **vacuum tubes**.

vacuum-tube oscillator An oscillator incorporating one or more **vacuum tubes**.

vacuum-tube rectifier A rectifier incorporating one or more **vacuum tubes**.

vacuum-tube voltmeter A voltmeter incorporating one or more **vacuum tubes**. Its abbreviation is **VTVM**.

vacuum ultraviolet Electromagnetic radiation in the ultraviolet region which is nearest to X-rays. It corresponds to wavelengths of approximately 4 to 200 nm, though the defined interval varies. Also called **extreme-ultraviolet radiation**.

valence Also spelled **valency**. 1. The ability of one chemical element to combine with another in specific proportions. It is expressed as the number of bonds that an atom of a given element forms when combined with other elements. Hydrogen is assigned a valence of one, so valence is the number of hydrogen atoms, or their equivalent, that an atom can combine with or replace when forming compounds. For example, in **HCl**, both chlorine and hydrogen have a valence of one, in H_2O oxygen has a valence of two, while in SiO_2 silicon has a valence of four. An element may have more than one valence. 2. The **valence** (1) of a molecule, ion, or radical.

valence band In a crystalline solid, such as a semiconductor, the most energetic energy band which can be filled with electrons.

valence electrons Electrons which are present in the outermost shell of an atom. Such electrons can move freely under the influence of an electric field, thus producing a net transport of electric charge. Valence electrons also determine the chemical properties of the atom. Also called **outer-shell electrons, peripheral electrons**, or **conduction electrons**.

valence shell The outermost shell of an atom. The **valence electrons** are contained in this shell.

valency Same as **valence**.

validity 1. The extent to which a result, such as that of a test, approximates the correct value. 2. The extent to which a test measures what it is intended to measure. 3. The extent to which data is free of errors. 4. The likelihood that a message, file, data, or the like has been unequivocally authenticated.

validity check The process of verifying the **validity** of a result, data, message, file, or the like. Also, a specific instance of such verification. Also called **validity checking**.

validity checking Same as **validity check**.

value 1. A numerical quantity or magnitude which is attained, calculated, assigned, or the like. Also, any given number, quantity, or magnitude. 2. In the Munsell system, the attribute of color that corresponds to the lightness of an object.

value-added network A communications network that provides additional services and/or capabilities, such as enhanced security, advanced error-detection and correction, resource management, message storing, and a guaranteed quality of service. Its acronym is **VAN**.

value-added reseller A business that purchases products made by other commercial entities, and sells them after purportedly adding something of value. For example, a company that buys from multiple vendors all the necessary hardware components for a computer, obtains an operating system and several applications, and sells the package to the public. Its acronym is **VAR**.

valve An active device consisting of a hermetically-sealed envelope within which electrons are conducted between electrodes. The cathode is the source of electrons, the positive electrode to which they travel is the anode or plate, while other electrodes that may be present include control grids and screen grids. Electron tubes may or may not contain a gas, and its presence and concentration affects the characteristics of said tubes. Such tubes have many applications, including their use in amplification, modulation, rectification, and oscillation. There are many examples, including CRTs, phototubes, pentodes, mercury-vapor tubes, and so on. When such a tube is evacuated to a degree that any residual gas present does not affect its electrical characteristics, it is called **vacuum tube**. Better known as **electron tube**, or **tube (1)**.

vampire tap A connection to a wire or cable in which one or more perforations are made through the protective jacket, so as to make contact with inner conductors. Used, for instance, to simplify ThickNet connections.

VAN Acronym for **value-added network**.

Van Allen belt Same as **Van Allen radiation belt**.

Van Allen radiation belt Each of the two regions of ionizing radiation which surround the planet earth, and which interact with its magnetic field. These belts are composed mostly of protons and electrons which originate from the solar wind which become trapped by the earth's magnetic field. These two belts extend from about 700 kilometers to beyond 50,000 kilometers, and their contained particles travel along the planet's lines of force. Also called **Van Allen belt**, or **radiation belt (2)**.

Van de Graaff generator An electrostatic generator in which a rapidly moving insulating belt collects electric charges, then discharges them inside a hollow metal sphere. Potentials of several million volts may be generated in this manner. Also called **belt generator**.

vanadium A silver-gray metallic element whose atomic number is 23. It is soft, ductile, and corrosion-resistant. It has about 15 known isotopes, of which one is stable. It is used mainly in steel alloys, providing additional strength and corrosion resistance. Its chemical symbol is **V**.

vanadium carbide Hard crystals whose chemical formula is **VC**. It is used in steel alloys, providing additional strength and corrosion resistance.

vanadium pentoxide A reddish-yellow powder whose chemical formula is V_2O_5. Used, for instance, in ceramics, and in glass which inhibits the transmission of UV.

vandal An executable file, such as that created using ActiveX, or an applet, which is intended to steal passwords, provoke a denial-of-service attack, or cause some other harm. Since it does not seek out other programs to replicate itself, it is not a virus. A vandal may arrive, for instance, in the form of an email attachment, or via content downloaded when accessing a Web page.

vane attenuator A waveguide attenuator in which a sheet or plate of dissipative material is placed through a non-radiating slot. As the sheet or plate is moved into and out of the slot, a variable amount of loss is introduced. Also called **flap attenuator**, **rotary-vane attenuator**, or **guillotine attenuator**.

vapor 1. A substance that is in the gaseous state, especially when it is liquid or solid under ordinary temperature and pressure conditions. 2. A substance that is in a gaseous phase at a temperature below its boiling point. 3. Water vapor, or another gas, which is present in air, or within an enclosure. 4. Same as **vaporize**.

vapor deposition A process, such as chemical-vapor deposition, in which a material in the vapor state is applied to a substrate.

vapor lamp A lamp which produces light when the gas it contains, usually at a low pressure, becomes ionized by an electric discharge passed through it. An example is a fluorescent lamp. Also known as **discharge lamp**, **electric-discharge lamp**, or **gas-discharge lamp**.

vapor-phase epitaxy A technique for growing epitaxial layers, in which the material to be deposited upon the substrate crystal is in vapor form. Used, for instance, in semiconductor heterostructure manufacturing. Its abbreviation is **VPE**.

vapor-phase soldering Soldering in which components, whose leads to be joined are coated with a paste or eutectic solder, are lowered into a chamber with boiling hydrocarbons. This melts the solder and forms good electrical connections. Used, for instance, in surface-mount technology, but does emit harmful hydrocarbons. Its abbreviation is **VPS**.

vapor pressure 1. The pressure exerted by a vapor which is in equilibrium with its liquid or solid form. The higher the vapor pressure, the more rapidly a liquid or solid will tend to evaporate. 2. The pressure exerted by water vapor molecules present in the atmosphere.

vaporize To become, or cause to become, a **vapor (1)**, **vapor (2)**, or **vapor (3)**. Also called vapor (4).

vaporware 1. Software which does not exist, but which is nonetheless hyped in the media so as to ascertain possible market interest, thwart the efforts of a competitor, or the like. When promoted in sales presentations, such as those using slides, usually called **slideware**. 2. Software which has been announced for release considerably sooner than the actual program becomes available. Usually utilized to refer to the deceptive practice of a vendor looking to prevent the purchase of an existing similar product offered by a competitor, or one that is looking to build a larger client base.

var 1. The SI unit for **reactive power**, which is the power in an AC circuit that cannot perform work. It is calculated by the following formula: $P = I \cdot V \cdot \sin\theta$, where P is the reactive power in watts, I is the current in amperes, V is the voltage in volts, and $\sin\theta$ is the sine of the angular phase difference between the current and the voltage. It is an acronym for volt-amperes reactive. 2. Abbreviation of **variable**.

VAR 1. Acronym for **value-added reseller**. 2. Abbreviation of **volt-amperes reactive**.

VAR-hour Abbreviation of volt-amperes reactive **hour**. A unit of reactive power equal to one VAR integrated over one hour. Also spelled **VARhour**.

VAR-hour meter An instrument or device which measures and indicates reactive power in **VAR-hours**. Also spelled **VARhour meter**.

VAR meter Same as **varmeter**.

varactor A pn-junction diode operated with reverse bias, so that it functions as a voltage-dependent capacitor. The p and n regions each act as the plates, with the depletion layer serving as the dielectric. Its capacitance varies as a function of its applied voltage, and is used, for instance, for electronic tuning. It is an acronym for **var**iable **reac**tor. Also called **varactor diode**, **varicap**, **varicap diode**, **variable-capacitance diode**, or **voltage-variable capacitor**.

varactor diode Same as **varactor**.

varactor tuning The use of one or more **varactors** for tuning. Used, for instance, in receivers or oscillators. Also called **varicap tuning.**

VARhour Same as **VAR-hour.**

VARhour meter Same as **VAR-hour meter.**

variable Its abbreviation is **var (2). 1.** A number, quantity, or magnitude which does not have a fixed value. Also, a symbol, such as **x**, representing such a variable. **2.** A number, quantity, or magnitude which can assume any of multiple values. Also, a symbol, such as **x**, representing such a variable. **3.** That which changes, or tends to change over time. Also, a symbol, such as **x**, representing such a variable. **4.** That which changes during a given process. Also, a symbol, such as **x**, representing such a variable.

variable attenuator A circuit or device which reduces the amplitude of a signal, ideally without introducing distortion, by a continuously-variable amount, or in steps.

variable bias A bias value that changes, or that can be adjusted. Automatic bias is an example.

variable bit rate In ATM, a minimum bandwidth which is guaranteed to be maintained. Utilized, for instance, for audio and/or video conferencing. Variable bit rate has real-time and non-real-time versions. Its abbreviation is **VBR.**

variable block A data block whose size or length may be varied, depending on the needs.

variable-capacitance diode Same as **varactor.**

variable capacitor A capacitor whose capacitance can be varied. This may be accomplished, for example, through the use of rotating plates. Also known as **variable condenser, adjustable capacitor**, or **adjustable condenser.**

variable-carrier modulation A type of amplitude modulation in which the amplitude of the carrier wave is varied according to the percentage of modulation, providing for an essentially constant modulation factor. Also called **controlled-carrier modulation**, or **floating-carrier modulation.**

variable component An electrical component, such as a capacitor or resistor, whose value which can be adjusted.

variable condenser Same as **variable capacitor.**

variable coupling Inductive coupling between circuits which can be varied. This may be accomplished, for instance, by adjusting the position of one coil with respect to another.

variable data printing Printing, such as that of brochures, in which each set of sheets has different content.

variable-depth sonar Sonar in which the depth of the underwater sound projectors and hydrophones can be varied. Its abbreviation is **VDS.**

variable field 1. A field, such as a dynamic magnetic field, whose intensity varies over time. **2.** Same as **variable-length field.**

variable frequency 1. A frequency, such as a modulated carrier frequency in FM, which varies over time. **2.** A frequency, such as that of a variable-frequency oscillator, which can be varied.

variable-frequency oscillator An oscillator whose frequency can be adjusted. This adjustment may occur in discrete steps, or through a continuous range. This contrasts with a **fixed-frequency oscillator**, whose frequency cannot be varied. Its abbreviation is **VFO.**

variable inductance Same as **variable inductor.**

variable inductor An inductor whose inductance can be adjusted. It may consist, for instance, of a coil whose inductance can be varied by using one or more moving elements such as taps or sliding contacts. This contrasts with a **fixed inductor**, which has a single inductance which cannot be varied. Also called **variable inductance.**

variable-length field A data field whose size in bits or bytes is determined by the amount of data stored. This contrasts with a **fixed-length field**, whose size can not be adjusted. Also called **variable field (2).**

variable-length record A data record composed of **variable-length fields**. This contrasts with a **fixed-length record**, which is composed of fixed-length fields.

variable-mu tube An electron tube whose mu, or amplification, factor can be varied.

variable output 1. An output, such as that of an audio amplifier, whose level can vary. **2.** An output whose level is varied.

variable reactor Same as **varactor.**

variable-reluctance motor A brushless synchronous motor that does not utilize a permanent magnet, and which does not have a coil in its rotor. Such a motor has pointed poles in both the stator and rotor, and the coils of the stator are fed the energizing current, which creates a magnetic pulling force that enables rotation. Reluctance motors are comparatively robust and efficient, and can be used when very high ambient temperatures are present. Its abbreviation is **VR motor**. Also called **reluctance motor**, or **switched-reluctance motor.**

variable-reluctance pickup A phonographic pickup that converts the movements of the stylus into the corresponding audio-frequency voltage output by varying the reluctance of an internal magnetic circuit. Also called **magnetic pickup**, or **magnetic cartridge.**

variable-reluctance transducer A transducer which converts a parameter, magnitude, signal, or phenomenon into a proportional change in the reluctance of a magnetic circuit, or vice versa.

variable-resistance transducer A transducer which converts a parameter, magnitude, signal, or phenomenon into a proportional change in the resistance of a circuit, or vice versa.

variable resistor Also known as **adjustable resistor. 1.** A resistor whose resistance may be varied by a mechanical device such as a sliding contact. Also called **rheostat. 2.** Same as **varistor. 3.** Any resistor whose resistance may be varied.

variable-speed motor A motor whose speed can be varied while under load. Also called **adjustable-speed motor.**

variable transformer A transformer whose output voltage can be varied between some minimum, or zero, and a maximum. This is generally accomplished by means of a sliding contact arm. In most cases, a variable transformer is an autotransformer. Also known as **adjustable transformer.**

variable voltage divider A voltage divider that utilizes adjustable resistors to vary the voltage. A potentiometer is an example. Also known as **adjustable voltage divider.**

Variac A **variable transformer** which is usually utilized for the adjustment of an AC voltage. Used, for instance, when a specific voltage above or below the line voltage is needed.

variance In statistics, the value obtained by adding the squares of the differences between each number and the mean value of a set, series, or population, and dividing by the number of members of said set. The square root of variance is called **standard deviation.**

varicap Same as **varactor.**

varicap diode Same as **varactor.**

varicap tuning Same as **varactor tuning.**

variometer A variable inductor incorporating two coils connected in series, one of which is rotated to adjust overall inductance values. Such coils are usually mounted concentrically.

varistor A semiconductor device which serves as a variable resistor. It resistance value is determined by its input voltage, and an example is a metal-oxide varistor. Also called **voltage-dependent resistor, variable resistor (2), voltage-controlled resistor**, or **adjustable resistor (2).**

varmeter An instrument which measures and indicates reactive power in vars. Also spelled **VAR meter**.

VB Abbreviation of **Visual BASIC**.

VBI Abbreviation of **vertical blanking interval**.

VBR Abbreviation of **variable bit rate**.

VBR-non-real time Abbreviation of variable bit rate-**non-real-time**. In ATM, variable bit rate service which does not support real-time applications such as audio and/or video conferencing. Its own abbreviation is **VBR-nrt**.

VBR-nrt Abbreviation of **VBR-non-real time**.

VBR-real time Abbreviation of variable bit rate-**real-time**. In ATM, variable bit rate service which supports real-time applications such as audio and/or video conferencing. Its own abbreviation is **VBR-rt**.

VBR-rt Abbreviation of **VBR-real time**.

VCA 1. Abbreviation of **voltage-controlled amplifier**. 2. Abbreviation of **voltage-controlled attenuator**.

vCalendar A specification or standard format pertaining to the exchange of scheduling information between applications and devices. Used, for instance, to synchronize such data between PDAs, PIMs, cell phones, and desktop computers.

vCard A specification or standard format for electronic business cards. Used, for instance, to share information such as name, job title, telephone numbers, fax number, emails, and physical address, between compatible devices, such as PDAs, PIMs, cell phones, and desktop computers.

VCCS Abbreviation of **voltage-controlled current source**.

VCD Abbreviation of **video CD**.

VCI Abbreviation of **Virtual Channel Identifier**.

VCL Abbreviation of **Virtual Channel Link**.

VCN Abbreviation of **virtual circuit number**.

VCO 1. Abbreviation of **voltage-controlled oscillator**. 2. Abbreviation of **Voice Carry Over**.

VCR 1. Abbreviation of **video cassette recorder**. 2. Abbreviation of **voltage-controlled resistor**.

VCSEL Acronym for Vertical-Cavity Surface-Emitting Laser. A laser diode that emits its coherent light from its surface, as opposed to its edge. Such lasers are simpler to manufacture and easier to test, in addition to being more efficient and durable. Since its emitted beam is narrower and more circular, it is especially well-suited for use with optical fibers.

VCVS Abbreviation of **voltage-controlled voltage source**.

VCXO Abbreviation of **voltage-controlled crystal oscillator**.

Vdc Abbreviation of **volts DC**.

VDC Abbreviation of **volts DC**.

VDD Abbreviation of **virtual device driver**.

VDR 1. Abbreviation of **video disc recorder**, or **video disk recorder**. 2. Abbreviation of **voltage-dependent resistor**.

VDS Abbreviation of **variable-depth sonar**.

VDSL Abbreviation of very high-speed digital subscriber line, or very high bit-rate digital subscriber line. A version of DSL utilizing copper twisted pairs, and which supports data transfer rates of up to over 100 Mbps upstream, and up to over 100 Mbps downstream. The transmission speed is affected significantly by the distance from the central office, and the shorter the distance, the greater the speed.

VDT Abbreviation of **video display terminal**, or **visual display terminal**.

VDU Abbreviation of **visual display unit**, or **video display unit**.

vector 1. Same as **vector quantity**. 2. A graphical representation of a **vector quantity**. 3. A one-directional array, such as a single column. 4. A course or direction of an aircraft, such as that which is guided by radio-transmitted instructions. 5. In computer graphics, a line whose starting and ending points are identified by specific coordinates, such as Cartesian coordinates.

vector diagram A graphical representation of one or more **vector quantities**.

vector display A display, such as that of a computer or oscilloscope, which draws **vectors (5)** on the screen.

vector drawing A drawing prepared using **vector graphics**.

vector font A font composed of **vectors (5)**, which lends itself to scaling. This contrasts, for instance, with a **bit-mapped font**, in which each character has its own stored bitmap.

vector graphics Images which are represented in the form of equations and graphics primitives. Used, for instance, in CAD and drawing programs. This contrasts with **bitmap graphics**, in which data structures composed of rows and columns of bits serve to represent the information of an image. Vector graphics are more flexible and are manipulated with greater ease than bitmap graphics, in addition to requiring less memory. Also called **object-oriented graphics**.

vector image An image prepared using **vector graphics**.

vector impedance meter An impedance meter which also provides phase angle readings.

Vector Markup Language Same as **VML**.

vector processor An array processor which performs data-parallel calculations on vectors.

vector quantity A quantity, such as force or magnetic field strength, which has a magnitude and direction. A **scalar quantity** has only magnitude. Also called **vector (1)**.

vector-sum excited linear predictive coding Same as **VSELP**.

vector-to-raster 1. To convert **vector graphics** into raster graphics, using a raster image processor. Also called **rasterize**. 2. The process of converting vector graphics into raster graphics, using a raster image processor. Also called **rasterization (2)**.

vector voltmeter A voltmeter which also provides phase angle readings.

vectorscope An oscilloscope specially designed to monitor, analyze, and tweak aspects of the color portion of a video signal. Used, for instance, to adjust color signals displayed in a TV transmission system.

vee antenna Same as **V-antenna**.

vee connection Same as **V connection**.

vehicle control The utilization of electrical, electronic, optical, magnetic, and/or mechanical components, devices, equipment, and systems, to monitor and control a vehicle. When such a vehicle is an aircraft or spacecraft, also called **flight control (1)**.

vehicle-control system A system utilized for **vehicle control**.

vel Abbreviation of **velocity**.

velocimeter 1. An instrument which measures and indicates the speed of sound in water. 2. An instrument which measures and indicates the flow of a fluid. Also called **flow-meter**, or **fluid meter**. 3. An instrument which measures and indicates **velocity**.

velocity A vector quantity describing the change in position of a body over time. Its magnitude is the speed of the body, while the direction is that in which it moves. Also, a specific velocity. Velocity is a vector quantity, while **speed** is a scalar quantity. Its abbreviation is v, or **vel**.

velocity coefficient Same as **velocity factor**.

velocity factor The ratio of the velocity of propagation of electromagnetic waves via a given transmission line or medium, to the velocity of propagation in free space. The ideal ratio is 1, or 100%. Also called **velocity coefficient**, or **velocity ratio**.

velocity microphone A microphone responds to the instantaneous particle velocity striking its diaphragm, which usually consists of a thin metal ribbon. Such microphones are highly directional, and the most common example is a ribbon microphone.

velocity-modulated tube A tube, such as a klystron, which has a cavity or region where electron velocities are modulated, so as to form electron bunches.

velocity modulation The variation of the velocities of the electrons in a stream, so as to form electron bunches. This usually occurs in an electron tube such as a klystron, with an RF signal helping form bunches by slowing down the faster electrons and speeding up the slower ones. Also, the process via which this occurs.

velocity of light The **speed of light** in a given direction.

velocity of propagation The **wave speed** in a given direction.

velocity of sound The **speed of sound** in a given direction.

velocity ratio Same as **velocity factor**.

velocity resonance Resonance in which the phase angle between the fundamental components of oscillation of a system and the applied signal is 90°. Also called **phase resonance**.

vendor-neutral That which is not dependent upon, based upon, dictated by, or otherwise favoring a specific vendor. Open systems are examples.

venetian-blind effect 1. An effect, often due to low-frequency electromagnetic fields, in which comparatively broad horizontal bars extend over an entire displayed TV picture. These bars are alternately black and white, and may or may not move up and down on the screen. Also called **horizontal hum bars**. 2. Any effect manifesting unwanted alternating lines similar to those in **venetian-blind effect (1)**.

Venn diagram A diagram in which two or more closed geometric shapes, usually circles, intersect with each other, with their position and overlap indicating their interrelationships.

vent 1. An opening which allows the escape of gases, fluids, or other substances which would create unwanted pressure, or whose presence would otherwise be undesired. For example, such a vent utilized to release unwanted gases from certain types of batteries or capacitors. 2. An opening which allows air or another fluid to flow in and/or out of an enclosure or area, so as to provide cooling or warming. Used, for instance, to regulate the temperature surrounding a device. 3. A carefully-dimensioned opening in a loudspeaker enclosure. It is employed for various purposes, but it is mainly used to increase and extend the reproduction of low frequencies. Also called **speaker port**, **port (6)**, **duct (5)**, or **ducted port**. 4. To release through a **vent (1)**, or **vent (2)**. 5. To provide with a **vent (1)**, **vent (2)**, or **vent (3)**.

vented cabinet A speaker enclosure which incorporates one or more **vents (3)**. Also called **vented enclosure**, or **ported cabinet**.

vented enclosure Same as **vented cabinet**.

vented loudspeaker Same as **vented speaker**.

vented speaker A speaker with a **vented cabinet**. Also called **vented loudspeaker**.

verification 1. The process of ascertaining or confirming the accuracy, correctness, or truth of something. For example, the verification that data has been entered correctly into a database. 2. The evidence of a **verification (1)**.

verify To ascertain or confirm the accuracy, correctness, or truth of something. For example, to verify that data has been entered correctly into a database

Verilog A hardware description language utilized to design and document electronic systems.

vernier 1. An auxiliary scale located alongside, or attached to, the main graduated scale of an instrument. This additional scale allows for the accurate measurement of fractional parts of the smallest divisions within the larger scale. Also called **vernier scale**. 2. A device or mechanism which assists in the fine tuning of a device or instrument.

vernier dial 1. A dial, such as a tuning dial, in which each complete rotation of the knob moves the main shaft a fraction of a revolution. Such a dial, for instance, may have a 2:1, 4:1, or an 8:1 turns ratio. 2. An instrument dial with a **vernier scale**.

vernier scale Same as **vernier (1)**.

Veronica Abbreviation of Very Easy Rodent-Oriented Net-wide Index to Computerized Archives, or Very Easy Rodent-Oriented Net-wide Index to Computer Archives. A Gopher search utility.

version A specific version of a given program or software package. For example *Program, release 1.01*. Also called **release version**.

version number The numbers, letters, and/or words which identify a specific version of a given program or software package. For example, *1.0*, *5.01a*, or *2004b beta build #1111*. Changes in whole numbers usually imply significant modifications, while fractions, letters, or builds often identify versions which correct bugs, provide minor enhancements, or the like. Also called **release version number**, or **program version**.

vert Abbreviation of **vertical**.

vertical Its abbreviation is **vert**. 1. Perpendicular to, or situated along a plane perpendicular to a horizon or a base line. Also, operating in a plane perpendicular to a horizon or a baseline. 2. At a right angle to a horizontal line. 3. Situated at the highest point.

vertical amplifier A circuit or device which amplifies the signals which produce a vertical deflection in an instrument such as an oscilloscope. Also called **Y-amplifier**.

vertical antenna An antenna, such as a dipole antenna, which consists of one or more radiators situated along a vertical plane.

vertical axis An axis which is considered to be parallel to, or situated along a vertical line, object, or the like. In a Cartesian coordinate system, for instance, it is usually considered to be the y-axis.

vertical beamwidth The angle between the points at which the intensity of an electromagnetic beam is at half of its maximum value, as measured in the vertical plane.

vertical blanking In a device such as a CRT, the cutting-off of the electron beam during **vertical retrace**. This prevents an extraneous line from appearing on the screen during this period. Also called **vertical retrace blanking**.

vertical blanking interval The interval within which **vertical blanking** occurs. Its abbreviation is **VBI**.

vertical blanking pulse A pulse which produces **vertical blanking**. This pulse occurs between the active vertical lines.

Vertical-Cavity Surface-Emitting Laser Same as **VCSEL**.

vertical centering The use of a **vertical centering control**.

vertical centering control In a CRT, a control utilized to center the image vertically on the screen. Also called **vertical positioning control**.

vertical convergence control In a color TV tube, a control which varies the voltage of the vertical dynamic convergence.

vertical deflection In a CRT, any vertical deflection of the electron beam.

vertical deflection coils In a CRT, a pair of coils which serve to deflect the electron beam vertically.

vertical deflection electrodes In a CRT, a pair of electrodes which serve to deflect the electron beam vertically.

vertical deflection oscillator In a TV receiver, an oscillator which produces signals controlling vertical sweeps. Also called **vertical oscillator**.

vertical deflection plates In a CRT, a pair of electrodes which serve to deflect the electron beam vertically.

vertical dipole A dipole antenna which consists of one or more radiators situated along a vertical plane. Also called **vertical dipole antenna**.

vertical dipole antenna Same as **vertical dipole**.

vertical directivity The directivity of an antenna or transducer along the vertical plane.

vertical dynamic convergence In a color TV tube with a three-color gun, the convergence of the three electron beams on a specified surface during vertical scanning.

vertical FET Abbreviation of **vertical field-effect transistor**. A field-effect transistor whose channel is perpendicular to the wafer surface, as opposed to being situated along the plane of said surface. This design features advantages such as increased packing density. Its abbreviation is **VFET**.

vertical field strength The field strength of an antenna measured in the vertical plane.

vertical flyback Same as **vertical retrace**.

vertical frequency In a CRT, the number of vertical sweeps or traces per second. Also called **vertical line frequency**.

vertical hold Same as **vertical-hold control**.

vertical-hold control A control which adjusts the frequency of the vertical deflection oscillator in a TV receiver, to maintain the displayed picture vertically steady. Also called **vertical hold**.

vertical industry portal Same as **vortal (1)**.

vertical line frequency Same as **vertical frequency**.

vertical linearity control In a CRT, a control which allows the vertical adjustment of that displayed, to reduce or eliminate distortion due to improper image linearity.

vertical magnetization Magnetization of a material or medium in a plane perpendicular to the plane of said material or medium. Multilayer materials may exhibit perpendicular magnetization. Also called **perpendicular magnetization**.

vertical MOS Abbreviation of **vertical metal-oxide semiconductor**. A metal-oxide semiconductor whose channel is perpendicular to the wafer surface, as opposed to being situated along the plane of said surface. An example is a vertical MOSFET. Its abbreviation is **VMOS**.

vertical MOSFET Abbreviation of **vertical metal-oxide semiconductor field-effect transistor**. A MOSFET whose channel is perpendicular to the wafer surface, as opposed to being situated along the plane of said surface. This design features advantages such as increased packing density. Its abbreviation is **VMOSFET**.

vertical movement In a robotic arm, up and down motion. This may be caused by moving the robot itself, or by moving the arm vertically.

vertical oscillator Same as **vertical deflection oscillator**.

vertical output stage In a TV receiver, an output amplifier following the vertical deflection oscillator.

vertical polarization **1.** Polarization of an electromagnetic wave in which the electric field vector is vertical. Thus, the magnetic field vector is horizontal. **2.** Radio-frequency transmissions utilizing waves with **vertical polarization (1)**. Under such circumstances, transmitting antenna elements are usually in the vertical plane, and receiving antenna elements are usually most sensitive in this plane.

vertical portal Same as **vortal (1)**.

vertical positioning control Same as **vertical centering control**.

vertical pulse Same as **vertical sync pulse**.

vertical radiator An antenna radiator which is situated along a vertical plane.

vertical recording Recording of a magnetic recording medium in which **vertical magnetization** is utilized. Also called **perpendicular recording**.

vertical redundancy check An error-checking method in which a parity bit is added to each transmitted byte. This bit is tested upon reception as a test of data integrity. Its abbreviation is **VRC**. Also called **vertical redundancy checking**.

vertical redundancy checking Same as **vertical redundancy check**.

vertical refresh rate Same as **vertical scanning frequency**.

vertical resolution In an image, such as that of a TV or fax, the number of picture elements in the vertical direction of scanning or recording. May be expressed, for instance, in pixels, or in dots per inch.

vertical retrace In a CRT, the return of the electron beam to its starting point after a vertical sweep or trace. Also called **vertical flyback**.

vertical retrace blanking Same as **vertical blanking**.

vertical retrace period The interval during which a **vertical retrace** occurs.

vertical scan Same as **vertical scanning**.

vertical scan frequency Same as **vertical scanning frequency**.

vertical scan rate Same as **vertical scanning frequency**.

vertical scanning Its abbreviation is **vertical scan**. **1.** In a CRT, the vertical movement of the electron beam. Also called **vertical sweep (1)**. **2.** In radars, the rotation of the antenna in the vertical direction.

vertical scanning frequency In a CRT, the number of vertical lines scanned by the electron beam per second. Its abbreviation is **vertical scan frequency**. Also called **vertical scanning rate**, **vertical scan rate**, or **vertical refresh rate**.

vertical scanning rate Same as **vertical scanning frequency**.

vertical scrolling Scrolling which is up or down within a computer document or program, while **horizontal scrolling** is to the left or to the right.

vertical signal A signal, such as a vertical blanking pulse, which determines a vertical parameter.

vertical sweep **1.** Same as **vertical scanning (1)**. **2.** In a CRT, to deflect the electron beam so that it has a vertical movement.

vertical sweep circuit A circuit producing a **vertical sweeps**.

vertical sync The scanning synchronization provided by a **vertical sync pulse**. It is an abbreviation of **vertical synchronization**. Its own abbreviation is **vsync**.

vertical sync pulse The pulse which synchronizes the vertical component of line-by-line scanning of a TV receiver with that of a TV transmitter, and triggers vertical retracing and vertical blanking. It is an abbreviation of **vertical synchronization pulse**, or **vertical synchronizing pulse**. Also called **vertical pulse**, **vertical synchronization signal**, **vertical synchronizing signal**, or **vertical sync signal**.

vertical sync signal Same as **vertical sync pulse**.

vertical synchronization Same as **vertical sync**. Its abbreviation is **vsync**.

vertical synchronization pulse Same as **vertical sync pulse**.

vertical synchronization signal Same as **vertical sync pulse**.

vertical synchronizing pulse Same as **vertical sync pulse**.

vertical synchronizing signal Same as **vertical sync pulse**.

vertically-polarized radiation Electromagnetic radiation in which the electric field vector is vertical. Thus, the magnetic field vector is horizontal.

vertically-polarized wave An electromagnetic wave in which the electric field vector is vertical. Thus, the magnetic field vector is horizontal.

Very Easy Rodent-Oriented Net-wide Index to Computer Archives Same as **Veronica**.

Very Easy Rodent-Oriented Net-wide Index to Computerized Archives Same as **Veronica**.

very-hard vacuum Same as **very-high vacuum**.

very high bit-rate digital subscriber line Same as **VDSL**.

very high frequency Same as **VHF**.

very high frequency antenna Same as **VHF antenna**.

very high frequency band Same as **VHF**.

very high frequency channel Same as **VHF channel**.

very high frequency omnidirectional radio range Same as **VOR**.

very high frequency omnidirectional range Same as **VOR**.

very high frequency omnirange Same as **VOR**.

very high frequency oscillator Same as **VHF oscillator**.

very high frequency range Same as **VHF**.

very high frequency receiver Same as **VHF receiver**.

very high frequency signal Same as **VHF signal**.

very high frequency television antenna Same as **VHF TV antenna**.

very high frequency television channel Same as **VHF TV channel**.

very high frequency transmitter Same as **VHF transmitter**.

very high frequency tuner Same as **VHF tuner**.

very high frequency TV antenna Same as **VHF TV antenna**.

very high frequency TV channel Same as **VHF TV channel**.

very high-level language A high-level computer language utilized in the context of object-oriented technology, relational databases, and the like. Its abbreviation is **VHLL**.

very high-rate digital subscriber line Same as **VDSL**.

very high-speed digital subscriber line Same as **VDSL**.

very high-speed DSL Same as **VDSL**.

Very High-Speed IC Same as **VHSIC**.

Very High-Speed Integrated Circuit Same as **VHSIC**.

very-high vacuum An environment or system whose pressure is below 10^{-6} torr, but above 10^{-9} torr. One torr equals 1 millimeter of mercury, or approximately 133.3 pascals. Also called **very-hard vacuum**.

Very Large Database A database whose content is greater than a given size, such as 1, 10, or 100 terabytes. Its abbreviation is **VLDB**.

Very Large Memory A memory system whose content is greater than a given amount, such as 4, 64, or 256 gigabytes. Its abbreviation is **VLM**.

very-large-scale integration In the classification of ICs, the inclusion of between 100,000 and 1,000,000 electronic components, such as transistors, on a single chip. This definition may require many more components in a few years to qualify for this level of integration, as miniaturization progresses. Its abbreviation is **VLSI**.

Very Long Instruction Word A CPU technology which combines a given number of instructions, such as four or more, for simultaneous execution. Also, a CPU architecture incorporating this technology. Its abbreviation is **VLIW**.

very long range Pertaining to **very long range radar**.

very long range radar Radar equipment which can effectively detect scanned objects whose range in greater than a given distance, such as 500 or 1,000 kilometers.

very low frequency A range of frequencies spanning from 3 kHz to 30 kHz. These correspond to wavelengths of 100 km to 10 km, respectively. Also, pertaining to this interval of frequencies. Its abbreviation is **VLF**. Also called **very low frequency band**.

very low frequency band Same as **very low frequency**.

very short range Pertaining to **very short range radar**.

very short range radar Radar equipment which can effectively detect scanned objects whose range is less than a given distance, such as 50 or 100 kilometers.

VESA Acronym for Video Electronics Standards Association. An organization that sets and promotes standards pertaining to video and multimedia interfaces, protocols, and devices.

VESA BIOS Extension An SVGA adapter interface defined according to **VESA** specifications.

VESA Bus Same as **VESA Local Bus**.

VESA DDC Abbreviation of **VESA D**isplay **D**ata **C**hannel. A **VESA** standard for the exchange of information between a monitor and a display adapter. For instance, a monitor can inform the graphics subsystem of the computer about its capabilities, such as maximum resolution. Also called **Display Data Channel**.

VESA Display Data Channel Same as **VESA DDC**.

VESA Display Power Management Signaling Same as **VESA DPMS**.

VESA DPMS Abbreviation of **VESA D**isplay **P**ower **M**anagement Signaling. A **VESA** standard pertaining to energy-conservation signaling for monitors. Also called **Display Power Management Signaling**.

VESA Local Bus A local bus architecture defined according to **VESA** specifications. Its abbreviation is **VL bus**, or **VLB**. Also called **VESA Bus**.

VESA standard A standard established and promoted by **VESA**.

vestigial-sideband filter A filter which serves to attenuate a carrier sideband.

vestigial-sideband transmission A radio signal transmission in which the amplitude modulation of one of the sidebands is transmitted as it is, while the other is attenuated substantially. Used, for instance, in TV transmissions. Also called **asymmetrical-sideband transmission**.

VFD Abbreviation of **vacuum fluorescent display**.

VFET Abbreviation of **vertical FET**.

VFO Abbreviation of **variable-frequency oscillator**.

VGA Abbreviation of Video Graphics Array, or Video Graphics Adapter. A video display standard which provided enhancements over **EGA**, but which has been superseded by **SVGA**. VGA can have a resolutions of up to 640 x 480 pixels supporting 16 colors, or 320 x 200 with 256 colors.

VHF Abbreviation of very high frequency. A range of radio frequencies spanning from 30 to 300 MHz. These correspond to wavelengths of ten meters to one meter, respectively, which are metric waves. Also, pertaining to this interval of frequencies. Also called **very high frequency range**, or **very high frequency band**.

VHF antenna Abbreviation of very high frequency **antenna**. An antenna that operates in the **VHF** range. Used, for instance, for TV reception.

VHF band Same as **VHF**.

VHF channel Same as **VHF TV channel**. Abbreviation of very high frequency **channel**.

VHF high band Within **VHF**, an interval spanning from about 148 to 174 MHz, although the defined interval varies.

VHF low band Within **VHF**, an interval spanning from about 30 to 50 MHz, although the defined interval varies.

VHF omnidirectional radio range Same as **VOR**.

VHF omnidirectional range Same as **VOR**.

VHF omnirange Same as **VOR**.

VHF oscillator Abbreviation of very high frequency **oscillator**. An oscillator which operates in the **VHF band**.

VHF range Same as **VHF**. Abbreviation of very high frequency **range**.

VHF receiver Abbreviation of very high frequency **receiver**. A receiver which responds to **VHF** signals.

VHF signal Abbreviation of very high frequency **signal**. An RF signal in the **VHF** range.

VHF television antenna Same as **VHF TV antenna**. Abbreviation of very high frequency **TV antenna**.

VHF television channel Same as **VHF TV channel**. Abbreviation of very high frequency **TV channel**.

VHF transmitter Abbreviation of very high frequency **transmitter**. A transmitter that emits **VHF** signals.

VHF tuner Abbreviation of very high frequency **tuner**. A tuner, such as that of a TV, utilized for reception of signals in the **VHF** range.

VHF TV antenna Abbreviation of very high frequency **TV antenna**. A **VHF antenna** utilized for TV reception.

VHF TV channel Abbreviation of very high frequency **TV channel**. A TV channel whose assigned frequency is in the **VHF** band. Also called **VHF channel**.

VHLL Abbreviation of **very high-level language**.

VHS Abbreviation of Video Home System. Also called **VHS video**. **1.** A standard videocassette format utilized for the recording and playing back of audiovisual content such as movies, broadcast programming, and the like. A VHS cassette has a long 0.5 inch magnetic tape which can provide from about 2 to 10 hours of play, depending on the length of the tape and the speed. It usually provides a resolution of about 250 lines per inch. There are improved versions, such as S-VHS. Also called **VHS format**. **2.** A cassette in **VHS** (1) format. Also called **VHS tape**, **VHS cassette**, or **VHS videotape**. **3.** A device which serves to record and playback **VHS** tapes. Also called **VHS recorder**, **VHS player**, **VHS deck**, **VHS video recorder**, or **VHS videocassette recorder**.

VHS cassette Same as **VHS** (2).

VHS deck Same as **VHS** (3).

VHS format Same as **VHS** (1).

VHS player Same as **VHS** (3).

VHS recorder Same as **VHS** (3).

VHS tape Same as **VHS** (2).

VHS video Same as **VHS**.

VHS video recorder Same as **VHS** (3).

VHS videocassette recorder Same as **VHS** (3).

VHS videotape Same as **VHS** (2).

VHSIC Abbreviation of Very High Speed Integrated Circuit. An IC with ultra-large-scale integration which operates at comparatively high speeds.

VI Abbreviation of **virtual interface**.

VI architecture Same as **virtual interface** (1). Abbreviation of **virtual interface architecture**.

VIA Same as **virtual interface** (1). Abbreviation of **virtual interface architecture**.

vibration **1.** A mechanical displacement of an object or body around a set point, a given value, an average value, or between two extreme values. For example, a swinging pendulum, a musical string which has been plucked, or a tuning fork which has been struck. **2.** The periodic variation of a physical quantity or object around a set point, a given value, an average value, or between two extreme values. For instance, AC, the movement of a pendulum, or the states of an astable multivibrator. Such vibration may be desired, as in that of a crystal utilized as a frequency standard, or unwanted, as in hunting. Also called **oscillation** (1). **3.** A single variation, movement, or cycle during **vibration** (1), or **vibration** (2). **4.** Oscillatory motion of particles such as atoms or molecules. For example, that which occurs when they absorb energy.

vibration alert An alert in which vibrations serve to convey information. For example, a properly equipped cell phone may vibrate when a call, message, or the like, is received.

vibration analyzer Same as **vibration meter**.

vibration isolation The use of a **vibration isolator** to protect components, circuits, devices, equipment, systems, materials, enclosures, or the like. Vibration isolation may also be enhanced through the design and/or placement of that which wishes to be isolated.

vibration isolator A device or material which serves to absorb or otherwise protect from **vibrations** (1).

vibration meter An instrument or device which utilizes one or more vibration sensors to measure and indicate **vibrations** (1). Used, for instance, to measure displacement, velocity, or acceleration of objects or bodies. Also called **vibrometer**, or **vibration analyzer**.

vibration pick-up Same as **vibration sensor**. Also spelled **vibration pickup**.

vibration pickup Same as **vibration sensor**. Also spelled **vibration pick-up**.

vibration sensor A transducer, such as that incorporating a laser or a piezoelectric crystal, which converts **vibrations** (1) into an electrical equivalent such as a voltage. Also called **vibration transducer**, or **vibration pickup**.

vibration table A platform used in conjunction with a device which delivers controlled vibrations. Utilized to test the ability of components, devices, equipment, systems, and materials, to withstand vibrations, shocks, jolts, and the like. Also called **vibrator** (4), **shake table**, or **shaking table**.

vibration transducer Same as **vibration sensor**.

vibrational Pertaining to, involving, requiring, or arising from **vibration**.

vibrator **1.** That, such as a vibration table, which serves to generate vibrations. **2.** An electromechanical device, such as a reed relay, which vibrates or produces vibrations. **3.** A **vibrator** (2) utilized to convert DC into AC. **4.** Same as **vibration table**.

vibrometer Same as **vibration meter**.

vibroscope An instrument or device which is utilized to monitor vibrations.

video **1.** Consisting of, or pertaining to images or sequences of images presented via a display device such as a TV, computer monitor, radar, or cell phone. The term usually includes any audio accompanying said images. Video consists of true motion which is divided into still frames, while **animation** is the simulation of movement using a properly sequenced set of pictures. **2.** Consisting of, or pertaining to images or sequences of images presented via a TV. Although the term usually refers to the visual portion of a TV signal or presentation, it may also refer to the accompanying sound. **3.** Pertaining to, utilizing, or resulting from signals at video frequencies. **4.** Data which is in a format suitable for displaying images or sequences of images via a device such as a computer monitor. **5.** Same as **videocassette**. **6.** Same as **videocassette player**.

video accelerator A video card which incorporates a chip, and which accelerates and enhances the video-handling performance of a system. A video accelerator significantly speeds up the updating of the images on a screen, which also frees the CPU to take care of other tasks. Especially useful, for instance, for displaying three-dimensional graphics.

Also called **video accelerator board, video accelerator card**, **graphics accelerator**, or **graphics accelerator card**.

video accelerator board Same as **video accelerator**.

video accelerator card Same as **video accelerator**.

video adapter Same as **video card**.

video amplifier A broadband amplifier which amplifies video frequencies and signals.

video animation Simulation of movement through the use of animated graphics. Also called **animation**.

video bandwidth A measure of the ability of a monitor to refresh its display screen. It can be calculated as the horizontal resolution multiplied by the vertical resolution multiplied by the refresh rate. For example, 1280 x 1024 x 75 = 98.304 MHz.

video board Same as **video card**.

video broadcast Same as **videocast**.

video buffer A buffer, usually consisting of a segment of memory in a video card, which holds the data that awaits to be sent to the display. Also called **screen buffer**, or **graphics buffer**.

video cable A cable, such as a coaxial cable, RCA cable, or a digital optical cable, utilized to carry video signals.

video cam 1. Abbreviation of **video camera**. **2.** Same as **videocam**.

video camera Its abbreviation is **video cam** (1). **1.** A device that converts images formed by lenses into electric signals. A photosensitive surface in the contained camera tube serves as the transducer which converts the optical image into electric video signals suitable for broadcasting, recording, or the like. Also known as **TV camera**, **camera** (2), or **telecamera**. **2.** A **video camera** (1) primarily designed for recording videotapes, as opposed to broadcasting live. Used, for instance, in camcorders.

video capture board An expansion board that enables a computer to capture frames from video sources such as TVs, VCRs, or digital cameras.

video card A computer card that enables it to display images. When a computer generates images, this card converts them into the electronic signals that serve as the input to the display device, such as the monitor. The graphics card determines factors such as the maximum resolution and number of colors that can be displayed. An appropriate monitor must be utilized to best benefit from the capabilities of such a card. Graphics cards usually incorporate their own memory, thus not occupying the RAM of the computer for preparing images. Also known by various other names, including **video adapter, video board, video controller, video graphics board, graphics card, graphics board, graphics adapter, display adapter, display card,** and **display board**.

video carrier A carrier, such as a picture carrier, which is modulated to convey video information. Also called **visual carrier**.

video cassette Same as **videocassette**.

video cassette deck Same as **videocassette player**. Also spelled **videocassette deck**.

video cassette format Same as **videocassette format**.

video cassette player Same as **videocassette player**.

video cassette recorder Same as **videocassette player**. Also spelled **videocassette recorder**. Its abbreviation is **VCR**.

video cassette tape Same as **videocassette tape**.

video CD A compact disc format which incorporates video, along with the audio. These may only be read by specific devices, such as a CD video player. Neither an audio CD player, nor a computer CD-ROM drive can access the information on discs in this format. Its abbreviation is **VCD**. Also called **CD video**.

video CD-ROM A CD-ROM which has video, or multimedia content including video, recorded.

video circuit A circuit carrying a **video signal**.

video codec 1. Abbreviation of **video co**der/dec**oder**. Hardware and/or software which converts analog video to digital video, and vice versa. **2.** Abbreviation of **video compressor/decompressor**. Hardware and/or software which compresses and decompresses video signals.

video coder/decoder Same as **video codec** (1).

video compression The encoding of video data so that it occupies less room. This helps maximize the use of storage space, facilitates faster transfer, and also reduces transmission bandwidth. Also called **digital video compression**.

video compressor/decompressor Same as **video codec** (2).

video conferencing Same as **videoconferencing**.

video controller Same as **video card**.

video converter Same as **video scan converter**.

video demodulator Same as **video detector**.

video detector In a TV, a circuit or device which demodulates a modulated video carrier, so as to extract video information. Also called **video demodulator**.

video digitizer A circuit, device, and/or program which converts an analog video input, such as that of a TV or VCR, into a digital video output. Used, for instance, to save an analog TV image to a hard disk or optical disc.

video disc player A device utilized to play back **videodiscs**. Also spelled **video disk player**.

video disc recorder A device utilized to record **videodiscs**. Also spelled **video disk recorder**. Its abbreviation is **VDR**.

video disk player Same as **video disc player**.

video disk recorder Same as **video disc recorder**. Its abbreviation is **VDR**.

video display Same as **video monitor**.

video display adapter Same as **video card**.

video display board Same as **video card**.

video display card Same as **video card**.

video display page The text, graphics, and so on which occupy the entire screen of a video monitor at any given moment. Also, the segment of video buffer or memory holding this data.

video display terminal A computer input/output device which incorporates a monitor, a video adapter, and a keyboard. It also usually includes a mouse. May be used, for instance, in networks. Its abbreviation is **VDT**. Also called **video display unit, video terminal, visual display terminal, visual display unit,** or **display terminal**.

video display unit Same as **video display terminal**. Its abbreviation is **VDU**.

video DRAM Video RAM consisting of dynamic RAM.

video driver A program which enables a computer operating system to communicate with, and control a monitor. Certain parameters, such as resolution and refresh rate can usually be set by the user using the video driver.

video editing The use of a video editor. Also, the changes so made.

video editing software Same as **video editor**.

video editor A program which is utilized to modify and manipulate video images and files. Used, for instance, to improve image quality, cut segments, change the sequences of frames, add transitions, and so on. Also called **video editing software**.

video effects The use of computers to enhance, modify, add, or remove images in movies, TV, and other multimedia presentations. Such effects are extensively used, and some of the techniques employed include motion capture and tweening. Also called **digital video effects**.

Video Electronics Standards Association Same as **VESA**.

video email Same as **video mail**.

video frequency **1.** An output frequency produced by a TV camera during recording, broadcasting, or the like. Such a signal may have a bandwidth extending from nearly zero Hz to over 4 MHz. **2.** A frequency utilized by, or producing, a video signal. The bandwidth of such signals can extend from nearly zero Hz to several GHz.

video gain control In a TV, a control which adjusts the gain of a **video signal**.

video game **1.** A computer program in which one or more users interact with the computer and/or other players, for amusement or competition. Such programs may be of varying complexity, and some feature faithfully realistic simulations of complicated activities such as aircraft piloting. Also called **computer game**, or **game**. **2.** A **video game (1)** with rich graphics, as opposed, for instance, to a game having mostly text or simple images.

Video Graphics Adapter Same as **VGA**.

Video Graphics Array Same as **VGA**.

video graphics board Same as **video card**.

video information In a device such as a TV, computer display, or radar monitor, the visual content presented. When referring specifically to TV, also known as **picture information**.

video mail Email which incorporates one or move videos. Its abbreviation is **v-mail**. Also called **video email**.

video memory Same as **video RAM**.

video mixer A circuit or device which mixes two or more video signals, such as those provided by TV cameras.

video mode A specific configuration of the settings of a computer monitor. For instance, the resolution, the number of colors that may be displayed, and whether it displays graphics and/or text. Also called **screen mode, monitor mode**, or **display mode**.

video modulation The modulation of a carrier wave, so as to carry video information.

video monitor Also called **video display**. **1.** A device utilized to display the images generated by a computer, TV, oscilloscope, radar, camera, or other similar device with a visual output. It incorporates the viewing screen and its housing. A specific example is a CRT. Also called **monitor (1), display (1)**, **display device**, or **display monitor**. **2.** A **video monitor (1)** utilized to present the output of a computer. Also called **monitor (2)**.

Video on Demand Its abbreviation is **VOD**. Cable or satellite programming, such as movies or TV programs, available to viewers whenever desired. A user accesses choices via an interactive menu displayed on the TV or other viewing device, and selects what wishes to be viewed, and the exact starting time. VOD may also provide other video content, such as games or training videos. All VOD services are fee-based.

video-out port A port, such as that providing an S-video connector, which enables a TV to be used as a computer display. Also called **video port (2)**, or **TV-out**.

video phone Same as **videophone**.

video player A device, such as a VCR or DVD, which plays back video content.

video port **1.** A port which serves to plug in a **video monitor (2)**. Also called **monitor port**, or **display port**. **2.** Same as **video-out port**.

video RAM Abbreviation of **video** random-access memory. Its own abbreviation is **VRAM**. Also called **video memory**. **1.** The memory utilized in a video card. It has to be fast enough to keep up with the screen refresh rate, usually consists of fast dynamic RAM, and allows for the CPU to write

to it at the same time said RAM refreshes the image being displayed at a given moment. When referring to its function as a buffer which stores the data contained in a single frame, also called **frame buffer**. **2.** Any memory utilized in the production, manipulation, or storage of video. The main memory of computer may be utilized, but is not optimized for such use, as **video RAM (1)** is.

video random-access memory Same as **video RAM**.

video recorder A device, such as a VCR or DVD recorder, which records video content.

video recording The process of producing a permanent, or semi-permanent, record of video content which is suitable for later reproduction. This includes recording upon magnetic tapes, magnetic disks, optical discs, and so on. Also, a specific recording. Also, that which has been recorded.

video scan conversion The conversion of one video scan format into another utilizing a **video scan converter**.

video scan converter A device which changes one video scan format into another. For example, that which converts the video output of a computer into an NTSC TV signal, or from an NTSC signal to a PAL signal, a PAL signal into an HDTV signal, or plan position indicator images into those which can be viewed by a computer monitor or TV. Also called **video converter**, or **scan converter**.

video server A network server specifically designed for and/or dedicated to the delivery of streaming video, especially video on demand.

video signal Also called **visual signal**. **1.** A signal occurring at a **video frequency**. **2.** A signal, such as a picture signal, which contains video information. **3.** A TV signal including the sound accompanying the images. **4.** A signal sent from a video adapter, or other source, in a format appropriate for viewing via a video monitor.

video streaming The transmission of streaming video over a computer network.

video tape Same as **videotape**.

video tape deck Same as **videocassette player**. Also spelled **videotape deck**.

video tape format Same as **videocassette format**. Also spelled **videotape format**.

video tape player Same as **videocassette player**. Also spelled **videotape player**.

video tape recorder Same as **videocassette player**. Also spelled **videotape recorder**. Its abbreviation is **VTR**.

video tape recording Same as **videotape recording**.

video teleconferencing Same as **videoconferencing**.

video telephone Same as **videophone**.

video terminal Same as **video display terminal**.

video toaster A full-featured video-editing program.

video transmitter Also called **visual transmitter**, or **picture transmitter**. **1.** A transmitter, such as a broadcast transmitter, which sends signals containing one or more pictures. Also, the equipment utilized for this purpose. **2.** The equipment utilized by a TV transmitter to send the video signal. This signal is combined with that of the sound transmitter for regular TV viewing.

video tuner A tuner, such as a TV tuner, utilized to select from given video frequencies.

video window A computer screen window displaying video.

videocam A video camera primarily intended for the transmission of images over the Internet. Such a camera may be utilized, for instance, for videoconferencing, in chat programs with video capabilities, for sending video email, or for monitoring a given area or environment. Also spelled **video cam (2)**. Also called **Webcam**, or **NetCam**.

videocassette A cassette, such as a VHS cassette, which contains a videotape, along with two floating reels, one for supply, the other for take-up. Also spelled **video cassette**. Also called **videotape (2)**, **videocassette tape (1)**, or **video (5)**.

videocassette deck   Same as **videocassette player**. Also spelled **video cassette deck**.

videocassette format   A specific **videocassette** format, such as VHS, or S-VHS. Also spelled **video cassette format**.

videocassette player   A device, such as a VHS player, which serves to record and playback **videocassettes**. Also spelled **video cassette player**. Its abbreviation is **VCR**. Also called **videocassette recorder**, **videocassette deck**, **videotape player**, **videotape recorder**, **videotape deck**, or **video (6)**.

videocassette recorder   Same as **videocassette player**. Also spelled **video cassette recorder**. Its abbreviation is **VCR**.

videocassette tape   Also spelled **video cassette tape**. **1.** Same as **videocassette**. **2.** Same as **videotape (1)**.

videocast   Abbreviation of **video** broad**cast**. An RF transmission, such as a TV broadcast, consisting of video signals intended for public or general reception.

videoCD   Same as **video CD**.

videoconferencing   Conferencing in which still or moving pictures are transmitted, usually along with voice and text, between users or entities at multiple locations. Such conferencing can use POTS for freeze-frame or jerky image transmission, although higher bandwidths can provide crisp real-time full-motion video. Also spelled **video conferencing**. Also called **video teleconferencing**.

videodisc   Also called **Laserdisc**. **1.** An optical disc format typically utilized for audio and video, such as movies and music. This format usually provides two analog and two digital audio channels, and its horizontal resolution is generally 425 lines. Also, a disc encoded in this format. Such discs are usually12 or 8 inches in diameter, and a laserdisc player is required to read them. **2.** Any optical disc format or disc.

videophone   Abbreviation of **video** tele**phone**. A telephone which in addition to providing speech, and other audible signals, has a display to present images. Such phones may provide still images which are periodically updated, or real-time video. Also spelled **video phone**.

videotape   Also spelled **video tape**. **1.** A magnetic tape, such as a VHS tape, which is suitable for recording and playing video signals. A video tape is contained within a **videocassette**. Also called **videocassette tape (2)**. **2.** Same as **videocassette**.

videotape deck   Same as **videocassette player**. Also spelled **video tape deck**.

videotape format   Same as **videocassette format**. Also spelled **video tape format**.

videotape player   Same as **videocassette player**. Also spelled **video tape player**.

videotape recorder   Same as **videocassette player**. Also spelled **video tape recorder**. Its abbreviation is **VTR**.

videotape recording   The use of a videocassette player to record TV programming or other video content. Also spelled **video tape recording**.

videotex   Same as **videotext**.

videotext   A service, such as teletext, in which text and graphics are presented on a TV screen, computer monitor, or the like, and which is used for shopping, or providing information such as weather or news. Also called **videotex**.

vidicon   A TV camera tube that is simpler than an image orthicon, and which is commonly utilized for closed-circuit TV.

view   **1.** To see what is displayed by a meter, instrument, computer monitor, TV, or the like. Also, that which is being viewed. **2.** To cause to be displayed on a screen, such as that of a TV or computer. **3.** A specific perspective of that which can be seen. For example, a given angle of a scene. **4.** To use a **viewer**. **5.** In a database management system, such as that of a relational database, a specific manner in which data is presented. For example, a view in which only the desired fields are displayed.

viewable image   The portion of a display screen, such as that of a TV or computer monitor, which has a usable viewing area. Computer CRTs, for instance, usually have a given space on the sides of the image, thus making the **screen size** larger than the size of viewable images. Usually measured diagonally.

viewer   A separate computer application which is called upon to interpret and view files and data which could otherwise not be seen by the current application. Also refers to multimedia files. For instance, a Web browser plug-in utilized to display video in a format that is not currently supported. Also called **external viewer**.

viewfinder   A device which serves to display an image corresponding to that which is being focused upon by a camera, such as a TV camera. Such a device may be optical and/or electronic. Also called **finder (2)**.

viewing screen   The viewing area of a display device such as a computer, TV, oscilloscope, or radar.

viewport   **1.** A specific portion of the text, graphics, or the like which occupy the entire screen of a computer monitor at any given moment. **2.** A specific portion of a window.

vignetting   A reduction in the brightness at the edges of an image, such as a photograph or slide. May be caused, for instance, by a poor lens design or a partial obstruction. Vignetting may also be utilized intentionally to highlight a given section of an image.

Villari effect   An effect which is the converse of **magnetostriction**. That is, when a magnetostrictive material is mechanically deformed, a magnetic property, such as magnetic induction or permeability, changes.

VINES   A UNIX-based network operating system. It is an acronym for **virtual n**etwork**i**ng **s**ystem, **virtual n**etworking software, **virtual n**etwork **s**ystem, or **virtual n**etwork software.

vinyl chloride   A flammable and colorless gas whose chemical formula is C_2H_3Cl. Widely utilized in the production of polymers such as polyvinyl chloride.

vinyl plastic   A plastic, such as polyvinyl chloride, made via the polymerization of **vinyl chloride**.

viral marketing   A marketing scheme intended to spread like a virus. For example, free email providers, such as Yahoo!, usually include an advertisement for their service in all emails sent, thus increasing its exposure each time an email is viewed.

virgule   In most computer operating systems, a character which serves to denote division, as seen, for instance, in: 22/7. This character also serves to separate elements within an internet address, as seen, for example, in: *http://www.yipeeee.com/whoo/yaah.html*. This character has a place on the keyboard, and is shown here between brackets: [/]. Also known as **forward slash**, or **slash**.

virtual   **1.** That which is not real, but is perceived as real, seemingly real, or which is otherwise simulated. Said especially of that which is computer-related or generated. For example, virtual reality, or virtual memory. **2.** That which is real, but appears or occurs elsewhere. Said especially of that which is computer-related. For example, a conversation between real people, each at their own location, but occurring in a chat room via a network. **3.** Pertaining to computer, online, and/or logical concepts. The term is frequently utilized interchangeably with related connotations of **electronic, digital, logical**, or **online**.

virtual address   In virtual memory, a memory location which is assigned to a place in a secondary storage device, such as a hard disk, as opposed to that in RAM.

virtual business Same as **virtual company**.

virtual card A digital greeting card sent over the Internet. There are many applications and Web sites which can be utilized to prepare such cards, and such greetings may incorporate text, still images, video, animation, audio, and so on, and can be chosen for virtually any occasion or purpose. Also called **virtual postcard, virtual greeting card, digital postcard,** or **electronic card**.

virtual cathode In an electron tube, a location within a space-charge region which can behave as if a cathode.

virtual channel 1. The path along which cells are transmitted in ATM. 2. Same as **virtual circuit**.

Virtual Channel Identifier In an ATM header, a field or tag that identifies a given **virtual channel** (1). Its abbreviation is **VCI**.

Virtual Channel Link A unidirectional connection between two ATM nodes or points. Its abbreviation is **VCL**.

virtual circuit Also called **virtual channel** (2), **virtual path** (2), **virtual route,** or **virtual connection**. 1. A logical connection between nodes of a communications network. Such a connection uses physical circuits, and may refer, for instance, to a permanent virtual circuit, or a switched virtual circuit. 2. A temporary path formed between nodes of a communications network. For example, that formed by a switched virtual circuit.

virtual circuit number In X.25, the number assigned to a virtual circuit. This number is attached to all packets during a call, differentiating them from other packets in other calls. Up to 4095 total VCNs in both directions are supported by X.25. Its abbreviation is **VCN**. Also called **logical channel number**.

virtual colocation The interconnection of the equipment of a customer or competitor to the facilities of a telecommunications company, without such equipment being within a physical location of the latter. This has benefits such as reducing the cost of operations, or being able to provide superior service. Used, for instance, by telephone companies or Internet providers. Also called **colocation** (2).

virtual community A collection of users of an online service or group. For example, all people who connect to the Internet on a regular basis, or the members of a newsgroup. Also called **online community**.

virtual company A business entity that relies primarily on computers and communication networks for most or all aspects of its function. Such an organization may have multiple employees which rarely or never see each other in person, instead employing telephone calls, videoconferencing, instant messages, emails, faxes, and so on for communication. Also called **virtual corporation,** or **virtual business**.

virtual connection Same as **virtual circuit**.

virtual corporation Same as **virtual company**.

virtual desktop A feature which enables a desktop to be larger than that which can be displayed by a computer screen at a given instant. A user utilizes left/right and/or up/down scrolling to view any areas hidden from view.

virtual device 1. A device which can be referenced, which is simulated, or does not otherwise exist physically. For instance, a virtual disk. 2. A device that exists in physical form, but which is made to appear as if in another form. For example, a virtual drive.

virtual device driver A device driver that enables direct communication between a hardware device and the operating system kernel. Its abbreviation is **VDD,** or **VxD**.

virtual disk A disk, such as a RAM disk, which is simulated.

virtual display A display, such as a head-mounted display, whose proximity to the eyes enables it to present the content equivalent to that of a regular computer display.

virtual drive A name given to a drive to facilitate identification by a given program or software system. This contrasts with a **physical drive**, which is the actual hardware. For instance, logical drives C: and D: may be a single partitioned physical drive. Also called **logical drive**.

virtual earth Same as **virtual ground**.

virtual environment An environment which is created via **virtual reality** (1). Also called **virtual reality environment**.

virtual greeting card Same as **virtual card**.

virtual ground A point in a circuit which is at zero potential, but is not connected to ground. Also called **virtual earth**.

virtual height The apparent height of an ionized atmospheric layer, as determined by the time interval that elapses between the transmission of a radio-frequency signal and the return of its ionospheric echo. The signal is assumed to be traveling at the speed of light. Also called **equivalent height**.

virtual host A **virtual server** which provides Web hosting.

virtual image 1. In an image-forming optical system, an image produced at an apparent source, from which the beam of light diverges. In a **real image**, the beam of light converges towards a given point, where the image is formed. 2. An image stored in computer memory. Such an image may larger than that which can be displayed on the monitor screen at any given moment.

virtual instrument An instrument, such as a multimeter or that utilized for playing or producing music, which is emulated utilizing computer hardware and software.

virtual interface Its abbreviation is **VI**. 1. An interface utilized for the high-speed transfer of data between multiple computers, such as servers, in which overhead is reduced by reducing or eliminating the breaking up of data into packets. Such an interface is usually platform-independent. Also, as standard for a virtual interface, or an architecture adhering to such a standard. Also called **virtual interface architecture**. 2. An interface, such as that incorporating goggles and special suits and/or a CAVE, utilized for interaction between a user and a **virtual reality**. Also called **virtual reality interface**.

virtual interface architecture Same as **virtual interface** (1). Its abbreviation is **VI interface,** or **VIA**.

virtual Internet service provider Same as **virtual ISP**.

virtual ISP Abbreviation of **virtual** Internet service provider. An ISP that utilizes the equipment, facilities, and services of another ISP. A virtual ISP basically offers that which the larger ISP provides, except that it uses its own brand name.

virtual LAN Abbreviation of **virtual** local-area network. Its own abbreviation is **VLAN**. A LAN which is created or subdivided using software, as opposed to rearranging the connecting wires and cables. This serves, for instance, to accelerate communications between users within the same workgroup, and for easily switching a user from one LAN to another.

virtual library 1. A collection of books, periodicals, documents, and other such materials which is available over the Internet. 2. All the books, periodicals, documents, and other such materials which are available over the Internet.

virtual local-area network Same as **virtual LAN**.

virtual machine Its abbreviation is **VM**. 1. A computer that is simulated by another computer. 2. A computer system which allows multiple operating systems and applications to be run simultaneously by two or more users. Such a system provides each user with what appears to be an independent computer. 3. A programming language interpreter, such as a Java Virtual Machine.

virtual memory Memory which is simulated to appear as larger than it really is. For example, memory management and secondary storage, such as a hard disk, may be utilized to provide such memory. The virtual memory so created is divided into segments called pages, and these are swapped to and from disk locations as needed, with the operating system keeping track of all mapping, addressing, and the like. Not only does this improve system performance and speed, it also enables running applications which would otherwise be too large. Its abbreviation is **VM**. Also called **virtual storage (2)**.

virtual network **1.** A communications network, such as a virtual LAN, that is created, interconnected, or subdivided using software, as opposed to rearranging the connecting wires and cables. **2.** Two or more interconnected networks which appear as a single network to a user.

virtual network software Same as **VINES**.

virtual network system Same as **VINES**.

virtual networking software Same as **VINES**.

virtual networking system Same as **VINES**.

virtual office Also called **virtual workplace**. **1.** A location or work environment where one or more members of a **virtual organization** perform work-related tasks. Such a setting may be virtual, with each individual contributing from a remote location. **2.** A location, real or virtual, where two or more members of a virtual organization meet, or where any combination of employees, clients, guests, or the like, meet. For instance, several people may meet via a videoconference.

virtual organization An organization, such as a business, which relies primarily on computers and communication networks for most or all aspects of its function. Such an organization may have multiple employees which rarely or never see each other in person, instead employing telephone calls, videoconferencing, instant messages, emails, faxes, and so on for communication.

virtual path **1.** The path along which cells are transmitted in ATM. **2.** Same as **virtual circuit**.

Virtual Path Identifier In an ATM header, a field or tag that identifies a given **virtual path (1)**. Its abbreviation is **VPI**.

virtual peripheral A peripheral which can be referenced, which is simulated, or which otherwise does not exist physically. For instance, a virtual printer.

virtual post card Same as **virtual card**.

virtual postcard Same as **virtual card**.

virtual printer A printer which can be referenced, which is simulated, or does not otherwise exist physically. For example, a section of a disk which accepts a file in a format suitable for printing while a printer becomes available.

virtual private network A network which has the appearance, functionality, and security of a private network, but which is configured within a public network, such as the Internet. The use of a public infrastructure while ensuring privacy using measures such as encryption and tunneling protocols, helps provide the security of a private network at a cost similar to that of a public network. Its abbreviation is **VPN**.

virtual private networking The creation, configuration, and use of a **virtual private network**.

virtual reality Its abbreviation is **VR**. **1.** A reality which is simulated, especially with the assistance of computers. A computer-assisted virtual reality may incorporate 3D graphics combined with special goggles which track the path of the user's vision, so that the virtual environment can be changed appropriately, plus sound effects, smells, changes in temperature, a body suit equipped with motion sensors, and so on, to faithfully replicate real surroundings and stimuli. Used, for instance, for training, education, or entertainment. **2.** The perception of reality experienced by a person or other being in a **virtual reality (1)**.

virtual reality environment Same as **virtual environment**.

virtual reality interface Same as **virtual interface (2)**.

Virtual Reality Modeling Language Same as **VRML**.

virtual root A directory that appears as one root directory to a user, but which actually points to another. May be seen, for instance, when a user connects to a network server, such as an FTP server. Also called **virtual root directory**.

virtual root directory Same as **virtual root**.

virtual route Same as **virtual circuit**.

virtual screen A viewing area which is larger than that which can be displayed by a computer screen at a given instant. A user utilizes left/right and/or up/down scrolling to view any areas hidden from view.

virtual server A server which works alongside other such servers within the same computer, as opposed to a **dedicated server**, which is a single computer working as a single server. Seen, for instance, in Web hosting, where one server can perform the tasks that would otherwise require multiple computers. In most cases this provides the same access, functionality, and performance of multiple servers. Each virtual server may or may not have its own IP address.

virtual storage Its abbreviation is **VS**. **1.** The storage of data over the Internet. Used, for instance, for backup, or for access from most any location in the world. **2.** Same as **virtual memory**.

virtual store A **virtual company** dedicated to selling goods and/or services. Also, a Web site where such an entity makes its offerings.

virtual terminal One type of terminal which is made to appear as if another. Used, for instance, to enable a user with a graphical-user interface to log-on to and enter commands into a text-based system such as an older mainframe. In this manner, such a terminal appears to the accessed mainframe as another mainframe.

virtual wallet Encryption software which serves to provide the virtual equivalent of a wallet. In it, may be contained digital cash, credit card information, shipping details, and a digital certificate for authentication of the wallet holder. Both merchants and customers benefit from the added security, expedience, and convenience. Also called **digital wallet**, or **electronic wallet**.

virtual workgroup A workgroup that communicates via teleconferencing, messaging, emails, file transfers, shared databases, telephone calls, and the like.

virtual workplace Same as **virtual office**.

virtual world An interactive environment, such as a CAVE, created via **virtual reality (1)**.

virus A computer program or programming code which replicates by seeking out other programs onto which to copy itself. It usually passes from computer to computer through the sharing of infected files, and goes unnoticed until it attacks, unless an antivirus program that recognizes the virus detects it first. The effects of a virus may be as slight as a friendly message appearing once on a screen, it might cause a system crash, or it may reach the extreme of destroying a hard drive. Also called **virus program**, or **computer virus**.

virus hoax A non-existent virus which is nonetheless hyped as a grave danger. Typically, credibility is simulated by including a bogus warning from a governmental agency such as the FBI, or another entity with purported credibility, such as Microsoft. The time, bandwidth, and other resources a virus hoax wastes can rival or exceed that of most real computer viruses.

virus program Same as **virus**.

virus scanner **1.** A computer program intended to identify, locate, isolate, and eliminate viruses. Such a program may also scan incoming files and data for viruses before a computer is exposed. Also called **antivirus program**. **2.** A utility, such as that found in a **virus scanner** (1), an email program, or via a Web site, which searches for, and identifies known viruses.

virus signature A unique bit string which each copy of a known virus has. Used, for instance, by an antivirus program for detection and removal of viruses.

VirusScan A proprietary virus scanner.

viscometer Abbreviation of **visco**sity **meter**. An instrument or device utilized to measure and indicate the **viscosity** of a fluid. Also called **viscosimeter**.

viscosimeter Same as **viscometer**.

viscosity A measure of the resistance of a fluid to flow. All fluids exhibit viscosity, with most gases having low viscosity and many oils having high viscosity.

viscosity meter Same as **viscometer**.

viscous damping The use of a fluid to reduce or limit the amplitude of a mechanical motion, such as a vibration. An example is the use of a ferrofluid for damping the motion of a high-frequency speaker. Also called **fluid damping**.

visibility factor The ratio of the minimum input signal that a receiver can detect, to the minimum output signal that can be detected by a user viewing the display of said receiver. May refer, for instance, of a radar receiver and its operator. Also called **display loss**.

visible horizon As seen from a given location, the junction at which the earth, or the sea, appears to meet the sky. Also called **visual horizon, apparent horizon**, or **horizon** (1).

visible light Same as **visible radiation**.

visible radiation Electromagnetic radiation whose wavelength enables it to be detected by an unaided human eye. The interval of wavelengths so detectable spans from approximately 750 nanometers (red) to approximately 400 nanometers (violet). Light is currently defined as traveling at 2.99792458×10^8 meters per second. Also called **visible light, light** (1), or **light radiation**.

visible spectrum The spectrum of frequencies, or range of wavelengths, encompassing **visible radiation**.

VisiBroker A proprietary Object Request Broker.

vision system A system which incorporates the hardware and software necessary to emulate functions of an eye. A charge-coupled device, for instance, may be used for the input of images, with the obtained information passed on to a processor. A vision system is not necessarily limited to the visible spectrum, and may also include interpretation of three-dimensional surfaces. Used, for instance, in many areas of robotics

visit An instance during which a **visitor** accesses a Web site.

visitor 1. A specific individual or entity which accesses a Web site, as distinguished from other such visitors. For example, a for a given period a Web site may have 100 hits, all by a single visitor, or 10 by 10 unique visitors, and so on. Individual users are usually distinguished by their IP number, but this may not be accurate as a single IP number may identify multiple users, and a single user may have a dynamic IP address. Also called **unique user**, or **unique visitor**. 2. A user that logs onto a network or system without having registered or otherwise established an account. Visitors typically have restricted access and/or privileges. Also called **guest**.

visitor account An account with restricted access and/or privileges provided to a **visitor** (2). Also called **guest account**.

visitor password A generic password assigned to a **visitor** (2). Also called **guest password**.

Visor A popular PDA.

Visual BASIC A version of BASIC utilized to build Windows applications. Its abbreviation is **VB**.

Visual C++ A version of C++ utilized to build Windows applications.

visual carrier Same as **video carrier**.

visual communication Any form of communication, such as flashing lights, which involves seeing the source.

visual display terminal Same as **video display terminal**. Its abbreviation is **VDT**.

visual display unit Same as **video display terminal**. Its abbreviation is **VDU**.

visual horizon Same as **visible horizon**.

visual programming Computer programming which is assisted by icons, buttons, menus, and other images, which allow for easy selection of components.

visual signal Same as **video signal**.

visual transmitter Same as **video transmitter**.

visualization 1. The showing of computer data in the form of graphics. For example, the presenting of data using 3D models as opposed to text or formulas. Also called **information visualization**, or **data visualization**. 2. The use of computers to represent, display, and manipulate scientific data and phenomena, such as the magnetic field around a point or object, or for the simulation of processes within a cell when metabolizing nutrients or when faced with a harmful substance. Enormous processing power must be combined with specifically tailored applications and display devices, and supercomputers are frequently utilized for such work. Since graphical data is analyzed by the brain at a more intuitive level than text data, more complex information can be more simply transmitted, helping provide even deeper insight into that being researched. Also called **scientific visualization**.

Viterbi algorithm A decoding algorithm which is utilized, for instance, in cellular telephony, speech recognition, and magnetic recording.

Viterbi decoding Decoding, such as that utilized in CDMA, using the **Viterbi algorithm**.

vitreous Of, pertaining to, characteristic of, similar to, derived from, or containing glass.

VL bus Same as **VESA Local Bus**.

VL local bus Same as **VESA Local Bus**.

VLAN Abbreviation of **virtual LAN**.

VLB Same as **VESA Local Bus**.

VLDB Abbreviation of **Very Large Database**.

VLF Abbreviation of **very low frequency**.

VLIW Abbreviation of **Very Long Instruction Word**.

VLM Abbreviation of **Very Large Memory**.

VLSI Abbreviation of **very large-scale integration**.

VM 1. Abbreviation of **virtual machine**. 2. Abbreviation of **virtual memory**.

VML Abbreviation of Vector Markup Language. An application of XML which defines the use of vector graphics in XML or HTML documents.

VMOS Abbreviation of **vertical MOS**.

VMOSFET Abbreviation of **vertical MOSFET**.

VMS Abbreviation of **voice messaging system**.

VOA Abbreviation of **volt-ohm-ammeter**.

VoATM Abbreviation of **Voice over ATM**.

vocoder Abbreviation of voice coder. 1. Same as **voice codec** (1). 2. Same as **voice codec** (2). 3. A device which analyzes and/or manipulates a voice input.

VOD Abbreviation of **Video on Demand**.

voder A **vocoder** (3) which utilizes a keyboard to the enter sounds to be analyzed and/or manipulated, as opposed to an input voice.

VoDSL Abbreviation of **Voice over DSL**.

VoFR Abbreviation of **Voice over Frame Relay**.

voice 1. The sounds produced and heard when a human, or other organism, makes sounds intended for communication. Humans, for instance, use the appropriate anatomical organs and conduits, such as the larynx and trachea, along with the necessary physiological mechanisms, such as forcing air through the trachea to vibrate the vocal cords located in the larynx, to producing such sounds. 2. A human **voice** (1). 3. Simulation of a **voice** (2) through the use of a device which incorporates a speaker and computer. Used, for example, in robotics, or to assist those with reduced speech ability. Also called **artificial voice**, or **artificial speech**. 4. Any sound resembling a **voice** (1).

voice-activated That which is actuated, operated, or controlled by a voice. Examples include properly equipped tape recorders and radio transmitters. Also called **voice-actuated**.

voice activation The initiating or activating of an event or process, such as recording, transmitting, or the entering of computer data, using a voice. Also called **voice actuation**.

voice-actuated Same as **voice-activated**.

voice actuation Same as **voice activation**.

voice amplifier An amplifier which is specially designed to amplify **voice frequencies**. Such an amplifier may be utilized, for instance, in a public-address or intercom system. Also called **speech amplifier**.

voice analyzer A circuit, device, piece of equipment, or system that evaluates various aspects of human voice. An example is a sound spectrograph. Used, for instance, in voice-recognition systems. Also called **speech analyzer**.

voice API Abbreviation of **voice application program interface**. An application program interface which facilitates tasks such as speech recognition and speech-to-text and text-to-speech conversions. Also called **speech API**.

voice band Same as **voice bandwidth**.

voice bandwidth The interval of frequencies encompassing **voice frequencies**. Also called **voice band**, or **speech bandwidth**.

voice beeper Same as **voice pager**.

Voice Carry Over A service, or a telephone unit which supports this service, which is similar to TDD, but which allows a user to speak responses as opposed to typing them. Such a unit may or may not have a keyboard. Its abbreviation is **VCO**.

voice channel 1. Same as **voice-grade channel**. 2. A channel carrying a **voice signal**.

voice chat A conversation via audioconferencing, a chat program with voice capabilities, Internet telephony, a telephone, or the like.

voice chip 1. A memory chip that is utilized to store voices. Used, for instance, in digital answering machines. 2. In a computer, a chip specifically designed for use with voice applications.

voice clipper A circuit or device which limits the amplitude of its output signal to a predetermined maximum, regardless of the variations of its input. Used, for example, for preventing component, equipment, or media overloads. Also called **speech clipper**.

voice clipping Also called **speech clipping**. 1. The action and effect of a **voice clipper**. 2. The unintended limiting of the amplitude of a speech signal, due to the exceeding of the capabilities of an amplifier or circuit.

voice codec Also called **voice coder**, or **speech codec**. 1. Abbreviation of **voice coder/decoder**. Hardware and/or software which converts speech to digital code, and vice versa. Also called **vocoder** (1). 2. Abbreviation of **voice compressor/decompressor**. Hardware and/or software which compresses and decompresses speech signals. Also called **vocoder** (2).

voice coder Same as **voice codec**. Its abbreviation is **vocoder**.

voice coil 1. In a dynamic speaker, a coil which is connected to the diaphragm, and through which an audio-frequency signal current is sent to move said coil in a piston-like manner, thus producing a sound-wave output. Also called **speaker voice coil**, or **loudspeaker voice coil**. 2. A motor that incorporates a coil and which is utilized to move an access arm in a disk drive. Such motors tend to be faster and more reliable than stepper motors utilized in this capacity.

voice compression The encoding of speech so that it occupies less room. This helps maximize the use of storage space, and also reduces transmission bandwidth. There are various speech-compression algorithms, including Code-Excited Linear Prediction. Also called **speech compression**.

voice control The operation or regulation of a component, circuit, device, piece of equipment, system, process, or mechanism, using a voice. For example, the use of voice commands to control navigation aids, temperature settings, audio selections, or seat adjustments in an automobile. Also called **speech control**.

voice-controlled A component, circuit, device, piece of equipment, system, process, or mechanism which is operated or regulated by a voice. Also called **voice-operated**, or **speech-controlled**.

voice-controlled computer A computer equipped with the proper interfaces, software, and processing capabilities to be fully and accurately controlled by voice commands. Used, for instance, by a surgical team during an operation. Also called **speech-controlled computer**.

voice digitization The conversion of a voice into digital form. This makes storage, processing, and transmission easier and more accurate. Also called **speech digitization**.

voice digitizer A device which converts a voice into digital form, such as that found in a compact disc. The larger the sample of the voice, and the higher the sampling rate, the more accurate the conversion. Also called **speech digitizer**.

voice enhancement The use of **voice processing** to augment the intelligibility of speech. Used, for instance, to help those with reduced hearing. Also called **speech enhancement**.

Voice Extensible Markup Language Same as **VXML**.

voice filter A filter which selectively transmits or rejects signals in one or more intervals of voice frequencies. Used, for instance, to improve intelligibility. Also called **speech filter**.

voice frequencies In the audio spectrum, the frequencies within which speech occurs. This interval varies by region, but for communications may be defined as spanning from approximately 200 Hz to 6000 Hz. Components, circuits, devices, transducers, equipment, and systems processing, converting, transmitting, or receiving speech must be able to properly handle this interval. Also called **speech frequencies**.

voice grade Pertaining to a communications channel or system whose bandwidth is sufficient to adequately support voice telephony.

voice-grade channel A communications channel, such as that provided by POTS, whose bandwidth is sufficient to adequately support voice telephony. Data transmission is limited to a bandwidth of a few kilohertz at best, and may be used, for instance for dial-up Internet access or fax. Also called **voice-grade circuit**, **voice-grade line**, or **voice channel** (1).

voice-grade circuit Same as **voice-grade channel**.

voice-grade communications Communications utilizing a **voice-grade channel**.

voice-grade line Same as **voice-grade channel**.

voice input The use of a natural or artificial voice to enter commands and/or data into a computer. Such a system makes use of voice recognition.

voice intelligibility The extent to which speech is capable of being understood. Usually used in the context of voice communications. The more intelligible, the clearer the reception of a voice transmission. Also called **speech intelligibility**.

voice interface Same as **voice user interface**.

voice inverter Same as **voice scrambler**.

voice mail A computerized system which receives and records incoming telephone calls, and which may also perform other functions, such as message forwarding. A caller hears a recorded greeting after a given number of rings, and may have the choice to leave a message, talk to an operator or other person, or, in the case of an account holder, access and manipulate messages after entering the appropriate code. Also spelled **voicemail**.

voice-mail notification A signal which indicates that one or more voice messages await. An example is a stutter dial tone heard when picking up a telephone to make a call.

voice menu A menu, such as that encountered when accessing an interactive voice response system, in which voice commands are utilized for selections. Such a menu may also respond to telephone keypad entries.

voice messaging The exchange of voice messages, as opposed to emails, SMSs, or the like. Also, a service enabling such messaging.

voice messaging system A system, such as that utilized for voice mail, which enables **voice messaging**. Its abbreviation is **VMS**.

voice modem A modem, which in addition to data-handling functions, also supports voice applications such as Internet telephony.

voice navigation 1. The use of voice commands to browse the World Wide Web. Used, for instance, with devices such as cell phones or PDAs, by for those with reduced motor function. 2. The use of voice commands to proceed through application menus.

voice network A network, such as a telephone network, via which speech is transmitted. Also called **speech network**.

voice-operated Same as **voice-controlled**.

voice-operated transmission Same as **VOX**.

voice-operated transmit Same as **VOX**.

voice output Computer output consisting of digital speech. Also called **speech output**.

Voice over ATM Its abbreviation is **VoATM**. 1. The process of transmitting voice traffic over an ATM-based network. Also, a protocol or standard for such transmissions. 2. A specific two-way speech or audio transmission using **Voice over ATM** (1).

Voice over DSL Its abbreviation is **VoDSL**. 1. The process of transmitting voice traffic over a DSL line. Many DSL services allow voice conversations to take place concurrently with rapid data transfers. Also, a protocol or standard for such transmissions. In addition, such lines permit multiplexing multiple voice channels, with a typical DSL line being able to handle hundreds of voice calls simultaneously. 2. A specific two-way speech or audio transmission using **Voice over DSL** (1).

Voice over Frame Relay Its abbreviation is **VoFR**. 1. The process of transmitting voice traffic over a Frame Relay network. Also, a protocol or standard for such transmissions. 2. A specific two-way speech or audio transmission using **Voice over Frame Relay** (1).

Voice over IP Its abbreviation is **VoIP**. 1. The process of transmitting voice traffic over an IP-based network. Also, a protocol or standard for such transmissions. 2. A specific two-way speech or audio transmission using **Voice over IP** (1).

Voice over Packet Its abbreviation is **VoP**. 1. The process of transmitting voice traffic over a packet-based network. Also, a protocol or standard for such transmissions. 2. A specific two-way speech or audio transmission using **Voice over Packet** (1).

voice pager A pocket-sized radio receiver or transceiver which serves to receive voice messages. These devices may emit a beep or vibrate when informing of a new message, making them suitable for those with reduced vision. Also called **voice beeper**.

voice port A device, located at the premises of a customer, which performs the proper conversions necessary for the transmission and reception of telephonic conversations through cable TV connections. Such a device may provide for multiple lines, and supports enhanced telephony features such as caller ID, call waiting, and the like, in addition to data services such as broadband data communications.

voice portal A Web portal that is accessed by voice. An example would be that allowing a user to dial in using a regular telephone, then checking email, news headlines, and the weather. The term may also be utilized to refer to other automated services or systems which are telephone-based, such as voice response systems. Its acronym is **vortal** (2).

voice power The power of an audio signal that is within the speech frequencies. Also called **speech power**, or **talk power**.

voice print Same as **voiceprint**.

voice processing Any form of processing performed on speech. These include digitization, amplification, compression, filtering, scrambling, and synthesis. Also called **speech processing**.

voice processor A circuit, device, piece of equipment, or system which serves for **voice processing**. For instance, a voice-recognition system. Also called **speech processor**.

voice recognition Also called **speech recognition**. 1. The ability of a device, piece of equipment, computer, or system, to detect and recognize spoken words. A speech-recognition system, for instance, must be able to properly identify all phonemes of a given language to properly process speech. A computer so enabled converts spoken words into commands, text, or the like. A computer so enabled converts spoken words into commands, text, or the like, for activating features, entering text, surfing the Internet, engaging in a voice chat, and so on. Also used, for instance, in an interactive voice response system. 2. A specific instance of **voice recognition** (1). 3. The technology employed in **voice recognition** (1). Also called **voice-recognition technology**.

voice-recognition application An application utilized for **voice recognition**. Used, for instance, for speech-to-text and text-to-speech conversions. Also called **speech-recognition application**.

voice-recognition device A device, such as a handheld computer, which is utilized for **voice recognition** (1). Also called **speech-recognition device**.

voice-recognition system A system, such as that incorporating the appropriate software and hardware, utilized for **voice recognition** (1). Also called **speech-recognition system**.

voice-recognition technology Same as **voice recognition** (3).

voice response Same as **voice response system**.

voice response system Any automated system, such as interactive voice response, in which a user accesses information which is presented via prerecorded sounds and/or the use of an artificial voice. Also called **voice response**, or **voice response unit**.

voice response unit Same as **voice response system**. Its abbreviation is **VRU**.

voice scrambler A circuit or device utilized for encoding a scrambled speech signal. Such a circuit or device, for instance, may divide the audio-frequency spectrum into bands which are displaced prior to transmission, and placed in their proper order by the receiving device. Also called **voice inverter**, **scrambler (1)**, or **speech scrambler**.

voice security 1. A biometric security measure which utilizes voiceprints or similar means for identification and authentication. **2.** The securing of the secrecy or privacy of voice communications utilizing techniques such as encryption, or devices such as voice scramblers.

voice-stress analyzer A device that monitors and records fluctuations in the voice of a subject which is being questioned. Such a device purportedly can detect when a person is lying.

voice synthesis Speech which is generated by a computer. The sounds are produced either by linking phonemes together, or by drawing from a database containing recorded words. Current technology provides for such speech to be almost completely natural. Its applications include assistance for those with reduced vision, or for retrieving email over any telephone. Also called **speech synthesis**, or **digital speech**.

voice synthesizer A device, piece of equipment, computer, or system, which is utilized for **voice synthesis**. Also called **speech synthesizer**.

voice telephony Telephone communication limited to the transmission and reception of voices and other sounds.

voice-to-text The conversion of a voice input into a text. Also called **voice-to-text conversion**, or **speech-to-text**.

voice-to-text conversion Same as **voice-to-text**.

voice user interface Its acronym is **VUI**. A user interface which utilizes voice recognition to enable a system to respond to voice commands and entries. Used, for instance, in computer systems, PDAs, or cell phones. The key difference between a VUI and an **interactive voice response system** is that the former accepts continuous speech and handles an extensive vocabulary, while the latter responds only specific words or short phrases. Also called **voice interface**, or **speech user interface**.

Voice XML Same as **VXML**.

voicemail Same as **voice mail**.

voiceprint A sound spectrogram resulting from the analysis of a voice sample. Used, for instance, in voice security. Also spelled **voice print**.

VoiceXML Same as **VXML**.

VoIP Abbreviation of **Voice over IP**.

vol Abbreviation of **volume**.

volatile 1. Said of a chemical substance which is readily evaporated at standard temperatures and pressures. That which is volatile has a high vapor pressure. **2.** Short-lived, unpredictable, and/or undergoing rapid or explosive change. **3.** Pertaining to, or characteristic of, **volatile memory**.

volatile memory Computer memory, such as most forms of RAM, which does not retain its content when the power is interrupted or turned off. Also called **volatile storage (2)**.

volatile RAM Abbreviation of **volatile** random-access memory. RAM, such as dynamic RAM, which does not retain its content when the power is interrupted or turned off.

volatile random-access memory Same as **volatile RAM**.

volatile storage 1. Computer storage which does not retain its content when the power is interrupted or turned off. This type of storage is very rare. **2.** Same as **volatile memory**.

volt The SI unit for potential difference, electric potential, and electromotive force. A volt may be defined as the potential difference between two points along a conductor when the current flowing between them is held at one ampere, and the power dissipated is one watt.

volt-ammeter Same as **voltammeter**.

volt-ampere A unit of apparent power. It is defined as the product of the RMS voltage, expressed in volts, by the RMS current, expressed in amperes, of an AC circuit. Also spelled **voltampere**. Its abbreviation is **VA**.

volt-ampere meter An instrument which measures and indicates apparent power, expressing readings in **volt-amperes**.

volt-amperes reactive The power in an AC circuit which cannot perform work. It is calculated by the following formula: $P = I \cdot V \cdot \sin\theta$, where P is the power in watts, I is the current in amperes, V is the voltage in volts, and $\sin\theta$ is the sine of the angular phase difference between the current and the voltage. Its abbreviation is **VAR**. Also called **reactive volt-amperes**, **reactive power**, **idle power**, **wattless power**, or **quadrature power**.

volt-ohm-ammeter An instrument which measures and indicates voltages, resistances, and currents in an electric circuit. Such a device usually provides an easy manner, such as a switch, to change between functions. It is the same as a **volt-ohm-milliammeter**, except that current readings are in amperes. Its abbreviation is **VOA**.

volt-ohm-milliammeter An instrument which measures and indicates voltages, resistances, and currents in an electric circuit. Such a device usually provides an easy manner, such as a switch, to change between functions. It is the same as a **volt-ohm-ammeter**, except that current readings are in milliamperes, or smaller units such as microamperes. Its abbreviation is **VOM**.

volt-ohmmeter An instrument which measures and indicates voltages and/or resistances in an electric circuit. Its abbreviation is **VOM**.

Volta effect A potential difference that is developed between contacts that are made of two dissimilar materials. This potential difference may be a few tenths of a volt. Also known as **contact emf**, or **contact potential**.

Volta pile Same as **voltaic pile**.

Volta's law When two conductors are placed in contact, each consisting of a different material, a potential difference is developed. Also called **Volta's principle**.

Volta's principle Same as **Volta's law**.

voltage The potential difference or electromotive force between two points, such as those in a circuit, expressed in volts, or in multiples of volts, such as kilovolts or millivolts. It has multiple symbols, which are E, **e**, V, and **v**. Also called **tension (1)**.

voltage amplification Also called **voltage gain**. **1.** For a component, circuit, or device, the ratio of the output voltage to the input voltage. Also called **voltage ratio**. **2.** The production of an output voltage which is greater than the input voltage.

voltage amplifier A component, circuit, or device whose output signal voltage is greater than its input signal voltage.

voltage antinode For a medium having standing waves, such as a transmission line or antenna, a point at which there is a maximum of voltage. Also called **voltage loop**.

voltage attenuation Also called **voltage loss (2)**. **1.** For a component, circuit, or device, the ratio of the input voltage to the output voltage. **2.** The production of an output voltage which is lesser than the input voltage.

voltage breakdown 1. A disruptive electrical discharge between the electrodes of an electron tube, produced by a voltage. **2.** A disruptive electrical discharge occurring in a gas, produced by a voltage. **3.** A disruptive electrical discharge occurring through an insulator, dielectric, or other material separating circuits, produced by a voltage. **4.** In a semiconductor, an avalanche breakdown produced by a voltage.

voltage calibrator   A source whose steady voltage level serves as a basis for calibrating instruments.

voltage coefficient 1. A coefficient depicting a voltage change resulting from a variation in another electrical parameter, such as current or resistance. **2.** A coefficient depicting a variation in another electrical parameter, such as current or resistance, resulting from a voltage change.

voltage comparator   An instrument or device, such as a differential amplifier, which compares two voltages.

voltage control   A component, circuit, or device which controls a voltage level. Used, for instance, in a power supply or generator.

voltage-controlled amplifier   An amplifier whose gain is regulated by its input voltage. Its abbreviation is **VCA.**

voltage-controlled attenuator   An attenuator whose reduction in the amplitude of a signal is regulated by its input voltage. Its abbreviation is **VCA.**

voltage-controlled capacitor   A capacitor, such as a varactor, whose capacitance is regulated by its input voltage. Also called **voltage-dependent capacitor.**

voltage-controlled crystal oscillator   A crystal oscillator whose frequency is regulated by its input voltage. Its abbreviation is **VCXO.**

voltage-controlled current source   A dependent source whose level of output current depends on its input voltage. An example is a transconductance amplifier. Its abbreviation is **VCCS.**

voltage-controlled device   A device, such as a switch or a voltage-controlled oscillator, whose function is controlled by an input voltage.

voltage-controlled oscillator   An oscillator whose frequency is regulated by its input voltage. Its abbreviation is **VCO.**

voltage-controlled resistor   Same as **varistor.** Its abbreviation is **VCR.**

voltage-controlled switch   A switch, such as a semiconductor device, whose switching action is determined by an input voltage.

voltage-controlled voltage source   A dependent source whose level of output voltage depends on its input voltage. An example is a voltage amplifier. Its abbreviation is **VCVS.**

voltage corrector   Same as **voltage regulator.**

voltage crest   Same as **voltage peak.**

voltage-current characteristic   For a component, circuit, or device, a curve plotting voltage as a function of current. Also called **voltage-current curve.**

voltage-current curve   Same as **voltage-current characteristic.**

voltage cut-off   Same as **voltage cutoff.**

voltage cutoff   Also spelled **voltage cut-off. 1.** A component, circuit, or device which prevents a voltage from exceeding a given limit. **2.** A component, circuit, or device, which prevents a voltage from dropping below a given limit.

voltage-dependent capacitor   Same as **voltage-controlled capacitor.**

voltage-dependent resistor   Same as **varistor.** Its abbreviation is **VDR.**

voltage detector   A component, circuit, or device which indicates the presence of a voltage.

voltage divider   A component, circuit, or device which is utilized to provide an output which is a desired fraction of the input voltage. For instance, a resistor, variable resistor, tapped resistor, or combination of resistors with taps, movable contacts, or junctions may be utilized for this purpose. Also called **potential divider.**

voltage doubler   A rectifier circuit whose output DC voltage is about twice the peak value of its input AC voltage. Such a circuit separately rectifies each half-cycle, then adds the rectified voltages. Also called **doubler (2).**

voltage drop   Also called **voltage loss (1),** or **potential drop. 1.** The voltage difference between any two points of a circuit or conductor, due to the flow of current. Also called **drop (3). 2.** The voltage difference between the terminals of a circuit element, due to the flow of current. Also called **drop (4).**

voltage end point   Also called **end point voltage (2). 1.** For a cell or battery, the voltage below which any device or piece of equipment connected to it will not operate, or below which operation is inadvisable. **2.** For a cell or battery, the voltage when discharge is complete.

voltage feed   The feeding of an antenna by connecting its transmission line at a **voltage antinode.**

voltage feedback   Feedback in which a proportion of the output voltage is fed back to the input.

voltage follower   An operational amplifier with unity gain. Thus, its output voltage is equal to its input voltage.

voltage gain   Same as **voltage amplification.**

voltage generator   A generator that provides a voltage to a component, circuit, device, piece of equipment, or system.

voltage gradient   For a given conductor or dielectric, the potential difference per unit length. Also called **potential gradient.**

voltage inverter   A component, circuit, or device whose output voltage has the opposite polarity or sign as the input voltage.

voltage jump 1. A sudden increase in voltage, especially that which is unwanted. **2.** A sudden increase in voltage in gas-discharge tube.

voltage lag   Within a circuit, a change in voltage which lags behind a change in current. For instance, in a capacitive circuit the current leads an applied voltage. This contrasts with **voltage lead.**

voltage lead   Within a circuit, a change in voltage which leads a change in current. For example, in an inductive circuit the current lags behind an applied voltage. This contrasts with **voltage lag.**

voltage level 1. The ratio of the voltage at a given point along a transmission system, to that of a given voltage reference, such as a zero level. **2.** A voltage value that is required for operation or that elicits a certain response. For example, a threshold voltage. **3.** A specific voltage, such as an AC line voltage of 115V. **4.** In digital logic, either a high-level voltage, or a low-level voltage.

voltage limit 1. The maximum output voltage a component, circuit, or device can safely provide. **2.** The maximum input voltage a component, circuit, or device can handle without damage.

voltage loop   Same as **voltage antinode.**

voltage loss 1. Same as **voltage drop. 2.** Same as **voltage attenuation.**

voltage magnitude   A given numerical value of **voltage.**

voltage maximum   Same as **voltage peak.**

voltage meter   Same as **voltmeter.**

voltage minimum 1. The lowest value of a voltage. **2.** The smallest value of the displacement from a reference position, such as zero, for a voltage. **3.** The **voltage minimum (1),** or **voltage minimum (2)** for a given time interval.

voltage multiplier 1. A component or circuit whose voltage output is a multiple of its input voltage. For example, a rectifier circuit which delivers a DC output voltage which is a multiple of the peak of its input AC voltage. Also called **multiplier (4). 2.** Same as **voltage-range multiplier (2).**

voltage node For a medium having standing waves, such as a transmission line or antenna, a point at which there is a minimum of voltage, or zero voltage.

voltage peak Also called **voltage maximum, voltage peak value,** or **voltage crest. 1.** The maximum value of a voltage. **2.** The maximum value of the displacement from a reference position, such as zero, for a voltage. **3.** The **voltage peak** (1), or **voltage peak** (2) for a given time interval.

voltage peak value Same as **voltage peak.**

voltage quadrupler A rectifier circuit whose output DC voltage is about four times the peak value of its input AC voltage. Also called **quadrupler** (2).

voltage quintupler A rectifier circuit whose output DC voltage is about five times the peak value of its input AC voltage. Also called **quintupler** (2).

voltage-range multiplier 1. A component or circuit which extends the range of an instrument. For example, a **voltage-range multiplier** (2). **2.** A precision resistor utilized in series, to extend the voltage range of an instrument such as a voltmeter. Also called **voltage multiplier** (2), **instrument multiplier** (2), or **multiplier** (5).

voltage rating Also called **working voltage. 1.** The maximum voltage that can be continuously applied to a conductor, material, or device, without compromising safety or reliability. **2.** The maximum voltage that a component, circuit, device, piece of equipment, or system can continuously produce without compromising safety or reliability.

voltage ratio Same as **voltage amplification** (1).

voltage reference A source of voltage whose steady value serves as a basis for comparison or operation. An example is a Weston standard cell. Also called **voltage standard,** or **voltage reference standard.**

voltage reference diode A diode whose steady output voltage value serves as a **voltage reference.**

voltage reference standard Same as **voltage reference.**

voltage-regulated power supply A power supply whose output voltage is essentially constant, despite variations in variables such as load resistance, line voltage, and temperature, so long as they are within a prescribed range. Also called **voltage-regulated supply, constant-voltage source,** or **constant-voltage power supply.**

voltage-regulated supply Same as **voltage-regulated power supply.**

voltage-regulating transformer A transformer which maintains the voltage of the output essentially constant, despite variations in variables such as load resistance, line voltage, and temperature, so long as they are within a prescribed range. Also called **voltage-regulator transformer,** or **regulating transformer.**

voltage regulation The maintenance of the voltage through a circuit essentially constant, utilizing a **voltage regulator.** Also called **voltage stabilization.**

voltage regulator A component, circuit, or device which maintains an output voltage within specified values, despite variations in variables such as load resistance, line voltage, or temperature, so long as they are within a prescribed range. Also called **voltage corrector, voltage stabilizer,** or **automatic voltage regulator.**

voltage-regulator diode A **voltage regulator** which incorporates a diode, such as a Zener diode.

voltage-regulator transformer Same as **voltage-regulating transformer.**

voltage relay A relay which is actuated at a specific voltage value, as opposed to a given current or power value.

voltage saturation In an electron tube, the condition in which the anode current can not be further increased, regardless of any additional voltage applied to it, since essentially all available electrons are already being drawn to said anode. Also called **plate saturation, current saturation, anode saturation,** or **saturation** (3).

voltage sensitivity 1. The minimum change in an input voltage that produces a change in the output of a component, circuit, device, piece of equipment, or system. **2.** The minimum change in a voltage that produces an observable change in the indication or output of a measuring instrument. **3.** The minimum voltage necessary to initiate a process, or to maintain proper operation.

voltage source Same as **voltage supply.**

voltage stabilization Same as **voltage regulation.**

voltage stabilizer Same as **voltage regulator.**

voltage standard Same as **voltage reference.**

voltage standing-wave ratio For a given transmission line, such as a coaxial cable or waveguide, the ratio of the maximum voltage to the minimum voltage. It is a measure of the impedance matching of the line. This ratio is equal to 1 when there is complete impedance matching, in which case the maximum possible RF power that reaches the load, such as an antenna. Its abbreviation is **VSWR.** Also called **standing-wave ratio** (1).

voltage supply A supply that provides a voltage to a component, circuit, device, piece of equipment, or system. Also called **voltage source.**

voltage surge A sudden and momentary increase in voltage. May be caused, for instance, by lightning, or faults in circuits. If protective measures are not employed, such a surge may bring about a failure or significant damage. Also called **surge voltage,** or **transient voltage.**

voltage-to-frequency converter A circuit or device whose output frequency is proportional to its input voltage. Used, for instance, for analog-to-digital conversions. This contrasts with a **frequency-to-voltage converter,** which converts a frequency input into a voltage output. Its abbreviation is **V/F converter.**

voltage to ground The potential difference between a given point in a circuit and a ground.

voltage transformer A transformer utilized to provide an output voltage different than the input voltage. It is a step-down transformer when the output voltage is lower than the input voltage, or a step-up transformer when the output is higher than the input voltage. Used, for instance, as an instrument transformer. Also called **potential transformer.**

voltage tripler A rectifier circuit whose output DC voltage is about three times the peak value of its input AC voltage. Also called **tripler** (3).

voltage tuning Tuning effected by varying a voltage, as occurs, for instance, in a voltage-controlled crystal oscillator.

voltage-variable capacitor Same as **varactor.**

voltaic Pertaining to the generation or flow of electricity, especially DC, as a result of chemical action. Also called **galvanic.**

voltaic battery A battery composed of **voltaic cells.**

voltaic cell A primary cell which converts chemical energy into electrical energy, and which consists of two electrodes immersed in an electrolyte. Each electrode is of a different metal, and DC is generated. It is a type of electrolytic cell. Also called **galvanic cell.**

voltaic couple Within a voltaic cell, two dissimilar conductors which generate a potential difference when immersed in the same electrolyte. Such conductors are usually metals, such as silver and zinc. Also called **galvanic couple.**

voltaic pile An early battery consisting of a series of alternated disks of dissimilar metals, usually zinc and copper, each separated by paper or cloth soaked in an electrolyte. Also called **Volta pile, galvanic pile,** or **pile** (1).

voltameter An instrument which measures electric charge, and expresses it in coulombs. It may consist, for instance, of an electrolytic cell in which the mass of a given substance liberated from a solution is correlated to coulombs of flowing current. Also called **coulometer, coulomb meter,** or **coulombmeter.**

voltammeter An instrument which measures and indicates voltages and/or currents in an electric circuit. Also spelled **volt-ammeter.** Also called **voltmeter-ammeter (1).**

voltampere Same as **volt-ampere.**

voltmeter Abbreviation of **volt**age **meter.** An instrument which measures and indicates **volt**ages. Examples include peak voltmeters, differential voltmeters, high-impedance voltmeters, electrostatic voltmeters, FET voltmeters, and picovoltmeters.

voltmeter-ammeter 1. Same as **voltammeter. 2.** A voltammeter which provides separate leads for each function.

voltmeter sensitivity The ratio of the total resistance of a voltmeter to its full-scale deflection, expressed in ohms per volt.

volts AC A specific value of an AC voltage, such as an AC line voltage. Its abbreviation is **VAC,** or **Vac.**

volts DC A specific value of a DC voltage, such as that of a battery. Its abbreviation is **VDC,** or **Vdc.**

volts per meter A unit of electric field strength. Its abbreviation is **V/m.**

volts RMS A specific RMS voltage, such as an RMS voltage rating. Its abbreviation is **VRMS.**

volume Its abbreviation is **vol,** and its symbol is **V. 1.** The amplitude of a sound, as perceived by a listener. It is a subjective measure of sound intensity, and as such varies from person to person. Volume is influenced by the absolute amplitude of the sound, its frequency, duration, and to a lesser extent other factors. Also called **loudness (1). 2.** The space a three dimensional object, entity, or region occupies. Usually expressed in cubic units, such as liters. **3.** A data storage unit such as file, a contiguous area of a disk, an area encompassing multiple disks, an optical disc, a tape cartridge, or a flash card. Also, the data contained in such a unit.

volume charge density The electric charge per unit volume. Usually expressed in coulombs per cubic meter. Also called **charge density (2).** Its symbol is ρ.

volume compression The limiting of the volume range of an audio-frequency signal. This may by accomplished, for instance, by using a volume compressor.

volume compressor A circuit that automatically limits the volume range of an audio signal. This is utilized, for example, by a radio transmitter to increase the average amount of modulation while avoiding overmodulation. Also called **automatic volume compressor.**

volume conductivity The reciprocal of **volume resistivity.**

volume control A control, such as that of a TV or stereo amplifier, which varies **volume (1).** Also called **loudness control.**

volume element Same as **voxel.**

volume expander A circuit or device which automatically increases the volume range of an audio-frequency signal. It works, essentially, to reverse the effect of volume compression, thus restoring the volume range of the original program. Also called **automatic volume expander,** or **expander (2).**

volume expansion The increasing of the volume range of an audio-frequency signal. This may by accomplished, for instance, by using a volume expander.

volume indicator Same as **VU meter.**

volume label The label or name assigned to a hard disk, DVD, floppy, data cartridge, or the like. Also called **volume name.**

volume-level meter Same as **VU meter.**

volume limiter A circuit or device which limits an output volume to a predetermined maximum, regardless of the variations in the input signal. Used, for example, for preventing component, equipment, or media overloads. An **automatic volume control** maintains the output volume at a constant level, despite variations in the input signal.

volume limiting The limiting of an output volume to a given value, through the use of a **volume limiter.**

volume magnetostriction The volume change of a magnetostrictive material when placed in a magnetic field.

volume meter Same as **VU meter.**

volume name Same as **volume label.**

volume pixel Same as **voxel.**

volume range 1. For a component, circuit, device, piece of equipment, system, or signal, the difference between the maximum and minimum volume levels. Usually expressed in decibels. **2.** A **volume range (1)** which can be safely handled, or which provides adequate performance. **3.** A **volume range (1)** or **volume range (2)** for a given time interval.

volume recombination rate In a semiconductor, the rate at which recombination occurs within the overall body. The **surface recombination rate** is that which occurs at the surface of a semiconductor material.

volume reference number Same as **volume serial number.**

volume resistance The ratio of the voltage applied between two electrodes placed within a given material, to the current flowing between said electrodes.

volume resistivity A measure of the inherent ability of a material to resist the flow of current, per unit volume. It is usually calculated as the resistance between opposite faces of a one centimeter cube, and expressed in ohm-centimeters. It is the reciprocal of **volume conductivity.** Also called **specific insulation resistance.**

volume serial number A unique number assigned to a hard disk, DVD, floppy, data cartridge, or the like. Such a number is not user selectable. Also called **volume reference number.**

Volume Table of Contents A file listing the contents of a disk. Its abbreviation is **VTOC.**

volume unit Its abbreviation is **VU. 1.** A unit of measurement of the power of a complex audio signal, and whose values are usually indicated by a VU meter. Volume units are usually specified relative to a reference power, such as 1 milliwatt or 4 dBm. **2.** A **volume unit (1)** utilized to measure a fluctuating AC or a complex electric wave.

volume-unit indicator Same as **VU meter.**

volume-unit meter Same as **VU meter.**

VOM 1. Abbreviation of **volt-ohmmeter 2.** Abbreviation of **volt-ohm-milliammeter.**

von Neumann architecture A computer architecture in which instructions and data addresses are stored in the same regions and are accessed through a single bus. Utilizing this architecture, instructions and data are analyzed and processed sequentially. This contrasts with **Harvard architecture,** where instructions and data addresses are stored in different regions and are accessed through separate buses.

VoP Abbreviation of **Voice over Packet.**

VOR Abbreviation of VHF **o**mnidirectional **r**ange, VHF omnirange, or VHF omnidirectional radio range. A ground-based radio navigation system in which each of its multiple stations transmits VHF signals which provide bearing information relative to the transmitter. Such a transmitter emits beacon signals 360° in azimuth along the horizontal plane, oriented from magnetic north. VOR frequencies are usually assigned between 108 and 118 MHz. There are enhanced systems, such as Doppler VOR. Used extensively by pilots and aircraft. Also called **VOR system**.

VOR receiver A device, located on an aircraft, which detects **VOR signals**.

VOR signal The VHF signals emitted by a **VOR station**.

VOR station Within a **VOR system**, a transmitting station. These are located at regular intervals to provide continuous coverage. Also called **VOR transmitter (1)**.

VOR system Same as **VOR**.

VOR transmitter 1. Same as **VOR station**. 2. The apparatus within a **VOR station** which emits the VHF signals.

vortal 1. Acronym for vertical p**ortal**. Abbreviation of vertical industry p**ortal**. A Web portal which is tailored to a specific industry, business, or the like. 2. Same as **voice portal**.

VOX Abbreviation of **voice-operated transmission**, or **voice-operated transmit**. 1. In a device such as a transmitter or transceiver, a feature which activates transmission only when a voice signal above a given level is present. Used, for instance, to eliminate the need for push-to-talk operation. 2. Radio communication utilizing **VOX (1)**.

voxel Acronym for **volume** pi**xel**. A unit of graphical information which defines a point in three-dimensional space, while a **pixel (1)** defines a point in two dimensions. A voxel has x, y, and z coordinates, while pixel has those for only x and y. It is also an abbreviation of **volume element**.

VPE Abbreviation of **vapor-phase epitaxy**.

VPI Abbreviation of **Virtual Path Identifier**.

VPN Abbreviation of **virtual private network**.

VPS Abbreviation of **vapor-phase soldering**.

VR Abbreviation of **virtual reality**.

VR motor Abbreviation of **variable-reluctance motor**.

VRAM Abbreviation of **video RAM**.

VRC Same as **vertical redundancy check**.

VRML Acronym for Virtual Reality Modeling Language. A computer language utilized to create interactive graphics which have a 3D feel to them. Used, for instance, to provide a visitor to a Web site 360° views of a given landscape, complete with the ability to move in any direction, triggering the appropriate changes in viewpoint.

VRMS Abbreviation of **volts RMS**.

VRU Abbreviation of **voice response unit**.

VS Abbreviation of **virtual storage**.

VSELP Abbreviation of vector-sum excited linear predictive coding. A speech-compression algorithm similar to Code-Excited Linear Prediction.

VSWR Same as **voltage standing-wave ratio**.

vsync Abbreviation of **vertical sync**, or **vertical synchronization**.

VTOC Abbreviation of **Volume Table of Contents**.

VTR Abbreviation of **video tape recorder**.

VTVM Abbreviation of **vacuum-tube voltmeter**.

VU Abbreviation of **volume unit**.

VU meter Abbreviation of volume-unit **meter**. An instrument which indicates the power of a complex audio signal, as expressed in volume units. Such a meter is usually calibrated to show a maximum recording or reproduction level, so as to maintain distortion within certain limits. Typically seen in recording equipment, and in some sound-reproduction devices such as audio amplifiers. Also called **volume indicator**, **volume-level meter**, **volume meter**, or **volume-unit indicator**.

VUI Abbreviation of **voice user interface**.

vulcanization A process that helps rubber becomes elastic, tough, and resistant to cold, heat, abrasions, water, and many chemicals. Although there are various processes utilized for this purpose, the most common involve heating rubber with sulfur.

vulnerable That which is susceptible to malicious actions, or which is especially exposed. Examples include data, operating systems, or communications networks which are not safeguarded against losses, damage, unwanted modifications, or unauthorized access.

VxD Abbreviation of **virtual device driver**.

VXML Abbreviation of **Voice XML**, which itself is an abbreviation of **Voice** Extensible Markup Language. An application of XML which enables a user to access the Internet via voice recognition. Thus, for instance, using an ordinary telephone and a voice-driven Web browser, a user can access email, get news, check on the schedule of an event, and so on.

W

w Symbol for **weight**.

W Symbol for **work (1)**.

W **1.** Symbol for **watt**. **2.** Chemical symbol for **wolfram** or **tungsten**. **3.** Symbol for **weight**.

W Band **1.** In communications, a band of radio frequencies extending from 75.00 to 110.00 GHz, as established by the IEEE. This corresponds to wavelengths of approximately 4.0 mm to 2.7 mm, respectively. **2.** In communications, a band of radio frequencies extending from 56 to 100 GHz. This corresponds to wavelengths of approximately 5.4 mm to 3.0 mm, respectively.

W-CDMA Abbreviation of **wideband CDMA**.

W/sr Abbreviation of **watt per steradian**.

W3 consortium Abbreviation of **World Wide Web Consortium**.

W3C Abbreviation of **World Wide Web Consortium**.

WAAS Abbreviation of **Wide-Area Augmentation System**.

wafer A slice of a semiconductor material which serves as a substrate for fabrication processes such as masking and etching. Also, to create such wafers. Also called **semiconductor wafer**, or **slice (2)**.

wafer charging The unwanted acquisition of a static charge by a **wafer** during processing.

wafer diameter The diameter of a **wafer** which has been cut into a circular form. Such diameters often range from a few millimeters to a few decimeters.

wafer fabrication The processes involved in the preparation of **wafers**. This involves steps such as cutting, grinding, polishing, and cleaning, so as to provide a wafer with the desired dimensions.

wafer sawing A technique for cutting a semiconductor material into dies or chips, in which high-speed precision saws, with diamond-coated blades which can be thinner than one micrometer, are used. Scribing is usually performed prior to sawing to facilitate cutting. Also called **wafer slicing**, or **sawing**.

wafer-scale integration The use of whole uncut **wafers** as components, as opposed to fabricating a wafer, cutting apart the chips, and connecting them together again. Used, for instance, to create multiple processors that will be used together in parallel processing. Its abbreviation is **WSI**.

wafer slicing Same as **wafer sawing**.

wafer switch A rotary switch with contacts arranged along one or more levels, each known as a wafer.

WAIS Acronym for **W**ide **A**rea **I**nformation **S**erver. A UNIX-based Internet search engine which accesses multiple databases across multiple servers.

wait state A delay which is introduced before a CPU can execute an instruction, while it waits for slower memory chips or devices to respond. This occurs, for instance, when the RAM works at a lower clock speed than the CPU. A wait state may be timed in different manners, such as a given number of clock cycles. The longer a wait state, the slower the performance of the CPU.

walk-up-and-use A system, such as a kiosk in an airport or museum, which needs to be designed in a manner that allows a user to simply approach and utilize it effectively without ever having used it before.

walkie-talkie A small and portable transceiver which is battery-powered.

wall box A location or enclosure which houses electrical and/or communications switches, connectors, circuit breakers, and the like. A wall box is usually recessed into a wall, or mounted upon one.

wall jack A jack, such as that of a telephone, mounted on a wall.

wall loudspeaker Same as **wall speaker**.

wall-mounted That, such as a wall box, a wall phone, or a wall outlet, which is recessed into, or mounted upon a wall.

wall outlet A power outlet located on a wall. Also called **wall socket (1)**, or **wall receptacle**.

wall phone Abbreviation of **wall** telephone. A telephone which is hung or otherwise mounted on a wall, as opposed, for instance, to a desktop phone.

wall plug A plug, such as that of CATV, mounted on a wall.

wall receptacle Same as **wall outlet**.

wall socket **1.** Same as **wall outlet**. **2.** A socket, such as that of a telephone, mounted on a wall.

wall speaker Also called **wall loudspeaker**. **1.** A usually lightweight speaker, such as a panel speaker, which is hung upon a wall. **2.** A speaker that is recessed into a wall. **3.** A large speaker with multiple drivers, often incorporating various woofers, midranges, and tweeters, which provides a virtual wall of sound.

wall telephone Same as **wall phone**.

walled garden The restricted access provided to a user by an Internet provider which takes measures to direct traffic to specific sites or content. Used, for instance, by ISPs, or entities, such as educational institutions, that provide Internet access and wish to route users to paid content.

wallpaper A picture, pattern, or other graphic which is displayed as a desktop surface or other screen background. In addition to available choices, which are countless, a user can designate or create most any image for this purpose.

WAN Acronym for **W**ide-**A**rea **N**etwork. A computer network which encompasses a large geographical area, such as a city or country, with some WANs, such as the Internet, covering the globe. A WAN may be a single large network, or consist of multiple LANs, with connections between nodes utilizing dedicated lines, existing telephony networks, satellites, or the like.

WAN analyzer Software and/or hardware which monitors the activity of a WAN, and which troubleshoots and performs tests such as the simulation of error conditions, so as to help it work smoothly.

WAN manager Software and/or hardware which monitors and controls a WAN, including its configuration, allocation of resources, and security.

wand A handheld optical reader, such as that shaped in the form of a pen, which is utilized to scan small amounts of data, such as bar codes at points of sale.

wander **1.** To move or fluctuate in a random or otherwise irregular manner. Said, for instance, of the movement of the needle of a dial. **2.** In radars, rapid fluctuations in the apparent location of a scanned object. Such variations oscillate around the mean position of said object. Also called **scintillation (5)**, **target glint**, or **target scintillation**.

WAP Abbreviation of **wireless application protocol**.

WAP gateway Abbreviation of **wireless application protocol gateway**.

WAP portal Abbreviation of **wireless application protocol portal**.

warble tone A tone with rapid but slight fluctuations in frequency. Used, for instance, for testing, or to produce a uniform sound field with no standing waves within a reverberation chamber.

warble-tone generator A circuit or device which generates **warble tones**.

warez Pirated software which is made available through Internet channels, such as FTP sites or bulletin board systems.

warm boot The restarting of a computer without powering down. This may be accomplished for instance, by entering an appropriate key sequence. If a program failure is not resolved through a warm boot, a **hard boot** may help. Also called **warm start**, **warm restart**, or **soft boot**.

warm restart Same as **warm boot**.

warm start Same as **warm boot**.

warm swap 1. A swap, such as that of a hard drive, where the computer is powered and functioning, but in which the bus the device connects to must be inactive at the moment of exchange. It is intermediate between a cold swap and a hot swap. 2. A hot swap performed when the computer is in an inactive state, such as a sleep mode.

warm up The process of becoming fully prepared for operation. Also, to become so prepared. Said, for instance, of devices which require a given internal operational temperature.

warm-up period Same as **warm-up time**. Also spelled **warmup period**.

warm-up time The time it takes a component, circuit, device, piece of equipment, system, or the like, to be fully prepared for proper operation when powered on. Also spelled **warmup time**. Also called **warm-up period**.

warmup period Same as **warm-up time**. Also spelled **warm-up period**.

warmup time Same as **warm-up time**.

warning An indication, such as a light or a sound, which signals an impending danger or other undesirable condition or state. Unless certain precautions are taken, there may be damage to a component, device, piece of equipment, or system, personnel may be harmed, and/or the surrounding environment may be endangered.

warning alarm An alarm, such as a flashing light or a siren, which provides **warning signals**.

warning beacon A beacon which informs of the proximity to a certain object or region.

warning bell A bell, such as that which is part of a smoke alarm, which provides **warning signals**.

warning buzzer A buzzer which provides **warning signals**.

warning device A device, such as a light or bell, which provides **warning signals**.

warning label A label placed in a visible location to inform of precautions that must be taken, so as to avoid serious damage to a component, device, piece of equipment, or system, or harm to personnel. Warning labels may be utilized, for instance, to advise of high-voltage installations.

warning lamp Same as **warning light**.

warning light A light, such as that on a control panel, which provides **warning signals**. Also called **warning lamp**.

warning message A message, such as that appearing on a computer display, which provides a **warning**.

warning signal A signal, such as flashing light or siren, which provides a **warning**.

warning sound A sound, such as a bell or siren, serves as a **warning signal**.

warning system A system, such as that incorporating a smoke detector and a smoke alarm, which provides **warning signals**.

warp 1. To become bent or twisted out of shape. Also, to be in such a state. 2. To deviate from a proper or desired course, or to otherwise affect in a manner that perturbs such a course. Also, such a deviation. Said, for instance, of a distortion in the amplitude values of an otherwise symmetrical wave.

washer A flat disk or ring, such as that made of rubber, plastic, felt, or metal, which is placed between mechanical components or parts, to prevent leakage, increase tightness, reduce friction, distribute pressure, or the like.

water-activated battery A battery which becomes functional only after being immersed in water, or having water added.

water analogy A model which facilitates the intuitive understanding of the manner in which electricity flows. In it, comparisons are made between the flow of water in a pipe or hose, and the flow of electric current through conductors. For example, water flows from a tank which is higher than another tank, the same way current flows from a point with a higher potential to that with a lower potential. The greater the difference in height between the tanks, the greater the flow, similar to the manner in which a greater potential difference produces an increased current flow. The thinner the pipe, the more restricted the water flow, analogous to the manner in which a thinner conductor has more resistance, among other examples. Of course, there are many other variables to consider, such as the differences in the resistance of dissimilar conductors leading to different rates of current flow, or the fact that electricity will not flow without a complete circuit path while water can simply fall from a hose onto a puddle, and so on. Thus, the analogy is not completely accurate, but is still helpful. Also called **water model**.

water calorimeter A device that measures the RF energy absorbed by a known volume of water, by measuring the change in temperature of said water.

water-cooled That, such as a tube or machine, which has heat dissipated via **water cooling**.

water-cooled tube An electron tube which is cooled by a stream of water which circulates near the outside of the anode.

water cooling The use of surrounding or circulating water for cooling objects or areas. Such water may be in direct contact with that which wishes to be cooled, or can be contained in channels such as pipes so as not to harm that which is damaged by moisture. Water, especially that which is refrigerated, can transfer much more heat away from objects than air.

water-flow alarm An alarm which is activated when a connected circuit or device detects a given water flow, such as that exceeding a specified rate.

water-flow indicator An instrument which indicates the flow rate of water.

water-flow meter An instrument which measures and indicates the flow rate of water. Such a meter may be ultrasonic, electronic, and so on.

water jacket A jacket, such as that of a cable, through which water flows so as to provide cooling.

water-level indicator A device, instrument, or system which indicates the level of water, such as that in a tank.

water load A waveguide termination consisting of an enclosure with stirred or flowing water. Used, for instance, as a test load.

water model Same as **water analogy**.

water-pipe ground A ground connection made by physically attaching a conductor to a water pipe. The cold water pipe is usually utilized, and it must be made of solid metal throughout, as opposed, for instance, to that having PVC sections and/or joints.

water power Electrical power generated through the conversion of energy contained in flowing or falling water. For example, a waterfall can be used to drive a water turbine coupled to a generator. Also called **hydropower, hydroelectric power,** or **hydroelectricity.**

water-pressure alarm An alarm which is activated when a connected circuit or device detects a given water pressure, such as that exceeding a specified value.

water-pressure gauge An instrument which indicates the pressure of water in an enclosure or pipe.

water-pressure meter An instrument which measures and indicates the pressure of water in an enclosure or pipe.

water rheostat A rheostat whose resistance is varied by raising, lowering, or otherwise moving electrodes within an aqueous solution.

waterfall model In product development, especially that of software, a model suggestive of a series of phases which progress sequentially without turning back, similar to manner in which the flow of a waterfall can not reverse its course as it approaches its basin. Such a model is very hard to implement effectively.

watermark Patterns of bits which are incorporated into digital intellectual property, such as voice, video, and/or text, to identify its source. The bits contain information such as the copyright holder, or its intended area of distribution. While ordinary watermarks are usually intended to be visible, digital watermarks are designed to be imperceptible. Such patterns, for instance, must be encoded into a CD or DVD without affecting sounds or images. Also called **digital watermark.**

WATS Acronym for Wide-Area Telephone Service. A long-distance service with fixed rates, and which is usually utilized by businesses. Inward WATS is configured only for receiving incoming calls, Outward WATS only allows outgoing calls to be placed, and there are services which combine both. A WATS line, for instance, may be charged a fixed monthly rate for usage up to a given number of hours, with additional fees applying after that. Also called **WATS service.**

WATS line A telephone line accessing a **WATS service.**

WATS service Same as **WATS.**

watt The SI unit of power. It is equal to the power rate of one joule of work per second. In terms of mechanical power one watt equals approximately 0.001341022 horsepower. Electrically it is equal to the power produced by a current of one ampere flowing through a potential of one volt. Its symbol is **W.**

watt current Within an AC circuit, the component of the current that is in phase with the voltage. Also known as **active current, in-phase current,** or **resistive current.**

watt-hour A unit of energy representing the power delivered at a continuous rate of one watt for an interval of one hour. This is equal to exactly 3.6×10^3 joules. Also spelled **watthour.** Its abbreviation is **Wh,** or **Whr.**

watt-hour capacity The power, expressed in watt-hours, that a storage battery can deliver under specified conditions.

watt-hour meter An instrument which measures and indicates, in watt-hours, or more commonly kilowatt-hours, electrical power consumption for a given or combined time interval. Also spelled **watthour meter.**

watt per steradian The SI unit of radiant intensity. Its abbreviation is **W/sr.**

watt-second A unit of energy equal to approximately 2.7777778×10^{-4} watt-hours, or exactly 1.0 joule. Its abbreviation is **Ws.**

wattage 1. Power, such as electrical power, expressed in watts or fractions/multiples of watts such as milliwatts or kilowatts. 2. Consumed power expressed in watts. For exam-

ple, that utilized at a given moment by an operating device, or the power consumed over a given time interval. 3. Same as **wattage rating.**

wattage rating Also called **wattage (3). 1.** A rated or named value stating the power, in watts, that a component, circuit, device, piece of equipment, or system can produce, consume, dissipate, or otherwise safely handle, when used in a given manner, or under specified conditions. For instance, the power rating of a speaker may indicate the RMS power that can be delivered to a dynamic speaker for extended periods without harming the voice coil or other components. The peak power rating would specify the maximum power level which can be briefly tolerated without harm. 2. The recommended power, expressed in watts, required for proper operation of a component, device, piece of equipment, or system.

watthour Same as **watt-hour.**

watthour meter Same as **watt-hour meter.**

wattless component In an AC circuit, the component of the current, voltage, or power which does not add power. These are, specifically, the wattless current, wattless voltage, or wattless power. Also called **idle component, reactive component,** or **quadrature component.**

wattless current The component of an alternating current which is in quadrature with the voltage. Such a component does not add power. Also called **idle current, reactive current,** or **quadrature current.**

wattless power The power in an AC circuit which cannot perform work. It is calculated by the following formula: $P = I \cdot V \cdot \sin\theta$, where P is the power in watts, I is the current in amperes, V is the voltage in volts, and $\sin\theta$ is the sine of the angular phase difference between the current and the voltage. Also called **reactive power, idle power, quadrature power, volt-amperes reactive,** or **reactive volt-amperes.**

wattless voltage The voltage component which is in quadrature with the current of an AC circuit. Such a component does not add power. Also called **reactive voltage, idle voltage,** or **quadrature voltage.**

wattmeter An instrument, such as a watt-hour meter, which measures and indicates electrical power consumption for a given or combined time interval, as expressed in watts or fractions/multiples of watts such as milliwatts or kilowatts. Its abbreviation is **WM.**

WAV A popular digital audio format. Also called **WAV format.**

WAV file A file in the **WAV format.**

WAV format Same as **WAV.**

wave A periodic disturbance which is propagated through a medium or space. Electromagnetic waves, for instance, are produced by the oscillation or acceleration of an electric charge, consist of sinusoidal electric and magnetic fields which are at right angles to each other and to the direction of motion, and propagate through a vacuum at the speed of light. There are other examples, including those caused by vibrations, such as acoustic waves. Characteristics which help describe waves include amplitude, frequency, waveform, wavelength, phase, and velocity. Waves can be classified in various manners. For instance, a wave may be longitudinal versus transverse, standing versus progressive, symmetrical versus non-symmetrical, and so on. As waves propagate, they may also undergo phenomena such as diffraction, refraction, reflection, and dispersion.

wave absorption The absorption by a surface, object, or region of all or part of the energy of a wave, or of certain frequencies.

wave amplitude For a wave, the maximum absolute value of the displacement from a reference position, such as zero.

wave analyzer An instrument which analyzes the content, such as frequency components and their amplitudes, of complex waves. Examples include spectrum, harmonic, and frequency analyzers. Also called **waveform analyzer**.

wave angle The angle at which electromagnetic radiation departs from, or is received by, an antenna. Such an angle is measured based on a reference plane, such as the horizon.

wave antenna A directional antenna consisting of one or more parallel horizontal conductors which are ½ to several wavelengths long. All conductors are suspended parallel to the ground, usually within a couple of meters of the surface. Also called **Beverage antenna**.

wave attenuation A reduction in the amplitude of a wave. For instance, a decrease in the amplitude of a wave as it propagates through a given medium.

wave beam **1.** A concentrated and essentially unidirectional stream of waves, such as a radio beam or a light beam. **2.** A **wave beam** (1) transmitted in a specific direction, such as that from an antenna or radio beacon.

wave celerity Same as **wave velocity**.

wave clutter In radars, unwanted echoes that appear on the display screen due to signal reflections from sea or ocean waves. Also called **wave return**.

wave converter A device, such as a mode transducer, which changes a given wave pattern or transmission mode into another.

wave crest Also called **wave peak**. **1.** The maximum instantaneous amplitude value of a wave. **2.** A **wave crest** (1) for a given time interval.

wave cycle One complete sequence of changes of a wave, such as those occurring during a cycle of AC.

wave direction The direction in which a wave travels. The sinusoidal electric and magnetic fields of electromagnetic waves, for instance, are each at right angles to each other, and to the direction of motion of said waves.

wave division multiplexing Same as **wavelength division multiplexing**. Its abbreviation **WDM**.

wave duct **1.** A pipe, tube, or channel through which waves are propagated. Also called **duct** (1). When referring to a pipe or tube with precise dimensions through which microwave energy is transmitted, also called **waveguide** (2). **2.** Within the troposphere, a layer which, depending on the temperature and humidity conditions, may act as a waveguide for the transmission of radio waves for extended distances. Also called or **duct** (3), **atmospheric duct**, or **tropospheric duct**. **3.** A narrow layer that forms under unusual conditions in the atmosphere or ocean, and which serves to propagate radio waves or sound waves. Also called **duct** (4).

wave envelope A curve whose points pass through the peaks of a wave, such as that showing the waveform of an amplitude-modulated carrier.

wave equation An equation which describes the behavior of a given type of wave or wave phenomena under specific conditions.

wave filter A filter, such as an electrical filter, which separates waves of different frequencies, or that which allows only certain transmission modes to pass, as does a mode filter.

wave front Same as **wavefront**.

wave function A function which describes the propagation of the wave associated with a given particle. It is a complex quantity representing the solution of the Schrödinger equation for such a particle. Multiple solutions depict the propagation of groups of particles.

wave group Two or more waves or wave trains traveling along the same path, whose frequencies, amplitudes, or the like, vary slightly.

wave heating The use of waves, such as those produced by a source of RF energy, for heating. Examples include high-frequency heating and dielectric heating.

wave impedance For a given point in an electromagnetic wave, the ratio of the electric field strength to the magnetic field strength.

wave interference **1.** The mutual effect two or more superimposed waves or vibrations have on each other. Also called **interference** (4). **2.** The variations in amplitude occurring in the wave resulting from the superimposition of waves from two or more coherent sources whose phase difference varies. As the phase difference approaches and reaches 0° there is constructive interference, resulting in an increase in amplitude, and as the phase difference approaches and reaches 180°, there is destructive interference, resulting in a reduction or cancellation in amplitude. Interference can occur in waves residing along any part of the electromagnetic spectrum. Also, the additive process by which this phenomenon occurs. Also called **interference** (3).

wave mechanics A branch of quantum theory which interprets physical phenomena based on the treatment of subatomic particles as matter waves.

wave motion Same as **wave propagation**.

wave movement Same as **wave propagation**.

wave normal A line or curve which is perpendicular to a wave front of an electromagnetic wave. A wave normal represents the propagation direction of such a wave.

wave number Same as **wavenumber**.

wave packet Multiple waves of different wavelengths which are superimposed and travel as a single disturbance.

wave-particle duality The dual nature of electromagnetic radiation, in which it has particle-like properties, as seen in the photoelectric effect, and wavelike properties, as evidenced by phenomena such as diffraction and interference. Certain particles such as electrons also exhibit this duality, as they have wavelike properties, as seen, for instance, in electron diffraction. Also called **particle-wave duality**.

wave path The path along which a wave, wave group, wave train, or the like is propagated.

wave peak Same as **wave crest**.

wave polarization A property of an electromagnetic wave which describes the time-varying direction of the electric field vector. Such polarization may be classified, for instance, as circular, elliptical, or linear, each of which describes the shaped traced by the electric field vector as a function of time. Polarization may also be defined in terms of the direction of the electric field with respect to a transmitting antenna or another reference, such as the surface of the earth, in which case it is said to have horizontal or vertical polarization. Also called **polarization** (2).

wave propagation The movement of a wave through a medium, such as the atmosphere, a vacuum, or a transmission line. For example, an electromagnetic wave traveling via a transmission line, or an acoustic wave traveling through water. Also, the process by which this occurs. Also called **wave motion**, **wave movement**, or **propagation** (2).

wave return Same as **wave clutter**.

wave shape Same as **waveform**. Also spelled **waveshape**.

wave soldering A automated method of soldering electronic components to circuit boards, in which molten solder is pumped from a reservoir through a spout to form a wave. The board is passed through the wave via an inclined conveyor. This technique minimizes the heating of the board. Also called **flow soldering**.

wave speed The speed at which a wave propagates through a given medium. It is the speed at which given points within it, such as crests and troughs, travel. In a vacuum it is equal to the product of the frequency and wavelength. In a dispersive medium the wave speed of electromagnetic waves varies as a function of the frequency, while in a non-dispersive medium the wave speed is independent of the frequency. Also called **wave celerity**, **celerity**, or **propagation speed**.

wave surface Same as **wavefront (1)**.

wave table synthesis Same as **wavetable synthesis**.

wave tail The falling portion of a waveform, as opposed to the **wavefront (2)** which is the rising portion.

wave train Multiple waves produced by the same disturbance, or a succession of waves emitted by the same source. Also spelled **wavetrain**.

wave trap A resonant circuit which is connected in series or in parallel with an antenna receiving system to help suppress unwanted signals at a given frequency. Also spelled **wavetrap**. Also called **trap (3)**.

wave trough 1. The minimum instantaneous amplitude value of a wave. 2. A **wave trough (1)** for a given time interval.

wave velocity The **wave speed** in a given direction. Also called **propagation velocity**, or **velocity of propagation**.

waveform The shape, or a graphical representation of such a shape, of a **wave**. A waveform follows the instantaneous values of the amplitude, or other periodically-varying quantity, of a wave as a function of time. Common waveforms include those of sine, square, and sawtooth waves, although waveforms may also be complex, aperiodic, and so on. Also called **waveshape**.

waveform-amplitude distortion In an amplifier, distortion that occurs when some frequencies are amplified more than others. Also known as **frequency distortion**, **amplitude distortion (1)**, or **amplitude-frequency distortion**.

waveform analyzer Same as **wave analyzer**.

waveform converter A circuit or device which converts an input with one waveform into an output with another waveform.

waveform distortion Any undesired changes in the waveform of a signal passing through a circuit, device, or transmission medium. Also, the extent of such changes. Examples include amplitude, harmonic, intermodulation, and phase distortion. Also called **distortion (2)**.

waveform generator A signal generator which can produce any of various selectable waveforms, such as those of sine, square, and sawtooth waves, over a wide range of frequencies. Also called **function generator (1)**.

waveform monitor An instrument, such as an oscilloscope, which serves to observe waveforms. Used, for instance, to monitor the quality of a broadcast signal.

waveform synthesizer A **waveform generator** capable of sophisticated waveform generation, as opposed to being limited to specific waveforms.

wavefront Also spelled **wave front**. 1. For a given wave, a surface which contains all the points having the same phase at a given instant. For an electromagnetic wave, for instance, such a surface is perpendicular to the direction of travel of the energy. Also called **wave surface**. 2. The rising portion of a waveform, as opposed to the **wave tail** which is the falling portion.

waveguide 1. A material medium whose physical boundaries confine and direct propagating electromagnetic waves. A waveguide, for instance, may be a hollow metallic conductor, a coaxial cable, a fiber-optic cable, or an atmospheric duct. Waveguides enable propagation of electromagnetic waves with very little attenuation. Also called **guide (2)**. 2. A **waveguide (1)** consisting of a hollow metal tube, and which is utilized primarily for propagating microwave energy. The cross-section of such a tube may have any of various shapes, the most common being rectangular, circular, and elliptical. The waves are propagated along the longitudinal axis. Also called **guide (3)**.

waveguide assembly A **waveguide (2)** complete with junctions, probes, plumbing, plungers, and the like.

waveguide attenuation The use and effect of a **waveguide attenuator**.

waveguide attenuator A device or object, such as a plate, which reduces the amplitude of waves traveling through a waveguide.

waveguide bend A change in the direction of the longitudinal axis of a waveguide, as seen, for instance, in an elbow bend. Also, the section which effects this change. Also called **bend (2)**.

waveguide cavity A **waveguide resonator** within a waveguide. A waveguide may have multiple such cavities.

waveguide components Components such as junctions, probes, plumbing, and plungers, which are used within or are connected to a waveguide.

waveguide connector Same as **waveguide junction**.

waveguide coupler Same as **waveguide junction**.

waveguide critical dimension In a waveguide, the dimension of the cross section, which in turn determines its cutoff frequency. Also called **critical dimension**.

waveguide cutoff For a given transmission mode in a waveguide, the frequency below which a traveling wave can not be maintained. That is, the waveguide functions efficiently only above this frequency. Also called **waveguide cutoff frequency**, **cutoff (4)**, or **cutoff frequency (3)**.

waveguide cutoff frequency Same as **waveguide cutoff**.

waveguide directional coupler Also called **directional coupler**. 1. A device which couples a primary waveguide system with a secondary waveguide system, so that the energy is transferred to the waves traveling in a specified direction, and not to those traveling in the other direction. Also called **directive feed**. 2. A waveguide device which extracts a small amount of the energy flowing in one direction, while ignoring that flowing in the other. Used, for instance, to monitor power output.

waveguide discontinuity An abrupt change in shape of a waveguide, which results in reflections. Also called **discontinuity (4)**.

waveguide elbow 1. A curved bend, usually of 90º, in a waveguide. Also called **elbow bend**, or **elbow**. 2. A waveguide connecter with a curved bend, usually of 90º.

waveguide filter In a waveguide, a filter, such as a ferrite isolator or a waveguide grating, utilized to selectively transmit or pass certain frequencies or specific types of waves.

waveguide flange A flange which serves to join one waveguide section with another.

waveguide gasket A gasket which serves to join one waveguide section with another.

waveguide grating In a waveguide, fine parallel lines which filter certain types of waves. In a circular waveguide, for instance, radial wires serve to block transverse electric waves. Also called **grating (1)**.

waveguide iris A diaphragm in a waveguide which introduces an inductance or capacitance. Such a diaphragm may consist of a conductive plate with an aperture, and serves for impedance matching. Also called **waveguide window**, or **iris (1)**.

waveguide junction A fitting which serves to join one waveguide section with another. Also called **waveguide coupler**, **waveguide connector**, or **junction (5)**.

waveguide mode For a guided electromagnetic wave, one of the possible manners of propagation. Such waves can propagate in one of three principal modes, which are as transverse electric waves, transverse magnetic waves, or transverse electromagnetic waves. Also called **propagation mode**.

waveguide phase shifter A device utilized to vary the phase relationship between the input signal and output signals of a waveguide.

waveguide plumbing The sections and devices, such as elbows and joints, utilized to channel energy in a waveguide. Also, the installation of such a system. Also called **plumbing (2)**.

waveguide plunger In a waveguide, a sliding cylinder, usually consisting of a metal or dielectric, which is utilized for tuning. Also called **piston (1)**, **plunger**, or **tuning piston**.

waveguide port An opening in a waveguide through which energy can be fed to, or taken from. For example, an opening utilized to take measurements. Also called **port (7)**.

waveguide post A post placed in a waveguide to introduce a reactance, susceptance, or the like. An example is a capacitive post. Also called **post (3)**.

waveguide probe A rod, pin, or wire which is inserted into a waveguide or cavity resonator, allowing energy to be transferred in or out. Also called **probe (2)**.

waveguide propagation 1. The movement of a wave through a waveguide. 2. The use of a waveguide for the propagation of radio waves.

waveguide radiator An opening in a waveguide through which energy is radiated into a reflector or into space.

waveguide resonator An enclosure which is able to maintain an oscillating electromagnetic field when suitably excited. The geometry of the cavity determines the resonant frequency, which is usually in the microwave range. Used, for instance, in klystrons, or magnetrons, or as a filter. Also known by various other names, including **cavity (2)**, **cavity resonator**, **resonant chamber**, **resonating cavity**, **microwave cavity**, **microwave resonance cavity**, and **rhumbatron**.

waveguide section A segment within a waveguide via which microwave energy is propagated. A waveguide may have one or more such sections, and these may have bends, twists, or the like.

waveguide stub In a waveguide, a stub which is usually utilized for tuning.

waveguide switch A switch that allows the positioning of a given waveguide section, so as to facilitate coupling with other such sections.

waveguide system A complete and functional installation of **waveguide plumbing**.

waveguide taper A change in the cross-section of a waveguide.

waveguide transformer A component in a waveguide which serves as an impedance transformer.

waveguide twist A waveguide section whose cross-section is rotated about its longitudinal axis.

waveguide window Same as **waveguide iris**.

wavelength For a periodic wave, such as an electromagnetic wave, the distance between any point having a given phase, and the point having the same phase in the next cycle. It is the length of a complete cycle, and is usually measured from peak to peak, or from trough to trough. In the case of electromagnetic radiation traveling through space, it is equal to the phase velocity divided by the frequency. Usually expressed in meters or fractions/multiples of meters such as millimeters or kilometers. Its symbol is λ.

wavelength constant The imaginary number component of the propagation constant. The real part is the **attenuation constant**. Usually expressed in radians per unit length. Also called **phase constant**.

wavelength division multiplexing Its abbreviation **WDM**. A data transmission technology in which multiple optical signals are multiplexed onto a single optical fiber. Each of the signals has a different wavelength, and with close spectral spacing, can increase the capacity of a single fiber to over 1 Tbps. In addition, WDM-based networks can carry different types of traffic at different speeds. For instance, ATM and

SONET simultaneously. Also called **wave division multiplexing**, or **dense wavelength division multiplexing**.

wavelength shifter 1. A component, circuit, or device whose input is at one wavelength, and whose output is at another. 2. A substance or material, such as certain dyes, which absorbs light at one wavelength and emits it at longer wavelength.

wavelet 1. A wave with short duration and amplitude. 2. A mathematic function which is utilized, for instance, for digital signal processing or image compression.

wavelet compression A lossy compression method which works on entire images as opposed to specific blocks, providing compression ratios which can exceed 300:1.

wavemeter An instrument which is utilized to measure the wavelength of a signal, especially that of a radio wave. Such an instrument may consist, for instance, of a tuned circuit incorporating a variable capacitor, and a resonance indicator. Examples include absorption, cavity, coaxial, and heterodyne wavemeters. A wavemeter can usually be utilized as a frequency meter, and vice versa, as either quantity can be obtained knowing the other, according to the following formula: $\lambda = c/f$, where λ is wavelength, c is the speed of light in a vacuum, and f is the frequency.

wavenumber For a given wave, the number of complete cycles per unit length. It is usually calculated by the following formula: $k = 1/\lambda$, where k is the wave number, and λ is the wavelength. Alternatively, it may be calculated as $k = 2\pi/\lambda$, where π is pi. Also spelled **wave number**.

waveshape Same as **waveform**. Also spelled **wave shape**.

wavetable synthesis The synthesizing of the sounds of musical instruments by storing digital information taken from recordings of the actual instruments. Such data can be mixed, combined, enhanced, and so on. Wavetable synthesis is more accurate than FM synthesis, and the higher the sampling rate, the more faithful the sound reproduction. Also spelled **wave table synthesis**.

wavetrain Same as **wave train**.

wavetrap Same as **wave trap**.

wax A smooth solid or semisolid substance featuring a low melting point and thermoplastic qualities. A wax may be natural or synthetic, and specific examples include paraffin wax, earth wax, and ceresin. Used, for instance as dielectrics, insulators, for waterproofing, or as a potting material.

wax paper A paper, such as kraft paper, which is impregnated with wax. Used, for instance, in paper capacitors.

Wb Symbol for **weber**.

Wb/m² Abbreviation of **weber per square meter**.

WBEM Abbreviation of **Web-Based Enterprise Management**.

WBN Abbreviation of **wireless broadband network**.

WCDMA Abbreviation of **wideband CDMA**.

WDM Abbreviation of **wavelength division multiplexing**, or **wave division multiplexing**.

weak coupling Also called **loose coupling**. 1. In a transformer, the arrangement of the primary and secondary windings in a manner which provides for a low degree of energy transfer. 2. In a transformer, the arrangement of the primary and secondary windings in a manner which provides minimum transfer. 3. A level of coupling which is below critical coupling. Also called **undercoupling**.

weak force The force which is responsible for certain nuclear decay processes, such as beta decay. Its force carriers are W and Z bosons. After the gravitational force, it is the weakest of the **four fundamental forces of nature**, the others being the electromagnetic force, and the strong force. Also called **weak nuclear force**.

weak nuclear force Same as **weak force**.

weak typing In programming, the allowing of certain exceptions in the enforcement of the data type of a given variable. Weak typing detects fewer errors when compiling than **strong typing**, which does not permit exceptions to data-typing rules.

wear-out failure Same as **wearout failure**.

wearable computer A computer whose components are all worn by a user. For example, a head-mounted device which incorporates a head-mounted display, a microphone and earphones for the input and output of data, and all other needed components such as the CPU, RAM, and disk drive. A wearable computer is well-suited, for instance, for the real-time analysis of data by scientists working under field conditions, rural doctors, people conducting business on the go, or to assist those with special needs, such as individuals with reduced vision or motor function. Also called **wearable PC**, or **hands-free computer**.

wearable computing The use of **wearable computers**.

wearable PC Same as **wearable computer**.

wearable personal computer Same as **wearable computer**.

wearout failure A failure of a component, device, piece of equipment, system, or material, occurring due to deterioration factors such as mechanical wear. It is a type of degradation failure. Also spelled **wear-out failure**.

weather protection Measures such as hermetic seals, special coatings, and rugged construction, which afford protection from harsh environmental conditions such as temperature extremes, heavy winds, and/or precipitation.

weather radar A radar specially designed to monitor weather conditions such as precipitation and clouds.

weather resistance The extent to which a component, device, piece of equipment, system, enclosure, or material can withstand harsh environmental conditions such as temperature extremes, heavy winds, and/or precipitation. Also, the resistance so possessed.

weather resistant Characterized by **weather resistance**.

weather satellite A satellite, such as a geostationary satellite, which monitors weather conditions such as upper atmosphere temperatures, humidity, precipitation, wind speeds, solar activity, and the like, relaying data and images to specially equipped receivers all over the world.

Web Abbreviation of **World Wide Web**.

Web accelerator An application which accelerates the retrieval of Web pages, utilizing techniques such as caching pages linked by the page being viewed at a given moment.

Web access The ability to access the Internet. Also, manner in which this is accomplished, such as utilizing the services of an Internet service provider. Also called **Internet access**.

Web address A sequence of characters which identifies a given Web page. It usually is the same as a **URL**. A bookmark may be set to easily return to any desired Web address.

Web administrator A person who is responsible for the key aspects of a Web site, such as its design, creation, supervision, management, traffic-monitoring, updating, maintenance, response to user feedback, and so on. A large organization may have multiple such individuals. Also called **Webmaster** or **Webmistress**, although these terms attempt to unnecessarily make reference to the gender of the administrator, which should be inconsequential in this context.

Web advertising Advertisements, such as banner, skyscraper, and pop-up ads, appearing over the Internet. Also, delivery of such ads. Also called **Internet advertising**, or **online advertising**.

Web Anonymizer A utility or service which allows anonymous Web surfing. Also called **Anonymizer**.

Web application Same as **Web-based application**.

Web-based application An application that is stored and maintained by a remote computer, and which is accessed or downloaded via the Web when needed. When such a program is accessed it is run on the remote computer, such as a server, and when downloaded such a program runs on the user's computer. Also called **Web application**, or **Web-based software**.

Web-based Distributed Authoring and Versioning Same as **WebDAV**.

Web-based e-mail Same as **Web-based email**.

Web-based email Email which is accessed via a Web browser, as opposed to using an email client. Web-based email offers benefits such as access to correspondence from virtually any location in the world, and can usually be utilized to download messages held on POP servers. Also called **Web-based mail**, **Web email**, **Web mail**, or **Internet email (1)**.

Web-Based Enterprise Management A set of technologies which help facilitate the management of the computing systems and networks of enterprises. Its abbreviation is **WBEM**.

Web-based mail Same as **Web-based email**.

Web-based software Same as **Web-based application**.

Web beacon Same as **Web bug**.

Web blocker Same as **Web filter**.

Web box **1.** A network computer used exclusively for access to the Internet, or designed for such use. Also called **Web computer (1)**, **Web PC (1)**, or **Internet box**. **2.** A computer or device, such as a **Web box (1)** or Internet appliance, designed primarily for accessing the Web, or which is utilized exclusively for this purpose.

Web browser A program which enables a computer user to browse, and perform other functions, such as downloading files and exchanging email, through the World Wide Web. This program locates and displays pages from the Web, and provides multimedia content. Most of the time there is interactive content on any given page. To have access to certain features, such as streaming video, specific plug-ins may be necessary, in addition to any hardware requirements. Also called **browser (2)**, or **Internet browser**.

Web bug A file, usually a GIF image, which works with Web cookies to surreptitiously gather user information such as their IP number, what Web pages they visit, what type of purchases they make, and so on. Used extensively by certain online services such as America Online. The GIF image utilized may be visible or transparent. Also called **Web beacon**.

Web cache An area of computer storage where the most recently downloaded Web pages may be temporarily placed, so that when one of these pages is returned to, it loads faster. Also called **browser cache**, **cache (2)**, or **Internet cache**.

Web cache server A server, such as a proxy server, which caches Web pages and other material downloaded from the Internet. If a page request can be satisfied by a cache server, the response time is faster, and bandwidth is economized. Pressing the browser's refresh icon usually accesses the specific location where the desired Internet resource is located. Also called **cache server**.

Web cam Same as **Webcam**.

Web camera Same as **Webcam**.

Web-centric Pertaining to applications, documents, services, or the like, which are specifically designed for, revolve around, or require the use of the Web.

Web client A client, such as a user's computer, served by a **Web server**. The term may also refer to an application, such as a browser, which is so served.

Web clipping A portion of the content of a given Web page. Used, for instance, to provide only the basic information needed when accessing the Internet via a low-bandwidth device such as certain cell phones or PDAs. Also called **clipping** (5).

Web cluster Same as **Web farm**.

Web computer 1. Same as **Web box** (1). **2.** A computer with the necessary hardware and software to connect to the Internet. Also called **Internet computer** (2).

Web conferencing The holding of conferences, or participation in conferencing, such as teleconferencing, videoconferencing, or data conferencing, via the Internet. Also called **Internet conferencing**.

Web content 1. Information which is accessed via the Internet, and which may consist of any combination of audio, video, files, data, or the like. Also called **Internet content**. **2.** The **Web content** (1) of a specific **Web page**. Also called **Web page content**. **3.** The **Web content** (1) of a specific **Web site**. Also called **Web site content**.

Web content filter Same as **Web filter**.

Web cookie A block of data prepared by a Web server which is sent to a Web browser for storage, and which remains ready to be sent back when needed. A user initially provides key information, such as that required for an online purchase, and this is stored in a cookie file. When a user returns to the Web site of this online retailer, the cookie is sent back to the server, enabling the display of Web pages that are customized to include information such as the mailing address, viewing preferences, or the content of a recent order. Also called **cookie, browser cookie**, or **Internet cookie**.

Web crawler A program which searches Web sites and organizes the located information. Also spelled **WebCrawler**. Also called **Web spider, crawler, spider** (1), or **bot** (2).

Web directory A directory which serves to index, organize, and simplify access to Web pages or sites.

Web e-mail Same as **Web-based email**.

Web email Same as **Web-based email**.

Web-enabled Hardware or software which is designed to be utilized in conjunction with the Web. For example, a cell phone which can retrieve email, or an application that prepares HTML documents.

Web farm Two or more Web servers which are housed at a single location. Used, for instance, to balance the overall load of a network, or for backup when certain pages have heavy traffic. Also called **Web cluster**, or **Web server farm**.

Web filter A program or utility which seeks to detect advertising and other bothersome or undesirable content before its loaded onto a Web page being accessed. Such a program, for instance, can filter Web page content, protect privacy, prevent pop-up ads from appearing, avert banner ads, eliminate certain JavaScript, stop animated GIFs, turn off ActiveX, disable Web bugs, and so on. Also called **Web blocker, Web content filter, content filter, Internet filter, net filter**, or **blocking software**.

Web filtering The use of **Web filter**.

Web host A business which provides server space for entities such as individuals or corporations. When an entity does not have its own servers, yet desires a presence on the Internet through one or more Web pages, a Web host provides the space. Hosting services may be fee-based, depending on the specific needs, such as the memory and bandwidth required, whether it is for commercial use, and so on. Also called **host** (2).

Web hosting The services a **Web host** provides. Also, the providing of said services. Also called **hosting** (2), or **host** (4).

Web index A Web site that has hyperlinks or hypermedia which facilitate locating information on given topics located elsewhere on the Web. Such an index may, for instance, consist of a list, directories, or may have a search engine available.

Web log Same as **Weblog**.

Web magazine Same as **Webzine**.

Web mail Same as **Web-based email**.

Web master Same as **Web administrator**. Also spelled **Webmaster**.

Web mistress Same as **Web administrator**. Also spelled **Webmistress**.

Web monitoring Also called **Internet monitoring**. **1.** Monitoring performed via the Internet, as in the use of a Webcam to observe a given area or environment. **2.** Monitoring performed on Internet usage. For example, the analysis of Internet traffic by time and/or region, or the examination of which users visit which Web sites.

Web pad Same as **Webpad**.

Web page A document within the World Wide Web. Such a document may be written, for instance, in HTML, and may require more or less space than that filling a single screen of a computer monitor at any given moment. Also called **page** (5).

Web page content Same as **Web content** (2).

Web page source The source code of a Web page, such as HTML code. Also called **Web page source code**, or **page source**.

Web page source code Same as **Web page source**.

Web payment Payment for goods and/or services via the Internet. Also called **electronic payment**, or **Internet payment**.

Web PC Abbreviation of **Web** personal computer. Also called **Internet PC**. **1.** Same as **Web box** (1). **2.** A personal computer with the necessary hardware and software to connect to the Internet.

Web PDA Abbreviation of **Web** personal digital assistant. **1.** A PDA enabled for Internet access, as most are. **2.** A PDA optimized and/or used exclusively for Internet access.

Web personal computer Same as **Web PC**.

Web personal digital assistant Same as **Web PDA**.

Web phone Abbreviation of **Web telephone**.

Web portal A Web site, such as *www.google.com*, which serves as a starting point to most any activity on the Internet. Web portals usually provide news, email, search engines, directories, chats, shopping, local interest topics, and so on. Also called **portal**, or **Internet portal**.

Web publishing The preparation of Web pages and Web sites, and their presentation via the Web.

Web radio The transmission of radio signals over the Internet. Also called **Internet radio**.

Web ring Same as **WebRing**.

Web search engine Software which examines or explores the Internet, looking through Web pages, documents, files, images, audio, multimedia, news, message boards, and the like, using selected keywords. A search engine may incorporate, for instance, a program which searches all available or applicable Web sites as selected by a proprietary algorithm, another program which creates an index or catalog of these sites, and another program which compares a given keyword search with all entries in the index and gives the user the results. A typical search can take just a fraction of second. A search engine may be limited to specific areas of interest, such as art or electronics, specific formats, such as mp3 or JPEG, and so on. Also called **search engine** (1).

Web search site A Web site, such as *www.google.com*, which provides one or more **Web search engines**. Also called **search site**.

Web server A server, such as an HTTP server, within the global system of servers which comprise the World Wide

Web. The Web pages users wish to access over the Internet are provided by Web servers. Also spelled **Webserver**.

Web server farm Same as **Web farm**.

Web site A given location within the World Wide Web. A site may have a home page which serves as the starting point for multiple pages, or may consist simply of a single Web page. Also spelled **Website**. Also called **site**.

Web site content Same as **Web content (3)**. Also spelled **Website content**.

Web site management The multiple tasks involved in the administration of a Web site, including hosting, monitoring, verification that all contained links are active, maintenance, and so on. Also spelled **Website management**. Also called **site management**.

Web site map A table, diagram, or other visual representation which summarizes the contents or pages of a Web site. A site map should provide a direct and intuitive presentation which enables a user to go directly to the desired content, when said user does not wish to navigate throughout the site. Such a map is frequently organized hierarchically, and the need for a site map increases proportionally to the size and complexity of the information offered. Also spelled **Website map**. Also called **site map**.

Web site search A Web search limited to a specific **Web site**. Also spelled **Website search**. Also called **site search**.

Web site usability The ease with which a specific **Web site** can be navigated. Also spelled **Website usability**.

Web spider Same as **Web crawler**. Also spelled **Webspider**.

Web surfing Exploring, researching, or otherwise spending time navigating from one Web site to the next. Also called **Internet surfing**, or **surfing (2)**.

Web switch A network switch that directs data across the Web.

Web telephone Its abbreviation is **Web phone**. Also called **Internet telephone**. **1.** Same as **Web telephony**. **2.** Software used for **Web telephony**.

Web telephony The use of the Internet for computer-to-computer, computer-to-telephone, or telephone-to-telephone calls. When making computer-to-computer calls, both systems must have, in addition to Internet access, a microphone, speakers or headphones, and compatible software. A computer-to-telephone call does not require the receiving party to have anything more than a phone line, while telephone-to-telephone calls use the Internet as the linking network. Also called **Web telephone (1)**, **Internet telephony**, or **net telephony**.

Web television Same as **Web TV**.

Web terminal A network terminal utilized to access the Internet. Also called **Internet terminal**.

Web TV Abbreviation of **Web television**. Also spelled **WebTV**. Also called **Internet TV**. **1.** The use of a TV for Internet access. This may be obtained utilizing Internet-ready TVs, or set-top boxes. **2.** The transmission of TV signals over the Internet.

Web wanderer Same as **Web crawler**.

Webcam Abbreviation of **Web camera**. A video camera primarily intended for the transmission of images over the Internet. Such a camera may be utilized, for instance, for videoconferencing, in chat programs with video capabilities, for sending video email, or for monitoring a given area or environment. Also spelled **Web cam**. Also called **videocam**, or **NetCam**.

WebCam Same as **Webcam**.

Webcamera Same as **Webcam**.

Webcast A specific program offered during **Webscasting**. Also, to transmit such content. Also called **Internet broadcast**, or **netcast**.

Webcasting Also called **Internet broadcasting**. **1.** The use of the Internet for transmission of programming intended for public or general reception. For example, a radio broadcaster may also provide a signal through the Internet, in addition to sending a signal which is received by antennas. **2.** The use of the Internet for transmission of programming intended for individuals and/or entities which have paid a fee or that are otherwise is entitled to such content.

WebCrawler Same as **Web Crawler**.

WebDAV Abbreviation of **Web**-based **D**istributed **A**uthoring and **V**ersioning. An IETF standard which sets a series of extensions to HTTP which allow users to edit, manage files, and otherwise collaborate from remote sites.

weber The SI unit of magnetic flux. It is the flux linking a circuit consisting of a single turn of wire which produces a potential of one volt when said flux is reduced uniformly to zero over one second. One weber is equal to 10^8 maxwells. It symbol is **Wb**.

weber per square meter A unit of magnetic flux density which is the equivalent of a tesla. Its abbreviation is **Wb/m²**.

Webisode A purported show, or episode of a show, which offers content dedicated to the promotion of something for sale. It is an abbreviation of **Web** episode.

Weblog Also spelled **Web log**. Its abbreviation is **blog**. **1.** A Web site, or a part of a Web site, where one or more individuals post thoughts, comments, or the like on a given topic, and which often serves as a personal diary that is available for public viewing. The contents of such a site are usually organized chronologically. **2.** To keep a **Weblog (1)**, and/or make entries. An individual may post to a Weblog several times a day, once a day, every few days, and so on.

Webmail Same as **Web-based email**.

Webmaster Same as **Web administrator**. Also spelled **Web master**.

Webmistress Same as **Web administrator**. Also spelled **Web mistress**.

Webpad A tablet computer utilized for Internet access. Also spelled **Web pad**.

WebPAD Same as **Webpad**.

Webring Same as **WebRing**.

WebRing A collection of related Web sites, all dealing with the same, or similar topics. There are hundreds of thousands of rings, and millions of sites are linked to each other via hyperlinks located on each page of such sites. Also spelled **Webring**, or **Web ring**.

Webserver Same as **Web server**.

Website Same as **Web site**.

Website content Same as **Web content (3)**. Also spelled **Web site content**.

Website management Same as **Web site management**.

Website map Same as **Web site map**.

Website search Same as **Web site search**.

Website usability Same as **Web site usability**.

Webspider Same as **Web crawler**. Also spelled **Web spider**.

WebTV Same as **Web TV**.

Webzine Abbreviation of **Web** magazine. A magazine, or similar publication, which is distributed or otherwise made available via the World Wide Web.

wedge 1. An object or material which is thicker at one end and tapers to a thinner edge at the opposite end. Also, that which is in such a shape. **2.** A waveguide termination consisting of a tapered object, such as a block, of a dissipative material such as carbon. **3.** In a TV, a test pattern or a part of a test pattern consisting of equally-spaced lines which help ascertain resolution.

wedge bonding A type of wire bonding in which the wire is shaped into the form of a wedge. This is usually accomplished using a wedge-shaped tool. Utilized to make very small electrical connections, such as those of semiconductors.

wedge filter An optical filter in which a property, such as thickness or density, varies progressively from one end to the other.

Wehnelt cylinder In a CRT, a cylindrically-shaped electrode which focuses the electrons emitted by the cathode.

weight Its symbol is **w**, or **W**, and its abbreviation is **wt**. 1. The force a mass experiences as a result of the gravitation of a celestial body, such as the earth. The weight of an object varies depending on the gravitational force exerted upon it, while its **mass (1)** does not. 2. In statistics, a factor or coefficient which helps represent the relative importance of a given term or value.

weight density The **weight (1)** of a given body or substance per unit volume.

weighting network An electric network that adjusts its frequency response in a predetermined manner, so as to minimize noise, improve transmission characteristics, adapt to a given loudness contour, or the like.

welcome page A Web page that serves as a starting point for navigation. A Welcome page may be that which is first loaded when a Web browser accesses the Internet, but usually refers to the first or main page within a group of pages, such as those of a business or individual. Also called **home page**, or **start page**.

weld 1. To join surfaces or parts via **welding**. 2. The spot or region which has been joined in a **weld (1)**.

weldability The capacity of a material to form a strong and reliable bond via **welding**. Also, a measure of said capacity.

welder A device, apparatus, or machine which performs **welding**.

welding The process of joining surfaces or parts by applying heat and/or pressure. A flux may or may not be used. The joined parts are usually metallic, and although heat is usually utilized, cold welding relies only on pressure. Other forms include arc, high-frequency, electron beam, laser, and resistance welding.

welding current Any current which is passed through given points during **welding**, as in resistance welding.

welding electrode An electrode through which a current or discharge is passed during **welding**, as in arc welding.

welding pressure Any pressure applied during **welding**, as in cold welding.

welding transformer A high-current, low-voltage transformer which is specially designed to supply electric current during **welding**.

well-behaved Said of a program which functions as expected within a given setting, or of that which adheres closely to a given standard. For example, an application that interacts as it should with given operating system, as opposed to bypassing it for certain operations. This contrasts with **ill-behaved**, in which the converse is true.

well-known port A port number which is widely utilized for a given purpose. Port 80, for instance, is commonly used over the Internet for HTTP traffic.

Weston cell Same as **Weston standard cell**.

Weston standard cell A standard cell used as a reference voltage source, in which the positive electrode is mercury, the negative electrode is an amalgam of cadmium and mercury, and the electrolyte is a solution of cadmium sulfate. It has a voltage of approximately 1.0186 at 20 °C. Also called **Weston cell**, or **cadmium cell**.

wet battery A battery composed of **wet cells**.

wet cell A primary cell in which the electrolyte is in liquid form. This contrasts with a **dry cell**, whose electrolyte is immobilized.

wet electrolytic capacitor An electrolytic capacitor whose electrolyte is in the form of a liquid. This contrasts with a **dry electrolytic capacitor**, whose electrolyte is a paste.

wet etching Etching in which the material to be removed is dissolved through immersion in an appropriate chemical solution. Once the etch is completed, the substrate is rinsed to remove any residual etchant which might otherwise continue etching or become a contaminant. Wet etching tends to be simpler and less precise than **dry etching**.

wetting 1. The coating of a surface, such as a contact, with a smooth adherent liquid film. 2. The coating of a surface to be soldered with a smooth adherent film of molten solder. This helps ensure proper heating, the minimization of the transfer of impurities, and the obtaining of a sound joint. Also called **solder wetting**.

wetware In the context of computers, a jocular term referring to the brain as a unit which works alongside hardware and software.

Wh Abbreviation of **watt-hour**.

What You See Is What You Get Same as **WYSIWYG**.

Wheatstone bridge A four-arm bridge circuit with a resistor in each arm, and which is used to measure unknown resistances. A Carey-Foster bridge, for instance is a type of Wheatstone bridge utilized to measure resistances which are nearly identical.

wheel mouse A mouse equipped with a scroll wheel. Also called **scroll mouse**.

wheel printer A printer that uses a spinning disk, with a given number of raised characters, which is spun so that a hammer may strike each character to be printed. Such printers are now obsolete. Also called **daisy-wheel printer**.

whip antenna An antenna consisting single element, in the form of a rod, which is usually flexible. A whip antenna may be telescoping, and is used, for instance, for portable communications devices such as transceivers, radio receivers, or cell phones.

Whirlwind An early digital computer utilizing magnetic core memory.

whisker 1. A single-crystal filament made of a metal, such as cobalt, or a refractory material, such as alumina. Such crystals feature very high tensile strength and temperature resistance. Also called **crystal whisker**. 2. A thin and flexible wire used to make electric contact. Used especially in semiconductors. Also called **contact wire (1)**. 3. In robotics, a contact sensor consisting of one or more thin wires protruding from an end-effector.

whistler A form of atmospheric interference, usually caused by lightning, which when heard by the appropriate radio receivers produces a sound whose pitch descends over one to several seconds.

White Book A document which details the specifications for the video CD format.

white box Also called **clear box**, or **open box**. 1. A component, device, or unit whose internal structure and function are known in great detail. Usually used in the context of its insertion into a system. 2. In computers, a hardware or software unit whose internal structure and function is known in great detail.

white-box testing Also called **structural testing**, **clear-box testing**, or **open-box testing**. 1. Any tests utilizing the assistance of a **white box (1)**. This contrasts with **black-box testing (1)**, in which the inner structure is not known, or is not necessary to be known. 2. Tests of software or hardware utilizing the assistance of a **white box (2)**. This contrasts with **black-box testing** (2), in which the inner structure is not known, or is not necessary to be known.

white compression In a TV signal, an attenuation in the gain at the levels which correspond to the light areas of the picture, so as to reduce the contrast in these areas. Also called **white saturation**.

white hat hacker A hacker which upon discovering a security vulnerability alerts the owner, administrator, or the like, of the breach. This contrasts with a **black hat hacker**, who uses such an intrusion to steal, damage, or otherwise harm. A **gray hat hacker** lies somewhere in between, and may, for instance, post security flaws on the Internet.

white level In a video signal, the amplitude level that corresponds to the maximum permitted white peaks, that is, to the screen being white. Also called **reference white level**.

white light Light which is perceived to be similar to daylight. Such light has a balanced mixture of the various colors across the visible spectrum. Also called **white radiation** (2).

white noise Noise whose sound energy or intensity per hertz is the same throughout a given frequency interval, such as the audible spectrum. The frequency response of such noise is flat, while that of **pink noise** drops off with frequency. Specifically, white noise passed through a filter with a 3 dB/octave rolloff results in pink noise. White noise is analogous to white light, as it contains equal amounts of all audible frequencies. An example is thermal noise.

white noise generator An instrument or device which generates **white noise**. Used, for instance, as a reference noise source, or for masking other noises. When used for purposes such as encryption, there should be no repetitive frequency or amplitude patterns.

white object An object that reflects and diffuses all light more or less equally across the visible spectrum.

white pages **1.** A Web-based directory service similar to the white pages of a telephone book, but which in addition may provide email addresses. Unlike regular white pages, these usually have flexible searching tools, including the ability to locate people by telephone number or address. **2.** A database of names and resources in an X.500 system. Also called **Directory Information Base**.

white paper A report, on a given topic or area of technical expertise, which is intended to be authoritative.

white radiation **1.** Electromagnetic or acoustic radiation whose energy or intensity per hertz is the same throughout a given frequency interval. An example is white noise. **2.** Same as **white light**.

white saturation Same as **white compression**.

white signal In fax communications, the amplitude of the signal produced when scanning the lightest portion of the transmitted material. Also called **picture white** (2).

white transmission A mode of fax transmission in which the maximum amplitude corresponds to the maximum lightness of the received material.

whiteboard An area on a display monitor that enables one or more users to draw, place images, text, or the like, on a screen. It is roughly the equivalent of a blackboard, and as such is used, for instance, for distance learning via the Internet.

whois An Internet directory service which enables users to find information, such as name and email address, pertaining to individuals or entities listed in a given database. One example is that utilized to obtain contact information for owners of given domain names.

Whr Abbreviation of **watt-hour**.

Wi-Fi Abbreviation of **Wireless Fidelity**. A device or network which is compatible with, or adheres to, IEEE 802.11 wireless standards and specifications. Utilized, for instance,

to help ensure compatibility between products from different manufacturers utilized in the same wireless network.

wicking The upward flow of a molten solder under the jacket of a wire.

Wide-Area Augmentation System A system utilized to improve the accuracy of GPS signals. Its abbreviation is **WAAS**.

Wide Area Information Server Same as **WAIS**.

Wide-Area Network Same as **WAN**.

Wide-Area Telephone Service Same as **WATS**.

wide band Same as **wideband**.

wide-band access Same as **wideband access**.

wide-band amplifier Same as **wideband amplifier**.

wide-band antenna Same as **wideband antenna**.

wide-band CDMA Same as **wideband CDMA**.

wide-band channel Same as **wideband channel**.

wide-band code-division multiple access Same as **wideband CDMA**.

wide-band communications Same as **wideband communications**.

wide-band connection Same as **wideband connection**.

wide-band electrical noise Same as **wideband electrical noise**.

wide-band FM Same as **wideband frequency modulation**. Also spelled **wideband FM**.

wide-band frequency modulation Same as **wideband frequency modulation**.

wide-band interference Same as **wideband interference**.

wide-band ISDN Same as **wideband ISDN**.

wide-band klystron Same as **wideband klystron**.

wide-band modem Same as **wideband modem**.

wide-band network Same as **wideband network**.

wide-band noise Same as **wideband noise**.

wide-band radar Same as **wideband radar**.

wide-band signal Same as **wideband signal**.

wide-band transmission Same as **wideband transmission**.

wide-format printer A printer, such as an inkjet printer, which can accommodate paper, or another appropriate medium, which is extra wide. Used, for instance, to print CAD drawings.

Wide SCSI A version of SCSI, such as Wide Ultra2 SCSI or Wide Ultra3 SCSI, which utilizes a 16-bit bus.

Wide Ultra SCSI A SCSI interface which utilizes a 16-bit bus and supports data rates up to 40 MBps. It uses a 68-pin cable.

Wide Ultra2 SCSI A SCSI interface which utilizes a 16-bit bus and supports data rates up to 80 MBps. It uses a 68-pin cable.

Wide Ultra3 SCSI A SCSI interface which utilizes a 16-bit bus and supports data rates up to 160 MBps. It uses a 68-pin cable. Also known as **Ultra3 SCSI**, or **Ultra160**.

wideband Also spelled **wide band**. Also called **broadband**. **1.** Operating at, or encompassing a wide range of frequencies. **2.** Operating at, or encompassing a wider range of frequencies than is ordinarily available. **3.** Communications in which multiple messages or channels are carried simultaneously over the same transmission medium. Cable TV, for instance, is broadband, as multiple channels are carried concurrently over a single coaxial cable. **4.** Communications at a speed of transmission higher than a determined amount. For example, that exceeding 1.544 Mbps. **5.** Communications at a speeds of transmission within two determined amounts. For example, greater than 64 kilobits per second, but no higher than 1.544 Mbps.

wideband access Access to a network, especially the Internet, via a wideband connection. Also spelled **wide-band access**. Also called **broadband access**.

wideband amplifier An amplifier, such as a video amplifier, that operates over a wide range of frequencies with an essentially flat frequency response. Also spelled **wide-band amplifier**. Also called **broadband amplifier**.

wideband antenna An antenna, such as a log-periodic antenna, that operates well over a wide range of frequencies. Also spelled **wide-band antenna**. Also known as **broadband antenna**.

wideband CDMA Abbreviation of **wideband code-division multiple access**. A third-generation wireless technology which has a significantly wider bandwidth than ordinary CDMA, and provides much greater throughput, which can exceed 2 Mbps. It is an ITU standard, and is used, for instance, in GSM as an improvement over TDMA. Also spelled **wide-band CDMA**. Its own abbreviation is **WCDMA**.

wideband channel Also spelled **wide-band channel**. Also called **broadband channel**. **1.** A communication channel that operates over a wide range of frequencies, or over a wider range of frequencies than is ordinarily available. **2.** A communication channel that can carry multiple messages simultaneously, or more than a certain number of messages simultaneously.

wideband code-division multiple access Same as **wideband CDMA**. Also spelled **wide-band code-division multiple access**.

wideband communications Also spelled **wide-band communications**. Also called **broadband communications**. **1.** Communications utilizing a wide range of frequencies, or over a wider range of frequencies than is ordinarily available. **2.** Communications in which multiple messages can be transmitted simultaneously, or in which more than a certain number of messages can be transmitted simultaneously. **3.** Communications at a speed of transmission higher than a determined amount. For example, that exceeding 1.544 Mbps.

wideband connection A connection to a communications network at a transmission speed exceeding a given rate, such as 1.544 Mbps. Also spelled **wide-band connection**. Also called **broadband connection**.

wideband electrical noise Electrical noise present over a wide range of frequencies. White noise is an example. Also spelled **wide-band electrical noise**. Also called **broadband electrical noise**.

wideband FM Abbreviation of **wideband frequency modulation**. Also spelled **wide-band FM**.

wideband frequency modulation FM in which the frequency deviation of the carrier wave is above a given minimum. Also spelled **wide-band frequency modulation**. Its abbreviation is **wideband FM**.

wideband interference Interference occurring over a wide range of frequencies. An electric motor may be a source. Also spelled **wide-band interference**. Also called **broadband interference**.

wideband ISDN Abbreviation of **wideband** Integrated Services Digital Network. An advanced version on ISDN based on ATM technology. It uses fiber-optic cables, and is capable of transmission speeds exceeding 1 Gbps. Also spelled **wide-band ISDN**. Also known as **BISDN**, or **broadband ISDN**.

wideband klystron A klystron with a wider range of frequencies than ordinarily available. Also spelled **wide-band klystron**. Also called **broadband klystron**.

wideband modem A modem for use on a **wideband network**. Also spelled **wide-band modem**. Also known as **broad-band modem**.

wideband network A LAN utilizing multiple transmission channels, each with its own carrier frequency, so that they do not interfere with each other. Such a network can simultaneously transmit data, voice, and video at very high speeds. Also spelled **wide-band network**. Also called **broadband network**.

wideband noise Noise present over a wide range of frequencies. May refer to electrical noise, thermal noise, radio noise, and so on. Also spelled **wide-band noise**. Also called **broadband noise**.

wideband radar A radar, such as an impulse radar, which utilizes a wide range of frequencies. Also spelled **wideband radar**.

wideband receiver A radio receiver that operates over a wide range of frequencies. Such a receiver, for instance, may be able to tune any frequency from a few kilohertz through several gigahertz.

wideband signal A signal encompassing a wide range of frequencies. For instance, such a signal sent over a broadband network. Also spelled **wide-band signal**. Also called **broadband signal**.

wideband transmission A transmission encompassing a wide range of frequencies. For instance, a transmission that simultaneously includes data, voice, and video. Also spelled **wide-band transmission**. Also called **broadband transmission**.

width **1.** The horizontal distance from a point or surface, such as an end, to another given point or surface, such as the other end. For example, the width of a displayed TV picture. **2.** For a given pulse, the time interval that elapses between the end of the rise time and the start of the decay time. The width may also be defined in other manners, such as the time interval between the 50% points of peak pulse amplitude. Also called **pulse duration** (1).

width control In a CRT, such as that of a TV or oscilloscope, a control which adjusts the width of that being displayed.

Wiedemann effect The twisting of a wire made of a magnetostrictive material which is placed within a longitudinal magnetic field, when a current flows through said wire.

Wiedemann-Franz Law A law stating that the ratio of the thermal conductivity to the electrical conductivity is proportional to the temperature of a metal. This relationship is a result of both forms of conductivity involving the free electrons in the metal, with increases in temperature raising thermal conductivity while decreasing electrical conductivity, and vice versa.

Wien bridge A four arm AC bridge utilized to measure unknown frequencies, capacitances, or inductances. One pair of adjacent arms of a Wien bridge has a resistor in each branch, while the other pair of adjacent arms has resistance-capacitance combinations in each branch, with the balancing of such a bridge being frequency-dependent. Also called **Wien bridge circuit**.

Wien bridge circuit Same as **Wien bridge**.

Wien bridge oscillator An oscillator whose frequency is determined by a **Wien bridge**.

Wien's displacement law A law which states that for a blackbody, the wavelength of maximum radiation is inversely proportional to its absolute temperature. Also called **Wien's law**.

Wien's law Same as **Wien's displacement law**.

WiFi Same as **Wi-Fi**.

Wikipedia A popular online encyclopedia which can be accessed free of charge.

wild card Same as **wildcard**.

wild card character Same as **wildcard**. Also spelled **wild-card character**.

wildcard A character or symbol, such as *, which can be utilized to represent any character, group of characters, value, or the like. For instance, using *quant** in a text search may provide results for multiple words starting with **quant-**, such as quantum, quantity, quantic, quantitative, and so on. A **?** is usually utilized to represent a single wild card character, as opposed to a string of them. Also spelled **wild card**. Also called **wildcard character**.

wildcard character Same as **wildcard**. Also spelled **wild card character**.

willemite A mineral whose chemical formula is Zn_2SiO_4, and whose color may be white, yellow, green, blue, brown, or red. It usually exhibits fluorescence under UV light, and may also show phosphorescence. Used, for instance, as a phosphor, and as a source of zinc. Also called **zinc silicate**.

Williams tube A CRT storage tube which held information on its screen in the form of electric charges drawn or written on its surface. It was used as an early data-storage unit.

Wilson cloud chamber A chamber which provides the proper pressure and humidity conditions to expose the tracks left by particles ionized by radiation. Such particles are made visible by a cloud of minute water droplets that temporarily surrounds said particles.

Wilson electroscope An electroscope which has a single hanging gold leaf.

Wimshurst machine A rotating machine consisting of two glass disks which rotate in opposite directions, and which can produce high voltages of static electricity. Each disk has sectors of tin foil, along with brushes to collect the charge. Used, for instance, to charge Leyden jars, or for demonstrations of discharges across a gap.

wind-driven generator A generator that utilizes a windmill, or similar apparatus, to convert mechanical energy produced by a blowing wind into electrical energy.

wind gage Same as **wind gauge**.

wind gauge An instrument for measuring and/or indicating wind and air flow speeds. Also spelled **wind gage**. Also called **anemometer**.

wind load Also spelled **windload**. Also called **wind loading**. **1.** A measure of the forces, such as torsion or bending, an antenna or similar structure is subjected to due to winds. **2.** A measure of the extent to which an antenna or similar structure can be subjected to a **wind load (1)** without damage or creating a safety hazard.

wind loading Same as **wind load**.

wind screen Same as **windscreen**.

wind shield Same as **windshield**.

winding **1.** One or more turns of one or more wires which form a continuous coil and which is utilized in devices and machines such as a transformers, generators, or motors. **2.** One complete turn within a **winding (1)**. **3.** Something wound around a core or other object.

winding wire A wire that is suitable for use in coils or windings, such as those of electromagnets or transformers. Such a wire is often composed of copper or aluminum, and is electrically insulated. Also called **magnet wire**.

windload Same as **wind load**.

Windom antenna A multiband antenna with a long horizontal radiator which is fed off-center by a vertical feed line which may also serve as a radiator.

window **1.** A time interval during which an activity may take place. Also, a time interval during which an activity must take place. **2.** An interval of radio frequencies which pass through or are propagated via the atmosphere. For example, a radio window. Also, an interval of frequencies which is transmitted through a given object or medium. **3.** A transparent section of a material, such as glass or plastic, through which the inner portion of a device or apparatus can be viewed. For example, such a window on certain radiation detectors. **4.** An area on a computer screen with defined boundaries, and within which information is displayed. Windows can be resized, maximized, minimized, placed side-by-side, overlaid, and so on. Each window can be that of a separate program, while a single program may have any number of windows open at a given moment. When two or more windows are open, only one is active, and the rest are inactive. Also called **application window**. **5.** A hole or opening between two cavities or waveguides, the bars of a transformer, or the like. **6.** A partial view of something, such as that of a dial or of a file. The viewed section can be varied, so as to see different portions at different moments. **7.** A material, such as foil, intended to interfere with radar operation.

window comparator A voltage comparator, such as that utilizing a voltage divider, that works within a given interval of values.

Windows A common operating system. There have been many versions, including 3.0, 95, and XP. As of Windows 95 the interface is similar to that utilized by Macintosh computers.

Windows Sockets Same as **Winsock**.

windscreen A covering placed on a microphone to minimize the sounds a wind would otherwise produce. Also spelled **wind screen**. Also called **microphone windscreen**.

windshield A protective material or enclosure which protects against any adverse effects a wind may produce. Also spelled **wind shield**.

Winsock Acronym for **Win**dows **Sock**ets. An interface between Windows and TCP/IP.

winterization Preparations made for winter weather. This may involve adding or replacing components, using a protective shield, or simply packing something up and storing it in a safe place.

WinZip A popular utility for zipping and unzipping files.

wipe To move smoothly over a surface while maintaining continuous contact, as with wiping contacts. Also called **slide**.

wiper Also called **slider**. **1.** An arm, brush, blade, strip, surface, or the like, which **wipes**. For example, a wiping contact. **2.** A movable contact that maintains continuous contact with a stationary contact.

wiping action The sliding contact made by **wiping contacts**.

wiping contacts Also called **sliding contacts**. **1.** Electrical contacts which slide or wipe past each other while maintaining smooth and continuous contact for added reliability. Such contacts may also be self-cleaning. **2.** Electrical contacts which wipe or slide past each other, resulting in a cleaning action when used. Also called **self-cleaning contacts**.

wire **1.** A single metallic strand, thread, or rod which is flexible and usually has a circular cross section. **2.** A **wire (1)** which serves as an electric conductor. Such a wire may or may not have insulation. Also called **electric wire**. **3.** One or more **wires (1)** bundled together and encased by a protective sheath. The contained conductors are insulated from each other. Also called **cable (1)**. **4.** To lay cables or connect components, devices, and equipment utilizing wires. **5.** An open telephone connection. **6.** To send a message via telegraph. Also, to send a message over a network using wires or cables. **7.** A message sent via **wire (6)**. **8.** To connect electrical components and devices directly with wires or cables, as opposed to utilizing intervening switches or radio links. Also called **hard-wire (1)**.

wire bonding **1.** In IC packaging, a method utilized to attach extremely fine wires between components or between components and the leads of a leadframe. Specific methods include ball bonding and wedge bonding. The wire utilized is usually high-purity gold. **2.** The joining of two wires or cables without soldering.

wire center A structure which houses one or more central offices.

wire communications A mode of transmission and/or reception of information, such as voice, video, data, or control signals, in which there are connecting wires or cables, as opposed to **wireless communications**.

wire control Remote control which utilizes connecting wires or cables, as opposed to the use of radio, infrared, or acoustic waves. Also called **wire remote control**.

wire drawing In wire manufacturing, the pulling of a metal through one or more dies to reduce its diameter to the desired value. Also spelled **wiredrawing**. Also called **drawing**.

wire duct A pipe, tube, or channel through which wires are run.

wire frame model Same as **wireframe model**.

wire frame modeling Same as **wireframe modeling**.

wire fuse A fuse, such as a link fuse, in which the melting conductor is a wire. This contrasts, for instance, with a strip fuse.

wire fusing current The level of current at which a given wire will melt. Also called **fusing current**.

wire gage Same as **wire gauge**.

wire gauge Also spelled **wire gage**. **1.** A scale or standard, such as Birmingham Wire Gauge, utilized for determining the diameter of wires. Also, a measurement expressed in such terms. For example, a 14-gauge wire. It is common for such scales or standards to also be utilized for determining thicknesses and diameters of tubing, sheets, rods, and the like. **2.** A device or instrument utilized to determine the gauge of a given wire. It is common for such a device or instrument to also be able to be used to determine thicknesses and diameters of tubing, sheets, rods, and the like.

wire harness A bundle of wires which is tied or otherwise attached together so as to be handled, installed, or removed as a unit.

wire leads Leads that consist of wires, as opposed, for instance, to strips, posts, or bars.

wire line A transmission line utilized for **wire communications**. Also called **wire link**.

wire link Same as **wire line**.

wire-link telemetry Telemetry in which signals are sent via wires and/or cables, as opposed to being sent over radio. Also called **hard-wire telemetry**.

wire pair A transmission line consisting of two similar conductors. For example, a two-wire circuit, or a two-wire telephone line. Also called **pair (3)**.

wire rate Same as **wire speed**.

wire recording An obsolete magnetic recording system in which audio or data is recorded upon a thin wire.

wire remote control Same as **wire control**.

wire speed In communications, the rate at which the hardware, such as switches and routers, and physical mediums, such as wires and cables, can transfer data across a network. When software, such as that performing encryption and decryption, works at the same rate as the hardware and physical mediums, it is said to run at wire speed. Also called **wire rate**.

wire splice The joining of two wires, usually by twisting their ends together, with or without soldering. Also, the act of so joining, and the place where such a joint occurs.

wire stripper A usually hand-held tool which cuts and removes the insulation of a wire without harming the inner conductors.

wire tap Same as **wiretap**.

wire-tapping Same as **wiretapping**.

wire telegraphy Telegraphy using connected wires, as opposed to radio waves, for transmission.

wire telephony Telephony using connected wires, as opposed to a radio link.

wire-wound resistor Same as **wirewound resistor**.

wire wrap The connection of a wire by wrapping several turns around a post, lug, pin, or terminal. This may be done by hand, or with a specialized tool. Also called **wire-wrap connection**.

wire-wrap connection Same as **wire wrap**.

wire-wrapping tool A tool specifically designed to make **wire-wrap connections**.

wiredrawing Same as **wire drawing**.

wireframe model A model illustrated using **wireframe modeling**. Also spelled **wire frame model**.

wireframe modeling Modeling, such as that utilized in CAD, in which 3D images or objects have the edges of each surface are represented by lines, as if the model were fashioned using strands of wire. Also spelled **wire frame modeling**.

wireless 1. A mode of transmission and/or reception of information, such as voice, video, data, or control signals, in which there are no connecting wires. Instead, communication is achieved by means of electromagnetic waves, such as radio-frequency waves or infrared waves, or via acoustic waves. There are many examples, including wireless telephones, wireless keyboards, and wireless networks. Also called **wireless communications**, or **wireless telecommunications**. **2.** A device, component, piece of equipment, or system which communicates in a **wireless (1)** manner.

wireless application protocol A protocol for wireless communications that enables wireless devices, such as properly equipped digital cellular telephones or PDAs, to access content from Internet, for uses such as retrieval of email, ecommerce, or researching via search engines. The Web pages displayed are usually stripped of complex graphics. Its abbreviation is **WAP**.

wireless application protocol gateway Software which performs the protocol conversions necessary to enable wireless devices, such as properly equipped digital cellular telephones or PDAs, to connect to another network, such as the Internet, which utilizes different protocols. Its abbreviation is **WAP gateway**.

wireless application protocol portal A Web portal which is accessed by wireless users and which tailors to their specific browsing needs, such as transmitting in a manner which minimizes the occupied bandwidth, and reducing the need for keyboard input. Its abbreviation is **WAP portal**. Also called **wireless portal**.

wireless bridge A device which serves to connect network nodes via radio-frequency waves, such as microwaves or infrared waves.

wireless broadband Broadband communications in which there are no connecting wires. Instead, communication is achieved by means of RF waves, such as microwaves waves or infrared waves. Also called **wireless broadband communications**, or **broadband wireless**.

wireless broadband communications Same as **wireless broadband**.

wireless broadband network A broadband network which provides high-speed communications, including data, voice, video, TV, Internet, videoconferencing, and so on, without connecting wires. Its abbreviation is **WBN**.

wireless cable Broadband cable TV, such as a Multichannel Multipoint Distribution Service, which does not utilize connecting wires. It is common for wireless cable to also offer Internet access and other data transmission services.

wireless communications Same as **wireless (1)**.

wireless connection A communications connection which is **wireless (1)**.

wireless data Information sent via **wireless data transmission**.

wireless data transmission The transmission of data in a **wireless (1)** manner.

wireless device A device which communicates in a **wireless (1)** manner.

Wireless Fidelity Same as **Wi-Fi**.

wireless headphones Headphones which do not have a cord. Such headphones receive signals from its base unit which is plugged into an audio amplifier, or similar device. Also called **cordless headphones**.

wireless intercom Also called **wireless intercom system**. **1.** An intercom system which does not utilize connecting wires, cords, or cables. **2.** An intercom system which utilizes the power-line wiring present in a given structure or location. Each station simply plugs into an available outlet.

wireless intercom system Same as **wireless intercom**.

wireless Internet Internet access via radio-frequency waves, such as microwaves. This type of access may be obtained, for instance, using a properly equipped cellular telephone or PDA. Its abbreviation is **wireless net (1)**.

wireless keyboard A computer keyboard which does not have a cord. Such a keyboard may use infrared or radio-frequency waves to communicate with the computer. Also called **cordless keyboard**.

wireless LAN Abbreviation of **wireless** local-area network. A LAN whose nodes communicate via radio-frequency waves, such as microwaves, or infrared waves. Useful, for instance, in settings where multiple nodes, such as computers, are constantly in motion. Its own abbreviation is **WLAN**. Also called **local-area wireless network**.

wireless local-area network Same as **wireless LAN**.

wireless local loop The use of RF waves, such as microwaves, for last mile connections between a telephone company's central office and the customers it serves. May be used, for instance, to provide telecommunications access to new communities without having to run wires or cables to each user. Its abbreviation is **WLL**. Also called **radio local loop**.

wireless loudspeakers Same as **wireless speakers**.

wireless markup language A markup language that is utilized to provide Internet content to mobile devices, such as properly equipped digital cellular telephones or PDAs. This format is tailored to the special needs of such devices, such as reducing the need for keyboard input. Its abbreviation is **WML**.

wireless medium A communications pathway which is not physically linked by cables or wires. For instance, the communications channels among the nodes of a wireless network.

wireless microphone A microphone which does not have a cord. It has its own power source, and transmits via infrared or radio-frequency signals. Also called **cordless microphone**.

wireless modem A modem which transmits data without utilizing connecting wires or cables. Used, for instance, in cells phones, PDAs, or wireless networks. Also called **radio modem**.

wireless mouse A computer mouse which does not have a cord. Such a mouse may use infrared or radio-frequency waves to communicate with the computer. Also called **cordless mouse**.

wireless net 1. Same as **wireless Internet**. **2.** Same as **wireless network**.

wireless network A network whose nodes communicate via radio-frequency waves, such as microwaves or infrared waves. Wireless LANs and wireless WANs are examples. Such networks feature benefits such as comparative ease of setting it up, and the removal of most mobility restrictions. Its abbreviation is **wireless net (2)**.

wireless networking The use of a **wireless network**.

wireless PDA Abbreviation of **wireless** personal digital assistant. A PDA which links to a network, peripheral, and/or another computer wirelessly.

Wireless Personal Area Network A network in which multiple devices are wirelessly interconnected when located within a given radius, such as ten meters, and which is suitable for a single user moving around employing PDAs, desktop computers, printers, and so on. Its abbreviation is **WPAN**. Also called **Personal Area Network**.

wireless personal digital assistant Same as **wireless PDA**.

wireless phone Abbreviation of **wireless** telephone. Also called **cordless phone**. A telephone which does not have a cord between the base unit and the handset, and which communicates via low-powered radio-frequency signals. In this context, although both **cordless telephones** and **cellular telephones** are wireless, the main difference between them is that the former plugs directly into a land telephone network, while the latter is linked to it via microwaves.

wireless portal Same as **wireless application protocol portal**.

wireless printer A printer that works wirelessly with properly equipped computers. Such printers and computers usually communicate via infrared signals. For best results, the communicating devices should be close to each other, and the signal should travel along an unobstructed path.

wireless reception The reception of wireless signals, such as radio-frequency waves or infrared waves, by a device such as a PDA or cellular telephone.

wireless remote control Remote control which utilizes radio, infrared, or acoustic waves, as opposed to connecting wires or cables.

wireless security system A security system whose components are not physically wired to each other, instead using radio-frequency and/or infrared signals to communicate. Such a system is ideally suited for structures that are already built, or in locations where a cabled arrangement would be impractical.

wireless service A service in which information, such as voice, video, or data, is transmitted and/or received without the use of connecting wires. Instead, signals are sent by means of electromagnetic waves such as radio-frequency waves or infrared waves. Cellular and satellite communications are two examples of such services.

wireless service provider A provider of wireless communications services, such as wireless Internet, or cellular telephony.

wireless speakers Speakers which do not have a cord. Such speakers incorporate their own amplifiers, and receive their signals from an audio amplifier, or similar device. Also called **wireless loudspeakers**.

wireless standard A standard or protocol, such as wireless application protocol, utilized for wireless communications.

wireless system A system which provides **wireless communications**. Such a system includes transmitting equipment, control mechanisms, the information-bearing signals so sent, the transmission path they follow, and receiving devices.

wireless technology Any technology, such as that utilized in cellular telephony, GPS, or wireless networks, which does not use connecting wires or cables for the exchange of information.

wireless telecommunications Same as **wireless (1)**.

wireless telegram A message sent via **wireless telegraphy**.

wireless telegraph Also called **radiotelegraph**. **1.** The transmission of **wireless telegrams**. **2.** A device utilized to send **wireless telegrams**.

wireless telegraphy Telegraphy in which radio waves are utilized, as opposed to wires. Also called **radiotelegraphy**.

wireless telephone Same as **wireless phone**.

wireless telephony The use of telephones for **wireless communications**.

wireless transmission The transmission of wireless signals, such as radio-frequency waves, by a device such as a microwave antenna.

wireless WAN Abbreviation of **wireless** wide-area network. A WAN whose nodes communicate via radio-frequency waves, such as microwaves. Such a network may span multiple countries. Its abbreviation is **WWAN**.

wireless Web Wireless access to the World Wide Web. A wireless portal, for instance, may be used for this.

wireless wide-area network Same as **wireless WAN**.

wirelessly Without the use of connecting wires, cables, or the like.

wiresonde An instrument similar to a radiosonde, except that the gathered data is transmitted via a connecting wire or cable.

wiretap Also spelled **wire tap**. **1.** A concealed device which serves to surreptitiously intercept a communication, especially a telephonic conversation. Also called **tap (4)**. **2.** To install a **wiretap (1)**, or make such a connection. Also called **tap (5)**. **3.** To monitor and/or record a communication using a **wiretap (1)**. Also called **tap (6)**.

wiretapping The use of a **wiretap (1)**. Also, the installation of a **wiretap (1)**. Also spelled **wire-tapping**.

wireway A specially designed channel through which cables and wires are run. A cable tray provides mechanical support, and protection which is tailored to specific needs. Such protection may include jackets and/or shields which safeguard against flames, high electrical noise, vibrations, crushing, and so on. Also called **cable tray**.

wirewound resistor A resistor in which the resistive element is a length of a resistance wire or strip, such as that made of nichrome, which is wound around insulating form or core. Also spelled **wire-wound resistor**.

wiring Also called **electric wiring**. **1.** The system of wires and/or conductors that connect electrical components, circuits, and devices together. For instance, the wires in a piece of electrical equipment, or the interconnections between components of an IC. **2.** The process of installing or manufacturing **wiring (1)**.

wiring board A patch panel, plugboard, printed-circuit board, or other panel or board serving for mounting and making connections between components and circuits.

wiring capacitance A capacitance which is not concentrated within a capacitor. Examples include the capacitance between the turns in a coil, or between adjacent conductors of a circuit. This contrasts with **lumped capacitance**, which is concentrated within a capacitor. Also called **self-capacitance**, **stray capacitance**, or **distributed capacitance**.

wiring closet A location or enclosure, such as a wall box or distribution frame, that houses communications wires and cables, and which provides terminals and connections.

wiring connector A device, object, or tool, such as a binding post or a wire-wrapping tool, which serves to join two or more conductors.

wiring diagram A graphical representation of the electrical elements in a circuit, and the way each is interconnected with each other. Each circuit element is represented by a symbol, while lines represent the wiring. Also called **wiring schematic**, **circuit diagram**, **schematic circuit diagram**, or **diagram (3)**.

wiring harness A group of insulated conductors which are bound together so as to facilitate the connection of their various terminals. Used, for instance, to run multiple automobile cables, each controlling individual items or systems, such as brake lights, turn indicators, trunk releases, and electrical signals enhancing braking.

wiring schematic Same as **wiring diagram**.

withstand voltage Same as **withstanding voltage**.

withstanding voltage The maximum voltage which can be applied to a dielectric without adverse effects, such as dielectric breakdown. Also called **withstand voltage**, or **dielectric withstanding voltage**.

wizard **1.** An interactive utility which provides help, such as that which may be needed during the installation or use of an application. Such a wizard usually provides guidance step-by-step. Also called **assistant**. **2.** An expert in some area of computing or networking, such as a hacker. **3.** A user with certain privileges in a given setting, such as MUDs.

WLAN Abbreviation of **wireless LAN**.

WLL Abbreviation of **wireless local loop**.

WM Abbreviation of **wattmeter**.

WML Abbreviation of **wireless markup language**.

WO Abbreviation of **write once**.

wobbulator A signal generator whose output is varied continuously between two limits. Used, for instance, to determine the frequency response of a circuit or device.

wolfram A lustrous silver-gray metal whose atomic number is 74. It is very hard and dense, ductile, has great corrosion resistance, and has the highest melting point and lowest vapor pressure of all known metals. At elevated temperatures it also has the highest tensile strength. It has almost 35 known isotopes, of which 3 are stable. Its applications include its use as bulb filaments, in electron tubes, in X-ray tubes, hard parts such as rocket nozzles, and in high-speed cutting devices and tools. Its chemical symbol is **W**. Also called **tungsten**.

Wollaston wire An extremely fine platinum wire which is used, for instance, in hot-wire instruments.

Wood's alloy Same as **Wood's metal**.

Wood's metal A silver-gray alloy whose composition is 50% bismuth, 25% lead, 12.5% tin, and 12.5% cadmium. It is used, for instance, as a solder with a low melting point. Also called **Wood's alloy**.

woofer A large speaker designed to reproduce frequencies below a given threshold, such as 1000 or 300 Hz. Depending on the design and components, a woofer may accurately reproduce frequencies below the limit of human hearing. Such a speaker unit is usually utilized with others, such as midranges and tweeters, for reproduction across the full audio spectrum. Also called **low-frequency speaker**.

Word A common word processing program.

word The fundamental unit of storage for a given computer architecture. It represents the maximum number of bits that can be held in its registers and be processed at one time. A word for computers with a 32-bit data bus is 32 bits, or 4 bytes. A word for computers with a 256-bit data bus is 256 bits, or 32 bytes, and so on. Also called **computer word**.

word processing The use of a **word processor**.

word processing application Same as **word processor**.

word processing program Same as **word processor**.

word processor A computer application utilized to produce, modify, display, and print text-based documents. Such programs usually are equipped with a heap of features and functions which simplify the preparation of documents, moving data between documents, and producing an output. These include including copying, pasting, inserting, deleting, and searching and replacing text, in addition to providing flexibility in the selection of fonts, paragraph and page layout, placement of headers, footers, and page numbers, plus spelling and grammar checking, thesaurus, file management, macros, the ability to incorporate graphics, insert tables, open multiple windows, display in WYSIWYG, and so on. Such a program may offer additional functionality, such as the handling of spreadsheets, or desktop publishing capabilities. Also called **word processing program**, or **word processing application**.

word stuffing A form of spamdexing in which the same word or words are placed dozens, hundreds, or even thousands of times on the same Web page, in order to deceive search engines into listing it at or near the top of search results through weighted relevance. Such words may be hidden from users, for instance, by employing a white text on a white background. Most search engines take measures to avoid rewarding such schemes.

WordPerfect A popular word processing program.

work 1. Its symbol is *W*. A transfer of energy from one body or system to another, which results from one body or system exerting a force which moves the other in the direction of said force. It calculated by the following formula: $W = Fd$, where *F* is force, and *d* is distance. Its SI unit it the joule, but various other units are often encountered, such as ergs, calories, kilogram-meters, or electronvolts. **2.** Same as **workpiece. 3.** To function or operate. Also, proper function or operation. **4.** To cause to function or operate.

work coil In an induction heater, an AC carrying coil which induces RF currents in the workpiece being heated. Also called **load coil (1)**.

work envelope The full range of motion available to a robot manipulator. Aside from the design of the manipulator, the workspace and position of the manipulator may restrict a work envelope.

work function The minimum energy required to remove an electron from a surface to an infinitely far point. For example, the minimum energy required by a photon to eject an electron from a given material is the photoelectric work function. Usually expressed in electronvolts. Its symbol is Φ.

workflow The automation of projects or processes, such as those in business, in which data, documents, tasks, and the like, are automatically passed from one participant to another following set procedures. Used, for instance, to help ensure that updates are properly tracked, applications utilized are coordinated, that documents and files are not lost, that projects are kept on schedule, and that they are indeed completed. Also called **workflow automation**.

workflow application An application that supports or handles a given activity, such as tracking updates and movements between participants, within **workflow automation**.

workflow automation Same as **workflow**.

workflow engine The software that implements the procedures necessary for **workflow automation** to proceed smoothly. Its responsibilities include the routing of projects at the appropriate time or stage to the appropriate individual or department.

workflow management The automatic routing of projects in **workflow automation**.

workflow management system A system utilized for defining, creating, and managing **workflow automation**.

workgroup Two or more individuals which work together on a project, and which share resources such as applications, files, and databases, and who usually collaborate via a LAN. Groupware is utilized to help in tasks such as emailing, scheduling, file transferring, application sharing, conferencing, the use of a whiteboard, and so on.

workgroup application Same as **workgroup software**.

workgroup computing The computer-assisted tasks and projects performed and worked on by **workgroups**.

workgroup software Software designed to facilitate groups of people, often in different locations, to work together on one more projects. Such software includes emailing, scheduling, file transferring, application sharing, conferencing, the use of a whiteboard, and so on. Also called **workgroup application**, **groupware**, or **teamware**.

working life 1. The time interval during which a component, circuit, device, piece of equipment, or system is calculated or expected to operate or otherwise work. **2.** The time interval during which a component, circuit, device, piece of equipment, or system actually operates or otherwise works.

working *Q* The value of the Q factor for a circuit or device with an external load. Also called **loaded *Q***.

working voltage Also called **voltage rating. 1.** The maximum voltage that can be continuously applied to a conductor, material, or device, without compromising safety or reliability. **2.** The maximum voltage that a component, circuit, device, piece of equipment, or system can continuously produce without compromising safety or reliability. **3.** The voltage required for the proper operation of a component, circuit, device, piece of equipment, or system. Also called **operating voltage**.

workpiece An object, device, or the like, which is shaped, heated, or otherwise worked on by a tool or machine during manufacturing. Said, for instance of a metal being worked on via induction heating. Also called **work (2)**.

Workplace Shell An OS/2 graphical user interface. Its abbreviation is **WPS**.

worksheet A matrix of interrelated cells which is manipulated by a spreadsheet program. Also called **spreadsheet (2)**.

workstation 1. A single-user computer which is especially powerful and fast, and which is utilized for applications such as CAD and scientific visualization. **2.** A network terminal which may or may not have its own processing power. **3.** A location or station with a terminal, where a single person works.

World Time An internationally agreed time standard based on time kept by atomic clocks, and which represents the local time at the 0° meridian, which passes through Greenwich, England. When utilizing World Time, each location on the planet has the same time, which is expressed utilizing a 24 hour clock, with a Z frequently appended. For example, 2100Z indicates 9 PM. Since the earth's rotation is gradually slowing, an extra second is added approximately once a year. It is based on International Atomic Time. Also called **Universal Coordinated Time, Universal Time Coordinated, Universal Time, Coordinated Universal Time, Zulu Time**, or **Z time**.

World Wide Web Its abbreviation is **WWW**, or **Web**. A global system of servers which support the same protocol, usually a version of HTTP, to exchange documents in specific formats, such as HTML, and which functions as the primary manner of exchanging information between the computers which comprise the **Internet**. The content of the WWW is usually accessed via a **Web browser**, and users proceed from Web page to Web page clicking on hyperlinks or icons, typing the address, accessing a bookmark, and so on. The amount of information available through the Web is extremely vast, with billions of pages already in existence, with any given page possibly having audio, pictures, streaming video, text, hypertext, and so on, or multimedia combinations.

World Wide Web Consortium An organization which develops and promotes standards and software for the World Wide Web. Its abbreviation is **W3C**, or **W3 consortium**.

worm 1. A shaft with a threaded surface utilized in a **worm gear**. 2. A computer program that replicates itself and spreads throughout a given computer or computer network, utilizing more and more resources and/or producing other damage such as system crashes. A **virus** also replicates itself, but must seek out other programs onto which to copy itself, while a worm does so autonomously.

WORM Acronym for **write once, read many**. An optical disc technology, such as DVD-R or CD-R, which allows for recording only once.

worm gear A gear which changes the direction of the axis of a rotary motion by utilizing a shaft with a threaded surface, called **worm** (1), and toothed wheel, called **worm wheel**.

worm wheel A toothed wheel utilized in a **worm gear**.

worst-case design A design, such as that of a component, circuit, or device, which provides sufficient margin for varying parameters and operational conditions, so as to enable adequate functioning under most any circumstances.

wow In sound reproduction, a form of distortion manifesting itself as slow fluctuations of pitch. It is caused by equipment speed variations occurring during recording, dubbing, or reproduction. When occurring at higher frequencies, it is called **flutter** (2).

wow and flutter In sound reproduction, a form of distortion manifesting itself as slow fluctuations of pitch. It is caused by equipment speed variations occurring during recording, dubbing, or reproduction. **Wow** describes the lower-frequency variations, while **flutter** (2) describes the higher-frequency fluctuations. Usually expressed as a percentage. In digital devices, such as CD players or DATs, wow and flutter values are drastically below the threshold of human detection.

wow factor The extent to which a product, function, or feature evokes admiration, amazement, or the like, in a user. Sometimes, however, a user may say "wow" in a mocking tone implying something along the lines of "what next?"

WPAN Abbreviation of **Wireless Personal Area Network**.

WPS Abbreviation of **Workplace Shell**.

wrist A set of joints between the arm and the end-effector of a robot, usually with the ability to move with various degrees of freedom.

wrist bend Same as **wrist pitch**.

wrist force sensor 1. In robotics, a sensor which measures the forces exerted by a wrist. 2. In robotics, a sensor which measures the forces exerted upon a wrist.

wrist pitch In a robot, up and down movement of a wrist. Also called **wrist bend**.

wrist rest A long pad, platform, or other device which serves to raise and rest the wrists of a person typing at a keyboard. Such a rest, when positioning the wrists correctly, may help reduce repetitive-strain injuries, such as carpal-tunnel syndrome. Also called **wrist support**.

wrist roll In a robot, rotation of a wrist in a clockwise or counterclockwise manner. Also called **wrist swivel**.

wrist strap A grounding strip which is typically wrapped around a wrist on one side, and attached to the chassis of the device or piece of equipment being worked on the other, so as to properly channel any static electricity a person may carry or generate. Used, for instance, to prevent a user from harming sensitive computer components when performing tasks such as the installation additional RAM. Also called **ESD wrist strap**, or **antistatic wrist strap**.

wrist support Same as **wrist rest**.

wrist swivel Same as **wrist roll**.

wrist yaw In a robot, side to side motion of a wrist.

write 1. To record, store, or otherwise transfer data to a memory location, storage device, magnetic stripe, or the like. Examples include copying data from RAM to hard disk, or from the CPU to RAM. The converse is a **read** (1). 2. Same as **write operation**.

write access The authorization to perform **write operations**.

write-back cache A type of cache memory in which changes to cached data are not simultaneously made to the original copy, but which are instead marked for later update. This contrasts with **write-through cache**, in which simultaneous changes are made.

write command A computer instruction to perform a **write operation**.

write cycle The complete sequence of events required to perform a **write operation**. Also, a single such cycle.

write cycle time The time interval that elapses between the starts of successive **write cycles**.

write error 1. An attempted **write operation** which is not successful. Possible causes include corrupted storage locations, program errors, or a malfunction of a read/write head. 2. A write operation in which there are one or more erroneous bits delivered or transferred.

write head A transducer which converts electrical signals into magnetic signals which can be recorded or stored on magnetic recording media, such as tapes or disks. Also called **writing head**, **recording head**, or **magnetic head** (1).

write once An optical technology or system which allows recording upon an optical disc once. Examples include DVD-R and CD-R. Also, pertaining to a disc, such as a DVD-R disc, which can only be written on once. Its abbreviation is **WO**.

write once, read many Same as **WORM**.

write operation The act or operation of performing a **write** (1). Also called **write** (2).

write protect 1. To prevent writing to a given data medium, such as a disk, or a file. Such protection may be afforded using software, or physically, using a device such as a write-protect tab. 2. A mode or state in which writing to a given data medium or file is prevented.

write-protect notch Same as **write-protect tab**.

write-protect tab A movable or removable object or device, such as a notch or a label, whose position or presence determines whether writing to a given medium is possible. Seen, for instance, on floppy disks. Also called **write-protect notch**.

write protection The preventing of writing to a given data medium, such as a disk, or to a file.

write-through cache A type of cache memory in which changes to cached data are simultaneously made to the original copy. This contrasts with **write-back cache**, in which simultaneous changes are not made, but are marked for later update.

write time The time interval that elapses between the initiation of a **write operation** and the delivery or transfer of the desired information.

writing The performance of a **write operation**.

writing gun In a storage tube, the electron gun which forms the displayed image. Said image is intensified by the electrons emitted by the flooding gun.

writing head Same as **write head**.

Ws Abbreviation of **watt-second**.

WSI Abbreviation of **wafer-scale integration**.

wt Abbreviation of **weight**.

WWAN Abbreviation of **wireless WAN**.

WWW Abbreviation of **World Wide Web**.

wye adapter Same as **wye connector**. Also spelled **Y-adapter**.

wye antenna A single-wire half-wave antenna whose impedance is matched to the characteristic impedance of the transmission line. The two leads of the transmission line are connected to the radiator, forming a Y shape, and since the antenna is not split, this also resembles a Greek letter Δ (delta). Also spelled **Y antenna**. Also called **wye-matched antenna**, or **delta-matched antenna**.

wye circuit Same as **wye network**. Also spelled **Y-circuit**.

wye connection Same as **wye network**. Also spelled **Y-connection**.

wye connector Also spelled **Y-connector**. Also called **wye adapter**. **1.** A connector that provides two outputs from a single input. Used, for instance, to split an audio signal to drive two outputs. **2.** A connector that provides a single output from two inputs. Used, for instance, to mix two signals.

wye-matched antenna Same as **wye antenna**. Also spelled **Y-matched antenna**.

wye network An electric network configuration in which three branches, such as impedance arms, are connected to a common central point or node. Also spelled **Y network**. Also called **wye circuit, wye connection**, or **T network**.

WYSIWYG Abbreviation of **What You See Is What You Get**. The ability to display on a monitor text and graphics exactly as it would appear if printed. In actuality this is only approximated, as printers tend to have much higher resolutions than monitors.

X

x Symbol for **X-coordinate**, or **abscissa**.

X Symbol for **reactance**.

X Symbol for **X-coordinate**, or **abscissa**.

x-amplifier Same as **X-amplifier**.

X-amplifier A circuit or device which amplifies the signals which produce a horizontal deflection in an instrument such as an oscilloscope. Also called **horizontal amplifier**.

x-axis **1.** In a two-dimensional Cartesian coordinate system, the horizontal axis, the other being the **y-axis**. Also, one of the three mutually-perpendicular axes in a three-dimensional Cartesian coordinate system, the others being the **y-** and **z-axes**. **2.** One of three reference axes in a quartz crystal, the others being the **y-** and **z- axes**. **3.** In a CRT, the axis representing the horizontal deflection of the electron beam.

X-axis Same as **x-axis**.

X Band **1.** In communications, a band of radio frequencies extending from 8.00 to 12.00 GHz, as established by the IEEE. This corresponds to wavelengths of approximately 3.8 to 2.5 cm, respectively. **2.** In communications, a band of radio frequencies extending from 5.2 GHz to 10.9 GHz. This corresponds to wavelengths of approximately 5.8 cm to 2.8 cm, respectively.

x-coordinate In a two-coordinate system, the horizontal coordinate. The **y-coordinate** is the vertical coordinate. Its symbol is **x**, or **X**. Also called **abscissa**.

X-coordinate Same **x-coordinate**.

X-cut A cut in a quartz crystal made in a manner that the x-axis is perpendicular to the faces of the resulting slab.

X-cut crystal A quartz crystal with an **X-cut**.

X-on/X-off Same as **Xon/Xoff**.

X/Open A consortium of international vendors which develops and promotes standards and specifications for open systems.

X-radiation Same as **X-rays**.

X-ray **1.** A single high-energy photon within **X-rays**. **2.** Pertaining to, consisting of, or arising from **X-rays**. **3.** Same as **X-ray image**. **4.** To produce an **X-ray image**.

X-ray absorption The absorption of **X-rays**. Such absorption, for instance, provides information on transitions between atomic energy levels.

X-ray absorption spectroscopy A spectroscopic technique which measures the absorption of X-rays by a sample. Used, for instance, to determine the structure of crystalline samples. Its abbreviation is **XAS**.

X-ray absorption spectrum A display or graph obtained through **X-ray absorption spectroscopy**.

X-ray analysis Any spectroscopic technique, such as X-ray diffraction, X-ray fluorescence, or X-ray absorption spectroscopy, utilizing X-rays.

X-ray astronomy Astronomy which studies the X-ray emissions of celestial bodies. Since the atmosphere absorbs X-rays, observations must be performed above a given altitude, such as 100 kilometers, using satellites or rockets. Balloons may be utilized closer to the surface of the earth, but this limits the detection of a significant part of the available X-ray spectrum.

X-ray crystallography The use of **X-ray diffraction** (1) to study the structure of crystals, using instruments such as Bragg spectrometers. Also called **X-ray diffraction (2)**.

X-ray diffraction Its abbreviation is **XRD**. **1.** The diffraction by a crystal of an incident X-ray beam. The angles at which the X-rays are diffracted are determined by the interatomic distances within the lattice, with the resulting characteristic pattern enabling the determination of the structure of the crystal. Such diffraction is the basis of **X-ray crystallography**. **2.** Same as **X-ray crystallography**.

X-ray diffraction pattern The characteristic pattern obtained via **X-ray diffraction**.

X-ray diffractometer An instrument utilized for **X-ray diffraction** analysis.

X-ray emission Same as **X-ray fluorescence**.

X-ray film A photographic film that has a radiosensitive emulsion coating one or both sides of a transparent base, and which is suitable for recording **X-ray images**.

X-ray fluorescence The emission of characteristic X-rays by a substance which is exposed to **X-rays**. Its abbreviation is **XRF**. Also called **X-ray emission**.

X-ray fluorescence spectroscopy An spectroscopic technique in which an electron microprobe directs a finely focused electron beam towards samples which then emit characteristic X-rays. Used, for instance, to determine the composition of materials. Its abbreviation is **XRF**.

X-ray hardness The relative readiness with which **X-rays** penetrate matter. The shorter the wavelength, the greater the hardness, and the more penetrating the rays.

X-ray image An image produced on a radiosensitive surface, such as a photographic film or plate, when exposed to penetrating radiation, especially X-rays. Typically, the object to be analyzed is placed between the X-ray source and the film or plate, with the relative opacity determining the intensity of the shadows produced. Used, for instance, as a medical diagnostic technique, for ascertaining product defects, or in the quantitative analysis of materials. Also called **X-ray (3)**, **X-ray photograph**, **X-ray picture**, **radiograph**, **radiogram (1)**, **roentgenogram**, **roentgenograph**, **shadowgraph**, or **skiagraph**.

X-ray inspection The use of **X-rays** to carefully examine below the surface of materials and bodies. Used, for instance, for non-destructive real-time inspection of printed-circuit boards and chips.

X-ray laser A laser which emits coherent **X-rays**.

X-ray lithography A lithographic technique in which **X-rays** are utilized to expose the resist, as opposed, for instance, to light or an electron beam. The short wavelengths minimize diffraction and provide for very high resolution, which can be in the nanometer range. The source of X-rays may be a synchrotron, and the mask is usually made of gold. Its abbreviation is **XRL**.

X-ray machine A device or instrument, such as that utilized for therapy, the taking of X-ray images, inspection of parts, or the determination of crystal structure, which utilizes **X-rays**. Such a machine includes an X-ray tube, and all components necessary for proper operation for the desired use. Also called **X-ray system**.

X-ray mask A mask utilized in **X-ray lithography**.

X-ray photoelectron spectroscopy A photoelectron spectroscopy technique in which the source of exciting radiation is **X-ray radiation**. Its abbreviation is **XPS**.

X-ray photograph Same as **X-ray image**.

X-ray photography The obtaining of **X-ray images**. Also, the science and techniques involved in obtaining such images, including reducing the exposure necessary for photographing, improvement of resolution, and the like. Also called **radiography**, **roentgenography**, **skiagraphy**, or **shadowgraphy**.

X-ray picture Same as **X-ray image**.

X-ray plate A photographic plate that has a radiosensitive emulsion coating one or both sides of a transparent base, and which is suitable for recording **X-ray images**.

X-ray radiation Same as **X-rays**.

X-ray region The region within the electromagnetic spectrum corresponding to **X-rays**.

X-ray source A source, such as an X-ray tube or the sun, of **X-rays**.

X-ray spectra A plural form of **X-ray spectrum**.

X-ray spectrograph An **X-ray spectrometer** equipped to produce a photographed or otherwise recorded output.

X-ray spectrometer An instrument utilized to obtain **X-ray spectrums**. Examples include those utilized in X-ray crystallography, X-ray fluorescence spectroscopy, or X-ray absorption spectroscopy.

X-ray spectrum **1.** A spectrum resulting from the diffraction, reflection, emission, or absorption of **X-rays**. Examples include spectrums resulting from X-ray diffraction, X-ray fluorescence spectroscopy, or X-ray absorption spectroscopy, or that of an X-ray source. **2.** The range of frequencies within the electromagnetic spectrum corresponding to X-rays.

X-ray system Same as **X-ray machine**.

X-ray telescope A telescope which detects, amplifies, and analyses the X-ray emissions of celestial bodies. Since the atmosphere absorbs X-rays, observations must be performed above a given altitude, such as 100 kilometers, using satellites or rockets, and orbits may range past 100,000 kilometers.

X-ray therapy The utilization of **X-rays** for therapeutic purposes. Used, for instance, to treat cancer.

X-ray topography An **X-ray diffraction** technique utilized especially to detect and characterize crystal defects.

X-ray tube An electron tube, usually a vacuum tube, which produces **X-rays**. In such a device, a beam of electrons emitted by a thermionic cathode is accelerated by an electrostatic field and directed towards the target electrode, which is usually the anode. The target electrode may be made, for instance, of tungsten or a tungsten alloy, and releases X-rays when bombarded by the electron beam, and in addition helps to dissipate the enormous heat created. The voltage across such a tube may range from a few thousand kilovolts to over a megavolt, depending on the application. Used, for instance, in X-ray spectrometers and X-ray machines.

X-rays Electromagnetic radiation whose wavelengths range between approximately 10^{-8} and 10^{-11} meter, which corresponds to frequencies of 10^{16} to 10^{19} Hz respectively, although the defined interval varies. X-rays are located between ultraviolet and gamma radiation. Such rays consist of high-energy photons, and penetrate living tissue and other matter to varying degrees, with less penetrating rays called soft X-rays, and more penetrating rays, whose energy and frequency are higher, known as hard X-rays. X-rays are produced, for instance, when matter is bombarded by electrons with sufficient energy, as occurs in an X-ray tube. X-rays are used as a medical diagnostic technique, for therapy, and in the study of the structure and quantitative analysis of materials. Also called **X-radiation**, **X-ray radiation**, or **roentgen-rays**.

X-wave Abbreviation of extraordinary **wave**. One of the two components into which a radio wave entering the ionosphere is split, under the combined influence of the earth's magnetic field and atmospheric ionization, the other being the **O-wave**. The electric vector of the X-wave rotates in the sense opposite of that of the ordinary wave. Also called **extraordinary component**, or **extraordinary-wave component**.

X window Same as **X window system**.

X window system A windowing system that works across multiple platforms, and which is in the public domain. Its abbreviation is **X window**.

x-y plotter A plotter which utilizes one pen, beam, or the like, to draw along the x-axis, and another to move along the y-axis. When utilized to chart the relationship between two variables, also called **x-y recorder**.

X-Y plotter Same as **x-y plotter**.

x-y recorder An **x-y plotter** utilized to chart the relationship between two variables.

X-Y recorder Same as **x-y recorder**.

X.25 An ITU and OSI standard protocol utilized for packet-switched networks. X.25 performs error-checking at each node, which makes it suitable for lower-speed data transfer across noisy lines.

X.400 An ITU and OSI standard protocol for email, which works at the application layer. X.400 supports TCP/IP, X.25, and dial-up lines, among others.

X.500 An ITU and OSI standard defining and managing online directories. Used, for instance, to support X.400 messaging systems.

X.509 An ITU and OSI standard pertaining to digital certificates.

X12 Abbreviation of **ANSI X12**.

XAS Abbreviation of **X-ray absorption spectroscopy**.

Xbase Any of various database languages similar to dBASE. Such languages are not necessarily compatible with dBASE.

Xbox A popular gaming system.

X_C Symbol for **capacitive reactance**.

xDSL Any of various DSL versions or technologies, such as ADSL, HDSL, IDSL, RADSL, SDSL, or VDSL.

Xe Chemical symbol for **xenon**.

XENIX A version of UNIX utilized in personal computers.

xenon A colorless and odorless noble gas whose atomic number is 54. Although it is in the noble gas group of the periodic table, it does has limited reactivity. It is present in very small amounts in the atmosphere, and has over 35 known isotopes, of which 8 are stable. It is used, for instance, in high-intensity arc lamps, flash lamps, radiation detectors, lasers, and semiconductors. Its chemical symbol is **Xe**.

xenon arc lamp An arc lamp whose ionized gas is **xenon**.

xenon flash lamp A flash tube whose ionized gas is **xenon**. Also called **xenon flash tube**, or **xenon tube**.

xenon flash tube Same as **xenon flash lamp**.

xenon lamp A lamp filled with **xenon** gas. When used as a pulsed light source it is a **xenon flash lamp**. Also called **xenon light**.

xenon light Same as **xenon lamp**.

xenon tube Same as **xenon flash lamp**.

xerography A method of reproducing images utilizing electricity and light. In it, a photoconductive surface, usually a drum, is positively charged with static electricity, then exposed to an optical image of that which is to be reproduced, forming a corresponding electrostatic pattern. A fine powder called toner, which has been negatively charged, is spread over this surface, adhering only to the charged areas of the drum. The toner is then fused to the paper, which has been positively charged, forming the final image. Used, for instance, in laser printers and copy machines. Also called **electrophotography**.

xeroradiograph An image obtained via **xeroradiography**.

xeroradiography A form of **xerography** in which images are produced through exposure to X-rays or gamma rays, as opposed to infrared, visible, or UV light.

XGA Abbreviation of Extended Graphics Array. An enhanced VGA standard supporting resolutions up to 1024 x 768.

XHTML Abbreviation of Extensible Hypertext Markup Language, or Extensible HTML. A markup language that combines HTML and XML, and which features greater portability and ease of extension and enhancement.

X_L Symbol for **inductive reactance**.

XLink Abbreviation of XML Linking Language. A markup language utilized to add hyperlinks to documents prepared utilizing XML.

XLISP A version of LISP which is in the public domain.

XLL Same as **XLink**. Abbreviation of **XML Linking Language**.

XLR connector An audio connector with three pins or leads, two for signals, and the third for grounding. Such connectors typically lock in place, and are often utilized to connect microphones.

XLR jack An XLR connector which has recessed openings into which an **XLR plug** with matching prongs is inserted. Also called **XLR socket**.

XLR plug An XLR connector with three pins or leads which is inserted into an **XLR jack** with matching recessed openings.

XLR socket Same as **XLR jack**.

Xmit Abbreviation of **transmit**.

XMIT Abbreviation of **transmit**.

XML Abbreviation of Extensible Markup Language. A specification for the format of documents and data to be used on the Web. It is a scaled-down version of SGML, and seeks to retain the comparative simplicity of HTML, yet offer greater flexibility in areas such as organization and presentation, while being fully compatible with both. Also called **XML format**.

XML document A document in **XML format**.

XML file A file in **XML format**.

XML format Same as **XML**.

XML Linking Language Same as **XLink**.

XML namespace A manner of uniquely identifying element types and attribute names in an **XML** document. A URI is usually utilized to identify each name.

XML parser A program or routine which identifies the structure and components of **XML** documents, such as the tags present.

XML Path Language Same as **XPath**.

XML Pointer Language Same as **XPointer**.

XML protocol A message-based protocol which utilizes **XML** syntax. Its abbreviation is **XMLP**.

XML schema Abbreviation of **XML schema** definition. A W3C specification utilized to describe the structure and content, such as data types, of an **XML** document.

XML schema definition Same as **XML schema**. Its abbreviation is **XSD**.

XMLP Abbreviation of **XML protocol**.

Xmodem A low-speed file-transfer protocol supported by many modems and communications applications.

XMT Abbreviation of **transmit**.

XNOR Abbreviation of exclusive **NOR**. A logical operation which is true if all of its elements are the same. For example, if all of its multiple inputs have a value of 0, then the output is 1. And, if all of its inputs have a value of 1, then the output is 1. Any other combination yields an output of 0. For such functions, a 1 is considered as a true, or high, value, and 0 is a false, or low, value. Also called **XNOR operation**.

XNOR circuit Same as **XNOR gate**.

XNOR gate Abbreviation of exclusive **NOR gate**. A circuit which has two or more inputs, and whose output is high only if all its inputs are the same. Also called **XNOR circuit**.

XNOR operation Same as **XNOR**.

XO Abbreviation of **crystal oscillator**.

xon/xoff Same as **Xon/Xoff**.

Xon/Xoff A protocol utilized for the synchronization of data transmitted using asynchronous communications. Xon signals that it is okay to transmit, while Xoff signals a stop, as may occur, for instance, when the buffer of the receiving computer or device is full. In this example, when more buffer room becomes available another Xon signal us sent.

XON/XOFF Same as **Xon/Xoff**.

XOR Abbreviation of exclusive **OR**. A logical operation which is false if all of its elements are the same. For example, if all of its multiple inputs have a value of 0, then the output is 0. And, if all of its inputs have a value of 1, then the output is 0. Any other combination yields an output of 1. For such functions, a 1 is considered as a true, or high, value, and 0 is a false, or low, value. Also called **XOR operation**.

XOR circuit Same as **XOR gate**.

XOR gate Abbreviation of exclusive **OR gate**. A circuit which has two or more inputs, and whose output is low if all of its inputs are the same. Also called **XOR circuit**.

XOR operation Same as **XOR**.

XP A common operating system.

XPath Abbreviation of XML **Path** Language. A language, based on XSLT and XPointer, which is utilized to define the parts of an XML document.

XPointer Abbreviation of XML **Pointer** Language. A language utilized to identify the structures within an XML document.

XPS Abbreviation of **ultraviolet photoelectron spectroscopy**.

XRD Abbreviation of **X-ray diffraction**.

XRF 1. Abbreviation of **X-ray fluorescence**. 1. Abbreviation of **X-ray fluorescence spectroscopy**.

XRL Abbreviation of **X-ray lithography**.

XSD Abbreviation of **XML schema definition**.

XSL Abbreviation of Extensible Stylesheet Language. Within **XML**, a standard defining stylesheets.

XSLT Abbreviation of Extensible Stylesheet Language Transformations. A language utilized for converting XML documents into other XML documents with different structures.

XTACACS Abbreviation of **Extended Terminal Access Controller Access Control System**.

xy plotter Same as **x-y plotter**.

XY plotter Same as **x-y plotter**.

xy recorder Same as **x-y recorder**.

XY recorder Same as **x-y recorder**.

Y

y **1.** Symbol for **yocto-**. **2.** Symbol for **y-coordinate**, or **ordinate**. **3.** Abbreviation of **year**.

Y Symbol for **admittance**.

Y **1.** Chemical symbol for **yttrium**. **2.** Symbol for **y-coordinate**, or **ordinate**. **3.** Symbol for **yotta-**.

Y-adapter Same as **Y-connector**. Also spelled **wye adapter**.

y-amplifier Same as **Y-amplifier**.

Y-amplifier A circuit or device which amplifies the signals which produce a vertical deflection in an instrument such as an oscilloscope. Also called **vertical amplifier**.

Y antenna A single-wire half-wave antenna whose impedance is matched to the characteristic impedance of the transmission line. The two leads of the transmission line are connected to the radiator, forming a Y shape, and since the antenna is not split, this also resembles a Greek letter Δ (delta). Also spelled **wye antenna**. Also called **Y-matched antenna**, or **delta-matched antenna**.

y-axis **1.** In a two-dimensional Cartesian coordinate system, the vertical axis, the other being the **x-axis**. Also, one of the three mutually-perpendicular axes in a three-dimensional Cartesian coordinate system, the others being the **x-** and **z-axes**. **2.** One of three reference axes in a quartz crystal, the others being the **x-** and **z-** axes. **3.** In a CRT, the axis representing the vertical deflection of the electron beam.

Y circuit Same as **Y network**. Also spelled **wye circuit**.

Y connection Same as **Y network**. Also spelled **wye connection**.

Y-connector Also spelled **wye connector**. Also called **Y-adapter**. **1.** A connector that provides two outputs from a single input. Used, for instance, to split an audio signal to drive two outputs. **2.** A connector that provides a single output from two inputs. Used, for instance, to mix two signals.

y-coordinate In a two-coordinate system, the vertical coordinate. The **x-coordinate** is the horizontal coordinate. Its symbol is **y (2)**, or **Y (2)**. Also called **ordinate**.

Y-coordinate Same as **y-coordinate**.

Y-cut A cut in a quartz crystal made in a manner that the y-axis is perpendicular to the faces of the resulting slab.

Y-cut crystal A quartz crystal with a **Y-cut**.

Y-matched antenna Same as **Y antenna**. Also spelled **wye-matched antenna**.

Y network An electric network configuration in which three branches, such as impedance arms, are connected to a common central point or node. Also spelled **wye network**. Also called **Y circuit**, **Y connection**, or **T network**.

y-parameters Same as **Y-parameters**.

Y-parameters Within an electric network, circuit parameters pertaining to admittance. It is an abbreviation of **admittance parameters**.

Y signal In color TV, the video signal which contains the brightness information. It is composed of 30% red, 59% green, and 11% blue. A black and white TV only displays Y signals. In a color TV, the sidebands of the chrominance carrier are added to this signal to convey the color information. Also called **luminance signal**.

YAG Abbreviation of **yttrium-aluminum-garnet**.

YAG laser Abbreviation of yttrium-aluminum-garnet **laser**. A solid-state laser whose active medium is a yttrium-aluminum-garnet crystal which is doped, for instance, with neodymium or erbium. Used, for example, in surgery, spectroscopy, welding, and communications.

Yagi Same as **Yagi antenna**.

Yagi antenna A highly directional antenna with multiple straight elements which lie along the same plane and are parallel to each other. The antenna typically has one driven element and two or more passive elements utilized as reflectors and directors, all approximately 0.5 electrical wavelengths. Widely used, for instance, for TV and radio signals. Also called **Yagi**, **Yagi-Uda antenna**, **Yagi array**, or **Yagi-Uda array**.

Yagi array Same as **Yagi antenna**.

Yagi-Uda antenna Same as **Yagi antenna**.

Yagi-Uda array Same as **Yagi antenna**.

Yahoo! A popular Web portal.

yard A unit of distance equal to exactly 0.9144 meter. Its abbreviation is **yd**.

yaw Side to side motion, such as that of a robot wrist. Also, the act of so moving.

Yb **1.** Chemical symbol for **ytterbium**. **2.** Abbreviation of **yottabit**.

YB Abbreviation of **yottabyte**.

Ybit Abbreviation of **yottabit**.

Ybyte Abbreviation of **yottabyte**.

YCbCr A color model utilized for color digital component video formats. **Y** represents the luminance signal, and **Cb** and **Cr** represent the two color-difference signals.

yd Abbreviation of **yard**.

year A unit of time representing the interval required for the earth to make a complete revolution around the sun. There are approximately 3.1556926×10^7 seconds in a year. Its abbreviation is **yr**, **y**, or **a**.

Yellow Book A document which details the specifications for the CD-ROM format.

yellow pages **1.** A Web-based directory service similar to the yellow pages of a telephone book, but which in addition may provide Web addresses, email addresses, maps, and so on, plus searching tools which are more flexible. **2.** A service that provides information on the users and/or resources of a communications network.

YHz Abbreviation of **yottahertz**.

yield The quantity or product obtained from a reaction or process. For example, the number of electrons released by a photocathode, per photon of incident radiation. Also, a proportion obtained in relation to a theoretical maximum. For example, the percentage of operational chips out of all chips manufactured.

yield point The point beyond which a material exhibits a given amount of permanent deformation.

yield strength The stress beyond which a material exhibits a given amount of permanent deformation.

yield stress The stress at which a material begins to exhibit permanent deformation. Also called **yield value**.

yield value Same as **yield stress**.

YIG Abbreviation of **yttrium-iron-garnet**.

YIQ Abbreviation of **YIQ** color model. A color model utilized for color NTSC broadcasting. **Y** represents the luminance signal, and **I** and **Q** represent the two color-difference signals.

YIQ color model Same as **YIQ**.

ym Abbreviation of **yoctometer**.

Ymodem An enhanced version of **Xmodem** which increases the transfer block size, enables sending multiple files at the same time, and provides improved error-checking.

yocto- A metric prefix representing 10^{-24}. For example, 1 yoctosecond is equal to 10^{-24} second. Its abbreviation is **y**.

yoctometer 10^{-24} meter. It is a unit of measurement of length. Its abbreviation is **ym**.

yoctosecond 10^{-24} second. It is a unit of time measurement. Its abbreviation is **ys**, or **ysec**.

yoke 1. Two or more magnetic recording heads, pole pieces, cores, or the like, which are physically joined together. **2.** In a CRT, a system of coils utilized for magnetic deflection of the electron beam. One possible arrangement consists of two sets of coils, one for horizontal deflection, and the other for vertical deflection. Also called **deflection yoke**, or **scanning yoke**.

yotta- A metric prefix representing 10^{24}. For example, yottahertz. Its symbol is **Y**.

yottabit 2^{80} bits, or approximately 1.2089×10^{24} bits. Often it is rounded to 1.0×10^{24}. Its abbreviation is **Yb**, or **Ybit**.

yottabyte 2^{80} bytes, or approximately 1.2089×10^{24} bytes. Often it is rounded to 1.0×10^{24}. Its abbreviation is **YB**, or **Ybyte**.

yottahertz 10^{24} Hz. This is within the cosmic-ray region of the electromagnetic spectrum. Its abbreviation is **YHz**.

Young's modulus The ratio of a longitudinal stress to the resulting longitudinal strain. It is a type of elasticity modulus.

YPbPr A color model utilized for color analog component video formats. **Y** represents the luminance signal, and **Pb** and **Pr** represent the two color-difference signals.

yr Abbreviation of **year**.

ys Abbreviation of **yoctosecond**.

ysec Abbreviation of **yoctosecond**.

ytterbium A lustrous silver-white metal whose atomic number is 70. It is soft, malleable, and ductile, and exhibits al- lotropy. It has about 30 known isotopes, of which 7 are sta- ble. Its applications include its use in lasers, semiconduc- tors, and as an X-ray source. Its chemical symbol is **Yb**.

ytterbium oxide A white solid whose chemical formula is Yb_2O_3. It is used, for instance, in ceramics and optics.

yttria Same as **yttrium oxide**.

yttrium A dark gray metal whose atomic number is 39. It is soft, malleable, and ductile, and exhibits allotropy. It has about 30 known isotopes, of which one is stable. Its applications include its use as a phosphor, in lasers, semiconductors, ceramics, and high-temperature alloys. Its chemical symbol is **Y**.

yttrium-aluminum-garnet A hard crystal which is composed of yttrium, aluminum, and garnet. In the context of YAG lasers, the term usually refers to such a crystal which is doped with another element, such as neodymium or erbium. Its abbreviation is **YAG**.

yttrium-iron-garnet A crystal which is composed of yttrium, iron, and garnet. It is utilized, for instance, in acoustic transducers and filters, lasers, and in microwave devices such as oscillators and ferrite isolators. Its abbreviation is **YIG**.

yttrium oxide A white powder whose chemical formula is Y_2O_3. It is used, for instance, as a phosphor, and as a mi- crowave filter. Also called **yttria**.

yttrium vanadate A white solid whose chemical formula is YVO_4. It is used, for instance, as a phosphor, and in lasers.

YUV Abbreviation of **YUV** color model. A color model which is utilized for color PAL and NTSC broadcasting. **Y** represents the luminance signal, and **U** and **V** represent the two color-difference signals.

YUV color model Same as **YUV**.

z 1. Symbol for **zepto-**.

Z 1. Symbol for **atomic number**. 2. Symbol for **zetta-**.

Z 1. Symbol for **impedance**.

z-amplifier Same as **Z-amplifier**.

Z-amplifier A circuit or device which amplifies the intensity of the electron beam in an instrument such as an oscilloscope. Also called **vertical amplifier**.

z-axis 1. One of the three mutually-perpendicular axes in a three-dimensional Cartesian coordinate system, the others being the **x-** and **y- axes**. 2. One of three reference axes in a quartz crystal, the others being the **x-** and **y- axes**. 3. In a CRT, the axis representing the intensity of the electron beam.

Z-axis modulation In a CRT, the modulation of the electron beam intensity according to the intensity of the signal source. Also called **Z-modulation**, or **intensity modulation (2)**.

Z-buffer A buffer, such as that in a 3D accelerator, which holds the Z, or depth, information of each pixel.

Z-cut A cut in a quartz crystal made in a manner that the z-axis is perpendicular to the faces of the resulting slab.

Z-cut crystal A quartz crystal with a **Z-cut**.

Z meter An instrument which measures and indicates impedances. Such an instrument usually provides a digital readout. It is an abbreviation of **impedance meter**.

Z-modulation Same as **Z-axis modulation**.

z-parameters Same as **Z-parameters**.

Z-parameters Within an electric network, circuit parameters pertaining to impedance. It is an abbreviation of **impedance parameters**.

Z time Same as **Zulu Time**.

z-transform An extension of Fourier and Laplace transforms utilized, for instance, to analyze and represent sampled data. Fourier and Laplace transforms are each special cases of z-transforms.

Z_0 Symbol for **characteristic impedance**, or **surge impedance**.

Z39.50 A protocol utilized for searching for documents located in online databases.

Zb Abbreviation of **zettabit**.

ZB Abbreviation of **zettabyte**.

Zbit Abbreviation of **zettabit**.

Zbyte Abbreviation of **zettabyte**.

Zeeman effect A splitting of spectral lines within a spectrum when the radiation source, such as light, is in the presence of a magnetic field. The greater the splitting, the stronger the influencing field.

Zeeman splitting A splitting of spectral lines due to the **Zeeman effect**.

Zener breakdown A form of avalanche breakdown occurring across reverse-biased pn junctions which have high doping on both sides of the boundary. Seen, for instance, in a Zener diode.

Zener diode A pn-junction diode with high doping on both sides of the boundary, and which utilizes **Zener breakdown** to provide a given maximum voltage across it. A Zener diode has a breakdown voltage of a few volts, up to about six, while an **avalanche diode** usually has a much higher breakdown voltage, sometime several hundred volts. Used, for instance, as a voltage-regulator diode.

Zener diode regulator Same as **Zener diode voltage regulator**.

Zener diode voltage regulator A voltage regulator which utilizes the constant-voltage characteristic and well-defined reverse-breakdown voltages of **Zener diodes**. Also called **Zener diode regulator**, or **diode voltage regulator**.

Zener effect The effect via which **Zener breakdown** occurs.

Zener voltage 1. The voltage at which **Zener breakdown** occurs. 2. The voltage at which avalanche breakdown occurs. Also called **breakdown voltage (4)**.

Zepp antenna Abbreviation of **Zeppelin antenna**. An end-fed half-wave horizontal antenna which provides multiband operation and uses a tuned feeder. Alternatively, its length may be multiple of a half-wavelength.

Zeppelin antenna Same as **Zepp antenna**.

zepto- A metric prefix representing 10^{-21}. For example, 1 zeptovolt is equal to 10^{-21} volt. Its abbreviation is **z**.

zeptofarad 10^{-21} farad. It is a unit of capacitance. Its abbreviation is **zF**.

zeptometer 10^{-21} meter. It is a unit of measurement of length. Its abbreviation is **zm**.

zeptosecond 10^{-21} second. It is a unit of time measurement. Its abbreviation is **zs**, or **zsec**.

zero 1. A number, whose symbol is **0**, which when added or subtracted from another number leaves the latter unchanged. When another number is multiplied by zero the result is zero, and division by zero is considered undefined with few exceptions. 2. A reading, point within a function or curve, position, input, condition, or the like, resulting in a zero value, magnitude, setting, output, state and so on. 3. To set an instrument or meter to a zero reading. Done, for instance, when calibrating or preparing for use. 4. To balance, or otherwise adjust, so as to obtain a zero reading or output. For example, to adjust the components of a bridge to obtain a zero output. 5. A number indicating the initial point, origin, or the like of something. For example, absolute zero. 6. Having a value that equals zero, or that which is too low to distinguish from zero.

zero adjuster A circuit, device, control, or mechanism which sets the reading of a meter or instrument to exactly zero when the measured quantity is actually at that value. Also called **zero set (1)**.

zero adjustment The use of a **zero adjuster**. Also, a specific instance of such use. Also called **zero set (2)**.

zero beat 1. The condition in which a beat note is not present or heard because the two frequencies being mixed in a nonlinear device are the same. 2. The condition in which a beat note is not present or heard because the two frequencies being mixed are the same. This may occur, for instance, during tuning.

zero-beat detector A circuit or device which detects a **zero beat** condition.

zero-beat reception Radio reception in which the incoming signal is combined with an internally generated signal of the same frequency. Also called **homodyne reception**.

zero bias 1. A lack of a voltage, current, capacitance, or other input which is applied to a component or device to establish a reference level for its operation. 2. A bias whose value is zero, as occurs, for instance, when the source and the gate of a FET have the same voltage.

zero compression 1. A file, file format, or data transfer in which no compression techniques are utilized. 2. Same as **zero suppression**.

zero condition Same as **zero state**.

zero-crossing detector A circuit, device, or instrument, such as a comparator, which detects the point at which a signal, such as a voltage, passes through a zero value in either direction.

zero-current switching Switching that occurs when a current amplitude passes through a zero value. This minimizes switching losses, especially at high frequencies.

zero-dispersion wavelength In a single-mode optical fiber, the wavelength at which dispersion is at a minimum. In silica-based optical fibers this usually occurs naturally at 1300 nm, while in doped-silica fibers the wavelength may be 1550 nm.

zero drift Also called **zero shift**. **1.** A gradual and undesired change in the zero reading or indication of a meter of instrument. This is usually corrected via a zero adjustment. **2.** A complete absence of drift in a component, circuit, device, instrument, piece of equipment, or system.

zero error 1. An error in the indication of a meter or instrument resulting from a non-zero reading when the measured quantity is actually zero. An index error is an example. **2.** A level of error that is equal to zero, or which is too low to distinguish from zero.

zero fill 1. To add non-significant zeroes when manipulating, storing, accessing, or transmitting data. For example, to add stuffing bits. **2.** To replace all storage locations on a disk with zeroes. Used, for instance, for testing, or to delete all content of a drive.

zero-frequency component A DC component within a complex wave or signal.

zero IF Abbreviation of **zero intermediate frequency**.

zero input An input, such as that of a flip-flop or logic gate, corresponding to a **zero state**. Also called **logic zero input**, or **logic low input**.

zero insertion force Same as **ZIF socket**. Its abbreviation is **ZIF**.

zero insertion force socket Same as **ZIF socket**.

zero intermediate frequency A lack of an intermediate frequency, as seen for instance, in direct-conversion receivers. Its abbreviation is **zero IF**.

zero level A specific level, such as that of a voltage, amplitude, sound pressure, or other magnitude, which is considered to be a reference, and upon which other levels are based. For example, 20 micropascals is considered to be the equivalent of 0 decibels, which is defined as the threshold, or zero level, of hearing. Also called **zero reference**.

zero meridian The meridian which has an assigned longitude of 0°. The Greenwich meridian is generally considered as the zero meridian of the planet earth. Also called **prime meridian**.

zero method A method of measurement in which a reading of zero is at the center of the scale. Indicated values may be greater than or less than the zero reading. One variation is the use of an audible signal for measurement, in which case the zero reading would be silent. Also called **balanced method**, or **null method**.

zero output An output, such as that of a flip-flop or logic gate, corresponding to a **zero state**. Also called **logic zero output**, or **logic low output**.

zero-point energy 1. The energy present in a substance at a temperature of absolute zero. This is a non-zero value. **2.** The energy present in a system in its lowest-energy, or ground, state. This is a non-zero value.

zero potential A potential arbitrarily considered to be zero. It may be the earth, or a large conducting body whose electric potential is also considered to be zero. Also called **zero voltage**, **earth potential**, or **ground potential**.

zero-power resistance For a given temperature, the resistance value of a thermistor at which no power is dissipated.

zero reference Same as **zero level**.

zero screw A screw utilized as a **zero adjuster**.

zero set 1. Same as **zero adjuster**. **2.** Same as **zero adjustment**.

zero shift Same as **zero drift**.

zero signal 1. An input signal which provides a reading of zero in an instrument or meter. **2.** A signal level which is considered to be a reference to which other signal levels are compared. **3.** A state in which there is no signal present. **4.** A state in which a signal too low to distinguish from zero is present.

zero stability 1. A complete lack of **zero drift**. **2.** A complete lack of stability.

zero state In digital logic, a level within the less positive of the two ranges utilized to represent binary variables or states. A low level corresponds to a 0, or false, value. Also called **zero condition**, **logic zero**, **logic 0**, **logic low**, or **low level**.

zero suppression The reduction or elimination of non-significant zeroes when manipulating, storing, accessing, or transmitting data. For example, to eliminate leading zeroes. Also called **zero compression** (2).

zero temperature The basis of a zero reading for a particular temperature scale. The Kelvin temperature scale, for, instance, is based on absolute zero, so its zero degree reading equals absolute zero. Another scale is the Celsius temperature scale, which is based on the freezing and boiling points of water under certain conditions. A reading of 0 °C is equal to approximately 273.15 K.

zero temperature coefficient A lack of a relationship between temperature and a given change in a physical quantity, phenomena, or parameter, such as resistance, current, expansion, reactivity, or accuracy. For example, a component which does not increase or decrease its resistance regardless of temperature variations.

zero time A specific instant of time which is considered to be a reference, and upon which the relative timing of other events are based. Used, for instance, for calibration, or for timing signals.

zero voltage Same as **zero potential**.

zero-voltage switching Switching that occurs when a voltage amplitude passes through a zero value. This minimizes switching losses, especially at high frequencies.

zero wait state 1. The execution of instructions with no wait states. SDRAM, for instance, is synchronized with the clock speed of the CPU it is running with, providing for no delays or wait states. **2.** RAM, such as SDRAM, which responds to the CPU fast enough to avoid wait states.

zeroth law of thermodynamics A law stating that if system A is in thermal equilibrium with system B, and that if system B is also in thermal equilibrium with system C, that system A and system C must be in thermal equilibrium. Measurements of temperature rely on this law.

zetta- A metric prefix representing 10^{21}. For example, zettahertz. Its symbol is **Z**.

zettabit 2^{70} bits, or approximately 1.1806×10^{21} bits. Often it is rounded to 1.0×10^{21}. Its abbreviation is **Zb**, or **Zbit**.

zettabyte 2^{70} bytes, or approximately 1.1806×10^{21} bytes. Often it is rounded to 1.0×10^{21}. Its abbreviation is **ZB**, or **Zbyte**.

zettahertz 10^{21} Hz. This is within the cosmic-ray region of the electromagnetic spectrum. Its abbreviation is **ZHz**.

zF Abbreviation of **zeptofarad**.

ZHz Abbreviation of **zettahertz**.

ZIF Same as **ZIF socket**. Abbreviation of **zero insertion force**.

ZIF socket Abbreviation of zero insertion force socket. A socket designed for the use of essentially zero force for insertion or removal of components, which virtually eliminates any mechanical stress. Such sockets usually have a lever or other device which assists insertion and removal, and are commonly utilized for placing and removing chips.

zig-zag Same as **zigzag**.

zigzag Also spelled **zig-zag**. 1. A line or path which has sharp turns in alternating directions. 2. One back and forth trace or interval within a **zigzag (1)**. 3. To move in a **zigzag (1)** pattern. 4. To have a **zigzag (1)** pattern.

zinc A lustrous bluish-white metallic chemical element whose atomic number is 30. Below 100 or 110 °C it is brittle, but above this temperature it becomes malleable and ductile. Zinc has been known since ancient times, can be refined to better than 99.999999%, and has about 25 known isotopes, of which 4 are stable. Its applications include its use for galvanization, in important alloys such as brass, in fuses, batteries, lasers, and semiconductors. Its chemical symbol is **Zn**.

zinc acetate White crystals used in ceramics, and for galvanizing.

zinc-air battery 1. Same as **zinc-air cell**. 2. A battery composed of two or more **zinc-air cells**.

zinc-air cell A cell in which the anode is zinc, and in which the cathode reactant is oxygen present in air. Such cells feature a comparatively flat discharge curve. Used, for instance, in button cells, such as those utilized in hearing aids. Also called **zinc-air battery (1)**.

zinc ammonium chloride A white powder used for galvanizing, and in batteries.

zinc bromide A white crystalline powder whose chemical formula is $ZnBr_2$. Used as a radiation viewing shield, and in batteries.

zinc-carbon cell A primary cell in which the positive electrode is carbon, and the negative electrode is zinc. It may be a wet cell or dry cell. When it is a dry cell with an ammonium chloride electrolyte, it is called **Leclanché cell**.

zinc chloride A white crystalline powder whose chemical formula is $ZnCl_2$. Used, for example, in batteries, soldering, and for galvanizing and etching.

zinc cyanide A white powder whose chemical formula is $Zn(CN)_2$. Used in plating.

zinc-mercuric-oxide cell A primary cell in which the anode is zinc, the cathode is mercuric oxide, and the electrolyte is a solution of potassium hydroxide. It produces an essentially constant output of 1.35V throughout its useful life, and is usually in the form of a small flat disk. It is used in hearing aids, watches, and other small devices, but is being replaced by other cells due to the toxicity of mercury. More commonly known as **mercury cell**.

zinc orthosilicate Same as **zinc silicate**.

zinc oxide A white crystalline powder whose chemical formula is ZnO. Used, for instance, for galvanizing, as a UV absorber, and in optics, lasers, semiconductors, ceramics, photography, and piezoelectric devices.

zinc selenide Yellow-red crystals whose chemical formula is $ZnSe$. Used as a phosphor, and in optics.

zinc silicate A mineral whose chemical formula is Zn_2SiO_4, and whose color may be white, yellow, green, blue, brown, or red. It usually exhibits fluorescence under UV light, and may also show phosphorescence. Used, for instance, as a phosphor, and as a source of zinc. Also called **zinc orthosilicate**, or **willemite**.

zinc-silver oxide cell A cell with a zinc anode, a silver oxide cathode, and a potassium hydroxide electrolyte. Usually encountered as button cells which feature high energy density, a comparatively flat discharge curve, and long shelf life. Also called **silver-oxide cell**.

zinc standard cell A standard cell used as a reference voltage source, in which the positive electrode is mercury, the negative electrode is an amalgam of zinc and mercury, and the electrolyte is a solution of zinc sulfate. It has a voltage of approximately 1.433 at 15 °C. Also called **Clark cell**.

zinc sulfide A yellow-white powder whose chemical formula is ZnS. Used, for instance, as a phosphor, and in lasers, semiconductors, and optics.

zinc telluride Reddish crystals whose chemical formula is $ZnTe$. Used, for instance, in lasers and semiconductors.

zip To compress files using a program such as WinZip or PKZIP.

zip cord 1. A two-wire cord, such as a lamp cord, which is adapted for another use. For example, a cut a lamp cord which is utilized to connect a speaker to an amplifier. 2. In optical communications, a two-fiber cable fashioned out of two single-fiber cables joined together.

Zip disk A 3.5 inch removable disk utilized with a **Zip drive**.

Zip drive A portable disk drive that uses 3.5 inch removable disks, and which is utilized, for instance, for archiving or for physically transferring files between locations as opposed to transmitting them over a network.

zip tone A short tone sent over a trunk which indicates that a call is arriving, readiness for an order, or to warn that a call is being monitored. Also called **beep tone**.

zipping The compression files using a program such as WinZip or PKZIP.

zirconia Same as **zirconium oxide**.

zirconium A lustrous silver-gray metal whose atomic number is 40. It is very strong, malleable, ductile, and heat and corrosion-resistant. When finely divided, it can ignite spontaneously in air, and is used in flash bulbs. It has about 25 known isotopes, of which 4 are stable. Its applications include its use in corrosion-resistant alloys, as a vacuum-tube getter, as a coating for nuclear rods, and in ceramics and welding. Its chemical symbol is **Zr**.

zirconium boride A gray metallic powder whose chemical formula is ZrB_2. Used in electrodes and thermocouples.

zirconium carbide A hard gray solid whose chemical formula is ZrC. Used, for instance, in ceramics, electrodes, lasers, and semiconductors.

zirconium hydride A gray-black powder whose chemical formula is ZrH. Used, for instance, as a nuclear moderator, in ceramics, and as a vacuum-tube getter.

zirconium nitride A hard yellow-red powder whose chemical formula is ZrN. Used, for instance, in ceramics.

zirconium oxide A white solid whose chemical formula is ZrO_2. Used, for instance, in ceramics, piezoelectric devices, induction coils, batteries, and optics. Also called **zirconia**.

zm Abbreviation of **zeptometer**.

Zmodem An enhanced version of **Xmodem** which increases the transfer block size, provides greater speed, and improved error-checking. Another feature provides for a transfer to resume at the point it was interrupted, which saves time and aggravation when using noisy lines to send or receive large files.

Zn Chemical symbol for **zinc**.

zone 1. One area or region, as opposed to another. Also, such a zone between other zones. For example, a skip zone. 2. A specific **zone (1)** when two or more such areas or regions are defined. For instance, a Fresnel zone, or the layers into which the atmosphere is divided. 3. A **zone (1)** where certain properties or characteristics are present. For example, a far-field region versus a near-field region. 4. In computers, an area designated for storage, or for storage with a specific purpose. May consist, for instance, of sectors, tracks, bands, or the like. Also, an area within a storage medium with a special purpose. For example, a landing zone.

zone file Within a Domain Name System, a file or database that contains the information necessary to correlate domain names and IP numbers.

zone leveling The use of **zone refining** to evenly distribute an impurity throughout a solid or crystalline material.

zone melting Same as **zone refining**.

zone of silence **1.** The region above an altitude of about 160 kilometers from the earth's surface, where the distance between air molecules is greater than the wavelength of sound, thus sound waves can not be propagated. This zone extends well beyond the planet into outer space. Also called **anacoustic zone**. **2.** A zone which runs from where ground waves become too weak for reception to the point where a sky wave returns to the surface of the earth. Also called **skip zone**, or **silent zone**.

zone purification Same as **zone refining**.

zone refining A method or process utilized to purify solid or crystalline materials, in which sections of the material to be refined are heated sequentially. Each section is heated until melting, at which time the impurities present concentrate in the liquid portion. As the heated region progresses through the bar or slab, the impurities continue to be dissolved in the molten zone, with only highly pure material being recrystallized. There may be multiple molten zones working simultaneously across a given sample, and the process can be repeated for even greater purity. Used, for instance, to remove impurities from semiconductors. Also called **zone melting**, **zone purification**, or **zone-refining purification**.

zone-refining purification Same as **zone refining**.

zone time The standard time utilized throughout a given **time zone**.

zoning The creation, assignment, determination, or division of something into **zones**.

zoom **1.** To use a **zoom lens** to change the size of an image. Also, to use such a lens to rapidly change the size of an image. **2.** Same as **zoom lens**. **3.** To change the size of the text and images appearing on a computer monitor, or within a given window. Zooming in increases or magnifies the size, while zooming out does the converse.

zoom lens A lens whose focal length can be varied, so as to change the size of an image. Such a lens enables such transitions to occur without defocusing. Also called **zoom (2)**.

Zr Chemical symbol for **zirconium**.

zs Abbreviation of **zeptosecond**.

zsec Abbreviation of **zeptosecond**.

Zulu Time An internationally agreed time standard based on time kept by atomic clocks, and which represents the local time at the 0° meridian, which passes through Greenwich, England. When utilizing Zulu Time, each location on the planet has the same time, which is expressed utilizing a 24 hour clock, with a Z frequently appended. For example, 2100Z indicates 9 PM. Since the earth's rotation is gradually slowing, an extra second is added approximately once a year. It is based on International Atomic Time. Also called **Z time, Universal Coordinated Time, Universal Time Coordinated, Universal Time, Coordinated Universal Time,** or **World Time**.

Greek Letters Utilized in the Dictionary

α (alpha) **1.** Symbol for **alpha**. **2.** Symbol for **attenuation constant**.

β (beta) Symbol for **beta**.

γ (gamma) rays Same as **gamma rays**.

Δ (delta) Symbol for **delta**.

ΔV Symbol for **potential difference**.

ϵ (epsilon) **1.** Symbol for **permittivity**. **2.** Symbol for **emissivity**.

ϵ_0 Symbol for **permittivity of free space**, or **electric constant**.

λ (lambda) **1.** Symbol for **lambda**. **2.** Symbol for **wavelength**. **2.** Symbol for **linear charge density**, or **charge density (3)**.

Λ (lambda) Symbol for **permeance**, or **magnetic permeance**.

μ (mu) **1.** Symbol for **mu**. **2.** Symbol for **micro-**. **3.** Symbol for **micron**. **4.** Symbol for **magnetic permeability** or **permeability (1)**.

μ factor Same as **amplification factor (2)**.

μ law Same as **mu-law**.

μ_0 Symbol for **permeability of free space**, or **magnetic constant**.

μA Abbreviation of **microampere**.

μA-h Abbreviation of **microampere-hour**.

μAh Abbreviation of **microampere-hour**.

μb Abbreviation of **microbar**.

μ_B Symbol for **Bohr magneton**, or **magneton (1)**.

μC Abbreviation of **microcoulomb**.

μcd Abbreviation of **microcandela**.

μCi Abbreviation of **microcurie**.

μF Abbreviation of **microfarad**.

μG Abbreviation of **microgauss**.

μg Abbreviation of **microgram**.

μGs Abbreviation of **microgauss**.

μH Abbreviation of **microhenry**.

μJ Abbreviation of **microjoule**.

μl Abbreviation of **microliter**.

μL **1.** Abbreviation of **microliter**. **2.** Abbreviation of **microlambert**.

μlm Abbreviation of **microlumen**.

μlx Abbreviation of **microlux**.

μm Abbreviation of **micrometer**.

μmol Abbreviation of **micromole**, or **micromol**.

μ_N Symbol for **nuclear magneton**, or **magneton (2)**.

μN Abbreviation of **micronewton**.

μOe Abbreviation of **microoersted**.

μph Abbreviation of **microphot**.

μR Abbreviation of **microroentgen**.

μrad **1.** Abbreviation of **microradian**. **2.** Abbreviation of **microrad**.

μrem Abbreviation of **microrem**.

μs Abbreviation of **microsecond**.

μS Abbreviation of **microsiemens**.

μsec Abbreviation of **microsecond**.

μT Abbreviation of **microtesla**.

μV Abbreviation of **microvolt**.

$\mu V/m$ Abbreviation of **microvolts per meter**.

μW Abbreviation of **microwatt**.

μW-h Abbreviation of **microwatt-hour**.

μW-hr Abbreviation of **microwatt-hour**.

$\mu \Omega$ Abbreviation of **microhm**.

ν (nu) Symbol for **frequency**.

π (pi) Symbol for **pi**.

π meson Same as **pi meson**.

π-network Same as **pi network**.

ρ (rho) **1.** Symbol for **volume charge density**, or **charge density (2)**. **2.** Symbol for **resistivity**.

σ (sigma) **1.** Symbol for **surface charge density**, or **charge density (1)**. **2.** Symbol for **conductivity**, or **electrical conductivity**. **3.** Symbol for **standard deviation**. **4.** Symbol for **Stefan-Boltzmann constant**.

ϕ (phi) Symbol for **phase (2)**.

Φ (phi) **1.** Symbol for **radiant flux**. **2.** Symbol for **work function**.

ψ (psi) Symbol for **electric flux**.

Ω (omega) Symbol for **ohm**.

Ω-cm Symbol for **ohm-centimeter**.

Ω-m Symbol for **ohm-meter**.

Symbols

∞ Symbol for **infinity**.

° Symbol for **degree**.

% Symbol for **percent**.

+ Symbol for **positive**.

– Symbol for **negative**.

' **1.** Symbol for **arcminute**, or **minute** (**2**).

" **1.** Symbol for **arcsecond**, or **second** (**2**).

↑ Symbol for **shift key**.

↵ Symbol for **enter key**.

← Symbol for **backspace key**.

? A character usually utilized to represent a single **wildcard** character, as opposed to a string of them, which may be represented by * (**1**).

* **1.** A character which may be utilized to represent a string of **wildcard** characters, as opposed to a single wild card, which is usually represented by ?. **2.** In most computer operating systems, a character which serves to denote multiplication, as seen, for instance, in: 7*3.

@ In an email address, the symbol, @, that separates a username from the domain name. Also called @ **sign**, or **at sign**.

@ **sign** Same as @.

Å Symbol for **angstrom**.

°C **1.** Symbol for **degree Celsius**. **2.** Symbol for **degree centigrade**.

.com Same as **dot com**.

.edu Same as **dot edu**.

°F Symbol for **degree Fahrenheit**.

°K Symbol for **degree Kelvin**.

.net Same as **dot net**.

.org Same as **dot org**.

°R Symbol for **degree Rankine**.

Numbers

0.3 to 3 GHz The interval encompassing **ultra high frequencies**.

0.3 to 3 kHz The interval encompassing **ultra low frequencies**, or **infralow frequencies**.

0.3 to 3 MHz The interval encompassing **medium frequencies**.

0.3 to 3 THz The interval encompassing **tremendously high frequencies**.

½ inch tape Abbreviation of **half-inch tape**.

1-2-3 A popular spreadsheet program.

1-bit DAC Abbreviation of **1-bit d**igital-to-analog **c**onverter. Digital-to-analog conversion in which each bit is converted serially.

1-pin connector A connector, such as an F connector, with 1 pin.

1G Abbreviation of **first-generation wireless**.

1G wireless Abbreviation of **first-generation wireless**.

1NF Abbreviation of **first normal form**.

1x Same as **1X**.

1X The usual or standard amount, speed, capacity, volume, or the like. For example, a 1X DVD player.

1xEV Abbreviation of **CDMA2000 1xEV**.

1xRTT Abbreviation of **CDMA2000 1xRTT**.

1.2M Same as **1.2MB**.

1.2MB An older floppy disk format whose capacity was 1.2 MB.

1.24416 Gbps A transmission rate of 1.24416 Gbps, as seen, for instance, in OC-24.

1.44M Same as **1.44MB**.

1.44MB An floppy disk format whose capacity is 1.44 MB.

1.544 Mbps A signaling rate of 1.544 Mbps, as seen, for instance, in T1 lines.

1.86624 Gbps A transmission rate of 1.86624 Gbps, as seen, for instance, in OC-36.

2-channel amplifier Abbreviation of **two-channel amplifier**, or **dual-channel amplifier**.

2-D Abbreviation of **two-dimensional**.

2-D array Abbreviation of **two-dimensional array**.

2-D graphics Abbreviation of **two-dimensional graphics**.

2-element beam Abbreviation of **two-element beam**.

2-input adder Abbreviation of **two-input adder**.

2-phase Abbreviation of **two-phase**.

2-phase AC Abbreviation of **two-phase AC**.

2-phase circuit Abbreviation of **two-phase circuit**.

2-phase commit Abbreviation of **two-phase commit**.

2-phase current Abbreviation of **two-phase current**.

2-phase generator Abbreviation of **two-phase generator**.

2-phase motor Abbreviation of **two-phase motor**.

2-phase power Abbreviation of **two-phase power**.

2-phase system Abbreviation of **two-phase system**.

2-phase transformer Abbreviation of **two-phase transformer**.

2-phase voltage Abbreviation of **two-phase voltage**.

2-pin connector A connector, such certain BNC connectors, with 2 pins.

2-port network Abbreviation of **two-port network**.

2's complement Abbreviation of **twos complement**.

2-state device Abbreviation of **two-state device**.

2-state logic Abbreviation of **two-state logic**.

2-tier client/server Abbreviation of **two-tier client/server**.

2-track recording Abbreviation of **two-track recording**.

2-track tape Abbreviation of **two-track tape**.

2-tuner picture-in-picture Abbreviation of **two-tuner picture-in-picture**.

2-tuner PIP Abbreviation of **two-tuner PIP**.

2-way beeper Abbreviation of **two-way beeper**.

2-way communication Abbreviation of **two-way communication**.

2-way intercom Abbreviation of **two-way intercom**.

2-way loudspeaker Abbreviation of **two-way loudspeaker**.

2-way pager Abbreviation of **two-way pager**.

2-way radio Abbreviation of **two-way radio**.

2-way radio communications Abbreviation of **two-way radio communications**.

2-way repeater Abbreviation of **two-way repeater**.

2-way speaker Abbreviation of **two-way speaker**.

2-way system Abbreviation of **two-way system**.

2-wire circuit Abbreviation of **two-wire circuit**.

2-wire line Abbreviation of **two-wire line**.

2-wire system Abbreviation of **two-wire system**.

2B1Q Abbreviation of **two-binary, one quaternary**.

2D Abbreviation of **two-dimensional**.

2D array Abbreviation of **two-dimensional array**.

2D graphics Abbreviation of **two-dimensional graphics**.

2G Abbreviation of **second-generation wireless**.

2G wireless Abbreviation of **second-generation wireless**.

2NF Abbreviation of **second normal form**.

2x Same as **2X**.

2X Two times the usual or standard amount, speed, capacity, volume, or the like. For example, a 2X DVD player, or 2x oversampling.

2.048 Mbps A signaling rate of 2.048 Mbps, as seen, for instance, in E1 lines.

2.48832 Gbps A transmission rate of 2.48832 Gbps, as seen, for instance, in OC-48.

2.5G Digital cellular communications providing greater bandwidth and features than **2G**.

3-channel sound Abbreviation of **three-channel sound**.

3-channel sound system Abbreviation of **three-channel sound system**.

3-channel stereo Abbreviation of **three-channel stereo**.

3-click rule Abbreviation of **three-click rule**.

3-D Abbreviation of **three-dimensional**.

3-D accelerator Same as **3D accelerator**.

3-D animation Abbreviation of **three-dimensional animation**.

3-D array Abbreviation of **three-dimensional array**.

3-D audio Abbreviation of **three-dimensional audio**.

3-D chat room Same as **3D chat room**.

3-D graphics Abbreviation of **three-dimensional graphics**.

3-D modeling Abbreviation of **three-dimensional modeling**.

3-D positional sound Abbreviation of **three-dimensional sound**.

3-D sound Abbreviation of **three-dimensional sound**.

3-D stereo enhancement Same as **3D stereo enhancement**.

3-D television Abbreviation of **three-dimensional television**.

3-gun CRT Abbreviation of **three-gun CRT**.

3-input adder Abbreviation of **three-input adder**.

3-phase Abbreviation of **three-phase**.

3-phase AC Abbreviation of **three-phase AC**.

3-phase bridge rectifier Abbreviation of **three-phase bridge rectifier**.

3-phase circuit Abbreviation of **three-phase circuit**.

3-phase current Abbreviation of **three-phase current**.

3-phase 4-wire Abbreviation of **three-phase four-wire**.

3-phase full-wave Abbreviation of **three-phase full-wave**.

3-phase generator Abbreviation of **three-phase generator**.

3-phase half-wave Abbreviation of **three-phase half-wave**.

3-phase meter Abbreviation of **three-phase meter**.

3-phase motor Abbreviation of **three-phase motor**.

3-phase power Abbreviation of **three-phase power**.

3-phase rectifier Abbreviation of **three-phase rectifier**.

3-phase system Abbreviation of **three-phase system**.

3-phase 3-wire Abbreviation of **three-phase three-wire**.

3-phase transformer Abbreviation of **three-phase transformer**.

3-phase voltage Abbreviation of **three-phase voltage**.

3-pin connector A connector, such as an XLR connector, with 3 pins.

3-prong electrical outlet Abbreviation of **three-prong electrical outlet**.

3-prong outlet Abbreviation of **three-prong outlet**.

3-state logic Abbreviation of **three-state logic**.

3-terminal regulator Abbreviation of **three-terminal regulator**.

3-tier client/server Abbreviation of **three-tier client/server**.

3 to 30 GHz The interval encompassing **super high frequencies**.

3 to 30 Hz The interval encompassing **extremely low frequencies**.

3 to 30 kHz The interval encompassing **very low frequencies**.

3 to 30 MHz The interval encompassing **high frequencies**.

3-way loudspeaker Abbreviation of **three-way loudspeaker**.

3-way speaker Abbreviation of **three-way speaker**.

3-way system Abbreviation of **three-way system**.

3-wire system Abbreviation of **three-wire system**.

3D Abbreviation of **three-dimensional**.

3D accelerator A graphics accelerator which is especially suited for handling 3D graphics.

3D animation Abbreviation of **three-dimensional animation**.

3D array Abbreviation of **three-dimensional array**.

3D audio Abbreviation of **three-dimensional audio**.

3D chat room A chat room which provides an environment in which 3D images, such as avatars, are incorporated.

3D graphics Abbreviation of **three-dimensional graphics**.

3D modeling Abbreviation of **three-dimensional modeling**.

3D positional sound Abbreviation of **three-dimensional sound**.

3D sound Abbreviation of **three-dimensional sound**.

3D stereo enhancement Circuitry which utilizes signal processing to attempt to simulate a 3-D listening environment. Such circuitry may be included, for instance, in a computer sound card.

3D Studio Max A popular program utilized for 3D modeling and animation.

3D television Abbreviation of **three-dimensional television**.

3D-TV Abbreviation of **three-dimensional television**.

3DES Abbreviation of **Triple DES**.

3DTV Abbreviation of **three-dimensional television**.

3G Abbreviation of **third-generation wireless**.

3G wireless Abbreviation of **third-generation wireless**.

3GL Abbreviation of **third-generation language**.

3NF Abbreviation of **third normal form**.

3.152 Mbps A signaling rate of 3.152 Mbps, as seen, for instance, in T1C lines.

4:1:1 In digital component video, a sampling ratio in which for every four luminance samples there is one for each of the color-difference signals.

4:2:2 In digital component video, a sampling ratio in which for every four luminance samples there are two for each of the color-difference signals.

4:3 The typical aspect ratio of a standard TV or computer monitor, which is equivalent to a 1.33:1 aspect ratio.

4-channel sound Abbreviation of **four-channel sound**.

4-channel sound system Abbreviation of **four-channel sound system**.

4-channel stereo Abbreviation of **four-channel stereo**.

4-pin connector A connector, such as a USB connector, with 4 pins.

4-point probe Abbreviation of **four-point probe**.

4-pole double-throw Abbreviation of **four-pole double-throw**.

4-terminal network Abbreviation of **four-terminal network**.

4-track recording Abbreviation of **four-track recording**.

4-track tape Abbreviation of **four-track tape**.

4-way loudspeaker Abbreviation of **four-way loudspeaker**.

4-way speaker Abbreviation of **four-way speaker**.

4-way system Abbreviation of **four-way system**.

4-wire circuit Abbreviation of **four-wire circuit**.

4-wire system Abbreviation of **four-wire system**.

4-wire wye Abbreviation of **four-wire wye**.

4CIF A **Common Intermediate Format** supporting a resolution of 704 x 576, at 30 frames per second.

4G Abbreviation of **fourth-generation wireless**.

4G wireless Abbreviation of **fourth-generation wireless**.

4GL Abbreviation of **fourth-generation language**.

4mm tape A tape, such as certain digital audio tapes, whose width is 4 mm.

4NF Abbreviation of **fourth normal form**.

4x Same as **4X**.

4X Four times the usual or standard amount, speed, capacity, volume, or the like. For example, a 4X DVD player, or 4x oversampling.

4X CD-ROM drive Abbreviation of **quad speed CD-ROM drive**.

4.97664 Gbps A transmission rate of 4.97664 Gbps, as seen, for instance, in OC-96.

5-pin connector A connector, such as certain DIN connectors, with 5 pins.

5GL Abbreviation of **fifth-generation language**.

5NF Abbreviation of **fifth normal form**.

5.1 Abbreviation of **5.1 surround sound**.

5.1 channel surround Same as **5.1 surround sound**.

5.1 channel surround sound Same as **5.1 surround sound**.

5.1 surround Abbreviation of **5.1 surround sound**.

5.1 surround sound A surround sound system providing six channels, which are the left front, left rear, center, right front, and right rear channels, plus the subwoofer. Since the bandwidth of the subwoofer is reduced, it is considered as the .1 portion. Its abbreviation is **5.1 surround**, or **5.1**. Also called **5.1 channel surround**, or **5.1 channel surround sound**.

6-DOF Abbreviation of **six degrees of freedom**.

6-pin connector A connector, such as that of a mouse port, with 6 pins.

6DOF Abbreviation of **six degrees of freedom**.

6x Same as **6X**.

6X Six times the usual or standard amount, speed, capacity, volume, or the like. For example, a 6X DVD player.

6.1 Abbreviation of **6.1 surround sound**.

6.1 channel surround Same as **6.1 surround sound**.

6.1 channel surround sound Same as **6.1 surround sound**.

6.1 surround Abbreviation of **6.1 surround sound**.

6.1 surround sound An enhancement of **5.1 surround sound** in which an additional channel, driving a rear center or back surround speaker, is added. Its abbreviation is **6.1 surround**, or **6.1**. Also called **6.1 channel surround**, or **6.1 channel surround sound**.

6.312 Mbps A signaling rate of 6.312 Mbps, as seen, for instance, in T2 lines.

7-bit ASCII Same as **standard ASCII**.

7-E-1 A common modem format in which there are **7** data bits, with even parity and **1** stop bit. Also called **E-7-1**.

7-pin connector A connector, such as that of certain flash cards, with 7 pins.

7-segment display Abbreviation of **seven-segment display**.

7.1 Abbreviation of **7.1 surround sound**.

7.1 channel surround Same as **7.1 surround sound**.

7.1 channel surround sound Same as **7.1 surround sound**.

7.1 surround Abbreviation of **7.1 surround sound**.

7.1 channel surround An enhancement of **6.1 channel surround** in which the additional channel is split into two, for an even greater surround sound experience. Its abbreviation is **7.1 surround**, or **7.1**. Also called **7.1 channel surround**, or **7.1 channel surround sound**.

8 A popular relational database.

8-bit Indicative of something dealing with single-byte, or 8-bit, increments. For example, 8-bit color, which allows for up to 256 possible colors to be displayed.

8-bit ASCII Same as **Extended ASCII**.

8-bit color Color in which the bit depth is 8, which allows for up to 256 possible colors to be displayed.

8-bit graphics Computer graphics utilizing **8-bit color**.

8-bit sound Sound based on 8-bit sound samples, providing for 256 possible increments. Such audio is of low quality.

8-N-1 A common modem setting in which there are **8** data bits, with no parity, and **1** stop bit. Also called **N-8-1**.

8-pin connector A connector, such as certain DIN connectors, with 8 pins.

8mm tape A tape, such as certain digital audio tapes, whose width is 8 mm.

8x Same as **8X**.

8X Eight times the usual or standard amount, speed, capacity, volume, or the like. For example, an 8X DVD player, or 8x oversampling.

8.448 Mbps A signaling rate of 8.448 Mbps, as seen, for instance, in E2 lines.

9-pin connector A connector, such as a DB-9 connector, with 9 pins.

9's complement Abbreviation of **nines complement**.

9-track Pertaining to a **9-track tape** or a **9-track tape drive**.

9-track tape A magnetic tape with nine parallel tracks, consisting of 8 data bits plus 1 parity bit.

9-track tape drive A tape drive used for reading or writing data on **9-track tapes**.

9.95328 Gbps A transmission rate of 9.95328 Gbps, as seen, for instance, in OC-192.

10/100 card An Ethernet card that supports both 10Base-T and 100Base-T.

10/100/1000 card An Ethernet card that supports 10Base-T, 100Base-T, and 1000Base-T.

10-GbE Same as **10-Gigabit Ethernet**.

10-Gigabit Ethernet An Ethernet technology which supports data-transfer rates of up to 10 gigabits, and which uses optical fibers. Its abbreviation is **10-GigE**, **10-Gig Ethernet**, or **10-GbE**.

10-GigE Same as **10-Gigabit Ethernet**.

10-Gig Ethernet Same as **10-Gigabit Ethernet**.

10 megabit Ethernet An Ethernet standard, such as 10Base-T, supporting data transfer rates of up to 10 Mbps.

10-pin connector A connector, such as that of certain flash cards, with 10 pins.

10's complement Abbreviation of **tens complement**.

10-turn potentiometer Abbreviation of **ten-turn potentiometer**.

10Base-F 10 megabit Ethernet that utilizes optical fibers, as opposed to **10Base-T**, which uses twisted pairs.

10Base-T 10 megabit Ethernet that utilizes twisted pairs, as opposed to **10Base-F**, which uses optical fibers. Also called **fast Ethernet**.

10Base2 An Ethernet standard using a coaxial cable which is about 0.5 centimeter in diameter, as opposed to **10Base5** whose cabling is about one centimeter in diameter. Also called **ThinNet**, **ThinWire**, **Thin Ethernet**, or **cheapernet**.

10Base5 An Ethernet standard using a coaxial cable which is about one centimeter in diameter, as opposed to **10Base2** whose cabling is about 0.5 centimeter in diameter. Also called **ThickNet**, **ThickWire**, or **Thick Ethernet**.

10BaseF Same as **10Base-F**.

10BaseT Same as **10Base-T**.

14.4K modem Abbreviation of **14.4Kbps modem**. A modem with a transmission rate of up 14,400 bps downstream.

14.4Kbps modem Same as **14.4K modem**.

15-pin connector A connector, such as that which attaches a network interface card and an Ethernet cable, with 15 pins.

16-bit Indicative of something dealing with two-byte, or 16-bit, increments. For example, 16-bit color, which allows for up to 65,536 possible colors to be displayed.

16-bit color Color in which the bit depth is 16, which allows for up to 65,536 possible colors to be displayed. Also called **high color**.

16-bit graphics Computer graphics utilizing **16-bit color**.

16-bit CPU A CPU which simultaneously processes two bytes at a time.

16-bit sound Sound based on 16-bit sound samples, providing for 65,536 possible increments. Such audio is approximately comparable to that of a standard CD.

16:9 The typical aspect ratio of an HDTV, which is equivalent to a 1.78:1 aspect ratio.

16CIF A **Common Intermediate Format** supporting a resolution of 1408 x 1152, at 30 frames per second.

16x Same as **16X**.

16X Sixteen times the usual or standard amount, speed, capacity, volume, or the like. For example, a 16X DVD player, or 16x oversampling.

21-pin connector A connector, such as a SCART connector, with 21 pins.

24-bit color Color in which the bit depth is 24, which allows for over 16.7 million possible colors to be displayed. 24 bits means that there are 8 for each of the primary colors, which are red, green, and blue. Each pixel is composed of a trio of color phosphor dots, one for each primary color. Also called **true color**.

24-bit graphics Computer graphics utilizing **24-bit color**.

24x Same as **24X**.

24X Twenty-four times the usual or standard amount, speed, capacity, volume, or the like. For example, a 24X DVD player, or a 24X CD-R.

25-pin connector A connector, such as a DB-25 connector, with 25 pins.

28.8K modem Abbreviation of **28.8K**bps **modem**. A modem with a transmission rate of up 28,000 bps downstream.

28.8Kbps modem Same as **28.8K modem**.

30-pin connector A connector, such as that for certain SIMM modules, with 30 pins.

30 to 300 Hz The interval encompassing **super low frequencies**.

30 to 300 GHz The interval encompassing **extremely high frequencies**.

30 to 300 kHz The interval encompassing **low frequencies**.

30 to 300 MHz The interval encompassing **very high frequencies**.

32-bit Indicative of something dealing with four-byte, or 32-bit, increments. For example, a 32-bit CPU, which simultaneously processes four bytes at a time.

32-bit CPU A CPU which simultaneously processes four bytes at a time.

32-bit color The same as **24-bit color**, plus an 8-bit alpha channel.

32-bit graphics Computer graphics utilizing **32-bit color**.

32x Same as **32X**.

32X Thirty-two times the usual or standard amount, speed, capacity, volume, or the like. For example, a 32X DVD player, or 32x oversampling.

33.6K modem Abbreviation of **33.6K**bps **modem**. A modem with a transmission rate of up 33,600 bps downstream.

33.6Kbps modem Same as **33.6K modem**.

34.368 Mbps A signaling rate of 34.368 Mbps, as seen, for instance, in E3 lines.

37-pin connector A connector, such as a DB-37 connector, with 37 pins.

39.81312 Gbps A transmission rate of 39.81312 Gbps, as seen, for instance, in OC-768.

44.736 Mbps A signaling rate of 44.736 Mbps, as seen, for instance, in T3 lines.

50-pin connector A connector, such as an RJ-21 connector, with 50 pins.

51.8 Mbps A transmission rate of 51.8 Mbps, as seen, for instance, in OC-1.

56K modem Abbreviation of **56K**bps **modem**. A modem with a transmission rate of up 56,000 bps downstream.

56Kbps modem Same as **56K modem**.

64-bit Indicative of something dealing with eight-byte, or 64-bit, increments. For example, a 64-bit CPU, which simultaneously processes eight bytes at a time.

64-bit CPU A CPU which simultaneously processes eight bytes at a time.

64K ISDN An ISDN line or service using a single 64 Kbps B channel.

64x Same as **64X**.

64X Sixty-four times the usual or standard amount, speed, capacity, volume, or the like. For example, a 64X DVD player, or 64x oversampling.

68-pin connector A connector, such as a PCMCIA connector, with 68 pins.

72-pin connector A connector, such as that for certain SIMM modules, with 72 pins.

80/20 rule A rule of thumb that applies in a variety of settings. For instance, roughly 20% of the features of a given application are used about 80% of the time, or about 80% of the failures or defects of a given product such as a chip are due to the top 20% or so of causes, among countless examples. Also called **Pareto Principle**.

95 A common operating system.

98 A common operating system.

100-GbE Same as **100-Gigabit Ethernet**.

100-Gigabit Ethernet An Ethernet technology which supports data-transfer rates of up to 100 gigabits, and which uses optical fibers. Its abbreviation is **100-GigE**, **100-Gig Ethernet**, or **100-GbE**.

100-Gig Ethernet Same as **100-Gigabit Ethernet**.

100-GigE Same as **100-Gigabit Ethernet**.

100 megabit Ethernet An Ethernet standard, such as 100Base-T, supporting data transfer rates of up to 100 Mbps.

100Base-FX 100 megabit Ethernet that utilizes optical fibers, as opposed to **100Base-TX**, which uses twisted pairs.

100Base-T An Ethernet standard supporting data transfer rates of up to 100 Mbps. Depending on the specific configuration, it may utilize two or four twisted-pair copper wires, or fiber-optic cables. Also called **fast Ethernet**.

100BaseFX Same as **100Base-FX**.

100BaseTX Same as **100Base-TX**.

100Base-TX 100 megabit Ethernet that utilizes twisted pairs, as opposed to **100Base-FX**, which uses optical fibers.

100BaseT Same as **100Base-T**.

123 A popular spreadsheet program.

128-bit Indicative of something dealing with sixteen-byte, or 128-bit, increments. For example, a 128-bit graphics accelerator, whose bus is 128 bits wide.

128-bit CPU A CPU which simultaneously processes sixteen bytes at a time.

128K ISDN An ISDN line or service using two 64 Kbps B channels.

128x Same as **128X**.

128X One hundred and twenty-eight times the usual or standard amount, speed, capacity, volume, or the like. For example, 128x oversampling.

139.264 Mbps A signaling rate of 139.264 Mbps, as seen, for instance, in E4 lines.

144-pin connector A connector, such as that for certain SO-DIMM modules, with 144 pins.

155.52 Mbps A transmission rate of 155.52 Mbps, as seen, for instance, in OC-3.

159.25248 Gbps A transmission rate of 159.25248 Gbps, as seen, for instance, in OC-3072.

168-pin connector A connector, such as that for certain DIMM modules, with 168 pins.

184-pin connector A connector, such as that for certain DDR SDRAM modules, with 184 pins.

200-pin connector A connector, such as that for certain DDR SDRAM modules, with 200 pins.

232-pin connector A connector, such as that for certain RIMM modules, with 232 pins.

256-bit Indicative of something dealing with 32-byte, or 256-bit, increments. For example, a 256-bit GPU.

256-bit CPU A CPU which simultaneously processes 256 bits at a time.

256-bit GPU A GPU which simultaneously processes 256 bits at a time.

274.176 Mbps A signaling rate of 274.176 Mbps, as seen, for instance, in T4 lines.

300 to 3000 GHz The interval encompassing **tremendously high frequencies**.

300 to 3000 Hz The interval encompassing **ultra low frequencies**, or **infralow frequencies**.

300 to 3000 kHz The interval encompassing **medium frequencies**.

300 to 3000 MHz The interval encompassing **ultra high frequencies**.

400.352 Mbps A signaling rate of 400.352 Mbps, as seen, for instance, in T5 lines.

404 Error Same as **Error 404**.

404 Not Found Same as **Error 404**.

466.56 Mbps A transmission rate of 466.56 Mbps, as seen, for instance, in OC-9.

488 bus Same as **IEEE 488 bus**.

488 standard Same as **IEEE 488 standard**.

512-bit Indicative of something dealing with 64-byte, or 512-bit, increments. For example, a 512-bit GPU.

512-bit CPU A CPU which simultaneously processes 512 bits at a time.

512-bit GPU A GPU which simultaneously processes 512 bits at a time.

565.148 Mbps A signaling rate of 565.148 Mbps, as seen, for instance, in E5 lines.

622.08 Mbps A transmission rate of 622.08 Mbps, as seen, for instance, in OC-12.

640x480 A computer monitor resolution in which 640 dots, or pixels, are displayed on each of the 480 lines of the display.

800x600 A computer monitor resolution in which 800 dots, or pixels, are displayed on each of the 600 lines of the display.

802 Same as **IEEE 802**.

802.1 Same as **IEEE 802.1**.

802.2 Same as **IEEE 802.2**.

802.3 Same as **IEEE 802.3**.

802.3z Same as **IEEE 802.3z**.

802.4 Same as **IEEE 802.4**.

802.5 Same as **IEEE 802.5**.

802.6 Same as **IEEE 802.6**.

802.7 Same as **IEEE 802.7**.

802.8 Same as **IEEE 802.8**.

802.9 Same as **IEEE 802.9**.

802.10 Same as **IEEE 802.10**.

802.11 Same as **IEEE 802.11**.

933.12 Mbps A transmission rate of 933.12 Mbps, as seen, for instance, in OC-18.

1000Base-CX Gigabit Ethernet which utilizes two twisted pairs.

1000Base-LX Gigabit Ethernet over optical fibers, using long-wavelength laser light.

1000Base-SX Gigabit Ethernet over optical fibers, using short-wavelength laser light.

1000Base-T Gigabit Ethernet using four copper twisted pairs.

1000BaseCX Same as **1000Base-CX**.

1000BaseLX Same as **1000Base-LX**.

1000BaseSX Same as **1000Base-SX**.

1000BaseT Same as **1000Base-T**.

1024-bit Indicative of something dealing with 128-byte, or 1024-bit, increments. For example, 1024-bit encryption.

1024x768 A computer monitor resolution in which 1024 dots, or pixels, are displayed on each of the 768 lines of the display.

1280x960 A computer monitor resolution in which 1280 dots, or pixels, are displayed on each of the 960 lines of the display.

1280x1024 A computer monitor resolution in which 1280 dots, or pixels, are displayed on each of the 1024 lines of the display.

1284 Same as **IEEE 1284**.

1284 cable Same as **IEEE 1284 cable**.

1284 compliant Same as **IEEE 1284 compliant**.

1284 printer cable Same as **IEEE 1284 printer cable**.

1284 standard Same as **IEEE 1284 standard**.

1394 serial bus Same as **IEEE 1394 serial bus**.

1394 standard Same as **IEEE 1394 standard**.

1600x1200 A computer monitor resolution in which 1600 dots, or pixels, are displayed on each of the 1200 lines of the display.

2000 A common operating system.

2048-bit Indicative of something dealing with 256-byte, or 2048-bit, increments. For example, 2048-bit encryption.